Analog Circuit Design

Volume 3

Design Note Collection

Analog Circuit Design
Volume 3

Design Note Collection

Edited by

Bob Dobkin

John Hamburger

AMSTERDAM • BOSTON • HEIDELBERG • LONDON • NEW YORK • OXFORD
PARIS • SAN DIEGO • SAN FRANCISCO • SINGAPORE • SYDNEY • TOKYO

Newnes is an imprint of Elsevier

Newnes is an imprint of Elsevier
225 Wyman Street, Waltham, MA 02451, USA
The Boulevard, Langford Lane, Kidlington, Oxford OX5 1GB, UK

First edition 2015

Library of Congress Cataloging-in-Publication Data
A catalog record for this book is available from the Library of Congress

British Library Cataloguing in Publication Data
A catalogue record for this book is available from the British Library

ISBN: 978-0-12-800001-4

For information on all Newnes publications
visit our web site at books.elsevier.com

Transferred to Digital Printing in 2015

Cover photo by Anne Hamersky

Dedicated to all the authors of these Design Notes.

And to analog engineers everywhere in hopes that we will continue to develop more.

For Sandra, Naomi, David and Sarah, the bright lights in my analog world.

Contents

Section 5 Switching Regulator Design: Boost Converters . . . 237

Publisher's Note

This book was compiled from Linear Technology Corporation's original Design Notes. These Design Notes have been renamed as chapters for the purpose of this book. However, throughout the text there are cross-references to different Design Notes, a few of which may not be in the book. For reference, this conversion table has been included; it shows both the book chapter numbers and the original Design Note numbers.

CHAPTER NUMBER	DESIGN NOTE
1	517
2	470
3	347
4	281
5	249
6	247
7	238
8	224
9	91
10	20
11	285
12	284
13	279
14	258
15	216
16	209
17	122
18	113
19	90
20	87
21	74
22	71
23	357
24	316

CHAPTER NUMBER	DESIGN NOTE
25	125
26	95
27	47
28	21
29	17
30	11
31	8
32	523
33	511
34	506
35	504
36	493
37	492
38	489
39	486
40	479
41	478
42	467
43	460
44	459
45	458
46	457
47	443
48	442

CHAPTER NUMBER	DESIGN NOTE
49	441
50	433
51	419
52	412
53	409
54	404
55	403
56	390
57	387
58	383
59	382
60	373
61	367
62	364
63	352
64	328
65	326
66	322
67	318
68	311
69	309
70	301
71	300
72	295
73	287
74	278
75	269
76	268
77	256
78	235
79	234
80	233
81	225

CHAPTER NUMBER	DESIGN NOTE
82	222
83	218
84	215
85	212
86	211
87	210
88	208
89	206
90	202
91	201
92	199
93	198
94	196
95	186
96	181
97	162
98	156
99	150
100	141
101	135
102	105
103	103
104	100
105	98
106	86
107	78
108	77
109	73
110	72
111	69
112	68
113	53
114	516

CHAPTER NUMBER	DESIGN NOTE
115	465
116	436
117	428
118	371
119	365
120	359
121	358
122	354
123	332
124	317
125	305
126	304
127	280
128	255
129	252
130	246
131	232
132	183
133	179
134	178
135	175
136	166
137	154
138	128
139	63
140	58
141	48
142	505
143	503
144	435
145	434
146	334
147	312

CHAPTER NUMBER	DESIGN NOTE
148	282
149	223
150	521
151	500
152	499
153	424
154	413
155	370
156	369
157	330A
158	307
159	109
160	108
161	52
162	49
163	507
164	314
165	292
166	220
167	172
168	157
169	119
170	115
171	82
172	44
173	33
174	32
175	524
176	518
177	488
178	474
179	469
180	446

CHAPTER NUMBER	DESIGN NOTE
181	438
182	430
183	411
184	385
185	377
186	299
187	261
188	59
189	444
190	386
191	363
192	356
193	93
194	76
195	496
196	472
197	471
198	420
199	415
200	395
201	393
202	380
203	342
204	336
205	320
206	283
207	277
208	275
209	271
210	250
211	244
212	242
213	239

CHAPTER NUMBER	DESIGN NOTE
214	194
215	188
216	170
217	160
218	144
219	124
220	121
221	120
222	111
223	54
224	41
225	491
226	483
227	310
228	243
229	189
230	142
231	509
232	410
233	344
234	296
235	260
236	158
237	31
238	18
239	498
240	487
241	485
242	450
243	484
244	405
245	497
246	495

CHAPTER NUMBER	DESIGN NOTE
247	466
248	437
249	421
250	402
251	397
252	360
253	353
254	346
255	319
256	265
257	263
258	253
259	226
260	217
261	200
262	197
263	155
264	139
265	519
266	425
267	361
268	350
269	338
270	427
271	408
272	401
273	391
274	389
275	384
276	372
277	321
278	290
279	272

CHAPTER NUMBER	DESIGN NOTE
280	270
281	149
282	508
283	501
284	490
285	462
286	461
287	449
288	445
289	440
290	422
291	417
292	406
293	392
294	388
295	376
296	349
297	345
298	340
299	325
300	315
301	303
302	267
303	264
304	231
305	164
306	133
307	101
308	99
309	520
310	512
311	481
312	464

CHAPTER NUMBER	DESIGN NOTE
313	455
314	453
315	452
316	398
317	378
318	374
319	396
320	394
321	343
322	327
323	205
324	92
325	81
326	79
327	65
328	57
329	55
330	522
331	494
332	477
333	468
334	463
335	456
336	400
337	379
338	368
339	341
340	274
341	259
342	257
343	236
344	237
345	219

CHAPTER NUMBER	DESIGN NOTE
346	207
347	184
348	180
349	177
350	165
351	159
352	153
353	146
354	138
355	116
356	104
357	88
358	66
359	60
360	448
361	431
362	337
363	214
364	167
365	131
366	127
367	96
368	297
369	112
370	106
371	38
372	35
373	26
374	24
375	19
376	13
377	10
378	5

CHAPTER NUMBER	DESIGN NOTE
379	2
380	1
381	329
382	289
383	228
384	203
385	193
386	191
387	176
388	174
389	168
390	161
391	130
392	129
393	102
394	94
395	80
396	75
397	64
398	39
399	34
400	30
401	29
402	27
403	22
404	14
405	4
406	510
407	426
408	323
409	302
410	173
411	51

CHAPTER NUMBER	DESIGN NOTE
412	45
413	40
414	513
415	502
416	473
417	454
418	451
419	429
420	414
421	399
422	355
423	333
424	331
425	308
426	306
427	298
428	286
429	266
430	254
431	241
432	230
433	182
434	171
435	163
436	148
437	140
438	136
439	132
440	107
441	89
442	84
443	83
444	70

CHAPTER NUMBER	DESIGN NOTE
445	61
446	56
447	50
448	46
449	43
450	42
451	36
452	28
453	25
454	23
455	15
456	12
457	3
458	423
459	407
460	348
461	227
462	190
463	351
464	229
465	145
466	324
467	313
468	291
469	276
470	251
471	245
472	221
473	195
474	169
475	147
476	67

CHAPTER NUMBER	DESIGN NOTE
477	37
478	16
479	9
480	7
481	248
482	185
483	137
484	123
485	293
486	262
487	362
488	339
489	288
490	515
491	514
492	482
493	480
494	476
495	447
496	439
497	432
498	418
499	381
500	375
501	366
502	335
503	273
504	240
505	187
506	152
507	134

Trademarks

These trademarks all belong to Linear Technology Corporation. They have been listed here to avoid repetition within the text. Trademark acknowledgment and protection applies regardless.

Linear Express, Linear Technology, LT, LTC, LTM, Burst Mode, Dust, Dust Networks, Eterna, FilterCAD, LTspice, Manager-on-Chip, OPTI-LOOP, Over-The-Top, PolyPhase, Silent Switcher, SmartMesh, SwitcherCAD, TimerBlox, µModule and the Linear logo are registered trademarks of Linear Technology Corporation. Adaptive Power, Bat-Track, BodeCAD, C-Load, ClockWizard, Direct Flux Limit, Direct-Sense, Easy Drive, EZSync, FilterView, FracNWizard, Hot Swap, isoSPI, LDO+, Linduino, LinearView, LTBiCMOS, LTCMOS, LTP, LTPoE++, LTpowerCAD, LTpowerPlanner, LTpowerPlay, Micropower SwitcherCAD, Mote-on-Chip, Multimode Dimming, No Latency ΔΣ, No Latency Delta-Sigma, No R_{SENSE}, Operational Filter, PanelProtect, PLL-Wizard, PowerPath, PowerSOT, PScope, QuikEval, RH DICE Inside, RH MILDICE Inside, SafeSlot, SmartMesh IP, SmartStart, SNEAK-A-BIT, SoftSpan, Stage Shedding, Super Burst, SWITCHER+, ThinSOT, Triple Mode, True Color PWM, UltraFast, Virtual Remote Sense, Virtual Remote Sensing, VLDO and VRS are trademarks of Linear Technology Corporation. All other trademarks are the property of their respective owners.

Acknowledgments

A project of this scale has many contributors. We want to thank most of all the amazing, talented writers who burned the midnight oil to write these Design Notes. And thanks to Linear Technology's graphic artists, particularly Gary Alexander, who did the precise layout of each note, Terri Yager, who helped deliver the notes in good form to the publisher, and Ron Sergi who provided proofing expertise.We also want to acknowledge our dedicated, talented partners at Elsevier/Newnes—Project Managers Charlie Kent and Pauline Wilkinson and our Editor Tim Pitts. Finally, I want to thank Bob Dobkin for his time, the care he put into this project and for the rare opportunity to work side by side with one of our analog gurus.

John Hamburger
Linear Technology Corporation

Introduction

All the Different Paths to Analog

Every design has a beginning and an end. The beginning is pretty easy to define; the end can be production or a scrap piece. The function of an analog circuit can take various alternative paths from the beginning to the end.

With digital circuits, it doesn't matter how you get there as long as the answer is correct. Digital is information, and as long as the computation ends correctly, the way it is achieved is less important. (Of course, there is timing and complexity, but these don't define whether the output number is correct.) So in digital circuits, the information once is it computed, is the important end result.

Analog circuits are different. The path you take to the end result affects the end result. Analog involves real world parameters, real world signals and real world measurements. The signal path from beginning to end operates on the input whether for amplification, detection, conversion, or any other function. The way we achieve that operation is important. Since it is the circuitry that operates on an analog signal, the circuitry leaves its stamp on the output signal.

Real world degradation of analog signals is easy to assess. Noise can increase, distortion can occur, voltage accuracy can worsen, and drifts can be introduced. Since the impact on the signal is a function of the circuitry used to operate on the signal,optimizing an analog circuit is very important.

It is relatively easy to see how an audio signal can be damaged by having the wrong circuitry. For example, the ultimate audio is a straight wire with gain. Next best is using really good op amps. If the circuit uses a low cost general purpose op amp with limited bandwidth, there is an increase in noise and distortion. Replace that op amp with a high speed, low noise amplifier and we get closer to the ideal.

But, less well known circuits are also subject to noise problems. For example, a temperature measurement bridge is made of three stable resistors and a temperature sensor. One might think that using a high gain, low noise amplifier is the only thing needed to give the best result. If we start to analyze the effects of the power supply, we see that power supply characteristics can be as important as the amplifier.

If the bridge is perfectly balanced, there is no signal at the output and small changes to the bridge's power supply don't affect the output. But as temperature changes, the sensor changes and the bridge is not always balanced.We amplify the electrical output from the bridge to obtain our temperature. This means that changes in the power supply affect the output of the bridge. We need to have a stable reference powering the bridge. Also the noise on the supply will appear at the output and the noise will be a function of how much the bridge is out of balance.

A real world example of this is using a bridge with a 24-bit analog-to-digital converter. The 24-bit analog-to-digital converter has a huge range on its input, so there is sufficient resolution of the temperature without offsetting and gaining the output of the bridge. But since we are looking at a small portion of the span—because we have enough resolution with 24-bit—we need to be very careful of the noise injected into the bridge. Our desired signal can be small and if there's too much noise on the power supply, it can show up on our output.

Experienced analog designers will realize some of the ancillary problems that can occur at the beginning of a design cycle. If the designer is inexperienced, it may take simulations and even breadboards to achieve a high-performance design. So having large numbers of finished analog circuits done by experienced designers is a great resource to providing well thought out starting points.

The teaching designs in this *Design Note Collection* help bring new designers up to speed and give experienced designers a starting point for even more sophisticated designs. This book has two purposes: to speed designs by presenting finished examples, as well as providing a teaching resource for designers. We hope this contributes in some way to future elegant analog circuit designs.

Bob Dobkin
Co-Founder, Vice President Engineering and
Chief Technical Officer
Linear Technology Corporation

Foreword

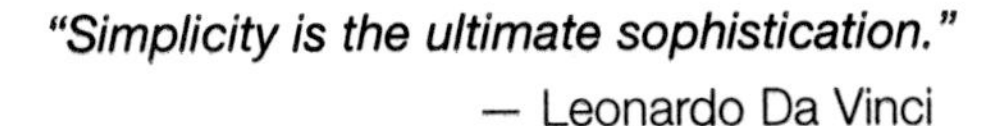

The *Design Note Collection* is the first effort to bring Linear's Design Notes into one volume. Design Notes were first published over 25 years ago, and after producing more than 500 notes, the genre is still going strong. One reason for their longevity is that while analog design is not a simple thing and doesn't follow simple rules, this format is succinct and readily applied.

In our present digital age, analog gurus are in short supply—and in great demand. That's where Design Notes come in. Since analog design requires example and practice, Design Notes were developed to help a growing community of designers better understand design via brief notes that explain specific circuit design challenges. Each Design Note is a pearl that forms around the grain of a design problem—to show an experienced designer how to solve a new problem or to explain to a budding engineer how to get around an age-old challenge.

That's why engineers have been tearing Design Notes out of their trade journals every month, filing them in binders and using them as a guide as they hammer out real world design problems on their lab benches.

Described as the "black art" of electronics, analog is studied, practiced and passed on by experienced gurus who teach the science and craft of analog in the lab.There it is still executed with probes, a soldering iron and oscilloscopes—the goal, not just to test the circuits, but to use and keep lab equipment in good tune.

It is our hope that this third volume of the *Analog Circuit Design* book series will be welcomed as enthusiastically as the first two volumes. The *Design Note Collection* has many authors, and it is to them we are grateful. You know some of their names—Jim Williams, Bob Dobkin, Carl Nelson, George Erdi, but there are a hundred more—talented analog designer/authors who have delivered these notes in a format that is familiar, easy to digest and can be tucked away for future reference.

We live in an analog world. That's why analog design is timeless. We hope that the *Design Note Collection*, organized logically by section topic, will provide good reading and a handy reference for many years to come.

John Hamburger
Linear Technology Corporation

PART 1

Power Management

High performance single phase DC/DC controller with power system management

1

Yi Sun

Introduction

The LTC3883 is a single phase synchronous step-down DC/DC controller featuring a PMBus interface for digital control and monitoring of key regulator parameters. It has integrated MOSFET gate drivers and can function either standalone or in a digitally managed system with other Linear Technology PMBus-enabled parts. The LTC3883 features:

- 4.5V to 24V input voltage range and 0.5V to 5.5V output voltage range
- ±0.5% output voltage accuracy over the full operational temperature range of −40°C to 125°C
- PMBus interface with programmable voltage, current limits, sequencing, margining, OV/UV thresholds, frequency synchronization and fault logging
- Telemetry readback parameters including V_{IN}, I_{IN}, V_{OUT}, I_{OUT}, temperature and faults
- External voltage dividers to set the chip address and default switching frequency and the output voltage
- Input current sensing and inductor DCR autocalibration

1.8V/30A single phase digital power supply with I_{IN} sense

Figure 1.1 shows a 7V to 14V input, 1.8V/30A output application that features inductor DCR current sensing. To improve the accuracy of the DCR current sense, the LTC3883 senses inductor temperature and compensates for the temperature coefficient of the winding resistance.

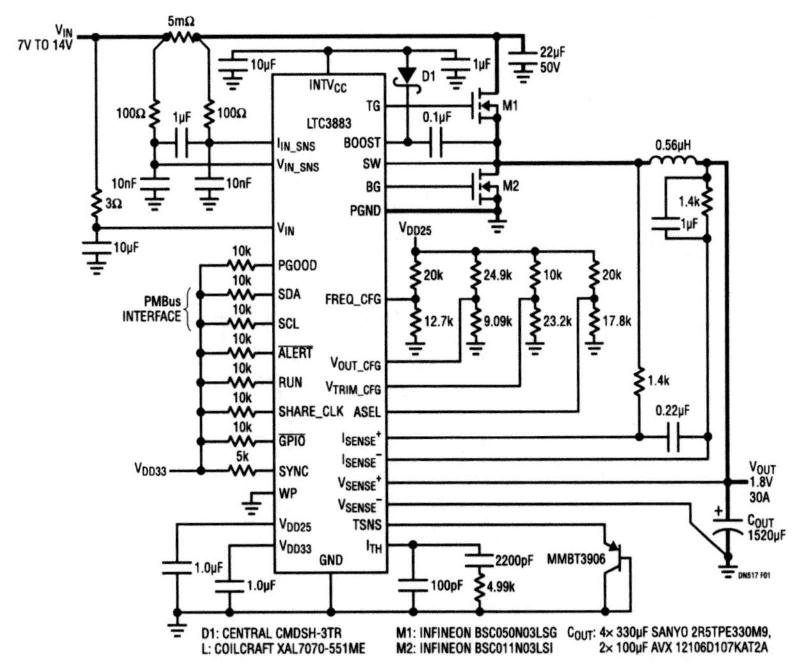

Figure 1.1 • 1.8V/30A Single Phase Digital Power Supply with I_{IN} Sense

Analog Circuit Design: Design Note Collection. http://dx.doi.org/10.1016/B978-0-12-800001-4.00001-6

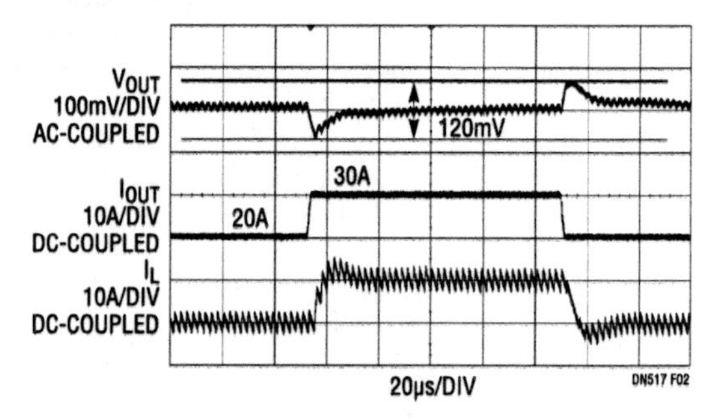

Figure 1.2 • Transient Performance for a 10A Load Step

This method ensures the accuracy of the readback current and overcurrent limit.

The LTC3883's control loop uses peak current mode control to achieve fast transient response and cycle-by-cycle current limit. Figure 1.2 shows the typical waveforms for a 10A load step transient, resulting in only a 60mV maximum deviation from nominal.

Input current sensing

The LTC3883 features input current sensing via a resistor in series with the input side of the buck converter—a 5mΩ sense resistor, as shown in Figure 1.1. The sense voltage is translated into a power stage input current by the LTC3883's 16-bit internal ADC. In addition, an internal IC sense resistor senses the chip's supply current at V_{IN}, so it can provide both the chip and the power stage's input current measurements.

Inductor DCR autocalibration

The problem with conventional inductor DCR current sensing is that the tolerance of the DCR can be as large as ±10%, greatly limiting the current read back accuracy. To solve this problem, the LTC3883 uses a proprietary inductor DCR autocalibration function. Figure 1.3 shows the simplified diagram of this circuit.

The LTC3883 accurately measures the input current (I_{IN}), the duty cycle (D) and the current sense voltage (V_{CS}) and calibrates the real DCR value based on the relation:

$$DCR_{CALIBRATED} = V_{CS} \cdot \frac{D}{I_{IN}}$$

With this autocalibration method, the output current read back accuracy is within 3% regardless of inductor DCR tolerance.

LTpowerPlay GUI

All power system management functions can be controlled by LTpowerPlay, a PC-based graphical user interface compatible with all of Linear's power system management products. With LTpowerPlay, designers can easily program and control the entire power system without writing a line of code. With this tool, it is easy to configure any chip on the bus, verify system status, read the telemetry, check fault status and control supply sequencing. LTpowerPlay can be downloaded at www.LTpowerPlay.com.

Conclusion

The LTC3883 combines a best in class analog DC/DC controller with complete power system management functions and precision data converters for unprecedented performance and control. Multiple LTC3883s can be used with other Linear Technology PMBus products to optimize multirail power systems. The powerful LTpowerPlay software simplifies the development of complex power systems. If dual outputs are needed, use the LTC3880 which shares common power system management features.

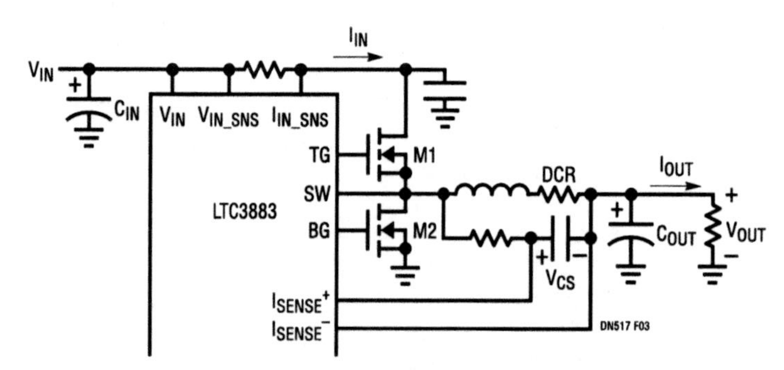

Figure 1.3 • DCR Autocalibration

One device replaces battery charger, pushbutton controller, LED driver and voltage regulator ICs in portable electronics

2

Marty Merchant

Introduction

The LTC3577/LTC3577-1 integrates a number of portable device power management functions into one IC, reducing complexity, cost and board area in handheld devices. The major functions include:

- Five voltage regulators to power memory, I/O, PLL, CODEC, DSP or a touch-screen controller
- A battery charger and PowerPath manager
- An LED driver for backlighting an LCD display, keypad and/or buttons
- Pushbutton control for debouncing the on/off button, supply sequencing and allowing end-users to force a hard reset when the microcontroller is not responding

By combining these functions, the LTC3577/LTC3577-1 does more than just reduce the number of required ICs; it solves the problems of functional interoperability—where otherwise separate features operate together for improved end-product performance. For instance, when the power input is from USB, the limited input current is logically distributed among the power supply outputs and the battery charger.

The LTC3577/LTC3577-1 offers other important features, including PowerPath control with instant-on operation, input overvoltage protection for devices that operate in harsh environments and adjustable slew rates on the switching supplies, making it possible to reduce EMI while optimizing efficiency. The LTC3577-1 features a 4.1V battery float voltage for improved battery cycle life and additional high temperature safety margin, while the LTC3577 includes a standard 4.2V battery float voltage for maximum battery run time.

Pushbutton control

The built in pushbutton control circuitry of the LTC3577/LTC3577-1 eliminates the need to debounce the pushbutton and includes power-up sequence functionality. A PB Status output indicates when the pushbutton is depressed, allowing the microprocessor to alter operation or begin the power-down sequence. Holding the pushbutton down for five seconds produces a hard reset. The hard reset shuts down the three bucks, the two LDOs and the LED driver, allowing the user to power down the device when the microprocessor is no longer responding.

Battery, USB, wall and high voltage input sources

The LTC3577/LTC3577-1 is designed to direct power from two power supply inputs and/or a Li-Ion/Polymer battery. The V_{BUS} input has selectable input current limit control,

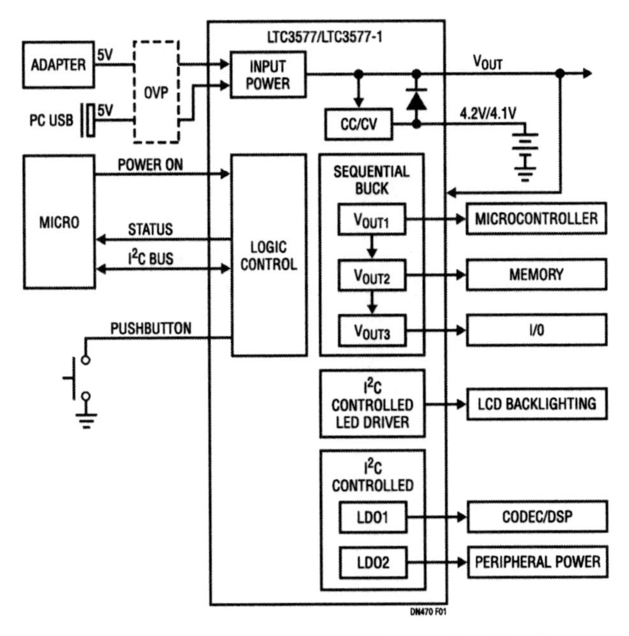

Figure 2.1 • Portable Device Power Distribution Block Diagram Featuring the LTC3577/LTC3577-1

Analog Circuit Design: Design Note Collection. http://dx.doi.org/10.1016/B978-0-12-800001-4.00002-8

designed to deliver 100mA or 500mA for USB applications, or 1A for higher power applications. A high power voltage source such as a 5V supply can be connected via an externally controlled FET. The voltage control (V_C) pin can be used to regulate the output of a high voltage buck, such as the LT3480, LT3563 or LT3505, at a voltage slightly above the battery for optimal battery charger efficiency.

Figure 2.1 shows a system block diagram of the LTC3577/LTC3577-1. An overvoltage protection circuit enables one or both of the input supplies to be protected against high voltage surges. The LTC3577/LTC3577-1 can provide power from a 4.2V/4.1V Li-Ion/Polymer battery when no other power is available or when the V_{BUS} input current limit has been exceeded.

Battery charger

The LTC3577/LTC3577-1 battery charger can provide a charge current up to 1.5A via V_{BUS} or wall adapter when available. The charger also has an automatic recharge and a trickle charge function. The battery charge/no-charge status, plus the NTC status, can be read via the I^2C bus. Since Li-Ion/Polymer batteries quickly lose capacity when both hot and fully charged, the LTC3577/LTC3577-1 reduces the battery voltage when the battery heats up, extending battery life and improving safety.

Three bucks, two LDOs and a boost/LED driver

The LTC3577/LTC3577-1 contains five resistor-adjustable step-down regulators: two bucks, which can provide up to 500mA each, a third buck, which can provide up to 800mA, and two LDO regulators, which provide up to 150mA each and are enabled via the I^2C interface. Individual LDO supply inputs allow the regulators to be connected to low voltage buck regulator outputs to improve efficiency. All regulators are capable of low-voltage operation, adjustable down to 0.8V.

The three buck regulators are sequenced at power-up (V_{OUT1}, V_{OUT2} then V_{OUT3}) via the pushbutton controller or via a static input pin. Each buck can be individually selected to run in Burst Mode operation to optimize efficiency or pulse-skipping mode for lower output ripple at light loads. A patented switching slew rate control feature, set via the I^2C interface, allows the reduction of EMI noise in exchange for efficiency.

The LTC3577/LTC3577-1 LED boost driver can be used to drive up to ten series white LEDs at up to 25mA or be configured as a constant voltage boost converter. As an LED driver, the current is controlled by a 6-bit, 60dB logarithmic DAC, which can be further reduced via internal PWM control. The LED current smoothly ramps up and down at one of four different rates. Overvoltage protection prevents the internal power transistor from damage if an open circuit fault occurs. Alternatively, the LED boost driver can be configured as a fixed voltage boost, providing up to 0.75W at 36V.

Many circuits require a dual polarity voltage to bias op amps or other analog devices. A simple charge pump circuit, as shown in Figure 2.2, can be added to the boost converter switch node to provide a dual polarity supply. Two forward diodes are used to account for the two diode voltage drops in the inverting charge pump circuit and provide the best cross-regulation. For circuits where cross-regulation is not important, or with relatively light negative loads, using a single forward diode for the boost circuit provides the best efficiency.

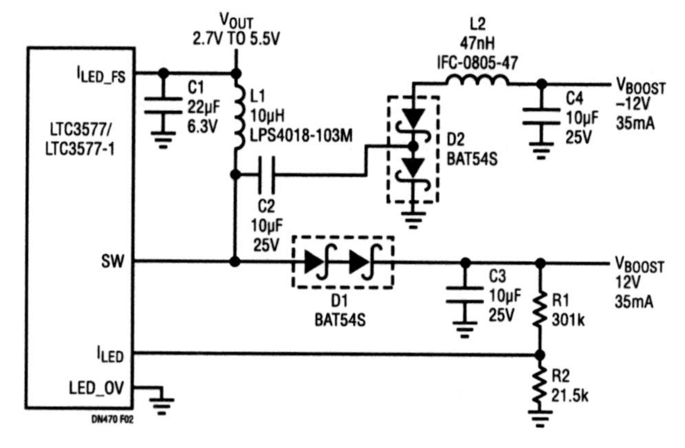

Figure 2.2 • Dual Polarity Boost Converter

Conclusion

The high level of integration of the LTC3577/LTC3577-1 reduces the number of components, required board real estate and overall cost; and greatly simplifies design by solving a number of complex power flow logic and control problems.

Simple circuit replaces and improves on power modules at less than half the price

3

Goran Perica

Introduction

Prefabricated power modules offer a simple but expensive solution to the complicated problem of efficiently converting high distribution voltages to the isolated low voltages required by digital ICs. Build-your-own solutions are much cheaper, but time constraints may drive even seasoned power supply designers to the power module shelf. The LT1952 changes that. It simplifies the design of a high performance synchronous single switch forward converter enough to make it a feasible and straightforward cost-cutting alternative to a power module.

A synchronous forward converter is a good choice for output voltages in the 1.2V to 26V range. In the forward converter family, among the several topologies that can achieve high efficiency, the single switch topology is the most cost-effective. It requires the least number of power MOSFETs and associated components. Until now, though, the single-switch forward converter required very complex PWM control circuitry, relegating it to the realm of power module makers.

In order for a single switch circuit to operate reliably, the PWM controller must provide soft-start, overcurrent protection and a volt-second clamp that ensures transformer

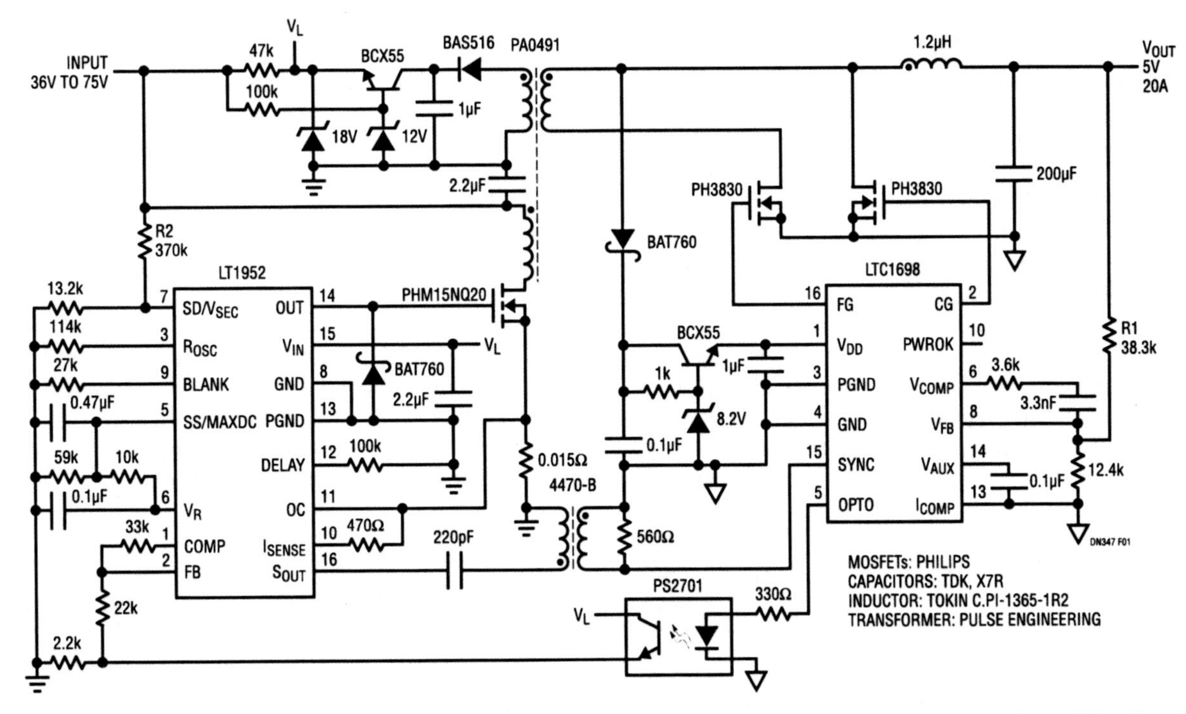

Figure 3.1 • High Performance 100W Forward Converter Replaces Power Modules at a Fraction of the Cost. This Circuit Is Easy to Modify for Any Output Voltage from 1.2V to 26V

Analog Circuit Design: Design Note Collection. http://dx.doi.org/10.1016/B978-0-12-800001-4.00003-X

core reset. Furthermore, the PWM controller must provide proper timing control for synchronous output rectification. The LT1952 PWM controller provides all of these functions and allows them to be adjusted for optimal performance. Additional adjustable features include undervoltage lockout (UVLO), leading edge blanking, current slope compensation and hysteretic start-up.

Perhaps the most important feature of the LT1952 controller is the programmable *adaptive* volt-second clamp. This function alone makes the LT1952 the easiest way to create a single switch forward converter. Single switch forward converters require a period of time when the primary MOSFET is turned off in order to reset the magnetic field inside the power transformer. Insufficient MOSFET off-time can saturate the transformer and lead to MOSFET failure. The LT1952 ensures that there is a sufficient amount of off-time for the transformer to reset under all input voltage conditions. This significantly reduces circuit size and cost as the transformer and MOSFET need not be oversized, as would be required without the *adaptive* volt-second clamp.

Some PWM controllers provide a fixed maximum duty cycle clamp for transformer reset. This method does not account for the increased energy left in the transformer at high input voltages during transients. Therefore, a fixed duty cycle clamp requires a larger, and more expensive, transformer to be used in order to withstand the higher volt-second product at high input voltage during transients. Furthermore, higher voltage MOSFETs have to be used in order to provide the appropriate reset voltage at high input voltage.

100W isolated synchronous forward converter in an eighth brick footprint

Figure 3.1 shows an example of a synchronous forward converter based on the LT1952. The converter in Figure 3.1 generates an isolated 5V, 20A output with a single set of power MOSFETs. Despite the simplicity of this circuit, performance is superior to most modules. Thanks to synchronous rectification and proper synchronous MOSFET timing, the efficiency is up to 93% (at 20A) and 95% (at 12A) as shown in Figure 3.2.

The converter in Figure 3.1 operates at 300kHz, utilizes only off-the-shelf components, fits into an eighth brick footprint area and has a bill of materials cost which is less than half of an equivalent power module.

Due to current mode operation, a fast feedback loop, and programmable current slope compensation, the converter uses ceramic output capacitors to achieve excellent transient response characteristics while requiring the smallest amount of output capacitance. Ceramic capacitors also ensure high reliability.

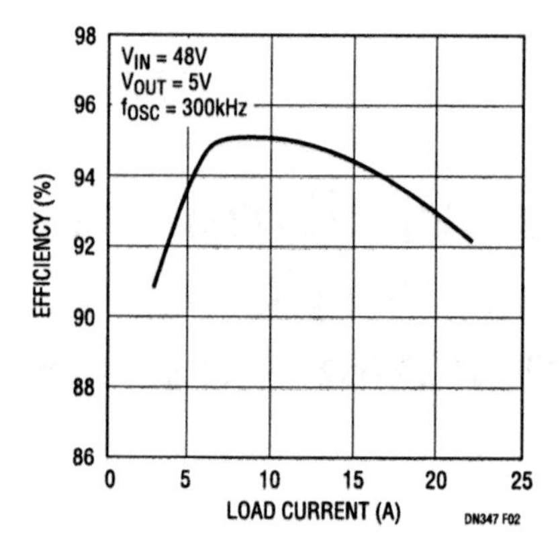

Figure 3.2 • Efficiency Is Very High for the Circuit in Figure 3.1

This circuit is flexible

The circuit in Figure 3.1 can be easily modified to fit a wide range of applications. For example, to change the output to 3.3V at 20A, simply change R1 to 21.5k. If higher current is required, add more MOSFETs in parallel and use a bigger transformer and output inductor. If lower power is required, smaller power components can be used. For lower input voltages use a lower voltage primary MOSFET, lower current sense resistor value, proportionally reduce the transformer turns ratio and lower the value of R2. Conversely, for higher input voltages, increase these values.

This circuit can also be modified to make a 12V, 20A bus converter. The accurate volt-second clamp of the LT1952 allows it to produce a more tightly regulated output over a wider input range than other bus converters. The LT1952 can regulate a 2:1 input voltage range to less than 10% regulation at the output. Because of the volt-second clamp, the optocoupler can be removed, and the LTC1698 circuit in Figure 3.1 can be replaced with a smaller LTC3900 circuit. The performance of the 12V bus converter is excellent with full overcurrent protection, gentle soft-start and efficiencies over 95%.

Conclusion

The LT1952 provides improved performance and makes it possible to replace expensive power modules at a fraction of the cost. It offers a simple, scalable high performance solution that reduces the amount of power converter design experience required to produce a reliable power supply.

4

Wide input range, high efficiency DDR termination power supply achieves fast transient response

Wei Chen

Introduction

Today's complicated computing and communication systems demand high system memory bandwidths. The emerging standard is double data rate (DDR) memory because of its higher data rate and relatively low cost. A typical DDR memory system needs at least two main power supplies: V_{DD} for the I/O power and V_{TT} for the termination power. To ensure good signal quality and fast data rate, the termination power supply (V_{TT}) must always track the V_{DD} with $V_{TT}=V_{DD}/2$. V_{TT} can be as low as 0.6V. Since the termination resistors can carry current in either direction, the V_{TT} power supply must be able to both source and sink current while tracking the V_{DD} supply. A family of new termination/tracking controllers, including the LTC3717, LTC3718 and LTC3831, satisfy these DDR requirements.

Overview of the LTC3717

The LTC3717 is a No R_{SENSE} current mode synchronous buck tracking controller that senses inductor current via the $R_{DS(ON)}$ of the bottom FET, eliminating the sense resistor and associated power loss. The LTC3717 implements a unique constant on-time architecture with on-time programmed by the input voltage and output voltage. This scheme allows a fairly constant switching frequency while achieving an extremely fast load transient response. In addition, the controller's minimum on-time is less than 100ns, allowing for a very small duty cycle at a very high switching frequency, thus minimizing the size of the inductor and capacitors. This circuit is suitable for high step-down applications such as 20V input and 1.25V output.

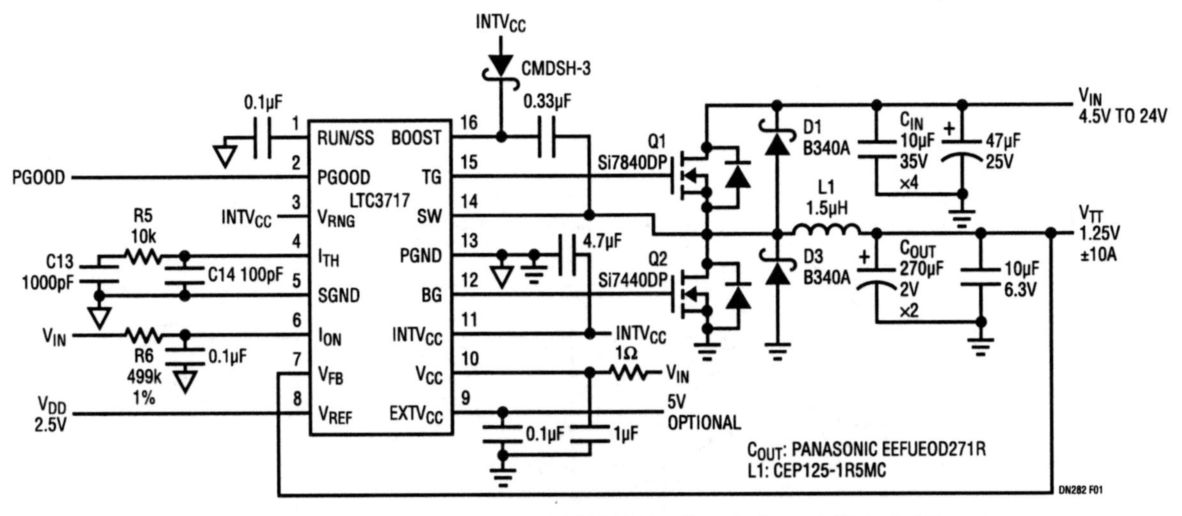

Figure 4.1 • High Efficiency ±10A LTC3717 V_{TT} Supply from 4.5V to 24V Input

Analog Circuit Design: Design Note Collection. http://dx.doi.org/10.1016/B978-0-12-800001-4.00004-1

The LTC3717 also includes a 5V internal LDO that can be used to drive an efficient logic-level power MOSFET. If an external 5V bias is available in the system, it can be applied to the EXTV$_{CC}$ pin to disable the internal 5V LDO and reduce the power loss of the controller at high input voltage. The reference tracking input of the LTC3717 is attenuated 50% internally, achieving a 0.65% regulation accuracy and eliminating the need for an external 1:1 resistor divider in the DDR termination power supply design.

Design example

Figure 4.1 shows a ±10A design using LTC3717. The input voltage can vary from 5V to 24V. The input voltage can be below 5V if an external 5V bias is available for powering the V$_{CC}$ pin of LTC3717. This design uses only two SO-8 PowerPak MOSFETs from Siliconix to deliver ±10A current. To achieve a higher output current, use an inductor with a higher current rating and lower $R_{DS(ON)}$ MOSFETs. This circuit achieves 84% efficiency at 250kHz switching frequency, 1.25V/10A output and 12V input, as shown in Figure 4.2.

The combination of the unique constant on-time current mode architecture of LTC3717 and the OPTI-LOOP compensation design produces an excellent load transient response. Figure 4.3 shows a typical load transient waveform. With only two SP output caps (270μF/2V), the output voltage variation is less than 100mV for a 10A load step (Figure 4.4).

Conclusion

The LTC3717 DDR termination power supply achieves high efficiency and fast load transient response for high input applications. If the available power sources are less than 5V, the LTC3718 can be used. The LTC3718 integrates an LTC3717 controller with a 1.2MHz boost regulator for the 5V MOSFET gate power. If the input voltage is between 3.3V and 8V, the LTC3831, a voltage mode synchronous buck tracking controller, can also be used.

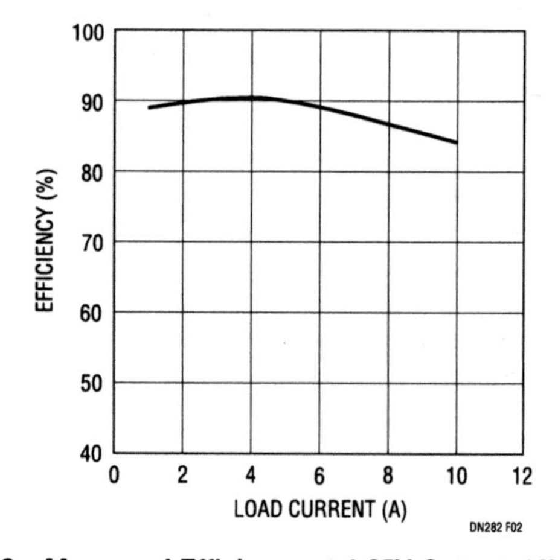

Figure 4.2 • Measured Efficiency at 1.25V Output, 12V Input

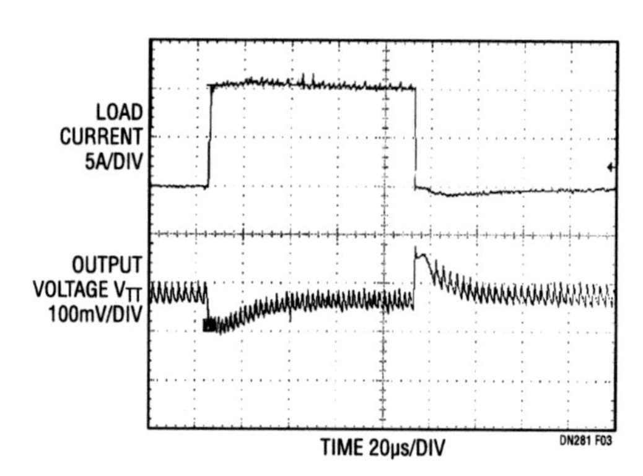

Figure 4.3 • Load Transient Waveforms

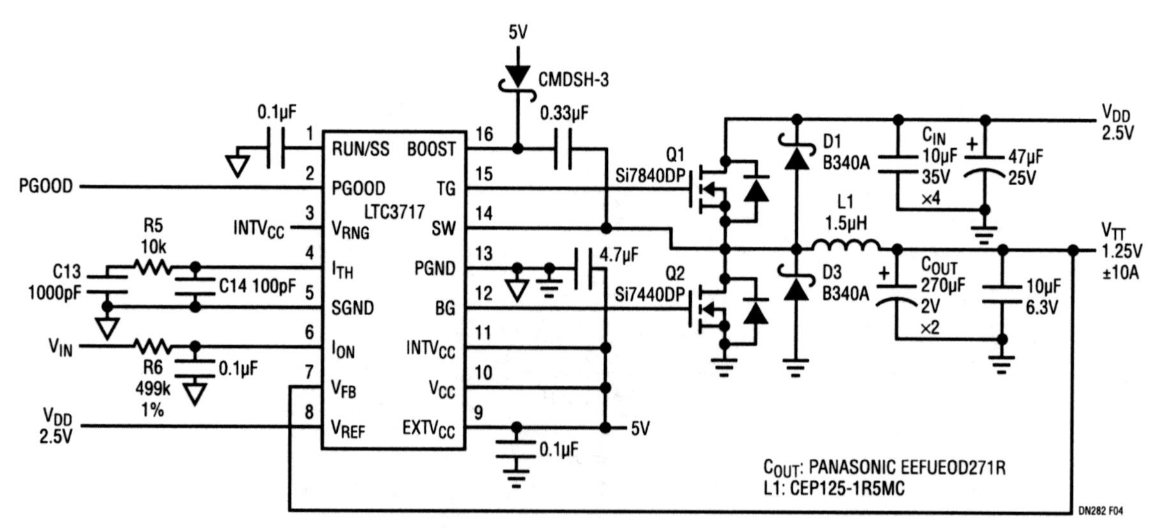

Figure 4.4 • High Efficiency ±10A LTC3717 V$_{TT}$ Supply from V$_{DD}$ Input

Minimize input capacitors in multioutput, high current power supplies

5

Wei Chen

Introduction

In broadband networking and high speed computing applications, multiple high current, low voltage power supplies are needed to power FPGAs, flash memory, DSPs and microprocessors. One such example calls for a maximum of 60A of current to power the CPU at 1.5V and up to 15A to power the memory at 2.5V from a 12V input. While a customized DC/DC module is usually expensive, the external circuitry for synchronization further increases the cost of individual supplies.

The LTC1628-SYNC PolyPhase controller can provide a simple and low cost solution. Compared to the LTC1628, the LTC1628-SYNC has a PLLIN pin that enables external synchronization. By combining the LTC1628-SYNC with the LTC1629, a true 3-phase circuit can be achieved for the 60A CPU supply and the second output of the LTC1628-SYNC circuit can be used to generate the memory power supply. Because the channel used for the memory power is interleaved out of phase from the other three channels used for CPU power, the net ripple current seen by the input bus is further reduced. In addition, the LTC1629's differential amplifier enables true remote sensing to ensure accurate voltage regulation at the CPU supply pins.

Design details

The block diagram and schematic diagrams are shown in Figures 5.1 and 5.2, respectively. With only twelve SO-8 MOSFETs (FDS7760A) and two SSOP-28 controllers, efficiencies of 85% and 88% are achieved for the 1.5V/60A and 2.5V/15A outputs, respectively.

Table 5.1 compares the input ripple current requirement of the PolyPhase design and a conventional single phase design. The PolyPhase technique reduces input capacitance by almost 60%.

Table 5.1 Comparison of Input Ripple Current and Input Capacitors for Single Phase and PolyPhase Configurations

PHASES	WORST-CASE INPUT RIPPLE CURRENT (A_{RMS})	NUMBER OF INPUT CAPACITORS: OS-CON 16SP270M AT 65°C
Single Phase	23.4	7
PolyPhase (LTC1629+LTC1628-SYNC)	10.2	3

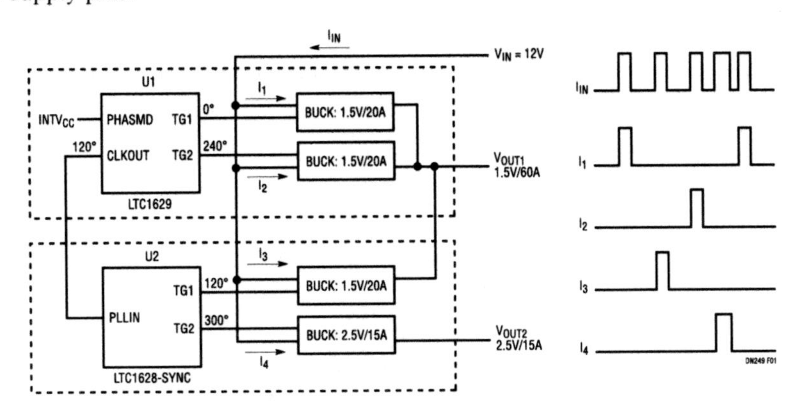

Figure 5.1 • Block Diagram of 2-Output Power Supplies

Analog Circuit Design: Design Note Collection. http://dx.doi.org/10.1016/B978-0-12-800001-4.00005-3

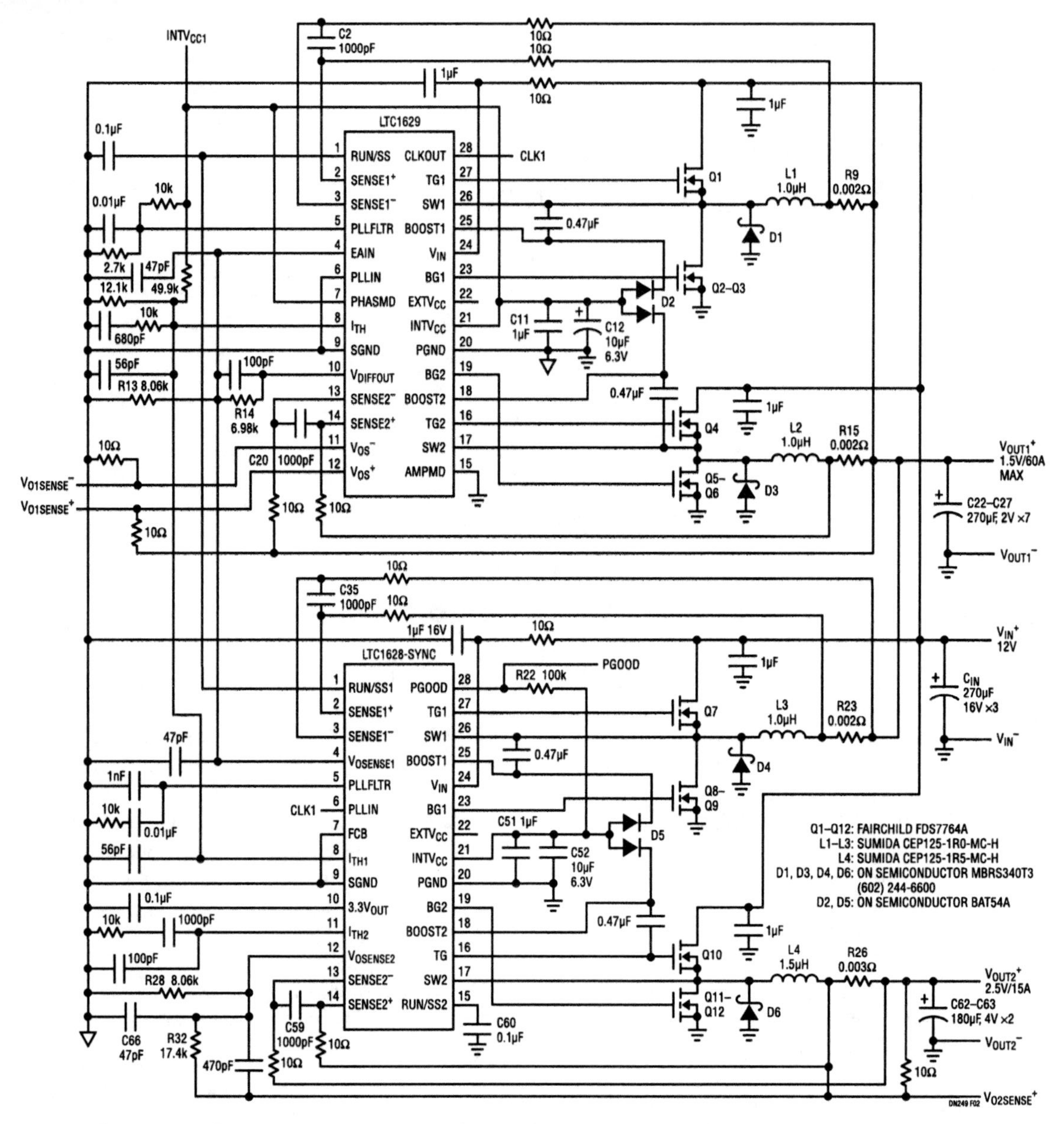

Figure 5.2 • Schematic Diagram of 2-Output Circuit: Input 12V; Outputs 1.5V/60A, 2.5V/15A

Conclusion

The synchronization capability of the LTC1628-SYNC helps reduce input capacitor requirements and prevents beat frequencies at the input bus. Combined with the LTC1629, it can effectively provide a 3-phase solution for multiple output applications and minimize the size and cost of the complete power supply.

For applications with more than two outputs, several LTC1628-SYNCs can be combined with the LTC1629 for multiphase operation. Refer to LTC1628-SYNC data sheet for more information.

Dual phase high efficiency mobile CPU power supply minimizes size and thermal stress

6

Wei Chen

Introduction

The increasing demand for computing power in notebook computers has significantly increased CPU clock frequencies and supply currents. Forthcoming mobile CPUs call for core currents as high as 25A to handle sophisticated computing tasks. Traditional single phase solutions have difficulty delivering such a high current. When the CPU supply voltage (0.7V to 1.8V) is converted directly from a high voltage (21V max) adapter input, single phase MOSFET drivers are not strong enough to efficiently drive high current MOSFETs without dV/dt shoot-through problems. The resulting excessive power loss in the MOSFETs increases thermal stress close to the CPU and reduces the battery run time. The physical size of a high current (25A) inductor becomes unacceptably large and more low ESR output capacitors are required to handle larger load steps. Also, current crowding in PCB traces near the inductor pads raises reliability concerns. As a result, the single phase solution is inefficient, bulky and can cause long-term reliability issues.

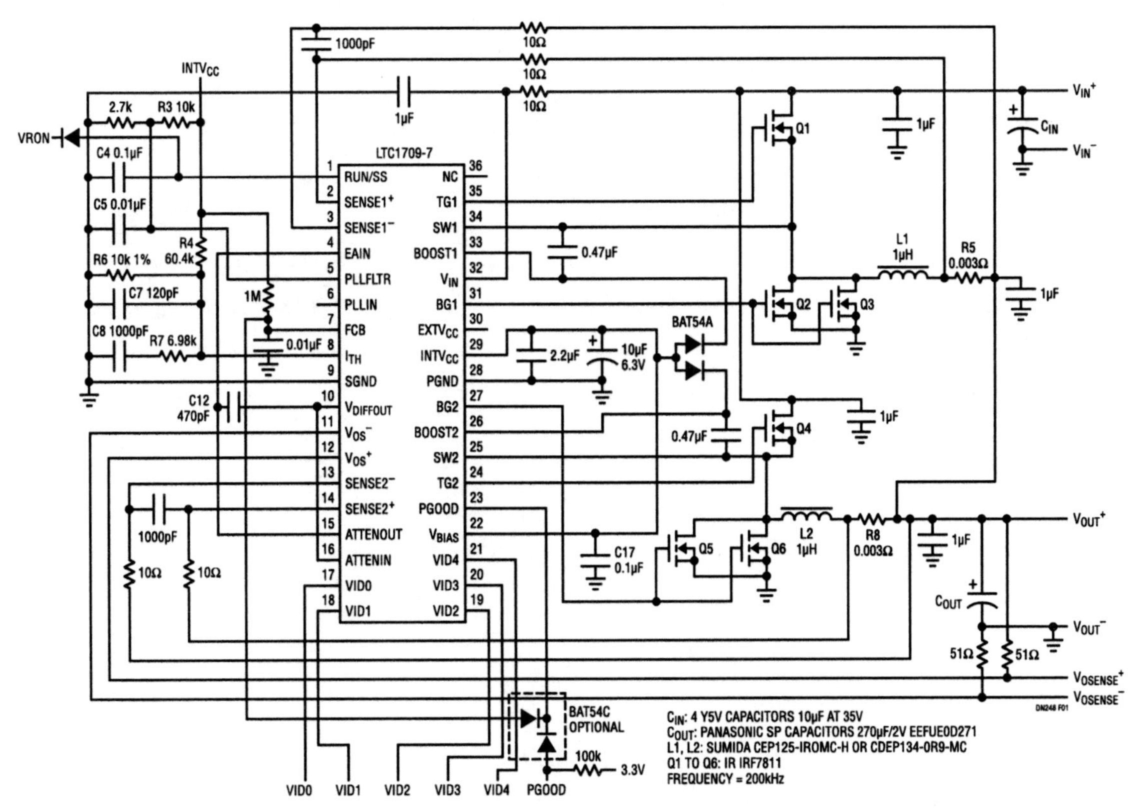

Figure 6.1 • Schematic Diagram of a 25A Mobile VRM

Analog Circuit Design: Design Note Collection. http://dx.doi.org/10.1016/B978-0-12-800001-4.00006-5

Dual phase switching is the best solution for this application. The LTC1709-7 is a dual phase controller that drives two synchronous buck stages 180 degrees out of phase, reducing input and output capacitors without increasing the switching frequency. The relatively low switching frequency and integrated high current MOSFET drivers help achieve high power conversion efficiency to maximize battery run time. Because of output ripple current cancellation, lower value, low profile inductors can be used, resulting in faster load-transient response and reduced component height. The LTC1709-7 also features discontinuous conduction mode and Burst Mode operation to minimize power losses when the CPU is in "sleep" mode. Because current is divided equally between two identical channels, heat is distributed uniformly and long-term PCB reliability is improved.

Design example

Figure 6.1 shows the schematic diagram of a 25A mobile CPU core power supply. With only one IC, six tiny SO-8 MOSFETs and two 1μH, low profile, surface mount inductors, an efficiency of about 85% is achieved for a 15V input and a 1.6V/25A output. Greater than 80% efficiency can be maintained throughout the load range from 5A to 25A.

Figure 6.2 shows the measured load transient waveform. The load current changes between 0A and 25A. The slew rate is 30A/μs. With only four SP capacitors (270μF/2V) at the output, the maximum output voltage variation during the load transient is less than 190mV$_{P-P}$. Resistors R4 and R6 provide the active voltage positioning with no loss in efficiency. Without the active voltage positioning, three more SP capacitors would be needed.

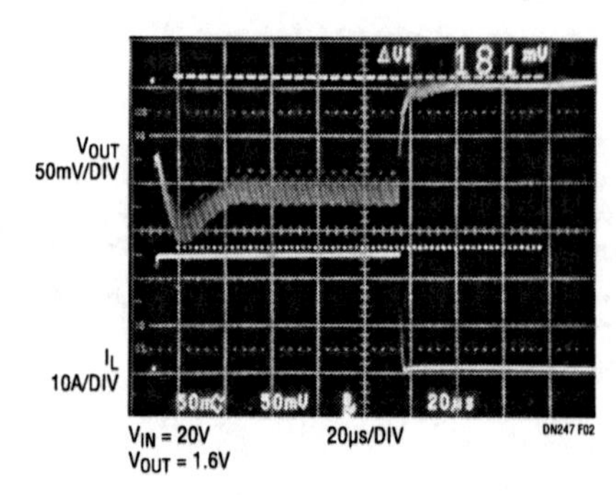

Figure 6.2 • Load Transient Waveforms at 25A Step and 30A/μs Slew Rate

Table 6.1 compares the performance and key component selections of the single phase and dual phase designs. The dual phase technique saves two 270μF SP output capacitors and two 10μF ceramic input capacitors. With the same number of MOSFETs and the same switching frequencies, the dual phase solution achieves much better efficiencies. The higher efficiency in the dual phase circuit, with more uniform current distribution, dramatically reduces the temperature rises in the MOSFETs and inductors.

Conclusion

Compared to the conventional single phase solution, the LTC1709-7 based, dual phase mobile VRM achieves higher efficiency, smaller size and lower solution cost. The dual phase solution extends battery life, minimizes thermal stress and improves long-term reliability.

Table 6.1 Comparisons of Single-Phase and Dual-Phase Solutions (Switching Frequency is 200kHz)

		SINGLE PHASE	DUAL PHASE
MOSFETs: IRF7811		Six (2 for Top, 4 for Bottom)	Six (per Phase, 1 for Top, 2 for Bottom)
Inductors and Sizes (L × W × H, Unit: mm)		One 1μH/25A (14.6 × 14.6 × 9)	Two 1μH/13A (12.5 × 12.5 × 4.9 Each)
Input Capacitors		Six 10μF/3V, Y5V Capacitors	Four 10μF/35V, Y5V Capacitors
Output Capacitors		Six SP Capacitors, 270μF/2V	Four SP Capacitors, 270μF/2V
Efficiency: $V_{IN}=20V$, $V_{OUT}=1.6V$, $I_{OUT}=25A$		80%	83%
Max Temperature*: $V_{IN}=21V$, $I_{OUT}=25A$, $V_{OUT}=1.6V$	Inductor	110°C	70°C
	MOSFETs	104°C	70°C

*Open air, after 20 minutes full-load operation. The temperatures are measured on the top surface of the components.

SOT-23 SMBus fan speed controller extends battery life and reduces noise

7

David Canny

Introduction

Battery run times for notebook computers and other portable devices can be improved and acoustic noise reduced by using Linear Technology's LTC1695 to optimize the operation of these products' internal cooling fans. The LTC1695 comes in a SOT-23 package and provides all the functions necessary for a system controller or microcontroller to regulate the speed of a typical 5V/≤1W fan via a 2-wire SMBus interface.

By varying the fan speed according to the system's instantaneous cooling requirements, the power consumption of the cooling fan is reduced and battery run times are improved. Acoustic noise is practically eliminated by operating the fan below maximum speed when the thermal environment permits. Designers also have the option of controlling the temperature in portable devices by using feedback from a temperature sensor to control the fan speed.

Figure 7.1 shows a typical application. Fan speed is easily programmed by sending a 6-bit digital code to the LTC1695 via the SMBus. This code is converted into an analog reference voltage that is used to regulate the output voltage of the LTC1695's internal linear regulator. The system controller can enable an optional boost feature that eliminates fan start-up problems by outputting 5V to the fan for 250ms before lowering the output voltage to its programmed value. Another important feature is that the system controller can read overcurrent and overtemperature fault conditions from information stored in the LTC1695. The part's SMBus address is hard-wired internally as 1110 100 (MSB to LSB, A6 to A0) and the data code bits D0 to D6 are latched at the falling edge of the SMBus Data Acknowledge signal (D6 is a Boost-Start Enable bit and D5 to D0 translate to a linearly proportional output voltage, 00–3F hex = 0V–5V). The LTC1694, which also appears in Figure 7.1, is a dual SMBus accelerator/pull-up device that may be used in conjunction with the LTC1695. Table 7.1 lists some 5V brushless DC fans suitable for typical LT1695 fan speed control applications.

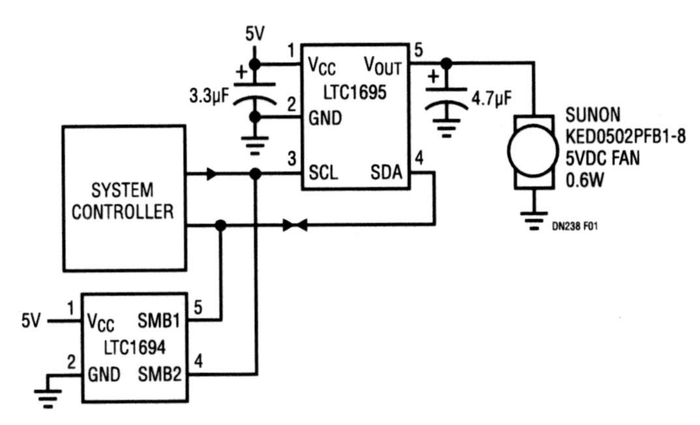

Figure 7.1 • SMBus Fan Speed Controller

Boost-start timer, thermal shutdown and overcurrent clamp features

A DC fan typically requires a starting voltage higher than its minimum stall voltage. For example, one Micronel 5V fan requires a 3.5V starting voltage, but once started, it will run until its terminal voltage drops below 2.1V (its stall voltage). Thus, the user must ensure that the fan starts up properly before programming the fan voltage to a value that is lower than the starting voltage. Monitoring the fan's DC current for stall conditions does not help because some fans consume almost the same amount of current at the same terminal voltage in both stalled and operating conditions. Another approach is to detect the absence of fan commutation ripple current. However, this is complex and requires customization for the characteristics of each brand of fan. The LTC1695 offers a simple and effective solution through the use of a boost-start timer. By setting the Boost-Start Enable bit high via the system controller, the LTC1695 outputs 5V for 250ms to the fan before lowering the voltage to its programmed value (see Figure 7.2 for the start-up voltage profile).

Analog Circuit Design: Design Note Collection. http://dx.doi.org/10.1016/B978-0-12-800001-4.00007-7

Table 7.1 Some 5V DC Fans' Characteristics

MANUFACTURER	PART NUMBER	AIRFLOW (CFM)	POWER (W)	SIZE (L • W • H) (mm³)
SUNON	KDE0501PFB2-8	0.65	0.50	20 • 20 • 10
ATC	AD0205HB-G51	0.80	0.45	25 • 25 • 10
SUNON	KDE0502PFB2-8	1.70	0.60	25 • 25 • 10
SUNON	KDE0503PFB2-8	3.20	0.60	30 • 30 • 10
SUNON	KDE0535PFB2-8	4.80	0.70	35 • 35 • 10
Micronel	F41MM-005XK-9	6.10	0.70	40 • 40 • 12

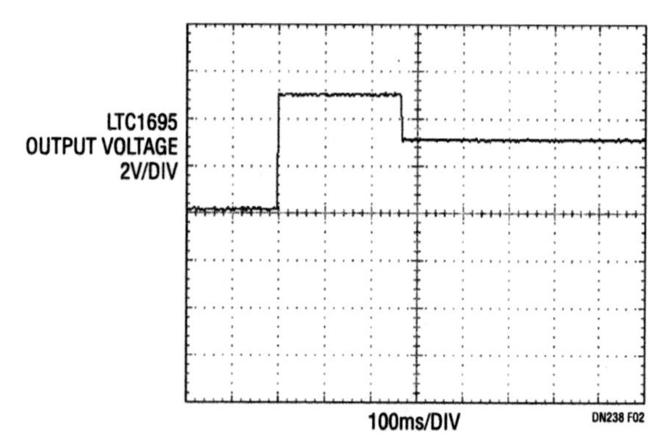

Figure 7.2 • Fan Start-Up Voltage Profile

During a system controller Read command, bits 6 and 7 in the data byte code are defined as the Thermal Shutdown Status (THE) and the Overcurrent Fault (OCF), respectively. The remaining bits of the data byte (0 to 5) are set low during host read back. The LTC1695 shuts down its PMOS pass transistor and sets the THE bit high if die junction temperature exceeds 155°C. During an overcurrent fault, the LTC1695's overcurrent detector sets the OCF bit high and actively clamps the output current to 390mA. This protects the LTC1695's PMOS pass transistor. Under dead short conditions ($V_{OUT}=0$), although the LTC1695 clamps the output current, the large amount of power dissipated on the chip will force the LTC1695 into thermal shutdown. These dual protection features protect both the IC and the fan, but more importantly, alert the host to system thermal management faults. During a fault condition, the SMBus logic continues to operate so that the host can poll the fault status data.

Conclusion

The LTC1695 improves battery run times and reduces acoustic noise in portable equipment. In addition, it provides important performance and protection features by controlling the operation of the equipment's cooling fan. It comes in a SOT-23 package and is easily programmed via the SMBus interface.

Active voltage positioning reduces output capacitors

8

Robert Sheehan

Introduction

Power supply performance, especially transient response, is key to meeting today's demands for low voltage, high current microprocessor power. In an effort to minimize the voltage deviation during a load step, a technique that has recently been named "active voltage positioning" is generating substantial interest and gaining popularity in the portable computer market. The benefits include lower peak-to-peak output voltage deviation for a given load step, without having to increase the output filter capacitance. Alternatively, the output filter capacitance can be reduced while maintaining the same peak-to-peak transient response.

Basic principle

The term "active voltage positioning" (AVP) refers to setting the power supply output voltage at a point that is dependent on the load current. At minimum load, the output voltage is set to a slightly higher than nominal level. At full load, the output voltage is set to a slightly lower than nominal level. Effectively, the DC load regulation is degraded, but the load transient voltage deviation will be significantly improved. This is not a new idea, and it has been observed and described in many articles. What is new is the application of this principle to solve the problem of transient response for microprocessor power. Let's look at some numbers to see how this works.

Assume a nominal 1.5V output capable of delivering 15A to the load, with a ±6% (±90mV) transient window. For the first case, consider a classic converter with perfect DC regulation. Use a 10A load step with a slew rate of 100A/μs. The initial voltage spike will be determined solely by the output capacitor's equivalent series resistance (ESR) and inductance (ESL). A bank of eight 470μF, 30mΩ, 3nH tantalum capacitors will have an ESR = 3.75mΩ and an ESL = 375pH. The initial voltage droop will be (3.75mΩ · 10A) + (375pH · 100A/μs) = 75mV. This leaves a 1% margin for set point accuracy. The voltage excursion will be seen in both directions, for the full load to minimum load transient and for the

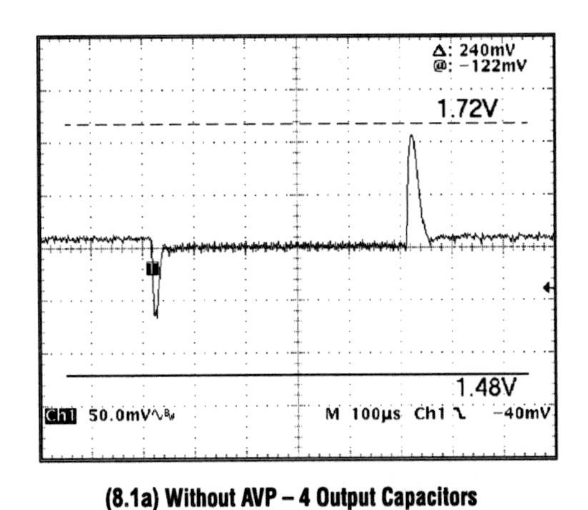

(8.1a) Without AVP – 4 Output Capacitors

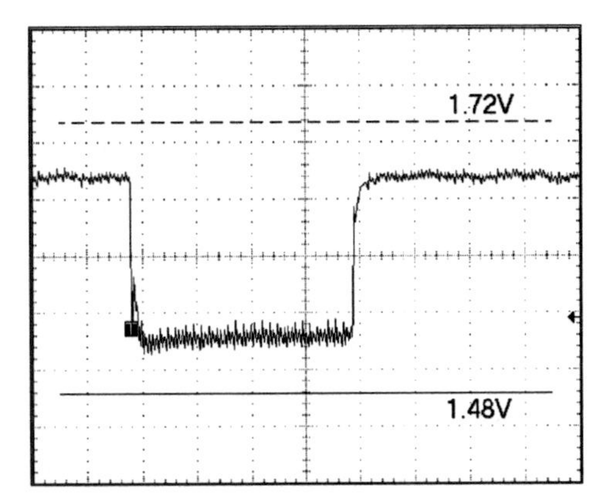

(8.1b) With AVP – 3 Output Capacitors

Figure 8.1 • Transient Response with Load Step from 0A to 12A

Analog Circuit Design: Design Note Collection. http://dx.doi.org/10.1016/B978-0-12-800001-4.00008-9

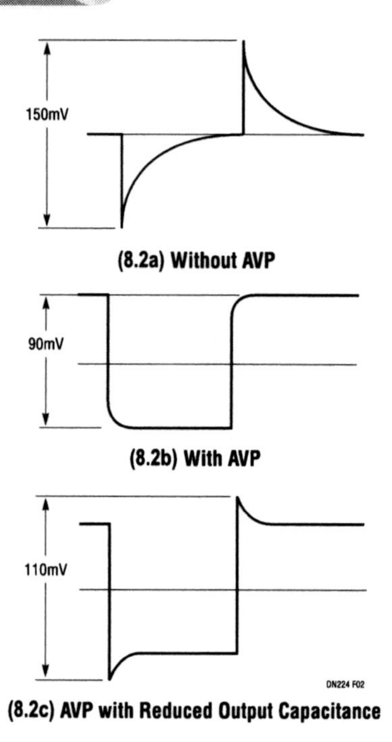

Figure 8.2 • Transient Response Comparison

minimum load to full load transient. The resulting deviation is $2 \cdot 75mV = 150mV$ peak-to-peak (Figure 8.2a).

Now look at the same transient using active voltage positioning. At the minimum load, purposefully set the output 3% (45mV) high. At full load, the output voltage will be set 3% low. During the minimum load to full load transient, the output voltage starts 45mV high, drops 75mV initially, and then settles to 45mV below nominal. For the full load to minimum load transient, the output voltage starts 45mV low, rises 75mV to 35mV above nominal, and settles to 45mV above nominal. The resulting deviation is now only $2 \cdot 45mV = 90mV$ peak-to-peak (Figure 8.2b). Now reduce the number of output capacitors from eight to six. The $ESR = 5m\Omega$ and $ESL = 500pH$. The transient voltage step is now $(5m\Omega \cdot 10A) + (500pH \cdot 100A/\mu s) = 100mV$. With the 45mV offset, the resultant change is ±55mV around center, or 110mV peak-to-peak (Figure 8.2c). The initial specification has been easily met with a 25% reduction in output capacitors.

An added benefit of voltage positioning is an incremental reduction in CPU power dissipation. With the output voltage set to 1.50V at 15A, the load power is 22.5W. By decreasing the output voltage to 1.47V, the load current is 14.7A and the load power is now 21.6W. The net saving is 0.9W.

Basic implementation

In order to implement voltage positioning, a method for sensing the load current is required. This information must then be used to move the output voltage in the correct direction. For a current mode controller, such as the LTC1736, a current sense resistor is already used. By controlling the error amplifier gain, we can achieve the desired result.

Current mode control example—LTC1736

Figure 8.3 shows the basic power stage and feedback compensation circuit for the LTC1736. In a non-AVP implementation, R_{A1} and R_{A2} are removed and C_C and R_C installed. The corresponding transient response with 20V input and 1.6V output is shown in Figure 8.1a. In order to implement voltage positioning, we will control the error amplifier gain at the I_{TH} pin. The internal g_m amplifier gain is equal to $g_m \cdot R_O$, where g_m is the transconductance in mmhos, and R_O is the output impedance in kohms. The voltage at the I_{TH} pin is proportional to the load current, where 0.48V = min load, 1.2V = half load, and 2V = full load in this application. $R_O = 600k\Omega$ and $g_m = 1.3mmho$. By setting a voltage divider to 1.2V from the 5V $INTV_{CC}$, the gain can be limited without affecting the nominal DC set point at half load. The Thevenin equivalent resistance is seen to be in parallel with the amplifiers R_O. Using the values shown in Figure 8.3 for R_{A1} and R_{A2}, the effective R_O will be $600k||91k||27k = 20.12k\Omega$. The voltage deviation at the amplifier input $\Delta V_{FB} = \Delta V_{ITH}/(g_m \cdot R_{Oeff})$. $\Delta V_{FB} = (2.0V - 0.48V)/(1.3mmho \cdot 20.12k\Omega) = 58mV$, which is ±29mV from the nominal half load set point.

Care should be taken to keep the amplifier input from being pulled more than ±30mV from its nominal value, or nonlinear behavior may result. The DC reference voltage at V_{FB} is 0.8V and V_{OUT} is set for 1.6V, so $\Delta V_{OUT} = 2 \cdot \Delta V_{FB} = 116mV$. The resulting transient response is shown in Figure 8.1b. The transient performance has been improved, while using fewer output capacitors.

The optimal amount of AVP offset is equal to $\Delta I \cdot ESR$. Figure 8.1b exhibits this condition.

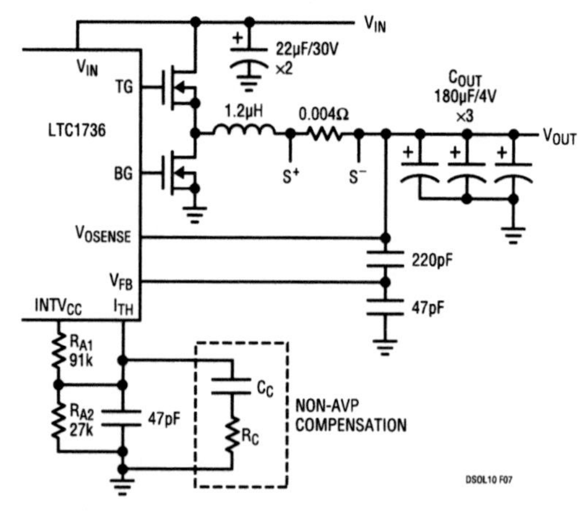

Figure 8.3 • LTC1736 with AVP

5V to 3.3V circuit collection

9

Richard Markell Craig Varga

High efficiency 3.3V regulator

The LTC1174-3.3 current mode DC/DC converter provides efficiencies better than 90% over a wide load current range while requiring only 1µA in shutdown.

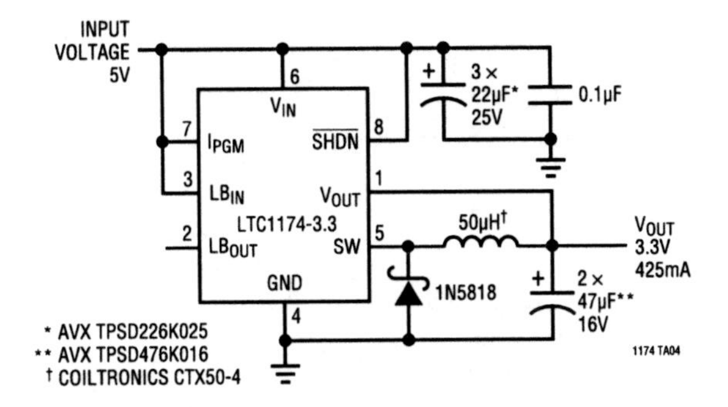

3.3V battery-powered supply with shutdown

The LT1121-3.3 low dropout linear regulator provides up to 150mA output current with 30µA quiescent current.

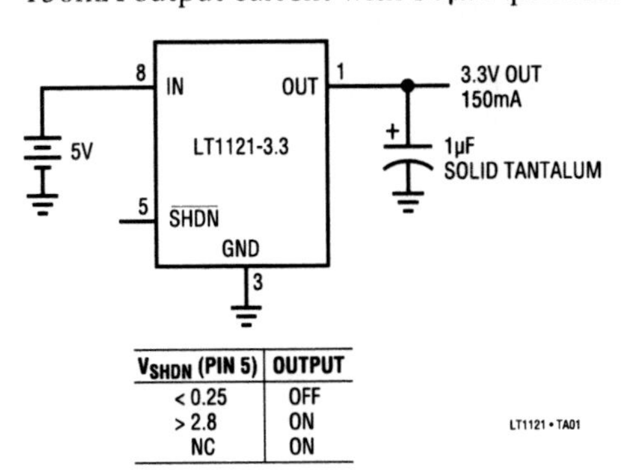

$V_{\overline{SHDN}}$ (PIN 5)	OUTPUT
< 0.25	OFF
> 2.8	ON
NC	ON

3.3V supply with shutdown

The LT1129-3.3 low dropout linear regulator provides more output current (to 700mA) with only a slight increase in quiescent current (50µA).

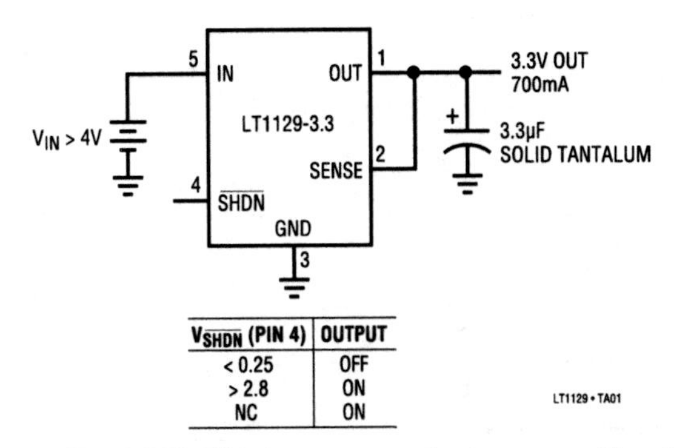

$V_{\overline{SHDN}}$ (PIN 4)	OUTPUT
< 0.25	OFF
> 2.8	ON
NC	ON

LT1585 linear regulator optimized for desktop Pentium processor applications

Linear regulator circuits provide simple solutions with superior transient performance for desktop resident Pentium processor-based systems.

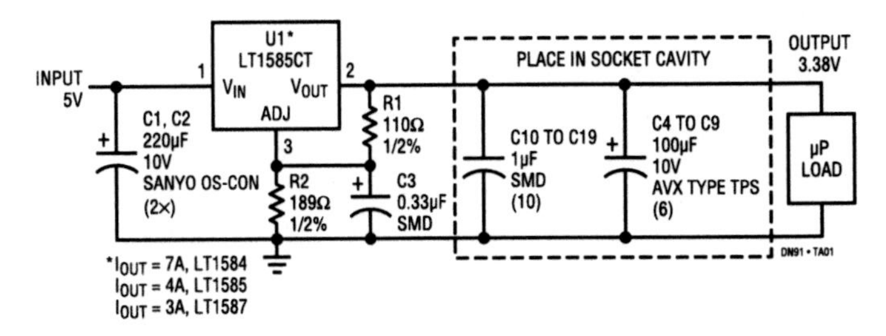

Analog Circuit Design: Design Note Collection. http://dx.doi.org/10.1016/B978-0-12-800001-4.00009-0

LTC1148 5V to 3.38V Pentium power solution 3.5A output current

This circuit achieves >90% efficiency using an LTC1148 synchronous switching regulator which consumes a mere 180μA quiescent current.

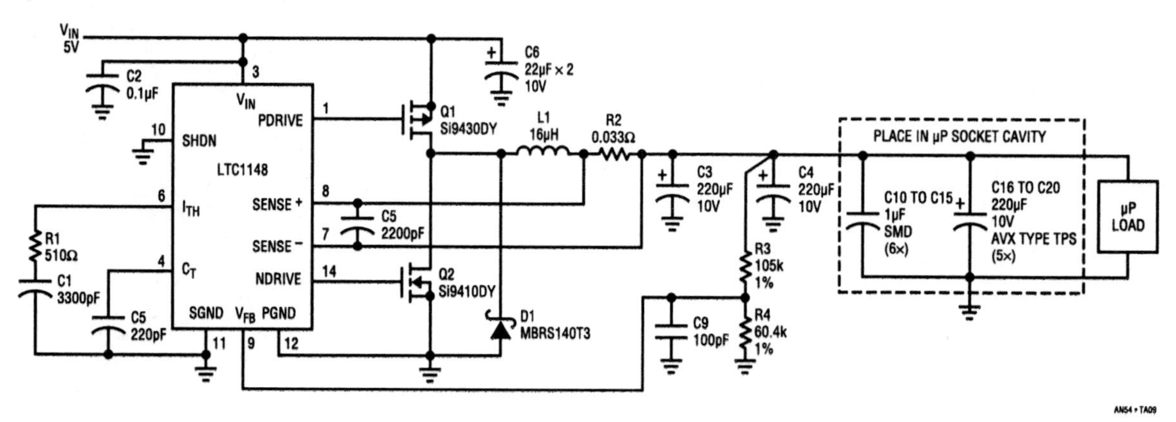

LTC1266 switching regulator converts 5V to 3.38V at 7A for Pentium and other high speed μPs

The LTC1266 drives N-channel MOSFETs directly and provides 7A output current at efficiencies greater than 90%.

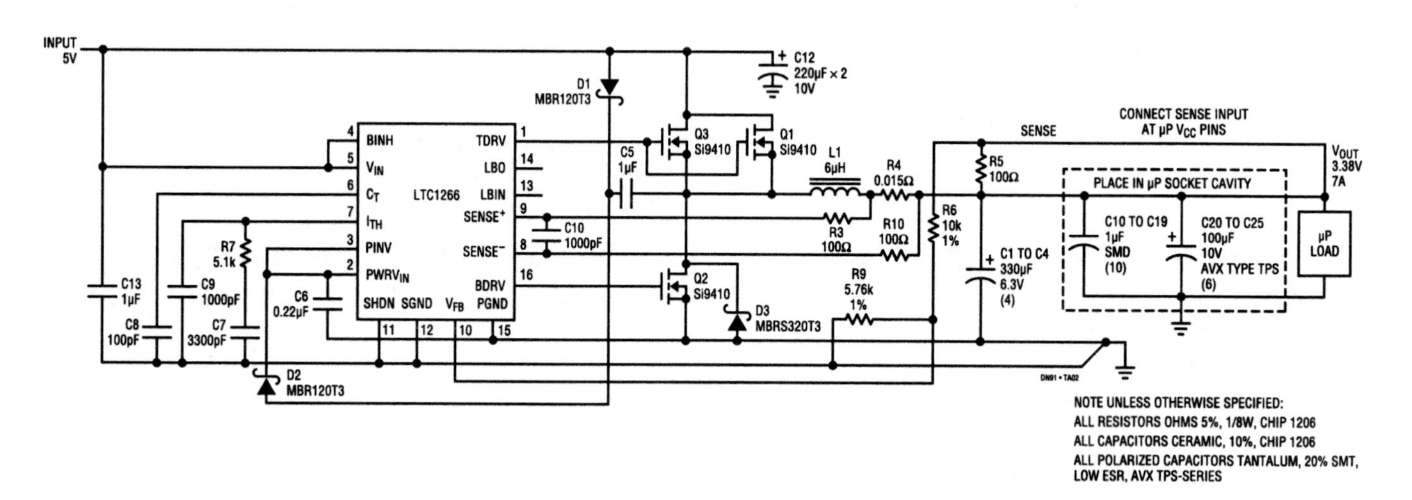

Hex level shift shrinks board space

10

Brian Huffman

Although simple in concept, interfacing digital levels between different logic families usually requires many parts and appreciable board space. Other applications that require some form of level shifting of the output swing have solutions just as complicated. A logic to CMOS analog switch (Figure 10.1) is just one example where a level shift must occur. A new device, the LTC1045, solves this and other level shift related problems conveniently. The LTC1045 is a hex level translator with a linear comparator on the front end. A latch and three-state output buffer are at the back. These features make it useful in other applications as a hex comparator or in interfacing to a data bus. Almost any input and output voltage requirements can be accommodated by simply setting the level of the appropriate power supply voltages.

The LTC1045 consists of six high speed comparators with output latches and three-state capability (see Figure 10.2). Each comparator's non-inverting input is brought out separately. The inverting inputs of comparators 1-4 are tied to V_{TRIP1} and 5-6 are tied to V_{TRIP2}. With these inputs the switching point of the comparators can be set anywhere within the common-mode range of V^- to ($V^+ - 2V$). There are four power supply pins on the LTC1045: V^+, V^-, V_{OH} and V_{OL}. V^+ and V^- power the comparator's front end, and V_{OH} and V_{OL} power the output drivers. Almost any combination of power supply voltages can be used. There are three restrictions: V_{OH} must be less than or equal to V^+; there must be a minimum differential voltage of 4.5V between V^+ and V^- and 3V between V_{OH} and V_{OL}; and the maximum voltage between any two pins must not exceed the 18V absolute maximum.

The supply current is programmed with an external resistor. The R_{SET} resistor allows trade-offs between speed and power consumption. The propagation delay, with the I_{SET} pin at V^- and a single 5V supply, is typically 100ns with a total supply current of 4.5mA. The quiescent current can be brought down to 100μA (15μA per comparator) with an R_{SET} of 1M and a propagation delay of only 1.2μs. In addition, the I_{SET} pin completely shuts off power and latches the translator output voltages. The DISABLE input sets the six outputs to a

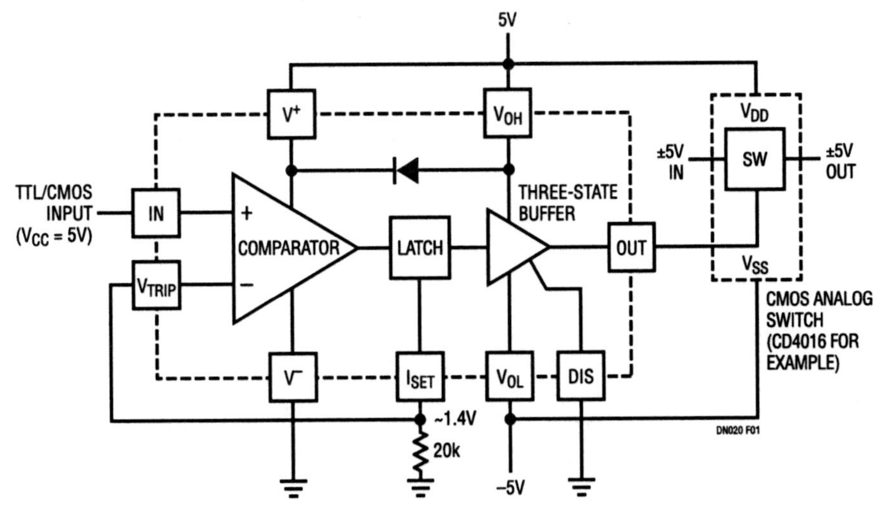

Figure 10.1 • TTL/CMOS Logic Levels to ±5V Analog Switch Driver

Analog Circuit Design: Design Note Collection. http://dx.doi.org/10.1016/B978-0-12-800001-4.00010-7

high impedance state allowing the LTC1045 to be interfaced to a data bus.

Figure 10.3 shows a simple way to build a battery-powered RS232 receiver. The input voltage may be driven ±30V without adverse effects because the 100k resistor prevents device damage. With a 1M R_{SET} the hex RS232 line receiver draws only 100μA of quiescent current and has a propagation delay of 1.2μs. Only a single supply is needed for operation.

Board space can be saved by using the LTC1045 level translator as a hex comparator—even though both comparator inputs are not available. Figure 10.4 shows the LTC1045 used as a power supply monitor. The outputs of three power supplies are tied to the positive inputs through an appropriate resistive voltage divider. The divider ratio is set so that the voltage into the comparator equals the reference on the inverting input when the power supply voltage is at a critical level.

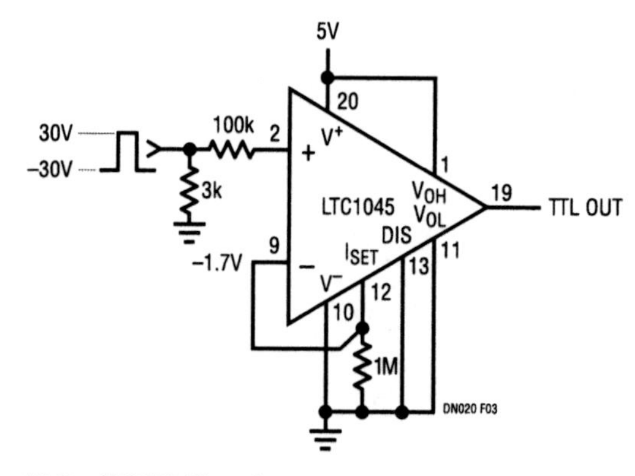

Figure 10.3 • RS232 Receiver

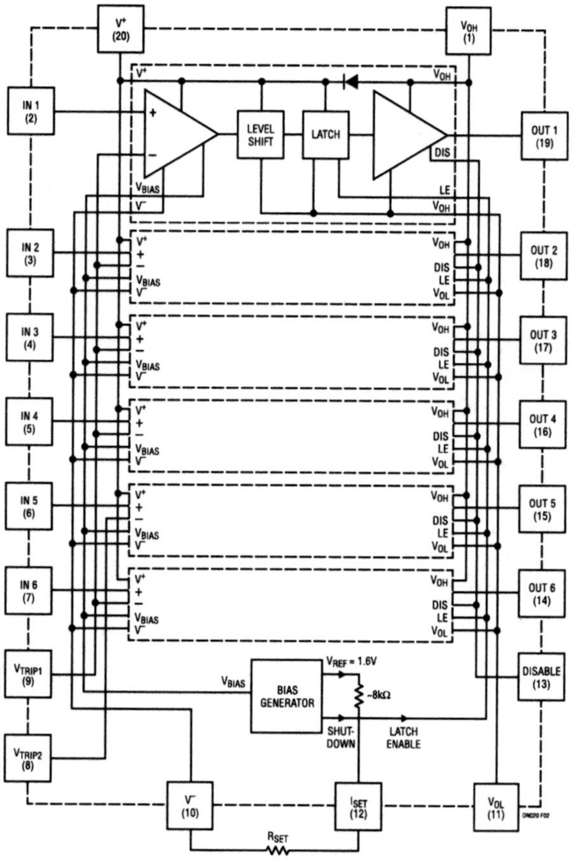

Figure 10.2 • LTC1045 Block Diagram

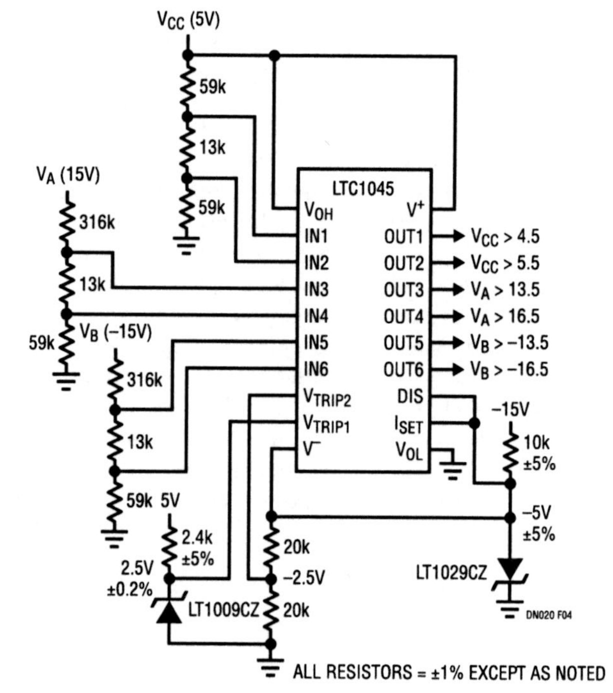

Figure 10.4 • Power Supply Monitor

Section 2

Microprocessor Power Design

Cost-effective, low profile, high efficiency 42A supply powers AMD Hammer processors

11

Henry J. Zhang   Wei Chen

Introduction

The new AMD Hammer processors require power supplies that can deliver more than 40A current at very low output voltages (0.8V to 1.55V). The LTC3719 PolyPhase controller meets these requirements. The LTC3719 is a current mode, 2-phase VID programmable controller that drives two synchronous buck stages 180 degrees out of phase of each other. The 2-phase architecture reduces the number of input and output capacitors without increasing the switching frequency. The relatively low switching frequency, along with integrated high current MOSFET drivers keeps efficiency high and solution size small.

Because of output current ripple cancellation, lower value inductors can be used, resulting in faster transient load responses and small inductor size. To further reduce the required input and output capacitance, the controller switching frequency can be synchronized to the external switching signal from another LTC PolyPhase controller, thus increasing the number of DC/DC converter phases. To ensure that current is equally shared among the output phases, the inductor current is sensed with current sensing resistors. Excellent current sharing eliminates the potential thermal overstress problem caused by unbalanced phase current. Therefore, the MOSFETs and inductors need not be oversized and the VRM can be operated reliably without heat sink.

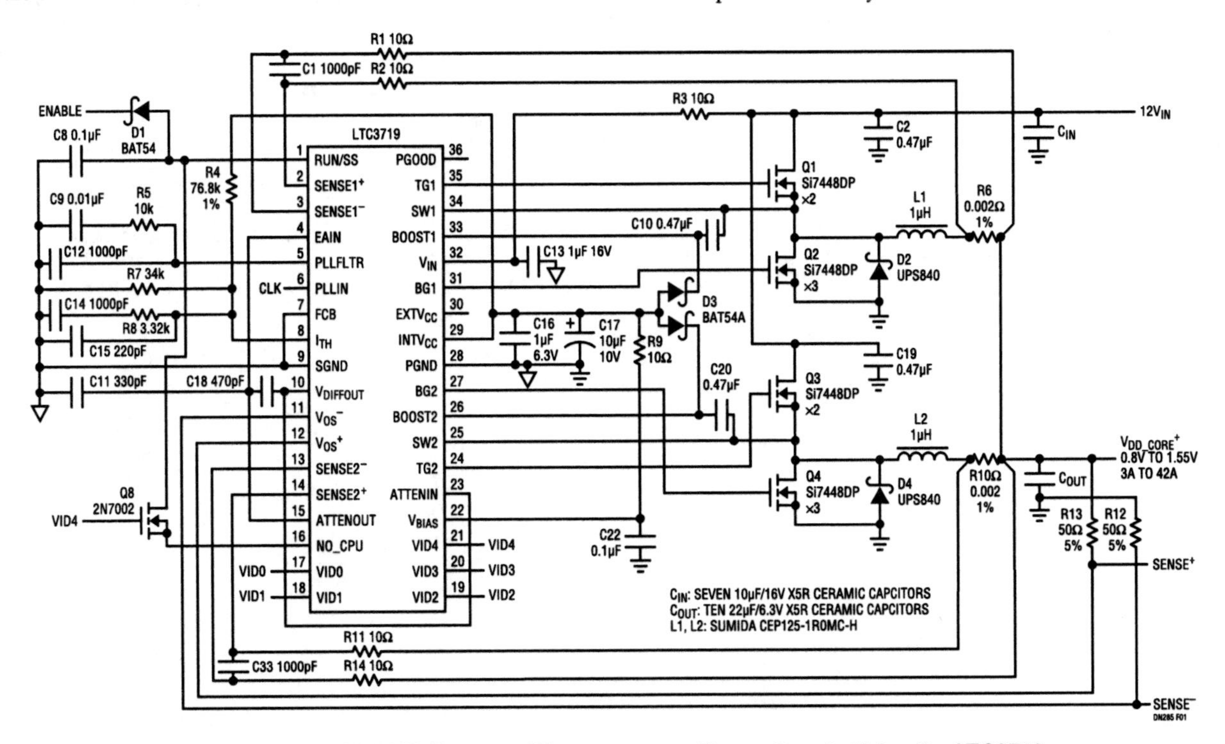

Figure 11.1 • 42A AMD Hammer Microprocessor Power Supply Using the LTC3719

Analog Circuit Design: Design Note Collection. http://dx.doi.org/10.1016/B978-0-12-800001-4.00011-9

An internal differential amplifier provides true remote sensing of the regulated supply's positive and negative output terminals at load point as required in high current applications. The internal 5-bit VID programmable attenuator complies with the AMD Hammer processor VID table (0.8V to 1.55V). The LTC3719 also incorporates two selectable light load operation modes: discontinuous conduction mode and Burst Mode operation to improve the light load efficiency. Power supply designs that use the LTC3719 are efficient, low profile and low cost.

Design example

Figure 11.1 shows a 2-phase 42A AMD Hammer processor power supply. With only one IC, ten tiny SO-8 size MOSFETs and two 1μH, low profile surface mount inductors, this supply is better than 86% efficient for a 12V input and a 1.55V/42A output. Figure 11.2 shows that efficiency remains high throughout a wide load range, from 3A to 42A. The highest components on the VRM are the output inductors which are less than 6mm high.

Figure 11.3 shows a measured load transient waveform. The load current step is 20A with a slew rate of 30A/μs. Output capacitor requirements are dominated by the total ESR of the output capacitor network on both the VRM and the transient load test circuit specified by the AMD Hammer Processor VRM Design Guide. Ten low profile ceramic capacitors (22μF/6.3V, X5R) are used on the VRM output. The maximum output voltage variation during the load transients is less than $100mV_{P-P}$. Active voltage positioning was employed in this design to reduce the number of output capacitors with no additional efficiency loss (refer to Design Solution 10 for more details on active voltage positioning).

Conclusion

The LTC3719, low voltage, high current power supply meets the AMD Hammer processor VRM requirement and allows low profile and high efficiency designs. The cost savings in the input and output capacitors, inductors and heat sinks help lower the cost of the overall power supply. For multiple CPU applications, several LTC1629-6s can be combined with an LTC3719 to increase the number of output phases, up to 12; the resulting supply current can be up to 200A. For more information about PolyPhase operation, see the LTC3719 data sheet and Application Note 77.

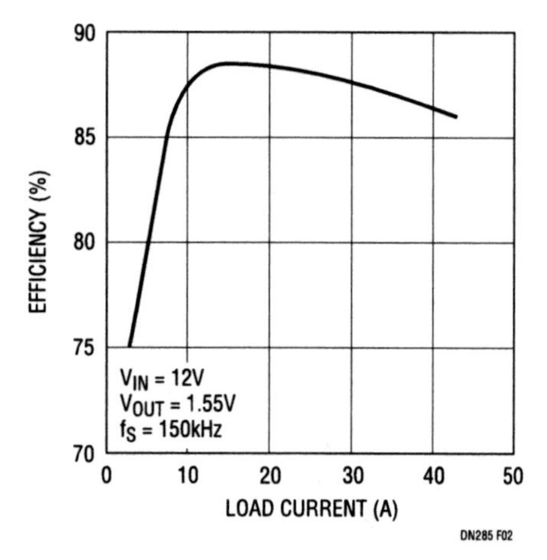

Figure 11.2 • Efficiency vs Load Current

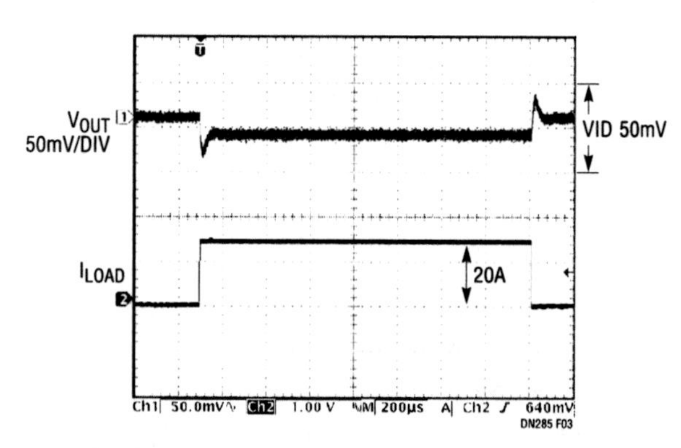

Figure 11.3 • Load Transient Waveforms for a 20A Step with a 30A/μs Slew Rate

Efficient, compact 2-phase power supply delivers 40A to Intel mobile CPUs

12

David Chen

Introduction

Notebook computers demand more power than ever before, and in turn, CPU supply current requirements have grown significantly to the point where the latest Intel mobile CPUs call for up to 40A. Such high currents reveal many design problems that are relatively insignificant at lower current levels. Today's notebook power supply designer must successfully manage the inevitable resulting power losses and thermal stresses within the already tight notebook space and still

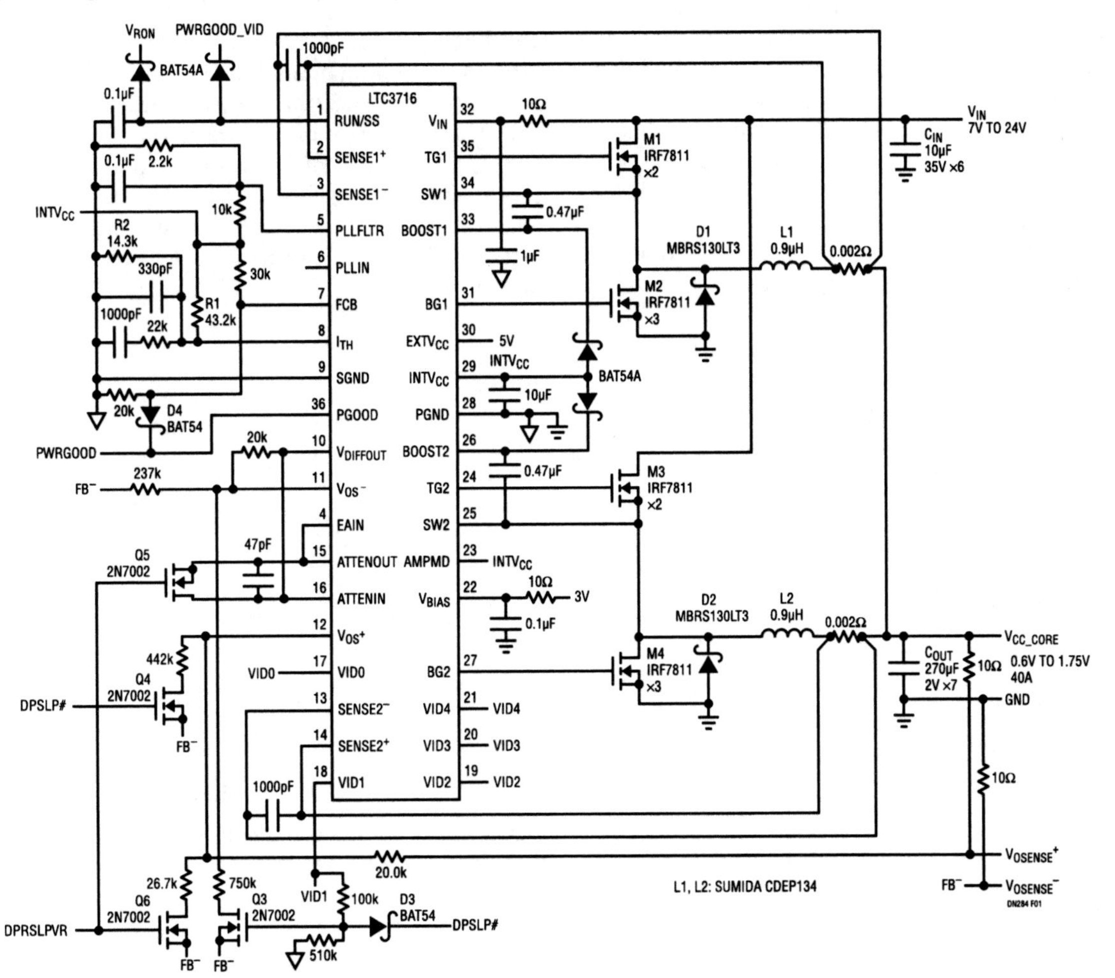

Figure 12.1 • 40A IMVP-III VR Power Supply Design for Intel Mobile Tualatin and Northwood Processors

Analog Circuit Design: Design Note Collection. http://dx.doi.org/10.1016/B978-0-12-800001-4.00012-0

maintain high efficiencies. The LTC3716 2-phase controller addresses these design issues by providing the following significant features in a compact package:

- Load current is distributed uniformly between two phases, reducing thermal stress on the PCB.
- Out-of-phase operation reduces both input and output capacitance requirements and improves overall efficiency.
- Integrated MOSFET gate drivers provide further space savings by precluding the need for separate driver ICs.

The VID table of the LTC3716 is compatible with the newest Intel IMVP specifications for Mobile Tualatin and Mobile Northwood processors.

Smaller inductors, simplified thermal management

The single phase, synchronous step-down regulator is the most common solution for low current CPU supplies. Nevertheless, when CPU current exceeds 20A, this simple topology is no longer suitable—the inductor is too large, power loss becomes excessive and the total solution cost becomes prohibitive.

A solution based on the LTC3716, on the other hand, allows for smaller footprint and lower profile inductors. Its high current gate drivers, each capable of driving up to three SO-8 MOSFETs, eliminate the need for separate driver ICs. With two phases sharing load current, the LTC3716 alleviates the current crowding problem that would occur in the single-phase solution. Heat is more uniformly distributed, simplifying thermal management. The out-of-phase operation of the LTC3716 decreases both input and output ripple currents, further improving efficiency.

Also, the transient performance of a 2-phase solution is superior to its single phase counterpart without sacrificing efficiency, due to two effects: the effective operating frequency is twice that of either phase and the equivalent inductance is half of either inductor.

Overall, 2-phase operation produces high current power supplies that are smaller, lower cost, more efficient and more reliable than single phase designs.

40A Intel IMVP-III voltage regulator

Figure 12.1 shows a 40A Intel IMVP-III voltage regulator design using the LTC3716. Q3, Q4 and Q6 provide the output offsets for battery mode, deep sleep mode and deeper sleep mode, respectively. D4 speeds up the high to low load transients while allowing Burst Mode operation, which maintains high efficiency at light loads. The resistor divider (R1/R2) implements the Intel Mobile Voltage Positioning (IMVP) requirement without additional power loss. Overall efficiency and transient waveforms are plotted in Figures 12.2 and 12.3, respectively. After 20 minutes of continuous full-load operation in an open air, 25°C ambient environment with no ventilation, both inductors measure 70°C and the MOSFETs measure 65°C.

Conclusion

The latest Intel mobile CPUs call for a supply current up to 40A. As an optimal combination of performance, cost and reliability, the LTC3716 2-phase solution delivers high current with high efficiency in a compact size. Other features of the LTC3716 include true current mode control, accurate current sharing, ±1% reference precision, adjustable soft-start, a wide (4V to 36V) input range and multiple circuit protection features. For more information, please refer to the LTC3716 data sheet.

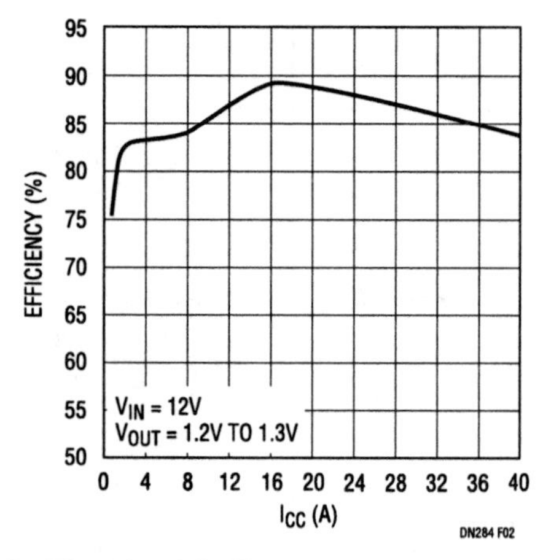

Figure 12.2 • The Circuit in Figure 12.1 Maintains High Efficiency Over a Wide Range of Load Currents

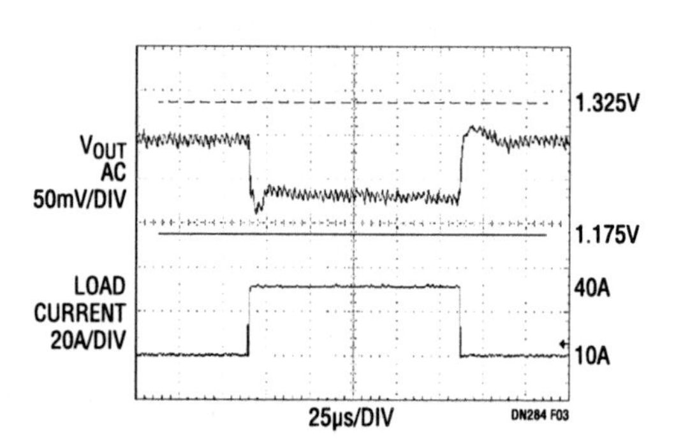

Figure 12.3 • With Seven SP Output Capacitors, the Output Excursion Is Within the Specification Limits with Plenty of Margin

Microprocessor core supply voltage set by I²C bus without VID lines

13

Mark Gurries

Introduction

Many modern CPUs run at two different clock speeds, where each speed requires a different core operating voltage to assure optimum performance. These voltages are documented in the manufacturer's VID (voltage identification) section of the CPU specification. Some new DC/DC converters (the LTC1909, for example) have built-in VID control that supports dual programmable output voltages, but many existing converters do not. The LTC1699 is a precision 2-state resistive divider that uses a simple SMBus interface to allow VID control on non-VID-enabled DC/DC converters. No dedicated VID lines are required.

How it works

DC/DC converters maintain consistent output voltage by comparing the output voltage, through an accurate voltage divider, against an internal reference and adjusting the output to compensate for differences. The LTC1699 is a 2-state resistive divider that replaces the fixed voltage divider in the feedback circuit, thus allowing the circuit to support two different voltage outputs. It is specifically designed to work with DC/DC PWM converters that use an internal 0.8V reference, such as the LTC1702A, LTC1628, LTC1735 and the LTC1778.

The LTC1699, on command, can choose between two programmable, precision output voltages. The two voltages are programmed via 5-bit VID words sent over the commonly used SMBus (System Management Bus) serial interface, precluding the need for dedicated VID control lines. The host system then has two methods to switch between the two voltages: digitally through the SMBus interface or via a logic signal at the select (SEL) pin. When CPU voltage is in regulation, the LTC1699 provides a power good signal that can be used to inform the CPU and satellite systems that power is up to specification. An enhanced version of the IC, the LTC1699EGN, expands the power sequencing control and status lines to coordinate multiple DC/DC converters that manage other CPU system voltages, such as those for the I/O and clock supplies (see Figure 13.2).

Since accurate CPU voltages are critical for reliable CPU operation, the voltage dividers in the LTC1699 are accurate to within ±0.35%. There are three versions of the LTC1699 to support different Intel CPUs and the unique voltage tables based on their 5-bit VID codes. The LTC1699-80 covers the Intel mobile specification while the desktop standards

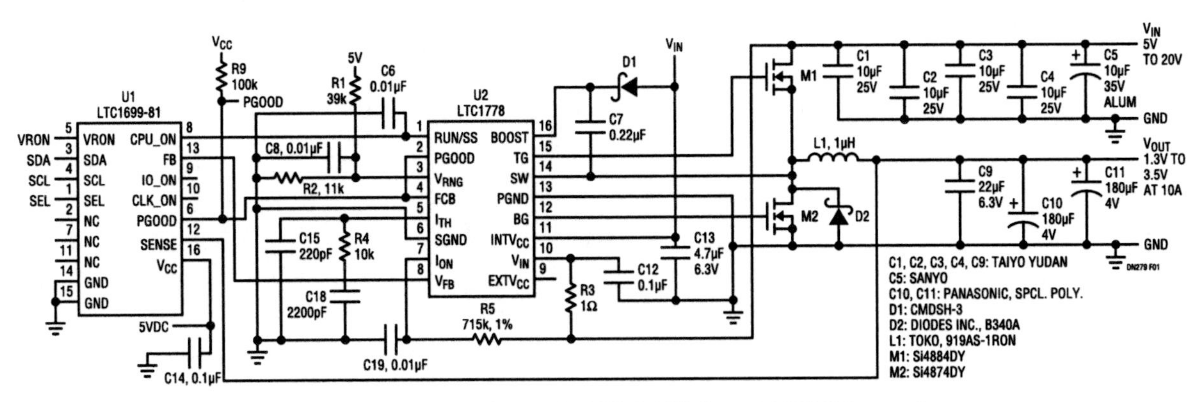

Figure 13.1 • SMBus Controlled High Efficiency DC/DC Converter

Analog Circuit Design: Design Note Collection. http://dx.doi.org/10.1016/B978-0-12-800001-4.00013-2

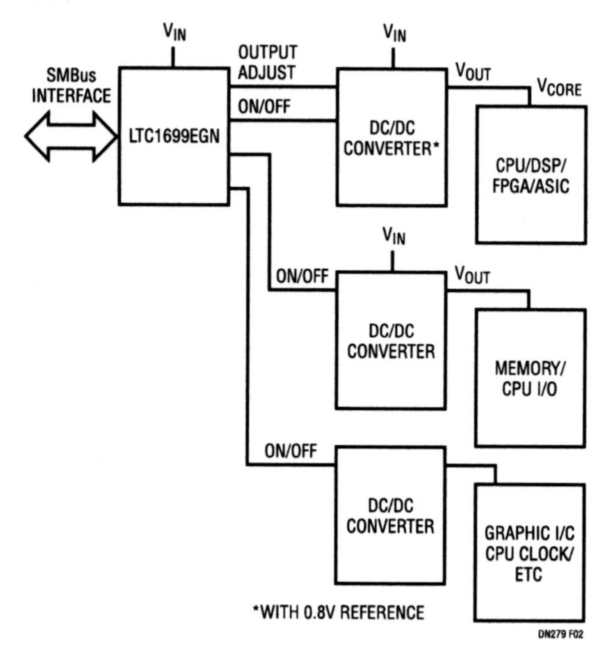

Figure 13.2 • SMBus Power Sequencing and Multiple DC/DC Converter Control with the 16-Pin SSOP Package of the LTC1699EGN

Table 13.1 Selection of V_{OUT} Ranges and VRMs with the Combination of the LTC1699 and an Appropriate DC/DC Converter

PART NUMBER	V_{OUT} RANGE	VRM COMPATIBILITY
LTC1699-80	0.9V to 2.0V	Mobile CPU
LTC1699-81	1.3V to 3.5V	VRM8.4
LTC1699-82	1.075V to 1.85V	VRM9.0

are covered by the LTC1699-81 for the VRM8.4 specification, and the LTC1699-82 for the VRM9.0 specification (see Table 13.1).

Why use an SMBus?

An SMBus is easy to implement and is growing as a system control standard. The SMBus was developed as a low power 2-wire serial interface to standardize the control and monitoring of the system support functions other than the CPU, originally defined for portable computers with intelligent rechargeable batteries. Most portable computers today use the SMBus for more than just battery control. It has evolved as the standard method of power flow control, system temperature monitoring and cooling control. It is now supported by popular operating systems and is integral to current PC design standards. Controlling the CPU voltage via SMBus is the next logical step, eliminating the need for proprietary control interfaces.

The SMBus does have some limitations. The SMBus version 1.0 standard has no error checking protocol, a potentially significant problem for modern CPUs that do not fare well when provided the wrong voltage. Although the newer SMBus v1.1 standard includes an optional error checking protocol, it is not widely used. Because most systems that traditionally use SMBus are error tolerant, upgrading current designs to the SMBus v1.1 protocol means a significant increase in communications software and hardware complexity. To address these issues and still take advantage of SMBus benefits, Linear Technology has developed several special protocol procedures and recommendations that provide ways to eliminate errors without use of v1.1 error checking.

The first is to allow the host to write and read the preprogrammed voltage values as often as needed to verify the value. The second is that a programmed value is not activated until the host sends two SMBus "ON" or "OFF" commands, one after the other. If any bit is out of place in the "ON" or "OFF" values, the preceding voltage programming command is rejected.

The LTC1699 also features two special lockout functions. The first is to ignore "ON" commands until the voltage registers are set up. In addition, when two valid "ON" command sequences are received, the VID registers are locked out to prevent changes while the power supplies are operating. Finally, the LTC1699 implements the new SMBus v1.1 logic levels for improved signaling integrity. Together these techniques offer robust and safe control of the CPU voltage using the popular SMBus.

Desktop/portable VID DC/DC converter

Figure 13.1 shows a typical implementation of a core voltage regulator using the LTC1778 controller and the LTC1699. The equivalent circuit is available as a monolithic IC in the LTC1909.

High efficiency I/O power generation for mobile Pentium III microprocessors

14

Wei Chen

The demand for higher performance notebook computers has fueled the development of faster and more power-hogging microprocessors. These microprocessors also require a faster input/output (I/O) bus and a faster clock. From a power management perspective, this means that core, I/O and clock power supplies should be able to handle more power. This requires that core and I/O DC/DC converters operate more efficiently and be as small as possible. Linear Technology recommends the LTC1778 to provide I/O power for the next generation Pentium III microprocessors. The I/O input voltage requirement is 1.25V; transient (AC) tolerance is ±9% and static (DC) tolerance is ±5%. Load current requirements are as follows:

Processor V_{TT}: 2.7A
830M chipset V_{TT}: 0.7A

The 830M chipset core has the following two possibilities:

1. Internal graphics using the 830M engine: 3.6A
2. External AGP graphics: 1.6A

Total maximum I/O current: 7.0A

The LTC1778 is a synchronous step-down switching regulator controller that provides synchronous drive for two external N-channel MOSFET switches. The true current

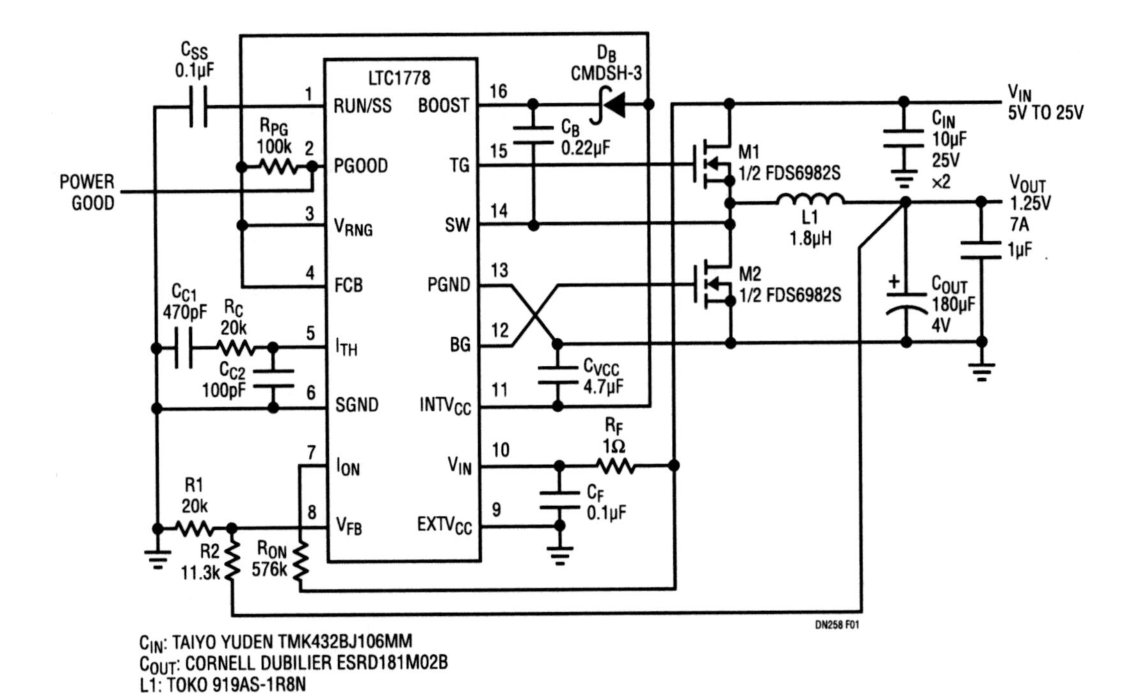

Figure 14.1 • LTC1778 Mobile Pentium III I/O Supply

Analog Circuit Design: Design Note Collection. http://dx.doi.org/10.1016/B978-0-12-800001-4.00014-4

mode control architecture has an adjustable current limit, can be easily compensated, is stable with ceramic output capacitors and does not require a power-wasting sense resistor. An optional discontinuous mode of operation increases efficiency at light loads. The LTC1778 operates over a wide range of input voltages from 4V to 36V and provides output voltages from 0.8 to 0.9 · V_{IN}. Switching frequencies up to nearly 2MHz can be chosen, allowing wide latitude in trading off efficiency for component size. Fault protection features include a power-good output, current limit foldback, optional short-circuit shutdown timer and an overvoltage soft latch. The LTC1778 is available in a 16-lead narrow SSOP package.

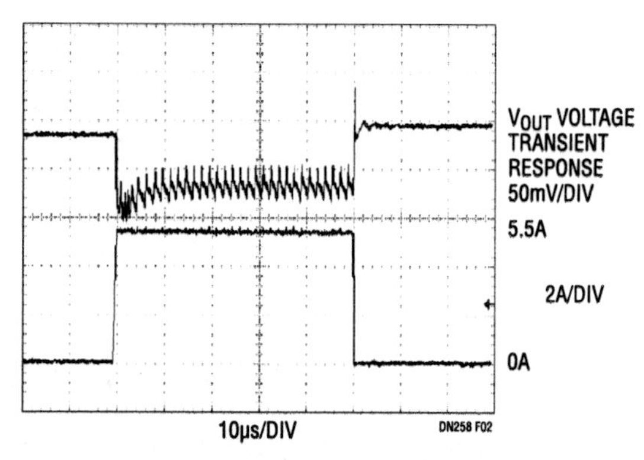

Figure 14.2 • Output Voltage Transient Response for $I_L = 0A$ to 5.5A with One SP Output Capacitor

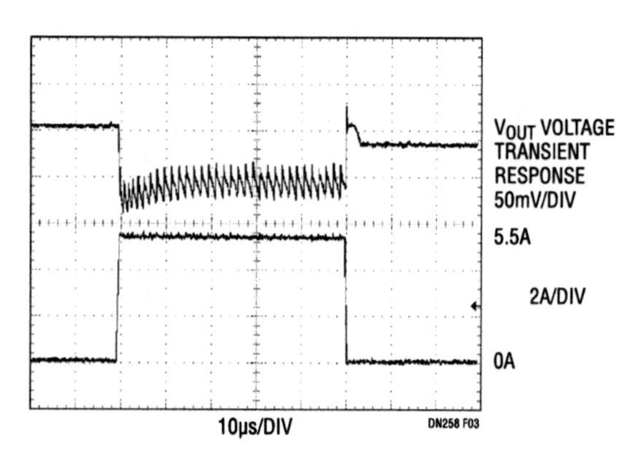

Figure 14.3 • Output Voltage Transient Response for $I_L = 0A$ to 5.5A with Two POSCAP Output Capacitors

Figure 14.1 shows a typical LTC1778 application schematic for a mobile Pentium III I/O supply. This circuit is optimized for small size and high efficiency from an input supply that varies from 5V to 24V. In order to reduce board space, it uses a single, dual N-channel FDS6982S MOSFET and only one 180µF (Panasonic SP) output capacitor. The superfast internal gate drivers with a typical rise time of 20ns help to minimize switching losses and the strong gate drivers help minimize conductive losses. Figure 14.2 shows the transient response for a 0A to 5.5A load step. It can be seen that the LTC1778 can easily meet the I/O transient and static specification with only one output capacitor. The LTC1778 allows the use of many different kinds of output capacitors such as aluminum electrolytic, tantalum, POSCAP, NEOCAP, SPs and ceramic, because of OPTI-LOOP compensation that allows the feedback loop to be compensated externally. Figure 14.3 shows the same output voltage transient with two 150µF POSCAP output capacitors. Note that the equivalent series resistance (ESR) of a POSCAP is 40mΩ, which is approximately twice that of an SP capacitor. Therefore, two POSCAPs are required to achieve the same output voltage transient response as one SP capacitor. Figure 14.4 shows typical efficiency curves for $V_{IN} = 5V$ and $V_{IN} = 15V$, $V_{OUT} = 1.25V$, $I_{LOAD} = 10mA$ to 7A. It can be seen that the efficiency is better than 85% for load currents up to 5A. The measured MOSFET case temperature is only 70°C for $V_{IN} = 12V$ at $I_{LOAD} = 7A$. This circuit can be implemented on a 0.5" × 1" board space.

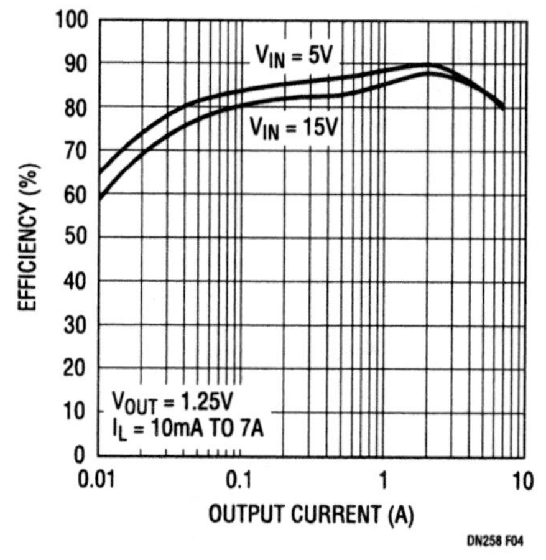

Figure 14.4 • Efficiency Curves for Figure 14.1

PolyPhase surface mount power supply meets AMD Athlon processor requirements with no heat sink

15

Craig Varga

Introduction

With the introduction of the AMD Athlon processor, the supply current requirement for a desktop PC's processor has, for the first time, surpassed the 40A mark. This, combined with low operating voltage (1.6V nominal) and very tight transient-response requirements, pushes conventional power supply design approaches to their limits. The LTC1929 PolyPhase current mode controller makes a slight break with convention. Instead of trying to deliver 40A with a single regulator, the load

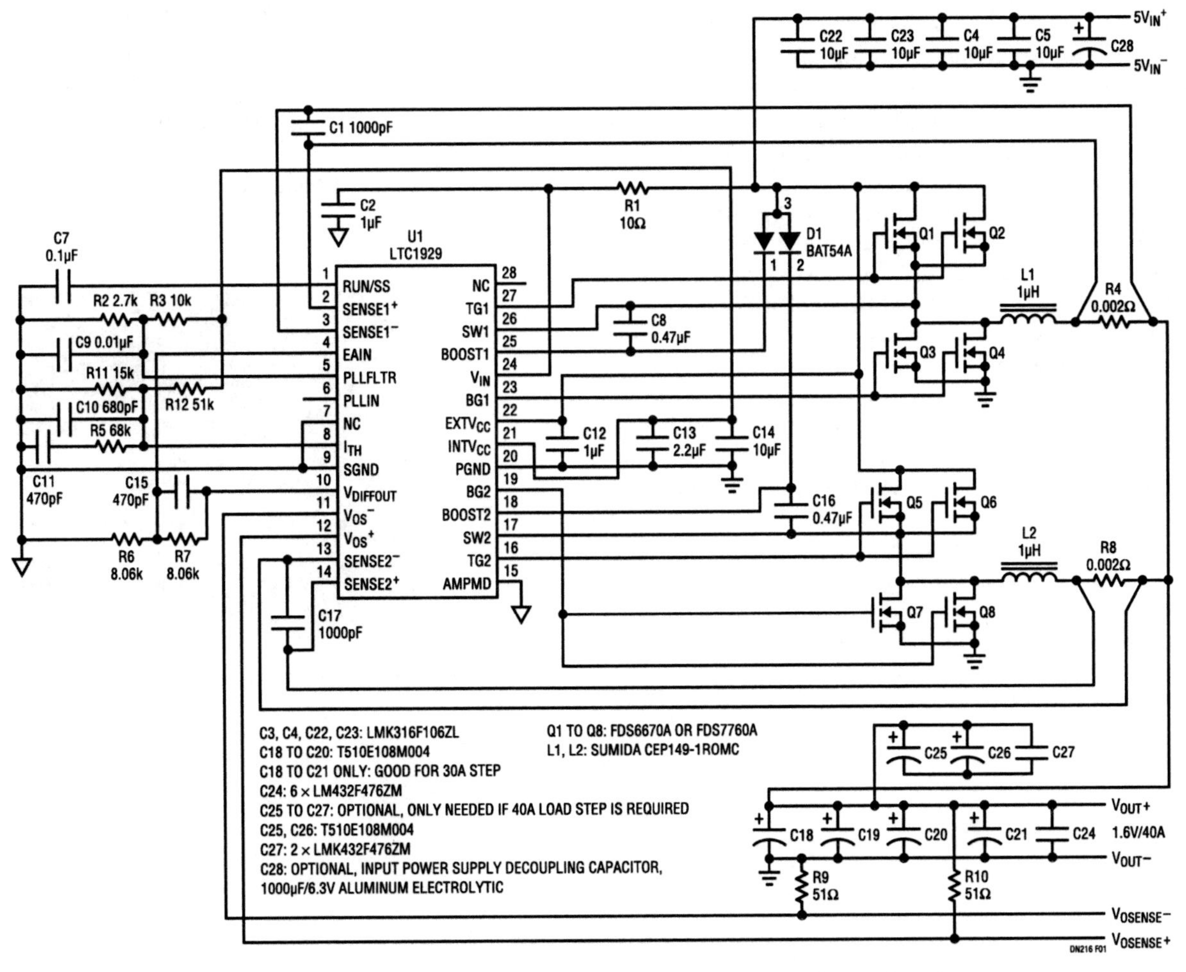

Figure 15.1 • Schematic for the AMD Athlon Processor Power Supply

Analog Circuit Design: Design Note Collection. http://dx.doi.org/10.1016/B978-0-12-800001-4.00015-6

is split in half by paralleling two regulators. The magic, however, is in the phase relationship of the two regulators' clocks.

PolyPhase architecture

The two synchronous buck regulators are connected in parallel and their clocks are synchronized 180° out of phase. This seemingly simple trick results in tremendous performance advantages as well as cost savings. A current mode architecture was chosen to reduce possible problems related to circulating currents that can appear in paralleled voltage mode regulators. Competing solutions use nonsynchronous regulators (using a diode for the low side switch) to eliminate this problem, but suffer a significant efficiency loss as a result.

The high side switch current waveform of a buck regulator is trapezoidal and varies between zero and approximately I_{OUT}. Since the input current is DC, the input capacitors must supply the difference between the instantaneous switch current and the average input current. This places a large ripple current burden on the input capacitor. By interleaving the two regulators, the peak current is halved as one stage tries to "fill the holes" left by the other. The net effect is a dramatic reduction in input capacitor ripple current.

The output ripple is reduced in a similar fashion. While one inductor's current is increasing, the other's is decreasing. There is also a significant reduction in the required inductor energy storage (approximately 75%). The inductor's volume, and therefore cost, are reduced as well. See Linear Technology's Application Note 77 for complete details.

In addition to a significant reduction in output ripple current, the power supply's maximum available current slew rate is dramatically increased. During a load step, the two inductors behave as if they were connected in parallel, while during steady-state operation, they appear to operate in series. The result is very low ripple and blazingly fast dynamic performance. The reduced ripple also takes a smaller bite out of the total error budget, leaving that much more margin for transient response. The bottom line is a significant reduction in the required output capacitance.

Figure 15.1 is the schematic of the AMD Athlon-optimized circuit. The design is quite straightforward: there are essentially two identical synchronous buck regulators connected in parallel. The controller drives them 180° out of phase. The LTC1929 has large gate drivers (approximately 1.5Ω) so it can drive large MOSFETs efficiently. There is also an accurate differential amplifier in the feedback path for easy remote sensing of both the output voltage and ground. The two PWM stages share a common error amplifier, which ensures that both channels provide the same amount of current to the load. Load sharing is therefore "open loop," eliminating oscillations in the share circuit that can occur with other approaches.

Figure 15.2 illustrates the efficiency for the circuit in Figure 15.1. The basic design will also operate with 12V as the main input source. The efficiency will be several points lower in this instance, but there may be advantages at the system-design level that need to be considered. Figure 15.3 shows the response of the regulator to a transient load step of 3A to 30A. To minimize PCB space, the design uses four 1000μF surface mount tantalum output capacitors. If lowest cost is an overriding objective, twelve 3900μF aluminum electrolytic capacitors can be substituted. If VID control is desired, the LTC1709 offers the same performance as the LTC1929 and includes a 5-bit VID DAC to program the output voltage from 1.3V to 3.5V.

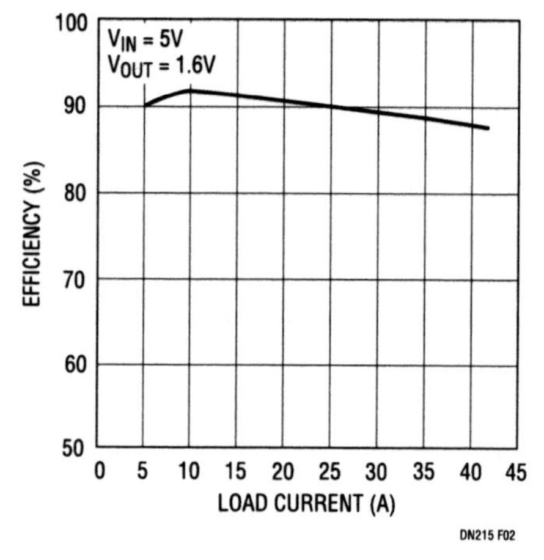

Figure 15.2 • Measured Efficiency

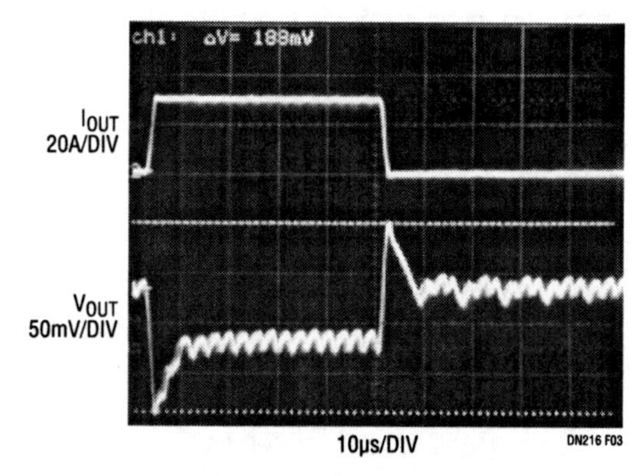

Figure 15.3 • Load Transient Response with Active Voltage Positioning

2-step voltage regulation improves performance and decreases CPU temperature in portable computers

16

John Seago

2-step regulation allows CPU power supply optimization in portable applications. Higher CPU clock frequencies mean lower core voltages, higher supply currents and more CPU power dissipation. Regulating CPU power from 5V reduces regulator switching losses permitting a higher switching frequency. Higher frequency operation results in smaller inductor size, fewer output capacitors and better transient response. OPTI-LOOP compensation and a 3.5V to 36V input range make the LTC1736 a good choice for either 1-step or 2-step configurations.

1-step vs 2-step power conversion

Traditionally, portable computer CPU voltage regulators operate over a wide range of input voltages. This 1-step regulation technique forces the power supply to operate from battery voltages as low as 8V to adapter voltages as high as 24V. This large input voltage range forces the designer to use a relatively large inductance value for the switch inductor. The large inductor value means more energy storage, so the overvoltage transient will be larger when the load current rapidly changes from high to low. Additional output capacitance may be required to meet transient specifications.

CPU core voltages are currently in the 1.5V region. Using the 1-step approach, regulating 24V down to 1.5V forces the regulator to regulate narrow “slivers” of input current to meet the transient requirements of high speed CPUs. A 24V wall adapter forces a 6.25% duty cycle when supplying a CPU voltage of 1.5V which means that the top MOSFET conducts for 0.2μs each cycle, at 300kHz.

Using 2-step regulation, CPU voltages are regulated from the 5V system supply. Decreasing the input voltage of the core voltage regulator reduces switching losses, allowing a higher switching frequency, a smaller inductor, decreased output capacitance, wider bandwidth, easier control loop optimization and higher efficiency. High efficiency and small inductor size are very important in portable computer applications since the core voltage regulator is normally located near the CPU. Temperature rise is a big problem around the CPU, and this area tends to be very crowded.

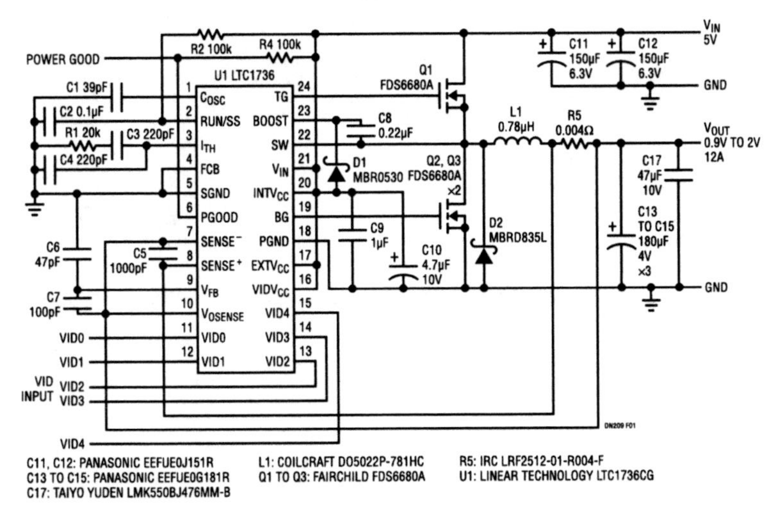

Figure 16.1 • CPU Core Voltage Regulator for 2-Step Application

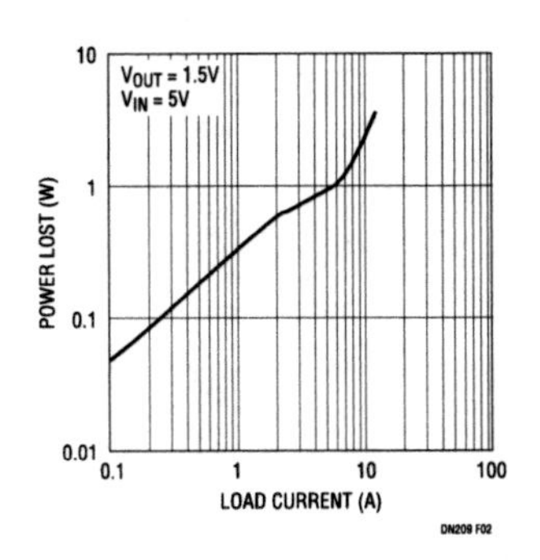

Figure 16.2 • Power Loss Curve for 2-Step Regulator

Analog Circuit Design: Design Note Collection. http://dx.doi.org/10.1016/B978-0-12-800001-4.00016-8

Circuit description

The circuit in Figure 16.1 shows the LTC1736 featuring a 5-bit, digitally controlled output voltage, from 0.9V to 2V at 12A from a 5V input. The LTC1736 is a constant frequency current mode, synchronous-buck controller, exactly like the popular LTC1735, but with 5-bit VID and a Power Good output. Only 540μF of output capacitance is required for a 0.1V transient response to a 0.2A to 12A load step from a 1.5V output. The circuit power loss curve is shown in Figure 16.2.

The circuit in Figure 16.3 shows the LTC1736 circuit redesigned for 8V to 24V inputs. 720μF of output capacitance is required for a 0.1V transient response for a 0.2A to 12A load step from a 1.5V output. The power loss curves in Figure 16.4 show the effect of increased switching losses caused by higher input voltages. Increased power dissipation caused by higher switching losses increases the ambient temperature around the CPU so the CPU temperature also increases.

The versatility of the LTC1736 can be seen by comparing the two schematics. The OPTI-LOOP architecture allows both frequency response and transient response to be optimized by adjusting the compensation component values connected to the I_{TH} pin. The V_{FB} pin provides access to the error amplifier so phase lead can be added to improve transient response and increase circuit stability. Capacitor C7 provides phase lead in the circuit of Figure 16.1. The LTC1735/6 is rated for input voltages from 3.5V to 36V, so inputs greater than 24V can be accommodated without concern for input voltage transients of 36V or less.

Regulator efficiency considerations

A common misconception is that the total efficiency of two circuits in series is the product of the efficiencies of each circuit. This is not true. Efficiency is defined as the total output power divided by the sum of the total output power plus all circuit losses.

2-step regulation takes advantage of the 5V regulator efficiency curve. The 5V supply has a peak efficiency of about 95% which is relatively flat over a wide range of load currents. The added current required to power the CPU supply causes the 5V regulator efficiency to decrease by about 1% but it also regulates the majority of the CPU power at about 94% efficiency. With a 5V input, the CPU regulator peaks at about 90% efficiency. Comparing 1-step and 2-step conversion efficiencies with a 12V input, the increased loss in the 5V regulator is about the same as the decreased loss in the CPU regulator so the overall system efficiency remains nearly constant.

The improved efficiency of 2-step regulation decreases the power dissipated near the CPU for all input voltages. The efficiency curves in Figure 16.4 show that more power would be dissipated near the CPU by a 1-step regulator when a higher voltage wall adapter powers the system.

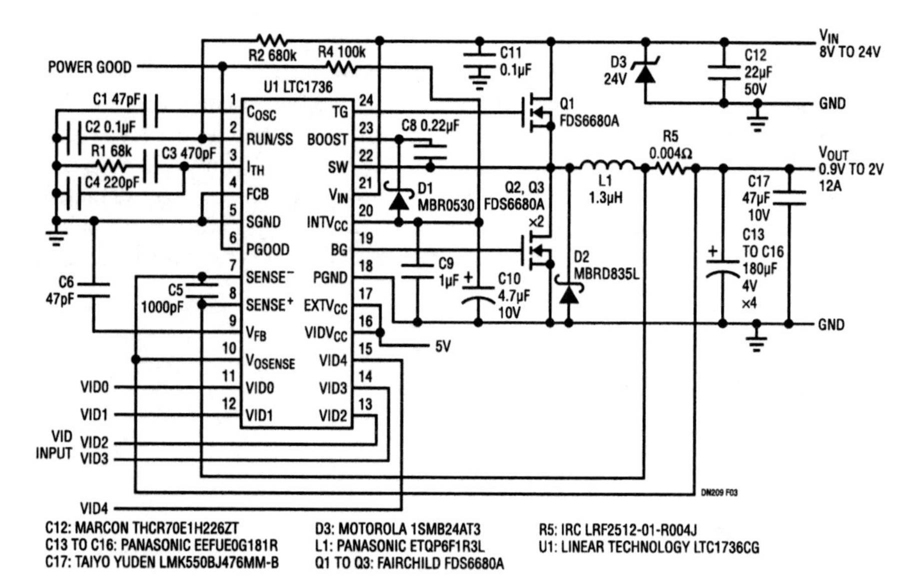

Figure 16.3 • CPU Core Voltage Regulator for 1-Step Application

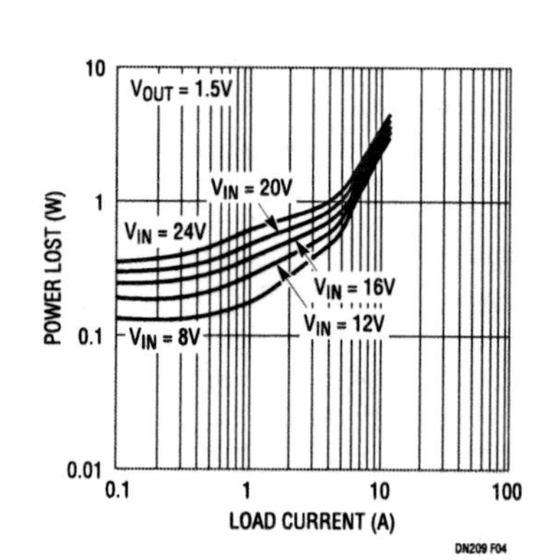

Figure 16.4 • Power Loss Curves for 1-Step Regulator

Dual regulators power Pentium processor or upgrade CPU

17

Craig Varga

Many manufacturers of Pentium processor-based motherboards have been searching for an economical solution to the problem of powering the present generation Pentium P54C and accommodating the upgrade processors that will soon become available. The existing processor uses a single supply for both the processor core and the I/O. For the highest frequency offerings, the supply required is 3.5V ±100mV (VRE specification). For the lower performance end of the clock frequency spectrum, a supply voltage of 3.3V ±5% is adequate. Recently, Intel respecified the standard 3.3V CPUs for operation at 3.5V. This allows designs for any clock frequency to be operated from a single 3.5V supply. The I/O ring and chipset should be powered by the same voltage as the CPU core, whether that is 3.3V or 3.5V.

The P55C upgrade processor, which will soon be available, requires separate supplies for the core and the I/O. The nominal core voltage is targeted at 2.500V ±5%, whereas the I/O supply is still nominally 3.3V. There is also a processor pin, V_{CC2DET}, at location AL1, that is bonded to ground on the P55C, but is open on the P54C. A significant complication is introduced by the core and I/O power pins of the P54C being shorted together on-chip. Figure 17.1 shows the system block diagram. If the core and I/O supplies do not deliver proportional currents, damage to the P54 metallization may occur. The LT1580/LT1587-based circuit shown in Figure 17.2 will automatically supply the required voltages to the CPU and the I/O circuitry based on the status of the V_{CC2DET} pin and share the load between the two regulators.

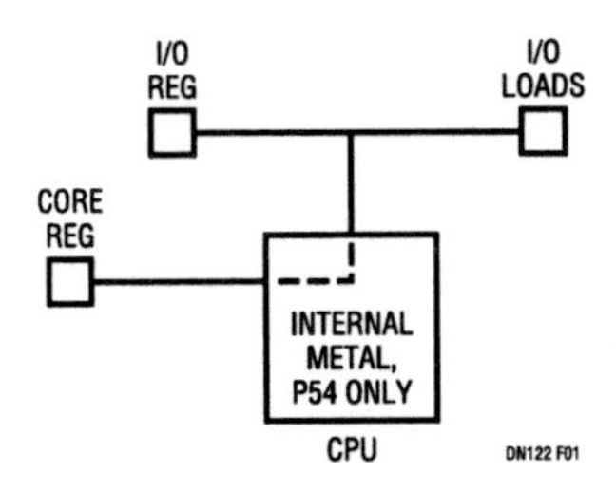

Figure 17.1 • System Configuration

A simple solution

This dual linear regulator circuit employs an LT1580 for the CPU core supply and an LT1587 for the I/O supply. The LT1580 has a precision reference, remote sense and exceptionally low dropout voltage. It is capable of meeting the stringent VRE voltage specification when subjected to the scrutiny of worst-case analysis. The LT1587 is rated at 3.0A maximum current and is adequate to power the I/O supply of most desktop systems. If more than 3A of I/O current is required—your design has a very large L2 cache, for

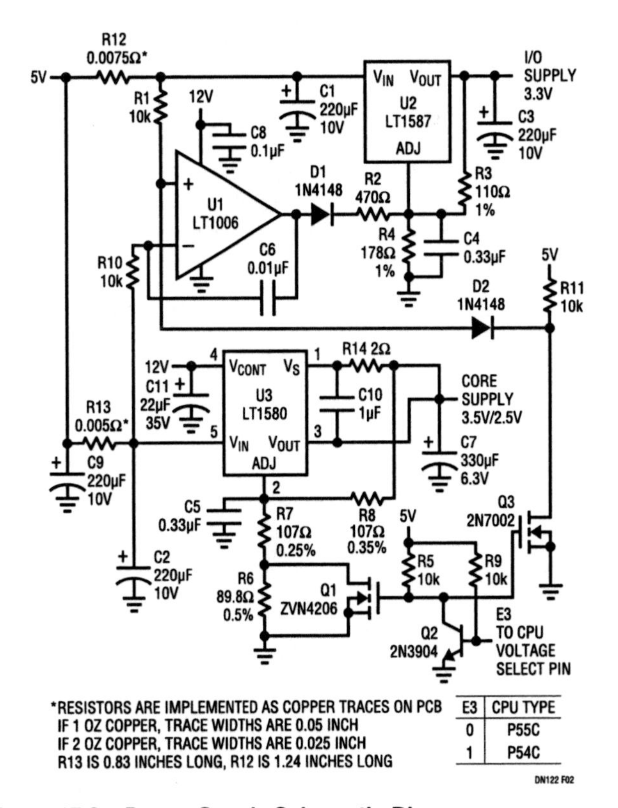

Figure 17.2 • Power Supply Schematic Diagram

Analog Circuit Design: Design Note Collection. http://dx.doi.org/10.1016/B978-0-12-800001-4.00017-X

example—an LT1585A, which is capable of 5.0A, may be substituted for the LT1587 by changing one resistor value (R12). See the design equations for details.

Op amp U1 forces the two regulators to share the load current when the outputs are shorted together by the CPU metalization. The load current is sensed by the two low value current sense resistors R12 and R13. These resistors are actually implemented as short traces on the PC board. The design does not depend on the sense resistors' absolute values being accurate; only ratiometric matching is required for the circuit to function properly. The resistance ratio will be very well-controlled across PC board production lots.

Amplifier U1 pulls up on the Adjust pin of U2, raising the output voltage of the I/O regulator until the proper current ratio between the two regulators is established. This condition is met when the voltage drop across the sense resistors is equal. The regulator currents are inversely proportional to the sense resistor values, and hence, to the resistor trace lengths. If a different current ratio is desired, just refigure the trace lengths per the equations given. The voltage drop across the resistors at full load is approximately 25mV. Of course, discrete resistors may be used if desired, but they are quite costly compared to a PC board trace.

Nonideal components will translate into errors in the current sharing ratio. With the components shown, the largest contributor to current-sharing errors is the error amplifier offset voltage. The very low offset of the economical LT1006CS8 (400μV max) ensures a worst-case share error of only 1.6%. If the through hole version of the LT1006 is used, this error drops by a factor of five. It is possible to further reduce the value of the sense resistors with this op amp.

If a user should upgrade to a P55C, E3 is now connected to ground. This turns off Q2, allowing Q1 and Q3 to turn on. Q1 shorts out part of the feedback divider of the LT1580, lowering its output to 2.500V. Q3 pulls the noninverting input of U1 low, forcing the op amp output to ground. D1 is now back biased, effectively disconnecting the op amp from the circuit, which causes the I/O regulator's output to drop to 3.3V. Resistor R11 pulls up the cathode of D2 when powering a P54 so that diode leakage current does not cause an error in current sharing.

The LT1580 permits remote sensing of the load voltage at the CPU. Also, by inserting a low value resistor in the sense line, a small intentional load regulation error is introduced, which, it can be shown, will reduce the peak-to-peak transient response of the regulator. The regulation error is well-controlled at the load and is not a function of any trace resistance or parasitics. This technique realizes a reduction in the amount of output capacitance required to control the core voltage transients.

Conclusion

With a small number of low cost components, it is possible to eliminate the need for replaceable power supply modules and still accommodate the desire to upgrade the microprocessor to improved technology. Moreover, the solution results in an "idiot proof" design, preventing the application of an inappropriate supply voltage, which could damage an expensive CPU.

Design equations

Assume V_S of approximately 25mV,

$$R13 = \frac{V_S}{I_{CORE}} \quad R12 = \frac{I_{CORE}}{I_{I/O}}(R13)$$

1oz copper thickness is 0.0036cm

2oz copper thickness is 0.0071cm

for 1oz copper PC board, use 0.127cm (0.050″) wide traces

for 2oz copper PC board, use 0.064cm (0.025″) wide traces

$$L = \frac{R}{R_S}(t)(w)$$

where L is the trace length in cm

R is the desired resistance

R_S is the specific resistivity of copper: 1.72μΩcm

t is the copper thickness of the PC board in cm

Big power for big processors: a synchronous regulator

18

Dave Dwelley

Linear Technology introduces the LTC1430 switching regulator controller for high power 5V step-down applications where efficiency, output voltage accuracy and board space requirements are critical. The LTC1430 is designed to be configured as a synchronous buck converter with a minimum of external components. It runs at a fixed switching frequency (nominally 200kHz) and provides all timing and control functions, adjustable current limit and soft-start, and level shifted output drivers designed to drive an all N-channel synchronous buck converter architecture. The switch driver outputs are capable of driving multiple, paralleled power MOSFETs with submicrosecond slew rates, providing high efficiency at very high current levels and eliminating the need for a heat sink in most designs.

The LTC1430 is usable in converter designs providing from a few amps to over 50A of output current, allowing it to supply 3.xV power to current hungry arrays of microprocessors. A novel "safety belt" feedback loop provides excellent large-signal transient response with the simplicity of a voltage feedback design. The LTC1430 also includes a micropower shutdown mode that drops the quiescent current to 1μA. An all N-channel synchronous buck architecture allows the use of cost-effective, high power N-channel MOSFETs. The on-chip output drivers feature separate power supply inputs and internal level shifters allowing the MOSFET gate drive to be tailored for logic level or standard devices. External component count in the high current path is minimized by eliminating low value current sense resistors. Voltage feedback eliminates the need for current sensing under normal operating conditions and output current limit is sensed by monitoring the voltage drop across the $R_{DS(ON)}$ of M1 during its ON state.

LTC1430 performance features

The LTC1430 uses a voltage feedback loop to control output voltage. It includes two additional internal feedback loops to improve high frequency transient response. A MAX loop responds within a single clock cycle when the output exceeds the set point by more than 3%, forcing the duty cycle to 0% and holding M2 on continuously until the output drops back into the acceptable range. A MIN loop kicks in when the output sags 3% below the set point, forcing the LTC1430 to 90% duty cycle until the output recovers. The 90% maximum ensures that charge pump drive continues to be supplied to the top MOSFET driver, preventing the gate drive to M1 from deteriorating

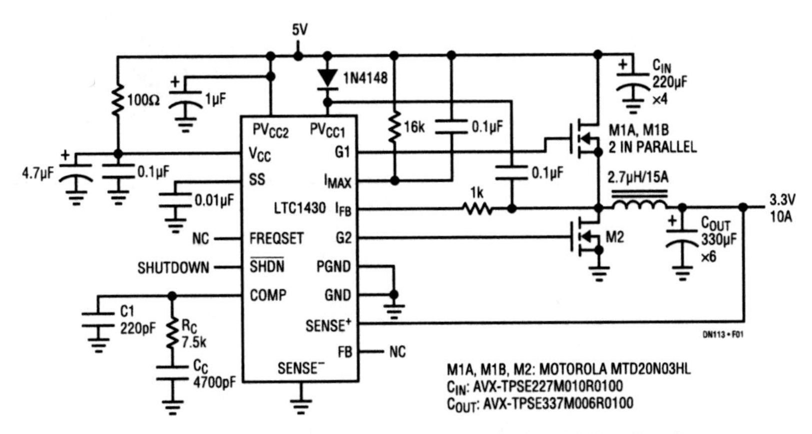

Figure 18.1 • Typical 5V to 3.3V, 10A LTC1430 Application

Analog Circuit Design: Design Note Collection. http://dx.doi.org/10.1016/B978-0-12-800001-4.00018-1

during extended transient loads. The MAX feedback loop is always active, providing a measure of protection even if the 5V input supply is accidentally shorted to the lower microprocessor supply. The MIN loop is disabled at start-up or during current limit to allow soft-start to function and to prevent MIN from taking over when the current limit circuit is active.

The LTC1430 includes an onboard reference trimmed to 1.265V ±10mV and an onboard 0.1% resistor divider string that provides a fixed 3.3V output. It specifies load regulation of better than 20mV and line regulation of 5mV, resulting in a total worst-case output error of ±1.7% when used with the internal divider or 0.1% external resistors. The internal reference will drift an additional ±5mV over the 0°C to 70°C temperature range, providing a ±2.1 total error budget over this temperature range.

The LTC1430 includes an internal oscillator that free runs at any frequency between 100kHz and 500kHz. The oscillator runs at a nominal 200kHz frequency with the FREQ pin floating. An external resistor from FREQ to ground will speed up the internal oscillator, whereas a resistor to VCC will slow the oscillator. The LTC1430 will shut down if the $\overline{\text{SHDN}}$ pin is low continuously for more than 50μs, reducing the supply current to 1μA.

A typical 5V to 3.3V application

The typical application for the LTC1430 is a 5V to 3.xV converter on a PC motherboard. The output is used to power a Pentium processor and the input is taken from the system 5V ±5% supply. The LTC1430 provides the precisely regulated voltage required without an external precision reference or trimming. Figure 18.1 shows a typical application with a 3.3V ±1% output voltage and a 12A output current limit. The power MOSFETs are sized so as not to require a heat sink under ambient temperature conditions up to 50°C. Typical efficiency is shown in Figure 18.2.

The 12A current limit is set by the 16k resistor R1 from PV_{CC} to I_{MAX} in conjunction with the 0.035Ω on resistance of the MOSFETs (M1a, M1b). The 0.1μF capacitor in parallel with R1 improves power supply rejection at I_{MAX}, providing consistent current limit performance with voltage spikes present at PV_{CC}. C_{SS} sets the soft-start time; the 0.01μF value shown provides a 3ms start-up time. The 2.7μH, 15A inductor allows the peak current to rise to the full current limit value without saturating the inductor core. This allows the circuit to withstand extended output short circuits. The inductor value is a compromise between peak ripple current and output current slew rate, which affects large-signal transient response. If the output load is expected to generate large output current transients (as large microprocessors tend to do) the inductor value will need to be quite low, in the 1μH to 10μH range.

Loop compensation is critical for obtaining optimum transient response with a voltage feedback system. The compensation components shown here give good response when used with the output capacitor values and brands shown (Figure 18.3). The output capacitor ESR has a significant effect on the transient response of the system; for best results, use the largest value, lowest ESR capacitors that will fit the budget and space requirements. Several smaller capacitors wired in parallel can help reduce total output capacitor ESR to acceptable levels. Input bypass capacitor ESR is also important to keep input supply variations to a minimum with $10A_{P-P}$ square wave current pulses flowing into M1. AVX-TPS series surface mount tantalum capacitors and Sanyo OS-CON organic electrolytic capacitors are recommended for both input and output bypass duty.

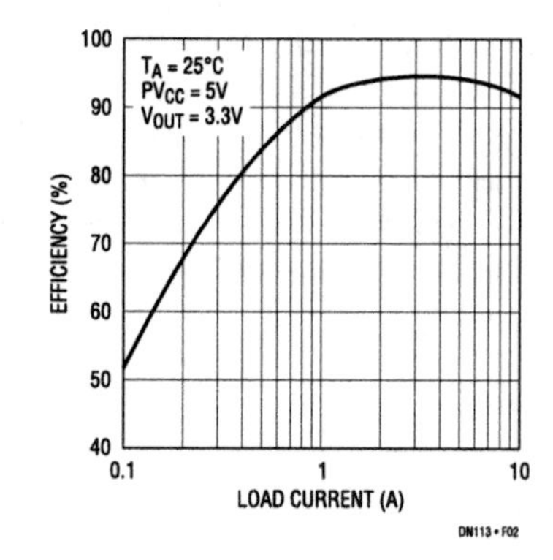

Figure 18.2 • Figure 18.1 Circuit Efficiency. Note That Efficiency Peaks at a Respectable 95%

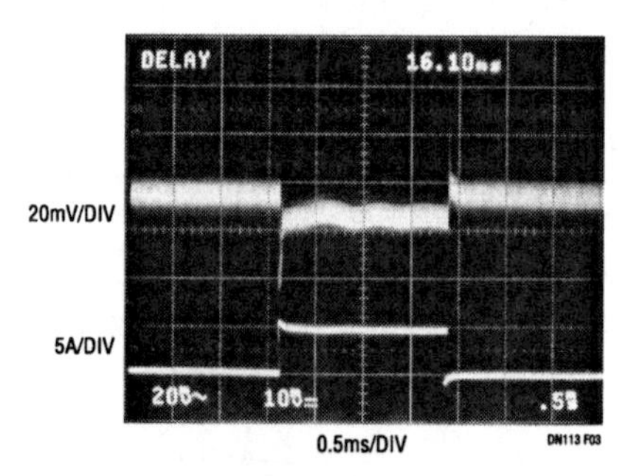

Figure 18.3 • Transient Response: 0A to 5A Load Step Imposed on Figure 18.1's Output

High efficiency power sources for Pentium processors

19

Craig Varga

In many applications, particularly portable computers, the efficiency of power conversion is critical both from the standpoint of battery life and thermal management. Desktop machines may also benefit from higher efficiency, particularly a "green PC." While linear regulators can offer low cost and high performance solutions, they can only offer 67% efficiency in 5V to 3.3V applications. Switching regulators are more efficient and minimize or even eliminate the need for heat sinks at a higher cost for the components. Efficiencies around 90% are routinely obtained with Linear Technology's best regulator designs (see Figure 19.2). The LTC1148-based circuit (Figure 19.1) meets the requirements of the P54-VR specification for output voltage transient response with the indicated decoupling network.

Selection of input source

Several options exist as to where to derive raw power for the regulator input. In most desktop systems a large amount of 5V power is available. Also, there is usually a reasonable source of 12V at hand. The 5V supply will most likely have the highest power output capability since it is called upon to power the bulk of the system logic. This logic can be sensitive to voltage changes outside of ±5%.

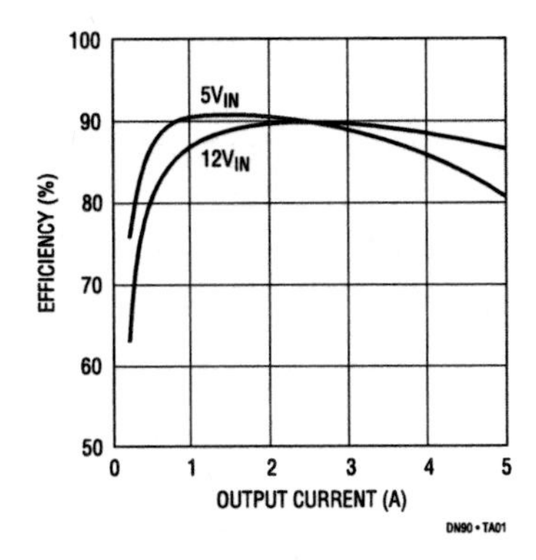

Figure 19.2 • Efficiency vs Load

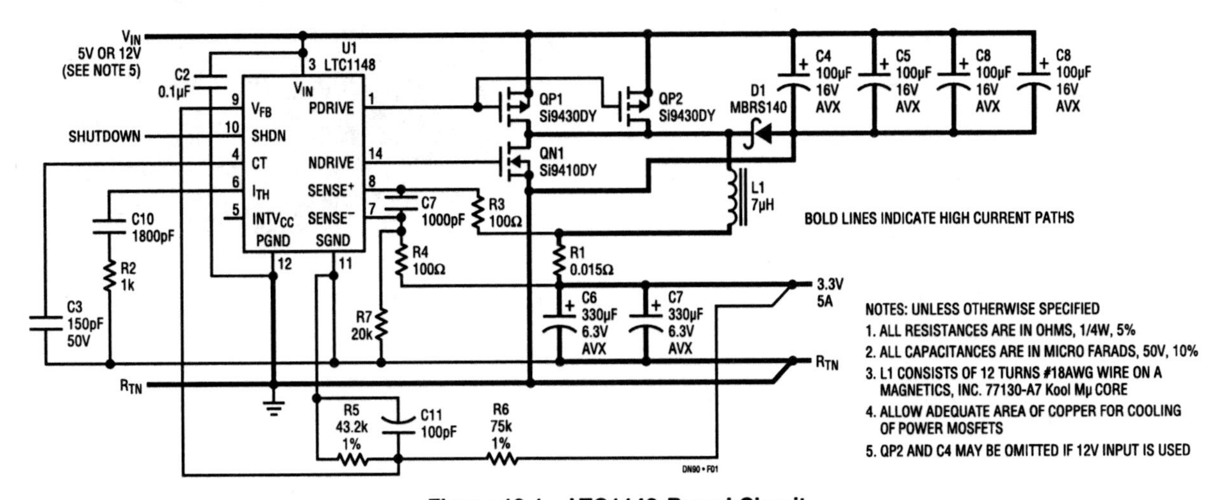

Figure 19.1 • LTC1148-Based Circuit

Analog Circuit Design: Design Note Collection. http://dx.doi.org/10.1016/B978-0-12-800001-4.00019-3

When the processor draws large transient currents, the 5V supply will be perturbed. In all "buck" type switching regulators there is an inductor in the path between the raw input supply and the load. This has the effect of limiting the rise time of the input currents and minimizing the disturbance to the 5V supply. However, the typical cheap off-line "brick" supply has terrible transient response, and the 5V supply may still be disturbed enough to cause logic problems. This is especially true as the load currents rise to the levels expected in multiprocessor systems.

If this is the case, using the 12V supply may prove advantageous. Since the 12V supply is not directly regulated, nothing that is terribly sensitive to voltage level is normally powered off the 12V bus. Moreover, with switching regulators, as a first order approximation, as the supply voltage rises the input current drops. As such, even though the input power is nominally the same whether running from a 5V or 12V supply, the current requirement is much lower if 12V is utilized for the input source.

The downside of 12V operation is lower light load efficiency than 5V operation. The efficiency with a 5V input powering a 3.3V switcher is likely to be several percentage points better than at 12V due to a reduction in switching losses. Every situation is somewhat different and a thorough analysis of the trade-offs must be undertaken to optimize the design. The schematic shown in Figure 19.1 offers the option to run from several supply choices. Each circuit was optimized for the specified input voltage, but will function well over a fairly wide range of supply voltages.

Transient response considerations

As with a linear regulator, the first several microseconds of a transient are out of the hands of the regulator and dropped squarely in the lap of the decoupling capacitor network. In the case of the switcher, the ultimate response of the regulator will be quite slow compared to a linear regulator. In the circuits shown, the approximate time required to ramp the regulator current to equal the high load condition is 11µs, about 2.4 times that of an LT1585 high speed linear regulator in the same application. This means, in layman's terms, that the LT1585 linear regulator requires less bulk capacitance than the LTC1148 switcher solution.

Circuit operation

Figure 19.1 is a schematic of the two regulators. For the 12V input, omit QP2 and C4. The design is a standard synchronous buck regulator that is discussed in detail in several Linear Technology Application Notes as well as the LTC1148 data sheet. Since the required output voltage is not the standard 3.3V, which is available factory set, an adjustable regulator is used. R5 and R6 set the output voltage to the desired level, in this case 3.38V. R7 is used to inhibit Burst Mode operation at light loads. If the system were permitted to operate in Burst Mode, the output voltage would rise by about 50mV at low load currents. If added low load efficiency is desired and the slightly higher low load output voltage can be tolerated, this resistor can be omitted.

To meet the transient requirements of the P54-VR, a fairly large amount of capacitance is needed beyond what is required to make the regulator function correctly. A viable decoupling scheme is to use ten each, 1µF surface mount ceramics and seven each, 220µF, 10V surface mount tantalums at the processor socket. In addition to the socket decoupling, two pieces of 330µF, 6.3V surface mount tantalums are required at the power supply.

The input capacitors were selected for their ability to handle the input ripple current. At a 5A load current this is a little over 4A with a 5V input and 2.6A for a 12V input. The capacitors are rated at slightly over 1A each at 85°C. If the input can be switched on very rapidly, the input capacitor voltage rating should be at least two times the supply voltage to prevent dV/dt failures.

By running the operating frequency at 150kHz, the small inductor used is sufficient. Also, since the design is synchronous, the ripple current may be permitted to get quite high without causing any problems for the regulator control loop. This would not be true in a non-synchronous design. A major advantage of high ripple current is the regulator's ability to ramp output current rapidly. The rate of rise of output current is directly proportional to input/output differential and inversely proportional to the inductor value. Using a small inductor aids in achieving fast response to transients.

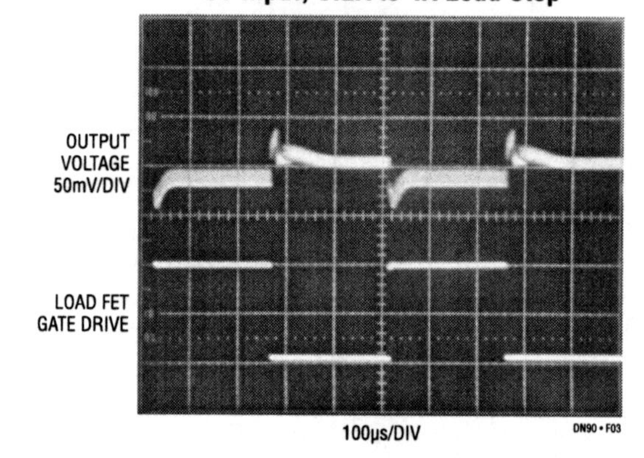

Fast regulator paces high performance processors

20

Mitchell Lee Craig Varga

New high performance microprocessors require a fresh look at power supply transient response. Pentium processors, for example, have current demands that go from a low idle mode of 200mA to a full load current of 4A in 20ns. A transition of the same magnitude occurs as the processor reenters its power saving mode. In addition, the overall supply tolerances have been narrowed significantly from the traditional ±5% for 5V supplies and include transient conditions. When all possible DC error terms are accounted for, the transient response of the power supply when subject to the load step mentioned above must be within ±46mV!

To address this problem Linear Technology has developed the LT1585 linear regulator. It features 1% initial accuracy, excellent temperature drift and load regulation, and virtually perfect line regulation. Complementing superb DC characteristics, the LT1585 exhibits extremely fast response to transients. The regulator is offered as an adjustable regulator requiring two resistors to set the operating point, as well as fixed versions which have been trimmed and optimized for 3.3V, 3.38V, 3.45V, and 3.6V outputs. Fixed versions are fully specified for worst-case DC error bounds; in adjustable designs the effects of the external voltage-setting resistors must be taken into account.

Transient response is affected by more than the regulator itself. Stray inductances in the layout and bypass capacitors, as well as capacitor ESR dominate the response during the first 400ns of transient. Figure 20.1 shows a bypassing scheme developed to meet all of the requirements for the Intel P54C-VR microprocessor. Multiple capacitors are required to reduce the total ESR and ESL, which affect the transient response.

Input capacitors C1 and C2 function primarily to decouple load transients from the 5V logic supply. The values used here are optimized for a typical 5V desktop computer "silver box" power supply input. C5 to C10 provide bulk capacitance at low ESR and ESL, and C11 to C20 keep the ESR and ESL low at high (>100kHz) frequencies. C4 is a damper and it minimizes ringing during settling.

A good place to locate the surface mount decoupling components is in the center of the Pentium socket cavity on the top side of the circuit board. Consider using concentric rings of power and ground plane on the top layer of the board within the socket center for bussing the capacitors together. Tie the main power and ground planes to these cavity planes with a minimum of two vias per capacitor. This will minimize parasitic inductance. The regulator and damper capacitor

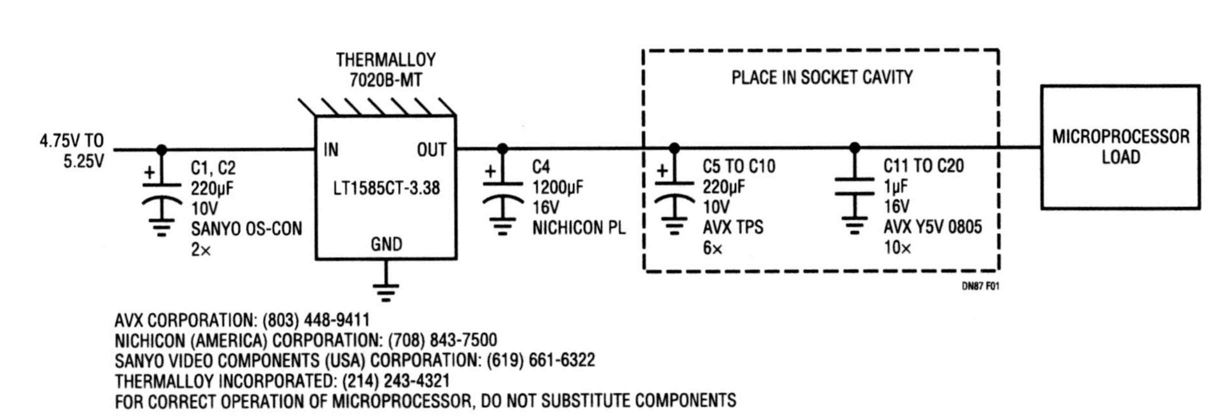

Figure 20.1 • Recommended Bypassing Scheme for Correct Transient Response

Analog Circuit Design: Design Note Collection. http://dx.doi.org/10.1016/B978-0-12-800001-4.00020-X

should be located close to (<1") the microprocessor socket to minimize circuit trace inductance.

Verifying the regulator and microprocessor layout can be accomplished with a controlled load such as the Power Validator manufactured by Intel. This device plugs directly into the microprocessor socket and simulates worst-case load transients conditions.

An oscilloscope photograph of the LT1585's response to a worst-case 200mA to 4A load step is shown in Figure 20.2. Trace C is the load current step, which is essentially flat at 4A with a 20ns rise time. Trace A is the output settling response at 20mV per division. Cursor trace B marks −46mV relative to the initial output voltage. At the onset of load current, the microprocessor socket voltage dips to −38mV as a result of inductive effects in the board and capacitors, and the ESR of the capacitors. The inductive effects persist for approximately 400ns. For the next 3μs the output droops as load current drains the bypass capacitors. The trend then reverses as the LT1585 catches up with the load demand, and the output settles after approximately 50μs.

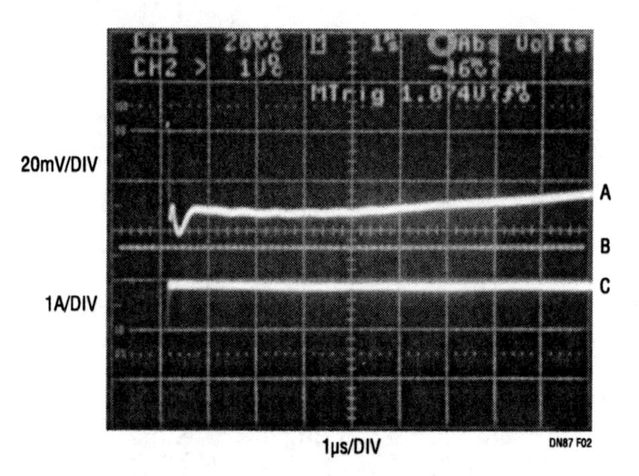

Figure 20.2 • Transient Response at Onset of 4A Load Current Step

Running 4A with a 1.7V drop, the regulator dissipates 6.8W. The heat sink shown in Figure 20.1, with 100ft/min air flow is adequate for worst-case operating conditions.

The adjustable version of the LT1585 makes it relatively easy to accommodate multiple mircoprocessor power supply voltage specifications (see Figure 20.3). To retain the tight tolerance of the LT1585 internal reference, 0.5% adjustment resistors are recommended. R1 is sized to carry approximately 10mA idling current (≤124Ω), and R2 is calculated from:

$$R2 = \frac{V_O - V_{REF}}{\frac{V_{REF}}{R1} + I_{ADJ}}$$

where:

$$I_{ADJ} = 60\mu A \text{ and } V_{REF} = 1.250V$$

Figure 20.3 shows the connections for R1 and R2. Note that C5 to C10 are reduced in value from Figure 20.1 without compromising the transient response. The addition of C3 makes this possible and also eliminates the need for C4.

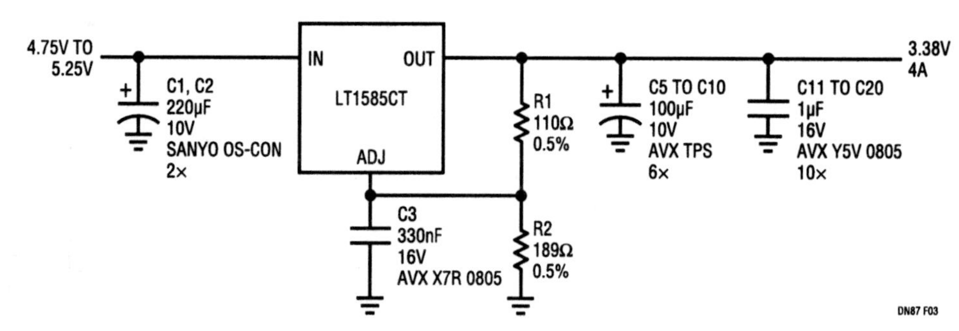

Figure 20.3 • Recommended Adjustable Circuit

Techniques for deriving 3.3V from 5V supplies

21

Mitchell Lee

Microprocessor chip sets and logic families that operate from 3.3V supplies are gaining acceptance in both desktop and portable computers. Computing rates and, in most cases, energy consumed by these circuits show a strong improvement over 5V technology. The main power supply in most systems is still 5V, necessitating a local 5V to 3.3V regulator.

Linear regulators are viable solutions at lower ($I_O \leq 1A$) currents, but they must have a low dropout voltage in order to maintain regulation with a worst-case input of only 4.5V. Figure 21.1 shows a circuit that converts a 4.5V minimum input to 3.3V with an output tolerance of only 3% (100mV). The LT1129-3.3 can handle up to 700mA in surface mount configurations, and includes both 16µA shutdown and 50µA standby currents for system sleep modes. Unlike other linear regulators, the LT1129-3.3 combines both low dropout and low voltage operation. Small input and output capacitors facilitate compact, surface mount designs.

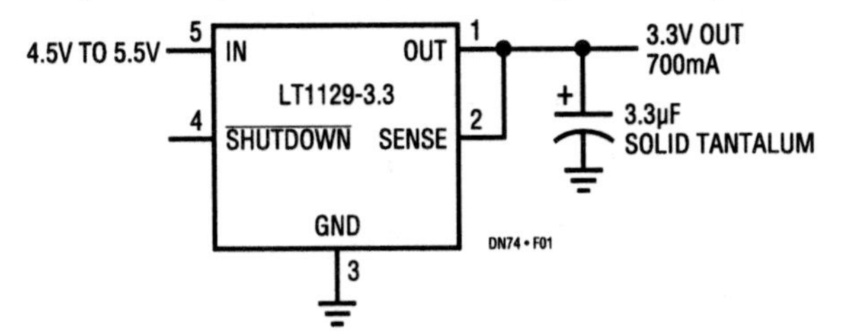

Figure 21.1 • Low Dropout Regulator Delivers 3.3V from 5V Logic Supply

For the LT1129-3.3, dissipation amounts to a little under 1.5W at full output current. The 5-lead surface mount DD package handles this without the aid of a heat sink, provided the device is mounted over at least 2500mm^2 of ground or power supply plane. Efficiency is around 62%.

Dissipation in linear regulators becomes prohibitive at higher current levels where they are supplanted by high efficiency switching regulators. A 2A, 5V to 3.3V switching regulator is shown in Figure 21.2. This synchronous buck

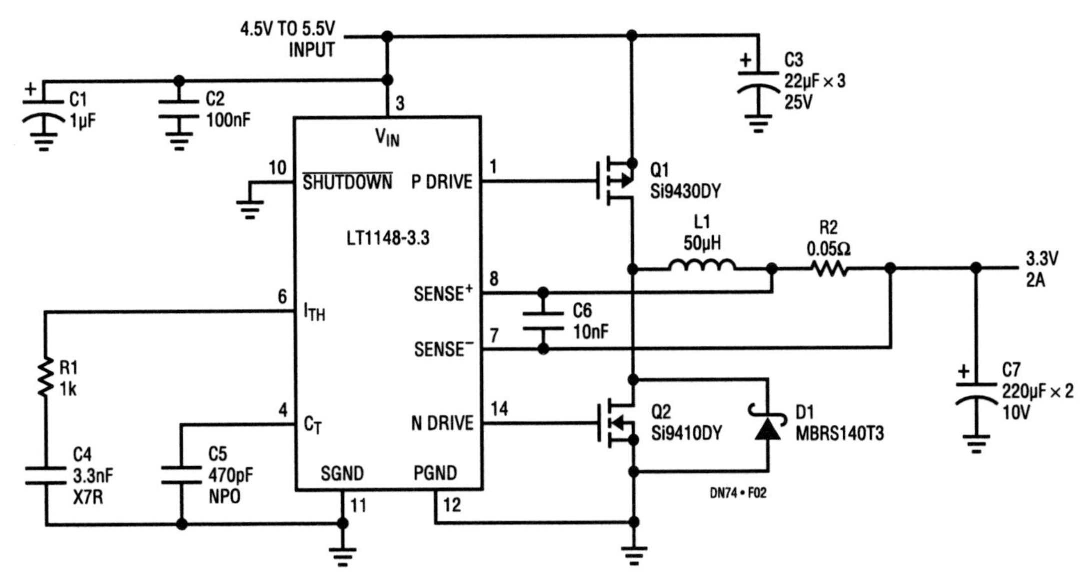

Figure 21.2 • 94% Efficiency Synchronous Buck Regulator Pumps Out 2A at 3.3V from 5V Logic Supply

Analog Circuit Design: Design Note Collection. http://dx.doi.org/10.1016/B978-0-12-800001-4.00021-1

converter is implemented with an LTC1148-3.3 converter. The LTC1148 uses both Burst Mode operation and continuous, constant off-time control to regulate the output voltage, and maintain high efficiency across a wide range of output

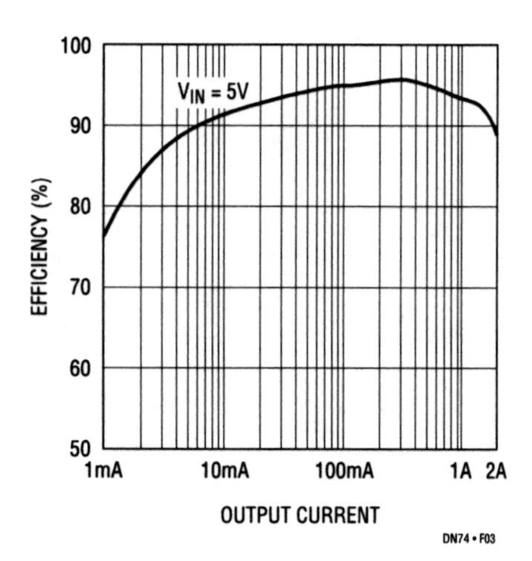

Figure 21.3 • LTC1148-3.3: Measured Efficiency

loading conditions. Efficiency as a function of output current is plotted in Figure 21.3.

All of the components used in the Figure 21.2 switching regulator are surface mount types, including the inductor and shunt resistor, which are traditionally associated with through hole assembly techniques.

Depending on the application, a variety of linear and switching regulator circuits are available for output currents ranging from 150mA to 20A. Choices in linear regulators are summarized in Table 21.1. There are some cases, such as in minicomputers and workstations, where higher dissipations may be an acceptable compromise against the circuit complexity and cost of a switching regulator, hence the >1A entries. Heat sinks are required.

Table 21.2 summarizes the practical current range of a number of switching regulators for 5V to 3.3V applications, along with their typical efficiencies.

A 5V to 3.3V converter circuit collection is presented in Application Note 55, covering the entire range of currents listed in Tables 21.1 and 21.2.

Table 21.1 Linear Regulators for 5V to 3.3V Conversion

LOAD CURRENT	DEVICE	FEATURES
150mA	LT1121-3.3	Shutdown, Small Capacitors
700mA	LT1129-3.3	Shutdown, Small Capacitors
800mA	LT1117-3.3	SOT-223
1.5A	LT1086	DD Package
3A to 7.5A	LT1083 LT1084 LT1085	High Current, Low Quiescent Current at High Loads
10A	2 × LT1087	Parallel, Kelvin Sensed

Table 21.2 Switching Regulators for 5V to 3.3V Conversion

LOAD CURRENT	DEVICE	EFFICIENCY	FEATURES
200mA to 400mA	LTC1174-3.3	90%	Internal P-Channel Switch, 1μA Shutdown
0.5A to 2A	LTC1147-3.3	92%	8-Pin SO, High Efficiency Converter
1A to 5A	LTC1148-3.3	94%	Ultra-High Efficiency Synchronous Converter
5A to 20A	LT1158	91%	Ultra-High Current Synchronous Converter

Regulator circuit generates both 3.3V and 5V outputs from 3.3V or 5V to run computers and RS232

22

David Dinsmore Richard Markell

Many portable microprocessor-based systems use a mix of 3.3V and 5V circuits. Some are still using only 5V and inevitably some systems will end up being solely 3.3V based. If accessories are to be plugged into, or connected to, any of these systems, a voltage conversion/power generation problem presents itself. The circuit shown in Figure 22.1 addresses the situation where *either* 5V or 3.3V power is available from the bus, but the accessory needs *both* 5V and 3.3V power.

The circuit consists of two sections, one being a DC/DC converter and the other being a pair of dual N-channel MOSFETs and their associated high side drivers that effectively form a DPDT switch.

When first powered up, a comparator inside of the LT1111 (IC2) determines the state of the circuit. The comparator's output (IC2, Pin 6) is wired to the input of the LTC1157 MOSFET driver (IC1). The LTC1157 internally generates a gate drive voltage which is 8.8V above the supply voltage and efficiently turns on and off the appropriate MOSFETs.

IC2 also forms a flying capacitor buck/boost DC/DC converter circuit. This topology is used so that no transformers are necessary. Q1 is used to control this section's voltage (V1). When V_{IN} is at 5V, Q1 is off, forcing this section to operate as a step-down converter. It produces 3.3V which is sent to the 3.3V output of the circuit through IC4B. In this state, 5V power is sent directly through IC3A while IC3B and IC4A are off.

When V_{IN} is at 3.3V, IC1 turns on Q1 shorting out the 140k resistor and forcing the DC/DC converter into step-up mode so that it generates 5V at V1 which is sent to the 5V output through IC3B, while 3.3V power is sent from input to output through IC4A. IC3A and IC4B are off.

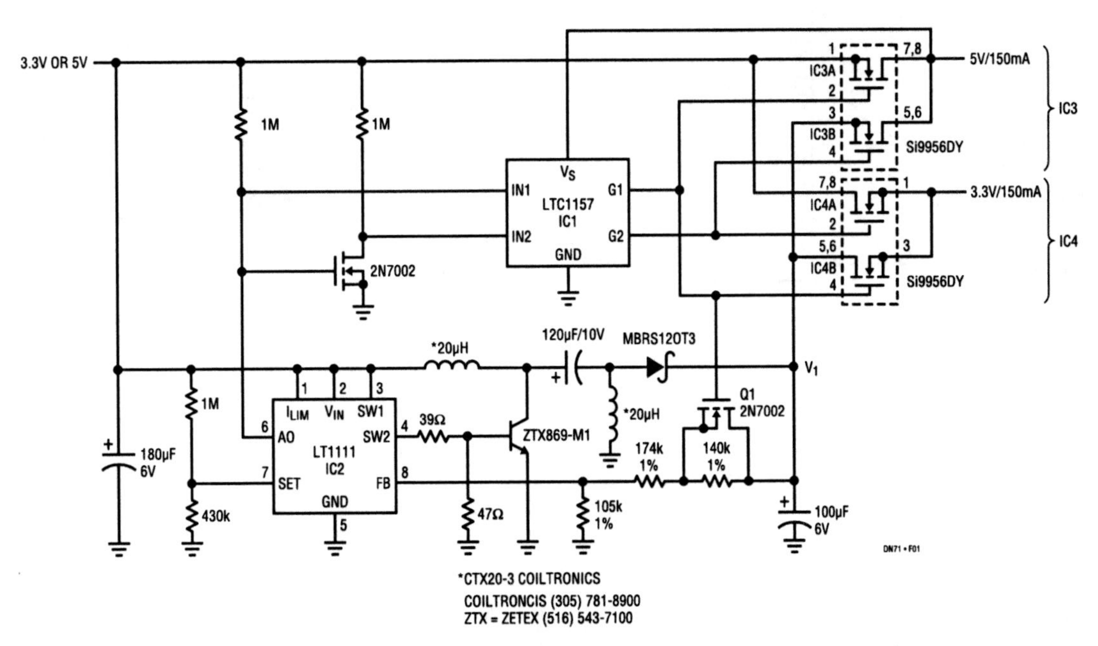

Figure 22.1 • Circuit Showing 3.3V or 5V Power

Analog Circuit Design: Design Note Collection. http://dx.doi.org/10.1016/B978-0-12-800001-4.00022-3

No load quiescent current is about 500μA. By replacing the LT1111 with the lower frequency LT1173 this could be reduced to 315μA, at the expense of a larger inductor size.

Overall efficiency of the circuit exceeds 80% with V_{IN}=3.3V and 86% with V_{IN}=5V. All components are available in surface mount.

Mixed 3.3V and 5V RS232 operation

Portable computers also require RS232 interfacing circuitry for inter-computer and mouse interfacing applications. Most portable computers now use a mix of 3.3V and 5V logic. Linear Technology offers a wide variety of interfacing circuits that can not only work with these voltages, but also upgrade to single 3.3V supplies when that is required.

Figure 22.2 shows the LT1330, a 3-driver/5-receiver, PC compatible, RS232 interface running on both 3.3V and 5V supplies. The LT1330's charge pump power is taken from the 5V supplies maximizing the RS232 transmitters load driving capability. The center trace of the photo demonstrates the ability of the transmitters to drive a 3000Ω/2500pF load at 120kBd. The drive level shown here are −6V to 7V when fully loaded.

The LT1330's receivers are powered by the 3.3V supply on Pin 14. This allows the logic levels to be compatible with either TTL or 3.3V logic since the output logic levels are typically 0.2V to 2.7V. Logic inputs to the transmitters respond to TTL levels, so they can be driven from either 3.3V or 5V logic families.

When the entire system can be operated on 3.3V, an LT1331 may be directly substituted for the LT1330. The LT1331 can be operated at 120kBd with the only limitation being transmitter output levels are −3.5V to 4V. While these levels are not RS232 compliant, they can be used to interface with all known RS232/RS562 systems. In all cases the LT1331 operated at 3.3V would provide a reliable communications link. The table below shows the details of 3-driver/5-receiver RS232 transceivers for 3.3V and mixed 5V/3.3V systems.

	LT1342	LT1330	LT1331	LTC1327
ESD Protection	±10kV	±10kV	±10kV	±10kV
3V Logic Interface	✓	✓	✓	✓
Power Supply	3V/5V	3V/5V	3V, 5V or 3V/5V	3V
Supply Current in SHUTDOWN	1μA	60μA	60μA	1μA
Receiver Active in SHUTDOWN		✓	✓	
Driver Disable	✓	✓	✓	
External Capacitors	0.1, 0.2μF	0.1, 0.2, 1μF	0.1, 0.2μF	0.1μF
Rx Output (Typ)	0.2V–2.7V	0.2V–2.7V	0.2V–2.7V	0V–3.3V
RS232 Tx Compliant	✓	✓		
RS232 Tx Compatible	✓	✓	✓	✓

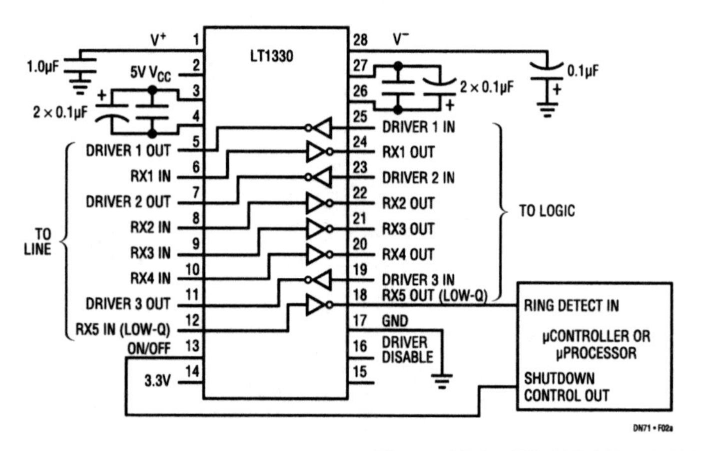

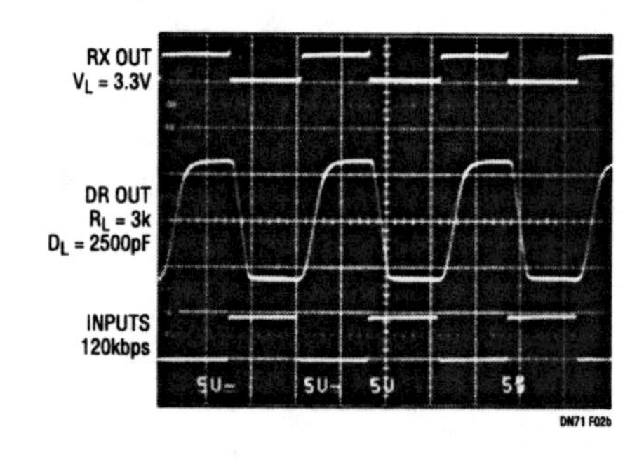

Figure 22.2 • LT1330 Mixed 5V/3V Operation

Section 3

Switching Regulator Basics

Tiny, highly flexible, dual boost/inverter tracks supplies

23

Jeffrey Huang

Introduction

Linear Technology's LT3471 is a space saving and extremely versatile dual switching regulator available in a tiny 3mm × 3mm DFN package. With a 2.4V to 16V input range, the device can easily implement any combination of boost, inverting and SEPIC topologies to generate two positive and/or negative outputs up to ±40V. It can readily make the two output voltages track each other with minimal additional circuitry, or it can independently regulate the outputs, thus saving space by replacing two separate ICs. Either way, the LT3471 fits into tight spaces and is easily adaptable for a wide range of applications.

LT3471 features

The LT3471 combines two independent, in-phase, 1.3A, 42V switches with error amplifiers that can sense to ground, providing boost and inverting capabilities. Both inputs of each error amplifier are available as high impedance feedback pins with a typical bias current of only 60nA. An accurate 1.00V reference is also pin accessible.

The LT3471 switches at a fixed 1.2MHz which allows for the use of tiny, low profile inductors and capacitors. Its internally compensated, current mode PWM architecture yields low, predictable output noise that is easy to filter. Additionally, the device provides a programmable soft-start function that requires only a simple external RC circuit to mitigate high inrush current.

Easy-to-implement ±15V dual tracking supplies

Figure 23.1 shows a Li-Ion to ±15V dual tracking supply application, highlighting the versatility of the LT3471 by taking advantage of the accessible error amplifier inputs and reference voltage. The circuit generates well-regulated ±15V

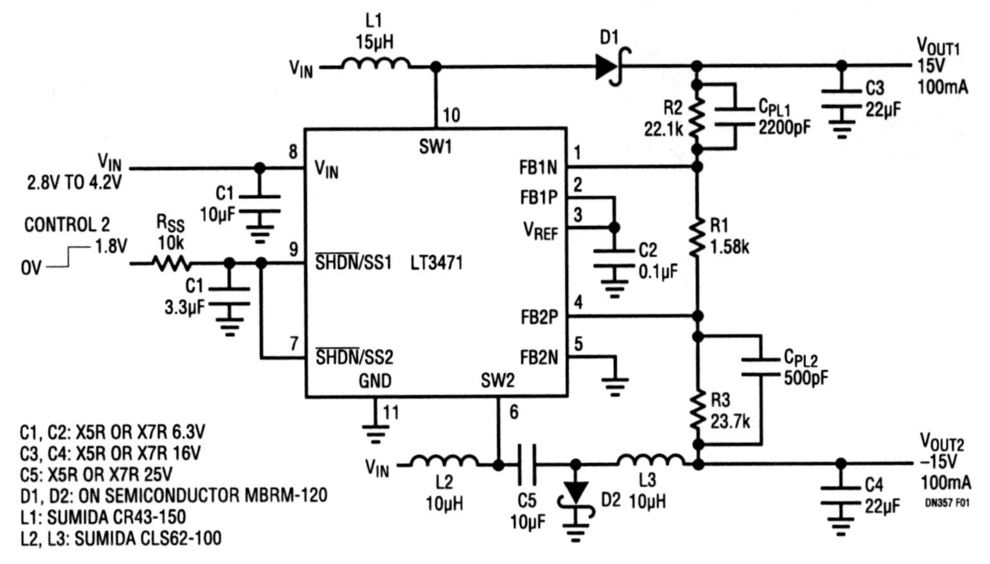

Figure 23.1 • Li-Ion to ±15V Dual Tracking Supplies

Analog Circuit Design: Design Note Collection. http://dx.doi.org/10.1016/B978-0-12-800001-4.00023-5

outputs that track each other with switchers 1 and 2 in boost and inverting configurations, respectively.

The two converters can talk to each other through their shared external feedback resistor network, specifically R1. When regulating, the control loop of switcher 1 servos FB1N to 1V while that of switcher 2 servos FB2P to ground. Selecting the value of R2 to be (14 • R1) and R3 to be (15 • R1) produces ±15V outputs. In this way, the magnitudes of V_{OUT1} and V_{OUT2} track each other; that is, a change in the +15V output causes a similar change in the −15V output, and vice versa. Figure 23.2 shows the circuit efficiency.

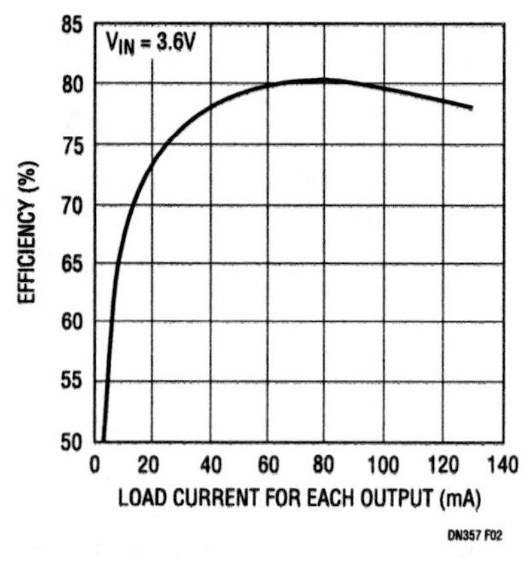

Figure 23.2 • Efficiency of the Circuit in Figure 23.1

The interaction between the two converters may compromise the stability of each. To prevent this, phase lead capacitors C_{PL1} and C_{PL2} improve the phase margin of both switchers. Figure 23.3 shows the transient response of the circuit under a constant 50mA load on V_{OUT1} and 50mA load step on V_{OUT2}. The upper waveform depicts the 15V output tracking the clean, damped response of the −15V output, the middle waveform. The output voltages remain within 0.3% of their nominal values during the step transient. V_{OUT2} similarly tracks V_{OUT1} under a load step on switcher 1 and constant load on switcher 2.

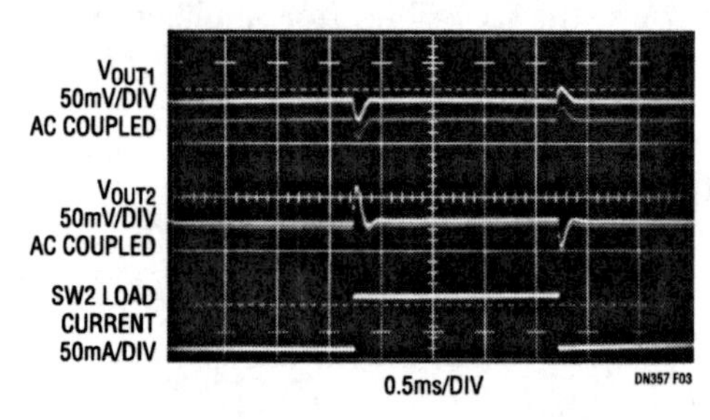

Figure 23.3 • Transient Response of the Circuit in Figure 23.1

Both switchers implement a programmable soft-start feature. The RC circuit at the SS1 and SS2 pins in Figure 23.1 sets the input current ramp rate. Figures 23.4 and 23.5 show the effect of soft-start on the output voltages (top two traces) and input current (bottom trace). Figure 23.4 depicts the start-up waveforms of the circuit without soft-start (C_{SS} removed). The output voltages reach ±15V very quickly but the inrush current sharply peaks above 2A. With soft-start, as shown in Figure 23.5, the input current peaks slightly to about 500mA, while the output voltages comfortably rise and settle to their steady-state values. In both cases, V_{OUT1} and V_{OUT2} track each other.

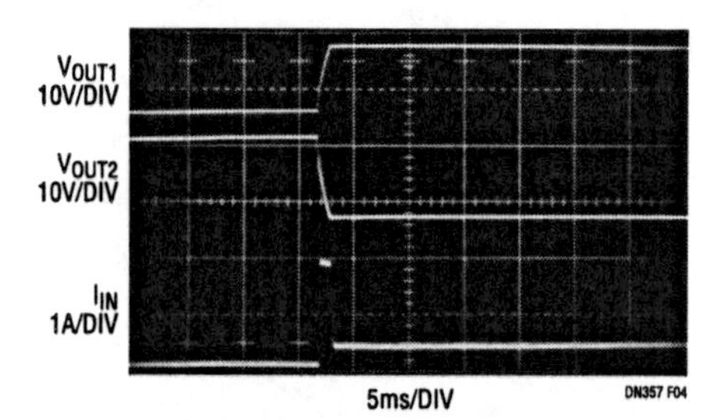

Figure 23.4 • Without Soft-Start (C_{SS} Removed in Figure 23.1), the Output Voltages Quickly Reach Regulation, but the LT3471 Draws Very High Inrush Current

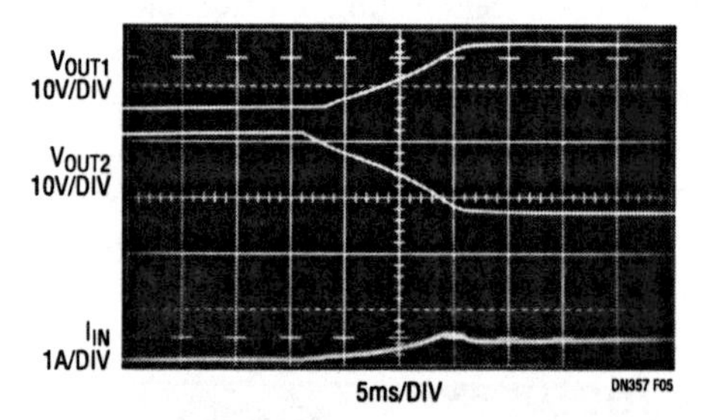

Figure 23.5 • With Soft-Start Enabled (via the $R_{SS}C_{SS}$ Network in Figure 23.1), the Output Voltages Slowly Reach Regulation, While the Controlled Inrush Current Safely Ramps Up with Little Peaking

Conclusion

The LT3471 is a tiny, dual switching regulator easily configurable for a broad array of applications, including organic and white LED drivers, TFT-LCD bias supplies and other portable device and medical diagnostic equipment functions. A flexible pinout makes it simple to generate any combination of two positive and/or negative outputs. The LT3471's small footprint and versatile, 2-switcher capability make it a great fit in a variety of power management solutions.

Ultralow noise switching power supplies simplify EMI compliance

24

David Canny

Introduction

Most electronic products must pass electromagnetic interference (EMI) compliance tests—such as the FCC Part 15 rules in the US—before they can be released into the market. When a company's product fails EMI compliance tests, the company loses control of the development process, shipping dates begin to slip, projected revenues diminish and expenses start piling up. Furthermore, excessive noise can interfere with sensitive electronics within the product itself. So it is essential that electronic products are designed from the beginning with EMI attenuation in mind. Often the EMI compliance problem lies in the product's switching power supply. In these cases, the best solution is prevention by using an ultralow noise switching power supply in the initial design, rather than adding shields and filters to the product after the fact.

Circuit description

Figure 24.1 shows the LT3439, one of several low noise switchers available from Linear Technology. Ultralow noise and low EMI are achieved by controlling both the output switch voltage and current slew rates. This switching architecture is particularly well suited to noise-sensitive systems such as industrial sensing and control, data conversion and wideband communications.

DC/DC push-pull topologies like the one shown in Figure 24.1 are commonly used in low noise systems because they offer relatively low input ripple current to the power

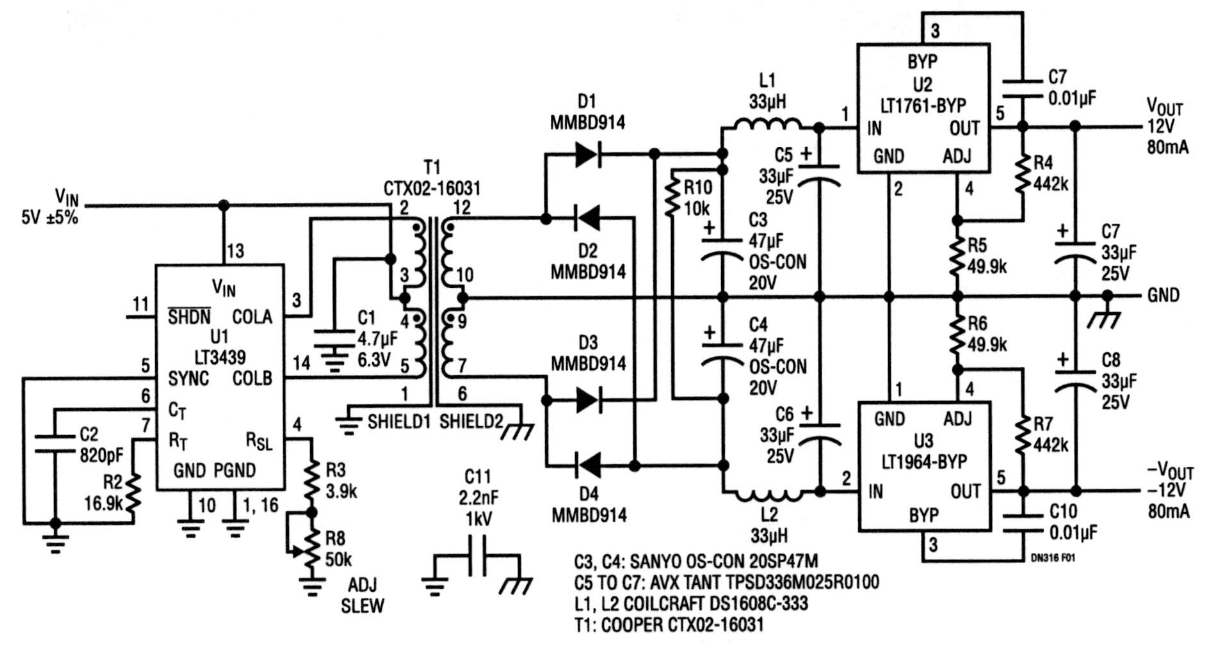

Figure 24.1 • 5V to ±12V Ultralow Noise Switching Power Supply

Analog Circuit Design: Design Note Collection. http://dx.doi.org/10.1016/B978-0-12-800001-4.00024-7

supply circuit. Nevertheless, DC/DC switcher topologies generate high frequency harmonics across the transformer primary windings (or inductor in the case of some topologies). The transformer provides little impedance to these differential mode harmonics as they are magnetically coupled across to the secondary winding. These high frequency harmonics on the secondary winding are difficult to suppress. Even with an output filter inductor, the high frequency harmonics pass through the filter inductor to the output via the inductor's parasitic capacitance. The best solution for attenuating the high frequency harmonics is to use a low noise switcher like the LT3439 to limit the voltage and current switching slew rates.

The LT3439 in Figure 24.1 drives the transformer T1 that produces isolated positive and negative output voltages. The two LT3439 internal switches are turned on out of phase at 50% duty cycles. Both of these switches operate at the oscillator frequency set by R2 and C2. During a switch's on time, V_{IN} is applied across the respective half primary side of the push-pull transformer. The voltage on the secondary side of the transformer is basically V_{IN} times the turns ratio of transformer T1. The diodes D1 to D4 rectify the secondary voltages. Capacitors C3 to C6 and inductors L1 and L2 filter the input voltage to the linear regulators.

The high frequency harmonics on the collector Pins 3 and 14 and throughout the circuit are reduced by as much as 40dB compared to a regular switcher by programming the output switch voltage and current slew rates using R8. The secondary winding of the center-tapped transformer can be wound to produce any desired output voltage based on the input voltage. The two linear regulators, U2 and U3, regulate the output voltages to ±12V. The transformer's Faraday shield blocks primary-to-secondary common mode noise while the 2.2nF capacitor C11 provides a low impedance return path for any primary-to-secondary currents through parasitic capacitances. Figure 24.2 shows the output noise on the $12V_{OUT}$ line (see Application Note 70 for low noise measurement techniques).

Conclusion

Ultralow noise switchers simplify product EMI compliance and help avoid last minute shipping delays. They reduce the high frequency harmonics associated with regular switchers by as much as 40dB while lower frequency harmonics are effectively suppressed by output filter inductors and capacitors.

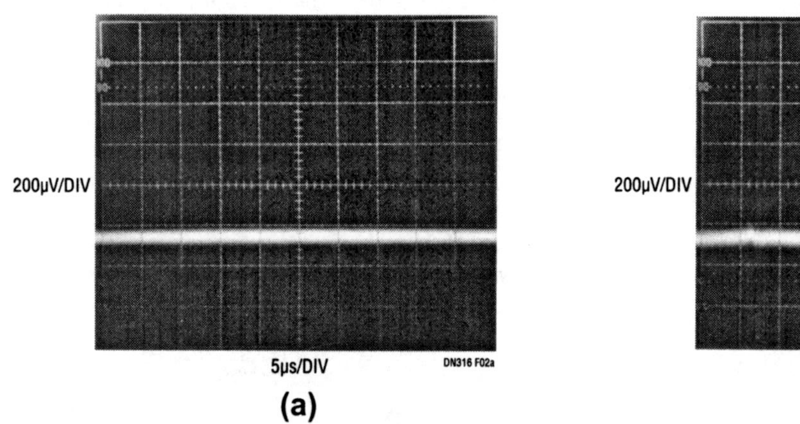

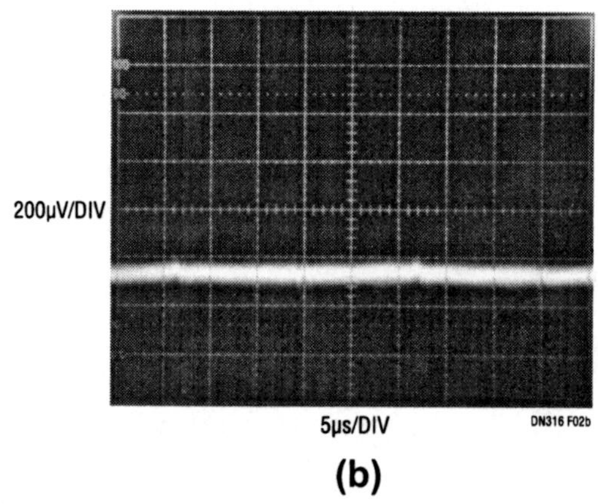

Figure 24.2 • (a) Test Setup Noise Floor. (b) $12V_{OUT}$ Noise at 80mA. Both Traces Were Captured with a 100MHz Measurement Bandwidth

Monolithic DC/DC converters break 1MHz to shrink board space

25

Mitchell Lee

In the never-ending quest for board space, operating frequency remains the most important variable in a DC/DC converter design. Higher frequency equates with smaller coils and capacitors. A new family of fast monolithic converters that allows circuit designers to reduce the size of their finished products is now available. Other improvements include quiescent currents well below those of slower converters and a new switch-drive technique that reduces dynamic losses by at least fourfold over previous methods, virtually eliminating these losses as a concern in efficiency calculations.

Table 25.1 shows the salient features of each member of the family. Each is configured as a grounded-switch step-up converter, but is equally useful in positive and negative high efficiency buck, SEPIC, inverting and flyback circuits. The converters are all based on the LT1372 design, which operates at 500kHz, draws 4mA quiescent current and contains a 1.5A, 0.5Ω switch. The LT1371 is designed for higher power applications, with a 3A, 0.25Ω switch. Supply current and operating frequency remain unchanged. Reduced quiescent current (1mA) makes the LT1373 useful in low power designs or in applications where the load current has a wide dynamic range. For the ultimate in miniaturization, the LT1377 features 1MHz operating frequency—especially helpful where post-filtering is employed.

Table 25.1 Family Characteristics

DEVICE	I_Q	SWITCH	FREE RUNNING FREQUENCY	SYNCHRONIZATION LIMIT
LT1371	4mA	3A	500kHz	800kHz
LT1372	4mA	1.5A	500kHz	800kHz
LT1373	1mA	1.5A	250kHz	360kHz
LT1377	4mA	1.5A	1MHz	1.6MHz

All devices share the same constant frequency PWM core, up to 90% duty cycle and 2.7V to 30V supply range with a maximum switch rating of 35V. Unique to these devices is a synchronization input that allows the internal oscillator to be overridden by an external clocking signal. The synchronization limit for each part is also shown in Table 25.1. Another unique feature is a second Feedback pin that allows direct regulation of negative outputs.

A simple boost converter using the LT1372 is shown in Figure 25.1. 350mA output current is available at 12V from a 5V input. Adaptive switch drive and a 0.5Ω collector resistance result in a peak efficiency of 87%, as shown in Figure 25.2.

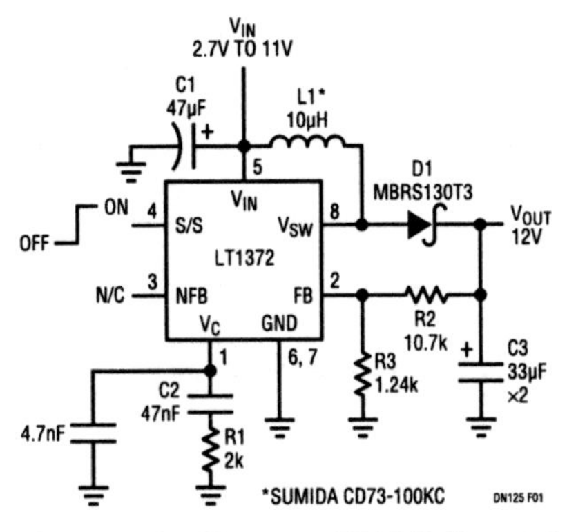

Figure 25.1 • Schematic Diagram: LT1372 Boost Converter

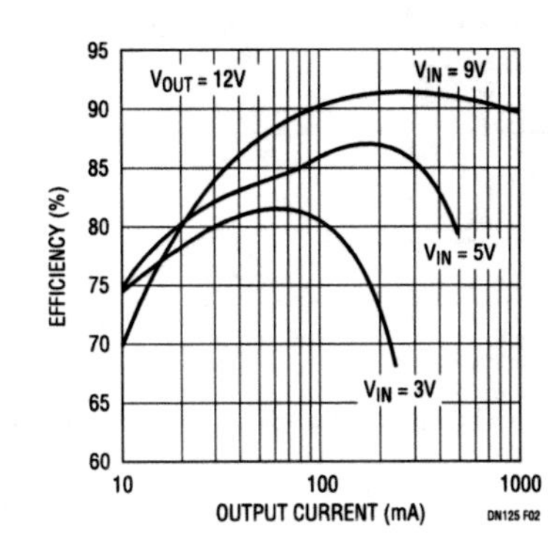

Figure 25.2 • Efficiency of Boost Converter Shown in Figure 25.1

Analog Circuit Design: Design Note Collection. http://dx.doi.org/10.1016/B978-0-12-800001-4.00025-9

Figure 25.3 shows a buck-boost (SEPIC) converter built around the 3-ampere LT1371. Inputs of 2.7V to 20V are converted to a 5V regulated output at up to 1.8A (see Figure 25.4). In spite of handling 9W output power, the 500kHz operating frequency of the LT1371 allows a 0.37-inch toroidal core to be used for the coupled inductor, with excellent efficiency. In shutdown, the output is completely disconnected from the input source.

The latest generation of disk drives has adopted magneto-resistive (MR) read-write heads. These operate with a low noise bias supply of −3V. Figure 25.5 shows a Cuk-configured LT1373 capable of generating −3V at 250mA. This converter topology exhibits inherently low output ripple and noise and uses a single-core coupled inductor. Operating at 250kHz allows the use of relatively small filter components. The speed-to-power ratio of the LT1373 is quite high; with only 1mA quiescent current, it maintains higher efficiency at light loads.

At the other end of the spectrum is the 1MHz LT1377. It has the same high speed-to-power ratio as the LT1373. In Figure 25.6 the LT1377 is used as a 50mA charger for four to six NiCd cells, operating from a 5V input. The charger is clamped against excessive output voltage at 11V, and maintains constant output current from 0V to 11V.

1MHz operation is also useful in radio applications where a 455kHz IF is present, as it gives one octave separation from that critical frequency. Figure 25.7 shows the LT1377 configured as a high efficiency buck converter, with a 5V, 1A output. A 20μH inductor is used in this application to maintain a low ripple current (10%), thus easing output filtering requirements.

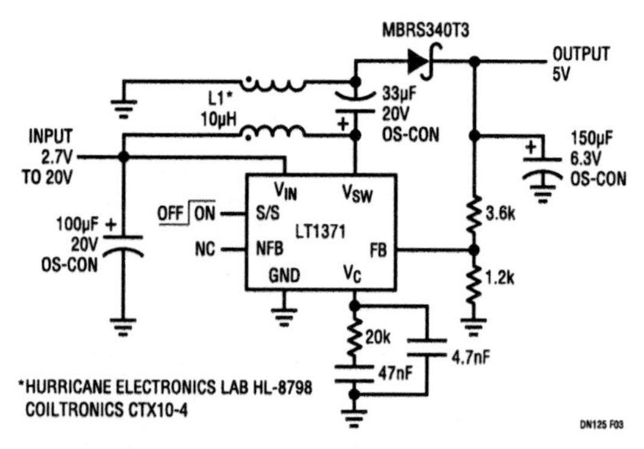

Figure 25.3 • 5V, 9W Converter Operates Over Wide Input Range with Good Efficiency

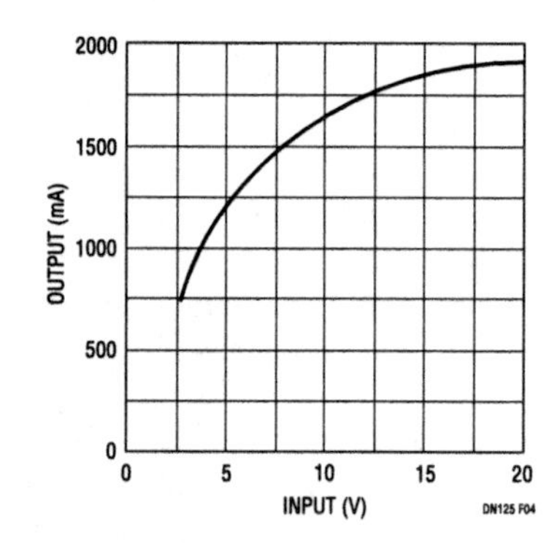

Figure 25.4 • Maximum Available Output Current of LT1371 9W Converter (Figure 25.3)

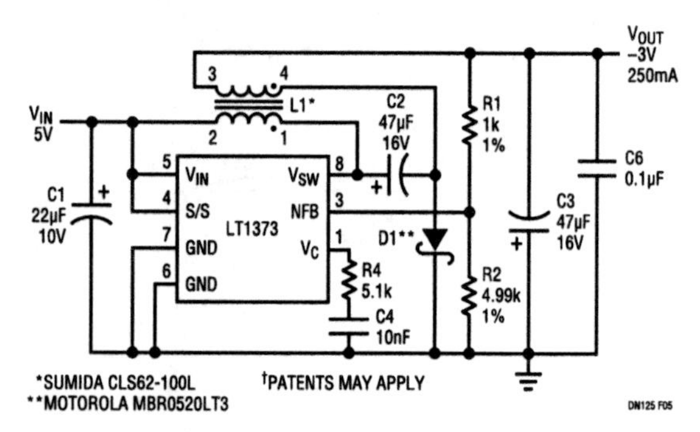

Figure 25.5 • Low Ripple 5V to −3V "Cuk"† Converter

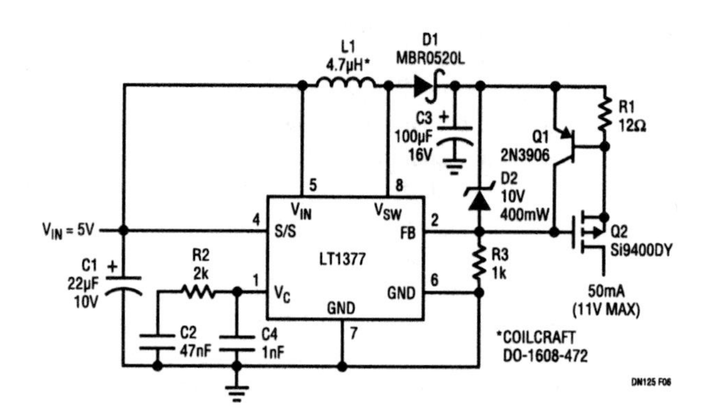

Figure 25.6 • Battery Charger

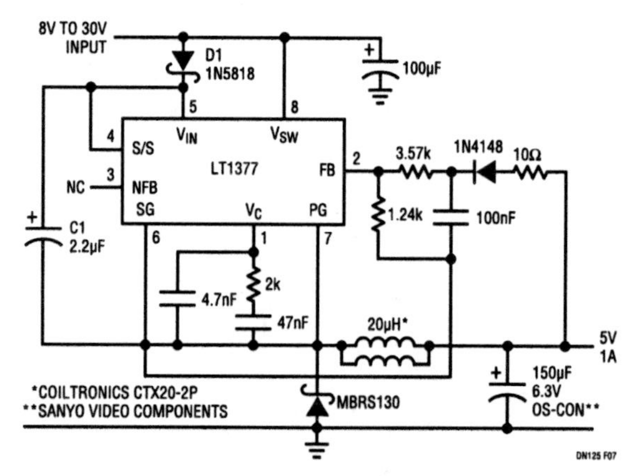

Figure 25.7 • 1MHz LT1377-Based Buck Converter

Capacitor and EMI considerations for new high frequency switching regulators

26

Carl Nelson   Bob Essaff

The LT1372 family of boost and flyback converters and the LT1376 buck converter form a new line of high frequency switching regulators offered by Linear Technology. Up to 10 times faster (250kHz to 1MHz) than older devices, these new converters are actually more efficient than older designs. Several design considerations for high frequency switchers are discussed here.

Capacitor technology considerations

At lower frequencies, magnetics dominates the size of DC-to-DC converters, but at frequencies above 200kHz the largest component is typically the input and output bypass capacitors. This makes it important to pick the right capacitor technology in critical size applications. The most important parameter in choosing the capacitor is cost versus size and impedance. The input and output capacitors must have low impedance at the switching frequency to limit voltage ripple and power dissipation.

It is also desirable to have low impedance at much higher frequencies because high amplitude narrow spikes are created if the capacitor has even a few nH of inductance. The amplitude of these spikes is determined by the rate-of-rise of current (dI/dt) fed to the capacitor. For a high speed converter, dI/dt may be as high as 0.5A/ns. Even 3nH of capacitor lead inductance will create $(0.5\ e^{9})\ (3\ e^{-9})=1.5$V spikes. The second part of this design note shows how to attenuate these spikes using parasitic PC board elements.

Figure 26.1 details the impedance characteristics of the most popular switching regulator capacitors. These are aluminum electrolytic, solid tantalum, OS-CON, and ceramic. Within one technology, impedance at lower frequencies tends to closely track physical volume of the capacitor, with larger volume giving lower impedance.

Aluminum electrolytics perform poorly and are rarely used at frequencies above 100kHz, but their low cost and higher voltage capability may cause them to be used for input bypass applications.

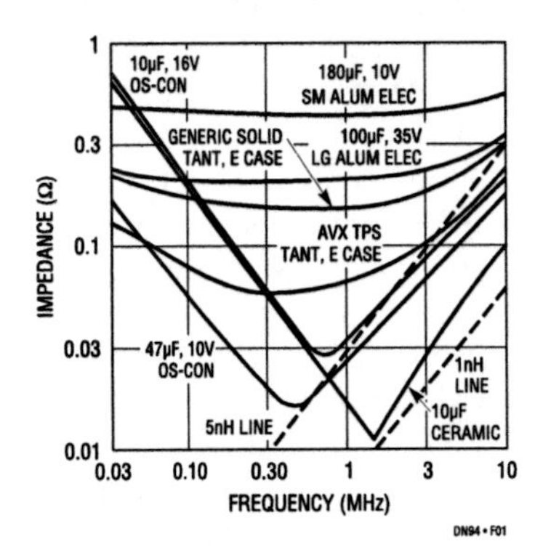

Figure 26.1 • Capacitor Impedance

Solid tantalum capacitors offer small size, low impedance, and low Q (to prevent loop stability problems). The downside of tantalum is limited voltage, typically 50V maximum, and a tendency for a small percentage of units to self-destruct when subjected to very high turn-on surge currents. AVX addresses the problem with their TPS line which features more rugged construction and special surge testing. Even TPS units must be voltage derated by 2:1 if high turn-on surges are expected, limiting practical operating voltage to 25V. Even with their shortcomings, solid tantalum seems to be the technology of choice for medium to high frequency DC-to-DC converters.

OS-CON capacitors from Sanyo and Marcon are made with a semiconductor-like dielectric that gives very low impedance per unit volume. They greatly outperform aluminum units, but have a few pitfalls of their own. Maximum voltage is 25V, height is significantly greater than solid tantalum, and most units are not surface mount. The very low ESR (effective series resistance) of OS-CON capacitors can also cause a problem with loop stability in switching regulators.

Ceramic capacitors offer the lowest impedance at 500kHz and beyond, but they suffer from high cost, large footprint, and limited temperature range. They are probably best suited

Analog Circuit Design: Design Note Collection. http://dx.doi.org/10.1016/B978-0-12-800001-4.00026-0

for input bypassing at higher voltages, and very high frequency (≥1MHz) applications.

Note that at high frequencies, most of the capacitors approach an inductive line of about 1nH to 5nH. Smaller units in parallel will reduce effective inductance, but cost goes up.

Controlling EMI: conducted and radiated

EMI strikes terror into system designers because its causes and cures are not well understood. There are two separate issues involved; meeting external FCC standards and avoiding internal system malfunctions. Higher frequency switching regulators would seem to make the whole problem much worse, but there are several mitigating circumstances that have allowed nearly all systems to use high frequency converters with few problems.

Contrary to popular misconceptions, older switching regulators rarely were a major contributor to overall system noise problems if the system contained reasonable amounts of logic chips. The noise created by a microprocessor, databusses, clock drivers, etc. usually swamps out switcher noise. This can be seen very simply by connecting a linear regulator temporarily in place of the switcher. Radiated noise using standard FCC methods, and supply line noise often show almost no change when the switcher is turned off.

The second reason for reduced high frequency switcher noise problems is that the components used are physically smaller. Radiated noise is proportional to radiating line length, so smaller, tightly packed components radiate significantly less.

Conducted EMI usually makes its appearance as ripple current fed back into the input supply, or as ripple voltage at the regulator output. Figure 26.2 shows how both of these effects can be virtually eliminated by using parasitic inductance in input or output lines. At 500kHz each inch of board trace represents nearly 0.1Ω of reactance, and for the higher harmonics the impedance is much higher. This impedance can be used in combination with existing capacitors to create "free" filter action.

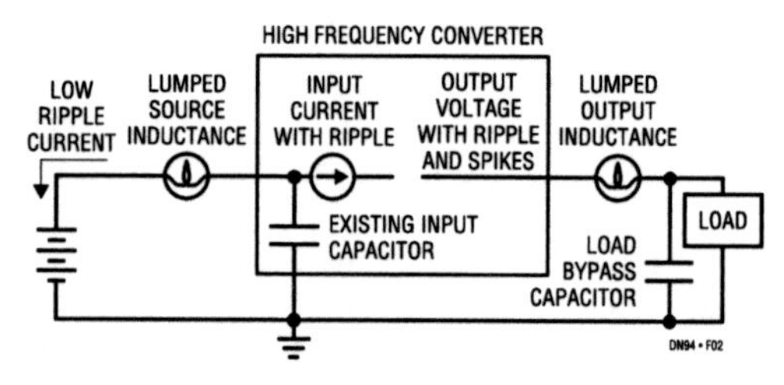

Figure 26.2 • Free High Frequency Filters Using Parasitics

At the input, as shown in Figure 26.3, the switching ripple current will all disappear into the input bypass capacitor if the reactance in the input lines is large compared to the capacitor impedance. For instance, an OS-CON 33μF, 20V capacitor has an impedance of about 0.02Ω at 500kHz. If the supply lines have an effective length of six inches (input plus return), the line reactance will be about 0.6Ω at 500kHz. The fundamental of the ripple current will be attenuated by the ratio of line impedance to capacitor impedance = 0.6/0.02 = 30:1. Higher harmonics will virtually disappear.

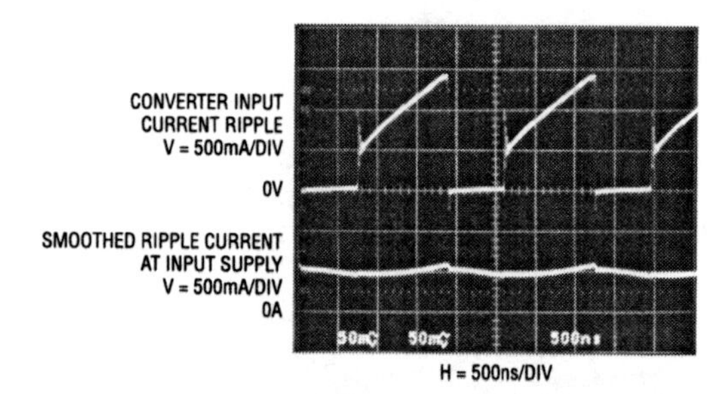

Figure 26.3 •

Output ripple can be filtered by utilizing the reactance of output lines in conjunction with load bypass capacitors as shown in Figure 26.4. With two inches of output trace (0.16Ω at 500kHz) and 0.15Ω capacitor impedance, fundamental ripple will be attenuated by two-to-one and output noise spikes virtually eliminated. The Fourier components of these spikes start above 25MHz where line reactance is 4Ω per inch.

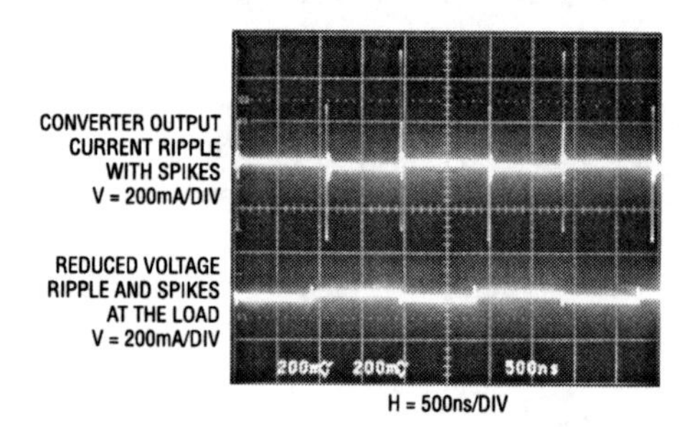

Figure 26.4 •

A common cause of radiated noise is the use of "open core" magnetics. The lowest cost high frequency inductors are wound on ferrite rods or barrels, which do not have a closed magnetic field path. This generates large local B fields that can play havoc with sensitive low level circuitry such as disk drives, A-to-D converters, and IF strips. Each application needs to be evaluated separately, but in situations where low signal levels exist in close proximity to the converter, closed core magnetics such as toroids or E cores should be used until extensive system level testing has proved out a cheaper solution.

Switching regulator generates both positive and negative supply with a single inductor

27

Brian Huffman

Many systems require ±12V from a 5V input. Analog or RS232 driver power supplies are obvious candidates. This requirement is usually solved by using a switcher with a multiple-secondary transformer or multiple switchers. These solutions can be complicated, requiring either transformer design or two inductors. An alternative approach, shown in Figure 27.1, uses a single inductor and charge pump to obtain the dual outputs. This solution is particularly noteworthy because is uses off-the-shelf components.

Figure 27.1 uses an LT1172 to generate both the positive and negative supply. The LT1172 is configured as a step-up converter to obtain the positive output. To generate the negative output a charge pump is used. The pump capacitor, C2, is charged up by the inductor when D2 is forward biased and discharges into C4 when the LT1172's power switch pulls the positive side of C2 to ground. The output capacitor provides current to the load during the charging cycle.

Figure 27.2 shows the regulator's operating waveforms. Since the LT1172 has a ground-referred power switch, the inductor has the input voltage applied across it when the switch is on. Trace A is the V_{SW} pin voltage and trace B is its current. The inductor current, trace C, rises slowly as the magnetic field builds up. The current rate of change is determined by the voltage applied across the inductor and its inductance. During this interval, energy is being stored in the inductor and no power is transferred to the +12V output. When the switch is turned off, energy is no longer transferred to the inductor, which causes the magnetic field to collapse. The collapsing magnetic field induces a change in voltage across the inductor causing the V_{SW} pin to rise until output diode D1 forward biases.

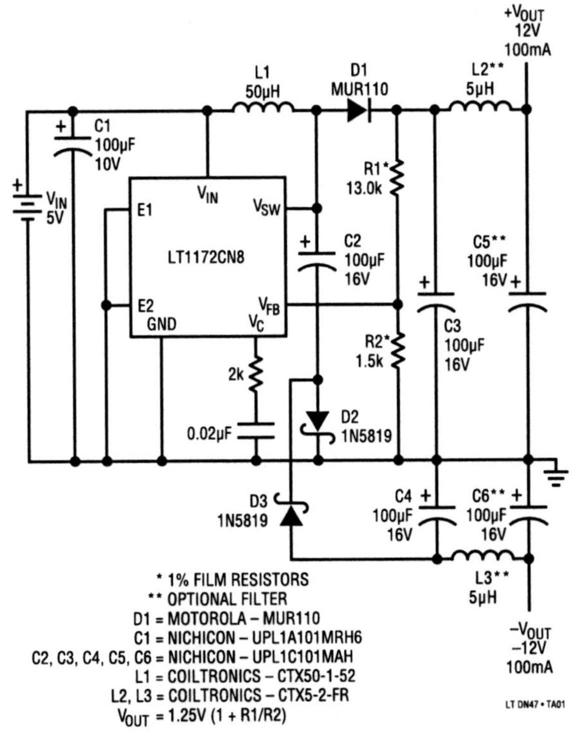

* 1% FILM RESISTORS
** OPTIONAL FILTER
D1 = MOTOROLA – MUR110
C1 = NICHICON – UPL1A101MRH6
C2, C3, C4, C5, C6 = NICHICON – UPL1C101MAH
L1 = COILTRONICS – CTX50-1-52
L2, L3 = COILTRONICS – CTX5-2-FR
V_{OUT} = 1.25V (1 + R1/R2)

Figure 27.1 • Inductor and Switch Capacitor Techniques Provide Bipolar Output

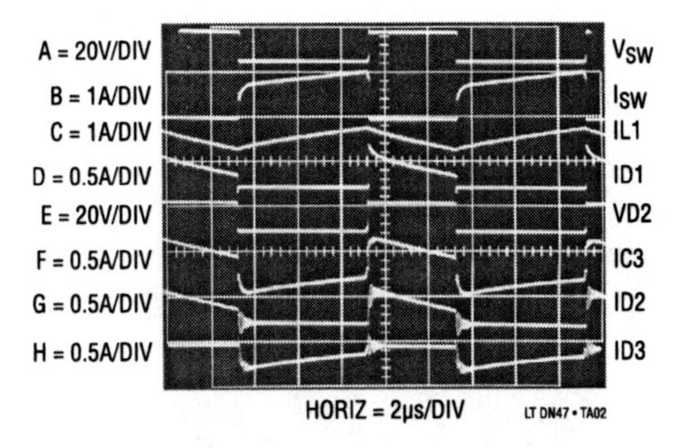

Figure 27.2 • Switching Waveforms for ±12V Output Converter

Analog Circuit Design: Design Note Collection. http://dx.doi.org/10.1016/B978-0-12-800001-4.00027-2

Trace D is the diode's current waveform. The diode provides a current path for the energy stored in the inductor to be transferred between the load and the output capacitor. When the diode is reverse biased, the output capacitor provides the load current. The LT1172's error amplifier compares the feedback pin voltage, from the 13kΩ–1.5kΩ divider, to its internal 1.24V reference and controls duty cycle. The output voltage can be varied by changing the R1–R2 divider ratio. An RC network at the V_C pin provides loop compensation.

A charge pump is used to invert the +12V output to a −12V output. When the LT1172's power switch turns off, the voltage on C2's positive side rises until D1 is forward biased. The inductor charges C2 when the voltage on C2's negative side rises enough to forward bias D2. Trace F shows C2's current waveform, trace E is D2's voltage waveform and trace G is its current. The voltage across C2 will be equal to a diode drop above $+V_{OUT}$ minus a Schottky diode drop. When the LT1172's power transistor turns on, the positive side of C2 is pulled to ground. During this period diode D3 is forward biased (trace H is its current waveform), and C4 is charged by C2. An optional LC filter is added to each output to attenuated output voltage ripple. Efficiency for this circuit generally exceeds 70%.

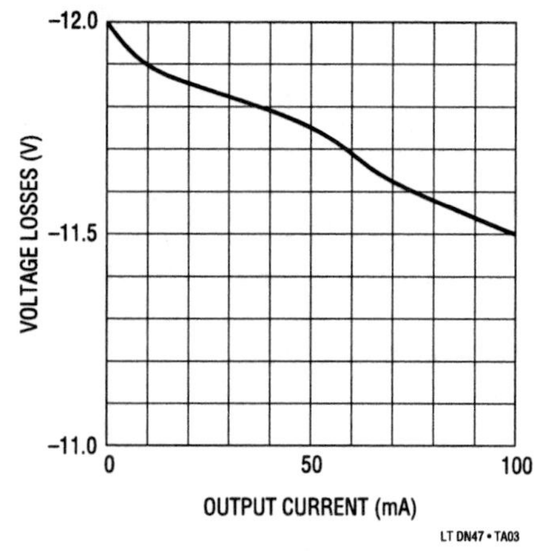

Figure 27.3 • Losses for Charge Pump Converter

Diode junction losses (D2 and D3) preclude ideal results, but performance is quite good. This circuit will convert $+V_{OUT}$ to $-V_{OUT}$ with losses as shown in Figure 27.3. Negative output load current should not exceed the positive output load by more than a factor of 5, otherwise the imbalance will cause the −12V transient response to suffer.

Figure 27.4 can be used for an LCD display contrast control. It is similar to the previous circuit except that all the load current is drawn from the negative output. This requires C3 to be small so negative output load fluctuations are quickly reflected to the positive output. Resistor R3 adjusts output voltage between −12V and −21V.

The LT1172 provides an elegant solution to power shutdown problems by integrating a shutdown feature, eliminating the need to place a power MOSFET in series with the input voltage. When the voltage of the V_C pin is pulled below 150mV, the IC shuts down pulling only 150µA. This is implemented by turning on Q1, reducing the circuit's quiescent current from 6mA to 150µA.

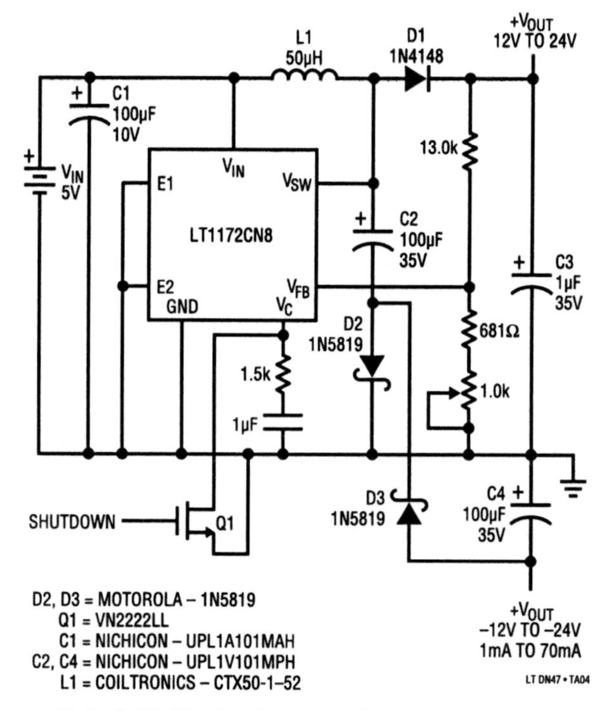

Figure 27.4 • LCD Display Contrast Control Power Supply

Floating input extends regulator capabilities

28

Brian Huffman

Many applications require circuit performance that is unachievable with conventional regulator design. This results in added complexity to the circuit. However, some problems can easily be solved by floating the input to the regulator. A floating input can either be a battery, or a secondary winding that is galvanically isolated from all other windings. With this method high efficiency negative voltage regulation, high voltage regulation, and low saturation loss positive buck switching regulator can all be achieved easily.

Low dropout negative voltage regulators are not currently available. This would seem to preclude high efficiency negative linear regulators. Such regulation is frequently desired in switching supply post regulators; however, if the secondary windings are isolated from one another, a low dropout positive voltage regulator can be used for negative regulation (Figure 28.1).

In this circuit the LT1086 servos the voltage between the output and the adjust pin to 1.25V. The positive regulation is accomplished by conventional regulator design. Negative voltage regulation is achieved by connecting the output of the positive voltage regulator to ground. The V_{IN} pin floats to 1.5V or greater, above ground. This technique can be used with any positive voltage regulator, although highest efficiency occurs with low dropout types.

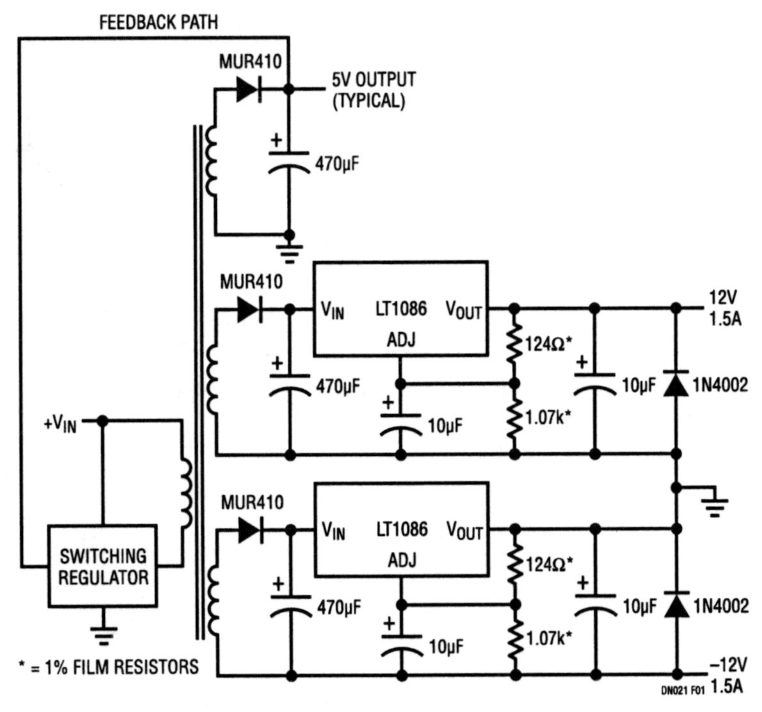

Figure 28.1 • High Efficiency Negative Voltage Regulation

Analog Circuit Design: Design Note Collection. http://dx.doi.org/10.1016/B978-0-12-800001-4.00028-4

Another example where floating a linear regulator can be useful is shown in Figure 28.2. In this case high voltage regulation can be handled if split secondary windings are available. This allows the regulators to be connected in series. Neither regulator exceeds its maximum differential voltage even under short-circuit conditions.

High current positive buck switching regulators can have excessive saturation losses since most switches are Darlingtons. As much as 2V can be dropped across a Darlington or composite PNP switching transistor. However, efficiency can be increased and power dissipation requirements greatly reduced if the input is allowed to float (Figure 28.3).

The circuit in Figure 28.3 uses an LT1070 to perform a buck conversion. The LT1070 is a current mode switching regulator. The V_{SW} pin output is a collector of a common emitter NPN, so current flows through it when it is low. The 40kHz repetition rate is set by the LT1070's internal oscillator. When the V_{SW} pins "on," current flows through the load, the inductor, and into the V_{SW} pin. During this time a magnetic field is built up in the inductor. When the switch is turned "off," the magnetic field collapses dumping energy into the load through D1. The input of the switching regulator floats to a potential set by the output.

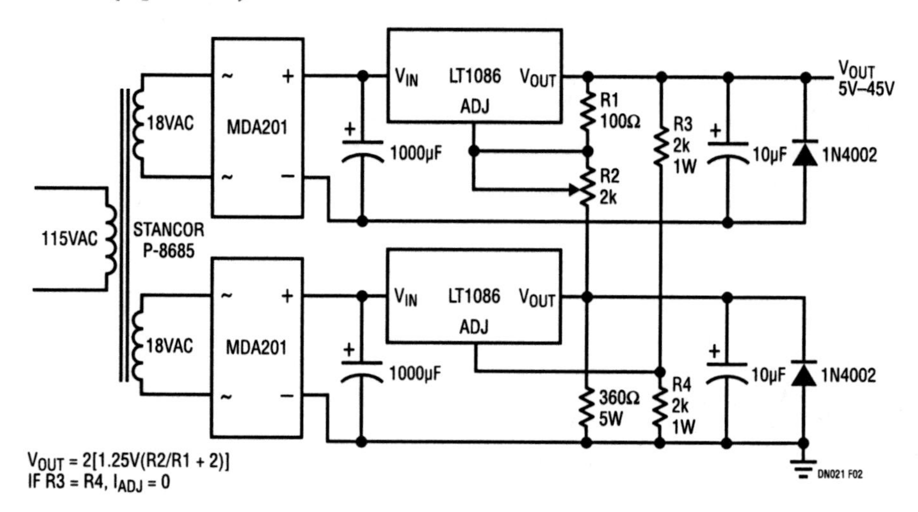

Figure 28.2 • High Voltage Regulation

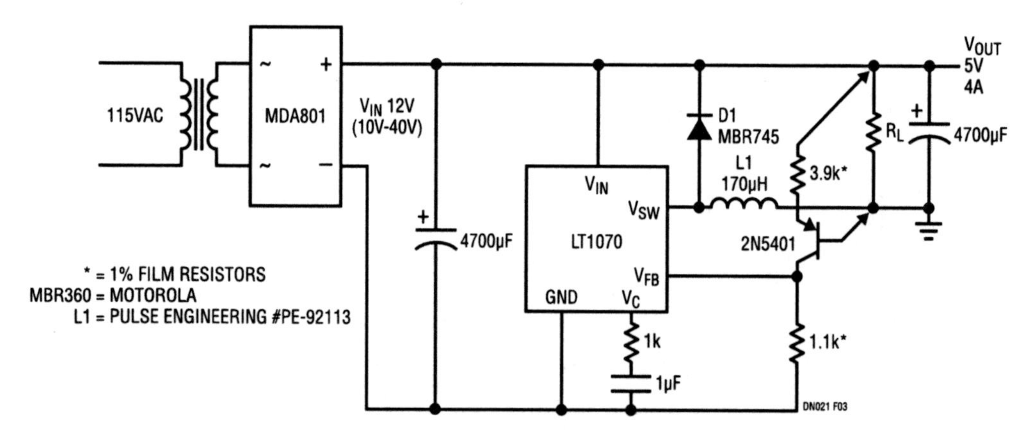

Figure 28.3 • Floating Input Low Saturation Loss Buck Regulator

Programming pulse generators for flash memories

29

Jim Williams

Recently introduced "flash" memories add electrical chip-erasure and reprogramming to established EPROM technology. These features make them a cost-effective and reliable alternative for updatable non-volatile memory. Utilizing the electrical program-erase capability requires linear circuitry techniques. The Intel 28F256 flash memory, built on the ETOX process, specifies programming operation with 12V or 12.75V (faster erase/program times) amplitude pulses. These "V_{PP}" amplitudes must fall within 1.6%, and excursions beyond 14.0V will damage the device.

Providing the V_{PP} pulse requires generating and controlling high voltages within the tightly specified limits. Figure 29.1's circuit does this. When the V_{PP} command pulse goes low (trace A, Figure 29.2) the LT1072 switching regulator drives L1, producing high voltage. DC feedback occurs via R1 and R2, with AC roll-off controlled by C1 and R3-C2. The result is a smoothly rising V_{PP} pulse (trace B) which settles to the required value. The specified R1 values allow either 12V or 12.75V outputs. The 5.6V Zener permits the output to return to 0V when the V_{PP} command goes high. It may be deleted in cases where a 4.5V minimum output is acceptable (see Intel 28F256 data sheet). The 0.1% resistors combine with the LT1072's tight internal reference to eliminate circuit trimming requirements. Additionally, this circuit will not spuriously overshoot during power-up or -down.

Figure 29.1's repetition rate is limited because the regulator must fully rise and settle for each V_{PP} command. Figure 29.3's circuit serves cases which require higher repetition rate V_{PP} pulses. Here, the switching regulator runs continuously, with the V_{PP} pulses generated by the A1-A2 loop. If desired, the "V_{PP} Lock" line can be driven, shutting down the regulator to preclude any possibility of inadvertent V_{PP} outputs. When V_{PP} Lock goes low (trace A, Figure 29.4) the LT1072 loop comes on (trace B), stabilizing at about 17V. Pulsing the V_{PP} command line low causes the 74C04 (trace C) to bias the LT1004 reference. The LT1004 clamps at 1.23V with A1 and A2 giving a scaled output (trace D). The 680pF capacitor controls loop slewing, eliminating overshoots. Figure 29.5 details the V_{PP} output. Trace A is the 74C04 output, with trace B showing clean V_{PP} characteristics. As in Figure 29.1, spurious V_{PP} outputs are suppressed during power-up or -down. The diode path around A2 prevents overshoot during short circuit recovery.

A good question might be "Why not set the switching regulator output voltage at the desired V_{PP} level and use a simple low resistance FET or bipolar switch?" Figure 29.6 shows that this is a potentially dangerous approach. Figure 29.6a shows the clean output of a low resistance switch operating directly at the V_{PP} supply. The PC trace run to the memory chip looks like a transmission line with ill-defined termination characteristics. As such, Figure 29.6a's clean pulse degrades

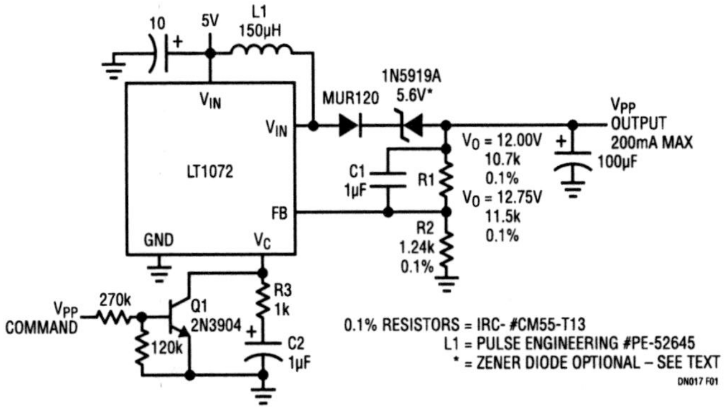

Figure 29.1 • Basic Flash Memory V_{PP} Pulse Generator

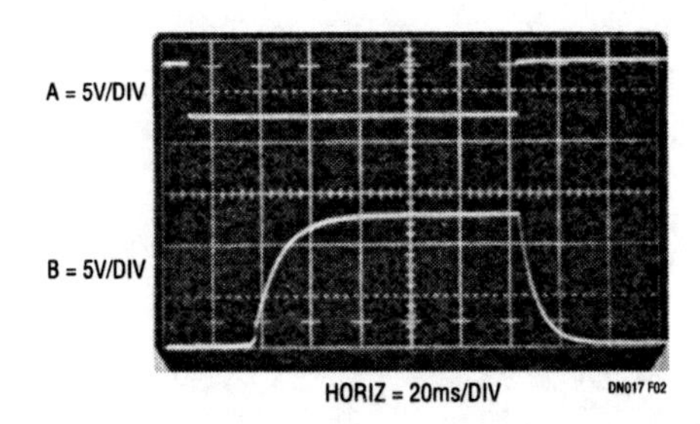

Figure 29.2 • Waveforms for Basic Flash Memory Pulser

Analog Circuit Design: Design Note Collection. http://dx.doi.org/10.1016/B978-0-12-800001-4.00029-6

and rings badly (Figure 29.6b) at the memory IC's pins. Overshoot exceeds 20V, well beyond the 14V destruction level. The controlled edge times of the circuits discussed eliminate this problem. Further discussion of these and other circuits appears in Application Note 31.

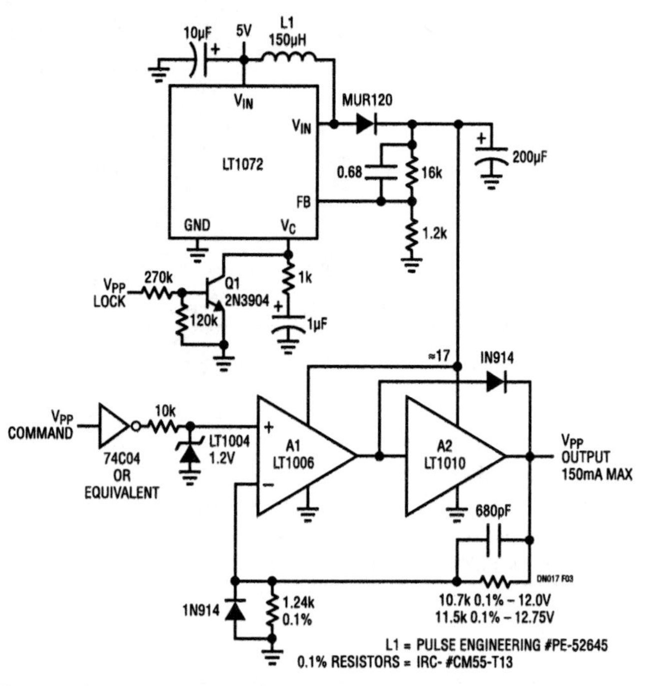

Figure 29.3 • High Repetition Rate V_{PP} Pulse Generator

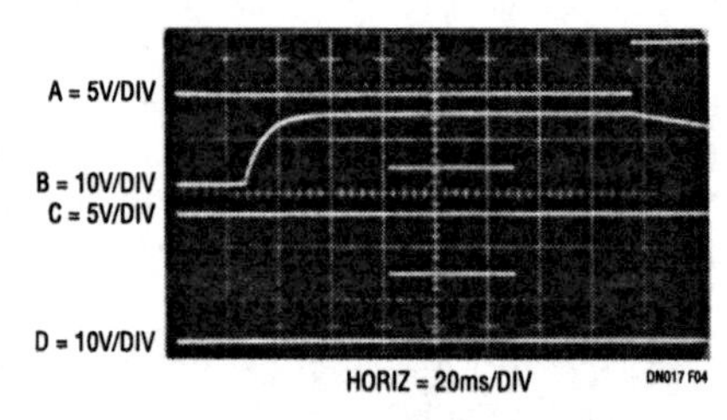

Figure 29.4 • Operating Details of High Repetition Rate Flash Memory Pulser

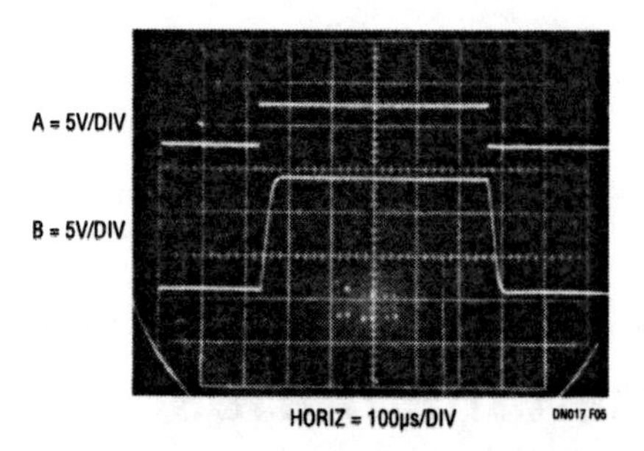

Figure 29.5 • Expanded Scale Display of Figure 29.3's V_{PP} Pulse. Controlled Risetime Eliminates Overshoots

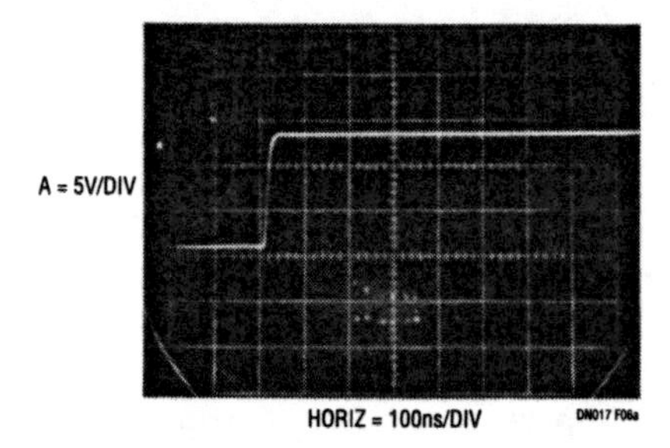

Figure 29.6a • An "Ideal" Flash Memory V_{PP} Pulse

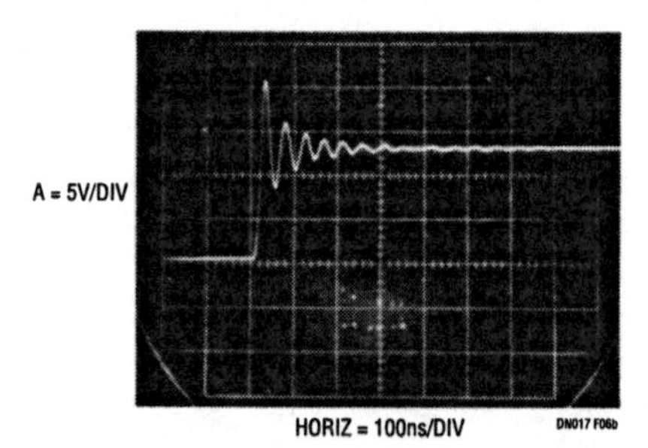

Figure 29.6b • Rings at Destructive Voltages After a PC Trace Run

Achieving microamp quiescent current in switching regulators

30

Jim Williams

Many battery-powered applications require very wide ranges of power supply output current. Normal conditions require currents in the ampere range, while standby or “sleep” modes draw only microamperes. A typical laptop computer may draw 1 to 2amperes running while needing only a few hundred microamps for memory when turned off. In theory, any switching regulator designed for loop stability under no-load conditions will work. In practice, a regulator's relatively large quiescent current may cause unacceptable battery drain during low output current intervals.

Figure 30.1 shows a typical flyback regulator. In this case the 6V battery is converted to a 12V output by the inductive flyback voltage produced each time the LT1070's V_{SW} pin is internally switched to ground. An internal 40kHz clock produces a flyback event every 25μs. The energy in this event is controlled by the IC's internal error amplifier, which acts to force the feedback (FB) pin to a 1.23V reference. The error amplifiers high impedance output (the V_C pin) uses an RC damper for stable loop compensation.

This circuit works well but pulls 9mA of quiescent current. If battery capacity is limited by size or weight this may be too high. How can this figure be reduced while retaining high current performance?

A solution is suggested by considering an auxiliary V_C pin function. If the V_C pin is pulled within 150mV of ground the IC shuts down, pulling only 50μA. Figure 30.2's special loop exploits this feature, reducing quiescent current to only 150μA. Here, circuitry is placed between the feedback divider and the V_C pin. The LT1070's internal feedback amplifier and reference are not used. Figure 30.3 shows operating waveforms under no-load conditions. The 12V output (trace A) ramps down over a period of seconds. During this time comparator A1's output (trace B) is low, as are the paralleled inverters. This pulls the V_C pin (trace C) low, putting the IC in its 50μA shutdown mode. The V_{SW} pin (trace D) is high, and no inductor current flows. When the 12V output drops about 20mV, A1 triggers and the inverters (74C04) go high, pulling the V_C pin up and turning on the regulator. The V_{SW} pin pulses the inductor at the 40kHz clock rate, causing the output to abruptly rise. This action trips A1 low, forcing the V_C pin back to shutdown. This “bang-bang” control loop keeps the 12V output within the 20mV ramp hysteresis window set by R3-R4. Diode clamps prevent V_C pin overdrive. Note that the loop oscillation period of 4–5seconds means the R6-C2 time constant at V_C is not a significant term. Because the LT1070 spends almost all of the time in shutdown, very little quiescent current (150μA) is drawn.

Figure 30.4 shows the same waveforms with the load increased to 3mA. Loop oscillation frequency increases to keep up with the loads sink current demand. Now, the V_C pin waveform (trace C) begins to take on a filtered appearance. This is due to R6-C2's 10ms time constant. If the load continues to increase, loop oscillation frequency will also increase. The R6-C2 time constant, however, is fixed. Beyond some frequency, R6-C2 must average loop oscillations to DC.

Figure 30.5 plots what occurs, with a pleasant surprise. As output current rises, loop oscillation frequency also rises until about 500Hz. At this point the R6-C2 time constant filters the V_C pin to DC and the LT1070 transitions into “normal” operation. With the V_C pin at DC it is convenient to think of A1 and the inverters as a linear error amplifier with a closed loop gain set by the R1-R2 feedback divider. In fact, A1 is still duty cycle modulating, but at a rate far above R6-C2's break frequency. The phase error contributed by C1 (which was selected for low loop frequency at low output currents) is dominated by the R6-C2 roll off and the R7-C3 lead into A1. The loop is stable and responds linearly for all loads beyond 80mA. In this high current region the LT1070 behaves like Figure 30.1's circuit.

Analog Circuit Design: Design Note Collection. http://dx.doi.org/10.1016/B978-0-12-800001-4.00030-2

The loop described provides a controlled, conditional instability to lower regulator quiescent current by a factor of 60 without sacrificing high power performance. Although demonstrated in a boost converter, it is readily exportable to other configurations, (e.g., multi-output flyback, buck, etc.) allowing LT1070 use in low quiescent power applications.

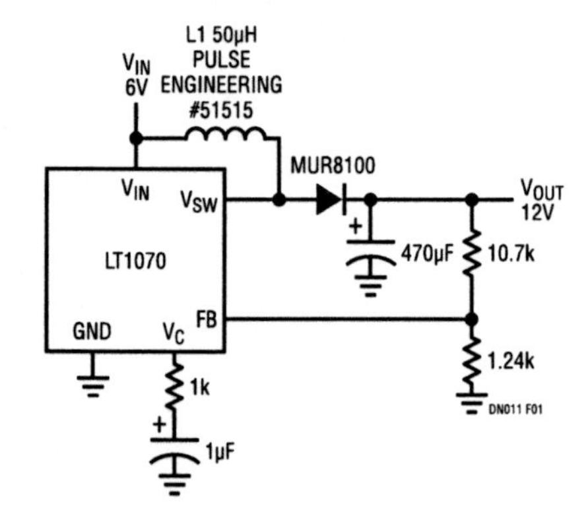

Figure 30.1 • Typical LT1070 Flyback Regulator

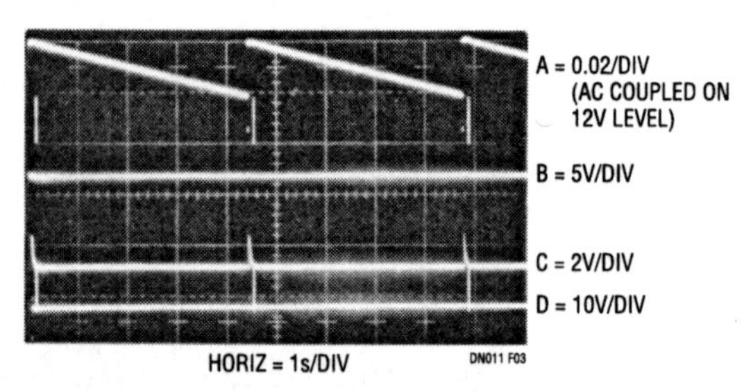

Figure 30.3 • Waveforms at No Load for Figure 30.2 (Traces B and D Retouched for Clarity)

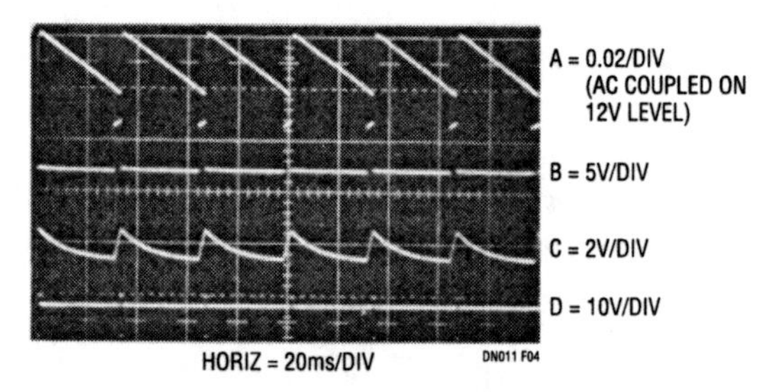

Figure 30.4 • Waveforms at 3mA Load for Figure 30.2

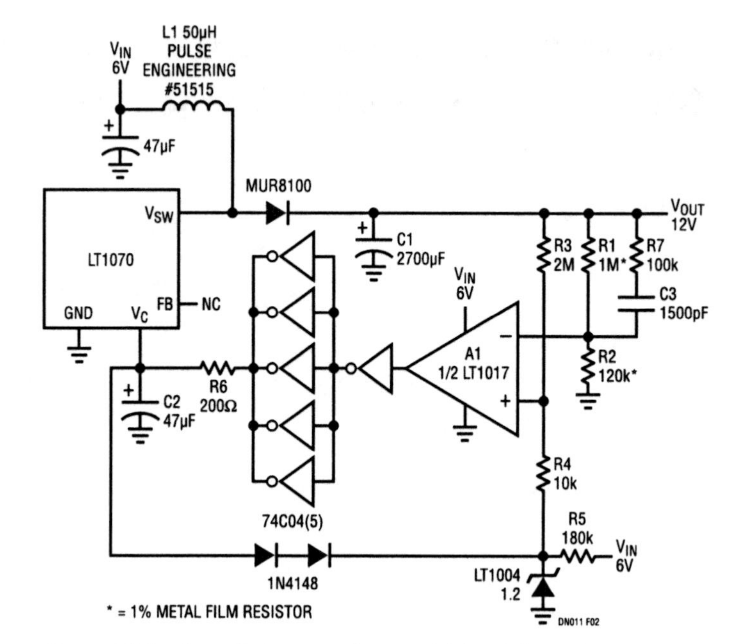

Figure 30.2 • Low Quiescent Current Flyback Regulator

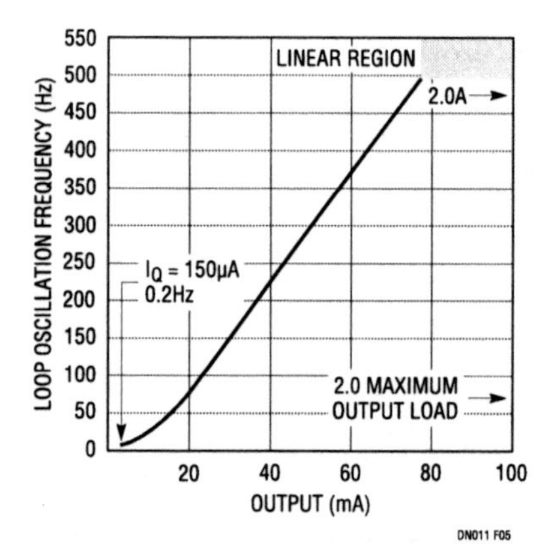

Figure 30.5 • Output Current vs Loop Oscillation Frequency for Figure 30.2

Inductor selection for switching regulators

31

Jim Williams

A common problem area in switching regulator design is the inductor, and the most common difficulty is saturation. An inductor is saturated when it cannot hold any more magnetic flux. As an inductor arrives at saturation it begins to look more resistive and less inductive. Under these conditions current flow is limited only by the inductor's DC copper resistance and the source capacity. This is why saturation often results in destructive failures.

While saturation is a prime concern, cost, heating, size, availability and desired performance are also significant. Electromagnetic theory, although applicable to these issues, can be confusing, particularly to the non-specialist.

Practically speaking, an empirical approach is often a good way to approach inductor selection. It permits real time analysis under actual circuit operating conditions using the ultimate simulator—a breadboard. If desired, inductor design theory can be used to augment or confirm experimental results.

Figure 31.1 shows a typical flyback regulator utilizing the LT1070 switching regulator. A simple approach may be employed to determine the appropriate inductor. A very useful tool is the #845 inductor kit[1] shown in Figure 31.2. This kit provides a broad range of inductors for evaluation in test circuits such as Figure 31.1.

Figure 31.3 was taken with a 450μH value, high core capacity inductor installed. Circuit operating conditions such as input voltage and loading are set at levels appropriate to the intended application. Trace A is the LT1070's V_{SWITCH}

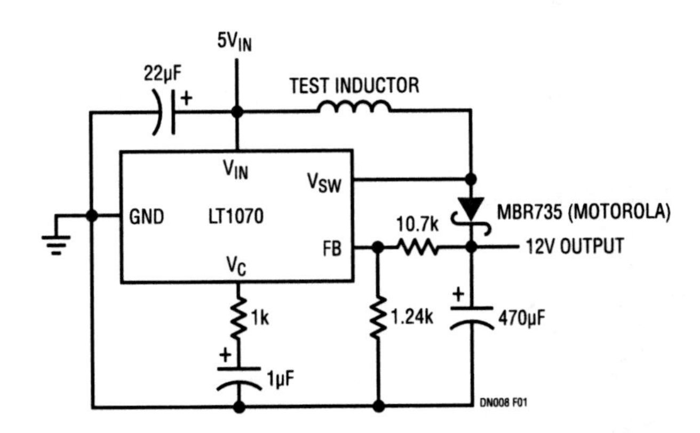

Figure 31.1 • Basic LT1070 Flyback Regulator Test Circuit

Note 1: Available from Pulse Engineering, Inc., P.O. Box 12235, San Diego, CA 92112, 619-268-2400.

Analog Circuit Design: Design Note Collection. http://dx.doi.org/10.1016/B978-0-12-800001-4.00031-4

pin voltage while trace B shows its current. When V_{SWITCH} pin voltage is low, inductor current flows. The high inductance means current rises relatively slowly, resulting in the shallow slope observed. Behavior is linear, indicating no saturation problems. In Figure 31.4, a lower value unit with equivalent core characteristics is tried. Current rise is steeper, but saturation is not encountered. Figure 31.5's selected inductance is still lower, although core characteristics are similar. Here, the current ramp is quite pronounced, but well controlled. Figure 31.6 brings some informative surprises. This high value unit, wound on a low capacity core, starts out well but heads rapidly into saturation, and is clearly unsuitable.

The described procedure narrows the inductor choice within a range of devices. Several were seen to produce acceptable electrical results, and the "best" unit can be further selected on the basis of cost, size, heating and other parameters. A standard device in the kit may suffice, or a derived version can be supplied by the manufacturer.

Using the standard products in the kit minimizes specification uncertainties, accelerating the dialogue between user and inductor vendor.

References

1. Williams, J., "Switching Regulators for Poets," Linear Technology Corporation, Application Note 25.
2. Nelson, C., "LT1070 Design Manual," Linear Technology Corporation, Application Note 19.

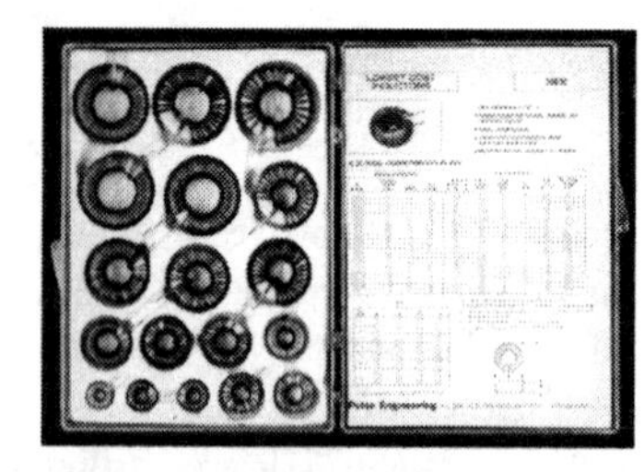

Figure 31.2 • Model 845 Inductor Selection Kit from Pulse Engineering, Inc. (Includes 18 Fully Specified Devices)

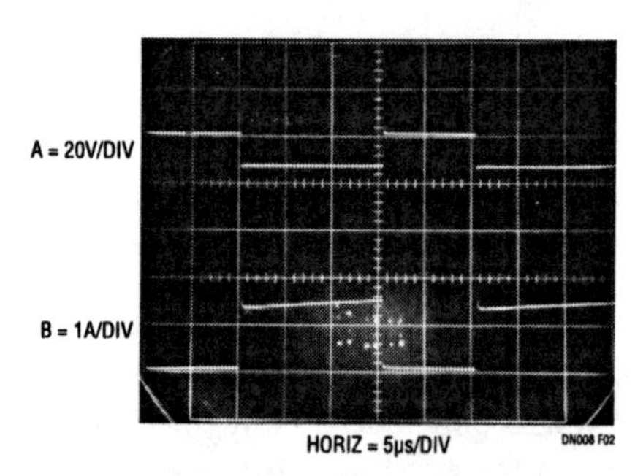

Figure 31.3 • Waveforms for 450μH, High Core Capacity Unit

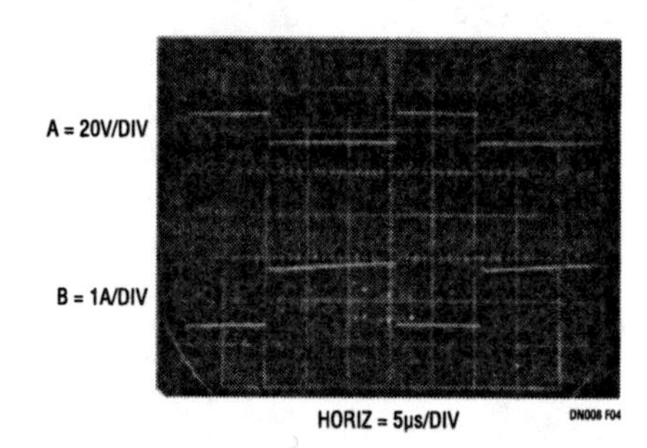

Figure 31.4 • Waveforms for 170μH, High Capacity Core Unit

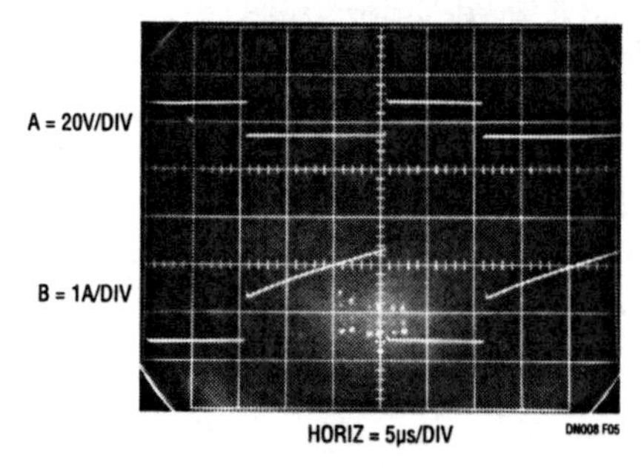

Figure 31.5 • Waveforms for 55μH, High Capacity Core Unit

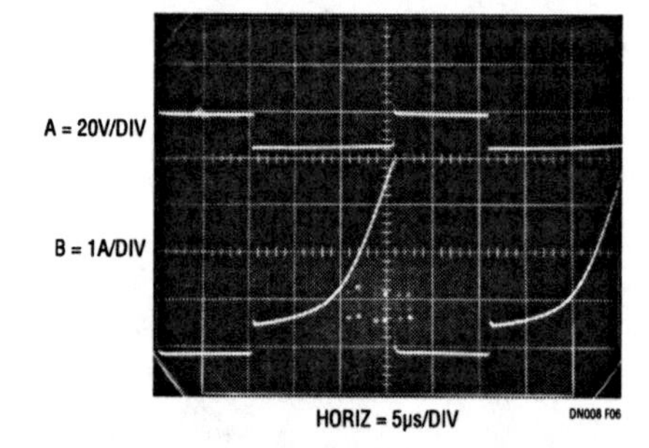

Figure 31.6 • Waveforms for 500μH, Low Capacity Core Inductor (Note Saturation Effects)

Section 4

Switching Regulator Design: Buck (Step-Down)

Inverting DC/DC controller converts a positive input to a negative output with a single inductor

32

David Burgoon

There are several ways to produce a negative voltage from a positive voltage source, including using a transformer or two inductors and/or multiple switches. However, none are as easy as using the LTC3863, which is elegant in its simplicity, has superior efficiency at light loads and reduces parts count compared to alternative solutions.

Advanced controller capabilities

The LTC3863 can produce a −0.4V to −150V negative output voltage from a positive input range of 3.5V to 60V. It uses a single-inductor topology with one active P-channel MOSFET switch and one diode. The high level of integration yields a simple, low parts-count solution.

The LTC3863 offers excellent light load efficiency, drawing only 70μA quiescent current in user-programmable Burst Mode operation. Its peak current mode, constant frequency PWM architecture provides positive control of inductor current, easy loop compensation and superior loop dynamics. The switching frequency can be programmed from 50kHz to 850kHz with an external resistor and can be synchronized to an external clock from 75kHz to 750kHz. The LTC3863 offers programmable soft-start or output tracking. Safety features include overvoltage, overcurrent and short-circuit protection, including frequency foldback.

−5.2V, 1.7A converter operates from a 4.5V to 16V source

The circuit shown in Figure 32.1 produces a −5.2V, 1.7A output from a 4.5V–16V input. Operation is similar to a flyback converter, storing energy in the inductor when the switch is on and releasing it through the diode to the output when the switch is off, except that with the LTC3863, no transformer is required. To prevent excessive current that can result from minimum on-time when the output is short-circuited, the controller folds back the switching frequency when the output is less than half of nominal.

The LTC3863 can be programmed to enter either high efficiency Burst Mode operation or pulse-skipping at light loads. In Burst Mode operation, the controller directs fewer, higher current pulses and then enters a low current quiescent state for a period of time depending on load. In pulse-skipping mode, the LTC3863 skips pulses at light loads. In this mode,

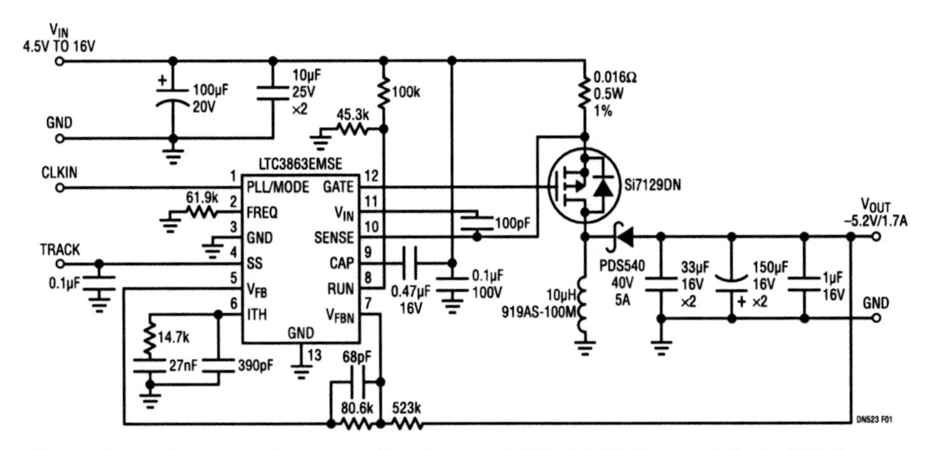

Figure 32.1 • Inverting Converter Produces −5.2V at 1.7A from a 4.5V to 16V Source

Analog Circuit Design: Design Note Collection. http://dx.doi.org/10.1016/B978-0-12-800001-4.00032-6

the modulation comparator may remain tripped for several cycles and force the external MOSFET to stay off, thereby skipping pulses. This mode offers the benefits of smaller output ripple, lower audible noise and reduced RF interference, at the expense of lower efficiency compared to Burst Mode operation. This circuit fits in about $0.5in^2$ ($3.2cm^2$) with components on both sides of the board.

Figure 32.2 shows the switch node voltage, inductor current and ripple waveforms at 5V input and −5.2V output at 1.7A. The inductor is charged (current rises) when the PMOSFET is on, and discharges through the diode to the output when the PMOS turns off. Figure 32.3 shows the same waveforms at 70mA out in pulse-skipping mode. Notice how the switch node rings out around 0V when the inductor current reaches zero. The effective period stops when the current reaches zero. Figure 32.4 shows the same load condition with Burst Mode operation enabled. Power dissipation drops by 31% at this operating point, and efficiency increases from 74% to 80.5%. At 12V input, the 45% reduction in dissipation is even more dramatic.

High efficiency

Figure 32.5 shows efficiency curves for both pulse-skipping and Burst Mode operation. Exceptional efficiency of 85.2% is achieved at 1.7A load and 12V input. Note how Burst Mode operation dramatically improves efficiency at loads less than 0.2A. Pulse-skipping efficiency at light loads is still much higher than that obtained from continuous conduction.

Conclusion

The LTC3863 simplifies the design of converters producing a negative output from a positive source. It is elegant in its simplicity, high in efficiency, and requires only a few inexpensive external components.

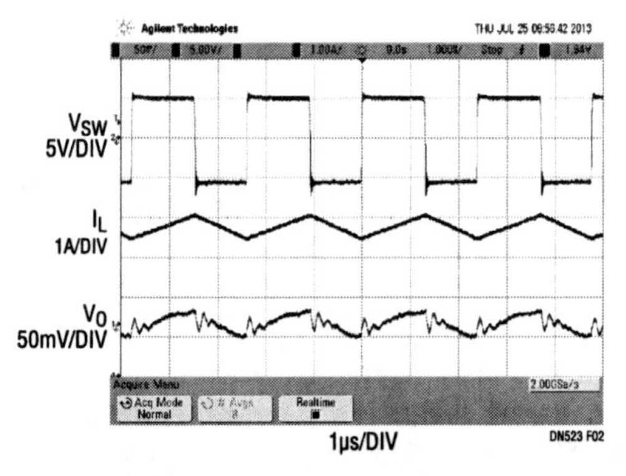

Figure 32.2 • Switch Node Voltage, Inductor Current and Ripple Waveforms at 5V Input and −5.2V Output at 1.7A

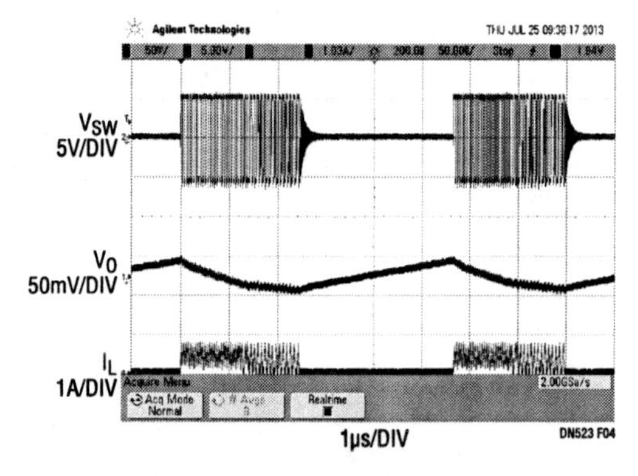

Figure 32.4 • Switch Node Voltage, Inductor Current and Ripple Waveforms at 5V Input and −5.2V Output at 70mA in Burst Mode Operation

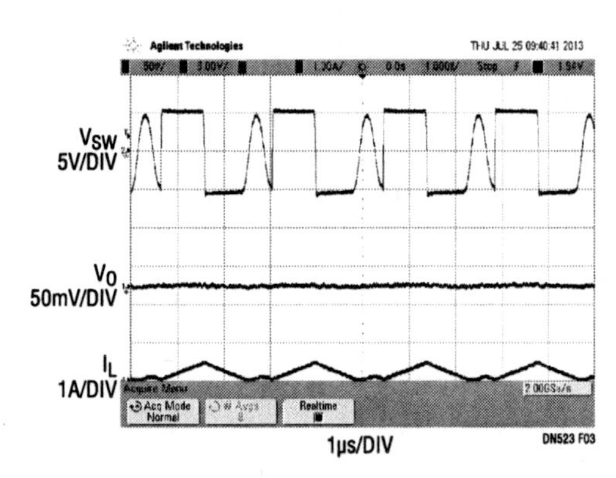

Figure 32.3 • Switch Node Voltage, Inductor Current and Ripple Waveforms at 5V Input and −5.2V Output at 70mA in Pulse-Skipping Mode

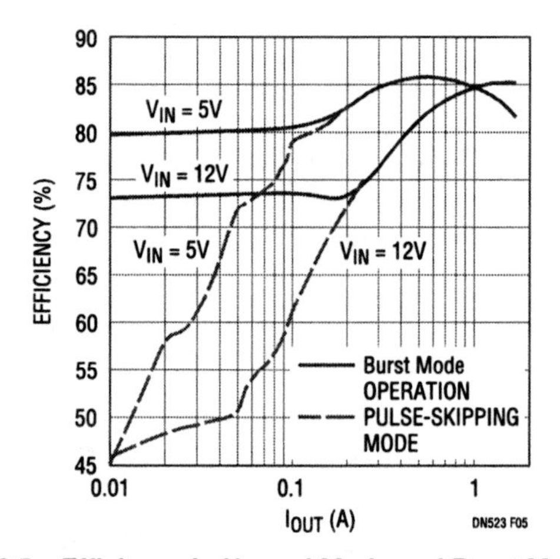

Figure 32.5 • Efficiency in Normal Mode and Burst Mode Operation of the Circuit in Figure 32.1

20V, 2.5A monolithic synchronous buck SWITCHER+ with input current, output current and temperature sensing/limiting capabilities

33

Tom Gross

Introduction

The LTC3626 synchronous buck regulator with current and temperature monitoring is the first of Linear's SWITCHER+ line of monolithic regulators. It is a high efficiency, monolithic synchronous step-down switching regulator capable of delivering a maximum output current of 2.5A from an input voltage ranging from 3.6V to 20V (circuit shown in Figure 33.1). The LTC3626 employs a unique controlled on-time/constant frequency current mode architecture, making it ideal for low duty cycle applications and high frequency operation, while yielding fast response to load transients (see Figure 33.2). It also features mode setting, tracking and synchronization capabilities. The LTC3626's 3mm × 4mm package has such low thermal impedance that it can operate without an external heat sink even while delivering maximum power to the load.

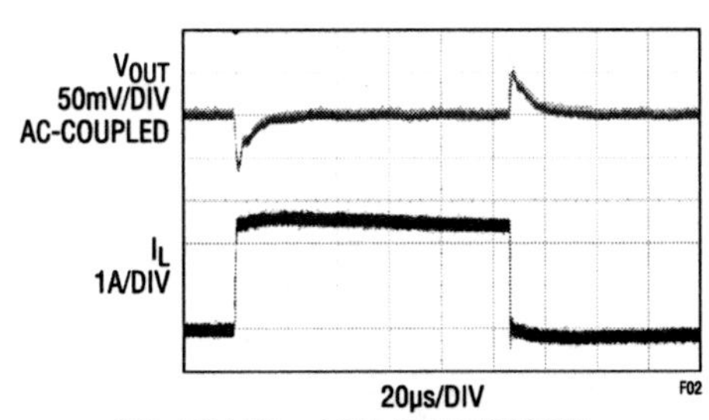

Figure 33.2 • Load Step Response for Figure 33.1 Circuit

Beyond its impressive regulator capabilities, the LTC3626's current and temperature monitoring functions stand out. They offer both monitoring and control capabilities with minimal additional components.

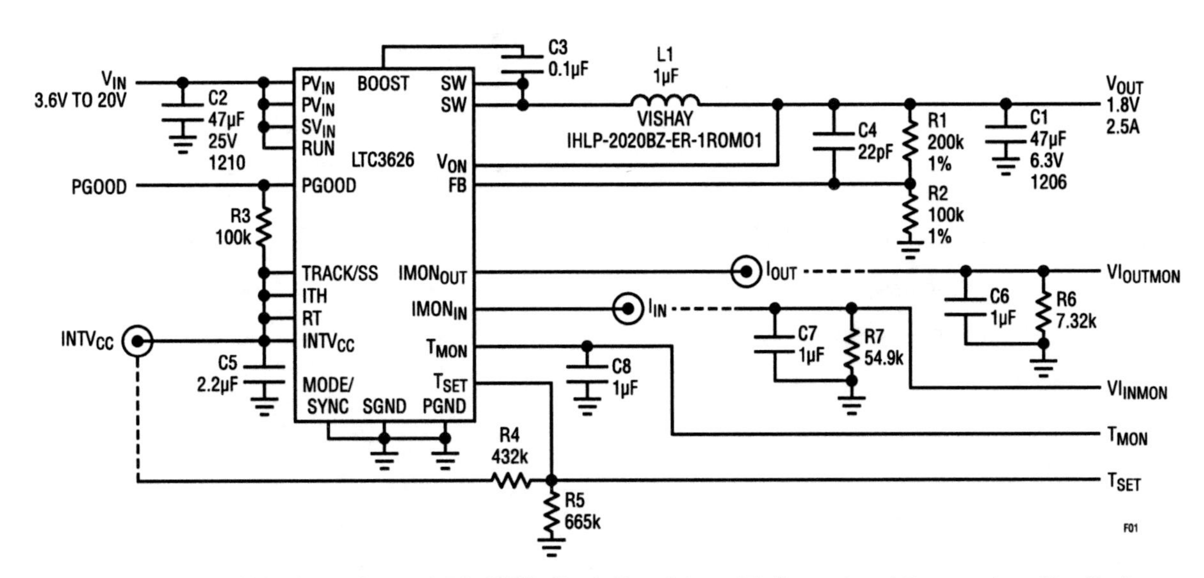

Figure 33.1 • 20V Maximum Input, 2.5A, 2MHz Buck Regulator with Current and Temperature Monitoring

Analog Circuit Design: Design Note Collection. http://dx.doi.org/10.1016/B978-0-12-800001-4.00033-8

Output/input current sensing

The LTC3626 senses the output current through the synchronous switch during the switch's on-time and generates a proportional current (scaled to 1/16000) at the $IMON_{OUT}$ pin. Figure 33.3 shows the accuracy of the $IMON_{OUT}$ output by comparing the measured output of the $IMON_{OUT}$ pin with calculated values. Error remains less than 1% over most of the output current range.

Likewise, this same sense current signal is combined with the buck regulator's duty cycle to produce a current proportional to the input current—again by 1/16000—at the $IMON_{IN}$ pin. A precision of better than 5% is achieved over a wide current range (see Figure 33.4).

Both current signals are connected to internal voltage amplifiers, referenced to 1.2V, that can shut down the part when tripped. So the input and output current limits are set by simply connecting a resistor to the $IMON_{IN}$ or $IMON_{OUT}$ pins, respectively, as shown in Figure 33.1. The relationship between the current limit and the resistor is:

$$I_{LIM} \cong \frac{1.2V \cdot 16000}{R_{LIM}}$$

For example, a 10k resistor sets a current limit of approximately 2A.

This simple scheme allows both monitoring and active control of the input and output current limits—the latter can be implemented via external control circuitry, such as a DAC with a few passive components.

Temperature sensing

The LTC3626 generates a voltage proportional to its own die temperature, which can be used to set a maximum temperature limit. The voltage at the temperature monitor pin (T_{MON}) is typically 1.5V at room temperature. To calculate the die temperature, T_J, multiply the T_{MON} voltage by the temperature monitor voltage-to-temperature conversion factor of 200K/V, and subtract the 273°C offset. The LTC3626 also has a temperature limit comparator fed by the temperature limit set pin, T_{SET}, and the T_{MON} pin. Hence, by applying a voltage to the T_{SET} pin, a maximum temperature limit can be set according to the following:

$$V_{TSET} = \frac{T_J + 273}{200°K/V}$$

Choosing a maximum temperature limit of 125°C equates to an approximate 2V setting on the T_{SET} pin—the IC will shut down once the die temperature T_J reaches this limit.

Conclusion

The LTC3626 combines current and temperature monitoring capabilities with a high performance buck regulator in a compact package. A microprocessor or other external control logic can supervise conditions via easy-to-use input and output current and temperature monitor pins, and it can shut itself down by setting a threshold voltage on the temperature set limit pin.

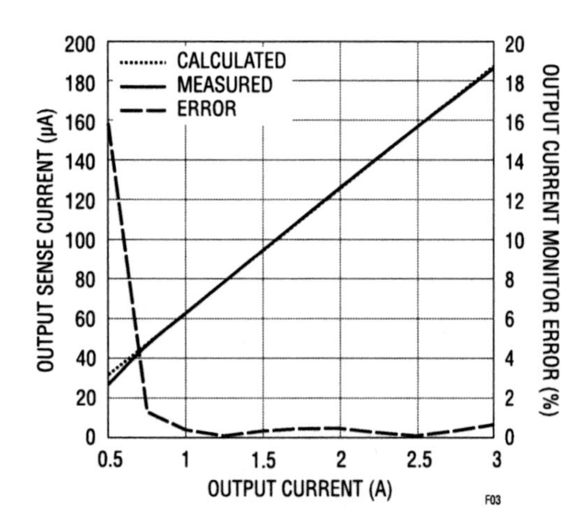

Figure 33.3 • Output Current vs Output Current Monitor

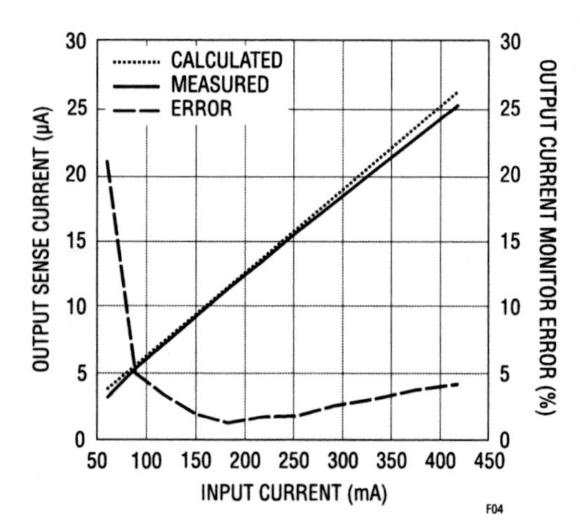

Figure 33.4 • Input Current vs Input Current Monitor

1.5A rail-to-rail output synchronous step-down regulator adjusts with a single resistor

34

Jeff Zhang

Introduction

A new regulator architecture the LTC3600 (first introduced with the LT3080 linear regulator) has wider output range and better regulation than traditional regulators. Using a precision 50µA current source and a voltage follower, the output is adjustable from "0V" to close to V_{IN}. Normally, the lowest output voltage is limited to the reference voltage. However, this new regulator has a constant loop gain independent of the output voltage giving excellent regulation at any output and allowing multiple regulators to be paralleled for higher output currents.

Operation

The LTC3600 is a current mode monolithic step-down buck regulator with excellent line and load transient responses. The 200kHz to 4MHz operating frequency can be set by a resistor or synchronized to an external clock. The LTC3600 internally generates an accurate 50µA current source, allowing the use of a single external resistor to program the reference voltage from 0V to 0.5V below V_{IN}. As shown in Figure 34.1, the output feeds directly back to the error amplifier with unity gain. The output equals the reference voltage at the I_{SET} pin. A capacitor can be paralleled with R_{SET} for soft start or to improve noise while an external voltage applied to the I_{SET} pin is tracked by the output.

Internal loop compensation stabilizes the output voltage in most applications, though the design can be customized with external RC components. The device also features a power good output, adjustable soft-start or voltage tracking and selectable continuous/discontinuous mode operation. These features, combined with less than 1µA supply current in shutdown, V_{IN} overvoltage protection and output overcurrent protection, make this regulator suitable for a wide range of power applications.

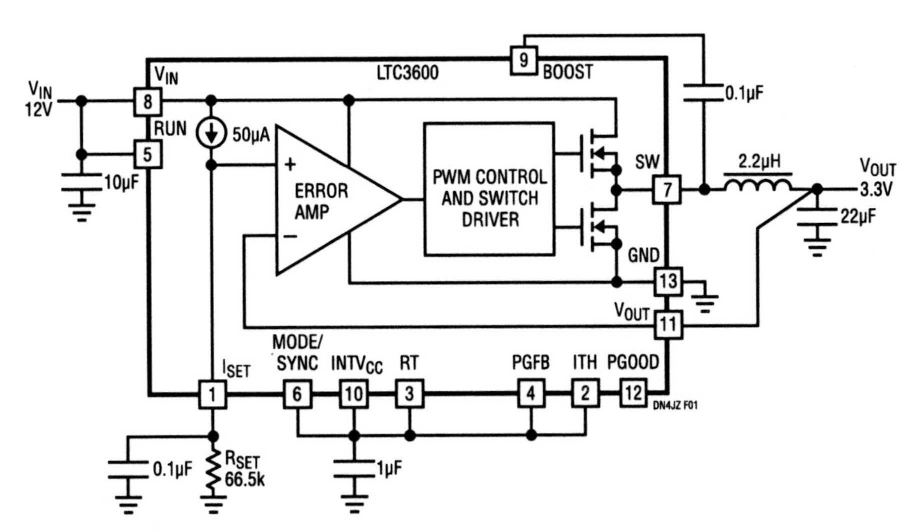

Figure 34.1 • High Efficiency, 12V to 3.3V 1MHz Step-Down Regulator with Programmable Reference

Analog Circuit Design: Design Note Collection. http://dx.doi.org/10.1016/B978-0-12-800001-4.00034-X

Applications

Figure 34.1 shows the complete LTC3600 schematic in a typical application that generates a 3.3V output voltage from 12V input. Figure 34.2 shows the load step transient response using internal compensation and with external compensation. Figure 34.3 shows the efficiency in CCM and DCM modes. Furthermore, the LTC3600 can be easily configured to be a current source, as shown in Figure 34.4. By changing the R_{SET} resistance from 0Ω to 3kΩ, the output current can be programmed from 0A to 1.5A.

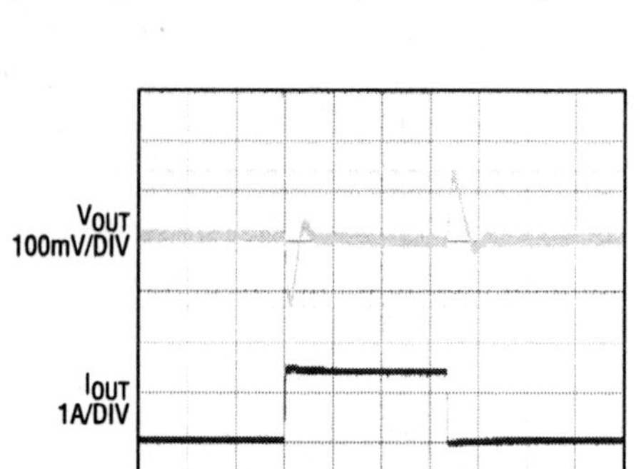

50µs/DIV
$12V_{IN}$ TO $3.3V_{OUT}$, INTERNAL COMPENSATION, ITH TIED TO $INTV_{CC}$

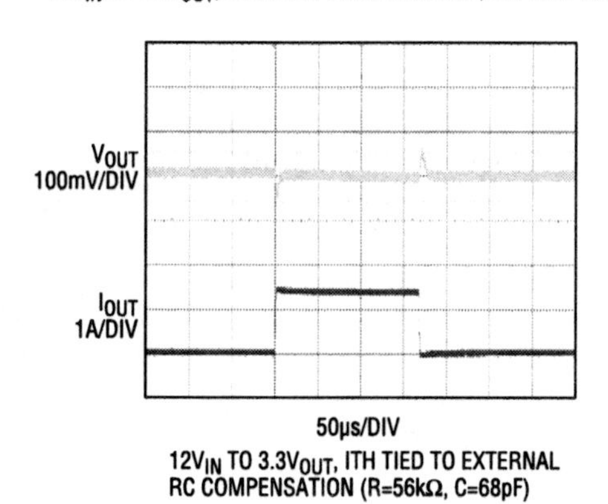

Figure 34.2 • 0A to 1.5A Load Step Response of the Figure 34.1 Schematic

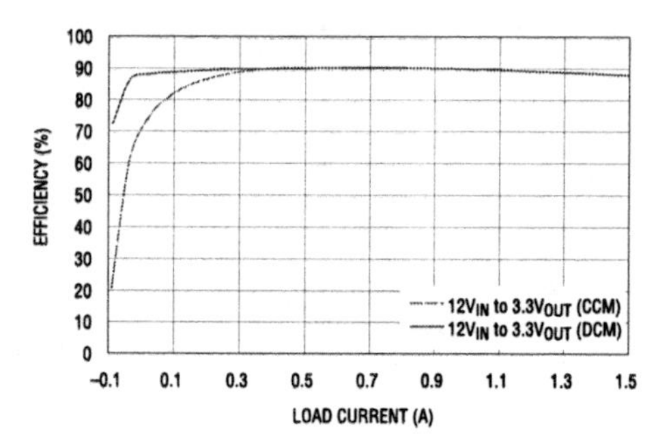

Figure 34.3 • Efficiency of 12V Input to 3.3V Output Regulator in CCM and DCM Mode

Conclusion

The LTC3600 uses an accurate internal current source to generate a programmable reference, expanding the range of output voltages. This unique feature gives the LTC3600 great flexibility, making it possible to dynamically change the output voltage, generate current sources, and parallel regulators for applications that would be difficult to implement using a standard DC/DC regulator configuration.

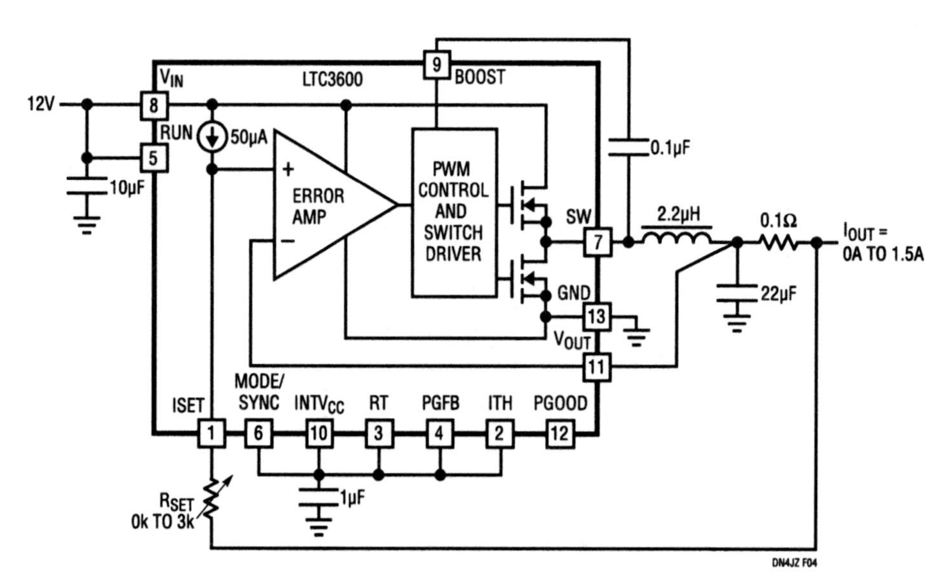

Figure 34.4 • The LTC3600 as a Programmable 0A to 1.5A Current Source

42V, 2.5A synchronous step-down regulator with 2.5μA quiescent current

35

Hua (Walker) Bai

Introduction

The LT8610 and LT8611 are 42V, 2.5A synchronous step-down regulators that meet the stringent high input voltage and low output voltage requirements of automotive, industrial and communications applications. To minimize external components and solution size, the top and bottom power switches are integrated in a synchronous regulator topology, including internal compensation. The regulator consumes only 2.5μA quiescent current from the input source even while regulating the output.

High efficiency synchronous operation

Replacing an external Schottky diode with an internal synchronous power switch not only minimizes the solution size, but also increases efficiency and reduces power dissipation. The efficiency improvement is significant in low output voltage applications where the voltage drop of the Schottky diode represents a relatively large portion of the output voltage. Figure 35.1 shows a 12V to 3.3V circuit. Figure 35.2 shows the efficiency of this circuit reaching 94%, which is 5% to 10% higher than a comparable nonsynchronous circuit.

Short-circuit robustness using small inductors

The LT8610 and LT8611 are specifically designed to minimize solution size by allowing inductor size to be selected based on the output load requirements of the application, rather than the maximum current limits of the IC. During overload or short-circuit conditions, the LT8610 and LT8611 safely tolerate operation with saturated inductors through the use of a high speed peak current mode architecture and a robust switch design. For example, an application that requires a maximum of 1.5A should use an inductor that has an RMS rating of >1.5A and a saturation current rating of >1.9A. This flexibility allows the user to avoid oversize inductors for applications requiring less than maximum output current.

Current sense and monitoring with the LT8611

The LT8611 includes a flexible current control and monitor loop using the ISN, ISP, IMON and ICTRL pins. The ISP and ISN pins connect to an external sense resistor that

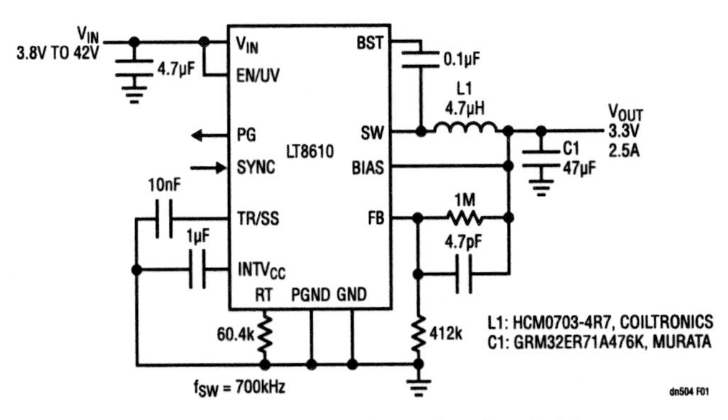

Figure 35.1 • LT8610 12V to 3.3V Application Achieves High Efficiency

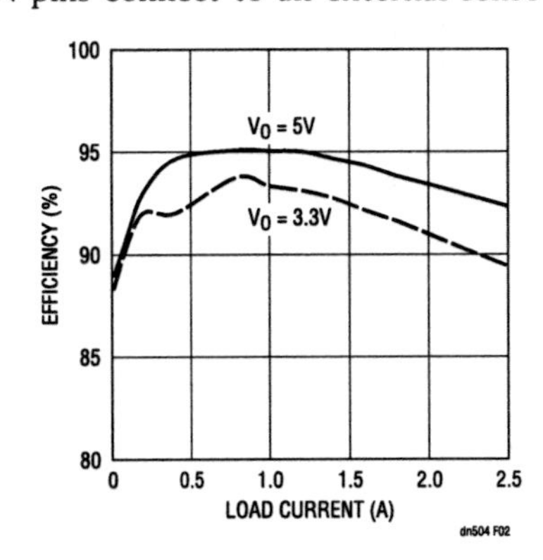

Figure 35.2 • Efficiency of the 12V to 3.3V Application (Circuit Shown in Figure 35.1)

Analog Circuit Design: Design Note Collection. http://dx.doi.org/10.1016/B978-0-12-800001-4.00035-1

may be in series with the input or output of the LT8611 or in series with other system currents. The current limit loop functions by limiting the LT8611 output current such that the voltage between the ISP and ISN pins does not exceed 50mV. The ICTRL pin allows the user to control this limit between 0mV and 50mV by applying 0V to 1V to the ICTRL pin. The IMON pin outputs a ground-referenced voltage that is 20·(ISP – ISN), which allows easy monitoring and may be used as an input to an A/D.

The LT8611 current sense and monitoring functionality may be used to limit short-circuit current or to create constant-current, constant-voltage (CCCV) supplies. Figure 35.3 shows well controlled current during a short-circuit event. The LT8611 can also be combined with a microcontroller with A/D and D/A to create sophisticated power systems. Typical apps include maximum power point tracking (MPPT) for solar charging and programmable LED current source.

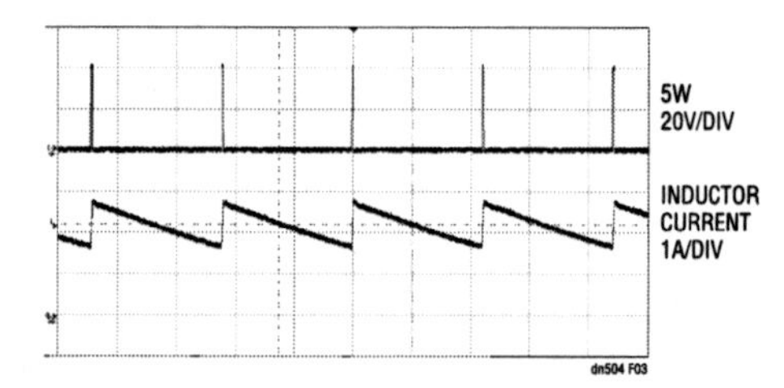

Figure 35.3 • Short-Circuit Current is Well Regulated at 42V with the LT8611

Wide input range operation at 2MHz

It is well known that higher switching frequencies allow for smaller solution sizes. In fact, a 2MHz switching frequency is often used in automotive applications to avoid the AM band and minimize solution footprint.

High switching frequencies, though, come with some trade-offs, including reduced ability to handle wide input voltage range commonly found in automotive and industrial environments. However, the LT8610 and LT8611 minimize these restrictions by allowing both high switching frequencies and high conversion ratios. This is due to their low minimum on-times (50ns typical) and low dropout, resulting in a wide input range, even at 2MHz. Figure 35.4 shows a 5V, 2A, 2MHz circuit that can accept 5.4V to 42V inputs. The circuit has a 2A output current limit.

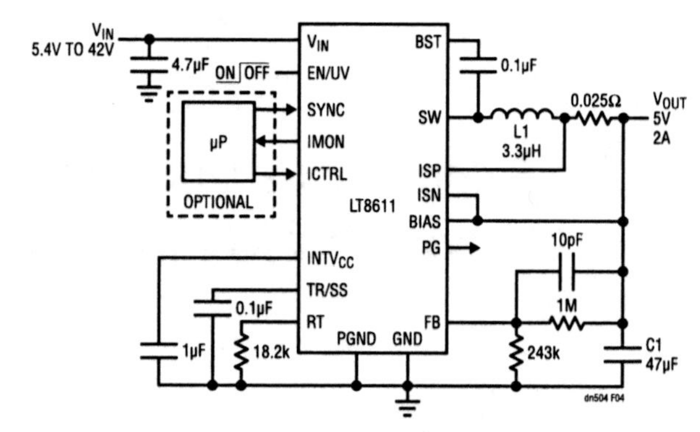

Figure 35.4 • LT8611 Running at 2MHz Reduces Solution Size, Avoids AM Band, and Still Allows High Duty Cycle

Low dropout operation

As the input voltage decreases toward the programmed output voltage, the LT8610 and LT8611 maintain regulation by skipping switch-off times and decreasing the switching frequency up to a maximum duty cycle of 99.8%. If the input voltage decreases further, the output voltage remains 450mV below the input voltage (at 2A load). The boost capacitor is charged during dropout conditions, maintaining high efficiency. Figure 35.5 shows the dropout performance.

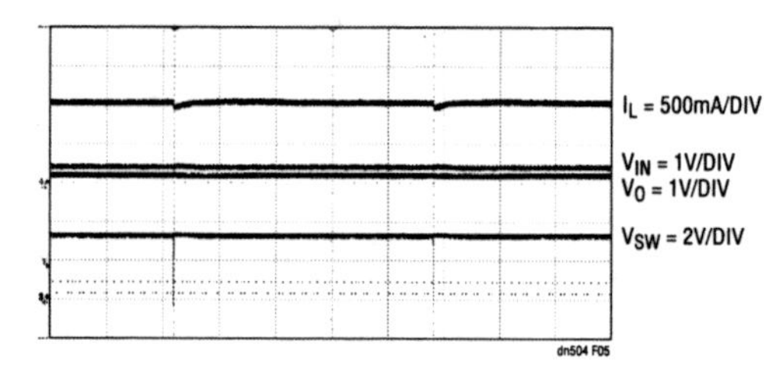

Figure 35.5 • LT8610/LT8611 Dropout Performance

Conclusion

LT8610 and LT8611 are 42V, 2.5A synchronous step-down regulators that offer 2.5µA quiescent current, high efficiency, fault robustness and constant current (LT8611 only), constant voltage operation in small packages. This combination of features makes them ideal for the harsh environment commonly found in automotive and industrial applications.

Bootstrap biasing of high input voltage step-down controller increases converter efficiency

36

Goran Perica Victor Khasiev

Introduction

High voltage buck DC/DC controllers such as the LTC3890 (dual output) and LTC3891 (single output) are popular in automotive applications due to their extremely wide 4V to 60V input voltage range, eliminating the need for a snubber and voltage suppression circuitry. These controllers are also well suited for 48V telecom applications where no galvanic isolation is required.

In a typical application for these controllers, the IC's supply voltage ($INTV_{CC}$) is provided by the on-chip LDO. This LDO produces 5V from input voltages up to 60V to bias control circuitry and provide power FET gate drive. Although simple, this built-in biasing scheme can be inefficient. Power losses can be significant in applications where the input voltage is consistently high, such as in 48V telecom applications. Reducing the power losses in the bias conversion can increase efficiency and also reduce the controller case operating temperature.

Employing $EXTV_{CC}$ to improve efficiency

One of the attractive features of the LTC3890 and LTC3891 controllers is the external power input ($EXTV_{CC}$). This is a second on-chip LDO, which can be used to bias the chip. When the input voltage is consistently high, it is more efficient to produce the biasing voltage by stepping down the converter's output voltage, which is fed into $EXTV_{CC}$, rather than generating 5V $INTV_{CC}$ from the high input voltage.

Figure 36.1 shows a block diagram for this scheme. The output can be directly connected to the $EXTV_{CC}$ pin of the chip as long as the output voltage is above 4.7V. However, extra circuitry (described in the following section) is required for outputs below 4.7V.

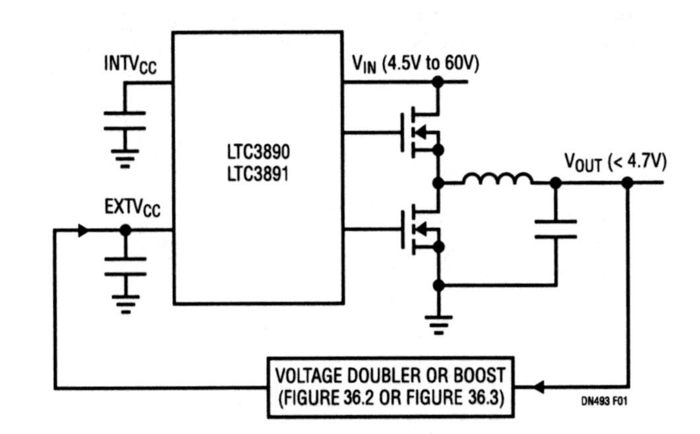

Figure 36.1 • Block Diagram Showing External Bias

Voltage doubler for output voltages below 4.7V

When the controller's output is below 4.7V, it must be stepped up to allow the built-in LDO to work. A simple voltage doubler solves this problem as long as the output is higher than 2.5V. Below 2.5V output, a multivibrator-based circuit can be used.

Analog Circuit Design: Design Note Collection. http://dx.doi.org/10.1016/B978-0-12-800001-4.00036-3

Figure 36.2 shows a simple, low cost solution for output voltages between 2.5V and 4.7V. This is a voltage doubler scheme based on small P-channel and N-channel MOSFETs, Q1 and Q2. The gates of these transistors are controlled by the bottom gate driver, BG of the controller. When BG is high, Q2 is on, Q1 is off and capacitor C1 charges from output voltage V_{OUT} through D1. When BG is low, Q2 is off, Q1 is on and capacitor C1 delivers a voltage close to $2 \cdot V_{OUT}$ to $EXTV_{CC}$.

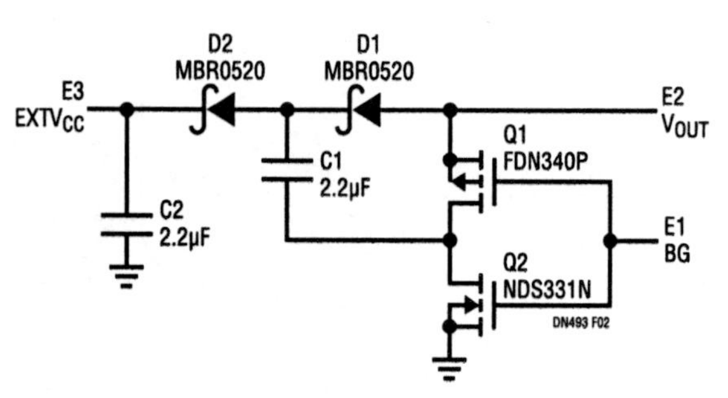

Figure 36.2 • Voltage Doubler Allows External Bias from V_{OUT} in the Range of 2.5V to 4.7V

Figure 36.3 shows a solution for voltages below 2.5V. The circuit consists of an astable multivibrator based on transistors Q1 and Q2, and a boost based on N-channel Q3 and inductor L1. Q1 and Q2 are biased from $INTV_{CC}$ and output voltage

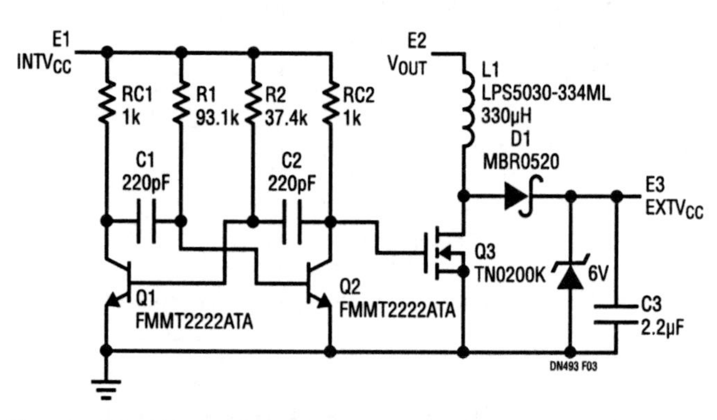

Figure 36.3 • Boost Controlled by Astable Multivibrator Is Used for V_{OUT} Lower than 2.5V

V_{OUT} is stepped up to 5V, which feeds $EXTV_{CC}$. The multivibrator frequency is set at 50kHz to minimize the EMI signature. The pulse width is defined by the ratio of resistors R1 and R2, as per the following expressions:

$$R1 = \frac{T \cdot (1 - D)}{0.7 \cdot C1}$$

$$R2 = \frac{T \cdot D}{0.7 \cdot C2}$$

$$D = \frac{EXTV_{CC} - V_{OUT}}{EXTV_{CC}}$$

$$T = \frac{1}{f}$$

Conclusion

The efficiency of high input voltage DC/DC controllers can be significantly improved by using the controller's output voltage to power the IC, instead of allowing the internal LDO to produce the bias voltage. For input voltages above 30V, efficiency improvements of 2% to 3% are realized when a voltage doubler circuit is used for a 3.3V at 5A output (see Figure 36.4). Similar efficiency improvements are shown for a 1.8V at 7A converter with a multivibrator-based circuit.

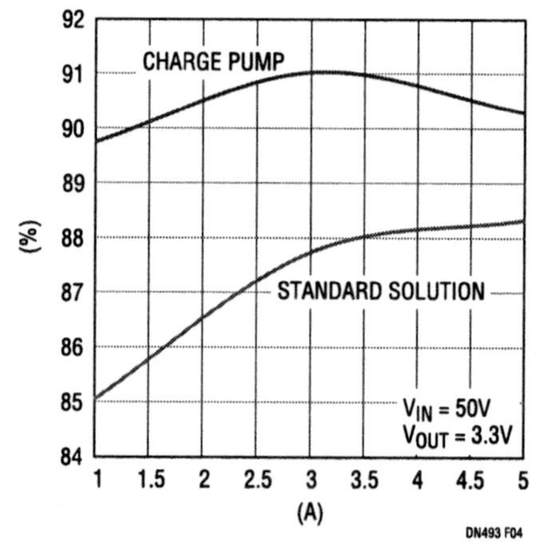

Figure 36.4 • LTC3890/LTC3891 Efficiency Improvement

36V, 3.5A dual monolithic buck with integrated die temperature monitor and standalone comparator block

37

Edwin Li

Introduction

Multioutput monolithic regulators are easy to use and fit into spaces where multichip solutions cannot. Nevertheless, the popularity of multioutput regulators is tempered by a lack of options for input voltages above 30V and support of high output currents. The LT3692A fills this gap with a dual monolithic regulator that operates from inputs up to 36V. It also includes a number of channel optimization features that allow the LT3692A's per-channel performance to rival that of multichip solutions.

The LT3692A is available in two packages: a 5mm × 5mm QFN and a 38-lead plastic TSSOP. Although both include the full feature set, the TSSOP package enhances the thermal performance of the dual buck.

High input voltage with high transient capability

The LT3692A can operate up to an input voltage of 36V and can sustain a transient voltage up to 60V for 1second, making it suitable for harsh operating environments such as those commonly found in automotive environments.

On-die temperature monitoring

The LT3692A provides an on-die temperature monitoring function which facilitates the application circuit design, debugging and package thermal optimization. The voltage at the T_J pin is directly proportional to the die temperature in Celsius (i.e., 250mV equals 25°C and 1.5V equals 150°C).

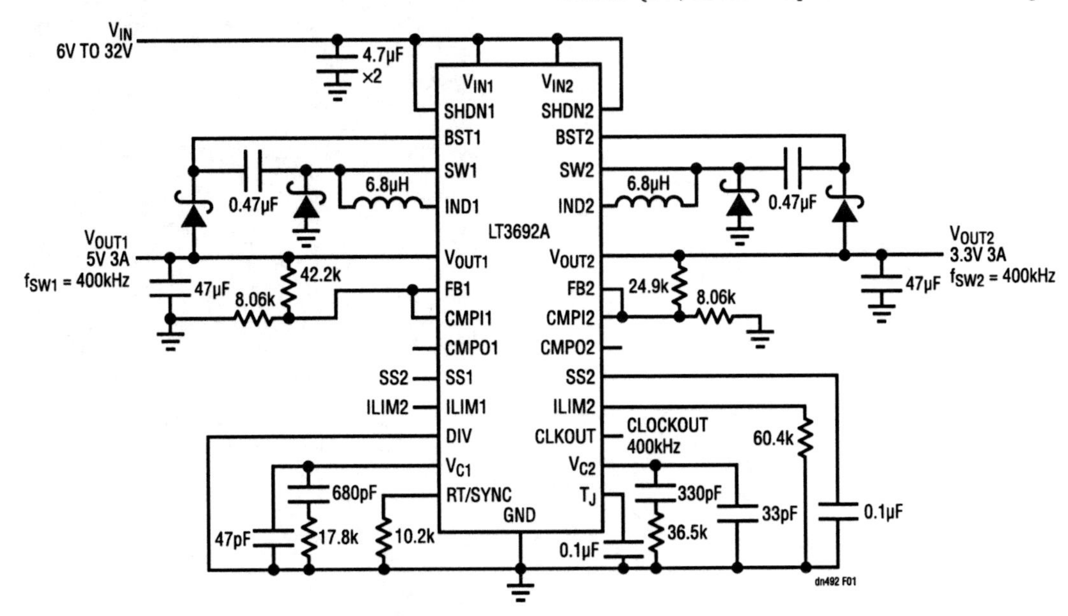

Figure 37.1 • Dual 5V/3A/400kHz, 3.3V/3A/400kHz Application Keeps Temperature Rise Low at a V_{IN} of 18V

Analog Circuit Design: Design Note Collection. http://dx.doi.org/10.1016/B978-0-12-800001-4.00037-5

The measured temperature of the LT3692A TSSOP die tops out at 80°C[1] with the two outputs each supporting 3A loads at 5V and 3.3V from an input voltage of 18V, with a switching frequency of 400kHz. Figure 37.1 shows the schematic of the measured application circuit. The same setup, but with 2.5A loads, drops the max die temperature to 68°C.[1]

Standalone comparator block

The LT3692A also includes a standalone comparator block, which provides a 720mV threshold with hysteresis and outputs an open-collector signal. This comparator can be configured as a power good flag signal by connecting the CMPI pin to the FB pin to monitor the output voltage. It can also be configured as a temperature flag, which gives a warning signal when the die temperature rises to a preset point. This function is realized together with the on-die temperature monitor. Figure 37.2 shows how to configure a 100°C temperature flag.

Other features

Independent adjustable current limit

The switch current limit on each output can be programmed from 2A to 4.8A. This expands the number of loading combinations that can be safely implemented without risking thermal overload of the package under extreme conditions, such as a short circuit. Likewise, the current limit can be used to protect the part in compact designs where the saturation margin on inductors is lowered to meet size constraints.

Independent synchronization

Independent synchronization allows any phase difference between the two outputs besides the standard 0° and 180°. The phase difference on the LT3692A is adjusted by controlling the duty cycle of the synchronization signal.

Frequency division

Frequency division makes it possible to tune the operating frequency of each channel to optimize overall performance and size. The frequency of channel 1 can be programmed to run at 1, 1/2, 1/4 or 1/8 the frequency of channel 2. Figure 37.3 shows the layout of a 3.3V/2.5A/550kHz channel and a 1.2V/1A/2.2MHz channel application. The relatively low 550kHz frequency of V_{OUT1} maximizes channel 1's input voltage to 36V while meeting minimum on-time requirements and keeping the efficiency high. The high 2.2MHz frequency of V_{OUT2} allows the use of smaller components for channel 2 as shown in Figure 37.3. Despite the reduction in size, electrical and thermal performance is uncompromised.

Conclusion

The LT3692A is a dual output monolithic regulator that combines the ease-of-use and compact solution size of typical monolithic regulators with the flexibility of discrete, multichip solutions. Its high transient voltage capability, die temperature monitor, standalone comparator block, adjustable current limit, adjustable switching frequency and frequency division function and independent synchronization enable the LT3692A to work in many applications that other monolithic chips cannot.

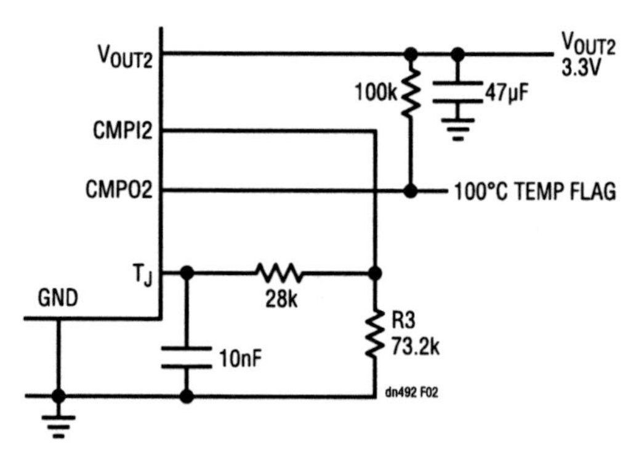

Figure 37.2 • Temperature Flag Using Comparator Block and Temperature Monitor

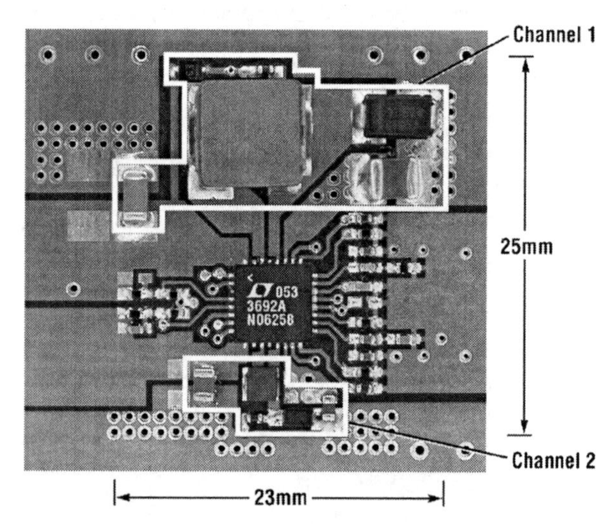

Figure 37.3 • Dual 3.3V/2.5A/500kHz, 1.2V/1A/2.2MHz Layout. Channel 2 Requires Half of the Area of Channel 1

Note1: The T_J pin reading on a standard demo board (DC1403A) running in a 25°C ambient temperature environment.

High efficiency, high density 3-phase supply delivers 60A with power saving Stage Shedding, active voltage positioning and nonlinear control for superior load step response

38

Jian Li Kerry Holliday

Introduction

The LTC3829 is a feature-rich single output 3-phase synchronous buck controller that meets the power density demands of modern high speed, high capacity data processing systems, telecom systems, industrial equipment and DC power distribution systems. The LTC3829's features include:

- 4.5V to 38V input range and 0.6V to 5V output range
- 3-phase operation for low input current ripple and output voltage ripple with Stage Shedding mode to yield high light load efficiency
- On-chip drivers in a 38-pin 5mm × 7mm QFN (or 38-pin FE) package to satisfy demanding space requirements

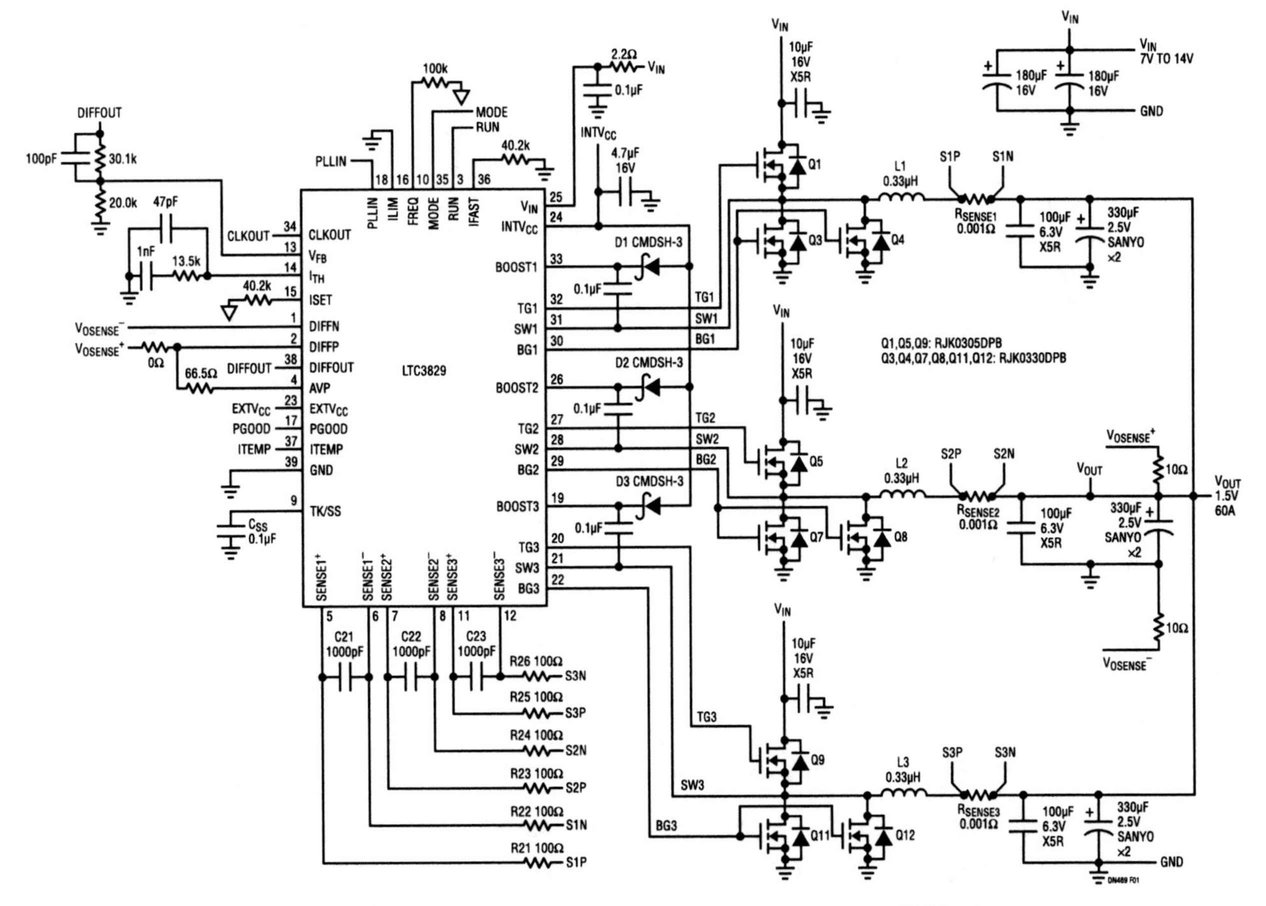

Figure 38.1 • A 1.5V/60A 3-Phase Converter Featuring the LTC3829

Analog Circuit Design: Design Note Collection. http://dx.doi.org/10.1016/B978-0-12-800001-4.00038-7

- Remote output voltage sensing and inductor DCR temperature compensation for accurate regulation
- Active voltage positioning (AVP) and nonlinear control ensure impressive load transient performance

1.5V/60A, 3-phase power supply

Figure 38.1 shows a 7V to 14V input, 1.5V/60A output application. The LTC3829's three channels run 120° out-of-phase, which reduces input RMS current ripple and output voltage ripple compared to single-channel solutions. Each phase uses one top MOSFET and two bottom MOSFETs to provide up to 20A of output current.

The LTC3829 includes unique features that maximize efficiency, including strong gate drivers, short dead times and a programmable Stage Shedding mode, where two of the three phases shut down at light load. Onset of Stage Shedding mode can be programmed from no load to 30% load. Figure 38.2 shows the efficiency of this regulator at over 86.5% with a 12V input and a 1.5V/60A output with Stage Shedding mode, dramatically increasing light load efficiency.

The current mode control architecture of the LTC3829 ensures that DC load current is evenly distributed among the three channels, as shown in Figure 38.3. Dynamic, cycle-by-cycle current sharing performance is similarly tight in the face of load transients.

A fast and controlled transient response is another important requirement for modern power supplies. The LTC3829 includes two features that reduce the peak-to-peak output voltage excursion during a load step: programmable nonlinear control or programmable active voltage positioning (AVP). Figure 38.4 shows the transient response without these features enabled. Figure 38.5 shows that nonlinear control improves peak-to-peak response by 17%. Figure 38.6 shows that AVP can achieve a 50% reduction in the amplitude of voltage spikes.

Conclusion

The LTC3829's tiny 5mm × 7mm 38-pin QFN package belies its expansive feature set. It produces high efficiency with a combination of strong integrated drivers and Stage Shedding/Burst Mode operation. It supports temperature compensated DCR sensing for high reliability. AVP and nonlinear control improve transient response with minimum output capacitance. Voltage tracking, multichip operation and external sync capability fill out its menu of features. The LTC3829 is ideal for high current applications such as telecom and datacom systems, industrial and computer systems.

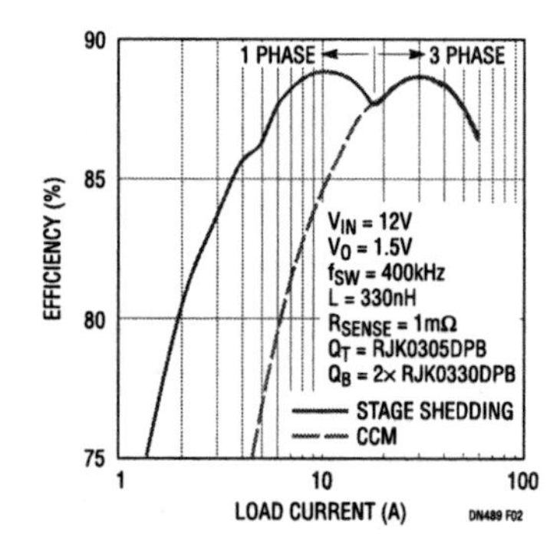

Figure 38.2 • Efficiency Comparison of Stage Shedding vs CCM

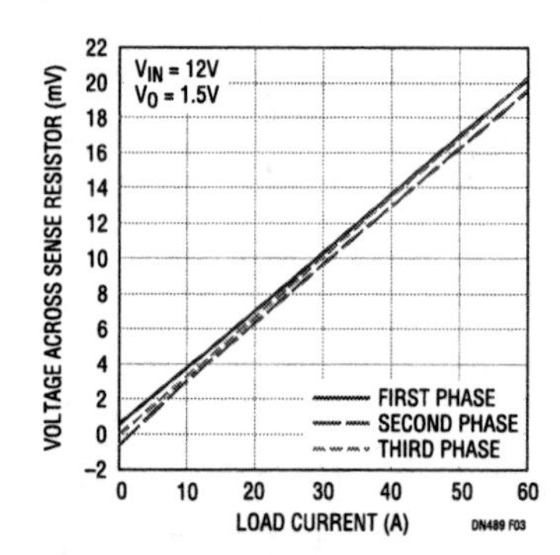

Figure 38.3 • Current Sharing Performance between Phases

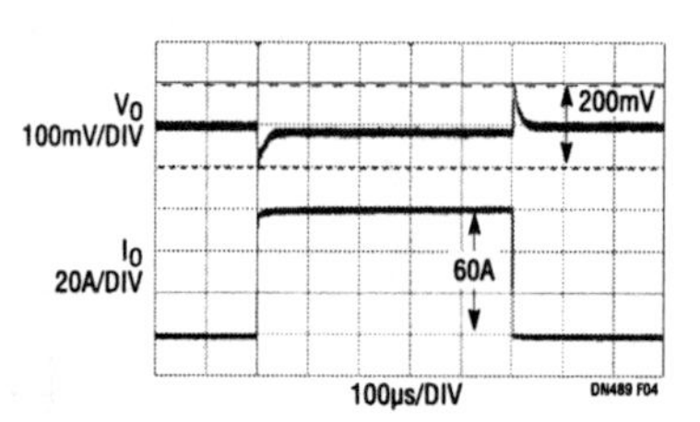

Figure 38.4 • Transient Performance without AVP and Nonlinear Control

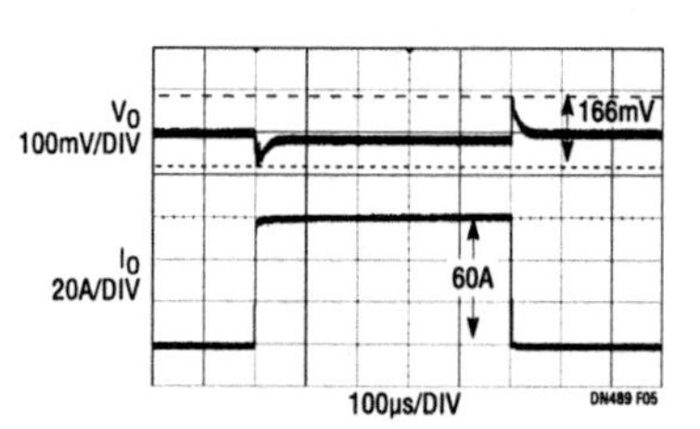

Figure 38.5 • Transient Performance with Nonlinear Control

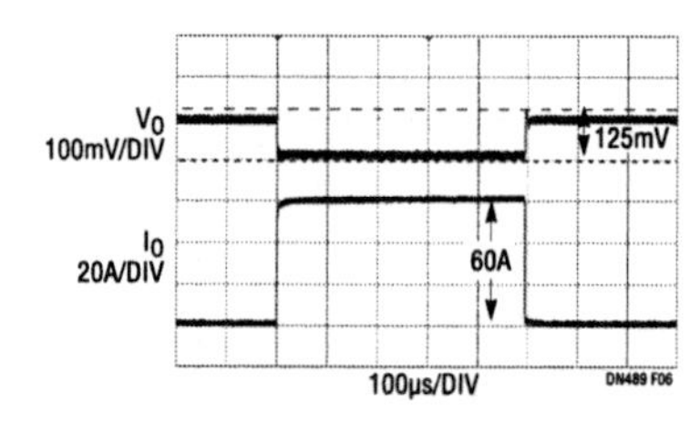

Figure 38.6 • Transient Performance with AVP

2-phase synchronous buck controller features light load Stage Shedding mode, active voltage positioning, low R_{SENSE} and remote V_{OUT} sensing

39

Charlie Zhao Jian Li

Introduction

Today's computer, datacom, and telecom systems demand power supplies that are efficient, respond quickly to load transients and accurately regulate the voltage at the load. For example, load current can be measured by using the inductor DCR, thus eliminating the need for a dedicated sense resistor. Inductor DCR sensing increases efficiency—especially at heavy load—while reducing component cost and required board space. The LTC3856 single-output 2-phase synchronous buck controller improves the accuracy of inductor DCR sensing by compensating for changes in DCR due to temperature.

DCR temperature compensation is just one of many performance enhancing features offered in the LTC3856. It also includes on-chip gate drivers, remote output voltage sensing, Stage Shedding mode for improved light load efficiency and adaptive voltage positioning for fast transient response. The LTC3856 can convert a wide input voltage range, 4.5V to 38V, to outputs from 0.6V to 5V. Despite the many features, the chip is small, available in 32-pin 5mm × 5mm QFN and 38-pin TSSOP packages.

High efficiency, 2-phase, 4.5V to 14V input, 1.5V/50A output converter

Figure 39.1 shows a typical LTC3856 application in a 4.5V to 14V input, 1.5V/50A output converter. The LTC3856's two channels operate out-of-phase, which reduces the input RMS current ripple and thus the required input capacitance. Up to six LTC3856s can be paralleled for up to 12-phase operation.

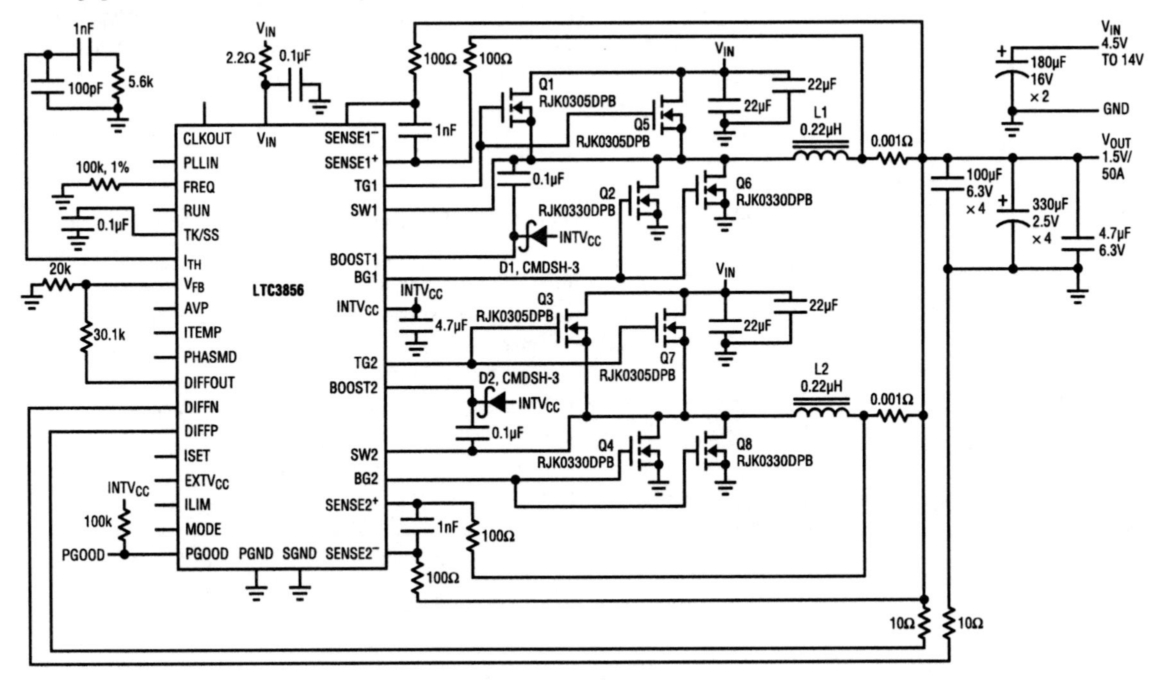

Figure 39.1 • 1.5V/50A, 2-Phase Synchronous Buck Converter Featuring the LTC3856

Analog Circuit Design: Design Note Collection. http://dx.doi.org/10.1016/B978-0-12-800001-4.00039-9

The LTC3856 has a phase-locked loop (PLL) and can be synchronized to an input frequency between 250kHz and 770kHz. Due to its peak current mode control architecture, the LTC3856 provides fast cycle-by-cycle dynamic current sharing plus tight DC current sharing, as shown in Figure 39.2.

Stage Shedding mode

At light loads, the LTC3856 can be programmed to operate in one of three modes: Burst Mode operation, forced continuous mode or Stage Shedding mode. With Stage Shedding mode, the LTC3856 can shut down one channel to reduce switching related loss which is the dominant loss at light loads. Stage Shedding mode is selected by simply tying the MODE pin to $INTV_{CC}$.

The efficiency improvements achieved by Stage Shedding mode are shown in Figure 39.3. Due to strong gate drivers and shorter dead-time, the LTC3856 can achieve 4%~5% higher efficiency than the LTC3729, a comparable single-output, 2-phase controller, over the whole load range. With Stage Shedding mode, significant efficiency improvement is further achieved at light load. At 5% load, the efficiency is improved by 13%.

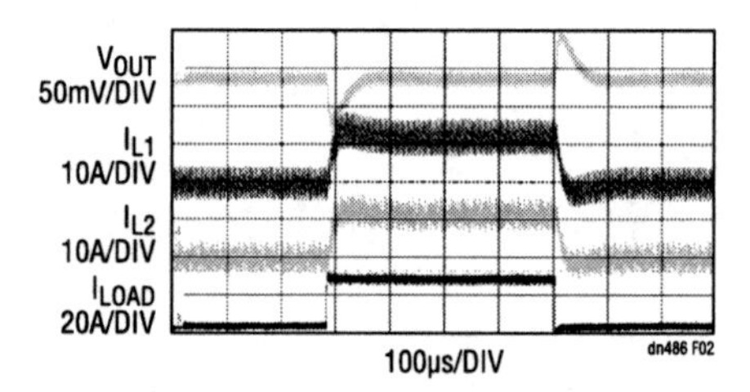

Figure 39.2 • Load Transient and Current Sharing: $V_{IN}=12V$, 25A to 50A Load Step

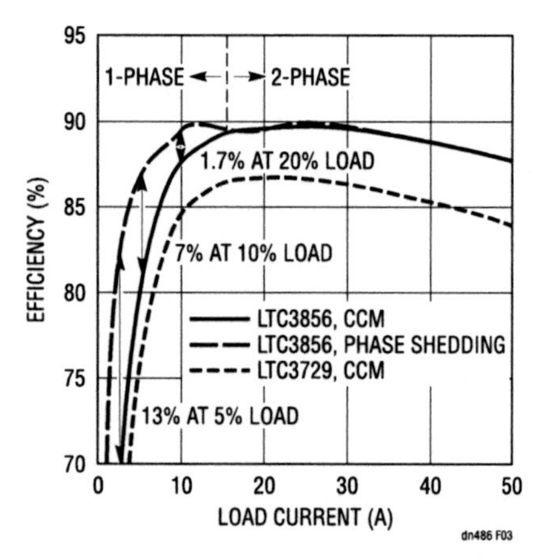

Figure 39.3 • Efficiency Comparison: $V_{IN}=12V$, $V_O=1.5V$, $F_{SW}=400kHz$, L=220nH, $R_{SENSE}=1m\Omega$, $Q_T=$RJK0305DPB, $Q_B=$2xRJK0330DPB

Current mode control allows the LTC3856 to transition smoothly from 2-phase to 1-phase operation and vice versa.

Active voltage positioning

User-selectable active voltage positioning (AVP) is another unique design feature of the LTC3856. AVP improves overall transient response and reduces required output capacitance by modifying the regulated output voltage depending on its current loading. With proper design, AVP can reduce load transient-induced peak-to-peak voltage spikes by 50%.

Inductor DCR sensing temperature compensation

Although not used here, inductor DCR sensing offers a lossless method of sensing the load current. The problem is that the DCR of the inductor typically has a positive temperature coefficient, causing the effective current limit of the converter to change with inductor temperature. The LTC3856 can sense the inductor temperature with an NTC thermistor, thereby adjusting the current limit based on the temperature. The result is a constant current limit over a broad temperature range. This improves inductor DCR sensing reliability in high current applications.

Output voltage remote sensing

For high output current, low voltage applications, board or wire interconnect resistance can cause a severe load regulation problem. To solve this problem, the LTC3856 includes a low offset, unity-gain, high bandwidth differential amplifier for true remote sensing. Common mode noise and ground loop disturbances can be rejected, and load regulation is greatly improved, especially when there are long trace runs between the load and the converter output.

Conclusion

The LTC3856 is a feature-rich single output, 2-phase synchronous step-down DC/DC controller. It achieves high efficiency in both heavy load and light load conditions, with temperature compensated DCR sensing and Stage Shedding mode or Burst Mode operation. AVP improves transient response even when the output capacitance is reduced. Remote sensing, a tight ±0.75% reference voltage accuracy over temperature, voltage tracking, strong on-chip drivers, multichip operation and external sync capability fill out its menu of features. The LTC3856 is ideal for high current applications and can meet the high standards of today's power supplies for telecom and datacom, industrial and computer applications.

Dual output high efficiency converter produces 3.3V and 8.5V outputs from a 9V to 60V rail

40

Victor Khasiev

Introduction

Among the many step-down DC/DC switching regulator controllers, the LTC3890 stands out because of its unique features. This 50μA quiescent current device can produce two output voltages ranging from 0.8V to 24V when powered from an input voltage of 4V to 60V.

Many high input voltage step-down DC/DC converter designs use a transformer-based topology or external high side drivers to operate from up to $60V_{IN}$. Others use an intermediate bus converter requiring an additional power stage. However, the LTC3890 simplifies design, with its smaller solution size, reduced cost and shorter development time compared to other design alternatives.

Feature rich

The LTC3890 is a high performance synchronous buck DC/DC controller with integrated N-channel MOSFET drivers. It uses a current mode architecture and operates from a phase-lockable fixed frequency from 50kHz to 900kHz. The device features up to 99% duty cycle capability for low voltage dropout applications, adjustable soft-start or voltage tracking and selectable continuous, pulse-skipping or Burst Mode operation with a no-load quiescent current of only 50μA. These features, combined with a minimum on-time of just 95ns, make this controller an ideal choice for high step-down ratio applications. Power loss and supply noise are minimized by operating the two output stages out-of-phase.

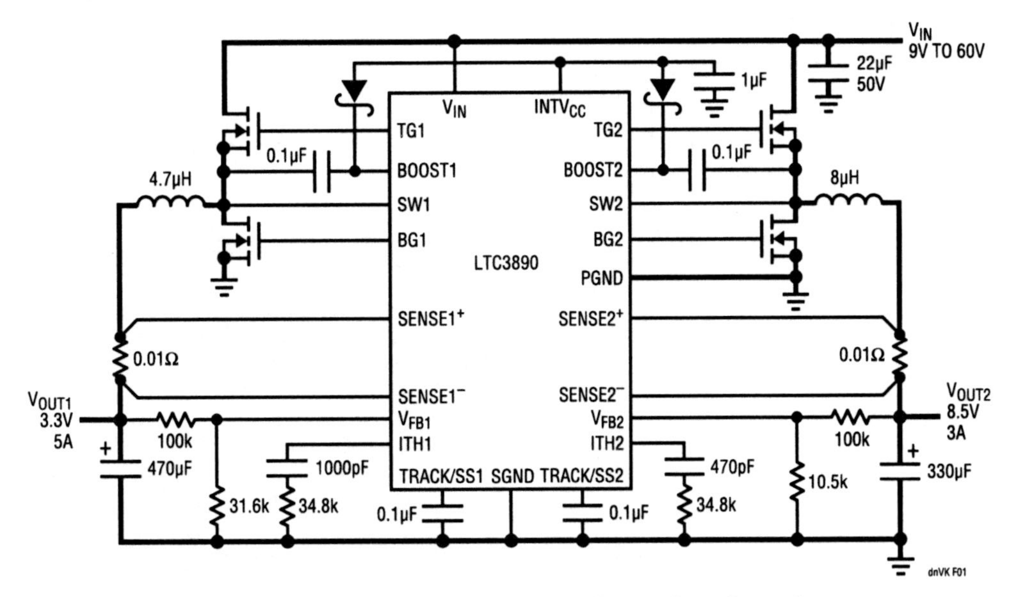

Figure 40.1 • High Efficiency Dual 8.5V/3.3V Output Step-Down Converter

Analog Circuit Design: Design Note Collection. http://dx.doi.org/10.1016/B978-0-12-800001-4.00040-5

Dual output application

Figure 40.1 shows the LTC3890 operating in an application that converts a 9V to 60V input into 3.5V/5A and 8.5V/3A outputs. The transient response for the 3.3V output with a 4A load step is less than 50mV (as shown in Figure 40.2).

Figure 40.3 shows the efficiency of the 8.5V channel with a 36V input voltage.

Single output application

The LTC3890 can also be configured as a 2-phase single output converter by simply connecting the two channels together. For example, a 9V to 60V input can be converted to an 8.5V output at 6A. Figure 40.4 shows the efficiency of this configuration at input voltages of 10V, 30V and 60V.

Current mode control provides good current balance between the phases. Less than 10% mismatch can be achieved, as shown in Figure 40.5.

Conclusion

Although there are many choices in dual output controllers, the LTC3890 brings a new level of performance with its high voltage operation, high efficiency conversion and ease of design.

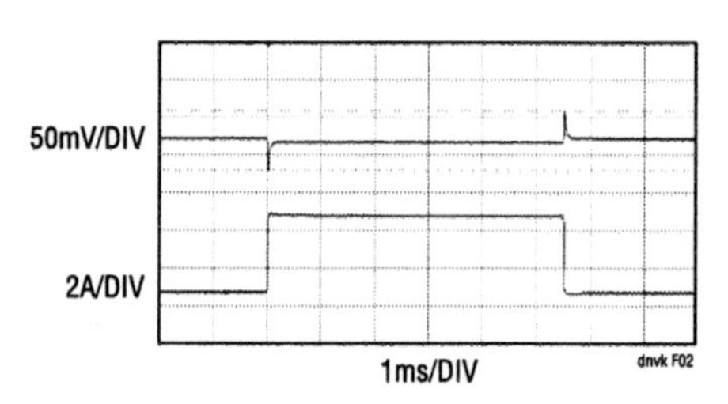

Figure 40.2 • Transient Response of 3.3V Channel (I_{OUT1}: 1A to 5A)

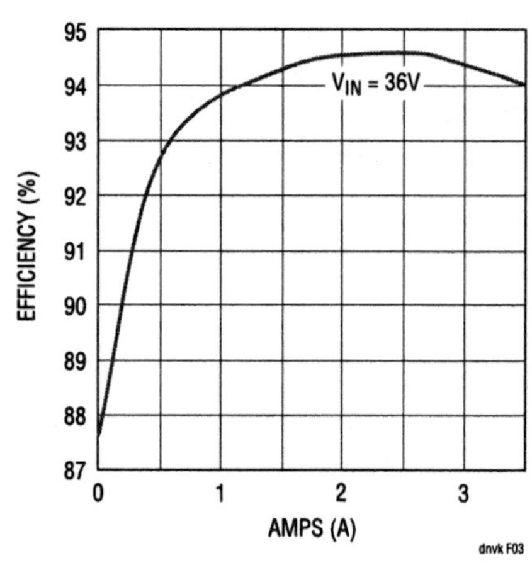

Figure 40.3 • Efficiency of the Converter in Figure 40.1 for the V_{OUT2A} 8.5V Channel

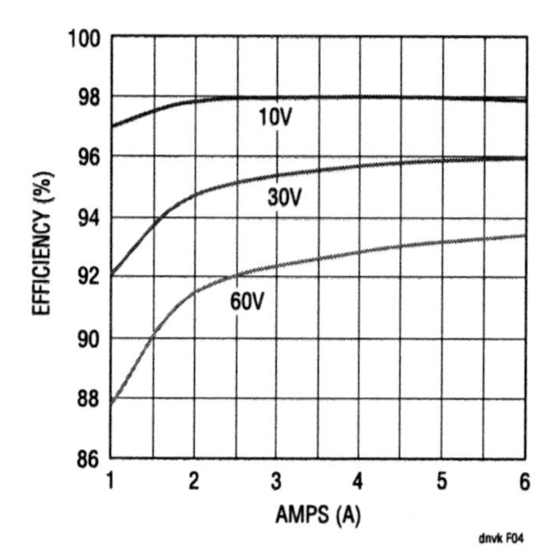

Figure 40.4 • Efficiency of the LTC3890 Configured as a 2-Phase Single Output of 8.5V at up to 6A from a 10V to 60V Input

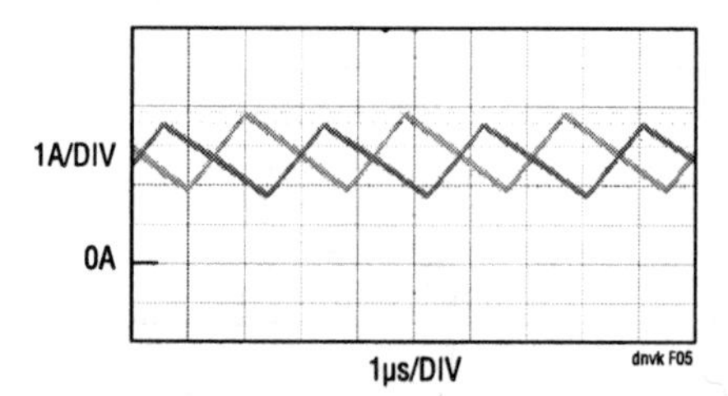

Figure 40.5 • The Inductor Current in a 2-Phase Single Output Converter. Currents in Both Inductors Shown with a 24V Input and 8.5V at 6A Output

Dual output step-down controller produces 10% accurate, efficient and reliable high current rails

41

Mike Shriver   Theo Phillips

Introduction

The LTC3855 makes it possible to generate high current rails with the accuracy and efficiency to satisfy the most demanding requirements of today's leading edge network, telecommunications and server applications. This 2-phase, dual output synchronous buck controller includes strong gate drivers that support operation with per-phase currents above 20A. The accurate 0.6V ±0.75% reference and its integrated differential amplifier (diff amp) allow remote sensing of the output of critical rails. This controller has an output voltage range from 0.6V to 12.5V when used without the diff amp and from 0.6V to 3.3V with the diff amp.

The LTC3855 uses the reliable peak current mode architecture to achieve a fast and accurate current limit and real time current sharing. Its current sense comparators are designed to sense the inductor current with either a sense resistor or with inductor DCR sensing. DCR sensing offers the advantage of reduced conducted power losses, since the current is measured using the voltage drop across the already-present inductor DC resistance—eliminating the losses incurred by adding a sense resistor. The trade-off is that DCR sensing is less accurate than a dedicated sense resistor because the DCR varies from part to part and over temperature. The LTC3855 uses an innovative scheme to improve the accuracy of DCR sensing by compensating for the DCR's variation with temperature.

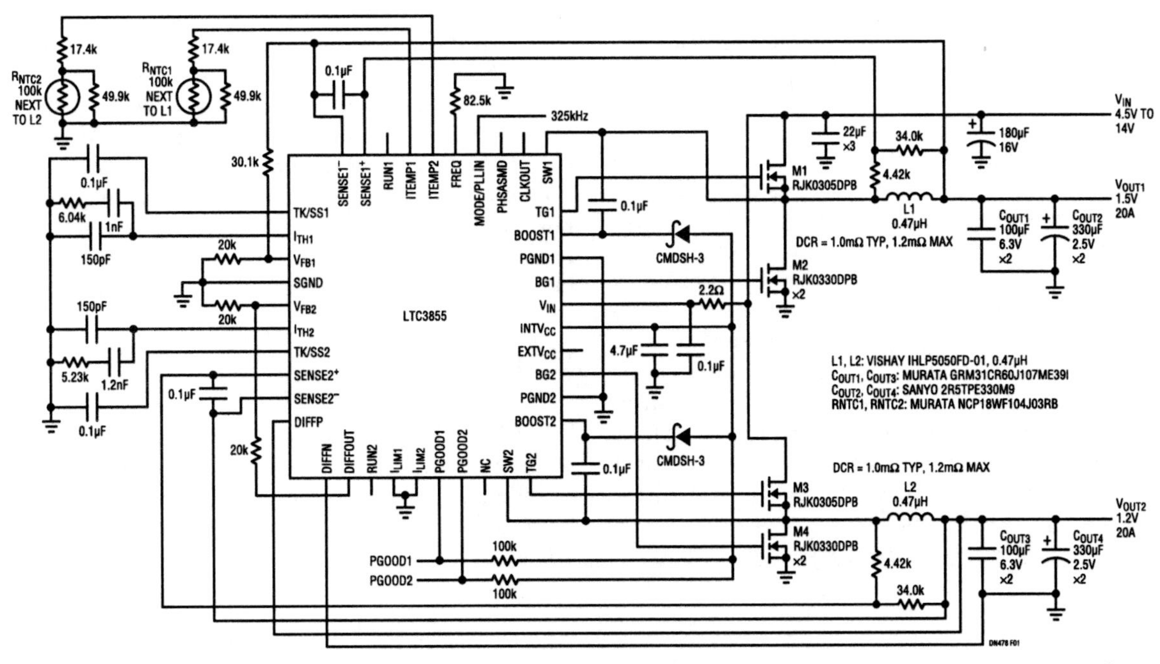

Figure 41.1 • Dual 1.5V/20A and 1.2V/20A Converter Operating at f_{SW}=325kHz. The Entire Circuit Fits within 1.7in^2 with Both Sides of the Board Populated

Analog Circuit Design: Design Note Collection. http://dx.doi.org/10.1016/B978-0-12-800001-4.00041-7

1.5V/20A and 1.2V/20A buck converter with remote sensing and NTC compensated DCR sensing

Figure 41.1 shows a 1.5V/20A and 1.2V/20A dual phase converter with DCR sensing, operating at 325kHz. High efficiency is achieved with the strong gate drivers, optimized dead-time and DCR sensing. The typical full load efficiency for the 1.5V and 1.2V rails is 89.5% and 87.8%, respectively (see Figure 41.2). The 1.2V output is remotely sensed with the diff amp. As a result, the 1.2V rail's output accuracy is unaffected by the voltage drops across the V_{OUT} and GND planes. The load step response for the 1.2V rail is shown in Figure 41.3.

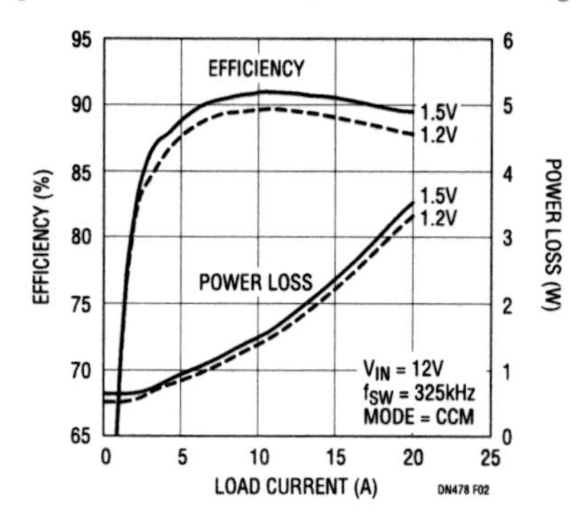

Figure 41.2 • Efficiency and Power Loss of the 1.5V/20A and 1.2V/20A Converter

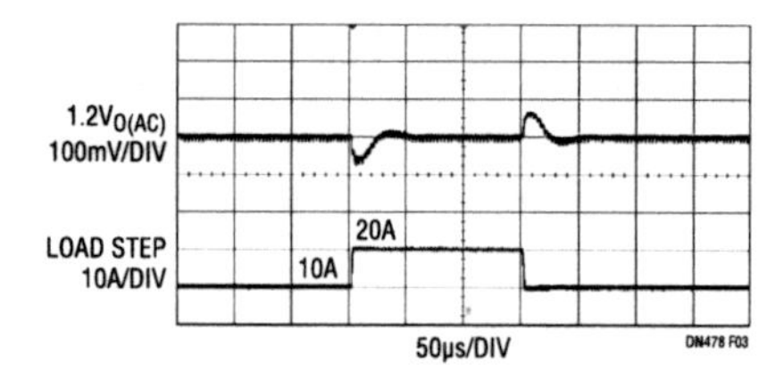

Figure 41.3 • 50% to 100% Load Step Response for the 1.2V Rail at V_{IN} = 12V

The LTC3855 features precise current limit thresholds of 30mV, 50mV and 75mV, selected via the I_{LIM} pins. The current limit threshold can be raised by biasing the ITEMP pins below 500mV. Since the ITEMP pins source 10μA of current, the peak current sense voltage can be increased by inserting a resistance of less than 50k from the ITEMP pin to ground. By placing an inexpensive NTC thermistor next to the inductor and connecting this thermistor to a linearization network from the ITEMP pin to ground, the current limit temperature coefficient can be greatly reduced. As Figure 41.4 illustrates, the compensated current limit is 20% higher than the uncompensated current limit at 110°C. Another use for the ITEMP pins is to increase the current limit for conventional DCR sense and R_{SENSE} applications.

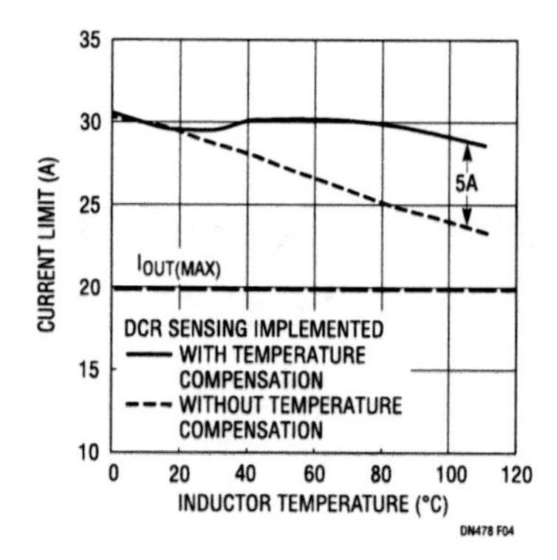

Figure 41.4 • Measured Current Limit of the 1.2V Rail Over Temperature with and without Temperature Compensation

PolyPhase operation

The LTC3855 provides inherently fast cycle-by-cycle current sharing due to its peak current mode architecture, plus very tight DC current sharing for single output PolyPhase applications. Up to 12-phase operation can be achieved by daisy chaining the CLKOUT and MODE/PLLIN pins and by programming the phase separation with the PHASMD pins. A major advantage of PolyPhase operation is the reduction of the required input and output capacitance due to ripple current cancellation. Also, single output PolyPhase applications have a faster load step response due to a smaller clock delay.

Other important features

The switching frequency of the LTC3855 can be programmed between 250kHz and 770kHz with a resistor placed from the FREQ pin to ground or synchronized to an external clock in this frequency range using its internally compensated phase lock loop. High efficiency at light load is achieved by selecting either Burst Mode operation or discontinuous mode operation, as opposed to continuous conduction mode. The LTC3855 can be used for inputs up to 38V, and its 100ns typical minimum on-time allows for high step-down ratios. The LTC3855 has a TK/SS pin for programmable soft-start or rail tracking, and dedicated RUN and PGOOD pins for each channel. The LTC3855 comes in either a 6mm × 6mm QFN or a thermally enhanced 38-lead TSSOP package.

Conclusion

The LTC3855 is a high performance dual output buck converter intended for low output voltage, high output current supplies. It provides the user with the benefits of a precise 0.6V 0.75% reference, an accurate current limit and high efficiency.

$15V_{IN}$, 4MHz monolithic synchronous buck regulator delivers 5A in 4mm × 4mm QFN

42

Tom Gross

Introduction

The LTC3605 is a high efficiency, monolithic synchronous step-down switching regulator that is capable of delivering 5A of continuous output current from input voltages of 4V to 15V. Its compact 4mm × 4mm QFN package has very low thermal impedance from the IC junction to the PCB, such that the regulator can deliver maximum power without the need for a heat sink. A single LTC3605 circuit can power a 1.2V microprocessor directly from a 12V rail—no need for an intermediate voltage rail.

The LTC3605 employs a unique controlled on-time/constant frequency current mode architecture, making it ideal for low duty cycle applications and high frequency operation. There are two phase-lock loops inside the LTC3605: one servos the regulator on-time to track the internal oscillator frequency, which is determined by an external timing resistor, and the other servos the internal oscillator to an external clock signal if the part is synchronized. Due to the controlled on-time design, the LTC3605 can achieve very fast load transient response while minimizing the number and value of external output capacitors.

The LTC3605's switching frequency is programmable from 800kHz to 4MHz, or the regulator can be synchronized to an external clock for noise-sensitive applications.

Furthermore, multiple LTC3605s can be used in parallel to increase the available output current. The LTC3605 produces an out-of-phase clock signal so that parallel devices can be interleaved to reduce input and output current ripple. A multiphase, or PolyPhase, design also generates lower high frequency EMI noise than a single phase design, due to the lower switching currents of each phase. This configuration also helps with the thermal design issues normally associated with a single high output current device.

$1.8V_{OUT}$, 2.25MHz buck regulator

The LTC3605 is specifically designed for high efficiency at low duty cycles such as $12V_{IN}$-to-$1.8V_{OUT}$ at 5A, as shown in Figure 42.1. High efficiency is achieved with a low $R_{DS(ON)}$ bottom synchronous MOSFET switch (35mΩ) and a 70mΩ $R_{DS(ON)}$ top synchronous MOSFET switch.

This circuit runs at 2.25MHz, which reduces the value and size of the output capacitors and inductor. Even with the high

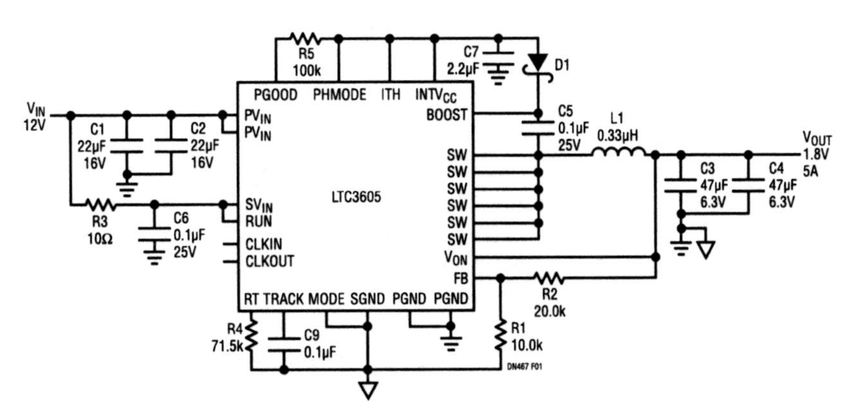

Figure 42.1 • 12V to 1.8V at 5A Buck Converter Operating at 2.25MHz

Analog Circuit Design: Design Note Collection. http://dx.doi.org/10.1016/B978-0-12-800001-4.00042-9

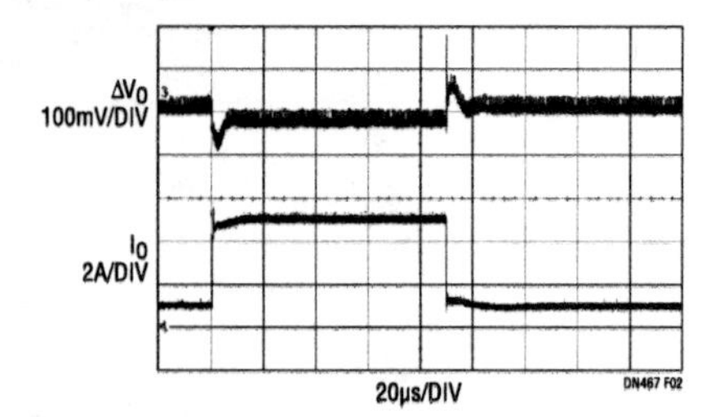

Figure 42.2 • Load Step Response of the Circuit in Figure 42.1

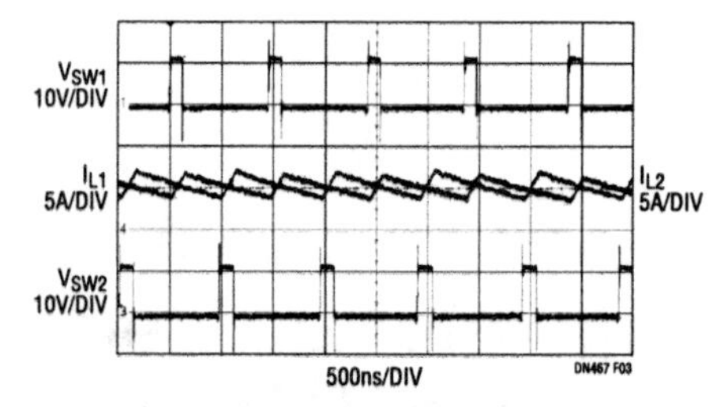

Figure 42.3 • Multiphase Operation Waveforms of the Circuit in Figure 42.4. The Switch Voltage and Inductor Ripple Currents Operate 180° Out of Phase with Respect to Each Other

switching frequency, the efficiency of this circuit is about 80% at full load.

Figure 42.2 shows the fast load transient response of the application circuit shown in Figure 42.1. It takes only 10μs to recover from a 4A load step with less than 100mV of output voltage deviation and only two 47μF ceramic output capacitors. Note that compensation is internal, set up by tying the compensation pin (ITH) to the internal 3.3V regulator rail ($INTV_{CC}$). This connects an internal series RC to the compensation point of the loop, while introducing active voltage positioning to the output voltage: 1.5% at no load and −1.5% at full load. The hassle of using external components for compensation is eliminated. If one wants to further optimize the loop, and remove voltage positioning, an external RC filter can be applied to the ITH pin.

1.2V_{OUT}, 10A, dual phase supply

Several LTC3605 circuits can run in parallel and out of phase to deliver high total output current with a minimal amount of input and output capacitance—useful for distributed power systems.

The 1.2V_{OUT} dual phase LTC3605 regulator shown in Figure 42.4 can support 10A of output current. Figure 42.3 shows the 180° out-of-phase operation of the two LTC3605s. The LTC3605 requires no external clock device to operate up to 12 devices synchronized out of phase—the CLKOUT and CLKIN pins of the devices are simply cascaded, where each slave's CLKIN pin takes the CLKOUT signal of its respective master. To produce the required phase offsets, simply set the voltage level on the PHMODE pin of each device to $INTV_{CC}$, SGND or $INTV_{CC}/2$ for 180°, 120° or 90° out-of-phase signals, respectively, at the CLKOUT pin.

Conclusion

The LTC3605 offers a compact, monolithic, regulator solution for high current applications. Due to its PolyPhase capability, up to 12 LTC3605s can run in parallel to produce 60A of output current. PolyPhase operation can also be used in multiple output applications to lower the amount of input ripple current, reducing the necessary input capacitance. This feature, plus its ability to operate at input voltages as high as 15V, make the LTC3605 an ideal part for distributed power systems.

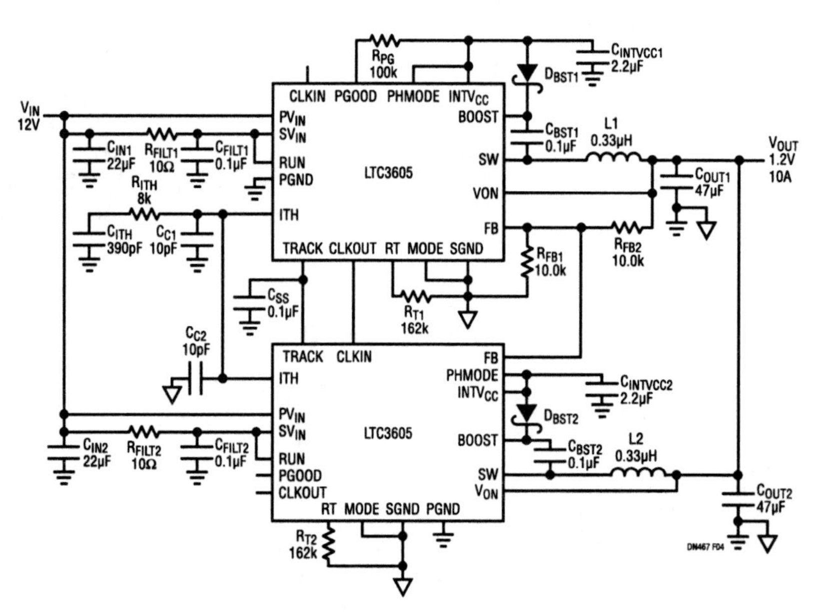

Figure 42.4 • 12V to 1.2V at 10A 2-Phase Buck Converter

Dual output buck regulator with current partitioning optimizes efficiency in space-sensitive applications

43

Johan Strydom

Introduction

The LTC3546 is a dual output current mode buck regulator with flexible output current partitioning. Beyond the advantages normally associated with dual output regulators (reduced size, cost, EMI and part count, with improved efficiency), the LTC3546's outputs can be partitioned for either 3A and 1A outputs, or two 2A outputs. This increases its application range and simplifies multiple supply rail designs. A configurable Burst Mode clamp for each output sets the current transition level between Burst Mode operation and forced continuous conduction mode to optimize efficiency over the entire output range. An adjustable switching frequency up to 4MHz and internal power MOSFET switches allow for small and compact footprints.

The LTC3546 utilizes a constant frequency current mode architecture that operates from an input voltage range of 2.25V to 5.5V—well suited to point-of-load (POL) conversion for intermediate power bus applications—and provides dual regulated output voltages as low as 0.6V.

The adjustable switching frequency can be set from 750kHz to 4MHz by an external resistor or synchronized to an external clock, allowing for a significant reduction in overall solution size through the reduction of the output capacitors and inductors. Furthermore, the 180 degree phase shift between outputs reduces input ripple current when compared with two independent regulators, as the ripple frequency is effectively doubled, thereby lowering EMI and allowing the use of a smaller input capacitor while also improving efficiency.

Additional features such as soft-start, supply sequencing and tracking, short-circuit protection and current foldback are all included in a thermally enhanced 28-lead 4mm × 5mm QFN package. An entire converter typically consumes less

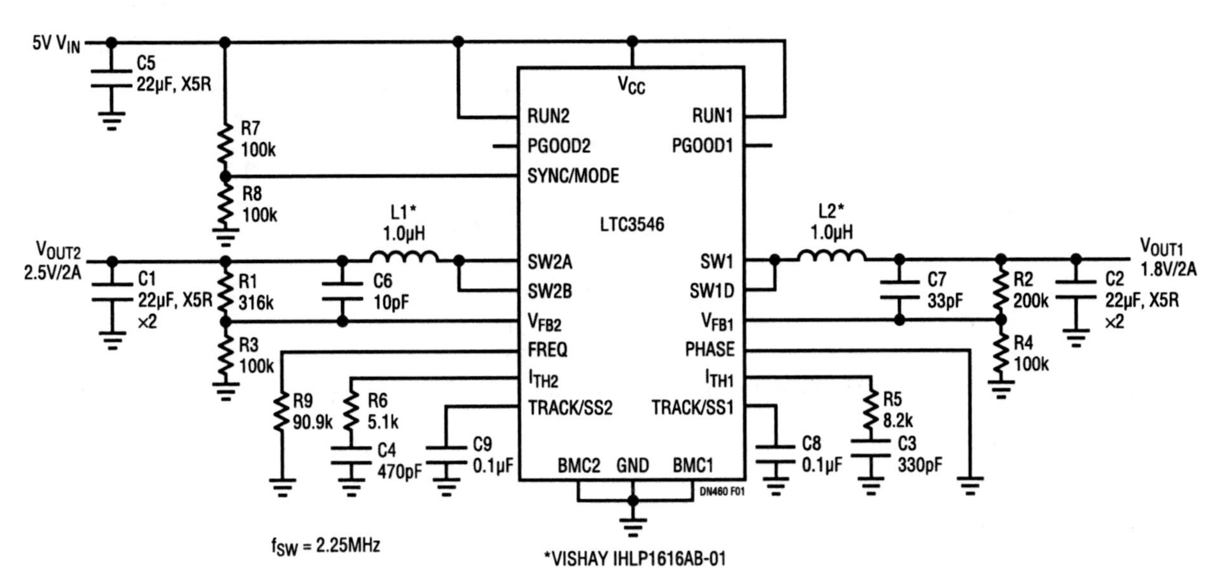

Figure 43.1 • A Low Profile Dual Output Converter for AM Frequency Sensitive Applications—2.5V/2A and 1.8V/2A Outputs

Analog Circuit Design: Design Note Collection. http://dx.doi.org/10.1016/B978-0-12-800001-4.00043-0

than 0.6 square inches of board real estate, single-sided, and is limited to 1.2mm in height.

Flexible current partitioning

A unique feature of the LTC3546 is its flexible current partitioning. The LTC3546 has two independently regulated outputs that can deliver up to 2A and 1A due to the 90mΩ and 180mΩ internal power MOSFET switches. An additional 1A output, with 180mΩ internal power MOSFET switches, can be paralleled to either of the outputs to produce either a 3A/1A or a 2A/2A dual output regulator. This external connection is internally detected and the dependent output is automatically gated in accordance to the connected output—nothing more is required.

Operation modes and efficiency

The LTC3546 can be configured for Burst Mode operation, forced continuous conduction (FCC) operation or pulse-skipping mode. In mobile applications where battery run time is of paramount importance, Burst Mode operation boosts efficiency by reducing gate charge losses at light loads, reducing supply current to just 125μA at no load. When noise control is important, forced continuous conduction operation may be preferred in order to trade efficiency for predictable, easily filtered constant frequency switching regardless of load current. Pulse-skipping mode provides a good compromise between light load efficiency and output voltage ripple. For optimum efficiency over all load conditions, the transition current between Burst Mode operation and forced continuous conduction mode can be set through the Burst Mode clamps (BMC) independently for each output.

Application examples

Figures 43.1 and 43.2 show the versatility of the LTC3546. Figure 43.1 shows a low profile converter with two independent outputs. It is configured for constant frequency (2.25MHz) operation regardless of load. This design is optimized for space sensitive applications where AM band (520kHz–1710kHz) interference could be a concern. Figure 43.2 shows a full featured dual POL converter with Burst Mode operation. The Burst Mode clamps are set for optimum efficiency over the entire load range of each output. This design is best suited for efficiency-sensitive applications with wide load current ranges.

Conclusion

The LTC3546 is a versatile dual output buck regulator suited for low to medium power applications. It delivers all the advantages of a dual output regulator with added flexible load current partitioning. The versatility of the LTC3546 makes it suitable for a wide range of applications that require compact, high efficiency power supplies.

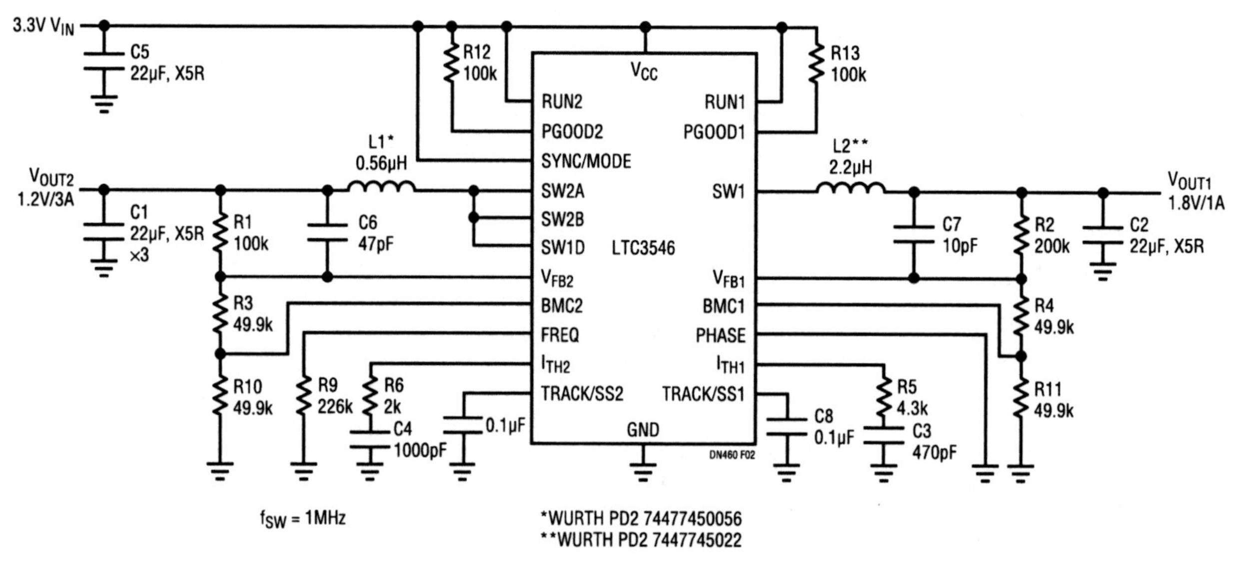

Figure 43.2 • A Dual Output Converter with High Efficiency Over the Entire Output Range—1.8V/1A and 1.2V/3A Outputs

Triple buck regulator features 1-wire dynamically programmable output voltages

44

Andy Bishop

Introduction

The LTC3569 is a compact power solution for handheld devices. Its tiny 3mm × 3mm QFN package includes three buck regulators with individually programmable output voltages. One regulator supports load currents up to 1200mA, while the other two support currents to 600mA. Two regulators can be paralleled for increased load capability. Each current mode regulator is internally compensated with excellent load and line regulation. A complete 2- or 3-output solution requires a minimum number of external passive components.

Three individually programmable bucks

The LTC3569's three output voltages are independently programmed by simply toggling their respective enable pins. Each time an enable pin sees a falling edge, a 4-bit counter is decremented. After a time-out delay of 120μs from the last rising edge at the enable pins, the counter state is latched into the feedback reference voltage DAC. In this way, the reference voltage can be programmed from 800mV (full scale) to 425mV in 25mV steps.

Configure parallel power stages for different loads

The LTC3569's buck regulators can be paralleled for higher load capability. By pulling the feedback pin of one of the two 600mA regulators up to the input supply voltage, that regulator's power stage is reconfigured as a slave, where switching is synchronized to its upstream master. Buck 2 can be a slave to Buck 1, or Buck 3 a slave to Buck 2. When operating in slave mode, the slave switch pin is tied in parallel with the master switch pin and the maximum output currents sum. This yields three possible combinations: three independent

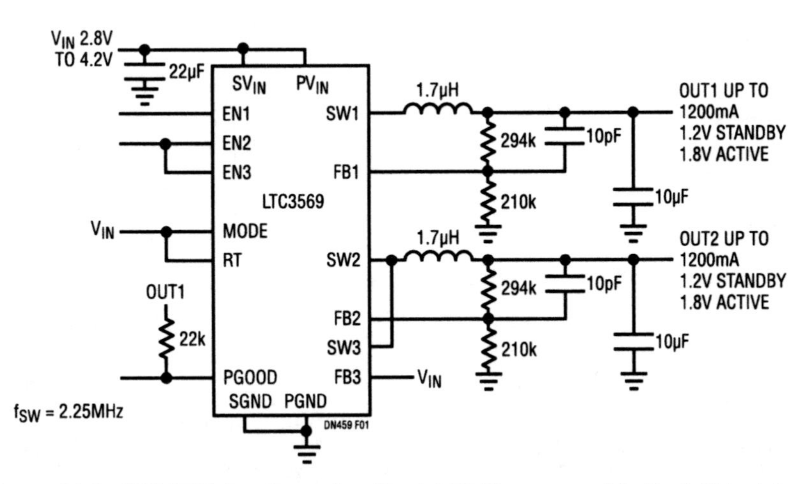

Figure 44.1 • LTC3569 Configured as Dual 1.2A Programmable Buck Regulators

Analog Circuit Design: Design Note Collection. http://dx.doi.org/10.1016/B978-0-12-800001-4.00044-2

regulators (a 1.2A buck and two 0.6A bucks), two independent 1.2A bucks, or two independent bucks of 1.8A and 0.6A.

Power good indicator

The LTC3569 has a PGOOD pin to indicate when any enabled regulator output voltage has risen to within 8% of the programmed value. If any of the enabled output voltages are lower than programmed, the PGOOD pin pulls low. If all of the regulators are off, the PGOOD pin pulls low and the LTC3569 enters a low power shutdown mode with <1μA of supply current.

Power saving operating modes

The LTC3569 offers two modes of operation (set via the MODE pin) that improve efficiency at light loads. Burst Mode operation is the most efficient at low load currents, while pulse-skipping mode produces lower ripple currents. At startup, until the end of the soft-start ramp, pulse-skipping mode is automatically selected.

Programmable clock frequency

The switching frequency is fixed at 2.25MHz by pulling RT up to the input supply, or the clock can be programmed to a frequency between 1MHz and 3MHz with a timing resistor to ground. If a clock signal is applied to the MODE pin the LTC3569's clock is injection locked to the external clock as long as the frequency is greater than that programmed using the RT pin. With injection locking, the operating mode is automatically set to pulse-skipping.

2-output, individually programmable 1.2A regulators

Figure 44.1 shows a 2-output application where each output can be reprogrammed at any time to a standby voltage of 1.2V or an active voltage of 1.8V. Both outputs provide up to 1.2A of load current from a Li-Ion battery voltage between 2.8V and 4.2V. Burst Mode operation is selected for high efficiency at light loads. Figures 44.2 through 44.5 show independent programming of the two output voltages via toggling of the respective enable pins while supplying a constant 625mA to each load.

Conclusion

The LTC3569 is a flexible solution for powering handheld Li-Ion battery applications. The ability to adjust or disable individual output voltages on the fly provides a simple solution to support energy saving operating modes in advanced microprocessor-based designs.

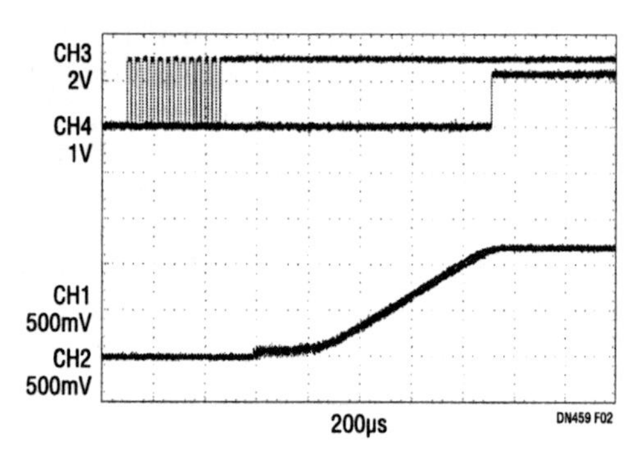

Figure 44.2 • Soft-Start Both Bucks into Standby. CH1 = OUT1, CH2 = OUT2, CH3 = EN1 = EN2, CH4 = PGOOD

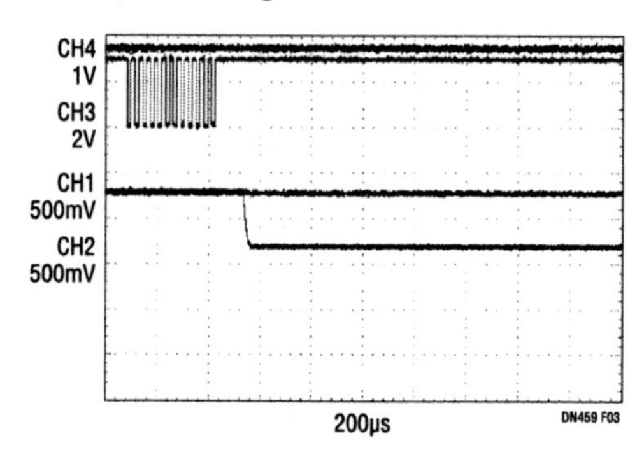

Figure 44.3 • Reprogram Buck 2 from Active to Standby with No Crosstalk on Buck 1 Output. CH1 = OUT1, CH2 = OUT2, CH3 = EN2, CH4 = PGOOD

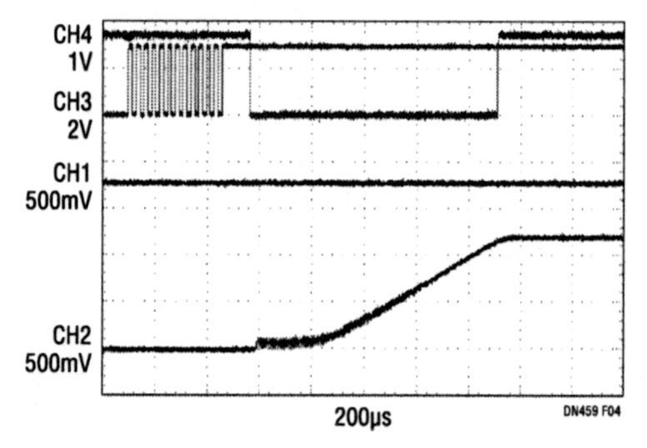

Figure 44.4 • Buck 1 Active, Buck 2 Soft-Start to Standby. CH1 = OUT1, CH2 = OUT2, CH3 = EN2, CH4 = PGOOD, No Crosstalk on Buck 1 Output

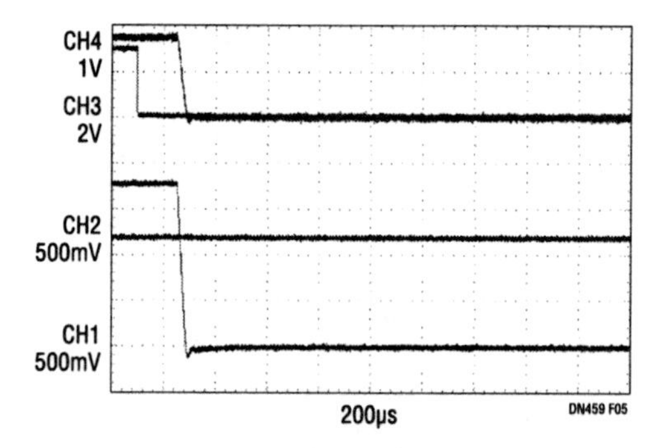

Figure 44.5 • Buck 1 Active to Shutdown, Buck 2 Standby. CH1 = OUT1, CH2 = OUT2, CH3 = EN2, CH4 = PGOOD, Note PGOOD Falls as It Is Tied to OUT1, No Crosstalk on Buck 2 Output

Buck converter eases the task of designing auxiliary low voltage negative rails

45

Victor Khasiev

Introduction

Many system designers need an easy way to produce a negative 3.3V power supply. In systems that already have a transformer, one option is to swap out the existing transformer with one that has an additional secondary winding. The problem with this solution is that many systems now use transformers that are standard, off-the-shelf components, and most designers want to avoid replacing a standard, qualified transformer with a custom version. An easier alternative is to produce the low negative voltage rail by stepping down an existing negative rail. For example, if the system already employs an off-the-shelf transformer with two secondary windings to produce ±12V, and a −3.3V rail is needed, a negative buck converter can produce the −3.3V output from the −12V rail.

Leave the transformer alone: $-3.3V_{OUT}$ from $-12V_{IN}$

Figure 45.1 shows a negative buck converter that generates −3.3V at 3A from a −12V rail. The power train (indicated by bold lines in Figure 45.1) includes an inductor L1, a diode D1 and a MOSFET Q1. The LTC3805-5 controller includes short-circuit protection (the current level can be precisely set), enable control and a programmable switching frequency. An internal shunt regulator simplifies biasing this IC directly from the input rail.

Despite the simplicity of this topology, there are some design hurdles. The first is that the feedback loop must control a negative output voltage via the controller's internal

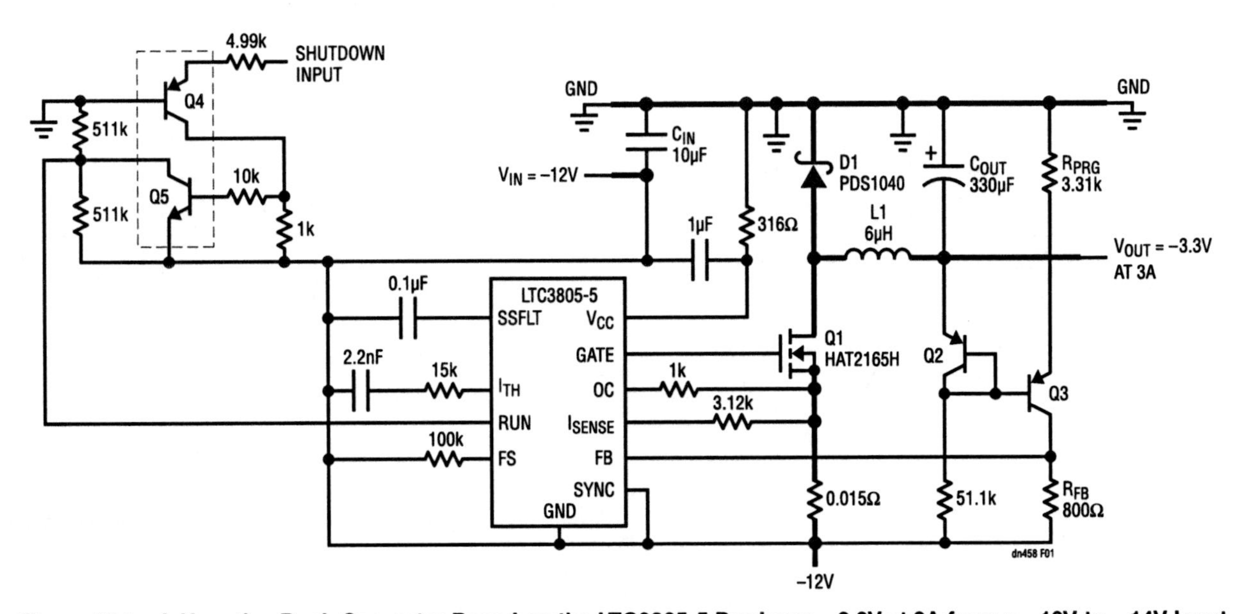

Figure 45.1 • A Negative Buck Converter Based on the LTC3805-5 Produces −3.3V at 3A from a −10V to −14V Input

Analog Circuit Design: Design Note Collection. http://dx.doi.org/10.1016/B978-0-12-800001-4.00045-4

positive reference. The second is that the on/off signal is referenced to the system ground.

To solve the output reference polarity problem, the regulation loop uses a current mirror based on transistors Q2 and Q3. Resistor R_{PRG} programs the current flowing into resistor R_{FB} which sets the output voltage. In this example, when the output voltage is at the desired −3.3V, the current through the 3.31k R_{PRG} resistor is 1mA. This current creates a 0.8V drop across resistor R_{FB}, which is equal to the reference voltage, V_{REF}, of the internal error amplifier:

$$V_{OUT} = \frac{V_{REF} \cdot R_{PRG}}{R_{FB}}$$

There is also an optional on/off circuit based on transistors Q4 and Q5. If 5V is applied to resistor R8, the LTC3805-5 shuts down. Both circuits are referenced to the system ground. The voltage stress on the power train components, the transfer function and other parameters are similar to positive input voltage buck converters.

This circuit operates at 90% efficiency, as shown in Figure 45.2. Figure 45.3 shows the progressive overcurrent protection as the load current increases. The output voltage drops at loads exceeding 4.5A, and at 5A the converter enters into a short-circuit protection state where the power is limited to 0.25W. The output voltage recovers after the short circuit is removed. In addition, the line and load regulation has a maximum deviation of less than 1%. Figures 45.4 and 45.5 show the start-up and transient response waveforms, respectively.

Conclusion

A negative buck converter is an easier way to generate an additional negative rail in systems that already have a larger negative voltage supply. This avoids undesirable replacement of standard transformers or modular components.

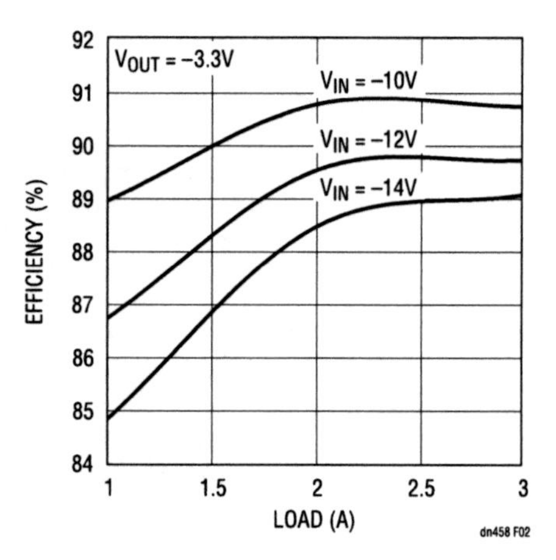

Figure 45.2 • Efficiency vs Input Voltage and Output Current for the Circuit in Figure 45.1

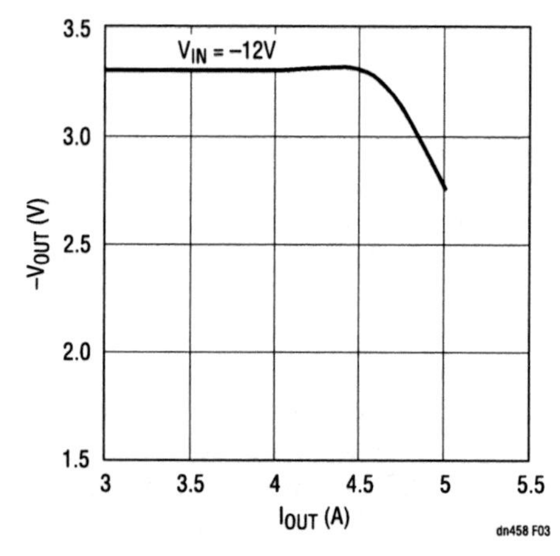

Figure 45.3 • Output Voltage vs Output Current, at −12V Input Voltage for the Circuit in Figure 45.1

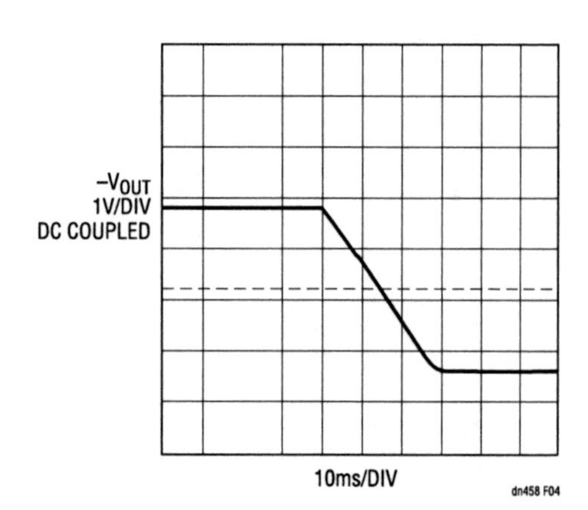

Figure 45.4 • Start-Up into Full Load

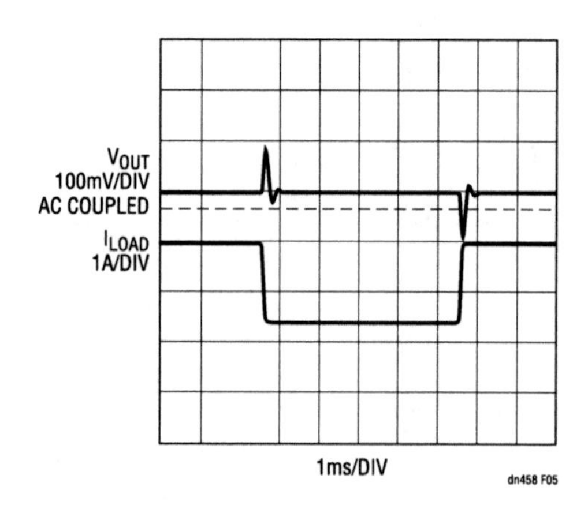

Figure 45.5 • Transient Response for a Load Current Step from 1A to 2.5A

Monolithic synchronous step-down regulator delivers up to 12A from a wide input voltage range

46

Charlie Zhao Henry Zhang

Introduction

The LTC3610 is a high power monolithic synchronous step-down DC/DC regulator that can deliver up to 12A of continuous output current from a 4V to 24V (28V maximum) input supply. It is a member of a high current monolithic regulator family (see Table 46.1) that features integrated low $R_{DS(ON)}$ N-channel top and bottom MOSFETs. This results in a high efficiency and high power density solution with few external components. This regulator family uses a constant on-time valley current mode architecture that is capable of operating at very low duty cycles at high frequency and with very fast transient response. All are available in low profile (0.9mm max) QFN packages.

Table 46.1 Comparison of LTC36XX Family Members

	LTC3608	LTC3609	LTC3610	LTC3611
INPUT VOLTAGE RANGE (V)	4 to 20	4.5 to 32	4 to 24	4.5 to 32
MAX LOAD CURRENT (A)	8	6	12	10
QFN PACKAGE SIZE (mm)	7 × 8 × 0.9	7 × 8 × 0.9	9 × 9 × 0.9	9 × 9 × 0.9

Typical application example

Figure 46.1 shows a typical application schematic of the LTC3608. This 7mm × 8mm regulator supplies 2.5V at 8A maximum load current from a 4.5V to 20V input source. This application switches at a nominal 650kHz, which allows the use of low profile inductor and capacitors while maintaining high efficiency. The switching frequency can be easily adjusted by a resistor (R_{ON} in Figure 46.1, connected to the I_{ON} pin). Figure 46.2 shows the operating efficiency.

The FCB pin is connected to ground to force continuous mode operation at light load for both low noise and small output ripple. The FCB pin can also be tied to $INTV_{CC}$ to enable

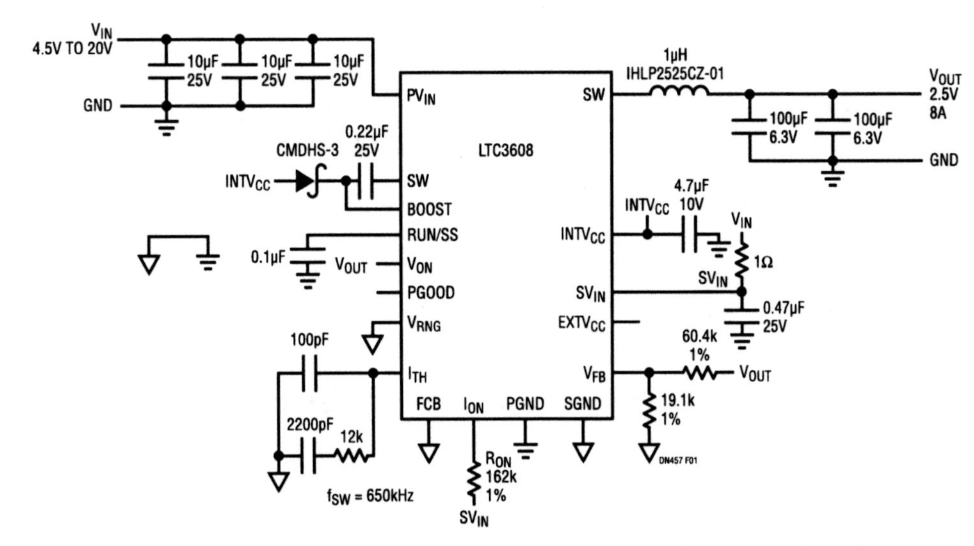

Figure 46.1 • 4.5V to 20V Input to 2.5V/8A High Density Step-Down Converter

Analog Circuit Design: Design Note Collection. http://dx.doi.org/10.1016/B978-0-12-800001-4.00046-6

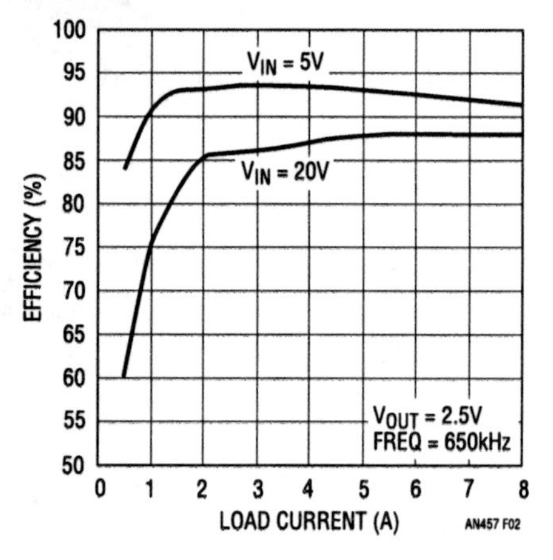

Figure 46.2 • Efficiency vs Load Current of Figure 46.1

discontinuous mode for higher efficiency at light load. Soft-start is programmable with a capacitor from the RUN/SS pin to ground. Forcing the RUN/SS pin below 0.8V shuts down the device.

The LTC36XX family are valley current mode regulators so they inherently limit the cycle-by-cycle inductor current. The inductor current is sensed using the $R_{DS(ON)}$ of the bottom MOSFET—no additional current sense resistor is required. The current limit is also adjustable with the voltage at the V_{RNG} pin. When the V_{RNG} pin is tied to ground, in the Figure 46.1 example, the current limit is set to about 16A.

An open-drain logic power good output voltage monitor (PGOOD) is pulled low when the output voltage is outside ±10% of the regulation point. In the case of overvoltage, the internal top MOSFET is turned off and the bottom MOSFET is turned on until the overvoltage condition clears. The LTC36XX also includes a foldback current limiting feature to further limit current in the event of a short circuit. If the output drops more than 25%, the maximum sense voltage is lowered to about one sixth of its original value.

Paralleling regulators for >12A

These parts can be easily paralleled for high output current applications. Figure 46.3 shows a 1.2V/24A application using two parallel LTC3610s. Because of the valley current mode control architecture, the paralleled regulators can operate at very low duty cycles with fast transient response and excellent load balance.

The current sharing is simple. Connect the I_{TH} pins together, since the I_{TH} pin voltage determines the cycle-by-cycle valley inductor current. The feedback pins of paralleled LTC3610s share a single voltage divider. The RUN/SS pins are connected so that the LTC3610s start up with same slew rate. The paralleled LTC3610s have excellent thermal balance due to good current sharing.

Conclusion

With broad input and output ranges, high current capability and high efficiency, these monolithic regulators provide small size, low external component count power solutions for many applications from communications infrastructure to industrial distributed power systems.

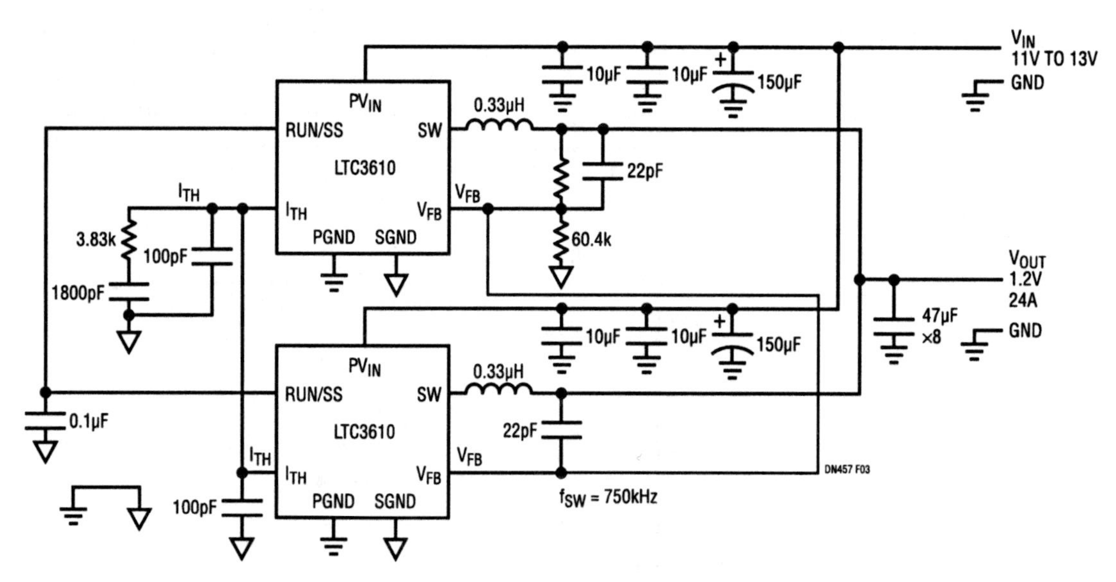

Figure 46.3 • Two LTC3610s in Parallel Can Provide 24A Output Current

Step-down synchronous controller operates from inputs down to 2.2V

47

Wei Gu

Introduction

Many telecommunications and computing applications need high efficiency step-down DC/DC converters that can operate from a very low input voltage. The high output power synchronous controller LT3740 is ideal for these applications, converting input supplies ranging from 2.2V to 22V to outputs as low as 0.8V with load currents from 2A to 20A. Applications include distributed power systems, point-of-load regulation and conversion of logic supplies.

A major challenge in designing a step-down converter with low V_{IN} is that the gate voltage for the N-channel MOSFETs is not readily available. The LT3740 solves this problem by integrating a DC/DC step-up converter for generating its own MOSFET gate drive voltage with a small inexpensive external inductor. This function permits the use of inexpensive off-the-shelf 5V gate-drive MOSFETs, offering up to 3% higher efficiency than sub-logic gate-drive MOSFETs and eliminating the need for a secondary supply.

The LT3740 operates at a fixed 300kHz frequency and employs valley current mode control to deliver excellent transient response and very low on-times. Furthermore, a power good signal is available to monitor the output voltage. The tracking capability built into XFER can be used to implement control of the output voltage when ramping up and ramping down. If less than 0.8V is applied to XFER, the LT3740 uses this voltage as its reference for regulation.

"Dying gasp" applications

The LT3740's low V_{IN} feature makes it a good fit in "dying gasp" applications for high reliability computers. In these systems, when the power gets cut, a large bank of electrolytic capacitors at the input, charged to 12V or so, serve as input to the power controller. The LT3740 can operate down to 2.2V, which allows it to run longer from the input capacitors' charged voltage than a controller requiring a higher

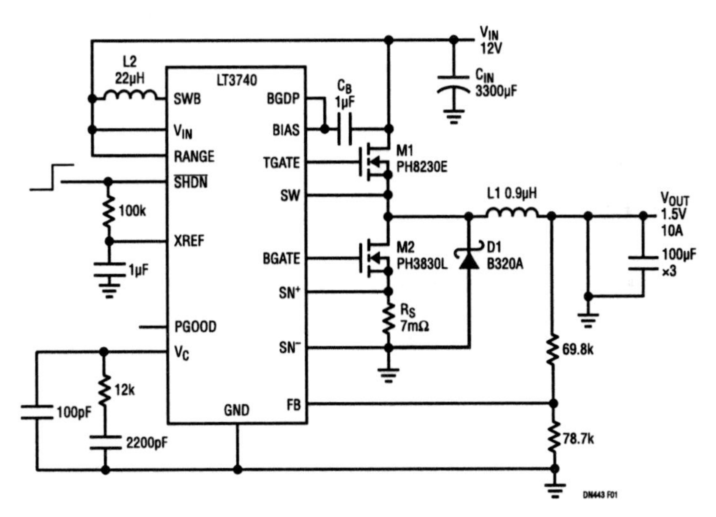

Figure 47.1 • "Dying Gasp" Application Circuit

Analog Circuit Design: Design Note Collection. http://dx.doi.org/10.1016/B978-0-12-800001-4.00047-8

input voltage. This effectively reduces the input capacitance requirements, which can save significant board space.

Figure 47.1 shows a typical application circuit. The LT3740 circuit, after its input power gets cut, is able to supply the load for 40ms as shown in Figure 47.2. In comparison, the output voltage would drop out after 25ms with a controller that shuts down when the input reaches 7V.

Generate a negative voltage from a low positive V_{IN}

Figure 47.3 shows a 2.4V to 14V input to −3.3V at 3A output converter. The LT3740 works particularly well in this application due to its wide input voltage range and ability to operate down to 2.2V. The LT3740 also operates with synchronous rectification, which allows the use of high efficiency MOSFETs, instead of less efficient switching diodes.

Wide input voltage range

The LT3740 offers high efficiency over a wide input voltage range (2.2V to 22V) and produces output voltages as low as 0.8V. The LT3740 employs valley current mode control to deliver excellent transient response and very low on-times. Figure 47.4 illustrates low duty cycle waveforms for a 22V input, 0.8V output application at a fixed frequency of 300kHz.

Conclusion

The LT3740 can operate from low input voltages, providing a space- and cost-saving solution over a wide input voltage range. The LT3740 is a versatile platform on which to build DC/DC converter solutions that use few external components and maintain high efficiencies over wide load ranges. The integrated step-up regulator facilitates true single supply operation with an input voltage as low as 2.2V.

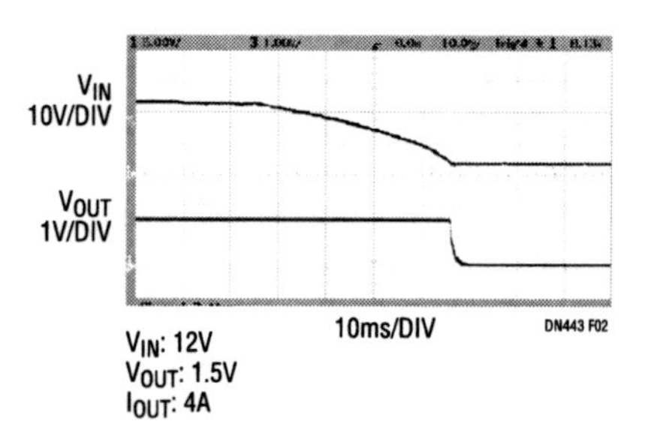

Figure 47.2 • Input and Output Voltages Waveforms When Input Power Is Removed

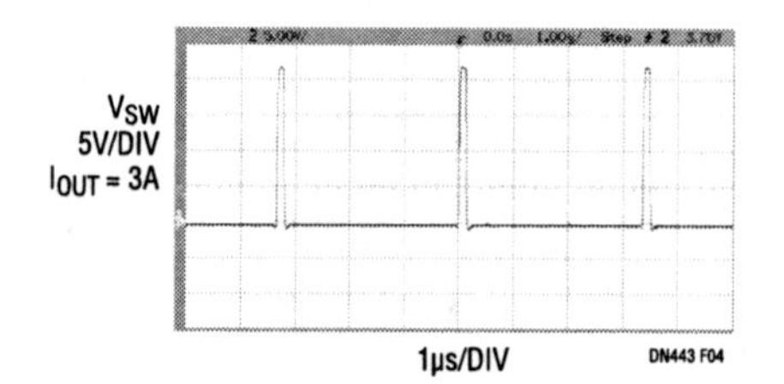

Figure 47.4 • LT3740 Low Duty Cycle Waveforms

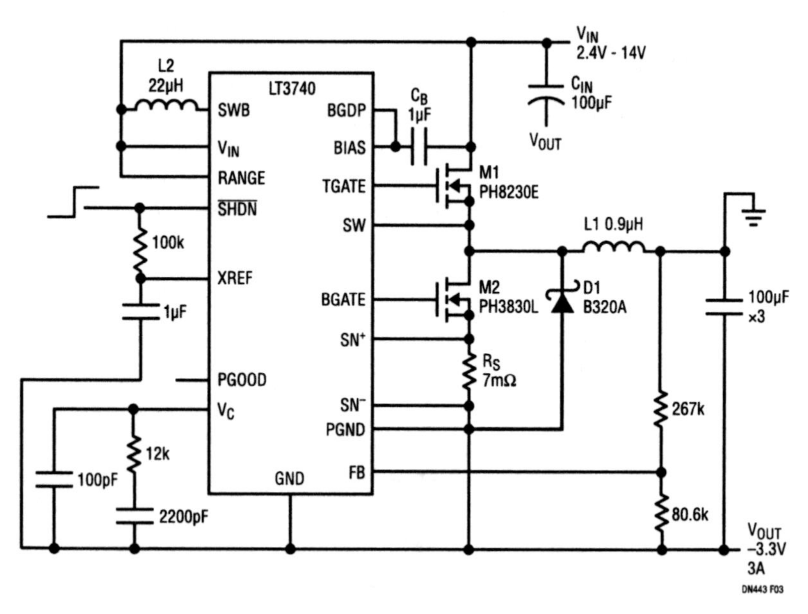

Figure 47.3 • A Positive to Negative Converter

Compact I²C-controllable quad synchronous step-down DC/DC regulator for power-conscious portable processors

48

Jim Drew

Introduction

The LTC3562 quad output step-down regulator is designed for multicore handheld microprocessor applications that operate from a single Li-Ion battery. Its four monolithic, high efficiency buck regulators support Intel's mobile CPU P-State and C-State energy saving operating modes. The output voltages are independently controllable via I²C, and each output can be independently started and shut down. Designers can choose from power saving pulse-skipping mode or Burst Mode operation, or select low noise LDO mode. The space-saving LTC3562 is available in a 3mm × 3mm QFN package and requires few external components.

Four I²C-controllable regulators

Two of the regulators provide up to 600mA of output current each while the other two provide up to 400mA each. All regulators are internally compensated, so no external compensation components are needed.

One of the 600mA regulators and one of the 400mA regulators (R600A and R400A) feature I²C-controllable feedback voltages, as shown in Figure 48.1. The output voltages of these "Type A" regulators are set by a combination of external programming resistors and I²C-adjustable feedback voltages—16 settings from 425mV to 800mV.

The "Type B" regulators (R600B and R400B) do not require external programming resistors because the resistors are integrated on-chip. The values of the internal feedback resistors are adjusted through the I²C port, resulting in 128 possible output voltages from 600mV to 3.775V in 25mV increments.

Inrush current limiting is provided by soft-start circuitry in all four regulators, as well as short-circuit protection and switch node slew rate limiting to reduce EMI.

Power saving operating modes

The LTC3562's step-down regulators offer four selectable modes of operation, which make it possible to balance low noise against efficiency. The four operating modes of the LTC3562 are shown in Figure 48.2.

At moderate to heavy loads, the constant frequency pulse-skipping mode provides the best output switching noise solution. At lighter loads, either Burst Mode operation or forced Burst Mode operation can be selected to maximize efficiency, though these modes produce higher ripple.

If the application calls for the lowest possible noise, LDO mode can be used for up to 50mA of load current. All four

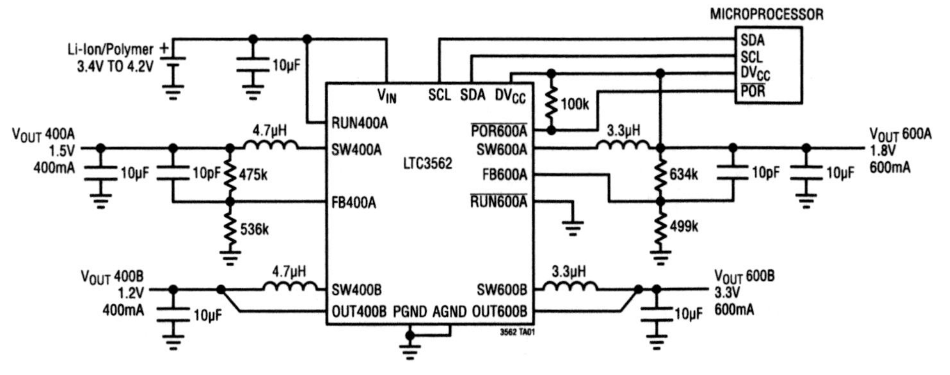

Figure 48.1 • High Efficiency Quad Step-Down Converter with I²C Control

Analog Circuit Design: Design Note Collection. http://dx.doi.org/10.1016/B978-0-12-800001-4.00048-X

converters support 100% duty cycle operation when the input voltage drops very close to the output voltage setting.

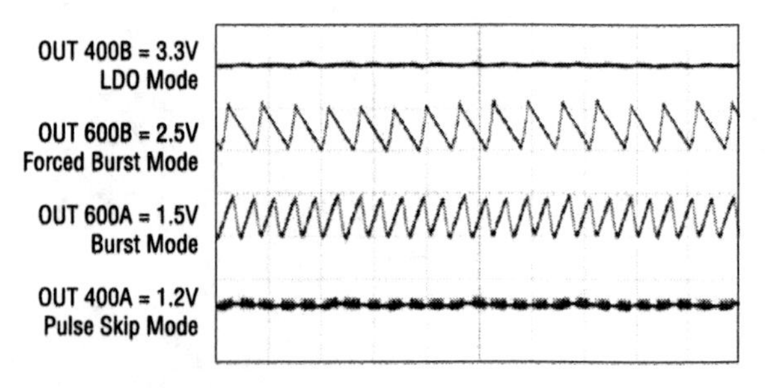

Figure 48.2 • Modes of Operation

I^2C programming of output voltages allows easy sequencing, tracking and margining

Each output can be programmed on the fly and independently enabled or disabled. These features taken together enable almost any sequencing or tracking scheme. A sequencing example is shown in Figure 48.3.

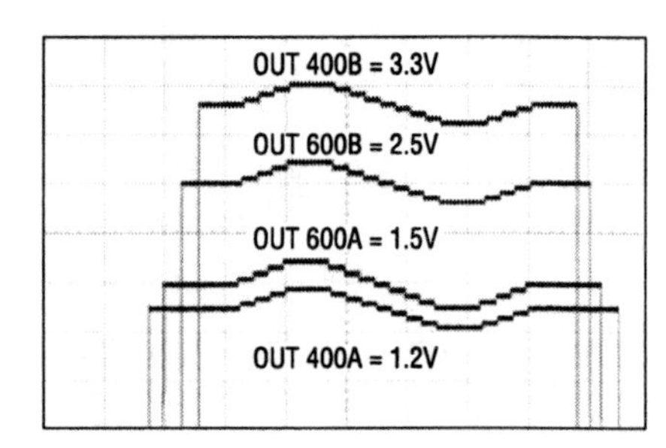

Figure 48.3 • LTC3562 Voltage Sequencing and Margining

A coincident voltage tracking example is shown in Figure 48.4. All of the outputs are ramped up together at power-up. At power-down, the highest output is incrementally ramped down until it reaches the value of the next higher voltage, which ramps down with the first. This is repeated until each output has tracked down to a minimum value and then disabled.

The ability to adjust the output voltage on the fly is also useful to margin the supplies for design evaluation or manufacturing quality audit testing. Voltage margining is applied to the nominal operating voltages in Figures 48.3 and 48.4.

Reducing the voltage or shutting down any output can reduce battery life or reduce energy usage in 'green' applications.

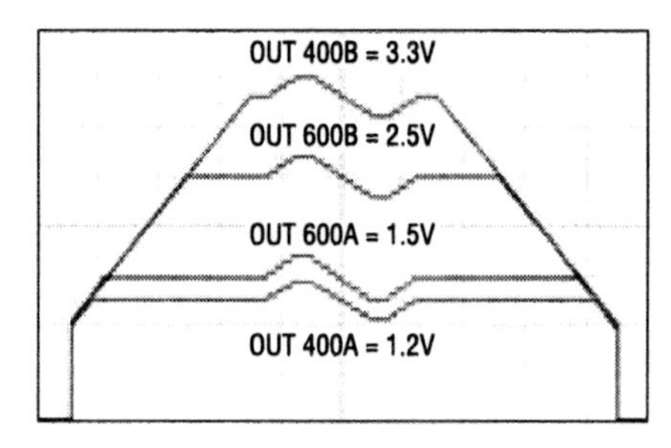

Figure 48.4 • LTC3562 Voltage Tracking and Margining

Conclusion

The LTC3562 is a versatile high efficiency quad output monolithic synchronous buck regulator controlled with an I^2C interface in a 3mm × 3mm QFN package. Four modes of operation allow the switching regulators to be tailored to the system's efficiency and noise requirements. This device is well suited for handheld microprocessor applications operating from a single Li-Ion battery where battery life is critical. The ability to use I^2C to adjust output voltages on the fly or disable output voltages supports Intel's mobile CPU P-State and C-State energy saving modes of operation and simplifies development and manufacturing tolerance testing.

Compact triple step-down regulator offers LDO driver and output tracking and sequencing

49

Tiger Zhou

Introduction

Typical industrial and automotive applications require multiple high current, low voltage power supplies to drive everything from disk drives to microprocessors. The LT3507 triple step-down converter fits easily into these applications. It is simple and compact compared to multi-chip solutions.

The LT3507 is a single IC current mode triple step-down regulator with internal power switches and a low dropout linear regulator driver. The switching converters are capable of generating one 2.4A output and two 1.5A outputs. All three converters are synchronized to a single oscillator, with the 2.4A output running antiphase to the other two converters, thereby reducing input ripple current. Each converter has independent shutdown and soft-start circuits and generates a power good signal when its output is in regulation, simplifying both supply sequencing and the interface with microcontrollers and DSPs. Separate input pins for each regulator offer additional flexibility; regulators can be cascaded to reduce circuit size, or each regulator can draw power from a different input source.

The switching frequency is set with a single resistor between 250kHz to 2.5MHz. High switching frequency allows the use of small inductors and capacitors resulting in a very compact triple output supply. The constant switching frequency, combined with low impedance ceramic capacitors, results in low output ripple. With its wide input voltage range of 4V to 36V, the LT3507 regulates a broad array of power sources including 5V logic rails, unregulated wall transformers, lead acid batteries and distributed power supplies.

6V to 36V input to four outputs—1.8V, 3.3V, 5V and 2.5V—one IC

The triple converter accommodates a 6V to 36V input voltage range and is capable of supplying up to 2.4A, 1.5A and 1.5A, respectively. The 20mA LDO driver output can drive an NPN transistor to provide a fourth low noise rail. Figure 49.1 shows

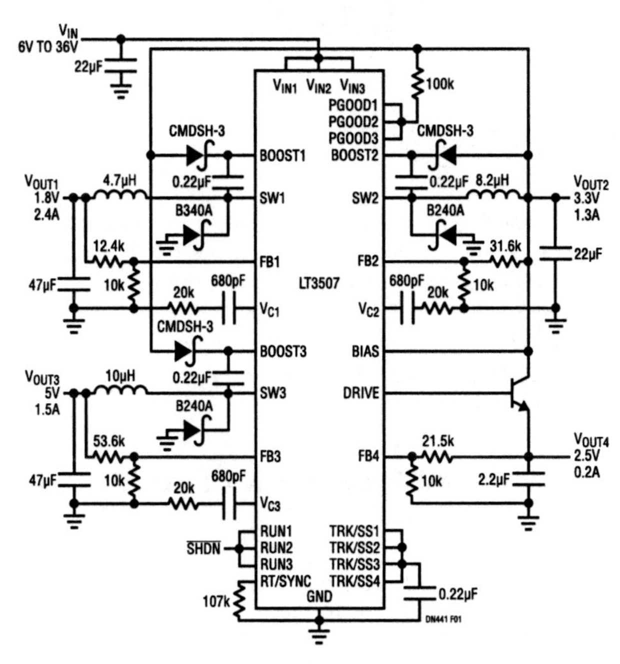

Figure 49.1 • A 4-Output Supply Including a Low Noise LDO

a typical application for four outputs—1.8V at 2.4A, 3.3V at 1.3A, 5.0V at 1.5A and 2.5V at 0.2A—from a 6V–36V input supply.

Low ripple high frequency operation even at high V_{IN}/V_{OUT} ratios

High frequency operation minimizes solution size, but one obstacle to high voltage (36V), high frequency (MHz) operation of monolithic step-down regulators is the minimum on-time constraint. Due to an internal logic propagation delay, a step-down regulator must remain on for a minimum time interval for proper operation. Otherwise, the converter operates in pulse-skipping mode at high input-to-output ratios, which

Analog Circuit Design: Design Note Collection. http://dx.doi.org/10.1016/B978-0-12-800001-4.00049-1

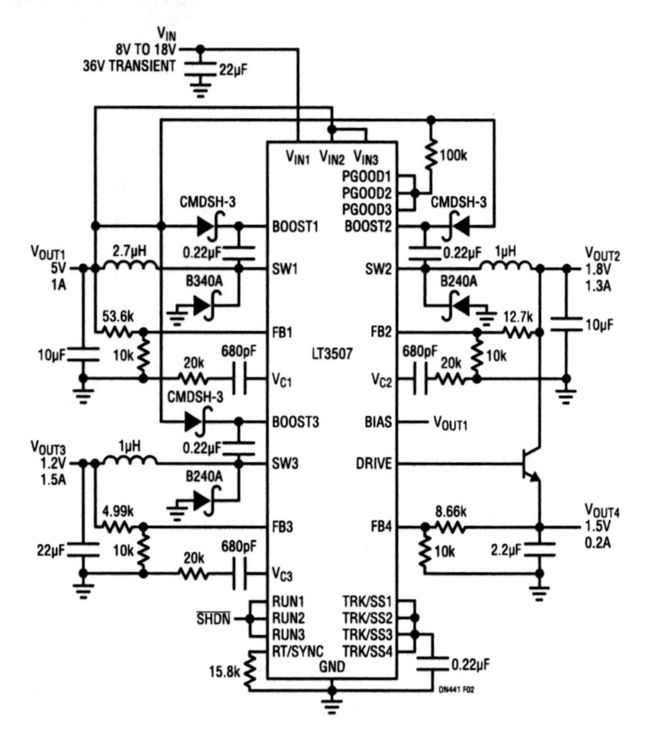

Figure 49.2 • Cascading Supplies Maintain High Frequency Operation Even with High V_{IN}/V_{OUT} Ratios

has the undesirable side effect of increasing output ripple. For example, the application in Figure 49.1 best operates at 450kHz when the input is 36V and the output is 1.8V.

However, the LT3507 has a built-in solution to this problem. By cascading the first converter and the other two, as shown in Figure 49.2, all three converters can be operated at 2MHz without pulse-skipping mode.

Input voltage lockout and sequencing

The LT3507's under- and overvoltage lockouts can be programmed with external resistors. When the schematic in Figure 49.2 is modified as in Figure 49.3, the LT3507 will accept V_{IN} up to 36V but operate only when V_{IN} is between 8V to 18V. This prevents the IC from operating during unintended or fault conditions, allowing the circuit designer to reduce the size of the external components. Figure 49.3 also shows a simple sequencing scheme: channel 1's power good indicator is tied to the tracking pins of the other three channels. Figure 49.4 shows the resulting start-up sequence, with channel 1 starting first and the remaining channels tracking during their start-up. Other sequencing and tracking examples can be found in the data sheet.

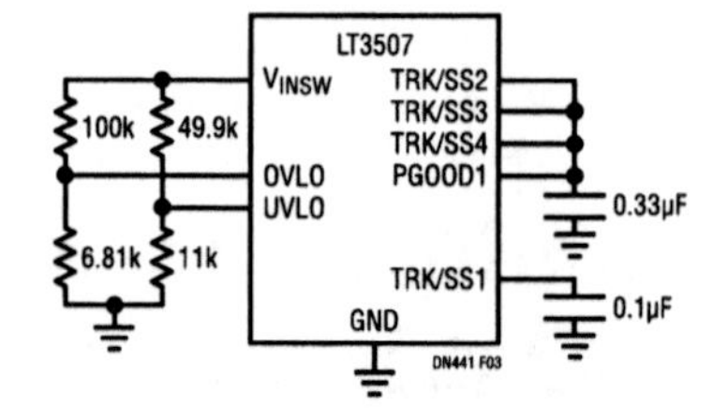

Figure 49.3 • External Resistors Program the Input Voltage Lockout; PGOOD1 Determines Sequencing and Tracking

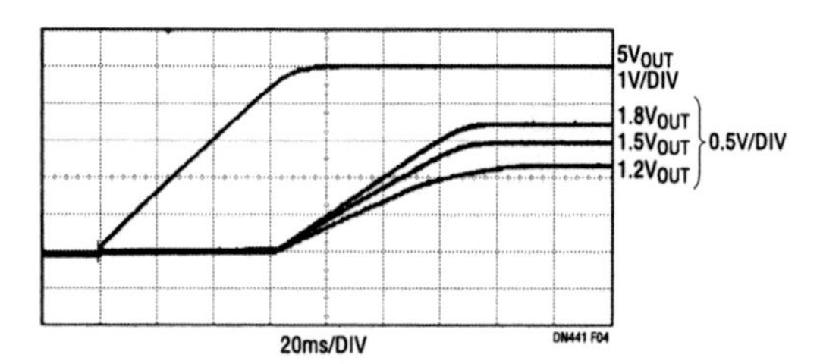

Figure 49.4 • Channel 1 Starts First, the Others Follow and Track

An additional feature of the LT3507 is the low noise LDO output. Figure 49.5 shows the LDO output ripple is reduced from the preregulated channel 2 output.

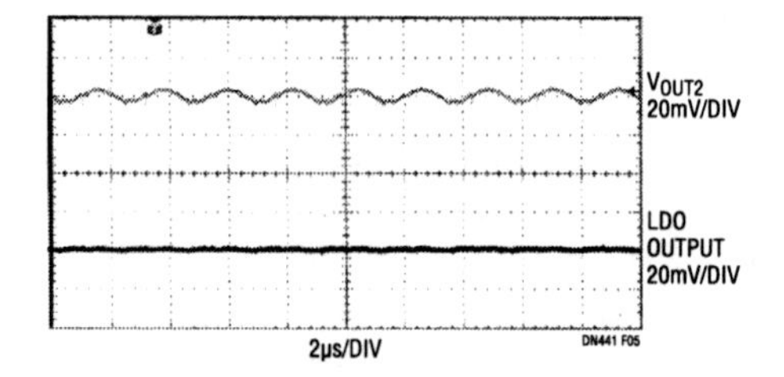

Figure 49.5 • Low Noise LDO Output: Top Trace Is Channel 2 Output, Bottom Trace Is Channel 4 LDO Output

Conclusion

The LT3507 integrates three buck regulators and an LDO driver in a QFN (5mm × 7mm) package, offering a compact solution for multiple-rail systems. Separate inputs for each converter offer wide design freedom, while separate PG indicators and TRK/SS pins further extend tracking and sequencing flexibility.

A positive-to-negative voltage converter can be used for stable outputs even with a widely varying input

50

Victor Khasiev

An obvious application of a positive-to-negative converter is generating a negative voltage output from a positive input. However, a not-so-obvious use is to produce a stable output voltage in an application that has a widely varying input. For example, a converter in a battery-powered device, which has an inherently variable input voltage, can produce a stable output voltage even if input voltage falls below the absolute value of the output voltage. However, an obvious drawback is reverse polarity, which can be easily overcome in this application. The supplied circuitry can use the negative output as the system ground and the negative battery terminal as the "positive" voltage source.

This topology is particularly useful when the input varies above or below the output. In such cases, a traditional step-down regulator would not be able to regulate once the battery voltage drops below the output, thus shortening the useful battery run time. Buck-boost solutions and other topologies such as a SEPIC solve this problem, but they tend to be more complicated and expensive. The positive-to-negative converter topology presented here combines the simplicity of a step-down converter and the regulation range of a buck-boost topology.

A new generation of Linear Technology high voltage synchronous step-down converters, such as the LT3845, make it possible to implement positive-to-negative conversions for a variety of applications.

Basic operation

Figure 50.1 shows a simplified block diagram of a positive-to-negative converter. Figure 50.2 shows an equivalent circuit, which helps in understanding the basic operation of the circuit in Figure 50.1. When transistor Q is on (Figure 50.2a), diode D is reverse biased and the current in inductor L increases. When Q is off (Figure 50.2b), inductor L changes polarity, diode D becomes forward biased, and current flows from inductor L to the load and capacitor C. The voltage across capacitor C and the load is negative, relative to system ground. Figure 50.3 shows a timing diagram.

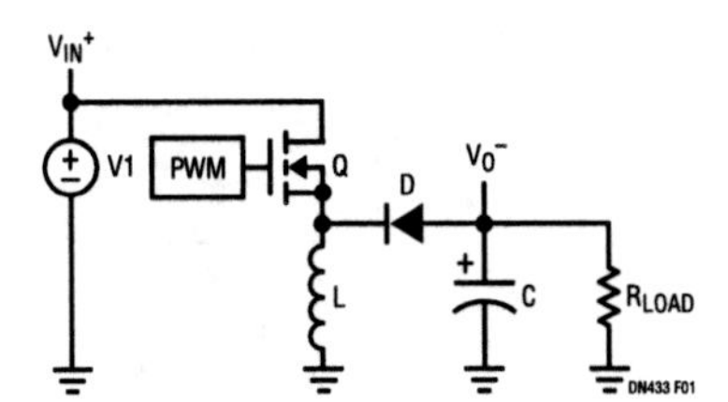

Figure 50.1 • Simplified Block Diagram of Positive-to-Negative Converter

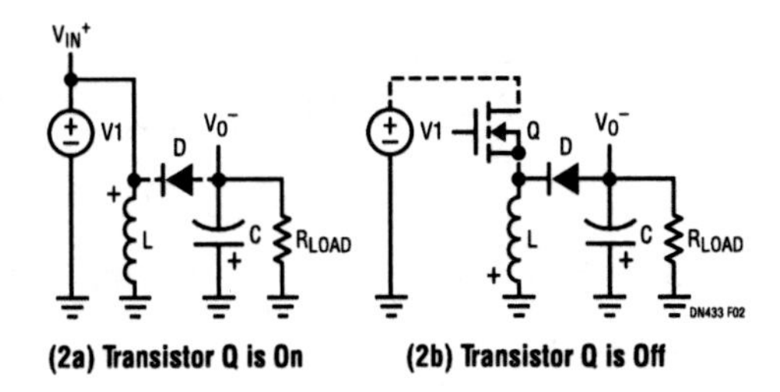

Figure 50.2 • Equivalent Circuits Show the Operation of the Positive-to-Negative Converter

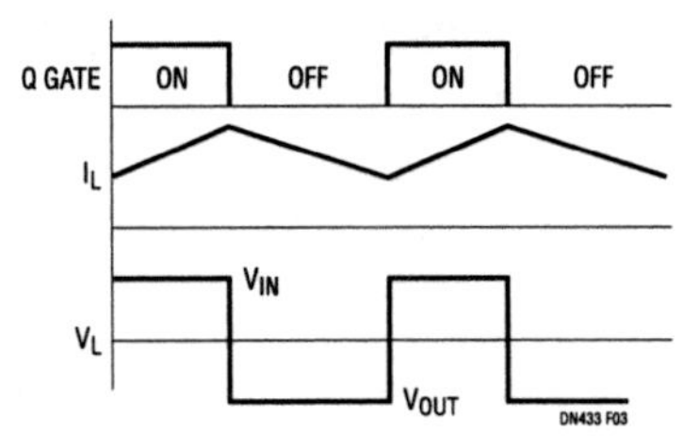

Figure 50.3 • Converter Timing Diagram

Analog Circuit Design: Design Note Collection. http://dx.doi.org/10.1016/B978-0-12-800001-4.00050-8

The duty cycle range can be found from following expression:

$$D = \frac{|V_O|}{V_{IN} + |V_O|}$$

$$D_{MAX} = \frac{|V_O|}{V_{IN(MIN)} + |V_O|}$$

$$D_{MIN} = \frac{|V_O|}{V_{IN(MAX)} + |V_O|}$$

Component stress in a positive-to-negative topology

V_{MAX} is the maximum voltage across transistor Q and diode D (Figure 50.2), where:

$$V_{MAX} = V_{IN(MAX)} + |V_O|$$

The maximum current, I_{MAX}, through transistor Q, inductor L and diode D can be derived based on the following equations, assuming continuous conduction mode:

$$I_L = \frac{I_O}{1 - D_{MAX}}, dI = \frac{V_{IN(MIN)} \cdot t \cdot D_{MAX}}{L}, I_{MAX} = I_L + \frac{dI}{2}$$

where t is a switching period.

Circuit description

Figure 50.4 shows a 9V to 15V input to −12V at 3A output converter. The high voltage LT3845 is used for several reasons, including the ability of its SW pin to withstand 65V, its integrated high side driver and differential current sense. The LT3845 can also provide synchronous rectification, which allows the use of efficient MOSFETs over less efficient switching diodes (Figures 50.5–50.7).

The entire converter power path contains the LT3845 high voltage PWM controller, MOSFETs Q1 and Q2, inductor L1, diode D1 and output filter capacitors C_{OUT1}–C_{OUT3}. Diode D2 is a bootstrap diode and diode D3 provides bias voltage for internal MOSFET drivers.

Conclusion

Very often electrical engineers have to design a negative voltage source supplied from a positive voltage rail. The positive-to-negative converter discussed in the article can be a good alternative to a flyback or a SEPIC approach.

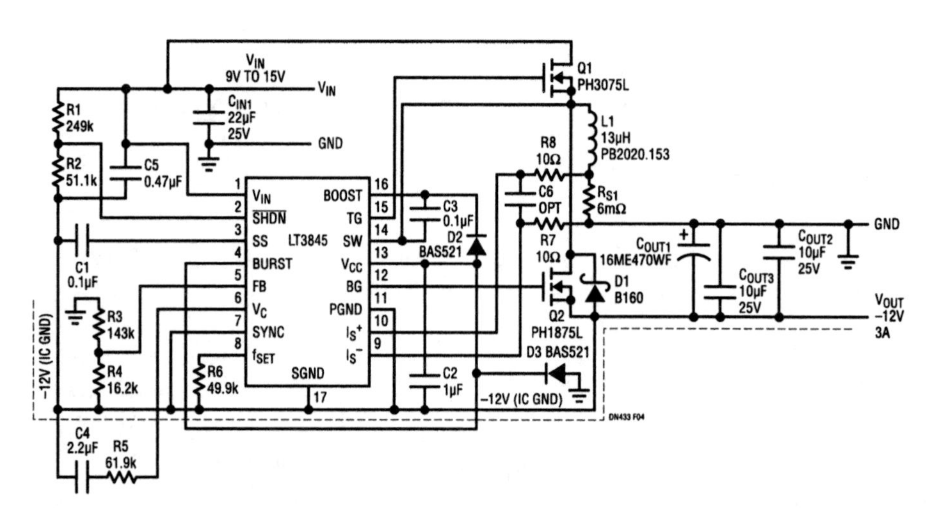

Figure 50.4 • Conversion of 9V–15V into −12V at 3A Based on the LT3845 High Voltage PWM Controller

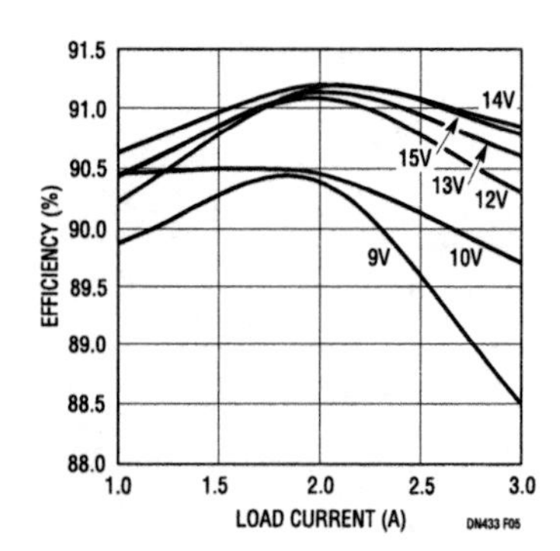

Figure 50.5 • Efficiency for the Figure 50.4 Circuit with Varying Input Voltage to a Fixed −12V Output

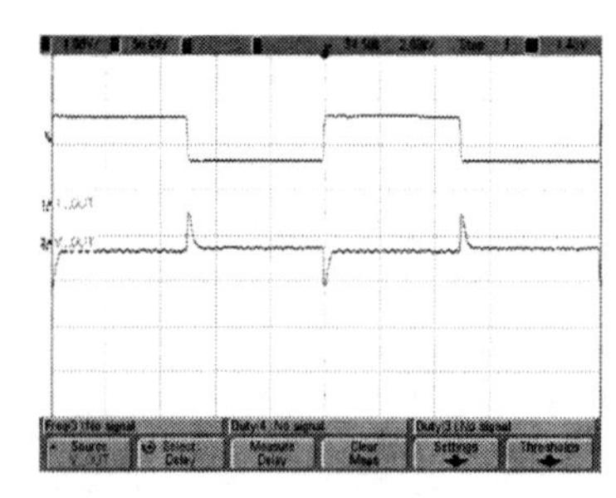

Figure 50.6 • Transient Response to an Output Load Step of 1A to 2A

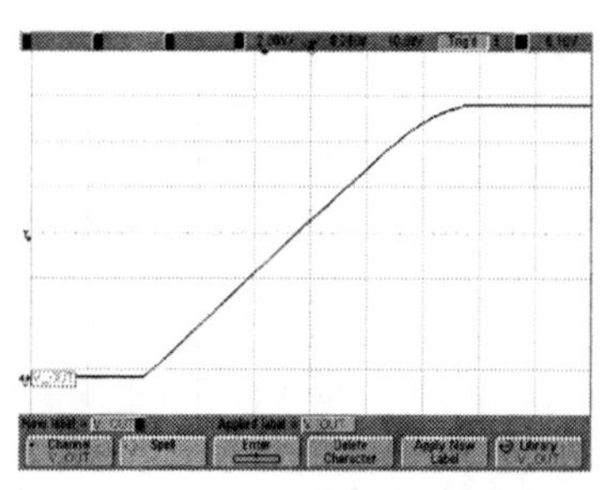

Figure 50.7 • Start-Up Waveform for the Circuit in Figure 50.4 with V_{IN} = 14V, V_{OUT} = −12V, I_{OUT} = 2A

51

One IC generates three sub-2V power rails from a Li-Ion cell

Frank Lee

Introduction

Shrinking geometries in IC technology have pushed the operating voltages of today's electronics well below 2V, presenting a number of design challenges. One common problem is the need for multiple supply voltages: for example, one voltage for a CPU core, another for I/O and still others for peripherals. Sensitive RF, audio and analog circuitry may require additional dedicated quiet supplies, separate from less noise-sensitive digital circuits. As the number of supplies increases, it becomes impractical to use a separate power supply IC for each voltage and special-requirements subsystem. Board area would be quickly consumed by power supplies. One solution to the space crunch is power supply integration, provided by a triple regulator like the LTC3446—three voltages from a single IC.

Triple supply in a tiny package

The LTC3446 combines a 1A synchronous buck regulator with two 300mA very low dropout (VLDO) linear regulators to provide up to three stepped-down output voltages from a single input voltage, all in a tiny 3mm × 4mm DFN. The 2.7V to 5.5V input voltage range is ideally suited for Li-Ion/Polymer battery-powered applications, and for powering low voltage logic from 5V or 3.3V rails. The output voltage range extends down to 0.4V for the VLDO regulators and 0.8V for the buck converter.

Each output is independently enabled or shut down via its own enable pin. When all outputs are shut down, V_{IN} quiescent current drops to 1µA or less, conserving battery power. The regulation voltage for each output is programmed by external resistor dividers. The buck regulator loop response can be tailored to the load by adjusting the RC network at the I_{TH} pin.

High efficiency and low noise

The 1A synchronous buck provides the main output with high efficiency, up to 90%. This buck converter features constant-frequency current mode operation at 2.25MHz, allowing small capacitors and inductor to be used. The two 300mA VLDO regulators can be connected to run off the buck output to provide two additional lower voltage outputs. This way,

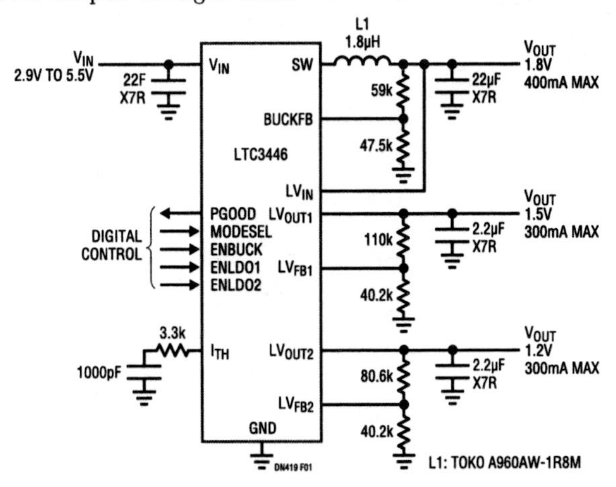

Figure 51.1 • Schematic Showing the LTC3446 Power Supply Configured to Deliver 1.8V from the 1A Buck, and 1.5V and 1.2V from the 300mA VLDO Regulators. The VLDO Regulators Are Powered from the Buck Output via the LV_{IN} Pin

Analog Circuit Design: Design Note Collection. http://dx.doi.org/10.1016/B978-0-12-800001-4.00051-X

the buck performs the bulk of the step-down at the high efficiencies typical of switching regulators, while the VLDO regulators provide additional lower voltages with good efficiency at the extremely low noise levels typical of linear regulators.

The schematic in Figure 51.1 shows the LTC3446 configured to deliver 1.8V from the buck, 1.5V from the first VLDO regulator, and 1.2V from the second VLDO regulator. Figure 51.2 shows the Figure 51.1 circuit assembled onto a printed-circuit board.

Selectable Burst Mode operation or pulse-skipping at light load

The LTC3446's buck regulator features Burst Mode operation for optimum efficiency when operating at light loads, at the cost of increased output ripple and the introduction of switching noise below the 2.25MHz clock frequency. Burst Mode operation can be defeated by bringing the MODESEL pin high, which commands the LTC3446 to continue to switch at the 2.25MHz clock frequency down to very light loads, whereupon pulses are skipped as needed to maintain regulation. Figure 51.3, which shows the efficiency of the buck regulator vs load current, also illustrates the typical efficiency gains from using Burst Mode operation at load currents below 100mA.

Very low dropout (VLDO) linear regulators

The VLDOs in the LTC3446 employ an NMOS source-follower architecture to overcome the traditional trade-off between dropout voltage, quiescent current and load transient response inherent in most PMOS- and PNP-based LDO regulator architectures. The V_{IN} pin (refer to Figure 51.1), supplies only the micropower bias needed by the VLDO control and reference circuits, typically at single-cell Li-Ion voltages. The actual load current is sourced from the LV_{IN} pin, which can be connected to the buck regulator output.

Each VLDO regulator provides a high accuracy output that is capable of supplying 300mA of output current with a typical dropout voltage of only 70mV from LV_{IN} to LV_{OUT}. V_{IN} should exceed the LV_{OUT} regulation point by 1.4V to provide sufficient gate drive to the internal NMOS pass device. Typical single-cell Li-Ion operating voltages extend down to 3.2V, supporting VLDO output voltages of up to 1.8V.

A single ceramic capacitor between 1μF and 2.2μF is all that is required for output bypassing. A low reference voltage of 400mV allows the VLDO regulators to be programmed to much lower voltages than are commonly available in LDO regulators.

Power good detection

The LTC3446 includes a built-in supply monitor. The PGOOD open-drain output pin is pulled low while any enabled output is more than ±8% from its regulation value. Once all enabled outputs are within this tolerance window, the PGOOD pin becomes high impedance. A microprocessor can monitor this open drain output pin to assess when a recently enabled output has completed start-up.

Conclusion

The LTC3446 packs an efficient 1A buck regulator and two 300mA VLDO regulators in a tiny 3mm × 4mm DFN package. With an output voltage range extending down to 0.4V for the VLDO regulators and 0.8V for the buck, and an input voltage range covering the single-cell Li-Ion range up to 5.5V, the LTC3446 is ideal for powering today's multi-voltage, sub-2V systems.

Figure 51.2 • The LTC3446 Triple Power Supply Assembled on a Printed Circuit Board

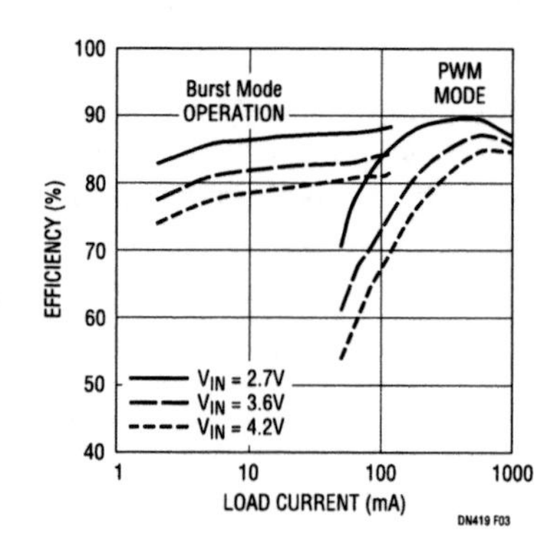

Figure 51.3 • Efficiency of the LTC3446's Buck Regulator vs Load

36V 2A buck regulator integrates power Schottky

52

David Ng

Introduction

Everyone wants more power in less space. However, the task of designing a power supply is easy to describe but difficult to execute. How does a designer select an optimal set of components that yields the best possible power supply in terms of size, cost and performance? Well, it is easier if the selection is reduced to only a handful of components.

For instance, the LT3681 reduces parts selection to only a few passive components by integrating all of the power semiconductors necessary to make a buck converter into a single package. Don't think that this high level of integration limits the usefulness of this part. The LT3681 accepts inputs from 3.6V to 34V, provides excellent line and load regulation and dynamic response, and offers a high efficiency solution over a wide load range while keeping the output ripple low during Burst Mode operation. Furthermore, its frequency is adjustable from 300kHz to 2.8MHz, enabling the use of small, low cost inductors and ceramic capacitors.

A small, simple solution

The LT3681 integrates a wide input voltage range, high performance buck controller, power switch, high side bootstrapping boost diode and a power Schottky diode. All of these attributes are crammed into a tiny but thermally efficient 14-pin 3mm × 4mm DFN package. So, all a designer needs to implement a full-featured buck converter is to add the output LC filter and a few passives.

The most obvious advantage of having the power Schottky diode integrated into the LT3681 is space savings, reducing the amount of board space required by the complete regulator by 15% or more. Moreover, the power Schottky diode has been optimized for the intended operation of the LT3681, so there is no need to agonize over finding the perfect form, fit and function diode for the application. Figure 52.1 shows a schematic of the LT3681 producing 5V at 2A from an input of 6.3V to 34V and Figure 52.2 shows the efficiency for a 12V input.

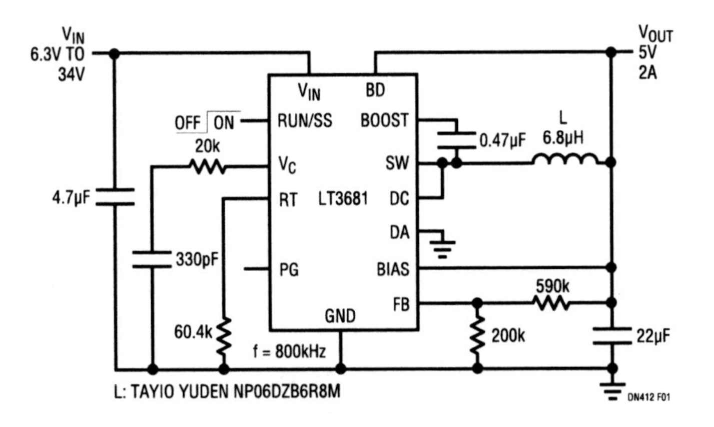

Figure 52.1 • The LT3681 Integrates All of the Power Semiconductors Necessary to Make a Simple 2A Buck Converter

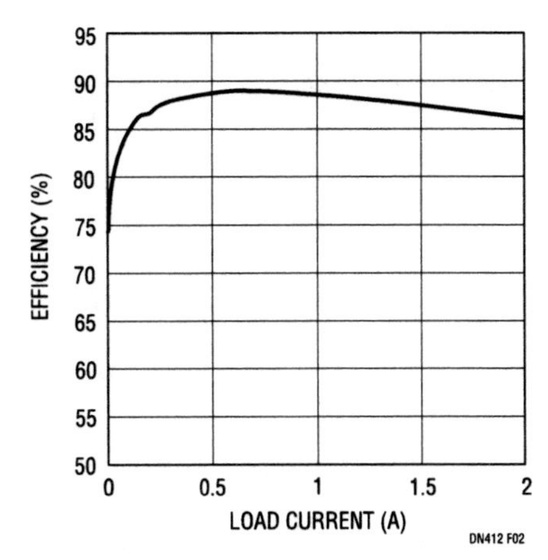

Figure 52.2 • The LT3681 Boasts High Efficiency (12V Input to 5V Output)

Analog Circuit Design: Design Note Collection. http://dx.doi.org/10.1016/B978-0-12-800001-4.00052-1

Low ripple and high efficiency solution over a wide load range

The LT3681 switching frequency can be programmed from 300kHz to 2.8MHz by using a resistor tied from the RT pin to ground. The LT3681 offers low ripple Burst Mode operation that maintains high efficiency at light loads while keeping the no load output voltage ripple below 15mV$_{P-P}$.

During Burst Mode operation, the LT3681 is able to deliver current in as little as one cycle to the output capacitor followed by sleep periods where all of the output power is delivered to the load by the output capacitor. Between bursts, all circuitry associated with controlling the output switch is shut down, reducing the input supply current to only 55μA. As the load current decreases toward no load, the percentage of time that the LT3681 operates in sleep mode increases and the average input current is greatly reduced, so high efficiency is maintained.

Figure 52.3 shows the low ripple and single cycle burst inductor current at no load for the 3.3V regulator shown in Figures 52.1 and 52.2. The LT3681 has a very low shutdown current (less than 1μA), significantly extending battery life in applications that spend long periods in sleep or shutdown mode.

For systems that rely on a well-regulated power source, the LT3681 provides a power good flag that signals when V_{OUT} reaches 90% of the programmed output voltage.

A resistor and capacitor on the RUN/SS pin programs the LT3681's soft-start, reducing the inrush current during start-up. In applications where the circuit is plugged into a live input source through long leads, an electrolytic input capacitor, which has higher ESR than a ceramic capacitor, is recommended to dampen the overshoot voltage. Refer to Application Note 88 for further details.

Frequency foldback saves chips

During short circuit, the LT3681 offers cycle-by-cycle current limit and frequency foldback, which decreases the switching frequency. This increases the off time, reducing the RMS current through the power switch and allowing the inductor current to safely discharge before the next switching cycle begins.

Conclusion

The robust design, small package and high level of integration of the LT3681 make it an excellent choice for a wide variety of step-down applications where a compact footprint and component optimization are critical. The high input voltage rating, high power switch capability and excellent package thermal conductivity adds to its versatility.

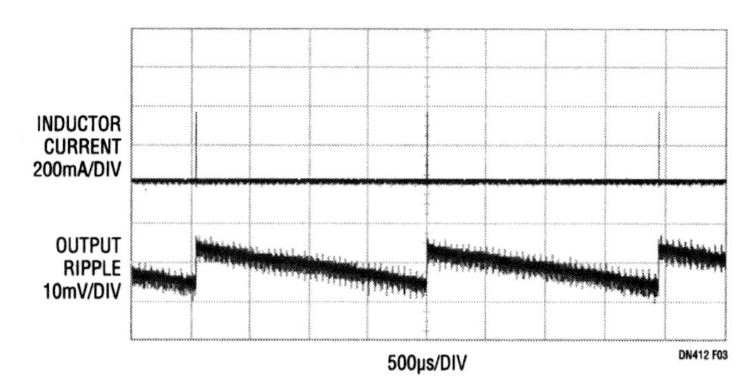

Figure 52.3 • This LT3681 Design Has Only 15mV of Output Ripple, Even at No Load Under Burst Mode Operation

Triple output 3-phase controller saves space and improves performance in high density power converters

53

Mike Shriver

Today's telecommunications, server and network applications require power from a multitude of voltage rails. Having more than ten rails ranging from 5V to 1V or less is common. These boards are typically crowded with heat-producing FPGAs or microprocessors, thus demanding power converters that are both compact and highly efficient. Furthermore, the converters may need to meet other requirements such as a fast load step response and rail tracking.

The LTC3773 switching regulator meets and even goes beyond the above requirements. This device is a 3-phase, triple output synchronous buck controller with built-in gate drivers packaged in either a 5mm × 7mm QFN or a 36-pin SSOP. Its switching frequency can be set to 220kHz, 400kHz or 560kHz, or it can be synchronized to an external clock between 160kHz and 700kHz. The controller can step down from input voltages as high as 36V and the output voltage can be programmed from 0.6V to 5V.

Figure 53.1 shows a high density triple output DC/DC converter with each output delivering up to 5A using the LTC3773 controller. Figure 53.2 shows the efficiency of each output versus load current; where up to 93% efficiency is achieved. Reductions in space are realized by the use of dual channel FETs and a switching frequency of 400kHz which permits the use of 7mm × 7mm ferrite inductors.

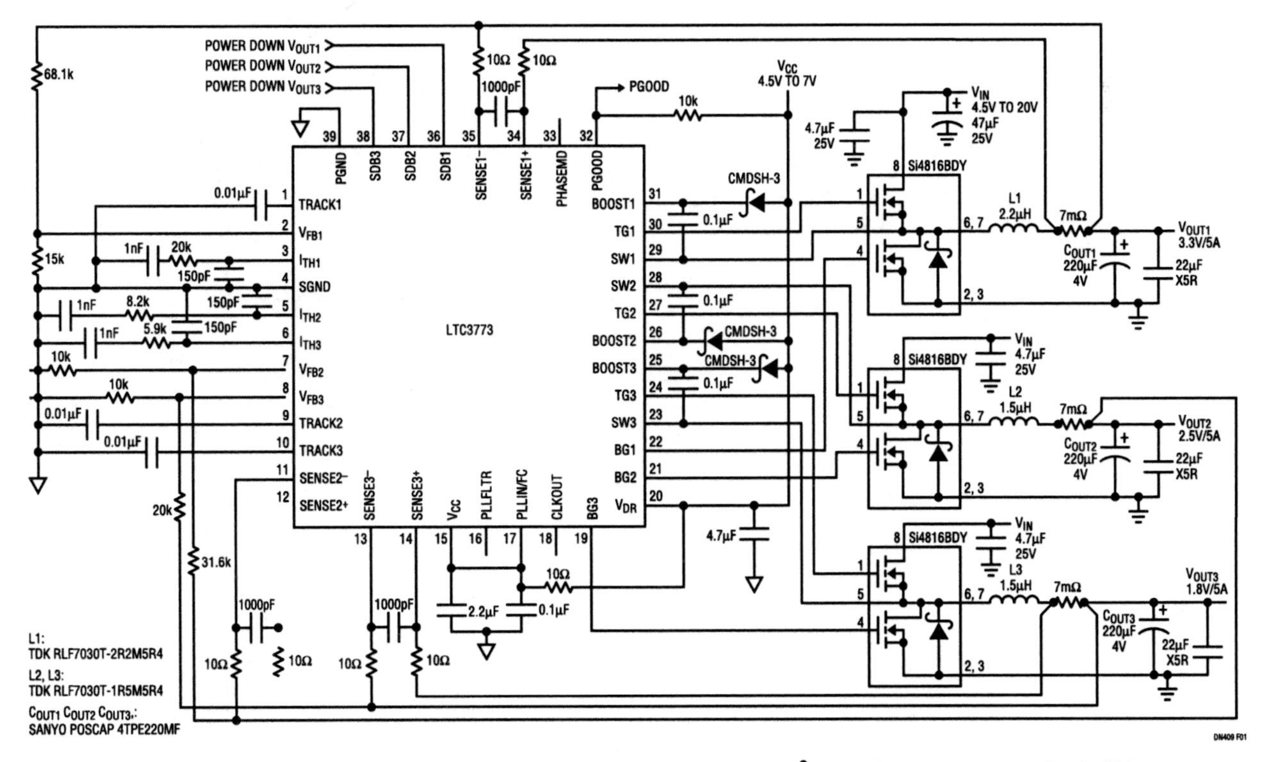

Figure 53.1 • High Density 5A Converter. Total Circuit Size = 1.5in², with Components on Both Sides

Analog Circuit Design: Design Note Collection. http://dx.doi.org/10.1016/B978-0-12-800001-4.00053-3

Switching the three rails out of phase results in improved performance and reduced cost. The use of triple phase operation instead of single phase can result in a reduction of the input capacitor ripple current by over 50% as shown in Figure 53.3, allowing the use of less input capacitance.

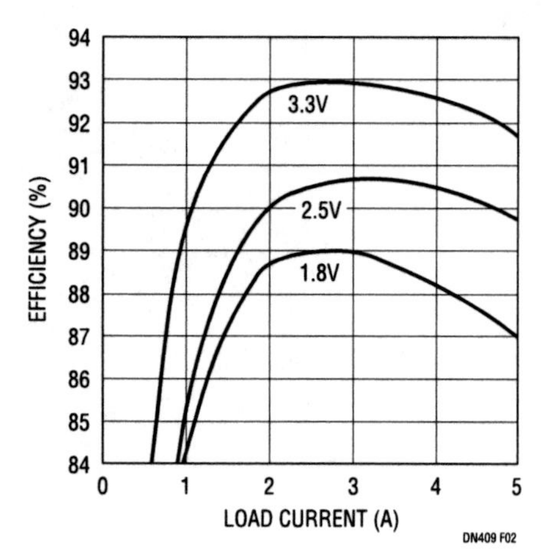

Figure 53.2 • Efficiency of the LTC3773 Converter at V_{IN} = 12V, f_{SW} = 400kHz. One Rail Enabled at a Time

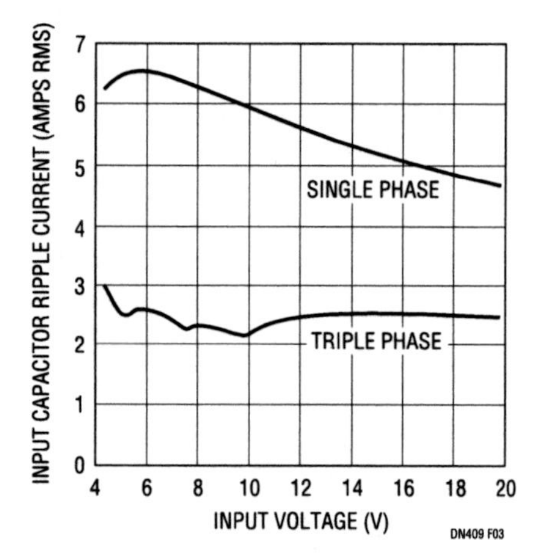

Figure 53.3 • Input Capacitor Ripple Current Comparison for Single Phase and Triple Phase Operation V_{OUT1} = 3.3V/5A, V_{OUT2} = 2.5V/5A, V_{OUT3} = 1.8V/5A
Single Phase: $\phi_{1,2,3}$ = 0°
Triple Phase: $\phi_{1,2,3}$ = 0°,120°, 240°

The outputs of two or more phases can be tied together which results in output ripple current reduction as well and a faster load step response. Up to six phases can be synchronized using the CLKOUT pin (on the QFN part only). Fast and accurate current sharing among the parallel phases is a result of the LTC3773's peak current mode architecture.

Compensation of each rail is achieved with an RC network on the I_{TH} pin (error amplifier output). The external I_{TH} compensation and the current mode topology allow the designer to easily stabilize a converter with the minimal amount of output capacitance using a variety of capacitor types including conductive polymer, tantalum and ceramic while still achieving a fast load step response (see Figure 53.4).

Other features of the LTC3773 include rail tracking and sequencing, a PGOOD signal, and three selectable light load operating modes (continuous conduction mode, Burst Mode operation and pulse skip mode).

Conclusion

Now designers have a clear and practical solution when they need a compact and cost-effective triple supply rail requirement in their telecom, server or network systems.

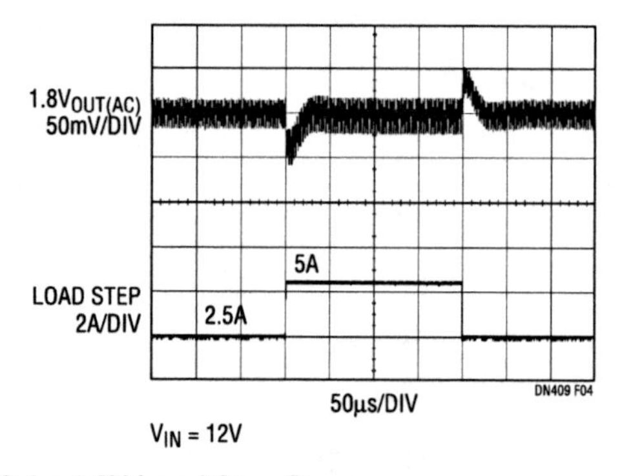

Figure 53.4 • 1.8V Load Step Response

Dual monolithic step-down switching regulator provides 1.6A outputs with reduced EMI and V_{OUT} as low as 0.8V

54

Hua (Walker) Bai

Introduction

Electronic devices are becoming smaller while the power requirements are increasing to satisfy ever more functionality. To preserve power and manage heat, switching regulators are desirable because of their high efficiency compared to linear regulators.

The LT3506 and the LT3506A are examples of how to squeeze power out of a supply without overheating the end product. These are dual 1.6A step-down monolithic regulators that simplify the lives of system engineers. They eliminate the need for external power switches, thereby reducing solution size and BOM cost. Their outputs can be as low as 0.8V, satisfying the needs of the latest DSPs. Integrated dual output channels reduce the part count, while anti-phase switching of the two channels maximizes efficiency and reduces input current ripple and EMI. Both the LT3506 and the LT3506A have 0.8V high accuracy voltage references.

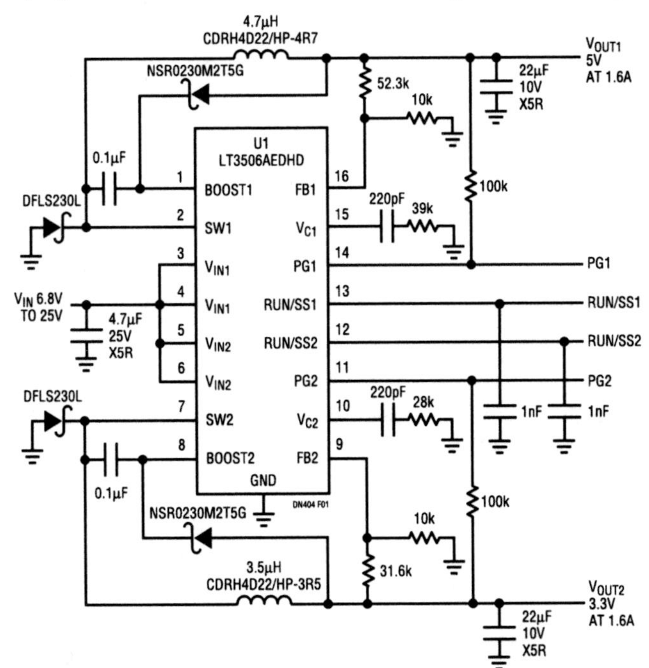

Figure 54.1 • LT3506A Application Circuit Provides Dual Outputs of 5V at 1.6A and 3.3V at 1.6A

Typical LT3506A and LT3506 applications

The primary difference between the LT3506A and the LT3506 is the switching frequency. The LT3506A's switching frequency is 1.1MHz while the LT3506's is 575kHz. Higher switching frequency allows smaller components. For lower output voltages, i.e., less than 3.3V, the LT3506 is recommended if $V_{OUT}/V_{IN(MAX)}$ is less than 15%. The lower switching frequency option usually results in higher efficiency due to the lower switching and inductor core losses.

The circuit in Figure 54.1 generates a 3.3V and a 5V output. The overall efficiency of the circuit (both channels) is shown in Figure 54.2. The circuit shown in Figure 54.3 generates a 1.2V and a 1.8V from a 3.6V to 21V input. The wide input voltage of both the LT3506 and the LT3506A

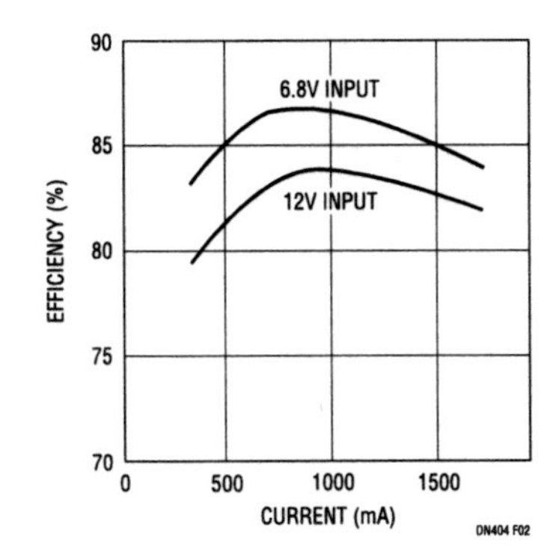

Figure 54.2 • Overall Circuit Efficiency (Both Outputs Loaded Identically)

Analog Circuit Design: Design Note Collection. http://dx.doi.org/10.1016/B978-0-12-800001-4.00054-5

(3.6V to 25V) can accept a variety of power sources, from lead-acid batteries and 5V rails to unregulated wall adapters and distributed power supplies.

The LT3506's high switching frequencies allow the use of small, low profile surface mount inductors and ceramic capacitors, resulting in smaller solution size and lower assembly cost. Furthermore, the low ESR of ceramic capacitors and high switching frequency results in very low, predictable output voltage ripple.

The internal supplies of both the LT3506 and LT3506A are powered from V_{IN1} pins. It is possible to supply power via the V_{IN2} pin from a different source, or it can be simply tied to V_{OUT1}, provided V_{OUT1} can supply sufficient current. The efficiency of V_{OUT2} can be improved when supplying V_{IN2} from a separate, lower supply voltage. Cascading V_{IN2} to V_{OUT1} also enables some low duty cycle applications for V_{OUT2}.

The thermally enhanced 16-lead DFN or TSSOP packages have an exposed ground pad on the bottom. This ground must be soldered to a ground pad on the PCB. Adding a dozen thermal vias to this pad improves thermal performance. A higher temperature grade part, I-grade, is also available.

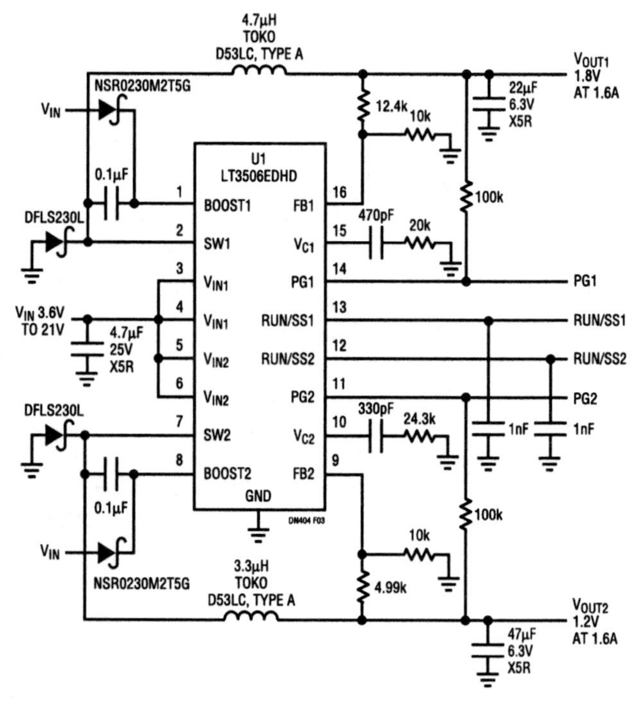

Figure 54.3 • LT3506 Application Circuit Provides Dual Outputs of 1.8V at 1.6A and 1.2V at 1.6A

Power sequencing without adding components

Supply sequencing is critical in many systems in order to prevent possible latch-up and to improve system reliability. Independent PG (Power Good) indicators, RUN/SS pins and V_C pins of the LT3506 and the LT3506A simplify supply sequencing. The PG pin remains low until the FB pin is within 10% of the final regulated output voltage. The easiest way to sequence outputs is to tie the PG1 pin to the V_{C2} pin. Then remove the pull-up resistor on PG1.This enables V_{OUT1} to come up before V_{OUT2} as shown in Figure 54.4.

2-phase switching eases EMI concerns

A step-down switching regulator draws current from its input supply and its input capacitor, resulting in a large AC current that can generate EMI. The LT3506's two regulators are synchronized to a single oscillator and switch out of phase by 180 degrees. Compared with two completely independent regulators, the input current ripple of the LT3506 is substantially reduced and its effective frequency doubled, thereby lowering EMI and allowing the use of a smaller input capacitor without reducing efficiency.

Conclusion

Both the LT3506 and LT3506A step down the output voltage to as low as 0.8V and provide 1.6A of current capability per channel. 2-channel anti-phase switching substantially reduces the input current ripple and eases EMI concerns. The PG, V_C, and RUN/SS pins make supply sequencing simple and straightforward. Two package options (leadless and leaded) and two switching frequency options allow optimal solutions for most applications.

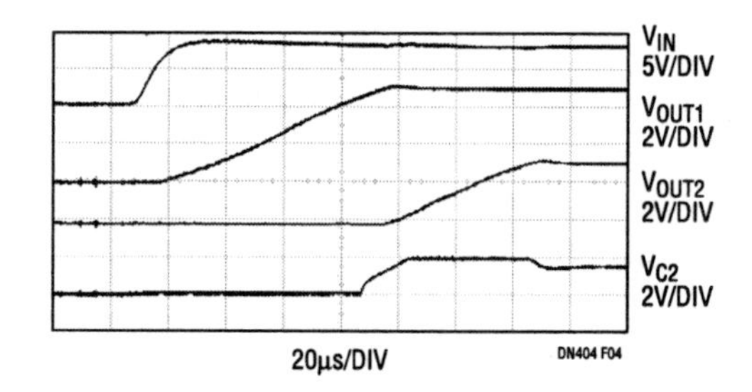

Figure 54.4 • Start-Up Waveforms of the Circuit in Figure 54.1 with Power Sequencing

A compact dual step-down converter with V_{OUT} tracking and sequencing

55

Tiger Zhou David Canny

Introduction

Typical industrial and automotive applications require multiple high current, low voltage power supply solutions to drive everything from disc drives to microprocessors. For many of these applications, particularly those that have size constraints, the LT3501 dual step-down converter is an attractive solution because it's compact and inexpensive compared to a 2-chip solution. The dual converter accommodates a 3V to 25V input voltage range and is capable of supplying up to 3A per channel. The circuit in Figure 55.1 produces 3.3V and 1.8V.

LT3501 dual converter features

- The LT3501 is feature-rich and comes with internal 3.5A switches and sense resistors to minimize solution size and cost.
- The LT3501 operates at a fixed frequency between 250kHz and 1.5MHz, programmed using a single resistor or synchronized to an external clock, allowing optimization of efficiency and solution size.
- A 180° phase relationship between the channels is maintained to reduce input voltage ripple and input capacitor size.
- Independent input voltage, feedback, soft-start and power good functions for each converter simplify the implementation of all the tracking and sequencing options available.
- Minimum input-to-output voltage ratios are extended by allowing the switch to stay on through multiple clock cycles resulting in a 95% maximum duty cycle regardless of switching frequency.
- The LT3501 automatically resets the soft-start function if the output drops out of regulation so that a short circuit or brownout event is graceful and controlled.

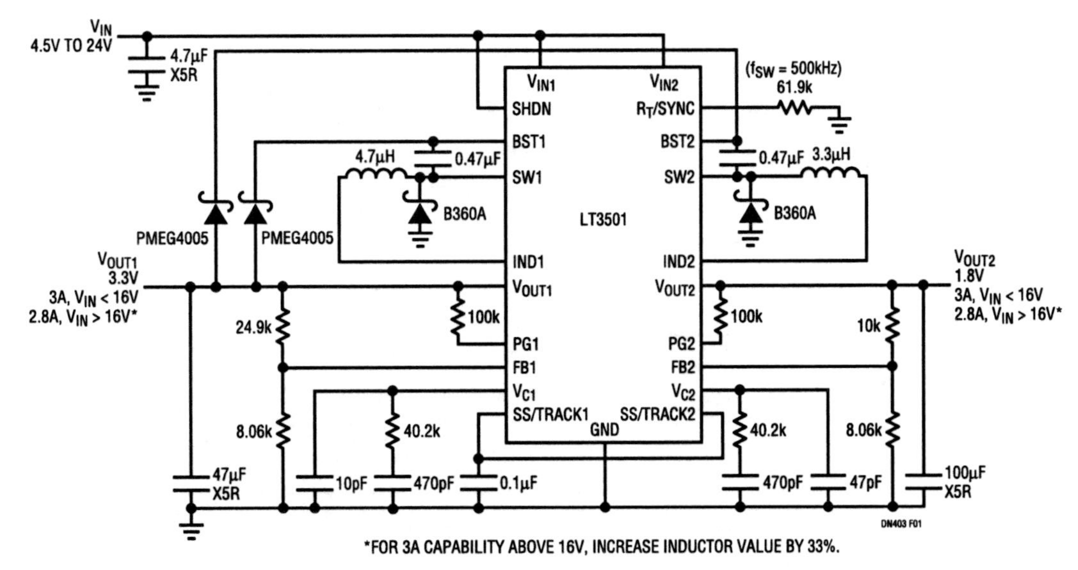

Figure 55.1 • Compact Dual 3A Step-Down Converter with Ceramic Capacitors

Analog Circuit Design: Design Note Collection. http://dx.doi.org/10.1016/B978-0-12-800001-4.00055-7

- One or both converters can be shutdown at any time if they're not being used, reducing input power drain.
- The LT3501 is available in a 20-pin TSSOP package with an exposed pad for low thermal resistance.

Output supply tracking and sequencing

Output voltage tracking and sequencing between channels can be implemented using the LT3501's soft-start and power good pins as shown in Figures 55.2a–c. Output sequencing can also be implemented as shown in Figure 55.2d.

High current single V_{OUT}, low ripple 6A output

The LT3501 can generate a single, low ripple 6A output as shown in Figure 55.3 with the dual converters sharing a single output capacitor. With this solution, ripple currents at the input and output are reduced, thus reducing voltage ripple and allowing the use of smaller, less expensive capacitors.

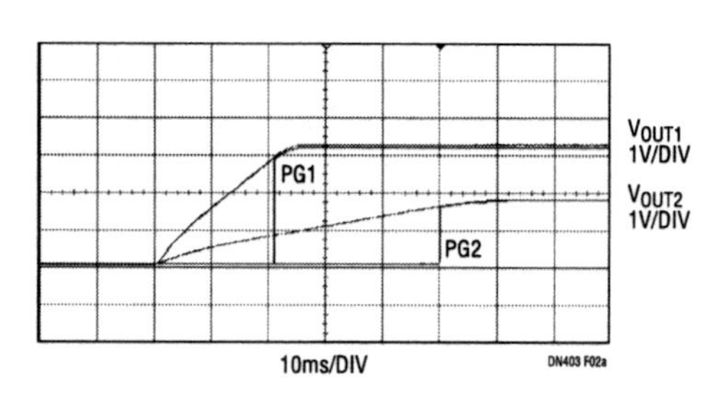

(55.2a) Independent

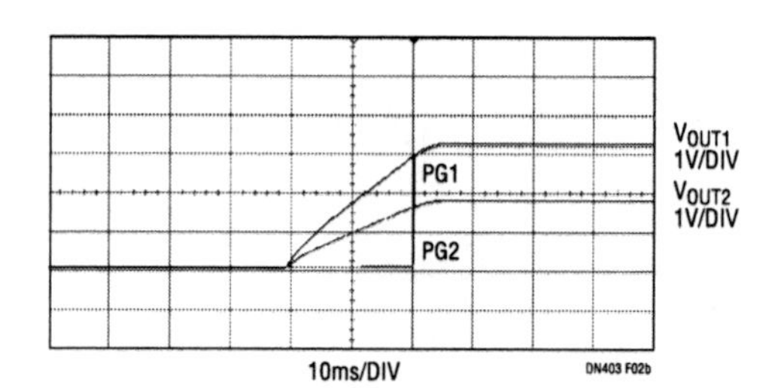

(55.2b) Ratiometric

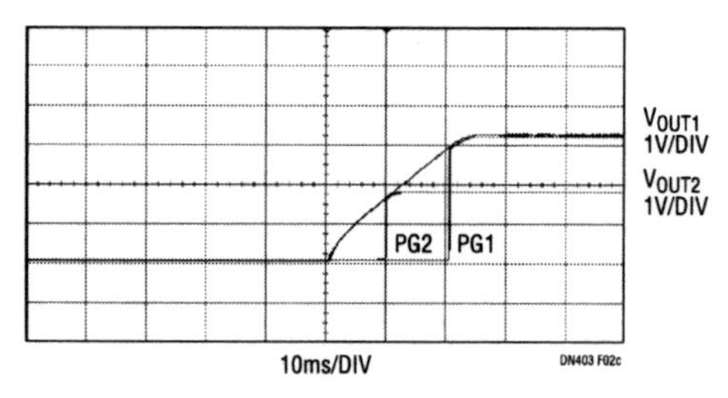

(55.2c) Absolute

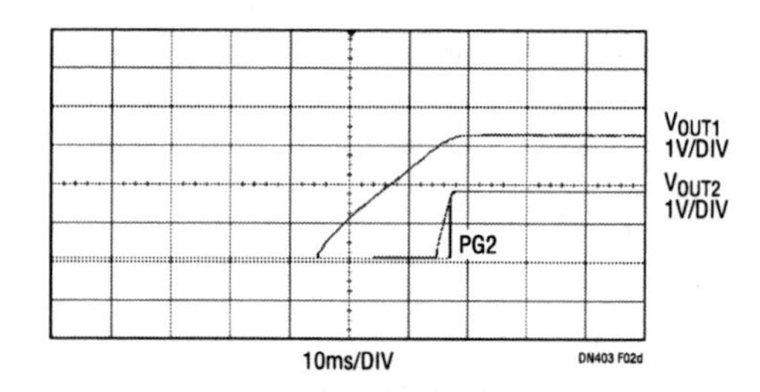

(55.2d) Output Sequencing

Figure 55.2 • Output Voltage Tracking and Sequencing

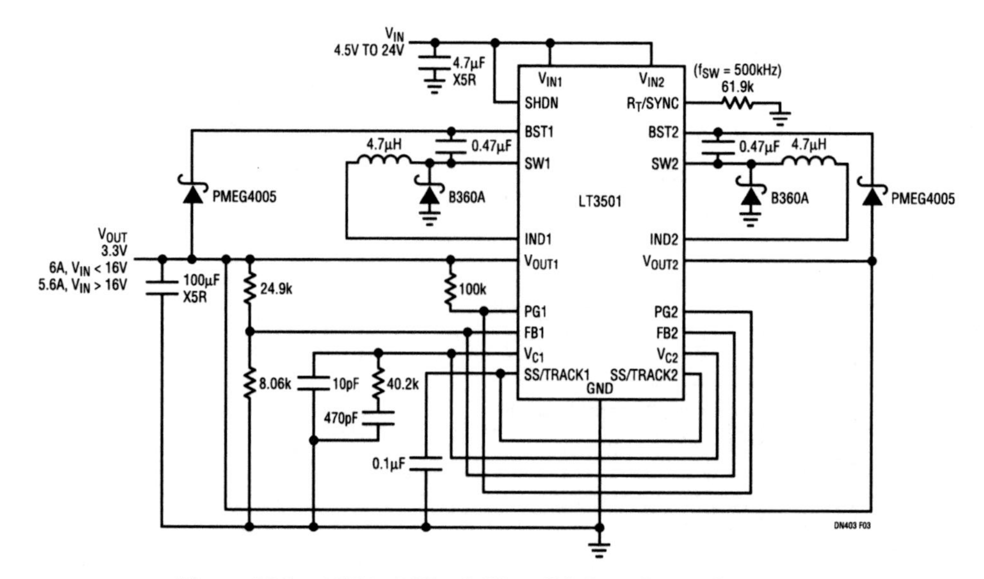

Figure 55.3 • 4.5V to $24V_{IN}$, $3.3V_{OUT}$/6A Step-Down Converter

Tiny monolithic step-down regulators operate with wide input range

56

Kevin Huang

Introduction

Automotive batteries, industrial power supplies, distributed supplies and wall transformers are all sources of wide-ranging high voltage inputs. The easiest way to step down these sources is with a high voltage monolithic step-down regulator that can directly accept a wide input range and produce a well-regulated output. The LT3493 accepts inputs from 3.6V to 36V and the LT3841 accepts inputs from 3.6V to 34V. Both provide excellent line and load regulation and dynamic response. The LT3481 offers a high efficiency solution over a wide load range and keeps the output ripple low during Burst Mode operation while the LT3493 provides a tiny solution with minimal external components. The LT3493 operates at 750kHz and the LT3481 has adjustable frequency from 300kHz to 2.8MHz. High frequency operation enables the use of small, low cost inductors and ceramic capacitors.

Low ripple and high efficiency solution over wide load range

The LT3481 is available in a 10-pin MSOP or a 3mm × 3mm DFN package with an integrated 3.8A power switch and external compensation for design flexibility. The switching frequency can be programmed from 300kHz to 2.8MHz by using a resistor tied from the RT pin to ground. Figure 56.1 shows the LT3481 producing 3.3V at 2A from an input of 4.5V to 34V. Figure 56.2 shows the circuit efficiency at 12V input.

The LT3481 offers low ripple Burst Mode operation that maintains high efficiency at light load while keeping the output voltage ripple below $15mV_{P-P}$. During Burst Mode operation, the LT3481 delivers single cycle bursts of current to the output capacitor followed by sleep periods when the output power is delivered to the load by the output capacitor. Between bursts, all circuitry associated with controlling the

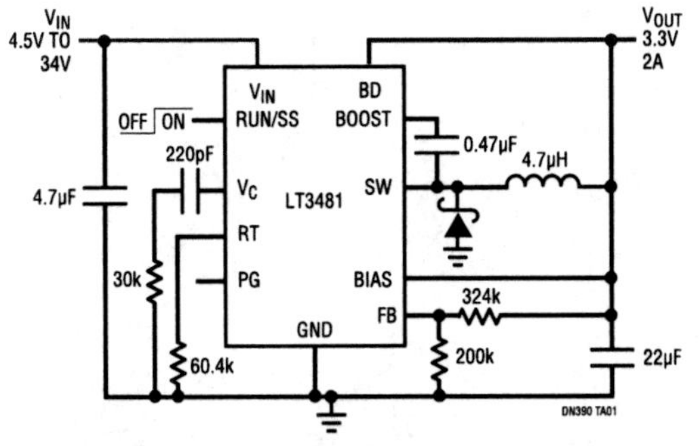

Figure 56.1 • 800kHz LT3481 DC/DC Converter Delivers 2A at 3.3V Output

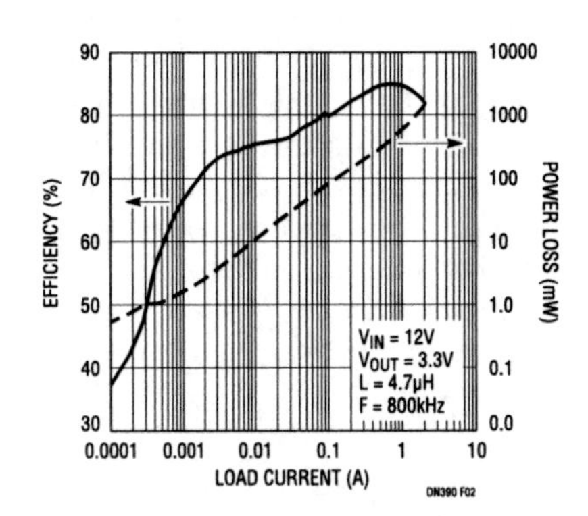

Figure 56.2 • Efficiency vs Load Current for Figure 56.1 Circuit

Analog Circuit Design: Design Note Collection. http://dx.doi.org/10.1016/B978-0-12-800001-4.00056-9

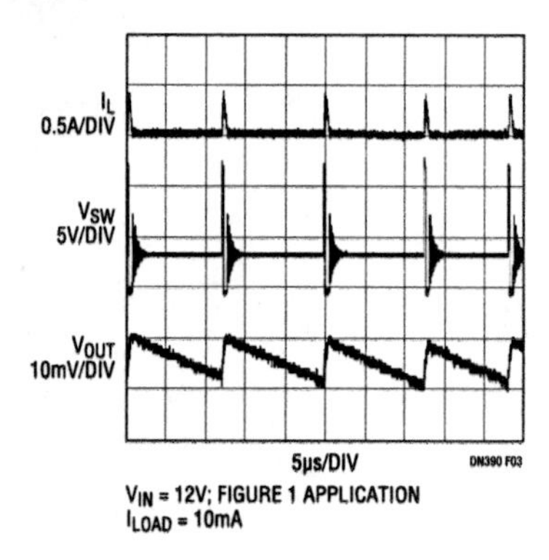

Figure 56.3 • LT3481 Burst Mode Operation at 10mA Load Current

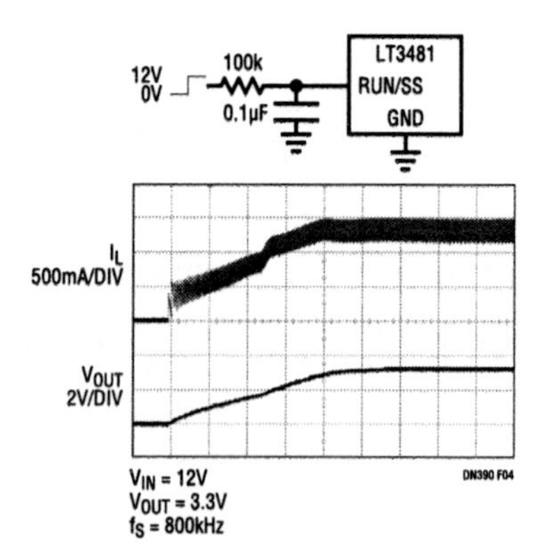

Figure 56.4 • Soft-Start of the LT3481

output switch is shut down reducing the input supply current to 50µA. Figure 56.3 shows the inductor current and output voltage ripple under single pulse Burst Mode operation from 12V input to 3.3V output. As the load current decreases to a no load condition, the percentage of time that the LT3481 operates in sleep mode increases and the average input current is greatly reduced resulting in high efficiency. The LT3481 has a very low shutdown current (less than 1µA) which significantly extends battery life in applications that spend long periods of time in sleep or shutdown mode.

The high side bootstrapping boost diode is integrated into the IC to minimize solution size and cost. When the output voltage is at least 2.8V, the anode of the boost diode can be connected to output. For output voltages lower than 2.5V, the boost diode can be tied to the input. For systems that rely on a well-regulated power source, the LT3481 provides a power good flag that signals when V_{OUT} reaches 90% of the programmed output voltage. A resistor and capacitor on the RUN/SS pin programs the LT3481's soft-start, reducing maximum inrush current during start-up. Figure 56.4 shows the circuit and start-up waveform.

Small solution size

The LT3493 includes an internal 1.75A power switch in a tiny 6-pin DFN package (2mm × 3mm). The current mode control circuit with its internal loop compensation eliminates

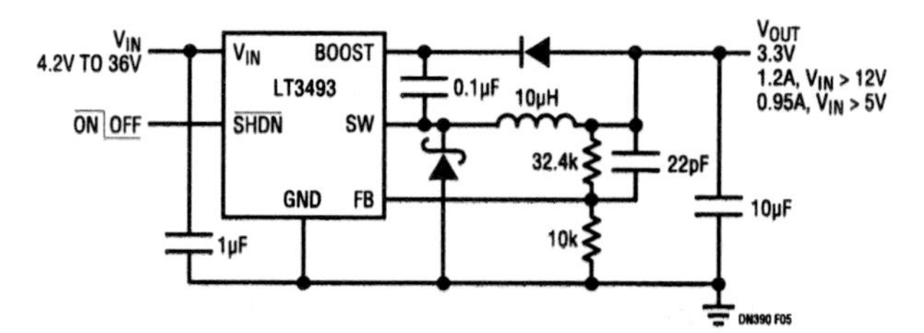

Figure 56.5 • LT3493 Wide Input Range DC/DC Converter Application to 3.3V

external compensation components, minimizing component count and reducing the PC board space to less than 50mm^2. The LT3493's reference voltage is 0.78V, making it suitable for applications with low output voltage. Figure 56.5 shows an application of the LT3493 switching at 750kHz. This circuit generates 3.3V from an input of 4.2V to 36V. In applications where the circuit is plugged into a live input source through long leads, a high ESR electrolytic capacitor at the input is recommended to damp the overshoot voltage. Refer to Application Note 88 for details. The $\overline{SHDN}$ pin can be driven through an external RC filter to soft-start the LT3493 (Figure 56.6).

Additional features of LT3481 and LT3493

During short circuit, both parts offer cycle-by-cycle current limit and frequency foldback which decreases the switching frequency when the output is low. The low frequency allows the inductor current to safely discharge.

Conclusion

The wide input ranges, small size and robust design of the LT3493 and LT3481 make them an excellent choice for a wide variety of step-down applications. Their high input voltage, high power switch capability and excellent package thermal conductivity add to their versatility.

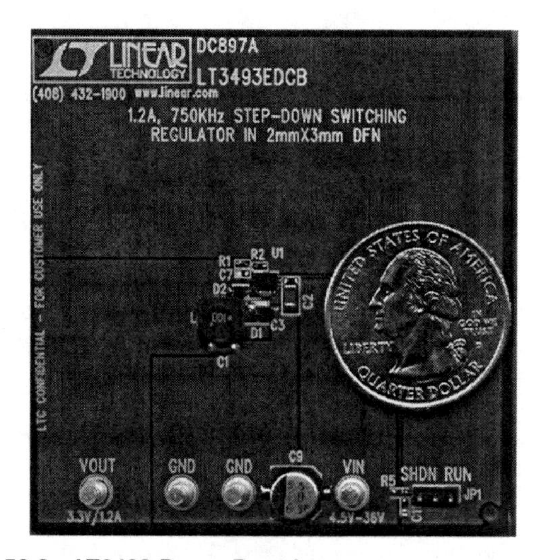

Figure 56.6 • LT3493 Demo Board

Cascadable 7A point-of-load monolithic buck converter

57

Peter Guan

Introduction

Easy-to-use and compact point-of-load power supplies are necessary in systems with widely distributed, high current, low voltage loads. The LTC3415 provides a compact, simple and versatile solution. It includes a pair of integrated complementary power MOSFETs (32mΩ top and 25mΩ bottom) and requires no external sense resistor. A complete design requires an inductor and input/output capacitors, and that's it. The result is a fast, constant frequency, 7A current mode DC/DC switching regulator.

Features

The overall solution is extremely compact since the LTC3415's 5mm × 7mm QFN package footprint is small and its high operating frequency of 1.5MHz allows the use of small low-profile surface mount inductors and ceramic capacitors. For loads higher than 7A, multiple LTC3415s can be cascaded to share the load while running mutually anti-phase, which reduces overall ripple at both the input and the output.

Other features include:

- Spread spectrum operation to reduce system noise
- Output tracking for controlled V_{OUT} ramp-up and ramp-down
- Output margining for easy system stress testing
- Burst Mode operation to lower quiescent current and boost efficiency during light loads
- Low shutdown current of less than 1μA
- 100% duty-cycle for low dropout operation
- Phase-lock-loop to allow frequency synchronization of ±50% of nominal frequency
- Internal or external I_{TH} compensation for ease of use or loop optimization, respectively

Operation

The LTC3415 offers several operating modes to optimize efficiency and noise reduction: Burst Mode operation, pulse-skipping mode or forced continuous mode. The mode is set by tying the Mode pin to SV_{IN}, $SV_{IN}/2$ or SGND, respectively. Burst Mode operation offers high efficiency at light load by shutting off the internal power MOSFETs as well as most of the internal circuitry between pulses. Forced continuous mode maintains a constant switching frequency throughout the entire load range, making it easier to filter switching noise for sensitive applications. Pulse-skipping mode allows constant frequency operation until the inductor current reaches zero, at which point it goes into discontinuous

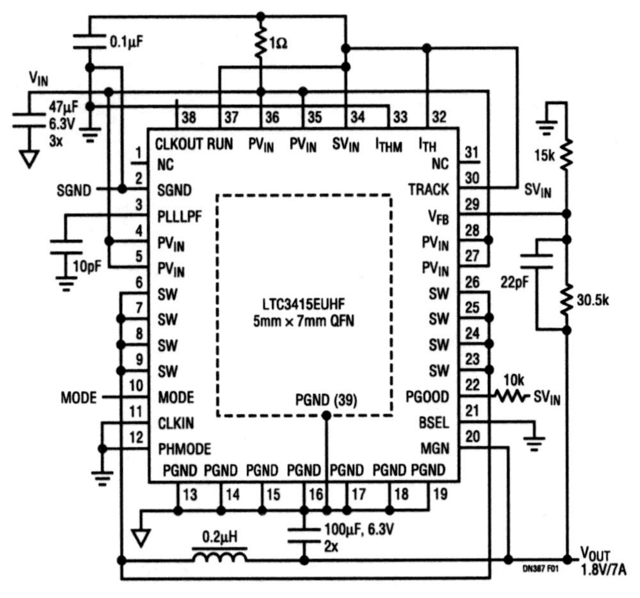

Figure 57.1 • 3.3V to 1.8V/7A Application

Analog Circuit Design: Design Note Collection. http://dx.doi.org/10.1016/B978-0-12-800001-4.00057-0

operation and finally it will skip cycles. Pulse-skipping mode offers low output voltage ripple while offering efficiency levels between Burst Mode operation and forced continuous mode.

Figure 57.1 shows an application of the LTC3415 in a 3.3V to 1.8V/7A step-down converter configuration. Figure 57.2 shows its efficiency and power loss vs load current in Burst Mode operation. Efficiency reaches as high as 92%. Figure 57.3 shows its fast transient response to a 5A load step. As shown, V_{OUT} recovers in 10μs with a dip of less than 100mV. Frequency can be changed easily from its nominal 1.5MHz to 1MHz or 2MHz by simply strapping the PLLLPF pin to SGND or SV_{IN}, respectively. Or if a particular frequency is desired, an external clock can be used to synchronize the operating frequency from 750kHz to 2.25MHz with the internal phase-lock-loop. Spread spectrum operation is available for EMI-sensitive applications by tying the CLKIN pin to SV_{IN}.

For applications that require controlled output voltage tracking between various outputs in order to prevent excessive current draw or even latch-up during turn-on and turn-off, the LTC3415 has a Track pin that allows the user to program how its output voltage ramps during start-up and shutdown. Figure 57.4 shows the output waveforms of two LTC3415s in track mode.

Greater than 7A outputs

By stacking multiple LTC3415s together, more output power is attained without increasing the number of input and output capacitors. Operating multiple LTC3415s out of phase not only allows accurate current sharing, but it also reduces the overall voltage ripple at both the input and the output, thus allowing fewer capacitors. Figure 57.5 shows an efficiency curve of the LTC3415 in 1-phase, 2-phase, 3-phase, 4-phase and 6-phase operation.

Conclusion

With its many operational features and compact total solution size, the LTC3415 is an ideal fit for today's point-of-load power supplies. It allows for accurate, compact, efficient and scalable power supplies with advanced features, including tracking and margining.

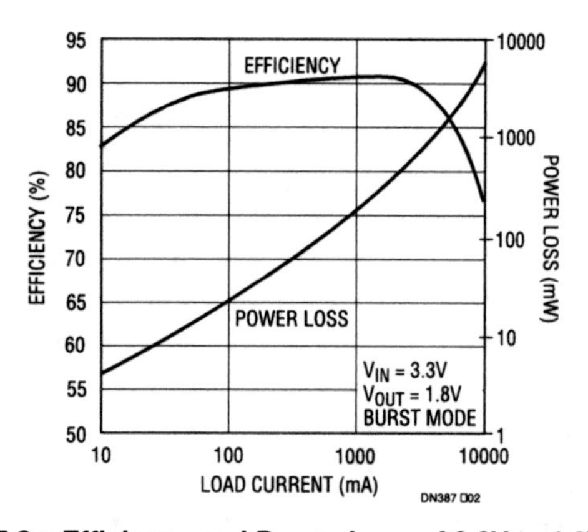

Figure 57.2 • Efficiency and Power Loss of 3.3V to 1.8V/7A Application in Figure 57.1

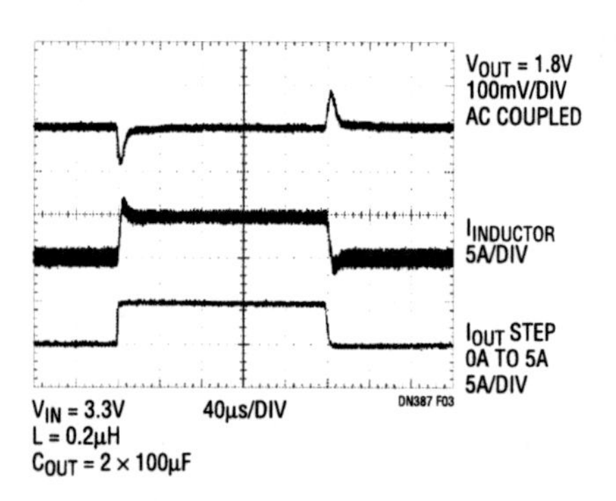

Figure 57.3 • V_{OUT} Transient Response to a 0A to 5A Load Step of the Circuit Shown in Figure 57.1

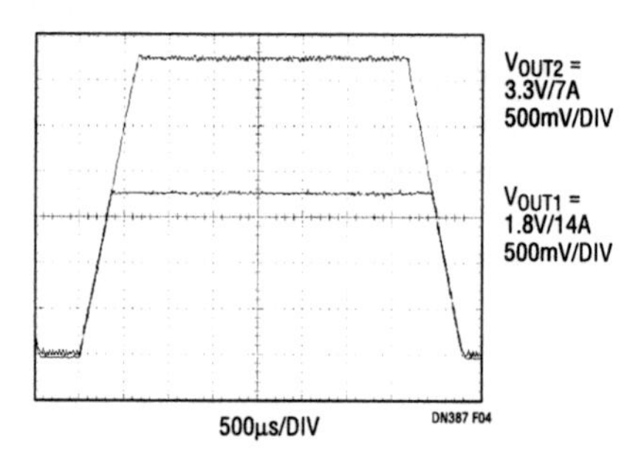

Figure 57.4 • Output Tracking of Two LTC3415s

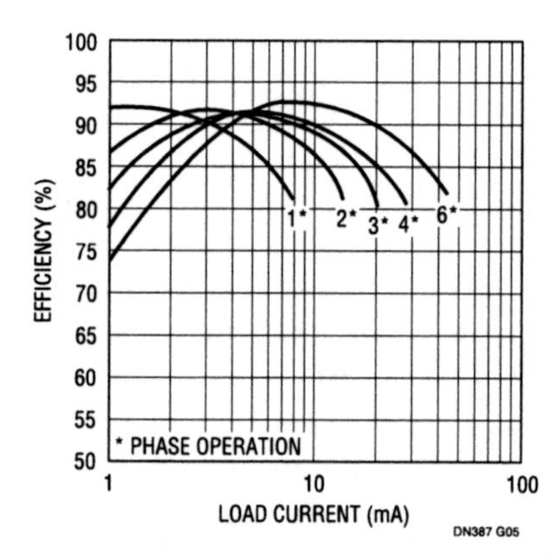

Figure 57.5 • Efficiency vs Load Current of LTC3415s in Multiphase Operation

High voltage current mode step-down converter with low power standby capability

58

Jay Celani

Introduction

Low power standby requirements are typically associated with battery-powered systems. Automotive systems, for example, commonly require power supplies to maintain output voltage regulation even under no-load conditions—while drawing minimal quiescent current to preserve battery life. Rising energy costs, however, have extended the need for low current standby operation to line-powered systems, such as small plugged-in appliances for home and business.

Designing a power supply that is very efficient at light loads is particularly difficult in systems where high input voltages and substantial load currents are required. A common approach in such high power systems is to add a secondary power path for low current operation—a potentially significant increase in the cost, board space and complexity of the power supply.

A better solution is to use the LT3800 as the core of a single supply synchronous DC/DC converter. The resulting power supply is simple and efficient. An LT3800-based converter requires few external components, maintains high conversion efficiencies over a wide load range and supports low power standby operation for compliance with system power-management requirements.

High efficiency at standby

The LT3800 is a $4V_{IN}$ to $60V_{IN}$, 200kHz fixed-frequency controller that uses synchronous operation and N-channel MOSFETs to maximize high current efficiency. Current mode operation with continuous high side inductor current sensing yields fast transient response and excellent line regulation. Low current standby requirements are met using Burst Mode operation. A reverse inductor current inhibit feature

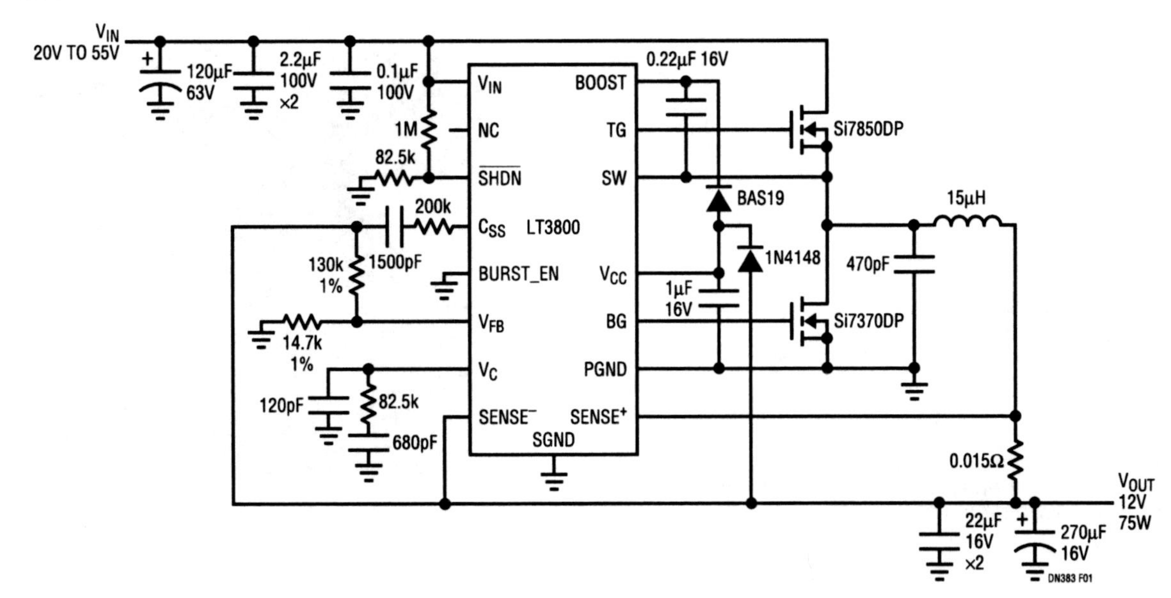

Figure 58.1 • LT3800 12V, 75W Buck DC/DC Converter with High Efficiency at Light Loads

Analog Circuit Design: Design Note Collection. http://dx.doi.org/10.1016/B978-0-12-800001-4.00058-2

also increases efficiency during light-load conditions. The LT3800 runs directly from the converter input supply, so there are no local supplies required to power the IC. The IC is also designed for easy use of output-derived power which further increases conversion efficiency. Cycle-by-cycle current limiting maintains the programmed current limit, even during instantaneous short-circuit fault conditions.

12V/75W synchronous buck DC/DC converter

Figure 58.1 shows a 12V, 75W DC/DC converter that can operate with input voltages from 20V to 55V. The 20V minimum input is set by a programmable UVLO function implemented using the precision hysteretic threshold of the LT3800 $\overline{\text{SHDN}}$ pin. The 55V upper bound is limited by switch FET margin. This converter provides full-load efficiencies above 95%, as shown in Figure 58.2, and can maintain a no-load output voltage with only 0.1mA of input supply quiescent current (see Figure 58.3).

This DC/DC converter incorporates a controlled dV/dt soft-start function which servos the converter output voltage to a programmed rising rate during start-up—in this case 1.3V/ms, yielding a start-up rise time of just under 10ms. The LT3800 automatically resets the soft-start function if the output drops out of regulation, so recovery from a short circuit or brownout event is graceful and controlled (see Figures 58.4 and 58.5).

The light load efficiency enhancement features of the LT3800 produce a supply that maintains high conversion efficiency across a 4-decade range. This DC/DC converter also features excellent heavy-load efficiency, producing peak conversion efficiencies as high as 96%.

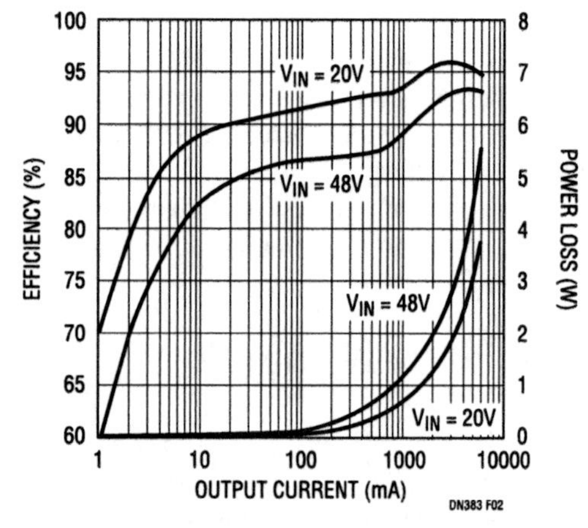

Figure 58.2 • The DC/DC Converter in Figure 58.1 Exhibits High Efficiency Across a Wide Load Range and Produces Peak Efficiencies Above 95%

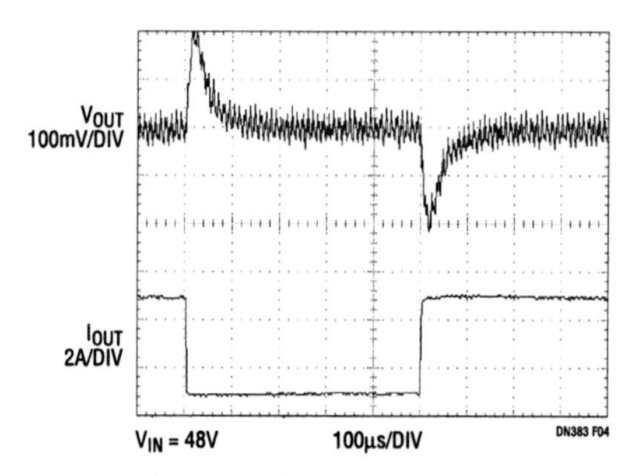

Figure 58.4 • 1A to 5A Load Step Generates <2% V_{OUT} Transient

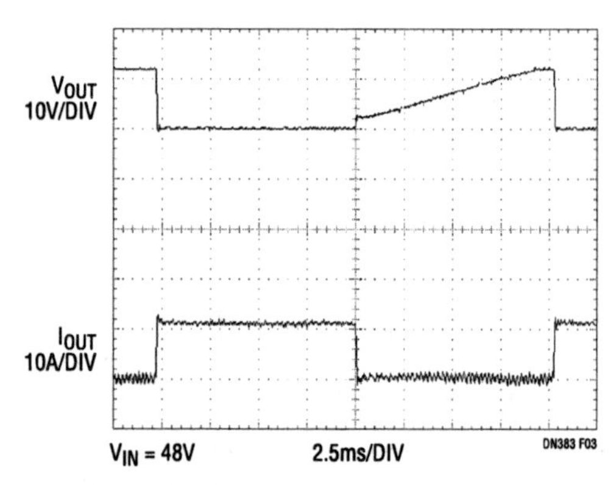

Figure 58.3 • Current Limit is Maintained during Instantaneous Short-Circuit and Auto-Reset Soft-Start Yields a Graceful Recovery from the Fault Event

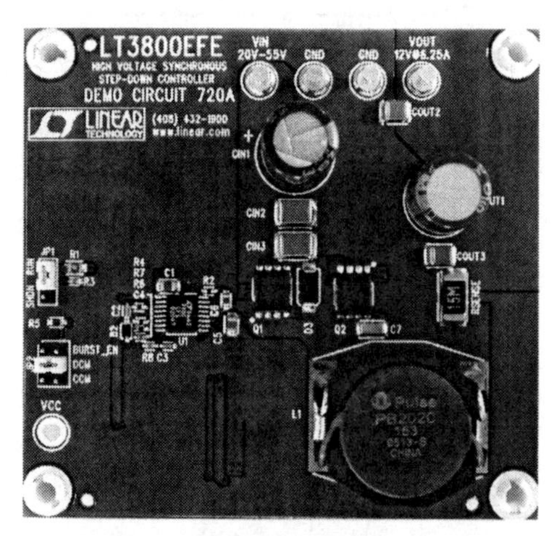

Figure 58.5 • LT3800 12V, 75W DC/DC Converter Layout

Low EMI synchronous DC/DC step-down controllers offer programmable output tracking

59

Charlie Zhao

Introduction

The LTC3808 synchronous DC/DC step-down controller packs numerous features required by the latest electronic devices into a low profile (0.75mm) 3mm × 4mm leadless DFN package or a leaded SSOP-16 package. Two similar parts, the LTC3809 and LTC3809-1, are even smaller, but less feature-rich versions of the LTC3808. The LTC3809 family is available in a 3mm × 3mm leadless DFN package or a 10-pin MSOP Exposed Pad package. All three parts can provide output voltages as low as 0.6V and output currents as high as 7A from a 2.75V to 9.8V input range, making them ideal devices for one or two lithium-ion cell inputs as well as distributed DC power systems.

The LTC3808 and LTC3809 also include important features for noise-sensitive applications, including a phase-locked loop (PLL) for frequency synchronization and spread spectrum frequency modulation to minimize generated electromagnetic interference (EMI). The adjustable operating frequency (300kHz to 750kHz) allows the use of small surface mount inductors and ceramic capacitors for compact power supply solutions.

Other features include:

- Low operating quiescent current to improve battery life and light load efficiency
- No R_{SENSE} current mode technology which senses the voltage across the main (top) power MOSFET to improve efficiency and reduce the size and cost of the solution
- Current mode control for excellent AC and DC line and load regulation
- Low dropout (100% duty cycle) for maximum energy extraction from a battery source
- Output overvoltage protection and short-circuit current limit protection
- Adjustable or fixed built-in soft-start timer
- Output voltage ramp control and the ability to track other voltage sources (LTC3808 and LTC3809-1)
- PowerGood voltage monitor (LTC3808)

Table 59.1 compares the features of these three parts.

Table 59.1 DC/DC Step-Down Controllers

	START-UP CONTROL	SPREAD SPECTRUM	ADJUSTABLE FREQ/PLL	POWER GOOD
LTC3808	Internal External Tracking	Yes	Yes	Yes
LTC3809	Internal	Yes	Yes	No
LTC3809-1	Internal External Tracking	No	No	No

Three choices for start-up control

The start-up of V_{OUT} for the LTC3808 and LTC3809-1 is based on the three different connections to the TRACK/SS pin. A typical application is shown in Figure 59.2. When TRACK/SS is connected to V_{IN}, the start-up of V_{OUT} is

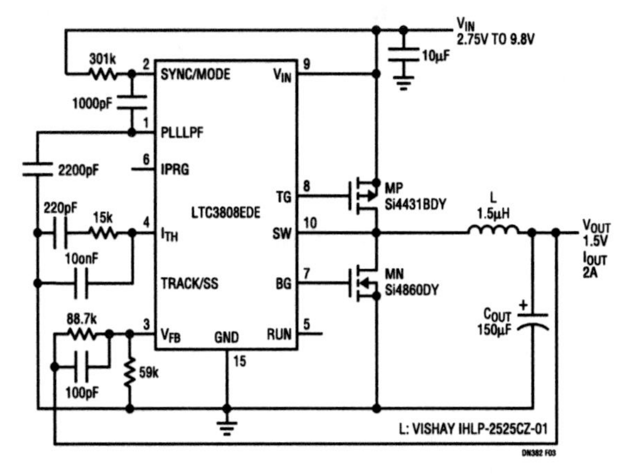

Figure 59.1 • Synchronous Converter with Spread Spectrum Frequency Modulation

Analog Circuit Design: Design Note Collection. http://dx.doi.org/10.1016/B978-0-12-800001-4.00059-4

controlled by the internal soft-start which ramps from 0V to (V_{FB}) in about 1ms. A second start up mode allows the 1ms soft-start time to increase or decrease by connecting an external capacitor C_{SS} between the TRACK/SS pin and ground. An internal 1µA current source and the value of C_{SS} control the ramp time of TRACK/SS from 0V to above 0.6V. In this case, the LTC3808 and LTC3809-1 regulate the VFB to the voltage at the TRACK/SS pin instead of the internal soft-start ramp. The third mode allows V_{OUT} of the LTC3808 and LTC3809-1 to track an external voltage, V_X, during start-up if a resistor divider from V_X is connected to the TRACK/SS pin. Figure 59.3 shows the start-up of V_{OUT} in these tracking modes for the circuit shown in Figure 59.2.

For simplicity, the LTC3809 only offers a 1ms internal soft-start.

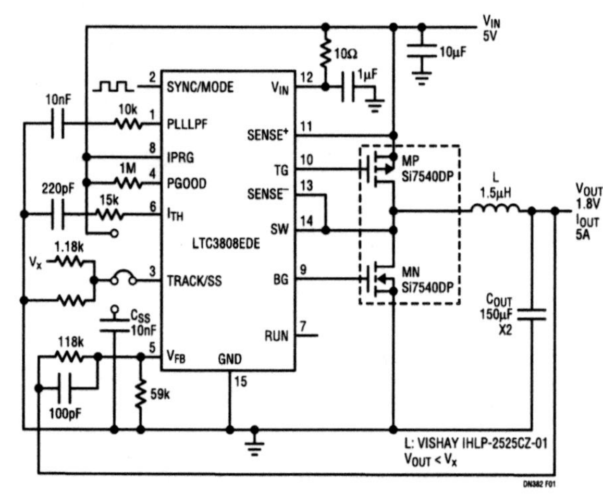

Figure 59.2 • The LTC3808 Offers the Flexibility of Start-Up Control Based on the Three Different Connections on the TRACK/SS Pin

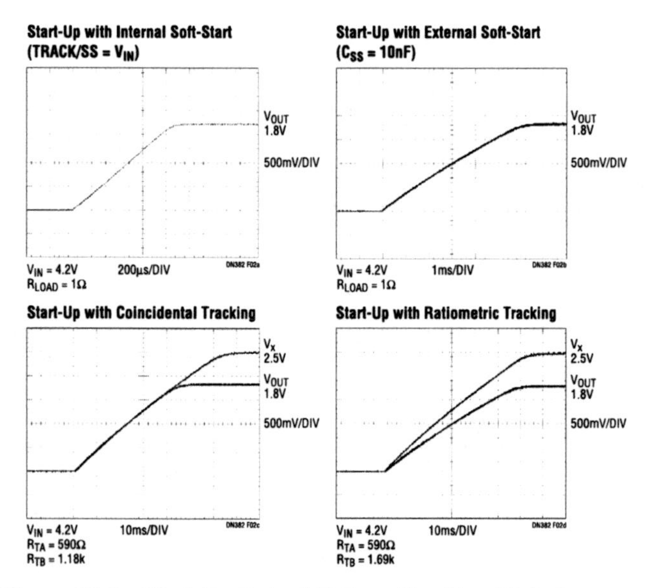

Figure 59.3 • Start-Up Output Voltage Tracking Plots for Circuit in Figure 59.2

Low EMI DC/DC conversion

The LTC3808 and LTC3809 minimize the need for EMI shields and filters in applications such as navigation systems, wireless LANs, data acquisition boards and industrial and military radio devices by optionally spreading the nominal operating frequency (550kHz) over a range of frequencies between 460kHz and 635kHz. Spread spectrum frequency modulation is enabled by biasing the SYNC/MODE pin to a DC voltage between 1.35V and ($V_{IN}-0.5V$). An internal 2.6µA pull-down current source at the SYNC/MODE pin can be used to set the DC voltage at this pin by tying a resistor with an appropriate value between SYNC/MODE and V_{IN}. Figure 59.1 shows the application circuit and Figure 59.4 shows the frequency spectral plots of the output (V_{OUT}) with and without spread spectrum modulation. Note the significant reduction in peak output noise (>20dBm) with spread spectrum enabled.

Conclusion

The LTC3808, LTC3809 and LTC3809-1 offer flexibility, high efficiency, low EMI and many other popular features in small thermally efficient packages. They offer excellent solutions for low voltage portable and distributed power systems that require a small footprint, high efficiency and low noise.

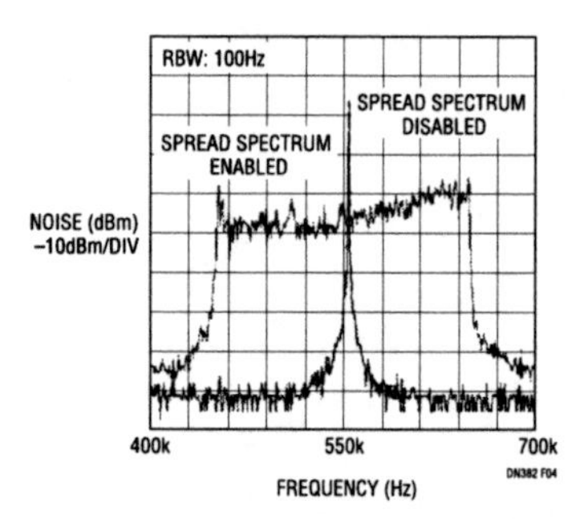

Figure 59.4 • Comparison of the V_{OUT} Spectrum with and without Spread Spectrum Modulation Enabled

ThinSOT micropower buck regulator has low output ripple

60

Keith Szolusha

Introduction

High voltage monolithic step-down converters simplify circuit design and save space by integrating the high side power switch into the device. In most cases, the switch is an n-type transistor (NMOS or NPN) with a boot-strapped drive stage, requiring an external boost diode and capacitor as well as the main catch diode, complicating the applications circuit.

The LT3470 is a 40V step-down converter with the power switch, catch diode and boost diode integrated in a tiny ThinSOT package. The boosted NPN power stage provides high voltage capability, high power density and high switching speed without the cost and space of external diodes.

The LT3470 accepts an input voltage from 4V to 40V and delivers up to 200mA to load. Micropower bias current and Burst Mode operation enable it to consume merely 26μA with no load and a 12V input. Hysteretic current mode control and single-cycle bursts result in very low output ripple and stable operation with small ceramic capacitors. The combination of small circuit size, low quiescent current and 40V input makes the LT3470 ideal for automotive and industrial applications.

Current mode control

The LT3470 uses a hysteretic current control scheme in conjunction with Burst Mode operation to provide low output ripple and low quiescent current while using a tiny inductor and ceramic capacitors. The switch turns on until the current ramps up to the level of the top current comparator, then turns off and the inductor current ramps down through the catch diode until the bottom current comparator trips and the minimum off-time has been met (Figure 60.1).

In continuous mode, the difference between the top and bottom current comparator levels is about 150mA. Since the switch only turns on when the catch diode current falls below threshold, switching frequency decreases, keeping switch current under control during start-up or short-circuit conditions (Figure 60.2).

If the load is light, the IC alternates between micropower and switching states to keep the output in regulation (Figure 60.3a). Hysteretic mode allows the IC to provide single switch-cycle bursts for the lowest possible light-

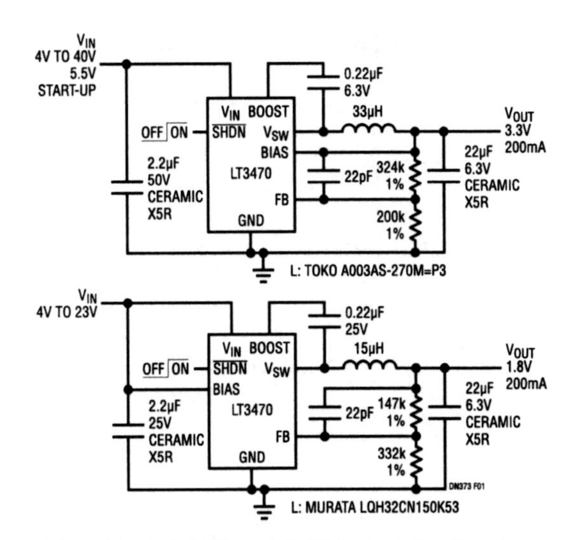

Figure 60.1 • Typical 3.3V and 1.8V Output Applications

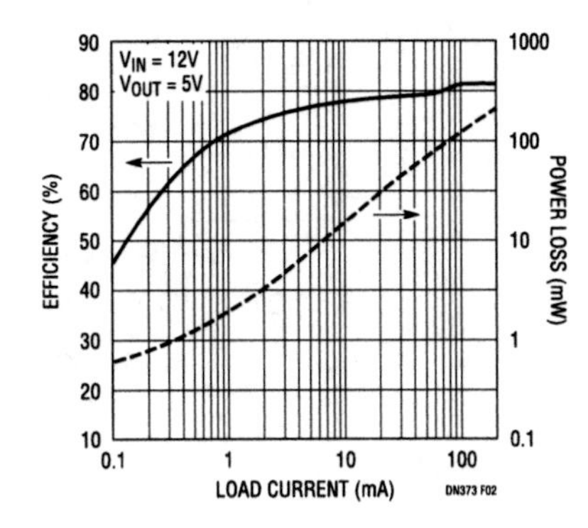

Figure 60.2 • Efficiency and Power Loss vs Load Current

Analog Circuit Design: Design Note Collection. http://dx.doi.org/10.1016/B978-0-12-800001-4.00060-0

load output voltage ripple (<20mV peak-to-peak from 12V to 3.3V at zero load). During continuous switching mode (Figure 60.3b) at higher current levels, the output voltage ripple is even smaller (<10mV peak-to-peak).

Design flexibility with integrated boost diode

A high side NPN power switch in a buck regulator design needs a driver voltage that is at least a few volts higher than the switch or input voltage. When there are no other high voltage lines available, a bootstrapping method of providing several volts of boost to the IC is required. When there is at least 2.5V on the output, the boost voltage can be most efficiently derived from the output. If the output voltage is too low, 1.8V for example, the boost voltage must be derived from the input.

Integration of the high side bootstrapping boost diode into the IC does not limit boost diode flexibility. Boost diode flexibility such as the ability to connect to various sources and/or the inclusion of a Zener blocking diode is needed for both high and low output voltages with and without wide input voltage ranges. The anode of the boost diode can be connected to different sources via the BIAS pin. In most cases, this is a simple connection to either the input, when the output voltage is below 2.5V, or the output, for output voltages above 2.5V. Additional Zener diode voltage drop in the boost diode path or a transistor bias supply as shown in Figure 60.4 protects the IC from BOOST pin overvoltage when there is a wide input voltage range.

Conclusion

The LT3470 is a wide input voltage range, hysteretic mode, fully integrated monolithic 300mA step-down DC/DC converter. The onboard high side NPN power switch, Schottky boost diode, and Schottky catch diode combined with the small ThinSOT package and high 40V input voltage make this a simple and versatile IC to use for many step-down applications with less than 200mA load current.

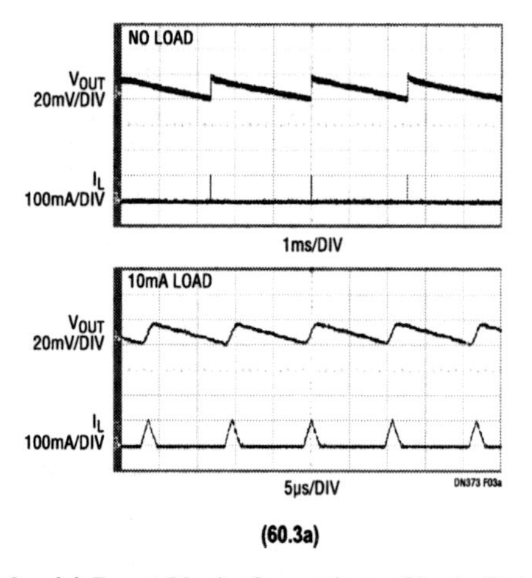

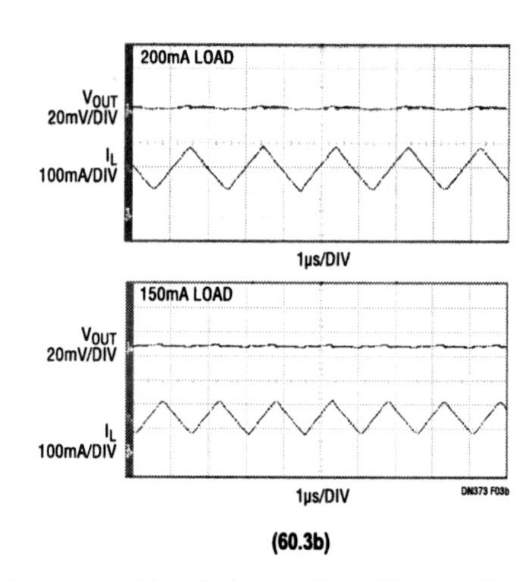

Figure 60.3 • (a) Burst Mode Operation—Single Pulse Burst Mode Operation Has Only 20mV$_{P-P}$ Ripple. (b) Continuous Operation—Extremely Low Output Voltage Ripple

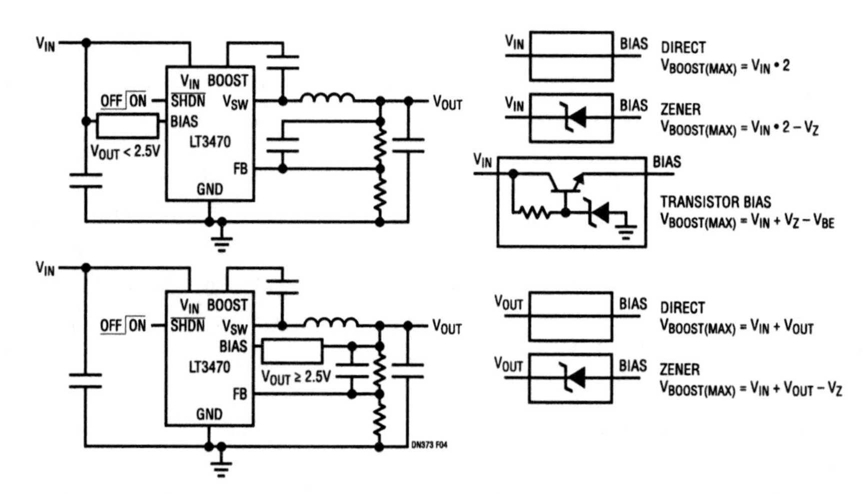

Figure 60.4 • BIAS and BOOST Pin Connection Variations Provide Input and Output Voltage Range Flexibility

Tiny versatile buck regulators operate from 3.6V to 36V input

61

Hua (Walker) Bai

Introduction

Linear Technology offers two new buck regulators that operate from a wide input voltage range (3.6V to 36V) and take so little space that they easily solve many difficult power supply problems. The LT1936 and LT1933 are perfect for applications with disparate power inputs or wide range input power supplies such as automotive batteries, 24V industrial supplies, 5V logic supplies and various wall adapters. Both parts are monolithic current mode PWM regulators which provide excellent line and load regulation and dynamic response. They operate at a 500kHz switching frequency, enabling the use of small, low cost inductors and ceramic capacitors, resulting in low, predictable output ripple.

Small size and versatility

The LT1936 regulator includes a 1.9A power switch in a tiny, thermally enhanced 8-lead MSOP. The LT1933 regulator includes an internal 0.75A power switch in a tiny 6-lead ThinSOT package, which occupies less than 0.15in^2 board space. The LT1936 offers the option of external compensation for design flexibility or internal compensation for compact solution size. Both parts offer soft-start via the $\overline{\text{SHDN}}$ pin, thus reducing maximum inrush currents during start-up. Both parts also have a very low, 2μA shutdown current which significantly extends battery life in applications that spend long periods of time in sleep or shutdown mode. During short circuit, both parts offer frequency foldback, where the switching frequency decreases by about a factor of ten. The lower frequency allows the inductor current to safely discharge, thereby preventing current runaway.

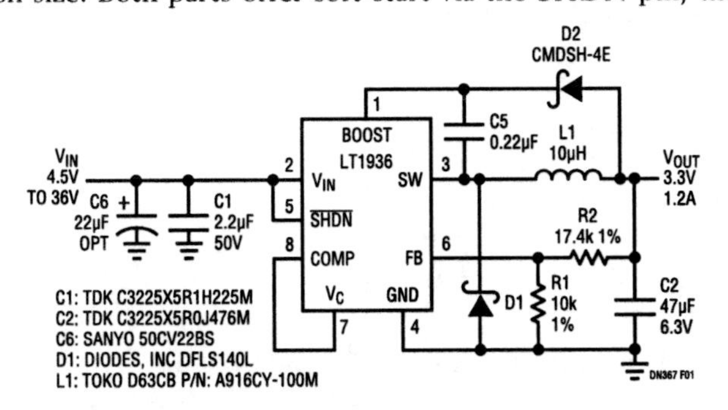

Figure 61.1 • Typical Application of LT1936 Accepts 4.5V to 36V and Produces 3.3V/1.2A

LT1936 produces 3.3V at 1.2A from 4.5V to 36V

Figure 61.1 shows a typical application for the LT1936. This circuit generates 3.3V at 1.2A from an input of 4.5V to 36V. With the same input voltage range, the LT1933 circuit can supply 500mA. The typical output voltage ripple of the Figure 61.1 circuit is less than 16mV while efficiency is as high as 89%. Excellent transient response is possible with either external compensation or the internal compensation; this circuit uses internal compensation to minimize component count. A high ESR electrolytic capacitor, C6 in Figure 61.1, is recommended to damp overshoot voltage in applications where the circuit is plugged into a live input source through long leads. For more information, refer to the LT1933 or LT1936 data sheet.

Producing a lower output voltage from the LT1936

In order to fully saturate the internal NPN power transistor of the LT1936, the BOOST pin voltage must be at least 2.3V above the SW pin voltage. A charge pump comprising D2 and C5 creates this headroom in Figure 61.1. Nevertheless, when the output voltage is less than 2.5V, different approaches are needed. Figure 61.2 shows one example. It allows V_{IN} to go up to 36V and generates 1.4A at 1.8V. In this circuit, Q2 serves as an inexpensive Zener. The emitter-base breakdown

Analog Circuit Design: Design Note Collection. http://dx.doi.org/10.1016/B978-0-12-800001-4.00061-2

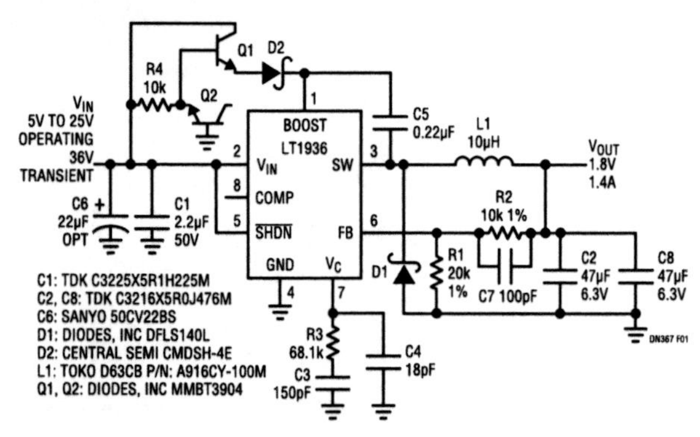

Figure 61.2 • This Circuit Generates Lower Output Voltage While Allowing Maximum Input Up to 36V

voltage of Q2 gives a stable 6V reference. The charging current for the BOOST capacitor, C5, passes through the follower, Q1. R4, Q1 and Q2 limit the BOOST pin voltage below its maximum rating of 43V. If the maximum V_{IN} in an application is less than 20V, simply tie V_{IN} to D2 to allow a lower minimum input voltage.

Negative output from a buck regulator

The circuit shown in Figure 61.3 can generate a negative voltage of −3.3V from a buck regulator such as the LT1933. This circuit effectively sets the ground reference of the LT1933 to −3.3V. The average inductor current of this circuit is the summation of the input and output current. The available output current is a function of the input voltage as shown in Figure 61.4.

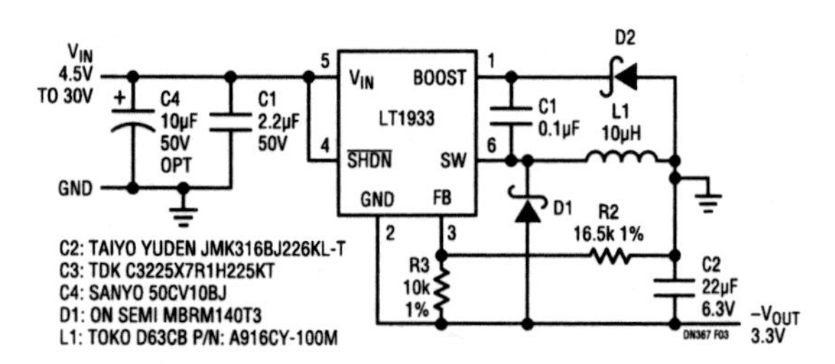

Figure 61.3 • This Circuit Produces −3.3V from 4.5V to 30V

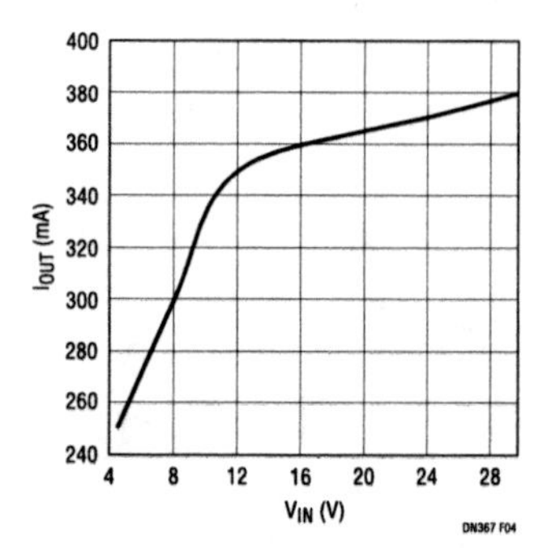

Figure 61.4 • Maximum Output Current of the Circuit in Figure 61.3 as a Function of the Input Voltage

Tiny circuit generates 3.3V and 5V from a minimum 4.5V supply

The circuit in Figure 61.5 is capable of generating two output voltages from a minimum 4.5V supply. One output is 3.3V at 300mA, the other 5V at 50mA. The circuit is especially useful in automotive cold crank conditions when the battery voltage drops below 5V but both the 3.3V and 5V outputs need to be alive. If more current is needed, the circuit can also be implemented using the LT1936. Even though the input of the LT1761-5 is unregulated, the 5V output is regulated by the LT1761-5 LDO. To maintain regulation, the 3.3V output current should be always well above the 5V output current, especially when V_{IN} is low.

Conclusion

The LT1933 and LT1936 step-down switching regulators accept a wide variety of input sources as well as offer compact, efficient and versatile solutions to many otherwise hard-to-solve problems.

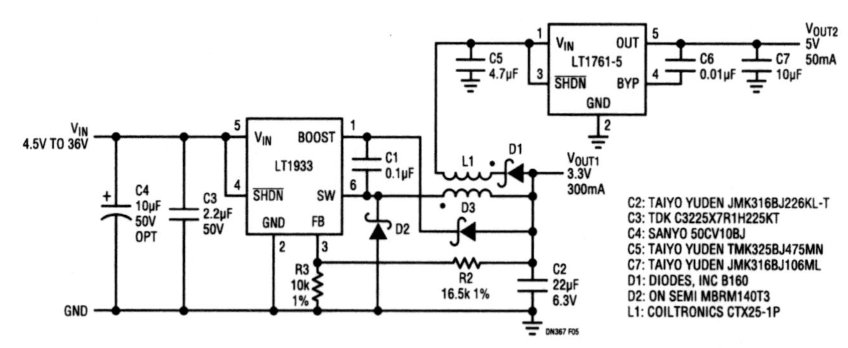

Figure 61.5 • From a Minimum 4.5V, This Circuit Produces Two Outputs at 3.3V/300mA and 5V/50mA

High accuracy synchronous step-down controller provides output tracking and programmable margining

62

Charlie Zhao

Introduction

Output voltage tracking and margining are two increasingly popular features in power supply designs for high performance server, ASIC and memory systems. These two functions are among the advanced features provided in the LTC3770, a wide operating range, high accuracy, synchronous step-down DC/DC controller. The LTC3770 operates with input voltages from 4V to 32V, generating output voltages down to 0.6V with its highly accurate ±0.67% 0.6V reference voltage. It has a constant on-time, valley current mode control architecture, which allows the LTC3770 to operate at very low duty cycles with fast transient response. The current sensing resistor is optional: leave it out for high efficiency, or in for most accurate current limit.

Figure 62.1 shows a 2.5V, 10A power supply with a 5V to 28V input range. The switching frequency of the converter can be selected by an external resistor, R_{ON}, and is compensated for variations in input supply voltage, or the controller can be synchronized to an external clock via an internal phase-lock loop. Figure 62.2 shows the efficiency of the circuit versus load current.

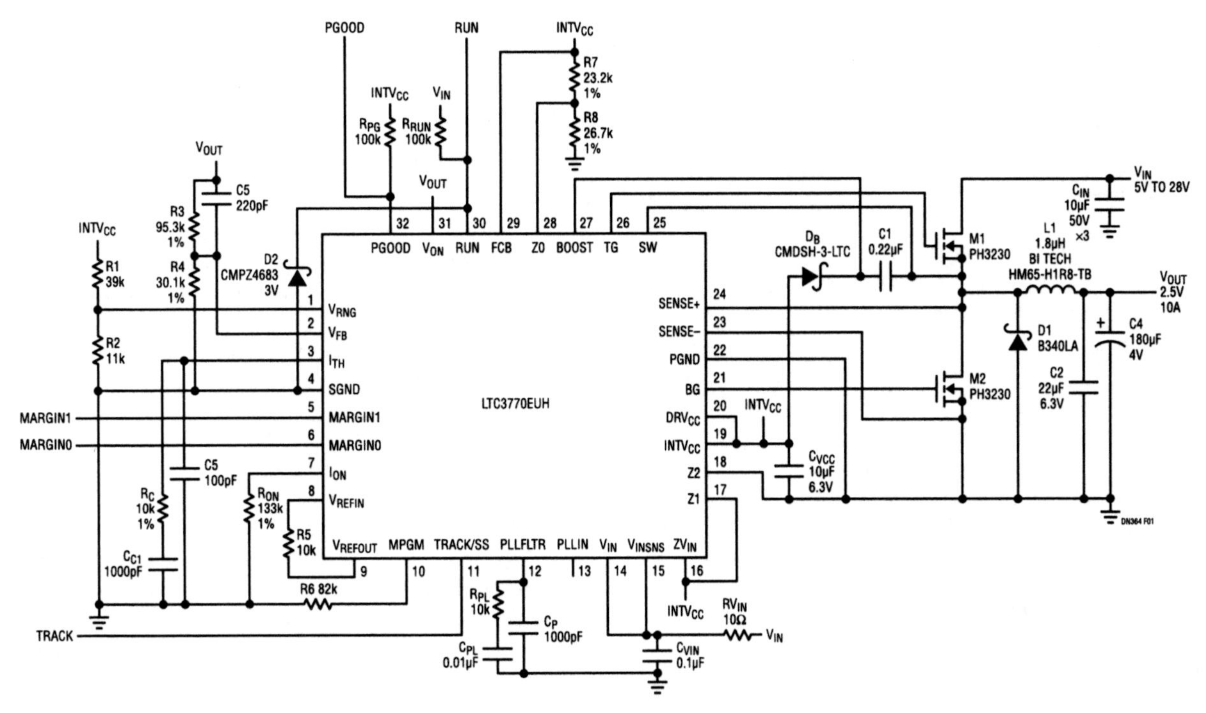

Figure 62.1 • High Efficiency 5V-28V_{IN} to 2.5V/10A Synchronous Buck Converter with Tracking and Margining

Analog Circuit Design: Design Note Collection. http://dx.doi.org/10.1016/B978-0-12-800001-4.00062-4

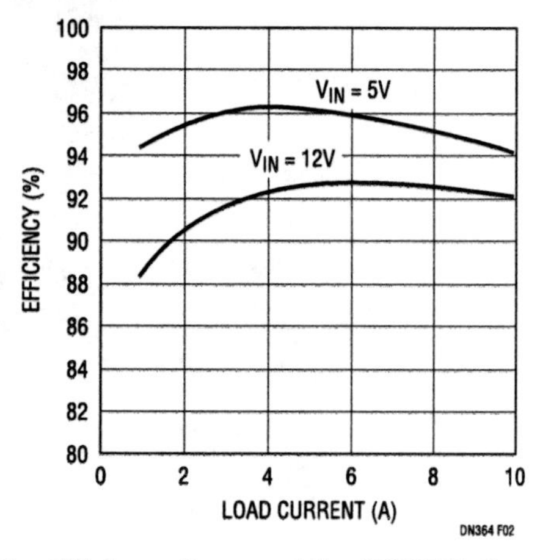

Figure 62.2 • Efficiency Curves of the LTC3770 Converter in Figure 62.1

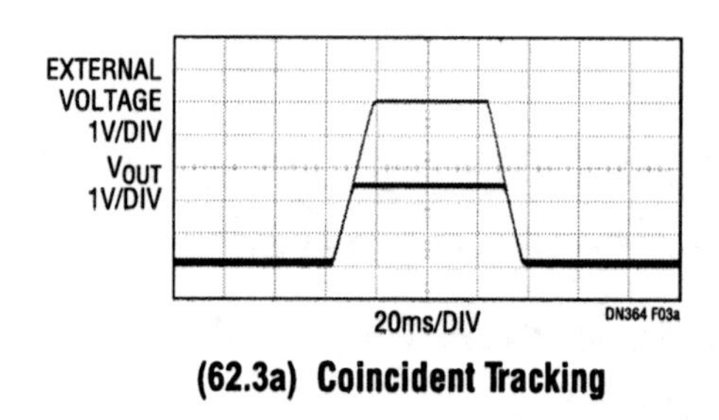

(62.3a) Coincident Tracking

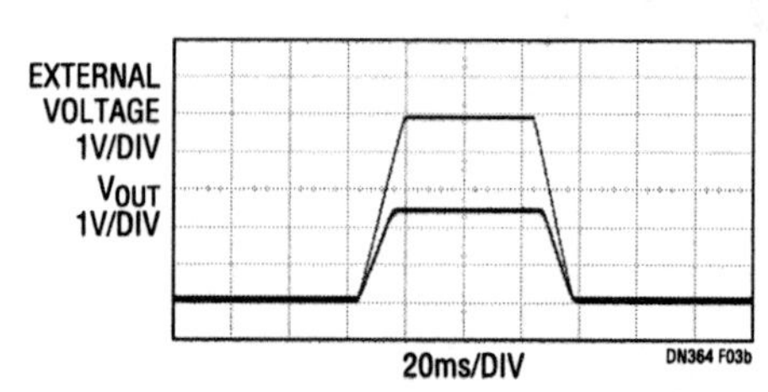

(62.3b) Ratiometric Tracking

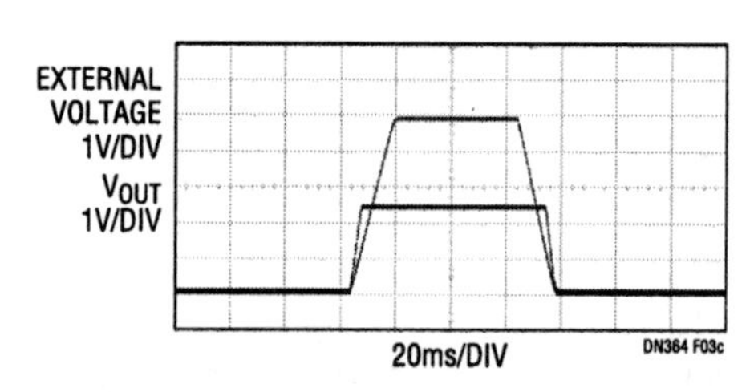

(62.3c) Special Ratiometric Tracking

Figure 62.3 • Up/Down Output Tracking. Upper Waveform: External Voltage; Lower Waveform: Output Voltage of the LTC3770 Converter

Start-up and shutdown output tracking

The LTC3770 output voltage can track another supply via the TRACK/SS pin. Tracking and sequencing functions allow the user to easily optimize the start-up and shutdown of multiple supplies, such as system core and I/O supplies. The tracking can be coincident or ratiometric, as shown in Figure 62.3.

Programmable voltage margining

Voltage margining is the dynamic adjustment of the output voltage to its worst-case operating value during production test in order to stress the load circuitry, verify control/protection functionality of the board and confirm system reliability. The LTC3770 has two logic control pins, MARGIN1 and MARGIN0, to enable margin up for higher output voltage or margin down for lower output voltage. Table 62.1 shows a summary of the configurations.

Table 62.1 Margining Function

MARGIN1	MARGIN0	MODE
Low	Low	No Margining
Low	High	Margin Up
High	Low	Margin Down
High	High	No Margining

The magnitude of the margin voltage shift is programmed by selecting the ratio of two resistors, R5 and R6 in Figure 62.1. When the margining function is enabled, the error amplifier reference voltage is adjusted to:

$$V_{REFIN} = 0.6V \pm \left(1.18V \cdot \frac{R5}{R6}\right)$$

For example, ±5% margining can be achieved by selecting R5 = 13k and R6 = 510k.

Additional features

The LTC3770 provides very strong gate drivers allowing up to 25A output currents, programmable current limit, output overvoltage protection and input undervoltage lockout. Other features include power good monitor, programmable soft-start, selectable discontinuous operation mode or forced continuous mode at light load and adjustable dead time between the top gate and bottom gate signals to optimize efficiency.

Conclusion

The LTC3770 has an accurate 0.6V reference voltage and a wide operating range. It includes advanced functions usually implemented by additional ICs—such as output tracking and programmable voltage margining. Available package options are a thermally enhanced 5mm × 5mm QFN package and a leaded 28-pin SSOP.

60V, 3A step-down DC/DC converter has low dropout and 100μA quiescent current

63

Keith Szolusha

Introduction

High voltage bipolar monolithic step-down converters are usually optimized for high efficiency at *high* output currents, often at the expense of light load efficiency and operation near dropout. The problem is that a 2mA quiescent current at zero load drains batteries in applications that spend long periods of time at minimum load current. One common solution for reducing quiescent current and improving battery runtime is a shutdown function, but shutdown drops the output voltage to zero. Shutdown is not acceptable in systems where a constant regulated output voltage is required for light load applications, system diagnostics and ready-to-use load transients. In some systems, a regulated output voltage is needed at very low input voltage (low dropout). Simply shutting down the converter to zero output to avoid output droop is not always an option.

The LT3434, 3A monolithic buck switching regulator is designed to optimize efficiency over *all* current and voltage levels, both high and low. Micropower bias current and Burst Mode operation enable it to consume merely 100μA at zero load and 12V input. The high efficiency bipolar NPN power switch (0.1Ω) provides up to 85% efficiency at a 2A load current. Combined with high duty cycle, the low dropout of the switch maintains a regulated 3.3V output down to 4V input at all load currents. This is important for automotive cold-crank operation.

The LT3434's 3.3V to 60V input voltage range makes it ideal for 14V and 42V automotive battery-fed applications with both 4V cold crank and high input voltage transients (up to 60V). The 3A switch current rating provides maximum load currents of up to 2.5A. The LT3434 maintains output regulation down to 4V input for 3.3V output and down to 3.3V input for 2.5V or lower output voltages.

The high input voltage and low quiescent current make this an ideal choice for many 48V nonisolated telecom applications, 40V FireWire peripherals and multisource battery-powered applications with autoplug adaptors. The LT3434 can survive load-dump input transients up to 60V that are common in these systems.

It also includes other important features to shrink solution size, simplify configuration and improve system robustness:

- Fixed 200kHz switching frequency provides low output ripple, high efficiency and the ability to provide wide input

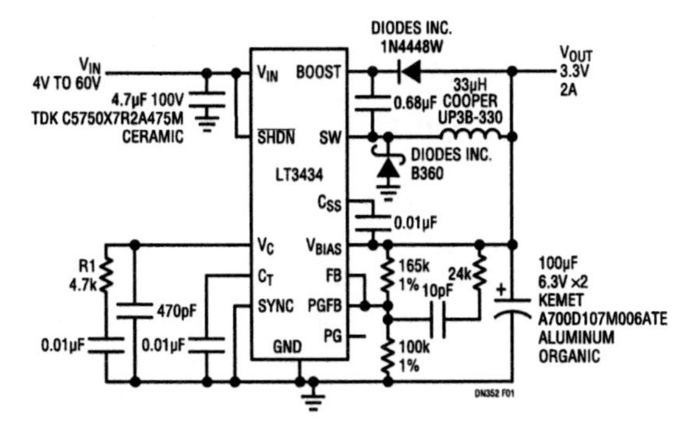

Figure 63.1 • LT3434 Wide Input Voltage Range DC/DC Converter Application to 3.3V Output at 2A Load Current Featuring Burst Mode Operation for Light Load Operation and 4V Low Dropout Operation

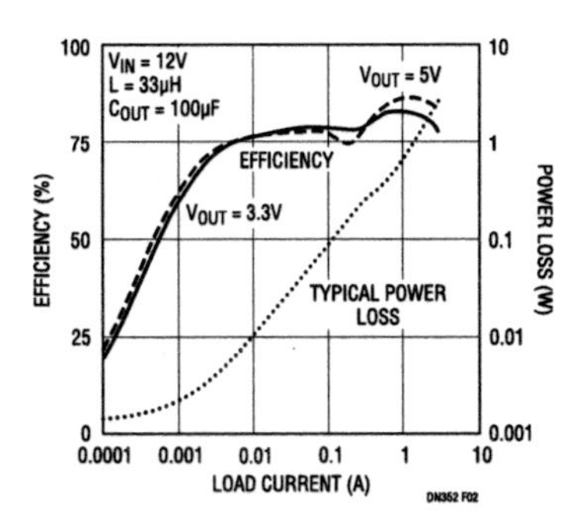

Figure 63.2 • The Efficiency of Figure 63.1 Is Typically Greater than 75%. At Light Loads, Supply Current Is Minimized with Burst Mode Operation

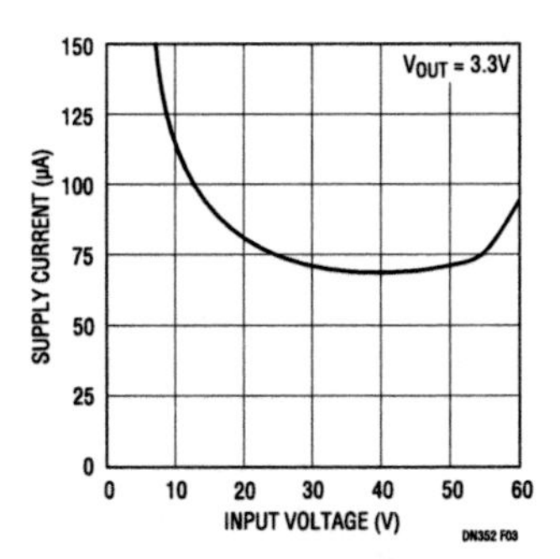

Figure 63.3 • With Zero Load Current, the Supply Current to the LT3434 Is Extremely Low, Typically Below 100μA ($V_{IN} \geq 12V$)

Analog Circuit Design: Design Note Collection. http://dx.doi.org/10.1016/B978-0-12-800001-4.00063-6

voltage range solutions. The LT3434 can be synchronized at frequencies up to 700kHz.

- The shutdown pin provides a 2.38V undervoltage lockout threshold as well as a 0.4V threshold for micropower shutdown (<1μA).
- A single capacitor provides soft-start capabilities and limits inrush current and output voltage overshoot in sensitive applications.
- A power good flag and power good comparator provide the system with an indication that the output voltage, the input voltage or some other line is above a desired voltage.
- The LT3434 is provided in a small 16-pin TSSOP thermally enhanced package for excellent thermal performance.

Burst Mode operation

Figure 63.1 shows a typical wide input voltage range step-down application to 3.3V output DC/DC converter. Burst Mode operation reduces light load quiescent current by disabling switching for a number of switch cycles and placing the part briefly in micropower shutdown until switching begins again. Bursts of switch pulses are enough to maintain output voltage regulation at light load. Figure 63.2 shows that the efficiency is high for nominal loads, between 100mA and 2A, and that at light load the quiescent current only sips from the battery during long periods of system inactivity. Figure 63.3 shows that for most typical input voltages, zero load quiescent current is below 100μA.

Low dropout

The LT3434 provides extremely low dropout with high maximum duty cycle (90%) and low power switch on-resistance (0.1Ω). Figure 63.4 demonstrates how the LT3434 maintains 3.3V output regulation with an input voltage down to 4V over the entire load current range. The minimum input voltage required to start up the output is slightly higher, as shown in Figure 63.4. Starting up the LT3434 at a duty cycle lower than maximum helps get the boost voltage high enough to run the power switch in low $V_{CE(SAT)}$ operation before entering extremely low dropout.

Soft-start

Only a single capacitor (C_{SS} in Figure 63.1) is required for soft-start. Soft-start avoids the problems created by large inrush currents at start-up, where switchers without

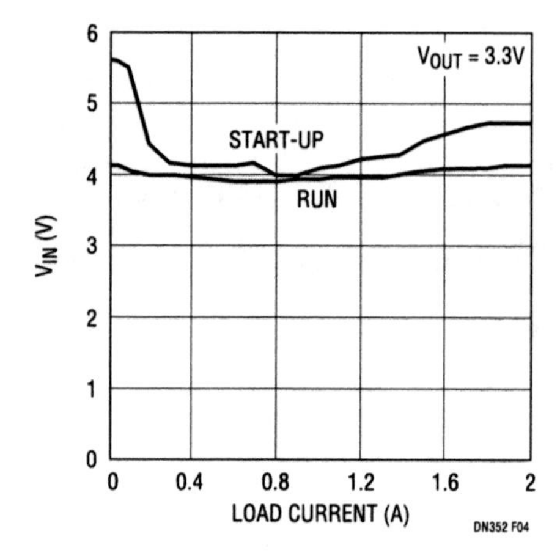

Figure 63.4 • Low Dropout Operation for 3.3V Output Is as Low as 4V Over the Entire Load Current Range. Start-Up Requires Slightly Higher Input Voltage

soft-start try to go from zero to regulation by consuming as much current as possible from the source and casting it into the output capacitor and load. This surge of current can both drag down a battery source voltage and cause overshoot in the output voltage. The soft-start capacitor for the LT3434 holds the peak current level clamp low, allowing it to slowly rise upon start-up. An external soft-start capacitor removes the inrush current surge and limits output voltage overshoot by controlling the output voltage ramp-up rate.

Power good

For systems that rely upon having a well-regulated power source or that follow a particular power-up sequence, the LT3434 provides a power good flag with programmable delay. The delay is programmed by C_T, starting when the power good feedback pin exceeds 90% of V_{REF} (1.25V). By tying the power good feedback pin (PGFB) directly to the feedback pin (FB), the power good comparator returns a "good" signal only when the output voltage has reached 90% and the C_T voltage exceeds its internal clamp. The power good feedback pin can also be tied to the input voltage, an external source or a resistor divider on any of these sources.

Conclusion

The LT3434 is a wide input voltage range, 200kHz, monolithic 3A, step-down DC/DC converter. High input voltage, high power switch capabilities, low quiescent current, low dropout and excellent package thermal conductivity make this an extremely useful and versatile IC that is simple to use in many step-down applications.

Monolithic synchronous regulator drives 4A loads with few external components

64

Joey M. Esteves

Introduction

The LTC3414 offers a compact and efficient voltage regulator solution for point of load conversion in electronic systems that require low output voltages (down to 0.8V) from a 2.5V to 5V power bus. Internal power MOSFET switches, with only 67mΩ on-resistance, allow the LTC3414 to deliver up to 4A of output current with efficiency as high as 94%. The LTC3414 saves space by operating with switching frequencies as high as 4MHz, enabling the use of tiny inductors and capacitors.

The LTC3414 employs a constant frequency, current mode architecture and can deliver up to 4A of output current. The switching frequency can be set between 300kHz and 4MHz by an external resistor; alternatively, it can be synchronized to an external clock where each switching cycle begins at the falling edge of the external clock signal. For improved thermal management, the LTC3414 is offered in a 20-lead TSSOP package with an exposed pad to facilitate heat transfer.

The LTC3414 can be configured for either Burst Mode operation, pulse-skipping or forced continuous operation. Burst Mode operation maximizes light-load efficiency and extends battery life by reducing gate charge losses at light loads—at no load, the LTC3414 consumes a mere 64μA of supply current. Forced continuous operation maintains a constant frequency throughout the entire load range, making it easier to filter the switching noise and reduce RF interference—important for EMI-sensitive applications. Pulse skip mode provides a good compromise between light load efficiency and output voltage ripple.

The LTC3414 provides for external control of the burst clamp current level, in effect allowing the burst frequency to be varied. Lower Burst Mode operating frequencies result in improved light load efficiencies, but there is a trade-off between light load efficiency and output voltage ripple—as the Burst Mode frequency decreases, the output ripple increases slightly. In the LTC3414, the burst clamp is adjusted by varying the DC voltage at the SYNC/MODE pin within a 0V to 1V range. The voltage level at this pin sets the minimum peak inductor current during each switching cycle in Burst Mode operation. Pulse skip mode is implemented by connecting the SYNC/MODE pin to zero volts. In pulse skip mode, the burst clamp is set to zero current and the minimum peak inductor current is determined by the minimum on-time of the control loop. Pulse-skipping minimizes the output voltage ripple by providing the lowest possible inductor current ripple.

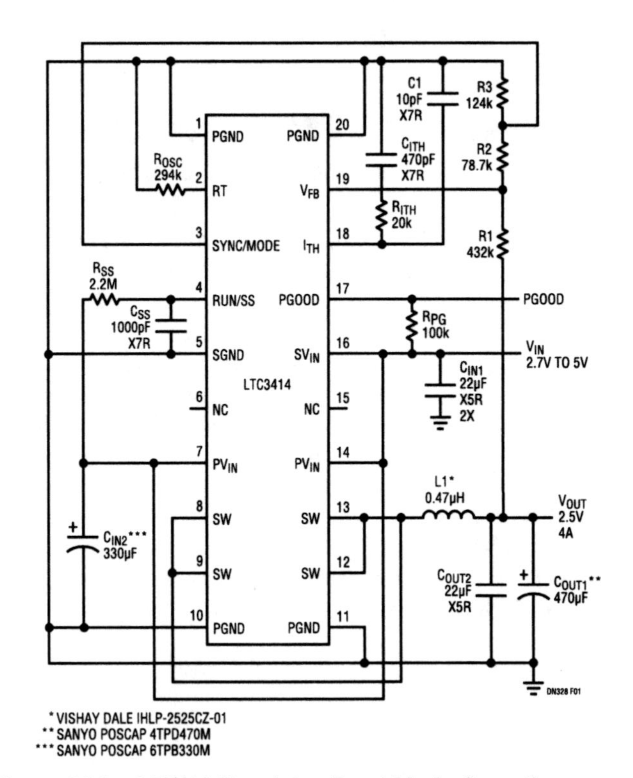

Figure 64.1 • 2.5V/4A Regulator, Burst Mode Operation

Analog Circuit Design: Design Note Collection. http://dx.doi.org/10.1016/B978-0-12-800001-4.00064-8

High efficiency 2.5V/4A step-down regulator

Figure 64.1 shows a 2.5V step-down DC/DC converter that is configured for Burst Mode operation. This circuit provides a regulated 2.5V output at up to 4A from a 2.7V to 4.2V input. Efficiency for this circuit, shown in Figure 64.2, is as high as 94% for a 3.3V input voltage. The switching frequency for this circuit is set at 1MHz by a single external resistor, R_{OSC}. Operating at frequencies this high allows the use of a lower valued (and physically smaller) inductor.

In this particular application, Burst Mode operation maintains the high efficiency at light loads. The burst clamp current is set by the R2 and R3 voltage divider, which generates a 0.49V reference at the SYNC/MODE pin. This corresponds to approximately 1.2A minimum peak inductor current, as shown in Figure 64.2.

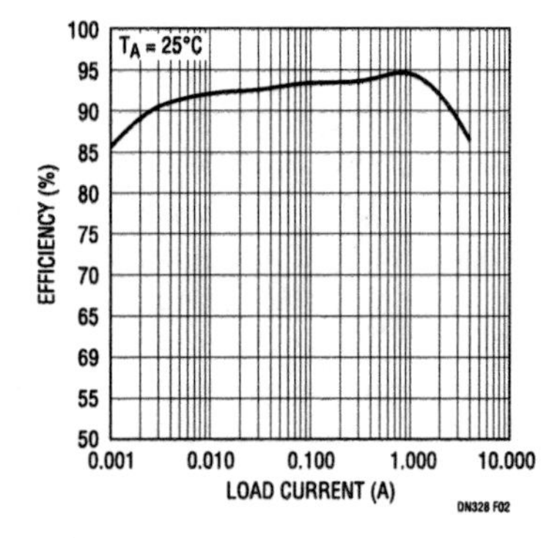

Figure 64.2 • Efficiency vs Load Current 3.3V to 2.5V Burst Mode Operation from Figure 64.1

High efficiency 3.3V/4A step-down regulator with all ceramic capacitors

Figure 64.3 shows a 3.3V step-down DC/DC converter using all ceramic capacitors. This circuit provides a regulated 3.3V output at up to 4A from a 5V input voltage. Efficiency for this circuit, shown in Figure 64.4, is as high as 93%. Ceramic capacitors offer low cost and low ESR, but many switching regulators have difficulty operating with them because the extremely low ESR can lead to loop instability. The phase margin of the control loop can drop to inadequate levels without the aid of the zero that is normally generated from the higher ESR of tantalum capacitors. The LTC3414, however, includes OPTI-LOOP compensation, which allows it to operate properly with ceramic input and output capacitors. The LTC3414 allows loop stability to be achieved over a wide range of loads and output capacitors with proper selection of the compensation components on the I_{TH} pin.

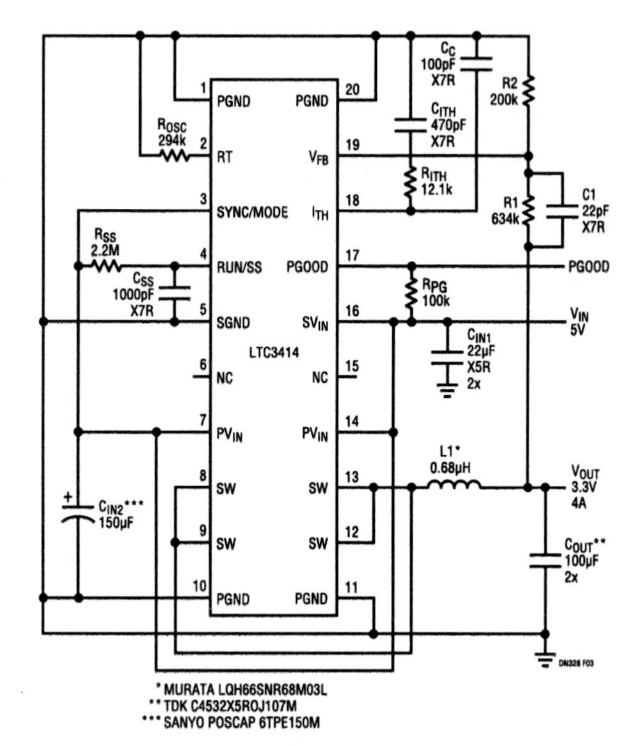

Figure 64.3 • 3.3V/4A Regulator, Forced Continuous Mode

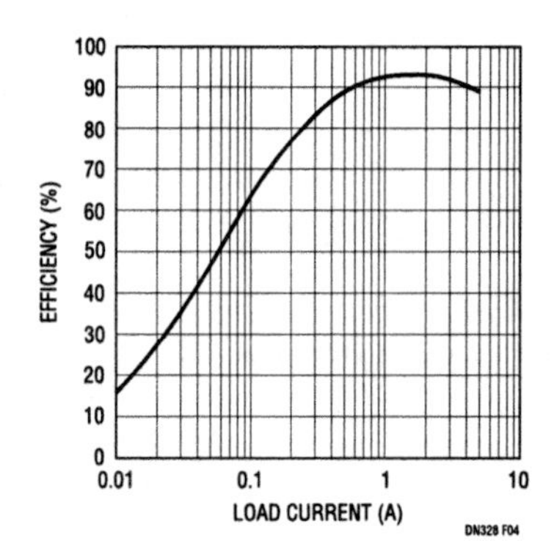

Figure 64.4 • Efficiency vs Load Current, 5V to 3.3V, Forced Continuous

Conclusion

The LTC3414 is a monolithic, synchronous step-down DC/DC converter that is well suited for applications requiring up to 4A of output current. Its high switching frequency and internal low $R_{DS(ON)}$ power switches make the LTC3414 an excellent choice for compact, high efficiency power supplies.

65

High performance power solutions for AMD Opteron and Athlon 64 processors

Henry J. Zhang Wei Chen

Introduction

AMD Opteron and Athlon 64 CPUs can draw 65A at very low voltages (0.8V to 1.55V). Such high currents put efficiency and related thermal issues at the top of a power supply designer's list of problems. The new LTC3733 PolyPhase synchronous buck controller solves these problems, and others, by enabling solutions that offer high efficiency, low profile and fast transient response.

The LTC3733 is a 3-phase current mode controller that drives three synchronous buck stages 120° out of phase. The three out-of-phase stages inherently self-cancel the current ripple and thus minimize the size of the input and output capacitors. Likewise, ripple cancellation and a high switching frequency (up to 600kHz) make it possible to use lower value output inductors, resulting in faster transient-load response and smaller physical inductor size. The LTC3733's integrated MOSFET drivers also simplify the design and minimize the overall supply footprint. The LTC3733 directly senses the inductor current of each phase to achieve excellent current sharing among phases, thus evenly distributing the thermal stress. As a result, overdesign of the MOSFETs and inductors is unnecessary and the circuit can be operated reliably without a heat sink. The total size and cost of the solution is minimized.

The LTC3733 comes in a 36-lead SSOP package and the LTC3733-1 comes in a low profile 7mm × 5mm QFN package with an exposed ground pad to minimize thermal impedance. An internal differential amplifier provides true remote

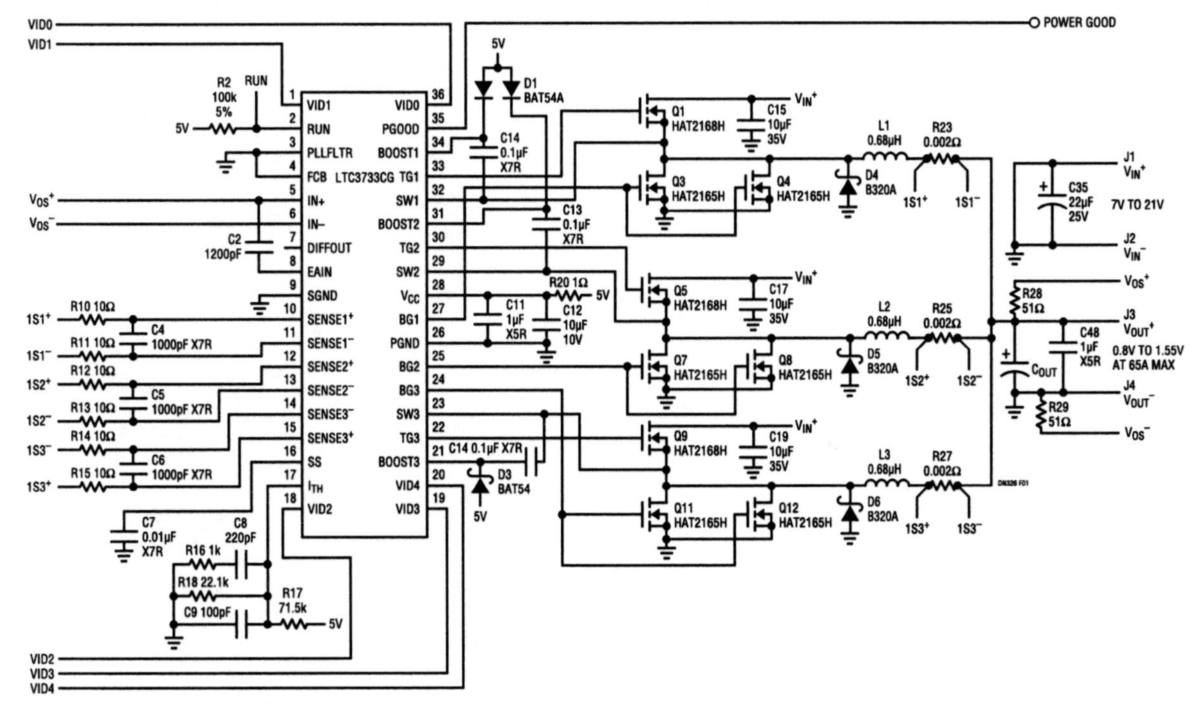

Figure 65.1 • 65A AMD Processor Power Supply Using the LTC3733

Analog Circuit Design: Design Note Collection. http://dx.doi.org/10.1016/B978-0-12-800001-4.00065-X

sensing of the regulated supply's positive and negative output terminals at the load, providing accurate output regulation in high current applications. The internal 5-bit VID programmable attenuator complies with the AMD Opteron and Athlon 64 processors' VID table (0.8V to 1.55V). The LTC3733 also incorporates two selectable light load operation modes: Burst Mode operation and Stage Shedding mode to significantly improve light load efficiency. With internal overvoltage and overcurrent protection features, the resulting LTC3733 solution provides superior performance for a wide range of applications, from enterprise CPU to mobile CPU applications.

3-phase, 65A AMD VRM design

Figure 65.1 shows a 3-phase 65A AMD processor power supply. This supply achieves better than 88% efficiency for a 7V input and a 1.55V/65A output, as shown in Figure 65.2. Even for a 21V input, efficiency peaks at more than 90%, remaining high throughout a wide 5A to 65A load range.

Figure 65.3 shows a measured load transient waveform. The load current step is 46A with a slew rate of 50A/μs. Ten low profile Sanyo POSCAP capacitors (330μF/2.5V) are used on the VRM output. The maximum output voltage variation during the load transients is just $88mV_{P-P}$.

Conclusion

The LTC3733-based low voltage, high current power supply meets the AMD processor VRM requirement. With 3-phase switching and superior light load operation modes (Stage Shedding and Burst Mode operation), this design can achieve high efficiency, small size and low overall solution cost. Accurate current sharing among the three phases improves thermal performance, and therefore, the long-term reliability of the CPU core power supply.

By combining the LTC3733 with one or more LTC3731s, a six or higher phase power supply can be easily obtained to power multiple CPUs. For more information, please contact Linear Technology Corporation.

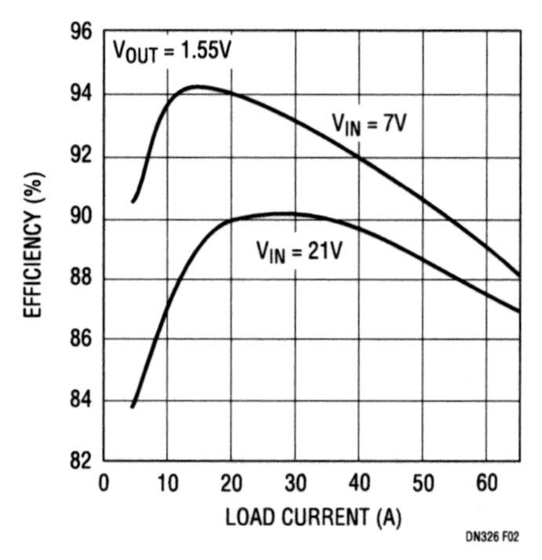

Figure 65.2 • Efficiency vs Load Current at 7V to 21V V_{IN}, 1.55V V_{OUT}

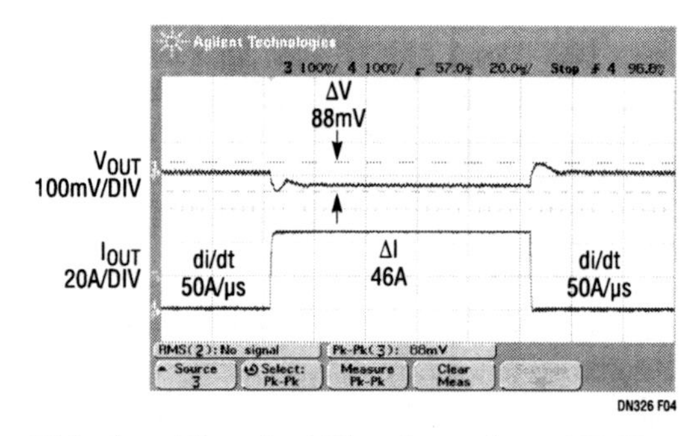

Figure 65.3 • Load Transient Waveforms for a 46A Step with a 50A/μs Slew Rate

High current step-down controller regulates to 0.6V output from 3V input

66

Charlie Zhao   Wei Chen

Introduction

The LTC3832 is a voltage mode, high efficiency, high power step-down switching regulator controller that can operate from input voltages as low as 3V. It integrates a 0.6V reference voltage, two powerful MOSFET drivers and other features which make it possible to build cost-effective, high efficiency, high current power supplies with output voltages as low as 0.6V.

The LTC3832 uses a synchronous switching architecture with external N-channel MOSFETs. An adjustable current limit is provided by sensing the current through the drain-source on-resistance of the top MOSFET, eliminating the need for a current sense resistor. The LTC3832 has a 300kHz free-running switching frequency that can be programmed or synchronized externally from 100kHz to 500kHz. All of these features promote a DC/DC converter solution with high efficiency and small size.

The LTC3832 also includes a thermal protection circuit which disables both gate drivers if the junction temperature reaches 150°C. The chip resumes normal operation when the temperature drops below 125°C.

Design examples

Figure 66.1 shows the schematic diagram and photo of a compact 1V/7A step-down DC/DC converter that accepts an input voltage of 3V to 8V. The design is based on the LTC3832-1, an SO-8 package option. Its efficiency is shown in Figure 66.2.

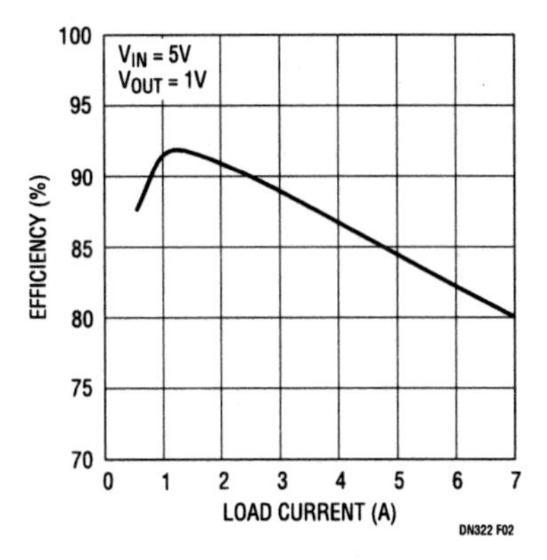

Figure 66.2 • Efficiency for the Circuit in Figure 66.1

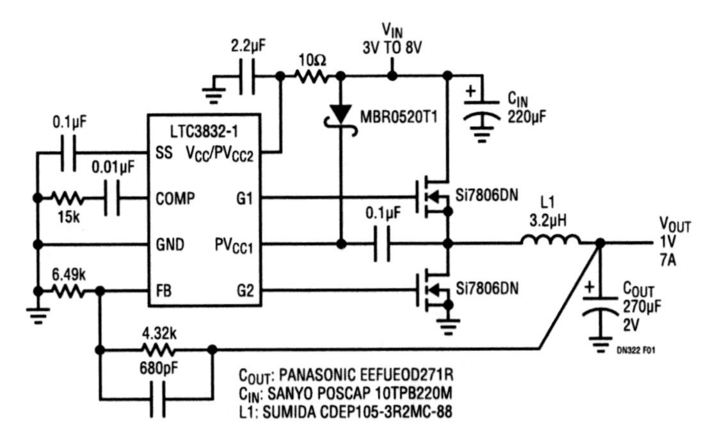

Figure 66.1 • A Compact 5V to 1V Power Supply at 7A

Analog Circuit Design: Design Note Collection. http://dx.doi.org/10.1016/B978-0-12-800001-4.00066-1

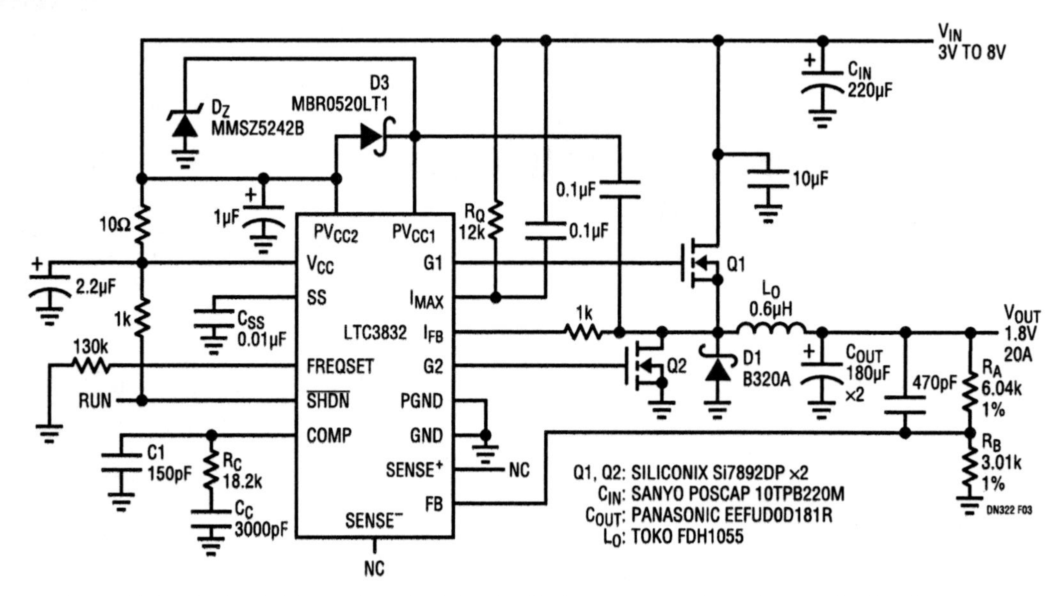

Figure 66.3 • High Efficiency 3.3V to 1.8V at 20A Synchronous Step-Down DC/DC Converter

Figure 66.3 shows the schematic diagram of a high efficiency 3V to 8V to 1.8V/20A synchronous step-down power supply. The switching frequency is set at 360kHz by the external resistor connected to the FREQSET pin.

For normal operation, a pull-up resistor brings the voltage high at the $\overline{SHDN}$ pin. A low voltage for more than 100µs at $\overline{SHDN}$ pin pulls the COMP and SS pins to ground and shuts Q1 and Q2 off. The current limit function is provided by connecting an external resistor R_Q from the I_{MAX} pin to the drain of the top MOSFET (Q1). The current limit threshold can be adjusted by the value of R_Q based on the $R_{DS(ON)}$ of Q1 and the load requirement. The 0.1µF decoupling capacitor across R_Q filters the switching noise.

The LTC3832 also includes a soft-start circuit. The soft-start voltage ramp rate in this circuit is set by C_{SS}.

Figure 66.4 shows the efficiency curve of the circuit. Up to 90% efficiency is obtained with a 3.3V input. For other output voltages, simply change the value of R_A. The output voltage can be as low as 0.6V (see Table 66.1). For higher output currents, parallel more MOSFETs and use a higher current inductor.

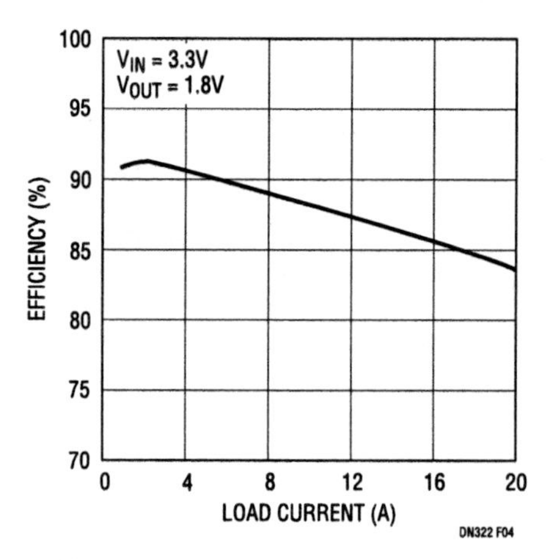

Figure 66.4 • Efficiency for the Circuit in Figure 66.3 with V_{IN}=3.3V

Table 66.1 R_A Values for Different Output Voltages

V_{OUT} (V)	R_A (Ω)
0.6	0
1.0	2.00k
1.2	3.01k
1.5	4.53k
1.8	6.04k
2.5	9.53k

Conclusion

The LTC3832 is a voltage mode controller optimized for high power, low voltage—as low as 0.6V—applications. It provides a set of features that promote high efficiency while keeping solution costs low. LTC3832 designs may be simulated with SwitcherCAD III which can be downloaded from www.linear.com.

Efficient dual polarity output converter fits into tight spaces

67

Keith Szolusha

Introduction

This design note describes a compact and efficient ±5V output dual polarity converter that uses a single buck regulator. The topology shown features 3mm maximum circuit height, high efficiency and low output voltage ripple on a 5V output—important considerations for many battery-powered, handheld and noise-sensitive devices. This combination of features is not easily achievable with other commonly used dual polarity topologies. For instance, one alternative topology, a flyback converter using a boost regulator, is relatively inefficient, requires a bulky (5mm or taller) transformer and generates high output voltage ripple. Another alternative, using two buck regulators, incurs both the cost of the additional regulator and the cost of the PCB real estate it occupies.

The single buck regulator topology shown here requires few components. To reduce the maximum circuit height, it uses two power inductors instead of a transformer. In the absence of the transformer core, the coupling capacitor allows energy to pass between the positive and negative sides of the circuit while maintaining a voltage potential between the two inductors, indirectly regulating the negative output.

12V input, ±5V output, only 3mm high

A dual polarity output converter uses a buck regulator, such as the LT1956 or LT3431. Both of these are 500kHz, 1.5A/3A peak switch current monolithic switchers. Figure 67.1 shows a 12V battery input (9V to 16V input with 36V transients) to ±5V_{OUT} dual polarity output converter using the LT1956EFE. Figure 67.2 shows the same circuit with twice the load current rating using the LT3431EFE.

Typical bucks with second, negative outputs

The dual polarity output configuration is similar to the typical buck regulator with a second, negative polarity output added to the circuit using a coupling capacitor and a second inductor, catch diode and output capacitor. The duty cycle remains the same as the typical buck regulator with the same V_{IN} and V_{OUT}. The positive 5V output maintains its low

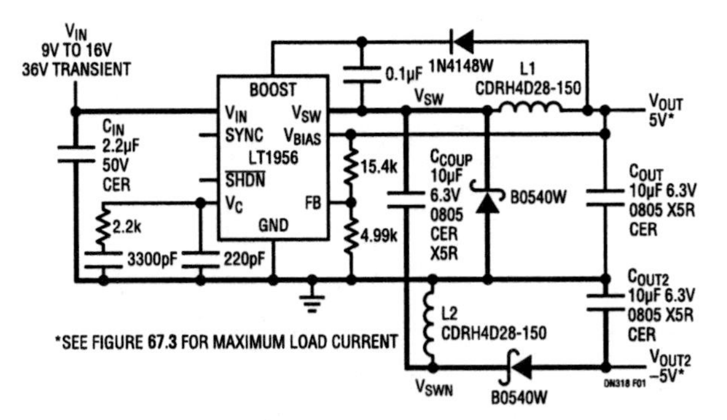

Figure 67.1 • LT1956 9V to 16V Input (with 36V Transients), ±5V Output, 3mm Height All Ceramic Dual Polarity Converter with High ΔI/Δt Crucial Layout Path Indicated in Bold

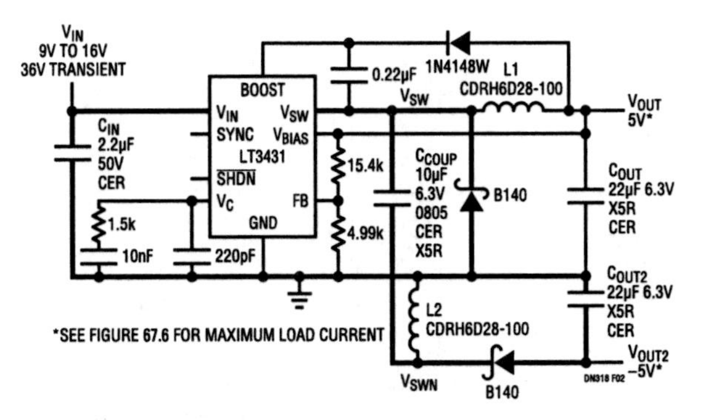

Figure 67.2 • LT3431 9V to 16V Input (with 36V Transients), ±5V Output, 3mm Height All Ceramic Dual Polarity Converter with High ΔI/Δt Crucial Layout Path Indicated in Bold

Analog Circuit Design: Design Note Collection. http://dx.doi.org/10.1016/B978-0-12-800001-4.00067-3

output voltage ripple characteristic from the buck regulator, but some of the current available for that output is rerouted through the coupling capacitor to the second output. The coupling capacitor charges up to and maintains a voltage equal to the output voltage (5V). This induces the same voltage and hence the same current ripple across both inductors. However, the average current in L1 is the 5V load current while the average current in L2 is the negative load current.

The maximum load current, shown in Figures 67.3 and 67.6, is reached when the sum of the peak inductor currents is equal to the peak switch current rating of the regulator, 1.5A (LT1956) or 3A (LT3430), or the negative output loses regulation. The peak switch current region is to the right of the peak in the curves. To the left of the peak in the curves, 1.5A and 3A cannot be reached by increasing the negative load current without losing over 3% regulation on V_{OUT2}.

$$I_{OUT(MAX)}[5V] = 1.5/3 - I_{OUT2}[-5V] - 2 \cdot IL_{P-P}/2$$
(for peak switch current region)

$$IL_{P-P} = (V_{IN} - 5V_{OUT}) \cdot DC/(L \cdot 500kHz)$$

Extremely low negative load currents can also cause a loss of regulation on V_{OUT2} as seen in Figures 67.4 and 67.7. In order to maintain relatively good regulation in low load applications, a preload resistor on V_{OUT2} of 12mA (LT1956) or 25mA (LT3430) may be required (Figure 67.5). Feedback is derived directly from V_{OUT} so that its load current can go to zero without a loss of regulation (see Figure 67.8).

Conclusion

The LT1956- and LT3431-based dual polarity output converters provide power for ±5V loads with a single buck regulator. This design offers size and efficiency advantages over other dual output designs, especially those that require a transformer.

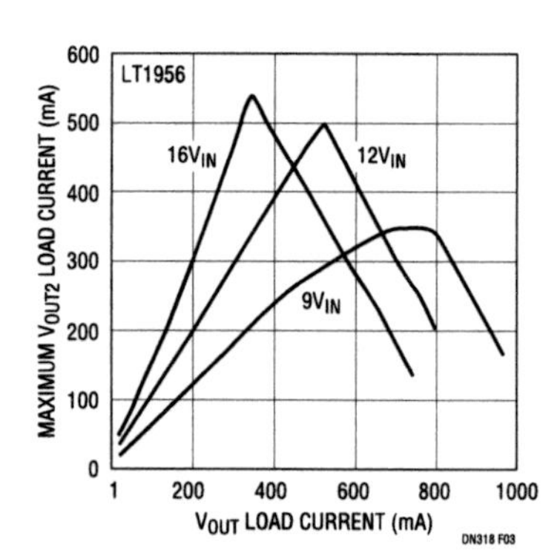

Figure 67.3 • Maximum Load Current Conditions for Figure 67.1

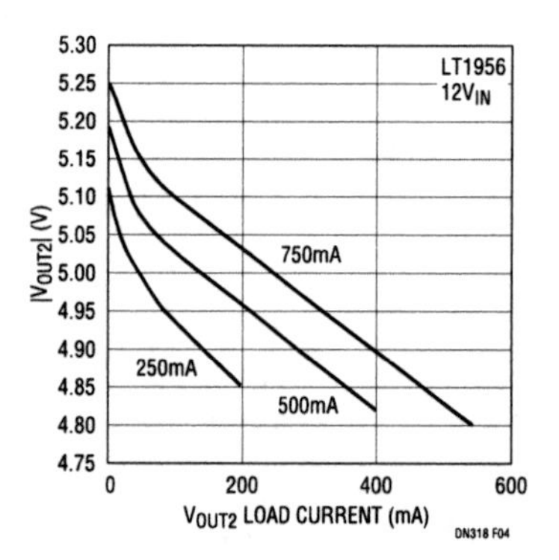

Figure 67.4 • The Negative Supply (V_{OUT2}) Maintains ±5% Regulation

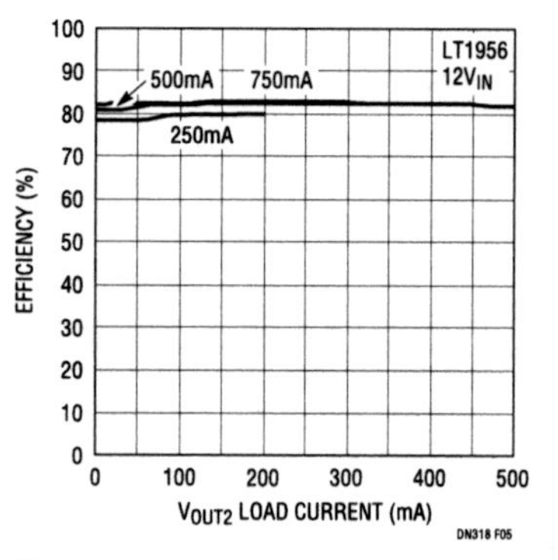

Figure 67.5 • The Efficiency of Figure 67.1

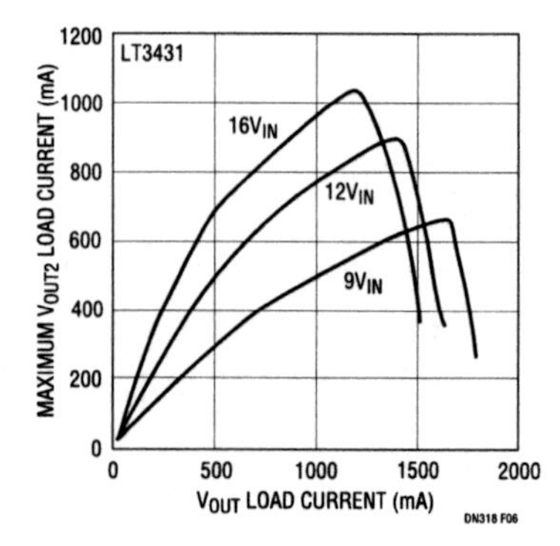

Figure 67.6 • Maximum Load Current Conditions for Figure 67.2

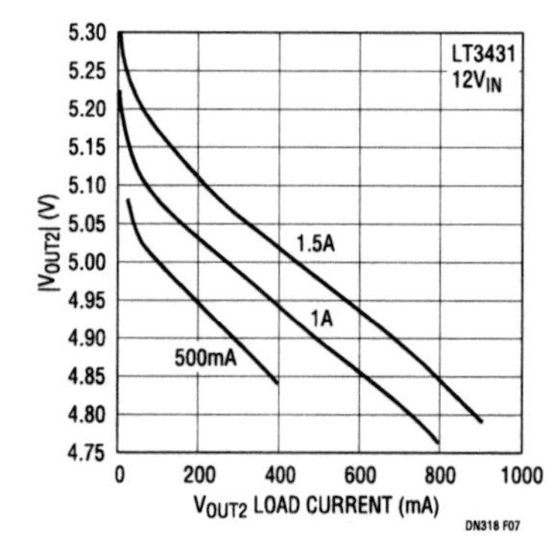

Figure 67.7 • The Negative Supply (V_{OUT2}) Maintains ±5% Regulation

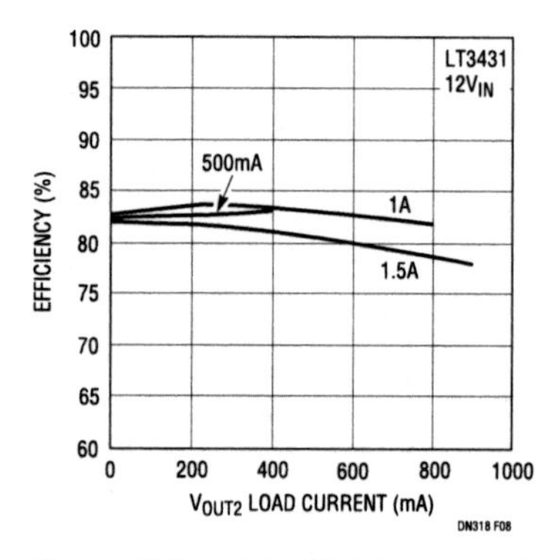

Figure 67.8 • The Efficiency of Figure 67.2

68

Dual output supply powers FPGAs from 3.3V and 5V inputs

Haresh Patel

Introduction

FPGAs often require multiple power supplies: one for the core voltage (usually 1.8V, but sometimes as low as 1.2V) and at least one more for the I/O circuitry (often 2.5V). The available input supply is either 3.3V or 5V. One way to provide the multiple step-down conversions is via multiple switcher-based power supplies; however, this may be more complicated and cumbersome than is warranted, especially if the I/O does not draw much current. In such instances, the dual output LTC1704 is a simple and space-saving option. It can supply two voltages with its versatile high frequency switcher and its space-saving LDO controller.

Circuit description

With a 5V input, the switcher channel can generate core voltages from 0.8V to 3.3V at currents up to 15A. This 550kHz switcher reduces required LC filter size while providing fast response to dynamic loads. Efficiency of the switcher is very high and features No R_{SENSE} technology, where the output current is sensed via the MOSFET's on resistance, to improve efficiency compared with regulators using a sense resistor.

The LDO uses an external pass transistor to regulate the I/O voltage. The LTC1704 provides base currents up to 30mA to control the output voltage under varying load conditions.

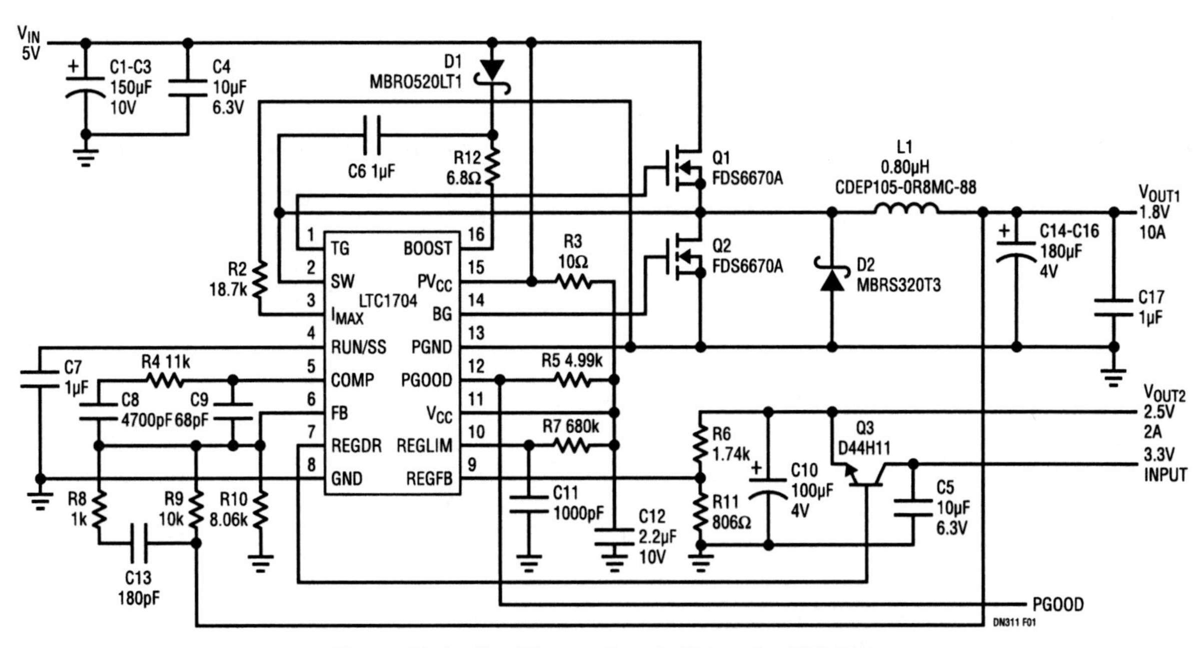

Figure 68.1 • Dual Power Supply Using the LTC1704

Analog Circuit Design: Design Note Collection. http://dx.doi.org/10.1016/B978-0-12-800001-4.00068-5

The circuit shown in Figure 68.1 provides 2.5V at 2A from a 3.3V supply.

The uncommitted collector and emitter of the external pass transistor gives the LDO versatility. It can convert 1.8V to 1.5V or 1.2V. At lower power levels, the input voltage can be as high as 5V. Output current would then be limited by thermal considerations.

Figure 68.1 shows a dual output 1.8V/10A and 2.5V/2A circuit using the LTC1704. This is a typical FPGA application where 1.8V is the core voltage and 2.5V is for I/O. In this case, the switcher supplies 1.8V and the LDO supplies 2.5V, taking power from either 3.3V or 5V for the external pass transistor.

The switcher channel uses all N-channel MOSFETs for improved efficiency and lower cost. R9 and R10 program the output voltage. Type III compensation—C9, R4, C8, R8 and C13—allows maximum flexibility in the choice of LC filter components. The current limit circuit uses the $R_{DS(ON)}$ of the bottom MOSFET to sense inductor current. A 10µA current from the I_{MAX} pin flowing into R2 produces the reference voltage for current limit. The current limit circuit discharges the soft-start capacitor C7 to control output current.

The linear regulator uses an external high gain low V_{CESAT} NPN series pass transistor Q3. The output voltage is $0.8V \cdot (1 + R6/R11)$. The maximum output voltage is limited to $(V_{CC} - V_{DRV} - V_{BE})$ and by $(V_{C(Q5)} - V_{CESAT})$. The maximum driver voltage drop (V_{DRV}) is 1.1V at 30mA. Limiting the base drive current provides short-circuit protection. R7 programs max base current drive. Pulling REGLIM down to below 0.8V turns off the LDO.

Figure 68.2 shows the efficiency of the 1.8V output over a 1A to 10A current range, over which the efficiency remains close to 90%. Figure 68.3 shows the load step response of the switcher to a 4A to 10A load step. At each edge of the load step, less than 50mV transient deviation occurs when using three 180µF 4V solid polymer capacitors.

Conclusion

The LTC1704 is suitable for applications requiring a high power switcher and a moderate power linear regulator where the cost and complexity of a second switcher would be unjustifiable. For applications that require more power from the second output than is practical with a linear regulator, the LTC1702A is a good choice with its two switchers that can deliver up to 15A each.

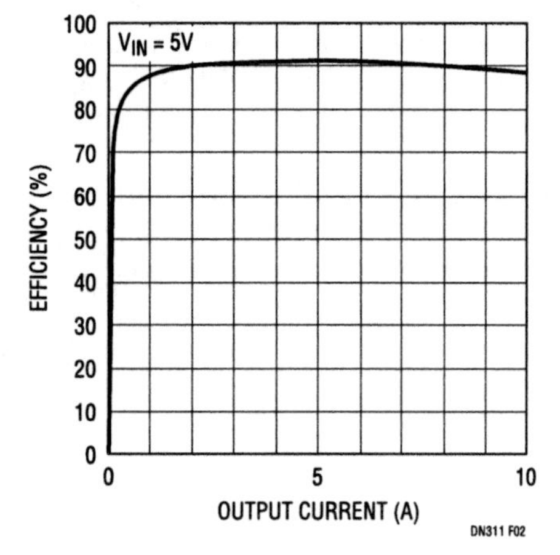

Figure 68.2 • 1.8V Efficiency

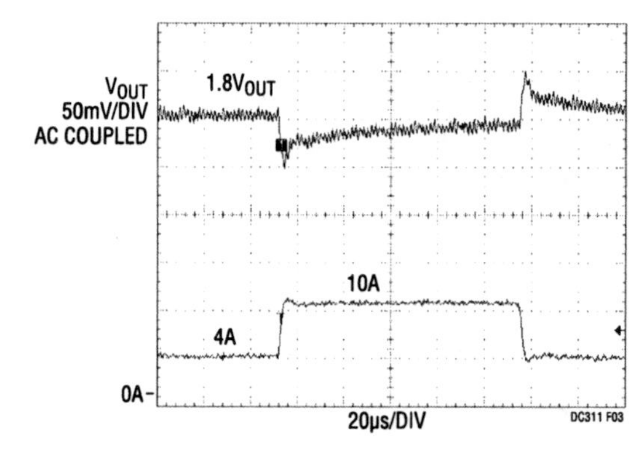

Figure 68.3 • Step Load Response of 1.8V Output

3A, 2MHz monolithic synchronous step-down regulator provides a compact solution for DDR memory termination

69

Joey M. Esteves

Introduction

The LTC3413 is a monolithic synchronous, step-down switching regulator that is capable of generating a bus termination voltage for DDR/DDR2 memory applications. While sourcing and sinking up to 3A of current, the LTC3413 allows switching frequencies as high as 2MHz. Increasing the switching frequency makes compact solutions possible by allowing the use of smaller inductors and capacitors. The internal power switches have a mere 85mΩ of on-resistance, making it possible to achieve efficiencies as high as 90% while generating an output voltage as low as 0.6V. For improved thermal handling, the LTC3413 is offered in a 16-lead TSSOP package with an exposed pad.

The LTC3413 utilizes a constant frequency, current mode architecture that operates from an input voltage range of 2.25V to 5.5V and provides a regulated output voltage equal to $V_{REF}/2$. The switching frequency can be set between 300kHz to 2MHz by a single external resistor. Output voltage ripple is inversely proportional to the switching frequency and the inductor value. Having the ability to increase the switching frequency as high as 2MHz allows lower inductor values to be used while still maintaining low output voltage ripple. Because smaller case sizes are usually offered for lower inductor values, the overall solution footprint can be reduced. An internal voltage divider halves the reference voltage, eliminating the need for external resistors to perform this task.

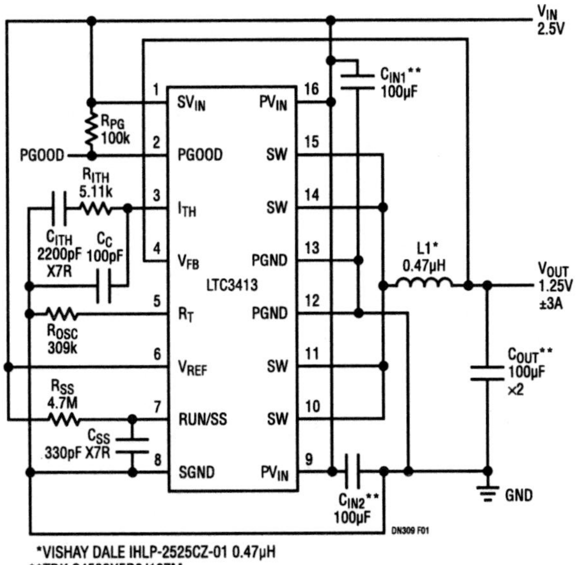

3A, 2.5V to 1.25V step-down DC/DC converter

Figure 69.1 illustrates a design solution for a 2.5V to 1.25V step-down DC/DC converter that is capable of sourcing and sinking up to 3A of output current. Efficiency for this circuit is as high as 90% as shown in Figure 69.2. Because of their low cost and low ESR, ceramic capacitors were selected for the input and output capacitors. Although many switching

*VISHAY DALE IHLP-2525CZ-01 0.47μH
**TDK C4532X5R0J107M

Figure 69.1 • 2.5V to 1.25V, ±3A DDR Memory Termination Supply

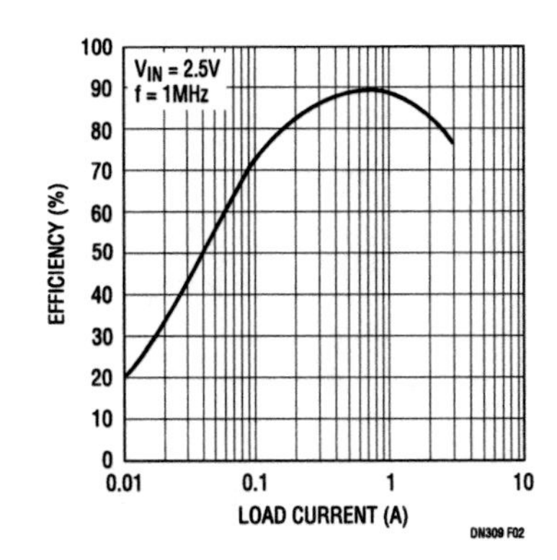

Figure 69.2 • Efficiency vs Load Current, $V_{IN}=2.5V$

Analog Circuit Design: Design Note Collection. http://dx.doi.org/10.1016/B978-0-12-800001-4.00069-7

regulators have difficulty operating with ceramic capacitors and rely on the zero that is generated by the larger ESR of tantalum capacitors, OPTI-LOOP compensation allows the LTC3413 to operate successfully with ceramic input and output capacitors. The frequency for this circuit is set at 1MHz by a single external resistor. Operating at frequencies this high allows the use of smaller external components such as the inductor and capacitors shown in Figure 69.1.

Many DDR termination applications require the bus termination voltage to be stepped down from a higher system voltage while tracking one-half of a reference voltage. This option is allowable in most systems since a reference voltage is typically available. Figure 69.3 shows a design solution for a 3.3V to 1.25V, ±3A DDR memory termination supply with a 2.5V external reference. Efficiency for this circuit is as high as 90% as shown in Figure 69.4. Figure 69.5 shows another design solution for a 3.3V to 0.9V, ±3A termination supply with a 1.8V external reference. Stepping down from a higher system voltage has the advantage of reducing the resistive losses due to the internal power switches, thus improving efficiency.

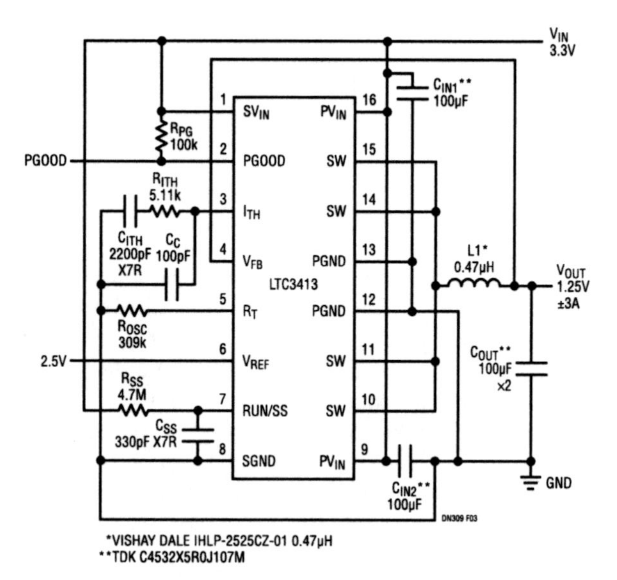

Figure 69.3 • 1.25V, ±3A DDR Memory Termination Supply from 3.3V

Conclusion

The LTC3413 is a monolithic, synchronous step-down DC/DC converter that is well suited for DDR memory termination applications requiring up to ±3A of output current. Its high switching frequency and internal low $R_{DS(ON)}$ power switches allow the LTC3413 to offer compact, high efficiency design solutions.

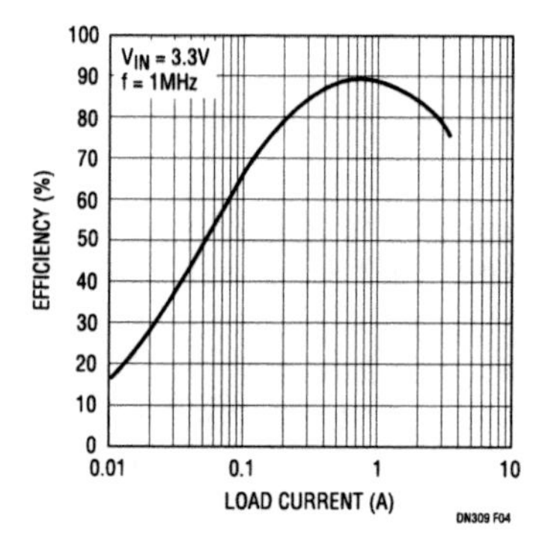

Figure 69.4 • Efficiency vs Load Current, V_{IN}=3.3V

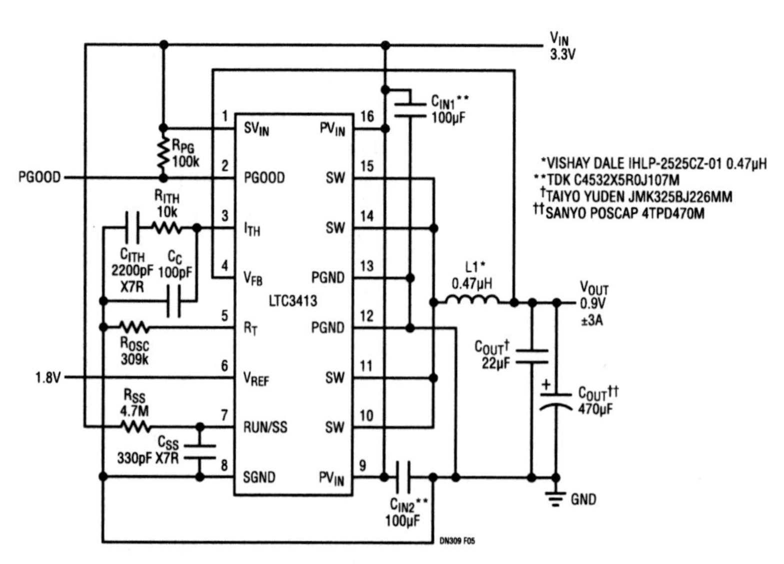

Figure 69.5 • 0.9V, ±3A HSTL Memory Termination Supply

60V/3A step-down DC/DC converter maintains high efficiency over a wide input range

70

Mark Marosek

Introduction

Today's high voltage applications—such as automotive, industrial and FireWire peripherals—place increasing demands on power supplies. They must provide high power, high efficiency and low noise, in a small space and over an ever-widening range of operational input voltages. Many high voltage DC/DC converter solutions can meet some of these conditions at high input voltages but they are unable to maintain high efficiencies at lower input voltages. Many of these same converters have frequency compensation schemes that require bulky input and output capacitors, which not only increase the size of the overall solution but also result in high output ripple voltage. The LT3430 is designed to alleviate all of these problems.

The LT3430 is a monolithic step-down DC/DC converter which utilizes a 3A peak switch current limit and has the ability to operate with a 60V input. The LT3430 runs at a fixed frequency of 200kHz and is housed in a small thermally enhanced 16-pin TSSOP package enabling it to save space while optimizing thermal management. Its 5.5V to 60V input range makes the LT3430 ideal for FireWire peripherals (typically 8V to 40V input), as well as automotive systems requiring 12V, 24V and 42V input voltages (with the ability to survive load dump transients as high as 60V). Furthermore, it was designed to maintain excellent efficiencies with both high and low input-to-output voltage differentials. Its current mode architecture adds flexible frequency compensation allowing the use of a ceramic output capacitor—resulting in small solutions with extremely low output ripple voltage (see Figures 70.3 and 70.4). Other features include a shutdown pin, which has an accurate 2.38V undervoltage lockout threshold and a 0.4V threshold for micropower shutdown (drawing only 25μA), and a SYNC pin, which allows the LT3430 to be synchronized up to 700kHz.

Efficiency

Monolithic step-down converters capable of operation at high input voltages are usually optimized for efficiency at high input-to-output voltage differentials, where the duty cycle is

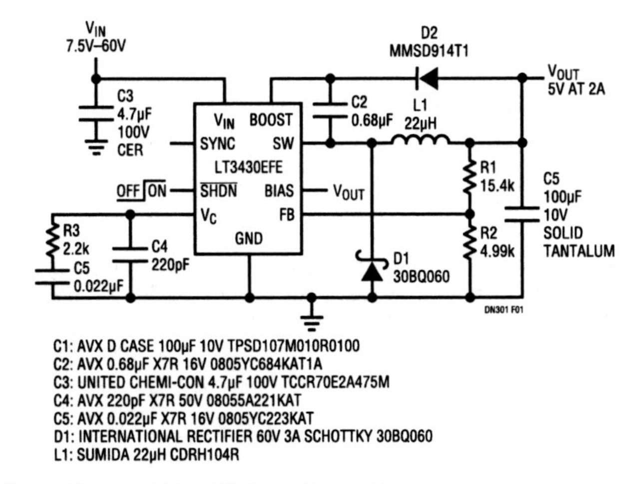

Figure 70.1 • 42V to 5V Step-Down Converter

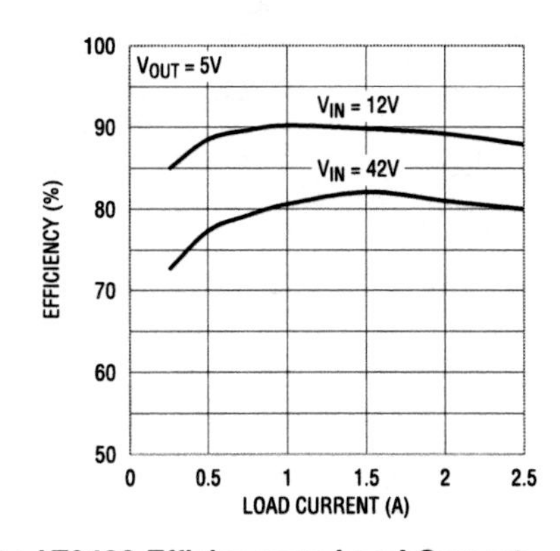

Figure 70.2 • LT3430 Efficiency vs Load Current

Analog Circuit Design: Design Note Collection. http://dx.doi.org/10.1016/B978-0-12-800001-4.00070-3

low. At low duty cycles, DC switch losses are small compared to the overall losses, so the switch design is often neglected, resulting in a switch resistance that can be as poor as 0.5Ω for some 3A converters. Such converters give up efficiency at high duty cycle operation and limit their minimum input voltage operating capability.

Figure 70.1 shows a 42V to 5V converter using the LT3430. To achieve high efficiency at high input voltages, the LT3430 provides fast output-switch edge rates. To further improve efficiency, the LT3430 provides a BIAS pin to allow internal control circuitry to be powered from the regulated output—light loads at high input voltages require minimal quiescent current to be drawn from the input. The peak efficiency for a 42V to 5V conversion is greater than 82%, as shown in Figure 70.2.

The LT3430 is also capable of excellent efficiencies at lower input voltages. The peak efficiency for a 12V to 5V conversion is greater than 90%, as shown in Figure 70.2. One key to achieving high efficiency for low input-to-output voltage conversions is to use a low resistance saturating switch. A pre-biased capacitor, connected between the BOOST and SW pins, generates a boost voltage above the input supply during switching. Driving the switch from this boost voltage allows the 100mΩ power switch to fully saturate. An output voltage as low as 3.3V is enough to generate the required boost supply.

Small size, low output ripple voltage (high switching frequency, all ceramic solution)

The high 200kHz switching frequency of the LT3430 keeps circuits small by minimizing the inductor value required to keep inductor ripple current low. The current mode architecture of the LT3430 allows for a small, low ESR ceramic capacitor to be used at the output—thus providing an extremely low output ripple voltage solution in a small space. Figure 70.3 shows a 5V/2A low profile (<3mm) solution for FireWire peripherals which uses a ceramic output capacitor. The output ripple voltage of this circuit is only $26mV_{P-P}$, much less than the $80mV_{P-P}$ incurred when a tantalum capacitor with an ESR of 80mΩ is used (see Figure 70.4).

Peak switch current (not your average current mode converter)

Most current mode converters have a reduced peak switch current limit at high duty cycles. This is a result of slope compensation, which is added to the converter's current sensing loop to prevent subharmonic oscillations for duty cycles above 50%. However, the LT3430 is able to maintain its peak switch current limit over the full duty cycle range. For applications that require high duty cycles, this offers significant advantages—including a lower inductor value, lower minimum V_{IN} and/or higher output current capability—over typical current mode converters with similar peak switch current limits.

Conclusion

The LT3430 features a 3A peak switch current limit, 100mΩ internal power switch and a 5.5V to 60V operating range, making it ideal for automotive, industrial and FireWire peripheral applications. It is highly efficient over the entire operating range and it includes important features to save space and reduce output ripple—including a 200kHz fixed operating frequency, a current mode architecture and availability in a small, thermally enhanced 16-pin TSSOP package.

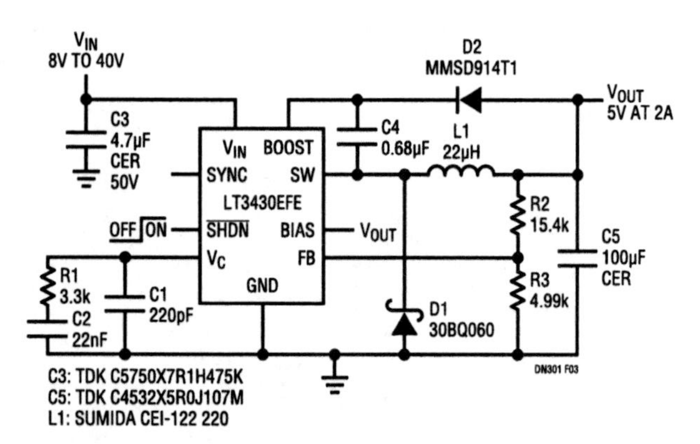

Figure 70.3 • Low Profile (Max Height of 3.0mm) Low Output Ripple Voltage Solution

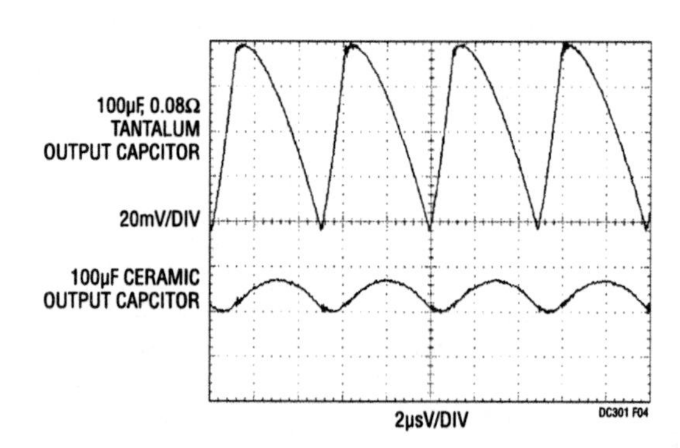

Figure 70.4 • Output Ripple Voltage Comparison for a Tantalum vs Ceramic Output Capacitor in the Circuit Shown in Figure 70.3 with V_{IN}=24V and I_{OUT}=2A

Monolithic synchronous step-down regulators pack 600mA current rating in a ThinSOT package

71

Jaime Tseng

Introduction

The new LTC3406, LTC3406-1.5, LTC3406-1.8, LTC3406B, LTC3406B-1.5 and LTC3406B-1.8 are the industry's first monolithic synchronous step-down regulators capable of supplying 600mA of output current in a 1mm profile ThinSOT package. These devices are designed to save space and increase efficiency for battery-powered portable devices. The LTC3406 series uses Burst Mode operation to increase efficiency at light loads, consuming only 20μA of supply current at no load. For noise-sensitive applications, the LTC3406B series disables Burst Mode operation and operates in a pulse-skipping mode under light loads. Both consume less than 1μA quiescent current in shutdown.

Space saving

Everything about the LTC3406/LTC3406B series is designed to make power supplies tiny and efficient. An entire regulator can fit into a 5 × 7mm board space. These devices are high efficiency monolithic synchronous buck regulators using a constant frequency, current mode architecture. Their on-chip power MOSFETs provide up to 600mA of continuous output current. Their internal synchronous switches increase efficiency and eliminate the need for an external Schottky diode. Internal loop compensation eliminates additional external components.

Versatile

These devices have a versatile 2.5V to 5.5V input voltage range, which makes them ideal for single cell Li-Ion or 3-cell NiCd and NiMH applications. The 100% duty cycle capability for low dropout allows maximum energy to be extracted from the battery. In dropout, the output voltage is determined by the input voltage minus the voltage drop across the internal P-channel MOSFET and the inductor resistance. The fixed voltage output versions—the LTC3406-1.5/LTC3406B-1.5 and LTC3406-1.8/LTC3406B-1.8, are fixed to 1.5V and 1.8V, respectively—require no external voltage divider for feedback, further saving space and improving efficiency. The adjustable voltage output versions—the LTC3406 and LTC3406B—allow the output voltage to be externally programmed with two resistors to any value above the 0.6V internal reference voltage.

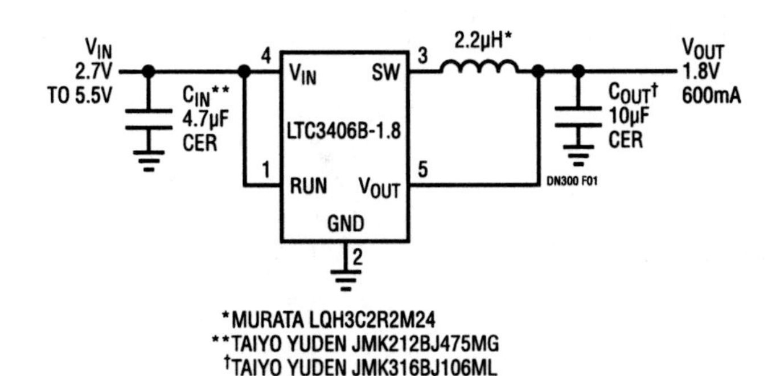

Figure 71.1 • 1.8V/600mA Step-Down Regulator Using All Ceramic Capacitors

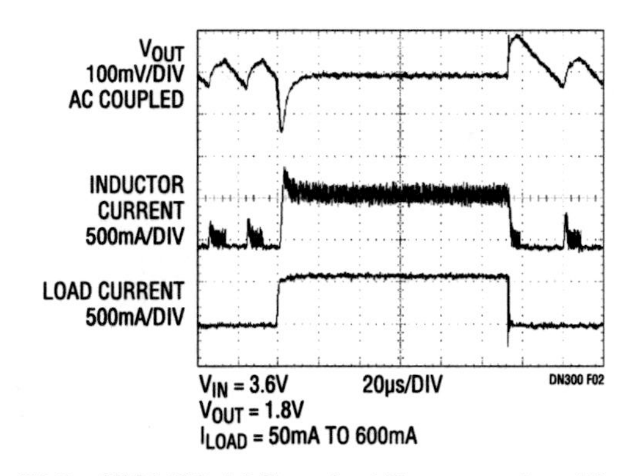

Figure 71.2 • LTC3406-1.8 Transient Response to a 50mA to 600mA Load Step (Burst Mode Operation at Light Load)

Analog Circuit Design: Design Note Collection. http://dx.doi.org/10.1016/B978-0-12-800001-4.00071-5

Fault protection

The LTC3406 and LTC3406B protect against output overvoltage, output short-circuit and power overdissipation conditions. When an overvoltage condition at the output (>6.25% above nominal) is sensed, the top MOSFET is turned off until the fault is removed. When the output is shorted to ground, the frequency of the oscillator slows to 210kHz to prevent inductor-current runaway. The frequency returns to 1.5MHz when V_{FB} is allowed to rise to 0.6V. When there is a power overdissipation condition and the junction temperature reaches approximately 160°C, the thermal protection circuit turns off the power MOSFETs allowing the part to cool. Normal operation resumes when the temperature drops to 150°C.

Efficient Burst Mode operation (LTC3406 series)

In Burst Mode operation, the internal power MOSFETs operate intermittently based on load demand (see Figure 71.2). Short burst cycles of normal switching are followed by longer idle periods where the load current is supplied by the output capacitor. During the idle period, the power MOSFETs and any unneeded circuitry are turned off, reducing the quiescent current to 20μA. At no load, the output capacitor discharges slowly through the feedback resistors resulting in very low frequency burst cycles that add only a few microamperes to the supply current.

Pulse-skipping mode (LTC3406B series) for low noise

Pulse-skipping mode lowers output ripple, thus reducing possible interference with audio circuitry. In pulse-skipping mode, constant frequency operation is maintained at lower load currents to lower the output voltage ripple. If the load current is low enough, cycle skipping eventually occurs to maintain regulation. Efficiency in pulse-skipping mode is lower than Burst Mode operation at light loads, but comparable to Burst Mode operation when the output load exceeds 50mA.

1.8V/600mA step-down regulator using all ceramic capacitors

Figure 71.1 shows an application of the LTC3406/LTC3406B-1.8 using all ceramic capacitors. This particular design supplies a 600mA load at 1.8V with an input supply between 2.5V and 5.5V. Ceramic capacitors have the advantages of small size and low equivalent series resistance (ESR), making possible very low ripple voltages at both the input and output. For a given package size or capacitance value, ceramic capacitors have lower ESR than other bulk, low ESR capacitor types (including tantalum capacitors, aluminum and organic electrolytics). Because the LTC3406/LTC3406B's control loop does not depend on the output capacitor's ESR for stable operation, ceramic capacitors can be used to achieve very low output ripple and small circuit size. Figures 71.2 and 71.3 show the transient response to a 50mA to 600mA load step for the LTC3406-1.8 and LTC3406B-1.8, respectively.

Efficiency considerations

Figure 71.4 shows the efficiency curves for the LTC3406-1.8 (Burst Mode operation enabled) at various supply voltages. Burst Mode operation significantly lowers the quiescent current, resulting in high efficiencies even with extremely light loads.

Figure 71.5 shows the efficiency curves for the LTC3406B-1.8 (pulse-skipping mode enabled) at various supply voltages. Pulse-skipping mode maintains constant-frequency operation at lower load currents. This necessarily increases the gate charge losses and switching losses, which impact efficiency at light loads. Efficiency is still comparable to Burst Mode operation at higher loads.

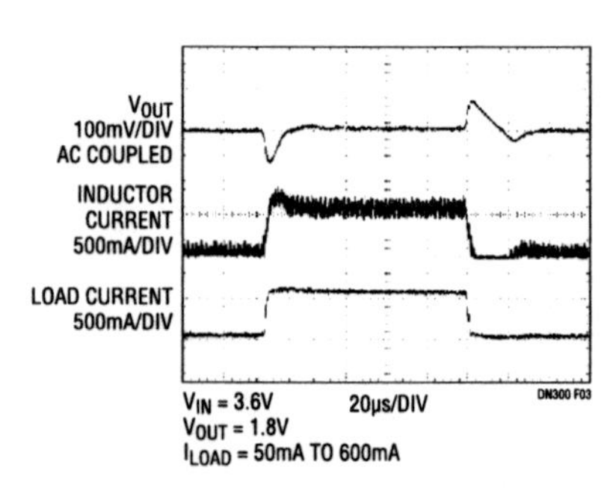

Figure 71.3 • LTC3406B-1.8 Transient Response to a 50mA to 600mA Load Step

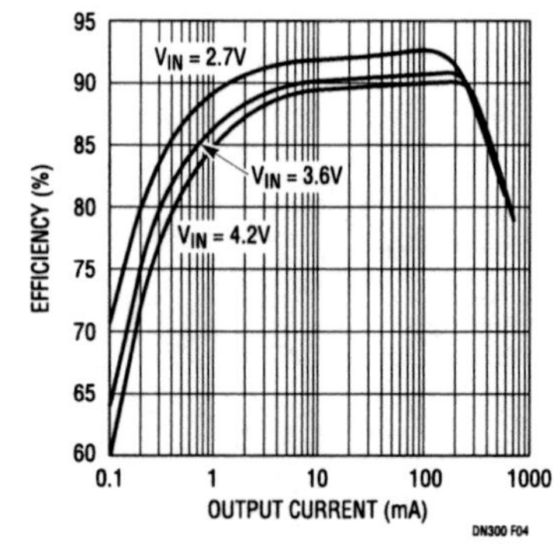

Figure 71.4 • Efficiency vs Load Current for LTC3406-1.8

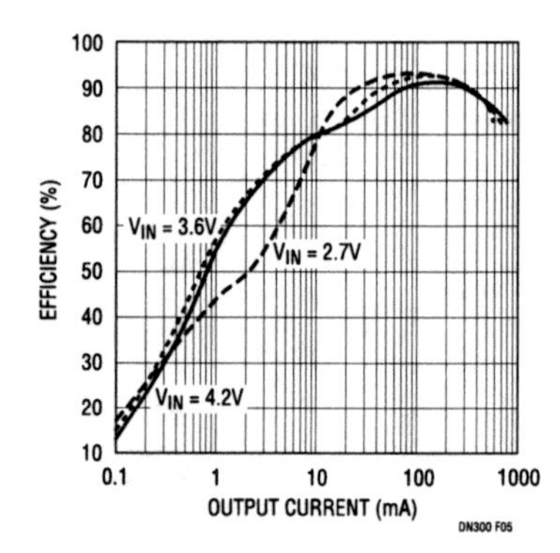

Figure 71.5 • Efficiency vs Load Current for LTC3406B-1.8

High efficiency adaptable power supply for XENPAK 10Gb/s Ethernet transceivers

72

Dongyan Zhou

Introduction

The XENPAK Multisource Agreement (MSA) defines a fiber optic module that conforms to the 10Gb/s Ethernet standard specified by IEEE 802.3ae. These modules require an adaptable power supply rail (APS) that has a variable output voltage from 0.9V to 1.8V, generated from a 3.3V input. The LTC1773 current mode synchronous buck regulator provides an efficient, cost-effective and space saving APS solution that exceeds the MSA standard as well as other similar standards.

The LTC1773 has features that make it a good match for this application including its 0.8V reference and 2.65V minimum input voltage. The reference is ±1.5% accurate over temperature which makes it easy to meet the ±4% APS tolerance specified in the MSA. The operating frequency is internally set at 550kHz, allowing the use of small surface mount components. Synchronous rectification increases efficiency and eliminates the need for an external Schottky diode, saving additional cost and space. Finally, the LTC1773 comes in a tiny 10-lead MSOP package, keeping the total board space for an entire APS solution under $0.4in^2$.

Adaptable power supply

Figure 72.1 shows an efficient APS design using the LTC1773. The APS SENSE pin provides for Kelvin sensing to compensate for PCB and connector voltage drops. The resistor R1, on the module side, programs the APS voltage. Table 72.1 lists R1 resistor values and corresponding APS voltages. Although Table 72.1 shows values for 1% resistors, higher accuracy resistors can be used to provide additional tolerance margin. In this design, the output voltage is set a little higher than specified to compensate for the ground plane drop.

The R4 pull-up resistor sets the output voltage at 0V when the APS SENSE lead is open. If the APS SENSE lead

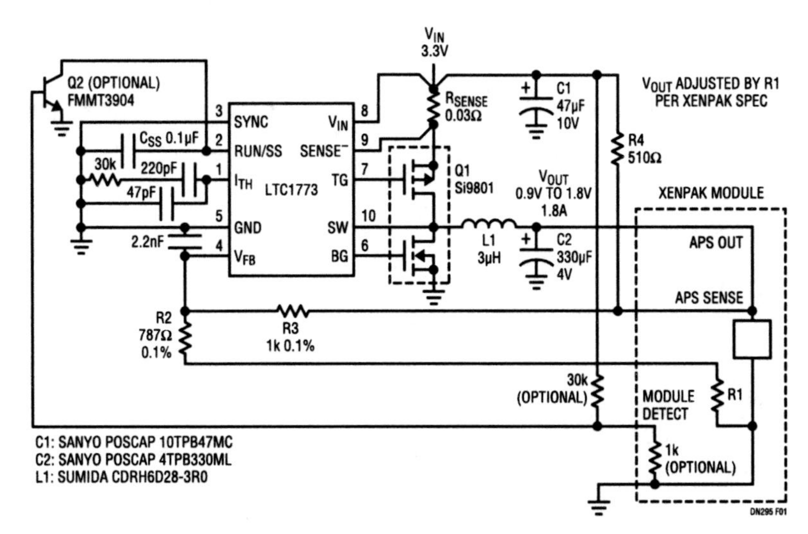

Figure 72.1 • Adaptable Power Supply Using the LTC1773

Analog Circuit Design: Design Note Collection. http://dx.doi.org/10.1016/B978-0-12-800001-4.00072-7

Table 72.1 APS Set Resistor Values

V_{NOM} (V)	R1 (Ω)
0.9	6810
1.0	3090
1.1	1820
1.2	1180
1.3	787
1.4	536
1.5	348
1.6	205
1.7	97.6
1.8	0

connects before the APS OUT and GND leads, the feedback pin of the LTC1773 is pulled low, resulting in an instantaneous high voltage from the APS. To prevent this, the APS OUT and GND leads are elongated in the XENPAK module pin description. If the APS SENSE lead is connected before the APS SET lead, the supply just defaults to the 0.8V feedback voltage. As an optional function, the MODULE DETECT pin can be used to turn on the APS. When the MODULE DETECT lead is open, the RUN/SS pin is pulled low by Q2 to shut down the converter. When the MODULE DETECT lead is connected, Q2 is turned off by the 1k pull-down resistor. The RUN/SS pin is then released to enable the converter. The LTC1773 also provides a soft-start function with a ramp rate controlled by the value of the soft-start capacitor C_{SS}. The sense resistor RSENSE sets an accurate current limit. The Si9801 complementary MOSFET saves cost and space.

Efficiency for this circuit peaks at 91% and stays above 78% across the output current and voltage range, as shown in Figure 72.2. Figure 72.3 shows the output voltage ripple measured at 15mV peak to peak (about 4.3mV_{RMS}) on the host side, which is well below the 40mV_{RMS} ripple requirement of the XENPAK MSA. The ripple measured at the input of the module is even lower due to the parasitic inductance and the input capacitor on the module.

Conclusion

The LTC1773 is a simple and efficient solution for XENPAK adaptable power supplies. A complete APS design requires few external components, saving space, cost and design time.

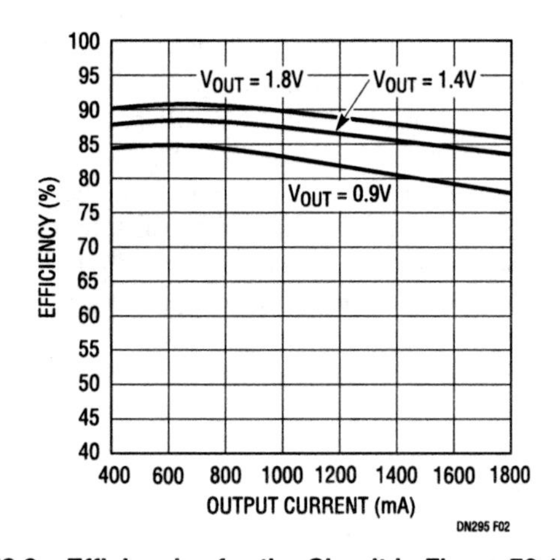

Figure 72.2 • Efficiencies for the Circuit in Figure 72.1 Where V_{IN}=3.3V

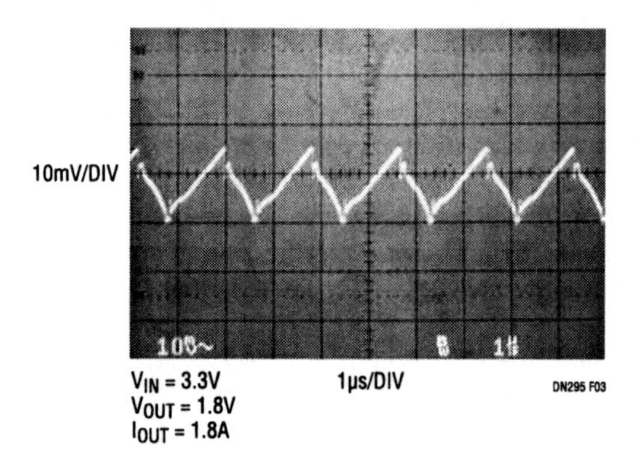

Figure 72.3 • Output Voltage Ripple for the Circuit in Figure 72.1

High voltage buck regulators provide high current, low profile power solutions for FireWire peripherals

73

Keith Szolusha

Introduction

Faster data transfer requirements between a personal computer and an increasing number of peripheral devices have boosted the popularity of the IEEE 1394 High Performance Serial Bus (FireWire technology), which enables speeds up to 400Mbps. FireWire technology is finding its way into real time data transfer and data intensive devices such as DVD players, digital video recorders, zip drives and CDRW drives. There is a growing need for cost-effective, low profile, low ripple DC/DC power supplies that provide 5V and 3.3V peripheral power directly from the FireWire cable at high currents (1A or greater). The LT1766 and LT3430 current mode regulators will satisfy these requirements.

Circuit descriptions

FireWire power sources, such as personal computers, provide 8V to 40V to the FireWire serial bus at currents of up to 1.5A and can power up to 16 peripherals while transferring data synchronously or asynchronously. Power sink peripherals, which do not supply power to the serial bus, receive power from the serial bus via the two power pins of the 6-pin FireWire port. Each peripheral must include an internal DC/DC converter to convert the bus's 8V to 40V input to a 5V or 3.3V output. Since many of these devices are portable consumer electronics, power consumption requirements are increasing while available space is decreasing. One way to decrease the total size and cost of a solution while keeping voltage ripple extremely low is to use ceramic input and output capacitors.

Figures 73.1 and 73.2 show two FireWire DC/DC converter solutions, both using ceramic input and output capacitors. Figure 73.1 shows a 3.3V, 1A output solution and Figure 73.2 shows a 5V, 2A output solution. Both have a 3.0mm maximum component height. The efficiencies and output voltage ripple for these two circuits are shown in Figures 73.3–73.5. The output voltage ripple is extremely low, less than $15mV_{P-P}$ for both solutions (see Figure 73.5). Higher output voltage yields overall higher efficiency since the forward voltage drop of the catch diode remains constant and is less significant.

The LT1766/LT3430 high voltage buck regulators used in these solutions provide high current for FireWire peripherals with several key advantages, such as 60V input capabilities,

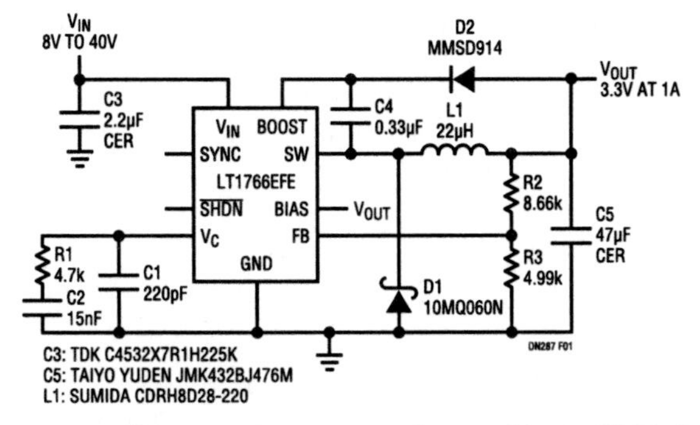

Figure 73.1 • This Low Component Count, 8V to 40V IN, 3.3V at 1A OUT DC/DC Converter Uses All Ceramic Capacitors (Max Height of 3.0mm)

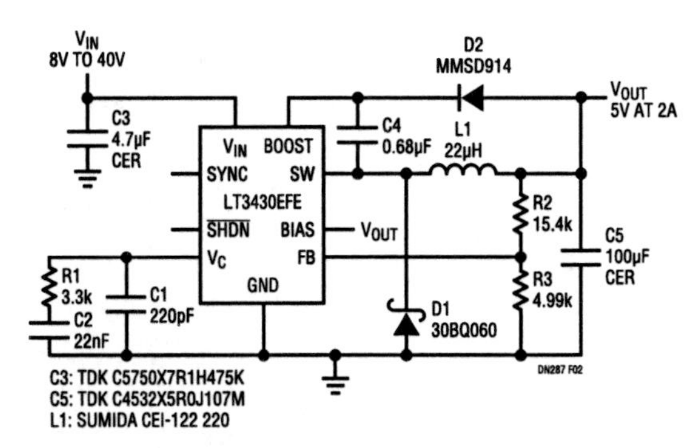

Figure 73.2 • This Low Component Count, 8V to 40V IN, 5V at 2A OUT DC/DC Converter Uses All Ceramic Capacitors (Max Height of 3.0mm)

Analog Circuit Design: Design Note Collection. http://dx.doi.org/10.1016/B978-0-12-800001-4.00073-9

200kHz operating frequency, high power internal 1.5A/3A power switch current limits, shutdown capability and thermally enhanced 16-pin TSSOP exposed lead-frame packages.

The external location of both the feedback and compensation components, combined with the current mode architecture of the LT1766 and LT3430 makes it possible to provide simple compensation solutions using ceramic output capacitors. The circuits in Figures 73.1 and 73.2 use a 3-element compensation scheme. Note that these designs are not limited to the input and output voltages and currents shown here. The transient response can be optimized for other input and output values by simple adjustments to the compensation network.

Current mode regulators have major advantages over voltage mode regulators which require higher ESR output capacitors or are difficult to design and have higher parts count compensation schemes. With a voltage mode buck regulator, it can be difficult to get proper phase and gain margins using ceramic output capacitors due to the capacitors' low ESR and lack of ripple voltage at the feedback node—a necessary evil for voltage mode controllers. Nonceramic, higher ESR capacitors supply the necessary increase in ripple voltage at the feedback node, but also have the undesirable side effect of increasing ripple at the output as well. Some manufacturers promote internal feedback or complicated compensation schemes that seem reasonable on the surface, but severely limit flexibility in choosing output capacitor sizes and types, input voltage ranges, and optimization of transient response and loop stability.

Conclusion

The LT1766 and LT3430 regulators provide simple, low profile power supply solutions for FireWire peripherals. They allow the use of ceramic input and output capacitors which minimizes the size, cost and ripple voltage of the supply while maintaining high efficiency and simplicity of design.

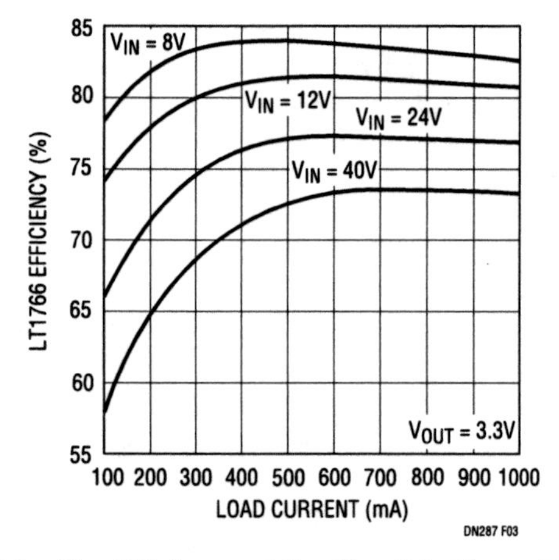

Figure 73.3 • The Efficiency of the Circuit in Figure 73.1 Is as High as 84% and Increases as Input Voltage Decreases

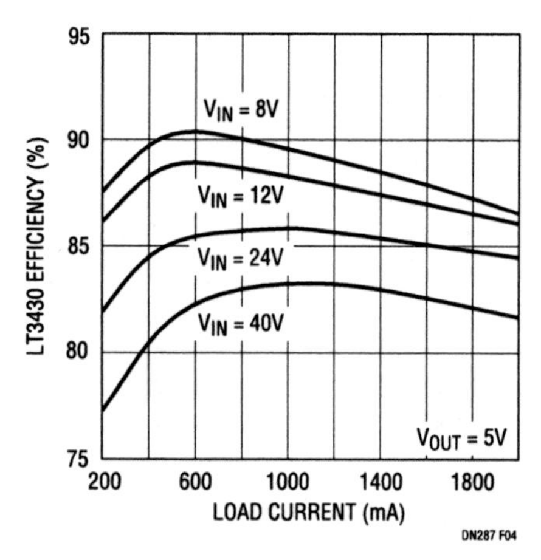

Figure 73.4 • The Efficiency of the Circuit in Figure 73.2 Is as High as 90%

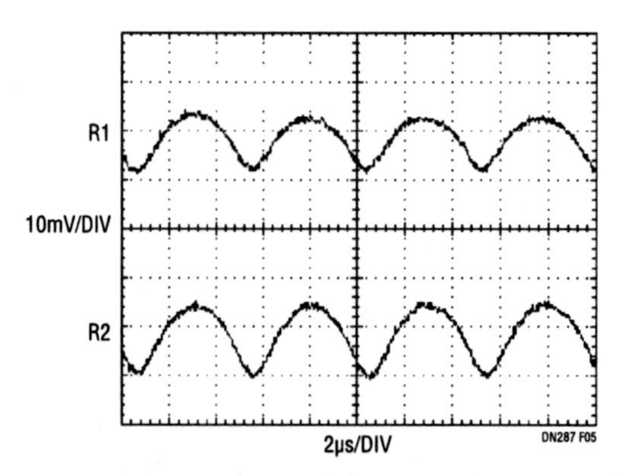

Figure 73.5 • Trace R1 Output Voltage Ripple of Figure 73.1 with 24V IN, 3.3V OUT, 1A OUT. Trace R2 Output Voltage Ripple of Figure 73.2 with 24V IN, 5V OUT, 2A OUT

Efficient DC/DC converter provides two 15A outputs from a 3.3V backplane

74

David Chen

Introduction

The 3.3V DC bus has become popular for broadband networking systems, where it is tapped for a variety of lower voltages to power DSPs, ASICs and FPGAs. These lower voltages range from 1V to 2.5V and often require high load currents. To maintain high conversion efficiency, power MOSFET conduction losses from the step-down converters must be minimized. The problem is that the 3.3V bus also brings with it frequent use of sub-logic level MOSFETs. Such MOSFETs have a relatively high $R_{DS(ON)}$, limiting the full-load efficiency of a converter to around 85%. A more efficient solution is to use logic-level MOSFETs, which have very low $R_{DS(ON)}$ but require a 5V supply. The LTC1876 allows the use of logic-level MOSFETs by combining a 1.2MHz boost regulator, which produces a 5V bias supply from a 3.3V input, with two step-down controllers, which provide the low voltage outputs. By integrating all three regulators in a single IC, the LTC1876 makes for efficient power supplies that can be small and inexpensive.

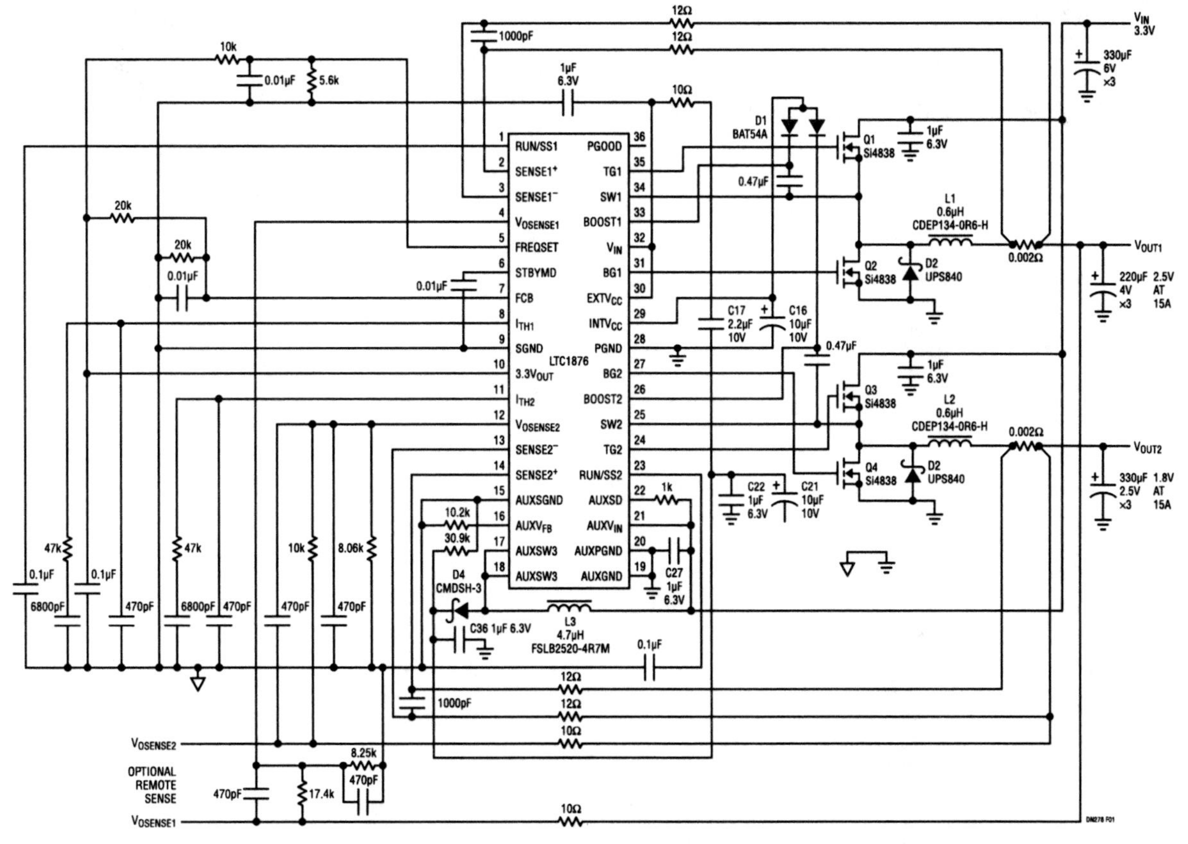

Figure 74.1 • An LTC1876 Design Converts 3.3V to 2.5V at 15A and 1.8V at 15A

Analog Circuit Design: Design Note Collection. http://dx.doi.org/10.1016/B978-0-12-800001-4.00074-0

Design example

Figure 74.1 shows a design that provides 2.5V/15A and 1.8V/15A from a 3.3V input. Because the LTC1876 provides a 5V bias for MOSFET gate drive, a very low $R_{DS(ON)}$ MOSFET Si4838 (2.4mΩ typical) can be used to achieve high efficiency. Figure 74.2 shows that the overall efficiency is above 90% over a wide range of loads.

Figure 74.2 also shows that the light load efficiency of this design is more than 84%. This is a direct benefit of the Burst Mode operation of the LTC1876. Further efficiency improvements come from operating the two step-down channels out-of-phase. The top MOSFET of the first channel is fired 180° out of phase from that of the second channel, thus minimizing the RMS current through the input capacitors. This significantly reduces the power loss associated with the ESR of input capacitors. Figure 74.3 shows detailed current waveforms of this operation.

Conclusion

The LTC1876 uses three techniques to efficiently power low voltage DSPs, ASICs and FPGAs from a low input voltage. The first technique uses an internal boost regulator to provide a separate 5V for the MOSFET gate drive. Secondly, its Burst Mode operation achieves high efficiency at light loads. Lastly is the out-of-phase technique which minimizes input RMS losses and reduces input noise. Complete regulator circuits are kept small and inexpensive, because all three switchers (one step-up regulator and two step-down controllers) are integrated into a single IC. For systems where a separate 5V is available or the input supply is greater than 5V, the internal boost regulator can be used to provide a third step-up output with up to 1A switch current.

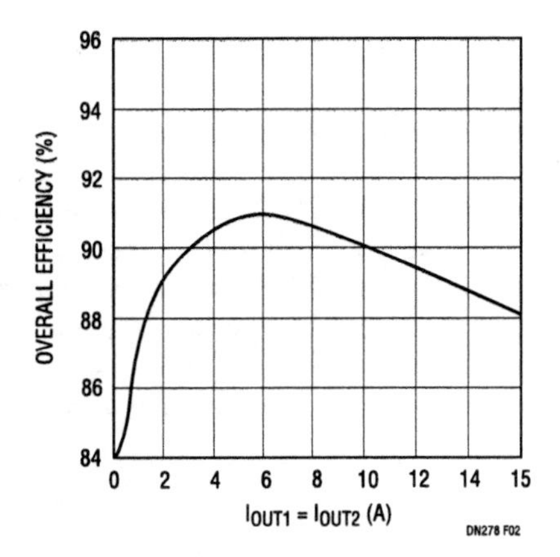

Figure 74.2 • High Efficiency of the Design in Figure 74.1

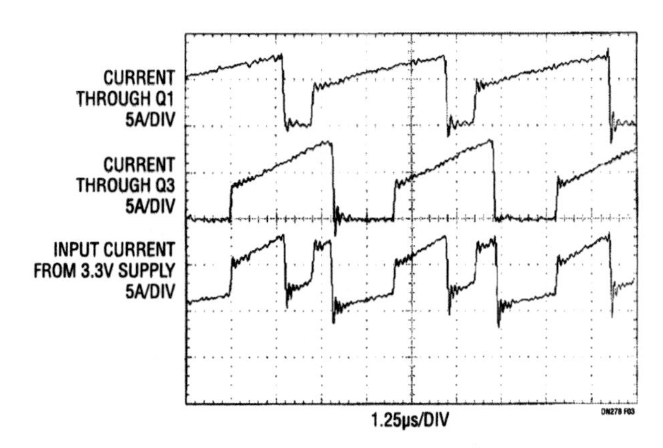

Figure 74.3 • Each Switcher Has 5A Peak Current but the Total Ripple at the Input Is Still Only 5A, Minimizing C_{IN} Requirements

60V step-down DC/DC converter maintains high efficiency

75

Mark W. Marosek

Introduction

Monolithic step-down converters capable of operation at high input voltages are usually optimized for efficiency at high input-to-output voltage differentials. At such low duty cycle operation where DC switch losses are less critical, the switch design is often neglected, resulting in a switch resistance that (for some 1.5A converters) can be as poor as 1Ω. Such converters sacrifice efficiency when lower input-to-output voltage differentials are required. The switch drop can also limit maximum duty cycle—putting a limit on the minimum input voltage for a given regulated output voltage.

The LT1766 is designed to optimize efficiency for both high and low input-to-output voltage differentials to support a wide input voltage range. In addition, the current mode topology used to provide fast transient response and good loop stability does not suffer from peak switch current fall off at low input-to-output voltage differentials, commonplace in most current mode converters.

The LT1766 is a 1.5A monolithic buck switching regulator. The 5.5V to 60V input voltage range makes the LT1766 ideal for 48V nonisolated telecom applications as well as 12V, 24V and (future) 42V automotive applications. These systems must survive load-dump input transients as high as 60V. Running at a fixed frequency of 200kHz, the LT1766 can be externally synchronized to clock frequencies up to 700kHz. A shutdown pin provides an accurate 2.38V undervoltage lockout threshold in addition to a 0.4V threshold for micropower shutdown (25μA). The LT1766 is provided in a small 16-pin SSOP (GN16) package with fused corner pins to improve thermal performance.

Efficiency

A typical high input voltage application, a 42V to 5V converter, is shown in Figure 75.1. To achieve high efficiency at high input voltages, fast output-switch edge rates are required; the LT1766 achieves edge rates of 1.2V/ns (rise)

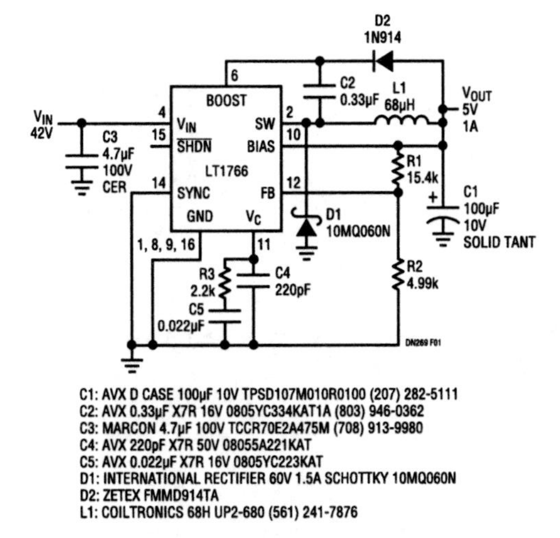

Figure 75.1 • 42V to 5V Step-Down Converter

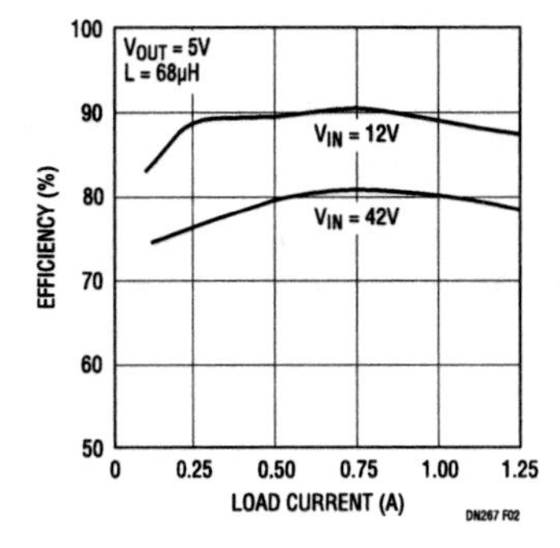

Figure 75.2 • LT1766 Efficiency

Analog Circuit Design: Design Note Collection. http://dx.doi.org/10.1016/B978-0-12-800001-4.00075-2

and 1.7V/ns (fall). In addition, light loads at high input voltages require minimal quiescent current to be drawn from the input. A BIAS pin allows the internal control circuitry to be supplied from the regulated output if greater than 3V. The peak efficiency for a 42V to 5V conversion is >80%.

The LT1766 is also capable of excellent efficiencies at lower input voltages. The peak efficiency for a 12V-to-5V conversion is >90% (Figure 75.2). One key to achieving high efficiency for low input-to-output voltage conversions is to use a low resistance saturating switch. A prebiased capacitor, connected between the BOOST and SW pins, generates a boost voltage above the input supply during switching. Driving the switch from this boost voltage allows the 200mΩ power switch to fully saturate. Furthermore, an output voltage as low as 3.3V is capable of generating the required boost supply (Figure 75.3).

Output ripple voltage

The output ripple voltage for the circuit in Figure 75.1, using a tantalum output capacitor, is approximately $35mV_{P-P}$ (Figure 75.4). Peak-to-peak output ripple voltage is the sum of a triwave (created by peak-to-peak ripple current in the inductor times the ESR of the output capacitor) and a square wave (created by the parasitic inductance (ESL) of the output capacitor times ripple current slew rate). A significant reduction in output ripple voltage to $12mV_{P-P}$ can be achieved using a ceramic output capacitor (Figure 75.4). With negligible ESR, the ceramic output capacitor reduces the portion of output ripple voltage generated by inductor ripple current times capacitor ESR. The useful feedback response zero provided by the tantalum output capacitor ESR for loop stabilization is now replaced by a capacitor inserted across R1 in the feedback resistor network.

Peak switch current

The LT1766 maintains peak switch current over the full duty cycle range. Although the LT1766 uses a current mode architecture to provide fast transient response and good loop stability, the LT1766 peak switch current does not fall off at high duty cycles, unlike most current mode converters. The fall off of peak switch current in most current mode converters is due to the addition of slope compensation to the current sensing loop of the converter in order to prevent subharmonic oscillations for duty cycles above 50%. The LT1766 uses patented circuitry to cancel the effect of slope compensation on peak switch current without affecting frequency compensation. For high duty cycle requirements, this is a significant benefit over typical current mode converters with similar peak switch current limits.

LT1766 features

- Wide Input Range: 5.5V to 60V
- 1.5A Peak Switch Current
- Small 16-Pin SSOP Package
- Constant 200kHz Switching Frequency
- 0.2Ω Saturating Switch
- Peak Switch Current Maintained Over Full Duty Cycle Range
- 25μA Shutdown Current
- 1.2V Feedback Reference
- Easily Synchronizable Up to 700kHz

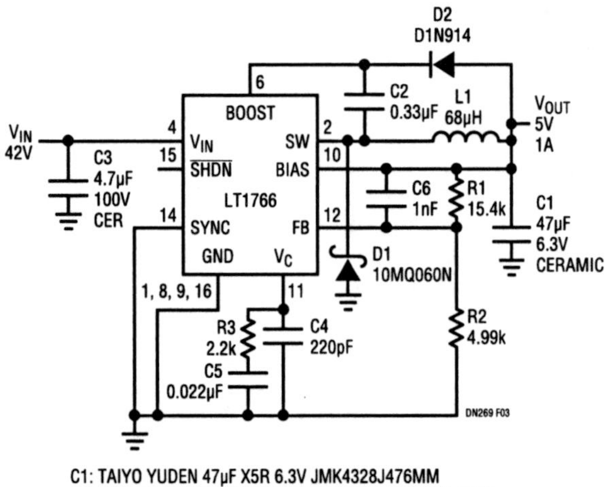

C1: TAIYO YUDEN 47μF X5R 6.3V JMK4328J476MM
C2: AVX 0.33μF X7R 16V 0805YC334KAT1A (803) 946-0362
C3: MARCON 4.7μF 100V TCCR70E2A475M (708) 913-9980
C4: AVX 220pF X7R 50V 08055A221KAT
C5: AVX 0.022F X7R 16V 0805YC223KAT
C6: AVX 1000pF X7R 50V 08055C102KAT
D1: INTERNATIONAL RECTIFIER 60V 1.5A SCHOTTKY 10MQ060N
D2: ZETEX FMMD914TA
L1: COILTRONICS 68H UP2-680 (561) 241-7876

Figure 75.3 • 42V to 5V (All Ceramic) Step-Down Converter with Low Output Ripple Voltage

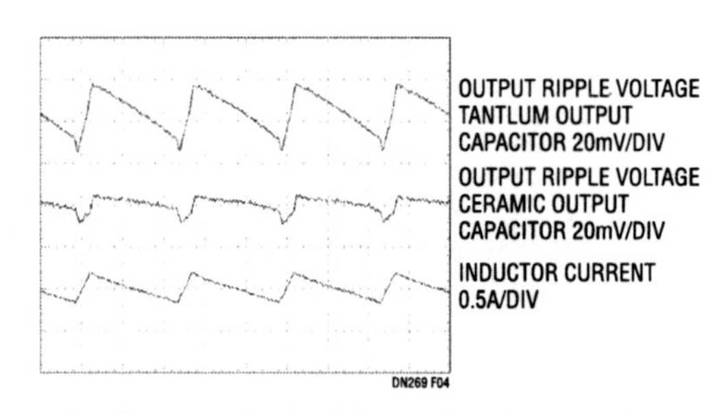

Figure 75.4 • Output Ripple Voltage Comparison (Tantalum vs Ceramic Output Capacitor)

Tiny buck regulator accepts inputs from 3.6V to 25V and eliminates heat sink

76

Jeff Witt

Introduction

The LT1616 is a complete fixed-frequency step-down switching regulator in a ThinSOT (1mm thick SOT-23) package. It meets the needs of circuit designers who require a large input voltage range or the smallest solution possible. The LT1616 accepts an input from 3.6V to 25V, produces a low voltage output at 400mA and occupies less than 0.15in^2 of board space. With this wide input range, the LT1616 can regulate a large variety of power sources, from 4-cell alkaline batteries to lead-acid automobile batteries, from 5V logic supplies to unregulated AC adapters. The LT1616 is an ideal replacement for bulky (and potentially hot) TO-220 linear regulators.

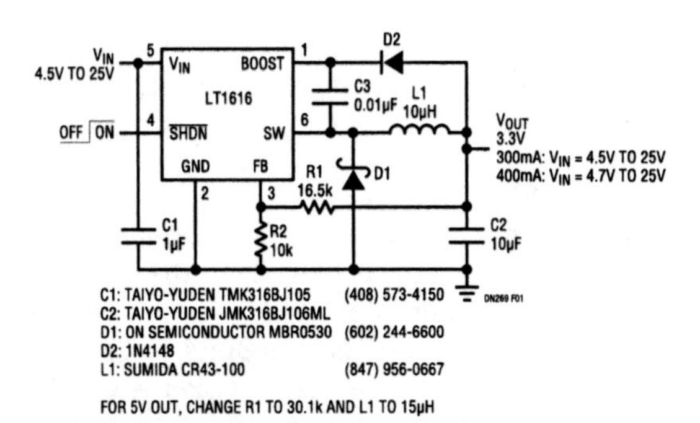

Figure 76.1 • The LT1616 Application Accepts an Input from 4.5V to 25V and Produces an Output of 3.3V at up to 400mA. The Circuit Is Easily Modified for 5V Output

Complete switcher in ThinSOT results in compact solution

Several features of the LT1616 enable this combination of small size and large voltage range. The high (1.4MHz) switching frequency allows the use of small inductors and capacitors. The current mode control circuit with its internal loop compensation eliminates additional components and handles a wide variety of output capacitors, including ceramic capacitors. The internal NPN power switch drops just 200mV at 300mA.

An external resistor divider programs the output voltage to any value above the part's 1.25V reference. The shutdown mode reduces the supply current to 1µA and disconnects the load from the input supply.

An internal 3.4V undervoltage lockout prevents switching at low input supply. The LT1616 will also withstand a shorted output. A fast current limit protects the circuit in overload and limits output power; when the output voltage is pulled to ground by a hard short, the LT1616 reduces its operating frequency to limit dissipation and peak switch current.

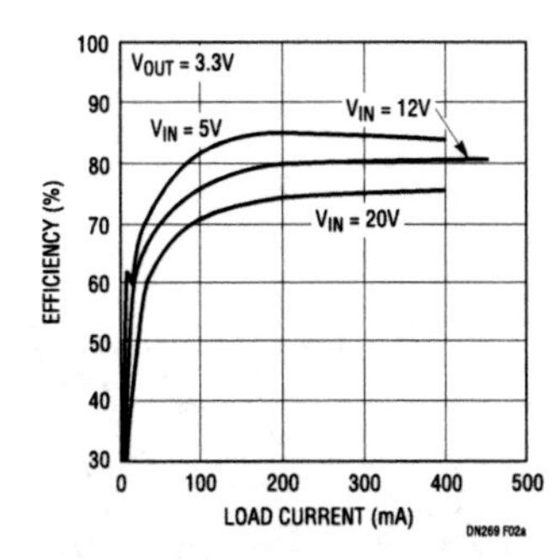

Figure 76.2a • Efficiency of Figure 76.1's Circuit, Output = 3.3V

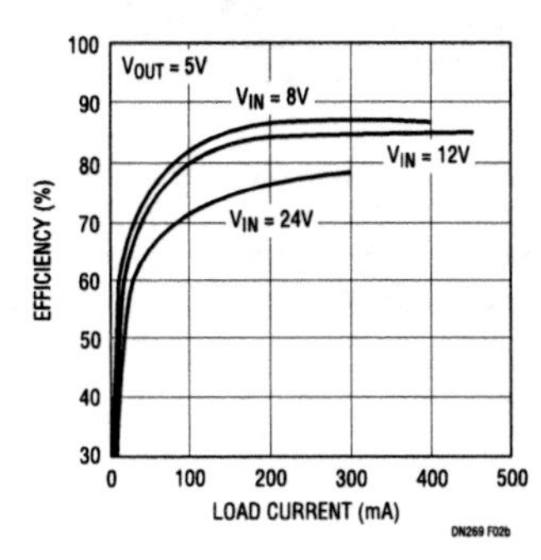

Figure 76.2b • Efficiency of Figure 76.1's Circuit, Output = 5V

Analog Circuit Design: Design Note Collection. http://dx.doi.org/10.1016/B978-0-12-800001-4.00076-4

The LT1616 produces 3.3V at 400mA

Figure 76.1 shows a typical application of the LT1616. This circuit generates 3.3V at 300mA from an input of 4.5V to 25V. From a slightly more restricted input range of 4.7V to 25V, it will supply 400mA to the load. Figure 76.2 shows the circuit's operating efficiency at several input voltages (it also shows the efficiency for a 5V output). This wide input range allows you to generate a local 3.3V logic supply from just about any source available.

Ceramic capacitors are best

The LT1616's ability to work with ceramic capacitors is a significant advantage. Where achieving low output ripple from a switching regulator is concerned, low equivalent series resistance (ESR) is the most important characteristic of a capacitor. For a given package size or capacitance value, a ceramic capacitor will have lower ESR than other bulk, low ESR capacitor types (including tantalum, aluminum and organic electrolytics). With its high switching frequency, the LT1616 requires less than 10μF of capacitance at the output. At this value, ceramics are both smaller and lower in cost than the competing low ESR capacitors.

To summarize, using ceramics results in low noise outputs and a small circuit size. Figure 76.3 shows the good transient response of the circuit in Figure 76.1. The output recovers from a load current step in less than 30μs, without ringing. Because the time scale of 50μs per division is much longer than the LT1616's switching period, the output ripple at the switching frequency is not directly visible. The ripple appears as a slight broadening of the upper trace and amounts to just $5mV_{P-P}$.

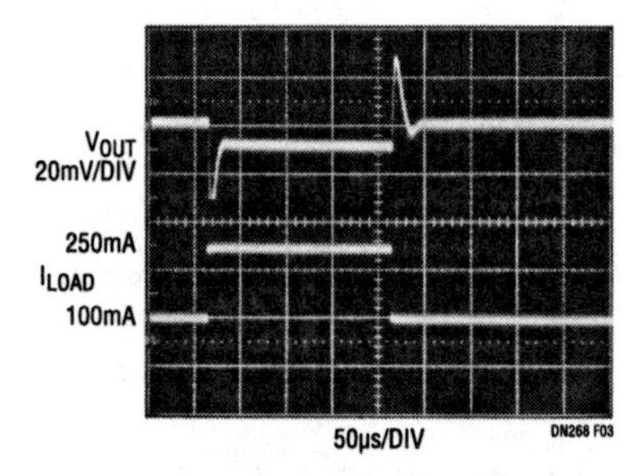

Figure 76.3 • The LT1616 Gets Along Fine with Ceramic Capacitors, Resulting in Good Transient Response and Low Output Ripple (~$5mV_{P-P}$). The Upper Trace Shows Output Voltage during a Stepped Load Current (Circuit of Figure 76.1 with V_{IN} = 10V)

Smaller than a TO-220

The small package size and high operating frequency of the LT1616 results in a very small circuit size. In most applications, the LT1616 circuit will occupy less space than a linear regulator performing the same task and will dissipate much less power. For example, an LT1616 circuit converting 12V to 3.3V at 300mA dissipates only 250mW. A linear regulator will dissipate 2.6W, requiring a TO-220 style package and either moving air or a heat sink to get rid of the heat. Figure 76.4 compares the size of the LT1616 solution with a TO-220 package. The circuit on the left is designed for a maximum input of 16V and an output of 350mA. The circuit on the right is designed for a maximum input of 25V (requiring a physically larger input capacitor) and uses a larger inductor to keep the efficiency high at its maximum load current of 400mA. Both circuits are low profile, with a maximum height of 2.2mm for the lower cost circuit on the left and 2mm for the circuit on the right.

Figure 76.4 • Tired of the Heat and Bulk of Linear Regulators? Switch! The Entire LT1616 Circuit Occupies Less Space than a TO-220

2.5V output

Figure 76.5 shows a 2.5V output circuit using the LT1616. The input range is limited on the low end by the undervoltage lockout (3.6V max) and on the high end by the voltage rating of the capacitors used and the maximum voltage rating of the BOOST pin.

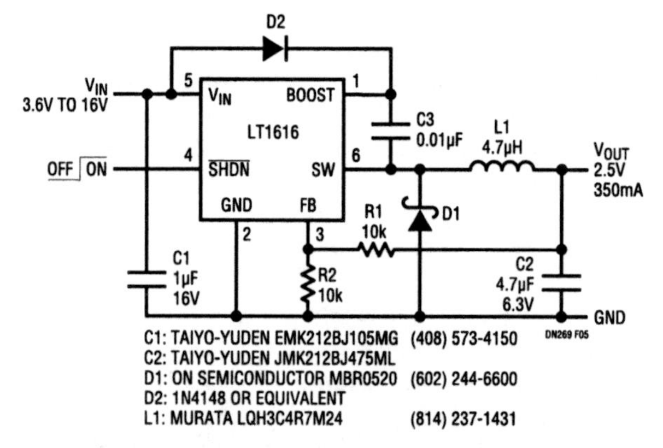

Figure 76.5 • This Circuit Produces 2.5V at 350mA from an Input Range of 3.6V to 16V

77

1.4MHz switching regulator draws only 10μA supply current

Jaime Tseng

Introduction

High switching frequency and low quiescent current are no longer conflicting requirements in the design of battery-powered products. Linear Technology's LTC3404 is the industry's first step-down switching regulator that operates at 1.4MHz while drawing only 10μA of supply current (using Burst Mode operation) at no load. This impressive feat allows for better than 90% efficiency over three decades of output load current while allowing the use of tiny external components. With the on-chip main and synchronous switches, minimal external components are necessary to make a complete, high efficiency (up to 95%) step-down regulator. Tiny external components and the LTC3404's MSOP package provide a minimum-area solution to meet the limited space requirements of today's portable applications.

LTC3404 features

The LTC3404 incorporates a constant frequency, current mode architecture that provides low noise and fast transient response. Its input voltage supply range of 2.65V to 6V and 100% duty cycle capability for low dropout make the LTC3404 ideal for moderate current (up to 600mA) battery-powered applications.

For maximum efficiency over the widest range of output load current, Burst Mode operation can be selected by driving SYNC/MODE pin HIGH with a logic-level signal or by tying it to VIN. For lower noise, pulse-skipping mode can be selected by driving the SYNC/MODE pin LOW with a logic-level signal or by tying it to ground. In this case, constant frequency operation is maintained at lower load currents together with minimum output ripple. If the load current is low enough, cycle skipping will eventually occur to maintain regulation. In this mode, the efficiency will be lower at very light loads, but becomes comparable to Burst Mode operation when the output load exceeds 50mA. For switching-frequency-sensitive applications, the LTC3404 can be externally synchronized to frequencies from 1MHz to 1.7MHz by applying an external clock signal to the SYNC/MODE pin. During synchronization, Burst Mode operation is inhibited and pulse-skipping mode is selected.

3.1V/600mA step-down regulator

Figure 77.1 shows a typical application suitable for a single Li-Ion cell or 3- to 4-cell NiCd or NiMH battery input. Note the small component values used in this application, made

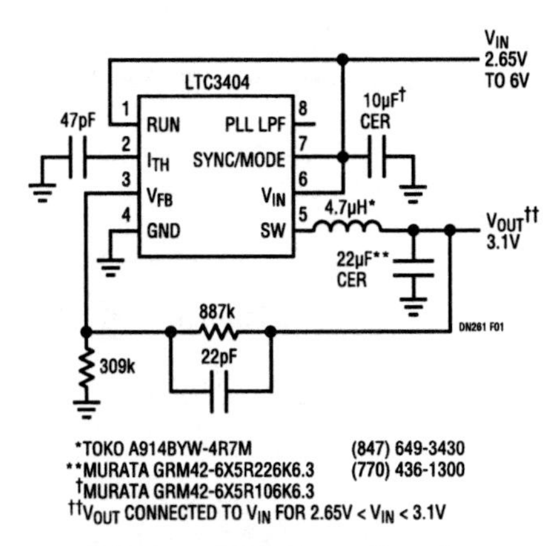

Figure 77.1 • 3.1V/600mA Step-Down Regulator

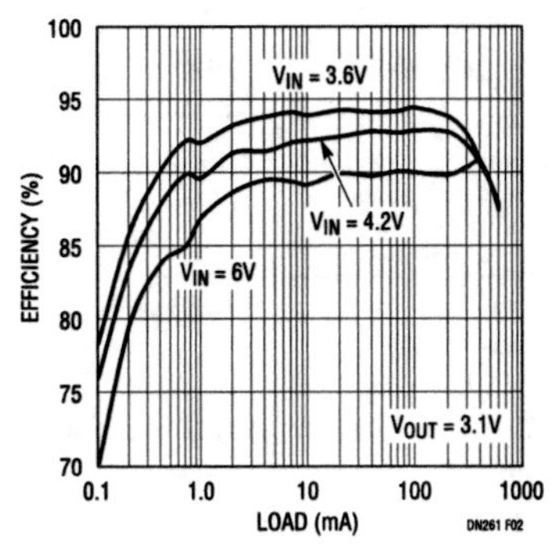

Figure 77.2 • Efficiency vs Load Current for Figure 77.1's Circuit (Burst Mode Operation Enabled)

Analog Circuit Design: Design Note Collection. http://dx.doi.org/10.1016/B978-0-12-800001-4.00077-6

possible by the high switching frequency of the LTC3404. Also, because of the part's internal synchronous switch, the Schottky diode normally seen on the SW pin is absent from Figure 77.1. This regulator occupies only 0.47″ × 0.31″ (0.146in²) of board area.

Figure 77.2 shows the efficiency for three different input voltages. The efficiency for a 3.6V input exceeds 90% over three decades of output current. The efficiency remains high down to loads as small as 100μA due to the LTC3404's ultralow quiescent current. Even though the quiescent current is very low, the transient performance is not compromised. Innovative new circuitry ensures that the error amplifier, while operating on less than 10μA at no load, can quickly respond to load changes. The oscilloscope photo in Figure 77.3 shows the part's outstanding transient performance when subjected to a 600mA load step. Note how the inductor current ripple is reduced as the load step forces continuous switching operation from Burst Mode operation. At maximum output load, the inductor current ripple is at its lowest due to the increased duty cycle caused by the IR drop of the main switch.

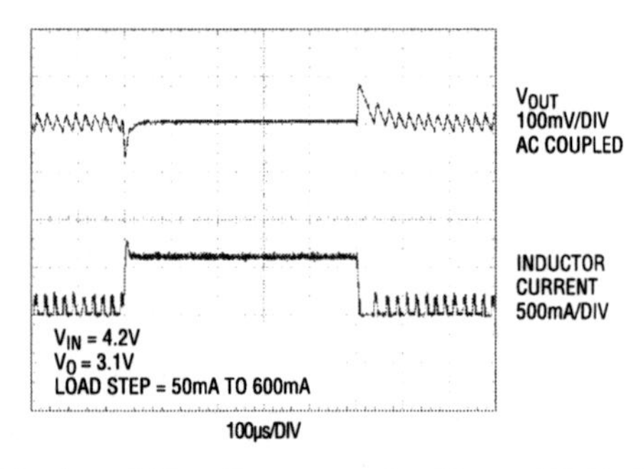

Figure 77.3 • Load Step Response for Figure 77.1's Circuit

Externally synchronized 3.1V/600mA step-down regulator

Figure 77.4 shows an application for low switching-frequency noise. The LTC3404 is synchronized to an external clock signal whereby Burst Mode operation is disabled automatically.

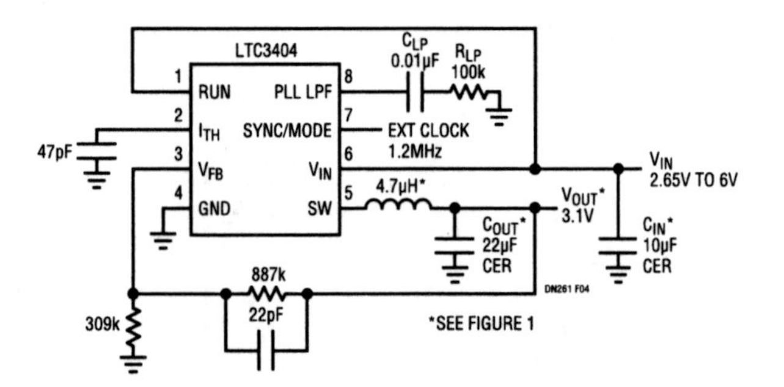

Figure 77.4 • Externally Synchronized 3.1V/600mA Step-Down Regulator

This provides constant frequency operation at lower load currents, reducing the ripple voltage. In this mode the efficiency is lower at light loads, as shown in Figure 77.5. However, the efficiency becomes comparable to that of Burst Mode operation when the output load exceeds 50mA.

The LTC3404 uses an internal phase-locked loop circuit to synchronize to an external signal. A voltage-controlled oscillator and a phase detector comprise the phase-locked loop. Filter components C_{LP} and R_{LP} smooth out the current pulses from the phase detector to provide a stable input to the voltage controlled oscillator. These components determine how fast the loop acquires lock. With the components shown in Figure 77.4, the loop acquires lock in about 100μs. When not synchronized to an external clock, the internal connection to the VCO is disconnected to prevent noise from altering the internal oscillator frequency. The oscilloscope photo in Figure 77.6 shows the transient performance with a 600mA load step.

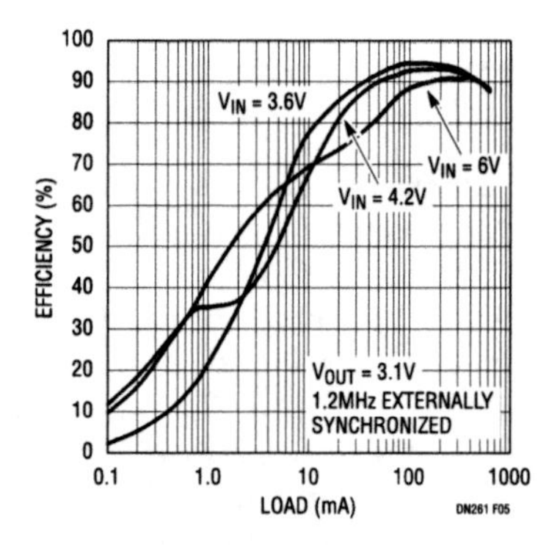

Figure 77.5 • Efficiency vs Load Current for Figure 77.4's Circuit (Burst Mode Operation Disabled)

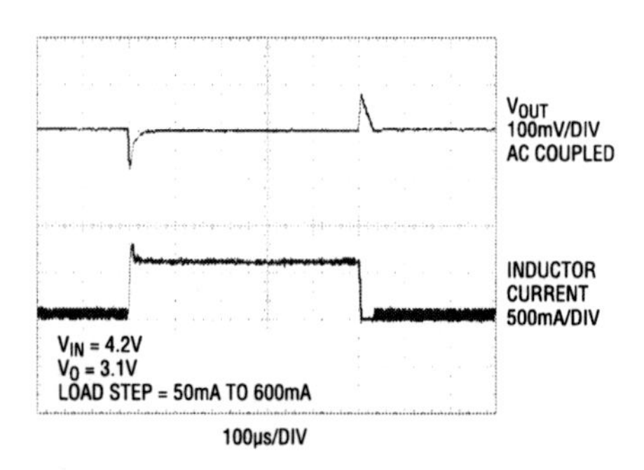

Figure 77.6 • Load Step Response for Figure 77.6's Circuit

Conclusion

The LTC3404 demonstrates that high switching frequency and low quiescent current can coexist, and thereby opens up a world of new possibilities in the design of battery-powered products.

10μA quiescent current step-down regulators extend standby time in handheld products

78

Greg Dittmer

Importance of low quiescent current

Many handheld products on the market today are used only occasionally but must be kept alive and ready all the time. When not being used, the circuitry is powered down to save battery energy, with a minimum amount of circuitry remaining on. Although the supply current is significantly reduced in this low power standby mode, the battery energy will still be slowly depleted to power the keep-alive circuitry and the regulator. If the device spends most of its time in this standby mode, the quiescent current of the regulator can have a significant effect on the life of the battery (see Figure 78.1).

To maximize the life of the battery in these types of products, Linear Technology has extended its family of 10μA step-down regulators to provide three new products. The LTC1771 is a constant off-time controller that drives an external P-channel MOSFET for output loads up to 5A. The LTC1877 and LTC1878 are monolithic regulators that provide constant frequency (550kHz) plus synchronous operation for loads up to 600mA. Due to their micropower architecture, the LTC1771, LTC1877 and LTC1878 require only 10μA of supply current to regulate their output at no load. These converters also feature micropower shutdown, current mode operation for excellent transient response and start-up behavior, short-circuit protection, 100% duty cycle for low dropout, availability in the tiny MSOP package and Burst Mode operation disable for low noise applications. The features of each of these parts are summarized in Table 78.1.

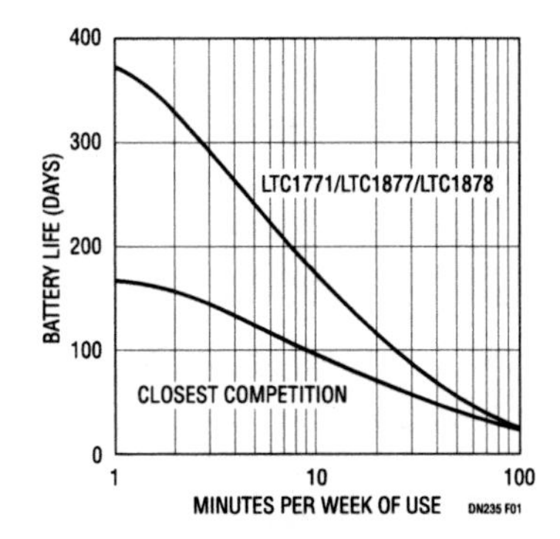

Figure 78.1 • 9V Battery Life Comparison for Load Requiring 100mA Operating and 100μA Standby Current at 3.3V

Table 78.1 Summary of Features of the LTC1771, LTC1877 and LTC1878

	LTC1771	LTC1877	LTC1878
No Load I_Q	10μA	10μA	10μA
Architecture	Controller	Monolithic	Monolithic
Maximum Load	Up to 5A (Programmable)	600mA (Fixed)	600mA(Fixed)
Input Supply Range	2.8V to 20V	2.65V to 10V	2.65V to 6V
Shutdown Current	2μA	<1μA	<1μA
Max Duty Cycle	100%	100%	100%
Package	MS8, SO-8	MS8	MS8
Frequency	3.5μs Off Time	550kHz	550kHz
Synchronous Rect	No	Yes	Yes
Synchronizable	No	Yes	Yes

LTC1878 single Li-Ion to 2.5V regulator

Figure 78.2 shows an application for converting a single cell Li-Ion to 2.5V at 600mA using the LTC1878. The internal synchronous switch, along with its 550kHz constant frequency operation, eliminates the need for an external Schottky diode and allows the use of small surface mount

Analog Circuit Design: Design Note Collection. http://dx.doi.org/10.1016/B978-0-12-800001-4.00078-8

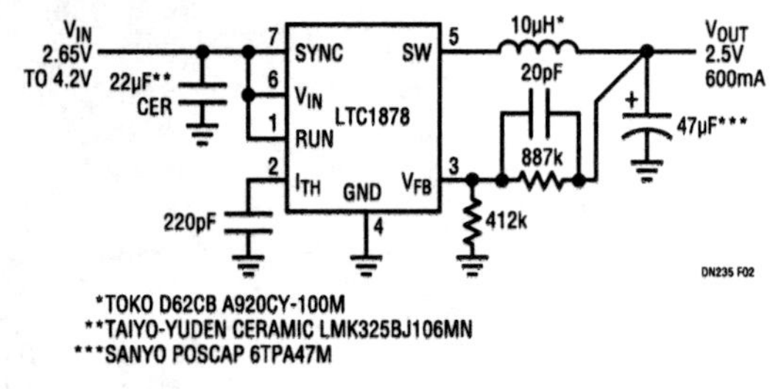

Figure 78.2 • LTC1878 2.5V/500mA Regulator

inductors and capacitors, saving external components and space. For higher input voltages such as dual Li-Ion applications, the LTC1877 can be used. The efficiency curves for this regulator are shown in Figure 78.3 and due to its ultralow quiescent current, the regulator provides outstanding efficiency down to loads as little as 100µA.

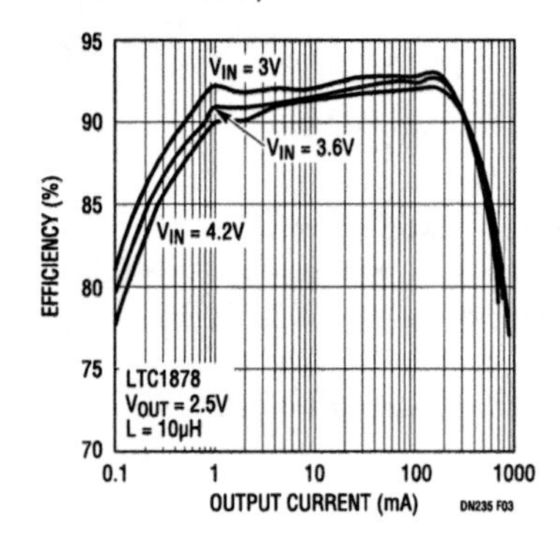

Figure 78.3 • Efficiency vs Load Current for Figure 78.2's Circuit

LTC1771 3.3V/2A regulator

For higher load currents or higher supply voltages with similar outstanding light load efficiency, the circuit shown in Figure 78.4 can be used. This circuit uses the LTC1771 controller which, when used with the appropriately sized external P-channel MOSFET, can provide output loads up to 5A. Due to the wide operating voltage of the LTC1771, the 3.3V/2A regulator shown in Figure 78.4 can operate with an input supply up to 18V. The 2A maximum load current is programmed with the 0.05Ω sense resistor. With a 15µH inductor and low ESR POSCAP output capacitor, the output ripple is less than 50mV. The Siliconix Si6447 or Si3443 are good MOSFET choices for the appropriate supply range due to their good compromise between low gate charge and low RDS(ON), while the Microsemi Powermite UPS5817 Schottky diode provides a good compromise between forward drop and reverse leakage. Low reverse leakage in the Schottky diode is important because the leakage current can potentially exceed the 10µA quiescent current of the LTC1771, thus significantly increasing the regulator's no load supply current. Unfortunately, Schottky diodes with lower reverse leakage tend to have higher forward drops and forward voltage drop affects efficiency at moderate to high loads. The efficiency curves for this regulator (Figure 78.5) show outstanding efficiency across more than four decades of load current.

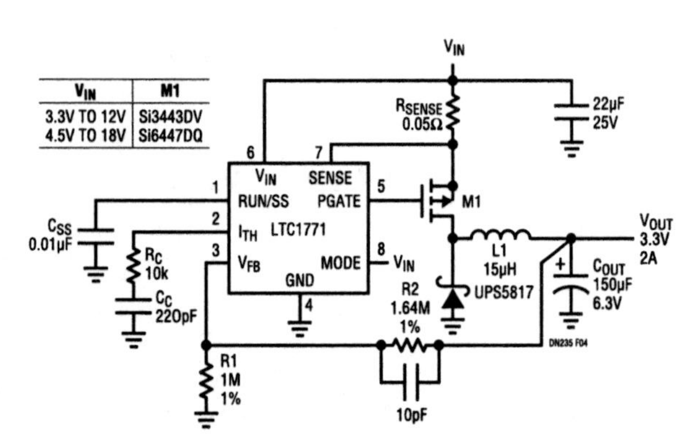

Figure 78.4 • LTC1771 3.3V/2A Regulator

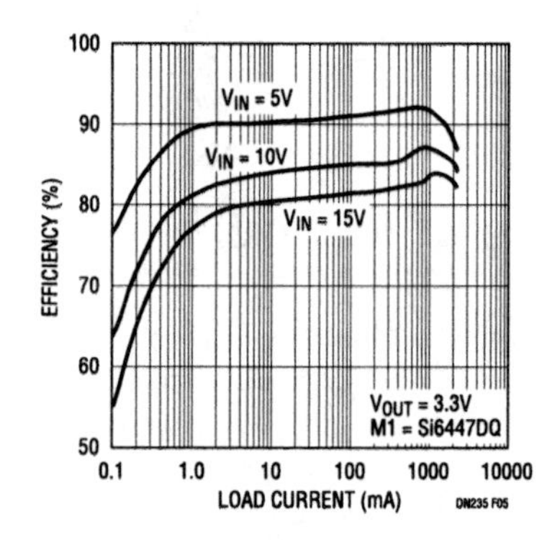

Figure 78.5 • LTC1771 Efficiency vs Load Current for Figure 78.4's Circuit

Low operating current without compromising transient response

Just because the LTC1771, LTC1877 and LTC1878 use so little supply current doesn't mean the transient performance is compromised. Innovative new circuitry ensures that the error amplifier, while operating on less than 10µA at no load, can quickly respond to load changes. The oscilloscope photo shown in Figure 78.6 shows the outstanding transient performance of the LTC1878 regulator of Figure 78.2 when subjected to a 500mA load step.

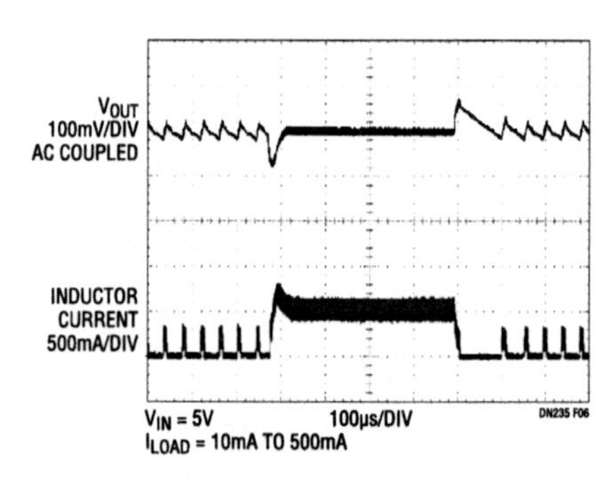

Figure 78.6 • Load Step Transient Response for Figure 78.2 Circuit

79

Low cost PolyPhase DC/DC converter delivers high current

Wei Chen

Introduction

The LTC1929/LTC1929-PG are current mode DC/DC controllers that drive two synchronous buck stages out of phase. The 2-phase architecture reduces the number of input and output capacitors without increasing the switching frequency. The relatively low switching frequency and integrated high current MOSFET drivers help obtain high power conversion efficiency for low voltage, high current applications. Because of output ripple current cancellation, lower value inductors can be used, resulting in faster load transient response. Power good indication can be added by using the LTC1929-PG. Typical applications include high current (up to 40A) low voltage (≤6V) power supplies for microprocessors, memory arrays and ASICs.

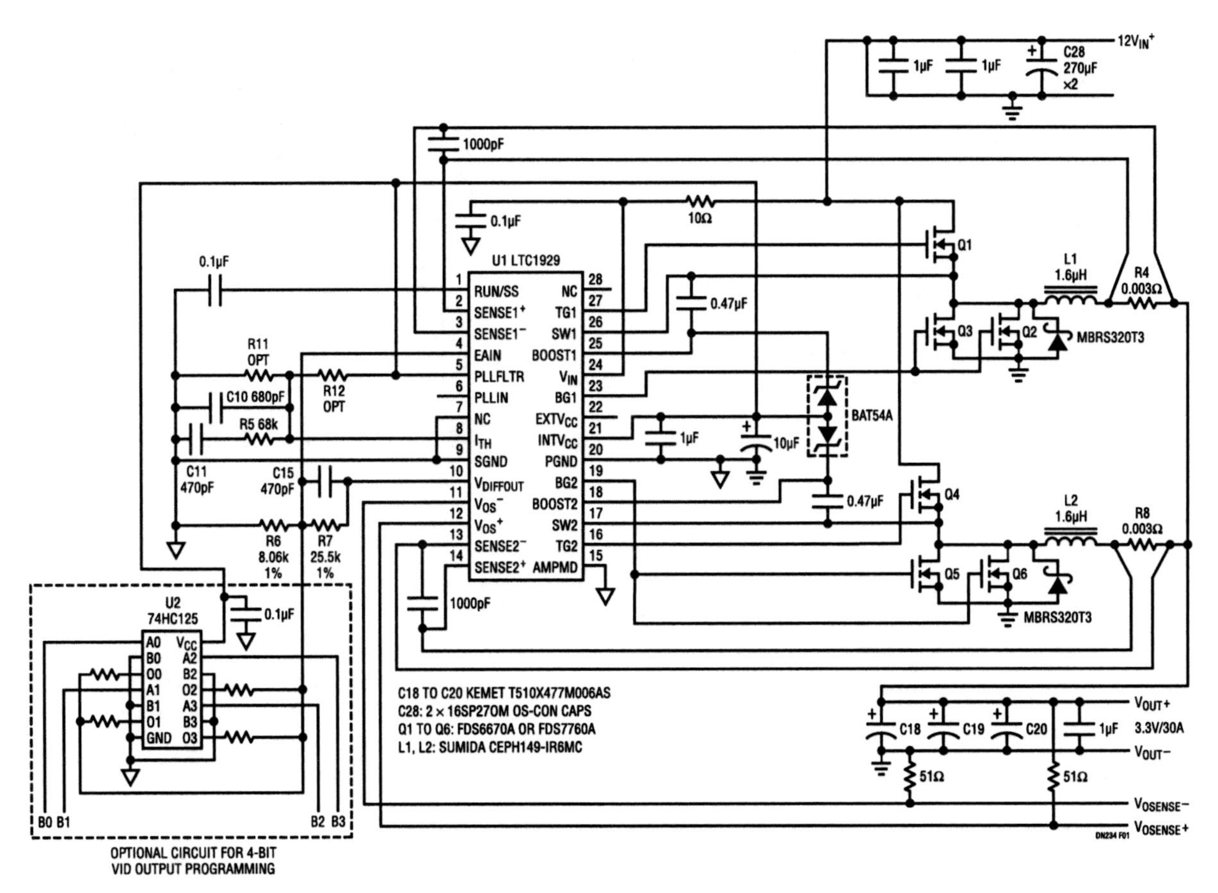

Figure 79.1 • Schematic Diagram of a 30A Power Supply

Analog Circuit Design: Design Note Collection. http://dx.doi.org/10.1016/B978-0-12-800001-4.00079-X

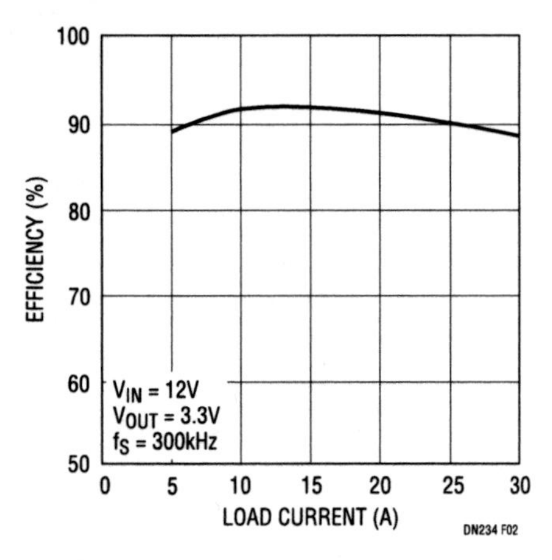

Figure 79.2 • Efficiency vs Load Current

Design example

Figure 79.1 shows the schematic diagram of a 30A power supply with 12V input and 3.3V output. With only one IC, six tiny SO-8 MOSFETs and two 1μH low profile surface mount inductors, greater than 88% efficiency can be maintained throughout the load range between 5A and 30A, as shown in Figure 79.2.

Active voltage positioning can be employed in this design to reduce the number of output capacitors (refer to Design Solutions 10 for more details on active voltage positioning). The optional resistors R11 and R12 provide active voltage positioning with no loss of efficiency.

Because of 2-phase operation, only two OS-CON input capacitors and three tantalum output capacitors are needed. If a single phase configuration were used, four OS-CON input capacitors and nine tantalum output capacitors would be needed.

Overcurrent limit

The LTC1929 has a built-in foldback current limit. When the output current exceeds the limit, the output voltage decreases. If the output voltage drops below 70% of its nominal output, the output current limit folds back. This feature protects the power supply from overheating as a result of abnormal operating conditions such as output short circuits. If the application requires a different output voltage threshold to activate foldback current limiting or to defeat foldback current limit entirely, the circuit shown in Figure 79.3 can be used.

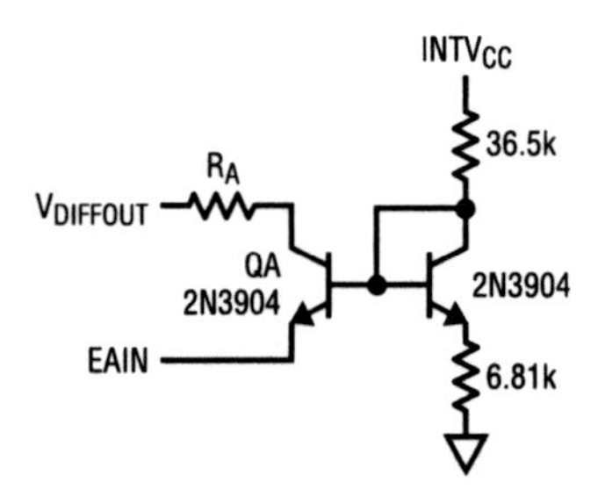

Figure 79.3 • Optional Circuit to Modify Foldback Current Limit

When the output voltage drops below 85% of the nominal value, QA will turn on to maintain a constant voltage on the EAIN pin even though the output voltage decreases. Consequently, a constant current limit is achieved as long as QA is not saturated. If the foldback current limit is triggered, the parallel combination of R_A and R7 will produce an EAIN voltage of about 0.65V. R_A can be adjusted to program the threshold of the foldback current limiting. If R_A connects to $INTV_{CC}$ instead of $V_{DIFFOUT}$, constant current limit will be maintained under the short-circuit condition.

Multiphase applications

LTC1929 provides 2-phase operation for output current up to 40A. In applications with very high output current (>40A) or multiple outputs, multiphase operation is usually preferred. The LTC1629 should then be used because its PHASMD and CLKOUT pins enable multiphase operation. The application information on the high current, single output design can be found in the LTC1629 data sheet and Design Note 215. Figure 79.4 shows an example of a 2-output power supply using LTC1629 and LTC1929. While each output sees a 2-phase ripple current, the common input sees 4-phase operation. The input ripple current is greatly reduced.

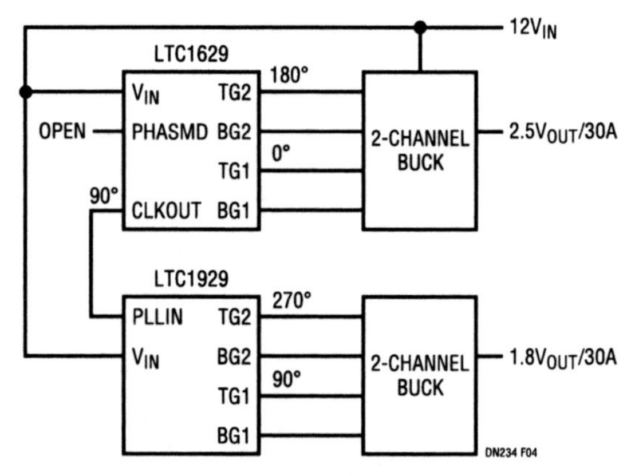

Figure 79.4 • Block Diagram of a 2-Output Power Supply

Conclusion

The LTC1929-PG provides a low cost, high efficiency power solution for low voltage, high current applications. By using multiple LTC1929s or LTC1629s, multiphase operation can be configured for either single output high current (>40A) applications or multioutput applications. Refer to the data sheets, Design Note 215 and Application Note 77 for more information.

Unique high efficiency 12V converter operates with inputs from 6V to 28V

80

Christopher B. Umminger

Generating an output DC voltage from an input voltage that can vary both above and below the regulation point is a challenging power supply problem with a variety of solutions. Flyback or forward converters work well, but are difficult to use if a small and efficient solution is desired at high output currents. A cascade of two converters, such as a boost followed by a buck or linear regulator, is another possibility. Alternately, one can use a switching regulator in a SEPIC configuration that requires two inductors and an intermediate capacitor. These solutions are complex and have a high cost in components, board area and efficiency. However, one can also use the LTC1625 No R_{SENSE} controller in a circuit that is capable of both up and down conversion and requires only a single inductor and no sense resistor.

12V output, single inductor, buck/boost converter

An example of such a circuit is shown in Figure 80.1 to provide a 12V output with inputs that can range from 6V to 18V. All of the circuitry to the left of the inductor is identical to a typical buck converter implemented with the LTC1625. However, now the output (right) side of the inductor is also switched using an additional MOSFET M3 and a diode D2. These two devices act like a boost converter stage. During the first portion of each cycle, switches M1 and M3 are on while M2 is off. The input voltage is applied across the inductor and

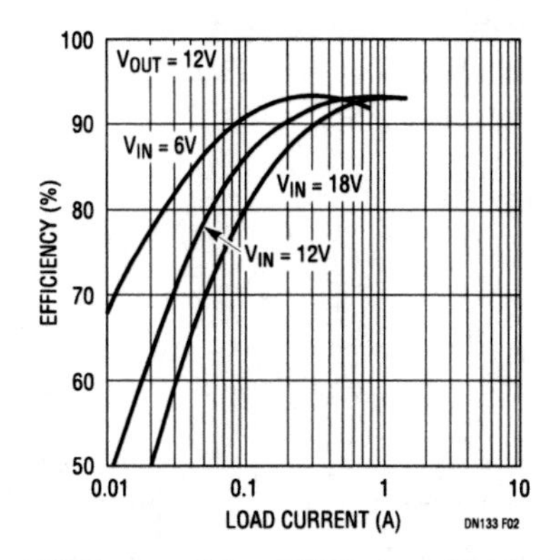

Figure 80.2 • Efficiency of the 12V Output Single Inductor Buck/Boost Converter

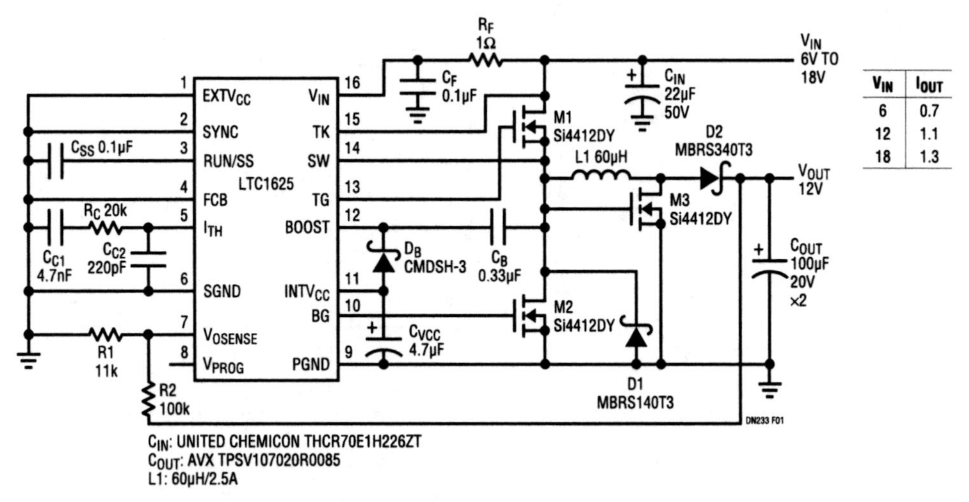

V_{IN}	I_{OUT}
6	0.7
12	1.1
18	1.3

Figure 80.1 • 12V Output Single Inductor Buck/Boost Converter

Analog Circuit Design: Design Note Collection. http://dx.doi.org/10.1016/B978-0-12-800001-4.00080-6

its current increases. After the LTC1625 current comparator trips, M1 and M3 are turned off and M2 and D2 conduct for the remainder of the cycle. During this time, current is delivered to the output while $-V_{OUT}$ is applied across the inductor and its current decreases.

The duty cycle for the Figure 80.1 circuit is equal to $V_{OUT}/(V_{IN}+V_{OUT})$. When V_{IN} is equal to V_{OUT}, a fifty percent duty cycle is required to balance the volt-seconds across the inductor. Both the input and output capacitors must filter a square pulse current in this topology. The average value of the inductor current is equal to the sum of the input and output currents. Since the LTC1625 uses MOSFET V_{DS} sensing to control the inductor current peaks, the output current limit depends upon the duty cycle and will vary with the input voltage. At $V_{IN}=12V$, the maximum output current is about 1.1A. Efficiency of the circuit is shown in Figure 80.2. Note that diode D2 prevents current reversal which causes cycle skipping at low load currents and improves the light load efficiency.

Synchronous circuit for higher power, higher V_{IN}

Several modifications can be made to the Figure 80.1 circuit to improve its operation as shown in Figure 80.3. In order to process more power, lower on-resistance MOSFET switches are used along with a higher current inductor. The number of input and output capacitors is also increased due to the higher RMS currents flowing through them.

At higher power levels, it is desirable to use a synchronous switch M4 across the output diode D2, allowing one to reduce the current rating of D2 and improve the converter efficiency. Gate drive for this switch is derived from the LTC1625 BG pin, buffered by the LTC1693-2 and then level shifted to the output with the network formed by C4, R4 and D4. Another change increases the allowed input voltage, which is limited in the Figure 80.1 circuit by the breakdown voltage of the M3 gate. This impediment is overcome using a clamp network formed by R5, C3 and Z1 to derive the turn-on signal for switch M3. The signal is buffered by the other half of the LTC1693-2 to drive M3. The efficiency of this circuit is shown in Figure 80.4.

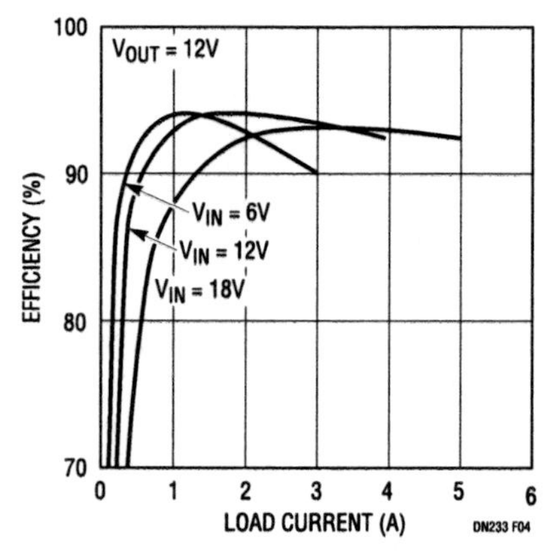

Figure 80.4 • Efficiency of the Synchronous Buck/Boost Circuit

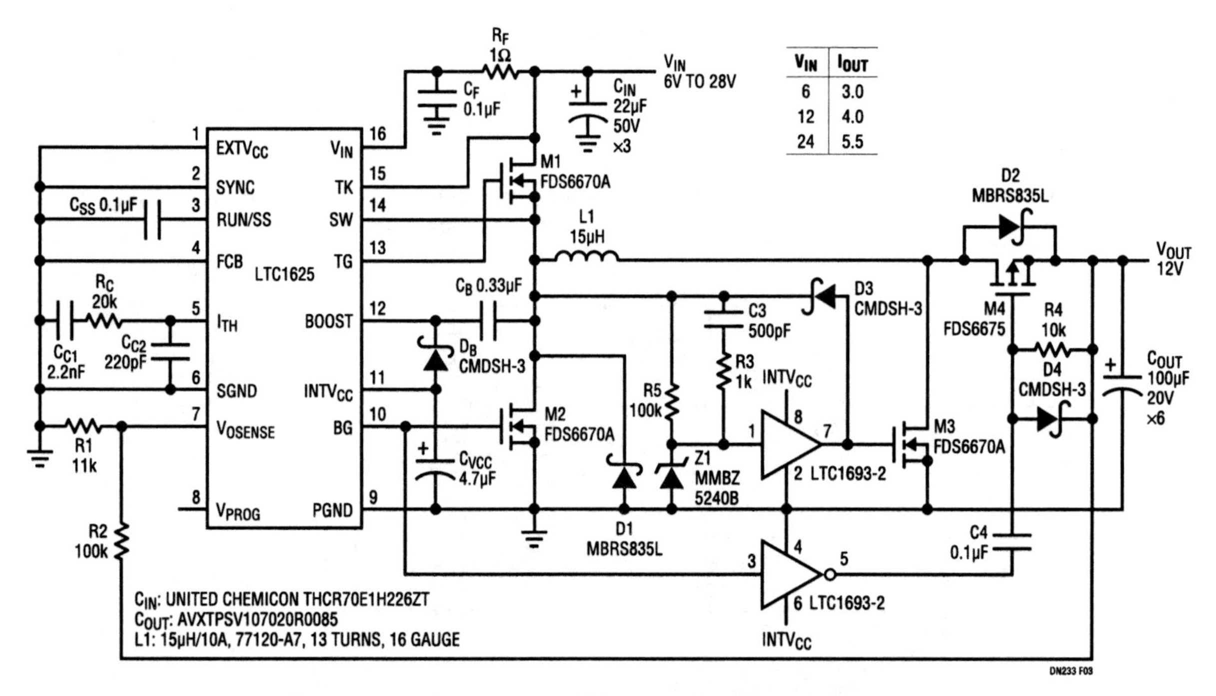

Figure 80.3 • Synchronous 12V Output Buck/Boost Converter

Low cost, high efficiency 42A DC/DC converter

81

Wei Chen

Introduction

The LTC1709-8 and the LTC1709-9 are dual, current mode, 2-phase controllers that drive two synchronous buck stages out of phase. This architecture reduces the number of input and output capacitors without increasing the switching frequency. The relatively low switching frequency and integrated high current MOSFET drivers help obtain high power conversion efficiency for low voltage, high current applications. Because of the output ripple current cancellation, lower inductance values can be used, resulting in faster load transient response. This, plus a 5-bit VID programmable attenuator, makes it particularly attractive for CPU power supply applications. Two VID tables are available to comply with VRM8.4 (LTC1709-8) and VRM9.0 (LTC1709-9).

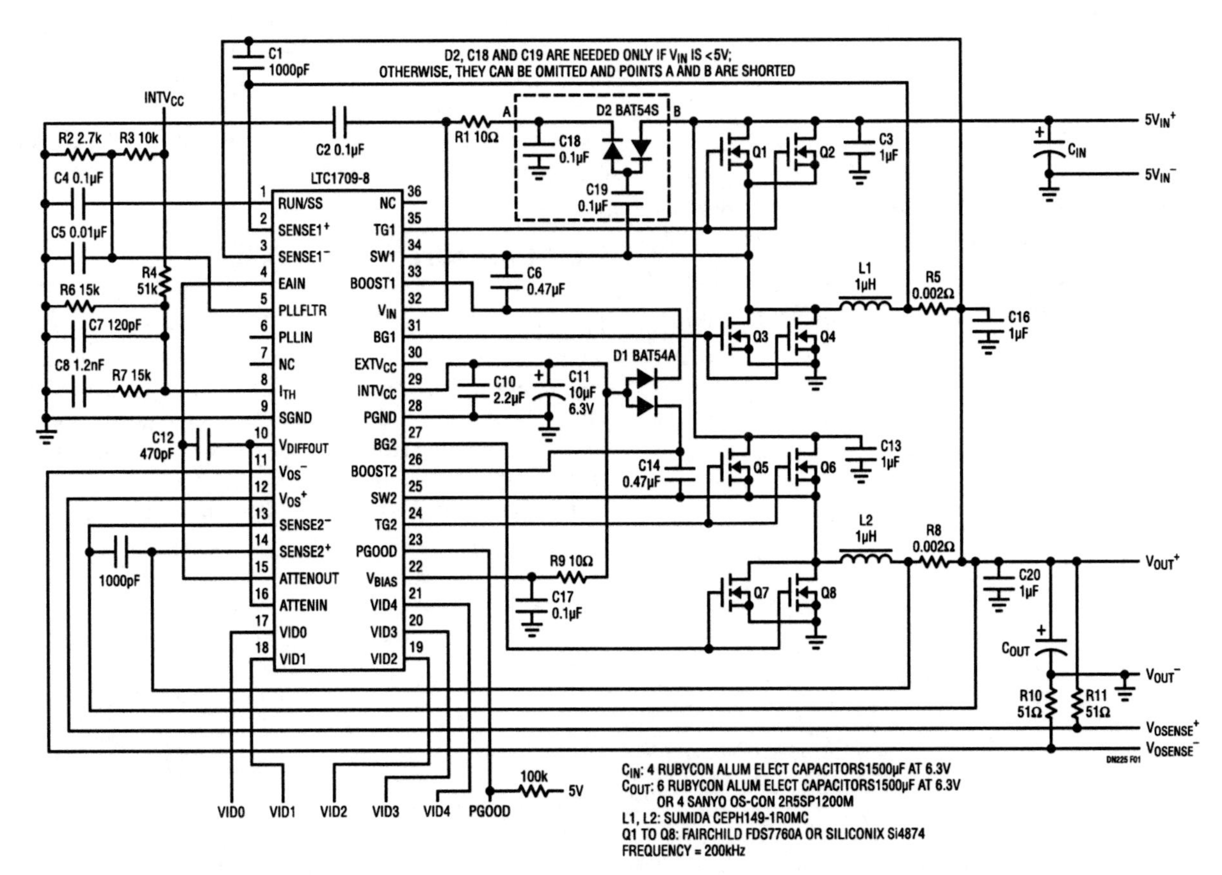

Figure 81.1 • Schematic Diagram of a 42A AMD Athlon Microprocessor Power Supply Using the LTC1709

Analog Circuit Design: Design Note Collection. http://dx.doi.org/10.1016/B978-0-12-800001-4.00081-8

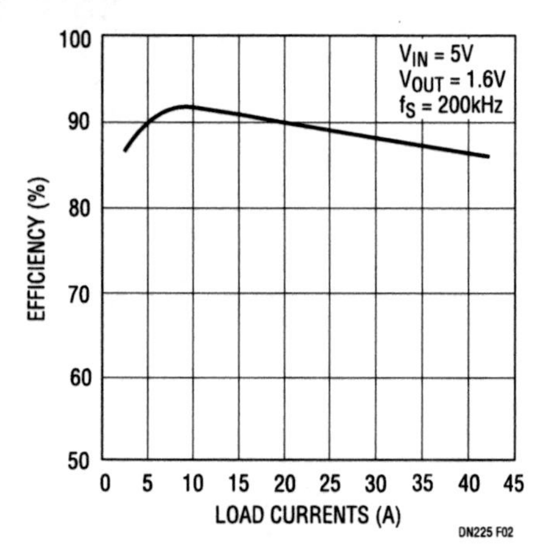

Figure 81.2 • Efficiency vs Load Currents

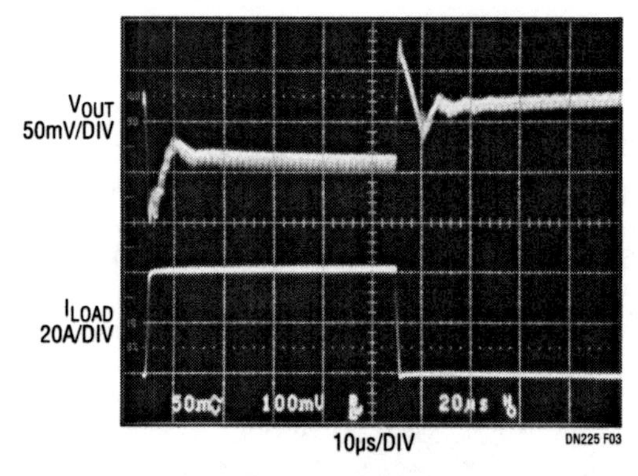

Figure 81.3 • Load Transient Waveforms at 40A Step and 30A/μs Slew Rate

Design example

Figure 81.1 shows the schematic diagram of a 42A power supply for the AMD Athlon microprocessor. With only one IC, eight tiny SO-8 MOSFETs and two 1μH low profile surface mount inductors, an efficiency of 86% is achieved for a 5V input and a 1.6V/42A output. Greater than 85% efficiency is maintained throughout the load range between 3A and 42A as shown in Figure 81.2. Because of the low input voltage, the reverse recovery losses in the body diodes of the bottom MOSFETs are not significant. No Schottky diodes are required in parallel with the bottom MOSFETs in this application.

Table 81.1 Comparisons of Input and Output Ripple Currents for Single Phase and Dual Phase Configurations (L = 1μH, f_S = 200kHz)

PHASES	INPUT RIPPLE CURRENT (A_{RMS})	OUTPUT RIPPLE CURRENT (A_{P-P})
1	19.7	10.9*
2	10.1	2.9

*Assumes that the single phase circuit uses two 1μH/21A inductors in parallel to provide 42A output.

Table 81.1 compares the input and output ripple currents for single phase and 2-phase configurations. A 2-phase converter reduces the input ripple current by 50% and the output ripple current by 75% compared to a single phase design. The resulting savings in input and output capacitors is significant.

Figure 81.3 shows the measured load transient waveform. The load current changes between 2A and 42A with a slew rate of about 30A/μs. Output capacitor requirements are dominated by the total ESR of the output capacitor network. Six low cost aluminum electrolytic capacitors (Rubycon, 1500μF/6.3V) are needed on the output to meet this requirement. The maximum output voltage variation during the load transients is less than 200mV_{P-P}. Active voltage positioning was employed in this design to reduce the number of output capacitors (refer to Design Solutions 10 for more details on active voltage positioning). R4 and R6 provide the desired load regulation with no efficiency loss. If OS-CON capacitors are used, four 1200μF/2.5V (2R51200M) capacitors will be sufficient.

Conclusion

The LTC1709-based, low voltage, high current power supply achieves high efficiency and small size simultaneously. The savings in the input and output capacitors, inductors and heat sinks help minimize the cost of the overall power supply. This LTC1709 circuit, with a few minor modifications, is also suitable for Intel VRM9.0 applications. Refer to Application Note 77 for more information on the PolyPhase technique.

High efficiency PolyPhase converter uses two inputs for a single output

82

Wei Chen Craig Varga

Introduction

As more functions are integrated into one IC, the power drawn by a single IC can easily exceed the capability of a single input power source. Redesigning the front-end power supply to increase the supply's capability will take time and money. Another solution is to use several available power sources to obtain the required output power, drawing some percentage of the total power from each source. The LTC1929 PolyPhase regulator provides a simple solution to this problem.

Design details

The LTC1929 is a PolyPhase dual current mode controller. It is capable of driving two synchronous buck channels 180 degrees out of phase to reduce output switching ripple current and voltage. One buck stage receives its input power from the 12V input and the other receives its power from the 5V input. In this 2-phase design, as the inductor current in the 5V circuit increases, the inductor current in the 12V circuit decreases. This results in a smaller net ripple current flowing into the output capacitor. Since there are two

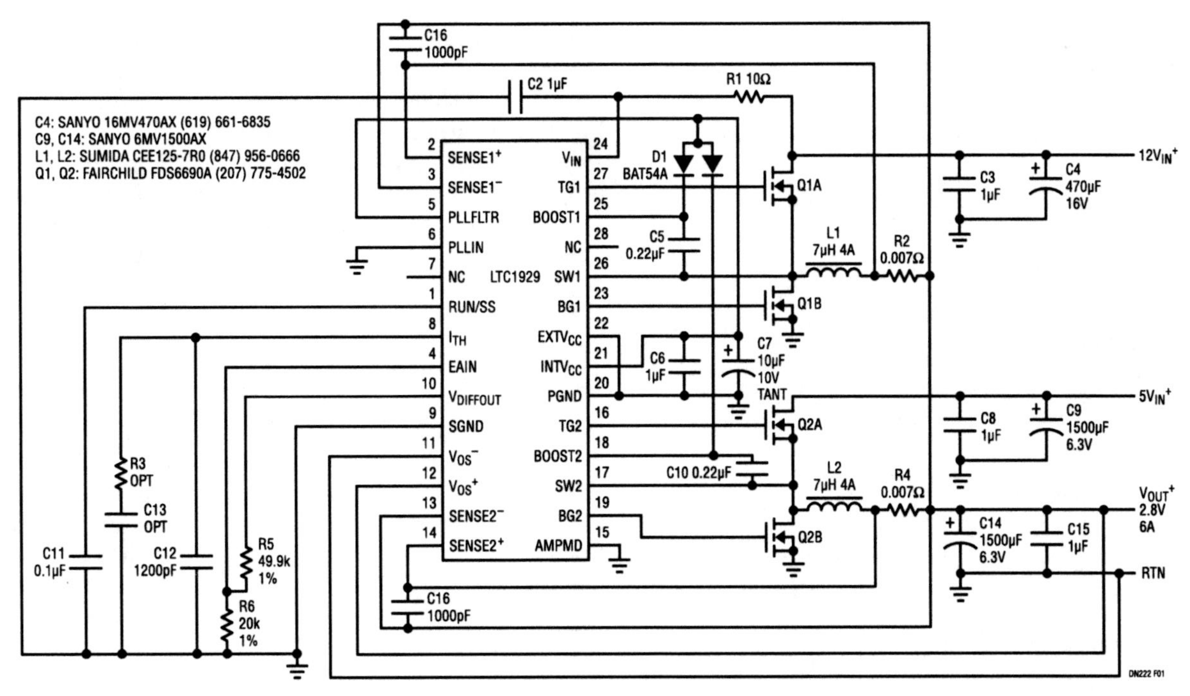

Figure 82.1 • Schematic Diagram Shows Both a 5V and 12V Input Supply for a High Current 2.8V Output

Analog Circuit Design: Design Note Collection. http://dx.doi.org/10.1016/B978-0-12-800001-4.00082-X

intervals in one switching period where ripple cancellation takes place, the output ripple voltage of the 2-phase design is much smaller than that of a single phase design, and fewer output capacitors can be used. Current mode operation provides inherent current sharing.

A typical application

The currents available from a PCI connector are limited to 2A for the 5V supply and 1A for the 12V supply. In the example shown here, the load can be as high as 6A or 16.8W at 2.8V. Neither the 5V nor the 12V source is capable of providing this power. Hence, it is desirable to design a power supply that can draw currents from both power sources, and whose maximum input currents from each source will not exceed the corresponding maximum limit. This design shows how to easily accomplish this using the LTC1929 PolyPhase controller. With only one IC, two dual FETs in SO-8 packages and two small inductors, a high efficiency, low noise power supply can be obtained.

Figure 82.1 shows the schematic diagram of the complete power supply. Since each buck circuit only supplies about 3.5A maximum, dual MOSFETs such as the Fairchild FDS6990A can be used. The switching frequency is about 300kHz per-channel for an effective output ripple frequency of 600kHz. The inductors in both stages are 7μH. The design uses Sumida CEE125-7R0 inductors, but any inductor with a similar inductance value and 4A or greater current rating should do the job. The current sense resistor is 0.007Ω for each channel.

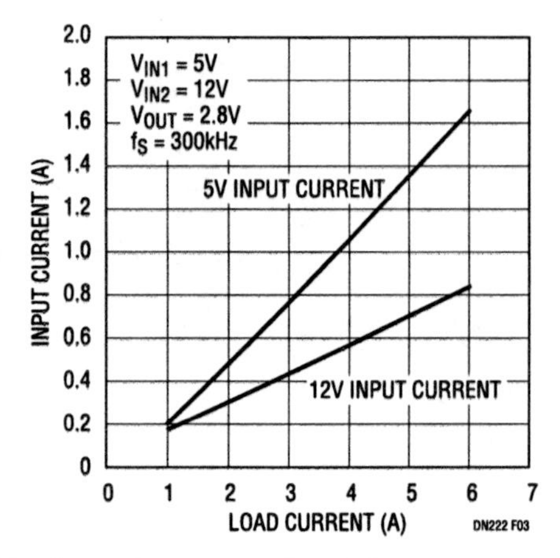

Figure 82.3 • Input Currents vs Load Currents

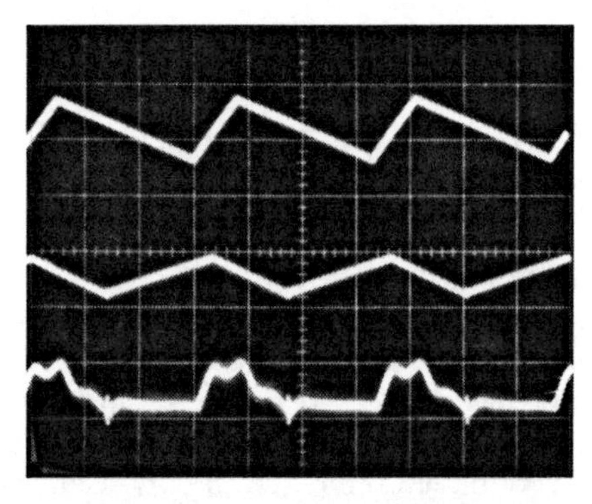

Figure 82.4 • Waveforms of Ripple Currents and Voltage (Top Trace: 12V Buck Inductor Current, 1A/DIV; Middle Trace: 5V Buck Inductor Current 1A/DIV; Bottom Trace: Output Ripple Voltage, 50mV/DIV)

Test results

Figure 82.2 shows the overall efficiency vs load currents. For most of the load range, the efficiency is above 90%. Figure 82.3 shows the distribution of the two input currents as the load current varies. The maximum input currents for the 5V and 12V sources are 1.66A and 0.84A, respectively, which are well below the PCI connector's current limits. Figure 82.4 shows the waveforms of the inductor ripple currents and output ripple voltages. Note the ripple cancellation phenomenon. The peak-to-peak switching ripple voltage at the output terminal is only 50mV$_{P-P}$ with one 1500μF/6.3V aluminum electrolytic capacitor. If two buck circuits are synchronized in phase, the ripple voltage will be 70mV$_{P-P}$, almost a 50% increase.

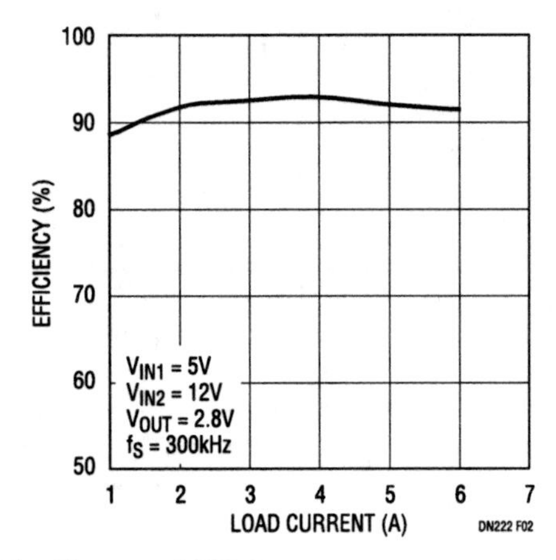

Figure 82.2 • Measured Efficiency

Conclusion

The PolyPhase technique reduces the output ripple voltage without increasing the switching frequency. High efficiency can be obtained for low output voltage applications. The LTC1929 PolyPhase controller provides a small, low cost solution for multi-input applications. If more than two inputs are needed, use the LTC1629 rather than the LTC1929. Multiple LTC1629s can be configured for 3-, 4-, 6- or even 12-phase operation.

High current dual DC/DC converter operates from 3.3V input

83

Ajmal Godil

In many telecommunication applications, isolated DC/DC "brick" modules are used to step 48V down to 3.3V. However, it may be necessary to generate 2.5V and 1.8V from the 3.3V supply to power a DSP, microprocessor and/or an ASIC. Formerly, linear regulators were commonly used for these applications, but since the current demanded by these loads has increased significantly (10A is not unusual), using a 54% efficient linear regulator (V_{IN}=3.3V, V_{OUT}=1.8V) would be a thermal nightmare. Real estate on these boards is extremely scarce, and dissipating 15W [(3.3V − 1.8V) · 10A] of power would require a big heat sink and lots of airflow. These high load currents require the use of high efficiency switching regulators that can step down from 3.3V.

The circuit in Figure 83.1 generates 1.8V at 12A and 2.5V at 5A from a 3.3V supply using the LTC1702. The LTC1702 is a 2-phase dual switching regulator controller optimized for high efficiency with input voltage ranging from 2.7V to 7.0V. It includes two complete, on-chip, independent switching regulator controllers, each designed to drive a pair of external N-channel MOSFETs in a voltage mode control, synchronous buck architecture. The LTC1702 uses a constant frequency, true PWM design that switches at 550kHz, minimizing external component size and cost and maximizing load transient performance. The LTC1702 also provides open-drain logic outputs (PGOOD1 and PGOOD2) that indicate whether either output has risen to within 5% of the final output voltage. An optional latching fault mode protects the load if the output rises 15% above the intended voltage. The dual output LTC1702 is packaged in a space saving 24-pin narrow SSOP.

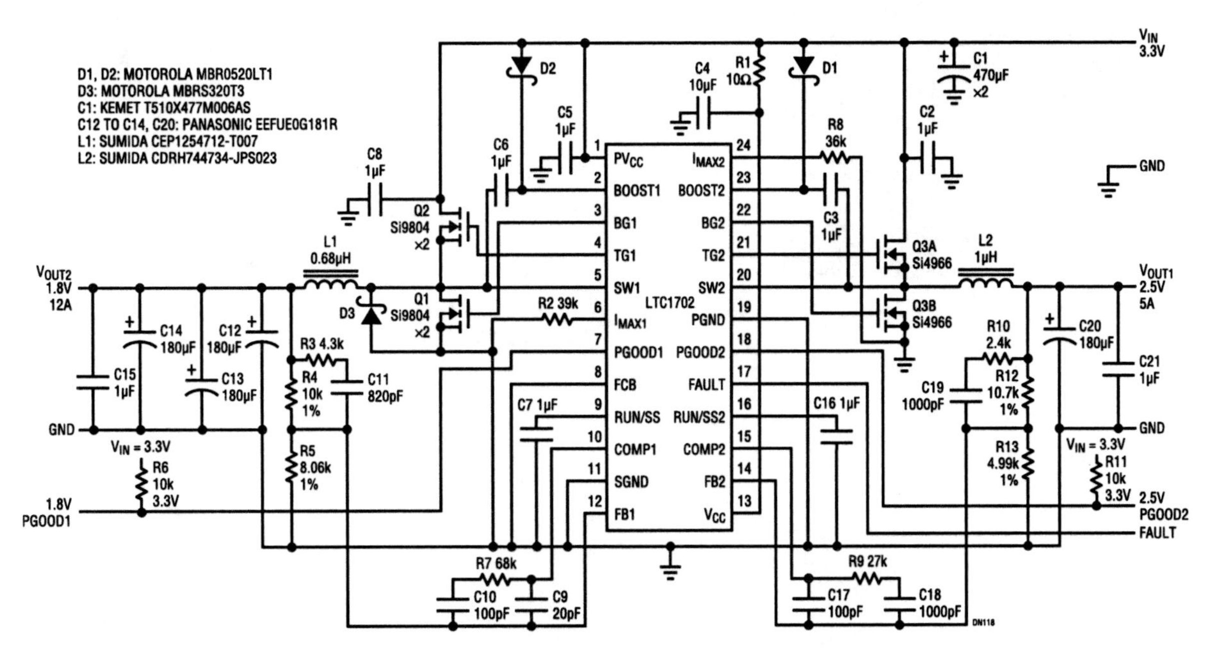

Figure 83.1 • Circuit Generating 1.8V at 12A and 2.5V at 5A from a 3.3V Supply with the LTC1702

Analog Circuit Design: Design Note Collection. http://dx.doi.org/10.1016/B978-0-12-800001-4.00083-1

The LTC1702 uses a true 25MHz gain bandwidth op amp as the feedback amplifier. This allows the use of an OPTI-LOOP compensation scheme that can precisely tailor the loop response by allowing the loop to be crossed over beyond 50kHz in most applications, while maintaining sufficient gain and phase margin for stable operation. Another feature of the LTC1702 significantly reduces the required input bulk capacitance. By running a single master clock that drives the two sides 180° out of phase, input ripple current cancellation is achieved. This technique is known as 2-phase switching and has the effect of doubling the frequency of the switching pulses seen by the input capacitor and significantly reducing their RMS value. The LTC1702 circuit in Figure 83.1 needs only two 470µF input capacitors, versus the four that would be required by a circuit using two individual switching regulators. This 50% reduction in input capacitors results in major cost savings. For more information about calculating RMS current carrying capability for the input capacitors, consult the LTC1702 data sheet and Application Note 77.

The 550kHz clock frequency and the low 3.3V input voltage allow the use of external inductors in the 1µH range while keeping the ripple current under control. The low inductance value helps in reducing the physical size of the core and raises the attainable dI/dt at the output of the circuit, decreasing the time that it takes for the circuit to correct for sudden changes in load current. This, in turn, reduces the amount of output capacitance required to support the output voltage during a load transient. Figure 83.2 shows the output voltage transient response for the 2.5V supply subjected to a 0A–5A load step with one 180µF capacitor. Figure 83.3 shows the output voltage transient response for the 1.8V supply subjected to a 0A–12A load step with three 180µF capacitors. Reduced input capacitance due to the LTC1702's 2-phase internal switching, combined with its 550kHz clock frequency, significantly reduces the total capacitance needed compared to a conventional switching regulator.

Figure 83.4 shows the typical efficiency curves for the LTC1702 circuit of Figure 83.1. It can be seen from Figure 83.4 that for the 2.5V supply, efficiency greater than 87% is achievable for load currents of up to 5A; for the 1.8V supply, efficiency greater than 85% is achievable for load currents of up to 12A with peak efficiencies of well over 90%. The circuit in Figure 83.1 can be laid out in less than 1.5 square inches, and can be easily modified to deliver higher or lower load currents by making a few minor changes to the power path components.

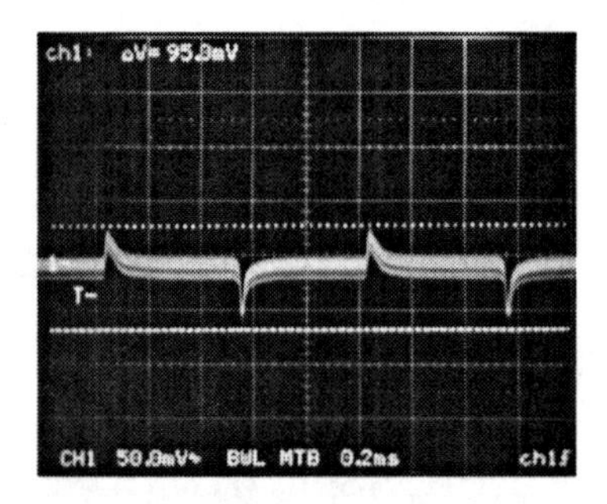

Figure 83.2 • LTC1702 Transient Response for I_{LOAD}=0A to 5A. V_{IN}=3.3V, V_{OUT1}=2.5V

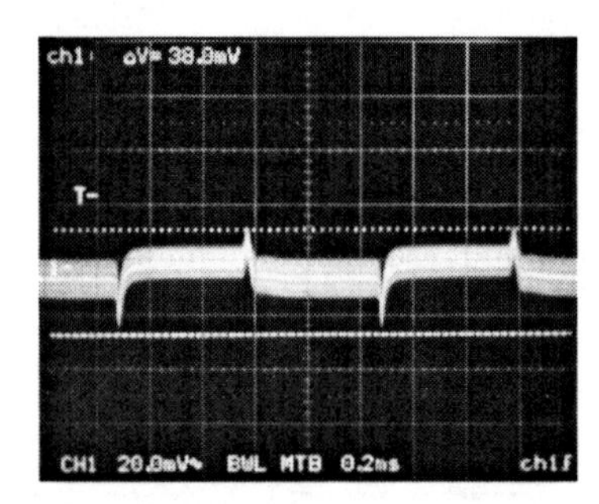

Figure 83.3 • LTC1702 Transient Response for I_{LOAD}=0A to 12A. V_{IN}=3.3V, V_{OUT2}=1.8V

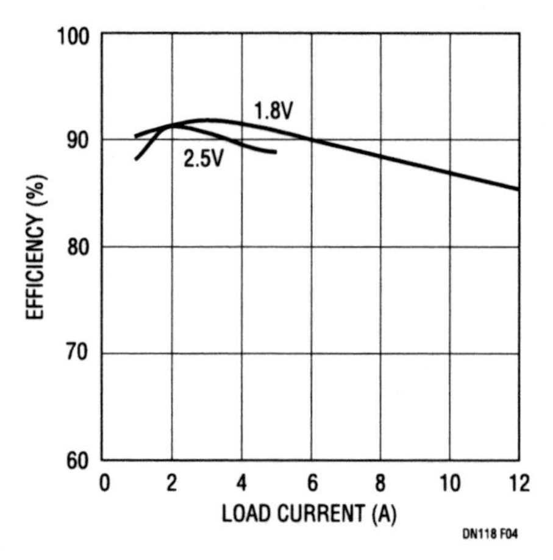

Figure 83.4 • LTC1702 Efficiency Curves. V_{IN}=3.3V, V_{OUT1}=2.5V, V_{OUT2}=1.8V

Low cost surface mount DC/DC converter delivers 100A

84

Wei Chen

Introduction

As computer systems get larger and more complex, their supply current requirements continue to rise. Systems requiring 100A at 3.3V are fairly common while next generation CPU current consumption is approaching 100A at slightly above 1V. Because few off-the-shelf standard power modules are capable of this current level, most system designers are forced to use several modules in parallel to obtain the required current. The resulting power solution is usually expensive, bulky and the performance is not always satisfactory.

The newly released LTC1629 is a dual current mode PolyPhase controller that provides a cost-effective solution for low voltage, high current applications. PolyPhase dramatically reduces input capacitor size and output switching ripple voltage by interleaving the clock signals of several paralleled power stages. Until now, multiphase designs have been difficult to implement because of complex timing and current sharing requirements. The introduction of the LTC1629 solves all of these problems. The advanced features of the LTC1629 include a differential amplifier for true remote sensing, high gate-drive capability, internal current sharing and selectable phasing control. Protection features include overvoltage protection, optional overcurrent latch-off and foldback current limit. This article illustrates a 6-phase power supply design using the LTC1629 with all surface mounted components. An efficiency of approximately 90% is obtained with an input voltage of 12V and an output of 3.3V at up to 100A.

Design details

Each LTC1629 is capable of driving two interleaved synchronous buck output stages. The PLL-based internal phasing circuitry enables 2-, 3-, 4-, 6- or 12-phase operation with a

Figure 84.1 • PolyPhase Converter Block Diagram

Table 84.1 Number of LTC1629s vs Output Current

OUTPUT CURRENT	<35A	35A TO 70A	70A TO 105A	105A TO 140A	140A TO 200A
No. of LTC1629s	1	2	3	4	6
Buck Stages	2	4	6	8	12

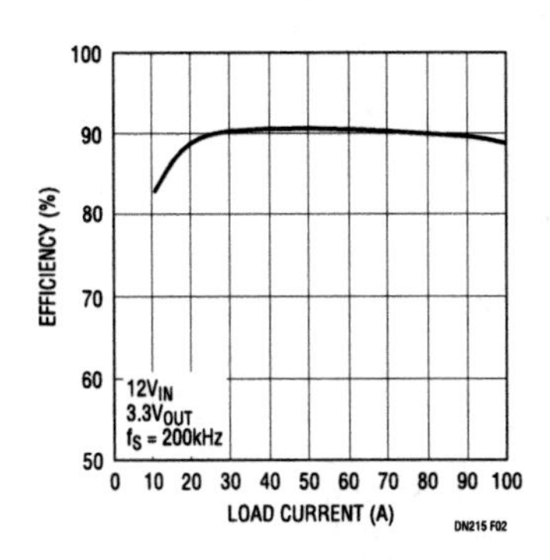

Figure 84.2 • Efficiency vs Load Current

Analog Circuit Design: Design Note Collection. http://dx.doi.org/10.1016/B978-0-12-800001-4.00084-3

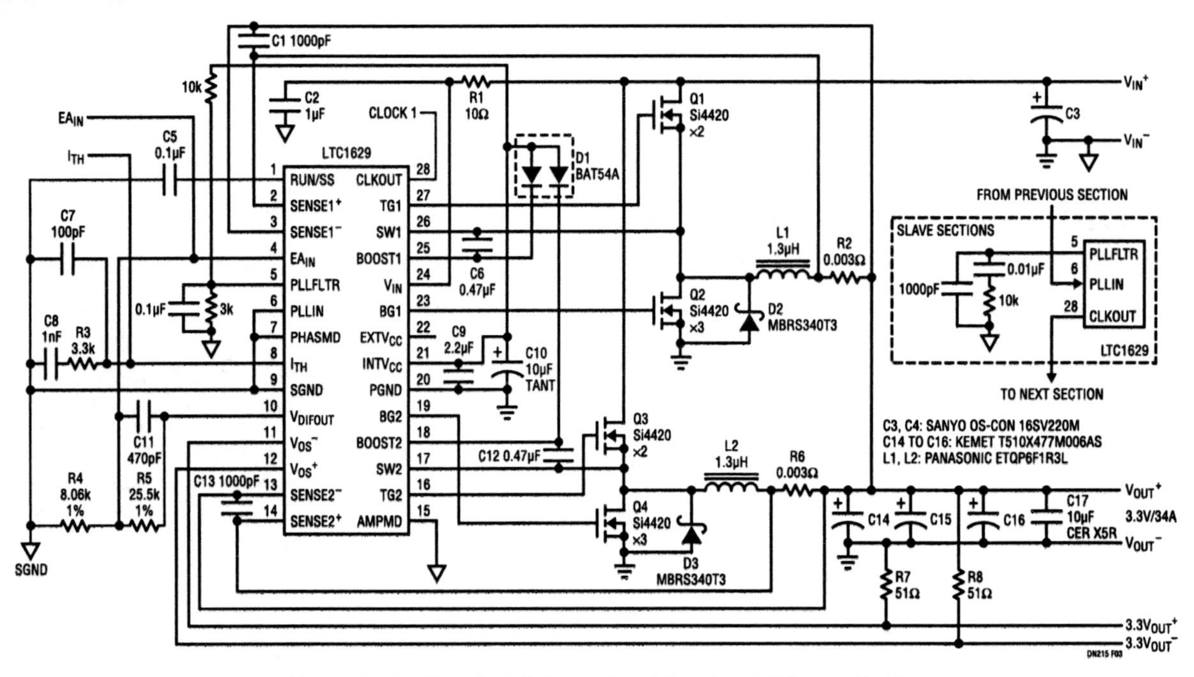

Figure 84.3 • Detailed Schematic of Section 1 (Master Unit)

Table 84.2 Comparison of Input and Output Ripple Current for Single Phase and 6-Phase Configurations. $L=1.3\mu H$, $f_S=200kHz$

NUMBER OF PHASES	INPUT RIPPLE CURRENT	OUTPUT RIPPLE CURRENT	NO. OF INPUT CAPACITORS SANYO OS-CON 16SV220M	NO. OF OUTPUT CAPACITORS FOR THE SAME OUTPUT RIPPLE VOLTAGE: KEMET T510X477M006AS
1	$48A_{RMS}$	$57A_{P-P}$*	13	248
6	$8A_{RMS}$	$2A_{P-P}$	3	9

*Assuming that the single phase circuit uses six 1.3μH/17A inductors in parallel to provide 100A output.

simple phase selection signal (high, low or open). This technique allows paralleling several LTC1629s to deliver output currents from 30A to 200A (see Table 84.1).

This particular design uses three LTC1629s to provide 6-phase operation. Figure 84.1 shows the block diagram of a complete power supply consisting of three almost identical sections: one master unit and two identical slave units. The slave units are the same as the master unit except that the V_{OS}^- and V_{OS}^+ pins of the slave LTC1629s are open. Only the master unit senses the output voltage. A total of 30 SO-8 MOSFETs and six small surface mounted inductors are used in this design (no heat sinks are required). The switching frequency is 200kHz. Figure 84.3 is a detailed schematic of section 1 (master unit).

Table 84.2 compares the input and output ripple current along with the required input and output capacitors for conventional single phase and 6-phase configurations. Compared to a single phase converter, the 6-phase circuit reduces the input ripple current by 83%. As a result, the 6-phase design requires only three OS-CON input capacitors (16SV220M), compared to the 13 required by a single phase circuit. The output ripple current reduction for the 6-phase circuit is even more significant: 96% lower than single phase. The output ripple voltage at 100A is only $5mV_{P-P}$ with only nine tantalum capacitors (T510X477M006AS). The resultant ripple frequency is 1.2MHz (six times the switching frequency). To achieve the same output ripple voltage a single phase design would require 248 tantalum capacitors. Figure 84.2 shows the measured overall efficiency (including the control circuit power loss) at close to 90% throughout most of the load range.

Conclusion

The LTC1629-based PolyPhase converter results in a simple and reliable design using standard manufacturing processes. This unique architecture achieves high efficiency, small size and low cost even at 100A and above. Refer to the URLs for more details.

High voltage, low noise buck switching regulator

85

Ajmal Godil

The LT1777 is a wide input range buck (step-down) switching regulator specially designed for low noise applications. The LT1777 can be beneficial in applications where low noise is critical, such as telecom, automotive cellular and GPS receiver power supplies. The schematic in Figure 85.1 highlights the capabilities of the LT1777.

The LT1777 can accept input voltages from 7.4V to 48V and has a nominal switching frequency of 100kHz. The monolithic die includes an onboard 700mA peak current switch, oscillator, control and protection circuitry. It uses current mode control that delivers excellent dynamic input supply rejection and short-circuit protection. In order to achieve low noise, the LT1777 is equipped with dI/dt limiting circuitry that is programmed via a small inductor (L_{SENSE} in Figure 85.1) in the power path. It also contains internal circuitry to limit dV/dt during switch turn-on and turn-off.

Figure 85.2 shows the V_{SW} node voltage and switch current for the low noise LT1777. Figure 85.3 shows the V_{SW} node voltage and switch current for the high voltage LT1676 buck regulator under the same test conditions (no slew rate limiting). It can be seen from Figures 85.2 and 85.3 that the switch node voltage and current waveforms for the LT1777 are more controlled and rise and fall more slowly than those of the LT1676 regulator. Conducted and radiated EMI are dramatically reduced by slowing down the sharp edges during

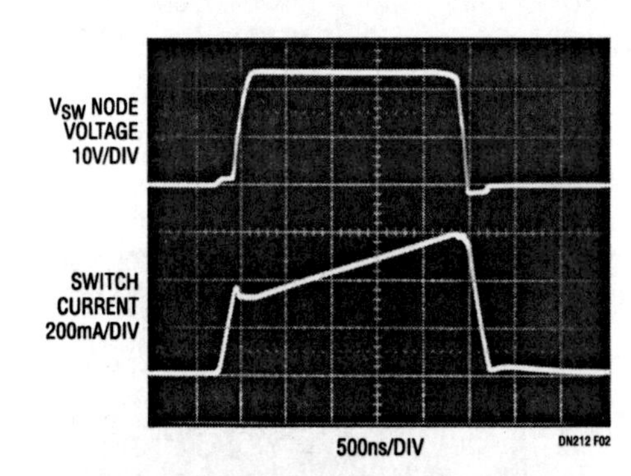

Figure 85.2 • V_{SW} Voltage and Switch Current for the LT1777

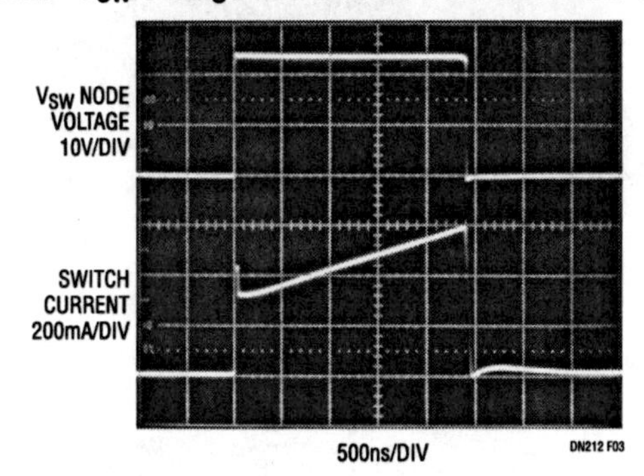

Figure 85.3 • V_{SW} Voltage and Switch Current for the LT1676 (No Slew Control)

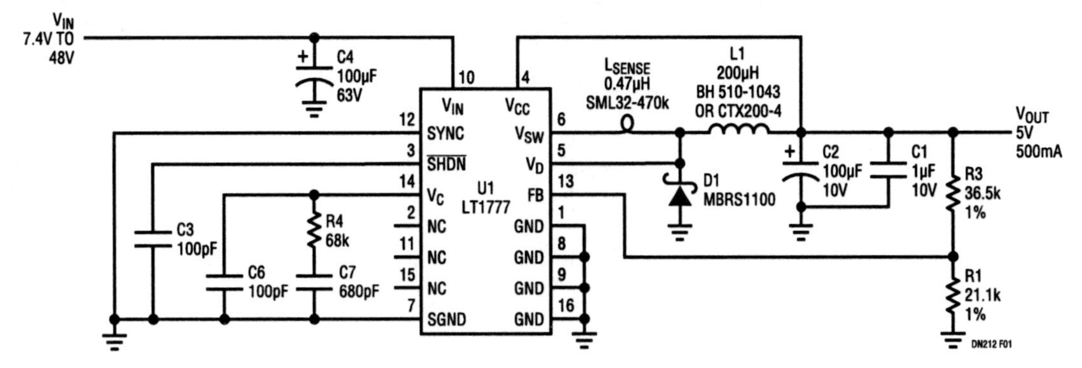

Figure 85.1 • 100kHz Low Noise Step-Down Switching Regulator

Analog Circuit Design: Design Note Collection. http://dx.doi.org/10.1016/B978-0-12-800001-4.00085-5

turn-on and turn-off of the power switch with only modest reduction in conversion efficiency.

Figure 85.4 shows a spectral analysis of the current wave forms for the LT1777 and the LT1676. The horizontal axis is 2MHz/DIV (0MHz to 20MHz), and the vertical axis is 10dB/DIV. It can be seen from Figure 85.4 that the LT1777 attenuates the high frequency noise by approximately −20dB compared to the LT1676.

The LT1777 can be disabled by connecting the shutdown ($\overline{SHDN}$) pin to ground, reducing input current to a few microamperes. For normal operation, decouple the $\overline{SHDN}$ pin with a 100pF capacitor to ground. The part also has a SYNC pin, used to synchronize the internal oscillator to an external clock, which can range from 130kHz to 250kHz. To use the part's internal oscillator, simply connect the SYNC pin to ground.

Since the LT1777 allows such a wide input range, the internal control circuitry draws power from the V_{CC} pin, which is normally connected to the output supply. During start-up, the LT1777 draws power from V_{IN}; after the switching supply output reaches 2.9V, the LT1777 uses the output voltage to power its internal control circuitry, thereby reducing quiescent power by hundreds of milliwatts when operating at high line voltages.

Figure 85.5 shows a typical efficiency curve for the LT1777 using sense inductor, $L_{SENSE} = 0.47\mu H$, $V_{IN} = 12V$ and $V_{OUT} = 5V$.

Generating low noise, dual-voltage supplies

The circuit in Figure 85.6 shows a cost-effective way to generate 5V and −5V low noise supplies from a single 10V to 28V supply using the LT1777 and a few off-the-shelf components.

L1A and L1B are two windings on a single core, used to generate ±5V. To minimize coupling mismatches between the two windings, C2 has been added, forcing the winding potentials to be equal and improving cross regulation. Total available current from both outputs is limited to 500mA. Maximum negative supply current is affected by the positive 5V load; a typical limit is one-half of the positive current.

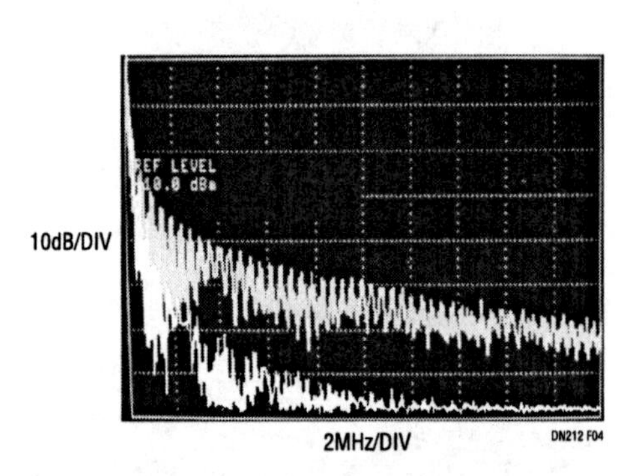

Figure 85.4 • Spectral Analysis of the Waveforms for the LT1676 and the Low Noise LT1777

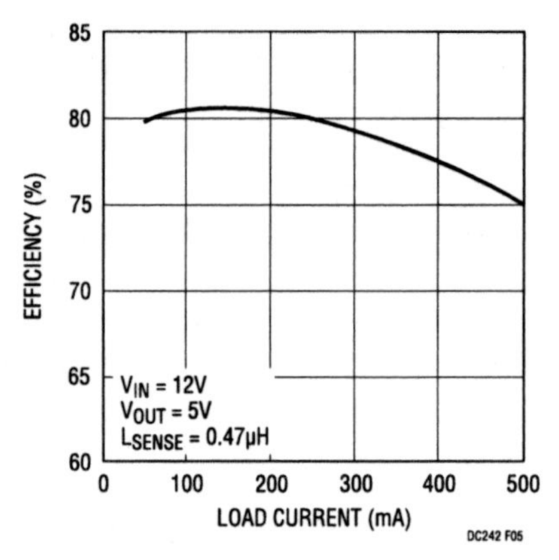

Figure 85.5 • LT1777 Output Efficiency

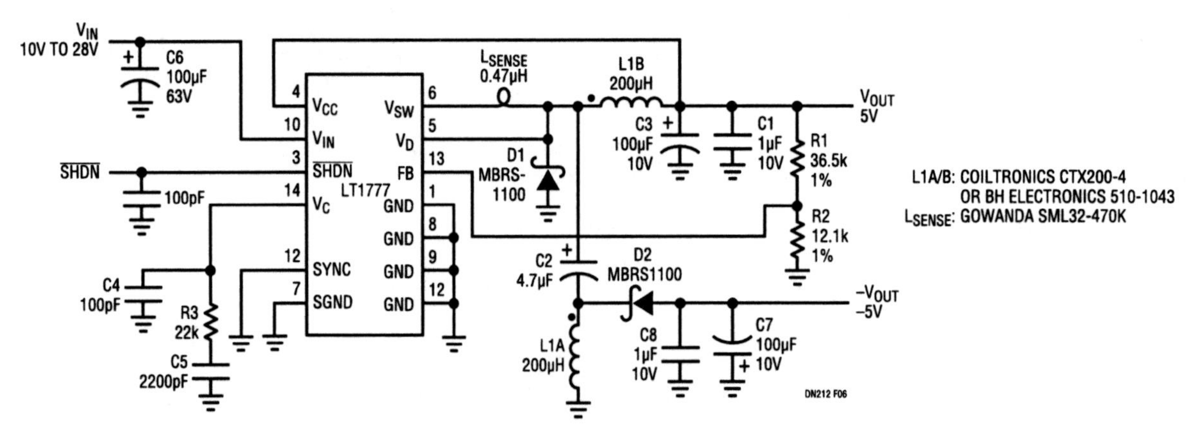

Figure 85.6 • This Cost-Effective Supply Generates ±5V from a 10V to 28V Input

Low cost, high efficiency 30A low profile PolyPhase converter

86

Wei Chen   Craig Varga

The growing need for very high current logic supplies often exceeds the capability of a single buck (step-down) converter. The solution is a PolyPhase converter where two or more buck sections work in parallel to deliver a single high current output.

Overview of the LTC1629

The LTC1629 is a current mode, PolyPhase controller that interleaves the clock signals of two synchronous buck stages, reducing input and output ripple currents without increasing the switching frequency. Because of the output ripple current cancellation, lower value inductors can be used, resulting in a faster load transient response. Lower current rating and decreased inductance also allow the use of smaller sized, low profile, surface mount inductors. The integrated high current MOSFET drivers are capable of driving low $R_{DS(ON)}$ MOSFETs efficiently. This LTC1629-based design achieves 90% efficiency with 12V input and 3.3V output voltage at 30A.

Each LTC1629-based regulator consists of two synchronous buck stages. Current mode control ensures current sharing allowing two or more such regulators to be paralleled directly. Moreover, the LTC1629 integrates proprietary phase locked loop-based clock phasing circuitry, enabling 2-, 3-, 4-, 6- or 12-phase operation with a simple phase selection signal (high, low or open). The LTC1629 includes an internal instrumentation amplifier, enabling true remote sensing. This is particularly useful for maintaining tight regulation actually at the point of load in high current applications.

The peak current mode control provides current sharing among the paralleled power stages. When multiple LTC1629-based regulators are used in parallel, the

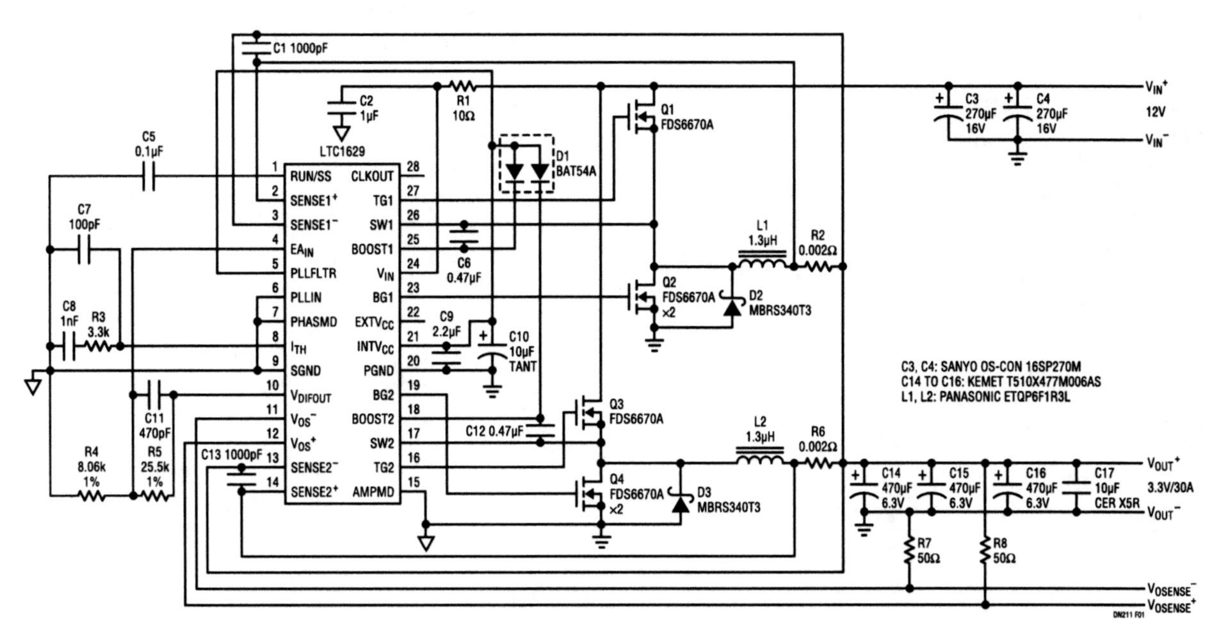

Figure 86.1 • Schematic Diagram of a 30A Power Supply Using LTC1629

Analog Circuit Design: Design Note Collection. http://dx.doi.org/10.1016/B978-0-12-800001-4.00086-7

differential amplifier of the master LTC1629 senses the output voltage and feeds a control voltage to the other slave LTC1629s for voltage regulation. The g_m error amplifier inside each LTC1629 allows direct paralleling of the I_{TH} pins (error amplifier outputs). The paralleled regulators share the same error voltage and source equal currents because the load current in a current mode regulator is proportional to the error voltage.

Design example: 30A 2-phase power supply

Figure 86.1 shows the schematic diagram of a 12V input, 3.3V/30A output PolyPhase power supply. Two synchronous buck output stages and one LTC1629 are employed to provide the 30A output current. Different output voltages may be achieved by changing R4 in the schematic diagram. The switching frequency is selected to be 300kHz. This results in the use of a 1.3μH/15A inductor for each buck circuit.

Table 86.1 compares the input and output ripple currents and input and output capacitors for single phase and 2-phase configurations. The single phase circuit employs only one buck converter. Compared to the single phase converter, the dual phase converter reduces the input ripple current by more than 45% and the output ripple current by more than 67%. The resulting reductions in the size and cost of capacitors are significant as shown in Table 86.1.

Figure 86.2 shows the measured output ripple voltage at full load. The output ripple voltage is less than $40mV_{P-P}$ and the ripple frequency is twice the switching frequency.

Figure 86.3 shows the measured efficiency as a function of load current. For most of the load range, the efficiency exceeds 90%. An overall efficiency of close to 90% was measured with a 3.3V/30A output. Note that only six SO-8 MOSFETs are used in the complete power supply.

Table 86.1 Comparisons of Input and Output Ripple Currents for Single Phase and Dual Phase Configurations ($L=1.3\mu H$, $f_S=300kHz$)

PHASES	INPUT RIPPLE CURRENT (A_{RMS})	OUTPUT RIPPLE CURRENT (A_{P-P})	NO. OF INPUT CAPACITORS: OS-CON 16P270M	NO. OF OUTPUT CAPACITORS FOR THE SAME OUTPUT RIPPLE VOLTAGE: KEMET T510X477M006AS
1	11.7	12.7[1]	4	9
2	6.4	4.2	2	3

[1]Assume that the single phase circuit uses two 1.3μH/15A inductors in parallel to provide 30A output.

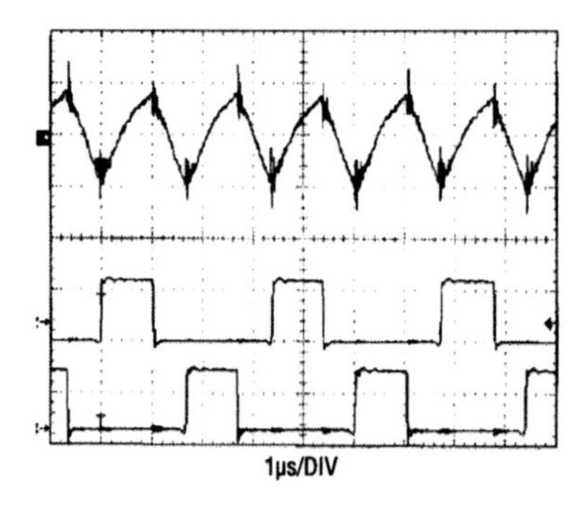

Figure 86.2 • Output Ripple Voltage Waveforms. Top Trace: Output Ripple Voltage, $20mV_{AC}$/DIV. Middle Trace: V_{DS} on Q2, 10V/Div. Bottom Trace: V_{DS} on Q4, 10V/DIV

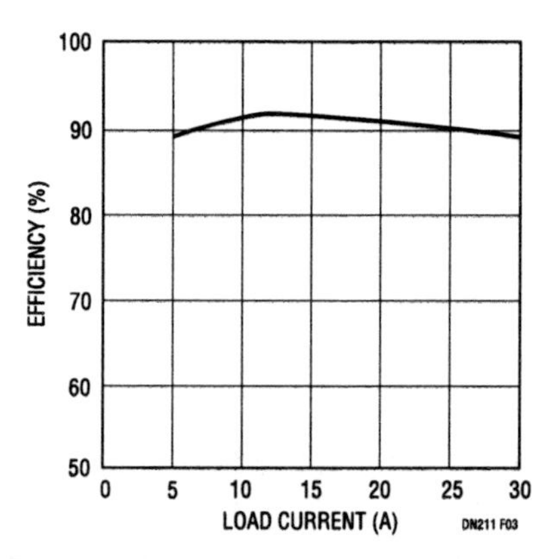

Figure 86.3 • Measured Efficiency vs Load Current

Conclusion

PolyPhase converters using the LTC1629 reduce the size and cost of the capacitors and inductors due to input and output ripple current cancellation. Lower output ripple voltage and smaller inductors help improve the circuit's dynamic performance during load transients. The LTC1629 helps minimize the external component count and simplifies the complete power supply design by integrating two PWM current mode controllers, true remote sensing, selectable phasing control, inherent current sharing capability, high current MOSFET drivers plus protection features (such as overvoltage protection, optional overcurrent latch-off and foldback current limiting) into one IC. The resulting manufacturing simplicity helps improve power supply reliability. High current MOSFET drivers allow the use of low $R_{DS(ON)}$ MOSFETs to minimize the conduction losses for high current applications. Lower current ratings on the individual inductors and MOSFETs also make it possible to use low profile, surface mount components. Therefore, an LTC1629-based PolyPhase high current converter can achieve high efficiency, small size and low profile simultaneously. The savings on the input and output capacitors, inductors and heat sinks minimizes the overall cost and size of the complete power supply.

2-phase switching regulator fits in tight places

87

Steve Hobrecht

Almost every week manufacturers present yet another component that has been reduced in size by 30% to 50% while maintaining similar performance characteristics. The size and weight requirements of the notebook and sub-notebook computer market have resulted in the development of increasingly smaller components. These consumer products have placed very stringent requirements on size, weight and run time. Because of component density, heat dissipation has also become one of the highest concerns for notebook computers and other portable equipment.

The third generation switching regulator controllers from Linear Technology address these new requirements. The LTC1628, a 2-phase, constant frequency, dual synchronous buck controller, has been specifically designed to minimize both the size and total cost of the power system while delivering the highest efficiency in the industry.

A 2-phase switching technique minimizes the peak and resultant RMS input current while doubling the effective frequency of the input current pulses. The input capacitor ESR requirement is reduced by 40% and the capacitance requirement is reduced by 75% when compared to a single phase controller solution! Ceramic input capacitors provide an ideal choice for the resultant requirements. Figure 87.1 contrasts the power loss due to the input RMS ripple current for a single and 2-phase solution over the 5V to 25V input voltage range for a 5V/3A and 3.3V/3A power supply. The input path resistive losses include the ESR of the input capacitor, internal battery resistance, PC trace/connector losses, fuse resistance and diode or MOSFET resistance that may be used in the power path for switching power sources.

The third generation controllers have also been designed to minimize the output capacitor requirements. External OPTI-LOOP compensation allows optimal loop compensation for the wide range of characteristics associated with switching power supply output capacitors. A unique Burst Mode architecture keeps the low current output voltage ripple down to a value close to the normal value for the continuous mode of operation, even when using output capacitors in the 47μF range. The output capacitor ESR requirement is determined by the expected load step and the capacitance value is determined by output current and speed of the feedback loop. A higher capacitance value cannot make up for high ESR and very low ESR cannot make up for insufficient capacitance.

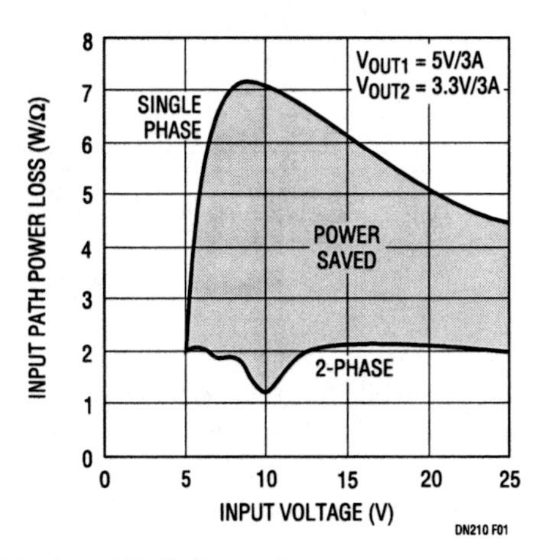

Figure 87.1 • Input Path Power Loss

A third key element in reducing solution size is an internal current foldback circuit that operates when the output voltage drops below 70% of the nominal value. This feature enables the use of MOSFETs properly sized for the output load current without the additional power dissipation margin required for short-circuit conditions.

The application shown in Figure 87.2 occupies less than two square inches of PC board area (components on one side only) and delivers 5V/3A, 3.3V/4A, and 12V/150mA—the three main system power supplies for a notebook computer. The application in this design note addresses very common requirements of small physical size and dynamic 1A step load changes. Single, low ESR capacitors were used for the input and each output. Dual-packaged MOSFETs are used to further minimize PC board space. The inductors were selected based upon the maximum inductance for a

Analog Circuit Design: Design Note Collection. http://dx.doi.org/10.1016/B978-0-12-800001-4.00087-9

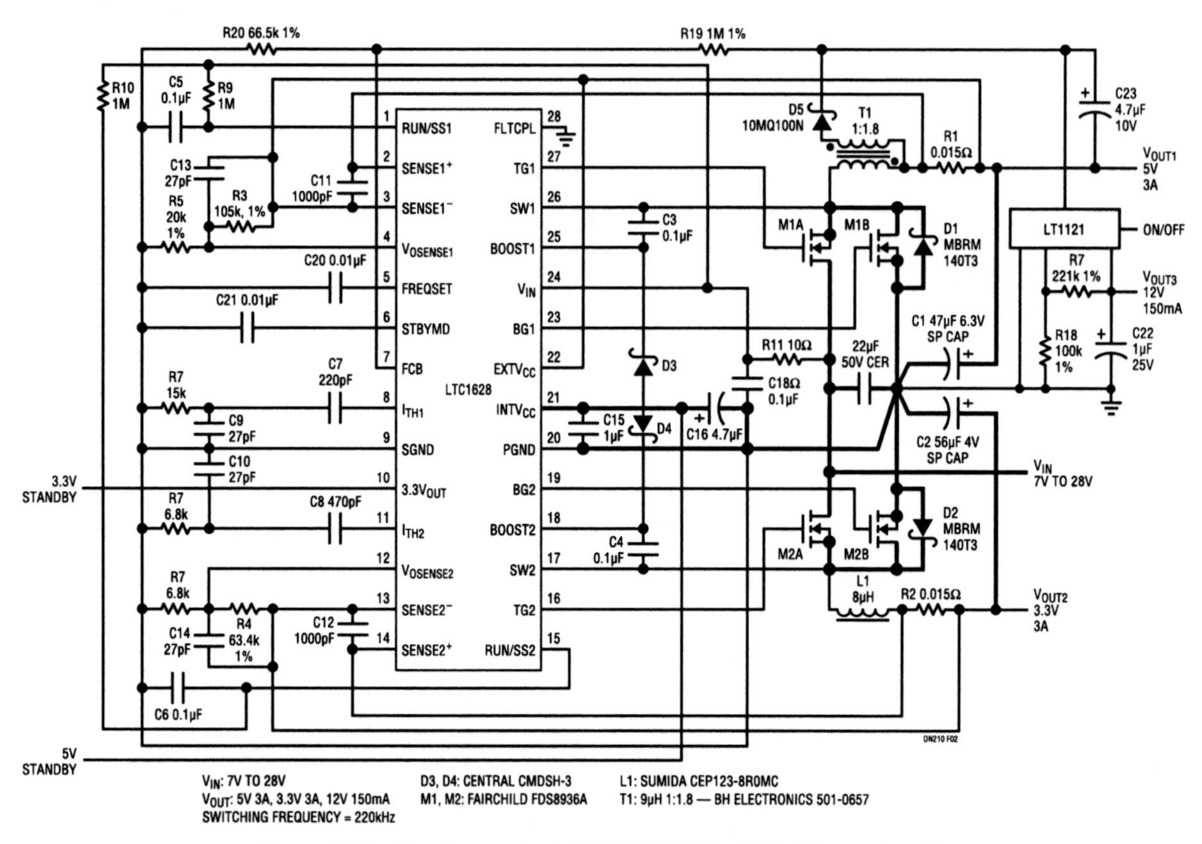

Figure 87.2 • Two Square Inch 5V/3.3V/12V Portable Power Supply

reasonable case size while meeting peak current requirements. The photo in Figure 87.3 shows normal output ripple voltage and transient response to a 1.5A to 2.5A load step in continuous mode (5V output). The output ripple voltage is 25mV$_{P-P}$ and the 1A load steps result in 60mV output voltage transients. The output ripple voltage in Burst Mode operation is 70mV$_{P-P}$. The efficiency of the power supply with a 15V input voltage and 5V/3A, 3.3V/3A and 12V/120mA output voltages measures 91%.

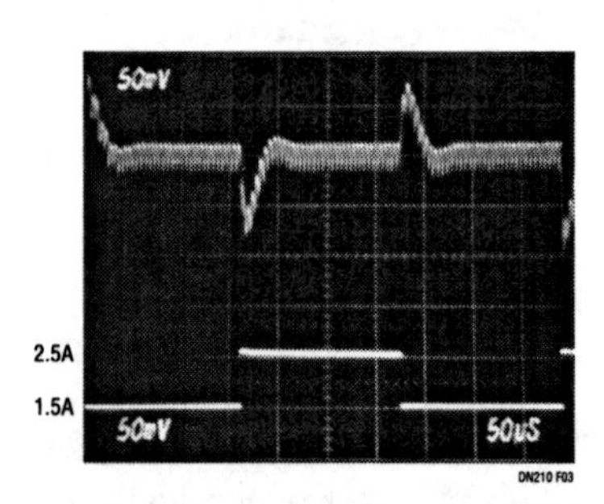

Figure 87.3 • Upper Trace: 5V Output, Transient Load Response (50mV/DIV) Bottom Trace: Load Current (1A/DIV)

Other features of the LTC1628 include:

- A 0.8V reference and 1% output voltage accuracy over line, load and temperature variations
- 7.5% overvoltage "soft-latch"
- Defeatable overcurrent latch-off timer
- Adjustable soft-start
- 25µA shutdown current
- 5V and 3.3V standby linear voltage regulators
- Loop bandwidth normally limited only by external inductor, output capacitor, switching frequency and the quality of the printed circuit board design

Low dropout 550kHz DC/DC controller operates from inputs as low as 2V

88

San Hwa Chee

The LTC1622 is a versatile high efficiency step-down controller that easily meets the size requirements of handheld portable applications with its small MSOP package. High frequency of operation (550kHz) allows the use of small magnetics, providing a complete power solution while occupying only a small amount of area. For even smaller magnetics, the LTC1622 can be synchronized up to 750kHz. Its wide operating input voltage range (2V to 8.5V) allows the part to be powered from a single or two lithium-ion batteries or 2- to 6-cell NiCd and NiMH battery packs.

The LTC1622 uses a constant frequency, current mode architecture, that provides excellent AC and DC load and line regulation. Its OPTI-LOOP compensation allows the transient response to be optimized while removing the constraints placed on C_{OUT}, such as limits on low ESRs. Burst Mode operation enhances efficiency at low load current while 100% duty cycle allows low dropout for extended operating time in battery-operated systems. Peak inductor current is set by an external sense resistor, allowing the design to be optimized for each specific application.

2.5V, 4A buck DC/DC converter

Figure 88.1 shows an application that can be used to power a handheld computer. The sense resistor has been selected to ensure low dropout operation while providing an output current of 4A. For short-circuit protection, a low cost diode, D2, is connected between the I_{TH} pin and V_{OUT} providing current foldback.

In addition, the LTC1622 operating frequency will be reduced to 110kHz when the voltage at the feedback pin drops below 0.3V. This feature ensures that all energy in the inductor will be completely dissipated at the end of each cycle, preventing inductor current runaway.

Due to a large amount of current that flows through the external P-channel MOSFET in this application, it is imperative that the MOSFET has proper heat sinking to thermally conduct heat away.

Figure 88.2 shows the efficiency curves. At 3V input, 90% efficiency is obtained between load currents of 0.5A to 2A.

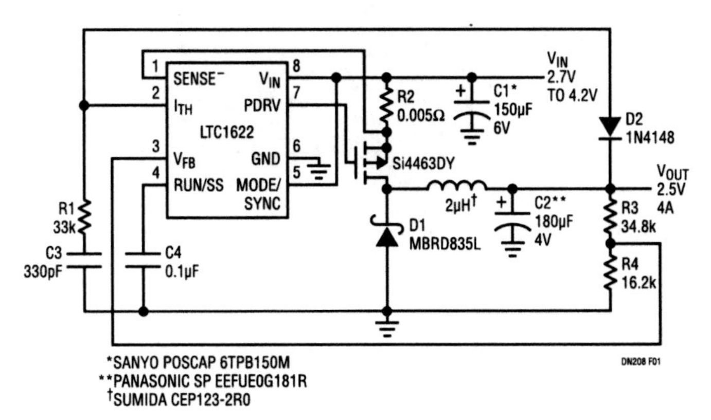

Figure 88.1 • 2.5V, 4A Step-Down Converter

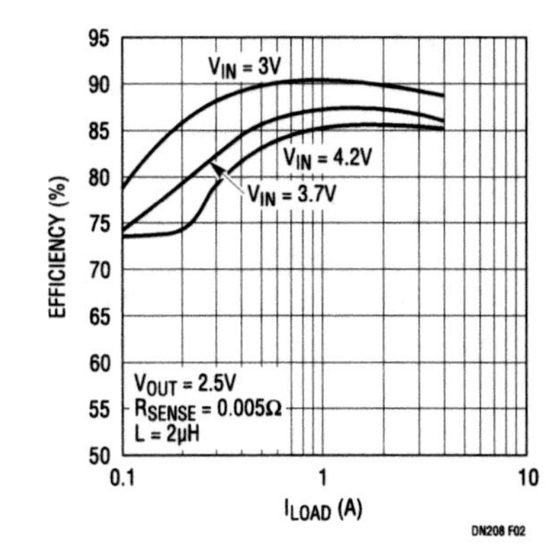

Figure 88.2 • Efficiency vs Output Load Current

Analog Circuit Design: Design Note Collection. http://dx.doi.org/10.1016/B978-0-12-800001-4.00088-0

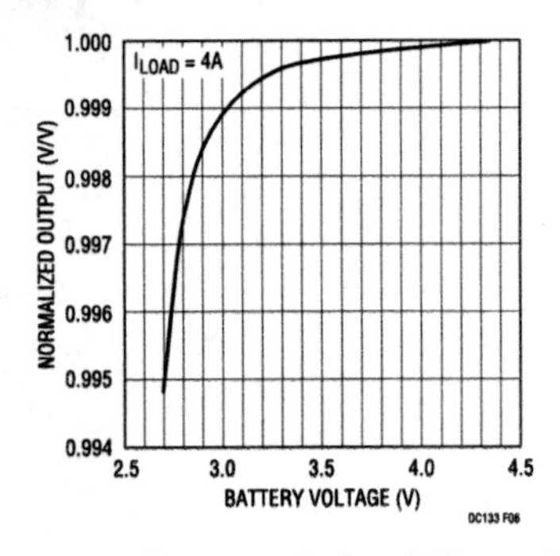

Figure 88.3 • Dropout Characteristic of Figure 88.1

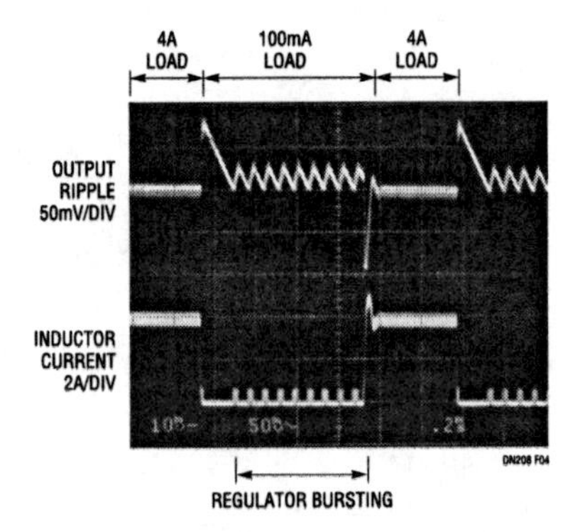

Figure 88.4 • Load Step Response (0.1A to 4A)

For load currents between 0.8A to 4A, the efficiency is at or above 85%.

Figure 88.3 shows the dropout characteristic of the circuit in Figure 88.1 and Figure 88.4 shows the load step response from 0.1A to 4A. To ensure the output voltage deviates less than 100mV during a load step, a low ESR output capacitor is chosen for this circuit. For a lower voltage deviation, a similar capacitor can be connected in parallel with the existing one.

"Zeta" step-up/step-down converter

For applications when the input voltage ranges between 4.2V to 2.7V (from a single lithium-ion) and the output is 3.3V, a zeta converter is needed. A zeta converter will step down the input when its input voltage is greater than the output voltage and step up the input voltage when input falls below the output voltage. Figure 88.5 shows a circuit with a capability of supplying 1A of output current. Note that inductors L1A and L1B are actually in a single package and they are connected in the polarity indicated. Again, diode D2 is used for limiting the short-circuit current. Once again, be sure to properly allow adequate copper area by the MOSFET to conduct heat away.

Figure 88.6 shows its efficiency for input voltages of 3V, 3.7V and 4.2V. Figure 88.7 shows the load step response from 50mA to 700mA. A single low ESR output capacitor is used to maintain low output voltage deviation during the load step.

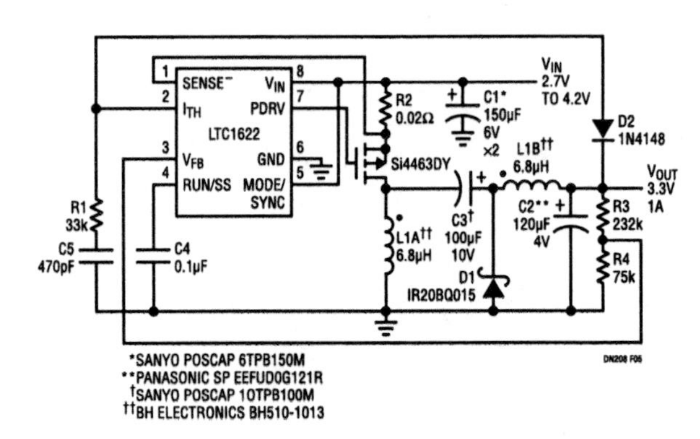

Figure 88.5 • Zeta Converter (3.3V, 1A)

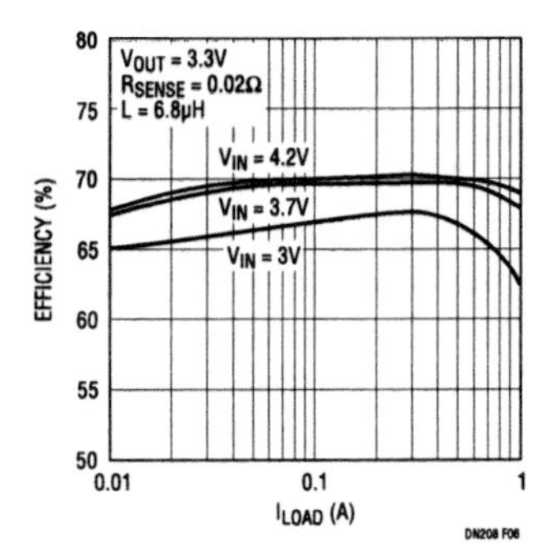

Figure 88.6 • Efficiency vs Output Load Current

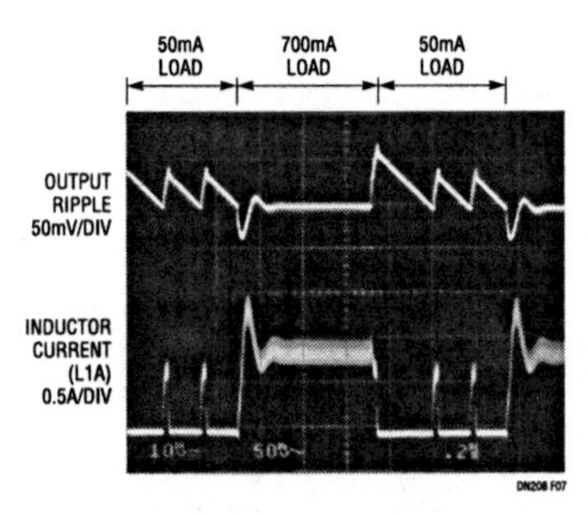

Figure 88.7 • Load Step Response (50mA to 700mA)

Switching regulator controllers set a new standard for transient response

89

Dave Dwelley

The LTC1702 is the first in a new family of low voltage, high speed switching regulator controllers. It is designed to operate from a standard 5V logic supply and generate two lower voltage, high current regulated outputs. Running at a fixed 550kHz switching frequency, each side of the LTC1702 features a voltage feedback architecture with a 25MHz gain-bandwidth op amp as the feedback amplifier, allowing loop crossover frequencies in excess of 50kHz. Powerful onboard MOSFET drivers allow the LTC1702 to drive large, high current external MOSFETs efficiently at 550kHz. The high feedback loop bandwidth maintains excellent transient response, and the high switching frequency allows the use of small external inductors and capacitors even as load currents rise beyond 10A. The dual output LTC1702 is packaged in a space-saving 24-pin narrow SSOP.

Mobile PCs using the most recent Intel Pentium III processors require LTC1702-level performance coupled with a DAC-controlled voltage at the core supply output. The LTC1703 is designed specifically for this application, and consists of a modified LTC1702 with an internal 5-bit DAC controlling the output voltage on side 1. The DAC conforms to the Intel Mobile VID specifications (Figure 89.1).

The LTC1702 and LTC1703 each consist of two independent switching regulator controllers in one package. Each controller is designed to be wired as a voltage feedback, synchronous step-down regulator, using two external N-channel MOSFETs per side as power switches (Figure 89.2). A small external charge pump (D_{CP} and C_{CP} in Figure 89.2) provides a boosted supply voltage to keep Q1 fully turned on. The switching frequency is set internally at

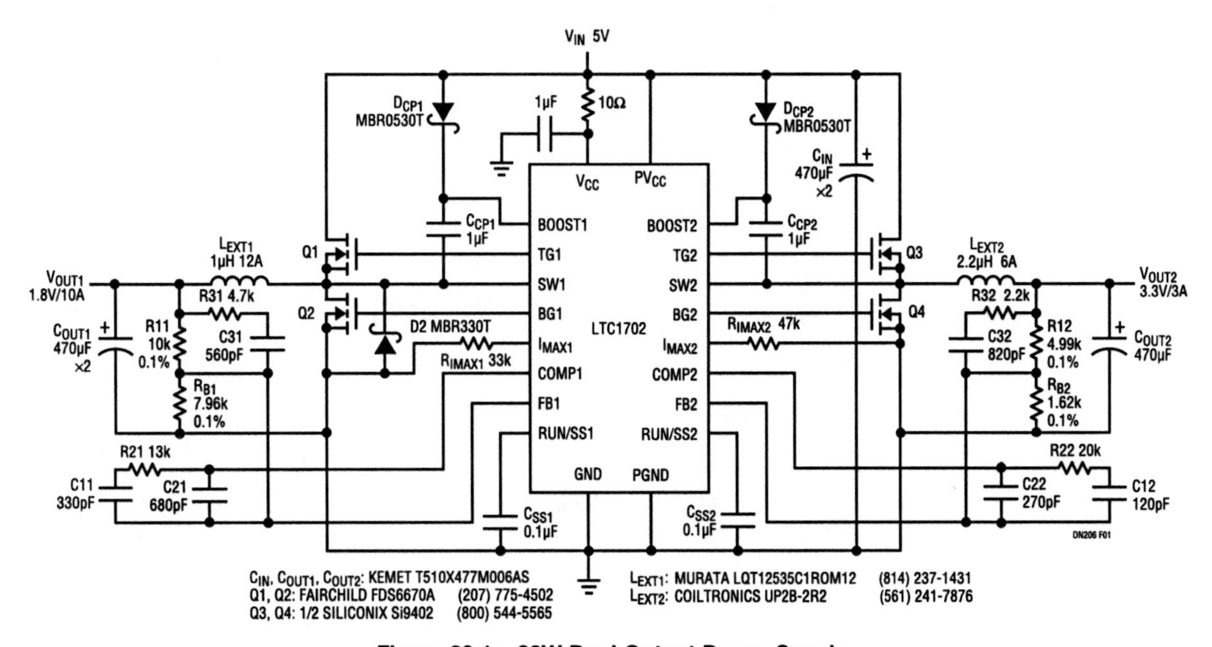

Figure 89.1 • 28W Dual Output Power Supply

Analog Circuit Design: Design Note Collection. http://dx.doi.org/10.1016/B978-0-12-800001-4.00089-2

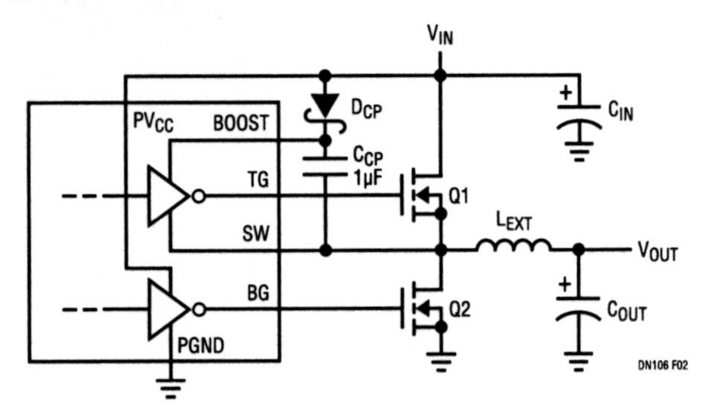

Figure 89.2 • LTC1702/LTC1703 Switching Architecture

550kHz. A user-programmable current limit circuit uses the synchronous MOSFET switch, Q2, as a current sensing element, eliminating the need for an external low value current sensing resistor.

Unlike conventional switching regulator designs, the LTC1702/LTC1703 uses a true 25MHz gain-bandwidth op amp as the feedback amplifier. This allows the use of an OPTI-LOOP compensation scheme that can precisely tailor the loop response. The high gain-bandwidth product allows the loop to be crossed over beyond 50kHz while maintaining good stability and significantly enhancing load transient response. Bias resistors, R_{B1} and R_{B2} (Figure 89.1) are used to set the DC output voltages along with two pole/zero pairs per side to compensate for phase shift caused by the inductor/output capacitor combinations.

Another feature of the LTC1702/LTC1703 reduces the required input capacitance with no performance penalty. The LTC1702/LTC1703 includes a single master clock that drives the two sides such that side 1 is 180° out of phase with side 2. This technique, known as 2-phase switching, has the effect of doubling the frequency of the switching pulses seen by the input capacitor and significantly reduces their RMS value. With 2-phase switching, the input capacitor is sized as required to support the larger of the two sides at maximum load current. As the load increases on the lower current side, it tends to cancel, rather than add to, the RMS current seen by the input capacitor; thus no additional capacitance needs to be added.

The 550kHz clock frequency and the low 5V input voltage allow the use of external inductors in the 1µH range while keeping ripple current under control. The low inductance value helps in two ways: it reduces the energy stored in the inductor during each switching cycle, reducing the physical core size required, and it raises the attainable dI/dt at the output of the circuit, decreasing the time that it takes for the circuit to correct for sudden changes in load current. This, in turn, reduces the amount of output capacitance required to support the output voltage during a load transient. Together with the reduced capacitance at the input due to the LTC1702/LTC1703 2-phase internal switching significantly reduces the amount of total capacitance needed compared to a conventional design running at 300kHz or less.

A typical LTC1702 application is shown in Figure 89.1. Input is taken from the 5V logic supply. Side 1 is set up to provide 1.8V at 10A, whereas side 2 is set to supply 3.3V at a lower 3A load level. System efficiency peaks at greater than 90% at each side. This circuit shows examples of both high power and lower power output designs possible with the LTC1702 controller. Side 1 uses a pair of ultralow $R_{DS(ON)}$ Fairchild SO-8 MOSFETs and a 1µH/12A Murata surface mount inductor. C_{IN} consists of two 470µF tantalum capacitors to support side 1 at full load, and C_{OUT1} uses two more 470µF devices to provide better than 3% regulation with 0A to 10A transients.

Side 2 uses a single SO-8 dual MOSFET and a smaller 2.2µH/6A inductor. C_{OUT2} is a single 470µF capacitor used to support 0A to 3A transients while maintaining better than 3% regulation. As the load current at side 2 increases, the LTC1702 2-phase switching actually reduces the RMS current in C_{IN}, removing the need for additional capacitance at the input beyond what side 1 requires. Both sides exhibit exceptional transient response (Figures 89.3 and 89.4). The entire 28W converter can be laid out in less than 1.5 square inches.

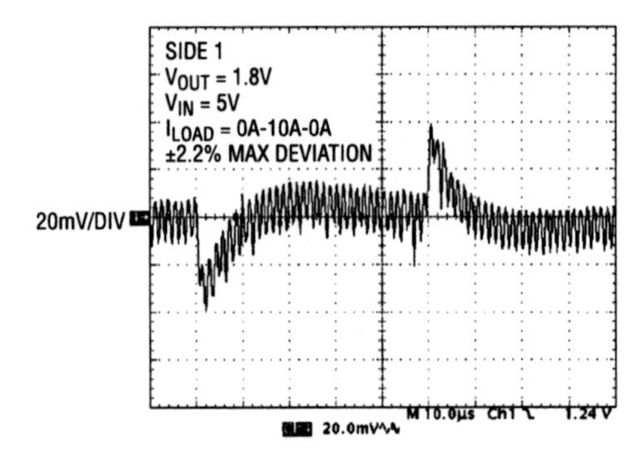

Figure 89.3 • 0A to 10A Transient Response (Figure 89.1, Side 1)

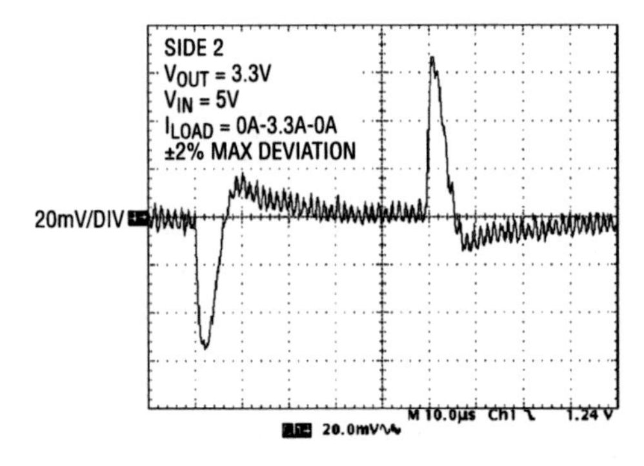

Figure 89.4 • 0A to 3A Transient Response (Figure 89.1, Side 2)

60V, high efficiency buck switching regulators in SO-8

90

Ajmal Godil

The LT1676 and LT1776 are wide input range, high efficiency buck (step-down) switching regulators in SO-8 packages. Typical applications include automotive DC/DC conversion, telecom 48V step-down and IEEE1394 (FireWire) step-down converters.

Both the LT1676 and the LT1776 can accept input voltages from 7V to 60V. The LT1676 has a nominal switching frequency of 100kHz whereas the LT1776 runs at 200kHz. For higher input voltage applications, the LT1676 is a better choice because it minimizes AC losses by switching at a lower frequency. For lower input voltage and cases where the input supply has high transient spikes, the LT1776's higher switching frequency allows a smaller inductor to be used. The LT1676 and the LT1776 each include an onboard 700mA peak current switch, oscillator, control and protection circuitry. These controllers use current mode control and exhibit high efficiency, low ripple and fast transient response. The circuit schematic in Figure 90.1 highlights the capabilities of the LT1676.

The LT1676 and the LT1776 can both be disabled by connecting the shutdown ($\overline{SHDN}$) pin to ground, reducing input current to a few microamperes. For normal operation, decouple the $\overline{SHDN}$ pin with a 100pF capacitor to ground. The part also has a SYNC pin, used to synchronize the internal oscillator to an external clock, which can range from 130kHz to 250kHz for the LT1676 and 250kHz to 400kHz for the LT1776. To use the part's internal oscillator, simply connect the SYNC pin to ground.

Since both parts allow such a wide input voltage range, they use two techniques to optimize efficiency. First, the internal control circuitry draws power from the V_{CC} pin, which is normally connected to the output supply. During start-up, the controllers draw power from V_{IN}. When the output voltage exceeds 2.9V, the bias power comes from the output. This reduces quiescent power consumption by hundreds of milliwatts when operating at high input voltage. Second, the switch circuitry maintains a fast rise time at high loads (see Figure 90.2). Both of these factors help in maximizing efficiency with high input voltages. At light loads, the switch rise time slows down (see Figure 90.3) to avoid pulse-skipping, thus maintaining a constant frequency from heavy

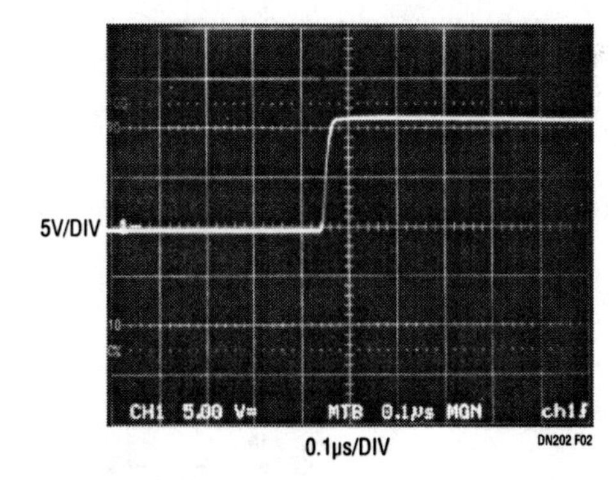

Figure 90.2 • Switch Rise Time at Heavy Load Climbs Quickly to Maximize Efficiency

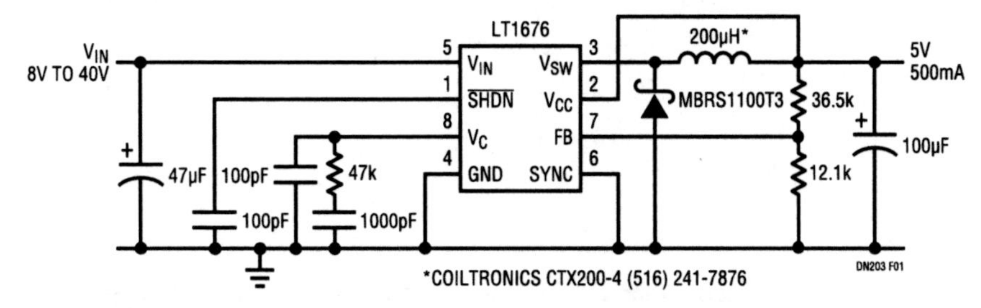

Figure 90.1 • LT1676 Application Circuit Generating 5V at 500mA from a FireWire Input (8V to 40V)

Analog Circuit Design: Design Note Collection. http://dx.doi.org/10.1016/B978-0-12-800001-4.00090-9

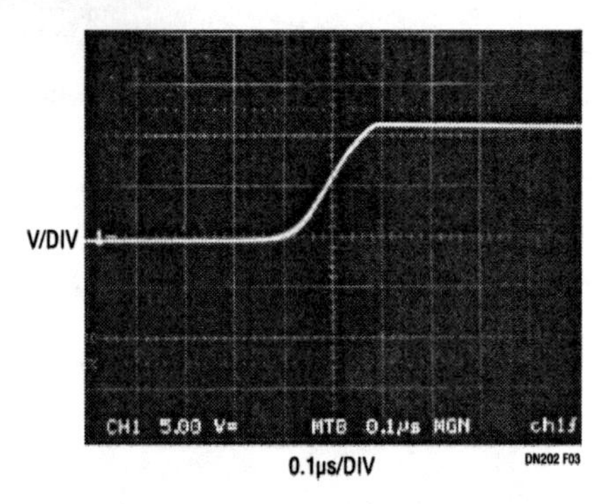

Figure 90.3 • Switch Rise Time at Light Load Slows Down to Avoid Pulse-Skipping

to light loads. This helps significantly in reducing the output ripple voltage and eliminates switching noise in the audio frequency spectrum.

Figure 90.4 shows the typical efficiency curves for various input voltages from 8V to 60V at an output voltage of 5V. Figure 90.5 shows the output voltage ripple with a steady-state load of 500mA. As can be seen in Figures 90.4 and 90.5, the LT1676 buck circuit has an efficiency of approximately 80% over a wide input voltage range, and only 16mV of output voltage ripple at a load current of 500mA.

Generating low cost, dual-voltage supplies

The circuit in Figure 90.6 shows a cost-effective way to generate 5V and −5V from a single 10V to 28V supply using the LT1776 and a few other off-the-shelf components.

L1A and L1B are two windings on a single core to generate ±5V. To minimize coupling mismatches between the two windings, C3 has been added, forcing the winding potentials to be equal and improving cross-regulation. The total available current is limited to 500mA. The maximum negative supply current is limited by the positive 5V load. A typical limit is one-half of the positive current.

Conclusion

The main benefits of the LT1676 and the LT1776 are their ability to handle high input voltages—up to 60V—and their control of switch turn-on time: slower at light loads to maintain constant switching frequency and faster at heavy loads to maximize efficiency.

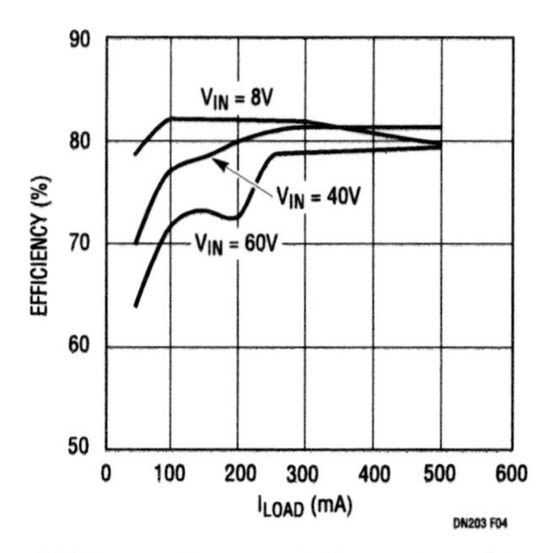

Figure 90.4 • Efficiency Curves of Figure 90.1 Circuit

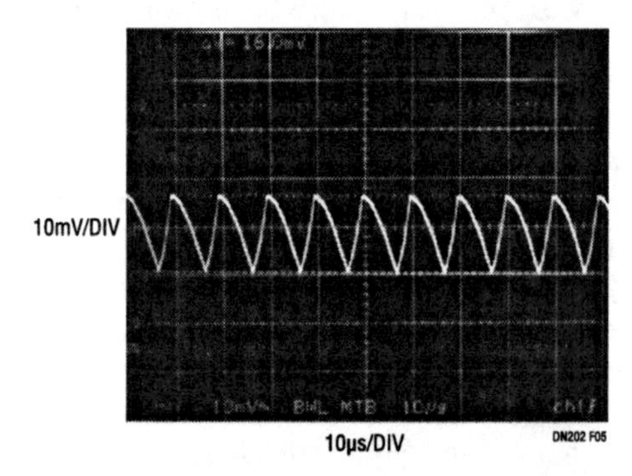

Figure 90.5 • Output Voltage Ripple for the LT1676 at $I_L = 500mA$

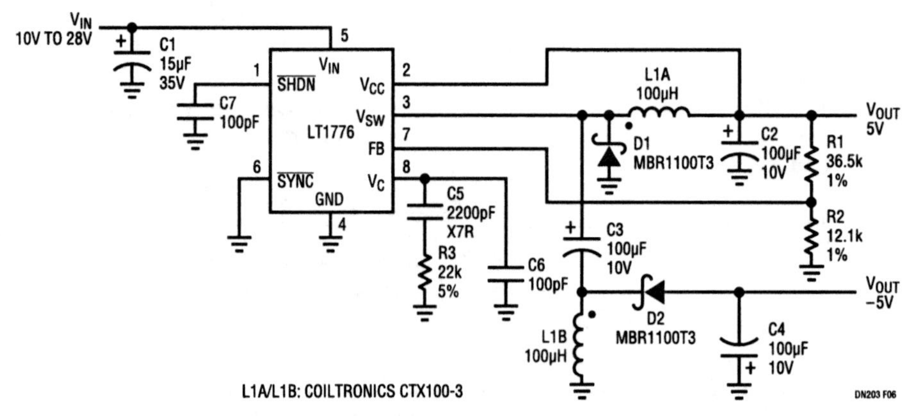

Figure 90.6 • Generating Dual Output Voltages from a Single Supply

High efficiency, monolithic synchronous step-down regulator works with single or dual Li-Ion batteries

91

Jaime Tseng

Li-Ion batteries, with their high energy density, are becoming the chemistry of choice for designers of many handheld products. In response to this growing trend, the LTC1627 incorporates features optimized for such chemistry. For instance, the LTC1627 integrates a precision undervoltage lockout circuit that shuts itself down when the supply voltage dips below 2.5V, drawing only 6μA of quiescent current. This prevents damaging a Li-Ion battery when it is nearing its end of charge. The operating supply range of the LTC1627 is from 2.65V to 8.5V; this accommodates one or two Li-Ion batteries and 3- to 6-cell NiCd and NiMH battery packs.

The LTC1627 is a current mode buck regulator using a fixed frequency architecture. It incorporates power saving Burst Mode operation and allows 100% duty cycle. The current mode architecture gives the LTC1627 excellent load and line regulation. Burst Mode operation provides high efficiency at low load currents. 100% duty cycle provides low dropout operation that extends operating time in battery-operated systems. The operating frequency is internally set at 350kHz, allowing the use of small surface mount inductors. For switching noise sensitive applications the LTC1627 can be externally synchronized at up to 525kHz by applying a clock signal of at least $1.5V_{P-P}$ to the SYNC/FCB pin.

Figure 91.1 shows a typical LTC1627 application circuit suitable for a single or dual Li-Ion battery input. Note that the Schottky diode normally seen on the SW pin is absent in Figure 91.1—the LTC1627 has no need for it, saving cost. Figure 91.2 shows the efficiency curves for three different input voltages. The efficiency with a 3.6V input exceeds 90% over a load range from 10mA to 600mA, making the LTC1627 attractive for all battery-operated products and efficiency sensitive applications.

Single Li-Ion applications

In single Li-Ion battery applications requiring a maximum output load current of 500mA, the top P-channel MOSFET's gate can be driven below ground to reduce its $R_{DS(ON)}$. This reduces the I^2R loss that dominates the efficiency loss in low V_{IN} applications. As V_{IN} drops, the duty cycle of the converter increases until, in the extreme case, it reaches dropout, where the P-channel MOSFET is on continuously. Figure 91.3 shows an application circuit suitable for a single Li-Ion cell that can deliver a 500mA load current down to $V_{IN}=3V$. The

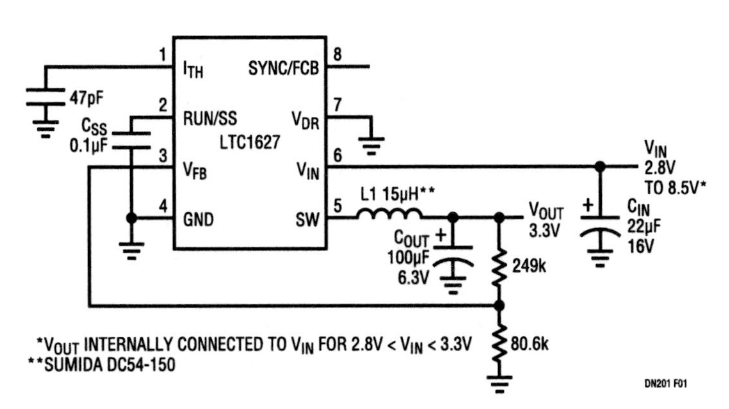

Figure 91.1 • High Efficiency Step-Down Converter

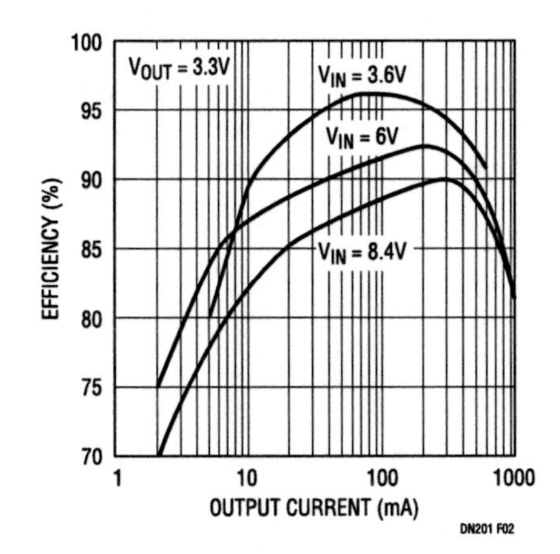

Figure 91.2 • Efficiency vs Output Load Current

Analog Circuit Design: Design Note Collection. http://dx.doi.org/10.1016/B978-0-12-800001-4.00091-0

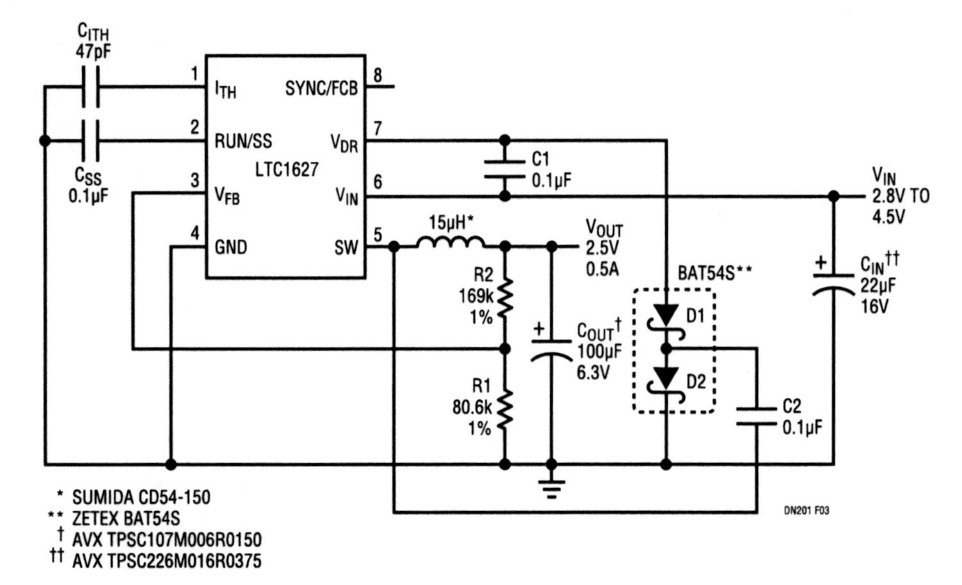

Figure 91.3 • Single Lithium-Ion to 2.5V/0.5A Regulator

top P-channel MOSFET driver makes use of a floating pin, V_{DR}, to allow biasing below ground. A simple charge pump bootstrapped to the SW pin realizes a negative voltage at the V_{DR} pin, as shown. Using the charge pump at $V_{IN} \geq 4.5V$ is not recommended, as this may cause $(V_{IN} - V_{DR})$ to exceed its absolute maximum value of 10V. If V_{IN} decreases to a voltage close to V_{OUT}, the loop may enter dropout and attempt to turn on the P-channel MOSFET continuously. A dropout detector counts the number of oscillator cycles that the P-channel MOSFET remains on, and periodically forces a brief off period to allow C1 to recharge. (100% duty cycle is allowed when V_{DR} is grounded as shown in Figure 91.1.)

Auxiliary winding control using SYNC/FCB pin

The SYNC/FCB pin controls the operation of the internal synchronous MOSFET by inhibiting Burst Mode operation and forcing continuous mode operation. When this pin drops below its ground referenced 0.8V threshold, continuous mode operation is forced. In continuous mode operation, the internal main and synchronous MOSFETs are switched continuously, regardless of the load on the main output. Synchronous switching removes the normal limitation that power must be drawn from the inductor primary winding in order to extract power from auxiliary windings. Synchronous operation allows power to be drawn from the auxiliary windings regardless of the primary output load.

The secondary output voltage is set by the turns ratio of the transformer in conjunction with a pair of external resistors returned to the SYNC/FCB pin, as shown in Figure 91.4. The secondary voltage V_{SEC} in Figure 91.4 is given by:

$$V_{SEC} \cong (N+1)V_{OUT} - V_{DIODE} > 0.8V\left(1 + \frac{R4}{R3}\right)$$

where N is the turns ratio of the transformer, V_{OUT} is the main output voltage sensed by V_{FB} and V_{DIODE} is the voltage drop across the Schottky diode.

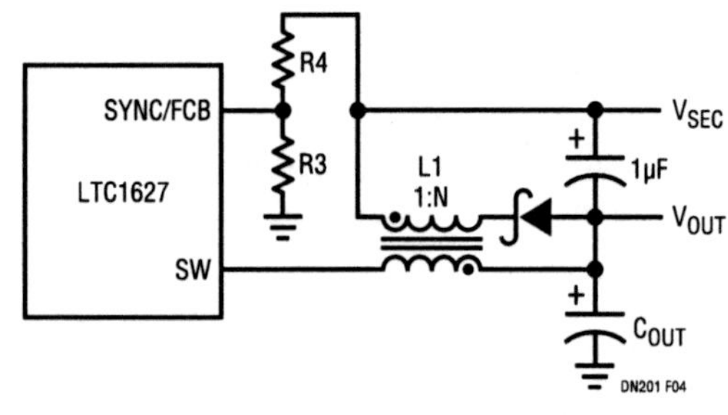

Figure 91.4 • Secondary Output Loop Connection

A low cost, efficient mobile CPU power

92

Randy G. Flatness

The LTC1735 is the newest member of Linear Technology's third generation of synchronous step-down controllers. This controller uses the same constant frequency, current mode architecture and Burst Mode operation as the industry standard LTC1435 and LTC1437 controllers but with improved features. With OPTI-LOOP compensation, new protection circuitry, tighter load regulation and stronger MOSFET drivers, the LTC1735 is ideal for the current and future generations of mobile CPUs. A companion part, the LTC1736, has all the features of the LTC1735 plus 5-bit VID voltage programming according to Intel mobile processor specifications.

The LTC1735 is pin compatible with the previous generation LTC1435 and LTC1437A controllers with only minor external component value changes. New protection features include internal foldback current limiting, output overvoltage crowbar and optional short-circuit shutdown. The 0.8V reference supports the low output voltages and 1% accuracy that will be demanded by future microprocessors. A constant operating frequency (synchronizable up to 500kHz) is set by an external capacitor, C_{OSC}, allowing maximum flexibility in optimizing efficiency.

The LTC1735's OPTI-LOOP compensation removes the constraints placed on C_{OUT} by other controllers (such as

Low cost dynamic VID for Pentium III processors

The circuit in Figure 92.1 generates CPU power (1.6V at 10A) from input voltages of 5V to 26V. Adding the low cost components shown in the insert of Figure 92.1 creates a dynamic VID two-level output voltage. This implementation produces output voltages of 1.3V with VG low and 1.5V with VG high. The LTC1735 has overvoltage protection that tracks the programmed output voltage, always protecting the CPU. If a power good output is needed, the LTC1735 can be replaced with the increased functionality LTC1735-1.

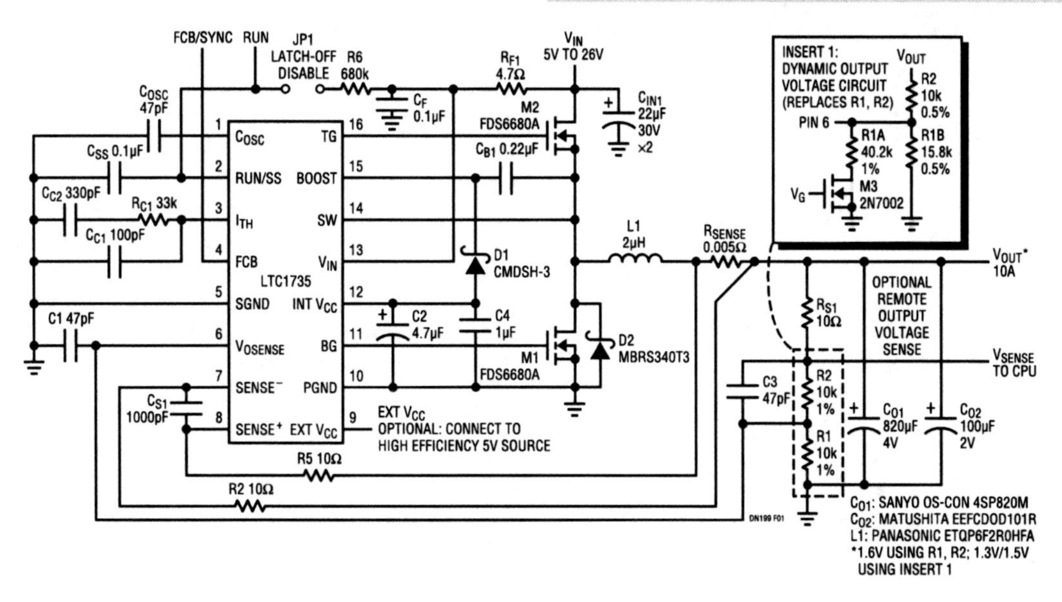

Figure 92.1 • Mobile CPU Power Supply

Analog Circuit Design: Design Note Collection. http://dx.doi.org/10.1016/B978-0-12-800001-4.00092-2

restrictions on very low ESR) for proper operation. A maximum duty cycle limit of 99% provides low dropout operation, which extends operating time in battery-operated systems. A wide input supply range allows operation from 3.5V to 30V (36V maximum).

The RUN/SS capacitor, C_{SS}, (refer to Figure 92.1) is used initially to turn on and limit the inrush current of the controller and as a short-circuit timer. If the output voltage falls to less than 70% of its nominal output voltage after C_{SS} is charged, it is assumed that the output is in a severe overcurrent and/or short-circuit condition and C_{SS} begins discharging. If the condition lasts for a long enough period, as determined by the size of C_{SS}, the controller will be shut down until the RUN/SS pin voltage is recycled. Jumper JP1 disables the overcurrent latch-off.

New internal protection features in the LTC1735 controller include foldback current limiting, short-circuit detection, short-circuit latch-off and overvoltage protection. These features protect the PC board, the MOSFETs and the CPU against faults.

Why should you defeat overcurrent latch-off? During the prototyping stage of a design, there may be a problem with noise pickup or poor layout causing the protection circuit to latch off. Defeating this feature will allow easy troubleshooting of the circuit and PC layout. The internal short-circuit detection and foldback current limiting remain active, thereby protecting the power supply system from failure. After the design is complete, you can decide whether to enable the latch-off feature.

The LTC1735 current comparator allows a maximum MOSFET current of $75mV/R_{SENSE}$. The use of a low loss sense resistor not only provides accurate current limiting, but allows the use of arbitrarily low output capacitor ESR values for outstanding load-step response. A 0.005Ω sense resistor programs a maximum load current of 10A, (higher currents may be set by lowering R_{SENSE}). The LTC1735 includes current foldback to help further limit load current when the output is shorted to ground. If the output falls by more than one-half, the maximum sense voltage is decreased to limit dissipation in the bottom MOSFET, resulting in a short-circuit current of 3.5A. Note that this function is always active and is independent of the short-circuit latch-off (see Figure 92.2).

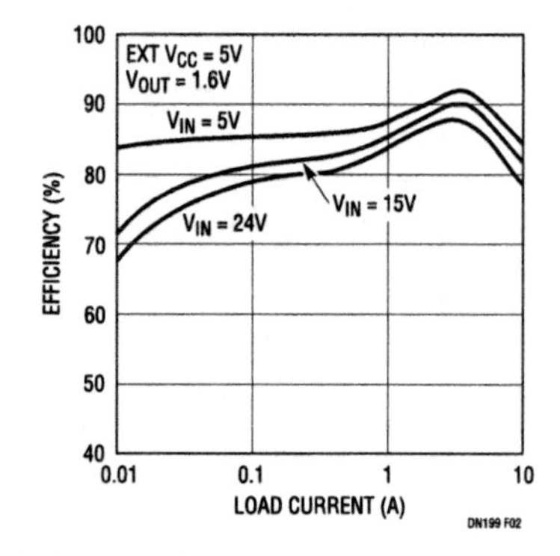

Figure 92.2 • Efficiency for Figure 92.1

The FCB pin is a multifunction pin that controls the operation of the synchronous MOSFET, is an input for external clock synchronization and reduces noise and RF interference by disabling Burst Mode operation. When the FCB pin drops below its 0.8V threshold (FCB = 0V), continuous mode operation is forced. In this case, the top and bottom MOSFETs continue to be driven synchronously regardless of the load on the main output.

The operating frequency is set to 270kHz by C_{OSC}. The LTC1735's internal oscillator can be synchronized to an external oscillator by applying a clock signal of at least $1.5V_{P-P}$ to the FCB pin. When synchronized to an external frequency, Burst Mode operation is disabled but cycle skipping occurs at low load currents because current reversal is inhibited. The bottom gate will come on every ten clock cycles to ensure that the bootstrap capacitor is kept refreshed and to keep the frequency above the audio range. The rising edge of an external clock applied to the FCB pin starts a new cycle. The range of synchronization with C_{OSC} = 47pF is from 240kHz to 400kHz.

The LTC1735 uses a new "soft latch" OVP circuit. Regardless of the operating mode, the synchronous MOSFET is forced on whenever the output voltage exceeds the regulation point by more than 7.5%. However, if the voltage then returns to a safe level, normal operation is allowed to resume, thereby preventing latch-off caused by noise or voltage reprogramming. This is important when dynamically changing output voltage since the overvoltage protection threshold tracks the new output voltage, always protecting the CPU.

Previous latching crowbar schemes for overvoltage protection have a number of problems. One of the most obvious, not to mention most annoying, is nuisance trips caused by noise or transients momentarily exceeding the OVP threshold. Each time that this occurs with latching OVP, a manual reset is required to restart the regulator.

The LTC1735 is designed to be used in higher current applications than the LTC1435 family. Stronger gate drives allow paralleling multiple MOSFETs or operating at higher frequencies. The LTC1735 has been optimized for low output voltage operation by reducing the minimum on-time to less than 200ns. Remember though, transition losses can still impose significant efficiency penalties at high input voltages and high frequencies. Just because the LTC1735 can operate at frequencies above 300kHz doesn't mean it should.

93

Optimizing a DC/DC converter's output capacitors

Ajmal Godil   Craig Varga

All DC/DC power supplies comprise closed-loop systems. In any closed-loop system, control theory dictates the need for adequate gain and phase margin for overall system stability. Trade-offs have to be made between phase and gain margin and transient response by adjusting the feedback gain for a given power stage.

Some power IC controller manufacturers have designed their products with internal loop compensation. Hence, the user is forced to select power stage components (mainly C_{OUT}) to meet stability criteria. The LTC1435A step-down DC/DC controller, on the other hand, allows the power stage component values to be chosen based simply on power

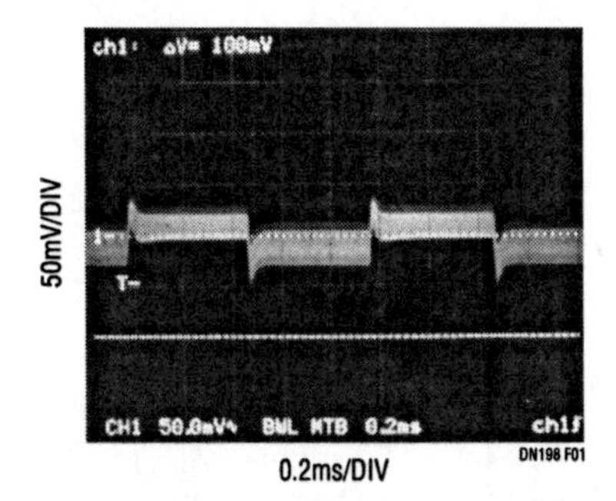

Figure 93.1 • Transient Response of LTC1435A. $C_{OUT}=2\times1500\mu F$ Sanyo VGX, $C_{C1}=100pF$, $C_{C2}=100pF$, $R_C=33k$, $I_L=0.5A$ TO 3A, $V_{IN}=12V$, $V_{OUT}=3.3V$

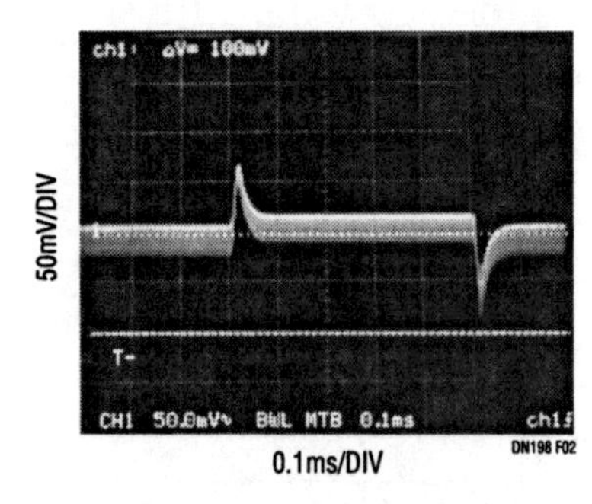

Figure 93.2 • Transient Response of LTC1435A. $C_{OUT}=2\times47\mu F$ OS-CON, $C_{C1}=470pF$, $C_{C2}=100pF$, $R_C=22k$, $I_L=0.5A$ TO 1.2A, $V_{IN}=12V$, $V_{OUT}=3.3V$

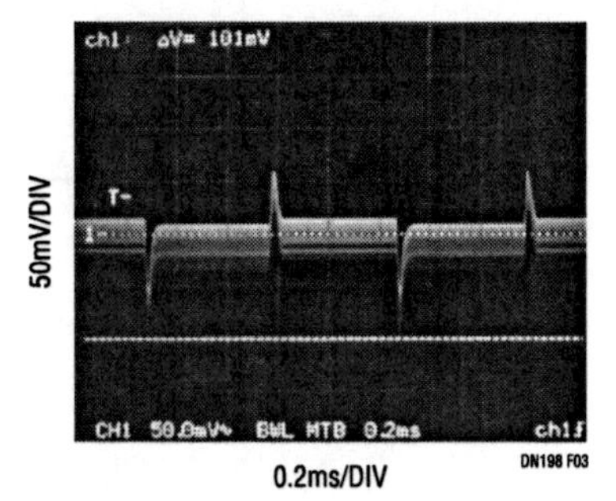

Figure 93.3 • Transient Response of LTC1435A. $C_{OUT}=1\times47\mu F$ OS-CON, $C_{C1}=1000pF$, $C_{C2}=100pF$, $R_C=15k$, $I_L=0.5A$ TO 1.2A, $V_{IN}=12V$, $V_{OUT}=3.3V$

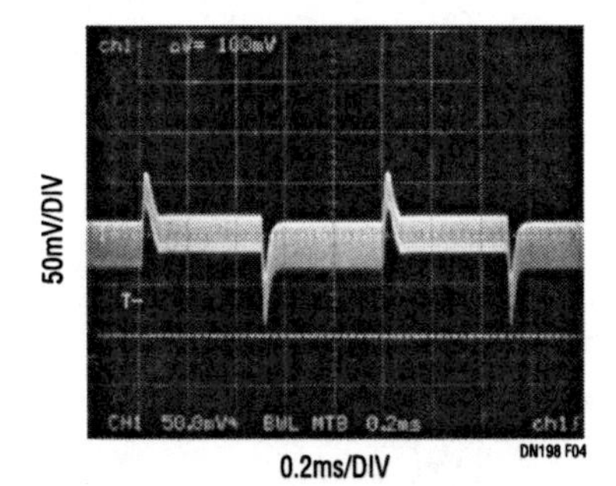

Figure 93.4 • Transient Response of LTC1435A. $C_{OUT}=2\times47\mu F$ POSCAP, $C_{C1}=1000pF$, $C_{C2}=100pF$, $R_C=22k$, $I_L=0.5A$ TO 1.8A, $V_{IN}=12V$, $V_{OUT}=3.3V$

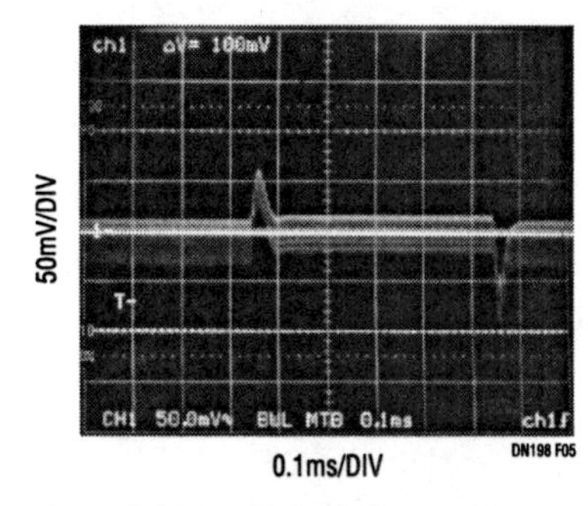

Figure 93.5 • Transient Response of LTC1435A. $C_{OUT}=1\times47\mu F$ Panasonic, SP, $C_{C1}=1000pF$, $C_{C2}=100pF$, $R_C=15k$, $I_L=0.5A$ TO 1.2A, $V_{IN}=12V$, $V_{OUT}=3.3V$

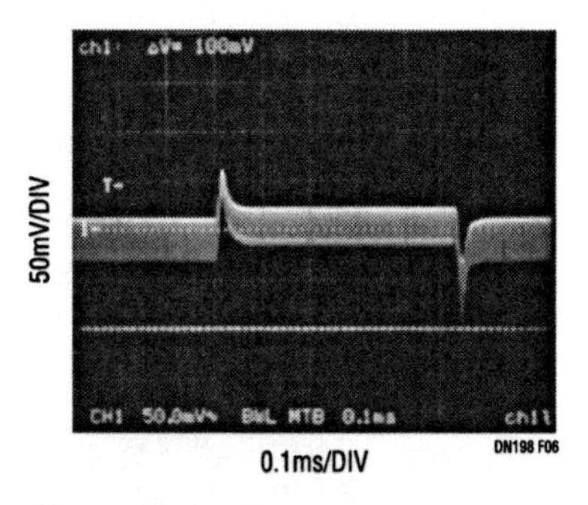

Figure 93.6 • Transient Response of LTC1435A. $C_{OUT}=2\times100\mu F$ NEOCAP, $C_{C1}=180pF$, $C_{C2}=100pF$, $R_C=47k$, $I_L=0.5A$ TO 2A, $V_{IN}=12V$, $V_{OUT}=3.3V$

Analog Circuit Design: Design Note Collection. http://dx.doi.org/10.1016/B978-0-12-800001-4.00093-4

requirements and allows feedback gain to be set independently, thus allowing minimization of expensive bulk output capacitors. This important design freedom results from the OPTI-LOOP architecture that makes the I_{TH} compensation point available in all Linear Technology DC/DC controllers.

Figure 93.1 through 93.6 show the output transient response of an LTC1435A circuit (application schematic in Figure 93.7) using various kinds of output bulk capacitors. The amplitude of the transient response is less than ±100mV. In each case, the number of output bulk capacitors used was chosen to produce 50mV or less of output voltage ripple at I_{LOAD} = 3.0A. Figure 93.8 shows the phase and gain margin for the application circuit in Figure 93.7 for C_{OUT} = 47µF × 2, 6.3V OS-CON capacitors. It can be observed from Figure 93.8 that the loop crosses 0dB at 21.8kHz and has a phase margin of 47.3°, which is more than enough for the loop to be unconditionally stable.

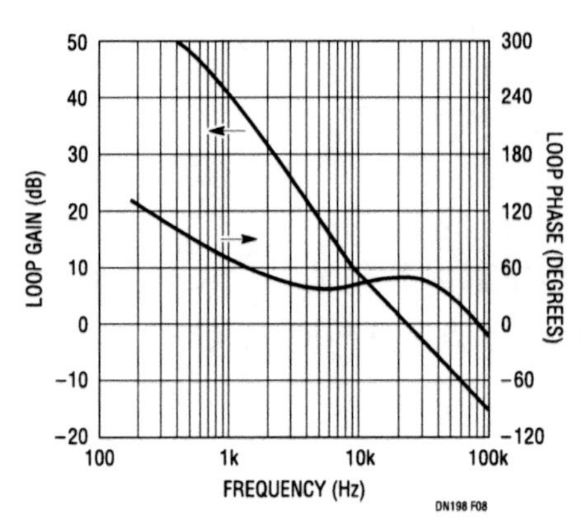

Figure 93.8 • Loop Gain and Phase vs Frequency

For each output capacitor type, the feedback loop compensation was adjusted for similar phase margin and dynamic performance. Note that the compensating resistor and capacitor values are not the same for each output capacitor configuration. Clearly, a fixed internal loop-compensation scheme does not allow optimization for all applications. For any general purpose power supply controller, the ability to tailor the feedback loop to the power path offers a significant advantage to the designer.

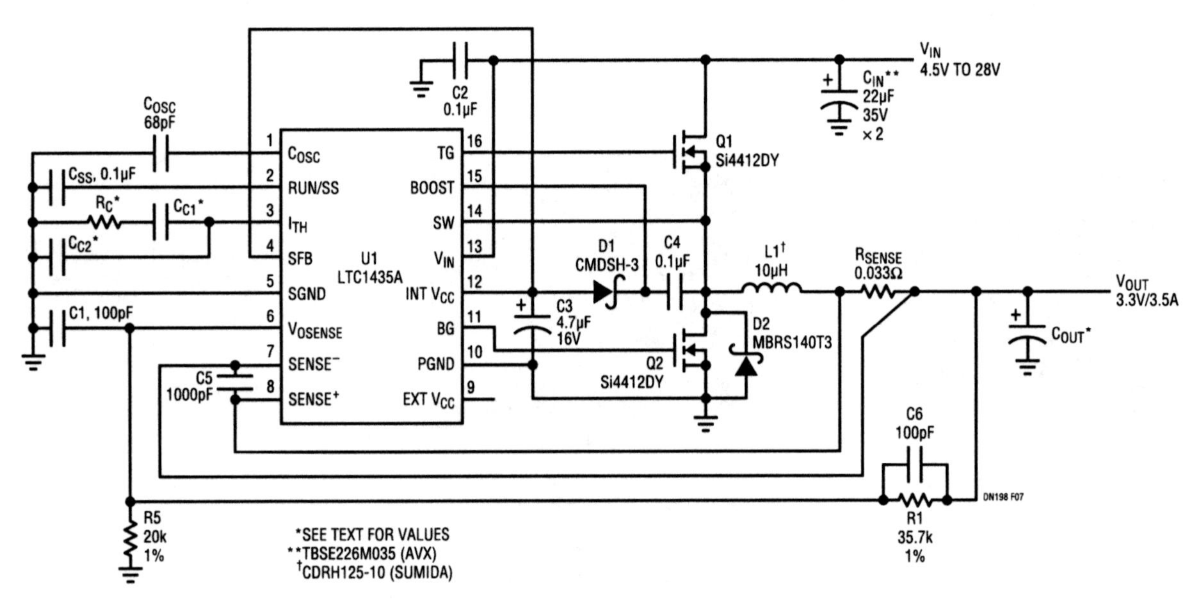

Figure 93.7 • LTC1435A Constant Frequency, High Efficiency Converter

Step-down converter operates from single Li-Ion cell

94

Tim Skovmand

Introduction

The LTC1626 is a low voltage, high efficiency, monolithic step-down DC/DC converter featuring an input supply voltage range of 2.5V to 6V, which makes it ideal for single-cell Li-Ion applications. A built-in low $R_{DS(ON)}$ switch provides high efficiency and allows up to 0.6A of output current. The LTC1626 incorporates automatic power saving Burst Mode operation to reduce gate-charge losses when the load current drops. With no load, the converter draws only 160μA and in shutdown it draws a mere 1μA, making it ideal for current-sensitive applications.

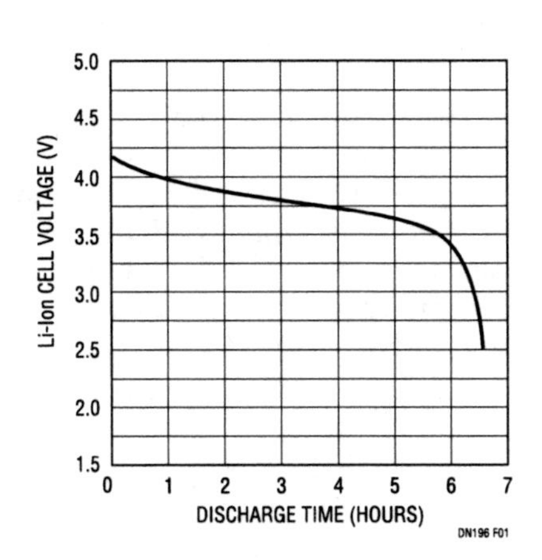

Figure 94.1 • Typical Single-Cell Li-Ion Discharge Curve

Single-cell Li-Ion operation

As shown in Figure 94.1, a fully charged single-cell Li-Ion battery begins the discharge cycle between 4.1V and 4.2V. During most of the discharge period, the cell produces between 3.5V and 4.0V. Toward the end of discharge, the cell voltage drops fairly quickly below 3V. The discharge is typically terminated somewhere around 2.5V (depending upon the manufacturer's specifications).

The LTC1626 is specifically designed to accommodate a single-cell Li-Ion discharge curve. For example, using the circuit shown in Figure 94.2, it is possible to produce a stable 2.5V/0.25A regulated output voltage with as little as a 2.7V from the battery, thus obtaining the maximum run time possible.

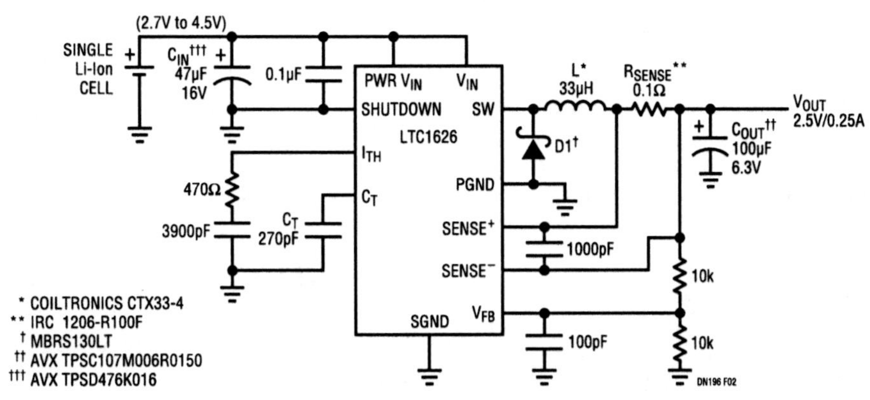

Figure 94.2 • Single-Cell Li-Ion Battery to 2.5V Converter

Analog Circuit Design: Design Note Collection. http://dx.doi.org/10.1016/B978-0-12-800001-4.00094-6

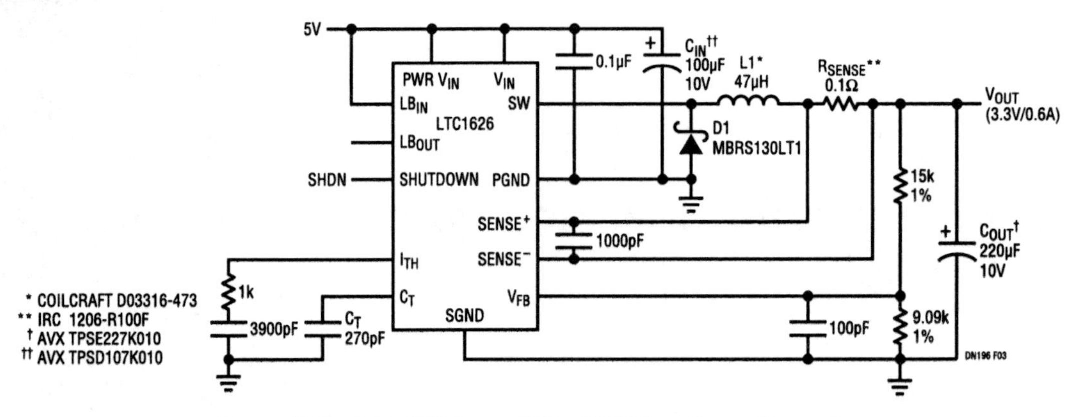

Figure 94.3 • High Efficiency 5V to 3.3V Step-Down Converter

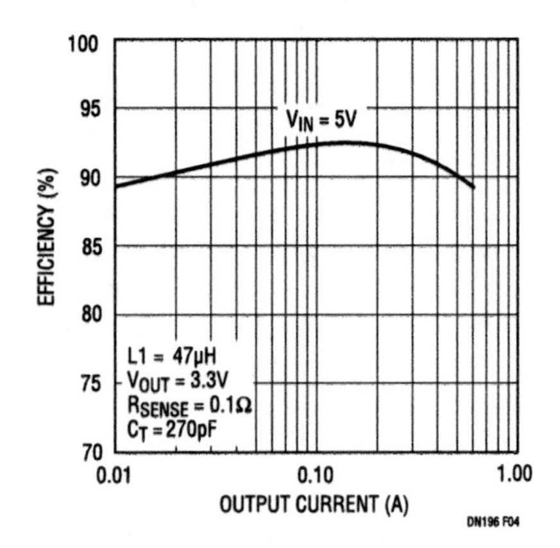

Figure 94.4 • Efficiency vs Load Current

100% duty cycle in dropout mode

As the Li-Ion cell discharges, the LTC1626 smoothly shifts from a high efficiency switch mode DC/DC regulator to a low dropout linear regulator (that is, 100% duty cycle). In this mode, the voltage drop between the battery input and the regulator output is limited only by the load current and the series resistance of the PMOS switch, the current sense resistor and the inductor. When the battery voltage rises again, the LTC1626 smoothly shifts back to a high efficiency DC/DC converter.

High efficiency 5V to 3.3V conversion

The circuit of Figure 94.3 shows the LTC1626 being used for board-level conversion of 5V to 3.3V at up to 0.6A. Although a linear regulator could also perform this function, it would result in an additional 1W of power loss. The high efficiency of the LTC1626 (Figure 94.4) reduces this loss to only 230mW.

Current mode architecture

The LTC1626 is a current mode DC/DC converter with Burst Mode operation. This results in a power supply that has very high efficiency over a wide load-current range, fast transient response and very low dropout characteristics. Further, the inductor current is predictable and well controlled under all operating conditions, making the selection of the inductor much easier.

Current mode control also gives the LTC1626 excellent start-up and short-circuit recovery characteristics. For example, when the output is shorted to ground, the off-time is extended to prevent inductor current runaway. When the short is removed, the output capacitor begins to charge and the off-time gradually decreases. The output returns smoothly to regulation without overshooting.

Low voltage low $R_{DS(ON)}$ switch

The integrated PMOS switch in the LTC1626 is designed to provide extremely low resistance at low supply voltages. Figure 94.5 is a graph of switch resistance versus supply voltage.

Note that the $R_{DS(ON)}$ is typically 0.32Ω at 4.5V and only rises to approximately 0.40Ω at 3.0V. This low switch resistance ensures high efficiency switching as well as low dropout DC characteristics at low supply voltages.

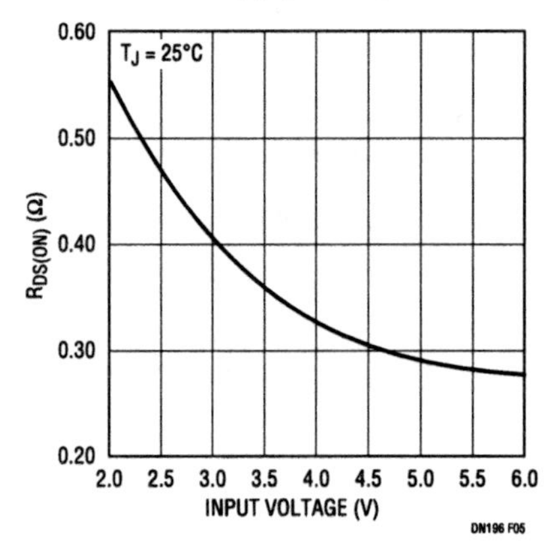

Figure 94.5 • PMOS Switch Resistance vs Input Supply Voltage

Conclusion

The LTC1626 is specifically designed to operate from a single-cell Li-Ion battery pack. With its low dropout, high efficiency and micropower operating modes, it is ideal for cellular phones and handheld industrial and medical instruments.

Optimized DC/DC converter loop compensation minimizes number of large output capacitors

95

John Seago

There is a trade-off between the cost of a few extra passive components and the flexibility that external loop compensation provides. Internal loop compensation is fixed, so it uses fewer passive parts but it also limits the designer's choice of output capacitors. The output capacitor should be chosen to meet the load requirements, not the regulator requirements. The external loop compensation provided by the LTC1435 family of parts allows the control loop to be optimized for the output capacitance required by the load.

External loop compensation can save money

By changing two or three passive component values, the LTC1435 allows the loop to be compensated for the output capacitor that meets the load requirements. External loop compensation allows the designer to optimize both the buck inductor and output capacitor for each application.

Although some loads have stringent transient requirements, many do not. The function of the output capacitor is to smooth the output voltage ripple and to source or sink output current until the regulator can respond to changes in load current. If the regulator can respond as quickly as the load current changes, very little output capacitance is required.

Figure 95.1 shows an LTC1435 configured for a 3.3V output with less than 50mV of output ripple and a 100mV transient response. The values for the primary loop-compensation components, C3 and R1, were selected by means of dynamic load testing, using the pulsed-load circuit shown in Figure 95.2. The load-pulser resistor values were selected to switch the load current between 1.5A and 3A at a 60mA/μs rate, to simulate actual load conditions. Figure 95.3 shows the output voltage transient waveform.

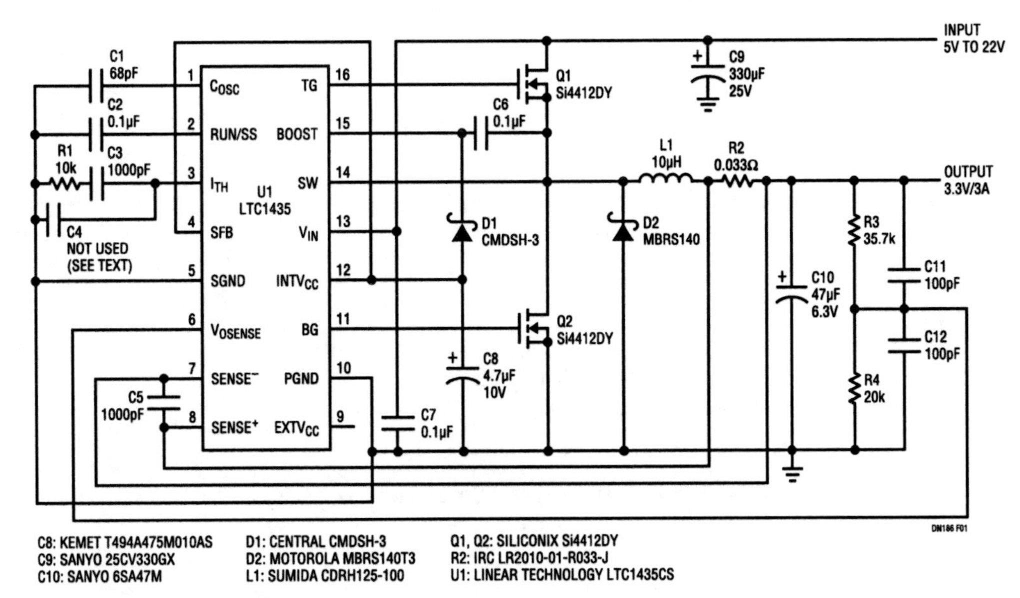

Figure 95.1 • Low Output Capacitance Voltage Regulator

Analog Circuit Design: Design Note Collection. http://dx.doi.org/10.1016/B978-0-12-800001-4.00095-8

Briefly, the values of C3, C4 and R1 determine the voltage gain and phase of the internal error amplifier at different frequencies. The value of C3 determines the low frequency gain, R1 determines the midband gain and C4 reduces gain at high frequencies. Generally, the values of C3 and C4 should be as small as possible and the value of R1 should be as large as possible.

Loop compensation using a dynamic load

Although many engineers consider control-loop theory difficult, most of the work is already done when optimizing a circuit for a particular load. The component values shown in the data sheet will provide stable operation under all static load conditions and most dynamic load conditions. The process of optimizing component values is not difficult. Using a dynamic load, or the pulsed-load circuit shown in Figure 95.2, select the appropriate output capacitor and adjust the values of C3, C4 and R1 in Figure 95.1 to minimize the overshoot and ringing on the output voltage waveform. Now, verify that the output voltage transient waveform is correct over the entire input voltage range.

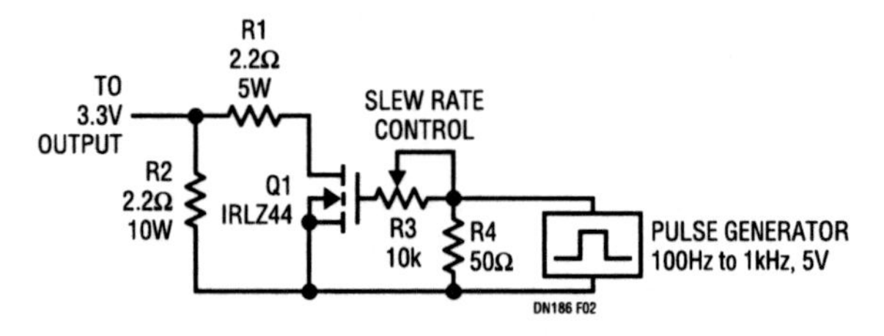

Figure 95.2 • Pulsed-Load Circuit

It is also important to verify that the control loop is stable over the required operating temperature range. It is common to use a heat gun and freeze spray to test temperature extremes but it is important to monitor the actual temperature to avoid overtesting the circuit. It is best to use a temperature-controlled chamber for all temperature testing.

Testing loop stability over temperature is even more important when using regulators with fixed compensation. It may be necessary to add even more output capacitance to ensure stable operation over temperature.

Most labs do not have a dynamic load for power supply testing. The circuit in Figure 95.2 shows an inexpensive way to test load-transient response. The value of R2 was selected to draw the nominal, pretransient load current, whereas the value of R1 was selected for the required load current step. Resistor R3 controls the slope of the load current step to better simulate actual load conditions.

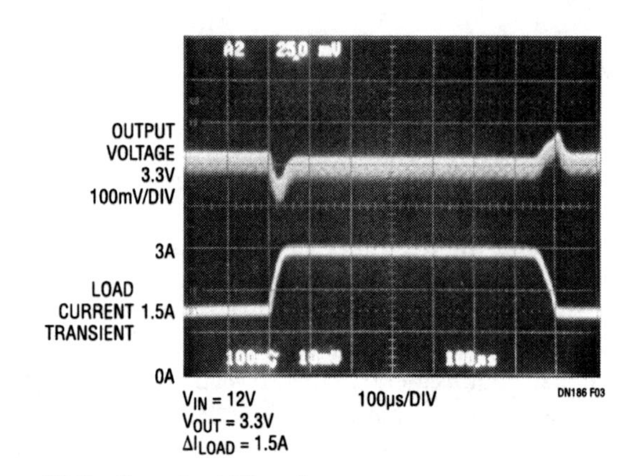

Figure 95.3 • Transient Waveforms

The advantage of adjustable loop compensation is simple: optimizing loop compensation components allows the lowest cost output capacitor to be used for a given load requirement. Adjustable loop compensation is available on all of the LTC1435 family of parts. As shown in Table 95.1, both single and dual versions are available with a variety of additional features.

Table 95.1 LTC1435 Related Parts

PART NUMBER	DESCRIPTION	COMMENTS
LTC1436/LTC1436-PLL/LTC1437	High Efficiency, Low Noise, Synchronous Step-Down Switching Regulator Controllers	Full-Featured Single Controllers
LTC1438	Dual Synchronous Controller with Power-On Reset and an Extra Comparator	Shutdown Current <30μA
LTC1439	Dual Synchronous Controller with Power-On Reset, Extra Linear Controller, Adaptive Power, Synchronization, Auxiliary Regulator and an Extra Uncommitted Comparator	Shutdown Current <30μA
LTC1538-AUX	Dual Synchronous Controller with AUX Regulator	5V Standby in Shutdown
LTC1539	Dual Synchronous Controller with the Same Features as the LTC1439	5V Standby in Shutdown

A high efficiency 500kHz, 4.5A step-down converter in an SO-8 package

96

Karl Edwards

Reducing board space and improving efficiency are key requirements in many systems, especially at higher currents, where component size and power losses generally increase. Linear Technology has addressed these issues with the new LT1374, a 500kHz, 4.5A monolithic buck converter designed to meet the needs of higher current applications. The LT1374 contains the power switch, logic, oscillator and all the control circuitry necessary to make a compact, high efficiency buck converter. The topology is current mode for fast transient response and good loop stability, with the added benefit of full cycle-by-cycle current limit.

The device is available in three package options: SO-8, DD and TO-220. For the most space-sensitive applications, the SO-8 retains the full 4.5A switch rating and is ideal for medium power applications with high peak loads. The DD package is intended for surface mount applications with continuous high current; the TO-220 is for high power, high ambient temperature systems.

A switching frequency of 500kHz allows the use of small, low value surface mount components to reduce board area. To further reduce power consumption, the LT1374 has two shutdown modes. A precise 2.38V threshold on the shutdown ($\overline{\text{SHDN}}$) pin keeps the internal reference alive but disables switching. This mode can be used as an accurate input undervoltage lockout, as shown in Figure 96.1. Grounding the $\overline{\text{SHDN}}$ pin takes the part into complete shutdown, reducing supply current to only 20μA.

For noise-sensitive applications, the $\overline{\text{SHDN}}$ pin can be replaced by SYNC (LT1374-SYNC), enabling the internal oscillator to be synchronized to an external system clock in the range of 580kHz to 1MHz. Both adjustable and fixed 5V output voltage parts are available. The LT1374, together with a minimum of small surface mount components, produces a 4.5A step-down regulator that is efficient in both power and board space.

High efficiency, 25V, 0.07Ω switch

High efficiency is the result of a fast bipolar process and a unique transistor layout that produces a high voltage switch with only 0.07Ω typical on-resistance. This permits the LT1374 to operate over an input voltage range of 5.5V to 25V with switch currents up to 4.5A. Figure 96.1 shows an example of the LT1374-5 in a typical 5V output step-down application. Efficiency for a 10V input is shown in Figure 96.2. Note that efficiency remains at over 88% from 0.5A up to the circuit's maximum 4A load current.

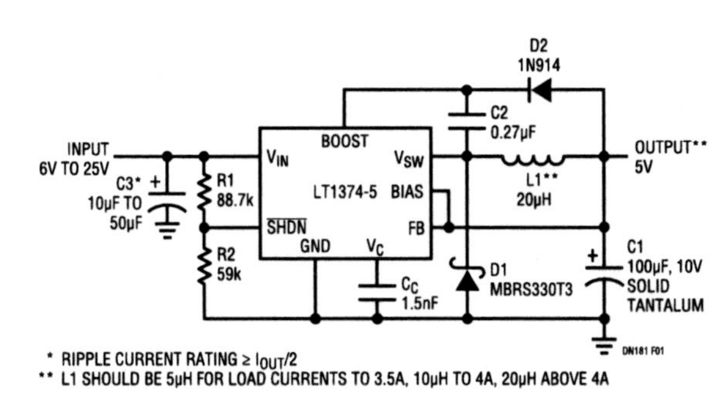

Figure 96.1 • 5V Buck Converter

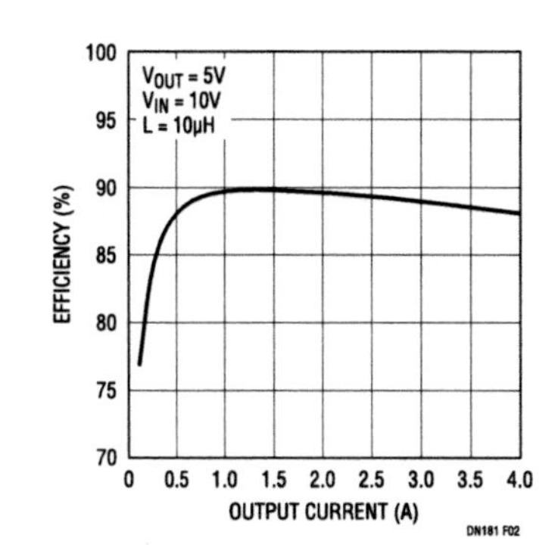

Figure 96.2 • 5V Efficiency vs Output Current

Analog Circuit Design: Design Note Collection. http://dx.doi.org/10.1016/B978-0-12-800001-4.00096-X

4.5A in an SO-8

The output switch of the LT1374 is designed to minimize power dissipation from both switch resistance and switch drive current. This allows the use of the SO-8 packaged LT1374 in applications that would have previously required a power package, especially when selection is defined by high dynamic load currents. Typical static and dynamic thermal characteristics for various load currents are shown in Figures 96.3 and 96.4. These measurements were made in still air with the LT1374 SO-8 placed on a $4in^2$ double-sided circuit board. Multiple vias conduct heat from the board's topside to a continuous copper plane on the bottom side. A typical application for the SO-8 package is supplying a motor driver. The motor may require 4A at start-up but only 2.5A when running. With a 60°C ambient temperature, the SO-8 package can provide 4A of load current for up to seven seconds, followed by 2.5A of continuous current. If 4A of continuous current were required, the surface mount DD package (θ_{JA}=30°C/W) could be used; for even higher power, use the TO-220 (θ_{JC}=4°C/W).

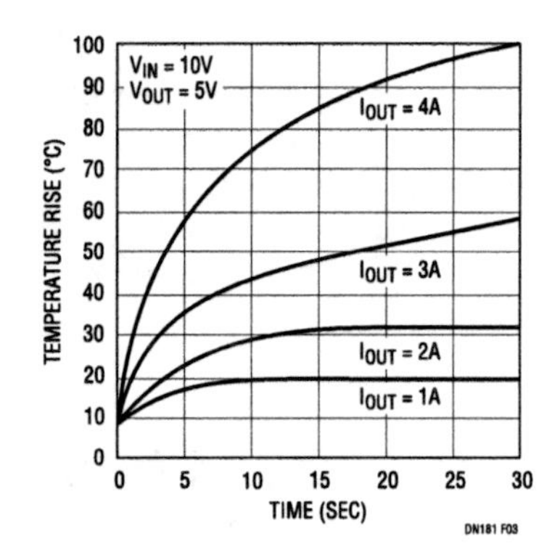

Figure 96.3 • Temperature Rise vs Time

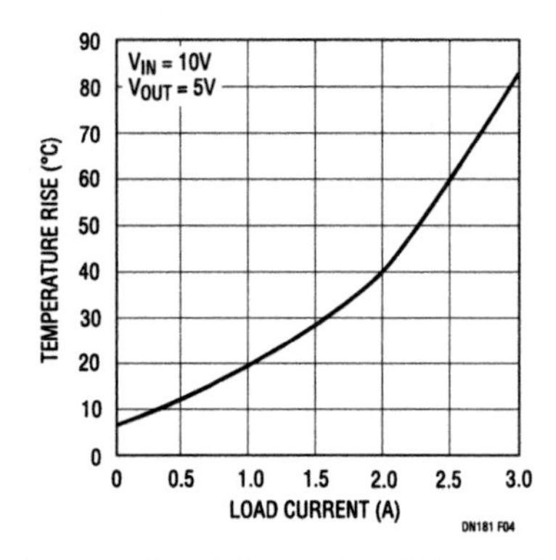

Figure 96.4 • Temperature Rise vs Load Current

Dual output SEPIC converter

The circuit in Figure 96.5 generates both positive and negative 5V outputs from two windings on a single core. The converter for the 5V output is a standard buck converter. The −5V topology would be a simple flyback winding coupled to the buck converter if C4 were not present. C4 creates a SEPIC (single-ended primary inductance converter) topology, which improves regulation and reduces ripple current in L1. Without C4, the voltage swing on L1B compared to L1A would vary due to relative loading and coupling losses. C4 provides a low impedance path to maintain an equal voltage swing in L1B, improving regulation. In a flyback converter, during switch on-time, all the converter's energy is stored in L1A only, since no current flows in L1B. At switch off, energy is transferred by magnetic coupling into L1B, powering the −5V rail. C4 pulls L1B positive during switch on-time, causing current to flow and energy to build in L1B and C4. At switch off, the energy stored in both L1B and C4 supplies the −5V rail. This reduces the current in L1A and changes L1B's current waveform from square to triangular.

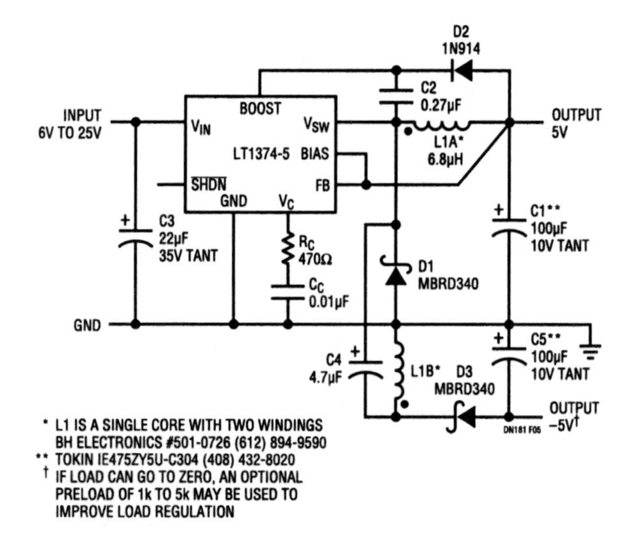

Figure 96.5 • Dual Output SEPIC Converter

High efficiency switching regulators draw only 10μA supply current

97

Greg Dittmer

Maximizing battery life, one of the key design requirements for all battery-powered products, is now easier with Linear Technology's new family of ultralow quiescent current, high efficiency step-down regulator ICs, the LTC1474 and LTC1475. The LTC1474 and LTC1475 are step-down regulators with on-chip P-channel MOSFET power switches. These regulators draw only 10μA supply current at no load while maintaining the output voltage. With the on-chip switch (1.4Ω at $V_{IN}=10V$), minimal external components are necessary to make a complete, high efficiency (up to 92%) step-down regulator. Low component count and the LTC1474/LTC1475's tiny MSOP packages provide a minimum-area solution to meet the limited space requirements of portable applications. Wide supply voltage range (3V to 18V) and 100% duty cycle capability for low dropout allow maximum energy to be extracted from the battery, making the LTC1474/LTC1475 ideal for moderate current (up to 300mA) battery-powered applications.

The peak inductor current is programmable via an optional current sense resistor to allow the design to be optimized for a particular application and to provide short-circuit protection and excellent start-up behavior. Other features include Burst Mode operation to maintain high efficiency over almost four decades of load current, an on-chip low-battery comparator and a shutdown mode to further reduce supply current to 6μA. The LTC1475 provides ON/OFF control with pushbutton switches for use in handheld products.

Inductor current control

Excessive peak inductor current can be a liability. Lower peak current offers the advantages of smaller voltage ripple ($\Delta V = I_{PEAK} \times ESR$), lower noise and less stress on alkaline batteries and other circuit components. Also, lower peaks allow the use of inductors with smaller physical size. The LTC1474/LTC1475 provides flexibility by allowing the peak switch/inductor current to be programmed with an optional sense resistor to provide just enough current to meet the load requirement. The sense resistor value required to set the desired peak inductor current is easily calculated from $R_{SENSE}=0.1/(I_{PEAK}-0.25)$. Without a sense resistor (that is, with Pins 6 and 7 shorted) the current limit defaults to its maximum of 400mA. Using the default current limit eliminates the need for a sense resistor and associated decoupling capacitor.

3.3V/250mA step-down regulator

A typical application circuit using the LTC1474-3.3 is shown in Figure 97.1. This circuit supplies a 250mA load at 3.3V with an input supply range of 4V to 18V (3.3V at no load).

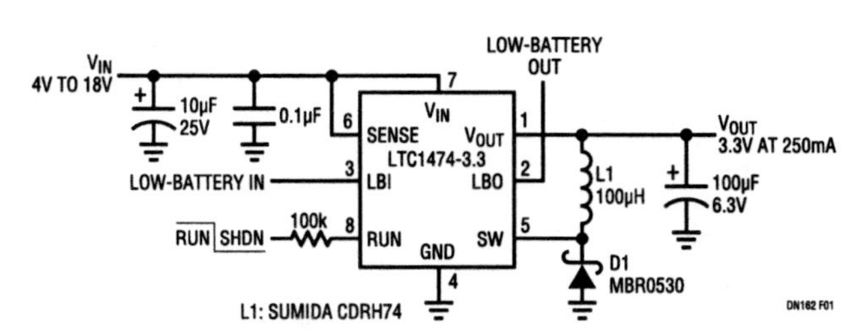

Figure 97.1 • High Efficiency Step-Down Converter

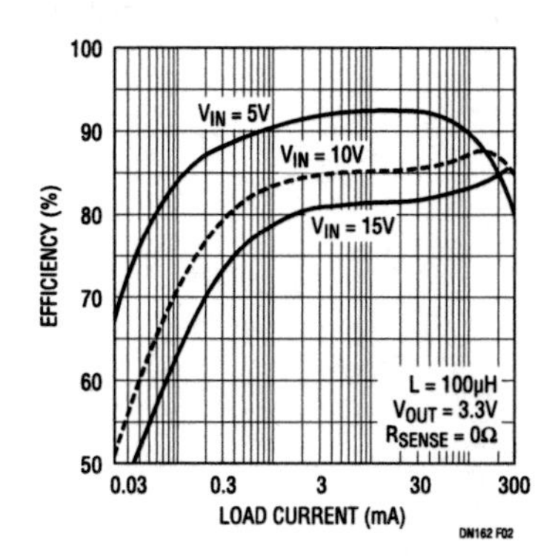

Figure 97.2 • Efficiency vs Load for Figure 97.1's Circuit

Analog Circuit Design: Design Note Collection. http://dx.doi.org/10.1016/B978-0-12-800001-4.00097-1

The SENSE pin is shorted to V_{IN} to set the peak inductor current to the 400mA maximum to meet the load requirement. Since the output capacitor dominates the output voltage ripple, an AVX TPS series low ESR (0.15Ω) output capacitor is used to provide a good compromise between size and low ESR. With this capacitor the output ripple is less than 60mV.

The efficiency curves for the 3.3V/250mA regulator at various supply voltages are shown in Figure 97.2. Note how the efficiency remains high down to extremely light loads. Efficiency at light loads depends on low quiescent current. The efficiency drops off as the load decreases below about 1mA because the non-load-dependent 10μA standby current loss then constitutes a more significant percentage of the output power. This loss is proportional to V_{IN} and thus its effect is more pronounced at higher V_{IN}.

Care must be used in selecting the catch diode to maximize both low and high current efficiency. Low reverse leakage current is critical for maximizing low current efficiency because the leakage can potentially approach the magnitude of the LTC1474/LTC1475 supply current. Low forward drop is critical for high current efficiency because loss is proportional to forward drop. These are conflicting parameters, but the MBR0530 0.5A Schottky diode used in Figure 97.1 is a good compromise.

3.3V/10mA regulator from a 4mA to 20mA loop

The circuit shown in Figure 97.3 is a 3.3V/10mA regulator that extracts its power from a 4mA to 20mA loop. This circuit demonstrates how an LTC1474/LTC1475-based regulator is easily optimized for such low current applications. The 2Ω sense resistor limits the peak inductor current to 40mA to minimize current ripple and provide good efficiency (84%). The 330μH inductor is a good value to use at this current level to keep the frequency low enough to avoid excessive switching losses without being so large that DCR losses are significant (see inductor section of the data sheet). The Zener diode at the input clamps the input voltage to 12V, which is then converted to 3.3V. This enables the 4mA (min) input current to be more than doubled at the output.

Pushbutton ON/OFF operation

The LTC1475 provides the option of pushbutton control of ON/OFF mode for handheld products. In contrast to the LTC1474's ON/OFF mode, which is controlled by a voltage level at the RUN pin (ground = OFF, open/high = ON), the LTC1475 ON/OFF mode is controlled by an internal S/R flip-flop that is set (ON) by a momentary ground at the ON pin and reset (OFF) by a momentary ground at the LBI/OFF pin. This provides simple ON/OFF control with two pushbutton switches. A simple implementation of this function is shown in Figure 97.4.

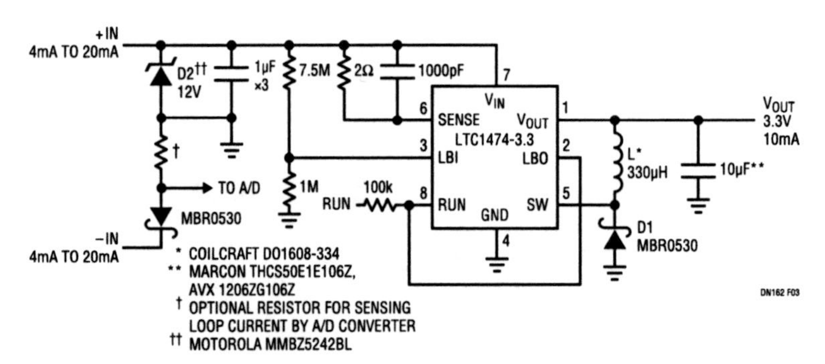

Figure 97.3 • High Efficiency 3.3V/10mA Output from 4mA to 20mA Loop

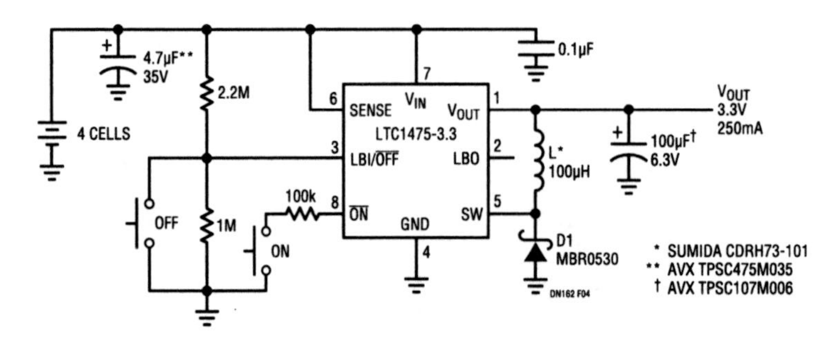

Figure 97.4 • Pushbutton ON/OFF 3.3V/250mA Regulator

High power synchronous buck converter delivers up to 50A

98

Dale Eagar

Introduction

The LT1339 is the buck/boost controller that needs no steroids. As a full-featured synchronous switching controller, the LT1339 incorporates the features needed for system level solutions. The unfortunate lack of such features in most PWM controllers forces designers to grope for handfuls of jellybean components. The LT1339 has an innovative slope-compensation function that allows the circuit designer freedom in controlling both the slope and offset of the slope-compensation ramp. Additionally, the LT1339 has an average current limit loop that yields a constant output current limit, regardless of input and/or output voltage. The LT1339 has a RUN pin that is actually the input to a precision comparator, giving the designer freedom to select an undervoltage lockout point and hysteresis appropriate for the design. The SYNC and SS (soft-start) pins allow simple solutions to system level design considerations. Like all Linear Technology controllers, the LT1339 has anti-shoot-through circuitry that ensures the robustness that is demanded in real world applications for medium and high power conversion.

For input voltages ranging from 12V to 48V and output voltages ranging from 1.3V to 36V, the LT1339 is a simple robust solution to your power conversion problems. The LT1339 is ideal for power levels ranging from tens of watts to tens of kilowatts. The LT1339 is straightforward and remarkably easy to use. This is one power controller that is not afraid of 20A, 50A or even 150A of load current.

Distributed power

Figure 98.1 details the typical low voltage buck converter. This circuit has a V_{IN} range of 10V to 18V with configurable output current and voltage. This simple circuit delivers 250W of

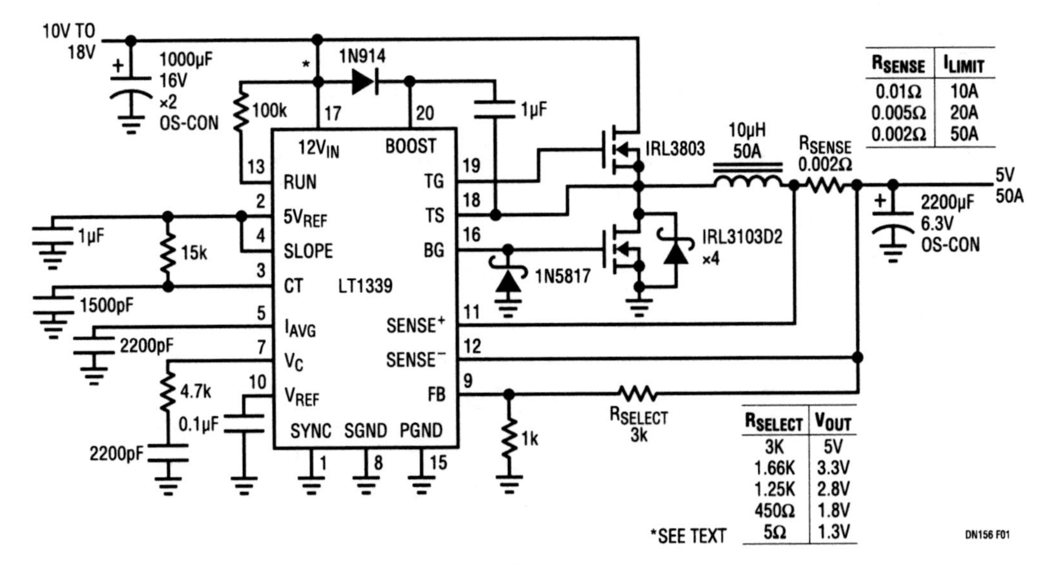

RSENSE	ILIMIT
0.01Ω	10A
0.005Ω	20A
0.002Ω	50A

RSELECT	VOUT
3K	5V
1.66K	3.3V
1.25K	2.8V
450Ω	1.8V
5Ω	1.3V

Figure 98.1 • $10V_{IN}$–$18V_{IN}$ to $5V_{OUT}$, 50A Converter

Analog Circuit Design: Design Note Collection. http://dx.doi.org/10.1016/B978-0-12-800001-4.00098-3

load power into a 5V load while maintaining efficiencies in the mid-nineties.

Higher input voltages

The circuit shown in Figure 98.1 is limited to 20V because of the maximum rating (Abs Max) of the LT1339 V_{IN} pin. The input voltage can be extended above 20V by inserting a 10V Zener diode in the circuit of Figure 98.1 where the asterisk (*) is shown. This will extend the input voltage of Figure 98.1's circuit up to 30V (the Abs Max ratings of the MOSFETs).

Blame it on the physicists

As the input voltage approaches 30V, the bottom MOSFETs will begin to exhibit "phantom turn-on." This phenomenon is driven by the instantaneous voltage step on the drain, the ratio of C_{MILLER} to C_{INPUT}, and yields localized gate voltages above V_t, the threshold voltage of the bottom MOSFETs. To defeat the physicists in this arena, in Figure 98.2 we add 3V of negative offset to the bottom gate drive. The physicists, however, having lost a battle, have not yet lost the war. From their dirty bag of tricks they pull the body diode effect of the bottom MOSFETs. In Figure 98.1, we use *FETKY* MOSFETs, FETs with internal Schottkys in parallel with the body diodes. Because *FETKY* MOSFETs are not available in the higher voltages, we must use external Schottkys. Because the inductance of the loop formed by the body diode and the external Schottky is correspondingly much higher, body-diode current is slow to move into the Schottkys. Our solution is to use smaller Schottkys and interdigitize them with the bottom MOSFETs to reduce inductance. Figure 98.2 details the victory over the physicists in the form of a $48V_{IN}$, $5V_{OUT}$, 50A synchronous buck converter. When this converter is configured for 24V output, it can deliver 960W of power from either a 36V or 48V input while maintaining 97% efficiency.

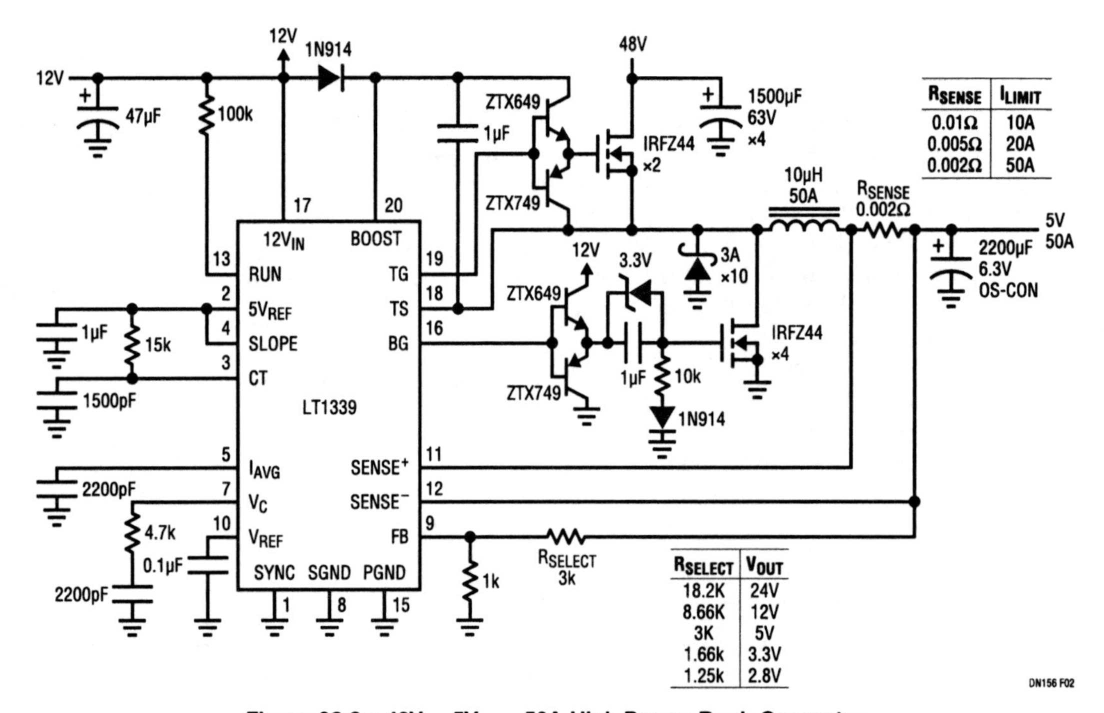

RSENSE	ILIMIT
0.01Ω	10A
0.005Ω	20A
0.002Ω	50A

RSELECT	VOUT
18.2K	24V
8.66K	12V
3K	5V
1.66k	3.3V
1.25k	2.8V

Figure 98.2 • $48V_{IN}$, $5V_{OUT}$, 50A High Power Buck Converter

Single IC, five output switching power supply system for portable electronics

Steve Hobrecht

The drive for higher performance portable electronic systems, concomitant with the need to remain compatible with existing interfacing standards and hardware, has caused the number of system power supply voltages to proliferate. This Design Note describes how to generate five separate output voltages using the LTC1538-AUX, a dual synchronous switching regulator controller with internal circuitry adaptable to many additional configurations. The five output voltages chosen in this example are:

- 5V ±4%/25mA linear regulator, which remains active regardless of the state of the switching regulator controllers
- 5V ±2%/3A synchronous switching buck regulator
- 3.3V ±2%/6A synchronous switching buck regulator
- 2.9V ±5%/3A peak low dropout linear regulator deriving power from the 3.3V output
- 12V ±5%/200mA synchronously rectified flyback output deriving power from the 5V output

The 100μA quiescent current, 5V standby linear regulator can efficiently power "wake-up" circuitry in portable systems and can deliver 25mA.

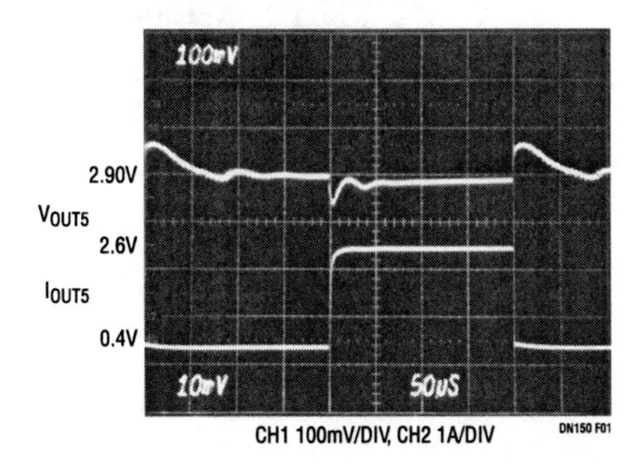

Figure 99.1 • 2.9V Output Transient Load Step Performance

The 5V and 3.3V switching regulators, independently activated by the two RUN/SS pins, provide very efficient, constant frequency operation. This is accomplished by using a synchronous-buck architecture at high currents and switching over to Burst Mode operation below approximately 10% to 15% of maximum current, as determined by the current sensing resistor for each controller.

The 5V output from the first controller, or any external voltage between 4.8V and 10V, can be tied to the $EXTV_{CC}$ pin. The power and voltage from this external source will supplant the internal 5V linear regulator if the applied voltage is greater than 4.8V. This technique improves efficiency by eliminating the power dissipated by the IC due to the current drawn through the internal linear regulator and the (V_{IN} – 5V) voltage drop.

The 2.9V linear regulator draws power from the 3.3V output and performs the function of a wide bandwidth, low dropout linear regulator having dynamic performance as illustrated in Figure 99.1.

The regulator's dominant pole is set by the output capacitance and the load resistance. The loop is stable, provided there is some ESR (0.02Ω to 0.1Ω) in the output capacitor in order to generate phase lead prior to the unity-gain crossover frequency. The AVX-TPS tantalum capacitor used here, or a Sanyo OS-CON type capacitor, has a complex impedance characteristic, providing close to zero phase shift at the unity-gain cross frequency of the amplifier.

The 12V/200mA output uses a synchronously driven, tightly coupled secondary winding on the first controller's primary winding to generate a highly efficient output with a ±5% tolerance under all load and line conditions. The output voltage is fed back to the SFB1 input pin and compared with the internal 1.19V reference. This feedback forces the primary controller into a "forced synchronous" mode, regulating the 12V output regardless of the loading of the primary 5V regulator. Tighter regulation and less ripple can be

Analog Circuit Design: Design Note Collection. http://dx.doi.org/10.1016/B978-0-12-800001-4.00099-5

attained with a slightly higher turns ratio and an additional linear regulator.

Figure 99.2 shows the overall efficiency for the circuit shown in Figure 99.3. The $I_{2.9}$ = Proportional curve documents efficiency when all outputs are loaded proportionally to their peak design loads of: 5V at 3A, 3.3V at 3A, 2.9V at 3A and 12V at 200mA. As an example, overall efficiency drops to 90% when the 2.9V output is loaded with 1A (33% of the designed maximum load). For this same efficiency point, the three other outputs are loaded as follows: 5V at 1A, 3.3V at 1A and 12V at 67mA. For the peak load of 36W, as specified above, efficiency drops to 85%.

The $I_{2.9} = 0$ curve shows efficiency when all outputs are loaded proportional to their peak loads *except* the 2.9V output which is not loaded. This curve documents a 100% design load of 27.3W. The circuit's efficiency peaks for this curve at 95% when the 5V and 3.3V outputs are loaded with 1A and the 12V output is loaded with 67mA.

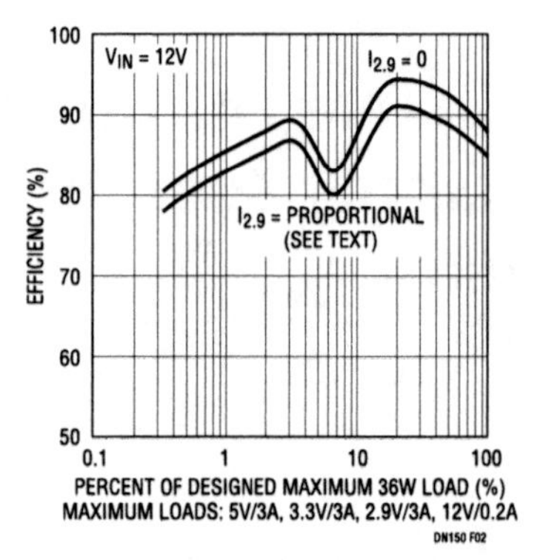

Figure 99.2 • Efficiency vs Percent of the Designed Maximum Load

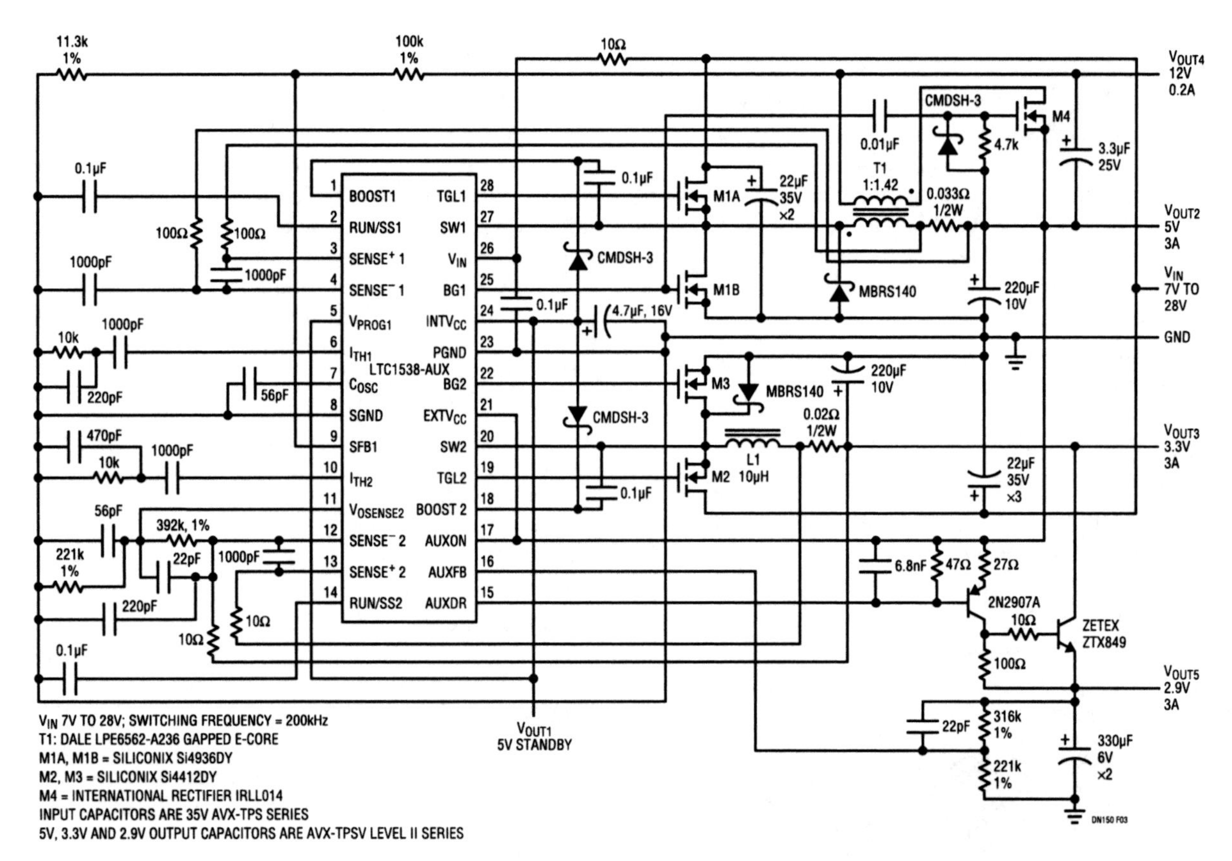

Figure 99.3 • Schematic Diagram of System

Low noise switching regulator helps control EMI

100

John Seago

Electromagnetic interference (EMI) is a potential problem for the circuit designer. Switching regulators can cause EMI in many products. Linear Technology has developed new techniques like spread-spectrum modulation, phase-locked synchronization and Adaptive Power mode that can reduce the amount of unwanted interference.

New IC solves old problems

The LTC1436-PLL is a constant frequency, current mode, synchronous step-down switching regulator that controls external N-channel MOSFETs for very efficient power conversion. It also features Adaptive Power mode, which provides constant frequency switching with good efficiency at light load currents. Figure 100.1a shows the audio frequencies generated by the 5V output of the circuit in Figure 100.2, while supplying 3mA of load current (0.1% of full load) using a cycle-skipping mode of operation. This mode may cause many cycles to be skipped between bursts of energy to the output capacitor. These energy bursts intrude into the audio band at sufficiently low output currents. Figure 100.1b shows that audio frequency noise is completely eliminated by the Adaptive Power mode under the same conditions.

Traditionally, efficiency is sacrificed to accomplish the audio frequency response shown in Figure 100.1b. Large synchronous MOSFETs are used to force continuous inductor current at the switching frequency, regardless of the load. The associated gate charge losses and losses caused by relatively large inductor ripple current result in very poor efficiency at light loads. The Adaptive Power mode uses only the small (SOT-23) MOSFET, Q3 and D2 in a conventional buck mode to allow constant frequency, discontinuous inductor current operation, which greatly decreases power loss. The gate-to-source capacitance of the small MOSFET is significantly less than either of the two large MOSFETs. Depending on component selection, there can be a 50-to-1 difference in gate-to-source capacitance between Q3 and Q1/Q2, so that Q3 requires only 2% of the gate drive power (loss) of Q1/Q2. This provides a substantial increase in efficiency at light loads.

New feature provides new EMI control

In addition to audio frequency suppression, the LTC1436-PLL has three additional RF EMI control mechanisms:

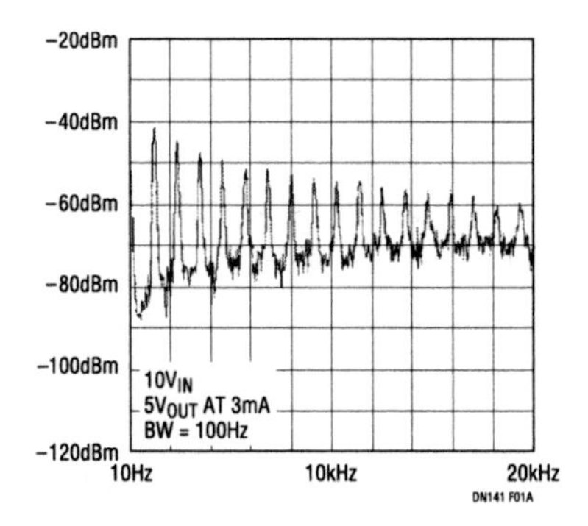

Figure 100.1a • Audio Frequency Generation in Cycle-Skipping Mode

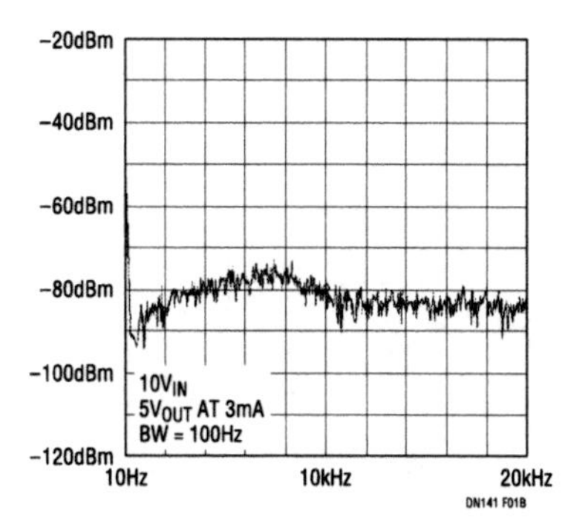

Figure 100.1b • Audio Frequency Amplitude with Adaptive Power Mode Operation

Analog Circuit Design: Design Note Collection. http://dx.doi.org/10.1016/B978-0-12-800001-4.00100-9

1. The LTC1436-PLL allows switching frequency modulation to spread the spectrum of switching noise. Through frequency modulation, peak energy is decreased and spread over a wide range of frequencies as shown in Figure 100.3. The normal 190kHz switching frequency and its harmonics are shown by the black trace. The colored trace shows the result of modulating the phase-locked loop low-pass filter (PLL LPF) pin with a 100Hz sawtooth waveform. Switching frequency energy is reduced by over 20dB when modulated; second and third harmonics are attenuated even more in this example. Figure 100.4 shows the spectrum out to 100MHz resulting from PLL LPF modulation under the same conditions as in Figure 100.3.
2. The switching frequency can be programmed anywhere from 50kHz to 400kHz by selecting the appropriate value of oscillator capacitor. This places harmonics away from sensitive frequencies like 455kHz.
3. The switching frequency can be phase-locked to an external system clock so that harmonics and sidebands of the switching frequency are common with those generated by the system. This phase lock can be maintained over a ±30% frequency range around f_{OSC}.

Additional features

The LTC1436-PLL provides a power-on reset timer function that flags an out-of-range output voltage condition, along with an auxiliary regulator that controls an external PNP transistor for an additional low noise, linear regulated output. The LTC1437 has all the features of the LTC1436-PLL plus an internal comparator with reference that can be used to detect a low-battery condition or provide other useful functions. The basic LTC1436 trades the phase-locked loop function for the additional comparator in the LTC1437.

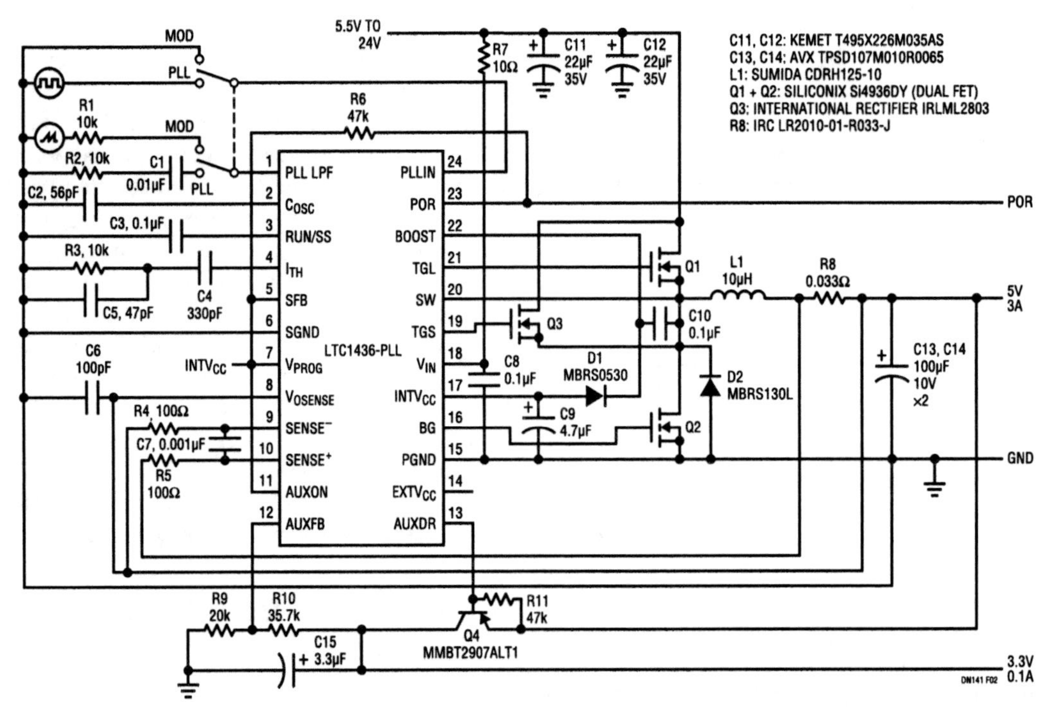

Figure 100.2 • Two Output LTC1436-PLL Circuit

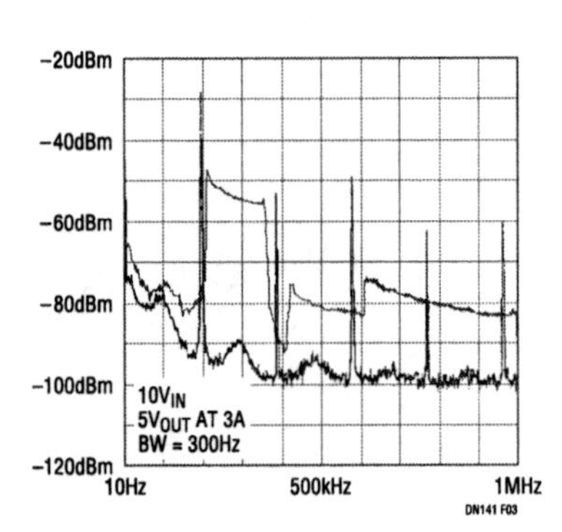

Figure 100.3 • Before and after Frequency Spreading

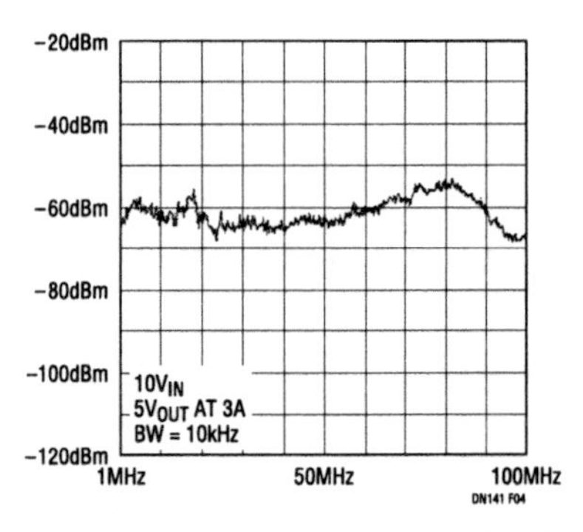

Figure 100.4 • High Frequency Response with Frequency Spreading

Efficient processor power system needs no heat sink

101

John Seago

New designs require more functionality in an ever decreasing package size. Compact, efficient high frequency power supplies are required and there is seldom room for a heat sink. Portable computers have some of the most demanding requirements. Powering the Pentium processor adds additional challenges for the power system designer.

New IC powers portable Pentium processor and much more

The LTC1435 is a current mode, constant frequency, synchronous step-down switching regulator that uses external N-channel MOSFETs for very high efficiency. With the wide input voltage range of 3.5V to 36V and low dropout (99% duty cycle) capability, the LTC1435 is a good choice for battery-powered circuits where the voltage of four NiCd cells can drop to 3.6V at the end of discharge. The ability to operate with 36V inputs allows application of a wide range of AC adapter voltages. The LTC1435 achieves battery-powered efficiencies of 90% to 95% and features Burst Mode operation for longest battery life under light load conditions. For constant frequency at all load conditions, the Burst Mode operation can be defeated easily on the LTC1435. The Adaptive Power mode, available on the LTC1436, provides constant frequency at light loads with greatly improved efficiency. With a "silver box" power supply or other high current voltage source, the LTC1435 can easily provide an output current of 12A.

The LTC1435's current mode architecture and 1% voltage reference provide a tightly controlled output voltage with excellent load and line regulation and outstanding set-point accuracy. The switching frequency can be selected between 50kHz and 400kHz, so total circuit cost, efficiency, component size and transient response can be properly balanced. The LTC1435 also features both logic-level on/off control and output current soft start. When the controller is in the shutdown mode, voltage is removed from the load and quiescent input current drops to a mere 15μA. The LTC1435 is available in the popular 16-pin SO and SSOP packages.

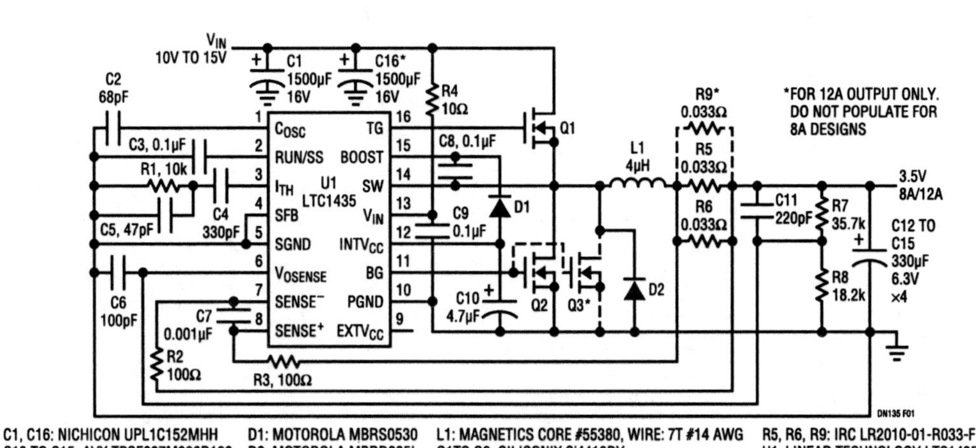

Figure 101.1 • 12V to 3.5V Regulator for 8A or 12A Applications

Analog Circuit Design: Design Note Collection. http://dx.doi.org/10.1016/B978-0-12-800001-4.00101-0

High performance Pentium processor power

The 150MHz Pentium load current changes from 0.2A to 4.5A in about 15ns. The core voltage must be maintained at 3.5V ±0.1V under all conditions.

The LTC1435 circuit shown in Figure 101.1 was used to power an Intel Power Validator to simulate the typical Pentium processor load transient of 0.2A to 4.5A. The transient waveform in Figure 101.2 shows a ±0.04V variation in output voltage with 700μF of local decoupling capacitance at the Power Validator and the 1300μF capacitance at the regulator output. The same base circuit was used to power both 8A and 12A static loads. By changing the value of R8, the output voltage can be set from 1.8V to 5V at a full 12A. Outputs of up to 9V are possible with some minor changes.

Portable Pentium processor power

The portable Pentium processor requires a core voltage of 2.9V ±0.165V and switches between 0.25A and 2.65A in about 30ns. This load was simulated by the Power Validator

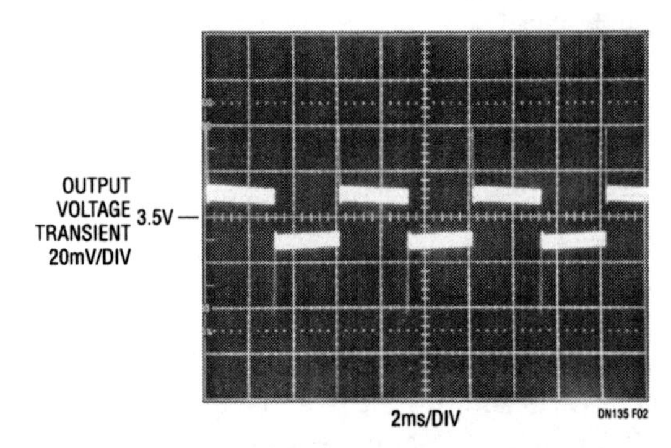

Figure 101.2 • High Performance Pentium Processor Load Transient Waveform

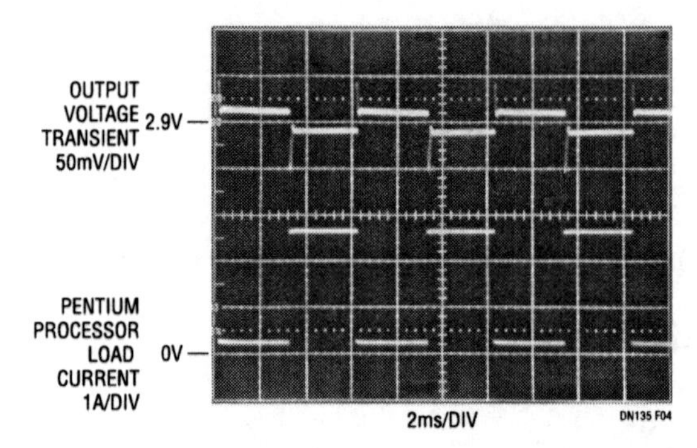

Figure 101.4 • Portable Pentium Processor Load Transient Waveform

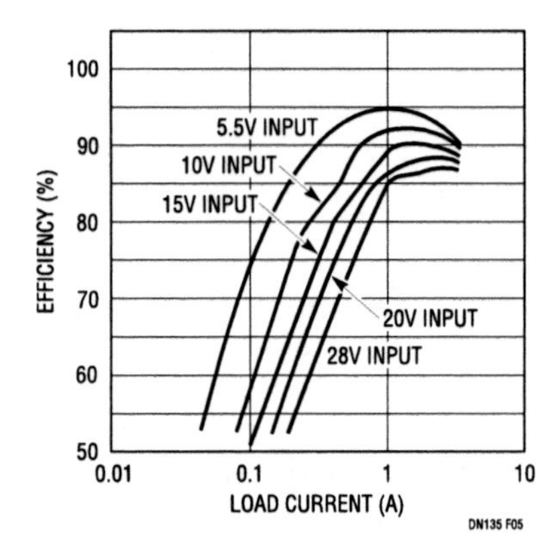

Figure 101.5 • LTC1435 Efficiency Curves for Different Input Voltages

and powered by the circuit of Figure 101.3. Although this circuit works very well over the input voltage range of 5.5V to 28V, it will continue to provide portable Pentium processor core voltage down to a 3.5V input by adding C13 and C14. Figure 101.4 shows the output voltage transient and load current waveforms with an input voltage of 3.5V. Figure 101.5 shows circuit efficiency over load and line conditions.

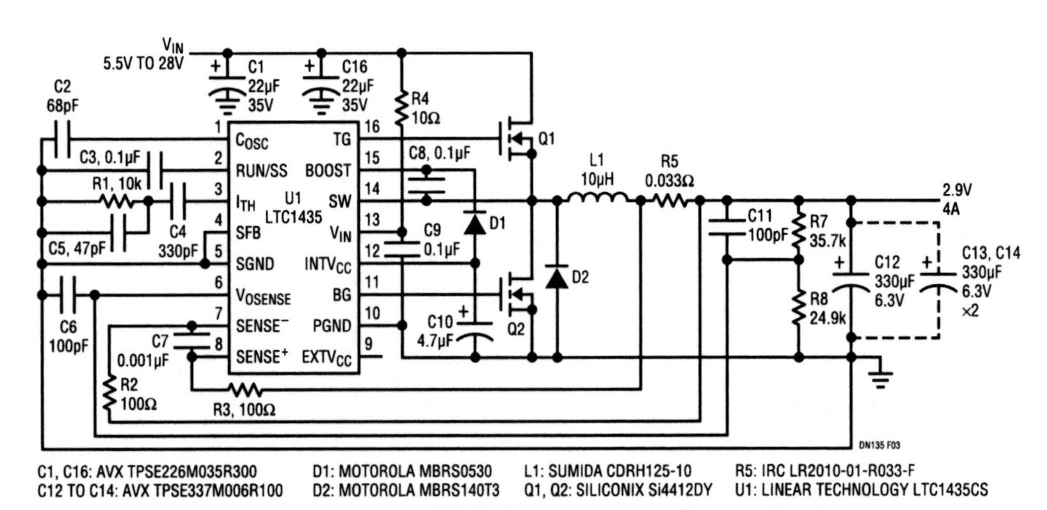

Figure 101.3 • 2.9V Regulator for Portable Pentium Processor

A new, high efficiency monolithic buck converter

102

San-Hwa Chee

The LTC1265 is a 14-pin SOIC step-down converter capable of operating at frequencies up to 700kHz. High frequency operation permits the use of small inductors for size sensitive applications. The LTC1265 has an internal 0.3Ω (at a supply voltage of 10V) P-channel power MOSFET switch, which is capable of supplying up to 1.2A of output current. With no load, the converter requires only 160μA of quiescent current; this decreases to a mere 5μA in shutdown conditions. In dropout mode, the internal P-channel power MOSFET switch is turned on continuously (at DC), thereby maximizing battery life. The part is protected from output shorts by its built-in current limiting. In addition to the features already mentioned, the LTC1265 incorporates a low-battery detector.

The LTC1265 is a current mode DC/DC converter with Burst Mode operation. The current mode architecture gives the LTC1265 excellent load and line regulation. Burst Mode operation results in high efficiency with both high and low load currents. The LTC1265 comes in three versions: LTC1265-5 (5V output), LTC1265-3.3 (3.3V output) and LTC1265 (adjustable). All versions operate down to an input voltage of 3.5V and up to an absolute maximum of 13V.

Efficiency

Figure 102.1 shows a typical LTC1265-5 application circuit. The efficiency curves for two different input voltages are shown in Figure 102.2. Note that the efficiency for a 6V input exceeds 90% over a load range from less than 10mA to 850mA. This makes the LTC1265 attractive for all battery-operated products and efficiency sensitive applications.

High frequency operation

Although the LTC1265 is capable of operation at frequencies up to 700kHz, the highest efficiency is achieved at an operating frequency of about 200kHz. As the frequency increases, losses due to the gate charge of the P-channel power MOSFET increase. In space sensitive applications, high frequency operation allows the use of smaller components at the cost of four to five efficiency points.

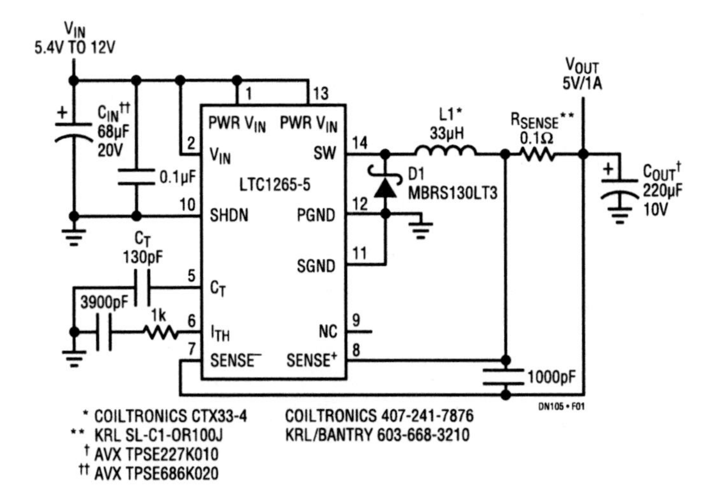

Figure 102.1 • High Efficiency 5V/1A Step-Down Converter

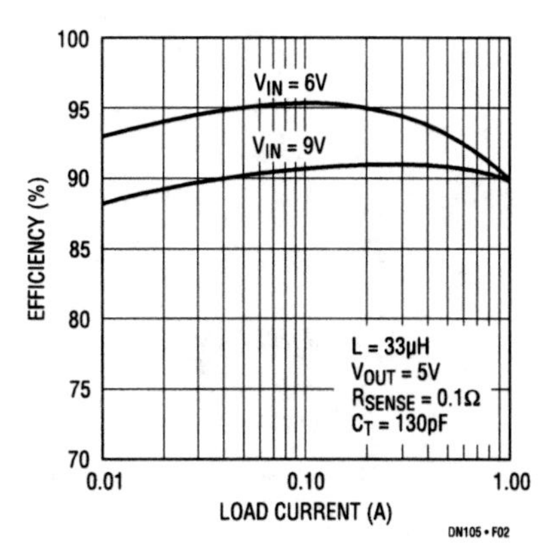

Figure 102.2 • Efficiency vs Load Current

Analog Circuit Design: Design Note Collection. http://dx.doi.org/10.1016/B978-0-12-800001-4.00102-2

Constant off-time architecture

The LTC1265 uses a constant off-time, current mode architecture. This results in a power supply that has very high efficiency over a wide load current range, fast transient response and very low dropout characteristics. The off-time is set by an external timing capacitor C_T and is constant whenever the output is in regulation. When the output is not in regulation, the off-time is inversely proportional to the output voltage. By using a constant off-time scheme, the inductor's ripple current is predictable and well controlled under all operating conditions, making the selection of the inductor much easier. The inductor's peak-to-peak ripple current is inversely proportional to the inductance in continuous mode. If a lower ripple current is desired, a larger inductor can be used for a given value of timing capacitor.

100% duty cycle in dropout mode

When the input voltage decreases, the switching frequency decreases. With the off-time constant, the on-time is increased to maintain the same peak-to-peak ripple current in the inductor. When the input-to-output voltage differential drops below 2.0V, the off-time is reduced. This prevents the operating frequency from dropping below 20kHz as the regulator approaches the dropout region. As the input voltage drops further, the P-channel switch is turned on for 100% of the cycle. The dropout voltage is governed by the switch resistance, load current and current sense resistor.

Good start-up and transient behavior

The LTC1265 exhibits excellent start-up behavior when it is initially powered on or recovering from a short circuit. This is achieved by making the off-time inversely proportional to the output voltage while the output is still in the process of reaching its regulated value. When the output is shorted to ground, the off-time is extended long enough to prevent inductor current runaway. When the short is removed, the output capacitor begins to charge and the off-time gradually decreases.

In addition, the LTC1265 has excellent load transient response. When the load current drops suddenly, the feedback loop responds quickly by turning off the internal P-channel switch. Sudden increases in output current will be met initially by the output capacitor, causing the output voltage to drop slightly. Tight control of the inductor's current, as mentioned above, means that output voltage overshoot and undershoot are virtually eliminated.

2.5mm typical height 5V-to-3.3V regulator

Figure 102.3 shows the schematic for a very thin 5V-to-3.3V converter. For the LTC1265 to be able to source 500mA output current and yet meet the height requirement, a small value inductor must be used. The circuit operates at a high frequency (typically 500kHz), increasing the gate charge losses.

Conclusion

The LTC1265, with its low dropout and high efficiency, is ideal for battery-operated products and efficiency sensitive applications. In addition, its ability to operate at high frequencies allows the use of small inductors for size sensitive applications.

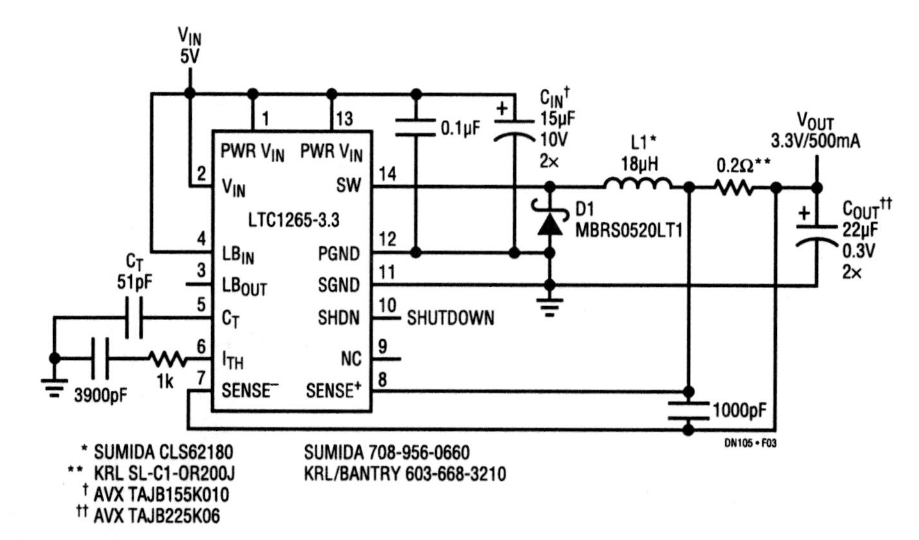

Figure 102.3 • 2.5mm High 5V-to-3.3V Converter (500mA Output Current)

Switching regulator provides high efficiency at 10A loads

103

Greg Dittmer

The new LTC1266 is a synchronous, step-down switching regulator controller that can drive two external, N-channel MOSFET switches. The superior performance of N-channel MOSFETs enables the LTC1266 to achieve high efficiency at loads of 10A or more with few additional components. Burst Mode operation provides high efficiency at light loads—efficiency is greater than 90% for loads from 10mA to 10A. The ability to provide 10A at high efficiency is critical for supplying power to Pentium processor applications.

The LTC1266 is based on the LTC1148 architecture and has most of the features of this successful product including constant off-time, current mode architecture with automatic Burst Mode operation. Pin selectable shutdown reduces the DC supply current to 40μA. The LTC1266 also provides pin selectable phase of the top-side driver which allows it to implement, in addition to an all N-channel step-down regulator, a low dropout regulator with high side P-channel or a boost regulator. Other new features of the LTC1266 include an on-chip low-battery comparator, pin-defeatable Burst Mode operation, a wider voltage supply range (3.5V to 20V), 1% load regulation and a higher maximum frequency of 400kHz.

N-channel vs P-channel

The key to the LTC1266's ability to drive large loads at high efficiencies is its ability to drive both top-side and bottom-side N-channel MOSFETs. The superiority of N-channel MOSFETs over P-channels at high currents is due to the lower $R_{DS(ON)}$ and lower gate capacitance of the N-channel parts. To compensate for the higher $R_{DS(ON)}$, the P-channel size is usually made larger, resulting in higher gate capacitance. Efficiency is inversely proportional to both $R_{DS(ON)}$ and gate capacitance. Higher $R_{DS(ON)}$ decreases efficiency due to higher I^2R losses and limits the maximum current the MOSFET can handle without exceeding thermal limitations. Higher gate capacitance increases losses due to the increased charge required to switch the MOSFETs on and off during each switching cycle.

Nonetheless, P-channel MOSFETs still have a home in lower current and low dropout applications due to the fact they can operate at 100% duty cycle. The LTC1266 offers the capability of driving either N-channel or P-channel.

Driving N-channel MOSFETs

P-channels have another distinct advantage—simplicity of the gate drive. Because of the negative threshold of the P-channel, the gate potential must decrease below the source (which is at V_{IN}) by at least $V_{GS(ON)}$ to turn it on. Hence, the top-side MOSFET can be gated between the available supply rail V_{IN} and ground.

Driving an N-channel top-side MOSFET is not so straightforward. When the top-side MOSFET is turned on, the source is pulled up to V_{IN}. Because the N-channel has a positive threshold voltage, the gate must be above the source by at least $V_{GS(ON)}$. Thus, the top-side drive must swing between ground and $V_{IN}+V_{GS(ON)}$. This requires a second, higher supply rail equal to at least $V_{IN}+V_{GS(ON)}$.

There are two ways to obtain this higher rail. The most straightforward way is to use a higher rail that is already available, as is the case in most desktop systems with 12V supplies. Note that the PWR V_{IN} input to the LTC1266 is dedicated to powering the internal drivers and is separate from the main supply input. The PWR V_{IN} voltage cannot exceed 18V (20V max), limiting the input voltage to $18V-V_{GS(ON)}$. For a converter with logic-level MOSFETs, this limits V_{IN} to about 13.5V. The PWR V_{IN} voltage must also meet its minimum requirement of $V_{IN}+V_{GS(ON)}$ (about 10V for a 5V-to-3.3V converter) in order not to burn up the high side MOSFET due to insufficient conductance at larger output loads. If a higher supply rail is not available, a charge-pump circuit can be used to pump V_{IN} to the required level.

Analog Circuit Design: Design Note Collection. http://dx.doi.org/10.1016/B978-0-12-800001-4.00103-4

Basic circuit configurations

Figures 103.1 and 103.2 show two basic circuit configurations for the LTC1266. Figure 103.1 show an LTC1266 in the charge pump configuration designed to provide a 3.3V/10A output. The Si4410s are new logic-level, surface mount N-channel MOSFETs from Siliconix that provide a mere 0.02Ω of on-resistance at $V_{GS}=4.5V$, and thus provide a 10A solution with minimal components. The efficiency plot shows that the converter still is close to 90% efficient at 10A. Because the charge pump configuration is used, PWR $V_{IN}=2\times V_{IN}$ plus any additional ringing on the switch node. Due to the high AC currents in this circuit, we recommend low ESR OS-CON or AVX input/output capacitors to maintain efficiency and stability.

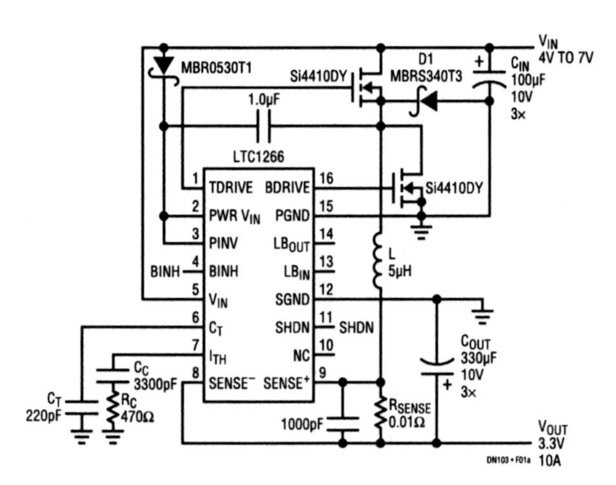

Figure 103.1a • All N-Channel Single Supply 5V to 3.3V/10A Regulator

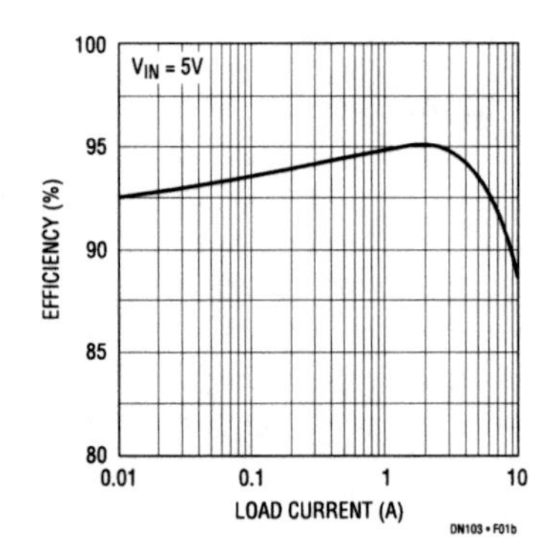

Figure 103.1b • Figure 103.1a Circuit Efficiency

The all N-channel, external PWR V_{IN} circuit shown in Figure 103.2 is a 3.3V/5A surface mount converter. The current sense resistor value is chosen to set the maximum current to 5A, according to the formula $I_{OUT}=100mV/R_{SENSE}$. With $V_{IN}=5V$, the 5µH inductor and 130pF timing capacitor provide an operating frequency of 175kHz and a ripple current of 1.25A. The $V_{GS(ON)}$ of the Si9410N-channel MOSFETs is 4.5V; thus the minimum allowable voltage at the external PWR V_{IN} is $V_{IN(MAX)}+4.5V$. At the other end, PWR V_{IN} should be kept under the maximum safe level of 18V, limiting V_{IN} to $18V-4.5V=13.5V$.

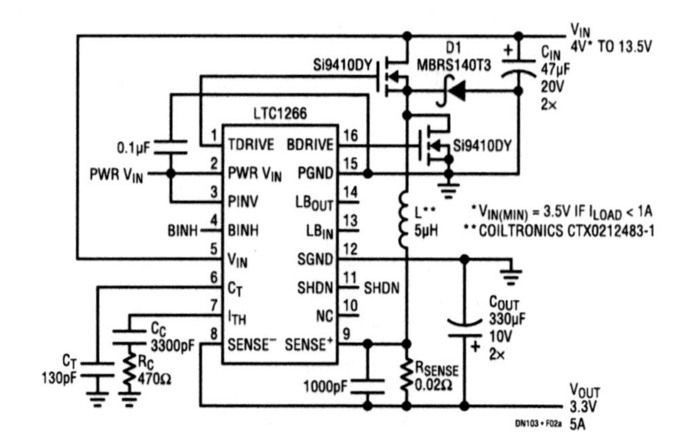

Figure 103.2a • All N-Channel 3.3V/5A Regulator with Drivers Powered from External Power V_{IN} Supply

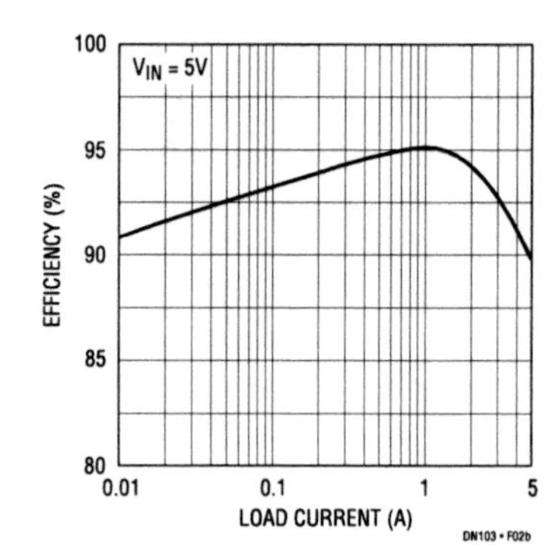

Figure 103.2b • Figure 103.2a Circuit Efficiency

The two application circuits demonstrate the fixed 3.3V version of the LTC1266. The LTC1266 is also available in fixed 5V and adjustable versions. All three versions are available in 16-pin narrow SOIC packages.

Conclusion

The new LTC1266 synchronous step-down regulator controller is the first Linear Technology synchronous controller with the ability to exploit the superior performance of N-channel MOSFETs to maximize efficiency and provide a low cost, compact solution for high current converters. The extra features provided in this product, Burst Mode inhibit and a low-battery comparator, make it ideal in a wide variety of applications.

Dual output regulator uses only one inductor

104

Carl Nelson

Many modern circuit designs still need a dual polarity supply. Communication and data acquisition are typical areas where both 5V and −5V are needed for some of the IC chips. It would be nice if a single switching regulator could supply both outputs with good regulation and a minimum of magnetic components. The circuit in Figure 104.1 is a good example of exploiting the best advantages of components and topologies to achieve a very small, dual output regulator with a single magnetic component.

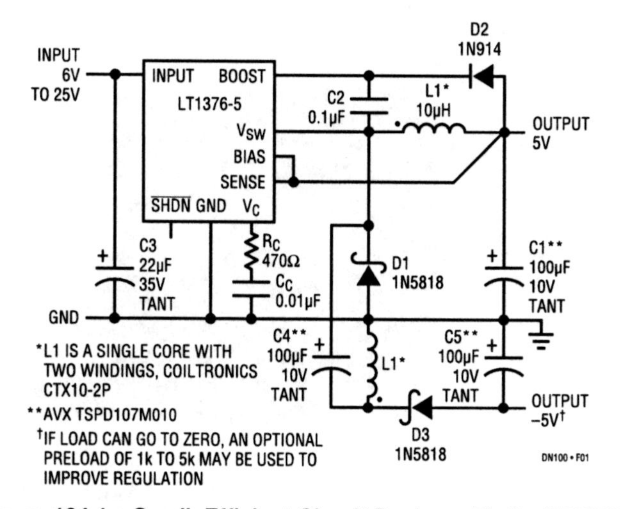

Figure 104.1 • Small, Efficient Circuit Design with the LT1376 Buck Converter

The 5V output is generated using the LT1376 buck converter. This device uses special design techniques and high speed processing to create a 500kHz design that is much smaller and more efficient than previous monolithic circuits. The current mode architecture and saturating switch design allow the LT1376 to deliver up to 1.5A load current from the tiny 8-pin SO package. L1 is a 10μH surface mount inductor from Coiltronics. It is manufactured with two identical windings that can be connected in series or parallel. One of the windings is used for the buck converter.

The second winding is used to create a negative output SEPIC (Single Ended Primary Inductance Converter) topology using D3, C4, C5, and the second half of L1. This converter takes advantage of the fact that the switching signal driving L1 as a positive buck converter is already the correct amplitude for driving a −5V SEPIC converter. During switch off time, the voltage across L1 is equal to the 5V output plus the forward voltage of D1. An identical voltage is generated in the second winding, which is connected to generate −5V using D3 and C5. Without C4, this would be a simple flyback winding connection with modest regulation. The addition of C4 creates the SEPIC topology. Note that the voltage swing at both ends of C4 is theoretically identical even without the capacitor. The undotted end of both windings goes to a zero AC voltage node, so the equal windings will have equal voltages at the opposing ends. Unfortunately, coupling between windings is never perfect, and load regulation at the negative output suffers as a result. The addition of C4 forces the winding potentials to be equal and gives much better regulation.

Regulation performance and efficiency

Figure 104.2 details the regulation performance of the circuit. The positive output combined load and line regulation is better than 1%, and this was considered good enough to forgo a graph. Negative output voltage is graphed as a function of negative load current for several values of positive load current. For best regulation, the negative output should have a preload of at least 1% of the maximum positive load.

Total output current of this circuit is limited by the maximum switch current of the LT1376. The following formula gives peak switch current, which cannot exceed 1.5A. This formula, in the spirit of simplicity, is simplified, so caution must be used if it indicates close to 1.5A peak current.

$$I_{PEAK} = 0.25A + (I^+) + (2)(I^-) \text{ (Must be less than 1.5A)}$$

Analog Circuit Design: Design Note Collection. http://dx.doi.org/10.1016/B978-0-12-800001-4.00104-6

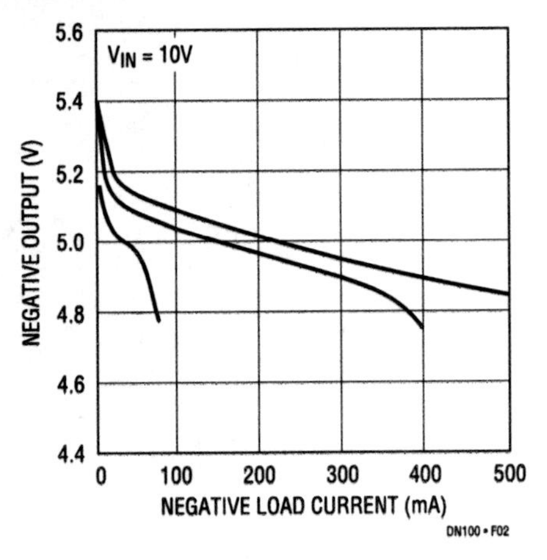

Figure 104.2 • −5V Regulation

Maximum *negative* load current is limited by the +5V load. A *typical* limit is one half of 5V current, but a more exact number can be found from:

$$\text{Max Negative Load} = (I^{+})(0.07)(V_{IN} - 2)$$

Note that as input voltage drops, less and less current is available from the negative output.

Efficiency of the switching regulator is not as good as a single buck converter, but still is very respectable, exceeding 80% over a wide current range. The inductor called out on the circuit schematic is a low cost off-the-shelf Coiltronics part made with a powdered iron core. Replacing that part with a Magnetics, Inc. Kool Mμ core #77130 with 13 turns of #24 wire will raise efficiency by 3% across most of the load current range (Figure 104.3).

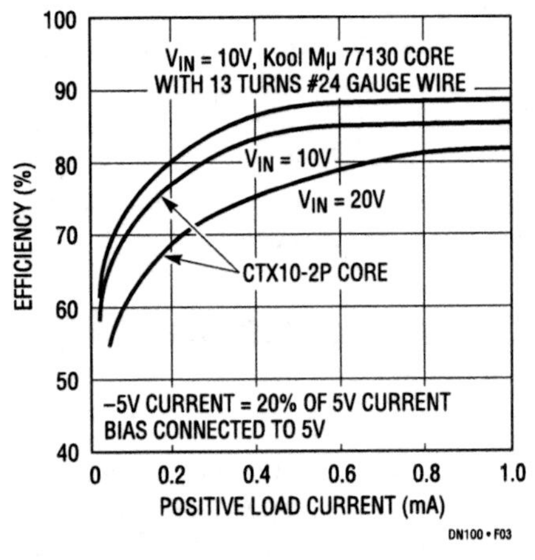

Figure 104.3 • Dual Output Efficiency

Output ripple voltage

Output ripple voltage is determined by the ESR of the output capacitors. The capacitors shown are AVX type TPS surface mount solid tantalum which are specially constructed for low ESR (<0.1Ω). Peak-to-peak ripple current into the +5V output capacitor is a triwave, typically $0.3A_{P-P}$, so an ESR of 0.1Ω in C1 will give $30mV_{P-P}$ output ripple. It is interesting to note that this ripple current is about one half of what would be expected for a buck converter. This occurs because the two windings are driven in parallel, so magnetizing current divides equally between the windings.

Ripple current peak-to-peak into the −5V output capacitor is approximately equal to twice the negative load current. The wave shape is roughly rectangular, and so is the resultant output ripple voltage. A 100mA negative load and 0.1Ω ESR output capacitor will have (2)(0.1A)(0.1Ω) = $20mV_{P-P}$ ripple. A word of caution, however; the current waveform contains fast edges, so the inductance of the output capacitor multiplied by the rate-of-rise of the current will generate very narrow spikes superimposed on the output ripple. With capacitor inductance of 5nH, and dI/dt = 0.05A/ns, the spike amplitude will be 250mV! Now for the good news. The effective bandwidth of the spikes is all above 20MHz, so it is very easy to filter them out. In fact, the inductance of the output PC board traces (20nH/in) coupled with load bypass capacitors will normally filter out the spikes. The only caveat is that if the load bypass capacitors are very low ESR types like ceramic, they should be paralleled with a larger tantalum capacitor to reduce the Q of the filter.

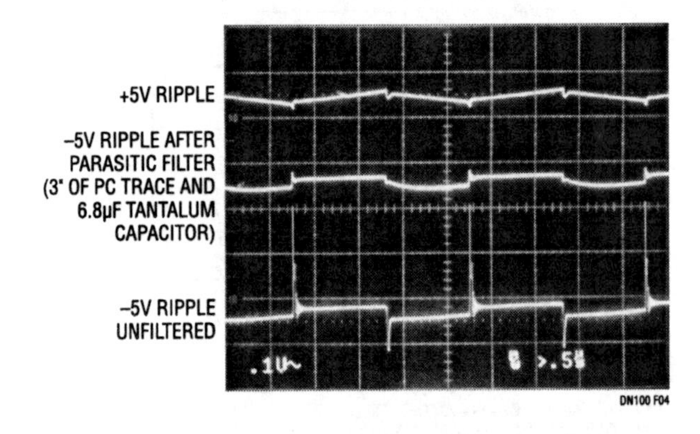

Both outputs can be shut down simultaneously by driving the LT1376 shutdown pin low. An undervoltage lockout function can also be implemented by connecting a resistor divider to the shutdown pin. See the LT1376 data sheet for details.

Highly integrated high efficiency DC/DC conversion

105

San-Hwa Chee   Howard Haensel

The LTC1574 and LTC1265 high efficiency step-down regulators minimize external components by using integrated low $R_{DS(ON)}$ P-channel switches. The LTC1574 goes one step further by including a low forward drop Schottky diode—an industry first. Both regulators also include on-chip low-battery detectors.

Burst Mode operation allows the LTC1574 and LTC1265 to achieve over 90% efficiency for load currents as low as 10mA. Current mode operation provides clean start-up, accurate current limit, and excellent line and load regulation. Inherent 100% duty cycle in dropout allows the user to extract maximum battery life. Both regulators can be shut down to a few microamperes.

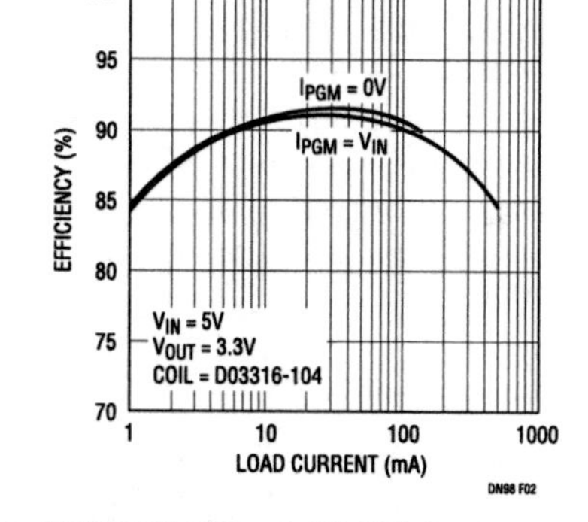

Figure 105.2 • LTC1574 5V to 3.3V Efficiency

LTC1574

The LTC1574 features the highest level of integration for a switching regulator. Besides an on-chip power MOSFET, it includes a low forward drop Schottky diode. The user needs only to provide an inductor and input/output filter capacitors for a complete high efficiency step-down converter. The current limit is pin selectable to either 340mA or 600mA, optimizing efficiency for a wide range of load currents.

Figure 105.1 shows a typical LTC1574 surface mount application requiring only three external components. It provides 3.3V at 150mA from an input voltage of 5V. Peak inductor current is limited to 340mA by connecting pin 6 (I_{PGM}) to ground. For applications requiring higher output current, connect pin 6 to V_{IN}. Under this condition the maximum load current is increased to 425mA. Efficiency curves for the two conditions on I_{PGM} are graphed in Figure 105.2. Note that all components remain the same for the two curves.

Low noise regulator

In some applications, it is important not to introduce any switching noise within the audio frequency range. Due to the Burst Mode nature of the LTC1574, there is a possibility that

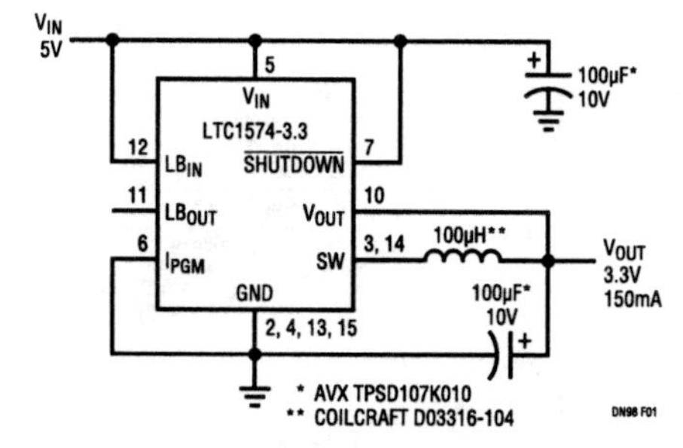

Figure 105.1 • LTC1574 3.3V, 150mA Surface Mount

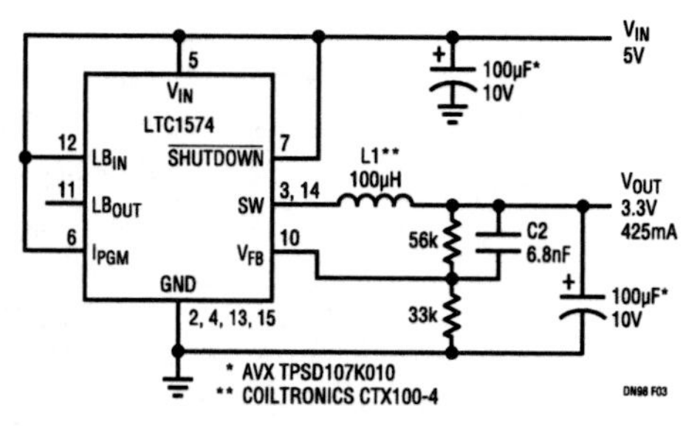

Figure 105.3 • Low Noise 5V to 3.3V Regulator

Analog Circuit Design: Design Note Collection. http://dx.doi.org/10.1016/B978-0-12-800001-4.00105-8

the regulator will introduce audio noise at some load currents. To circumvent this problem, a feed-forward capacitor can be used to shift the noise spectrum up and out of the audio band. Figure 105.3 shows the low noise connection with C2 being the feed-forward capacitor. The peak-to-peak output ripple is reduced to 30mV over the entire load range. A toroidal surface mount inductor L1 is chosen for its excellent self-shielding properties. Open magnetic structures such as drum and rod cores are to be avoided since they inject high flux levels into their surroundings. This can become a major source of noise in any converter circuit.

LTC1265

Whereas the LTC1574 can only supply a load current up to 425mA, the LTC1265 can source up to 1.2A. It features a low 0.3Ω (V_{IN} = 10V) internal P-channel MOSFET to provide high efficiency at high load current. The inductor current is user programmable via an external current sense resistor. Operation up to 700kHz permits the use of small surface mount inductors and capacitors. The LTC1265 employs an external Schottky diode.

Unlike the LTC1574 which always operates in Burst Mode, the LTC1265 only operates in Burst Mode at light loads and switches to continuous operation at heavier loads. For the LTC1265 to operate in Burst Mode, the load current has to be less than 15mV/R_{SENSE}.

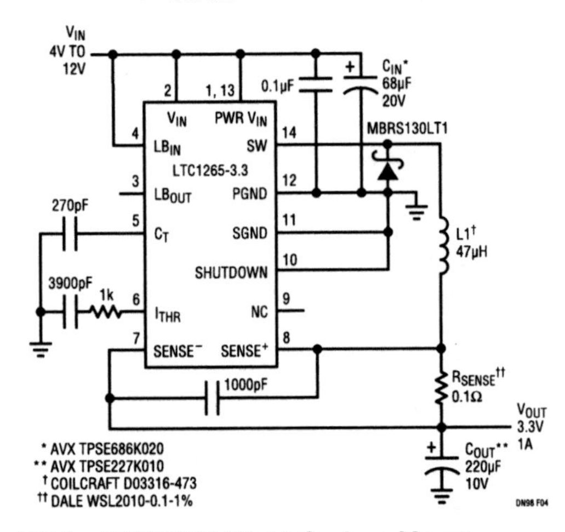

Figure 105.4 • LTC1265 3.3V, 1A Surface Mount

Figure 105.4 shows a typical LTC1265 surface mount application. It provides 3.3V at 1A from an input voltage range of 4V to 12V. Efficiency at various input voltages is plotted in Figure 105.5. Here the sense resistor is chosen as 0.1Ω, therefore the LTC1265 will go into continuous mode operation for load currents greater than 150mA. The peak efficiency approaches 93% at mid-current levels.

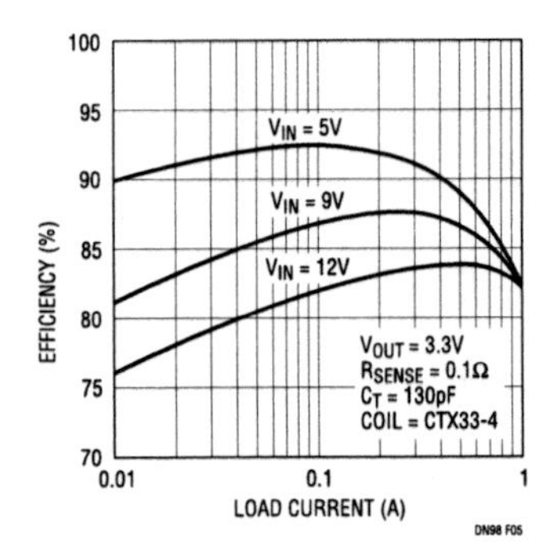

Figure 105.5 • LTC1265 5V to 3.3V Efficiency

Battery charger application

In Figure 105.6, the LTC1265 is configured as a battery charger for a four-NiCd stack. It has the capability of performing a fast charge of 1A, a trickle charge of 100mA or the charger can be shut off. In shut-off, diode D1 serves two purposes. First, it prevents the LTC1265 circuitry from drawing battery current and second, it eliminates "back powering" the LTC1265 which avoids a potential latch condition at power-up.

LTC1574 or LTC1265?

The LTC1574 and LTC1265 are differentiated by both the output current level and operating mode. For loads less than 425mA, the LTC1574 is the ideal choice because of its simplicity and ease of use. However, for applications requiring continuous mode operation, or more than 425mA output current, the LTC1265 must be used. Both devices can be tailored to meet a wide range of requirements.

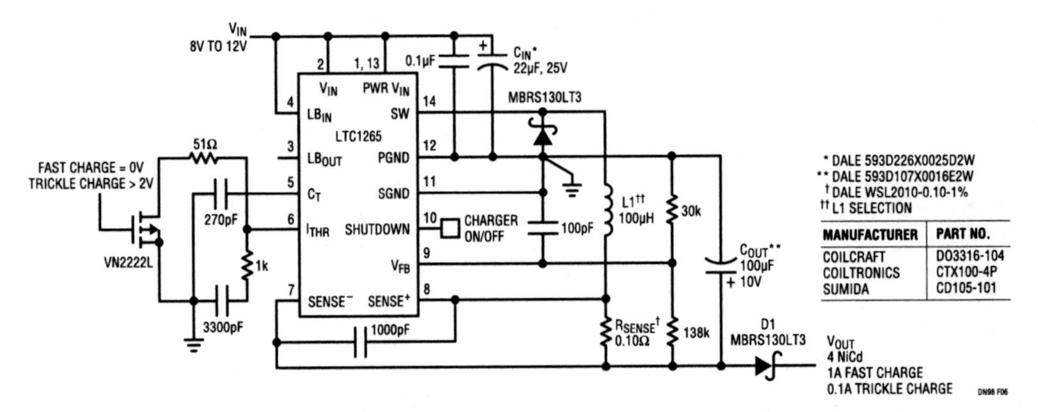

MANUFACTURER	PART NO.
COILCRAFT	D03316-104
COILTRONICS	CTX100-4P
SUMIDA	CD105-101

Figure 105.6 • NiCd Battery Charger

Ultralow power, high efficiency DC/DC converter operates outside the audio band

106

Mitchell Lee

Portable communications products are densely packed with signal processing, microprocessor, radio frequency, and audio circuits. Digital clock noise must be eliminated not only from the audio sections, but also from the antenna which, by the very nature of the product, is located only inches from active circuitry. If a switching regulator is used in the power supply, it becomes another source of noise. The LTC1174 step-down converter is designed specifically to eliminate noise at audio frequencies while maintaining high efficiency at low output currents.

Figure 106.1 shows an all surface mount solution for a 5V, 120mA output derived from five to seven NiCd or NiMH cells. Small input and output capacitors that are capable of handling the necessary ripple currents help conserve space. In applications where shutdown is desired this feature is available (otherwise short this pin to V_{IN}).

The LTC1174's internal switch, connected between V_{IN} and V_{SW}, is current controlled at a peak of approximately 340mA. Low peak switch current is one of the key features that allows the LTC1174 to minimize system noise compared to other chips which carry significantly higher peak currents, easing shielding and filtering requirements and decreasing component stresses. Output currents of up to 450mA are possible with this device by connecting the I_{PGM} pin to V_{IN}. This increases the peak current to 600mA, allowing for a high average output current.

To conserve power and maintain high efficiency at light loads, the LTC1174 uses Burst Mode operation. Unfortunately, this control scheme can also generate audio frequency noise at both light and heavy loads. In addition to electrical noise, acoustical noise can emanate from capacitors and coils under these conditions. A feed-forward capacitor (C2) shifts the noise spectrum up and out of the audio band, eliminating these problems. C2 also reduces peak-to-peak output ripple to approximately 30mV over the entire load range.

A toroidal surface mount inductor (L1) is chosen for its excellent self-shielding properties. Open magnetic structures such as drum and rod cores are to be avoided since they inject high flux levels into their surroundings. This can become a major source of noise in any converter circuit.

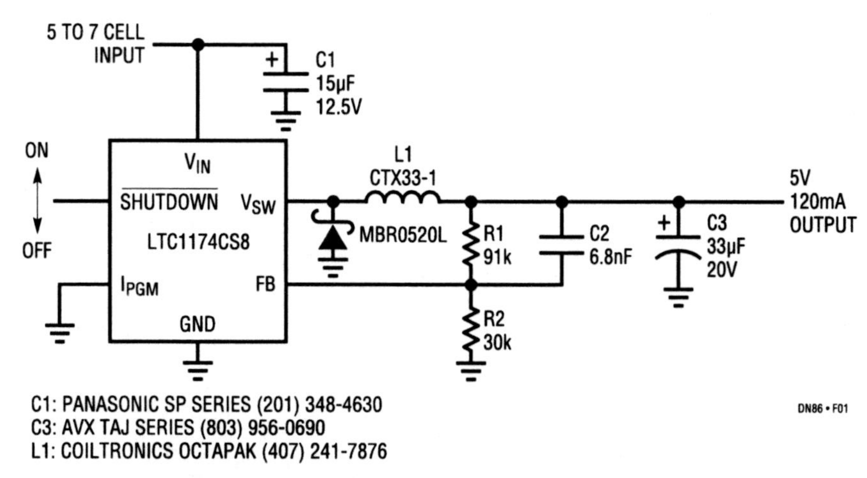

Figure 106.1 • Low Noise, High Efficiency Step-Down Regulator for Personal Communications Devices

Analog Circuit Design: Design Note Collection. http://dx.doi.org/10.1016/B978-0-12-800001-4.00106-X

The interactions of load current, efficiency, and operating frequency are shown in Figure 106.2. High efficiency is maintained even at low current levels, dropping below 70% at around 800μA. No load supply current is less than 200μA, dropping to approximately 1μA in shutdown. The operating frequency rises above the telephony bandwidth of 3kHz at a load of 1.2mA. Most products draw milliampere range load currents only in standby with the audio circuits squelched, when low frequency noise is not an issue.

The frequency curve depicted in Figure 106.2 was measured with a spectrum analyzer, not a counter. This ensures that the lowest frequency noise peak is observed rather than a faster switching frequency component. Any tendency to generate subharmonic noise is quickly exposed using this measurement method.

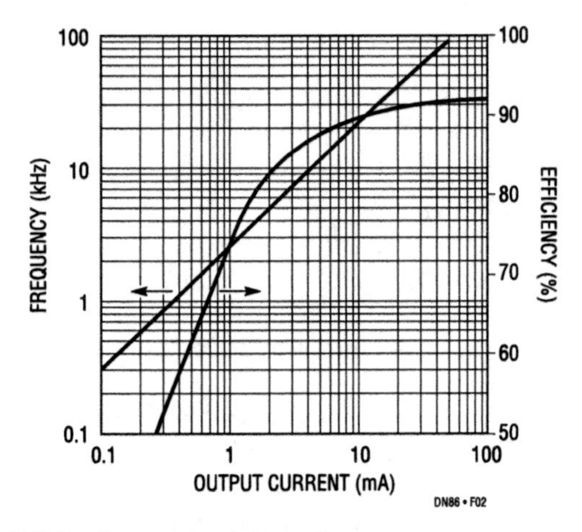

Figure 106.2 • Parameter Interaction

A spectrum analysis of noise from 100kHz to 10MHz is shown in Figure 106.3. The fundamental switching component in this test was approximately 85kHz, and the second harmonic shows up at twice that frequency. It measures approximately $3mV_{RMS}$. Harmonics of the 85kHz fundamental disappear into a 10μV "mud" between 1MHz and 2MHz. Noise in the critical 455kHz region ranges from 10μV to 300μV, depending on operating frequency. At 10.7MHz, an important and sensitive intermediate frequency, the noise is broadband and well below $10\mu V_{RMS}$.

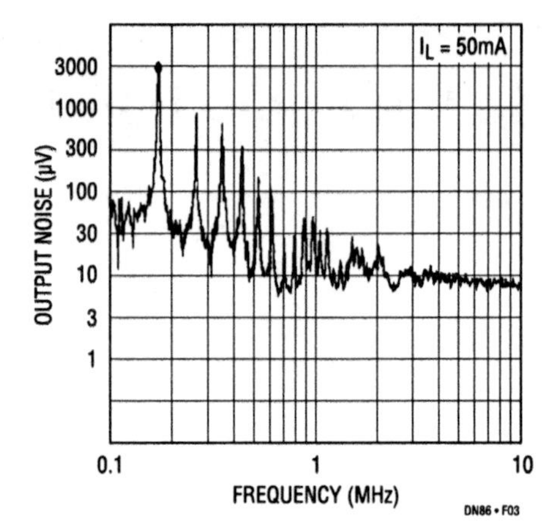

Figure 106.3 • Noise in the 100kHz to 10MHz Band

Further noise reduction is possible by adding an output filter (see Figure 106.4). A small surface mount ferrite bead is placed in series with the 5V output, close to the LTC1174 and bypassed by a 1μF surface mount ceramic capacitor. Noise attenuation at 10MHz exceeds 20dB.

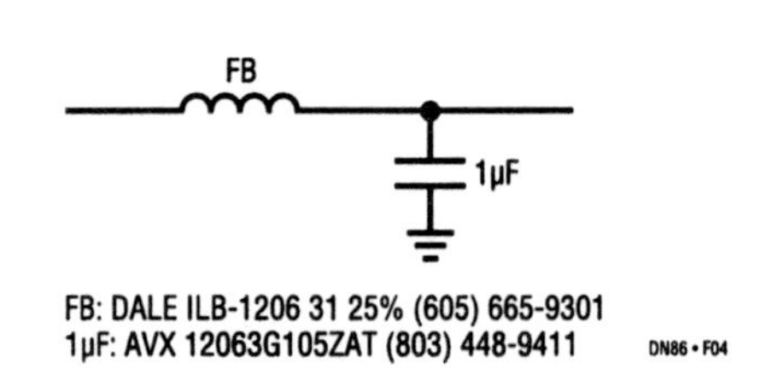

Figure 106.4 • An Effective Filter for Attenuating Noise Components above 1MHz

Triple output 3.3V, 5V, and 12V high efficiency notebook power supply

107

Randy G. Flatness

The new LTC1142 is a dual 5V and 3.3V synchronous step-down switching regulator controller featuring automatic Burst Mode operation to maintain high efficiencies at low output currents. Two independent regulator sections, each driving a pair of complementary MOSFETs, may be shut down separately to less than 20μA/output. This feature is an absolute necessity to maximize battery life in portable applications. Additionally, the input voltage to each regulator section can be individually connected to different potentials (20V maximum) allowing a wide range of novel applications.

The operating current levels for both regulator sections are user programmable, via external current sense resistors, to set current limit. A wide input voltage range for the LTC1142 allows operation from 4V to 16V. The LTC1142HV extends this voltage range to 20V, permitting operation with up to 12-cell battery packs.

Both regulator blocks in the LTC1142 and LTC1142HV use a constant off-time current mode architectures with Burst Mode operation. This results in a power supply that has very high efficiency over a wide load current range, fast transient response, and very low dropout. The LTC1142 is ideal for applications requiring 5V and 3.3V output voltages with high conversion efficiencies over a wide load current range in a small amount of board space.

The application circuit in Figure 107.2 is configured to provide output voltages of 3.3V, 5V, and 12V. The current capability of both the 3.3V and 5V outputs is 2A (2.5A peak). The logic controlled 12V output can provide 150mA (200mA peak), which is ideal for flash memory applications. The operating efficiency shown in Figure 107.1 exceeds 90% for both the 3.3V and 5V sections.

The 3.3V section of the circuit in Figure 107.2 is comprised of the main switch Q4, synchronous switch Q5, inductor L1, and current shunt R_{SENSE3}. The current sense resistor R_{SENSE} monitors the inductor current and is used to set the output current according to the formula $I_{OUT}=100mV/R_{SENSE}$. Advantages of current control include excellent line and load transient rejection, inherent short-circuit protection

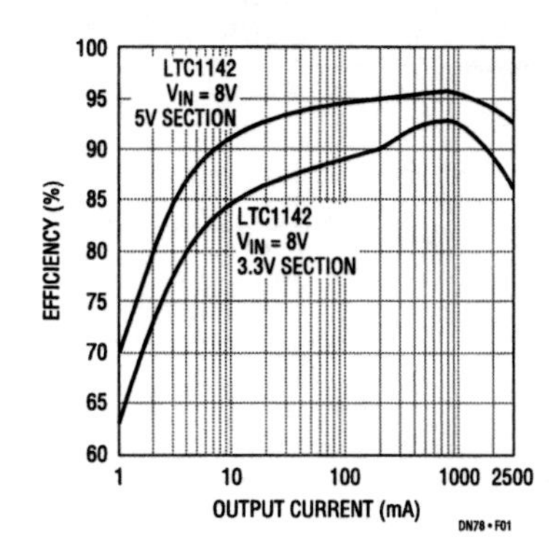

Figure 107.1 • LTC1142 Efficiency

and controlled start-up currents. Peak inductor currents for L1 and T1 of the circuit in Figure 107.2 are limited to $150mV/R_{SENSE}$ or 3.0A and 3.75A respectively.

When the output current for either regulator section drops below approximately $15mV/R_{SENSE}$, that section automatically enters Burst Mode operation to reduce switching losses. In this mode the LTC1142 holds both MOSFETs off and sleeps at 160μA supply current while the output capacitor supports the load. When the output capacitor discharges 50mV, the LTC1142 briefly turns this section back on, or "bursts" to recharge the output capacitor. The timing capacitor pins, which go to 0V during the sleep interval, can be monitored with an oscilloscope to observe burst action. As the load current is decreased the circuit will burst less and less frequently.

The timing capacitors C_{T3} and C_{T5} set the off-time according to the formula $t_{OFF}=1.3 \times 10^4 \times C_T$. The constant off-time architecture maintains a constant ripple current while the operating frequency varies with input voltage. The 3.3V section has an off-time of approximately 5μs resulting in a operating frequency of 120kHz at 8V input voltage. The 5V section has an off-time of 3.5μs and a switching frequency of 107kHz at 8V input voltage.

Analog Circuit Design: Design Note Collection. http://dx.doi.org/10.1016/B978-0-12-800001-4.00107-1

The operation of the 5V section is identical to the 3.3V section with inductor L1 replaced by transformer T1. The 12V output voltage is derived from an auxiliary winding on the 5V inductor T1. The output from this additional winding is rectified by diode D3 and applied to the input of an LT1121 regulator. The 12V output voltage is set by resistors R3 and R4. A turns ratio of 1:1.8 is used for T1 to ensure that the input voltage to the LT1121 is high enough to keep the regulator out of dropout while maximizing efficiency.

The LTC1142 synchronous switch removes the normal limitation that power must be drawn from the primary 5V inductor winding in order to extract power from the auxiliary winding. With synchronous switching the auxiliary 12V output may be loaded without regard to the 5V primary output load providing that the loop remains in continuous mode operation.

When the 12V output is activated by a TTL high (6V maximum) on the 12V enable line, the 5V section of the LTC1142 is forced into continuous mode. A resistor divider composed of R1, R5 and switch Q1 forces an offset subtracting from the internal 25mV offset at pin 14. When this external offset cancels the built-in 25mV offset Burst Mode operation is inhibited.

For additional high efficiency circuits see Application Note 54.

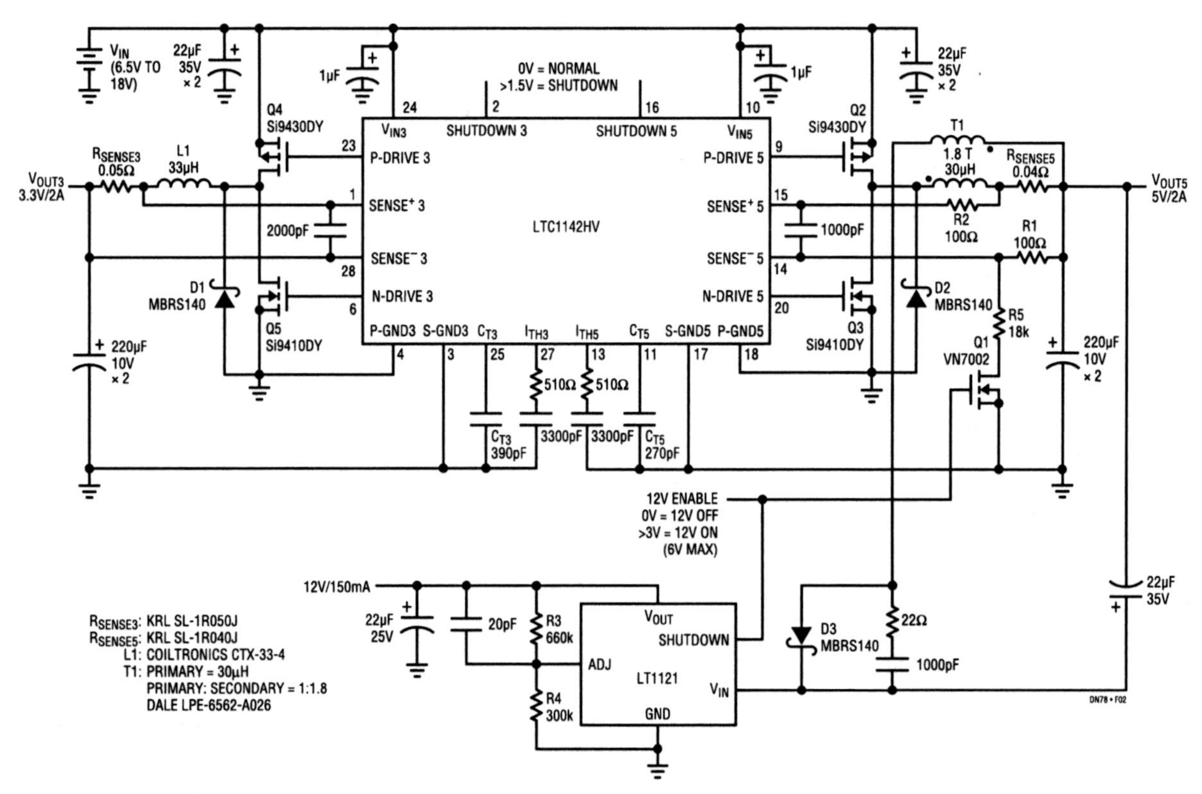

Figure 107.2 • LTC1142 Triple Output High Efficiency Power Supply

Single device provides 3.3V and 5V in surface mount

108

Peter Schwartz

This Design Note describes a circuit which uses one LTC1149 to regulate both a 3.3V and a 5V output with a 17W capability. The circuit presented is an improved version of the one detailed in Design Note 72 (DN72). Enhancements include an emphasis upon the use of surface mount components and an extended input voltage range (6V to 24V vs 8V to 24V).

The schematic diagram is given in Figure 108.3. For the principles of operation, please refer to DN72 (copies are available from any LTC representative). One significant difference between this circuit and that of DN72 is that QN2's gate drive is AC-coupled, ensuring full enhancement of QN2 even at low input voltages. The circuit as shown operates down to $V_{IN}=7V$. Adding two 220µF capacitors in parallel with C3/C4 extends minimum V_{IN} to 6V or less.

The assembled circuit (Demo Circuit 027A) measures only 2.15″ × 1.63″ (Figure 108.1). This compact and inexpensive design provides excellent efficiency, generally approaching and often exceeding 90% (Figure 108.2).[1] Cross-regulation between the two outputs is also quite good (Table 108.1).[1] Additionally, network D1/D2/C7 ensures that the 3.3V and 5V outputs both reach their rated voltages at the same time following power-up.

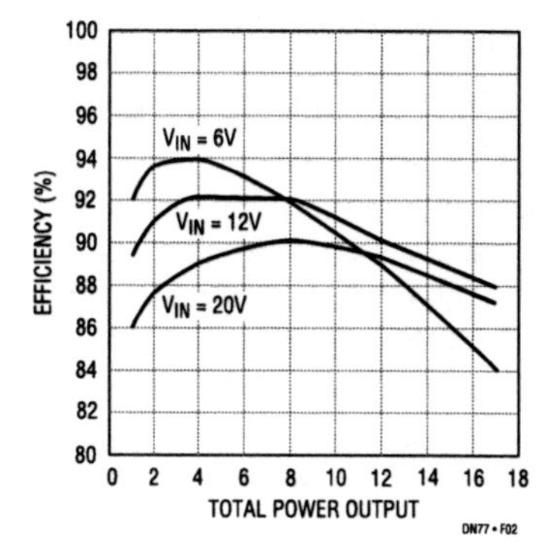

Figure 108.2 • Efficiency vs P_{OUT} and V_{IN}

Table 108.1 Cross-Regulation vs V_{IN} and I_{OUT}

V_{IN}	$I_{3.3V}$	$V_{3.3V}$	I_{5V}	V_{5V}
6.00V	0mA	3.36V	0mA	5.03V
	5A	3.23V	0mA	5.06V
	2.12A	3.41V	2A	4.80V
	0mA	3.55V	3A	4.68V
8.00V	0mA	3.36V	0mA	5.04V
	5A	3.24V	0mA	5.07V
	2.12A	3.39V	2A	4.90V
	0mA	3.47V	3A	4.81V
24.00V	0mA	3.38V	0mA	5.14V
	5A	3.24V	0mA	5.12V
	2.12A	3.31V	2A	4.98V
	0mA	3.36V	3A	4.93V

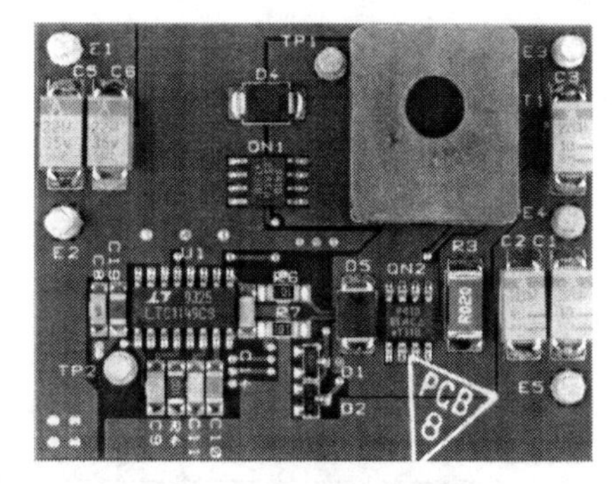

Figure 108.1 • Demonstration Circuit Board

Note 1: Data at $V_{IN}=6V$ taken with $C_{5V}=(4\times 220\mu F)$.

Analog Circuit Design: Design Note Collection. http://dx.doi.org/10.1016/B978-0-12-800001-4.00108-3

Customizing the circuit

The circuit of Figure 108.3 is the result of a significant R&D effort by LTC, providing a 3.3V/5V power solution combining performance and price benefits with manufacturability and reliability. At the same time the circuit is flexible enough to accommodate a number of variations. Some of these are:

1. **Peak Power > 17W:** Useful for starting disk drives and other "surge" loads, increased peak power is obtained by lowering the value of R3, and if necessary adding capacitance to the 3.3V output to meet equivalent series resistance requirements. Under most conditions, the total ripple current rating of the 3.3V output capacitance is determined by maximum **continuous** power (capacitor current ratings are determined by I^2R heating and have an associated thermal time constant). For additional details and assistance, contact the factory.
2. **Lower Power Output:** When the full 17W capability featured here is not needed, some input and output filter capacitors can be removed from the circuit. Frequently QP2 can be deleted as well, and a smaller transformer used. Please consult LTC for further information.
3. **Lowest Cost:** Circuit cost can be reduced by using aluminum electrolytic capacitors for C1 to C6 and C15 to C18. The Nichicon "PL" or United Chemi-Con "LXF" series are good choices. On the outputs, Sanyo 10SA220MOS-CON capacitors (220µF, 10V) provide excellent performance in a small case size. Deleting D5 will save area and cost with only a slight efficiency reduction. In low voltage applications, the LTC1148 can be substituted for the LTC1149, with quiescent current and price advantages. For applications where $V_{IN} \geq 12V$, QP2 can often be removed with little or no effect on efficiency.

Construction notes

1. Figure 108.3 shows several ground lines. These should be run separately (single point ground). Heavy line widths in the schematic indicate wide power and ground traces on the PC board.
2. Pin 10 of the LTC1149 is sensitive to switching noise. The PCB layout should take this into account.
3. The Demonstration Board uses tantalum input filter capacitors (C5/C6 and C17/C18) for space reasons. For best life, specific voltage and current derating criteria apply to tantalum devices. **If these capacitors are to be subjected to voltages in excess of 18V DC, contact the capacitor vendor. For applications where the input will be subjected to high dV/dt or high dI/dt surges (e.g., switch closure to a battery pack), aluminum electrolytic input capacitors are definitely preferred due to their higher reliability under such conditions.**

Other

Linear Technology has a Gerber file of this Demonstration Board (DC027A) available along with a complete parts list. For this Demonstration Board, a Hurricane Electronics Lab throughhole transformer (HL-8700) was used to reduce overall height. Beckman Industrial Corporation has developed a very low profile, surface mount transformer suitable for applications where $V_{IN} \geq 9V$ (HM00-93839). Capacitors C1 to C6 and C15 to C18 are AVX "TPS" series capacitors and should not be casually substituted. For more information, Hurricane can be reached at (801) 635-2003, Beckman at (714) 447-2656, and AVX Application Assistance at (800) 282-4975.

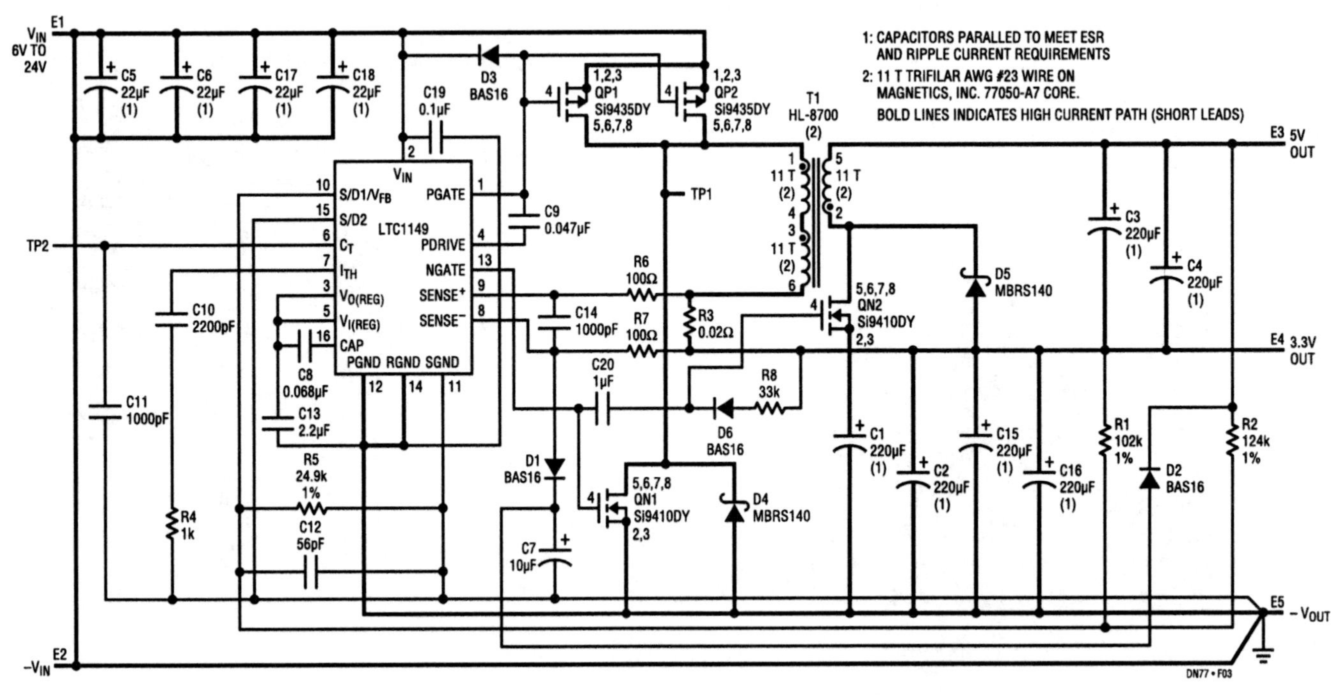

Figure 108.3 • Single LTC1149 Provides 3.3V and 5V in SMT

A simple high efficiency, step-down switching regulator

109

San-Hwa Chee

The new LTC1174 requires only four external components to construct a complete high efficiency step-down regulator. Using Burst Mode operation, efficiency of 90% is achievable at output currents as low as 10mA. The LTC1174 is protected against output shorts by an internal current limit which is pin selectable to either 340mA or 600mA. This current limit also sets the inductor's peak current. This allows the user to optimize the converter's efficiency depending upon the output current requirement.

To help the user get the most out of their battery source, the internal 0.9Ω (at supply voltage of 9V) power P-channel MOSFET switch is turned on continuously (DC) at dropout. In addition, an active low shutdown pin is included to power down the LTC1174, reducing the no load quiescent current from 130μA to just 1μA. An on-chip low battery detector is also included, with the trip point set by two external resistors.

Figure 109.1 shows a typical LTC1174 surface mount application. It provides 5V at 175mA from an input voltage range of 5.5V to 12.5V. Figure 109.2 shows the circuit's efficiency approaching 93% at an input voltage of 9V. Peak inductor current is limited to 340mA by connecting Pin 7 (I_{PGM}) to ground. The advantages of controlling the inductor's current include: excellent line and load transient response, short-circuit protection and controlled start-up current.

For applications requiring higher output current, connect Pin 7 (I_{PGM}) to V_{IN}. Under this condition, the maximum load current is increased to 425mA. Figure 109.3 shows the resulting circuit. Note that all components remain the same as in Figure 109.1. The new efficiency curve is shown in Figure 109.4.

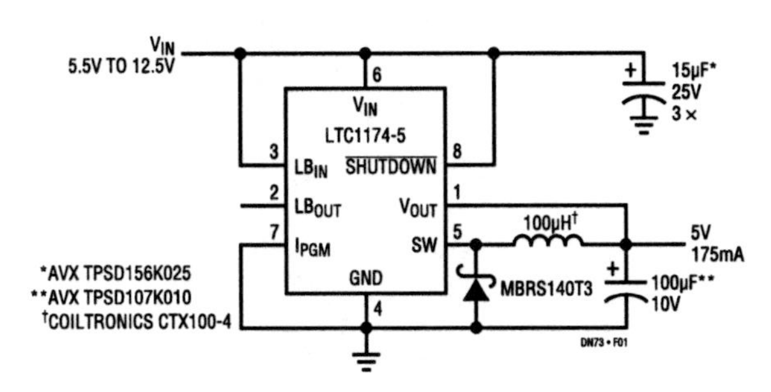

Figure 109.1 • LTC1174 5V, 175mA Surface Mount

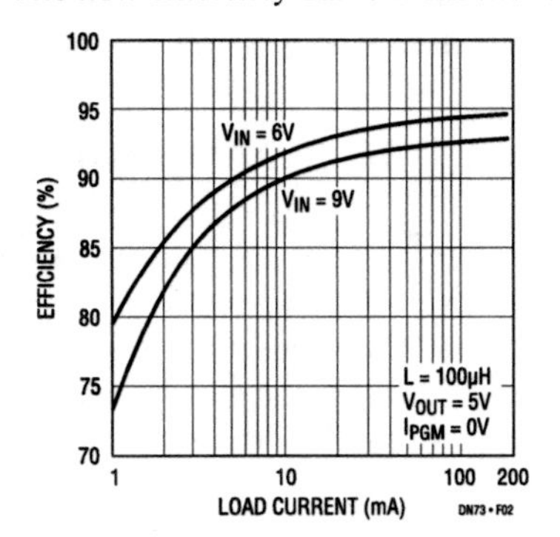

Figure 109.2 • LTC1174 5V, 175mA Efficiency

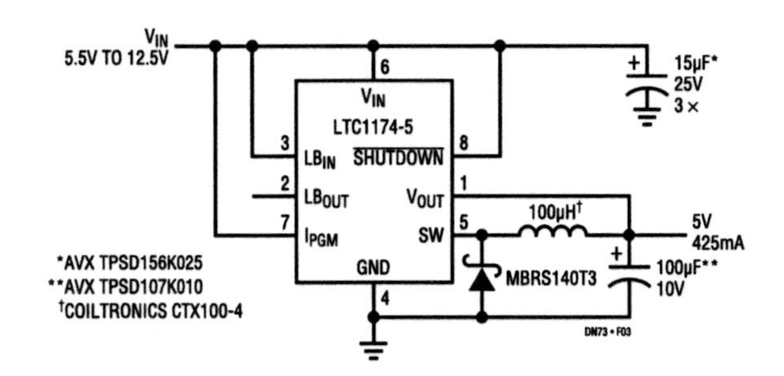

Figure 109.3 • LTC1174 5V, 425mA Surface Mount

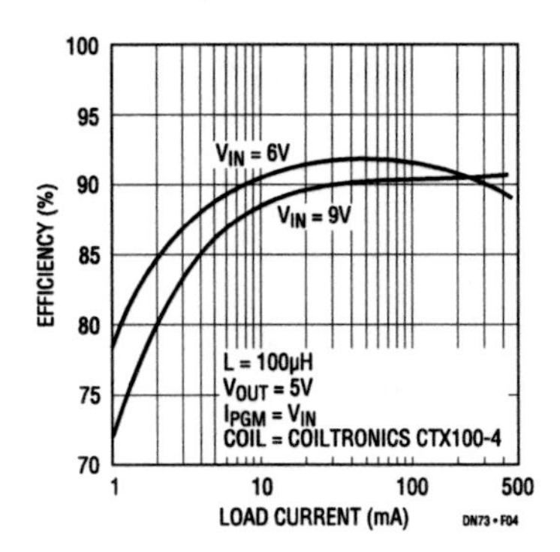

Figure 109.4 • LTC1174 5V, 425mA Efficiency

Analog Circuit Design: Design Note Collection. http://dx.doi.org/10.1016/B978-0-12-800001-4.00109-5

To have good control of inductor ripple current, a constant off-time architecture is used for the LTC1174. This scheme allows the ripple current to remain constant while the input voltage varies, easing the inductor's selection. However, the switching frequency is a function of input voltage. For an input voltage range of 6V to 12V with an output voltage of 5V, the operating frequency varies from about 42kHz to 146kHz. Figure 109.5 shows a normalized plot of the switching frequency as a function of the differential input/output voltage. The normalized value of 1 is equivalent to 111kHz.

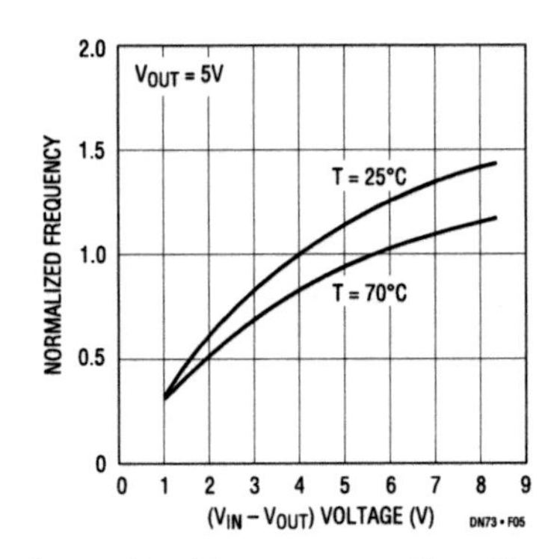

Figure 109.5 • Operating Frequency vs $V_{IN} - V_{OUT}$

100% duty cycle in dropout

When the input voltage decreases, the switching frequency decreases. With the off-time constant, the on-time is increased to maintain the same peak-to-peak ripple current in the inductor. Ultimately, a steady-state condition will be reached where Kirchoff's Voltage Law determines the dropout voltage. When this happens, the P-channel power MOSFET is turned on DC (100% duty cycle). The dropout voltage is then governed by the load current multiplied by the total DC resistance of the MOSFET, inductor, and the internal 0.1Ω current sense resistance. Figure 109.6 shows the dropout voltage as a function of load current.

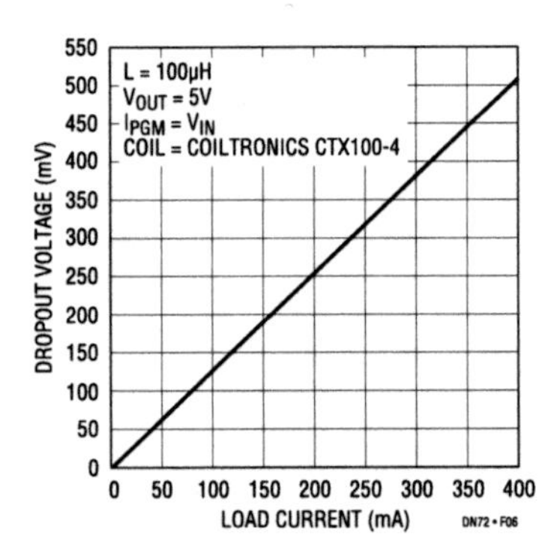

Figure 109.6 • Dropout Voltage vs Output Current

Positive-to-negative converter

The LTC1174 can easily be set up for a negative output voltage. If −5V is desired, the LTC1174-5 is ideal for this application as it requires the least components. Figure 109.7 shows the schematic for this application including low battery detection capability. The LED will turn on at input voltages less than 4.9V. The corresponding efficiency curve is shown in Figure 109.8.

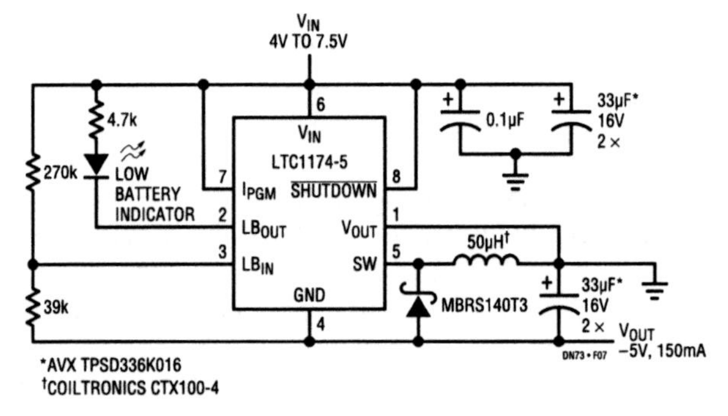

Figure 109.7 • Positive to −5V Converter with Low Battery Detection

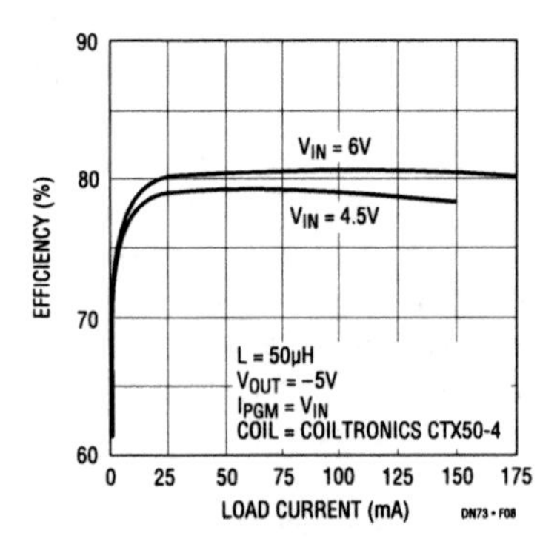

Figure 109.8 • Efficiency vs Load Current for a −5V Output Regulator

Delivering 3.3V and 5V at 17W

110

Peter Schwartz

This design note shows how one LTC1149 synchronous switching regulator can deliver both 3.3V and 5V outputs. The design's simplicity, low cost, and high efficiency make it a strong contender for portable, battery-powered applications. The circuit described accepts input voltages from 8V to 24V, to power **any** combination of 3.3V and 5V loads totalling 17W or less. For input voltages in the 8V to 16V range, the LTC1148 may be used, reducing both quiescent current and cost. For operation at input voltages below 8V, please contact the factory.

For convenience, the test circuit was built using mostly through-hole components. A follow-on design note will give details on building this circuit with surface mount parts.

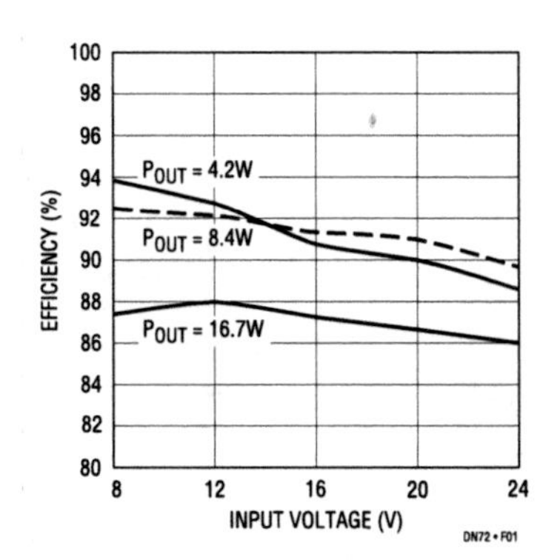

Figure 110.1 • Efficiency vs V_{IN} and P_{OUT}

Table 110.1 Cross-Regulation vs V_{IN} and I_{OUT}

V_{IN}	$I_{3.3V}$	$V_{3.3V}$	I_{5V}	V_{5V}
8V	0mA	3.43V	0mA	5.14V
	5A	3.27V	0mA	5.19V
	2A	3.42V	2A	4.95V
	0mA	3.52V	3A	4.84V
24V	0mA	3.42V	0mA	5.14V
	5A	3.26V	0mA	5.12V
	2A	3.32V	2A	4.97V
	0mA	3.42V	3A	4.93V

Performance

Efficiency of this circuit is excellent, generally approaching and frequently exceeding 90% (Figure 110.1). The cross-regulation between the two outputs (a measure of their interdependency) is quite good (Table 110.1). At low power levels, the LTC1149 cleanly enters Burst Mode operation with a quiescent current of only 0.7mA.

Theory of operation

The complete circuit is shown in Figure 110.2. To develop the 3.3V output, the LTC1149 acts as a synchronous step-down (buck) converter. L1A and L1B in series form the 3.3V buck inductor, and the C1/C2 combination is the 3.3V output filter capacitor. When Q1/Q2 are ON, the current through L1 ramps up. When Q1/Q2 turn OFF, Q3 is turned ON to provide the current in L1 with a low resistance recirculation path. This use of Q3 as a synchronous rectifier increases efficiency by virtually eliminating conduction voltage drop.

The 5V output is produced by L1, L2, Q4, and C3/C4. Since Q3 has essentially zero voltage drop when turned ON, the voltage across L1 is fixed at 3.3V during that time. With the voltage across L1 known, transformer action develops a predictable voltage across L2. If Q4 is turned on for the same interval as Q3 (forming a second synchronous rectifier), current will flow from L2 into C3/C4. Using a turns ratio of 2:1 between L1A/L1B and L2, C3/C4 will charge to a total voltage of $(0.5 \cdot 3.3V) + 3.3V = 5V$. Feedback to the LTC1149's

Analog Circuit Design: Design Note Collection. http://dx.doi.org/10.1016/B978-0-12-800001-4.00110-1

error amplifier comes from the 3.3V and 5V outputs through R1 and R2 (this "split feedback" enhances cross-regulation).

In addition to simplicity, this topology offers some more subtle advantages over other dual output techniques:

1) The 3.3V and 5V outputs are inherently synchronous to each other.

2) Both outputs achieve their rated voltage at the same time after power-up or after a short circuit.

3) A short to ground on either output will automatically disable the other output. This is difficult to achieve with techniques employing two independent control loops.

Circuit particulars

There are three areas of this circuit which require special attention. They are the transformer (L1A, L1B, L2), the input and output capacitors, and the layout.

The transformer must be trifilar-wound. Trifilar winding is a standard production technique in which three wires are wound at the same time on the same magnetic core. The three resulting coils form a transformer with excellent magnetic coupling. In this circuit these attributes improve cross-regulation and efficiency. Two of the three coils are connected in series to form L1. The third coil becomes the boost winding, L2. This inherently provides the required 2:1 turns ratio between L1 and L2. The test transformer was made by using three windings of ten turns #23 wire, on a Kool Mμ 77050-A7 toroid (finished size: 0.625″ diameter × 0.25″ high). If an off-the-shelf transformer is desired, Coiltronics, Inc. and Hurricane Labs both carry suitable parts. Coiltronics can be reached at (305) 781-8900; Hurricane's number is (801) 635-2003.

The values and sizes of the input and output capacitors are determined by ESR and ripple current ratings. The following lists critical parameters. Specific vendors and types are suggested in Figure 110.2.

C1, C2: Total parallel ESR ≤ 0.035Ω
Total IRMS rating ≥ 2.5A
C3, C4: Total IRMS rating ≥ 2.5A
C5, C6: Total IRMS rating ≥ 1.6A

In general, layout practices should follow those for other switching power supplies. Some examples are: keep separate types of grounds separate (e.g., signal ground, main power ground), and return the various grounds to a single common point. Power and ground leads should be kept short and isolated as much as possible from signal traces. Details for the successful layout of circuits using the LTC1148/LTC1149 can be found on the data sheets for these parts, which should be consulted for routing recommendations. As noted above, a surface mount layout for this circuit will appear in the follow-on design note.

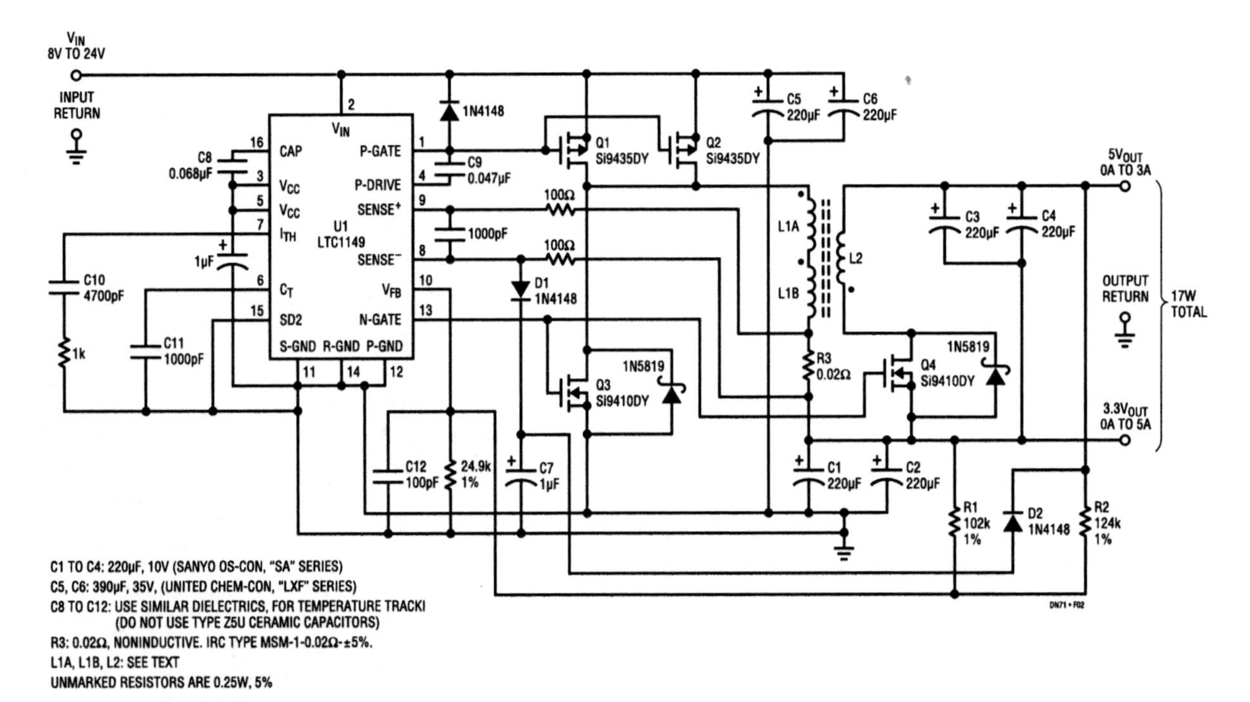

Figure 110.2 • Dual Output LTC1149 Supply Provides High Efficiency at Low Cost

Low parts count DC/DC converter circuit with 3.3V and 5V outputs

111

Ron Vinsant

This design note describes a simple low cost dual output step-down converter circuit based on the LT1076 five terminal switching regulator.

Performance

Input voltages can range from 8V to 30V. The load range on the 5V is 0.05A to 0.5A while the 3.3V load range is 0.1A to 1A. The circuit is self-protecting under no load conditions; it will "burp" in the same fashion as many off-line flyback power supplies.

Output voltage regulation is excellent. Over all load and line conditions, including cross-regulation, the 3.3V output varies from 3.25V to 3.27V. The 5V output varies from 4.81V to 5.19V under the same conditions.

Performance Table

V_{IN}	V_{OUT}, OUTPUT 1 (5V)	V_{OUT}, OUTPUT 2 (3.3V)
	At I_{OUT} = 0.4A	At I_{OUT} = 1A
8V	4.81	3.26
30V	5.07	3.26
	At I_{OUT} = 0.05A	At I_{OUT} = 1A
8V	5.14	3.25
30V	5.19	3.25
	At I_{OUT} = 0.4A	At I_{OUT} = 0.1A
8V	4.81	3.26
30V	5.02	3.27
	At I_{OUT} = 0.05A	At I_{OUT} = 0.1A
8V	5.07	3.26
30V	5.11	3.26

In a typical application of 0.5A on the 3.3V and 0.25A on the 5V, efficiency is typically 76%. With an input voltage of 30V and full load, the efficiency drops to 66%. In normal operating regions efficiency is always better than 70%.

The 5V ripple is less than 75mV and the 3.3V ripple less than 50mV over all line and load conditions.

This design can help save both parts and cost by the elimination of a second regulator. Only a few additional parts are required to make the second output. They are: two resistors, a Schottky diode, a small ceramic capacitor and a filter electrolytic; only five additional components! The normal single winding inductor has one small winding added to create the additional output.

The circuit has been built in our lab and has only been evaluated for room temperature performance. No stability analysis has been done.

Inductor

The inductor is based on an EP-13 ferrite core which is available from a number of vendors. In our breadboard we used a Ferroxcube core in 3C81 material gapped to 6 mils (center gap). The 3.3V winding is 22 turns of #25 AWG while the 5V winding is 13 turns of #28 AWG. Any magnetics vendor should be able to wind this device. The inductor has only a 14°C temperature rise. Coiltronics at (305) 781-8900, or Hurricane Labs at (801) 635-2003 can supply this inductor off the shelf.

Capacitors

Ripple current in the output capacitors C2 and C3 is $250mA_{RMS}$ total with the input voltage at 30V and maximum load on the two outputs.

The input capacitor (C1), which undergoes higher stress, has a ripple current of 830mA maximum at 14V input and maximum load.

Analog Circuit Design: Design Note Collection. http://dx.doi.org/10.1016/B978-0-12-800001-4.00111-3

The input and output capacitors have been chosen primarily for ESR, not for voltage. The 50V rating of the capacitors is not due to stress from the circuit but from the fact that the lowest ESR for a particular can size occurs at 50V, in this series of capacitors, from this specific manufacturer.

The capacitors in the frequency compensation network should be at least X7R ceramic, never Z5U, and if broad temperature operation is expected, polyester or polycarbonate film caps should be used.

Layout

In order to achieve proper performance it is important to lay out the circuit as shown in Figure 111.1. Use a single point ground at the output of the converter as shown. The term "short" indicates that the trace should be as short as possible between the two points shown. These traces should have a minimum width of 0.2″ in 2oz copper for a length of less than 1.5″. Traces longer than this should be avoided on heavier lines of the schematic.

Heat sinking

Any heat sink of 30°C/W or lower will keep the LT1076 at an acceptable temperature up to a 70°C ambient. See the LT1076 data sheet for further information.

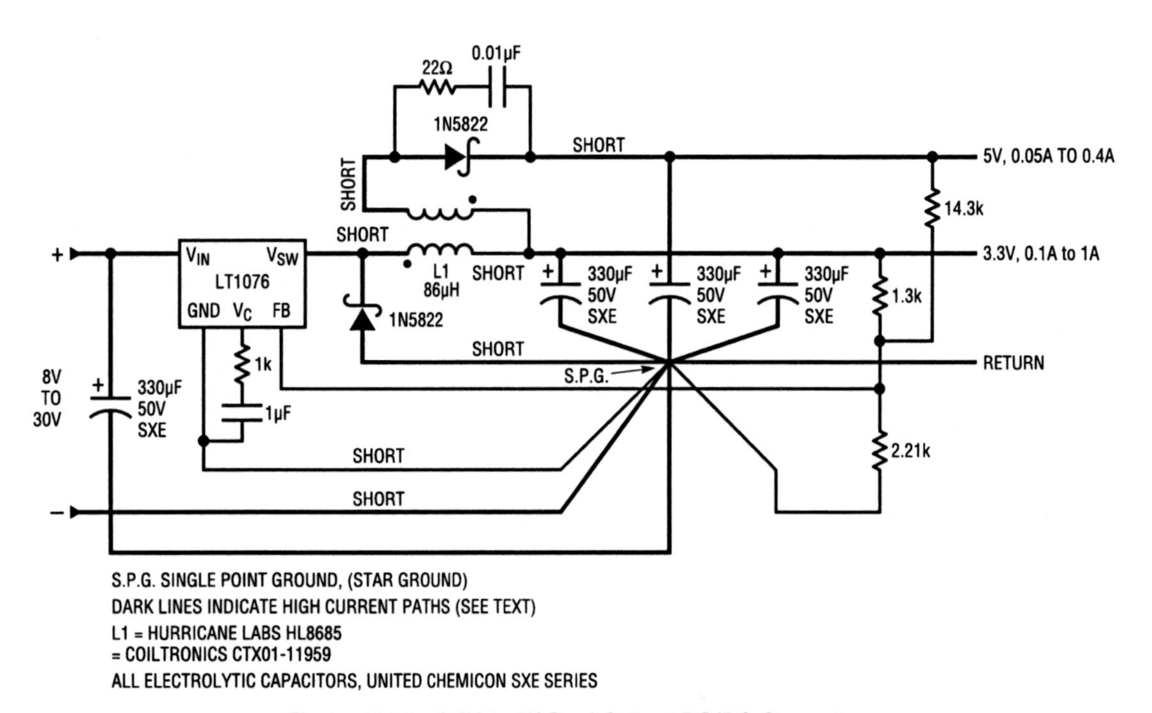

Figure 111.1 • 3.3V to 5V Dual Output DC/DC Converter

New synchronous step-down switching regulators achieve 95% efficiency

112

Brian Huffman   Milt Wilcox

The new LTC1148 and LTC1149 synchronous switching regulator controllers make high efficiency DC/DC conversion possible in a wide range of applications. These controllers share a current-mode architecture which combines synchronous switching for maximum efficiency at high currents with an automatic low current operating mode, called Burst Mode operation, which makes 90% efficiencies possible at output currents as low as 10mA.

Figure 112.1 shows a typical LTC1148 surface mount application providing 5V at 2A from an input voltage of 5.5V to 13.5V. The operating efficiency, shown in Figure 112.2, peaks at 97% and exceeds 90% from 10mA to 2A with a 10V input. Q1 and Q2 comprise the main switch and synchronous switch, respectively, while inductor current is measured via the voltage drop across current shunt R_{SENSE}. R_{SENSE} is the key component used to set the output current capability according to the formula $I_{OUT} = 100mV/R_{SENSE}$. Advantages of current control include excellent line and load transient rejection, inherent short-circuit protection and controlled start-up currents. Peak inductor current is limited to 150mV/R_{SENSE} or 3A for the Figure 112.1 circuit.

The timing capacitor, C_T, sets the off-time according to the formula $t_{OFF} = 1.3 \cdot 104 \cdot C_T$. The constant off-time architecture maintains a constant inductor ripple current, while the operating frequency varies with input voltage. The Figure 112.1 circuit has an off-time of approximately 6μs, resulting in an operating frequency which varies from 60kHz to 90kHz over an 8V to 12V input range.

When the output current drops below approximately 15mV/R_{SENSE}, the LTC1148 automatically enters Burst Mode operation to reduce switching losses. In this mode, the LTC1148 holds both MOSFETs off and sleeps at 200μA supply current, while the output capacitor supports the load.

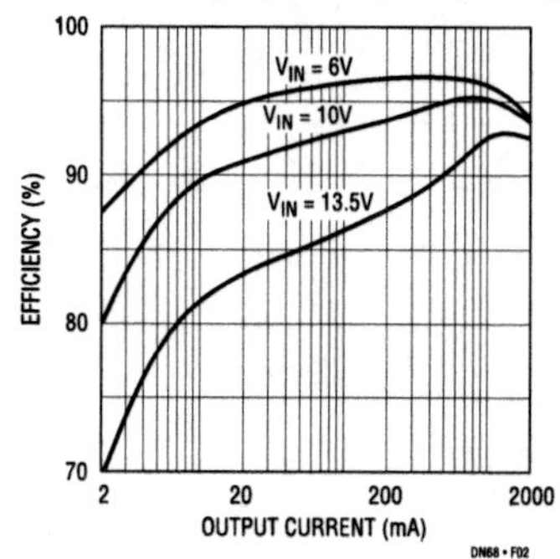

Figure 112.2 • LTC1148-5: 5.5V to 13.5V Efficiency

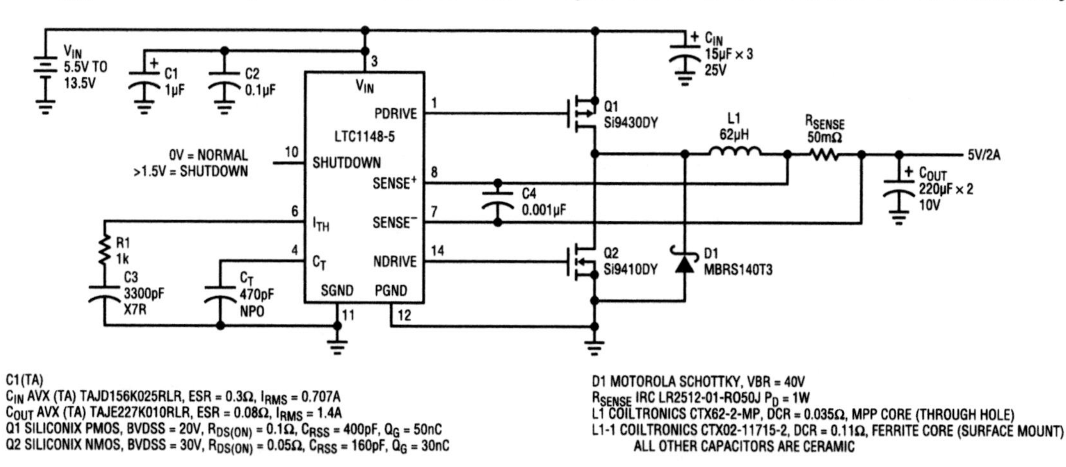

Figure 112.1 • LTC1148 (5.5V–13.5V to 5V/2A) Surface Mount

Analog Circuit Design: Design Note Collection. http://dx.doi.org/10.1016/B978-0-12-800001-4.00112-5

When the output capacitor discharges 50mV, the LTC1148 briefly turns back on, or "bursts," to recharge the capacitor. The timing capacitor Pin 4, which goes to 0V during the sleep interval, can be monitored with an oscilloscope to observe burst action. You will observe the circuit bursting less and less frequently as the load current is reduced. Complete shutdown reduces the supply current to only 10μA.

For applications which require greater than 13.5V, the higher voltage LTC1149 includes all of the operating features of the LTC1148 plus an internal regulator and a gate drive level shift circuit which allow operation up to V_{IN} = 48V. The design and performance of an LTC1149 based circuit is similar to that of the Figure 112.1 LTC1148 circuit, with a slight increase in sleep current (600μA) and shutdown current (150μA) due to the additional LTC1149 high voltage circuitry.

Although highly efficient at output currents of under 2A, P-channel MOSFETs can become a dominate loss element at higher output currents, limiting overall circuit efficiency. Consequently, N-channel MOSFETs are better suited for use in high current applications because they have a substantially lower ON resistance. The circuit shown in Figure 112.3 utilizes the low loss characteristics of N-channel MOSFETs, providing efficiency in excess of 90% at an output current of 5A.

Figure 112.3's operation is similar to that of the Figure 112.1 circuit, but it utilizes an LTC1149, which accommodates higher input voltages, and has been modified to drive the top N-channel MOSFET. The circuit operation is as follows: the LTC1149 provides a PDRIVE output (Pin 4) that swings between ground and 10V which turns Q3 on and off. While Q3 is on, the N-channel MOSFET (Q4) is off because its gate is pulled low by Q3 through D2. During this interval, the NGATE output (Pin 13) turns the synchronous switch (Q5) on, creating a low resistance path for the inductor current.

In order to turn Q4 on, its gate must be driven above the input voltage. This is accomplished by bootstrapping capacitor C2 off the source of Q4. The LTC1149 V_{CC} output (Pin 3) supplies a regulated 10V output that is used to charge C2 through D1 while Q4 is off. With Q4 off, C2 charges to 5V for the first cycle in Burst Mode operation and 10V thereafter.

When Q3 turns off, the N-channel MOSFET is turned on by the SCR connected NPN-PNP (Q1 and Q2) network. Resistor R2 supplies Q2 with enough base drive to trigger the SCR. Q2 then forces Q1 to turn on which supplies more base drive to Q2. This regenerative process continues until both transistor are fully saturated. During this period, the source of Q4 is pulled to the input voltage. While Q4 is on, its gate source voltage is approximately 10V, fully enhancing the N-channel MOSFET.

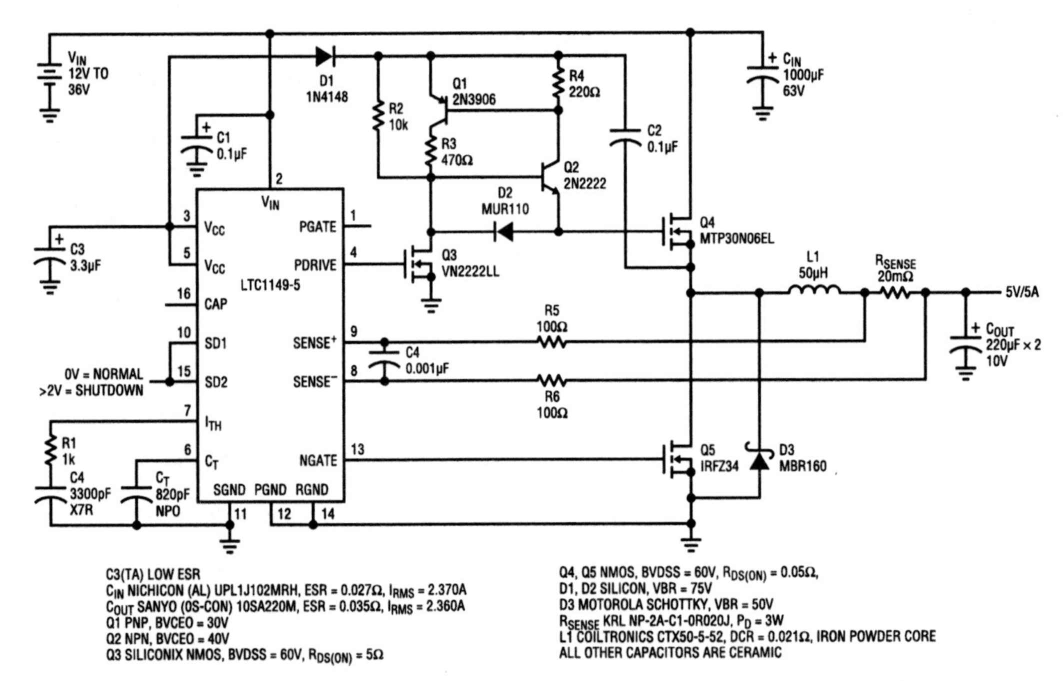

Figure 112.3 • LTC1149-5 (12V–36V to 5V/5A) Using N-Channel MOSFETs

High performance frequency compensation gives DC-to-DC converter 75μs response with high stability

113

Ron Vinsant

This design note describes four high performance, low cost, 1.75A step-down converter circuits based on the LT1076 five terminal switching regulator. All four circuits have exceptional transient response; indeed, it is superior to most three terminal linear regulators. Transient response is important to loads that are switched on and off or that require high peak currents. Examples are digital circuits that are turned on and off, disk drive motors, stepper motors and linear amplifiers. The frequency compensation schemes shown in this design note, when compared to the usual R and C technique, allow greater variation in output capacitor ESR without causing stability problems. This is important in applications where wide temperature variations occur (which change capacitor ESR) such as industrial control, automotive and military, and when the use of multiple capacitor vendors with different capacitor specifications is required.

Phase margin is always more than 50° and gain margin is a minimum of 18dB. Bode plots are available from the factory upon request.

The efficiency of these circuits is typically 80% with output ripple less than 50mV. Input voltages can be as high as 45V. Input ripple rejection is an exceptional 60dB due to the feed-forward architecture of the LT1076. These circuits use a small number of external parts that are available off-the-shelf.

Many of the problems associated with five terminal switching regulators have been addressed by these circuits. Start-up overshoot is less than 5% with the optional soft start circuit. On recovery from a short circuit, a 10% overshoot is realized.

For a 15V output, line regulation is typically 0.06% (10mV) for a 20V to 40V input voltage change. Load regulation is difficult to measure; in fact, it is only 1mV to 2mV at the point of regulation. This applies to all output voltages (see Tables 113.1 and 113.2).

Each circuit has been built in our lab and evaluated for stability, temperature, component life and tolerance. Two circuit options are shown: a simple soft start circuit and an output voltage adjustment (see Figure 113.1).

Inductors

The inductors shown in Table 113.3 are designed around two different core materials. The first is powdered iron based for low cost. The second is tape wound steel for smaller size and higher efficiency but at greater cost. For rapid evaluation of these circuits, powdered iron cores are available in sample quantities from Micrometals at 1-800-356-5977. Completed inductors are available from Coiltronics at 305-781-8900.

Capacitors

Ripple current in the output capacitor is 150mA maximum with the input voltage at 40V and maximum load. At 35°C ambient estimated life-time with the specified capacitor and full load is 28 years.

The input capacitor, which undergoes higher stress, has a ripple current of 830mA maximum at 14V input and maximum load. The life-time of this capacitor is 14 years at 35°C. If the ambient temperature is higher, the life of the capacitor will be cut in half for every 10°C increase. The ESR specification affects the output ripple as well as frequency compensation. Its value of capacitance is not critical.

The capacitors in the frequency compensation network should be at least X7R ceramic, never Z5U, and, if broad temperature operation is expected, polyester or polycarbonate film caps should be used.

Manufacturing technologies must also be taken into account. If an IR furnace is used for soldering, use only ceramic capacitors. A wave or hand soldering operation is suggested for both film and electrolytic capacitors. This is an area of continuing development so be sure to contact the capacitor manufacturers for temperature profiles.

Analog Circuit Design: Design Note Collection. http://dx.doi.org/10.1016/B978-0-12-800001-4.00113-7

Layout

In order to achieve proper performance it is important to lay out the circuit as the schematic indicates. Use a single point ground at the output of the converter as shown. The term "short" indicates that the trace should be as short as possible between the two points shown. These traces should have a minimum width of 0.2 inches in 2oz. copper for a length of less than 1.5 inches. Traces longer than this should be avoided on the heavily shaded portions of the schematic.

Output adjustment

A potentiometer can be added to the output divider string, provided the string does not change its overall resistance value. A table showing resistance values is shown with the schematic.

Heat sinking

Any heatsink of 30°C/W (~2 square inches) or lower will keep the LT1076 at an acceptable temperature up to a 70°C ambient. See LT1076 data sheet for further information.

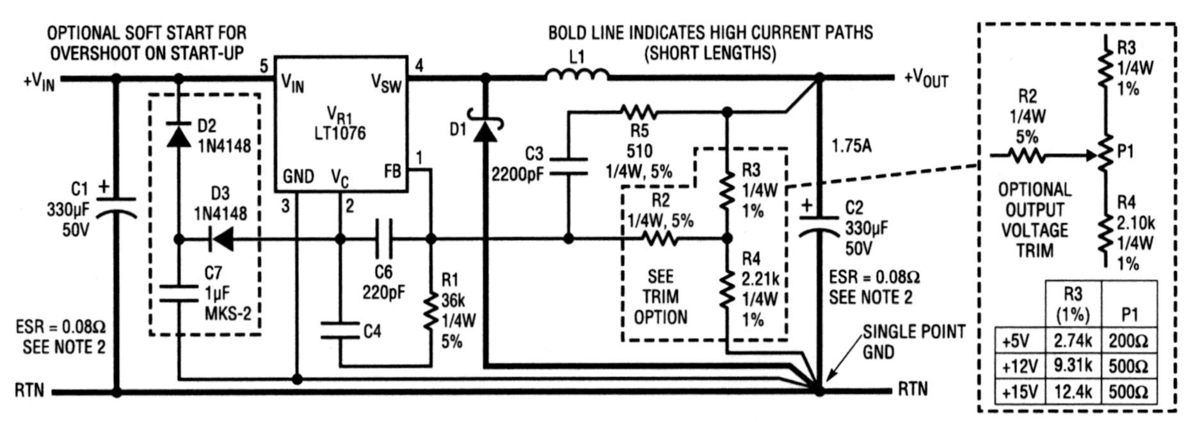

Figure 113.1 • High Performance DC-to-DC Converter

Table 113.1 Components

#	V_{IN}	V_{OUT}	E (%) AT V_{IN}	L (µH)	D1	R2 (5%)	R3 (1%)	C4 (10%)
1	8V–20V	+5V	83% at 10	75	MBR330P	1.5k	2.80k	0.0068µF
2	8V–40V	+5V	76% at 24	91	MBR350	1.5k	2.80k	0.0068µF
3	15V–40V	+12V	86% at 24	180	MUR415	1.2k	9.79k	0.01µF
4	18V–40V	+15V	86% at 24	240	MUR415	1.2k	12.7k	0.01µF

Table 113.2 Performance

#	V_{OUT}	MIN LOAD	REGULATION (MIN TO MAX)		RIPPLE REJECTION 50Hz–400Hz	OUTPUT RIPPLE
			LOAD	LINE		
1	+5V	0.200	0.1%	15mV	60dB	50mV
2	+5V	0.175	0.1%	15mV	60dB	50mV
3	+12V	0.175	0.1%	15mV	60dB	50mV
4	+15V	0.175	0.1%	15mV	60dB	50mV

Note 1: V_{IN} = 24V except #1 at 14V.
Note 2: Temperature = 25°C.
Note 3: Periodic and random deviation (PARD). With optional adjustment = ±2.5%. Without optional adjustment = ±4.5%.

Table 113.3 Inductor

L (µH)	NUMBER TURNS	CORE	COILTRONICS P/N	SMALLER TOROID
75	37 #18	T68-52A	CTX75-2-52	CTX75-2-KM
91	38 #18	T80-52B	CTX91-2-52	CTX91-2-KM
180	53 #18	T80-52B	CTX180-2-52	CTX180-2-KM
240	61 #18	T80-52B	CTX240-2-52	CTX240-2-KM

Note 1: ΔL with DC current is 20% max.

Section 5

Switching Regulator Design: Boost Converters

$1\mu A$ I_Q synchronous boost converter extends battery life in portable devices

114

Goran Perica

Introduction

Boost converters are regularly used in portable devices to produce higher output voltages from lower battery input voltages. Common battery configurations include two to three alkaline or NiMH cells or, increasingly, Li-Ion batteries, yielding a typical input voltage between 1.8V and 4.8V.

The 12V output converter shown in Figure 114.1 is designed to run from any typical small battery power source. This design centers around the LTC3122 boost converter, which can efficiently generate a regulated output up to 15V from a 1.8V to 5.5V input. The LTC3122 includes a 2.5A internal switch current limit and a full complement of features to handle demanding boost applications, including switching frequency programming, undervoltage lockout, Burst Mode operation or continuous switching mode, and true output disconnect. The integrated synchronous rectifier is turned off when the inductor current approaches zero, preventing reverse inductor current and minimizing power loss at light loads.

This unique output disconnect feature is especially important in applications that have long periods of idle time. While idling, the part can be shut down, leaving the output capacitor fully charged and standing by for quick turn-on. In shutdown, the part draws less than $1\mu A$ from the input source.

Because the batteries used in portable devices are usually as small as possible, they present high internal impedance under heavy loads, especially close to the end of their discharge cycle. Unlike other boost converters that struggle with high source impedance at start-up, the LTC3122 prevents high surge currents at start-up.

1.8V to 5.5V input to 12V output boost regulator

The circuit in Figure 114.1 is designed for high efficiency and small size. The LTC3122 operates at 1MHz to minimize the size of the filter capacitors and boost inductor, and uses Burst Mode operation to maintain high efficiency at light loads, as shown in Figure 114.2. At heavier loads, the converter can operate in constant frequency mode, resulting in lower input and output ripple. Constant frequency operation can result in lower EMI and is easier to filter.

Efficiency can be increased by running the LTC3122 at a relatively low switching frequency. Figure 114.3 shows the results of reducing the switching frequency from 1MHz to 500kHz.

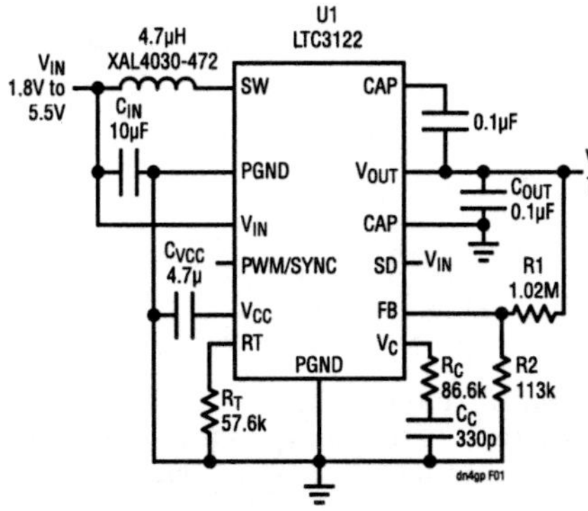

Figure 114.1 • The 1MHz Operating Frequency and Small Inductor Make This Converter Suitable for Demanding Portable Battery-Powered Applications

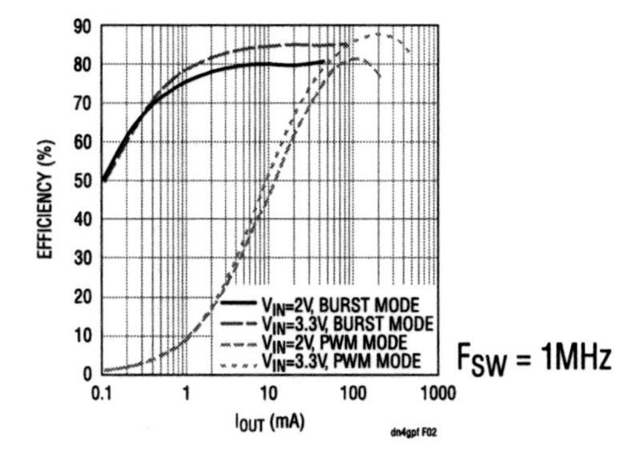

Figure 114.2 • The High Efficiency of the LTC3122 Boost Converter Extends Battery Life in Portable Applications

Analog Circuit Design: Design Note Collection. http://dx.doi.org/10.1016/B978-0-12-800001-4.00114-9

Efficiency can be improved further by increasing the inductor size. Figure 114.4 shows the increase in efficiency achieved by replacing the 4mm × 4mm boost inductor (XAL4030-472) with a 7mm × 7mm inductor (744-777-910 from Würth). The 90% efficiency at 10mA is 5% higher than the efficiency shown in Figure 114.3.

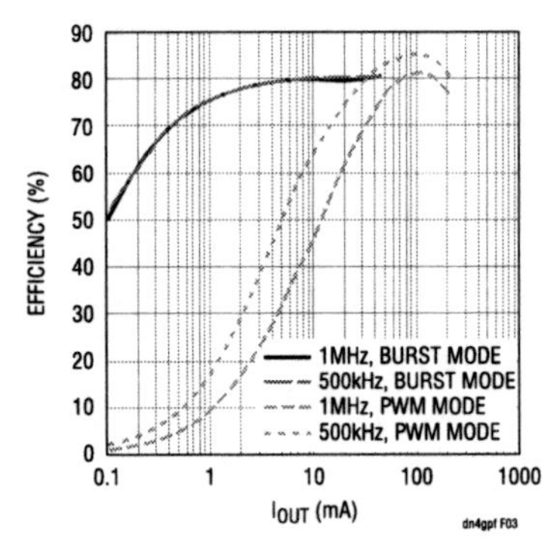

Figure 114.3 • The Efficiency Is Greatly Affected by the Operating Frequency. At 100mA Load an Additional 4% Can Be Gained by Reducing the Switching Frequency from 1MHz to 500kHz.

Battery size should be taken into account when considering inductor size. Using a relatively small inductor running at a high frequency may necessitate a correspondingly higher capacity battery to achieve the same run time at relatively lower efficiency. In other words, space gains achieved with a smaller inductor may be replaced by the need for a bigger battery.

Output disconnect

Typical boost converters cannot disconnect the output from the input because of the boost diode. Current always flows from the input through the inductor and boost diode to the output. Therefore the output can not be shorted or disconnected from the input, a significant problem in many applications, especially in shutdown. In contrast, the LTC3122 includes an internal switch that disconnects the boost MOSFET body diode from the output. This also allows for inrush current limiting at turn-on, minimizing the surge currents seen by the input power source.

Figure 114.5 shows the output of the LTC3122 disconnected in shutdown. The output voltage is pulled to zero by the load following shutdown, and the LTC3122 consumes less than 1µA of current.

Start-up inrush current limiting

To simulate a real battery-operated application, the circuit in Figure 114.1 was tested with 1Ω of equivalent series resistance (ESR) placed between the power source and the LTC3122 circuit. Once the LTC3122 is enabled, it controls the start-up so that the input power source can lift the output rail to regulation. The input current slowly ramps up. The input current overshoot required to charge the output capacitor is limited to only 200mA and the input power source voltage droop is limited to 0.5V, as shown in Figure 114.5.

Conclusion

The LTC3122 boost converter serves the needs of battery-operated applications that require low standby quiescent current and high efficiency. Unlike many other boost converters, it includes features, enabling operation from batteries near full discharge when battery ESR becomes high. Its very low quiescent and shutdown currents, combined with output disconnect, extend battery run time in applications with long idle periods. The LTC3122 includes a complete set of features for high performance battery operated applications and comes in a small, thermally enhanced 3mm×4mm package.

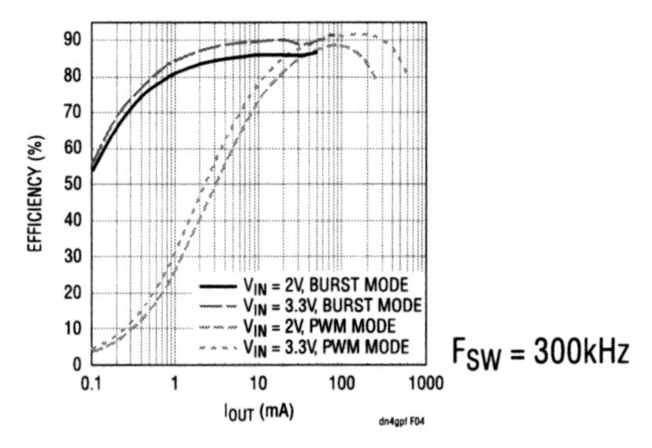

Figure 114.4 • With a Lower Switching Frequency and a Larger Inductor, a Smaller Battery Can Be Used. Efficiency Gain Up to 30% in the 1mA to 10mA Load Range (in PWM Mode) Can Significantly Improve Applications That Operate with Light Loads.

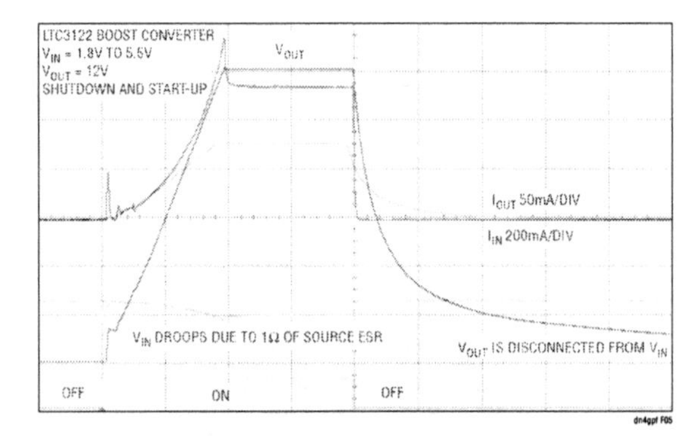

Figure 114.5 • Inrush Current Limiting at Turn-On Minimizes Surge Currents Seen by the Input Source. The Output Is Disconnected from Input During Shutdown.

Ultralow power boost converters require only 8.5μA of standby quiescent current

115

Xiaohua Su

Introduction

Industrial remote monitoring systems and keep-alive circuits spend most of their time in standby mode. Many of these systems also depend on battery power, so power supply efficiency in standby state is very important to maximize battery life. The LT8410/-1 high efficiency boost converter is ideal for these systems, requiring only 8.5μA of quiescent current in standby mode. The device integrates high value (12.4M/0.4M) output feedback resistors, significantly reducing input current when the output is in regulation with no load. Other features include an integrated 40V switch and Schottky diode, output disconnect with current limit, built in soft-start, overvoltage protection and a wide input range, all in a tiny 8-pin 2mm × 2mm DFN package.

Application example

Figure 115.1 details the LT8410 boost converter generating a 16V output from a 2.5V-to-16V input source. The LT8410/-1 controls power delivery by varying both the peak inductor current and switch off time. This control scheme results in low output voltage ripple as well as high efficiency over a wide load range. Figures 115.2 and 115.3 show efficiency and output peak-to-peak ripple for Figure 115.1's circuit. Output ripple voltage is less than 10mV despite the circuit's small (0.1μF) output capacitor.

The soft-start feature is implemented by connecting an external capacitor to the V_{REF} pin. If soft-start is not needed, the capacitor can be removed. Output voltage is set by a resistor divider from the V_{REF} pin to ground with the center tap connected to the FBP pin, as shown in Figure 115.1. The FBP pin can also be biased directly by an external reference.

The $\overline{SHDN}$ pin of the LT8410/-1 can serve as an on/off switch or as an undervoltage lockout via a simple resistor divider from V_{CC} to ground.

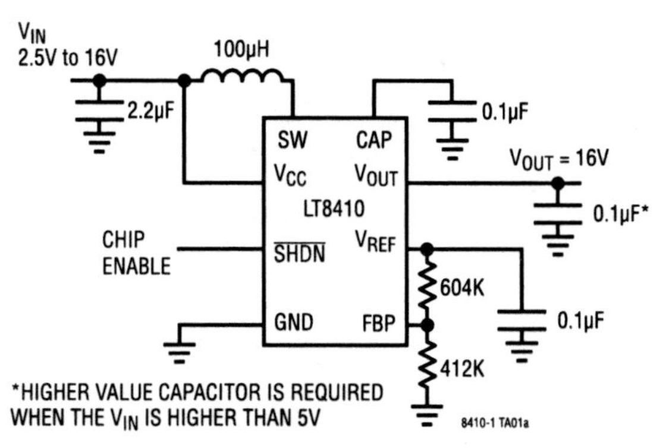

Figure 115.1 • 2.5V–16V to 16V Boost Converter

Ultralow quiescent current boost converter with output disconnect

Low quiescent current in standby mode and high value integrated feedback resistors allow the LT8410/-1 to regulate a 16V output at no load from a 3.6V input with about 30μA of average input current. Figures 115.4–115.6 show typical quiescent and input currents in regulation with no load.

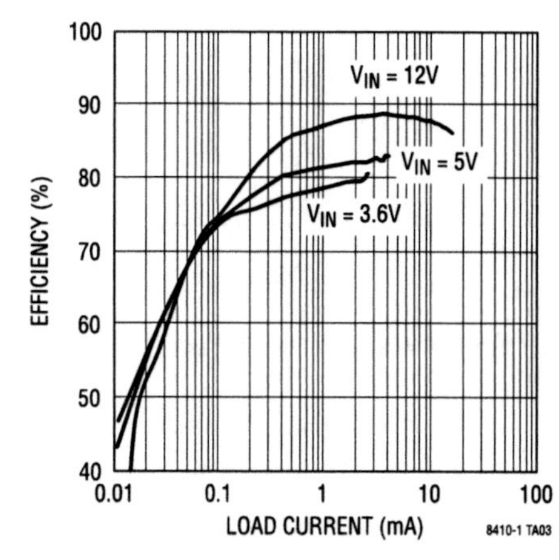

Figure 115.2 • Efficiency vs Load Current for Figure 115.1 Converter

Analog Circuit Design: Design Note Collection. http://dx.doi.org/10.1016/B978-0-12-800001-4.00115-0

The device also integrates an output disconnect PMOS, which blocks the output load from the input during shutdown. The maximum current through the PMOS is limited by circuitry inside the chip, allowing it to survive output shorts.

Compatible with high impedance batteries

A power source with high internal impedance, such as a coin cell battery, may show normal output on a voltmeter, but its voltage can collapse under heavy current demands. This makes it incompatible with high current DC/DC converters. With very low switch current limits (25mA for the LT8410 and 8mA for the LT8410-1), the LT8410/-1 can operate very efficiently from high impedance sources without causing inrush current problems. This feature also helps preserve battery life.

Conclusion

The LT8410/-1 is a smart choice for applications which require low standby quiescent current and/or require low input current, and is especially suited for power supplies with high impedance sources. The ultralow quiescent current and high value integrated feedback resistors keep average input current very low, significantly extending battery operating time. The LT8410/-1 is packed with features without compromising performance or ease of use and is available in a tiny 8-pin 2mm × 2mm package.

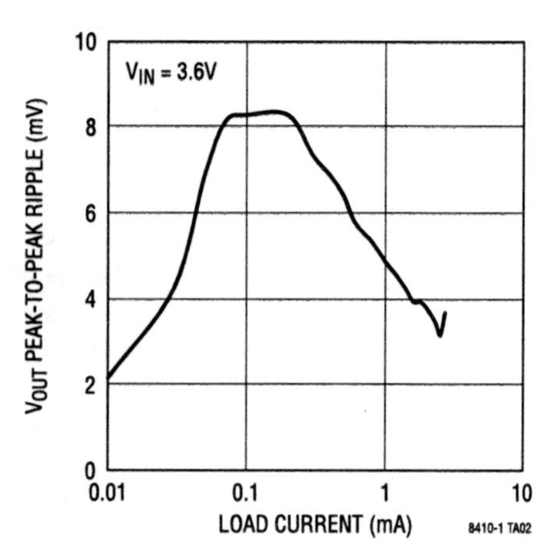

Figure 115.3 • Output Peak-to-Peak Ripple vs Load Current for Figure 115.1 Converter at 3.6V

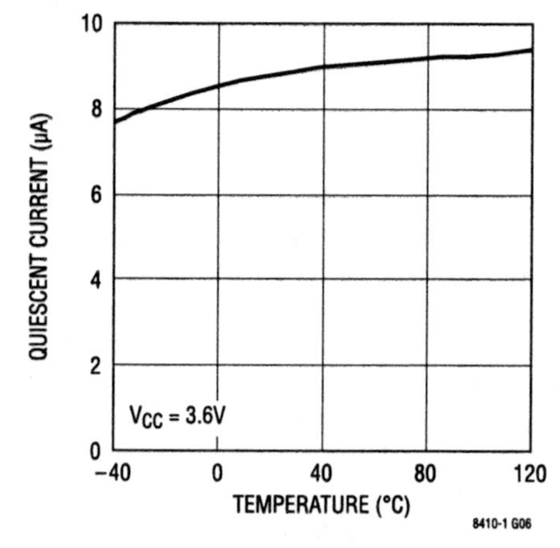

Figure 115.4 • Quiescent Current vs Temperature (Not Switching)

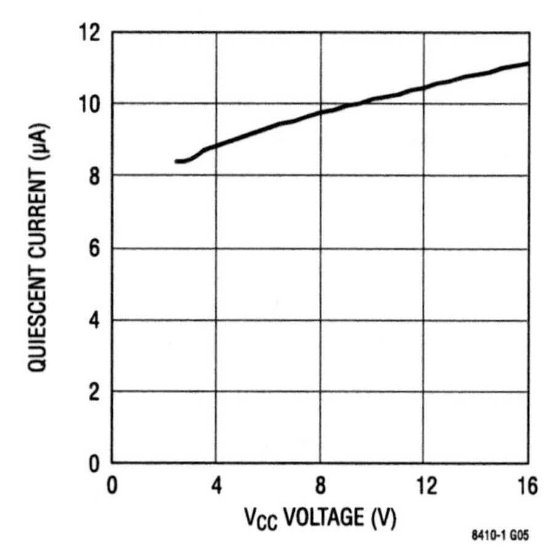

Figure 115.5 • Quiescent Current vs V_{CC} Voltage (Not Switching)

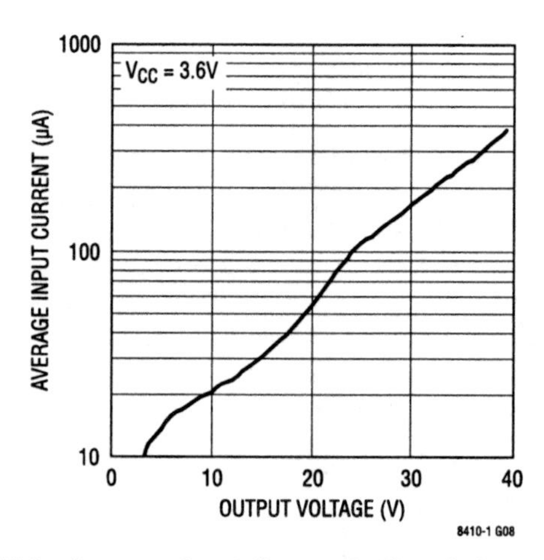

Figure 115.6 • Average Input Current in Regulation with No Load

Tiny dual full-bridge Piezo motor driver operates from low input voltage

116

Wei Gu

Introduction

Piezoelectric motors are used in digital cameras for autofocus, zooming and optical image stabilization. They are relatively small, lightweight and efficient, but they also require a complicated driving scheme. Traditionally, this challenge has been met with the use of separate circuits, including a step-up converter and an oversized generic full-bridge drive IC. The resulting high component count and large board space are especially problematic in the design of cameras for ever shrinking cell phones. The LT3572 solves these problems by combining a step-up regulator and a dual full-bridge driver in a 4mm × 4mm QFN package. Figure 116.1 shows a typical LT3572 Piezo motor drive circuit. A step-up converter is used to generate 30V from a low voltage power source such as a Li-Ion battery or any input power source within the part's wide input voltage range of 2.7V to 10V. The high output voltage of the step-up converter, adjustable up to 40V, is available for the drivers at the V_{OUT} pin. The drivers operate in a full-bridge fashion, where the OUTA and OUTB pins are the same polarity as the PWMA and PWMB pins, respectively, and the $\overline{OUTA}$ and $\overline{OUTB}$ pins are inverted from PWMA and PWMB, respectively.

Figure 116.2 • Typical Layout for the Figure 116.1 Converter

The step-up converter and both Piezo drivers have their own shutdown control. Figure 116.2 shows a typical layout.

Single driver application

Each full-bridge Piezo driver can be independently enabled and disabled by controlling the $\overline{SHDNA}$ and $\overline{SHDNB}$ pins. When held below 0.3V, $\overline{SHDNA}$ and $\overline{SHDNB}$ prevent

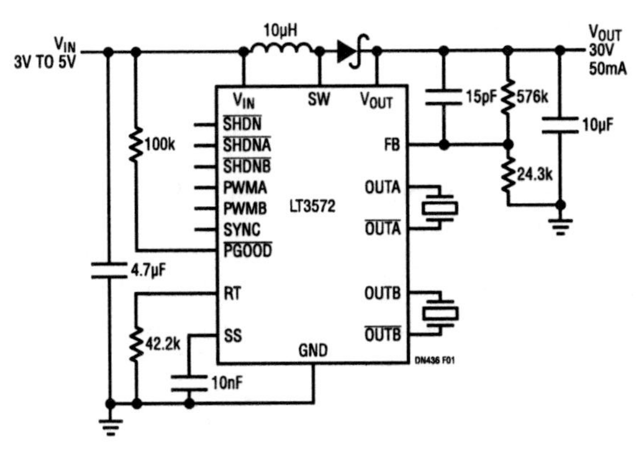

Figure 116.1 • Typical Circuit

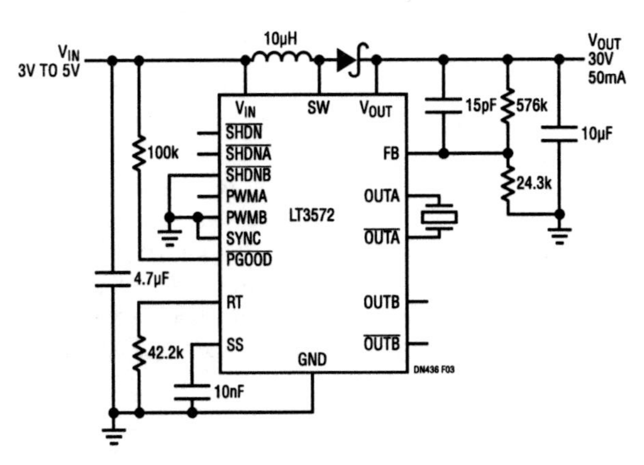

Figure 116.3 • Single Driver Application Circuit

Analog Circuit Design: Design Note Collection. http://dx.doi.org/10.1016/B978-0-12-800001-4.00116-2

the drivers from switching and keep the outputs in a high impedance state.

In applications where only one driver is used, the unused driver can be simply turned off without wasting any power by tying either $\overline{SHDNA}$ or $\overline{SHDNB}$ pin to the GND. Figure 116.3 shows a typical single driver application circuit where only driver A is enabled. The input pin PWMB is tied to GND.

Using external power supply

The high output voltage of the step-up converter, adjustable up to 40V, is available for the drivers at the V_{OUT} pin. For some multiple Piezo motor applications with multiple LT3572s, all the full-bridge drivers are powered by an external high voltage power supply. In this case, the integrated step-up converter can be simply disabled and only the dual drivers are used. In Figure 116.4, the $\overline{SHDN}$ pin is tied to the ground so the step-up regulator is prevented from switching. The SW pin, RT pin, SS pin and $\overline{PGOOD}$ pin are left open. The V_{IN} pin should be connected to a voltage source between 2.7V and 10V and FB pin to any voltage between 1.3V and 3V. In this example, the V_{IN} pin and FB pin are connected together, and both drivers are fully functional while the step-up converter is not running. The V_{IN} current is normally below 10mA.

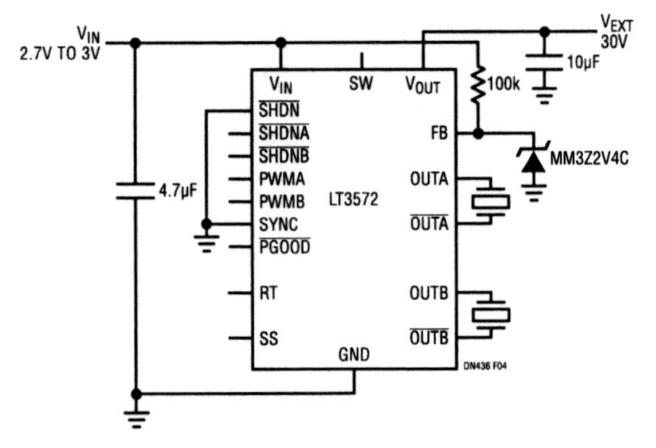

Figure 116.4 • Using External Power Supply with Integrated Step-Up Converter Disabled

Operating Piezo motor with long wires

In some cases, the Piezo motors are physically located far away from the driver. The parasitic inductance of the long connecting wires and capacitive Piezo motor form a high Q resonant LC tank. If the oscillation is not properly dampened, the driver pins would see large negative voltages, possibly causing spurious operation of the IC. Schottky diodes can be added at the OUTA and OUTB pins to prevent ICs from seeing large negative voltage. Another way to solve this problem is to add a resistor between the driver and the Piezo motor,

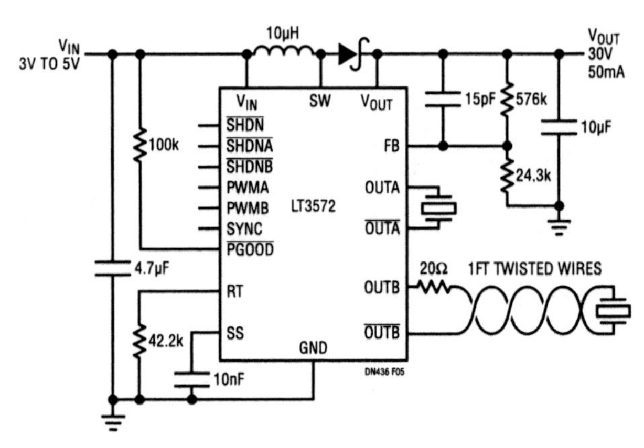

Figure 116.5 • Adding a Resistor when Operating with Long Wires

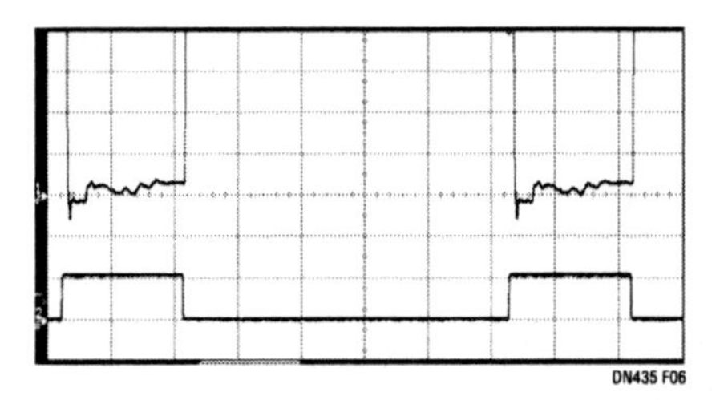

Figure 116.6 • $\overline{OUTB}$ Voltage Without the Resistor. Top Trace: $\overline{OUTB}$ Voltage (2V/Div), Bottom Trace: PWMB Voltage (2V/Div)

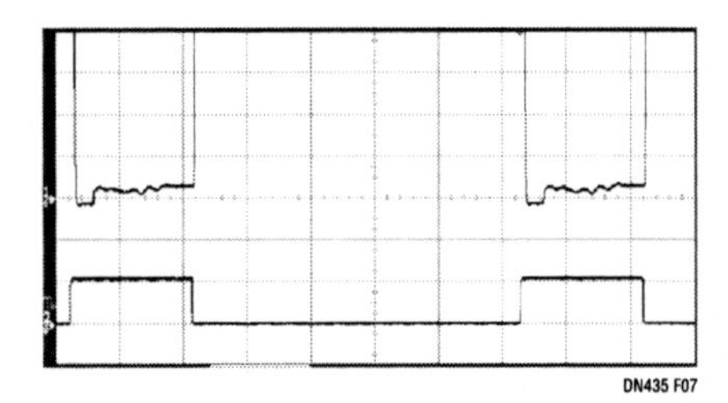

Figure 116.7 • $\overline{OUTB}$ Voltage with the Resistor. Top Trace: $\overline{OUTB}$ Voltage (2V/Div), Bottom Trace: PWMB Voltage (2V/Div)

as shown in Figure 116.5, to slow down the driving speed and dampen the oscillation. In this example, the connecting wires are 1-foot long twisted wires and the resistor is 20Ω. The voltage waveforms of the $\overline{OUTB}$ pin are shown in Figure 116.6 without the resistor, and Figure 116.7 with the resistor.

Conclusion

The LT3572 is a complete Piezo motor drive solution with a built-in high efficiency internal switch and integrated dual full-bridge drivers. Its fixed frequency, soft-start function, internal compensation and small footprint make the LT3572 a very simple and small solution to drive Piezo motors.

Tiny synchronous step-up converter starts up at 700mV

117

Dave Salerno

Introduction

Alkaline batteries are convenient because they're easy to find and relatively inexpensive, making them the power source of choice for portable instruments and devices used for outdoor recreation. Their long shelf life also makes them an excellent choice for emergency equipment that may see infrequent use but must be ready to go on a moment's notice. It is important that the DC/DC converters in portable devices operate over the widest possible battery voltage range to extend battery run time, and thus save the user from frequent battery replacement.

Single-cell alkaline batteries, with a 1.6V to 0.9V range, present a special challenge to DC/DC converters because of their low voltage and the fact that their internal resistance increases as the battery discharges. Thus, a DC/DC converter that can both start up and operate efficiently at low input voltages is ideally suited for single-cell alkaline products.

The LTC3526L is a 1MHz, 550mA synchronous step-up (boost) converter with a wide input voltage range of 0.7V to 5V and an output voltage range of 1.5V to 5.25V. Housed in a 2mm × 2mm DFN package, the LTC3526L has a typical start-up voltage of just 700mV, with operation down to 400mV once started.[1] Despite the LTC3526L's tiny solution size, it includes many advanced features, including output disconnect, short-circuit protection, low noise fixed frequency operation, internal compensation, soft-start, thermal shutdown and Burst Mode operation for high efficiency at light load. For low noise applications, the LTC3526LB offers fixed frequency operation at all load currents. With an output voltage range that extends down to 1.5V, the LTC3526L and LTC3526LB can even be used in applications previously requiring a boost converter followed by a buck converter.

A typical single-cell boost application is shown in Figure 117.1. In this example the LTC3526LB is used to generate 1.8V for a Bluetooth radio application. The LTC3526LB was selected for its small size, minimal external component count and low-noise, fixed frequency operation at all load currents. A graph of output current capability versus input voltage is shown in Figure 117.2. Note that the converter starts up at 700mV at no load and once running, can deliver 25mA

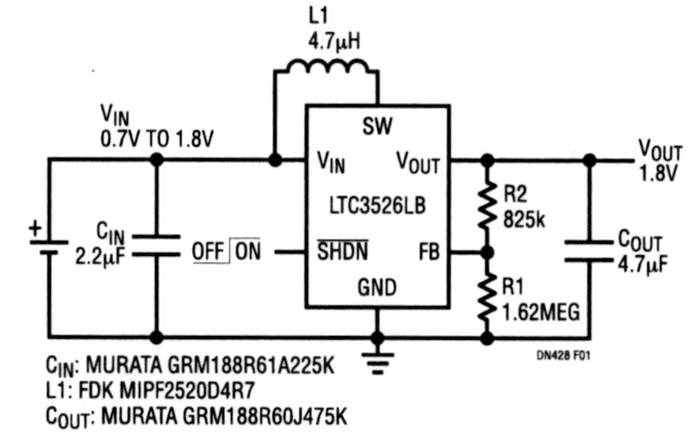

Figure 117.1 • Single-Cell 1.8V Boost Converter for a Bluetooth Radio Application Features a Low Start-up Voltage and Uses a Monolithic Chip Inductor for a Maximum Component Height of Just 1mm

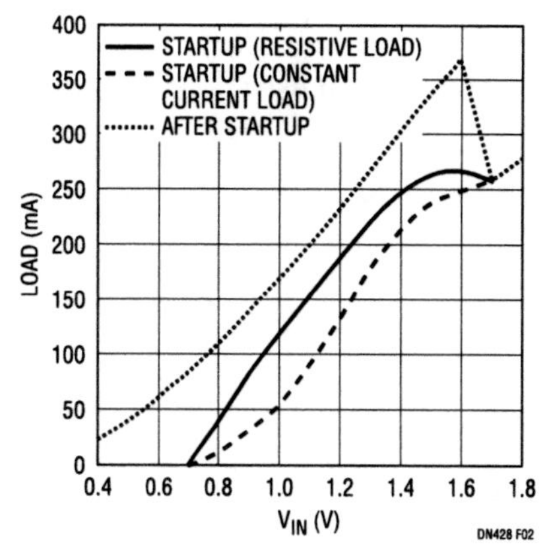

Figure 117.2 • Maximum Load Capability during and after Start-up for the Circuit in Figure 117.1

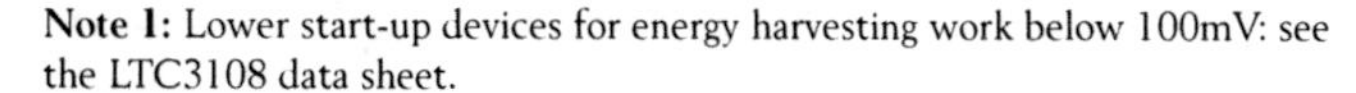

Note 1: Lower start-up devices for energy harvesting work below 100mV: see the LTC3108 data sheet.

Analog Circuit Design: Design Note Collection. http://dx.doi.org/10.1016/B978-0-12-800001-4.00117-4

of output current with an input voltage of only 400mV. The 1MHz switching frequency allows the use of small, low profile inductors, such as the monolithic chip inductor shown in this application. This provides a complete solution with a footprint that's just $36mm^2$ with a 1mm profile.

Many new battery types are available to the consumer, some of which are aimed at high-tech, high power applications. One of these is the disposable lithium AA/AAA battery, which offers a significant improvement in run time over traditional alkaline batteries. Furthermore, in applications that see infrequent use, the long shelf life of lithium batteries gives them a performance edge over nickel-based rechargeable batteries, which have a high self-discharge rate.

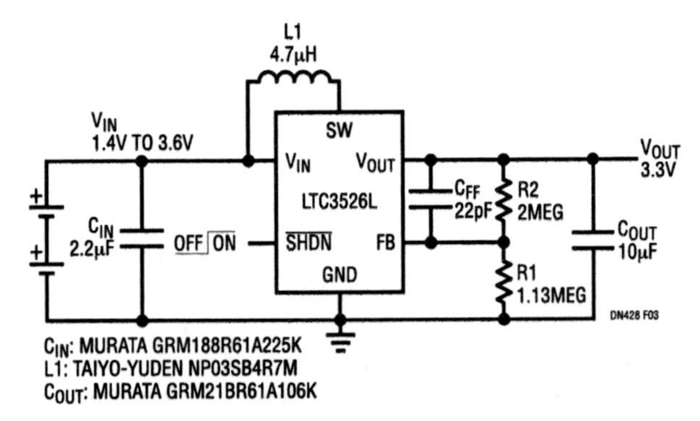

Figure 117.3 • Two AA Lithium Cell to 3.3V Boost Converter with 250mA Load Capability Maintains High Efficiency Over Three Decades of Load Current and Operates with $V_{IN} \geq V_{OUT}$

One characteristic of the lithium battery is that its voltage can be as high as 1.8V when the battery is fresh, compared to 1.6V for a typical alkaline battery. This is a problem for 2-cell alkaline applications that use a traditional boost converter to produce a 3.3V output from an alkaline 3.2V max input. Most boost converters cannot maintain regulation when the input is higher than the output, as it is with two fresh lithium batteries (3.6V).

The LTC3526L solves this problem by maintaining regulation even when the input voltage exceeds the output voltage. An example of a 2-cell to 3.3V boost converter using the LTC3526L is shown in Figure 117.3. A small feed-forward capacitor has been added across the upper divider resistor to reduce output ripple in Burst Mode operation. Efficiency vs load curves are shown in Figure 117.4. These curves demonstrate the high efficiency at light load made possible by the low 9µA quiescent current of Burst Mode operation. The curve in Figure 117.5 illustrates the efficiency at input voltages above and below the output voltage.

Conclusion

The LTC3526L is a highly integrated step-up DC/DC converter in a 2mm × 2mm package designed to easily fit a wide variety of battery-powered applications. Low start-up and operating voltages extend run time in single-cell applications. It even regulates in step-down situations where the fresh battery voltage (V_{IN}) may exceed V_{OUT}. For high efficiency at light loads, or low noise operation, it offers a choice of Burst Mode or fixed frequency operation.

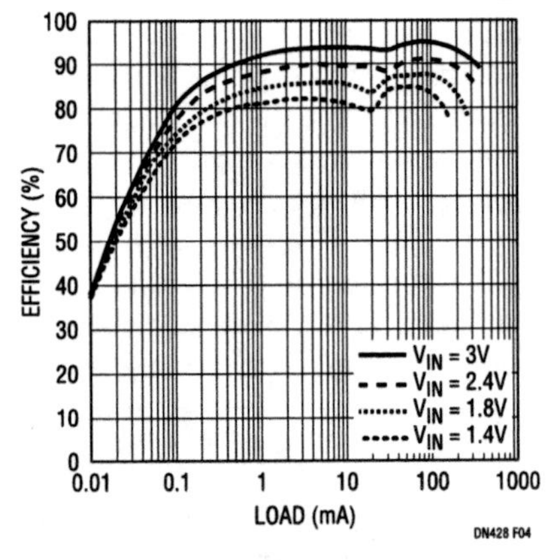

Figure 117.4 • Efficiency vs Load for the Circuit in Figure 117.3

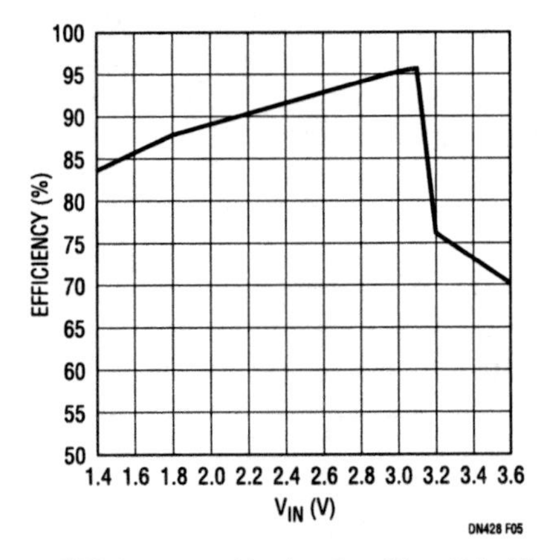

Figure 117.5 • Efficiency vs V_{IN} for the Circuit in Figure 117.3 (at 100mA Load Current)

High efficiency 2-phase boost converter minimizes input and output current ripple

118

Goran Perica

Introduction

Many automotive and industrial applications require higher voltages than is available on the input power supply rail. A simple DC/DC boost converter suffices when the power levels are in the 10W to 50W range, but if higher power levels are required, the limitations of a straightforward boost converter become quickly apparent. Boost converters convert a low input voltage to a higher output voltage by processing the input current with a boost inductor, power switch, output diode and output capacitor. As the output power level increases, the currents in these components increase as well. Switching currents also increase proportional to the output-to-input voltage conversion ratio, so if the input voltage is low, the switching currents can overwhelm a simple boost converter and generate unacceptable EMI.

For example, consider Figure 118.1, a 12V input to 24V, 10A output switching converter operating at 300kHz. The currents processed by the converter in Figure 118.1 are shown in the first row of Table 118.1. The relatively high current levels in the switcher are reflected in high input and output ripple currents, which results in increased EMI.

The circuit shown in Figure 118.2 performs the same DC/DC conversion, but with greatly reduced input and

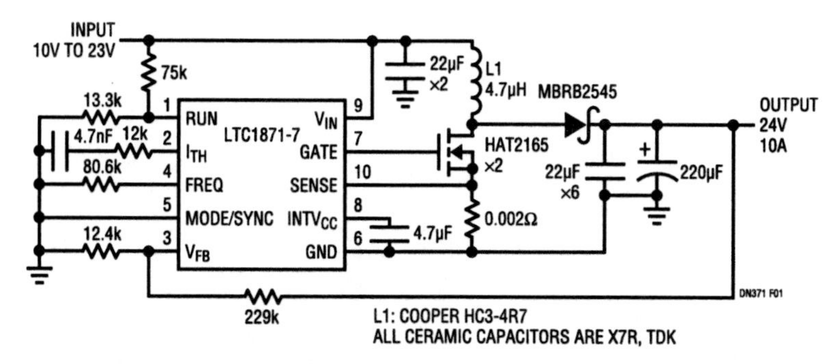

Figure 118.1a • Single Phase Boost Converter: Can Be Used to Convert 12V Input to 24V, 10A Output

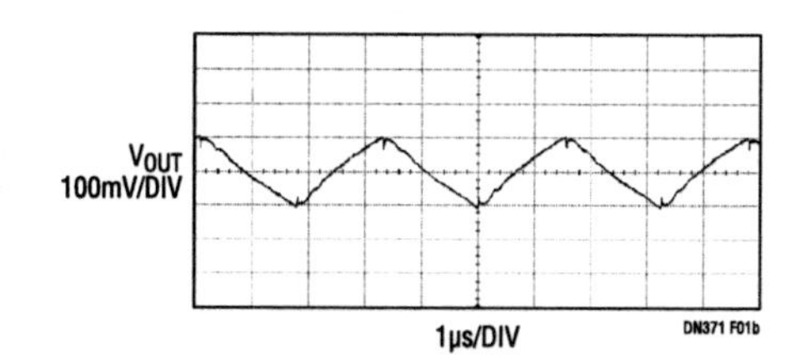

Figure 118.1b • Single Phase Boost Converter Output Voltage Ripple

Table 118.1 Dual Phase Boost Converter Has Lower Input and Output Ripple Currents and Voltages Than Single Phase Boost Converter

	INPUT RMS CURRENT	INPUT RIPPLE CURRENT	MOSFET RMS DRAIN CURRENT	OUTPUT DIODE RMS CURRENT	OUTPUT CAPACITOR RMS CURRENT	OUTPUT CAPACITOR FREQUENCY	OUTPUT VOLTAGE RIPPLE
Single Phase Boost Converter	21.1A	$4.2A_{P-P}$	15.4A	14.4A	10.5A	300kHz	212mV
Dual Phase Boost Converter	20.7A	**$0.17A_{P-P}$**	2 × 7.4A	2 × 7.2A	**1.9A**	600kHz	**65mV**

Analog Circuit Design: Design Note Collection. http://dx.doi.org/10.1016/B978-0-12-800001-4.00118-6

output ripple, significantly reducing EMI, and at a higher effective switching frequency, which allows the use of two 22μF output capacitors versus six 22μF output capacitors required in Figure 118.1.

The trick is the 2-phase boost topology, which interleaves two 180° out-of-phase output channels to mutually cancel out input and output ripple current—the results are shown in the second row of Table 118.1. Each phase operates at 50% duty cycle and the rectified output currents from each phase flow directly to the load—namely the low inductor ripple current—so only a small amount of output current (shown in Table 118.1) is handled by the output capacitors.

The centerpiece of the design in Figure 118.2 is the LT3782 2-phase current mode PWM controller. Current mode operation ensures balanced current sharing between the two power converters resulting in even power dissipation between the power stages.

The efficiency of the dual phase converter, shown in Figure 118.3, is high enough that it can be built entirely with surface mount components—no need for heat sinks. In a 240W boost supply application, the power dissipation of 12.9W is relatively easy to manage in a well laid out, large multilayer PCB with some forced airflow.

Conclusion

The simple LT3782 dual phase switching boost converter improves on single phase alternatives by allowing high power output with lower ripple currents, reduced heat dissipation and a more compact design.

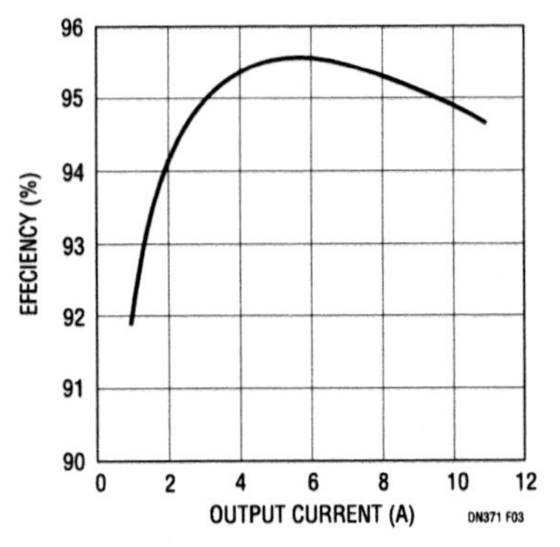

Figure 118.3 • 12V Input to 24V Output Dual Phase Boost Converter Efficiency

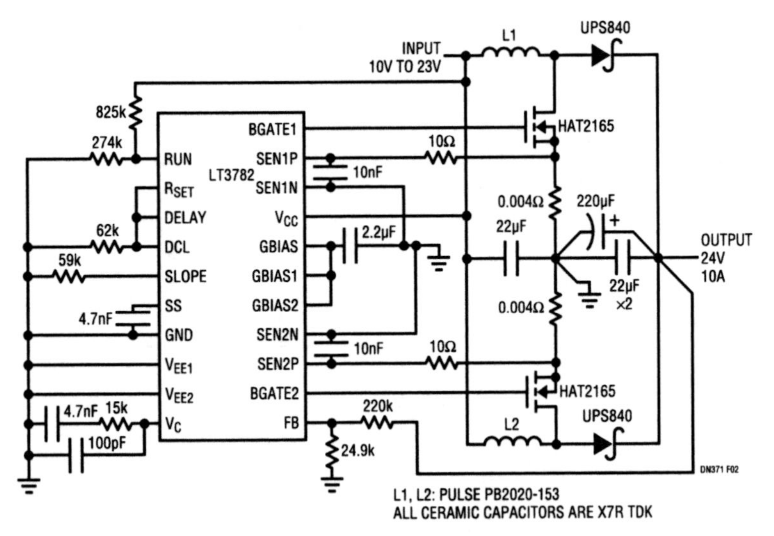

Figure 118.2a • Dual Phase Boost Converter Reduces EMI and Ripple Currents with a Minimum Input and Output Filtering

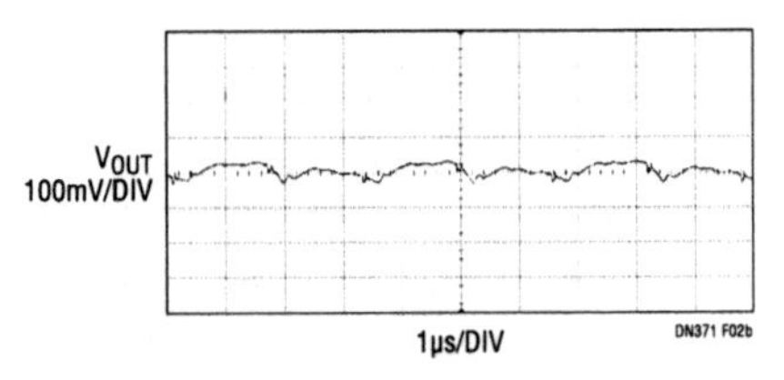

Figure 118.2b • Dual Phase Boost Converter Output Voltage Ripple

ThinSOT switching regulator controls inrush current

119

Jesus Rosales

Introduction

Inrush current can be troublesome, especially when regulators are operated from batteries or a current-limited source. In such cases, excessive inrush current can cause the input source to collapse and prevent the converter from ever starting up. Inrush current can also trip input line fuses or cause the output to overshoot. There are many ways to deal with this problem, but most of them are difficult to implement or have undesirable side effects, often making the cure worse than the cause.

A simple solution

Figure 119.1 shows a solution to this problem using the LT3467 in a 10V–16V input to 12V/300mA output SEPIC converter. The LT3467 ThinSOT switching regulator integrates a soft-start function with a built-in 42V, 1.1A switch. Figures 119.2 and 119.3 illustrate the effect that the LT3467's soft-start feature has on inrush currents.

Figure 119.2 shows the start-up waveforms for the Figure 119.1 circuit with the SS feature disabled (by removing C_{SS} from the SS pin). The inrush current in this case is limited only by the maximum duty cycle or maximum switch current of the LT3467. The top trace shows input current and the bottom trace shows output voltage. It takes only about 100µs for the output to go from 0V to 12V, requiring an input current of about 0.85A peak, thus exceeding the nominal steady-state level.

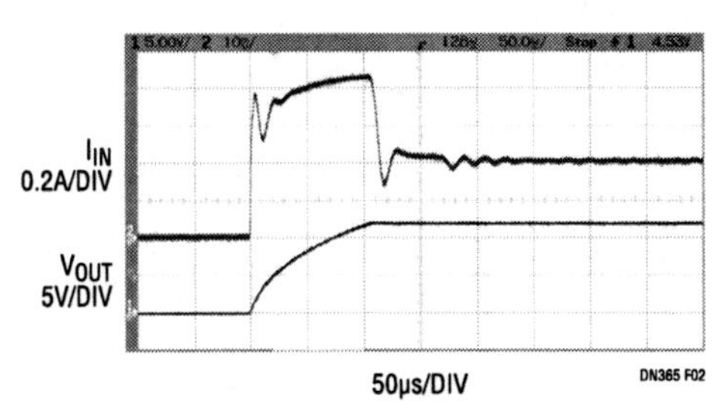

Figure 119.2 • Input Current and Output Voltage at Turn-On Without C_{SS} Capacitor for Figure 119.1 Circuit

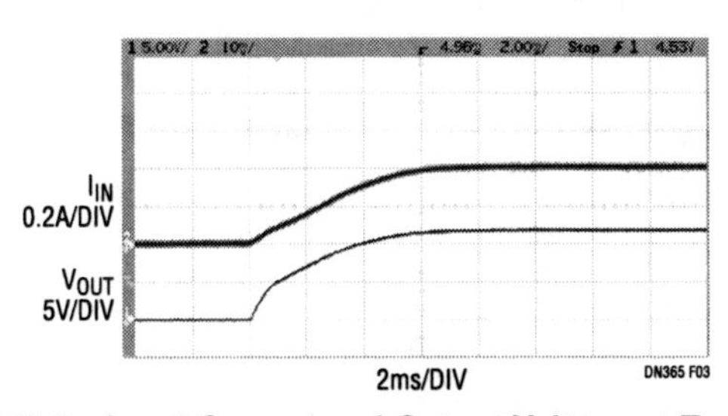

Figure 119.3 • Input Current and Output Voltage at Turn-On with C_{SS} Capacitor for Figure 119.1 Circuit

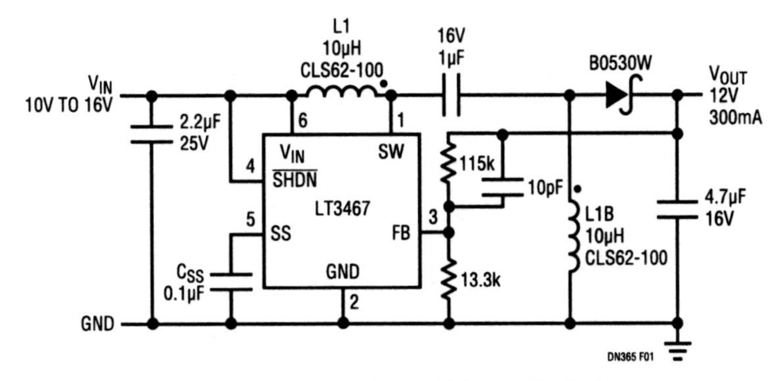

Figure 119.1 • $10V_{IN}$–$16V_{IN}$ to 12V at 300mA SEPIC Converter

Analog Circuit Design: Design Note Collection. http://dx.doi.org/10.1016/B978-0-12-800001-4.00119-8

Figure 119.3 shows the same waveforms with a 0.1μF capacitor connected to the SS pin. Now, both input current and output voltage rampup in a controlled fashion and settle comfortably into their respective steady-state levels. By using only one small capacitor between the SS pin and ground, inrush current has been eliminated. Figure 119.4 shows an efficiency curve for the SEPIC converter of Figure 119.1.

Figure 119.5 shows a boost converter that provides 12V at 300mA from a 5V input source. Figures 119.6 and 119.7 show the input and output currents, with and without the soft-start feature enabled. Figure 119.6 shows that without soft-start, even though the output voltage follows the input by conducting through the inductor and diode, there is an inrush of current which exceeds 2A at the point where the converter begins to switch. Figure 119.7 shows the improved start-up results with soft-start enabled. Figure 119.8 shows the efficiency of the circuit.

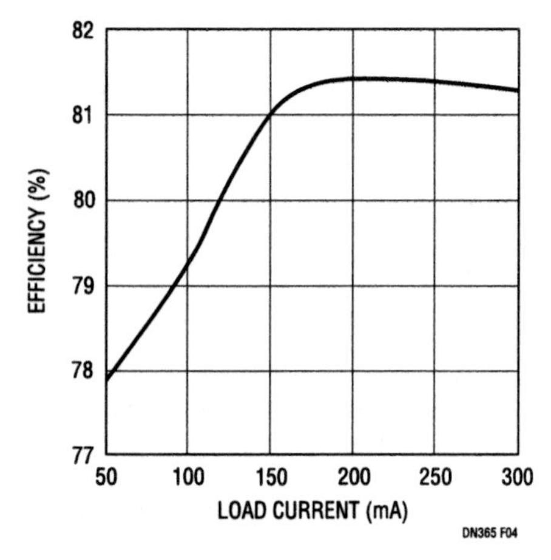

Figure 119.4 • SEPIC Converter Efficiency for Figure 119.1 Circuit

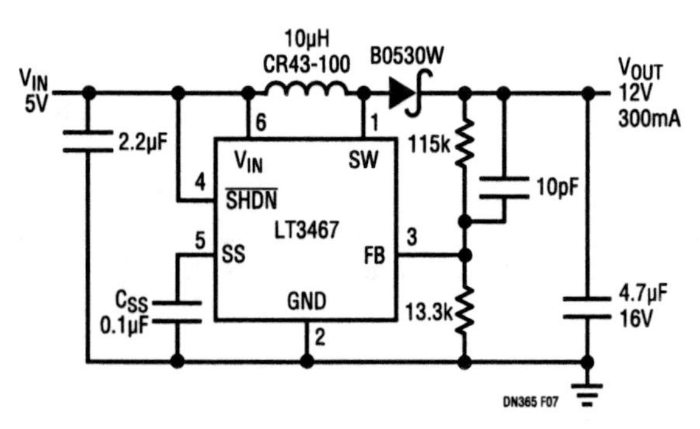

Figure 119.5 • 5V_{IN} to 12V at 300mA BOOST Converter

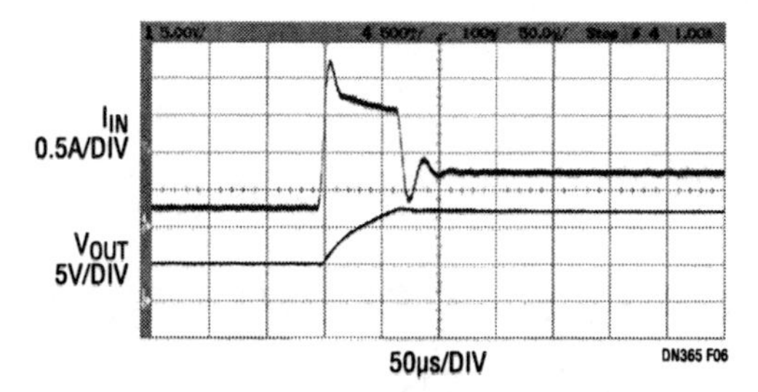

Figure 119.6 • Input Current and Output Voltage at Turn-On Without C_{SS} Capacitor for Figure 119.5 Circuit

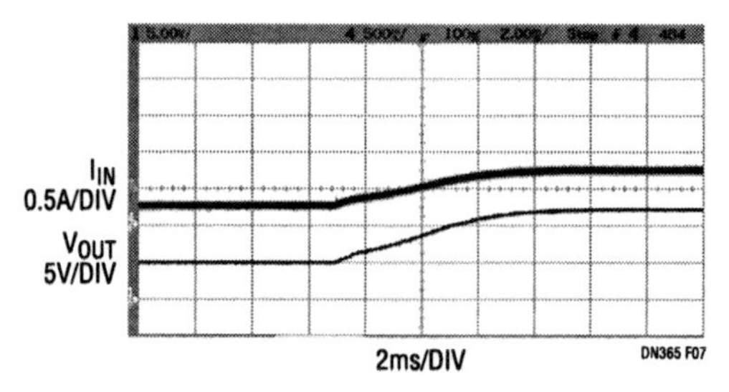

Figure 119.7 • Input Current and Output Voltage at Turn-On with C_{SS} Capacitor for Figure 119.5 Circuit

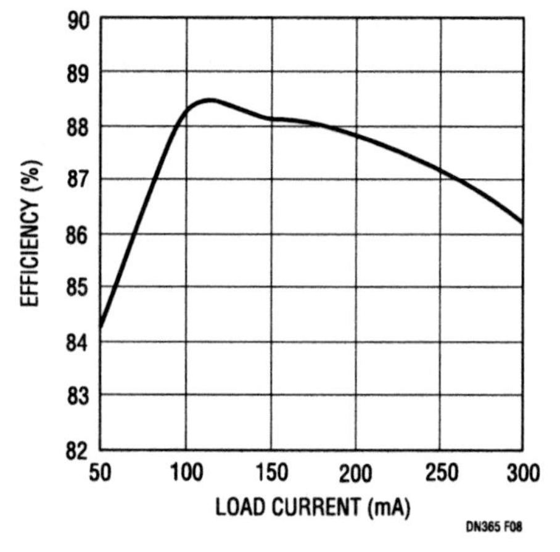

Figure 119.8 • BOOST Converter Efficiency for Figure 119.5 Circuit

Conclusion

Countless hours can be spent looking for ways to control inrush current or just getting the output to come up. The LT3467 provides a simple solution. The addition of a single, tiny, low cost capacitor can make the difference between a smooth design cycle or a laborious one.

Dual DC/DC converter with integrated Schottkys generates ±40V outputs and consumes only 40μA quiescent current

120

David Kim

Introduction

As portable devices become more sophisticated and require higher display resolution, there is an increased demand for accurate, high voltage bias supply solutions with wide input and output voltage ranges. The traditional methods of using arrays of capacitors to implement a charge pump or a bulky and expensive transformer no longer meet the accuracy and size requirements of today's portable devices.

The LT3463 is both accurate and compact. It fits both a positive output converter plus a negative output converter into a tiny (3mm × 3mm) DFN package, including Schottky diodes and switches capable of 250mA (400mA on the negative channel of the LT3463A).

The LT3463 works in a wide range of applications due to its 2.3V to 15V input voltage range and output capability to ±40V. Each converter is designed to operate with a quiescent current of only 20μA, which drops to less than 1μA in shutdown, making the LT3463 solution ideal for battery-powered portable applications.

Dual output ±20V converter

Figure 120.1 shows a ±20V LCD bias voltage supply using the LT3463. This circuit generates both positive and negative 20V outputs from a Li-Ion battery. Low profile inductors and capacitors keep the circuit under 9mm × 9.5mm × 1.2mm, making this circuit ideal for small wireless devices such as cellular phones or DSCs.

This design can produce ±20V at 9mA from a 2.7V input and up to 20mA from a 5V input. The efficiency shown in Figure 120.2 exceeds 70% over a wide load current range reaching 75% at 20mA. The LT3463's constant off-time architecture allows 20μA quiescent current operation for each output, making the LCD bias circuit efficient even at 100μA load current. This circuit can accommodate different load voltages as the LCD bias voltage varies for different manufacturers (typically 9V ~ 25V).

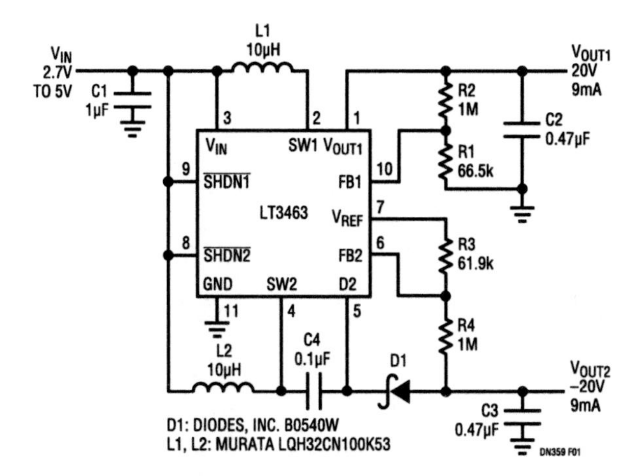

Figure 120.1 • Dual Output ±20V Converter

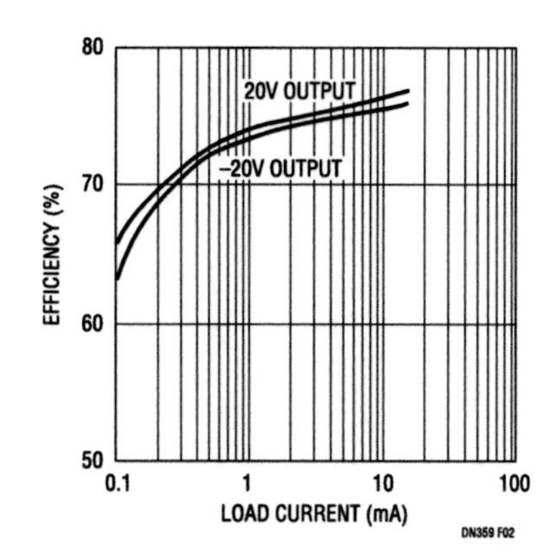

Figure 120.2 • Efficiency of Circuit in Figure 120.1 at V_{IN} = 3.6V

Analog Circuit Design: Design Note Collection. http://dx.doi.org/10.1016/B978-0-12-800001-4.00120-4

Dual output (±40V) converter

The circuit in Figure 120.3 demonstrates the impressive input and output voltage range of the LT3463. As shown, the 42V internal switches allow up to ±40V output without a transformer or an array of diodes and capacitors. The output voltages can be easily changed by adjusting the values of R1 and R3. The circuit is designed to operate from a Li-Ion battery or two alkaline cells (down to 2.4V).

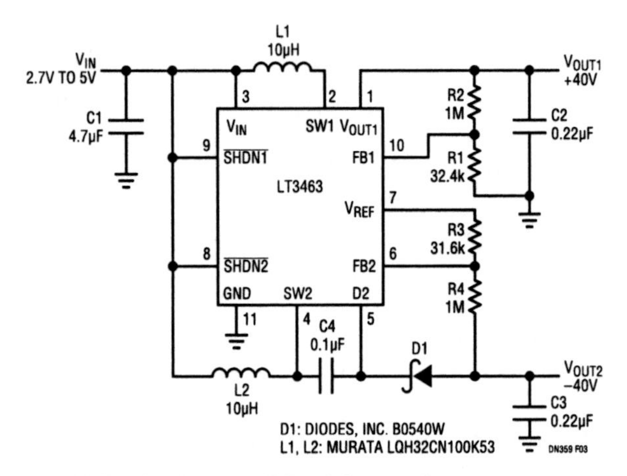

Figure 120.3 • 2.7V to ±40V Dual Output Converter

CCD sensor bias supply

The circuit in Figure 120.4 shows a CCD sensor bias supply for a cellular camera phone application. The two outputs, 15V and −8V, are generated from a Li-Ion battery input. With a minimum input voltage of 3.3V, the circuit is designed to output 15V at 10mA and −8V at 40mA to accommodate the maximum current consumption of the CCD sensor. The low power consumption of the LT3463 and its small circuit size also make this solution ideal as a general-purpose TFT display bias supply for portable devices. Figure 120.5 shows the efficiency and power loss data for the circuit.

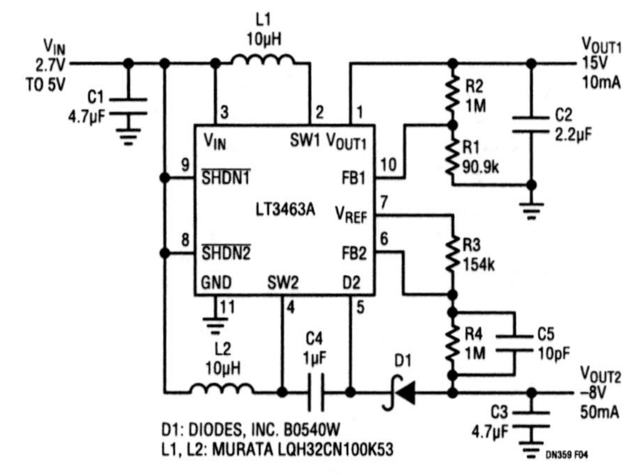

Figure 120.4 • CCD Sensor Bias Supply

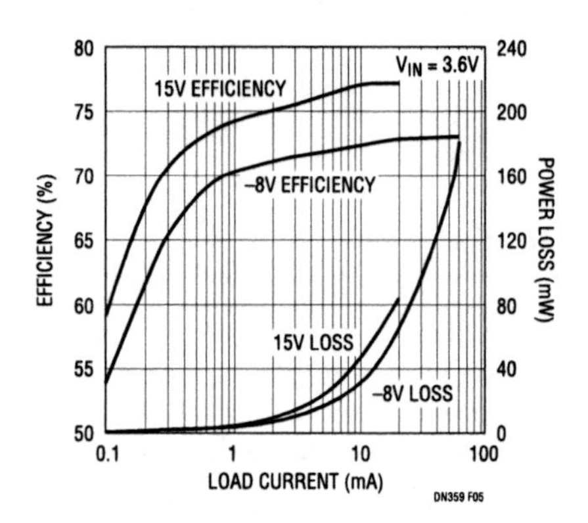

Figure 120.5 • Efficiency and Power Loss for the Circuit in Figure 120.4

Conclusion

The LT3463 and LT3463A are ideal solutions for high resolution portable display applications requiring multiple (positive and negative) high output voltages, wide input voltage range, low quiescent current, small circuit size and accurate output regulation.

Compact step-up converter conserves battery power

121

Mike Shriver

Introduction

The LT3464 is an ideal choice for portable devices which require a tiny, efficient and rugged step-up converter. The device, housed in a low profile (1mm) 8-lead ThinSOT package, integrates a Schottky diode, NPN main switch and PNP output disconnect switch. For light load efficiency, Burst Mode operation is used to deliver power to the load. This results in high efficiency and minimal battery current draw over a broad range of load current. Quiescent current is only 25μA. While in shutdown, the output disconnect switch separates the load from the input, further increasing battery run time. This same feature reduces the fault current to 45mA (typ) when the output is shorted to ground, a feature that few boost converters offer.

Another advantage of the LT3464 is its small solution size. A constant off-time architecture is used with fixed peak current limit switching. The low current limit of 115mA and an off-time of 250ns enable the use of tiny surface mount inductors and capacitors, while an internal phase lead capacitor reduces output voltage ripple.

The LT3464 provides the designer with much flexibility. Output voltages up to 34V can be attained with the 36V rating of the main switch, while external control of the output voltage can be accomplished via the control pin. Its wide input voltage range of 2.3V to 10V allows for a variety of input voltage sources including one or two lithium-ion battery cells.

16V bias supply

Figure 121.1 shows a 16V bias supply that can provide 6.5mA at an efficiency of 77% from a lithium-ion battery (V_{IN}=3.6V) as shown in Figure 121.2. The circuit uses a 22μH surface mount chip inductor with a 1210 footprint and a 0.33μF output capacitor with an 0805 footprint. The entire circuit occupies an area of 51mm^2. Smaller components can be used to further reduce the circuit area at the expense of efficiency. This supply can be used to bias small LCD panels and small passive organic LED (OLED) panels as well.

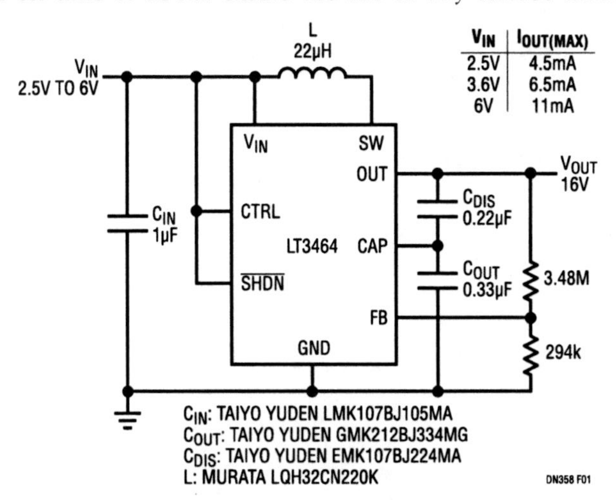

Figure 121.1 • Efficient 16V Bias Supply Using 0805 Output Capacitor and 1210 Choke

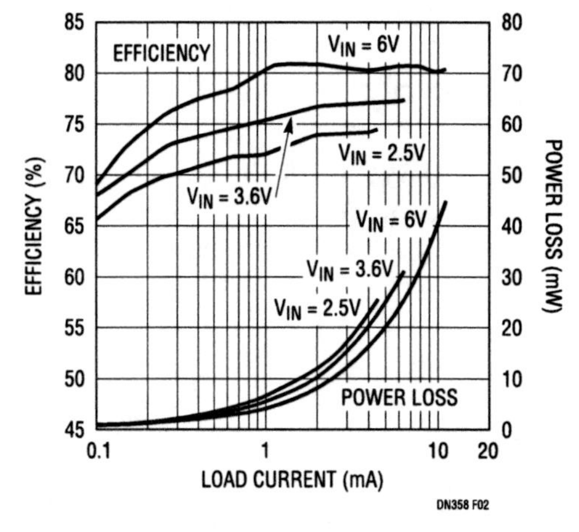

Figure 121.2 • Efficiency and Power Loss of 16V Output LT3464 Boost Converter (Figure 121.1 Circuit)

Analog Circuit Design: Design Note Collection. http://dx.doi.org/10.1016/B978-0-12-800001-4.00121-6

20V bias supply with variable output voltage

Manual adjustment of the bias voltage is required in some LCD applications in order to vary the contrast. The LT3464 CTRL pin eases this task. When a DC voltage of 1.25V or less is applied to the control pin, the internal reference is overridden, allowing external control of the bias voltage from (V_{IN} – 0.8V) to nominal V_{OUT}. Figure 121.3 shows a DAC-controlled bias supply. Other methods of driving the control pin include using a filtered PWM signal or a potentiometer.

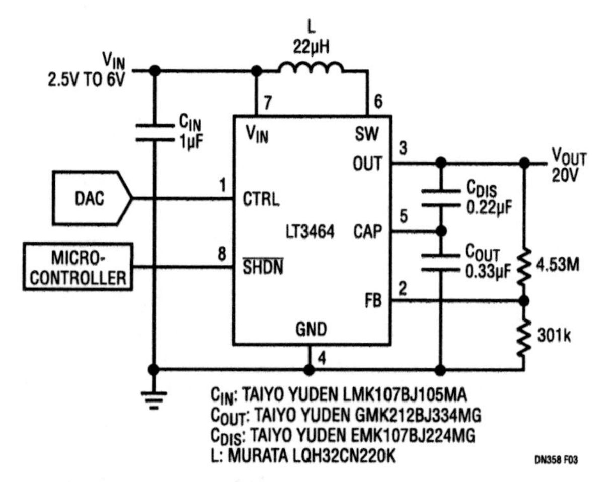

Figure 121.3 • 20V Nominal Bias Supply with DAC Controlled Output Voltage and Shutdown

±20V bias supply

A dual, ±20V bias supply is shown in Figure 121.4. The +20V rail is regulated and an inverting charge pump tapped from its switch node forms a quasi-regulated –20V output. For a 10:1 difference in load currents, the two outputs are regulated within 5% of each other. The full load efficiency is 77% at an input voltage of 3.6V. One benefit of this circuit is that both outputs are isolated from the input during shutdown; the positive output is isolated by the internal disconnect switch and the negative output is isolated by the charge pump capacitor C_{CP}.

34V bias supply

The 36V rating of the main switch allows output voltages up to 34V as illustrated in Figure 121.5. The 34V bias supply shown can supply 3.5mA at 76% efficiency from a 3.6V input.

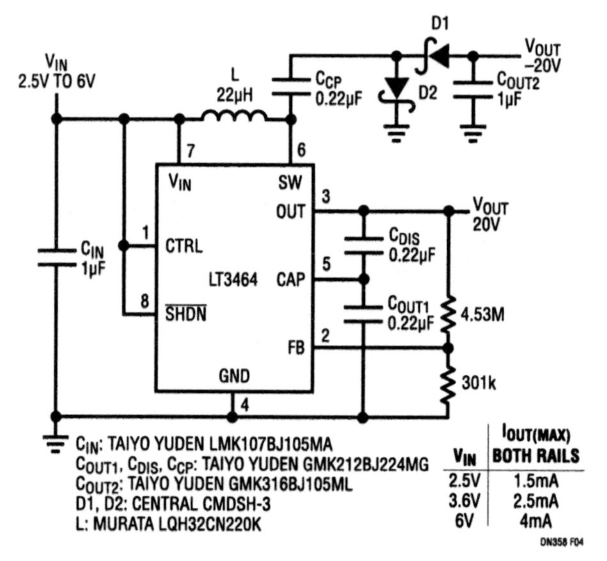

V_{IN}	$I_{OUT(MAX)}$ BOTH RAILS
2.5V	1.5mA
3.6V	2.5mA
6V	4mA

Figure 121.4 • ±20V Bias Supply with Output Disconnect on Both Rails

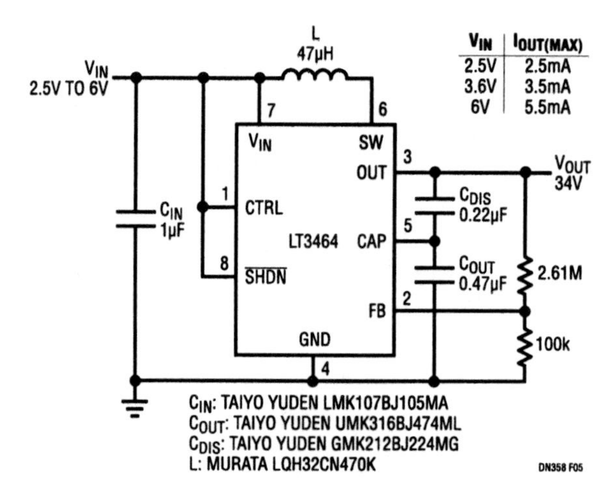

V_{IN}	$I_{OUT(MAX)}$
2.5V	2.5mA
3.6V	3.5mA
6V	5.5mA

Figure 121.5 • 34V Bias Supply Using the LT3464

The 47μH inductor has a 1210 footprint and the 0.47μF output capacitor has a 1206 footprint. The circuit occupies an area of only $55mm^2$.

Conclusion

The LT3464 provides a compact, complete solution for generating high voltage, low current bias supplies.

2-phase boost converter delivers 10W from a 3mm × 3mm DFN package

122

Jesus Rosales

Introduction

Small size, high efficiency, low noise and simplicity are all key features for battery-powered applications and point-of-load converters in low voltage systems. The LTC3428 is well suited for these applications because it offers these features as well as minimum output ripple and component count. It can start up with as little as 1.5V and operates with inputs up to 4.5V. Its dual phase architecture allows for an effective switching frequency of 2MHz (1MHz/phase), which minimizes inductor and capacitor size.

Dual phase converter reduces output ripple

The LT3428 incorporates two internal 93mΩ, N-channel MOSFET switches—enabling it to supply 2A of current at 5V from an input of 3.3V.

Figure 122.1 shows a dual phase 5V, 2A design. This converter switches at 1MHz per phase. The two phases are 180° out of phase, effectively doubling the output ripple frequency. This reduces the peak-to-peak output ripple current, which in turn makes it easier to filter out switching frequency ripple and noise. Input ripple current is also reduced, which minimizes stress on other components and reduces required input capacitance. The circuit in Figure 122.1 produces only $20mV_{P-P}$ of output voltage ripple, as shown in Figure 122.2.

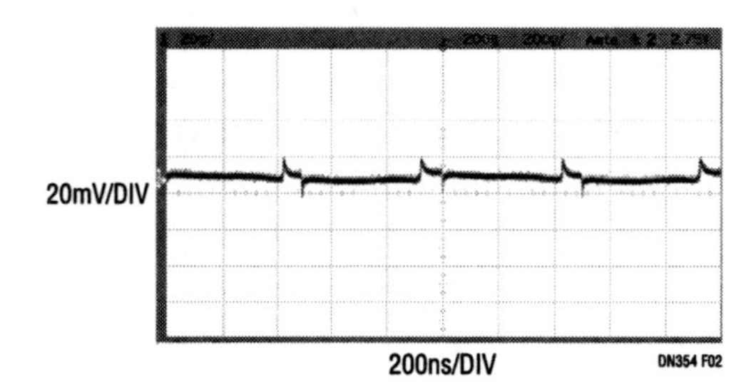

Figure 122.2 • Output Ripple for the 5V Dual Phase Boost Converter in Figure 122.1

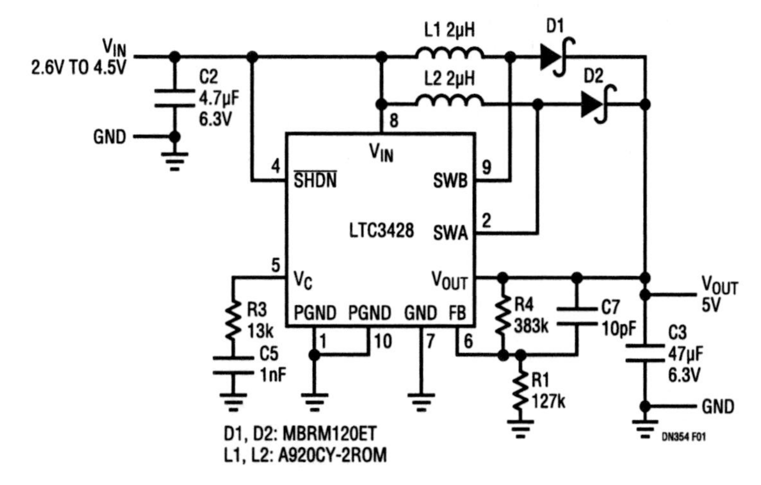

Figure 122.1 • A Dual Phase, 5V at 2A Output Boost Converter

Analog Circuit Design: Design Note Collection. http://dx.doi.org/10.1016/B978-0-12-800001-4.00122-8

Smaller layout is possible by reducing the number of external components

The LTC3428 requires very few external components for a complete boost circuit (see Figure 122.1). This, combined with its 3mm × 3mm footprint and 0.75mm profile make for an extremely compact, but feature rich converter. It can provide as much as 10W of power at $3.3V_{IN}$, and includes integrated features including internal soft-start and thermal shutdown. Figure 122.3 shows a photo of a typical layout, while Figure 122.4 shows the efficiency of this circuit at $3.3V_{IN}$.

Figure 122.3 • Typical Layout for a 5V Dual Phase Boost Converter

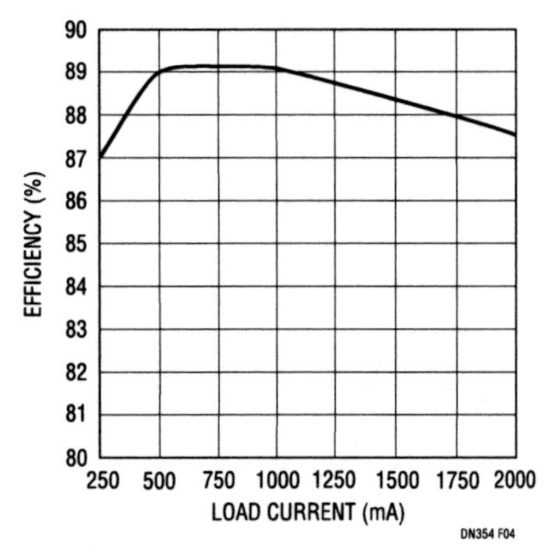

Figure 122.4 • Efficiency for the 5V Dual Phase Boost Converter in Figure 122.1 with $3.3V_{IN}$

Antiringing feature in discontinuous operation

During discontinuous mode operation, the inductor current is discharged to zero before the end of the switching period. Once the diode is turned off, there is high frequency ringing (caused by the inductor and parasitic capacitance) on the switch node, which can cause EMI radiation. The LTC3428 features an antiringing circuit that significantly reduces the discontinuous operation ringing. Figure 122.5 shows a switch waveform of a converter with antiringing control and Figure 122.6 shows one without.

Conclusion

The LTC3428's dual phase architecture reduces input and output ripple when compared to a single phase design, while providing high efficiency for up to 2A at 5V from a 3.3V input. Its 3mm x 3mm footprint and integrated features keep the circuit layout simple and small.

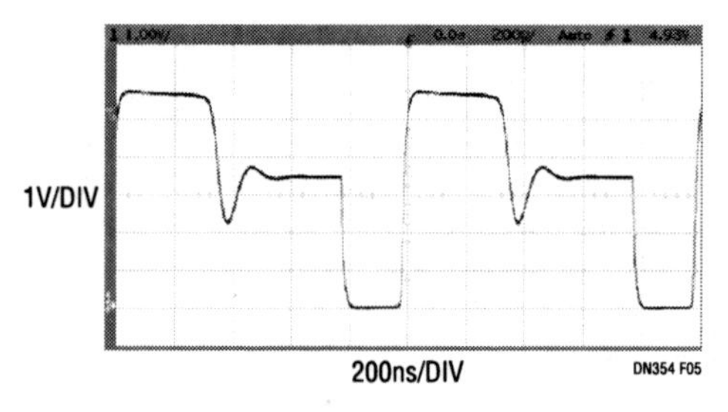

Figure 122.5 • Switch Waveform for the 5V Dual Phase Boost Converter in Discontinuous Mode with Antiringing Circuit

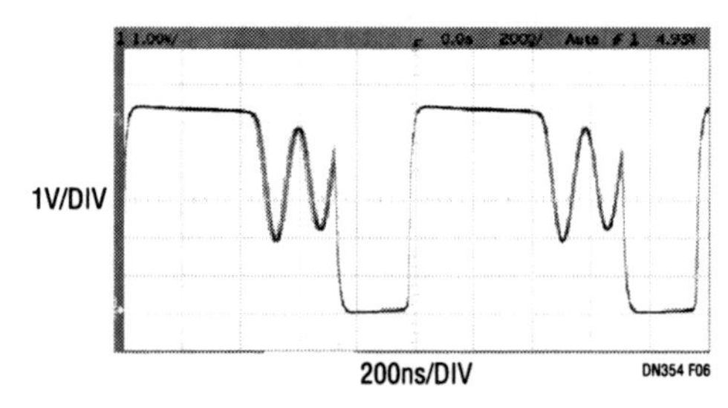

Figure 122.6 • Switch Waveform for a 5V Single Phase Boost Converter in Discontinuous Mode

4-phase monolithic synchronous boost converter delivers 2.5A with output disconnect in a 5mm × 5mm QFN package

123

David Salerno

Introduction

The LTC3425 is the industry's first 4-phase, monolithic synchronous boost converter. It can start up with as little as 1V and operate with inputs up to 4.5V. The output voltage range is 2.4V to 5.25V, making it well suited for battery-powered applications as well as point-of-load regulation in low voltage systems. The 4-phase architecture allows for an effective switching frequency of up to 8MHz, which in turn reduces output ripple current and peak inductor current by a factor of four (over an equivalent single phase circuit). This allows the use of small, low cost, low profile inductors and ceramic capacitors even at high load currents.

Integrated output disconnect allows V_{OUT} to go to 0V in shutdown while eliminating the high inrush current typical of traditional boost converters during start-up. With 5A peak current capability and an effective switch $R_{DS(ON)}$ of 40mΩ (NMOS) and 50mΩ (PMOS), the LTC3425 is capable of efficiently delivering 2.5A load current from a 0.8mm maximum profile, 32-pin 5mm × 5mm QFN package.

Multiple operating modes optimize performance in different applications

The LTC3425 can be configured for automatic Burst Mode operation, fixed frequency mode with forced continuous conduction or fixed frequency mode with pulse skipping. Programmable automatic Burst Mode operation is ideal for portable applications, where the load current can vary over a wide range and efficiency is paramount. A quiescent current of just 12µA in Burst Mode operation extends battery life

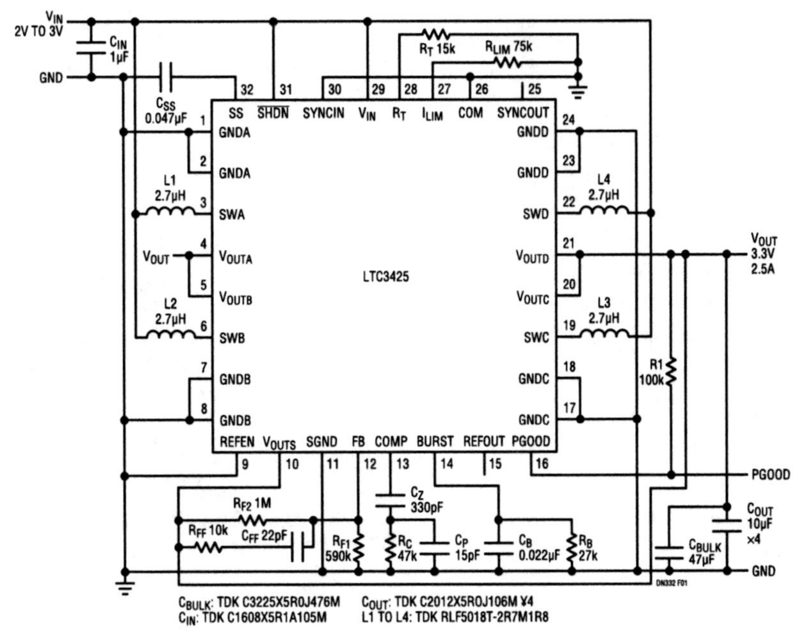

Figure 123.1 • 2-Cell to 3.3V/2.5A Boost Converter

Analog Circuit Design: Design Note Collection. http://dx.doi.org/10.1016/B978-0-12-800001-4.00123-X

during light load operation. For noise-sensitive applications fixed frequency mode can be selected, with either forced conduction for low noise at light load or pulse skipping for improved light load efficiency. Quiescent current in shutdown is less than 1μA.

Fault protection

The LTC3425 includes short-circuit protection, programmable peak current limit and thermal shutdown.

High power and high efficiency in a small package

Figure 123.1 shows an LTC3425 application using all ceramic capacitors and low profile inductors to deliver 2.5A load current at 3.3V from a 2-cell input (V_{IN} of 2V to 3V). Maximum component profile is only 2.5mm. In this example, the oscillator frequency is programmed for 4MHz (1MHz/phase), resulting in 10mV_{P-P} output ripple at full load (see Figure 123.2). The burst resistor is selected to automatically transition from Burst Mode operation to fixed frequency mode when the load exceeds 100mA for optimal efficiency (peaking at 96%) over the load range (see Figure 123.3). No-load input current is only 39μA at 2.4V.

The current mode architecture results in excellent transient response during a load step (see Figure 123.4). The use of a small feed-forward RC network across the top feedback resistor provides improved transient response and reduced output ripple in Burst Mode operation, especially when using low output capacitor values. The feed-forward capacitor also compensates for the effect of stray capacitance at the FB pin when using large value feedback resistors (Figure 123.5).

The open-drain PGOOD output goes low when V_{OUT} is 11.5% below its regulated value and goes high when V_{OUT} is within 9% of its regulated value.

Conclusion

The LTC3425's multiphase architecture makes it possible to deliver high power in a small, low profile package with very low output ripple. It offers features that are demanded in high performance portable applications, including output disconnect, automatic Burst Mode and high efficiency over a wide load range. It also offers flexible design features, including its programmable soft-start, current limit and oscillator frequency, external compensation and multiple available operating modes.

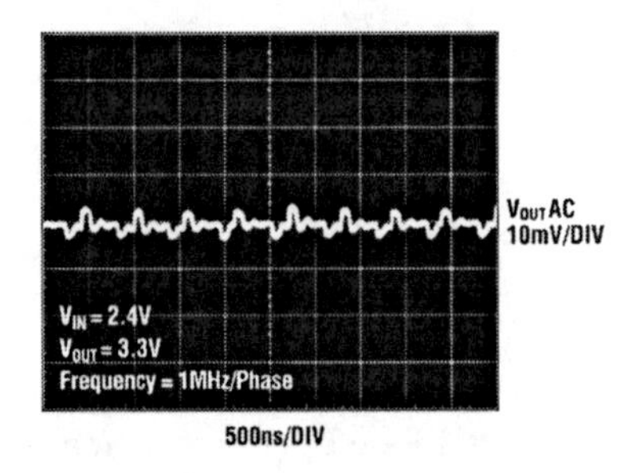

Figure 123.2 • Output Voltage Ripple at 2.5A Load for Converter of Figure 123.1

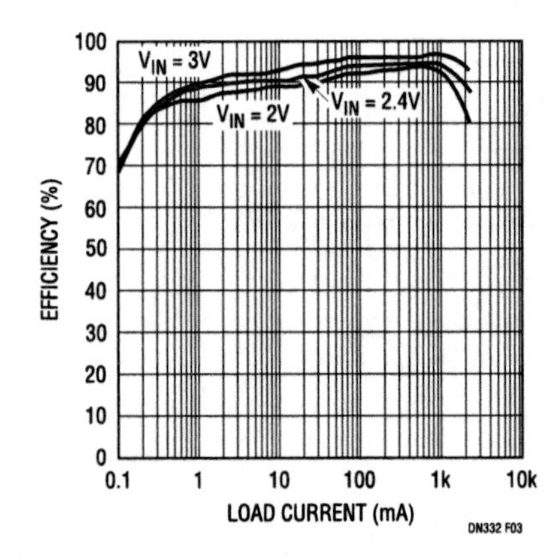

Figure 123.3 • Efficiency vs Load of the Converter in Figure 123.1

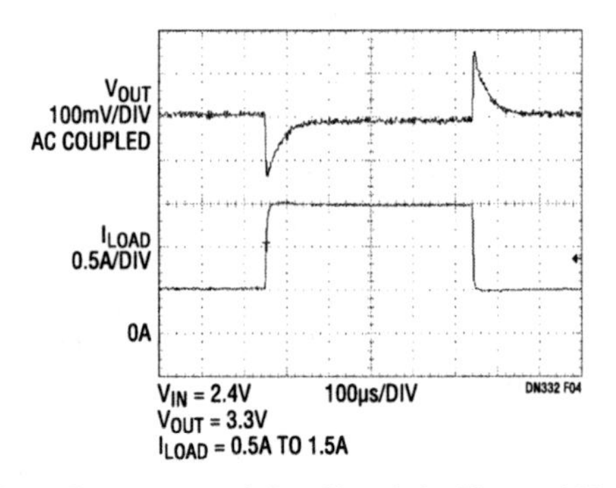

Figure 123.4 • Response of the Circuit in Figure 123.1 to a 1A Load Step

Figure 123.5 • The LTC3425 Can Deliver 10W in a Low Profile 0.7in² Footprint

Boost regulator makes low profile SEPIC with both step-up and step-down capability

124

Keith Szolusha

Introduction

Automotive, distributed power and battery-powered applications often operate at a voltage that is derived from a widely variable bus voltage. Frequently the operating voltage falls somewhere in the middle of the bus voltage range, such as a 12V automotive operating voltage, from a 4V to 18V bus. These applications require a DC/DC converter that can step up or step down, depending on the voltage present on the bus. Flyback and SEPIC designs are commonly used single-switch solutions for this problem, but both of these solutions typically use a transformer which poses layout and height problems for applications where space is at a premium.

One alternative to a transformer-based topology is to use two low profile inductors and a SEPIC coupling capacitor which transfers the energy between the two inductors much like the core of a transformer. The coupling capacitor provides a low impedance path for the inductor currents to pass either from the input (primary) inductor through the catch diode and to the output, or from the output (secondary) inductor back through the switch to ground. Both inductors act continuously and independently, making their selection easier than selecting the transformer for a flyback or a typical SEPIC circuit. The inductors are not restricted to having the same inductance and can be individually picked for peak currents and allowable ripple.

3V to 20V input, 5V output, 3mm maximum height SEPIC

Figures 124.1 and 124.2 shows a 3V to 20V input, 5V output 3mm maximum height SEPIC using the LT1961, a 1.25MHz, current mode, monolithic, 1.5A peak switch current, boost converter. The output current capability of this circuit varies with input voltage (see Figure 124.3). At 3V input, the converter can supply up to 410mA of load current and as high as 830mA of load current at 20V input. The tiny coupling

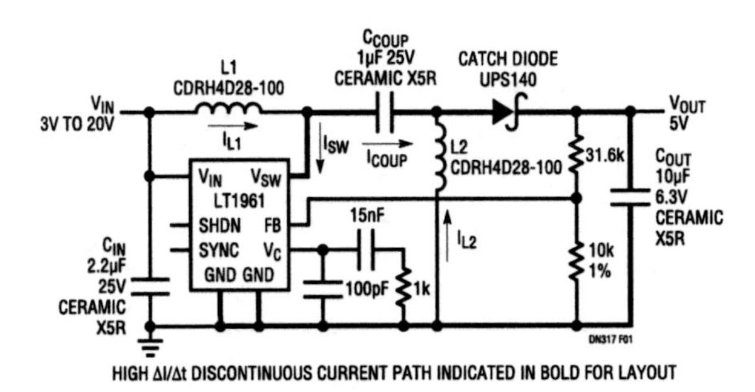

Figure 124.1 • LT1961 in a 3V to 20V Input to 5V Output All Ceramic SEPIC (3mm Maximum Height)

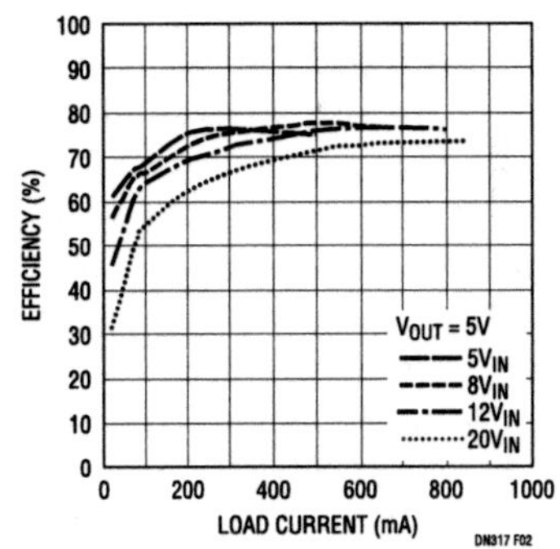

Figure 124.2 • Efficiency of the Circuit in Figure 124.1

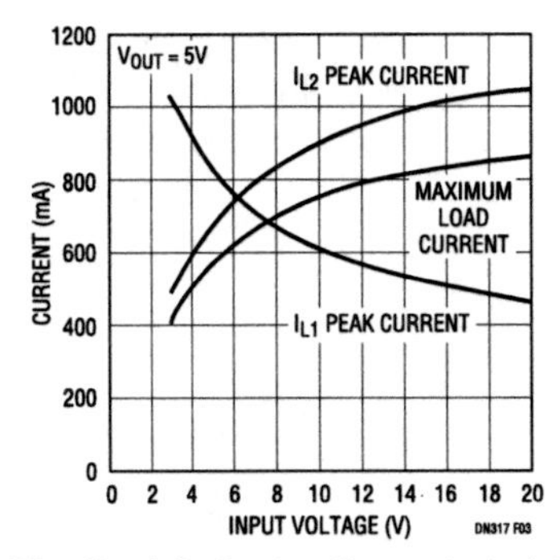

Figure 124.3 • The Peak Inductor Currents in L1 and L2 Sum to 1.5A, the Peak Switch Current. Maximum Output Current Is the Average Current in L2 at Peak Switch Current

Analog Circuit Design: Design Note Collection. http://dx.doi.org/10.1016/B978-0-12-800001-4.00124-1

capacitor used here is large enough to handle the RMS ripple current transferring between the primary and secondary sides of the circuit, and to maintain a voltage equal to the input voltage in order to provide good regulation and maximum output power. The current mode control topology of the LT1961 and the small 10μF ceramic output capacitor provide excellent transient response over the wide input voltage range.

4V to 18V input, 12V output, 3mm maximum height SEPIC

12V buses are often derived from sources with a wide input voltage range. For instance, automotive solutions can have steady-state operating voltages as high as 18V and as low as 4V for cold-crank conditions. Figure 124.4 shows a simple, low cost and low profile (≤3mm) solution that avoids the high cost of using both a boost and buck converter and maintains 12V system power during cold-crank conditions.

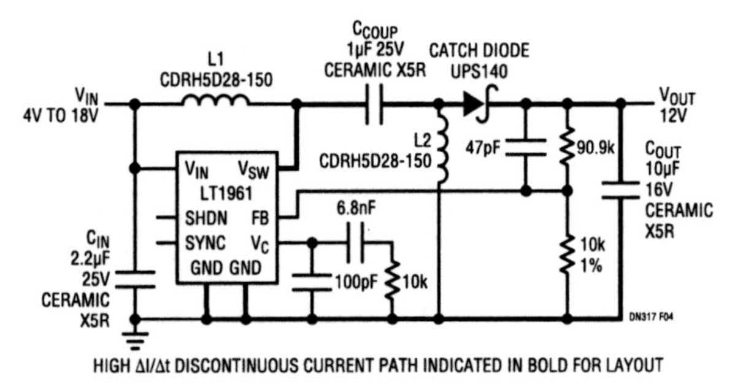

Figure 124.4 • LT1961 in a 4V to 18V Input to 12V Output 3mm Maximum Height All Ceramic SEPIC

The catch diode has a 40V reverse breakdown voltage rating in order to handle the voltage induced across it during the switch off-time which is equal to the output voltage plus the input voltage. The 35V maximum switch voltage rating of the LT1961 allows the input voltage to go up as high as 18V. With a DC voltage equal to the input voltage, the coupling capacitor raises the voltage at the switch node to a level equal to the input voltage plus the output voltage. Tiny voltage spiking present on the switch node of any switching converter requires a few volts of headroom between the maximum switch voltage rating and the sum of the input and output voltages. The switching spikes are reduced to a minimum by keeping the high ΔI/Δt discontinuous current path (indicated in bold in Figures 124.1 and 124.4) as short as possible. The placement of the two power inductors is not crucial which makes it easier to create a power supply layout that fits confined spaces.

Efficiency, as shown in Figure 124.5, is typically greater than 75% and as high as 80%. This is better than average for 12V SEPICs and not much less than a similarly priced and sized 12V buck converter solution which is limited to greater than 14V input. Maximum load current increases with input voltage, as shown in Figure 124.6. 500mA load current is possible at 12V input and up to 600mA at 18V. The maximum switch current of the LT1961 is 1.5A and is the sum of the peak current in L1 and L2. Higher output voltage raises the current in the input inductor.

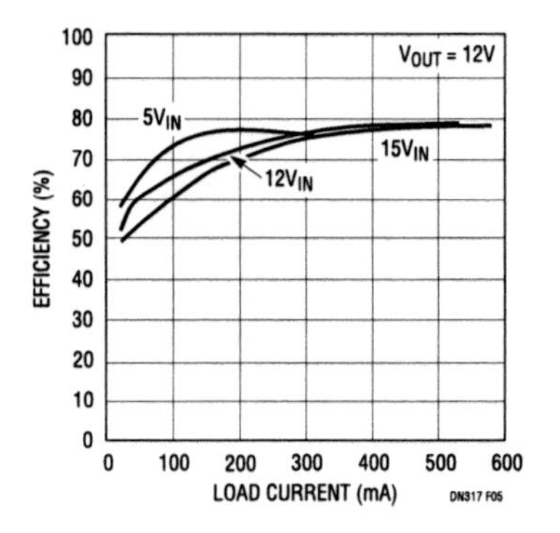

Figure 124.5 • Efficiency of the Circuit in Figure 124.4

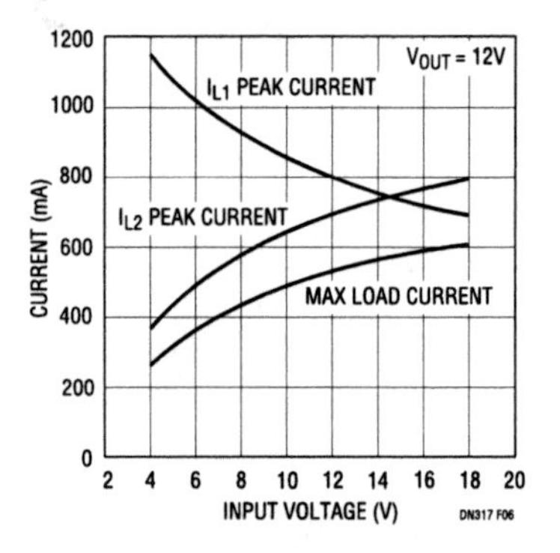

Figure 124.6 • The Peak Inductor Currents and Maximum Load Current of the Circuit in Figure 124.4

Conclusion

The LT1961 fits into SEPIC solutions for applications with wide input voltage ranges. The solutions are small, simple and low profile. All ceramic capacitors and tiny components help keep power supply costs to a minimum. The 2-inductor SEPICs shown here eliminate the use of a tall transformer and offer layout flexibility to fit tight design constraints.

Dual monolithic buck regulator provides two 1.4A outputs with 2-phase switching to reduce EMI

125

Jeff Witt

Introduction

Advanced electronic systems that use a single voltage supply are a thing of the past. Today's electronic systems require several regulated voltages; even relatively simple subsystems need a minimum of two supplies. Microprocessors and DSPs, for example, might require both a 1.8V core supply and a 3.3V supply for I/O and memory. Many board level systems require both 3.3V and 5V. Added to this mix is the ever-increasing packaging density of electronic products. Voltage regulators must fit in a small space in close proximity to sensitive circuits, meaning they must be small, efficient and low noise. Linear regulators may generate too much heat, or be too bulky for some applications, and switching regulators present possible EMI problems.

Circuit description

The LT1940 is a dual step-down switching regulator that solves these problems for systems requiring two or more regulated voltages. Its wide input range, 3.6V to 25V, accepts a variety of power sources. The low profile 16-lead TSSOP package has an exposed metal backside, improving thermal performance and allowing the LT1940 to produce two 1.4A outputs without additional heat sinks. High frequency, 2-phase switching minimizes ripple and EMI while each channel has independent soft-start and power good indicators. These features make it possible to design small, low noise supplies that interface easily with existing systems.

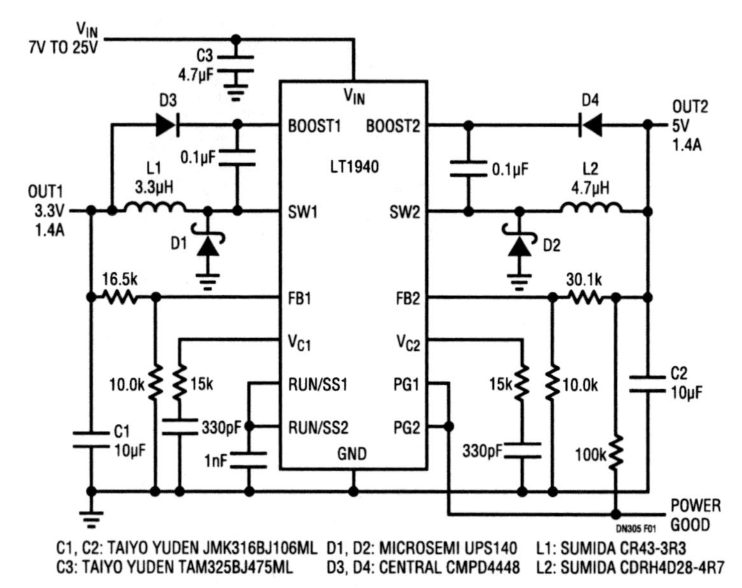

Figure 125.1 • The LT1940 Produces Two Low Noise Outputs Using Small Ceramic Capacitors

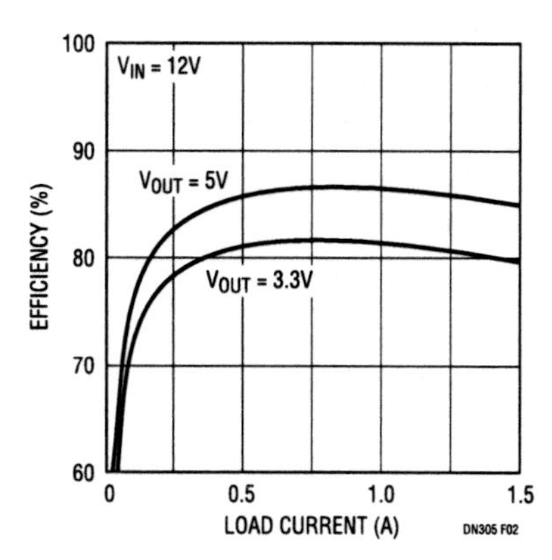

Figure 125.2 • The Efficiency of the Circuit in Figure 125.1 Remains High, Even at Full Load

Analog Circuit Design: Design Note Collection. http://dx.doi.org/10.1016/B978-0-12-800001-4.00125-3

The circuit in Figure 125.1 generates 3.3V and 5V from an input of 7V to 25V. In this circuit, the two RUN/SS pins are tied together and a single capacitor programs the soft-start. Also, the two power good pins are tied together providing a single power good signal that goes high when both outputs are in regulation (see Figure 125.2).

High frequency, current mode switching minimizes component size

The LT1940's high 1.1MHz switching frequency and current mode control allow the use of small components including low profile inductors and ceramic capacitors. Because the control loop can be easily compensated even with a high loop bandwidth, the output capacitors can have relatively low values and still provide fast and stable transient performance. The high switching frequency, combined with the low ESR of the ceramic capacitors, results in a very low output ripple (<$5mV_{P-P}$).

2-phase switching eases EMI concerns

A buck regulator draws pulses of current from its input supply, resulting in large AC currents that can cause EMI problems. The LT1940's two regulators are synchronized to a single oscillator and switch out of phase by 180°. This substantially reduces the input ripple current, thereby lowering EMI and allowing the use of a single input capacitor. Synchronization also eliminates the audible noise that can occur when two switchers run at slightly different frequencies.

Soft-start and power good pins simplify supply sequencing

Multisupply systems often require output sequencing. For example, a microprocessor's core supply should be in regulation before power is applied to the I/O circuits. Figure 125.3 shows a simple way to sequence the two outputs of the LT1940. Channel 1 produces the 1.8V core supply. Its power good pin pulls V_{C2} low, disabling channel 2 until the 1.8V output is in regulation (see Figure 125.4).

Conclusion

The LT1940 has the right set of features to implement a high performance, dual output power supply. High switching frequency, 2-phase operation and all ceramic capacitors produce a small, low ripple, low EMI circuit that interfaces easily to any system.

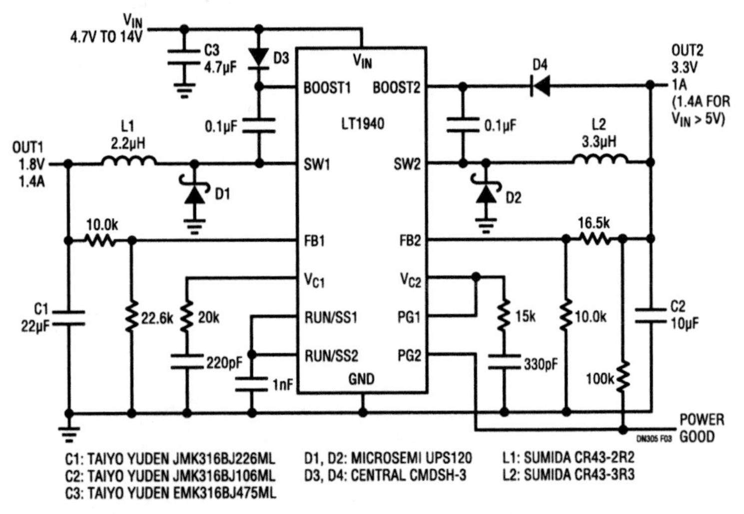

Figure 125.3 • This 1.8V/3.3V Circuit Uses the Power Good Output of Channel 1 to Sequence the Two Outputs (Channel 1 Starts First)

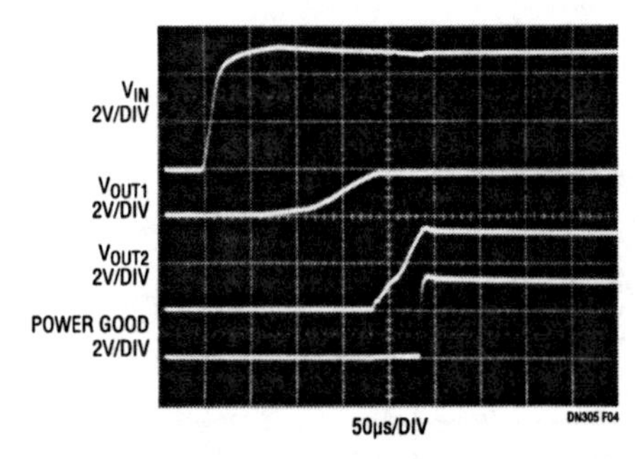

Figure 125.4 • Start-Up Waveforms of the Circuit in Figure 125.3

4MHz monolithic synchronous step-down regulators bring high efficiency to space-sensitive applications

126

Joey M. Esteves

Introduction

The LTC3411 and LTC3412 provide compact and efficient power supply solutions for portable electronics such as cell phones, PDAs and notebook computers. These two monolithic, synchronous step-down regulators provide DC/DC conversion from either a 3.3V or 5V system voltage to outputs as low as 0.8V. They also offer switching frequencies as high as 4MHz, allowing the use of tiny inductors and capacitors. Both devices save additional space by integrating the power switches into their monolithic architecture. The LTC3412's built-in power switches have an 85mΩ on-resistance, enabling it to deliver up to 2.5A of output current with efficiencies as high as 95%. The LTC3411 is optimized for lower power applications. Its 110mΩ power switches allow output currents as high as 1.25A.

The LTC3411 and LTC3412 both utilize a constant frequency, current mode architecture that operates from an input voltage range of 2.5V to 5.5V and provide an adjustable regulated output voltage from 0.8V to 5V. The switching frequency for either part can be set from 300kHz to 4MHz by an external resistor or synchronized to an external clock. The ability to increase the switching frequency as high as 4MHz allows for lower inductor values while still maintaining low output voltage ripple since output voltage ripple is inversely proportional to the switching frequency and the inductor value. Because smaller case sizes are usually offered for lower inductor values, the overall solution size is reduced. The LTC3411 is offered in an MSOP package to further reduce the footprint. For optimal thermal handling, the LTC3412 is offered in a 16-lead TSSOP package with an exposed pad.

Multiple operating modes allow optimization of efficiency and noise suppression

Both the LTC3411 and LTC3412 can be configured for either Burst Mode operation or forced continuous mode, while the LTC3411 also offers pulse-skipping mode. Burst Mode operation provides high efficiency and extends battery life by reducing gate charge loss at light loads. Forced continuous

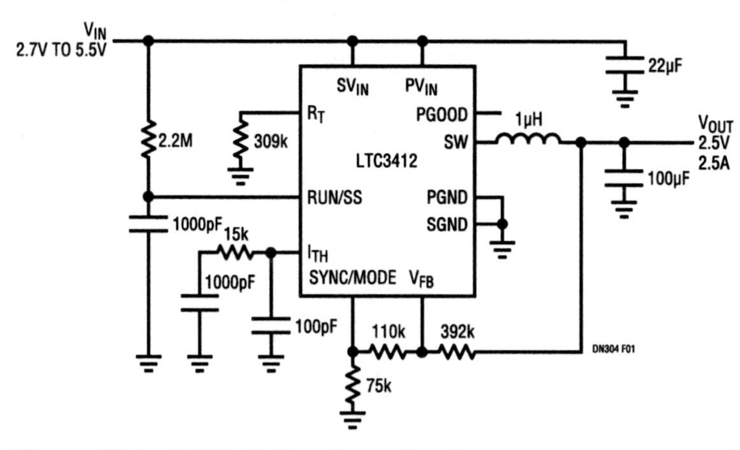

Figure 126.1 • 2.5V/2.5A Step-Down Regulator

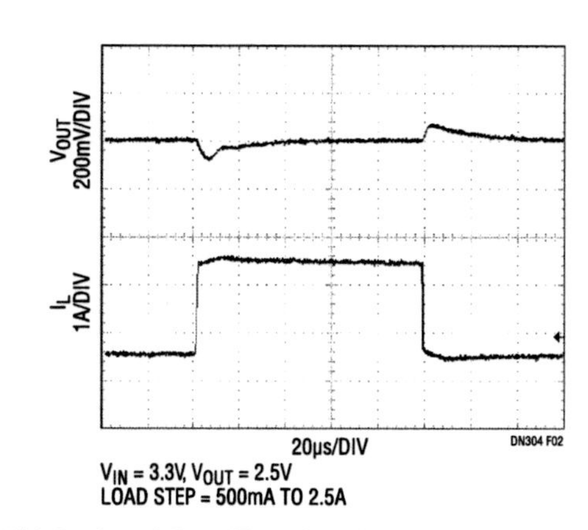

Figure 126.2 • Load Step Transient Response

Analog Circuit Design: Design Note Collection. http://dx.doi.org/10.1016/B978-0-12-800001-4.00126-5

mode is not as efficient at light loads but it offers advantages in noise-sensitive applications. Pulse-skipping mode is a compromise between the two. With no load, the LTC3411 and LTC3412 consume only 62μA of supply current.

In Burst Mode operation, there is a trade-off between output voltage ripple and efficiency at light loads. With the LTC3412, external control of the burst clamp level allows the burst frequency to be varied. A lower burst frequency increases efficiency at light loads due to lower gate charge losses but this also slightly increases output voltage ripple. The burst clamp level can be adjusted on the LTC3412 by varying the voltage at the SYNC/MODE pin in the range of 0V to 1V.

For applications in which noise suppression is a priority, both devices offer forced continuous (frequency) mode, in which constant frequency is maintained regardless of output load. The LTC3411 also offers pulse-skipping mode. In pulse-skipping mode, the LTC3411 continues to switch at a constant frequency down to very low output currents, minimizing the ripple voltage and ripple current at the output (see Figure 126.2).

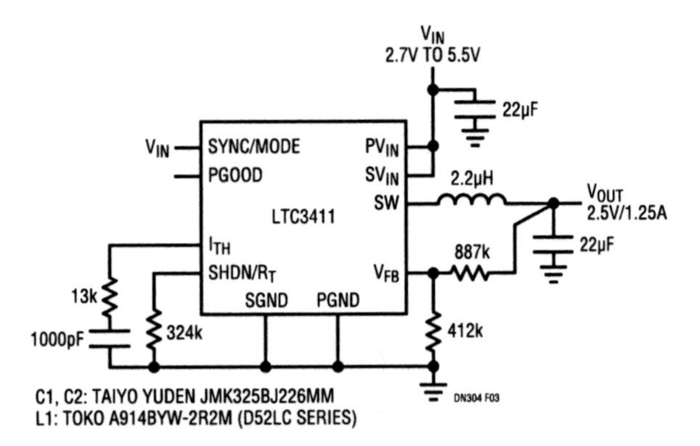

Figure 126.3 • 2.5V/1.25V Step-Down Regulator

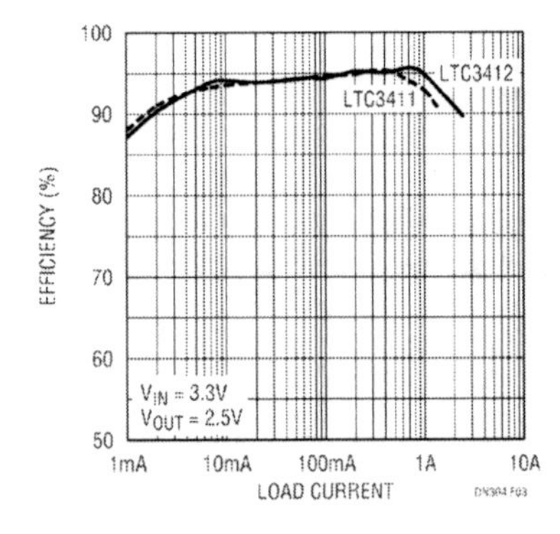

Figure 126.4 • Efficiencies for the Circuits Shown in Figures 126.1 and 126.3

Two 2.5V step-down converters

Figure 126.1 shows a design using the LTC3412 for a 2.5V step-down DC/DC converter that is capable of sourcing up to 2.5A of output current. Figure 126.3 shows a design using the LTC3411 for a 2.5V step-down converter that is capable of sourcing up to 1.25A of output current. Efficiencies for these circuits are as high as 95% for a 3.3V input as shown in Figure 126.4. The input and output capacitors are ceramic, which are desirable because of their low cost and low ESR. Many switching regulators have difficulty operating with ceramic capacitors because they rely on the feedback response zero that is generated by the larger ESR of tantalum capacitors. The LTC3412 and LTC3411, however, feature loop compensation, which allows them to operate successfully with ceramic capacitors. The frequencies for these particular demonstration circuits are set at 1MHz by a single external resistor, allowing for small inductors and capacitors, as illustrated in Figures 126.1 and 126.3.

Conclusion

The LTC3411 and LTC3412 are high performance monolithic, synchronous step-down DC/DC converters that are well suited for applications requiring up to 1.25A and 2.5A of output current, respectively. Their high switching frequency and internal low $R_{DS(ON)}$ power switches allow the LTC3411 and LTC3412 to offer compact, high efficiency power supply solutions for any application, as can be seen in Figure 126.5.

Figure 126.5 • The LTC3411 Is a Space-Saving Step-Down Regulator

Tiny and efficient boost converter generates 5V at 3A from 3.3V bus

127

Dongyan Zhou

Introduction

Circuits that require 5V remain popular despite the fact that modern systems commonly supply a 3.3V power bus, not 5V. The tiny LTC1700 is optimized to deliver 5V from the 3.3V bus at very high efficiency, though it can also efficiently boost other voltages. The small MSOP package and 530kHz operation promote small surface mount circuits requiring minimal board space, perfect for the latest portable devices. By taking advantage of the synchronous rectifier driver, the LTC1700 provides up to 95% efficiency. To keep light load efficiency high in portable applications, the LTC1700 draws only 180µA in sleep mode. The LTC1700 features a start-up voltage as low as 0.9V, adding to its versatility.

The LTC1700 uses a constant frequency, current mode PWM control scheme. Its No R_{SENSE} feature means the current is sensed at the main MOSFET, eliminating the need for a sense resistor. This saves cost, space and improves efficiency at heavy loads. For noise-sensitive applications, Burst Mode operation can be disabled when the SYNC/MODE pin is pulled low or driven by an external clock. The LTC1700 can be synchronized to an external clock ranging from 400kHz to 750kHz.

3.3V input, 5V/3A output boost regulator

Figure 127.1 shows a 3.3V input to 5V output boost regulator which can supply up to 3A load current. Figure 127.2 shows that the efficiency is greater than 90% for a load current range of 200mA to 3A and stays above 80% all the way down to a 3mA load.

C2 is a tantalum capacitor providing bulk capacitance to compensate for possible long wire connections to the input supply. In applications where the regulator's input is connected very close to a low impedance supply, this capacitor is not needed.

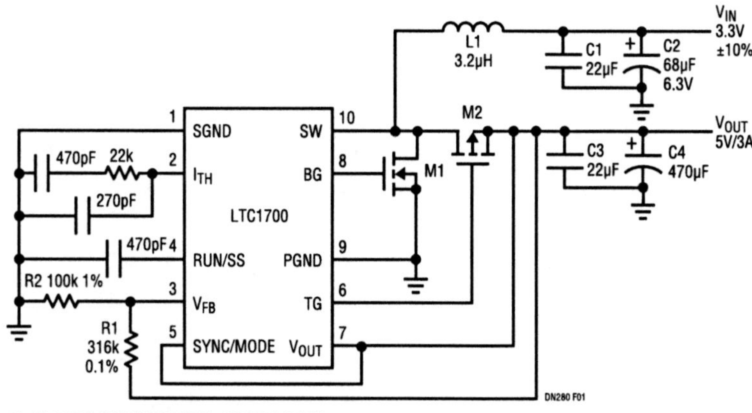

C1, C3: TAIYO YUDEN CERAMIC JMK325BJ226M
C2: AVX TAJB686K006R
C4: SANYO POSCAP 6TPB470M
L1: SUMIDA CEP1233R2
M1: INTERNATIONAL RECTIFIER IR7811W
M2: SILICONIX Si9803

Figure 127.1 • 3.3V to 5V, 3A Boost Regulator

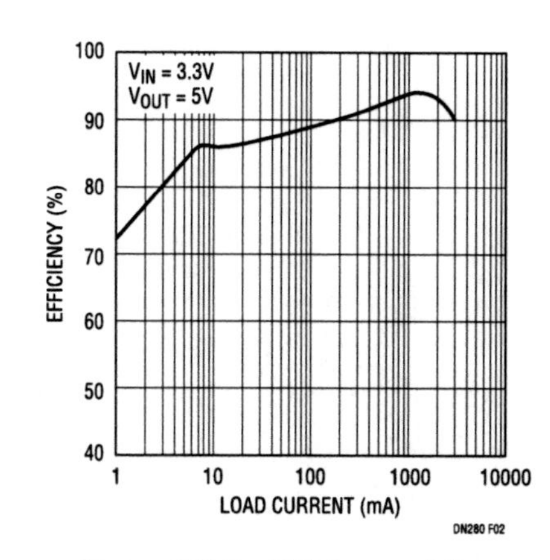

Figure 127.2 • Efficiency of the Figure 127.1 Circuit

Analog Circuit Design: Design Note Collection. http://dx.doi.org/10.1016/B978-0-12-800001-4.00127-7

2-cell input, 3.3V/1A output regulator

In digital cameras and other battery-powered devices, the LTC1700 makes for a high efficiency boost regulator in a small package. Figure 127.3 shows a 2-alkaline cell to 3.3V output circuit. This circuit can supply 1A maximum output current. Figure 127.4 shows the efficiency at different battery voltages. Efficiency of this circuit peaks at 93%. If a lower $R_{DS(ON)}$ MOSFET (such as Si6466) is used for M1, the maximum output current can be increased to 1.4A with about a 2% reduction in efficiency due to the increased gate capacitance. MOSFETs with lower than 2.5V gate threshold voltages are recommended. The LTC1700 is also an ideal device for single cell Li-Ion battery to 5V applications.

Conclusion

The LTC1700 boost controller brings high efficiency and small size to low voltage applications. Its features are ideally suited to both battery-powered and line-powered applications.

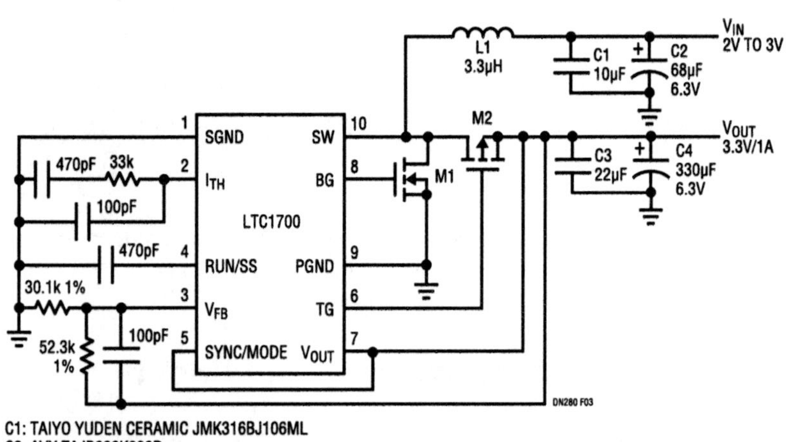

C1: TAIYO YUDEN CERAMIC JMK316BJ106ML
C2: AVX TAJB686K006R
C3: TAIYO YUDEN CERAMIC JMK325BJ226M
C4: SANYO POSCAP 6TPB330M
L1: MURATA LQN6C
M1: SILICONIX Si9804
M2: SILICONIX Si9803

Figure 127.3 • 2-Cell to 3.3V, 1A Boost Regulator

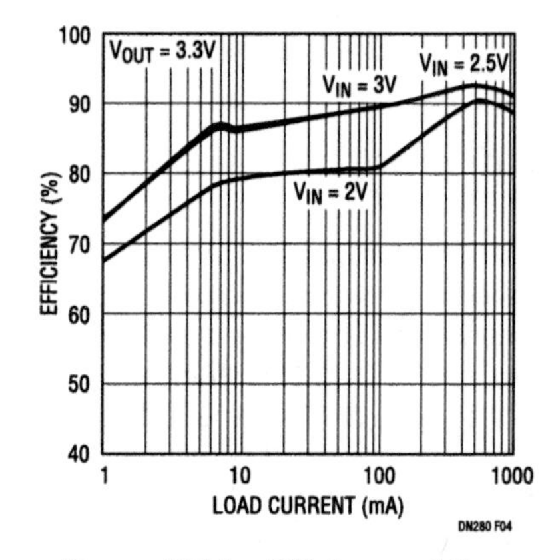

Figure 127.4 • Efficiency of the Figure 127.3 Circuit

Tiny boost controller provides efficient solutions for low voltage inputs

128

Keith Szolusha

Introduction

The expanding world of low voltage portable and microprocessor electronics has created the need for small, highly efficient, low cost, low voltage boost controllers with high current capabilities. The new LT1619, available in a tiny MS8 package, is a high current, low input voltage boost controller with a powerful rail-to-rail MOSFET driver and ultralow current-sense voltage (53mV). These features, used in the common boost or versatile SEPIC topologies, provide low cost, efficient and tiny DC/DC solutions for many applications, including 3.3V to 5V converters and automotive-range 12V to 5V converters.

The LT1619 provides a complete solution for low input voltage applications that require low side MOSFET drive. It is a 300kHz, current mode PWM controller capable of operating from inputs ranging from 1.9V to 18V. The rail-to-rail, 1A MOSFET driver is capable of driving an external MOSFET gate to within 350mV of the supply rail and to within 100mV of ground. Bootstrapping the driver supply pin (DRV) to the output enables the power supply to operate from input voltages as low as 1.9V yet still drive the MOSFET gate voltage high enough for full enhancement. The 53mV low side current limit threshold improves efficiency by reducing the sense resistor's power dissipation. At light loads, the controller automatically switches to Burst Mode operation to conserve power. In shutdown, the LT1619 requires only 15μA of quiescent current.

3.3V to 5V converters

Figure 128.1 shows a 3.3V to 5V/2.2A boost supply using the LT1619. Low parts count, small size and high efficiency (greater than 90%) make it a perfect solution when a moderate amount of 5V power is required in a predominately 3.3V system. The output voltage can be returned to the DRV pin, further enhancing M1.

Low current-sense voltage, although more efficient, can be more susceptible to switching noise. However, the internal current sense amplifier is blanked for 280ns to prevent spurious switching spikes (and therefore PWM jitter) across the sense resistor caused by the gate charging current at switch turn-on. Although this blanking sets a minimum switch

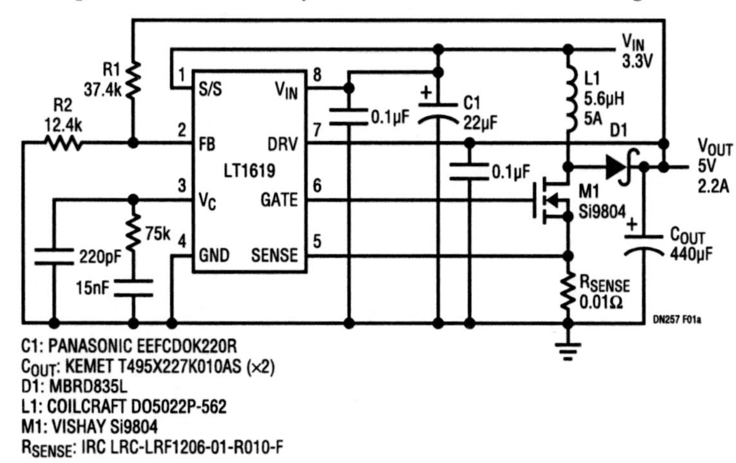

Figure 128.1a • 3.3V to 5V Step-Up Converter Rated for 2.2A

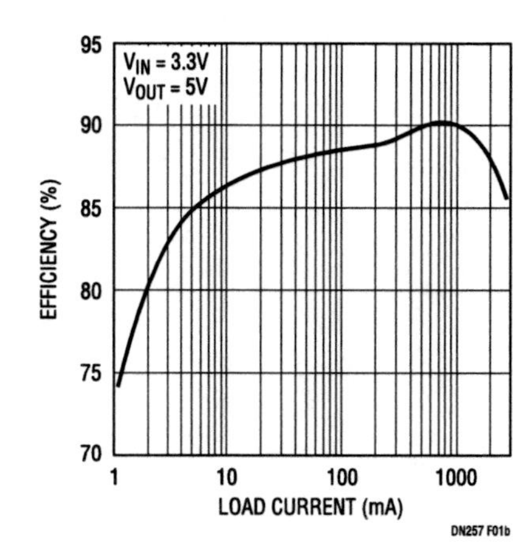

Figure 128.1b • Efficiency of Figure 128.1

Analog Circuit Design: Design Note Collection. http://dx.doi.org/10.1016/B978-0-12-800001-4.00128-9

on-time, the controller is capable of skipping cycles at light load with Burst Mode operation disabled.

Choosing the MOSFET

The LT1619 is designed to drive an N-channel MOSFET with up to 60nC of total gate charge (Q_G). Recently, significant advances have been made in low voltage (<30V) power MOSFETs. 10mΩ, low voltage, low threshold FETs with less than 60nC of gate charge are readily available. MOSFETs with less than 60nC of gate charge can be driven directly by the LT1619, resulting in a simple, low cost design.

Automotive supply

Figure 128.2 shows a 5V, 1A SEPIC (single-ended primary inductance converter) designed to operate from a 12V battery. The bias supply on VIN and DRV limits the voltage to about 4V maximum at start-up, limiting the amount of quiescent power lost and maintaining a high efficiency. The LT1619 is powered from the output through D2 after regulation is achieved, increasing efficiency.

The low input voltage threshold of the LT1619 allows the battery voltage to drop to 3V or less during normal operation (cold-crank support). With a sublogic-level MOSFET, this converter can still be started at voltage levels as low as 3V without having to add extra components. The power supply will also work well with a battery voltage as high as 18V, which provides margin for battery charging voltage (15V) plus several volts of inductive spiking.

SEPIC capacitor C2 provides a path for continuous input current and directs T1's leakage energy to the output. Using C2 increases efficiency and reduces input capacitor ripple current requirements. The LT1619's 300kHz operating frequency allows for smaller magnetics (0.45″ × 0.45″ × 0.25″h) and smaller capacitors than required by lower frequency controllers.

Conclusion

The LT1619 solves many of the problems associated with low input voltage source DC/DC converters. Its numerous features make it an ideal choice for a wide range of applications requiring low side MOSFET power transistors, high efficiency, low cost and high currents in very little board space.

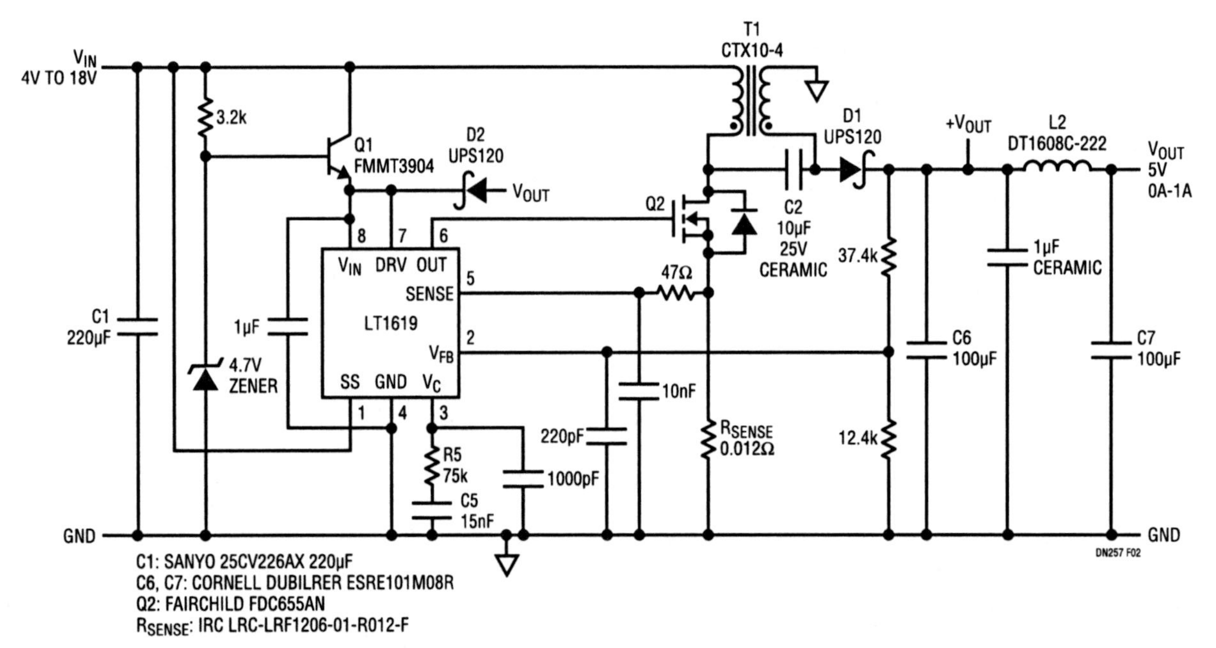

Figure 128.2 • SEPIC Conversion Delivers 5V at 1A from a 12V Battery

Current-limited DC/DC converter simplifies USB power supplies

129

Bryan Legates

Many portable universal serial bus (USB) devices power themselves from the USB host or hub power supply when plugged into the USB port. Several requirements must be met to ensure the integrity of the bus: the USB specification dictates that the input capacitance of a device must be less than 10μF to minimize inrush currents when the device is plugged into the USB port; when first plugged in, the device must draw less than 100mA from the port; and for high power devices, the current drawn from the port can increase to 500mA only after it is given permission to do so by the USB controller. These requirements can be easily met using the LT1618 DC/DC converter, which provides an accurate input current control ideal for USB applications. The LT1618 combines a traditional voltage feedback loop with a unique current feedback loop to operate as a constant-current, constant-voltage source.

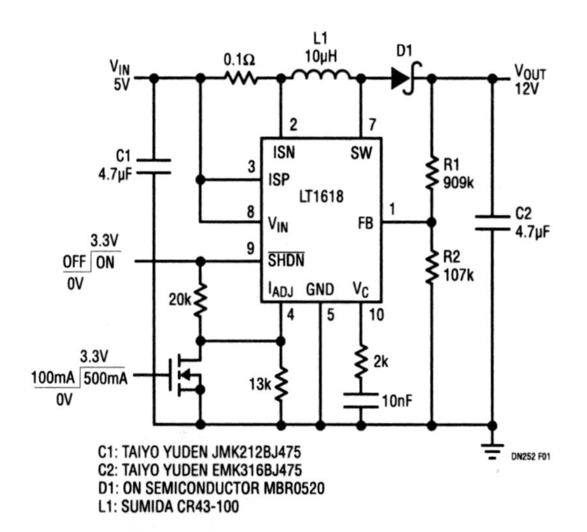

Figure 129.1 • USB to 12V Boost Converter (with Selectable 100mA/500mA Input Current Limit)

USB to 12V boost converter

Figure 129.1 shows a 5V to 12V boost converter ideal for USB applications. The converter has a selectable 100mA/500mA input current limit, allowing the device to be easily switched between the USB low and high power modes. Efficiency, shown in Figure 129.2, exceeds 85%. If the load demands more current than the converter can provide with the input current limited to 100mA (or 500mA), the output voltage will simply decrease and the LT1618 will operate in constant-current mode. For example, with an input current limit of 100mA, around 35mA can be provided to the 12V output. If the load increases to 50mA, the output voltage will reduce to around 8V to maintain a constant 100mA input current.

USB to 5V SEPIC DC/DC converter with short-circuit protection

The single-ended primary inductance converter (SEPIC) shown in Figure 129.3 is ideal for applications where the output must reduce to zero during shutdown. The input current

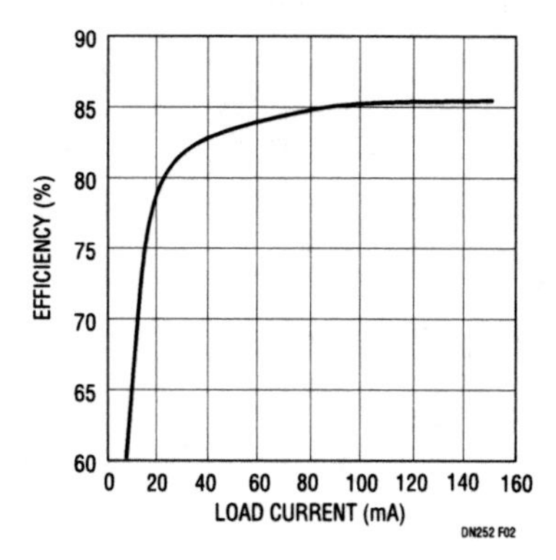

Figure 129.2 • USB to 12V Boost Efficiency

Analog Circuit Design: Design Note Collection. http://dx.doi.org/10.1016/B978-0-12-800001-4.00129-0

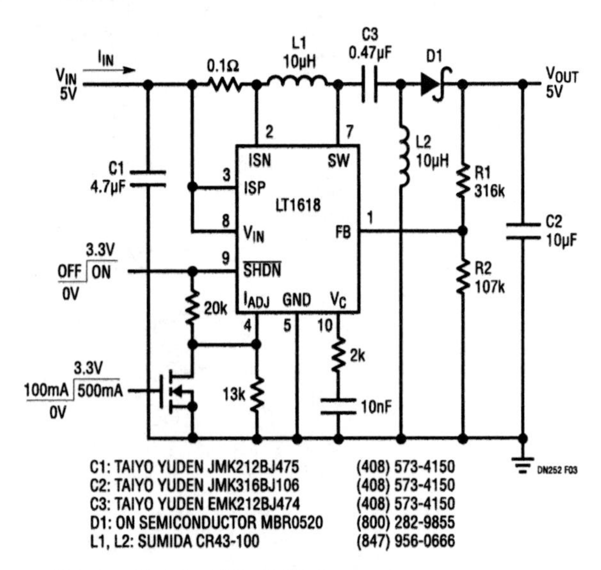

Figure 129.3 • USB to 5V SEPIC Converter

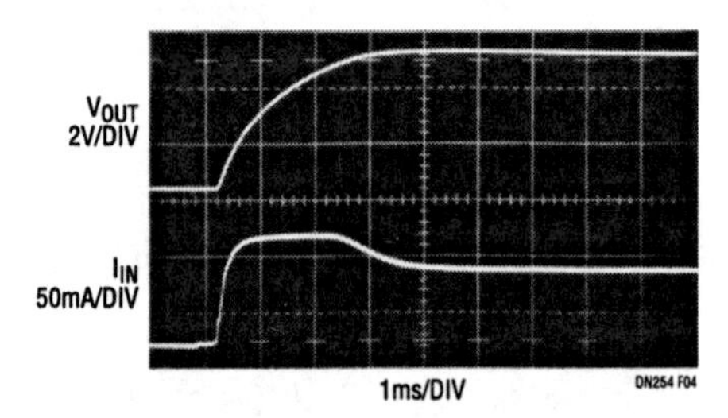

Figure 129.4 • USB SEPIC during Start-Up

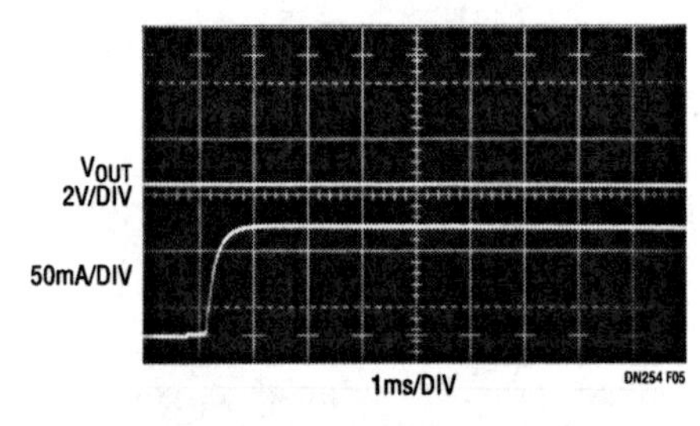

Figure 129.5 • USB SEPIC Start-Up with Output Shorted

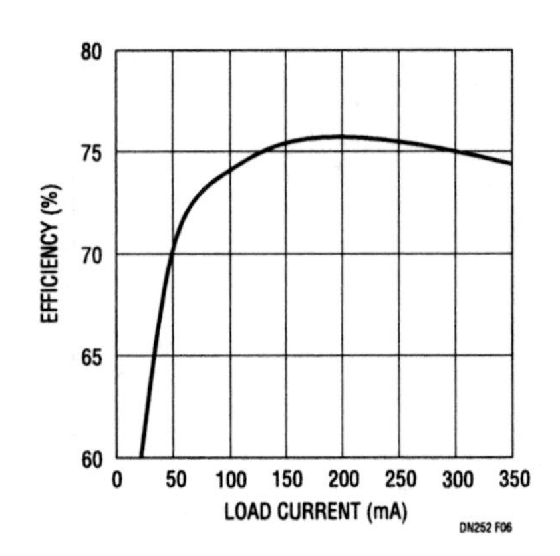

Figure 129.6 • USB to 5V SEPIC Efficiency

limit not only helps soft-start the output but also provides short-circuit protection, ensuring USB device compliance even under output fault conditions. Figure 129.4 shows the start-up characteristic of the SEPIC with a 50mA load. By limiting the input current to 100mA, the output is soft-started, smoothly increasing and not overshooting its final 5V value. Figure 129.5 shows that the input current does not exceed 100mA even with the output shorted to ground (thus the flat output voltage waveform in the picture). Efficiency for this SEPIC converter is shown in Figure 129.6.

Li-Ion white LED driver

In addition to providing an accurate input current limit, the LT1618 can also be used to provide a regulated output current for current-source applications. White LED drivers are one such application for which the LT1618 is ideally suited. With an input voltage range of 1.6V to 18V, the LT1618 can provide LED drive from a variety of input sources, including two or more alkaline cells, or one or more Li-Ion cells. The circuit in Figure 129.7 is capable of driving six white LEDs from a single Li-Ion cell. LED brightness can be adjusted using a pulse width modulated (PWM) signal, as shown, or by using a DC voltage to drive the I_{ADJ} pin directly, without the R3/C3 low-pass filter. If brightness control is not needed, simply tie the I_{ADJ} pin to ground. Typical output voltage with the LEDs shown is around 22V, and the R1, R2 output divider sets the maximum output voltage to around 26V to protect the LT1618 if the LEDs are disconnected. The LT1618's constant-current loop regulates 50mV across the 2.49Ω sense resistor, setting the LED current to 20mA.

Efficiency for this circuit exceeds 70%, which is significantly higher than the 30% to 50% efficiency obtained when using a charge pump for LED drive. When the LT1618 is turned off, no current flows in the LEDs. Their high forward voltages prevent them from turning on, ensuring a true low current shutdown with no excess battery leakage or light output.

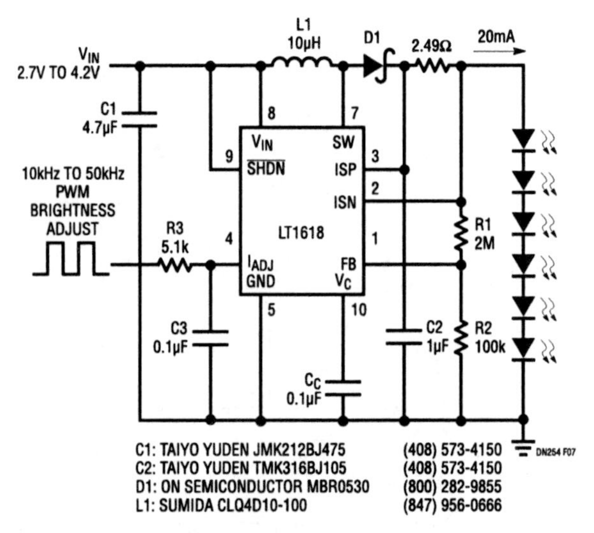

Figure 129.7 • Li-Ion White LED Driver

3MHz micropower synchronous boost converters deliver 3W from two cells in a tiny MSOP package

130

Mark Jordan

Portable electronic devices operating from one or more alkaline or nickel-metal-hydride cells require a boosted supply that is both small and efficient. Linear Technology's new LTC3401 and LTC3402 are high frequency micropower synchronous boost converters that operate from an input voltage below 1V and deliver up to 1A of output current from two cells. DC/DC efficiency can be as high as 97%. The operating frequency, output voltage and Burst Mode operation are all user programmable, allowing these products to fit in various applications where size considerations must be balanced against efficiency. Furthermore, all of this functionality is packed into a small, thermally enhanced MSOP-10 package.

The LTC3401 is optimized for applications requiring less than 1.5W of total output power from two alkaline cells, whereas the LTC3402 is optimized for applications requiring 3W or less from two cells. High efficiency is achieved through internal features such as lossless current sensing, low gate charge, low $R_{DS(ON)}$ synchronous power switches (0.16Ω NMOS, 0.18Ω PMOS) and fast switching transitions to minimize power loss. An external Schottky diode is not required, but may be used to maximize efficiency.

The current mode control architecture, along with OPTI-LOOP compensation and adaptive slope compensation, allows the transient response to be optimized over a wide range of loads, input voltages and output capacitors. The IC remains in fixed-frequency mode until the user forces the IC to enter Burst Mode operation by driving the MODE/SYNC pin high. The IC consumes only 38μA of quiescent current in Burst Mode operation, maximizing efficiency at light loads. The part can also be commanded to shut down by pulling the $\overline{\text{SHDN}}$ pin low when the part draws less than 1μA of quiescent current. The PGOOD pin provides an open-drain output flag that pulls low when the output voltage is more than 9% below the regulation voltage.

All-ceramic-capacitor, 2-cell to 3.3V, 1A converter

A 3W, 2-cell alkaline application using the LTC3402 is shown in Figure 130.1. The operating frequency is set at 1MHz for this application, providing a good compromise between size and efficiency. The area of the converter is less than $0.25in^2$. The efficiency of the circuit of Figure 130.1 is shown in Figure 130.2. The efficiency peaks at 96% at 300mW output power and is greater than 85% up to 3W. The Burst Mode efficiency is 80% with a 500μA load, making it ideal for applications that power down for extended periods of time.

In many applications, the output filter capacitance can be reduced for the desired transient response by having the device commanding the change in load current (i.e., system microcontroller) inform the power converter of the changes as they occur. Specifically, a "load-feed-forward" signal coupled into the V_C pin gives the inner current loop a head start in providing the change in output current. The transconduct-

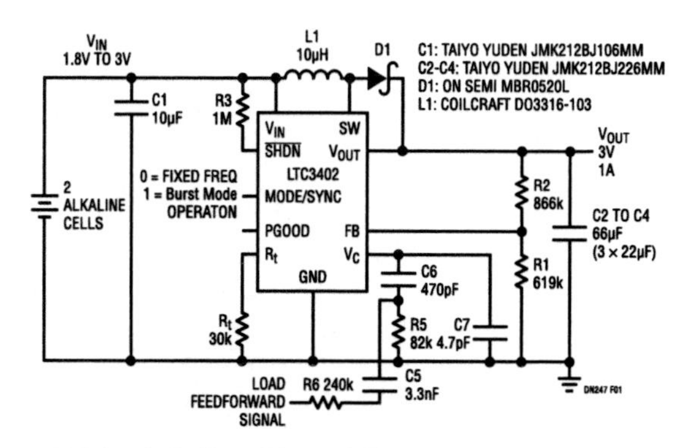

Figure 130.1 • 2-Cell to 3V at 1A Boost Converter Utilizing the LTC3402

Analog Circuit Design: Design Note Collection. http://dx.doi.org/10.1016/B978-0-12-800001-4.00130-7

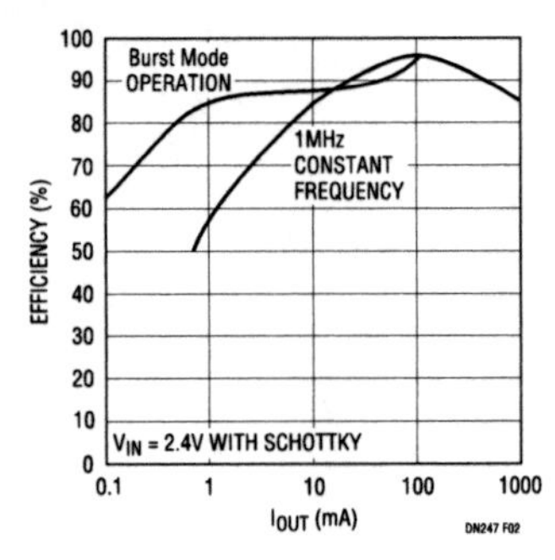

Figure 130.2 • Efficiency of the Circuit in Figure 130.1

ance of the LTC3402 converter at the V_C pin with respect to the inductor current is typically 170mA/100mV, so the amount of signal injected is proportional to the anticipated change of inductor current with load. The outer voltage loop performs the remainder of the correction, but because of the load feedforward signal, the range over which it must slew is greatly reduced. This results in an improved transient response. The load transient response of the circuit in Figure 130.1 for a 100mA to 1A load step is shown in Figure 130.3. A logic-level feed-forward signal is coupled through components C5 and R6. The peak-to-peak output voltage ripple for just 66μF of output capacitance is 330mV. To achieve this ripple without the load feedforward signal, the output capacitance would have to increase to over 150μF.

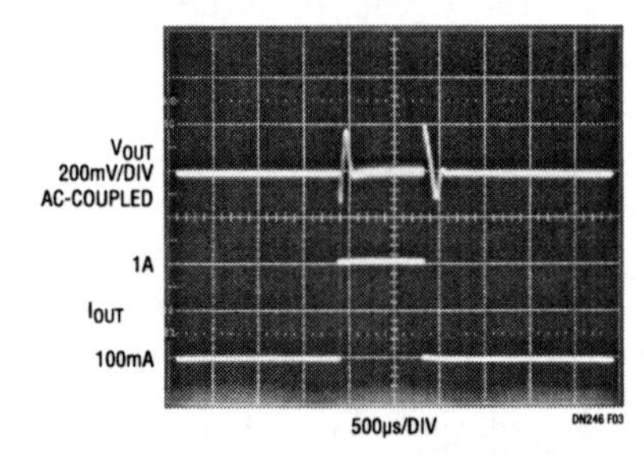

Figure 130.3 • Transient Response of DC/DC Converter with 100mA to 1A Load Step

High efficiency Li-Ion CCFL backlight application

Small portable applications with CCFL backlights, such as PDAs, require a highly efficient backlight converter solution to maximize operating time before recharging. A high efficiency Li-Ion CCFL supply is shown in Figure 130.4. The LTC3401 is set up as a current regulator rather than a voltage regulator and provides the tail current to the self-oscillating resonant Royer circuit that generates the high voltage sinusoidal wave to the lamp. The lamp dimming is provided by means of a control voltage, but alternate dimming techniques can be used.

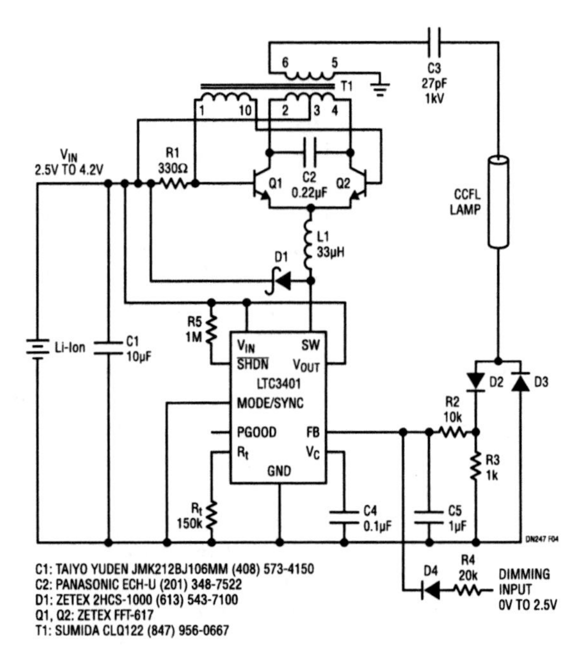

Figure 130.4 • High Efficiency, Compact CCFL Supply with Remote Dimming

SOT-23 switching regulator with integrated 1A switch delivers high current outputs in a small footprint

131

Albert Wu

Linear Technology's LT1930 is the industry's highest power SOT-23 switching regulator. The device contains an internal 1A, 36V switch and is pin compatible with both the low power LT1613 and the micropower LT1615. The LT1930 provides a simple upgrade path for users of these parts who need more power for new designs. In addition to portable applications requiring higher output currents, the device also fits well in nonbattery-operated equipment. Multiple output power supplies can now use a separate regulator for each output voltage, replacing cumbersome quasi-regulated approaches using a single regulator and custom transformers. The LT1930 utilizes a constant frequency, internally compensated, current mode PWM architecture that results in low, predictable output noise that is easy to filter. Its 1.2MHz switching frequency allows the use of tiny, low cost capacitors and low profile inductors. With an input voltage range of 2.6V to 16V, the LT1930 is a good fit for a variety of applications.

5V local supply

Figure 131.1 shows a typical 3.3V to 5V boost converter using the LT1930. The circuit can provide an impressive output current of 480mA while occupying less than 0.3″ by 0.35″ of board area (less than 0.105in^2). The efficiency, shown in Figure 131.2, remains above 83% over a wide load current range of 60mA to 450mA, reaching 86% at 200mA. The maximum output voltage ripple of this circuit is 40mV$_{P-P}$, which corresponds to less than 1% of the nominal 5V output. Figure 131.3 is an oscillograph of the transient response. The lower waveform depicts a load step from 200mA to 300mA, the middle waveform shows the inductor current and the upper waveform shows the output voltage. The output voltage remains within 1% of the nominal value during the transient steps and displays a nice damped response with little ringing.

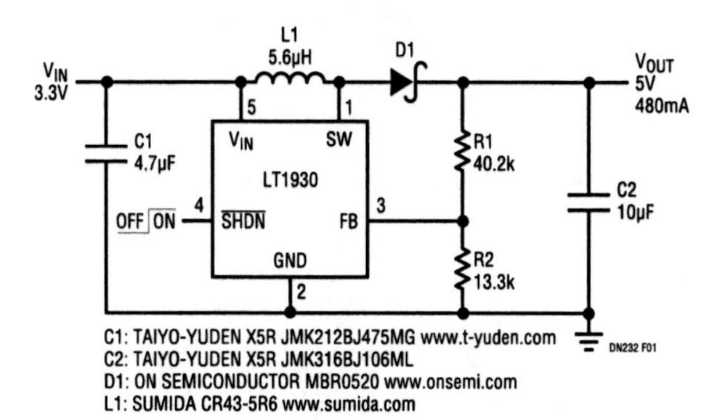

Figure 131.1 • 3.3V to 5V Boost Converter

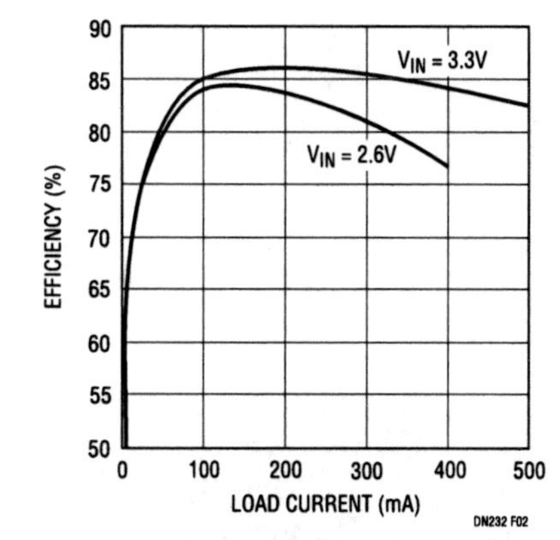

Figure 131.2 • 3.3V to 5V Boost Converter Efficiency

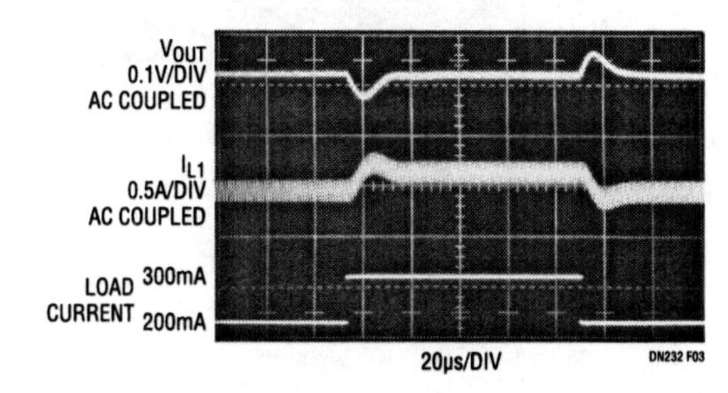

Figure 131.3 • 3.3V to 5V Boost Converter Transient Response

Analog Circuit Design: Design Note Collection. http://dx.doi.org/10.1016/B978-0-12-800001-4.00131-9

12V local supply

Another typical application is a 5V to 12V boost converter as shown in Figure 131.4. This circuit can provide 300mA of output current and achieves efficiencies of up to 87% as shown in Figure 131.5. The maximum output voltage ripple of this circuit is $60mV_{P-P}$, which corresponds to 0.05% of the nominal 12V output. As seen in Figure 131.6, the output voltage remains within 1% of the nominal value during a 50mA load step.

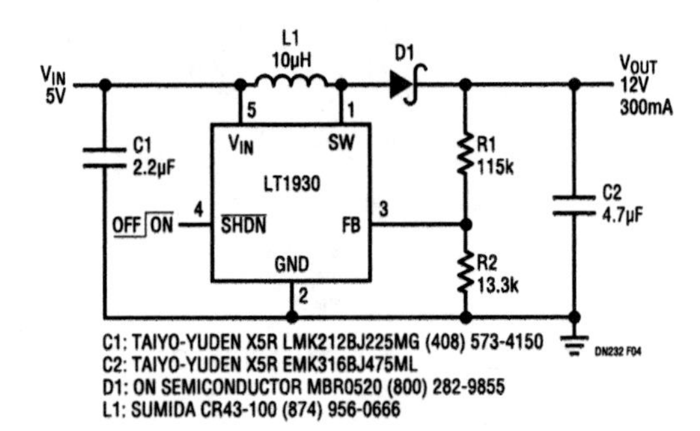

Figure 131.4 • 5V to 12V Boost Converter

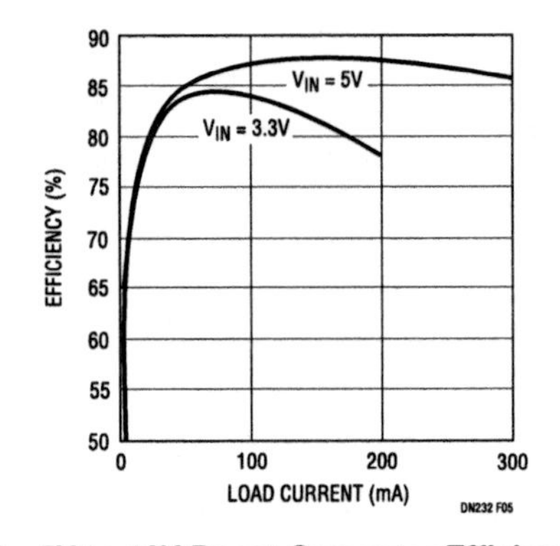

Figure 131.5 • 5V to 12V Boost Converter Efficiency

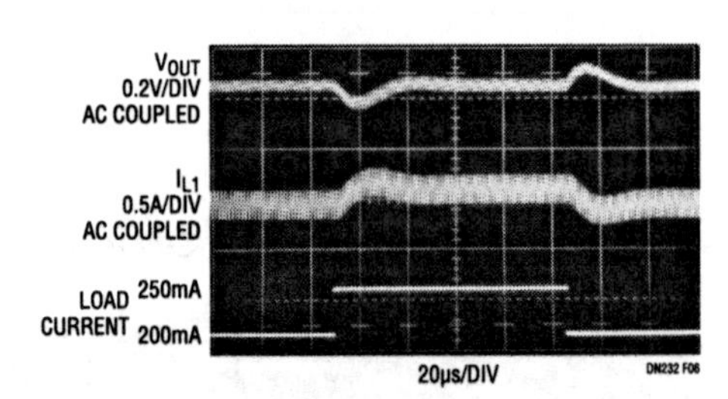

Figure 131.6 • 5V to 12V Boost Converter Transient Response

±15V dual output converter with output disconnect

A ±15V dual output converter using the LT1930 is shown in Figure 131.7. Both outputs are developed using charge pumps, so both are disconnected from the input when the LT1930 is turned off. Since the supplies are generated in the same manner, this circuit features excellent cross-regulation. For a 5× difference in output currents, the positive and negative output voltages differ less than 1%; for a 10× difference, they differ less than 2%. Both outputs of this circuit can each supply up to 70mA of current. The efficiency plot for this circuit is shown in Figure 131.8.

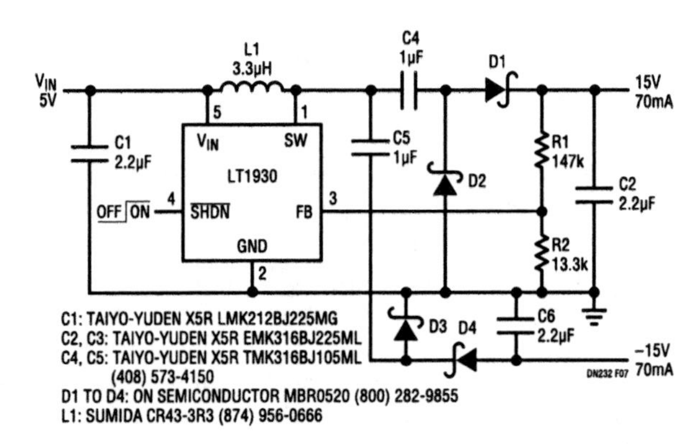

Figure 131.7 • ±15V Dual Output Converter with Output Disconnect

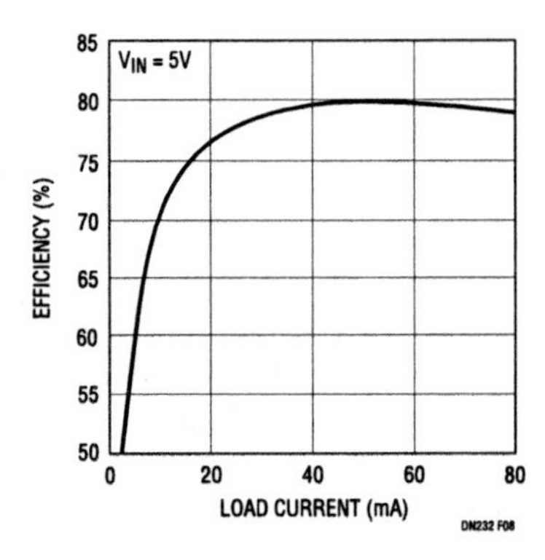

Figure 131.8 • ±15V Dual Output Converter Efficiency

A 500kHz, 6A monolithic boost converter

132

Karl Edwards

Complementing and expanding on the current LT1371/LT1372 family of 500kHz switchers, Linear Technology introduces the LT1370, a 6A boost converter. A high efficiency switch is included on the die, along with all the oscillator, control and protection circuitry necessary for a complete switching regulator. This part combines the convenience and low parts count of a monolithic solution with the switching capabilities of a discrete power device and controller. At 0.065Ω on-resistance, 42V maximum switch voltage and 500kHz switching frequency, the LT1370 can be used in a wide range of output voltage and current applications. Only a few surface mount components are needed to complete a small, high efficiency DC/DC converter. LT1370 features include current mode operation, external synchronization and low current shutdown mode (12μA typical).

Circuit description

The LT1370 is a current mode switcher. This means that switch duty cycle is directly controlled by the switch current rather than by the output voltage. This technique has several advantages: immediate response to input voltage variations, greatly simplified closed-loop frequency compensation, and pulse-by-pulse current limiting, which provides maximum switch protection. An internal low dropout regulator provides a 2.3V supply to all control circuitry. This low dropout design allows the input voltage to vary from 2.7V to 30V with virtually no change in device performance. An internal 500kHz oscillator is the basic clock for all timing. A bandgap provides the reference for the feedback error amplifier.

As with the LT1371, error amplifier circuitry allows the LT1370 to directly regulate negative output voltages. The NFB pin regulates at −2.48V, while the amplifier's output internally drives the FB pin to 1.245V. The error amplifier is a current output (g_m) type, so its output voltage, present on the V_C pin, can be externally clamped to lower the current limit. A capacitor-coupled external clamp provides soft start.

The S/S pin has two functions: synchronization and shutdown. The internal oscillator can be synchronized to a higher frequency by applying a TTL square wave to this pin. This allows the part to be synchronized to a system clock. If the S/S pin is held low, the LT1370 will enter shutdown mode. In this mode, all internal circuitry is disabled, reducing supply current to 12μA. An internal pull-up ensures start-up when the S/S pin is left open circuit.

5V to 12V boost converter

Figure 132.1 shows a typical 5V to 12V boost application. The high 6A switch rating permits the circuit to deliver up to 24W. Figure 132.2 shows the overall converter efficiency. Notice that peak efficiency is 90%; efficiency stays above 86% at the circuit's maximum 2A output current. The inductor

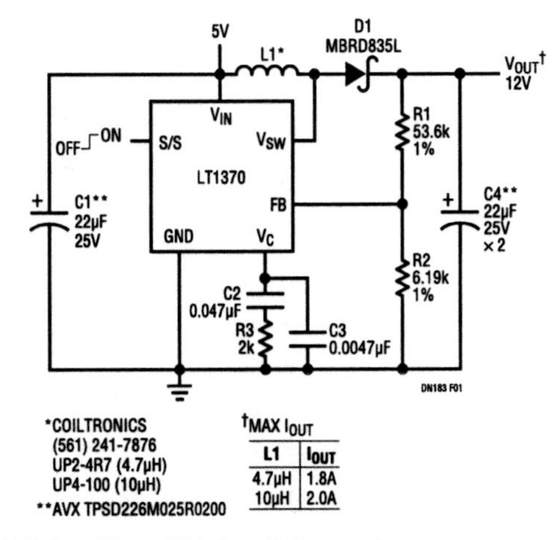

Figure 132.1 • 5V to 12V Boost Converter

Analog Circuit Design: Design Note Collection. http://dx.doi.org/10.1016/B978-0-12-800001-4.00132-0

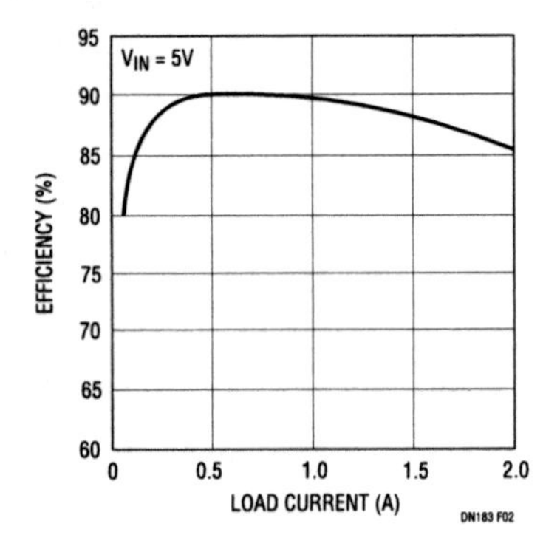

Figure 132.2 • 12V Output Efficiency

needs to be chosen carefully to meet peak current values. The output capacitor can see high ripple currents—often, as in this application, higher than the ripple rating of a single capacitor. This requires the use of two surface mount tantalums in parallel; both capacitors should be of the same value and manufacturer. The input capacitor does not have to endure such high ripple currents and a single capacitor will normally suffice. The catch diode, D1, must be rated for the output voltage and average output current. The compensation capacitor, C2, normally forms a pole in the 2Hz to 20Hz range, with a series resistor, R3, to add a zero at 1kHz to 5kHz. The S/S pin in this example is driven by a logical on/off signal, a low input forcing the LT1370 into its 12μA shutdown mode.

Positive to negative converter

The negative feedback (NFB) pin, enables negative output regulators to be designed with direct feedback. In the circuit shown in Figure 132.3, a 2.7V to 13V input, −5V output converter, the output is monitored by the NFB pin and a simple divider network. No complex level shifting or unusual grounding techniques are required. The S/S pin is used to synchronize the switching frequency to a 600kHz external clock signal.

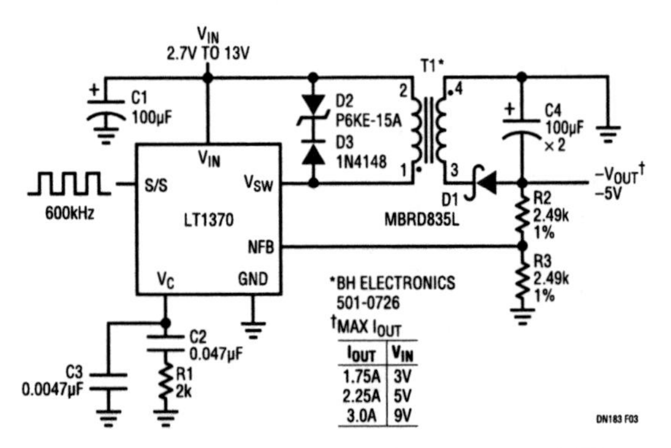

Figure 132.3 • Positive to Negative Converter with Direct Feedback

The switch clamp diodes, D2 and D3, prevent the leakage spike from the transformer, T1, from exceeding the switch's absolute maximum voltage rating. The Zener voltage of D2 must be higher than the output voltage, but low enough that the sum of input voltage and clamp voltage does not exceed the switch-voltage rating.

5V SEPIC converter

Figure 132.4 is an example of a SEPIC converter. The SEPIC topology has the advantage of an input voltage range that extends both above and below the output voltage. In Figure 132.4, the batteries can be at a charge level from 9V to below 4V while maintaining a fixed 5V output. Also, there is no direct path from input to output. When the S/S pin is grounded, forcing the LT1370 into shutdown, there is no leakage into the output. In shutdown, battery current is reduced to 12μA, the input current of the LT1370. The magnetic coupling of inductors L1A and L1B is not critical for operation, but generally they are wound on the same core. C2 couples the inductors together and eliminates the need for a switch snubber network.

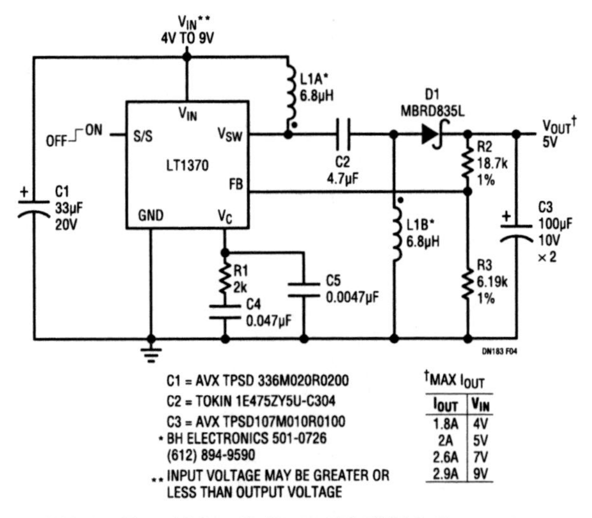

Figure 132.4 • Two Li-Ion Cells to 5V SEPIC Converter

Conclusion

With its low resistance switch, 6A operating current and 500kHz operation, the LT1370 is ideal for small, low parts count, high current applications. Its high switching frequency removes the need for large bulky magnetics and capacitors. Compared to a separate control device and power switch, the LT1370's monolithic approach simplifies the design effort, allows operation at lower input voltages and reduces the board space required to implement a complete DC/DC converter.

Micropower 600kHz step-up DC/DC converter delivers 5V at 1A from a Li-Ion cell

133

Steve Pietkiewicz

Linear Technology introduces a new micropower DC/DC converter designed to provide high output power from a single cell or higher input voltage. The LT1308 features an onboard switch capable of handling 2A with a voltage drop of 300mV and operates from an input voltage as low as 1V. The LT1308 features Burst Mode operation at light load; efficiency is 75% or better for load currents of 1mA. The device switches at 600kHz; this high frequency keeps associated power components small and flat; additionally, troublesome interference problems in the sensitive 455kHz IF band are avoided. The LT1308 is intended for generating power on the order of 2W to 5W. This is sufficient for RF power amplifiers in GSM terminals or for digital camera power supplies. The LT1308 is available in the 8-lead SO package.

Single Li-Ion cell to 5V/1A DC/DC converter for GSM

GSM terminals have emerged as a worldwide standard. A common requirement for these products is an efficient, compact, step-up converter that develops 5V from a single Li-Ion cell to power the RF amplifier. The LT1308 performs this function with a minimum of external components. The circuit is detailed in Figure 133.1. Many designs use a large aluminum electrolytic capacitor (1000µF to 3300µF) at the DC/DC converter output to sustain the output voltage during the transmit time slice, since the amplifier can require more than 1A. The output capacitor, along with the LT1308

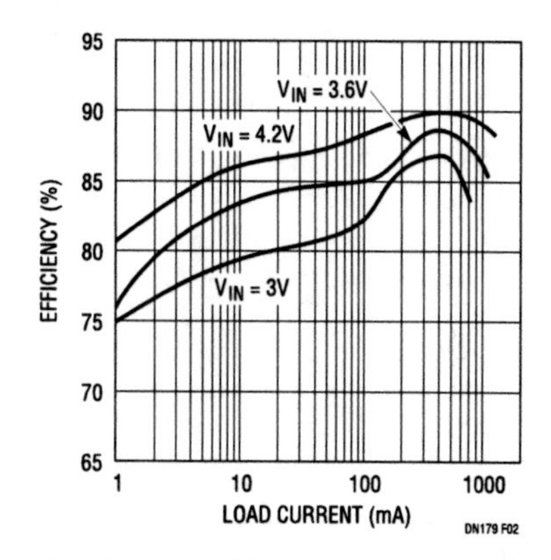

Figure 133.2 • Efficiency of Figure 133.1's Circuit Reaches 90%

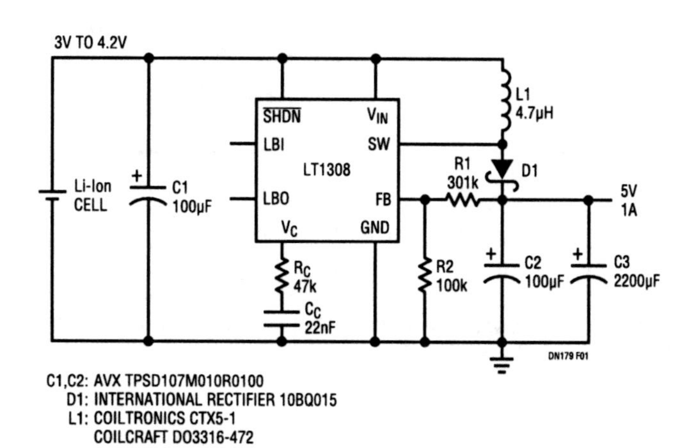

Figure 133.1 • Single Li-Ion Cell to 5V/1A DC/DC Converter

Analog Circuit Design: Design Note Collection. http://dx.doi.org/10.1016/B978-0-12-800001-4.00133-2

compensation network, serves to smooth out the input current demanded from the Li-Ion cell. Efficiency, which reaches 90%, is shown in Figure 133.2. Transient response of a 0A to 1A load step with typical GSM profiling (1:8 duty cycle, 577μs pulse duration) is depicted in Figure 133.3. Voltage droop (top trace) is 200mV. Inductor current (bottom trace) increases to 1.7A peak; the input capacitor supplies some of this current, with the remainder drawn from the Li-Ion cell.

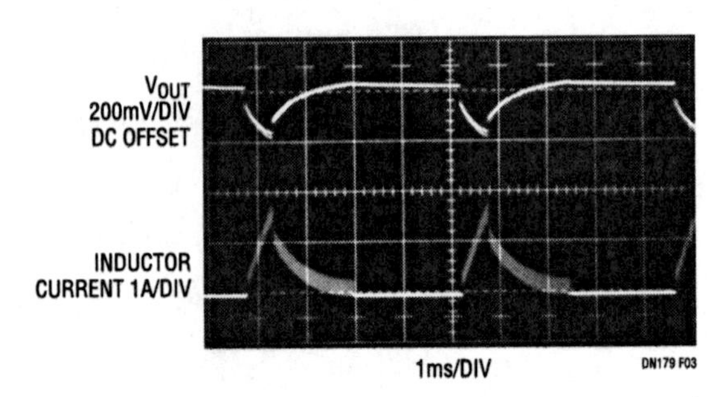

Figure 133.3 • Transient Response of DC/DC Converter: V_{IN} = 3V, 0A to 1A Load Step

2-cell digital camera supply produces 3.3V, 5V, 18V and −10V

Power supplies for digital cameras must be small and efficient while generating several voltages. The DSP and logic need 3.3V, the ADC and LCD display need 5V and biasing for the CCD element requires 18V and −10V. The power supplies must also be free of low frequency noise, so that post filtering can be done easily. The obvious approach, to use a separate DC/DC converter IC for each output voltage, is not cost-effective.

A single LT1308, along with an inexpensive transformer, generates 3.3V/200mA, 5V/200mA, 18V/10mA and −10V/10mA from a pair of AA or AAA cells. Figure 133.4 shows the circuit. A coupled-flyback scheme is used, actually an extension of the SEPIC (single ended primary inductance converter) topology. The addition of capacitor C6 clamps the SW pin, eliminating a snubber network. Both the 3.3V and 5V outputs are fed back to the LT1308 FB pin, a technique known as split feedback. This compromise results in better overall line and load regulation. The 5V output has more influence than the 3.3V output, as can be seen from the relative values of R2 and R3. Transformer T1 is available from Coiltronics, Inc. (561-241-7876). Efficiency vs input voltage for several load currents on both 3.3V and 5V outputs is pictured in Figure 133.5. The CCD bias voltages are loaded with 10mA in all cases.

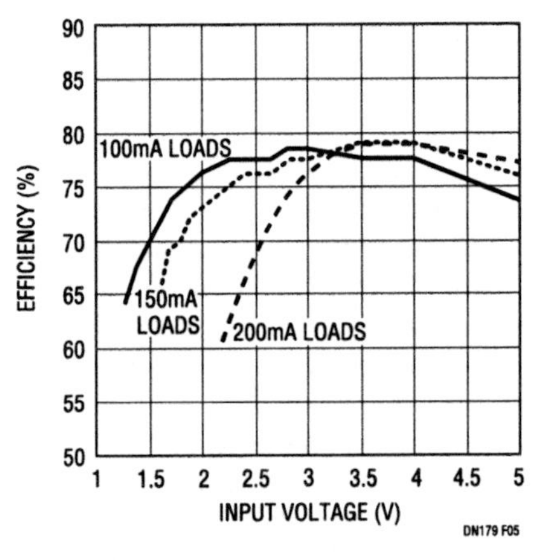

Figure 133.5 • Efficiency vs Input Voltage for 100mA, 150mA and 200mA Loads on 3.3V and 5V Outputs

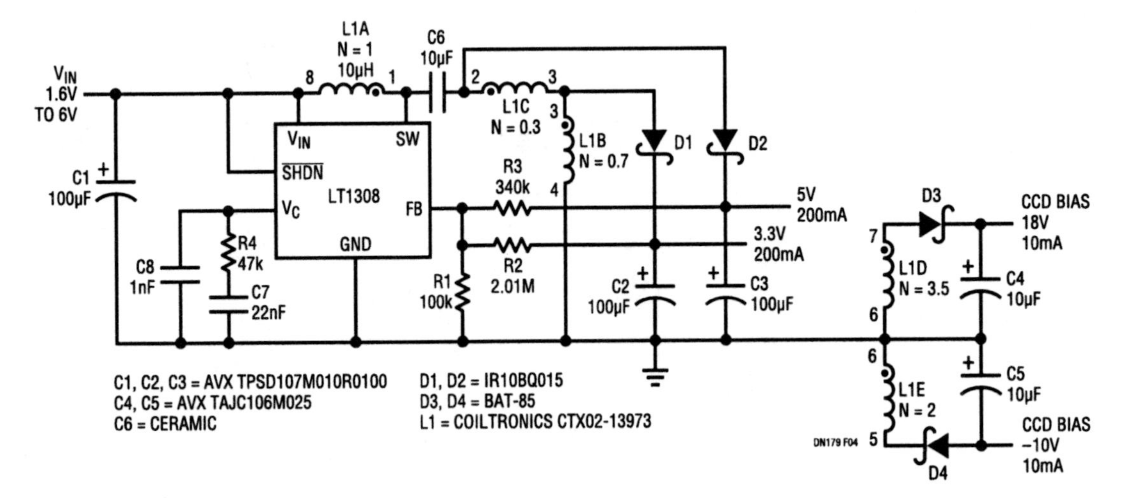

Figure 133.4 • This Digital-Camera Power Supply Delivers 5V/200mA, 3.3V/200mA, 18V/10mA and −10V/10mA from Two AA Cells

Ultralow noise switching regulator controls EMI

134

Jeff Witt

Today's circuit designer is often challenged to assemble a high performance system by combining sensitive analog electronics with potentially noisy DC/DC converters. Requirements for a small, efficient, cost-effective solution are in conflict with acceptable noise performance—noisy switching regulators call for filtering, shielding and layout revisions that add bulk and expense. Most electromagnetic interference (EMI) problems associated with DC/DC converters are due to high speed switching of large currents and voltages. To maintain high efficiency, these switch transitions are designed to occur as quickly as possible. The result is input and output ripple that contains very high harmonics of the switching frequency. These fast edges also couple through stray magnetic and electric fields to nearby signal lines, making efforts to filter the supply lines ineffective.

The LT1534 ultralow noise switching regulator provides an effective and flexible solution to this problem. Using two external resistors, the user can program the slew rates of the current through the internal 2A power switch and the voltage on it. Noise performance can be evaluated and improved with the circuit operating in the final system. The system designer need sacrifice only as much efficiency as is necessary to meet the required noise performance. With the controlled slew rates, system performance is less sensitive to layout, and shielding requirements can be greatly reduced; expensive layout and mechanical revisions can be avoided.

The LT1534's internal oscillator can be programmed over a broad frequency range (20kHz to 250kHz) with good initial accuracy. It can also be synchronized to an external signal placing the switching frequency and its harmonics away from sensitive system frequencies.

Low noise boost regulator

In Figure 134.1, the LT1534 boosts 3.3V to supply 650mA at 5V with its oscillator synchronized to an external 50kHz clock. The circuit relies on the low ESR of capacitor C2 to keep the output ripple low at the fundamental frequency; slew rate control reduces the high frequency ripple. Figure 134.2 shows waveforms of the circuit as it delivers 500mA. The top trace shows the voltage on the collector of the internal bipolar power switch (the COL pins), and the middle trace shows the switch current. The lowest trace is the output ripple. The slew rates are programmed to their fastest here, resulting in good efficiency (83%), but also generating excessive high frequency ripple. Figure 134.3 shows the same waveforms with the slew rates reduced. The large high frequency transients have been eliminated.

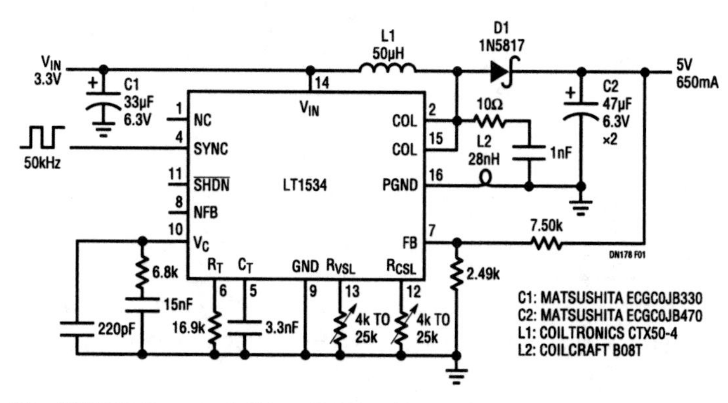

Figure 134.1 • The LT1534 Boosts 3.3V to 5V. The Resistors on the R_{VSL} and R_{CSL} Pins Program the Slew Rates of the Voltage on the Power Switch (COL Pins) and the Current Through It

Analog Circuit Design: Design Note Collection. http://dx.doi.org/10.1016/B978-0-12-800001-4.00134-4

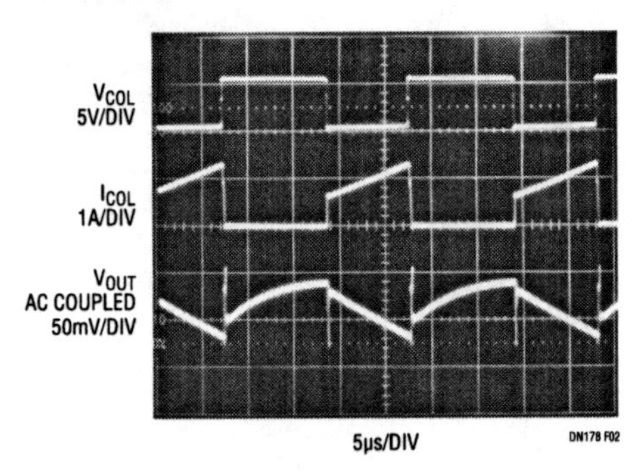

Figure 134.2 • High Slew Rates ($R_{CSL}=R_{VSL}=4k$) Result in Good Efficiency but Excess High Frequency Ripple

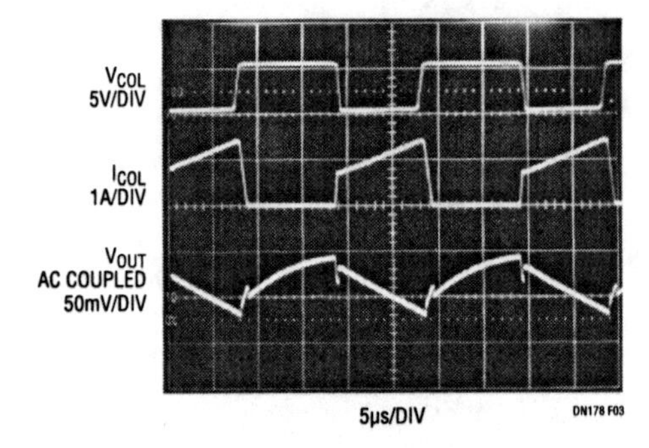

Figure 134.3 • Low Slew Rates ($R_{CSL}=R_{VSL}=24k$) Result in an Output without Troublesome High Frequency Transients

Low noise bipolar supply

Many high performance analog systems require quiet bipolar supplies. This circuit (Figure 134.4) will generate ±5V from a wide input range of 3V to 12V, with a total output power of 1.5W. By using a 1:1:1 transformer, the primary and secondary windings can be coupled using capacitors C2 and C3, allowing the LT1534 to control the switch transitions at the output rectifiers as well as at the switch collector. Secondary damping networks are not required.

Additional LT1534 features

The LT1534 is a complete, low noise switching regulator with an internal 2A power switch, packaged in a 16-lead narrow plastic SO. The current mode architecture provides fast transient response and cycle-by-cycle current limit. Undervoltage lockout and thermal shutdown provide further protection. The large input range (2.7V to 23V) and high switch voltage (25V), combined with a 12μA shutdown mode, result in a very flexible part suitable for battery-powered operation. The LT1534 can directly regulate either positive or negative output voltages.

The LT1533, closely related to the LT1534, provides two slew rate-controlled 1A power switches. Optimized for push-pull topologies, the LT1533 provides even greater opportunity for reducing DC/DC converter noise. For further applications, consult the LT1533 and LT1534 data sheets and Application Note 70.

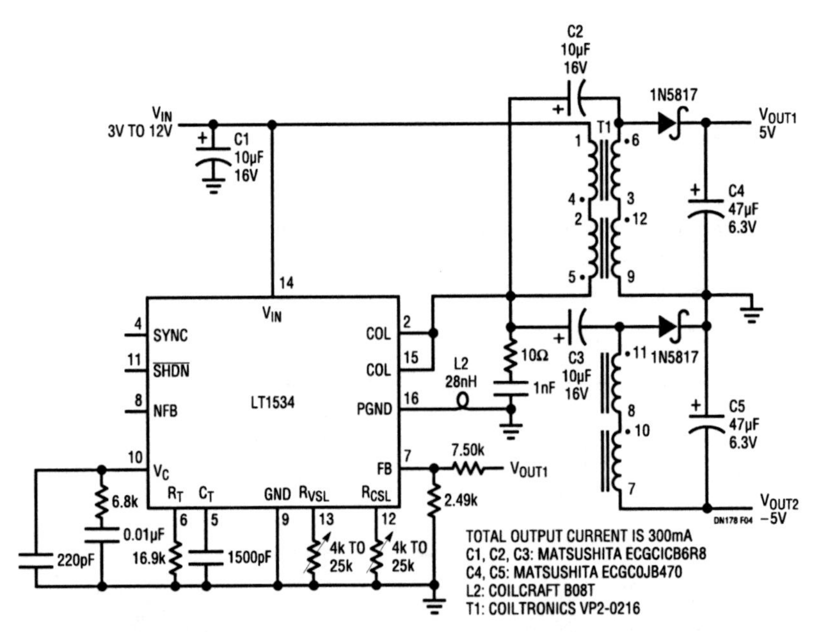

Figure 134.4 • A Low Noise, Wide Input Range ±5V Supply

Off-line low noise power supply does not require filtering to meet FCC emission requirements

135

Jim Williams

Introduction

Off-line power supplies require input filtering components to meet FCC emission requirements. Additionally, board layout is usually quite critical, requiring considerable experimentation even for experienced off-line supply designers. These considerations derive from the wideband harmonic energy generated by the fast switching of traditional off-line supplies. A new device, the LT1533 low noise switching regulator, eliminates these issues by continuous, closed-loop control of voltage and current

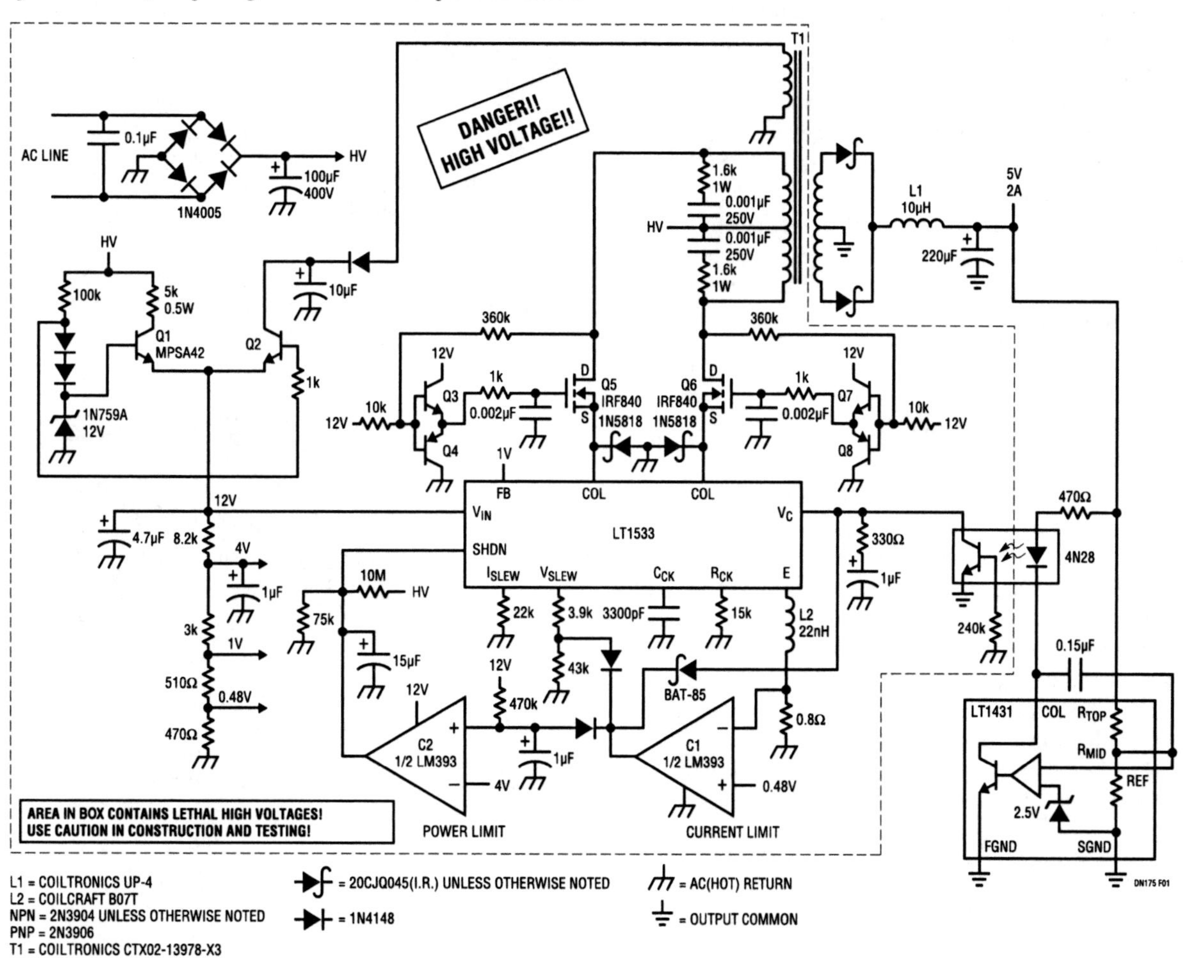

Figure 135.1 • 10W Off-Line Power Supply Passes FCC Emission Requirements without Filter Components

Analog Circuit Design: Design Note Collection. http://dx.doi.org/10.1016/B978-0-12-800001-4.00135-6

switching times.[1] Additionally, the device's push-pull output drive eliminates the flyback interval of conventional approaches. This further reduces harmonics and smooths input current drain characteristics. Although intended for DC/DC conversion, the LT1533 adapts nicely to off-line service, while eliminating emission, filtering, layout and noise concerns.

Circuitry details

Figure 135.1 shows the supply. Q5 and Q6 drive T1, with a rectifier filter, the LT1431 and the optocoupler closing an isolated loop back to the LT1533. The LT1533 drives Q5 and Q6 in cascode fashion to achieve high voltage switching capability. It also continuously controls their current and voltage switching times, using the resistors at the I_{SLEW} and V_{SLEW} pins to set transition rates. FET current information is directly available, although FET voltage status is derived via the 360k–10k dividers and routed to the gates via the NPN-PNP followers. The source wave shapes, and hence the voltage slewing information at the LT1533 collector terminals, are nearly identical in shape to the drain waveforms.

Q1, Q2 and associated components provide a bootstrapped bias supply, with start-up transistor Q1 turning off once T1 begins supplying power to Q2. The resistor string at Q2's emitter furnishes various "housekeeping" bias potentials. The LT1533's internal 1A current limit is too high for effective overcurrent protection. Instead, current is sensed via the 0.8Ω shunt at the LT1533's emitter pin (E). C1, monitoring this point, goes low when current limit is exceeded. This pulls the V_C pin low and also accelerates voltage slew rate, resulting in fast limiting while minimizing instantaneous FET stress. Prolonged short-circuit conditions result in C2 going low, putting the circuit into shutdown. Once this occurs, the C1–C2 loop oscillates in a controlled manner, sampling current for about a millisecond every second or so. This action forms a power limit, preventing FET heating and eliminating heat sink requirements.

Performance characteristics

Figure 135.2 shows waveforms for the power supply. Trace A is one FET source; traces B and C are its gate and drain waveforms, respectively. FET current is trace D. The cascoded drive maintains waveshape fidelity, even as the LT1533 tightly regulates voltage and current transition rates. The wideband harmonic activity typical of off-line supply waveforms is entirely absent. Power delivery to T1 (center screen, trace C) is particularly noteworthy. The waveshapes are smoothly controlled, and no high frequency content is observable.

Figure 135.3, a 30MHz wide spectral plot, shows circuit emissions well below FCC requirements. This data was taken with no input filtering LC components and a nominally non-optimal layout.

Output noise is composed of fundamental ripple residue, with essentially no wideband components. Typically, the low frequency ripple is below 50mV. If additional ripple attenuation is desired a 100μH–100μF LC section permits <100μV output noise. Figure 135.4 shows this in a 100MHz bandpass. Ripple and noise are so low that the oscilloscope requires a 40dB low noise preamplifier to even register a display (see footnote 1).

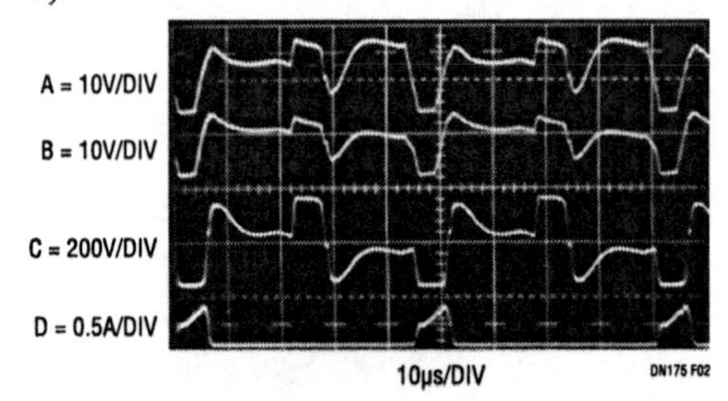

Figure 135.2 • Waveforms for One of the Power Supplies' FETs Show No Wideband Harmonic Activity. LT1533 Provides Continuous Control of Voltage and Current Slewing. Result Is Smoothly Controlled Waveshapes for FET Source (A), Gate (B) and Drain (C). FET Current is Trace D

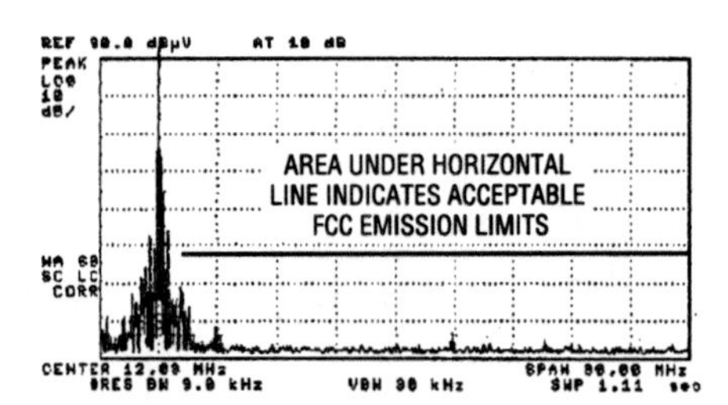

Figure 135.3 • 30MHz Wide Spectral Plot Shows Circuit Emissions Well Below FCC Requirements Despite Lack of Traditional Filter Components

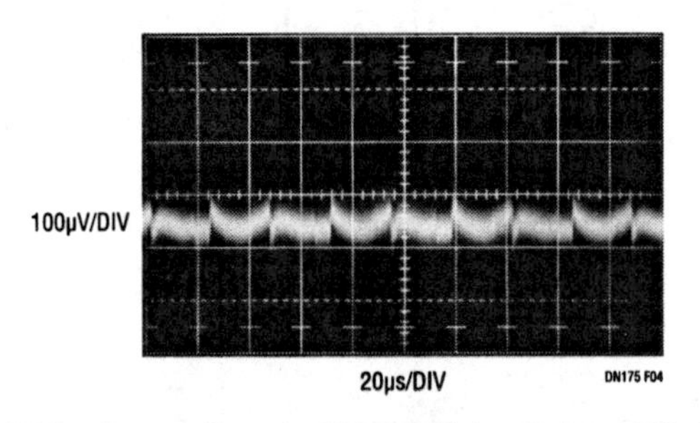

Figure 135.4 • Power Supply Output Noise Below 100μV (100MHz Measurement Bandwidth) Is Obtainable Using Additional Output LC Section. Without LC Section Wideband Harmonic Is Still Absent, Although Fundamental Ripple Is 50mV

Note 1: In depth coverage of this device, its use and performance verification appears in LTC Application Note 70, "A Monolithic Switching Regulator with 100μV Output Noise," by Jim Williams.

"LCD bias" and "backup supply" applications for a micropower DC/DC converter

136

Gary Shockey Jeff Witt

Some step-up DC/DC converter functions require input current limiting because of high source impedance or limited capability of power components. The LT1316, a micropower step-up DC/DC converter with peak switch current control, meets these needs. The device draws 33μA quiescent current and contains a 0.6Ω, 30V switch that can be programmed for a maximum peak current between 30mA and 600mA with an external resistor. It also has a low-battery detector that remains active in shutdown, where quiescent current drops to 3μA. The two circuit examples below illustrate how the LT1316's current limit function allows realization of difficult converter circuits.

2-cell, low profile LCD bias generator fits in small places

Portable electronic products with LCDs are getting thinner, resulting in severe restrictions on component height. LCD bias generators placed in or near the display housing need to use low profile (under 2mm) components to meet height restrictions. These low profile inductors and capacitors have somewhat higher parasitic resistance than their higher profile equivalents; hence, switching regulator peak current must be controlled to keep the inductor from saturating and to keep output voltage ripple under control. The LT1316, with its programmable current limit function, is ideal for use as an LCD bias generator.

Figure 136.1's circuit delivers 5mA at up to 28V from a 2-cell battery, using components that are under 2mm high. Peak current is limited to 350mA by 10k resistor R3 at the R_{SET} pin. The parallel combination of a 1μF, 35V tantalum and a 0.47μF, 50V ceramic keep output ripple voltage to 180mV, less than 1% of the output voltage. Output voltage and inductor current waveforms at an input voltage of 2V and load current of 4mA are detailed in Figure 136.2. The 28V output can be varied by changing the value of R2 or by summing a current into the LT1316 FB pin.

Higher output current can be generated if a higher input supply voltage is available. Table 136.1 shows output current for supply voltages of 2V, 3.3V and 5V. Up to 20mA at 28V can be generated from a 5V supply. Efficiency using these low profile components is a few points lower than it would be with larger components, but it is still above 74%.

Table 136.1 Output Current for Input Voltages of 2V, 3.3V and 5V

V_{IN}	L	PEAK CURRENT	R_{SET}	OUTPUT CURRENT
2V	22μH	350mA	10k	5mA
3.3V	22μH	550mA	7.5k	15mA
5V	47μH	350mA	10k	20mA

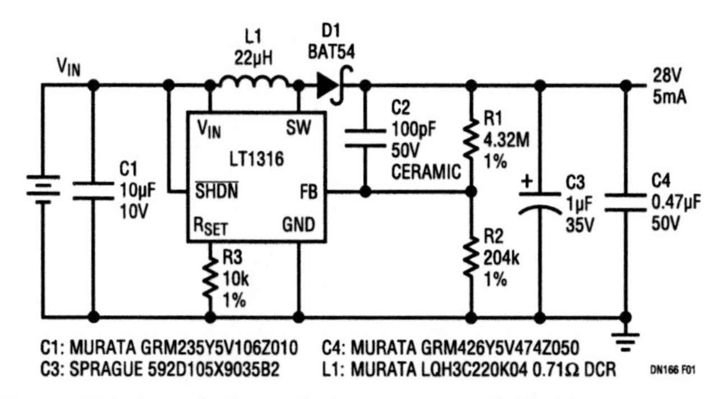

Figure 136.1 • 2-Cell to 28V Converter for LCD Bias Generators Uses Components Under 2mm Tall

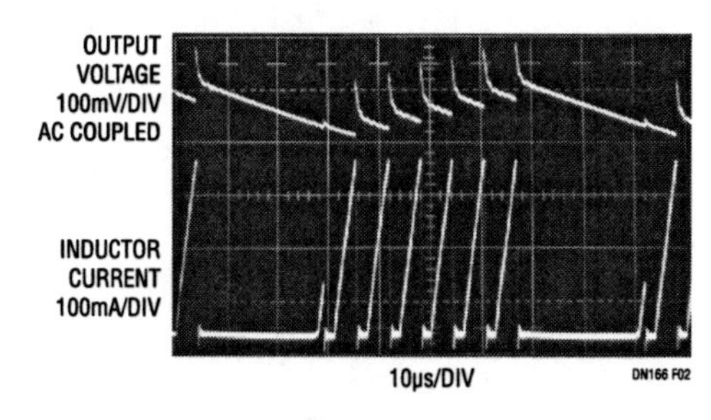

Figure 136.2 • Controlled, Low Peak Current Keeps Output Voltage Ripple Under 180mV$_{P-P}$

Analog Circuit Design: Design Note Collection. http://dx.doi.org/10.1016/B978-0-12-800001-4.00136-8

Supercapacitor-powered backup supply

Typical backup supplies for low power (several μW) logic systems operate from a lithium battery or a high energy density capacitor (a "supercap"). Some systems may require a higher power backup: for example, a "last gasp" write to flash memory might require several mW for several seconds. There are obstacles to efficient operation at higher loads from these power sources. Both the long-life lithium batteries and supercaps have large series resistances that result in reduced efficiency at high RMS currents and poor regulation due to IR drop. In addition, the supercap output voltage, in contrast to a battery's, decreases continuously as power is drawn and the capacitor must be substantially discharged to obtain its stored energy. A micropower switching regulator is required, and the LT1316, with its ability to precisely control peak switch current, is ideally suited to such high impedance energy sources.

Figure 136.3 shows a 5V, 6mA backup supply operating from a 0.1F, 5.5V, 75Ω supercap. The supercap, C_{SUP}, is charged through R1 from a normally present 5V. The charge state is monitored with the LT1316's low-battery detector; the READY line is high when C_{SUP} is near full charge. When a power loss is detected, the system can pull the RUN line high to turn on the backup supply. The LT1316 operates as a simple boost regulator, generating 5V power until C_{SUP} has discharged to 1.5V. R_{SET} programs the peak switch current of the LT1316. Figure 136.4 shows the input and output voltage as the circuit supplies a fixed 6mA load. The output remains regulated at 5V as the input voltage drops. With peak switch current programmed to ~500mA (R_{SET}=5.1k), output regulation is maintained for 9.6 seconds. Also plotted are the results with a peak current of 100mA (R_{SET}=33k), enough switch current to satisfy the 6mA load current at the lowest input voltage. The benefit is obvious; the lower peak current results in lower RMS current from the supercap, reducing losses and extending backup time by 22% to 11.7 seconds.

The accurate control of peak switch current also allows the designer to better match the inductor to the power demands of the application, reducing system size and cost. Figure 136.5 shows the circuit operation under identical operating conditions, with a smaller CD43 series inductor substituted for the larger CD54. At higher peak currents, the additional inductor loss lowers operating time by 5%. With a low peak switch current, there is essentially no penalty for using the small inductor.

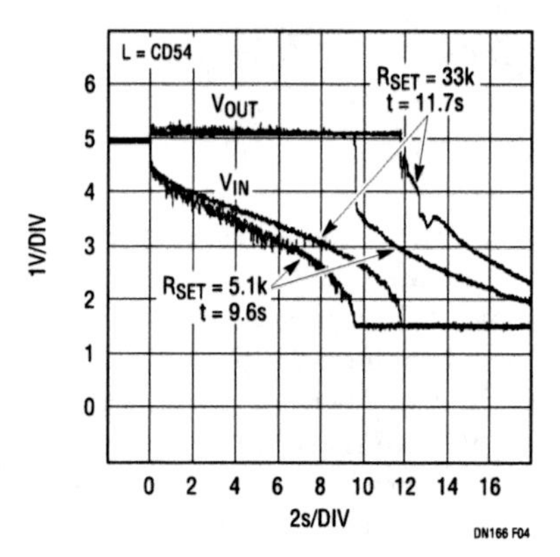

Figure 136.4 • Lower Peak Current Results in Longer Operating Time

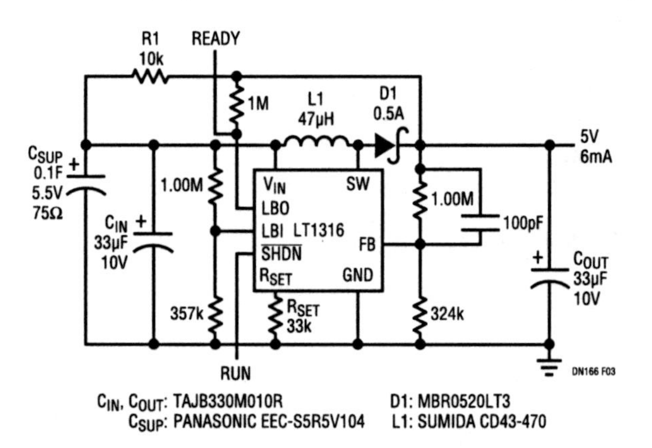

Figure 136.3 • Supercap Backup Supply

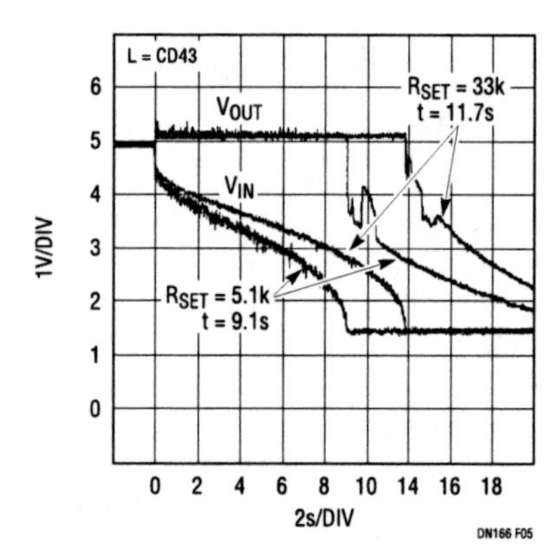

Figure 136.5 • Lower Peak Current Allows the Use of Smaller Inductors

Short-circuit protection for boost regulators

137

Jeff Witt

The basic boost regulator topology provides no short-circuit protection. When the output is pulled low, a large current can flow from the input through the inductor and catch diode, limited only by the series resistance of these parts. The result may be damage to the boost regulator, the load or the power source. This design note presents several solutions to this potential problem.

Short-circuit protection and load disconnect with the LTC1477

The LTC1477 protected high side switch contains an internal, low loss NMOS power switch. It provides protection to external circuits in two ways: switch current is limited to 0.85A, 1.5A or 2A (depending on configuration) and thermal shutdown circuitry turns the switch off when power dissipation raises the die temperature to 130°C. In Figure 137.1, the LTC1477 protects an LT1304 micropower boost regulator and its load from excessive currents. An increasing load current will first be limited to 0.85A; as more power is dissipated in the switch, the LTC1477 will cycle the switch to limit average current and dissipation to an acceptable level. The enable pin can be used to disconnect the output. In this circuit the LT1304 uses its low-battery detector to shut itself down and disconnect the load when the battery voltage drops below 2.7V.

For higher power applications, the V_{IN2} and V_{IN3} pins of the LTC1477 should also be connected to the input. This raises the current limit to 2A; switch resistance will be just 0.07Ω. The LTC1477 operates from 2.7V to 5.5V and is packaged in an SO-8. The LTC1478 is a dual version available in a 16-lead SO.

Current-limited boost regulator

It may be desirable to limit, rather than interrupt, output current to a heavy load. Figure 137.2 shows a 2-cell to 5V boost converter with output current limited to 150mA. The LT1304 generates 5V at the source of Q1, which, with its gate grounded through R1 and R2, turns on to supply current to the output. As the load current increases, the voltage across R_{SENSE} reaches 0.6V, Q2 turns on and current through R2 raises the voltage at the feedback pin; the LT1304 begins to regulate the output current to 150mA. When the output

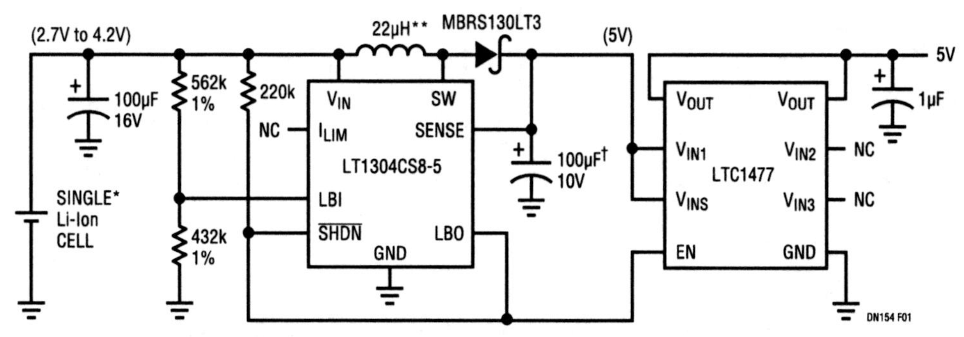

Figure 137.1 • Boost Converter with Short-Circuit Protection and Automatic Load Disconnect

Analog Circuit Design: Design Note Collection. http://dx.doi.org/10.1016/B978-0-12-800001-4.00137-X

voltage is pulled lower than the input voltage, the LT1304 is no longer able to control the output current, and Q2 regulates the current by pulling up on Q1's gate. Note that Q1 must dissipate approximately 0.4W when the output is shorted to ground. An extra transistor (Q3) will disconnect the load when the LT1304 is shut down. Although R_{SENSE} lowers the efficiency near full load, the additional circuitry has essentially no effect on the excellent light load efficiency of the LT1304.

Short-circuit protection at higher power

At higher currents, low $R_{DS(ON)}$ transistors and low voltage current sensing are necessary to maintain high efficiency and manageable power dissipation. The LTC1153 circuit breaker IC drives an external high side N-channel FET and will turn off the FET when the voltage across a current sense resistor exceeds 100mV. Trip delay and reset times can be adjusted with external components.

The LT1270 boost converter can generate 2A at 12V from a 5V input (Figure 137.3). Protection against a shorted output is provided with an LTC1153 programmed to trip at 2.5A. At start-up, the LTC1153 drives the gate of Q1 through a filter (R1 and C1). This limits dV/dt at the output, controlling the inrush current to the capacitive load. When the LTC1153 senses a voltage drop across R_{SENSE} exceeding 100mV, the gate of Q1 is grounded through diode D1. The FET will remain off for a period determined by the capacitor tied to the C_T pin. With this reset period set longer than the trip delay (60μs max) and the turn-on time of Q1 (~R1 C1), the average output current will be much lower than the peak current; this keeps power dissipation of the FET and the load at a safe level. An open collector at the STATUS pin of LTC1153 indicates the state of the circuit breaker. The output can be disconnected by setting the IN pin low. The LTC1153 operates from 4.5V to 18V and is available in an 8-lead PDIP or SO.

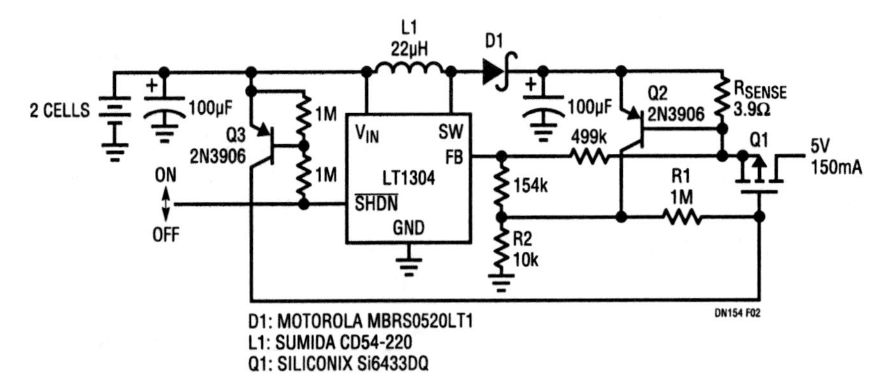

Figure 137.2 • Current Limited Boost Converter with Load Disconnect

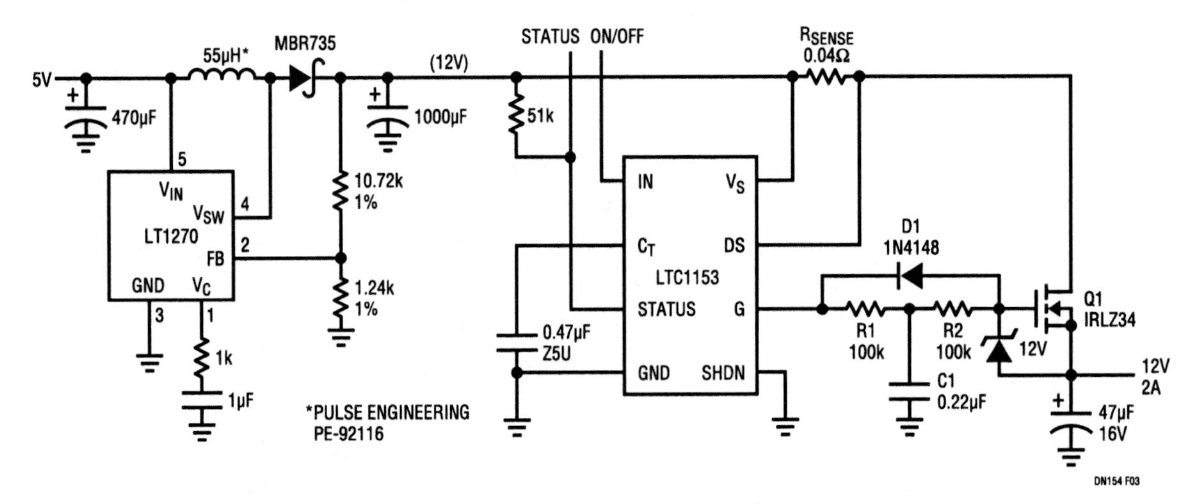

Figure 137.3 • 2A/12V Step-Up Regulator with Circuit Breaker (Post Regulator), Breaker Status Feedback and Ramped Output

Single-cell micropower fixed-frequency DC/DC converter needs no electrolytic capacitors

138

Steve Pietkiewicz

Today's low power boost converter ICs have been rejected by designers of products incorporating RF communications for two reasons. First, the converters use some form of variable-frequency control to maintain acceptable efficiency during periods of light load. Significant spectral energy in the sensitive 455kHz band can occur, introducing difficult interference problems with the system's IF amplifier. Second, large output capacitors are required to keep output ripple voltage at an acceptable level. Most battery-powered products have neither the space nor budget for the D-case size tantalum capacitor usually required. The LT1307 current mode PWM switching regulator eliminates these concerns by using small, low cost ceramic capacitors for both input and output and by employing fixed frequency 575kHz operation to keep spectral energy out of the 455kHz band. Dense high speed bipolar process technology enables the LT1307 to fit in the subminiature MSOP package. The LT1307 consumes just 60μA at no load and includes a low-battery detector comparator with a 200mV reference voltage. The internal power switch is rated at 500mA with a V_{CESAT} of 300mV.

Single-cell boost converter

A complete single cell to 3.3V converter is shown in Figure 138.1. The circuit generates 3.3V at up to 75mA from a 1V input. The 10μF ceramic output capacitor can be obtained from several vendors. Efficiency, detailed in Figure 138.2, exceeds 70% over the 1:500 load range of 200μA to 100mA at a 1.25V input. Figure 138.3 shows output voltage and inductor current as the load current is stepped from 5mA to 55mA. The oscillograph reveals substantial detail about the operation of the LT1307. With a 5mA load, V_{OUT} (top trace) exhibits a ripple voltage of 60mV at

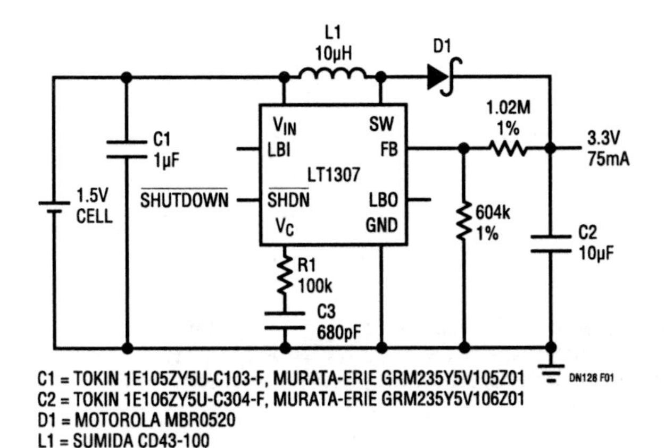

Figure 138.1 • Single Cell to 3.3V Boost Converter Delivers 75mA at a 1V Input

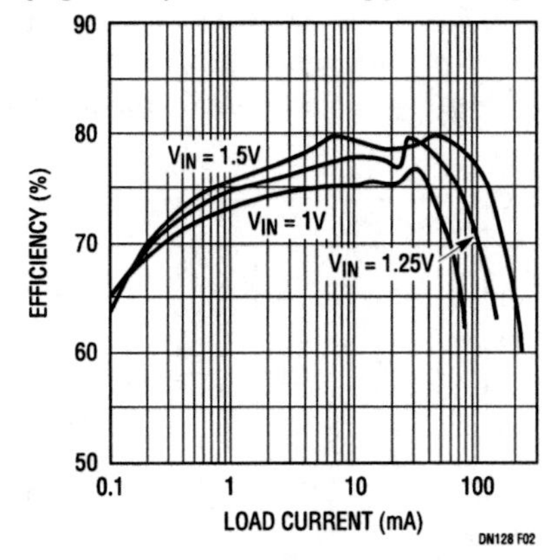

Figure 138.2 • 3.3V Converter Efficiency

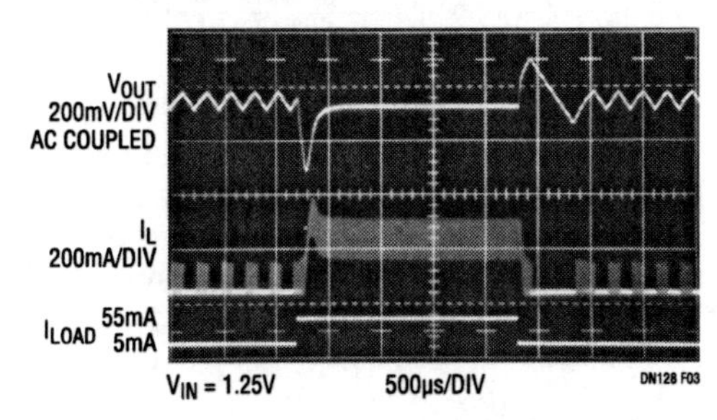

Figure 138.3 • Transient Response with 5mA to 55mA Load Step

Analog Circuit Design: Design Note Collection. http://dx.doi.org/10.1016/B978-0-12-800001-4.00138-1

4kHz. The device is in Burst Mode operation at this output current level. Burst Mode operation enables the converter to maintain high efficiency at light loads by turning off all circuitry inside the LT1307 except the reference and error amplifier. When the LT1307 is not switching, quiescent current decreases to 60μA. When switching, inductor current (middle trace) is limited to approximately 100mA. Switching frequency inside the "bursts" is 575kHz. As the load is stepped to 55mA, the device shifts from Burst Mode operation to constant switching mode. Inductor current increases to about 300mA peak and the low frequency Burst Mode ripple goes away. R1 and C3 stabilize the loop.

455kHz noise considerations

Switching regulator noise is a significant concern in many communication systems. The LT1307 is designed to keep noise energy out of the 455kHz band at all load levels while consuming only 60μW to 100μW at no load. At light load levels, the device is in Burst Mode operation, causing low frequency ripple to appear at the output. Figure 138.4 details spectral noise directly at the output of Figure 138.1's circuit in a 1kHz to 1MHz bandwidth. The converter supplies a 5mA load from a 1.25V input. The Burst Mode fundamental at 5.1kHz and its harmonics are quite evident, as is the 575kHz switching frequency. Note, however, the absence of significant energy at 455kHz. Figure 138.5's plot reduces the frequency span from 255kHz to 655kHz with a 455kHz center. Burst Mode low frequency ripple creates sidebands around the 575kHz switching fundamental. These sidebands have low signal amplitude at 455kHz, measuring – 55dBmV$_{RMS}$. As load current is further reduced, the Burst Mode frequency decreases. This spaces the sidebands around the switching frequency closer together, moving spectral energy further away from 455kHz. Figure 138.6 shows the noise spectrum of the converter with the load increased to 20mA. The LT1307 shifts out of Burst Mode operation eliminating low frequency ripple. Spectral energy is present only at the switching fundamental and its harmonics. Noise voltage measures –5dBmV$_{RMS}$ or 560μV$_{RMS}$ at the 575kHz switching frequency, and is below –60dBmV$_{RMS}$ for all other frequencies in the range. By combining Burst Mode operation with fixed frequency operation the LT1307 keeps noise away from 455kHz making the device ideal for RF applications where the absence of noise in the 455kHz band is critical.

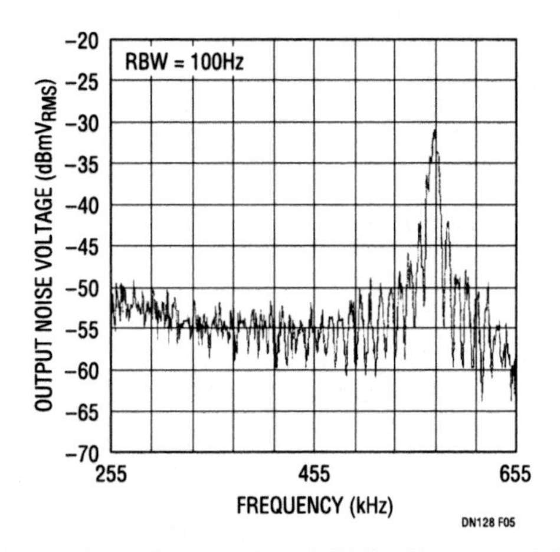

Figure 138.5 • Span Centered at 455kHz Shows –55dBmV$_{RMS}$ (1.8μV$_{RMS}$) at 455kHz. Burst Mode Operation Creates Sidebands 5.1kHz Apart about Switching Fundamental of 575kHz

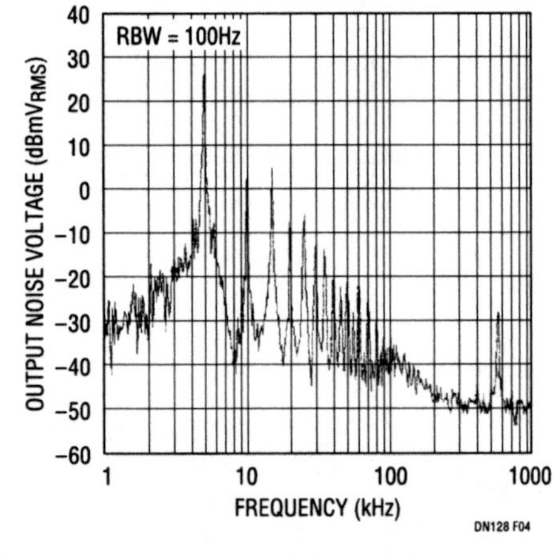

Figure 138.4 • Spectral Noise Plot of 3.3V Converter Delivering 5mA Load. Burst Mode Operation at 5.1kHz Is 23dBmV$_{RMS}$ or 14mV$_{RMS}$. Switching Fundamental at 575kHz Is –31dBmV$_{RMS}$ or 28μV$_{RMS}$

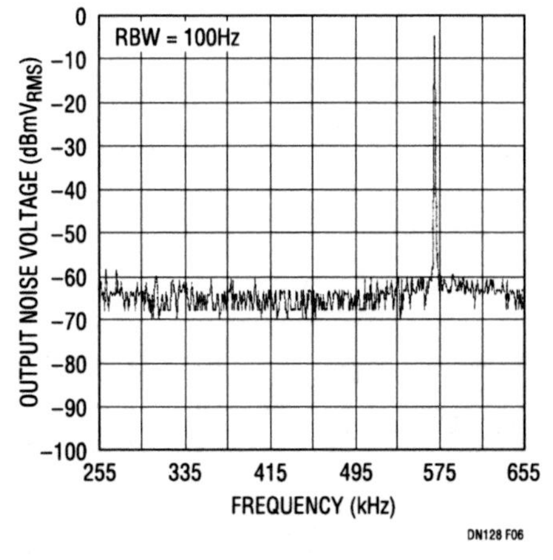

Figure 138.6 • With Converter Delivering 20mA, Low Frequency Sidebands Disappear. Noise Is Present Only at the 575kHz Switching Frequency and its Harmonics

2 AA cells replace 9V battery, extend operating life

139

Steve Pietkiewicz

Operating life is an important feature in many portable battery-operated systems. In many cases the power source is the ubiquitous 9V "transistor" battery. 5V generation is accomplished with a linear regulator. Significant gains in battery life can be obtained by replacing the 9V/linear regulator combination with 2 AA cells and a step-up switching regulator. Two (alkaline) AA cells occupy 1.3cubic inches, the same as a 9V battery, but contains 6WH of energy, compared to just 4WH in an alkaline 9V battery. Two AA cells also cost less than a 9V battery.[1] The additional energy in the AA cells provides longer operating life when compared to a 9V battery based solution.

An evaluation of the three approaches with a 30mA load illustrates the differences in battery life. An HP7100B strip chart recorder provides a nonvolatile record of circuit performance. The linear regulator circuit shown in Figure 139.1 uses an LT1120 micropower low dropout regulator IC. A minimum of external components are required. No inductors or diodes are needed; however, the linear step-down process is inherently inefficient. The step-down switcher shown in Figure 139.2 uses an LT1173 configured in step-down mode driven from an alkaline 9V battery. In Figure 139.3 the step-up circuit uses an LT1173 configured in step-up mode driven from a pair of alkaline AA cells. The two switching circuits require an external inductor, diode and output capacitor in addition to the IC.

Circuit operation of the switching step-down regulator is straightforward. A comparator inside the LT1173 senses output voltage on its "sense" pin. When V_{OUT} drops below 5V, the on-chip switch cycles. As current ramps up and ramps down in L1, it flows into C1 and the load, raising output voltage.

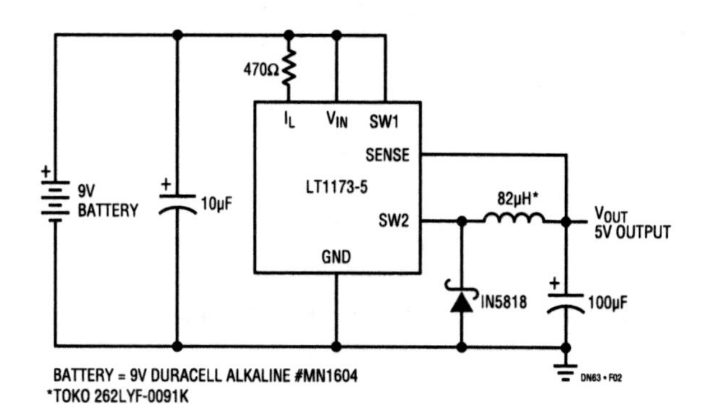

Figure 139.2 • 9V to 5V Step-Down Regulator

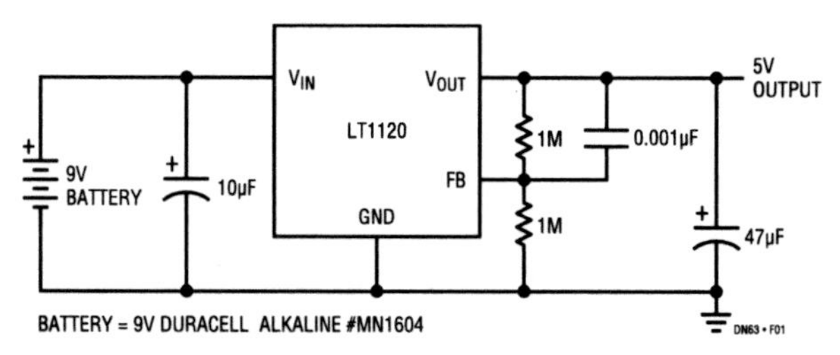

Figure 139.1 • 9V to 5V Linear Regulator

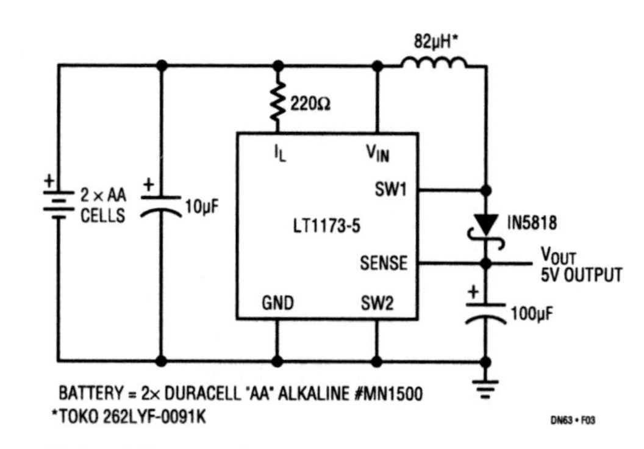

Figure 139.3 • 3V to 5V Step-Up Regulator

Note 1: A quick check at the local drugstore yielded $2.99 for a 4-pack of alkaline AA cells and $2.49 for a single 9V battery (after $1.00 mail-in rebate).

Analog Circuit Design: Design Note Collection. http://dx.doi.org/10.1016/B978-0-12-800001-4.00139-3

When V_{OUT} rises above 5V, the cycling action stops and the regulator goes into a standby mode, pulling 110μA from the supply. C1 is left to supply energy to the load. These "bursts" of cycles occur as needed to keep the output voltage at 5V. 50mV of hysteresis at the sense pin eliminates the need for frequency compensation. The step-up regulator operates in a similar fashion, although in this case the inductor current flows into the load only on the discharge half of the switch cycle. Output voltage is regulated in a similar manner.

Efficiency curves for the three circuits are shown in Figures 139.4 and 139.5. The linear regulator circuit has efficiency of 52% with a fresh battery. As the input-output differential decreases, the efficiency increases and at end of battery life exceeds 90%. Regulator ground current limits efficiency at drop-out. The switch-mode step-down circuit has almost constant efficiency, ranging from 84% at 6.3V input to 82% at 9.5V input. Minimum V_{IN} is set by the drop of the emitter follower switch inside the LT1173. Performance for the step-up converter is shown in Figure 139.5. At higher inputs, the switch drop is a lower percentage of supply, resulting in higher efficiency.

The three regulators show substantial differences in operating life. The linear regulator operates for 16.5 hours, as shown

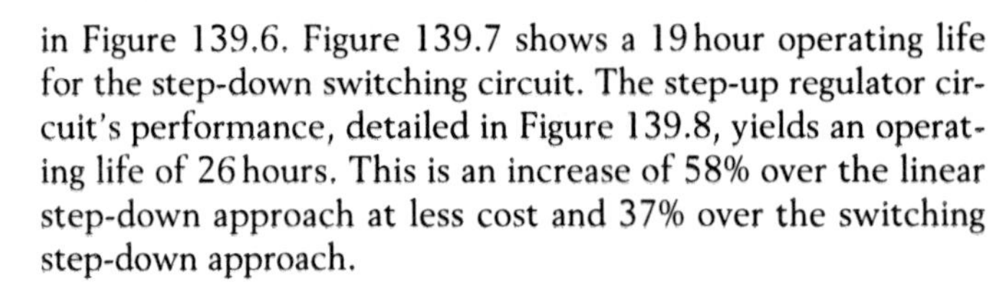
in Figure 139.6. Figure 139.7 shows a 19 hour operating life for the step-down switching circuit. The step-up regulator circuit's performance, detailed in Figure 139.8, yields an operating life of 26 hours. This is an increase of 58% over the linear step-down approach at less cost and 37% over the switching step-down approach.

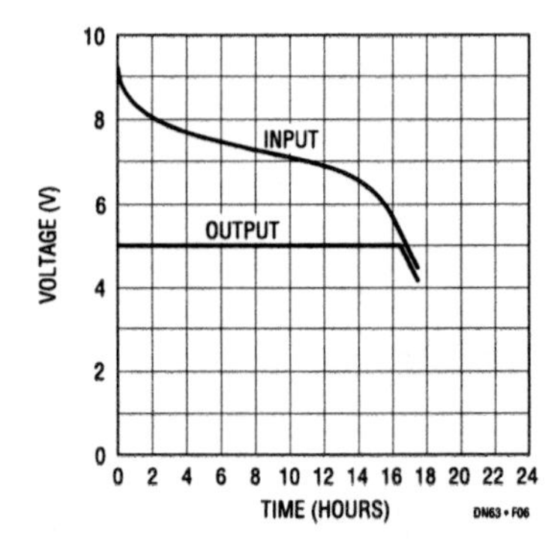

Figure 139.6 • 9V to 5V Step-Down Linear—LT1120, 30mA Load

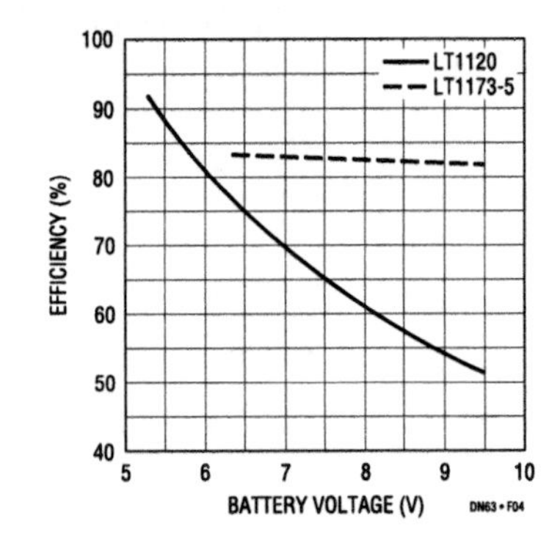

Figure 139.4 • Step-Down Conversion Efficiency—5V Output, 30mA Load

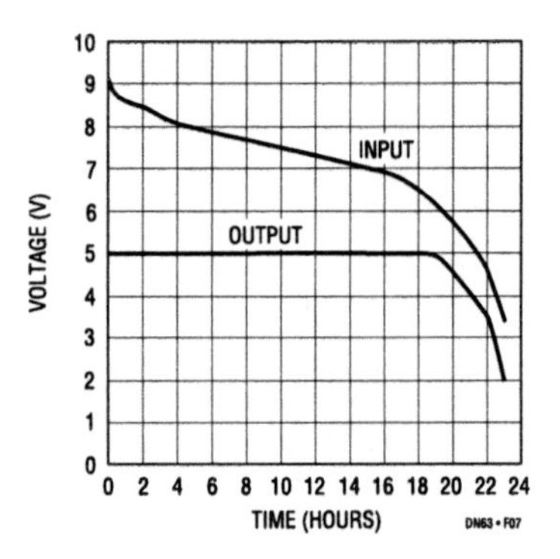

Figure 139.7 • 9V to 5V Step-Down Switcher—LT1173-5, 30mA Load

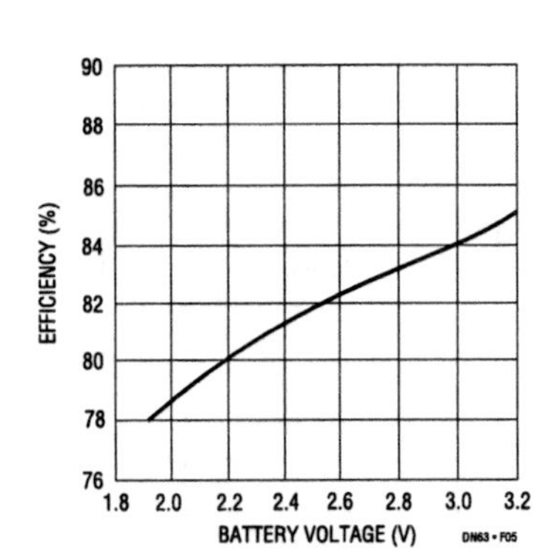

Figure 139.5 • Step-Up Conversion Efficiency—5V Output, 30mA Load

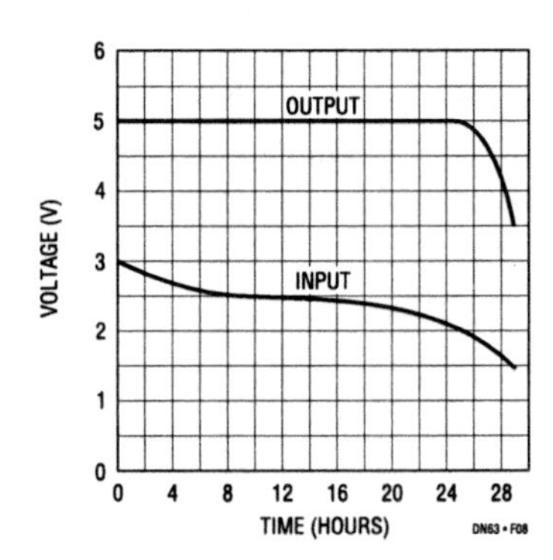

Figure 139.8 • 3V to 5V Step-Up Switcher—LT1173-5, 30mA Load

A simple, surface mount flash memory Vpp generator

140

Steve Pietkiewicz Jim Williams

"Flash" type memories add electrical chip-erasure and reprogramming to established EPROM technology. These features make them a cost effective and reliable alternative for updatable non-volatile memory. Utilizing the electrical program-erase capability requires linear circuitry techniques. Intel flash memory, built on the ETOX process, specifies programming operation with 12V amplitude pulses. These "Vpp" amplitudes must fall within tight tolerances, and excursions beyond 14.0V will damage the device.

Providing the Vpp pulse requires generating and controlling high voltages within the tightly specified limits. Figure 140.1's circuit does this. When the Vpp command pulse goes high (Trace A, Figure 140.2) the LT1109 switching regulator drives L1, producing high voltage. DC feedback occurs via the regulator's sense pin. The result is a smoothly rising Vpp pulse (Trace B) which settles to the required value. Trace C, a time and amplitude expanded version of Trace B, details the desired settling to 12V. Artifacts of the switching regulator's action are discernible, although no overshoot or poor dynamics are displayed.

This circuit is well suited for providing Vpp power to flash memory. All associated components, including the inductor, are surface mount devices. As such, the complete circuit occupies very little space (see Figure 140.3). In the shutdown mode the circuit pulls only 300μA. Output voltage goes to V_{CC} minus a diode drop when the converter is in shutdown mode. This is an acceptable and specified condition for flash memories and does not harm the memory. A 0V output is possible by placing a 5.6V Zener diode in series with the output rectifier (Figure 140.4a). An alternative configuration, suggested by J. Dutra of LTC, AC couples the output to achieve a 0V output (Figure 140.4b). Both of these methods add component count, decrease efficiency and slightly limit available output current. They are unnecessary unless the user desires a 0V output on the Vpp line.

A good question might be; "Why not set the switching regulator output voltage at the desired Vpp level and use a simple low resistance FET or bipolar switch?" This is a potentially dangerous approach. Figure 140.5 shows the clean output of a low resistance switch operating directly at

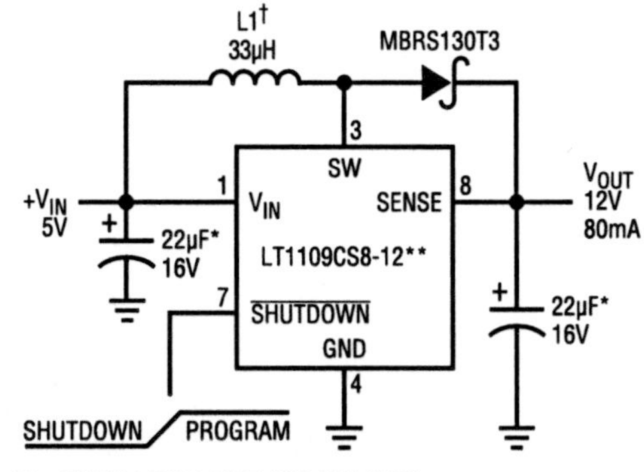

Figure 140.1 • All Surface Mount Flash Memory Vpp Generator

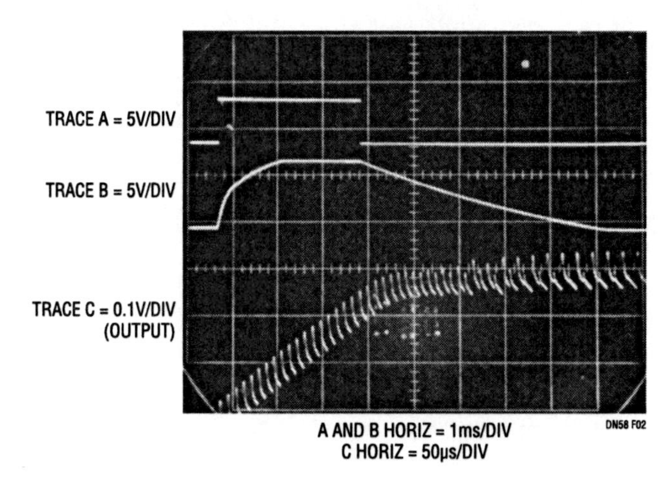

Figure 140.2 • Waveforms for the Flash Memory Pulser Show No Overshoot

Analog Circuit Design: Design Note Collection. http://dx.doi.org/10.1016/B978-0-12-800001-4.00140-X

the Vpp supply. The PC trace run to the memory chip looks like a transmission line with ill-defined termination characteristics. As such, Figure 140.5's clean pulse degrades and rings badly (Figure 140.6) at the memory IC's pins. Overshoot exceeds 20V, well beyond the 14V destruction level. The controlled edge times of the circuit discussed eliminate this problem. Further discussion of this and other circuits appears in LTC Application Note 31, "Linear Circuits for Digital Systems" and LTC Demo Manual DC019, "Flash Memory Vpp Generator."

Figure 140.3 • Simple Flash Memory Pulser Uses all Surface Mount Components

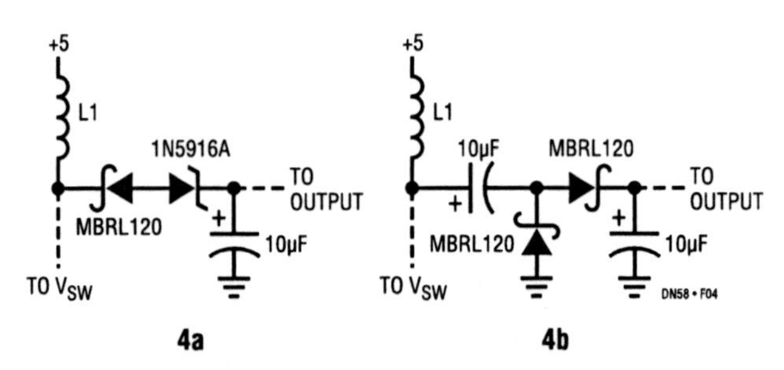

Figure 140.4 • Alternative 0V Output Solutions

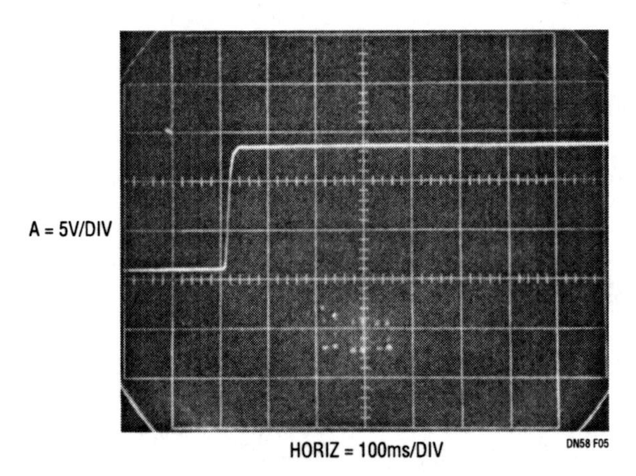

Figure 140.5 • An "Ideal" Flash EPROM Vpp Pulse

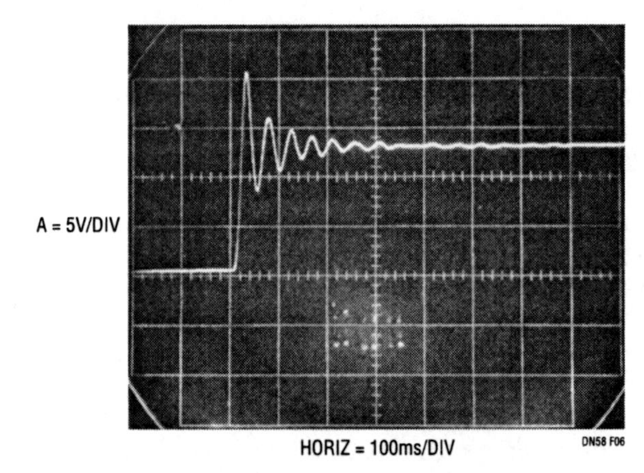

Figure 140.6 • Rings at Destructive Voltages after a PC Trace Run

No design switching regulator 5V, 5A buck (step-down) regulator

141

Ron Vinsant

Introduction

This simple, no design regulator, is a step-down DC/DC converter designed to convert an 8V to 40V input to a regulated 5V output. The 5V output is capable of sourcing up to 5A of output current.

This converter is based on the Linear Technology LT1074 switching regulator IC. This device needs only a few external parts to make up a complete regulator including thermal protection and current limit. This design uses off-the-shelf parts for low cost and easy availability of components. Specifications for the circuit are in Table 141.1.

Circuit description

Figure 141.1 shows the schematic of the circuit. For the purpose of this explanation assume that the output is at a constant +5V DC and that the input voltage is greater than +8V DC.

At intervals of ≈10μs (100kHz) the control portion of the LT1074 turns on the switch transistor between the V_{IN} and V_{SW} pins impressing a voltage across the inductor, L1. This causes current to build up in the inductor while also supplying current to the load and capacitor C1.

The control circuit determines when to turn off the switch during the 10μs interval to keep the output voltage at +5V DC. When the switch transistor turns off, the magnetic field in the inductor collapses and the polarity of the voltage across the inductor changes to try and maintain the current in the inductor. This current in the inductor is now directed (due to the change in voltage polarity across the inductor) by the diode, D1, to the load. The current will flow from the inductor until the switch turns on again, (continuous operation) or until the inductor runs out of energy (discontinuous operation).

Referring back to Figure 141.1, the divider circuit of R1 and R2 is used to set the output voltage of the supply against an internal voltage reference of 2.21V DC.

R3 and C2 make up the frequency compensation network used to stabilize the feedback loop.

Conclusion

This Design Note demonstrates a fully characterized step-down converter circuit that is both simple and low cost. This design can be taken and reliably used in a production environment without the need for any custom components. A PC board layout and FAB drawing are available from Linear Technology.

Table 141.1 Performance Summary (Operating Temperature Range 0°C to 50°C)

INPUT VOLTAGE RANGE			**+8.0V TO +40.0V DC**
Output	Output Voltage (±0.15V DC)		+5.00V DC
	Max Output Current V_{IN} = 8.0V to 40.0V		5.0A DC
	Typical Output Ripple at I_{OUT} = 4.0A DC at Switching Frequency	With Optional Filter (L2 and C4)	$5mV_{P-P}$
		Without Optional Filter (L2 and C4)	$50mV_{P-P}$
	Load Regulation V_{IN} = 8V	At I_{OUT} = 0.5A DC to I_{OUT} = 5.0A DC	0.5%
	Line Regulation I_{OUT} = 5A	At V_{IN} = +8.0V DC to V_{IN} = +40.0V DC	0.5%

Analog Circuit Design: Design Note Collection. http://dx.doi.org/10.1016/B978-0-12-800001-4.00141-1

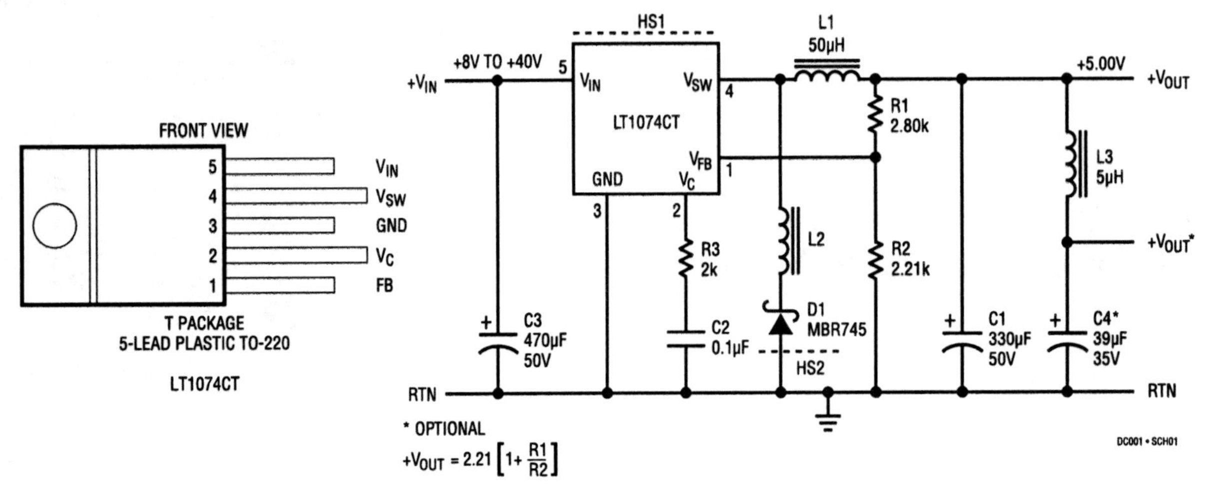

Figure 141.1 • Package and Schematic Diagrams

Table 141.2 Parts List

REFERENCE DESIGNATOR	QUANTITY	PART NUMBER	DESCRIPTION	VENDOR
PCB	1	001A	PCB FAB, Buck Switching Regulator	LTC
D1	1	MBR745	Diode, Schottky, 7A, 45V	Motorola
HS2	1	6038B-TT	Heat Sink	Thermalloy
L2	1	2664000101	Shield Bead	Fair-Rite
VR1	1	LT1074CT	Switching Regulator, 100kHz	LTC
HS1	1	7020B-MT	Heat Sink	Thermalloy
C1	1	SXE50VB331M12X20LL	Cap, Alum Elect, 330µF, 50V	United Chemicon
C2	1	CK06BX104K	Cap, Ceramic, 0.1µF, 50V	AVX
C3	1	UPL1H471MRH	Cap, Alum Elect, 470µF, 50V	Nichicon
C4	1	UPL1V390MAH	Cap, Alum Elect, 39µF, 35V	Nichicon
L1	1	CTX50-5-MP	Inductor, 50µH, 5A	Coiltronics
L3	1	CTX5-5-FR	Inductor, 5µH, 5A	Coiltronics
R1	1	MF 1/8W 2.80kΩ	Res, MF, 1/8W, 1%, 2.80k	
R2	1	MF 1/8W 2.21kΩ	Res, MF, 1/8W, 1%, 2.21k	
R3	1	CF 1/4W 2kΩ	Res, CF, 1/4W, 5%, 2k	

Section 6

Switching Regulator Design: DC/DC Controllers

Dual controller provides 2μs step response and 92% efficiency for 1.5V rails

142

Mike Shriver

Introduction

The LTC3838 is a dual output, dual phase buck controller that employs a controlled constant on-time, valley current mode architecture to provide fast load step response, high switching frequency and low duty cycle capability. The switching frequency range is 200kHz to 2MHz—its phase-locked loop keeps the frequency constant during steady-state operation and can be synchronized to an external clock. The LTC3838 accepts a wide input range, 4.5V to 38V, and can produce 0.6V to 5.5V outputs.

The remotely sensed V_{OUT1} has a voltage regulation accuracy of 0.67%, from 0°C to 85°C, even with a voltage difference of ±0.5V between local ground and remote ground. The current sense comparators are designed to sense the inductor current with either a sense resistor for high accuracy or with the inductor DCR directly for reduced power losses and circuit size.

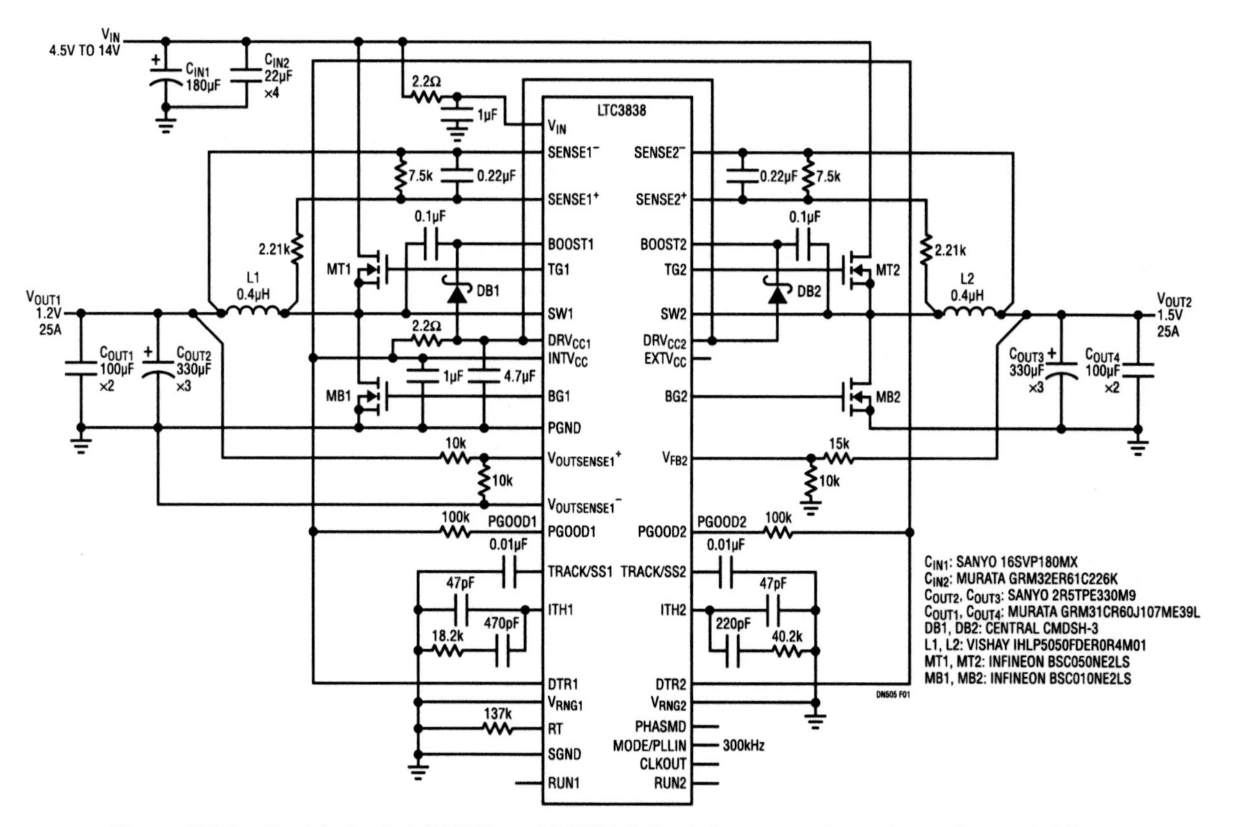

Figure 142.1 • Dual Output, 1.5V/25A and 1.2V/25A Buck Converter Operating at F_{SW} = 300kHz

Analog Circuit Design: Design Note Collection. http://dx.doi.org/10.1016/B978-0-12-800001-4.00142-3

1.5V/25A and 1.2V/25A buck converter

Figure 142.1 shows a dual 25A output buck converter synchronized to an external 300kHz clock. The controlled constant on-time valley current mode architecture allows the switch node pulses to temporarily compress when a 5A to 25A load step is applied to the 1.2V rail, resulting in a voltage undershoot of only 58mV (see Figure 142.2).

The full load efficiency for the 1.5V and 1.2V rails is 91.8% and 90.8%, respectively, as shown in Figure 142.3. The high efficiency is realized by the strong gate drivers, optimized dead time and DCR sensing.

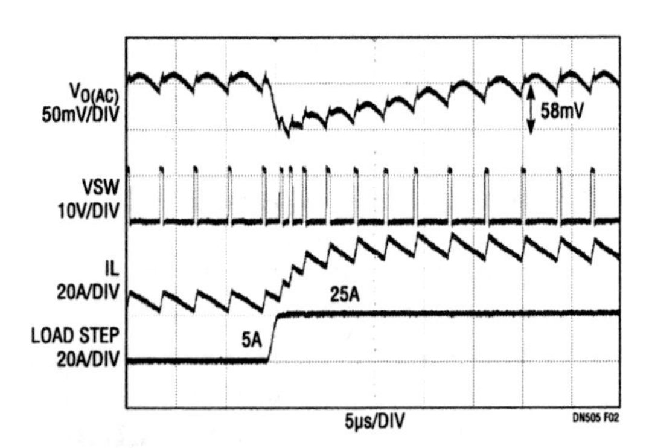

Figure 142.2 • 20% to 100% Step Load Response of the 1.2V Rail at V_{IN}=12V, F_{SW}=300kHz, Mode=FCM

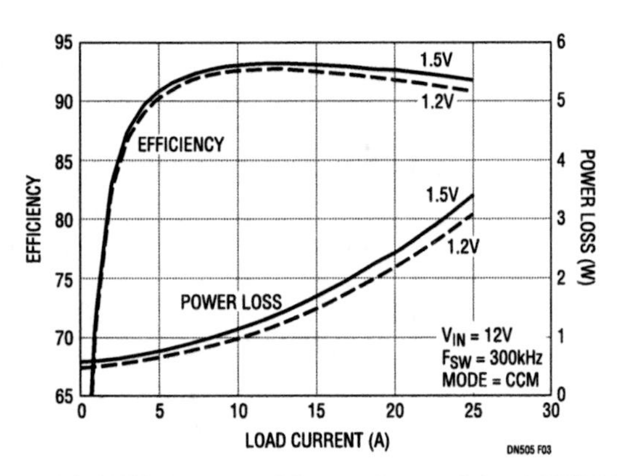

Figure 142.3 • Efficiency and Power Loss of the 1.5V/25A and 1.2V/25A Converter

The two channels operate 180° out-of-phase, which permits the use of fewer input capacitors due to input capacitor ripple current cancellation. For higher current applications, two or more phases can be tied together to form a single output, PolyPhase converter. The benefits include a faster load step response, reduced input and output capacitance and reduced thermal dissipation.

Detect transient feature further speeds up transient response

An innovative feature of the LTC3838 is the load release transient detect feature. The DTR pin indirectly monitors the output voltage by looking at the AC-coupled ITH signal. If the inferred overshoot exceeds a user set value, the bottom FET turns off. This allows the inductor current to slew down at a faster rate, which in turn reduces the overshoot. Per Figure 142.4, a 32% reduction in the overshoot is realized on the 1.2V rail. Greater improvements occur at lower output voltages.

Conclusion

The LTC3838 is a dual output buck controller ideal for applications that require a fast load step response, high switching frequency, high efficiency and accurate output voltages. Other features include selectable operating modes: forced continuous mode (FCM) for fixed frequency operation or discontinuous mode (DCM) for higher efficiency at light load, programmable current limit thresholds, soft-start, rail tracking and individual PGOOD and RUN pins. The LTC3838 comes in a 5mm × 7mm QFN package or a thermally enhanced 38-lead TSSOP package.

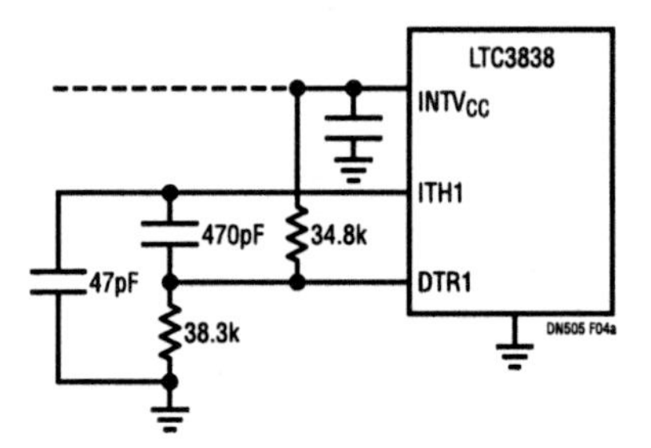

Figure 142.4a • Implementation of the Detect Transient Feature on the 1.2V Rail

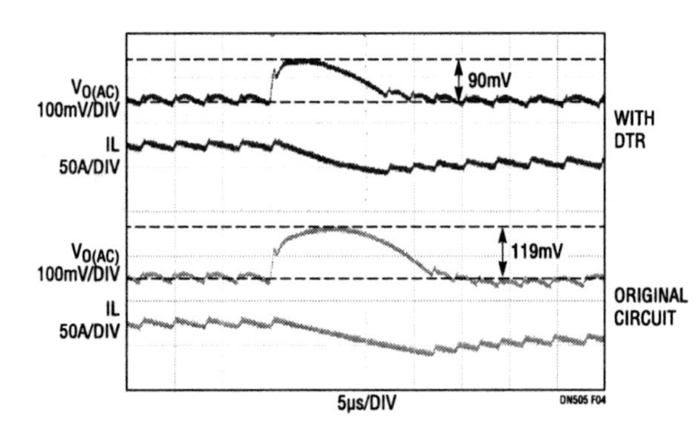

Figure 142.4b • 100% to 20% Step Load Response of the 1.2V Rail with and without the Detect Transient Feature, V_{IN}=12V, F_{SW}=300kHz, Mode=FCM

Dual DC/DC controller for DDR power with differential V_{DDQ} sensing and ±50mA V_{TT} reference

143

Ding Li

Introduction

The LTC3876 is a complete DDR power solution, compatible with DDR1, DDR2, DDR3 and DDR4 lower voltage standards. The IC includes V_{DDQ} and V_{TT} DC/DC controllers and a precision linear V_{TT} reference. A differential output sense amplifier and precision internal reference combine to offer an accurate V_{DDQ} supply. The V_{TT} controller tracks the precision VTTR linear reference with less than 20mV total error. The precision VTTR reference maintains 1.2% regulation accuracy, tracking one-half V_{DDQ} over temperature for a ±50mA reference load.

The LTC3876 features controlled on-time, valley current mode control, allowing it to accept a wide 4.5V to 38V input range, while supporting V_{DDQ} outputs from 1.0V to 2.5V, and V_{TT} and VTTR outputs from 0.5V to 1.25V. Its phase-locked loop (PLL) can be synchronized to an external clock between 200kHz and 2MHz. It also features voltage-tracking soft-start, PGOOD and fault protection.

High efficiency, 4.5V to 14V input, dual output DDR power supply

Figure 143.1 shows a DDR3 power supply that operates from a 4.5V to 14V input. Figure 143.2 shows efficiency curves for discontinuous and forced continuous modes of operation.

Load-release transient detection

As output voltages drop, a major challenge for switching regulators is to limit the overshoot in V_{OUT} during a load-release transient. The LTC3876 uses the DTR pin to monitor the

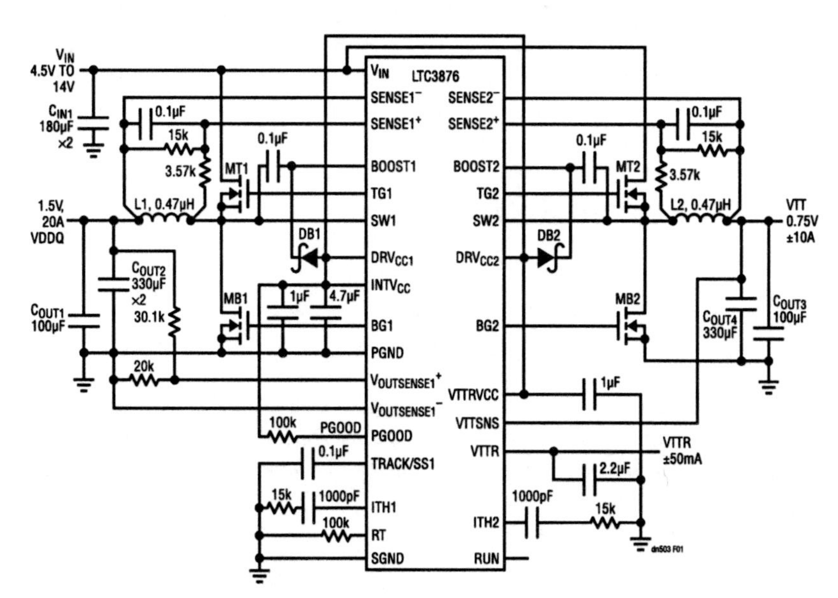

Figure 143.1 • 1.5V V_{DDQ}/20A 0.75V V_{TT}/10A DDR3 Power Supply

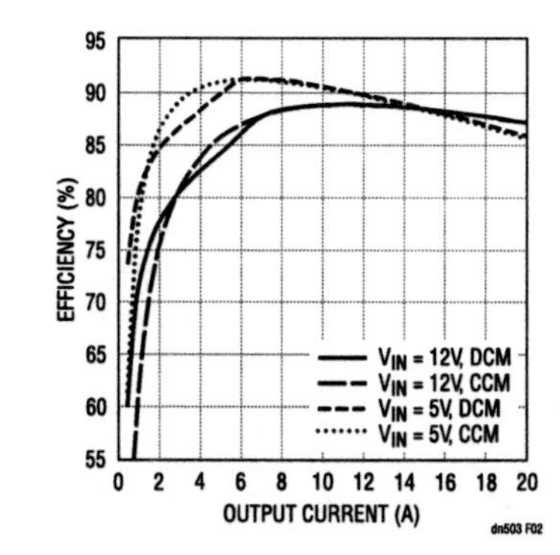

Figure 143.2 • Efficiency of Circuit in Figure 143.1 (V_{DDQ} = 1.5V, f_{SW} = 400kHz, L = 470nH)

Analog Circuit Design: Design Note Collection. http://dx.doi.org/10.1016/B978-0-12-800001-4.00143-5

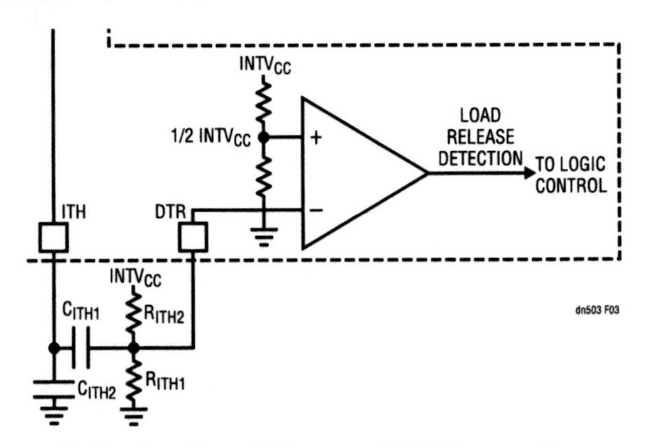

Figure 143.3 • Functional Diagram of DTR Connection for Load Transient Detection

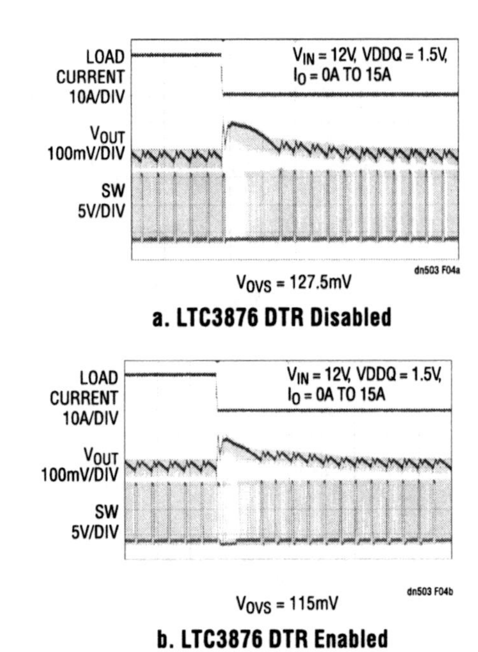

Figure 143.4 • Load Release Comparison

first derivative of the ITH voltage to detect load release transients. Figure 143.3 shows how this pin is used for transient detection.

The two R_{ITH} resistors establish a voltage divider from $INTV_{CC}$ to SGND, and bias the DC voltage on the DTR pin (at steady-state load or ITH voltage) slightly above half of $INTV_{CC}$. For a given C_{ITH1}, this divider does not change compensation performance as long as R_{ITH1}/R_{ITH2} equals R_{ITH} that would normally be used in conventional single-resistor OPTI-LOOP compensation.

The divider sets the RC time constant needed for the DTR duration. The DTR sensitivity can be adjusted by the DC bias voltage difference between DTR and half $INTV_{CC}$. This difference could be set as low as 100mV, as long as the ITH ripple voltage with DC load current does not trigger the DTR. If the load transient is fast enough that the DTR voltage drops below half of $INTV_{CC}$, a load release event is detected. The bottom gate (BG) is turned off, so that the inductor current flows through the body diode in the bottom MOSFET.

Note that the DTR feature causes additional losses on the bottom MOSFET, due to its body diode conduction. The bottom FET temperature may be higher with a load of frequent and large load steps—an important design consideration. Test results show a 20°C increase when a continuous 100%-to-50% load step pulse chain with 50% duty cycle and 100kHz frequency is applied to the output.

V_{TT} reference (VTTR)

The linear V_{TT} reference, VTTR, is specifically designed for large DDR memory systems by providing superior accuracy and load regulation for up to ±50mA output load. VTTR is the buffered output of the V_{TT} differential reference resistor divider. VTTR is a high output linear reference, which tracks the V_{TT} differential reference resistor divider and equals half of the remote-sense V_{DDQ} voltage.

Connect VTTR directly to the DDR memory VREF input. Both input and output supply decoupling are important to performance and accuracy. A 2.2μF output capacitor is recommended for most typical applications. It is suggested to use no less than 1μF and no more than 47μF on the VTTR output. The VTTR power comes from the VTTRVCC pin. The typical recommended input VTTRVCC RC decoupling filter is 2.2μF and 1Ω. When VDDQSNS is tied to INTVCC, the VTTR linear reference output is 3-stated and VTTR becomes a reference input pin, with voltage from another LTC3876 in a multiphase application (see Figure 143.4).

V_{TT} supply

The V_{TT} supply reference is connected internally to the output of the VTTR V_{TT} reference output. The V_{TT} supply operates in forced continuous mode and tracks V_{DDQ} in start-up and in normal operation regardless of the MODE/PLLIN settings. In start-up, the V_{TT} supply is enabled coincident with the V_{DDQ} supply. Operating the V_{TT} supply in forced continuous mode allows accurate tracking in start-up and under all operating conditions.

Conclusion

The LTC3876 is a complete high efficiency and high accuracy solution for DDR memory power supplies. The unique controlled on-time architecture allows extremely low step-down ratios while maintaining a fast, constant switching frequency. The wide input voltage range of 4.5V–38V and programmable, synchronizable switching frequency from 200kHz to 2MHz gives designers the flexibility needed to optimize their systems.

Single resistor sets positive or negative output for DC/DC converter

144

Jesus Rosales

Introduction

Many electronic subsystems, such as VFD (vacuum fluorescent display), TFT-LCD, GPS or DSL applications, require more than just a simple step-down or step-up DC/DC converter. They may require inverting, noninverting converters or both. Designers usually resort to different regulator ICs to control various polarity outputs, thus increasing the inventory list. The LT3580 solves this problem by controlling either positive or negative outputs using the same feedback configuration. It contains an integrated 2A, 42V switch and packs many popular features such as soft-start, adjustable frequency, synchronization and a wide input range into a small footprint. The LT3580 comes in an 8-pin 3mm × 3mm DFN or MSOP packages and can be used in multiple configurations such as boost, SEPIC, flyback and Cuk topologies.

Sensing output voltage has never been easier

The LT3580 has a novel FB pin architecture that simplifies the design of inverting and noninverting topologies. Namely, there are two internal error amplifiers; one senses positive outputs and the other negative. Additionally, the LT3580 has integrated the ground side feedback resistor to minimize component count. To illustrate the benefits, notice how the schematics in Figures 144.1, 144.3 and 144.5 need only one feedback resistor.

A single sense resistor simply connects to the FB pin on one side and to the output on the other regardless of the output polarity, eliminating the confusion associated with positive or negative output sensing and simplifying the board layout. A user decides the output polarity he needs, the topology he wants to use and the LT3580 does the rest.

Adjustable/synchronizable switching frequency

It is often necessary to operate a converter at a particular frequency, especially if the converter is used in an RF communications product that is sensitive to spectral noise in certain frequency bands. Also, if the area available for a converter is limited, operating at higher frequencies allows the use of tiny component sizes, reducing the real estate required and the output ripple. If power loss is a concern, switching at a lower frequency reduces switching losses, improving efficiency. The switching frequency can be set from 200kHz to 2.5MHz via a single resistor from the RT pin to ground. The device can also be synchronized to an external clock via the SYNC pin.

Soft-start and undervoltage lockout

To alleviate high inrush current levels during start-up, the LT3580 includes a soft-start feature which controls the ramp rate of the switch current by the use of a capacitor from SS to ground.

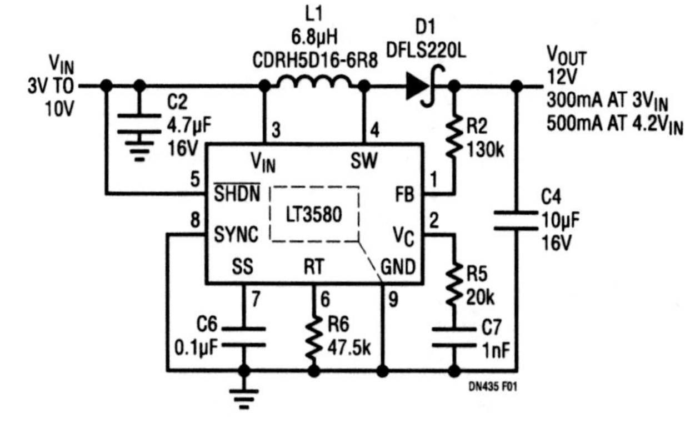

Figure 144.1 • 3V–10V to 12V, 300mA Boost Converter

Analog Circuit Design: Design Note Collection. http://dx.doi.org/10.1016/B978-0-12-800001-4.00144-7

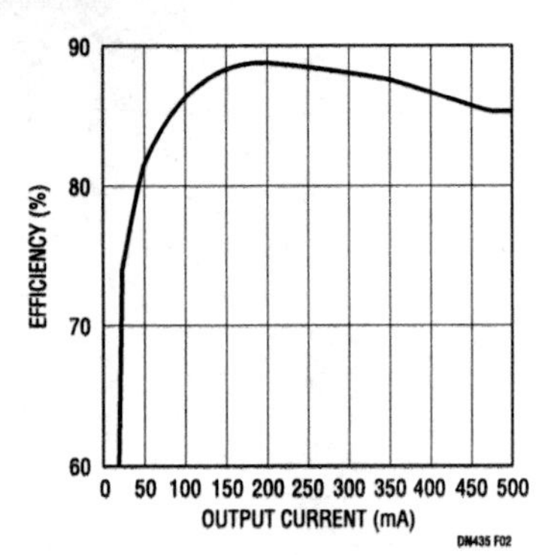

Figure 144.2 • Efficiency for the Figure 144.1 Converter at 4.2V_{IN}

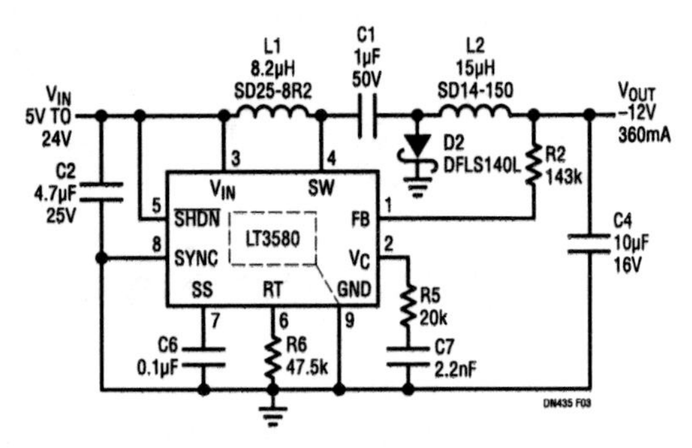

Figure 144.3 • 5V–24V to –12V, 350mA Cuk Converter

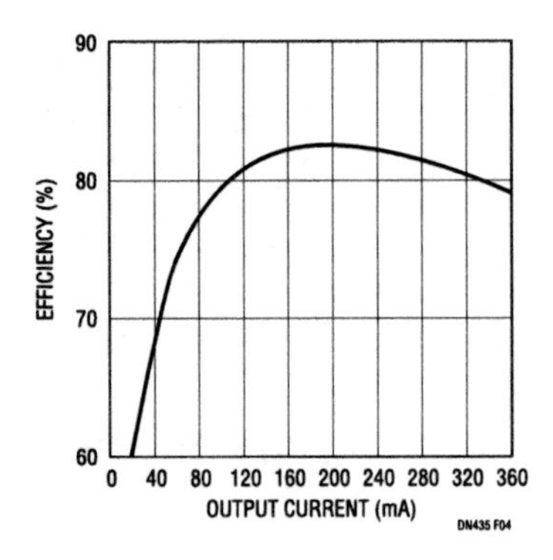

Figure 144.4 • Efficiency for the Figure 144.3 Converter at 5V_{IN}

The $\overline{SHDN}$ pin in the LT3580 serves two purposes. Tying it high or low turns the converter on or off. In situations where the input supply is current limited, has a high source impedance or ramps up/down slowly, the $\overline{SHDN}$ pin can be configured to provide undervoltage lockout through a simple resistor divider from V_{IN} to ground.

Boost converters

A boost converter, shown in Figure 144.1, produces a positive output voltage always higher than its input. Figure 144.2 shows the efficiency graph for the boost converter in Figure 144.1 at a 4.2V input.

Cuk converter

Figure 144.3 shows a schematic for a Cuk converter, which produces a negative output with no DC path to the source. The output can be either higher or lower in amplitude than the input. The Cuk converter has output short-circuit protection, which is made more robust by the frequency foldback feature in the LT3580. Figure 144.4 shows the efficiency graph for the Cuk converter in Figure 144.3 at a 5V input.

SEPIC converters

Figure 144.5 shows a SEPIC converter. A SEPIC converter is similar to the Cuk in that it can step up or step down the input; it offers output disconnect and short-circuit protection but produces a positive output. Figure 144.6 shows the switch waveform of the SEPIC converter during an output short-circuit event. Notice how the switching frequency folds back to one-fourth of the regular frequency as soon as the output voltage is shorted to ground. This feature enhances short-circuit performance for both Cuk and SEPIC converters.

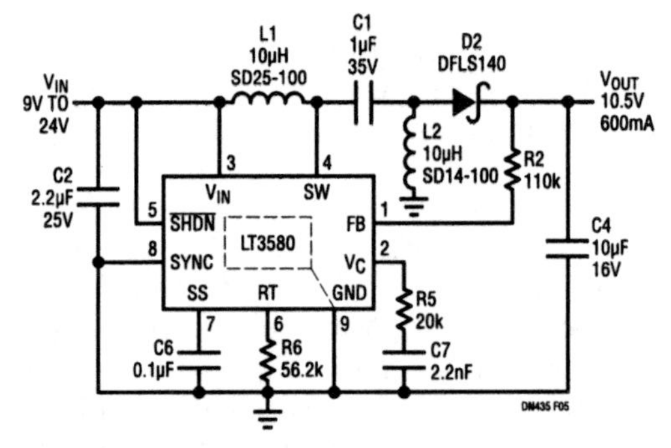

Figure 144.5 • 9V–24V to 10.5V, 600mA SEPIC Converter

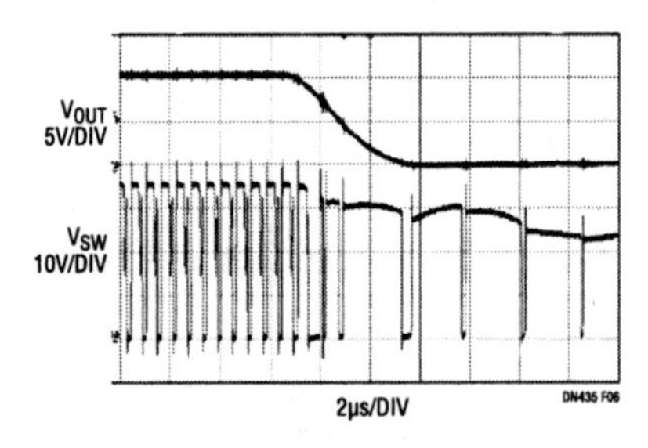

Figure 144.6 • Short-Circuit Event for the Figure 144.5 Converter at 24V_{IN}

Conclusion

The LT3580 features a unique feedback architecture that allows it to be configured as an inverting or noninverting converter. Now, the same device can be used to produce regulated voltages of either polarity, allowing for a reduction in inventory count. Its many additional features such as soft-start, adjustable switching frequency, shutdown, synchronizing capability, configurable undervoltage lockout, frequency foldback, external compensation and wide input range simplify the design of inverting and noninverting converters.

Multiphase DC/DC controller pushes accuracy and bandwidth limits

145

Tick Houk

Introduction

Speed and accuracy do not always go hand-in-hand in DC/DC converter systems—that is, until now. The LTC3811 is a dual output, fixed frequency current mode DC/DC switching regulator controller designed for one of today's most demanding power supply applications: high current, low voltage processor core supplies.

With supply current requirements in excess of 100A and supply voltages as low as 1V, every milliohm of PCB resistance and every millivolt of IR drop count. The LTC3811 has an output voltage tolerance of ±0.5% over temperature, giving power supply designers unprecedented flexibility when making component value and board layout choices.

In addition to high accuracy, the LTC3811's low minimum on-time (typically 65ns) allows users to convert a 12V input to a 1V output at switching frequencies up to 750kHz, optimizing load transient response and reducing the solution size.

A dual output, 2-phase supply with differential remote sensing and inductor DCR sensing

Figure 145.1 illustrates a dual output supply using the LTC3811. The 1.5V, 15A output is regulated using the integrated differential remote sense amplifier and tracks the output of channel 1 during start-up. Both outputs use DCR

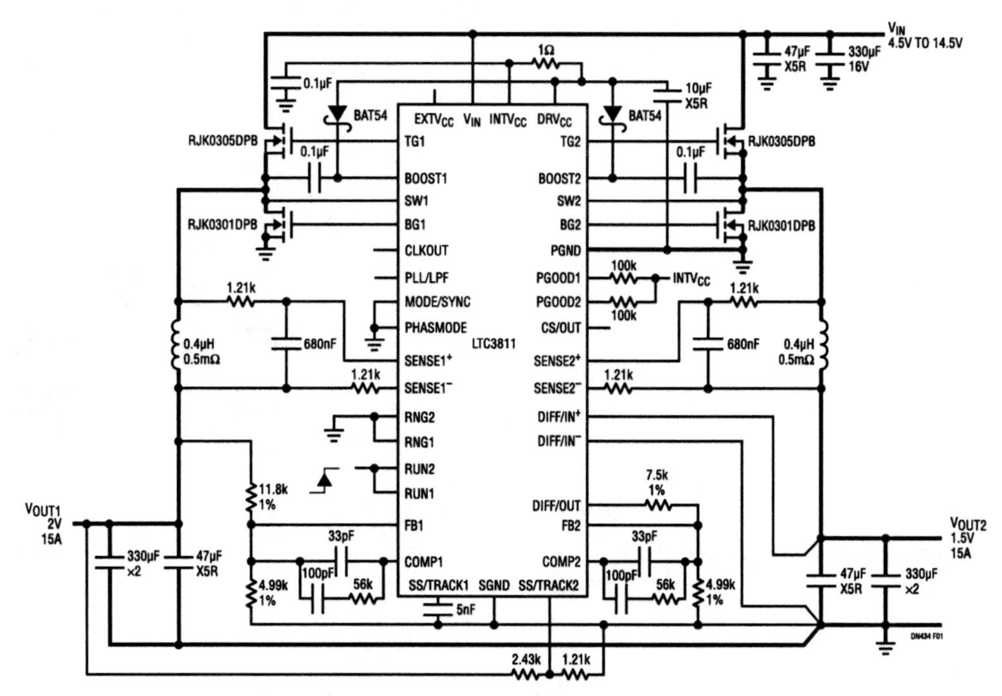

Figure 145.1 • Dual Output, 2-Phase Supply with Differential Remote Sensing and Inductor DCR Sensing

Analog Circuit Design: Design Note Collection. http://dx.doi.org/10.1016/B978-0-12-800001-4.00145-9

sensing in order to maximize efficiency and operate 180° out of phase in order to reduce the size of the input capacitor. Figure 145.2 illustrates the load step response for channel 2. Figure 145.3 illustrates low duty cycle waveforms for a 20V input, 1.2V output application.

For noise-sensitive applications, where the switching frequency needs to be synchronized to an external clock, the LTC3811 contains a PLL with an input range of 150kHz to 900kHz. In addition, the MODE/SYNC, PHASEMODE and CLKOUT pins allow multiple LTC3811s to be daisy-chained in order to produce a single high current output. The LTC3811 can be configured for 2-, 3-, 4-, 6- or 12-phase operation, extending the load current range to beyond 200A.

A tried-and-true architecture

The fixed frequency, peak current mode control architecture was chosen for its excellent channel-to-channel current matching and its robust cycle-by-cycle current limit. Current sensing can be done using either a resistor in series with the inductor or by sensing the DCR of the inductor with an RC filter. This gives the user a choice between optimum control of the maximum inductor current and maximum efficiency.

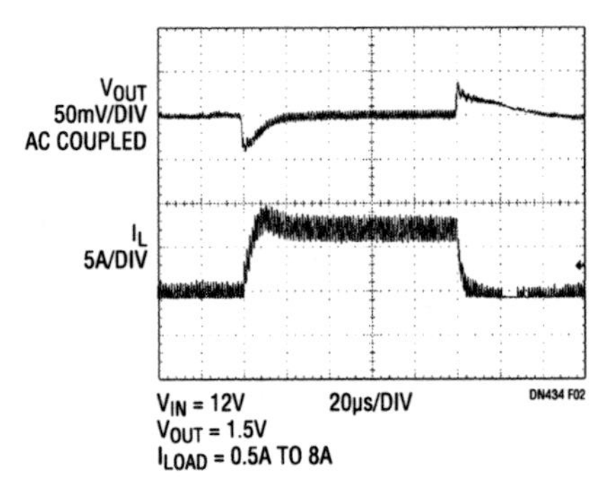

Figure 145.2 • Load Step Response

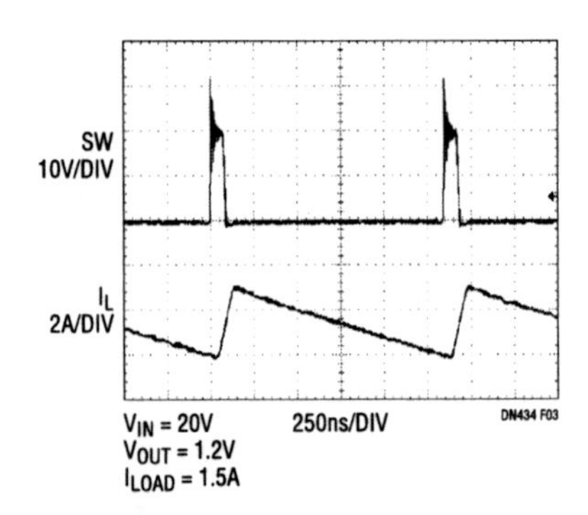

Figure 145.3 • LTC3811 Low Duty Cycle Waveforms

In order to accommodate the use of low DCR inductors and still maintain good control over the maximum output current, the current sense voltage for each channel is programmable from 24mV to 85mV using the RNG pins.

The LTC3811 has a 4.5V to 30V input voltage range and is available in two package options: a 38-pin 5mm×7mm QFN and a 36-pin SSOP.

Load step improvement with voltage positioning

For single-output multiphase applications, the LTC3811 contains an amplifier for voltage positioning purposes. The current sense input voltages are converted to an output current by a multiple-input, single-output transconductance amplifier, so that an error voltage proportional to the load current can be introduced at the input of the differential amplifier. This transconductance amplifier allows the user to program an output load line, improving the DC and AC output accuracy in the presence of load steps. Figure 145.4 illustrates the load step response for a 2-phase, single output power supply using the LTC3811.

Conclusion

The LTC3811 is a versatile, high performance synchronous buck controller optimized for low voltage, high current supply applications. With an output accuracy of ±0.5% and a remote sensing differential amplifier, it represents a new benchmark for DC/DC converters. It can easily be configured for either single or dual output supplies, inductor DCR sensing or a sense resistor, and it takes advantage of Linear Technology's proprietary PolyPhase current sharing architecture. The combination of a very low minimum on-time and a fixed frequency peak current mode control architecture no longer force the power supply designer to trade off performance for protection.

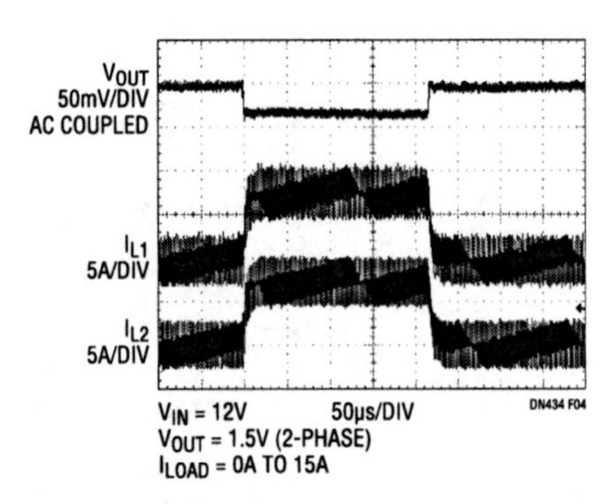

Figure 145.4 • Load Step Response for a 2-Phase, Single Output Supply with Voltage Positioning (Forced Continuous)

2-phase DC/DC controller makes fast, efficient and compact power supplies

146

David Chen

High efficiency, fast transient response and small size are often at odds in power supply designs. Fortunately, the 2-phase LTC3708 PWM controller makes it possible to simultaneously achieve all three. Figure 146.1 show the LTC3708 in a compact, high efficiency dual-output power supply that has excellent transient response.

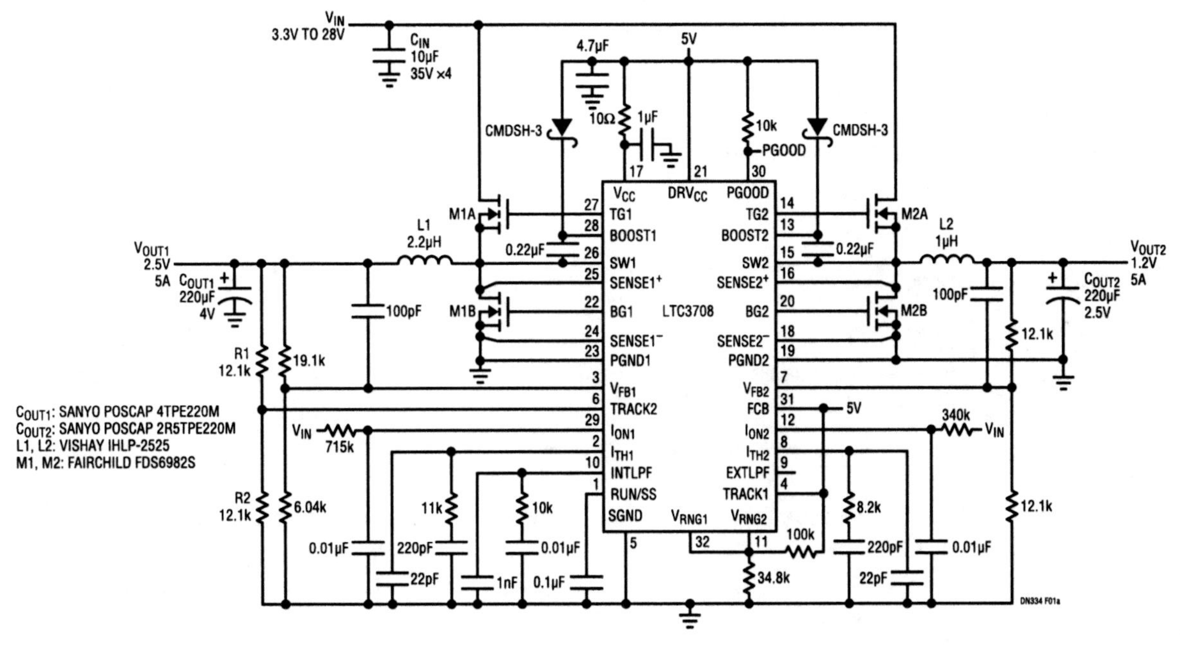

Figure 146.1a • Dual Power Supply with V_{OUT2} Tracking V_{OUT1}

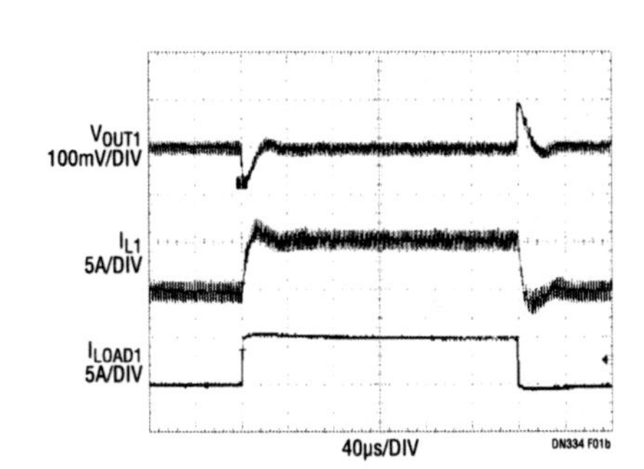

Figure 146.1b • Load Step on V_{OUT1}

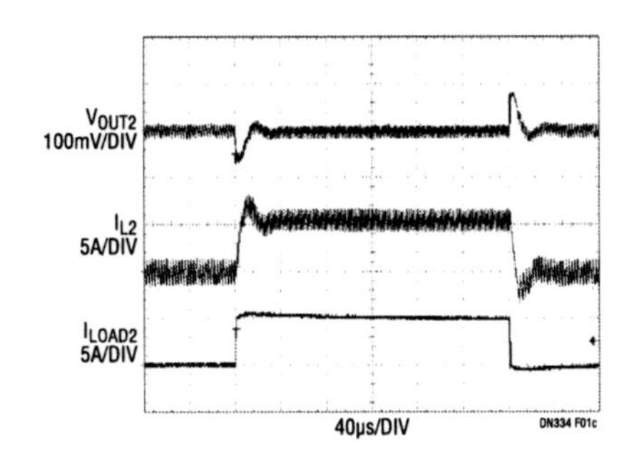

Figure 146.1c • Load Step on V_{OUT2}

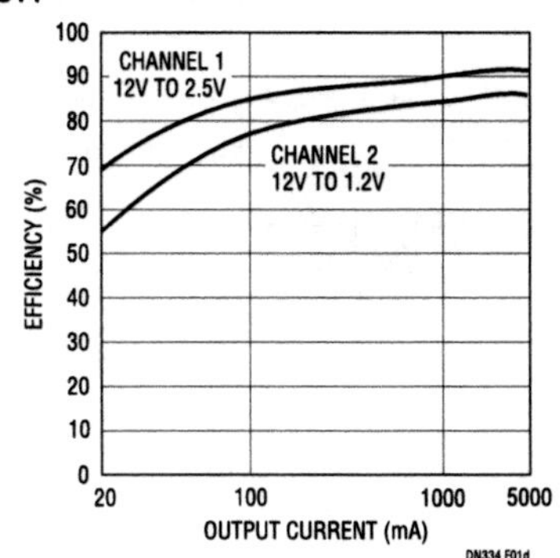

Figure 146.1d • Efficiency

Analog Circuit Design: Design Note Collection. http://dx.doi.org/10.1016/B978-0-12-800001-4.00146-0

The device uses a combination of design features to make all of this possible. High efficiency is attained through a combination of features: a No R_{SENSE} current sensing technique, 2-phase operation mode, onboard high current synchronous MOSFET gate drivers and a pulse skipping function that reduces the switching and gate driving losses at light loads. Fast transient response comes from the constant on-time control architecture with a very narrow pulse width (minimum $t_{ON} < 85ns$). Compact solution size is achieved because of the LTC3708's high frequency capability, minimized input and output capacitance requirements and high levels of circuit integration. All control circuitry and MOSFET gate drivers are incorporated within a small 5mm × 5mm QFN package. The LTC3708 also provides accurate voltage tracking over the entire output range at both ramp-up and ramp-down transitions (Figure 146.2). Table 146.1 explains these features in more detail.

The LTC3708 hosts other features that make it an ideal controller for high performance power management applications. The input voltage can be as high as 36V and the output regulates down to 0.6V. The current limit is user programmable to accommodate the variation of MOSFET $R_{DS(ON)}$ values. Protection functions include cycle-to-cycle current limit, an overvoltage crowbar and an optional short-circuit timer. When either output is out of regulation, a Power Good indicator falls low after 100μs masking—a technique that prevents transitory glitches and noise from falsely triggering system protection.

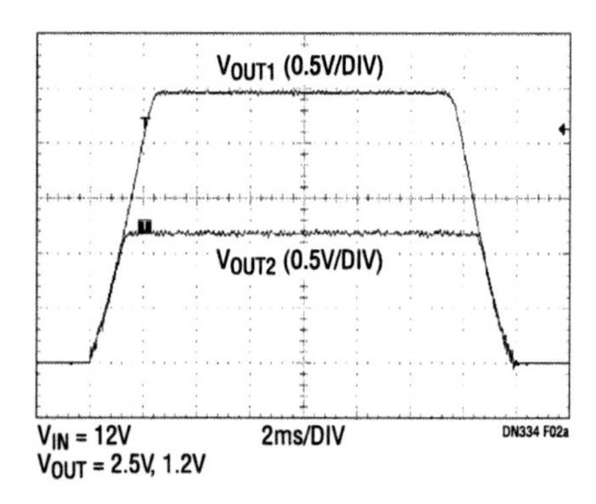

Figure 146.2a • Coincident Tracking (R1 = R2 = 12.1k)

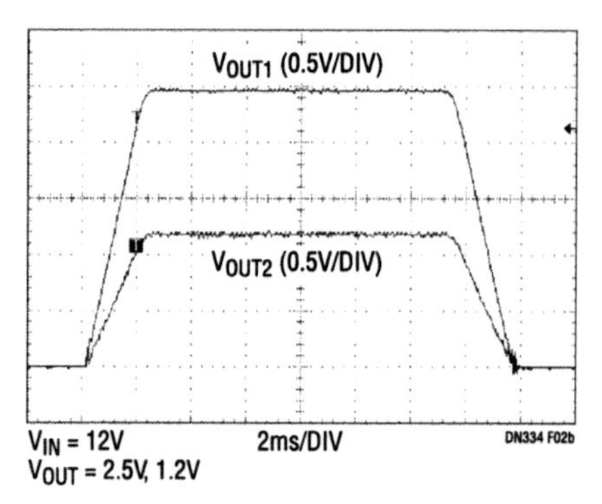

Figure 146.2b • Ratiometric Tracking (R1 = 19.1k, R2 = 6.04k)

Table 146.1 LTC3708 Design Features

FEATURES	FUNCTIONS	BENEFITS
Output Tracking	Various Modes of Tracking and Sequencing can be Programmed: Coincident, Ratiometric, etc	Simplifies the Timing Design of Multiple Supply Systems
No R_{SENSE}	Output Current is Sensed through the Synchronous MOSFET	Improves the Efficiency of Low Output Applications ($V_{OUT} \leq 5V$)
2-Phase Operation	Two Output Channels Operate at Same Frequency with 180° Phase Shift	Reduces Input RMS Current and EMI Noises; Minimizes Input Capacitance
Constant On-Time Control Architecture	The Top MOSFETs can be Turned on Immediately Without Clock Latency	Expedites Transient Response and Reduces Output Capacitance
Minimum $t_{ON} < 85ns$	This is the Minimum Duration that the Top MOSFETs Need to be On	Expedites Transient Response and Enables High Frequency Designs
Frequency Synchronization	The Switching Frequency can be Synchronized to an External Clock	Maintains Constant Frequency Operation and Synchronizes all Switching Regulators in a System
Pulse Skipping at Light Loads	The Switching Period Extends at Light Loads and Reverse Current is Inhibited	Improves Light Load Efficiency with Minimum Switching and Gate Driving Loss

High performance 3-phase power supply delivers 65A and high efficiency over the entire load range

147

Wei Chen

Introduction

CPUs used in notebook computers and other mobile applications now draw more than 30A of current and may draw as much as 65A in the near future. For these heavy loads, high efficiency power supplies are required to protect the system from excessive thermal stress. With this in mind, PolyPhase switching DC/DC converters have become the standard for CPU power supplies because they are extremely efficient at these heavy loads. Nevertheless, for portable applications that spend much of the time in sleep or standby modes, *light load* efficiency has also grown in importance, where conserving energy for maximum battery run-time is also a priority. Unfortunately, traditional PolyPhase converters, although exceptional performers at heavy loads, do not yield comparable efficiencies at light load. Linear Technology Corporation offers a solution to this problem with a new family of PolyPhase controllers that allow the design of converters that are efficient over the entire CPU load range.

These new 3-phase controllers, the LTC3730, LTC3731 and LTC3732, operate efficiently at both heavy loads and light loads. These new controllers introduce Stage Shedding operation to improve light load efficiency. Like LTC's 2-phase controllers, these new 3-phase controllers provide true

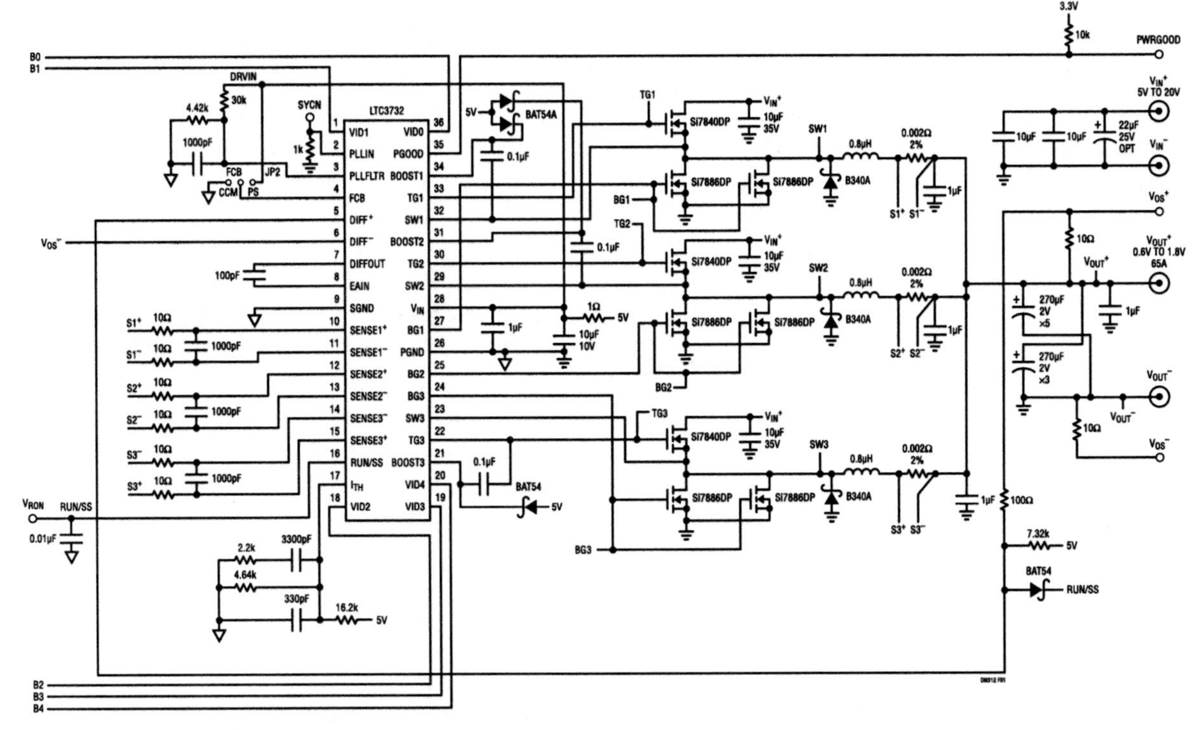

Figure 147.1 • Schematic Diagram of a 3-Phase LTC3732 65A VRM9.x Power Supply

Analog Circuit Design: Design Note Collection. http://dx.doi.org/10.1016/B978-0-12-800001-4.00147-2

remote sensing on both the positive and negative output rails to ensure tight output regulation at high output currents, and Kelvin sensing (positive and negative) on the pads of each current sense resistor to achieve accurate current sharing, even if the layout of parallel power stages is not symmetrical. All controllers feature integrated high current MOSFET drivers for up to 600kHz switching frequency, thus minimizing the overall power supply size and component count.

The LTC3731 is a versatile 3-phase controller that generates a 30 or 60 degree phase-shifted clock output based on the voltage level of the PHASMD pin. This feature allows several LTC3731s to be paralleled for up to a 12-phase operation. The output voltage is programmed by external resistors. The LTC3730 is a dedicated 3-phase controller with 5-bit VID output programming that is compatible with IMVP2 and IMVP3. The internal op amp can be used to program voltage offsets for different CPU operation modes. The LTC3732 is another 3-phase controller with a 5-bit VID output programming that is compatible with Intel's VRM9.x specs. All three controllers are available in space saving 36-lead SSOP packages, while the LTC3731 is also available in a much smaller, thermally enhanced 5mm × 5mm QFN package.

Stage Shedding operation

In high current applications, low $R_{DS(ON)}$ MOSFETs are typically chosen to minimize conduction losses at full loads. At light loads, though, the high gate charge and parasitic capacitance of these MOSFETs often cause power losses associated with gate driving and switching. Also, the core loses of inductors dominate the overall inductor power losses at light loads. Since the switching losses, gate driving losses and inductor core losses do not decrease with load currents, light load efficiency suffers.

Another cause of low efficiency at light loads is the circulating current among the paralleled stages. In a PolyPhase synchronous buck converter, the inductor current in each synchronous buck stage can reverse at light loads due to synchronous rectification. In a practical PolyPhase design, a current sharing error always exists because of tolerances in the sense resistors and slight differences between paralleled channels within the controllers. Any current sharing error among the paralleled stages introduces circulating currents that introduce additional power losses.

To reduce these light load power losses, Stage Shedding operation shuts down all but one channel, completely eliminating the circulating currents. Furthermore, this mode eliminates the gate drive losses, MOSFET switching losses and inductor core losses of the unused channels. The result is much higher efficiency at light loads and since the controller maintains the basic regulation loop, Stage Shedding operation has no effect on output regulation accuracy.

3-phase high efficiency VRM9.x power supplies for Pentium 4 CPU

Figure 147.1 shows a 3-phase VRM9.x power supply for a Pentium 4 microprocessor. It uses the LTC3732 to drive nine small PowerPak SO-8 MOSFETs for 65A output current. To provide higher output currents, simply use lower $R_{DS(ON)}$ MOSFETs and higher current rated inductors. R3 and R4 implement a lossless active voltage positioning (AVP) technique to minimize the output capacitor size. For more detailed technical information about AVP, see the LTC1736 data sheet or Design Note 224. Figure 147.2 shows the measured efficiency under different load conditions. The input is 12V, the output voltage is 1.4V and the switching frequency is 450kHz. Figure 147.2 shows the efficiencies measured when Stage Shedding operation is enabled and for conventional PolyPhase operation. As the chart shows, Stage Shedding operation significantly improves efficiencies at light loads (≤10A). For instance, at 1% of full load (0.6A), Stage Shedding operation improves efficiency by more than 25%.

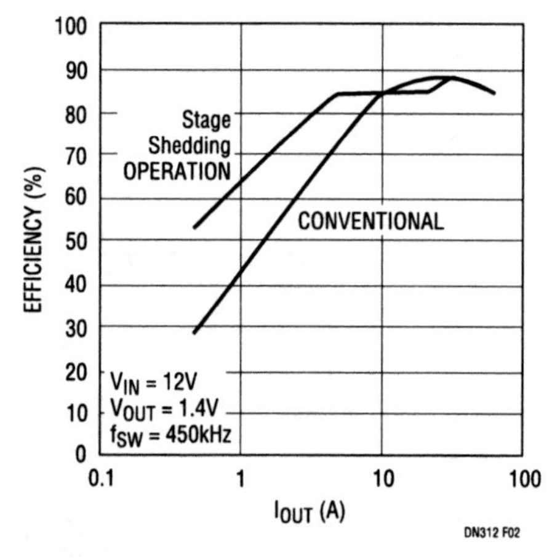

Figure 147.2 • Measured Efficiency (Stage Shedding Operation vs Conventional Operation). Load Range is 0.6A to 55A

Reduce component count and improve efficiency in SLIC and RF power supplies

148

Tick Houk

Introduction

Nonisolated power supplies with negative output voltages require op amps and other "glue" circuitry in order to level-shift the feedback voltage to the error amplifier. In applications such as SLIC (subscriber line interface circuit) power supplies, this extra circuitry adds unwanted cost and requires more space on the PC board, with no increase in system performance.

The LTC3704 is a high performance, single-ended current mode DC/DC controller IC that allows the user to directly sense a negative output voltage using the NFB pin. A unity-gain inverting amplifier within the IC, combined with a precision bandgap voltage reference, give users a −1.230V input that can be directly connected to a resistor divider between the negative output voltage and ground. Additional features, such as No R_{SENSE} current mode control, programmable operating frequency (50kHz to 1MHz), synchronization capability and a wide input range (2.5V to 36V) improve efficiency and greatly ease design. The LTC3704 is housed in a small 10-lead MSOP package.

A dual output SLIC supply with simplified feedback using the LTC3704

Figure 148.1 illustrates a dual output telecom power supply designed for use in subscriber line interface circuits or SLICs. The input to the SLIC power supply is some form of battery (lead-acid or lithium-ion, for example) so that extended talk-time can be provided to POTS (plain old telephone system) phones during an AC line failure (or rolling blackout). The output voltage is typically proportional to the distance the

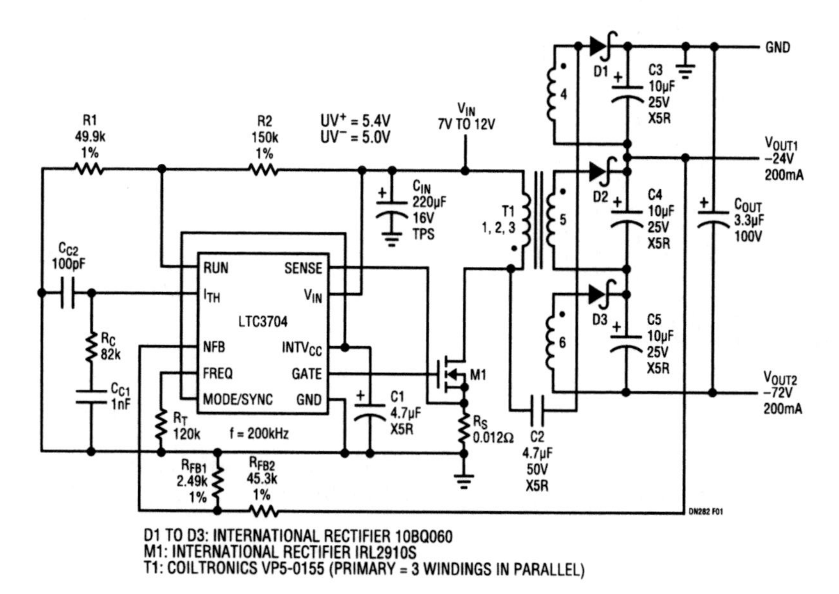

Figure 148.1 • Dual Output SLIC Supply with the LTC3704 Simplifiers Feedback

Analog Circuit Design: Design Note Collection. http://dx.doi.org/10.1016/B978-0-12-800001-4.00148-4

subscriber line runs from the local hub to a house or office, in order to compensate for the impedance of the loop. Multiple output supplies are used to supply groups of users at different distances from the hub.

This SLIC power supply takes advantage of the negative feedback input (NFB) on the LTC3704 and therefore, eliminates the need for an additional op amp. The −24V output is regulated directly and the −72V output is obtained by stacking additional windings on the −24V output. The −24V output of this supply uses one secondary winding in a SEPIC-type configuration in order to reduce the input ripple current, whereas the −72V output uses the other two windings in a conventional flyback mode. A standard, 6-winding VERSA-PAC transformer (VP5-0155) was used, with three windings connected in parallel on the primary in order to satisfy the high current demand (the peak switch current can be almost 4A at an input of 7V). A 5.2V LDO within the LTC3704 provides a regulated supply for the gate driver which is capable of driving very large power MOSFETs (50nC to 100nC).

Improved battery protection using the LTC3704's programmable undervoltage lockout

With most lead-acid or lithium-ion chemistries, deep discharging of the battery cells can severely reduce the life of the battery. As a result, it is important to monitor the battery and turn the converter off before the battery voltage reaches an unsafe level. In Figure 148.1, the RUN pin on the LTC3704 monitors the battery voltage using resistors R1 and R2 and turns off the power supply when the input supply drops below 5V. A hysteresis level of 8% is provided in order to increase noise immunity (UV^+ is 5.4V).

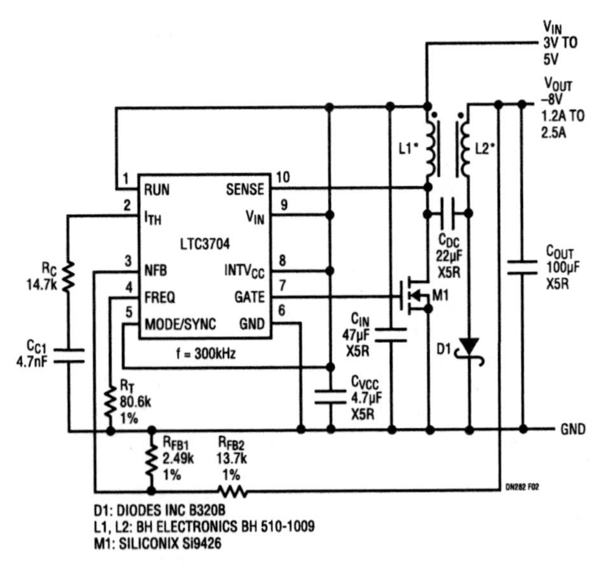

Figure 148.2 • A High Efficiency −8V RF Power Supply

A current mode, −8.0V, 1.2A RF power supply with no current sense resistor

The design illustrated in Figure 148.2 takes advantage of LTC's proprietary No R_{SENSE} technology to provide true current mode control without a discrete current sense resistor. The voltage drop across the power MOSFET is sensed during the on-time, thereby providing the control loop with a "lossless" method of measuring the switch current. This technique provides the maximum efficiency possible for a single-ended current mode converter, reduces board space and reduces the overall cost of the power supply in applications where the drain of the power MOSFET is less than 36V (the absolute maximum rating of the SENSE pin). It should be noted that the output voltage and maximum output current of this supply can easily be scaled by the choice of the components around the chip without modifying the basic design.

Figures 148.3 and 148.4 illustrate the maximum output current vs input voltage for the supply and the efficiency vs load current at input voltages of 3V and 5V, respectively.

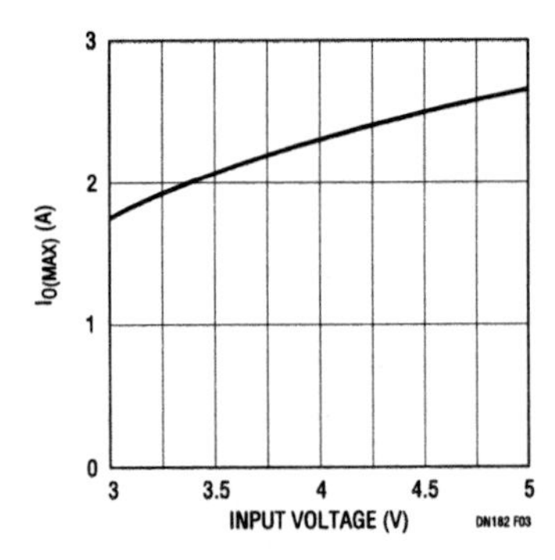

Figure 148.3 • Maximum Output Current vs Input Voltage for the −8V RF Power Supply

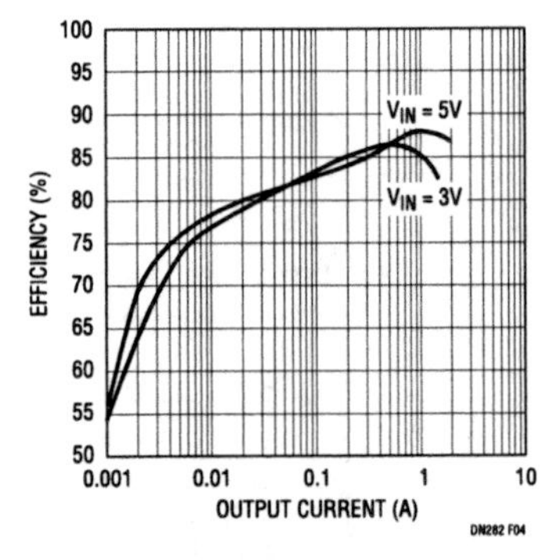

Figure 148.4 • Efficiency vs Output Current for the −8V RF Power Supply

SOT-23 DC/DC converters generate up to ±35V outputs and consume only 20µA of quiescent current

149

Bryan Legates

Today's portable devices need small power supply solutions that operate with a minimum of supply current. To meet these needs, Linear Technology introduces the LT1615, LT1615-1, LT1617 and LT1617-1 micropower SOT-23 DC/DC converters. With an input voltage range as low as 1V and an output voltage range as high as ±35V, these devices provide considerable power supply design flexibility.

The LT1615 and LT1615-1 are designed to regulate positive output voltages, whereas the LT1617 and LT1617-1 are designed to directly regulate negative output voltages without the need for feedback level-shifting circuitry. The LT1615 and LT1617 have a 350mA current limit and a minimum input voltage of 1.2V, whereas the LT1615-1 and LT1617-1 have a lower, 100mA current limit and a minimum input voltage of 1V. All four converters use tiny, low profile inductors and capacitors to minimize the overall system footprint and cost. With a quiescent current of only 20µA and a shutdown current of 0.5µA, these devices squeeze the most life out of any battery-powered application.

±20V dual output converter with output disconnect

Today, most portable devices use a liquid crystal display (LCD). Different manufacturers require substantially different bias voltages for their LCDs. Typically, a single 9V to 25V supply is needed (either positive or negative), but some LCDs require both a positive and negative supply. Figure 149.1 shows a ±20V dual output converter ideally suited for LCD bias applications needing both supplies. Both outputs are developed using charge pumps, so both are disconnected from the input when the LT1617 is turned off. Because the supplies are generated in the same manner, this circuit features excellent cross-regulation: for a 5× difference in output currents, the positive and negative output voltages differ less than 1%; for a 10× difference, they differ less than 2%. A similar circuit can be implemented using the LT1615 if the regulation of the positive output is more important. As shown in Figure 149.2, efficiency reaches 78% with a fresh 4-cell alkaline battery.

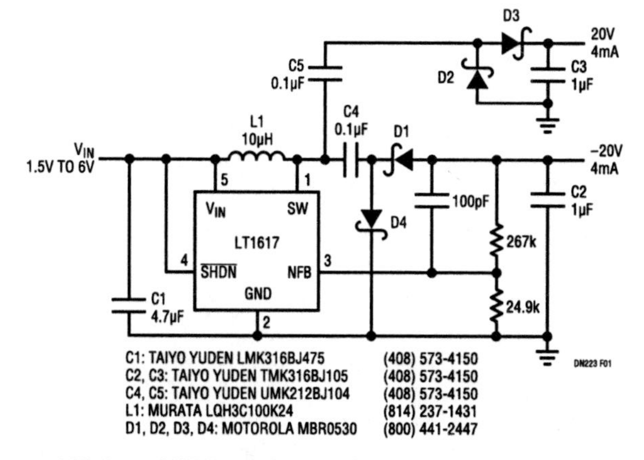

Figure 149.1 • ±20V Dual Output Converter with Output Disconnect

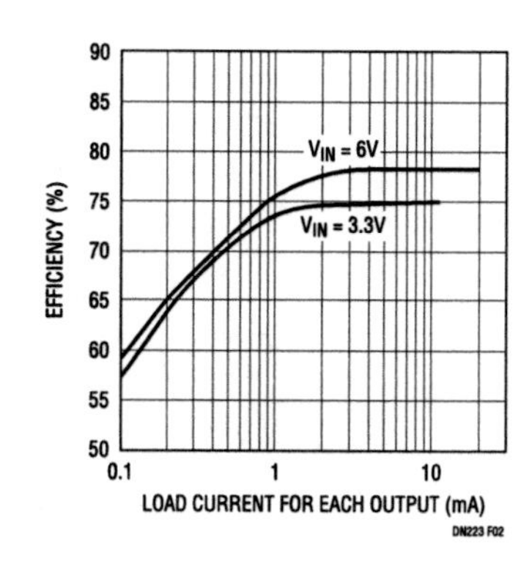

Figure 149.2 • ±20V Dual Output Converter Efficiency

Analog Circuit Design: Design Note Collection. http://dx.doi.org/10.1016/B978-0-12-800001-4.00149-6

24V boost converter

Figure 149.3 shows a circuit ideal for LCD applications needing only a positive bias voltage. This 24V boost converter delivers 10mA from a nearly discharged single Li-Ion cell. An input voltage as low as 1.5V can be used with this converter, but the output current capability reduces to 5mA. Converter efficiency is shown in Figure 149.4.

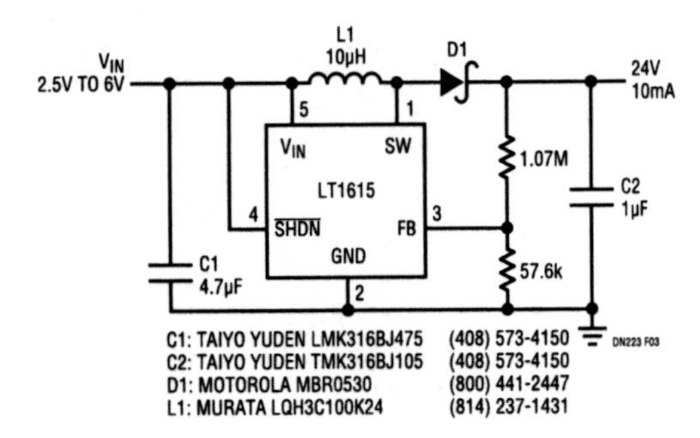

Figure 149.3 • 24V Boost Converter

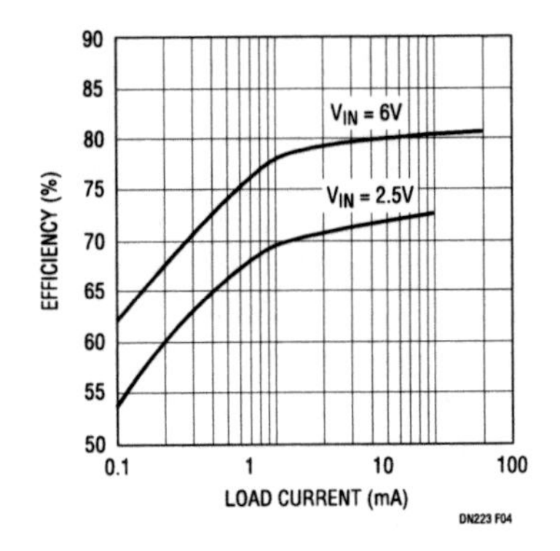

Figure 149.4 • 24V Boost Converter Efficiency

1V to 35V boost converter

The circuit in Figure 149.5 shows the impressive input and output voltage range of the LT1615-1. As shown, the circuit will work from one to four alkaline cells or a single cell Li-Ion battery. The maximum input voltage for this circuit is limited by the 6.3V voltage rating on the input capacitor, C1. The LT1615-1 can operate with an input voltage as high as 15V. The output current is limited by the 1V minimum input (this converter can provide 2mA with a 3V input). If a larger output current is needed, but operation from a 1V input is not required, use an LT1615 in place of the LT1615-1 to obtain a 3× increase in maximum output current.

1-cell to 3V boost converter

A 1-cell alkaline to 3V boost converter using the LT1615-1 is shown in Figure 149.6. Capable of providing 15mA of output current, this converter occupies a board area less than 1/4″ by 5/16″ (less than 0.078inches^2). See Figure 149.7 for converter efficiency, which reaches 75% with a fresh 1-cell alkaline battery.

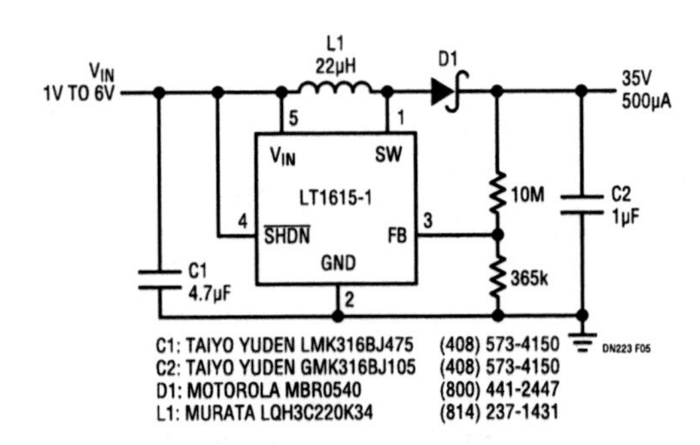

Figure 149.5 • 1V to 35V Boost Converter

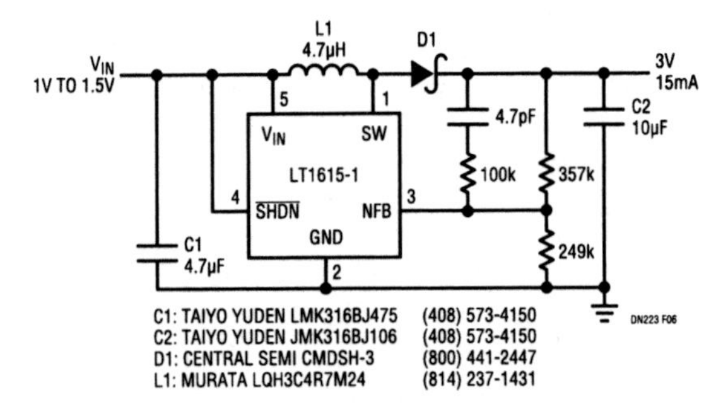

Figure 149.6 • 1-Cell Alkaline to 3V Boost Converter

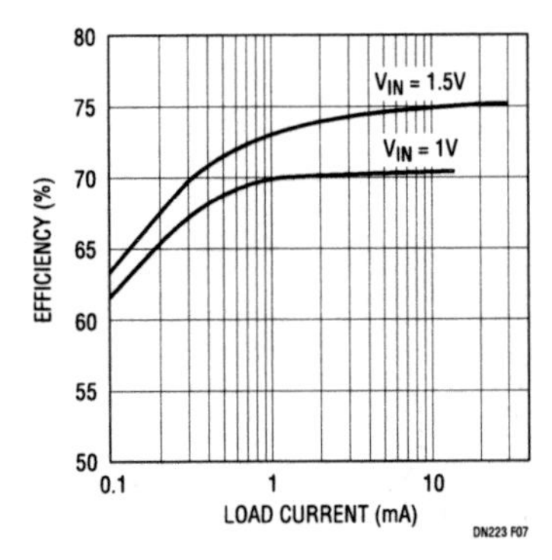

Figure 149.7 • 1-Cell to 3V Boost Converter Efficiency

Section 7

Switching Regulator Design: Buck-Boost Controllers

80V synchronous 4-switch buck-boost controller delivers hundreds of watts with 99% efficiency

150

Keith Szolusha Tage Bjorklund

Introduction

A DC/DC converter's efficiency and component temperature are important considerations in high power applications where high current could overheat the catch diode used in an asynchronous buck or boost topology. Replacing the catch diode with a synchronous switch can significantly improve overall converter efficiency and eliminate much of the heat that would be otherwise generated in the non-synchronous catch diode.

The advantages offered by a synchronous buck or boost topology can also be applied to a buck-boost topology, where the converter's output voltage falls within its input range. In this case, a synchronous 4-switch buck-boost converter using a single inductor offers the same advantages as a 2-switch synchronous buck or boost.

The LT8705 is a synchronous 4-switch buck-boost controller IC that can deliver hundreds of watts with high efficiency for constant voltage or constant current applications from wide-ranging inputs (up to 80V). It uses a robust synchronous switch topology and adds the versatility of four servo loops (voltage and current at input and output), making it possible to design high power battery chargers and solar panel converters with minimum component count. These are only two examples of the many high power, high current telecom, automotive and industrial solutions the LT8705 can produce.

240W 48V 5A telecom power supply

In telecom applications the input voltage has a wide (36V to 72V) range. Power converters that deliver a stabilized 48V DC voltage to the loads are commonly used. The LT8705 can easily handle hundreds of watts at 48V output voltage, at efficiency as high as 99%. Figure 150.1 shows one example with 5A (240W) output (Figure 150.2).

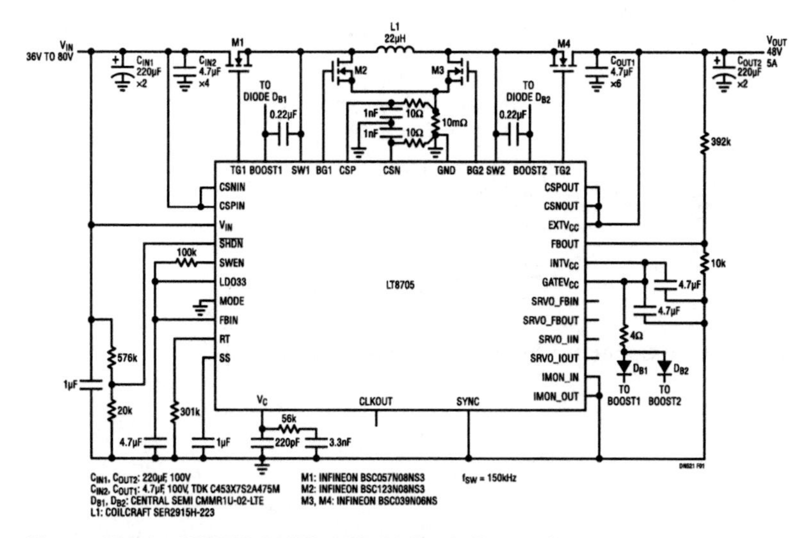

Figure 150.1 • LT8705 240W, 48V, 5A Buck-Boost Converter for Telecom Voltage Stabilization

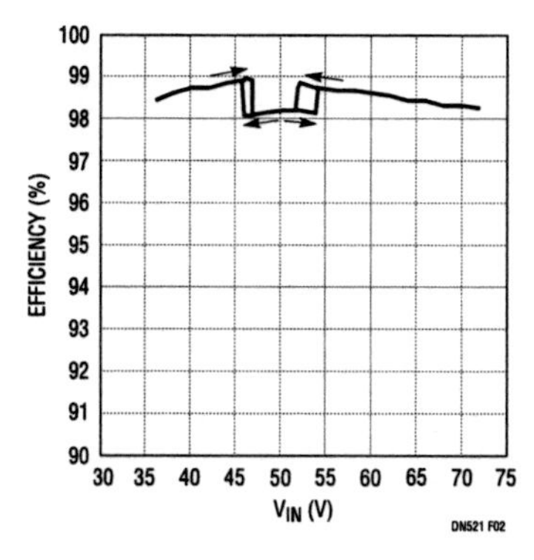

Figure 150.2 • The Efficiency of the LT8705, 48V Converter Is as High as 99%

Analog Circuit Design: Design Note Collection. http://dx.doi.org/10.1016/B978-0-12-800001-4.00150-2

500W charger for 12S LiFePO4 battery

Figure 150.3 shows a circuit for charging a lithium iron phosphate battery from a 48V (±10%) input voltage. The battery has 12 cells in series, so the maximum charge voltage is 44V. This means that the circuit will operate in step-down (buck) mode most of the time, but at lowest input voltage it must operate in buck-boost mode.

At a 48V input voltage the circuit has an efficiency of 99% at full load. The efficiency is high because only the input stage (M1, M2) is switching at high duty cycle and M4 is on continuously. The efficiency drops slightly if the input voltage is reduced to the minimum input voltage (43.2V), as the LT8705 circuit then must operate in buck-boost mode when all MOSFETs are switching.

An external microcontroller can be used for the charge algorithm, and for controlling current and voltage from the LT8705 power converter.

Four servo loops and wide voltage range

The LT8705's 2.8V to 80V input and its 1.3V to 80V output ranges, combined with its four servo loops, allow it to easily solve a number of traditionally complex problems. The four servo loops can be used to control input and output voltages and currents. For instance, the input voltage and current can be regulated along with the output voltage and current for maximum power point solar panel applications.

The IC outputs a flag for each servo loop, indicating which is in control at any given time. This is particularly useful information for microcontrollers in battery chargers and solar panel converters.

Conclusion

The LT8705 is an 80V synchronous 4-switch buck-boost controller, that can provide hundreds of watts at up to 99% efficiency with a single inductor. Its four servo-loops allow it to regulate current and/or voltage for both the input and output.

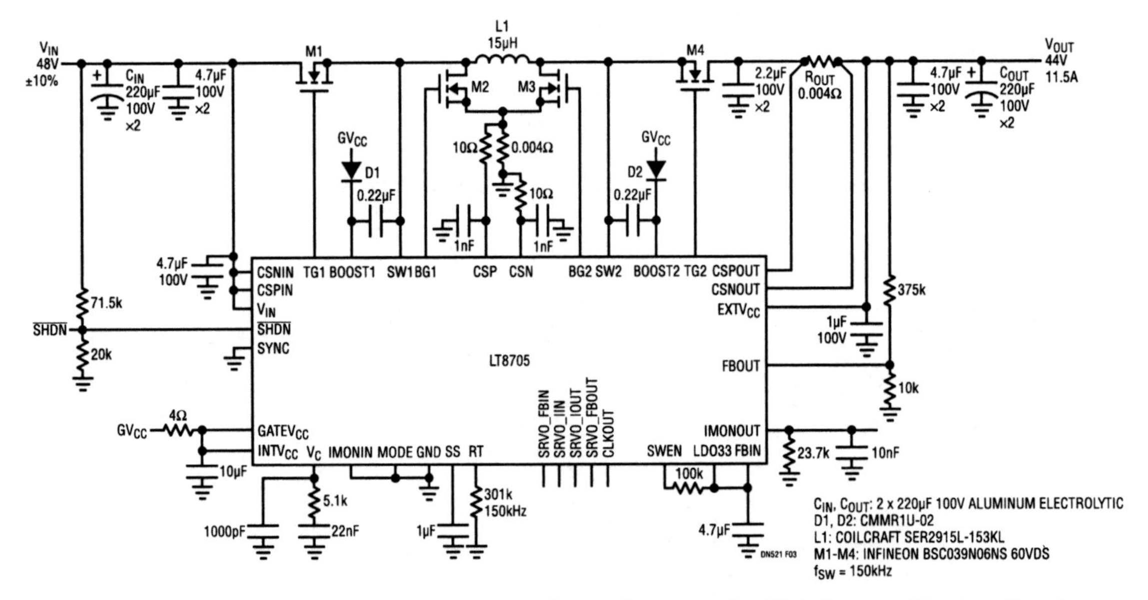

Figure 150.3 • LT8705 500W, 44V, 11.5A Buck-Boost Converter for High Powered Battery Supply

Wide input voltage range boost/inverting/SEPIC controller works down to an input voltage of 1.6V

151

Zhongming Ye

Introduction

Many of today's electronic devices require an inverting or noninverting converter or sometimes both. They also need to operate from a variety of power sources including USB, wall adapters, alkaline and lithium batteries. To produce various polarity outputs from variable input voltages, power supply designers often use a variety of regulator ICs, which makes for a long inventory list.

The LT3759 operates over an input voltage range from 1.6V to 42V and controls either positive or negative outputs using the same feedback pin, thus shortening the inventory list and simplifying design. It also packs many popular features such as soft-start, adjustable frequency and synchronization into a small footprint. The LT3759 comes in a 5mm × 4mm 12-pin MSE package and can be used in multiple configurations such as boost, SEPIC, flyback and Cuk topologies.

Wide input voltage range with internal LDO

The LT3759's wide input range simplifies the design of power supplies that must be compatible with a wide array of power sources. There is no need to add an external regulator or a slow-charge hysteretic start scheme because the LT3759 includes two internal low dropout (LDO) voltage regulators powered from V_{IN} and DRIVE respectively, allowing simple start-up and biasing. The LT3759's internal $INTV_{CC}$ current limit function protects the IC from excessive on-chip power dissipation.

Sensing output voltage made easier

LT3759 features a novel FBX pin architecture that simplifies the design of inverting and noninverting converters. It contains two internal error amplifiers—one senses positive

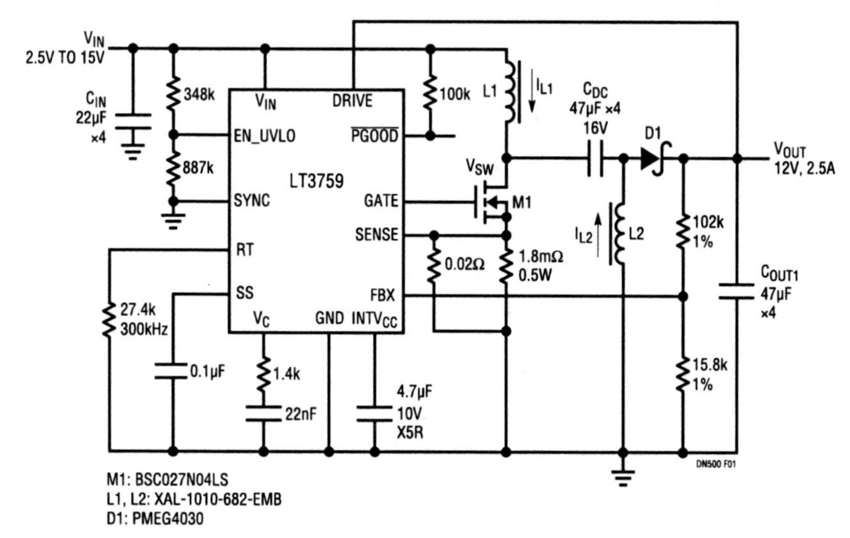

Figure 151.1 • SEPIC Converter Produces 12V Output from 2.5V to 15V Inputs

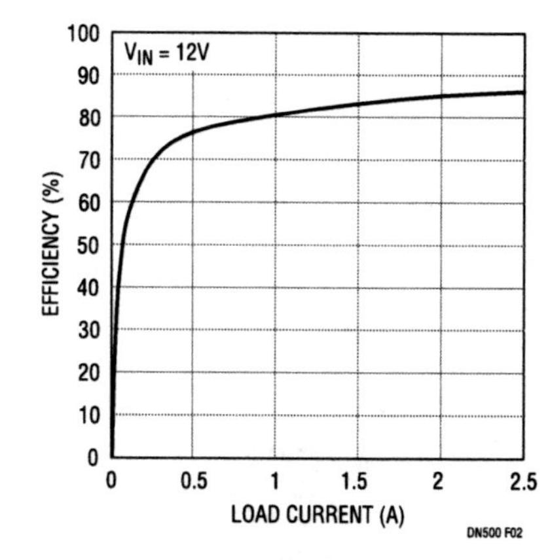

Figure 151.2 • Efficiency for the Converter in Figure 151.1

Analog Circuit Design: Design Note Collection. http://dx.doi.org/10.1016/B978-0-12-800001-4.00151-4

outputs and the other negative—allowing the FBX pin to be connected directly to a divider from either a positive output or a negative output, eliminating any confusion associated with positive or negative output sensing and simplifying the board layout. Simply decide the output polarity and topology and the LT3759 does the rest.

Adjustable/synchronizable switching frequency

It is often necessary to operate a converter at a particular frequency, especially if the converter is used in an RF communications product that is sensitive to spectral noise in certain frequency bands. Also, if the area available for a converter is limited, operating at higher frequencies allows the use of smaller component sizes, reducing the real estate required and the output ripple. If power loss is a concern, switching at a lower frequency reduces switching losses, improving efficiency. The switching frequency can be set from 100kHz to 1MHz via a single resistor from the RT pin to ground. The device can also be synchronized to an external clock via the SYNC pin.

Precision UVLO and soft-start

Input supply undervoltage (UVLO) for sequencing or start-up overcurrent protection is easily achieved by driving the UVLO with a resistor divider from the V_{IN} supply. The divider output produces 1.22V at the UVLO pin when V_{IN} is at the desired UVLO rising threshold voltage. The UVLO pin has an adjustable input hysteresis, which allows the IC to ignore a settable input supply droop before disabling the converter. During a UVLO event, the IC is disabled and V_{IN} quiescent current drops to 1μA or lower.

The SS pin provides access to the soft-start feature, which reduces the peak input current and prevents output voltage overshoot during start-up or recovery from a fault condition. The SS pin reduces the inrush current by lowering the switch peak current. In this way soft-start allows the output capacitor to charge gradually toward its final value.

A 2.5V to 15V to 12V SEPIC converter

Figure 151.1 shows a 2.5V to 15V input, 12V/2.5A output SEPIC power supply using the LT3759. The typical efficiency for this converter is shown in Figure 151.2. Figure 151.3 shows the switch waveform during an output short-circuit event. Notice how the switching frequency folds back to one-third of the regular frequency as soon as the output voltage is shorted to ground. This feature enhances short-circuit performance of both Cuk and SEPIC converters.

A 1.8V to 4.5V to 5V/2A boost converter

Figure 151.4 shows a 5V, 2A output converter that takes an input of 4.5V to as low as 1.8V. The LT3759 is configured as a boost converter for this application, where the converter output voltage is higher than the input voltage. The 500kHz operating frequency allows the use of small inductor and output capacitors.

Conclusion

The LT3759 is a versatile IC that integrates a rich set of unique features in a tiny 5mm × 4mm 12-pin MSE package. It accepts wide input voltage range from 1.6V to 42V, with low shutdown current and frequency foldback at output short-circuit. The LT3759 is ideal for wide input voltage applications from single-cell, lithium-ion powered systems to automotive, industrial and telecommunications power supplies. The high level of integration yields a simple, low parts-count solution for boost, SEPIC and Inverting converters.

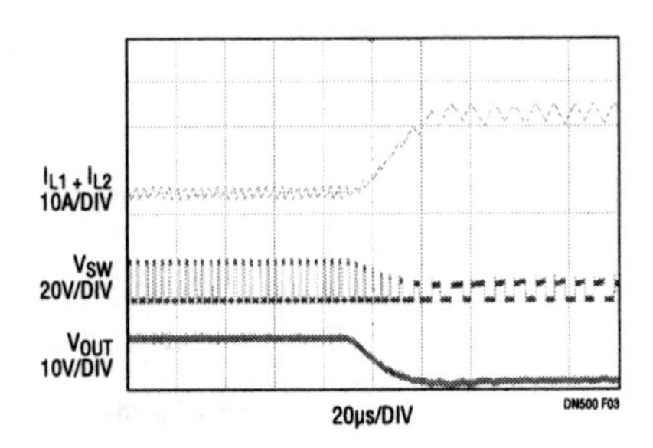

Figure 151.3 • Short-Circuit Event for the Converter in Figure 151.1

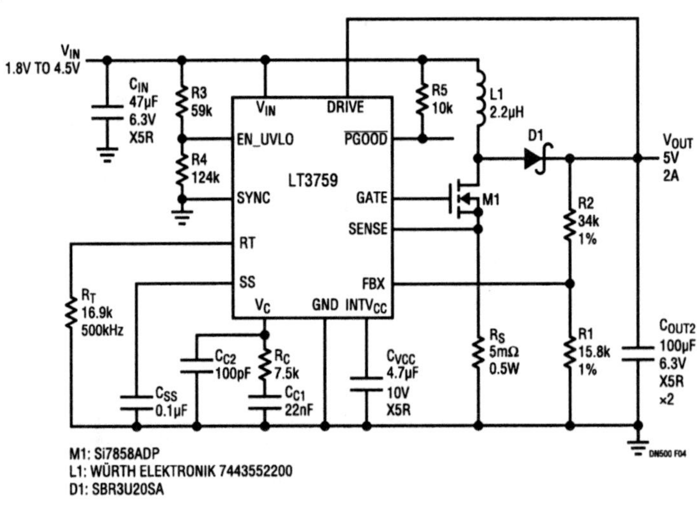

Figure 151.4 • Boost Converter Produces 5V/2A Output from 1.8V to 4.5V Input

High efficiency 4-switch buck-boost controller provides accurate output current limit

152

Tage Bjorklund

Introduction

A 4-switch buck-boost converter (Figure 152.1) is often a better alternative to a transformer-based topology when a converter's input voltage can be above or below the regulated output voltage and isolation is not required. A buck-boost converter provides a wider input voltage range, better efficiency and has no need of a bulky transformer. A buck-boost converter is much more efficient than a comparable SEPIC converter.

The LTC3789 is a buck-boost switching regulator controller that operates in current mode at a constant switching frequency. Current mode control simplifies loop compensation and yields excellent load and line transient response with only small output and input capacitance. Furthermore, an accurate inductor current limit allows use of a small size inductor.

LTC3789 features

The internal oscillator can be programmed by a resistor, applied voltage, or phase-locked to an external clock within a 200kHz to 600kHz frequency range. The LTC3789's wide 4V to 38V (40V maximum) input and output range and seamless, low noise transitions between operating regions is ideal for automotive, telecom and battery-powered systems. Higher input voltage is easily enabled by adding a high voltage gate driver for the input side MOSFETs, for example the LTC4444-5.

Figure 152.1 shows a simplified schematic of a buck-boost converter. The LTC3789 controls four external N-channel MOSFETs (A, B, C, D). The power stage uses one inductor (L), and a current sense resistor (R_{SENSE}) for current mode control and inductor current limit. Figures 152.2–152.4 show the switch waveforms for buck, buck-boost, and boost modes of operation.

The LTC3789 is available in low profile 28-pin 4mm × 5mm QFN and narrow SSOP packages.

12V, 5A output from a 4V to 38V input

Figure 152.5 shows a buck-boost regulator that takes a 4V to 38V input and converts it to a fixed 12V.

The LTC3789 includes two internal series regulators that provide 5.5V for gate drivers and control circuitry. During start-up the current is drawn from V_{IN} (40V maximum rating), but

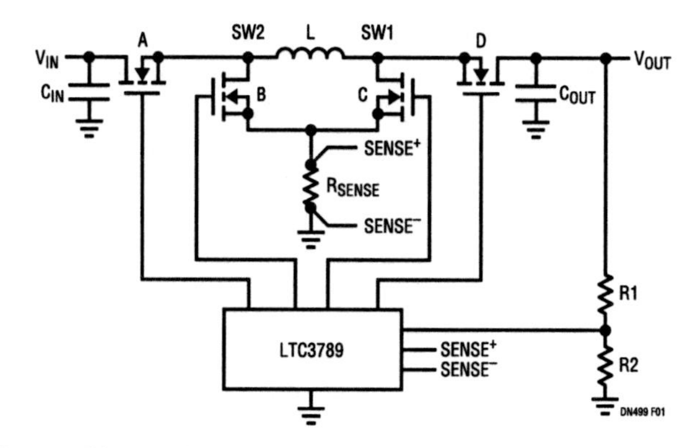

Figure 152.1 • 4-Switch Buck-Boost Converter Using the LTC3789

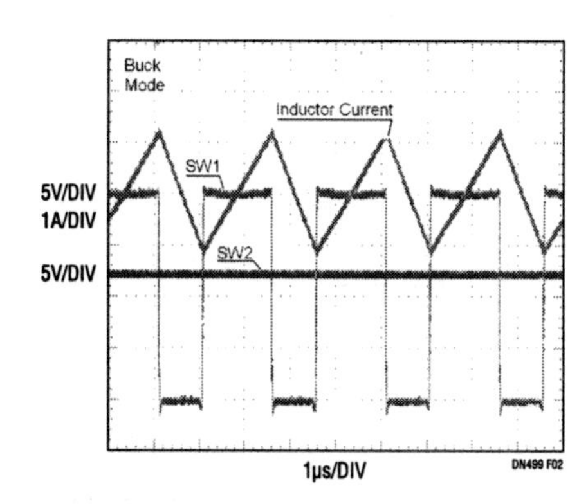

Figure 152.2 • Buck Mode [V_{IN} = 20V, V_{OUT} = 12V]

Analog Circuit Design: Design Note Collection. http://dx.doi.org/10.1016/B978-0-12-800001-4.00152-6

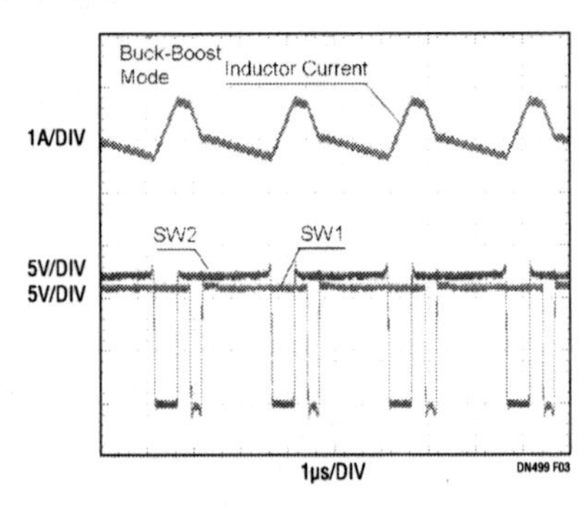

Figure 152.3 • Buck-Boost Mode [V_{IN} = 11V, V_{OUT} = 12V]

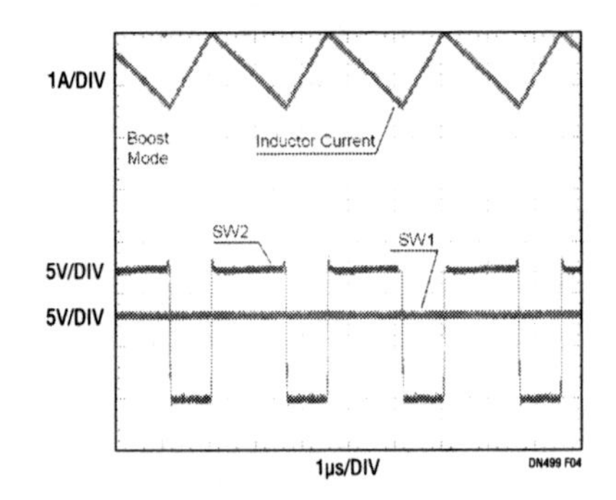

Figure 152.4 • Boost Mode [V_{IN} = 8V, V_{OUT} = 12V]

switches to $EXTV_{CC}$ (14V maximum rating) as soon as the voltage on $EXTV_{CC}$ exceeds 4.8V, to decrease power loss.

This 4-switch topology allows the output to be disconnected from the input during shutdown. LTC3789 features adjustable soft-start, no reverse current during start-up and 1% output voltage accuracy.

The MODE/PLLIN pin can be used to select between pulse-skipping mode and forced continuous mode operation, or allows the IC to be synchronized to an external clock. Pulse-skipping mode offers the lowest ripple at light loads, while forced continuous mode operates at a constant frequency for noise-sensitive applications.

A power good output pin (PGOOD) indicates when the output is within 10% of its set point.

Accurate output (or input) current limit

The optional output current feedback loop shown in Figure 152.6 provides support for battery charging and other constant current applications, where an accurate output current limit is essential. The output current limit function provides a constant current characteristic—the output current is limited to a constant level even as the output voltage is pulled down by an overcurrent condition. Alternatively the sense resistor can simply be moved to the input side to accurately control the maximum input current.

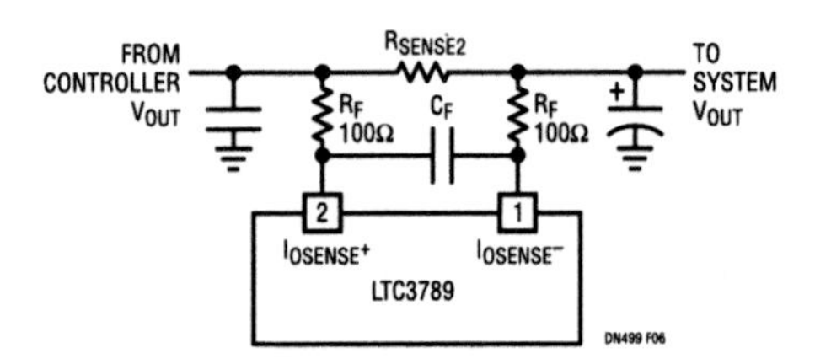

Figure 152.6 • Optional Output Current Limit

Conclusion

The LTC3789 is a constant frequency current mode buck-boost switching regulator controller that accepts a wide 4V to 38V input voltage range. The single inductor topology yields high power density and high efficiency in a compact footprint. Its optional current-limit function is useful for applications such as battery charging, where accurate current control is necessary.

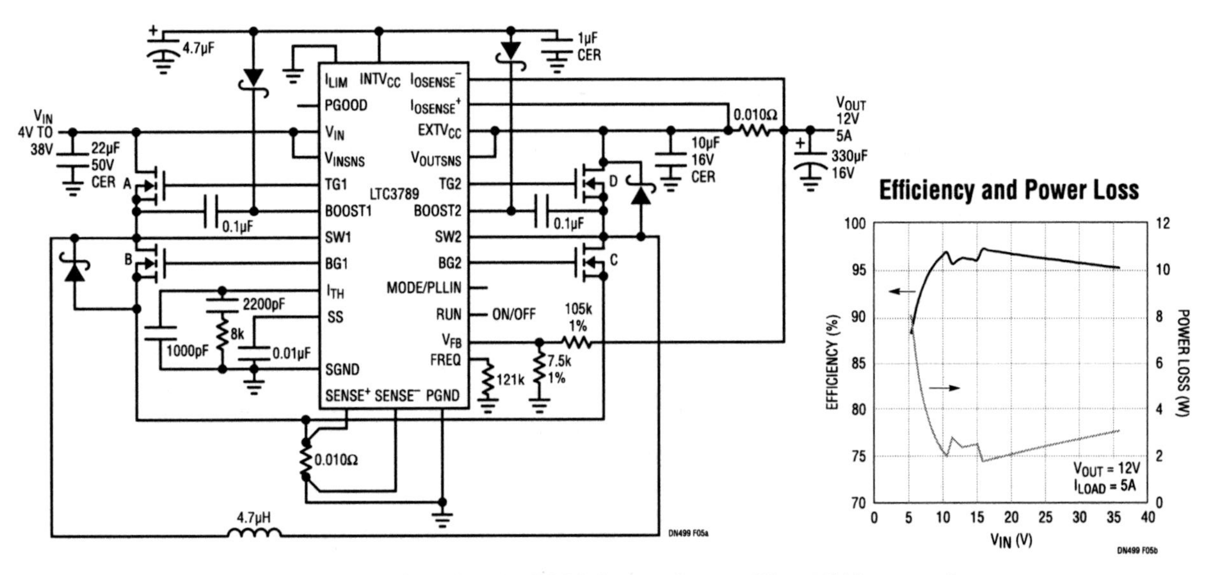

Figure 152.5 • Regulated 12V, 5A Output from a 4V to 38V Input and Associated Efficiency and Power Loss Curves

Buck-boost controller simplifies design of DC/DC converters for handheld products

153

David Burgoon

Introduction

A number of conventional solutions have been available for the design of a DC/DC converter where the output voltage is within the input voltage range—a common scenario in Li-Ion battery-powered applications—but none were very attractive until now. Conventional topologies, such as SEPIC or boost followed by buck, have numerous disadvantages, including low efficiency, complex magnetics, polarity inversion and/or circuit complexity/cost. The LTC3785 buck-boost controller yields a simple, efficient, low parts-count, single-converter solution that is easy to implement, thus avoiding the drawbacks associated with traditional solutions.

High efficiency controller capabilities

The LTC3785 serves applications requiring input and output voltages in the range of 2.7V to 10V—ideal for applications powered from one or two Li-Ion cells or multiple cell NiMH, NiCad or alkaline batteries. It supports a single inductor, 4-switch, buck-boost topology, which is ideal for an output voltage that is within the input voltage range. The high level of integration yields a simple, low parts-count solution. The LTC3785 provides for current-limit and shutdown in all modes of operation (which is not normally available with boost converters).

Very high efficiency is achieved by synchronous rectification, high side drive (allowing the use of N-channel

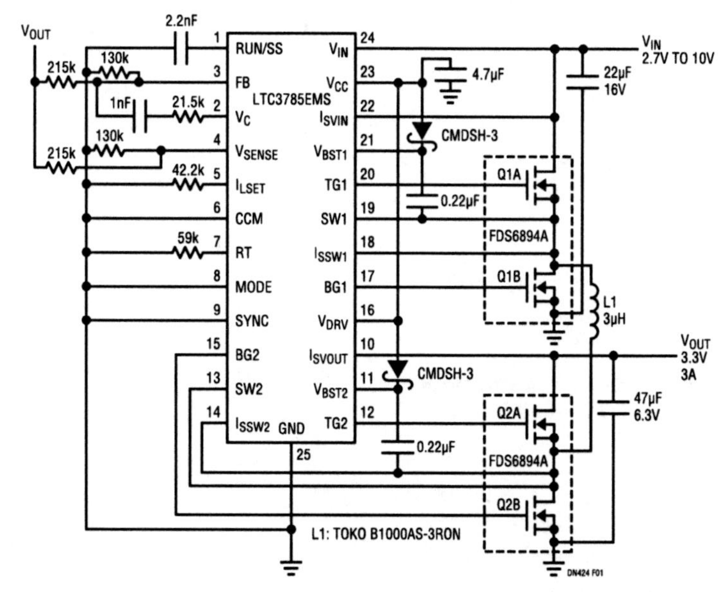

Figure 153.1 • Schematic of Buck-Boost Converter Using the LTC3785 to Provide 3.3V at 3A from a 2.7V to 10V Source

Analog Circuit Design: Design Note Collection. http://dx.doi.org/10.1016/B978-0-12-800001-4.00153-8

MOSFETs), $R_{DS(ON)}$ current sensing, and Burst Mode operation for efficient light load operation. Protection features include soft-start, overvoltage, undervoltage and foldback current-limit with burp-mode or latch-off for extended faults.

3.3V, 3A converter operates from 2.7V–10V source

The circuit shown in Figure 153.1 utilizes the LTC3785 controller to produce a synchronous, 4-switch, buck-boost design. It provides a fixed 3.3V, 3A output from a 2.7V–10V input. It is short-circuit protected: the controller offers a choice of recycling or latch-off protection for severe overload faults.

The circuit produces seamless operation throughout the entire input voltage range, operating as a synchronous buck converter, synchronous boost converter, or a combination of the two through the transition region. At input voltages well above the output, the converter operates in buck mode. Switches Q1A and Q1B commutate the input voltage, and Q2A stays on, connecting L1 to the output. As the input voltage is reduced and approaches the regulated output voltage, the converter approaches maximum duty cycle on the input (buck) side of the bridge, and the output (boost) side of the bridge starts to switch, thus entering the buck-boost or 4-switch region of operation. As the input is reduced further, the converter enters the boost region at the minimum boost duty cycle. Switch Q1A stays on, connecting the inductor to the input while switches Q2A and Q2B commutate the output side of the inductor between the output capacitor and ground. In boost mode, this converter has the ability to limit input current, and also to shut down and disconnect the source from the output—two desirable features that a conventional boost converter cannot provide. Figures 153.2, 153.3 and 153.4 show input side and output side switch waveforms along with inductor current for buck (10V input), boost (2.7V input) and buck-boost (3.8V input) modes of operation.

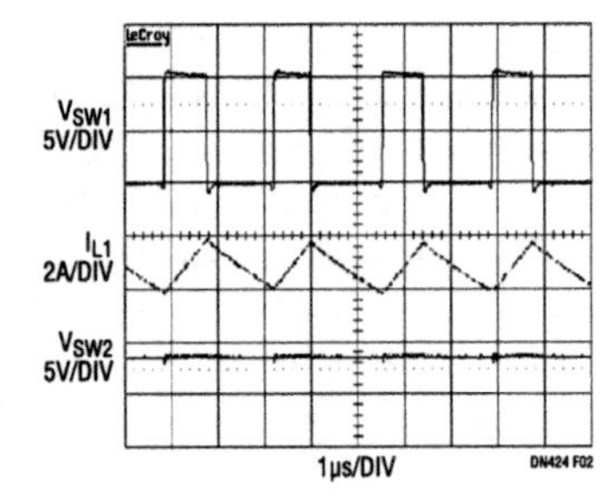

Figure 153.2 • Input Side and Output Side Switch Waveforms Along with Inductor Current for Buck (10V Input) Mode Operation for the Circuit in Figure 153.1

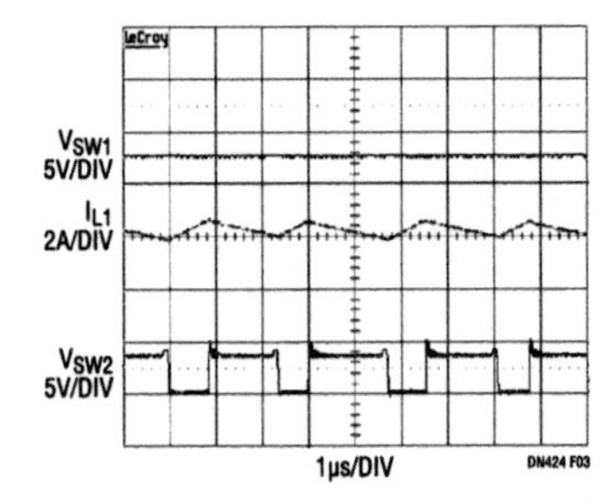

Figure 153.3 • Input Side and Output Side Switch Waveforms Along with Inductor Current for Boost (2.7V Input) Mode Operation for the Circuit in Figure 153.1

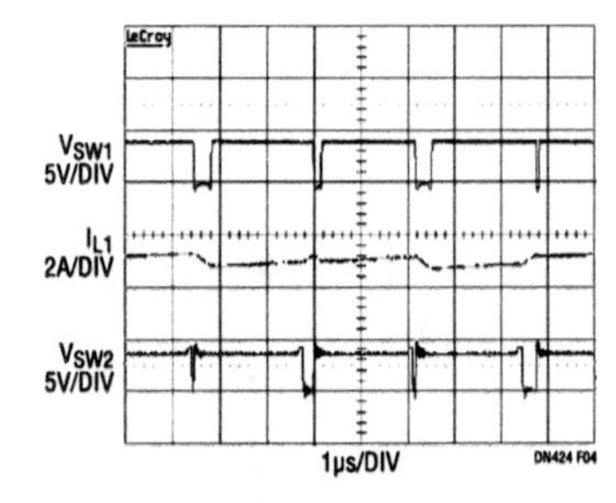

Figure 153.4 • Input Side and Output Side Switch Waveforms Along with Inductor Current for Buck-Boost (3.8V Input) Mode Operation for the Circuit of Figure 153.1

95% efficiency

Figure 153.5 shows efficiency curves for both normal (not forced continuous conduction) and Burst Mode operation. Exceptional efficiency of 95% is achieved at typical loads, resulting from sophisticated controller features including high side drivers and $R_{DS(ON)}$ current sensing. Even higher efficiencies are possible by using a larger ferrite inductor. This circuit easily fits in 0.6in^2 with components on both sides of the board. The curves show how Burst Mode operation improves efficiency at extremely light loads—an important determinant of battery run time.

Conclusion

The LTC3785 is the latest addition to the class of buck-boost converters developed by Linear Technology to satisfy the requirements of battery-powered applications, specifically those requiring an output voltage that is within the input voltage range. A topology based on the LTC3785 controller overcomes the deficiencies of conventional designs. It is elegant in its simplicity, high in efficiency, and requires only a small number of inexpensive external components.

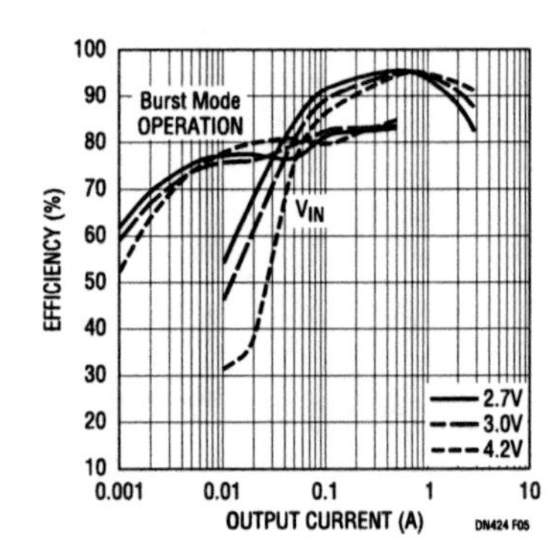

Figure 153.5 • Efficiency in Normal and Burst Mode Operation for the Circuit in Figure 153.1

Wide input voltage range buck-boost converter simplifies design of variable input supplies

154

John Canfield

Introduction

Many of today's portable electronic devices require the ability to operate from a variety of power sources including USB, wall adapters, and alkaline and lithium batteries. Designing a power conversion solution that is compatible with this wide array of power sources can be daunting. The LTC3530 monolithic synchronous converter simplifies the task by operating in both buck and boost modes over an extended input voltage range of 1.8V to 5.5V. No complicated topology is required to account for varying inputs that can be above, below or equal to the output.

The LTC3530 utilizes a proprietary switching algorithm that provides seamless transitions between buck and boost modes while simultaneously optimizing efficiency over all operating conditions. Using this advanced control algorithm, the LTC3530 is capable of high efficiency, fixed frequency operation with input voltages that are above, below, or equal to the output voltage, while requiring only a single inductor. This capability makes the LTC3530 well suited for lithium-ion/polymer and 2-cell alkaline or NiMH applications which require a supply voltage that is within the battery voltage range. In such cases, the high efficiency and extended input operating range of the LTC3530 offer greatly improved battery run-time, as much as 25% in some cases, over alternative solutions.

At 3.3V output, a load current of up to 600mA can be supported over the entire lithium-ion input voltage range; 250mA of load current is supported when the input is 1.8V. The output voltage is user programmable from 1.8V to 5.25V via an external resistor divider. The LTC3530 includes a soft-start circuit to minimize the inrush current transient during power-up. The duration of the soft-start period can be programmed via the time constant of an external resistor and capacitor.

The switching frequency of the LTC3530 is user programmable via a single external resistor, allowing the converter to be optimized to meet the space and efficiency requirements of each particular application. An external resistor and capacitor provide compensation of the feedback loop, enabling the frequency response to be adjusted to suit a wide array of external components. This flexibility allows for rapid output voltage transient response regardless of inductor value and output capacitor size.

The LTC3530 features an automatic transition to Burst Mode operation at a user programmable current level to improve light load efficiency. For noise sensitive applications, the LTC3530 can be forced into fixed frequency operation

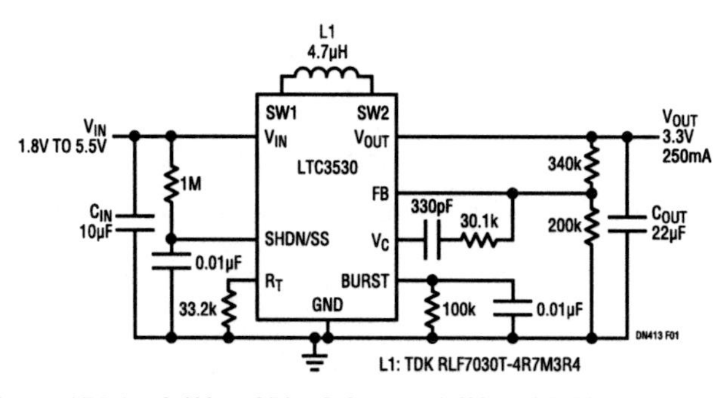

Figure 154.1 • 3.3V at 250mA from a 1.8V to 5.5V Input

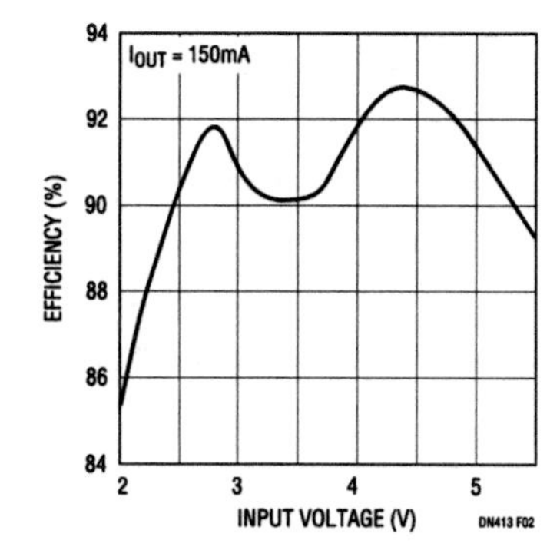

Figure 154.2 • Efficiency vs Input Voltage of the Circuit in Figure 154.1

Analog Circuit Design: Design Note Collection. http://dx.doi.org/10.1016/B978-0-12-800001-4.00154-X

at all load currents by connecting the BURST pin to V_{IN}. The LTC3530 also features short-circuit protection and overtemperature shutdown. Internal reverse current limiting circuitry prevents damage to the part should the output voltage be pulled above regulation through an external path.

Efficiency

Figure 154.1 illustrates a typical LTC3530 application circuit configured with a 1MHz switching frequency, which represents a good compromise between the PCB area and efficiency for most applications. The efficiency curve versus input voltage for this application circuit is shown in Figure 154.2. The LTC3530 achieves greater then 85% efficiency with an input voltage greater than or equal to 2V. These high levels of efficiency in combination with its wide input voltage range make the LTC3530 an attractive solution for battery-operated products and other efficiency-sensitive applications.

Programmable Burst Mode operation

The LTC3530 provides automatic Burst Mode operation, which greatly improves efficiency at light load currents. Burst Mode operation reduces the operating current of the LTC3530 to only 40μA in order to improve light load efficiency and extend battery run time. The LTC3530 automatically transitions to Burst Mode operation when the average output current falls below a user programmable level set via an external resistor. When the load current rises above the Burst Mode threshold, the part automatically returns to fixed frequency PWM operation.

The precise control circuitry of the LTC3530 allows the Burst Mode threshold to be set at load currents as low as 20mA. In addition, the LTC3530 directly monitors the average load current thereby providing a Burst Mode transition threshold that is independent of input voltage, output voltage, and inductor value, unlike other devices that rely on the level of peak inductor current.

In noise sensitive applications, the LTC3530 can be forced into fixed frequency PWM operation at all load currents by simply connecting the BURST pin to V_{IN}. In addition, the BURST pin can be driven dynamically in the application to provide low noise performance during critical phases of operation, or to reduce voltage transients during periods of expected large load transitions.

1.27mm profile Li-Ion to 3.3V regulator

The high switching frequency and advanced buck-boost switching algorithm of the LTC3530 allow the use of small external components. Figure 154.3 shows a circuit that is optimized to reduce the total application size. The entire converter has a maximum height of 1.27mm and occupies a PCB area of only 0.135 square inches making it ideal for height constrained applications such as PC cards. Figure 154.4 shows the efficiency versus input voltage for this area-optimized application circuit. This converter is able to support a 600mA load for output voltages above 2.4V and obtains greater than 86% efficiency over the entire Li-Ion input voltage range.

Conclusion

The LTC3530 with its high efficiency, wide input voltage range, and tiny circuit size is well suited to a variety of battery-operated products and other efficiency-sensitive applications. With the IC's array of programmable features, the circuit can be customized to meet the needs of any application, while still maintaining a compact total solution footprint.

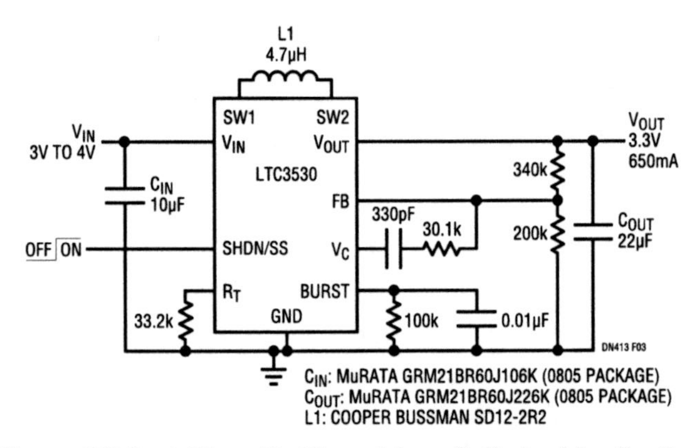

Figure 154.3 • 1.27mm Profile and Area Optimized Application Circuit

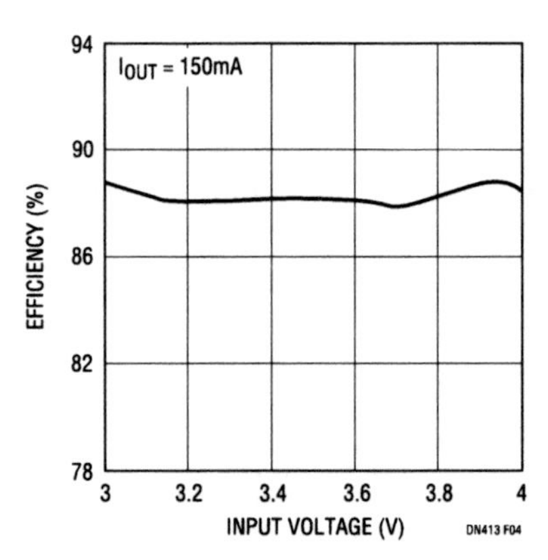

Figure 154.4 • Efficiency vs Input Voltage of Figure 154.3 Circuit

Buck or boost: rugged, fast 60V synchronous controller does both

155

Greg Dittmer

Introduction

Automotive, telecom and industrial systems are harsh, unforgiving environments that demand robust electronic systems. For example, an automotive battery system may be a nominal 12V, 24V or 42V, but load dump conditions can generate transients up to 60V. The LTC3703-5 is a synchronous switching regulator controller that can directly step down input voltages up to 60V and withstand transients up to 80V, making it ideal for harsh environments. The ability to step down the high input voltage directly allows a simple single inductor topology, resulting in a compact high performance power supply—in contrast to the low side drive topologies that require bulky, expensive transformers.

Feature rich controller

The LTC3703-5 drives external logic-level N-channel MOSFETs using a constant frequency, voltage mode architecture. A high bandwidth error amplifier and patented line feed forward compensation provide very fast line and load transient response. Strong 1Ω gate drivers minimize switching losses—often the dominant loss component in high voltage supplies—even when multiple MOSFETs are used for high current applications. Other features include:

- Low minimum on-time (200ns) for low duty cycle applications
- Precise 0.8V ±1% reference
- Programmable current limit utilizing the voltage drop across the synchronous MOSFET to eliminate the need for a current sense resistor
- Programmable operating frequency (100kHz to 600kHz)
- Low shutdown current (25μA), external clock synchronization input and selectable pulse skip mode operation
- Packaged in a 16-pin narrow SSOP or a 28-pin SSOP if high voltage pin spacing is desired.

High efficiency 48V to 3.3V/6A power supply

The circuit shown in Figure 155.1 provides direct step-down conversion of a typical 48V telecom input rail to 3.3V at 6A. The circuit can handle input transients up to 60V without

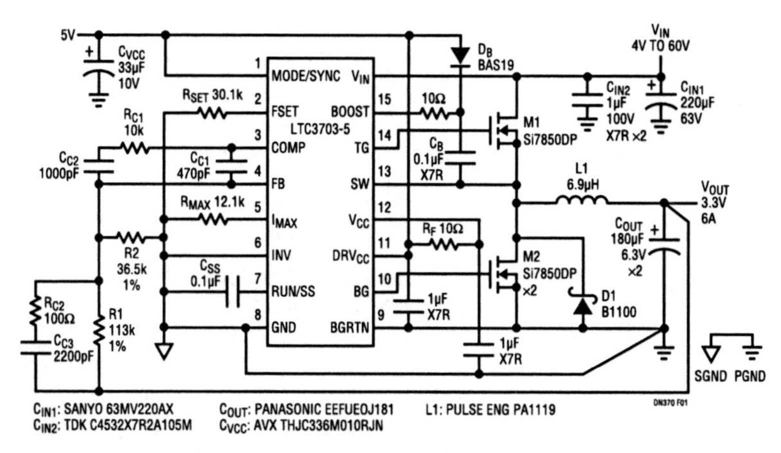

Figure 155.1 • Buck: 48V to 3.3V/6A Synchronous Step-Down Converter

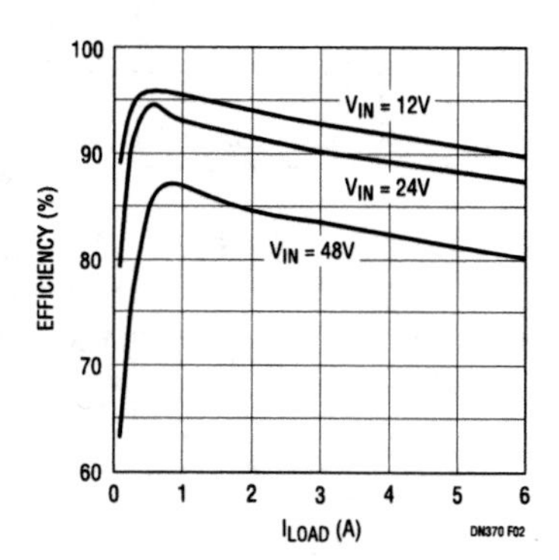

Figure 155.2 • Efficiency of Figure 155.1's Circuit

Analog Circuit Design: Design Note Collection. http://dx.doi.org/10.1016/B978-0-12-800001-4.00155-1

requiring protection devices or 80V if appropriate MOSFETs are used. The frequency is set to 250kHz to optimize efficiency and output ripple. Figure 155.2 shows a mid-range efficiency of over 90% at 24V input and 83.5% at 48V input.

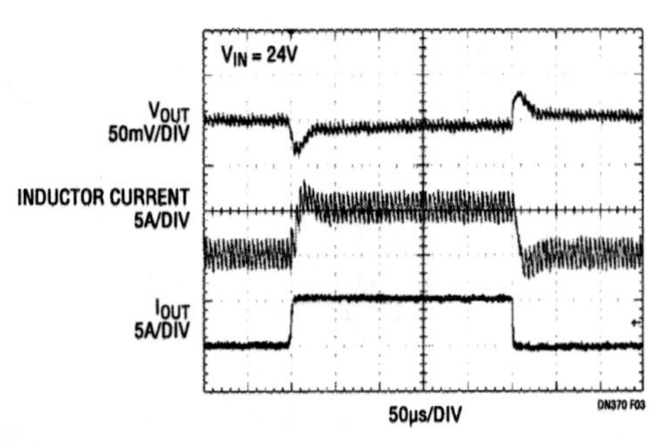

Figure 155.3 • Load Transient Performance of Figure 155.1 Circuit Shows 20μs Response Time to 5A Load Step

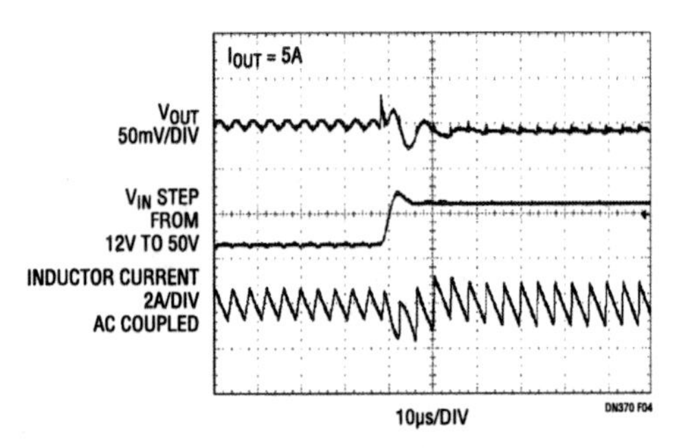

Figure 155.4 • Line Transient Performance of Figure 155.1 Circuit Shows Almost Complete Rejection of 12V to 50V Supply Transient

The loop is compensated for a 50kHz crossover frequency which provides 20μs response time to load transients (see Figure 155.3). The outstanding line transient performance is shown in Figure 155.4. The 12.1k R_{MAX} resistor value is chosen to limit the inductor current to about 12A during a short-circuit condition.

High efficiency 12V to 24V/5A synchronous step-up fan power supply

Synchronous boost converters have a significant advantage over non-synchronous boost converters in higher current applications due to the low power dissipation of the synchronous MOSFET compared to that of the diode in a non-synchronous converter. The high power dissipation in the diode requires a much larger package (e.g. D^2PAK) than the small S8-size package required for the synchronous MOSFET for the same output current.

Figure 155.5 shows the LTC3703-5 implemented as a synchronous step-up converter for generating 24V/5A from 12V—a common voltage for driving fans. This supply achieves a peak efficiency over 96% (see Figure 155.6). The LTC3703-5 is set to operate as a synchronous boost converter by simply connecting the INV pin to greater than 2V. In boost mode, the BG pin becomes the main switch and TG becomes the synchronous switch. Aside from this phase inversion, boost mode operation is similar to buck mode. In boost mode, the LTC3703-5 can produce output voltages as high as 60V.

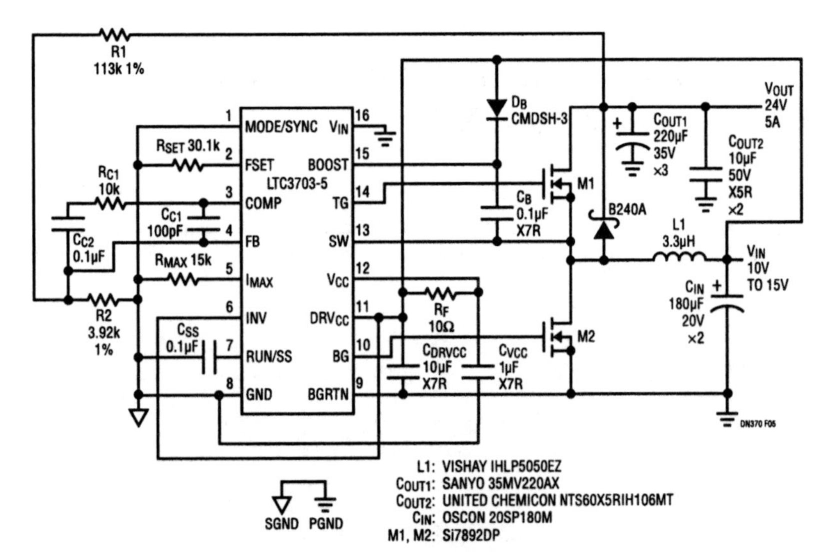

Figure 155.5 • Boost: 12V to 24V/5A Synchronous Step-Up for Fan Power Supply

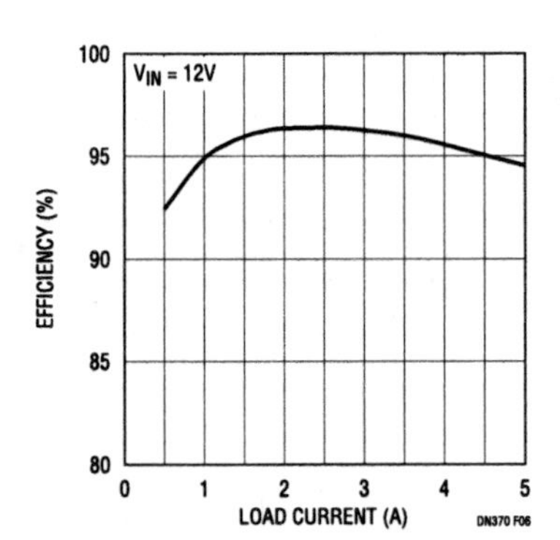

Figure 155.6 • Efficiency of Figure 155.5's Circuit

Industry's first 4-switch buck-boost controller achieves highest efficiency using a single inductor

156

Wilson Zhou   Theo Phillips

Introduction

One of the most common DC/DC converter problems is generating a regulated voltage that falls somewhere in the middle of a wide range of input voltages. When the input voltage can be above, below or equal to the output voltage, the converter must perform step-down and step-up functions. Unlike solutions requiring bulky transformers, the LTC3780 meets these requirements in the most compact and efficient manner, using just one off-the-shelf inductor and a single current sense resistor.

The LTC3780 uses a constant frequency current mode architecture which allows seamless transitions between buck, boost and buck-boost modes with a wide 4V to 30V (36V maximum) input and output range. Burst Mode operation and skip cycle mode provide high efficiency operation at light loads, while forced continuous mode and discontinuous mode reduce output voltage ripple by operating at a constant frequency. A soft-start feature reduces output overshoot and inrush currents during start-up. Overvoltage protection, current foldback and on-time limitation provide protection for fault conditions, including short circuit, overvoltage and inductor current runaway. The LTC3780 is available in low profile 24-pin TSSOP and 32-lead 5mm × 5mm QFN packages.

High efficiency 4-switch buck-boost converter

Figure 156.1 shows a simplified LTC3780 4-switch buck-boost converter. When V_{IN} exceeds V_{OUT}, the LTC3780 operates in buck mode. With switch D on and switch C off, switches A and B turn on and turn off alternately, as they would in a typical synchronous buck regulator. Conversely, when V_{IN} is lower than V_{OUT}, the LTC3780 operates in boost mode. With switch A on and synchronous switch B off, switch C and synchronous switch D turn on and turn off alternately, behaving as a typical synchronous boost regulator.

When V_{IN} is close to V_{OUT}, the controller is in buck-boost mode. Switches A and D are on for most of each period. Brief connections between V_{IN} and ground, and V_{OUT} and ground, are made through the inductor and switches B-D and A-C to regulate the output voltage. In buck-boost mode, inductor peak-to-peak current is much lower than that of SEPIC converters and traditional buck/boost converters. Figure 156.2 shows the inductor current and switch node waveforms.

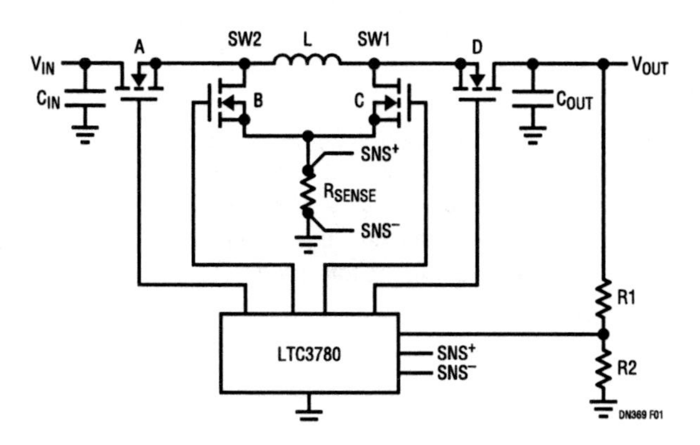

Figure 156.1 • 4-Switch Buck-Boost Converter

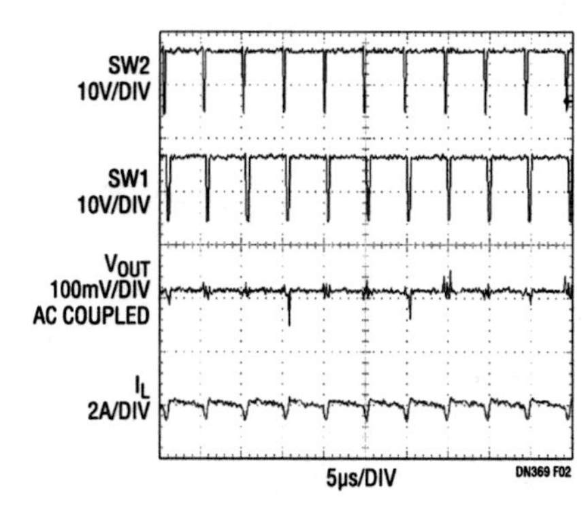

Figure 156.2 • Switch Nodes and Inductor Current Waveforms ($V_{IN} = V_{OUT} = 12V$)

Analog Circuit Design: Design Note Collection. http://dx.doi.org/10.1016/B978-0-12-800001-4.00156-3

Low inductor ripple current and the use of synchronous rectifiers allow the LTC3780 to achieve very high efficiency over a wide V_{IN} range. When the input and output voltages are both 12V, the 4-switch buck-boost has 99% efficiency at 2A load and 98% at its maximum 5A load (Figure 156.3). With its current mode control architecture, the converter has excellent load and line transition response, minimizing the required filter capacitance and simplifying loop compensation. As a result, very little filter capacitance is required. The single sense resistor structure dissipates little power (compared with multiple resistor sensing schemes) and provides consistent current information for short-circuit and overcurrent protection.

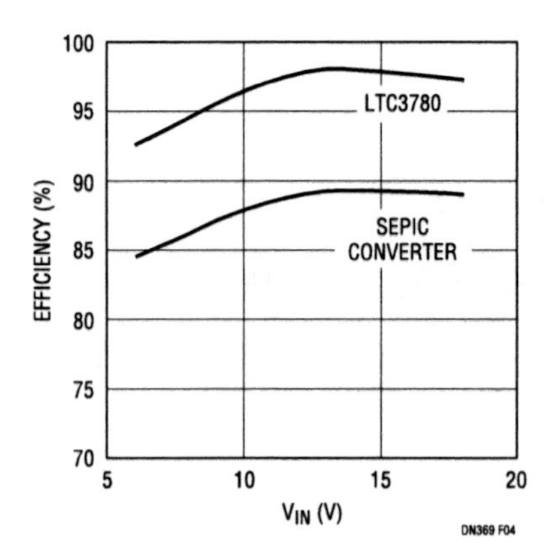

Figure 156.4 • Efficiency Comparison between the LTC3780 and a SEPIC Converter (V_{OUT} = 12V, I_{LOAD} = 5A)

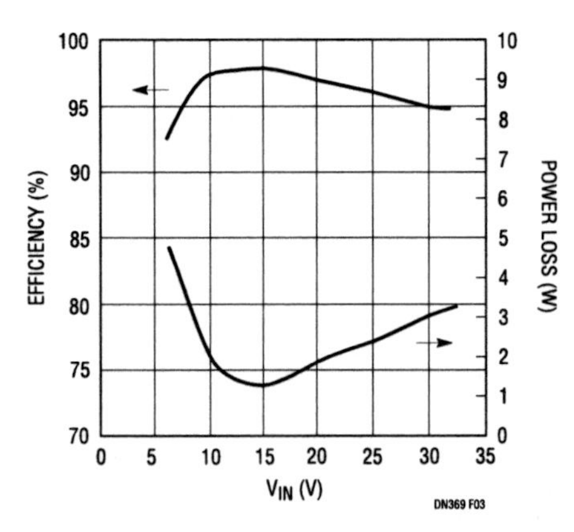

Figure 156.3 • Efficiency and Power Loss (V_{OUT} = 12V, I_{LOAD} = 5A)

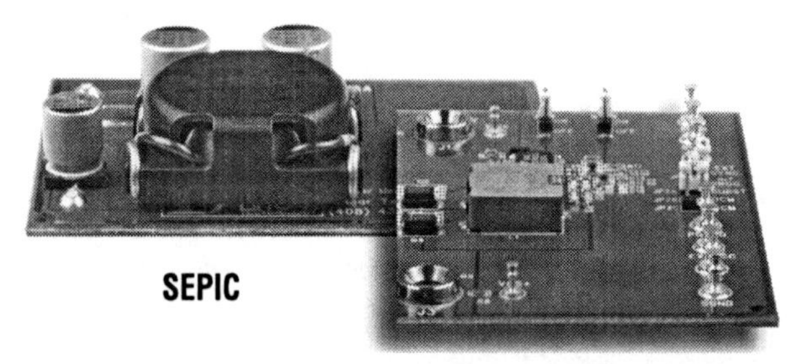

Figure 156.5 • Inductor Size Comparison between the LTC3780 5A/12V Converter (Right, 12.7mm x 12.7mm x 4mm) and a Typical SEPIC (Left, 21mm x 21mm x 10.8mm)

Replacing a SEPIC converter

This single inductor buck-boost approach has high power density and high efficiency. Compared with a coupled inductor SEPIC converter, its efficiency can be 8% higher. Figure 156.4 shows the efficiency comparison between the LTC3780 4-switch buck-boost and a typical SEPIC converter. Note that a SEPIC converter has a maximum switch voltage equal to the input voltage plus the output voltage. So for a given maximum input voltage, a SEPIC would dictate the use of a higher voltage external switch than is required with the LTC3780. Moreover, the typical inductor occupies about 1/5th of a SEPIC transformer's footprint, less than 1/15th the volume and less than one-half the profile, as shown in Figure 156.5.

Protection for boost operation

The basic boost regulator topology provides no short-circuit protection. When the output is pulled low, a large current can flow from the input to the output. Without shutting down the whole circuit, the LTC3780 circumvents this problem by forcing the converter into buck mode and using current foldback to limit the inductor current.

Simplify

For certain applications such as those requiring low current or not requiring current sinking, Switch D can be replaced with a Schottky diode. This simplified topology has approximately 2% lower efficiency.

Conclusion

The LTC3780 is a constant frequency current mode buck-boost switching regulator controller that allows the input voltage to be above, below or equal to the output voltage. Its high efficiency, high power density and single inductor topology make this product ideal for automotive, telecom, medical and battery-powered systems.

High input voltage monolithic switcher steps up and down using a single inductor

157

Jay Celani

Introduction

Ultrawide input voltage requirements are a common design problem for DC/DC converter applications, but when that range includes voltages both above and below the output voltage, the converter must perform both step-up and step-down functions. The LT3433 is a high voltage monolithic DC/DC converter that incorporates two switch elements, allowing for a unique topology that accommodates both step-up and step-down conversion using a single inductor.

The LT3433 uses a 200kHz constant frequency, current mode architecture and operates with input voltages from 4V to 60V. An internal 1% accurate voltage reference allows programming of precision output voltages up to 20V using an external resistor divider. Burst Mode operation improves efficiencies during light-load conditions, reducing the device's quiescent current to 100μA during no-load conditions. A soft-start feature reduces output overshoot and inrush currents during start-up, and both current limit foldback and frequency foldback are employed to control inductor current runaway during start-up and short-circuit conditions. The LT3433 is available in a 16-pin fused TSSOP exposed pad package which provides a small footprint and excellent thermal characteristics.

When the converter input voltage is significantly higher than the output voltage, the LT3433 operates as a modified buck converter using a boosted-drive high side switch. If the converter input voltage becomes close enough to the output voltage to require a duty cycle greater than 75% in buck mode, the LT3433 automatically enables a second switch. This second switch pulls the output side of the switched inductor to ground during the "switch on" time, creating a bridged switching configuration.

During bridged switching, the LT3433 merges the elements of buck and boost DC/DC converters as shown in Figure 157.1. In the simplest terms, a buck DC/DC converter switches the V_{IN} side of the inductor, while a boost converter switches the V_{OUT} side of the inductor. Combining the elements of both topologies achieves both step-up and step-down functionality using a single inductor, so voltage conversion can continue when V_{IN} approaches or is less than V_{OUT}.

4V–60V input to 5V output DC/DC automotive converter

A 4V–60V to 5V DC/DC converter is shown in Figure 157.2. This converter is well suited for 12V automotive battery applications, maintaining output voltage regulation with battery line voltages from 4V cold crank through 60V load dump. The threshold for bridged mode operation is about 8V, so the converter will operate primarily in buck mode except during a cold crank condition. During buck operation, this converter can provide load currents up to 350mA with input voltages up to 60V. Operating with a nominal 13.8V input,

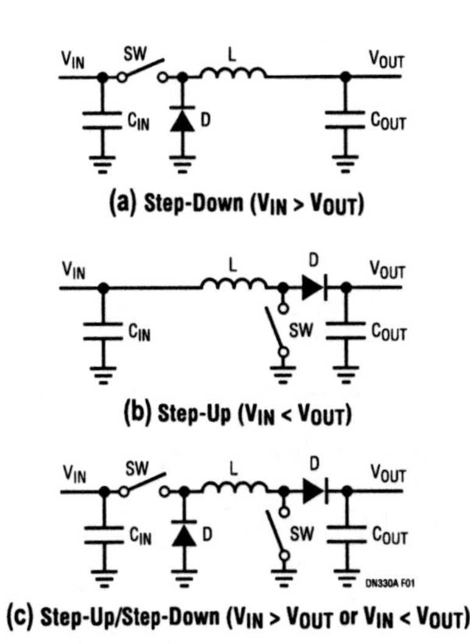

Figure 157.1 • The LT3433 Merges the Elements of Step-Up and Step-Down DC/DC Converters

Analog Circuit Design: Design Note Collection. http://dx.doi.org/10.1016/B978-0-12-800001-4.00157-5

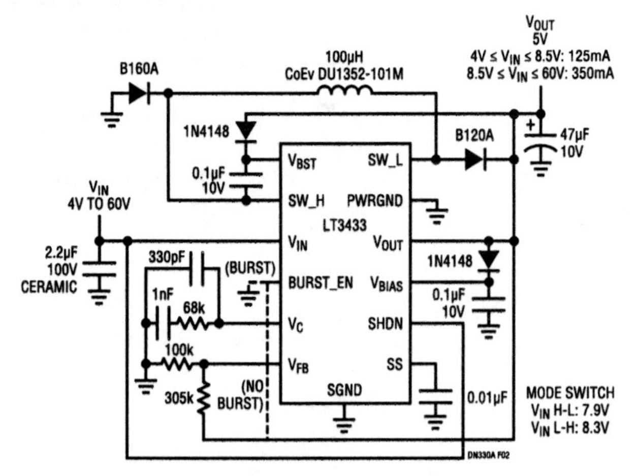

Figure 157.2 • 4V–60V to 5V DC/DC Converter

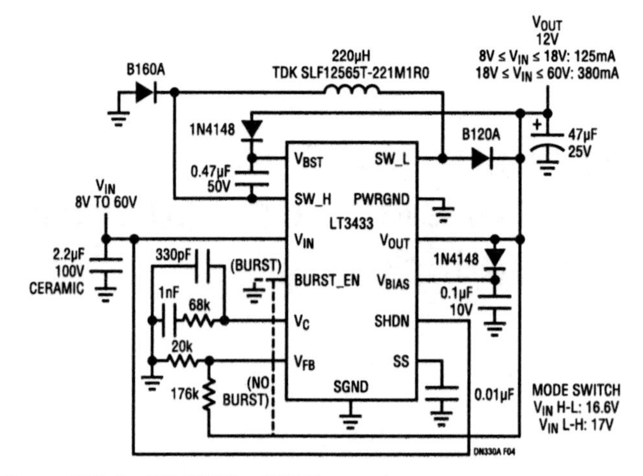

Figure 157.4 • 8V–60V to 12V Converter

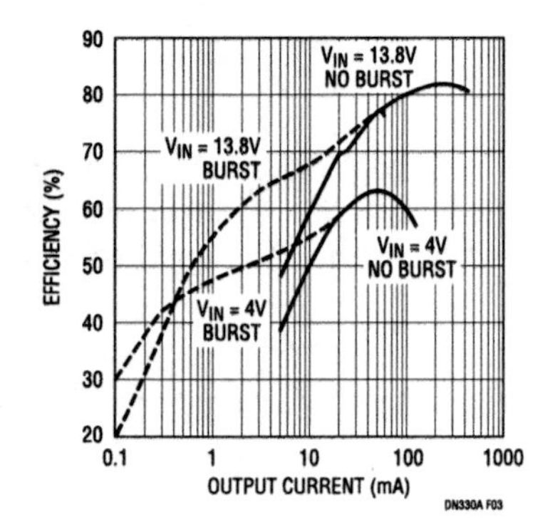

Figure 157.3 • 4V–60V to 5V Conversion Efficiency

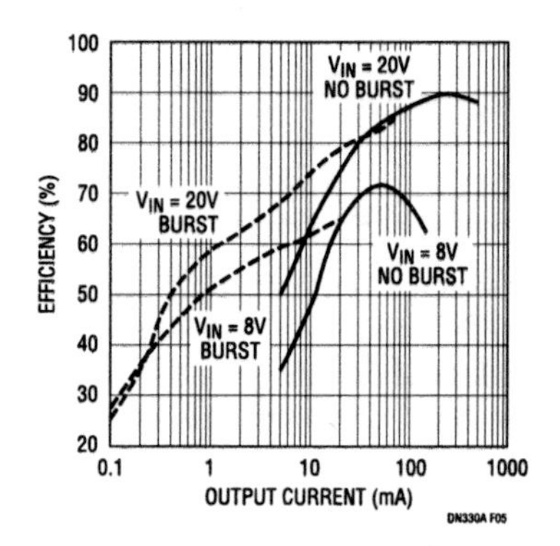

Figure 157.5 • 8V–60V to 12V Conversion Efficiency

this LT3433 converter accommodates loads of 400mA and produces efficiencies up to 82%.

When the input voltage drops below 8V, the converter switches into bridged operation to maintain output voltage regulation. Because the LT3433 switch current limit is fixed, converter load capability is reduced while operating in bridged mode. With an input of 4V, the converter accommodates loads up to 125mA. Not only does this LT3433 converter operate across a large range of DC input voltages, but it also maintains tight output regulation during input transients. When subjected to a 1ms 13.8V to 4V input transition to simulate a cold crank condition, regulation is maintained to 1% with a 125mA load Figure 157.3.

8V–60V input to 12V output DC/DC converter

As converter output voltages increase, switch current and duty-cycle limitations prevent operation with V_{IN} at the extreme low end of the LT3433 operational range. The 12V output converter shown in Figure 157.4 can provide load current up to 125mA with an input voltage as low as 8V. This is suitable for 12V automotive applications without cold-crank requirements, as well as many other applications such as those powered by inexpensive wall adapters. This converter operates in buck mode with input voltages above 17V, accommodating loads up to 380mA. This converter accommodates loads up to 435mA and produces efficiencies above 89% at 20V input Figure 157.5.

Conclusion

The LT3433 simplifies ultrawide input range DC/DC voltage conversion, enabling simple and inexpensive solutions to a variety of design problems. Automatic transitioning between buck and bridged modes of operation provides seamless output regulation for wide input voltage ranges and input voltage transients. The use of a small footprint TSSOP package, a single inductor and few external components reduce board space requirements, increase efficiency and improve thermal characteristics.

Supply 2A pulses for GSM transmission from 500mA USB or PCMCIA ports

158

Dongyan Zhou

Introduction

GSM modems have become popular as a wireless data transfer solution; however, they frequently require large bursts of current that exceed the maximum input current available to the voltage regulator. Many can be powered from a USB port (4.5V to 5.5V input) or a PCMCIA port (3V to 3.6V input), where 3.xV is required to supply the RF power amplifier that draws current pulses up to 2A. Since the input from the USB or PCMCIA port is current limited to 500mA, a high efficiency buck-boost converter with input current limit provides the best power supply solution. High efficiency maximizes the average output power, while a bulk output capacitor is used to maintain the voltage during the high current pulses.

The LTC3440 buck-boost converter is a compact and high efficiency solution for GSM modems that can be powered from USB or PCMCIA. It has four integrated switches to provide synchronous buck-boost operation and requires only one inductor, saving cost and board space. The IC is available in a tiny 10-pin MSOP package and operates up to 2MHz, allowing the use of tiny surface mount components. Quiescent current is only 25μA in Burst Mode operation. The LTC3440 also disconnects the output the from the input during shutdown, which is required for many USB applications.

Powering GSM modems from USB or PCMCIA

Figure 158.1 shows a buck-boost converter powered from USB or PCMCIA. The input current limit is implemented by using half of the LT1490A dual micropower op amp. The op amp is configured as a current source which limits the input current to 500mA. The other half of the LT1490A is used as a buffer. Efficiency at the given pulsing load is shown in Figure 158.2.

The magnitude and duration of the pulsing current, together with the ripple voltage specification, determine the choice of the output capacitor. Both the ESR of the capacitor and the charge stored in the capacitor each cycle contribute to the output voltage ripple. The ripple due to the charge is approximately:

$$V_{RIPPLE\_BULK} = \frac{(I_{PULSE} - I_{STANDBY}) \cdot t_{ON}}{C_{OUT}}$$

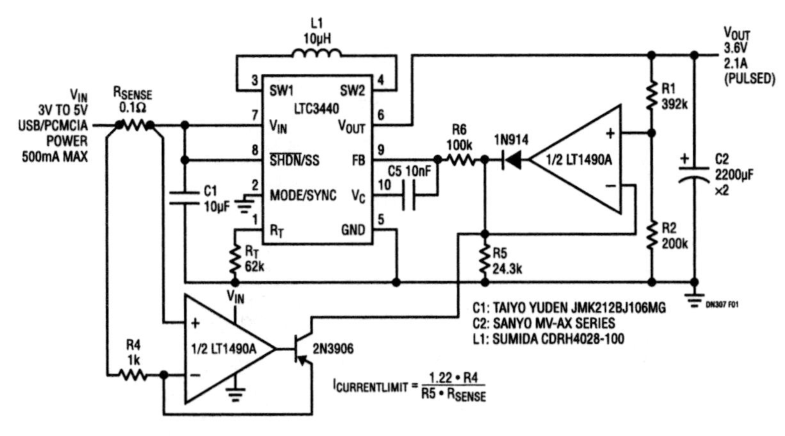

Figure 158.1 • Converter Powering GSM Modem from USB or PCMCIA

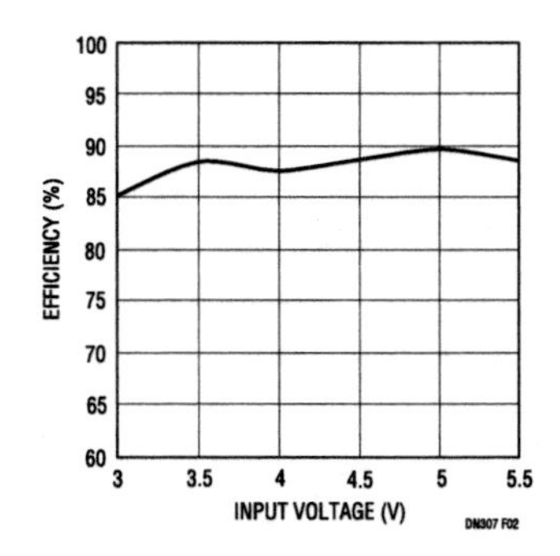

Figure 158.2 • Efficiency of the Converter in Figure 158.1 (Pulsing Load: 1.2A for 1.15ms, 80mA for 3.45ms, within a Period of 4.6ms)

Analog Circuit Design: Design Note Collection. http://dx.doi.org/10.1016/B978-0-12-800001-4.00158-7

where I_{PULSE} and t_{ON} are the peak current and on time during transmission burst and $I_{STANDBY}$ is the current in standby mode. The above is a worst-case approximation assuming all the pulsing energy comes from the output capacitor.

The ripple due to the capacitor ESR is:

$$V_{RIPPLE\_ESR} = (I_{PULSE} - I_{STANDBY}) \cdot ESR$$

Low ESR and high capacitance are critical to maintain low output ripple. In this application, two 2200μF SANYO electrolytic capacitors are used. Each capacitor has less than 38mΩ ESR. For applications requiring a very low profile, the BestCap series from AVX and PowerStor Aerogel Capacitors from COOPER offer very high capacitance and low ESR in 2mm height packages.

The GSM standard specifies 575μs transmission burst within a 4.6ms period (1/8 duty factor). The converter in Figure 158.1 can provide up to 2.1A during each transmission burst and 100mA in standby for input voltages as low as 3V with a maximum of 500mA input current. If the minimum input voltage is higher or the 500mA input current limit is only required for USB input (4.5V to 5.5V), more power is available at output. Figure 158.3 shows the output voltage ripple along with input and output currents. Other standards (such as GPRS) define a higher data rate. One popular requirement is 1.15ms transmission bursts within a 4.6ms period (1/4 duty factor). The converter can deliver 1.2A pulsing current with 80mA standby current, again assuming 3.0V minimum input voltage with a 500mA current limit. The output ripple is similar to Figure 158.3 except that the ripple caused by ESR is less due to the lower load step.

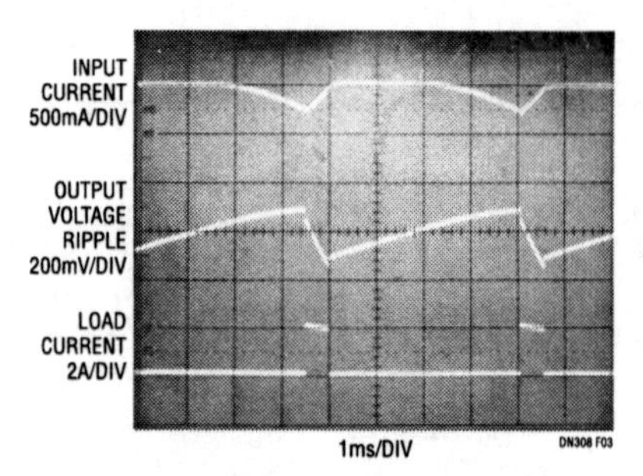

Figure 158.3 • Waveforms Showing Input Current (Top) and Output Voltage Ripple (Middle) at Pulsing Load (Bottom)

5V converter in USB On-The-Go devices

As portable devices using USB increase in popularity, there is a growing need for them to communicate directly with each other when a PC is not available. The result is USB On-The-Go (OTG). USB OTG is a new supplement to the USB 2.0 specification that augments the capability by adding a host function for connection to USB peripherals. These USB OTG dual-role devices need a 5V converter to supply USB peripherals. The LTC3440 provides a compact solution with high efficiency, low quiescent current and shutdown disconnect. Accurate current limit can again be achieved by using the LT1490A dual op amp. Figure 158.4 shows a converter with 100mA current limit. Efficiency is plotted in Figure 158.5.

To optimize light load efficiency, the MODE/SYNC pin can be pulled high to enable Burst Mode operation where quiescent current is only 25μA.

Conclusion

The LTC3440 synchronous buck-boost converter provides an optimal solution for GSM modems and USB OTG devices. Compared with the traditional SEPIC or Boost cascaded with an LDO converter, the LTC3440 converter provides much higher efficiency with fewer components and smaller size.

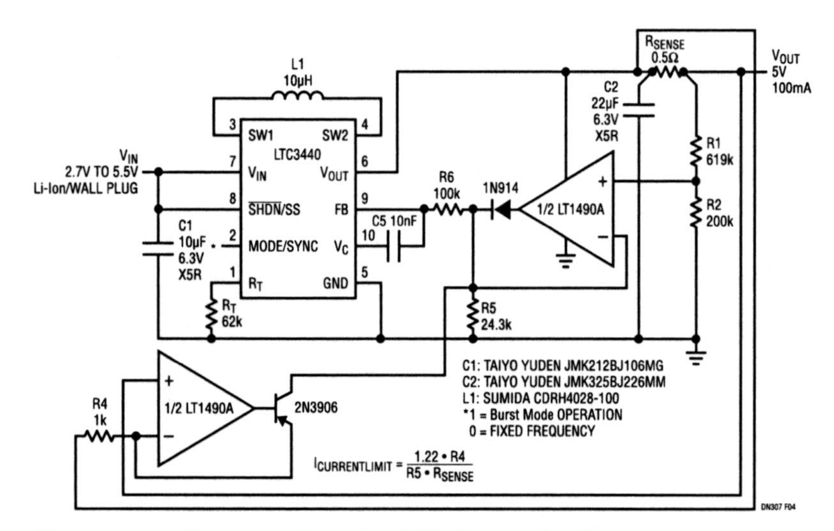

Figure 158.4 • Converter Powering USB Compatible Devices

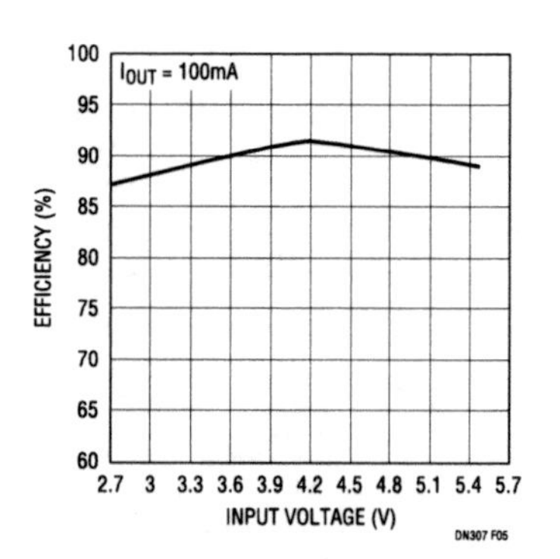

Figure 158.5 • Efficiency of the Converter in Figure 158.4

Micropower buck/boost circuits: converting three cells to 3.3V

159

Mitchell Lee

Two combinations of cell count and output voltage are to be strictly avoided: three cells converted to 3.3V and four cells converted to 5V. These combinations are troublesome because no ordinary regulator (boost, buck or linear) can accommodate a situation where the input voltage range overlaps the desired output voltage.

This design note presents four circuits capable of solving the 3-cell dilemma. The LT1303 and LT1372 high efficiency DC/DC converters are used throughout, giving a fair comparison of each topology's efficiency. The LT1303 is optimized for battery operation and includes a low-battery detector which is required to implement one of the topologies. The LT1372 500kHz converter is used for compact layouts at higher current levels.

You can expect 200mA output from LT1303 based circuits and 300mA from the LT1372 circuit without modification. All of the circuits feature output disconnect; in shutdown the outputs fall to 0V. The input range of LT1303 based

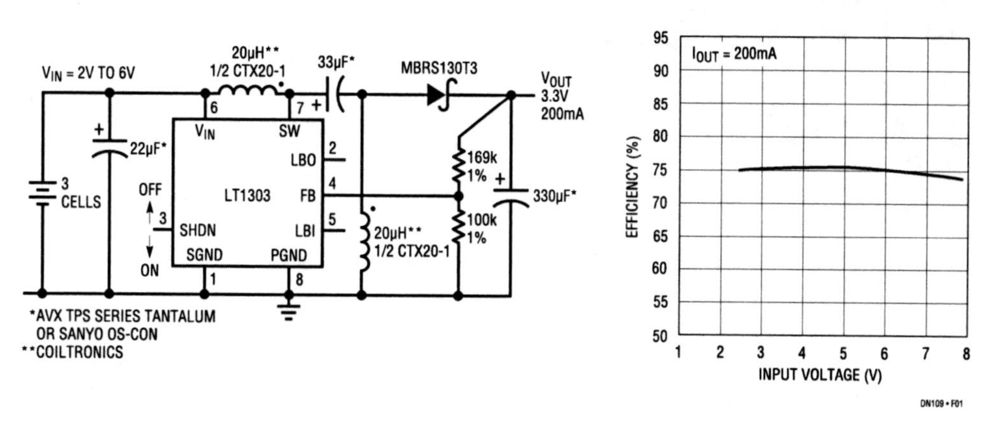

Figure 159.1 • 3-Cell to 3.3V SEPIC

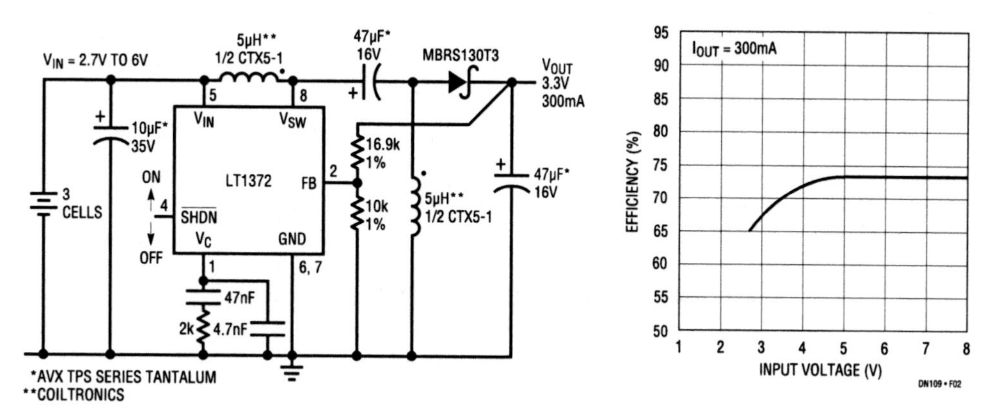

Figure 159.2 • 3-Cell to 3.3V SEPIC

Analog Circuit Design: Design Note Collection. http://dx.doi.org/10.1016/B978-0-12-800001-4.00159-9

converters extends well beyond the 3-cell source shown. These function at 1.8V, and although not fully characterized for efficiency, can accept inputs of up to 10V. The LT1372 converter operates from 2.7V to 10V.

The circuits in Figures 159.1 and 159.2 are based on the SEPIC (Single-Ended Primary Inductance Converter) topology. Although not stellar, the efficiency is quite consistent over a wide input voltage range. Peculiar to the SEPIC topology is its use of two inductors. These, however, are wound together on a single core and consume no more space than a simple 2-terminal inductor of similar rating. A wide selection of stock 2-winding, 4-terminal inductors are available from Coiltronics and other magnetics vendors.

Peak efficiency improves in Figure 159.3 using a bipolar buck/boost topology. This circuit is essentially a boost converter with a linear post regulator. For $V_{IN} < V_{OUT}$, the LT1303 boosts the input driving the bipolar emitter just high enough to maintain the desired output voltage—the transistor is saturated. For $V_{IN} > V_{OUT}$, the LT1303 drives the emitter to a value just higher than the *input voltage* sufficient to develop the base current necessary to support any load current. In this condition the transistor serves as a linear post regulator, cascoding the output of the boost converter and dissipating power as would any linear regulator.

Highest peak efficiency is obtained with the circuit in Figure 159.4 using a MOSFET buck/boost converter. For $V_{IN} < V_{OUT}$, the circuit operates as a boost converter and the MOSFET, driven by the LT1303's low-battery detector/amplifier, is held 100% ON. The output voltage is developed and controlled by the boost converter.

For $V_{IN} > V_{OUT}$, the boost function can no longer control the output voltage and it begins to rise. Staggered feedback (R3, R4, R5) allows the low-battery detector/amplifier to take control using the MOSFET as a linear pass element. Because the MOSFET requires no base drive, and because it has such a low ON resistance, the efficiency peaks at well over 90%. Furthermore, the efficiency peak occurs in the vicinity of a NiCd's nominal terminal voltage of 3 × 1.25 = 3.75V, right where the efficiency is needed most.

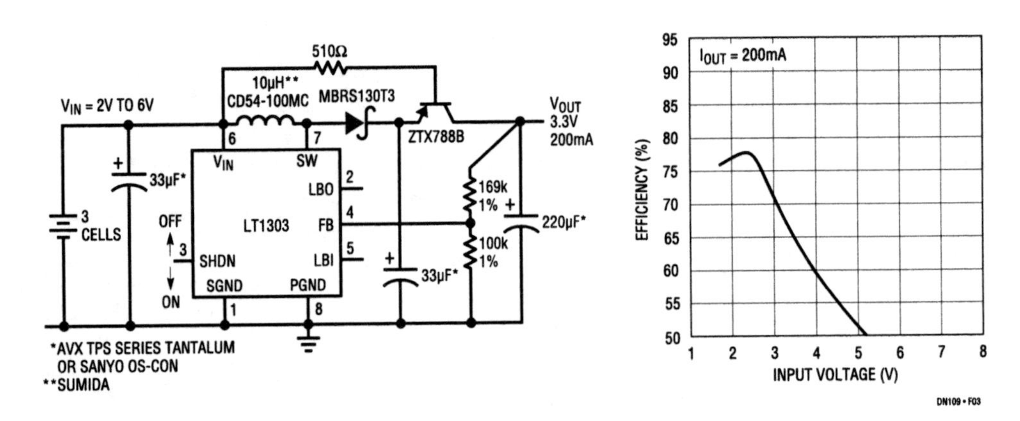

Figure 159.3 • 3-Cell to 3.3V Bipolar Buck/Boost

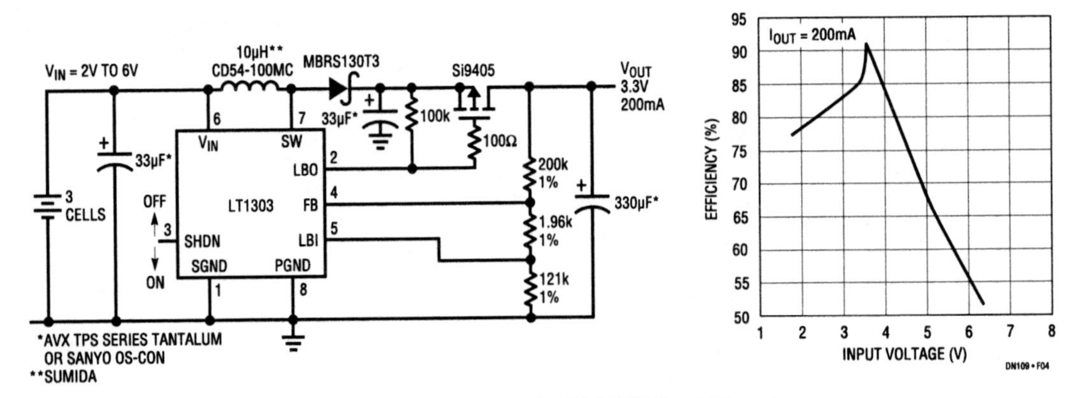

Figure 159.4 • 3-Cell to 3.3V MOSFET Buck/Boost

250kHz, 1mA IQ constant frequency switcher tames portable systems power

160

Bob Essaff

DC-to-DC power conversion remains one of the toughest tasks for portable system designers. Dealing with various battery technologies and output voltage requirements dictates the need for creative circuit solutions. The two circuits discussed are tailored for operation from a single lithium-ion (Li-Ion) cell. These new batteries are finding widespread use due to their high energy storage capabilities. The first circuit has a 3.3V output and the second circuit has both 5V and −5V outputs for applications requiring dual supplies.

At the heart of each circuit is the LT1373 current mode switching regulator. Guaranteed to operate down to 2.7V, this part allows the full energy storage capacity of a single Li-Ion battery to be used. The LT1373 draws only 1mA of quiescent current for high efficiency at light loads and has a low resistance 1.5A switch for good efficiency at higher loads. Switching at 250kHz saves space by reducing the size of the magnetics, and the fixed-frequency switching also reduces the noise spectrum generated. To avoid sensitive system frequencies the part can be externally synchronized to a specific frequency from 300kHz to 360kHz. The LT1373 can also be shut down where it draws only 12μA supply current.

3.3V SEPIC converter

Generating a 3.3V output from a single Li-Ion cell is not straight forward because at full charge the battery voltage is above the output voltage and when discharged, the battery voltage is below the output voltage. A conventional buck or boost regulator topology will not work. The circuit in Figure 160.1 uses the SEPIC (single-ended primary inductance converter) topology which allows the input voltage to be higher or lower than the output voltage. The circuit's two inductors, L1A and L1B, are actually two identical windings on the same inductor core, though two individual inductors can be used. The topology is essentially identical to a 1:1 transformer-flyback circuit except for the addition of capacitor C2 which forces identical AC voltages across both windings. This capacitor performs three tasks. First, it eliminates the power loss and spikes created by flyback-converter leakage inductance. Secondly, it forces the input current to be a triangular waveform riding on top of a DC component instead of forming a large amplitude square wave. Finally, it eliminates the voltage spike across the output diode when

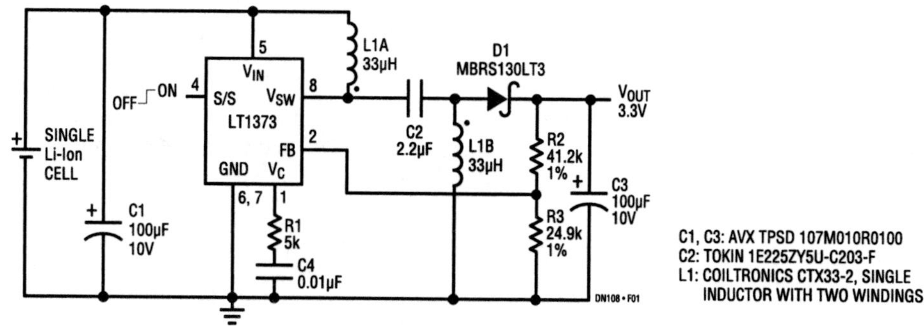

Figure 160.1 • Single Li-Ion Cell to 3.3V SEPIC Converter

Analog Circuit Design: Design Note Collection. http://dx.doi.org/10.1016/B978-0-12-800001-4.00160-5

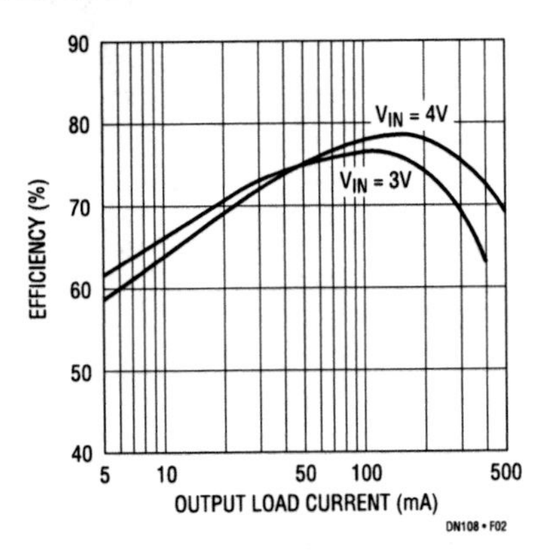

Figure 160.2 • 3.3V Efficiency

the switch turns on. Another feature of the SEPIC topology is that, unlike a typical boost converter, there is no DC path from the input to the output. This means that when the LT1373 is shut down, the load is completely disconnected from the input power source. Figure 160.2 shows that the 3.3V SEPIC converter maintains reasonable efficiency over two decades of output load current even though 3.3V circuits typically have low efficiency due to catch diode losses.

Dual output converter

Many portable systems still require a negative bias voltage to operate interface or other circuitry, where the voltage accuracy is not critical. Using a single inductor, the circuit in Figure 160.3 generates a regulated 5V output and a quasi-regulated −5V output for such applications. The circuit first converts a single Li-Ion cell input voltage to a well-regulated 5V output. It then takes advantage of the switching waveform on the V_{SW} pin to generate the −5V output in a charge pump fashion. The voltage on the V_{SW} pin is 5V plus D1's forward voltage when V_{SW} is high. At this time, C3 charges to the V_{SW} voltage minus D2's forward voltage or about 5V. When the V_{SW} pin goes low, the minus side of C3 goes to −5V which turns on D3 and charges C5 to −5V. This generates a −5V supply which is only quasi-regulated due to diode drop and switch saturation losses. Figure 160.4 shows the regulation of the negative output for various positive output load currents. As shown in Figure 160.5, the dual output converter has high efficiency over two decades of load current.

For more information, please consult the LT1373 data sheet. For parts similar to the LT1373 with higher switching frequencies (500kHz and 1MHz), consult the LT1372/LT1377 data sheet.

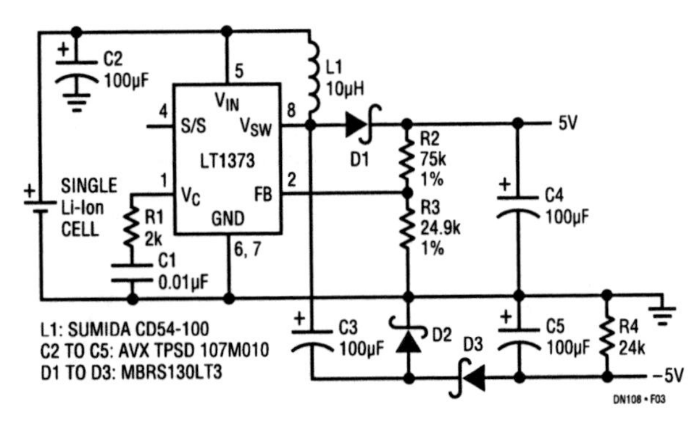

Figure 160.3 • Single Li-Ion to ±5V

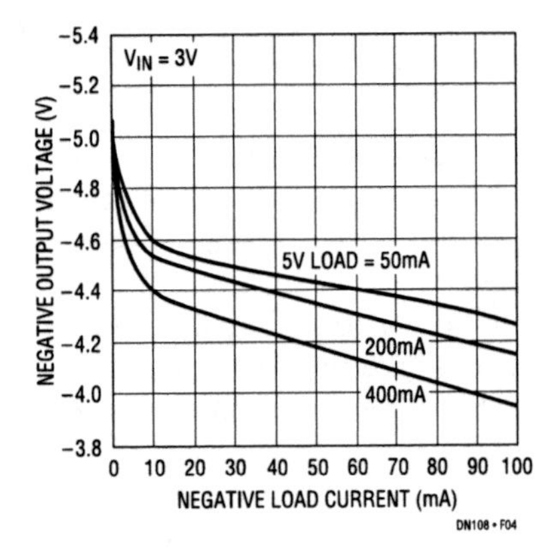

Figure 160.4 • −5V Regulation

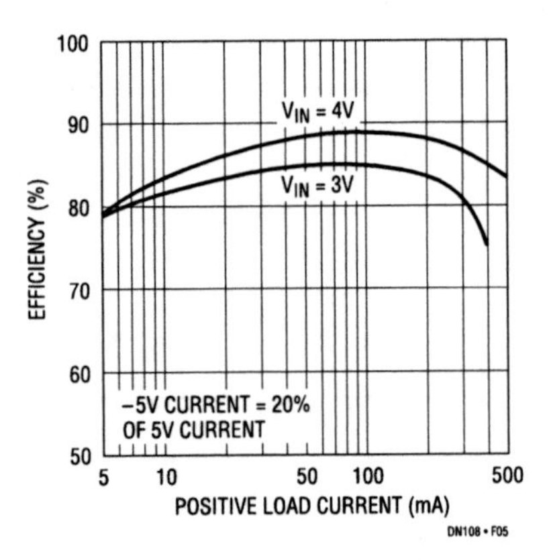

Figure 160.5 • Dual Output Efficiency

DC/DC converters for portable computers

161

Steve Pietkiewicz   Jim Williams

Portable computers require simple and efficient converters for 5V power and display driving. A regulated 5V supply can be generated from two AA cells using the circuit shown in Figure 161.1. U1, an LT1073-5 micropower DC/DC converter, is arranged as a step-up, or "boost" converter. The 5V output, monitored by U1's SENSE pin, is internally divided down and compared to a 212mV reference voltage inside the device. U1's oscillator turns on when the output drops below 5V, cycling the switch on and off at a 19kHz rate. This action alternately causes current to build up in L1, then dump into C1 through D1, increasing the output voltage. When the output reaches 5V, the oscillator turns off. The gated oscillator provides the mechanism to keep the output at a constant 5V. R1 invokes the current limit feature of the LT1073, limiting peak switch current to 1A. U1 limits switch current by turning off the switch when the current reaches the programmed limit set by R1. Switch "on" time, therefore, decreases as V_{IN} is increased. Switch "off" time is not affected. This scheme keeps peak switch current constant over the entire input voltage range, allowing maximum energy transfer to occur at low battery voltage without exceeding L1's maximum current rating at high battery voltage.

The circuit delivers 5V at 150mA from an input range of 3.5V to 2.0V. Efficiency measures 80% at 3.0V, decreasing to 70% at 2.0V for load currents in the 15mA to 150mA range. Output ripple measures 170mV$_{P-P}$ and no-load quiescent current is just 135μA.

A −24V LCD bias generator is shown in Figure 161.2. In this circuit U1 is an LT1173 micropower DC/DC converter. The 3V input is converted to +24V by U1's switch, L1, D1, and C1. The switch pin (SW1) then drives a charge pump composed of C2, C3, D2, and D3 to generate −24V. Line regulation is less than 0.2% from 3.3V to 2.0V inputs. Load regulation, although it suffers somewhat since the −24V output is not directly regulated, measures 2% from a 1mA to 7mA load. The circuit will deliver 7mA from a 2.0V input at 73% efficiency.

If greater output power is required, Figure 161.2's circuit can be driven from a 5V source. R1 should be changed to 47Ω and C3 to 47μF. With a 5V input, 40mA is available at 75%

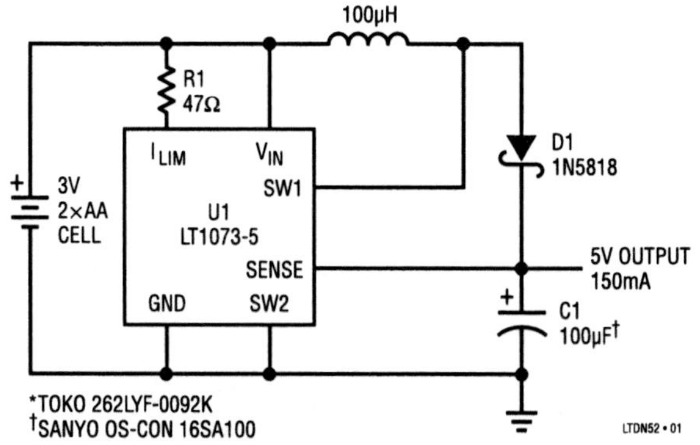

Figure 161.1 • Two AA Cell to 5V Step-Up Converter Delivers 150mA

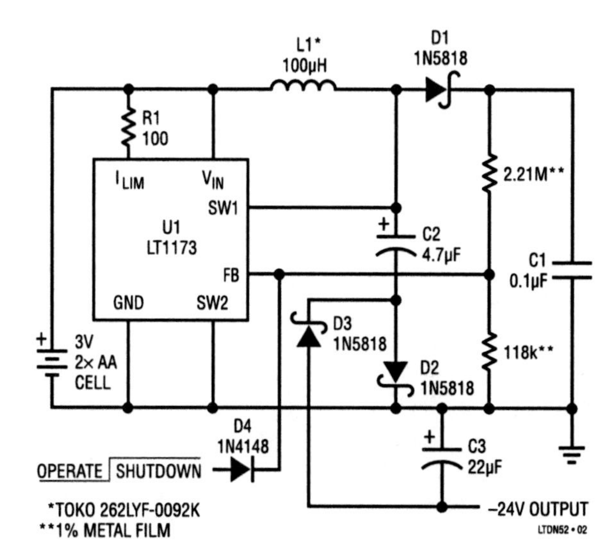

Figure 161.2 • DC/DC Converter Generates −24V from 3V to 5V

Analog Circuit Design: Design Note Collection. http://dx.doi.org/10.1016/B978-0-12-800001-4.00161-7

efficiency. Shutdown is accomplished by bringing the anode of D4 to a logic high, forcing the feedback pin of U1 to go above the internal reference voltage of 1.25V. Shutdown current is 110μA from the input source and 36μA from the shutdown signal.

Current generation portables require back-lit LCD displays using cold cathode fluorescent lamps (CCFLs). Figure 161.3 provides 78% efficiency with full control over lamp brightness. 82% efficiency is possible if the LT1072 is driven from a low voltage (e.g., 3V to 5V) source. Additional benefits include a 4.5V to 20V supply range and low radiated power due to sine wave-based operation.

L1 and the transistors comprise a current driven Royer class converter which oscillates at a frequency primarily set by L1's characteristics and the 0.02μF capacitor. LT1072-driven L2 sets the magnitude of the Q1-Q2 tail current, and hence L1's drive level. The 1N5818 diode maintains current flow when the LT1072 is off.

The 0.02μF capacitor combines with L1's characteristics to produce sine wave voltage drive at the Q1 and Q2 collectors. L1 furnishes voltage step-up, and about 1400V_{P-P} appears at its secondary. Current flows through the 33pF capacitor into the lamp. On negative waveform cycles the lamp's current is steered to ground via D1. Positive waveform cycles are directed, via D2, to the ground referred 562Ω to 50k potentiometer chain. The positive half-sine appearing across these resistors represents one-half the lamp current. This signal is filtered by the 10k-1μF pair and presented to the LT1072's feedback pin. This connection closes a control loop which regulates lamp current. The 2μF capacitor at the LT1072's V_C pin provides stable loop compensation. The loop forces the LT1072 to switch-mode modulate L2's average current to whatever value is required to maintain a constant current in the lamp. The constant current's value, and hence lamp intensity, may be varied with the potentiometer. The constant current drive allows full 0–100% intensity control with no lamp dead zones or "pop-on" at low intensities. Additionally, lamp life is enhanced because current cannot increase as the lamp ages. Detailed information on this circuit appears in Application Note 45.

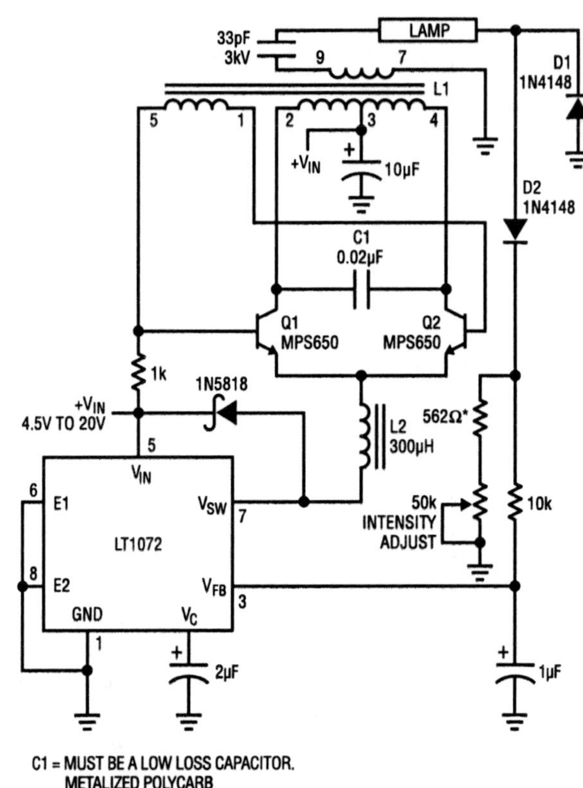

Figure 161.3 • Cold Cathode Fluorescent Lamp Power Supply

No design switching regulator 5V buck-boost (positive-to-negative) regulator

162

Ron Vinsant

Introduction

This simple, no design regulator, operates with an input between 4.5V DC and 40V DC. It provides a −5V output at a maximum output current of 1A to 3A depending on input voltage.

This converter is based on the Linear Technology LT1074 switching regulator IC. This device needs only a few external parts to make up a complete regulator including thermal protection and current limit. This design uses off-the-shelf parts for low cost and easy availability of components. Specifications for the circuit are in Table 162.1.

Circuit description

Figure 162.1 shows the schematic of the circuit. For the purpose of this explanation assume that the output is at a constant −5V DC and that the input voltage is greater than +4.5V DC.

At intervals of ≈10μs (100kHz) the control portion of the LT1074 turns on the switch transistor between the V_{IN} and V_{SW} pins impressing a voltage across the inductor, L1. This causes current to build up in the inductor.

The control circuit determines when to turn off the switch during the 10μs interval to keep the output voltage at −5V DC. When the switch transistor turns off, the magnetic field in the inductor collapses and the polarity of the voltage across the inductor changes to try and maintain the current in the inductor. This current in the inductor is now directed (due to the change in voltage polarity across the inductor) by the diode, D1, to the load. The current will flow from the inductor until the switch turns on again, (continuous operation) or until the inductor runs out of energy (discontinuous operation).

C2 is a low ESR-type electrolytic capacitor that is used in conjunction with L1 as the output filter. C5 and L2 form a post filter that reduces output ripple further.

Referring back to Figure 162.1, the divider circuit of R1, R2, R3 and R4 is used to set the output voltage of the supply against an internal voltage reference of 2.21V DC.

R3, R4, C3 and C4 make up the frequency compensation network used to stabilize the feedback loop (Table 162.2).

Conclusion

This design note demonstrates a fully characterized positive to negative converter circuit that is both simple and low cost. This design can be taken and reliably used in a production environment without the need for any custom magnetics. A PC board layout and FAB drawing are available from Linear Technology.

Table 162.1 Performance Summary (Operating Temperature Range 0°C to 50°C)

Input Voltage Range			+4.5V to +40.0V DC
Output	Output Voltage (±0.15V DC)		−5.00V DC
	Max Output Current at V_{IN} = 4.5V DC		1.0A DC
	Max Output Current at V_{IN} = 40.0V DC		3.5A DC
	Typical Output Ripple at I_{OUT} = 2.5A DC at Switching Frequency	With Optional Filter (L2 and C5) Without Optional Filter (L2 and C5)	$50mV_{p-p}$ $300mV_{p-p}$
	Load Regulation V_{IN} = 4.5V DC	At I_{OUT} = 0.1A DC to 1.0A DC	0.6%
	Line Regulation I_{LOAD} = 1A	At V_{IN} = 4.5V DC to 40.0V DC	0.2%

Analog Circuit Design: Design Note Collection. http://dx.doi.org/10.1016/B978-0-12-800001-4.00162-9

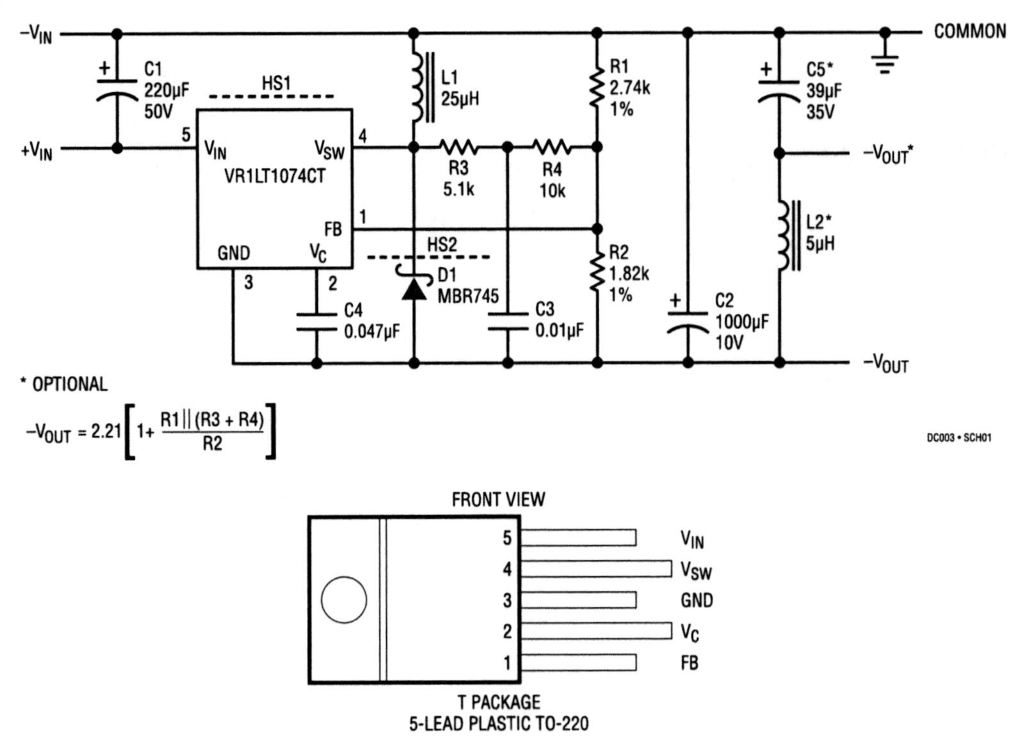

Figure 162.1 • Package and Schematic Diagrams

Table 162.2 Parts List

REFERENCE DESIGNATOR	QUANTITY	PART NUMBER	DESCRIPTION	VENDOR
PCB	1	003A	PCB FAB, Buck-Boost Converter	LTC
D1	1	MBR745	Diode, Schottky, 7A, 45V	Motorola
HS2	1	6038B-TT	Heat Sink	Thermalloy
VR1	1	LT1074CT	Switching Regulator, 100kHz	LTC
HS1	1	7020B-MT	Heat Sink	Thermalloy
C1	1	UPL1H221MPH	Cap, Alum Elect, Low ESR, 220μF, 50V	Nichicon
C2	1	LXF10VB272M12X30LL	Cap, Alum Elect, Low ESR, 1000μF, 10V	United Chemicon
C3	1	CK06BX103K	Cap, Ceramic, 0.01μF, 100V	AVX
C4	1	CK05BX473K	Cap, Ceramic, 0.047μF, 100V	AVX
C5	1	UPL1V390MAH	Cap, Alum Elect, Low ESR, 39μF, 35V	Nichicon
L1	1	CTX 25-5-52	Inductor, 25μH, 5A	Coiltronics
L2	1	CTX5-5-FR	Inductor, 5μH, 5A	Coiltronics
R1	1	MF 1/8W 2.74kΩ	Res, MF, 1/8W, 1%, 2.74kΩ	
R2	1	MF 1/8W 1.82kΩ	Res, MF, 1/8W, 1%, 1.82kΩ	
R3	1	CF 1/4W 5.1kΩ	Res, CF, 1/4W, 5%, 5.1kΩ	
R4	1	CF 1/4W 10kΩ	Res, CF, 1/4W, 5%, 10kΩ	

Section 8

Linear Regulator Design

High voltage inverting charge pump produces low noise positive and negative supplies

163

Marty Merchant

Introduction

Dual-polarity supplies are commonly needed to operate electronics such as op amps, drivers, or sensors, but there is rarely a dual-polarity supply available at the point of load. The LTC3260 is an inverting charge pump (inductorless) DC/DC converter with dual low noise LDO regulators that can produce positive and negative supplies from a single wide input (4.5V to 32V) power source. It can switch between high efficiency Burst Mode operation and low noise constant frequency mode, making it attractive for both portable and noise-sensitive applications. The LTC3260 is available in a low profile 3mm × 4mm DFN or a thermally enhanced 16-lead MSOP, yielding compact solutions with minimal external components. Figure 163.1 shows a typical 12V to ±5V application featuring the LTC3260.

Inverting charge pump

The LTC3260 can supply up to 100mA from the inverted input voltage at its charge pump output, V_{OUT}. V_{OUT} also serves as the input supply to a negative LDO regulator, LDO⁻. The charge pump frequency can be adjusted between 50kHz to 500kHz by a single external resistor. The MODE pin is used to select between a high efficiency Burst Mode operation or constant frequency mode to satisfy low noise requirements.

Constant frequency mode

A single resistor at the RT pin sets the constant operating frequency of the charge pump. If the RT pin is grounded, the charge pump operates at 500kHz, where the open-loop output resistance (R_{OL}) and the output ripple are optimized, allowing maximum available output power with only a few millivolts peak-to-peak output ripple.

Light load efficiency can be increased by reducing the operating frequency, as shown in Figure 163.2, but at the expense of increased output ripple. The lower operating frequency produces a higher effective open-loop resistance (R_{OL}), but the reduced switching rate also reduces the input current, resulting in increased efficiency at light loads. Furthermore, at relatively heavy loads, the increased R_{OL} reduces the effective

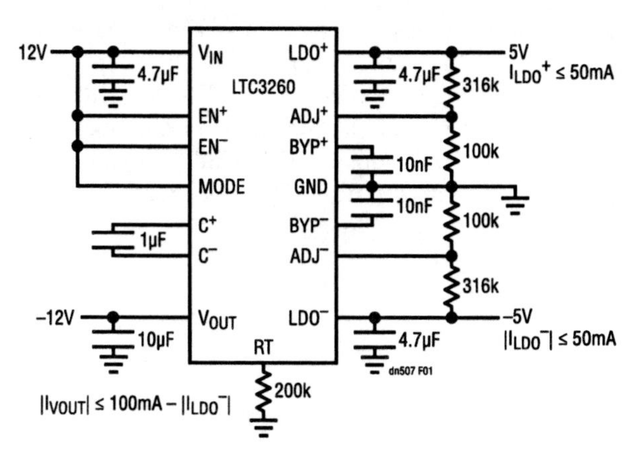

Figure 163.1 • Typical 12V to ±5V Supply

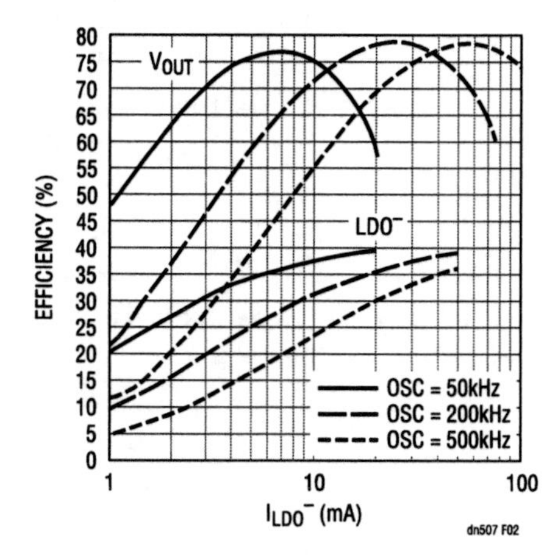

Figure 163.2 • LTC3260 V_{IN} to V_{OUT} and V_{IN} to LDO⁻ Efficiency vs Frequency for the Circuit in Figure 163.1

Analog Circuit Design: Design Note Collection. http://dx.doi.org/10.1016/B978-0-12-800001-4.00163-0

difference between V_{OUT} and LDO^- —decreasing the power dissipation in the negative LDO. The cumulative result is higher overall efficiency with high input voltages and/or light loads.

Reducing the frequency increases the output ripple as shown by the expression below and in Figure 163.3.

$$V_{RIPPLE(PK-PK)} \approx \frac{I_{OUT} \cdot t_{OFF}}{C_{OUT}}$$

$$\text{where } t_{OFF} = \left(\frac{1}{f_{OSC}} - 1\mu s\right)$$

In general, constant frequency mode is suitable for applications requiring low output ripple even at light loads, but further gains in light load efficiency can be gained by using Burst Mode operation, described below.

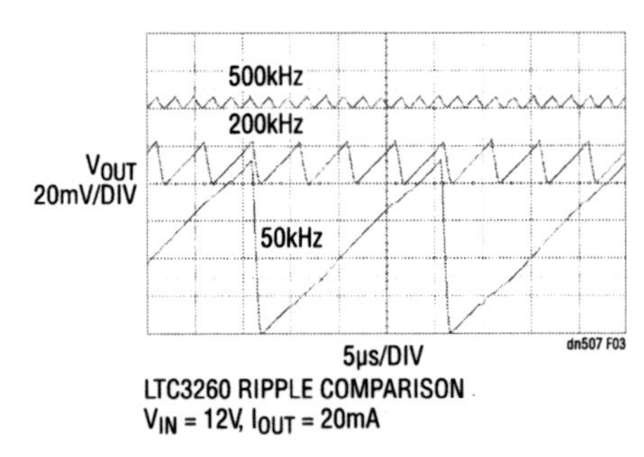

Figure 163.3 • V_{OUT} Constant Frequency Ripple Comparison at 500kHz, 200kHz and 50kHz at 20mA Load

Burst Mode operation

Figure 163.4 shows the light-load efficiency of the charge pump in Burst Mode operation. Burst Mode operation increases the output ripple over constant frequency mode, but the increase in ripple is only a small percentage of V_{IN}, as shown in Figure 163.5.

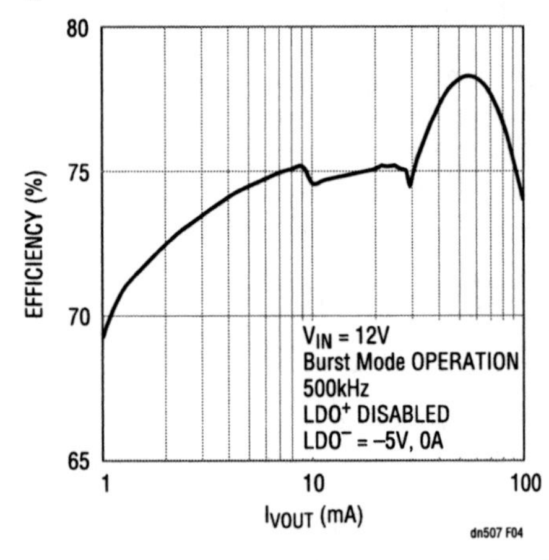

Figure 163.4 • LTC3260 Burst Mode Operation Efficiency

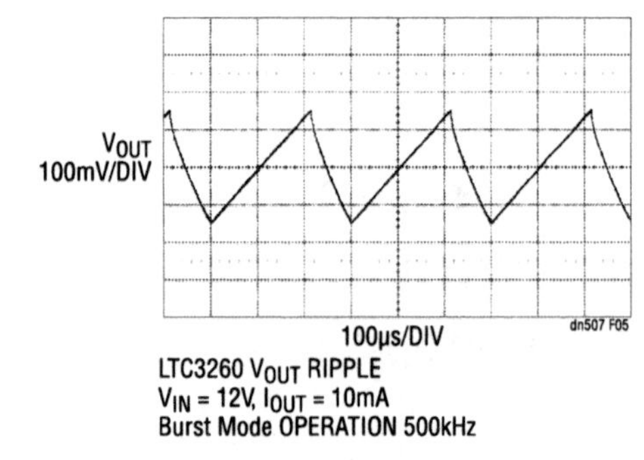

Figure 163.5 • V_{OUT} Ripple in Burst Mode Operation

Burst Mode operation is implemented by charging V_{OUT} close to $-V_{IN}$. The LTC3260 then enters a low quiescent current sleep state, about 100μA with both LDO regulators enabled, until the burst hysteresis is reached. Then the charge pump wakes up and the cycle repeats. The average V_{OUT} is approximately $-0.94V_{IN}$. As the load increases, the charge pump runs more often to keep the output in regulation. If the load increases enough, the charge pump automatically switches to constant frequency mode in order to maintain regulation.

Dual LDOs

Both of the LTC3260's LDOs—the positive LDO regulator supplied from V_{IN}, and the negative LDO regulator supplied from V_{OUT}—are capable of supporting 50mA loads. Each LDO has a dropout voltage of 300mV with a 50mA output and has an adjust pin, allowing the output voltage to be set by a simple resistor divider. The LDO regulators can be individually enabled. The EN^- pin enables both the inverting charge pump and LDO^-. When both regulators are disabled, the part shuts down with only 2μA of quiescent current. The LDO references can be filtered by adding a capacitor on each of the bypass pins to further reduce noise at the LDO regulator outputs.

Conclusion

The LTC3260 produces low noise positive and negative supplies from a single positive power source. The LTC3260 features optional Burst Mode operation for light-load efficiency in battery-powered devices, or low noise constant frequency mode for noise-sensitive applications. The LTC3260's combination of inverting charge pump and dual LDO regulators yields elegant solutions to applications with 4.5V to 32V inputs.

164

80V linear regulator is micropower

Todd Owen

Introduction

Industrial, automotive and telecom applications pose tough design challenges due to harsh operating environments and large voltage transients on already high (12V to 48V) rails. Some switching power supplies are robust enough to provide localized low voltage/high current power from high voltage input rails, but switch-mode supplies are overly complex for low power keep-alive circuits that typically consume only a few milliamps of current. For most of these low power circuits, a wide input range linear regulator is an ideal solution.

Introducing the LT3010 high voltage LDO

The LT3010 is a high voltage, micropower, low dropout linear regulator in a thermally enhanced 8-lead MSOP package. It can provide up to 50mA of output current from input supplies ranging from 3V to 80V. At 50mA of output current, dropout voltage on the LT3010 is only 300mV. The LT3010 operates normally on only 30μA of quiescent current. It can also be put into a low power shutdown state by pulling the $\overline{\text{SHDN}}$ pin low, bringing quiescent current down to just 1μA. For standard operation, the $\overline{\text{SHDN}}$ pin can be pulled as high as 80V (regardless of input voltage) or left floating. The LT3010 is offered in both adjustable and fixed 5V output versions.

Figure 164.1 shows a typical application for the LT3010, illustrating how easy it is to design a low current supply running from a high voltage rail. The only external components required are input and output bypass capacitors, and the input bypass is not required if the device is located close enough to the main supply bypass capacitor. Internal frequency compensation on the LT3010 stabilizes the output for a wide range of capacitors. A minimum of 1μF output capacitance is required for stability, and almost any type of output capacitor can be used. Small ceramic capacitors with low ESR can be used without requiring extra series resistance as is sometimes required with other regulators.

Protection features are incorporated in the LT3010, safeguarding itself and sensitive load circuits. If the input voltage reverses—from a backwards battery or fault on the line—no current flows into the device and no negative voltage is seen at the load. No external protection diodes are necessary when using the LT3010. With a reverse voltage from output to input, the LT3010 acts as though it has a diode in series with its output and prevents reverse current flow. For dual supply applications where the regulator load is returned to a negative supply, the output can be pulled below ground by several volts while still allowing the device to start and operate. The LT3010 also provides current limiting and thermal limiting features.

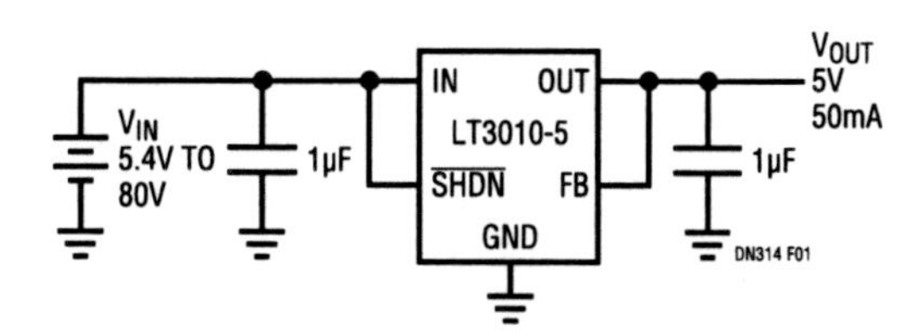

Figure 164.1 • LT3010 Typical Application

Analog Circuit Design: Design Note Collection. http://dx.doi.org/10.1016/B978-0-12-800001-4.00164-2

A versatile and rugged regulator

The LT3010 provides an optimum solution for harsh conditions. Long wire runs for high voltage rails can have transient voltage spikes as loads are switched on and off. Existing automotive applications run from 12V while some new systems are transitioning to 42V, but both can have transients greater than 60V. Telecom applications typically run from a 48V supply that may extend as high as 72V. Industrial applications can span even wider input voltage ranges. Reverse input or reverse voltages from output to input are also possible.

Figure 164.2 shows a typical automotive or telecom application for the LT3010 that takes advantage of its micropower quiescent current. In an automotive application, this might be an always-on circuit that runs whether the ignition is on or not—common for many modern automotive subsystems. The total current consumed by all always-on subsystems must be no more than several milliamps to prevent excessive battery drain. The LT3010 can also be placed into the shutdown state whenever the subsystem is not needed.

Other features of the LT3010 make it ideal for automotive applications. The small size of the LT3010 and its associated external components keep board space and height to a minimum. Power connections to the LT3010 can come directly from a battery because input transients will not damage the regulator or the load. Even reversed battery connections present no worry since the LT3010 prevents reverse current from flowing and damaging sensitive load circuits. Above all, the LT3010's input limit of 80V saves design time and costs by allowing subsystems to migrate directly from 12V to 42V (or anywhere in between) without redesign.

For telecom applications, the 48V rail powers a keep-alive circuit for monitoring or other purposes. The quiescent current is important, especially when battery back-up must kick in to keep the output alive when a fault occurs on the input. Should a fault on the 48V rail occur, the battery back-up takes over and the internal protection of the LT3010 prevents current flow from the output back to the input, removing the need for protection diodes. Component size can still be a concern, depending upon application constraints. The 48V input rail in telecom applications can have transient voltages as high as 72V. The LT3010 can handle these transients without the need for preregulation or protection devices. Finally, the thermally enhanced 8-lead MSOP package provides a very compact, thermally efficient solution footprint. With a θ_{JA} of only 40°C/W, it is able to dissipate heat from high power transients found in these applications.

Conclusion

The LT3010 offers exceptional performance in a small package. It can supply low power from high voltage rails in applications that previously required external pre-regulation schemes or complex switching supplies. Low quiescent current minimizes the power consumption that can be dropped even further by placing the part into shutdown. Stable output voltage is available with a wide range of output capacitors, including small ceramics. Internal protection circuitry in the LT3010 eliminates the need for external protection diodes. The thermally enhanced 8-lead MSOP package provides low thermal resistance making it easy to design the part into harsh environments.

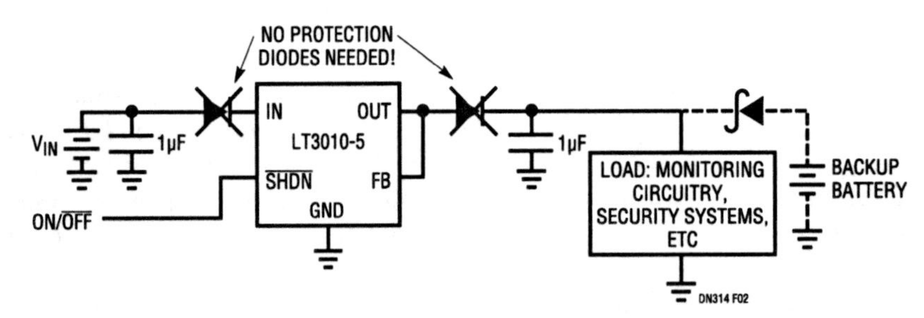

Figure 164.2 • LT3010 Automotive or Telecom Application

Very low dropout (VLDO) linear regulators supply low voltage outputs

165

Tom Gross

Introduction

With each new generation of computing systems, total power continues to increase while system voltages fall. CPU core voltages and logic supplies below 1.8V are now commonplace. Power supplies must not only regulate low output voltages but they must also operate from low input voltages. A low voltage, low dropout linear regulator is an attractive conversion option for applications where output currents are in the several ampere range. Component count and cost are low in comparison to switching regulator solutions, and with low input-to-output voltage differentials, efficiencies are comparable.

VLDO circuit descriptions

The LT1580 monolithic low dropout linear regulator, the LT1573 LDO PNP driver and the LT1575 LDO MOSFET controller/driver are devices well suited to deliver lower output voltages from a supply of 1.8V or lower. Each device offers excellent line/load regulation, temperature performance and transient load step response.

Figure 165.1 illustrates the LT1580 delivering 1.3V at 3A (maximum), from a 1.8V input supply. This particular configuration requires a higher voltage supply to bias the control circuitry. Specifically, the control voltage must be 1V above the

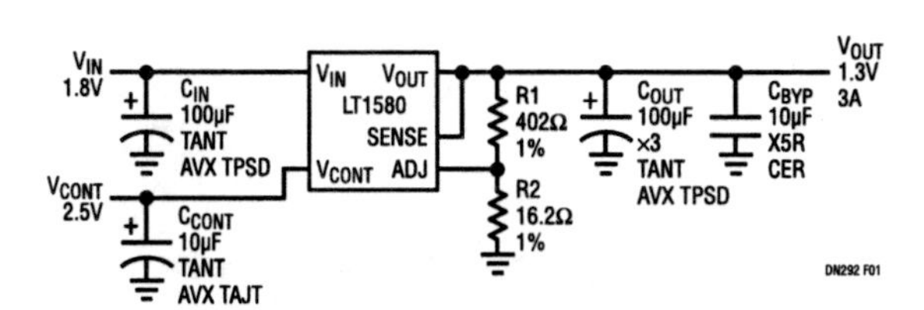

Figure 165.1 • LT1580 Fast Transient Response Low Dropout Linear Regulator

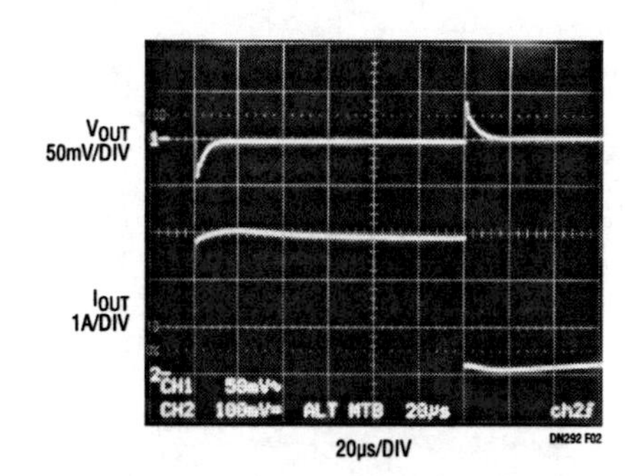

Figure 165.2 • Load Transient Response for Circuit in Figure 165.1

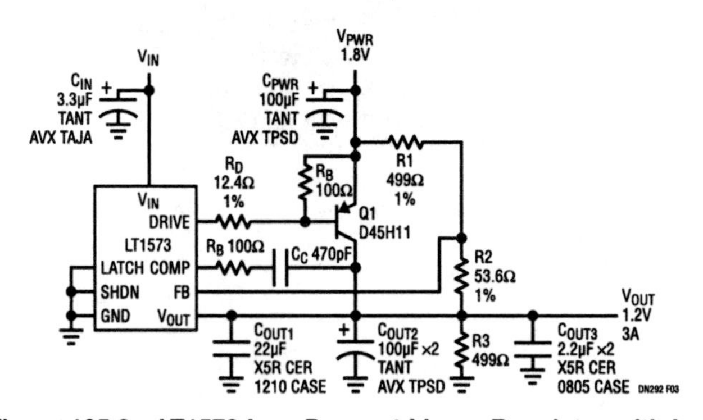

Figure 165.3 • LT1573 Low Dropout Linear Regulator with Low Output Voltage

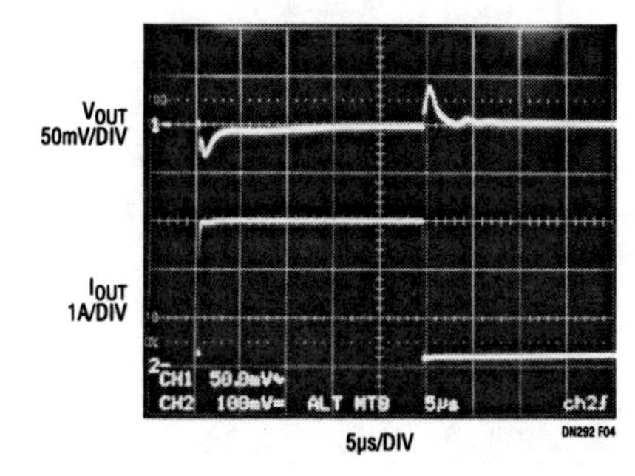

Figure 165.4 • 3A Load Transient Response for Circuit in Figure 165.3

Analog Circuit Design: Design Note Collection. http://dx.doi.org/10.1016/B978-0-12-800001-4.00165-4

output voltage for proper operation or 2.3V minimum in this case. In current systems, a 2.5V supply is typically available and is used here as the control supply voltage. The dropout voltage from input-to-output is 300mV. The load step transient response is shown in Figure 165.2. With a 3A load step, the output voltage deviation is less than 50mV and the output voltage recovers within 20 microseconds.

Figure 165.3 shows a circuit where the regulated output voltage is less than the feedback reference voltage. The circuit consists of an LT1573 linear regulator generating 1.2V at 3A from a 1.8V supply. As in the previous circuit, a second 3.3V input voltage is required for the control circuitry. A resistor divider connected from the 1.8V supply to the output biases the feedback pin above the regulated output voltage by 65mV. This allows the feedback pin to regulate at 1.265V with a 1.2V output. R3's value is chosen such that Q1 must be biased in order for the feedback pin to reach its regulated voltage. This method of generating the feedback voltage is acceptable when the input voltage is regulated. If necessary, an external voltage reference can be used to acquire a tighter output voltage tolerance. Figure165.4 shows the transient response for a 3A load step. Like the previous circuit, the minimum input voltage is 1.6V.

A linear regulator based on the LT1575 MOSFET driver can handle higher output power and very low dropout requirements. Figure 165.5 shows the LT1575 controller driving an external N-channel MOSFET. The regulator converts 1.8V to 1.5V, capable of delivering 4A maximum, using a logic-level Siliconix Si4410 MOSFET as the pass element. The LT1613 boost converter, which is able to operate with input voltages down to 1.1V, generates the appropriate gate drive for the MOSFET. Note that the input capacitors used, Panasonic SP capacitors (part number EEFUE0E221R), were chosen because they represent a typical output capacitor network of microprocessor power supplies. High frequency, low ESR tantalum capacitors, such as AVX TPS capacitors can be substituted for these capacitors.

Figure165.6 depicts the LT1575 load step transient response of less than 50mV output voltage deviation and under 100µs response. The efficiency of this linear regulator is 83%, primarily due to the low input-to-output voltage differential. An external MOSFET with lower $R_{DS(ON)}$ can make for even lower dropout performance and potentially higher efficiency. The LT1575 linear regulator controller offers the lowest dropout voltage performance of any commercially available linear regulator. For instance, the dropout voltage for this circuit is less than 100mV.

Conclusion

Low output voltage, low dropout voltage linear regulators are practical alternatives to switching regulators in the current and future generations of computer systems. All the circuits described above offer viable solutions for the power supply designer. For systems with bus voltages less than 2.5V, the LT1580 linear regulator circuit provides the necessary power conversion with the fewest external components. The LT1573 driver allows the use of a high current PNP pass transistor. The LT1575 linear regulator combines very low dropout voltage performance with high output current capability.

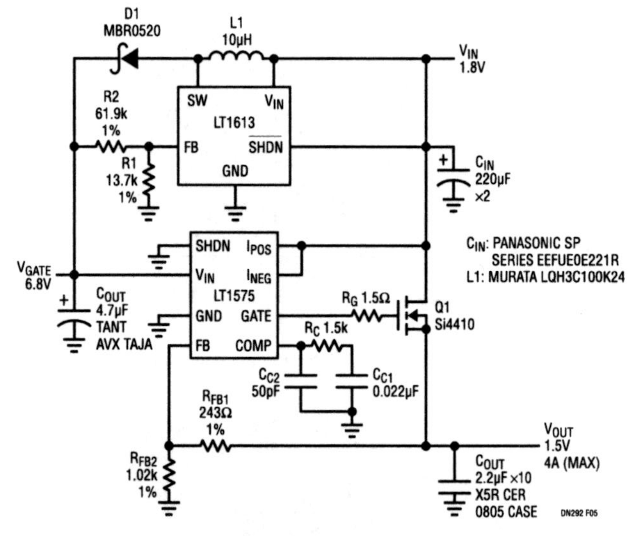

Figure 165.5 • LT1575/LT1613 Low Dropout, High Current Linear Regulator

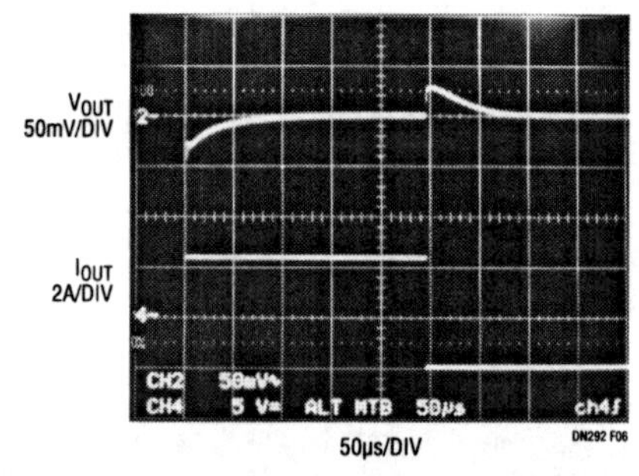

Figure 165.6 • Transient Response for a 4A Step in the Circuit of Figure 165.5

Lowest noise SOT-23 LDOs have 20μA quiescent current, 20μV_{RMS} noise

166

Todd Owen Jim Williams

Telecom and instrumentation applications often require a low noise voltage regulator. Frequently this requirement coincides with the need for low regulator dropout and small quiescent current. LTC recently introduced a family of devices to address this problem. Table 166.1 shows a variety of packages, power ranges and features in three basic regulator types. The SOT-23 packaged LT1761 has only 20μV_{RMS} noise with 300mV dropout at 100mA. Quiescent current is only 20μA.

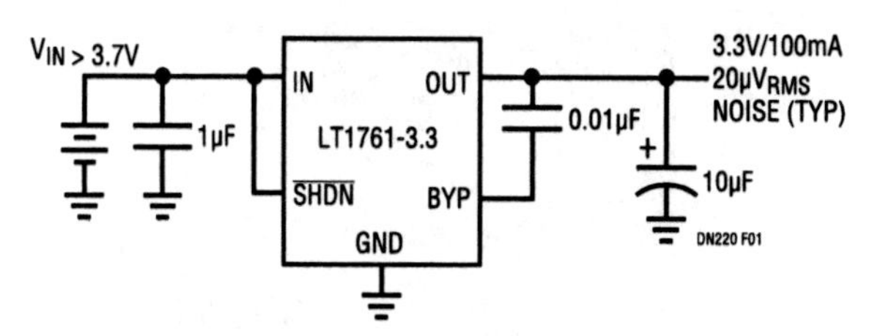

Figure 166.1 • Applying the Low Noise, Low Dropout, Micropower Regulator. Bypass Pin and Associated Capacitor Are Key to Low Noise Performance

Applying the regulators

Applying the regulators is simple. Figure 166.1 shows a minimum parts count, 3.3V output design. This circuit appears similar to conventional approaches with a notable exception: a bypass pin (BYP) is returned to the output via a 0.01μF capacitor. This path filters the internal reference's output, minimizing regulator output noise. It is the key to the 20μV_{RMS} noise performance. A shutdown pin ($\overline{SHDN}$), when pulled low, turns off the regulator output while keeping current drain inside 1μA. Dropout characteristics appear in Figure 166.2. Dropout scales with output current, falling to less than 100mV at low currents.

Noise performance

Noise performance is displayed in Figure 166.3. This measurement was taken in a 10Hz to 100kHz bandwidth with a "brick wall" multipole filter.[1] The photo's trace, applied to a

Table 166.1 Low Noise LDO Family Short-Form Specifications. Quiescent Current Scales with Output Current Capability, Although Noise Performance Remains Constant

REGULATOR TYPE	OUTPUT CURRENT	RMS NOISE (10Hz TO 100kHz) C_{BYP}=0.01μF	PACKAGE OPTIONS	FEATURES	QUIESCENT CURRENT	SHUTDOWN CURRENT
LT1761	100mA	20μV	SOT-23	Shutdown, Reference Bypass, Adjustable Output. SOT-23 Package Mandates Selecting Any Two Features	20μA	<1μA
LT1762	150mA	20μV	MS8	Shutdown, Reference Bypass, Adjustable Output	25μA	<1μA
LT1763	500mA	20μV	SO-8	Shutdown, Reference Bypass, Adjustable Output	30μA	<1μA

Note 1: Noise measurement and specification of regulators requires care and will be comprehensively treated in a forthcoming LTC Application Note.

Analog Circuit Design: Design Note Collection. http://dx.doi.org/10.1016/B978-0-12-800001-4.00166-6

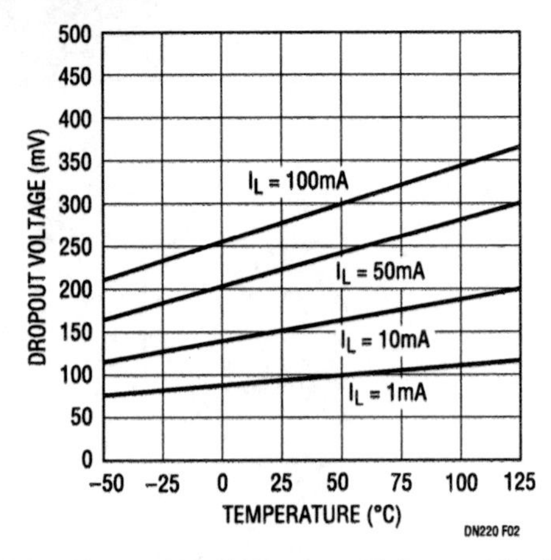

Figure 166.2 • Figure 166.1's Dropout Voltage at Various Currents

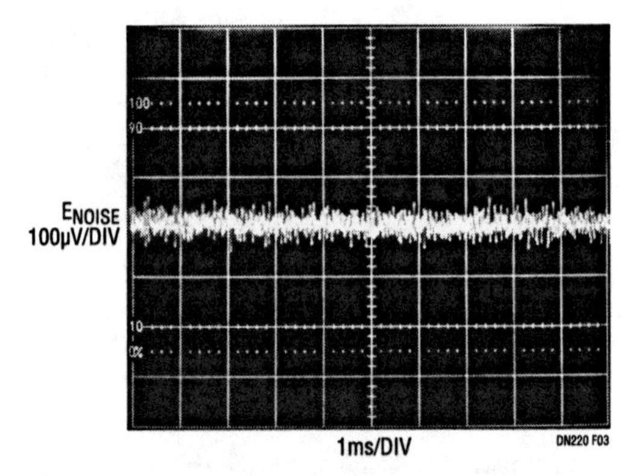

Figure 166.3 • LT1761 Output Voltage Noise in a 10Hz to 100kHz Bandwidth. 20µV_{RMS} Noise Is the Lowest Available in an LDO

thermally responding RMS voltmeter, contains less than 20µV_{RMS} noise. Figure 166.4 shows noise in the frequency domain with noise power falling with increasing frequency.

Other advantages

The LT1761 family is stable (no output oscillation) even when used with low ESR ceramic output capacitors. This is in stark contrast to LDO regulators from other manufacturers that often oscillate with ceramic capacitors.

The unique internal architecture provides an added bonus in transient performance when adding a 0.01µF noise capacitor. Transient response for a 10mA to 100mA step with a 10µF output capacitor is shown in Figure 166.5. Figure 166.6 shows the same setup with the addition of a 0.01µF bypass capacitor. Settling time and amplitude are markedly reduced.

Conclusion

These devices provide the lowest available output noise in a low dropout regulator without compromising other parameters. Their performance, ease of use and versatility allow use in a variety of noise-sensitive applications.

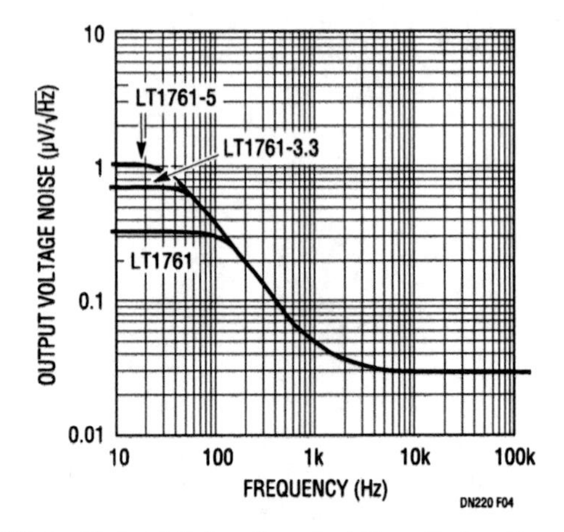

Figure 166.4 • Output Noise Spectral Density for Figure 166.1's Circuit. Curves for Three Output Versions Show Dispersion Below 200Hz

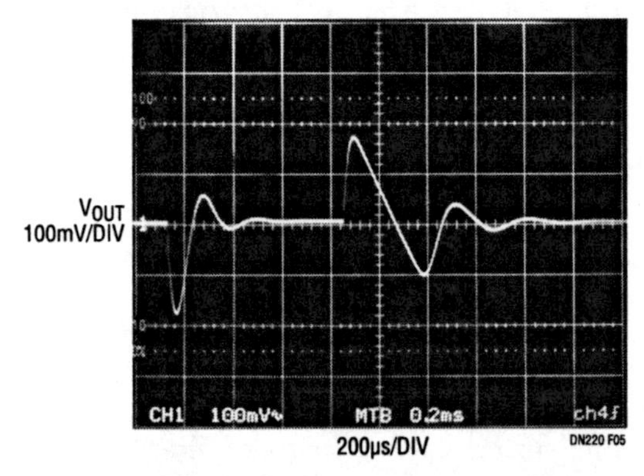

Figure 166.5 • Transient Response with No Noise Bypass Capacitor

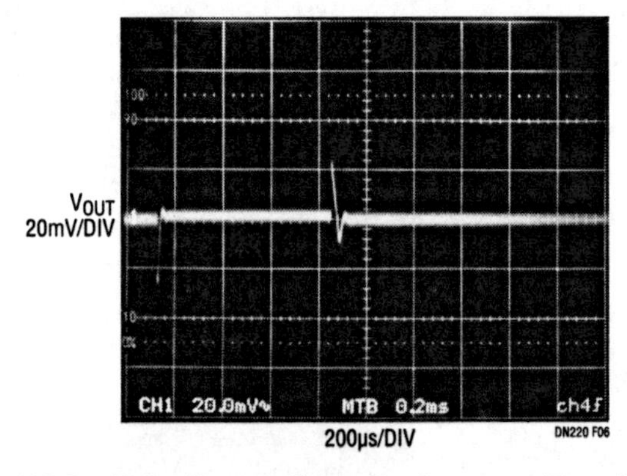

Figure 166.6 • Noise Bypass Capacitor Improves Transient Response. Note Change in Voltage Scale

High efficiency linear and switching solutions for splitting a digital supply

167

Dave Dwelley Gary Maulding

It can be inconvenient to generate a split supply in a typical digital system. The classic solution is to use a pair of resistors between 5V and GND to create a 2.5V "ground" for analog circuitry (Figure 167.1). Unfortunately, the resultant "ground" has a painfully high impedance and the resistors draw a large amount of supply current. The output can be buffered with an op amp to lower the impedance, but a specialized op amp is required to handle any significant bypass capacitance at the output. This Design Note presents two alternate methods of creating a split supply that can provide good transient response while conserving supply current.

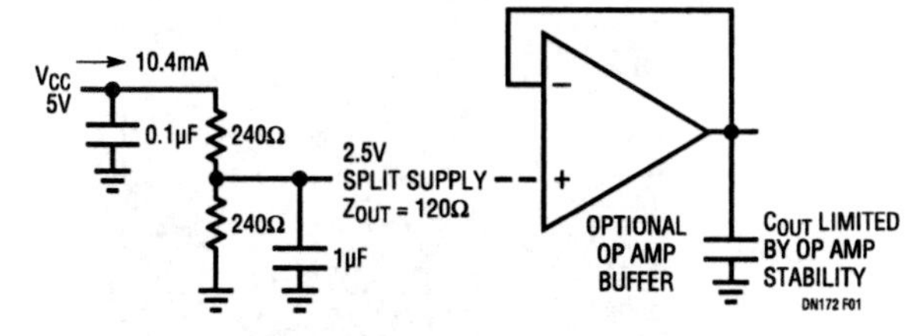

Figure 167.1 • Resistor Divider Supply Splitter

The LT1118 is a specialized linear regulator designed to source or sink current as necessary to keep its output in regulation. It can handle output capacitors of arbitrarily large size, improving output transient response. Available with a fixed 2.5V output (ideal for splitting 5V supplies), it draws only 600µA quiescent current typically and can source 800mA or sink 400mA, enough to satisfy most analog subsystems.

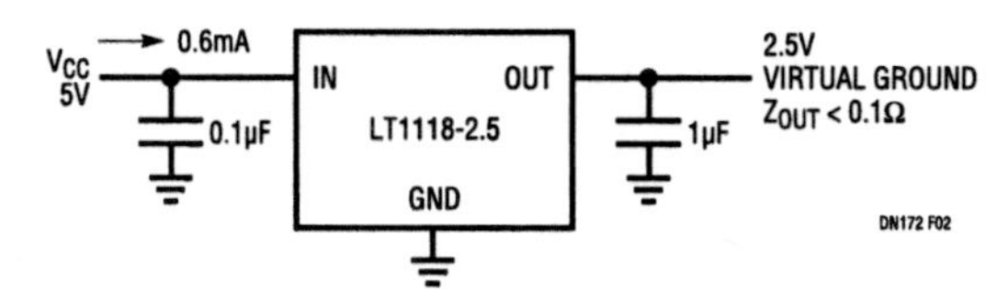

Figure 167.2 • LT1118-2.5 Supply Splitter

The LT1118 requires only two external components (Figure 167.2) and features a DC output impedance below 0.1Ω under all loading conditions, far better than any practical resistor divider solution. The LT1118 draws only enough supply current to meet the demands of the load at the split supply, providing nearly 50% power efficiency over a wide range of load currents (Figure 167.3). Load transient response is excellent, with less than 5µs recovery time from a ±400mA current load step (Figure 167.4). At low current levels, the LT1118 is the optimum solution for splitting a digital supply.

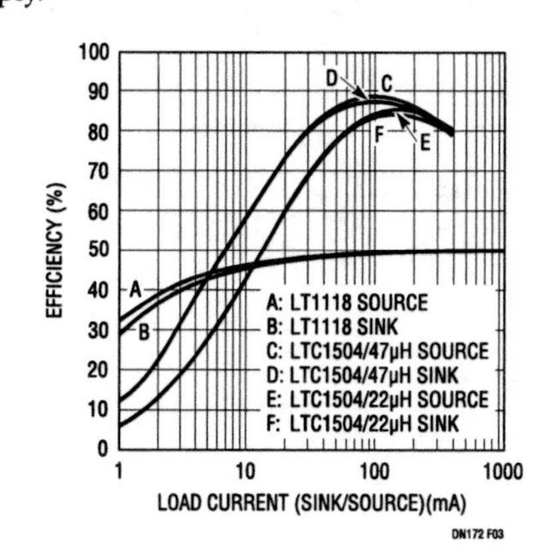

Figure 167.3 • Effiency vs Load Current for Linear and Switching Circuits

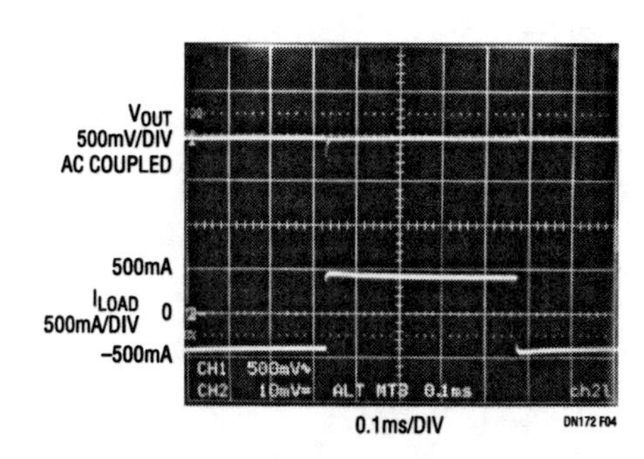

Figure 167.4 • LT1118 Transient Response

Analog Circuit Design: Design Note Collection. http://dx.doi.org/10.1016/B978-0-12-800001-4.00167-8

At higher power levels, the 50% efficiency of the LT1118 can become a liability in power-sensitive or battery-powered systems since half of all the power drawn from the split supply is wasted heating up the LT1118. The LTC1504 addresses this situation by providing as much as 90% efficiency while sourcing or sinking up to 500mA. The LTC1504 is a synchronous switching regulator with on-board power switches. The continuous conduction, synchronous buck architecture inherently sinks current as well as sourcing it, making the circuit an effective supply splitter. Quiescent current is 3mA with typical components. This penalizes efficiency at low current levels when compared to the LT1118, but the intrinsic power conversion abilities of the inductor-based switching architecture allow power efficiencies approaching 90% above 100mA (Figure 167.3 again). A typical LTC1504 circuit will draw only 56mA from the 5V supply while sourcing 100mA from the 2.5V output—magic!

The switching architecture of the LTC1504 requires a few more external components than the LT1118 (Figure 167.5), and generates a small amount of output noise at the 200kHz switching frequency. Transient recovery is controlled primarily by the value of the external inductor. With a 47μH inductor, switching noise is minimal and the circuit recovers from a ±400mA output load step in 30μs (Figure 167.6). Switching to a 22μH inductor brings transient recovery time down to 15μs (Figure 167.7), but output ripple and quiescent current increase. The LTC1504 features a shutdown pin that drops quiescent current below 10μA when the split supply is not required.

Both the LT1118 and the LTC1504 provide superior supply splitting when compared to simple resistor- or regulator-based circuits. The LT1118 fits best where impedance requirements are critical at low current levels, or where low output noise is paramount. The LTC1504 is the best solution where efficiency, especially at high current levels, is the overriding concern. Both devices can also be used in similar applications where source/sink capability is important, such as SCSI or positive ECL supplies.

C_{OUT}: AVX TAJC476M016R
L_{EXT}: SUMIDA CDRH73-470 (47μH: LOW RIPPLE, HIGH EFFICIENCY)
SUMIDA CDRH73-220 (22μH: FAST TRANSIENT RESPONSE) DN172 F05

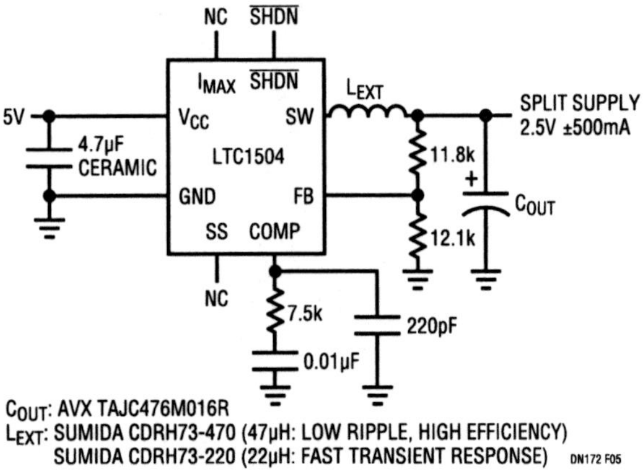

Figure 167.5 • LTC1504 Supply Splitter

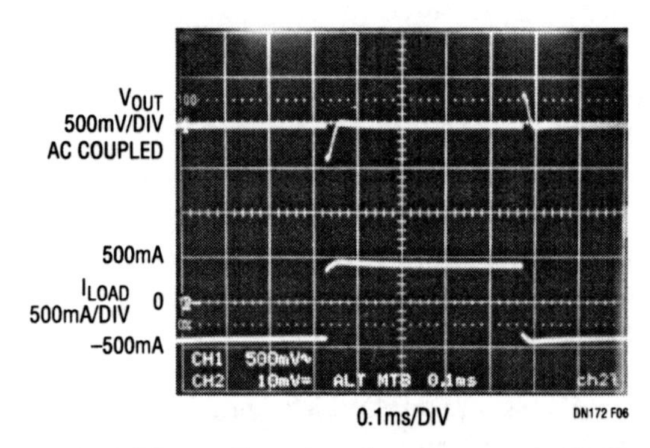

Figure 167.6 • LTC1504 Transient Response with 47μH Inductor

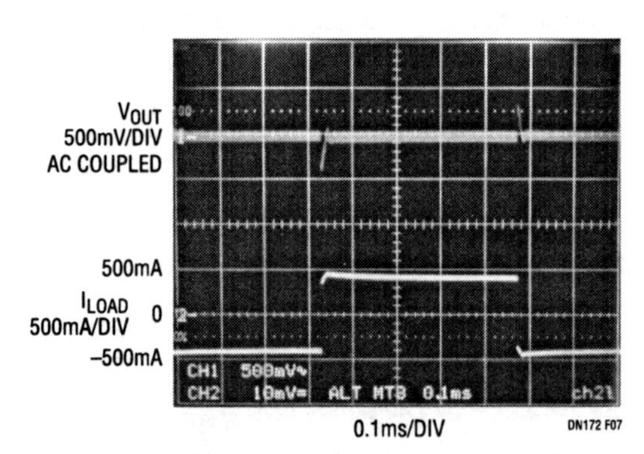

Figure 167.7 • LTC1504 Transient Response with 22μH Inductor

UltraFast linear regulator eliminates all bulk tantalum and electrolytic output capacitors

168

Anthony Bonte

Introduction

Powering 200+MHz microprocessors requires high current, tight tolerance, fast transient response power supplies. Fast load transients mandate bulk output capacitance to maintain regulation and thus, cost increases. Surface mount tantalum capacitors are expensive and require voltage derating for reliable performance. Electrolytic capacitors are physically large and exhibit increased ESR with age. Therefore, transient response and regulation performance degrade.

To improve profit margins, some manufacturers reduce output capacitance and ignore the true regulation requirements. Many power supplies are deemed reliable if Windows95 boots up more than once. Most motherboards are only warranted for 90 days. LTC believes that many system crashes (blamed on software) are attributable to poor power supply regulation. To address these issues, Linear Technology has introduced the LT1575/LT1577.

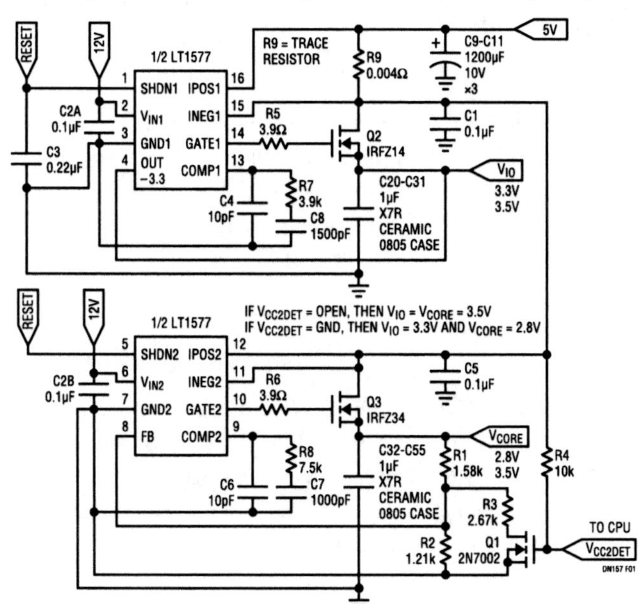

Figure 168.1 • LT1577 P54C/P55C Pentium Processor Autoselect Circuit

New LTC regulator controllers

The LT1575/LT1577 family of controller ICs drives discrete N-channel MOSFETs and produces low dropout, UltraFast transient response regulators. These ICs feature 1% typical performance over all DC tolerances. **Superior transient load performance eliminates all bulk output capacitors.** An LT1577 based P55C Pentium processor power supply operates with only 24 high frequency decoupling, 1μF ceramic capacitors required for the microprocessor core.

Adjustable and fixed voltage versions accommodate any microprocessor voltage. MOSFET $R_{DS(ON)}$ selection allows custom dropout voltage performance. The controllers also provide current limiting, on/off control and overvoltage protection or thermal shutdown. The single LT1575 package is an 8-pin SO or PDIP and the dual LT1577 package is a 16-pin narrow body SO.

Figure 168.1 illustrates an LT1577 application with a fixed 3.3V and an adjustable voltage regulator for a P54C/P55C Pentium processor autoselect circuit. The P54C Pentium processor core and I/O circuitry operate from 3.5V. The P55C Pentium processor I/O operates from 3.3V and the core operates from 2.8V.

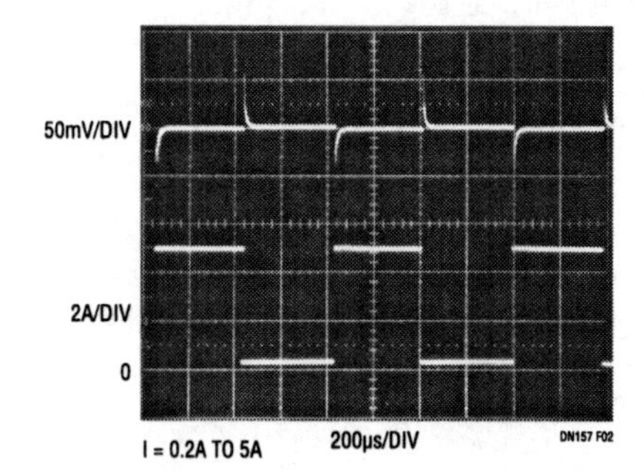

Figure 168.2 • Transient Response for 0.2A to 5A Output Load Step

Analog Circuit Design: Design Note Collection. http://dx.doi.org/10.1016/B978-0-12-800001-4.00168-X

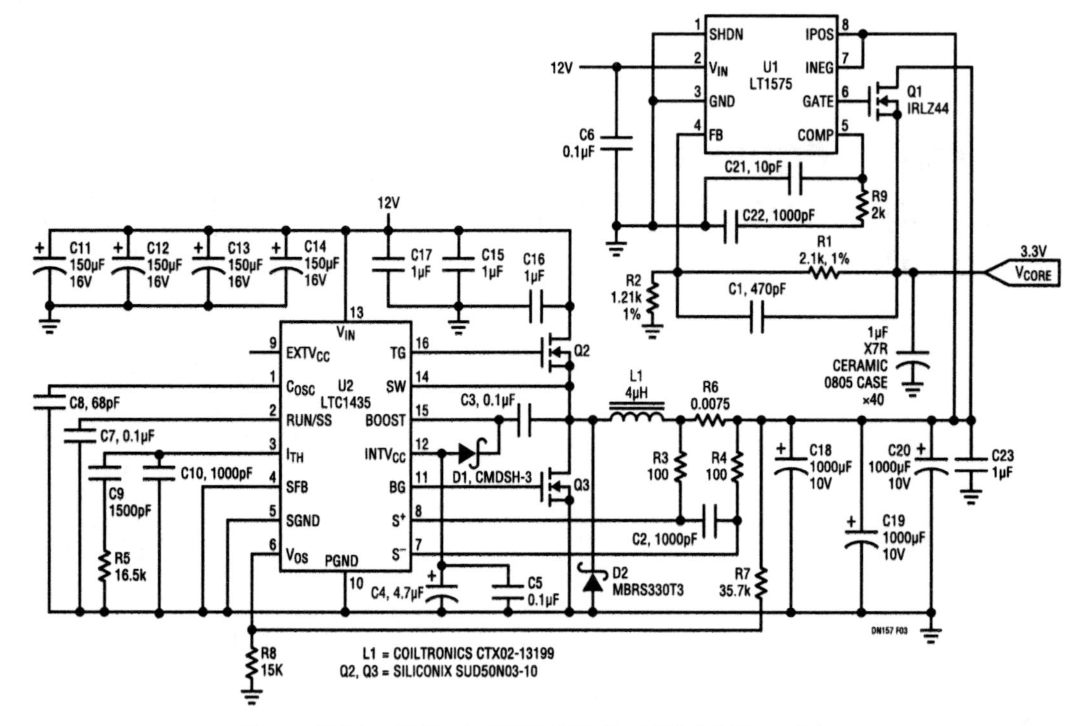

Figure 168.3 • 12V to 3.3V/9A (14A Peak) Hybrid Regulator

V_{CC2DET}'s signal determines circuit operation. In a P54C circuit, V_{CC2DET} is open and the core and I/O supply planes connect together. Q1 turns on and the Q3 (IRFZ34) regulator controls its output to 3.5V. The Q2 (IRFZ14) regulator attempts to control its output to 3.3V, but its feedback pin (Pin 4) senses 3.5V and turns Q2 off. Q3 supplies all core and I/O power.

In a P55C circuit, V_{CC2DET} is grounded and the core and I/O supply planes are separate. Q2 controls the I/O voltage to 3.3V and Q3 controls the core voltage to 2.8V. The I/O circuitry's lower current requirement permits a lower cost MOSFET for Q2 and reduced output capacitance.

The current limit sense resistor is made of "free" PCB trace. Q2's and Q3's common-drain connection permits common heat sink mounting. The COMP pin components adjust frequency compensation for each regulator relative to the MOSFET and output capacitors used.

Figure 168.2 shows the core regulator transient response for a 4.8A load current step in a P55C setup. Compensation limits overshoot/undershoot to 50mV. The ±100mV tolerance for a VRE processor is easily met. The autoselect concept is easily extended to the multiplicity of voltages required by various processors. Consult LTC for details.

Figure 168.3 shows a 3.3V, 14A logic supply that uses an LT1575 as a post-regulator on an LTC1435 synchronous buck regulator, generating 3.3V from 12V with an overall efficiency of 72%. The LT1575 uses an IRLZ44 as the pass transistor, allowing <550mV dropout voltage. The switching regulator's output is set to 4V.

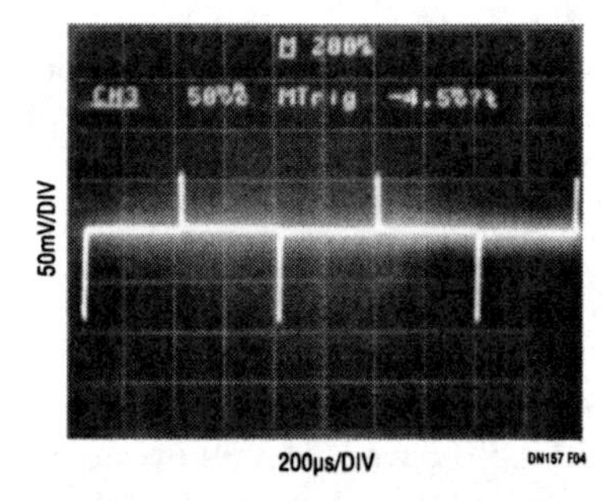

Figure 168.4 • Transient Response for Figure 168.3's Circuit to a 10A Load Step

Figure 168.4 shows the transient response for a 10A, 50ns rise/fall time load step. The only output capacitors are 40, 1μF surface mount ceramic capacitors. The circuit eliminates about a dozen low ESR tantalum capacitors, which would be required without the linear regulator.

Conclusion

The LT1575/LT1577 combine the benefits of low dropout voltage, precision performance, UltraFast transient response and significant output capacitance cost savings. The LT1575/LT1577 controller ICs step to the next performance level required by motherboard designers.

Fast response low dropout regulator achieves 0.4 dropout at 4A

169

Craig Varga

Low dropout regulators have become more common in desktop computer systems as microprocessor manufacturers have moved away from 5V only CPUs. A wide range of supply requirements exists today with new voltages just over the horizon. In many cases, the input-output differential is very small, effectively disqualifying many of the low dropout regulators on the market today. Several manufacturers have chosen to achieve lower dropout by using PNP-based regulators. The drawbacks of this approach include much larger die size, inferior line rejection and poor transient response.

Enter the LT1580

The new LT1580 NPN regulator is designed to make use of the higher supply voltages already present in most systems. The higher voltage source is used to provide power for the control circuitry and supply the drive current to the NPN output transistor. This allows the NPN to be driven into saturation, thereby reducing the dropout voltage by a V_{BE} compared to a conventional design. Applications for the LT1580 include 3.3V to 2.5V conversion with a 5V control supply, 5V to 4.2V conversion with a 12V control supply, or 5V to 3.6V conversion with a 12V control supply. It is easy to obtain dropout voltages as low as 0.4V at 4A, along with excellent static and dynamic specifications.

The LT1580 is capable of 7A maximum with approximately 0.8V input-to-output differential. The current requirement for the control voltage source is approximately 1/100 of the output load current or about 70mA for a 7A load. The LT1580 presents no supply-sequencing issues. If the control voltage comes up first, the regulator will not try to supply the full-load demand from this source. The control voltage must be at least 1V greater than the output to obtain optimum performance. For adjustable regulators, the adjust-pin current is approximately 60 μA and varies directly with absolute temperature. In fixed regulators, the ground pin current is about 10mA and stays essentially constant as a function of load. Transient response performance is similar to that of the LT1584 fast-transient-response regulator. Maximum input voltage from the main power source is 7V and the absolute maximum control voltage is 13V. The part is fully protected from overcurrent and over-temperature conditions. Both fixed voltage and adjustable voltage versions are available. The adjustables are packaged in 5-pin TO-220s, whereas the fixed-voltage parts are 7-pin TO-220s.

The LT1580 brings many new features

Why so many pins? The LT1580 includes several innovative features that require additional pins. Both the fixed and adjustable versions have remote-sense pins, permitting very accurate regulation of output voltage at the load, where it counts, rather than at the regulator. As a result, the typical load regulation over a range of 100mA to 7A with a 2.5V output is approximately ±1mV. The Sense pin and the V_{CONT} pin, plus the conventional three pins of an LDO regulator, give a pin count of five for the adjustable design. The fixed voltage part adds a GND pin for the bottom of the internal feedback divider, bringing the pin count to six. Pin 7 is a no connect.

Note that the Adjust pin is brought out even on the fixed voltage parts. This allows the user to greatly improve the dynamic response of the regulator by bypassing the feedback divider with a capacitor. In the past, using a fixed regulator meant suffering a loss of performance due to the lack of such a bypass. A capacitor value of 0.1 μF to approximately 1 μF will generally provide optimum transient response. The value chosen depends on the amount of output capacitance in the system.

In addition to the enhancements already mentioned, the reference accuracy has been improved by a factor of two, with a guaranteed 0.5% tolerance. Temperature drift is also very well controlled. When combined with ratiometrically accurate internal divider resistors, the part can easily hold 1% output

Analog Circuit Design: Design Note Collection. http://dx.doi.org/10.1016/B978-0-12-800001-4.00169-1

accuracy over temperature, guaranteed, while operating with an input/output differential of well under 1V.

In some cases, a higher supply voltage for the control voltage will not be available. If the Control pin is tied to the main supply, the regulator will still function as a conventional LDO and offer a dropout specification approximately 70mV better than conventional NPN-based LDOs. This is the result of eliminating the voltage drop of the on-die connection to the control circuit that exists in older designs. This connection is now made externally, on the PC board, using much larger conductors than are possible on the die.

Circuit example

Figure 169.1 shows a circuit designed to deliver 2.5V from a 3.3V source with 5V available for the control voltage. Figure 169.2 shows the response to a load step of 200mA to 4.0A. The circuit is configured with a 0.33 μF adjust-pin bypass capacitor. The performance without this capacitor is shown in Figure 169.3. This difference in performance is the reason for providing the Adjust pin on the fixed-voltage devices. A substantial savings in expensive output decoupling capacitance may be realized by adding a small ceramic capacitor at this pin.

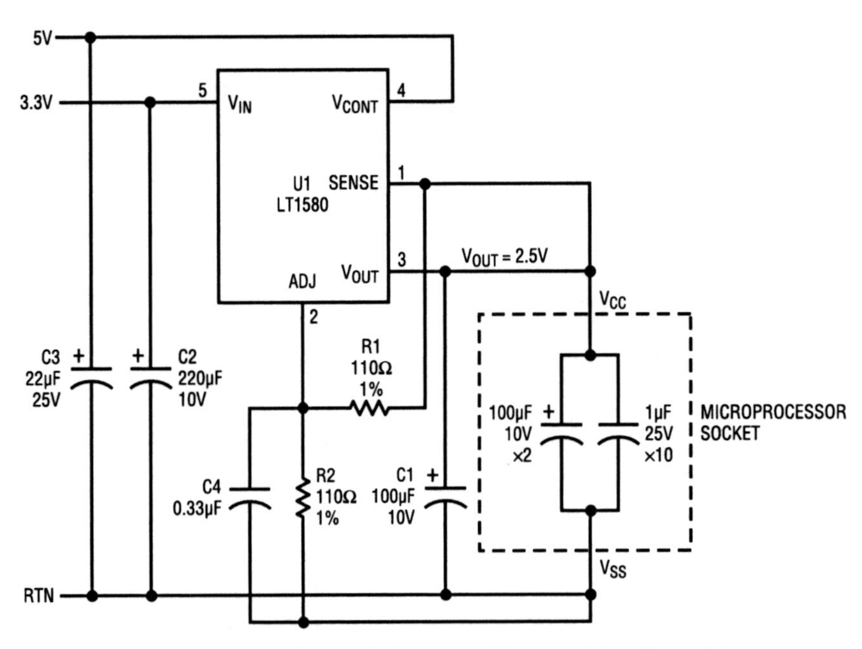

Figure 169.1 • LT1580 Delivers 2.5V from 3.3V at Up to 6A

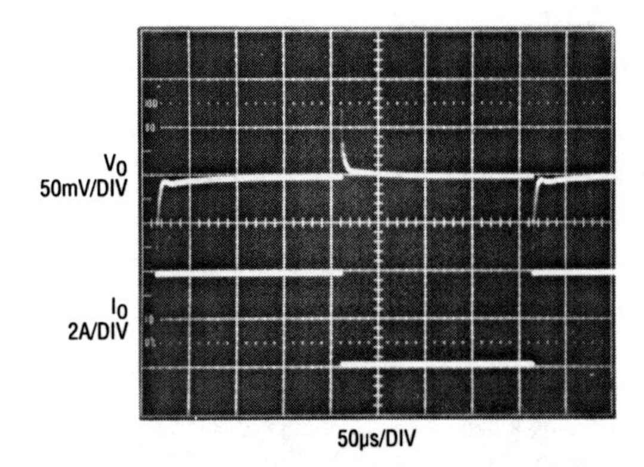

Figure 169.2 • Transient Response of Figure 169.1's Circuit with Adjust-Pin Bypass Capacitor. Load Step is from 200mA to 4 Amps

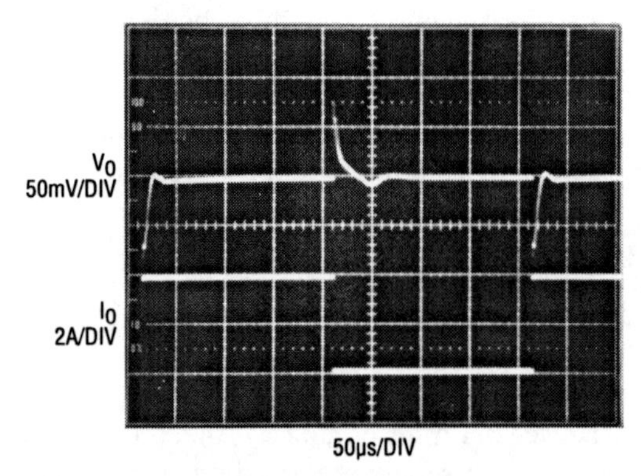

Figure 169.3 • Transient Response Without Adjust-Pin Bypass Capacitor. Otherwise, Conditions are the Same as in Figure 169.2

Create a virtual ground with a sink/source voltage regulator

170

Gary Maulding

Analog signal processing functions embedded in dominantly digital systems usually require the addition of a negative power supply. Although many single supply analog components are available, optimum performance characteristics are often available only in devices that require both positive and negative power supplies. Even the use of single supply components does not solve all of the problems of single 5V operation of analog functions. Signal swings are restricted to a positive direction and may approach but not reach ground potential. The LT1118-2.5 source/sink voltage regulator, used as a supply splitter, provides the designer with an alternative to adding a negative supply. Referencing the signal levels to the regulator's 2.5V output allows symmetrical signal swings and also allows the selection of op amps, active filters and other components that do not have common mode range to ground.

Resistive dividers are frequently used as supply splitters to establish a virtual ground. This approach is often effective, but the circuit designer must work around the resistive divider's severe performance limitations. A fundamental limitation of a resistive divider is the high power dissipation required to obtain any suitable output impedance. A divider made from 240Ω resistors, as in Figure 170.1a, pulls over 10mA of quiescent current and has a DC output impedance of 120Ω. In contrast, the LT1118-2.5 splitter shown in Figure 170.1b achieves an output impedance of less than 0.025Ω with a quiescent current of 600μA. The resistive divider's high output impedance restricts its use to high impedance, low frequency signal sources; otherwise, the output impedance of the splitter will cause gain errors and AC response degradation in all op amp configurations except unity-gain buffers. Use of large feedback network resistors will limit frequency response and noise performance.

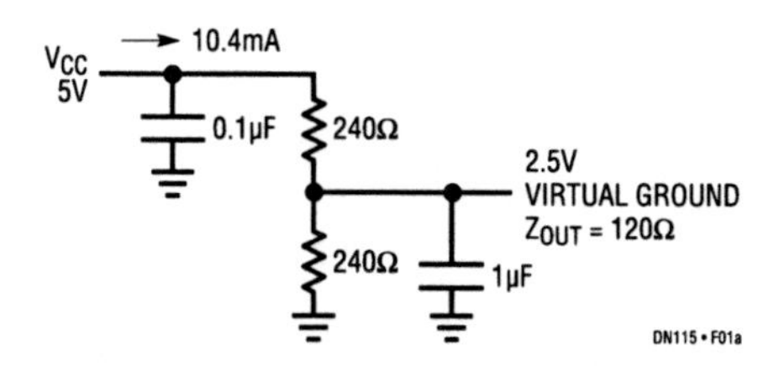

Figure 170.1a • Resistive Divider Supply Splitter

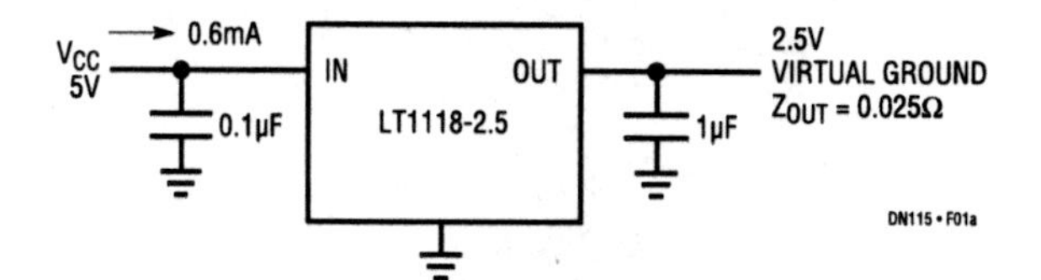

Figure 170.1b • LT1118-2.5 Supply Splitter

Figure 170.2 shows the LT1118-2.5 output impedance vs frequency for several output current loads. Examination of this plot reveals several operating characteristics of the regulator. At low frequencies, the output impedance is below 0.025Ω at all operating points. No passive divider can approach this low impedance. The quiescent current condition has the maximum impedance, with both sink or source current loads causing further output impedance reduction. This level of impedance variation is so low however, that no measurable distortion is introduced in the signal.

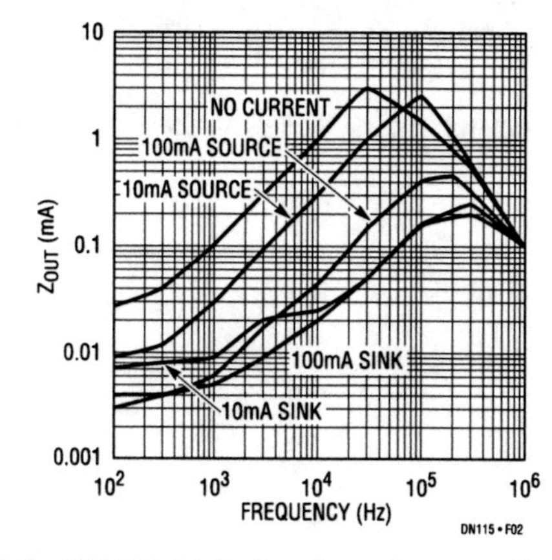

Figure 170.2 • LT1118-2.5 Output Impedance vs Frequency

Analog Circuit Design: Design Note Collection. http://dx.doi.org/10.1016/B978-0-12-800001-4.00170-8

The low output impedance also allows low feedback resistance video amplifiers to operate from the virtual ground without gain errors or distortion. As frequencies increase, the output impedance of the regulator increases, until the output capacitor's impedance dominates and rolls the impedance off to lower values again. With the 1μF capacitor used in this example, maximum output impedance is 3Ω at 30kHz. Larger output capacitors will reduce this maximum impedance by intersecting the regulator's output impedance curve at lower frequencies. Only in the most demanding applications should a larger capacitor be required.

The LT1118 family of regulators are all capable of sourcing 800mA and sinking 400mA of load current. This high level of current will handle almost any signal levels encountered in supply splitter applications. Figure 170.3 shows the fast (less than 5μs) load transient settling characteristics of the LT1118-2.5. This response was obtained using a 1μF output capacitor with 800mA to −400mA load steps. The regulator is stable with any output capacitance greater than 0.2μF, with larger capacitance reducing the output voltage transient amplitude. Competing supply splitter circuits do not exhibit well-behaved transient behavior. This is illustrated in Figure 170.4, which shows a competitor's load transient response from −25mA to 25mA, the circuit's full output capability. With 1μF of output capacitance the regulator is almost unstable.

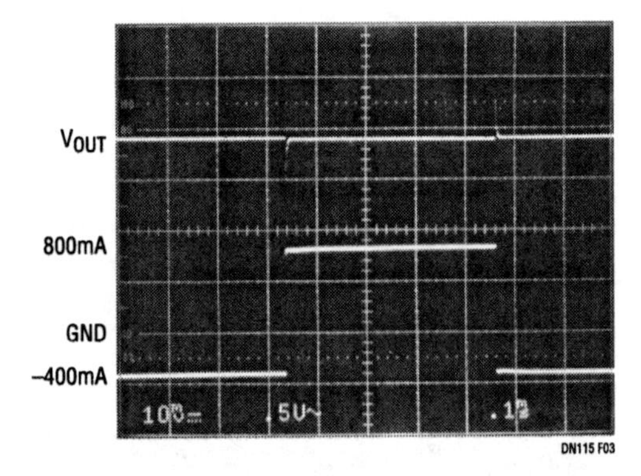

Figure 170.3 • LT1118-2.5 Load Transient Response

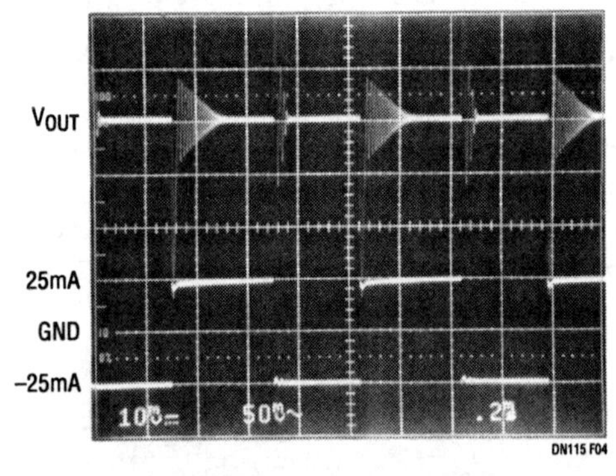

Figure 170.4 • Competitor's Supply Splitter Load Transient

The LT1118-2.5, like the LT1118-2.85/LT1118-5, is offered in both SOT-223 and SO-8 packages. The SO-8 version includes a Shutdown pin, which turns the output high impedance and drops supply current to zero. The SO-8 versions can be used to minimize power consumption when the analog functions are not in use. All members of the LT1118 family share the performance features of fast settling and excellent transient response for loads between 400mA sinking to 800mA sourcing.

The LT1118-5 is a useful regulator for applications requiring very fast settling to full load transients. The current sinking output limits output voltage overshoot compared to conventional regulators.

The LT1118-2.85 is an ideal choice for SCSI terminator applications. The increasing use of active negation drivers to improve noise immunity in fast SCSI systems causes conventional voltage regulators to lose regulation. The current sinking output mode of the LT1118-2.85 maintains a solid 2.85V output with the maximum 24 of 27 data lines negated. The use of one LT1118-2.85 and twenty-seven 110Ω surface mount resistors provides a more economical, higher performance solution for terminating fast wide SCSI data busses than SCSI terminators with on-chip resistors. Terminators with on-chip resistors have serious deficiencies in power dissipation and output capacitance compared to an LT1118-2.85 solution. Up to 3.5W of power is dissipated terminating a 27-line SCSI bus. Terminators with on-chip resistors cannot dissipate this much power from a surface mount IC package, so multiple chips must be used to terminate the entire bus. One LT1118-2.85 with external resistors can do the whole job, since 2W is dissipated in the resistors and only 1.5W in the IC. The LT1118-2.85 plus external resistors also provides better noise immunity than terminators with on-chip resistors. The pin capacitance of the IC can be as high as 2pF. Capacitance on the SCSI data lines causes reflections and loss of noise immunity at high data rates. With the LT1118-2.85 plus external resistor solution, the data line capacitance is minimized to the PC board stray capacitance.

171

5V to 3.3V regulator with fail-safe switchover

Mitchell Lee

Newer microprocessors designed for replacing existing 5V units operate from lower voltage supplies. In the past a processor swap was simply a matter of removing one IC and replacing it with an updated version. But now the upgrade path involves switching from a 5V chip to one that requires 3.xxV.

One means of changing supply voltage from 5V to 3.xxV is to clip a jumper that bypasses a local 3.xxV regulator. This is not a good solution since it leaves too much to chance. Failure to remove the jumper can result in the instant destruction of the new microprocessor upon application of power. A means of automatically sensing the presence of a 3.xxV or 5V processor is necessary.

Intel microprocessors include a special pin called "VOLDET" which can be used to determine whether or not a particular chip needs 3.xxV or 5V. Figure 171.1 shows a simple circuit that takes advantage of this pin to automatically "jumper out" a 3.xxV regulator whenever a 5V processor is inserted into the socket. VOLDET is pulled low on 3.xxV processors; it is buffered by transistor Q2 which grounds the gate of a bypassing switch (Q1). Q1 is turned off leaving the LT1085 to regulate the microprocessor's V_{CC} line.

For 5V microprocessors the VOLDET pin is high; Q2 is turned off allowing Q1's gate to pull up to 12V, turning itself on. With the LT1085 shorted from input to output by the MOSFET, 5V flows directly to the microprocessor. No service intervention is required to ensure correct V_{CC} potential.

The circuit in Figure 171.1 is fine for cases where 12V is available to enhance the MOSFET switch. However, in portable applications, 12V is frequently not available or available only on an intermittent basis. Figure 171.2 shows a second solution using a high side gate driver to control the MOSFET. A VOLDET pull-up resistor is required in both figures because in some cases VOLDET is an open circuit or a shorting link, and in other cases it is an open-drain output.

VOLDET is pulled up to 5V in both circuits. This could pose a problem for 3.xxV processors with open-drain

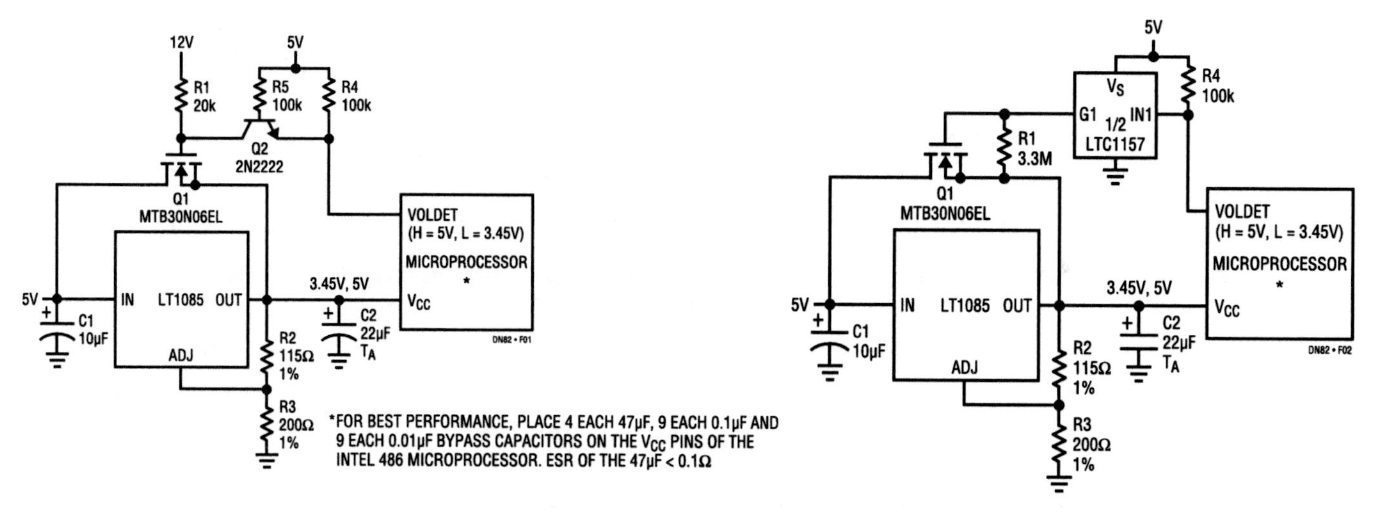

Figure 171.1 • Bypass Circuit for 3.xxV and 5V Microprocessor Swaps Using Transistor Buffer

Figure 171.2 • Bypass Circuit for 3.xxV and 5V Microprocessor Swaps Using High Side Gate Driver (No 12V Supply Required)

Analog Circuit Design: Design Note Collection. http://dx.doi.org/10.1016/B978-0-12-800001-4.00171-X

VOLDET pins, but for 3.xxV devices VOLDET is always pulled low and 5V never reaches it. The 5V reaches VOLDET only on 5V devices.

For certain families of microprocessors, 3.3V is required. The circuits shown in Figures 171.1 and 171.2 are fully compatible with 3.3V applications by simply substituting a fixed 3.3V version of the regulator (use an LT1085-3.3). Higher current operation is also possible. The LT1085 is suitable for 3A applications; use an LT1084 and an MTB50N06EL for up to 5A. Table 171.1 shows the wide range of linear regulators available at currents of up to 10A.

In some applications the complexity of a high efficiency switching regulator may be justified for reasons of battery life. Figure 171.3 shows a switcher that not only converts 5V to 3.45V but also acts as its own bypass switch for applications where a 5V output is required. An open drain or collector pulling the V_{FB} pin low causes the top-side P-channel MOSFET to turn on 100%, effectively shorting the output to the 5V input. If the open collector is turned off, the LTC1148 operates as a high efficiency buck mode power converter, delivering a regulated 3.45V to the load. For 3.3V applications a fixed 3.3V version of the LTC1148 is available.

Table 171.1 Linear Regulators for 5V to 3.3V Conversion

LOAD CURRENT	DEVICE	FEATURES
150mA	LT1121-3.3	Shutdown, Small Capacitors
700mA	LT1129-3.3	Shutdown, Small Capacitors
800mA	LT1117-3.3	SOT-223
1.5A	LT1086	DD Package
3A to 7.5A	LT1083 LT1084 LT1085	High Current, Low Quiescent Current at High Loads
10A	2×LT1087	Parallel, Kelvin Sensed

The topic of powering low voltage microprocessors in a 5V environment is covered extensively in Application Note 58, available on request. Both linear and switching solutions are discussed.

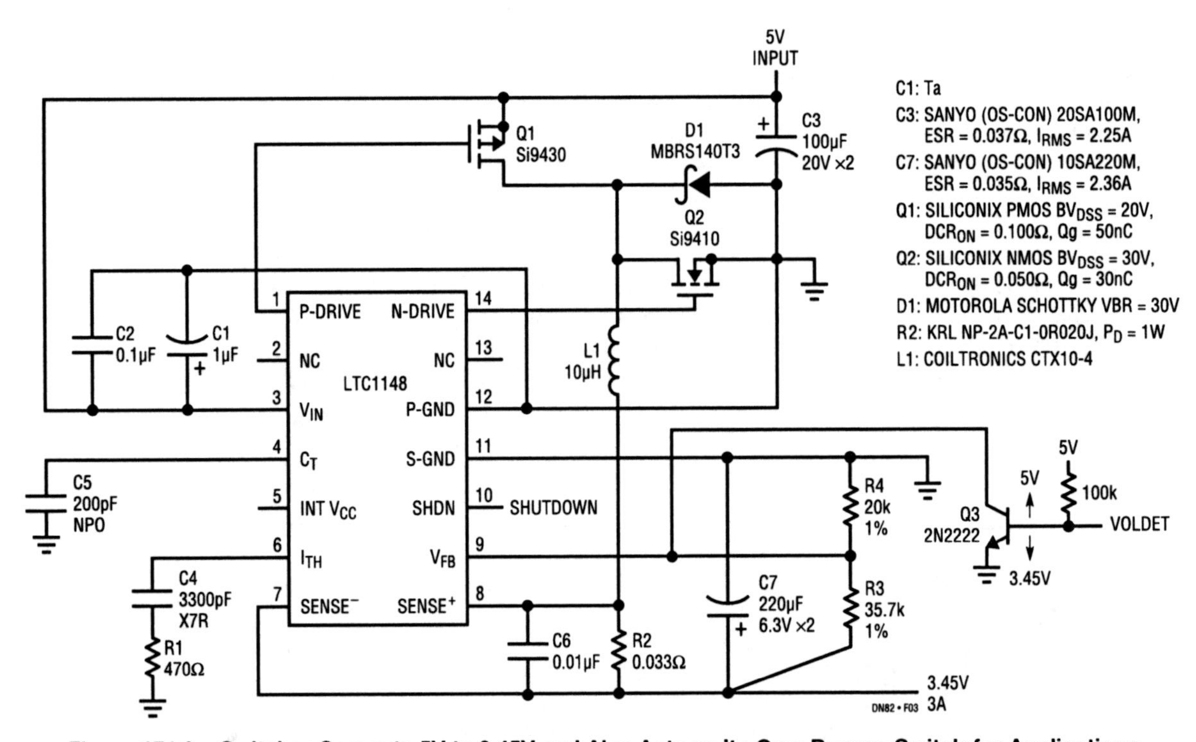

Figure 171.3 • Switcher Converts 5V to 3.45V and Also Acts as its Own Bypass Switch for Applications Where a 5V Output Is Required

A simple ultralow dropout regulator

172

Jim Williams

Single Devices Now Provide This Function

Switching regulator post regulators, battery powered apparatus, and other applications frequently require low drop-out linear regulators. Often, battery life is significantly affected by the regulator's dropout performance. Figure 172.1's simple circuit offers lower dropout voltage than any monolithic regulator. Dropout is below 50mV at 1A, increasing to only 450mV at 5A. Line and load regulation are within 5mV, and initial output accuracy is inside 1%. Additionally, the regulator is fully short circuit protected, and has a no load quiescent current of 600μA.

Circuit operation is straightforward. The 3-pin LT1123 regulator (TO-92 package) servo controls Q1's base to maintain its feedback pin (FB) at 5V. The 10μF output capacitor provides frequency compensation. If the circuit is located more than six inches from the input source the optional 10μF capacitor should bypass the input. The optional 20Ω resistor limits LT1123 power dissipation and is selected based upon the maximum expected input voltage (see Figure 172.2).

Normally, configurations of this type offer unpredictable short circuit protection. Here, the MJE1123 transistor shown has been specially designed for use with the LT1123. Because of this, beta based current limiting is practical. Excessive output current causes the LT1123 to pull down harder on Q1 until beta limiting occurs. Under these conditions the controlled pull down current combines with Q1's beta and safe operating area characteristics to provide reliable short circuit limiting. Figure 172.3 details current limit characteristics for 30 randomly selected transistors.

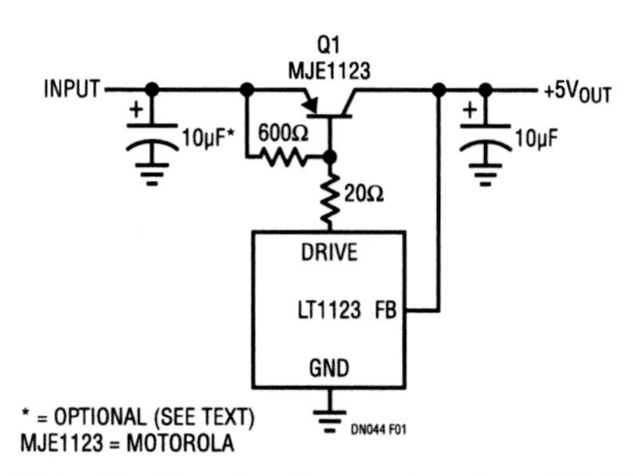

Figure 172.1 • The Ultra-Low Dropout Regulator. LT1123 Combines with Specially Designed Transistor for Lowest Dropout and Short Circuit Protection

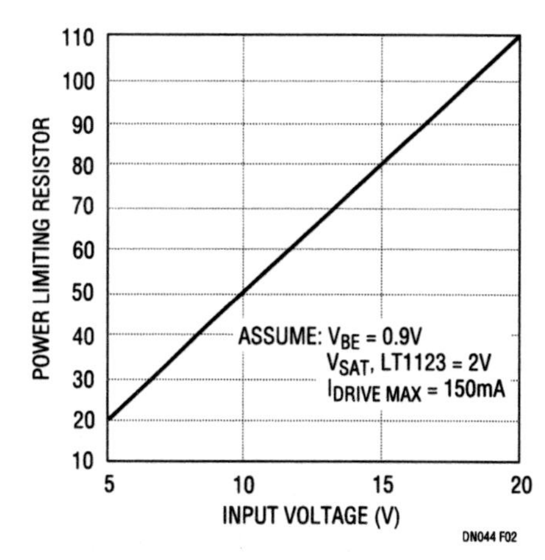

Figure 172.2 • LT1123 Power Dissipation Limiting Resistor Value vs Input Voltage

Analog Circuit Design: Design Note Collection. http://dx.doi.org/10.1016/B978-0-12-800001-4.00172-1

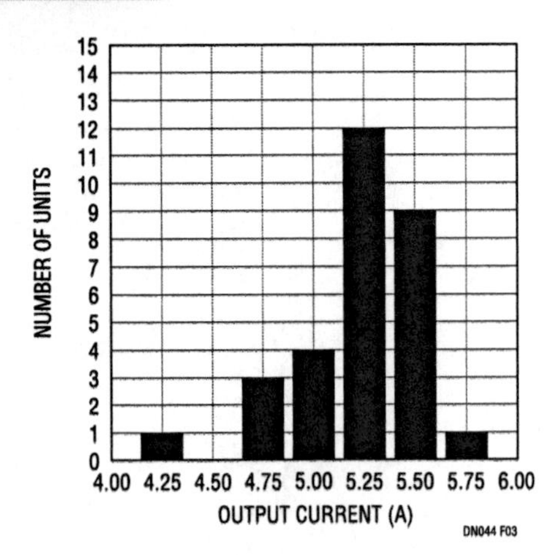

Figure 172.3 • Short Circuit Current for 30 Randomly Selected MJE1123 Transistors at $V_{IN}=7V$

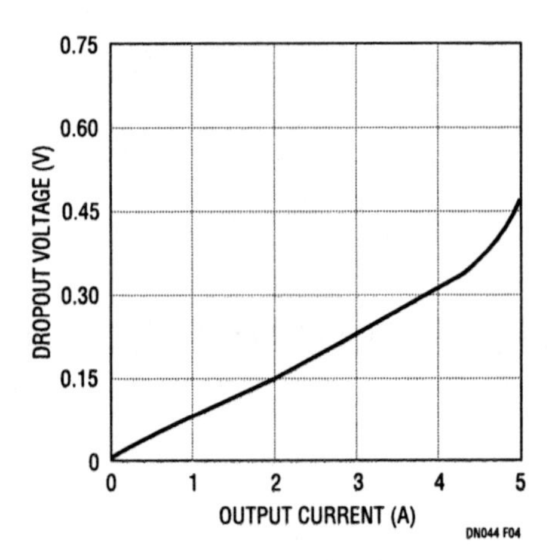

Figure 172.4 • Dropout Voltage vs Output Current

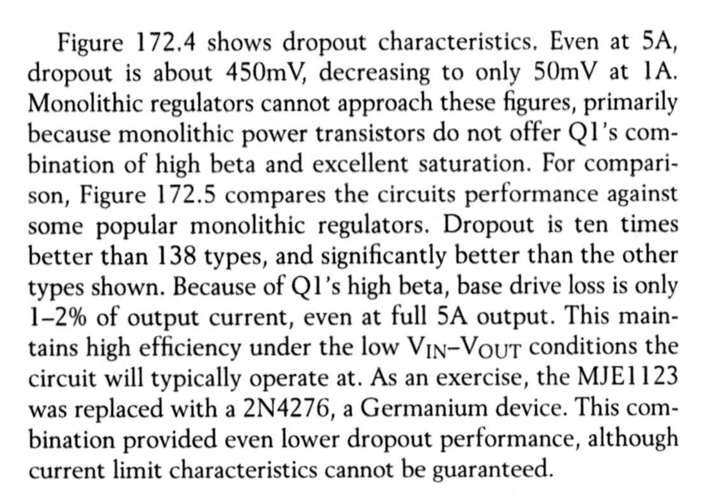

Figure 172.4 shows dropout characteristics. Even at 5A, dropout is about 450mV, decreasing to only 50mV at 1A. Monolithic regulators cannot approach these figures, primarily because monolithic power transistors do not offer Q1's combination of high beta and excellent saturation. For comparison, Figure 172.5 compares the circuits performance against some popular monolithic regulators. Dropout is ten times better than 138 types, and significantly better than the other types shown. Because of Q1's high beta, base drive loss is only 1–2% of output current, even at full 5A output. This maintains high efficiency under the low $V_{IN}–V_{OUT}$ conditions the circuit will typically operate at. As an exercise, the MJE1123 was replaced with a 2N4276, a Germanium device. This combination provided even lower dropout performance, although current limit characteristics cannot be guaranteed.

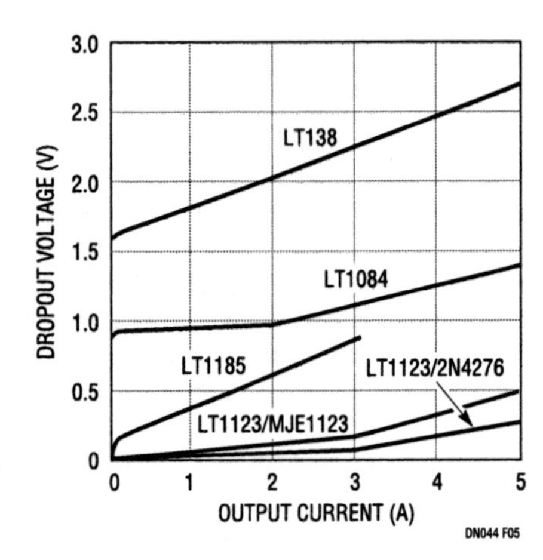

Figure 172.5 • Dropout Voltage vs Output Current for Various Regulators

Powering 3.3V digital systems

173

Dennis O'Neill

The new generation of high density digital devices requiring 3.3V power supplies impose some unique constraints on power supply designers. In nearly all cases the computers using these devices already have a 5V system supply. Deriving the 3.3V supply from the existing 5V rail permits system upgrades with a simple on-card solution. In many cases the 5V rail is the only supply available, mandating this approach. The first decision to be made is whether to use a switching regulator or a linear regulator? Switchers have a clear efficiency advantage when there is a large difference between the input and output voltage, but that advantage diminishes as the input voltage approaches the output voltage.

Simple calculations show that the efficiency of a switcher is marginally better in this application. Assuming a nominal input voltage of 5.0V and an output voltage of 3.3V, the efficiency of a linear regulator, (LT1083 type Figure 173.1A), independent of output current, is simply 3.3V/5.0V = 66%. For a switcher (Figure 173.1B) the efficiency is tougher to

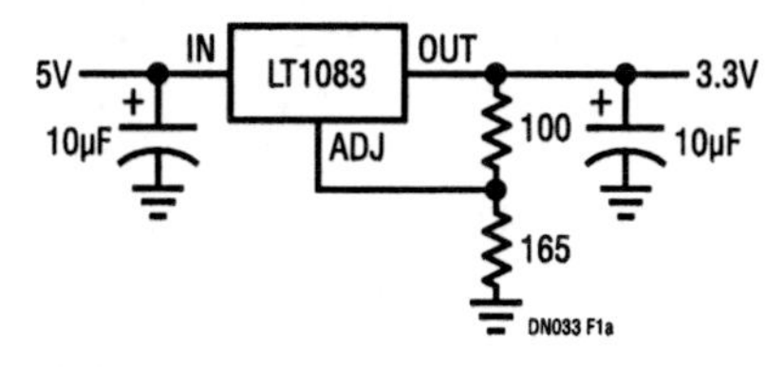

Figure 173.1A • Linear Regulator

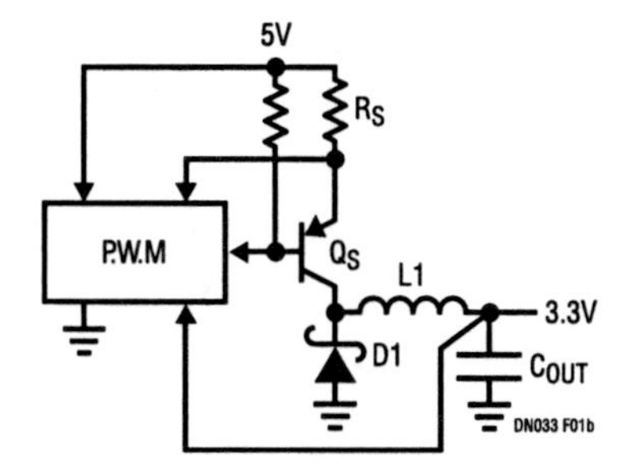

Figure 173.1B • Buck Switching Regulator

calculate. With only 5V available a PNP switch must be used, a MOSFET is not practical due to its gate drive requirements. The average inductor current will be equal to the load current. The duty cycle is determined by:

$$DC = (V_{OUT} + V_D)/(V_{IN} - V_{SAT} - V_D)$$

where $V_{OUT} = 3.3V$, $V_{IN} = 5.0V$, $V_{SAT} = V_{CE}$ sat. of Q_S, V_D = forward voltage of D1.

Assuming V_{CE} sat. of Q_S to be 0.6V at a forced Beta of 10 and the forward voltage of D1 to be 0.6V, (using a Schottky diode; a silicon diode would be closer to 1V at rated current), the formula indicates an 80% duty cycle. Significant power losses are listed below.

Switch Saturation Voltage	(0.6V)(1.0A)(80%) = 0.48W
Switch Base Current	(5.0V)(0.1A)(80%) = 0.40W
Diode Forward Voltage	(0.6V)(1.0A)(20%) = 0.12W
Inductor Voltage	(0.1V)(1.0A)(100%) = 0.10W
Switching Transients	≈0.10W
R_{SENSE} Voltage	(0.1V)(1.0A)(80%) = 0.08W
P.W.M. Circuit	(5.0V)(0.02A)(100%) = 0.10W
Total	1.38W

The efficiency is Power Out/Power In = 3.3W/(1.38W + 3.3W) ≈ 70%. This says that the switcher could be more efficient, but by a small margin (4%). Other considerations, such as noise filtering, further decrease the switchers efficiency. In addition, circuit design becomes complex. For example, short circuit sensing in the emitter lead of Q_S might require generating another supply greater than 5V to power the sense amplifier. Also, some form of adaptive base drive is needed to maintain efficiency at light loads.

When the small efficiency gain of a switcher is balanced against the advantages of a linear regulator (superior transient

Analog Circuit Design: Design Note Collection. http://dx.doi.org/10.1016/B978-0-12-800001-4.00173-3

response, low noise, and ease of design) it becomes clear that the linear regulator is the best choice in this application.

Regulator design

Figure 173.1A shows a basic linear regulator circuit utilizing an LT1083 adjustable low dropout regulator. These deices, specified for dropout voltage at several points over their operating current range, are ideal for this application. Nominal tolerance on the 5V rail in most systems is ±5% (4.75V to 5.25V). If the regulator dropout voltage is at the upper extreme of its specification (1.5V at maximum current and temperature for LT1083 family) it would still be able to supply 3.25V to the memory devices when the 5V rail is at the low end of its specification (4.75V). This is well within the allowable digital supply voltage range of 3.3V ±10% (3.0V–3.6V).

LT1083 family behavior in dropout is benign. Once the device enters dropout the output simply follows the input. There is no increase in quiescent current during dropout as is common in PNP type regulators. The basic regulator circuit shown in Figure 173.1A supplies currents up to 7.5A. The device has all the normal protection features associated with high current supplies, such as thermal and short circuit protection.

At currents greater than 7.5A several LT1087's are used in parallel (Figure 173.2). The LT1087 is the newest member of the LT1083 family. The device is a version of the LT1084 with two additional sense pins and is available in a 5-pin TO-220 package. When tied together the sense pins are used to Kelvin sense the output voltage. When used separately they form the inputs to a differential amplifier whose output changes the devices 1.2V reference voltage.

In Figure 173.2 the master LT1087's sense pins are tied together and connected to point A. This device senses and controls the output voltage using the Kelvin sense feature. R3 and C1 filter the voltage fed back to the sense pins. At low frequencies the output pin voltage of the master LT1087 is forced positive by the internal loop to maintain point A at the desired 3.3V value. This voltage is set by the ratio of R2/R1 according to the formula in Figure 173.2. The voltage across R_M is proportional to the load current.

The slave unit operates differently. This device senses the voltage across R_M and adjusts the voltage across R_S to be equal, effectively forcing this device to output a current equal to the master unit. The differential gain from the sense pins to the output is low (11), so to make the devices share current equally, R_M and R_S need to be scaled $R_M/R_S = 1.0/0.9$.

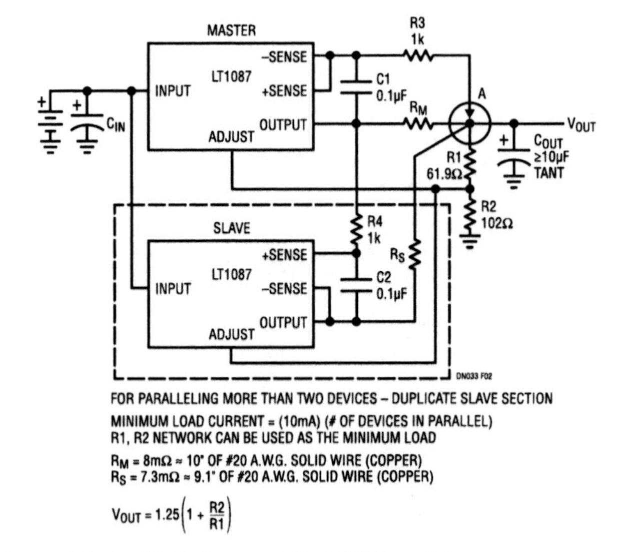

Figure 173.2 • High Current 5V–3.3V Regulator

R4 and C2 filter the feedback to the slave unit. The minimum load current for the total circuit is 10mA per device. R1 and R2 can be scaled to guarantee that the minimum load current spec will be met, and minimizes the output voltage error due to the adjust pin currents (50µA per device). The 10µF output capacitor is the minimum value needed to ensure stability. Larger output capacitors will improve transient response. R_M and R_S are chosen so that the voltage drop across them, at full load current (5A per device) is 40mV. This value is chosen to be large enough to ensure proper current sharing without significantly degrading dropout voltage. As noted in Figure 173.2, R_M and R_S will be in the 7mΩ–8mΩ range. These values are small enough to be made from either short lengths of wire or from carefully laid out PC traces. The absolute value of these resistors is not critical, but the ratio of R_M:R_S should be maintained.

The circuit as shown will source 10A. This capability can be increased, in increments of 5A, by duplicating the slave unit and properly sizing R1 and R2 to sink the additional 10mA per device. C_{OUT}'s value may require adjustment upwards to maintain optimum transient response.

Circuits using up to five units have been tested, but there is no limit to the number of slave units, other than limits on space. These regulators offer a simple solution to powering 3.3V digital devices with almost no efficiency compromise. Low cost, space savings, and easy design make them attractive for this application.

A simple ultralow dropout regulator

174

Jim Williams

Linear voltage regulators with low dropout characteristics are a frequent requirement, particularly in battery powered applications. It is desirable to maintain regulation until the battery is almost entirely depleted. Regulator dropout limits significantly impact useful battery life, and as such should be minimized. Figure 174.1 shows dropout characteristics for a monolithic regulator, the LT1085. The <1.5V dropout performance is about twice as good as standard monolithic regulators. In many cases this device will serve nicely, but applications requiring lower dropout mandate a different approach.

Figure 174.2's simple regulator has only 85mV dropout at 2.5A–a 13x improvement. At lower currents dropout decreases to vanishingly small values. This circuit is particularly applicable in battery driven lap top computers, where multi-output power supplies are used. In operation, the LT1431 shunt regulator adjusts its output ("collector") to whatever value is required to force circuit output to 5V. The LT1431's internal trimming eliminates the usual feedback resistors and trimpots. Q1, the pass element, runs as a voltage overdriven source follower. This configuration offers the lowest possible dropout voltage,[1] although it does require a +12V bias source for Q1's gate. This +12V source is commonly present in lap top computers and similar devices because of disc drive and peripheral power requirements. Power drain on the +12V supply is a few milliamperes.

Providing short circuit protection without introducing significant loss requires care. A1 achieves this by sensing across a 0.002Ω shunt (1.5″ of #23 wire). This introduces only 6mV of drop at the circuits 3A current limit threshold. A 6mV current limit trip point is derived by grounding A1's offset pin 5. The 6mV input offset generated at A1 by doing this is stable

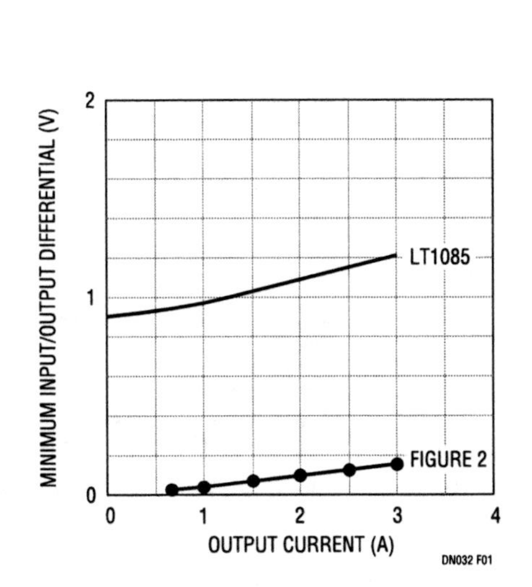

Figure 174.1 • Dropout Performance for a Low Dropout Monolithic Regulator vs Figure 174.2

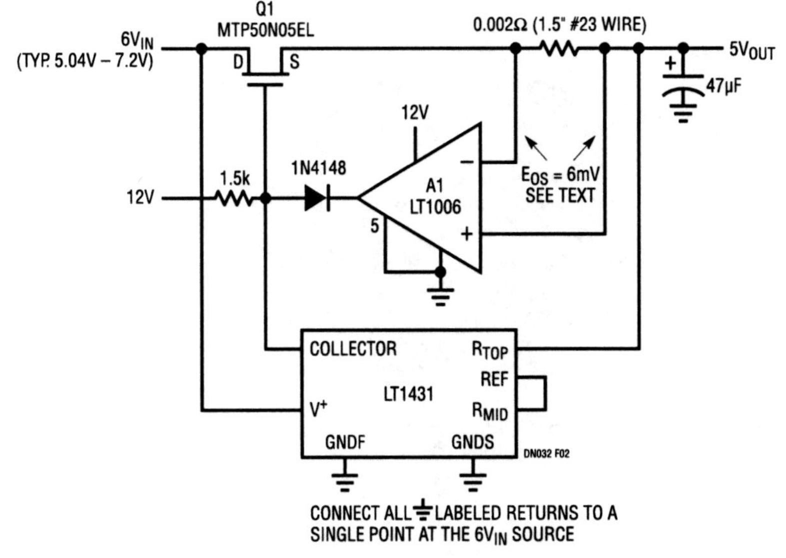

Figure 174.2 • Ultra-Low Dropout Regulator

Note 1: A detailed discussion of various methods for achieving low dropout appears as Appendix A ("Achieving Low Dropout") in LTC Application Note 32, "High Efficiency Linear Regulators."

Analog Circuit Design: Design Note Collection. http://dx.doi.org/10.1016/B978-0-12-800001-4.00174-5

over time, temperature and unit-unit variation, and substitutions for A1 are not advisable. Currents beyond 3A cause A1 to pull low, stealing Q1's gate drive and shutting off the regulators output. Under overload conditions A1 and Q1 from a well controlled linear current control loop with smooth limiting. Figure 174.3 details dropout characteristics. Results for the MTP50N05EL MOSFET specified for Q1 show only 85mV dropout, decreasing to just 8mV at 0.25A. For comparison, data for some higher resistance transistors also appears.

Q1's source follower connection makes regulator dynamics quite good compared to common source/emitter approaches. Figure 174.4 shows no load (Trace A low) to full load (Trace A high) response. Regulator output (Trace B) dips only 200mV and recovers quickly with clean damping. The positive slew recovery time is due to the 1.5kΩ bias resistor acting against Q1's input capacitance (Trace C is Q1's gate). Quicker response is possible by a reduction in this value, although current drain from the +12V supply will increase. The value used represents a good compromise. Transient recovery for load removal is also well controlled.

This regulator offers a simple solution to applications requiring extremely low dropout over a range of output currents. The performance, low parts count and lack of trimming make it an attractive alternative to other approaches. For reference, pertinent information on construction of wire shunts appears in Figures 174.5 and 174.6.

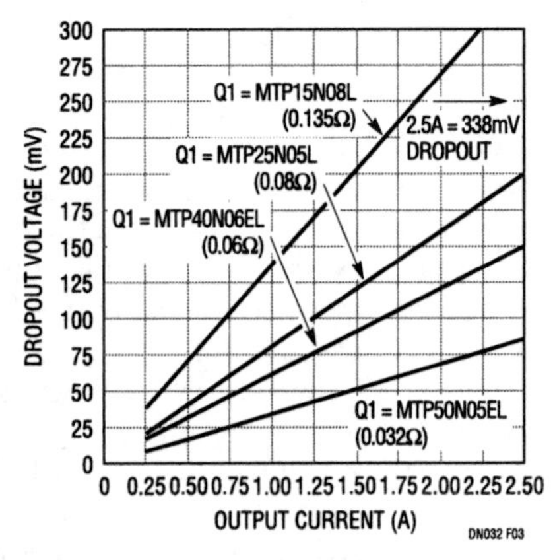

Figure 174.3 • Dropout Characteristics for Figure 174.2. Q1's Saturation Directly Influences Performance

WIRE GAUGE	μΩ/INCH
10	83
11	100
12	130
13	160
14	210
15	265
16	335
17	421
18	530
19	670
20	890
21	1000
22	1300
23	1700
24	2100
25	2700

Figure 174.5 • Resistance vs Size for Various Copper Wire Types

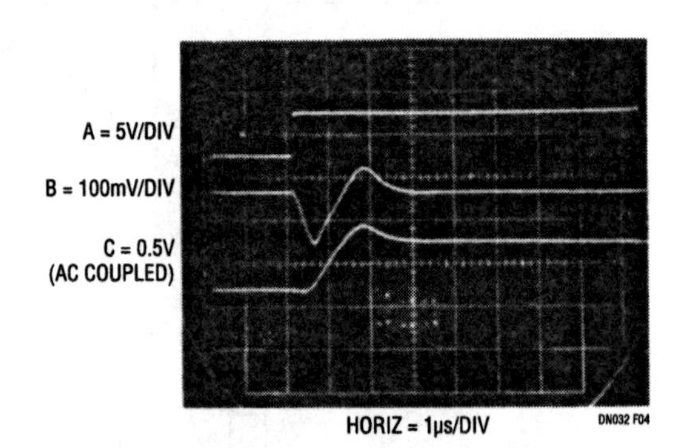

Figure 174.4 • Transient Response for a Full Load Step. Follower Connection Provides Clean Dynamics

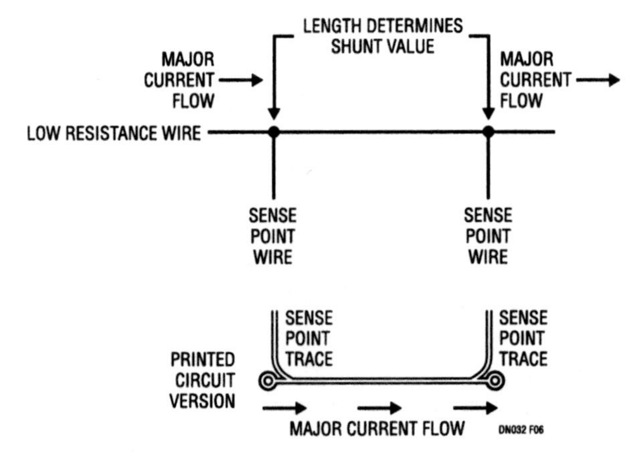

Figure 174.6 • Detail of a Low Resistance Current Shunt

Section 9

Micromodule (μModule) Power Design

Dual 13A μModule regulator with digital interface for remote monitoring & control of power

175

Jian Li Gina Le

Digital power system management: set, monitor, change and log power

Managing power and implementing flexibility in a high rail count circuit board can be challenging, requiring hands-on probing with digital voltmeters and oscilloscopes, and often rework of PCB components. To simplify power management, especially from a remote location, there is a trend to configure and monitor power via a digital communications bus. Digital power system management (PSM) enables on-demand telemetry capability to set, monitor, change and log power parameters.

Dual μModule regulator with precision READ/WRITE of power parameters

The LTM4676 is a dual 13A output constant frequency switching mode DC/DC μModule (micromodule) regulator (Figure 175.1). In addition to delivering power at a point-of-load, the LTM4676 features configurability and telemetry-monitoring of power and power management parameters over PMBus—an open standard I^2C-based digital serial interface protocol. The LTM4676 combines best-in-class analog switching regulator performance with precision mixed signal data acquisition. It features ±1% maximum DC output voltage error and ±2.5% current read back accuracy over temperature ($T_J = -40°C$ to 125°C), and integrated 16-bit delta-sigma ADC and EEPROM.

The LTM4676's 2-wire serial interface allows outputs to be margined, tuned and ramped up and down at programmable slew rates with sequencing delay times. Input and output currents and voltages, output power, temperature, uptime and peak values are readable. The device is comprised of fast, dual analog control loops, precision mixed signal circuitry, EEPROM, power MOSFETs, inductors and supporting components housed in a 16mm × 16mm × 5.01mm BGA (ball grid array) package.

The LTM4676 operates from a 4.5V to 26.5V input supply and steps down V_{IN} to two outputs ranging from 0.5V to 5.4V. Two outputs can current share to provide up to 26A (i.e., 13A + 13A as one output).

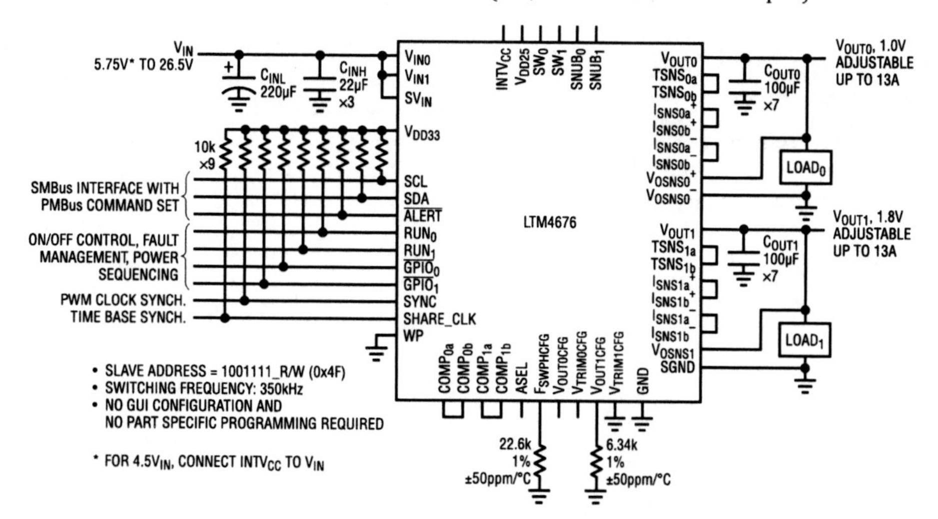

Figure 175.1 • LTM4676: Dual 13A Output μModule Regulator with PMBus Interface

Analog Circuit Design: Design Note Collection. http://dx.doi.org/10.1016/B978-0-12-800001-4.00175-7

Internal or external compensation

The LTM4676 offers both internal or external compensation, which can optimize the transient response over a wide operating range. Figure 175.2 shows that the peak-to-peak output voltage is only 94mV with a 50% load step.

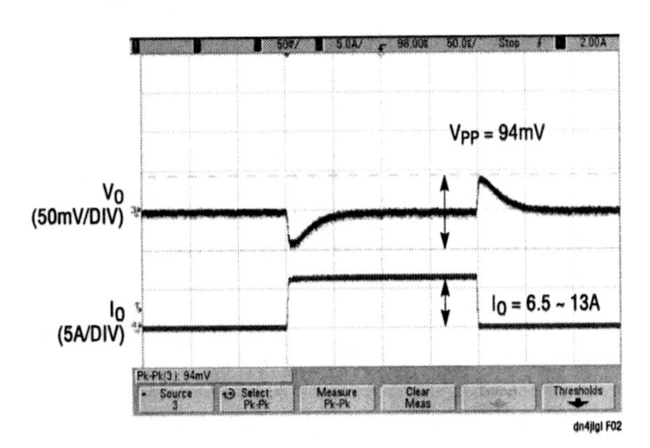

Figure 175.2 • Transient Response of the LTM4676 in Figure 175.1 at V_{IN} = 12V, V_{OUT1} = 1.8V, I_O = 6.5A ~ 13A

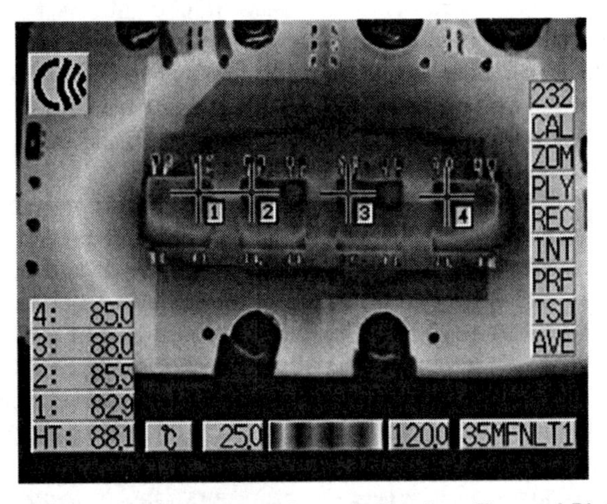

Figure 175.3 • Four LTM4676 Current Sharing: Thermal Picture at V_{IN} = 12V, V_{OUT} = 1.0V/100A, 300LFM Airflow

Current share for up to 100A at $1V_{OUT}$

The LTM4676 uses a constant frequency peak current mode control architecture, which offers a cycle-by-cycle current limit and easy current sharing among multiple phases. Paralleling modules can achieve much higher output current capability. For example, four LTM4676 μModule regulators can be paralleled to provide up to 100A output current. Figure 175.3 shows the thermal picture. With 300LFM of airflow, the hot spot temperature rise is only 64.3°C. The even thermal distribution among modules is due to excellent current sharing performance. Figure 175.4 is a photo of the demo board with four LTM4676 μModule regulators assembled to provide 100A at 1V.

Figure 175.4 • Four LTM4676, Each in a 16mm x 16mm x 5.01mm LGA Package Deliver 100A at $1V_{OUT}$

Conclusion

Linear Technology's digital power system management (PSM) products provide users with critical power-related data. One can access load current, input current, output voltages, compute power consumption, efficiency, and access other power management parameters via a digital bus. This enables predictive analytics, minimizes operating costs, increases reliability and ensures smart energy management decisions can be made.

On-demand digital control and monitoring of system power with the LTM4676 eliminates PCB layout and circuit component manipulation and accelerates system characterization, optimization and data mining during prototyping, deployment and field operation.

For demo kits, free download of the LTPowerPlay GUI and PSM training videos, please visit www.linear.com and type "PSM" in the SEARCH box.

36V input, low output noise, 5A μModule regulator for precision data acquisition systems

176

Jaino Parasseril

Introduction

Low output noise, fast transient response and high efficiency are just a few of the stringent power supply demands made by applications featuring high data rate FPGA I/O channels and high bit count data converters. The power supply designer faces the difficult task of meeting all of these requirements with as few components as possible, since no single topology easily meets all three.

For instance, high performance linear regulators achieve the required low output noise and fast transient response, but tend to dissipate more power than a switching topology, resulting in thermal issues. Switching regulators, on the other hand, are generally more efficient and run cooler than linear regulators, but generate significantly more output noise and cannot respond as quickly to transients. Power supply designers often resort to combining the two topologies, using a switching regulator to efficiently step down a relatively high bus voltage, followed by a linear post regulator to produce a low noise output. Although it is possible to produce a low noise supply in this way, it requires careful design to achieve high efficiency and fast transient response.

An easier way to reap the benefits of both a linear regulator and a switching regulator is to use the LTM8028, which achieves low noise, fast transient response and high efficiency by combining both regulators into a single part.

Integrated switching and linear regulators

The LTM8028 is a $36V_{IN}$, 5A μModule regulator that combines a synchronous switching converter and low noise linear regulator in a 15mm × 15mm × 4.92mm BGA package. It operates from an input range of 6V to 36V with an output voltage that can be programmed between 0.8V and 1.8V. The combination of the two converters results in tight tolerance of line and load regulation over the −40°C to 125°C temperature range.

The switching frequency can be adjusted between 200kHz and 1MHz with the RT resistor, or the SYNC pin can synchronize the internal oscillator to an external clock. The 5A current limit can be reduced by utilizing the IMAX pin. The PGOOD pin can be used to detect when the output voltage is within 10% of the target value.

PCB trace voltage compensation using SENSEP

The resistance of PCB traces between the μModule regulator and the load can result in voltage drops that cause a load regulation error at the point of load. As the output current increases, the voltage drop increases accordingly. To eliminate this voltage error, the LTM8028's SENSEP pin can be connected directly to the load point.

Programmable output voltage

The output voltage can be digitally programmed in 50mV increments by controlling the LTM8028's 3-state inputs: VO0, VO1 and VO2. Additionally, the MARGA pin can be used for output margining via analog control that adjusts the output voltage by up to ±10%.

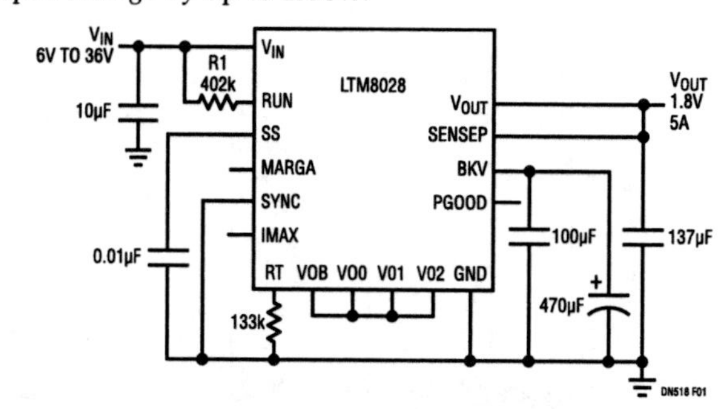

Figure 176.1 • μModule Regulator Takes a Wide Ranging 6V to 36V Input and Produces a Low Noise 1.8V Output with Up to 5A Output Current

Analog Circuit Design: Design Note Collection. http://dx.doi.org/10.1016/B978-0-12-800001-4.00176-9

DC1738A highlights the LTM8028 capabilities

A 1.8V output application is shown in Figure 176.1. The LTM8028 comes in a 15mm × 15mm × 4.92mm BGA package and is featured in the demonstration circuit DC1738A, shown in Figure 176.2.

Noise test comparison using LTC2185 ADC

When powering high speed analog-to-digital converters (ADCs), it is important to use a power supply that is as clean as possible. Any switching spurs that are present on the power supply rail will translate into AM modulation in the ADC output spectrum. The noise performance of the LTC2185, a 16-bit ADC, was evaluated to see the difference between using (1) a typical LDO, (2) a typical switching regulator, and (3) the LTM8028 low noise µModule regulator. A simplified schematic of the test is shown in Figure 176.3, where the DUT is represented by either of the configurations.

Figure 176.4 shows the FFT plots using the three different methods of powering the LTC2185 when sampling a 70MHz tone at 100Msps. The LDO provides a clean power supply, achieving a SINAD of 76.22dB. However, when powered by a typical 250kHz switching regulator, there are spurs around the fundamental with an offset frequency of 250kHz. These are switching regulator spurs that are AM modulated around the carrier frequency. The sampling process produces 250kHz spurs at baseband. As a result, the SINAD drops to 71.84dB, around 4dB compared to an LDO. This reduces the LTC2185 to nearly 12-bit performance. In demanding applications where tenths of dBs are significant, losing 4dB of SINAD because of a noisy regulator is unacceptable. In addition to degrading the SINAD of the ADC, these spurs may land on neighboring channels or on other signals of interest, making it impossible to receive meaningful data from those channels. With the LTM8028, only a few extraneous spurs exist near the desired frequency and the SINAD performance is only 0.03dB worse than the LDO baseline. The spurious content that was very pronounced in the spectrum of the switching regulator is virtually eliminated. As a result, there will not be any performance degradation of the LTC2185 when using a LTM8028 regulator.

Conclusion

The LTM8028 µModule regulator combines a linear regulator and a switching regulator to form a DC/DC converter with minimal power loss, low noise and UltraFast transient response, all in a 15mm × 15mm × 4.92mm BGA package.

Figure 176.2 • The LTM8028 Makes it Possible to Build a Minimal Component-Count Regulator that Meets Stringent Noise, Efficiency and Transient Response Requirements

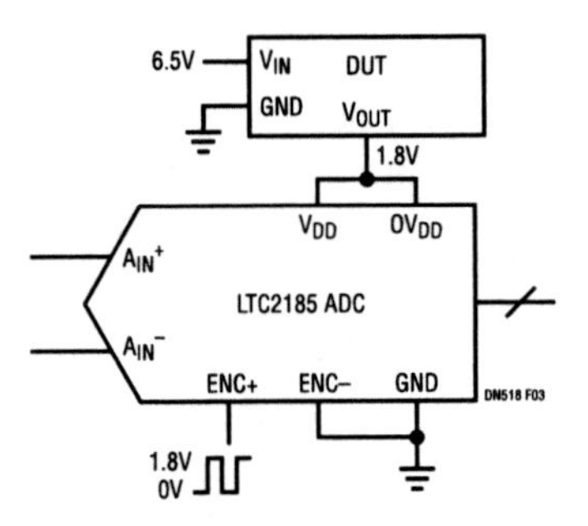

Figure 176.3 • Noise Test Schematic Using Different Supplies to Power 16-Bit LTC2185 ADC

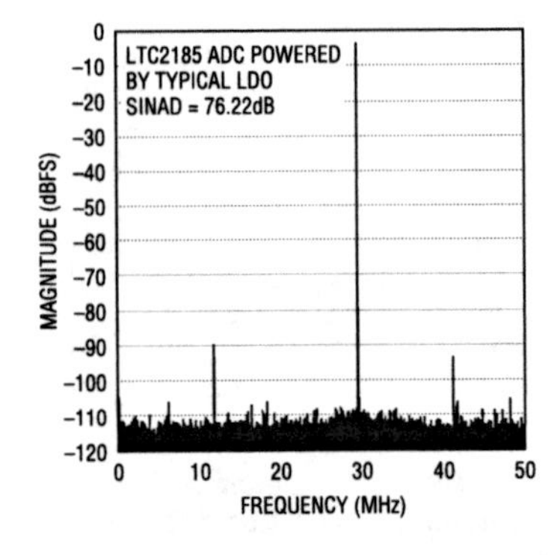

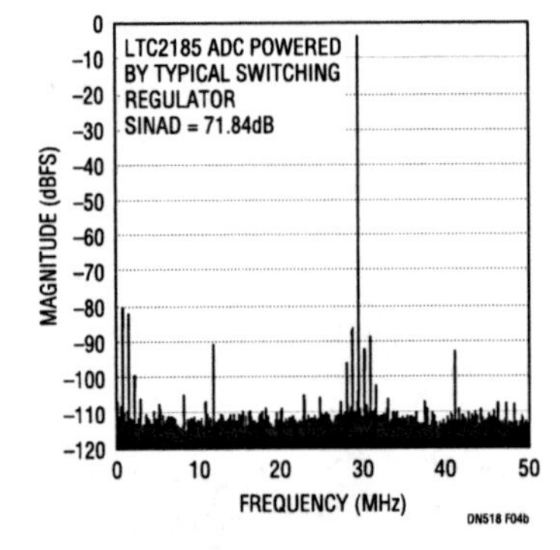

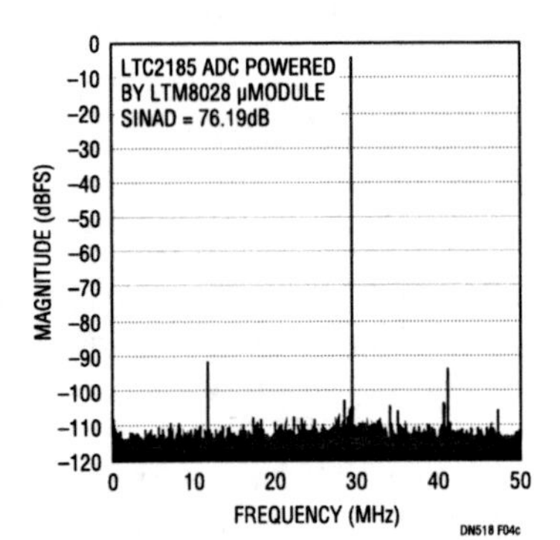

Figure 176.4 • 32k-Point FFT, f_{IN} = 70.3MHz, –1dBFs, 100Msps, Using CMOS Clock Drive

Step-down μModule regulator produces 15A output from inputs down to 1.5V—no bias supply required

177

1.5V to 5.5V input, 0.8V to 5V output from a 15mm × 15mm × 4.32mm LGA package

Alan Chern Jason Sekanina

15A high efficiency output from a low input voltage

The LTM4611 is a switch mode, step-down DC/DC μModule regulator in a compact 15mm × 15mm × 4.32mm LGA surface mount package. The switching controller, MOSFETs, inductor and supporting components are housed in the package. With a built-in differential remote sense amplifier, the LTM4611 can tightly regulate its output voltage from 0.8V to within 300mV of V_{IN} and deliver 15A output efficiently from 1.5V to 5.5V input.

Only a handful of components are needed to create a complete point-of-load (POL) solution with the LTM4611 (see Figure 177.1). The C_{SS} capacitor provides smooth start-up on the output and limits the input surge current during power-up. C_{FF} and C_P set the loop-compensation for fast transient response and good stability. The output voltage, 1.5V, is set by a single resistor, R_{SET}.

Efficiency is exceptional, even down to the lowest input voltages, as shown in Figure 177.2.

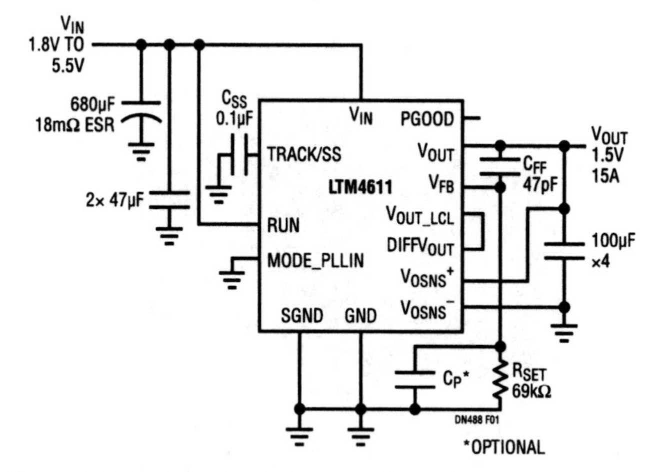

Figure 177.1 • 1.8V_{IN} to 5.5V_{IN} to 1.5V_{OUT} with 15A Output Load Current

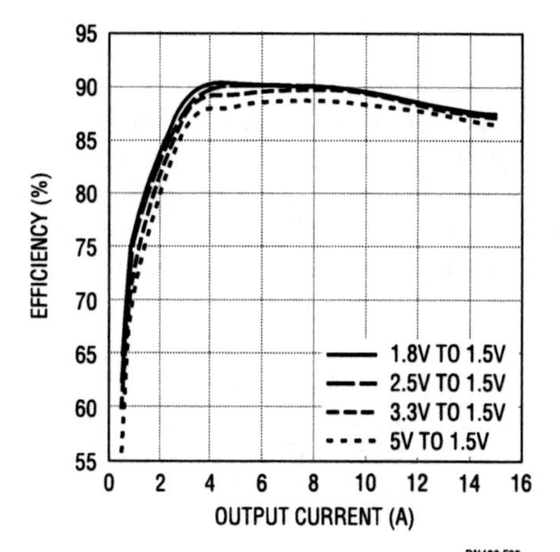

Figure 177.2 • Efficiency of Figure 177.1 Circuit

VIDEO AT video.linear.com/56

Input and output ripple

Output capacitors should have low ESR to meet output voltage ripple and transient requirements. A mixture of low ESR polymer and/or ceramic capacitors is sufficient for producing low output ripple with minimal noise and spiking. Output capacitors

Scan this code with your smart phone to view informative videos.

VIDEOS AT
m.linear.com/4611

Some phones may require that you download a QR Code scanner/reader application.

Analog Circuit Design: Design Note Collection. http://dx.doi.org/10.1016/B978-0-12-800001-4.00177-0

are chosen to optimize transient load response and loop stability to meet the application load-step requirements by using the Excel-based LTpowerCAD design tool. (Table 5 of the LTM4611 data sheet provides guidance for applications with 7.5A load-steps and 1μs transition times.) For this design example, four 100μF ceramic capacitors are used. Figures 177.3 and 177.4 show input and output ripple at 15A load with 20MHz bandwidth-limit. View the associated videos to see the test methodology, as well as ripple waveforms without bandwidth limiting.

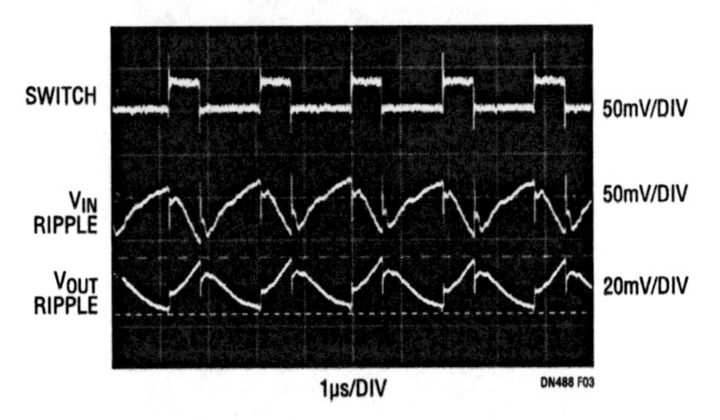

Figure 177.3 • 5V_{IN} to 1.5V_{OUT} at 15A Output Load

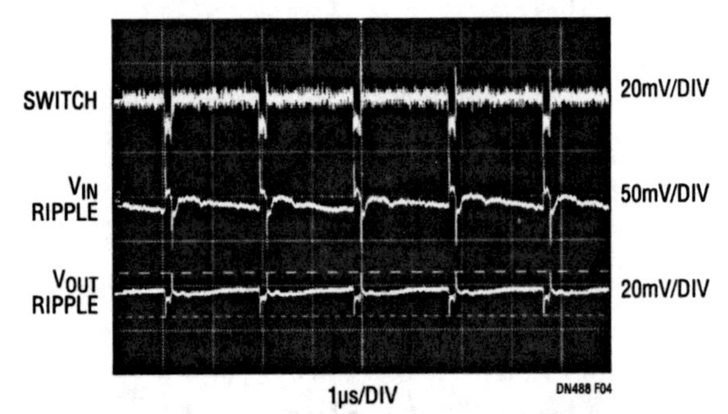

Figure 177.4 • 1.8V_{IN} to 1.5V_{OUT} at 15A Output Load

VIDEO AT
video.linear.com/57

For this design, the choice of input capacitors is critical due to the low input voltage range. Long input traces can cause voltage drops, which could nuisance-trip the μModule regulator's undervoltage lockout (UVLO) detection circuitry. Input ripple, typically a non-issue with higher input voltages, may fall a significant percentage below nominal—close to UVLO—at lower input voltages. In this case, input voltage ripple should be addressed since input filter oscillations can occur due to poor damping under heavy load current. This design uses a large 680μF POSCAP and two 47μF ceramic capacitors to compensate for meter-long input cables used during bench testing.

Thermally enhanced packaging

The device's LGA packaging allows heat sinking from both the top and bottom, facilitating the use of a metal chassis or a BGA heat sink. This form factor promotes excellent thermal dissipation with or without air flow. Figure 177.5 shows the top view thermal imaging of the LTM4611 at a power loss of 3.5W with no air flow, when converting 5V to 1.5V.

Internal self-heating of the LTM4611 remains quite low even at a low 1.8V input voltage due to its micropower bias generator that enables strong gate drive for its power MOSFETs. Figure 177.6 shows a power loss of 3.2W with hot spots slightly changed from their positions with a 5V input—the nominal surface temperature is 60°C. Watch the associated videos to see the test set-up and watch 200 LFM of air flow cool the unit by 10°C.

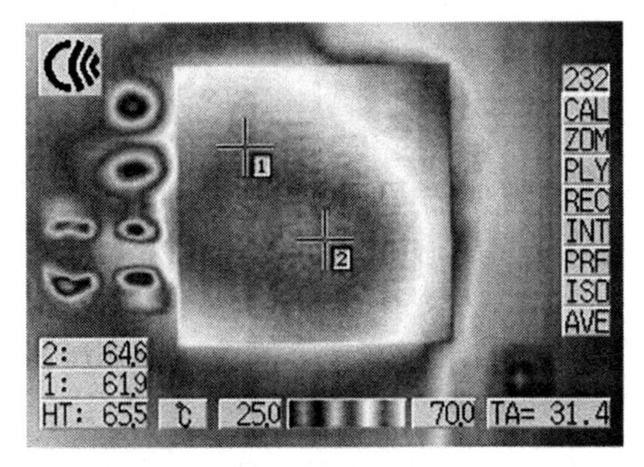

Figure 177.5 • 5V_{IN} to 1.5V_{OUT} at 15A Output Load. 3.5W Power Loss with 0LFM and 65°C Surface Temperature Hot Spot

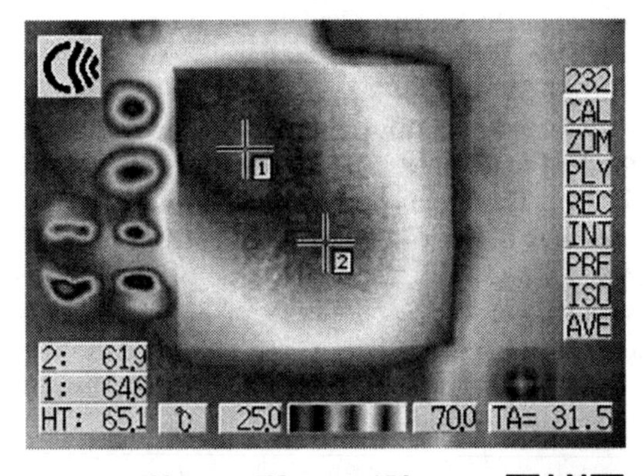

Figure 177.6 • 1.8V_{IN} to 1.5V_{OUT} at 15A Output Load. 3.2W Power Loss with 0LFM and 65°C Surface Temperature Hot Spot

VIDEO AT
video.linear.com/55

Conclusion

The LTM4611 is a step-down μModule regulator that easily fits into POL applications needing high output current from low voltage inputs—from 1.5V to 5.5V. Efficiency and thermal performance remain high across the entire input voltage range, simplifying electrical, mechanical and system design in data storage, RAID, ATCA, and many other applications.

Dual μModule DC/DC regulator produces high efficiency 4A outputs from a 4.5V to 26.5V input

178

Alan Chern

Dual system-in-a-package regulator

Systems and PC boards that use FPGAs and ASICs are often very densely populated with components and ICs. This dense real estate (especially the supporting circuitry for FPGAs, such as DC/DC regulators) puts a burden on system designers who aim to simplify layout, improve performance and reduce component count. A new family of DC/DC μModule regulator systems with multiple outputs is designed to dramatically reduce the number of components and their associated costs.

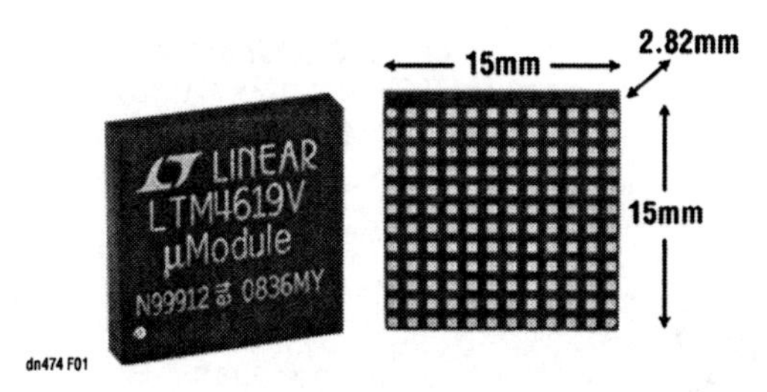

Figure 178.1 • The LTM4619 LGA Package is Only 15mm x 15mm x 2.82mm and Houses Dual DC/DC Switching Circuitry, Inductors, MOSFETs and Support Components

These regulators are designed to eliminate layout errors and to offer a ready-made complete solution. Only a few external components are needed since the switching controllers, power MOSFETs, inductors, compensation and other support components are all integrated within the compact surface mount 15mm × 15mm × 2.82mm LGA package. Such easy layout saves board space and design time by implementing high density point-of-load regulators (Figure 178.1).

The LTM4619 switching DC/DC μModule converter regulates two 4A outputs from a single wide 4.5V to 26.5V input voltage range. Each output can be set between 0.8V and 5V with a single resistor. In fact, only a few components are needed to build a complete circuit (see Figure 178.2).

Figure 178.2 shows the LTM4619 μModule regulator in an application with 3.3V and 1.2V outputs. The output voltages can be adjusted with a value change in R_{SET1} and R_{SET2}. Thus, the final design requires nothing more than a few resistors and capacitors. Flexibility is achieved by pairing outputs, allowing the regulator to form different combinations such as single input/dual independent outputs or single input/parallel single output for higher maximum current output.

The efficiency of the system design for Figure 178.2 is shown in Figure 178.3 and power loss is shown in Figure 178.4, both at various input voltages. Efficiency at light

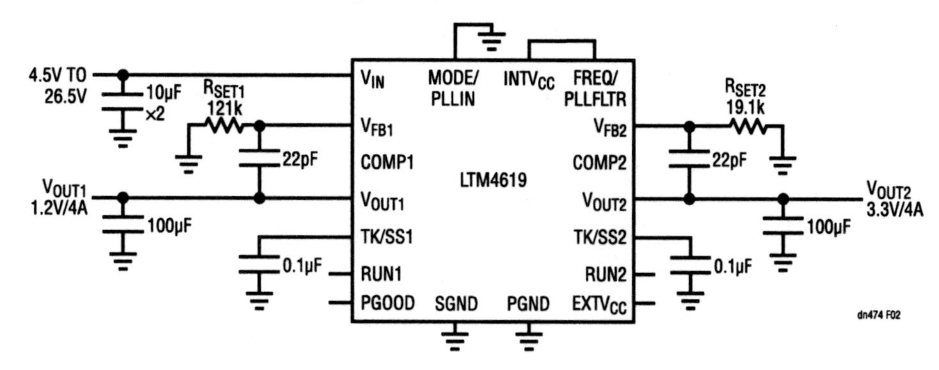

Figure 178.2 • 4.5V to 26.5V Input to Dual 3.3V and 1.2V Outputs with 4A Maximum Output Current Each

Analog Circuit Design: Design Note Collection. http://dx.doi.org/10.1016/B978-0-12-800001-4.00178-2

load operation can be improved with selective pulse-skipping mode or Burst Mode operation by tying the mode pin high or leaving it floating.

Multiphase operation for four or more outputs

For a 4-phase, 4-rail output voltage system, use two LTM4619s and drive their MODE_PLLIN pins with a LTC6908-2 oscillator, such that the two μModule devices are synchronized 90° out of phase. Reference Figure 21 in the LTM4619 data sheet. Synchronization also lowers voltage ripple, reducing the need for high voltage capacitors whose bulk size consumes board space. The design delivers four different output voltage rails (5V, 3.3V, 2.5V and 1.8V) all with 4A maximum load.

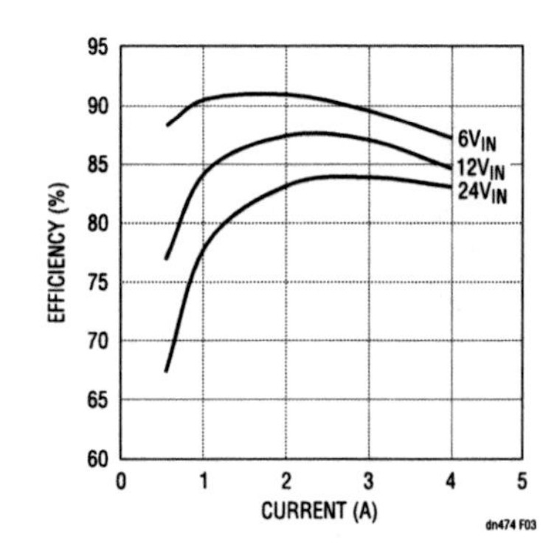

Figure 178.3 • Efficiency of the Circuit in Figure 178.2 at Different Input Voltage Ranges for 3.3V and 1.2V Outputs

Thermal performance

Exceptional thermal performance is shown in Figure 178.5 where the unit is operating in parallel output mode; single $12V_{IN}$ to a single $1.5V_{OUT}$ at 8A. Both outputs tied together create a combined output current of 8A with both channels running at full load (4A each). Heat dissipation is even and minimal, yielding good thermal results. If additional cooling is needed, add a heat sink on top of the part or use a metal chassis to draw heat away.

Conclusion

The LTM4619 dual output μModule regulator makes it easy to convert a wide input voltage range (4.5V to 26.5V) to two or more 4A output voltage rails (0.8V to 5V) with high efficiency and good thermal dissipation. Simplicity and performance are achieved through dual output voltage regulation from a single package, making the LTM4619 an easy choice for system designs needing multiple voltage rails.

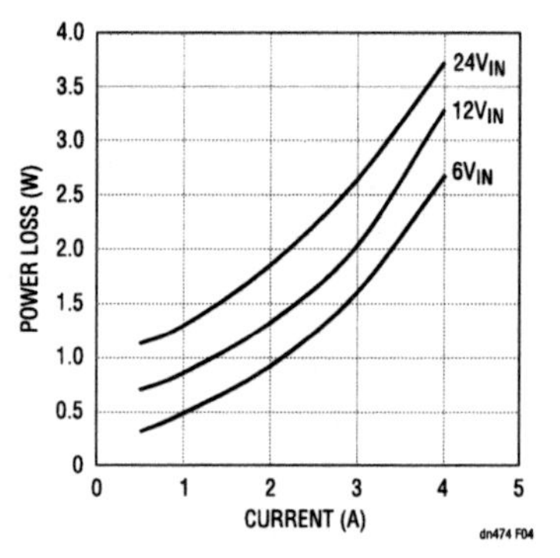

Figure 178.4 • Power Loss of the Circuit in Figure 178.2 at Different Input Voltages for 3.3V and 1.2V Outputs

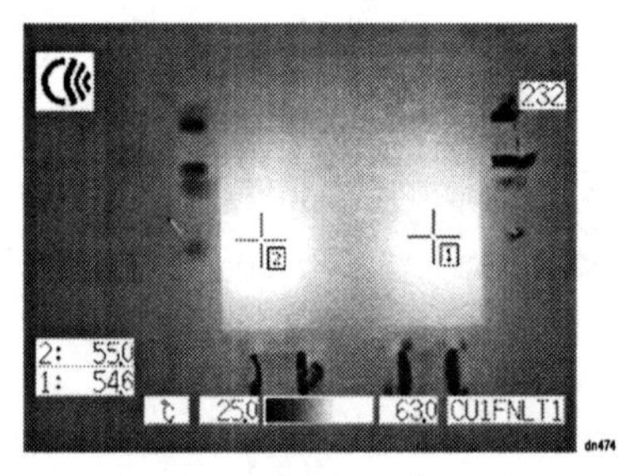

Figure 178.5 • LTM4619: Exceptional Thermal Performance of a Paralleled Output μModule Regulator ($12V_{IN}$ to Paralleled $1.5V_{OUT}$ at 8A Load)

Triple output DC/DC μModule regulator in 15mm × 15mm × 2.8mm surface mount package replaces up to 30 discrete components

179

Eddie Beville Alan Chern

Introduction

When space and design-time are tight in multivoltage systems, the solution is a multioutput DC/DC regulator IC. For more space and time constraint systems, a better solution is an already-fabricated compact multioutput DC/DC system that includes not only the regulator ICs but the supporting components such as the inductors, compensation circuits, capacitors and resistors.

Dual switching 4A and 1.5A VLDO regulators

The LTM4615 offers three separate power supply regulators in a 15mm × 15mm × 2.8mm LGA surface mount package: two switching DC/DC regulators and one very low dropout VLDO linear regulator (Figure 179.1). MOSFETs, inductors, and other support components are all built in. Each power supply can be powered individually or together, to form a single input, three output design. Moreover, for an otherwise complex triple output circuit design, the task is eased to designing with only one device while the layout is as simple as copying and pasting the LTM4615's package layout. One LTM4615 replaces up to 30 discrete components when compared to a triple-output high efficiency DC/DC circuit.

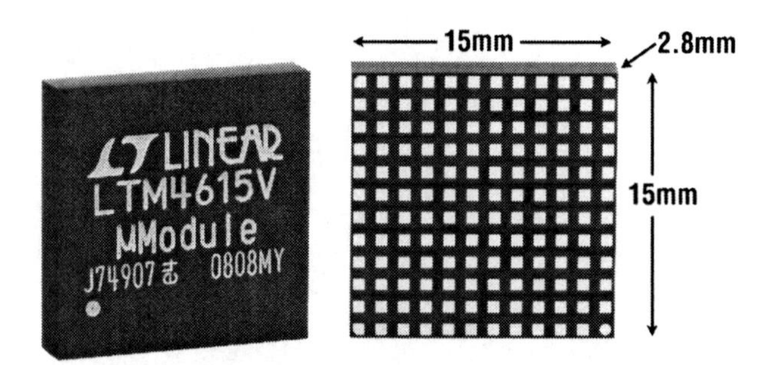

Figure 179.1 • Three DC/DC Circuits in One Package

The two switching regulators, operating at a 1.25MHz switching frequency, accept input voltages between 2.35V to 5.5V and each delivers a resistor-set output voltage of 0.8V to 5V at 4A of continuous current (5A peak). The output voltages can track each other or another voltage source. Other features include, low output voltage ripple and low thermal dissipation.

The VLDO regulator input voltage (1.14V to 3.5V) is capable of up to 1.5A of output current with an adjustable output range of 0.4V to 2.6V, also via a resistor. The VLDO regulator has a low voltage dropout of 200mV at maximum load. The regulator can be used independently, or in conjunction with either of the two switching regulators to create a high efficiency, low noise, large-ratio step-down supply—simply tie one of the switching regulator's outputs to the input of the VLDO regulator.

Multiple low noise outputs

The LTM4615 is capable of operating with all three regulators at full load while maintaining optimum efficiency. A typical LTM4615 design (Figure 179.2) for a 3.3V input to three outputs has the VLDO input driven by V_{OUT2}. The efficiency of this design is shown in Figure 179.3.

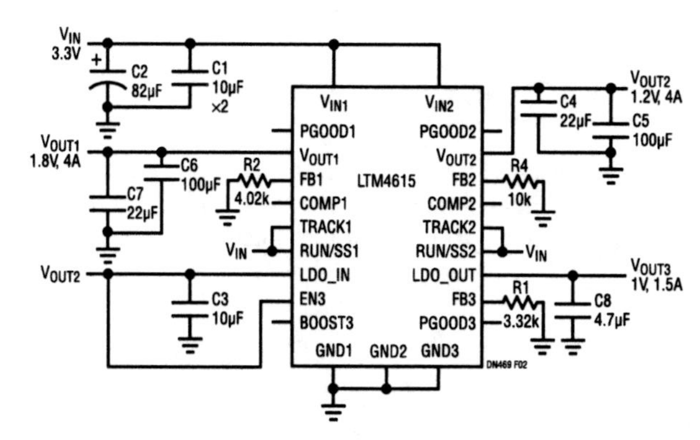

Figure 179.2 • Triple Output LTM4615: 3.3V Input, 1.8V (4A), 1.2V (4A), 1.0V (1.5A)

Analog Circuit Design: Design Note Collection. http://dx.doi.org/10.1016/B978-0-12-800001-4.00179-4

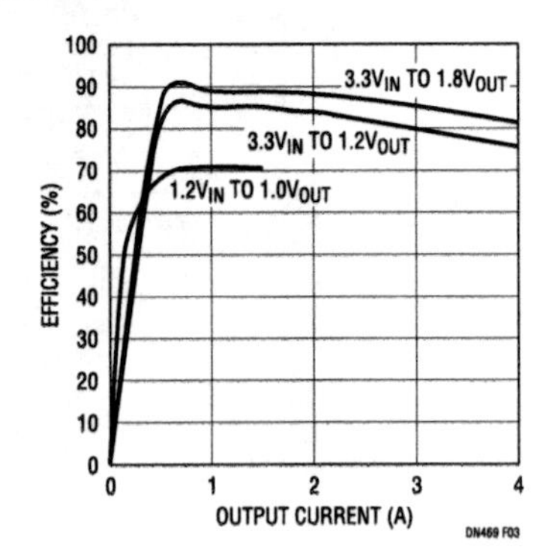

Figure 179.3 • Efficiency of the Circuit in Figure 179.2, 1.8V, 1.2V and 1.0V (VLDO)

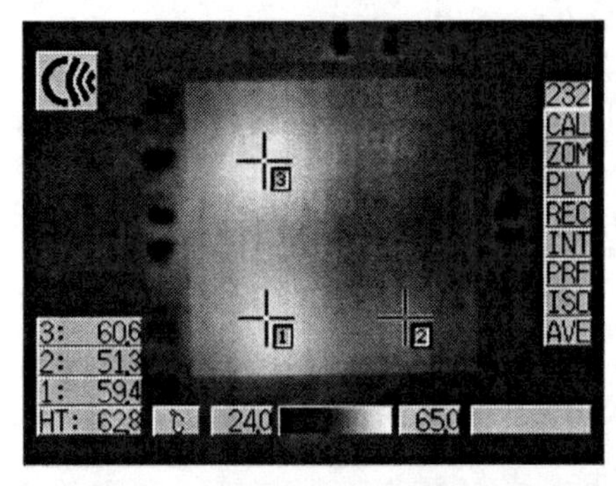

Figure 179.5 • Top View Thermal Imaging of the Unit at Full Load in Ambient Temperature with No Airflow. Even Temperatures (Cursors 1 and 3) indicate Balanced Thermal Conductivity between the Two Switching Regulators. 3.3V Input, 1.8V (4A) and 1.2V (4A).

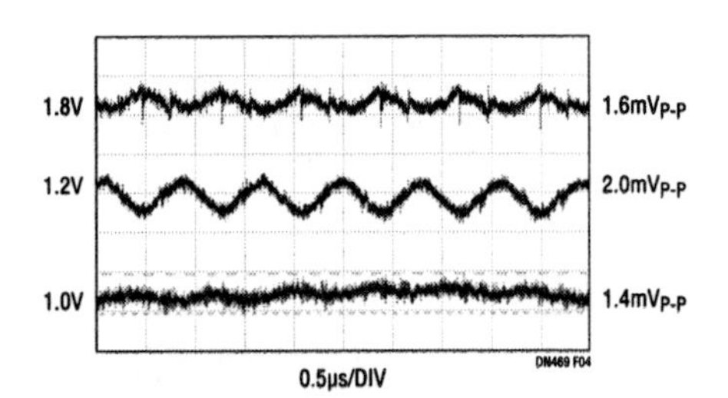

Figure 179.4 • Low Output Voltage Ripple (3.3V Input)

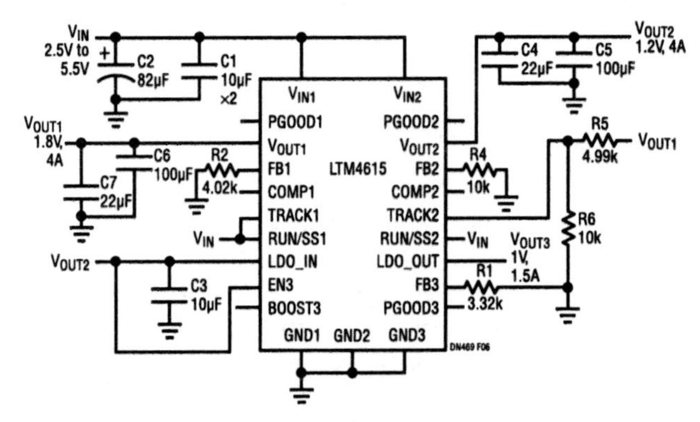

Figure 179.6 • Output Voltage Tracking Design V_{OUT2} (1.2V) Tracks V_{OUT1} (1.8V)

The LTM4615 comes prepackaged with ceramic capacitors and additional output capacitors are only needed under full 4A load and if the input source impedance is compromised by long inductive leads or traces.

The VLDO regulator provides a particularly low noise 1.0V supply as it is driven by the output of the 1.2V switching regulator (V_{OUT2}). The low output voltage ripple for all three outputs is shown in Figure 179.4.

Thermally enhanced packaging

The LGA packaging allows heatsinking from both the top and bottom. This design utilizes the PCB copper layout to draw heat away from the part and into the board. Additionally, a heat sink can be placed on top of the device, such as a metal chassis, to promote thermal conductivity. Thermal dissipation is well balanced between the two switching regulators (Figure 179.5).

Output voltage tracking

A tracking design (Figure 179.6) and output (Figure 179.7) can be programmed using the TRACK1 and TRACK2 pins. Divide down the master regulator's output with an external resistor divider that is the same as the slave regulator's feedback divider on the slave's TRACK pin for coincident tracking.

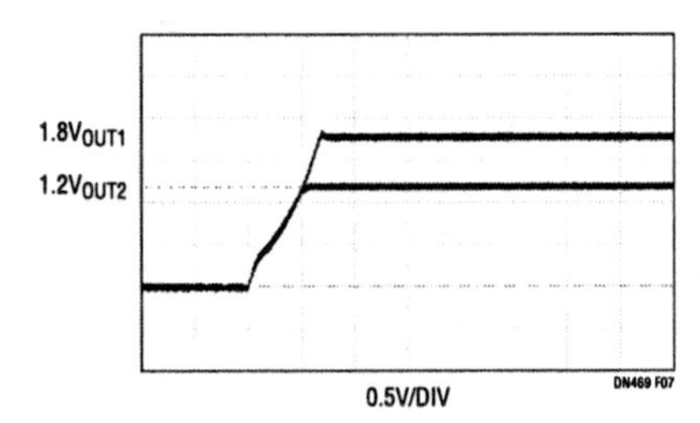

Figure 179.7 • Start-Up Voltage for Figure 179.5 Circuit V_{OUT1} (1.8V) Coincidentally Tracks V_{OUT2} (1.2V) for Coincident Tracking

Dual 8A DC/DC μModule regulator is easily paralleled for 16A

180

Eddie Beville Alan Chern

Two independent 8A regulator systems in a single package

The LTM4616 is a dual input, dual output DC/DC μModule regulator in a 15mm × 15mm × 2.8mm LGA surface mount package. Only a few external components are needed since the switching controller, MOSFETs, inductor and other support components are integrated within the tiny package.

Both regulators feature an input supply voltage range of 2.375V to 5.5V and an adjustable output voltage range of 0.6V to 5V with up to 8A of continuous output current (10A peak). For higher output current designs, the LTM4616 can operate in a 2-phase parallel mode allowing the part to deliver a total output current of 16A. The default switching frequency is set to 1.5MHz, but can be adjusted to either 1MHz or 2MHz via the PLLLPF pins. Moreover, CLKIN can be externally synchronized from 750kHz to 2.25MHz. The device supports output voltage tracking for supply rail sequencing. Safety features include protection against short circuit, overvoltage and thermal shutdown conditions.

Simple and efficient

The LTM4616 can be used as completely independent dual switching regulators with different inputs and outputs or paralleled to provide a single output. Figure 180.1 shows a typical design for a 5V common input and two independent outputs, 1.8V and 1.2V. Figure 180.2 shows the efficiency of the circuit at both 5V and 3.3V inputs.

Few external components are needed since the integrated output capacitors can accommodate load steps to the full 8A. Each output voltage is set by a single set resistor from FB1 (or FB2) to GND. In parallel operation, the FB pins can be tied together with a single resistor for adjustable output voltage.

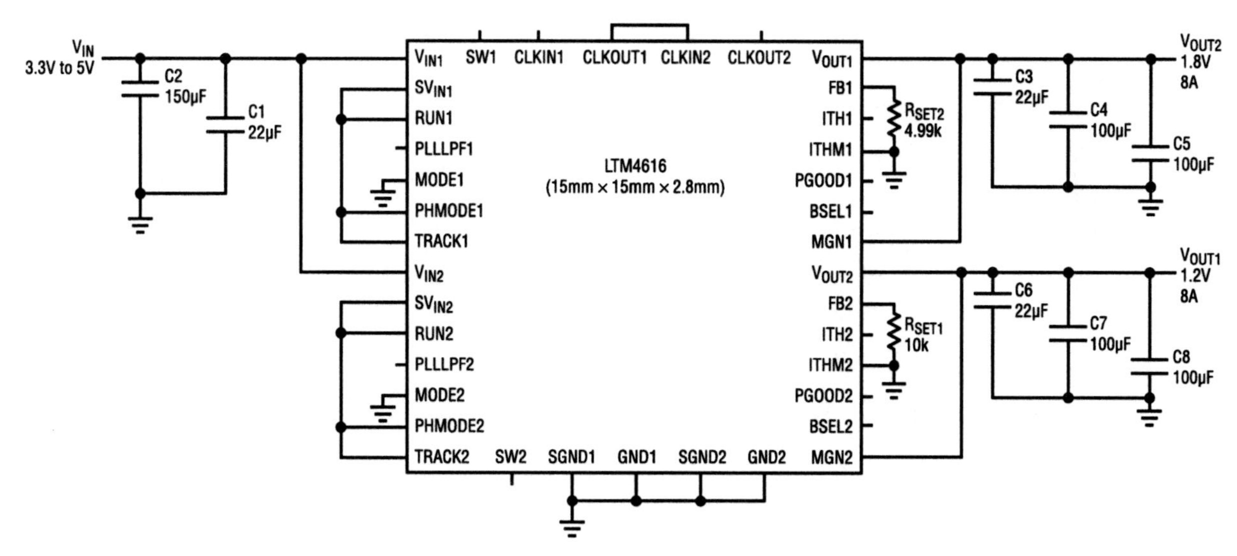

Figure 180.1 • Dual Output LTM4616 for a Single 3.3V to 5V Input, Independent 1.8V and 1.2V Outputs at 8A Each

Analog Circuit Design: Design Note Collection. http://dx.doi.org/10.1016/B978-0-12-800001-4.00180-0

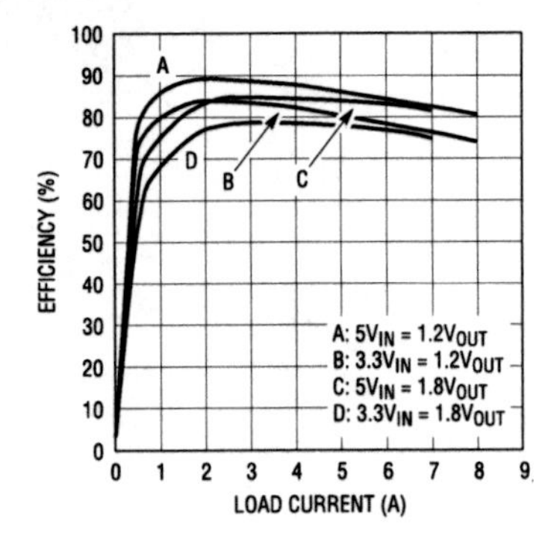

Figure 180.2 • LTM4616 Efficiency: Dual Output

Parallel operation for increased output current

You can double the maximum output current to 16A by running the two outputs in parallel as shown in Figure 180.3. Note that the FB pins share a single voltage-set feedback resistor that is half the value of the feedback resistor in the usual two output configuration. This is because the internal 10kΩ top feedback resistors are in parallel with one another, making the top value 5kΩ.

It is preferred to connect CLKOUT1 to CLKIN2 when operating from a single input voltage. This minimizes the input voltage ripple by running the two regulators out of phase with each other. If more than 16A output current is required, then multiple LTM4616 regulators can be configured for multiphase operation with up to 12 phases via the PHMODE pin. Figure 180.4 shows the expected efficiency of the parallel system at 5V and 3.3V inputs to 1.8V output. Note that the two regulators drive equal output current even during soft-start, as shown in Figure 180.5.

Conclusion

Whether you require a single 16A high current output or dual 8A outputs with sequencing, the LTM4616 provides a simple and efficient solution.

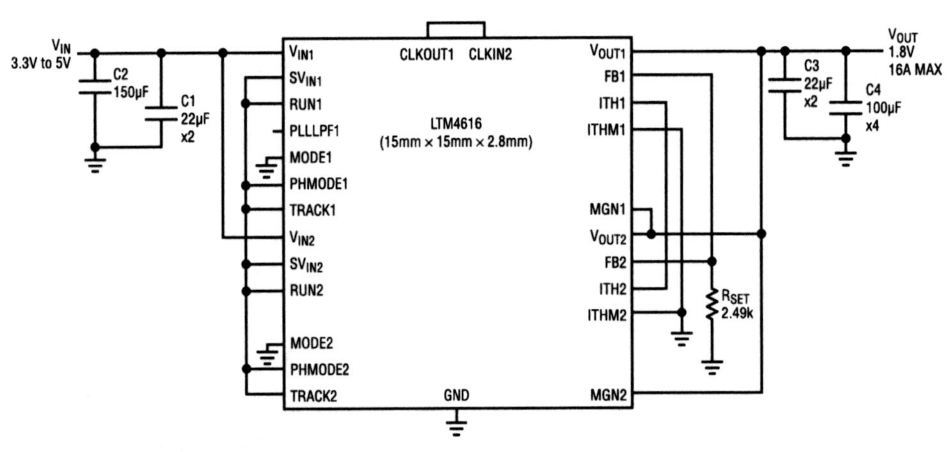

Figure 180.3 • LTM4616 with 16A Parallel Operation

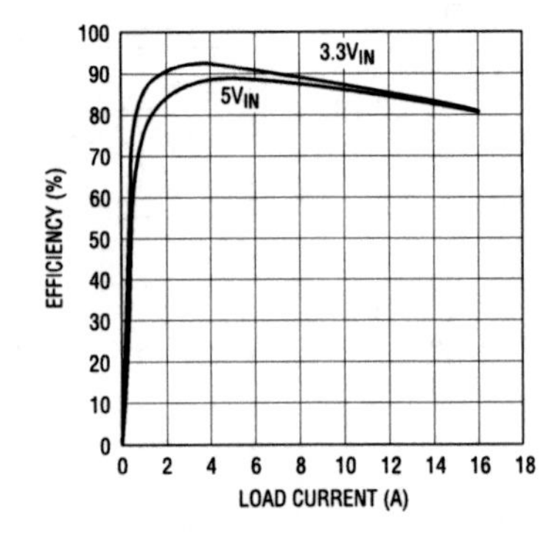

Figure 180.4 • LTM4616 Efficiency: Single 1.8V Output

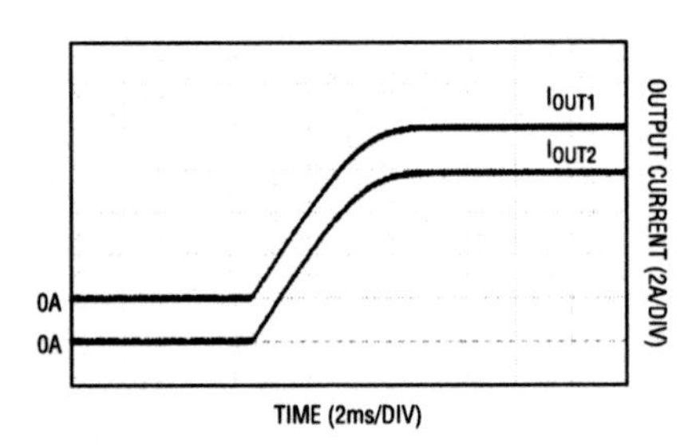

Figure 180.5 • Balanced Current Sharing for Even Heat Dissipation [$5V_{IN}$ to $1.8V_{OUT}$ at 16A]

μModule buck-boost regulators offer a simple and efficient solution for wide input and output voltage range applications

181

Jian Yin Eddie Beville

Introduction

An increasing number of applications require DC/DC converters that produce an output that falls somewhere within the input voltage range. The problem is that conventional buck-boost converter topologies, such as SEPIC or boost followed by buck, are complex, inefficient and consume a relatively large board area. Linear Technology offers 4-switch-topology buck-boost regulators that significantly improve efficiency and save space, but a complete regulator design still requires a number of external components and meticulous board layout decisions related to electrical and thermal considerations. The next clear step to simplify the design is a modular approach—a buck-boost regulator system in an IC form factor. The LTM4605 and LTM4607 μModule buck-boost regulators take that approach. Each requires only one external inductor and a single sensing resistor to produce a compact, high performance, high efficiency buck-boost regulator with exceptional thermal performance.

High efficiency

The LTM4605 and LTM4607 are high efficiency switch mode buck-boost power supply modules. The LTM4605 can operate over an input voltage range of 4.5V to 20V and support any output voltage within the range of 0.8V to 16V, set by a single resistor. As shown in Figure 181.1, the LTM4607 supports 4.5V to 36V inputs and outputs of 0.8V to 16V. Both can provide 92% to 98% efficiency over the wide input range. This high efficiency design delivers up to 5A continuous current in boost mode (12A in buck mode). Only the inductor, sensing resistor, and bulk input and output capacitors are needed to finish the design. Figure 181.2 shows a typical LTM4605 application with an output of 12V at 5A. An optional RC snubber is added here to reduce switching noise for applications where radiated EMI noise is a concern.

Low profile solution

These power modules are offered in a space saving and thermally enhanced 15mm × 15mm × 2.8mm LGA package. This low profile package can fit the back side of PC boards for many high density point-of-load applications. Their high switching frequency and current mode architecture enable a fast transient response to line and load changes without sacrificing stability. Both can be frequency synchronized with an external clock to reduce undesirable frequency harmonics. Fault protection comes in the form of overvoltage protection and foldback current protection.

Smooth transition and circuit simplicity

Both the LTM4605 and LTM4607 include the switching controller, four power FETs, compensation circuitry and support components. The 4-switch topology provides high efficiency in

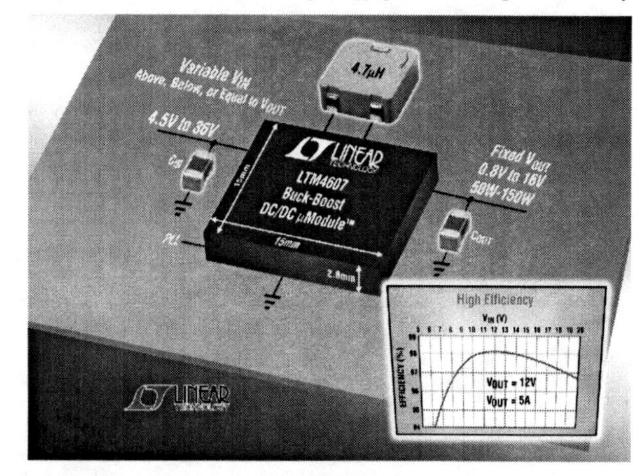

Figure 181.1 • There is No Easier Way to Design a High Efficiency, High Power Density Buck-Boost Regulator than with the LTM4605 or LTM4607

Analog Circuit Design: Design Note Collection. http://dx.doi.org/10.1016/B978-0-12-800001-4.00181-2

all three modes of operation—buck, buck-boost and boost—with a smooth transition between each. Figure 181.2 shows an actual buck-boost design with external components chosen to satisfy the boost mode's 5A maximum load current. For buck-only applications, the maximum load current can be 12A at $12V_{OUT}$ with the same external components. For instance in a buck-only configuration, such as in Figure 181.3, the load current can be increased up to 7A at $12V_{OUT}$ for 168W capability. This application can achieve better than 98% efficiency as shown in Figure 181.4.

Excellent thermal performance

The low profile LGA package has a low thermal resistance from junction to pin (4°C/W), thus maintaining an acceptable junction temperature even when satisfying high power requirements. Typically, operation in room temperature ambient conditions requires no special heat sinking or added airflow, but for warmer ambient environments or high loads, simply add a heat sink to the top of the case for 2-sided cooling and add air flow to significantly lower the thermal resistance from junction to ambient. The data sheet provides more details about adding heat sinks and air flow considerations.

Conclusion

There is no easier way to design an efficient high-density buck-boost converter than with the LTM4605 or LTM4607 μModule regulator. No design tricks are necessary to achieve efficiencies up to 98%—only one inductor, a single sensing resistor and bulk capacitance are required to complete a design. Low profile LGA packages fit on the back side of PCBs and have good thermal performance, enabling a 168W power output in an 8cm × 8.4cm 4-layer PCB. These devices are ideal for automotive, telecom, medical, motor drive and battery-powered applications.

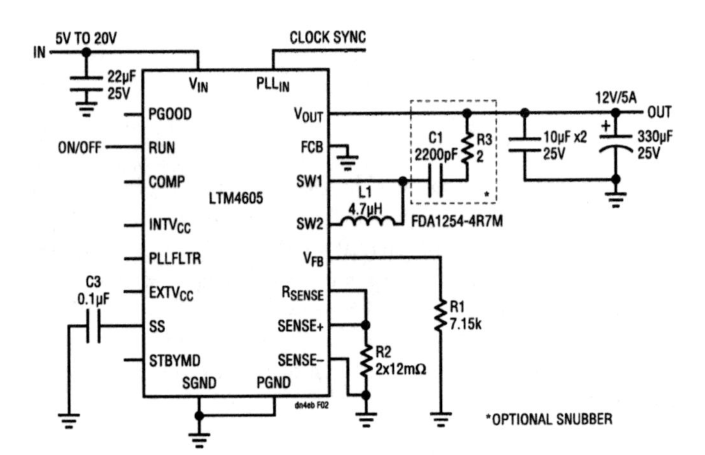

Figure 181.2 • Buck-Boost Converter Produces $12V_{OUT}$ at 5A from a 5V to 20V Input Range

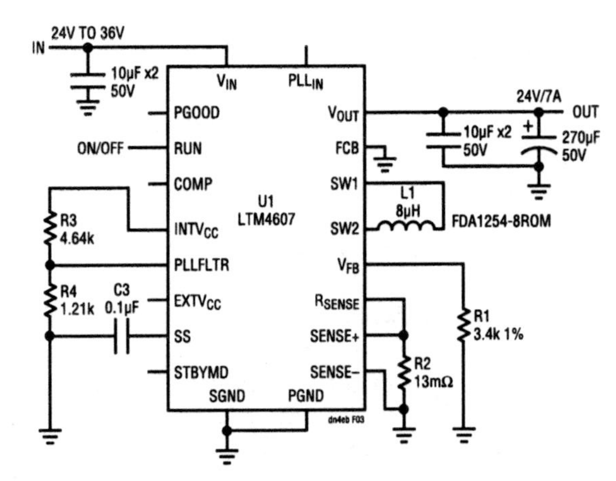

Figure 181.3 • Buck Converter Produces a 24V Output with 168W Capability

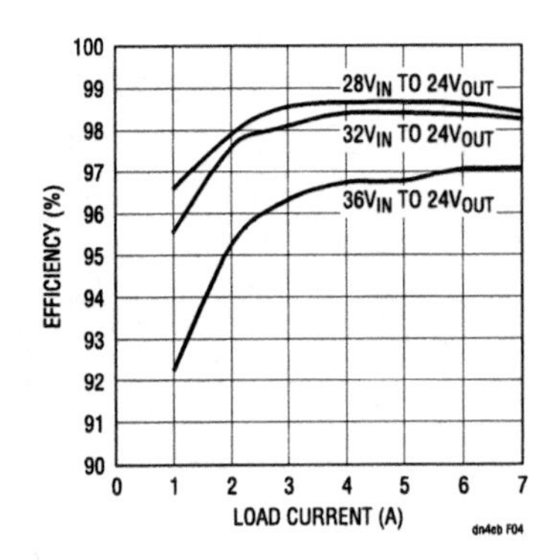

Figure 181.4 • Efficiency for the $24V_{OUT}$ Converter in Figure 181.3

8A low voltage, low profile DC/DC μModule regulator in 9mm × 15mm package weighs only 1g

182

Eddie Beville Alan Chern

Introduction

In communications, industrial and other high power systems, board-mounted point-of-load (POL) DC/DC power supplies simplify thermal management and offer high performance. An ideal POL power supply module takes a minimal amount of space and mounts on the board much like other surface mount ICs without special tooling. It should also demonstrate exceptional thermal performance with excellent efficiency and low power dissipation.

8A DC/DC μModule regulator in an IC form factor

The LTM4608 μModule regulator is a complete high density power supply in a low profile (15mm × 9mm × 2.8mm) LGA surface mount package (Figure 182.1). Its small form factor houses the switching controller, MOSFETs, inductor and all support components, and weighs only 1g. At this size, it can be mounted on the back side of a system board, taking advantage of otherwise unused space.

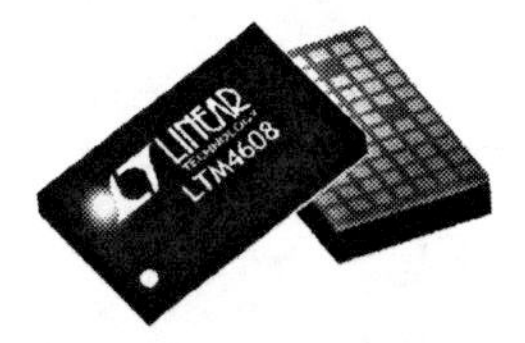

Figure 182.1 • The LTM4608 Offers High Power Density in a 9mm x 15mm x 2.8mm LGA Package

The LTM4608 operates from an input supply range of 2.375 to 5.5V, and a single resistor is all that is needed to set the output voltage within a 0.6V to 5V output range. Its high efficiency design and low thermal impedance package delivers up to 8A continuous current.

Wealth of features

The LTM4608's 1.5MHz switching frequency and current mode architecture allow it to react quickly to line and load transients without sacrificing stability. Cycle-by-cycle current mode control also enables excellent current sharing for parallel operation. The integrated clock enables multiphase operation and frequency synchronization, and a frequency spread spectrum feature can also be activated to further reduce switching noise harmonics. The device supports output voltage tracking or simpler supply rail sequencing. Programmable output voltage margining is supported for ±5%, ±10%, and ±15% levels. Fault protection features include over voltage protection, over current protection, and thermal shutdown.

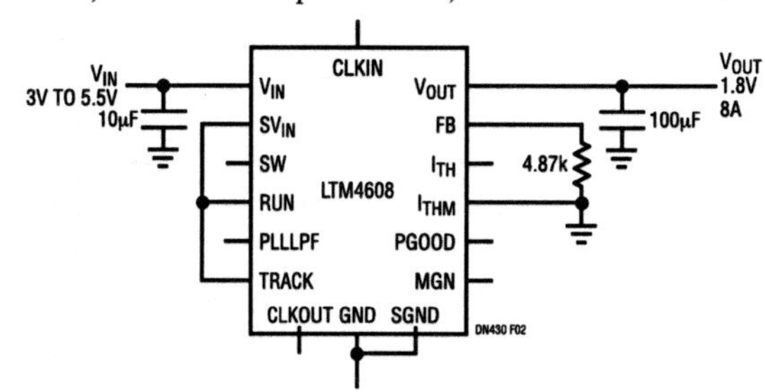

Figure 182.2 • Few Components Are Required for a 1.8V/8A Application

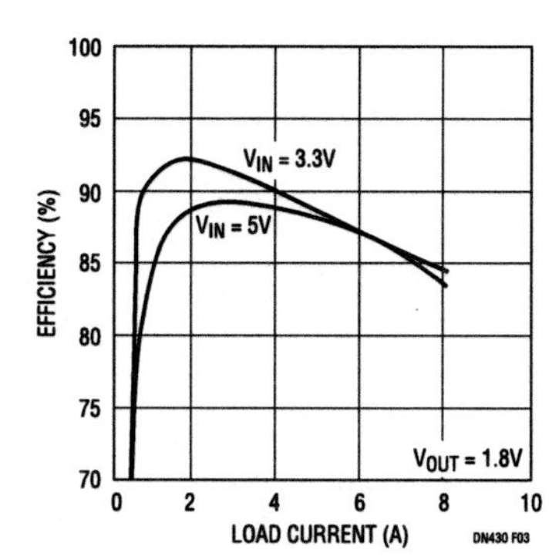

Figure 182.3 • Efficiency of the Application in Figure 182.2

Analog Circuit Design: Design Note Collection. http://dx.doi.org/10.1016/B978-0-12-800001-4.00182-4

Quick and easy design

Figure 182.2 shows a typical 1.8V output design; its efficiency is shown in Figure 182.3. Because the LTM4608 includes two integrated 10μF ceramic capacitors, additional input capacitors are only needed for large load steps up to the full 8A level. Linear Technology provides a μModule Power Design tool and SwitcherCAD simulation tool to calculate the necessary capacitance for any particular design (www.linear.com/micromodule).

For low output voltage ripple and low droop during load transients, low ESR capacitors should be used. A low ESR polymer or ceramic capacitor is sufficient. Typical ranges are 100μF to 200μF. The output voltage is set with an external resistor from the FB pin to ground.

Thermally enhanced packaging

The LTM4608's package has a low thermal resistance of 7°C/W junction-to-pin and 25°C/W junction-to-ambient when mounted on a four-layer board with no airflow. The device's unique packaging allows simple heat sinking from both the top and bottom, making it possible to use a metal chassis as a heat sink.

Output voltage tracking

Output voltage tracking is programmed via the Track pin. The slave output can be tracked up and down with another regulator's output.

Current sharing: 8A+8A=16A

Two or more LTM4608 μModule regulators can be paralleled to provide multiples of 8A of load current. Because of the LTM4608's current mode architecture, the output current and power are evenly and safely distributed across each LTM4608. Figure 182.4 shows a 16A design that also operates 180° out-of-phase to reduce input and output ripple current.

Fault conditions: overcurrent limit and thermal shutdown

The LTM4608's current mode control inherently limits cycle-by-cycle inductor current, not only in steady state operation, but also in transient. The LTM4608 device has over temperature shutdown protection that inhibits switching operation above 150°C.

Conclusion

Weighing 1g, occupying 135mm^2, and standing only 2.8mm tall, the LTM4608 is a complete and efficient point-of-load DC/DC system that eases circuit and layout challenges by fitting in the tightest spaces even on the bottom of the PCB. With the LTM4608, the design of an 8A switchmode regulator is as simple as a linear regulator. This DC/DC μModule regulator is rich in features and provides circuit protection as well as capability to current share for applications requiring more than 8A.

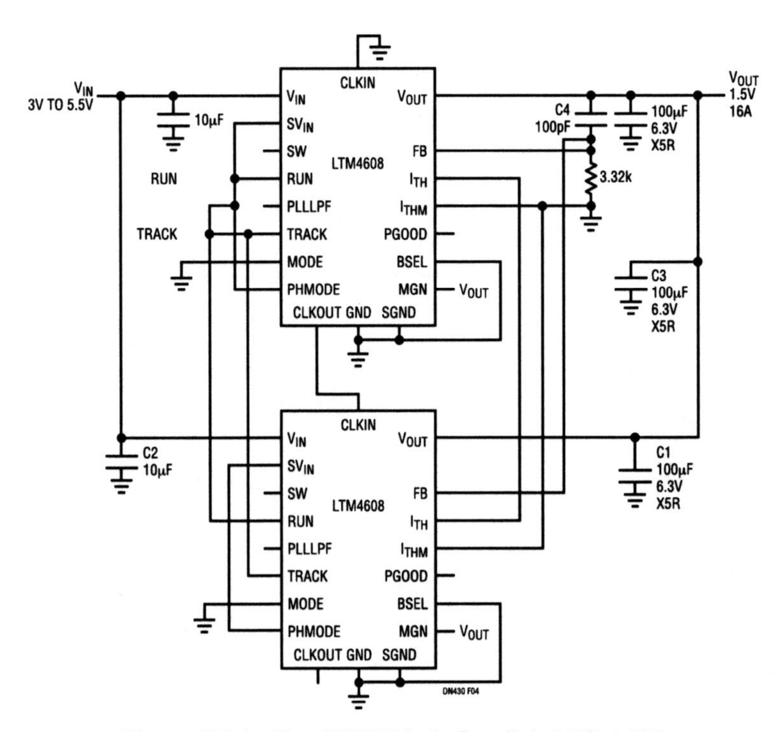

Figure 182.4 • Two LTM4608s in Parallel, 1.5V at 16A

Simple and compact 4-output point-of-load DC/DC μModule system

183

Jian Yin Eddie Beville

Introduction

Advancements in board assembly, PCB layout and digital IC integration have produced a new generation of densely populated, high performance systems. The board-mounted, point-of-load (POL) DC/DC power supplies in these systems are subject to the same demanding size, performance and power requirements as other subsystems—demands that are difficult to meet with traditional power modules or controller/regulator ICs. The LTM4601 DC/DC μModule converter meets these demands by shrinking an entire solution to the size of a low profile IC. Its frequency synchronization and voltage tracking features allow multiple LTM4601s to be easily and quickly configured for multioutput applications.

4-output DC/DC converter power system

Figure 183.1 shows the photo of a 4-output DC/DC supply using four μModule converters with frequency synchronization and output tracking. The operating waveforms of the four outputs are interleaved with a 90° relative phase difference, thus reducing the effective input current ripple. This in turn significantly reduces the bulk capacitance of the circuit and the circuit size.

Figure 183.2 presents the efficiency of each output in Figure 183.1. With 12V input voltage, each output is tested up to 12A by disabling the other three outputs. The high efficiencies up to 92% guarantee low losses in the circuit board, thus leading to a reduced system profile.

Figure 183.3 shows the simplified block diagram for Figure 183.1. For a detailed schematic, please refer to page 22 of the LTM4601 data sheet. An intermediate bus input of 8V–16V is converted to four different outputs: 1.5V at 12A, 1.8V at 12A, 2.5V at 12A and 3.3V at 10A. The output voltages are set by resistances on the LTM4601 VFB pins. A 4-phase oscillator LTC6902 generates 90° interleaved clock signals. Moreover, spread spectrum frequency modulation (SSFM) can be activated by adding an external resistor from the LTC6902 MOD pin to V+.

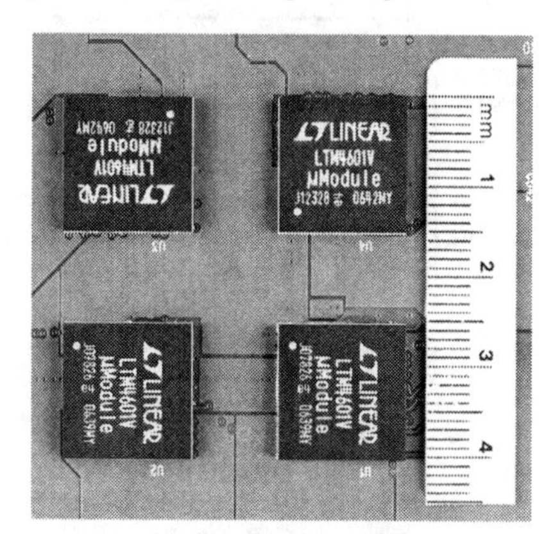

Figure 183.1 • A 4-Output 103W DC/DC System Can Fit This Tiny Space (Each LTM4601 μModule DC/DC Converter Contains an Inductor, MOSFETs, Bypass Capacitors, etc.)

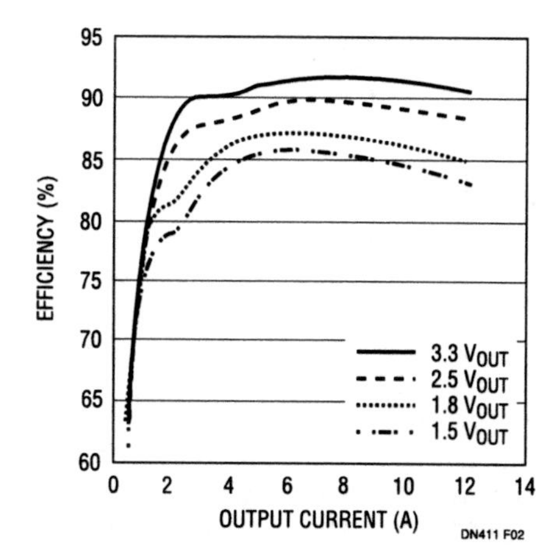

Figure 183.2 • Efficiency of Each Output for the Circuit in Figure 183.1

Analog Circuit Design: Design Note Collection. http://dx.doi.org/10.1016/B978-0-12-800001-4.00183-6

Output tracking

The output voltage of the LTM4601 can track another converter's output ratiometrically or coincidently. The circuit in Figure 183.1 implements coincident tracking by connecting the 3.3V output (master) to the TRACK/SS pins of the other μModule converters (slaves) via resistive dividers. For coincident tracking, the master must have a higher output voltage than the slaves. The soft-start capacitor on the TRACK/SS pin of the 3.3V master supply sets the ramp rate of the start-up voltage. Figure 183.4 shows the start-up waveforms of the four outputs with output tracking.

Frequency synchronization

The operating frequency of the LTM4601 can be synchronized with an external clock to reduce undesirable frequency harmonics, and its operation can be interleaved with other LTM4601s. Figure 183.5 shows the input current ripple of the 180° phase-interleaved 1.8V and 3.3V outputs of the circuit in Figure 183.1. The input current ripple of the 3.3V output is synchronized with its PLLN signal in Figure 183.5. Therefore, with four 90° interleaved inputs, the input current ripples are partially cancelled, reducing the required input capacitance.

Conclusions

The synchronization and tracking features of the LTM4601 allow interleaved phases in the 4-output solution, thus reducing input capacitance and producing a compact design. High efficiency and excellent thermal performance make it possible to handle the total maximum power of 103W in a 4-layer PCB at 11cm × 11cm.

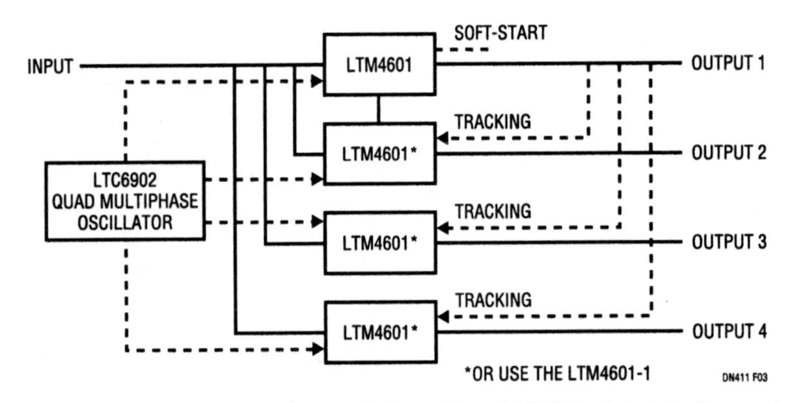

Figure 183.3 • Simplified Schematic of a Compact 4-Output Point-of-Load DC/DC μModule Converter Solution. The LTC6902 Interleaves the Operating Waveforms of the Four μModule Converters, so That the Ripple Currents Cancel Each Other. This Significantly Reduces the Size of the Required Input Capacitors. Start-Up and Shutdown Voltage Tracking Is Simply Accomplished by Connecting the Output of the 3.3V μModule (Output 1) to the Other μModule Converters

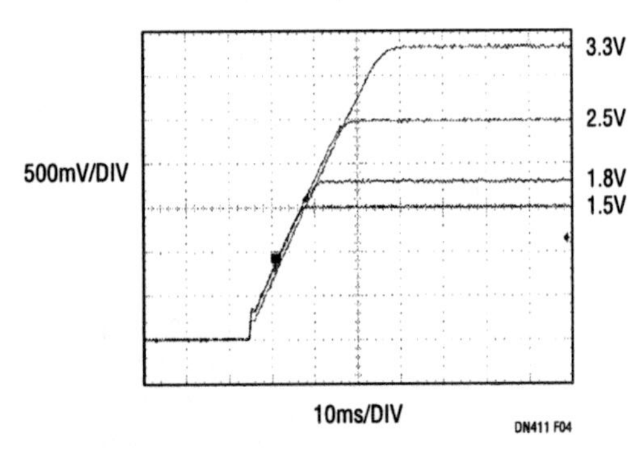

Figure 183.4 • Start-Up Voltage Waveforms of the Circuit in Figure 183.1 Show Coincident Tracking of the Outputs

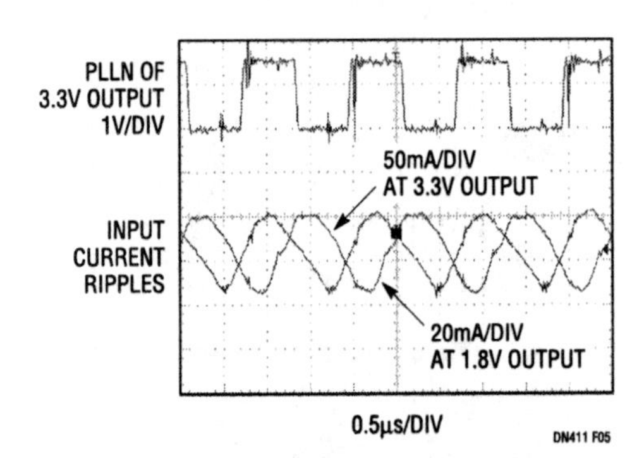

Figure 183.5 • Input Current Ripple Is Reduced by Interleaving the Operation of the Supplies Using Frequency Synchronization

10A high performance point-of-load DC/DC μModule regulator

4.5V to 28V input, 0.6V to 5V output in a 15mm × 15mm × 2.8mm package

184

Eddie Beville

Introduction

Advancements in board assembly, PCB layout and digital IC integration have produced a new generation of densely populated, high performance systems. The board-mounted point-of-load (POL) DC/DC power supplies in these systems are subject to the same demanding size, high power and performance requirements as other subsystems. The rigorous new POL demands are difficult to meet with traditional controller or regulator ICs, or power modules.

For such demanding applications, an ideal POL power supply must meet high performance specifications while simplifying board assembly—mounting similar to other surface mount ICs on the board without requiring special tooling. Such POL DC/DC regulators must also demonstrate exceptional thermal performance with innovative packaging technology. Power density increases without the danger of overheating and shortened device life. The LTM4600 μModule regulator does all of these things.

10A DC/DC μModule regulator in IC form factor

The LTM4600 μModule regulator is a complete power supply point-of-load DC/DC regulator with a low profile IC form-factor. The controller, onboard inductor, MOSFETs and compensation circuitry are all housed in a 15mm × 15mm × 2.8mm LGA surface mount package which weighs only 1.73g (Figure 184.1). These size parameters allow the LTM4600 to be mounted on the back side of a system board, taking advantage of the otherwise unused space. The μModule regulator switches at a nominal 800kHz in a synchronous topology to offer very high efficiency in a small form factor and low profile.

The μModule regulator is offered in two versions. The LTM4600EV operates from an input supply range of 4.5V to 20V; the LTM4600HVEV operates from 4.5V to 28V. Both offer adjustable output voltages from 0.6V to 5V and output currents of 14A peak and 10A continuous. Fault protection features include overvoltage protection and overcurrent protection.

Quick and easy design

Figure 184.2 shows a typical LTM4600EV design for a 2.5V output and Figure 184.3 shows the efficiency of the circuit. Although bulk capacitors on the input and output suffice in most applications, this design uses two low ESR 10μF 25V ceramic capacitors to reduce input RMS ripple. The output voltage is set with an external resistor from the V_{OSET} pin to ground. The output capacitors are selected for low ESR to maintain an initial voltage droop of the output voltage to approximately $\Delta V_{OUT} = I_{LOADSTEP} \cdot R_{ESR}$ in a transient step.

Figure 184.1 • The LTM4600 Offers Unprecedented Power Density in a Small Package

Analog Circuit Design: Design Note Collection. http://dx.doi.org/10.1016/B978-0-12-800001-4.00184-8

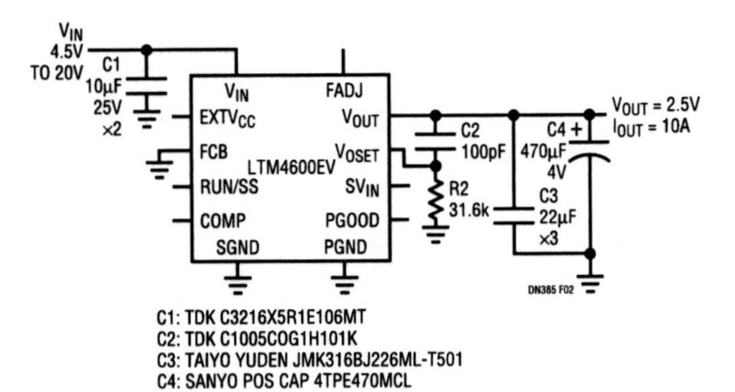

Figure 184.2 • Few Components are Required in this 2.5V/10A Application

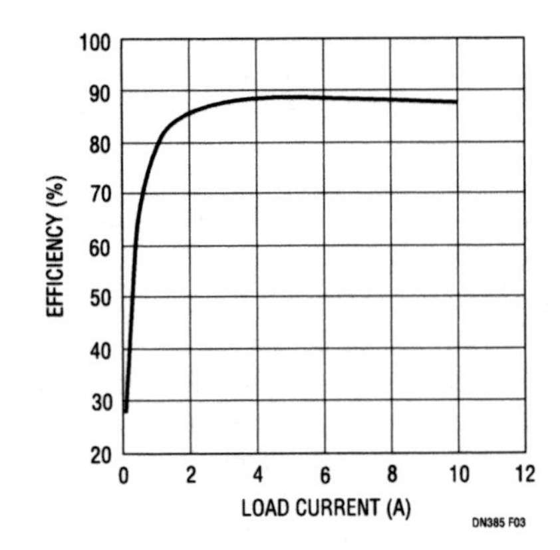

Figure 184.3 • Efficiency of the Application in Figure 184.2

Thermally enhanced packaging

The μModule regulator package has extremely low thermal resistance of 6°C/W and 15°C/W junction-to-case and junction-to-ambient, respectively. It allows heat-sinking from both the top and bottom of the device. Figure 184.4 shows the top view thermal imaging of the LTM4600 at full throttle with no airflow and heat sink. Refer to Application Note 103 for detailed thermal analysis and measurements.

Fast transient response

A unique feature of the LTM4600 is its no-clock-latency valley current mode architecture. This feature allows very fast loop response to rapid load transients with minimum output capacitance. Typically, the output voltage turns around in 4 to 6

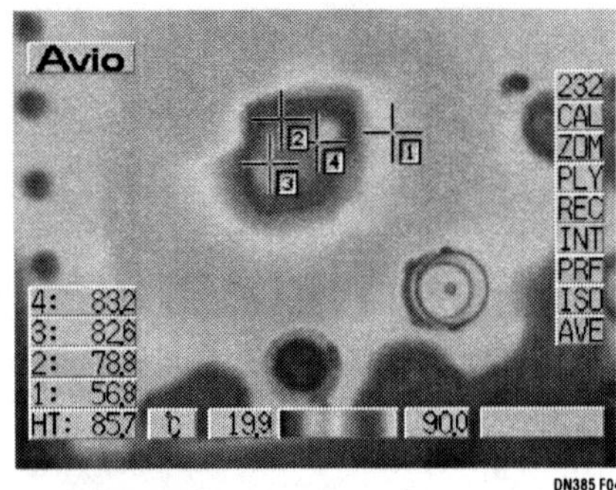

Figure 184.4 • The LTM4600 Exhibits Impressive Thermal Performance, Even Without Air Flow and Heat Sink (24V to 3.3V at 10A, Top View). For a Color Representation, Download the File at www.linear.com

microseconds and fully recovers in 20 to 25 microseconds. Figure 184.5 shows the transient deviation of only 55mV on a 2.5V output with a 5A load step. The 6μs of turnaround is achieved with only a 470μF POS cap and the three 22μF ceramics.

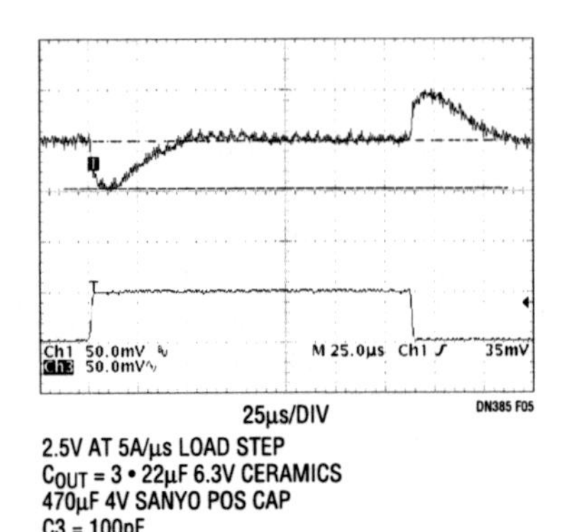

Figure 184.5 • Load Transient Response for the Application in Figure 184.2

Paralleling the μModule regulator for 20A output

Two LTM4600 μModule regulators can be used in parallel to double the output current. The current mode architecture and precision current limiting allow two modules to equally share the output current, thus maximizing efficiency and equally distributing the heat.

Section 10

Switching Regulators for Isolated Power Design

Isolated converters have buck simplicity and performance

185

Kurk Mathews

Buck converter designers have long benefited from the simplicity, high efficiency and fast transient response made possible by the latest buck controller ICs, which feature synchronous rectification and PolyPhase interleaved power stages. Unfortunately, these same features have been difficult or impossible to implement in the buck converter's close relative, the forward converter, often used in isolated industrial and telecom applications. That is, until now. The LTC3706 secondary-side synchronous controller and its companion smart gate driver, the LTC3725, make it possible to create an isolated forward converter with the simplicity and performance of a buck converter.

Simple isolated 3.3V, 30A forward converter

Many isolated supplies place the controller IC on the input (primary) side and rely on indirect synchronous rectifier timing and optoisolator feedback to control the output (secondary). The circuit shown in Figure 185.1 offers a more direct approach using fewer components. The LTC3706 controller is used on the secondary and the LTC3725 driver with self-starting capability is used on the primary. When an input voltage is applied, the LTC3725 begins a controlled soft-start

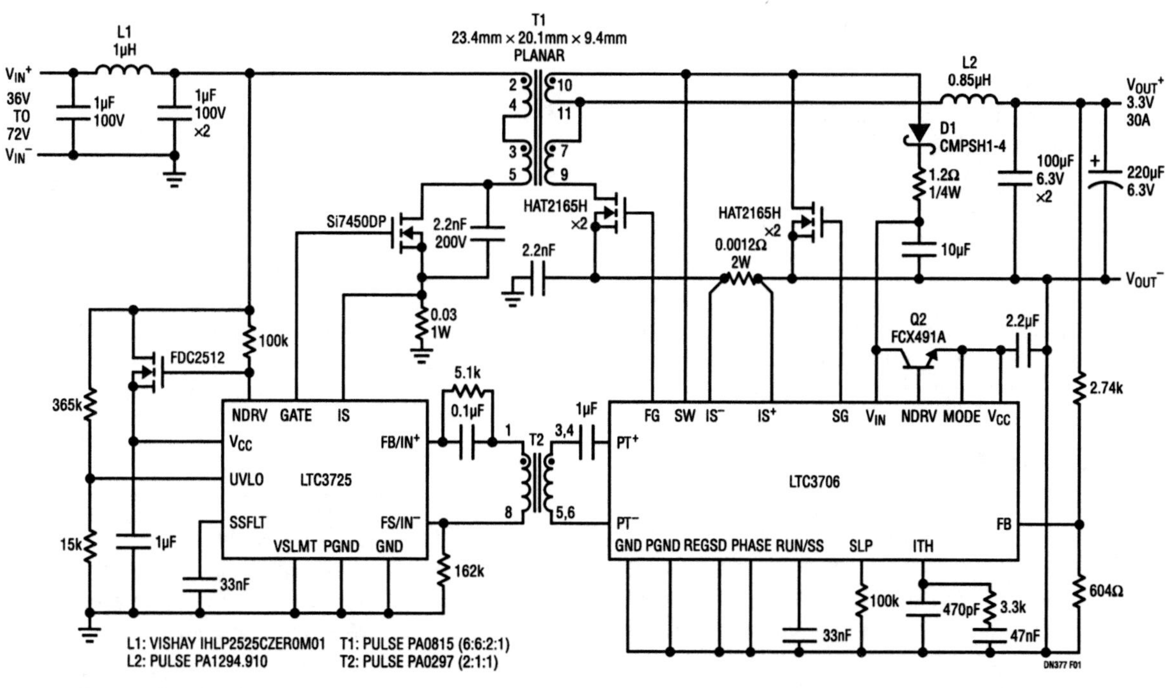

Figure 185.1 • Complete 100W High Efficiency, Low Cost, Minimum Part Count Isolated Telecom Converter. Other Output Voltages and Power Levels Require Only Simple Component Changes

Analog Circuit Design: Design Note Collection. http://dx.doi.org/10.1016/B978-0-12-800001-4.00185-X

of the output voltage. As the output voltage begins to rise, the LTC3706 secondary controller is quickly powered up via T1, D1 and Q2. The LTC3706 then assumes control of the output voltage by sending encoded PWM gate pulses to the LTC3725 primary driver via signal transformer, T2. The LTC3725 then operates as a simple driver receiving both input signals and bias power through T2.

The transition from primary to secondary control occurs seamlessly at a fraction of the output voltage. From that point on, operation and design simplifies to that of a simple buck converter. Secondary sensing eliminates delays, tames large-signal overshoot and reduces output capacitance. The design shown in Figure 185.1 features off-the-shelf magnetics and high efficiency (see Figure 185.2).

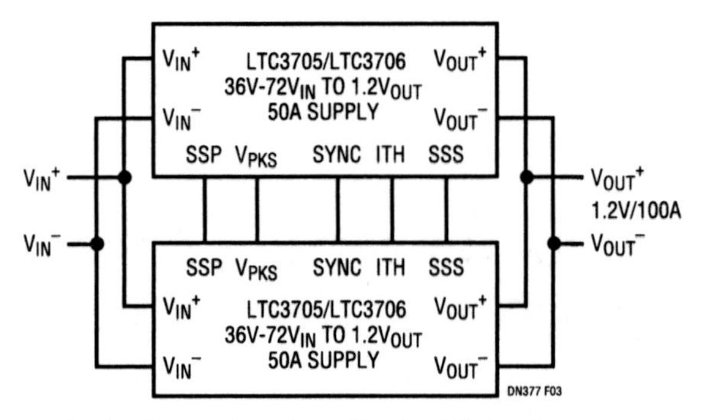

Figure 185.3 • Paralleling Supplies for Higher Power Operation

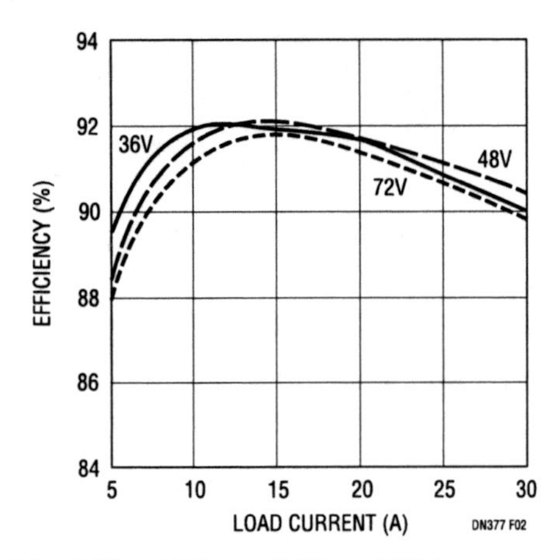

Figure 185.2 • 36V to 72V_{IN} to 3.3V_{OUT} Efficiency

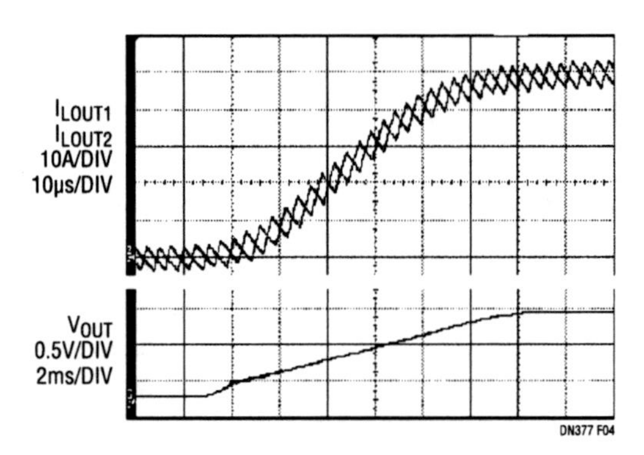

Figure 185.4 • 1.2V, 100A Load Current Step (Top Trace) and Start-Up (Bottom Trace)

PolyPhase design ups power limit

The LTC3706 defies typical forward converter power limits by allowing simple implementation of a PolyPhase current share design. PolyPhase operation allows two or more phase-interleaved power stages to accurately share the load. The advantages of PolyPhase current sharing are numerous, including much improved efficiency, faster transient response and reduced input and output ripple.

The LTC3706 supports standard output voltages such as 5V, 12V, 28V and 52V as well as low voltages down to 0.6V. Figure 185.3 shows how easy it is to parallel two 1.2V supplies to achieve a 100A supply. Figure 185.4 shows the excellent output inductor current tracking during a 0A to 100A load current step and the smooth handoff during start-up to secondary-side control at 0.5V output.

Related products

The LTC3705 is a dual switch forward driver version of the LTC3725 single switch forward driver. The LTC3705 includes an 80V (100V transient) high side gate driver. The 2-switch topology eliminates transformer reset concerns, further simplifying design.

The 16-pin LTC3726 secondary controller is an option to the 24-pin LTC3706. The LTC3726 does not include the remote voltage sensing or the linear regulator features found in the LTC3706, so it is suitable in a single phase design or as a PolyPhase slave device. Both controllers may be used without the primary driver for nonisolated applications.

Features

These ICs include features that provide robust performance with few external parts and a simple feedback loop. For example, the LTC3725 primary driver includes a linear regulator controller and internal rectifier, eliminating the need for a primary bias supply. The LTC3725 also includes a volt-second and primary current limit. The LTC3706 controller includes a synchronous rectifier crowbar and remote voltage sensing.

Conclusion

The new LTC3706 controller and LTC3725 driver bring an unprecedented level of simplicity and performance to the design of isolated supplies. These two devices work together to offer high efficiency, low cost solutions using off-the-shelf external components.

Multiple output isolated power supply achieves high efficiency with secondary side synchronous post regulator

186

Charlie Zhao Wei Chen

Introduction

The newly released LT3710, a secondary side synchronous post regulator controller, provides tight regulation for all outputs of multiple output isolated power supplies. Solutions that use the LT3710 are efficient, use minimal space and provide fast transient response.

The LT3710 is a voltage mode controller with leading edge modulation, programmable current limit and dual MOSFET drivers. It generates a tightly regulated second output directly from the transformer secondary winding, thus minimizing the size of the output inductor and capacitor of the main output stage. The use of synchronous MOSFETs significantly improves efficiency, making it suitable for low output voltage applications.

Design example

Figures 186.1a and 186.1b show a dual output high efficiency isolated DC/DC power supply with 36V to 72V input range and two outputs of 3.3V/10A and 2.5V/10A. The basic power stage topology is a 2-switch forward converter with synchronous rectification. The primary side controller uses the LT3781; a current mode 2-switch forward controller with built-in MOSFET drivers. On the secondary side a synchronous rectifier controller, the LTC1698, provides the voltage feedback for the main 3.3V output as well as the gate drive for the synchronous MOSFETs. The LT3710 circuit precisely regulates the 2.5V output. Total load and line regulation for the 2.5V output is better than 0.2%. From a 48V input, the overall efficiency is about 87% with full load on both outputs.

Conclusion

The LT3710 is a high efficiency secondary side synchronous post regulator controller that is designed to generate a tightly regulated secondary output for multiple output isolated power supplies. It can be used with any buck derived single-ended or dual-ended isolated topologies, such as forward, push-pull, half-bridge and full-bridge converters.

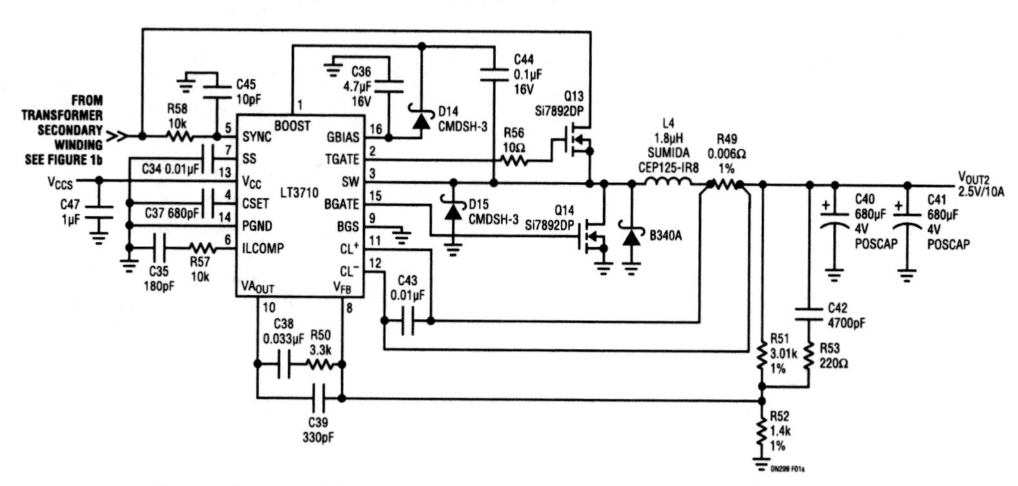

Figure 186.1a • 36V to 72V DC to 3.3V/10A and 2.5V/10A Dual Output Isolated Power Supply (Part 1 of 2: Post Regulator Circuit)

Analog Circuit Design: Design Note Collection. http://dx.doi.org/10.1016/B978-0-12-800001-4.00186-1

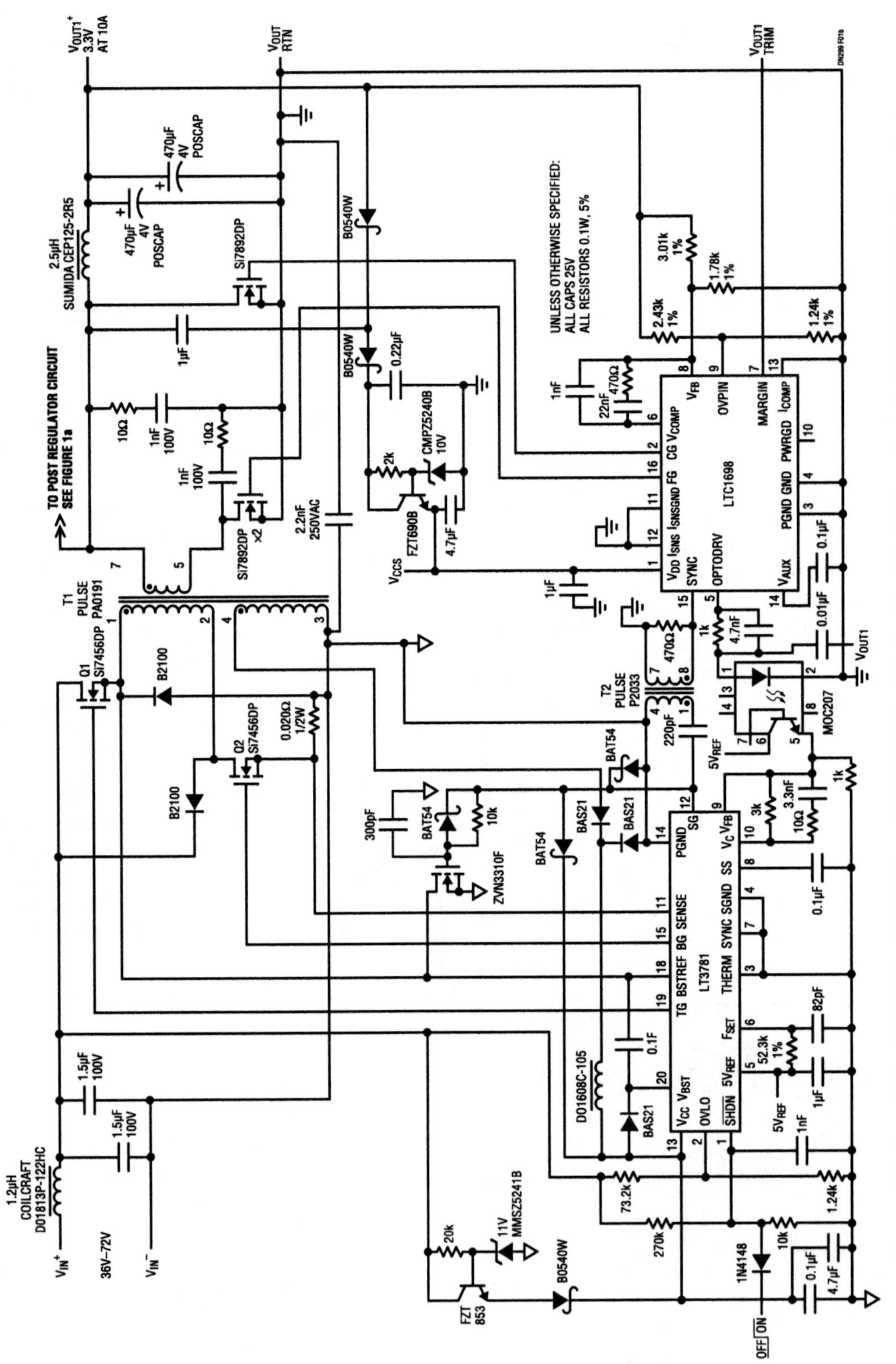

Figure 186.1b • 36V to 72V DC to 3.3V/10A and 2.5V/10A Dual Output Isolated Power Supply (Part 2 of 2: Main Converter)

Chip set offers low cost alternative to 48V telecom modules

187

Kurk Mathews

The demand for high performance 48V input telecom modules has never been higher. The latest generation modules offer high current, high efficiency alternatives to traditional low voltage potted supplies. Although these drop-in solutions are attractive, their higher cost can quickly add up as product volumes increase. Replacing a module with an on-board solution based on the new LT1681/LTC1698 chip set significantly reduces system cost and takes full advantage of board area, minimizing the need to derate output power at elevated ambient temperatures.

The new LT1681 controller contains all the necessary functions for a synchronous forward converter and works together with the LTC1698 secondary controller for isolated applications. The LT1681 features include current-mode operation up to 250kHz, 75V high side driver, leading edge blanking, input under/overvoltage protection, synchronous gate output and a thermal shutdown pin. The LTC1698 interfaces directly with the LT1681 providing an error amplifier, current limit and output overvoltage protection in addition to synchronous gate drivers.

Figure 187.1 • 3.3V/20A Supply in Half-Brick Footprint

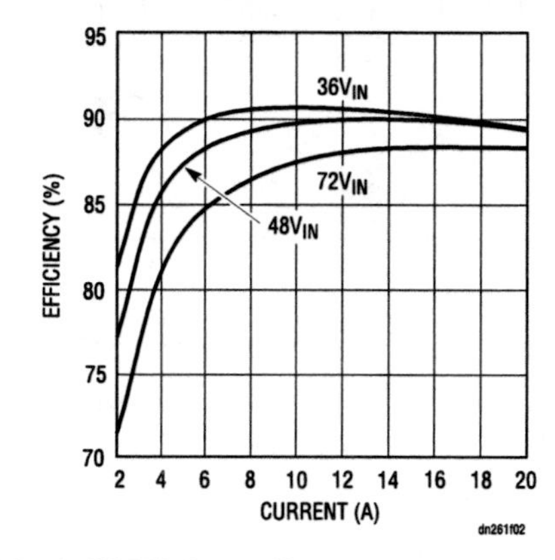

Figure 187.2 • 3.3V Efficiency (See Figure 187.3)

Isolated 48V to 3.3V supply

The schematic in Figure 187.3 presents a complete $36V_{IN}$ to $72V_{IN}$ to $3.3V_{OUT}/20A$ power supply. Total component cost is typically below $30 (50k pcs). The 2-transistor forward converter utilizes low voltage primary switches and provides for recovery of the transformer's mutual and leakage energy. The LTC1698 secondary controller synchronizes with the LT1681 via a small pulse transformer and drives secondary synchronous MOSFETs. Efficiency is shown in Figure 187.2. The LTC1698 includes an error amplifier and optocoupler drive buffer, eliminating the output feedforward path and loop compensation issues associated with '431 type references. A margin pin allows the output voltage to be adjusted ±5%. Other output voltages can be realized by substituting components into the same basic circuit. Contact the factory for more information including half-brick layout information (see Figure 187.1).

Conclusion

The LT1681 and LTC1698 combine to provide a low cost, discrete alternative to telecom modules. The LT1681/LTC1698 features reduce circuit complexity. High efficiency synchronous operation combined with the thermal advantages of an on board supply make the LT1681/LTC1698 the ideal choice for high current, low cost isolated converters.

Analog Circuit Design: Design Note Collection. http://dx.doi.org/10.1016/B978-0-12-800001-4.00187-3

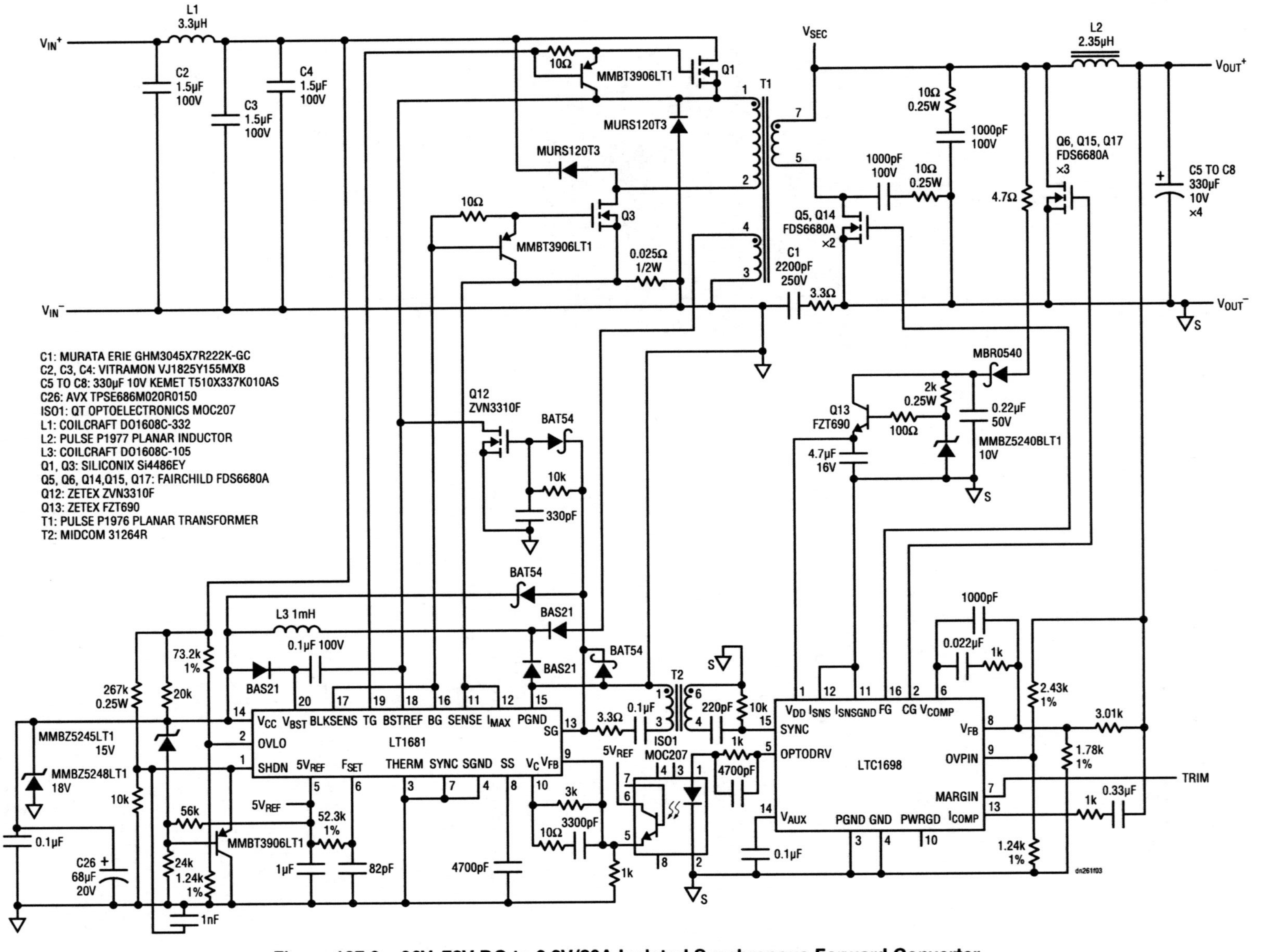

Figrue 187.3 • 36V–72V DC to 3.3V/20A Isolated Synchronous Forward Converter

5V high current step-down switchers

188

Ron Vinsant Milton Wilcox

Low cost high efficiency (80%), high power density DC/DC converter

The LT1241 current mode PWM control IC can be used to make a simple high frequency step-down converter. This converter also has low manufacturing costs due to simple magnetic components. This circuit exhibits a wide input range of 30V to 60V while maintaining its 12A 5V output. It has short-circuit protection and uses minimal PC board area due to its 300kHz switching frequency.

Figure 188.1 shows the LT1241 being used to drive the switching transistor Q1 through a ferrite pulse transformer T2. This transformer is built on a high μ material resulting in an 11 turn bifilar wound toroid that is only 0.15 inches in diameter and can be surface mounted. T1 acts as a current sense transformer whose volt·second balance is assured by the duty cycle limit of 50% inherent in the LT1241. The output inductor (L1) is made of Magnetics Kool-Mu material and is only 0.7 inches in diameter.

Short-circuit protection is provided through bootstrap operation of the LT1241. If the output is shorted the LT1241

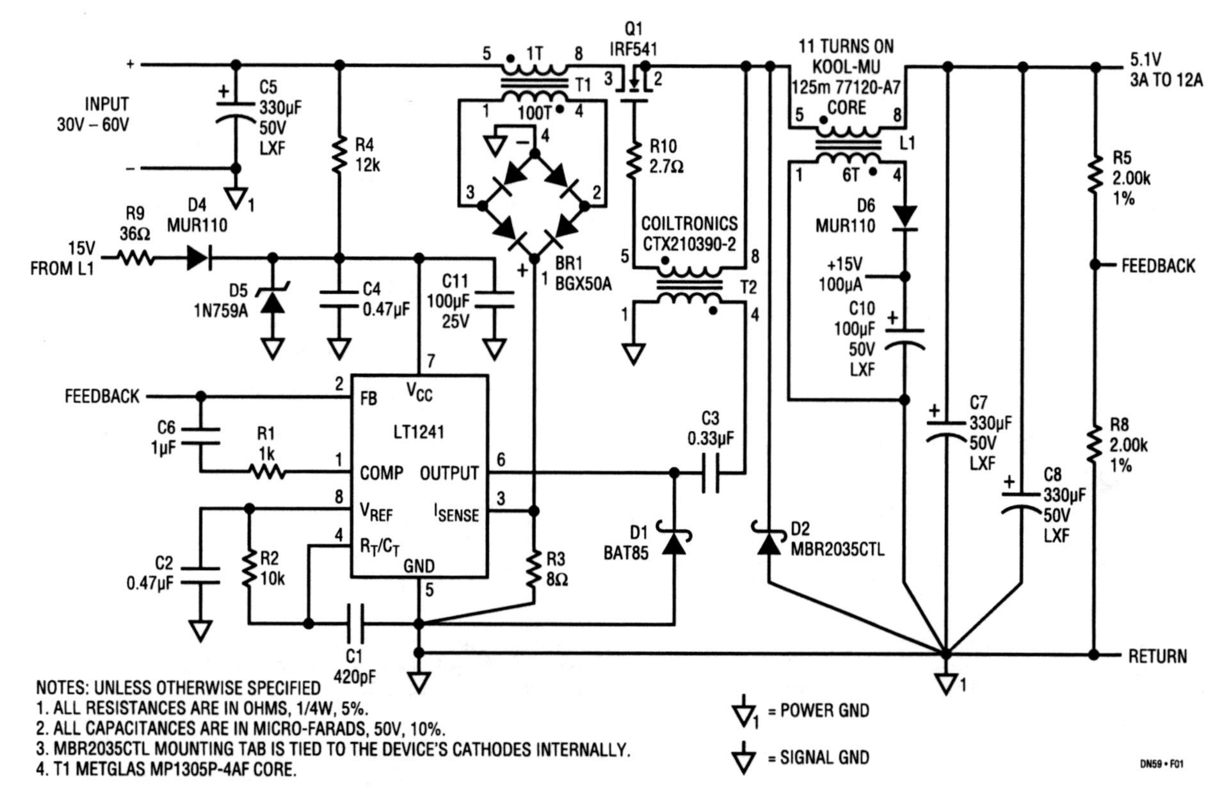

Figure 188.1 • The LT1241 Driving the Switching Transistor Q1 Through a Ferrite Pulse Transformer T2

Analog Circuit Design: Design Note Collection. http://dx.doi.org/10.1016/B978-0-12-800001-4.00188-5

limits its pulse width to ≤250ns. Because there is not enough current supplied to make the aux winding on the output inductor 15V, the LT1241 stops operation. It will then try to start by C11 charging through R4. If the output is still shorted it will stop again. Thus in a short, the circuit starts and stops, protecting itself from overload.

Synchronous switching eliminates heat sinks in a 50W DC/DC converter

The new LT1158 half-bridge N-channel power MOSFET driver makes an ideal synchronous switch driver to improve the efficiency of step-down (buck) switching regulators. The diode losses in a conventional step-down regulator become increasingly significant as V_{IN} is increased. By replacing the high current Schottky diode with a synchronously-switched power MOSFET, efficiencies well over 90% can be realized (see Figure 188.2).

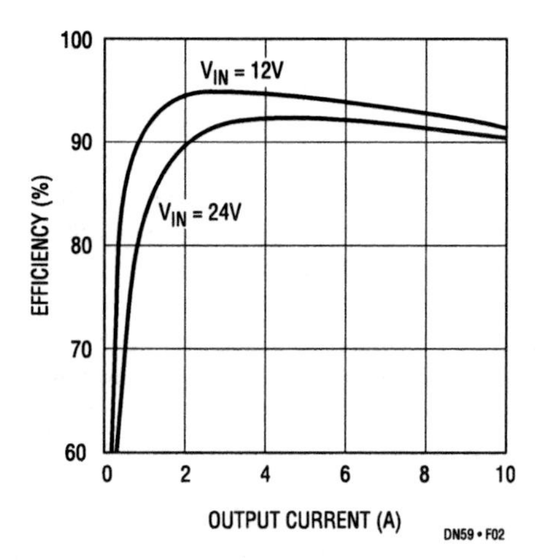

Figure 188.2 • Operating Efficiency for Figure 188.3 Circuit

In the Figure 188.3 circuit an LT3525 provides a voltage mode PWM to drive the LT1158 input pin. The LT1158 drives (2) 28mΩ power MOSFETs for each switch, reducing individual device dissipation to 0.7W worst case. This eliminates the need for heat sinks for operation up to 10A at a temperature of 50°C ambient. The inductor and current shunt losses for the Figure 188.3 circuit are 1.2W and 0.7W respectively at 10A.

An additional loss potentially larger than those already mentioned results from the gate charge being delivered to multiple large MOSFETs at the switching frequency. This driver loss can only be controlled by running the oscillator at as low a frequency as practical—in the case of the Figure 188.3 circuit, 25kHz. A very low ESR (<20mΩ) output capacitor is used to limit output ripple to less than 50mVp-p with 2.5Ap-p ripple current.

The LT1158 also provides current limit for DC/DC converter applications. When the voltage across R_S exceeds 110mV, the LT1158 $\overline{\text{fault}}$ pin conducts, and assumes control of the PWM duty cycle. This provides true current mode short-circuit protection with soft recovery. The Figure 188.3 regulator current limit is set at 15A which raises the dissipation in each bottom MOSFET to 1.7W during a short. Therefore 30°C/W heat sinking must be added for the bottom side MOSFETs if continuous short-circuit operation is required. Care should also be taken when routing the sense+ and sense− leads to prevent coupling from the inductor.

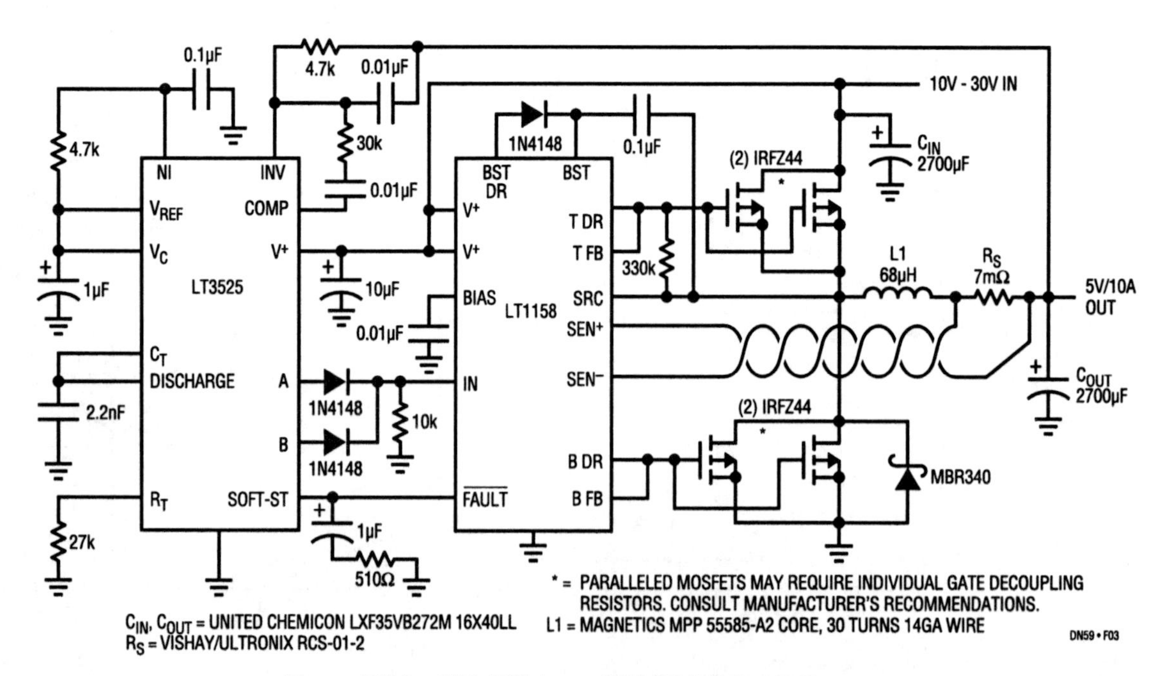

Figure 188.3 • High Efficiency 50W DC/DC Converter

Section 11

Power Control & Ideal Diode Design

Ideal diodes protect against power supply wiring errors

189

Meilissa Lum

Introduction

High availability systems often employ dual feed power distribution to achieve redundancy and enhance system reliability. ORing diodes join the feeds together at the point of load, most often using Schottky diodes for low loss. MOSFET-based ideal diodes can be used to replace Schottky diodes for a significant reduction in power dissipation, simplifying the thermal layout and improving system efficiency. Figure 189.1 shows the LTC4355 and LTC4354 combining the inputs and returns in a −48V, 5A dual feed application. This solution reduces the power dissipation from 6W using Schottky diodes to just 1.1W with MOSFETs.

With two supply sources and four supply connections there are plenty of ways to incorrectly connect the wires. Although the likelihood of a wiring error is small, the cost is high if downstream cards are not designed to tolerate such errors. Wiring errors could include reverse polarity or cross-feed connections. Knowing this, circuit designers are accustomed to using discrete diode solutions to protect against such mishaps. It is important that active ideal diodes give similar protection.

Types of misconnections

Figure 189.2 shows the correct power supply connections. RTNA and RTNB are close in potential by virtue of the common connections to safety ground represented by R_{GND}.

Figure 189.3 shows a reversed input connection with RTNA and NEGA swapped. The associated ideal diodes are reverse biased, making the wiring error transparent to the load with BATTERY B providing power.

Figure 189.4 shows another misconnection with RTNB and NEGA swapped, so one power supply is connected across the RTN inputs of the LTC4355 and the other supply across the −48V inputs of the LTC4354. In this case, the reverse input protection network of three diodes shown in Figure 189.1 prevents damage to the LTC4355. The load operates from BATTERY B, but only after the current has passed through the ground wiring.

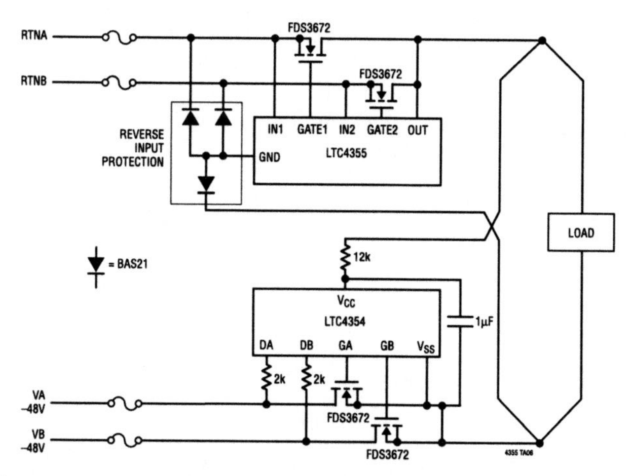

Figure 189.1 • −48V Ideal Diode-OR

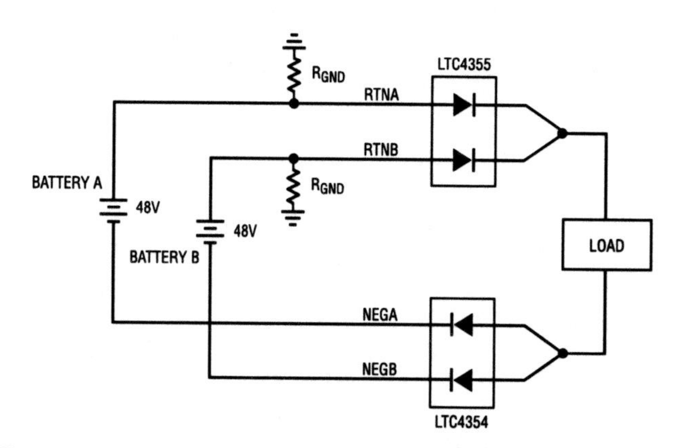

Figure 189.2 • Correct Power Supply Connections

Analog Circuit Design: Design Note Collection. http://dx.doi.org/10.1016/B978-0-12-800001-4.00189-7

Figure 189.5 shows BATTERY B installed incorrectly. The reversed battery has no effect on the load because the diode connected to NEGB is reverse biased. The voltage across the LTC4354 can exceed 100V and an external clamp may be added to protect its DRAIN pin.

Figures 189.2–189.5 have the correct safety ground connections to RTNA and RTNB. Damage can occur if there is a large potential difference between RTNA and RTNB. Figure 189.6 shows the safety ground, R_{GND}, mistakenly connected to NEGB instead of RTNB. This connects the power supplies in series and the voltage seen across the load nears 100V which can cause damage, a situation no different than encountered with a discrete diode solution. A TransZorb placed across the output protects the load, until a fuse on the input opens to isolate the high voltage from the load.

Conclusion

In dual feed applications, the supply connections can be erroneously wired, potentially causing damage to the load. An ideal diode solution using the LTC4355 and LTC4354 provides protection similar to Schottkys, but with much lower power dissipation. The end result is a compact layout and improved efficiency.

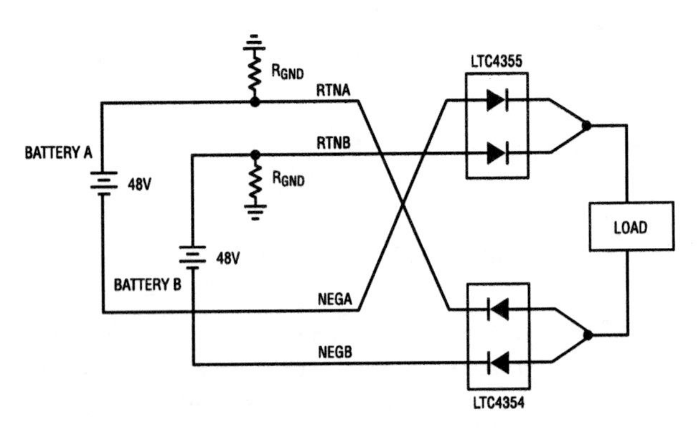

Figure 189.3 • Reversed RTNA and NEGA Connection

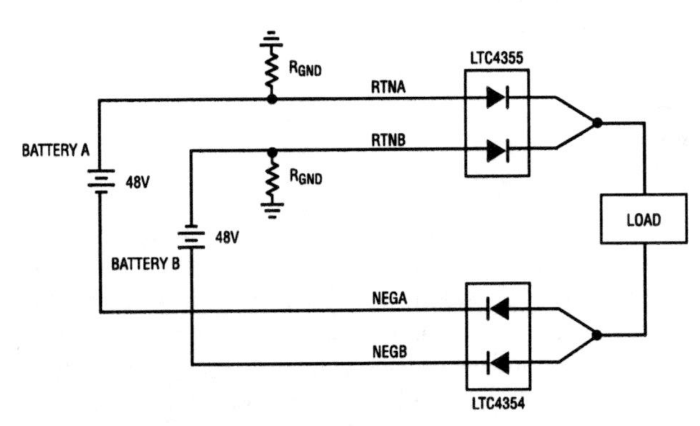

Figure 189.5 • Reversed BATTERY B

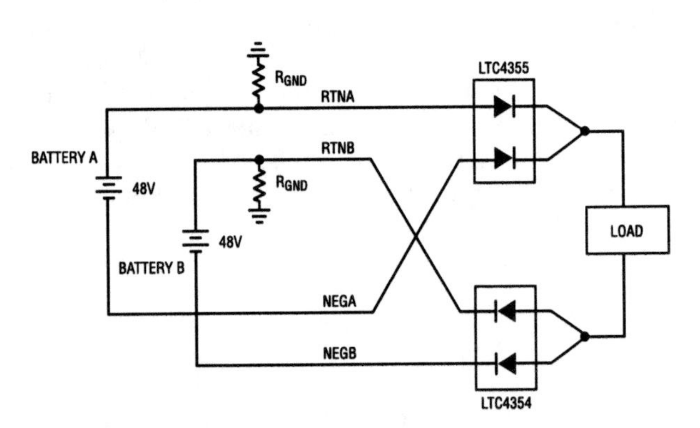

Figure 189.4 • RTNB and NEGA Swapped

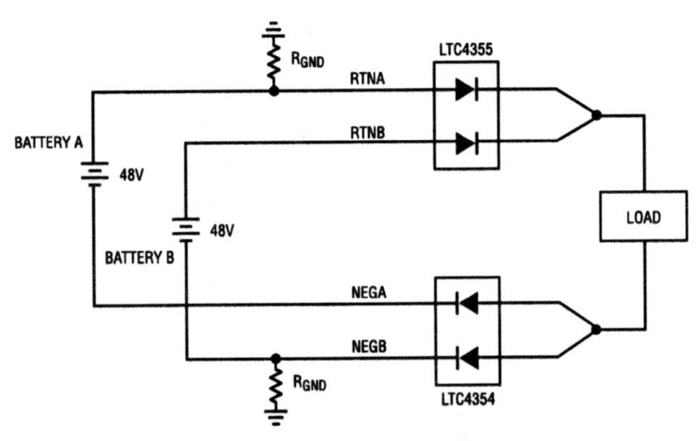

Figure 189.6 • GND Misconnected to NEGB

Ideal diode controller eliminates energy wasting diodes in power OR-ing applications

190

David Laude

Introduction

Many modern electronic devices need a means to automatically switch between power sources when prompted by the insertion or removal of any source. The LTC4412 simplifies PowerPath management and control by providing a low loss, near ideal diode controller function. Any circuit that could otherwise use a diode OR to switch between power sources can benefit from the LTC4412. The forward voltage drop of an LTC4412 ideal diode is far less than that of a conventional diode, and the reverse current leakage can be smaller for the ideal diode as well (see Figure 190.1). The tiny forward voltage drop reduces power losses and self-heating, resulting in extended battery life. Features include:

- Voltage drop across the controlled external MOSFET of only 20mV (typical)
- Low component count helps keep overall system cost low
- 6-pin ThinSOT package permits a compact design solution
- Wide supply operating range of 2.5V to 28V (36V absolute maximum)

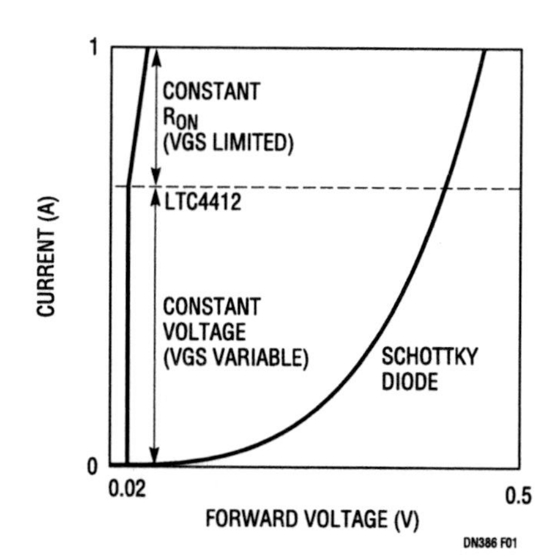

Figure 190.1 • LTC4412 Ideal Diode Controller vs Schottky Diode Characteristics

- Protection of MOSFET from excessive gate-to-source voltage with VGS limiter
- Low quiescent current of 11μA with a 3.6V supply, independent of the load current
- A status pin that can be used to enable an auxiliary MOSFET power switch or to indicate to a microcontroller that an auxiliary supply, such as a wall adapter, is present
- A control input pin for external control, such as from a microcontroller

Applications include anything that takes power from two or more inputs:

- Cellular phones
- Portable computers
- PDAs
- MP3 players and electronic video and still cameras
- USB peripherals
- Wire-ORed multipowered equipment
- Uninterruptible power supplies for alarm and emergency systems
- Systems with standby capabilities
- Systems that use load sharing between two or more batteries
- Multibattery charging from a single charger
- Logic controlled power switches

Automatic power switching between two power sources

Figure 190.2 illustrates an application circuit for the automatic switchover of load between two power sources, in this example a wall adapter and a battery. While the wall adapter is absent, the LTC4412 controls the gate of Q1 to regulate the voltage drop across the MOSFET to 20mV, thus wasting negligible battery power. The STAT pin is an open circuit while the battery provides power. When a wall adapter or other

Analog Circuit Design: Design Note Collection. http://dx.doi.org/10.1016/B978-0-12-800001-4.00190-3

supply connected to the auxiliary input is applied, the SENSE pin voltage rises. As the SENSE pin voltage rises above $V_{IN} - 20mV$, the LTC4412 pulls the GATE voltage up to turn off the P-channel MOSFET. When the voltage on SENSE exceeds $V_{IN} + 20mV$, the STAT pin sinks 10μA of current to indicate that an AC wall adapter is present. The system is now in the reverse turn-off mode, where power to the load is delivered through the external diode and no current is drawn from the battery. The external diode is used to protect the battery against some auxiliary input faults such as a short to ground. Note that the external MOSFET is wired so that the drain to source diode will reverse bias and not deliver current to the battery when a wall adapter input is applied.

Load sharing

Figure 190.3 shows a dual battery load sharing application with automatic switchover of power between the batteries and a wall adapter. In this example, the battery with the higher voltage supplies all of the power until it has discharged to the voltage of the other battery. Once both batteries have the same voltage, they share the load with the battery with the higher capacity providing proportionally higher current to the load. In this way, the batteries discharge at a relatively equal rate, maximizing battery run time.

When a wall adapter input is applied, both MOSFETs turn off and no load current is drawn from the batteries. The LTC4412's STAT pins provide information as to which input is supplying the load current. The ganging of the LTC4412s can be applied to as many power inputs as are needed.

Conclusion

The LTC4412 provides a simple means to implement a low loss ideal diode controller that extends battery life and reduces self-heating. The low external parts count results in low implementation cost and with its ThinSOT 6-pin package, a compact design as well. Its versatility is useful in a variety of applications (see the LTC4412 data sheet for additional applications).

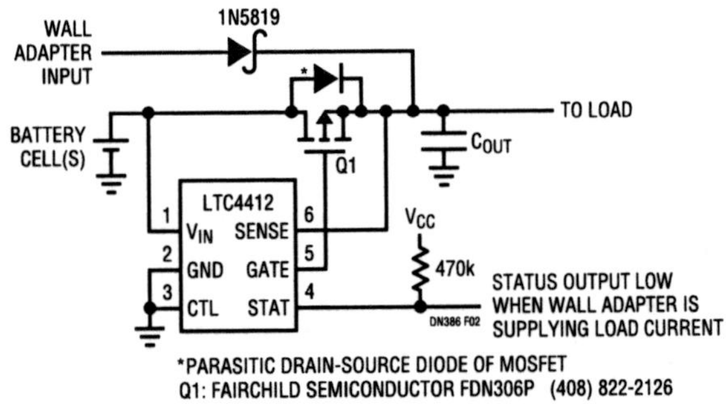

*PARASITIC DRAIN-SOURCE DIODE OF MOSFET
Q1: FAIRCHILD SEMICONDUCTOR FDN306P (408) 822-2126

Figure 190.2 • Automatic Power Switching between a Battery and a Wall Adapter

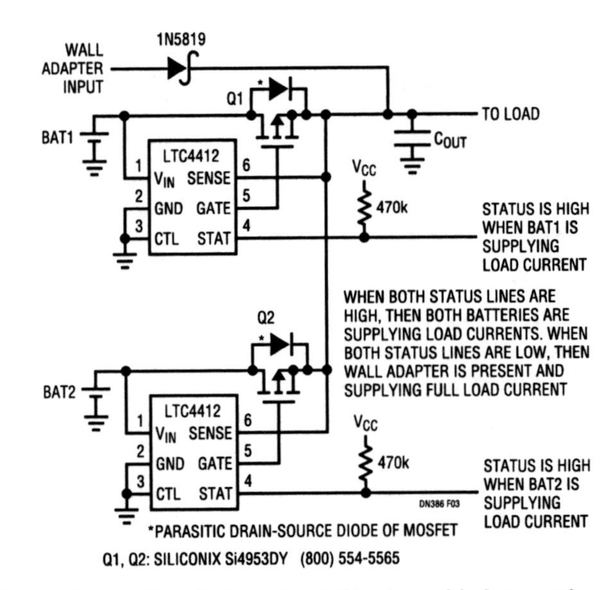

Figure 190.3 • Dual Battery Load Sharing with Automatic Switchover of Power from Batteries to Wall Adapter

Replace ORing diodes with MOSFETs to reduce heat and save space

191

James Herr Mitchell Lee

Introduction

High availability telecom systems employ redundant power supplies or battery feeds to enhance system reliability. Discrete diodes are commonly used to combine these power sources at the point of load. The disadvantage of this approach is the significant forward voltage drop and resulting power dissipation, even with Schottky diodes. This drop also reduces the available supply voltage, which is sometimes critical at the low end of the input operating range. A circuit with "ideal" diode behavior overcomes the dissipation and voltage loss problems by eliminating the forward drop.

The LTC4354 negative voltage diode-OR controller realizes near-ideal diode behavior. External N-channel MOSFETs, actively driven by the LTC4354, act as pass transistors to replace the diodes. The device maintains a small 30mV forward voltage drop across the MOSFETs at light load; under heavy load the voltage drop becomes a function of $R_{DS(ON)}$. For example, an 18mΩ MOSFET and 5A load current produce a drop of 90mV, representing a more than fivefold improvement in drop and power dissipation over a Schottky diode, which exhibits a 500mV drop under the same operating conditions. Lower power dissipation conserves board space and saves the cost of heat sinks. At the same time, 410mV of input operating range is added—a critical factor when the system is running on hold-up capacitors with only a few volts of headroom.

Ideal −48V ORing diode

Figure 191.1 shows a comparison of power dissipation for a diode and a MOSFET driven by the LTC4354. At 10A, the voltage drop across a 100V Schottky diode (MBR10100) is around 620mV; a heat sink is required to handle resulting 6.2W of power dissipation. Using an LTC4354 driving a 100V N-channel MOSFET (IRFR3710Z), the dissipation is only 1.8W due to the low 18mΩ (max) $R_{DS(ON)}$ of the MOSFET. This is easily dissipated in the circuit board with no additional heat sinking.

The LTC4354 implements two ideal diodes, simultaneously controlling two external N-channel MOSFETs with the source pins tied together, as shown in Figure 191.2. This common source node is connected to the V_{SS} pin, the negative supply of the device. Its positive supply is derived from −48V_RTN through an external current limiting resistor. The LTC4354 includes an internal shunt to regulate the V_{CC} pin at 11V.

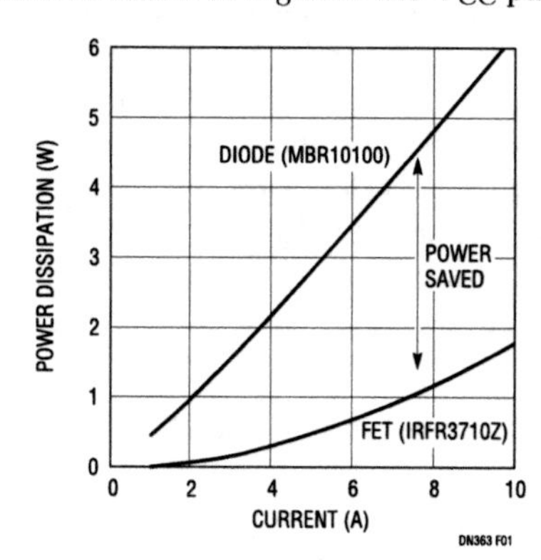

Figure 191.1 • Comparison of Power Dissipation for a Diode and a MOSFET Driven by the LTC4354

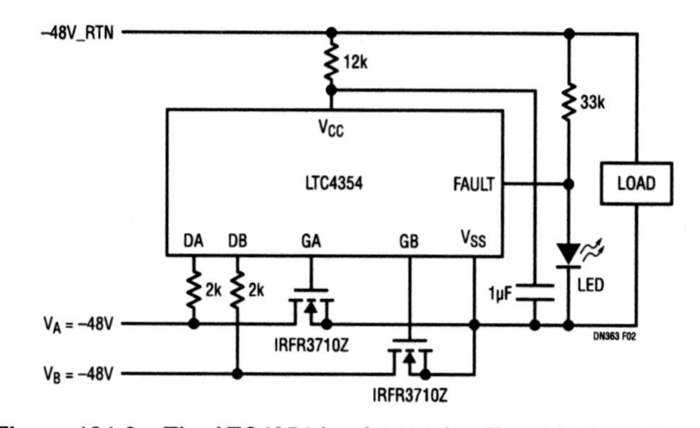

Figure 191.2 • The LTC4354 Implementing Two Ideal Diodes, Controlling Two N-Channel MOSFETs with the Source Pins Tied Together

Analog Circuit Design: Design Note Collection. http://dx.doi.org/10.1016/B978-0-12-800001-4.00191-5

At power-up, the initial load current flows through the body diode of the MOSFET and returns to the supply with the lower (most negative) voltage. The associated gate pin immediately ramps up and turns on the MOSFET. The LTC4354 tries to regulate the voltage drop across the source and drain terminals to 30mV. If the load current causes more than 30mV of drop, the gate is driven higher to further enhance the MOSFET. Eventually the MOSFET is driven fully on and the voltage drop increases as dictated by $R_{DS(ON)} \cdot I_{LOAD}$.

If the power supply voltages are nearly equal, this regulation technique ensures that the load current is smoothly shared between them without oscillation. The current level flowing through each MOSFET depends on the $R_{DS(ON)}$ of the MOSFETs, the output impedance of each supply and distribution resistance.

In the case of supply failure, such as an input supply short to −48V_RTN, a potentially large reverse current could flow from the −48V_RTN through the MOSFET that is on. The LTC4354 detects this condition as soon as it appears and turns off the MOSFET in less than 1μs. This fast turn-off prevents reverse current from reaching a damaging level, exhibiting a behavior not unlike a discrete diode with a recovery time measured in hundreds of nanoseconds.

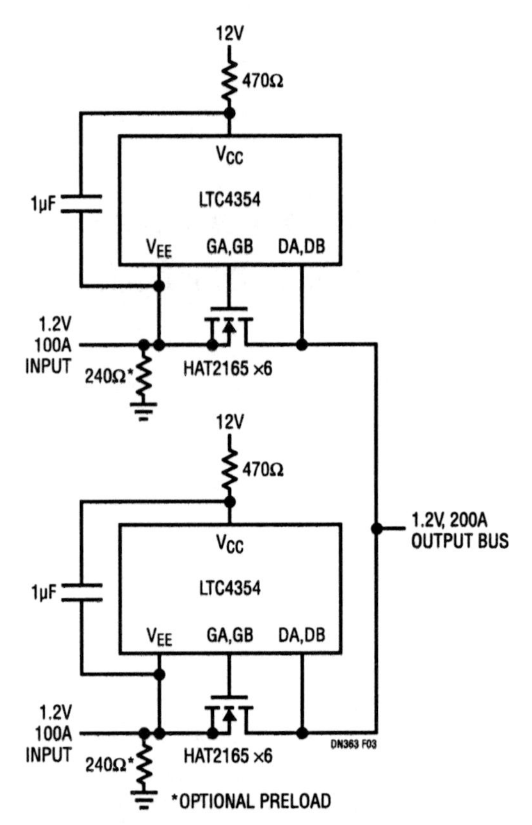

Figure 191.3 • Positive Low Voltage Diode-OR Combines Multiple Switching Converters

Fault output detects damaged MOSFETs and fuses

The LTC4354 monitors each MOSFET and reports any excessive forward voltage that is indicative of an overcurrent fault. When the pass transistor is fully on but the voltage drop across it exceeds the 260mV fault threshold, the open-drain FAULT pin goes high. This allows an LED or optocoupler to turn on and flag the system controller. It is important to recognize excessive voltage drop in the MOSFETs because extra heat is being dissipated. If the condition persists the system controller can take action and shut down the load.

Positive low voltage ideal diodes

LTC4354 is also suited for positive, low voltage applications, as shown in Figure 191.3. With this circuit the outputs of multiple high current switching converters can be combined, without concern about back feeding or supply failure shorting out the common bus. One diode "channel" comprises the LTC4354 and six parallel MOSFETs, supplying 100A to a 1.2V load. The circuit is easily adapted to any supply voltage between 0V and 5V, provided there is a path for up to 4mA V_{EE} current to ground at either the input or the output. Most high current switching converters can easily sink 4mA and no preload is necessary. No circuit changes are necessary for different operating voltages.

Conclusion

The trend in today's telecom infrastructure is toward higher current and smaller module space. Traditional diode ORing is increasingly cumbersome. The LTC4354 provides an improved solution by controlling low $R_{DS(ON)}$ N-channel MOSFETs to reduce power dissipation and save board space and heat sinks, on both sides of the isolation barrier. Furthermore, the LTC4354 monitors and reports fault conditions, information not provided by a traditional diode-OR circuit.

Dual monolithic ideal diode manages multiple power inputs

192

Andy Bishop

Introduction

The LTC4413 dual monolithic ideal diode helps reduce the size and improve the performance and reliability of handheld battery-operated devices. The LTC4413 is a single-chip solution that can automatically select between and isolate up to three power sources; such as a wall adapter, an auxiliary supply and a battery. It provides a low loss automatic PowerPath management solution for demanding applications that may require short-circuit protection, thermal management and system-level power management and control.

The LTC4413 contains two isolated low voltage (2.5V to 5.5V) monolithic ideal diodes. Each ideal diode channel provides a low forward voltage drop (typically as low as 40mV when conducting low current) and a low $R_{DS(ON)}$ (below 100mΩ) when conducting high current—features that are important to extend battery life and reduce heat in portable applications. Furthermore, each channel is capable of providing up to 2.6A of continuous current from a small 3mm × 3mm, 10-pin DFN package. Current-limit and thermal shutdown features further enhance system reliability.

Triple supply power management

Figure 192.1 shows a schematic with the LTC4413 configured to automatically switchover from a battery to either a USB supply or to a wall adapter. This circuit provides uninterrupted power to the load—while isolating all three power sources—allowing the user to remove the battery without affecting load voltage when either of the two other supplies is present. This circuit exploits the LTC4413 to provide low loss uninterrupted power while automatically prioritizing which source should be connected to the load.

Referring to Figure 192.1, if a wall adapter is applied in the absence of a USB supply, the body diode in MP1 forward biases, pulling the output voltage above the battery voltage and turning off the ideal diode connected between INA and OUTA. This causes the STAT voltage to fall, turning on MP1.

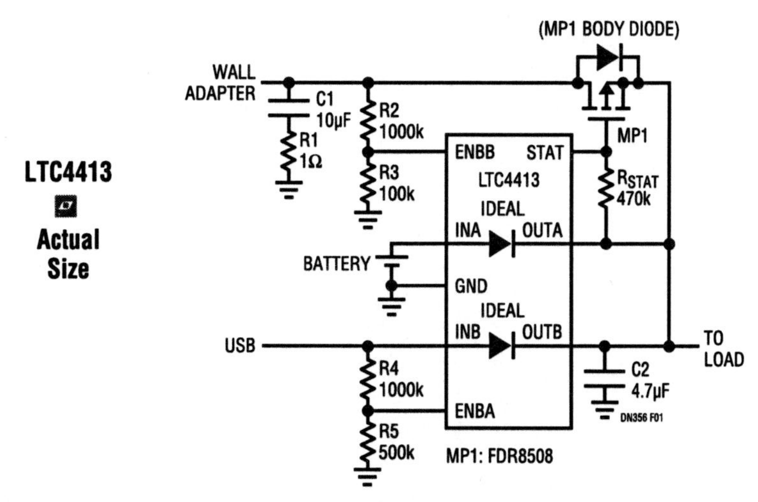

Figure 192.1 • Automatic Switchover from a Battery to a USB Supply or Wall Adapter

Analog Circuit Design: Design Note Collection. http://dx.doi.org/10.1016/B978-0-12-800001-4.00192-7

The load then draws current from the wall adapter and the battery is disconnected from the load. If the USB is present when the wall adapter is removed, the output voltage falls until the USB voltage exceeds the output voltage causing the STAT voltage to rise, disabling the external PFET; the USB then provides power to the load. In the absence of a USB supply when the wall adapter is removed, the load voltage droops until the battery voltage exceeds the load voltage, likewise causing the STAT voltage to rise and disabling MP1; the battery then provides the load power.

When a USB supply is applied, the voltage divider at ENBA disables the power path from battery to OUT. The USB then provides load current, unless a wall adapter is present as described above.

Automatic switchover between a battery and a wall adapter with a battery charger

Figure 192.2 illustrates an application where the LTC4413 performs the function of automatically switching a load over from a battery to a wall adapter, while controlling an LTC4059A battery charger. When no wall adapter is present, the LTC4413 connects the load to the Li-Ion battery. In this configuration, the STAT voltage is high, thereby disabling the battery charger. If a wall adapter is connected, the load voltage rises as the ideal diode from INB to OUTB conducts. As soon as the load voltage exceeds the battery voltage, the battery is disconnected from the load and the STAT voltage falls, turning on the LTC4059A battery charger and beginning a charge cycle.

When the wall adapter is removed, the load voltage collapses until it is below the battery voltage. The battery reverts to supplying power to the load and the STAT pin falls disabling the battery charger.

Conclusion

The LTC4413 performs automatic PowerPath management functions in high performance battery-powered applications. The applications described herein have the added benefit that while an alternate supply powers the load, the battery may be replaced without disturbing load voltage. This feature demonstrates the benefit of the LTC4413 as compared with alternative solutions that do not allow the battery to be replaced without impacting load power.

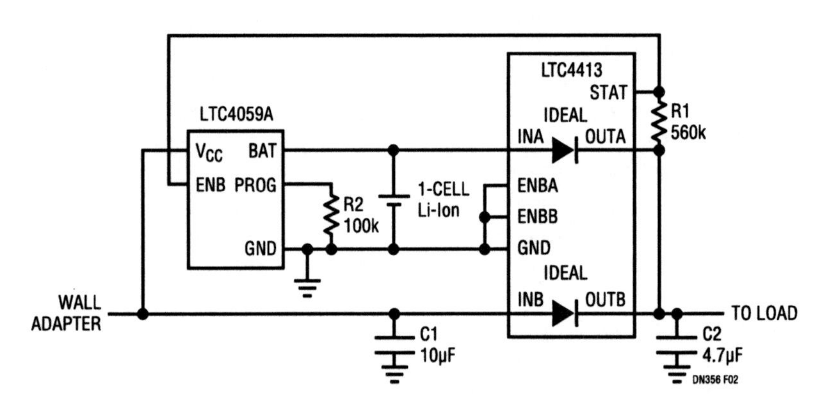

Figure 192.2 • Automatic Switchover from a Battery to a Wall Adapter with a Battery Charger

193

PCMCIA socket voltage switching

Why your portable system needs SafeSlot protection

Doug La Porte

Introduction

Most portable systems have built-in PCMCIA sockets as the sole means of expansion. The requirements of the PCMCIA specification have led to some confusion among system designers. This Design Note will attempt to lessen the confusion and highlight other practical system issues.

Host power delivery to the PC card socket flows through two paths: the main V_{CC} supply pins and the VPP programming pins. Both supplies are switchable to different voltages to accommodate a wide range of card types. The V_{CC} main card supply must be capable of delivering up to 1A at either 3.3V or 5V. The 1A rating is an absolute maximum derived from the contact rating of 500mA per pin for both V_{CC} pins and assumes that both pins are in good condition and current is shared equally. One of the most stringent actual current requirements is during hard drive spin-up. Present hard drives require 5V at 600mA to 800mA for a short duration during spin-up. Current draw drops to 300mA to 420mA during read and write operations. A low switch resistance on the 3.3V switch is critical to assure that the specified 3.0V minimum is maintained. The VPP supply must source 12V at up to 120mA and 3.3V or 5V at lesser currents. The VPP supply is intended solely for flash memory programming. The 120mA current requirement allows writing to flash devices and simultaneously erasing two other parts as required by many flash drives.

The host PCMCIA socket designer also has several other practical aspects of the design to consider. The exposed socket pins are vulnerable to being shorted by foreign objects such as paper clips. In addition the users will attempt to install damaged cards. In short, once in the hands of the consumer, the designer and manufacturer have little control over use and abuse. To ensure a robust system and a satisfied customer, switch protection features such as current limiting

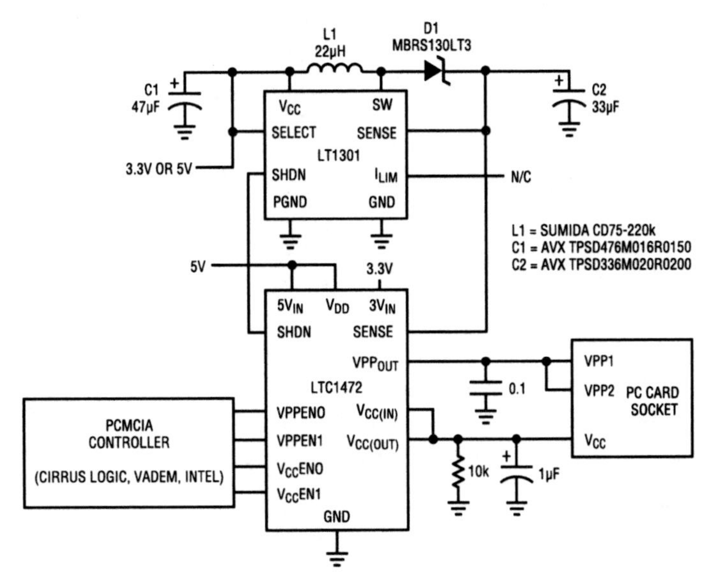

Figure 193.1 • Typical LTC1472 Application with the LT1301 3.3V Boost Regulator

Analog Circuit Design: Design Note Collection. http://dx.doi.org/10.1016/B978-0-12-800001-4.00193-9

and thermal shutdown are a necessity. The nature of the PC cards and portable systems requires the card being powered on and off as needed to conserve power. Many PC cards have large input capacitance and draw over 2W. The power up/down sequencing can put demanding transient requirements on your system power supply. To make the transient response of the system supply manageable, the PCMCIA switch should have break-before-make switching, controlled rise and fall times and current limiting. The slowed rise time coupled with current limiting are critical in controlling the immense in-rush current difficulties seen when charging the large input capacitance of many cards.

LTC1472: complete V_{CC} and VPP PCMCIA switch matrix with SafeSlot protection

The LTC1472 is a complete, fully integrated V_{CC} and VPP switch matrix that addresses all of the PCMCIA socket switching needs. Figure 193.1 shows a typical LTC1472 application used in conjunction with the LT1301 to supply 12V for flash memory programming. The LTC1472's logic inputs allow direct interfacing with both logic high and logic low industry standard controllers without any external glue logic.

Table 193.1 LTC1472 Truth Table

VCC Switch		
V_{CC}ENO	**V_{CC}EN1**	**$V_{CC(OUT)}$**
0	0	Off
1	0	5V
0	1	3.3V
0	1	Off
VPP Switch		
VPPENO	**VPPEN1**	**VPP_{OUT}**
0	0	OV
0	1	$V_{CC(IN)}$
1	0	VPP_{IN}
1	1	Hi-Z

The LTC1472 is available in the space saving narrow 16-pin SOIC package. The V_{CC} switch's $R_{DS(ON)}$ is 0.14Ω to support the 1A current requirement. The V_{CC} output is switched between 3.3V, 5V and high impedance. The VPP output pin is switched between 0V, V_{CC}, 12V and high impedance. Table 193.1 shows the V_{CC} and VPP truth tables.

The LTC1472 features SafeSlot protection. The built-in SafeSlot current limiting and thermal shutdown features are vital to ensuring a robust and reliable system. The V_{CC} current limit is above 1A to maintain compatibility with all existing cards yet provide protection. The VPP current limit is above 120mA to also maintain compatibility. All switches are break-before-make type with controlled, slowed rise and fall times for minimal system supply impact. In-rush current, from even the largest card input capacitance, is kept under control by the LTC1472's slowed rise-time switching and current limiting.

The LTC1472 has on-chip charge pumps for switch driving. For this reason, the device does not require a continuous 12V source. Most of the time the LT1301 is in shutdown mode, consuming only 10μA. The LT1301 becomes operational only during flash memory programming. The LTC1472 itself conserves power by going to a low 1μA standby mode when V_{CC} and VPP outputs are switched off. The use of the LT1301 is optional. Any suitable 12V supply can be directly connected to the VPP_{IN} pin. Caution should be exercised when using a general purpose 12V supply; make certain that it does not have spikes or transients exceeding the flash memory 14V maximum voltage rating and that the regulation is within the 5% flash memory tolerance.

Conclusion

PCMCIA sockets are the preferred method of expansion in portable systems. As these devices proliferate to less sophisticated users, there will be greater opportunity for abuse. To counter this trend the portable system design must take safeguards to protect the system. The high level of integration, SafeSlot protection features and controlled rise and fall switching make the LTC1472 the ideal solution for portable systems.

Linear Technology has a family of PCMCIA socket voltage control products to suit a broad range of customer's needs. Table 193.2 lists the present part offerings. For assistance on your specific design needs, call Linear Technology.

Table 193.2 Linear Technology's PCMCIA Host Socket Voltage Control Products

PART NUMBER	REMARKS
LT1312	VPP Linear Regulator
LT1313	Dual Slot VPP Linear Regulator
LT1314	Low Cost V_{CC} and VPP Switch Matrix
LT1315	Dual Low Cost V_{CC} and VPP Switch Matrix
LTC1470	Complete SafeSlot Protected V_{CC} Switch Matrix
LTC1471	Dual Complete SafeSlot Protected V_{CC} Switch Matrix
LTC1472	Complete SafeSlot Protected V_{CC} and Switch Matrix

PC card power management techniques

194

Tim Skovmand

Most portable computers have sockets built in to accept small PC cards for use as extended memories, data/fax modems, network interfaces, wireless communicators, etc. The Personal Computer Memory Card International Association (PCMCIA) has released specifications, 1.0 and 2.0, which outline the general voltage and power requirements for these cards.

Power is provided by the host computer to the PC card through the card socket via the main V_{CC} supply pin(s) and the VPP programming supply pins. Both supplies can be switched to different voltages to accommodate a wide range of card types and applications.

The V_{CC} supply can be switched from 5V to 3.3V and must be capable of supplying upwards of 1A for short periods of time and hundreds of milliamps continuously. Three low resistance MOSFET switches are typically used to select the card V_{CC} power as shown in Figure 194.1. The LTC1165 inverting triple MOSFET driver accepts active-low logic commands directly from a common PCMCIA controller and generates gate drive voltages above the positive rail to fully enhance low $R_{DS(ON)}$ N-channel MOSFET switches. Two back-to-back MOSFET switches, Q2 and Q3, isolate the parasitic body diode in Q2.

The LTC1165 drives the three MOSFET gates at roughly the same rate producing a smooth transition between supply voltages. Further, the LTC1165 provides a natural break-before-make action due to the asymmetry between the turn-on times and the turn-off times; i.e., no external delays are required. A noninverting version, the LTC1163, is also available. Both devices are available in 8-lead surface mount packaging.

The second path for card power is via the two VPP programming pins on each card socket which are typically tied together. These two pins were originally intended for programming

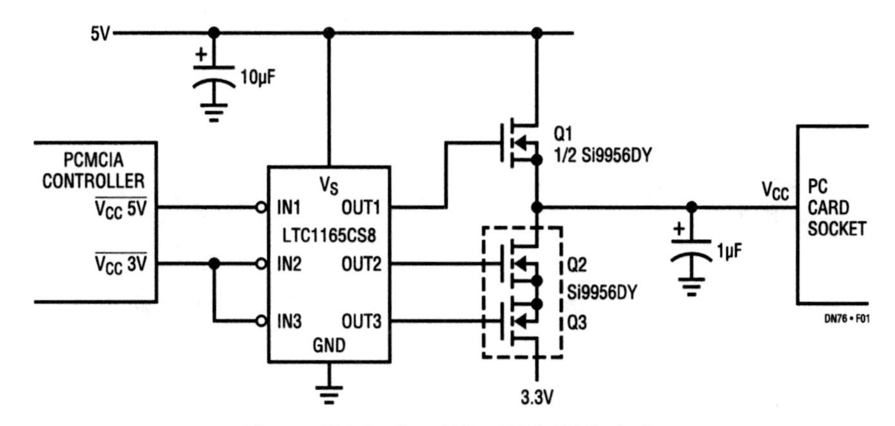

Figure 194.1 • Card V_{CC} 5V, 3.3V Switch

Analog Circuit Design: Design Note Collection. http://dx.doi.org/10.1016/B978-0-12-800001-4.00194-0

flash memories but are sometimes used as an alternate power source for the card. The VPP supply voltage is therefore capable of being switched between four operating states: 12V, V_{CC}, 0V and Hi-Z. Figures 194.2 and 194.3 are two different approaches to solving the four state output problem. Figure 194.2 shows a circuit that produces 12V "locally" by converting the incoming V_{CC} supply through a step-up converter for programming flash memories, etc. This circuit has the unique capability of supplying up to 500mA when the VPP pin is programmed to V_{CC}. Figure 194.3 is a switched output voltage linear regulator which is powered from a auxiliary 13V to 20V unregulated supply. This circuit supplies 120mA at 12V and protects the card slot from overcurrent damage. The supply currents are listed in the truth table to the right of each schematic. All components shown, including the integrated circuits, are available in surface mount packaging.

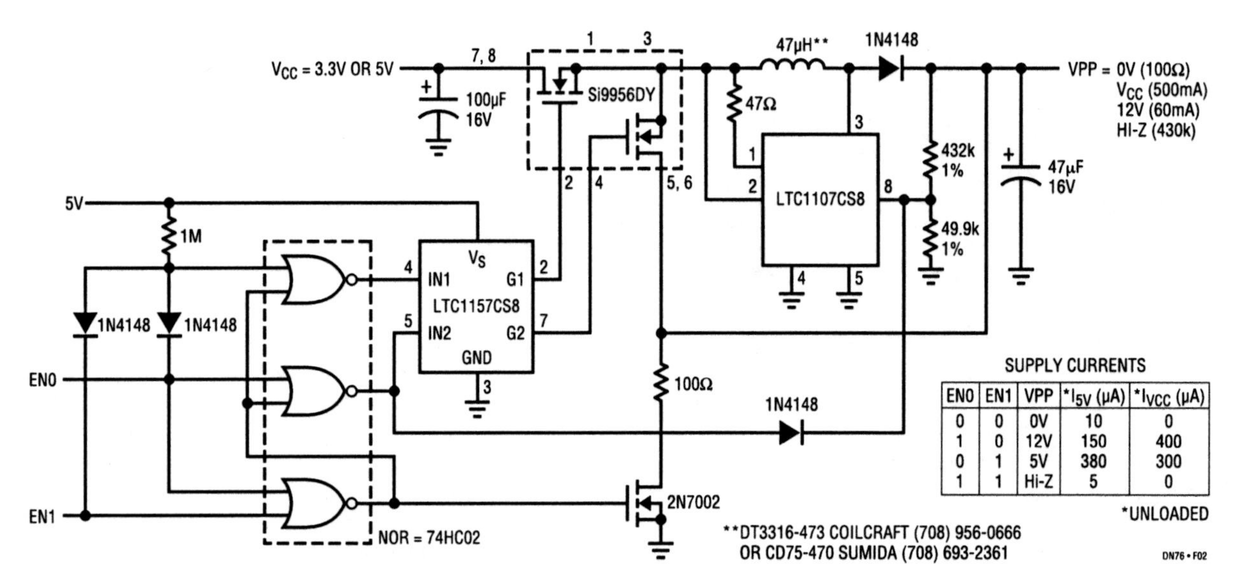

Figure 194.2 • Step-Up Regulator VPP Power Management

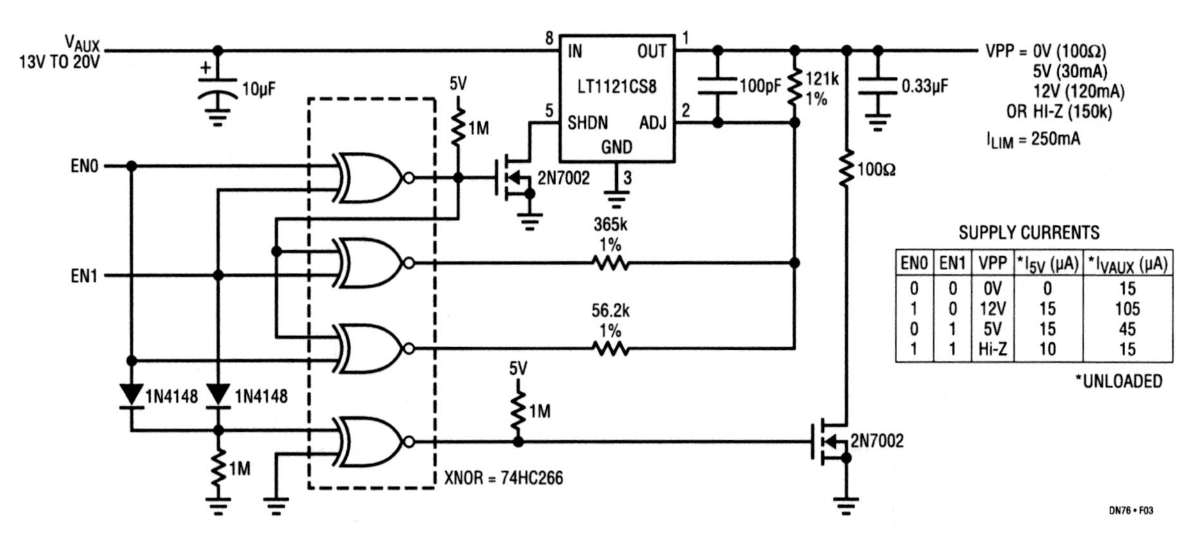

Figure 194.3 • Current-Limited Linear Regulator VPP Power Management

Section 12

Battery Management

Complete battery charger solution for high current portable electronics

3.5A charger for Li-Ion/LiFePO$_4$ batteries multiplexes USB and wall inputs

195

George H. Barbehenn

Introduction

The LTC4155 and LTC4156 are dual multiplexed-input battery chargers with PowerPath control, featuring I^{2}C programmability and USB On-The-Go for systems such as tablet PCs and other high power density applications. The LTC4155's float voltage (V_{FLOAT}) range is optimized for Li-Ion batteries, while the LTC4156 is optimized for lithium iron phosphate (LiFePO$_4$) batteries, supporting system loads to 4A with up to 3.5A of battery charge current. I^{2}C controls a broad range of functions and USB On-The-Go functionality is controlled directly from the USB connector ID pin.

Input multiplexer

A distinctive feature of the LTC4155/LTC4156 PowerPath implementation is a dual-input multiplexer using only N-channel MOSFETs controlled by an internal charge pump. The input multiplexer has input priority selection and independent input current limits for each channel.

Applications include any device with a high capacity Li-Ion or LiFePO$_4$ battery that can be charged from a high current wall adapter input or from the USB input—such as a tablet PC. Figure 195.1 shows a typical application and Figure 195.2 shows its efficiency.

This dual input multiplexer implementation allows the use of inexpensive, low $R_{DS(ON)}$ N-channel MOSFETs. The

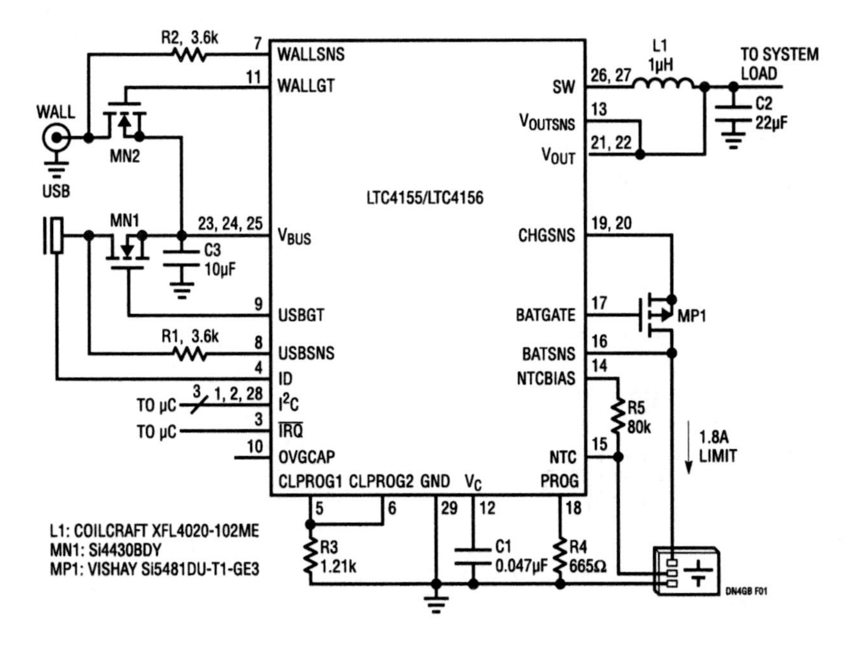

Figure 195.1 • Typical Application Using a Simple Input Multiplexer with No Backdrive Protection

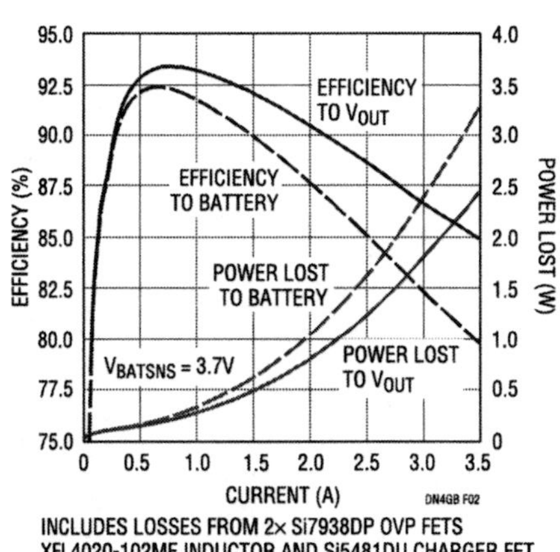

Figure 195.2 • Switching Regulator Efficiency

Analog Circuit Design: Design Note Collection. http://dx.doi.org/10.1016/B978-0-12-800001-4.00195-2

MOSFETs also provide overvoltage protection (OVP) and if desired, backdrive blocking and reverse-voltage protection (RVP). Backdrive blocking prevents voltage on the wall input from backdriving the USB, or vice-versa. Backdrive blocking can be implemented on one or both inputs. Reverse-voltage protection prevents a negative voltage applied to the protected input from reaching downstream circuits.

Dual high current input application

Figure 195.3 shows a dual 3.5A input application, featuring OVP, RVP and backdrive protection. The FDMC8030 MOSFETs provide ±40V of OVP and RVP protection.

0V ~ 6V input on either WALL or USB

In the circuit of Figure 195.3, when a 0V ~ 6V input is present on either input, the corresponding gate signal rises to approximately twice V_{IN}, enabling the two series N-channel MOSFETS and connecting the input to V_{BUS}. The undervoltage lockout (UVLO) is approximately 4.35V on each channel.

The LTC4155/LTC4156 have an input priority bit, which defaults to WALL. If a valid voltage is present on both inputs, only WALLGT is activated. The input priority bit can be changed via I^2C to make the USB channel preferred when both inputs are present.

>6V input on either WALL or USB

When either input goes above 6V, the corresponding WALLGT or USBGT pin is pulled low, shutting off the corresponding MOSFETs and disconnecting the input. If both inputs have 5V on them, and the input that is enabled by the input priority bit goes above 6V, the LTC4155/LTC4156 automatically and seamlessly switches to the other input—with no disturbance on V_{OUT}.

The diode-connected NPNs (Q3 and Q4) serve to prevent excess V_{GS} on the MOSFET closest to the input of the corresponding channel current from the input flows through the diode, through the B-C junction of the NPN bipolar transistor, and pulls the gate up through the 5M resistor. This prevents the gate from dropping below the source by more than 2V.

A voltage greater than 6V on one input does not prevent the other input from operating normally.

<0V input on either WALL or USB

The USBSNS and WALLSNS pins ignore any negative inputs, but clamp the pins to $-V_F$ (about −0.6V). The negative voltage forward-biases the base-emitter junction of the NPN bipolar transistor, shorting the gate to the input and ensuring that the gate is never more than about 0.5V above the source.

A negative voltage on one input does not prevent the other input from operating normally.

OTG operation

The LTC4155/LTC4156 drive the USBGT pin high when USB On-The-Go operation is enabled, connecting V_{BUS} to the USB input and enabling up to 500mA to be sourced.

Conclusion

The LTC4155/LTC4156 implement an overvoltage and reverse-voltage protected, prioritized input multiplexer for products that need to support multiple system power or battery charging functionality. Optional backdrive blocking prevents the appearance of voltages at an unconnected input.

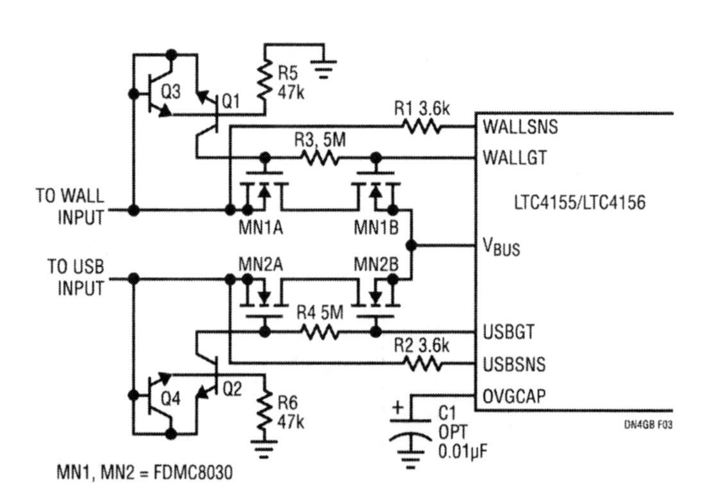

Figure 195.3 • LTC4155/LTC4156 Input Multiplexer with OVP, RVP and Backdrive Blocking

196

Battery conditioner extends the life of Li-Ion batteries

George H. Barbehenn

Introduction

Li-Ion batteries naturally age, with an expected lifetime of about three years. But, that life can be cut very short–to under a year–, if the batteries are mishandled. It turns out that the batteries are *typically* abused in applications where intelligent conditioning would otherwise significantly extend the battery lifetime. The LTC4099 battery charger and power manager contains an I²C controlled battery conditioner that maximizes battery operating life, while also optimizing battery run time and charging speed (see Figure 196.1).

The underlying aging process in Li-Ion batteries

Modern Li-Ion batteries are constructed of a graphite negative terminal, cobalt, manganese or iron phosphate positive terminal and an electrolyte that transports the lithium ions.

The electrolyte may be a gel, a polymer (Li-Ion/Polymer batteries) or a hybrid of a gel and a polymer. In practice, no suitable polymer has been found that transports lithium ions effectively at room temperature. Most 'pouch' Li-Ion/Polymer batteries are in fact hybrid batteries containing a combination of polymer and gel electrolytes.

The charge process involves lithium ions moving out of the negative terminal material, through the electrolyte and into the positive terminal material. Discharging is the reverse process. Both terminals either release or absorb lithium ions, depending on whether the battery is being charged or discharged.

The lithium ions do not bond with the terminals, but rather enter the terminals much like water enters a sponge; this process is call 'intercalation.' So, as is often the case with charge-based devices such as electrolytic capacitors, the resulting charge storage is a function of both the materials used and the physical structure of the material. In the case of the electrolytic capacitor, the foil is etched to increase its surface area. In the case of the Li-Ion battery the terminals must have a sponge-like physical makeup to accept the lithium ions.

The choice of positive terminal material (cobalt, manganese or iron phosphate) determines the capacity, safety and aging properties of the battery. In particular, cobalt provides superior capacity and aging characteristics, but it is relatively unsafe compared to the other materials. Metallic lithium is flammable and the cobalt positive terminal tends to form metallic lithium during the discharge process. If several safety

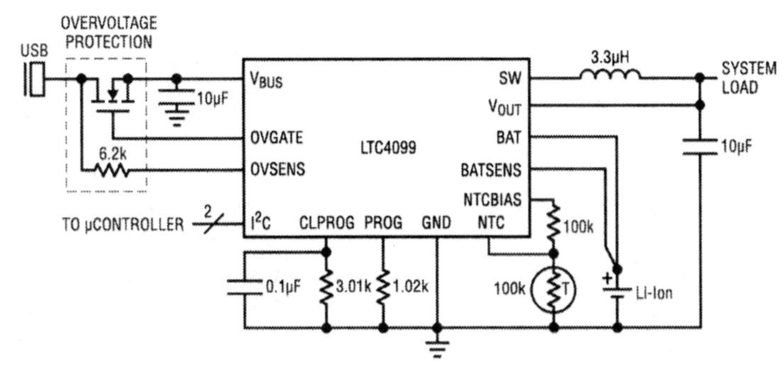

Figure 196.1 • The LTC4099 with I²C Controlled Battery Conditioner

Analog Circuit Design: Design Note Collection. http://dx.doi.org/10.1016/B978-0-12-800001-4.00196-4

measures fail or are defeated, the resulting metallic lithium can fuel a "vent with flame" event.

Consequently, most modern Li-Ion batteries use a manganese or iron phosphate-based positive terminal. The price for increased safety is slightly reduced capacity and increased aging.

Aging is caused by corrosion, usually oxidation, of the positive terminal by the electrolyte. This reduces both the effectiveness of the electrolyte in lithium-ion transport and the sponge-like lithium-ion absorption capability of the positive terminal. Battery aging results an increase of the battery series resistance (BSR) and reduced capacity, as the positive terminal is progressively less able to absorb lithium ions.

The aging process begins from the moment the battery is manufactured and cannot be stopped. However, battery handling plays an important role in how quickly aging progresses.

Conditions that affect the aging process

The corrosion of the positive terminal is a chemical process and this chemical process has an activation energy probability distribution function (PDF). The activation energy can come from heat or the terminal voltage. The more activation energy available from these two sources the greater the chemical reaction rate and the faster the aging.

Li-Ion batteries that are used in the automotive environment must last 10 to 15years. So, suppliers of automotive Li-Ion batteries do not recommend charging the batteries above 3.8V. This does not allow the use of the full capacity of the battery, but is low enough on the activation energy PDF to keep corrosion to a minimum. The iron phosphate positive terminal has a shallower discharge curve, thus retaining more capacity at 3.8V.

Battery manufacturers typically store batteries at 15°C (59F) and a 40% state of charge (SoC), to minimize aging. Ideally, storage would take place at 4% or 5% SoC, but it must never reach 0%, or the battery may be damaged. Typically, a battery pack protection IC prevents a battery from reaching 0% SoC. But pack protection cannot prevent self-discharge and the pack protection IC itself consumes some current. Although Li-Ion batteries have less self-discharge than most other secondary batteries, the storage time is somewhat open-ended. So, 40% SoC represents a compromise between minimizing aging and preventing damage while in storage (see Figure 196.2).

In portable applications, the reduction in capacity from such a reduced SoC strategy is viewed negatively in marketing specifications. But it is sufficient to detect the combination of high ambient heat and high battery SoC to implement an algorithm that minimizes aging while ensuring maximum capacity availability to the user.

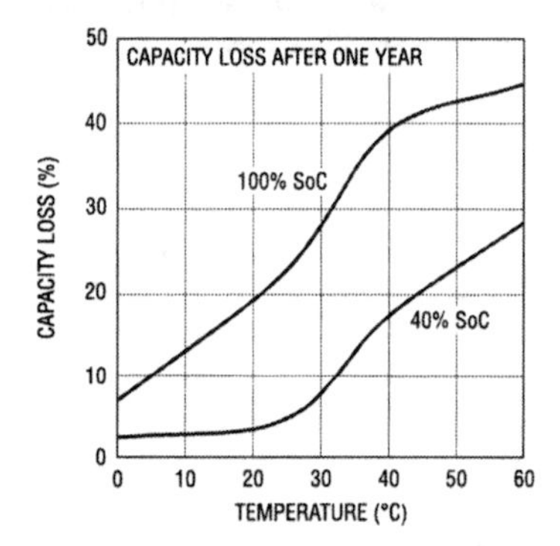

Figure 196.2 • Yearly Capacity Loss vs Temperature and SoC for Li-Ion Batteries

Battery conditioner avoids conditions that accelerate aging

The LTC4099 has a built-in battery conditioner that can be enabled or disabled (default) via the I^2C interface. If the battery conditioner is enabled and the LTC4099 detects that the battery temperature is higher than ~60°C, it gently discharges the battery to minimize the effects of aging. The LTC4099 NTC temperature measurement is always on and available to monitor the battery temperature. This circuit is a micropower circuit, drawing only 50nA while still providing full functionality.

The amount of current used to discharge the battery follows the curve shown in Figure 196.3, reaching zero when the battery terminal voltage is ~3.85V. If the temperature of the battery pack drops below ~40°C and a source of energy is available, the LTC4099 once again charges the battery. Thus, the battery is protected from the worst-case battery aging conditions.

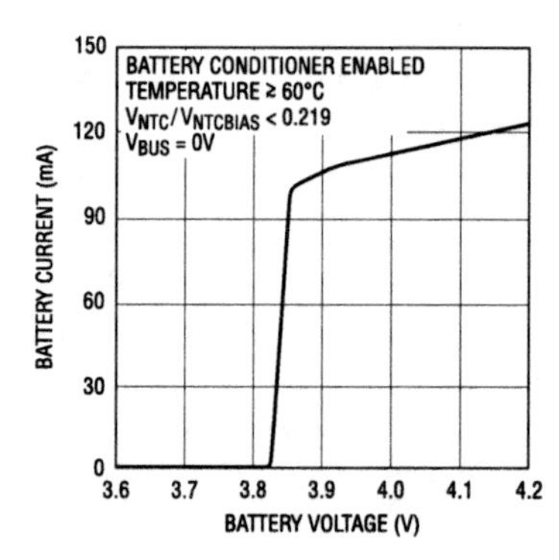

Figure 196.3 • Battery Discharge Current vs Voltage for the LTC4099 Battery Conditioning Function

Conclusion

Although the aging of Li-Ion batteries cannot be stopped, the LTC4099's battery conditioner ensures maximum battery life by preventing the battery-killing conditions of simultaneous high voltage and high temperature. Further, the micropower, always on NTC monitoring circuit ensures that the battery is protected from life-threatening conditions at all times.

Simple calibration circuit maximizes accuracy in Li-Ion battery management systems

197

Jon Munson

Introduction

In Li-Ion battery systems it is important to match the charge condition of each cell to maximize pack performance and longevity. Cell life improves by avoiding both deep discharge and overcharge, so typical systems strive for operation between 20% and 80% states of charge (SOC). Detection and correction of charge imbalances assures that all cells track within the desired SOC window, preventing premature aging of some cells that could compromise the entire pack capacity. Highly accurate measurements are required to determine SOC with

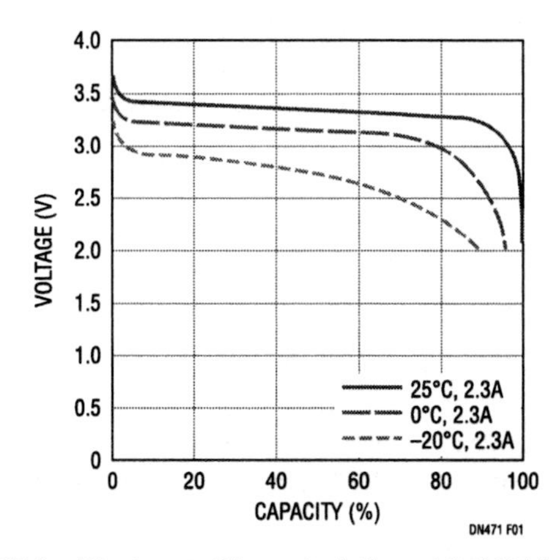

Figure 197.1 • Discharge Characteristics of 3.3V Li-Ion Cell

Li-Ion cells due to their exceptionally flat discharge characteristics, particularly with the lower voltage chemistries (see example in Figure 197.1).

Although the popular LTC6802 Battery Stack Monitor offers high accuracy analog-to-digital conversion, some applications demand accuracy that is only attainable with a dedicated voltage reference IC. The LT1461 is especially

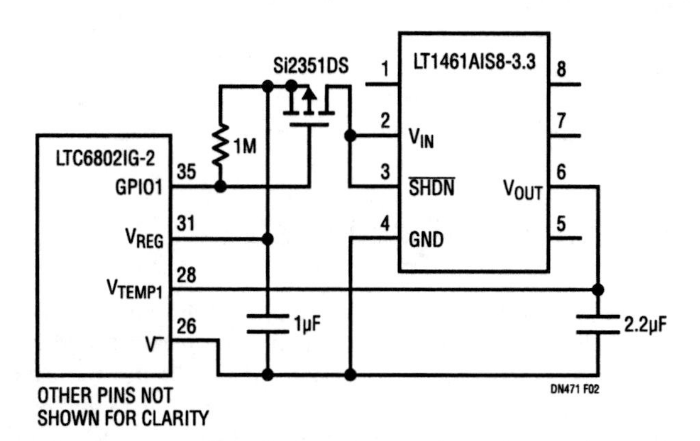

Figure 197.2 • LT1461 as an External Calibration Source for an LTC6802 Li-Ion Battery Monitor

suited as a high performance calibration source, available in the small SO-8 package. Figure 197.2 illustrates this configuration. The calibration reference is measured with an ADC channel normally intended for temperature measurement. A programmable I/O bit controls power to the reference.

Accounting for the error sources

Fundamentally, there are several key characteristics that comprise an overall accuracy specification:

- Quantization error of the ADC
- Initial accuracy of the ADC (or calibration reference)
- Variation from channel to channel
- Variation with temperature
- Hysteresis effects, primarily that of the soldering process
- Variation with operating time (long-term drift)

The maximum specified error in the data sheet for the LTC6802IG-2 includes the first four items and is ±0.22%; about ±7mV when measuring 3.3V, the most demanding

Analog Circuit Design: Design Note Collection. http://dx.doi.org/10.1016/B978-0-12-800001-4.00197-6

region of the discharge curve. The spec budgets ±3.3mV (±0.1%) as the maximum variation over the −40°C to 85°C operating temperature. Since the differential non-linearity (DNL) of the ADC is about ±0.3 LSB, the quantization error contribution is about ±0.8 LSB, or ±1.2mV. Typical channel-to-channel variation is minimal, under ±1mV, leaving about ±1.5mV for trim resolution and accuracy in the IC manufacturing process. Thermal hysteresis is specified as 100ppm, and an additional approximately ±0.1% error may develop from the shift of the printed-circuit soldering process.

Projected typical long-term drift is under $60ppm/\sqrt{khr}$. If the practical vehicle battery system active life cycle is targeted at 5khr (about 15years or 150,000miles), an uncertainty of around ±0.5mV could develop. This is a relatively small contribution to total error.

The LT1461AIS8-3.3 voltage reference IC has an output tolerance of ±0.04% and less than ±1.2mV of change over temperature with its exemplary 3ppm/°C worst-case stability. The LT1461 exhibits a long-term drift of under $60ppm/\sqrt{kHr}$ and thermal hysteresis of 75ppm. Solder reflow shift is expected to be under 250ppm (±0.8mV).

Since a significant portion of the LTC6802 ADC error accumulates after the initial delivery of the IC, an external calibration technique improves accuracy in a finished product.

Examining calibration strategies

There are a number of options to improve system accuracy, at the expense of additional complexity. With the simple circuit of Figure 197.2, several options are available that take advantage of the external calibration reference. Accuracy projections of several methods are tabulated in Table 197.1 and described below.

The simplest scheme (method 1) involves no local memory or measurements at production. This method takes readings of the nominal 3.300V calibration voltage periodically and normalizes all ADC readings with the same computed correction factor. The tolerance and drift of the reference and channel-to-channel variations are left uncorrected but the net uncertainty would be improved by almost a factor of 2, to ±6.2mV.

A slightly more complex technique (method 2) involves storage of a single correction factor that accounts for the true reference voltage as measured with high accuracy test-fixture instrumentation. This then eliminates the initial error of the LT1461, improving the overall accuracy to ±4.1mV, nearly a 3× total improvement.

While small, there is still some channel-to-channel variation that can be calibrated out with a method that uses more initial test-fixture measurements (method 3). This is similar to method 2, but with high accuracy measurements of every channel taken (including the reference) and the saving of individual correction factors for each. This further reduces the error to ±3.1mV (almost a 4× total improvement).

Conclusion

A precision voltage reference, such as the LT1461, can improve the accuracy of an LTC6802-based battery management system to about ±3mV worst-case. The reference is a simple addition to the highly integrated LTC6802 Li-Ion monitoring solution, thanks to the spare general-purpose ADC channels available. The low operating current of the LT1461 voltage reference also makes it ideal for this and other battery-powered applications.

Reference

"Battery Stack Monitor Extends Life of Li-Ion Batteries in Hybrid Electric Vehicles," Linear Technology Magazine, Vol. 19, No. 1, 2009, page 1.

Table 197.1 Accuracy of Calibration Methods Described for 3.3V Measurements

EXTERNAL CALIBRATION METHOD (ALL TOLERANCES SHOWN IN ±MV)	QUANTIZATION	FACTORY TRIM	SOLDERING SHIFT	CHANNEL MATCH	THERMAL VARIATION	THERMAL HYSTERESIS	LONG-TERM DRIFT	TOTAL ERROR
LTC6802 Without External Calibration	1.2	1.5	3.3	1.0	3.3	0.3	0.5	11.1
1: Calibration with LT1461, No Stored Information	1.2	1.3	0.8	1.0	1.2	0.2	0.5	6.2
2: Calibration with LT1461, Store Calibration Values for Reference Voltage	1.2	–	–	1.0	1.2	0.2	0.5	4.1
3: Calibration with LT1461, Store Calibration Values for Reference Voltage and Each Input	1.2	–	–	–	1.2	0.2	0.5	3.1

USB power solution includes switching power manager, battery charger, three synchronous buck regulators and LDO

198

Brian Shaffer

Introduction

Linear Technology offers a variety of parts to simplify the task of extracting power from a battery or a USB cable. These devices seamlessly manage the power flow between an AC adapter, USB cable and Li-Ion battery, all while maintaining USB power specification compliance. As battery capacities rise, battery chargers must keep pace by steadily improving efficiency to minimize thermal concerns and charge times. A USB-based battery charger must squeeze as much power from the USB as possible, and do it efficiently to meet the stringent space and thermal constraints of today's power-intensive applications.

The LTC3555 combines a USB switching power manager and battery charger with three synchronous buck regulators and an LDO to provide a complete power supply solution in one small (4mm×5mm) package (Figure 198.1). The constant-current, constant-voltage Lithium-Ion/Polymer charger utilizes a Bat-Track feature to maximize the efficiency of the battery charger by generating an input voltage that automatically tracks the battery voltage (described below). An I^2C serial interface affords the system designer complete control over the charger and the DC/DC bucks for ultimate adaptability to changing operating modes in a wide range of applications.

Switching PowerPath controller maximizes available power to the system load

The LTC3555 improves over earlier generations of USB battery chargers with the addition of several new features. It uses a proprietary switching power manager to extract power from a current-limited USB port with the highest possible efficiency, while maintaining average input current compliance. It minimizes power lost in the linear charger with its Bat-Track feature.

First generation USB applications implemented a current-limited battery charger directly between the USB port and the battery, where the battery voltage powers the system.

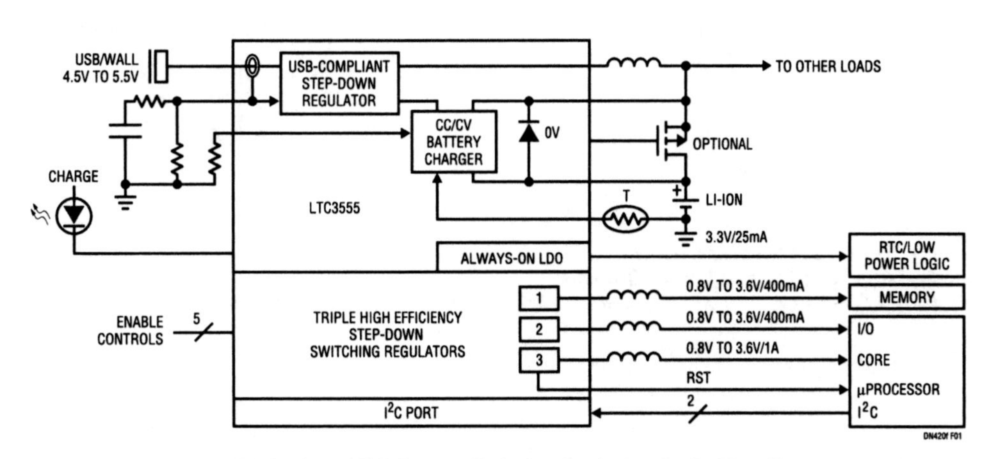

Figure 198.1 • All-in-One USB Power Solution Includes Switching Power Manager, Battery Charger, Three Synchronous Buck Regulators and LDO

Analog Circuit Design: Design Note Collection. http://dx.doi.org/10.1016/B978-0-12-800001-4.00198-8

This is referred to as a battery-fed system. In a battery-fed system, the available system power is $I_{USB} \cdot V_{BAT}$ because V_{BAT} is the only voltage available to the system load. When the battery is low, nearly half of the available power is lost to heat within the linear battery charger element.

Second generation USB chargers developed an intermediate voltage between the USB port and the battery. This intermediate bus voltage topology is referred to as a PowerPath system. In PowerPath ICs, a current-limited switch is placed between the USB port and the intermediate voltage. The intermediate voltage, V_{OUT}, powers the linear battery charger and the system load. By using the intermediate bus voltage topology, the battery is decoupled from the system load and charging may be carried out opportunistically. PowerPath systems have the added benefit of being "instant-on" because the intermediate voltage is available for system loads as soon as power is applied to the circuit, independent of the state of the battery. In a PowerPath system, more of the 2.5W available from the USB port is made available to the system load as long as the input current limit has not been exceeded. PowerPath systems offer improvements over battery-fed systems, but significant power may still be lost in the linear battery charger element if the battery voltage is low.

The LTC3555 is the first IC in the third generation of USB PowerPath chargers. These PowerPath devices produce an intermediate bus voltage from a USB-compliant step-down regulator that is regulated to a fixed amount over the battery voltage (a Bat-Track feature). The regulated intermediate voltage is just high enough to allow proper charging through the linear charger. By tracking the battery voltage in this manner, power lost in the linear battery charger is minimized, efficiency increases and power available to the load is maximized.

Figure 198.2 provides an efficiency comparison and power savings between chargers with switching vs linear PowerPath systems. The amount of power saved while charging large batteries can make the difference between a device that runs in thermal limit and one that runs cool.

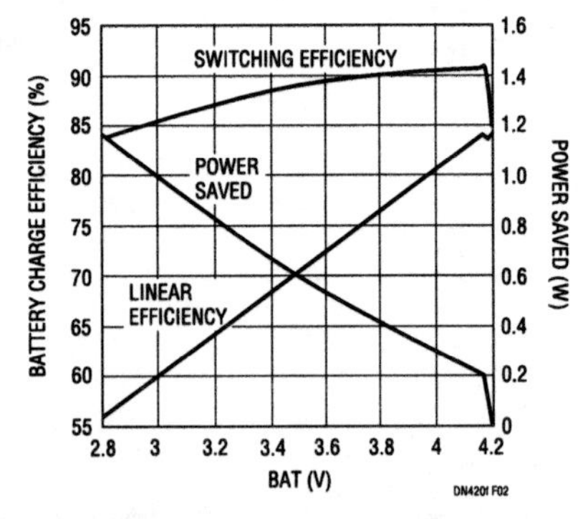

Figure 198.2 • Switching PowerPath Battery Charger Efficiency and Power Savings Relative to a Linear Charger. (V_{BUS}=5V, 5X MODE, R_{CLPROG}=2.94k, R_{PROG}=1k, I_{BAT}=0.7A at V_{BAT}=2.8V)

Complete power solution in a single IC

The LTC3555 also contains three user configurable step-down DC/DC converters capable of delivering 0.4A, 0.4A and 1A. Regulator 1 has a fixed reference voltage of 0.8V while regulators 2 and 3 may have their reference voltage changed via the I^2C interface between 0.8V and 0.425V. All of the converters operate at a switching frequency of 2.25MHz, allowing the use of small passive components while maintaining efficiencies up to 92% for output voltages greater than 1.8V (see Figure 198.3). All three regulators may be programmed to operate in pulse-skipping mode, Burst Mode operation or LDO mode via the I^2C port or through I/O pins. In Burst Mode operation the output ripple amplitude is slightly increased and the switching frequency varies with the load current to improve efficiency at light loads. If noise is a concern, all of the regulators may be set to operate in LDO mode or pulse-skipping mode. The device also provides an always-on 3.3V output capable of delivering 25mA for system needs such as a real-time clock or pushbutton monitor.

Conclusion

The LTC3555 is an advanced and complete power solution in a single chip. The third generation PowerPath management technology with its reduction in both heat generation and battery charge time is ideally suited for tomorrow's high density, feature rich battery-powered products. By integrating three I^2C-controlled, highly efficient step-down DC/DC converters, the LTC3555 allows the system designer complete flexibility to adapt to changing demands and operating modes.

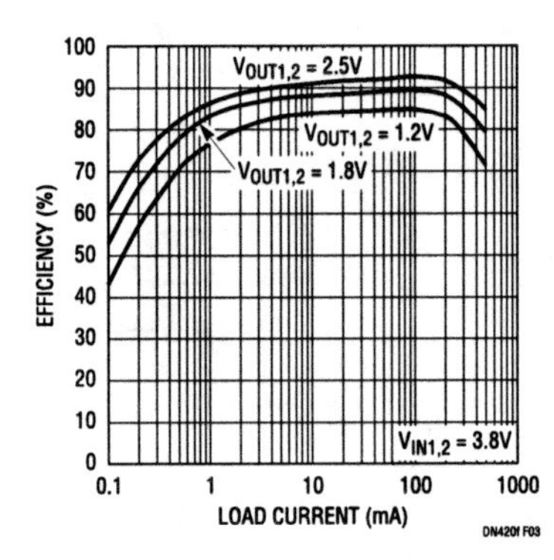

Figure 198.3 • Efficiency of Switching Regulators 1 and 2 with Burst Mode Operation

Switching USB power manager with PowerPath control offers fastest charge time with lowest heat

199

Dave Simmons

Introduction

Lithium-Ion and Lithium Polymer batteries are common in portable consumer products because of their relatively high energy density—they provide more capacity than other available chemistries within given size and weight constraints. USB battery charging is also becoming commonplace, as many portable devices require frequent interfacing with a PC for data transfer.

As portable products become more complex, the need for higher capacity batteries increases, with a corresponding need for more advanced battery chargers. Larger batteries require either higher charging current or additional time to charge to their full capacity. Most consumers look for shorter charge times, so increasing the charge current seems obviously preferable, but increasing charge current presents two major problems. First, with a linear charger, increased current creates additional power dissipation (i.e., heat). Second, the charger must limit the current drawn from the 5V USB bus to either 100mA (500mW) or 500mA (2.5W) depending on the mode that the host controller has negotiated.

PowerPath controllers deliver more power to the system load

There are two methods commonly used to extract power from a USB port. The first method uses a current limited battery charger directly between the USB port and the battery. This is referred to as a Battery Fed System because the system load is powered directly from the battery. Available power is given by $I_{USB} \cdot V_{BAT}$ because V_{BAT} is the only voltage available to the system load. When the battery is low, nearly half of the available power can be lost within the linear battery charger element. In low battery voltage protection mode, as little as 5% of the available power may be usable.

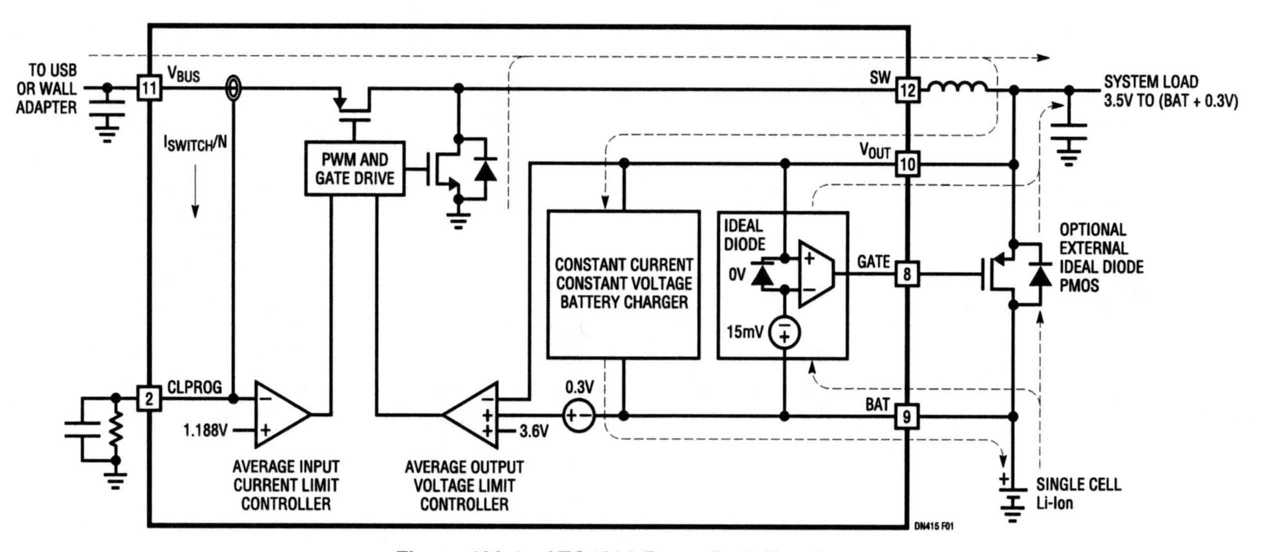

Figure 199.1 • LTC4088 PowerPath Topology

Analog Circuit Design: Design Note Collection. http://dx.doi.org/10.1016/B978-0-12-800001-4.00199-X

The second method develops an intermediate voltage between the USB port and the battery. This intermediate voltage bus topology is referred to as a PowerPath System. In PowerPath ICs, a current limited switch is placed between the USB port and the intermediate voltage. The intermediate voltage, V_{OUT}, then powers both a linear battery charger as well as the entire portable product. By using the intermediate voltage bus topology, the battery is decoupled from the system load and charging can be carried out opportunistically. During charging with a PowerPath system, the full 2.5W from the USB port is made available to the system load as long as the input current limit has not been exceeded. In this case V_{OUT} is just under the input voltage (5V for example). However, since the battery voltage is much lower than the 5V input, significant power is still lost to the linear battery charger element.

LTC4088 makes charging more efficient

The LTC4088 replaces the current limited switch in traditional PowerPath systems with a 2.25MHz buck mode synchronous switching regulator, as shown in Figure 199.1. The intermediate voltage, V_{OUT}, is regulated to just above the battery voltage. Because power is conserved in a switching regulator, the available output current is higher than the input current.

LTC4088 reduces USB charge time

This additional current can be used to power the portable product and charge the battery more quickly. Figure 199.2 shows the typical improvement in charge current versus a linear charger when powered from a 500mA USB port.

LTC4088 eases thermal constraints

The second benefit of the switching regulator is heat reduction. Power lost by inefficient charging can cause the external case of a portable product to become uncomfortably warm, and in extreme cases, it can cause thermal limiting of the battery charger. Figure 199.3 shows the typical efficiency and power savings of the LTC4088 relative to a linear charger when connected to a 500mA USB port.

The LTC4088 also includes a mode designed for use with AC powered wall adapters, in which the maximum input current is limited to 1A. Available current to the system load and battery charger ranges somewhere between 1A and 1.8A, depending on the battery voltage. Many higher capacity batteries are capable of charging at these higher rates, but with a volt or more difference between the wall adapter and the battery, the accompanying dissipative heating cannot be tolerated. Until now, these applications simply had to settle for a lower than optimal charge rate, and accompanying longer charge time.

Conclusion

The LTC4088 offers a dramatic advancement in battery charging and power path management technology, with its reduction in both heat generation and battery charge time. Designed specifically for portable applications, its high switching frequency and internal compensation require only a small inductor and output capacitor. Only the LTC4088's unique topology of a buck mode switching regulator working in tandem with a linear battery charger can give this unparalleled performance.

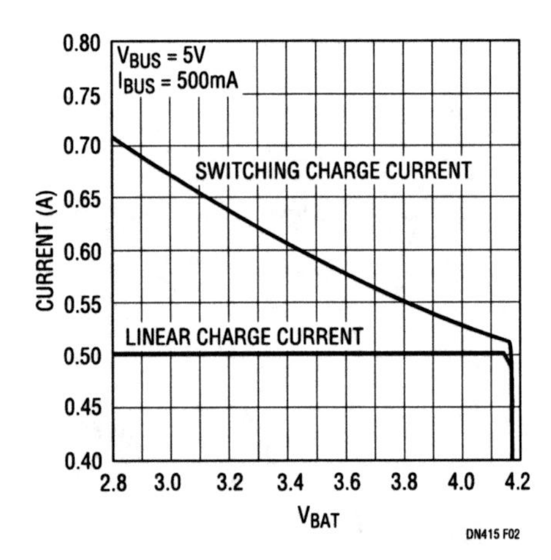

Figure 199.2 • Typical Charge Current for LTC4088 vs Linear Charger When Powered from a 500mA USB Port

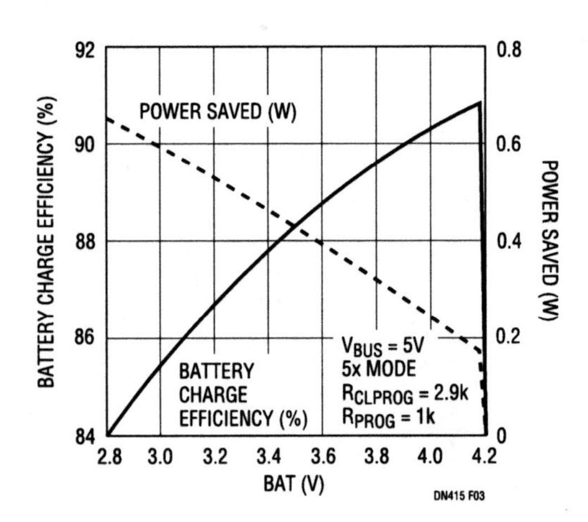

Figure 199.3 • Battery Charger Efficiency and Power Savings Relative to a Linear Charger When Charging from a USB Port

Universal Li-Ion battery charger operates from USB and 6V to 36V input in just 2cm^2

200

Liu Yang

Introduction

There are a number of advantages to offering USB and high input voltage power and battery-charging capability in handheld devices such as GPS navigators, PDAs, digital still cameras, photo viewers and MP3 players. For instance, charging and operation from USB offers the obvious convenience of not requiring a travel adapter. High voltage sources, such as Firewire and 12V to 24V adapters are even better, since they provide faster charging than USB and allow charging in more places, such as in the car. Nevertheless, there is an important design consideration with high voltage power sources: the voltage difference between the high voltage source and the battery in the handheld is very large. Since a linear charger cannot handle the power dissipation, a switching charger is required.

The LTC4089 and LTC4089-5 (see Figure 200.1) conveniently integrate a high voltage and wide input range (6V to 36V with 40V absolute maximum) monolithic 1.2A buck switching regulator and a USB power manager/charger into a compact thermally enhanced 3mm × 6mm DFN package. The LTC4089's buck regulator output voltage tracks the battery voltage to within 300mV. This Bat-Track feature minimizes overall power dissipation. The LTC4089-5 has a fixed 5V at OUT when power is applied at HVIN. When power is supplied from the USB port, the power manager maximizes the available power to the system load; up to the full USB available power of 2.5W. It automatically adjusts the Li-Ion battery charge current with respect to the system load current to maintain the total input current compliance within the USB limits. The total solution size is less than 2cm^2 with all components on one side of the PCB.

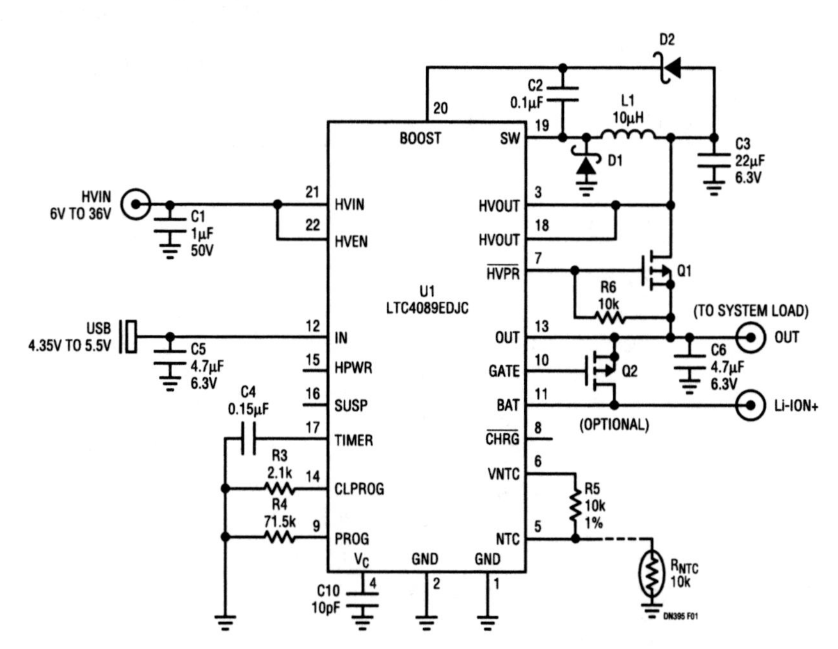

Figure 200.1 • The LTC4089 Schematic Illustrates Multiple Input Voltage Capability

Analog Circuit Design: Design Note Collection. http://dx.doi.org/10.1016/B978-0-12-800001-4.00200-3

Adaptive high voltage buck minimizes total power loss

The LTC4089's buck converter output voltage V_{OUT} tracks the battery voltage V_{BAT}. It is always 0.3V higher than V_{BAT}, so that the battery can be charged quickly while minimizing overall power dissipation. Figure 200.2 shows the overall efficiency at various input voltages, where the total power dissipation is less than 1.1W. Furthermore, if the battery is excessively discharged and V_{BAT} falls too low, the minimum V_{OUT} is 3.6V to ensure continuous system operation.

USB power manager maximizes power available to the system

In a traditional dual input device, the input charges the battery and the system's power is directly taken from the battery. This creates a number of problems. One of these is that the system's available power is reduced by the low-battery voltage when there is USB power present. For example, when $V_{BAT} = 3.3V$, the available power to the system is only 1.65W while the USB itself supplies 2.5W. The balance is dissipated as heat. The LTC4089 successfully solves this problem by providing an intermediate voltage V_{OUT} to power the system load. This V_{OUT} is independent of the battery voltage and equal to the USB voltage, thus the full USB power is available to the system load. Table 200.1 shows the advantages of the LTC4089 power manager over the traditional dual input configuration.

Small footprint

With all the necessary components on the same side of the PCB, the total solution size is less than $2cm^2$ (11.3mm × 17.5mm) as shown in Figure 200.3.

Summary

The LTC4089 integrates a high voltage wide input monolithic switching regulator, USB power manager and Li-Ion battery charger into a 3mm × 6mm DFN package and improves the functionality of USB-based and multiple power input portable devices.

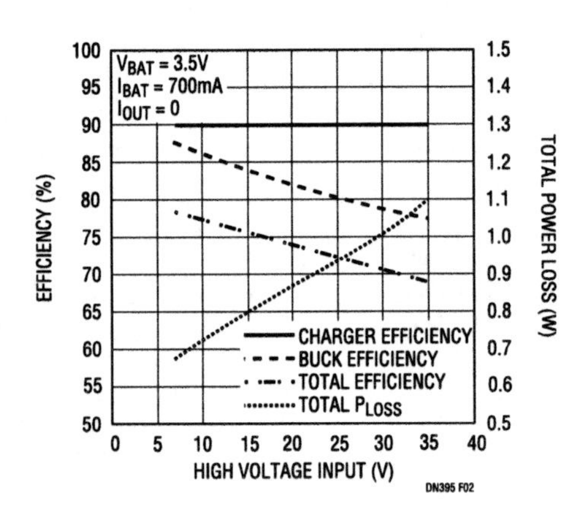

Figure 200.2 • The LTC4089 High Voltage Charger Efficiency and Total Power Loss

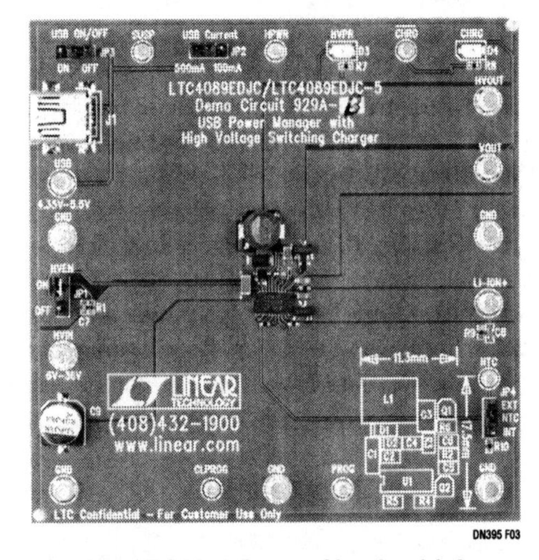

Figure 200.3 • The LTC4089 Demo Circuit with Layout in Bottom Right Corner

Table 200.1 Comparison of Traditional Dual Input Charger and Linear Technology's LTC4089 Power Manager/Charger for USB Charging

SCENARIO	TRADITIONAL DUAL INPUT CHARGER	LTC4089 POWER MANAGER/CHARGER
Battery Voltage is Below Trickle Charging Voltage	Available Current to System is Only Trickle Charge Current (50mA to 100mA), Which may not be Sufficient to Start the System	Full Adapter/USB Power is Available to System, Although Battery is in Trickle Charge
Battery is not Present	Most Chargers Consider this as a Fault. The System Cannot Start	Full Adapter/USB Power is Available to System
$V_{BAT} = 3.3V$ at USB Input	Available Power to System is Only 1.65W. The System Power cannot be Greater Than this	Full 2.5W USB Power is Available to the System
System Consuming Close to the Input Power Limit	Cannot Distinguish the Available Charging Current. Charger Timer Runs Out before the Battery is Fully Charged	Charger Time Proportionally Increases with Less Available Charge Current. The Battery is Always Fully Charged

Handheld high power battery charger

201

Mark Gurries

Introduction

As the performance of many handheld devices approaches that of laptop computers, design complexity also increases. Chief among them is thermal management—how do you meet increasing performance demands while keeping a compact and small product cool in the user's hand?

For instance, as battery capacities inevitably increase, charge currents will also increase to maintain or improve their charge times. Traditional linear regulator-based battery chargers will not be able to meet the charge current and efficiency demands necessary to allow a product to run cool. What is needed is a switching-based charger that takes just about the same amount of space as a linear solution—but without the heat.

Small PCB footprint

Figure 201.1 shows how simple a feature-laden LTC4001-based charger solution can be. This switching-based charger only requires the IC, a small 1.5μH inductor, two small 1206-size 10μF ceramic capacitors, and a few other tiny components. Furthermore, even simpler configurations are possible (see Figure 201.2). This monolithic 2A, 1.5MHz synchronous PWM standalone battery charger is packaged in a 4mm × 4mm 16-pin QFN package, which contains the built-in switching MOSFETs and charge termination controller. Figure 201.3 shows an actual PCB solution.

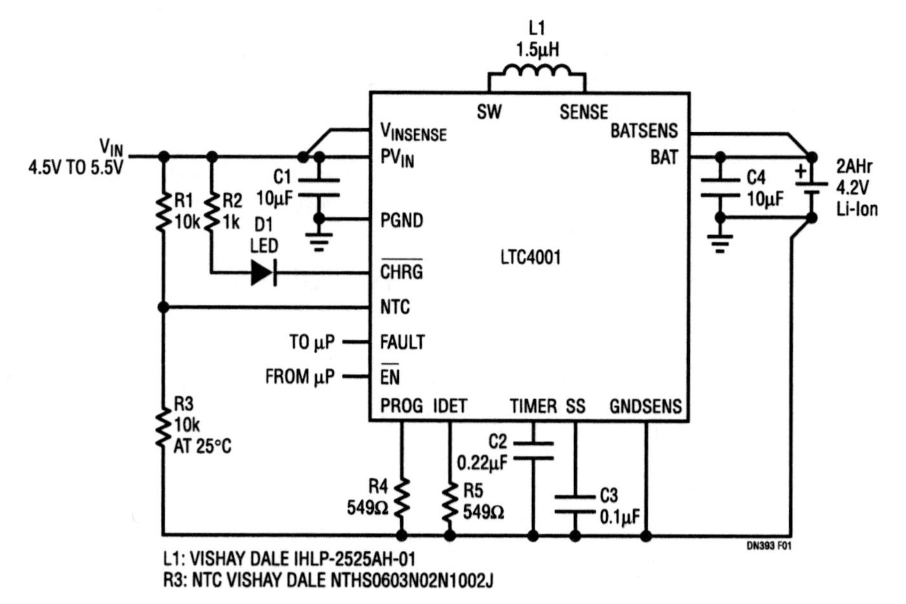

Figure 201.1 • Li-Ion Battery Charger with 3-Hour Timer, Temperature Qualification, Soft-Start, Remote Sensing and C/10 Indication

Analog Circuit Design: Design Note Collection. http://dx.doi.org/10.1016/B978-0-12-800001-4.00201-5

Advanced features and functions

One of the more unique features of the LTC4001 is its full remote voltage sense capability which permits faster charge rates by bypassing voltage drops in narrow PCB traces, EMI filters or current sense resistors for gas gauge related support; this is placed on the system side of the battery connector. Eliminating these losses in the sense circuit can significantly shorten the constant voltage phase of the overall charge time.

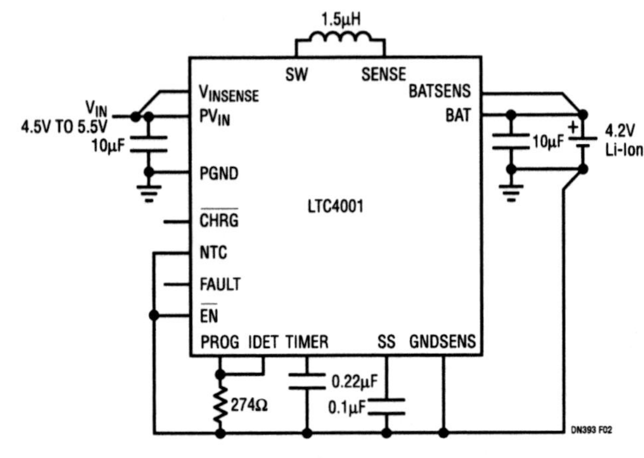

Figure 201.2 • Simple 2A Battery Charger

Figure 201.3 • Actual LTC4001 Demo Board Showing a Compact Footprint (Height ≤ 1.8mm)

Another important feature is programmable soft-start, which requires only a small ceramic capacitor on the SS pin. Soft-start saves design time and cost by simplifying the power source requirements, precluding the need to handle fast start-up load transients commonly found with switching power supplies.

Other advanced features include: 50mA trickle charge recovery of over-discharged batteries below 3V/cell; adjustable charge timer via a single capacitor on the TIMER pin; and automatic restart of a charge cycle when the battery voltage falls below 100mV of the full charge voltage.

Two signals provide status. First is the FAULT pin, which in conjunction with the LTC4001 thermistor circuit, reports an out-of-range temperature situation. When a temperature fault occurs, the charge process is stopped immediately. Charging a battery when it is out of its normal temp range can damage it. Second, the $\overline{CHRG}$ pin shows three states relating to the charge state of the battery or charger. In addition to the normal OFF indication, it also indicates when the battery is below its user programmable I_{DET} threshold or when in bulk charge mode.

The $\overline{EN}$ (Enable) pin allows for shutdown of the charger, thus reducing its V_{IN} quiescent current below 50µA and the battery drain current to less than 3µA. Shutdown also occurs automatically if V_{IN} falls to less than 250mV above the current battery voltage.

Flexible options

The LTC4001 provides a number of flexible options in its small package. Bulk charge current is programmable via the PROG pin and a simple resistor, from 2A and below. A separate resistor on the I_{DET} pin is all that is required to set the full charge current termination or indication threshold independently of the bulk charge current setting. Typically the I_{DET} threshold is set to 1/10 (C/10) of the bulk charge current, which equates to a battery being about 95% to 98% full. Raising the I_{DET} current trip threshold significantly reduces charge time by having a full charge indication occur sooner in exchange for a slightly lower full state of charge. Likewise, increasing the trip threshold extends the timer to approach a 100% state of charge if there are no serious time constraints. The type of charge termination is also flexible. In addition to timer-based termination, the charge can be terminated when the I_{DET} threshold is reached or charge termination can be defeated all together to allow an external power manager to decide.

Conclusion

The LTC4001 is the charger of choice for the next generation of handheld devices with its tiny solution size, unmatched power capability, high efficiency, protection features and flexible options.

Fast, high efficiency, standalone NiMH/NiCd battery charging

202

Fran Hoffart

Introduction

Although recent popular attention is focused on Lithium Ion batteries, one must not forget that other battery chemistries, such as Nickel Cadmium (NiCd) and Nickel Metal Hydride (NiMH) have advantages in rechargeable power systems. Nickel-based batteries are robust, capable of high discharge rates, have good cycle life, do not require special protection circuitry and are less expensive than Li-Ion. Among the two, NiMH batteries are rapidly replacing NiCd because of their higher capacity (40% to 50% more) and the environmental concerns of the toxic cadmium contained in NiCd batteries.

The LTC4010 and LTC4011 are NiCd/NiMH battery chargers that simplify Nickel-based battery charger design and include power control and charge termination for fast charging up to 16 series-connected cells using a synchronous buck topology. The LTC4011 provides a full feature set in a 20-lead TSSOP while the LTC4010 comes in a 16-lead TSSOP. The LTC4010 removes the PowerPath control output, top-off charge indicator, DC power sense input and provides limited thermistor options.

NiCd/NiMH battery charging basics

Batteries come in many sizes and capacity ratings. When specifying charge current, it is commonly related to a battery's capacity, or simply "C". The letter "C" is a term used to indicate the manufacturers' stated battery discharge capacity which is measured in milliamp-hours (mAh). This capacity rating becomes important when fast charging because it determines the required charge current for proper charge termination.

There are several commonly used methods for charging Nickel batteries. They are all related to the length of the charge cycle which determines the recommended charge current. A slow charge (or low rate charge) consists of a

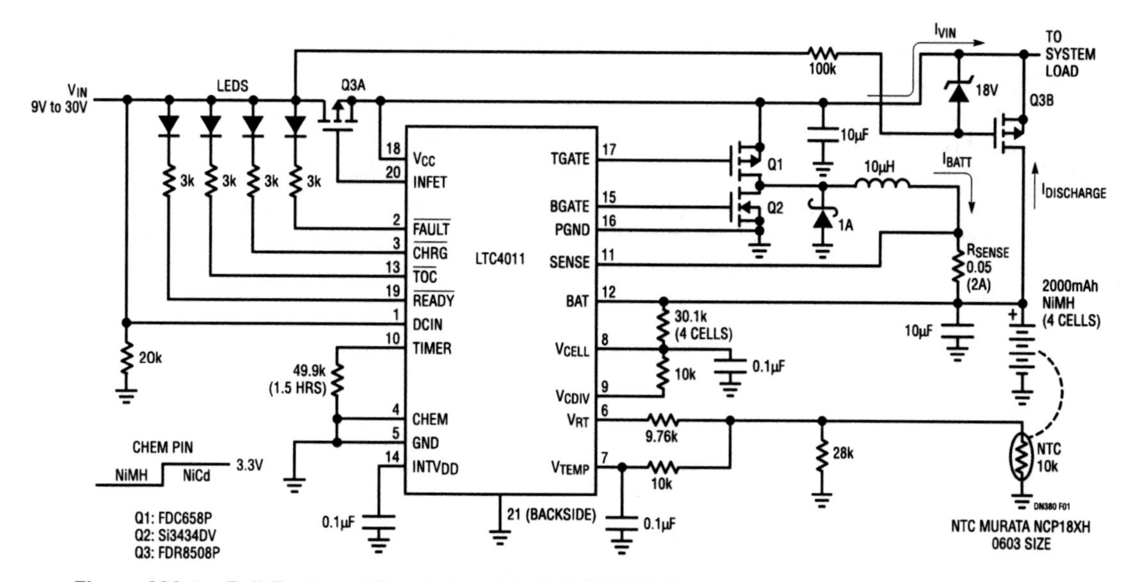

Figure 202.1 • Full Featured Standalone 2A, 4-Cell NiMH Fast Charger with PowerPath Control

Analog Circuit Design: Design Note Collection. http://dx.doi.org/10.1016/B978-0-12-800001-4.00202-7

relatively low charge current, typically 0.1C, applied for approximately 14hours set by a timer. A quick charge applies a constant current of approximately 0.3C to the battery while a fast charge applies a constant current of 1C or higher. Both quick and fast charge cycles require that the charge current terminate when the battery becomes fully charged.

During a fast charge cycle, a constant current is applied to the battery while allowing the battery voltage to rise to the level required (within limits) to force this current. As the battery accepts charge, the battery voltage and temperature slowly rise. As the battery approaches full charge, the voltage rises faster, reaches a peak, then begins to drop ($-\Delta V$); at the same time, the battery temperature begins to quickly rise ($\Delta T/\Delta t$). Most fast or quick charge termination methods use one or both of these conditions to end the charge cycle.

Complete 4-cell NiMH battery charger

Figure 202.1 shows a fast, 2A charger featuring the high efficiency LTC4011 550kHz synchronous buck converter. The LTC4011 simplifies charger design by integrating all of the features needed to charge Ni-based batteries, including constant current control circuitry, charge termination, automatic trickle and top off charge, automatic recharge, programmable timer, PowerPath control and multiple status outputs. Such a high level of integration lowers the component count, enabling a complete charger to occupy less than $4cm^2$ of board area.

Initial battery qualification verifies that sufficient input voltage is present for charging and that the battery voltage and battery temperature are within an acceptable range before charging at full current. For deeply discharged batteries, a low current trickle charge is applied to raise the battery voltage to an appropriate level before applying full charge current. When qualification is complete, the full programmed constant-current begins.

Standalone charge termination

The charge termination methods used by the LTC4010 and LTC4011 utilize battery voltage and battery temperature changes to reliably indicate when full charge is reached as a function of the charge current selected. The charge current must be sufficiently high (between 0.5C and 2C) for the battery to exhibit the voltage and temperature profile required for proper charge termination. Figure 202.2 shows a typical fast-charge profile displaying charge current, battery temperature and per cell voltage. This profile indicates that the charge cycle terminated due to the rate of temperature rise or $\Delta T/\Delta t$.

The $-\Delta V$ charge termination algorithm begins shortly after the full charge current starts flowing. A fixed delay time prevents false termination due to battery voltage fluctuations from batteries that are deeply discharged or haven't been charged recently. For batteries that are near full charge, the $-\Delta V$ termination sequence begins immediately to prevent overcharging.

During the charge cycle, both the $-\Delta V$ and $\Delta T/\Delta t$ termination methods are active. For NiMH batteries, the $-\Delta V$ termination requires that the single cell battery voltage drop 10mV from the peak voltage or the rate-of-temperature rise ($\Delta T/\Delta t$) be greater than 1°C/minute. The measurements are taken every 30seconds and the results must be consistent for four measurements for termination to take place. Typically the $\Delta T/\Delta t$ termination method occurs earlier in the charge cycle. If this occurs, the LTC4010/4011 adds a top-off charge at a reduced charge current for 1/3 of the programmed time. Top-off only occurs when charging NiMH batteries.

After the charge cycle has ended, the charger continues monitoring the battery voltage. If the voltage drops below a fixed threshold level, due to an external load on the battery or self-discharge, a new charge cycle begins with the charge termination algorithms immediately enabled.

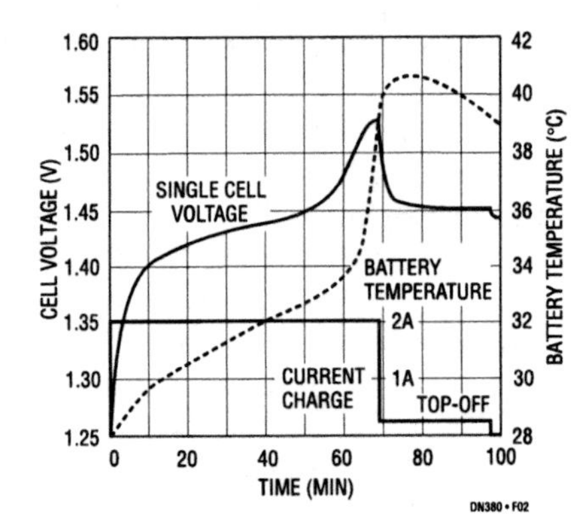

Figure 202.2 • Typical NiMH Fast Charge Profile

Conclusion

The LTC4010 and LTC4011 provide complete standalone solutions for reliable, robust and safe fast charging of NiCd and NiMH batteries. Proper charging is critical to not only obtain maximum battery capacity but to also avoid high temperatures, overcharge and other conditions which adversely affect battery life.

Dual Smart Battery charger simplifies battery backup for servers

203

Mark Gurries

Introduction

Smart Batteries are an increasingly popular choice for more than just traditional compact consumer electronic devices. For example, Smart Batteries are being used as battery backup for products such as blade servers, where knowing battery status is very important.

LTC1760 dual smart battery charger

Figure 203.1 shows a typical dual battery charger. This circuit can charge batteries with up to 4A and switch continuously down to zero load currents. This circuit takes advantage of ceramic capacitors' space saving features without producing any audible noise. The high 300kHz switching frequency allows the use of small low cost 10μH inductors.

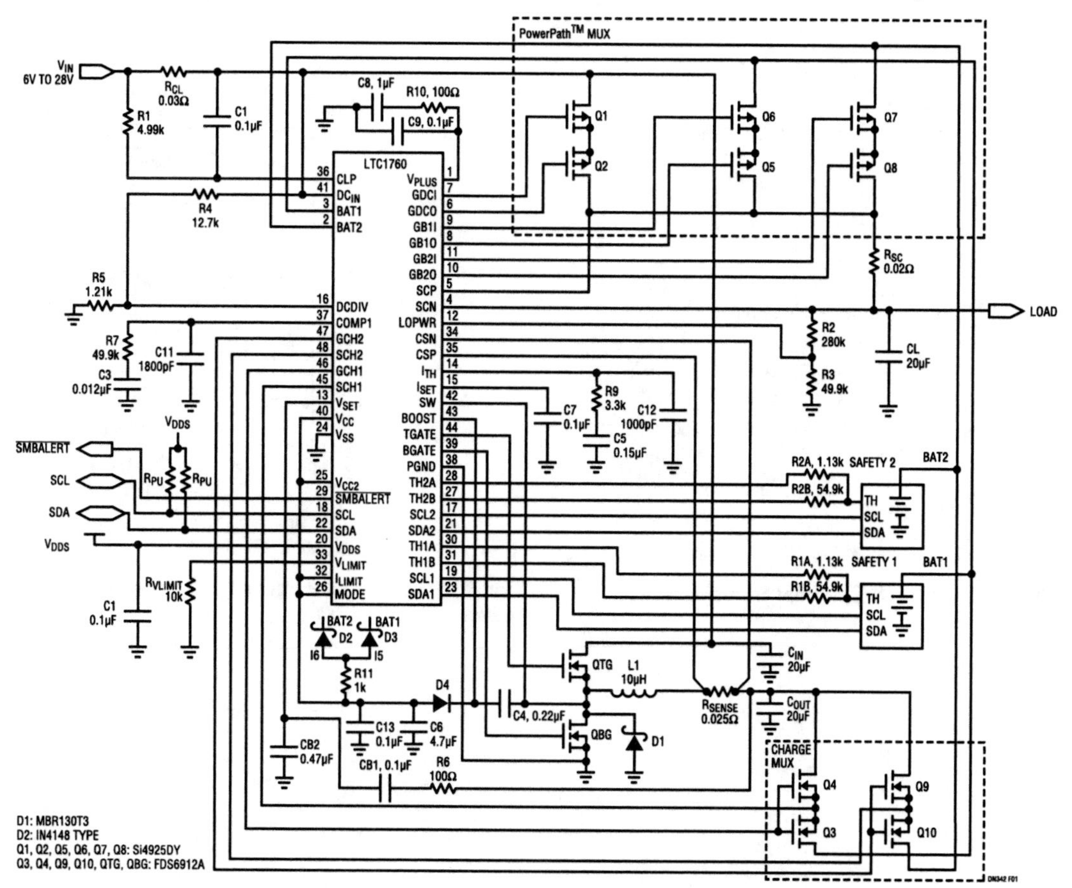

Figure 203.1 • 4A Dual Battery System

Analog Circuit Design: Design Note Collection. http://dx.doi.org/10.1016/B978-0-12-800001-4.00203-9

The LTC1760 complies with the Smart Battery System Manager (SBSM) specification V1.1. It has a very wide input and charge output voltage range of 6V to 28V. Current and voltage accuracies of 0.2% of the reported values provide precision charge capability. Low dropout is achieved with 99% maximum duty cycle while maintaining efficiency greater than 95%. The LTC1760 also offers many unique features, including a special current limit and voltage limit system that prevents SMBus data corruption errors from generating false charge values, which could harm the battery. An SMBus accelerator increases data rates in high capacitance traces while preventing bus noise from corrupting data.[1]

Other features include: an AC present signal with precision 3%-accurate user adjustable trip points; a safety signal circuit that rejects false thermistor tripping due to ground bounce caused by the sudden presence of high charge currents, and an ultrafast overvoltage comparator circuit that prevents voltage overshoots when the battery is suddenly removed or disconnects itself during charge. Last but not least is an input current limit sensing circuit that limits charge current to prevent wall adapter overload as the system power increases.[2]

LTC1760 power management

Dual battery systems are traditionally used to simply extend system battery run time by allowing a sequential battery drain—drain battery 1, then battery 2. New server applications are also using batteries and demand drain currents beyond the capability of a single battery.

The LTC1760 addresses this need by allowing the safe parallel discharge of two batteries. Parallel discharge offers more than just increased current capability. It reduces I^2R losses and improves voltage regulation under extremely high load conditions, both of which can improve total discharge time over a sequential solution. Figure 203.2 compares discharge times for equivalent parallel and sequential solutions. In high current, rapid discharge applications, quick recharging of the batteries is a priority. Again the LTC1760 goes beyond the simple sequential solution and offers *parallel charging*, which, depending on the battery chemistry, can result in significant charge time reductions over a sequential solution, as shown in Figure 203.3.

Safely managing the charge and discharge states of multiple batteries and the DC input power source presents a significant power management issue that historically has involved a host processor running custom written application software. The LTC1760 simplifies this task by operating in a standalone Level 3 Bus Master mode. It autonomously controls simultaneous battery charging and discharging, full dual battery conditioning support and ideal diode PowerPath switching between two batteries and a wall adapter *without requiring any host processor.*

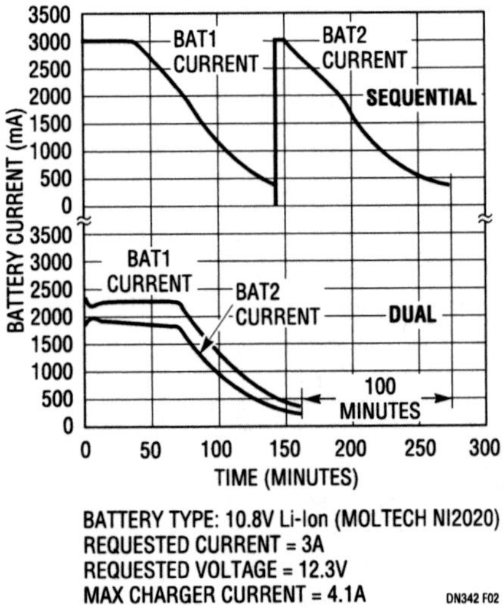

BATTERY TYPE: 10.8V Li-Ion (MOLTECH NI2020)
REQUESTED CURRENT = 3A
REQUESTED VOLTAGE = 12.3V
MAX CHARGER CURRENT = 4.1A
DN342 F02

Figure 203.2 • Dual Battery vs Sequential Battery Charge Time

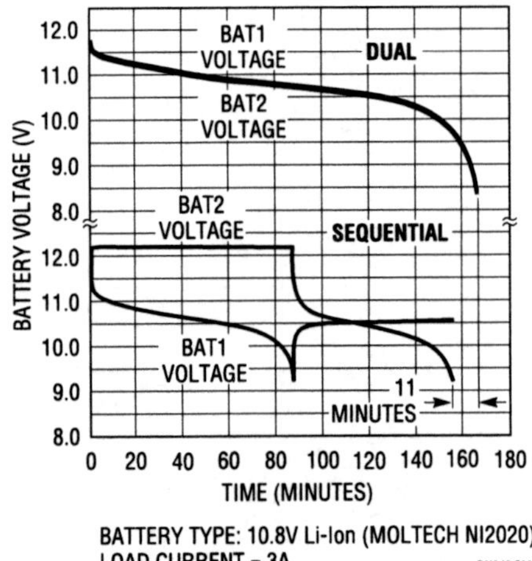

Figure 203.3 • Dual Battery vs Sequential Battery Discharge Time

Note 1: U.S. Patent number 6650174.

Note 2: U.S. Patent number 5723970.

Advanced topology USB battery charger optimizes power utilization for faster charging

204

John Shannon

Linear Technology offers a variety of parts to simplify the task of extracting power from a USB cable, including the new, easy-to-use LTC4055 high performance Li-Ion charger and power controller. The LTC4055 seamlessly manages power flow between an AC adapter, USB cable and Li-Ion battery, all while maintaining USB power specification compliance.

Thanks to its sophisticated intermediate voltage bus topology, the LTC4055 charges batteries faster with less heat generation than a traditional charger-fed topology. Better yet, a typical LTC4055-based USB charger solution is compact and requires only 10 external components and 100mm^2 of PCB real estate.

Benefits of the LTC4055

In order to fully appreciate the benefits of the LTC4055, let us first analyze the difference between the intermediate voltage bus topology and a charger fed topology. Figure 204.1 is a simplified block diagram of the two power topologies. In the intermediate bus voltage topology, the output called V_{MAX} is derived from one of the three available power sources—wall adapter, USB or battery. The system load, usually consisting of DC/DC converters, LED drivers or disk drives, is powered from V_{MAX}.

In order to simplify the analysis, the voltage drop of the input current source is assumed to be zero. The current drawn by the system is then the power required divided by the voltage input to the system. In the case of the LTC4055, that voltage is the highest of the adapter, USB or battery. Excess current, beyond what is required to power the system loads, is available to charge the battery.

In contrast, charger fed systems place the system loads in parallel with the battery. The voltage input to the system load is the battery voltage. The current drawn by the system is the power requirement divided by the battery voltage. Like in the intermediate bus voltage topology, excess current not required by the system is available to charge the battery.

Figure 204.1 shows these two topologies and compares their power losses at typical system loads and battery voltages. It is clear that the intermediate voltage bus topology has advantages over the charger fed topology. In the presence of

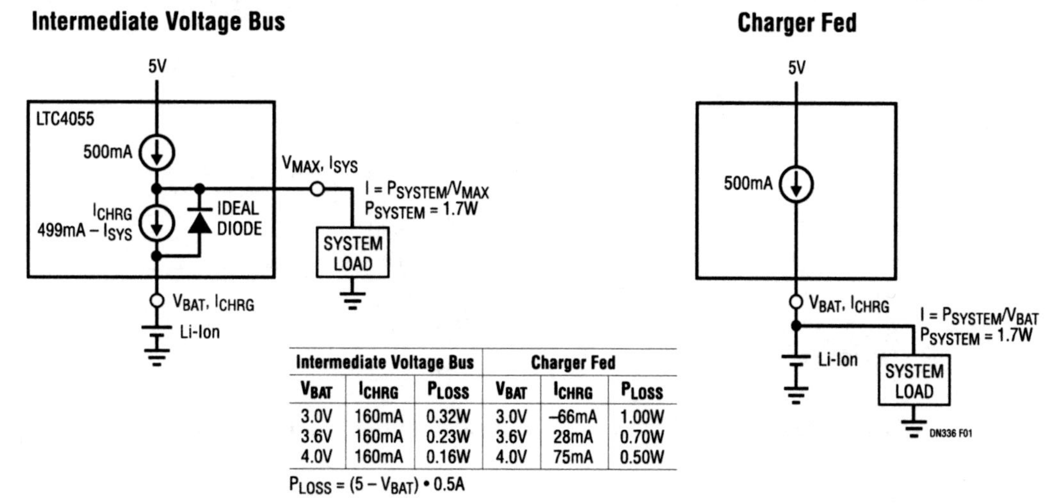

Intermediate Voltage Bus			Charger Fed		
V_{BAT}	I_{CHRG}	P_{LOSS}	V_{BAT}	I_{CHRG}	P_{LOSS}
3.0V	160mA	0.32W	3.0V	–66mA	1.00W
3.6V	160mA	0.23W	3.6V	28mA	0.70W
4.0V	160mA	0.16W	4.0V	75mA	0.50W

$P_{LOSS} = (5 - V_{BAT}) \cdot 0.5A$

Figure 204.1 • LTC4055's Topology Optimizes Power Utilization by Reducing Power Loss

Analog Circuit Design: Design Note Collection. http://dx.doi.org/10.1016/B978-0-12-800001-4.00204-0

system loads, the intermediate bus voltage topology is able to charge in situations where the charger fed topology would actually be discharging the battery. The intermediate bus voltage topology also reduces power losses in the LTC4055, which means less heat generation in the system enclosure. In addition to these clear advantages, the intermediate bus voltage topology offers other more subtle benefits. For instance, the battery is isolated from the rest of the system—as long as system current draw is within limits, the system can run even if the battery is missing or deeply discharged.

Simple circuit automatically selects the best power source

Figure 204.2 shows a typical LTC4055 implementation. This charger provides power to the OUT (V_{MAX}) pin from the best of the three available power sources: wall, battery or USB.

Operation with wall adapter present

If a wall adapter is present, it powers the intermediate V_{MAX} bus providing power to the system load and the battery charger. The battery, if present, is charged via the LTC4055 with a constant current, constant voltage, timer terminated charger. The charge current, when powered by an adapter, is programmable up to 1A. When the adapter is present, the $\overline{\text{ACPR}}$ status pin is pulled low. Pulling the SHDN pin high disables charging. The $\overline{\text{CHRG}}$ status pin is pulled low while the battery is being charged (timer has not terminated).

Operation with no wall adapter, but USB available

If wall power is not available, but USB power is, then the USB power is switched through to OUT via a current limiting circuit in order to enforce USB current limit compliance. Like the case where a wall adapter is present, the voltage at V_{MAX} is used to power the system load and charge the battery. If the load does not completely use all the available USB power, then the extra power is used to charge the battery. As the system load increases, the battery charging current decreases enough to maintain USB input current compliance. If the load current exceeds the allowed USB current, battery charging ceases and the battery discharges through an ideal diode, internal to the LTC4055, into OUT (V_{MAX}). In this way, USB power provides what current it can, while the battery shoulders the rest of the load.

The maximum current drawn from the USB bus is simply programmed via a resistor. Grounding the HPWR pin reduces the maximum current draw by a factor of five, for compliance with the low power USB mode. Grounding the SUSP pin further reduces USB current consumption to 200μA, to comply with the USB suspend requirements.

Unplugged operation

During unplugged operation, the LTC4055 minimizes power losses by supplying the system load through an "ideal" diode function. The device also maximizes battery run time by powering V_{MAX} through an ideal diode connected to the Li-Ion cell.

Conclusion

The LTC4055 uses an intermediate voltage bus topology to yield faster charging and less heating than less sophisticated charger fed systems. Further, its use greatly simplifies PowerPath control in handheld USB compatible products. The typical schematic of an optimized complete three-supply (wall/USB/battery) selection/charger system is shown in Figure 204.2.

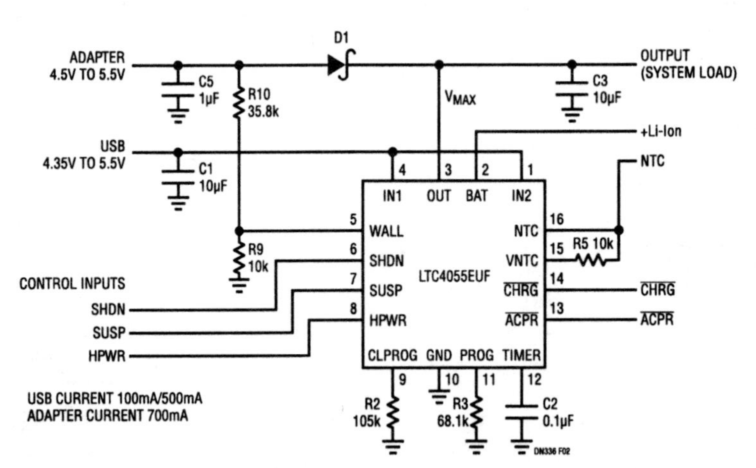

Figure 204.2 • The LTC4055 Implementation is Compact and Simple

Simplify battery charging from the USB

205

John Shannon

Introduction

The USB is fast becoming the most popular way to attach peripherals to personal computers. An important feature of the USB standard is that it allows for power transmission over the USB cable, making it possible for a PDA or other handheld device to charge its battery while it is connected to a host PC.

The USB specification defines load current limits that USB peripherals can draw from the USB. Three operating modes are allowed:

- Suspend mode, where the USB peripheral draws less than 500μA from the USB cable
- Low power mode, where the current consumption can be up to 100mA
- High power mode, where the current is allowed to be up to 500mA

The USB specification also states that USB peripherals are not allowed to backfeed into the USB cable.

Figure 205.1 shows a minimum component solution to the USB battery charger problem. In this case, only five components including the LTC4053, are required to achieve a fully compliant USB charger.

The charger is based on the LTC4053—a standalone timer-terminated constant current/constant voltage linear charger. The LTC4053 features a thermal loop which maximizes the charging rate under all conditions by regulating the maximum die temperature. Less sophisticated linear chargers must run lower charging rates or risk overheating, thus increasing charge times.

Operation of the LTC4053 is simple. The charge current is set by a resistor on the PROG pin, the termination timer period is set by a capacitor on the timer pin and float voltage is preset at 4.2V.

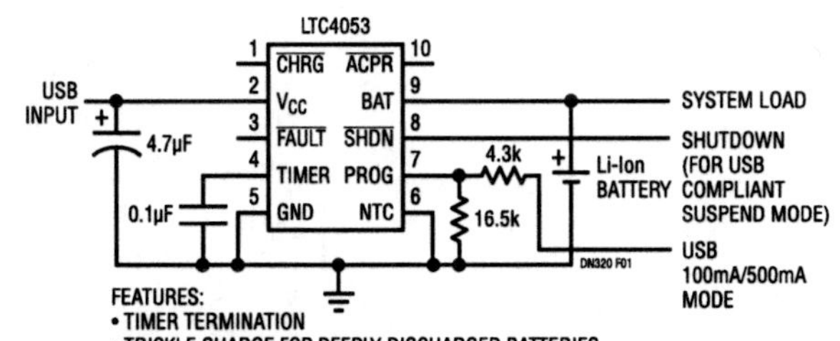

Figure 205.1 • Minimum Component Count USB Compliant Li-Ion Charger

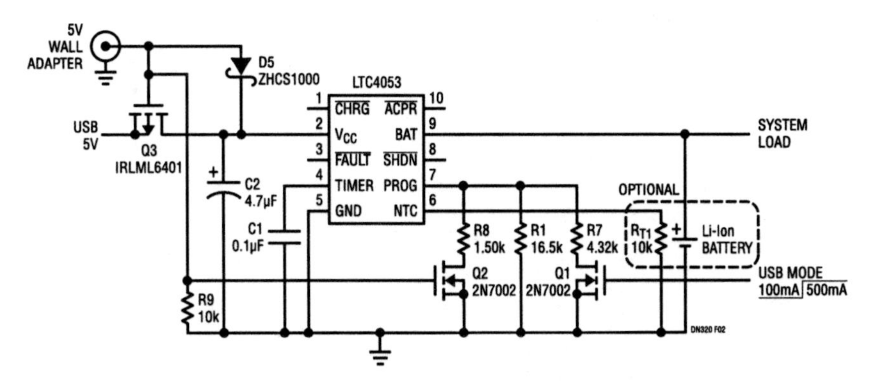

Figure 205.2 • Dual Input Charger: 1A Fast Charging from Wall Adapter or 100mA/500mA from USB

Analog Circuit Design: Design Note Collection. http://dx.doi.org/10.1016/B978-0-12-800001-4.00205-2

Charging from USB or a wall adapter

USB charge current is limited to 500mA, but the LTC4053 is capable of higher output currents. This capability can be exploited by adding some additional components around the LTC4053 to allow for a higher charge rate from an alternate power source such as a wall adapter.

The circuit shown in Figure 205.2 requires only 12 components, is compliant with the USB specification and provides high rate charging from an adapter. The NTC feature of the LTC4053 is also implemented so that charging is inhibited if the battery is too hot or too cold.

Two things must happen to add high rate adapter-based charging. The first is that the USB input must be isolated from the adapter to prevent adapter power feeding into the USB. The second is that the higher charging rate must be selectable by changing the resistive load on the PROG pin.

A P-channel MOSFET, Q3, is used to select between USB or adapter power. Q3 is on if USB power is present and the adapter is absent since its gate is at ground (through R9) and its source is at 5V. As the wall adapter power comes up, the gate of Q3 is pulled high and Q3 turns off. A 1A charging current flows from the adapter through diode D5. The 1A charge rate is programmed by turning on Q2 when the adapter is present which alters the resistive load on the PROG pin.

Faster charging with system in full operation

Both of these simple circuits draw system power directly from the battery. This makes for a simple solution but also reduces the total available power from the USB port. The USB port is capable of 2.5W (5V · 500mA). Placing the device system bus on the battery directly (normally 3.6V) only allows the system to extract 1.8W (3.6V · 500mA) from the USB cable before the 500mA limit is reached and the battery must supply any excess power. The lower the battery voltage, the less power is available to the system load. The solution is to power the system load directly from the USB, measure the system current consumption and back off the charger current as required to maintain USB compliance.

Figure 205.3 shows a compact seven component solution using the LTC4410 USB power manager. The system load is supported directly by the USB when the USB is present. This allows a maximum of 2.5W for the system load. The current flowing from the USB port is measured and the battery charging current is reduced in order to maintain USB compliance. When USB power is absent, the load is supported by the battery through the switch Q1. Compared to the solution shown in Figure 205.2, this solution more fully exploits the power available at the USB port, thus resulting in faster charging.

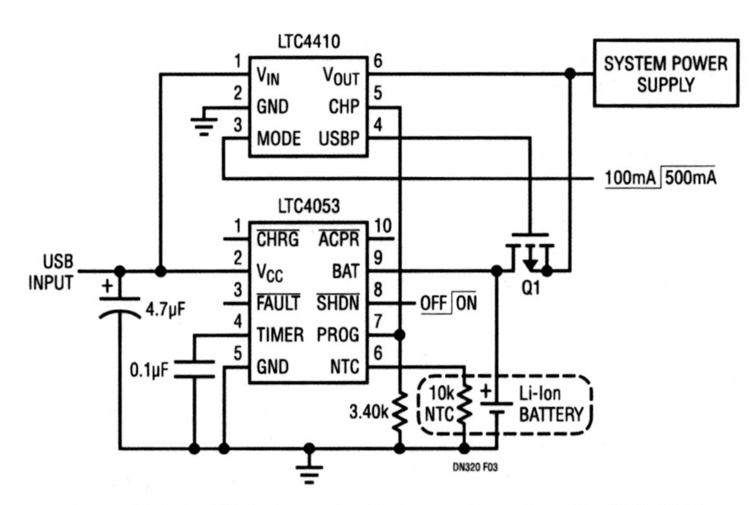

Figure 205.3 • USB Compliant Charger Based on the LTC4410 Fully Utilizes the Power Available from the USB Input

Li-Ion linear charger allows fast, full current charging while limiting PC board temperature to 85°C

206

Fran Hoffart

Introduction

Linear battery chargers are typically smaller, simpler and less expensive than their switcher-based counter parts, but they have one major disadvantage: excessive power dissipation when the input voltage is high and the battery voltage is low (discharged battery). Typically, such conditions are temporary—as the battery's voltage rises with its charge—but one must consider this worst-case situation when determining the maximum allowable values for charge current and IC temperature. One simple solution to this overheating problem is to decrease the charge current for the entire constant current part of the charging process. The problem with this method is a corresponding increase in charge time. A better option is to use the LTC1733 Li-Ion single cell linear charger which overcomes any overheating problem while maintaining fast charge times. A unique thermal feedback loop within the IC allows full current, fast charging under nominal conditions without overheating under worst-case conditions (including high ambient temperature, high input voltage or low battery voltage situations).

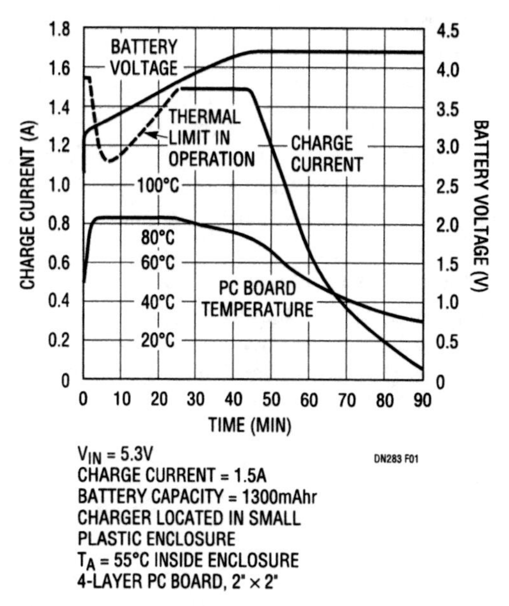

Figure 206.1 • LTC1733 Li-Ion Battery Charge Cycle for High Ambient Temperature Conditions

Thermal feedback loop limits IC temperature

A thermal feedback loop limits the maximum junction temperature of the LTC1733 to approximately 105°C, well below the maximum allowable junction temperature of 125°C. As the junction temperature approaches 105°C, the on-chip temperature sensor begins to smoothly decrease the charge current to a level that will limit the maximum junction temperature to 105°C (see Figure 206.1). Unlike ICs that simply shut down at 160°C to protect themselves, the LTC1733 can operate in this temperature control mode indefinitely. Devices with a 160°C thermal shutdown temperature could begin switching on and off at the thermal limit or might not operate correctly as a charger. Thermal shutdown is not a healthy mode of operation, it is rather intended to protect the IC from failure when overstressed.

Charge cycle with thermal limit in operation

Figure 206.1 shows a typical single cell Li-Ion charge cycle for a worst-case temperature condition. The curves show battery voltage, charge current and PC board temperature vs time.

A charge cycle begins when the input power is applied with the battery connected and the program resistor connected to ground. Deeply discharged batteries are trickle charged at 10% of full current until the battery voltage reaches 2.48V at which point the charger switches to full current.

Analog Circuit Design: Design Note Collection. http://dx.doi.org/10.1016/B978-0-12-800001-4.00206-4

At the start of the charge cycle, the charge current quickly rises to the programmed value of 1.5A, resulting in the battery voltage rising to 3.2V. With an input voltage of 5.3V, the 3.2W of power dissipated by the LTC1733 raises the junction temperature to approximately 105°C with the 2″×2″ PC board temperature (heat sink) reaching approximately 85°C in approximately 1.5 minutes. The thermal feedback loop limits any additional temperature rise by reducing the charge current. As the battery voltage rises, the LTC1733 temperature begins to drop allowing the charge current to rise back up to the programmed current level of 1.5A. The charging continues at the 1.5A constant current level until the battery voltage reaches 4.2V, at which time the constant voltage portion of the charge cycle begins. This continues with the charge current continually dropping until the 3-hour timer ends the charge cycle (Figure 206.1 shows the first 90 minutes).

Thermally enhanced package dramatically improves power dissipation

A special low profile (1.1mm) 10-pin MSOP package with an exposed bottom-side metal pad allows the IC to be soldered directly to the PC board copper to greatly reduce the junction-to-case thermal resistance. A good thermal layout allows the LTC1733 to dissipate up to 2.5W continuously using a 2″×2″ 4-layer PC board at a 25°C ambient temperature.

A good thermal layout consists of PC board copper directly below the package spreading out to copper areas and feedthrough thermal vias to internal and backside copper layers. For surface mount devices, the PC board copper can become an effective heat sink.

It is also important to solder the entire area of the IC's metal pad to the board to assure good heat conduction. Tests have shown that with a large initial power of 4.5W applied to a package, an improperly soldered package will reach the thermal feedback temperature in just seconds, while a good solder attachment will take over a minute.

Complete standalone charger

The LTC1733 is a complete constant current, constant voltage, power limiting linear charger for a single cell Li-Ion battery as illustrated in Figure 206.2. The IC includes a 1.5A power MOSFET, current sense resistor, programmable charge current, programmable timer, selectable charge voltage and thermistor input to monitor battery temperature for charge qualification. There are three status outputs capable of driving LEDs to indicate "AC power good," "charge" and "fault." There is also an output to monitor charge current. Input voltage requirements are 4.5V to 6.5V with manual shutdown and a micropower sleep mode when the input voltage is removed. No input blocking diode is required because of the internal MOSFET construction.

Conclusion

The LTC1733 is a standalone Li-Ion battery linear charger IC that allows the charge current to be programmed for nominal conditions of V_{IN}, $V_{BATTERY}$ and ambient temperature without the excessive temperatures associated with certain temporary charge conditions. This allows for higher charge currents (resulting in a faster charge) with the assurance that an occasional worst-case scenario will not overheat the system.

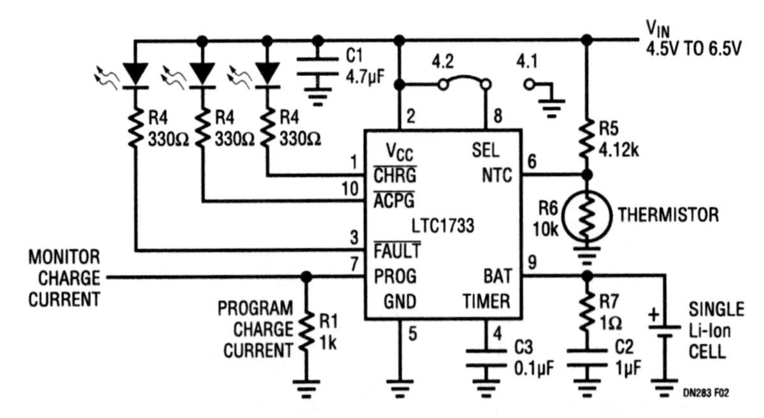

Figure 206.2 • Complete 1.5A Single Cell Li-Ion Charger for 4.1V or 4.2V Cells (No External MOSFET, Blocking Diode or Sense Resistor Required)

Dual battery power manager increases run time by 12% and cuts charge time in half

207

Mark Gurries

Introduction

To save space and provide longer battery run times, many high performance notebook computers support dual, swappable batteries, where each battery bay can hold a battery or an optional peripheral. Two batteries are obviously better than one, providing increased run time over a single battery, but how much better are they? The answer is that two batteries can produce performance that is better than twice that of a single battery. The trick lies in simultaneously charging and discharging the two batteries, as opposed to the traditionally easier, sequential method.

Although a multibattery simultaneous charge and discharge system can be more difficult to implement than a sequential system, paralleling the charging and discharging of dual batteries significantly reduces charge time and extends run time. The LTC1960 solves many design complexities by placing all of the hard-to-design charge and discharge control functions in a single package, making implementation of dual battery management systems possible for a wide variety of applications. In addition to controlling simultaneous battery charging and discharging, it controls all PowerPath switching between the two batteries, the wall adapter and the equipment's DC/DC converters, plus it includes many circuit protection functions. Figure 207.1 shows a block diagram of a typical application.

The LTC1960 comprises two controllers: the PowerPath controller manages the delivery of power from the two batteries and a DC input supply while the charge controller manages charging the batteries.

The heart of the PowerPath controller is the ideal diode circuit that allows precise voltage tracking between the batteries. The ideal diode circuit uses the same MOSFET transistors that turn power on and off, and makes them act like diodes, but without the power loss or variation in voltage drop as a function of current. Voltage loss, and hence power loss, is typically reduced by a factor of 30 over a Schottky diode. High speed comparators monitor reverse current conditions and shut off the MOSFETs in microseconds. An undervoltage detector watches for sudden loss of voltage at the load and turns on all the power sources in 10μs with no host intervention required. The host can also shut down the PowerPath in an emergency via a high speed shutdown input in the case of CPU overvoltage conditions or any other system-level crisis. Finally, there is a combined time and current based short-circuit protection system that protects the power path MOSFETs from destruction in the event of a short.

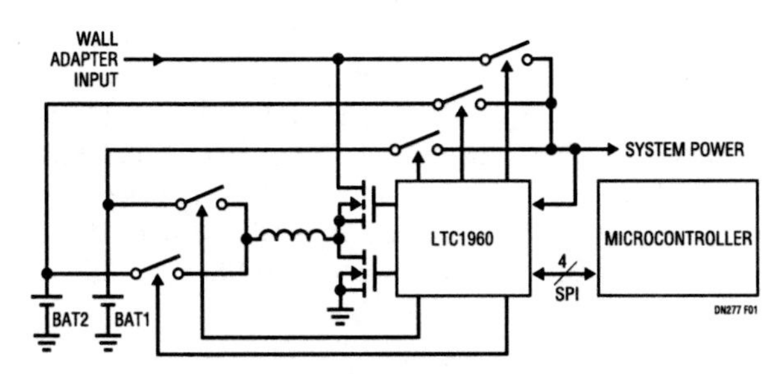

Figure 207.1 • The LTC1960 System Architecture

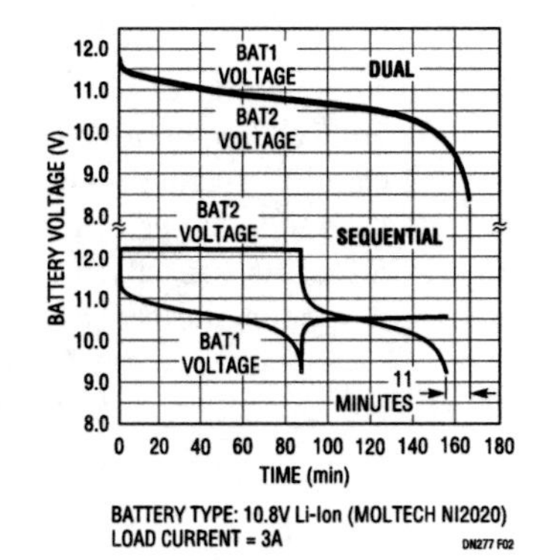

Figure 207.2 • Dual Discharge Extends Run Time

Analog Circuit Design: Design Note Collection. http://dx.doi.org/10.1016/B978-0-12-800001-4.00207-6

The charger controller uses synchronous rectification with a 0.5V low dropout capability and a 99% max duty cycle. An 11-bit voltage DAC with a system level accuracy of ±0.8% is provided along with a 5% system accurate 10-bit current DAC. With the ability to program from milliamps to amps, maintaining good current accuracy at low current is a challenge. The LTC1960 charger solves this problem by providing an optional pulse charge in the low current mode. A patented input current limit maximizes charge rate while the end product is operating without overloading the wall adapter. The IC's 5% precision current limit permits the user to accurately size the wall adapter and avoid overdesign and higher cost. An overvoltage comparator detects a sudden battery disconnect and shuts off the charger until the overvoltage condition is cleared.

Automatic current sharing

The LTC1960 does not control the current flowing into and out of each battery as they are charged and discharged. The ideal diode feature helps optimize battery charge times by allowing the batteries themselves to control current sharing. This is because the capacity or Amp-Hour rating of each battery determines how the current is shared. The current simply divides according to the ratio of the batteries' capacity ratings. Automatic steering of current allows both batteries to reach their full charge or full discharge points at the same time.

Simultaneous discharge increases run time

In high current drain applications, discharging two batteries in parallel more than doubles the run time over that of a single battery (see Figure 207.2). When two batteries share the load current equally, the current is halved in each battery, so internal battery I^2R power losses are reduced by one fourth. This reduction in internal battery power loss can lead to longer run times with increases of 12% or more.

Faster charge times with a second battery

It is possible with the LTC1960 to charge two batteries in the time it would take to charge one, without having to create separate charge circuits. Batteries that use a Constant Voltage (CV) mode during charge termination take a long time to reach their full capacity relative to batteries that use a Constant Current (CC) mode. Specifically, a Li-Ion battery charges to about 85% of its capacity in the first half of a charge time cycle and spends the second half filling the remaining 15%.

If two batteries receive charge current at the same time in the CV phase, the result can be a 25% reduction of total charge time. Another 25%, or more, time savings can come in the CC phase of charge, where the automatic current sharing of the batteries allows charging at a higher current rate relative to a single battery. In sum, you can reduce the charge time by about 50% relative to the sequential method, as shown in Figure 207.3.

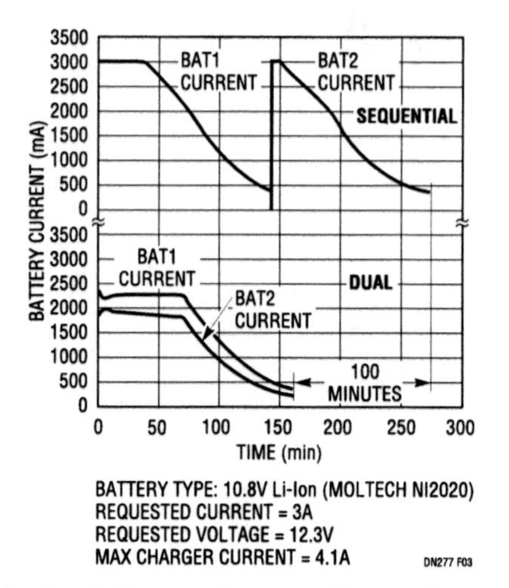

Figure 207.3 • Dual Charge Shortens Charge Time

Automatic crisis power management

An important feature of PowerPath control is the ability to handle sudden loss of power to the load. The LTC1960 manages power by monitoring the voltage on the power summing point of all three power sources comprising the wall adapter, battery 1 and battery 2. If there is a loss of power to the load, a programmable voltage comparator detects it and immediately connects all three sources to the load before it fails. This state is called 3-diode mode (3DM). The power source with the highest voltage will pick up the load, with multisource current sharing possible. The ideal diode circuit prevents energy transfer from any power source to any other power source. The system can remain in 3DM mode continuously making it possible to plug the LTC1960 into a circuit, without having to worry about control interfaces or programming—just plug and play.

Conclusion

The LTC1960 represents the first complete dual battery charge-discharge system solution on a chip. It reduces solution cost, development time, PCB space and part count while at the same time provides more control, safety, and automatic crisis management relative to any other solution available today. Combined with a host microcontroller, it has the flexibility to work in both user proprietary and Smart Battery based applications. The limits of what can be accomplished with LTC1960 are solely dependent on the software controlling the IC.

Single inductor, tiny buck-boost converter provides 95% efficiency in lithium-ion to 3.3V applications

208

Mark Jordan

Introduction

In portable applications powered by a single lithium-ion cell, the input voltage can typically change from 4.2V initially, down to 2.5V at end of life. It is a challenging task to provide a regulated voltage within the range of the battery. Until now, the most popular solution has been the SEPIC converter, but its mediocre efficiency and requirement of both a coupled inductor and a high current flyback capacitor make it a less than optimal solution. Another option is to cascade a boost converter with either an LDO or a buck converter, but the additional area and cost of the extra components, as well as low efficiency, are major drawbacks. Linear Technology's new LTC3440 buck-boost converter provides the most compact solution with the highest efficiency, thereby reducing cost, increasing battery life and saving precious real estate.

The LTC3440 incorporates a patent-pending control technique to efficiently regulate an output voltage above, below or equal to the input source voltage with a single inductor by properly phasing the four internal switches. Efficiencies well over 90% are achieved for the entire battery range without the use of Schottky rectifier diodes. The low $R_{DS(ON)}$ (0.19Ω NMOS, 0.22Ω PMOS), low gate charge synchronous switches, along with minimal break-before-make times, provide high frequency, low noise operation with high efficiency.

For light loads, the part offers user controlled Burst Mode operation to maximize battery life, drawing only 25μA of quiescent current. The operating frequency can be programmed from 300kHz to 2MHz by changing the value of the timing resistor on the R_T pin. Users can synchronize the operating frequency by connecting an external clock to the MODE/SYNC pin. The part can also be commanded to shut down by pulling the $\overline{SHDN}$/SS pin low. In shutdown, the part draws less than 1μA of quiescent current and disconnects the output from the input supply. During start-up, the ramp rate of the output voltage is controlled by the external soft-start components. This controlled ramp rate provides for inrush current limiting. Housed in a thermally enhanced 10-lead MSOP package, the LTC3440 is ideal for portable power applications requiring less than 2W of output power.

All ceramic capacitor, single inductor, 2W Li-Ion to 3.3V converter

An all-ceramic capacitor, lithium-ion to 3.3V application at 600mA is shown in Figure 208.1. The operating frequency is programmed to 1MHz and soft-start is incorporated with R4 and C3. The efficiency curves versus load current for the Li-Ion battery range are shown in Figure 208.2. With Burst

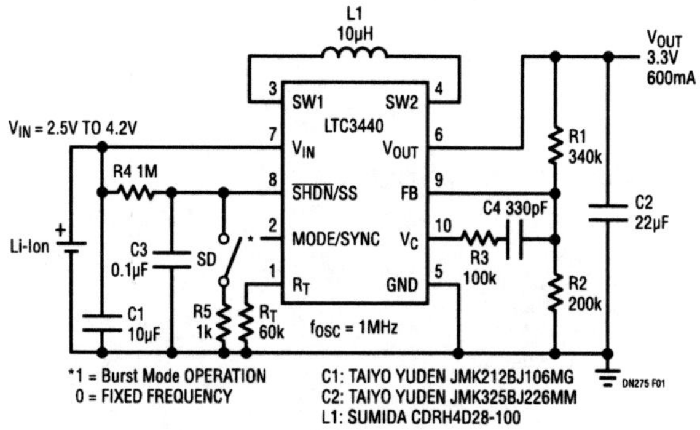

Figure 208.1 • Simple Lithium-Ion to 3.3V Converter at 600mA

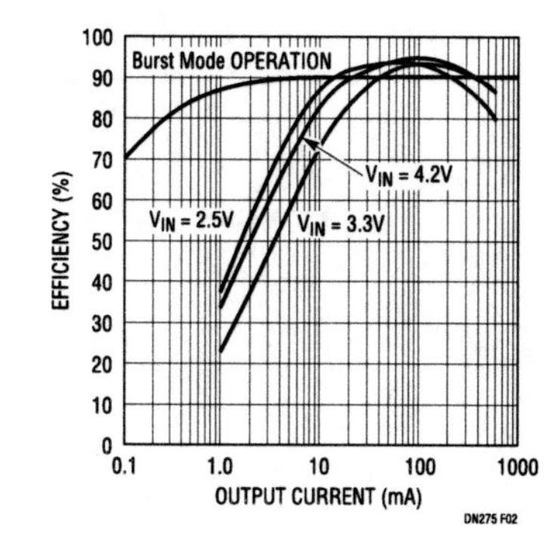

Figure 208.2 • Li-Ion to 3.3V Efficiency

Analog Circuit Design: Design Note Collection. http://dx.doi.org/10.1016/B978-0-12-800001-4.00208-8

Mode operation enabled at light loads, efficiencies of over 85% are achieved for more than three decades of load current. At 200μA, the efficiency remains above 70%, primarily due to the low 25μA quiescent current in Burst Mode operation. In many applications the decreased load demand on the converter is known by the application and the converter can be commanded to enter power saving Burst Mode operation by driving the MODE/SYNC pin high.

WCDMA dynamically controlled power amp power supply

For the new third generation (3G) cellular phones, the high speed data transmission imposes a stringent power demand on the battery. Maximum overall efficiency and operation over the entire battery voltage range are required to maximize run time. A 2W, dynamically controlled power supply for a WCDMA cell phone power amplifier (PA) is shown in Figure 208.3. By adjusting the voltage across the PA, the overall efficiency to the antenna is improved, and a linear PA can be utilized. At peak power, the PA requires the highest programmed voltage, typically 3.4V to 4V depending on the PA. At the lowest power level, when only voice is transmitted and the user is close to the base station, the PA draws less than 100mA and requires a lower voltage, typically between 0.4V and 2V. Since the LTC3440 can regulate an output voltage above, below or equal to the battery voltage, the maximum transmit power can be maintained over the entire voltage range. For applications requiring a program voltage below 2V, a Schottky diode is required from the SW2 to V_{OUT} pins to provide a low impedance power path since the internal synchronous switch loses gate drive at low output voltage. Figure 208.4 demonstrates the efficiency of the converter versus input voltage at various load currents. The transient response of the power converter for a 1.5V output voltage change, commanded by the DAC, is shown in Figure 208.5.

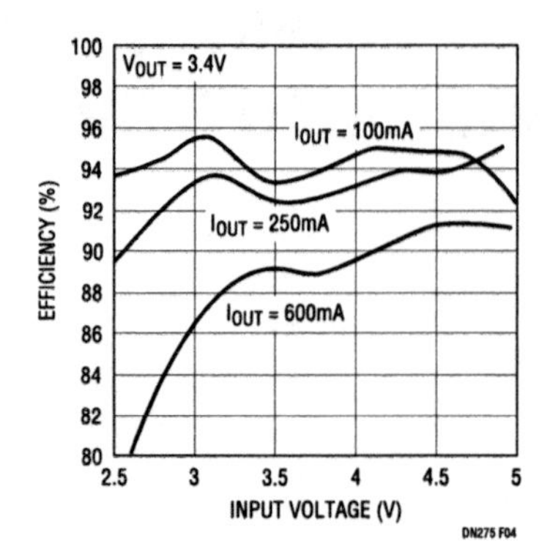

Figure 208.4 • Efficiency of the WCDMA Power Amp Power Supply

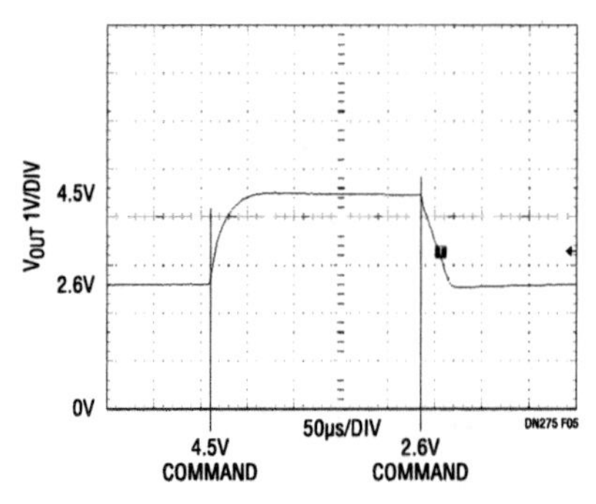

Figure 208.5 • Output Voltage Transient Response of the WCDMA Power Supply

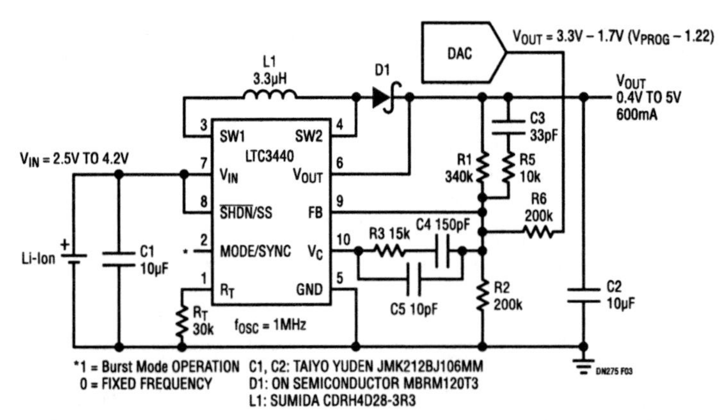

Figure 208.3 • WCDMA Power Amp Power Supply with Dynamic Voltage Control

Tiny step-up/step-down power supply delivers 3.3V at 1.3A in battery-powered devices

209

Keith Szolusha

Introduction

Many of today's cellular phones, PDAs, MP3 players and other portable devices require that a consistent 3.3V power supply be delivered from a single, rechargeable lithium-ion battery. The problem is that a Li-Ion battery at full charge has a voltage somewhat higher than 3.3V and loses voltage over the life of a charge to less than 3.3V. A power supply that depends on a Li-Ion battery requires a voltage regulator that can maintain consistent 3.3V output from both high and low inputs. Since typical Li-Ion applications are portable mass-market devices where the focus is on short time-to-market, long battery life, small size and low cost, the regulator must take minimal space, be highly efficient and use inexpensive off-the-shelf components wherever possible.

Figure 209.1 shows a simple and efficient SEPIC converter that provides up to 1.3A output at 3.3V with input from a single 2.7V to 4.2V battery. Its simplicity, low cost, efficiency and small component size (no component is higher than 3.1mm) satisfy many of the size and power-consumption requirements of battery-powered portable devices.

Regulated output voltage from a range of inputs

Although a SEPIC is only one of many possible configurations for a DC/DC converter, it has a major advantage over other choices when applied in a modern portable device. The SEPIC in Figure 209.1 delivers 3.3V output throughout the input voltage range, 2.7V to 4.2V. One alternative to the SEPIC configuration is a simple step-down converter. A step-down converter delivers 3.3V as long as the battery voltage remains above 3.6V, but it begins to drop out when the battery voltage drops below 3.6V, producing output voltage a few hundred millivolts below the input voltage. The output voltage will drop all the way down below 2.5V as the input voltage drops to 2.7V.

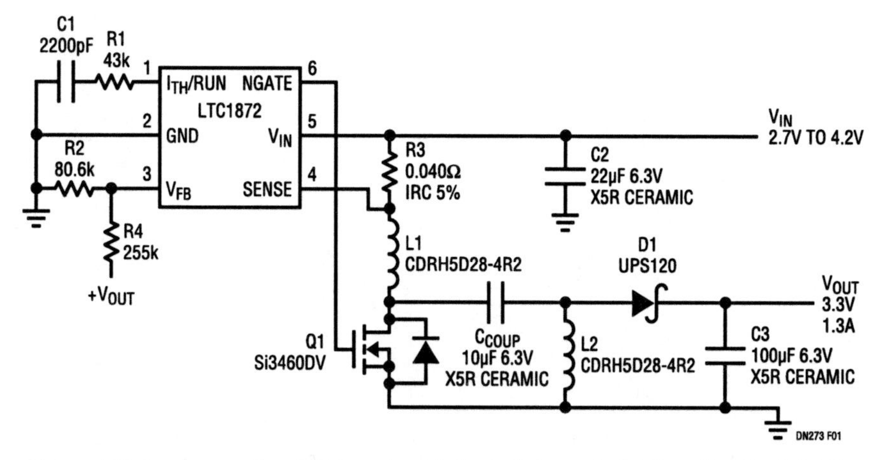

Figure 209.1 • Converting Lithium-Ion Battery Voltage to 3.3V, 1.3A with SOT-23 DC/DC Converter Has Minimal Components Count and 3.1mm Maximum Height

Analog Circuit Design: Design Note Collection. http://dx.doi.org/10.1016/B978-0-12-800001-4.00209-X

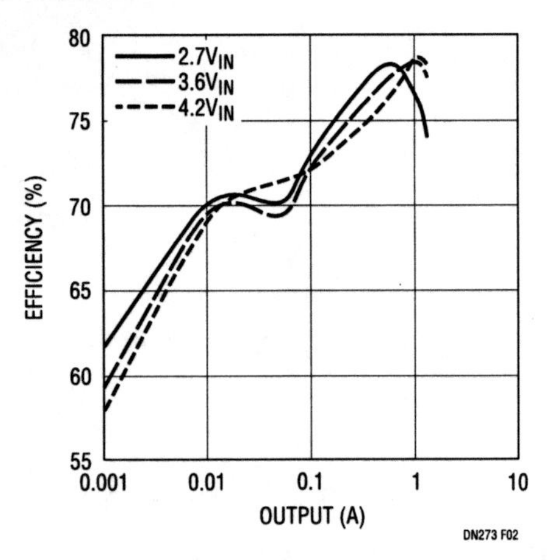

Figure 209.2 • Typical Efficiency of Lithium-Ion Battery to 3.3V, 1.3A LTC1872 DC/DC Converter

Figure 209.3 • Two Inductor Design Offers Better Performance than a Design Using a Transformer

Highly efficient

This design can provide DC/DC conversion efficiencies of up to 78%, important for maximizing battery life in portable devices. Figure 209.2 shows the efficiency curve for various input voltages and output currents. Note that above 10mA output the circuit is greater than 70% efficient and climbs to 78% efficiency at 1A output. It remains greater than 60% for output currents as low as 1mA, due mainly to the low quiescent current of the LTC1872. If a shutdown mode is desired, the LTC1872's extremely low 22μA (maximum) shutdown current also prolongs battery life.

The simplicity of this circuit minimizes cost, board space and design headaches. This 550kHz current mode SOT-23 controller drives a single TSOP-6 N-channel MOSFET. The 10μF ceramic coupling capacitor has extremely high RMS ripple current capabilities for its small size and cost. The two small 4.2μH, 2.2A inductors (L1 and L2) are no more than 3.0mm high and need not be placed next to each other as in the alternative magnetics choice, a single bigger and costlier transformer with maximum height well above 3mm. In general, uncoupled power inductors are available from more sources and come in more inductance and DC current ratings than equivalent transformers. A transformer also limits the flexibility of this layout. Figure 209.3 demonstrates how two separate inductors (with a combined cost still less than an equivalent transformer) are placed to minimize the circuit size and form the shortest possible high frequency AC switching path and thus the most noise immune layout possible. The most likely choice for a transformer replacement of the two inductors is shown for reference. The transformer has a maximum height of 6.0mm, well above the 3.1mm maximum height of the two inductor circuit. The bulky 4-pin transformer increases the size of the circuit and makes layout more difficult.

The 100μF ceramic output capacitor has a high ripple current rating and extremely low ESR resulting in limited output voltage ripple. This particular design requires only a 22μF ceramic input capacitor because SEPICs have low input ripple current (due to the continuous current in inductor L1). The tiny Schottky diode with 2A current rating takes very little space and the whole board can be kept very small by using the LTC1872 in its thin SOT-23 package.

The LTC1872 provides a 2.5V undervoltage lockout feature that prevents current runaway at low input voltages, particularly important for Li-Ion battery-powered devices.

A very low cost SOT-23 Li-Ion battery charger requires little area and few components

210

David Laude

The LTC1734 is a low cost, single cell Li-Ion battery charger with constant voltage and constant current control. The small quantity and low cost of the external components results in a very low overall system cost, and with the IC's 6-pin SOT-23 package, provides a compact design solution. Previous products usually required an external current sensing resistor and blocking diode, but these functions are now provided within the LTC1734. Other features include:

- 1% accurate 4.1V or 4.2V float voltage
- Programmable constant current range of 200mA to 700mA
- Charging current monitor and manual shutdown for use with a microcontroller
- Automatic shutdown with no battery drain after wall adapter removal

Applications include portable devices such as cellular phones, digital cameras and handheld computers. The LTC1734 can also be used as a general purpose current source or for charging nickel-cadmium and nickel-metal-hydride batteries.

A simple low cost Li-Ion charger

A battery charger programmed for 300mA in the constant current mode with a charge current monitoring function is shown in Figure 210.1. The PNP is needed to source the charging current and resistor R1 is used to program the maximum charging current. The I_{SENSE} and BAT pins are used to monitor charge current and voltage, respectively, while the DRIVE pin controls the PNP's base. Note that no external current sense resistor or diode to block reverse current is required. For most other chargers a blocking diode, in series with the supply, is required to prevent draining the battery should the unpowered supply input become a low impedance. When the supply is opened or shorted to ground, the charger shuts down and only a few nanoamps of leakage current flows from the battery to the charger. This feature extends battery life, especially if the portable device is off for long periods of time. The supply voltage can range from 4.75V to 8V, but power dissipation of the PNP may become excessive near the higher end, especially at higher charging current levels. The PNP's power dissipation will require attention to adequate heat sinking. Refer to the PNP manufacturer's data sheet for heat sinking requirements.

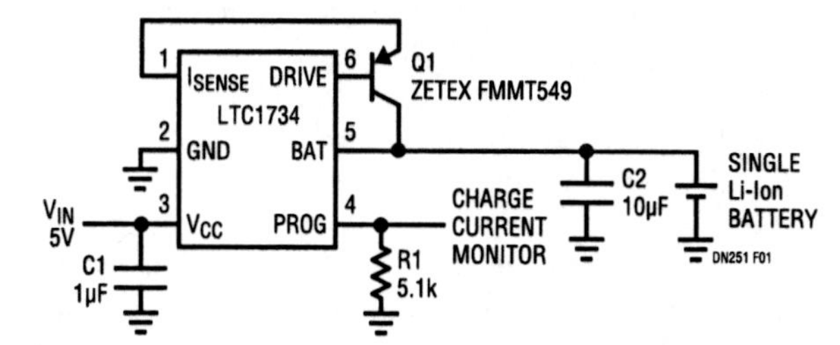

Figure 210.1 • Low Cost Li-Ion Charger Programmed for 300mA

With the supply voltage near its low end, the PNP's saturation voltage becomes important. In this case a low V_{CESAT} transistor, such as those shown in the figures, may be required to prevent the PNP from heavily saturating and demanding excessive base current from the DRIVE pin.

To maintain good AC stability in the constant voltage mode, a capacitor is required across the battery to compensate for inductance in the wiring to the battery. This capacitor (C2) may range from 4.7µF to 100µF, and its ESR can range from near zero to several ohms depending on the inductance to be compensated. In general, compensation is best with a capacitance of 4.7µF to 22µF and an ESR of 0.5Ω to 1.5Ω. In the constant current mode, good AC stability is realized by keeping capacitance on the PROG pin to less than 25pF. Higher capacitive loading, such as from a low-pass input filter to an ADC, can be easily tolerated by isolating the capacitance with at least 1kΩ of resistance.

If the input supply is hot plugged, a ceramic input capacitor (C1) should be avoided because its high Q can cause volt-

Analog Circuit Design: Design Note Collection. http://dx.doi.org/10.1016/B978-0-12-800001-4.00210-6

age transients of up to twice the DC supply level and possibly damage the charger. If using such a low ESR capacitor, adding a resistance of 1Ω to 2Ω in series with C1 will sufficiently dampen these transients.

The programming pin (PROG) accomplishes several functions. It is used to set the current in the constant current mode, monitor the charging current and manually shut down the charger. In the constant current mode, the LTC1734 maintains the PROG pin at 1.5V. The program resistor value is determined by dividing 1.5V by R1's desired current while in the constant current mode. The charging current is always 1000 times the current through R1 and is therefore proportional to the PROG pin voltage. The PROG pin voltage drops below 1.5V as the constant voltage mode is entered and charging current drops off. At 1.5V the charging current is the full 300mA, while at 0.15V the current is 1000 · (0.15/5100) or about 30mA. If the grounded side of R1 is pulled above 2.15V or is allowed to float, the charger enters the manual shutdown mode and charging ceases. These features support charging the battery to its full capacity by allowing a microcontroller to monitor the charging current and shut down the charger at the appropriate time. An internal 3μA pull-up current will pull the floated PROG pin up. By design, this current adds no error, but does set a minimum current through the program resistor of 3μA.

While charging in the constant voltage mode, currents produced by active dynamic loads may create excessive transient levels on the PROG pin. If desired, these transients can be filtered with a simple RC low-pass filter. Connect a 1k resistor to the PROG pin with its opposite end connected to a 0.1μF capacitor with its other end grounded. Monitor the filtered PROG voltage at the RC common node. Load transients are not reflected onto the PROG pin if the charger remains in the constant current mode.

A programmable constant current source

An example of a programmable current source is shown in Figure 210.2. To insure that only the constant current mode is activated, the BAT pin is tied to ground to prevent the constant voltage control loop from engaging. Control inputs (CONTROL 1, CONTROL 2) either float or are pulled to ground. This can be achieved by driving them from the drains of NMOS FETs or from the collectors of NPNs. When both inputs are floating, manual shutdown is entered. Connecting Control 1 to ground causes 500mA of current while Control 2 results in 200mA of current. When both control inputs are grounded the current is 700mA. The choice of the PNP depends upon its power dissipation. A voltage DAC, connected to the PROG pin through a resistor, could also be used to control the current. A PWM source, connected to a control input, can be used to modulate the current. Pulse width modulation is useful for wide range or fine control of the average current and can be used to extend the constant current range to below 200mA. Applications include charging nickel-cadmium or nickel-metal-hydride batteries, driving LEDs or biasing bridge circuits.

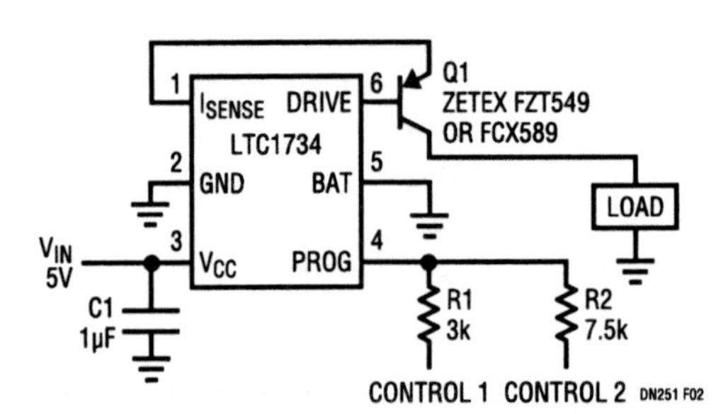

Figure 210.2 • Programmable Current Source with Output Current of 0mA, 200mA, 500mA or 700mA

Simple Li-Ion charge termination using the LT1505

211

Mark Gurries

Li-Ion batteries are normally charged with a current limited constant voltage for a fixed length of time. At the end of this time period, the voltage must be removed to prevent internal chemistry changes in the battery. At a minimum, a timer is needed to terminate the charging process after the maximum amount of time required to fully charge the battery. However, if the battery is never fully discharged, this overall time period is too long, forcing a longer charge cycle than is really needed. One of the ways to determine when a Li-Ion battery is in a fully charged state is to monitor the charge current flowing into the battery during the constant voltage phase of the charging cycle. When the charge current falls below a preset level, the battery is fully charged and the charger can be disabled. The problem is how to detect this low level of current so as to terminate the charge cycle. The LT1505 produces a signal to allow this to happen.

The LT1505 provides a logic signal output called "FLAG" that is intended to help implement a Li-Ion charge termination mechanism. In the constant voltage portion of the charge cycle, the battery continues to accept charge and the charge current continues to drop. When the charge current drops below 20% of the full-scale value, the FLAG output goes low.

Normally, the FLAG function is used to let the system know that the battery is nearly fully charged. However, the flag can also be used to change the time scale of the overall charge-cycle timer and convert it to a top-off charge timer. Thus, the user gets a battery that is almost fully charged, but the charger remains on so that the remaining time can be used to charge the battery to 100%, after which the charger will turn off.

The nature of Li-Ion battery charging is such that the battery spends about half of the total charge time receiving about 85% of its capacity and the remaining half to receive the last 15%. The LT1505's 20% current threshold represents a compromise between charge time and a fully charged battery. It represents around 85% of a typical battery's charge capacity (see Figure 211.1).

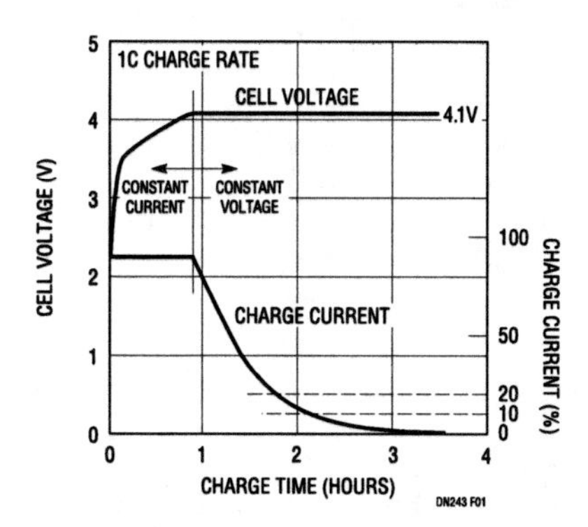

Figure 211.1 • Typical Lithium-Ion Charge Characteristics

The circuit in Figure 211.2 uses a CD4541 programmable timer IC controlled by the LT1505. When a low-true reset start pulse, created by the pulse circuit around Q1, is sensed by the MR(L) pin of the CD4541, the internal counters are reset and begin counting at a rate of 3Hz for 2^{16} (65,536) counts, or approximately three hours. If the FLAG pin of the LT1505 goes low, a new reset pulse is generated and the count is changed to 2^{13} (8192) counts or approximately 45 minutes.

Another popular current trip point is 10% of full scale for about 90% or more battery capacity charge. The trade-off is charge time. Since the current falls asymptotically toward zero, there is a significantly longer time period needed to reach the 10% point relative to the 20% point. The choice of 20% or 10% is a system design issue. To rescale the trip point, a resistor, R_{CAP}, is added between the LT1505 CAP pin and ground in parallel with the existing 0.1µF capacitor. For a 4A charger, the R_{CAP} resistor is 68k for a 400mA threshold. Threshold values as low as 7.5% of full scale are possible. However, below this level the repeatable accuracy becomes poor due to analog offset voltage inside the LT1505. Consult Linear Technology's Application department for more information.

Analog Circuit Design: Design Note Collection. http://dx.doi.org/10.1016/B978-0-12-800001-4.00211-8

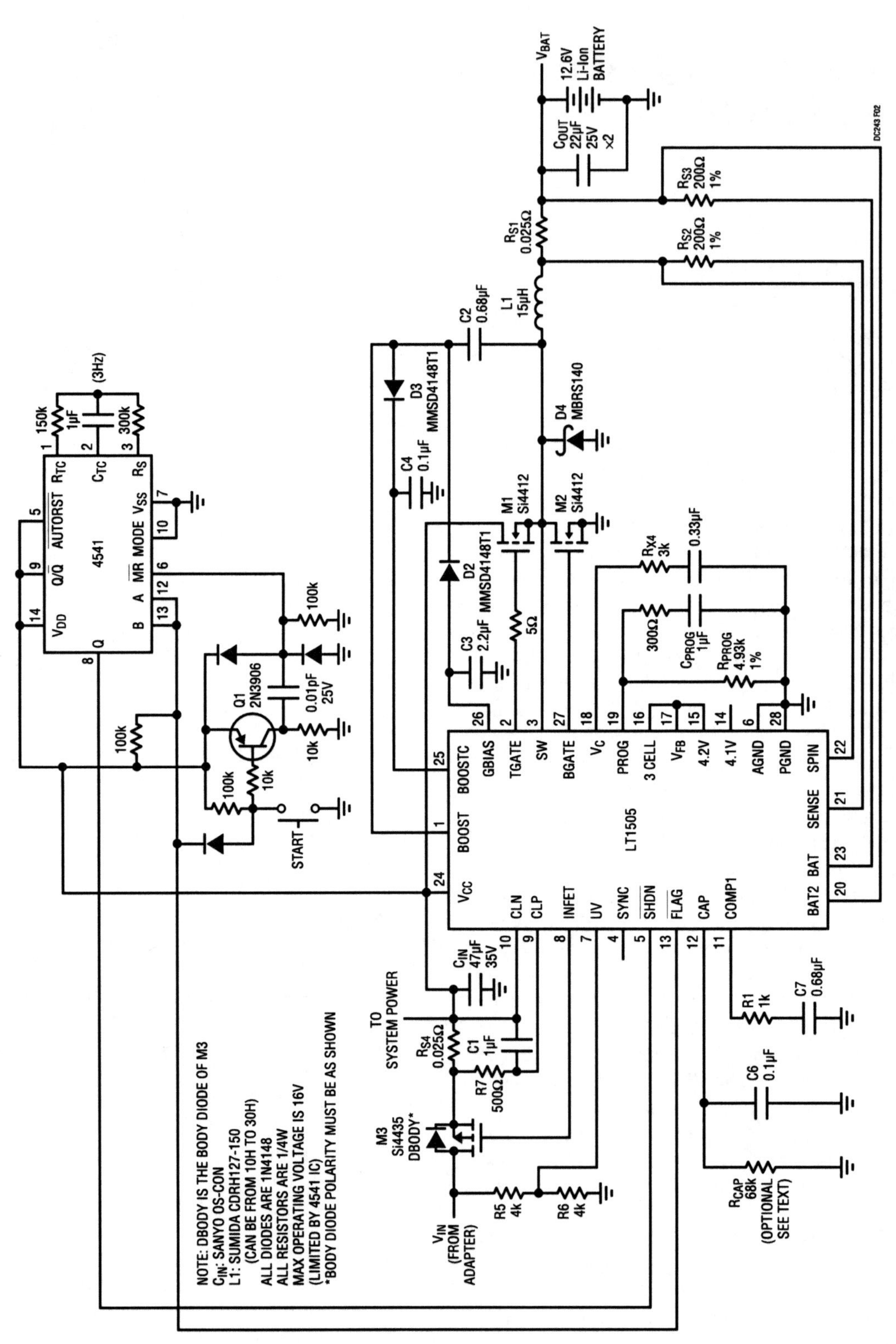

Figure 211.2 • LT1505 Li-Ion Charger with Termination

Li-Ion charge termination IC interfaces with PWM switchers

212

Fran Hoffart

Rechargeable lithium-ion batteries are rapidly becoming the battery of choice for many battery-powered products. These products include notebook computers, PDAs, video camcorders, digital cameras, cellular phones, portable test equipment and others. Compared to other rechargeable power sources, Li-Ion batteries have higher energy density for both weight and volume and provide longer run time between charges.

Charging Li-Ion batteries is a relatively simple process. Apply a current-limited (at a 1**C** rate) constant voltage (±1% tolerance) for approximately three hours, then stop charging. (**C** is a battery term used to indicate the ampere-hour capacity of a battery). A complete charge cycle may also include precharge qualification for battery temperature and precharge qualification for deeply discharged batteries.

Battery pack protection

Because of the high energy associated with lithium-ion cells and their sensitivity to abuse, many battery manufacturers require protective devices inside the battery pack for both safety and performance reasons. These devices often consist of poly fuses, thermal fuses and bimetallic breakers, which protect the battery from overtemperature and overcurrent conditions. In addition, battery packs may also contain back-to-back MOSFET switches that disconnect the battery if an overcurrent condition exists for either charge or discharge, or if an overvoltage or undervoltage condition exists. Many of these precautions are needed because lithium-ion cells are easily damaged by both overcharge and overdischarge conditions.

LT1510 battery charger IC

The LT1510 is a high efficiency switching regulator power IC designed specifically for battery charging applications. A step-down current mode 200kHz or 500kHz PWM topology is used. Included on the die is a 2A switch along with programmable current and voltage control circuitry. Available in a 16-pin SO and the tiny MSOP surface mount package, the LT1510 is capable of providing up to 1.5A of charge current

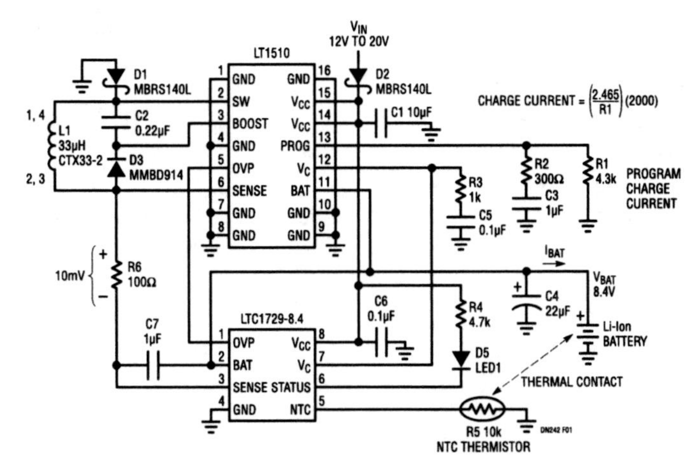

Figure 212.1 • Complete 1.3A Battery Charger for Two Li-Ion Cells

Analog Circuit Design: Design Note Collection. http://dx.doi.org/10.1016/B978-0-12-800001-4.00212-X

in many situations. Although the LT1510 provides many charger functions, it lacks a timer and requires precision resistors to program the charge voltage. This is when the LTC1729 should be added.

LTC1729 Li-Ion charge termination IC

This 8-pin IC interfaces with the LT1510 (as well as other LTC charger products, such as the LT1511, LT1769, LT1505, LT1512 and LT1513) to provide a complete Li-Ion charger solution. The LTC1729 provides a precision voltage divider for programming the charge voltage and includes a preconditioning trickle charge for deeply discharged cells. Battery temperature is monitored using a thermistor; a 3-hour timer ends the charge cycle. Also included is a status output pin that provides a signal when the charge current drops below a programmable threshold level, indicating a near-full-charge condition. This signal can be used to drive an LED, to provide charge indication to other circuitry or to terminate the charge when the charge current drops below the threshold level.

Complete 2-cell Li-Ion charger

A complete constant-current/constant-voltage Li-Ion 2-cell charger is shown in Figure 212.1. The LT1510 provides the charge current and the LTC1729 provides the charge termination. R1 allows the charge current to be easily programmed.

The charge cycle

A typical charge cycle of the circuit in Figure 212.1 is as follows: with the input voltage applied and no battery connected, the charger output is pulled high by an internal 200μA current source in the LT1510 and clamped at 9V by the LTC1729. Connecting a discharged 2-cell battery to the charger will pull the charger output (BAT pin) down to the battery voltage, starting the charge cycle. For temperature qualification, the voltage on the NTC (thermistor) pin must be between 0.405V and 2.79V, indicating that the temperature is between 0°C and 50°C; otherwise the charge cycle is put on hold until the temperature is within this range. For deeply discharged batteries with voltages below 5.2V, a preconditioning 16mA trickle charge begins and continues until the battery voltage exceeds 5.2V.

After the qualification and preconditioning is completed, the constant-current portion of the charge cycle begins (see curves in Figure 212.2). As the battery accepts charge, the battery voltage rises and approaches the programmed voltage of 8.4V, at which time, the constant-voltage portion of the charge cycle begins. With the battery voltage held constant, the charge current will drop exponentially, eventually reaching tens of mA before the 3-hour timer expires, thus ending the charge cycle.

The open-drain STATUS pin is pulled low when the battery is installed and the charge current is greater than 100mA. When the charge current drops below 100mA, a 50μA current source pulls this pin low and, after the 3-hour timer has timed out, this pin is open circuit. The 100mA threshold level is programmable by changing R6.

Board layout and testing

Although the LT1510 can provide charge current in excess of 1A, a good thermal layout of the PC board is required. Wide copper traces for the ground pins, feedthrough vias and generous amounts of copper on both sides of the board are all necessary to minimize the IC temperature rise.

When testing the charger, use either a battery or a battery simulator for a load (a conventional electronic load presents a high impedance, unlike a battery). A simple battery simulator consists of an adjustable lab power supply with a load resistor across the output. Select a resistor that will result in approximately twice the rated charge current flowing through it. The power supply can now be used in place of the battery for testing purposes. A fully discharged to a fully charged battery can be simulated by varying the power supply voltage.

For additional information, please consult the LT1510 and LTC1729 data sheets.

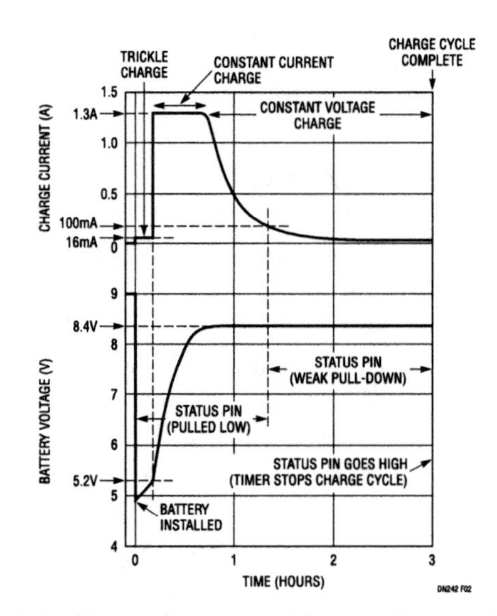

Figure 212.2 • Charge Current and Battery Voltage for a Typical Charge Cycle

A miniature, low dropout battery charger for lithium-ion batteries

213

James Herr

Introduction

Lithium-ion (Li-Ion) batteries are becoming the power source of choice for today's small handheld electronic devices due to their light weight and high energy density. However, there are a number of important issues associated with the charging of these batteries. As an example, if they are overcharged, they can be potentially hazardous to the user.

The LTC1731 is a constant-current/constant-voltage linear charger controller for single cell lithium-ion batteries. Its output accuracy of ±1% (max) over the −40°C to 85°C range prevents the possibility of overcharging. The output float potential is internally set to either 4.1V or 4.2V, without the use of an expensive external 0.1% resistor divider. Furthermore, the charging current is user programmable with ±7% accuracy.

At the beginning of the charging cycle, if the battery voltage is below 2.457V, the LTC1731 will precharge the battery with only 10% of the full-scale current to avoid stressing the depleted battery. Once the battery voltage reaches 2.457V, normal charging can commence. Charging is terminated by a user-programmed timer. After this timer has completed its cycle, the charging can be restarted by removing and then reapplying the input voltage source, or by shutting down the part momentarily. A built-in end-of-charge (C/10) comparator indicates that the charging current has dropped to 10% of the full-scale current. The output of this comparator can also be used to stop battery charging before the timer completes its cycle.

The LTC1731 is available in the 8-pin MSOP and SO packages.

Operation and circuit description

Figure 213.1 shows a detailed schematic of a 500mA single cell Li-Ion battery charger using the LTC1731-4.2. The charge current is programmed by the combination of a program resistor (R_{PROG}), from the PROG pin to ground, and a sense resistor (R_{SENSE}), between the V_{CC} and SENSE pins. R_{PROG} sets a program current through an internal, trimmed 800Ω resistor, setting up a voltage drop from V_{CC} to the input of the current amplifier (CA). The current amplifier controls the gate of an external P-channel MOSFET (Q1) to force an equal voltage drop across R_{SENSE}, which in turn, sets the charge current. When the potential at the BAT pin approaches the preset float voltage, the voltage amplifier starts sinking current, which decreases the required voltage drop across R_{SENSE}, thereby reducing the charge current.

Charging begins when the potential at the V_{CC} pin rises above 4.1V. At the beginning of the charge cycle, if the battery voltage is below 2.457V, the charger goes into trickle charge mode. The trickle charge current is 10% of the full-scale current. If the battery voltage stays low for one quarter of the total programmed charge time, the charge sequence is terminated.

The charger goes into the fast charge, constant-current mode after the voltage on the BAT pin rises above 2.457V. In constant-current mode, the charge current is set by the combination of R_{SENSE} and R_{PROG}. When the battery approaches the final float voltage, the voltage loop takes control and the charge current begins to decrease. When the current drops to

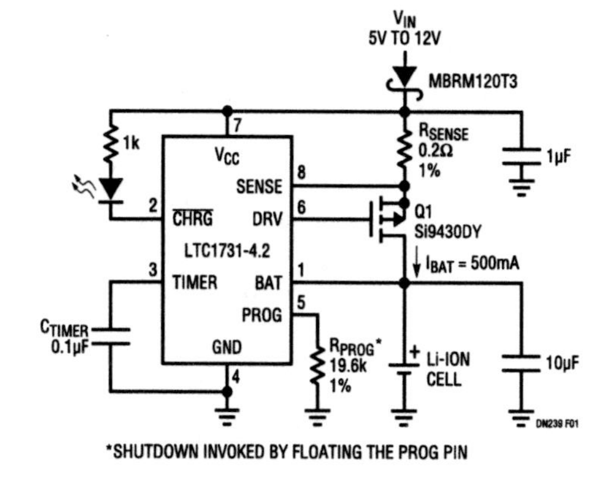

Figure 213.1 • Single Cell 500mA Li-Ion Battery Charger

Analog Circuit Design: Design Note Collection. http://dx.doi.org/10.1016/B978-0-12-800001-4.00213-1

10% of the full-scale charge current, an internal comparator turns off the pull-down N-channel MOSFET at the $\overline{CHRG}$ pin and connects a weak current source to ground to indicate an end-of-charge (C/10) condition.

An external capacitor on the TIMER pin (C_{TIMER}) sets the total charging time. Once the timer cycle completes, the charging is terminated immediately and the $\overline{CHRG}$ pin is forced to a high impedance state. To restart the charge cycle, simply remove the input supply and then reapply it, or alternatively, float the PROG pin momentarily.

For batteries such as lithium-ion that require accurate final float potential, the internal 2.457V reference, voltage amplifier and the resistor divider provide regulation with better than ±1% accuracy. For NiMH and NiCd batteries, the LTC1731 can be turned into a current source by simply connecting the TIMER pin to V_{CC}. When in the constant-current only mode, the voltage amplifier, timer and the trickle charge function are all disabled.

When the input voltage is not present, the charger goes into a sleep mode, dropping I_{CC} to 7μA. This greatly reduces the current drain on the battery and increases the standby time. The charger can always be shut down by floating the PROG pin. An internal current source will pull this pin's voltage high and clamp it at 3.5V.

Programming charge current

The formula for the battery charge current is:

$$I_{BAT} = (I_{PROG}) \cdot (800\Omega/R_{SENSE}) = (2.457V/R_{PROG}) \cdot (800\Omega/R_{SENSE})$$

where R_{PROG} is the total resistance from the PROG pin to ground.

For example, if a 500mA charge current is needed, select a value for R_{SENSE} that will drop 100mV at the maximum charge current. $R_{SENSE} = 0.1V/0.5A = 0.2\Omega$, then calculate:

$$R_{PROG} = (2.457V/500mA) \cdot (800\Omega/0.2\Omega) = 19.656k$$

For best accuracy over temperature and time, 1% resistors are recommended. The closest 1% resistor value is 19.6k.

Typical application

1.5A single cell battery charger

The LTC1731 can also be connected as a switcher-based battery charger for higher charging current applications (see Figure 213.2). As in the linear charger, the charge current is set by R3 and R5. The $\overline{CHRG}$ pin output will indicate an end-of-charge (C/10) condition when the average current drops down to 10% of the full-scale value. A 100μF bypass capacitor is required at the BAT pin to keep the ripple voltage low.

Conclusion

The LTC1731 makes a very compact, low parts count and low cost multiple battery chemistry charger. The onboard programmable timer provides charge termination without interfacing to a microprocessor.

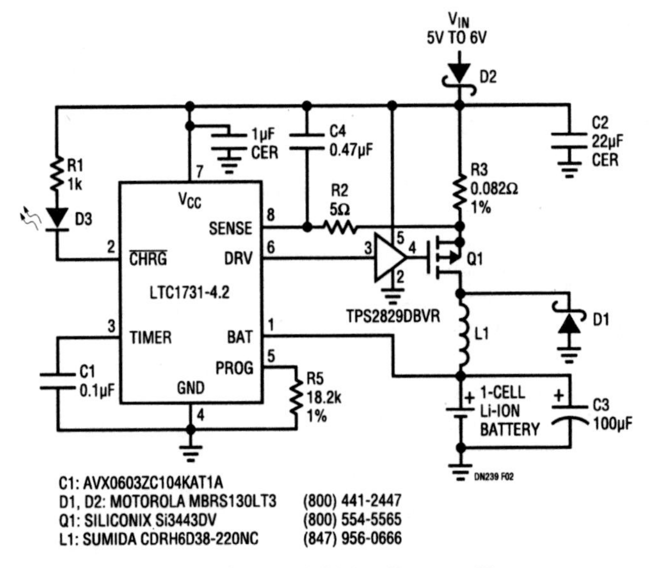

Figure 213.2 • Single Cell 1.5A Li-Ion Battery Charger

New charger topology maximizes battery charging speed

214

Fran Hoffart

Introduction

Battery charging in notebook computers and other portable products generally involves compromises. A notebook computer's AC adapter is usually sized to charge the battery at its maximum rate when the computer is off. In this condition, the computer draws essentially no power, so the full capacity of the adapter can be used for charging the battery. However, when the computer is turned on, charging current is usually reduced to a low rate to avoid overloading the AC adapter. You can use your computer or fast charge your battery, but not both at the same time.

Linear Technology has developed a new battery charger topology that maximizes the battery charging rate, even when the computer is on, without increasing the size or capacity of the AC adapter power source. Instead of simply reducing the charging current to an arbitrarily low level, the charger monitors the current drawn from the AC adapter and automatically reduces the charging current only when necessary to avoid overloading the AC adapter. At all other times, the charging current can be at the maximum programmed value if the battery demands it. Since average notebook power consumption is considerably lower than peak demand, battery charging can now continue at nearly the same rate whether the computer is on or off. This translates to faster battery charging and, in many cases, a smaller, less expensive AC adapter.

LT1511 battery charger IC

The block diagram in Figure 214.1 shows the basic functions performed by a battery charger IC using this patented topology.[1] The LT1511 is a high efficiency 200kHz switching regulator IC in a step-down configuration suitable for charging lithium-ion batteries. It contains multiple feedback loops for constant charge voltage, constant charge current and input

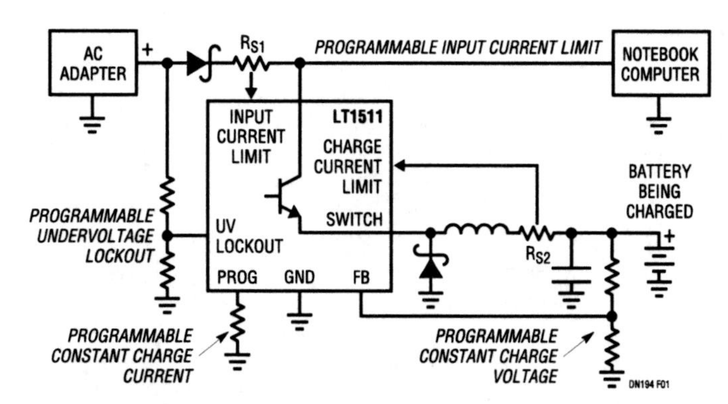

Figure 214.1 • Block Diagram of LT1511 Step-Down Battery Charger Illustrating Input Current and Charge Current Limit Functions

Note 1: US patent number 5,723,970.

Analog Circuit Design: Design Note Collection. http://dx.doi.org/10.1016/B978-0-12-800001-4.00214-3

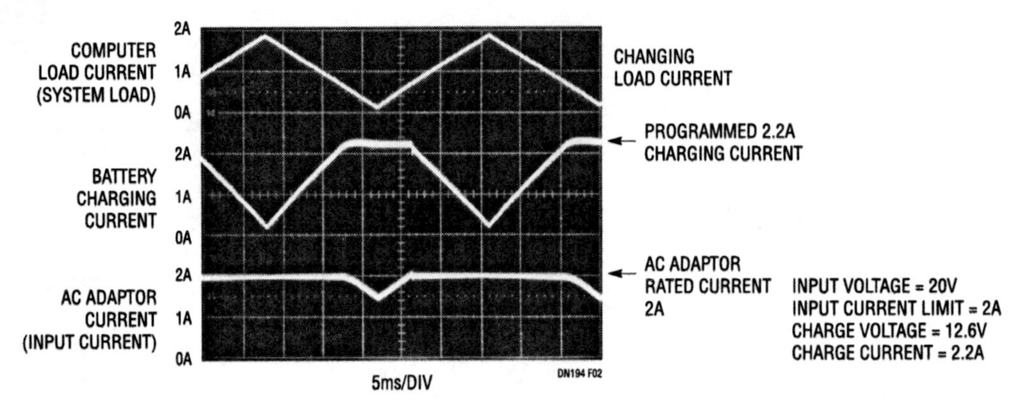

Figure 214.2 • Current Waveforms Show How Charging Current (Middle) Drops When Laptop Computer Current (Top) Rises to Ensure That AC Adapter Current (Bottom) Does Not Exceed Programmed Limit

current limit. Low value resistors (R_{S1} and R_{S2}) are used to sense the charge current and the current drawn from the input power source (AC adapter). The input current limit control loop allows the input power supply or AC adapter to provide current to power notebook circuitry and simultaneously charge a battery without overloading the input power supply. As the notebook current requirements increase, the LT1511 begins adjusting the battery charging current downward to keep the input power supply current below a predetermined limit.

Also included on the die are a 4A switching transistor, precision 0.5% voltage reference, adjustable undervoltage lockout and autoshutdown control (3μA battery drain when input power is removed).

The oscilloscope photo in Figure 214.2 illustrates how the charging current (center) decreases as the load current (top) increases, so as not to exceed the AC adapter current limit (bottom). Note: for this photo, the changing computer load current is shown as a triangle waveform. The actual computer load current waveform will be much different.

All surface mount lithium-ion charger

The circuit shown in Figure 214.3 is a CC/CV charger that can be used to charge up to five series lithium-ion cells at currents up to 3A. Charge current is easily adjustable using a single resistor, a control voltage, a PWM signal or a DAC output. The circuit values shown are for 12.6V out (3 cells) and 2.2A charge current. The input current limit is set at 2A by the 0.05Ω current sense resistor, R4, which develops the required 100mV sense voltage. Resistors R1 and R2 program undervoltage lockout, which keeps the charger off until the input voltage reaches 11V.

Battery manufacturers recommend terminating the charge after a fixed amount of time has passed. Using an external timer, the charging can be stopped by programming the charge current to zero by using the program pin (Pin 19), pulling the program pin low, or pulling the V_C pin low. For additional circuit information, please consult the LT1511 data sheet, Design Note 124 or Design Note 144.

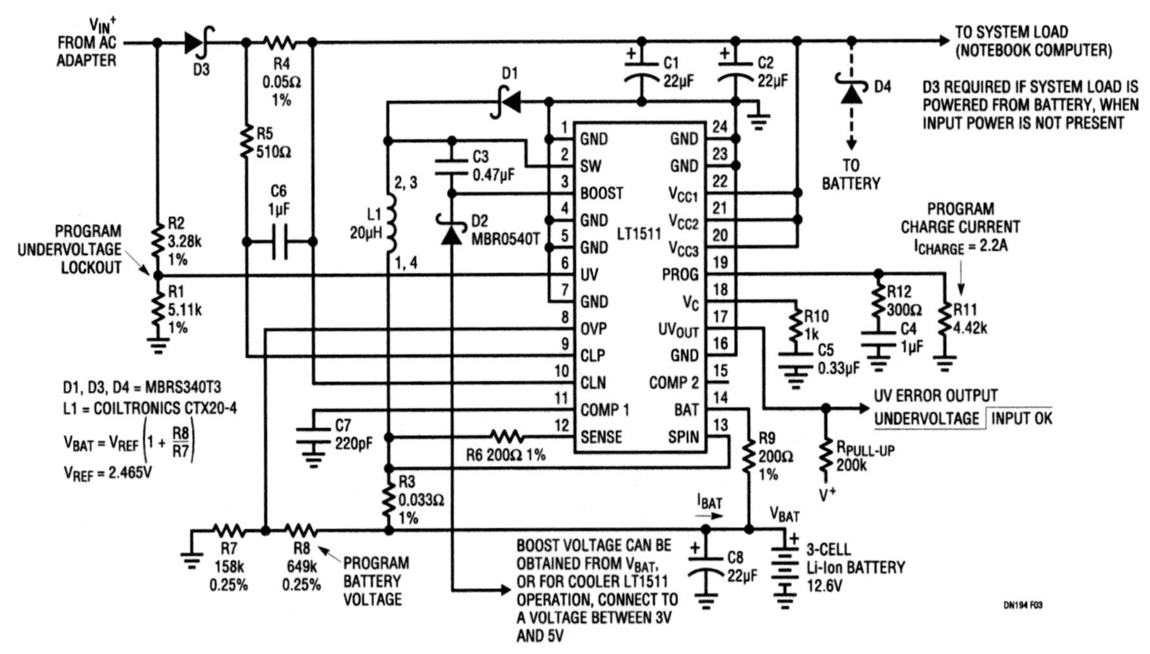

Figure 214.3 • Programmable Constant-Current/Constant-Voltage Battery Charger with Input Current Limit

Inexpensive circuit charges lithium-ion cells

215

David Bell

Introduction

A single lithium-ion cell is often the battery of choice for portable equipment because of its high energy density. The 3V to 4.1V provided by a lithium-ion cell is also a good match for modern low voltage circuits, often simplifying the power supply. Despite these advantages, designers are often frustrated when attempting to design precision charging circuitry that meets battery manufacturers' specifications. Figure 215.2 is a simple, cost-effective linear charger that satisfies these precision lithium-ion charging requirements.

Lithium-ion cell manufacturers generally recommend a constant-current/constant-voltage (CC/CV) charging technique. Although conceptually simple, charging a lithium-ion cell requires very accurate control of the float voltage to obtain high capacity with long cycle life. If the voltage is too low, the cell will not be fully charged; if the voltage is too high, the cycle life is significantly degraded. Excessive voltage to the cell can also result in venting and other hazardous conditions (specific hazards depend upon the cell's construction and chemistry).

Figure 215.1 depicts CC/CV charging characteristics for a typical lithium-ion cell. A fully discharged cell will initially be charged by a constant current, since the cell's voltage is below the 4.1V constant-voltage limit. (Most lithium-ion cells require either a 4.1V or 4.2V float voltage, depending on the cell's chemistry.) Once the cell's voltage rises to the float voltage of 4.1V, the charger limits further rise in terminal voltage

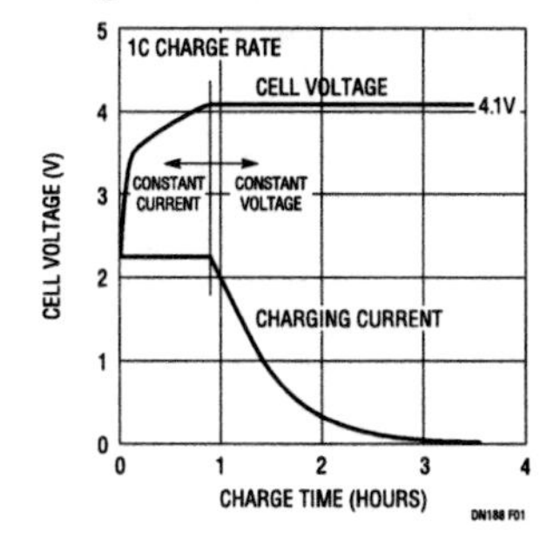

Figure 215.1 • Typical Lithium-Ion Charge Characteristics

and the charging current naturally begins to fall off. Most battery manufacturers recommend that charging be terminated roughly one hour after the current has fallen to 10% of its peak value. Alternatively, a timer can be started when charging begins, with time-out used to suspend charging once sufficient time has elapsed to charge a fully depleted cell.

Circuit description

Figure 215.2 depicts a simple and inexpensive linear charger that can be used to charge a single lithium-ion cell. The circuit provides constant-current/constant-voltage (CC/CV) charging from an inexpensive, unregulated 6V wall adapter. The charger is built around a single LTC1541, which contains a voltage reference, op amp and comparator. The high accuracy voltage reference (±0.4%) regulates the battery float voltage to ±1.2%, as required by most lithium-ion battery manufacturers. Even tighter accuracy can be obtained by specifying higher accuracy for feedback divider resistors R6 and R7. The charger may be configured to float at either 4.1V or 4.2V by changing the value of R6. Use 252k for a 4.1V float voltage; use 261k for a 4.2V float voltage.

Transistor Q1 is used to regulate battery charging current. Q1's base current is controlled by the op amp output (Pin 1) and buffered by transistor Q2 for additional current gain. Diode D1 is needed to prevent reverse current flow when the wall adapter is unplugged or unpowered. Because this is a linear regulator, the designer must consider power dissipation in Q1. As shown with a 6V wall adapter, Q1 dissipates a maximum of about 1W and can be heat sinked directly to the printed circuit board. Higher current levels or higher input voltages will increase dissipation and additional heat sinking will be needed.

Battery charging current is sensed by R11 and fed to the op amp's noninverting input via R10. IC1's internal 1.2V reference voltage is divided to 44mV by R4 and R2 and connects to the op amp's inverting input. The op amp compares the current sense voltage against the 44mV reference and adjusts

Analog Circuit Design: Design Note Collection. http://dx.doi.org/10.1016/B978-0-12-800001-4.00215-5

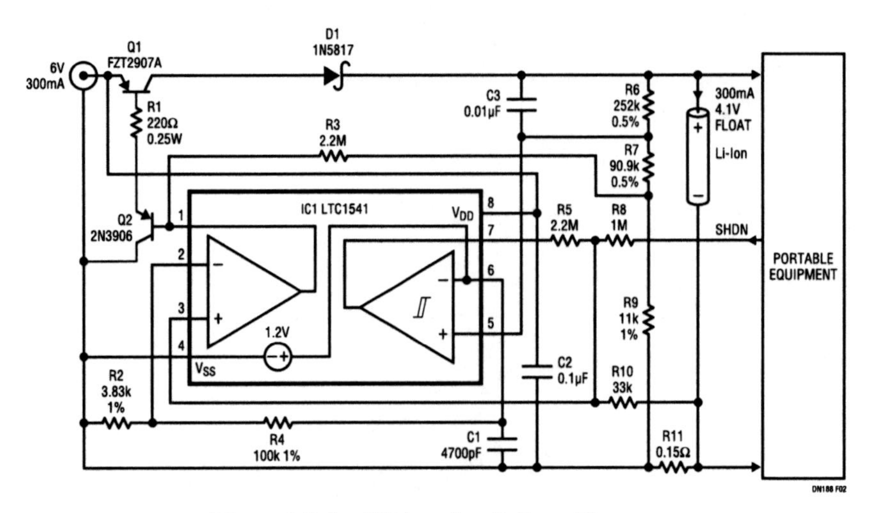

Figure 215.2 • Lithium-Ion Battery Charger

the base drive to Q1 as needed to regulate current to 300mA. The op amp's ±1.25mV maximum input offset voltage guarantees accurate charge current regulation while dropping only 44mV across the sense resistor.

Once the battery charges to 4.1V, the voltage loop begins to reduce charging current to maintain the desired float voltage. A resistor divider comprising R6, R7 and R9 generates a feedback voltage to IC1's comparator (Pin 5). Once the voltage at this node reaches 1.2V, the comparator output goes high, pulling the current sense signal high via R5. During voltage regulation, the comparator output (Pin 7) is a pulsed waveform; however, the low slew rate of the micropower op amp smoothes this signal to a small amplitude triangle wave at Pin 1. In fact, the voltage at Pin 7 may be monitored by a microprocessor to detect the onset of constant-voltage regulation.

Current sense resistor R11 is in series with the battery charging path and would normally result in voltage regulation beginning at around 4.05V instead of the desired 4.1V (44mV is dropped across the sense resistor). However, the addition of R3 and R9 compensates for this voltage drop, and results in activation of the constant-voltage loop at 4.1V. In essence, R3 and R9 create a negative output impedance from the regulator that cancels the 0.15Ω resistance of R11. By carefully selecting the values of R3, R7 and R9, one can produce an even larger negative output resistance and compensate for the internal resistance of the battery and its internal protection circuitry. The result is faster recharge times without exceeding the 4.1V limit within the cell itself.

Because the current sense loop actually monitors the total current drawn from the wall adapter, the charger will automatically "load share" with the portable equipment. In other words, when the equipment draws no power, all of the 300mA is available to charge the battery. However, any current drawn by the equipment will simply subtract from the battery charging current, keeping the wall adapter load limited to 300mA. The charger may also be shut down by logic control, as shown in Figure 215.2. Pulling high on the shutdown signal forces the current feedback signal above the 44mV threshold, thereby turning off Q1. The charger operates normally when the shutdown pin is a high impedance state—the default state on most microprocessor port pins. R8 may be eliminated if the shutdown feature is not needed.

Other charging options

The simple linear charger described above is suitable for many handheld portable products, where total charging current is modest. For higher current or multicell applications, a high efficiency switching regulator charger, such as the LT1510 or LT1511, may be appropriate. The LT1510 is a CC/CV charger capable of delivering up to 1.25A of charging current from an SO-8 package. The LT1511 can deliver up to 3A and includes an input current limiting feature.

The CC/CV charging technique is not limited to lithium-ion batteries; sealed lead-acid (SLA) batteries may also be charged using similar circuitry. The constant-voltage control loop can easily be converted to a "hysteretic charger" for optimal SLA charging. Contact Linear Technology Applications Engineering for additional details.

Battery backup regulator is glitch-free and low dropout

216

Mitchell Lee Todd Owen

A new class of linear regulator has been developed for battery backup applications. It eliminates both the losses associated with steering diodes, the glitches and battery-to-battery cross conduction inherent in MOSFET switching schemes and the poor regulation inherent in dual regulator schemes previously used for this purpose. See the comparison detailed in Table 216.1.

Figure 216.1 shows a simplified block diagram of the LT1579 dual input regulator. The regulator features 300mA output capability and low dropout. Two batteries, or a battery and an AC-derived power source, are connected to V_{IN1} and V_{IN2}. The relative voltage of these two sources plays no role in determining which input supplies power to the load: the primary input (V_{IN1}) is normally used to power the output, and the secondary input (V_{IN2}) takes over as a backup when the primary source fails. Unlike diode steering circuits, this allows batteries of any voltage to serve as primary or backup.

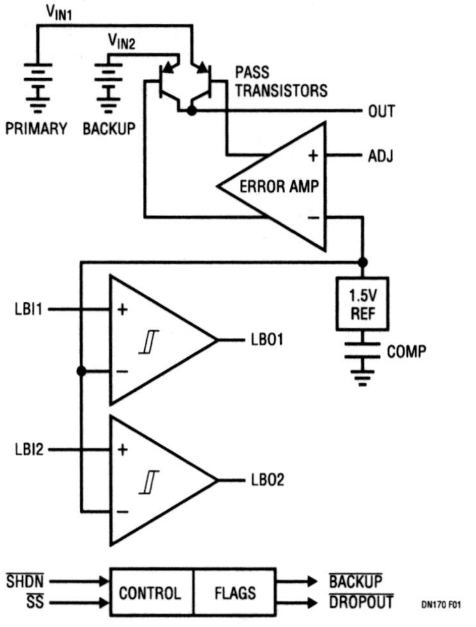

Figure 216.1 • LT1579 Block Diagram

Either battery can be removed and replaced without disturbing the output voltage.

Several other important features are included to simplify integration of the LT1579 into a battery-backed system. Again referring to Figure 216.1, two logic flags, $\overline{\text{BACKUP}}$ and $\overline{\text{DROPOUT}}$, are useful for monitoring the status of the regulator. $\overline{\text{BACKUP}}$ goes low when V_{IN1} fails and V_{IN2} takes over, whereas $\overline{\text{DROPOUT}}$ indicates that both V_{IN1} and V_{IN2} have failed.

Two comparators independently monitor the condition of the batteries. In contrast to $\overline{\text{BACKUP}}$ and $\overline{\text{DROPOUT}}$, the low-battery detectors give advance warning of impending battery failure. Secondary Select ($\overline{\text{SS}}$) overrides V_{IN1}'s priority over V_{IN2}, forcing the regulator to abandon the primary battery in favor of the backup.

The primary battery normally supplies the load until its terminal voltage is nearly equal to the output voltage; however, some battery types may be damaged if discharged this far. $\overline{\text{SS}}$, used in conjunction with a low-battery detector,

Table 216.1 Backup Method Comparison

	LT1579	STEERING DIODES	MOSFET SWITCHING	TWO REGULATORS
Guaranteed Battery-to-Battery Isolation	✓	✓		
Prioritized Inputs	✓		✓ (Needs Circuitry)	
Seamless Switching	✓	✓		✓
Seamless Regulation	✓			
Logic Override	✓		✓	
Battery Disconnect	✓		✓	✓

Analog Circuit Design: Design Note Collection. http://dx.doi.org/10.1016/B978-0-12-800001-4.00216-7

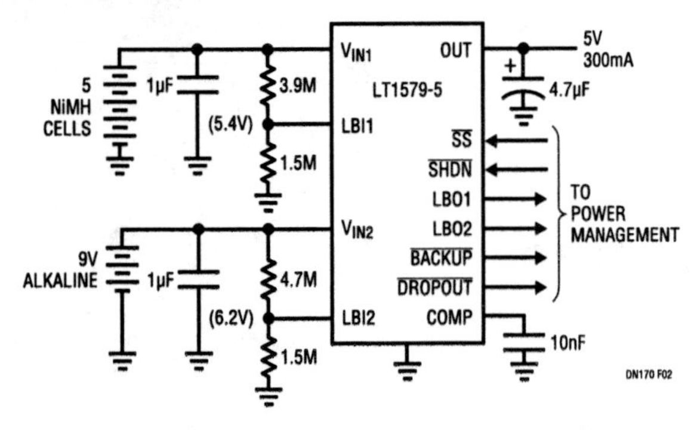

Figure 216.2 • A 9V Battery Backs Up Five NiMH Cells

allows the regulator to abandon the primary battery at a higher, nondamaging end-of-discharge voltage.

One last feature is $\overline{\text{SHUTDOWN}}$; this turns the regulator off and reduces total drain from both inputs to less than 7µA.

Figure 216.2 shows a typical application of the LT1579 with primary power supplied by five NiMH cells and backup provided by a 9V alkaline. Both low-battery comparators are used; they report the condition of the primary and backup batteries to a microprocessor. The $\overline{\text{BACKUP}}$ and $\overline{\text{DROPOUT}}$ flags keep the microprocessor apprised of the regulator's status. In addition to the fixed 5V version shown, 3V, 3.3V and adjustable versions are also available.

9V snap terminals are easily reversed by the end-user during installation of the battery. No harm is done to the LT1579 because both inputs are reverse-battery protected. No current is drawn from the reversed battery and no excess current is drawn from the adjacent battery. Best of all, the load never knows the difference. The regulator continues to deliver the correct output voltage throughout the entire event.

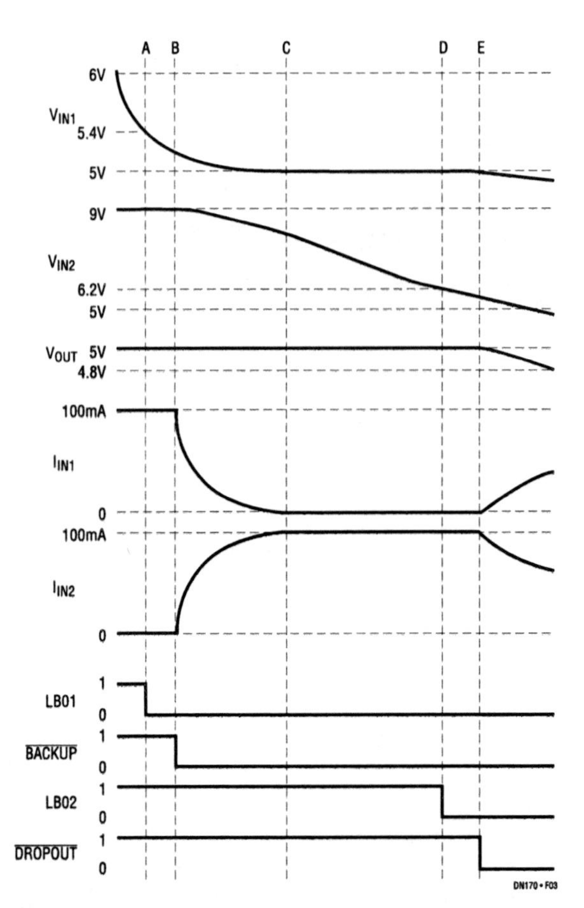

Figure 216.3 • Typical Event Sequence for the Circuit of Figure 216.2. The Time Scale Is Distorted for Purposes of Illustration

Figure 216.3 shows a typical sequence of events for the circuit of Figure 216.2. Initially, both batteries are fully charged and all status flags (LBO1, LBO2, $\overline{\text{BACKUP}}$ and $\overline{\text{DROPOUT}}$) are high. Load current, assumed to be 100mA, is drawn from the primary battery at V_{IN1}. After a period of time, V_{IN1} begins to falter. At point A, $V_{IN1} = 5.4V$ and LBO1 goes low, predicting the eventual depletion of the primary battery. When V_{IN1} enters dropout (B), $\overline{\text{BACKUP}}$ goes low and the regulator begins to gradually transfer the load to the backup battery at V_{IN2}. By time C, the primary battery has dropped below the point where it can deliver useful current to the output and all load current is supplied by the backup battery.

The backup battery reaches its low voltage threshold at point D, signaling impending doom. This is the last chance for the system to alert the user, store critical data, shed load and find ways to survive until the batteries are replaced or recharged. In Figure 216.3 the relentless load continues unabated and discharges the backup battery. The regulator can no longer maintain its 5V output at point E and a logic low appears on the $\overline{\text{DROPOUT}}$ pin. Now the output falls below 5V and some current is once again drawn from the primary battery.

The LT1579 integrates a complex current steering function into one chip and can significantly reduce board space and design time while adding a number of useful power management features. It is ideally suited for battery- or line-operated equipment that relies on a backup battery for reliability and uninterrupted service. The output is unaffected by battery changes or power cycling, and status flags allow implementation of a very complete power management system with minimal external components.

Dual PowerPath controller simplifies power management

217

Jaime Tseng

As the demand for portable electronics with multiple batteries continues to grow, so does the need for simple and efficient solutions for switching between batteries. The LTC1473 simplifies the design of circuitry for switching between two batteries or a battery and an AC adapter.

The LTC1473 is designed to drive two sets of low loss, back-to-back N-channel MOSFET switches to route power where needed, typically to the input of the main system switching regulator. An internal micropower boost regulator supplies the gate drive for the N-channel MOSFET switches.

The LTC1473 simplifies the designer's task by incorporating a number of protection features. Short-circuit protection shuts off the gate drive when a fault condition exceeds the user-programmable time limit, inrush current limiting limits the current flowing into or out of the system bypass capacitor and a diode default mode allows power to flow from the highest potential until the inputs can be defined.

Automatic switchover between battery and AC adapter

A protected automatic switchover between a battery and a power source connected at DCIN can be constructed using the circuit in Figure 217.1. Under normal conditions, this circuit will route the voltage at DCIN to the output. If the voltage at DCIN drops below 9.75V, DCIN is deselected and the battery voltage is routed to the output. If the battery voltage

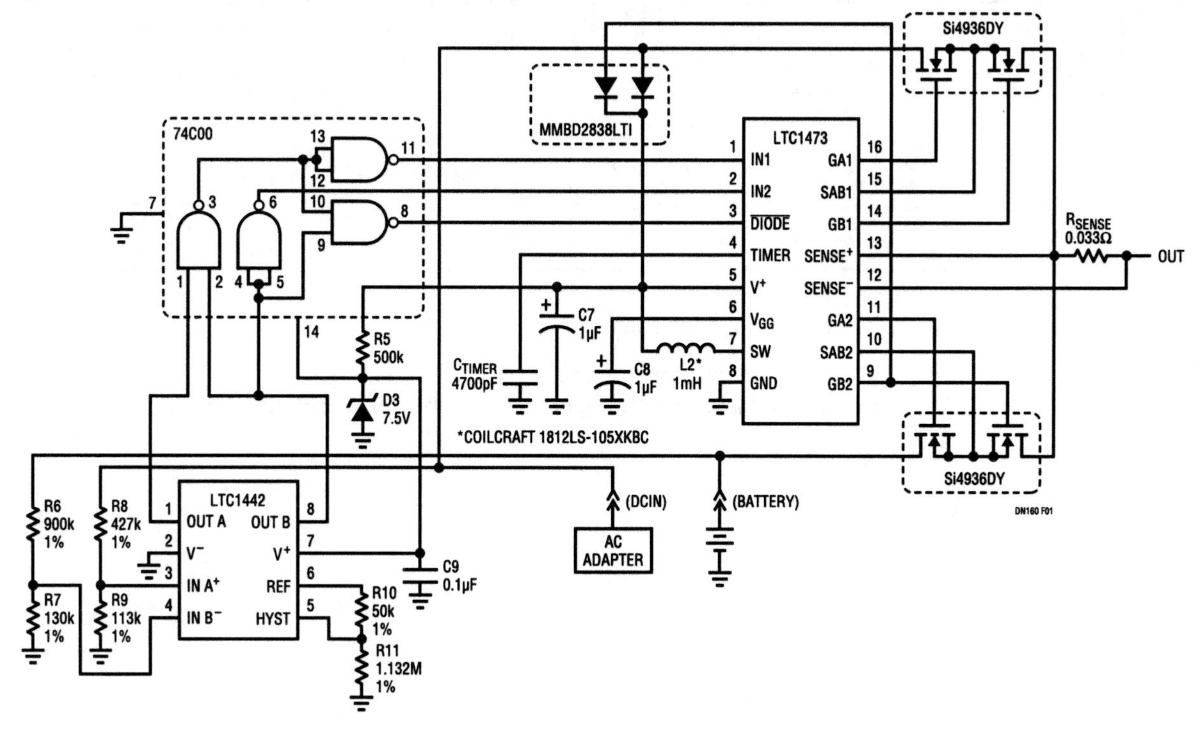

Figure 217.1 • Protected Automatic Switchover between Battery and AC Adapter

Analog Circuit Design: Design Note Collection. http://dx.doi.org/10.1016/B978-0-12-800001-4.00217-9

is less than 5.9V, each switch is made to mimic a diode, allowing power to flow from the highest potential source to the output. In this "2-diode" mode, the first half of each PowerPath switch pair is turned on, and the second half is turned off. Thus, two diodes are formed by the body diodes of the MOSFET switches that are turned off.

The inrush current limit of 6A is selected with a 0.033Ω R_{SENSE} resistor. The fault timer is set to 1.1ms with a 4700pF C_{TIMER} capacitor. If a MOSFET switch is in current limit for more than 1.1ms, an internal latch in the LTC1473 is set and the MOSFET switch is turned off.

The LTC1442 shown in Figure 217.1 is an ultralow power dual comparator with a precision 1.182V reference. This comparator monitors the voltage at DCIN and the battery voltage and selects which MOSFET switch to turn on. Simple logic, comprising CMOS NAND gates, decodes the comparator outputs to control the inputs of the LTC1473. A 7.5V Zener shunt regulator in series with a 500k resistor supplies power for both the CMOS NAND gates and the LTC1442.

Power routing circuit for microprocessor controlled dual battery systems

The microprocessor controlled dual battery system shown in Figure 217.2 uses two LTC1473s to provide input power routing and battery charging multiplexing. The two batteries can be of different chemistries. One LTC1473 is used to connect the output of the charger to the battery; the other connects the battery to the input of the system switching regulator.

The power-management microprocessor provides overall control of the power management system in concert with the two LTC1473s and the auxiliary power-management systems. The microprocessor decides which battery to connect to the input of the system switching regulator and which battery is in need of recharging. To charge a battery, the microprocessor selects the charging algorithm for that particular battery chemistry.

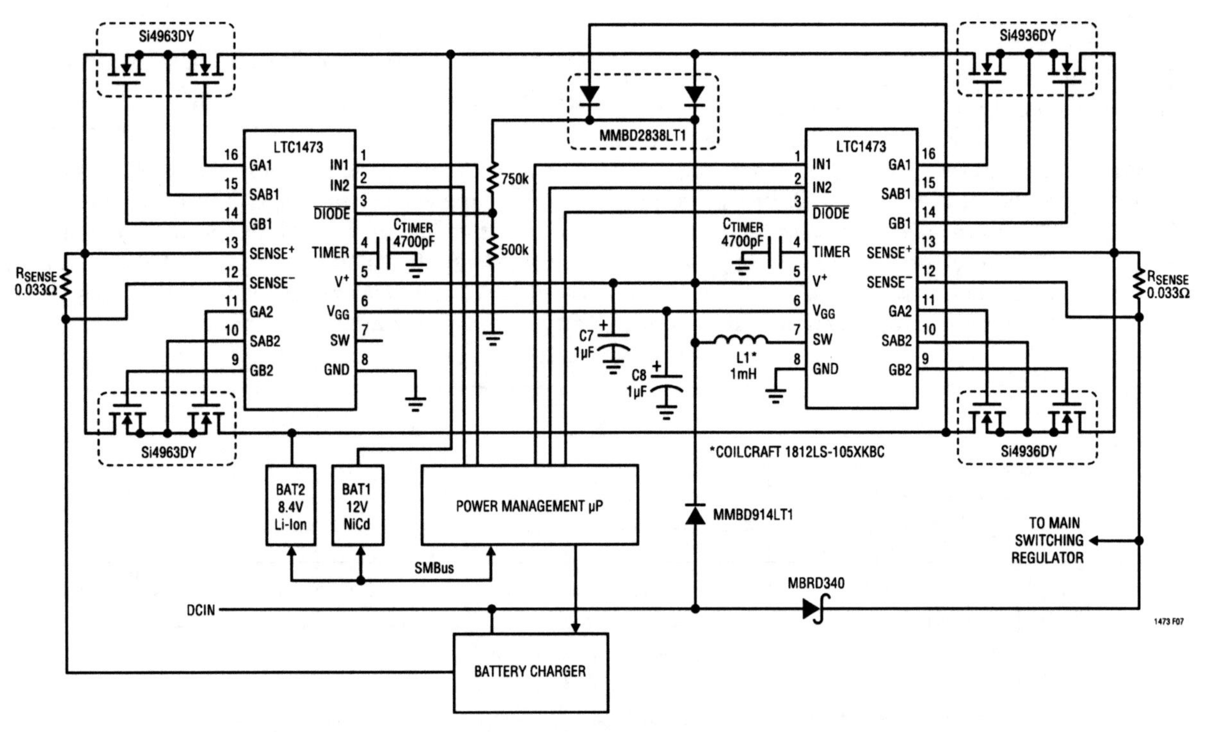

Figure 217.2 • Power Routing Circuit for Microprocessor Controlled Dual Battery System

Low dropout, constant-current/constant-voltage 3A battery charger

218

Chiawei Liao

Introduction

The LT1511 current mode PWM battery charger is the simplest, most efficient solution for fast charging modern rechargeable batteries that require constant-current and/or constant-voltage charging including lithium-ion (Li-Ion), nickel-metal-hydride (NiMH) and nickel-cadmium (NiCd).

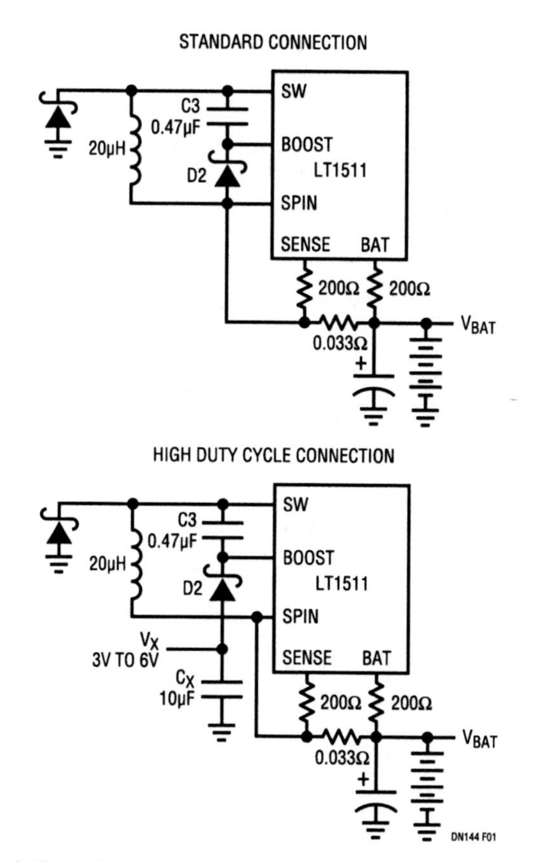

Figure 218.1 • High Duty Cycle

Higher duty cycle for the LT1511 battery charger

Maximum duty cycle for the LT1511 constant-current/constant-voltage battery charger is typically 90%, but this may be too low for some applications. For example, if an 18V ±3% adapter is used to charge ten NiMH cells, the charger must put out 15V maximum. A total of 1.6V is lost in the input diode, switch resistance, inductor resistance and parasitics, so the required duty cycle is 15/16.4 = 91.4%. As it turns out, duty cycle can be extended to 93% by restricting boost voltage to 5V instead of using V_{BAT}, as is normally done. This lower boost voltage also reduces power dissipation in the LT1511. Connect an external source of 3V to 6V at the V_X node in Figure 218.1[1] with a 10μF C_X bypass capacitor.

Enhancing dropout voltage

For even lower dropout and/or to reduce heat on the board, the input diode can be replaced with a FET (see Figure 218.2). It is straightforward to connect a P-channel FET across the input diode and connect its gate to the battery so that the FET commutates off when the input goes low. The problem is that the gate must be pumped low so that the FET is fully turned on even when the input is only a volt or two above the battery voltage. There is also a turnoff speed issue. To avoid large current surges from the battery back through the charger into the FET, the FET should turn off instantly when the input is dead shorted. Gate capacitance slows turnoff, so a small P-channel (Q2) is used to discharge the gate capacitance quickly in the event of an input short. The body diode of Q2 creates the necessary pumping action to keep the gate of Q1 low during normal operation. Note the Q1 and Q2 have a V_{GS} spec limit of 20V. This restricts V_{IN} to a maximum of 20V.

Note 1: Figures 218.1, 218.2 and 218.4 circuits are for 3A charging current. See the LT1511 data sheet for other design values.

Analog Circuit Design: Design Note Collection. http://dx.doi.org/10.1016/B978-0-12-800001-4.00218-0

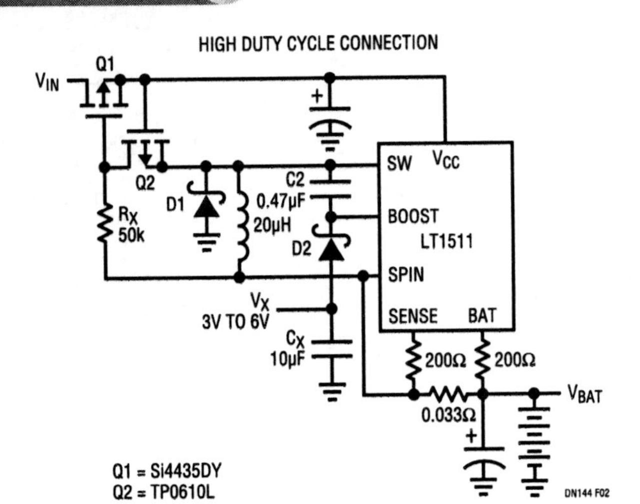

Figure 218.2 • Replacing the Input Diode

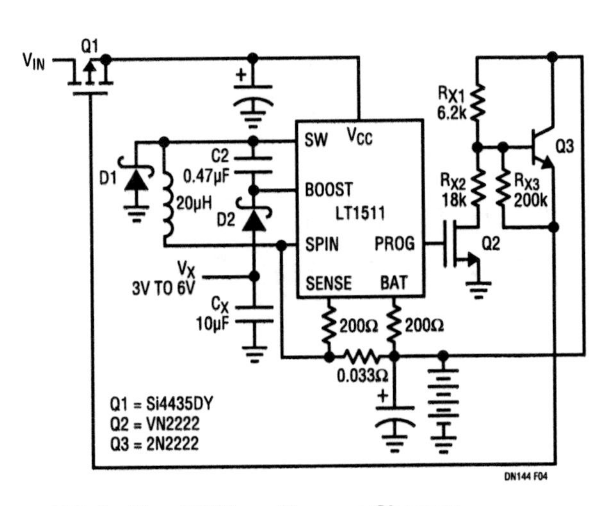

Figure 218.4 • V_{IN}>20V Low Dropout Charger

Figure 218.3 is a 3A complete constant-current/constant-voltage charger for 15V batteries. Input is from an 18V ±3% adapter. For adapter current limit and undervoltage lockout functions, please see the LT1511 data sheet.

For V_{IN} > 20V, the circuit in Figure 218.4 can be used to clamp V_{GS} to <20V. R_{X1} and R_{X2} are chosen to draw 1mA or less to ensure that the 2.5V on PROG is sufficient to turn Q2 on. This gives a value for R_{X1} of about 6.2k and R_{X2} is calculated from:

$$R_{X2} = R_{X1}\left(\frac{V_{BAT}}{V_{GS} - V_{IN} + V_{BAT}} - 1\right)$$

V_{BAT} = Highest battery voltage

V_{GS} = Minimum Q1 gate drive

V_{IN} = Lowest input voltage

If R_{X1} = 6.2k, V_{BAT} = 19.5V, V_{GS} = 7V, V_{IN} = 21.5V, then R_{X2} = 18k.

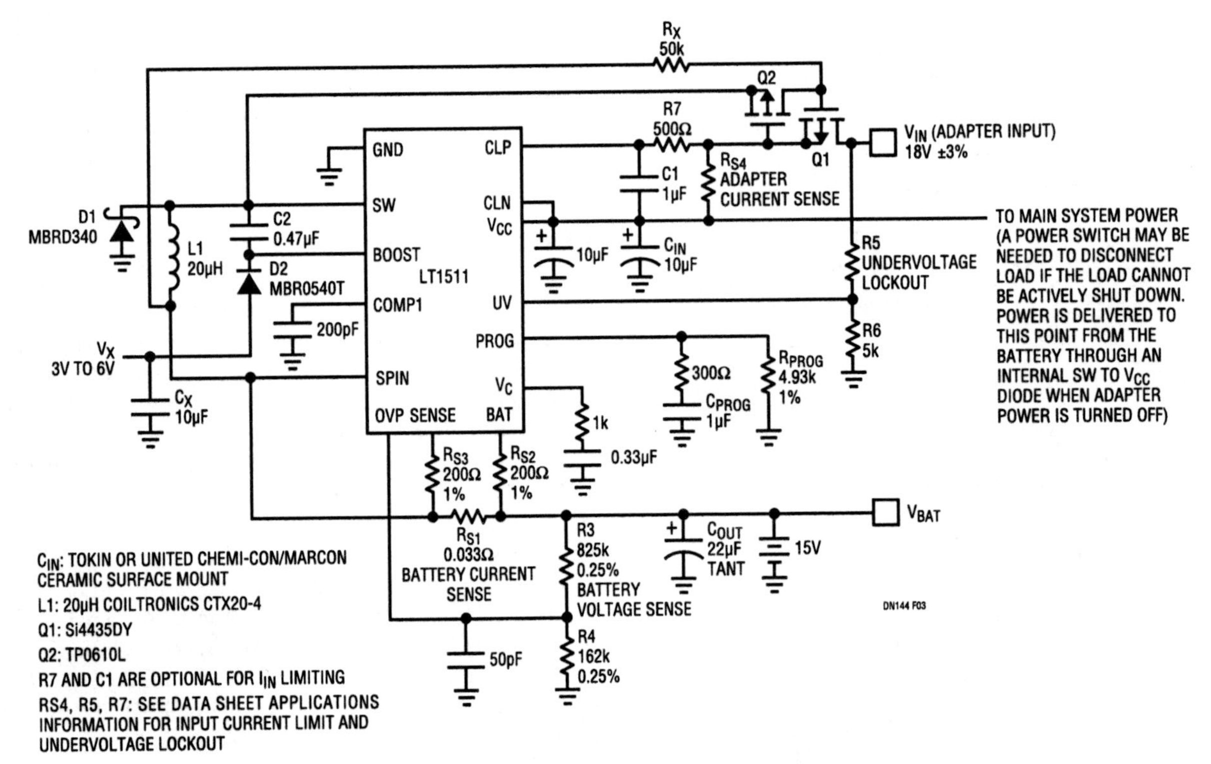

Figure 218.3 • 3A Constant-Current/Constant-Voltage Low Dropout Battery Charger

Fused lead battery charger ICs need no heat sinks

219

Carl Nelson

The LT1510 and LT1511 are high efficiency battery-charging chips capable of output powers up to 25W and 50W, respectively. These chips are intended for fast-charging batteries in modern portable equipment where space is at a premium. They are therefore packaged in low profile surface mount packages and run at a fairly high frequency (200kHz) to minimize the height and footprint of the complete battery charger. To minimize thermal resistance, these packages are specially constructed with the die-attach paddle connected (fused) directly to multiple package leads.

When operating these chips near their maximum power levels, extra care should be taken to minimize chip power dissipation and to keep the overall thermal resistance of the package-board combination as low as possible. Figure 219.1 shows a simple way to reduce power dissipation with no loss in performance. The LT1510 and LT1511 use an external diode (D2) and capacitor (C_X) to generate a voltage that is higher than the input voltage. This voltage is used to supply the base drive to the internal NPN power switch to allow it to saturate with a forced h_{FE} of about 50. Switching speed and on-resistance losses are minimized with this technique. The *required* boost voltage is only 3V, but with the normal connection of D2, the resulting boost voltage is equal to the battery voltage. Unnecessarily high base drive losses result with high battery voltages. Connecting D2 to a 3.3V or 5V supply (V_X) instead of to the battery reduces chip dissipation by approximately:

$$\text{Power reduction} = (V_{BAT} - V_X)(I_{CHRG})(V_{BAT})/(50)(V_{IN})$$

With a 20V adapter charging a 12.6V battery at 3A, and $V_X = 3.3V$, chip power is reduced by 0.35W.

Fused-lead packages conduct most of their heat out the leads. This makes it very important to provide as much PC board copper around the leads as is practical. Total thermal resistance of the package-board combination is dominated by the characteristics of the board in the immediate area of the package. This means both lateral thermal resistance across the board and vertical thermal resistance through the board to other copper layers. Each layer acts as a thermal heat spreader

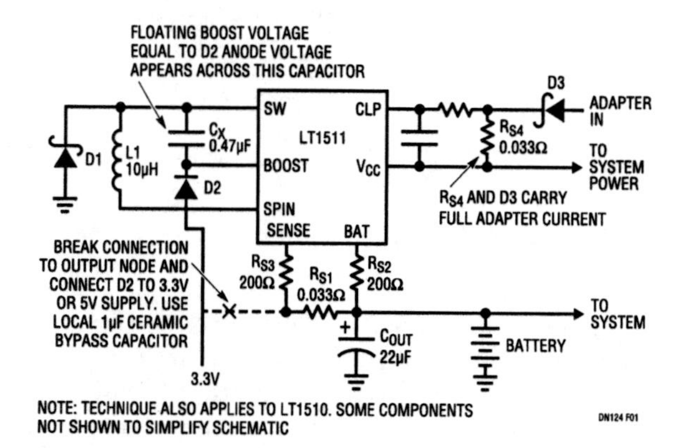

Figure 219.1 • A Simple Way to Reduce Power Dissipation with No Loss in Performance

that increases the heat sinking effectiveness of extended areas of the board.

Total board area becomes an important factor when the area of the board drops below about 20 square inches. The graphs in Figures 219.2 and 219.3 show thermal resistance versus board area for 2-layer and 4-layer boards. Note that 4-layer boards have significantly lower thermal resistance, but both types show a rapid increase for reduced board areas. Figures 219.4 and 219.5 show actual measured lead temperatures for chargers operating at full current. Battery voltage and input voltage will affect device power dissipation, so the data sheet power calculations must be used to extrapolate these readings to other situations.

Vias should be used to connect board layers together. Planes under the charger area can be cut away from the rest of the board and connected with vias to form both a low thermal resistance system and to act as a ground plane for reduced EMI.

Glue-on, chip-mounted heat sinks are effective only in moderate power applications where the PC board copper cannot be used, or where the board size is small. They offer very little improvement in a properly laid out multilayer board of reasonable size.

Analog Circuit Design: Design Note Collection. http://dx.doi.org/10.1016/B978-0-12-800001-4.00219-2

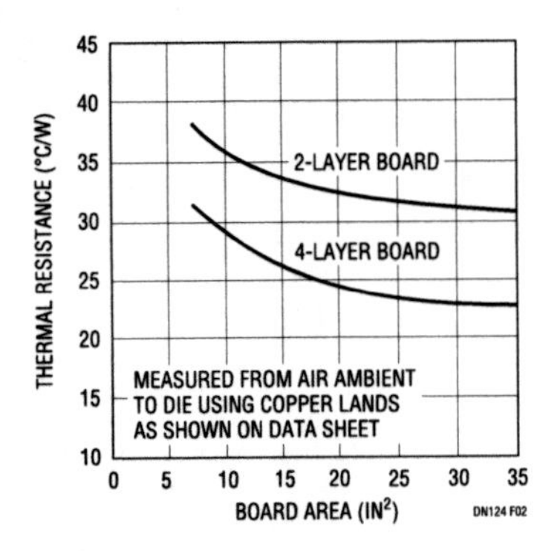

Figure 219.2 • LT1511 Thermal Resistance

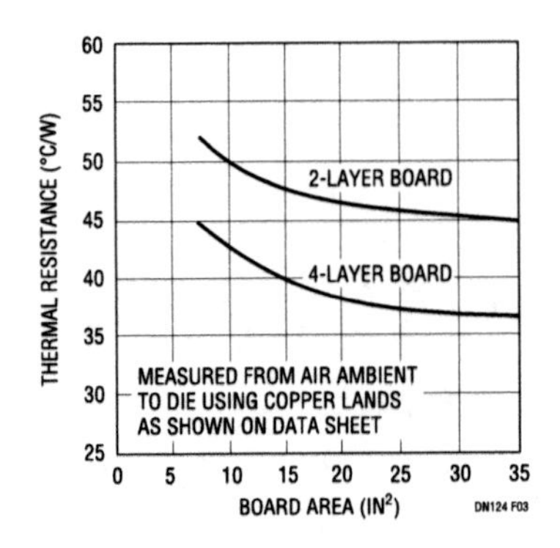

Figure 219.3 • LT1510 Thermal Resistance

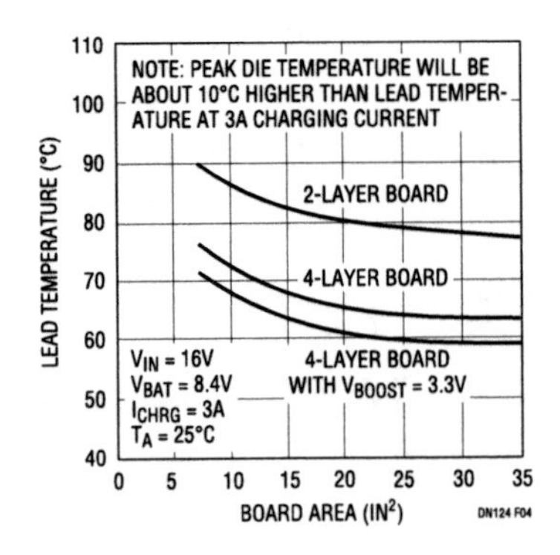

Figure 219.4 • LT1511 Lead Temperature

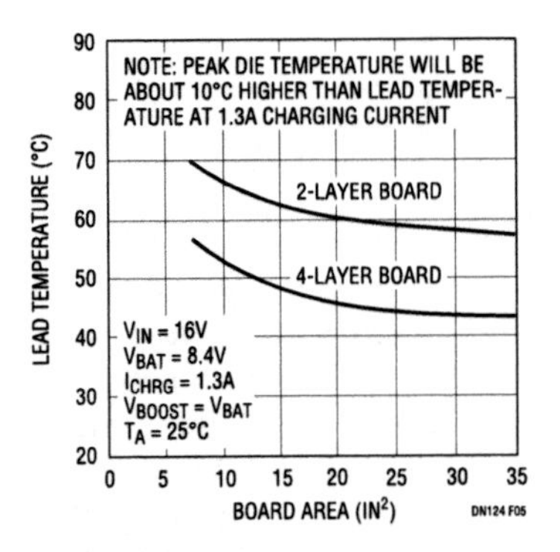

Figure 219.5 • LT1510 Lead Temperature

The suggested methods for a final check on chip operating temperature are to solder a small thermocouple on top of one of the IC ground leads or to use an IR sensor on top of the package. Using either method, the temperature measured either way will be about 10°C lower than actual peak die temperature when the charger is delivering full current. The 125°C rating of these charger chips means that lead temperature readings as high as 100°C (with maximum ambient temperature) are still comfortably within the device ratings.

Another consideration is the power dissipation of other parts of the charger and the surrounding circuitry used for other purposes. The catch diode (D1) will dissipate a power equal to:

$$P_{DIODE} = (I_{CHRG})(V_F)(V_{IN} - V_{BAT})/V_{IN}$$

$V_{IN}=16V$, $V_{BAT}=8.4V$, $V_F=0.45V$, and $I_{CHRG}=3A$, the diode dissipates 0.64W. Unfortunately, it must be located very close to the charger chip to prevent inductive spikes on the Switch pin. Increase in charger chip temperature due to power dissipation in D1 is about 12°C/W. D3 is used for input protection and can also dissipate significant power but it can be located away from the charger. The current sense resistors used with the LT1511 dissipate power equal to:

$$P(R_{S1}) = R_{S1}(I_{CHRG})^2$$

$$P(R_{S4}) = R_{S4}(I_{ADPT})^2$$

R_{S4} power depends only on charger output current, but R_{S1} carries full adapter current. These resistors, which typically dissipate a total of about 0.5W, will also increase chip temperature at about 12°C/W, assuming that they are close to the charger chip.

It is assumed that L1 contributes only slightly to chip temperature rise because its power dissipation is normally fairly low compared to its heat sinking ability. This will be true if a low loss core is used (Kool Mμ, etc.), and the winding resistance is less than $0.2V/I_{CHRG}$. If the charger is located near other high power dissipation circuitry, direct temperature testing may be the only accurate way to ensure safe device temperatures.

Finally, don't forget about losses in PC board trace resistance. A 100mil wide trace is huge by modern standards, but a pair of these traces six inches long, in 1/2oz copper, delivering 3A, would have a resistance of $\approx 0.12\Omega$, and a power loss of 1.1W!

New micropower, low dropout regulators ease battery supply designs

220

Mitchell Lee John Seago

Three new linear regulators simplify the design of battery-operated equipment. The LT1521 is a 300mA, positive low dropout regulator with micropower quiescent current and shutdown. The LT1175 is a 500mA negative complement with adjustable current limit. A third product, the LT1118, has the unique capability of maintaining output regulation while sourcing or sinking load current.

The LT1521 contains all of the features associated with battery-operated applications. In designs where memory must be powered continuously, the LT1521's 12μA quiescent current eliminates the need for a separate micropower backup supply. In shutdown the quiescent current drops to 6μA. Figure 220.1 shows an example application using the LTC1477 as a means of disconnecting all circuitry except for memory and ON/OFF control logic. The LTC1477 protected high-side switch draws only 10nA in shutdown, eliminating itself as well as its load as a factor in battery shelf life.

In battery-backed memory applications, the output of the LT1521 can be held up by the backup battery while in shutdown or even with the input power removed. No series output diode is required as reverse current flow is internally limited to about 5μA.

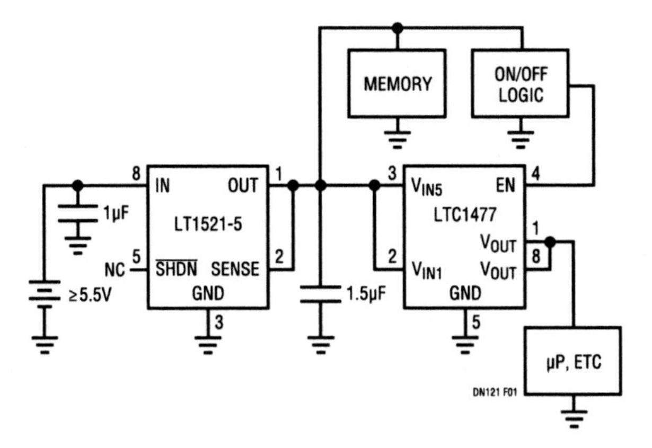

Figure 220.1 • The LT1521's 12μA Standby Current Eliminates the Need for a Separate Memory Backup Supply

A common problem in portable equipment is the chance of installing the batteries backwards, thereby destroying the electronics contained within. The LT1521 needs no reverse protection diode to guard against this condition, as it is designed to withstand up to −20V input while also protecting the load. Low dropout is preserved.

Most important for battery applications is low dropout, a characteristic not neglected in the design of the LT1521. At 300mA the dropout is just 500mV, dropping to approximately 290mV at a 50mA load. This enables the LT1521 to maintain regulation while draining the last drop of power from the battery.

Like the LT1521, the LT1175 also features low dropout, running 500mV at 500mA load current. Quiescent current is 45μA dropping to 10μA in shutdown.

The LT1175 offers several unique features not available in other negative regulators. First, the current limit is adjustable by pin strapping to 200mA, 400mA, 600mA or 800mA. This allows the current limit to be tailored to suit normal load requirements while not exceeding the maximum safe current drain from the battery during a short-circuit fault.

Shutdown on a negative regulator could be a mixed blessing, particularly if positive logic was used to control a negative input. The LT1175 solves this problem by accepting either positive or negative shutdown signals. Holding the Shutdown pin within 0.8V of ground disables the LT1175. If pulled to 2.5V or more *positive or negative* with respect to ground, the regulator is enabled. If shutdown is not desired, either float or connect the Shutdown pin to V_{IN} and the regulator will be enabled whenever power is applied.

The LT1521 and LT1175 work well together as the basis for a split supply as shown in Figure 220.2. Low output capacitance requirements allow the use of ceramic units instead of larger, more expensive electrolytic or tantalum capacitors.

Owing to the LT1175's unique Shutdown pin, the Shutdown pins of both devices can be joined together as shown and driven from a positive control logic signal. Behavior of the outputs relative to shutdown control is shown in Figure 220.3.

Analog Circuit Design: Design Note Collection. http://dx.doi.org/10.1016/B978-0-12-800001-4.00220-9

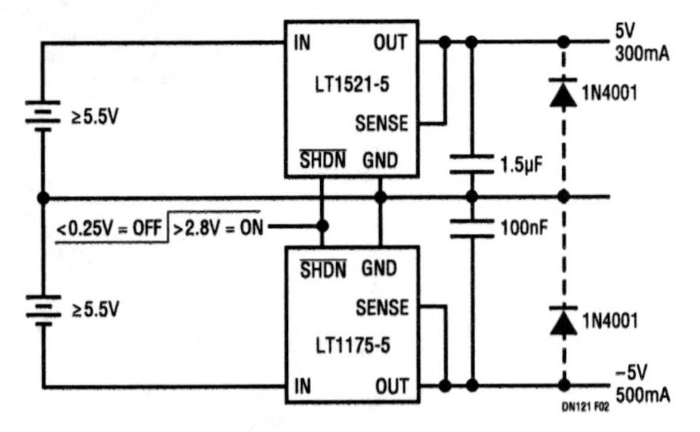

Figure 220.2 • In a Split Supply Application the Shutdown Pins May Be Commanded in Parallel Using Positive Logic

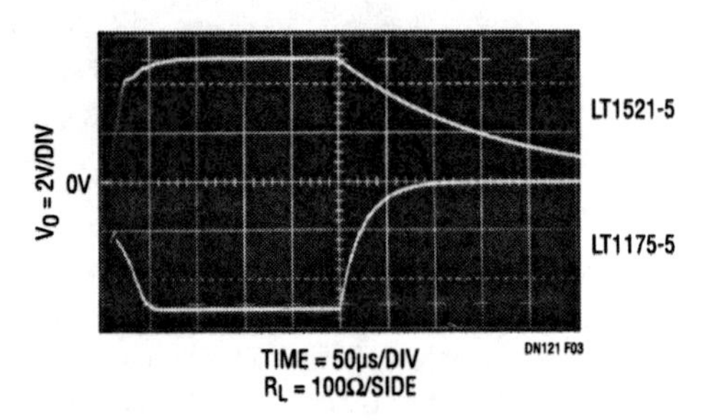

Figure 220.3 • Clean Start-Up and Shutdown Is Assured by Utilizing Ganged Shutdown Control of the LT1521 and LT1175.

Although the LT1521-5 can tolerate up to −20V forced output potential with respect to its input, supply reversal diodes (1N4001) are often required to protect both linear and digital load circuitry from damage under transient start-up or fault conditions. The LT1175 is designed to withstand up to +2V forced output voltage. For both devices, start-up and recovery from short circuit or thermal shutdown is guaranteed under these conditions.

As anyone who has designed and built a "single supply" op amp circuit can attest, few can be implemented without the use of a mid-supply bias point or resistive divider providing that same function. The LT1118 serves as a low power means of obtaining a regulated, low dropout bias point for critical applications. This device features the ability to both source and sink load current (+800mA, −400mA), and exhibits an output impedance of about 16µH across a wide range of frequencies. The output remains stable irrespective of any bypass capacitance of 220nF or more. An output impedance of less than 3Ω can be achieved across a 10MHz bandwidth with the addition of a 1µF bypass; less than 1Ω with a 10µF bypass.

The LT1118 is available in 5V, 2.85V and 2.5V versions. Where the 5V version might serve as a standalone regulator, the 2.5V version is a good choice for splitting an existing 5V rail (see Figure 220.4). In addition to greatly reducing power consumption, the DC output impedance is less than 0.025Ω—unmatched by any resistive divider solution. A separate Enable pin shuts off the LT1118, reducing its supply current to 1µA. Figure 220.5 shows typical output impedance under a variety of operating conditions.

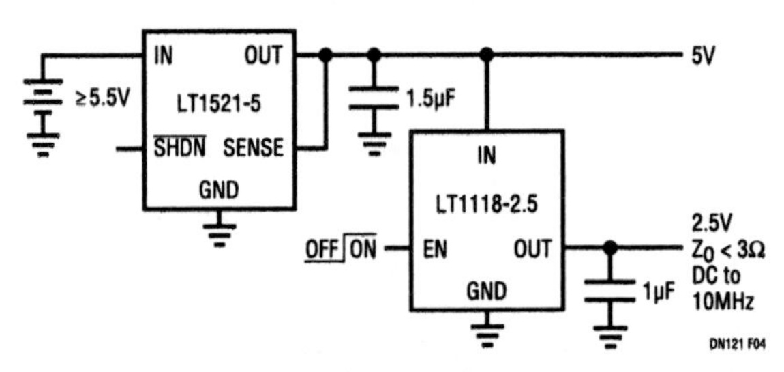

Figure 220.4 • Splitting the Supply Saves Power and Holds Bias Point DC Resistance to Less Than 0.025Ω

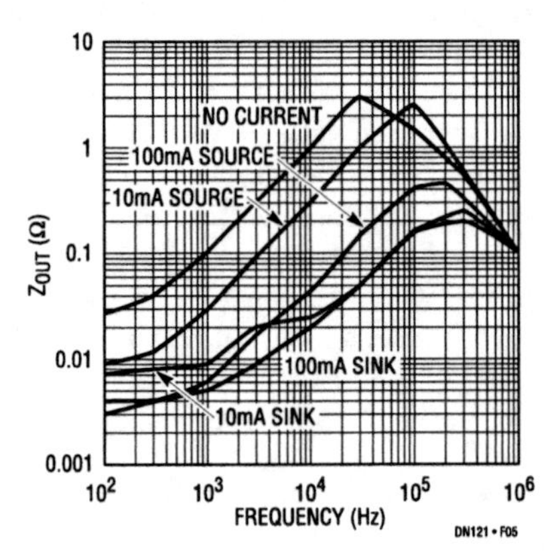

Figure 220.5 • LT1118-2.5 Output Impedance vs Frequency

Micropower DC/DC converter with independent low-battery detector

221

Steve Pietkiewicz

In the expanding world of low power portable electronics, a 2- or 3-cell battery remains a popular power source. Designers have many options for converting the 2V to 4V battery voltage to 5V, 3.3V and other required system voltages using low voltage DC/DC converter ICs. The LT1304 offers users a micropower step-up DC/DC converter featuring Burst Mode operation and a low-battery detector that stays alive when the converter is shut down. The device consumes only 125μA when active, yet can deliver 5V at up to 200mA from a 2V input. High frequency operation up to 300kHz allows the use of tiny surface mount inductors and capacitors. When the device is shut down the low-battery detector draws only 10μA. An efficient internal power NPN switch handles 1A with a drop of 500mV. Up to 85% efficiency is obtainable in 2-cell to 5V converter applications. The fixed output LT1304-5 and LT1304-3.3 versions have internal resistor dividers that set the output voltage to 5V or 3.3V, respectively.

A 2-cell to 5V converter

A compact 2-cell to 5V converter can be constructed using the circuit of Figure 221.1. The LT1304-5 fixed output device eliminates the need for external voltage setting resistors, lowering component count. As the battery voltage drops, the circuit continues to function until the LT1304's undervoltage lockout disables the part at approximately V_{IN}=1.5V. Up to 200mA output current is available at a battery voltage of 2V. As the battery voltage decreases below 2V, cell impedance starts to quickly increase. End-of-life is usually assumed to be around 1.8V, or 0.9V per cell. Burst Mode micropower operation keeps efficiency above 70% even for load current below 1mA. Efficiency, detailed in Figure 221.2, reaches 85% for a 3.3V input. Load transient response is illustrated in Figure 221.3. Since the LT1304 uses a hysteretic comparator in place of the traditional linear feedback loop, the circuit responds immediately to changes in load current. Figure 221.4 details

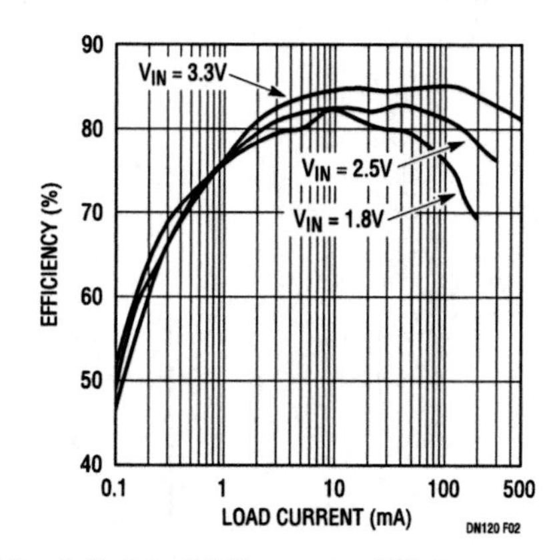

Figure 221.2 • 2-Cell to 5V Converter Efficiency

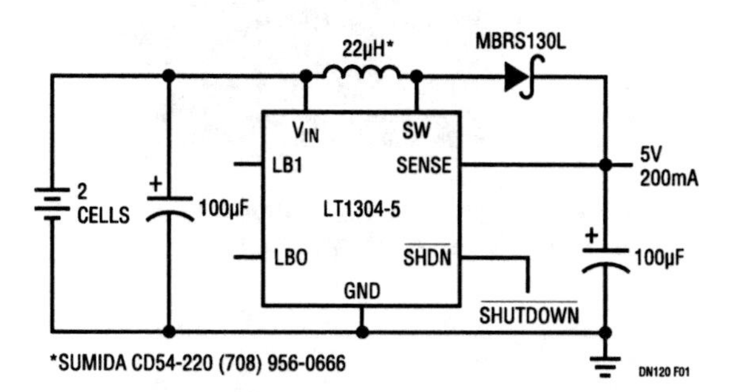

Figure 221.1 • 2-Cell to 5V/200mA Boost Converter Requires Only Four External Parts

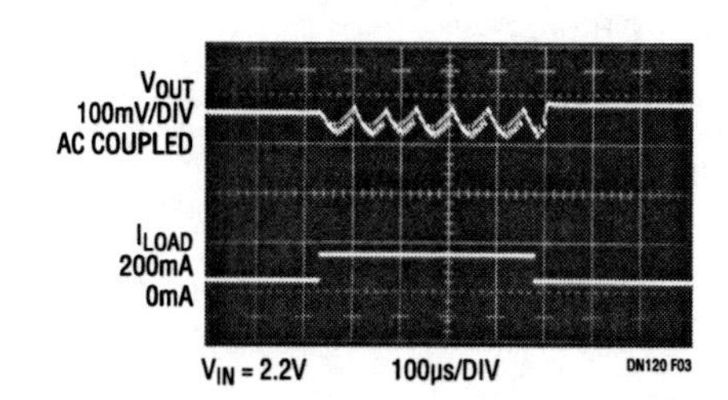

Figure 221.3 • Boost Converter Load Transient Response

Analog Circuit Design: Design Note Collection. http://dx.doi.org/10.1016/B978-0-12-800001-4.00221-0

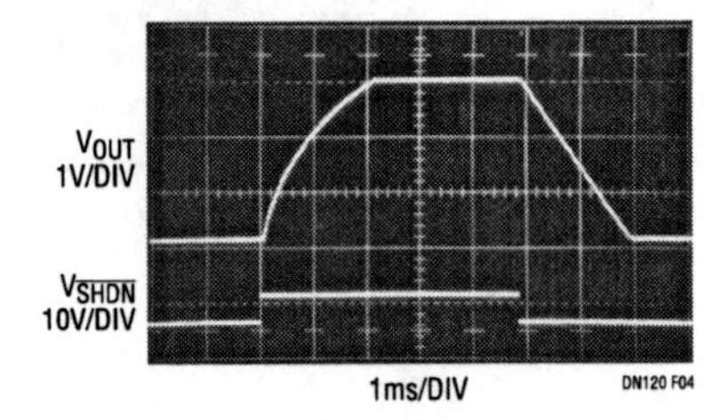

Figure 221.4 • Start-Up Response into 200mA Load. V_{OUT} Reaches 5V in Just Over 2ms

start-up behavior. After the device is enabled, output voltage reaches 5V in approximately 2ms while delivering 200mA.

Super Burst Mode operation: 5V/80mA DC/DC with 15μA quiescent current

The LT1304's low-battery detector can be used to control the DC/DC converter. The result is a reduction in quiescent current by almost an order of magnitude. Figure 221.5 details this Super Burst Mode operation circuit. V_{OUT} is monitored by the LT1304's LBI pin via resistor divider R1/R2. When LBI is above 1.2V, LBO is high, forcing the LT1304 into shutdown mode and reducing current drain from the battery to 15μA. When V_{OUT} decreases enough to overcome the low-battery detector's hysteresis (about 35mV) LBO goes low. Q1 turns on, pulling $\overline{SHDN}$ high and turning on the rest of the IC. R3 limits peak current to 500mA; it can be removed for higher output power. Efficiency is shown in Figure 221.6. The converter is approximately 70% efficient at a 100μA load, 20 points higher than the circuit of Figure 221.1. Even at a 10μA load, efficiency is in the 40% to 50% range, equivalent to

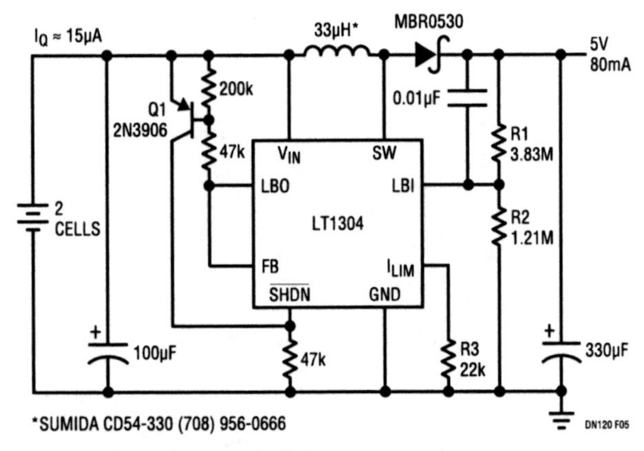

Figure 221.5 • Super Burst Mode Operation 2-Cell to 5V DC/DC Converter Draws Only 15μA Unloaded. Two AA Alkaline Cells Will Last for Years

100μW to 120μW total power drain from the battery. In contrast, Figure 221.1's circuit consumes approximately 300μW to 400μW unloaded.

An output capacitor charging cycle or "burst" is shown in Figure 221.7 with the circuit driving a 50mA load. The slow response of the low-battery detector results in the high number of individual switch cycles or "hits" within the burst.

Figure 221.8 depicts output voltage at the modest load of 100μA. The burst repetition rate is around 4Hz. With the load removed, the repetition rate drops to approximately 0.2Hz, or one burst every 5 seconds. Systems that spend a high percentage of operating time in sleep mode can benefit from the greatly reduced quiescent power drain of Figure 221.5's circuit.

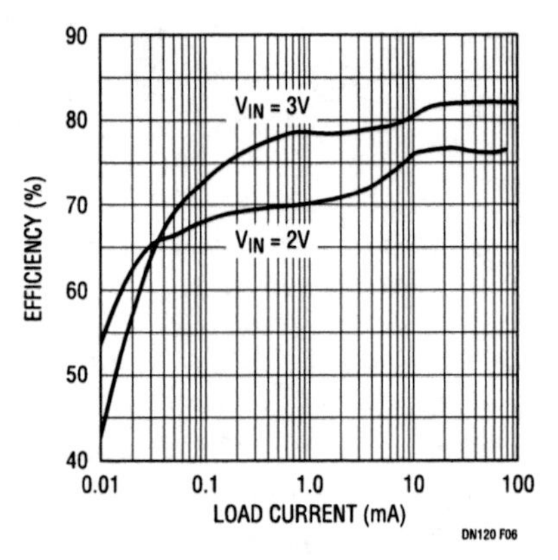

Figure 221.6 • Super Burst Mode Operation DC/DC Converter Efficiency

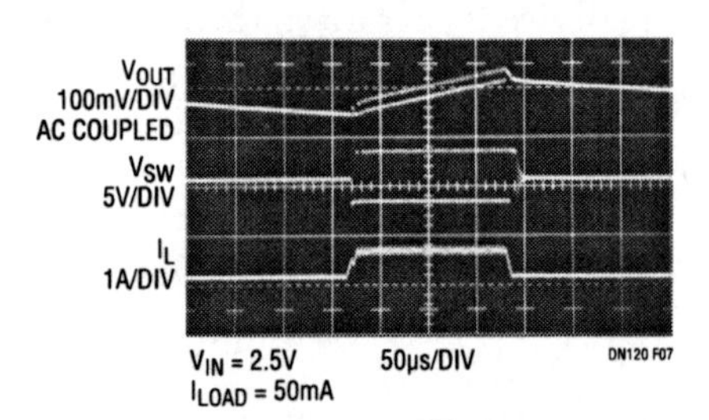

Figure 221.7 • Super Burst Mode Operation in Action

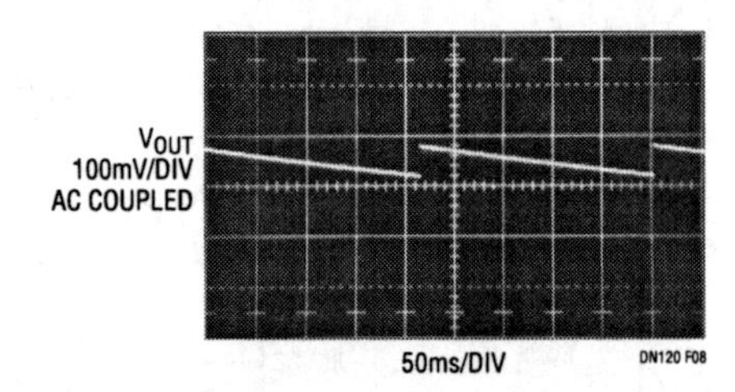

Figure 221.8 • Figure 221.5's Circuit, 100μA Load. Burst Occurs Approximately Once Every 240ms

High efficiency lithium-ion battery charger

222

Chiawei Liao

The LT1510 current mode PWM battery charger is the simplest, most efficient solution for fast charging modern rechargeable batteries including lithium-ion (Li-Ion), nickel-metal-hydride (NiMH) and nickel-cadmium (NiCd) that require constant current and/or constant voltage charging. The internal switch is capable of delivering 1.5A DC current (2A peak current). The onboard current sense resistor (0.1Ω) makes the charge current programming very simple. One resistor (or a programming current from a DAC) is used to set the charging current to within 5% accuracy. With 0.5% reference voltage accuracy, the LT1510 16-lead S package meets the critical constant voltage charging requirement for lithium cells.

The LT1510 can charge batteries ranging from 1V to 20V. A blocking diode is not required between the chip and the battery because the chip goes into sleep mode and drains only 3μA when the wall adaptor is unplugged. Soft-start and shutdown features are also provided.

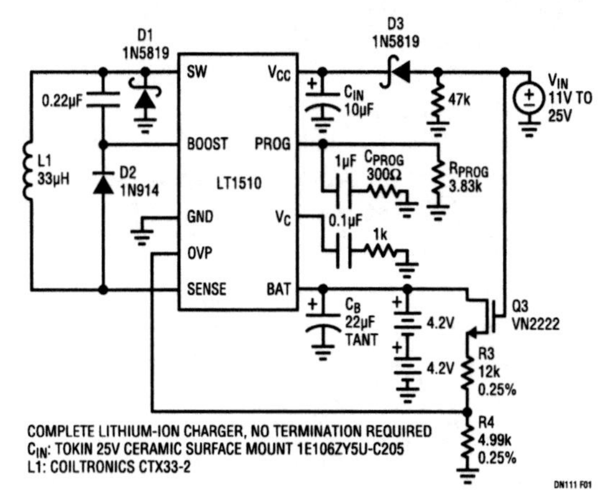

Figure 222.1 • Charging Lithium-Ion Batteries (Efficiency at 1.3A = 86%)

Lithium-ion battery charger

The circuit in Figure 222.1 uses the 16-lead LT1510 to charge lithium-ion batteries at a constant 1.3A until battery voltage reaches 8.4V set by R3 and R4. The charger will then automatically go into a constant voltage mode with current decreasing toward near zero over time as the battery reaches full charge. This is the normal regimen for lithium-ion charging, with the charger holding the battery at "float" voltage indefinitely. In this case, no external sensing of full charge is needed. Figure 222.2 shows typical charging characteristics.

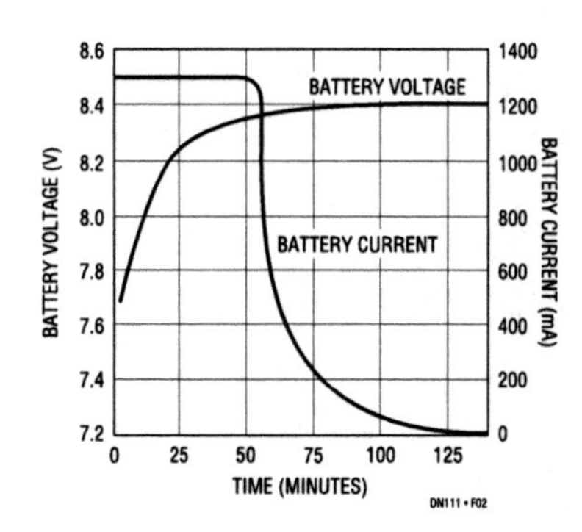

Figure 222.2 • Battery Charging Characteristics

The battery DC charging current is programmed by a resistor R_{PROG} (or a DAC output current) at the PROG pin. High DC accuracy is achieved with averaging capacitor C_{PROG}. The basic formula for full charging current is:

$$I_{BAT} = (I_{PROG})(2000) = (2.465/R_{PROG})(2000)$$
$$= (2.465/3.83k)(2000) = 1.3A$$

Approximately 0.25mA flows out of the BAT pin at all times when adapter power is applied. Therefore, to ensure a regulated output even when the battery is removed, the voltage divider current should be set at 0.5mA. Q3 is used to eliminate this current drain when adapter power is off, with a 47k resistor to pull its gate low.

Analog Circuit Design: Design Note Collection. http://dx.doi.org/10.1016/B978-0-12-800001-4.00222-2

With divider current set as 0.5mA, R4=2.465/0.5mA=4.93k, let R4=4.99k:

$$R3 = R4\left(\frac{V_{BAT}}{2.465} - 1\right) = 4.99k\left(\frac{8.4}{2.465} - 1\right) = 12k$$

V_{IN} has to be at least 3V higher than battery voltage and between 8.5V to 25V.

Lithium-ion batteries typically require float voltage accuracy of 1% to 2%. The LT1510 OVP voltage has 0.5% accuracy at 25°C and 1% over full temperature. This may suggest that very accurate (0.1%) resistors are needed for R3 and R4. Actually, in float mode the charging currents have tapered off to a low value and the LT1510 will rarely heat up past 50°C, so 0.25% resistors will provide the required level of overall accuracy.

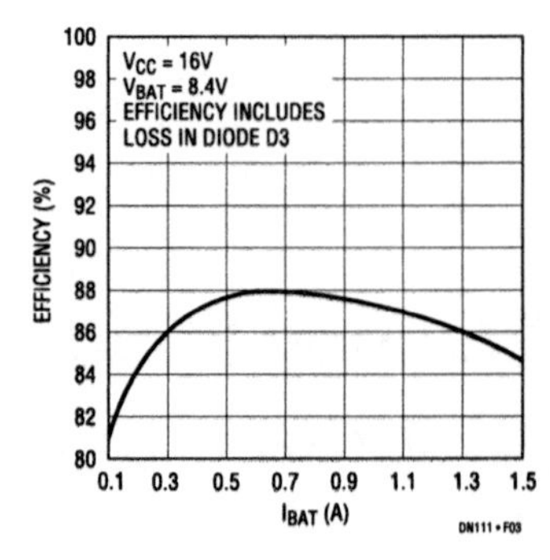

Figure 222.3 • Efficiency of Figure 222.1 Circuit

Thermal calculations

Although the battery charger achieves efficiency of approximately 86% at 1.3A, a thermal calculation should be done to ensure that junction temperature will not exceed 125°C. Power dissipation in the IC is caused by bias and driver current, switch resistance, switch transition losses and the current sense resistor. The 16-lead SO, with a thermal resistance of 50°C/W, can provide a full 1.5A charging current in many situations. Figure 222.3 shows the efficiency for charging currents up to 1.5A.

$$P_{BIAS} = (3.5mA)(V_{CC}) + (1.5mA)(V_{BAT}) + \frac{(V_{BAT})^2(7.5mA + 0.012 \cdot I_{BAT})}{V_{CC}}$$

$$P_{DRIVE} = \frac{(I_{BAT})(V_{BAT})^2}{50(V_{CC})}$$

$$P_{SWITCH} = \frac{(I_{BAT})^2(R_{SW})(V_{BAT})}{V_{CC}} + (T_{OL})(V_{CC})(I_{BAT})$$

$$P_{SENSE} = (0.18\Omega)(I_{BAT})^2$$

R_{SW} = Switch on resistance ≈ 0.35Ω

T_{OL} = Effective switch overlap time ≈ 10ns

$V_{CC} = V_{IN} - 0.4V$

Example: $V_{IN} = 16V, V_{BAT} = 8.4V, I_{BAT} = 1.3A$

$$P_{BIAS} = (3.5mA)(15.6) + (1.5mA)(8.4) + \frac{(8.4)^2(7.5mA + 0.012 \cdot 1.3)}{15.6} = 0.10W$$

$$P_{DRIVE} = \frac{(1.3)(8.4)^2}{50(15.6)} = 0.12W$$

$$P_{SWITCH} = \frac{(1.3)^2(0.35)(8.4)}{15.6} + (10^{-8})(15.6)(1.3)(200kHz) = 0.36W$$

$$P_{SENSE} = (0.18)(1.3)^2 = 0.30W$$

Total power in the IC is $0.1 + 0.12 + 0.36 + 0.30 = 0.88W$

Temperature rise in the IC will be:

$$(50°C/W)(0.88W) = 44°C$$

Some battery manufacturers recommend termination of constant voltage float mode 30 to 90 minutes after charging current has dropped below a specified level (typically 50mA to 100mA). Check with the manufacturers for details. The circuit in Figure 222.4 will detect when charging current has dropped below 75mA. This logic signal is used to initiate a timeout period, after which the LT1510 can be shut down by pulling the VC pin low with an open collector or drain. Some external means may be used to detect the need for additional charging or the charger may be turned on periodically to complete a short float voltage cycle. The current trip level is determined by the battery voltage, R1 through R3, and the internal LT1510 sense resistor (≈0.18Ω pin-to-pin). D2 generates hysteresis in the trip level to avoid multiple comparator transitions. R2 and R3 are chosen to total about 1M to minimize battery loading. D2 is assumed to be off during high current charging when the comparator output is high. To ensure this, the ratio of R2 to R3 is chosen to make the center node voltage less than the logic supply. R4 is somewhat arbitrary and does not affect trip point. R1 is adjusted to set the trip level:

$$R1 = \frac{(I_{TRIP})(R2 + R3)(0.18\Omega)}{V_{BAT}} = \frac{(75mA)(560k + 430k)(0.18)}{8.4V} = 1.6k$$

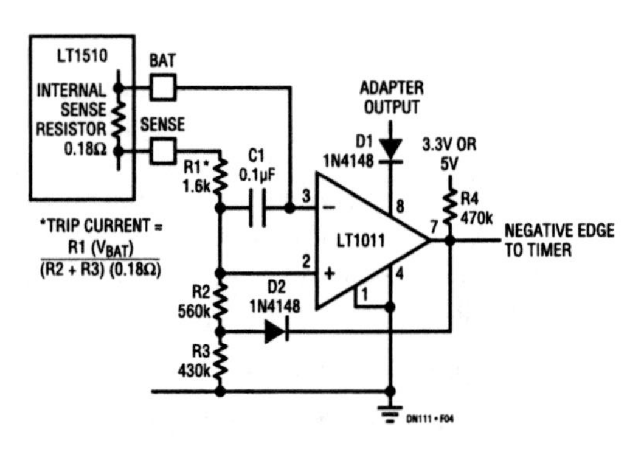

Figure 222.4 • Current Comparator for Initiating Float Timeout

A 4-cell NiCd regulator/charger for notebook computers

223

Tim Skovmand

The new LTC1155 dual power MOSFET driver delivers 12V of gate drive to two N-channel power MOSFETs when powered from a 5V supply with no external components required. This ability, coupled with its micropower current demands and protection features, makes it an excellent choice for high side switching applications which previously required more expensive P-channel MOSFETs.

A notebook computer power supply system is a good example of an application which benefits directly from this high side driving scheme. A 4-cell, NiCad battery pack can be used to power a 5V notebook computer system. Inexpensive N-channel power MOSFETs have very low on resistance and can be used to switch power with low voltage drop between the battery pack and the 5V logic circuits.

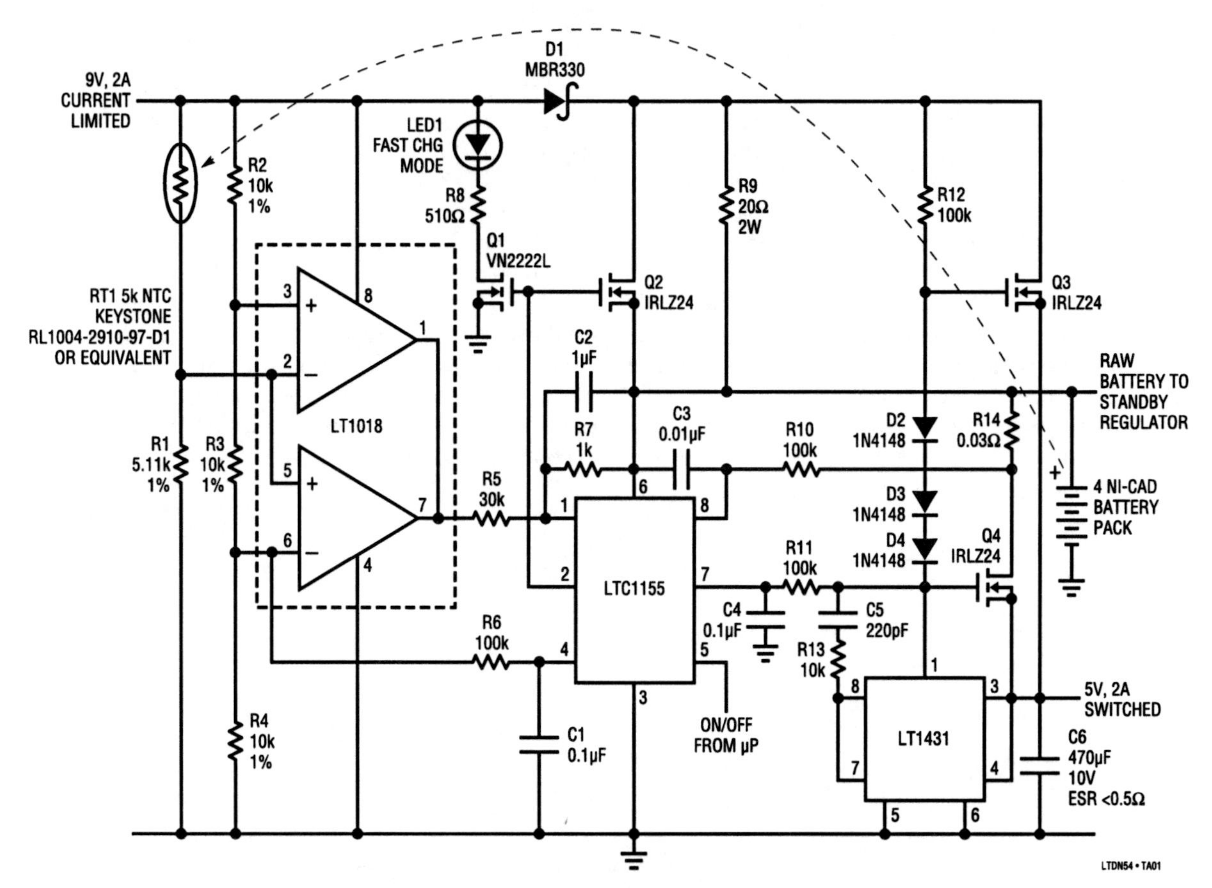

Figure 223.1 • The LTC1155 Dual MOSFET Driver Provides Gate Drive and Protection for a 4-Cell NiCd Charger and Regulator

Analog Circuit Design: Design Note Collection. http://dx.doi.org/10.1016/B978-0-12-800001-4.00223-4

Figure 223.1 shows how a battery charger and an extremely low voltage drop 5V regulator can be built using the new LTC1155 and three inexpensive power MOSFETs.

Quick charge battery charger

One-half of the LTC1155 dual MOSFET driver controls the charging of the battery pack. The 9V, 2A current limited wall unit is switched directly into the battery pack through an extremely low resistance MOSFET switch, Q2. The gate drive output, Pin 2, generates about 13V of gate drive to fully enhance Q1 and Q2. The voltage drop across Q2 is only 0.17V at 2A and, therefore, can be surface mounted to save board space.

An inexpensive thermistor, RT1, measures the battery temperature and latches the LTC1155 off when the temperature rises to 40°C by pulling low on Pin 1, the drain sense input. The window comparator also ensures that battery packs which are very cold (<10°C) are not quick charged.

Q1 drives an indicator lamp during quick charge to let the computer operator know that the battery pack is being charged properly. When the battery temperature rises to 40°C, the LTC1155 latches off and the battery charge current flowing through R9 drops to 150mA.

Extremely low voltage drop regulator

A 4-cell NiCd battery pack produces about 6V when fully charged. This voltage will drop to about 4.5V when the batteries are nearly discharged. The second half of the LTC1155 provides gate voltage drive, Pin 7, for an extremely low voltage drop MOSFET regulator. The LT1431 controls the gate of Q4 and provides a regulated 5V output when the battery is above 5V. When the battery voltage drops below 5V, Q4 acts as a low resistance switch between the battery and the regulator output.

A second power MOSFET, Q3, connected between the 9V supply and the regulator output "bypasses" the main regulator when the 9V supply is connected. This means that the computer power is taken directly from the AC line while the charger wall unit is connected. The LT1431 provides regulation for both Q3 and Q4 and maintains a constant 5V at the regulator output. The diode string made up of diodes D2-D4 ensure that Q3 conducts all the regulator current when the wall unit is plugged in by separating the two gate voltages by about 2V.

R14 acts as a current sense for the regulator. The regulator latches off at 3A when the voltage drop between the second drain sense input, Pin 8, and the supply, Pin 6, rises above 100mV. R10 and C3 provide a short delay. The μP can restart the regulator by turning the second input, Pin 5, off and then back on.

The regulator is switched off by the μP when the battery voltage drops below 4.6V. The standby current for the 5V, 2A regulator is less than 10μA. The regulator is switched on again when the battery voltage rises during charging.

Very low power dissipation

The power dissipation in the notebook computer is very low. The current limited wall unit dissipates the bulk of the power created by quick charging the battery pack. Q2 dissipates less than 0.5W. R9 dissipates about 0.7W. Q4 dissipates about 2W for a very short period of time when the batteries are fully charged and dissipates less than 0.5W as soon as the battery voltage drops to 5V. The three integrated circuits shown are micropower and dissipate virtually no power.

Cost-effective and efficient power system

The circuit shown in Figure 223.1 consumes very little board space. The LTC1155 is available in a 8-pin SO package and the three power MOSFETs can also be housed in SO packaging. Q4 must be heat sinked properly for the short period of time that the battery voltage is above 5.5V (consult the MOSFET manufacturer data sheet for SO heat sink recommendations).

The LTC1155 allows the use of inexpensive N-channel MOSFET switches to directly connect power from a 4-cell NiCd battery pack to the charger and the load. This technique is very cost-effective and is also very efficient. Nearly all the battery power is delivered directly to the load to ensure maximum operating time from the batteries.

Switching regulator allows alkalines to replace NiCds

224

Brian Huffman

In many applications it is desirable to substitute non-rechargeable batteries for chargeable types. This capability is necessary when the NiCds cannot be recharged or long charge times are unacceptable. alkaline batteries are an excellent choice in this situation. They are readily available and have reasonable energy density. Compared to alkalines, NiCds provide a more stable terminal voltage as they discharge. NiCds decay from 1.3V to 1.0V, while alkalines drop from 1.5V to 0.8V. Replacing NiCds with alkalines can cause unacceptable low supply voltage, although available energy is adequate. A boost type switching regulator obviates this problem, allowing alkaline cells to replace NiCds. The circuit shown in Figure 224.1 accommodates the alkaline cells widely varying terminal voltage while providing a constant output voltage.

This circuit is a step-up boost type switching regulator. It maintains a constant 6V output as battery voltage fails. The inductor accumulates energy from the battery when the LT1270 switch pin (V_{SW}) switches to ground and dumps its stored energy to the output when the switch pin (V_{SW}) goes off. The feedback pin (V_{FB}) samples the output from the 6.19k–1.62k divider. The LT1270's error amplifier compares the feedback pin voltage to its internal 1.24V reference and controls the V_{SW} pin switching current, completing a control loop. The output voltage can be varied by changing the resistor divider ratio. The RC damper on the V_C pin provides loop frequency compensation. The minimum start-up voltage for this circuit is 3V. If a 3.3V start-up voltage is permissible R1 and Q1 can be removed with D2 replaced by a short.

Bootstrapping the V_{IN} pin off the output voltage allows the battery voltage to drop below the minimum start-up voltage, while maintaining circuit operation. For example, with three C cells the battery voltage is initially 4.5V and operates down to 2.4V. With this bootstrapped technique the circuit provides a constant output voltage over the battery's complete operating range, maximizing battery life.

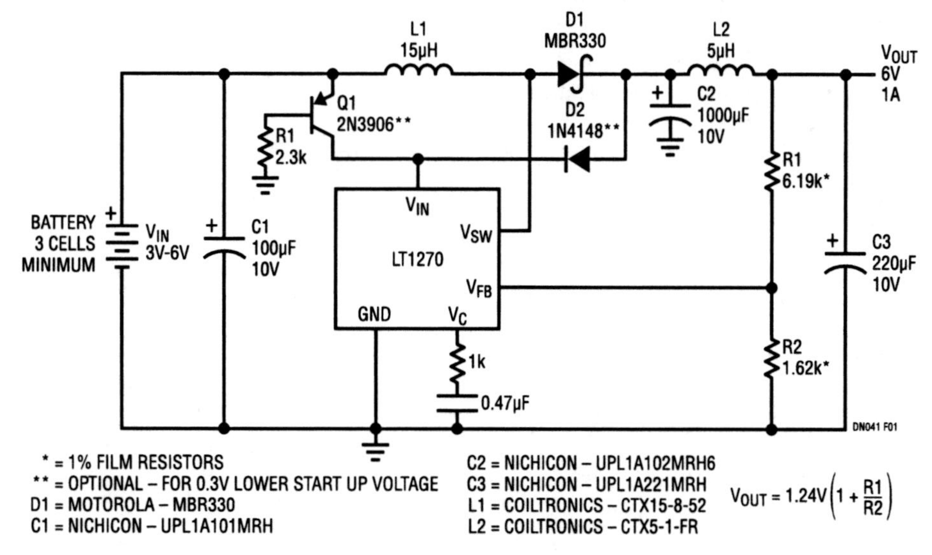

Figure 224.1 • Low Voltage Circuit Provides Constant Output Voltage as Battery Discharges

Analog Circuit Design: Design Note Collection. http://dx.doi.org/10.1016/B978-0-12-800001-4.00224-6

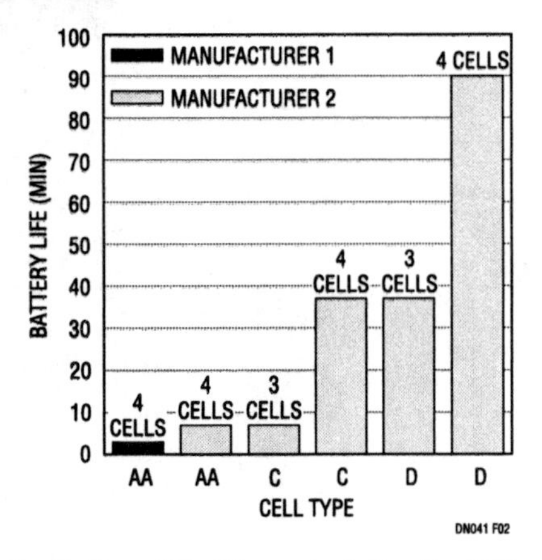

Figure 224.2 • Battery Life Characteristics for Different Batteries for a 6W Load

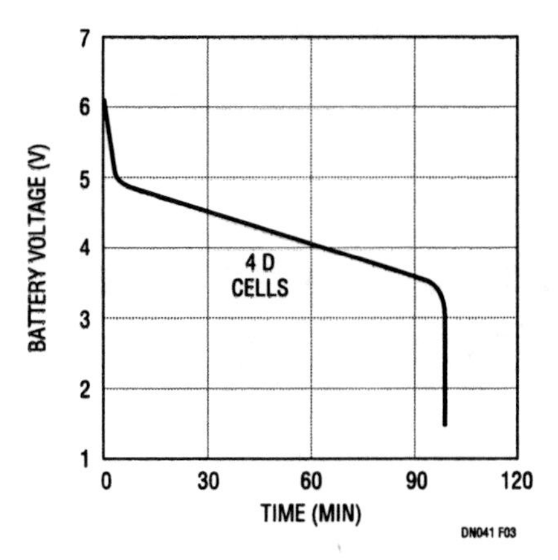

Figure 224.3 • alkaline Battery Discharge Characteristic with 6W Load

Battery life characteristics are different for various cell types. Figure 224.2 compares battery life between AA, C, and D cells with a 6W load. In this application the power drain from the battery remains relatively constant. As the battery voltage decreases the battery current increases. The AA types discharge quicker than the C or D cells. They are physically smaller than the other cells, and therefore store less energy. The AA cells are 3 times smaller than the C cells and 6 times smaller than the D cells.

Current drain also influences cell life. Battery life significantly decreases at high current discharge. Slightly higher battery stack voltages permit surprising battery life increases. The higher voltage means lower current drain for a constant power load. Operating at just 33% less current the four C cells last 5 times longer than three C cells.

Battery life characteristics vary widely between manufacturers. Some manufacturers' cells are optimized to operate more efficiently at lower current levels, making it wise to consult the battery manufacturer's discharge characteristics.

Figure 224.3 shows alkaline battery discharge characteristics for four D cells. A fresh cell measures 1.5V and operates down to 0.8V before the cell dies. The battery stack voltage drops quickly and then stabilizes until it reaches 3.2V; 0.8V per cell. There is no usable battery life beyond this point.

Figure 224.4 shows efficiency exceeding 85%. The diode and LT1270 switch are the two main loss elements. The Schottky diode introduces a relatively constant 7% loss, while the LT1270 switch loss varies with battery voltage. As battery voltage decreases, switch current and duty cycle increase. This has a dramatic effect on switch loss, because switch loss is proportional to the square of switch current multiplied by duty cycle. Therefore, at low input voltages efficiency is degraded because this loss is a higher percentage of the battery power drain.

If lower output current is desired, an LT1170, LT1171, or LT1172 can be used.

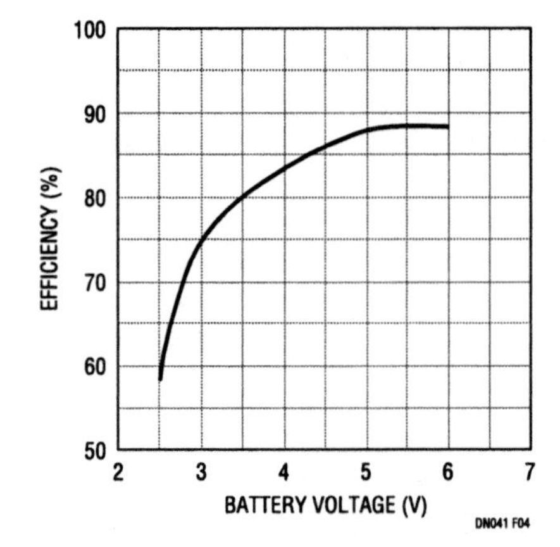

Figure 224.4 • Efficiency for Various Battery Voltages

Section 13

Energy Harvesting & Solar Power Circuits

225

Tiny 2-cell solar panel charges batteries in compact, off-grid devices

Fran Hoffart

Introduction

Advances in low power electronics now allow placement of battery-powered sensors and other devices in locations far from the power grid. Ideally, for true grid independence, the batteries should not need replacement, but instead be recharged using locally available renewable energy, such as solar power. This Design Note shows how to produce a compact battery charger that operates from a small 2-cell solar panel. A unique feature of this design is that the DC/DC converter uses power point control to extract maximum power from the solar panel.

The importance of maximum power point control

Although solar cells or solar panels are rated by power output, a panel's available power is hardly constant. Its output power depends heavily on illumination, temperature and on the load current drawn from the panel. To illustrate this, Figure 225.1 shows the V-I characteristic of a 2-cell solar panel at a constant illumination. The I-vs-V curve features a relatively constant-current characteristic from short circuit (at the far left) to around 550mA load current, at which point it bends to a constant-voltage characteristic at lower currents, approaching maximum voltage at open circuit (far right). The panel's power output curve shows a clear peak in power output around 750mV/530mA, at the knee of the I-vs-V curve. If the load current increases beyond the power peak, the power curve quickly drops to zero (far left). Likewise, light loads push power toward zero (far right), but this tends to be less of an issue.

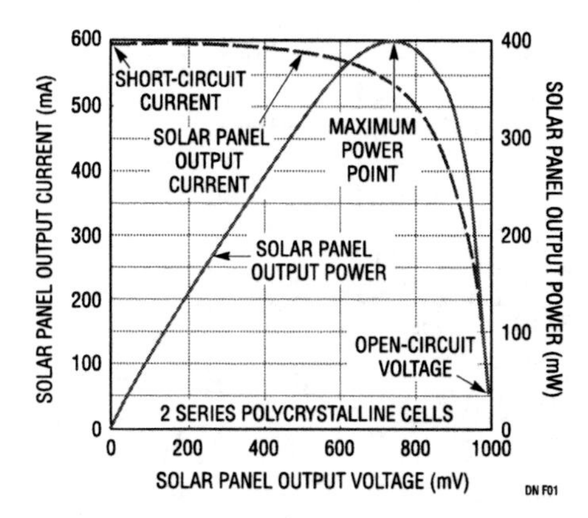

Figure 225.1 • Solar Panel Output Voltage, Current and Power

Of course, panel illumination affects available power—less light means lower power output; more light, more power. Although illumination directly affects the *value* of peak power output, it does not do much to affect the peak's *location* on the voltage scale. That is, regardless of illumination, the panel output voltage at which peak power occurs remains relatively constant. Thus, it makes sense to moderate the output current so that the solar panel voltage remains at or above this peak power voltage, in this case 750mV. Doing so is called maximum power point control (MPPC).

Figure 225.2 shows the effects of varying sunlight on the charge current, with maximum power point control and without. The simulated sunlight is varied from 100% down to approximately 20%, then back up to 100%. Note that as the sunlight intensity drops about 20%, the solar panel's output voltage and current also drop, but the LTC3105 maximum power point control prevents the panel's output voltage from dropping below the programmed 750mV. It accomplishes this by reducing the LTC3105 output charge current to prevent the solar panel from collapsing to near zero volts, as is shown in the plot on the right side of Figure 225.2. Without power point control, a small reduction in sunlight can completely stop charge current from flowing.

LTC3105 boost converter with input power control

The LTC3105 is a synchronous step-up DC/DC converter designed primarily to convert power from ambient energy sources, such as low voltage solar cells and thermoelectric

Analog Circuit Design: Design Note Collection. http://dx.doi.org/10.1016/B978-0-12-800001-4.00225-8

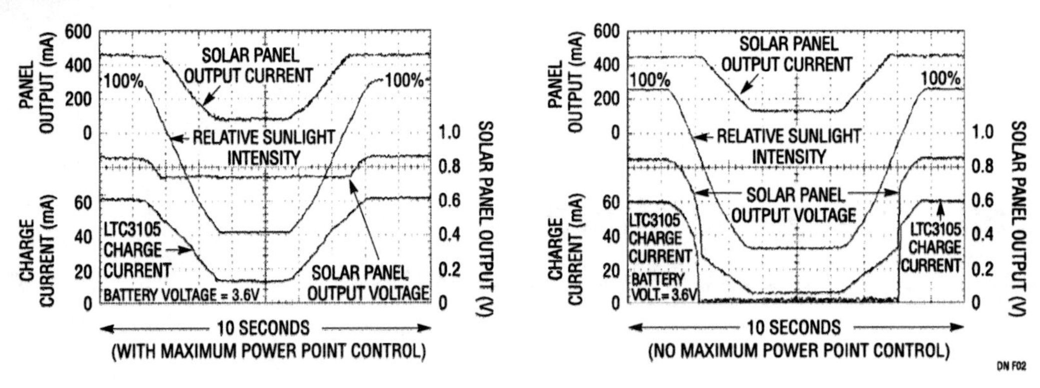

Figure 225.2 • Changing Sunlight Intensity Effects on Charge Current

generators, to battery charging power. The LTC3105 uses MPPC to deliver maximum available power from the source. It accomplishes this by reducing the LTC3105 output current to prevent the solar panel from collapsing to near zero volts. The LTC3105 is capable of starting up with an input as low as 250mV, allowing it to be powered by a single solar cell or up to nine or ten series-connected cells.

Output disconnect eliminates the isolation diode often required with other solar powered DC/DC converters and allows the output voltage to be above or below the input voltage. The 400mA switch current limit is reduced during start-up to allow operation from relatively high impedance power sources, but still provides sufficient power for many low power solar applications once the converter is in normal operation. Also included are a 6mA adjustable output low dropout linear regulator, open-drain power good output, shutdown input and Burst Mode operation to improve efficiency in low power applications.

Solar-powered Li-Ion battery charger

Figure 225.3 shows a compact solar-powered battery charger using a LTC3105 as a boost converter and an LTC4071 as a Li-Ion shunt charger. A 2-cell 400mW solar panel provides the input power to the LTC3105 to produce over 60mA of charge current in full sunlight. Maximum power point control prevents the solar panel voltage from dropping below the 750mV maximum power point, as shown in Figure 225.1. The converter's output voltage is programmed for 4.35V, slightly above the 4.2V float voltage of the Li-Ion battery. The LTC4071 shunt charger limits the voltage across the battery to 4.2V. Grounding the FBLDO pin programs the low dropout regulator to 2.2V, which powers the "charging" LED. This LED is on when charging and off when the battery voltage is within 40mV of the float voltage, indicating near full charge. An NTC thermistor senses battery temperature and lowers the LTC4071 float voltage at high ambient temperatures for increased battery safety. To prevent battery damage from over-discharge, the low battery disconnect feature disconnects the battery from the load if the battery drops below 2.7V.

Conclusion

Although the circuit described here produces only a few hundred milliwatts, it can provide enough power to keep a 400mAhr Li-Ion battery fully charged under most weather conditions. The low input voltage, combined with input power control, makes the LTC3105 ideal for low power solar applications. In addition, the LTC4071 shunt charging system complements the LTC3105 by providing the precision float voltage, charge status and temperature safety features to assure long battery life in outdoor environments.

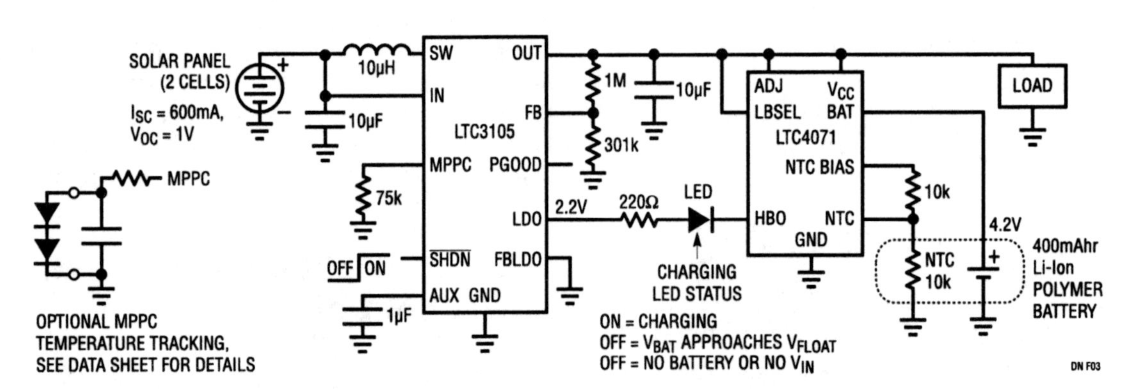

Figure 225.3 • 2-Cell Solar Panel Li-Ion Battery Charger

Energy harvester produces power from local environment, eliminating batteries in wireless sensors

226

Jim Drew

Introduction

Recent advances in ultralow power microcontrollers have produced devices that offer unprecedented levels of integration for the amount of power they require to operate. These are systems on a chip with aggressive power-saving schemes, such as shutting down power to idle functions. In fact, so little power is needed to run these devices that many sensors are going wireless, since they can readily run from batteries. Unfortunately, batteries must be regularly replaced, which is a costly and cumbersome maintenance project. A more effective wireless power solution may be to harvest ambient mechanical, thermal, or electro-magnetic energy in the sensor's local environment.

The LTC3588-1 shown in Figure 226.1 is a complete energy harvesting solution optimized for high impedance sources such as piezoelectric transducers. It contains a low loss full wave bridge rectifier and a high efficiency synchronous buck converter, which transfer energy from an input storage device to an output at a regulated voltage capable of supporting loads up to 100mA. The LTC3588-1 is available in 10-lead MSE and 3mm × 3mm DFN packages.

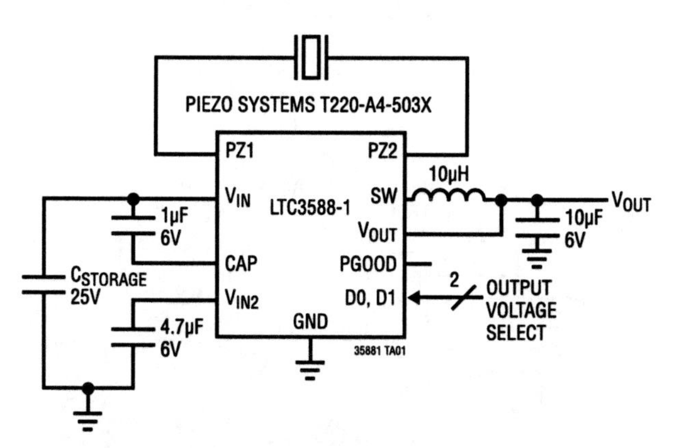

Figure 226.1 • Complete Energy Harvesting Solution Optimized for High Impedance Sources Such as Piezoelectric Transducers

Ambient energy sources

Ambient energy sources include light, heat differentials, vibrating beams, transmitted RF signals or any other source that can produce an electrical charge through a transducer. For instance:

- Small solar panels have been powering handheld electronic devices for years and can produce 100s of mW/cm^2 in direct sunlight and 100s of $\mu W/cm^2$ in indirect light.
- Seebeck devices convert heat energy into electrical energy where a temperature gradient is present. Sources of heat energy vary from body heat, which can produce 10s of $\mu W/cm^2$, to a furnace exhaust stack where surface temperatures can produce 10s of mW/cm^2.
- Piezoelectric devices produce energy by either compression or deflection of the device. Piezoelectric elements can produce 100s of $\mu W/cm^2$ depending on their size and construction.
- RF energy harvesting is collected by an antenna and can produce 100s of pW/cm^2.

Successfully designing a completely self-contained wireless sensor system requires power saving microcontrollers and transducers that consume minimal electrical energy from low energy environments. Now that both are readily available, the missing link is the high efficiency power conversion product capable of converting the transducer output to a usable voltage.

Figure 226.2 shows an energy harvesting power system that includes the energy source/transducer, an energy storage element and a means to convert this stored energy into a useful regulated voltage. There may also be a need for a voltage rectifier network between the energy transducer and the energy storage element to prevent energy from back-feeding into the transducer or to rectify an AC signal in the case of a piezoelectric device.

Analog Circuit Design: Design Note Collection. http://dx.doi.org/10.1016/B978-0-12-800001-4.00226-X

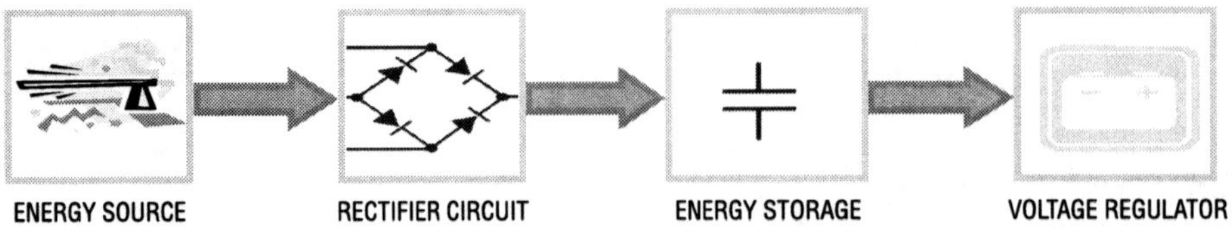

ENERGY SOURCE | RECTIFIER CIRCUIT | ENERGY STORAGE | VOLTAGE REGULATOR

Figure 226.2 • Energy Harvesting System Components

Application examples

The LTC3588-1 requires the output voltage of the transducer to be above the undervoltage lockout rising threshold limit for the specific output voltage set at the D0 and D1 input pins. For maximum energy transfer, the energy transducer must have an open circuit voltage of twice the input operating voltage and a short-circuit current of twice the input current required. These requirements must be met at the minimum excitation level of the source to achieve continuous output power.

Piezoelectric transducer application

Figure 226.3 shows a piezoelectric system that, when placed in an air stream, produces 100μW of power at 3.3V. The deflection of the piezoelectric element is 0.5cm at a frequency of 50Hz.

Seebeck transducer application

Figure 226.4 shows an energy harvesting system that uses a Seebeck transducer from Tellurex Corporation. A heat differential produces an output voltage that supports a 300mW output load. Connecting the transducer to the PZ1 input prevents reverse current from flowing back into the Seebeck device when the heat source is removed. The 100Ω resistor provides current limiting to protect the LTC3588-1 input bridge.

Harvest energy from the EM field produced by standard fluorescent lights

This application requires some outside-the-box thinking. Figure 226.5 shows a system that harvests energy from the electric fields surrounding high voltage fluorescent tubes. Two 12″ × 24″ copper panels are placed 6″ from a 2′ × 4′ fluorescent light fixture. The copper panels capacitively harvest 200μW from the surrounding electric fields and the LTC3588-1 converts that power to a regulated output.

Conclusions

The LTC3588-1 allows remote sensors to operate without batteries by harvesting ambient energy from the surrounding environment. It contains all the critical power management functions: a low loss bridge rectifier, a high efficiency buck regulator, a low bias UVLO detector that turns the buck converter on and off, and a PGOOD status signal to wake up the microcontroller when power is available. The LTC3588-1 supports loads up to 100mA with just five external components.

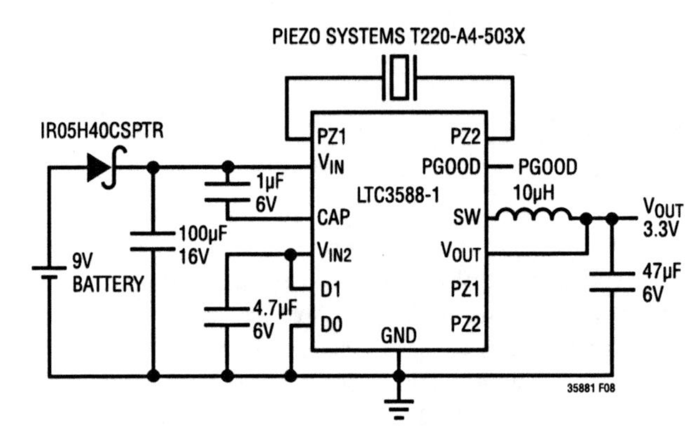

Figure 226.3 • Piezoelectric Energy Harvester

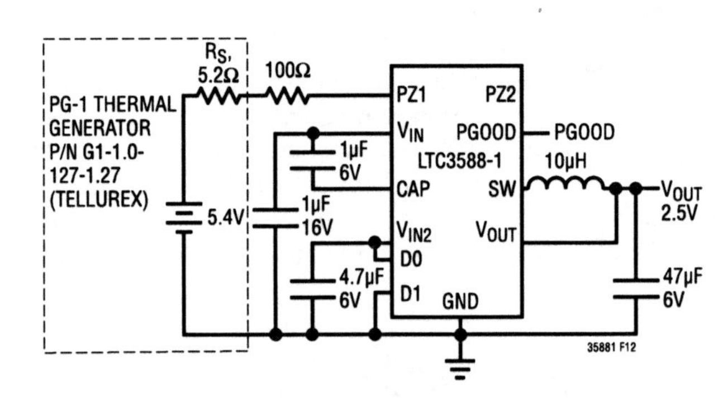

Figure 226.4 • Seebeck Energy Harvester

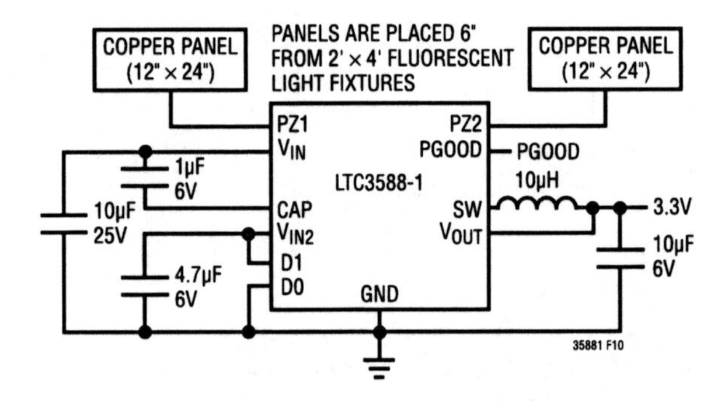

Figure 226.5 • Electric Field Energy Harvester

Section 14

Charge Pump DC/DC Converter Design

Step-down charge pumps are tiny, efficient and very low noise

227

William Walter

Introduction

Inductorless charge pump DC/DC converters are popular in space-constrained applications with 10mA to 500mA load currents. Such converters are available in small packages, operate with very low quiescent current and require minimal external components. However, the downside for most charge pumps is noise. Noise that is generated at the power input can interfere with RF transmission and reception in wireless applications, and noise at the output can couple into sensitive circuits or even create audible noise. The new LTC3250 and LTC3251 step-down charge pumps solve this problem with a new switching architecture that mitigates noise without sacrificing any of the space saving, high efficiency benefits typical of charge pump converters.

The LTC3250-1.5 is optimized for applications that require up to 250mA of current from a fixed 1.5V output, and where space is at an absolute premium. The LTC3251 provides more flexibility in a slightly larger footprint: it features an adjustable output voltage (0.9V to 1.6V), up to 500mA output, ultralow input and output noise and spread spectrum operation. Both ICs use 2-to-1 switched capacitor fractional conversion to improve efficiency by 50% over a linear regulator. Both parts require only 35μA of operating current.

Efficient low noise fixed 1.5V output charge pump with ultrasmall footprint

The LTC3250-1.5 switched capacitor step-down DC/DC converter squeezes into the tightest spaces while providing 1.5V at 250mA from a single 3.1V to 5.5V supply. To keep the converter footprint small, the LTC3250 operates at high frequency (1.5MHz) making it possible to use only three tiny low cost ceramic capacitors for operation. The LTC3250 is available in a 6-pin ThinSOT package making it possible to build a complete converter in an area of less than $0.04in^2$ (see Figure 227.1).

The LTC3250's constant frequency architecture achieves regulation by sensing the output voltage and regulating the amount of charge transferred per cycle. This method of regulation provides much lower input and output voltage ripple than that of conventional switched capacitor charge pumps

Figure 227.1 • The LTC3250 Is Available in a Tiny 6-Pin ThinSOT Package Making It Possible to Fit a Complete Converter in Less Than 0.04in²

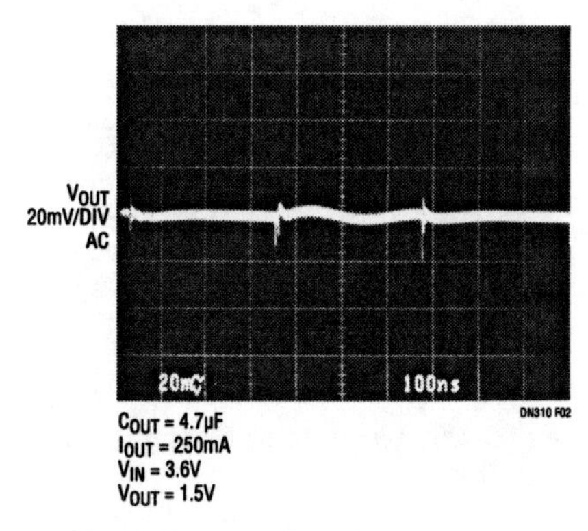

Figure 227.2 • Output Voltage Ripple

Analog Circuit Design: Design Note Collection. http://dx.doi.org/10.1016/B978-0-12-800001-4.00227-1

that regularly have 50mV or more of ripple. The constant, high frequency charge transfer of the LTC3250 makes filtering input and output noise less demanding than traditional switched capacitor charge pumps where switching frequencies depend on load current and can range over several orders of magnitude. Figure 227.2 shows the low output ripple for the LTC3250-1.5 with a 250mA load.

Ultralow noise adjustable charge pump with spread spectrum operation

Switching regulators are commonly used to provide power conversion inside handheld devices such as cellular phones and PDAs. Such devices, particularly those which provide RF communication, tend to be very sensitive to noise and electromagnetic interference (EMI). Switching regulators operate on a cycle-by-cycle basis to transfer power to an output. For conventional step-down regulators, the first half cycle the input is supplying current to the output and charging the storage element (either capacitor or inductor), and the other half cycle the output current is supplied via the storage element from ground, and no current is supplied via the input. This rectangular wave of current at the input can cause a large ripple voltage with harmonics extending to very high frequencies (see Figure 227.3). Since the operating frequency of conventional step-down regulators is either fixed or variable, the output will still have a large component of noise at the frequency of operation and some harmonics (see Figure 227.4).

The LTC3251 significantly reduces input noise through the use of a dual phase spread spectrum charge pump. The dual phase architecture works by supplying current on both clock phases, thus drawing a constant input current that is half the output current. Additionally, with spread spectrum operation the internal oscillator of the LTC3251 produces a clock pulse whose period is random on a cycle-by-cycle basis, but fixed between 1MHz and 1.6MHz. This has the benefit of spreading the switching noise over a range of frequencies, thus significantly reducing harmonic noise. This architecture achieves extremely low output and input noise. Figure 227.5 shows the virtual elimination of input harmonics using only a tenth of the input capacitance of a conventional step-down regulator, and Figure 227.6 shows the significant reduction in peak output noise using only half of the output capacitance.

Versatility

The LTC3251 has four modes of operation which are selected by the mode pins MD0 and MD1. The modes are: Continuous Spread Spectrum for low noise at all operating currents; Burst Mode operation for high efficiency at light loads; Super Burst mode for ultralow operating current at very light loads (9μA typical at no load); and shutdown.

Conclusion

The LTC3251 is available in a 10-pin thermally enhanced MSOP package; the LTC3250 is available in a tiny ThinSOT package. Their small size, relatively high current outputs, and low noise make them ideally suited for space-constrained battery-powered applications.

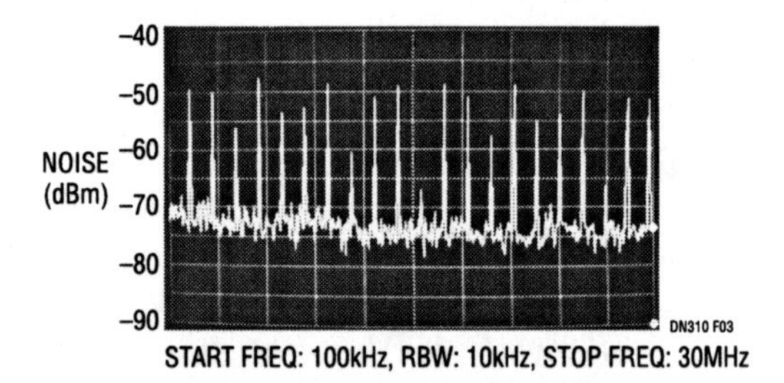

Figure 227.3 • Conventional Step-Down Regulator Input Noise Spectrum with 10μF Input Capacitor (I_O = 500mA)

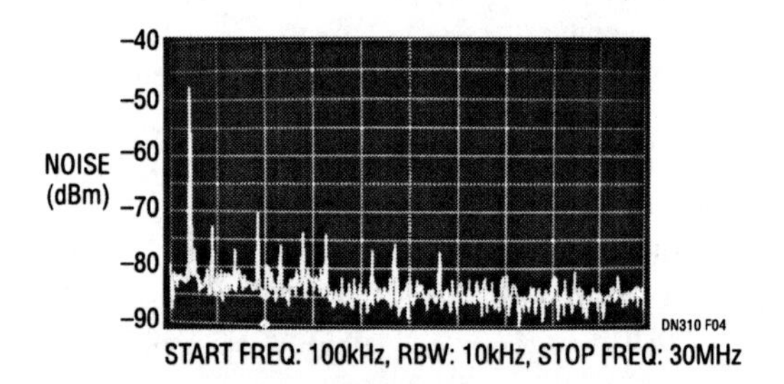

Figure 227.4 • Conventional Step-Down Regulator Output Noise Spectrum with 22μF Output Capacitor (I_O = 500mA)

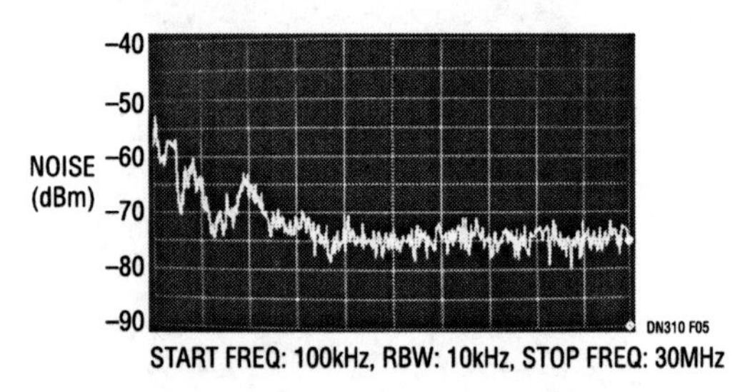

Figure 227.5 • LTC3251 Input Noise Spectrum with 1μF Input Capacitor (I_O = 500mA)

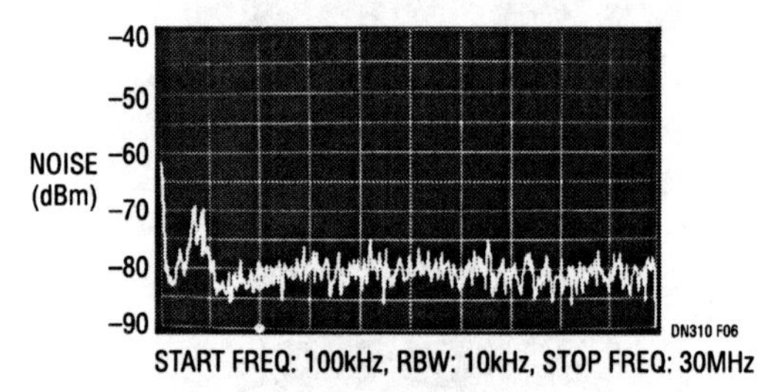

Figure 227.6 • LTC3251 Output Noise Spectrum with 10μF Output Capacitor (I_O = 500mA)

New charge pumps offer low input and output noise

228

Sam Nork

Charge pump (inductorless) DC/DC converters are quite popular in space-constrained applications where low to moderate load currents must be supplied. Such converters are available in small packages, operate with very low quiescent current and require minimal external components. However, noise generation is one undesirable characteristic of most charge pumps.

Unwanted noise can create a variety of problems. Noise generated at the power input can interfere with RF transmission and reception in wireless applications. Noise at the output can couple onto sensitive circuits or even create audible noise. The new LTC3200 family of boost charge pumps employs a new architecture designed to minimize noise at the input and output to mitigate such unwanted behavior.

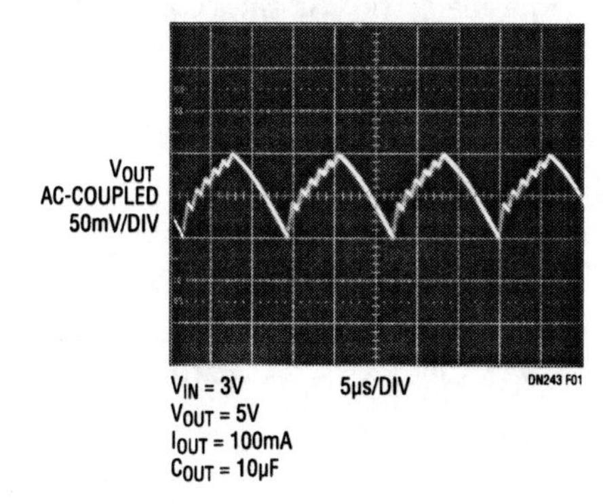

Figure 228.1 • Typical Burst Mode Output Ripple

Burst Mode operation vs constant frequency

Most regulating charge pump DC/DC converters operate using a Burst Mode architecture. Such regulator architectures provide the lowest quiescent current, but generate the highest levels of both input and output noise. With Burst Mode parts, the charge pump switches are either delivering maximum current to the output or are turned off completely. A hysteretic comparator and reference control the turn-on and -off of the charge pump to provide output regulation. Low frequency ripple appears at the output and is required for regulation (see Figure 228.1). This bursting on and off also results in large input ripple current that must be supplied by the input source. Any impedance in the input source creates voltage noise at the input. This noise must then be rejected by the rest of the circuitry powered from the same source.

The LTC3200 and LTC3200-5 have been designed to minimize both input and output noise. These parts are regulating boost charge pumps that can supply up to 100mA of output current. The LTC3200-5 produces a regulated 5V output and is available in a 6-lead SOT-23 package. The LTC3200 produces an adjustable output voltage and is available in an 8-lead MSOP package. Both parts use a constant frequency architecture that eliminates low frequency output noise. Charge pump switching is continuous, even with no load, and a linear control loop regulates the amount of charge transferred to the output on each clock cycle. Since the output regulation loop is linear, the peak-to-peak output ripple can be approximated as $V_{RIPPLE} = (I_{LOAD}/C_{OUT})/(2 \cdot f_{OSC})$, with no additional ripple due to regulator hysteresis.

The parts' 2MHz oscillator frequency allows low output ripple to be achieved even with small output capacitors. Figure 228.2 illustrates the output ripple achievable with the

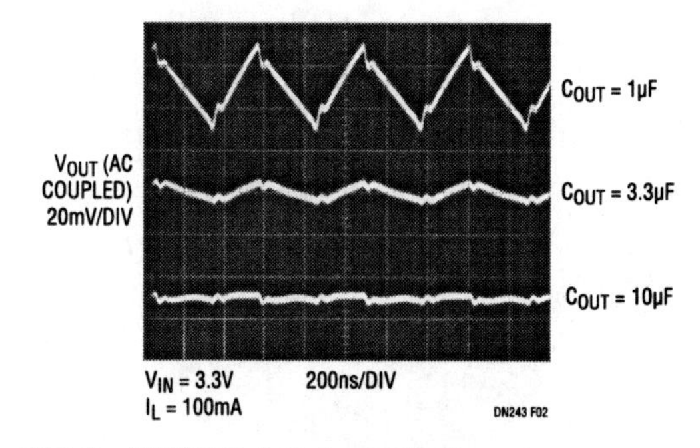

Figure 228.2 • LTC3200-5 Output Ripple

Analog Circuit Design: Design Note Collection. http://dx.doi.org/10.1016/B978-0-12-800001-4.00228-3

LTC3200-5 supporting a 100mA load with different values of output capacitance.

Input noise reduction

Although constant frequency generation alone provides substantial input noise improvement, the LTC3200 family goes one step further. A unique internal control circuit regulates the input current on both phases of the charge pump clock. This technique prevents RC current decay during one or both half-clock cycles of the charge pump oscillator, thereby minimizing the input-referred ripple due to changing input current. Figure 228.3 shows the difference in input noise between the LTC3200 and a typical Burst Mode charge pump. Both parts are shown producing a regulated 5V output at 100mA of output current from a 3.6V input. 0.1Ω of input impedance is used for testing purposes. The typical Burst Mode part uses 10μF ceramic capacitors at both the input and the output. The LTC3200 uses 1μF ceramic caps of the same dielectric. As shown in Figure 228.3, significant improvements in input noise are achieved with the LTC3200—even with one-tenth the bypass capacitance.

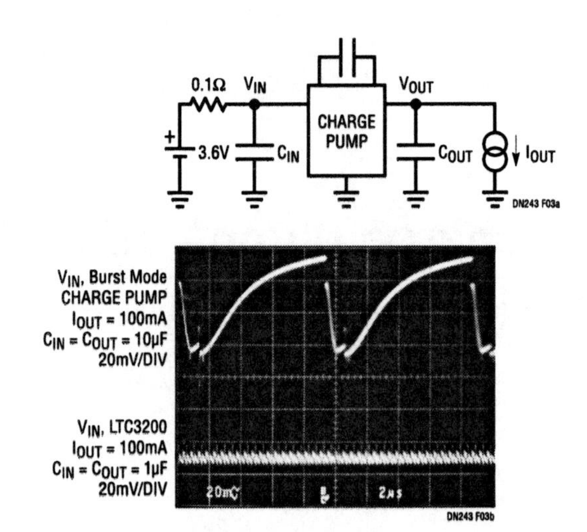

Figure 228.3 • Input Noise Test Circuit

Typical applications

Charge pumps are commonly used to provide low power boost conversion inside handheld devices such as cellular phones and PDAs. Such devices, particularly those which contain RF communication, tend to be very sensitive to noise. A popular application for a low noise charge pump in such products is powering white LEDs used for backlighting a small color LCD display. The circuit shown in Figure 228.4 produces a low noise boosted supply for driving up to six white LEDs. The LTC3200's FB pin is used to regulate the LED current flowing through each ballasted LED. By using the LTC3200, the user can provide boosted power to the backlight circuit directly from the battery without the cumbersome problem of filtering low frequency noise.

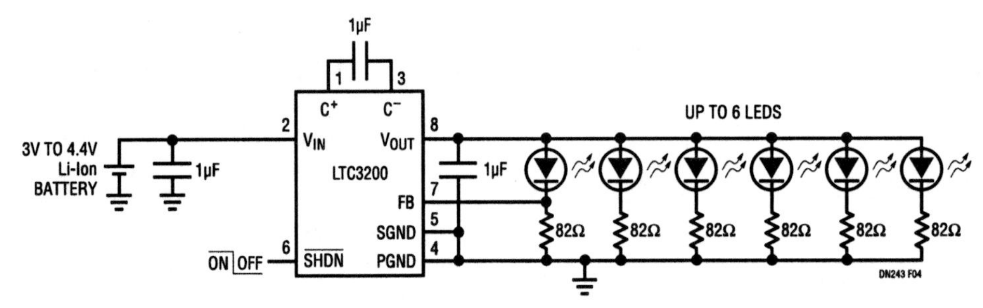

Figure 228.4 • Low Noise White LED Driver with LED Current Control

Step-up/step-down DC/DC conversion without inductors

229

Sam Nork

Introduction

Many applications require a regulated supply from an input source that may be greater than or less than the desired output voltage. Such applications place unique constraints on the DC/DC converter and, as a general rule, add complexity and cost to the power supply. A typical example is generating 5V from a 4-cell NiCd battery. When the batteries are fully charged, the input voltage is around 6V. When the batteries are near end of life, the input voltage may be as low as 3.6V. Maintaining a regulated 5V output for the life of the batteries typically requires an inductor-based DC/DC converter (for example, a SEPIC converter) or a complex, hybrid step-up/step-down solution. The LTC1514/LTC1515 family of switched capacitor DC/DC converters handles this task using only three external capacitors (Figure 229.1).

A unique architecture allows the parts to accommodate a wide input voltage range (2.0V to 10V) and adjust the operating mode as needed to maintain regulation. As a result, the parts can be used with a wide variety of battery configurations and/or adapter voltages (Figure 229.2). Low power consumption ($I_Q = 60\mu A$ typical) and low external parts count make the LTC1514 and LTC1515 well suited for space-conscious, low power applications, such as cellular phones, PDAs and portable instruments. The parts come in adjustable and fixed output voltages and include additional features such as power-on reset capability (LTC1515 family) and an uncommitted comparator that is kept alive during shutdown (LTC1514 family).

Regulator operation

The parts use a common internal switch network to implement both step-up and step-down DC/DC conversion. The action of the switch network is controlled by internal circuitry that senses the voltage differential between V_{IN} and V_{OUT}. When the input voltage is lower than the output voltage, the switch network operates as a step-up voltage doubler with a free-running frequency set by the internal oscillator (650kHz typ). When the input voltage is greater than the output, the switch network operates as a step-down gated switch. Regulation is achieved by comparing the divided output voltage to the internal reference voltage. When the divided output drops below the reference voltage, the switch network is enabled to boost the output back into regulation. The net result is a stable, tightly regulated output supply that can tolerate widely varying input voltages and load transients (Figures 229.3 and 229.4).

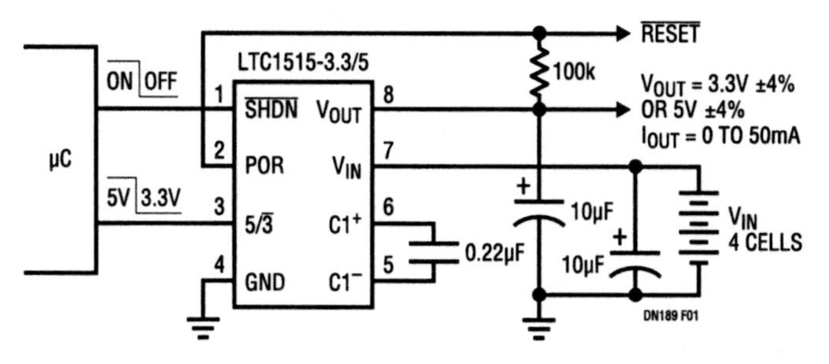

Figure 229.1 • Programmable 5V/3V Power Supply with Power-On Reset

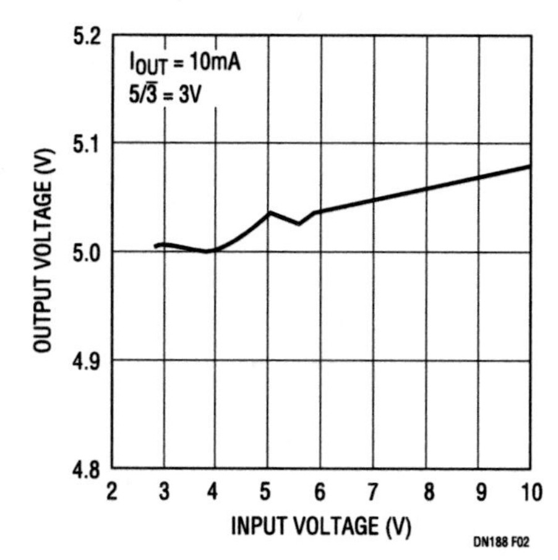

Figure 229.2 • LTC1515-X 5V Output Voltage vs Input Voltage

Analog Circuit Design: Design Note Collection. http://dx.doi.org/10.1016/B978-0-12-800001-4.00229-5

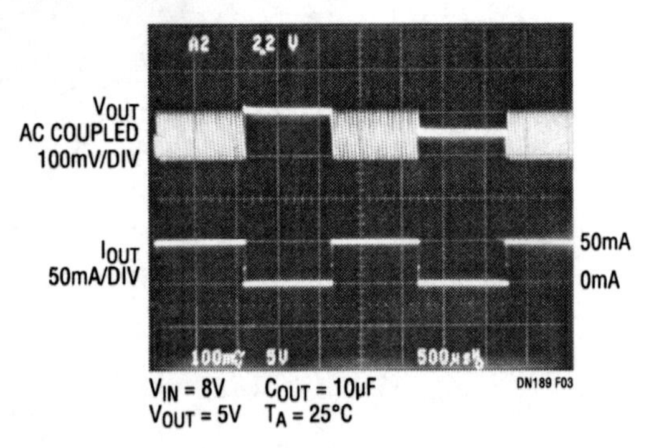

Figure 229.3 • LTC1515-X Step-Down Mode 5V Load Transient Response

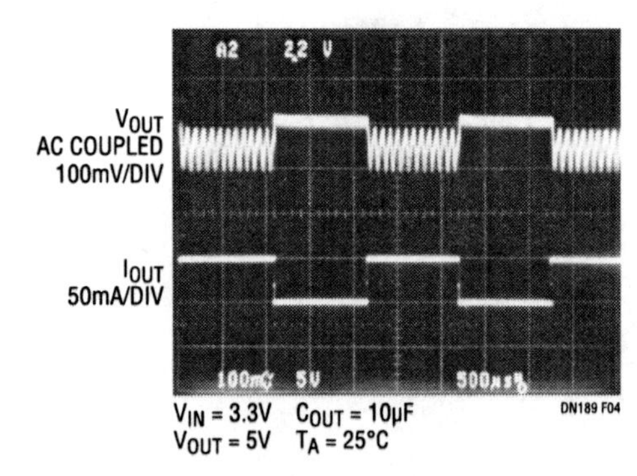

Figure 229.4 • LTC1515-X Step-Up Mode 5V Load Transient Response

Dual output supply from a 2.7V to 10V input

The circuit shown in Figure 229.5 uses the low-battery comparator to produce an auxiliary 3.3V regulated output from the VOUT of the LTC1514-5. A feedback voltage divider formed by R2 and R3 connected to the comparator input (LBI) establishes the output voltage. The output of the comparator (LBO) enables the current source formed by Q1, Q2, R1 and R4. When the LBO pin is low, Q1 is turned on, allowing current to charge output capacitor C4. Local feedback formed by R4, Q1 and Q2 creates a constant current source from the 5V output to C4. Peak charging current is set by R4 and the V_{BE} of Q2, which also provides current limiting in the case of an output short to ground. With the values shown in Figure 229.5, the auxiliary regulator can deliver up to 50mA before reaching its current limit. However, the combined output current from the 5V and 3.3V supplies may not exceed 50mA. Since the regulator implements a hysteretic feedback loop in place of the traditional linear feedback loop, no compensation is needed for loop stability. Furthermore, the high gain of the comparator provides excellent load regulation and transient response.

Conclusion

With low operating current, minimal external parts count and robust protection features, the LTC1514 and LTC1515 offer a simple and cost-effective solution to low power step-up/step-down DC/DC conversion. The shutdown, POR and low-battery-detect features provide additional functionality. The ease of use and versatility of these parts make them ideal for low power DC/DC conversion applications.

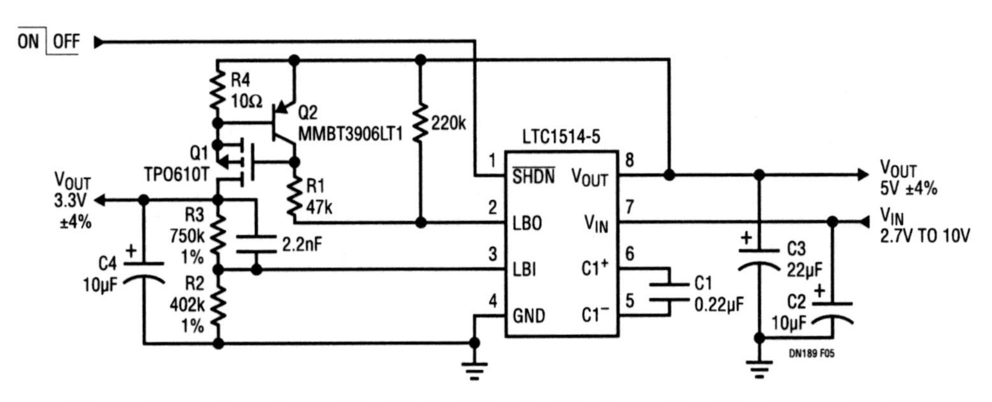

Figure 229.5 • Low Power Dual Output Supply (Maximum Combined I_{OUT} = 50mA)

Ultralow quiescent current DC/DC converters for light load applications

230

Sam Nork

In lightly loaded battery applications that require regulated power supplies, the quiescent current drawn by the DC/DC converter can represent a substantial portion of the average battery current drain. In such applications, minimizing the quiescent current of the DC/DC converter becomes a primary objective because this results in longer battery life and/or an increased power budget for the rest of the circuitry. The following two circuits provide regulated step-up and step-down DC/DC conversion and consume extremely low quiescent current.

2-cell to 5V conversion with $I_Q = 12\mu A$

The circuit in Figure 230.1 produces a regulated 5V output from a 2V to 5V input and consumes only 12μA (typical) of supply current. The LTC1516 is a charge pump DC/DC converter that uses Burst Mode operation to provide a regulated 5V output.

This circuit achieves ultralow quiescent current by disabling the internal charge pump when the output is in regulation. The charge pump is enabled only when the output load forces the voltage on C_{OUT} to droop by approximately 80mV. External capacitors C1 and C2 are then used to transfer charge from V_{IN} to V_{OUT} until the output climbs back into regulation. This regulation method results in approximately 100mV of voltage ripple at the output.

The circuit is capable of providing up to 50mA of output current (for $V_{IN} \geq 3V$). As shown in Figure 230.2, typical efficiency exceeds 70% with load currents as low as 50μA (see Figure 230.3).

The low quiescent current of the LTC1516 may render shutdown of the 5V supply unnecessary because the 12μA quiescent current is lower than the self-discharge rate of many batteries. However, the part is also equipped with a 1μA shutdown mode for additional power savings.

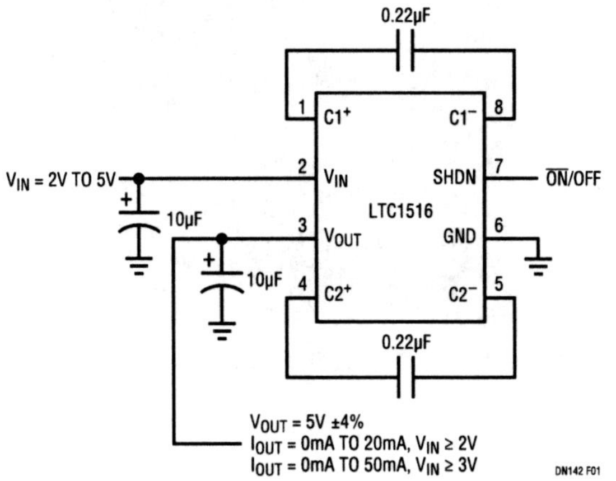

Figure 230.1 • Regulated 5V Output from a 2V to 5V Input

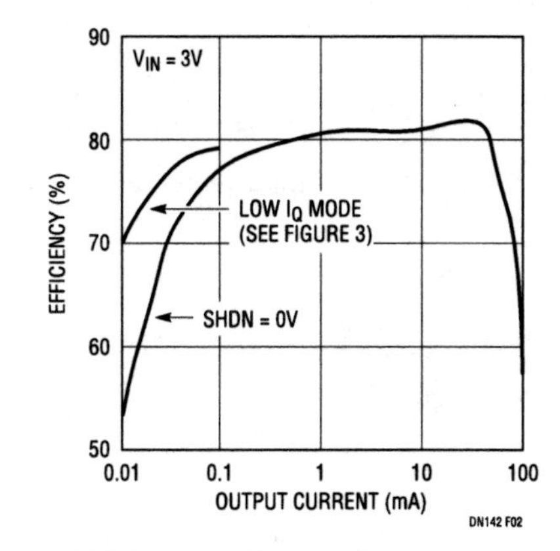

Figure 230.2 • Efficiency vs Output Current

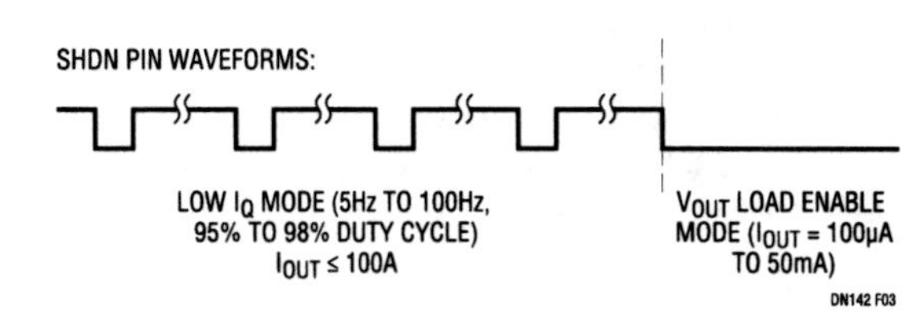

Figure 230.3 • SHDN Pin Waveforms for Ultralow Quiescent Current Supply

Analog Circuit Design: Design Note Collection. http://dx.doi.org/10.1016/B978-0-12-800001-4.00230-1

Ultralow quiescent current (I_Q<5μA) regulated supply

The LTC1516 contains an internal resistor divider that draws only 1.5μA (typ) from V_{OUT}. During no-load conditions, the internal load causes a droop rate of only 150mV per second on V_{OUT} with C_{OUT}=10μF. Applying a 5Hz to 100Hz, 95% to 98% duty cycle signal to the SHDN pin ensures that the circuit of Figure 230.1 comes out of shutdown frequently enough to maintain regulation during no-load or low-load conditions. Since the part spends nearly all of its time in shutdown, the no-load quiescent current (see Figure 230.4) is approximately equal to $(V_{OUT})(1.5\mu A)/(V_{IN})(\text{Efficiency})$.

The LTC1516 must be out of shutdown for a minimum duration of 200μs to allow enough time to sense the output and keep it in regulation. As the V_{OUT} load current increases, the frequency with which the part is taken out of shutdown must also be increased to prevent V_{OUT} from drooping below 4.8V during the OFF phase. A 100Hz 98% duty cycle signal on the SHDN pin ensures proper regulation with load currents as high as 100μA. When load current greater than 100μA is needed, the SHDN pin must be forced low, as in normal operation. The typical no-load supply current for this circuit with V_{IN}=3V is only 3.2μA.

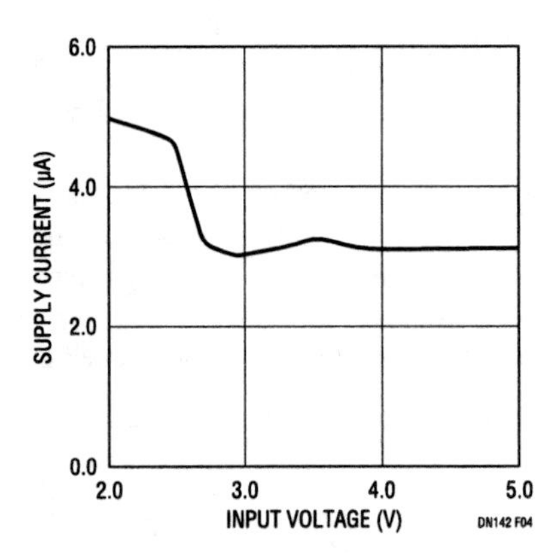

Figure 230.4 • No-Load I_{CC} vs Input Voltage for Low I_Q Mode

Micropower LDO regulator consumes <5μA

The micropower linear regulator shown in Figure 230.5 delivers a regulated 3.3V output using less than 5μA quiescent current. With such low operating current, a standard 9V alkaline battery can power this regulator for 10 years.

Circuit operation is very straightforward. The LTC1440's internal reference connects to one input of the feedback comparator. A feedback voltage divider formed by R2 and R3 establishes the output voltage. The output of the comparator enables the current source formed by Q1, Q2, R1 and R4. When LTC1440's output is low, Q1 is turned on, allowing current to charge output capacitor C4. Local feedback formed by R4, Q1 and Q2 creates a constant current source from V_{IN} to C4. Peak charging current is set by R4 and the V_{BE} of Q2, which also provides current limiting in case of an output short to ground. With the values shown in Figure 230.5, the regulator is guaranteed to deliver at least 10mA output current with inputs as low as 4.8V (that is, from a fully discharged 9V battery).

Because the regulator implements a hysteretic feedback loop in place of the traditional linear feedback loop, no compensation is needed for loop stability. Furthermore, the extremely high gain of the comparator provides excellent load regulation and transient response. However, as with the LTC1516, the comparator hysteresis necessarily produces a small amount of output ripple. Output ripple can be reduced to 10mV–20mV peak-to-peak with feed-forward capacitor C3 (see Figure 230.6), but no-load quiescent current increases by approximately 1.5μA. Without C3 the quiescent current is about 4.5μA, but output ripple is 50mV to 100mV peak-to-peak.

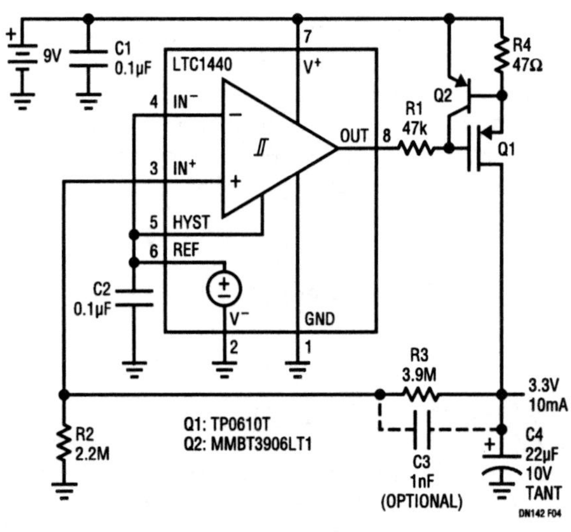

Figure 230.5 • Micropower LDO Regulator

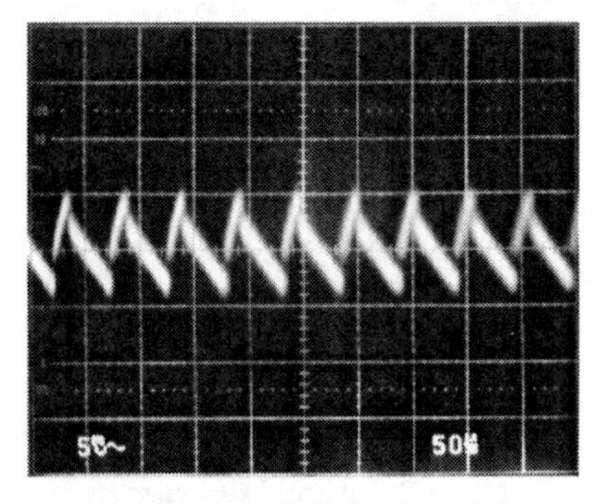

Figure 230.6 • Typical Output Ripple Using 1nF Feed Forward Capacitor

Section 15

Flyback Converter Design

Micropower isolated flyback converter with input voltage range from 6V to 100V

231

Zhongming Ye

Introduction

Flyback converters are widely used in isolated DC/DC applications because of their relative simplicity and low cost compared to alternative isolated topologies. Even so, designing a traditional flyback is not easy—the transformer requires careful design, and loop compensation is complicated by the well known right-half plane (RHP) zero and the propagation delay of the optocoupler.

Linear Technology's no-opto flyback converters, such as the LT3573, LT3574, LT3575, LT3511 and LT3512, simplify the design of flyback converters by incorporating a primary-side sensing scheme and running the converter in boundary mode. The LT8300 high voltage monolithic isolated flyback converter further simplifies flyback design by integrating a 260mA, 150V DMOS power switch, an internal compensation network and a soft-start capacitor. The LT8300 operates with input supply voltages from 6V to 100V and delivers output power of up to 2W with as few as five external components.

The LT8300 operates in boundary mode and offers low ripple Burst Mode operation, enabling the design of converters that feature high efficiency, low component count and minimal power loss in standby.

Simple and accurate primary-side voltage sensing

The LT8300 eliminates the need for an optocoupler by sensing the output voltage on the primary side when the output diode current drops to zero during the primary switch-off period. This greatly improves the load regulation since the voltage drop is zero across the transformer secondary winding and any PCB traces. This allows an LT8300-based flyback converter to produce ±1% typical load regulation at room temperature. Figure 231.1 shows the schematic and Figure 231.2 the load regulation curves of a flyback converter with a 5V output.

Very small size, low component count solution

The LT8300 integrates a 260mA, 150V DMOS power switch along with all high voltage circuitry and control logic into a 5-lead TSOT-23 package. The isolated output voltage is set via a single external resistor with compensation and soft-start

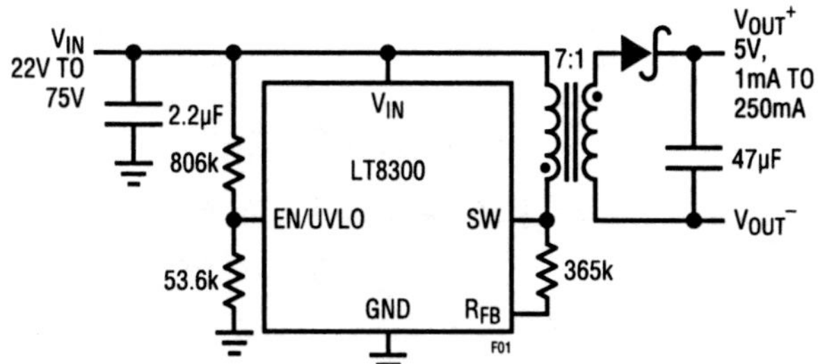

Figure 231.1 • A Complete 5V Flyback Converter for a 22V to 75V Input

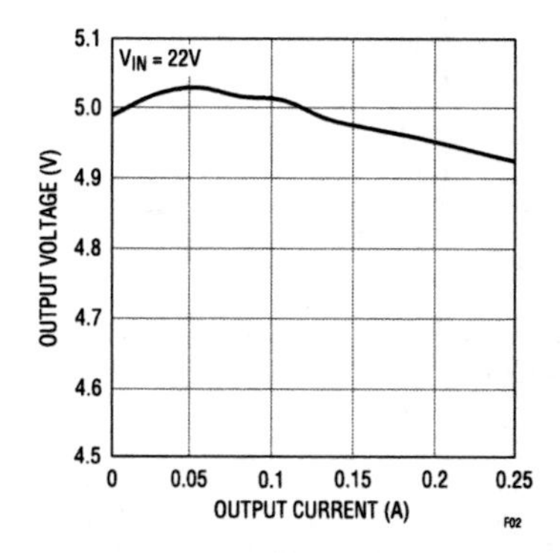

Figure 231.2 • Regulation of a 22V to 75V Input to 5V Flyback Converter of Figure 231.1

Analog Circuit Design: Design Note Collection. http://dx.doi.org/10.1016/B978-0-12-800001-4.00231-3

Figure 231.3 • Demonstration Circuit of 22V to 75V to 5V/0.3A Converter (See Figure 231.1)

circuitry integrated in the IC. Low ripple Burst Mode operation maintains high efficiency at light loads while minimizing the output voltage ripple.

The converter turns on the internal switch immediately after the secondary diode current reduces to zero, and turns off when the switch current reaches the predefined current limit; the diode has no reverse-recovery loss. Furthermore, since the switch is turned on with zero current, switching losses are minimized. The reduction in power losses allows the converter to operate at a relatively high switching frequency, which in turn, allows the use of a smaller transformer than would be required at a lower operating frequency. Overall, the LT8300 significantly reduces converter size compared to other solutions.

Figure 231.3 shows the standard demo circuit DC1825A for an isolated flyback using a small EP7 core transformer. The six key components are the input and output capacitors (C2, C3), output diode (D1), feedback resistor (R3), transformer (T1) and the LT8300. For the same application, a traditional flyback circuit would require, at minimum, 11 additional components, plus complicated start-up and bias power circuits in both the primary and secondary sides.

Low I_Q, small preload and high efficiency

As the load lightens, the LT8300 reduces the switching frequency until the minimum current limit is reached, and the converter then runs in discontinuous mode. The LT8300 features an accurate minimum current limit and very small propagation delay. At very light loads, it further reduces the loss by running in low ripple Burst Mode operation, where the part switches between sleep mode and active mode. The typical quiescent current is 70µA in sleep mode and 330µA in switching mode, reducing the effective quiescent current.

The typical minimum switching frequency is about 7.5kHz, with the circuit requiring a very small preload (typical 0.5% of full load). Therefore, LT8300 power losses in standby mode are very low—important for applications requiring high efficiency in always-on applications. Figure 231.4 shows a solution that produces 20mA at 24V from a 12V input. Efficiency peaks at 87%, and remains high at 84% with a 20mA load, as shown in Figure 231.5.

Conclusion

The LT8300 is an easy-to-use flyback converter with a rich set of unique features integrated in a small 5-lead TSOT-23 package. It accepts a wide input voltage range, from 6V to 100V, with very low shutdown current and standby power consumption. Boundary mode operation reduces switching loss, shrinks converter size, simplifies system design and offers superior load regulation. Other features, such as internal soft-start, accurate current limit, undervoltage lockout and internal loop compensation further facilitate an easy flyback converter design.

The LT8300 is ideal for a broad range of applications, from battery-powered systems to automotive, industrial, medical, telecommunications power supplies and isolated auxiliary/housekeeping power supplies. The high level of integration yields a simple, low parts-count solution for low power flyback converters.

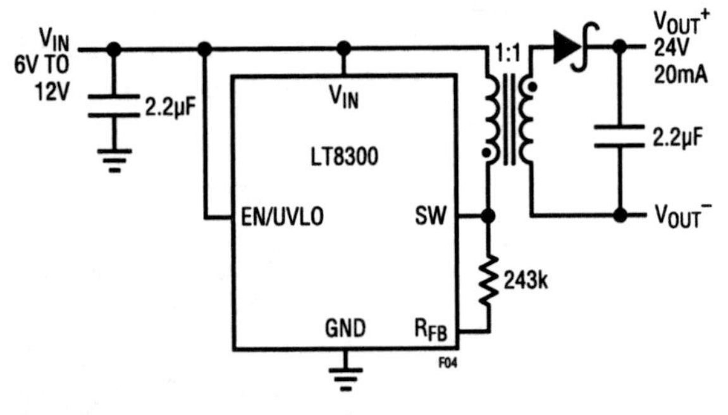

Figure 231.4 • Flyback Converter Optimized for Low Standby Power (6V–12V to 24V/20mA)

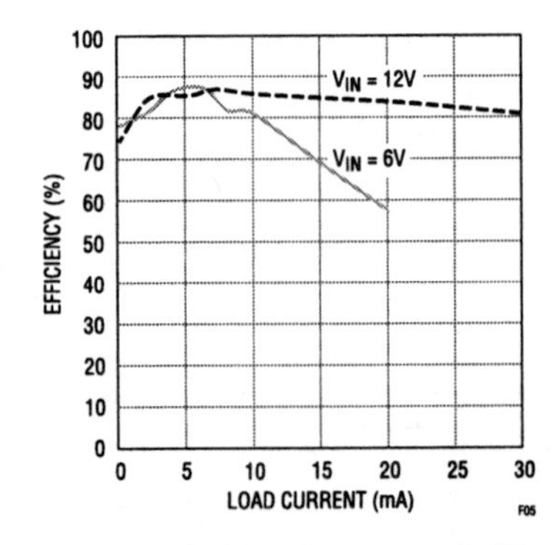

Figure 231.5 • Efficiency of the Converter in Figure 231.4

Flyback controller simplifies design of low input voltage DC/DC converters

232

David Burgoon

Introduction

Small, high efficiency DC/DC converters are critical to the design of leading-edge electronics. Achieving high accuracy and efficiency traditionally means adding extra components, complexity, and size. Not so with the LT3837. This flyback controller serves 10W to 60W isolated applications with high performance, simplicity, small size, and a minimum component count.

High efficiency controller capabilities

The LT3837 operates from a 4.5V to 20V input, but the converter input range can be extended upwards by using a V_{CC} regulator and/or a bias winding on the transformer. It also provides a synchronous rectifier output with adjustable timing to optimize efficiency and enhance cross-regulation in multiple-output supplies.

The LT3837 eliminates the need for the traditional secondary-side reference, error amplifier, and opto-isolator circuits by sampling the flyback voltage on a primary-side winding. Accuracy is enhanced with output resistance compensation. Current mode control with leading edge blanking yields a high performance loop that is easy to compensate.

The operating frequency is adjustable from 50kHz to 250kHz or can be synchronized to an external clock. Soft-start provides well-controlled start-up with limited inrush current. Protection features include current limit with soft-start cycling for severe overloads, undervoltage lockout, and thermal shutdown.

3.3V, 10A converter operates from a 9V to 18V source

The circuit shown in Figure 232.1 is a flyback design for a 3.3V, 10A output from a 9V to 18V input with a minimum of external components. The LT3837 samples the voltage on the

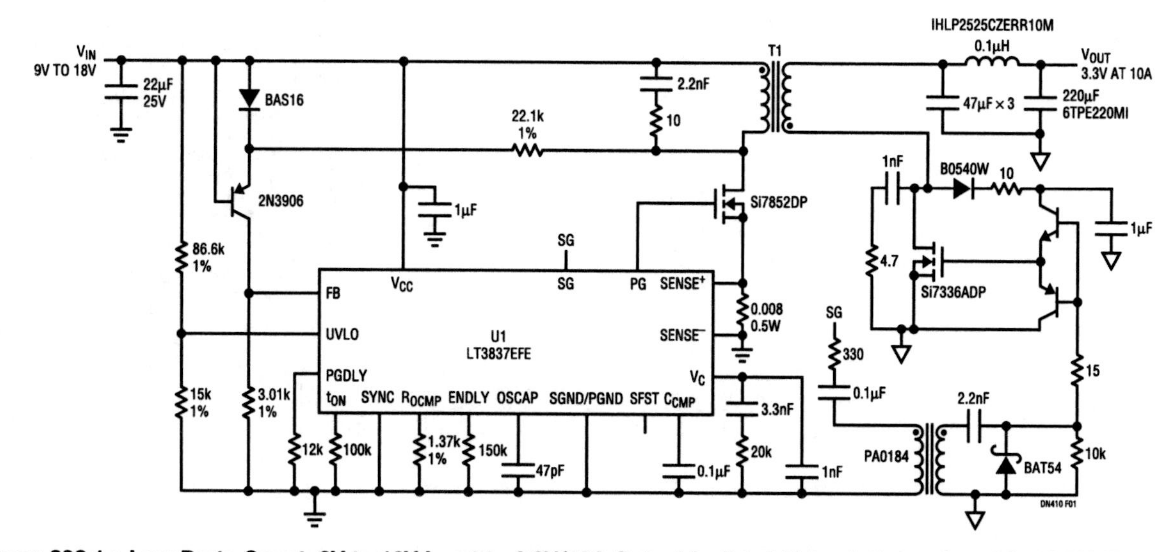

Figure 232.1 • Low Parts Count, 9V to 18V Input to 3.3V/10A Output Isolated Flyback Converter with ±0.7% Regulation

Analog Circuit Design: Design Note Collection. http://dx.doi.org/10.1016/B978-0-12-800001-4.00232-5

primary winding during the flyback interval to provide superb regulation. Figure 232.2 shows ruler-flat regulation at 9V input, and a tight regulation window of ±0.7% over line and load. Synchronous rectification with adjustable timing yields excellent efficiency—88% over a wide range of operating conditions—as shown in Figure 232.3.

3.3V, 10A converter operates from a 9V to 36V source

Figure 232.4 shows an enhanced circuit that extends the input operating range of the LT3837 to 9V to 36V. Operation is converted to hysteretic start-up for efficient wide-range operation. Q1 provides a low-drop current source for start-up, and Q2 creates a suitable undervoltage circuit for V_{CC}. These circuits, together with the V_{CC} winding on the transformer, result in low V_{CC} power at higher input voltages and low dissipation cycling when operating into a short circuit. This circuit is implemented in a 1.5in^2 footprint. This circuit exhibits excellent regulation of ±1.2% over line and load, and efficiency of 88% over much of its operating range.

Conclusion

The LT3837 is part of a new class of flyback controllers developed by Linear Technology to satisfy the demand for economical, high performance power converters. It provides synchronous rectifier drive and eliminates the need for secondary regulation circuits and opto-isolators. The LT3837 makes it easy to implement high performance flyback designs that are cost-effective, small and efficient.

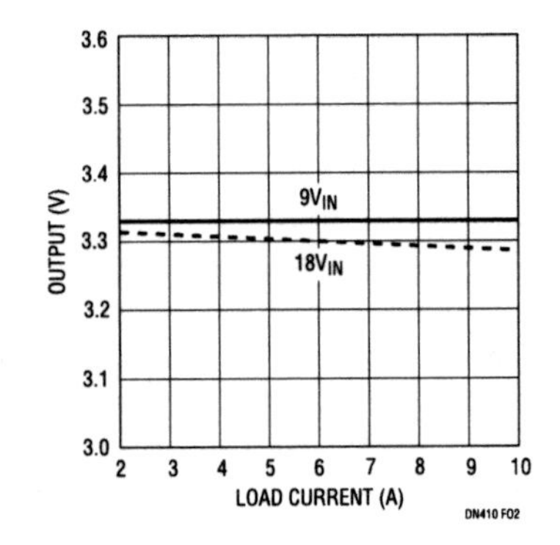

Figure 232.2 • Regulation of the Converter in Figure 232.1

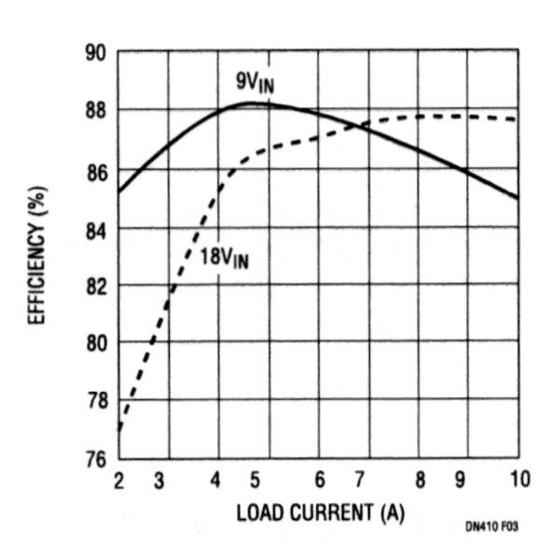

Figure 232.3 • Efficiency of the Converter in Figure 232.1

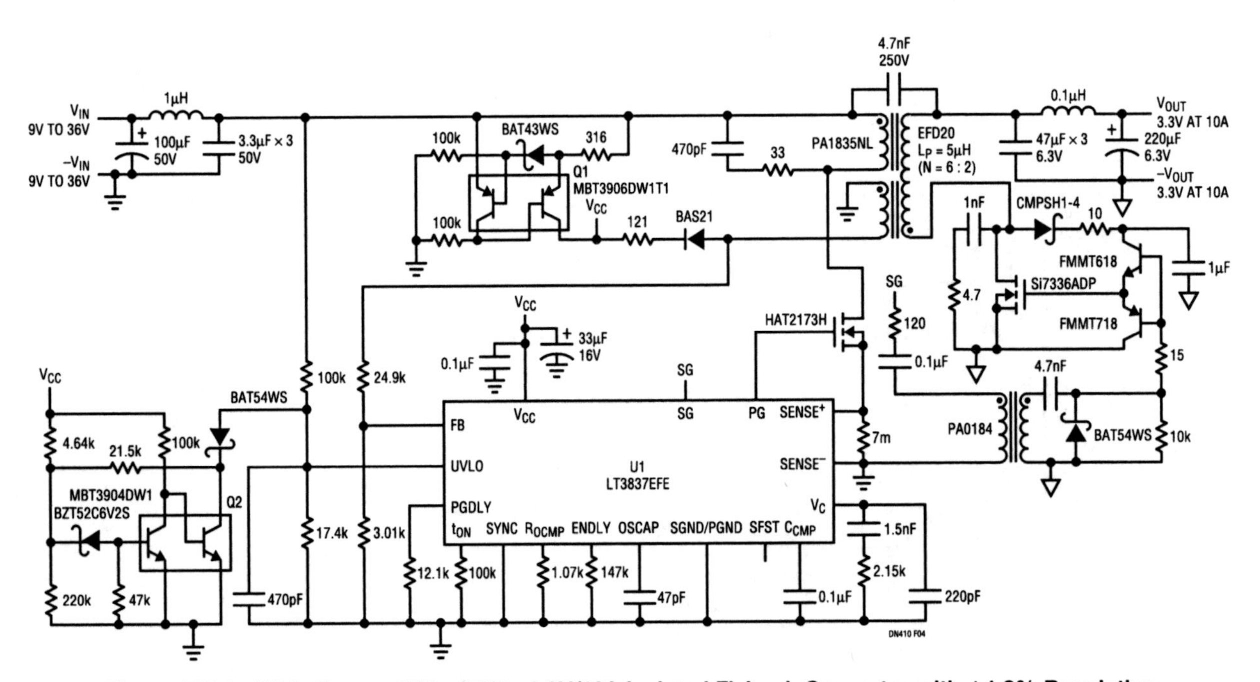

Figure 232.4 • Wide Range, 9V to 36V to 3.3V/10A Isolated Flyback Converter with ±1.2% Regulation

Flyback controller improves cross regulation for multiple output applications

233

Tom Hack

Introduction

Flyback converters are often used in power supplies requiring low to medium output power at several output voltages. With a flyback, multiple outputs incur little additional cost or complexity—each additional output requires only another transformer winding, rectifier and output filter capacitor. Power over Ethernet (PoE) applications may require several low to medium power outputs and are good candidates for multiple output flyback converters.

While multiple output flyback converters are simple and inexpensive, they often suffer from poor cross regulation. Usually, one output is tightly regulated via feedback, while the other outputs are controlled via less accurate transformer action. Schottky diode voltage drops, transformer leakage inductances and transformer winding resistances degrade regulation. Also, outputs remain uncoupled when the rectifiers are off, which contributes to poor cross regulation.

The LTC3806 is a new flyback controller that improves load and cross regulation for multiple output flyback converters. Multi-output systems that previously used post regulators, multiple switching supplies or other methods to provide tight output voltage tolerances can now use lower cost flyback converters. The LTC3806 uses forced continuous conduction operation improving cross regulation

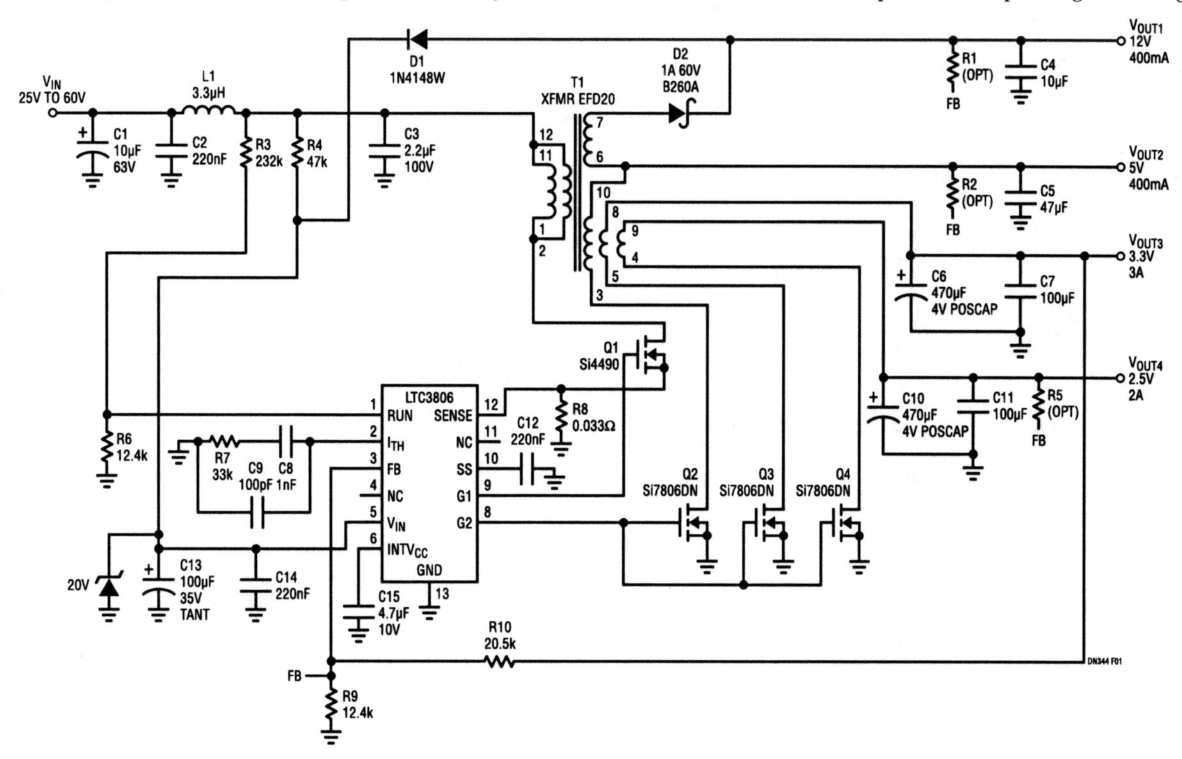

Figure 233.1 • Multiple Output Synchronous Flyback Converter

Analog Circuit Design: Design Note Collection. http://dx.doi.org/10.1016/B978-0-12-800001-4.00233-7

by keeping the rectifiers on and the outputs coupled (via transformer action) for a larger fraction of each power conversion cycle. The LTC3806 also uses synchronous rectification to reduce rectifier voltage drops, improving converter efficiency and load regulation.

Improved load and cross regulation

Figure 233.1 shows a multiple output flyback converter that takes full advantage of the features of the LTC3806. Performance of this design was compared against a non-synchronous version created by replacing Q2 with an SL13 Schottky rectifier, and Q3 and Q4 with B540C Schottky rectifiers.

Load regulation measurements were taken with one output swept from no load to maximum output current while all other outputs delivered maximum output current. Table 233.1 shows how load regulation is improved in the synchronous design. Load regulation for output 1 and output 4 in the non-synchronous design are given only from 20% to 100% of maximum output current because load regulation rapidly deteriorates below 20% of maximum output current.

Cross regulation measurements were taken under the same conditions as the load regulation measurements. Table 233.2 shows the superior cross regulation performance of the synchronous design.

Table 233.1 Load Regulation (% V_{OUT})

	OUTPUT 1	OUTPUT 2	OUTPUT 3	OUTPUT 4
Synchronous	7.86%	6.31%	1.39%	9.47%
Nonsynchronous	8.88% (20%–100% of I_{MAX})	6.42%	2.43%	10.32% (20%–100% of I_{MAX})

Table 233.2 Cross-Regulation (% V_{OUT}) (Nonsynchronous Results in Parenthesis)

OUTPUT 1	OUTPUT 2	OUTPUT 3	OUTPUT 4
Swept*	0.87% (2.02%)	0.22% (0.42%)	0.29% (0.43%)
0.31% (0.94%)	Swept	0.16% (0.23%)	0.17% (0.30%)
−5.49% (−8.52%)	−4.88% (−8.20%)	Swept	−4.43% (−8.61%)
−0.63% (−0.36%)	−0.12% (0.13%)	0.16% (0.39%)	Swept*

*20% to 100% of I_{MAX} for nonsynchronous

The case for using synchronous rectification is even stronger when load and cross-regulation measurements are taken with one output swept, and all other outputs unloaded. Synchronous rectification provides acceptable load and cross-regulation while Schottky rectification does not. To get reasonable regulation in a non-synchronous design, all outputs require preloading of several percent of maximum I_{OUT}.

Efficiency

Peak efficiency for this design is 87.6% at 60% of maximum output power (Figure 233.2). Such high efficiency is possible because of the significant reduction in conduction loss afforded by the synchronous rectification, particularly with low voltage outputs. At full load (21.4W total output power), supply dissipation is only 3.38W. For the non-synchronous design, peak efficiency is 81.9% and power dissipation is 5.31W. If light load regulation is required, this degrades to 78.9% due to preloading.

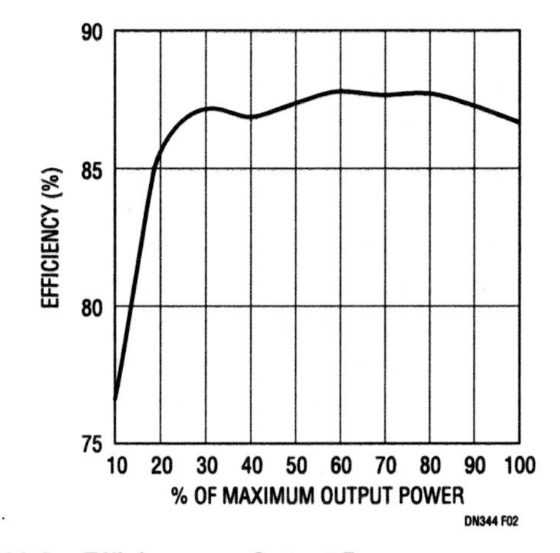

Figure 233.2 • Efficiency vs Output Power

Composite feedback provides additional design flexibility

In Figure 233.1, only output 3 is controlled via feedback while the remaining outputs are set via transformer action. To further improve load and cross regulation on these other outputs (though at the expense of degraded load and cross-regulation on output 3), composite feedback may be used by adding optional resistors R1, R2, and R5.

Conclusion

The LTC3806 makes cost-effective multiple output flyback converters attractive for low and medium power applications such as Power over Ethernet devices. Improved load and cross regulation eliminates the need for additional switching supplies or post regulators. Synchronous rectification improves efficiency and lowers power dissipation.

No R_{SENSE} controller is small and efficient in boost, flyback and SEPIC applications

234

Tick Houk

Introduction

The increase in demand for high bandwidth signal processing in telecom systems has driven a corresponding increase in demand for space-efficient boost, flyback and SEPIC power supplies. The LTC1871 addresses these specific needs by providing a high performance, single-ended current mode DC/DC controller IC in a small, 10-lead MSOP package. The LTC1871 is efficient and easy to use, mainly due to features such as No R_{SENSE} current mode control, a user-programmable operating frequency (50kHz to 1MHz), a strong gate driver, programmable undervoltage lockout, synchronization capability and a wide input range (2.5V to 36V).

A high efficiency 5V, 2A networking logic supply

Figure 234.1 illustrates a small, high efficiency networking 5V logic supply that can operate from either a 2.5V or a 3.3V input supply. This design takes advantage of LTC's proprietary No R_{SENSE} technology to provide true current mode control without an external current sense resistor. The voltage drop across the power MOSFET is sensed during the on-time, thereby providing the control loop with a "lossless" method of measuring the switch current. This technique provides the maximum efficiency possible for a single-ended current mode converter. It also saves board space and reduces the cost of the power supply in applications where the drain of the power MOSFET is less than 36V (the absolute maximum rating of the SENSE pin). It should be noted that the output voltage and maximum output current of this supply can easily be scaled by the choice of the components around the chip without modifying the basic design.

An operating frequency of 300kHz is programmed using a single resistor to ground, allowing the use of a small, inexpensive 1.8μH inductor from Toko. A Siliconix/Vishay SO-8 power MOSFET (Si9426) and an International Rectifier surface mount diode (30BQ015) were chosen for the 2A output current level, and a low ESR ceramic output capacitor keeps the output ripple below $60mV_{P-P}$.

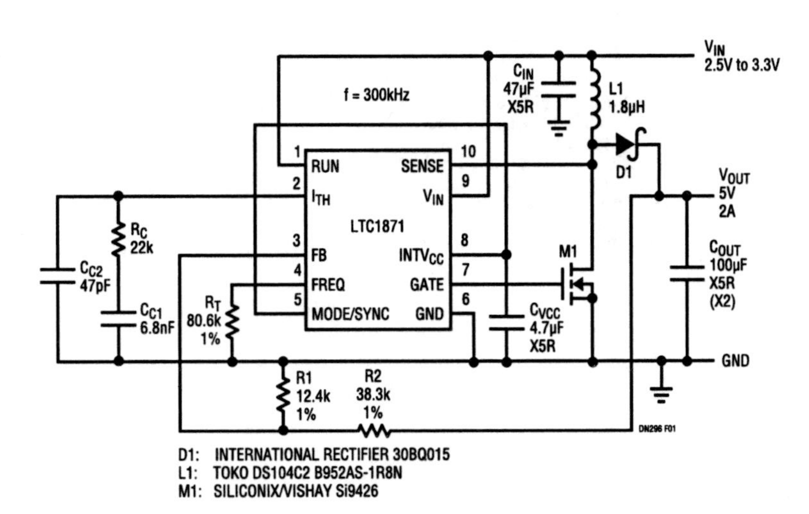

Figure 234.1 • A High Efficiency 5V, 2A Networking Logic Supply

Analog Circuit Design: Design Note Collection. http://dx.doi.org/10.1016/B978-0-12-800001-4.00234-9

A 2 square inch, 12V non-isolated flyback housekeeping supply for telecom applications

The LTC1871 is also ideally suited for telecom input applications (36V to 72V input), where a single-switch flyback topology results in low component count and small size. Figure 234.2 illustrates a 12V, 0.4A housekeeping power supply which occupies only two square inches of board space. The converter operates in discontinuous mode (DCM) at 200kHz and provides power to other analog and digital ICs on the PC board. A standard, 6-winding VERSA-PAC transformer (VP1-0076) is connected with three windings in series on the primary and three in parallel on the secondary. A small SuperSOT-6 MOSFET from Fairchild provides the primary switch, along with a surface mount diode from International Rectifier.

Due to the high input voltage, a pre-regulator—consisting of transistor Q1, diode D1 and resistor R5—is needed to provide start-up power to the LTC1871. Once the 12V output is in regulation, D2 provides DC power to the IC and Q1 turns off. This bootstrapping technique ensures that the gate drive power is provided by the switcher itself, thus maintaining the highest efficiency possible. A 5.2V LDO within the LTC1871 provides a regulated supply for the gate driver, which is capable of driving MOSFETs with gate charges up to 100nC.

Programmable undervoltage lockout provides clean start-up and power-down

In order to provide well controlled start-up and power-down, the RUN pin on the LTC1871 monitors the input voltage—at resistive divider R1-R2—and turns off the converter when the input drops below 29.5V. An 8% hysteresis level provides increased noise immunity (UV^+ is 31.8V). The optional capacitor C1 can be used to provide "ride-through" capability for short-duration input dropouts.

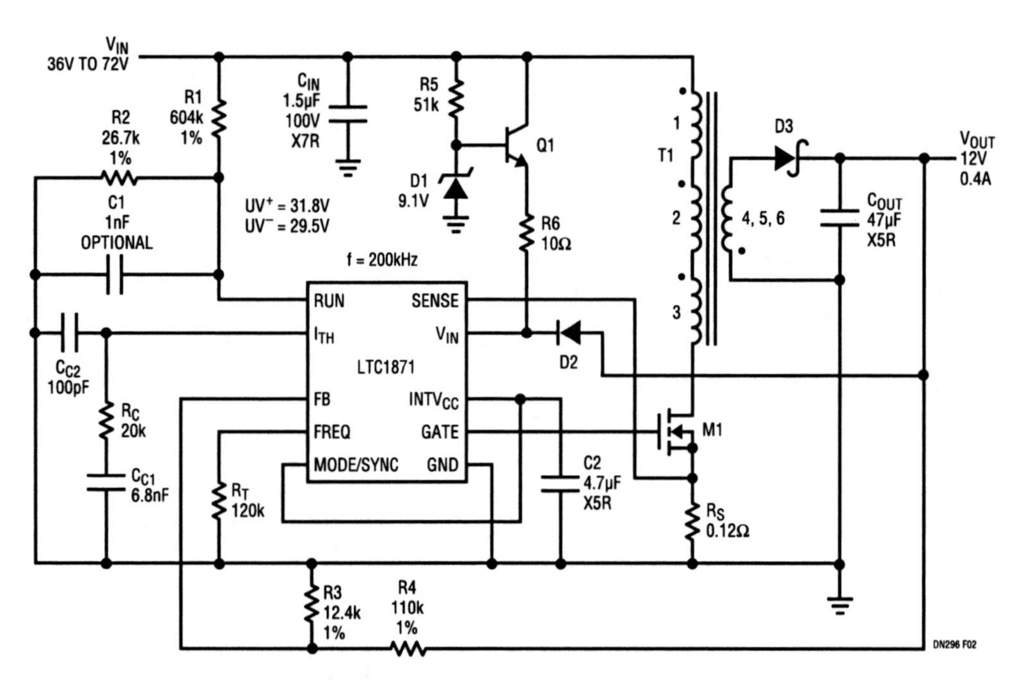

T1: COILTRONICS VP1-0076 (3 WINDINGS IN SERIES ON PRIMARY, 3 IN PARALLEL ON SECONDARY)
M1: FAIRCHILD FDC2512 (150V, 0.5 OHM)
Q1: ZETEX FZT605 (120V)
D1: ON SEMICONDUCTOR MMBZ5239BLT1 (9.1V)
D2: ON SEMICONDUCTOR MMSD4148T11
D3: INTERNATIONAL RECTIFIER 10BQ060

Figure 234.2 • A Small, Non-Isolated 12V Flyback Telecom Housekeeping Supply

Isolated flyback converter regulates without an optocoupler

235

Robert Sheehan

Introduction

Designing an isolated power supply can be a daunting task. Optocouplers, secondary-side error amplifiers and frequency compensation can send you scurrying to your distributor catalog to find a module. Fortunately, there are new IC solutions that make isolated power much simpler and less expensive. Here is a gem of a circuit that takes the headaches out of your isolated power supply design task and at a small fraction of a module's cost.

The design criteria

- Input Voltage: 100V DC to 300V DC
- Output Voltage: 5V
- Output Current: 100mA to 1A
- Regulation: <±3% total line and load
- Efficiency: >70% at full load
- Isolation: 500V DC
- Size: 1.5″ × 2″ × 0.5″
- Cost: Under $10 for total solution

Circuit description

A flyback topology was chosen that allows for wide variations of input voltage and output current. At the heart of the circuit is the LT1725 isolated flyback controller (Figure 235.1). This is a fixed frequency, current mode PWM controller. It drives the gate of an external MOSFET and is powered from the third (bias) winding of the flyback transformer.

The bias winding also provides output voltage feedback information. This is done directly through a resistive divider, without any rectification or filtering. The unique features of the LT1725 allow regulation to be maintained well into discontinuous mode operation (light load).

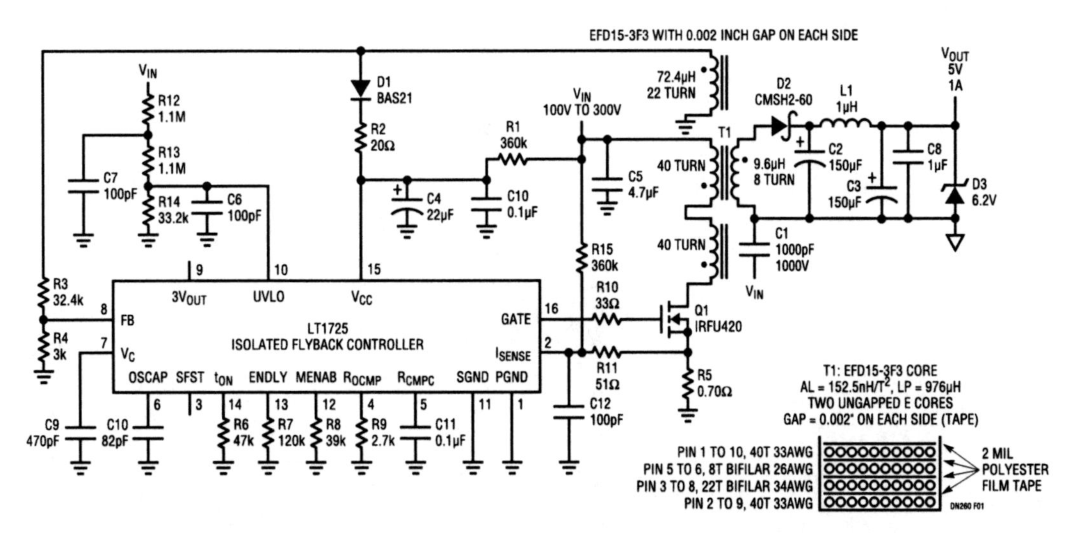

Figure 235.1 • $100V_{IN}$ to $300V_{IN}$, $5V_{OUT}$ at 1A Isolated Flyback Power Supply

Analog Circuit Design: Design Note Collection. http://dx.doi.org/10.1016/B978-0-12-800001-4.00235-0

The transformer is an EFD15 using a gapped ferrite core. The primary inductance is 976µH and the turns ratio is 10:1. The input and output capacitors are aluminum electrolytic for low cost. This provides a high degree of damping at the input and easy frequency compensation for the output. The standard level MOSFET is a D-PAK IRFU420 rated for 500V. It is chosen for low gate charge to minimize switching losses. The output rectifier is a CMSH2-60 rated for 2A at 60V and is sized to handle continuous short-circuit current.

Circuit operation

Upon application of input power, R1 trickle charges C4, applying a voltage to the V_{CC} pin of the LT1725. When the voltage at V_{CC} reaches 15V, the LT1725 turns on and begins switching the gate of Q1. With each switch cycle, energy builds in T1 and the output voltage begins to rise. The voltage across the output winding is reflected into the bias winding by the transformer turns ratio. When this voltage reaches about two thirds of the final output voltage, D1 conducts and provides power for the LT1725 from the bias winding. C4 is sized to provide the power for the LT1725 while the output voltage is ramping up. Too low a value for C4 results in the converter cycling on and off with a sawtooth voltage on the output and across C4. An added benefit of this type of trickle-charge start-up circuit is lower power dissipation during short circuits. With a sustained output short, the converter harmlessly cycles on and off at a low frequency until the short is removed, restoring normal operation. R13 and R14 set input undervoltage lockout at 85V. This prevents operation at a lower input voltage that could overload the input supply.

The operating frequency is set to 120kHz by the 82pF capacitor at the OSCAP pin. This is a reasonable frequency to keep the switching losses relatively low when operating at higher input voltages.

R3 and R4 sample output voltage feedback from the bias winding. The feedback amplifier inside the LT1725 looks at the flyback signal only while the output rectifier, D2, is conducting. This ensures a good sample of the output voltage, maintaining accuracy over a wide range of operating conditions. Adjusting the value of R9 compensates output voltage load regulation. Resistors R7 and R8 allow the timing of the feedback amplifier sampling to be tailored for specific applications. Frequency compensation is accomplished by a single 470pF capacitor at the V_C pin.

Current is sensed at the MOSFET source, using a ground referenced signal. Leading edge spikes on the current sense signal are ignored due to the current sense amplifier's blanking time, set by R6. R15 introduces a DC offset that is proportional to the input voltage. This minimizes variation in current limit over line conditions. R11 and C12 create a high frequency filter to keep out any switching noise that may be fed from the input voltage through R15. This filter is not always required if the line compensation is removed.

The flyback topology is characterized by pulsating currents in the input and output capacitors. This can result in a relatively high output voltage ripple on C2. L1 and C3 are used to further attenuate the output ripple, resulting in a clean DC voltage. D3 is used to clamp the output voltage in case the load drops below the 100mA minimum level.

C1 provides an AC return path for any common mode current generated in the transformer. Since it bridges the isolation barrier, a voltage rating greater than the isolation voltage is required.

Conclusion

The flyback circuit represents a simple, low cost solution for isolated power. The versatility of the LT1725 allows the circuit to be tailored to specific applications. Accurate sampling of the output voltage from the bias winding also eliminates the need for an optocoupler circuit.

236

Isolated DC/DC conversion

Kurk Mathews

The LT1425 is designed for applications requiring well regulated, isolated voltages, such as isolation amplifiers, remote sensors and telecommunication interfaces. A unique feedback amplifier eliminates the need to cross the isolation barrier twice, resulting in a simpler, lower parts count supply. The LT1425, available in 16-pin SO, is a 275kHz current mode controller with an integral 1.5A switch, designed primarily to provide well regulated, isolated voltages from 3V to 20V sources.

Figure 236.1 shows a typical flyback LAN supply, including an alternate transformer for a complete PCMCIA type II height solution. Load regulation is ±1% for output currents of 0mA to 250mA. Feedback is accomplished by averaging the flyback voltage on the primary side of T1. The internal switch is located between the V_{SW} and P_{GND} pins. The R_{FB} pin is internally biased to V_{IN}. During the switch off-time, a feedback current proportional to V_{OUT}/n (n is the transformer's turn ratio) is developed into the R_{FB} pin (via R4). Flyback voltage on T1 is not present during the switch on-time or when the secondary current decays to zero (discontinuous flyback mode). Collapse-detect and blanking circuitry ensure that the feedback amplifier ignores information during these times.

Resistor R3 provides additional load compensation, necessary to compensate for winding resistance and output diode voltage drop. It generates a current proportional to the average switch current (and therefore, to load current). This current subtracts from the feedback signal, compensating for the

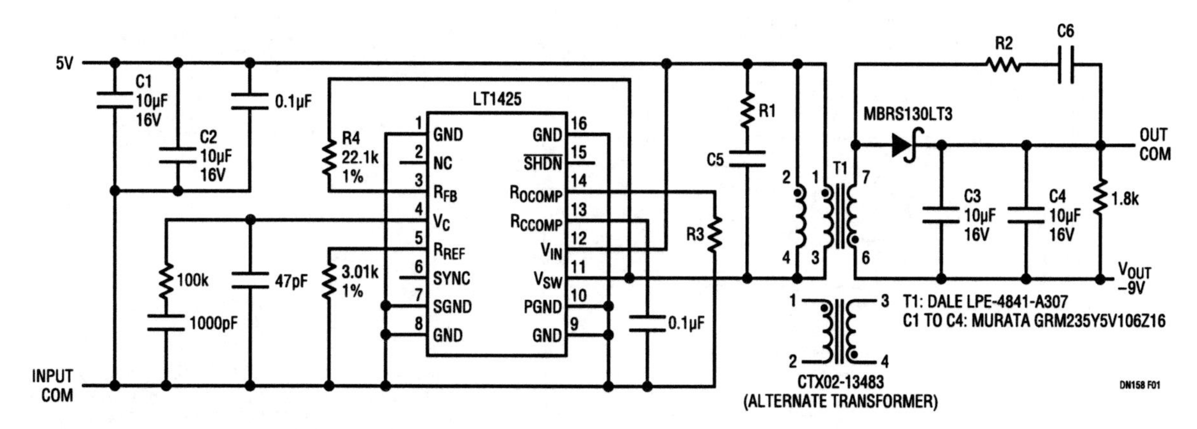

Transformer T1

	LPRI	RATIO	ISOLATION	(L × W × H)	I_{OUT}	EFFICIENCY	R1, R2	C5, C6	R3
DALE LPE-4841-A307	36μH	1:1:1	500VAC	10.7 × 11.5 × 6.3mm	250mA	76%	47Ω	330pF	13.3k
COILTRONICS CTX02-13483	27μH	1:1	500VAC	14 × 14 × 2.2mm	200mA	70%	75Ω	220pF	5.9k

Figure 236.1 • 5V to −9V/250mA Isolated LAN Supply

Analog Circuit Design: Design Note Collection. http://dx.doi.org/10.1016/B978-0-12-800001-4.00236-2

parasitic voltage drops that tend to lower the output voltage with increasing load.

The result of this feedback method is excellent load regulation and fast dynamic response not found in similar isolated flyback schemes. Referring again to Figure 236.1, the −9V output changes only 300mV during a 50mA to 250mA load transient.

Figure 236.2 shows a ±5V supply with 1.5kV of isolation. The sum of line/load/cross-regulation is better than ±3%. Full load efficiency is between 72% (V_{IN} = 5V) and 80% (V_{IN} = 15V). The isolation voltage is ultimately limited only by bobbin selection and transformer construction.

In Figure 236.3, an external cascoded 200V MOSFET is used to extend the LT1425's 35V maximum switch voltage limit. The input voltage range (36V to 72V) also exceeds the LT1425's 20V maximum input voltage, so a bootstrap winding is used. D1, D2, Q2, Q3 and associated components form the necessary start-up circuitry with hysteresis. When C1 charges to 15V, switching begins and the bootstrap winding begins to supply power before C1 has a chance to discharge to 11V. Feedback voltage is fed directly through a resistor divider to the R_{REF} pin. The load compensation circuitry is bypassed, resulting in ±5% load regulation.

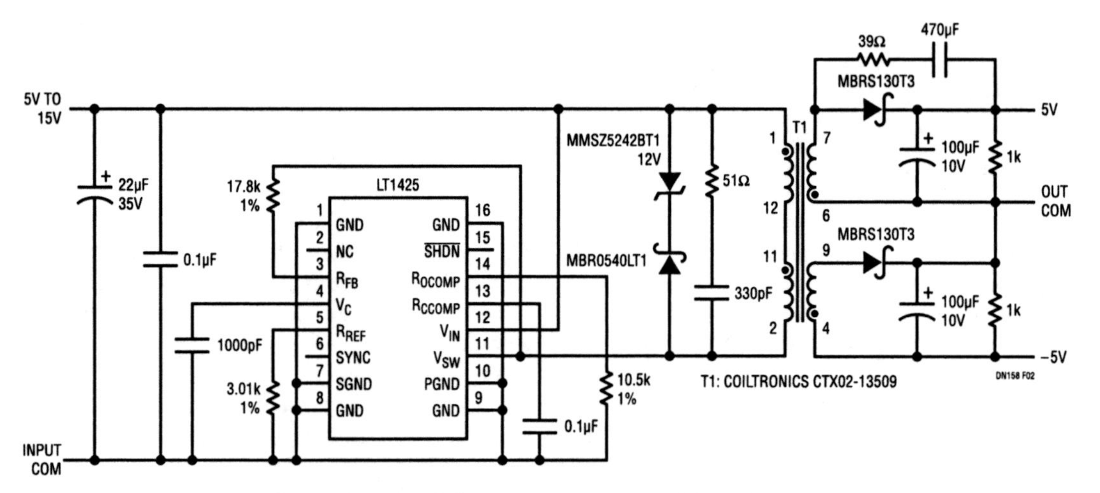

Figure 236.2 • 5V Fully Isolated ±5V, ±220mA Supply

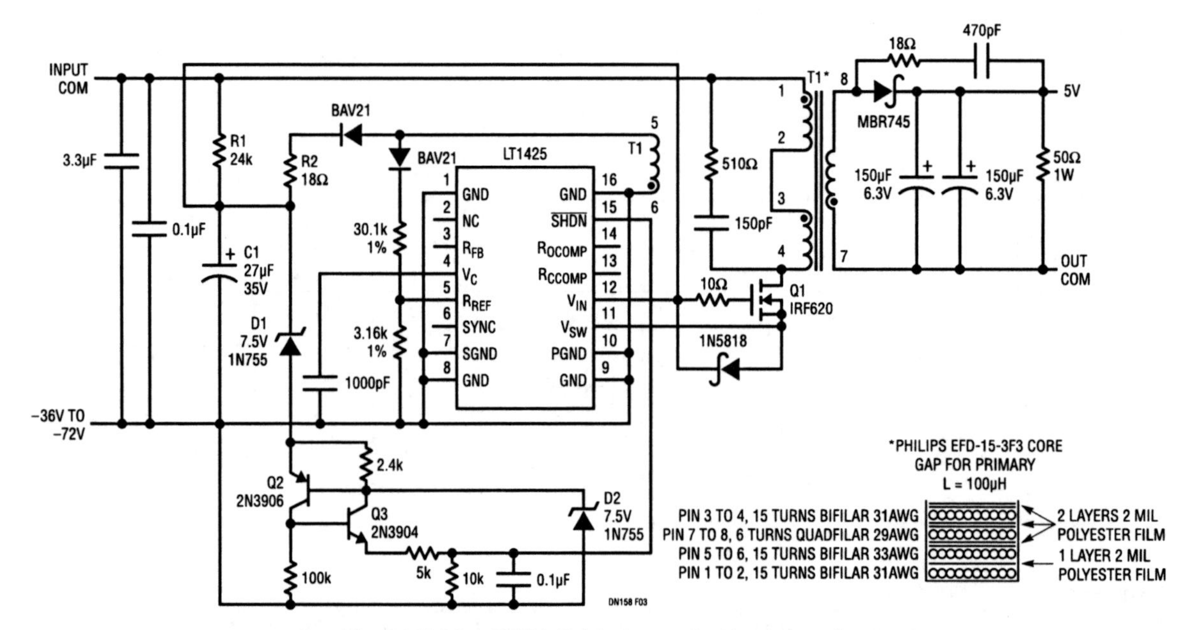

Figure 236.3 • 5V/2A Telecommunications Supply

Isolated power supplies for Local Area Networks

237

Sean Gold

Introduction

Local Area Networks such as Ethernet or Cheapernet, require low cost isolated power supplies with modest line and load regulation. Table 237.1 summarizes the objective design specifications based on IEEE 802.3 and ECMA 200-V. The LT1072 high efficiency switching regulator can be used in isolated flyback mode to satisfy these requirements with minimal support circuitry.[1]

Table 237.1 Power Supply Specifications for Figure 237.1

PARAMETER	VALUE	COMMENTS
V_{OUT}	−9V	Ethernet $11.4 < V_{IN} < 12.6V$
		Cheapernet $4.55V < V_{IN} < 5.45V$
Ripple	$V_n < 10mV_{p-p}$	
I_{LOAD}	150mA	40mA Min, 250mA Max
Load Reg	5%	
Line Reg	5%	
Efficiency	e > 70%	
Isolation	3000V	Ethernet
	500V	Cheapernet

Circuit design

Figure 237.1 illustrates the design approach. In isolated flyback mode, the LT1072 has no electrical connection to the load; instead, the regulator obtains a feedback signal from the transformers flyback voltage during the switch off-time. The voltage sense occurs after a 1.5μs delay, which prevents the internal error amplifier from regulating the voltage spike due to transformer leakage inductance. The LT1072 compares the feedback signal with a reference voltage, which is set at the feedback pin with a resistor to ground. The primary voltage is regulated to $16V + (V_{FB}/R_{FB})7k$. The feedback pin voltage V_{FB}, clamps to about 400mV, and the term $(V_{FB}/R_{FB})7k$ is nominally set to 2V, making the total flyback voltage 18V.

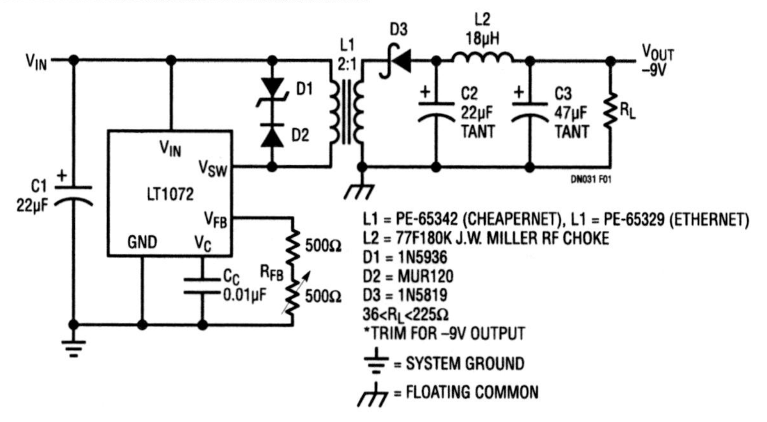

Figure 237.1 • Isolated Switching Regulator for LAN

Note 1: LTC's Application Note 19, the LT1070 Design Manual, presents a detailed discussion of isolated flyback mode and general information on switching regulator design.

Analog Circuit Design: Design Note Collection. http://dx.doi.org/10.1016/B978-0-12-800001-4.00237-4

The circuit is programmed for −9V output by setting the transformer turns ratio to 2 to 1. The feedback resistor R_{FB}, includes a 500Ω trim to take into account variations in the clamp voltage and gain within the LT1072.

A snubber network consisting of a fast turn-on, high breakdown diode and a 36V Zener diode, limits the magnitude of the leakage inductance spike. This snubber configuration improves efficiency because it minimizes the duration of the inductance spike. A Schottky diode in the secondary reduces the voltage loss to the output and increases efficiency.

Specifications for power supply filters are application dependent. When noise levels of 150mV are tolerable, a single 100μF tantalum capacitor is a suitable supply filter. When output noise below 10mV is required, the use of large output capacitors is often impractical. An LC filter is an appropriate recourse. The optional LC filter in Figure 237.1 contains an RF choke L2, and tantalum filter capacitors C1 and C2. These components have low effective series resistance (ESR) which helps maintain 5% load regulation.

Figure 237.2 shows the voltage on the switch pin, trace A, and the current flowing through the inductor, trace B. Trace C is a magnified view of trace A, which more clearly shows regulation of the primary voltage after the switch off-time. Figure 237.3 shows the voltage and current noise at the output.

Transformer design

The circuit design for 12V to −9V (Ethernet) and 5V to −9V (Cheapernet) circuits are identical except for the transformer specifications. Both circuits develop a regulated 18V primary voltage, but the available input voltage determines the required primary inductance.

$$L_{PRI} = \frac{V_{IN}}{(\Delta I)(f)(1 + V_{IN}/V_{PRI})} = \frac{5V}{(0.3A)(40kHz)(1 + 5/18)} = 326\mu H(\text{Minimum})$$

where,

ΔI = Magnetizing Current
f = Switching Frequency
V_{IN} = Input Voltage
V_{PRI} = Primary Voltage

Ethernet requires a larger primary inductance than Cheapernet, which implies a larger transformer. Increased isolation also mandates a larger core to accommodate additional insulation. The transformers used in both applications are shown in Figure 237.4. The PE-65329 for Ethernet (right) achieves 3700V isolation, while the PE-65342 for Cheapernet (left) provides 500V isolation.[2] These transformers are constructed with low loss core material and low resistance wire, to further improve efficiency.

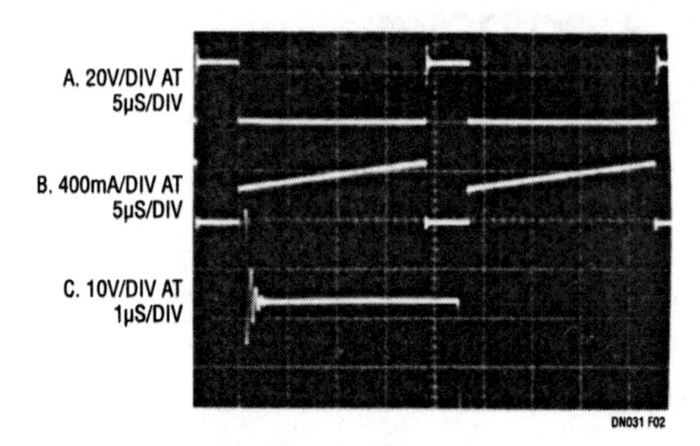

Figure 237.2 • Switching Waveforms

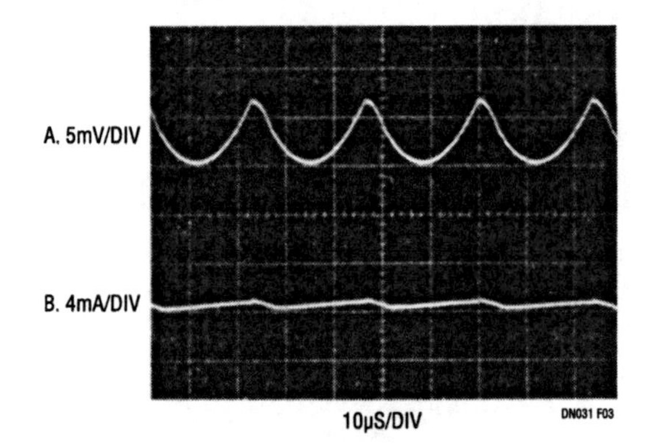

Figure 237.3 • Voltage and Current Noise

Figure 237.4 • LAN Transformers, PE-65342 (Left), PE-65329 (Right)

Note 2: A 500V version of the Ethernet transformer (PE-65330) is available in the 0.5inch package in Figure 237.4.

A battery-powered laptop computer power supply

238

Brian Huffman

Most battery-powered laptop computers require regulated multiple output potentials. Problems associated with such a supply include magnetic and snubber design, loop compensation, short-circuit protection, size and efficiency. Typical output power requirements include 5V @ 1A for memory and logic circuitry and ±12V @ 300mA to drive the analog components. Primary power may be either a 6V or 12V battery. The circuit in Figure 238.1 meets all these requirements. The LT1071 simplifies the power supply design by integrating most of the switching regulator building blocks. Also, the off-the-shelf transformer eliminates all the headaches associated with the magnetic design.

The circuit is a basic flyback regulator. The transformer transfers the energy from the 12V input to the 5V and ±12V outputs. Figure 238.2 shows the voltage (trace A) and the current (trace B) waveforms at the V_{SW} pin. The V_{SW} output is a collector of a common emitter NPN, so current flows through it when it is low. The circuit's 40kHz repetition rate is set by the LT1071's internal oscillator. During the V_{SW} (trace A) "on" time, the input voltage is applied across the primary winding. Notice that the current in the primary (trace C) rises slowly as the magnetic field builds up. The magnetic field in the core induces a voltage on the secondary windings. This voltage is proportional to the input voltage times the turns ratio. However, no power is transferred to the outputs because the catch diodes are all reversed biased. The energy is stored in the magnetic field. The amount of energy stored in the magnetic field is a function of the current level, how long the current flows, the primary inductance and the core material. When the switch is turned "off" energy is no longer transferred to the core, causing the magnetic field to collapse. The voltage on the transformer windings is proportional to time-rate-of-change of the magnetic field. Hence, the collapsing magnetic field causes the voltages on the windings to change. Now the catch diodes are forward biased and the energy is transferred to the outputs. Trace D is the voltage seen on the 5V secondary and trace E is the current flowing through it. The energy transfer is controlled by the LT1071's internal error amplifier, which acts to force the feedback (FB) pin to a 1.24V reference. The error amplifiers high impedance output (V_C pin) uses an RC damper for stable loop compensation.

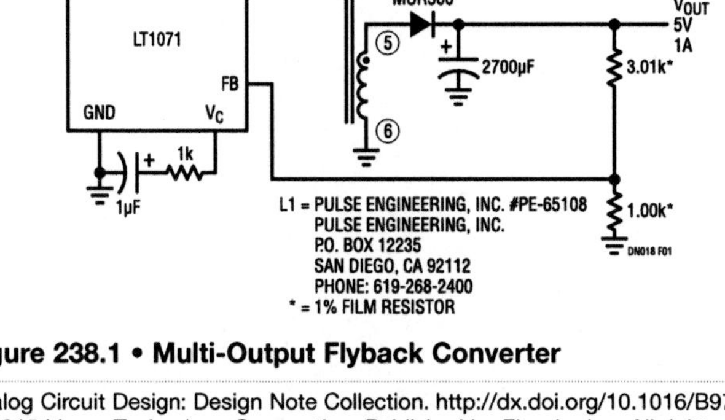

L1 = PULSE ENGINEERING, INC. #PE-65108
PULSE ENGINEERING, INC.
P.O. BOX 12235
SAN DIEGO, CA 92112
PHONE: 619-268-2400
* = 1% FILM RESISTOR

Figure 238.1 • Multi-Output Flyback Converter

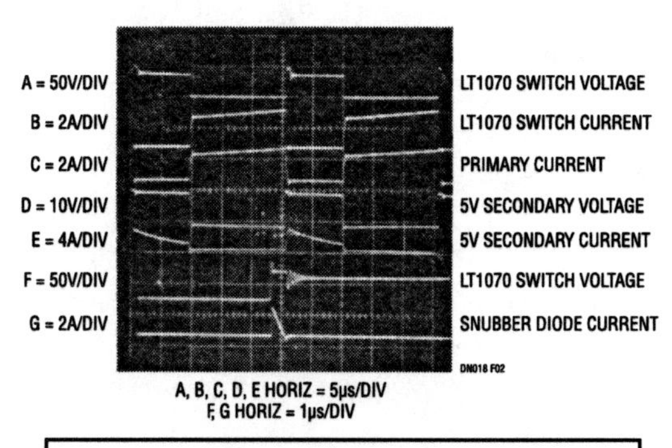

Use LT1171 and LT3080 for Higher Efficiency

Figure 238.2 • Waveforms for Continuous Mode Operation

Analog Circuit Design: Design Note Collection. http://dx.doi.org/10.1016/B978-0-12-800001-4.00238-6

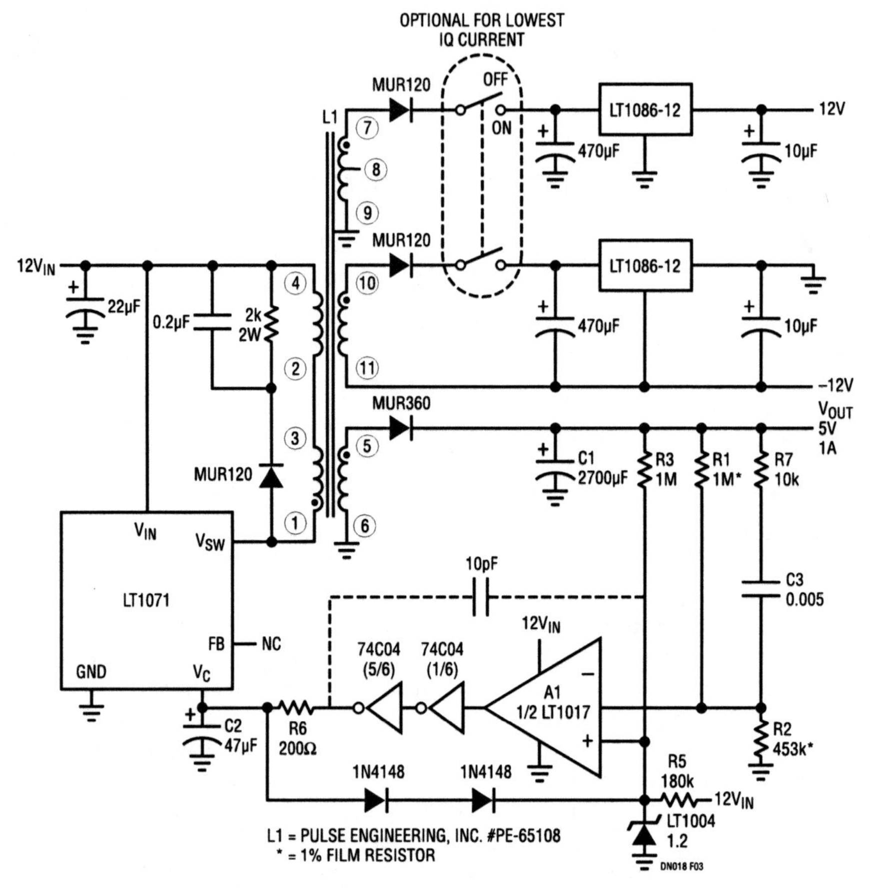

Figure 238.3 • Multi-Output, Transformer Coupled Low Quiescent Current Converter

If a 6V input is desired, use just one primary winding and an LT1070.

This is not an ideal transformer so not all the energy is coupled into the secondary. The energy left in the primary winding causes the overvoltage spike seen on the V_{SW} pin (trace F). This phenomenon is modeled by a leakage inductance term placed in series with the primary winding. When the switch is turned "off" current continues to flow in the inductor, causing the snubber diode to conduct (trace G). The snubber network clamps the voltage spike, preventing excessive voltage at the LT1071's V_{SW} pin. When the snubber diode current reaches zero, the V_{SW} pin voltage settles to a potential related to the turns ratio, output voltage and input voltage.

Post regulators are needed to the unregulated outputs if the cross-regulation error is too great. Such error can be as much as 20% depending upon output loading conditions. Note that the floating secondaries allow a −12V output to be obtained with a positive voltage regulator. The isolation allows the input of the regulator to float above ground. The LT1086 positive voltage regulators maintain both positive and negative outputs with 1%.

If battery capacity is limited by size or weight this circuit's 9mA quiescent current may be too high. Figure 238.3's modification offers output current in the ampere range with only microamps of quiescent drain. Further information about this circuit can be found in the LTC Application Note 29 "Some Thoughts on DC-DC Converters," page 8.

By using standard magnetics and a simplified switching regulator the design time needed to implement this power supply is greatly reduced. Although these circuits demonstrated in flyback topology, the LT1070/LT1071/LT1072 can easily handle other configurations including buck, boost, forward and inverting. Examples are given in the LTC Application Notes; AN19 "LT1070 Design Manual," AN25 "Switching Regulators for Poets," and AN29 "Some Thoughts on DC-DC Converters."

Section 16

Supercapacitor Charging

Supercapacitor-based power backup system protects volatile data in handhelds when power is lost

239

Jim Drew

Introduction

Handheld electronic devices play a key role in our everyday lives. Because dependability is paramount, handhelds are carefully engineered with lightweight power sources for reliable use under normal conditions. But no amount of careful engineering can prevent the mistreatment they will undergo at the hands of humans. For example, what happens when a factory worker drops a bar code scanner, causing the battery to pop out? Such events are electronically unpredictable, and important data stored in volatile memory would be lost without some form of safety net—namely a short-term power holdup system that stores sufficient energy to supply standby power until the battery can be replaced or the data can be stored in permanent memory.

Supercapacitors are compact, robust, reliable and can support the power requirements of a backup system for short-term power-loss events. Like batteries, they require careful charging and power regulation at the output. The LTC3226 is a 2-cell series supercapacitor charger with a PowerPath controller that simplifies the design of backup systems. Specifically, it includes a charge pump supercapacitor charger with programmable output voltage and automatic cell voltage balancing, a low dropout regulator and a power-fail comparator for switching between normal and backup modes. Low input noise, low quiescent current and a compact footprint make the LTC3226 ideal for compact, handheld, battery-powered applications. The device comes in a 3mm × 3mm 16-lead QFN package.

Backup power application

Figure 239.1 shows a power holdup system that incorporates a supercapacitor stack with the capacity to provide standby power of 165mW for about 45 seconds in the absence of battery power. An LDO converts the output of the supercapacitor stack to a constant voltage supply during backup mode.

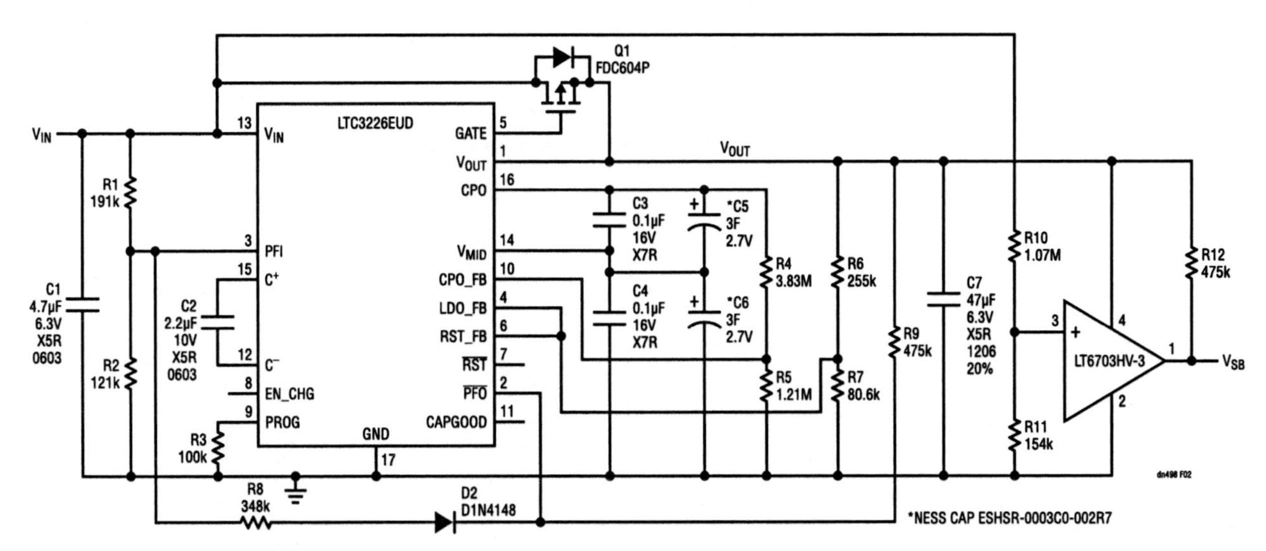

Figure 239.1 • A Typical Power Backup System Using Supercapacitors

Analog Circuit Design: Design Note Collection. http://dx.doi.org/10.1016/B978-0-12-800001-4.00239-8

Designing a power backup system is easy with the LTC3226. For example, take a device that has an operating current of 150mA and a standby current (I_{SB}) of 50mA when powered from a single-cell Li-Ion battery. To ensure that a charged battery is present, the power-fail comparator (PFI) high trigger point is set to 3.6V. The device enters standby mode when the battery voltage reaches 3.15V and enters backup mode at 3.10V ($V_{BAT(MIN)}$), initializing holdup power for a time period (t_{HU}) of about 45 seconds.

The standby mode trigger level is controlled by an external comparator circuit while the backup mode trigger level is controlled by the PFI comparator. While in backup mode, the device must be inhibited from entering full operational mode to prevent overly fast discharge of the supercapacitors.

The design begins by setting the PFI trigger level. R2 is set at 121k and R1 is calculated to set the PFI trigger level at the PFI pin (V_{PFI}) to 1.2V.

$$R1 = \frac{V_{BAT(MIN)} - V_{PFI}}{V_{PFI}} \cdot R2 = 191.6k\Omega$$

Set R1 to 191k.

The hysteresis on the V_{IN} pin needs to be extended to meet the 3.6V trigger level. This can be accomplished by adding a series combination of a resistor and diode from the PFI pin to the PFO pin. $V_{IN(HYS)}$ is 0.5V, $V_{PFI(HYS)}$ is 20mV and V_f is 0.4V.

$$R8 = \frac{V_{PFI} + V_{PFI(HYS)} - V_f}{V_{IN(HYS)} - \frac{V_{PFI(HYS)}}{R2} \cdot (R1 + R2)} \cdot R1 = 349.3k\Omega$$

Set R8 to 348k.

Set the LDO backup mode output voltage to 3.3V by setting R7 to 80.6k and calculating R6. $V_{LDO(FB)}$ is 0.8V.

$$R6 = \frac{V_{OUT} - V_{LDO(FB)}}{V_{LDO(FB)}} \cdot R7 = 251.9k\Omega$$

Set R6 to 255k.

The fully charged voltage on the series-connected supercapacitors is set to 5V. This is accomplished with a voltage divider network between the CPO pin and the CPO_FB pin. R5 is set to 1.21M and R4 is calculated. $V_{CPO(FB)}$ is 1.21V.

$$R4 = \frac{V_{CPO} - V_{CPO(FB)}}{V_{CPO(FB)}} \cdot R5 = 3.78M\Omega$$

Let R4 equal 3.83M.

As the voltage on the supercapacitor stack starts to approach V_{OUT} in backup mode, the ESR of the two supercapacitors and the output resistance of the LDO must be accounted for in the calculation of the minimum voltage on the supercapacitors at the end of t_{HU}. Assume that the ESR of each supercapacitor is 100mΩ and the LDO output resistance is 200mΩ, which results in an additional 20mV to $V_{OUT(MIN)}$ due to the 50mA standby current. $V_{OUT(MIN)}$ is set to 3.1V, resulting in a discharge voltage (ΔV_{SCAP}) of 1.88V on the supercapacitor stack. The size of each supercapacitor can now be determined.

$$C_{SCAP} = 2 \cdot \frac{I_{SB} \cdot t_{HU}}{\Delta V_{SCAP}} = 2.39F$$

Each supercapacitor is chosen to be a 3F/2.7V capacitor from Nesscap (ESHSR-0003C0-002R7).

Figure 239.2 shows the actual backup time of the system with a 50mA load. The backup time is 55.4 seconds due to the larger 3F capacitors used in the actual circuit.

Conclusion

High performance handheld devices require power backup systems that can power the device long enough to safely store volatile data when the battery is suddenly removed. Supercapacitors are compact and reliable energy sources in these systems, but they require specialized control systems for charging and output voltage regulation. The LTC3226 makes it easy to build a complete backup solution by integrating a 2-cell supercapacitor charger, PowerPath controller, an LDO regulator and a power-fail comparator, all in a 3mm × 3mm 16-lead QFN package.

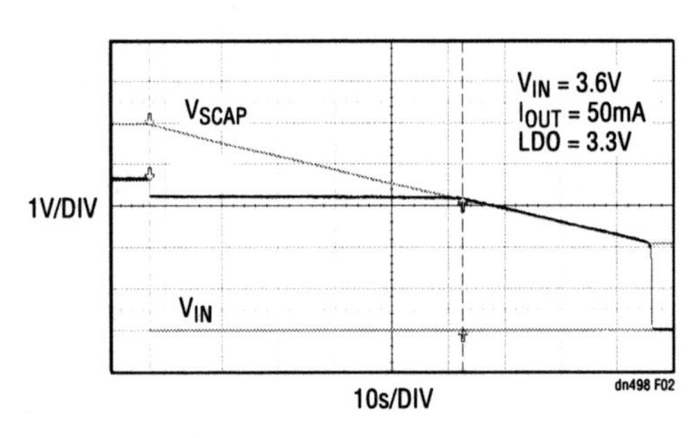

Figure 239.2 • Backup Time Supporting 50mA Load

Supercapacitor-based power backup prevents data loss in RAID systems

240

Jim Drew

Introduction

Redundant arrays of independent disks, or RAID, systems by nature are designed to preserve data in the face of adverse circumstances. One example is power failure, thereby threatening data that is temporarily stored in volatile memory. To protect this data, many systems incorporate a battery-based power backup that supplies short-term power—enough watt-seconds for the RAID controller to write volatile data to nonvolatile memory. However, advances in flash memory performance such as DRAM density, lower power consumption and faster write time, in addition to technology improvements in supercapacitors such as lower ESR and higher capacitance per unit volume, have made it possible to replace the batteries in these systems with longer lasting, higher performance and "greener" supercapacitors. Figure 240.1 shows a supercapacitor-based power backup system using the LTC3625 supercapacitor charger, an automatic power crossover switch using the LTC4412 PowerPath controller and an LTM4616 dual output µModule DC/DC converter.

The LTC3625 is a high efficiency supercapacitor charger ideal for small profile backup in RAID applications. It comes in a 3mm × 4mm × 0.75mm 12-lead DFN package and requires few external components. It features a programmable average charge current up to 1A, automatic cell voltage balancing of two series-connected supercapacitors and a low current state that draws less than 1µA from the supercapacitors.

Backup power applications

An effective power backup system incorporates a supercapacitor stack that has the capacity to support a complete data transfer. A DC/DC converter takes the output of the supercapacitor stack and provides a constant voltage to the data recovery electronics. Data transfer must be completed before the voltage across the supercapacitor stack drops to the minimum input operating voltage (V_{UV}) of the DC/DC converter.

To estimate the minimum capacitance of the supercapacitor stack, the effective circuit resistance (R_T) needs to be determined. R_T is the sum of the ESR of the supercapacitors, distribution losses (R_{DIST}) and the $R_{DS(ON)}$ of the automatic crossover's MOSFETs:

$$R_T = ESR + R_{DIST} + R_{DS(ON)}$$

Allowing 10% of the input power to be lost in R_T at V_{UV}, $R_{T(MAX)}$ may be determined:

$$R_{T(MAX)} = \frac{0.1 \cdot V_{UV}{}^2}{P_{IN}}$$

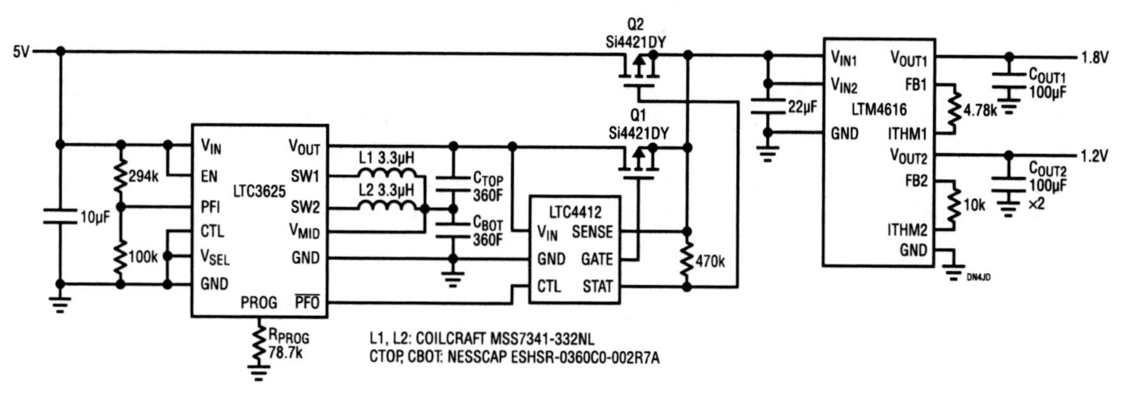

Figure 240.1 • Supercapacitor Energy Storage System for Data Backup

Analog Circuit Design: Design Note Collection. http://dx.doi.org/10.1016/B978-0-12-800001-4.00240-4

The voltage required across the supercapacitor stack ($V_{C(UV)}$) at V_{UV}:

$$V_{C(UV)} = \frac{V_{UV}^2 + P_{IN} \cdot R_T}{V_{UV}}$$

The minimum capacitance (C_{MIN}) requirement can now be calculated based on the required backup time (t_{BU}) to transfer data into the flash memory, the initial stack voltage ($V_{C(O)}$) and ($V_{C(UV)}$):

$$C_{MIN} = \frac{2 \cdot P_{IN} \cdot t_{BU}}{V_{C(0)}^2 - V_{C(UV)}^2}$$

C_{MIN} is half the capacitance of one supercapacitor. The ESR used in the expression for calculating R_T is twice the end-of-life ESR. End of life is defined as when the capacitance drops to 70% of its initial value or the ESR doubles.

The Charge Profile into Matched SuperCaps graph in the LTC3625 data sheet shows the charge profile for two configurations of the LTC3625 charging a stack of two 10F supercapacitors to 5.3V with R_{PROG} set to 143k. This graph, combined with the following equation, is used to determine the value of R_{PROG} that would produce the desired charge time for the actual supercapacitors in the target application:

$$R_{PROG} = 143k \cdot \frac{10F}{C_{ACTUAL}} \cdot \frac{5.3V - V_{C(UV)}}{V_{OUT} - V_{C(UV)}} \cdot \frac{t_{RECHARGE}}{t_{ESTIMATE}}$$

$V_{C(UV)}$ is the minimum voltage of the supercapacitors at which the DC/DC converter can produce the required output. V_{OUT} is the output voltage of the LTC3625 in the target application (set by V_{SEL} pin). $t_{ESTIMATE}$ is the time required to charge from $V_{C(UV)}$ to the 5.3V, as extrapolated from the charge profile curves. $t_{RECHARGE}$ is the desired recharge time in the target application.

Design example

For example, say it takes 45 seconds to store the data in flash memory where the input power to the DC/DC converter is 20W, and the V_{UV} of the DC/DC converter is 2.7V. A $t_{RECHARGE}$ of ten minutes is desired. The full charge voltage of the stack is set to 4.8V—a good compromise between extending the life of the supercapacitor and utilizing as much of the storage capacity as possible. The components of R_T are estimated: $R_{DIST} = 10m\Omega$, $ESR = 20m\Omega$ and $R_{DS(ON)} = 10m\Omega$.

The resulting estimated values of $R_{T(MAX)} = 36m\Omega$ and $R_T = 40m\Omega$ are close enough for this stage of the design. $V_{C(UV)}$ is estimated at 3V. C_{MIN} is 128F. Two 360F capacitors provide an end-of-life capacitance of 126F and ESR of 6.4mΩ. The crossover switch consists of the LTC4412 and two P-channel MOSFETs. The $R_{DS(ON)}$, with a gate voltage of 2.5V, is 10.75mΩ (max). An R_T of 26.15mΩ is well within $R_{T(MAX)}$. The value for R_{PROG} is estimated at 79.3k. The nearest standard 1% resistor is 78.7k. The data sheet suggests a 3.3μH value for both the buck and boost inductors.

The LTC3625 contains a power-fail comparator, which is used to monitor the input power to enable the LTC4412. A voltage divider connected to the PFI pin sets the power fail trigger point (V_{PF}) to 4.75V.

Figure 240.2 shows the actual backup time of the system with a 20W load. The desired backup time is 45 seconds, whereas this system yields 76.6 seconds. The difference is due to a lower R_T than estimated and an actual V_{UV} of 2.44V. Figure 240.3 shows the actual recharge time of 685 seconds compared to the 600 seconds used in the calculation, a difference due to the lower actual V_{UV}.

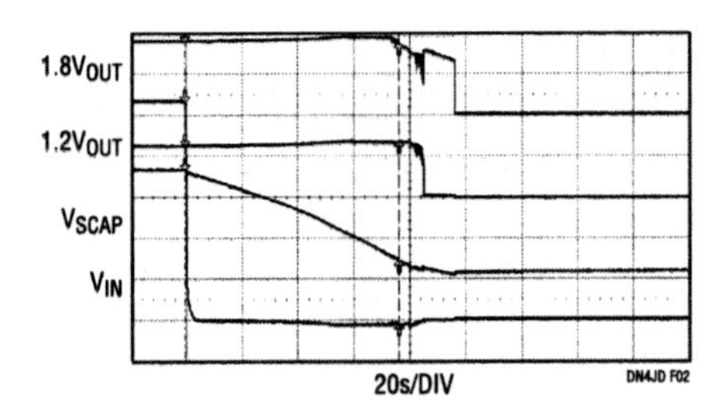

Figure 240.2 • Supercapacitor Backup Time Supporting a 20W Load

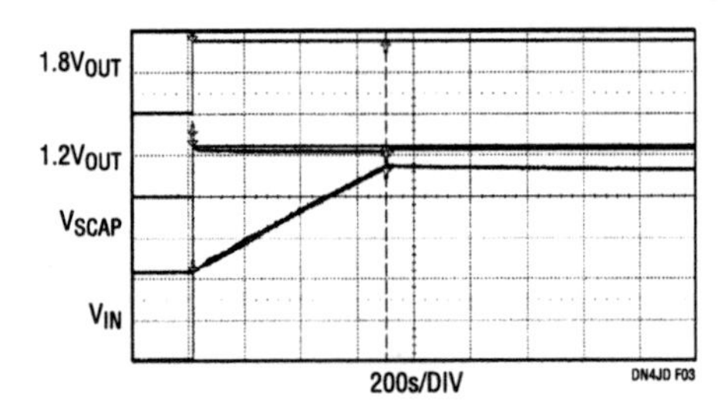

Figure 240.3 • Recharge Time After Backup

Conclusion

Supercapacitors are replacing batteries to satisfy green initiative mandates for data centers. The LTC3625 is an efficient 1A supercapacitor charger with automatic cell balancing that can be combined with the LTC4412 low loss PowerPath controller to produce a backup power system that protects data in storage applications.

Complete energy utilization improves run time of a supercap ride-through application by 40%

241

George H. Barbehenn

Introduction

Many electronic systems require a local power source that allows them to ride through brief main power interruptions without shutting down. Some local power sources must be available to carry out a controlled shutdown if the main power input is abruptly removed.

A battery backup can supply power in the event of a mains shutdown, but batteries are not well suited to this particular application. Although batteries can store significant amounts of energy, they cannot deliver much power due to their significant source impedance. Also, batteries have finite lives of ~2 to 3 years, and the maintenance required for rechargeable batteries is substantial.

Supercapacitors are well suited to such ride-through applications. Their low source impedance allows them to supply significant power for a relatively short time, and they are considerably more reliable and durable than batteries.

Complete energy utilization maximizes run time of supercap ride-through application

Figure 241.1 shows a complete 3.3V/200mA ride-through application that maximizes the amount of power extracted from the supercap to support the load.

The main components of the ride-through application include:

- The LTC4425 complete 2A supercapacitor charger. It clamps the individual cell voltages to ensure that the cells do not overvoltage during charging and balances the cells throughout charge and discharge.

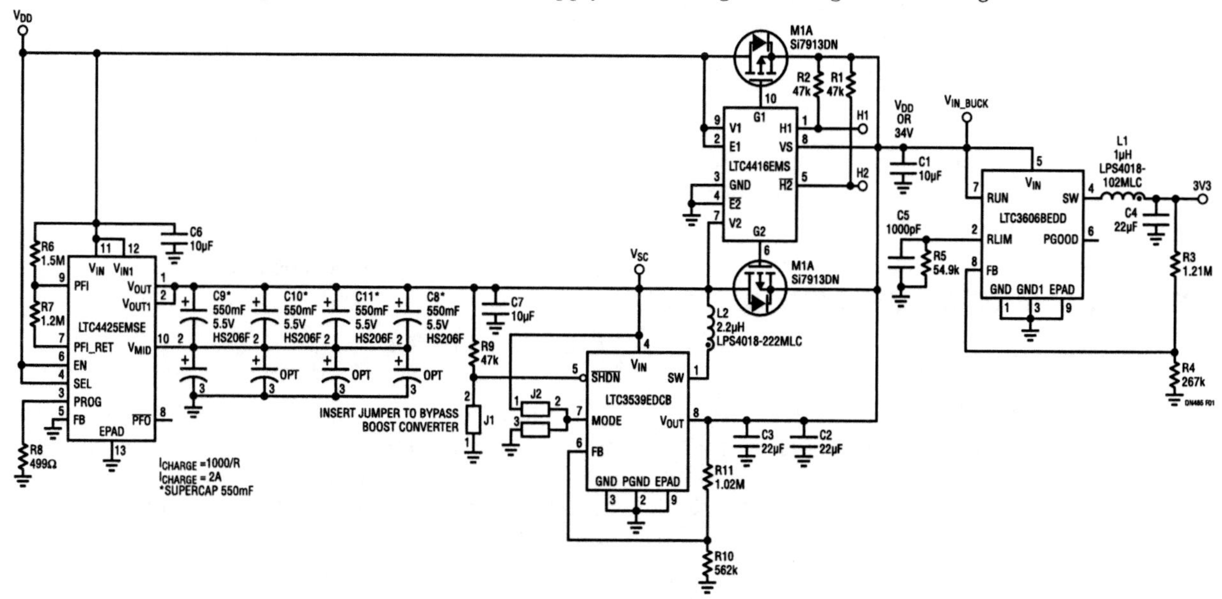

Figure 241.1 • This Supercap-Based Power Ride-Through Circuit Maximizes Run Time Using an Energy Scavenging Scheme

Analog Circuit Design: Design Note Collection. http://dx.doi.org/10.1016/B978-0-12-800001-4.00241-6

- The LTC3606 micropower buck regulator produces the regulated 3.3V output.
- The LTC4416 dual ideal diode switches the supercap in and out depending on need.
- The LTC3539 micropower boost regulator with output disconnect recovers nearly all the energy in the supercap and it keeps the input to the LTC3606 above dropout as the supercap voltage drops. This boost regulator operates down to 0.5V.

40% improvement in run time

Figure 241.2 shows the waveforms if the LTC3539 boost circuit is disabled. Run time from input power off to output regulator voltage dropping to 3V is 4.68 seconds. Figure 241.3 shows the waveforms if the LTC3539 boost circuit is operational. Run time from input power off to the output regulator dropping to 3V is 7.92 seconds. Note in Figure 241.3 that the output is a steady 3.3V voltage with a sharp cutoff.

How it works

When the LTC3539 boost regulator is disabled, as soon as input power falls, the LTC4416 ideal diodes switch the input energy supply for the LTC3606 buck regulator to the supercap. In Figure 241.2, the voltage across the supercap (V_{SC}) is seen to linearly decrease due to the constant power load of 200mA at 3.3V on the buck regulator output (3V3).

In Figure 241.3, when the LTC3539 boost regulator is enabled, the voltage across the supercap (V_{SC}) is seen to linearly decrease due to the constant power load of 200mA at 3.3V on the buck regulator. When the voltage at V_{SC} reaches 3.4V, the regulation point of the boost regulator, the boost regulator begins switching. This shuts off the ideal diode and disconnects the buck regulator from the supercapacitor. The energy input to the buck regulator is now the boost regulator's output of 3.4V.

Because the input of the buck regulator remains at 3.4V, its output remains in regulation. When the boost regulator reaches its input UVLO and shuts off, its output immediately collapses, and the buck regulator shuts off.

Maximizing usage of the energy in the supercap

Because each power conversion lowers the overall efficiency, the boost circuit should be held off as long as possible. Therefore, set the boost regulator output voltage as close to the buck regulator input dropout voltage as possible, in this case, 3.4V.

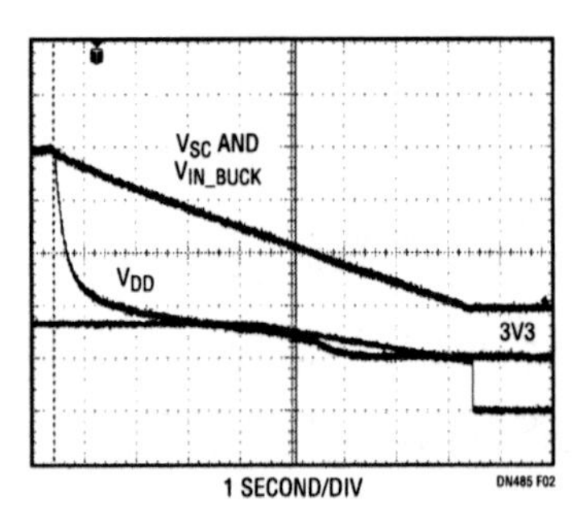

Figure 241.2 • Power Ride-Through Application Results without Boost Circuit

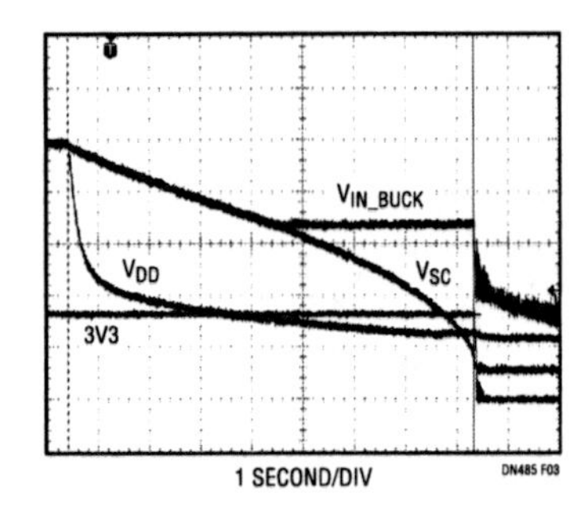

Figure 241.3 • Power Ride-Through Application Results with Boost Circuit Enabled. The Boost Circuit Yields a 40% Improvement in Run Time

If the supercapacitor is initially charged to 5V, then the energy in the supercapacitor is 6.875J:

$$\frac{1}{2}CV^2 = \frac{1}{2}0.55F \cdot 5^2 = 6.875J$$

$$0.67W(3.33 \cdot 0.2A)$$

The output power is $3.33V \cdot 0.2A = 0.67W$, so the percentage of energy extracted from the full supercap when the boost regulator is disabled is 45.1%:

$$\frac{\varepsilon_{LOAD}}{\varepsilon_{CAP}} = \frac{0.67 \cdot 4.68s}{6.875} = 45.1\%$$

The percentage of the energy extracted from the supercap's available storage when the boost regulator is enabled is 77%:

$$\frac{\varepsilon_{LOAD}}{\varepsilon_{CAP}} = \frac{0.67 \cdot 7.92s}{6.875} = 77\%$$

This represents a 40% improvement in ride-through run time—significant when seconds count.

Conclusion

The run time of any given supercapacitor-based power ride-through system can be extended by 40% if energy is utilized from the discharging supercap. This is particularly relevant if the supercapacitor charge voltage is reduced to ensure high temperature reliability.

Supercapacitors can replace a backup battery for power ride-through applications

242

Jim Drew

Introduction

Supercapacitors (or ultracapacitors) are finding their way into an increasing number of applications for short-term energy storage and applications that require intermittent high energy pulses. One such application is a power ride-through circuit, in which a backup energy source cuts in and powers the load if the main power supply fails for a short time. This type of application has typically been dominated by batteries, but electric double layer capacitors (EDLCs) are fast making inroads as their price-per-farad, size and effective series resistance per capacitance (ESR/C) continue to decrease.

Figure 242.1 shows a 5V power ride-through application where two series-connected 10F, 2.7V supercapacitors charged to 4.8V can support 20W for over a second. The LTC3225, a new charge pump-based supercapacitor charger, is used to charge the supercapacitors at 150mA and maintain cell balancing while the LTC4412 provides automatic switchover between the supercapacitor and the main supply. The LTM4616 dual output DC/DC μModule regulator creates the 1.8V and 1.2V outputs. With a 20W load, the output voltages remain in regulation for 1.42 seconds after the main power is removed.

Supercapacitor characteristics

A 10F, 2.7V supercapacitor is available in a 10mm × 30mm 2-terminal radial can with an ESR of 25mΩ. One advantage supercapacitors offer over batteries is their long lifetime. A capacitor's cycle life is quoted as greater than 500,000 cycles, whereas batteries are specified for only a few hundred cycles. This makes the supercapacitor an ideal "set and forget" device, requiring little or no maintenance.

Two critical parameters of a supercapacitor in any application are cell voltage and initial leakage current. Initial leakage current is really dielectric absorption current, which disappears after some time. The manufacturers of supercapacitors rate their leakage current after 100 hours of applied voltage

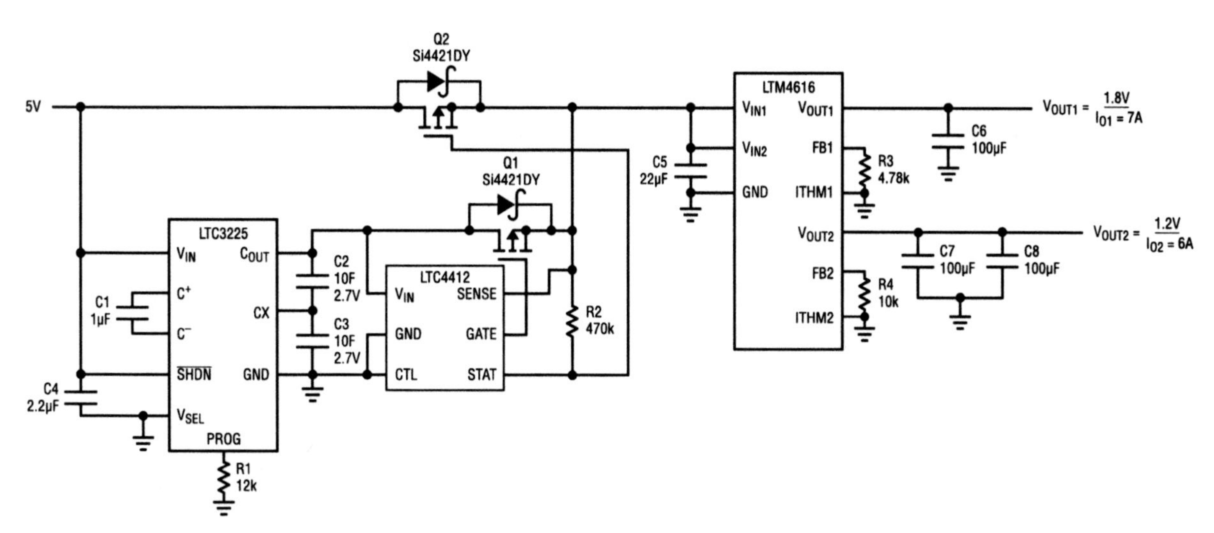

Figure 242.1 • 5V Ride-Through Application Circuit Delivers 20W for 1.42 Seconds

Analog Circuit Design: Design Note Collection. http://dx.doi.org/10.1016/B978-0-12-800001-4.00242-8

while the initial leakage current in those first 100 hours may be as much as 50 times the specified leakage current.

The voltage across the capacitor has a significant effect on its operating life. When used in series, the supercapacitors must have balanced cell voltages to prevent overcharging of one of the series capacitors. Passive cell balancing, where a resistor is placed across the capacitor, is a popular and simple technique. The disadvantage of this technique is that the capacitor discharges through the balancing resistor when the charging circuit is disabled. The rule of thumb for this scheme is to set the balancing resistor to 50 times the worst case leakage current, estimated at 2μA/Farad. Given these parameters, a 10F, 2.5V supercapacitor would require a 2.5k balancing resistor. This resistor would drain 1mA of current from the supercapacitor when the charging circuit is disabled.

A better alternative is to use a non-dissipative active cell balancing circuit, such as the LTC3225, to maintain cell voltage. The LTC3225 presents less than 4μA of load to the supercapacitor when in shutdown mode and less than 1μA when input power is removed. The LTC3225 features a programmable charging current of up to 150mA, charging two series supercapacitors to either 4.8V or 5.3V while balancing the individual capacitor voltages.

To provide a constant voltage to the load, a DC/DC converter is required between the load and the supercapacitor. As the voltage across the supercapacitor decreases, the current drawn by the DC/DC converter increases to maintain constant power to the load. The DC/DC converter drops out of regulation when its input voltage reaches the minimum operating voltage (V_{UV}).

To estimate the requirements for the supercapacitor, the effective circuit resistance (R_T) needs to be determined. R_T is the sum of the capacitors' ESRs plus the circuit distribution resistances, as follows:

$$R_T = ESR + R_{DIST}$$

Assuming 10% of the input power is lost in the effective circuit resistance when the DC/DC converter is at the minimum operating voltage, the worst case R_T is:

$$R_{T(MAX)} = \frac{0.1 \cdot V_{UV}^2}{P_{IN}}$$

The voltage required across the supercapacitor at the minimum operating voltage of the DC/DC converter is:

$$V_{C(UV)} = \frac{V_{UV}^2 + P_{IN} \cdot R_T}{V_{UV}}$$

The required effective capacitance can then be calculated based on the required ride-through time (T_{RT}), and the initial voltage on the capacitor ($V_{C(0)}$) and $V_{C(UV)}$ shown by:

$$C_{EFF} = \frac{2 \cdot P_{IN} \cdot T_{RT}}{V_{C(0)}^2 - V_{C(UV)}^2}$$

The effective capacitance of a series-connected bank of capacitors is the effective capacitance of a single capacitor divided by the number of capacitors while the total ESR is the sum of all the series ESRs.

The ESR of a supercapacitor decreases with increasing frequency. Manufacturers usually specify the ESR at 1kHz, while some manufacturers publish both the value at DC and at 1kHz. The capacitance of supercapacitors also decreases as frequency increases and is usually specified at DC. The capacitance at 1kHz is about 10% of the value at DC. When using a supercapacitor in a ride-through application where the power is being sourced for seconds to minutes, use the effective capacitance and ESR measurements at a low frequency, such at 0.3Hz. Figure 242.2 shows the ESR effect manifested as a 180mV drop in voltage when input power is removed.

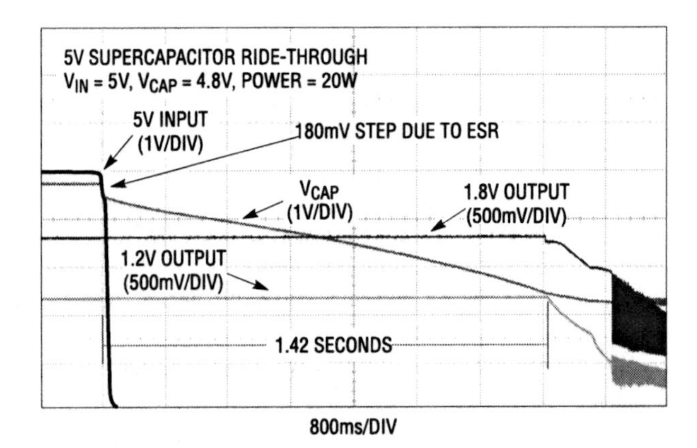

Figure 242.2 • 5V Ride-Through Application Timing

Conclusion

Supercapacitors can meet the needs of power ride-through applications where the time requirements are in the seconds to minutes range. Supercapacitors offer long life, low maintenance, light weight and environmentally friendly solutions when compared to batteries. To this end, the LTC3225 provides a compact, low noise solution for charging and cell balancing series-connected supercapacitors, without degrading performance.

Section 17

Current Source Design

Convert temperature to current at high linearity with current source

243

Todd Owen

Electronics 101

One of the first lessons in a basic electronics course covers the symbols for resistors, capacitors, inductors, voltage sources and current sources. Although each symbol represents a functional component of a real-world circuit, only some of the symbols have direct physical counterparts. For instance, the three discrete passive devices—resistors, capacitors, inductors—can be picked off a shelf and placed on a real board much as their symbolic analogs appear in a basic schematic. Likewise, while voltage sources have no direct 2-terminal analog, a voltage source can be easily built with an off-the-shelf linear regulator.

The black sheep of basic electronics symbols has long been the 2-terminal current source. The symbol shows up in every basic electronics course, but Electronics 101 instructors must take time to explain away the lack of a real-world equivalent. The symbol presents a simple electronics concept, but building a current source has, until now, been a complex undertaking.

A real 2-terminal current source

With the introduction of the LT3092, it is now as easy to produce a 2-terminal current source as it is to create a voltage source. Figure 243.1 shows how the LT3092 uses an internal current source and error amplifier, together with the ratio of two external resistors, to program a constant output current at any level between 0.5mA and 200mA. The flat temperature coefficient of the internal reference current (highlighted in Figure 243.2) is as good as many voltage references. Low TC resistors do not need to be used; the temperature coefficients of the external resistors need only match one another for optimum results.

No frequency compensation or supply bypass capacitors are needed. Frequency compensation is internal and the internal reference circuitry is buffered to protect it from line changes.

No input-to-output capacitors are required. While extensive testing has been done to ensure stable operation under the widest possible set of conditions, complex load impedance conditions could provoke instability. As such, testing in situ with final component values is highly recommended. If stability issues occur, they can be resolved with small capacitors or series RC combinations placed on the input, output, or from input to output.

The LT3092 offers all the protection features expected from a high performance product: thermal shutdown, overcurrent protection, reverse-voltage and reverse-current protection. Because a simple resistor ratio sets the current, a wide variety of techniques can be utilized to adjust the

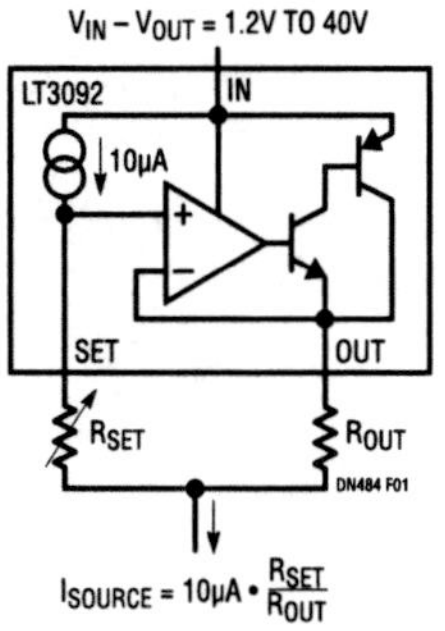

$V_{IN} - V_{OUT} = 1.2V$ TO 40V

LT3092 IN 10μA SET OUT R_{SET} R_{OUT}

$I_{SOURCE} = 10\mu A \cdot \frac{R_{SET}}{R_{OUT}}$

Figure 243.1 • 2-Terminal Current Source Requires Only Two Resistors to Program

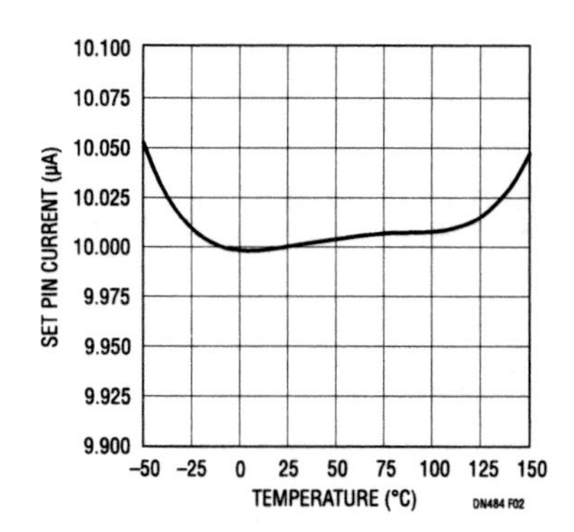

Figure 243.2 • SET Pin Current vs Temperature

Analog Circuit Design: Design Note Collection. http://dx.doi.org/10.1016/B978-0-12-800001-4.00243-X

current on the fly. The LT3092 can also be configured as a linear regulator without output capacitors for use in intrinsic safety environments.

The LT3092 as a T-to-I converter

Omega's 44200 series linear thermistor kits[1] include thermistors and resistors that together create a linear response to temperature when appropriately configured. These kits generate either a voltage or resistance proportional to temperature with high accuracy; the #44201 kit is listed for the 0°C to 100°C temperature range with 0.15°C accuracy.

Obviously, these kits easily satisfy the needs of a wide variety of applications, but problems arise when the thermistor must be placed at the end of a long wire—application information from Omega suggests no more than 100 feet of #22 wire for thermistor kit #44201. Wire impedance interferes with the thermistor resistance and defeats the accuracy inherent in the kit.

By adding the LT3092 to the thermistor kit along with three 0.1% accuracy resistors and one final trim, a very accurate 2-terminal temperature-to-current converter can be built. This circuit measures 700μA operating current at 0°C, dropping by 2μA every degree until 100°C, at which point the current measures 500μA. The obvious advantage to this T-to-I converter over a T-to-V converter is that current remains constant regardless of the wire length—as long as there is sufficient voltage to meet the compliance of the LT3092 circuit while not exceeding its absolute maximum. Electronics 101: Kirchoff's laws dictate conservation of current in the wire runs as long as there are no nodes for current to leak along the run.

Figure 243.3 shows the schematic for linear thermistor kit #44201 from Omega with the LT3092 and the additional resistor values. The formulas under the figure allow for substitution of other thermistor kit values and determination of appropriate complementary resistors to fit the application.

Once the initial circuit is built, any initial tolerance, variations, and offsets are easily trimmed out by connecting a voltmeter from node A to node B and trimming the potentiometer to measure 302mV (for this design). This voltage remains constant regardless of temperature.

Now, one wire runs out and back for temperature sensing at significant distances. By providing input voltage above the compliance level of the LT3092 (less than 2V for this circuit and resistor combination) and sensing the resultant current (use a 1k resistor and DVM) one can measure temperature. Figure 243.4 shows the current output from the circuit across temperature and the difference between measured and calculated response.

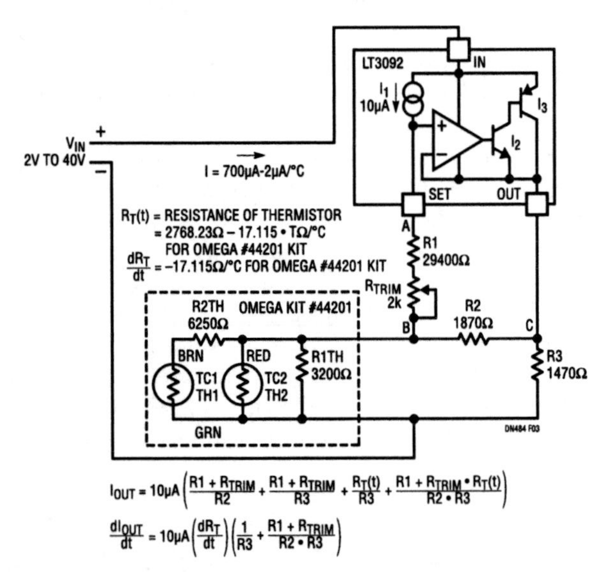

$$I_{OUT} = 10\mu A\left(\frac{R1 + R_{TRIM}}{R2} + \frac{R1 + R_{TRIM}}{R3} + \frac{R_T(t)}{R3} + \frac{R1 + R_{TRIM} \cdot R_T(t)}{R2 \cdot R3}\right)$$

$$\frac{dI_{OUT}}{dt} = 10\mu A\left(\frac{dR_T}{dt}\right)\left(\frac{1}{R3} + \frac{R1 + R_{TRIM}}{R2 \cdot R3}\right)$$

Figure 243.3 • 2-Terminal Temperature-to-Current Thermometer Suitable for Use at the End of Long Wire Runs

Conclusion

The LT3092 requires only two external resistors to produce a 2-terminal current source that references to input or ground, or sits in series with signal lines.

A 2-terminal current source enables a number of applications, especially those involving long wire runs, as Kirchoff's laws dictate the conservation of current over long wire distances—distances where a voltage signal would be corrupted. The example presented here uses the LT3092 and a linear thermistor kit to convert temperature to current, creating a 2-terminal current output thermometer. Placing this in series with long distances of wire maintains accuracy despite the distance of wire used.

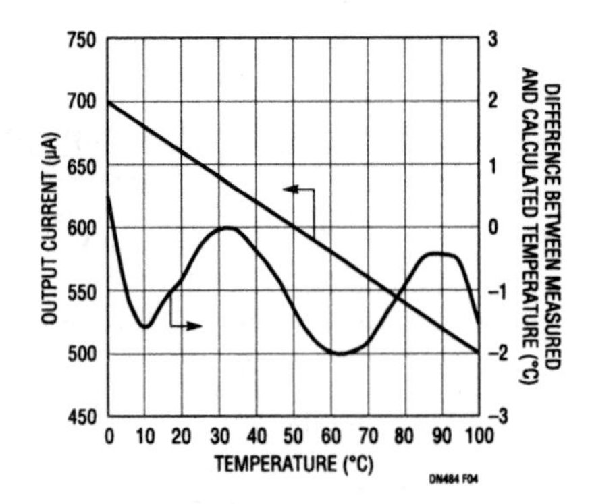

Figure 243.4 • Calculated vs Measured Performance of the Thermometer in Figure 243.3.

Note 1: Available from www.omega.com.

Versatile current source safely and quickly charges everything from large capacitors to batteries

244

David Ng

Introduction

The LT3750 is a current mode flyback controller optimized to easily and efficiently provide a controlled current to charge just about any capacitive energy storage device. The LT3750's simple but flexible feature set allows it to handle a wide variety of charging needs. These include large high voltage capacitors for professional photoflash equipment and emergency beacons, small capacitors that are charged and discharged thousands of times a second, and batteries for long term energy needs.

Safe, small and flexible

All of the control and feedback functions of the LT3750 are referred to the charger's input. The target voltage is set by just two resistors in a simple, low voltage network that monitors the flyback voltage of the transformer. When charging a capacitor to a high voltage, there is no need to connect any components to the hazardously high output potential. The charging current is a triangle wave whose amplitude is set by an external sense resistor and the flyback transformer turns ratio.

The LT3750 operates in boundary mode, at the edge of continuous and discontinuous conduction, which significantly reduces switching losses. This in turn allows for high frequency operation, and a correspondingly small flyback transformer size. The LT3750 is itself tiny, available in a 10-lead MSOP package.

The LT3750 is also compatible with a wide range of control circuitry. It is equipped with a simple interface consisting of a CHARGE command input bit and an open-drain $\overline{\text{DONE}}$ status flag. Both of these signals are compatible with most digital systems, yet tolerate voltages as high as 24V. The LT3750 operates from 3V to 24V DC.

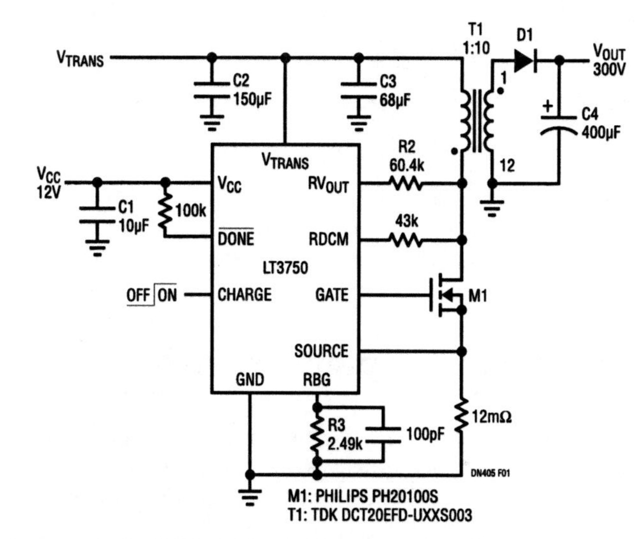

Figure 244.1 • LT3750 Circuit Charges 400μF Capacitor to 300V. Danger High Voltage—Operation by High Voltage Trained Personnel Only

Simple strobe capacitor charger

Figure 244.1 shows a LT3750 circuit that charges a 400μF strobe capacitor to 300V. This capacitor and voltage combination is typical of professional photoflash systems, security devices and automotive light strobes. The target voltage is set by the two resistors R2 and R3, which together monitor the MOSFET drain voltage. This voltage, when referenced to the input rail, is directly proportional to the output potential while power is being transferred to the output capacitor. The LT3750 compares this to an internal reference and terminates the charge cycle when the output has reached the desired target voltage, after which the LT3750 sets the $\overline{\text{DONE}}$ bit to signal the system microcontroller that the charge cycle is complete.

Analog Circuit Design: Design Note Collection. http://dx.doi.org/10.1016/B978-0-12-800001-4.00244-1

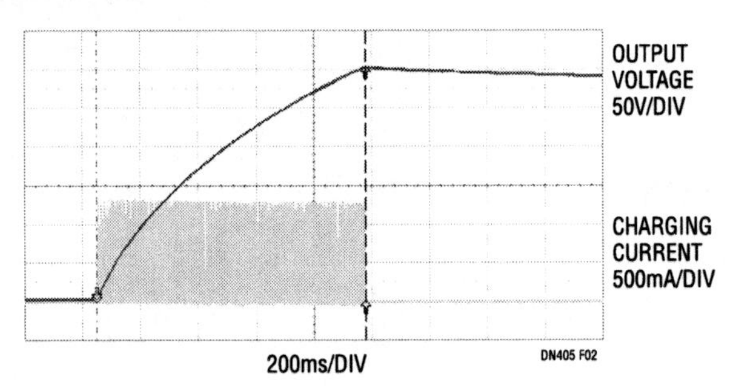

Figure 244.2 • LT3750 Charges 400μF to 300V in 0.92 Seconds

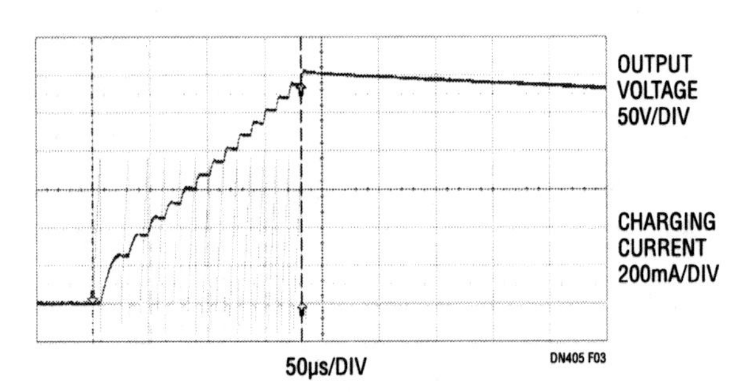

Figure 244.3 • LT3750 Charges 0.1μF to 300V in 180μs

As shown in Figure 244.2, the LT3750 charges the 400μF to 300V in about 0.92 seconds when the circuit is powered from a 12V source. Note that the output current amplitude is constant throughout the charge cycle.

Charge small capacitors fast

Many devices need to provide energy to a transducer multiple times per second, such as diagnostic equipment and device testers. Figure 244.3 shows that, for the same circuit as in Figure 244.1, the LT3750 is capable of charging a 0.1μF capacitor to 300V in just 180μs. The only change in the circuit is the replacement of the 400μF output cap with one that is much smaller. The performance of the circuit is essentially the same, other than the charge time. As far as the output device is concerned, the LT3750 circuit is a current source.

Charge batteries too

Another type of system that needs a controlled current source is a fast charger for a lead-acid battery. A fast charger for a lead-acid battery differs from the capacitor charging applications in that it needs to charge at high current, but at a much lower voltage. Figure 244.4 shows a circuit that charges at 6A until the lead-acid battery potential reaches the 14V float voltage. Again, the circuit is remarkably similar to the previous two designs—the transformer turns ratio is now 1:1 and the R2 set resistor has been changed to set the target float voltage to 14V. Other float voltages may be accommodated by simply changing R2 to the appropriate value.

When the battery voltage reaches 14V, the LT3750 sets the $\overline{\text{DONE}}$ bit. This can then be used to signal the system microcontroller, which can then enter a "trickle-charge" mode by setting the CHARGE bit at a fixed, low frequency interval.

Conclusion

The LT3750 is an easy-to-use controller that is ideal for applications where there is a need to charge an energy storage device to a predetermined target voltage. Its unique architecture allows it to be used in just about any application where a controlled current source is needed, with almost no limitation on the output voltage.

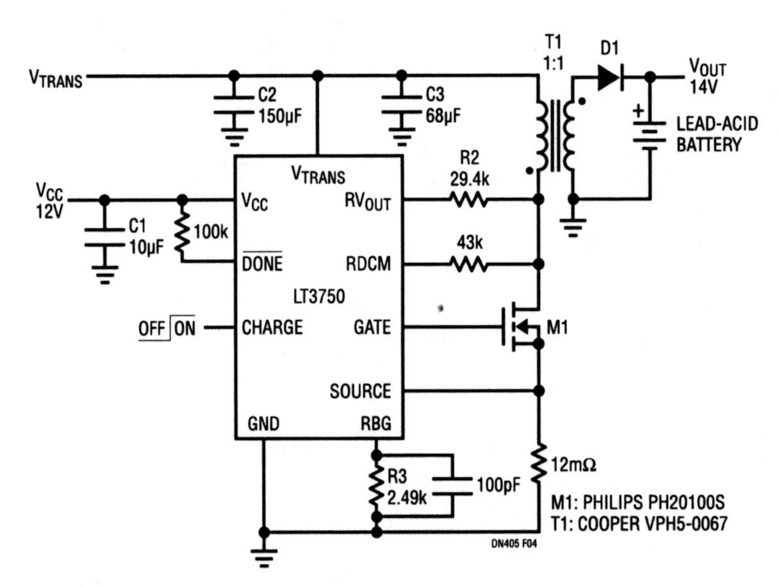

Figure 244.4 • LT3750 Battery Charger with Microcontroller Interface for Variable Current Charging

Section 18

Hot Swap and Circuit Protection

Protect sensitive circuits from overvoltage and reverse supply connections

245

Victor Fleury

Introduction

What would happen if someone connected 24V to your 12V circuits? If the power and ground lines were inadvertently reversed, would the circuits survive? Does your application reside in a harsh environment, where the input supply can ring very high or even below ground? Even if these events are unlikely, it only takes one to destroy a circuit board.

To block negative supply voltages, system designers traditionally place a power diode or P-channel MOSFET in series with the supply. However, diodes take up valuable board space and dissipate a significant amount of power at high load currents. The P-channel MOSFET dissipates less power than the series diode, but the MOSFET and the circuitry required to drive it increases costs. Both of these solutions sacrifice low supply operation, especially the series diode. Also, neither protects against voltages that are too high—protection that requires more circuitry, including a high voltage window comparator and charge pump.

Undervoltage, overvoltage and reverse supply protection

The LTC4365 is a unique solution that elegantly and robustly protects sensitive circuits from unpredictably high or negative supply voltages. The LTC4365 blocks positive voltages as high as 60V and negative voltages as low as −40V. Only voltages in the safe operating supply range are passed along to the load. The only external active component required is a dual N-channel MOSFET connected between the unpredictable supply and the sensitive load.

Figure 245.1 shows a complete application. A resistive divider sets the overvoltage (OV) and undervoltage (UV) trip points for connecting/disconnecting the load from V_{IN}. If the input supply wanders outside this voltage window, the LTC4365 quickly disconnects the load from the supply.

The dual N-channel MOSFET blocks both positive and negative voltages at V_{IN}. The LTC4365 provides 8.4V of enhancement to the gate of the external MOSFET during

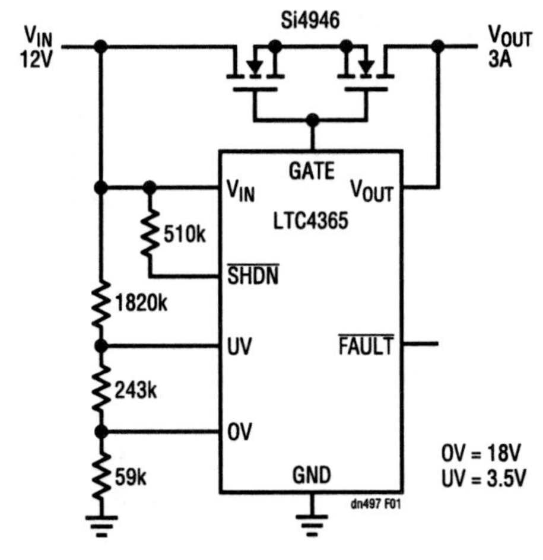

Figure 245.1 • Complete 12V Automotive Undervoltage, Overvoltage and Reverse Supply Protection Circuit

Analog Circuit Design: Design Note Collection. http://dx.doi.org/10.1016/B978-0-12-800001-4.00245-3

normal operation. The valid operating range of the LTC4365 is as low as 2.5V and as high as 34V—the OV to UV window can be anywhere in this range. No protective clamps at V_{IN} are needed for most applications, further simplifying board design.

Accurate and fast overvoltage and undervoltage protection

Two accurate (±1.5%) comparators in the LTC4365 monitor for overvoltage (OV) and undervoltage (UV) conditions at V_{IN}. If the input supply rises above the OV or below the UV thresholds, respectively, the gate of the external MOSFET is quickly turned off. The external resistive divider allows a user to select an input supply range that is compatible with the load at V_{OUT}. Furthermore, the UV and OV inputs have very low leakage currents (typically <1nA at 100°C), allowing for large values in the external resistive divider.

Figure 245.2 shows how the circuit of Figure 245.1 reacts as V_{IN} slowly ramps from −30V to 30V. The UV and OV thresholds are set to 3.5V and 18V, respectively. V_{OUT} tracks V_{IN} when the supply is inside the 3.5V to 18V window. Outside of this window, the LTC4365 turns off the N-channel MOSFET, disconnecting V_{OUT} from V_{IN}, even when V_{IN} is negative.

Novel reverse supply protection

The LTC4365 employs a novel negative supply protection circuit. When the LTC4365 senses a negative voltage at V_{IN}, it quickly connects the GATE pin to V_{IN}. There is no diode drop between the GATE and V_{IN} voltages. With the gate of the external N-channel MOSFET at the most negative potential (V_{IN}), there is minimal leakage from V_{OUT} to the negative voltage at V_{IN}.

Figure 245.3 shows what happens when V_{IN} is hot-plugged to −20V. V_{IN}, V_{OUT} and GATE start out at ground just before the connection is made. Due to the parasitic inductance of the V_{IN} and GATE connections, the voltage at V_{IN} and GATE pins ring significantly below −20V. The external MOSFET must have a breakdown voltage that survives this overshoot.

The speed of the LTC4365 reverse protection circuits is evident by how closely the GATE pin follows V_{IN} during the negative transients. The two waveforms are almost indistinguishable on the scale shown. Note that no additional external circuits are needed to provide reverse protection.

There's more! AC blocking, reverse V_{IN} Hot Swap control when V_{OUT} is powered

After either an OV or UV fault has occurred (or when V_{IN} goes negative), the input supply must return to the valid operating voltage window for at least 36ms in order to turn the external MOSFET back on. This effectively blocks 50Hz and 60Hz unrectified AC.

LTC4365 also protects against negative V_{IN} connections even when V_{OUT} is driven by a separate supply. As long as the breakdown voltage of the external MOSFET is not exceeded (60V), the 20V supply at V_{OUT} is not affected by the reverse polarity connection at V_{IN}.

Conclusion

The LTC4365 controller protects sensitive circuits from overvoltage, undervoltage and reverse supply connections using back-to-back MOSFETs and no diodes. The supply voltage is passed to the output only if it is qualified by the user-adjustable UV and OV trip thresholds. Any voltage outside this window is blocked, up to 60V and down to −40V.

The LTC4365's novel architecture results in a rugged, small solution size with minimal external components, and it is available in tiny 8-pin 3mm × 2mm DFN and TSOT-23 packages. The LTC4365 has a wide 2.5V to 34V operating range and consumes only 10μA during shutdown.

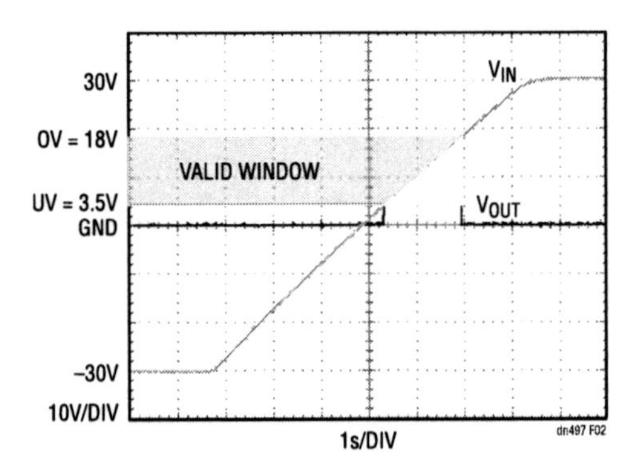

Figure 245.2 • Load Protection as V_{IN} Is Swept from −30V to 30V

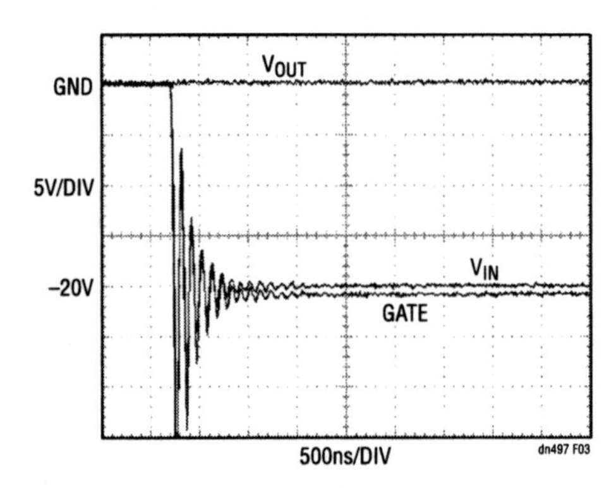

Figure 245.3 • Hot Swap Protection from V_{IN} to −20V

Simple energy-tripped circuit breaker with automatic delayed retry

246

Tim Regan

Introduction

A circuit breaker protects sensitive load circuits from excessive current flow by opening the power supply when the current reaches a predetermined level. The simplest circuit breaker is a fuse, but blown fuses require physical replacement. An electronic circuit breaker provides the same measure of circuit protection as a fuse without the single-use problem. Nevertheless, an electronic circuit breaker with a fixed trip current threshold, while effective for protection, can become a nuisance if tripped by short duration current transients—even if the circuit breaker self-resets.

One way to minimize nuisance breaks is to employ a slow-blow technique, which allows relatively high levels of current for short intervals of time without tripping the breaker. Ideally, the breaker's trip threshold would be a function of total transient energy, instead of just current. This article describes an electronic circuit breaker, combining current sensing with timing to create an energy-tripped breaker, which protects sensitive circuits while minimizing nuisance trips.

Higher currents permitted for shorter time intervals

The circuit of Figure 246.1 has three distinct parts—circuit breaking, current sensing and timing.

The circuit breaking function can be any type of electronically controlled relay or solid state switch, properly sized for voltage and current ratings of the load being protected.

Load current sensing is achieved via an LT6108-2 current sense amplifier with built-in comparator. The LT6108-2 converts the voltage drop across a small valued sense resistor to a ground-referenced output voltage that is directly proportional to the load current. The trip threshold is created by scaling the output voltage via resistor divider and feeding the result to the integrated comparator with a precision 400mV voltage reference. The comparator changes state when the load current exceeds the threshold.

To prevent short duration transients from causing nuisance trips, an LTC6994-2 TimerBlox delay timer is added between

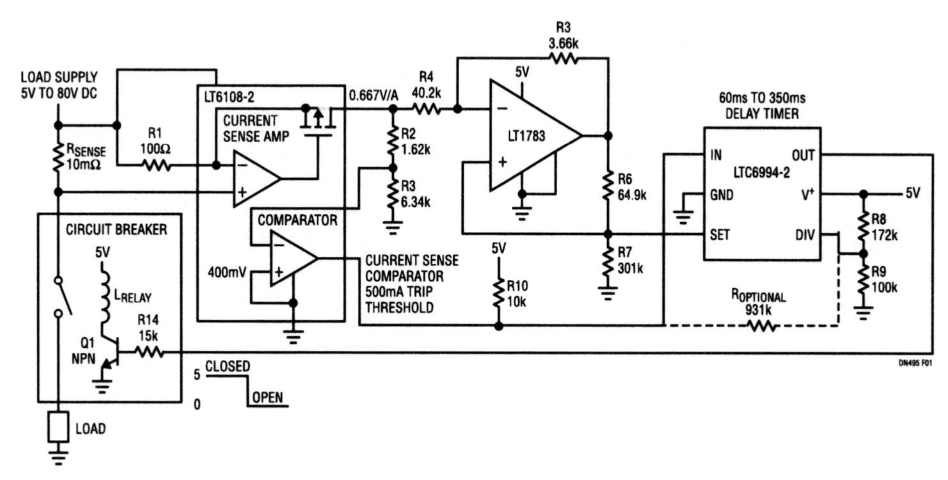

Figure 246.1 • Energy-Tripped Circuit Breaker Trips After a Time Interval that Varies as a Function of Sensed Load Current

Analog Circuit Design: Design Note Collection. http://dx.doi.org/10.1016/B978-0-12-800001-4.00246-5

the comparator output and the circuit breaker. Once tripped, the comparator falling edge starts a variable time delay interval, which, if allowed to complete, signals the circuit breaker to open. Nothing happens if the transient duration is shorter than the delay.

A current-controlled delay interval

The LTC6994-2 delays from an edge appearing at its IN pin by a time ranging from 1μs to 33s. The delay time is controlled by the current sourced by the SET pin, which programs an internal oscillator frequency, while the bias voltage on the DIV pin selects a frequency divide ratio.

The LT1783 op amp circuit takes the output voltage from the current sense amplifier and adjusts the SET pin current, thereby making the delay time a function of the load current (see Figure 246.2). As shown, the current sense comparator trip threshold is 500mA. A current of 500mA creates a falling edge and starts a time delay of 350ms. Should the load current drop below 500mA before the delay time expires, the timer output remains high and the circuit breaker does not trip.

Higher load currents correspond to higher current sense amplifier output voltages, which in turn reduce the delay time interval (Figure 246.2). For instance, a 5A load current trips the circuit breaker in only 60ms. Depending on the average load current in excess of the 500mA threshold, the delay interval or trip time will fall somewhere between 30ms and 400ms.

Once tripped, the load current drops to zero. This resets the current sense comparator high. This rising edge is also delayed by the LTC6994-2. The minimum current sense output voltage stretches this delay to a maximum time of ~1.3s. After this delay the circuit breaker closes and reapplies power to the load. This automatic retry function requires no additional components.

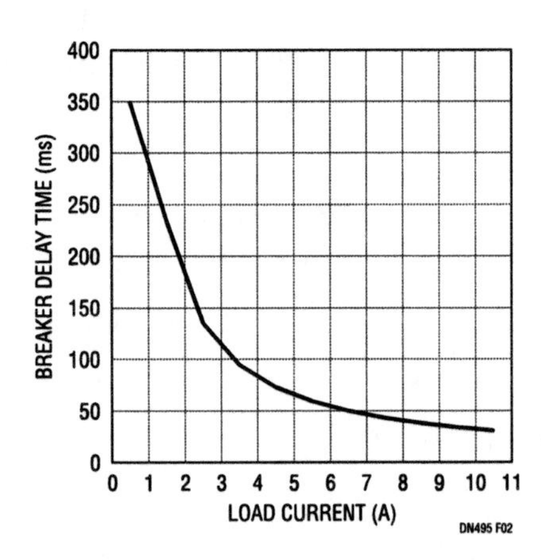

Figure 246.2 • Low Current Transients Must Last Relatively Longer to Trip the Breaker. Higher Currents Trip the Circuit Breaker in Less Time

The response of the circuit to a 5A load current spike and automatic retry is shown in Figure 246.3. If the load current remains too high, the trip/retry cycle repeats continually. A current surge is fairly common when the circuit breaker is first closed and can trip the comparator. If the duration is less than the timer delay, the breaker remains closed, thus avoiding an endless loop of self-induced nuisance trips.

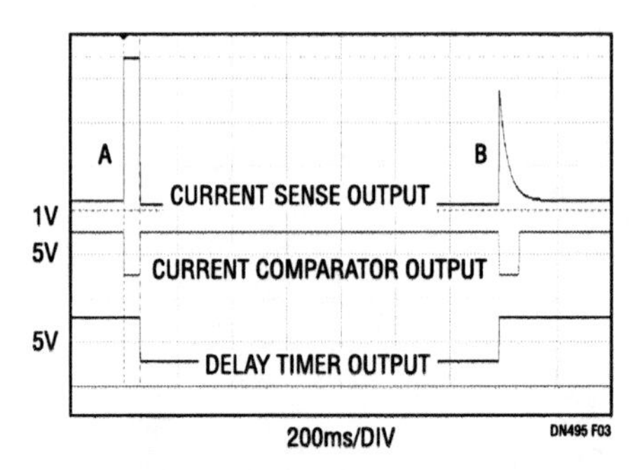

Figure 246.3 • An Example Trip and Retry Sequence. At Time Point A, the 5A Load Current Spike Trips the Comparator and 60ms Later the Breaker Is Opened. At Time B, After a Delay Time of 1.3 sec, the Timer Closes the Breaker. The Resulting Short Duration Spike of Start-Up Current Is Not Large Enough or Long Enough in Duration to Trip the Breaker Again

Extending the retry time interval

The LTC6994-2 delay timer has eight divider settings for a wide range of timing intervals. Adding the single optional resistor shown in Figure 246.1 shifts the delay block to a new setting, increasing the retry time interval if desired. This can give any fault condition more time to subside. The circuit breaker response time interval is not affected.

For the values shown, when the circuit breaker trips and the current drops to zero, the comparator high level biases the DIV pin to a higher voltage level, resulting in a longer retry delay time of 10 seconds.

Conclusion

The circuit shown here can be easily modified to different timing requirements with a few resistor value changes. Other current sense devices such as the LT1999 can also be used to monitor bidirectional load currents with variable breaker timing functionality.

Hot Swap controller, MOSFET and sense resistor are integrated in a 5mm × 3mm DFN for accurate current limit and load current monitoring in tight spaces

247

Vladimir Ostrerov

Introduction

In general, a Hot Swap controller provides two important functions to boards that can be plugged and unplugged from a live backplane:

- It limits the potentially destructive inrush current when a board is plugged in.
- It acts as a circuit breaker, with the maximum current and maximum time at that current factored into the breaker function.

Two of the components required to implement these functions, the power MOSFET and sense resistor, tend to dominate the board real estate taken by the Hot Swap circuit. The LTC4217 saves space by combining these two components with a Hot Swap controller in a 16-pin 5mm × 3mm DFN package (or 20-lead TSSOP). This 2A integrated Hot Swap controller fits easily onto boards operating in the voltage range from 2.9V to 26.5V. A dedicated 12V version, LTC4217-12, is also available, which contains preset 12V specific thresholds. Figure 247.1 shows how little space is required for a complete Hot Swap circuit.

LTC4217 features

Figure 247.2 shows a simplified block diagram of the LTC4217. The controller provides inrush current control and a 5% accurate 2A current limit with foldback. For soft-start, an internal current source charges the gate of the N-channel MOSFET with 300V/s slew rate. Lower soft-start output voltage slew rates can be set by adding an external gate capacitor.

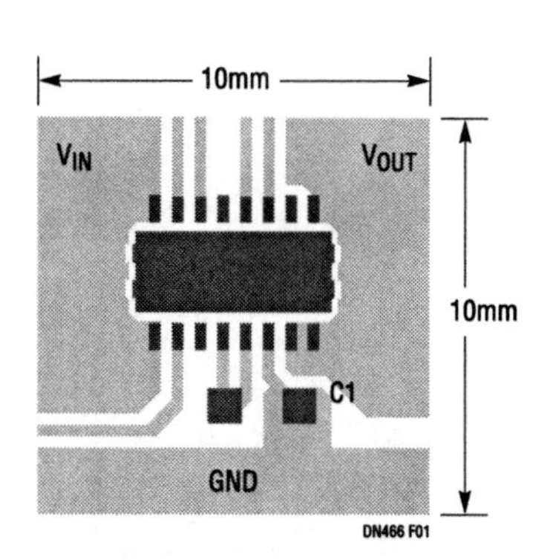

Figure 247.1 • Tiny Integrated Controller Package Results in a Small Footprint

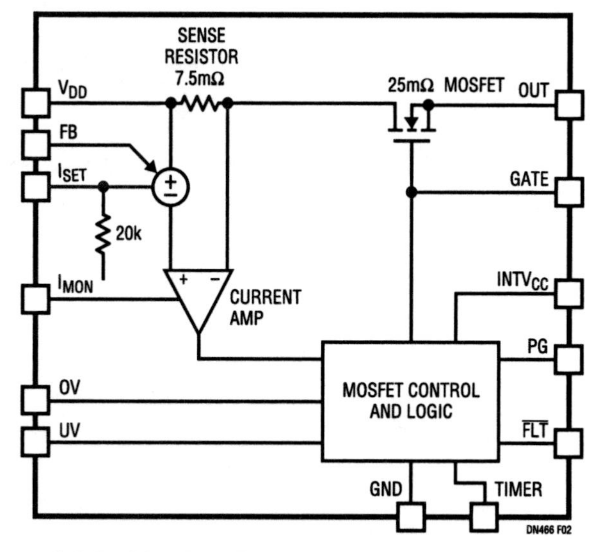

Figure 247.2 • Simplified Block Diagram of the LTC4217

Analog Circuit Design: Design Note Collection. http://dx.doi.org/10.1016/B978-0-12-800001-4.00247-7

Integrated MOSFET and sense resistor

The LTC4217 integrates a 25mΩ MOSFET and 7.5mΩ current sense resistor. The default value of the active current limit is 2A, which can be adjusted to a lower value by adding an external resistor. Using an external analog switch for the connection of this resistor allows a start-up current to be larger than the maximum load current in steady state operation.

Adjustable current limit

The voltage at the I_{SET} pin determines the active current limit, which is 2A by default. This pin is driven by a 0.618mV voltage reference through a 20k resistor. An external resistor placed between the I_{SET} pin and ground forms a resistive divider with the internal 20k resistor. The divider acts to lower the voltage at the I_{SET} pin and therefore lower the current limit threshold.

The I_{SET} pin voltage increases linearly with temperature, with a slope of 3.2V/°C when no external resistor is present. This compensates for the temperature coefficient of the sense resistor—important for applications that must maintain monitoring accuracy over a wide temperature range. This also provides a convenient means to monitor the MOSFET temperature. If the die temperature exceeds 145°C, the MOSFET is turned off. It is turned on again when the temperature drops below 125°C.

Voltage and current monitoring

The LTC4217 protects the load from overvoltage and undervoltage conditions with a 2% accurate comparator threshold. The LTC4217 also features an adjustable current limit timer, a current monitor output and a fault output.

The adjustable current limit timer sets the time duration for current limit before the MOSFET is turned off. The current monitor produces a voltage signal scaled to the load current. The fault output is an open drain that pulls low when an overcurrent fault has occurred and the circuit breaker trips.

Typical application

The LTC4217 application circuit shown in Figure 247.3 operates with a 100ms auto-retry time and a 2ms overcurrent condition when the load current reaches 2A. It also produces a voltage signal for an ADC to monitor load current.

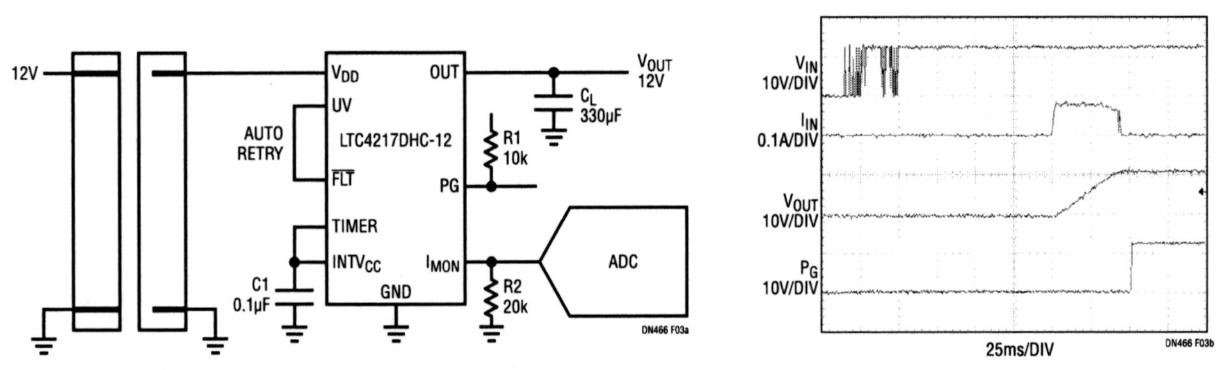

Figure 247.3 • Typical Application with Current Monitored by an ADC

Hot Swap solution meets AMC and MicroTCA standards

248

Vladimir Ostrerov

Introduction

The LTC4223 is a dual Hot Swap controller that meets the power requirements of the Micro Telecommunication Computing Architecture (MicroTCA) specification recently ratified by the PCI Industrial Computer Manufacturers Group (PICMG).

The LTC4223 includes an internal pass FET for the 3.3V auxiliary supply and a driver for an external N-channel pass FET for the 12V payload supply. Inrush current for both supplies is controlled: the auxiliary supply has a fixed 240mA active current limit and the 12V ramp rate is controlled by an external capacitor. A timed circuit breaker and fast current limit protect both supplies against severe overcurrent faults. It also features an adjustable analog current limit with a circuit breaker timeout for the 12V supply.

The LTC4223 monitors 12V load current by sensing voltage across an external resistor and outputs a ground-referenced voltage (at the 12IMON pin) proportional to the load current. It also provides separate power-good outputs for the two supplies and a single, common fault output. Additional features include card detection and independent control of the two supplies. The LTC4223-1 latches off after a circuit breaker fault timeout expires while the LTC4223-2 provides automatic retry after a fault. A fault on the 12V supply shuts down only the 12V path, leaving the 3.3V auxiliary power available for system management functions. A fault on the 3.3V AUX supply shuts down both supplies.

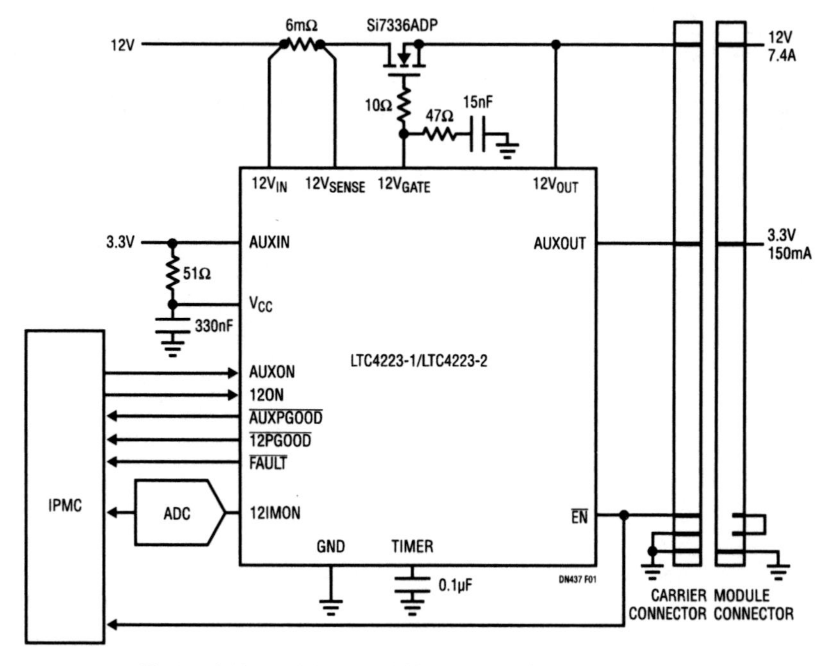

Figure 248.1 • Advanced Mezzanine Card Application

Analog Circuit Design: Design Note Collection. http://dx.doi.org/10.1016/B978-0-12-800001-4.00248-9

Advanced Mezzanine Card application

Figure 248.1 shows a typical MicroTCA application. The current limit on the 12V rail is 7.6A, determined by the 6mΩ sense resistor. Auxiliary rail current limit is internally set to 240mA. (Section 4.2.1 of the MicroTCA specification details the requirements for payload and auxiliary voltage, current, and protection.) Figure 248.2 shows the power-up transients when a card is inserted. Figures 248.3 and 248.4 show the two modes of overcurrent protection on the 12V supply. In Figure 248.3, the load current is increased above the analog current limit (ACL) threshold. The LTC4223 responds by reducing the current to the ACL threshold, allowing the card to ride out short overcurrent faults. If the fault persists, the timer expires and power is turned off. In the event of a severe overcurrent fault, load current is reduced to the ACL limit in 8μs as shown in Figure 248.4. Once again, if the fault persists and the timer expires, power is turned off entirely.

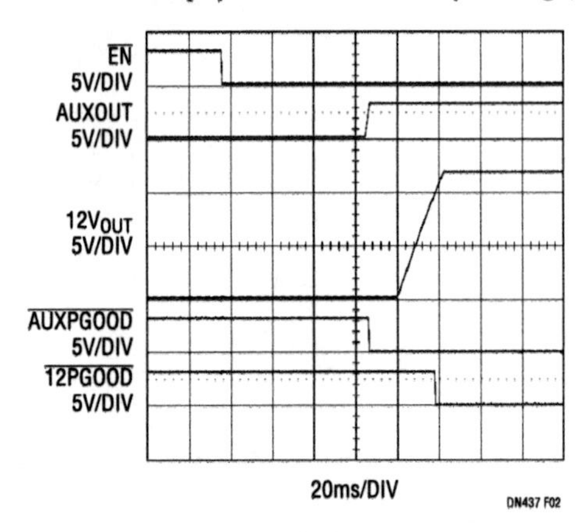

Figure 248.2 • Normal Power-Up Waveform

Conclusion

The LTC4223 aims to simplify Hot Swap control for Advanced Mezzanine Cards in ATCA and MicroTCA systems. It succeeds by meeting all MicroTCA requirements for controlling both payload and auxiliary power with only a 5mm × 4mm DFN package and a few minimal external components. Individual card monitoring and control functions are further simplified by LTC4223's status and control lines.

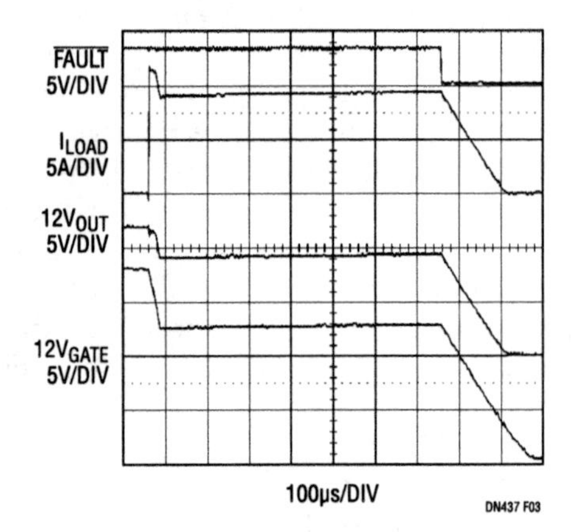

Figure 248.3 • Overcurrent Fault on 12V Output

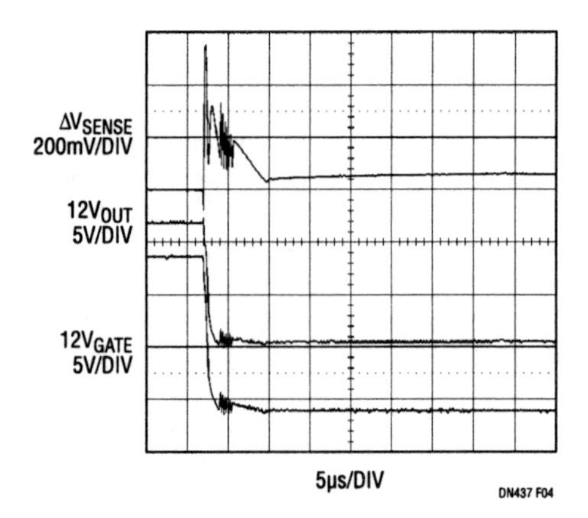

Figure 248.4 • Short-Circuit Fault on 12V Output

An easy way to add auxiliary control functions to Hot Swap cards

249

Mark Thoren

Introduction

A Hot Swap controller is essential to any system in which boards are inserted into a live backplane. The controller must gently ramp up the supply voltage and current into the card's bypass capacitors, thus minimizing disturbances on the backplane and to other cards. Likewise, it must disconnect a faulty card from the backplane if it draws too much current. The controller also monitors undervoltage and overvoltage conditions on the backplane supply, ensuring reliable operation of the card's circuitry. The LTC4215-1 takes the obvious next step and integrates three general purpose I/O (GPIO) lines and an accurate ADC into the Hot Swap controller to provide quantitative information on board voltage and current. Upgrading to the LTC4215-1 is analogous to replacing a car's venerable "Check Engine" light with a modern dashboard information display.

Additional control

There are many functions on a card that are considered part of the "power gateway," apart from the actual function of the board (telecommunications, data acquisition, etc.). These include sequencing power supplies, providing supply status information, monitoring pushbuttons, etc. The LTC4215-1 GPIO pins are well suited to these functions. Tying the ON pin high turns on the pass FET after a 100ms power-on delay. Grounding the ON pin enables software control of the FET. The state of the GPIO pins can be set before enabling the FET, ensuring a known state when downstream power is enabled. GPIO1 defaults high on power-up, and can sink 5mA. GPIO2 defaults high and can sink 3mA. GPIO3 defaults low and can sink 100μA.

For instance, Figure 249.1 shows an application that monitors a "request to remove card" pushbutton and lights an

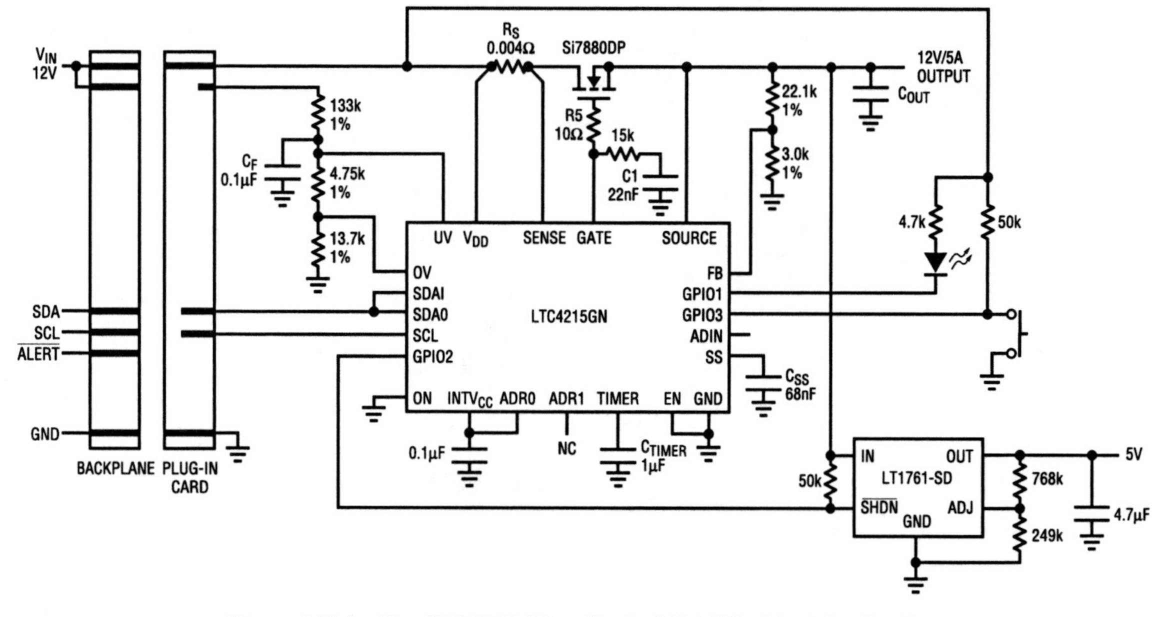

Figure 249.1 • The LTC4215-1 in a Typical Card Resident Application

Analog Circuit Design: Design Note Collection. http://dx.doi.org/10.1016/B978-0-12-800001-4.00249-0

"okay to remove" LED when the card is ready for removal. This permits graceful shutdown of the card. For example, it can transfer collected data before shutting down so that it is not lost. GPIO1, which defaults high, controls the LED. GPIO3 is reprogrammed as an input that monitors the state of the pushbutton. The GPIO2 pin controls the operation of an onboard regulator. This is important in mixed signal circuits, where analog circuitry may need to be powered up before digital signals are enabled.

Figure 249.2 uses a GPIO pin to control an LTC4210-1 Hot Swap controller, which in turn controls a 3.3V rail. Once again, this is useful for sequencing supplies and may eliminate the need for additional sequencing circuits.

Figure 249.3 uses all three GPIO pins to light one of eight LEDs using a 74HC138 decoder. These can indicate system status or power consumption. Other possible functions include issuing a microprocessor reset, adding additional channels to the ADC using the GPIO pins to control a multiplexer, or interfacing with an advanced power supply sequencer such as the LTC2928.

Conclusion

The LTC4215-1 is a smart power gateway for Hot Swap circuits. It provides fault isolation, closely monitors the health of the power path, and provides an unprecedented level of control over the inrush current profile. The three general purpose I/O pins and a spare ADC channel allow further control of power path and system initialization/shutdown related functions.

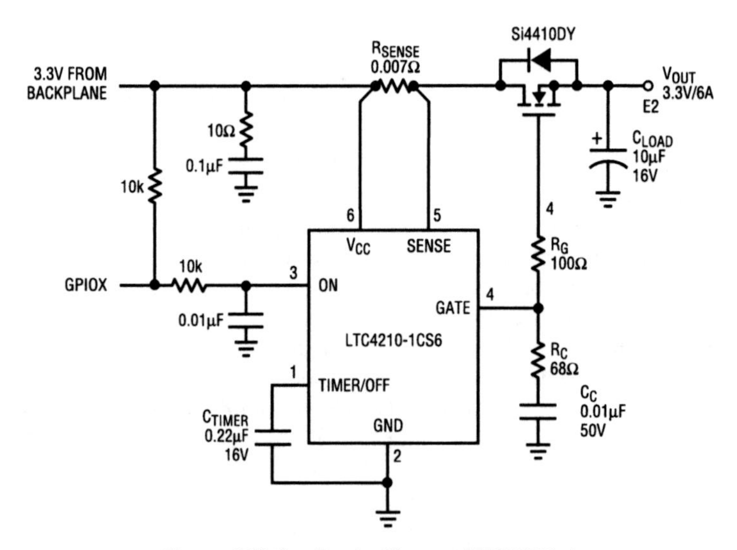

Figure 249.2 • Controlling an LTC4210-1

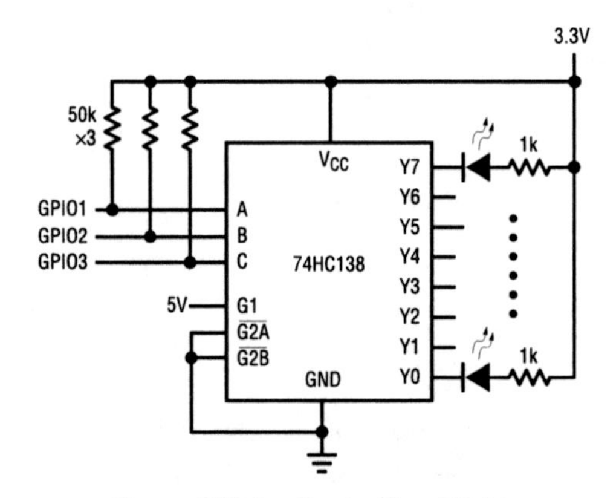

Figure 249.3 • Controlling Eight Status LEDs

Electronic circuit breaker in small DFN package eliminates sense resistor

250

S. H. Lim

Introduction

Traditionally, an electronic circuit breaker (ECB) comprises a MOSFET, a MOSFET controller and a current sense resistor. The LTC4213 does away with the sense resistor by using the $R_{DS(ON)}$ of the external MOSFET. The result is a simple, small solution that offers a significant low insertion loss advantage at low operating load voltage. The LTC4213 features two circuit breaking responses to varying overload conditions with three selectable trip thresholds and a high side drive for an external N-channel MOSFET switch.

Overcurrent protection

The SENSEP and SENSEN pins monitor the load current via the $R_{DS(ON)}$ of the external MOSFET and serve as inputs to two internal comparators—SLOWCOMP and FASTCOMP—with trip points at V_{CB} and $V_{CB(FAST)}$, respectively. The circuit breaker trips when an overcurrent fault causes a substantial voltage drop across the MOSFET. An overload current exceeding $V_{CB}/R_{DS(ON)}$ causes SLOWCOMP to trip the circuit breaker after a 16μs delay. In the event of a severe overload or short-circuit current exceeding $V_{CB(FAST)}/R_{DS(ON)}$, the FASTCOMP trips the circuit breaker within 1μs, protecting both the MOSFET and the load.

Both of the comparators have a common mode input voltage range from ground to $V_{CC}+0.2V$. This allows the circuit breaker to operate as the load supply turns on from 0V.

Flexible overcurrent setting

The LTC4213 has an I_{SEL} pin to select one of these three overcurrent settings:

I_{SEL} at GND, $V_{CB}=25mV$ and $V_{CB(FAST)}=100mV$
I_{SEL} left open, $V_{CB}=50mV$ and $V_{CB(FAST)}=175mV$
I_{SEL} at V_{CC}, $V_{CB}=100mV$ and $V_{CB(FAST)}=325mV$

Overvoltage protection

The LTC4213 can provide load overvoltage protection (OVP) above the bias supply. When $V_{SENSEP}>V_{CC}+0.7V$ for 65μs, an internal OVP circuit activates with the GATE pin pulling low and the external MOSFET turning off. The OVP circuit protects the system from an incorrect plug-in event where the V_{IN} load supply is much higher than the V_{CC} bias voltage.

Typical electronic circuit breaker (ECB) application

Figure 250.1 shows the LTC4213 in a dual supply ECB application. An input bypass capacitor is recommended to prevent transient spikes when the V_{IN} supply powers up or the ECB responds to overcurrent conditions. Figure 250.2 shows a normal power-up sequence. The LTC4213 exits reset mode once the V_{CC} pin is above the internal under voltage lockout threshold and the ON pin rises above 0.8V (see trace 1 in Figure 250.2). After an internal 60μs de-bounce cycle, the GATE pin capacitance is charged up from ground by an internal 100μA current source (see trace 2). As the GATE pin and the gate of MOSFET charges up, the external MOSFET turns

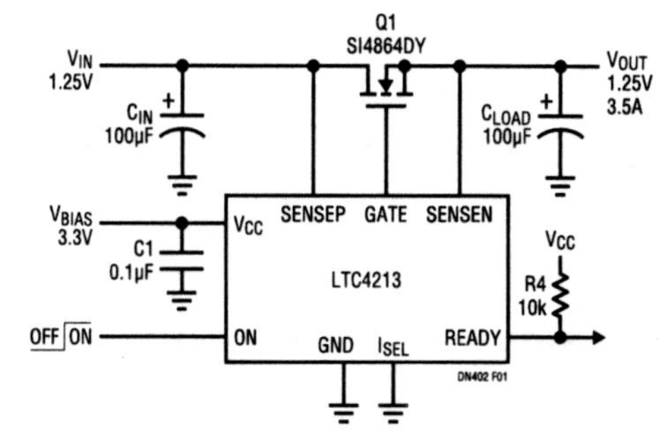

Figure 250.1 • The LTC4213 in an Electronic Circuit Breaker Application

Analog Circuit Design: Design Note Collection. http://dx.doi.org/10.1016/B978-0-12-800001-4.00250-7

on when V_{GATE} exceeds the MOSFET's threshold. The circuit breaker is armed when V_{GATE} exceeds ΔV_{GSARM}, a voltage at which the external MOSFET is deemed fully enhanced and $R_{DS(ON)}$ minimized. Then, 50μs after the circuit breaker is armed, the READY pin goes high (see trace 3) and signals the system to power up V_{IN}. Trace 4 shows the related V_{OUT} waveform when V_{IN} powers up. In order to not trip the circuit breaker during start-up, the load current must be lower than V_{CB}/R_{SENSE}. If needed, the I_{SEL} pin can be stepped dynamically for a higher overcurrent threshold at start-up and a lower threshold when the load current has stabilized.

Accurate ECB with sense resistor

The $R_{DS(ON)}$ voltage drop sensing method trades the circuit breaker accuracy for system simplicity. The majority of sensing inaccuracy is due to the external MOSFET's $R_{DS(ON)}$ varied by operating temperature and under different V_{GS} bias condition. The MOSFET vendors also do not specify the $R_{DS(ON)}$ distribution tightly due to manufacturing variation. If an external tight tolerance resistor is employed for current sensing instead, the LTC4213 reveals its ±10% circuit breaker accuracy. Figure 250.3 shows a tolerable R_{SENSE} resistor voltage drop and the LTC4213 is used for accurate ECB applications.

High side switch for N-channel logic level MOSFET

Logic level N-channel MOSFET applications usually requires a minimum gate drive voltage of 4.5V. Figure 250.4 shows the LTC4213 in a high side switch application. The LTC4213's internal charge pump boosts the GATE above the logic level gate drive requirement and ensures the MOSFET is fully enhanced for $V_{CC} \geq 3V$. The typical gate drive versus bias supply voltage curve is shown in Figure 250.5.

Conclusion

The LTC4213 is a small package, No R_{SENSE} electronic circuit breaker that is ideally suited for low voltage applications with low MOSFET insertion loss. It includes selectable dual current level and dual response time circuit breaker functions. The circuit breaker has wide operating input common-mode-range from ground to V_{CC}.

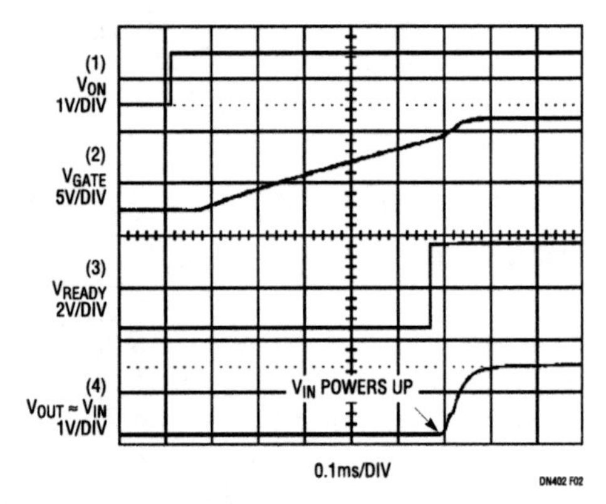

Figure 250.2 • Normal Power-Up Sequence

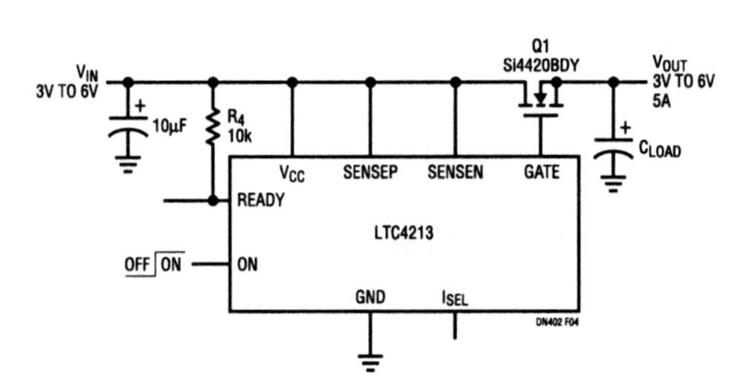

Figure 250.4 • High Side Switch for Logic Level N-Channel MOSFET, $V_{CC} > 3V$

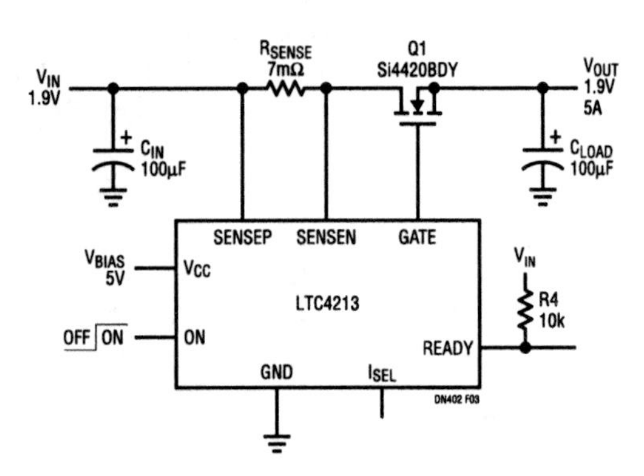

Figure 250.3 • Accurate ECB with High Side Sense Resistor

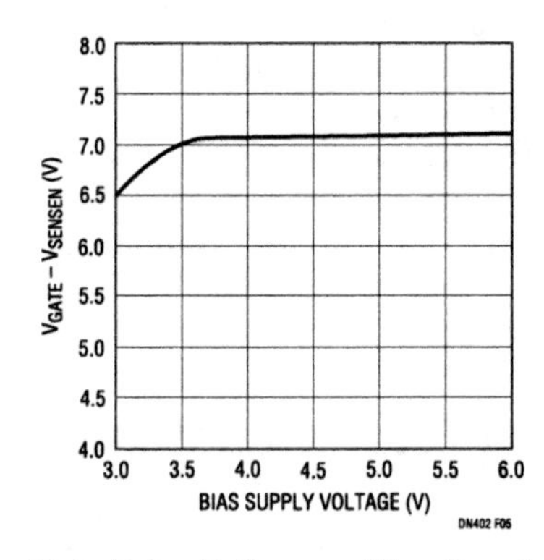

Figure 250.5 • Gate Drive Voltage vs Bias Supply Voltage

AdvancedTCA Hot Swap controller monitors power distribution

251

Mitchell Lee

Introduction

AdvancedTCA is a modular computing architecture developed by the PCI Industrial Computer Manufacturers Group for use in central office telecom environments. PICMG 3.0 defines, among other things, the electrical and mechanical attributes of the backplane, connectors and removable cards in these −48V systems.

Each removable card, or front board, is designed for live insertion into a working system. A power draw of up to 200W per front board is allowed, placing the maximum load current in the 4A to 5A range.

Card-centric inrush limiting and quantitative current and voltage monitoring are highly desirable to sanitize the incoming battery feeds, minimize power plane disturbances, allow for budgeting power consumption and permit failure prediction in an otherwise functional system. The LTC4261 Hot Swap controller provides these features. Also included is a digital interface for controlling the functions of the LTC4261, and for reading the current and voltage measurement registers.

Circuit solutions

Figure 251.1 shows a complete circuit designed to handle up to the maximum available power. The LTC4261's accurate current limit is set to provide at least 5.5A under all conditions, a comfortable margin for 200W, yet trips off just under 7A to preserve fuse integrity in the presence of nuisance overloads. At insertion the LTC4261 allows contact bounce to settle, then soft-starts the load using a ramped current. Inrush current is increased gradually to a few hundred milliamperes and held there until the MOSFET is fully on.

Current is monitored by the SENSE pin and an 8mΩ shunt resistor. Direct measurement of the current is possible via the I^2C port, with 10-bit resolution and 8A full scale.

Cutting diode dissipation

ATCA's redundant −48V power feeds are combined on-card with ORing diodes. At 5A current consumption even Schottky rectifiers present a serious problem in terms of both voltage drop and power dissipation: a conducting pair drop more than 1V and dissipate 6W. Following the diode manufacturer's recommendations, 8 square inches of board area are needed to satisfy the heat sinking requirements.

Diode dissipation, voltage loss and board area is reduced in Figure 251.1 by using MOSFETs as active rectifiers with the LTC4354 diode OR driver. Total dissipation is cut to less than 1W for two conducting "diodes" at maximum load.

Zero Volt Transient

The so-called zero volt transient requirement is a legacy of earlier telecom equipment standards stipulating uninterrupted system operation during the course of a 5ms input voltage dropout. An energy of 1J is needed to sustain a 200W load during this interval.

The accepted method of energy storage to satisfy the 1J requirement is a bulk reservoir capacitor which is charged through resistors. This technique dictates the use of bulky high voltage storage capacitors, such as 100V (or rare 80V) rated units which can handle the maximum input voltage of 75V. Since the zero volt transient test commences at 44V, nothing is gained by storing a higher voltage. Compact 50V capacitors are used instead, by limiting the charging voltage with a simple zener-transistor circuit.

The ATCA connector pin configuration presents a special design challenge. Here extraction is inferred from the difference between each ENABLE and its associated VRTN, thereby ignoring input dropouts. A PNP transistor pulls up on $\overline{EN}$ in the event of an ENABLE disconnect, shutting down the LTC4261 and permitting safe extraction with no connector damage. During a zero volt transient, no signal reaches the $\overline{EN}$ pin; power flows uninterrupted to the load when the input voltage recovers.

Analog Circuit Design: Design Note Collection. http://dx.doi.org/10.1016/B978-0-12-800001-4.00251-9

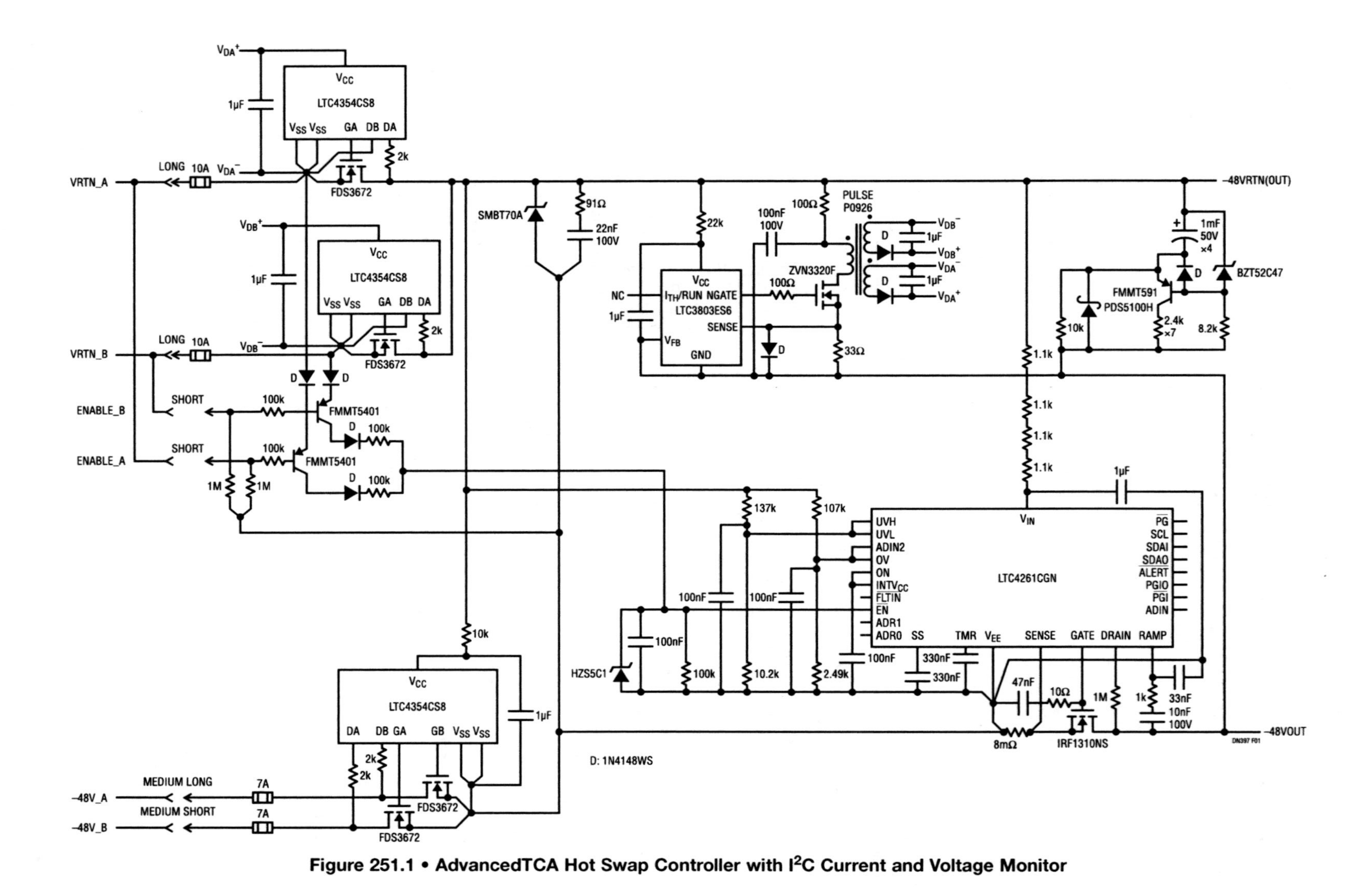

Figure 251.1 • AdvancedTCA Hot Swap Controller with I²C Current and Voltage Monitor

Protecting and monitoring hot swappable cards in high availability systems

252

David Soo

Introduction

Mission critical systems, such as those used by financial institutions and internet providers, require high availability computing equipment. These systems duplicate or make redundant certain functional blocks in the system. This allows the system to operate on the backup elements when blocks fail. High availability systems typically place these blocks into separate cards that can be swapped out if they fail or show signs of trouble. This requires sophisticated monitoring to determine when a functional block has failed and if a back-up is operating.

Redundant power

Often the reliability of the power distribution is improved by providing two supplies to each card. Each supply is fuse protected and then diode OR'ed to form one supply rail for the card. To further increase reliability, the voltage and current information is monitored for power usage and usage history. This allows the system to determine if the card is using its allotted power or operating abnormally. Cards operating at abnormal currents are likely sources of failure and could be flagged for service before something actually fails. Such information can also be monitored remotely by a maintenance provider. In addition to measuring the board current and voltages, the fuses are monitored for faults. This information warns the user that a blown fuse has compromised the redundant power path to the card.

Monitoring power through a Hot Swap controller

Before considering the method for current and voltage monitoring, we need to first discuss the function of a Hot Swap controller. For the sake of discussion, let's assume the system

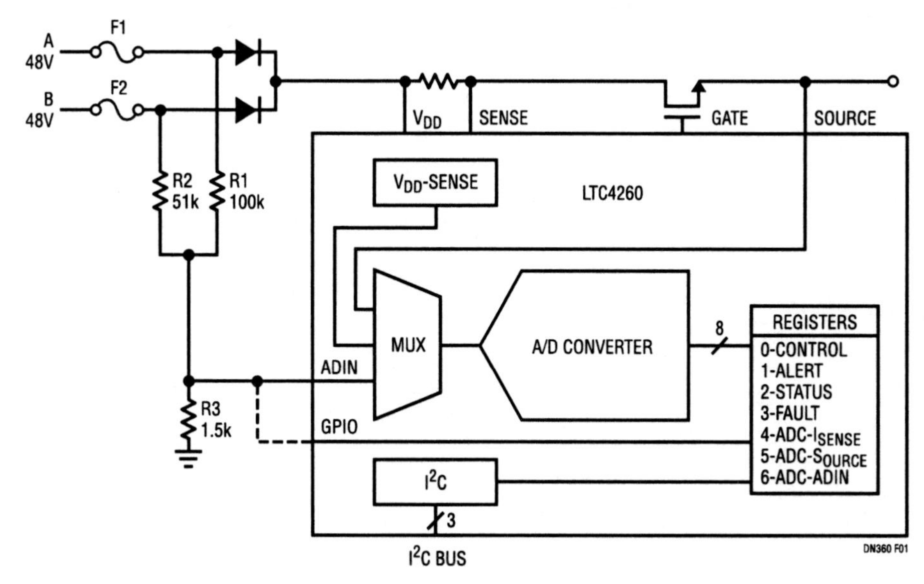

Figure 252.1 • Fuse Monitoring Using the LTC4260

Analog Circuit Design: Design Note Collection. http://dx.doi.org/10.1016/B978-0-12-800001-4.00252-0

backplane provided two well regulated 48V (±5%) supplies to each card. One 48V supply provides power in the event the other supply fails. This requires that upon failure the bad supply is removed and then replaced while the system remains powered. Installing a new supply could lead to a 10% step up in voltage if the lower supply is augmented with a new supply at the high end of the 48V ±5% range. Such a high voltage step would cause large inrush currents in downstream capacitors on the cards. Similarly, large inrush currents occur when an unpowered card is inserted into a live backplane.

A Hot Swap controller uses a power switch and sense resistor in series with the power rail to eliminate this inrush current. When the current through the sense resistor reaches a limiting threshold, the Hot Swap controller uses the power switch to limit the current. This is useful during power-up and also during fault conditions such as the voltage step described above. The current limit feature develops a controlled rise in the voltage across the downstream card capacitance and protects the backplane power bus against disturbances.

The power switch with sense resistor also acts as a gateway for power to the card and it becomes an ideal location to monitor and collect power supply data. Recognizing this, Linear Technology developed the LTC4260, a Hot Swap controller with power monitoring features. Included in the LTC4260 is an 8-bit ADC capable of measuring the current via the voltage drop across the sense resistor. It also measures the voltage at the load side of the power switch and the voltage at an external general purpose ADC input pin. The part detects faults in the power path such as overcurrent, input overvoltage or undervoltage, output undervoltage and a shorted power switch. The LTC4260 records the occurrence of faults plus the three ADC measurements in registers that are accessible via an I^2C bus.

Adding fuse detection

If a fuse in series with an ORing diode opens, then one of the supplies is lost. The system continues to operate on the remaining supply but without the assurances of redundancy. Additional circuitry is required to detect an open fuse in this configuration.

Adding fuse detection could be accomplished a couple of ways. An external resistor taps off between the fuses and diode to connect to a common point. That common point is terminated through a grounded resistor and feeds the GPIO pin. The GPIO pin is used as a general purpose digital input that is accessible through the I^2C bus. If a fuse is blown, the GPIO pin voltage drops. The resulting logic low indicates one of the redundant paths is open.

Of course it would be better to know *which* fuse failed. The solution to this problem uses the general purpose ADC input available on the LTC4260, as shown in Figure 252.1. By sizing the pull-up resistor to one fuse different from the pull-up to the other fuse, it is possible to detect which fuse is open-circuit. When both fuses F1 and F2 are present, the equivalent resistance of R1 in parallel with R2 is 34k. The R1||R2, R3 resistive divider results in a voltage of 2.04V on the ADIN pin. If one of the fuses fails, the voltage at the common point connected to the ADIN pin drops to a value set by the remaining resistive divider to the remaining 48V supply. If fuse A fails then the R2/R3 voltage divider yields 1.37V on the ADIN pin. Likewise an open on fuse B pulls the pin to 0.71V.

Summary

In order to increase the reliability of high availability systems, it is prudent to monitor the health and integrity of the power distribution network. These systems commonly use a Hot Swap controller to limit inrush current to cards. Since all card power flows through the inrush control circuitry, it is the natural place to add this monitoring capability. In addition to monitoring power, these circuits can also be used to check the integrity of redundant supply networks.

AdvancedTCA Hot Swap controller eases power distribution

253

Mitchell Lee

Introduction

AdvancedTCA is a new modular computing architecture developed by the PCI Industrial Computer Manufacturers Group for use in central office telecom environments. PICMG 3.0 defines, among other things, the electrical and mechanical attributes of the backplane, connectors and removable cards in these systems.

System power is supplied by the −48V dual battery feed typical of telecom installations, and ATCA borrows many of its related specifications from established telecom standards.

Power requirements

Each removable card or front board, is designed for live insertion into a working system. A power draw of up to 200W per front board is allowed, placing the maximum load current in the 4A to 5A range.

As is common in these types of systems, card-centric inrush limiting and current and voltage monitoring are highly desirable to sanitize the incoming battery feeds and minimize power plane disturbances. The LTC4252A Hot Swap controller is a good match for −48V, 0W to 200W applications.

Circuit solutions

Figure 253.1 shows a complete circuit designed to handle up to the maximum available power. The LTC4252A's accurate current limit is set to provide at least 5.5A under all conditions, a comfortable margin for 200W, yet trips off just under 7A to preserve fuse integrity in the presence of nuisance overloads.

Over- and undervoltage monitoring are both implemented in this circuit. UV thresholds are set at −37V turning on and −33.3V turning off, as measured after the ORing diodes. OV turns off at −74.7V and back on at −73.2V, again as measured after the ORing diodes. This assures operation over

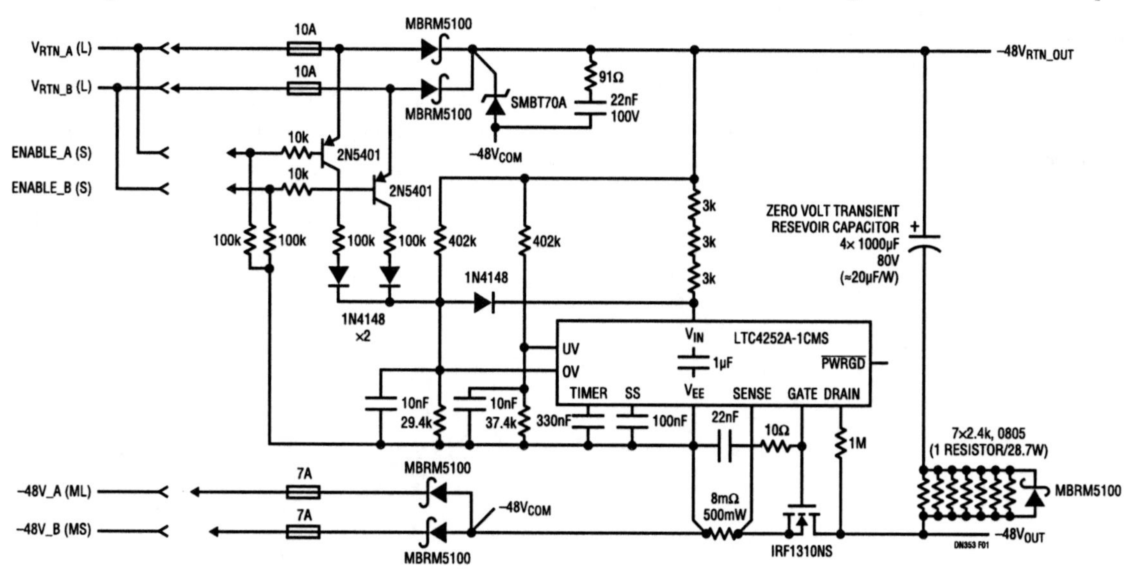

Figure 253.1 • 200W AdvancedTCA Hot Swap Controller Circuit

Analog Circuit Design: Design Note Collection. http://dx.doi.org/10.1016/B978-0-12-800001-4.00253-2

the full range of −43V to −72V, as well as under conditions of input surge to −75V and transients to −100V in accordance with ATCA specifications.

Once insertion is detected, the LTC4252A pauses for 230ms to allow for contact bounce, then soft-starts the load using a ramped current scheme. Inrush current is increased gradually until the MOSFET is fully on.

Current overloads detected by the SENSE pin and 8mΩ shunt are dealt with by three distinct levels of response. If a small, sustained overload of 7A or more is detected, the TIMER pin delays for 5.7ms before shutting down. If the overload exceeds 7.5A the LTC4252A throttles back the MOSFET and holds the current to that value. Again, after a 5.7ms delay the circuit shuts down. If the overload is severe, a strong and very fast amplifier quickly corrects the MOSFET gate voltage, bringing it down to near threshold for the device. Then the LTC4252A's current limit circuit takes over and maintains 7.5A for the duration of the 5.7ms TIMER delay period.

The LTC4252A also monitors the voltage drop across the MOSFET and reduces the TIMER delay to as little as 1.8ms as the voltage stress increases. This keeps the MOSFET comfortably within its safe operating area in the presence of hard faults.

Zero Volt Transient

The so-called zero volt transient requirement is a legacy of earlier telecom equipment standards stipulating uninterrupted system operation during the course of a 5ms input voltage dropout. An energy of 1J is needed to sustain a 200W load during this interval.

In addition to the energy storage requirement, the connector pin configuration presents a special design challenge. Front board insertion is detected by two short pins ENABLE_A and ENABLE_B. Rather than simply looping through the backplane, these pins connect to V_{RTN_A} and V_{RTN_B} which complicates their use for insertion detection.

If ENABLE_A and ENABLE_B were used to detect insertion by directly driving the LTC4252A's UV pin resistors, the MOSFET would turn off at the first sign of a voltage drop, regardless of the cause. The LTC4252A would initiate a new start-up cycle when the input was restored, complete with debounce and soft-start. If recovering from a zero volt transient, the 1J storage mechanism would deplete long before the MOSFET recovered, interrupting front board operation.

Here extraction is inferred from the difference between each ENABLE and its associated V_{RTN}, thereby ignoring input dropouts. A PNP transistor pulls up on OV in the event of an ENABLE disconnect while V_{RTN} is powered, shutting down the LTC4252A and permitting safe extraction with no connector damage. In contrast the base and emitter terminals remain shorted during a zero volt transient; no signal reaches the OV pin and the MOSFET remains on. Power flows uninterrupted to the load when the input voltage is restored.

Energy storage

The accepted method of energy storage to satisfy the 1J requirement is a bulk reservoir capacitor which is charged through resistors. Several benefits arising from the use of the resistors are not immediately obvious. First, transient overvoltages are blocked from the capacitors allowing the use of diminutive, 80V rated units. Second, an energy equal to that stored in the capacitors is dissipated in the charging path. If Q1 alone shouldered this burden a much larger and more expensive high SOA device would be necessary. Instead this charging energy is dissipated in a few resistors.

Computing energy

The bulk storage reservoir capacitance is calculated knowing a minimum input voltage at the start of a zero volt transient (specified as −43V at the input of the front board), decaying to a worst-case UV detection voltage of −34V. Diode losses result in a load voltage of −41V decaying to −34V over the 5ms dropout period.

In reckoning the necessary capacitance, a common mistake is to apply the familiar $E=(1/2)CV^2$ energy equation, but using ΔV for the voltage term. This is incorrect and leads to an alarmingly big capacitance of over 40,000μF. Fortunately the value is far smaller and using the correct formula:

$$E = \frac{1}{2}C\left(V1^2 - V2^2\right) \qquad (1)$$

is found to be just 3,800μF, for 1J of available energy. Reduce this value in direct proportion to the load in lower power applications.

PCI Express power and MiniCard solutions

254

Mitchell Lee Vladimir Ostrerov

Introduction

PCI Express is the third generation of PCI (Peripheral Component Interconnect) technology used to connect I/O peripheral devices in computer systems. It is intended as a general purpose I/O device interconnect that meets the needs of a wide variety of computing platforms such as desktop, mobile, server and communications. It also specifies the electrical and mechanical attributes of the backplane, connectors and removable cards in these systems.

Power requirements

Each add-in card connector requires two power rails: 12V and 3.3V, along with a third, optional 3.3V auxiliary rail. If the designer elects to provide the auxiliary 3.3V rail in his system or if the platform supports the WAKE# signal, the $3.3V_{AUX}$ rail must be supplied to all connectors.

In PCI Express, as in PCI, the Hot Swap controller is resident on the platform side of the system, delivering power to the connectors only after a card is inserted. This way, empty and otherwise unused connectors are safely unpowered, reducing the risk of inadvertent power supply faults, connector damage and simplifying power-up when a card is inserted into the connector.

Table 254.1 shows the PCI Express power supply rail specifications per connector, based on the number of connectors in the system.

PCI Express Mini Card also provides two power sources: 3.3V and 1.5V. The auxiliary voltage ($3.3V_{AUX}$) is sourced over the same pins as the primary 3.3V supply and is available during the standby/suspend state to support wake-event processing on the Mini Card. Table 254.2 summarizes PCI Express Mini Card power specifications.

There is no requirement for supply sequencing in either PCI Express or PCI Express Mini Card; the supplies may come up or go down in any order.

Table 254.1 Summary of PCI Express Power Supply Requirements

POWER RAIL	×1 CONNECTOR	×4/×8 CONNECTOR	×16 CONNECTOR
12V			
Supply Current	0.5A	2.1A	4.4A (up to 5.5A)
Capacitive Load	300μF	1000μF	2000μF
3.3V			
Supply Current	3A	3A	3A
Capacitive Load	1000μF	1000μF	1000μF
$3.3V_{AUX}$			
Supply Current	0.375A	0.375A	0.375A
Capacitive Load	150μF	150μF	150μF

Table 254.2 Summary of PCI Express Mini Card Supply Requirements

POWER RAIL	PRIMARY POWER	AUXILIARY POWER
3.3V		
Peak Supply Current	1A	–
1.5V		
Peak Supply Current	0.5A	–
$3.3V_{AUX}$		
Peak Supply Current	0.33A	0.25A

Note: All values are maximums.

Analog Circuit Design: Design Note Collection. http://dx.doi.org/10.1016/B978-0-12-800001-4.00254-4

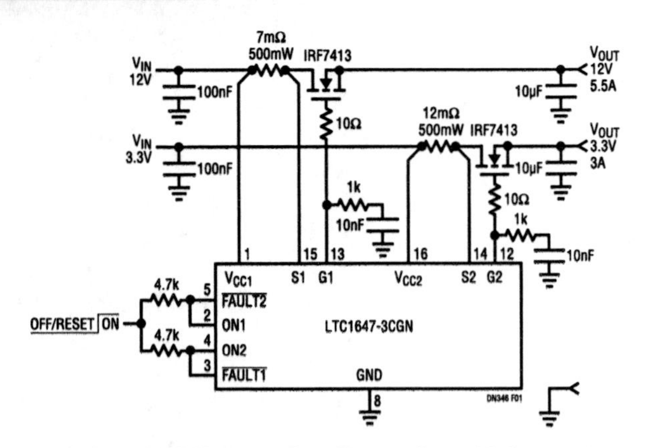

Figure 254.1 • PCI Express Port Power for ×16 Connector Satisfies dV/dt Specifications, Soft-Starts Full Load Capacitance and Guards against Short Circuits and Overloads

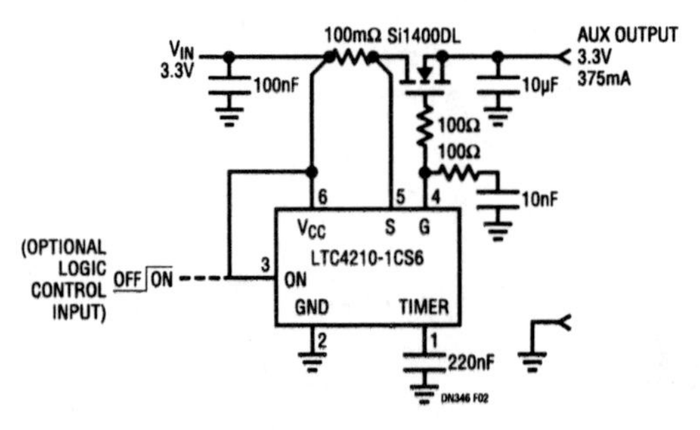

Figure 254.2 • Optional 3.3V$_{AUX}$ Port Power with Self-Resetting, Current-Limited Circuit Breaker Operates Autonomously or Under Logic Control

Circuit solutions

Figure 254.1 shows a circuit for a ×16 connector PCI Express system, implemented with an LTC1647 dual Hot Swap controller. As shown, port power is gated by a control signal applied through 4.7k resistors to the ON pins. The $\overline{\text{FAULT}}$ outputs are cross-coupled with the ON pins such that if an overcurrent fault occurs on either supply, the unaffected rail also shuts down. Shutdown is reset by toggling the control signal into the off state for at least 100μs.

The supply rails slew at less than 1.5V/ms, well under any of the PCI Express limits. Thus the only alteration necessary to scale the design for ×1, ×4 and ×8 connector systems is to replace the 12V supply sense resistor with 70mΩ (×1) or 15mΩ (×4/×8) as appropriate.

Figure 254.2 shows a current limited circuit breaker for the optional 3.3V$_{AUX}$ rail, using the LTC4210 Hot Swap controller in a 6-lead ThinSOT package. With the ON pin (3) hard-wired to 3.3V, the LTC4210 operates autonomously, featuring a timed current limit to protect against overloads. Should the circuit breaker trip in the presence of a sustained fault, the LTC4210 automatically restarts after a time delay, requiring no intervention on the part of the PCI Express port manager.

PCI Express Mini Card

Mini Card circuit requirements differ from PCI Express in several ways, but most importantly the Hot Swap controller resides on the card itself. In this configuration, it is important for the Hot Swap controller to monitor the incoming power and switch on only when those voltages are fully established.

Figure 254.3 shows a Mini Card Hot Swap control circuit implemented with an LTC1645, with undervoltage and overcurrent monitoring on both the 3.3V and 1.5V supplies. A fault on either supply rail causes both channels to shut down simultaneously. An open-drain signal at the ON pin permits optional logic control over the LTC1645. Similar to the LTC1647, pulsing ON low resets the LTC1645 after a fault.

For 3.3V$_{AUX}$, simply add the circuit of Figure 254.2 in parallel at the 3.3V output. Note that with both circuits enabled and providing power to the 3.3V output, the available current is greater than 1.5A.

A complete definition of the PCI Express power delivery requirements is found in the PCI Express Card Electromechanical Specification and PCI Express Mini Card Electromechanical Specification. Further information is available at www.pcisig.com.

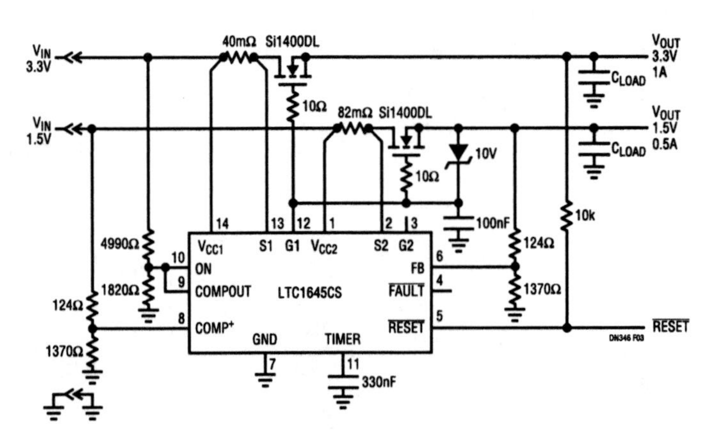

Figure 254.3 • PCI Express Mini Card Hot Swap Circuit Resides on Mini Card

Low voltage Hot Swap controller ignores backplane noise and surges

255

Mitchell Lee

First generation single supply Hot Swap controllers, such as the LTC1422, combine the features of various discrete inrush limiting circuits into a single IC, including an electronic circuit breaker, adjustable power-up rate, reset output, high-side MOSFET gate drive, undervoltage lockout and a wide operating voltage range. The LTC4211 advances these features by adding a dual level, dual response time electronic circuit breaker, adjustable soft-start with inrush current limiting, fault detection, faster response time to severe overloads and operation from 2.5V to 16.5V—all in a small 10-lead MSOP package.

Control 25W with a 10-lead MS package

Figure 255.1 shows a 5V, 5A Hot Swap application designed for inclusion on a removable circuit board. As is customary, the circuit board ground and V_{CC} planes are wired to long connector pins, eliminating reliability problems associated with pin-to-receptacle arcing. Until the PC board is firmly seated in the backplane, the MOSFET switch remains safely off and isolates C_{LOAD} (typically 100µF or more) from the backplane. Once the connector is fully seated, the R1-R2 divider which serves both as a connection sense and an undervoltage lockout, activates the LTC4211. When activated, the LTC4211 waits one TIMER pin period as set by C_{TIMER} and then turns on the MOSFET. $\overline{RESET}$ signals a successful start-up as measured by divider R3-R4 and the FB pin.

Dual level current control

Advanced current control features separate the LTC4211 from other Hot Swap controllers. The SENSE pin monitors load current and shuts off the MOSFET in the event the current exceeds $50mV/R_{SENSE}$. Instead of using a simple circuit breaker approach, the SENSE pin has two thresholds: a slow 50mV trip point whose timing is governed by C_{FILTER} at the FILTER pin, and a fast 150mV (3×) trip point which triggers in just 300ns to interrupt the flow of current in the event a catastrophic fault occurs on the output. Thus, the LTC4211 ignores temporary surges and overloads but responds quickly when a genuine fault is detected.

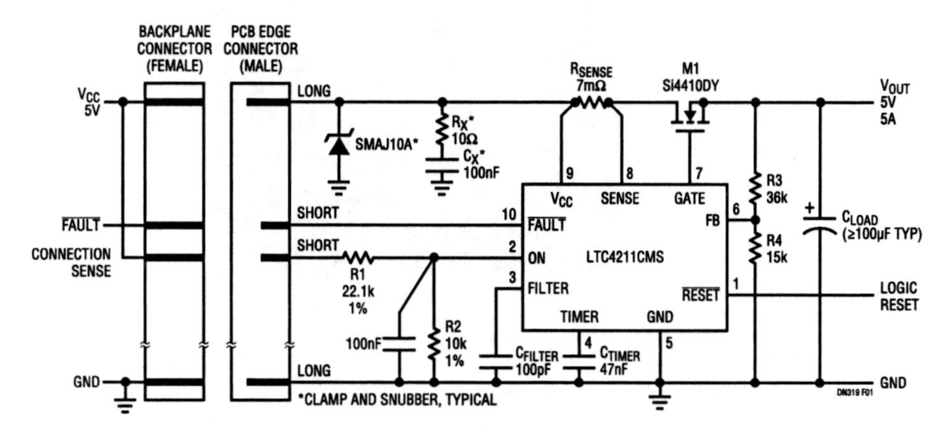

Figure 255.1 • Single Channel 5V Hot Swap Controller in 10-Lead MS with Dual Level Current Control

Analog Circuit Design: Design Note Collection. http://dx.doi.org/10.1016/B978-0-12-800001-4.00255-6

Inrush limiting

Conspicuous by its absence in Figure 255.1 is a gate capacitor which might otherwise define the ramp rate and therefore the inrush current of the output during start-up. Instead, the 50mV SENSE pin threshold is used to servo the inrush current to a value of 50mV/R_{SENSE}. The 50mV threshold circuit breaker function is suspended during this critical period, but the 150mV SENSE threshold remains active to catch catastrophic faults. Once the soft-start period is over, the 50mV circuit breaker is armed.

There are several advantages to dispensing with the usual gate capacitor including the elimination of an external component and the elimination of the turn-off delay inherent in discharging that capacitor during a fault. Even so, the LTC4211 is not restricted to using its internal current control mode at start-up—an external capacitor can still be employed in the ordinary way, integrating the 10μA GATE pin current to produce a well-controlled soft-start ramp. An 8-pin version, the LTC4211CS8 is available as a backwards-compatible upgrade to the LTC1422CS8 in applications demanding current limited start-up.

Adaptive response to overloads

During start-up, the LTC4211 operates in the aforementioned 50mV/R_{SENSE} current limit mode (regardless of whether a gate capacitor is included in the design) at any time the 50mV threshold is exceeded. For example, if the LTC4211 attempts to start up into a short circuit, the current is first limited to 50mV/R_{SENSE} and then cut off after the expiration of one TIMER pin cycle.

After successful start-up, the timed circuit breaker function takes over. The 50mV threshold still applies, but instead of instantly tripping on an overcurrent condition, the FILTER circuit delays turn-off, thereby rejecting temporary surges and spikes. This prevents minor backplane disturbances from interrupting delivery of power to critical subsystems and memory.

Figure 255.2 illustrates the action of the FILTER pin. An overload of approximately 8A is drawn from the 5V output which exceeds the 50mV/7mΩ SENSE pin threshold. C_{FILTER} delays shutdown 100μs. A shorter duration overload would have been rejected. Extreme overloads must be recognized and cleared immediately to prevent collateral damage. The 150mV SENSE pin threshold reacts to these overloads in just 300ns, bypassing the FILTER pin delay.

Figure 255.2 also shows the importance of input clamping in high current applications. Readily recognized is the dI/dt dip at the input, V_{CC}, precipitated by the ≈1A/μs load current slew rate coupled with a backplane/wiring harness inductance of nearly 3μH. There is no input bypassing. Upon commutation of the 8A load current, a potentially destructive consequence of the inductive feedpoint impedance rears its ugly head: a voltage spike limited only by the input clamp. The LTC4211's high absolute maximum V_{CC} rating of 17V eases selection of an appropriate clamp, as these devices tend to have wide tolerances.

Recovery from faults

Once the circuit breaker has tripped, for whatever reason, the LTC4211 safely latches off and helps guard against any damage to the MOSFET or affected circuitry. The $\overline{FAULT}$ pin then alerts the system controller for further action. The circuit breaker can be reset by either cycling the ON pin low under microprocessor control or by allowing the chip to reset itself (by tying $\overline{FAULT}$ back to the ON pin).

$\overline{FAULT}$ also serves as an input. If $\overline{FAULT}$ is pulled low externally by another open drain logic signal, the LTC4211 circuit breaker trips and turns off the output. This feature allows multiple supplies to shut down simultaneously when a short circuit occurs on any one output. The $\overline{FAULT}$ pin of each LTC4211 is wired to a common point, so that a fault on one is communicated to the others.

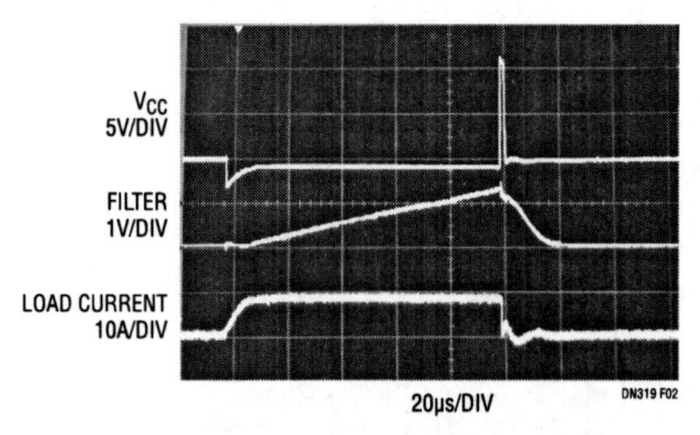

Figure 255.2 • Temporary Surges and Spikes Are Rejected during the Time Set by the FILTER Pin. A Persistent Overload of ~8A Trips the Circuit Breaker After 100μs FILTER Delay

Hot Swap circuit meets InfiniBand specification

256

Pat Madden

An InfiniBand backplane distributes bulk power at 12V DC to a number of plug-in modules, each having a DC/DC converter that steps the 12V supply down to the voltages required locally. A pass transistor on each module isolates the converter's input capacitors during module plug-in, allowing modules to be added without disturbing others on the backplane. The requirements governing module hardware design are found in the *InfiniBand Architecture Specification, Volume 2, Release 1.0.* This article describes a circuit fully implementing the specification for sequencing bulk power. The power conversion problem is not discussed here.

Figure 256.1 shows the circuit. Power is supplied from the backplane at 12V DC between VB_In and VB_Ret. There are two control signals, VBxEn_L (active low) and Local Power Enable (active high). The LTC1642 Hot Swap controller IC performs four functions in this circuit. It controls pass transistor Q1's gate in response to both signals; it limits Q1's drain current; it monitors VB_In for both overvoltage and undervoltage conditions; and starts the DC/DC converter downstream. The LTC1642 was selected because it can respond to two control inputs and its voltage range suits the application's requirements. Although the backplane supplies 12V nominally, the InfiniBand specification requires steady-state operation to 14V DC and up to 16V for 1ms transients; this is satisfactory because the LTC1642 is fully specified for operation up to 16.5V. Its 33V maximum rating is an added benefit: capacitive bypassing at VB_In is limited to 5nF by the specification, so high frequency ringing to voltages

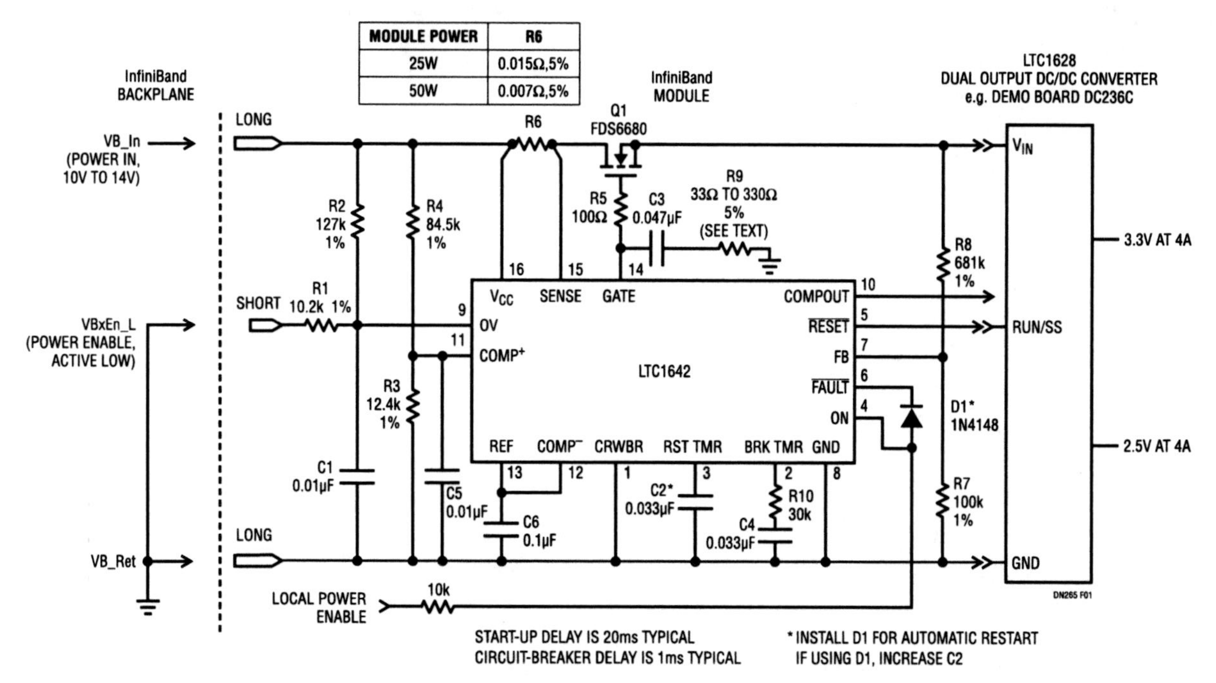

Figure 256.1 • LTC1642 InfiniBand Hot Swap Circuit

Analog Circuit Design: Design Note Collection. http://dx.doi.org/10.1016/B978-0-12-800001-4.00256-8

substantially higher than 16V after "hot" module removal or a load short is to be expected.

During module plug-in, the circuit operates as follows. The longest pins, VB_In and VB_Ret, mate first with the backplane. The LTC1642's OV (overvoltage) pin is pulled high by R2, holding Q1 off. OV is pulled low by R1 when the shorter VBxEn_L pin mates with the backplane, where it is shorted to VB_Ret. VBxEn_L is only asserted if the voltage at OV is less than 1.223V (±1%), so Q1 remains off if VB_In exceeds 16V DC; C1 filters out ringing. The chip also monitors VB_In for undervoltage conditions through the R3–R4 divider, holding the COMPOUT pin low when VB_In is less than 10V. COMPOUT can be tied to the ON pin to lock out undervoltages on VB_In. Local Power Enable is driven by an InfiniBand Management Device, powered separately from a 5V auxiliary, also supplied by the backplane. It is asserted at voltages greater than 1.223V (±1%). Tie it high through a 10k current limiting resistor if there is no management device on the board. After assertion of both control signals, there is a delay of 20ms, which is proportional to C2, before Q1's source begins ramping. The ramp rate, 0.5V/ms (inversely proportional to C3), safely limits the charging current drawn by the DC/DC converter's input capacitors. R5 prevents self-oscillations in Q1, which operates as a source follower while ramping. The R7–R8 divider monitors Q1's source voltage, with the $\overline{RESET}$ output holding the DC/DC converter off until Q1's source voltage reaches 10V.

R6 senses Q1's drain current. When its voltage drop reaches a limiting value, a control loop in the LTC1642 servos Q1's gate voltage to prevent further increases in drain current. The loop regulates in 5μs to 10μs after activation, safely within the 20μs specification. R9 works with C3 to compensate this loop. Its value depends on the parasitic inductance in series with V_{CC}. If it's less than 1μH, use 330Ω; if more, use 33Ω. Q1's maximum allowed drain current increases as its source voltage, sensed by the R7–R8 divider, increases. The voltage across R6 is limited to 23mV when the LTC1642's FB pin is grounded, increasing gradually to 53mV when FB is 1V or more. This "foldback" current limiting tends to equalize Q1's power dissipation during "hard" and "soft" shorts. If Q1 remains in current limit too long, it is latched off to keep it from overheating, and the $\overline{FAULT}$ output is asserted. The delay before latch-off is 1ms, proportional to C4. R10 is included to ensure that Q1 eventually latches off after repetitive, transient current limiting individually lasting less than 1ms. Optional diode D1 is particularly useful on boards lacking an InfiniBand Management Device; it automatically restarts the LTC1642 if it should latch off. Increase C2 if using it, to increase Q1's off time and keep it from overheating during repeated start-up cycles. To restart the circuit if D1 is omitted, either an InfiniBand Management Device must detect the fault and toggle Local Power Enable or the bulk power to the module must be cycled off then on.

The circuit was tested with a 50W DC/DC converter, LTC1628. Figure 256.2 shows an oscillograph captured during module plug-in. The delay from VBxEn_L assertion to the converter's output coming into regulation is 73ms, safely within the 500ms InfiniBand specification.

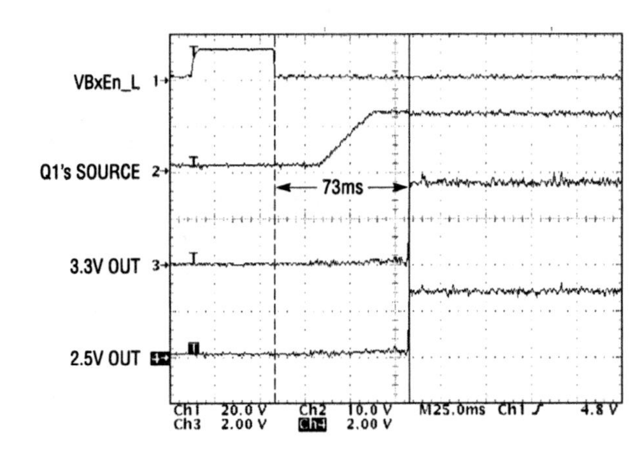

Figure 256.2 • Waveforms during InfiniBand Module Plug-In

Hot Swap and buffer I²C buses

257

John Ziegler

As server systems have grown, the number and complexity of input/output (I/O) cards that contain control circuitry to monitor the servers have grown in proportion. Zero down time systems require users to insert I/O cards into a live backplane. While many IC vendors have developed chips to safely Hot Swap the power supply and ground lines, until now no one has developed a monolithic solution to "hot swap" the system data (SDA) and clock (SCL) lines in I²C and SMBus systems. Expansion of these systems have made rise and fall time specifications difficult to meet, as the SDA and SCL capacitances of each I/O card add directly to those of the backplane. The LTC4300-1 allows the user to plug I/O cards into a live backplane without corrupting the data transaction taking place on the backplane. It also provides bidirectional buffering, keeping the backplane and card capacitances isolated.

Figure 257.1 shows an application of the LTC4300-1 that safely hot swaps the SDA and SCL lines. The LTC4300 resides on the edge of a peripheral card, with the SCLOUT pin connected to the card's SCL bus, and the SDAOUT connected to the card's SDA bus. When the card is plugged into a live backplane via a staggered connector, ground makes connection first, followed by V_{CC}.

After ground and V_{CC} connect, SDAIN and SCLIN make connection with the backplane SDA and SCL lines. During this time, the 1V precharge circuitry is active and forces 1V through 100k nominal resistors to the low capacitance (less than 10pF) SDA and SCL pins, minimizing the worst-case voltage differential these pins will see at the moment of connection. These precharge and low capacitance features result in minimal disturbances on the backplane SDA and SCL busses during hot swapping.

The voltages on the SDA backplane bus and LTC4300-1 SDAIN pin during card insertion are shown in Figure 257.2. A 100pF capacitor to ground is used to emulate the equivalent SDA bus capacitance. Just before the insertion, the backplane SDA bus is approaching 4V, while the LTC4300-1's SDAIN pin is precharged to 1V. Due to the high resistance and low capacitance of the SDAIN pin, the voltage on the pin rises up to the backplane voltage at the moment of insertion, whereas the backplane voltage is barely affected. The two signals are now shorted together.

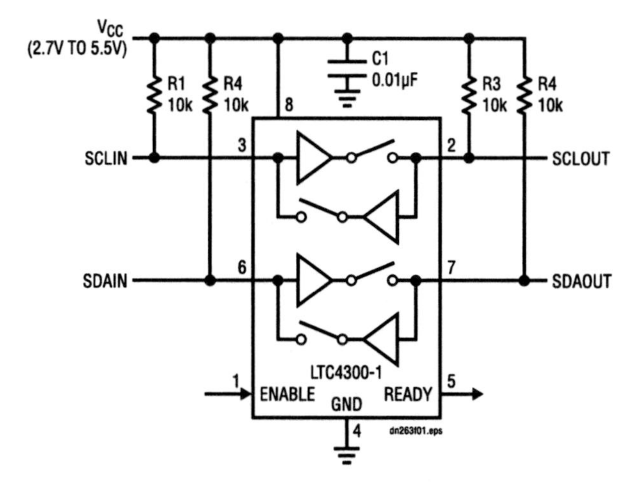

Figure 257.1 • Using the LTC4300-1 to Hot Swap SDA and SCL Lines

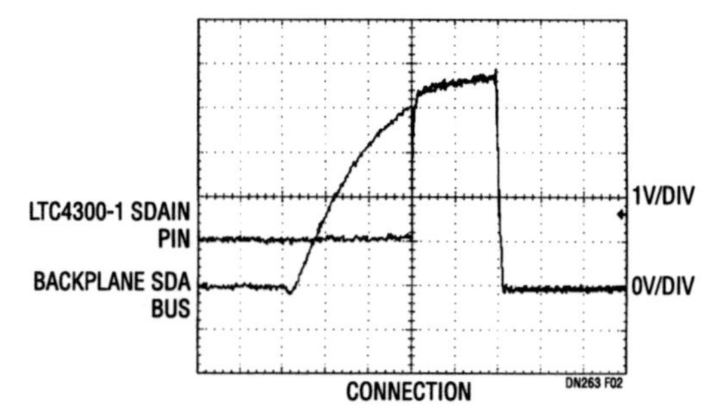

Figure 257.2 • LTC4300-1 SDAIN Pin Connecting to Backplane SDA Bus

Analog Circuit Design: Design Note Collection. http://dx.doi.org/10.1016/B978-0-12-800001-4.00257-X

Once a Stop Bit or Bus Idle occurs on the backplane without bus contention on the card, the LTC4300-1 disables the precharge circuitry and activates input-to-output connection circuitry, joining the backplane SDA and SCL busses with those on the card.

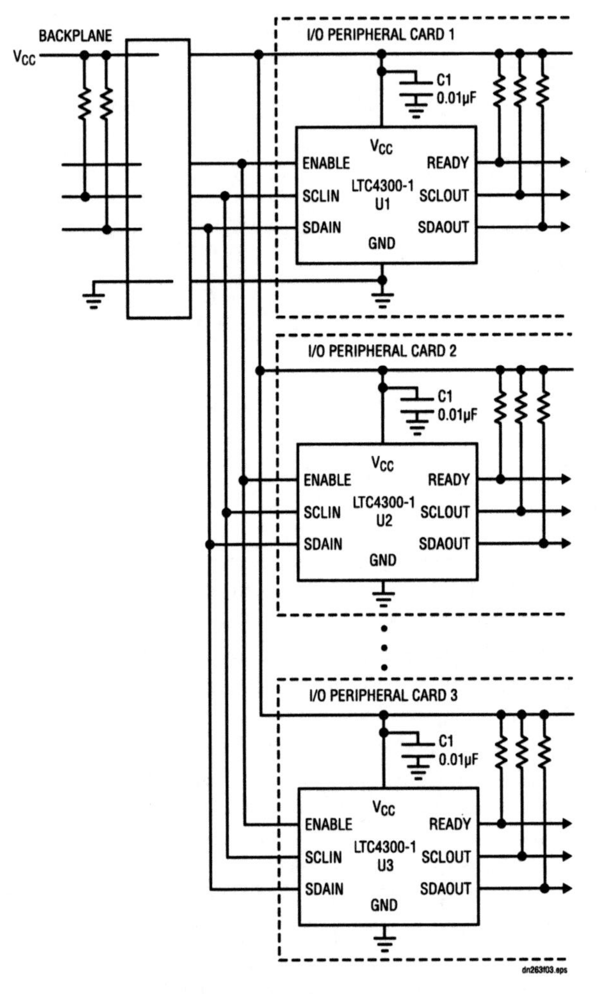

Figure 257.3 • Multiple I/O Cards Plugging into a Backplane

Capacitance buffering and rise time accelerator features

The key feature of the input-to-output connection circuitry is that it provides bidirectional buffering. Figure 257.3 shows an application that takes advantage of this feature. If the I/O cards were plugged directly into the backplane, all of the backplane and card capacitances would add directly together, making rise and fall time requirements difficult to meet. Placing an LTC4300-1 on the edge of each card, however, isolates the card capacitance from the backplane. For a given I/O card, the LTC4300-1 drives the capacitance on the card, and the backplane must drive only the low capacitance of the LTC4300-1. The LTC4300-1 further aids in meeting system rise time requirements by providing rise time accelerator circuits on all four SDA and SCL pins. Figure 257.4 shows the improved rise time provided by the accelerators for 10pF and 100pF equivalent bus capacitances.

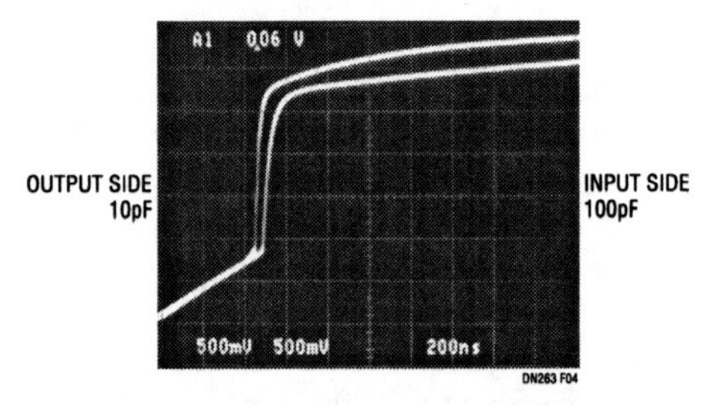

Figure 257.4 • Rise Time Accelerators Pulling Up on 10pF and 100pF Capacitance

Conclusion

The LTC4300-1 allows users to plug an I/O card into a live backplane without corrupting the backplane SDA and SCL signals in 2-wire bus systems. In addition, the connection circuitry provides bidirectional buffering, keeping the backplane and card capacitances isolated. Rise time accelerator circuits provide additional help in meeting rise time requirements.

Power supply isolation controller simplifies hot swapping the CompactPCI bus for 5V-/3.3V-only applications

258

Andy Gardner

Although ±12V supplies are provided on the CompactPCI backplanes, many plug-in boards only require 5V and 3.3V. The LTC1646 is the ideal power supply isolation controller for these applications.

The LTC1646 Hot Swap controller is designed to meet the power supply isolation requirements found in the CompactPCI hot swap specification PICMG 2.1 for plug-in boards requiring 5V and/or 3.3V supplies. The chip will turn a board's supply voltages on and off in a controlled manner, allowing the card to be safely inserted or removed without causing glitches on the power supplies and causing other boards in the system to reset. It also protects against short circuits, precharges the bus I/O connector pins during insertion and extraction and reports on the state of the supply voltages via the HEALTHY# signal.

LTC1646 feature summary

The LTC1646 features can be summarized as follows:

- Controls 5V and/or 3.3V CompactPCI supplies.
- Dual level, programmable circuit breakers: this feature is enabled after power-up is complete. If either supply exceeds current limit for more than 20μs, the circuit breaker will trip and the chip latches off. In the event that either supply exceeds three times the set current limit, all supplies are disabled and the chip latches off without delay.
- Current limited power-up: the supplies are allowed to power up in current limit. This allows the chip to power up boards with widely varying capacitive loads without

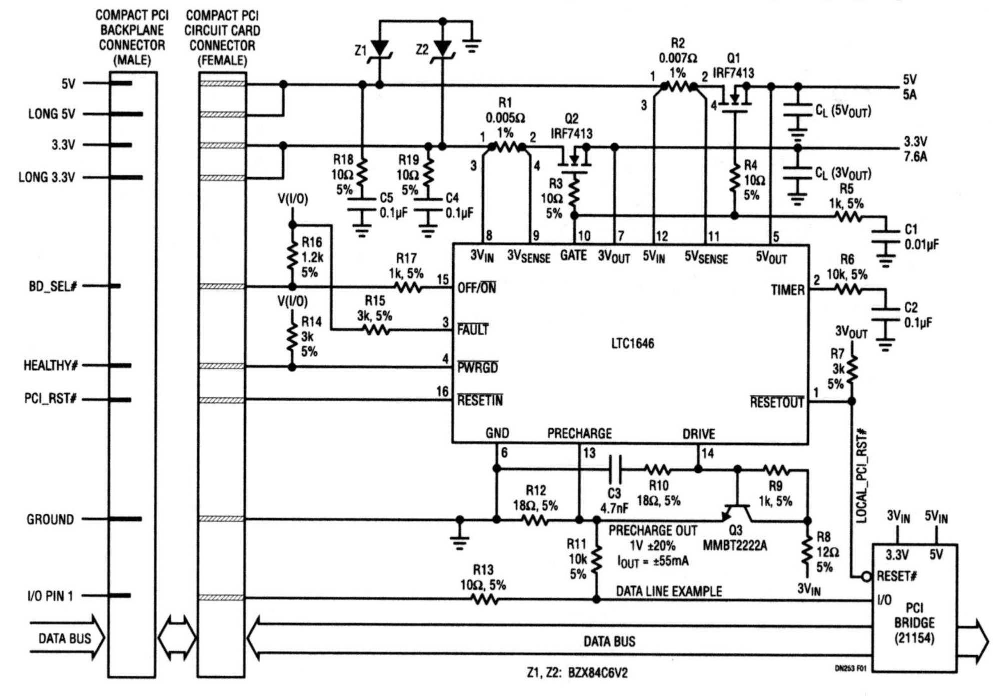

Figure 258.1 • CompactPCI Application with Only 3.3V and 5V Supplies

Analog Circuit Design: Design Note Collection. http://dx.doi.org/10.1016/B978-0-12-800001-4.00258-1

tripping the circuit breaker. The maximum allowable power-up time is programmable using the TIMER pin and an external capacitor.

- Programmable foldback current limit: a programmable analog current limit with a value that depends on the output voltage. If the output is shorted to ground during the power-up cycle, the current limit drops to keep power dissipation and supply glitches to a minimum.
- Precharge output: on-chip reference and amplifier provide 1V for biasing bus I/O connector pins during card insertion and extraction.
- BD_SEL#, HEALTHY#, PCI_RST# and LOCAL_PCI_RST# signals are supported.
- Space saving 16-pin SSOP package.

Typical application

Figure 258.1 shows a typical application using the LTC1646.

The main 3.3V and 5V inputs to the LTC1646 come from the medium-length power pins. The long 3.3V and 5V connector pins are shorted to the medium length 5V and 3.3V connector pins on the plug-in card and provide early power for the LTC1646's precharge circuitry, the V(I/O) pull-up resistors and the PCI bridge chip. The BD_SEL# signal is connected to the OFF/$\overline{\text{ON}}$ pin while the $\overline{\text{PWRGD}}$ pin is connected to the HEALTHY# signal. The HEALTHY# signal is combined on chip with the PCI_RST# signal to generate the LOCAL_PCI_RST# signal which is available at the $\overline{\text{RESETOUT}}$ pin.

The power supplies are controlled by placing external N-channel pass transistors Q1 and Q2 in the 3.3V and 5V power paths. Resistors R1 and R2 sense overcurrent conditions, and R5 and C1 provide current control-loop compensation. Resistors R3 and R4 prevent high frequency oscillations in Q1 and Q2.

Power-up sequence

Figure 258.2 shows a typical power-up sequence.

When the CompactPCI card is inserted, the long 5V, 3.3V and GND connector pins make contact first. The LTC1646's precharge circuit biases the bus I/O pins to 1V during this stage of the insertion. The 5V and 3.3V medium length pins make contact during the next stage of the insertion, but the slot power is disabled as long as the OFF/$\overline{\text{ON}}$ pin is pulled high by the 1.2k pull-up resistor to V(I/O). During the final stage of the board insertion, the BD_SEL# short connector pin makes contact and the OFF/$\overline{\text{ON}}$ pin can be pulled low. This enables the pass transistor to turn on and a 5μA current source is connected to the TIMER pin. Each supply is then allowed to power up at the rate $dV/dt = 13\mu A/C1$ or as determined by the current limit and the load capacitance, whichever is slower. Current-limit faults are ignored while the TIMER pin voltage is less than 1.25V. Once both supplies are within tolerance, the HEALTHY# signal is pulled low and the LOCAL_PCI_RST# signal goes high.

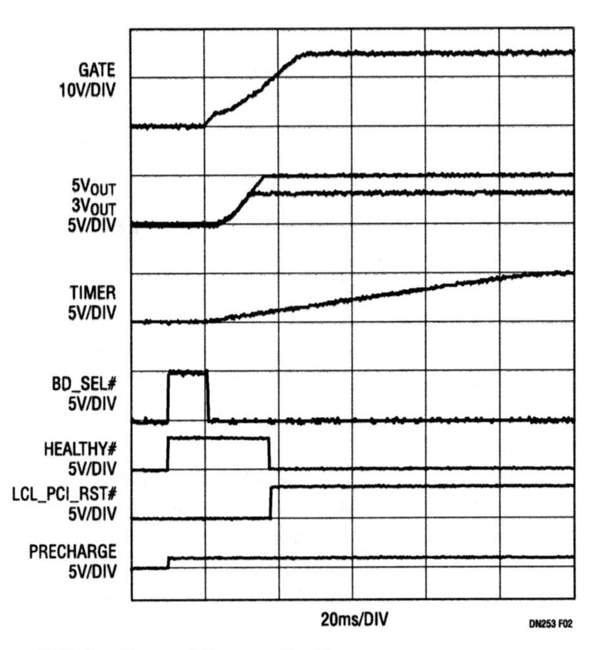

Figure 258.2 • Normal Power-Up Sequence

Conclusion

Using the LTC1646, a CompactPCI board can be made hot swappable so the system power can remain uninterrupted while the board is being inserted or removed. With the LTC1646, safe hot swapping becomes as easy as hooking up an IC, a couple of power FETs and a handful of resistors and capacitors.

259

A 24V/48V Hot Swap controller

Robert Reay

As supply voltages continue to drop, designers face the difficult task of minimizing the voltage drops through a distributed power system. Designers are leaning toward distributing power at a higher voltage, commonly 24V/48V, then stepping down the voltage on each board in the system to the final desired value using power modules.

Most power modules require an input bypass capacitor that is typically hundreds of microfarads in value. When the board is hot-plugged into a live power rail, the input capacitor can draw huge inrush currents as it charges. The inrush current can cause permanent damage to the board's components and cause glitches on the system power supply that can make the system function improperly.

The LT1641 provides a simple and flexible solution to the hot swapping problems. The chip allows a board to be safely inserted into or removed from a live backplane with an operating supply voltage range from 9V to 80V. It features a programmable analog foldback current limit, programmable undervoltage lockout and overvoltage protection, automatic restart or latched operation and direct power module enable control.

Typical application

Placing an external N-channel pass transistor (Q1) in the power path as shown in Figure 259.1, controls power supply on a board. Resistor R_S provides current detection and capacitor C1 provides control of the GATE slew rate. C2 is the timing capacitor. Resistor R6 provides current control loop compensation and R5 prevents high frequency oscillations in Q1. Resistors R1 and R2 provide undervoltage sensing. D1 clamps the V_{GS} of Q1 at 18V and thus protects Q1 from breaking down if the DRAIN node is shorted to ground. After the power pins first make contact, transistor Q1 is turned off.

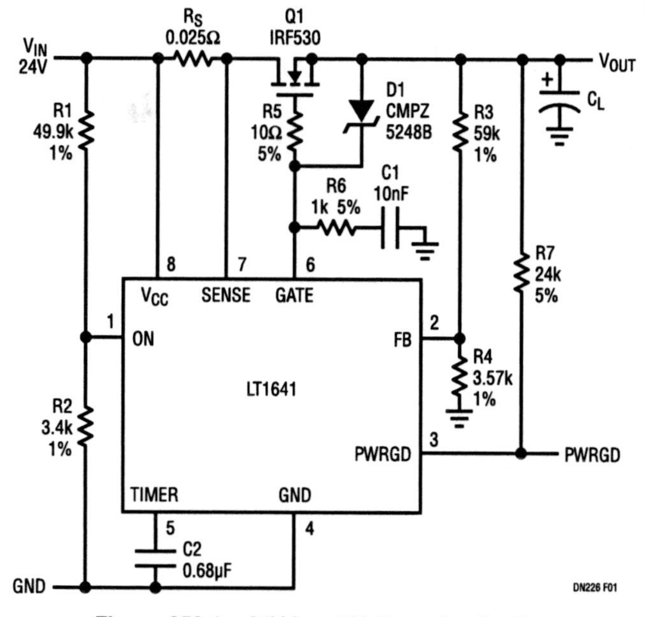

Figure 259.1 • 24V Input Voltage Application

Analog Circuit Design: Design Note Collection. http://dx.doi.org/10.1016/B978-0-12-800001-4.00259-3

If the chip detects that the input voltage is high enough, it charges the GATE pin high with 10μA. The supply inrush current is set at $I_{INRUSH} = 10\mu A \cdot C_L/C1$. Once the voltage at the output has reached its final value, as sensed by resistors R3 and R4, the PWRGD pin goes high. The waveforms are shown in Figure 259.2.

The LT1641 features a programmable foldback current limit with an electronic circuit breaker that protects against short circuits or excessive supply currents. The current limit level is set by placing a sense resistor between the V_{CC} pin and SENSE pin. When the load exceeds the current limit, the LT1641 regulates the GATE pin voltage to keep the current through the sense resistor at a constant value. In the meantime, C2 is charged with an 80μA pull-up current. If the voltage at the TIMER pin reaches 1.233V and an overcurrent condition still exists, the LT1641 turns off the pass transistor.

To prevent excessive power dissipation in the pass transistor and to prevent voltage spikes on the input supply during short-circuit conditions at the output, the current folds back as a linear function of the output voltage, which is sensed at the FB pin. The foldback ratio is 4:1.

Automatic restart

To force the LT1641 to automatically restart after an overcurrent fault, the bottom plate of capacitor C1 can be connected to the ON pin. When an overcurrent condition occurs, the GATE pin is driven to maintain a constant voltage across the sense resistor. Capacitor C2 begins to charge. When the voltage at the TIMER pin reaches 1.233V, the GATE pin is immediately pulled to GND and transistor Q1 is turned off. Capacitor C1 momentarily pulses the ON pin to low and clears the internal fault latch. When the voltage at the TIMER pin decreases to 0.5V, the LT1641 turns on again. If the short-circuit condition at the output still exists, the cycle will repeat itself indefinitely with 3.75% on-time duty cycle and prevent Q1 from overheating. The waveforms are shown in Figure 259.3.

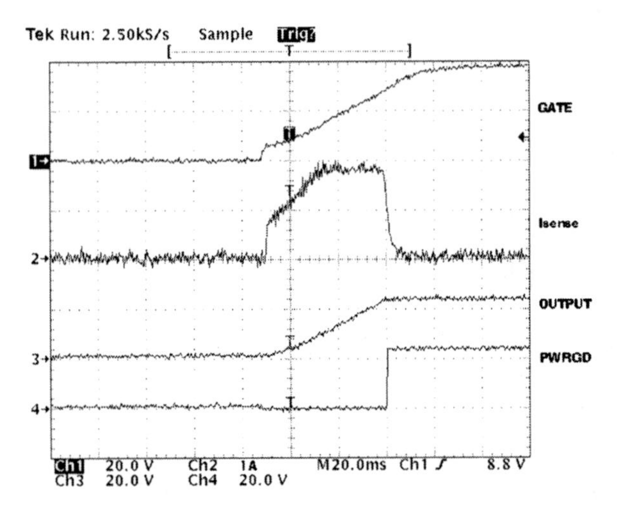

Figure 259.2 • Power-Up Waveforms

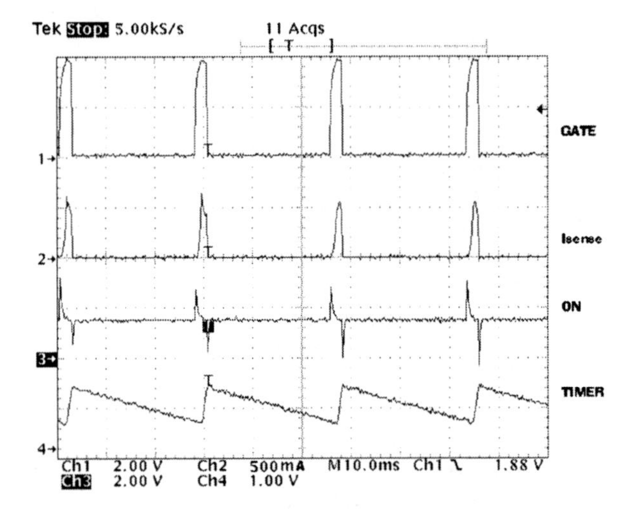

Figure 259.3 • Automatic Restart Waveforms

Dual channel Hot Swap controller/ power sequencer allows insertion into a live backplane

260

Bill Poucher

The LTC1645 is a two-channel Hot Swap controller and power sequencer that allows a board to be safely inserted and removed from a live backplane. When a board is hot swapped, the supply bypass capacitors on the board can draw huge transient currents from the backplane power bus as they charge. These transient currents can cause permanent damage to the capacitors, connector pins and board traces and can disrupt the system supply, causing other boards in the system to reset.

Using external N-channel pass transistors, supply voltages from 1.2V to 12V can be ramped together or separately at a programmable rate. Programmable electronic circuit breakers protect against shorts at either output. The LTC1645 is available in 14- and 8-pin SO packages. The 8-pin version includes a control input, dual gate drives and dual circuit breakers. The 14-pin version additionally provides a system reset signal and a spare comparator to indicate when board supply voltages drop below programmable levels. It also has a fault signal to indicate an overcurrent condition, as well as a timer pin to create a delay before ramping up the supply voltages and before deasserting the system reset signal.

Basic operation

The LTC1645 controls a board's power supplies with external N-channel pass transistors in the power paths as shown in Figures 260.1 and 260.3. The LTC1645 contains an internal charge pump to provide high side drive to the N-channel FETs. R_{SENSE1} and R_{SENSE2} provide current fault detection, and R1 and R2 prevent high frequency oscillation. By ramping the gates of the pass transistors up and down at a controlled rate, the transient surge current $(I = C \cdot dV/dt)$ drawn from the main backplane supply, is limited to a safe value when the board makes connection.

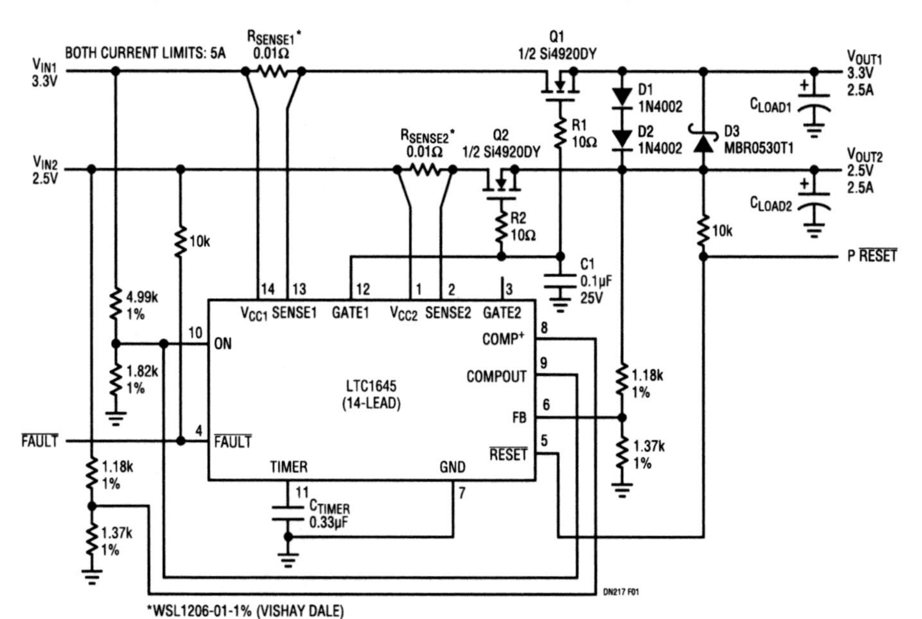

Figure 260.1 • Ramping 3.3V and 2.5V Up and Down Together

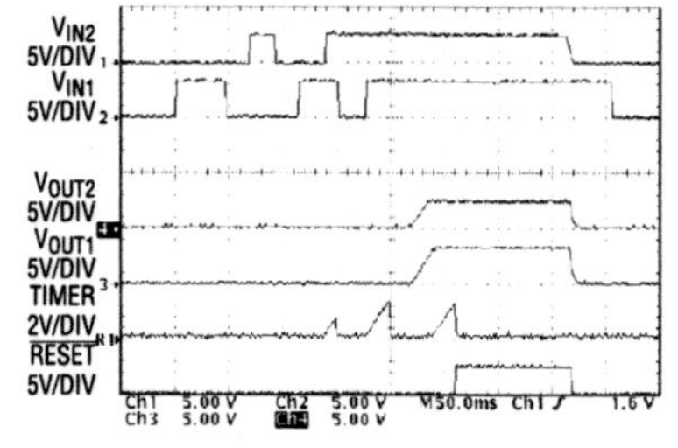

Figure 260.2 • Supply Tracking Waveforms

Analog Circuit Design: Design Note Collection. http://dx.doi.org/10.1016/B978-0-12-800001-4.00260-X

Taking the ON pin above 0.8V turns on GATE1 after one timing cycle, while taking the ON pin above 2V turns GATE2 on if the ON pin has been above 0.8V for at least one timing cycle. The circuit breaker trips whenever the voltage across either sense resistor is greater than 50mV for more than 1.5μs. When it trips, both GATE pins are immediately pulled to ground, the external FETs are quickly turned off and (in the 14-pin version) $\overline{FAULT}$ is asserted.

The 14-pin version of the LTC1645 provides two open-drain output comparators for monitoring input or output voltage levels. One releases $\overline{RESET}$ one timing cycle after the FB pin exceeds 1.238V, while the other releases COMPOUT immediately whenever COMP+ is above 1.238V.

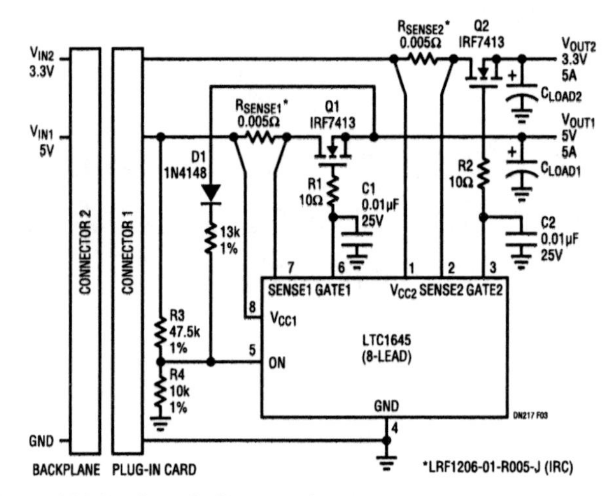

Figure 260.3 • Supply Sequencing

Power supply tracking and sequencing

In addition to general purpose hot swapping, the LTC1645 helps simplify power supply tracking and sequencing circuits. Some applications require that the difference between two power supplies never exceed a certain voltage. This requirement applies during power-up and power-down as well as during steady-state operation, often to prevent latch-up in a dual-supply ASIC. Other systems require one supply to come up after another; for example, when a system clock needs to start before a block of logic. Typical dual supplies or backplane connections may come up at arbitrary rates depending on load current, capacitor size, soft-start rates, etc. Traditional solutions can be cumbersome or require complex circuitry to meet the necessary requirements.

Figure 260.1 shows an application ramping V_{OUT1} and V_{OUT2} up and down together. The ON pin must reach 0.8V to ramp up V_{OUT1} and V_{OUT2}. The spare comparator pulls the ON pin low until V_{IN2} is above 2.3V, and the ON pin cannot reach 0.8V before V_{IN1} is above 3V. Thus, both input supplies must be within regulation before a timing cycle can start. At the end of the timing cycle, the output voltages ramp up together. If either input supply falls out of regulation or if an overcurrent condition is detected, the gates of Q1 and Q2 are pulled low together. Figure 260.2 shows an oscilloscope photo of the circuit in Figure 260.1.

On power-up, V_{OUT1} and V_{OUT2} ramp up together. On power-down, the LTC1645 turns off Q1 and Q2 simultaneously. Charge remains stored on C_{LOAD1} and C_{LOAD2} and the output voltages will vary depending on the loads. D1 and D2 turn on at ≈1V (≈0.5V each), ensuring that V_{OUT1} never exceeds V_{OUT2} by more than 1.2V. D3 guarantees V_{OUT2} is never greater than V_{OUT1} by more than 0.4V. Barring an overvoltage condition at the input(s), the only time these diodes might conduct current is during a power-down event, and then only to discharge C_{LOAD1} or C_{LOAD2}. In the case of an input overvoltage condition that causes excess current to flow, the circuit breaker will trip if the current limit level is set appropriately.

Figure 260.3 shows the LTC1645 in a Hot Swap application configured to ramp up V_{OUT1} before V_{OUT2}. V_{OUT1} is initially discharged and D1 is reverse-biased, thus the voltage at the ON pin is determined only by V_{CC1} through the resistor divider R3 and R4. If V_{CC1} is above 4.6V, the voltage at the ON pin exceeds 0.8V and V_{OUT1} ramps up. As V_{OUT1} ramps up, D1 forward-biases and pulls the ON pin above 2V when $V_{OUT1} \approx 4.5V$. This turns on GATE2 and V_{OUT2} ramps up. Use the 14-pin version if additional voltage monitoring is desired.

Conclusion

Designing hot insertion systems normally requires a significant effort by an experienced analog designer. With the LTC1645, safe and reliable hot swapping becomes as easy as hooking up an IC, a couple of power FETs and a handful of resistors and capacitors.

261

Hot swapping the CompactPCI bus

Robert Reay

One of the reasons for the increase in popularity of the CompactPCI bus is its ability to be hot swapped into and out of a live backplane without turning off the system power. One of the key elements required for hot swapping a CompactPCI board is the power supply isolation controller.

The LTC1643L is designed to meet the power supply isolation requirements found in the CompactPCI Hot Swap Specification PICMG 2.1. The chip will turn a board's supply voltages on and off in a controlled manner, allowing the board to be safely inserted or removed without causing glitches on the power supplies and causing other boards in the system to reset. It also protects against short circuits and reports the state of the supply voltages via the HEALTHY# signal.

LTC1643 feature summary

The LTC1643 features can be summarized as follows:

1. Controls all four CompactPCI supplies: −12V, 12V, 3.3V and 5V.
2. Programmable foldback current limit: a programmable analog current limit with a value that depends on the output voltage. If the output is shorted to ground, the current limit drops to keep power dissipation and supply glitches to a minimum.

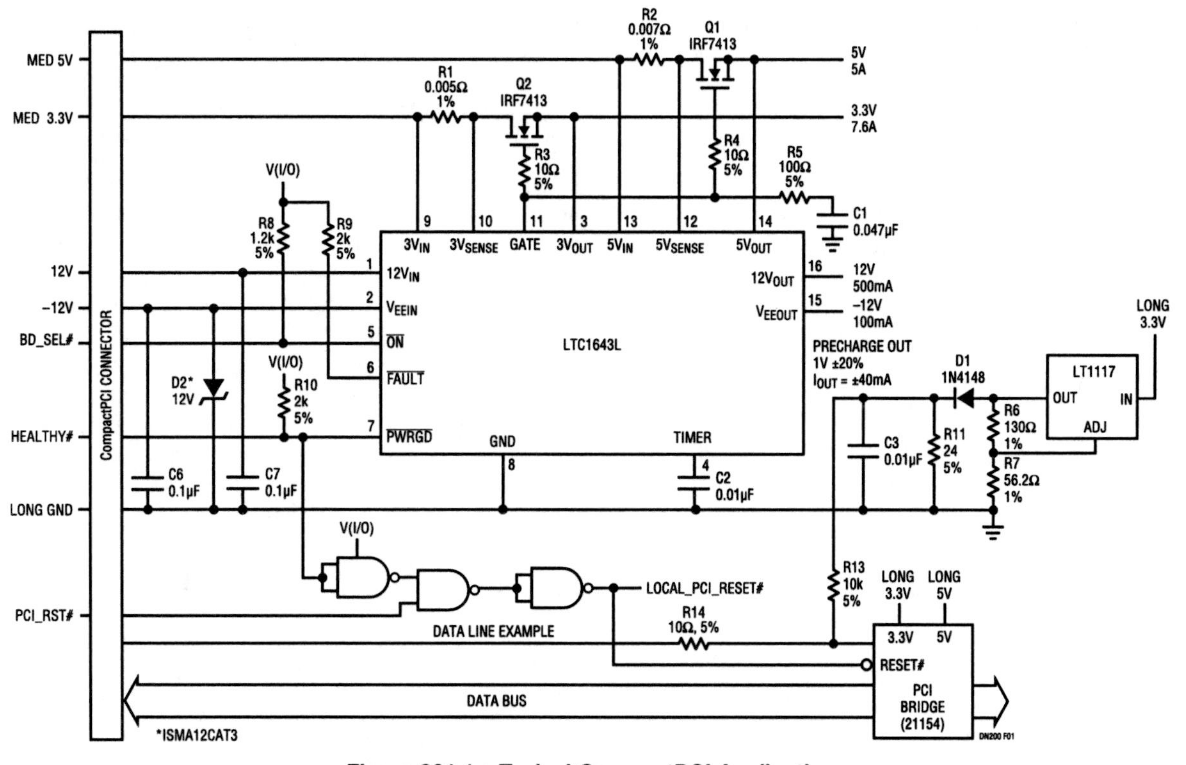

Figure 261.1 • Typical CompactPCI Application

Analog Circuit Design: Design Note Collection. http://dx.doi.org/10.1016/B978-0-12-800001-4.00261-1

3. Programmable circuit breaker: if a supply remains in current limit too long, the circuit breaker will trip, the supplies will be turned off and the FAULT# pin will be pulled low.
4. Current limited power-up: the supplies are allowed to power up in current limit. This allows the chip to power up boards with large capacitive loads without tripping the circuit breaker. The maximum power-up time is programmable using the TIMER pin.
5. −12V and 12V power switches on chip.
6. BD_SEL# and HEALTHY# signals.
7. Space-saving 16-pin SSOP package.

Typical application

Figure 261.1 shows a typical application using the LTC1643L.

The power supplies are controlled by placing external N-channel pass transistors Q1 and Q2 in the 5V and 3.3V power paths. Internal pass transistors control the 12V and −12V power paths. Resistors R1 and R2 sense overcurrent conditions and R5 and C1 provide current control loop compensation. Resistors R3 and R4 prevent high frequency oscillations in Q1 and Q2. Capacitor C2 provides the power-up timing and C6 and C7 provide chip bypassing on the 12V and −12V inputs. Diode D2 protects the part from voltage surges below −15V on the −12V supply.

The 3.3V, 5V, 12V and −12V inputs of the LTC1643 come from the medium length power pins on the CompactPCI connector. The long 3.3V, 5V and V(I/O) pins power up the bus-precharge circuit, the PCI bridge chip and the LOCAL_PCI_RESET# logic. The BD_SEL# signal is connected to the $\overline{\text{ON}}$ pin and the HEALTHY# signal is connected to the $\overline{\text{PWRGD}}$ pin. The HEALTHY# signal is combined with the PCI_RESET# signal to generate the LOCAL_PCI_RESET# signal.

The 1V precharge voltage for the data bus lines is generated by an LT1117 low dropout regulator. The output of the LT1117 is set to 1.8V, then the voltage is dropped by an 1N4148 diode to generate 1.0V. The precharge circuit is capable of sourcing and sinking 40mA.

Power-up sequence

Figure 261.2 shows a typical power-up sequence.

When the BD_SEL# is pulled low, the pass transistors are allowed to turn on and a 20μA current source is connected to the TIMER pin. The current in each pass transistor increases until it reaches the current limit for that supply. Each supply is then allowed to power up at the rate dV/dt = 50μA/C1 or as determined by the current limit and the load capacitance, whichever is slower. Current-limit faults are ignored while the TIMER pin voltage is ramping up. Once all four supply voltages are within tolerance, the HEALTHY# signal will pull low.

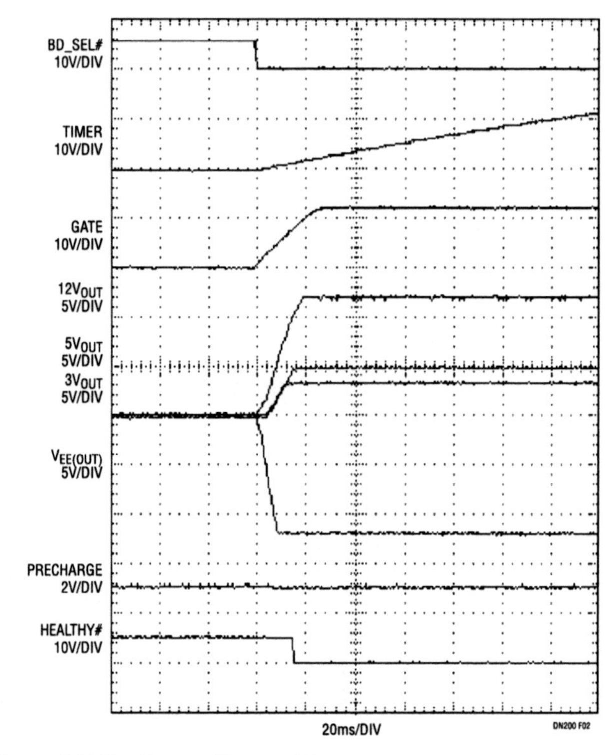

Figure 261.2 • Normal Power-Up

Conclusion

Using the LTC1643L, a CompactPCI board can be made hot swappable so the system power can remain on when the board is inserted or removed. With the LTC1643L, safe hot swapping becomes as easy as hooking up an IC, a couple of power FETs and a handful of resistors and capacitors.

Power solutions for the Device Bay

262

Ajmal Godil

"Device Bay" is an industry specification defining a mechanism for easily adding and upgrading PC peripheral devices to a live system without opening the chassis or shutting off the power. This requires Hot Swap capability for the Device Bay to ensure the integrity of users' data and applications. For example, a CD-R drive could be added to provide a large storage medium for digital imaging or a DVD drive could be added to enable DVD video playback. The Device Bay specification applies to all classes of computers, including desktop, mobile, home and server computers. Device Bay comes in three form factors:[1] DB32, DB20 and DB13. The DB32 is designed for desktops with a maximum power consumption of 25W, DB20 is designed for laptops and desktops with a maximum power consumption of 4W, and DB13 is designed for laptops with a maximum power consumption of 4W.

Device Bay power requirements

According to the Device Bay Interface Specification,[1] Revision 0.85, the DB32, DB20 and DB13 form factors all require an identification voltage (Vid_3.3V), which needs to be provided by the host system. It is recommended that this voltage be increased slowly when the device is inserted into the system, because the supply bypass capacitors on the device can draw huge transient currents from the system power supply as they charge. The transient currents can cause permanent damage to the connector pins and glitches on the system supply. These glitches may force other boards in the system to reset.

Table 262.1 Maximum Allowable DB32 Device Current

VOLTAGE	TRANSIENT <100µs (A)	PEAK CURRENT 100µs to 30s (A)	ELECTRICAL CONTINUOUS CURRENT 30s to 300s[1] (A)	THERMAL CONTINUOUS CURRENT >300s[1] (A)
Vid_3.3V	0.91	0.45	0.45	0.45
12V	5	3.75	2.5	2
5V	4	3	2	2
3.3V	7	5.25	3.5	3.5

Table 262.2 Maximum Allowable DB20 and DB13 Device Current

VOLTAGE	TRANSIENT <100µs (A)	PEAK CURRENT 100µs to 30s (A)	ELECTRICAL CONTINUOUS CURRENT 30s to 300s[1] (A)	THERMAL CONTINUOUS CURRENT >300s[1] (A)
Vid_3.3V	0.91	0.45	0.45	0.45
5V	2	1.6	0.8	0.8
3.3V	3.03	2.42	1.21	1.21

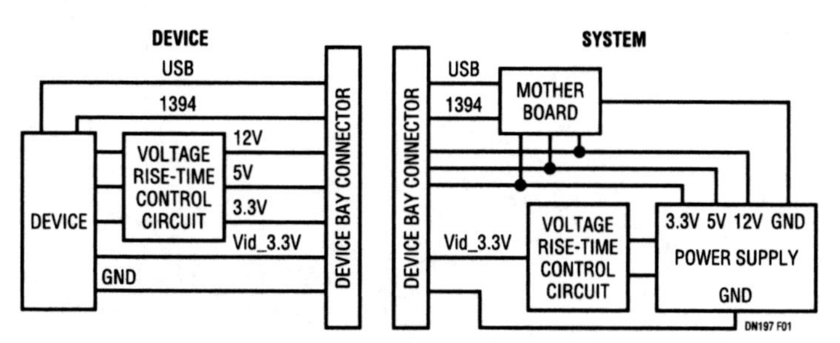

Figure 262.1 • Block Diagram for the Device Bay

Note 1: Device Bay Interface Specification, Revision 0.85, February 6 1998.

Analog Circuit Design: Design Note Collection. http://dx.doi.org/10.1016/B978-0-12-800001-4.00262-3

The device uses the identification voltage to power the 1394 bus or the USB interface. Additionally, the DB32 form factor needs 12V, 5V and 3.3V supplies. The DB20 and DB13, on the other hand, only need 5V and 3.3V supplies. It is recommended that these supply voltages also increase slowly to minimize current spikes. Figure 262.1 shows a block diagram of a circuit that controls the rise time of the various voltage sources on the device side and the system side for the DB32 form factor. The maximum allowable currents for various Device Bay form factors are given in Table 262.1 and 262.2.

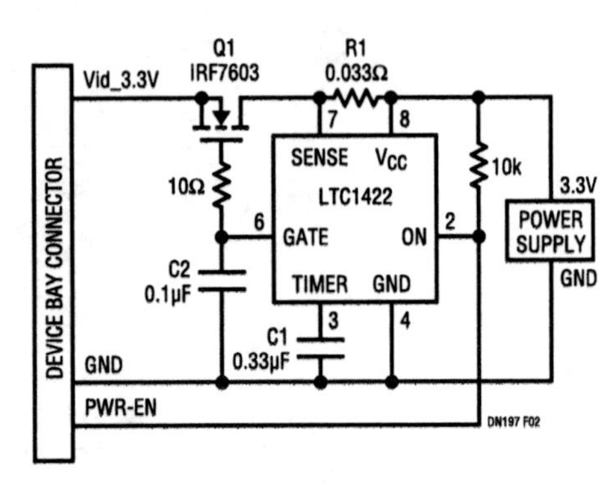

Figure 262.2 • Vid_3.3V Supply Solution for DB32, DB20 and DB13 Form Factors

Power solution for Vid_3.3V on the system side

Figure 262.2 shows the power solution for DB32, DB20 and DB13's Vid_3.3V supply using the LTC1422. The LTC1422 is an 8-pin Hot Swap controller that allows safe insertion of an external peripheral into the Device Bay slot. By increasing the voltage on the MOSFET Q1's gate at a controlled rate, the transient surge current drawn by the device's bulk capacitor is limited to a safe value when the device is connected to the bay.

When power is first applied to the bay system, the gate of Q1 is pulled low. After the PWR-EN pin is released from GND by the Device Bay connector and the software controlled PWR_CTL bit, the voltage on the LTC1422 ON pin changes to a logic high. After the ON pin is held high for at least one timing cycle ($t_1 = 1.232 \cdot C1/2\mu A$), the charge pump turns on. The voltage at the gate begins to rise with a slope equal to $10\mu A/C2$, where C2 is the external capacitor connected between the GATE pin and GND. The rise time for the supply is equal to $3.3V \cdot C2/10\mu A$ (Figure 262.3).

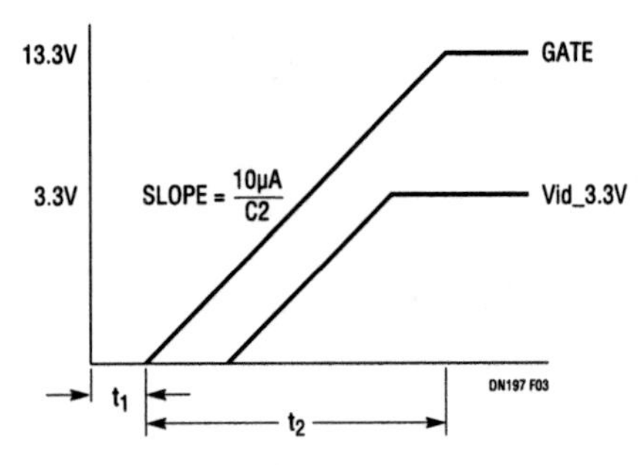

Figure 262.3 • Supply Turn On

The LTC1422 features an electronic circuit breaker function that protects against short circuits or excessive currents from the supply. By placing a sense resistor, R1, between the supply input and the SENSE pin, the circuit breaker will be tripped whenever the voltage across the sense resistor is greater than 50mV for more than 10µs. The sense resistor should be sized to allow 150% of the maximum load current to flow before the circuit breaker trips. Therefore, by choosing $R1 = 0.033\Omega$, approximately 1.5A (versus Device Bay specification of 0.91A) would flow into the device before the circuit breaker trips. When the circuit breaker trips, the GATE pin is immediately pulled to GND and the external MOSFET Q1 is quickly turned off.

Power solutions for DB32, DB20 and DB13 form factors on the device side

Figure 262.4 shows the application circuit for DB32, DB20 and DB13 form factors using the LTC1422 on the device side. These circuits operate in the same fashion as the circuit in Figure 262.2, except that a single LTC1422 is used to control the gates of multiple MOSFETs to allow the respective voltage supplies to rise simultaneously. Also, since a single LTC1422 controls multiple supplies, the circuit breaker feature in Figure 262.4 has been eliminated by shorting Pins 7 and 8 together.

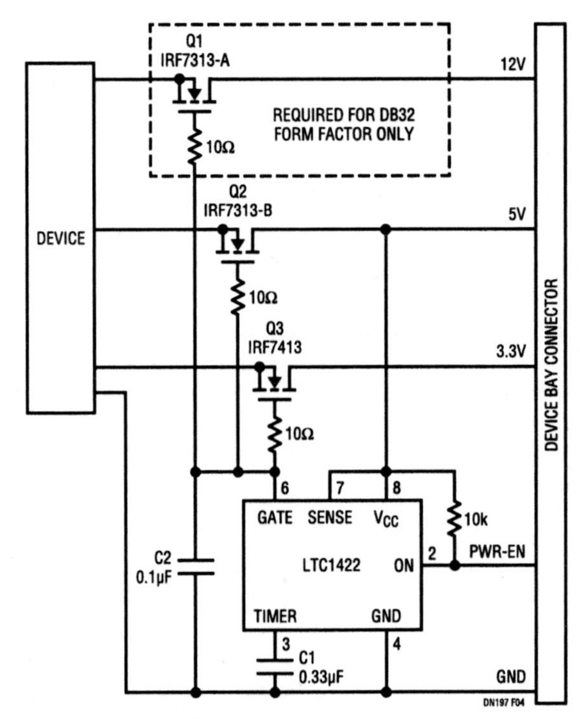

Figure 262.4 • Device Side Voltage Ramp-Up Circuit for DB32, DB20 and DB13 Form Factors

263

Hot swapping the PCI bus

James Herr Paul Marshik Robert Reay

The Peripheral Component Interconnect (PCI) bus has become widely used in high volume personal computers and single-board computer designs. With a 32-bit data path and a bandwidth of up to 133Mbps, PCI offers the throughput demanded by the latest I/O and storage peripherals. Unfortunately, the original PCI specification does not require the bus to be hot swappable, so the system power must be turned off when a peripheral is inserted into or removed from a PCI slot.

With the migration of the PCI bus into servers, industrial computers and computer telephony systems, the ability to plug a peripheral into a live PCI slot becomes mandatory. By using the LTC1421 to control the power supplies and a QuickSwitch bus switch to buffer the data bus, a peripheral can be inserted into a PCI slot without turning off the system power.

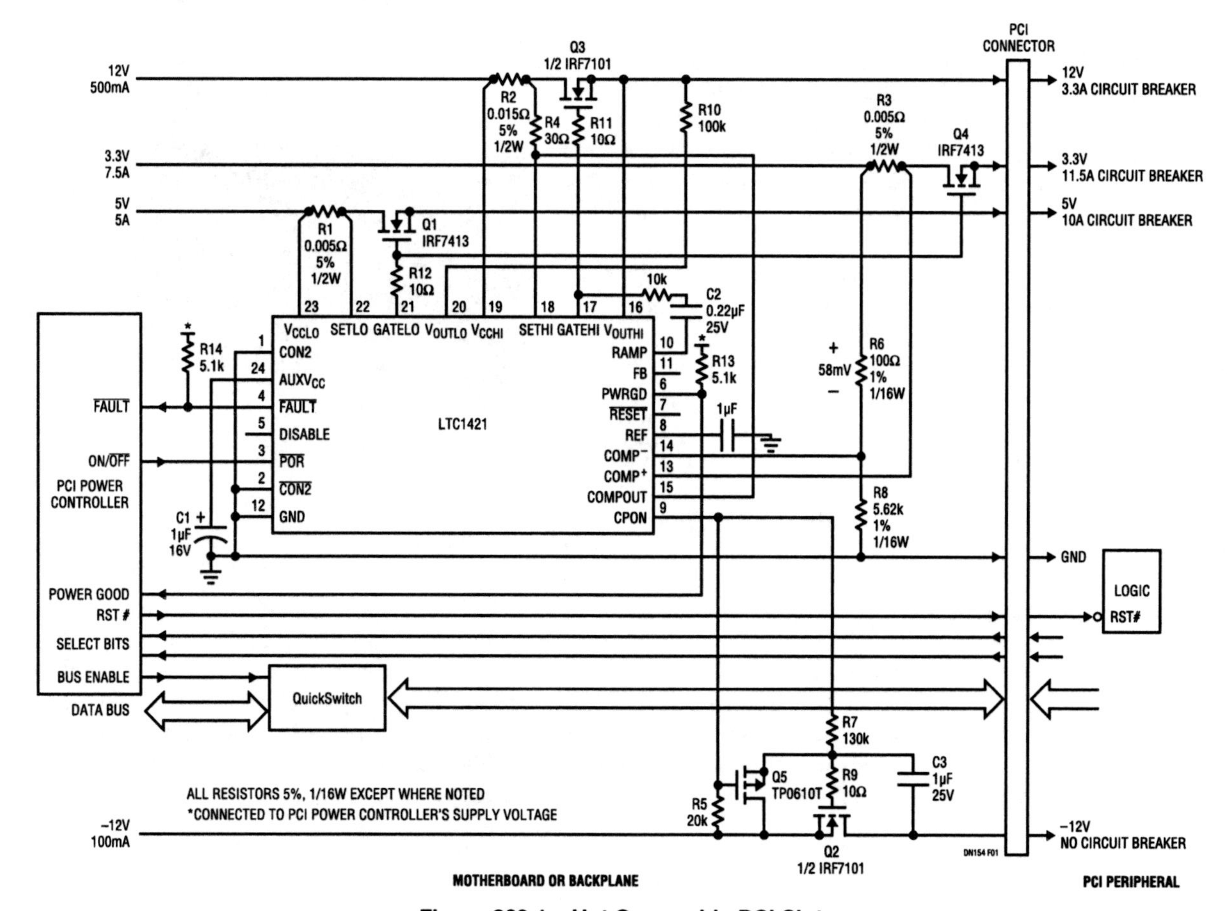

Figure 263.1 • Hot Swappable PCI Slot

Analog Circuit Design: Design Note Collection. http://dx.doi.org/10.1016/B978-0-12-800001-4.00263-5

Inrush current and data bus problems

When the peripheral is inserted, the supply bypass capacitors on the peripheral can draw huge transient currents from the PCI power bus as they charge up. The transient currents can cause permanent damage to the connector pins and board traces, while causing glitches on the system supply that force other peripherals in the system to reset.

The second problem involves the diodes to V_{CC} at the input or output of most logic families. With the peripheral initially unpowered, the V_{CC} input to the logic gate is at ground potential. When the data bus pins make contact, the bus lines are clamped to ground through the diodes to V_{CC} and the data is corrupted. With current flowing into the V_{CC} diode, the logic gate may latch up and destroy itself when power is applied.

Hot swappable PCI slot using the LTC1421

The circuitry for a hot swappable PCI slot on a motherboard or backplane is shown in Figure 263.1.

The power supplies for each PCI slot are controlled by an LTC1421 and four external FETs, and the data bus is buffered by several QS3384 QuickSwitches or equivalent. A PCI power control ASIC, FPGA, microprocessor or the like, controls all of the slots within the system.

The 12V, 5V, 3.3V and −12V supplies are controlled by placing external N-channel pass transistors, Q1 to Q4, in the power path. By increasing the voltage on the gate of the pass transistors at a controlled rate, the transient surge current ($I = C \cdot dV/dt$) drawn from the PCI supplies can be limited to a safe value. The ramp rate for the positive supplies is set by $dV/dt = 20\mu A/C2$. The −12V supply ramp rate is set by R7 and C3, while resistor R5 and transistor Q5 help turn off transistor Q2 quickly. Resistors R9, R11 and R12 prevent potential high frequency FET oscillations. Resistors R13 and R14 pull up PWRGD and $\overline{\text{FAULT}}$ to the proper logic level.

Sense resistors R1, R2 and R3 provide current fault protection. When the voltage across R1 and R2 is greater than 50mV for more than 10µs, the LTC1421 circuit breaker is tripped. All of the FETs are immediately turned off and the $\overline{\text{FAULT}}$ pin is pulled low. The circuit breaker is reset by cycling the $\overline{\text{POR}}$ pin. The current fault protection for the 3.3V supply is provided by resistive divider R6 and R8 and the uncommitted comparator in the LTC1421. Because the current levels on the −12V supply are so low, overcurrent protection is not necessary.

The QuickSwitch bus switch contains a low resistance N-channel placed in series with the data bus. The switch is turned off when the board is inserted and then enabled after the power is stable. The switch inputs and outputs do not have a parasitic diode back to V_{CC} and have very low capacitance.

System timing

The system timing is shown in Figure 263.2. The PCI power controller senses when a board has been inserted into the PCI via the power-select bits. Alternatively, the user can inform the controller that a board has been inserted via a front panel or keyboard. The PCI controller holds the RST# pin low and disables the QuickSwitch bus switches, then turns on the LTC1421 via the $\overline{\text{POR}}$ pin. The power supplies turn on at a controlled rate and when the 12V supply is within 10% of its final value, the PWRGD signal pulls high. The PCI power controller waits one reset time-out period and then pulls RST# high and enables the QuickSwitch devices.

When the board is turned off, RST# is pulled low, the QuickSwitch bus switches are disabled and the LTC1421 is turned off by pulling the $\overline{\text{POR}}$ pin low. After a 20ms delay, the external FETs are turned off and the supply voltages collapse.

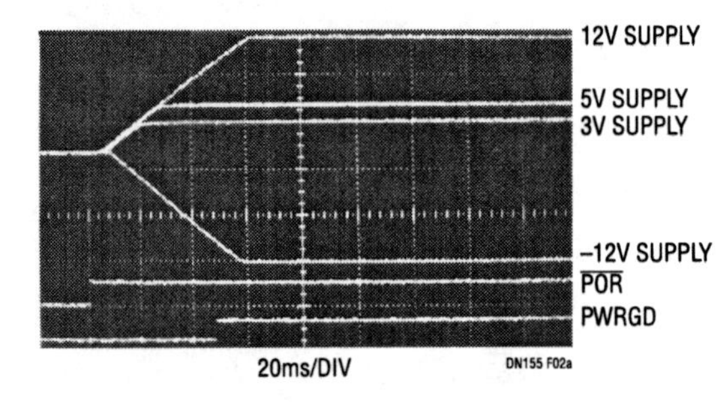

Figure 263.2a • Power-Up

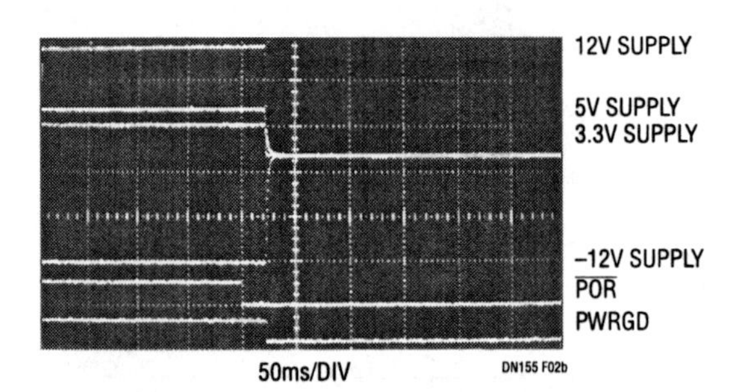

Figure 263.2b • Power-Down

Conclusion

Using the LTC1421 and a QuickSwitch bus switch, a PCI slot can be made hot swappable so the system power can remain on when a peripheral is inserted or removed. Up to now, the design of Hot Swap circuitry has required the talents of an analog guru. With the LTC1421, safe hot swapping becomes as easy as hooking up an IC, a couple of power FETs and a handful of resistors and capacitors.

264

Safe hot swapping

James Herr Robert Reay

When a circuit board is inserted into a live backplane, the large bypass capacitors on the board can draw huge inrush currents from the backplane power bus as they charge. The inrush current, on the order of 10A to 100A, can destroy the board's bypass capacitors, metal traces or connector pins. The inrush current can also cause a glitch on the backplane power bus, which could force all of the other boards in the system to reset. In addition, the system data bus can be disrupted when the board's data pins make or break contact.

The LTC1421 can turn on two positive and one negative board supply voltages at a programmable rate, allowing a board to be safely inserted in, or removed from, a live backplane. The device provides internal charge pumps for driving the gates of external N-channel pass transistors, board connection sensing, flexible supply voltage monitoring, power on reset output, short-circuit protection and soft or hard reset via software control.

Typical application

Figure 264.1 shows a typical application using the LTC1421.

The LTC1421 works best with a staggered, 3-level connector. Ground makes connection first to discharge any static

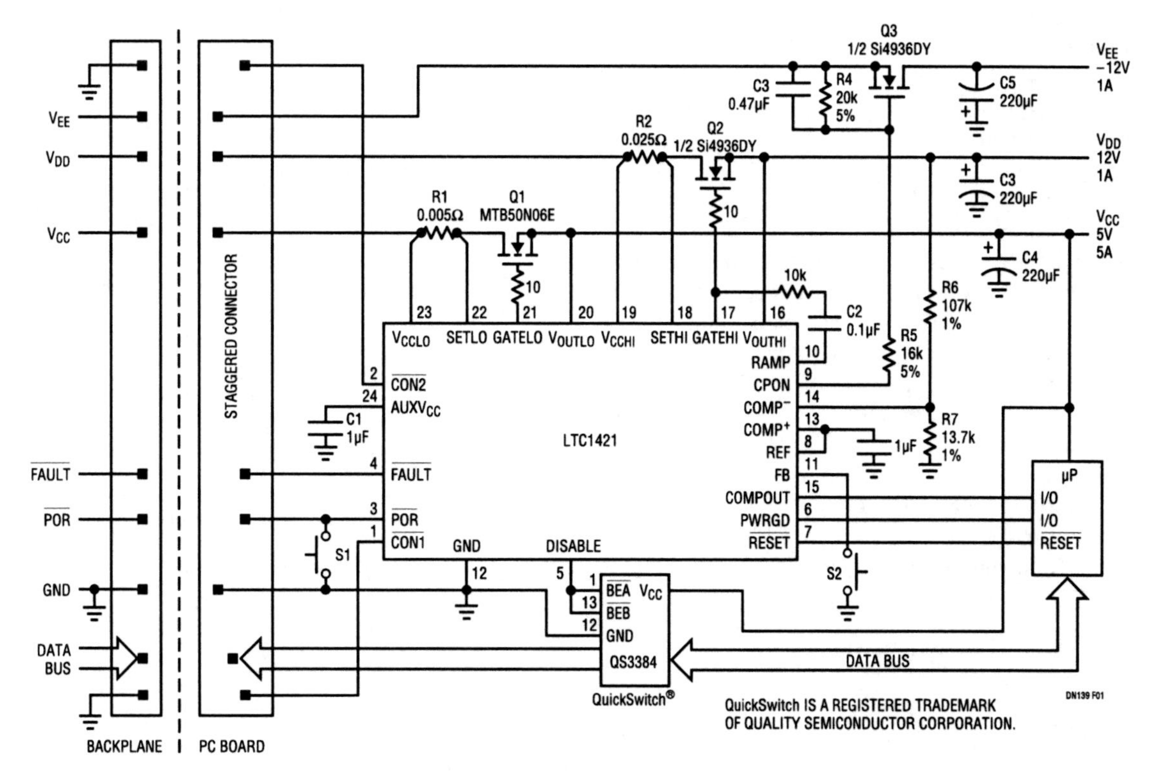

Figure 264.1 • LTC1421 Typical Application

Analog Circuit Design: Design Note Collection. http://dx.doi.org/10.1016/B978-0-12-800001-4.00264-7

build-up. V_{CC}, V_{DD} and V_{EE} make connection second and the data bus and all other pins last. The connection sense pins $\overline{CON1}$ and $\overline{CON2}$ are located on opposite ends of the connector to allow the board to be rocked back and forth during insertion.

The power supplies on the board are controlled by placing external N-channel pass transistors Q1, Q2 and Q3 in the power path for V_{CC}, V_{DD} and V_{EE}, where V_{CC} and V_{DD} can range from 3V to 12V, and V_{EE} from −5V to −12V. By ramping up the voltage on the pass transistors' gates at a controlled rate, the transient surge current [I=(C)(dV/dt)] drawn from the main backplane supply will be limited to a safe value. The ramp rate is set by the value of capacitor C2.

The board's data bus is buffered by a QS3384 QuickSwitch from Quality Semiconductor. Disabling the QuickSwitch via the DISABLE pin during board insertion and removal prevents corruption of the system data bus.

Resistors R1 and R2 form an electronic circuit breaker function that protects against excessive supply current. When the voltage across the sense resistor is greater than 50mV for more than 20μs, the circuit breaker trips, immediately turning off Q1 and Q2 while the $\overline{FAULT}$ pin is pulled low. The device will remain in the tripped state until the $\overline{POR}$ pin is pulsed low or the power on V_{CCLO} and V_{CCHI} is cycled. The circuit breaker can be defeated by shorting V_{CCLO} to SETLO and V_{CCHI} to SETHI.

The $\overline{RESET}$ signal is used to reset the system microcontroller. When the voltage on the V_{OUTLO} pin rises above the reset threshold, PWRGD immediately goes high and $\overline{RESET}$ goes high 200ms later. When the V_{OUTLO} supply voltage drops below the reset threshold, PWRGD immediately goes low, and $\overline{RESET}$ goes low 60μs later, allowing the PWRGD signal to be used as an early warning that a reset is about to occur. When the FB is left floating, the reset threshold is 4.65V; when the FB pin is tied to V_{OUTLO}, the reset threshold is 2.90V.

The uncommitted comparator and internal voltage reference, along with resistors R6 and R7, are used to monitor the 12V supply. When the supply drops below 10.8V, the COMPOUT pin will go low. The comparator can be used to monitor any voltage in the system.

Pushbutton switches S1 and S2 are used to generate a hard and soft reset, respectively. A hard or soft reset may also be initiated by a logic signal from the backplane. Pushing S1 shorts the $\overline{POR}$ pin to ground, generating a hard reset that cycles the board's power. Pass transistors Q1 to Q3 are turned off and V_{OUTLO} and V_{OUTHI} are actively pulled to ground. When V_{OUTLO} discharges to within 100mV of ground, the LTC1421 is reset and a normal power-up sequence is started.

Pushing S2 shorts the FB pin to ground, generating a soft reset that does not cycle the board's power. PWRGD immediately goes low, followed 64μs later by $\overline{RESET}$. When S2 is released, PWRGD immediately goes high, followed 200ms later by $\overline{RESET}$.

Board insertion timing

When the board is inserted, the GND pin makes contact first, followed by V_{CCHI} and V_{CCLO} (Figure 264.2, time point 1). DISABLE is immediately pulled high, so the data bus switch is disabled. At the same time $\overline{CON1}$ and $\overline{CON2}$ make contact and are shorted to ground on the host side (time point 3). When $\overline{CON1}$ and $\overline{CON2}$ are both forced to ground for more than 20ms, the LTC1421 assumes that the board is fully connected to the host and power-up can begin. When V_{CCLO} and V_{CCHI} exceed the 2.45V undervoltage lockout threshold, the 20μA current reference is connected from RAMP to GND, the charge pumps are turned on and CPON is forced high (time point 4). V_{OUTHI} and V_{OUTLO} begin to ramp up. When V_{OUTLO} exceeds the reset threshold voltage, PWRGD will immediately be forced high (time point 5). After a 200ms delay, $\overline{RESET}$ will be pulled high and DISABLE will be pulled low, enabling the data bus (time point 6).

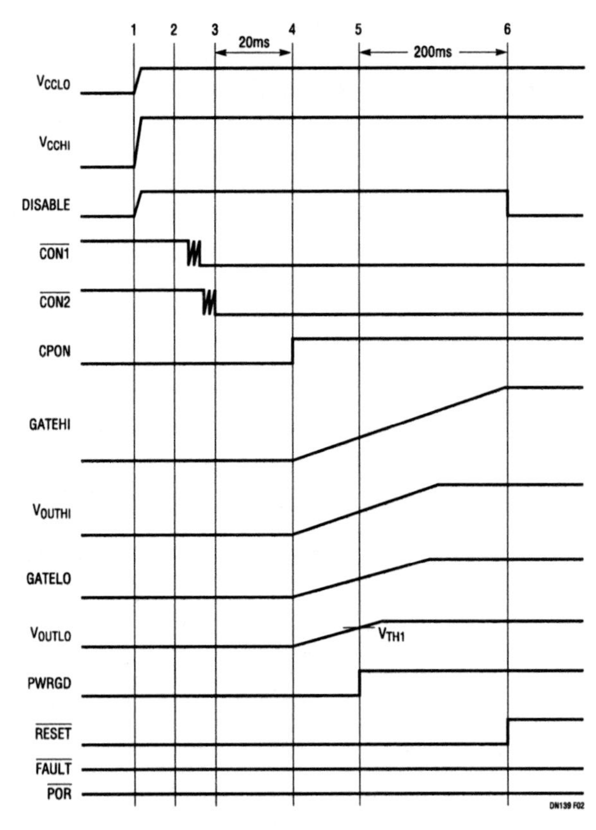

Figure 264.2 • Board Insertion Timing

Section 19

Power over Ethernet

Active bridge rectifiers reduce heat dissipation within PoE security cameras

265

Ryan Huff

Introduction

Power over Ethernet (PoE) has been embraced by the video surveillance industry as a solution to an age-old problem: complicated cabling. For instance, a basic, traditional fixed-view security camera requires two cables: one for power (10W to 15W from a 24V AC or 12V DC), and a separate, coax cable for the video signal. With PoE, a single Ethernet cable carries both video data and power. Everything is simplified. Right?

Not quite. To meet compatibility with existing systems, camera manufacturers must produce PoE-enabled cameras that are also compatible with legacy power sources—they must accept PoE 37V to 57V DC from an RJ-45 jack or 24V AC, +12V DC, or −12V DC from an auxiliary power connector.

The old way loses power

Figure 265.1 shows the power architecture used by many PoE camera manufacturers to solve this problem. A full-bridge diode rectifier after the auxiliary (old-school) input produces positive DC power from either 24V AC, +12V DC or −12V DC. The resulting DC power and the PoE inputs are diode-ORed with the winning supply fed to a wide input voltage isolated switching power supply, which in turn powers the camera electronics.

This power architecture presents a few challenges. When the camera is powered from the auxiliary input, three diodes (circled in Figure 265.1) fall into the power path. In addition to the inefficiency of this design and possible heat problems from the power dissipated by the diodes, the three diodes lead to a significant voltage drop at the input to the switching power supply. With a 10W to 15W camera, these challenges are easily surmountable, but the latest security cameras have doubled this power consumption. Features like pan/tilt/zoom (PTZ) and camera lens heaters for outdoor operation have made this power architecture unsuitable for this new wave of cameras.

To illustrate the architecture's deficiencies, consider a 26W camera. For a 12V DC auxiliary input (assumed to actually be 9V DC due to use of unregulated wall warts/AC adapters) and three 0.5V drop Schottky diodes, the input voltage of

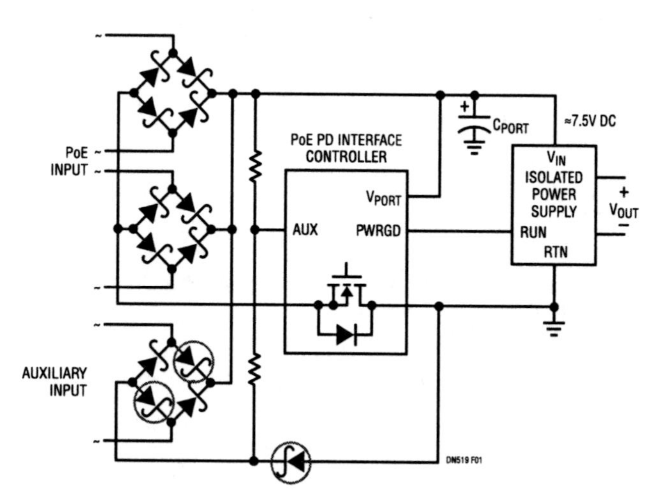

Figure 265.1 • Auxiliary Input and PoE Power Architecture

Analog Circuit Design: Design Note Collection. http://dx.doi.org/10.1016/B978-0-12-800001-4.00265-9

the switching power supply is 7.5V (9V – 3·0.5V). The input current for this camera is approximately 3.5A (26W/7.5V). The resultant power dissipation of the three Schottky diodes in the power path is 5.2W (3.5A·3·0.5V). This power dissipation leads to higher temperature within the camera, which is difficult, time consuming and expensive to mitigate.

Improve performance with ideal diodes

Figure 265.2 shows a way to counter this shortcoming. Here, the two diodes of the full-bridge rectifier are replaced by ideal diodes, circled (black) in Figure 265.2. Ideal diodes are simply MOSFETs controlled to behave like regular diodes. The advantage of an ideal diode is that one can use MOSFETs with low channel resistance ($R_{DS(ON)}$), thus reducing the forward voltage drop ($I_{DS} \cdot R_{DS(ON)}$) to much less than a Schottky diode drop. The LT4320 ideal diode bridge controller enables the control of four MOSFETs in a full-bridge configuration. The third diode drop due to the diode-OR in Figure 265.1 is eliminated by the LT4275 LTPoE++/PoE+/PoE PD controller. Its topology allows the use of a few small-signal diodes, circled together in Figure 265.2, for auxiliary input sensing. These diodes are not in the power path as in the traditional architecture, so they do not contribute any additional voltage drop or heat issues.

Results

The power architecture shown in Figure 265.2 significantly reduces overall power losses when compared to that of Figure 265.1. To quantify, the LT4320 combined with low channel resistance MOSFETs results in a 20mV drop across each ideal diode bridge MOSFET. This produces an input at the isolated supply of 8.96V (9V – 2·20mV). The higher input voltage drops the required input current to only 2.9A (26W/8.96V) versus the original 3.5A.

The resulting power dissipation of the improved architecture is now a scant 116mW (2.9A·2·20mV), versus 5.2W for the original architecture—a 45× reduction! Additionally, the lower input current further reduces power dissipation in the isolated power supply's power components (i.e., input filter inductor, power transformer and switching MOSFETs) due to the reduction of their I^2R power losses. A simple calculation puts this reduction at 31% ($100\% - 2.9A^2/3.5A^2$).

Conclusion

Adding the LT4320 and LT4275 to the auxiliary and PoE inputs of a PoE-enabled security camera recovers more than 5W (5.2W – 116mW) of power dissipation over traditional full-bridge/diode-OR designs. This reduction of power eases the thermal design time and complexity of PoE security cameras.

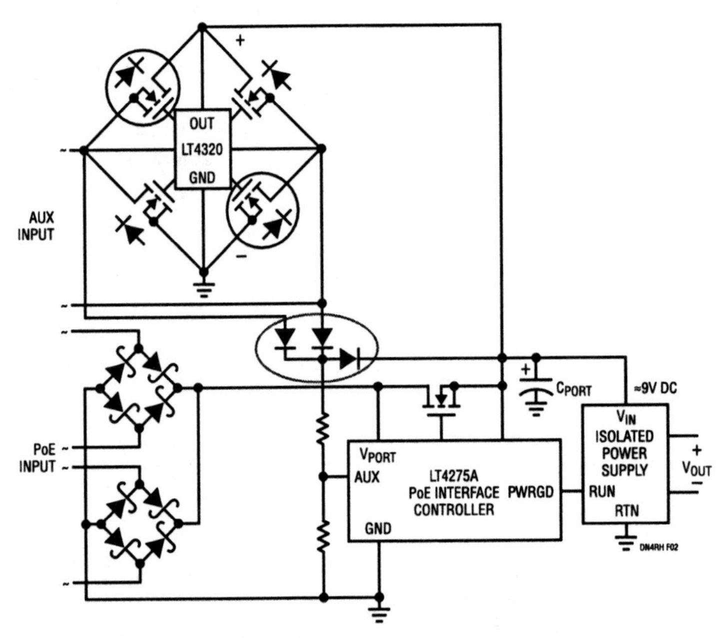

Figure 265.2 • Improved Power Architecture with No Diode Drops in PowerPath

High power PoE PD interface with integrated flyback controller

266

Dilian Reyes

Introduction

To this day, Power over Ethernet (PoE) continues to gain popularity in today's networking world. The 12.95W delivered to the Powered Device (PD) input supplied by the Power Sourcing Equipment (PSE) is a universal supply. Each PD provides its own DC/DC conversion from a nominal 48V supply, thus eliminating the need for a correct voltage wall adapter. However, higher power devices cannot take advantage of standard PoE because of its power limitations, and must rely on a large wall adapter as their primary supply. The new LTC4268-1 breaks this power barrier by allowing for power of up to 35W for such power-hungry 2-pair PoE applications. The LTC4268-1 provides a complete solution by integrating a high power PD interface control with an isolated flyback controller.

PD interface controller

The PD interface controller provides the same 25k signature detection resistance defined in the standard PoE. An extended optional class can be read by a customized PSE that looks for such a class. Once a PSE detects and classifies the PD, it fully powers on the device. The LTC4268-1 provides a low inrush current limit, allowing load capacitance to ramp up to the line voltage in a controlled manner without interference from the PSE current limit. After the load capacitance is charged up, the LTC4268-1 switches to the high input current limit and provides a power good signal to its switching regulator indicating that it can start its operation. During this time, the LTC4268-1 remains in its high current limit state allowing for up to 35W delivered to the load.

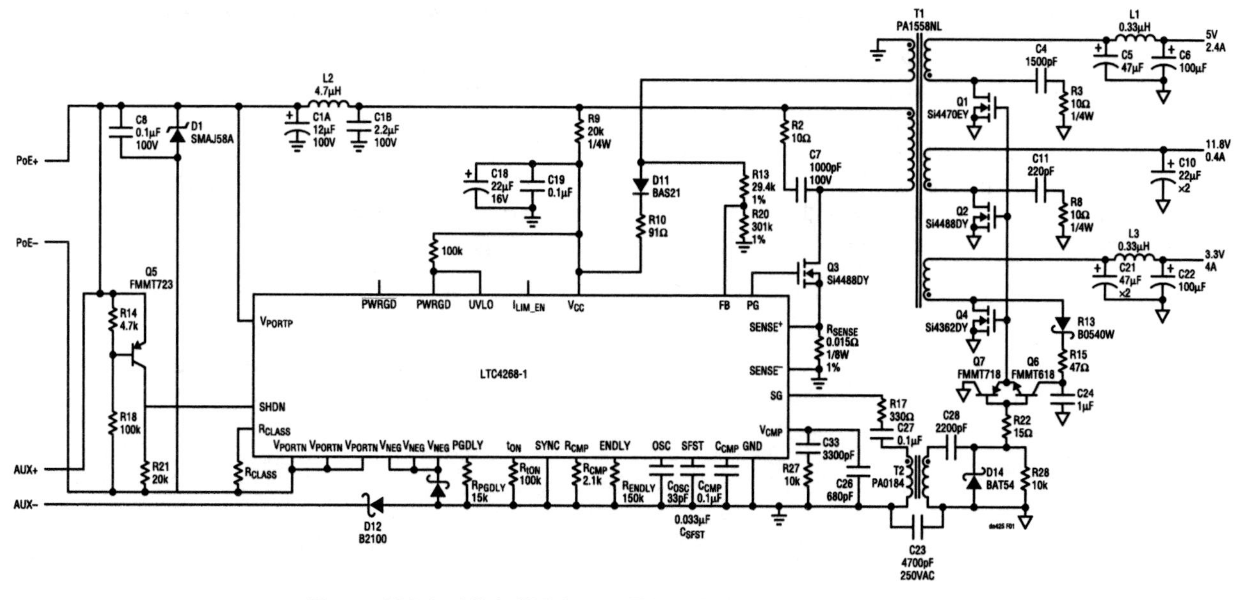

Figure 266.1 • High Efficiency, Triple Output, High Power PD

Analog Circuit Design: Design Note Collection. http://dx.doi.org/10.1016/B978-0-12-800001-4.00266-0

Synchronous flyback controller

Once power is switched over to the synchronous flyback controller, the LTC4268-1 regulates the output voltages by sensing the average of all the output voltages via a transformer winding during the flyback time. This allows for tight output regulation without the use of an opto-isolator, providing improved dynamic response and reliability. Synchronous rectification increases the conversion efficiency and cross-regulation effectiveness above a conventional flyback topology. No external driver ICs or delay circuits are needed to achieve synchronous rectification; a single resistor is all that is needed to program the synchronous rectifier's timing.

High efficiency, triple output, high power PD

Figure 266.1 shows a design using the LTC4268-1 in a high power, triple output PD. A high power PSE connects through an Ethernet cable to the RJ45 connector. PSE detection and power is passed through the data pairs' high power Ethernet transformer or directly to the spare pairs in this 2-pair 10/100BaseT PoE system. The PSE power is then controlled by the LTC4268-1 PD interface and forwarded on to its switching regulator. An auxiliary supply option can also be connected to bypass and disable the PD interface which gives the auxiliary priority in power supply over PoE. Power conversion is then from the auxiliary supply down to the output voltages.

The small supply of the LTC4268-1 utilizes an isolated flyback topology with synchronous rectification that requires no opto-isolator, lowering the parts count. This circuit gives efficiencies at full load of 83% when powered from a PSE and over 85% power sourced from an auxiliary supply.

PSE and auxiliary supplies

Standard PSEs are capable of providing as low as 15.4W at the port output. This would not be sufficient power for a high power PD operating at full load. Here, a customized PSE capable of delivering higher power must be used, or a PSE controller designed for high power such as an LTC4263-1 single port PSE controller. In cases where a high power PSE is not available, an auxiliary supply can be used.

2-pair vs 4-pair PD

2-pair power is used today in IEEE 802.3af systems. One pair of conductors is used to deliver the current and a second pair is used for the return while two conductor pairs are not powered. This architecture offers the simplest implementation method but suffers from higher cable loss than an equivalent 4-pair system.

4-pair power delivers current to the PD via two conductor pairs in parallel allowing for an even higher level of power. This lowers the cable resistance but raises the issue of current balance between each conductor pair. Differences in resistance of the transformer, cable and connectors along with differences in diode bridge forward voltage in the PD can cause an imbalance in the currents flowing through each pair. Using two independent LTC4268-1s (Figure 266.2) allows for interfacing and power from two independent PSEs, and independent DC/DC converters resolve the current imbalance.

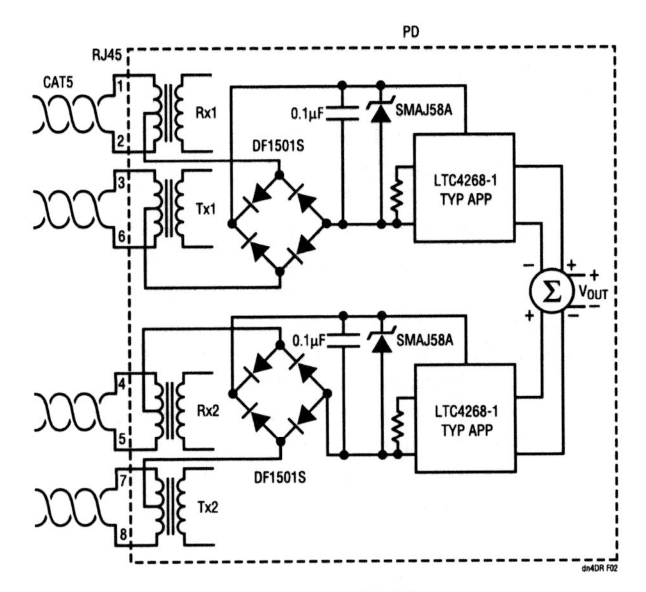

Figure 266.2 • 4-Pair, High Power PD Diagram

Conclusion

The LTC4268-1 is a highly integrated solution for the next generation of PD products. It offers PoE PD functionality with control for efficient high power delivery to the output load.

Simple battery circuit extends Power over Ethernet (PoE) peak current

Mark Gurries

Introduction

Power over Ethernet (PoE) is a new development that allows for the delivery of power to Ethernet-based devices via standard Ethernet CAT5 cable, precluding the need for wall adapters or other external power sources. The PoE specification defines a hardware detection protocol where Power Sourcing Equipment (PSE) is able to identify PoE Powered Devices (PDs), thus allowing full backwards compatibility with non-PoE-aware (legacy) Ethernet devices. The PoE specification also sets an upper limit on the power that can

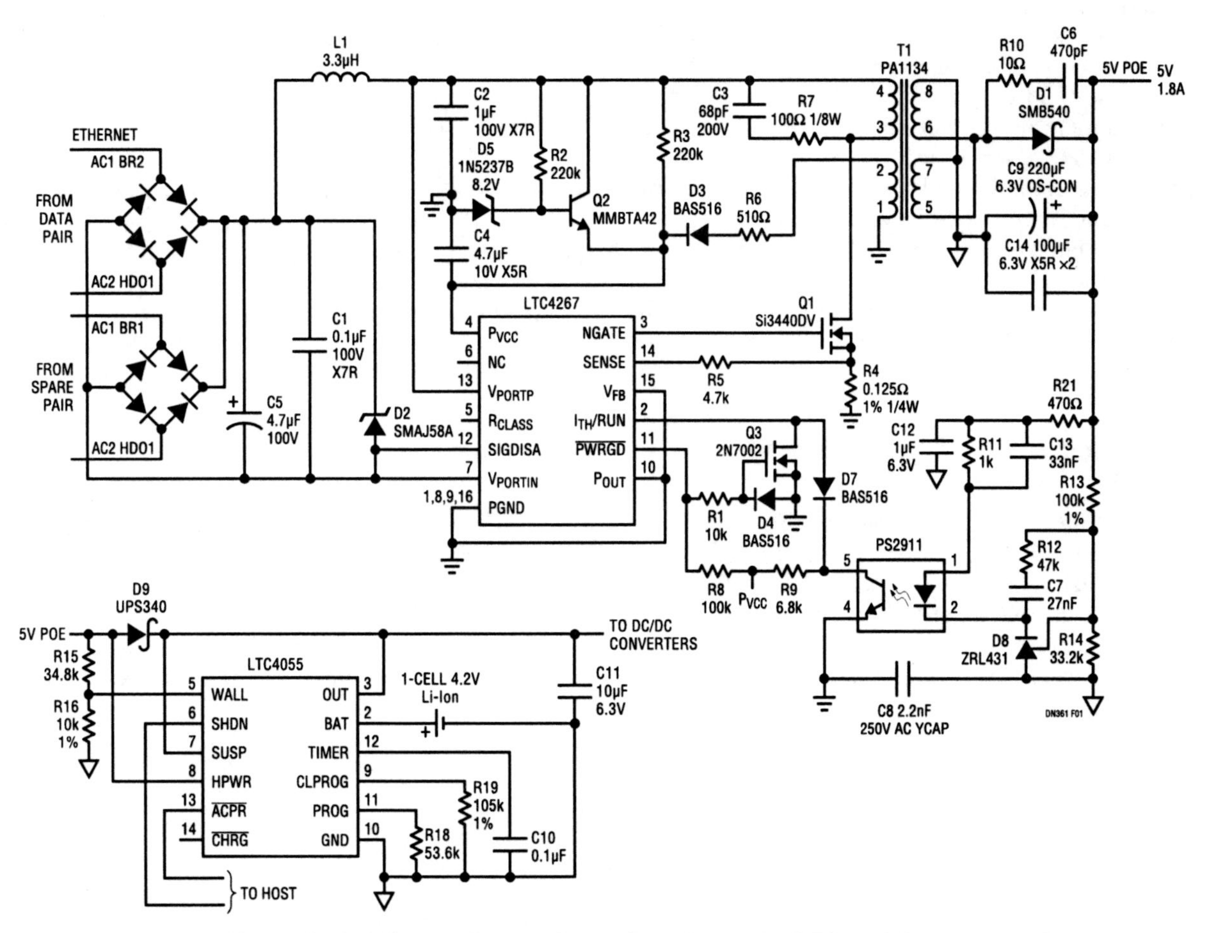

Figure 267.1 • Simple Battery Charger/PowerPath Controller (LTC4055) Augments PoE Regulator's (LTC4267) Peak Output Power to Overcome PoE Power Constraints

Analog Circuit Design: Design Note Collection. http://dx.doi.org/10.1016/B978-0-12-800001-4.00267-2

be drawn by a PD. The problem is: what happens when a PD must draw more power than allowed by the PoE standard? Examples may be the spin up of a disk drive or a period of sustained transmission of data from an RF transmitter. If the *average* power load of these applications is less than the available PoE power, one solution is to store power in the PD when power consumption is low and then tap the reserve to augment PoE power when needed. For many applications, a rechargeable battery fits the bill.

Of course, one cannot just throw a battery and a battery charger into the mix. The power path must be able to change seamlessly, on the fly, from PoE-powers-device-and-charges-battery, to PoE-and-battery-power-device, to battery-powers-device. Figure 267.1 shows a complete and compact solution.

The PoE circuit

By default, power over the Ethernet is *not* available. The standard calls for a protocol to be implemented that allows the Ethernet hub to identify the device needing power. The LTC4267 simplifies the design of PDs by providing wholesale implementation of the protocol and power management functions.

PoE power comes in the form of −48V at 350mA. If the PoE current is allowed to exceed 400mA, the standard calls for the PSE to break the circuit. This is a problem for devices that occasionally need a little more juice than PoE will offer. Another problem is that −48V does not easily convert to commonly used positive voltage supply rails. Designers are forced to provide DC isolation along with the inverted down conversion to a more usable voltage. To meet these requirements, the LTC4267 used in Figure 267.1's circuit implements an input current limited DC input isolated flyback converter, providing a user-settable regulated low voltage.

The LTC4267 circuit in Figure 267.1 supplies 5V at 1.8A. 5V is a popular supply voltage to run logic, interface with other devices such as USB, and of primary concern in this application, to charge a single Li-Ion cell to its target termination voltage of 4.2V.

PowerPath and charger circuit

In Figure 267.1, the LTC4055 provides triple PowerPath control and Li-Ion battery charging. One path is created by connecting an external Schottky diode to the LTC4055's OUT pin and the built-in wall adapter detection circuits. In this case, the "wall adapter" power comes from the LTC4267 5V power supply called 5V PoE. The second path is for USB power, not used in this application. The third path is the battery discharge path. When the 5V PoE power goes away or drops out of regulation, the LTC4055 automatically switches the battery power over to the OUT pin using its internal ideal diode circuit. There is no delay in the switchover, so power is never lost.

When 5V PoE power is restored, the battery is disconnected from the load and charging is permitted. The LTC4055 charge current is adjustable and in Figure 267.1, the circuit is limited to 900mA which is drawn from the OUT pin. That leaves 900mA to run the system while charging. Powered devices connected to the OUT pin must be compatible with the Li-Ion voltage range. The $\overline{\text{ACPR}}$ pin of the LTC4055 can be used to indicate which power source is providing power, allowing the PD to configure itself accordingly.

High transient load or continuous current load operation

When the power limit of the 5V PoE supply is reached, the voltage drops and the battery charger shuts down to relieve the PSE of the charge current load. If the voltage continues to collapse, the battery automatically is placed into parallel operation with the 5V PoE power supply, thus increasing the available peak load current. The LTC4055 $\overline{\text{ACPR}}$ signal is active high during the overload. Battery charging automatically resumes once the overload goes away and the 5V PoE voltage has risen enough to show recovery.

Optimization options

If sustained currents approaching 1.8A are expected from the 5V PoE and there are thermal management issues related to the diode's heat dissipation, the diode D9 can be replaced with the LTC4411 ideal diode for more efficient operation. Recommended DC/DC converters to generate logic supplies in this application include the LTC3443 buck-boost and/or the LTC3407-2 dual buck regulators.

Conclusion

The highly integrated LTC4267 and LTC4055 simplify the design of compact, simple and complete battery-based power systems that run from Ethernet power. More importantly, seamless PowerPath control enables circuits that can use a battery to augment Ethernet power when an application momentarily demands more than the PoE standard allows.

Fully autonomous IEEE 802.3af Power over Ethernet midspan PSE requires no microcontroller

268

Dilian Reyes

Introduction

The IEEE802.3af Power over Ethernet (PoE) standard defines how power will be delivered over CAT5 lines. Despite the differences between legacy devices and those that adhere to the new standard, there is no need to completely replace existing systems. The nominal 48V required by powered devices (PDs) can be delivered by midspan power sourcing equipment (PSE), which is connected to the front end in series with legacy routers and switches. The LTC4259A is a quad PSE controller designed for both endpoint and midspan PSEs that integrates PD signature detection, power level classification, AC and DC disconnect detection and current limit without the need for a microcontroller.

A PSE's duties

The responsibilities of the PSE are to correctly detect if a compliant PD has been connected to a port, optionally classify the PD and properly apply power to the PD while

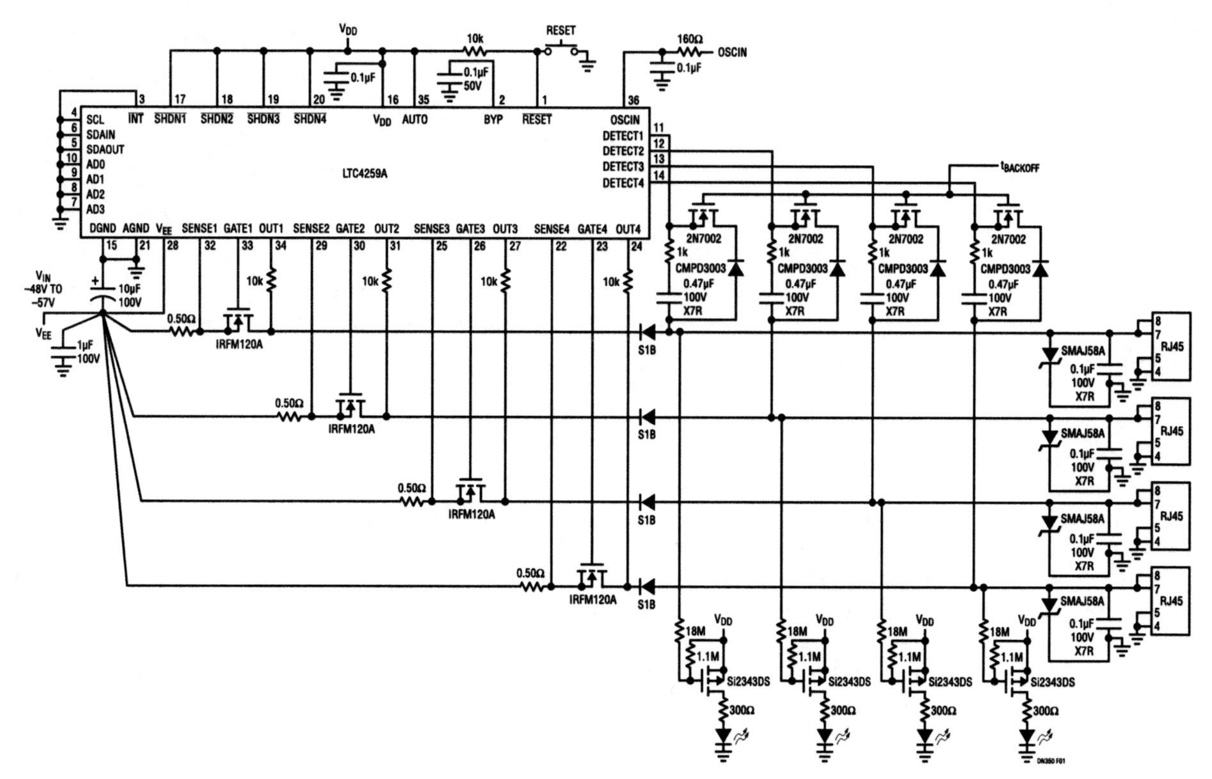

Figure 268.1 • Autonomous 4-Port Power over Ethernet Midspan PSE

Analog Circuit Design: Design Note Collection. http://dx.doi.org/10.1016/B978-0-12-800001-4.00268-4

protecting the port from fault conditions. Once a PD is powered on, the PSE monitors a PD's presence and switches off power when the device is removed. A PSE must also provide overcurrent protection to prevent damage to the PSE and PD.

Traditional PSE solutions use a microcontroller to perform the detection measurements and calculations and control additional circuitry that switches power to a PD. The LTC4259A in Figure 268.1, by contrast, requires no microcontroller and runs autonomously carrying out signature detection. It automatically interprets the loading conditions and powers on a valid PD.

The midspan PSE also must not interfere with an endpoint's operation. An endpoint PSE applies power on either the signal pairs or the spare pairs of the CAT5 cable, while the midspan PSE must apply power to only the spare pairs. To avoid conflict if the two were to be connected at the same time, the circuit in Figure 268.2 implements an LTC1726 watchdog timer to periodically disable the LTC4259A's detection scheme for two seconds. Midspan devices are required to have a backoff capability after a failed attempt of detection to allow for a potentially present endpoint PSE to detect and power on a port.

After the backoff interval is complete, LTC4259A detection is re-enabled for at least one full detection cycle. If a midspan or endpoint PSE is able to detect a valid signature 25kΩ (R_{SIG}) and power up the PD, a compliant PD would no longer display the R_{SIG} to prevent any further good signature detects and power ups from a second PSE. Hardware implementation of the backoff timer eliminates the need for a microcontroller software timing routine.

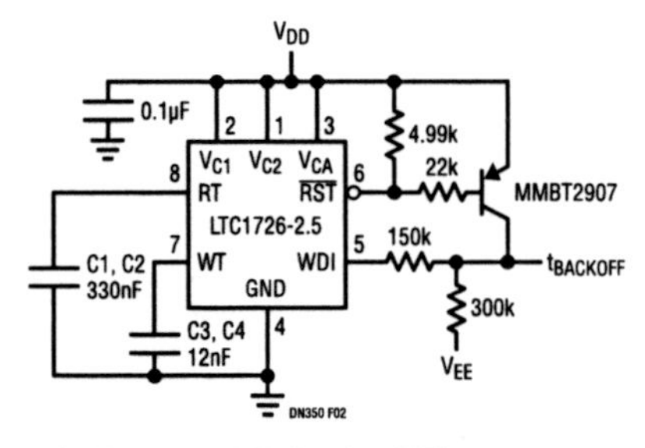

Figure 268.2 • Midspan PSE Backoff Timer

Disconnect detection

When a PD is unplugged from a powered port, the IEEE standards specify that a PSE must implement at least one of two power disconnect detections modes for port power removal: DC disconnect and/or AC disconnect.

DC disconnect measures a minimum current drawn from a port to determine if a PD is present and requires power. While this is easier to implement, AC disconnect is considered a more accurate detection of a PD's presence. AC disconnect measures the PD impedance and would keep a port powered even for PDs that idle at low power.

The LTC4259A auto mode uses the AC disconnect method by default. The LT1498 in Figure 268.3 is a dual rail-to-rail op amp used to output a sine wave to drive OSCIN of the LTC4259A. The LTC4259A applies the AC signal to the lines and detects its presence when a PD has been removed and the port power is to be switched off.

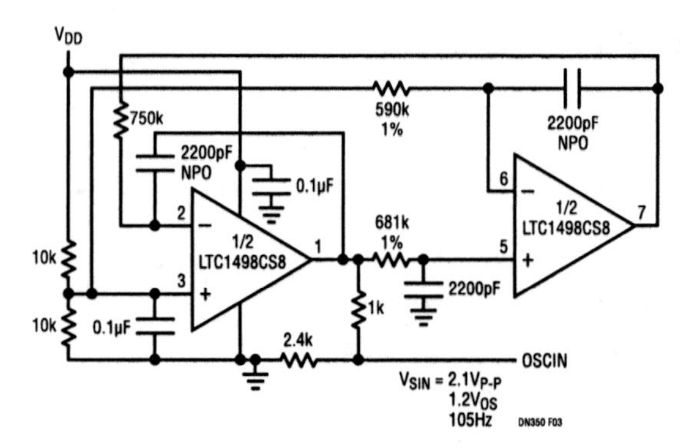

Figure 268.3 • Sine Wave Circuit for AC Disconnect

Supplying 3.3V from −48V

A 3.3V supply powers the digital portion of the LTC4259A. The LTC3803 circuit in Figure 268.4 converts −48V to 3.3V eliminating the need for a second power supply. This boost regulator circuit achieves a tight 2% regulation and outputs 400mA, enough for up to 12 LTC4259As and port indicator LEDs in a 48-port application.

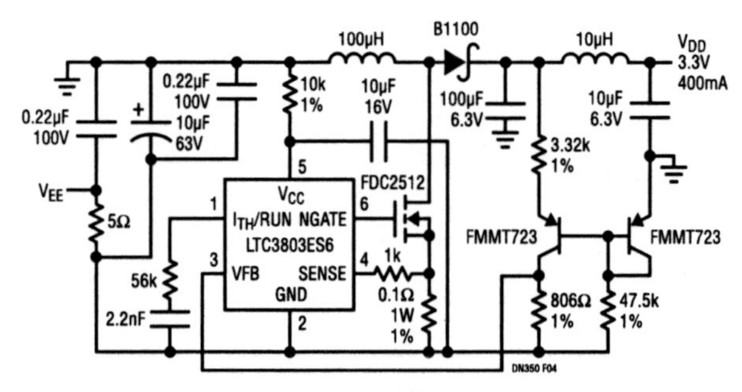

Figure 268.4 • −48V to 3.3V Boost Converter

LTC4259A options

The LTC4259A also allows flexibility when designing an endpoint or midspan PSE. Internal registers can be accessed via I^2C for additional control and settings, including the option of DC disconnect. The LTC4259A aids in IEEE-compliant power management by providing PD classification—a better method than guessing via monitoring current—of devices that present a class, such as an LTC4257 PD Interface Controller.

Power over Ethernet isolated power supply delivers 11.5W at 90% efficiency

269

Jesus Rosales

When powering IP telephones, wireless access points, PDA charging stations and other PDs (Powered Devices) from an Ethernet cable, designers have at most 12.95W of available power per the IEEE 802.3af standard. Increased demands for power mandate a very efficient power converter especially for class 3 devices (consuming between 6.49W and 12.95W). The more power lost in the converter, the less power available for the PD.

The voltage available from the PSE (Power Sourcing Equipment) ranges from 44V to 57V, but PDs need to operate with as much as 20Ω of series wire resistance. A PD can never draw more than 350mA or 12.95W continuously. With the maximum input current of $350mA_{RMS}$, the input voltage can droop as much as 7V bringing the lower side of the input range to 37V. To avoid interfering with the classification signature impedance measurement, a PD must not draw significant current below 30V.

There are many topologies to choose from when designing an isolated DC/DC converter, but for PoE (Power over Ethernet) applications, the choices are few. When trying to maximize efficiency every milliwatt counts. MOSFET gate driving losses become significant so the fewer switches to turn on and off the better. A push-pull converter could be used, but the additional complexity is not justified at this power level. A single transistor forward converter is another option, but requires an additional output inductor and rectifier. A flyback converter is the simplest choice. Flyback converters are thought to be less efficient than forward and push-pull converters, but that changes when the output is synchronously rectified (a MOSFET is used instead of a diode to rectify the output).

The LT1725 switching regulator controller greatly simplifies the design of PoE supplies. The LT1725 is specifically designed for the isolated flyback topology and includes features that make it a good match for PoE supplies, including programmable input undervoltage lockout, hysteretic start-up and a patented feedback circuit that eliminates the need for an optocoupler while providing excellent output regulation.[1]

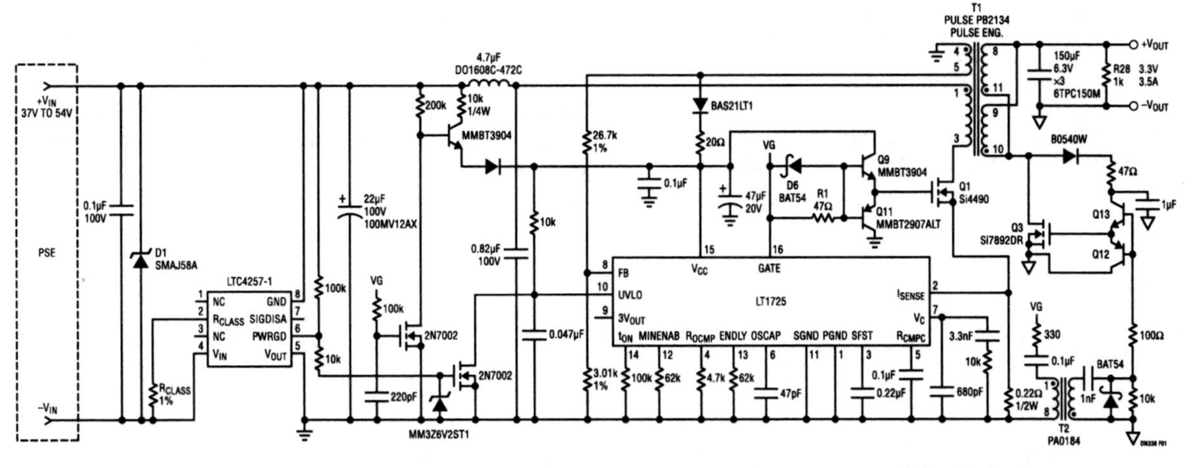

Figure 269.1 • 36V–72V Input to 3.3V at 3.5A Output, Isolated Synchronous Flyback Converter

Note 1: U.S. Patent No. 05438499, 05305192, 0584163.

Analog Circuit Design: Design Note Collection. http://dx.doi.org/10.1016/B978-0-12-800001-4.00269-6

In order to maximize power to the load, the converter chosen must have synchronous rectification. Diodes will simply dissipate too much power. Figure 269.1 shows a schematic for an isolated synchronous flyback converter using the LT1725 that achieves efficiencies as high as 90% at 11.5W out. Figure 269.2 shows an efficiency curve for this synchronous converter and a second curve for the same converter with a 6CWQ06FN diode used in place of Q3 (making the converter non-synchronous). While the synchronous converter needs a few extra components to control the rectifying MOSFET Q3, the resulting efficiency gain (approximately 10%) is significant. An additional benefit of the lower power dissipation with a MOSFET versus a diode is that a heat sink is no longer necessary, allowing for a dramatic reduction in board space. The other 5% efficiency gain comes from the elimination of preload. A synchronous converter does not need any preload to keep the output voltage in regulation while the non-synchronous converter does. The output of a non-synchronous converter can float up, uncontrolled, without a preload.

Another advantage of a synchronous converter is tighter load regulation as shown by Figure 269.3. The main reason is that the forward voltage in a rectifier diode changes with load current while the voltage drop in a MOSFET remains consistent and low.

This circuit was designed primarily to provide 3.3V at 3.5A from an input of 37V to 54V. Nevertheless, converter operation is seamless over the full 36V to 72V input range (remove D1 if operated at $72V_{IN}$). Operation of this circuit is straightforward. MOSFET Q1 turns on and energy is stored in the transformer T1. Energy is then delivered to the output during the time MOSFET Q1 is off. MOSFET Q3 turns on whenever MOSFET Q1 turns off, providing output rectification. Secondary MOSFET Q3 is driven by transistor drivers Q12 and Q13. T2 inverts the LT1725 gate signal and drives the common bases of Q12 and Q13. R1, D6, Q9 and Q11 buffer the primary side gate signal and provide a small delay so that Q1 and Q3 are never on at the same time.

Complementing this converter is the LTC4257-1, which provides complete signature and interface functions for a PD operating in an IEEE 802.3af PoE system.

Conclusion

To get the most out of the power available for a PoE supply, a converter needs to have synchronous rectification. The flyback topology offers the simplest solution and can operate synchronously with small incremental cost. The power consumed by the Q3 MOSFET in this design is roughly one tenth of that of a rectifying diode. The 1.4W of total converter dissipation is evenly distributed among switching MOSFETs, power transformer and controller IC, allowing the designer to compact board layout. The LT1725 greatly simplifies circuit design because of its patented feedback circuit which eliminates the need for an optocoupler and a secondary reference without sacrificing load regulation.

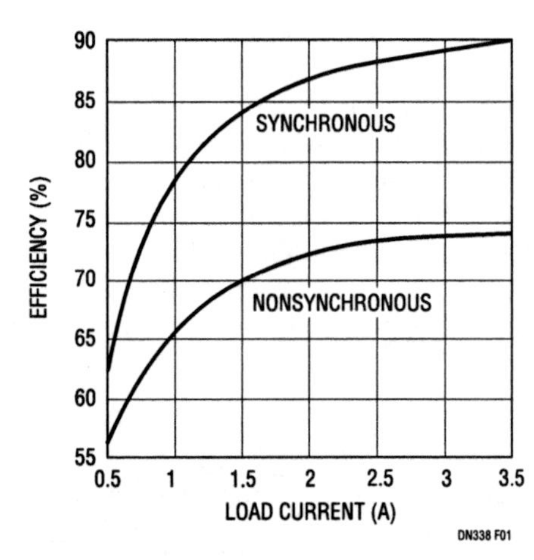

Figure 269.2 • Efficiency Comparison between a Synchronous and Non-synchronous Converter Using the LT1725

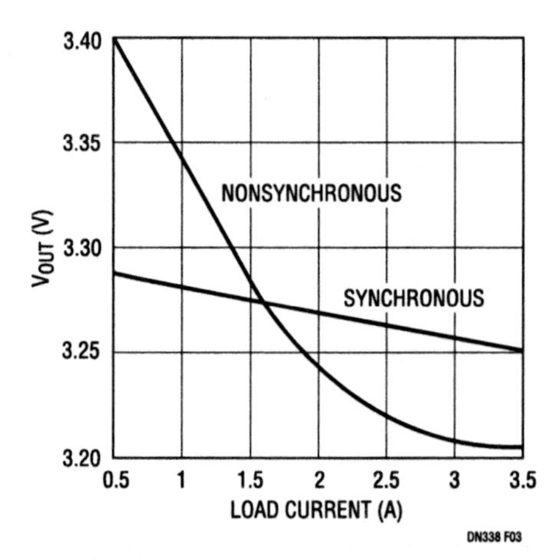

Figure 269.3 • Load Regulation Comparison Between a Synchronous and Non-synchronous Converter Using the LT1725

Section 20

System Monitoring and Control

Pushbutton on/off controller with failsafe voltage monitoring

270

Victor Fleury

Introduction

Have you had the exasperating experience of a laptop or PDA defiantly not responding to your commands? You frantically press key after key, but to no avail. As hope turns to anger (but just before you throw the company's laptop through the window) you slam your finger against the on/off power button. Ten seconds later, your laptop finally surrenders and the screen goes black in a high pitched whimper.

The unresponsive pushbutton was likely the result of an unresponsive µP or system logic—as evidenced by the crash. By pressing and holding the on/off pushbutton, the LTC2953 provides the user with the ability to force system power off, even under fault conditions. This long pushbutton command works independently of system logic and automatically shuts off power after the adjustable timer expires. The length of time the pushbutton must be held low in order to force a power-down is adjustable with an external capacitor on the PDT pin.

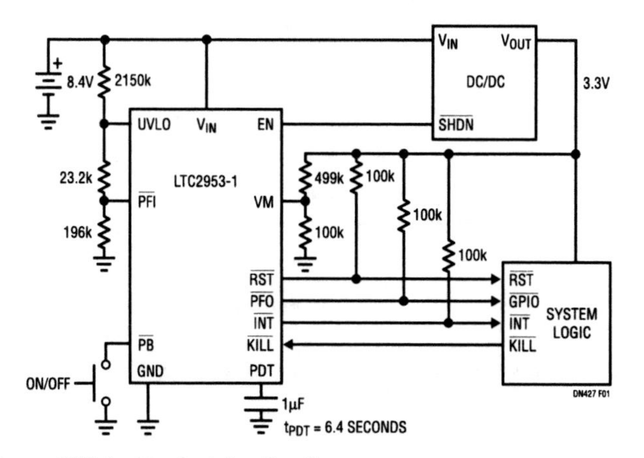

Figure 270.1 • Typical Application

Pushbutton challenges

The on/off pushbutton of electronic devices presents the system designer with a unique set of challenges. The circuits that monitor the pushbutton translate the chattering pushbutton signal into a clean voltage step that enables a DC/DC converter or turns on a power switch. These circuits communicate with system logic to make sure that power turns on and turns off in an orderly manner. Additionally, failsafe features should disable system power if there is a problem with either the input or output power supply. The pushbutton monitor must also be rugged: absorb high levels of electrostatic discharge, tolerate voltage transients below ground and operate at high voltage levels.

The LTC2953 pushbutton on/off controller with voltage monitoring addresses all of these issues by providing a complete solution for interfacing to the on/off pushbutton of electronic devices. This tiny IC integrates the timing circuitry needed to clean up the pushbutton chatter and provides a simple communication protocol for orderly system power turn on and turn off. The LTC2953 includes a deglitched lockout comparator that prevents the system from drawing power from a dead battery or low supply. The device also provides a single adjustable supply reset monitor with 200ms delay.

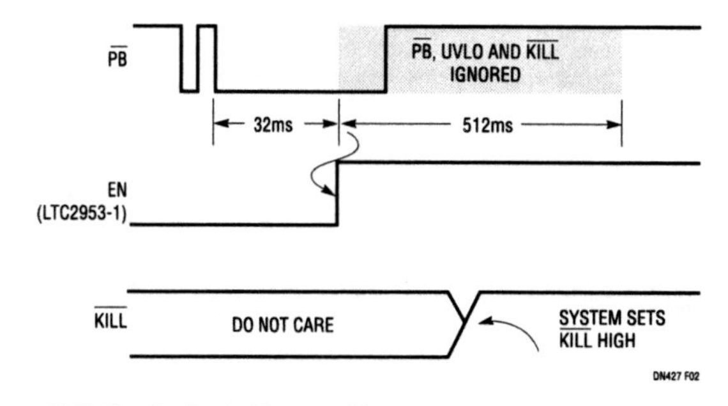

Figure 270.2 • Orderly Power-On

Analog Circuit Design: Design Note Collection. http://dx.doi.org/10.1016/B978-0-12-800001-4.00270-2

The LTC2953's wide input voltage range (2.7V to 27V) is designed to operate from single-cell to multicell battery stacks, thus eliminating the need for a high voltage LDO. The part's feature set allows the system designer to turn off power to all circuits except the LTC2953, whose low quiescent current (14µA typical) extends battery life. The device is available in a space saving 12-lead 3mm × 3mm DFN package.

Orderly power-on

The rugged pushbutton input of the LTC2953 connects directly to the electronic device's noisy, chattering mechanical on/off switch. To turn on system power, the LTC2953 asserts the enable output 32ms after detecting the end of pushbutton chatter. Once power has been enabled, the system must set the $\overline{\text{KILL}}$ input high within 512ms. This 512ms timeout period is a failsafe feature that prevents the user from turning on the electronic device when there is a faulty DC/DC converter or an unresponsive microprocessor. The LTC2953 turns off power if $\overline{\text{KILL}}$ is not set high during this time window. See Figure 270.1's application circuit and Figure 270.2's timing diagram.

Orderly power-off: short interrupt pulse

Under normal conditions, an electronic device is turned off by pulsing the on/off power switch. To turn off system power, the LTC2953 asserts the interrupt output 32ms after detecting the end of pushbutton chatter. Upon noticing this interrupt signal, system logic performs power-down and housekeeping tasks and asserts $\overline{\text{KILL}}$ low when done. The LTC2953 subsequently releases the enable output, thus turning off system power (see Figure 270.3's timing diagram).

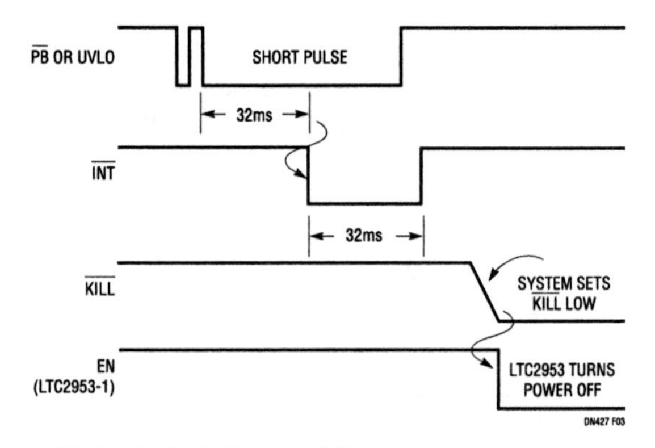

Figure 270.3 • Orderly Power-Off

Failsafe features

The LTC2953 provides three comparators for voltage monitoring: UVLO, Power Fail and Reset. The UVLO comparator detects three types of aberrant behavior at the input supply. If the supply glitches for longer than 32ms, the LTC2953 will issue an interrupt signal. If the supply falls and stays below the user adjustable level, the LTC2953 will turn off system power after the user-adjustable timer expires. Additionally, the UVLO comparator prevents a user from turning on system power if the input supply is too low (see Figure 270.4). The power fail is a general purpose uncommitted comparator, useful for distinguishing between a $\overline{\text{PB}}$ interrupt and a low supply interrupt. The reset comparator is a single adjustable voltage monitor with fixed 200ms delay.

Conclusion

The LTC2953 is a low power, wide input voltage range (2.7V to 27V) pushbutton on/off controller with input and output voltage monitoring. The LTC2953 provides a simple and complete solution for manually toggling power of many types of systems. Desirable features include a power fail comparator that issues an early warning of a decaying supply, along with a UVLO comparator that prevents a user from turning on a system with a low supply or dead battery. The LTC2953 provides even greater system reliability by integrating an adjustable single supply supervisor. Two versions of the part accommodate either positive or negative enable polarities. The device is available in a space saving 3mm×3mm DFN package.

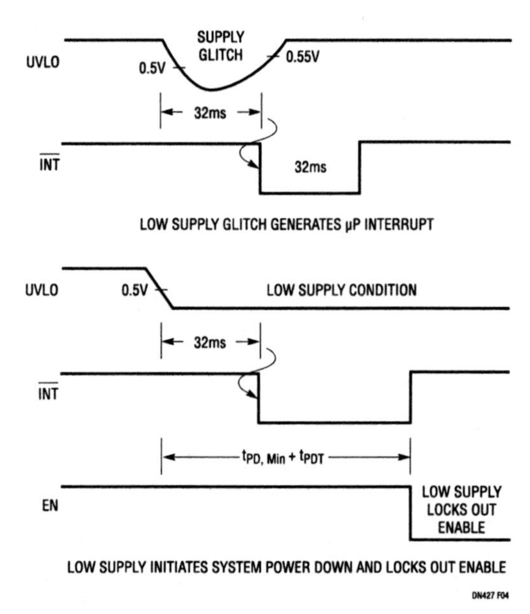

Figure 270.4 • Multifunction UVLO Comparator

Versatile voltage monitors simplify detection of overvoltage and undervoltage faults

271

Scott Jackson

Introduction

Many modern electronic systems have strict power supply operating ranges—requiring accurate monitoring of each supply. Some systems must know that all supplies are present and stable before start-up and some must know if the supplies deviate from safe operating conditions.

The LTC2912, LTC2913, and LTC2914 supervisors respectively monitor single, dual, and quad power supplies for undervoltage and overvoltage with tight 1.5% threshold accuracy over temperature. All monitors in the multiple-monitor devices share a common undervoltage output and a common overvoltage output with a timeout period that is adjustable or disabled. Each monitor has input glitch rejection to ensure reliable reset operation without false or noisy triggering.

Each part has at least two options: one with capability to latch the overvoltage output and one with capability to externally disable both outputs. The LTC2912 has a third option with latching capability and a non-inverted overvoltage output. Each part has an internal 6.5V shunt regulator allowing the device to be used in a system with any supply level.

Basic operation

Figure 271.1 shows a typical application for the LTC2914. Each monitored input is compared to a 0.5V threshold. Any channel can be configured to monitor both undervoltage and overvoltage conditions using a 3-resistor divider. When monitoring a positive voltage, the VH input of the channel is connected to the high-side tap of the resistive divider and triggers an undervoltage condition while the VL input is connected to the low-side tap of the resistive divider and triggers an overvoltage condition. When an undervoltage condition is detected, the $\overline{UV}$ output asserts low.

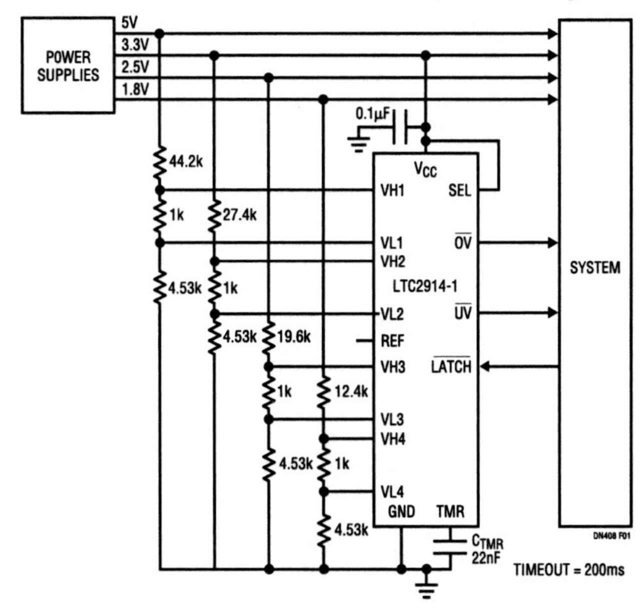

Figure 271.1 • Quad UV/OV Supply Monitor

Analog Circuit Design: Design Note Collection. http://dx.doi.org/10.1016/B978-0-12-800001-4.00271-4

Once all undervoltage conditions clear, the $\overline{UV}$ output remains asserted until a timeout period has elapsed. This timeout period is set by a capacitor between the TMR and GND pins. The timeout period can be disabled by tying the TMR pin to V_{CC}. Figure 271.2 shows the timeout period versus TMR capacitance. The $\overline{OV}$ output behaves in a similar manner. On parts with latching capability, the $\overline{OV}$ output latches when asserted until cleared by the $\overline{LATCH}$ pin. Holding the $\overline{LATCH}$ pin high bypasses the overvoltage latch.

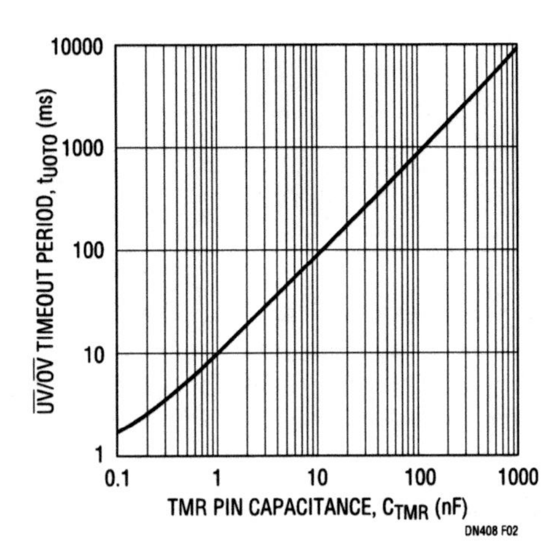

Figure 271.2 • Timeout Period vs Capacitance

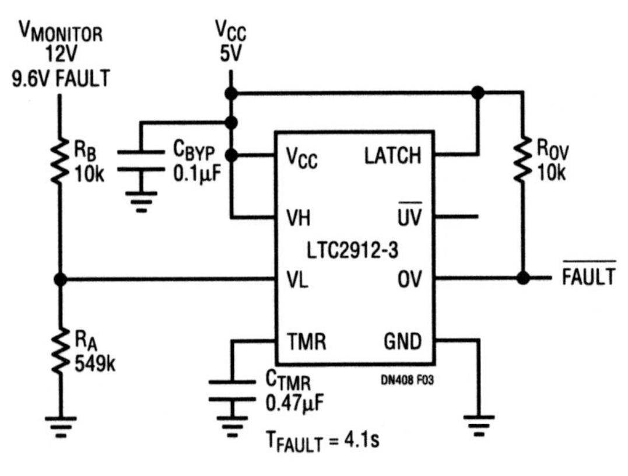

Figure 271.3 • Fault Detection Circuit for a 4.1s Undervoltage Condition

Minimum fault length monitor

The LTC2912-3 can be used to detect an undervoltage condition with a minimum duration by using the VL input and the non-inverted OV output. For example, an automobile system may need to monitor the 12V power supply during a power-up condition. During the initial cranking of the automobile, the power supply droops. If the supply droop exists for an extended period, the system may need to disconnect various circuits from the supply for protection or disconnect circuits to reduce the load. This is accomplished by the circuit shown in Figure 271.3.

The timeout function of the LTC2912 typically starts when a fault clears. However, because the VL input is used in this case to monitor an undervoltage instead of an overvoltage, the timeout function occurs at the beginning of an undervoltage condition and the OV output remains high until the period has elapsed. If the fault still exists when this timeout period elapses, the OV output ($\overline{FAULT}$) pulls low until the fault clears. Choosing a 0.47μF timing capacitor produces a 4.1s timeout delay. Therefore, any supply droop lower than 9.6V and longer than 4.1s asserts FAULT. Figure 271.4 shows the resulting waveform of a supply fault condition of 9V for 5 seconds.

Conclusion

The LTC2912, LTC2913, and LTC2914 simplify power supply monitoring of any voltage level by offering superior performance and flexibility. Only a few resistors are needed to configure monitoring of multiple voltages for both undervoltage and overvoltage conditions. The LTC2914 offers these features in a 16-lead SSOP and 16-lead (5mm×3mm) DFN package, the LTC2913 in a 10-lead MSOP and 10-lead (3mm×3mm) DFN package, and the LTC2912 in a tiny 8-lead ThinSOT and 8-lead (3mm×2mm) DFN package.

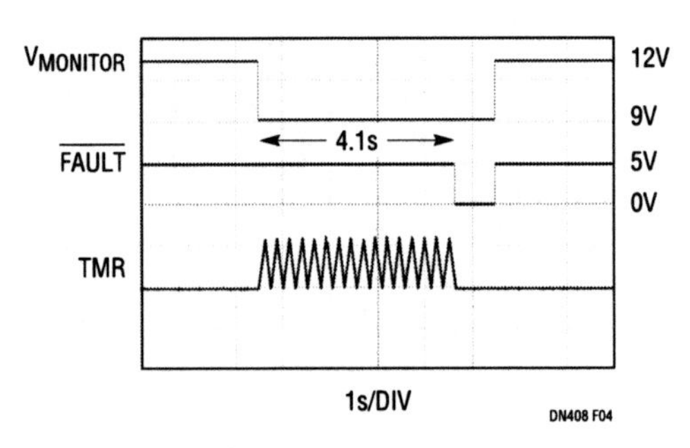

Figure 271.4 • Fault Detection Waveform of a 4.1s Undervoltage Condition

Power supply sequencing made simple

272

Bob Jurgilewicz

Introduction

System designers face a number of problems when it comes to controlling multiple power supplies. Turn-on and turn-off characteristics, supply monitoring, fault management and reset generation are a few issues that affect both the short-term system performance and long-term reliability. Design is further complicated by a process that often puts final decisions about supply requirements at the end of the design phase. So, a good supervisor/control solution allows for easy design and adjustment anywhere in the design process.

Firmware solutions place a daunting hurdle directly in the critical path of the design. Every change involves software engineers, a load of testing and worst of all, waiting. Loading code during production is time consuming and costly.

A better solution uses hardware, but easy-to-change, relatively inexpensive hardware. How about generic re-usable circuit blocks that are added early in the system design with little regard to the final specific power requirements? The existing blocks are left unfinished, simply waiting for passive component values to be determined. When final decisions about the power supplies' operating specifications are determined, calculate the values for a few passive components and populate the empty spaces in the circuit. Fortunately, such a solution exists.

The LTC2928 is a 4-channel cascadable power supply sequencer and high accuracy supervisor. Multiple LTC2928s can be easily connected to sequence an unlimited number of power supplies. Cascade action is via a single pin connection and is functional during sequence-up and sequence-down operations. Sequencing thresholds, order and timing are configured with just a few external components, eliminating the need for PC board layout or software changes during system development. Sequence outputs control supply enable pins or N-channel pass gates. Precision input comparators with individual outputs monitor power supply voltages to 1.5% accuracy. Supervisory functions include under- and overvoltage monitoring and reporting as well as reset generation. The reset output may be forced high to complement margin testing.

Application faults, whether generated by the LTC2928 or communicated by a host, can shut down all controlled supplies. The type and source of faults are reported for diagnosis. Individual channel controls are available to independently exercise enable outputs and supervisory functions. A high voltage input allows the LTC2928 to be powered from voltages as high as 16.5V. A buffered reference output permits negative power supply sequencing and monitoring operations.

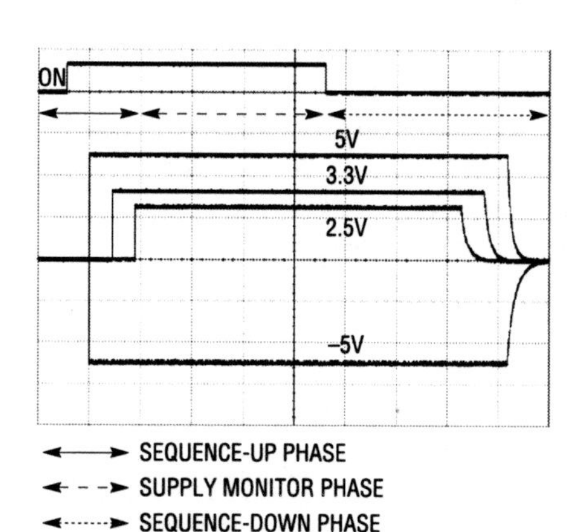

Figure 272.1 • Sequencing Application Waveforms

Three phases of the power management cycle

A complete power management cycle is divided into three phases as shown in Figure 272.1. The sequence-up phase initiates by transitioning the ON pin above threshold with a logic signal or power supply. The controlled supplies sequence-up with user-configured order and timing. All supplies must exceed a user-defined sequence-up threshold within the configured "power-good" time. If any supply fails to turn on properly, a sequence fault occurs and all controlled supplies are

Analog Circuit Design: Design Note Collection. http://dx.doi.org/10.1016/B978-0-12-800001-4.00272-6

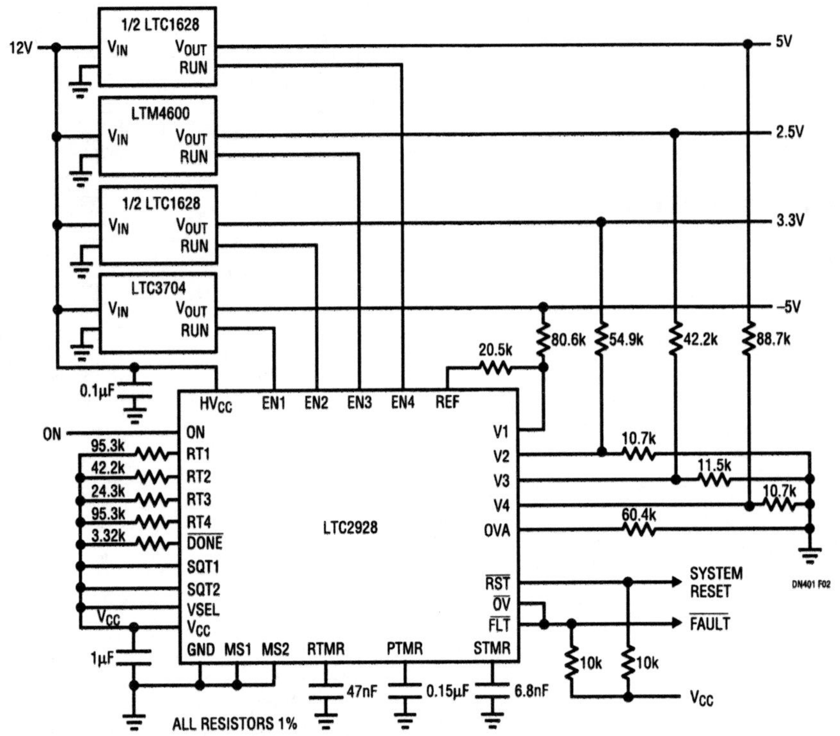

Figure 272.2 • LTC2928 Application Schematic (Component Values Calculated with the LTC2928 Configurator Tool)

shut down. Once all supplies reach their sequence-up threshold, the supply monitor phase begins (see Figure 272.2).

During the supply monitor phase, the input signals are continuously compared against user configured undervoltage and overvoltage thresholds. The comparators filter out minor glitches coupled to their inputs. If any supply is out of compliance with sufficient magnitude and duration, the reset output ($\overline{RST}$) and/or overvoltage output ($\overline{OV}$) pulls low. Once all inputs are within compliance, the respective monitor output pulls high after the user defined reset delay. Users may select whether or not a fault is generated on the basis of under- or overvoltage events. A generated fault shuts down all controlled supplies. Shutting down upon an undervoltage fault is often a critical operation. For example, consider a temporary short on one supply due to a probe slip. Once the short is removed, the supply may recover, but it might do so out of sequence if the other supplies are unaffected. A reset fault shuts off all the supplies, allowing for a new in-sequence start-up procedure.

The sequence-down phase initiates by transitioning the ON pin below threshold with a logic signal or power supply. The controlled supplies sequence-down with user-configured order and timing. All supplies must fall below the user-defined sequence-down threshold within the configured "power-good" time. If any supply fails to turn off properly, a sequence fault occurs and all controlled supplies are shut down.

LTC2928 configuration software designs it for you

To make life truly simple, Linear Technology offers free configuration software that calculates all resistor values, capacitor values and required logic connections. The tool also generates schematics and a passive element bill-of-materials. All you need to know are your supply parameters and sequence order. Contact Linear Technology for details.

Conclusion

The LTC2928 greatly reduces the time and cost of power management design by eliminating the need to develop, verify and load firmware at back end test. System control issues such as sequence order, timing, reset generation, supply monitoring and fault management are all handled with the LTC2928.

Pushbutton on/off controller simplifies system design

273

Victor Fleury

Introduction

Handheld designers often grapple with ways to de-bounce and control the on/off pushbutton of portable devices. Traditional de-bounce designs use discrete logic, flip-flops, resistors and capacitors. Other designs include an onboard microprocessor and discrete comparators which continuously consume battery power. For high voltage multicell battery applications, a high voltage LDO is needed to drive the low voltage devices. All this extra circuitry not only increases required board space and design complexity, but also drains the battery when the handheld device is turned off. Linear Technology addresses this pushbutton interface challenge with a pair of tiny pushbutton controllers.

The LTC2950 integrates all the flexible timing circuits needed to de-bounce the on/off pushbutton of handheld devices. The part also provides a simple yet powerful interface that allows for controlled power-up and power-down of the handheld device. The LTC2951 offers an adjustable timer for applications that require more time during power-down.

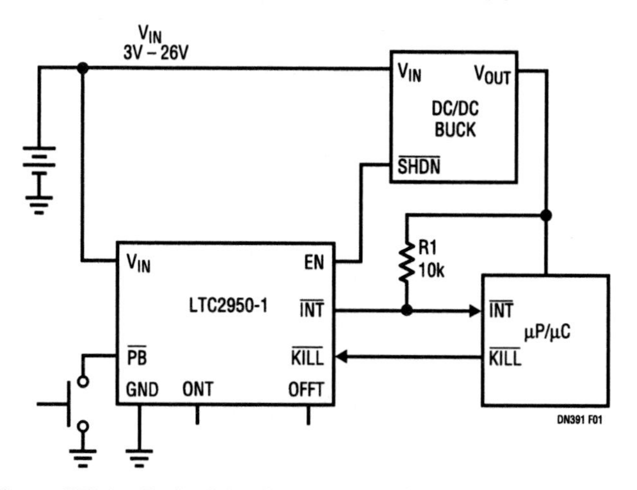

Figure 273.1 • Typical Application with One External Component

These two micropower, high voltage (2.7V to 26V) parts are offered in space-saving 8-pin 3mm×2mm DFN and TSOT-8 packages.

De-bounces turn-on

The circuit in Figure 273.1 provides manual control of the shutdown pin of a DC/DC converter. To turn on the converter, the LTC2950 first de-bounces the pushbutton input and then releases the low leakage enable (EN) output. The turn-on de-bounce time defaults to 32ms and is extendable by placing an optional capacitor on the ONT pin. This allows the handheld designer to adjust the length of time the user must hold down the pushbutton before turning on power to the device. The timing of Figure 273.2 illustrates performance with a noisy $\overline{PB}$ pin.

Protect against faults at power-up

The LTC2950 starts a 512ms blanking timer after it enables the DC/DC converter. If the $\overline{KILL}$ input is not driven high within this time period, the part automatically shuts off the

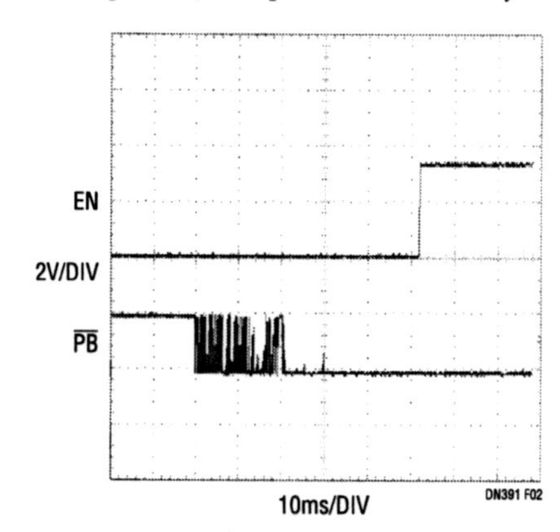

Figure 273.2 • Turn-On De-Bounce Timing

Analog Circuit Design: Design Note Collection. http://dx.doi.org/10.1016/B978-0-12-800001-4.00273-8

converter. This failsafe feature prevents the user from turning on the handheld device when there is a faulty power converter or an unresponsive microprocessor.

Controlled power-down

To turn off the handheld device, the LTC2950 first de-bounces the pushbutton input and then asserts the interrupt output (see Figure 273.3). The turn-off de-bounce time defaults to 32ms and is extendable by placing an optional capacitor on the OFFT pin.

The LTC2950 then starts an internal 1024ms blanking timer that allows the microprocessor to perform its power-down housekeeping functions. At the end of the timer period, the part shuts down power to the handheld device by turning off the DC/DC converter. Additionally, the LTC2951 provides an extendable power-down blanking timer (optional KILLT external capacitor) that accommodates lengthier microprocessor housekeeping tasks. Note that the LTC2950/LTC2951 de-bounces both the rising and falling edges of the pushbutton.

Operation without μP

The LTC2950 is easily adapted for applications that do not use a μP or μC. Simply connect the $\overline{\text{INT}}$ and $\overline{\text{KILL}}$ pins to the output of the DC/DC converter. When the user presses the pushbutton to turn off system power, the interrupt output asserts the $\overline{\text{KILL}}$ input, which then shuts off the converter. See Figure 273.4.

High voltage, micropower

The LTC2950 operates from a wide 2.7V to 26.4V input voltage range to accommodate a wide variety of input power supplies. This eliminates the need for a high voltage, low power LDO.

The LTC2950 is ideally suited for maximizing the battery life of a handheld device. When power is turned off to the handheld device, the LTC2950's very low quiescent current (6μA typical) is an insignificant drain on the battery.

Conclusion

The LTC2950 and LTC2951 provide simple, low power, small footprint solutions to the de-bounce problem. The LTC2950 integrates adjustable turn-on and turn-off timing, plus a fixed 1024ms power-down housekeeping timer. Alternatively, the LTC2951 provides a fixed 128ms turn-on timer, an adjustable turn-off timer and an adjustable power-down housekeeping timer. A simple microprocessor interface protects against faults at power-up and allows for graceful power-down.

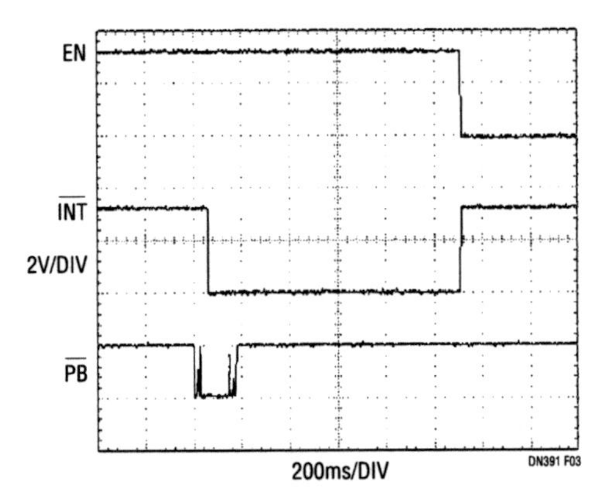

Figure 273.3 • Turn-Off De-Bounce Timing

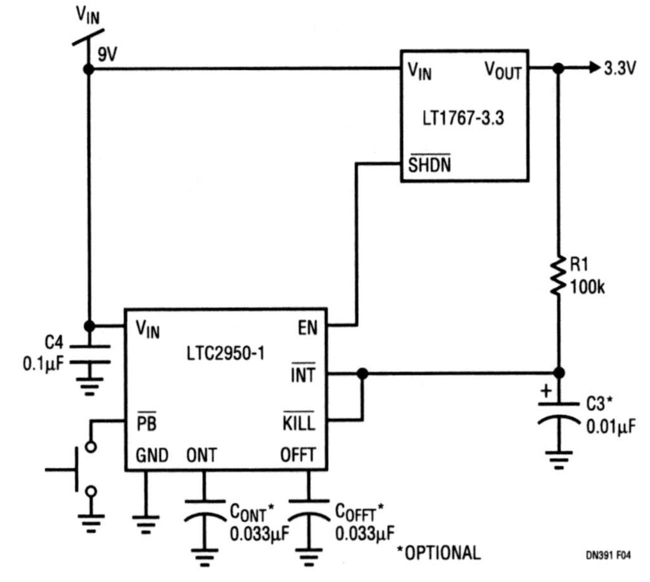

Figure 273.4 • No μP Application

Tracking and sequencing made simple with tiny point-of-load circuit

274

Scott Jackson

Introduction

Multiple-voltage electronics systems often require complex supply voltage tracking or sequencing, which if not met, can result in system faults or even permanent failures in the field. The design difficulties in meeting these requirements are often compounded in distributed-power architectures where point-of-load (POL) DC/DC converters or linear regulators are scattered across PC board space, sometimes on different board planes. The problem is that power supply circuitry is often the last circuitry to be designed into the board, and it must be shoehorned into whatever little board real estate is left. Often, a simple, drop-in, flexible solution is needed to meet these requirements.

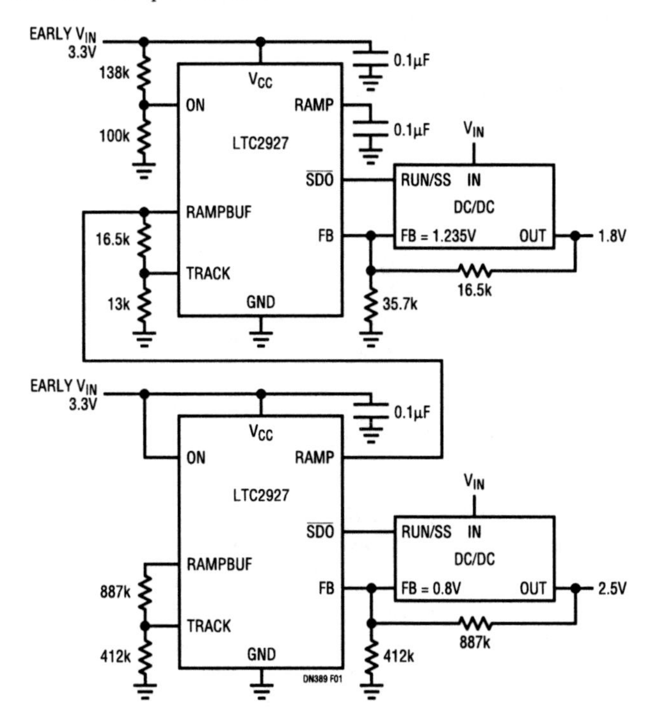

Figure 274.1 • Dual Supply Tracking Application

The LTC2927 provides a simple and versatile solution in a tiny footprint for both tracking and sequencing without the drawbacks of series MOSFETs. Furthermore, power supply stability and transient response remain unaffected because the LTC2927 offsets the output voltage of the regulator without altering the power supply control loop dynamics.

Basic operation

Each POL converter that must be tracked or sequenced can have a single LTC2927 placed at point-of-load as shown in Figure 274.1. By selecting a few resistors and a capacitor, a supply is configured to ramp up and ramp down with a variety of voltage profiles. The choice of resistors can cause a slave supply to track the master signal exactly or with a different ramp rate, voltage offset, time delay, or combination of these.

Figure 274.2 shows a 4-supply tracking and sequencing profile that highlights the flexibility of the LTC2927. A master signal is generated by tying a capacitor from the RAMP pin to ground or by supplying another ramping signal to be tracked. This ramping signal can be a master signal generated by another LTC2927 or another tracking controller such as the LTC2923. Likewise, another supply voltage can be used as the master signal. If an external ramping signal is used, it

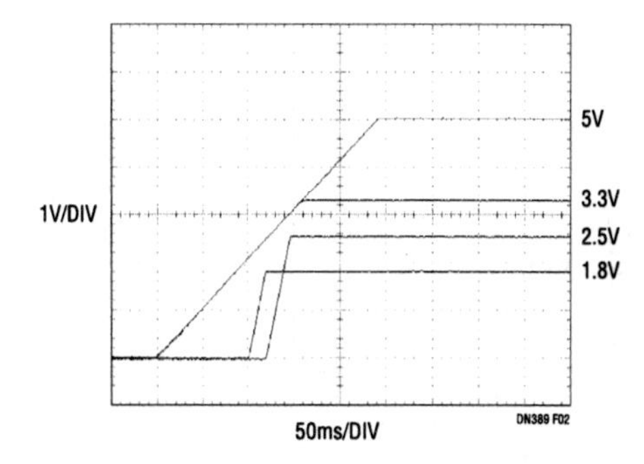

Figure 274.2 • Output Profile of a 4-Supply System Showing Tracking, Sequencing and Ramp Rate Control

Analog Circuit Design: Design Note Collection. http://dx.doi.org/10.1016/B978-0-12-800001-4.00274-X

can be connected directly to the RAMP pin or to the resistive divider connected to the TRACK pin.

For applications that require master control of the shutdown or RUN/SS pins of the slave supplies, the LTC2927 provides an SDO output. SDO pulls low when the ON pin is below 1.23V and the RAMP pin is below 200mV.

Negative supply tracking

The LTC2927 can also be used to track negative voltage regulators. Figure 274.3 shows a tracking example using an LT3462 inverting DC/DC converter to produce a −5V supply. This converter has a ground-based reference, which allows current to be pulled from a node where R_{FA} has been divided. To properly pull current from the LT3462 FB network, a current mirror must be placed between the LTC2927 and the converter. Figure 274.4 shows the tracking profile of Figure 274.3 with a ramp rate of 100V/s. V_{MASTER} is positive, but the inverse is shown for clarity. The −5V slave does not pull all the way up to 0V at V_{MASTER} = 0V because the ground referenced current mirror cannot pull its output all the way to ground. If the converter has an FB reference voltage greater than 0V or if a negative supply is available for the current mirror, the offset can be removed. Figure 274.5 shows the resulting waveform.

Conclusion

The LTC2927 simplifies power supply tracking and sequencing by offering superior performance in a tiny point-of-load footprint. Only a few resistors are needed to configure simple or complex supply behaviors. Series MOSFETs are eliminated along with their parasitic voltage drops and power consumption. The LTC2927 offers all of these features in tiny 8-lead ThinSOT and 8-lead (3mm × 2mm) DFN package.

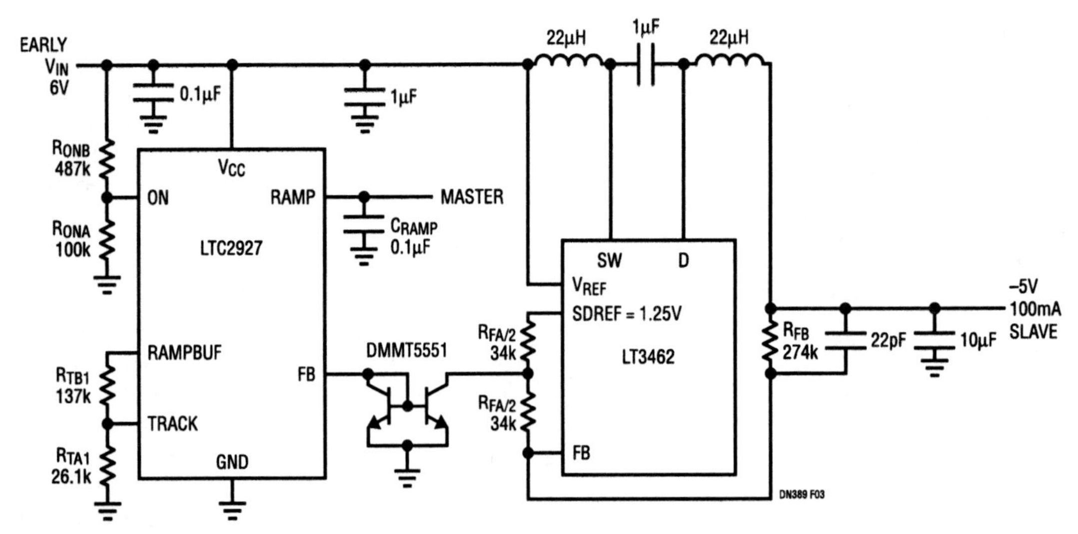

Figure 274.3 • Supply Tracking of a GND Referenced Negative Regulator

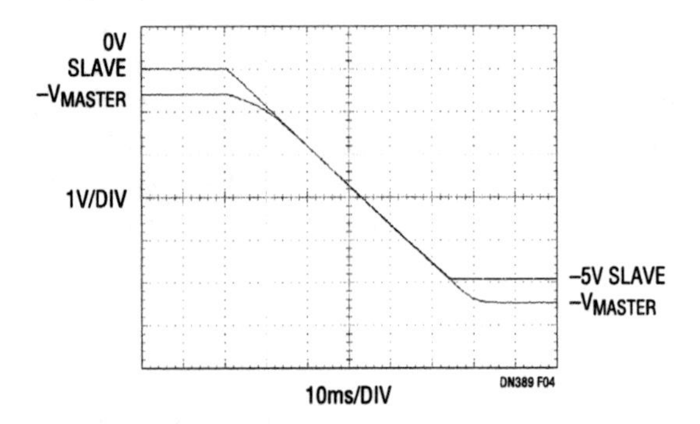

Figure 274.4 • Tracking Profile of the Negative Regulator Application in Figure 274.3

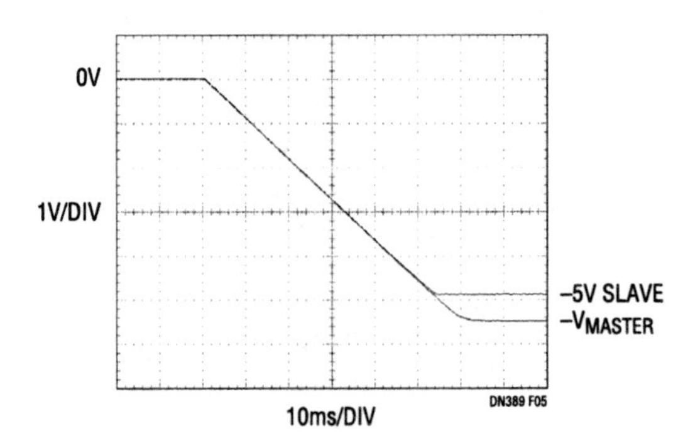

Figure 274.5 • Tracking Profile of the Negative Regulator Application without the Current Mirror Pull-Down Limitation

Accurate power supply sequencing prevents system damage

275

Jeff Heath Akin Kestelli

Introduction

Many complex systems—such as telecom equipment, memory modules, optical systems, networking equipment, servers and base stations—use FPGAs and other digital ICs that require multiple voltage rails that must start up and shut down in a specific order, otherwise the ICs can be damaged. The LTC2924 is a simple and compact solution to power supply sequencing in a 16-pin SSOP package (see Figures 275.1 and 275.2).

How it works

Four power supplies can be easily sequenced using a single LTC2924 and multiple LTC2924s can be just as easily cascaded to sequence any number of power supplies. With slightly reduced functionality, six power supplies can be sequenced with a single LTC2924.

The LTC2924 controls the start-up and shutdown sequence and ramp rates of four power supply channels via output pins (OUT1 to OUT4). Each OUT pin uses a 10µA current source connected to an internal charge pump and a low resistance switch to GND. This combination makes the outputs flexible enough to connect them directly to many power supply shutdown pins or to external N-channel MOSFET switches.

The LTC2924 monitors the output voltage of each sequenced power supply via four input pins (IN1 to IN4). These inputs use precision comparators and a trimmed bandgap voltage reference to provide better than 1% accuracy. The power ON and power OFF voltage thresholds are set using resistive dividers for each of the four channels. The power ON threshold and the power OFF threshold are individually selectable on a channel by channel basis (see "Selecting the

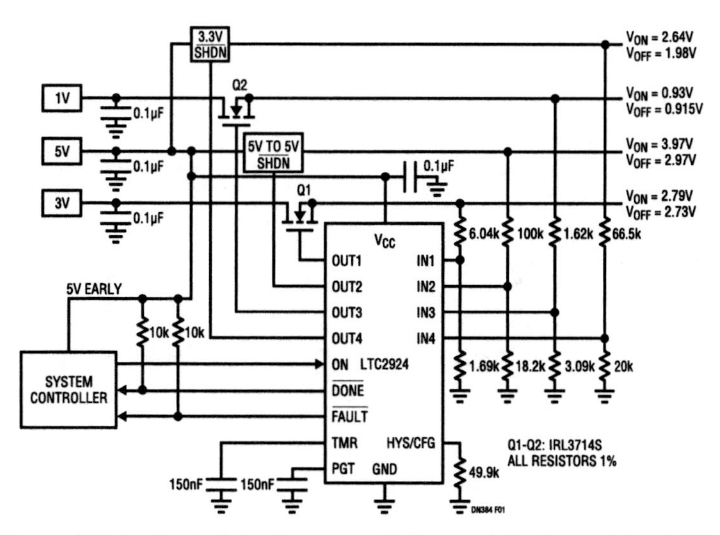

Figure 275.1 • Typical Application with External N-Channel MOSFETs

Analog Circuit Design: Design Note Collection. http://dx.doi.org/10.1016/B978-0-12-800001-4.00275-1

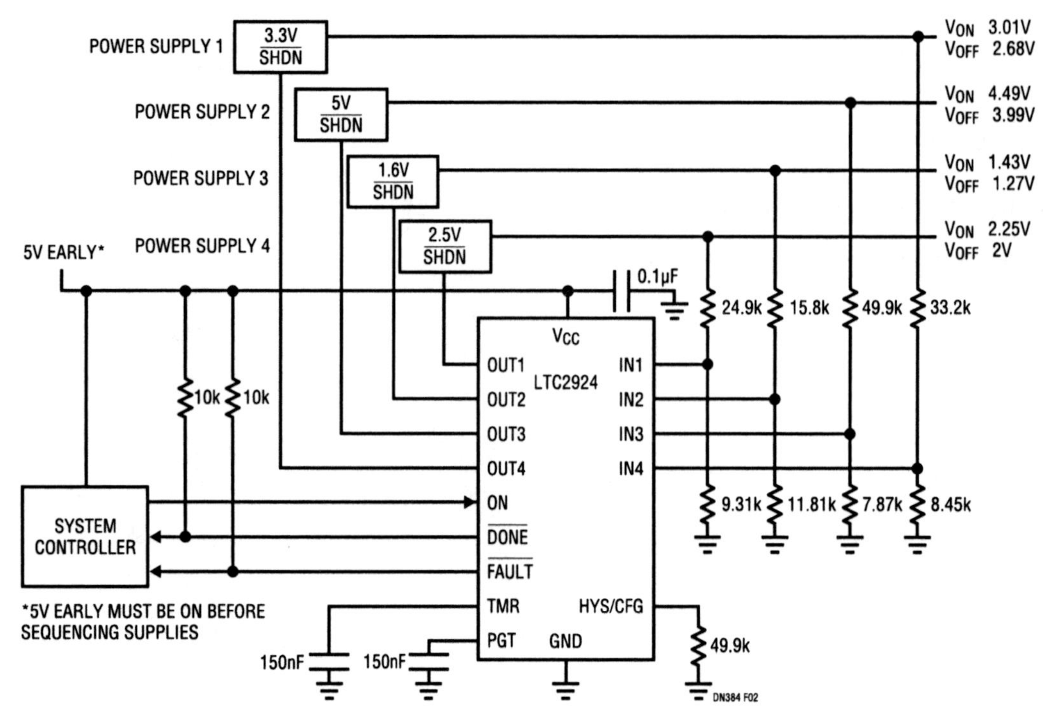

Figure 275.2 • Four Power Supply Sequencer Using Shutdown Pins

Hysteresis Current and IN Pin Feedback Resistors" in the data sheet for details).

The LTC2924 timer pin (TMR) is used to provide an optional delay between the completion of start-up of one supply and the start-up of the next power supply. The delay time is selected by placing a capacitor between the TMR pin and ground (delay = 200ms/μF), whereas floating the TMR pin removes any delay. The start-up delay can be different than the shutdown delay. Figure 275.3 shows a simple circuit where the shutdown delay is half the start-up delay.

The LTC2924 also includes a power good timer (PGT). The LTC2924 starts the PGT as each individual power supply is enabled. If any power supply fails to reach its nominal specified voltage within the allotted time interval, a power ON fault is detected.

Conclusion

The LTC2924 fits into a wide variety of power supply sequencing and monitoring applications. With very few external components and a 16-pin narrow SSOP, an LTC2924-based sequencing solution requires very little board space.

The power supply enable pins require no configuration by the designer, yet are versatile enough to directly drive shutdown pins or external N-channel MOSFETs. Soft-start of power supplies can be achieved simply by adding a capacitor. If the sequencing of more than four power supplies is required, the LTC2924 can be cascaded to sequence a virtually unlimited number of power supplies. Tailoring the LTC2924 to a specific application requires no software and designs can be fine tuned during system integration simply by changing resistor and capacitor values. Ease of design, low component cost, and a small footprint make the LTC2924 an excellent choice for power supply sequencing and monitoring.

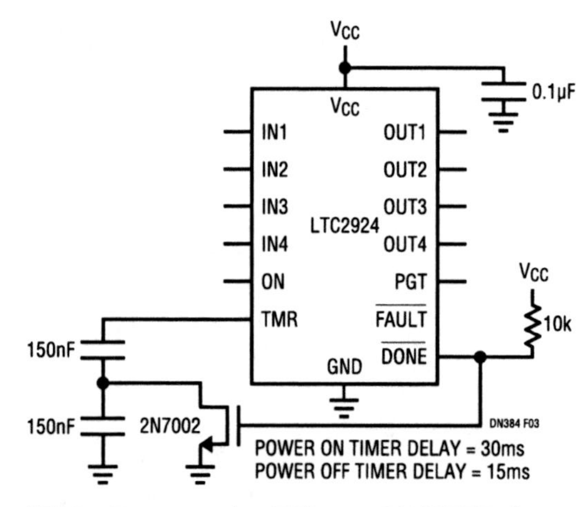

Figure 275.3 • Programming Different ON/OFF Delays

Power supply tracker can also margin supplies

276

Dan Eddleman

Power supply margining is a technique commonly used to test circuit boards in production. By adjusting power supply output voltages, electrical components are tested at the upper and lower supply voltage limits specified for a design. The LTC2923 power supply tracking controller can be used to margin supplies in addition to its usual task of tracking multiple power supplies.

The LTC2923 uses the simple tracking cell shown in Figure 276.1 to control the ramp-up and ramp-down behavior of multiple supplies. This cell servos the TRACK input to 0.8V and mirrors the current supplied by that pin at the FB output pin. The FB pin connects to the feedback node of the slave power supply. Normally, a resistive divider connects the master signal to the TRACK pin. By selecting the appropriate resistor values, R_{TA} and R_{TB}, the relationship of the slave power supply is configured relative to the master signal.

The supply margining application uses an LTC2923 tracking cell to margin a supply high and low under the control of a three-state I/O pin.

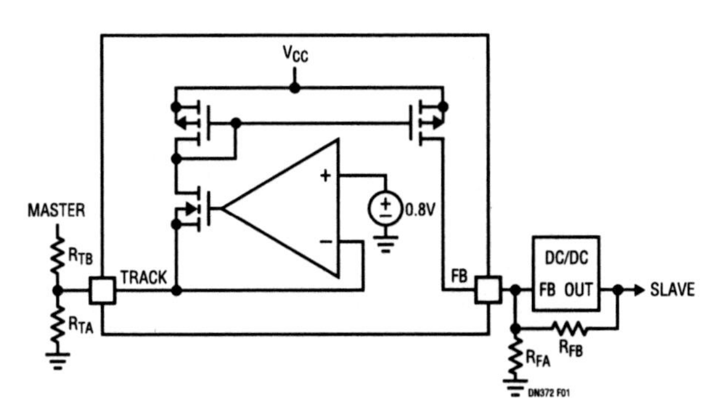

Figure 276.1 • Simplified Tracking Cell

In the circuit shown in Figure 276.2, the supply is margined to its high, low and nominal output voltages by driving the I/O pin to its high, low and high impedance states respectively. This example shows calculated resistor values rather than standard resistor values for ease of illustration. If the feedback voltage, V_{FB}, of the power supply is 0.8V, solve for the value of R_{FM1} that must be added in parallel with R_{F1} of the existing design to produce the desired high margin output.

In Figure 276.2, the feedback resistors R_{F2} and R_{F1} produce an output voltage of 2.5V. To margin 10% high to 2.75V requires a 54.4k resistor, R_{FM1}, in parallel with R_{F1}. Now connect a resistor, R_{TM1}, whose value is equal to R_{FM1} between the TRACK pin and ground. If the output will be margined low by the same voltage that it was margined high, then connect another resistor, R_{TM2}, equal to R_{FM1}, between the TRACK pin and the three-state I/O pin.

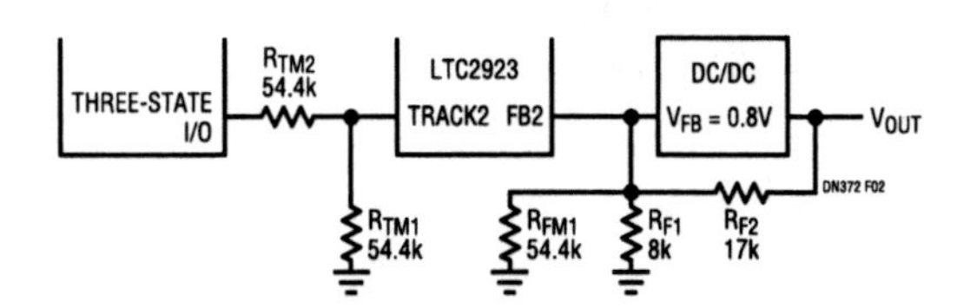

Figure 276.2 • The LTC2923 Margins the Output of a 2.5V Supply 10% High or Low Under the Control of a Three-State I/O

In the circuit shown in Figure 276.3, an LTC2923 ramps up a 3.3V supply through a series FET, tracks a 2.5V supply to that 3.3V supply, and margins the 2.5V supply up and down by 10%. The first tracking cell connected to pins TRACK1 and FB1 causes the 2.5V supply to track this 3.3V supply during power-up and power-down as shown in Figure 276.4. The tracking cell connected to TRACK2 and FB2 is used to margin the 2.5V supply up and down by 10%.

The operation of the circuit in Figure 276.3 is simple. To margin high, the I/O pin is pulled above 1.6V. This pulls the TRACK2 pin above 0.8V so that no current is sourced into the feedback node of the power supply. The supply then defaults to its margined high output of 2.75V. For a nominal output, the I/O is high impedance. Now, no current flows

Analog Circuit Design: Design Note Collection. http://dx.doi.org/10.1016/B978-0-12-800001-4.00276-3

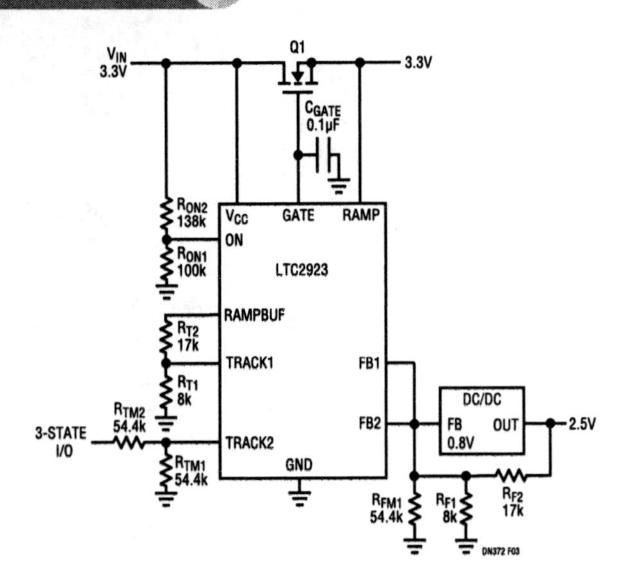

Figure 276.3 • The 2.5V Supply Tracks the 3.3V Supply and Can Be Margined High or Low by 10% Under Control of a Three-State I/O

through R_{TM2} but 14.7µA flows through R_{TM1} and is mirrored at the feedback node of the power supply. This forces the output voltage down by 250mV to 2.50V. For a margined low output, the I/O pin is pulled to ground. Now, 14.7µA flows through R_{TM2} in addition to the 14.7µA flowing through R_{TM1}. This current is mirrored at the power supply feedback node, and drives the output down by an extra 250mV from nominal.

Note that the ability to configure a current driven into the feedback node with R_{TM1} often allows the nominal output voltage to be closer to the ideal value than is possible with a single pair of standard value resistors, R_{F1} and R_{F2}, in the power supply feedback network.

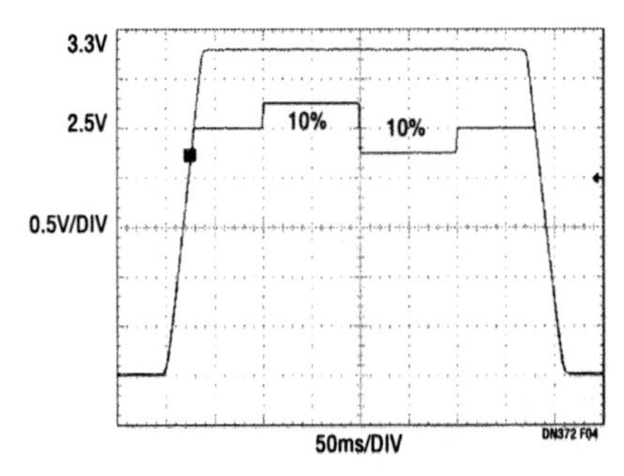

Figure 276.4 • Output of Circuit in Figure 276.3. The 2.5V Supply Tracks the 3.3V Supply and Is Margined High and Low by 10%

If the desired high and low voltage margins, ΔV_{HIGH} and ΔV_{LOW}, are not equal simply adjust R_{TM2}. In this case, choose R_{FM1} as above to configure the high margin, and set $R_{TM1} = R_{FM1}$. Scale the voltage step ΔV_{LOW} relative to the voltage step ΔV_{HIGH} by choosing R_{TM2} by $R_{TM1}/R_{TM2} = \Delta V_{LOW}/\Delta V_{HIGH}$. For example, to change the margins in the above example to 10% high and 20% low, leave R_{FM1} and R_{TM1} unchanged at 54.4k, but reduce R_{TM2} is to 27.2k.

If the feedback voltage, V_{FB}, of the power supply is not 0.8V then the values of R_{TM1} and R_{TM2} are scaled by $0.8V/V_{FB}$. If the feedback voltage in the above example were 1.23V, then R_{TM1} and R_{TM2} would be scaled so that $R_{TM1} = R_{TM2} = R_{FM1} \cdot 0.8V/1.23V = 35.4k$.

Conclusion

The LTC2923's primary application is tracking power supplies, but its versatile architecture is suited to other functions as well. The application described here allows a three-state I/O to control supply margining using a few resistors and an LTC2923 tracking cell.

Dual micropower comparator with integrated 400mV reference simplifies monitor and control functions

277

Jon Munson

Introduction

The LT6700 dual comparator incorporates features to reduce part count in space-critical designs, including a trimmed on-chip 400mV bandgap derived reference and internal hysteresis mechanisms. The LT6700 also features low voltage micropower single supply operation (1.4V to 18V, 7μA typical) and Over-The-Top I/O capability to maximize versatility and provide solutions especially useful in portable battery-powered applications. The outputs are open collector to permit logical wire-AND functionality, and can drive relatively heavy loads (up to 40mA) such as relays or LED indicators.

The LT6700 supports a wide range of design configurations, but still offers a minimum pin count package (ThinSOT, 6-lead). This is made possible by offering the LT6700 in three different versions, each with a different input configuration. The LT6700-1 provides the designer with one inverting and one noninverting input, especially useful in window detection functions; the LT6700-2 provides two inverting inputs; and the LT6700-3 offers two noninverting inputs. The internal reference is connected to one of the inputs of each comparator section, as shown in Figure 277.1, and the remaining two connections are brought out for signal sensing by the user.

"Gas gauge" battery monitor

It is easy to create a simple and accurate battery monitor using the LT6700, thanks to the accurate internal reference (±2% over temperature). Figure 277.2 shows an implementation of a 2-threshold "alkaline-cell" battery monitor. For the resistor values shown, the Pin 1 output goes low when the pack voltage falls below 2V (1V per cell) which corresponds to about 30% capacity remaining. The Pin 6 output goes low as well at 1.6V (0.8V per cell) as the battery pack reaches its rated end-of-life voltage. The number of threshold points may easily be increased by extending the resistor-divider chain and using additional comparators.

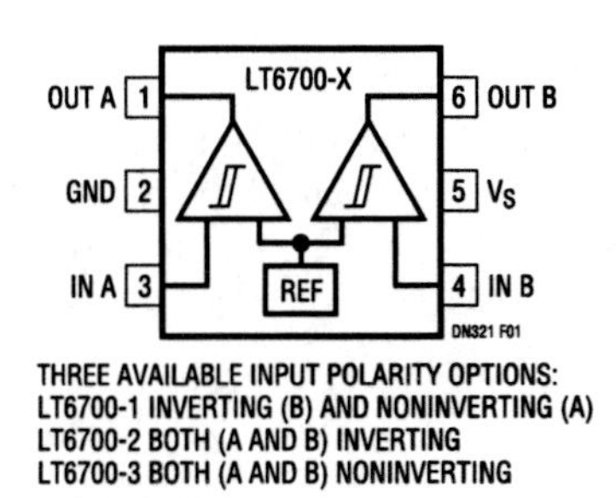

Figure 277.1 • Pin Functions of the LT6700 Family

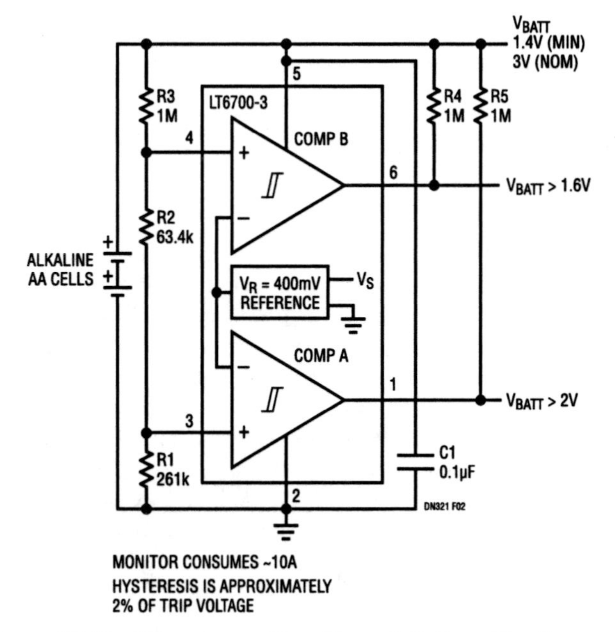

Figure 277.2 • Micropower "Gas Gauge" Battery Monitor

Analog Circuit Design: Design Note Collection. http://dx.doi.org/10.1016/B978-0-12-800001-4.00277-5

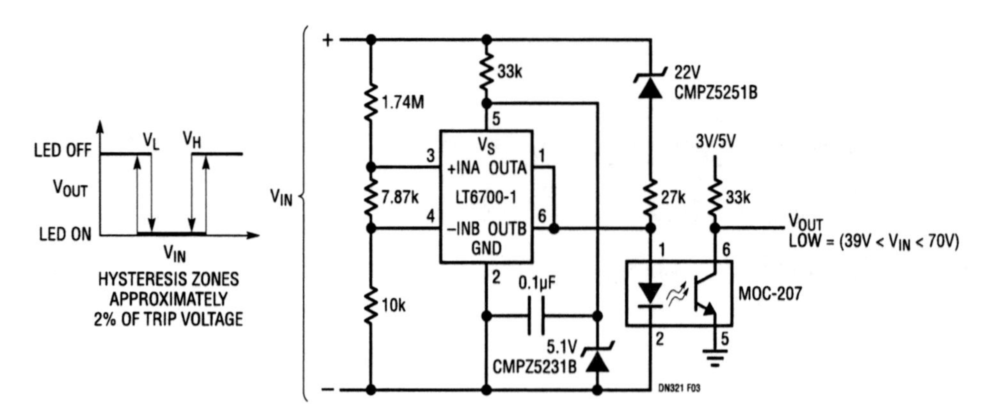

Figure 277.3 • 48V Power Bus Status Monitor

Simple window-function status monitor

The LT6700-1 lends itself nicely to window comparison applications, where the output wire-AND feature can be exploited. Figure 277.3 shows a 48V power bus monitor that provides an opto-isolated alarm indication when voltage limits are exceeded. The micropower operation of the circuit allows it to derive operating power directly from the monitored voltage using simple Zener diode techniques.

During bus operation within the normal voltage range, neither comparator output is active, so the LED is on and the alarm output is low (alarm clear). If the bus voltage deviates sufficiently, then one of the comparators shunts away the LED drive resulting in assertion of the alarm. Notice that any failure mode that causes an open connection to the phototransistor or prevents the LED from operating (i.e., other open-circuit conditions), produces a fail-safe alarm indication at the destination logic input. The ability of the LT6700 to operate down to very low voltage assures correct alarm indication even during deep sags in bus potential (the 22V Zener further eliminates the possibility of false-fault indication during a bus power-down transition by disabling the LED).

Micropower thermostat/temperature alarm

Though the 400mV reference is not directly available to the circuit designer, an inverting comparator section can be used to scale an external voltage in proportion to the reference by implementing a simple "bang-bang" servo. This technique is shown in Figure 277.4, where the multiplier is set to two by an equal-resistor feedback path. The inverting comparator steers current to hold the voltage on a capacitor such that the feedback "bangs" between the input hysteresis points. The LT6700 hysteresis is nominally 6.5mV so this circuit has about $13mV_{P-P}$ ripple on the servo capacitor.

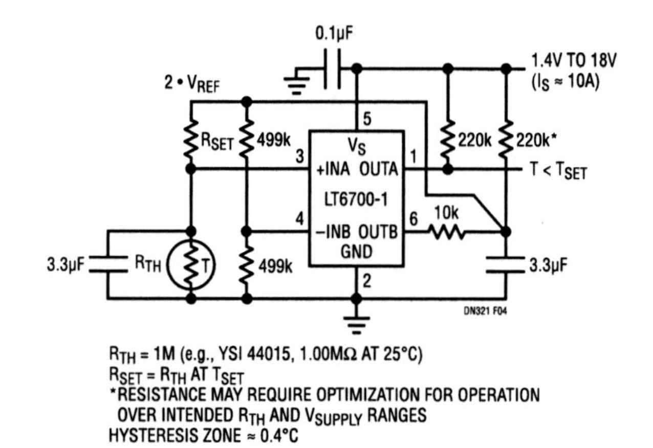

Figure 277.4 • Micropower Thermostat/Temperature Alarm

The other comparator section actually performs the alarm decision which is simply based on the imbalance of a resistor half-bridge. In this circuit, the thermistor resistance is balanced against a known resistor, so the temperature threshold is established easily by selecting R_{SET} from the thermistor table for the temperature of interest. Since the resistance varies about −4.4%/°C for the thermistor shown, the temperature hysteresis of the output signal is about 0.4°C, suitable for most environmental control applications. The capacitor in parallel with the thermistor filters away the ripple of the reference multiplier circuit. The micropower consumption of this circuit permits over two years of continuous operation from a common 3V coin-cell (i.e., CR2032).

Conclusion

The LT6700 provides compact, micropower solutions for threshold-based status and control functions. The extra-wide supply range and Over-The-Top features offer performance ideal for portable, battery-powered products as well as industrial applications.

Monitor network compliant –48V power supplies

278

Brendan Whelan

Introduction

Reliability is a top priority for the designers of modern telephone and communication equipment. Designers take extra care to protect circuitry from failure-causing temperature and voltage changes, employing redundancy whenever possible, especially for power supplies. The power supplies are monitored to obtain early warning of impending failure. Often complicated circuitry that can include a voltage reference, comparators, voltage regulator and several precision resistor dividers is used. Designers may also use discrete components to monitor and indicate the state of power supply fuses. The resulting circuitry can be expensive in terms of component cost, board space and engineering time. The LTC1921 replaces the complicated monitoring circuitry with a simple integrated precision monitoring system contained entirely in an 8-lead MSOP package.

Features

The LTC1921 is the only integrated solution that can monitor two independent −48V power supplies, plus associated fuses and drive up to three opto-isolators or LEDs to indicate status. The required external components are three resistors and optocouplers or LEDs, as shown by the simple circuit in Figure 278.1. The LTC1921 can withstand ±100V DC at the supply and fuse input pins and tolerates ±200V transients.

The LTC1921 monitors supply voltages by dividing the voltage internally and comparing to an internal precision reference. Since no critical precision external components are required, component cost, board space and design time are minimized while accuracy is maximized. The LTC1921 comes with telecom industry accepted preset voltage thresholds, including undervoltage (−38.5V), undervoltage recovery (−43V) and overvoltage (−70V). The overvoltage threshold

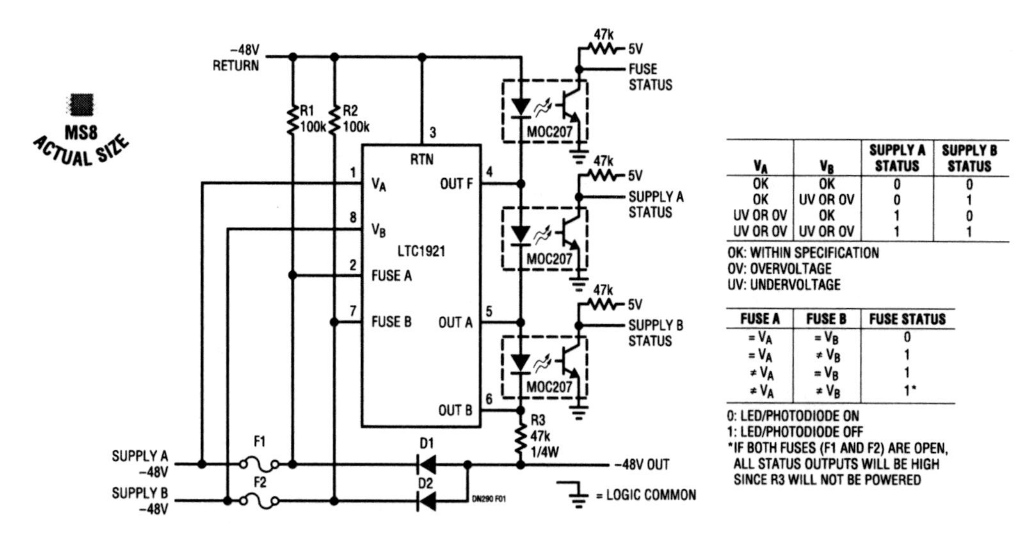

V_A	V_B	SUPPLY A STATUS	SUPPLY B STATUS
OK	OK	0	0
OK	UV OR OV	0	1
UV OR OV	OK	1	0
UV OR OV	UV OR OV	1	1

OK: WITHIN SPECIFICATION
OV: OVERVOLTAGE
UV: UNDERVOLTAGE

FUSE A	FUSE B	FUSE STATUS
$= V_A$	$= V_B$	0
$= V_A$	$\neq V_B$	1
$\neq V_A$	$= V_B$	1
$\neq V_A$	$\neq V_B$	1*

0: LED/PHOTODIODE ON
1: LED/PHOTODIODE OFF
*IF BOTH FUSES (F1 AND F2) ARE OPEN, ALL STATUS OUTPUTS WILL BE HIGH SINCE R3 WILL NOT BE POWERED

Figure 278.1 • The LTC1921 Requires Few External Components for Monitoring Two Supplies

Analog Circuit Design: Design Note Collection. http://dx.doi.org/10.1016/B978-0-12-800001-4.00278-7

has a 1.3V hysteresis that defines the overvoltage recovery threshold. These thresholds are trimmed to meet exacting requirements. This eliminates the messy worst-case threshold tolerance error calculation required when using discrete comparators, resistors and a separate voltage reference.

The LTC1921 is designed to indicate proper supply status across a wide variety of conditions. In order to accomplish this, the internal architecture is symmetrical. The LTC1921 is powered via the supply monitor input pins, V_A and V_B. Supply current can be drawn from either or both pins so the device can operate properly as long as one supply is within the operating range. Since power is not drawn from a combined supply (such as would be available with a diode OR), the LTC1921 will function properly even if the fuses or diodes are not functional. In addition, the LTC1921 has a low voltage lockout. If both supply voltages are very low, all three outputs of the LTC1921 lock into a fault indication state, thus communicating to supervisory systems that there is a problem even though there is not enough power for the LTC1921 to maintain accuracy.

The device monitors fuses by comparing the voltage potentials on each side of each fuse. If a significant difference (about 2V) is sensed, the LTC1921 signals that a fuse has opened. The voltage difference across the damaged fuse may be reduced by diode reverse leakage, making it difficult to detect a damaged fuse. Weak pull-up resistors (R1 and R2 in Figure 278.1) ensure that the LTC1921 can detect an open fuse circuit. The value of the pull-up resistors used is a function of the reverse leakage current of the OR'ing diodes used.

The LTC1921 can communicate supply and fuse status by controlling external optocouplers or LEDs. This allows for intelligent system monitoring despite high isolation voltage requirements. Control of the LEDs or optocouplers is accomplished by connecting the LTC1921 outputs in parallel with the LEDs or photodiodes. During normal supply and fuse conditions, the LTC1921 outputs are high impedance; current flows through the external diodes continuously. If a fuse opens or a supply voltage falls outside of the allowed window, then the proper LTC1921 output shunts the current around the diode, thus indicating a fault. *The outputs may be ORed to reduce the number of required opto-isolators.*

Application example

Figure 278.2 shows an LTC1921 and an LT4250 Hot Swap controller comprising a complete power system solution. The LTC1921 monitors both −48V supply inputs from the power bus, as well as the supply fuses. The status signals may be wired off the board via opto-isolators to an isolated microprocessor or microcontroller to control system performance and warning functions. Resistors R9 and R10 pull up the fuse pins so that damaged fuses can be detected. The LT4250L controls the combined −48V supply during hot swapping and low supply conditions, and monitors the combined supply voltage. The $\overline{\text{PWRGD}}$ pin drives an opto-isolator signaling the status of the LT4250's switched output.

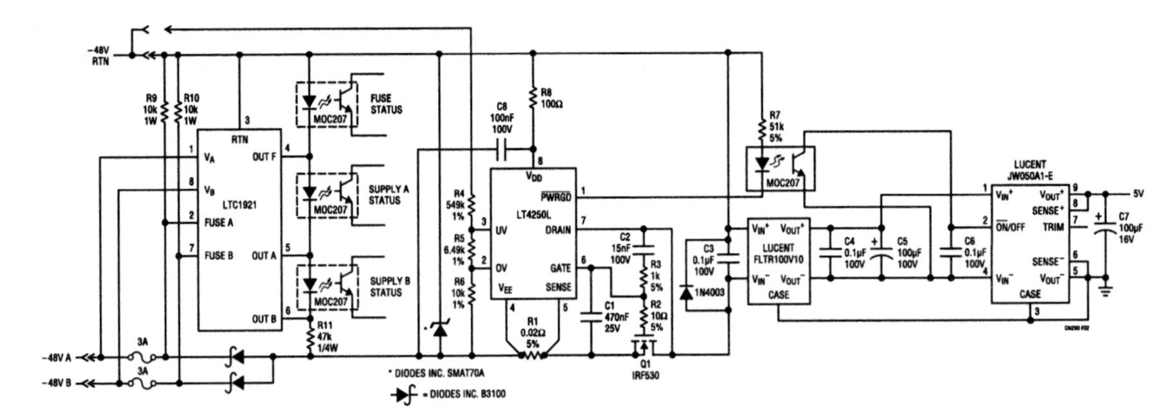

Figure 278.2 • Network Switch Card Monitor with Hot Swap Control

Multiple power supplies track during power-up

279

Vladimir Ostrerov

Introduction

Many modern circuits require multiple power supplies that must turn on in a certain order to avoid damaging sensitive components. In many cases, forcing the supplies to ramp up together is the preferred solution. Unfortunately, this can be difficult when the supplies are generated from multiple sources, each with its own power-up timing and transient response. However, there is a simple solution for up to five supplies ramping up simultaneously.

The circuit shown in Figure 279.1 solves this problem by placing an N-channel MOSFET between the output of each power supply and the load. When power is first applied to the circuit, the MOSFETs are turned off and each power supply is allowed to power up at its own rate. Once each power supply has settled, the common gates of the MOSFETs are ramped up, forcing the outputs to ramp up simultaneously as shown in Figure 279.2.

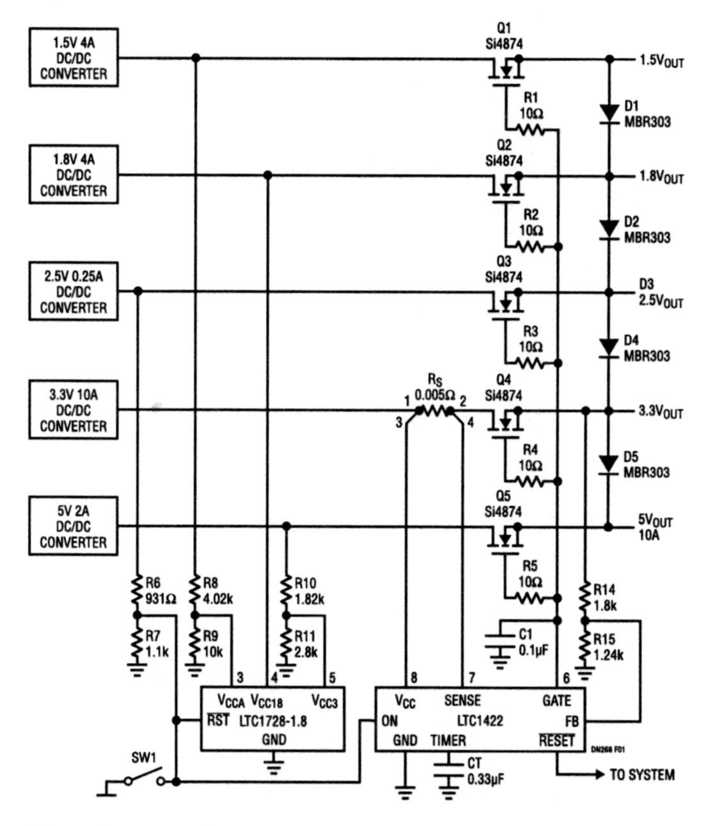

Figure 279.1 • 5-Supply Voltage Tracker

Five supply voltage tracker circuit

The key components of the circuit in Figure 279.1 are the LTC1728-1.8 triple supply monitor and the LTC1422 single Hot Swap controller. The LTC1728-1.8 directly monitors three supply outputs: 5V, 1.8V and 1.5V. The outputs of the 3.3V and 2.5V supplies are monitored by the LTC1422. Short-circuit protection for the 3.3V supply is provided by the LTC1422 using sense resistor R_S, but all other voltages rely on their individual power supply's current limit.

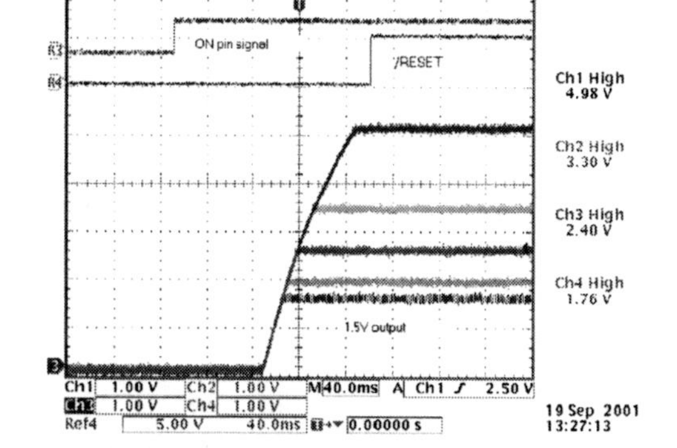

Figure 279.2 • Circuit Waveforms

Analog Circuit Design: Design Note Collection. http://dx.doi.org/10.1016/B978-0-12-800001-4.00279-9

When all three supplies monitored by the LTC1728-1.8 are in compliance, the open-drain pull-down on the $\overline{RST}$ pin turns off after a 200ms delay. The 2.5V resistor divider monitor connected to the LTC1422 ON pin is then enabled. When the 2.5V supply is within tolerance as measured by the ON pin, and the 3.3V supply exceeds the LTC1422 undervoltage lockout threshold, the LTC1422 turns on.

After one timing cycle (set by C2 at the TIMER pin), the voltage at the GATE pin begins to ramp up, turning on transistors Q1 to Q5. The slope of the voltage rise is set by the total capacitance at the GATE pin (C_G) and 0.1µA GATE pull-up current:

$$\frac{dV_{GATE}}{dt} = \frac{I_{GATE}}{C_G}$$

Capacitance C_G is equal to the sum of capacitor C1 and the total MOSFET gate capacitance. Because each MOSFET is connected as a source follower, the inrush current into each load capacitance is limited according to:

$$I_{INRUSH} = \frac{C_{LOAD} \cdot I_{GATE}}{C_G}$$

Once the 3.3V output is within tolerance as measured by the FB pin, the $\overline{RESET}$ pin open-drain pull-down turns off after a timing cycle. A complete timing diagram is shown in Figure 279.3.

Power-down can be initiated by forcing the ON pin signal low with the switch SW1, or by turning off any of the power supplies. The GATE pin is pulled low immediately, disconnecting the loads from the power supplies, and the loads start to discharge at the rate determined by the load capacitance and load current. Diodes D1 to D5 are included to insure worst-case differential levels between supplies during power-down and catastrophic fault conditions.

For better performance, use low drop power MOSFETs and adjust the preliminary power supply voltage to account for the voltage drop across the transistor.

Conclusion

Although the circuit of Figure 279.1 controls five supplies, it can be easily modified to accommodate fewer supplies. Unused monitor inputs can be tied off to a higher supply voltage and the unused MOSFETs removed. Different supply voltages can be accommodated by selecting the appropriate voltage option of the LTC1728-1.8 and changing resistor values. In sum, with only a handful of components, the circuit solves the tricky problem of controlling the power-up of multiple supplies in a complex system.

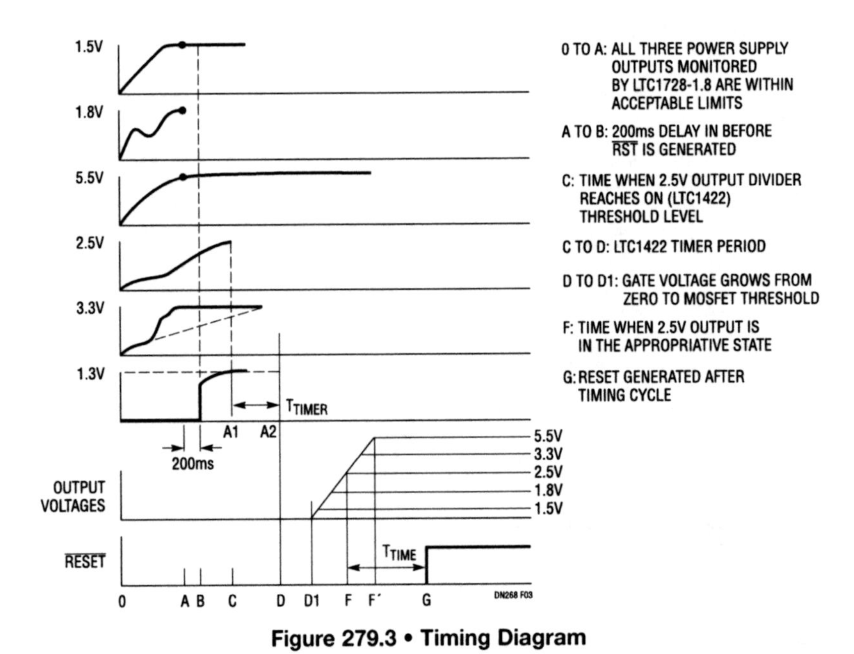

Figure 279.3 • Timing Diagram

I^2C fan control ensures continuous system cooling

280

Dilian Reyes

Introduction

Linear Technology's LTC1840 is a dual fan speed controller for high availability servers and other rack-based network and telecom equipment. The LTC1840 offers advanced control and monitoring capabilities, accessed via an I^2C and SMBus compatible 2-wire serial interface. In addition to two fan speed control channels, the LTC1840 also includes a fan tachometer and fault monitoring, nine slave addresses and four general-purpose programmable I/O pins in a 16-pin SSOP package. Adjusting fan speed to match instantaneous cooling requirements increases energy efficiency and reduces noise. By operating at reduced speeds, fan bearings are subjected to less wear, increasing fan life and reliability.

Figure 280.1 shows a block diagram for a fan speed control system using the LTC1840. The LTC1840 contains two current DACs used to gain full control over fan speed. The scaled currents individually adjust the fan-driving output voltage of a switching regulator. V_O increases as the current I_{DAC} is increased under command of the serial interface. The number of fans controlled by one DAC is limited only by the switching regulator output power.

The TACH of the LTC1840 monitors the speed of fans that include a tachometer output. Internal logic accumulates a maximum of 255 counts between the fan tachometer's rising edges. The rate of the counter is determined by a divisor (2, 4, 8 or 16 chosen via the serial interface) from the 50kHz internal oscillator. Fans slowing down due to worn bearings or halted from a jam will cause an overflow in the internal counter and a corresponding bit is set low in the fault register. The system controller can then take action, shutting down the faulty fan and summoning maintenance.

The chip contains four general purpose input/output (GPIO) pins that are configured independently. As open-drain outputs, they can be set high, low or to pulse at a 1.5Hz rate. The outputs are rated at 10mA sink current for compatibility with LEDs. Configured as inputs, the GPIO pins can monitor thermal switches, pushbuttons and switching regulator and Hot Swap controller fault or power good outputs. State changes are detected and flagged in a fault register.

Internal data registers are read and programmed via I^2C by specifying device address and register address. DACA and DACB registers control the 100μA current outputs on a 255-step scale. The STATUS register allows the user to enable the TACHA and TACHB fault data and set the divisor for the internal counter frequency. The internal count, which is inversely proportional to tachometer speed, is stored in the TACHA and TACHB registers. Unmasked faults set the $\overline{FAULT}$ pin high as an instant hardware alert. The GPIO setup and GPIO data registers configure the GPIO pins, assign output and fault status and read input state.

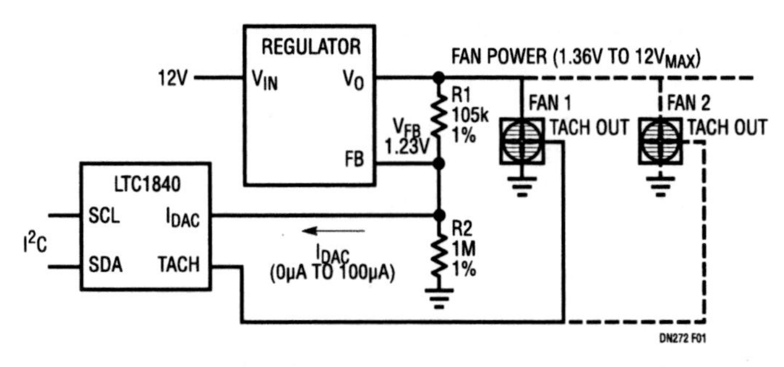

Figure 280.1 • LTC1840 Fan Speed Control Block Diagram

Analog Circuit Design: Design Note Collection. http://dx.doi.org/10.1016/B978-0-12-800001-4.00280-5

Continuous system cooling and tachometer monitoring

The circuit in Figure 280.2 demonstrates the capabilities of the LTC1840. Power for up to four 12V, 420mA fans is supplied by each LTC1771 high efficiency step-down regulator. As shown, the upper LTC1771 drives a single fan backed up by an idle, redundant fan. In the event the primary fan fails, GPIO3 turns off the LTC1771 and simultaneously activates the backup fan which runs at full speed. These fans operate one at a time, so the tachometer outputs are wire ORed and only one input (TACHA) is required to monitor their speed.

Two fans are driven in parallel by the lower LTC1771 and alternately monitored by TACHB. These fans operate simultaneously, so their tachometer outputs are muxed by a quad NAND gate. GPIO2 operates in pulsing mode and serves to clock the mux.

Additional features

For applications requiring multiple fan controllers, the LTC1840's three-state (high, low, no connect) address programming inputs support nine user-selectable slave addresses. The FAULT output bypasses the serial interface and brings immediate attention to fault conditions detected by the LTC1840, including slow downs in the tachometer and changes in GPIO logic state.

If the BLAST pin is high at start-up or presented with a high to low transition at anytime, the DAC output currents are forced instantaneously to full scale and the chip awaits commands from the serial bus. In addition, when BLAST is set high the LTC1840 guards against system controller crashes with an internal watchdog timer. If the device is not accessed for a period of more than 1.5 minutes, both DAC outputs are set to full scale to guarantee adequate system cooling.

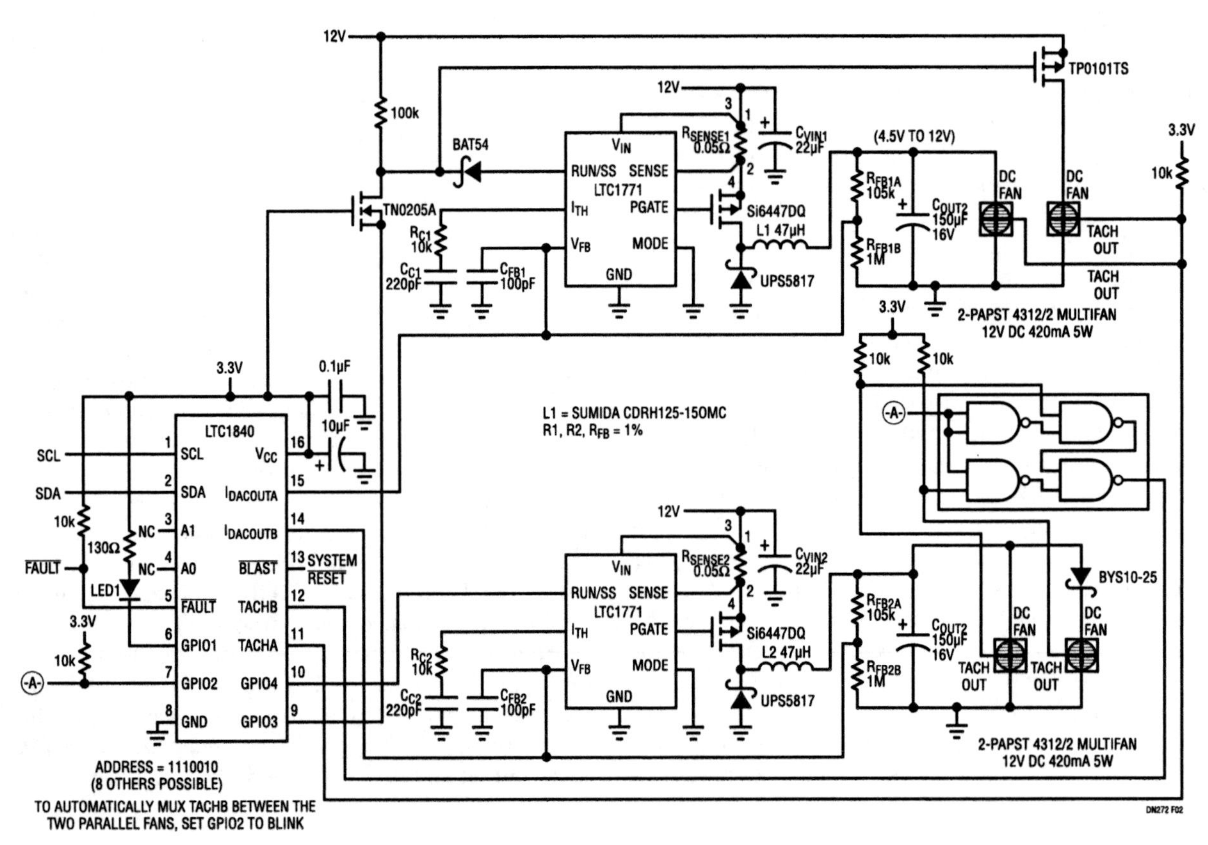

Figure 280.2 • Controlling Various Fan Operations with the LTC1840

Monitor system temperature and multiple supply voltages and currents

281

Kevin R. Hoskins Alan Rich

Fault-tolerant systems ensure their users uninterrupted service and prevent data loss through acquisition of accurate supply voltage, load current and temperature information. By comparing this information to predetermined operating profiles, the system decides if the measured data is within nominal ranges or if a fault condition is looming. If a fault condition is imminent, the system takes corrective action before data is lost.

The measured data can also be used to establish preventive maintenance schedules. The system develops stress profiles

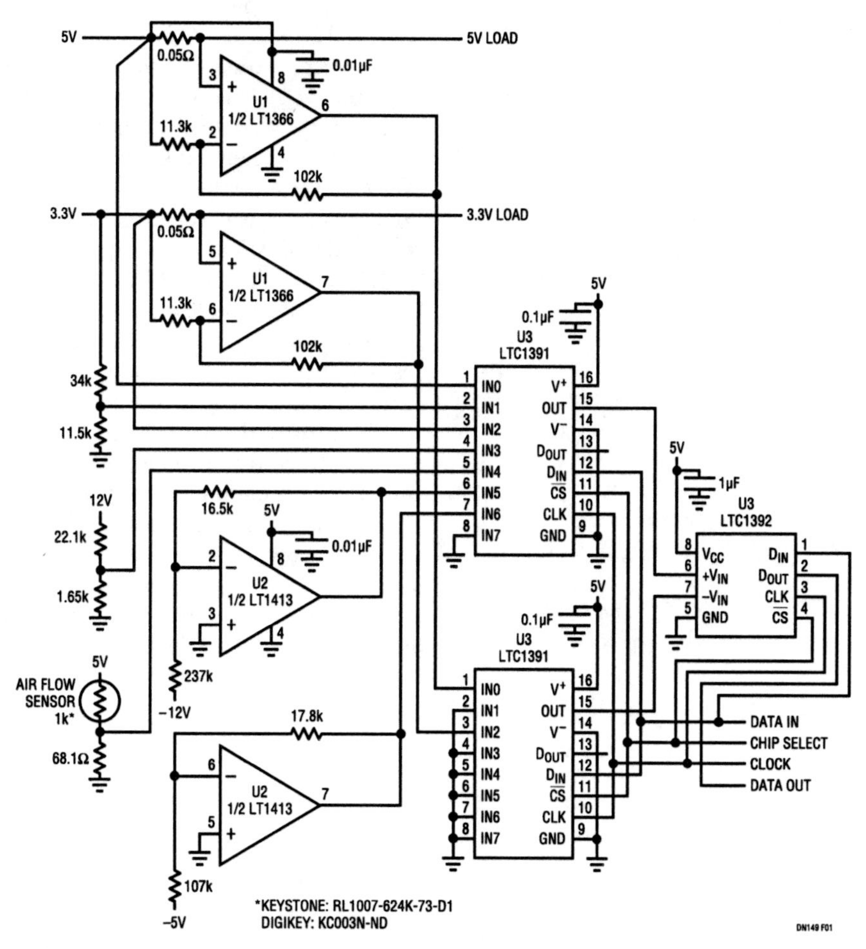

Figure 281.1 • The LTC1392, with the Help of a Few Friends, Helps Prevent Data Loss and Down Time by Providing Feedback as It Monitors a Fault-Tolerant Computer System

Analog Circuit Design: Design Note Collection. http://dx.doi.org/10.1016/B978-0-12-800001-4.00281-7

based on operating time and the measured values of the different parameters. These profiles suggest a need for increased preventive maintenance when the measured values indicate high sustained peak operation over a given time period or a relaxed schedule during times of reduced activity.

The circuit shown in Figure 281.1 monitors the multiple supply voltages, supply currents and temperatures of a fault-tolerant system. The heart of the circuit is the LTC1392. This 8-pin system monitor looks at supply voltage, load current and temperature. It operates on a single 5V (typ) supply and draws a nominal 350μA while converting or just 0.2μA when idle. It includes an onboard temperature sensor, differential sample-and-hold, bandgap reference and a 10-bit ADC. The LTC1392 communicates with a host system through a serial interface.

Multitude of measurements

As configured, the circuit in Figure 281.1 monitors a system's 3.3V, ±5V and ±12V supply voltages. It also captures the 3.3V and 5V supplies' output current magnitude. Lastly, it acquires two temperatures: ambient from the LTC1392's built-in sensor and a remote from an external sensor.

The monitoring circuit selects the current, voltage and temperature signals using two LTC1391 serially programmed multiplexers. Multiple LTC1391s are programmed either by daisy-chaining a device's serial data output pin (D_{OUT}) to the data input pin (D_{IN}) of the next device or by connecting together the data input pins of each part. Daisy-chaining multiple LTC1391s works best for applications that require unique channel selection on each MUX or need simultaneous selection of different channel combinations across multiple MUXs. This latter connection allows an ADC with differential inputs to convert the difference between different combinations of signals.

Figure 281.1's circuit shows the connection that selects the same channel on each LTC1391. This connection simplifies the software because only one channel selection data byte is created and applied to each MUX simultaneously. By pairing the same channel on each LTC1391, the circuit converts signals that are either across the current sense resistors or ground referred. The signals across the current sense resistors are applied to the LTC1392's differential input by selecting the same channel of each LTC1391. Ground-referred signals are applied to one LTC1391 while the corresponding channel on the other LTC1391 is grounded.

The current sensing resistors in the 5V and 3.3V supply outputs are kept small, minimizing their voltage drop and dissipation. A dual LT1366 op amp amplifies the small drop across the sense resistors by a gain of 10. The amplifiers' outputs are selected with the LTC1391 multiplexer and applied to the LTC1392's input. The LTC1392's inputs have a common mode input range that includes ground and the voltage applied to the V_{CC} pin. For the 0.05Ω series resistor value shown, the full-scale current range is 2A, with a resolution of 19.5mA.

The LTC1392 measures the 5V supply voltage through the on-chip V_{CC} supply line with an accuracy of ±39mV over a $-40°C \leq T_A \leq 85°C$ range. This accuracy is guaranteed over a supply voltage monitor range of 4.5V to 6V.

The 3.3V, −5V and ±12V are scaled and measured using the LTC1392's 1V input range. The chosen scaling factor allows for as much as 20% supply output fluctuation without exceeding the ADC's full-scale input range.

The application circuit measures ambient temperature and cooling air flow using two different temperature sensors. One of the temperature sensors is built into the LTC1392 system monitor and measures ambient temperature. The monitor's small, SO-8 size makes it unobtrusive and easily placed in tight quarters. It communicates with a host system using a serial interface that requires just three connections. Together, the small size and minimum interface allow remote placement. The LTC1392 is specified over a −40°C to 85°C temperature range with 0.25°C/count. The output code varies from zero at −130°C to full scale at 125°C.

The second temperature sensor, an NTC 1kΩ thermistor, and a 68.1Ω series resistor were chosen to induce self-heating in the thermistor. The thermistor is placed in a cooling fan's air stream and helps the host system, using the "wind-chill" effect, determine the fan's performance. As air moves across the thermistor, it is cooled, increasing both its resistance and voltage drop. At the same time, the voltage drop across the 68.1Ω resistor decreases and is measured by the LTC1392. Low data values indicate that the thermistor is cooled by a properly working fan. Degradation in the fan's performance (caused by a clogged air filter, blocked air flow, a coffee cup or an LTC databook sitting on the air inlet) reduces cooling, which, in turn, decreases the thermistor's resistance. As this occurs, the voltage drop across the 68.1Ω resistor increases, with a corresponding increase in LTC1392 output code. The 68.1Ω resistor also sets the LTC1392's maximum input voltage of 1V to correspond to approximately 50°C. Because of many possible variations in the mechanical and physical layout of different systems, the thermistor and series resistor values may require custom selection.

Section 21

Powering LED Lighting & Other Illumination Devices

60V, synchronous step-down high current LED driver

282

Hua (Walker) Bai

Introduction

The meaning of the term "high power LED" is rapidly evolving. Although a 350mA LED could easily earn the stamp of "high power" a few years ago, it could not hold a candle to the 20A LED or the 40A laser diodes of today. High power LEDs are now used in DLP projectors, surgical equipment, stage lighting, automotive lighting and other applications traditionally served by high intensity bulbs. To meet the light output requirements of these applications, high power LEDs are often used in series. The problem is that several series-connected LEDs require a high voltage LED driver circuit. LED driver design is further complicated by applications that require fast LED current response to PWM dimming signals.

The LT3763 is a 60V synchronous, step-down DC/DC controller designed to accurately regulate LED current at up to 20A with fast PWM dimming. It is a higher voltage version of its predecessor, the LT3743. It can be applied in a number of other applications thanks to its three additional regulation loops:

1) An output voltage regulation loop enables constant output voltage operation. This can be used to provide open LED protection or charging termination for a battery charger.
2) A second current regulation loop can be used to set an input current limit.
3) An input voltage regulation loop can be used for maximum power tracking (MPPT) in solar-powered applications.

48V input to 35V output, 10A LED driver optimized for efficiency

Figure 282.1 shows a design that delivers 350W output power to drive up to seven LEDs in series from a 48V source. At this high power level, dissipated power is a major concern, so

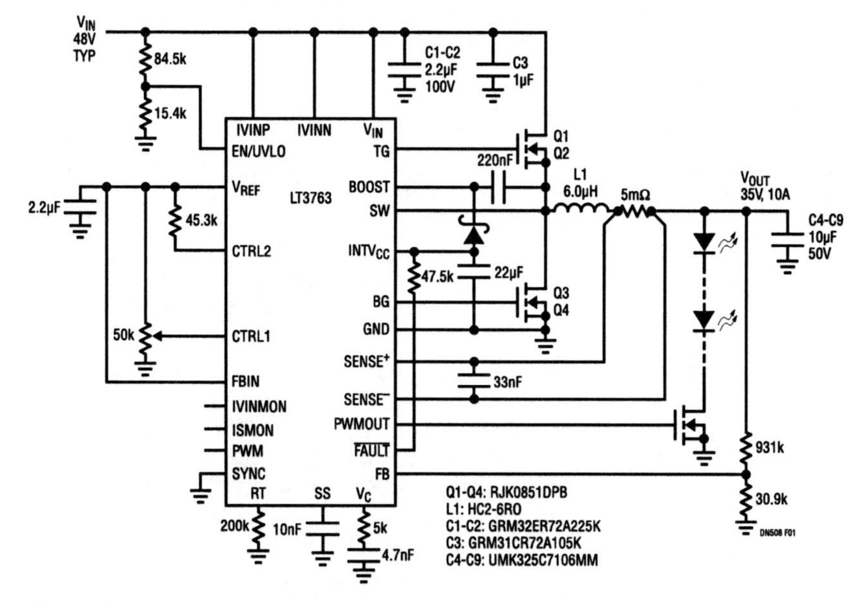

Figure 282.1 • 48V Input to 35V Output, 10A

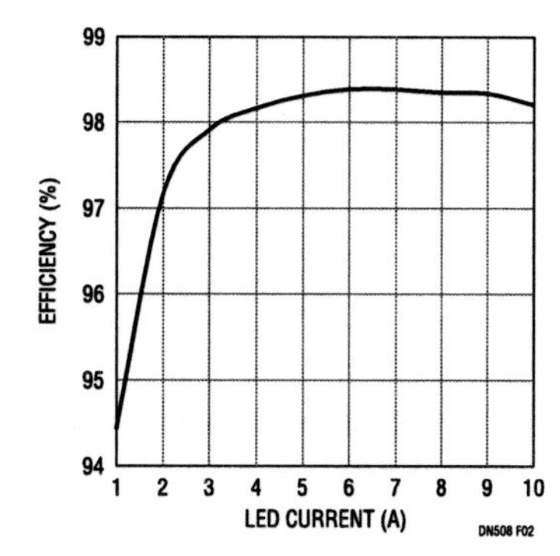

Figure 282.2 • Efficiency of the 48V Input to 35V Output Circuit

Analog Circuit Design: Design Note Collection. http://dx.doi.org/10.1016/B978-0-12-800001-4.00282-9

high efficiency is critical. Each 1% of efficiency improvement reduces the loss by 3.5W—significant if the total power loss budget is less than 7W. This circuit is optimized to operate with 98.2% efficiency at full load—Figure 282.2 shows the efficiency reaching 98% when LED current is above 3A and peaking at 98.4% at ~6A.

At high voltage, the switching losses of the MOSFETs and the inductor outweigh conduction losses. The switching frequency is set to 200kHz to minimize switching losses while maintaining small solution size. Running at full load, this circuit's hot spot occurs at the top MOSFETs, which settles at less than 50°C temperature rise—a very comfortable range for the MOSFETs.

36V input to 20V output, 10A LED driver with fastest PWM dimming

PWM LED dimming is the standard dimming method for high power, high performance lighting applications. Fast LED current response to a PWM signal is important in image-producing applications, such as DLP projectors. Figure 282.3 shows the LT3763 in an application optimized for fast LED PWM dimming.

To achieve fast LED current response to the PWM signal, the LT3763 includes many innovative features. For a given input voltage, the smaller the inductance, the faster the inductor current ramps up, which translates to faster LED current response. This circuit takes only a few microseconds to reach full LED current from zero current when a PWM dimming signal is turned on. Figure 282.4 shows the performance in the PWM dimming application. Efficiency is 97% at full load.

Solar-powered battery charger

The LT3763 can also regulate its input voltage by adjusting its output current. This is useful for applications that must track peak input power such as in a solar-powered battery charger.

Every solar panel has a point of maximum output power that depends on panel illumination, voltage and output current of the panel. In general, peak power is achieved by maintaining the panel voltage in a small range by reducing output current when needed to prevent the panel voltage from moving out of this range. This is called maximum power point tracking (MPPT).

The LT3763's input voltage regulation loop keeps the panel voltage in maximum power point range by adjusting output current. The constant current, constant voltage (CCCV) operation and C/10 function make the part a natural fit for battery charger applications.

Conclusion

The LT3763 is a 60V, synchronous, high current step-down LED driver controller that can be used to drive the latest high power LEDs, with fast PWM dimming response if needed. The LT3763 is not limited to LED driver applications, due to its three additional voltage and current regulation loops and a number of other powerful features.

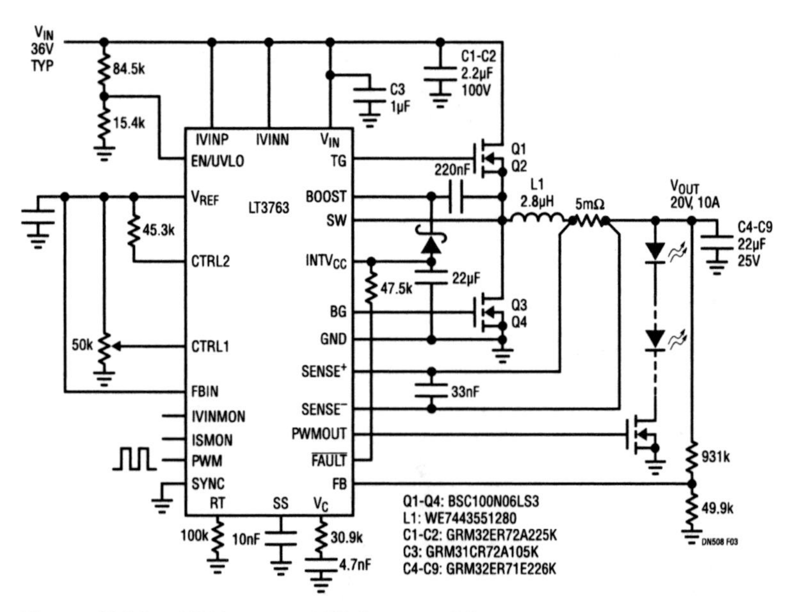

Figure 282.3 • 36V Input to 20V Output, 10A

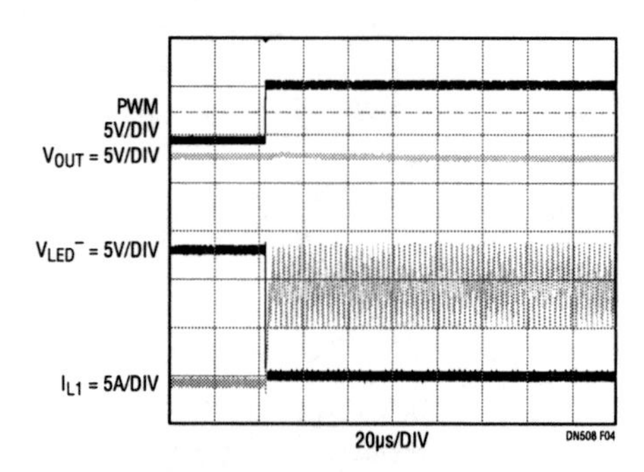

Figure 282.4 • PWM Dimming Performance of Figure 282.3 Circuit

60V buck-boost controller drives high power LEDs, charges batteries and regulates voltage with up to 98.5% efficiency at 100W and higher

283

Keith Szolusha

Introduction

The LT3791 is a 4-switch synchronous buck-boost DC/DC converter that regulates both constant-current and constant-voltage at up to 98.5% efficiency with a single inductor. It can deliver hundreds of watts and features a 60V input and output rating, making it ideal for driving high power LED strings and charging high voltage batteries when both step-up and step-down conversion is needed. It can also be used as a constant-voltage buck-boost regulator with current limiting and monitoring for both input and output.

Buck-boost controller drives 100W LED string for airplane and truck lights

Airplanes and big trucks with 24V batteries need powerful, efficient and robust headlights and spotlights. Figure 283.1 shows a 33.3V, 3A (nine Luminus SSR-90 LEDs) buck-boost LED driver that runs from 15V to 58V input with up to 98.5% efficiency.

The 4-switch synchronous topology drives high power LEDs with minimal switch power loss (and minimal temperature rise). Unlike other topologies, the LT3791 buck-boost can be shorted from LED+ to both LED− and GND and, as a feature, can be programmed to latch off or keep trying to turn back on if the short is removed. Diagnostic output flags report both short-circuit and open-LED conditions.

The solution in Figure 283.1 features up to 100:1 PWM dimming at 100Hz for accurate brightness adjustment without color shift and analog LED dimming when a PWM oscillator is not present (Figure 283.2).

36V, 2.5A SLA battery charger

The buck-boost converter shown in Figure 283.3 charges a 36V 12Ah SLA battery at 44V with 2.5A DC from a 9V to 58V input.

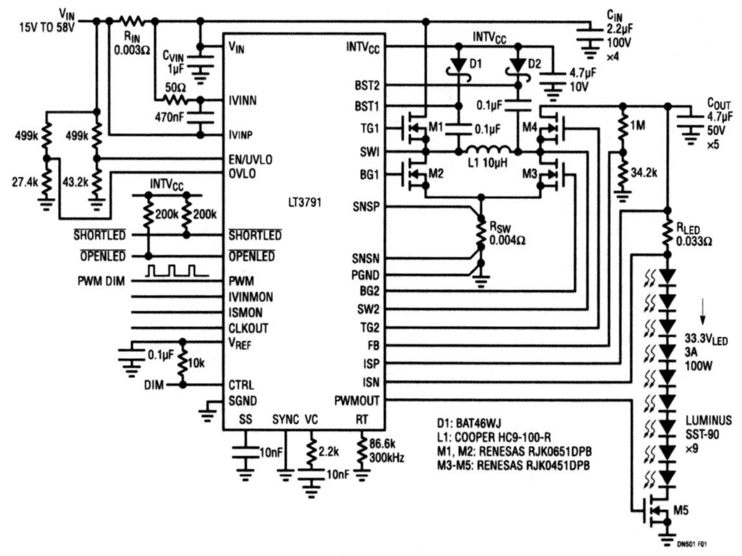

Figure 283.1 • 15–58V_{IN} to 33.3V 3A LED Driver with up to 98.5% Efficiency

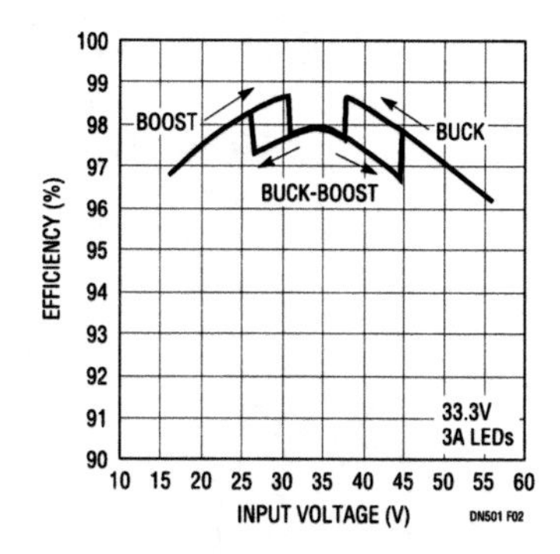

Figure 283.2 • Efficiency of Figure 283.1

Analog Circuit Design: Design Note Collection. http://dx.doi.org/10.1016/B978-0-12-800001-4.00283-0

Specially integrated C/10 current sensing and battery voltage detection drops the battery voltage from its charging voltage (44V) to its float voltage (41V) when the battery is near full charge. The $\overline{\text{OPENLED}}$ flag is used to change the state of the charger from charge to float. When the battery voltage drops far enough, voltage feedback returns the charger to its charge state.

The LT3791 can be tailored to charge a range of battery chemistries and capacities from a variety of input sources regardless of the voltage relationship between them. An external microcontroller can be programmed and used to create a maximum power point tracking device to charge the battery from a solar panel. The output diagnostics and dimming input pins make this a simple interface for high power solar panel applications design.

120W, 6V to 55V voltage regulator

The LT3791 can also be used in high power constant-voltage buck-boost applications. The FB pin doubles as both the main voltage feedback for the buck-boost, and the overvoltage protection detection and regulation when used as an LED driver.

Figure 283.1 can be easily turned into a 100W voltage regulator with V_{OUT} between 6V and 55V by placing a 100k resistor between the SS and V_{REF} pins to defeat soft-start reset during a fault.

The voltage regulator is short-circuit proof. The $\overline{\text{SHORTLED}}$ flag reports when there is a short circuit on the output. The input current monitor can still be used in constant-voltage regulation to protect any system that is input-current limited. R_{LED} sets the output current limit for short circuit, but can be removed to provide a $\overline{\text{PGOOD}}$ flag using $\overline{\text{OPENLED}}$ by shorting ISP to ISN.

Conclusion

The LT3791 synchronous buck-boost controller delivers 100W and higher power at up to 98.5% efficiency to a number of different loads. The wide, 4.7V to 60V input and 0V to 60V output voltage range make it both powerful and versatile, and its short-circuit capability makes this a robust choice for many applications in potentially hazardous environments. This seemingly limitless IC can be used in applications where a typical buck or boost converter cannot because of the crossover of input and output voltage ranges.

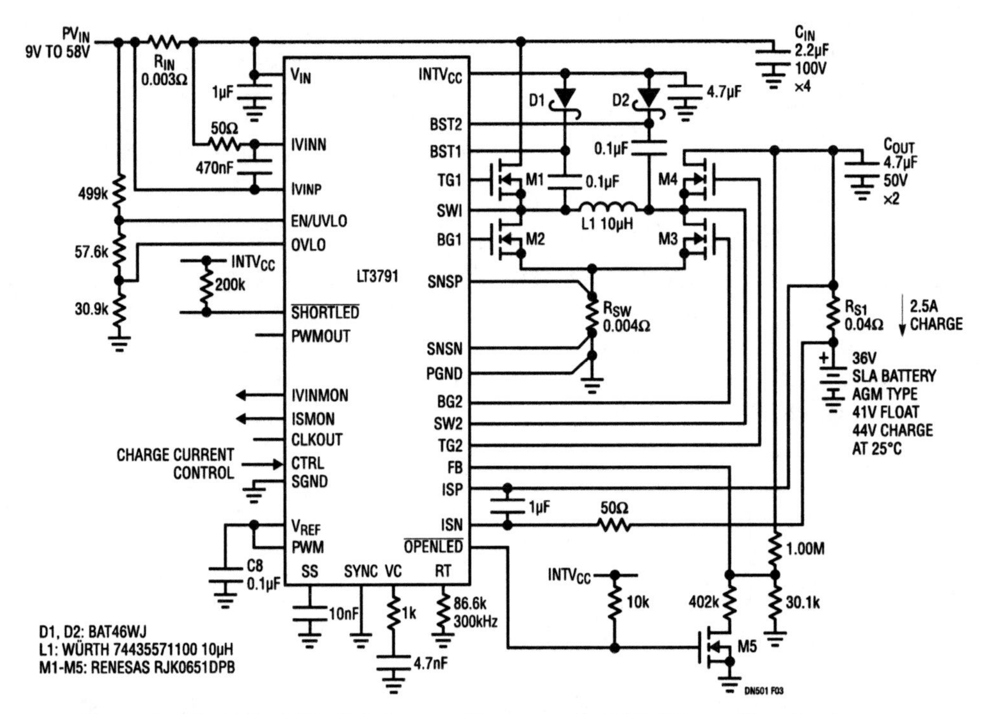

Figure 283.3 • 36V, 2.5A SLA Battery Charger with C/10 Charge Termination

Offline LED lighting simplified: high power factor, isolated LED driver needs no opto-isolators and is TRIAC dimmer compatible

284

Wei Gu

Introduction

As environmental concerns over traditional lighting increase and the price of LEDs decreases, high power LEDs are fast becoming a popular lighting solution for offline applications. In order to meet the requirements of offline lighting—such as high power factor, high efficiency, isolation and TRIAC dimmer compatibility—prior LED drivers used many external discrete components, resulting in cumbersome solutions. The LT3799 solves complexity, space and performance problems by integrating all the required functions for offline LED lighting.

The LT3799 controls an isolated flyback converter in critical conduction (boundary) mode, suitable for LED applications requiring 4W to over 100W of LED power. Its novel current sensing scheme delivers a well-regulated output current to the secondary side without using an optocoupler. Its unique bleeder circuit makes the LED driver compatible with TRIAC dimmers without additional components. Open- and shorted-LED protection ensures long term reliability.

No-opto operation

Figure 284.1 shows a complete LED driver solution. The LT3799 senses the output current from the primary side switch current waveform. For a flyback converter operating in boundary mode, the equation for the output current is:

$$I_{OUT} = 0.5 \cdot I_{PK} \cdot N \cdot (1 - D)$$

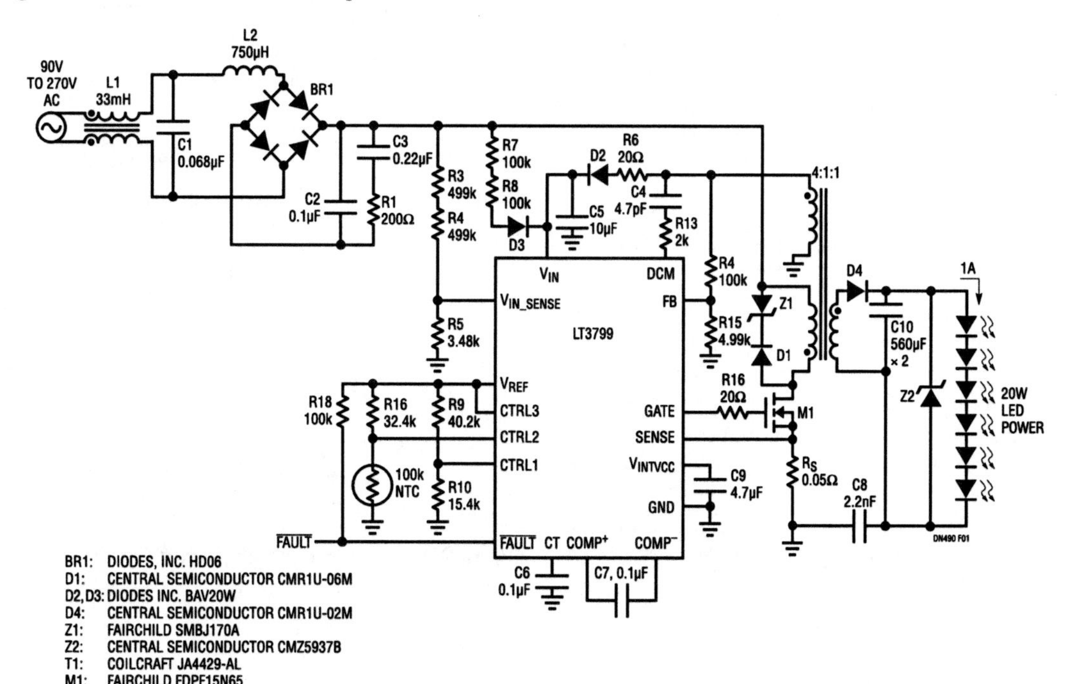

Figure 284.1 • TRIAC Dimmable 20W Offline LED Driver Using the LT3799

Analog Circuit Design: Design Note Collection. http://dx.doi.org/10.1016/B978-0-12-800001-4.00284-2

I_{PK} is the peak switch current, N is the primary to secondary turns ratio and D is the duty cycle. The IC regulates the output current by adjusting the peak switch current and the duty cycle through a novel feedback control. Unlike other primary side sensing methods that need to know input power and output voltage information, this new scheme provides much better output current regulation since the accuracy is barely affected by transformer winding resistance, switch $R_{DS(ON)}$, output diode forward voltage drop and LED cable voltage drop.

High power factor, low harmonics

By forcing the line current to follow the applied sine-wave voltage, the LT3799 achieves high power factor and complies with IEC61000-3-2, Class C lighting equipment Harmonics Requirement. A power factor of one is achieved if the current drawn is proportional to the input voltage. The LT3799 modulates the peak switch current with a scaled version of the input voltage. This technique provides power factors of 0.97 or greater. A low bandwidth feedback loop keeps the output current regulated without distorting the input current.

TRIAC dimmer compatible

When the TRIAC dimmer is in the off state, it's not completely off. There is considerable leakage current flowing through its internal filter to the LED driver. This current charges up the input capacitor of the LED driver, causing random switching and LED flicker. Prior solutions added a bleeder circuit, including a large, expensive high voltage MOSFET. The LT3799 eliminates the need for this MOSFET or any other extra components by utilizing the transformer primary winding and the main switch as the bleeder circuit. As shown in Figure 284.2, the MOSFET gate signal is high and the MOSFET is on when the TRIAC is off, bleeding off the leakage current and keeping the input voltage at 0V. As soon as the TRIAC turns on, the MOSFET seamlessly changes back into a normal power delivery device.

Open- and shorted-LED protection

The LED voltage is constantly monitored through the transformer third winding. The third winding voltage is proportional to the output voltage when the main switch is off and the output diode is conducting current. In the event of overvoltage or open-LED, the main switch turns off and the capacitor at the CT pin discharges. The circuit enters hiccup mode as shown in Figure 284.3.

In a shorted LED event, the IC runs at minimum frequency before the V_{IN} pin voltage drops below the UVLO threshold as the third winding cannot provide enough power to the IC. The IC then enters its start-up sequence as shown in Figure 284.4.

CTRL pins and analog dimming

The LT3799's output can be adjusted through multiple CTRL pins. For example, the output current would follow a DC control voltage applied to any CTRL pin for analog dimming. Overtemperature protection and line brownout protection can also be easily implemented using these CTRL pins.

Conclusion

The LT3799 is a complete offline LED driver solution featuring standard TRIAC dimming, active PFC and well-regulated LED current with no optocoupler. This high performance and feature-rich IC greatly simplifies and shrinks offline LED driver solutions.

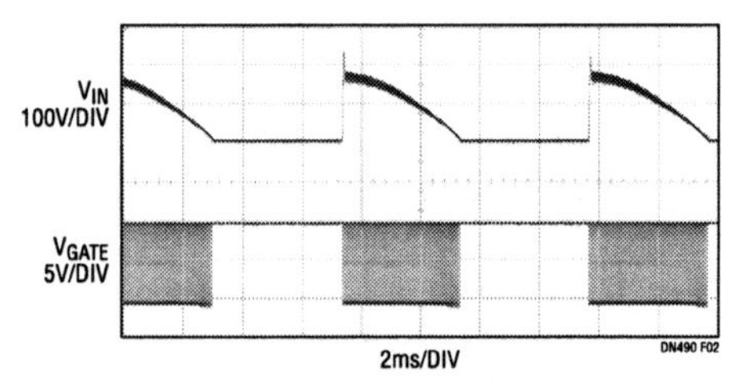

Figure 284.2 • MOSFET Gate Signal and V_{IN}

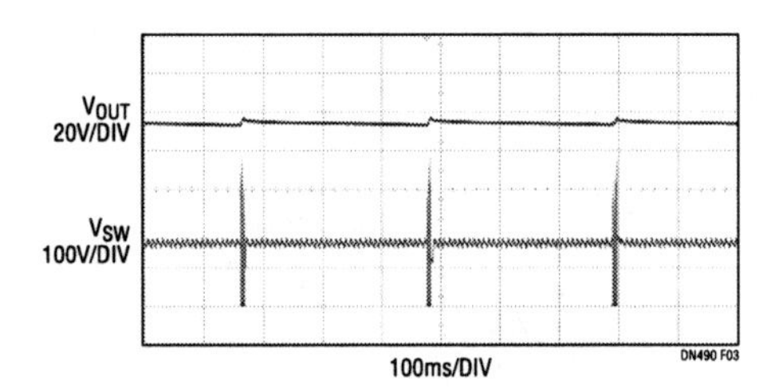

Figure 284.3 • Output Open-Circuit Event

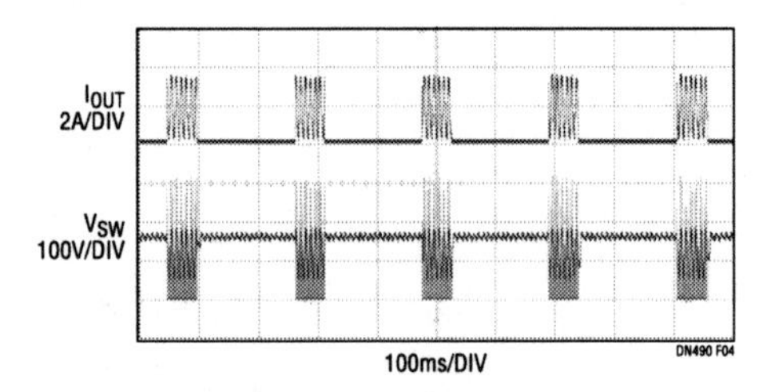

Figure 284.4 • Output Short-Circuit Event

Reduce the cost and complexity of medium LCD LED backlights with a single inductor LED driver for 60 LEDs

285

Daniel Chen

Introduction

One inductor, one IC, one string of LEDs. This is the conventional way to build a boost LED driver for LCD display backlights. Although this is a perfectly acceptable solution for small LCD displays that only require a few strings, in larger displays the number of control ICs and inductors multiplies quickly, as do the expense and PCB real estate requirements. This is a major hurdle in the race to replace CCFLs with robust, spectrally superior LEDs in medium sized, bright displays.

A better driver is needed to bring the cost and complexity of LED backlights in line with CCFLs. The LT3598 answers the call by driving six strings of ten LEDs at up to 30mA per string. It also has a built-in power switch to save space and design time. Efficiency is optimized via an adaptive feedback loop that monitors all LED pin voltages to provide an output voltage just high enough to light all LED strings. The LED current is regulated even when V_{IN} is greater than V_{OUT}. The LED current can be derated based on programmed LED temperature through an NTC resistor divider or by programming die junction temperature.

Typical application

Figure 285.1 shows the LT3598's six channels driving 60 LEDs, with each string programmed at 20mA. The CTRL pin and PWM pin provide analog and digital dimming, respectively. True Color PWM dimming delivers constant LED color with a 3000:1 dimming ratio. Figure 285.2 shows the typical ±0.5% current matching between strings, which yields the uniform light distribution that is so important in large backlight applications.

Need more current?

For applications that demand more than 30mA per string, multiple channels of the LT3598 can be easily combined for higher LED current. Figure 285.3 shows a configuration that drives two strings at up to 90mA per string. The 1000:1 PWM dimming waveform at 125°C junction temperature (worst case) is shown in Figure 285.4.

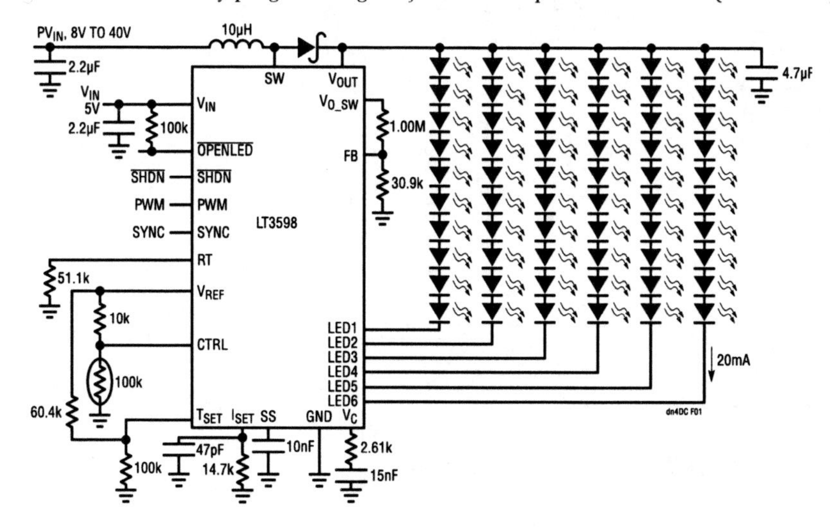

Figure 285.1 • LED Driver for 60 x 20mA LEDs

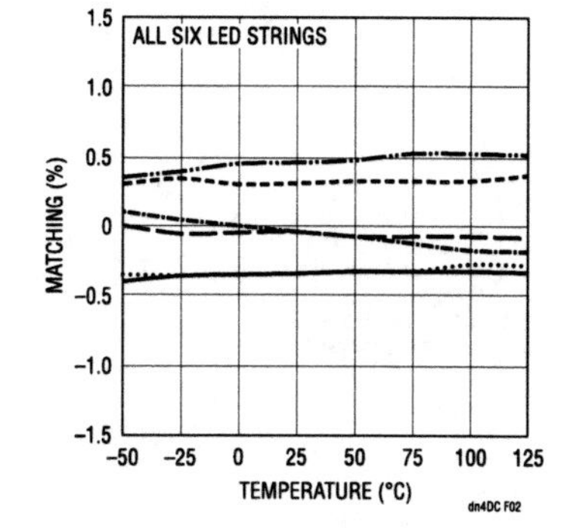

Figure 285.2 • Current Matching for Figure 285.1

Analog Circuit Design: Design Note Collection. http://dx.doi.org/10.1016/B978-0-12-800001-4.00285-4

T_{SET} pin for thermal protection

The T_{SET} pin voltage can be programmed to limit the internal junction temperature of the LT3598. Once this temperature is reached, the LED current will linearly decrease if the junction temperature keeps increasing, as shown in Figure 285.5. This thermal regulation feature provides important protection at high ambient temperatures, and allows a given application to be optimized for typical, instead of worst-case, ambient temperatures.

Channel disable capability

Unused LED pins can be tied to V_{OUT} to disable them, so no current flows into the disabled channels. Fault detection ignores any channels tied to V_{OUT}. Figure 285.6 shows an application with two disabled channels that yields efficiency as high as 90%.

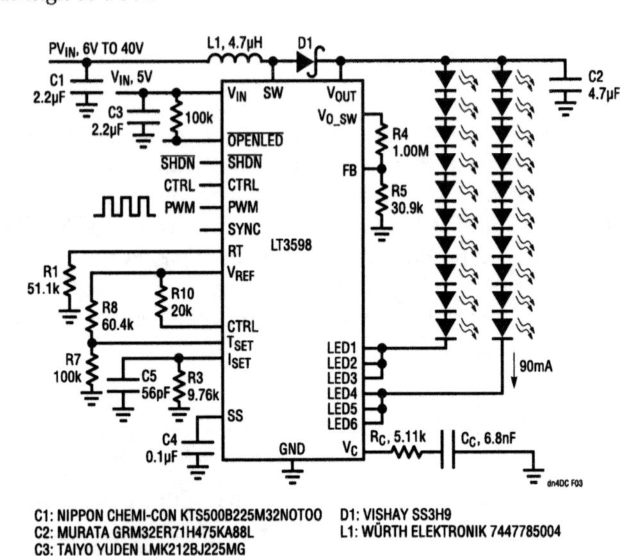

Figure 285.3 • LED Driver for Two Strings of 90mA LEDs

Conclusion

LT3598 is a versatile LED driver with a built-in power switch for multiple LED strings. High PWM dimming is possible even with its robust fault detection. Furthermore, a voltage loop regulates the output voltage when all LED strings are open.

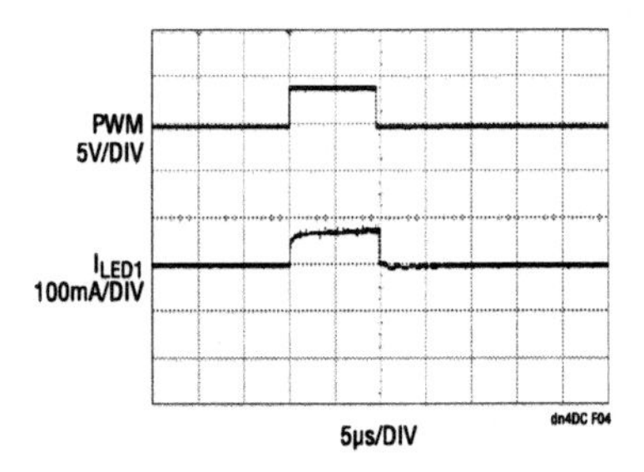

Figure 285.4 • 1000:1 PWM Dimming for Figure 285.3 at 125°C

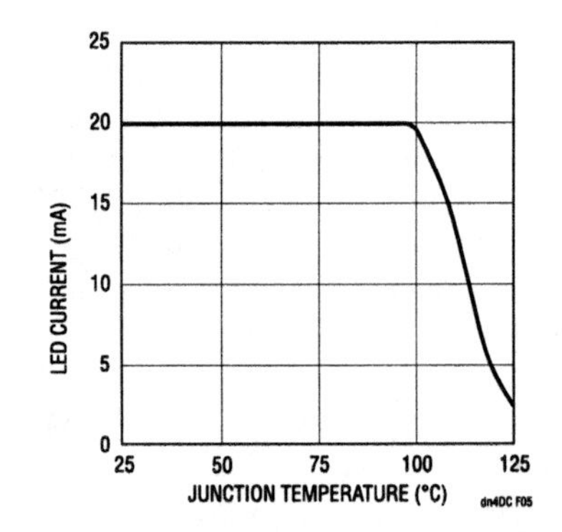

Figure 285.5 • T_{SET} Function Reduces LED Current at High Temperatures

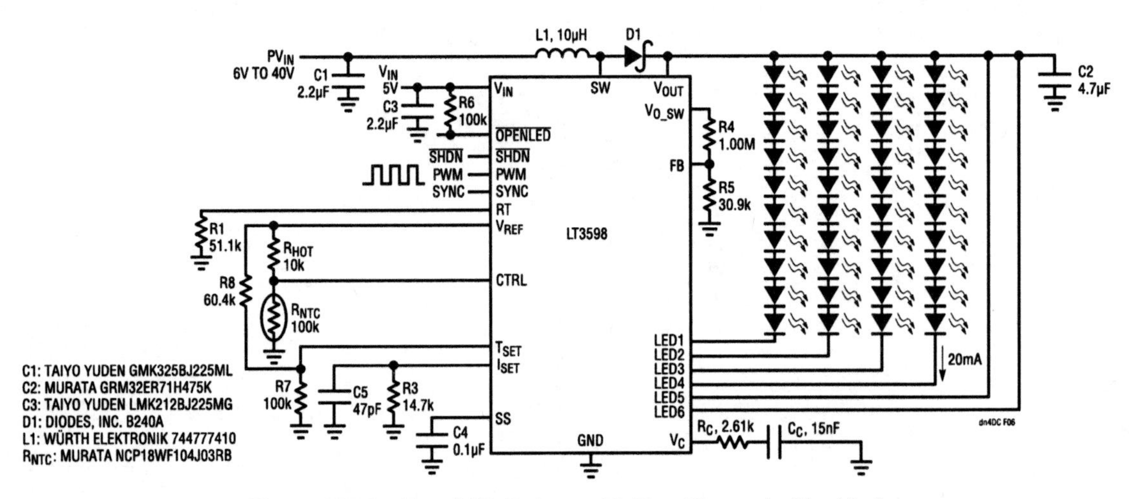

Figure 285.6 • Four LED Strings with Two Channels Disabled

100V controller drives high power LED strings from just about any input

286

Keith Szolusha

Introduction

Strings of high power solid-state LEDs are replacing traditional lighting technologies in large area and high lumens light sources because of their high quality light output, unmatched durability, relatively low lifetime cost, constant-color dimming and energy efficiency. The list of applications grows daily, including LCD backlights and projection, industrial and architectural lighting, automotive lights, streetlights, billboards and stadium lights.

As the list expands, so does the range of V_{IN} for the LED drivers. LED drivers must be able to handle wide ranging inputs, including transient voltages of automotive batteries, a wide range of other batteries and wall wart voltages. For LED lighting manufacturers, applying a different LED driver for each application means stocking, testing and designing with a number of controllers. It would be better to use just one that can be applied to many solutions.

The LT3756 high voltage LED driver features a unique topological versatility that allows it to be used in boost, buck-boost mode, buck mode, SEPIC, flyback and other topologies. Its high power capability provides potentially hundreds of watts of LED power over a wide input voltage range. Its 100V floating LED current sense inputs provide accurate LED current sensing. Excellent PWM dimming architecture produces high dimming ratios.

A number of features protect the LEDs and surrounding components. Shutdown and undervoltage lockout, when combined with analog dimming derived from the input, provide the standard ON/OFF feature as well as a reduced LED current should the battery voltage drop to unacceptably low levels. Analog dimming is accurate and can be combined with PWM dimming for a wide range of brightness control. Soft-start prevents spiking inrush currents. The $\overline{\text{OPENLED}}$ pin informs of open or missing LEDs and the SYNC (LT3756-1) pin can be used to sync switching to an external clock. The FB voltage loop limits the max V_{OUT} to protect the converter in the case of open LEDs.

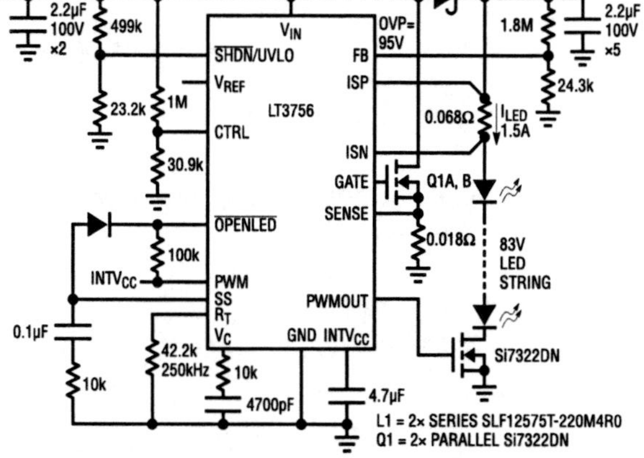

Figure 286.1 • A 125W, 83V at 1.5A, 97% Efficient Boost LED Driver for Stadium Lighting

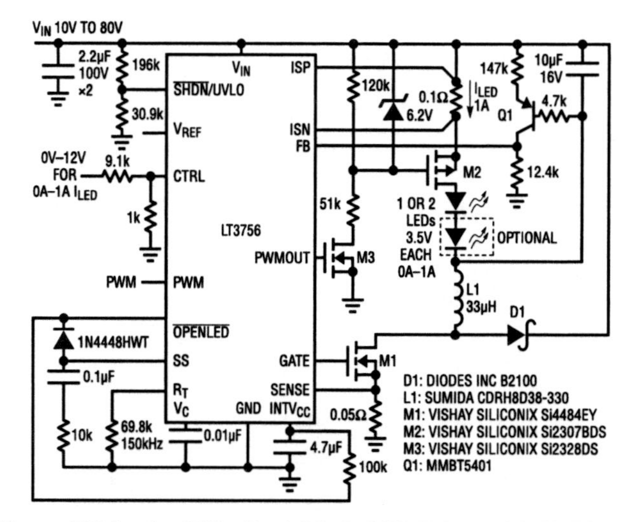

Figure 286.2 • An $80V_{IN}$ Buck Mode LED Driver with PWM Dimming for Single or Double LEDs

Analog Circuit Design: Design Note Collection. http://dx.doi.org/10.1016/B978-0-12-800001-4.00286-6

The 16-pin IC is available in a tiny QFN (3mm × 3mm) and an MSE package, both thermally enhanced. For lower input voltage requirements, the $40V_{IN}$, $75V_{OUT}$ LT3755 LED controller is a similar option.

Boost

Lighting systems for stadiums, spotlights and billboards require huge strings of LEDs running at high power. The LT3756 controller drives up to 100V LED strings. The 125W LED driver in Figure 286.1 has a 40V–60V input.

The high power gate driver switches two 100V MOSFETs at 250kHz. This switching frequency minimizes the size of the discrete components while maintaining high 97% efficiency, producing a less-than-50°C discrete component temperature rise—more manageable than the heat produced by the 125W LEDs.

Even if PWM dimming is not required, the PWMOUT MOSFET is useful for LED disconnect during shutdown. It prevents current from running through the string of LEDs.

If the LED string is removed, the FB constant-voltage loop takes over and regulates the output at 95V. Without overvoltage protection, the LED sense resistor would see zero current and the output cap voltage would go over 100V, exceeding several max ratings. While in OVP $\overline{\text{OPENLED}}$ goes low.

Buck mode

When V_{IN} is higher than V_{LED}, the LT3756 can serve equally well as a buck mode LED driver. The buck mode LED driver in Figure 286.2 operates with a wide 10V-to-80V input range to drive one or two LEDs at 1A.

PWM dimming requires a level-shift from the PWMOUT pin to the high-side LED string. The max PWM dimming ratio increases with higher switching frequency, lower PWM dimming frequency, higher V_{IN} and lower LED power. In this case, a 100:1 dimming ratio is possible with a 100Hz dimming frequency and a 48V input. Although higher switching frequency is possible, the duty cycle has its limits. Generous minimum on-time and minimum off-time restrictions require a frequency on the lower end of its range (150kHz) to meet both the harsh high-V_{IN}-to-low-V_{LED} ($80V_{IN}$ to one 3.5V LED) and low-V_{IN}-dropout requirements ($10V_{IN}$ to $7V_{LED}$).

OVP of the buck mode LED driver has a level shift as well. Without the level-shifted OVP network tied to FB, an open LED string would result in the output capacitor charging up to V_{IN}. Although the buck mode components will survive this scenario, the LEDs may not survive being plugged into a potential equal to V_{IN}.

Buck-boost mode

A common LED driver requirement is that the ranges of both the LED string voltage and the input voltage are wide and overlapping. In fact, some designers prefer to use the same LED driver circuit for several different battery sources and several different LED strings. Such a versatile configuration trades some efficiency, component cost, and board space for design simplicity, and time-to-market.

The buck-boost mode driver in Figure 286.3 uses a single inductor. It accepts inputs from 9V to 36V to drive 10V–50V LED strings at 400mA.

The inductor current is the sum of the input current and the LED string current; the peak inductor current is equal to the peak switching current. Below 9V input, CTRL analog dimming scales back the LED current to keep the inductor current under control. UVLO turns off the LEDs below $6V_{IN}$. C_{OUT}, DI and MI can see voltages as high as 95V here.

Conclusion

The LT3756 controller is a versatile high power LED driver. It has all the features required for large (and small) strings of high power LEDs. Its high voltage rating, optimized LED driver architecture, high performance PWM dimming, host of protection features and accurate high side current sensing make the LT3756 a single-IC choice for a variety of lighting systems.

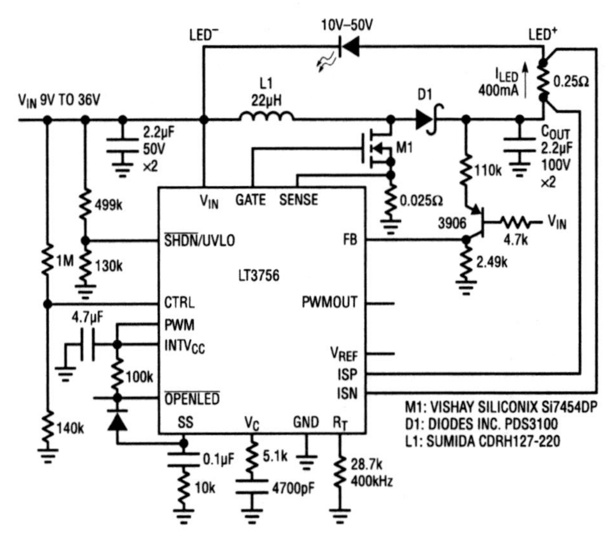

Figure 286.3 • A Buck-Boost Mode LED Driver with Wide-Ranging V_{IN} and V_{LED}

Triple LED driver in 4mm × 5mm QFN supports LCD backlights in buck, boost or buck-boost modes and delivers 3000:1 PWM dimming ratio

287

Hua (Walker) Bai

Introduction

By integrating three independent LED drivers, the LT3496 offers a highly efficient, compact and cost-effective solution to drive multiple LED strings. All three drivers have independent on/off and PWM dimming control, and can drive different numbers or types of LEDs. High side current sensing and built-in gate drivers for PMOS LED disconnect allow the LT3496 to operate in buck, boost, SEPIC or buck-boost modes with up to 3000:1 True Color PWM dimming ratio.

The LT3496 is offered in a single 4mm × 5mm QFN or FE28 package. The efficiency of each driver can exceed 95%.

Integrated PMOS drivers improve PWM dimming ratio to 3000:1

A high PWM dimming ratio is critical in many display applications, especially in high end LCD panels. Beware, though, the definition of dimming ratio varies among suppliers. When comparing dimming ratios, pay close attention to the PWM dimming frequency and linearity of the LED average current at different PWM duty cycles. For instance, the LT3496's high 3000:1 PWM dimming ratio can be achieved at a 100Hz PWM frequency—high enough to keep the display flicker-free over the entire dimming range.

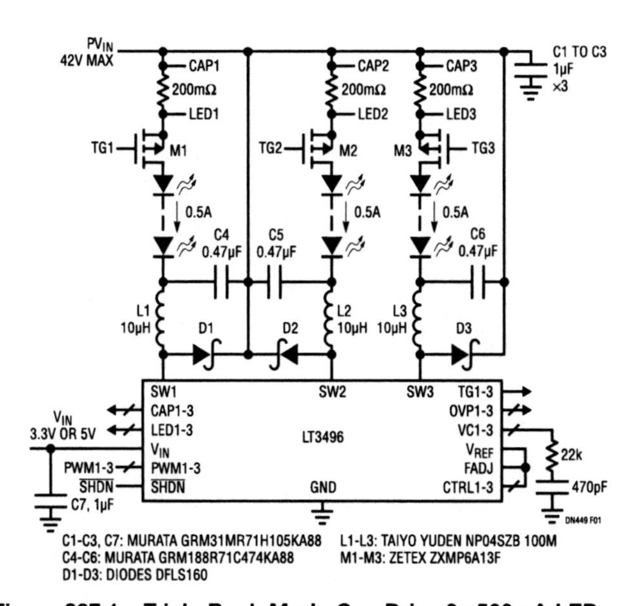

Figure 287.1 • Triple Buck Mode Can Drive 3x 500mA LED Strings

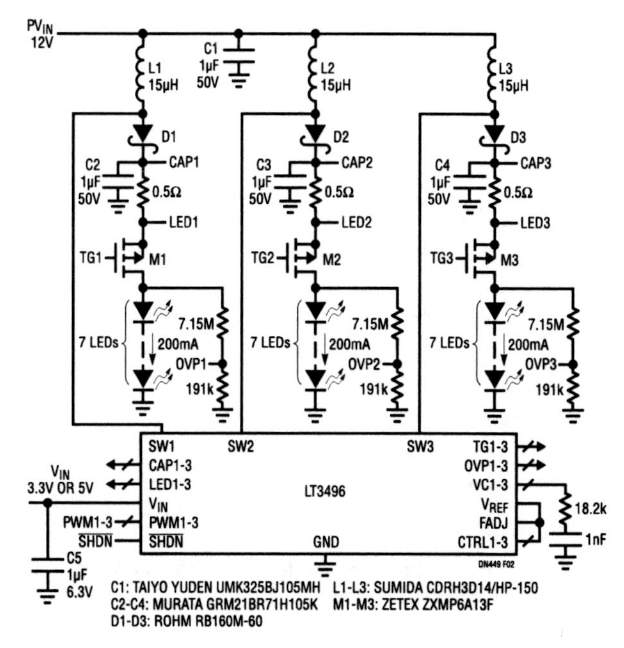

Figure 287.2 • Triple Boost Mode Can Drive 200mA LEDs

Analog Circuit Design: Design Note Collection. http://dx.doi.org/10.1016/B978-0-12-800001-4.00287-8

Buck mode circuit drives three 500mA LED strings

Figure 287.1 shows a triple buck mode LED driver. Each channel drives 500mA of current to its LEDs. Each string can have from eight to twelve LEDs, depending on type. The 2.1MHz switching frequency minimizes the solution size by allowing the use of low profile inductors and capacitors. The overall size of the circuit is less than 16mm×16mm, with a maximum height of 1.5mm.

Efficiency can be above 95% for a LT3496 buck mode driver. A further reduction in the parts count is possible by removing M1, M2 and M3. However, the dimming ratio drops without those MOSFETs. To improve the efficiency, the V_{IN} pin should be biased from a 3.3V or 5V supply. Energy to the LEDs is supplied by PV_{IN}. OVP protection is omitted in Figure 287.1.

Boost mode circuit drives three 200mA LED strings

Figure 287.2 shows a triple boost mode driver that delivers 200mA to each LED string from a regulated 12V. Figure 287.3 shows the superior PWM dimming performance of the circuit. The LED current reaches a programmed 200mA in less than 500ns. The efficiency of this circuit is 90% at a 2.1MHz switching frequency. Unlike the buck mode driver, the boost mode and buck-boost mode drivers always require an OVP circuit at the output for open LED protection.

Buck-boost mode circuit survives load dump events

In automotive applications, load dump is a condition under which an IC is expected to experience 40V transient. In such applications, the LED string voltage often falls in the middle of the 8V to 40V input supply range, thus requiring buck-boost mode.

In a buck-boost circuit, the switch voltage is the sum of the input voltage and the LED voltage. Therefore, it is necessary to turn off the internal power switch before the input voltage gets too high. The LT3496 circuit in Figure 287.4 drives four LEDs, at 200mA per channel. The circuit monitors the Schottky diodes' cathode voltage (V_{SC}). The OVP logic turns off the main switch when V_{SC} is above 38V, preventing the switch voltage from rising further. Since no IC pin experiences absolute maximum voltage, the circuit survives the load dump event.

Conclusion

Multiple output LED drivers, such as the LT3496, offer excellent current matching, efficiency and space savings. The flexibility to operate in buck, boost or buck-boost mode makes the LT3496 feasible in many rugged applications.

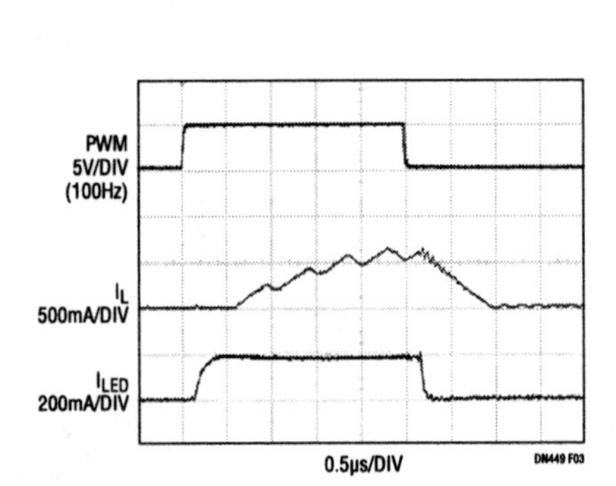

Figure 287.3 • Achieving Greater Than 3000:1 PWM Dimming Ratio with a PMOS Disconnect

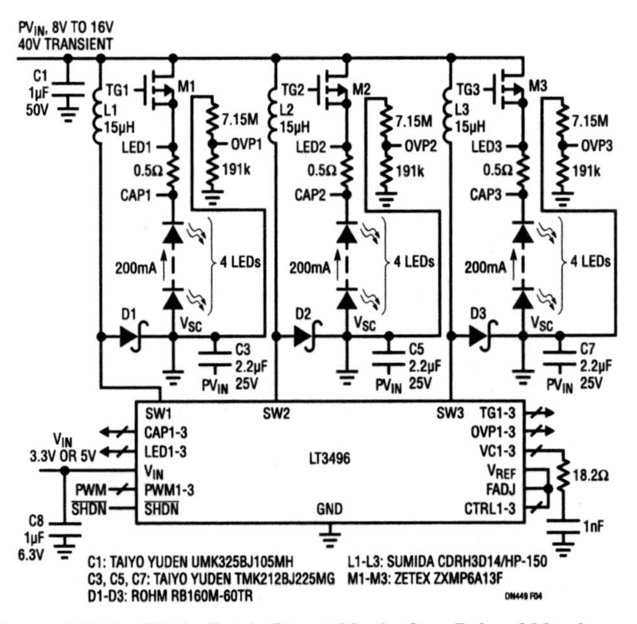

Figure 287.4 • Triple Buck-Boost Mode Can Drive 200mA LEDs While Surviving Load Dump

μModule LED driver integrates all circuitry, including the inductor, in a surface mount package

288

David Ng

Introduction

Once relegated to the hinterlands of low cost indicator lights, the LED is again in the spotlight of the lighting world. LED lighting is now ubiquitous, from car headlights to USB-powered lava lamps. Car headlights exemplify applications that capitalize on the LED's clear advantages—unwavering high quality light output, tough-as-steel robustness, inherent high efficiency—while a USB lava lamp exemplifies applications where *only* LEDs work. Despite these clear advantages, their requirement for regulated voltage *and* current make LED driver circuits more complex than the venerable light bulb, but some new devices are closing the gap. For instance, the LTM8040 μModule LED driver integrates all the driver circuitry into a single package, allowing designers to refocus their time and effort on the details of lighting design critical to a product's success.

A superior LED driver

The LTM8040 is a complete step-down DC/DC switching converter system that can drive up to 1A through a string of LEDs. Its 4V to 36V input voltage range makes it suitable for a wide range of power sources, including 2-cell lithium-ion battery packs, rectified 12VAC and industrial 24V. The LTM8040 features both analog and PWM dimming, allowing a 250:1 dimming range. The built-in 14V output voltage clamp prevents damage in the case of an accidental open LED string. The default switching frequency of the LTM8040 is 500kHz, but switching frequencies to 2MHz can be set with a resistor from the RT pin to GND.

Easy to use

The high level of integration in the LTM8040 minimizes external components and simplifies board layout. As shown in Figure 288.1, all that is necessary to drive an LED string up to 1A is the LTM8040 and an input decoupling capacitor. Even with all this built-in functionality, the LTM8040 itself is small, measuring only 15mm×9mm×4.32mm.

Rich feature set

The LTM8040 features an ADJ pin for precise LED current amplitude control. The ADJ pin accepts a full-scale input voltage range of 0V to 1.25V, linearly adjusting the output LED current from 0A to 1A. Figure 288.2 shows the ratiometric response of the output LED current versus the ADJ

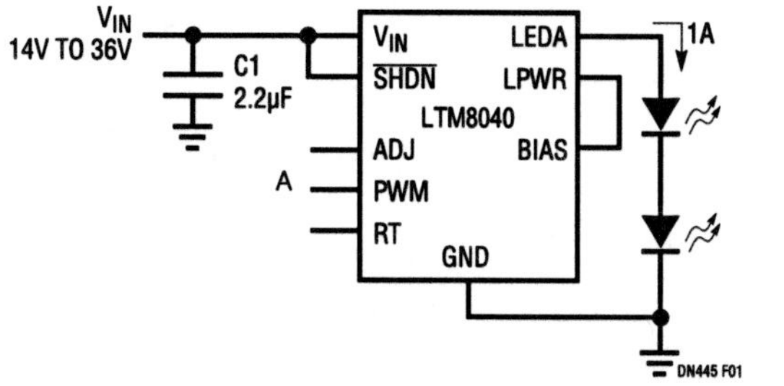

Figure 288.1 • Driving an LED String with the LTM8040 Is Simple—Just Add the Input Capacitor and Connect the LED String

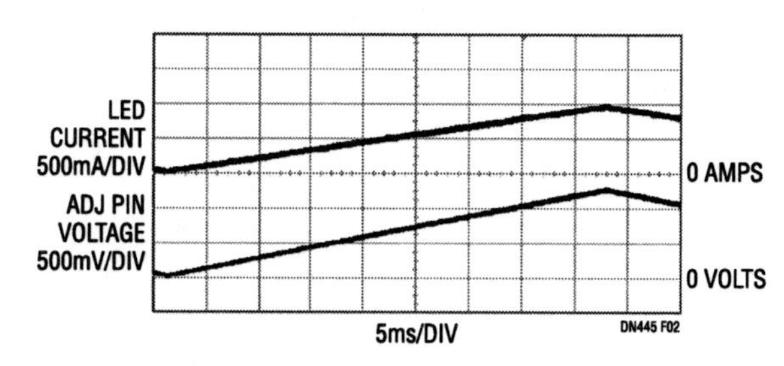

Figure 288.2 • Drive a 0V to 1.25V Voltage into the ADJ Pin to Control the LED Current Amplitude

Analog Circuit Design: Design Note Collection. http://dx.doi.org/10.1016/B978-0-12-800001-4.00288-X

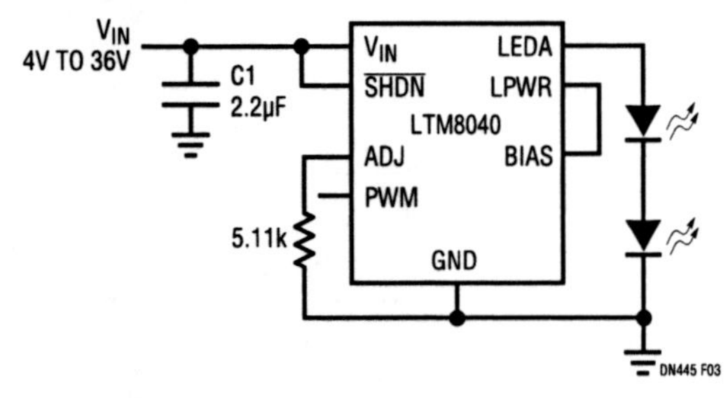

Figure 288.3 • Control the LED Current with a Single Resistor from ADJ to Ground

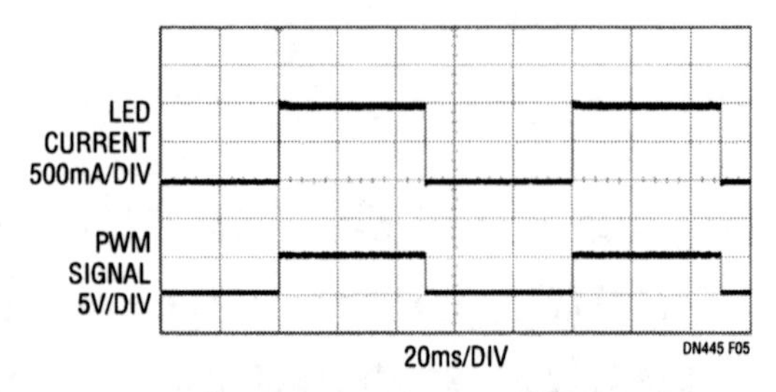

Figure 288.5 • The LTM8040 Can PWM LED Current with Minimal Distortion, Even at Frequencies as Low as 10Hz

voltage. The ADJ pin is internally pulled up through a 5.11k precision resistor to an internal 1.25V reference, so the output LED current can also be adjusted by applying a single resistor from ADJ to ground, as shown in Figure 288.3.

The PWM control pin allows high dimming ratios. With an external MOSFET in series with the LED string as shown in Figure 288.4, the LTM8040 can achieve dimming ratios in excess of 250:1. As seen in Figure 288.5, there is little distortion of the PWM LED current, even at frequencies as low as 10Hz. The 10Hz performance is shown to illustrate the capabilities of the LTM8040—this frequency is too low for practical pulse-width modulation, being well within the discrimination range of the human eye.

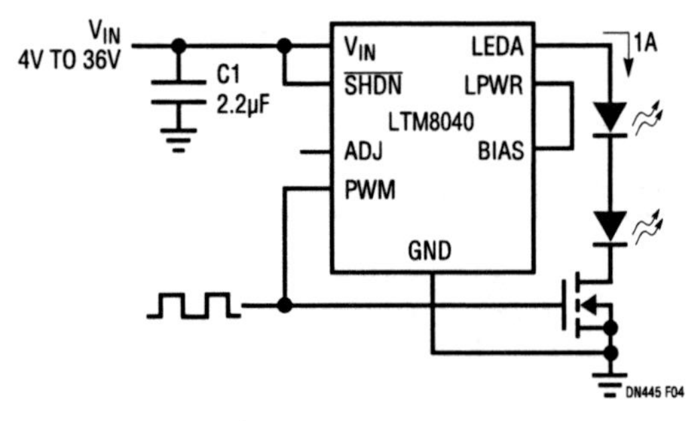

Figure 288.4 • The LTM8040 Can PWM Its LED String with an External MOSFET

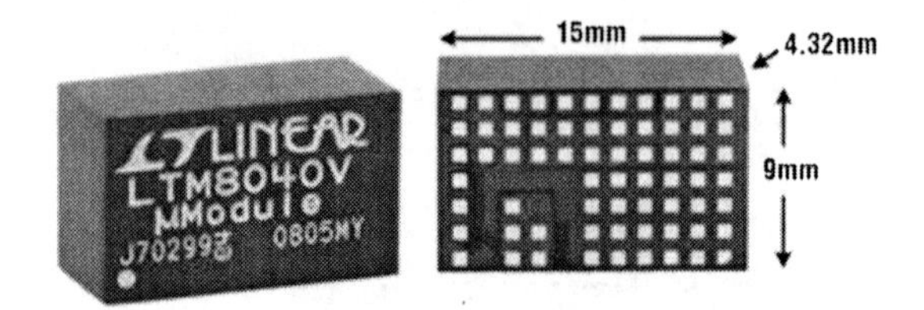

Figure 288.6 • Only 9mm x 15mm x 4.32mm, the LTM8040 LED Driver Is a Complete System in an LGA Package

The LTM8040 also features a low power shutdown state. When the $\overline{SHDN}$ pin is active low, the input quiescent current is less than 1µA.

Conclusion

The LTM8040 µModule LED driver makes it easy to drive LEDs. Its high level of integration and rich feature set, including open LED protection, analog and PWM dimming, save significant design time and board space.

Versatile TFT LCD bias supply and white LED driver in a 4mm × 4mm QFN

289

Eddy Wells

Introduction

The makers of handheld medical, industrial and consumer devices use a wide variety of high resolution, small to medium sized color TFT LCD displays. The power supply designers for these displays must contend with shrinking board area, tight schedules, and variations in display types and feature requirements. The LTC3524 simplifies the designer's job by combining a versatile, easily programmed, TFT LCD bias supply and white LED backlight driver in a low profile 4mm × 4mm QFN package.

The LTC3524's 2.5V to 6V input supply range is ideally suited for portable devices powered from Li-Ion or multiple alkaline or nickel cells. Both the LCD and LED drivers operate at 1.5MHz, allowing the use of tiny, low cost, inductors and capacitors.

The TFT bias portion of the circuit consists of a synchronous boost converter, adjustable between 3V and 6V, providing the main analog V_{OUT} for the TFT. Low current gate drive voltages (VH and VN) are generated using integrated charge-pump circuits. These low noise outputs are programmable to ±20V, allowing optimal bias for multiple display types and makers. The TFT outputs are sequenced at power-up and discharged at power-down as shown in Figure 289.1.

A second non-synchronous boost converter generates the voltage required to regulate one or two LED strings at up to 25mA each. LED current can be adjusted by either analog or digital means, optimizing the TFT display for varying ambient light conditions. Each string is independently enabled and can contain one to five LEDs in series. Internal circuitry maintains equal current in the strings, even when the forward voltage drops of the LEDs do not match. Open LED protection is provided to prevent the output from exceeding 24V.

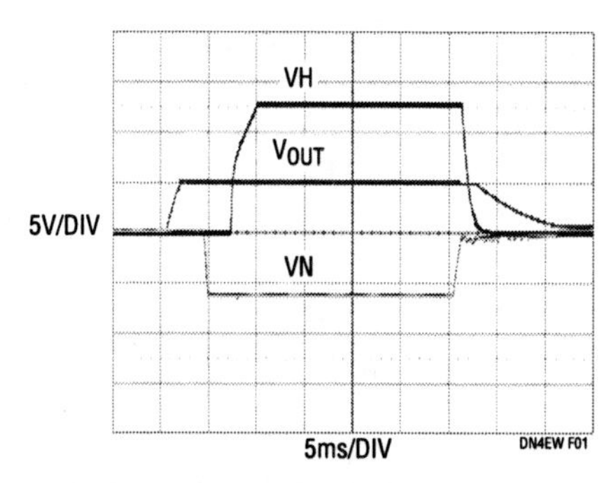

Figure 289.1 • LTC3524 TFT LCD Supply Sequencing at Power-Up and Power-Down

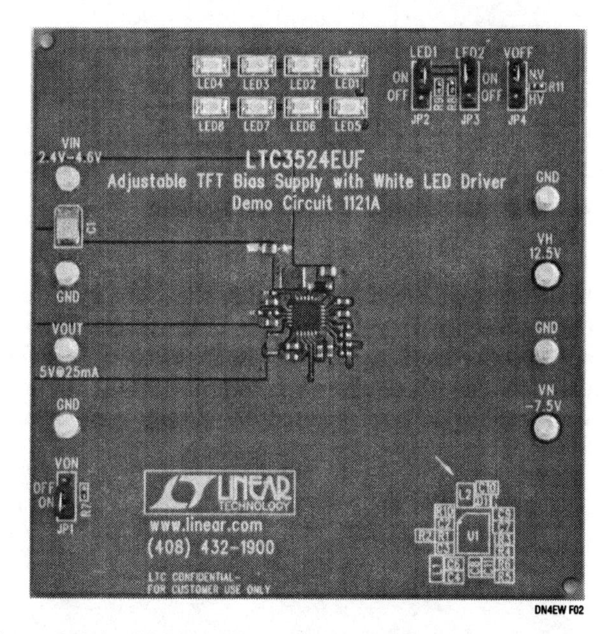

Figure 289.2 • LTC3524-Based LCD and White LED Supply

Analog Circuit Design: Design Note Collection. http://dx.doi.org/10.1016/B978-0-12-800001-4.00289-1

3-output TFT supply with digitally dimmed LED backlight

A LTC3524-based TFT and backlight solution for a 4 to 6 inch LCD is shown in Figure 289.2. High frequency operation of the power components and the QFN package shrinks the total converter footprint to approximately 120mm^2 (single sided).

The circuit schematic is shown in Figure 289.3. The TFT bias portion of the circuit provides a 5V, 25mA output for the TFT drivers as well as 12.5V and −7.5V outputs with up to 2mA for the gate bias. These voltages are programmed using the FBVO, FBH, and FBN pins respectively.

As shown in Figure 289.1, these outputs are sequenced with V_{OUT}, VN, then VH powered, as required by most displays. The outputs are actively discharged when ELCD is brought low, removing voltage from the display.

The white LED backlight for the Figure 289.3 circuit consists of two strings with four series LEDs. The LEDs are driven from the high side with the LTC3524, allowing the strings to terminate at ground, reducing the number of wires required to power the display. With R_{PROG} = 100k, each LED is regulated to 20mA. Maximum power for the backlight is approximately 600mW, assuming a forward voltage around 3.6V per element.

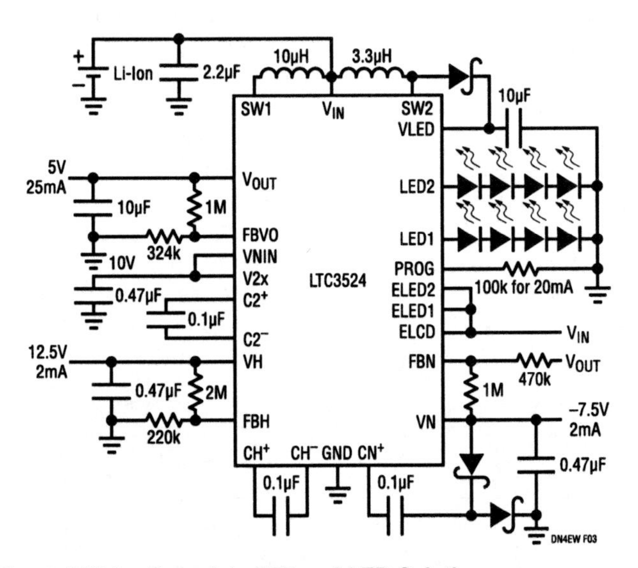

Figure 289.3 • Complete TFT and LED Solution

Dimming is achieved by changing the duty cycle of a 200Hz power signal applied to the LED strings. The frequency is high enough to prevent visually detectable flickering, but low enough to allow a better than 100:1 dimming range. Dimming is implemented by simply connecting a

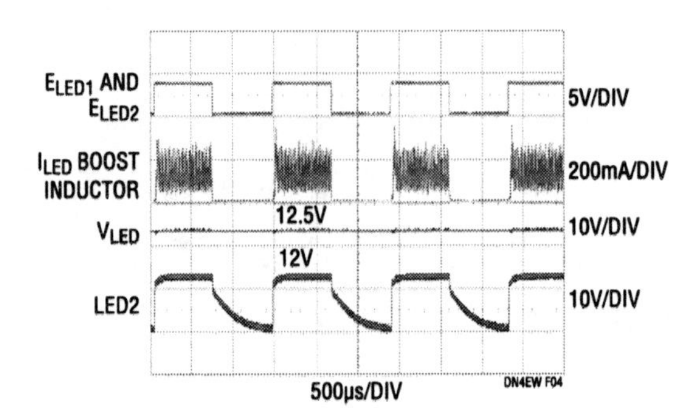

Figure 289.4 • Burst Dimming Waveforms

microprocessor controlled port to ELED1 and ELED2. Scope waveforms at 50% duty cycle are shown in Figure 289.4.

Efficiency results for this design are given in Figure 289.5 with a 3.6V input. The LCD efficiency curve shows the performance of the synchronous boost converter with V_{OUT} at 5V and varying load current. This curve includes the no load quiescent current of the charge-pumps, which are powered from V_{OUT}.

Analog dimming of the LEDs can be implemented by adjusting the current through the PROG pin. Efficiency for analog dimming is shown in Figure 289.5. Efficiency with PWM dimming would remain close to 78% over a wide dimming range.

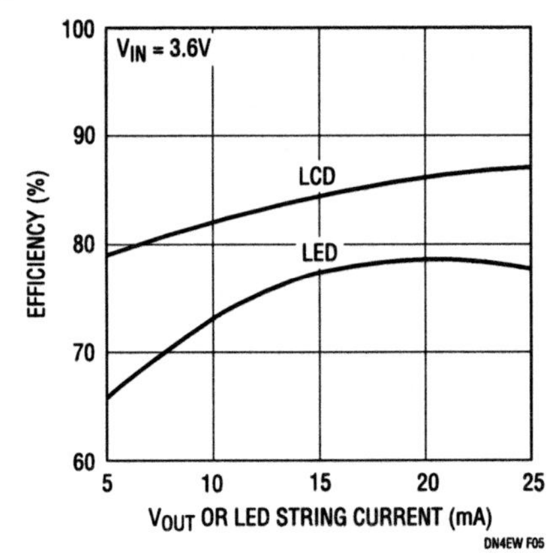

Figure 289.5 • LCD Bias and LED Efficiency

Conclusion

The LTC3524 shrinks and simplifies the design of small to medium sized TFT LCDs by combining the LCD supply and LED driver in a single compact package. LCD bias voltages and LED currents are programmable, making it possible to simplify parts stock by using the LTC3524 for a wide variety of displays.

Tiny universal LED driver can gradate, blink or turn on nine individual LEDs with minimal external control

290

Marty Merchant

Introduction

LEDs are the lighting workhorse of cell phones, MP3 players and diagnostic lights in telecom systems. Their uses are many, from utilitarian backlighting to eye-catching aesthetic effects such as slowly pulsing multicolor indicators. As device designers strive to differentiate their products on the shelf, the number and complexity of lighting effects grows. It would seem that each new effect requires significant additional hardware, and/or complex software, right? Actually, no, there is a way to apply these effects to a number of LEDs with only a single driver IC.

The LTC3219 9-output universal LED (ULED) driver can be programmed to individually gradate, blink or turn on nine individual LEDs using internal logic and circuitry to drive nine 6-bit DAC-controlled LED current sources. Because the gradation and blinking features are controlled internally, effects can be realized without adding ICs, extensively tying up the I^2C bus or filling valuable memory space with complex programming subroutines. Any feature on any 0mA to 28mA output can be configured to activate via the external enable (ENU) pin or I^2C interface.

The LTC3219 operates from a 2.9V to 5.5V input. The charge pump provides up to 250mA output current, and to optimize efficiency, it automatically changes charge pump mode to 1x, 1.5x or 2x depending on the output current requirement. Any of these modes can also be forced.

Blinking and gradation modes

Each output can be set to blink each output with a 156ms or a 625ms on time and a 1.25s or a 2.5s period. Blink mode can be initiated and ended via the I^2C interface or using the ENU pin. Once blinking has been initiated, the LED(s) continue to blink without any interaction from the I^2C interface or the ENU pin. This allows the controlling interface device to shutdown and save battery power until needed.

The LTC3219 can gradually turn on, or gradually turn off, any number of the LED channels. The gradation ramps up from 0mA to the programmed LED intensity with ramp times of 240ms, 480ms or 960ms (likewise for turn off). Like blinking mode, gradation mode can be implemented via minimal I^2C interaction or by the ENU pin as shown in Figure 290.1.

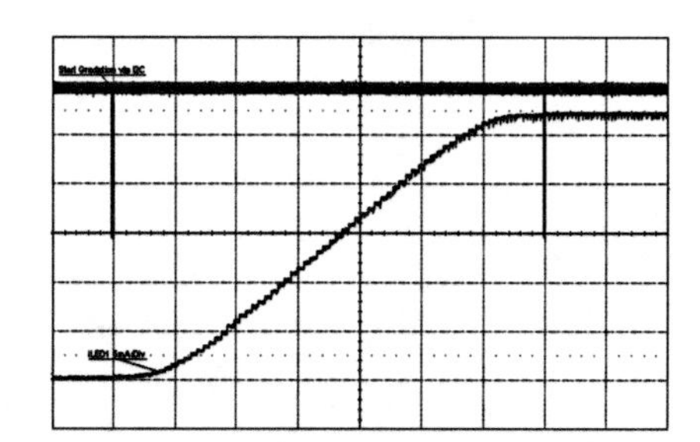

Figure 290.1 • The LTC3219 Gradating an LED from 0mA to 28mA in 960ms. Prior to the Gradation Ramp, the Gradation Timer, Up Bit and ULED Registers Are Set. A Stop Bit on the Last I^2C Write Starts the Gradation Ramp. After the Gradation Ramp Has Finished, Gradation Is Disabled with the LED Set at Full Intensity

Single IC drives cell phone backlight, new message/missed call/battery charger indicator, and RGB function select button

The circuit in Figure 290.2 illustrates a flip cell phone lighting circuit with four white LEDs for backlighting the keypad, a multicolor indicator, and a function select button illuminated by an RGB LED. The multicolor indicator consists of a red

Analog Circuit Design: Design Note Collection. http://dx.doi.org/10.1016/B978-0-12-800001-4.00290-8

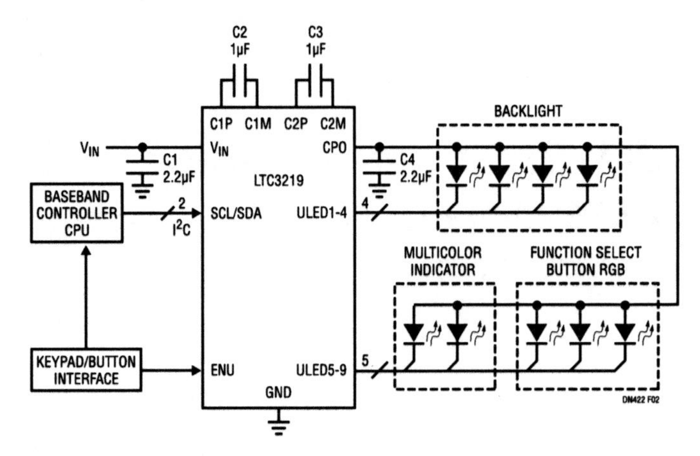

Figure 290.2 • A Single IC, Multilighting Cell Phone Application. The LTC3219 Comes in a 3mm x 3mm 20-Lead QFN Package and Only Requires Five External Components

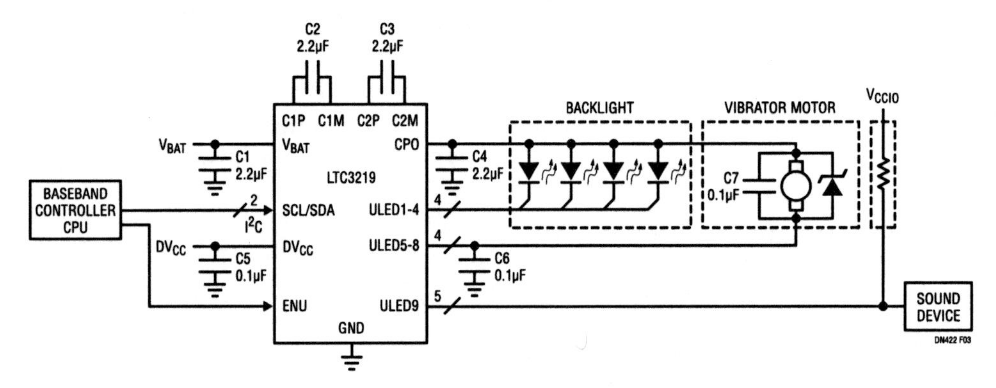

Figure 290.3 • Cell Phone with Backlighting, Vibrator Motor and Sound Controller

and a green LED. The RGB LED provides full color gamut, including white by varying its individual LED intensities.

When the cell phone is powered on or flipped open, the keypad and the function select button gradually illuminate to an intensity set by the baseband controller and CPU using the gradation feature of the LTC3219. The Function Select button may also gradually change colors using the gradation feature. After an idle period or during power off, the LEDs gradually turn off using the gradation feature. When a call is missed, the baseband controller and CPU set the multicolor indicator to blink red to indicate a missed call or blink green if the caller left a message. Once the multicolor indicator is blinking, the baseband controller hands off control to the external enable pin and shuts down to save battery power. The keypad and button interface holds the ENU pin high until the cell phone user takes action to turn off the blinking indicator.

Control for cell phone backlight, vibrator motor and sound

Cell phones use various combinations of vibration, sound and light to alert the users of an incoming call or message. Figure 290.3 illustrates a cell phone with four backlighting LEDs, a vibrator motor and a logic controlled sound device. A single logic pin, ENU, turns on all simultaneously.

If the vibrator motor requires more than 100mA, simply gang-up the ULED outputs to provide enough current. A small ceramic capacitor may be needed across the motor terminals and between the ULED output pins and ground to reduce inductive spikes and to prevent false dropout.

The speed and current in the motor is proportional to the voltage across the motor, so the voltage across the motor must be controlled in order to control the motor speed and current. One voltage-control method is to connect a shunt Zener diode across the motor. Use a Zener diode that provides the desired voltage across the motor with minimal Zener current for maximum efficiency.

Conclusion

The LTC3219 is an LED driver and charge pump which can independently control nine outputs. Special features such as gradation, blinking, and GPO modes require minimal I^2C communication. The LTC3219 is an ideal device for many applications that use multiple lighting, logic or other current controlled devices.

291

Drive large TFT-LCD displays with a space-saving triple-output regulator

Jesus Rosales

Introduction

The power appetite of large TFT-LCDs appears to be insatiable. Power supplies must feed increasing numbers of transistors and improved display resolutions, and do so without taking much space.

The triple output supply shown in Figure 291.1 shows a compact design based on the LT3489, which is optimized for driving large TFT panel displays. The main output provides 8V at 600mA while the 23V and −8V outputs provide 10mA and 20mA respectively, all from a 3.3V input. Even though TFT converters generally run from a regulated 3.3V or 5V source, this converter can operate seamlessly from a lithium-ion battery, delivering 5W when the battery is drained to 3.3V and 8W when it is at 4.2V.

The LT3489 squeezes a 2.5A, 0.12Ω, 40V switch into a tiny 3mm × 3mm MS8E footprint. It offers external or internal compensation, an internal soft-start, a 2MHz switching frequency, and it is also pin compatible with the popular LT1946. The design process with the LT3489 is easy and predictable.

The circuit in Figure 291.2 operates as a SEPIC converter, allowing the output to be higher or lower than the input.

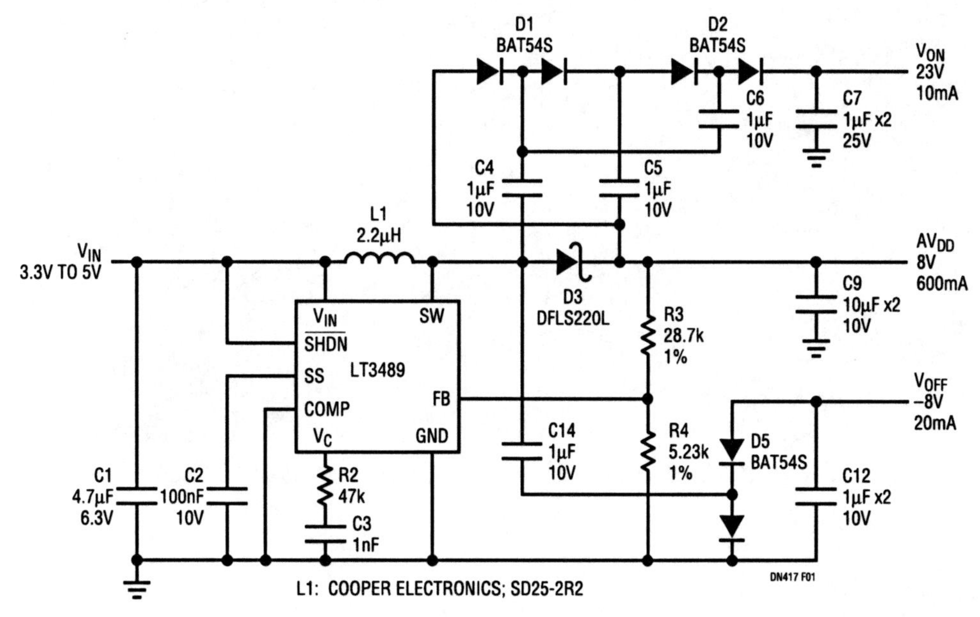

Figure 291.1 • A 3.3V Input, Triple-Output—8V, 23V and −8V—2MHz, TFT Converter

Analog Circuit Design: Design Note Collection. http://dx.doi.org/10.1016/B978-0-12-800001-4.00291-X

Pulling the $\overline{SHDN}$ pin to ground sets the output at 0V even while the voltage source is connected to the input.

Both applications take full advantage of the soft-start feature in which a single capacitor programs the voltage ramp rate of the output at start-up. The 2MHz switching frequency makes possible the use of small surface mount inductors and ceramic capacitors, which reduce the total footprint of the design. Figure 291.3 shows how small the triple-output TFT circuit in Figure 291.1 can be.

Conclusion

For large TFT-LCD panel displays, local bias supplies, DSL modems or portable devices, the LT3489 delivers big power from a small 3mm × 3mm MS8E package. Its rugged 2.5A, 0.12Ω, 40V internal switch, soft-start feature, fixed frequency and flexible compensation simplify the design and improve the performance of many applications.

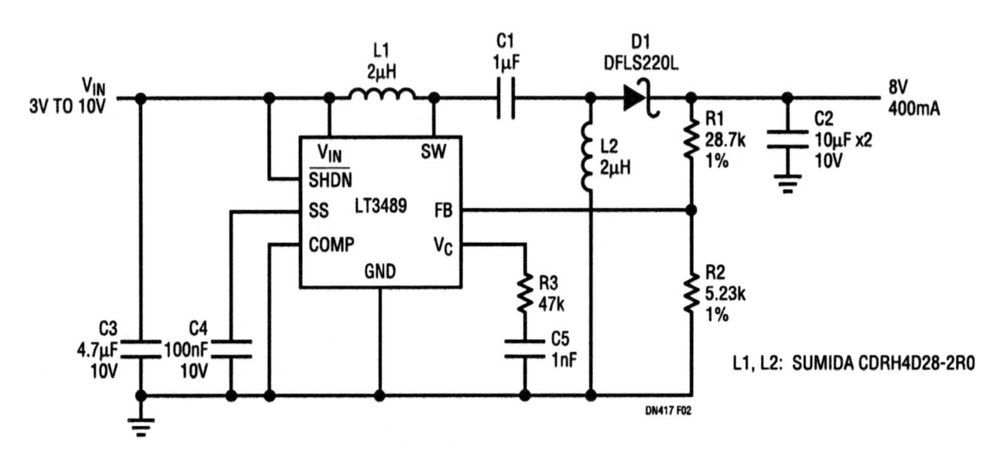

Figure 291.2 • A 2MHz, 3V to 10V Input to 8V at 400mA to 900mA with Output Disconnect Converter

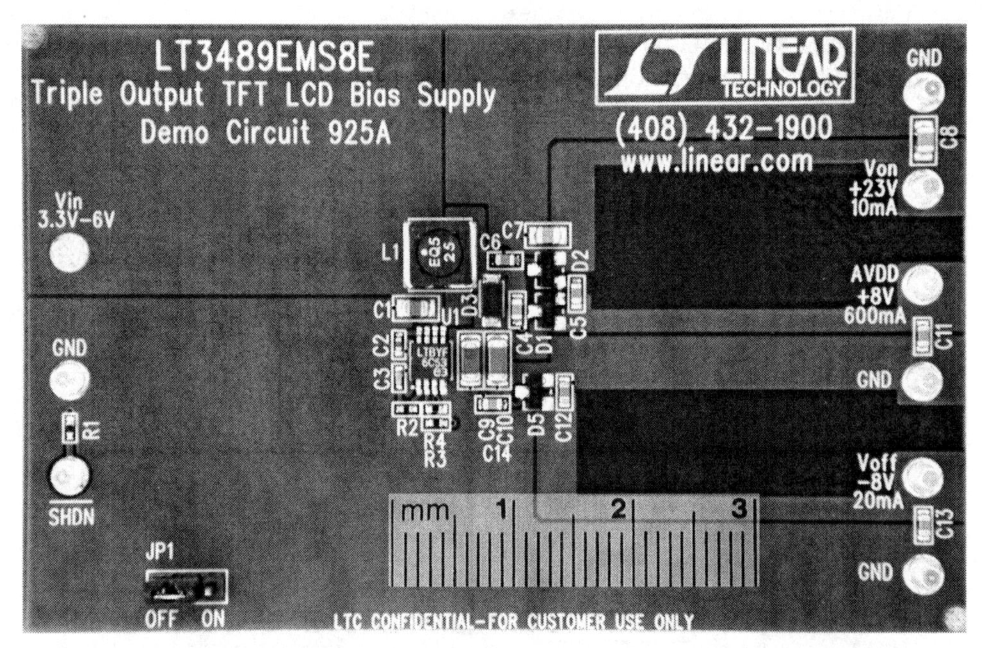

Figure 291.3 • LT3489 Demo Circuit with Layout for the Figure 291.1 Schematic

Versatile high power LED driver controller simplifies design

292

Ryan Huff

Introduction

The increased popularity of high power LEDs over the last several years has challenged electronic engineers to come up with accurate and efficient, yet simple drive solutions. The task is more difficult as the market for LEDs enters the realm of high-powered lights, such as those for automobile headlights or large LCD backlights. High light-output solutions usually involve large arrays of individual LEDs stacked in series. Conventionally, driving high power strings with accurate current is at odds with simplicity and efficiency—typically involving an inefficient linear regulator scheme or a more complicated, multiple IC switching regulator configuration. There is a simpler and better way via a low parts count, single IC solution for driving high power LED strings. At the heart of this highly efficient, simple and accurate solution is the LTC3783 controller IC.

Fully integrated, high power LED driver controller

The LTC3783 has all of the functions that are normally required to run an LED string: an accurate current regulation error amplifier, a switch mode power supply (SMPS) controller with FET drivers, and two different ways to control the brightness of the LED string.

The current regulating error amplifier uses the voltage drop across a sense resistor in series with the LED string to precisely regulate the LED current. The SMPS control portion of the LTC3783 takes advantage of current mode operation to easily compensate the loop response of the many possible topologies such as boost, buck, buck-boost, flyback and SEPIC. The integrated FET drivers allow fast switching of the power MOSFETs that are needed to efficiently convert input

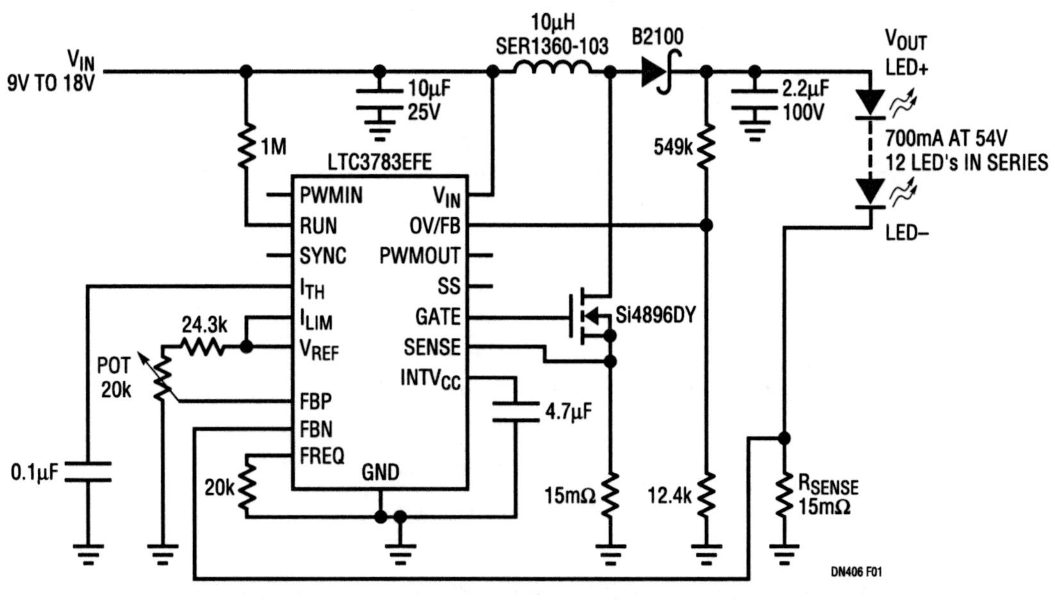

Figure 292.1 • LTC3783 in a Boost Configuration to Drive 12 LEDs in Series

Analog Circuit Design: Design Note Collection. http://dx.doi.org/10.1016/B978-0-12-800001-4.00292-1

power to LED power without having to add external gate drive ICs.

LED dimming

Two different ways of controlling LED brightness are included. Analog dimming varies the LED current from a maximum value down to about 10% of this maximum (a 10:1 dimming range). Since an LED color spectrum is related to current, this approach is not appropriate for some applications. However, PWM, or digital dimming, switches between zero current and the maximum LED current at a rate fast enough that visual flicker is not apparent, typically greater than 100Hz. The duty cycle changes the effective average current. This method allows up to a 3000:1 dimming range, limited only by the minimum duty cycle. Because the LED current is either maximum or off, this method also has the advantage of avoiding LED color shifts that come with the current changes associated with analog dimming.

Boost circuit

Figure 292.1 shows a boost configuration using all off-the-shelf components. The input voltage, which ranges from 9V to 18V, is boosted to an LED string voltage of 30V to 54V. The LED string can consist of twelve 700mA LEDs of any color in series for a total of up to 38W of LED power. At an input voltage of 18V and an LED string voltage of 54V, this circuit achieves an astounding power efficiency of over 95%! This high efficiency results in no greater than a 25°C temperature rise for any circuit component.

Buck-boost circuit

Figure 292.2 shows a buck-boost solution that can be used when the input voltage range overlaps the LED string voltage. Here the input voltage ranges from 9V to 36V and the LED string ranges from 18V to 37V. This 8-LED series string runs up to 1.5A. At the nominal input voltage of 14.4V and an LED string voltage of 36V at 1.5A (54W output power), the efficiency is almost 93%. Again, this was achieved using exclusively off-the-shelf components.

LED protection and other features

The LTC3783 can operate from a wide 3V to 36V (or higher) input voltage supply range. A programmable undervoltage lockout ensures that too low an input voltage is ignored by the chip. If an LED string is inadvertently left open, an overvoltage protection feature ensures that the output voltage does not exceed a programmable level. A soft-start function is included in order to limit the in-rush of current from the input supply during start-up. The switching frequency can be set by a single resistor to any value between 20kHz and 1MHz, or it can be synchronized to an external clock.

Conclusion

Driving high power LED strings with the LTC3783 yields a highly efficient, low parts count and flexible solution. Furthermore, being able to use standard off-the-shelf components helps to simplify the design without sacrificing performance.

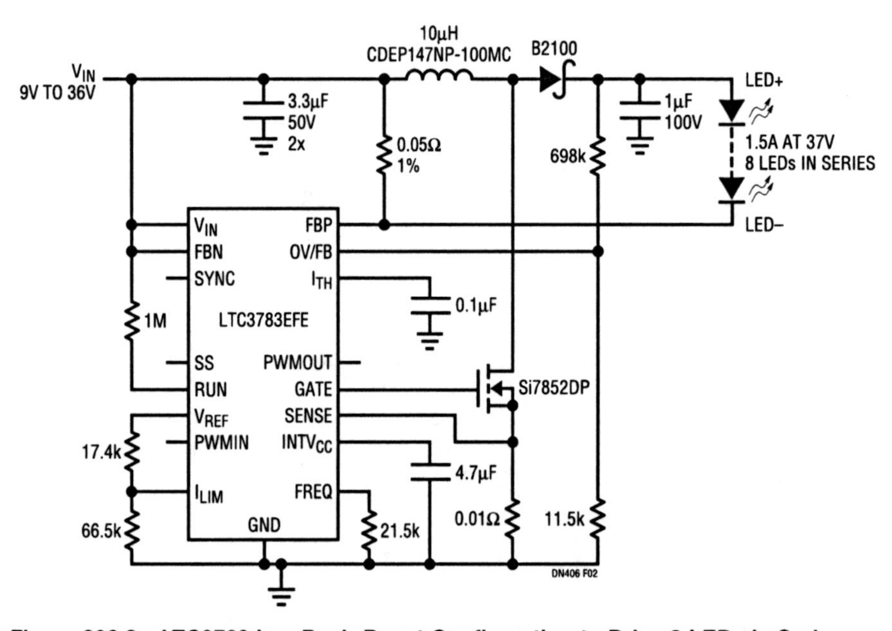

Figure 292.2 • LTC3783 in a Buck-Boost Configuration to Drive 8 LEDs in Series

High voltage buck converters drive high power LEDs

293

Keith Szolusha

Introduction

High power LEDs continue to replace traditional bulbs in new automotive, industrial, backlight display and architectural detail lighting systems. LEDs excel in a wide range of performance and cost parameters, including excellent spectral performance, long life, robustness, falling manufacturing cost and relatively safe materials. Linear Technology offers a large and growing family of high voltage DC/DC converters tailored specifically to drive high-powered LEDs.

The LT3474 and LT3475, for example, are high voltage, high current, single- and dual-channel buck LED converters with wide PWM dimming ratios that can drive one or more LEDs up to 1A and 1.5A for 80 lumens to 120 lumens per LED (or more as higher output LEDs become available). These dedicated LED drivers have onboard high voltage NPN power switches and internal sense resistors to minimize board space, reduce component count and simplify design.

With their high side sense resistors, the LT3474 and LT3475 can drive LEDs tied to ground, an important advantage in many systems. Current mode control and a precise reference voltage optimize loop dynamics for a well regulated, low ripple constant LED current. Thermally enhanced exposed pad packages keep the junction temperature low during high power operation in stressful environments. A PWM pin uses the dimming MOSFET gate signal to extend the dimming ratio of the converter by maintaining constant output capacitor voltage and control loop state during PWM dimming off-time. Shutdown and external analog current adjust pins provide simple interface for further LED light and current control flexibility in any system.

Single buck 1A LED driver

The LT3474 buck converter 1A LED driver shown in Figure 293.1 has features that suit it to automotive applications (and other battery-powered applications) or to industrial applications with limited board space, high voltage and high ambient temperature. This scheme uses a high side integrated 100mΩ sense resistor for true LED current sensing and regulation, superior to the common and less efficient method of biasing LEDs with a constant voltage and a power wasting bias resistor.

The 4V to 36V input voltage range makes it ideal to use with little-to-no input transient protection circuitry in automotive, industrial and avionic applications where long cables from the battery result in very high input spikes.

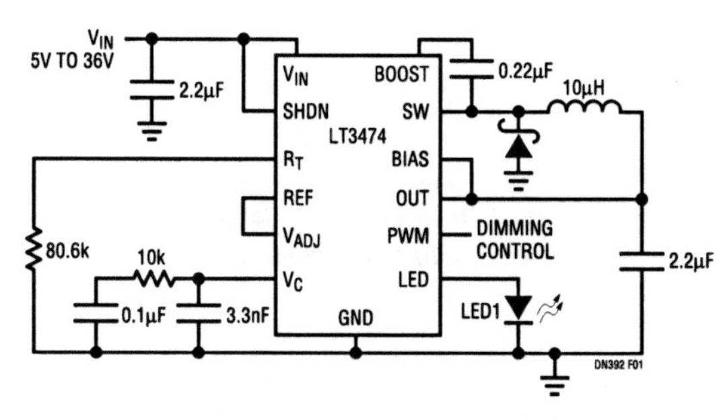

Figure 293.1 • LT3474 High Voltage Buck LED Driver Regulates 1A

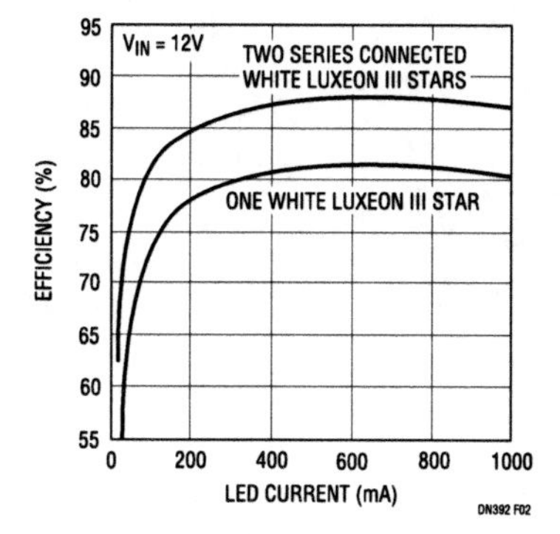

Figure 293.2 • LT3474 Buck Drives Single or Multiple LEDs with High Efficiency

Analog Circuit Design: Design Note Collection. http://dx.doi.org/10.1016/B978-0-12-800001-4.00293-3

The boosted NPN power switch results in high efficiency for both 1- and 2-LED applications (Figure 293.2). The boost diode is integrated to further reduce component count. Driving the shutdown pin to ground turns off the LEDs and reduces the input current to less than 2μA for battery longevity.

LED brightness is controlled by either the 400:1 True Color PWM dimming with an external MOSFET driver or with an analog 25:1 (or filtered PWM) signal on the V_{ADJ} pin. Applications can be optimized for highest efficiency or smallest component size via an external resistor that programs the switching frequency from 200kHz to 2MHz.

The maximum output voltage of the LT3474 is clamped at 13.8V which protects the LT3474 output from LED open circuit. Short-circuit protection is the final detail that makes the LT3474 a bulletproof converter in the case of all types of LED failures.

Dual buck 1.5A LED driver

Figure 293.3 shows a dual-channel 1.5A buck converter LED driver using the LT3475 which is essentially two LT3474 converters combined in a single IC with a few additional features. This simple solution is ideal for automotive applications where two overhead or dashboard lights are needed in the same system. Both light channels (each a single or a string of LEDs) have separate V_{ADJ} voltages and PWM signals for independent operation, but a single shutdown pin further improves battery-saving micropower operation by reducing the total battery drain of the circuit to 2μA in shutdown.

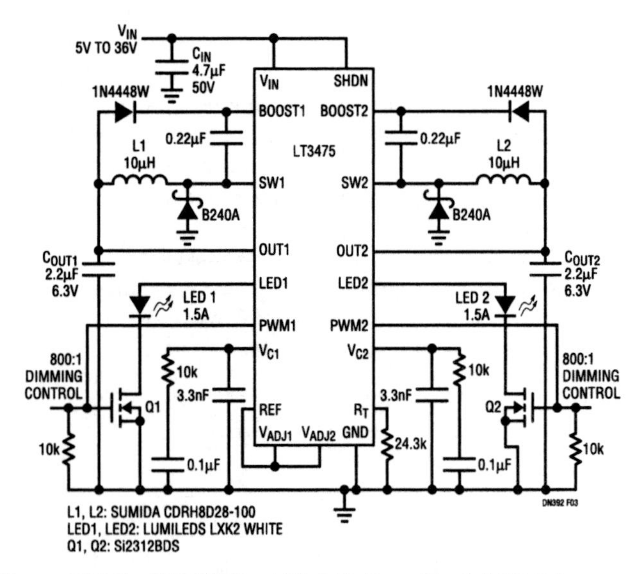

Figure 293.3 • LT3475 Dual High Voltage Buck LED Driver Regulates 1.5A

Each of the dual outputs can be driven as high as 1.5A for more powerful LEDs or LEDs that require higher current and less forward voltage such as red and amber brake and signal LEDs. Although the maximum output voltage is clamped at the same level as the LT3474 at 13.8V, the maximum power output capability of the LT3475 is three times higher. The PWM dimming ratio is also greater—1200:1 or higher with the extended dimming ratio circuit in Figure 293.4. Improvements in PWM dimming techniques with lower minimum dimming on-time requirements help this IC achieve extreme automotive and nighttime dimming levels while maintaining the same true color as 100% duty cycle. Independent analog V_{ADJ} dimming ratio is 30:1 (50mA LED current) for each channel. To reduce internal power dissipation, the boost diode for each channel is left out of the IC.

Compared to the LT3474, the LT3475 offers three times the power capability, the same shutdown current, the same switching frequency range, a slightly higher input voltage (36V operating, 40V maximum), a higher dimming ratio, a higher LED current and only a slightly bigger package (20-pin versus 16-pin exposed thermal pad TSSOP)—making it a great choice for higher power solutions. In addition, the anti-phase switching of the two channels in the LT3475 reduces the input ripple seen by the source and limits the need for extra high voltage input capacitors.

Conclusion

The LT3474 and LT3475 are excellent choices for high voltage, high current, buck LED drivers in automotive, industrial, backlight display and architectural display lighting systems. The heavily integrated ICs reduce component count and board space, while still providing flexible features such as adjustable LED current, PWM dimming and adjustable operating frequency. Accurate LED current regulation makes these ICs superior to other DC/DC voltage regulators or LED drivers. Efficiency as high as 88% combined with less than 2μA of shutdown current save battery power and extend lifetime.

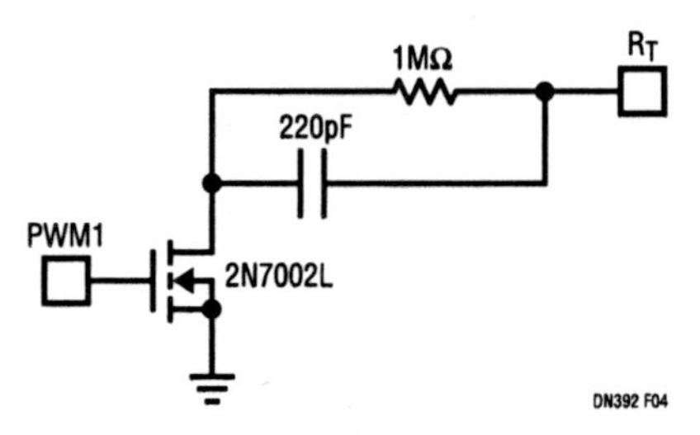

Figure 293.4 • LT3475 Extended Dimming Range Circuit Provides 1200:1 PWM Dimming Ratio When Added to Figure 293.3, and up to 3000:1 at 1.4MHz

Wide input range 1A LED driver powers high brightness LEDs with automotive and $12V_{AC}$ supplies

294

John Tilly Awo Ashiabor

Introduction

Today's ultrabright LEDs far exceed the performance of incandescent bulbs in both efficiency and lifetime. Taking full advantage of these features requires a correspondingly efficient and reliable LED driver, such as the LT3474. The LT3474 is a step-down 1A LED driver that supports a variety of power sources, has a wide 4V to 36V input voltage range and is programmable to deliver LED current from 35mA to 1A at up to 88% efficiency. It requires minimal external circuitry and is available in a space saving 16-lead TSSOP package.

Automotive LED driver

Figure 294.1 shows the configuration of the LT3474 operating from a 12V automotive battery input. As shown, the circuit can tolerate voltage swings from 4V to 36V, common in an automotive environment. With an integrated NPN switch, boost diode and sense resistor, the LT3474 cuts the external component count to a minimum. The high side sense allows a grounded cathode connection, easing wiring constraints. Both PWM and analog dimming are available with minor circuit modification; see the LT3474 data sheet for details.

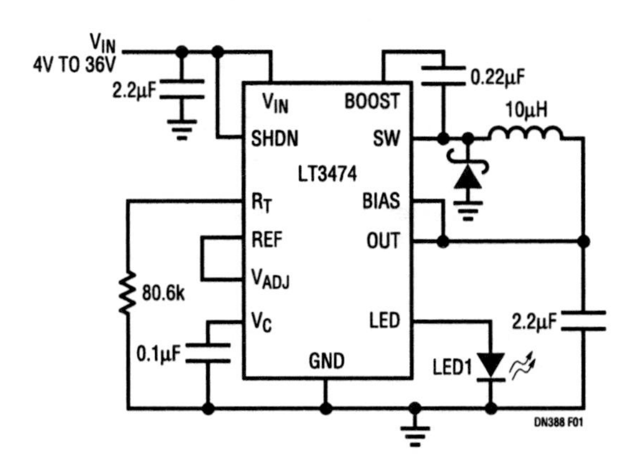

Figure 294.1 • 4V–36V Input Voltage 1A LED Driver Requires Few Components

Driving LEDs from $12V_{AC}$ input

The LT3474 directly regulates LED current, maintaining constant LED current over changing V_{IN}. The wide input range of the LT3474 allows direct connection to a rectified $12V_{AC}$ input. Using a small input capacitor, as shown in Figure 294.2,

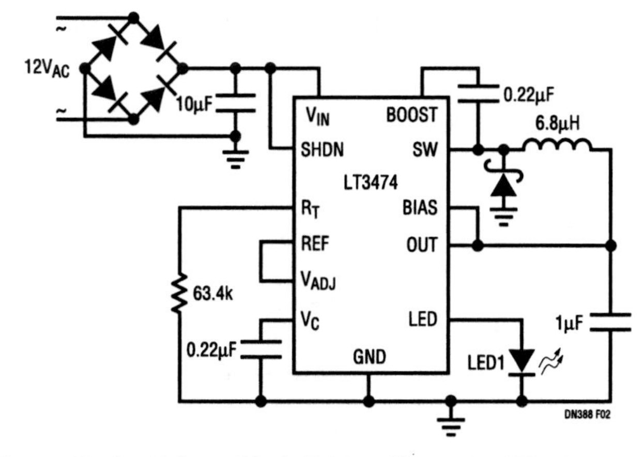

Figure 294.2 • Using a Diode Bridge Allows the LT3474 to Drive an LED from a $12V_{AC}$ Input

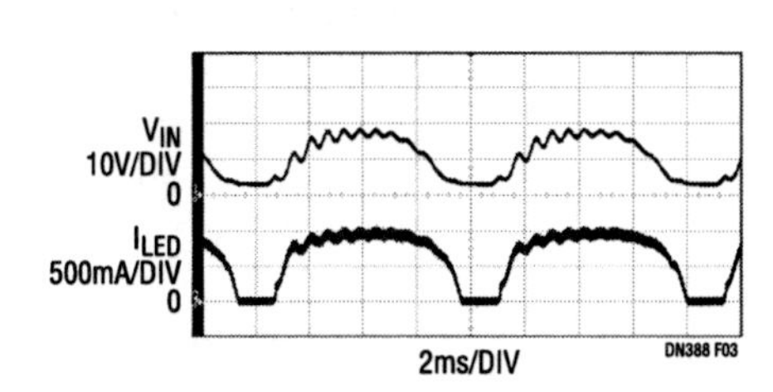

Figure 294.3 • Using a 10µF Input Capacitance, the LT3474 Delivers Nearly 1A of LED Current with Smallest Board Size

Analog Circuit Design: Design Note Collection. http://dx.doi.org/10.1016/B978-0-12-800001-4.00294-5

minimizes size. In this case, the LT3474 delivers nearly 1A of LED current as shown in Figure 294.3. Adding more capacitance to the input, as shown in Figure 294.4, holds the input voltage above the LED voltage. In this case, the LT3474 can deliver a constant LED current even with significant 120Hz ripple on the input as shown in Figure 294.5.

Thermal regulation

The issue of heat management is at the core of many LED applications. A reliable solution maintains the longevity of the LED by keeping the LED junction temperature below the recommended limit. One answer to this problem is to mount massive heat sinks, wasting space and money. Figure 294.6 shows a better solution. The temperature of the LED is sensed by the thermistor mounted near the LED and is translated into a voltage signal to the V_{ADJ} pin. The V_{ADJ} pin reduces the current through the LED appropriately to meet the power derating specified by the Luxeon III Star manufacturer. Only slight modifications to the resistor values are required to adjust the circuit for use with other high brightness LEDs (see Figure 294.7).

Conclusion

High power white LEDs are fast becoming the lighting of choice in architectural, automotive, museum and avionic systems due to their efficiency, high quality light and long lifetimes. The LT3474 makes it easy to create compact, efficient, robust and versatile LED drivers from a variety of power supplies. Designers can now focus their time on creating imaginative new LED applications, instead of on LED drivers.

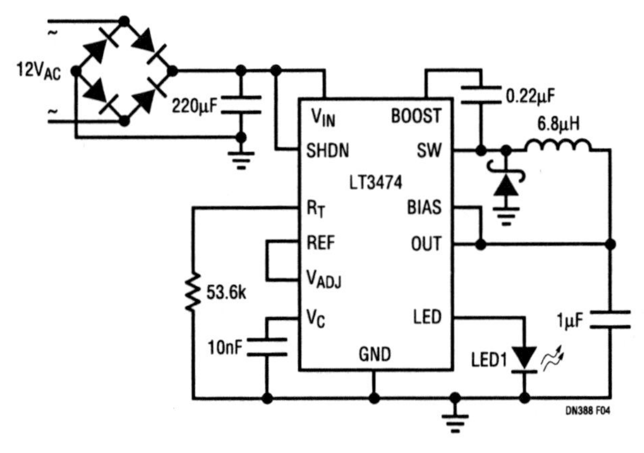

Figure 294.4 • With a 220µF Input Capacitor, the LT3474 Supplies a Constant 1A Current to the LED

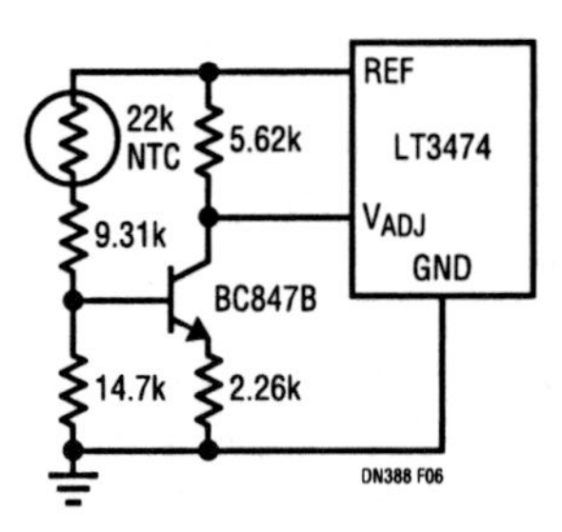

Figure 294.6 • Compact, Economical Thermo-Regulating Circuit. The NTC and NPN, Mounted Close to the LED, Monitor the LED's Temperature

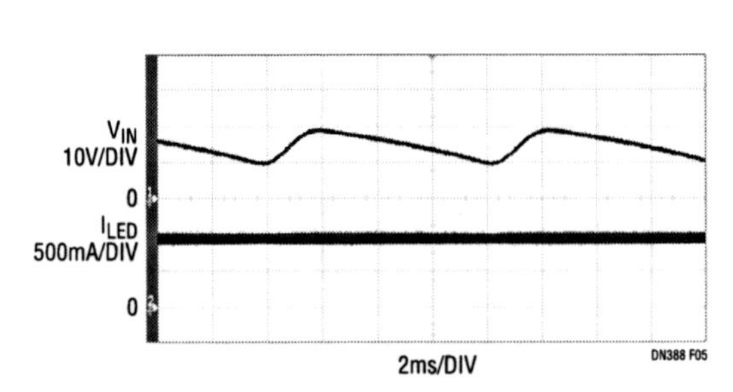

Figure 294.5 • With a 220µF Input Capacitor, the LT3474 Delivers Constant 1A LED Current with Changing Input Voltage

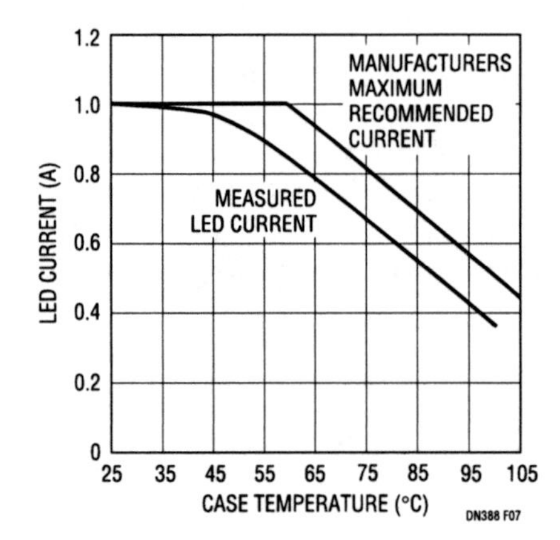

Figure 294.7 • LED Current Safely Lies within Specified Limits for the Luxeon III Star Power

Monolithic converter drives high power LEDs

295

Keith Szolusha

Introduction

High power LEDs are quickly expanding their reach as the light source of choice for flat panel computer and TV monitors, TV projection, signage, portable lights and automotive interior, trim and brake lighting. The input voltage and LED voltage combinations across these applications are as diverse as the applications themselves, precluding the ability for a single topology to satisfy the needs of them all. Nevertheless, all LED drivers, whether buck, boost, buck-boost or SEPIC must regulate a constant LED current, regardless of input and output voltages. Now a single switching regulator with the ability to be configured in a large variety of topologies for high power constant LED current is available.

The LT3477 can drive high power LEDs at constant current in any of the topologies stated above. It is a current mode, 3A DC/DC step-up converter that incorporates dual rail-to-rail current sense amplifiers and an internal 3A, 42V switch. It combines a traditional voltage feedback loop and two unique current feedback loops to operate as a constant-current and/or constant-voltage source. The floating rail-to-rail current sense amplifiers allow for both ground-referenced and floating LED solutions in different topologies, along with the added benefit of inrush current or short-circuit protection.

Both current sense voltages are 100mV and can be adjusted independently using the I_{ADJ1} and I_{ADJ2} pins. Efficiencies of up to 91% can be achieved in typical applications. The LT3477 features a programmable soft-start function to limit inductor current during start-up. Both inputs of the error amplifier are available externally, allowing positive and negative output voltages. The switching frequency is programmable from 200kHz to 3.5MHz through an external resistor. It comes in two thermally enhanced packages: a 20-pin (4mm × 4mm) QFN and a 20-pin TSSOP.

Boost driver

The LT3477's internal ground-referenced 3A NPN power switch is most commonly used for boost applications. Figure 295.1 shows a 5V to four 1W LED boost converter with open LED protection and 330mA constant LED current.

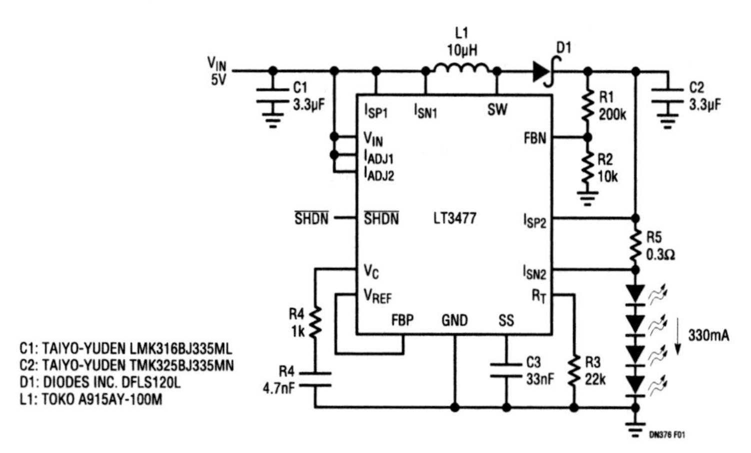

Figure 295.1 • 330mA Boost LED Driver with Open LED Protection

Analog Circuit Design: Design Note Collection. http://dx.doi.org/10.1016/B978-0-12-800001-4.00295-7

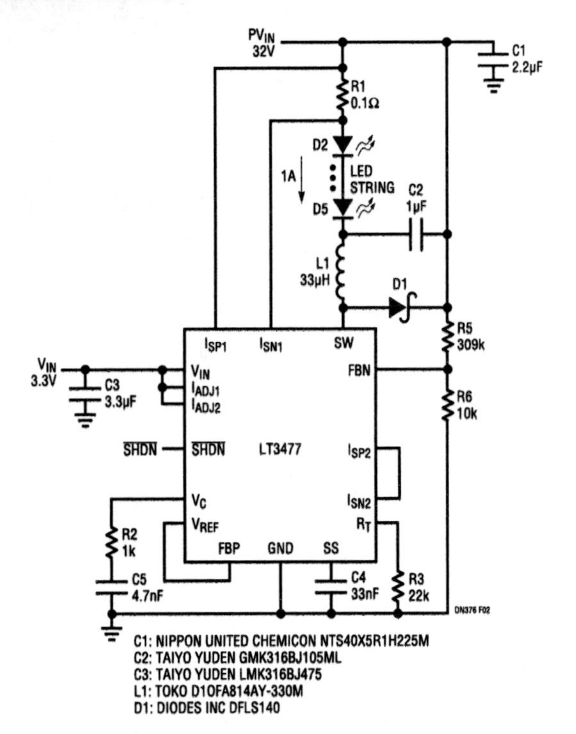

Figure 295.2 • 1A Buck LED Driver

The constant current is regulated using a current sense amplifier and a 0.3Ω sense resistor. The feedback voltage amplifier is only used for overvoltage protection in case the LEDs are removed from the circuit.

The forward voltage of the four LEDs ranges from 12V to 16V. The input voltage range is 2.5V (minimum LT3477 input voltage) up to just below the LEDs' forward voltage. The LT3477 can drive more LEDs as long as the peak switch current remains below 3A.

Buck driver

LEDs are best driven with a constant current source, but unlike most system loads, they do not have to be ground referenced. Therefore, the LT3477 with its floating current sense amplifiers can be converted into a buck LED driver with low LED ripple current. Figure 295.2 shows a 32V input voltage buck converter driving a string of 1A LEDs. The LEDs are tied to the input source through a sense resistor and the typical output Schottky catch diode is tied back to V_{IN}, converting the boost IC to a monolithic buck LED driver. Once again, the feedback voltage amplifier is only used for overvoltage protection—this time on the input—to prevent damage to the

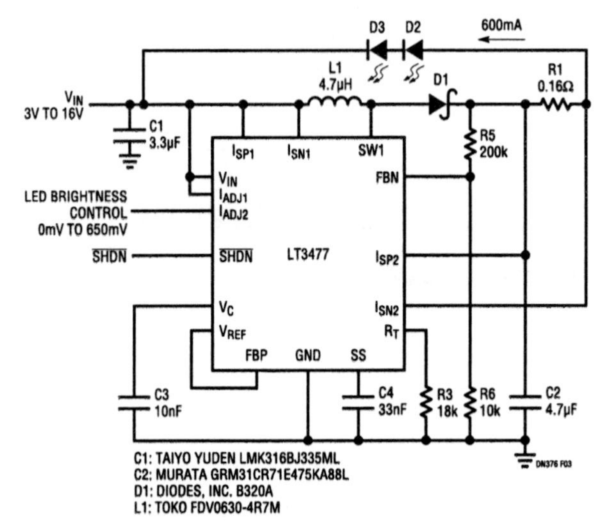

Figure 295.3 • Buck-Boost LED Driver

42V switch. In buck mode, the V_{IN} pin need only be tied to a 3.3V or 5V source for maximum efficiency, and to keep it below its absolute maximum rating of 25V. The dimming I_{ADJ} pin functions work the same in buck mode as boost mode, reducing the LED current proportionally (see Figure 295.3).

Buck-boost driver

If the battery range lies both above and below the forward voltage range of the LEDs, the LT3477 can simply be converted into a buck-boost converter as floating current sense amplifiers allow the LED string and its series sense resistor to be tied anywhere in the circuit. By altering the typical boost LED driver application so that the LED string returns to V_{IN} as opposed to ground, step-up/step-down capability is provided.

Conclusion

The LT3477 is a versatile, monolithic boost, buck, and buck-boost or SEPIC LED driver with a high power 3A, 42V switch. It can also be used for boost or SEPIC voltage converters requiring inrush or short-circuit protection. Two floating current sense amplifiers and a ground-referenced voltage feedback amplifier help give the LT3477 its high level of versatility. The externally programmable switching frequency, a shutdown pin, LED current dimming adjustment and a single soft-start capacitor satisfy the additional requirements of LED drivers and boost/SEPIC regulators.

Quad output switching converter provides power for large TFT LCD panels

296

Dongyan Zhou

Introduction

The LT1943 is a highly integrated, 4-output regulator designed to power large TFT LCD panels. The LT1943 employs switching regulators—instead of linear regulators—to minimize power dissipation and accommodate a wide input voltage range. The wide input range, 4.5V to 22V, allows it to accept a variety of power sources, including the commonly used 5V, 12V and 19V AC adaptors. The first buck regulator provides a logic voltage with up to 2A of current. The other three switching regulators provide the three bias voltages, AV_{DD}, V_{ON} and V_{OFF}, required by LCDs.

All four regulators are synchronized to a 1.2MHz internal clock, allowing the use of small, low cost inductors and ceramic capacitors. Since different types of panels may require different bias voltages, all output voltages are adjustable for maximum flexibility. Programmable soft-start

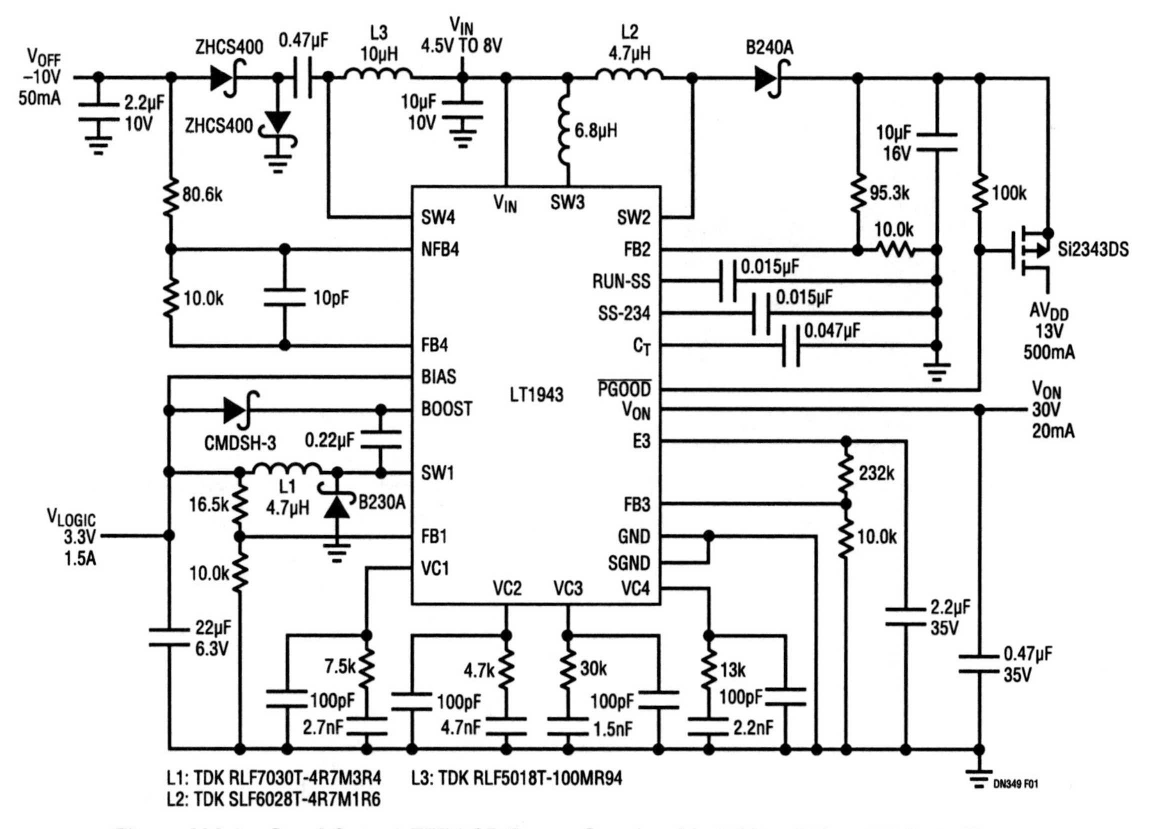

Figure 296.1 • Quad Output TFT LCD Power Supply with 4.5V to 8V Input Voltage Range

Analog Circuit Design: Design Note Collection. http://dx.doi.org/10.1016/B978-0-12-800001-4.00296-9

capability is included in all outputs to limit inrush current. The LT1943 has a built-in start-up sequence and panel protection feature. The LT1943 is available in a low-profile 28-pin TSSOP package.

4-output supply with soft-start

Figure 296.1 shows a 4-output TFT LCD power supply with a 4.5V to 8V input range. The first output provides a 3.3V, up to 1.5A, logic supply using a buck regulator. The second output employs a boost converter to generate a 13V, 500mA AV_{DD} bias supply. Another boost converter and an inverter generate V_{ON} and V_{OFF}.

When power is first applied to the input, the RUN/SS pin starts charging. When its voltage reaches 0.7V, switcher 1 is enabled. The capacitor at RUN/SS pin controls the V_{LOGIC} ramping rate and inrush current in L1.

Switchers 2, 3 and 4 are controlled by the BIAS pin, which is usually connected to V_{LOGIC}. When the BIAS pin is higher than 2.8V, the SS-234 pin begins charging to enable switchers 2, 3 and 4. When AV_{DD} reaches approximately 90% of its programmed voltage, the PGOOD pin is pulled low. When AV_{DD}, V_{OFF} and E3 all reach 90% of their programmed voltages, the C_T timer is enabled and a 20μA current source begins to charge C_T. When the C_T pin reaches 1.1V, an output PNP turns on, enabling V_{ON}. Since V_{ON} has to be present to turn on the LCD panel, the V_{ON} turn-on delay gives the column drivers and digital circuitry in the LCD panel time to get ready, preventing high currents from flowing into the panel. Figure 296.2 illustrates the start-up sequencing of the 4-output power supply in Figure 296.1. Figure 296.3 gives the overall efficiency for the circuit in Figure 296.1.

If one of the regulated voltages, V_{LOGIC}, AV_{DD}, V_{OFF} or E3 drops more than 10%, the internal PNP turns off to shut down V_{ON}. This action protects the panel in a fault condition. The PGOOD pin is used to drive an optional PMOS device at the output of the AV_{DD} boost regulator to disconnect AV_{DD} from the input during shutdown.

The converter uses all ceramic capacitors. X5R- or X7R-type ceramic capacitors are recommended, as these materials retain their capacitance over a wide temperature range.

Wide input range supply

If the input voltage may be higher than the AV_{DD} set value, a SEPIC regulator can be used in place of a boost regulator to generate the AV_{DD} output. This covers the commonly used 12V and 19V inputs. Details for this are covered in the LT1943 data sheet.

Conclusion

The LT1943 simplifies and shrinks power supplies for TFT LCD panels. Its four integrated switching regulators enable a wide input voltage range and reduce power dissipation. All regulators have a 1.2MHz switching frequency and allow the exclusive use of ceramic capacitors to minimize circuit size, cost and output ripple.

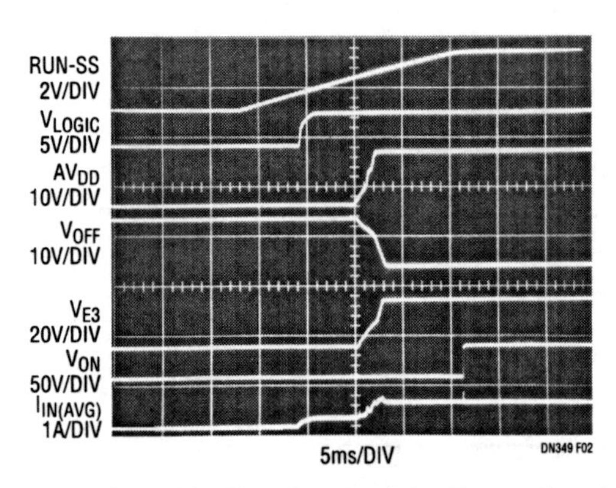

Figure 296.2 • Start-Up Waveforms of the Power Supply in Figure 296.1

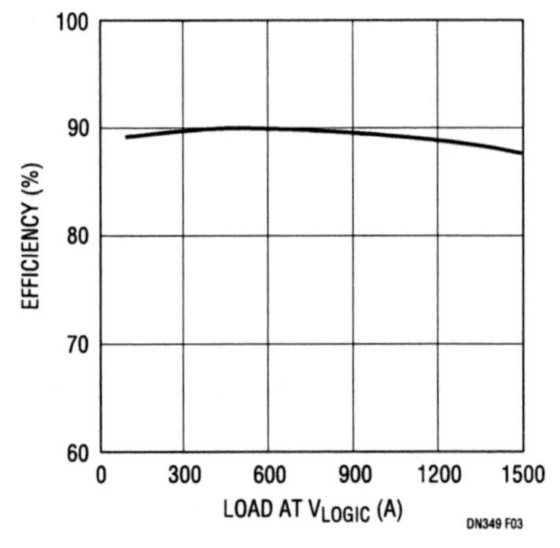

Figure 296.3 • Total Circuit Efficiency of the Power Supply in Figure 296.1 (Load at AV_{DD}: 500mA)

Basic flashlamp illumination circuitry for cellular telephones/cameras

297

Jim Williams

Introduction

Next generation cellular telephones will include high quality photographic capability. Flashlamp-based lighting is crucial for good photographic performance. A previous full-length Linear Technology publication detailed flash illumination issues and presented flash circuitry equipped with "red-eye" reduction capability.[1,2] Some applications do not require this feature; deleting it results in an extremely simple and compact flashlamp solution.

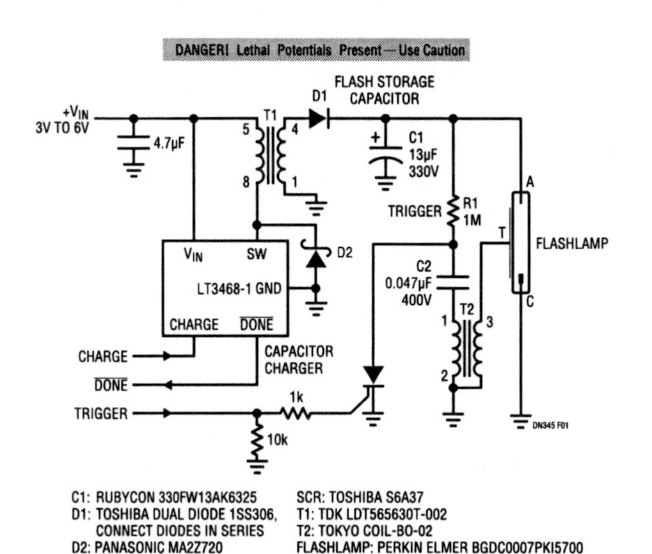

Figure 297.1 • Complete Flashlamp Circuit Includes Capacitor Charging Components, Flash Capacitor C1, Trigger (R1, C2, T2, SCR) and Flashlamp. TRIGGER Command Biases SCR, Ionizing Lamp via T2. Resultant C1 Discharge Through Lamp Produces Light

Note 1: See LTC Application Note 95, "Simple Circuitry for Cellular Telephone/Camera Flash Illuminaton" by Jim Williams and Albert Wu, March 2004.

Note 2: "Red-eye" in a photograph is caused by the human retina reflecting the light flash with a distinct red color. It is eliminated by causing the eye's iris to constrict in response to a low intensity flash immediately preceding the main flash.

Flashlamp circuitry

Figure 297.1's circuit consists of a power converter, flashlamp, storage capacitor and an SCR-based trigger. In operation the LT3468-1 charges C1 to a regulated 300V at about 80% efficiency. A "trigger" input turns the SCR on, depositing C2's charge into T2, producing a high voltage trigger event at the flashlamp. This causes the lamp to conduct high current from C1, resulting in an intense flash of light. LT3468-1 associated waveforms, appearing in Figure 297.2, include trace A, the "charge input," going high. This initiates T1 switching, causing C1 to ramp up (trace B). When C1 arrives at the regulation point, switching ceases and the resistively pulled-up "$\overline{\text{DONE}}$" line drops low (trace C), indicating C1's charged state. The "TRIGGER" command (trace D), resulting in C1's discharge via the lamp, may occur any time (in this case ≈600ms) after "$\overline{\text{DONE}}$" goes low. Normally, regulation feedback would be provided by resistively dividing down the output voltage. This

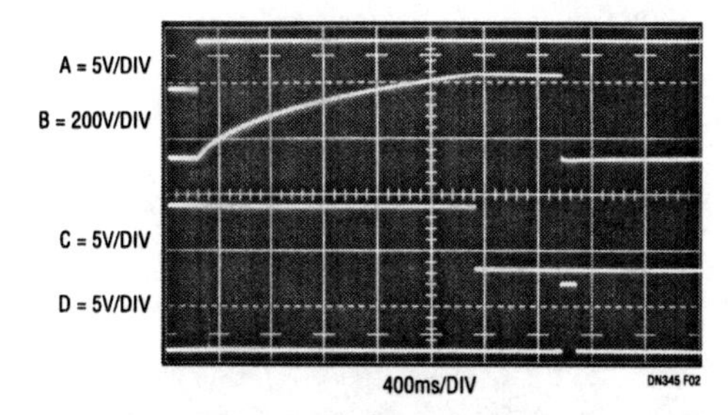

Figure 297.2 • Capacitor Charging Waveforms Include Charge Input (Trace A), C1 (Trace B), $\overline{\text{DONE}}$ Output (Trace C) and TRIGGER Input (Trace D). C1's Charge Time Depends upon Its Value and Charge Circuit Output Impedance. TRIGGER Input, Widened for Figure Clarity, May Occur Any Time After $\overline{\text{DONE}}$ Goes Low

Analog Circuit Design: Design Note Collection. http://dx.doi.org/10.1016/B978-0-12-800001-4.00297-0

approach is not acceptable because it would require excessive switch cycling to offset the feedback resistor's constant power drain. While this action would maintain regulation, it would also drain excessive power from the primary source, presumably a battery. Regulation is instead obtained by monitoring T1's flyback pulse characteristic, which reflects T1's secondary amplitude. The output voltage is set by T1's turns ratio. This feature permits tight capacitor voltage regulation, necessary to ensure consistent flash intensity without exceeding lamp energy or capacitor voltage ratings. Also, flashlamp energy is conveniently determined by the capacitor value without any other circuit dependencies.

Figure 297.3 shows high speed detail of the high voltage trigger pulse (trace A), the flashlamp current (trace B) and the light output (trace C). Some amount of time is required for the lamp to ionize and begin conduction after triggering. Here, 3µs after the 4kV$_{P-P}$ trigger pulse, flashlamp current begins its ascent to over 100A. The current rises smoothly in 3.5µs to a well defined peak before beginning its descent. The resultant light produced rises more slowly, peaking in about 7µs before decaying. Slowing the oscilloscope sweep permits capturing the entire current and light events. Figure 297.4 shows that light output (trace B) follows lamp current (trace A) profile, although current peaking is more abrupt. Total event duration is ≈200µs with most energy expended in the first 100µs.

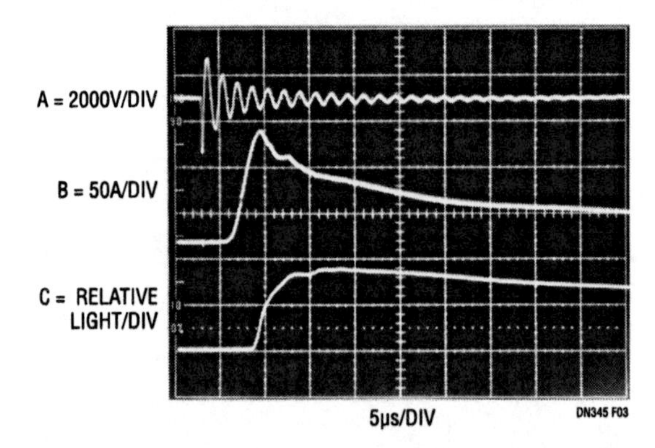

Figure 297.3 • High Speed Detail of Trigger Pulse (Trace A), Resultant Flashlamp Current (Trace B) and Relative Light Output (Trace C). Current Exceeds 100A After Trigger Pulse Ionizes Lamp

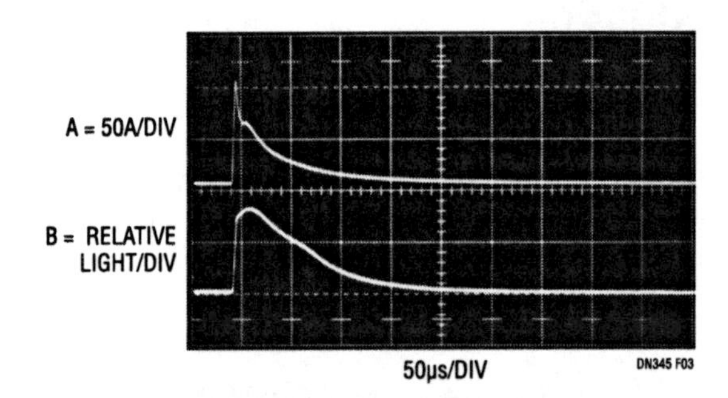

Figure 297.4 • Photograph Captures Entire Current (Trace A) and Light (Trace B) Events. Light Output Follows Current Profile Although Peaking Is Less Defined. Waveform Leading Edges Enhanced for Figure Clarity

Conclusion

The circuit presented constitutes a basic, but high performance, flash illumination solution. Its low power, small size and few components suit cellular telephone/camera applications where size and power drain are important. It provides a practical, readily adaptable path to accessing flashlamp-based illumination's photographic advantages.

DC/DC converter drives white LEDs from a variety of power sources

298

Keith Szolusha

Introduction

LEDs are usually driven with a constant DC current source in order to maintain constant luminosity. However, most DC/DC converters are designed to deliver a constant voltage by comparing a feedback voltage to an internal reference via an internal error amplifier. The easiest way to turn a simple DC/DC converter into a constant current source is to use a sense resistor to turn the output current into a voltage and use it as the feedback. The problem with this method is its reduced efficiency—700mA across a 1.2V (typical reference voltage) drop produces an 840mW power loss. One solution is to use an external op amp to amplify the voltage of a low value resistor to the given reference voltage. This saves converter efficiency, but significantly increases the cost and complexity of a simple converter with the additional components and board space.

A superior solution is to use the LT1618 constant current, constant voltage converter, which combines a traditional voltage feedback loop and a unique current feedback loop to operate as a constant-voltage, constant-current DC/DC converter. No external op amps are required for this extremely compact solution. The I_{ADJ} (current adjustment) pin provides the capability to dim the LED during normal operation by varying the resistor setting or injecting a PWM signal. Access to both the positive and negative inputs of the internal constant-current amplifier allows the sense resistor to be placed anywhere in the converter's output or input path and provides constant output or input current. Without access to both inputs, either a ground referenced sense resistor or some additional level-shifting transistors or operational amplifier would be required.

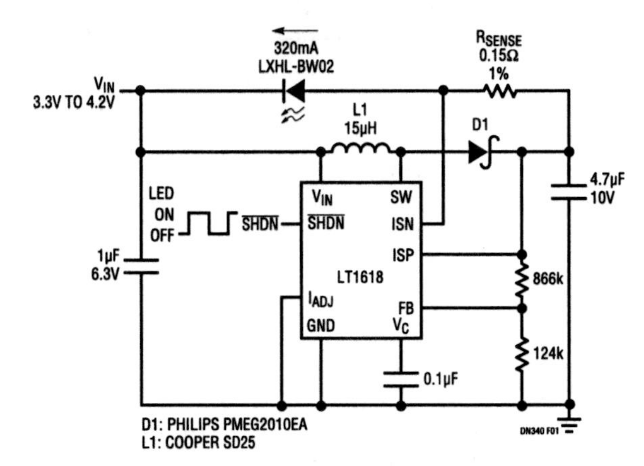

Figure 298.1 • The LT1618 Powers the LXHL-BW02 White LED from a Single Lithium-Ion Battery with 70% Efficiency

Lithium-ion source (3.3V to 4.2V)

The LT1618EDD is a 1.4MHz constant-current, constant-voltage boost converter in a tiny 10-pin thermally enhanced DFN package. The monolithic (onboard) low side switch has a maximum peak current limit of 1.5A. This enables extremely compact high-current solutions for portable and battery-powered applications. The high switching frequency allows the input and output capacitors and the inductor to be extremely small.

Although the LT1618 is conventionally used as a high frequency boost converter with the load being driven between V_{OUT} and ground, the unique method, shown in Figure 298.1, of tying the load from V_{OUT} back to V_{IN} allows it to be used to drive the LXHL-BW02 1W white LED from a lithium-ion battery input. Tying the load back to V_{IN} allows the forward voltage of the LED (the load voltage) to be either above or below the input voltage as the battery voltage changes. This topology avoids the need for an additional inductor as would be required in other buck-and-boost topologies such as SEPIC or flyback.

The single inductor used here is extremely small and low cost, matching the tiny all-ceramic capacitors and low-profile IC. Tying the load back to V_{IN} increases the inductor current by summing both the input and output currents. Due to increased switch losses, the overall efficiency of the solution

Analog Circuit Design: Design Note Collection. http://dx.doi.org/10.1016/B978-0-12-800001-4.00298-2

is approximately 70% over the input voltage range. Nevertheless, at this efficiency, it is difficult to match the compactness and low cost of this solution. The LED is turned off by grounding the $\overline{SHDN}$ pin and input or output disconnect is not required.

The constant-current sensing resistor, R_{SENSE}, is only 150mΩ. At 320mA, it dissipates about 15mW. R_{SENSE} is tied directly to the positive and negative input pins of the LT1618 and is placed in the path of the LED, returning to V_{IN}. The LT1618's access to both the positive and negative pins keeps this circuit simple and compact.

2-alkaline cell source (1.8V to 3.0V)

Powered from two alkaline cells, the circuit in Figure 298.2 can illuminate a 1W LXHL-BW02 Lumileds white LED with a constant 320mA current at 85% efficiency. The forward voltage of the 1W LED, 3.6V typical, requires a boosted voltage from the 2-cell input. In order to turn off the LED, an input disconnect switch is required. Typically in handheld lighting applications, this is a simple pushbutton or switch. Without input disconnect, the shutdown pin would only prevent the IC from switching, still providing a direct path of current from the input to the output through the inductor and catch diode.

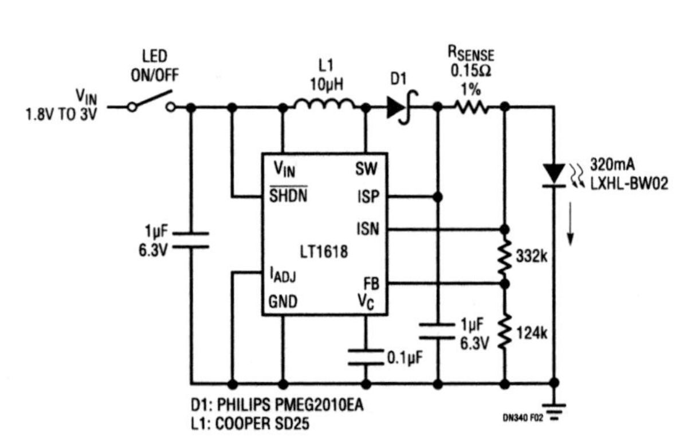

Figure 298.2 • The LT1618 Powers the LXHL-BWO2 White LED from Two Alkaline Cells with 85% Efficiency

Automotive power source (9V to 16V)

With floating sense resistor inputs, the constant-current LED can be placed in DC/DC converter circuits where one would not normally think to place a load. For example, Figure 298.3 shows the LXHL-PW09 white LED placed in the input path of what appears to be a boost converter with V_{OUT} connected to V_{IN}. In fact, the forward voltage drop across the LED allows the boost topology to provide what would be an output of V_{IN} while it appears to see an input of $V_{IN}-V_F$ (3.6V typical forward voltage). However, the load that is driven is the white LED with a constant 700mA current at approximately 70% efficiency. This circuit is therefore a buck that uses a boost converter. One advantage of this topology is that only a single input/output capacitor is required, but an increased inductor size is needed to limit the ripple seen by both the LED and the input/output capacitor. Once again, the $\overline{SHDN}$ pin can be used to turn the LED on and off without the need for an input or output disconnect switch.

Conclusion

The LT1618 has many features that make it an ideal IC for providing constant current to Lumileds white LEDs from a variety of power sources.

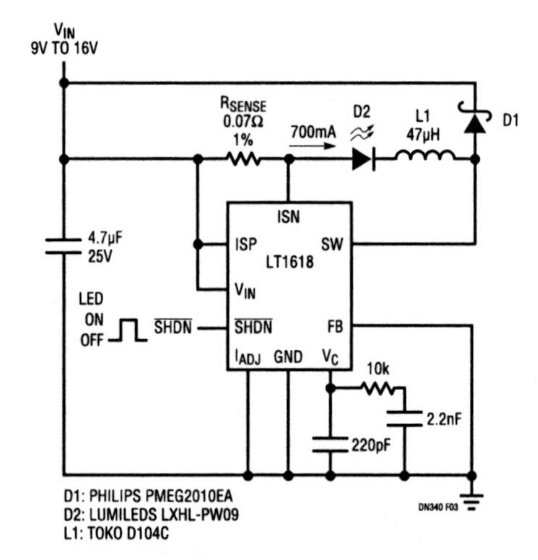

Figure 298.3 • The LT1618 Powers the LXHL-PW09 White LED from an Automotive Battery with 70% Efficiency

High efficiency ThinSOT white LED driver features internal switch and Schottky diode

299

David Kim

Introduction

The LT3465 white LED driver is ideal for backlight circuits in small, battery-powered portable devices—such as cellular phones, PDAs and digital cameras. The LT3465 includes important features such as automatic soft-start to prevent large inrush current, open LED protection and an integrated Schottky diode in a low profile (<1mm) ThinSOT package to save space. The LT3465 is optimized for color display backlight applications with two to four white LEDs and a Li-Ion battery input. Even so, its internal 30V switch is capable of driving up to six LEDs in series.

The LT3465 uses a constant current, step-up architecture which directly regulates the LED current and guarantees a consistent light intensity and color in each LED, regardless of the differences in their forward voltage drops. The constant 1.2MHz switching frequency allows the use of tiny external components and minimizes both input and output ripple for applications requiring low input and output noise. The internal compensation of the LT3465 reduces output capacitor requirements to a single 0.22μF ceramic capacitor—a significant space and cost saving over compensation schemes that have more strict output capacitance requirements.

The 200mV feedback voltage, high efficiency internal power switch and internal Schottky diode minimize power losses in the LT3465, resulting a typical efficiency of 80%. The LT3465 also comes in a 2.7MHz switching frequency version (LT3465A), allowing the use of even smaller components, such as chip inductors.

Li-Ion-powered driver for four white LEDs

The compact white LED driver circuit shown in Figure 299.1 is designed to fit into small wireless devices such as cellular phones or PDAs. The efficiency of this circuit is higher than that of switched-capacitor based drivers in a parallel architecture due to the constant current step-up series LED architecture.

This design supplies 15mA of constant current, driving four LEDs in series from a Li-Ion battery or a 5V adapter input. The integrated Schottky diode, internal soft-start and open LED protection simplify the circuit and improve performance. The 1.2MHz constant frequency and integrated optimized compensation allow the use of small components, including tiny 0603-size ceramic input and output capacitors and a tiny inductor (an even smaller chip inductor can be

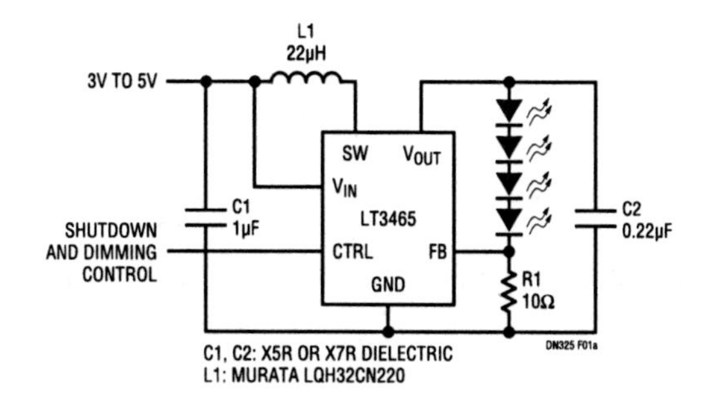

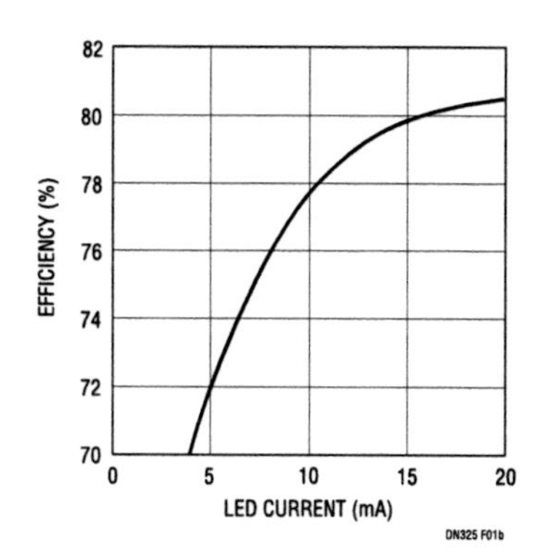

Figure 299.1 • Li-Ion-Powered Driver for Four White LEDs

Analog Circuit Design: Design Note Collection. http://dx.doi.org/10.1016/B978-0-12-800001-4.00299-4

Table 299.1 R1 Resistor Value Selection

FULL I_{LED} (mA)	R1 (Ω)
5	40.2
10	20.0
15	13.3
20	10.0

used with the LT3465A). The LED current is programmed with resistor R1 at the feedback pin by the simple formula (see Table 299.1):

$$I_{LED} = 200mV/R1 \text{ or } R1 = 200mV/I_{LED}$$

Precision resistors (1%) are recommended for applications that require highly accurate LED current.

Dimming control

The LT3465 features single pin shutdown and dimming control. By applying a DC voltage (Figure 299.2a) of 0.2V to ~1.5V at the CTRL pin (Pin 4), the feedback voltage changes from 25mV to 200mV. For a 20mA LED current application (R_{FB} = 10Ω), changing the CTRL pin voltage from 0.2V~1.5V yields LED current from 2.5mA to 20mA. A CTRL voltage below 50mV will shut down the device. The curve in Figure 299.3 shows the correlation between V_{FB} vs V_{CTRL}.

Three dimming methods are shown in Figure 299.2 via a DC voltage, a filtered PWM signal and a logic signal. The first two methods adjust LED brightness by varying voltage at the CTRL pin and the logic signal method adjusts the brightness by changing the feedback voltage directly.

The filtered PWM dimming shown in Figure 299.2b works similar to DC voltage dimming except the V_{DC} input comes from a filtered PWM signal. The filter is the 5kΩ, 100nF RC circuit which filters the PWM signal to a DC voltage proportional to the duty cycle of the PWM signal. In this case, the LED current increases proportionally with the duty cycle of the PWM signal. A 100% duty cycle corresponds to full LED current, and a 0% duty cycle corresponds to zero LED current. The frequency recommended for filtered PWM dimming is 3kHz with a magnitude of 2V.

The logic signal dimming method (Figure 299.2c) uses an N-channel MOSFET and R_{INC} to reduce the value of R1 when a logic signal turns on the MOSFET. The value of R1 sets the minimum LED current and the parallel combination of R_{INC} and R1 sets the higher LED current. This simple circuit allows for a single dimming step. For more than two LED current levels, add additional MOSFET-R_{INC} circuits in parallel. Keep in mind that a separate logic signal is needed for each current setting and that the LED current calculation involves the parallel resistances of each $R_{INC}(n)$. With n switches on, $I_{LED}(n) = 200mV\ [R1||R_{INC}(1)...||R_{INC}(n)]$. The CTRL pin can still be used as different resistors are switched in to provide greater control of LED current.

Conclusion

The LT3465 is a white LED driver optimized for driving two to four LEDs from a Li-Ion input. It features a 36V, 1.2MHz internal power switch, internal Schottky diode, automatic soft-start, open LED protection and optimized internal compensation. The LT3465 is ideal for wireless devices requiring small circuit size, high efficiency and matching LED brightness.

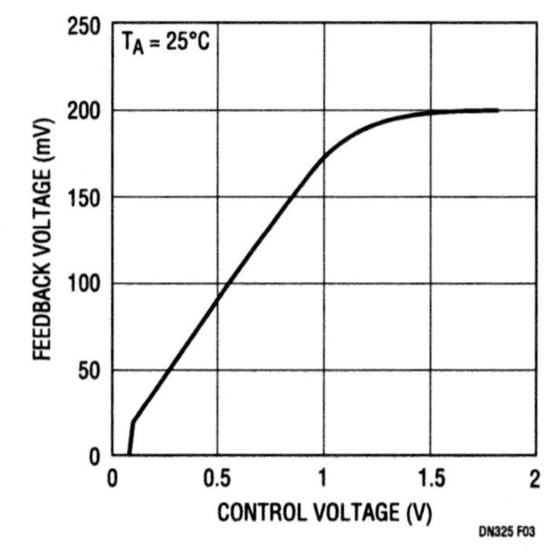

Figure 299.3 • V_{FB} vs V_{CTRL} Correlation Curve

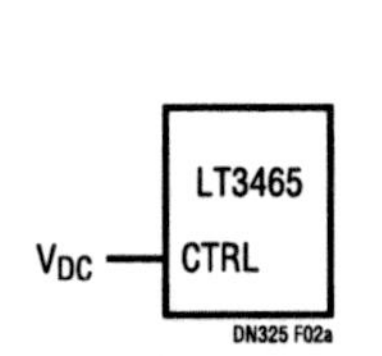

Figure 299.2a • DC Voltage Dimming

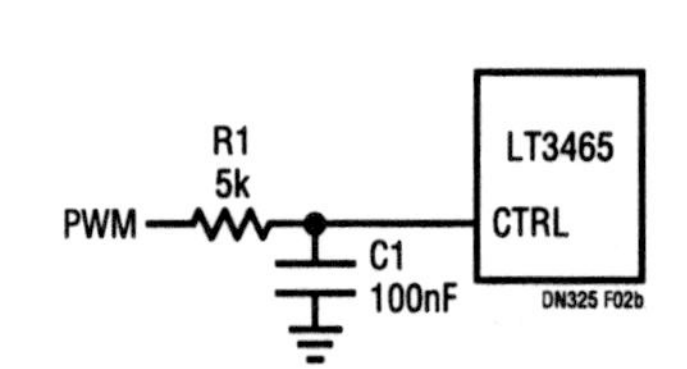

Figure 299.2b • Filtered PWM Dimming

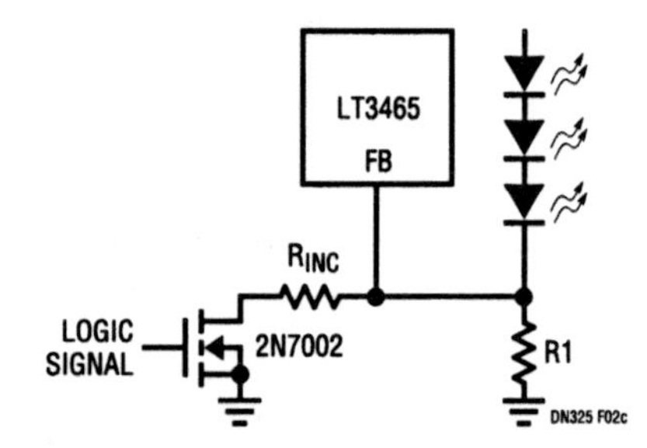

Figure 299.2c • Logic Signal Dimming

White LED driver in tiny SC70 package delivers high efficiency and uniform LED brightness

300

David Kim

Introduction

The LT1937 step-up white LED driver is an ideal solution for small battery-powered portable devices such as cellular phones, PDAs and digital cameras. The LT1937 features an internal 36V switch that is capable of driving up to eight LEDs in series, but it is optimized for Li-Ion powered color display backlight applications that use two to four white LEDs. The LT1937 guarantees a constant light intensity and color in each LED, regardless of differences in their forward voltage drops due to a constant current step-up architecture that directly regulates the LED current. The constant 1.2MHz switching allows for the use of tiny external components and minimizes input and output ripple voltage—meeting the noise level requirements of products with sensitive wireless circuitry. The superior internal compensation of LT1937 lowers the output capacitor requirement to a single 0.22µF ceramic, saving space and cost. The low 95mV feedback voltage and an efficient internal switch minimize power losses in the LT1937. The result is a typical efficiency of 84%. The LT1937 is available in the tiny SC70 or 1mm tall ThinSOT package.

Li-Ion-powered driver for three white LEDs

Figure 300.1 shows a white LED driver circuit that is intended for small wireless devices. The constant current step-up series LED architecture of this circuit has much better efficiency than the alternative switched capacitor based parallel LED architecture. The circuit is designed to provide 15mA of constant current to drive three LEDs in series from a Li-Ion battery or 5V adapter input. The 1.2MHz constant frequency and superior internal compensation results in 0603 size ceramic input and output capacitors and a tiny ferrite

Table 300.1 R1 Resistor Value Selection

I_{LED} (mA)	R1 (Ω)
5	19.1
10	9.53
12	7.87
15	6.34
20	4.75

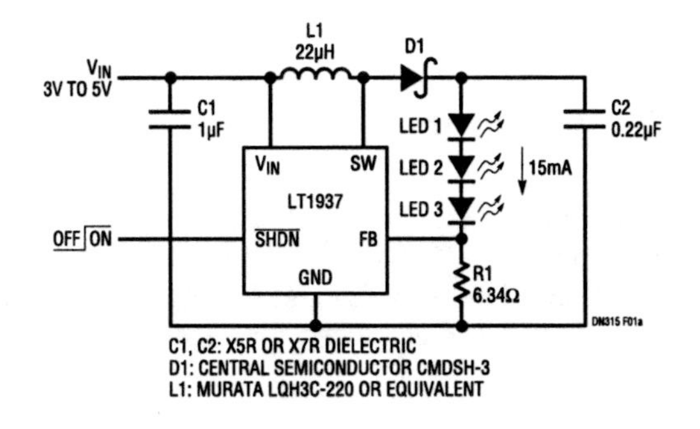

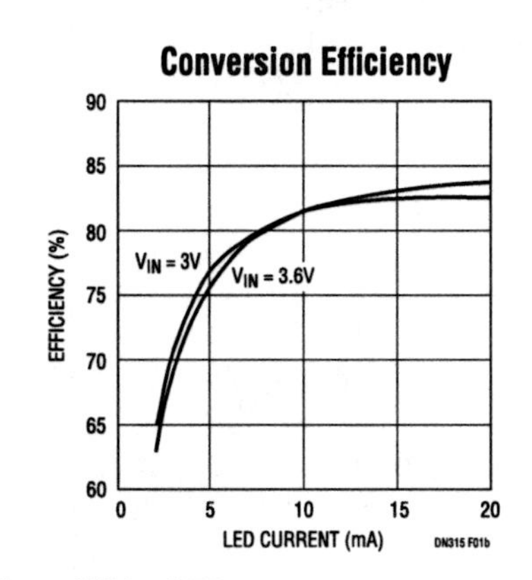

Figure 300.1 • Li-Ion-Powered Driver for Three White LEDs

Analog Circuit Design: Design Note Collection. http://dx.doi.org/10.1016/B978-0-12-800001-4.00300-8

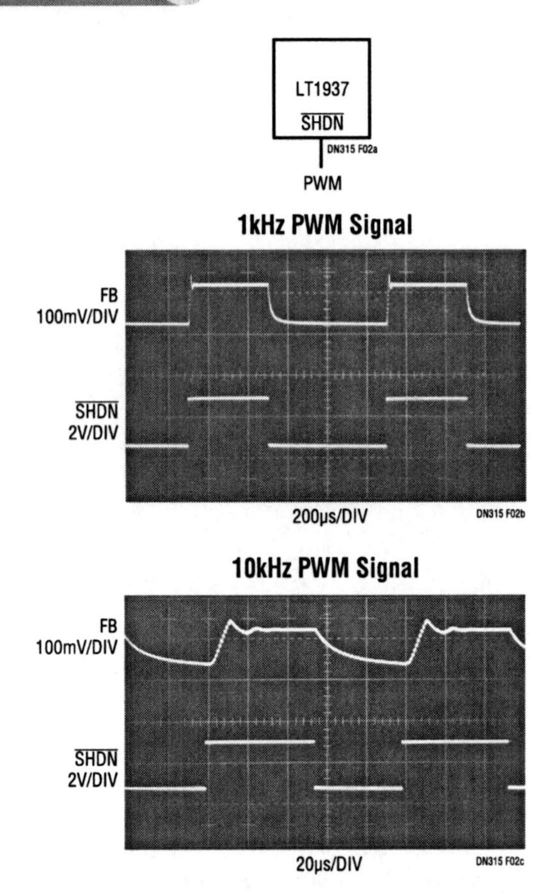

Figure 300.2 • PWM Dimming Control Using the $\overline{SHDN}$ Pin

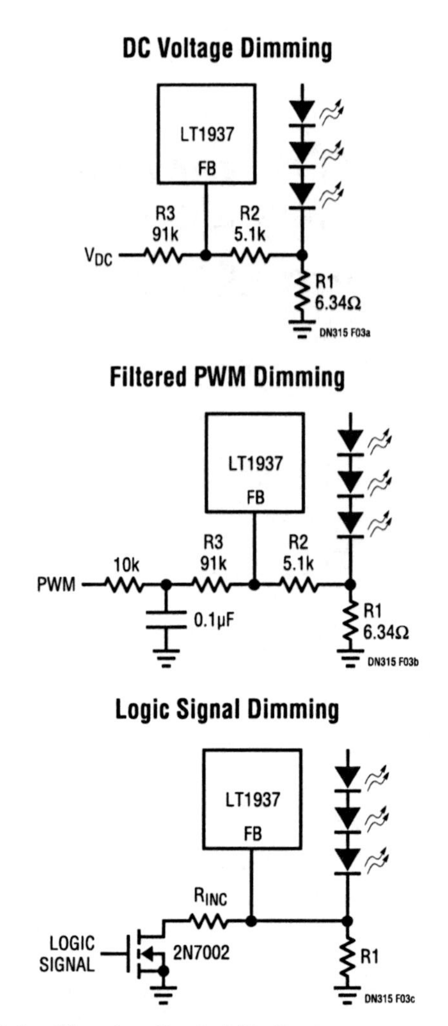

Figure 300.3 • Dimming Control Methods

core inductor (a chip inductor can be used to save even more space). The constant LED current is set with the R1 resistor at the feedback pin. By using a simple LED current calculation, $I_{LED} = 95mV/R1$ or $R1 = 95mV/I_{LED}$, a resistor value selection table is easily calculated (see Table 300.1). For accurate LED current, high precision (1%) resistors are needed.

Easy dimming control

The brightness of the LED can be adjusted using a PWM signal, a filtered PWM signal, a logic signal or a DC voltage. The brightness control using the PWM signal to $\overline{SHDN}$ pin and PWM dimming waveforms are shown in Figure 300.2. With the PWM signal applied to the $\overline{SHDN}$ pin, the LT1937 is turned on or off by this signal. The average LED current increases proportionally with the duty cycle of the PWM signal, where 0% duty cycle sets the LED current to zero and a 100% duty cycle sets it to full current. The typical frequency range recommended for PWM dimming is for a 1kHz to 10kHz signal with at least a 1.5V amplitude.

Figure 300.3 shows alternative LED brightness control methods using a DC voltage, filtered PWM signal and logic signal. The DC voltage dimming control shown in Figure 300.3 is designed to control LED current from 0mA to 15mA using the 0V to 2V DC voltage at the V_{DC} input. As the voltage at the V_{DC} input increases, the voltage drop on R2 increases and voltage drop on R1 decreases, resulting in a decrease of LED current. The filtered PWM dimming works the same way except that the V_{DC} input now comes from a filtered PWM signal. The 10k, 0.1μF RC filters the PWM signal so that it is close to DC and the duty cycle of the PWM signal changes the DC voltage level. The LED current can also be adjusted in discrete steps using a logic signal dimming method shown in Figure 300.3. R1 sets the minimum LED current when the NMOS is off and R_{INC} increases LED current by reducing the resistor value when the NMOS is turned on.

Conclusion

The LT1937 is a white LED driver optimized for driving two to four LEDs from a Li-Ion battery input. With its 36V, 1.2MHz internal switch and superior internal compensation, the LT1937 is well suited for small wireless devices requiring very small circuit size, high efficiency and uniform LED brightness.

Photoflash capacitor charger has fast efficient charging and low battery drain

301

Albert Wu

Introduction

The LT3420 is designed to charge large-valued capacitors—such as those used for the strobe flashes of digital and film cameras—to high voltages. These photoflash, or strobe, capacitors range in value from a hundred microfarads to over a millifarad, with target output voltages above 300V. Traditional strobe capacitor charging methods are either inefficient or require software overhead. The LT3420 provides a compact, simple to use and efficient charger solution that requires no software, saving space, battery life, design time and cost.

LT3420 charger circuits typically achieve efficiencies greater than 75%. The LT3420 includes important features such as automatic capacitor charge refresh and control/indicator pins which make it highly flexible and easy to use. Its versatility allows it to be used in applications that require a simple standalone photoflash charger as well as applications where it is completely controlled by a microprocessor (described below). No voltage divider is needed on the high voltage output.

Figure 301.1a shows a typical LT3420 circuit that can charge a 220μF photoflash capacitor to 320V in 3.5 seconds from a 5V input. Figure 301.1b shows the charge time as a function of battery voltage. In Figure 301.1a, the circuitry to the right of C4 shows a typical method to generate the light pulse once the photoflash capacitor is charged. When the SCR is fired, the flying lead along the glass envelope of the Xenon bulb reaches many kilovolts in potential. This ionizes the gas inside the bulb forming a low impedance path across the bulb. The energy stored in the photoflash capacitor quickly flows through the Xenon bulb, producing the burst of light needed for flash photography.

Features

The LT3420 includes an integrated 1.4A power switch and utilizes a patent-pending control technique. Precise control of the switching current is achieved by sensing both the primary and secondary currents of transformers, a method which prolongs battery life. Figure 302.2 shows the relevant waveforms when the output has reached 300V in the circuit of

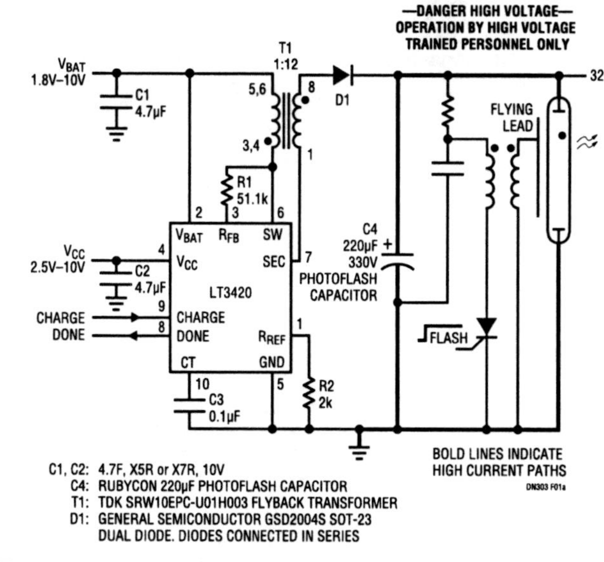

—DANGER HIGH VOLTAGE—
OPERATION BY HIGH VOLTAGE
TRAINED PERSONNEL ONLY

C1, C2: 4.7F, X5R or X7R, 10V
C4: RUBYCON 220μF PHOTOFLASH CAPACITOR
T1: TDK SRW10EPC-U01H003 FLYBACK TRANSFORMER
D1: GENERAL SEMICONDUCTOR GSD2004S SOT-23
DUAL DIODE. DIODES CONNECTED IN SERIES

BOLD LINES INDICATE
HIGH CURRENT PATHS

Figure 301.1a • 320V Photoflash Capacitor Charging Circuit

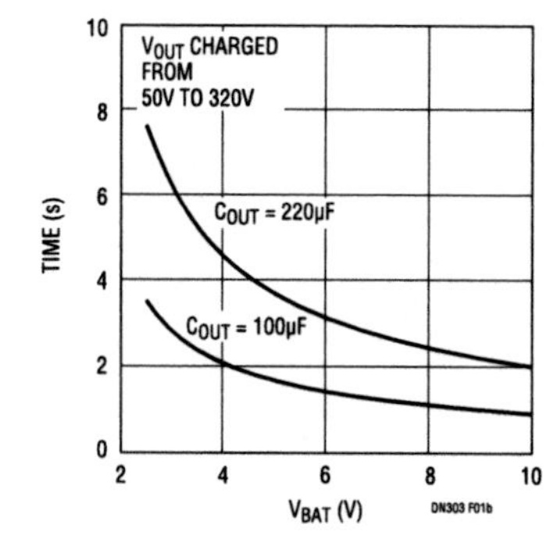

Figure 301.1b • Charge Time

Analog Circuit Design: Design Note Collection. http://dx.doi.org/10.1016/B978-0-12-800001-4.00301-X

Figure 301.1a. The peak primary current is limited to 1.4A (typical), while the primary current when the power switch turns on is 480mA (typical). By operating the part in Continuous Conduction Mode (CCM), charge time is minimized. The output voltage is detected via the flyback waveform on the primary of the transformer—V_{SW} in Figure 301.2. The target output voltage is controlled by two resistors, R1 and R2. This flyback detection scheme removes the need for a resistor divider string from the high voltage output to ground, thus eliminating the associated power loss found in many competing flash modules.

Once the target output voltage is reached, the device enters a refresh mode where the quiescent current of the device is reduced to 90μA (typical). The LT3420 has a user programmable refresh timer built in. The value of C3 determines the time period after which the part comes out of the refresh mode and recharges the output to the target voltage. This process repeats to maintain the output at the desired voltage. Figure 301.3 shows the different modes of the LT3420 from shutdown, to charging and finally refresh.

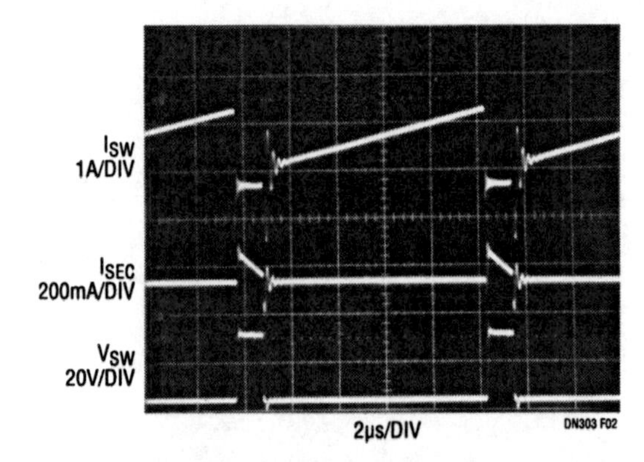

Figure 301.2 • Switching Waveforms with V_{OUT}=300V, V_{CC}=V_{BAT}=3.3V

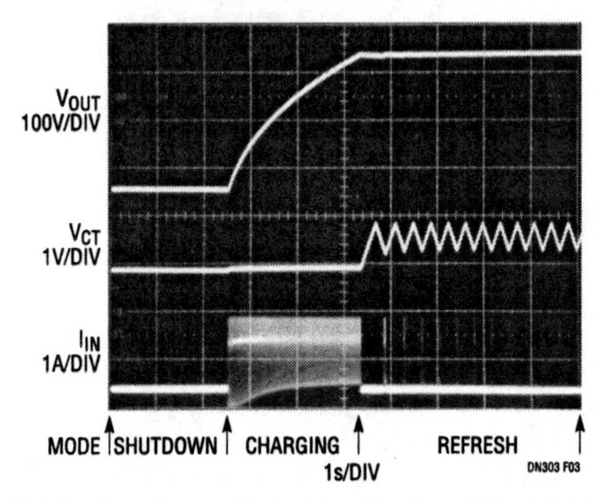

Figure 301.3 • The Three Operating Modes of the LT3420: Shutdown, Charging and Refresh of the Photoflash Capacitor

Interfacing to a microcontroller

The LT3420 can be easily interfaced to the microcontroller found in digital cameras. The CHARGE and DONE pins are the control and mode indicator pins, respectively, for the part. By utilizing these pins, the LT3420 can be selectively disabled and enabled at any time. Figure 301.4 shows the LT3420 circuit being selectively disabled when the CHARGE pin is driven low midway through the charge cycle. This might be necessary during a sensitive operation in a digital camera. Once the CHARGE pin is returned to the high state, the charging continues from where it left off.

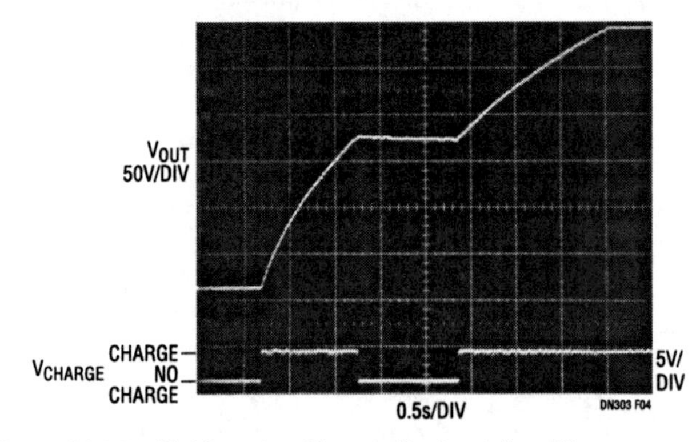

Figure 301.4 • Halting the Charge Cycle at Any Time

Conclusion

The LT3420 provides a highly efficient and integrated standalone solution for charging photoflash capacitors. Many important features are incorporated into the device, including automatic refresh, tightly controlled currents and an integrated power switch, thus reducing external parts count. The LT3420 comes in a small, low profile, MSOP-10 package, making a complete solution that takes significantly less PC board space than traditional methods.

High efficiency white LED driver guarantees matching LED brightness

302

Dave Kim

Introduction

White LEDs are widely used in small LCD backlight applications for simplicity, high reliability and low cost. However, due to variations in the forward voltage drop of white LEDs, matching the brightness becomes a major design consideration for multiple LED applications. Since LED brightness is proportional to current rather than voltage, a constant-current source driving LEDs in series should be used to ensure the same illumination of each LED. The LT1932 uses a constant-current step-up architecture that directly regulates the LED current and guarantees a constant light intensity in each LED, regardless of differences in their forward voltage drops. Its unique internal current source accurately regulates LED current even when the input voltage is higher than the LED voltage, greatly simplifying the battery/adapter power designs. The internal 36V switch is capable of driving up to eight LEDs in series with 20mA of LED current.

The LT1932 comes in the tiny 1mm, ThinSOT package. It operates at a constant 1.2MHz switching frequency, permitting the use of tiny, low profile chip inductors and capacitors to minimize circuit size and cost in space-conscious portable applications such as cellular telephones and handheld computers.

Li-Ion LED driver for four white LEDs

The white LED driver shown in Figure 302.1 is an ideal solution for Li-Ion powered, 4-LED backlight applications. The LT1932 and five external components require less than 0.65cm^2 of printed circuit board space.

The constant-frequency step-up topology provides low noise, high efficiency and matched LED brightness along with adjustable current control. The LT1932 permits no LED current during shutdown, resulting in less than 1µA of battery

Table 302.1 R_{SET} Resistor Values

I_{LED} (mA)	R_{SET} VALUE (Ω)
40	562
30	750
20	1.13k
15	1.50k
10	2.26k
5	4.53k

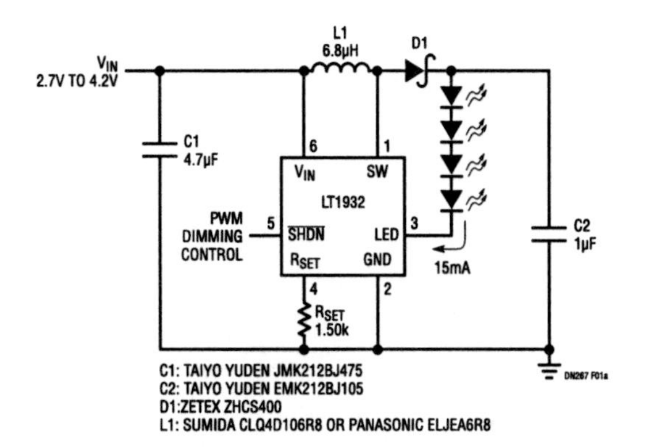

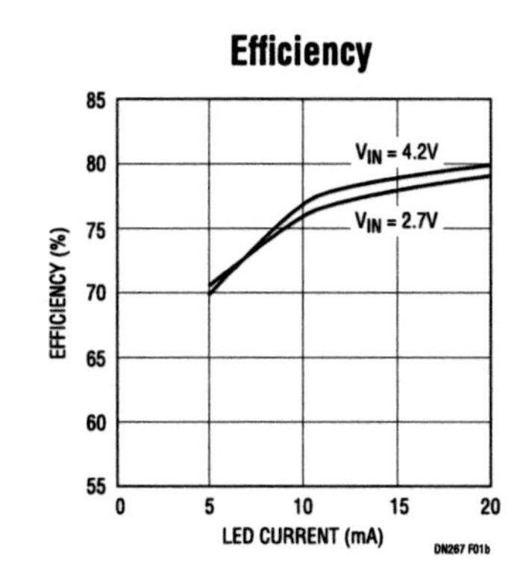

Figure 302.1 • Li-Ion (or 5V) LED Driver for Four White LEDs

Analog Circuit Design: Design Note Collection. http://dx.doi.org/10.1016/B978-0-12-800001-4.00302-1

current in the standby mode. The LED current is easily programmed over a range of 5mA to 40mA by selecting the value of R_{SET} as shown in Table 302.1. By applying an additional DC voltage or a pulse-width modulated (PWM) signal to the R_{SET} pin, the LED current can be adjusted for dimming or brightness control.

Dimming control

The brightness of the LED can be easily adjusted using a PWM signal, a filtered PWM signal, a logic signal or a DC voltage. Five LED dimming schemes are shown in Figure 302.2. The LT1932 using PWM brightness control provides the widest dimming range and the purest white LED color over the entire dimming range. This results in better than a 20:1 dimming ratio without the undesirable blue tint common to white LED backlights. PWM controlled LED current is shown in Figure 302.3. Average LED current changes with duty cycle by switching between full current and zero current. This ensures that when the LEDs are on, they can be driven at the appropriate current to give the purest white light (at 15mA or 20mA) while the light intensity changes with the PWM duty cycle. See the LT1932 data sheet for more information about the dimming control methods shown in Figure 302.2.

Conclusion

For multiple white LED backlight applications, a constant current, series LED driver is required to ensure the light intensity matching in each LED. The constant current topology and 36V internal switch make the LT1932 an ideal solution for multiple white LED driver applications. The LT1932 features the purest white LED color dimming control, less than 1μA standby mode quiescent current, selectable current level, guaranteed LED brightness matching and extremely small circuit size making it well suited for portable cellular phone and handheld computer applications.

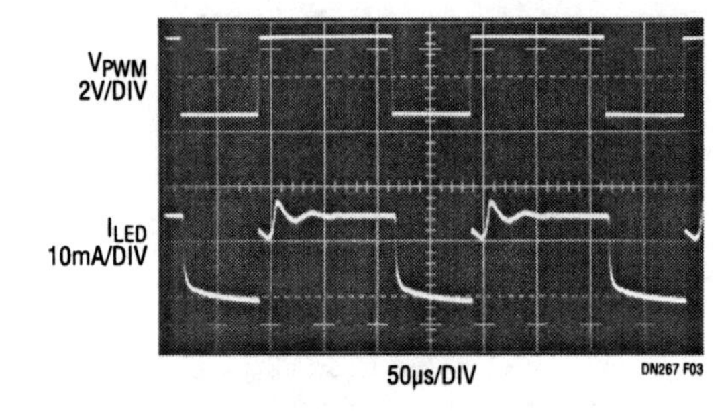

Figure 302.3 • PWM Dimming Using the SHDN Pin

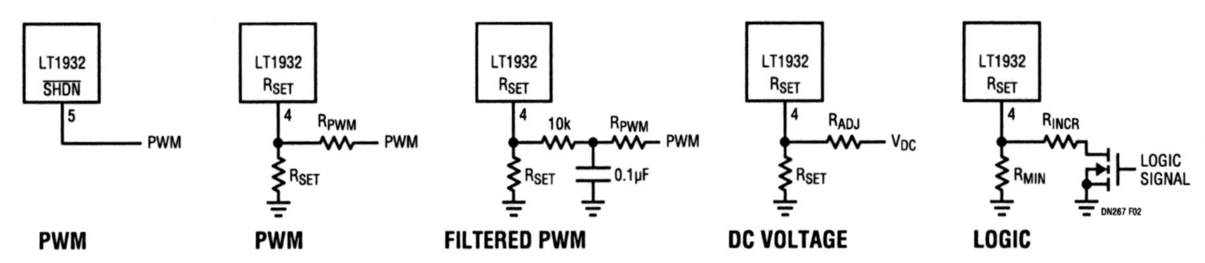

Figure 302.2 • Five Methods of LED Dimming

High power desktop LCD backlight controller supports wide dimming ratios while maximizing lamp lifetime

303

Rich Philpott

Introduction

Liquid crystal displays (LCDs), long standard in laptop computers and handheld instruments, are gaining in popularity as desktop computer displays. Larger displays require multiple high power cold cathode fluorescent lamps (CCFLs). The lamps must have a dimming range and life expectancy comparable to those of previous generations of desktop displays. To achieve maximum lamp lifetime and dimming range while maintaining efficiency, CCFL drive should be sinusoidal, contain zero DC component, and not exceed the CCFL manufacturer's current ratings. Providing a low crest-factor sinusoidal CCFL drive also maximizes current-to-light conversion efficiency, reduces display flicker and minimizes EMI and RFI emissions. The LT1768 high power CCFL controller, with its unique Multimode Dimming, provides the necessary drive to enable a wide dimming range, while maximizing lamp lifetime.

LT1768 dual CCFL backlight inverter

The circuit in Figure 303.1 is a dual, grounded-lamp backlight inverter that operates from an input of 9V to 24V, delivers current from 0mA to 9mA per CCFL and has a dimming ratio greater than 100:1. The LT1768 in the circuit is a 350kHz fixed frequency, current mode, pulse-width modulator that provides the lamp-current control function.

The CCFL current is controlled by a DC voltage on the PROG pin of the LT1768. This voltage feeds the LT1768's Multimode Dimming block, which converts it to a current and feeds it to the V_C pin. As the V_C pin voltage rises, the LT1768's GATE pin is pulse-width modulated at 350kHz.

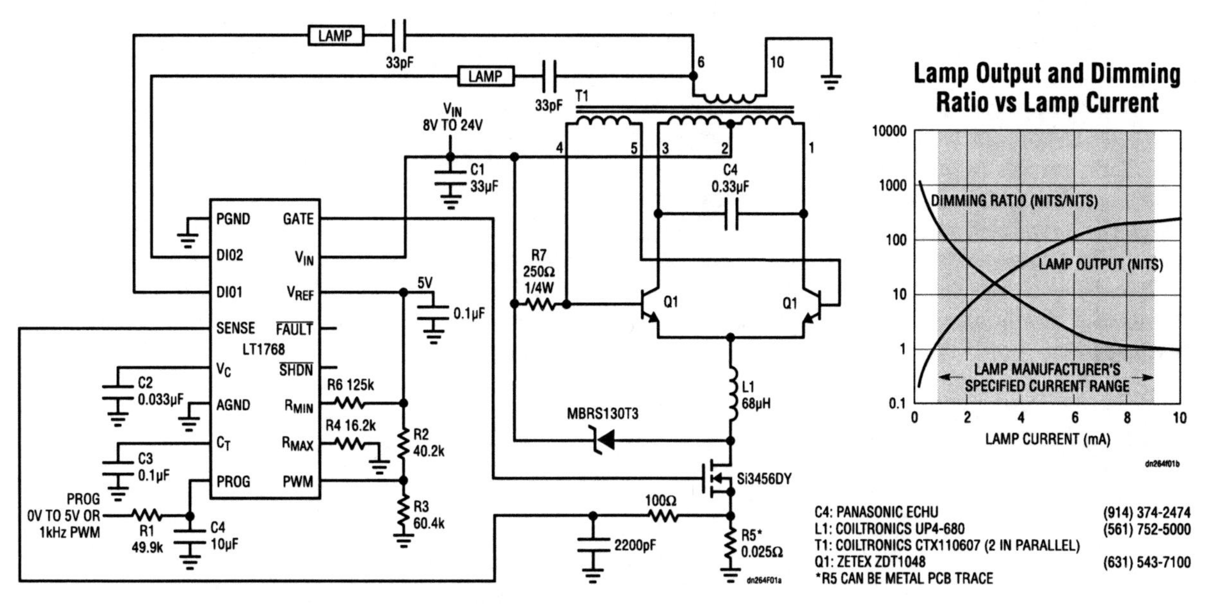

Figure 303.1 • 14W CCFL Supply Produces a 100:1 Dimming Ratio While Maintaining Minimum and Maximum Lamp-Current Specifications

Analog Circuit Design: Design Note Collection. http://dx.doi.org/10.1016/B978-0-12-800001-4.00303-3

The pulse-width modulation produces an average current in inductor L1 proportional to the voltage on the V_C pin. The CCFLs are driven by the Royer-class converter comprised of T1, C4 and Q1. The Royer converter produces a 90% efficient, zero DC component, 60kHz sinusoidal waveform based on the average current in L1. Sinusoidal currents from both CCFLs are returned to the LT1768 through the DIO1/DIO2 pins. A fraction of the CCFL current pulls against the V_C pin closing the loop. A single capacitor on the V_C pin provides loop compensation and CCFL current averaging, which results in constant CCFL current regardless of line and load conditions. Varying the value of the V_C current source via the Multimode Dimming block varies the CCFL current and resultant light intensity.

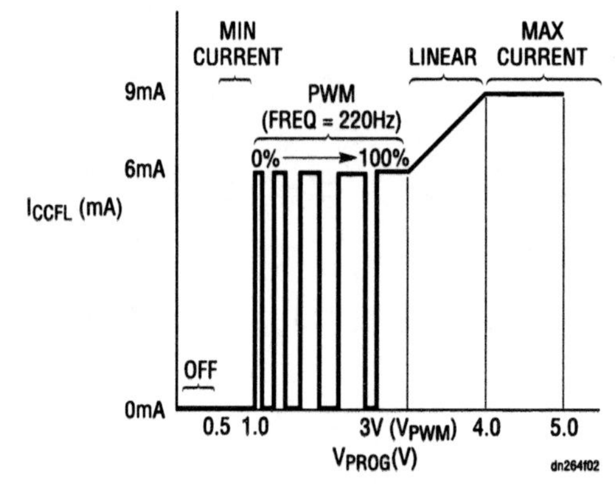

Figure 303.2 • Lamp Current vs PROG Voltage

Multimode Dimming

Previous solutions used intensity control schemes that were limited to either linear or PWM control. Linear control schemes provide the highest efficiency circuits but either limit dimming range or violate lamp specifications to achieve wide dimming ratios. PWM control schemes offer wide dimming range but produce high crest-factor waveforms detrimental to CCFL life and waste power at higher currents. The LT1768's patented Multimode Dimming combines the best of both control schemes to extend CCFL life while providing the widest possible dimming range.

The circuit in Figure 303.1 accepts either a 0V to 5V DC voltage, or a 0V to 5V, 1kHz PWM waveform and converts to a DC voltage. The filtered input voltage is sent to the LT1768 PROG pin, which controls lamp intensity by placing the LT1768 into one of five distinct modes of operation. Referring to Figure 303.2, which mode is in use is determined by the voltages on the PROG and PWM pins and by the currents that flow out of the R_{MAX} and R_{MIN} pins.

Off mode ($V_{PROG} < 0.5V$) sets the CCFL current to zero.

Minimum current mode ($0.5V < V_{PROG} < 1.0V$) sets the CCFL current to a precise minimum level set by the R_{MIN} resistor. This mode determines the minimum lamp current and intensity.

Maximum current mode ($V_{PROG} > 4V$) sets the CCFL current to a precise maximum level set by the R_{MAX} resistor. Setting the CCFL current in this mode to the manufacturers maximum rating achieves maximum intensity and ensures no degradation in the lamp lifetime.

In linear mode ($V_{PWM} < V_{PROG} < 4V$), CCFL current is controlled linearly with the voltage on the PROG pin. Linear mode provides the best current-to-light conversion and highest efficiency.

In PWM mode ($1V < V_{PROG} < V_{PWM}$), the CCFL current is modulated between the minimum CCFL current and the value for CCFL current in linear mode with $V_{PROG} = V_{PWM}$. The PWM frequency is set by a single capacitor on the C_T pin. The PWM duty cycle is set by the voltage on the PROG pin with 1V equal to 0%, and 100% (linear mode) equal to V_{PWM}. The LT1768's PWM mode enables wide dimming ratios while reducing the high crest factor found in PWM-only dimming solutions.

When combined, these five modes of operation allow the creation of a DC-controlled CCFL current profile that can be tailored to enable the widest possible dimming ratio while maximizing CCFL lifetime.

LT1768 fault modes

The LT1768 also has fault detection to ensure that lamp current and Royer transformer ratings are not exceeded under fault conditions. If one CCFL lamp is open, the LT1768 activates a fault flag and adjusts the current in the remaining so that it never exceeds the maximum current set by the R_{MAX} resistor. If both lamps are open circuit, the LT1768 shuts down the Royer section to avoid any hazardous high voltage conditions.

Additional features

The LT1768 also provides a temperature-compensated 5V reference, an undervoltage lockout feature, thermal shutdown and a logic-compatible shutdown pin that reduces supply current when activated. The LT1768 is available in a 16-pin SSOP package.

Tiny regulators drive white LED backlights

304

Dave Kim

Introduction

The emergence of color LCD displays in handheld information appliances has created the need for a small, bright white backlight. Fortunately, the recent commercialization of high intensity white LEDs provides the perfect solution. These tiny LEDs are capable of delivering ample white light without the fragility problems and costs associated with fluorescent backlights commonly used in notebook computers. However, they do pose a problem—the forward voltage of white LEDs can be as high as 4V, precluding powering them directly from a single lithium-ion cell.

This design note describes several different circuits that may be used to boost and regulate the Li-Ion battery voltage to power white LEDs. These circuits provide sufficient power to drive multiple white LEDs and are small enough to easily fit within cellular telephones and handheld computers.

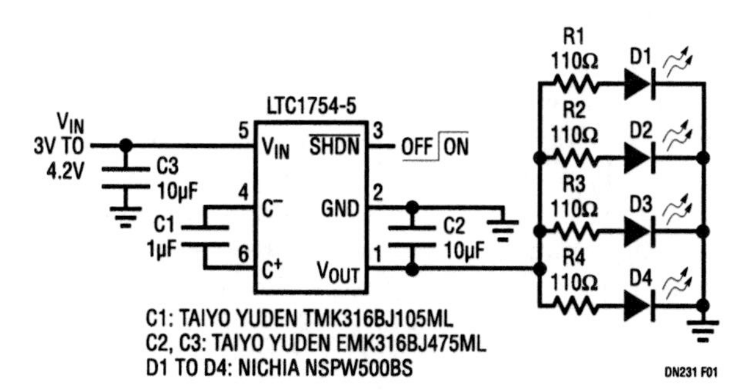

Figure 304.1 • LTC1754 White LED Driver

Circuit descriptions

Figure 304.1 depicts a very small charge pump DC/DC converter that is capable of powering four white LEDs at 15mA each (the typical forward current used for backlighting). The LTC1754-5 used in this application is a tiny SOT-23 device that delivers a regulated 5V output without any inductors. The entire circuit occupies less than 0.1in^2 of board space (excluding the LEDs) and may be powered directly from a lithium-ion cell.

A constant-current backlight supply may be constructed using the LTC1682. This architecture has several advantages. First, it directly controls the LED current regardless of the forward voltage drop across the LED. Additionally, the LTC1682's output is regulated by an on-chip linear regulator to deliver a very low noise output to the LEDs. Output ripple is less than 4mV and eliminates the risk of RF interference in sensitive cellular phone applications. The LTC1682's high current output is capable of powering five parallel LEDs at 15mA each from a single lithium-ion cell.

Applications requiring more white LEDs or higher efficiency can use an LT1615 boost converter to drive a series connected string of LEDs. The high efficiency circuit shown in Figure 304.3 can provide a constant-current drive for up

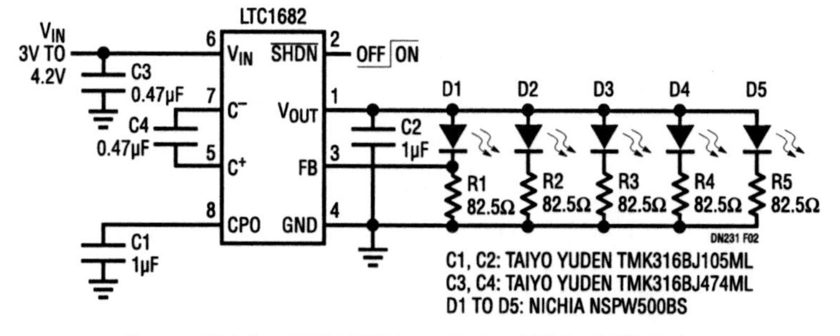

Figure 304.2 • LTC1682 Low Noise White LED Driver

Analog Circuit Design: Design Note Collection. http://dx.doi.org/10.1016/B978-0-12-800001-4.00304-5

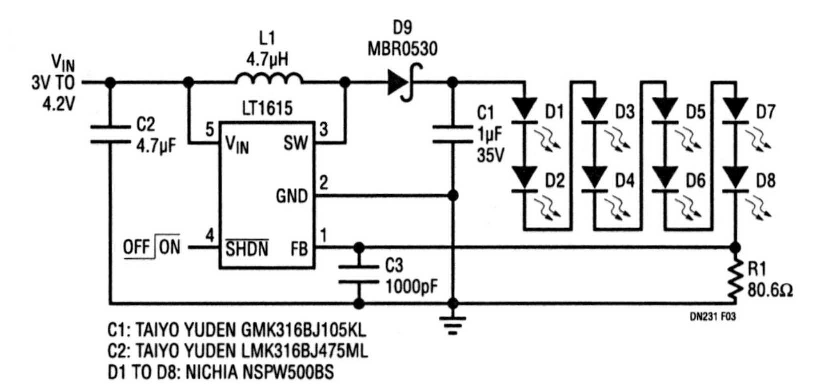

Figure 304.3 • LT1615 White LED Driver

to eight LEDs. Driving eight white LEDs in series requires approximately 29V at the output and is possible due to the internal 36V, 350mA switch in the LT1615. The constant-current design of the circuit guarantees the same LED current through all series LEDs, regardless of the forward voltage differences between the LEDs. Although this circuit is designed to operate from a single Li-Ion battery (2.5V to 4.2V), the LT1615 is also capable of operating from inputs as low as 1V with commensurate output power reductions.

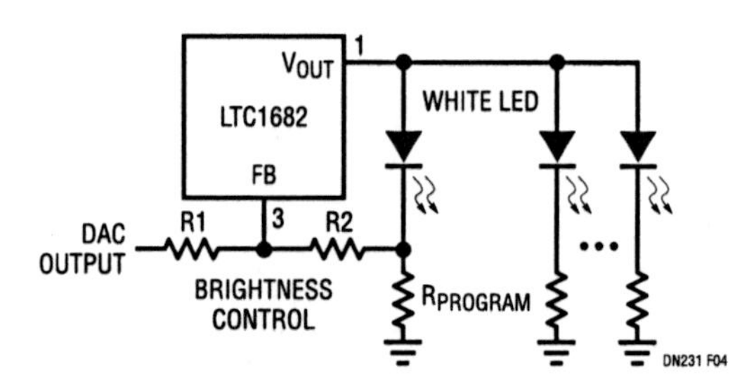

Figure 304.4 • Brightness Control Using DAC Output

Brightness control

The brightness of the LED can be controlled by applying a PWM signal to the SHDN pin on any of the backlight circuits shown as long as a couple of precautions are taken. Because of the "soft-start" circuitry incorporated in these DC/DC converters, the output voltage will not immediately rise to full output after the SHDN pin is taken high. Consequently, a PWM signal in the range of 200Hz is recommended—much faster and brightness control will be nonlinear; much slower and flicker may be observed. It may also be desirable to place a resistor between the DC/DC converter output and ground (in parallel with the LED load) to discharge the output during shutdown. Select a resistor that will draw approximately 1mA when the DC/DC converter is operational. (A parallel resistor is not required with the LTC1682 because it contains internal discharge circuitry.)

As an alternative to using PWM control, a DAC output can also be used to control the brightness of the LEDs in Figures 304.2 and 304.3. As depicted in Figure 304.4, the DAC output controls the brightness of the LED by varying the voltage across $R_{PROGRAM}$. Since the regulator holds the feedback voltage constant, varying the DAC voltage will affect the current flowing through the LEDs. A lower DAC voltage will result in higher brightness, while a high DAC voltage will result in lower brightness.

Summary

The circuits shown are several examples of very tiny step-up regulators that are suitable for driving white LEDs. These circuits contain such desirable features as constant-current LED drive, low noise (important for cellular telephone applications), brightness control and low voltage input (LT1615). Consult Linear Technology for additional applications assistance with white LED circuits.

High power CCFL backlight inverter for desktop LCD displays

305

Jim Williams

Large LCD (liquid crystal display) displays designed to replace CRTs (cathode ray tubes) in desktop computer applications are becoming available. The LCD's reduced size and power requirements allow much smaller product size, a highly desirable feature.

CRT replacement requires a 10W to 20W inverter to drive the CCFL (cold cathode fluorescent lamp) that illuminates the LCD. Additionally, the inverter must provide the wide dimming range associated with CRTs, and it must have safety features to prevent catastrophic failures.

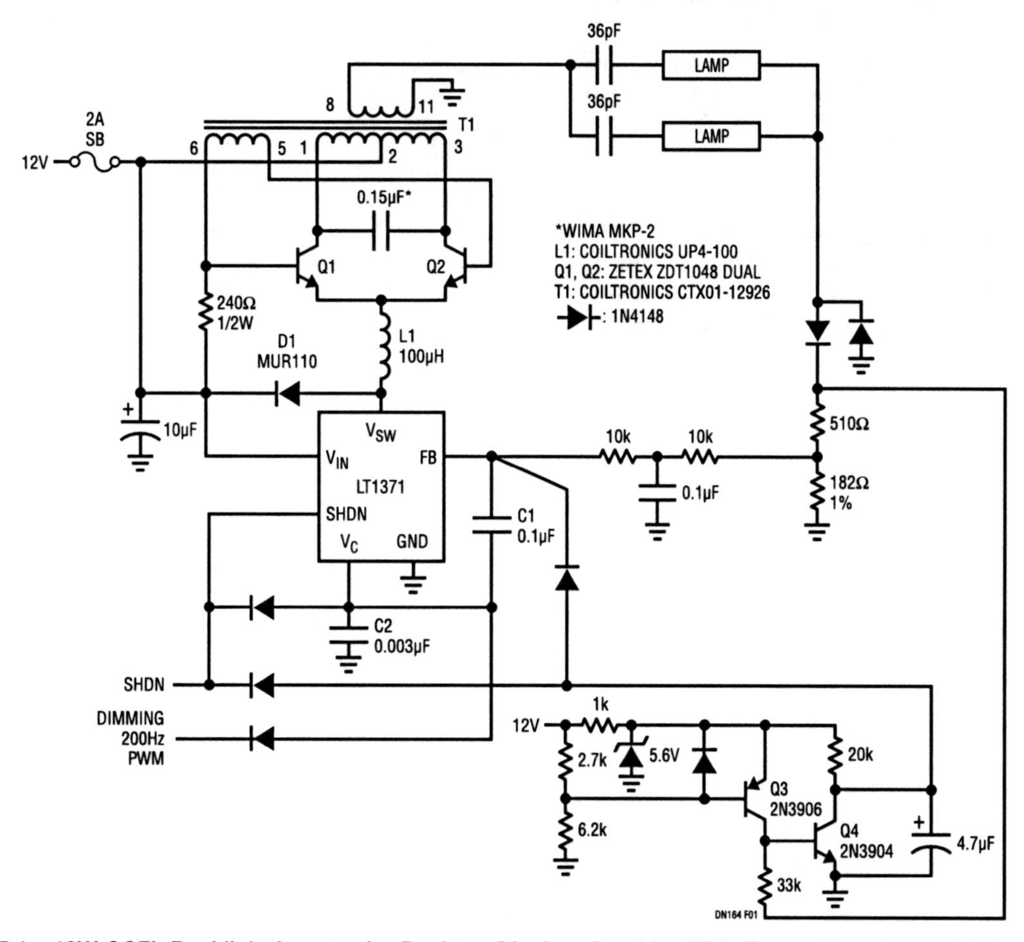

Figure 305.1 • 12W CCFL Backlight Inverter for Desktop Displays Provides Wide Range Dimming and Safety Features

Analog Circuit Design: Design Note Collection. http://dx.doi.org/10.1016/B978-0-12-800001-4.00305-7

Figure 305.1's circuit meets these requirements. It is a modified, high power variant of an approach employed in laptop computer displays.[1] T1, Q1, Q2 and associated components form a current fed, resonant Royer converter that produces high voltage at T1's secondary. Current flows through the CCFL tubes and is summed, rectified and filtered, providing a feedback signal to the LT1371 switching regulator. The LT1371 delivers switched mode power to the L1–D1 node, closing a control loop around the Royer converter. The 182Ω resistor provides current-to-voltage conversion, setting the lamp current operating point. The loop stabilizes lamp current against variations in time, supply, temperature and lamp characteristics. The LT1371's frequency compensation is set by C1 and C2. The compensation responds quickly enough to permit the 200Hz PWM input to control dimming over a 30:1 range with no degradation in loop regulation. Applicable waveforms appear in Figure 305.2.

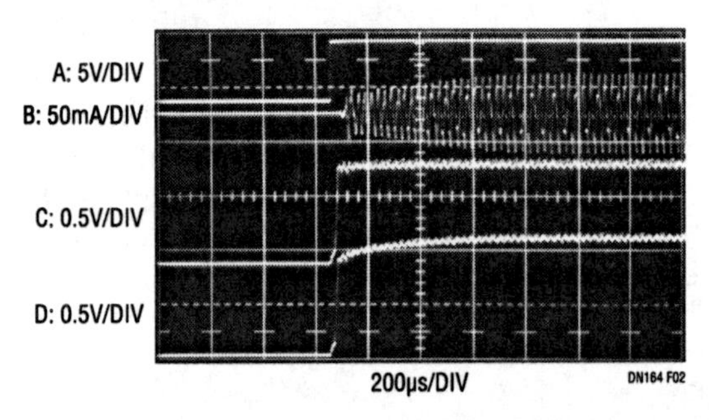

Figure 305.2 • Fast Loop Response Maintains Regulation at 200Hz PWM Rate. Waveforms Include PWM Command (A), Lamp Current (B), LT1371 Feedback (C) and Error Amplifier V_C (D) Pins. Loop Settling Occurs in 500μs

Q3 and Q4 shut down the circuit if lamp current ceases (open or shorted lamps or leads, T1 failure or similar malfunction). Normally, Q4's collector is held near ground by the lamp-current-derived base biasing. If lamp current ceases, Q4's collector voltage increases, overdriving the feedback node and shutting down the circuit. Q3 prevents unwanted shutdown during power supply turn-on by driving Q4's base until supply voltage is above about 7V.

Figure 305.3 shows the shutdown circuit reacting to the loss of lamp feedback. When lamp feedback ceases, the voltage across the 182Ω current sense resistor drops to zero (visible between Figure 305.3's second and third vertical graticule lines, trace A). The LT1371 responds to this open-loop condition by driving the Royer converter to full power (Q1's collector is trace B). Simultaneously, Q4's collector (trace C) ramps up, overdriving the LT1371's feedback node in about 50ms. The LT1371 stops switching, shutting off the Royer converter drive. The circuit remains in this state until the failure has been rectified.

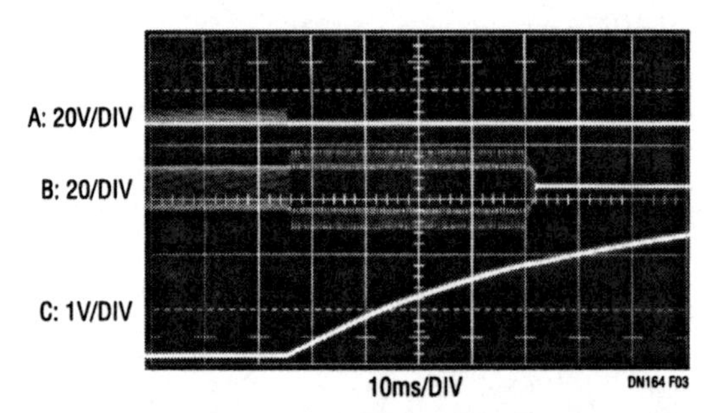

Figure 305.3 • Safety Feature Reacts to Lamp Feedback Loss by Shutting Down Power. Lamp Current Dropout (Trace A) Allows Monitoring Circuit to Ramp Up (Trace C), Shutting Off Drive (Trace B)

This circuit's combination of features provides a safe, simple and reliable high power CCFL lamp drive. Efficiency is in the 85% to 90% range. The closed-loop operation ensures maximum lamp life while permitting extended dimming range. The safety feature prevents excessive heating in the event of malfunction and the use of off-the-shelf components allows ease of implementation.

Note 1: See LTC Application Note 65, A Fourth Generation of LCD Backlight Technology.

Low input voltage CCFL power supply

306

Fran Hoffart

Cold cathode fluorescent lamps (CCFLs) are often used to illuminate liquid crystal displays (LCDs). These displays appear in laptop computers, gas pumps, automobiles, test equipment, medical equipment and the like. The lamps themselves are small, relatively efficient and inexpensive, but they must be driven by specialized power supplies. High AC voltage (significantly higher than the operating voltage) is needed to start the lamp. A sinusoidal waveform is desired, the current must be regulated, efficiency should be high and the power supply must be self-protecting in the event of an open-lamp condition.

CCFL power supplies consisting of a Royer class, self-oscillating sine wave converter driven by an LT1513 switching regulator are shown in Figures 306.1 and 306.2. These circuits are especially suited for low voltage operation, with guaranteed operation for input voltages as low as 2.7V and as high as 20V. High voltage output regulated Royer converters, although capable of 90% efficiency, are not well-suited for low input voltage operation and have difficulty operating with input voltages below 5V. The circuits shown here overcome this limitation while providing efficiency exceeding 70%.

The LT1513 is a 500kHz current mode switching regulator featuring an internal 3A switch and unique feedback circuitry. In addition to the Voltage Feedback pin (V_{FB}), a second feedback node (I_{FB}) provides a simple means of controlling output current in a flyback or SEPIC (single-ended primary inductance converter) topology.

Two CCFL driver circuits are shown. The first (Figure 306.1) drives one end of the lamp, with the other end effectively grounded. Lamp current is directly sensed at

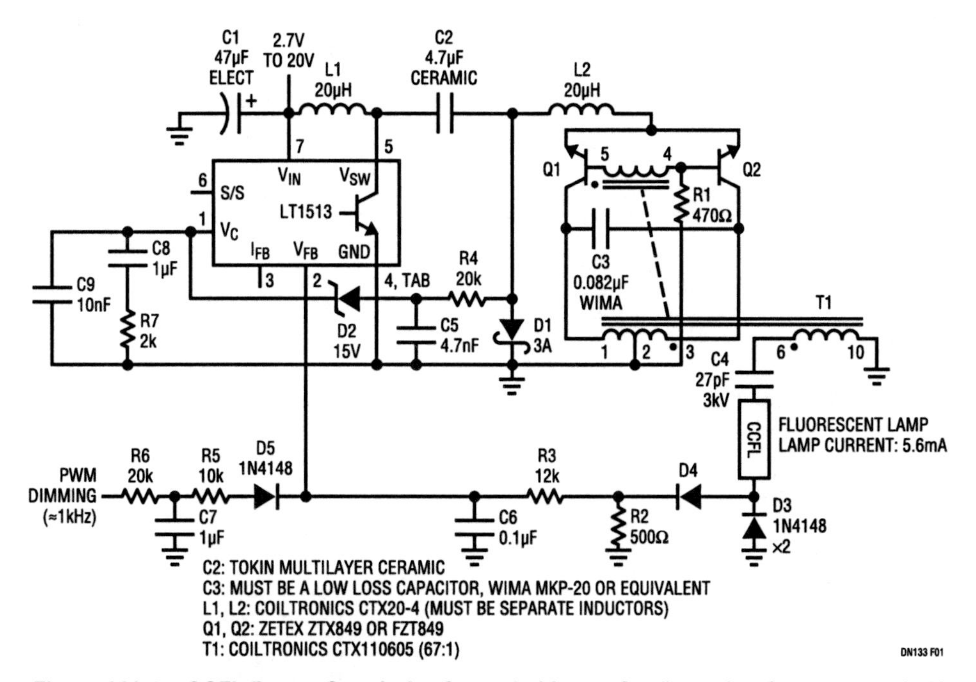

Figure 306.1 • CCFL Power Supply for Grounded Lamp Configuration Operates on 2.7V

Analog Circuit Design: Design Note Collection. http://dx.doi.org/10.1016/B978-0-12-800001-4.00306-9

the low side of the lamp, half-wave rectified by D4, and then used to develop a feedback voltage across R2. This voltage, filtered by R3 and C6, drives the V_{FB} pin (2) to complete the feedback loop. The RMS lamp current is tightly regulated and is equal to 2.82V/R2.

Because of the high voltage 60kHz lamp drive used by the CCFL lamps, any stray capacitance from the lamp and lamp leads to ground will result in unwanted parasitic current flow, thus lowering efficiency. The lamp and display housing often have relatively high stray capacitance, which can dramatically lower the overall circuit efficiency.

In some displays that exhibit high capacitance, a floating lamp drive can provide much higher overall efficiency. The operation of the floating lamp circuit shown in Figure 306.2 is similar to that of the grounded lamp, except for the transformer secondary and the feedback method used. In this circuit, the lamp current is controlled by sensing and regulating the Royer input current. This current is sensed by R2, filtered by R3 and C6 and fed into the I_{FB} pin (3) of the LT1513, thus completing the feedback loop. The sense voltage required at the I_{FB} pin is −100mV. Because the Royer input current rather than the actual lamp current is regulated, the regulation of Figure 306.2 is not as tight as that of Figure 306.1.

There are three considerations to keep in mind when laying out a PC board for these supplies. The first is related to high frequency switcher characteristics. The 500kHz switching frequency allows very small surface mount components to be used, but it also requires that PC board traces be kept short (especially the input capacitor, Schottky diode and LT1513 ground connections). The second item is the high voltage section, which includes T1's secondary, the ballasting capacitor C3 and the lamp wiring. Lamp starting voltages can easily exceed 1000V, which can cause a poorly designed board to arc, resulting in catastrophic failure. Board leakages can also increase dramatically with time, resulting in destructive field failures. Third, surface mount components rely on the PC board copper to conduct the heat away from the components and dissipate it to the surrounding air. Good thermal PC board layout practices are necessary.

In both circuits, lamp current can be adjusted downward to provide lamp dimming. A 5V pulse-width modulated (PWM) 1kHz signal or an adjustable DC voltage can provide a full range of dimming. In Figure 306.1, 100% duty cycle represents minimum lamp brightness, whereas in Figure 306.2 100% duty cycle is maximum lamp brightness.

The lamp drive is a constant current, and without protection circuitry, voltages could become very high in the event of an open-lamp connection, causing transformer arcing or LT1513 failures. Open-lamp or high input voltage fault protection is provided by R4, C5 and the 15V Zener D2, which limit the maximum voltage available for the Royer converter.

See Application Notes 49 and 65, Design Note 99 and the LT1513 Data Sheet for additional information on driving CCFL lamps.

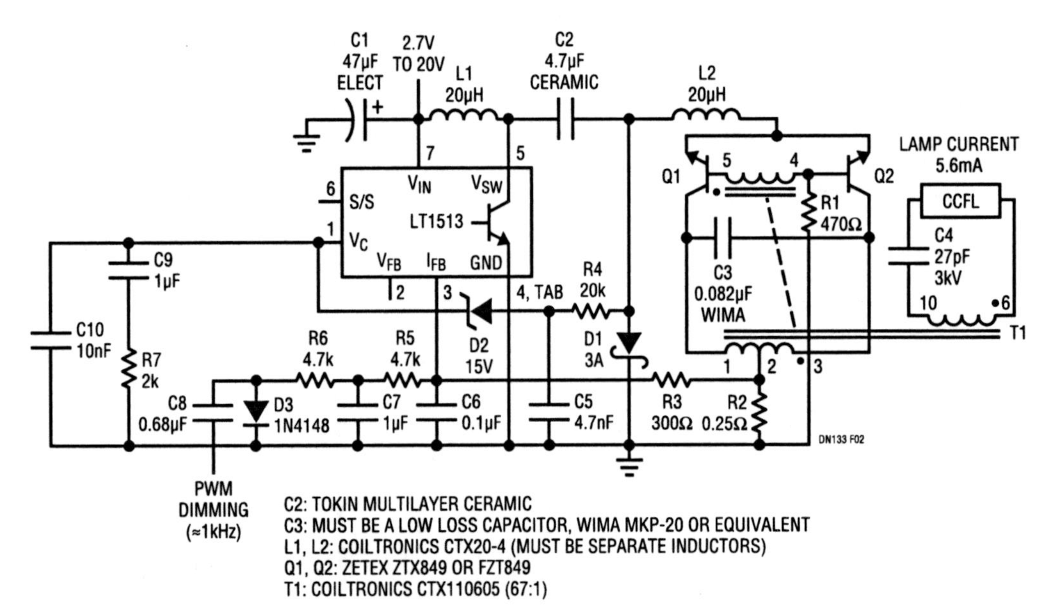

Figure 306.2 • CCFL Power Supply for Floating Lamp Configuration Operates on 2.7V

A precision wideband current probe for LCD backlight measurement

307

Jim Williams

Evaluation and optimization of cold cathode fluorescent lamp (CCFL) performance requires highly accurate AC current measurement. CCFLs, used to backlight LCD displays, typically operate at 30kHz to 70kHz with measurable harmonic content into the low MHz region.[1] Accurate determination of RMS operating current is important for electrical and emissivity efficiency computations and to ensure long lamp life. Additionally, it is desirable to be able to perform current measurements in the presence of high common-mode voltage (>1000V_{RMS}). This capability allows investigation and quantification of display and wiring induced losses, regardless of their origins in the lamp drive circuitry.

Current probe circuitry

Figure 307.1's circuitry meets the discussed requirements. It signal conditions a commercially available "clip-on" current probe with a precision amplifier to provide 1% measurement accuracy to 10MHz. The "clip-on" probe provides convenience, even in the presence of the high common voltages noted. The current probe biases A1, operating at a gain of about 3.75. No impedance matching is required due to the probe's low output impedance termination. Additional amplifiers provide distributed gain, maintaining wide bandwidth with an overall gain of about 200. The individual amplifiers avoid any possible crosstalk-based error that could be introduced by a monolithic quad amplifier. D1 and Rx are selected for polarity and value to trim overall amplifier offset. The 100Ω trimmer sets gain, fixing the scale factor. The output drives a thermally based, wideband RMS voltmeter. In practice, the circuit is built into a 2.25" × 1" × 1" enclosure which is *directly* connected, via BNC hardware, to the voltmeter. No cable is used. The result is a "clip-on" current probe with 1% accuracy over a 20kHz to 10MHz bandwidth. Figure 307.2 shows response for the probe-amplifier as measured on a Hewlett-Packard HP-4195A network analyzer.

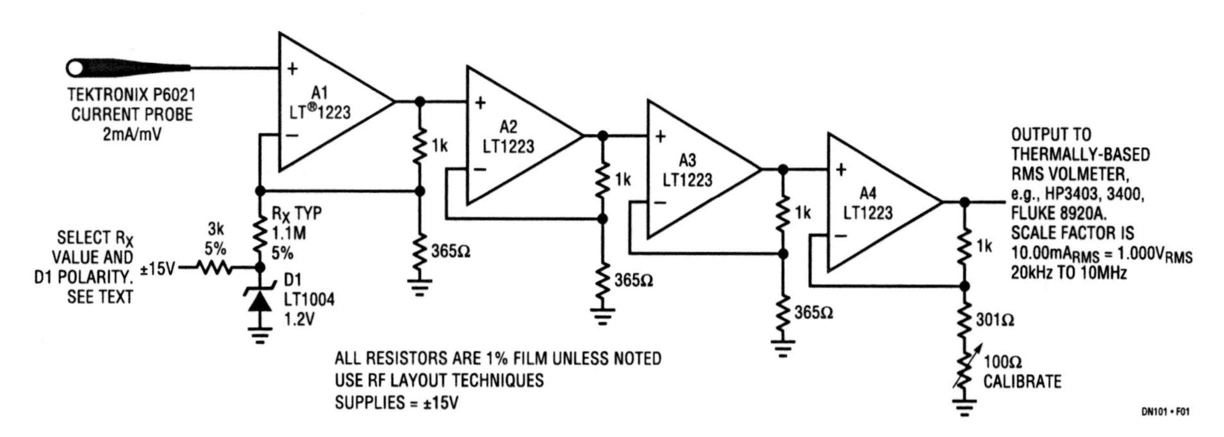

Figure 307.1 • Precision "Clip-On" Current Probe for CCFL Measurements Maintains 1% Accuracy over 20kHz to 10MHz Bandwidth

Note 1: Williams, Jim, "Techniques for 92% Efficient LCD Illumination." Linear Technology Corporation AN55, August 1993.

Analog Circuit Design: Design Note Collection. http://dx.doi.org/10.1016/B978-0-12-800001-4.00307-0

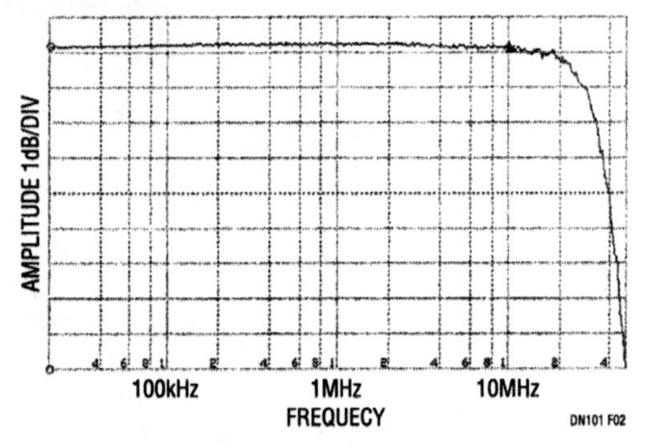

Figure 307.2 • Amplitude vs Frequency Output of HP4195A Network Analyzer. Current Probe-Amplifier Maintains 1% (0.1dB) Error Bandwidth from 20kHz to 10MHz. Small Aberrations between 10MHz and 20MHz Are Test Fixture Related

Current calibrator

Figure 307.3's circuit, a current calibrator, permits calibration of the probe-amplifier and can be used to periodically check probe accuracy. A1 and A2 form a Wein bridge oscillator. Oscillator output is rectified by A4 and A5 and compared to a DC reference at A3. A3's output controls Q1, closing an amplitude stabilization loop. The stabilized amplitude is terminated into a 100Ω, 0.1% resistor to provide a precise 10.00mA, 60kHz current through the series current loop. Trimming is performed by altering the nominal 15k resistor for exactly 1.000V_{RMS} across the 100Ω unit.

In use, this current probe has shown 0.2% baseline stability with 1% absolute accuracy over one year's time. The sole maintenance requirement for preserving accuracy is to keep the current probe jaws clean and avoid rough or abrupt handling of the probe.[2]

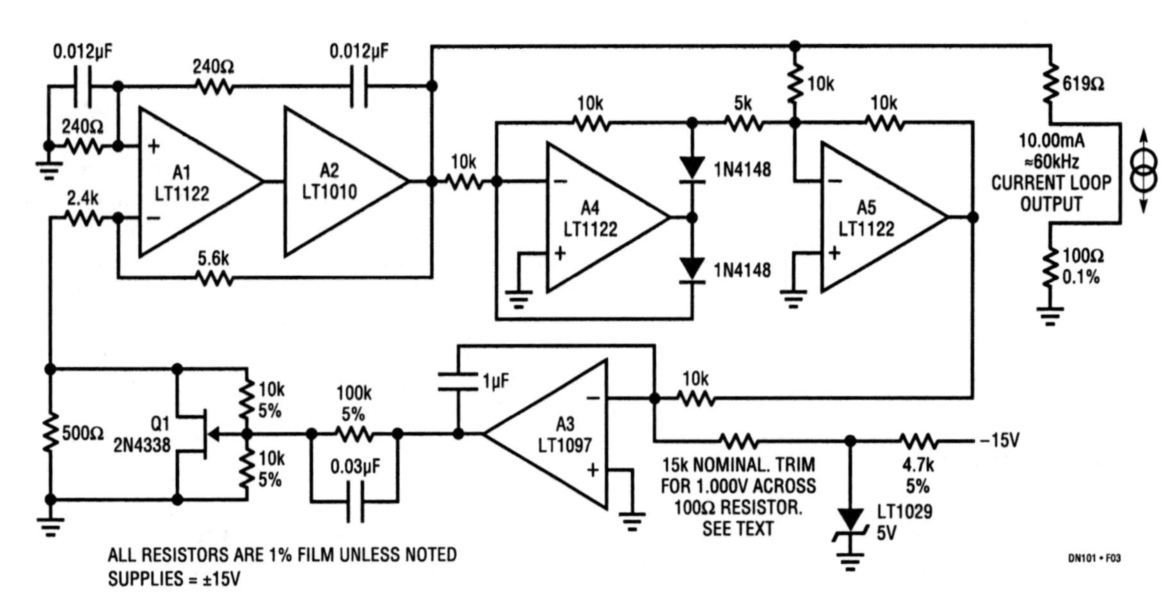

Figure 307.3 • Current Calibrator for Probe Trimming and Accuracy Checks. Stabilized Oscillator Forces 10.00mA through Output Current Loop at 60kHz

Note 2: Private Communication. Tektronix, Inc.

Floating CCFL with dual polarity contrast

308

Anthony Bonte

Current generation portable computers and instruments use backlit liquid-crystal displays (LCDs). Cold cathode fluorescent lamps (CCFLs) provide the highest available efficiency for backlighting the display. The lamp requires high voltage AC to operate, mandating an efficient, high voltage DC/AC converter. The LCD also requires a bias supply for contrast control. The supply's output must regulate and provide adjustment over a wide range.

Manufacturers offer a wide array of monochrome and color displays. These displays vary in size, lamp drive current, contrast voltage polarity, operating voltage range and power consumption. The small size and battery-powered operation associated with LCD-equipped apparatus dictate low component count and high efficiency. Size constraints place limitations on circuit architecture and long battery life is a priority. All components, including PC board and hardware, must fit within the LCD enclosure with a height restriction of 5mm to 10mm.

Linear Technology addresses these requirements by introducing the LT1182/LT1183/LT1184F/LT1184. The LT1182 and LT1183 are dual fixed frequency, current mode switching regulators that provide the control function for cold cathode fluorescent lighting and liquid-crystal display contrast. The LT1184F/LT1184 provides only the CCFL function.

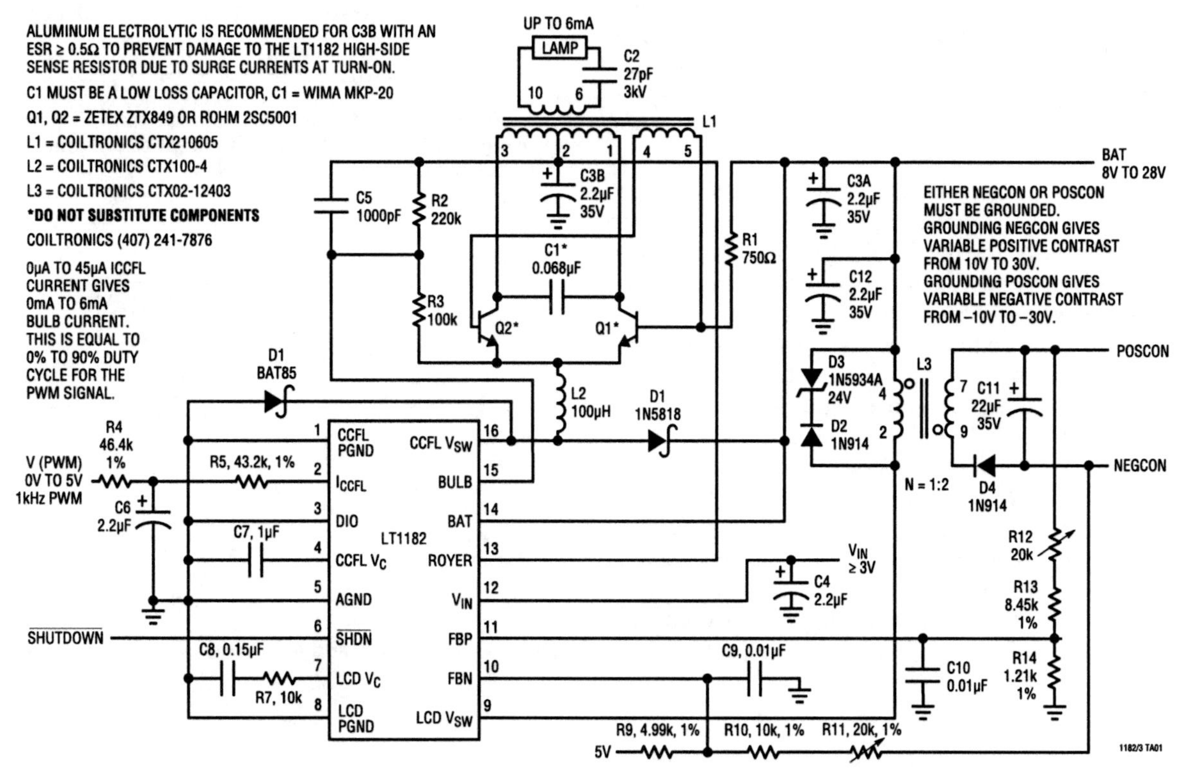

Figure 308.1 • LT1182 Floating CCFL Configuration with Variable Positive/Variable Negative LCD Contrast

Analog Circuit Design: Design Note Collection. http://dx.doi.org/10.1016/B978-0-12-800001-4.00308-2

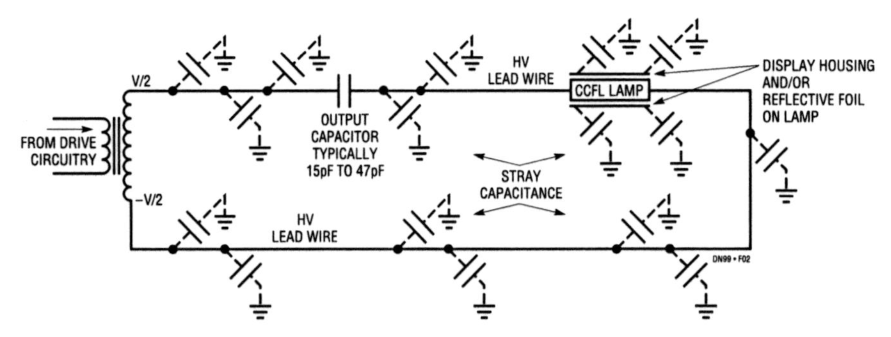

Figure 308.2 • Loss Path due to Stray Capacitance in a Floating LCD Installation. Differential, Balanced Lamp Drive Reduces This Loss Term and Improves Efficiency

The ICs include high current, high efficiency switches, an oscillator, a reference, output drive logic, control blocks and protection circuitry. All of the devices supports grounded lamp or floating lamp configurations using a unique lamp current control circuit. The LT1182/LT1183 supports negative voltage or positive voltage LCD contrast operation with a new dual polarity error amplifier. In short, this new family reduces system power dissipation, requires fewer external components, reduces overall system cost and permits a high level of system integration for a backlight/LCD contrast solution.

Figure 308.1 is a complete floating CCFL circuit with variable negative/variable positive contrast voltage capability based on the LT1182. Lamp current is programmable from 0mA to 6mA using a 0V to 5V 1kHz PWM signal at 0% to 90% duty cycle. LCD contrast output voltage polarity is determined by which side of the transformer secondary (either POSCON or NEGCON) the output connector grounds. In either case, LCD contrast output voltage is variable from an absolute value of 10V to 30V. The input supply voltage range is 8V to 28V. The CCFL converter is optimized for photometric output per watt of input power. CCFL electrical efficiency up to 90% is possible and requires strict attention to detail. LCD contrast efficiency is 82% at full power.

Achieving high efficiency for a backlight design requires careful attention to the physical layout of the lamp, its leads and the construction of the display housing. Parasitic capacitance from any high voltage point to DC or AC ground creates paths for unwanted current flow. This parasitic current degrades electrical efficiency. The loss term is related to $1/2CV^2f$ where C is the parasitic capacitance, V is the voltage at any point on the lamp and f is the Royer operating frequency. Losses up to 25% have been observed in practice. Figure 308.2 indicates the loss paths present in a typical LCD enclosure for a floating lamp configuration. Layout techniques that increase parasitic capacitance include long high voltage lamp leads, reflective metal foil around the lamp and displays supplied in metal enclosures.

Lossy displays are the primary reason to use a floating lamp configuration. Providing symmetric, differential drive to the lamp reduces the total parasitic loss term by one-half in comparison to a grounded lamp configuration. As an added benefit, floating lamp configurations eliminate field imbalance along the length of the lamp. Figure 308.3 illustrates this effect. Eliminating field imbalance improves the illumination range from about 6:1 for a grounded lamp configuration to 30:1 for a floating lamp configuration. Figure 308.4 is a graph of normalized nits/watts versus lamp current for a typical manufacturer's display with a 6mA lamp. Performance for the display is compared in a floating lamp configuration versus a grounded lamp configuration. The benefit of reduced parasitic loss is readily apparent.

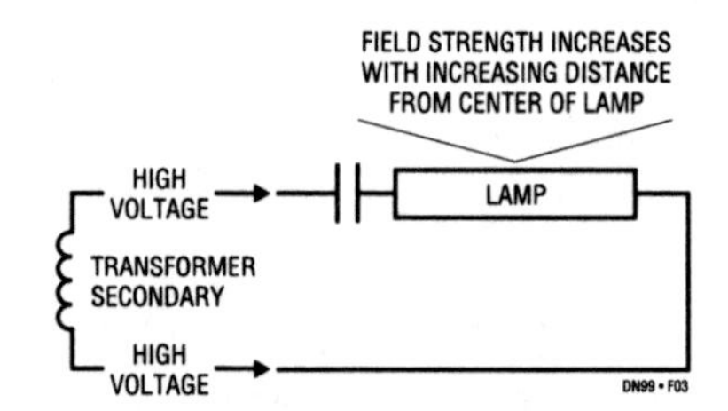

Figure 308.3 • Field Strength vs Distance for a Floating Lamp. Improving Field Imbalance Permits Extended Illumination Range at Low Levels

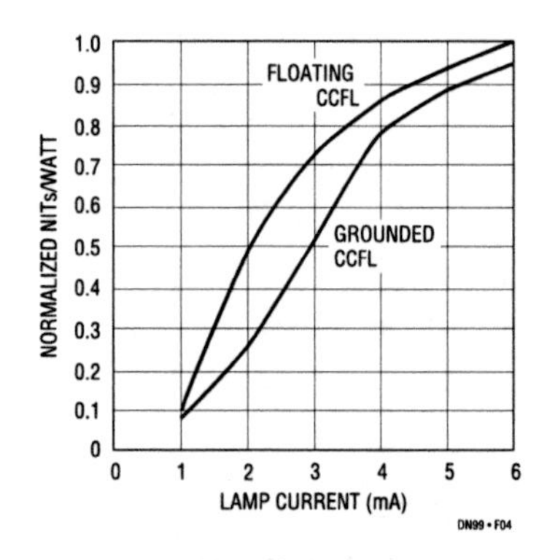

Figure 308.4 • Normalized Nits/Watts vs Lamp Current

Section 22

Automotive and Industrial Power Design

Versatile industrial power supply takes high voltage input and yields from eight 1A to two 4A outputs

309

Martin Merchant

Introduction

AQ1

Today's industrial electronic systems contain many of the same components as consumer electronics—microcontrollers, FPGAs, system-on-chip ASICs and other electronics—requiring multiple low voltage rails at widely varied load currents. Industrial applications can also demand a pushbutton interface, an always-on supply for a real-time clock (RTC) or memory and the ability to take input power from a high voltage supply. Other required features may be a watchdog timer (WDT), a kill or reset button, software adjustable voltage levels and error reporting of low input/output voltages and high die temperature.

The LTC3375 is a highly configurable multioutput step-down power converter that offers the features often required by industrial electronics while providing the flexibility to configure various outputs with maximum currents ranging from 1A to 4A.

Configurable maximum output current

The LTC3375's eight 1A channels can be combined to produce various combinations of 1A, 2A, 3A and 4A buck regulators, as shown by the 15 different output current configurations in Table 309.1.

Connecting the feedback pin of a given channel to its V_{IN} pin configures that channel as a slave to the adjacent channel. The switch pins of the two channels are connected together to share a single inductor and output capacitor. Master/slave channels are enabled via the master's enable pin and regulate to the master's feedback network.

Output current can be increased to 3A or 4A by connecting additional adjacent channels. The circuit in Figure 309.1 shows the LTC3375 configured with a 3A output, a 1A output, two 2A outputs and an always-on LDO. It also illustrates how the LTC3375 can be connected to control the start-up of an upstream external buck controller via the on-chip pushbutton interface to supply input power to the LTC3375 buck regulators.

Table 309.1 LTC3375 Maximum Current Configurations

NUMBER OF BUCKS	OUTPUT CONFIGURATION
8	1A, 1A, 1A, 1A, 1A, 1A, 1A, 1A
7	1A, 1A, 1A, 1A, 1A, 1A, 2A
6	1A, 1A, 1A, 1A, 1A, 3A
6	1A, 1A, 1A, 1A, 2A, 2A
5	1A, 1A, 1A, 1A, 4A
5	1A, 1A, 1A, 2A, 3A
5	1A, 1A, 2A, 2A, 2A
4	1A, 1A, 2A, 4A
4	1A, 1A, 3A, 3A
4	1A, 2A, 2A, 3A
4	2A, 2A, 2A, 2A
3	1A, 3A, 4A
3	2A, 2A, 4A
3	2A, 3A, 3A
2	4A, 4A

External V_{CC} LDO and external input power supply start-up control

The LTC3375 can control an external LDO pass device to supply its V_{CC} power and any other low current electronics such as an RTC. The V_{CC} powers the internal pushbutton circuitry, WDT, internal registers and open-drain pull-ups. The external LDO in Figure 309.1 creates a 3.3V supply from the 24V rail.

Analog Circuit Design: Design Note Collection. http://dx.doi.org/10.1016/B978-0-12-800001-4.00309-4

When the pushbutton is pressed, the ON pin is released and the RUN pin is pulled high on the LTC3891, supplying input power to the buck regulators of the LTC3375. When the LTC3891 achieves regulation, the PGOOD pin is released, enabling EN1 of the LTC3375 and turning on the 2A regulator. The remaining regulators can be enabled with the precision threshold enable pins or via software-controlled I^2C commands. Pressing the pushbutton again for 10 seconds or more, or pulling $\overline{KILL}$ low for 50ms or more, causes the ON pin to be pulled low, disabling all of the buck regulators.

Unique power control and features

The I^2C interface allows extensive control of regulator operation. Each regulator may be set to a high efficiency Burst Mode operation to save power at light loads or set to forced continuous mode for lower output ripple voltage. Each regulator can also have the switching cycle phase shifted by 0°, 90°, 180°, or 270° with respect to the reference clock to allow a lower input ripple current when multiple outputs are supplying large loads. Another feature is the ability to manipulate each output voltage up or down by adjusting the feedback reference voltage from the default 725mV setting in 25mV steps (ranging from 425mV to 800mV). The I^2C interface is also used for reporting error conditions for each regulator.

The LTC3375 has a reset ($\overline{RST}$) pin and an interrupt request ($\overline{IRQ}$) pin, which can be programmed to report when any regulator's output voltage has dropped below 92.5% of the regulation point. The $\overline{IRQ}$ pin can also be programmed to report when the input voltage drops below the undervoltage lockout (UVLO) threshold or when the die temperature has reached a set temperature threshold. The regulator's PGOOD and UVLO status, the die temperature warning and the measured die temperature can be monitored by the microprocessor via the I^2C interface.

One problem with microprocessors is that a software bug can cause the program to hang. The LTC3375 includes a watchdog timer input (WDI) pin to monitor the SCL pin or some other pin to determine if the software is still running. If the software has stopped running, the watchdog timer output (WDO) pin can be used to reset the microprocessor or power down the HV buck and the LTC3375 buck regulators. Connecting the WDO pin to the $\overline{RST}$ pin of a microprocessor causes the microprocessor to reset when the WDT is not satisfied. Connecting the WDO pin to the $\overline{KILL}$ pin causes the ON pin to go low, disabling the HV buck and all LTC3375 regulators. The $\overline{KILL}$ pin can be pulled low by a pushbutton "paper clip" switch to power down all the regulators as a last resort.

Conclusion

The LTC3375 can be configured with multiple regulated 1A to 4A outputs totaling up to 8A, and includes many features required by today's industrial electronics.

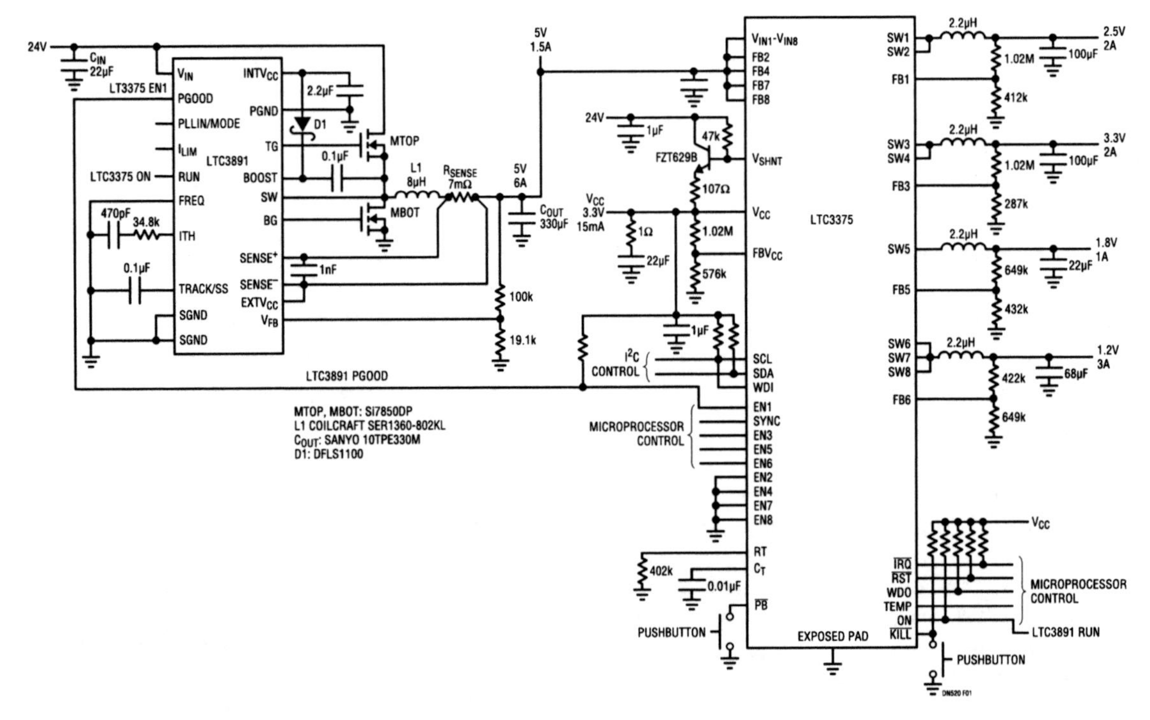

Figure 309.1 • Low Voltage Power Supply with Pushbutton Control of Upstream HV Buck and Always-On LDO

65V, 500mA step-down converter fits easily into automotive and industrial applications

310

Charlie Zhao

Introduction

The trend in automobiles and industrial systems is to replace AQ1 mechanical functions with electronics, thus multiplying the number of microcontrollers, signal processors, sensors and other electronic devices throughout. The issue is that 24V truck electrical systems and industrial equipment use relatively high voltages for motors and solenoids while the microcontrollers and other electronics require much lower voltages. As a result, there is a clear need for compact, high efficiency step-down converters that can produce very low voltages from the high input voltages.

65V input, 500mA DC/DC converter with an adjustable output down to 800mV

The LTC3630 is a versatile Burst Mode synchronous step-down DC/DC converter that includes three pin-selectable preset output voltages. Alternatively, the output can be set via feedback resistors down to 800mV. An adjustable output or input current limit from 50mA to 500mA can be set via a single resistor. The hysteretic nature of this topology provides inherent short-circuit protection. Higher output currents are possible by paralleling multiple LTC3630s together and connecting the FBO of the master device to the VFB pin of a slave device. An adjustable soft-start is included. A precision RUN pin threshold voltage can be used for an undervoltage lockout function.

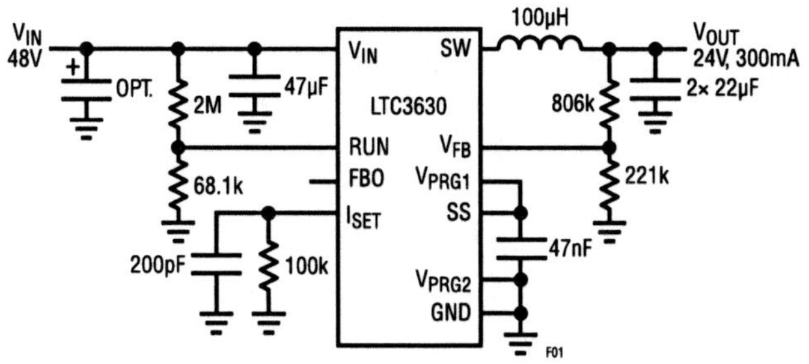

Figure 310.1 • High Efficiency 24V Regulator with Undervoltage Lockout and 300mA Current Limit

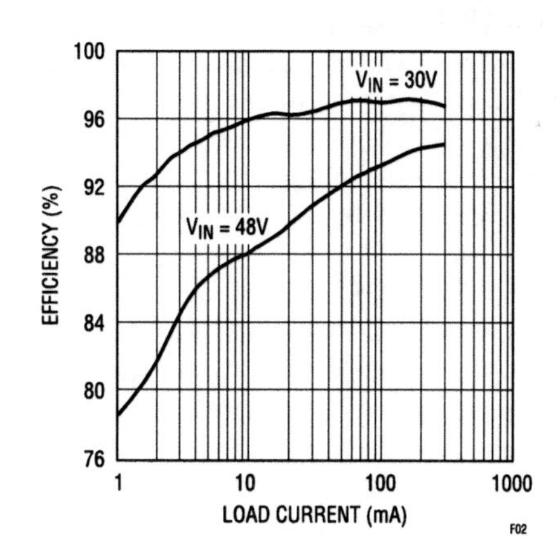

Figure 310.2 • Efficiency of Circuit in Figure 310.1

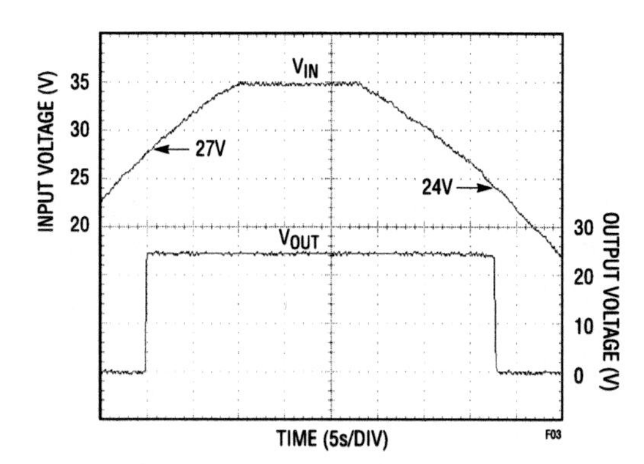

Figure 310.3 • Input Voltage Sweep vs Output Voltage Showing Undervoltage Lockout Threshold Levels

Analog Circuit Design: Design Note Collection. http://dx.doi.org/10.1016/B978-0-12-800001-4.00310-0

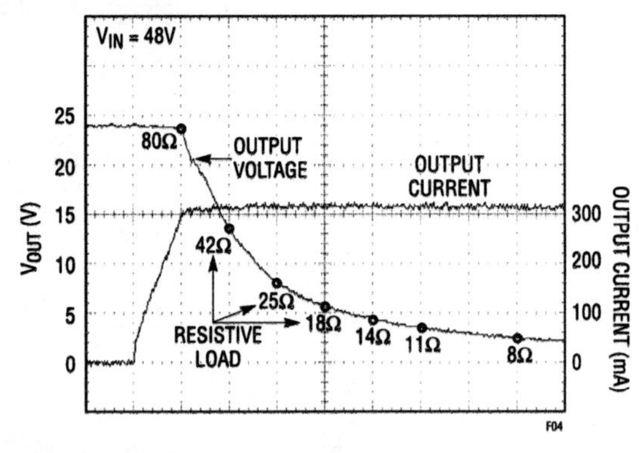

Figure 310.4 • Resistive Load Sweep vs Output Current vs Output Voltage with Output Current Limit Set to 300mA

24V regulator with 300mA output current limit and input undervoltage lockout

Figure 310.1 shows a 48V to 24V application that showcases several of the LTC3630's features, including the undervoltage lockout and output current limit. Operational efficiencies are shown in Figure 310.2.

The RUN pin is programmed for V_{IN} undervoltage lockout threshold levels of 27V rising and 24V falling. Figure 310.3 shows V_{OUT} vs V_{IN}. This feature assures that V_{OUT} is in regulation only when sufficient input voltage is available.

The 24V output voltage can be programmed using the 800mV 1% reference or one of the preset voltages. This circuit uses the 5V preset option along with feedback resistors to program the output voltage. This increases circuit noise immunity and allows lower value feedback resistors to be used.

Although the LTC3630 can supply up to 500mA of output current, the circuit in Figure 310.1 is programmed for a maximum of 300mA. An internally generated 5μA bias out of the I_{SET} pin produces a voltage across an I_{SET} resistor, which determines the maximum output current. Figure 310.4 shows the output voltage as a resistive load is varied from approximately 100Ω down to 8Ω while maintaining the output current near the programmed value of 300mA. In addition, the hysteretic topology used in this DC/DC converter provides inherent short-circuit protection.

Input current limit

Another useful feature of the LTC3630 is shown in Figure 310.5. In this 5V circuit, the current limit is set by a resistive divider from V_{IN} to I_{SET}, which produces a voltage

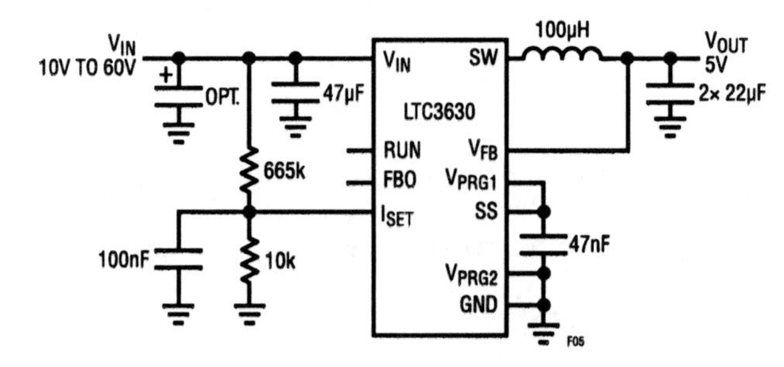

Figure 310.5 • 5V Regulator with 55mA Input Current Limit

on the I_{SET} pin that tracks V_{IN}. This allows V_{IN} to control output current which determines input current.

An increased voltage on I_{SET} increases the converter's current limit. Figure 310.6 shows the steady-state input current vs input voltage and the available output current before the output voltage begins to drop out of regulation. For the values shown in Figure 310.5, the input current is limited to approximately 55mA over a 10V to 60V input voltage range.

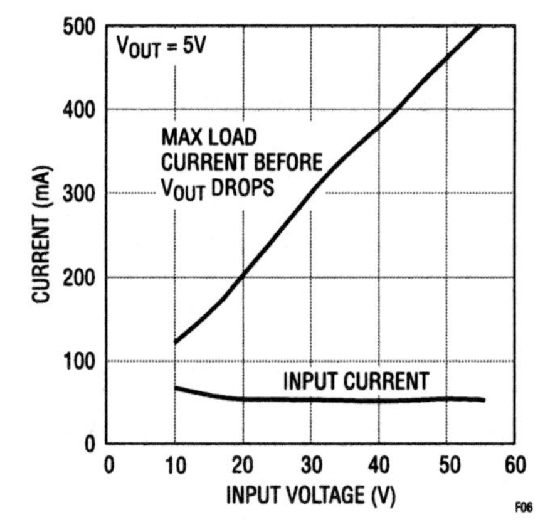

Figure 310.6 • Input Voltage vs Load Current and Input Current with Input Current Limit Circuit Shown in Figure 310.5

Conclusion

The LTC3630 offers a mixture of features useful in high efficiency, high voltage applications. Its wide output voltage range, adjustable current capabilities and inherent short-circuit tolerant operation makes this DC/DC converter an easy fit in demanding applications.

2-phase, dual output synchronous boost converter solves thermal problems in harsh environments

311

Goran Perica

Introduction

Boost converters are regularly used in automotive and industrial applications to produce higher output voltages from lower input voltages. A simple boost converter using a Schottky boost diode (Figure 311.1) is often sufficient for low current applications. However, in high current or space-constrained applications, the power dissipated by the boost diode can be a problem especially in high ambient temperature environments. Heat sinks and fans may be needed to keep the circuit cool, resulting in high cost and complexity.

To solve this problem, the Schottky output rectifier can be replaced by a synchronous MOSFET rectifier (Figure 311.2). If MOSFETs with very low $R_{DS(ON)}$ are used, the power dissipation can be reduced to the point where no heat sinks or active cooling is required, thus reducing costs and saving space.

Advantages of synchronous rectification

Consider the power dissipation of the single output circuit in Figure 311.1. The output diode D1 carries 6.7A of RMS current to produce 3A of output current from a 5V input. At

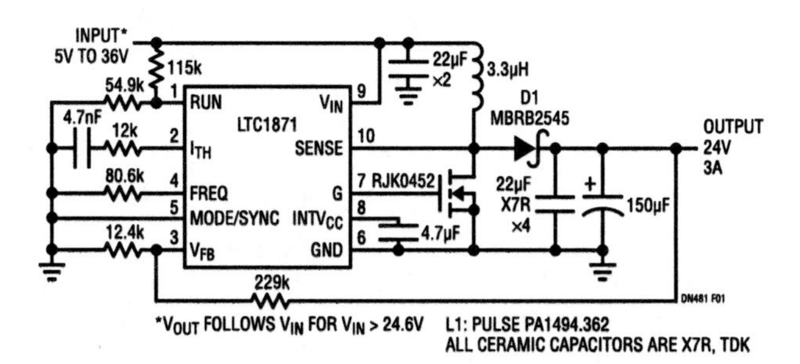

Figure 311.1 • Although This Simple Circuit Is Capable of 3A of Output Current, Beware of Power Dissipation in the Output Diode D1

this current level, diode D1's voltage drop is 0.57V, resulting in 1.6W of power lost as heat. Dissipating 1.6W in an 85°C (or higher) automotive operating environment is not trivial. To keep the circuit cool, heat sinks, cooling fans and multilayer printed circuit boards must be used. This, of course, adds complexity, cost and size to an ostensibly simple boost converter.

A far better solution (featured in a dual output configuration) is shown in Figure 311.2, where a synchronous power MOSFET rectifier replaces the output diode. Under the same conditions, the voltage drop across output synchronous MOSFET Q2 is only 42mV or 7.4% of the voltage drop in the diode D1. The resulting power dissipation of 115mW in Q2 is relatively trivial. Another advantage of using a MOSFET as the output rectifier is the elimination of leakage current, about 10mA in the case of the MBR2545 diode—an additional 240mW of power dissipation in the application of Figure 311.1.

Dual output automotive boost converter

Figure 311.2 illustrates a typical automotive boost application with a 5V to 36V input voltage range. Here, the converter produces a 12V output for generic automotive loads such as entertainment systems, and a 24V output for circuits such as high power audio amplifiers. The two outputs are completely independent and can be controlled separately.

Because the circuit in Figure 311.2 is a boost converter, the output voltage can be regulated only for input voltages that are lower than the output voltage. The output voltage regulation versus input voltage is shown in Figure 311.3. When the input voltage is higher than the preset output voltage, synchronous MOSFETs Q2 and Q4 are turned continuously ON and boost MOSFETs are not switching. This feature allows the converter to be used in applications that require boosting only during load transients such as cold-cranking of a car

Analog Circuit Design: Design Note Collection. http://dx.doi.org/10.1016/B978-0-12-800001-4.00311-2

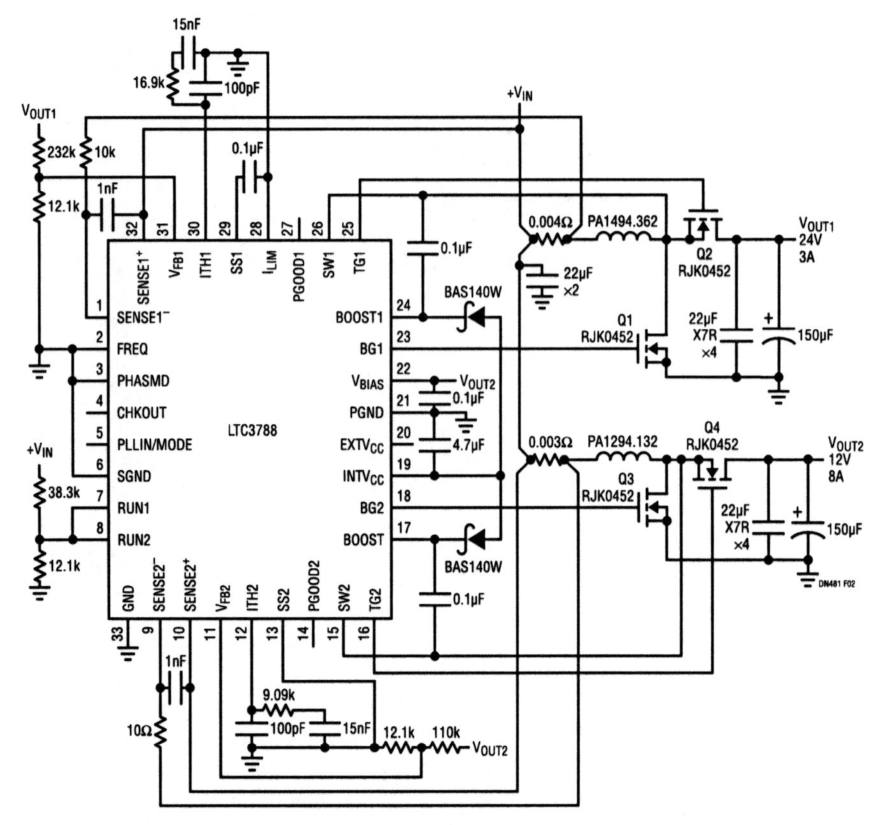

Figure 311.2 • The LTC3788 Converter Is over 95% Efficient Even under Worst-Case Conditions. When $V_{IN} > V_{OUT(SET)}$, Efficiency Approaches 100% as Shown in Figure 311.4

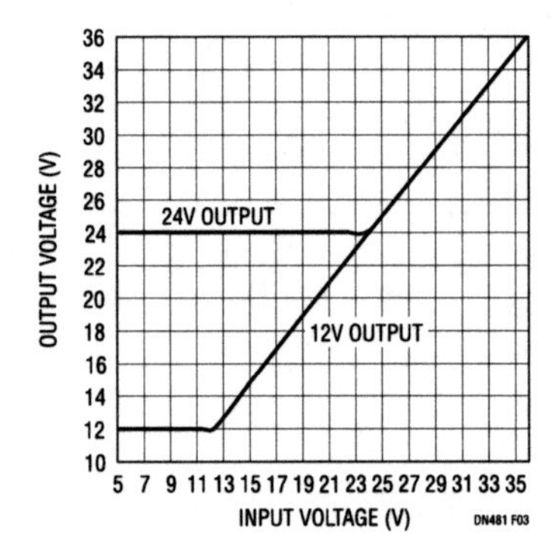

Figure 311.3 • The Output Voltage Follows the Input Voltage When $V_{IN} > V_{OUT(SET)}$

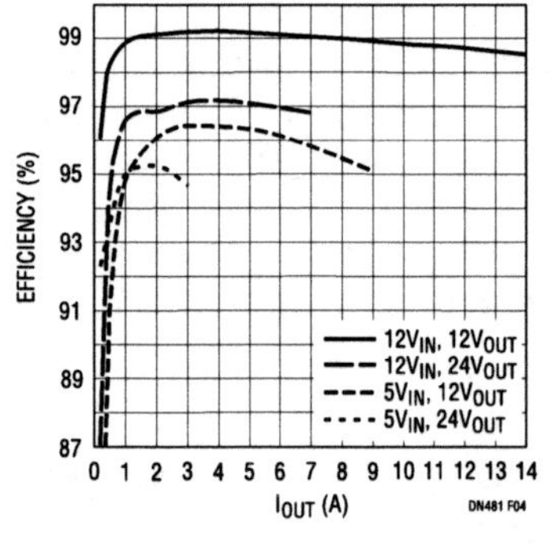

Figure 311.4 • The Converter in Figure 311.2 Peaks at 95% Efficiency When Operating from a 5V Input

engine. In this case, the LTC3788 circuit's input voltage could be as low as 2.5V.

The efficiency of this converter (Figure 311.4) is high enough that it can be built entirely with surface mount components, requiring no heat sinks. A multilayer PCB with large copper area may be sufficient to dissipate the small amount of heat resulting from the MOSFETs' DC resistance, even at high ambient temperatures.

If higher output currents are required, or if lower output ripple voltage is desired, the two LTC3788 channels can be combined for a single current-shared output. Simply connect the two outputs and short the respective FB, ITH, SS and RUN pins. Because the two channels operate out of phase, output ripple currents are greatly reduced—nearly canceling out at 50% duty cycle. Thus, smaller output capacitors can be used with lower output ripple currents and voltages.

Conclusion

The LTC3788 dual synchronous boost controller is a versatile and efficient solution for demanding automotive and industrial applications. By minimizing power losses in the output rectifier, this converter can be designed in a very small footprint and operate safely at elevated ambient temperatures.

High efficiency USB power management system safely charges Li-Ion/Polymer batteries from automotive supplies

312

George H. Barbehenn

Introduction

Automotive power systems are unforgiving electronic environments. Transients to 90V can occur when the nominal voltage range is 10V to 15V (ISO7637), along with battery reversal in some cases. It is fairly straightforward to build automotive electronics around this system, but increasingly end-users want to operate portable electronics, such as GPS systems or music/video players, and to charge their Li-Ion batteries from the automotive battery. To do so requires a compact, robust, efficient and easy-to-design charging system.

Complete USB/battery charging solution for use in large transient environments

Figure 312.1 shows such a design. This complete PowerPath manager and battery charger system seamlessly charges the Li-Ion battery from a wide ranging high voltage or USB source.

In this circuit, the LTC4098 USB power manager/Li-Ion battery charger controls an LT3480 HV step-down regulator. The LTC4098's Bat-Track feature provides a high efficiency,

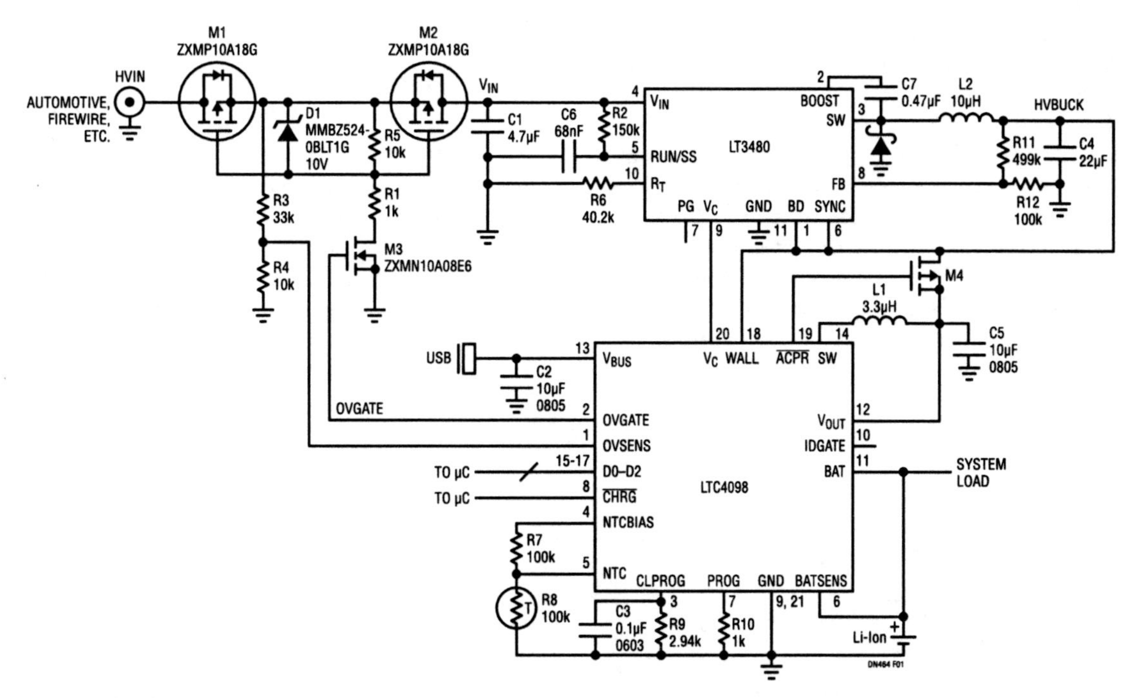

Figure 312.1 • LTC4098 USB Power Manager/Li-Ion Battery Charger Works with an LT3480 HV Buck Regulator to Accept Power from an Automotive Environment or FireWire System. Overvoltage Protection Protects Both ICs and Downstream Circuits

Analog Circuit Design: Design Note Collection. http://dx.doi.org/10.1016/B978-0-12-800001-4.00312-4

low power dissipation battery charger from low and high voltages alike. The Bat-Track feature controls an internal input current-limited switching regulator to regulate V_{OUT} to approximately $V_{BAT}+0.3V$ which maximizes battery charger efficiency, and thus minimizes power dissipation by operating the battery charger with minimal headroom. Furthermore, the Bat-Track feature reduces charge time by allowing a charge current greater than the USB input current limit—the switching regulator behaves like a transformer exchanging output voltage for output current.

The LTC4098 can extend the Bat-Track concept to an auxiliary regulator via the WALL and V_C pins. When sufficient voltage is present on WALL, Bat-Track takes control of the auxiliary regulator's output via the V_C pin, maintaining the regulator's output at $V_{BAT}+0.3V$.

The LTC4098 also includes an overvoltage protection function—important in volatile supply voltage environments. Overvoltage protection shuts off a protection N-channel MOSFET (M2) when the voltage at the OVSENS pin exceeds approximately 6V. The upper limit of voltage protection is limited only by the breakdown voltage of the MOSFET, and by the current flowing into the OVSENS pin.

Overvoltage protection covers the entire battery charger/power manager system

The overvoltage protection function of the LTC4098 can protect any part of the circuit. In Figure 312.1, the protection has been extended to the LT3480 V_{IN} input. The overvoltage shutdown threshold has been set to 24V. This threshold provides ample margin against destructive overvoltage events without interfering with normal operation.

In Figure 312.1, M1 is a P-channel MOSFET that provides reverse voltage protection, whereas M2 is the overvoltage protection MOSFET, and M3 level-shifts the OVGATE output of the LTC4098.

If the HVIN voltage is less than zero, the gate and source voltages of both M1 and M2 are held at ground through R3, R4, and R5, ensuring that they are off. If the HVIN voltage is between 8V and approximately 24V, the gate of M3 is driven high via the LTC4098's OVGATE pin. This turns on M1 and M2 by pulling their gates 7V to 10V below their sources via M3, D1, R1 and R5. With M1 and M2 on current flows from HVIN to V_{IN} and the system operates normally.

If the HVIN input exceeds approximately 24V, the LTC4098 drives the gate of M3 to ground, which allows R5 to reduce the V_{GS} of M1 and M2 to zero, shutting them off and disconnecting HVIN from V_{IN}.

M1, M2 and M3 have a BV_{DSS} of 100V, so that this circuit can tolerate voltages of approximately −30V to 100V. It will operate normally from 8V to approximately 24V. This combination is ideal for the harsh automotive environment, providing a robust, low cost and effective solution for Li-Ion battery charging from an automotive power system.

Finally, setting the OVSENS resistor divider requires some care. For an OVSENS voltage between approximately 2V and 6V, $V_{OVGATE}=1.9 \cdot V_{OVSENS}$. OVSENS is clamped at 6V and the current into (or out of) OVSENS should not exceed 10mA. The chosen resistor divider attenuates HVIN by a factor of 4, so M3 has sufficient gate voltage to turn on when HVIN exceeds approximately 8V. When HVIN = 100V, the current into OVSENS is just 2.25mA—well below the 10mA limit.

As shown in Figure 312.2, V_{IN} is only present when HVIN is in the 8V to 24V region. Figure 312.3 shows a close-up centered on the load dump ramp. The ISO7637 test ramp rises from 13.2V to 90V in 5ms. There is a 220µs turn-off delay—OVGATE going low to the gates of M1 and M2—which results in an overshoot on V_{IN}. The maximum value of this overshoot is 3.5V ($V_{VIN(MAX)} \approx 27.5V$). The magnitude of this overshoot can be calculated for different ramp rates, such that

$$V_{OVERSHOOT} = \Delta V/\Delta t \cdot t_{DELAY}$$

where $\Delta V=(90V-13.6V)$, $\Delta t=5ms$, and $t_{DELAY}=220\mu s$, so, $V_{OVERSHOOT}=3.36V$.

If less delay, and thus less overshoot, is desired, an active turn-off circuit can reduce the delay from OVGATE to the gates of M1 and M2 to a few microseconds.

Conclusion

The LT3480 high voltage step-down regulator and LTC4098 Li-Ion/Polymer battery charger, combined with a few external components, produce a robust high performance Li-Ion charger suitable for portable electronics plugged into an automotive power source and maintain compatibility with USB power. The circuit provides all the functionality that customers expect, along with voltage protection from battery reversal and load dump transients.

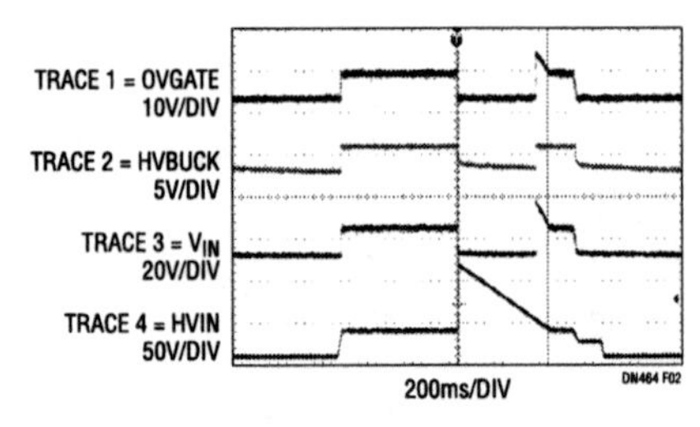

Figure 312.2 • Overvoltage Protection through Input Transients per ISO7637 Standards

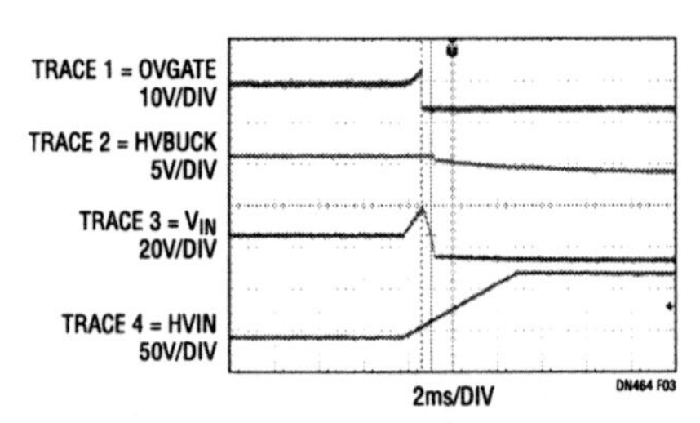

Figure 312.3 • Closeup of Figure 312.2 Waveforms Showing Overshoot on HVIN

Low profile synchronous, 2-phase boost converter produces 200W with 98% efficiency

313

Victor Khasiev

Introduction

Automotive audio amplifiers require a high power boost converter that is both efficient and compact. High efficiency is essential to keep dissipated heat low and avoid bulky and expensive heat sinks. The LT3782A is a 2-phase synchronous PWM controller, making it possible to produce a low profile, high power boost supply that achieves 98% efficiency.

A 24V output boost converter at 8.5A (continuous), 10.5A (peak) from a car battery

Figure 313.1 shows a boost converter that generates 24V from an input voltage range of 8.5V to 18V. Output power is 200W continuous and 250W for short pulse loads, corresponding to 8.5A continuous current and 10.5A pulsed current.

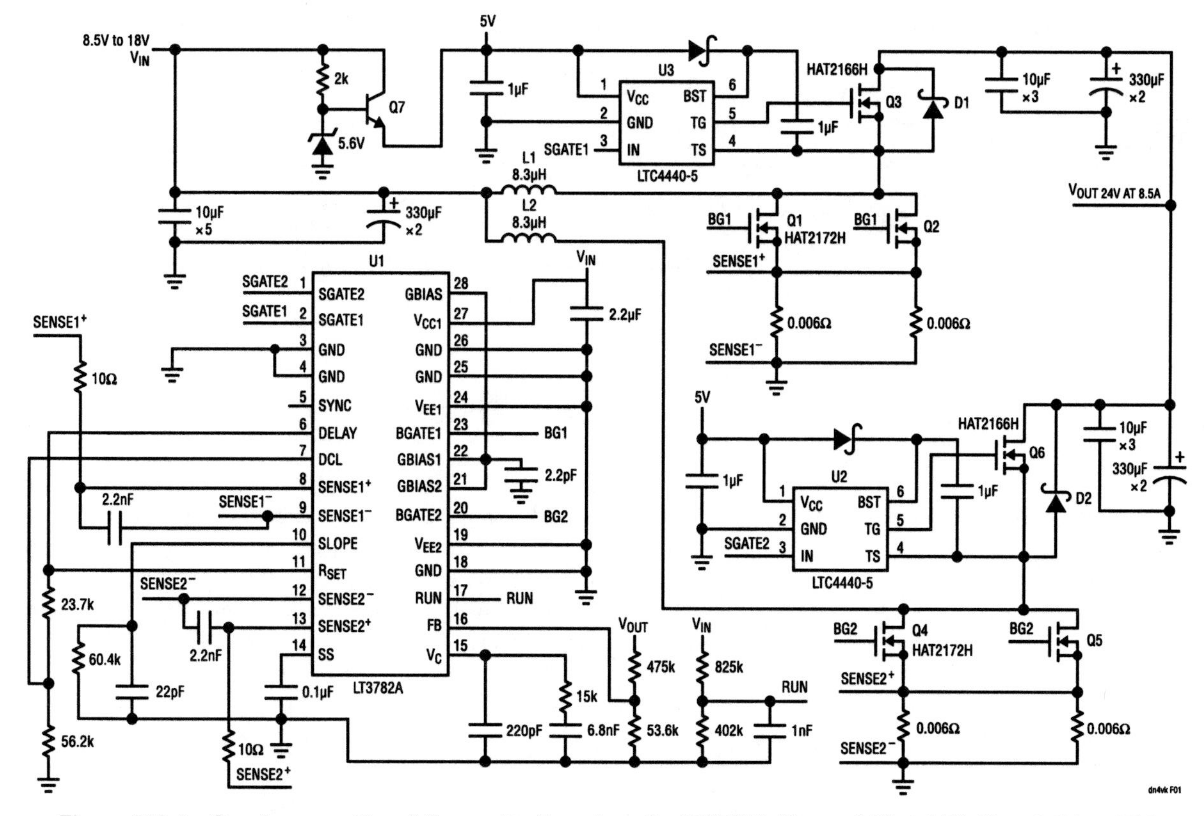

Figure 313.1 • Synchronous Boost Converter Based on the LT3782A (V_{OUT}=24V at 8.5A, V_{IN}=8.5V to 18V)

Analog Circuit Design: Design Note Collection. http://dx.doi.org/10.1016/B978-0-12-800001-4.00313-6

This circuit comprises three major sections. Two are the phase-interleaved power trains, and the third is the control circuit.

Each power train includes an inductor, two switching MOSFETs, a synchronous MOSFET and an output capacitive filter. The output filters are connected together in parallel. Schottky diodes D1 and D2 increase efficiency during the dead time.

The control circuitry can be further divided into three parts: the PWM functionality based on the LTC3782A (U1) and two high side drivers around the LTC4440-5 (U2 and U3). A linear pre regulator based on Q7 generates the required bias voltage for U2 and U3. This approach allows the use of logic-level MOSFETs to minimize gate losses.

The centerpiece of the control circuit is the LT3782A. This 2-phase PWM controller features low side gate signals and corresponding synchronous signals for high side gate control. Control signals are interleaved by running the two stages 180° out-of-phase. The 2-phase approach with accurate channel-to-channel current sharing minimizes electrical and thermal stress on power train components, and reduces EMI. Using the LTC4440-5 as a high side driver allows for high frequency switching.

Performance results

This converter aims for high efficiency and low profile, and it succeeds in reaching both goals with efficiency reaching 98% (Figure 313.2) and a maximum component height of 10.5mm. Output voltage regulation over the full input voltage and output current ranges is better than 2%. Figure 313.3 shows the transient response with a 3A step load.

Basic calculations and component selection

This section shows how to make a preliminary selection of inductors and MOSFETs. Detailed calculations of the losses and converter efficiency evaluation can be found in Robert W. Erickson's *Fundamentals of Power Electronics*, 2nd edition.

For CCM operation, the maximum duty cycle at low line can be found from the following expression:

$$D_{MAX} = \frac{V_{OUT} - V_{IN(MIN)}}{V_{OUT}}$$

Average inductor current and peak current can be calculated as follows:

$$I_{L(AVG)} = \frac{I_{OUT}}{2 \cdot (1 - D_{MAX}) \cdot \eta}; I_{L(PEAK)} = I_{L(AVG)} + \frac{\Delta I}{2}$$

The peak current through the switching MOSFET is equal to $I_{L(PEAK)}$, and the RMS value of the MOSFET current is:

$$I_{SW(RMS)} = I_{L(AVG)} \cdot \sqrt{D_{MAX}} \cdot \sqrt{1 + \frac{1}{3} \cdot \left(\frac{\Delta I}{I_{L(AVG)}}\right)^2}$$

The peak current through the synchronous MOSFET is equal to $I_{L(PEAK)}$, and the RMS MOSFET current value is:

$$I_{SR(RMS)} = I_{L(AVG)} \cdot \sqrt{1 - D_{MAX}} \cdot \sqrt{1 + \frac{1}{3} \cdot \left(\frac{\Delta I}{I_{L(AVG)}}\right)^2}$$

The MOSFETs should be rated to handle the output voltage plus 20% to 30% of headroom.

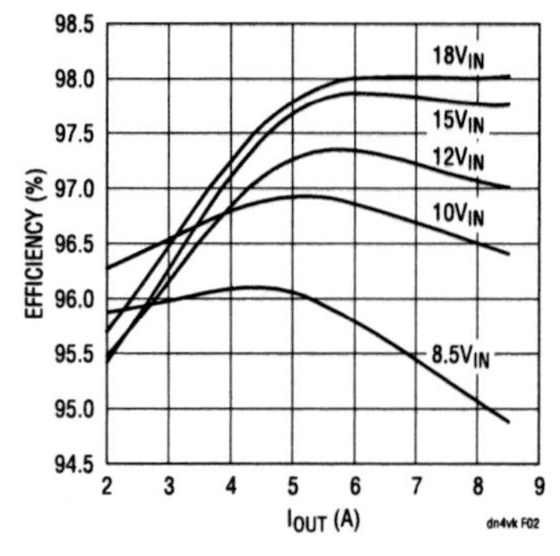

Figure 313.2 • Efficiency vs I_{OUT} (from Circuit in Figure 313.1)

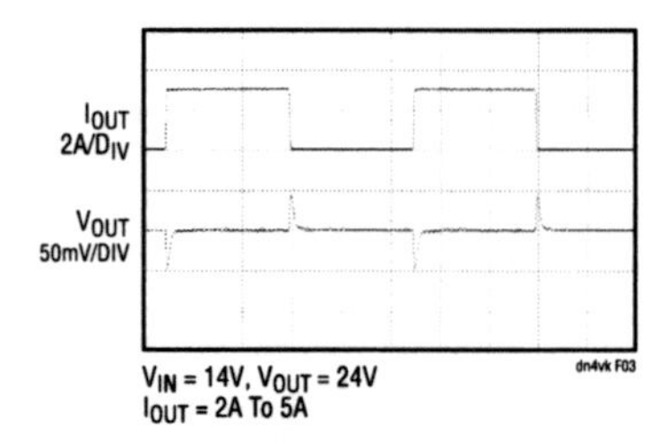

Figure 313.3 • Transient Response of the Circuit in Figure 313.1 for a 3A Loadstep

Conclusion

The LT3782A-based 2-phase synchronous boost converter provides high efficiency, excellent transient response and excellent line and load regulation over a wide input voltage range. High power, high efficiency and a low component profile allow this converter to fit into tight spaces commonly found in automotive environments.

4-phase boost converter delivers 384W with no heat sink

314

Victor Khasiev

Introduction

High power boost converters are becoming increasingly popular among designers in the automotive, industrial and telecom industries. When power levels of 300W or more are required, high efficiency (low power loss) is imperative in the power train components to avoid the need for bulky heat sinks and forced-air cooling. Interleaving power stages (multiphase operation) improves efficiency and reduces ripple voltage and currents in both the input and output capacitors, allowing the use of smaller filter components.

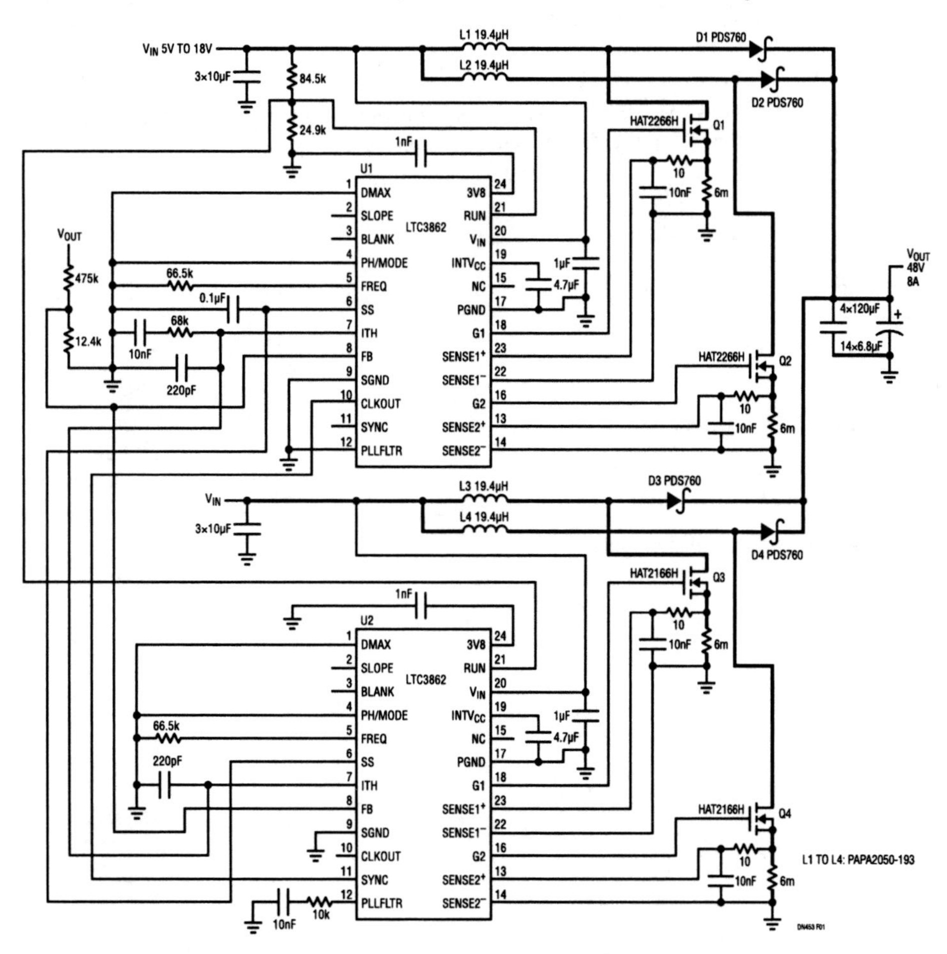

Figure 314.1 • A 4-Phase Boost Converter Based on the LTC3862 Produces 48V at 8A from a 5V to 18V Input

Analog Circuit Design: Design Note Collection. http://dx.doi.org/10.1016/B978-0-12-800001-4.00314-8

384W boost converter

Figure 314.1 shows a 4-phase boost converter using two 2-phase LTC3862 current mode controllers configured in a master-slave configuration (as described in the LTC3862 data sheet). U1 is the master controller; it generates the clock signal that serves to synchronize the two controllers. Synchronization is achieved by connecting U1's CLKOUT pin to U2's SYNC pin and terminating the PLLFLTR pin of U2 with a lowpass filter. Each controller operates with a 180° phase shift between its two channels, and there is a 90° phase shift between the two controllers as defined by the state of U1's PHASEMODE pin, to form an interleaved 4-phase system as shown in Figure 314.2.

The power train includes four inductors L1-L4, four MOSFETs Q1-Q4, four diodes D1-D4, along with the input and output filter capacitors. The ITH, FB, SS and RUN pins of the two controllers are connected together, which forces current share balancing and consistent start-up timing between the phases, and forces both controllers to turn on at the same input voltage.

This converter can deliver 8A at 48V continuously from a 12V to 18V input. It can even support the 48V output with input voltages down to 5V, at a reduced output current. Figure 314.3 shows the converter efficiency above 96% for much of the load range. The transient response to a 3A load step as shown in Figure 314.4 has only a 100mV deviation from nominal.

Up to 12 power stages can be paralleled and clocked out-of-phase for even higher power applications. The LTC3862 has an input voltage range of 4V to 36V and an output voltage that is dependent upon the choice of external components, making it an excellent choice for 12V automotive high power boost converter applications such as audio amplifiers and fuel injection systems.

Conclusion

The LTC3862 2-phase controller is a powerful building block for multiphase boost converter applications that demand high efficiency, low power loss, reduced ripple voltage and currents, and have a small solution size without the need for forced-air cooling or heat sinks.

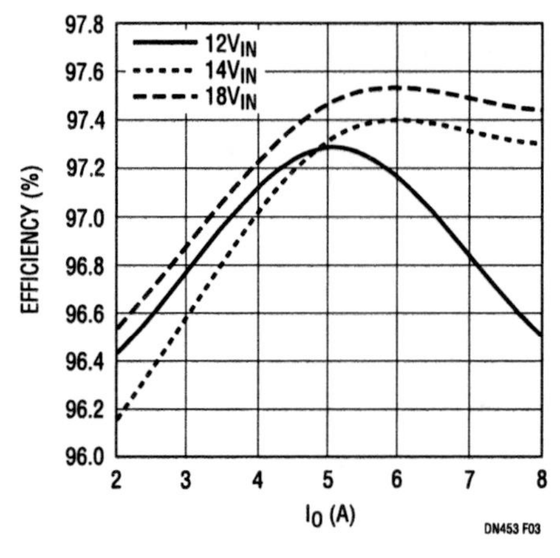

Figure 314.3 • Efficiency vs Output Current Input Voltage for Multiple V_{IN}

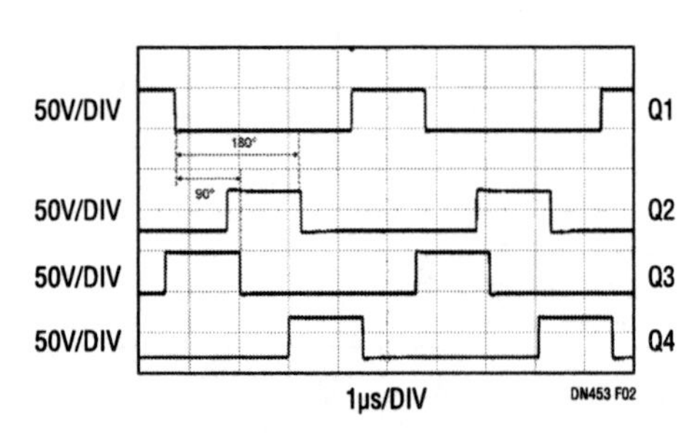

Figure 314.2 • Timing Diagram Showing 4-Phase Operation

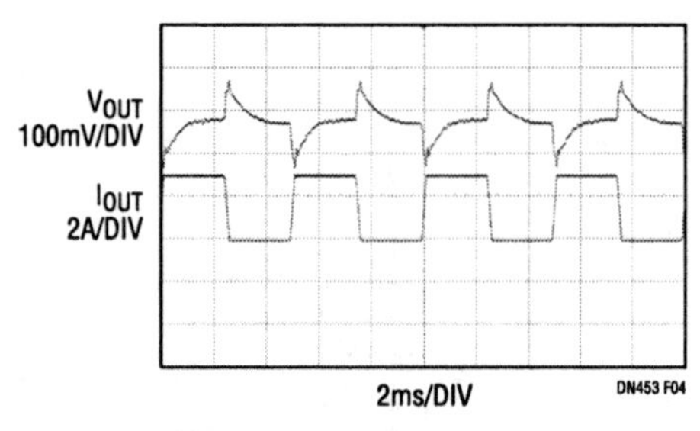

Figure 314.4 • Transient Response for a Current Load Step from 2A to 5A

Power monitor for automotive and telecom applications includes ADC and I²C interface

315

Dilian Reyes

Introduction

The LTC4151 is a high side power monitor that includes a 12-bit ADC for measuring current and voltage, as well as the voltage on an auxiliary input. Data is read through the widely used I^2C interface. An unusual feature in this device is its 7V to 80V operating range, allowing it to cover applications from 12V automotive to 48V telecom.

Automotive power monitoring

Automobile batteries serve more systems than ever before, many of which operate when the battery is not charging, such as information/entertainment systems or devices plugged into the accessory socket.

The high input voltage of the LTC4151 is a good fit for monitoring power in high transient environments such as automotive. Figure 315.1 shows the LTC4151 monitoring up to 16A through a 5mΩ sense resistor at an accessory socket, and feeding data via I^2C to a microcontroller.

A portable GPS unit is used to illustrate the principle. In this case it is powered up and charging its own internal battery, drawing 396mA from the 12.1V supply. The 4.8W of power is relatively low and thus calls for no immediate need for alarm. However, a higher power device such as a built-in DVD player with dual LCD displays or an external 60W thermoelectric cooler plugged into the accessory socket would drain the battery considerably faster than the GPS. The digital information from the LTC4151 high resolution and accurate 12-bit ADC can be interpreted and displayed on an in-dash screen, or used by the host system to shut down the channel to avoid fully draining the battery.

Telecom power monitoring with PoE

One major advantage of the wide range input voltage of the LTC4151 is the ability to monitor higher voltage applications such as those used in telecommunications. The emerging IEEE802.3af Power over Ethernet (PoE) standard has gained much interest in the past few years.

In Figure 315.2 the LTC4151-1 monitors the isolated 48V power supply to the LTC4263 single port Power Sourcing Equipment (PSE) controller. Communication across isolation through optocouplers to a microcontroller is simplified with the LTC4151-1's split bidirectional SDA line to separate data

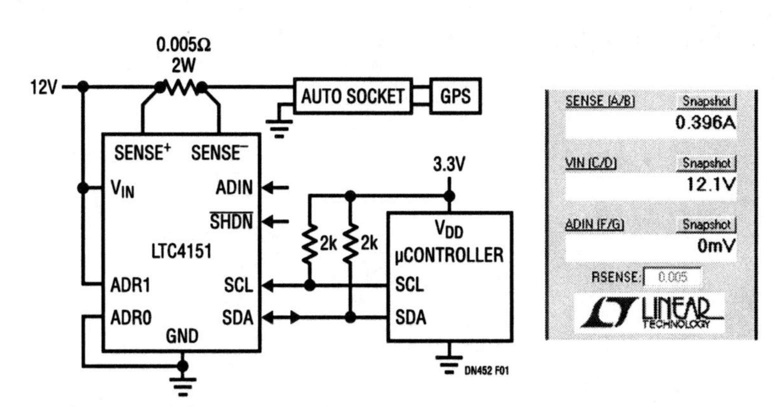

Figure 315.1 • The LTC4151 Monitoring Voltage and Current of an Auto Socket with a GPS Unit

Analog Circuit Design: Design Note Collection. http://dx.doi.org/10.1016/B978-0-12-800001-4.00315-X

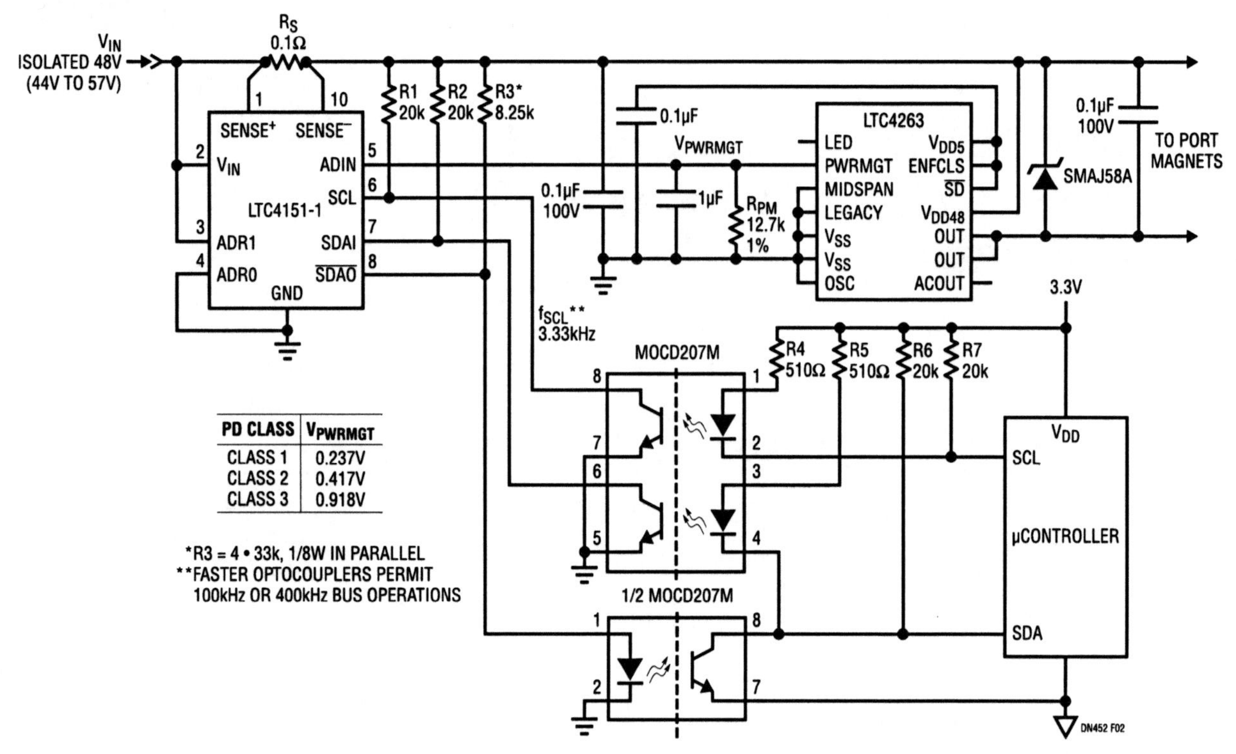

PD CLASS	V_{PWRMGT}
CLASS 1	0.237V
CLASS 2	0.417V
CLASS 3	0.918V

Figure 315.2 • The LTC4151-1 in a PoE Single Port PSE with the LTC4263. I²C Communication to an Isolated Microcontroller

in and data out. Pull-up resistors tie directly to the 48V supply for pins SCL and SDAI, which are internally clamped to 6V, and inverted SDAO is configured to be clamped by an optocoupler diode. With the low speed optocouplers shown, the LTC4263 generating its own 5V supply, and the LTC4151 high voltage protected I²C pins, a separate digital supply is not needed on the PoE side, just the single isolated 48V supply.

Optional power classes (4W, 7W and 15.4W) categorize the power requirements of a Powered Device (PD) on the cable end. The LTC4263 outputs a current to a power management resistor that is proportional to the power class of the plugged-in PD. The LTC4151-1 measures the resulting voltage through its auxiliary ADC input and reports this to the microcontroller, which in turn interprets what power class is present. The microcontroller can then read the current being drawn at the port to determine if the PD is abiding by its power class, and confirm that the supply voltage to the PSE controller meets PoE standards.

The LTC4151 has a configurable address so multiple LTC4151s can operate on the same bus, allowing for a multiport solution with monitoring at each port. This assists with the controller power management function, which utilizes the available power from an optimized supply to the individual ports.

Additional benefits of the LTC4151 are the integrated current sense amplifier, input voltage resistor divider, precise ADC reference voltage and channel select MUX. These improve accuracy versus variances of external components and also save on costs of discrete parts.

Conclusion

The LTC4151 is an easy to use but feature-rich power monitoring device suitable for a wide variety of automotive, telecom and industrial applications. It provides accurate voltage and current monitoring of a positive supply rail from 7V to 80V via a simple I²C interface.

Direct efficient DC/DC conversion of 100V inputs for telecom/automotive supplies

316

Greg Dittmer

Introduction

Automotive, telecom and industrial systems have harsh, unforgiving environments that demand robust electronic systems. In telecom systems the input rail can vary from 36V to 72V, with transients as high as 100V. In automotive systems the DC battery voltage may be 12V, 24V or 42V with load dump conditions causing transients up to 60V or more. The LTC3810 is a current mode synchronous switching regulator controller that can directly step down input voltages up to 100V, making it ideal for these harsh environments. The ability to step down the high input voltage directly allows a simple single inductor topology, resulting in a compact high performance power supply—in contrast to the low side drive topologies that require bulky, expensive transformers.

Feature-rich controller

The LTC3810 drives two external N-channel MOSFETs using a synchronizable constant on-time, valley current mode architecture. A high bandwidth error amplifier provides fast line and load transient response. Strong 1Ω gate drivers minimize switching losses—often the dominant loss component in high voltage supplies—even when multiple MOSFETs are used for high current applications. The LTC3810 includes an internal linear regulator controller to generate a 10V IC/driver supply

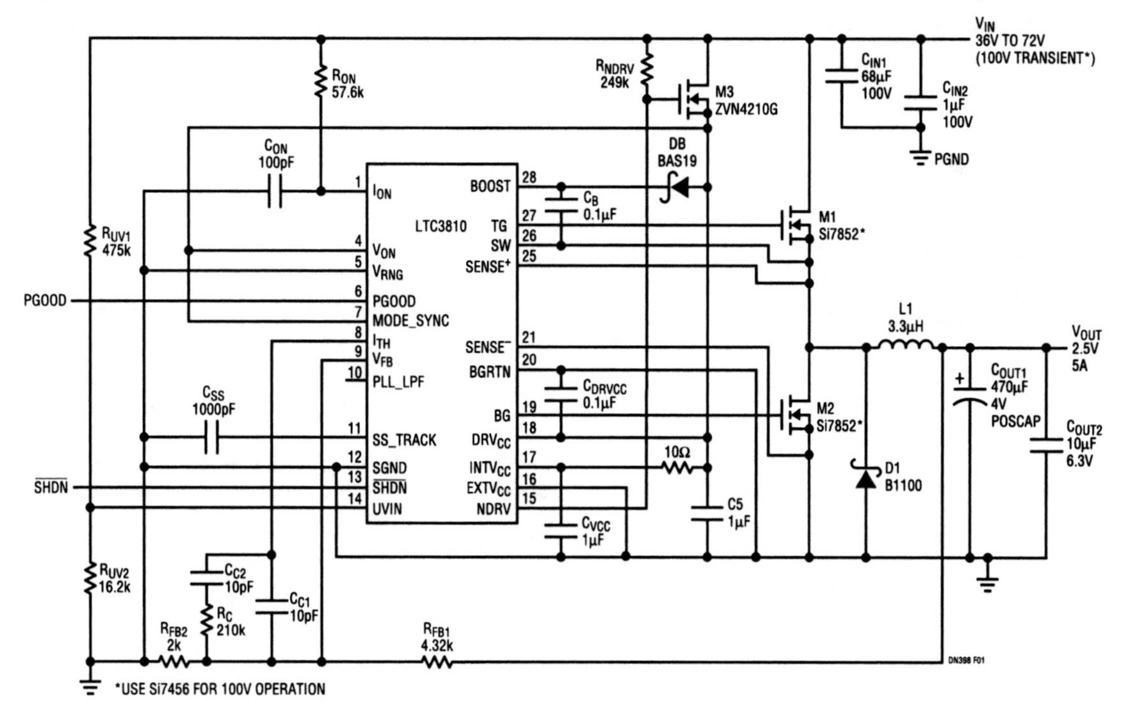

Figure 316.1 • Compact 36V–72V to 2.5V/6A Synchronous Step-Down Converter

Analog Circuit Design: Design Note Collection. http://dx.doi.org/10.1016/B978-0-12-800001-4.00316-1

from the high voltage input supply with a single external SOT23 MOSFET. When the output voltage is above 6.7V, the 10V supply can be generated from the output, instead of the input, for higher efficiency. Other features include:

- Programmable cycle-by-cycle current limit, with tight tolerances, provides control of the inductor current during a short-circuit condition. No R_{SENSE} current sensing utilizes the voltage drop across the synchronous MOSFET to eliminate the need for a current sense resistor.
- Low minimum on-time (<100ns) for low duty cycle applications. The on-time is programmable with an external resistor and is compensated for changes in input voltage to keep switching frequency relatively constant over a wide input supply range.
- Precise 0.8V ±0.5% reference over the operating temperature range of 0°C to 85°C.
- Phase-locked loop for external clock synchronization, selectable pulse-skip mode operation, tracking, programmable undervoltage lockout and power good output voltage monitor.
- 28-pin SSOP package with high voltage pin spacing.

High efficiency 36V–72V to 2.5V/6A power supply

The circuit shown in Figure 316.1 provides direct step-down conversion of a typical 48V telecom input rail to 2.5V at 5A. With the 100V maximum DC rating of the LTC3810 and 80V for the MOSFETs, the circuit can handle input voltages of up to 80V without requiring protection devices (up to 100V if appropriate MOSFETs are used). This circuit demonstrates how the low minimum on-time of the LTC3810 enables high step-down ratio applications: 2.5V output from a 72V input at 250kHz is a 140ns on-time.

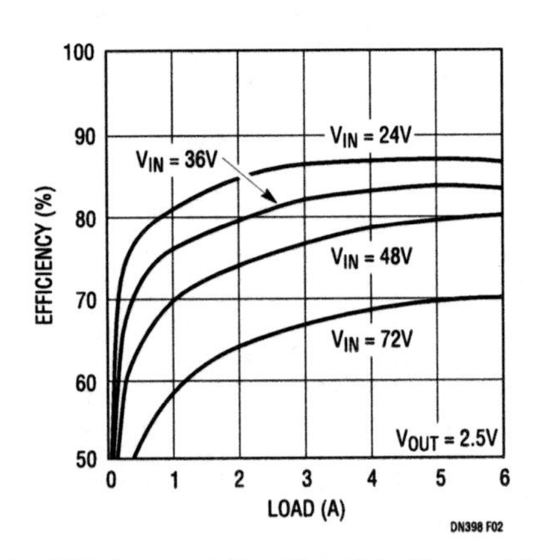

Figure 316.2 • Efficiency of the Circuit in Figure 316.1

The frequency is set to 250kHz with the R_{ON} resistor to optimize efficiency while minimizing output ripple. Figure 316.2 shows mid-range efficiencies of 80% to 84% at 36V input and 65% to 70% at 72V input. Type II compensation is used to set the loop bandwidth to about 75kHz, which provides a 20μs response time to load transients (see Figure 316.3).

The V_{RNG} pin is set to 0V to set the current limit to about 8A (3A after foldback) during a short-circuit condition (see Figure 316.4). The resistor divider (R_{UV1}, R_{UV2}) sets the input supply undervoltage lockout to 24V, keeping the LTC3810 shut down until the V_{IN} >24V.

The LTC3810's internal linear regulator controller generates the 10V IC/driver supply ($INTV_{CC}$, DRV_{CC} pins) from the input supply with a single external MOSFET, M3. For continuous operation the power rating of M3 must be at least $(72V - 10V) \cdot (0.02A) = 1.2W$. If another low voltage supply (between 6.2V and 14V) capable of supplying the ~20mA IC/driver current is available, this supply could be connected to $INTV_{CC}/DRV_{CC}$ pins to increase efficiency by up to 10% at loads above 1A.

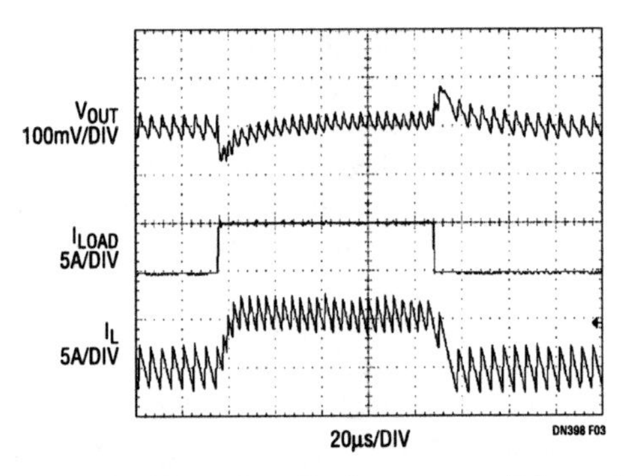

Figure 316.3 • Load Transient Performance of Figure 316.1 Circuit Shows 20μs Response Time to a 5A Load Step

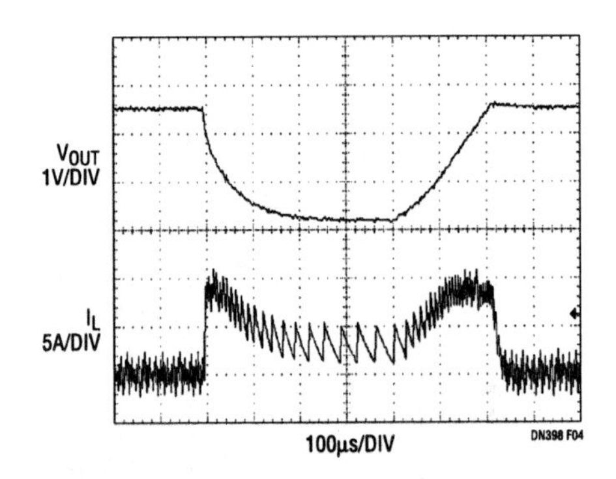

Figure 316.4 • Short-Circuit Condition in Figure 316.1 Circuit Shows Tight Control of Inductor Current and Foldback

Monolithic step-down regulator withstands the rigors of automotive environments and consumes only 100μA of quiescent current

317

Rich Philpott

Introduction

Automobile electronic systems place high demands on today's DC/DC converters. They must be able to precisely regulate an output voltage in the face of wide temperature and input voltage ranges—including load dump transients in excess of 60V and cold crank voltage drops to 4V. The converter must also be able to minimize battery drain in always-on systems by maintaining high efficiency over a broad load current range. Similar demands are made by many 48V nonisolated telecom applications, 40V FireWire peripherals and battery-powered applications with auto plug adaptors. The LT3437's best in class performance meets all of these requirements in a small thermally enhanced 3mm × 3mm DFN package.

Features of the LT3437

The LT3437 is a 200kHz fixed frequency, 500mA monolithic buck switching regulator. Its 3.3V to 80V input voltage range makes the LT3437 ideal for harsh automotive environments. Micropower bias current and Burst Mode operation help to maintain high efficiency over the entire load range and result in a no load quiescent current of only 100μA for the circuit in Figure 317.1. The LT3437 has an undervoltage lockout and a shutdown pin with an accurate threshold for a <1μA quiescent current shutdown mode.

External synchronization can be implemented by driving the SYNC pin with a logic-level input. The SYNC pin also doubles as Burst Mode defeat for applications where lower output ripple is desired over light load efficiency. A single capacitor provides soft-start capability which limits inrush current and output voltage overshoot during start-up and recovery from brown-out situations. The LT3437 is available in either a low profile 3mm × 3mm 10-pin DFN or 16-pin TSSOP package, both with an exposed pad leadframe for low thermal resistance.

Brutal input transients

Figure 317.2 shows the LT3437's reaction to the severe input transients that are possible in an automotive environment. Here, the input voltage rises from a nominal 12V to 80V in a 100ms load dump pulse, then drops to 4V in a 150ms cold crank pulse. The 200kHz fixed frequency and current mode topology of the LT3437 allow it to take it all in its stride—response to the input transients are less than 1% of the regulated voltage.

The fuzziness seen on the output voltage is due to the ESR of the output capacitor and the change in inductor current

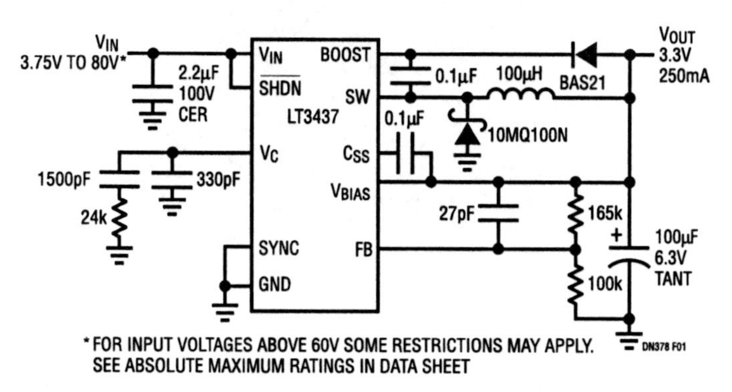

Figure 317.1 • 14V to 3.3V Step-Down Converter with 100μA No Load Quiescent Current

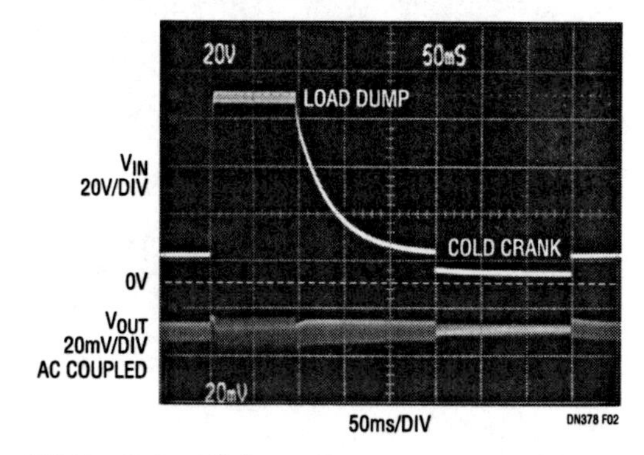

Figure 317.2 • Output Voltage Response to Load Dump and Cold Crank Input Transients

Analog Circuit Design: Design Note Collection. http://dx.doi.org/10.1016/B978-0-12-800001-4.00317-3

ripple as the input voltage transitions between levels. This ripple can be eliminated by changing the output capacitor type from tantalum to ceramic.

Low quiescent currents

Many of today's automotive applications are migrating to always-on systems which require low average quiescent current to prolong battery life. Loads are switched off or reduced during low demand periods, then activated for short periods. Quiescent current for the application circuit in Figure 317.1 is less than 1μA in shutdown mode, and a mere 100μA (Figure 317.3) for an input voltage of 12V under a no load condition. The LT3437 provides excellent step response from a no-load to load situation as shown in Figure 317.4. Automatic Burst Mode operation ensures efficiency over the entire load range as seen in Figure 317.5. Burst Mode operation can be defeated if lower ripple is desired over light load efficiency by pulling the SYNC pin high or driving it with an external clock.

Soft-start capability

The rising slope of the output voltage is determined by the output voltage and a single capacitor. Initially, when the output voltage is close to zero, the slope of the output is determined by the soft-start capacitor. As the output voltage increases, the output slope is increased to full bandwidth near the regulated voltage. Since the circuit is always active, inrush current and voltage overshoot are minimized for start-up and recovery from overload (brown-out) conditions. Figure 317.6 illustrates the effect of several soft-start capacitor values.

Conclusion

The LT3437's wide input range, low quiescent current, robust design and small thermally enhanced packages make it an ideal solution for all automotive and wide input voltage, low quiescent current applications.

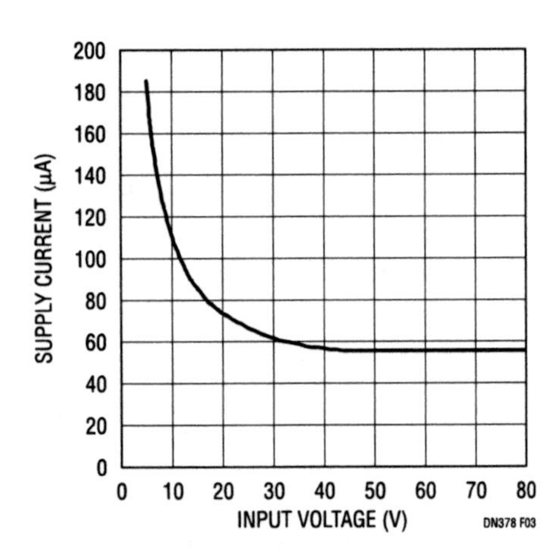

Figure 317.3 • Supply Current vs Input Voltage for Circuit in Figure 317.1

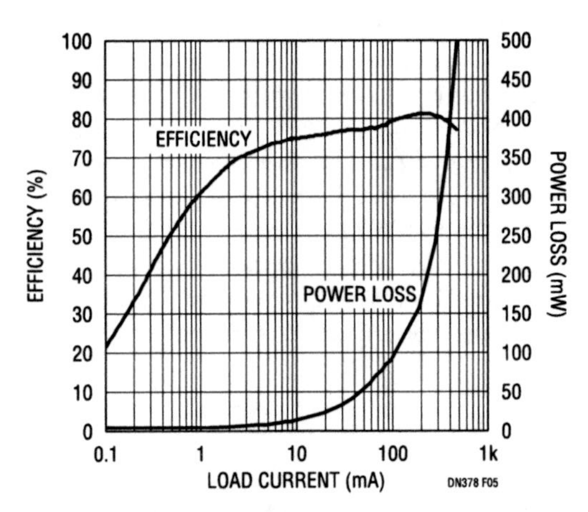

Figure 317.5 • Efficiency and Power Loss vs Load Current for the Circuit in Figure 317.1

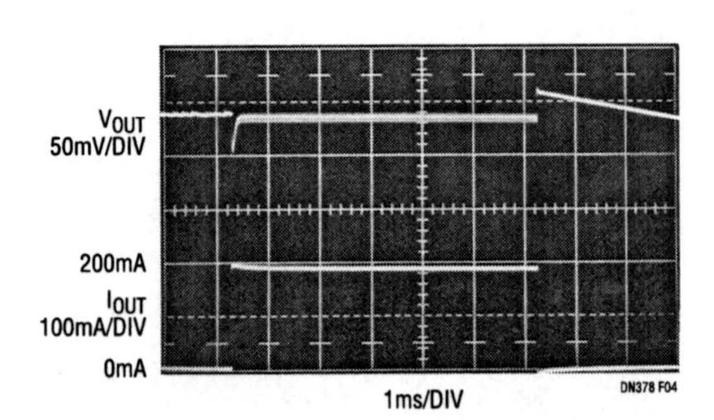

Figure 317.4 • Output Voltage Response for 0mA to 200mA Load Step

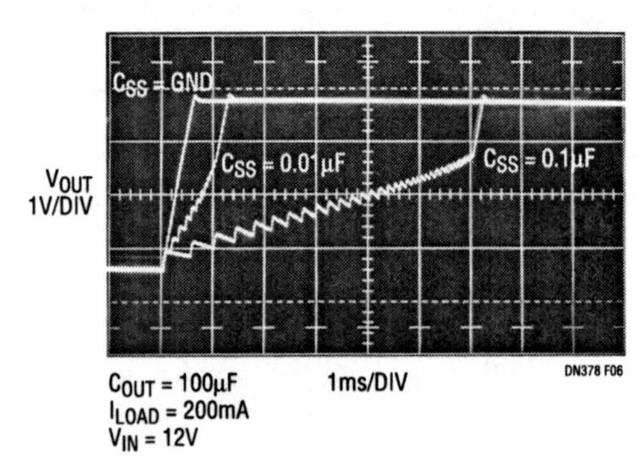

Figure 317.6 • Output Voltage Soft-Start

Monitor and protect automotive systems with integrated current sensing

318

Jon Munson

Introduction

An automobile is an unforgiving environment for integrated circuits, where under-the-hood operating temperatures run from −40°C to 125°C and large transient excursions on the battery voltage bus are expected. In the past, electronics were part of the well-protected and centralized Engine Control Unit (ECU), but the trend is toward more distributed electronics. Electrically driven accessories and fault-protection monitoring functions are leaving the protective umbrella of the ECU and migrating directly into vehicle subsystems.

For example, many functions formerly driven by the engine—via belt and pulley or hydraulics—are now electrically driven (motorized), such as water pumps, steering mechanisms, brake actuators and various body controls. These functions can become a safety risk if they are not continually monitored for operational readiness and/or have a backup mode of operation. In either case, real-time monitoring becomes necessary and generally involves accurately measuring the current draw of each subsystem.

Simple current monitoring solutions

The LT6100 and LTC6101 are high side current-sense amplifiers that have been developed specifically to address the automotive designers' needs. These parts require a minimum number of support components to operate in the harsh automotive battery-bus environment.

Figure 318.1 shows a basic high side current monitor using the LTC6101. The selection of R_{IN} and R_{OUT} establishes the desired gain of this circuit, powered directly from the battery bus. The current output of the LTC6101 allows it to be located remotely to R_{OUT}. Thus, the amplifier can be placed directly at the shunt, while R_{OUT} is placed near the monitoring electronics without ground drop errors.

This circuit has a fast 1μs response time that makes it ideal for providing MOSFET load switch protection. The switch element may be the high side type connected between the sense resistor and the load, a low side type between the load and ground or an H-bridge. The circuit is programmable to produce up to 1mA of full-scale output current into R_{OUT}, yet draws a mere 250μA supply current when the load is off.

Figure 318.2 shows the LT6100 used as a combination current sensor and fuse monitor. This part includes on-chip

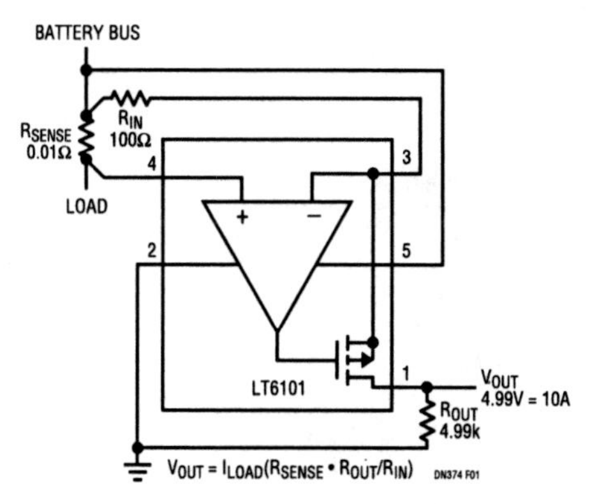

Figure 318.1 • Simple LTC6101 High Side Current Sense Amplifier

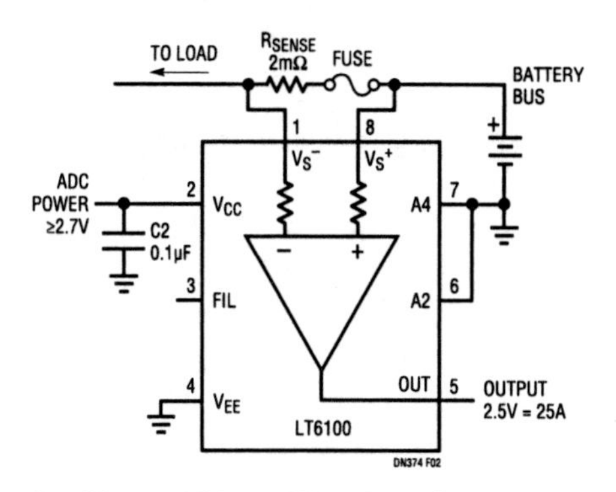

Figure 318.2 • Simple LT6100 High Side Current Sense Amplifier and Fuse Monitor

Analog Circuit Design: Design Note Collection. http://dx.doi.org/10.1016/B978-0-12-800001-4.00318-5

output buffering and was designed to operate with the low supply voltage (≥2.7V), typical of vehicle data acquisition systems, while the sense inputs monitor signals at the higher battery bus potential. The LT6100 inputs are tolerant of large input differentials, thus allowing the blown-fuse operating condition (this would be detected by an output full-scale indication). The LT6100 can also be powered down while maintaining high impedance sense inputs, drawing less than 1µA max from the battery bus.

Solving the H-bridge problem

Many of the newer electric drive functions, such as steering assist, are bidirectional in nature. These functions are generally driven by H-bridge MOSFET arrays using pulse-width modulation (PWM) methods to vary the commanded torque. In these systems, there are two main purposes for current monitoring. One is to monitor the current in the load, to track its performance against the desired command (i.e., closed-loop servo law), and another is for fault detection and protection features.

A common monitoring approach in these systems is to amplify the voltage on a "flying" sense resistor, as shown in Figure 318.3. Unfortunately, several potentially hazardous fault scenarios go undetected, such as a simple short to ground at a motor terminal. Another complication is the noise introduced by the PWM activity. While the PWM noise may be filtered for purposes of the servo law, information useful for protection becomes obscured. The best solution is to simply provide two circuits that individually protect each half-bridge and report the bidirectional load current. In some cases, a smart MOSFET bridge driver may already include sense resistors and offer the protection features needed. In these situations, the best solution is the one that derives the load information with the least additional circuitry.

Figure 318.4 shows a differential load measurement for an ADC using twin unidirectional sense measurements. Each LTC6101 performs high side sensing that rapidly responds to fault conditions, including load shorts and MOSFET failures. Hardware local to the switch module (not shown in the diagram) can provide the protection logic and furnish a status flag to the control system. The two LTC6101 outputs taken differentially produce a bidirectional load measurement for the control servo. The ground-referenced signals are compatible with most ΔΣ ADCs. The ΔΣ ADC circuit also provides a "free" integration function that removes PWM content from the measurement. This scheme also eliminates the need for analog-to-digital conversions at the rate needed to support switch protection, thus reducing cost and complexity.

Conclusion

The LT6100 and LTC6101 high side current-sense amplifiers simplify designs in the automotive environment. High transient voltage tolerance (105V for the LTC6101HV) and ground-referenced outputs make it possible to improve robustness and substantially reduce the parts count over traditional solutions.

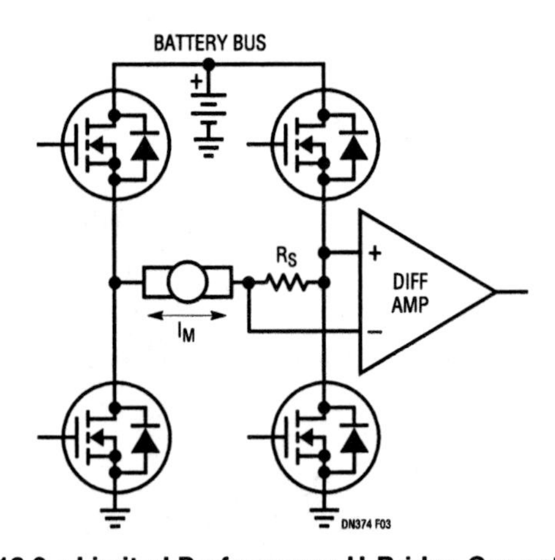

Figure 318.3 • Limited Performance H-Bridge Current Monitor

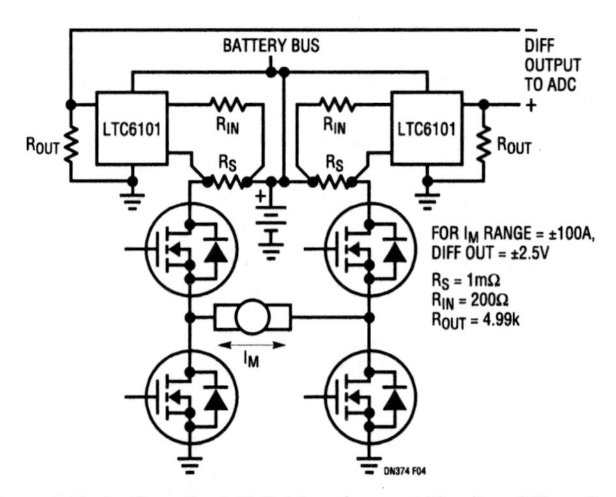

Figure 318.4 • Practical H-Bridge Current Monitor Offers Fault Detection and Bidirectional Load Information

Section 23

Video Design Solutions

High resolution video solutions using single 5V power

319

Jon Munson

Introduction

Video cable driver amplifier output stages traditionally require a supply voltage of at least 6V in order to provide the required output swing. This requirement is usually met with 5V supplies by adding a boost regulator or a small local negative rail, say via the popular LT1983-3. Such additional circuitry is unnecessary in typical $1V_{P-P}$ video connections, such as HD component video, if the cable driver amplifiers simply offer near rail-to-rail output capability when powered from 5V.

Standard definition and SVGA (800 × 600 pixel) low voltage devices have been available from Linear Technology for some time (see Chapter 322), but a number of recent device developments have made it possible to produce high resolution video devices that operate on a single 5V power supply. Some parts that fit this mold include the LT6556, a UXGA-resolution (1600 × 1200 pixel) RGB 2:1 input-port buffered multiplexer (MUX); the LT6557 and LT6558 UXGA fixed gain triple amplifiers that include on-chip biasing to minimize external part count; and the LT6559 triple

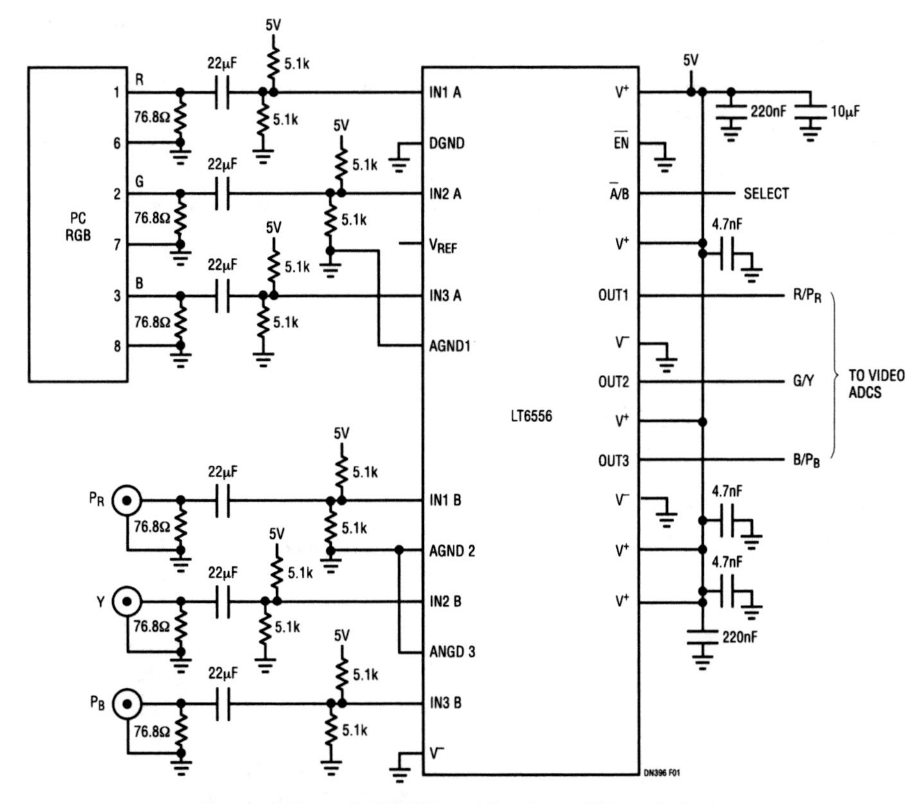

Figure 319.1 • LT6556 Provides Input Video Selector and ADC Driver for Multimedia Display System

Analog Circuit Design: Design Note Collection. http://dx.doi.org/10.1016/B978-0-12-800001-4.00319-7

amplifier that provides flexible, cost-effective solutions in SXGA (1280 × 1024 pixel) products.

High resolution video input-port multiplexer

High performance multimedia video display systems usually include a multiple-input feature to select between a VESA-compliant D-type PC connection and consumer component video that uses RCA jacks. The incoming video signal is at most $1V_{P-P}$ nominal (Y-channel, $1.5V_{P-P}$ worst-case when AC-coupled) and the required gain is unity for digitizing by an analog-to-digital converter set (ADC) or other signal routing. This input selection function is readily implemented with the LT6556 on 5V as shown in Figure 319.1, supporting all video resolutions including UXGA by virtue of its 750MHz bandwidth and 6.5ns settling time. The part is available in either SSOP-24 or QFN-24 packaging and includes layout-friendly flow through pinouts. For the AC-coupling shown, the outputs swing approximately ±0.7V about the mid-supply level, within an available range of $2.6V_{P-P}$. Though not explicitly shown in Figure 319.1, coupling to the ADC inputs usually involves series resistances to reduce capacitive loading of the amplifiers to preserve the smoothest frequency response and optimal settling.

High resolution single-supply cable driver

The LT6557 is a triple video amplifier specifically engineered to provide UXGA level performance on a single 5V power supply. A quasi rail-to-rail output stage and an almost slew-unlimited 400MHz large signal bandwidth make this the part of choice for the most performance-critical applications.

The LT6557 has internal gain-setting resistors to establish a nominal gain of two and incorporates a single-resistor-programmable input biasing system to eliminate the usual input divider resistors used in single supply applications. As seen in Figure 319.2, the entire cable driver function is largely reduced to the IC and blocking capacitors. The internal input biasing may be defeated for DC-coupled applications, such as with direct digital-to-analog converter (DAC) output applications where a software controlled offset is introduced to set the signal dynamics. A unity gain version, the LT6558, is also available.

Economical SXGA/HD cable driver

In cost sensitive applications like consumer video-playback equipment, the LT6559 provides excellent bang-for-the-buck in a tiny QFN-16 (3mm × 3mm) package. As a basic triple current-feedback op amp (CFA) with individual channel enables, the LT6559 offers great flexibility in forming various multiplexer, cable driver, and ADC driver functions at low cost. Even though the LT6559 is not a true rail-to-rail output device, there remains approximately $3V_{P-P}$ of available output swing on 5V due to its high performance output stage design. Figure 319.3 shows a typical AC-coupled application as an economical HD or SXGA-grade triple cable driver (one channel shown for brevity). As a general purpose CFA, the feedback resistor value (301Ω) optimizes the frequency response. This circuit is ideal as an output buffer/driver for following passive reconstruction filters such as for the increasingly popular 1080p HD format (i.e. 60MHz lowpass).

Conclusion

As system designers continue to reduce the number of supply voltages used within their products, pressure to maximize analog performance on available 5V logic supplies has led to the need for viable low voltage high performance video solutions. For high resolution applications, Linear Technology offers the LT6556 buffered MUX, the LT6557/LT6558 AC-coupled amps, and the economical LT6559 triple CFA—all well suited to operate in the 5V environment.

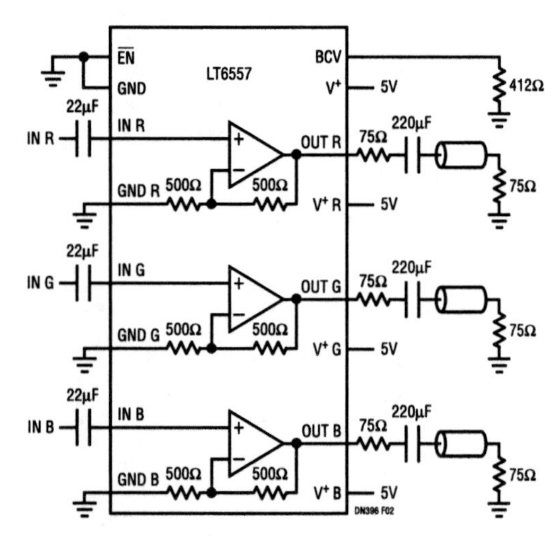

Figure 319.2 • LT6557 Provides Low Part Count UXGA-Resolution Cable Driver on 5V Supply

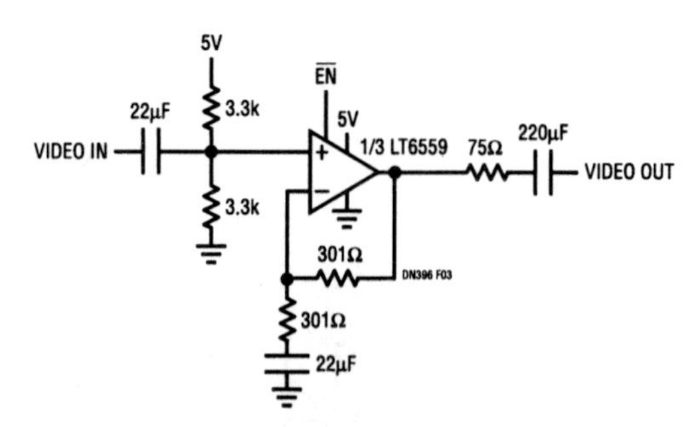

Figure 319.3 • HD Video Cable Driver Using Economical LT6559 (Depicting One Channel of Three)

Pass HDMI compliance tests with ease

320

Bill Martin

Introduction

The high definition multimedia interface (HDMI) is fast becoming the de facto standard for passing digital audio and video data in home entertainment systems. This standard includes an I²C type bus called a display data channel (DDC) that is used to pass extended digital interface data (EDID) from the sink device (such as a digital TV) to the source device (such as a digital A/V receiver). EDID includes vital information on the digital data formats that the sink device can accept. The HDMI specification requires that devices have less than 50pF of input capacitance on their DDC bus lines, which can be very difficult to meet. The LTC4300A's capacitance buffering feature allows devices to pass the HDMI DDC input capacitance compliance test with ease.

AQ1

LTC4300A-1 bus buffer

The LTC4300A-1 is a 2-wire bus buffer that includes capacitance buffering between input and output, an enable pin for input-to-output connection control through hardware and rise time accelerators to provide for swift bus transitions through the bus logic thresholds. Due to the sub-10pF input capacitance of the LTC4300A-1, the capacitance buffering right at the HDMI connector interface allows the component to easily pass the DDC input capacitance test limit of 50pF even if the internal capacitance of the channel is substantially higher. The HDMI cable connector must see the OUT side of the LTC4300A-1 for the input capacitance compliance testing to be accurate (see Figure 320.1).

In HDMI, the sink pulls the hot plug detect (HPD) signal high to tell the source that it is ready to accept commands through the DDC. This signal can be controlled by the READY pin of the LTC4300A-1 to prevent the possibility of erroneous attempts by the source to contact the sink before the sink is ready to return EDID. The READY pin only goes high after 5V is applied and the LTC4300A-1 ENABLE pin is pulled high by the HDMI receiver IC, a controller in the sink, or the 5V line itself.

The rise time accelerators in the LTC4300A-1 compress transition times on rising signal edges, minimizing the chance of interrupted data transfer due to noise and allowing the DDC to meet I²C timing requirements. That is, HDMI specification allows for 800pF of load; enough that the DDC

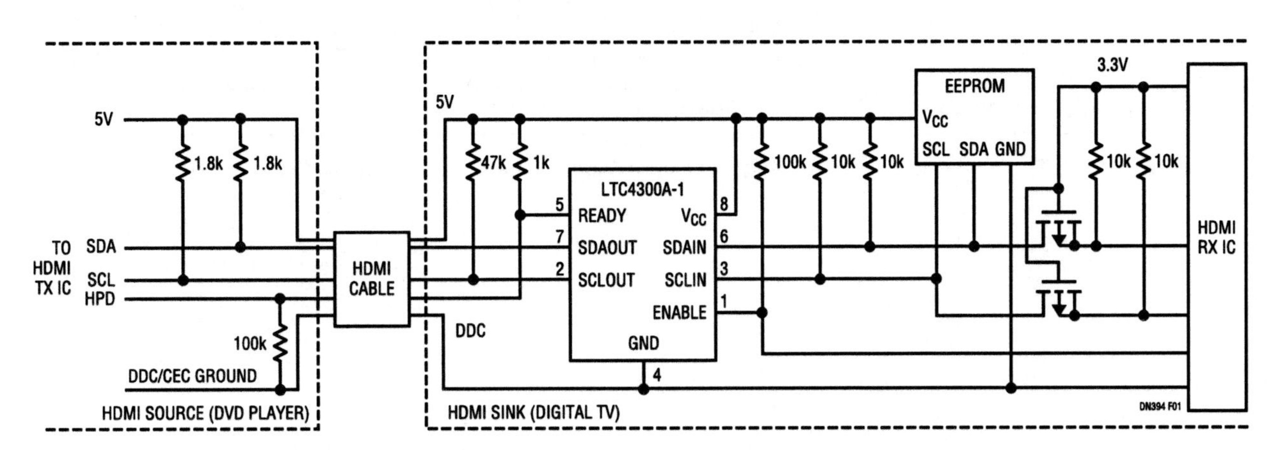

Figure 320.1 • LTC4300A-1 in HDMI Capacitance Buffering Application

Analog Circuit Design: Design Note Collection. http://dx.doi.org/10.1016/B978-0-12-800001-4.00320-3

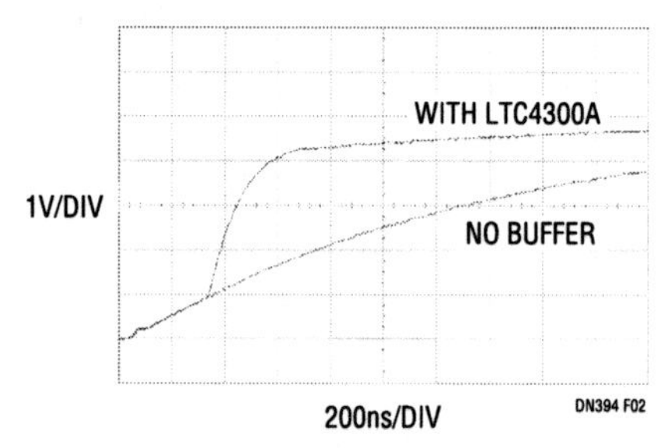

Figure 320.2 • The LTC4300A Provides Capacitance Buffering for the DDC While Improving Bus Timing

cannot be guaranteed to meet the required 100kHz I^2C 1μs rise time specification with the allowed DDC pullup resistance values. Rise time accelerators allow this timing requirement to be met even with capacitances well above 800pF.

If the 5V supply for the DDC is changed to 3.3V in future versions of the HDMI specification, the LTC4300A-1 can remain in the design as is, for it can work with supply voltages from 2.7V to 5.5V. The LTC4300A-1 will transparently support new and legacy equipment in the case of an HDMI specification change.

Figure 320.2 shows how the LTC4300A-1 provides capacitance buffering at the cable interface while improving the rise time of the heavily loaded 5V bus (750pF in this example). Without the LTC4300A-1, the signal is failing the I^2C 1μs rise time specification (measured between $0.3V_{CC}$ and $0.7V_{CC}$). In the DDC capacitance test, only the capacitance of the connector, the traces to the LTC4300A-1 and the less than 10pF input capacitance of the LTC4300A-1 will be measured.

LTC4300A-3 level shifting buffer

The LTC4300A-3 level shifting I^2C buffer is also a good solution for this application. Figure 320.3 shows the LTC4300A-3 being used for capacitance buffering and 5V to 3.3V level shifting. In this application, the EEPROM is powered by a backup 3.3V supply that is available when the component is turned off. The EDID in the EEPROM should be available for reading even when a component's power is off. The level shifting between the 5V and 3.3V bus segments is accomplished by having separate supply pins for the two segments.

Having two supply pins also allows the LTC4300A-3 to provide rising edge acceleration on the 3.3V and 5V bus segments. This is a useful feature for the bus segment that is inside the component, but cable capacitance values of well over the 700pF HDMI spec will be encountered in the up to 30m HDMI cables that are being used for home theaters, so rise time acceleration is a most valuable feature on the cable side bus segment.

Although the applications shown are for HDMI receive channels, the LTC4300A-1 and LTC4300A-3 can also be used in HDMI transmit channels with equal success.

Conclusion

The LTC4300A-1 and LTC4300A-3 solve the DDC capacitance testing problem in HDMI while also substantially improving the timing performance of the bus and providing a high level of ESD protection.

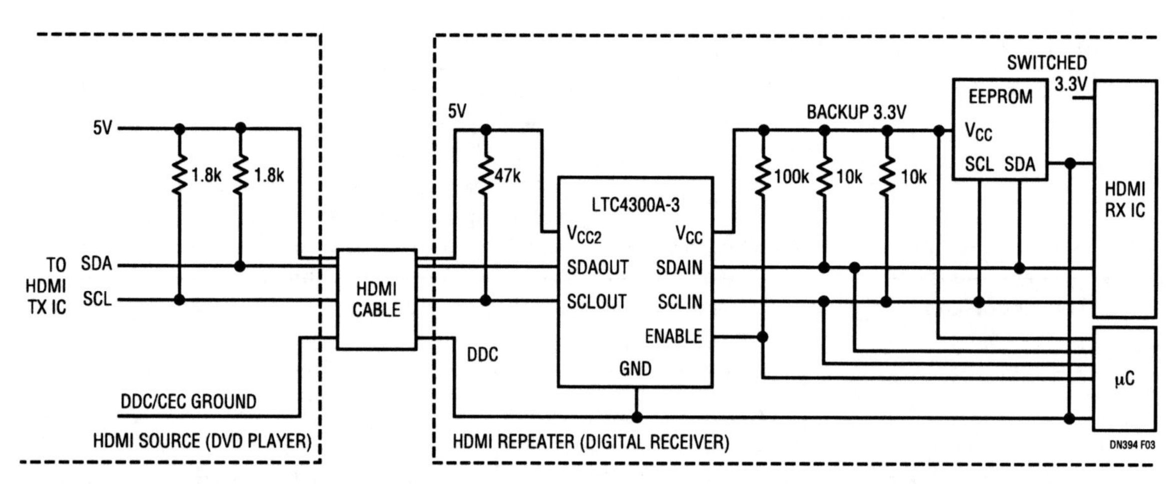

Figure 320.3 • LTC4300A-3 in a Level Shifting and Capacitance Buffering HDMI Application with Backup 3.3V

Video difference amplifier brings versatility to low voltage applications

321

Jon Munson

Introduction

The LT6552 is a specialized dual-differencing 75MHz operational amplifier ideal for rejecting common mode noise as a video line receiver. The input pairs are designed to operate with equal but opposite large-signal differences and provide exceptional high frequency common mode rejection (CMRR of 65dB at 10MHz), thereby forming an extremely versatile gain block structure that minimizes component count in most situations. The dual input pairs are free to take on independent common mode levels, while the two voltage differentials are summed internally to form a net input signal. The LT6552 is optimized for low supply voltages, ranging from 3V to 12V between the rails. In single supply operation at 3.3V, the input range includes ground and an output that can swing within 400mV of either supply rail when driving a 150Ω load. Additionally, a shutdown feature is provided to allow easy power management in supply current-sensitive applications.

Dual input pair zaps common mode noise pickup

The basic application of the LT6552 is that of a high CMRR difference amplifier as shown in Figure 321.1. Only two resistors are required to set the difference gain (Gain = 2 in this example). For unity gain, no additional components are required. Single supply operation usually requires a means of output offset control so that signals are not clipped. The dual input structure of the LT6552 makes this easy (see $V_{DC(ADJ)}$ in Figure 321.1). The circuit shown in Figure 321.1 is especially effective in removing ground noise from video signals in vehicular and industrial applications, as shown in Figure 321.2, killing common mode noise by over a factor of 1000 and providing a lower solution cost than other op amp based topologies because of the fewer required components.

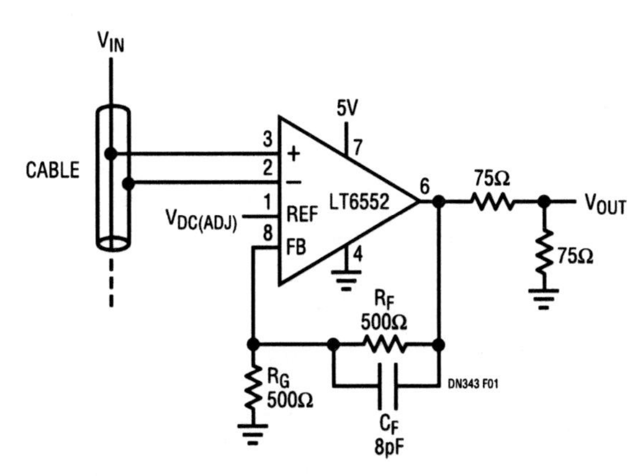

Figure 321.1 • Noise Rejecting Sense Amplifier with DC Adjust

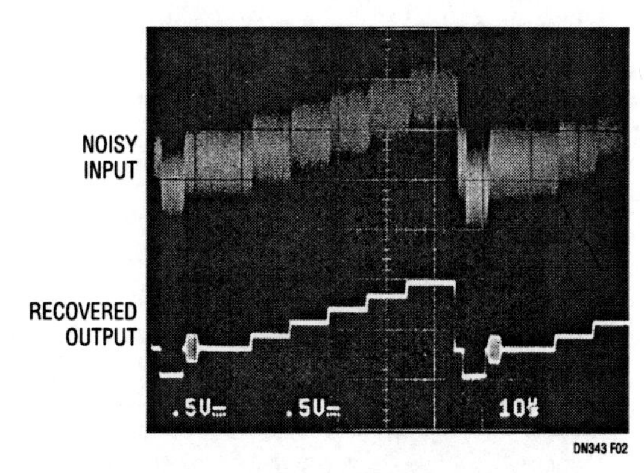

Figure 321.2 • Common Mode Noise Eliminated from Video Signal

Analog Circuit Design: Design Note Collection. http://dx.doi.org/10.1016/B978-0-12-800001-4.00321-5

Perform video rate analog arithmetic

Because of its dual differencing input structure, the LT6552 is able to readily process both additive (noninverting) and subtractive (inverting) variables without the need for complicated resistor networks. Figure 321.3 shows that the output is a function of three input variables with the overall scaling set by the feedback resistors. This property provides a useful means of differential to single-ended conversion with offset control or performing multivariate functions such as the YP_BP_R to RGB video adapter application shown in Figure 321.4. The circuit of Figure 321.4 uses the fewest possible amplification stages to accomplish the needed matrix conversion functions while operating on the lowest supply voltages (±3V). Input black levels should be near 0V for best results and sync on Y is mapped to all three outputs, though typically only needed on G.

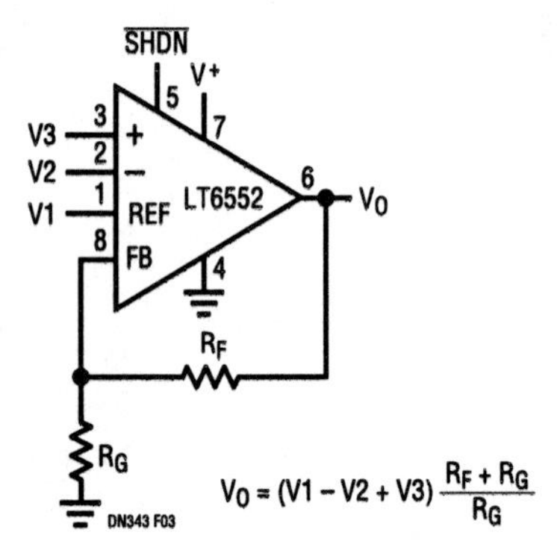

Figure 321.3 • General Purpose Arithmetic Block

Conclusions

By virtue of its dual differencing input structure and ability to operate from low supply voltages, the LT6552 provides a versatile and high performing gain block for modern analog applications. Of particular value is the part's ability to recover difference signals in the presence of strong common mode interference, such as in vehicular and closed-circuit video applications.

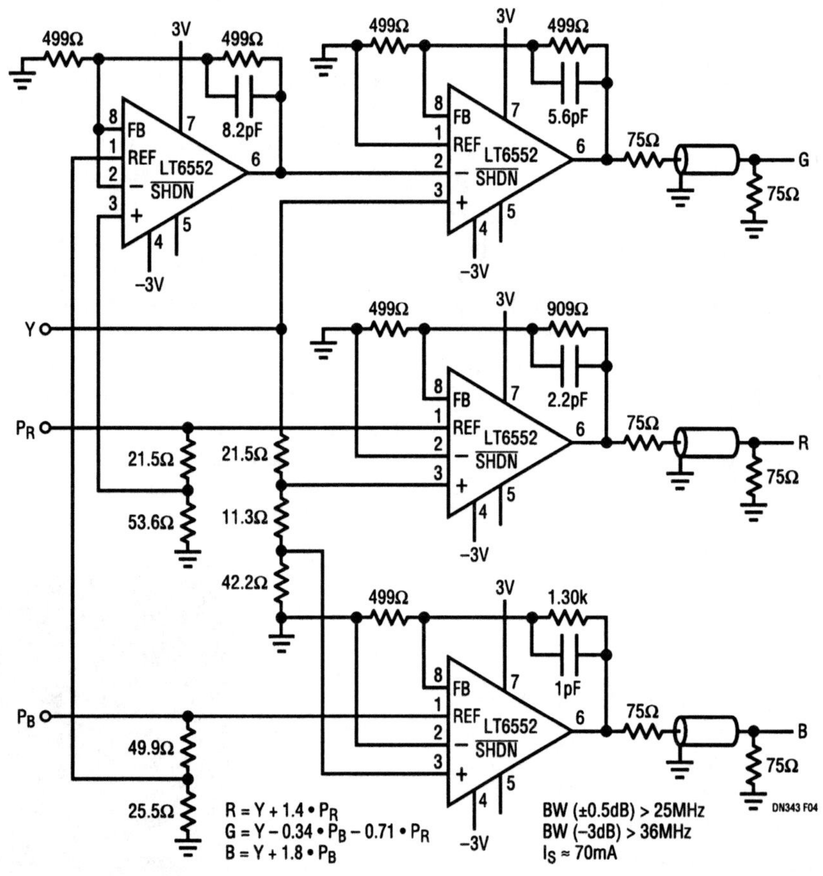

Figure 321.4 • YP_BP_R to RGB Component Video Converter

322

Video signal distribution using low supply voltage amplifiers

Jon Munson

Introduction

Video designs are often pressed to operate on the lowest possible rail voltages—a simple result of the trend towards lower voltage logic and the advantages of sharing supply potentials wherever possible. Video designs are further complicated by the need to accommodate the dynamic offset variation inherent in AC-coupled designs as picture content varies. Traditional op amps require relatively large amounts of output-swing headroom and are therefore impractical for AC coupling at even 5V. Linear Technology offers a new family of video op amps which addresses these issues and offers the ability to operate down to 3.3V in most instances. This family includes the LT6205 (single), LT6206 (dual), LT6207 (quad), LT6550 (triple, fixed gain of 2) and LT6551 (quad, fixed gain of 2).

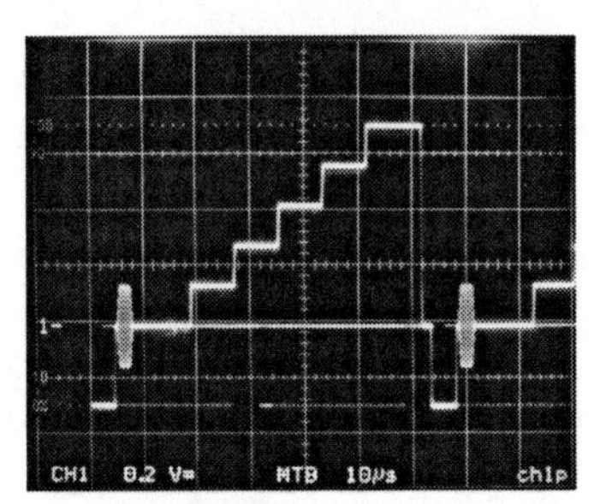

Figure 322.1 • Typical $1V_{P-P}$ Video Waveform (Several Lines Overlaid)

Video signal characteristics

To determine the minimum video amplifier supply voltage, we must first examine the nature of the signal. Composite video is the most commonly used signal in broadcast-grade products and combines Luma (or luminance, the intensity information), Chroma (the colorimetry information) and Sync (vertical and horizontal raster timing) elements into a single signal, NTSC and PAL being the common formats.

Typical video waveforms are specified to have a nominal $1.0V_{P-P}$ amplitude, as shown in Figure 322.1. The lower 0.3V is reserved for sync tips that carry timing information. The black level (zero intensity) of the waveform is at (or set up slightly above) this upper limit of the sync information. Waveform content above the black level is intensity information, with peak brightness represented at the maximum signal level. The sync potential represents blacker-than-black intensity, so scan retrace activity is invisible on a CRT. In the case of composite video, the modulated color subcarrier is superimposed on the waveform, but the dynamics generally remain inside the $1V_{P-P}$ limit (a notable exception is the chroma ramp used for differential gain and differential phase measurements, which can reach $1.15V_{P-P}$).

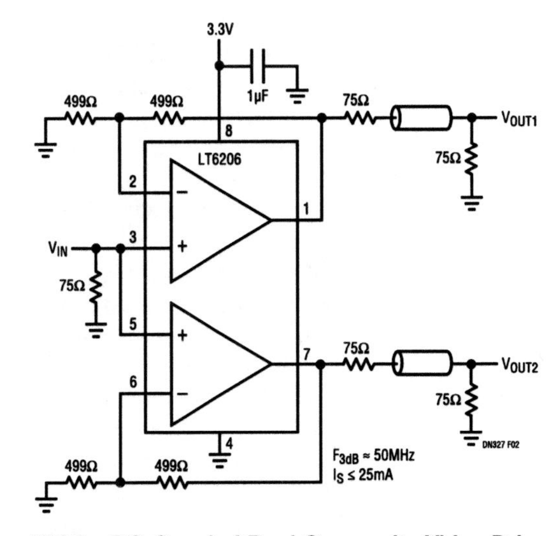

Figure 322.2 • DC-Coupled Dual Composite Video Driver Powered from 3.3V

Analog Circuit Design: Design Note Collection. http://dx.doi.org/10.1016/B978-0-12-800001-4.00322-7

Amplifier considerations

Most video amplifiers drive cables that are series terminated (back terminated) at the source and load terminated at the destination with resistances equal to the cable characteristic impedance, Z_O (usually 75Ω). This configuration forms a 2:1 resistor divider in the cabling that must be corrected in the driver amplifier by delivering $2V_{P-P}$ output into an effective $2 \cdot Z_O$ load (e.g. 150Ω). Driving the cable can require in excess of 13mA while the output is approaching the saturation limits of the amplifier output. The absolute minimum supply is $V_{MIN} = 2 + V_{OH} + V_{OL}$, where the V_O values represent the minimum voltage drops that an amplifier can be guaranteed to develop with respect to the appropriate supply rail.

For example, the LT6206 dual operating on 3.3V in Figure 322.2, with exceptionally low $V_{OH} \leq 0.5V$ and $V_{OL} \leq 0.35V$, provides a design margin of 0.45V, enough to cover supply variations and DC bias accuracy for the DC-coupled video input.

Handling AC-coupled video signals

Unfortunately, one cannot always be assured that source video has the appropriate DC content to satisfy the amplifier involved, so other design solutions are frequently required. AC-coupled video inputs are intrinsically more difficult to handle than those with DC-coupling because the average signal voltage of the video waveform is affected by the picture content, meaning that the black level at the amplifier wanders with scene brightness. By analyzing the worst-case wander, we can determine the AC-coupled constraint.

Figure 322.3 shows two superimposed AC-coupled waveforms, the higher trace being black field and the lower trace being white field. The wander is measured as 0.56V for the $1V_{P-P}$ NTSC waveform shown, so an additional 1.12V allowance must be made in the amplifier supply (assuming gain of 2, so $V_{MIN} = 3.12 + V_{OH} + V_{OL}$). The amplifier output (for gain of 2) must swing 1.47V to −1.65V around the DC-operating point, so the biasing circuitry needs to be designed accordingly for optimal fidelity. For example, an LT6551 operating on 5V, with excellent $V_{OH} \leq 0.8V$ and $V_{OL} \leq 0.2V$, has a healthy design margin of 0.88V for a composite signal.

A popular method of further minimizing supply requirements with AC-coupling is to employ a simple clamping scheme as shown in Figure 322.4. In this circuit, the LT6205 is able to operate from 3.3V by having the sync-tips control the charge on the coupling capacitor, thereby reducing the black-level input wander to ≈0.07V. A minor drawback to this circuit is the slight sync-tip compression (≈0.025V at input) due to the diode conduction current, though the picture content retains full fidelity. This circuit has nearly the design margin of its DC-coupled counterpart, at 0.31V (for this circuit, $V_{MIN} = 2.14 + V_{OH} + V_{OL}$).

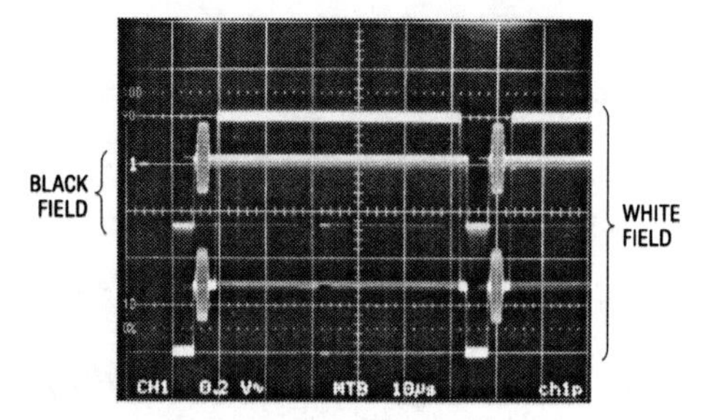

Figure 322.3 • Video Offset Shift due to Picture Content with Conventional AC Coupling

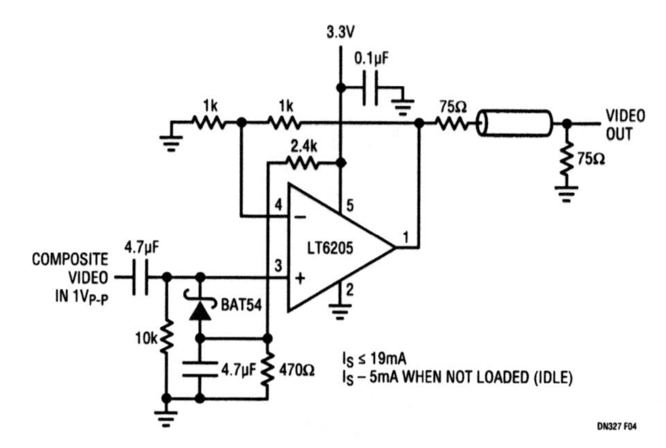

Figure 322.4 • AC-Coupled Clamped Video Amplifier Powered from 3.3V

Conclusion

With the industry's lowest output saturation characteristics, the low voltage video amplifiers, including the LT6205 (single), LT6206 (dual), LT6207 (quad), LT6550 (triple, fixed gain of 2) and LT6551 (quad, fixed gain of 2), offer the video designer the ability to share reduced supply voltages along with the logic circuitry. This ability to share supply voltages helps save space and cost by reducing power dissipation and power converter complexity.

Tiny RGB video multiplexer switches pixels at 100MHz

323

Frank Cox John Wright

Introduction

The LT1675, a new 3-channel, 2-input 250MHz video multiplexer, is designed for pixel switching, video graphics and RGB routing. The complete circuit squeezes into a 16-lead SSOP package and uses only 0.25in^2 of PC board area (Figure 323.1). The LT1675 features a fixed gain of 2 for driving double-terminated cables. By incorporating internal feedback resistors, the circuit simplifies PC board layout and boosts performance by eliminating stray capacitance. A single channel 2:1 MUX, the LT1675-1 is available in the small MSOP package. Table 323.1 summarizes the major performance specifications of this new multiplexer, and Figure 323.2 shows a typical application: switching between two RGB sources.

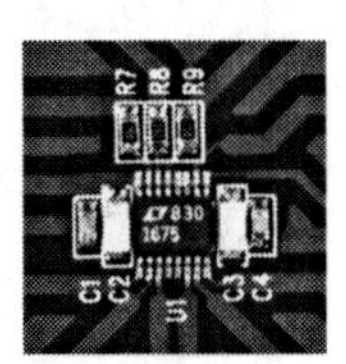

Figure 323.1 • Board Photo Actual Size

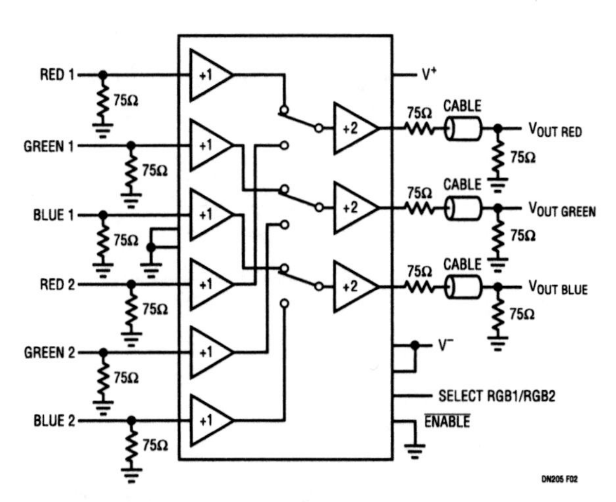

Figure 323.2 • LT1675 Typical Application: Switching between Two RGB Sources and Driving Three Cables

Table 323.1 LT1675 Performance, VS = ±5V

PARAMETER	CONDITIONS	TYPICAL VALUES
−3dB Bandwidth	$R_L = 150\Omega$	250MHz
0.1dB Gain Flatness	$R_L = 150\Omega$	70MHz
Crosstalk	Between Active Channels at 10MHz	−60dB
Slew Rate	$R_L = 150\Omega$	1100V/μs
Differential Gain	$R_L = 150\Omega$	0.07%
Differential Phase	$R_L = 150\Omega$	0.05°
Channel Select Time	$R_L = 150\Omega$, $V_{IN} = 1V$	2.5ns
Enable Time	$R_L = 150\Omega$	10ns
Output Voltage Swing	$R_L = 150\Omega$	±3V
Output Offset Voltage		20mV
Supply Current	All Three Channels Active	30mA
Supply Current Disabled		1μA

Expanding inputs does not increase power dissipation

In video-routing applications, where the ultimate in speed is not mandatory as it is in pixel switching, it is possible to expand the number of MUX inputs by shorting the LT1675 outputs together and switching between the two LT1675s with the $\overline{\text{ENABLE}}$ pins. This technique, shown in Figure 323.3, does not increase the power dissipation because LT1675s draw virtually zero current when disabled.

Analog Circuit Design: Design Note Collection. http://dx.doi.org/10.1016/B978-0-12-800001-4.00323-9

Add your own logo

The circuit of Figure 323.4 highlights a section of picture under control of a synchronous key signal. The technique is used for adding the logo you see in the bottom corner of commercial television pictures or any type of overlay signal, such as a crosshair or a reticule. The key signal has two bits of control, so there can be four levels of highlighting: unmodified video, video plus 33% white, video plus 66% white and 100% white. Two LT1675s are configured as a 2-bit DAC, and resistors on the output set the relative bit weights. The output of the LT1675, labeled B, is one-half the weight of the A device. To properly match the 75Ω video cable, the output resistors are selected so the parallel combination of the two is 75Ω. The output will never exceed peak white, which is 0.714V for this NTSC-related RGB video. The reference white signal is adjusted to a lower level than peak white to make the effect less intrusive.

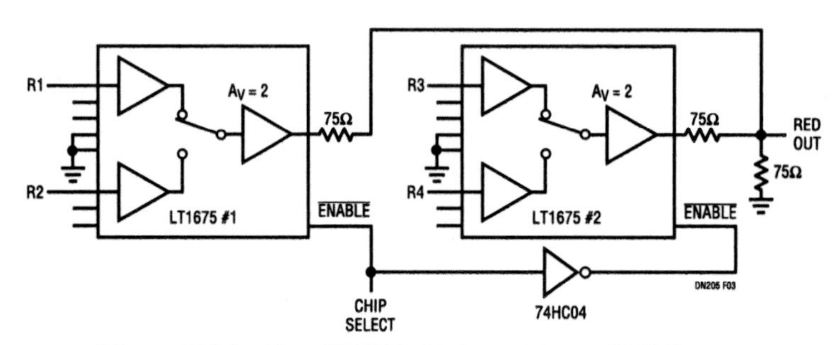

Figure 323.3 • Two LT1675s Make a 4-Input RGB Router

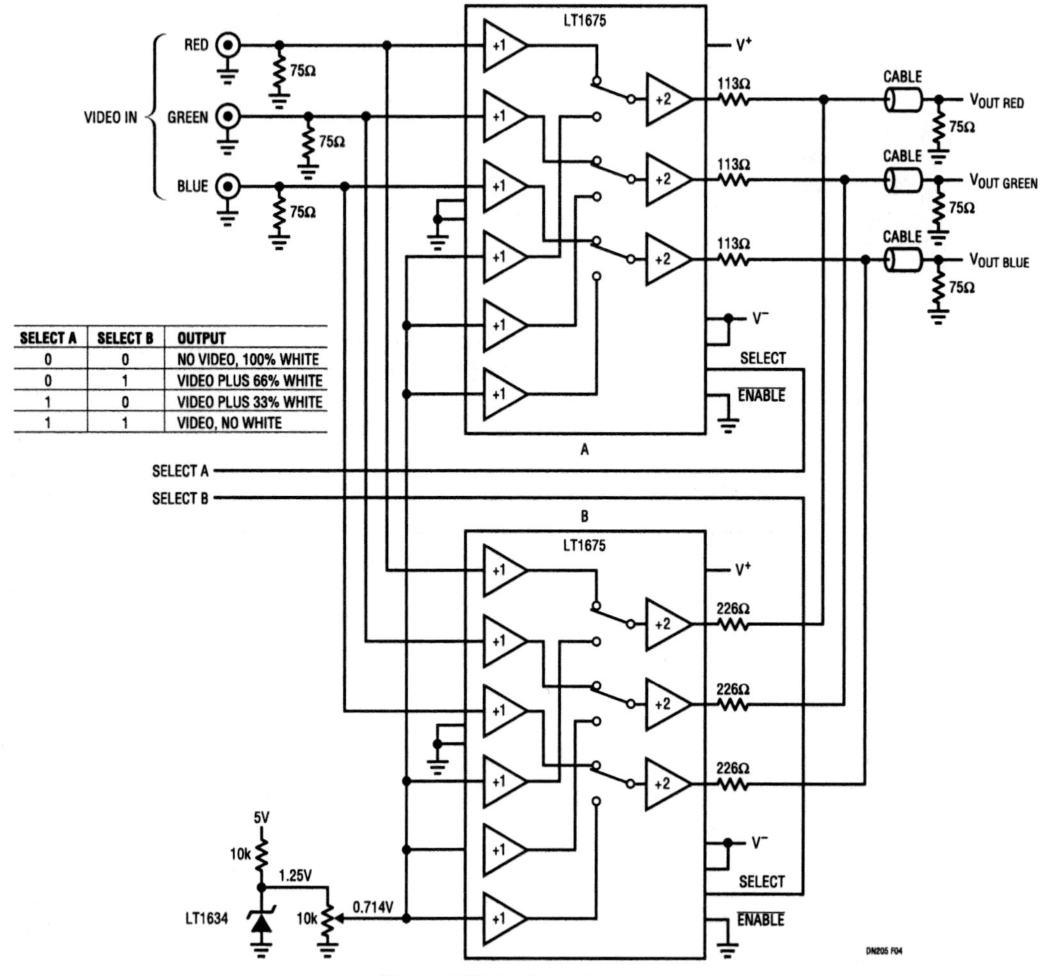

SELECT A	SELECT B	OUTPUT
0	0	NO VIDEO, 100% WHITE
0	1	VIDEO PLUS 66% WHITE
1	0	VIDEO PLUS 33% WHITE
1	1	VIDEO, NO WHITE

Figure 323.4 • Logo Inserter

324

An adjustable video cable equalizer

Frank Cox

This design note presents a voltage controlled cable equalizer based on the LT1256 video fader. The circuit features ease of adjustment, simplicity and the capability for remote control. The amount of equalization can be adjusted continuously from the maximum allowed by the passive components to none at all. While the example shows video, this high performance equalizer can be used in any system using long runs of coaxial cable or twisted pair to transmit analog signals.

A voltage or current controlled equalizer is essential in systems which automatically set cable compensation. In systems where cable equalization is set manually, a voltage controlled equalizer is still preferred as it does not require routing the signal path to the control. Instead, only a DC control voltage passes from the front panel to the equalizer.

Automatic equalization is possible for properly characterized video cables. Maximum equalization is set to coincide with the maximum length of cable expected; the equalizer is controlled by a servo loop. One method of generating the necessary control voltage is to sample the color burst amplitude and compare this with a reference voltage using a summing integrator. Since the frequency roll-off of the cable is known (and fixed for a given cable) only the amount of equalization needs to be adjusted.

In many applications color video is transmitted down long runs of coaxial or twisted-pair cable. Losses in the cable increase with signal frequency and cable length. The type of cable will determine the rate of high frequency loss. Color information in NTSC video is contained primarily in the high frequency portion of the spectrum. Besides causing a loss of detail in the picture, excessive high frequency loss will make reliable decoding of the composite color signal more difficult or impossible. Most commercial distribution amplifiers have provisions for equalizing the cable losses, but many times these units come at a high cost.

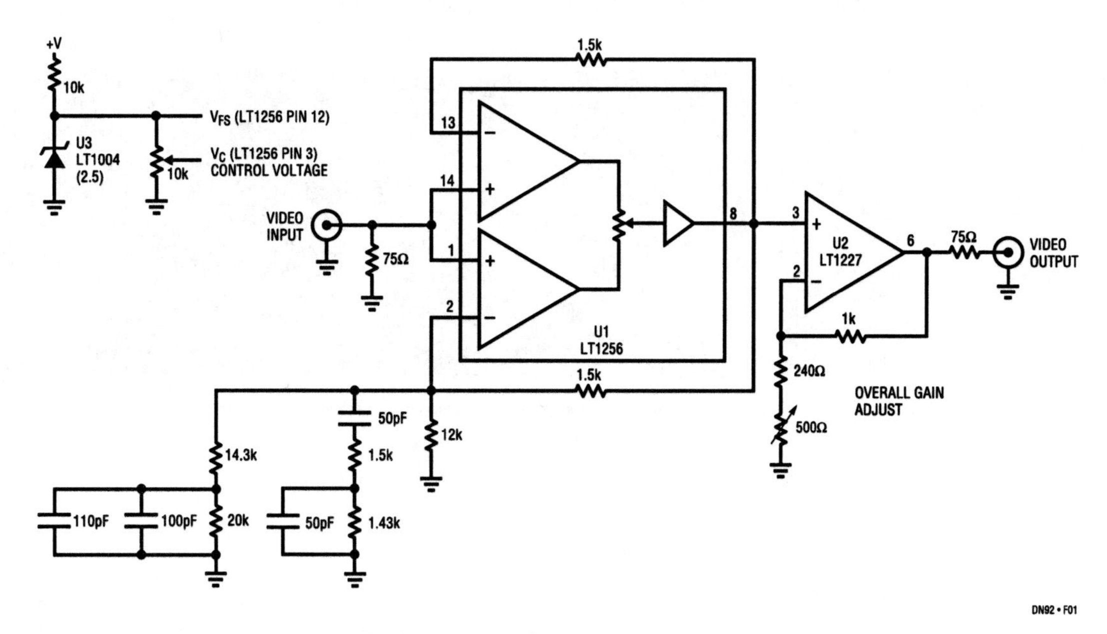

Figure 324.1 • LT1256 Cable Equalizer

Analog Circuit Design: Design Note Collection. http://dx.doi.org/10.1016/B978-0-12-800001-4.00324-0

Figure 324.1 is a complete schematic of the cable equalizer. The LT1256 (U1) is a two input, one output 40MHz current feedback amplifier with a linear control circuit that sets the amount each input contributes to the output. One amplifier (input pins 13 and 14) of the LT1256 is configured as a gain of one with no frequency equalization. The other amplifier (input pins 1 and 2) has frequency equalizing components in parallel with the 12k gain resistor. The equalization components for this demonstration circuit were chosen empirically as no data on the cable was available. As the control voltage is varied the output contains a summation of the two separate input channels; one containing the input video with no compensation and the other with the maximum depth of equalization. By adjusting this mix it is possible to smoothly adjust the equalizer depth. An additional amplifier (U2, LT1227) is used to set the overall gain. Two amplifiers were used here to make setting the gain a single adjustment, but in a production circuit the LT1256 can be configured to have the necessary gain and the whole function can be done with one chip.

In this demonstration a spool of over 250 feet[1] of good quality coax was used to transmit NTSC video. The LT1256 equalizer is placed at the receive end[2] and is adjusted with the use of a test pattern and a video waveform monitor (or oscilloscope). Figure 324.2 shows the video after transiting the cable and without equalization. Three standard video test signals were used; the multiburst, the 2T and the 12.5T. The 2T and the 12.5T test signals are sensitive indicators of phase and amplitude distortion in the video signal. The effect of the equalization circuit is shown in Figure 324.3. The resultant frequency response is flat and the time domain behavior is also excellent. Network analyzer plots of gain vs frequency for various settings of equalizer depth are given in Figure 324.4.

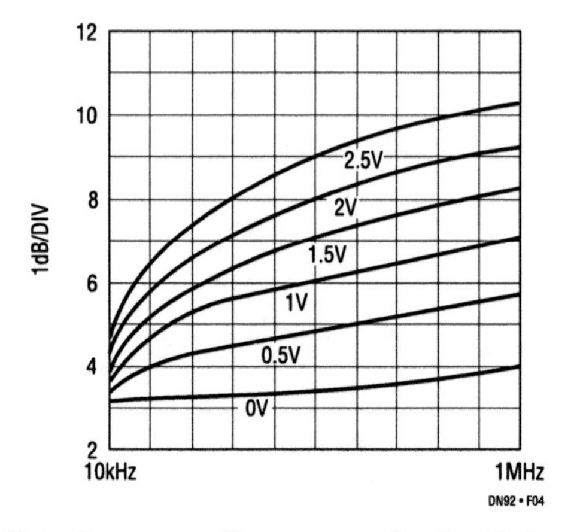

Figure 324.4 • Frequency Response vs Control Voltage

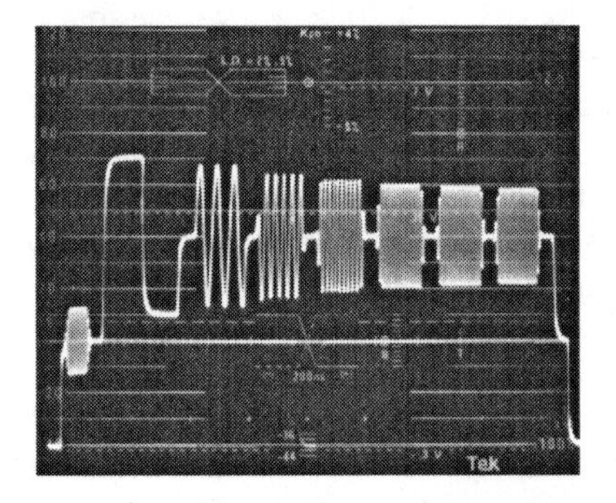
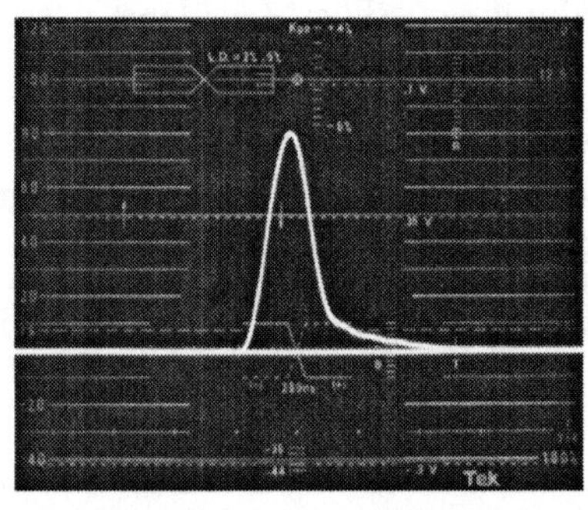
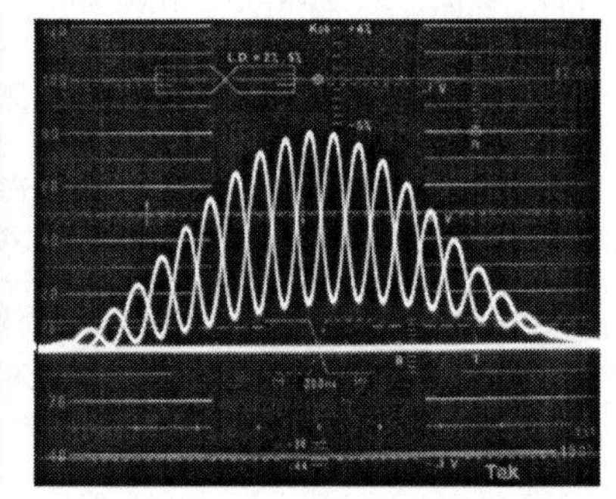

Figure 324.2 • Multiburst, 2T and 12.5T After 250 Feet of Coax

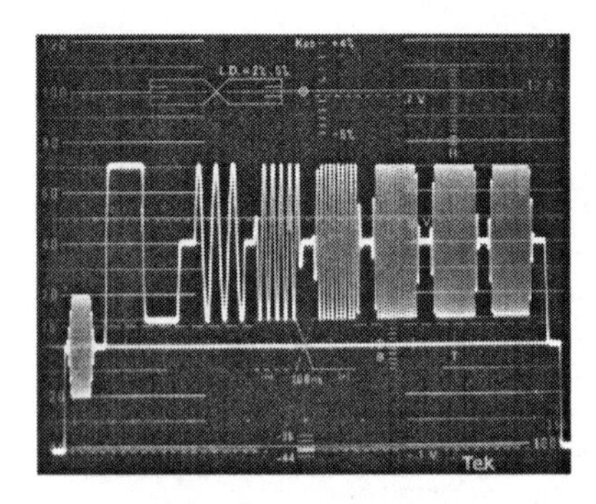
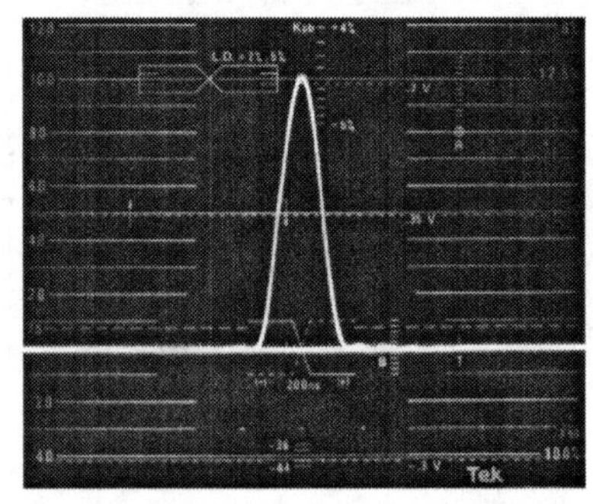
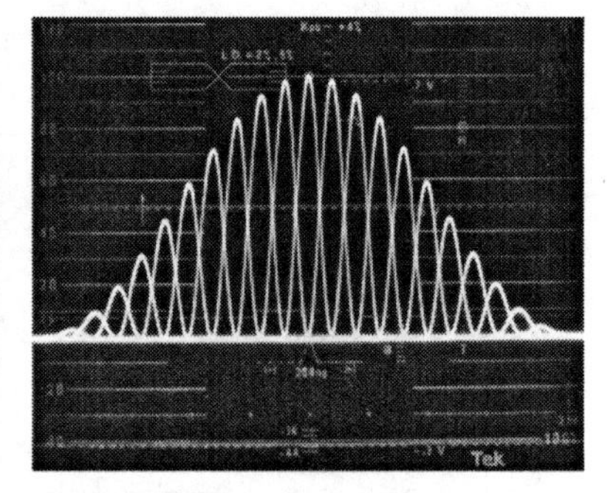

Figure 324.3 • Multiburst, 2T and 12.5T After 250 Feet of Coax and Equalization, Circuit in Figure 324.1

Note 1: It should also be noted that the LT1256 can be used to equalize much longer coaxial and twisted-pair cable than the one in this demonstration.

Note 2: Since the equalizer provides gains at high frequencies there is a possibility of overload if the circuit is placed at the transmit end of the cable rather than the receive end of the cable, especially if there is a need for a great deal of equalization. For example, 2000 feet of Belden 8281 precision video cable will have about 11dB of loss at 5MHz. This requires the driving amp to swing 3.5 times normal if the transmit end is boosted to compensate for the cable.

4 × 4 video crosspoint has 100MHz bandwidth and 85dB rejection at 10MHz

325

John Wright

4 × 4 crosspoint

The compact high performance 4 × 4 crosspoint shown in Figure 325.1 uses four LT1205s to route any input to any or all outputs. The complete crosspoint uses only six SO packages, and less than six square inches of PC board space. The LT1254 quad current feedback amplifier serves as a cable driver with a gain of 2. A ±5V supply is used to ensure that the maximum 150°C junction temperature of the LT1254 is not exceeded in the SO package. With this supply voltage the crosspoint can operate at a 70°C ambient temperature and drive 2V (peak or DC) on to a double-terminated 75Ω video cable. The feedback resistors of these output amplifiers have been optimized for this supply voltage. The −3dB bandwidth of the crosspoint is over 100MHz with only 0.8dB of peaking. All Hostile Crosstalk Rejection is 85dB at 10MHz when a shorted input is routed to all outputs. Keys to attaining this high rejection include:

1. Mount the feedback resistors for the surface mount LT1254 on the backside of the PC board.
2. Keep the (−) input traces of LT1254 as short as possible.
3. Route V^+ and V^- for the LT1205s on the component (top) side and under the devices (between inputs and outputs).
4. Use the backside of the PC board as a solid ground plane. Connect the LT1205 device grounds and bypass capacitor's grounds as vias to the backside ground plane.
5. Surround the LT1205 output traces by ground plane and route them away from (−) inputs of the other three LT1254s.

Each pair of logic inputs labeled SELECT LOGIC OUTPUT is used to select a particular output. The truth table is used to select the desired input, and is applied to each pair of logic inputs. For example, to route Channel 1 Input to Output 3, the fourth pair of logic inputs labeled SELECT LOGIC OUTPUT 3 is coded A = Low and B = High. To route Channel 3 Input to all outputs, set all eight logic inputs High. Channel 3 is the default input with all logic inputs open. To shut off all channels, a pair of LT1259s can be substituted for the LT1254. The LT1259 is a dual current feedback amplifier with a shutdown pin that reduces the supply current to 0μA.

4 × 4 Crosspoint All Hostile Crosstalk

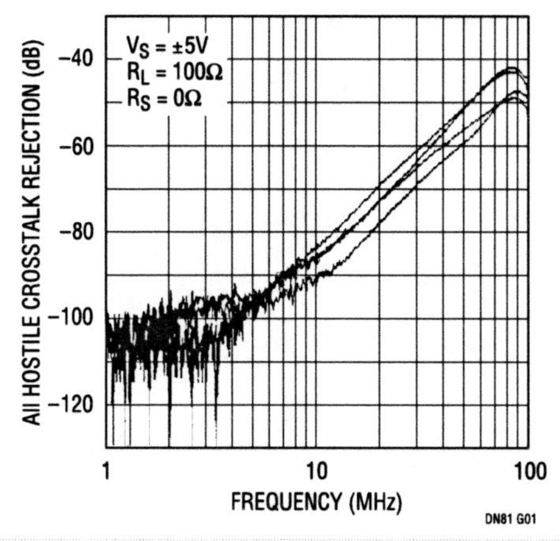

4 × 4 Crosspoint Response

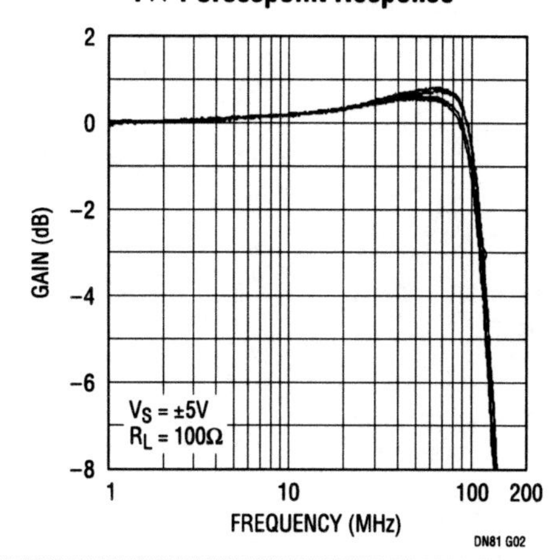

Analog Circuit Design: Design Note Collection. http://dx.doi.org/10.1016/B978-0-12-800001-4.00325-2

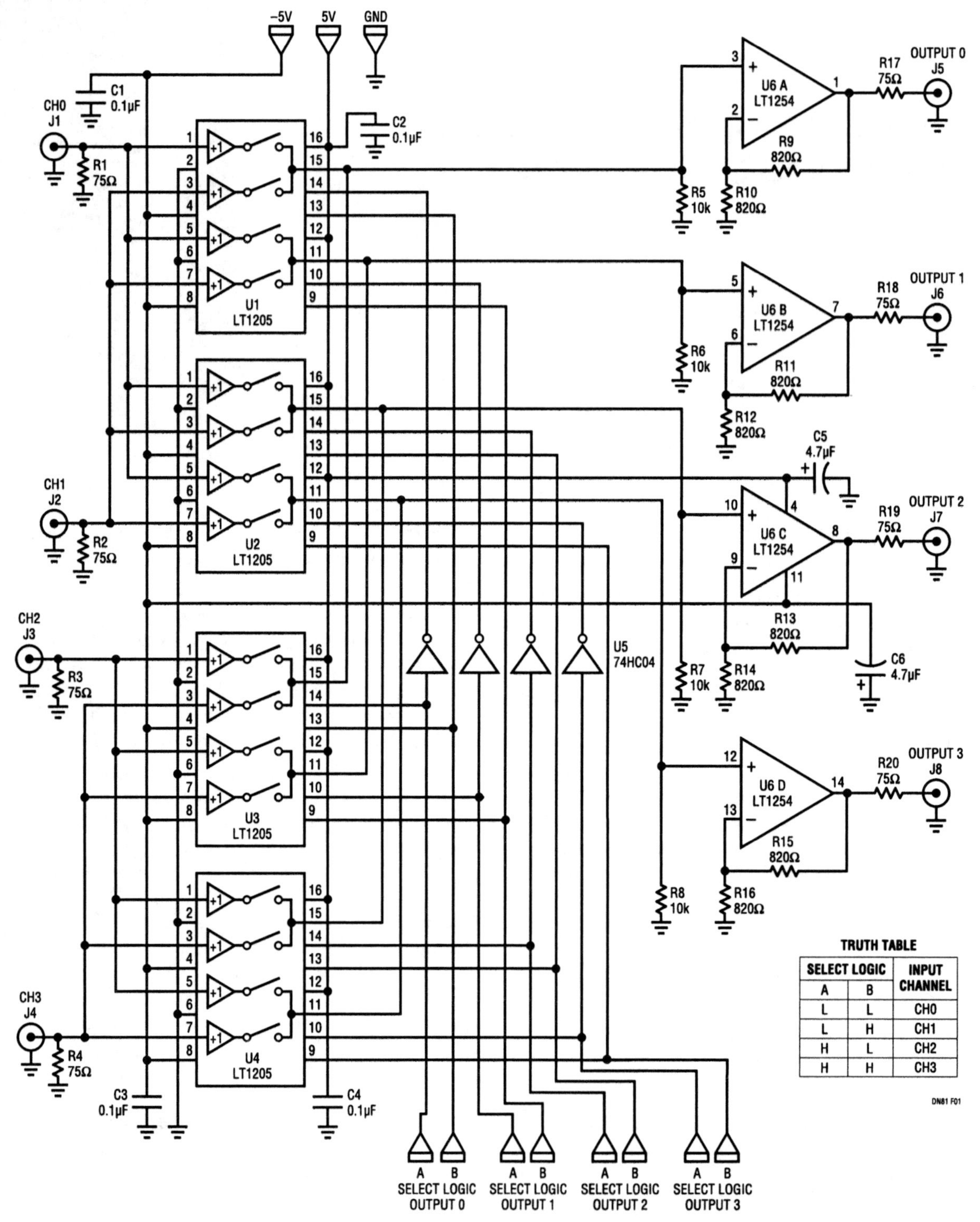

TRUTH TABLE

SELECT LOGIC		INPUT CHANNEL
A	B	
L	L	CH0
L	H	CH1
H	L	CH2
H	H	CH3

Figure 325.1 • 4 × 4 Crosspoint and Truth Table

Single 4-input IC gives over 90dB crosstalk rejection at 10MHz and is expandable

326

John Wright

Introduction

Professional video systems need to multiplex between many signals without interference from adjacent video sources that are not selected. Final system crosstalk rejection of all non-selected or "Hostile" signals of 72dB is regarded as "professional quality." This level of isolation is very difficult to achieve because every doubling in the number of inputs degrades the crosstalk by 6dB. In the past because no single IC was good enough, cascades of discrete switches and amplifiers were used to achieve the necessary isolation. An additional requirement of some video multiplexers is the ability to switch quickly and cleanly so the sources can be changed in picture without visible lines or distortion. New emerging multimedia systems require the performance of professional systems in the PC environment.

The new LT1204 4-input video multiplexer IC speeds the design of high performance video selection products. It features easy input expansion, and over 90dB crosstalk rejection on a PC board up to 10MHz even when expanded to 16 inputs. Additionally, this new multiplexer has low switching transients and includes a 75MHz current feedback amplifier to drive 75Ω cables. Figure 326.1 shows the LT1204 in a typical application.

Expanding the number of inputs

To expand the number of MUX inputs LT1204s can be paralleled by shorting their outputs together. The Disable feature ensures that amplifier outputs that are not selected do not alter the cable termination. When the LT1204 is disabled (pin 11 low), the output stage is turned off and the feedback resistors are bootstrapped, effectively removing them from the circuit. This has the effect of raising the "true" output impedance to about 25k in Figure 326.1. The LT1204 disable logic has been designed to prevent shoot-through current when two or more amplifiers have their outputs shorted together. The LT1204 also has a logic controlled shutdown (pin 12 low)

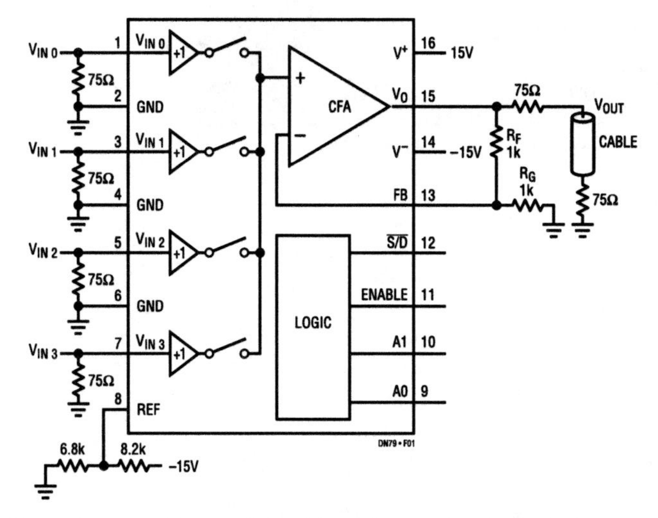

Figure 326.1 • 4-Input Video Multiplexer with Cable Driver

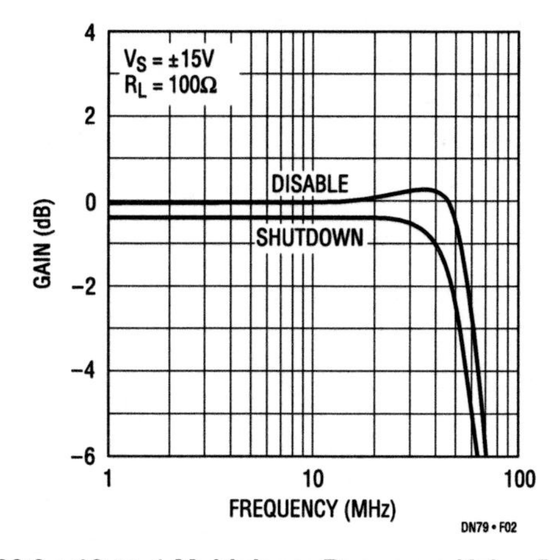

Figure 326.2 • 16-to-1 Multiplexer Response Using Disable Feature vs Shutdown Feature

Analog Circuit Design: Design Note Collection. http://dx.doi.org/10.1016/B978-0-12-800001-4.00326-4

that drops the supply current from 19mA to 1.5mA. When shut down, the feedback resistors load the output because the bootstrapping is inoperative. Figure 326.2 shows this loading effect for a 16-to-1 MUX made with four LT1204s using the Disable feature vs the Shutdown feature.

PC board layouts

Crosstalk is a strong function of the IC package, the PC board layout, as well as the IC design. Layout of a PC board that has over 90dB crosstalk rejection at 10MHz is not trivial. PC boards have been fabricated to show the component and ground placement required to attain this level of performance. It has been found empirically from these PC boards that capacitive coupling across the package of greater than 3fF (0.003pF) will diminish the rejection. Keys to the layout are: placing ground plane between inputs, minimizing the feedback pin trace length, putting feedback resistors on the back side of the surface mount PC board, and guarding pin 13 with ground plane.

Crosstalk in P-DIP and SOL vs Frequency

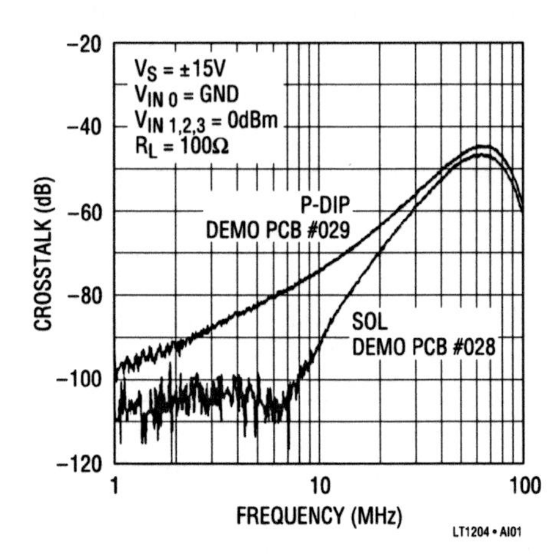

Switching transients

Multimedia systems switch active video "in-picture" to create special effects and this requires fast clean transitions with low "glitch" energy. In the past video source selection was made during the blanking period and switching transients were not visible. The LT1204 has input buffers that isolate the internal make-before-break switches. These buffers ensure glitches are minimized at the inputs. This is important because loop-through connections send these glitches to other equipment. When two channels are on momentarily the more positive voltage passes through; if both are equal, there is only a 40mV error at the input of the CFA. The time of this 40mV error can be reduced by adjusting the voltage on the Reference (pin 8). The Reference pin is used to trade off positive input voltage range for switching time. On ±15V supplies, settling the voltage on pin 8 to −6.8V reduces the switching transient to a 50ns duration, and the positive input range reduces from 6V to 2.35V. The negative input range remains unchanged at −6V. Included are photos of the switching transients for the new LT1204 as well as competitive CMOS and bipolar MUXs.

LT1204 Output Switching Transients

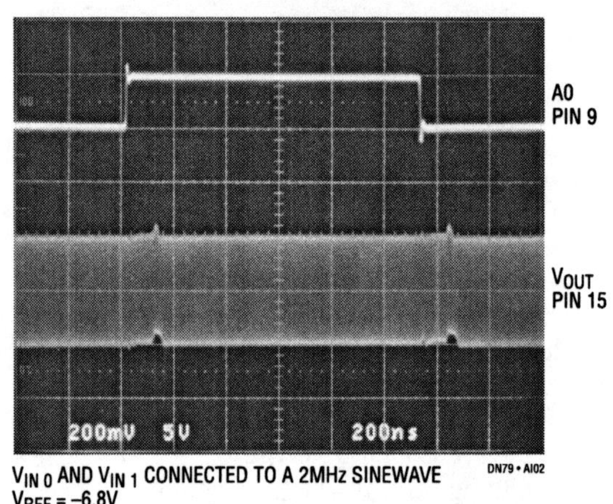

$V_{IN\,0}$ AND $V_{IN\,1}$ CONNECTED TO A 2MHz SINEWAVE
V_{REF} = −6.8V

Competitive MUXs

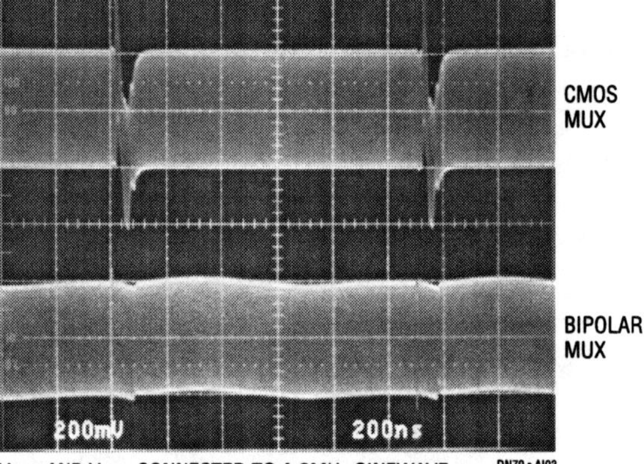

$V_{IN\,0}$ AND $V_{IN\,1}$ CONNECTED TO A 2MHz SINEWAVE

Send color video 1000 feet over low cost twisted-pair

327

John Wright

It is now possible to send and receive color composite video signals appreciable distances on a low cost twisted-pair. This technique is similar to the push toward twisted-pair cables in EtherNet systems to replace costly coaxial cable connections. The cost advantage of these techniques is significant. Standard 75Ω RG-59/U coaxial cable costs between 25¢ and 50¢ per foot, but a PVC twisted-pair is only pennies per foot. This means hundreds of dollars are saved in installations as short as 1000 feet, easily paying for additional electronics. The system also provides for "drops" or receiver taps along the twisted-pair.

This bidirectional "video bus" consists of the low cost LT1190 op amp and the LT1193 video difference amplifier shown in Figure 327.1. A pair of LT1190s at TRANSMIT 1 is used to generate differential signals to drive the line which is back-terminated in its characteristic impedance. These amplifiers have high 50mA load driving ability, while maintaining a very high 450V/µs slew rate and 50MHz gain-bandwidth. The twisted-pair receiver is an LT1193 video difference amplifier at RECEIVE 1, and it converts signals from differential to single-ended. The LT1193 offers features unavailable with other op amp configurations. In addition to speed and load driving

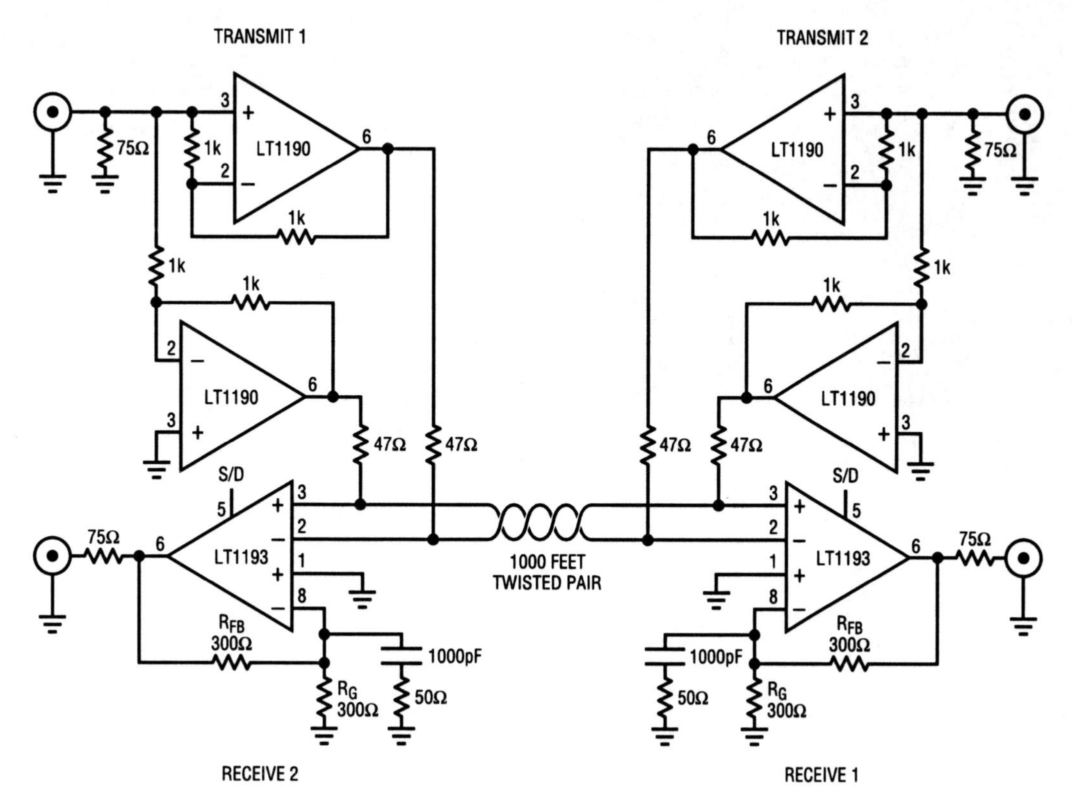

Figure 327.1 • Bidirectional Video Bus

Analog Circuit Design: Design Note Collection. http://dx.doi.org/10.1016/B978-0-12-800001-4.00327-6

ability, the LT1193 provides high input impedance (+) and (−) inputs, and common-mode rejection in excess of 40dB at 10MHz. Because of the LT1193's unique topology, it is possible to provide cable compensation at the amplifier's feedback node as shown. In this case, 1000 feet of twisted-pair is compensated with 1000pF and 50Ω to boost the −3dB bandwidth of the system from 750kHz to 4MHz. The effect of this compensation can be seen in Figures 327.2 and 327.3. This bandwidth is adequate to pass 3.58MHz chroma subcarrier, and the 4.5MHz sound subcarrier. Attenuation in the cable can be compensated by lowering the gain-set resistor R_G in Figure 327.1. At TRANSMIT 2, another pair of LT1190s serves the dual function of providing cable termination via low output impedance, and generating differential signals for TRANSMIT 2.

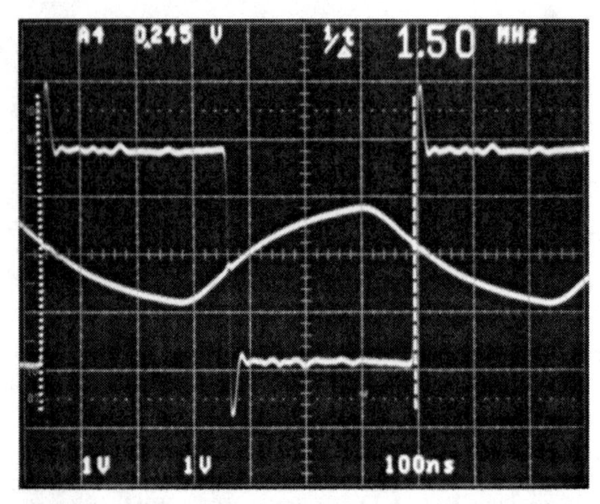

Figure 327.2 • 1.5MHz Square Wave Input and Unequalized Response through 1000 Feet of Twisted-Pair

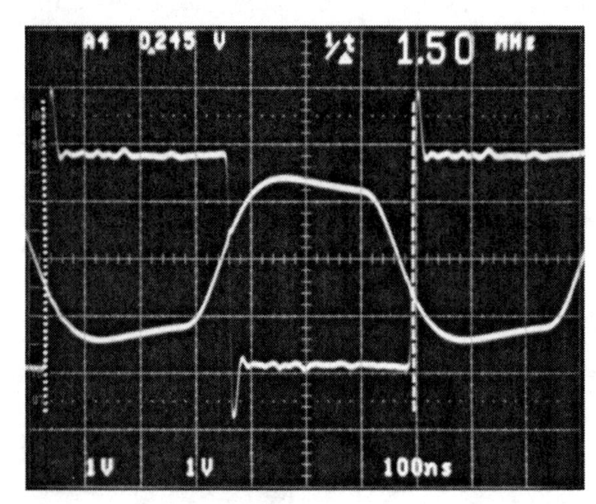

Figure 327.3 • 1.5MHz Square Wave Input and Equalized Response through 1000 Feet of Twisted-Pair

A good indication of the system's ability to pass color composite video is shown in Figure 327.4. This multiburst pattern was passed through 1000 feet of low cost PVC twisted-pair, and it contains a 3.58MHz chroma subcarrier and a 4.5MHz sound subcarrier. Although the scope photo shows these frequencies to be attenuated about 3dB, a clean picture is present at the end of the twisted-pair. Additional receiver taps can be added along the twisted-pair. The trick is to leave the taps unterminated and limit their length. Longer drops cause the 3.58MHz chroma subcarrier to reflect and interfere with the phase of the transmitted subcarrier. The effect is color smudge or in the limit ghosting.

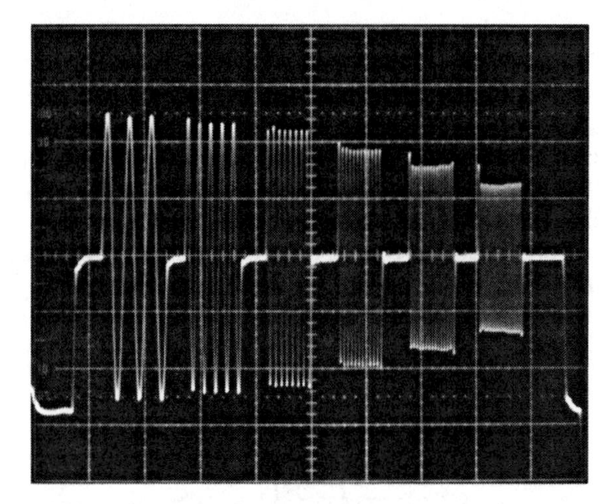

Figure 327.4 • Multiburst Pattern Passed through 1000 Feet of Twisted-Pair

The LT1190 and LT1193 include a shutdown feature that drops their dissipation to only 15mW when not in use. This feature is useful for shutdown of unused receivers; however, shutdown of the drivers will result in mistermination of the twisted-pair. If power consumption is a major concern, the LT1190 and LT1193 can be replaced with the low power LT1195 and LT1187, resulting in only a slight performance degradation.

Video circuits collection

William H. Gross

Introduction

This note shows how to make several different video circuits using high speed op amps. All of these circuits work with composite, RGB and monochrome video. For best results, bypass the power supply pins of these amplifiers with 1μF to 10μF tantalum capacitors in parallel with 0.01μF disc capacitors. It is important to terminate both ends of video cables to preserve frequency response. When properly terminated, the cable looks like a resistive load of 150Ω.

Lots of Inputs Video MUX Cable Driver (LT1227)

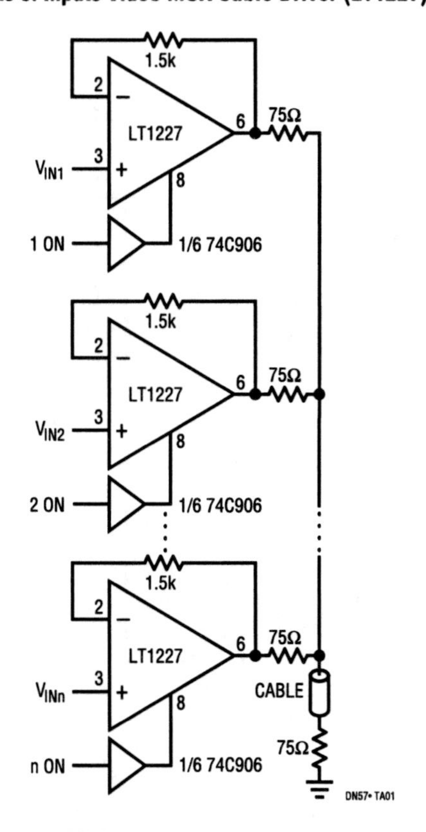

Multiplex amplifiers

Often it is desirable to select one of several signals to send down a cable. Connecting the outputs of several amplifiers together and using the amplifier's shutdown pin to disable all but one accomplishes this goal. The LT1190, LT1191, LT1192, and LT1193 are shutdown by pulling Pin 5 to the negative supply.

The LT1223 and LT1227 current feedback amplifiers are shutdown by pulling Pin 8 to ground. During normal operation Pin 8 is open and at the positive supply potential. An easy way to interface Pin 8 to logic is with a logic level N-channel FET or a 74C906 (open-drain hex buffer).

Two Input Video MUX Cable Driver (LT1190)

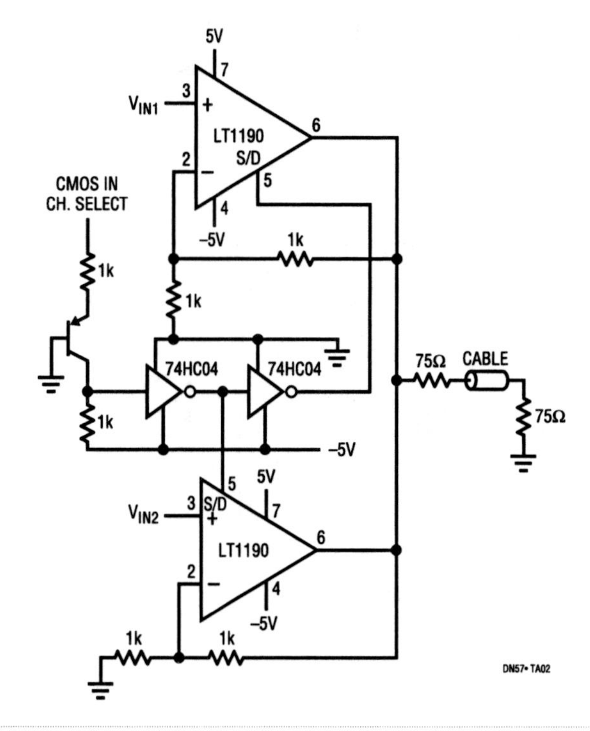

Analog Circuit Design: Design Note Collection. http://dx.doi.org/10.1016/B978-0-12-800001-4.00328-8

Differential Gain and Phase of Several Amplifiers

	DIFFERENTIAL			
	LOAD = 1kΩ		LOAD = 150Ω	
PART NUMBER	GAIN	PHASE	GAIN	PHASE
LT1190*	0.05	0.02	0.23	0.16
LT1191*	0.03	0.01	0.09	0.07
LT1192**	0.10	0.01	0.23	0.15
LT1193*	0.20	0.08	0.20	0.08
LT1194**	0.20	0.08	0.20	0.08
LT1223	0.01	0.02	0.12	0.26
LT1227	0.01	0.01	0.01	0.01
LT1228	0.01	0.01	0.04	0.10
LT1229	0.01	0.01	0.04	0.10

$V_S = \pm15V, A_V = 2$
*$V_S = \pm8V, A_V = 2$
**$V_S = \pm8V, A_V = 10$

Loop through cable receivers

Most video instruments require high impedance differential input amplifiers that will not load the cable even when the power is off.

Differential Input Video Loop Through Amplifier Using a Video Difference Amplifier (LT1194)

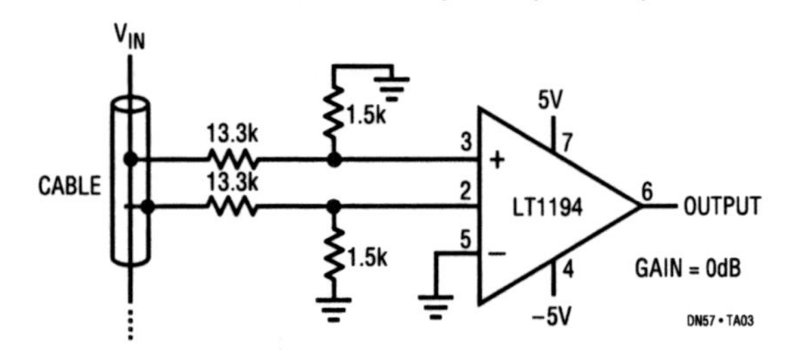

Electronically Controlled Gain, Video Loop Through Amplifier (LT1228)

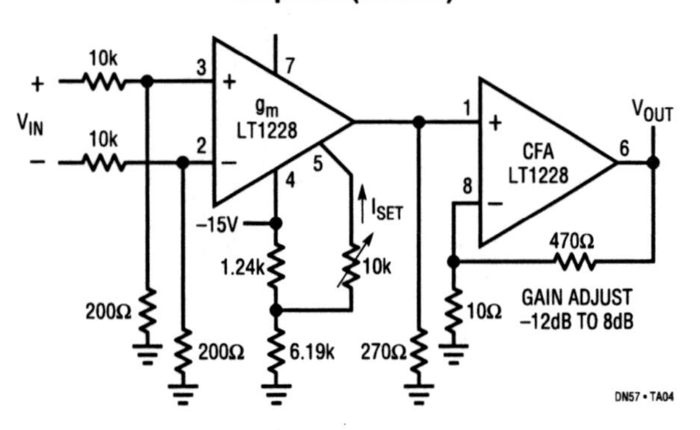

DC restore circuits

The following circuit restores the black level of a monochrome composite video signal to 0V at the beginning of every horizontal line. This circuit is also used with CCD scanners to set the black level.

Video DC Restore (Clamp) Circuit (LT1228)

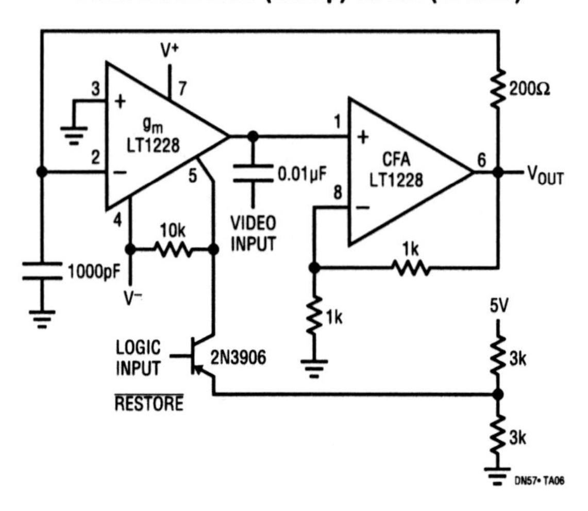

Fader circuits

Using two LT1228 transconductance amplifiers in front of a current feedback amplifier forms a video fader. The ratio of the set currents into Pin 5 determines the ratio of the inputs at the output.

Video Fader (LT1228, LT1223)

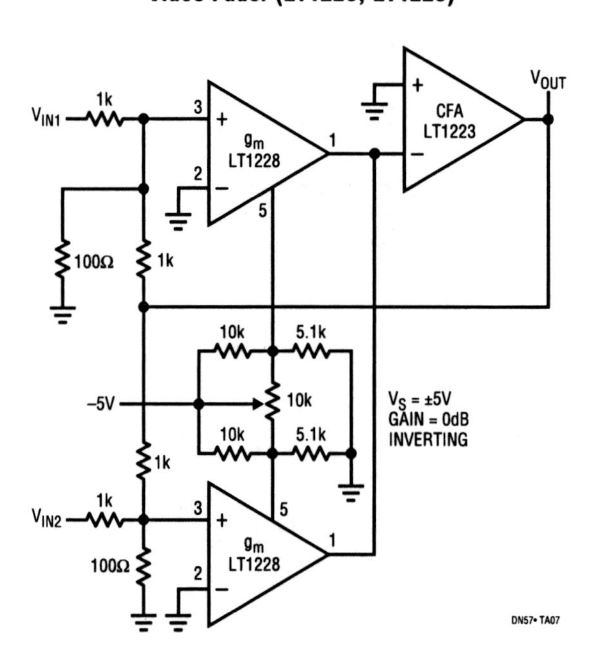

Low cost differential input video amplifiers simplify designs and improve performance

329

John Wright

The LT1190 is a family of high speed amplifiers optimized for video performance on ±5V or single +5V supplies. The family includes three voltage feedback op amps and two video difference amplifiers. All amplifiers slew at 450V/μs, and deliver ±50mA output current for driving cables. The LT1193 video difference amplifier features uncommitted high input impedance (+) and (−) inputs, and can be used in differential or single-ended configurations. In addition, the LT1193 has an adjustable gain of 2 or greater, with a −3dB bandwidth of 80MHz.

Wideband voltage controlled amplifier

The LT1193 video difference amplifier combined with an MC1496 balanced modulator make a low cost 50MHz Voltage Controlled Amplifier (VCA), shown in Figure 329.1. The input signal of the MC1496 at Pin 1 is multiplied by the Control Voltage on Pin 10, and appears as a differential output current at Pins 6 and 12. The LT1193 acts to level shift the differential signal and convert it to a single-ended output. Resistor R_B is used to set the bias current in the MC1496 to 1mA in each 200Ω R_L, while R_{CM} is used to shift the differential output into the common mode range of the LT1193. Resistors R1 through R4 bias the MC1496 inputs so that the Control Voltage V_C can be referenced to 0V. Positive VC causes positive gain; negative V_C gives a phase inversion ($-A_V$), while 0V on V_C gives maximum attenuation (within the V_{OS} of the MC1496 control inputs). The value of R_e is chosen by knowing the maximum input signal:

$$R_e = (V_{IN}\ \max)/1mA$$

For the example shown the maximum input signal is 100mV peak, therefore, $R_e = 100\Omega$. At this maximum input signal there is significant distortion from the remodulation of the input pair. Linearity can be improved by increasing R_e at the expense of gain. The maximum voltage gain of the VCA is:

$$\frac{V_0}{V_i} = (A_V \text{ of LT1193})\left(\frac{2\,R_L}{R_e + 2\,r_e}\right)$$

$$V_0/V_{IN} = 2(2 \times 200)/(100 + 52) = 5.26 = 14.42\text{dB}$$

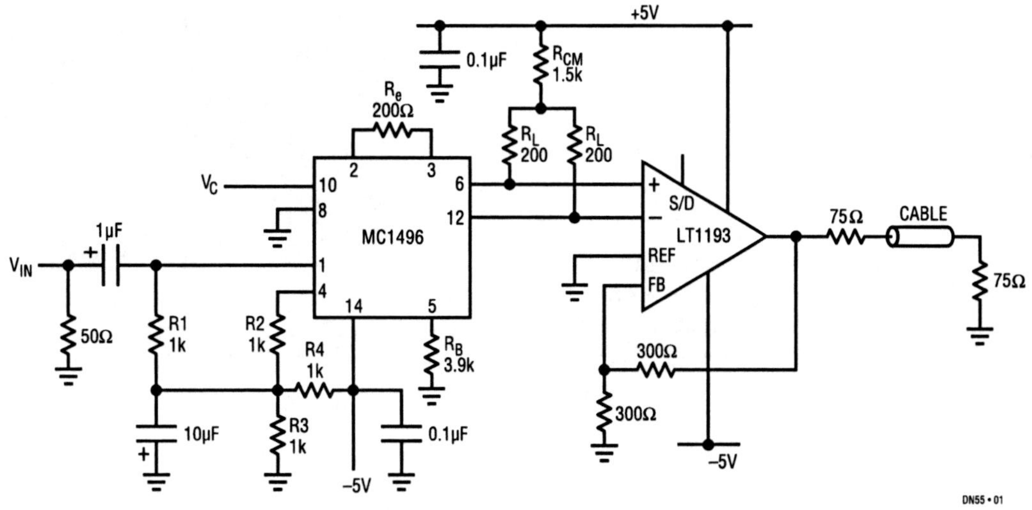

Figure 329.1 • Low Cost 50MHz Voltage Controlled Amplifier

Analog Circuit Design: Design Note Collection. http://dx.doi.org/10.1016/B978-0-12-800001-4.00329-X

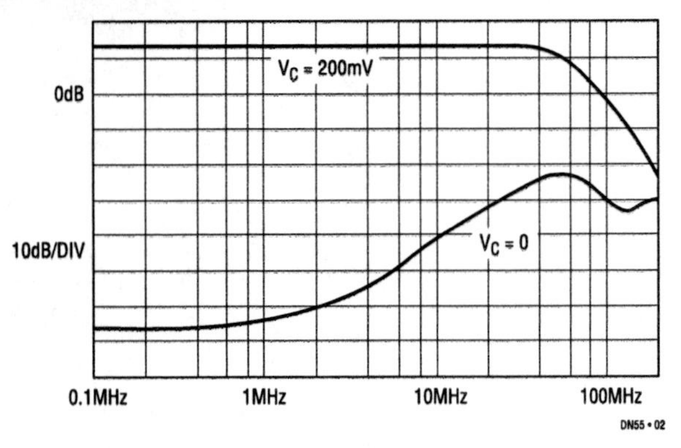

Figure 329.2 • Gain and Attenuation of VCA

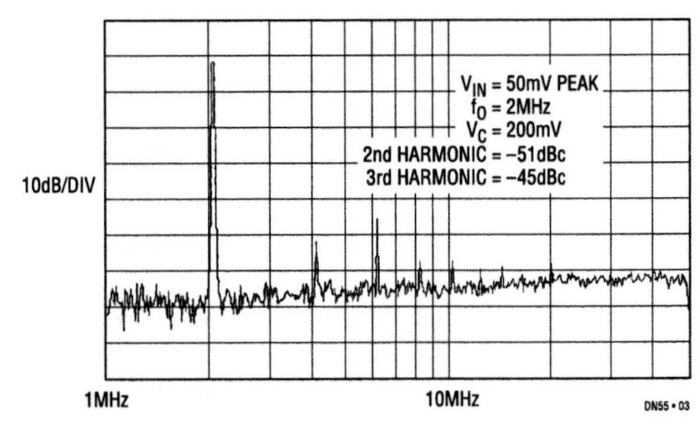

Figure 329.3 • VCA Output Spectrum

Figures 329.2 and 329.3, show the frequency response and harmonic distortion of the VCA.

The voltage gain of the VCA can be increased at the expense of bandwidth by changing the value of load resistors R_L. Shorting RCM and increasing R_L to 2k will increase the maximum gain by 20dB and the −3dB bandwidth will drop to approximately 10MHz.

The LT1193 has a shutdown feature that reduces its power dissipation to only 15mW and forces a three-state output. The three-state output occurs when Pin 5 is taken to V^-. The high Z state, dominated by the impedance of the feedback network, is useful for multiplexing several amplifiers on the same cable. The impedance of the feedback resistors should not be raised above 1kΩ because stray capacitance on the (−) input can cause instability.

Extending the input range on the LT1193

Figure 329.4 shows a simplified schematic of the LT1193. In normal operation the REF pin 1 is grounded or taken to a DC offset control voltage, while differential signals are applied between Pins 2 and 3. The LT1193 has been optimized for gains of 2 or greater, and this means the input stage must handle fairly large input signals. The maximum input signal occurs when the input differential amplifier is tilted over hard in one direction, and 1.2mA flows through the 1kΩ R_e. The maximum input swing is therefore $1.2V_P$ or $2.4V_{P-P}$. The second differential pair is running at slightly larger current so that when the first input stage limits, the second stage remains biased to maintain the feedback.

Occasionally it is necessary to handle signals larger than $2.4V_{P-P}$ at the input. The LT1193 input stage can be tricked to handle up to $4.8V_{P-P}$. To do this, it is necessary to ground Pin 3 and apply the differential input signal between Pin 1 and 2. The input signal is now applied across two 1k resistors in series. Since the input signal is applied to both input pairs, the first pair will run out of bias current before the second pair, causing the amplifier to go open loop. This effect is shown in Figure 329.5 for the amplifier operating in a closed loop gain of 1. The LT1193 has a unity gain phase margin of only 40 degrees, so when operating at unity gain, care must be taken to avoid instability.

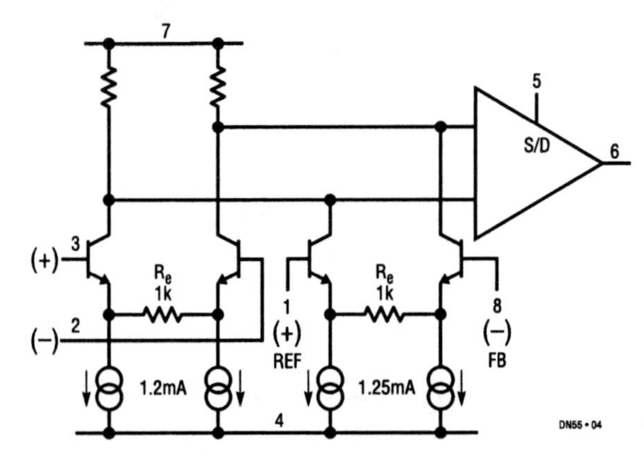

Figure 329.4 • LT1193 Simplified Schematic

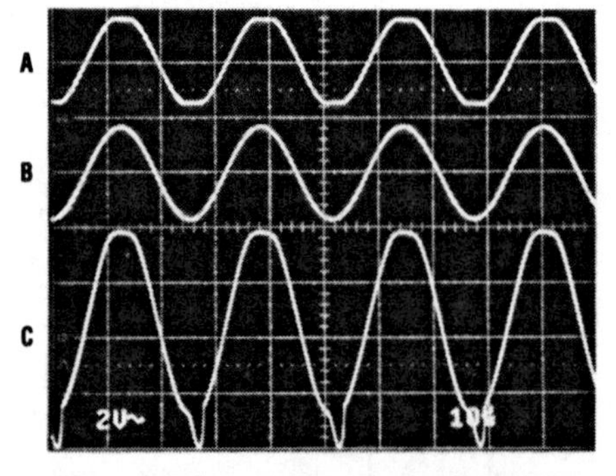

Figure 329.5 • LT1193 in Unity Gain. (A) Standard Inputs, Pins 2 to 3, $V_{IN}=3.6V_{P-P}$. (B) Extended Inputs, Pins 1 to 2, $V_{IN}=3.6V_{P-P}$ (C) Extended Inputs, Pins 1 to 2, $V_{IN}=7.0V_{P-P}$

PART 2

Mixed Signal

Section 1

Data Conversion: Analog-to-Digital

330

Generating a ±10.24V true bipolar input for an 18-bit, 1Msps SAR ADC

Guy Hoover

Introduction

The LTC2338 is an 18-bit fully differential SAR ADC that is remarkably easy to drive. This 1Msps ADC operates from a single 5V supply and achieves ±4LSB INL maximum with −111dB THD and 100dB SNR. Its fully differential ±20.48V true bipolar input range minimizes the need for range scaling, and its 2kΩ resistive input greatly reduces the charge kickback from the internal sampling capacitor.

ADCs claiming similar performance require scaling to exceed what is typically a 0V to V_{REF} input range, resulting in low impedance inputs or an additional buffer stage requirement. To band limit noise and minimize disturbances reflected into the buffer from sampling transients, other ADCs require filter circuitry composed of expensive film or C0G capacitors at the driver output. In contrast, the simple driver circuit presented here requires only a dual precision op amp and two resistors to drive the LTC2338-18. Layout strategies for this circuit are also shown.

Simple driver circuit

The circuit of Figure 330.1 uses only the LT1469 dual precision op amp and two metal film resistors to form a single-ended to differential driver for the LTC2338-18. This circuit takes a single-ended ±10.24V input voltage and converts it to the ±20.48V fully differential signal, which is required for proper operation of the LTC2338-18.

Typical offset for the driver portion of this circuit is less than the equivalent of 1LSB (156μV) for the LTC2338-18. Typical AC performance for this circuit includes THD of −110dB and SNR of 100dB. This performance can be seen in the FFT of Figure 330.2. The THD and SNR performance are similar to the typical performance numbers found in the

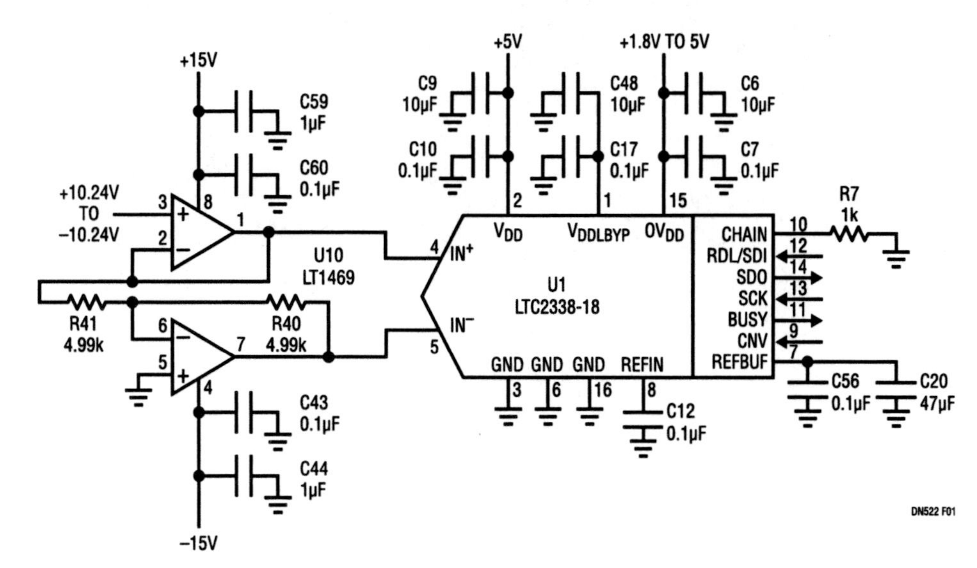

Figure 330.1 • Single-Ended to Differential Driver for LTC2338 18-Bit SAR ADC with a ±10.24V Input Range

Analog Circuit Design: Design Note Collection. http://dx.doi.org/10.1016/B978-0-12-800001-4.00330-6

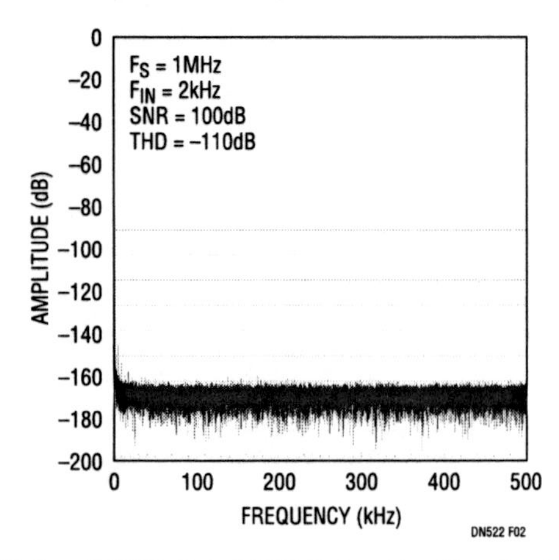

Figure 330.2 • 131072-Point FFT Using the Circuit of Figure 330.1

LTC2338-18 data sheet—this simple driver produces negligible performance degradation.

Layout is important

PC board layout can have a significant effect on the performance of a high speed 18-bit ADC. When considering layout, keep the following in mind:

- A ground plane should always be used—a solid ground plane just below the component layer is recommended.
- Keep traces as short as possible.
- Keep bypass capacitors as close to the supply pins as possible, and each bypass capacitor should have its own low impedance return to ground.
- The analog input traces should be screened by ground.
- The layout involving the ADC analog inputs should be as symmetrical as possible, so that parasitic elements cancel each other out.
- The reference bypass capacitors should be as close to the REFBUF and REFIN pins as possible.

Figure 330.3 shows a close up of the layout connecting the LT1469 and the LTC2338-18 on a demonstration board. Device, pin and component numbers shown in the photograph of Figure 330.3 correspond to the numbers shown in the schematic of Figure 330.1. See the DC1908 demo board manual and PCB files available at www.linear.com/demo/ for the complete DC1908 schematic and layout.

Conclusion

The LTC2338-18, with its large true bipolar input voltage range and resistive input, greatly simplifies the task of driving an 18-bit fully differential SAR ADC. Using the simple driver circuit presented here, consisting of only a dual precision op amp and two resistors, it is possible to maintain the good AC and DC specifications of this ADC. PCB layout is an important consideration in achieving this level of performance. Proper use of a ground plane, keeping bypass capacitors near pins being bypassed and symmetrical layout around the analog inputs help ensure a high level of performance.

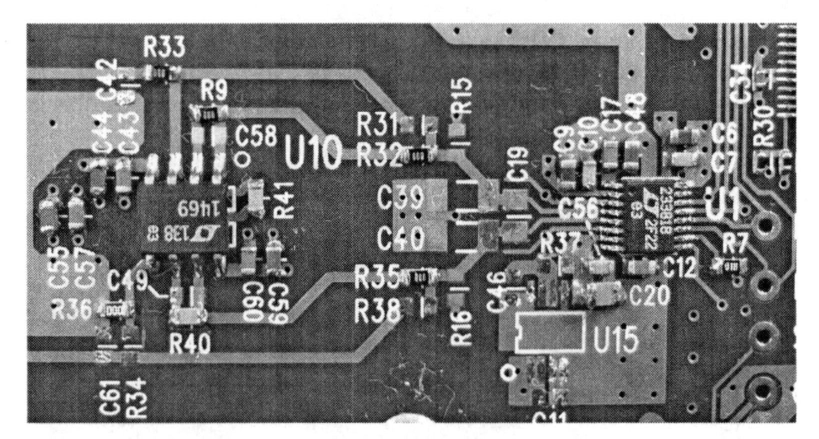

Figure 330.3 • Close Up of Demonstration Circuit DC1908 Shows Important Layout Considerations to Achieve High Level Performance

Driving a low noise, low distortion 18-bit, 1.6Msps ADC

331

Guy Hoover

Introduction

The LTC2379-18 is an 18-bit, 1.6Msps SAR ADC with an extremely high SNR of 101dB and THD of −120dB. It also features a unique digital-gain compression function, which eliminates the need for a negative supply in the ADC driver circuit.

Designing a driver circuit to get the best possible performance from the LTC2379-18 is not difficult. The two circuits presented here demonstrate differential and single-ended solutions using dual and single supplies. Note that the components used here have been carefully chosen with the ADC's accuracy and acquisition time requirements in mind, so any modifications should be thoroughly tested.

Fully differential driver

The circuit of Figure 331.1 converts a fully differential ±5V signal to a fully differential 0V to 5V signal—the normal input range for the LTC2379-18. This circuit is useful for sensors that produce a fully differential output.

Filter networks R3, R5, C6 and R4, R6, C7 limit the input bandwidth to approximately 100kHz. Matching on these networks is important to achieve the lowest distortion, as a mismatch in delay results in the development of a common mode signal. The filter network comprising R1, R2, C1, C2 and C3 minimizes the noise contribution of the LT6203 and

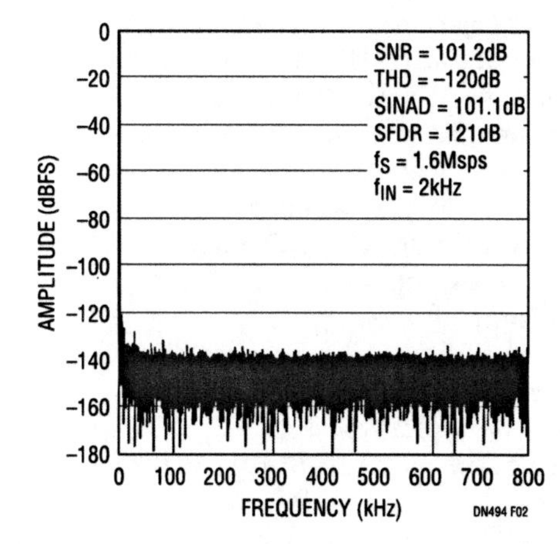

Figure 331.2 • 32k Point FFT Using the Circuit of Figure 331.1

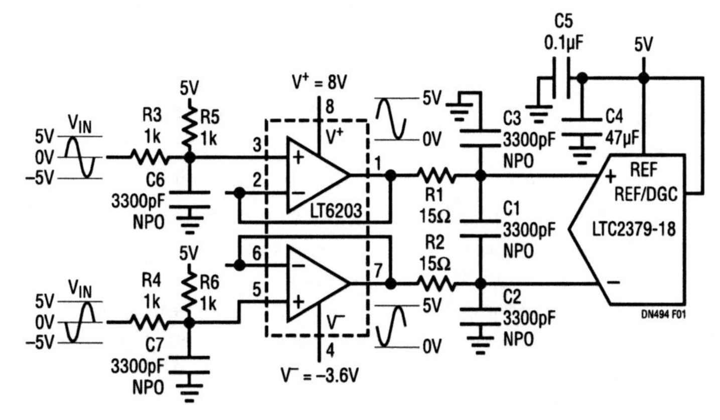

Figure 331.1 • An LTC2379-18 Fully Differential ±5V Driver Using the LT6203

Analog Circuit Design: Design Note Collection. http://dx.doi.org/10.1016/B978-0-12-800001-4.00331-8

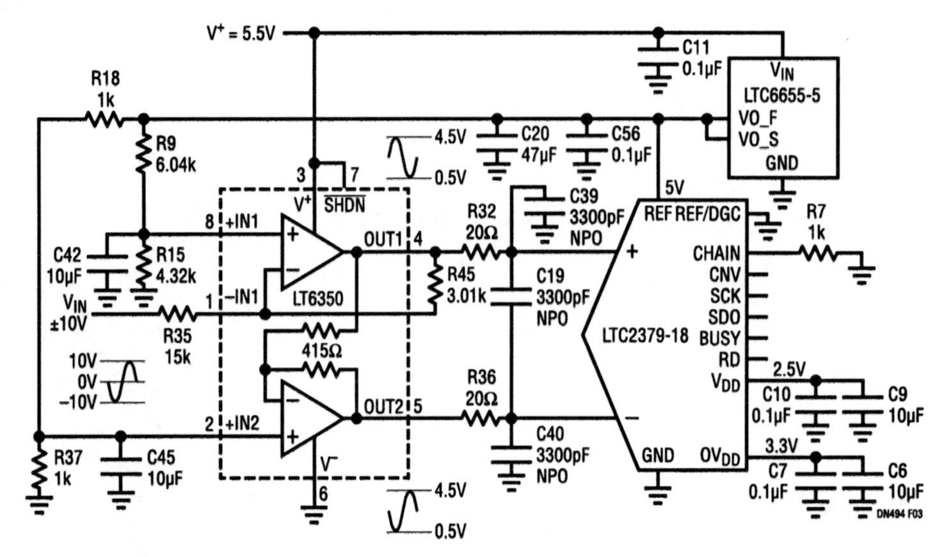

Figure 331.3 • LTC2379-18 Single Supply, ±10V Single-Ended Driver Using the LT6350

minimizes disturbances reflected into the LT6203 from sampling transients. The 32k point FFT in Figure 331.2 shows the performance of the LTC2379-18 in the circuit of Figure 331.1.

Single supply driver

The circuit of Figure 331.3 uses the digital-gain compression feature of the LTC2379-18, which defines the ADC full-scale input swing to be 10% to 90% of the reference voltage. This means that for a 5V reference the full-scale swing is 0.5V to 4.5V. This is sufficient headroom for the LT6350, so a negative supply is not needed. This not only saves the cost and complexity of providing a negative supply, it also reduces the overall power consumption of the ADC driver portion of the circuit by a factor of two.

By using the LTC6655-5 precision low noise reference, which only requires a supply 0.5V above its output, the entire circuit can be operated from a single 5.5V supply.

This circuit accepts a ±10V single-ended input voltage and converts it to a 0.5V to 4.5V fully differential signal. SNR for this circuit is 99dB due to the reduced input swing and THD is still a very good −95dB.

Layout considerations

When dealing with a high speed 18-bit ADC, PC board layout must be carefully considered. Always use a ground plane. Keep traces as short as possible. Keep bypass capacitors as close to the supply pins as possible. Each bypass capacitor should have its own low impedance return to ground. The analog input traces should be screened by ground. The layout involving the ADC analog inputs should be as symmetrical as possible so that parasitic elements cancel each other out. The output of the reference and the REF pin bypass capacitors should be as close to the REF pin as possible.

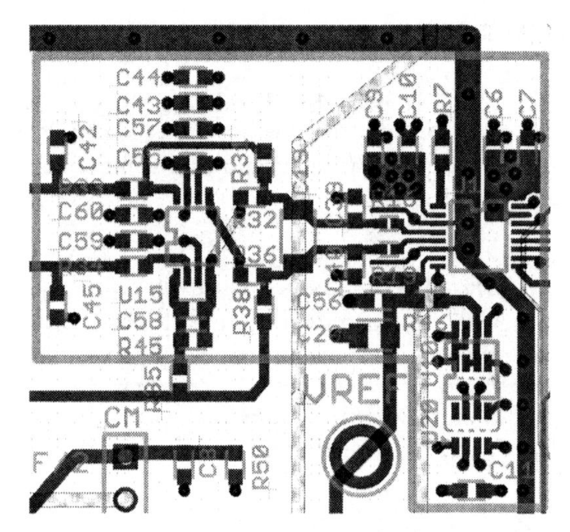

Figure 331.4 • LTC2379-18 Sample Layout

Figure 331.4 shows a sample layout for the LTC2379-18. Figure 331.4 is a composite of the top, ground, bottom and silk screen layers. Component numbers used in the circuit of Figure 331.3 refer to layout of Figure 331.4. See the DC1783A demo board manual available at www.linear.com for a complete LTC2379-18 layout example.

Conclusion

Driving the LTC2379-18 is not difficult. Using the simple circuits described here, the LTC2379-18 can be driven over a variety of input voltage ranges with fully differential or single-ended inputs. With its unique digital-gain compression function, the LTC2379-18 can be driven with a single supply, which saves power while reducing cost and complexity.

Driving lessons for a low noise, low distortion, 16-bit, 1Msps SAR ADC

332

Guy Hoover

Introduction

Designing an ADC driving topology that delivers uncompromising performance is challenging, especially when designing around an ultralow noise SAR ADC such as the 1Msps LTC2393-16. For both single-ended and differential applications, a well thought out driving topology can fully realize the ultralow noise and low distortion performance required in your data acquisition system.

The LTC2393-16 is the first in a family of high performance SAR ADCs from Linear Technology that utilizes a fully differential architecture to achieve an excellent SNR of 94.2dB and THD of −105dB. And in order to take full advantage of the ADC performance, we present driving solutions for both single-ended and differential applications. Both topologies fully demonstrate the ultralow noise and low distortion capabilities of the LTC2393-16.

Single-ended to differential converter

The circuit of Figure 332.1 converts a single-ended 0V to 4.096V signal to a differential ±4.096V signal. This circuit is useful for sensors that do not produce a differential signal. Resistors R1, R2 and capacitor C2 limit the input bandwidth to approximately 100kHz.

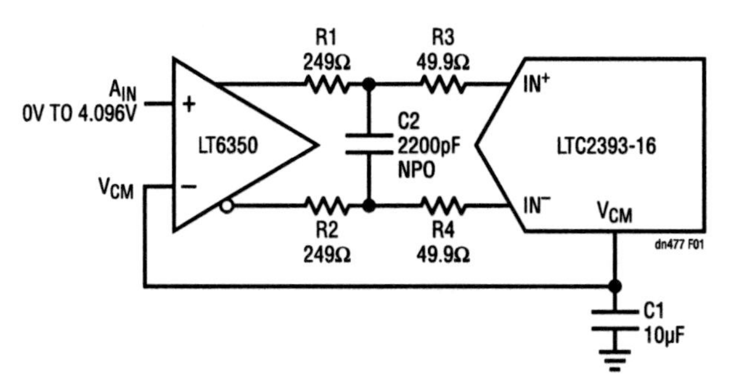

Figure 332.1 • Single-Ended to Differential Converter

When driving a low noise, low distortion ADC such as the LTC2393-16, component choice is essential for maintaining performance. All of the resistors used in this circuit are relatively low values. This keeps the noise and settling time low. Metal film resistors are recommended to reduce distortion caused by self-heating. An NPO capacitor is used for C2 because of its low voltage coefficient, which minimizes distortion. The excellent linearity characteristics of NPO and silver mica capacitors make these good choices for low distortion applications. Finally, the LT6350 features low noise, low distortion and a fast settling time.

The 16k-point FFT in Figure 332.2 shows the performance of the LTC2393-16 in the circuit of Figure 332.1. The measured SNR of 94dB and THD of −103dB match closely with the typical data sheet specs for the LTC2393-16, showing that little, if any, degradation of the ADC's specifications result from inserting the single-ended to differential converter into the signal path.

Fully differential drive

The circuit of Figure 332.3 AC-couples and level shifts the sensor output to match the common mode voltage of the ADC. The lower frequency limit of this circuit is about 10kHz. The lower frequency limit can be extended by increasing the values of C3 and C4. This circuit is useful for sensors with low impedance differential outputs.

The circuit of Figure 332.1 could be AC-coupled in a similar manner. Simply bias A_{IN} to V_{CM} through a 1k resistor and couple the signal to A_{IN} through a 10μF capacitor.

PCB layout

The circuits shown are quite simple in concept. However, when dealing with a high speed 16-bit ADC, PC board layout must also be considered. Always use a ground plane. Keep traces as short as possible. If a long trace is required for a bias

Analog Circuit Design: Design Note Collection. http://dx.doi.org/10.1016/B978-0-12-800001-4.00332-X

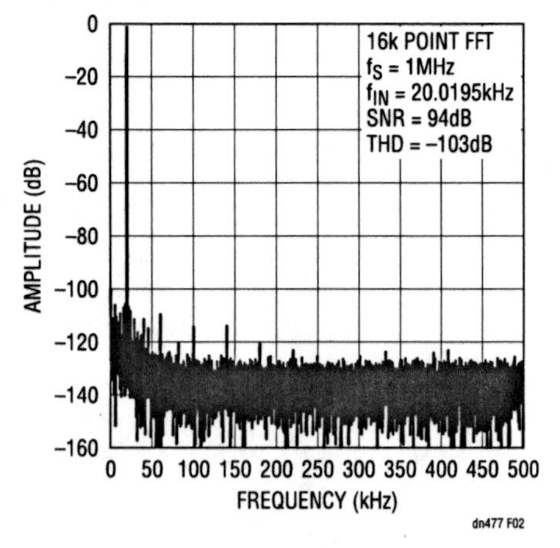

Figure 332.2 • LTC2393-16 16k Point FFT Using Circuit of Figure 332.1

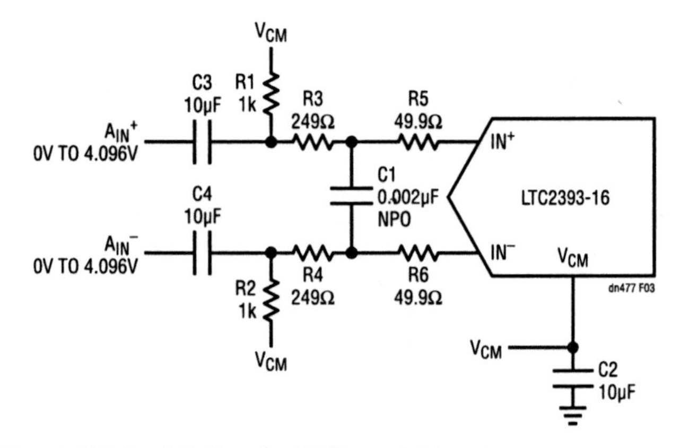

Figure 332.3 • AC-Coupled Differential Input

node such as V_{CM}, use additional bypass capacitors for each component attached to the node and make the trace as wide as possible. Keep bypass capacitors as close to the supply pins as possible. Each bypass capacitor should have its own low impedance return to ground. The analog input traces should be screened by ground. The layout involving the analog inputs should be as symmetrical as possible so that parasitic elements cancel each other out.

Figure 332.4 shows a sample layout for the LTC2393-16. Figure 332.4 is a composite of the top metal, ground plane and silkscreen layers. See the DC1500A Quick Start Guide available at www.linear.com for a complete LTC2393-16 layout example.

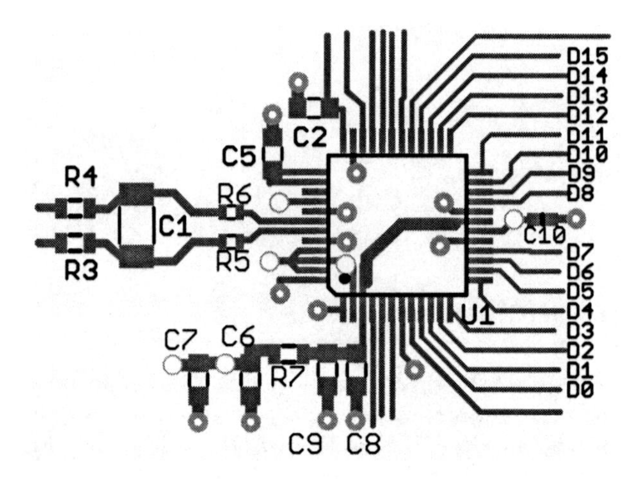

Figure 332.4 • Sample Layout for LTC2393-16

Conclusion

The LTC2393-16 with its fully differential inputs can improve SNR by as much as 6dB over conventional differential input ADCs. This ADC is well suited for applications that require low distortion and a large dynamic range. Realizing the potential low noise, low distortion performance of the LTC2393-16 requires combining simple driver circuits with proper component selection and good layout practices.

Maximize the performance of 16-bit, 105Msps ADC with careful IF signal chain design

333

Clarence Mayott Derek Redmayne

Introduction

Modern communication systems require an ADC to receive an analog signal and then convert it into a digital signal that can be processed with an FPGA. The job of a mixed signal engineer is to optimize the signal at the input of the ADC to maximize overall system performance. This usually requires a signal chain comprised of multiple gain and filtering sections. An ADC is only as good as the signal it is measuring.

For instance, the LTC2274 provides excellent AC performance with an appropriate IF signal chain. The LTC2274 is a 16-bit, 105Msps ADC that serially transmits 8B/10B encoded output data compliant with the JESD204 specification. It uses a single differential transmission line pair to reduce the number of IO lines required to transmit output data. The LTC2274 has 77dB of SNR, and 100dB of spurious free dynamic range.

Signal chain topology

Figure 333.2 details a signal chain optimized for a 70MHz center frequency and a 20MHz bandwidth driving the LTC2274. The final filter and circuitry around the ADC are shown in detail. The earlier stages of the chain can be changed to suit a target application.

The first stage of amplification in the chain uses an AH31 from TriQuint Semiconductor. This GaAs FET amplifier offers a low noise figure and high IP3 point, which minimizes distortion caused by the amplifier stage. It provides 14dB of gain over a wide frequency region. The high IP3 prevents intermodulation distortion between frequencies outside the passband of the surface acoustic wave (SAW) filter.

A SAW filter follows the amplification stage for band selection. The SAW filter offers excellent selectivity and a flat passband if matched correctly. Gain before the SAW must not be higher than the maximum input power rating of the SAW; otherwise it leads to distortion. A digitally controlled step attenuator may be required in the signal path to control the power going into the SAW filter.

The second stage of amplification is used to recover the loss in the SAW filter. The insertion loss of the SAW filter is about −15dB, so the final amplifier should have at least this much gain, plus enough gain to accommodate the final filter. By splitting the gain between two amplifiers, the noise and distortion can be optimized without overdriving the SAW filter. It also allows for a final filter with better suppression of noise from the final amplifier, improving SNR and selectivity.

The output stage of the final filter needs to be absorptive to accommodate the ADC front end. This suppresses glitches reflected back from the direct sampling process.

This signal chain will not degrade the performance of the LTC2274. When receiving a 4-channel WCDMA signal with a 20MHz bandwidth, centered at 70MHz, the ACPR is 71.5dB (see Figure 333.1).

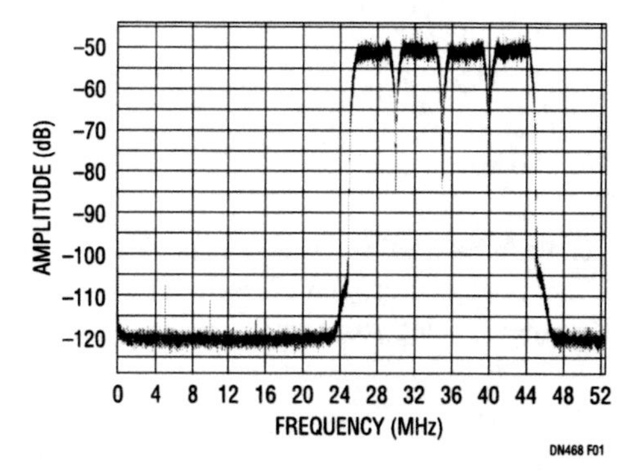

Figure 333.1 • Typical Spread Spectrum Performance

Analog Circuit Design: Design Note Collection. http://dx.doi.org/10.1016/B978-0-12-800001-4.00333-1

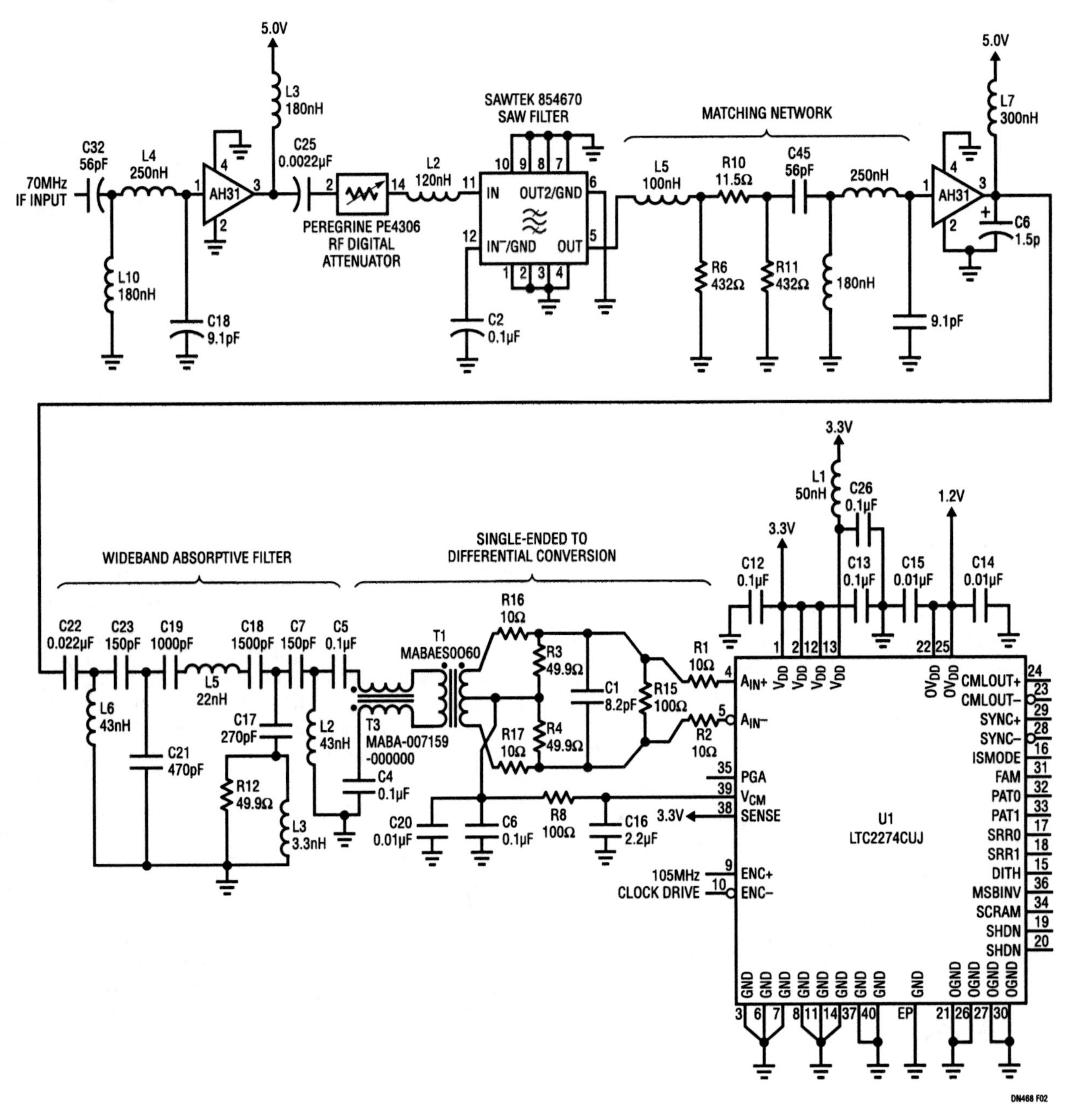

Figure 333.2 • IF Receiver Chain

Conclusion

The LTC2274 can be used to receive high IF frequencies, but getting the most out of this high performance ADC requires a carefully designed analog front end. The performance of the LTC2274 is such that it is possible to dispense with the automatic gain control and build a receiver with a low fixed gain. The LTC2274 is a part of a family of 16-bit converters that range in sample rate from 65Msps to 105Msps. For complete schematics of this receiver network, visit www.linear.com.

Upgrade your microcontroller ADC to true 12-bit performance

334

Guy Hoover

Introduction

Many 8-bit and 16-bit microcontrollers feature 10-bit internal ADCs. A few include 12-bit ADCs, but these often have poor or nonexistent AC specifications, and certainly lack the performance to meet the needs of an increasing number of applications. The LTC2366 and its slower speed versions offer a high performance alternative, as shown in the AC specifications in Table 334.1. Compare these guaranteed specifications with the ADC built into your current microcontroller.

This family's DC specifications are equally impressive. INL and DNL are guaranteed to be less than ±1LSB. Operating from a single 2.5V, 3V or 3.3V supply, the current draw on these parts is a maximum of 4mA during a conversion. This can be reduced to less than 1μA by placing the part into SLEEP mode during periods of inactivity, which greatly reduces the average supply current at lower sample rates.

These ADCs are available in tiny 6-lead and 8-lead TSOT-23 packages. The 8-lead devices have adjustable V_{REF} and OV_{DD} pins. The adjustable V_{REF} pin allows the input span to be reduced to 1.4V. This, combined with the high ADC input impedance, can eliminate the need for gain or buffer stages in many applications. The OV_{DD} pin, which controls the digital output level, can be adjusted from 1V to 3.6V, simplifying communication with different logic families. For applications that do not require an adjustable reference or adjustable output levels, the 6-lead device with $V_{REF} = OV_{DD} = V_{DD}$ should suffice.

The SPI interface requires only three wires to communicate with the microcontroller, keeping the overall solution size small in low power, high speed applications.

Application circuits

Figure 334.1 shows a single supply AC-coupled amplifier driving the LTC2366. This circuit is useful in applications where the sensor output level is too low to achieve full SNR performance from the ADC. The output of the LT6202 swings rail-to-rail. This feature maximizes the circuit's dynamic range

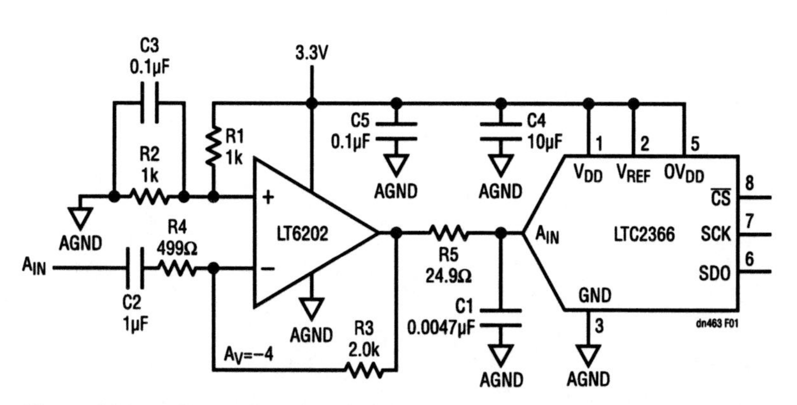

Figure 334.1 • Single Supply AC-Coupled Amplifier Level Shifts Input for Maximum Dynamic Range

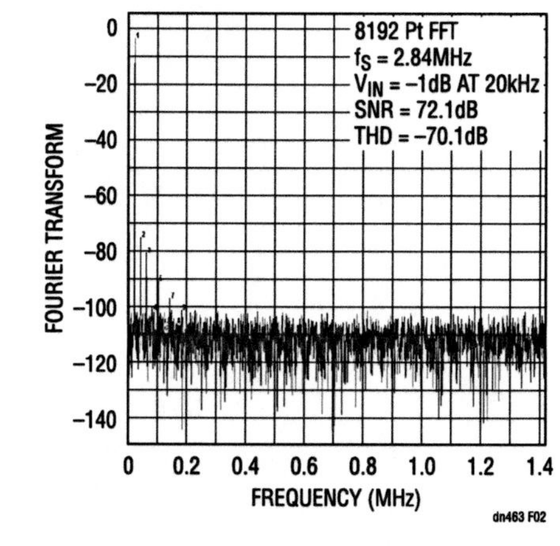

Figure 334.2 • FFT Shows Low Noise and Distortion of Figure 334.1 Circuit

Analog Circuit Design: Design Note Collection. http://dx.doi.org/10.1016/B978-0-12-800001-4.00334-3

when the op amp output is level shifted to the center of the ADC's swing. The FFT of Figure 334.2 demonstrates the low noise and distortion of this circuit.

In Figure 334.3, a single supply DC amplifier with a programmable gain of 0 to 4096 drives the LTC2360. With a maximum offset of 10μV and a DC to 10Hz noise of 2.5μV_{P-P}, the LTC6915 is a good choice for high gain applications. This circuit is useful for very low level signals or for applications with a wide range of input levels.

Conclusion

The 12-bit ADCs in the LTC236x family guarantee AC specifications that most built-in microcontroller ADCs cannot meet, thus improving performance when used in place of on-chip ADCs. The LTC236x family is easily interfaced to most microcontrollers via its SPI interface. A wide range of sample rates, an external reference pin and a separate OV_{DD} pin provide additional flexibility.

Table 334.1 LTC236x ADC Family AC Specifications

PART NUMBER	SAMPLE RATE	SINAD	SNR	THD	FULL LINEAR BANDWIDTH
LTC2366	3Msps	68dB (Min)	69dB (Min)	−72dB (Max)	2.5MHz (Typ)
LTC2365	1Msps	68dB (Min)	70dB (Min)	−72dB (Max)	2.0MHz (Typ)
LTC2362	500ksps	72dB (Typ)	73dB (Typ)	−85dB (Typ)	1.0MHz (Typ)
LTC2361	250ksps	72dB (Typ)	73dB (Typ)	−85dB (Typ)	1.0MHz (Typ)
LTC2360	100ksps	72dB (Typ)	73dB (Typ)	−85dB (Typ)	1.0MHz (Typ)

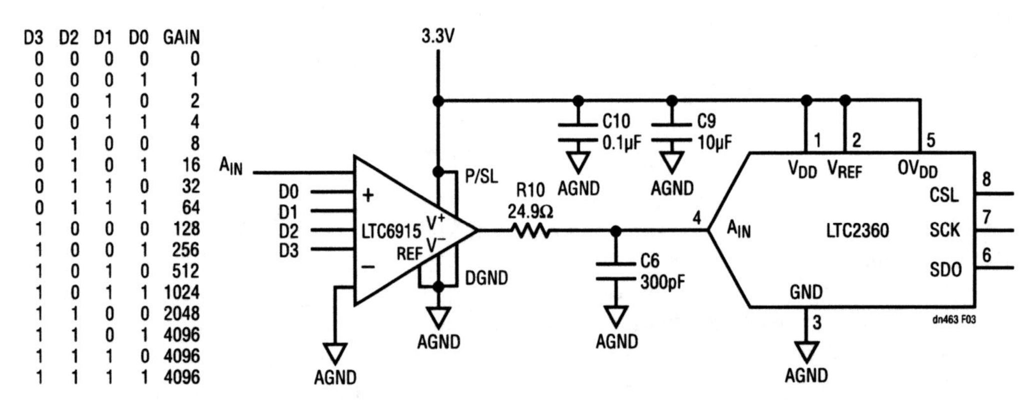

Figure 334.3 • Single Supply DC Amplifier Provides Programmable Gain from 0 to 4096

Digitize a $1000 sensor with a $1 analog-to-digital converter

335

Mark Thoren

Introduction

The LTC2450 is a 16-bit, single-ended input, delta-sigma ADC in a 2mm × 2mm DFN package, but don't let its small size and low cost fool you. The LTC2450 has impressive DC specs, including 2 LSB INL, 2 LSB offset and a 0.01% gain error, making it a perfect match for many high-end industrial sensors, as well as a wide variety of data acquisition, measurement, control and general purpose voltage monitoring applications.

Digitize an accurate sensor with an accurate ADC

The Setra Model 270 is an exquisitely accurate barometric pressure sensor often used in weather stations and semiconductor manufacturing. A 600 to 1100 millibar input range corresponds to a 0V to 5V output. Despite its high accuracy, it does not require an expensive ADC to take full advantage of this sensor's performance. The LTC2450's DC specifications are more than adequate for the 270's 0.05% accuracy. Figure 335.1 shows the basic connections, and Figure 335.2 shows the change in barometic pressure from walking up and down a flight of stairs, taking a short break on each stair. After sample 2500, the sensor is resting on the bottom stair measuring only the change in ambient pressure. The spikes at the end of the graph are from a door opening and closing. The Setra 270 must be treated as a 4-terminal device, as the negative output is nominally 5V above the negative excitation. Figure 335.3 shows an isolated supply for the sensor's excitation, and the 4.7μF capacitor keeps switching noise from affecting the measurement.

Not so obvious features

Mixed signal designers often try to extract more resolution from an ADC by averaging samples. Averaging reduces the signal bandwidth and improves resolution, but it consumes processing resources, complicates firmware and necessitates the use of a fast ADC even though the application may require a much slower data rate. It also does not improve accuracy. The LTC2450 is inherently very accurate, and it does the averaging for you. The front-end sample rate of the LTC2450 is

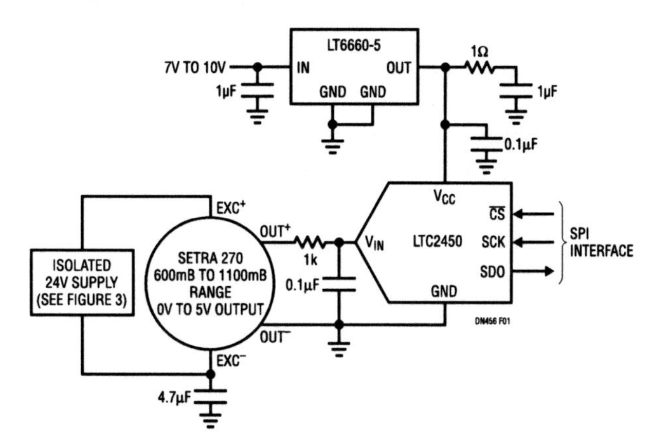

Figure 335.1 • 16-Bit Barometric Pressure Measurement

Analog Circuit Design: Design Note Collection. http://dx.doi.org/10.1016/B978-0-12-800001-4.00335-5

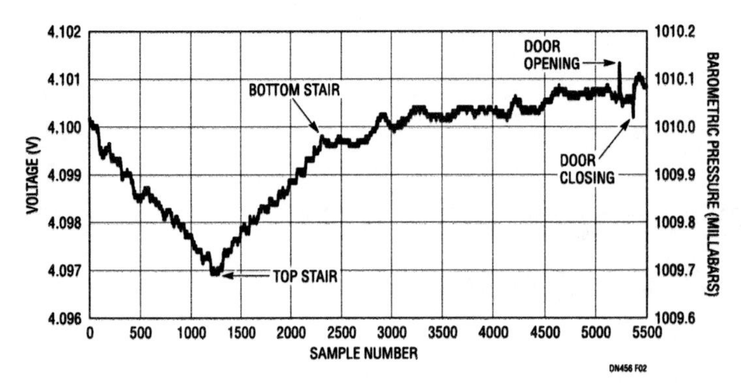

Figure 335.2 • Stairway Pressure

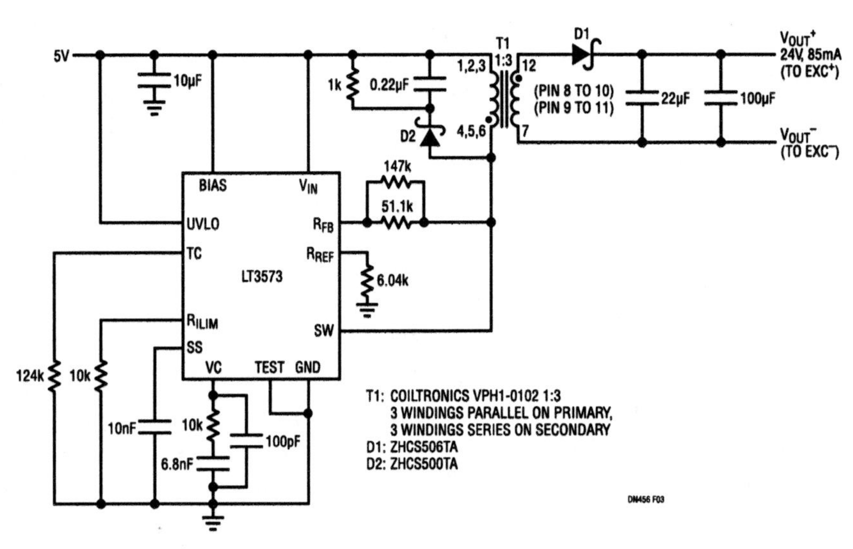

Figure 335.3 • 24V Isolated Excitation Supply

3.9Msps, which is decimated to 30 samples per second by a Sinc digital filter. The filter's effective bandwidth is approximately 30Hz, which means noise between 30Hz and 3.9MHz is greatly attenuated. When combined with a simple 1-pole filter, wideband noise is generally not a concern.

A breakthrough feature of the LTC2450 is the proprietary modulator switching scheme that reduces the average input current by orders of magnitude compared to ADCs with similar specifications. Ordinarily, an RC filter produces offset and gain errors due to the ADC's average input current flowing through the resistor. The very low 50nA average sampling current of the LTC2450 produces less than a 1LSB error with a 1k, 0.1μF filter. This filter is more than adequate for limiting the wideband noise of most active devices driving the ADC, as well as preventing sampling current "spikes" from the modulator from affecting the source.

Conclusion

With manufacturers claiming up to 32 bits of resolution on their precision ADCs, and a confusing array of choices between 8 and 32 bits, often with highly obfuscated data sheets, where does a "medium speed, medium resolution" part like the LTC2450 fit in? It fits pretty much anywhere where you would use a 4-1/2 digit (40,000 count) digital voltmeter to test your circuit. At the other end of the performance spectrum, the LTC2450 is more economical than many 12-bit ADCs. The next time you are searching for an ADC, consider the LTC2450. It may be just what you need for performance, while occupying a tiny amount of board space that matches its tiny price ($1.15 each in 1000-piece quantities).

True rail-to-rail, high input impedance ADC simplifies precision measurements

336

Mark Thoren

Introduction

High input impedance and a wide input range are two highly desirable features in a precision analog-to-digital converter, and the LTC2449 delta-sigma ADC has both. With just a few external components, the LTC2449 forms an exceptional measurement system with very high input impedance and an input range that extends 300mV beyond the supply rails.

A designer may trade off the LTC2449's 200nV resolution for faster conversion rates, but otherwise the LTC2449 requires few to no performance trade-offs. It simultaneously achieves 1ppm linearity (Figure 336.2), 200nV input resolution and a 5V input span. Ten filter oversample ratios are available, providing data rates from 6.8 samples per second to 3500 samples per second. Normal mode rejection of 50Hz and 60Hz is better than 87dB in the 6.8sps mode. *All DC specifications hold for all speeds*—only the resolution changes. Such persistent high performance simplifies the design of otherwise challenging applications, such as 6-digit voltmeters, sensor interfaces, and industrial control. In addition, the LTC2449 digital interface and timing are extremely simple, and the No Latency architecture eliminates concerns about filter settling when scanning multiple input channels.

Solving common issues

One unique feature of the LTC2449 is that the analog inputs are routed to the MUXOUT pins, and an external buffer isolates these signals from the switched capacitor ADC inputs (See Figure 336.1). The external buffer yields high impedance through the multiplexer and back to the analog inputs. This has a distinct advantage over integrated buffers because

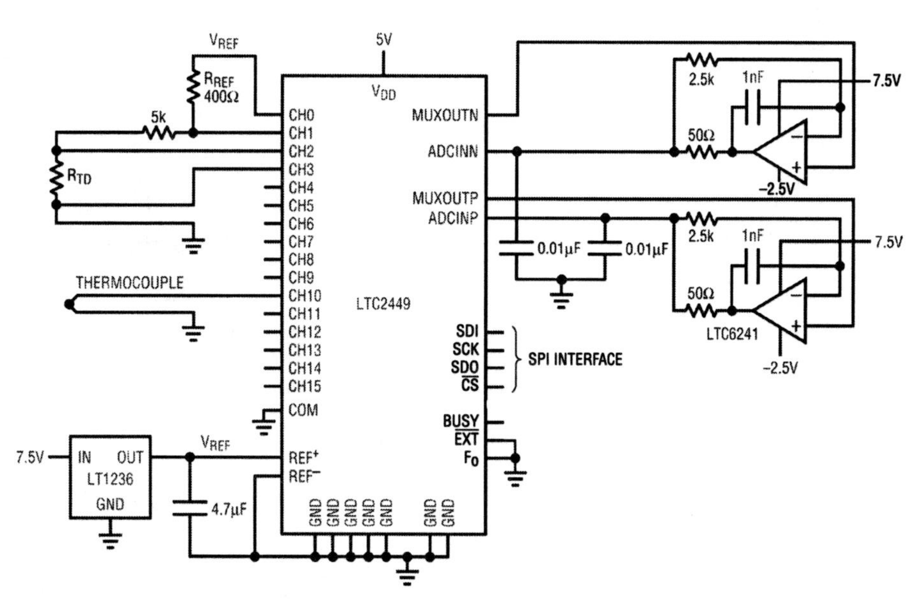

Figure 336.1 • Temperature Sensing Application Example

Analog Circuit Design: Design Note Collection. http://dx.doi.org/10.1016/B978-0-12-800001-4.00336-7

the analog inputs are truly rail-to-rail, and slightly beyond, with appropriate buffer supply voltages.

The LTC6241 is a precision CMOS amplifier with 1pA bias current and impressive DC specifications: the maximum offset is 125μV and the open loop gain is 1.6 million, typical. While the offset is not important in this application because it is removed by the LTC2449's multiplexer switching technique, the high open loop gain ensures that the 10ppm typical gain error of the LTC2449 does not degrade. Figure 336.1 shows proper interfacing of the LTC6241 to the LTC2449. The amplifier's 0.01μF capacitive load and compensation network provides the LTC2449 with a charge reservoir to average the ADC's sampling current while the 2.5k feedback resistor maintains DC accuracy.

The LTC6241 has a rail-to-rail output stage, and an input common mode range from the negative supply to 1.5V lower than the positive supply. Since no rail-to-rail amplifier can actually pull its outputs to the rails, an LT3472 boost/inverting regulator is used to create the −2.5V and 7.5V op amp supplies from the 5V supply as shown in Figure 336.3. This regulator can provide enough current for several amplifiers and other circuitry that *really* needs to swing to the rails. In addition, the LT3472's 1.1MHz switching frequency is close to the middle of the LTC2449 digital filter stopband. The center of the stopband is 900kHz when using the internal conversion clock and is independent of the selected speed mode.

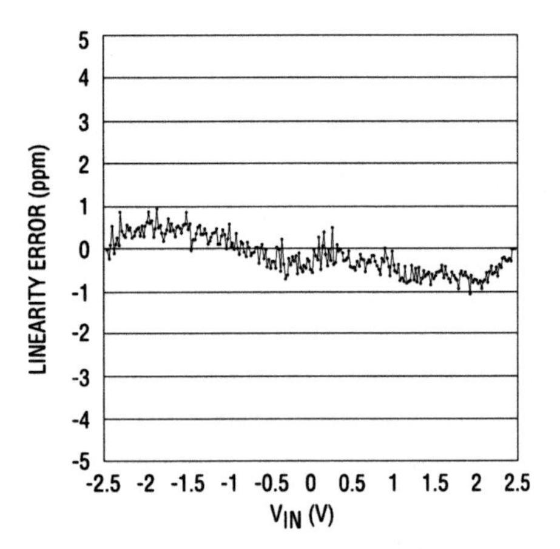

Figure 336.2 • LTC2449 Integral Nonlinearity

Applications

The LTC2449 is commonly used with thermocouples and RTDs as shown in Figure 336.1. Thermocouple outputs produce very small changes (tens of microvolts per degree C) and the output will be negative if the thermocouple is colder than the "cold junction" connection from the thermocouple to the copper traces on the PCB. The RTD is measured by comparing the voltage across the RTD to the voltage across a reference resistor. This provides a very precise resistance comparison and it does not require a precise current source. Grounding the sensors as shown is a good first line of defense for reducing noise pickup; however, the ADC must accommodate input signals that are very close to or slightly outside the supply rails. The LTC2449 handles these signals perfectly.

Conclusion

The LTC2449 solves many of the problems that designers encounter when trying to apply delta-sigma ADCs in demanding applications. High impedance, rail-to-rail inputs and a very simple serial interface simplify both hardware and software design.

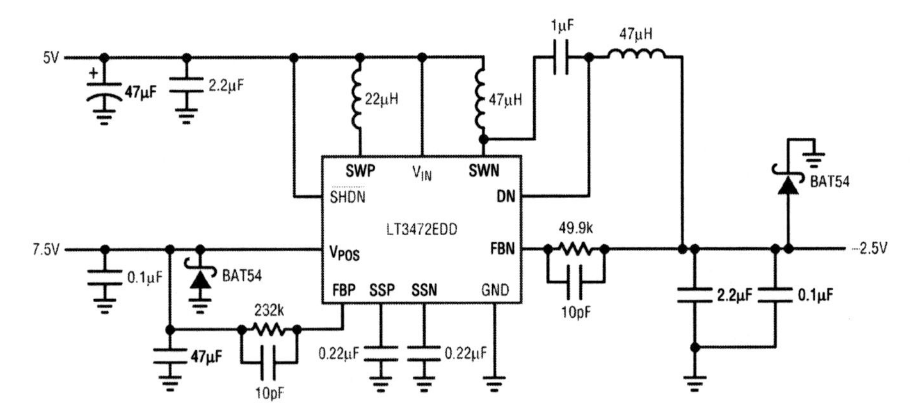

Figure 336.3 • Power Supply for Buffers

Easy Drive ADCs simplify measurement of high impedance sensors

337

Mark Thoren

Delta-sigma ADCs, with their high accuracy and high noise immunity, are ideal for directly measuring many types of sensors. Nevertheless, input sampling currents can overwhelm high source impedances or low-bandwidth, micropower signal conditioning circuits. The LTC2484 family of delta-sigma converters solves this problem by balancing the input currents, thus simplifying or eliminating the need for signal conditioning circuits.

A common application for a delta-sigma ADC is thermistor measurement. Figure 337.1 shows the LTC2484 connections for direct measurement of thermistors up to 100kΩ. Data I/O is through a standard SPI interface and the sampling current in each input is approximately:

$$\frac{\left(\frac{V_{REF}}{2}\right) - V_{CM}}{1.5M\Omega}, \text{ where } V_{CM} = \frac{V_{IN^+} + V_{IN^-}}{2}$$

or about 1.67µA when V_{REF} is 5V and both inputs are grounded.

Figure 337.2 shows how to balance the thermistor such that the ADC input current is minimized. If reference resistors R1 and R4 are exactly equal, the input current is zero and no errors result. If the reference resistors have a 1% tolerance, the maximum error in the measured resistance is 1.6Ω due to the slight shift in common mode voltage; far less than the 1% error of the reference resistors themselves. No amplifier is required, making this an ideal solution in micropower applications.

It may be necessary to ground one side of the sensor to reduce noise pickup or simplify wiring if the sensor is remote. The varying common mode voltage produces a 3.5kΩ full-scale error in the measured resistance if this circuit is used without buffering.

Figure 337.3 shows how to interface a very low power, low bandwidth op amp to the LTC2484. The LT1494 has excellent DC specs for an amplifier with 1.5µA supply current—the maximum offset voltage is 150µV and the open loop gain

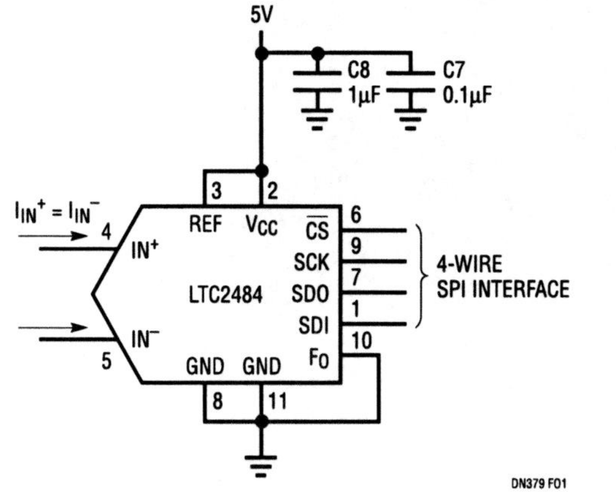

Figure 337.1 • LTC2484 Connections

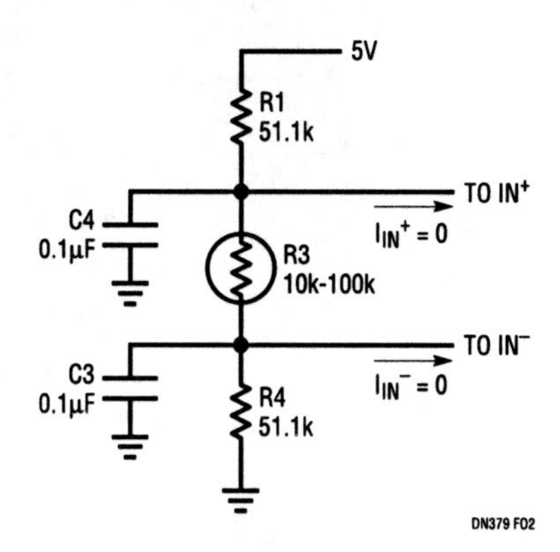

Figure 337.2 • Centered Sensor

Analog Circuit Design: Design Note Collection. http://dx.doi.org/10.1016/B978-0-12-800001-4.00337-9

is 100,000—but its 2kHz bandwidth makes it unsuitable for driving conventional delta-sigma ADCs. Adding a 1kΩ, 0.1μF filter solves this problem by providing a charge reservoir that supplies the LTC2484's instantaneous sampling current, while the 1kΩ resistor isolates the capacitive load from the LT1494. Don't try this with an ordinary delta-sigma ADC—the sampling current from ADCs with specifications similar to the LTC2484 family would result in a 1.4mV offset and a 0.69mV full-scale error in the circuit shown in Figure 337.3. The LTC2484's balanced input current allows these errors to be easily cancelled by placing an identical filter at IN^- (Figures 337.4 and 337.5).

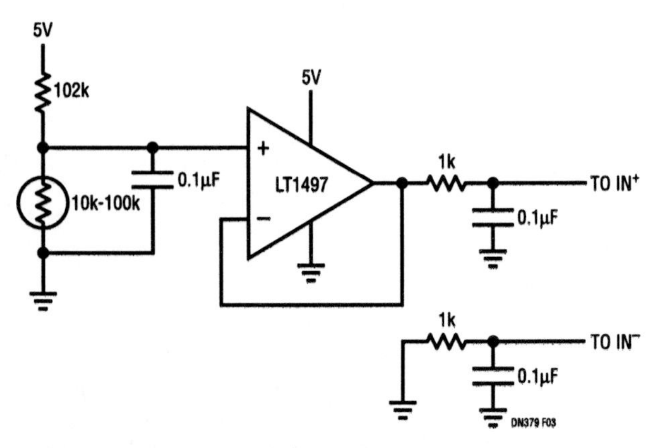

Figure 337.3 • Grounded, Buffered Sensor

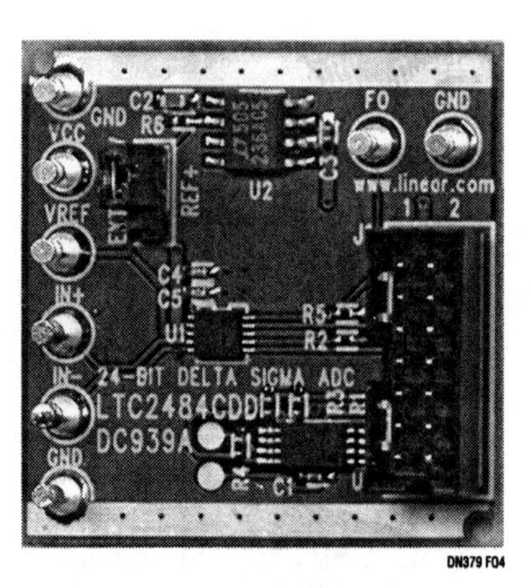

Figure 337.4 • LTC2484 Demo Board

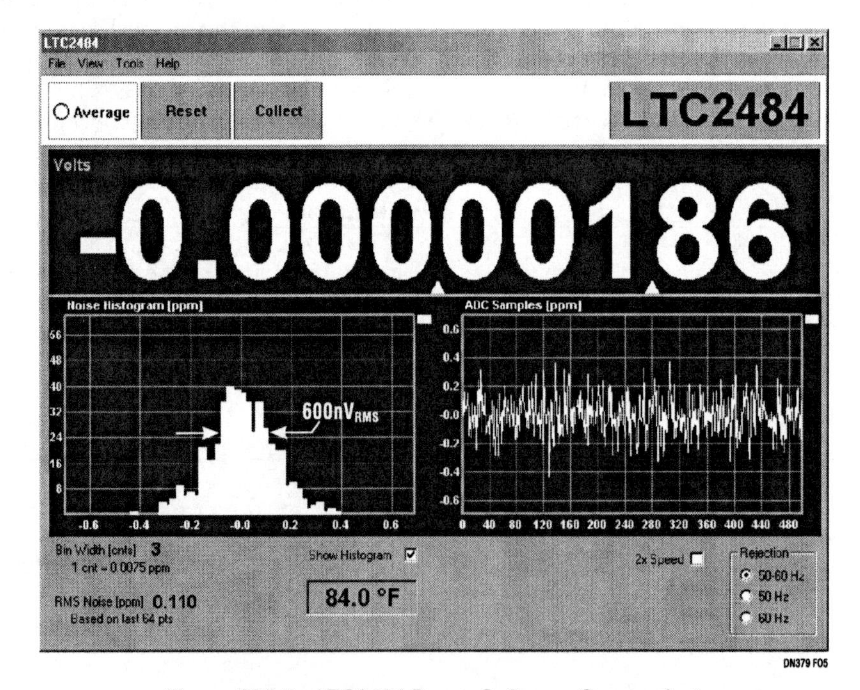

Figure 337.5 • LTC2484 Demo Software Screenshot Showing Microvolt Offset and $600nV_{RMS}$ Noise

Easy Drive delta-sigma analog-to-digital converters cancel input current errors

338

Mike Mayes

Introduction

It is now possible to place large RC networks directly in front of high resolution ΔΣ analog-to-digital converters without degrading their DC accuracy (see Figure 338.1). The LTC248x family of converters solves this problem with Easy Drive technology, a fully passive sampling network that automatically cancels the differential input current. Easy Drive technology does not use on-chip buffers, which compromise performance (see What is Wrong with On-Chip Buffers?), but instead uses a new architecture that maintains 0.002% full-scale error with input RC networks up to 100kΩ and 10μF. This new technology offers many advantages over previous generation ΔΣ ADCs:

- Rail-to-rail common mode input range
- Direct digitization of high impedance sensors
- Elimination of sampling spikes seen at the ADC input pins
- Simple external lowpass filtering
- Noise/power reduction
- Cancellation of external RC settling errors
- Easy interface to external amplifiers
- Removal of transmission line effects for remote sensors

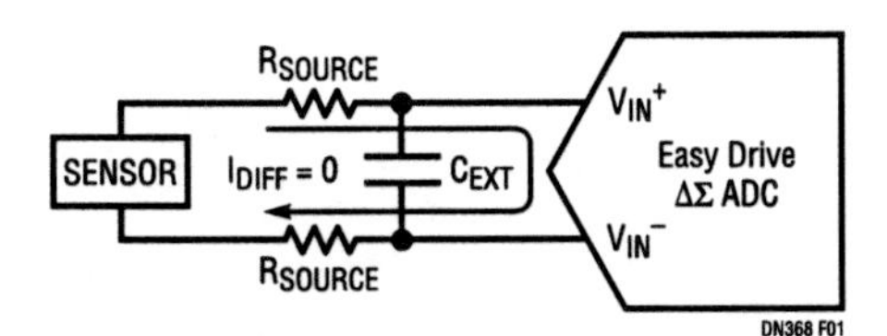

Figure 338.1 • Easy Drive Technology Automatically Cancels Differential Input Current, Thus Allowing Direct Digitization of Large External RC Networks

How does it work?

Delta-Sigma converters achieve high resolution by combining many low resolution conversions into one high resolution result. Most commercially available ΔΣ converters combine hundreds or even thousands of 1-bit conversions into a single 16-, 20- or 24-bit result. The obvious advantage is that it's much easier to implement a 1-bit converter than a 24-bit converter. In order to achieve high resolution, the input is sampled many times during the conversion cycle.

The problem is that the input structure of ΔΣ converters is a switched capacitor network. Capacitors are rapidly switched (up to 10MHz) between the input, reference and ground as a function of the final output code. Each time these capacitors are switched to the ADC input, a current pulse is generated. A pattern of charging/discharging pulses is seen at the input pin of the ADC. This pattern is a complex function of the input and reference voltages. External RC networks that do not completely settle during each sample period cause large DC errors.

The trick to solving this problem is to take advantage of the oversampling properties of ΔΣ converters. The front-end capacitor switching on a per sample basis is identical to conventional ΔΣ converter sampling. An innovative front-end sampling architecture controls the switching pattern of the capacitor array. When summed over the entire conversion cycle, the total differential input current is zero, independent of the differential input voltage, common mode input voltage, reference voltage or output code. The common mode input current is constant and proportional to the difference between the input common mode voltage and reference common mode voltage.

RC networks placed in front of ΔΣ ADCs significantly improve their performance and ease-of-use while providing lowpass and anti-alias filtering. External RC networks applied to the input of the LTC248x simply integrate (average) the input current spikes generated by the ADC.

Analog Circuit Design: Design Note Collection. http://dx.doi.org/10.1016/B978-0-12-800001-4.00338-0

Since the average differential input current is zero, the total error introduced by the external RC network is zero if the resistance tied to the plus/minus inputs of the ADC is balanced. Resistances up to 100k, combined with capacitors up to 10μF, may be placed in front of the ADC with less then 0.002% full-scale error (20ppm), while conventional ΔΣ ADCs with the same input network have greater than 10% full-scale errors (100,000ppm). Furthermore, no errors are introduced even if the external resistances are not balanced, as long as the common mode input voltage is equal to the common mode reference voltage. Even if the common mode input voltage does not match the common mode reference voltage, the differential input current remains zero and the common mode input current results in an offset voltage which may be removed through system calibration.

Direct digitization of external sensors with impedances up to 100kΩ is now possible without the need for external or on-chip amplifiers (see Figure 338.2). Bridges, RTDs, thermocouples and other sensors may tie directly to the ADC input. The addition of external capacitors reduces the charge kickback spikes seen at the input of the ADC. An external 1μF capacitor reduces a 1V spike to 18μV. This improves the noise performance of systems where the sensor cannot be placed near the ADC input and eases the drive requirements in applications where external amplifiers are used. The addition of a large resistor between the amplifier output and the ADC input isolates the amplifier from the large bypass capacitor, thus improving its stability.

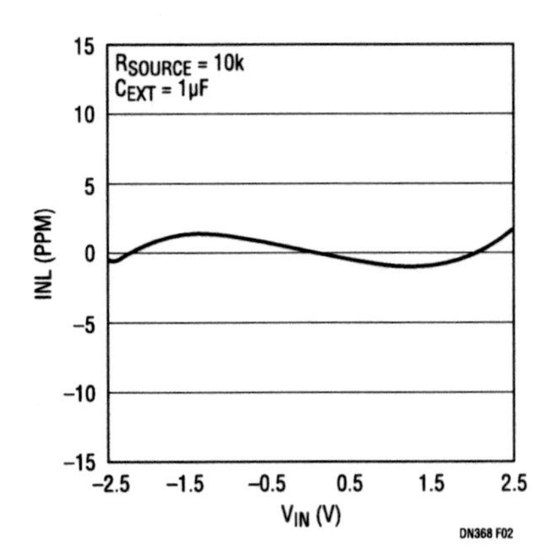

Figure 338.2 • Easy Drive Technology Directly Digitizes Large External RC Networks without Degrading Linearity

What is wrong with on-chip buffers?

One historical solution to the input current settling problem is to integrate a buffer amplifier on the same chip as the ΔΣ ADC. This isolates the ADC input from the switched capacitor array making the ADC input appear high impedance. While this solution looks good on paper, the fact is data converters using on-chip buffers suffer from the limitations of those amplifiers. The common mode input range can no longer swing rail-to-rail. Input signals need to be shifted at least 50mV above ground and a volt or more below V_{CC}. Amplifier offset errors, offset drift, PSRR, CMRR and noise are combined directly with the input signal and result in reduced converter performance. Additionally, on-chip amplifiers require significant power in order to drive the high speed capacitive sampling network. For these reasons, most manufacturers of ΔΣ ADCs using this technology offer a mode to shut off and bypass on-chip amplifiers.

Another solution is coarse/fine input sampling. During the first half of the sampling period (coarse), the input voltage is sampled through an on-chip buffer amplifier, thus isolating the ADC input from the charging capacitor. During the second half of the sampling period (fine), the buffer is switched off and the capacitor is tied directly to the input. While this decreases the magnitude of the spikes seen at the input of the ADC, it results in nonlinear settling errors as a function of op amp offset voltage, CMRR, input signal level and external RC time constants. For these reasons, manufacturers of ΔΣ ADCs using this technology bypass coarse/fine sampling for input signal levels below 100mV.

Conclusion

New Easy Drive technology simplifies the drive requirements of ΔΣ ADCs. The solution lies in a purely passive input current cancellation algorithm that enables rail-to-rail inputs without the added power requirements of on-chip buffer amplifiers and the errors they introduce. Easy Drive technology enables ΔΣ ADCs to directly interface to high impedance sensors, lowpass filters and input bypass capacitors without degrading the DC performance.

Devices using the Easy Drive technology are currently available in 16- and 24-bit versions with an on-chip temperature sensor, no latency conversions for simple multiplexing, on-chip oscillators with guaranteed line frequency rejection, precise DC specifications and the ease-of-use common to all of Linear's ΔΣ ADC converters.

16-bit ADC simplifies current measurements

339

Mark Thoren

Introduction

The LTC2433-1 is a high performance 16-bit delta-sigma ADC for DC measurements. With an input noise floor of 1.45μV_{RMS} and a reference range of 100mV to V_{CC}, the input resolution and range can be optimized for a wide variety of applications. The flexible SPI interface can be configured to self-clock, simplifying isolation or level shifting of digital signals in applications where the ADC must be referenced to a different potential than the data acquisition system.

Data transfer

Figure 339.1a shows a −48V telecom supply current monitor with a 5.4A full scale. The LTC2433-1 serial interface is configured for internal serial clock, continuous conversion mode. This mode is selected by tying chip select low and pulling SCK high at power-up. In this mode, the LTC2433-1 continuously converts at 6.8 samples per second and clocks out its data at a serial data rate of 17.5kHz ±2%, the tolerance of the internal oscillator.

The LTC2433-1 serial data format lends itself to asynchronous reception. While a conversion is taking place, SDO is high. At the end of a conversion, SDO goes low for two clock cycles ($\overline{EOC}$ and DUMMY bits) and then continues outputting the remaining data bits. Thus the $\overline{EOC}$ bit can be used as a start bit as in standard asynchronous communication schemes such as RS232. Unfortunately, the internal oscillator tolerance makes it risky to receive all 19 bits asynchronously. One solution is to apply a crystal-controlled clock signal to the F_O pin, but there is a simpler (and cheaper) way.

Exclusive ORing the SDO and SCK signals produces a serial data signal with embedded clock information similar to Manchester encoding that can easily be decoded by a microcontroller or FPGA. The data format is shown in Figure 339.2.

While a conversion is taking place, both SCK and SDO are high, so the XOR output is low. At the end of a conversion,

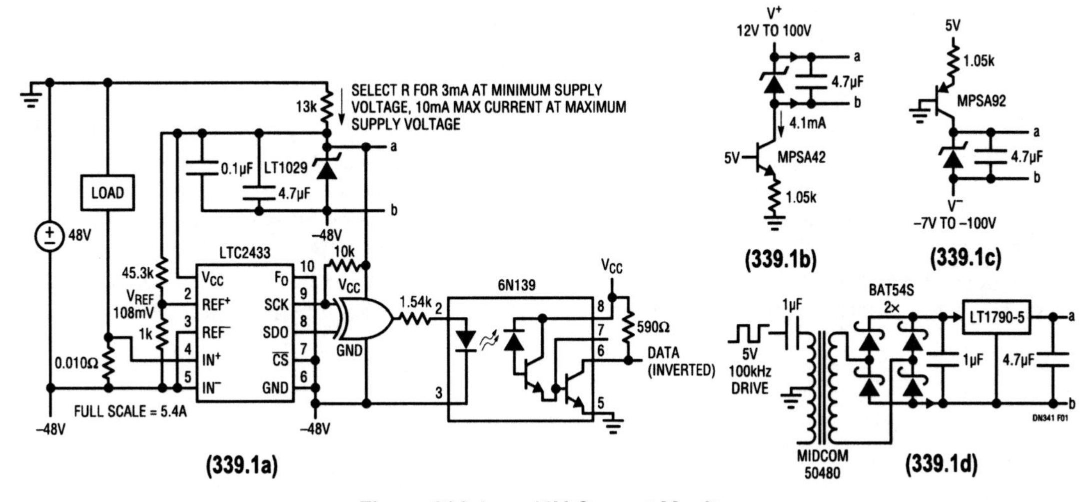

Figure 339.1 • −48V Current Monitor

Analog Circuit Design: Design Note Collection. http://dx.doi.org/10.1016/B978-0-12-800001-4.00339-2

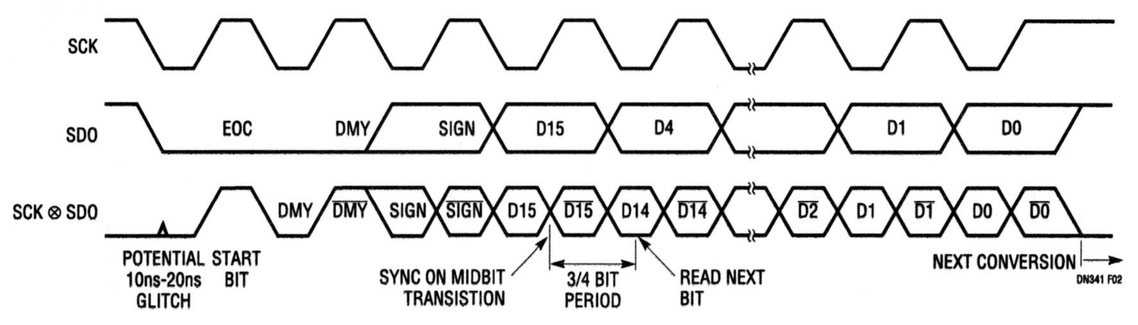

Figure 339.2 • Timing Diagram

both SDO and SCK fall, which may produce a glitch of up to 10ns as these edges are separated only by internal gate delays. The receiving device should look for a high level for at least 20ns to ensure that it is the start bit and not a glitch. (The optocoupler circuit shown does not respond to a pulse narrower than 500ns, so the glitch is not a problem.) The next rising edge is the center of the DUMMY bit, which synchronizes the sampling of the SIGN bit three quarters of one bit period later. After sampling SIGN, the next transition starts another three-quarter bit period delay to synchronize the sampling of D15. The procedure continues until all data bits are received.

This data reception technique tolerates a total timing error of −50% to 33% including errors due to differences between the optocoupler rise and fall times, timing error of the receiving device and the 2% error of the LTC2433-1 internal oscillator.

Data reception pseudocode

The following pseudocode can be ported to an appropriate microcontroller or used to design a state machine in a programmable logic device.

1. Wait for data high state for more than 20ns.
2. Wait for low. This is the end of the start bit.
3. Wait for transition (middle of dummy bit).
4. Wait three quarters of a clock period.
5. Sample SIGN, wait for transition.
6. Wait three quarters of a clock period.
7. Sample D15, wait for transition.
8. Wait three quarters of a clock period.
9. Sample D14, wait for transition.
10. Continue until all bits are read.

The circuit was tested using a PIC microcontroller running at 20MHz. Code should be thoroughly tested for adequate timing margins. Also, good programming dictates that code should have timeouts in case an edge is missed, as might occur if the data reading procedure is pre-empted by an interrupt. This can be as simple as aborting a read if it takes more than double the theoretical time for all 19 bits to be clocked out.

Power and analog inputs

Power and reference in Figure 339.1a are derived from an LT1029 precision shunt reference. The series resistor should be chosen such that the LT1029 current is at least 1mA at all times. While a conversion is taking place, the LTC2433-1 draws 200μA. During the data output phase, ADC current drops to 4μA and the 6N139 optocoupler draws 2mA at 50% duty cycle. The 6N139 meets the low input current and medium speed requirements of this application. Data inversion is required to keep the LED off while a conversion is taking place.

The 5V reference is divided down to 108mV for current measurements, giving a differential input range of ±54mV to match standard 50mV output current shunts with 4mV of over range capacity. For voltage monitoring applications, the 5V reference can be used directly and the input can be divided to accommodate the resulting ±2.5V input range.

This circuit can be adapted to a wide variety of applications. Figure 339.1b is suitable for high side current sensing up to 100V (limited by dissipation in the current source transistor). Figure 339.1c is for low side sensing of negative supplies. Figure 339.1d is a fully isolated supply using a small telecom transformer and an LT1790-5 series reference for both power and reference voltage.

Conclusion

The LTC2433-1 is a simple and cost-effective solution to challenging DC monitoring problems. It is possible to simplify applications that once required complex (and inaccurate) analog level shifting by placing this highly accurate ADC “at the source”—all that is needed is a creative, but simple use of the differential input and reference, along with the flexible SPI interface offered by the LTC2433-1.

12-bit ADC with sequencer simplifies multiple-input applications

340

Jeff Huehn

The LTC1851 is a new 12-bit, 1.25Msps ADC with an 8-channel input multiplexer, a programmable gain sample-and-hold and an internal reference. This design note describes a novel feature of the LTC1851: a programmable sequencer that can automatically control the input mux and sample-and-hold.

New ADC automatically converts multiple inputs with different spans at different rates

Let's imagine a hypothetical application with the following inputs: input A has a range of 0V to 4.096V and needs to be sampled at 400ksps, input B needs 400ksps with a range of 0V to 2.048V, input C has a range of ±2.048V around 2.5V and needs 200ksps, input D needs 100ksps with a range of ±1.024V and is truly differential with a common mode of 2V and input E needs 100ksps with a range of 1V to 3.048V. There are both single-ended and differential, unipolar and bipolar inputs, two different spans and different required sampling rates.

The solution is the LTC1851 sequencer, which allows the user to program a repeating pattern of up to 16 independent mux addresses and configurations and allocates the bandwidth of the LTC1851 as needed. The LTC1851 can easily be programmed to read all five of these inputs continuously and automatically.

Table 340.1 shows a 12-step sequence sampling input A every third conversion, input B every third conversion, input C every sixth conversion, and inputs D and E once every 12 conversions. This will result in effective sampling rates of 1.25Msps · 4/12 = 416.67ksps for inputs A and B, 208.33ksps for input C and 104.167ksps for inputs D and E. The LTC1851 handles the channel selection and input configuration and will cycle through these 12 steps automatically as conversions are performed (see Figure 340.1).

Writing and reading the sequencer

To write to the sequencer, $\overline{RD}$ must be high and M0 taken low (see Figure 340.2). The falling edge of $\overline{WR}$ enables the configuration control inputs (DIFF, A2, A1, A0, $\overline{UNI}$/BIP and PGA) and the rising edge latches the data and advances to the next location. Subsequent $\overline{WR}$ low pulses will write up to 16 locations. After the last desired location is written, M0 should be taken high.

To confirm the integrity of the programmed sequence, the user can read the contents of the Sequencer. $\overline{WR}$ must be

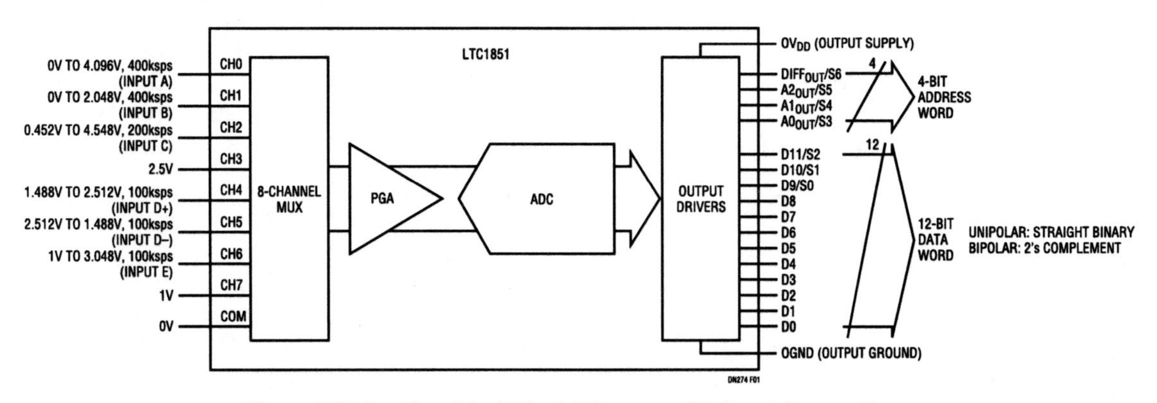

Figure 340.1 • Simplified Block Diagram with Input Connections

Analog Circuit Design: Design Note Collection. http://dx.doi.org/10.1016/B978-0-12-800001-4.00340-9

Table 340.1 12-Step Sequence Provides Effective Sampling Rate of 416ksps for Inputs A and B, 208ksps for Input C and 104ksps for Inputs D and E

STEP	INPUT	+INPUT PIN	−INPUT PIN	INPUT RANGE (V)	MEMORY LOCATION	DIFF	A2	A1	A0	$\overline{\text{UNI}}$/BIP	PGA
1	A	CH0	COM	0 TO 4.096	0000	0	0	0	0	0	1
2	B	CH1	COM	0 TO 2.048	0001	0	0	0	1	0	0
3	C	CH2	CH3	±2.048	0010	1	0	1	0	1	1
4	A	CH0	COM	0 TO 4.096	0011	0	0	0	0	0	1
5	B	CH1	COM	0 TO 2.048	0100	0	0	0	1	0	0
6	D	CH4	CH5	±1.024	0101	1	1	0	0	1	0
7	A	CH0	COM	0 TO 4.096	0110	0	0	0	0	0	1
8	B	CH1	COM	0 TO 2.048	0111	0	0	0	1	0	0
9	C	CH2	CH3	±2.048	1000	1	0	1	0	1	1
10	A	CH0	COM	0 TO 4.096	1001	0	0	0	0	0	1
11	B	CH1	COM	0 TO 2.048	1010	0	0	0	1	0	0
12	E	CH6	CH7	1 TO 3.048	1011	1	1	1	0	0	1

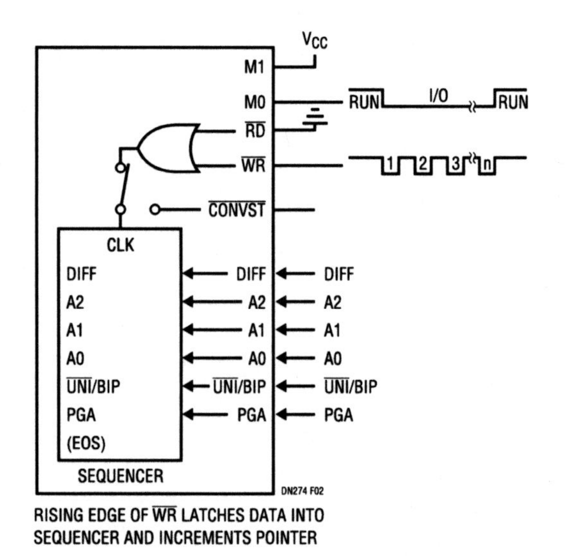

Figure 340.2 • Writing the Sequencer

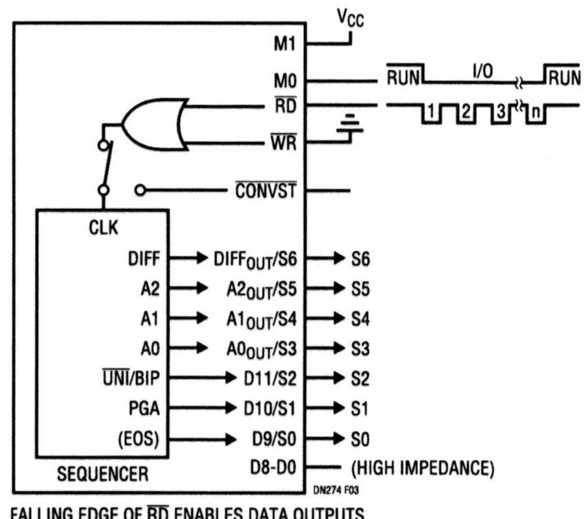

Figure 340.3 • Reading the Sequencer

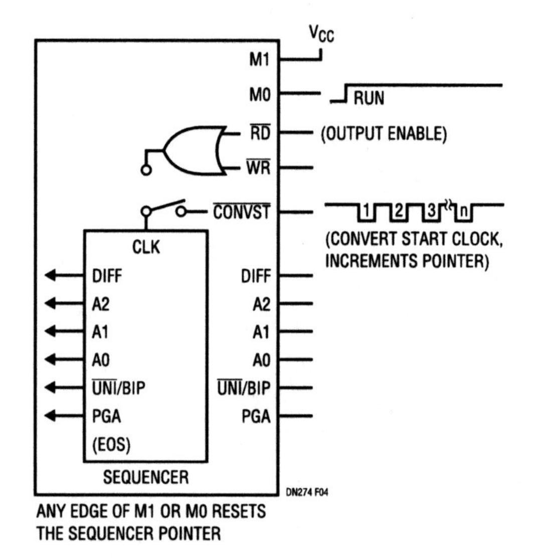

Figure 340.4 • Running the Sequencer

high and M0 taken low (see Figure 340.3). An $\overline{\text{RD}}$ low pulse will output the first sequencer location on the seven status word output pins (S6 to S0). The rising edge of $\overline{\text{RD}}$ will return the output pins to a high impedance state and advance to the next location. Subsequent $\overline{\text{RD}}$ low pulses will read through all 16 locations. The last location in the sequence will be indicated by a logic 1 on the S0 pin.

Running the sequencer

The M0 pin must be returned high, which will reset the pointer to location 0000 (see Figure 340.4). The LTC1851 will begin acquiring the input signal using the configuration stored in that location. The falling edge of $\overline{\text{CONVST}}$ will sample the inputs and begin a conversion. After the conversion, the LTC1851 will begin acquiring the next input using the configuration stored in location 0001 and the 12-bit data output word (D11 to D0), along with the 4-bit mux address ($DIFF_{OUT}$, $A2_{OUT}$, $A1_{OUT}$, $A0_{OUT}$), will be available on the data output pins. (The 12-bit output word will automatically switch format for unipolar and bipolar inputs.) When the last programmed location is reached, the sequencer will start over at location 0000. The program stored in the sequencer memory is retained as long as power is continuously applied to the part.

Conclusion

The LTC1851 greatly simplifies the task of continuously converting multiple inputs. It can be programmed to handle a wide variety of inputs and automate channel selection and input configuration. For more information see the November 2001 issue of *Linear Technology* magazine or the LTC1851 data sheet.

A-to-D converter does frequency translation

341

Derek Redmayne

The need to characterize frequency sources, both in the laboratory and in the field, is increasingly important. The circuit in Figure 341.1 offers some interesting attributes in a compact and relatively inexpensive scheme. It uses an LTC1420 ADC to undersample a higher frequency, driving an LTC1668 DAC, followed by a filter to perform a down conversion. The output of the filter is subsequently resampled to produce a manageable sample rate for a single-chip microcontroller. In addition to characterizing the carrier in an IF strip or the output of a local oscillator, this technique is also useful for characterizing ADCs, DACs, clock sources, signal sources or the effects of logic devices or phase-locked loops on phase noise.

Frequency conversion or translation is usually performed by a diode mixer or a Gilbert cell mixer. Down conversion is most often encountered in radio receivers; up conversion is more commonly used in transmitters. The common superheterodyne receiver usually involves one conversion to produce a fixed intermediate frequency (IF). Spectrum analyzers, cellular base stations, cable modems, microwave and satellite receivers, radar and optical communications systems all include frequency conversion blocks.

Down conversion with an ADC

It may not be commonly known that down conversion can be performed using an ADC, by undersampling a signal frequency. The resulting output signal frequency is the difference between the sample frequency (f_S) (or a multiple of f_S) and the incoming frequency. An ADC may be used to undersample any frequency that is within its full linear bandwidth.

As in the case of a mixer, the result of this operation is a sum and a difference frequency. The sum frequency, however, ends up at the same apparent frequency as the difference frequency in a discrete time sampled system. Essentially, only the difference frequency remains.

The major constraint in an undersampled system is that the bandwidth of the incoming signals must not fall outside the Nyquist zone in use. (A Nyquist zone extends over a bandwidth of $f_S/2$, above or below an integral multiple of the sample frequency.) Any signal falling outside the desired Nyquist zone wraps back into the DC-to-$f_S/2$ zone. The above constraint can be relaxed if subsequent bandpass filtering in the digital domain limits the frequency range of interest. So long as an unwanted

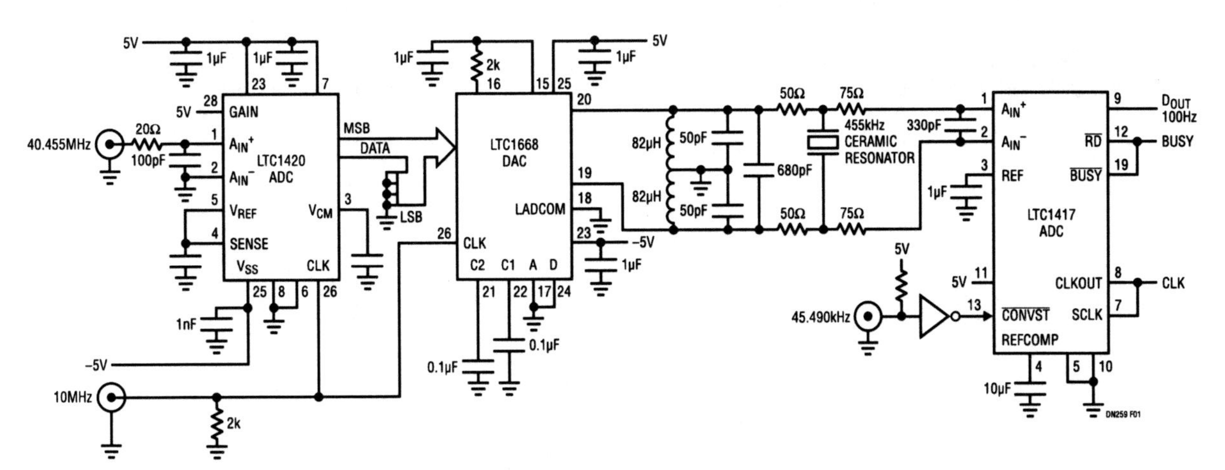

Figure 341.1 • Undersampling 40MHz Performs 2-Stage Frequency Translation to 100Hz

Analog Circuit Design: Design Note Collection. http://dx.doi.org/10.1016/B978-0-12-800001-4.00341-0

signal does not wrap back into the frequency range of interest, its effect on the spectrum of interest is negligible.

In Figure 341.1, the 10Msps LTC1420 translates the 40.455MHz input signal to 455kHz at its output. When a high speed DAC is used to reproduce the 455kHz signal, a subsequent analog bandpass filter adds little cost or power dissipation. One advantage of the analog filter is that it does not exhibit mathematical artifacts if the signal frequency is not coherent with the sample rate. In fact, this scheme allows the intermediate frequency to be tailored to suit the conversion rate of the resampling ADC.

An incentive for using a high speed infinite sample-and-hold in this fashion is the benefit of a high sample rate, without the need to process samples at that rate. The data rate delivered by a high speed ADC can be too fast for a low power processor to handle and data rate decimation may reduce signal-to-noise ratio too much. The use of an analog filter after the DAC may seem old fashioned, but the filter characteristics available from ceramic resonators, active filters or tuned LC filters may be hard to match in a digital filter. The use of a higher resolution ADC following the initial 12-bit quantization allows details to be resolved if the original signal contains a few LSB of noise (dither), as well as improving frequency measurement capability.

The 455kHz intermediate signal was chosen to allow the use of readily available 455kHz ceramic resonators or LC filters. Note that the LTC1560-1 monolithic 5th order elliptic lowpass filter could also be used in this application. The LTC1668, as it is a current output device, can drive a tank circuit tuned to the desired frequency. The subsequent resampling of this signal at a submultiple of 455KHz–100Hz (45,490sps) produces a sinusoid at 100Hz.

In Figure 341.2, the resampled output of the DAC is shown. Figure 341.3 is the result of an FFT performed on of the output of the LTC1417.

As mentioned earlier, the output of the DAC is not only the difference frequency of 455kHz. The DAC acts like a mixer and produces in addition to the fundamental (455kHz), the sum and the difference frequencies of the 10MHz conversion clock and the 455kHz signal.

The lower of these unwanted frequencies, 9.545MHz (10MHz–455kHz), is approximately 20 times the 455kHz or 4.4 octaves above the carrier. The signal level in these components without filtering is approximately 25dB below the carrier; hence, a lowpass or bandpass filter is required. A 2nd order LPF with a 12dB/octave roll-off in the transition region will reduce these unwanted components to approximately 77dB below the carrier, the region of other harmonic and noise components. If the signal under scrutiny is a single tone, a lowpass filter is adequate.

These techniques can also be used on the bench to evaluate the performance of signal generators and clock sources and of course, ADCs and DACs, as well as performing monitoring functions in the field.

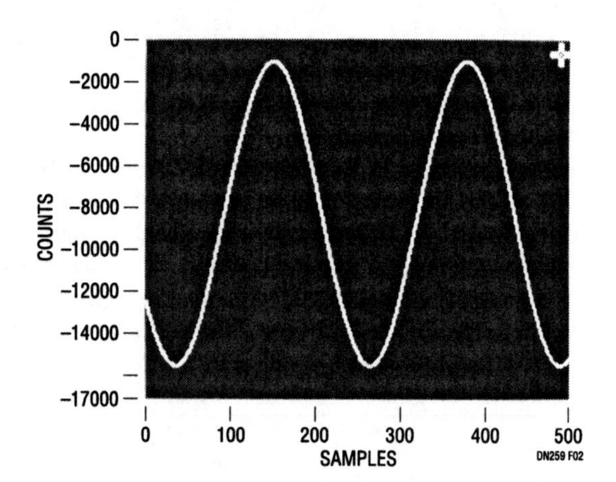

Figure 341.2 • The Downconverted 100Hz Output Exaggerates Phase or Frequency Variation in Original Signal

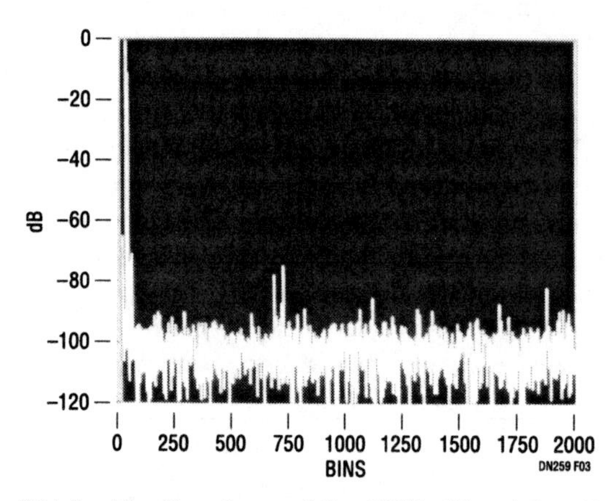

Figure 341.3 • The Spectrum of the 100Hz Signal Can Be Processed to Determine Characteristics of the Original Signal

Resolving very small temperature differences with a delta-sigma ADC

342

Derek Redmayne

The LTC2402, a 2-channel 24-bit, No Latency Delta-Sigma ADC, is well-suited to differential temperature sensing using a number of different sensors. Extremely good matching between channels allows two absolute temperature measurements to be compared in order to determine the differential or gradient between two points. In the circuits shown below, the LTC2402, with averaging, can resolve a temperature differential of a few millidegrees, although in the case of a single-shot pair of readings, a peak error of 0.05°C is possible. Independently adjustable full (FS_{SET}) and zero scale (ZS_{SET}) facilitate 3- or 4-wire measurements with platinum RTDs and the automatic sequencing of the two channels makes measurements across an isolation barrier easier. The no latency feature of these ADCs also resolves dilemmas associated with pulsed excitation. The use of pulsed excitation minimizes or eliminates parasitic thermocouple effects that can compromise RTD measurements. In addition, pulsed excitation can reduce standby or average power consumption in battery-powered applications.

Platinum RTDs

Platinum RTDs are the most stable and accurate temperature sensors available on the market. However, their low output levels and self-heating effects require attention to detail in both physical placement and electrical design. RTDs, although more linear than thermocouples and thermistors, are somewhat nonlinear; linearization should therefore be used. In the past, this has often been done in a first-order approximation by feeding back a portion of the amplified output voltage to the excitation, thereby providing a multiplication factor. With a processor-driven design, hardware linearization is unnecessary.

Self-heating effects

The self-heating effect that results from the excitation current is dependent on the thermal resistance of the RTD element to its environment. If this thermal resistance is a constant (as it would be if the sensor is bonded to a solid mass), the self-heating effect can be calculated based on the resistance of the RTD at temperature, then subtracted from the reading. This assumes that the ambient medium undergoes no phase changes in the temperature range of interest. For absolute temperature measurements, self-heating effects can be problematic; however, if temperature differential is the issue, the use of a higher excitation current can be an advantage. If the thermal resistance of the two sensors to their local environment matches reasonably well, the self-heating effect will have an effect on the absolute measurement only. The absolute temperature measurement is needed only for purposes of determining the slope of the resistance change, and hence is not very critical, possibly eliminating the need for precise linearization.

At a rather high excitation current of 1mA, a 100Ω RTD (European curve) will produce 385μV/°C of output signal in the neighborhood of 25°C (290μV/°C at 850°C). Averaging 30 readings will resolve the temperature differential to a couple of millidegrees, although a single channel reading will exhibit approximately 0.05°C peak-to-peak noise.

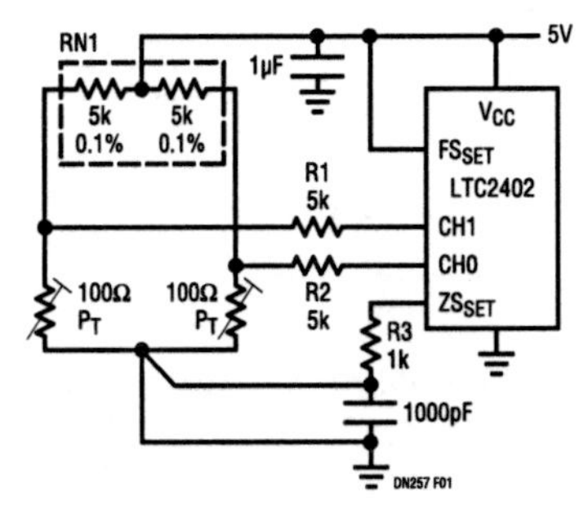

Figure 342.1 • 2-Channel ADC Measures Differential Temperature from Bridge Configured RTDs, ZS_{SET} Input Eliminates Drop in Connecting Wire

Analog Circuit Design: Design Note Collection. http://dx.doi.org/10.1016/B978-0-12-800001-4.00342-2

Bridge connection of RTDs

Figure 342.1 shows the use of the two channels of the LTC2402 to sense two RTDs in essentially the same manner. If the two sensors are physically close together, remote sensing of the common return by virtue of the connection to ZS_{SET} eliminates the effect of the voltage drop across the return wiring. The wiring leading to the upper terminal of the sensors is only responsible for a small gain error. For example, 20 feet of 26-gauge copper wire would only produce 0.068% gain error (in contrast, if remote sensing of the common return were not used, the error associated with individual returns would be approximately 20°C).

The tolerance of the two reference resistors can contribute a much greater error, by virtue of producing a gain error. One-percent (1%) resistors would produce a gain error of up to 2%, resulting in an error of 5°C. However, the gain error can be recorded and eliminated in software if conditions producing zero temperature gradient can be produced or identified. It is recommended that resistors R1 and R2 be either a matched divider pair or very low temperature coefficient devices such as Vishay S102 or Z201 series. Temperature tracking in integrated pairs can be as tight as 0.1ppm/°C.

Series connection of RTDs

Figure 342.2 shows a means by which the gain variation between the two channels can be reduced. As there is only one excitation current, there is no need for precisely matched resistors. The tolerance of the reference resistor (R4) only significantly affects the absolute measurement and if the quantity that is desired is the differential temperature, the absolute temperature needs only to be known to nominal accuracy. A 5% resistor, in this case, may give adequate results. The only error mechanism, other than noise, is nonlinearity in the ADC between the nominal 0mV to 400mV range and the 400mV to 800mV range. As linearity does not exhibit any abrupt discontinuities over the full input range, this is not more than a ppm of full scale over this more restricted range.

Pulsed excitation

Figure 342.3 is an example of how pulsed excitation current can be used. When the analog switch (SW1) is turned off, the full reference voltage is still applied to the reference terminal, but there is no appreciable excitation applied to the RTDs. The voltages measured at the inputs are now determined by offset and parasitic thermocouple voltages present in all the connections between the RTD and the ADC. These measurements should be subtracted from the voltages measured with excitation current flowing. Alternation of measurements with and without excitation will suppress 1/f noise, if amplification were introduced to amplify the difference voltage. The amplifier shown within dashed lines is an instrumentation amplifier such as the LT1167. If the amplifier is used, SW3, R6 and R7 must be used to allow signals to remain within the input range of the amplifier.

The use of copper or tungsten RTDs, in conjunction with the LTC2402 and the techniques described above, would allow very subtle temperature gradients to be measured in numerous industrial, consumer or automotive applications.

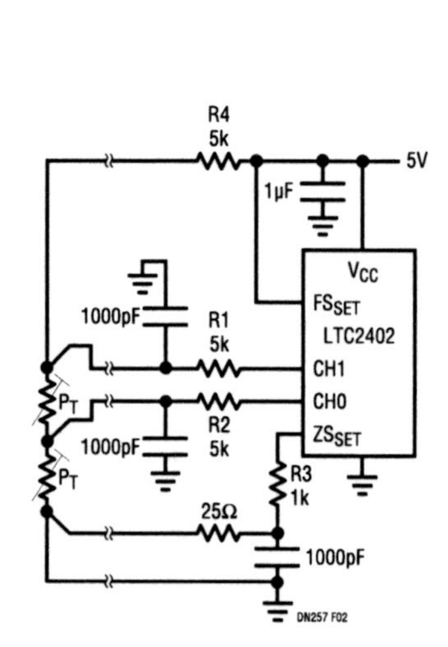

Figure 342.2 • Stacked Half-Bridge Eliminates Half the Excitation Current and the Matching Requirements of Figure 342.1.

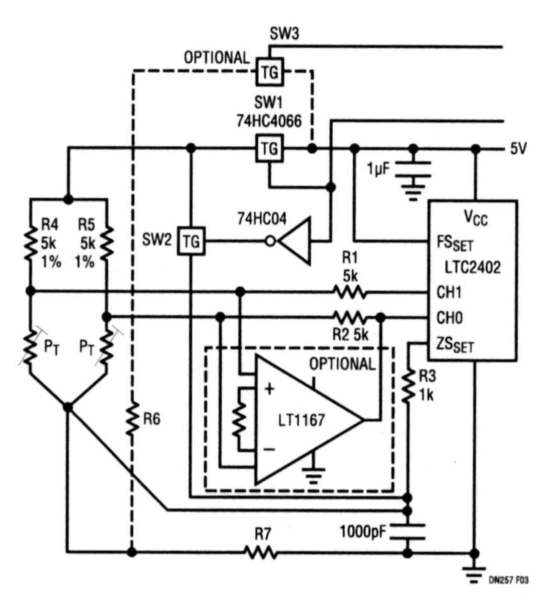

Figure 342.3 • Switched Excitation Reduces Standby Current and Cancels Offset, Thermocouples and Low Frequency Noise

1- and 2-channel No Latency ΔΣ 24-bit ADCs easily digitize a variety of sensors, part 1

343

Michael K. Mayes

Since its introduction, the LTC2400 has transformed the method of designing analog-to-digital converters into a variety of systems. Some key features separating the LTC2400 from conventional high resolution ADCs are:

1. Ultralow offset (1ppm), offset drift (0.01ppm/°C), full-scale error (4ppm) and full-scale drift (0.02ppm/°C) without user calibration
2. Absolute accuracy less than 10ppm total (linearity + offset + full scale + noise) over the full operating temperature range
3. Ease-of-use (eight pins, no configuration registers, internal oscillator and latency-free conversion)
4. Low noise and wide dynamic range (0.3ppm RMS with $V_{REF} = V_{CC} = 5V$)

This two part design note introduces two new products leveraging the technology used in the LTC2400. Both parts are in tiny 10-pin MSOP packages. They include full-scale and zero-scale set inputs for removing systematic offset/full-scale errors. The first part (LTC2401) is a single-ended 1-channel device. The second part is a 2-channel device with automatic ping-pong channel selection (LTC2402).

The absolute accuracy and near zero drift of these devices enable several novel applications, of which two are presented in part 1 and two in part 2. The first application uses the full-scale set (FS_{SET}) and zero-scale set (ZS_{SET}) inputs of the 1-channel device (LTC2401) to digitize a half-bridge sensor. The second combines the LTC2402's ping-pong channel selection, absolute accuracy and excellent rejection into a pseudo-differential bridge digitizer. The third (part 2) is a thermocouple digitizer with a digital cold junction compensation scheme using the automatic ping-pong channel selection of the LTC2402 for simplified optical isolation. The final application (in part 2) uses the LTC2402 to digitize an RTD temperature sensor and remove voltage drop errors due to long lead lengths using the second channel and the LTC2402's underrange capabilities.

Single-ended half-bridge digitizer with reference and ground sensing

Many types of sensors convert real world phenomena (temperature, pressure, gas levels, etc.) into a voltage. Typically, this voltage is generated by passing an excitation current through the sensor. This excitation current also flows through parasitic resistors R_{P1} and R_{P2} (see Figure 343.1). The voltage drop across these parasitic resistors leads to systematic offset and full-scale errors.

In order to eliminate the errors associated with these parasitic resistors, the LTC2401/LTC2402 includes a full-scale set input (FS_{SET}) and a zero-scale set input (ZS_{SET}). As shown in Figure 343.1, the FS_{SET} pin acts as a high impedance full-scale sense input. Errors due to parasitic resistance R_{P1} in series with the half-bridge sensor are removed by the FS_{SET} input to the ADC.

The absolute full-scale output of the ADC (data out = $FFFFFF_{HEX}$) will occur at $V_{IN} = V_B = FS_{SET}$. Similarly, the offset errors due to R_{P2} are removed by the ground sense input ZS_{SET}. The absolute zero output of the ADC

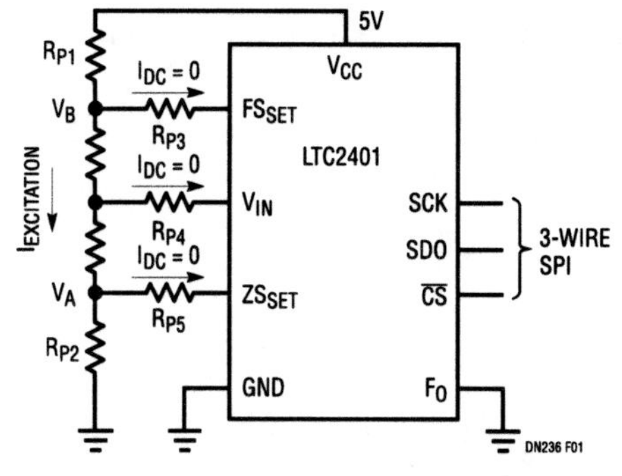

Figure 343.1 • Half-Bridge Digitizer with Zero-Scale and Full-Scale Sense

Analog Circuit Design: Design Note Collection. http://dx.doi.org/10.1016/B978-0-12-800001-4.00343-4

(data out = 000000_{HEX}) occurs at $V_{IN} = V_A = ZS_{SET}$. Parasitic resistors R_{P3} to R_{P5} have negligible errors due to 1nA (typ) leakage current at the FS_{SET}, ZS_{SET} and V_{IN} pins. The wide dynamic input range (−300mV to 5.3V) and low noise (0.6ppm RMS) enable the LTC2401 to directly digitize the output of the bridge sensor.

Pseudo-differential applications

Generally, designers choose fully differential topologies for several reasons. First, the interface to a 4- or 6-wire bridge is simple (it is a differential output). Second, they require good rejection of line-frequency noise. Third, they typically look at a small differential signal sitting on a large common mode voltage and need accurate measurements of the differential signal independent of the common mode input voltage. Many applications currently using fully differential analog-to-digital converters for any of the above reasons may migrate to a pseudo-differential conversion using the LTC2402.

The LTC2402 interfaces directly to a 4- or 6-wire bridge, see Figure 343.2. Like the LTC2401, the LTC2402 includes a FS_{SET} and a ZS_{SET} for sensing the excitation voltage directly across the bridge. This eliminates errors due to excitation currents flowing through parasitic resistors. The LTC2402 also includes two single-ended input channels that can tie directly to the differential output of the bridge. The two conversion results may be digitally subtracted yielding the differential result.

Additionally, the measurements from the individual channels represent the common mode output of the sensor. This information can be used to characterize and digitally compensate the temperature drift of the sensor.

Noise rejection

The LTC2402's single-ended rejection of line frequencies (50Hz/60Hz ±2%) and harmonics is better than 110dB. Since the device performs two independent single-ended conversions each with >110dB rejection, the overall common mode and differential rejection is much better than the 80dB rejection typically found in other differential delta-sigma converters.

In addition to excellent rejection of line frequency, the LTC2402 also exhibits excellent single-ended noise rejection of a wide range of frequencies due to its fourth order sinc filter, see Figure 343.3. Each single-ended conversion independently rejects high frequency noise (>60Hz). Care must be taken to insure noise at frequencies below 15Hz and at multiples of the ADC sample rate (15,600Hz) are not present. For this application, it is recommended the LTC2402 is placed in close proximity to the bridge sensor in order to reduce the noise applied to the ADC input. By performing three successive conversions (CH0-CH1-CH0) the drift and low frequency noise can be measured and compensated for digitally.

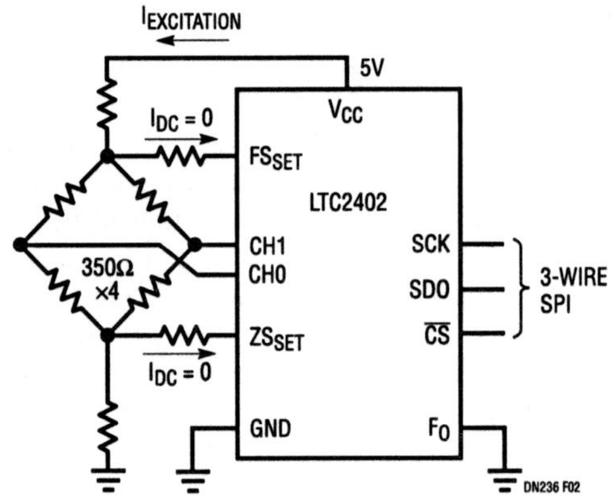

Figure 343.2 • Pseudo-Differential Strain Gauge Application

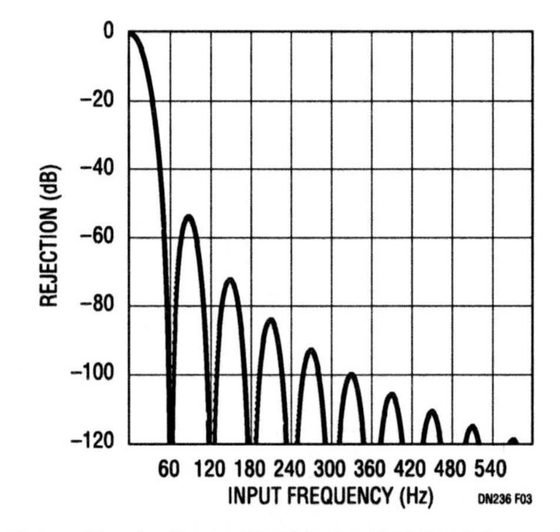

Figure 343.3 • Single-Ended LTC2401/LTC2402 Input Rejection

1- and 2-channel No Latency ΔΣ 24-bit ADCs easily digitize a variety of sensors, part 2

344

Michael K. Mayes

Introduction

This design note is a continuation of application ideas using the LTC2401/LTC2402 high accuracy, tiny (MSOP), 24-bit delta-sigma converters. Part 1 introduced the two new devices and bridge digitizer circuits utilizing the unique features of these devices. It showed how to remove offset/full-scale errors due to excitation currents using the full-scale and zero-scale set inputs. Furthermore, part 1 showed pseudo-differential application circuits for directly digitizing 4- and 6-terminal bridge circuits.

This article introduces two new applications using the 2-channel LTC2402. The first is a digital cold junction compensation circuit for remote measurements. The second shows how to digitize an RTD sensor and remove voltage drop errors digitally by using the ADC's second channel and under-range capabilities.

Digital cold junction compensation

In order to measure absolute temperature with a thermocouple, cold junction compensation must be performed. The LTC2402 enables simple digital cold junction compensation. One channel measures the output of the thermocouple while the other measures the output of the cold junction sensor (diode, thermistor, etc.); see Figure 344.1.

The selection between CH0 (thermocouple) and CH1 (cold junction) is automatic. The LTC2402 alternates conversions between the two input channels and outputs a bit corresponding to the selected channel in the serial data output word. This simplifies the user interface by eliminating a channel select input pin. As a result, the LTC2402 is ideal for systems performing isolated measurements; it only requires two opto-isolators (one for serial data out and one for the serial data output clock).

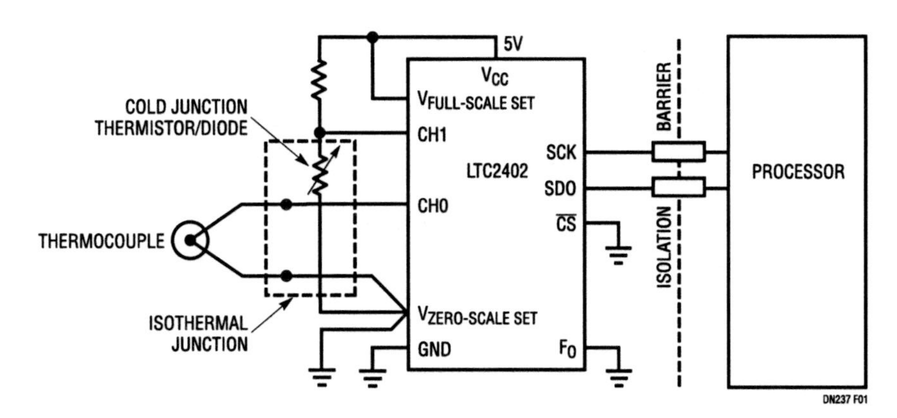

Figure 344.1 • Digital Cold Junction Compensation Circuit

Analog Circuit Design: Design Note Collection. http://dx.doi.org/10.1016/B978-0-12-800001-4.00344-6

Alternating conversions between two input channels is difficult with conventional delta-sigma analog-to-digital converters. These devices require a 3–5 conversion cycle settling every time the input channel is switched. On the other hand, the LTC2402 uses a completely different architecture than previous delta-sigma converters. This results in latency free, single cycle settling. The LTC2402 enables continuous conversion between two alternating channels without the added complexity associated with other delta-sigma A/D converters.

RTD temperature digitizer

RTDs used in remote temperature measurements often have long lead lengths between the ADC and RTD sensor. These long lead lengths result in voltage drops due to excitation current in the interconnect to the RTD. This voltage drop can be measured and digitally removed using the LTC2402; see Figure 344.2.

The excitation current (typically 200μA) flows through a long lead length to the remote temperature sensor (RTD). This current is applied to the RTD, whose resistance changes as a function of temperature (100Ω to 400Ω for 0°C to 800°C). The same excitation current flows back to the ADC ground and generates another voltage drop across the return lead. In order to get an accurate measurement of the temperature, these voltage drops must be measured and removed from the conversion result. Assuming the resistance is approximately the same for the forward and return paths (R1=R2=R), the auxiliary channel on the LTC2402 can measure this drop. These errors are then removed with simple digital correction.

As in the previous example, the LTC2402 alternately converts CH0 and CH1. The result of the first conversion on CH0 corresponds to an input voltage of $V_{RTD} + R \cdot I_{EXCITATION}$. The result of the second conversion (CH1) is $-R \cdot I_{EXCITATION}$. Note, the LTC2402's input range is not limited to the supply rails, it has underrange capabilities. The device's input range is −300mV to V_{REF}+300mV. Adding the two conversion results together, the voltage drop across the RTD leads are cancelled and the final result is V_{RTD}.

Conclusion

Linear Technology has introduced two new converters to its 24-bit delta-sigma converter family. The family consists of the LTC2400 (1-channel 8-pin SO), LTC2404/LTC2408 (4- and 8-channel 24-bit ADCs) and the LTC2401/LTC2402 shown here.

Also recently introduced are the LTC2420 reduced cost 20-bit delta-sigma ADC with a 100 samples-per-second turbo mode, as well as 4- and 8-channel versions, the LTC2424 and LTC2428.

Additionally, the LTC2410, a fully differential input/reference device in a GN16 is available as well as a pin-compatible LTC2413 featuring simultaneous 50Hz and 60Hz rejection. These devices feature 800nV$_{RMS}$ noise over a wide input range of 5V, near zero offset error, full-scale error and linearity drift.

If board space is an issue, a fully differential input/reference device (LTC2411) is available in a 10-lead MSOP package.

Each device features absolute accuracy, ease-of-use and near zero drift. The LTC2401/LTC2402 also include full-scale set and zero-scale set inputs for removing errors due to systematic voltage drops. The performance, features and ease-of-use of these devices warrant designers to rethink their future industrial system and instrumentation designs.

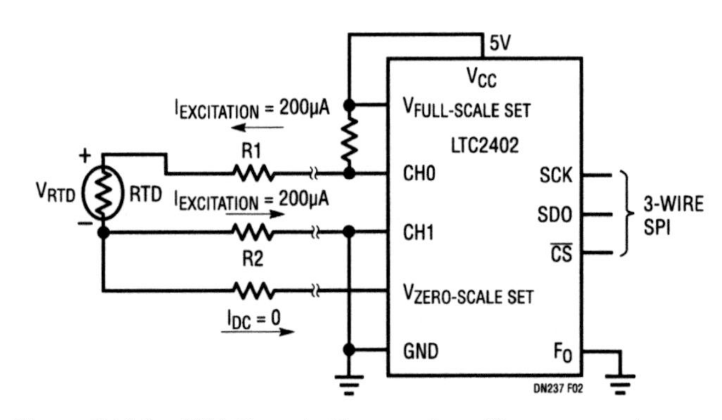

Figure 344.2 • RTD Remote Temperature Measurement

24-bit ADC measures from DC to daylight

345

Derek Redmayne

The need to measure a wide variety of physical phenomena is increasingly important in intelligent sensors. The following circuit gives eight examples of the flexibility of the LTC2408 for sensing real-world phenomena. The eight inputs of the LTC2408 are used here in a variety of ways that would not be practical without the dynamic range of the LTC2408 (see Figure 345.1).

All of the examples shown use single-ended sensing and an absolute minimum of external circuitry. Inputs to the circuit shown range from high voltage DC to ultraviolet light. Output

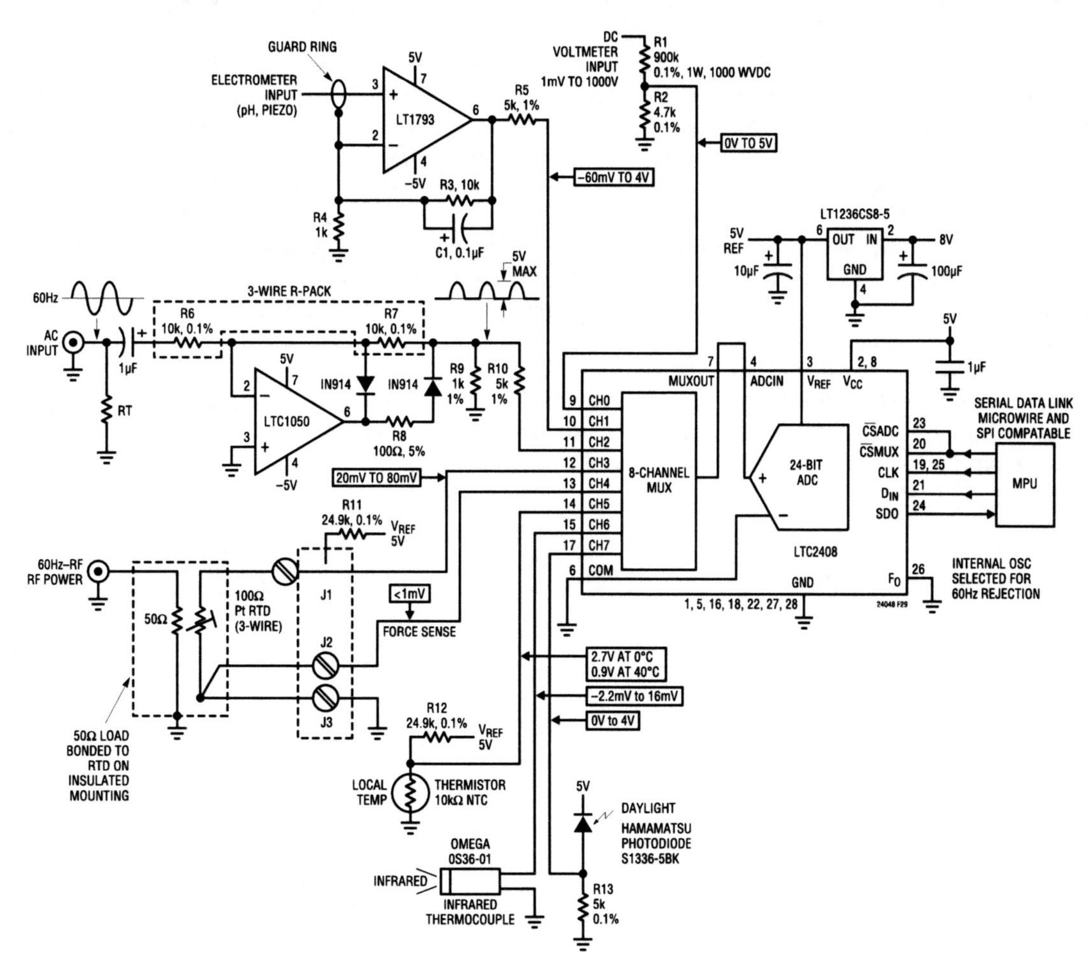

Figure 345.1 • Measure DC to Daylight Using the LTC2408

Analog Circuit Design: Design Note Collection. http://dx.doi.org/10.1016/B978-0-12-800001-4.00345-8

data represents amplitude or power levels for signals in all the AC inputs.

CH0, with an appropriate 1W resistor rated to withstand 1000V, is able to measure DC voltages over a single range of −60V to 1000V. No autoranging is required.

CH1 shows an LT1792 FET input amplifier as an electrometer for low frequency applications, such as pH. Physical phenomena with very high source impedance cannot drive a switched capacitor converter directly; hence, some form of buffering is necessary.

CH2 shows a precision rectifier that relies on the LTC2408's $sinc^4$ digital filter for integration of the resulting half-wave rectified signal. This circuit can be used at 60Hz, 120Hz, and from 400Hz to 1000Hz with good results. Above 1000Hz, amplifier overshoot and gain/bandwidth will start to compromise results. The dynamic range is limited by the LTC1050's offset voltage. The dynamic range of the system is approximately five orders of magnitude as limited by noise. A stable signal source allows parameters such as magnetic reluctance, permeability or eddy current loss to be measured. A second precision rectifier on another channel can be used to provide ratiometric operation.

CH3 and CH4 are used for RTD temperature sensing with a resolution of approximately 0.03°C. CH4 senses the voltage drop on the force lead for 3-wire RTDs. Subtract 2 × CH4's reading from CH3's reading to give the true reading at the sensor. If a 2-wire RTD is located in close proximity to the LTC2408, CH4 can be used for another signal.

The use of an RTD to sense the temperature of a 50Ω load resistor allows true RMS/RF input power to be measured thermally with a fair degree of accuracy from audio to GHz frequencies.

In practice, the RTD must be bonded to an appropriate resistor [noninductive, low TC and able to survive the same temperature range as the RTD (850°C)]; the assembly must then be mounted inside an insulating enclosure that exhibits only direct thermal transfer.

Convection and radiation should be eliminated. The insulating material, filler, adhesives and any substrate material must withstand cycling through these temperature extremes. Resolution in the center of the power range is approximately 1 part in 1000. The maximum temperature of the resistor or sensor, whichever is less, limits the maximum power level that can be measured. The minimum level that can be resolved is limited by LTC2408 noise ($1.5\mu V_{RMS}$). This basic approach to sensing wide band AC can be tailored for high or low power levels. Physical implementation determines the results.

It is best implemented with RTDs and, with less precision, with thermocouples or thermistors. The dynamic range of this technique is not wide, as power is measured, and once translated back to signal amplitude, exhibits a range that is the square root of the dynamic range at the converter.

CH5 shows a half-bridge connected thermistor, used to sense the case temperature of the RTD-based thermal power measurement scheme on channels 3 and 4. Thermistors give very good resolution over a limited temperature range. Resolution of 0.001°C is possible, although accuracy is limited by self-heating effects and thermistor characteristics.

CH6 is directly connected to an infrared thermocouple allowing noncontact measurement of temperature or, alternatively, high levels of infrared light. The resolution of this type of sensor, as used with the LTC2408, is comparable to that of a conventional thermocouple. Resolution with a Type J is ≈0.03°C. The temperature range of these devices is more restricted than a conventional thermocouple and they are tailored for limited ranges. Note that the output impedance of the sensor is 3kΩ. As a result, open-detection schemes typically used with thermocouples cannot be used. These devices do not need cold junction compensation. In addition, conventional thermocouples can be connected directly to the LTC2408 (not shown) and cold junction sensing can be provided by another temperature sensor on a different channel or by using an LT1025 for cold junction compensation.

CH7 is used to sense daylight via photodiode current. With a resolution of 300pA, the optical dynamic range of this circuit covers six orders of magnitude.

This application is intended to demonstrate the mix and match capabilities of the LTC2408, which allow very low level signals to be handled alongside high level signals with a minimum of complexity.

Many other single-ended sensing schemes can be connected directly to the LTC2408. Differential signals can also be accommodated via instrumentation amplifiers, also available from LTC.

High accuracy differential to single-ended converter for ±5V supplies

Differential to single-ended converter has very high uncalibrated accuracy and low offset and drift

346

Kevin R. Hoskins Derek V. Redmayne

Introduction

The circuit in Figure 346.1 is ideal for low level differential signals in applications that have a ±5V supply and need high accuracy without calibration. The circuit achieves 19.6-bit resolution and 18.1-bit accuracy. These and other specifications are summarized in Table 346.1.

Operation

The circuit in Figure 346.1 combines an LTC1043 and LTC1050 as a differential to single-ended amplifier that has an input common mode range that includes the power supplies. It uses the LTC1043 to sample a differential input voltage, holds it on C_S and transfers it to a ground-referred capacitor, C_H. The voltage on C_H is applied to the LTC1050's noninverting input and amplified by the gain set by resistors R1 and R2 (101 for the values shown). The amplifier's output is then converted to a digital value by the LTC2400.

The LTC1043 achieves its best differential to single-ended conversion when its internal switching frequency operates at a nominal 300Hz, as set by the 0.01µF capacitor C1 and when 1µF capacitors are used for C_S and C_H. C_S and C_H should be a film type such as mylar or polypropylene. Conversion accuracy is enhanced by placing a guard shield around C_S and connecting the shield to Pin 10 of the LTC1043. This minimizes nonlinearity that results from stray capacitance transfer errors associated with C_S. To minimize the possibility of PCB leakage currents introducing an error source into C_H, an optional

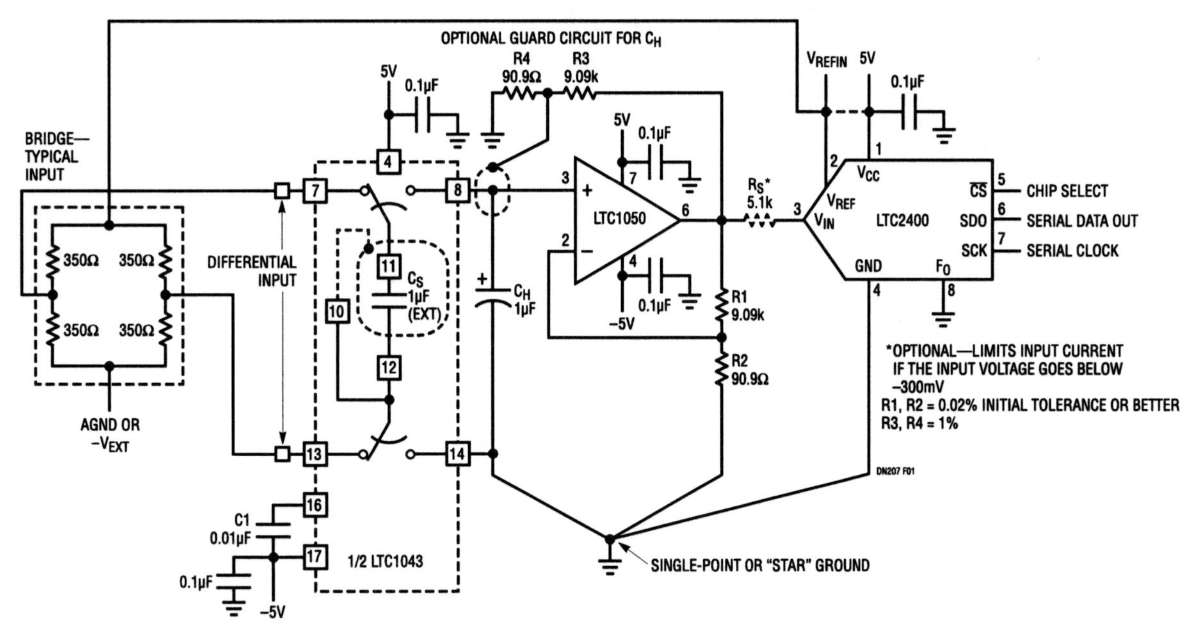

Figure 346.1 • Differential to Single-Ended Converter for Low Level Inputs, Such as Bridges, Maintains the LTC2400's High Accuracy

Analog Circuit Design: Design Note Collection. http://dx.doi.org/10.1016/B978-0-12-800001-4.00346-X

guard circuit could be added as shown. The common point of these two resistors produces the potential for the guard ring. Consult the LTC1043 data sheet for more information. As is good practice in all high precision circuits, keep all lead lengths as short as possible to minimize stray capacitance and noise pickup.

The LTC1050's closed-loop gain accuracy is affected by the tolerance of the ratio of the gain-setting resistors. If cost considerations preclude using low tolerance resistors (0.02% or better), the processor to which the LTC2400 is connected can be used to perform software correction. Operated as a follower, the LTC1050's gain and linearity error is less than 0.001%.

As stated above, the LTC1043 has the highest transfer accuracy when using 1.0μF capacitors. For example, 0.1μF will typically increase the circuit's overall nonlinearity tenfold to 0.001% or 10ppm.

A source of errors is thermocouple effects that occur in soldered connections. Their effects are most pronounced in the circuit's low level portion, before the LTC1050's output. Any temperature changes in any of the low level circuitry's connections will cause linearity perturbations in the final conversion result. Their effects can be minimized by balancing the thermocouple connections with reversed redundant connections and by sealing the circuit against moving air.

A subtle source of error arises from ground lead impedance differences between the LTC1043 circuit, the LTC1050 preamplifier and the LTC2400. This error can be avoided by connecting Pin 14 of the LTC1043, the bottom end of R2 and Pin 4 of the LTC2400 to a single-point "star" ground.

The circuit's input current consists of common mode and differential mode components. The differential mode input current can be as much as ±25nA when $V_{IN(CM)}$ is equal to the + or −5V common mode limit, and V_{DIFF} is 40mV (typ). The ratio of common mode to differential input current is nominally 3:1. Under balanced input conditions ($V_{IN(CM)}=0$), the total input current is typically 1nA. The values may vary from part to part. Figure 346.1's input is analogous to a 2μF capacitor in parallel with a 25MΩ connected to ground. The LTC1043's nominal 800Ω switch resistance is between the source and the 2μF capacitance.

The circuit schematic shows an optional resistor, R_S. This resistor can be placed in series with the LTC2400's input to limit current if the input goes below −300mV. The resistor does not degrade the converter's performance as long as any capacitance, stray or otherwise, connected between the LTC2400's input and ground is less than 100pF. Higher capacitance will increase offset and full-scale errors.

Other differential to single-ended conditioning circuits for the LTC2400 are available at www.linear.com.

Table 346.1 Performance Specifications for Figure 346.1's Circuit. $V_{CC}=V_{REF}=$ LT1236-5; $V_{FS}=40mV$, $R_{SOURCE}=175\Omega$ (Balanced)

PARAMETER	CIRCUIT (MEASURED)	LTC2400	TOTAL (UNITS)
Input Voltage Range	−3 to 40		mV
Zero Error	12.7	1.5	μV
Input Current	See Text		
Nonlinearity	±1	4	ppm
Input-Referred Noise (without averaging)	0.3*	1.5	μV_{RMS}
Input-Referred Noise (averaged 64 readings)	0.05*		μV_{RMS}
Resolution (with averaged readings)	19.6		Bits
Overall Accuracy (uncalibrated**)	18.1		Bits
Supply Voltage	±5	5	V
Supply Current	1.6	0.2	mA
CMRR	120		dB
Common Mode Range	±5		V

*Input-referred noise with a gain of 101.
**Does not include gain setting resistors.

Micropower MSOP 10-bit ADC samples at 500ksps

347

Guy Hoover Marco Pan Kevin R. Hoskins

Introduction

The LTC1197/LTC1199 10-bit serial ADCs offer small size, low power operation and fast sample rates with good AC and DC performance. These parts are ideal for low power, high speed and/or compact designs. This design note discusses the features and performance that make the LTC1197/LTC1199 excellent choices for such new designs.

Features

Smallest size (MSOP)

The LTC1197/LTC1199 are among the smallest ADCs available. The LTC1197/LTC1199's serial interface allows the use of 8-pin MSOP and SO packages. Although the SO package consumes little area, the MSOP package reduces the small footprint even further. These are some of the first ADCs available in the MSOP package, which is about half the size of the SO-8.

3V or 5V supplies

The LTC1197/LTC1199 are 5V parts (V_{CC} = 4V to 9V for the LTC1197 and 4V to 6V for the LTC1199). Also available for use in 3V systems are the LTC1197L and the LTC1199L (V_{CC} = 2.7V to 4V). Designed for use in mixed-supply systems, these devices operate flawlessly even when the digital input is greater than the V_{CC} voltage. This is useful in systems where the ADC is running at a lower supply voltage than the processor. If the ADC is running at a higher supply voltage than the processor, the ADC serial data output voltage can easily be decreased to a level appropriate for the processor.

Performance

Micropower performance with auto shutdown at full speed

Running continuously, the LTC1197L consumes only 2.2mW at the maximum sampling rate (25mW for the LTC1197). The power consumption drops dramatically, as shown in Figure 347.1, at lower sampling rates. The formula for calculating power consumption is:

$$P_D = V_{CC} \cdot I_{CC} \cdot t_{CONV} \cdot f_S$$

where P_D is the power consumption, V_{CC} is the supply voltage, I_{CC} is the supply current while the conversion is occurring, t_{CONV} is the conversion time and f_S is the sample rate. As you can see from the formula, lowering f_S reduces the power consumption linearly. It is also important to minimize t_{CONV} by clocking the ADC at its maximum rate during the conversion. In this way, the total power is less because the device is on for a shorter period of time.

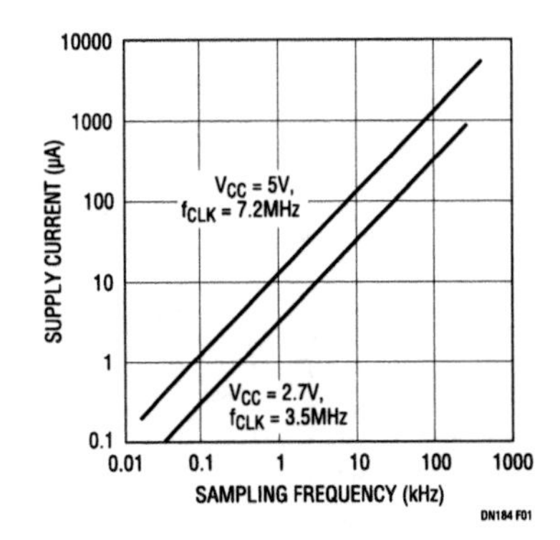

Figure 347.1 • The LTC1197/LTC1199 Reduce Their Supply Current Consumption and Save Power When Operating at Lower Sample Rates

Analog Circuit Design: Design Note Collection. http://dx.doi.org/10.1016/B978-0-12-800001-4.00347-1

High speed capability

Even though the LTC1197/LTC1199 are capable of micropower operation, they are able to sample at rates of up to 500kHz. These parts can also digitize fast input signals up to the Nyquist frequency (250kHz for the LTC1197) with over nine effective number of bits (ENOBs).

Good DC and AC specs

The DC specifications of these parts are very good. Linearity (both INL and DNL) is typically 0.3 LSB with a maximum spec of 1 LSB. Offset is specified at 2 LSBs (max) and gain error is specified at 4 LSBs (max). These specifications are guaranteed over the full temperature range of the part. Both commercial and industrial temperature range versions are available.

AC performance is equally impressive. S/(N+D) is typically 60dB (58dB for the L version). THD is typically −64dB (−60dB for the L version) and the peak harmonic or spurious noise is typically −68dB (−63dB for the L version). An FFT of the LTC1197's conversion performance is shown in Figure 347.2.

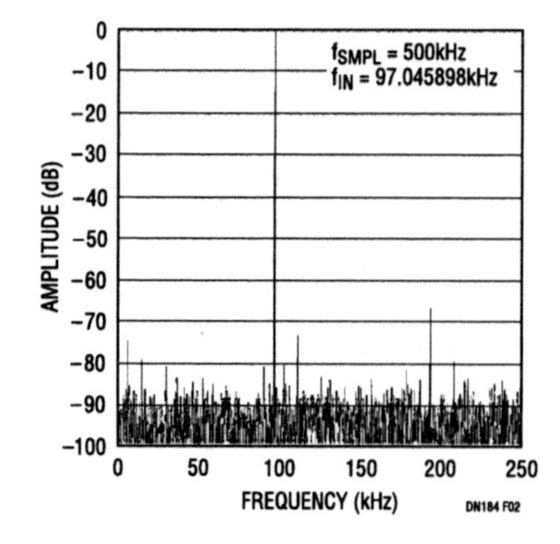

Figure 347.2 • The LTC1197's Typical 60dB SINAD Shown in the FFT Curve Is Among the Best and Translates into 9.7 Effective Bits

Flexible inputs

The LTC1197 has a single differential input and a wide range reference input. The reference input allows the full scale to be reduced to as low as 200mV. This translates to an LSB size of only 200μV. Combined with the high impedance of the analog input, this allows direct digitization of low level transducer outputs, which can save board space and the cost of a gain stage.

With its software-selectable 2-channel MUX, the LTC1199 is capable of measuring either one differential or two single-ended input signals. Both parts have a built-in sample-and-hold.

Serial I/O

The LTC1197/LTC1199 are hardware and software compatible with SPI and MICROWIRE protocols using either 3- or 4-wire serial interfaces. This compatibility is achieved with no additional circuitry, allowing easy interface to many popular processors.

Battery current monitor

Figure 347.3 shows a 2.7V to 4V battery current monitor that draws only 45μA from the battery it monitors. Supply current is conserved by sampling at 1Hz and using the LTC1152's shutdown pin to keep the op amp off between conversions. The LTC1197 automatically shuts down after a conversion. The circuit can be located near the battery, serially transmitting data to the microprocessor.

Conclusion

Conserving space and power, the LTC1197/LTC1199 have a small footprint and are capable of micropower operation. They have a versatile, SPI/MICROWIRE compatible serial interface. The adjustable reference input, 2-channel, software-selectable MUX and 5V or 3V operation increase this ADC family's versatility. When this versatility is combined with the high conversion rate and good DC and AC performance, you can see why these ADCs are good choices for low power, high speed and/or compact designs.

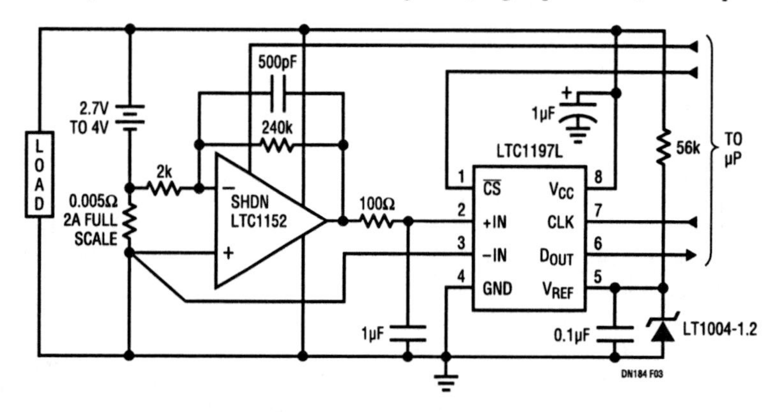

Figure 347.3 • This 0A to 2A Battery Current Monitor Draws Only 45μA from a 3V Battery

16mW, serial/parallel 14-bit ADC samples at 200ksps

348

Chi Fai Yeung Kevin R. Hoskins

Introduction

The new LTC1418 14-bit ADC samples at 200ksps, consuming only 16mW from a single 5V supply. Its pin configurable serial or parallel interface is designed to be versatile, easy to use and adaptable, requiring little or no support circuitry in a wide variety of applications. Some of the key features of this new device include:

- 200ksps Throughput
- Low Power: 16mW
- Parallel and Serial Data Output Modes
- 1.25LSB INL Max and 1LSB DNL Max
- Two Input Ranges: 0V to 4.096V with 5V Supply; ±2.048V with ±5V Supply
- NAP and SLEEP Power Shutdown Modes
- Small Package: 28-Pin SSOP

High performance without high power

As shown in Figure 348.1, the LTC1418 includes a high performance differential sample-and-hold circuit, an ultra-efficient successive approximation ADC, an on-chip reference, and a digital interface that allows easy serial or parallel interface to a microprocessor, FIFO or DSP. The LTC1418's factory calibrated 14-bit performance is achieved without requiring any autocalibration cycles. The result is a 1LSB (max) differential linearity error, no missing codes and 1.25LSB (max) integral linearity error guaranteed over temperature.

For AC applications the dynamic performance of the LTC1418 is exceptional. The low distortion, differential sample-and-hold acquires input signals at frequencies up to 10MHz. At the 100kHz Nyquist frequency, the spurious free

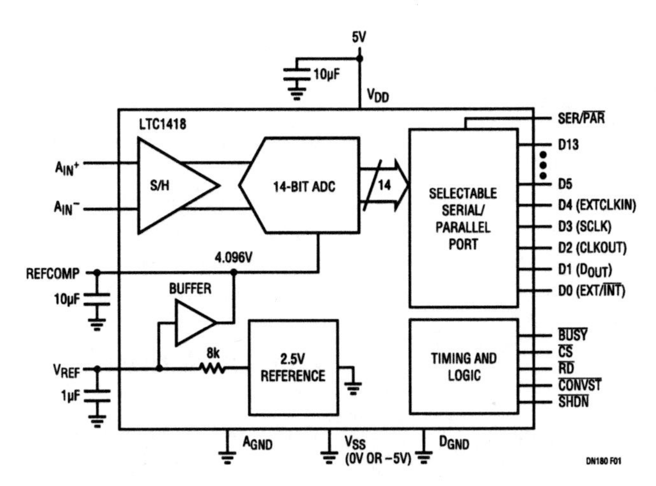

Figure 348.1 • LTC1418 Block Diagram

Analog Circuit Design: Design Note Collection. http://dx.doi.org/10.1016/B978-0-12-800001-4.00348-3

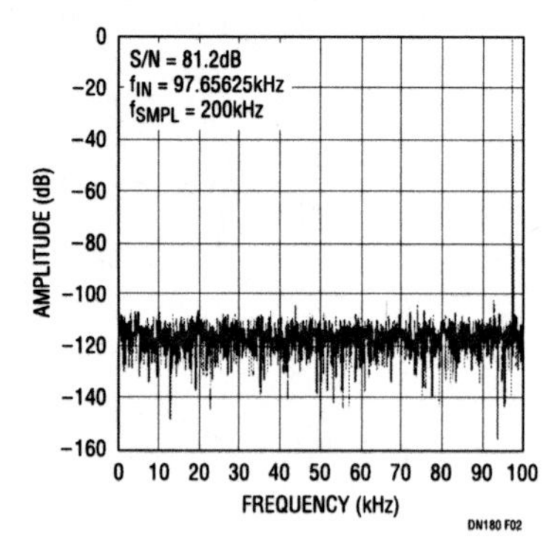

Figure 348.2 • LTC1418 FFT

dynamic range is typically 95dB. The noise is also low with a signal-to-noise ratio (SNR) of 82dB from DC to well beyond Nyquist and a SINAD of 81.2dB as shown in Figure 348.2.

The LTC1418's superior AC and DC performance is achieved while dissipating just 16mW at 200kHz and 10mW at sample rates below 50kHz, the lowest power of any 14-bit ADC. Two shutdown modes, NAP and SLEEP, make it possible to cut dissipation further at lower sample rates.

NAP mode reduces dissipation by 80% and SLEEP mode reduces dissipation to 10μW.

Differential inputs with wideband CMRR

The LTC1418's differential input has excellent common mode rejection, eliminating the need for most input conditioning circuitry. As shown in Figure 348.3, the LTC1418's CMRR remains excellent at high frequencies unlike that of op amps or instrumentation amplifiers whose rejection degrades substantially.

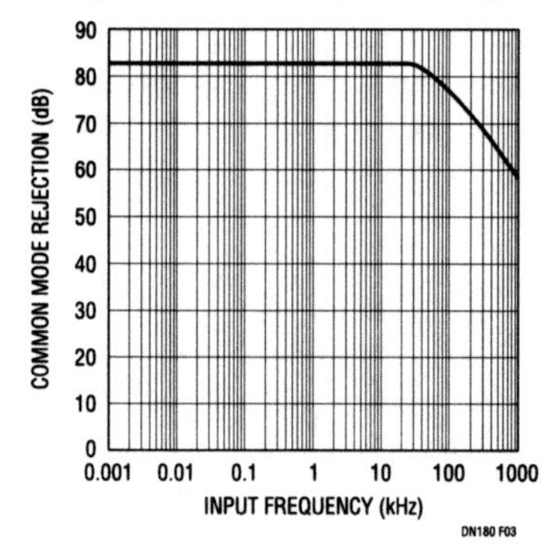

Figure 348.3 • Input Common Mode Rejection vs Input Frequency

Single supply or dual supply operation

The LTC1418 can operate with single or dual supplies. Dual supplies allow direct coupling of signals that swing below ground. The ADC is equipped with circuitry that automatically detects when −5V is applied to the V_{SS} pin. With a −5V supply the ADC operates in bipolar mode and the full-scale range becomes ±2.048V for A_{IN}^+ with respect to A_{IN}^-. With a single supply (V_{SS}=0V), the ADC operates in unipolar mode with an input range of 0V to 4.096V.

On-chip reference

The on-chip 2.5V reference is available on the V_{REF} pin (Pin 3). Internally the reference's 2.5V output is amplified to 4.096V, setting the full-scale span for the ADC. The 4.096V output is available on the REFCOMP pin (Pin 4) and may be used as a reference for other external circuitry. With a temperature coefficient of 10ppm/°C, both V_{REF} and REFCOMP are suitable as the master reference for the system. If, however, an external reference is required, it can easily overdrive either reference output.

Parallel or serial data output

The parallel output mode of the LTC1418 allows the lowest digital overhead. A microcontroller can initiate a conversion and perform other tasks while the conversion is running. The ADC can then signal the microcontroller after the conversion is complete with the $\overline{BUSY}$ signal, at which time valid data is available on the parallel output bus. $\overline{BUSY}$ may also be used to clock latches or a FIFO directly since data is guaranteed to be valid with the rising edge of $\overline{BUSY}$.

The serial output mode of the LTC1418 is simple and flexible, requiring just three pins for data transfer: a data out pin, a serial clock pin and a control pin. Serial data can be clocked with either the internal shift clock or an external shift clock. Data can be clocked out during or after the conversion. The serial I/O is SPI and MICROWIRE compatible.

Perfect for telecom: wide dynamic range

The wide dynamic range required by telecommunications is satisfied by the LTC1418's low noise, low distortion and extremely wide dynamic range over its entire Nyquist bandwidth. Spurious free dynamic range is typically 95dB to beyond the Nyquist frequency. The sample-and-hold circuit's ultralow, $5ps_{RMS}$ jitter keeps the low SNR constant from DC to 1MHz, ideal for undersampling applications.

The LTC1418 is designed to have ultralow bit error rates, another important requirement for telecom systems. The error rate is so low that it is immeasurable. This, combined with excellent AC and DC specification, makes the LTC1418 ideal for a wide variety of applications.

Conclusions

The new LTC1418 low power, 14-bit ADC will find uses in many types of applications from industrial instrumentation to telephony. The LTC1418's adaptable analog input and digital interface reduces the need for expensive support circuitry and can result in a lower cost, smaller system.

16-bit, 333ksps ADC achieves 90dB SINAD, –100dB THD and no missing codes

349

Marco Pan Kevin R. Hoskins

Fastest 16-bit sampling ADC

Linear Technology's recently introduced LTC1604 is the fastest, highest performance 16-bit sampling ADC on the market. This device samples at 333ksps and delivers excellent DC and AC performance. The LTC1604 operates on ±5V supplies and typically dissipates just 220mW. It has a fully differential input sample-and-hold and an onboard reference. Two power shutdown modes, NAP and SLEEP, reduce power consumption to 7mW and 10μW, respectively. At 333ksps, this 16-bit device not only offers performance superior to that of the best hybrids, but does so with low power, the smallest size, an easy-to-use parallel interface and the lower cost of a monolithic part. It is available in a tiny 36-pin SSOP package. Some of the key features of this new device include:

- 333ksps throughput
- 16bits with no missing codes and ±2LSB INL
- Low power dissipation and power shutdown (10μW in SLEEP mode)
- Excellent AC performance: 90dB SINAD and –100dB THD
- Small 36-pin SSOP package

These features of the LTC1604 can simplify, improve and lower the cost of current data acquisition systems and open up new applications that were not previously possible because no similar parts were available.

Outstanding DC and AC performance

As shown in Figure 349.1, the LTC1604 combines a high performance differential sample-and-hold circuit with an extremely fast successive-approximation ADC and an on-chip reference. Together, they deliver an excellent combination of DC and AC performance.

The DC specifications include 16-bit with no missing codes and ±2LSB integral nonlinearity error, all guaranteed over temperature. The ADC includes an on-chip, curvature corrected bandgap reference. Figures 349.2a and b show the LTC1604's exceptional INL and DNL error.

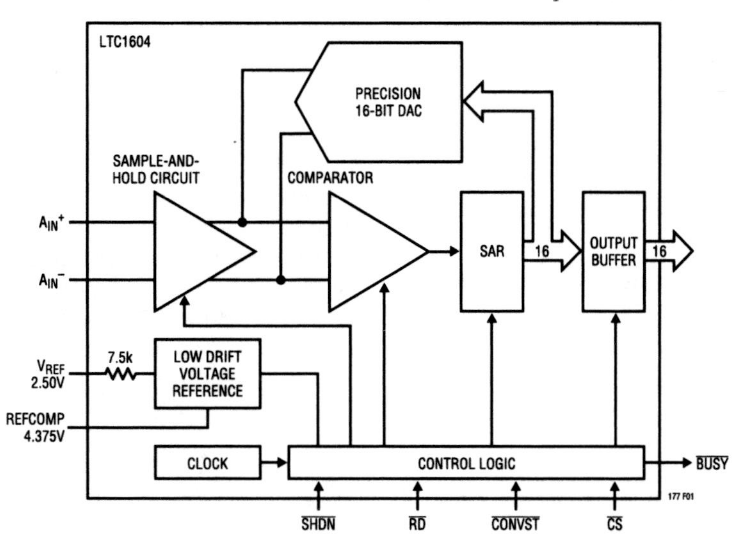

Figure 349.1 • The 333ksps, 16-Bit ADC Features a True Differential S/H with Excellent Bandwidth and CMRR

Analog Circuit Design: Design Note Collection. http://dx.doi.org/10.1016/B978-0-12-800001-4.00349-5

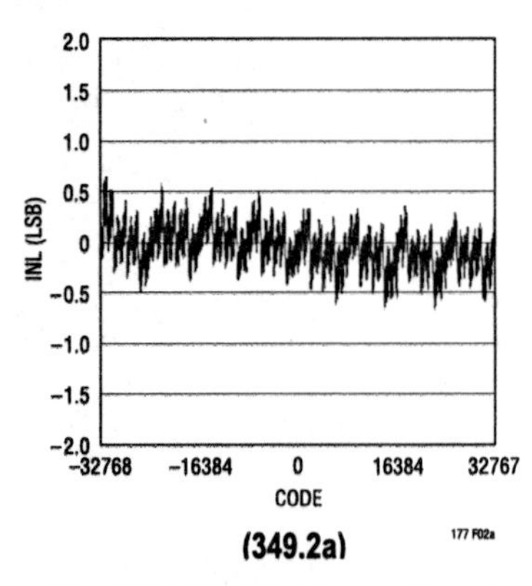

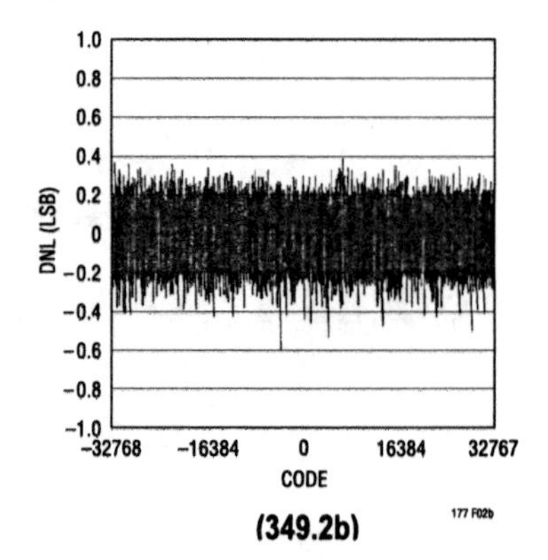

Figure 349.2 • The LTC1604 Achieves Excellent INL (349.2a) and DNL (349.2b) without Cumbersome Autocalibration

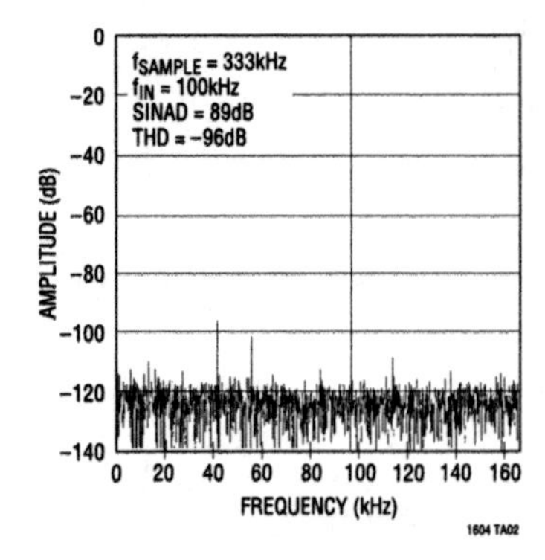

Figure 349.3 • SINAD Is over 90dB and THD Is −100dB at Low Input Frequencies. Even with 100kHz Inputs, SINAD Remains 89dB and THD Is −96dB as Shown

In addition to its outstanding linearity, the LTC1604 provides exceptional spectral purity at 333ksps; better than 90dB SINAD and −100dB THD for a 20kHz input and 89dB SINAD and −96dB THD for a 100kHz input (Figure 349.3). Figure 349.4 shows how well the converter's signal-to-noise plus distortion ratio (SINAD) holds up as the input frequency is increased.

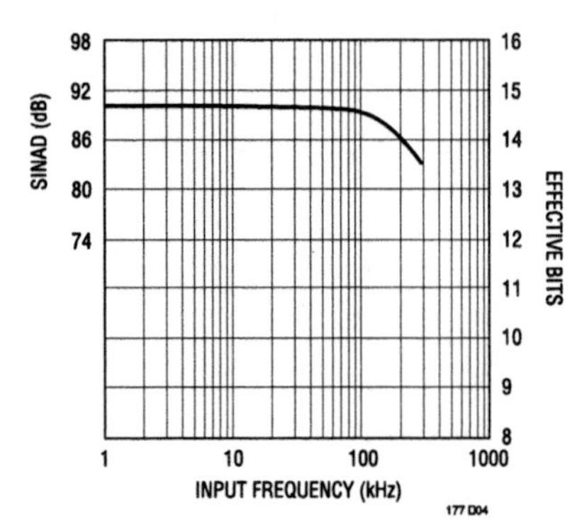

Figure 349.4 • The Wideband S/H Captures Signals Well Beyond Nyquist

Differential inputs reject common mode noise

Getting a clean signal to the input of an ADC, especially a 16-bit ADC, is not an easy task in many systems. In a single-ended sampling system accuracy and dynamic range can be limited by ground noise. When a single-ended signal is applied to an ADC's input, the ground noise adds directly to the applied signal. Although a filter can reduce this noise, this does not work for in-band noise or common mode noise at the same frequency as the input signal. Figure 349.5 shows how the LTC1604 provides relief. Because of its excellent CMRR, the LTC1604's differential inputs reject ground noise, even at the frequency of the desired input signal. Further, the LTC1604's wideband CMRR can eliminate high frequency noise up to 1MHz and beyond.

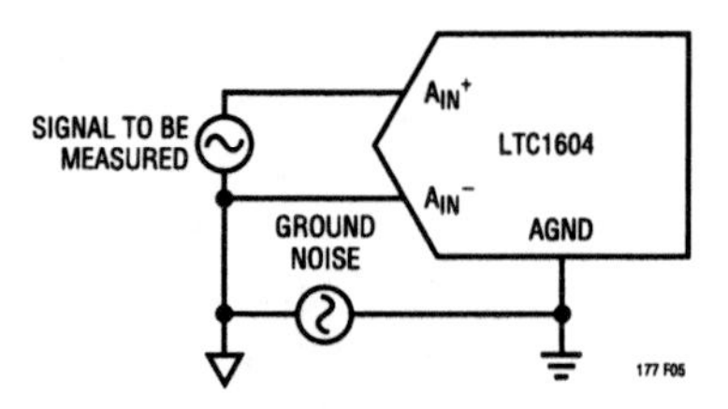

Figure 349.5 • The LTC1604's Differential Inputs Reject Common Mode Noise by Measuring Differentially

Applications

The performance of the LTC1604 makes it very attractive for a wide variety of applications, such as digital signal processing, PC data acquisition cards, medical instrumentation and high resolution or multiplexed data acquisition.

With its excellent dynamic performance and linearity, and high sample rate the LTC1604 is the ideal ADC for high speed, 16-bit DSP and PC data acquisition card applications.

Applications such as single-channel or multiplexed high speed data acquisition systems benefit from the LTC1604's high sample rate and high impedance inputs. The high sample rate allows designers to multiplex more channels of a given bandwidth than slower 16-bit ADCs while meeting the demands of a low power budget.

16-bit, 100ksps A/D converter runs on 5V supply

350

Sammy Lum   Kevin R. Hoskins

The LTC1605 is a new 16-bit, 100ksps ADC from Linear Technology. Its outstanding DC accuracy and ±10V analog input range are ideal for industrial control and instrumentation applications. Its simple I/O, low power and high performance make it easy to design into applications requiring wide dynamic range and high resolution.

Product features

- No missing codes and ±2LSB max INL over temperature
- Single 5V supply with 55mW typical power dissipation
- ±10V analog input with ±20V overvoltage protection for harsh environments
- Complete ADC contains sample-and-hold and reference
- 28-pin PDIP, SO and SSOP packages

Circuit description

The LTC1605 converts ±10V analog input signals while operating on a 5V supply. A resistor network is used to attenuate the input signal. This reduced internal signal is digitized by a differential, switched-capacitor, 16-bit SAR ADC at a rate of up to 100ksps. The differential architecture provides high immunity to power supply and other external noise sources. The trimmed 2.5V bandgap reference can be overdriven with an external reference if desired.

The digital interface is simple. Conversions are started using the read-convert (R/C) pin. Data is available as a 16-bit word or two 8-bit bytes (Figure 350.1).

AC and DC performance

Figure 350.2 shows the fast Fourier transform (FFT) of a ±10V 1kHz sine wave signal digitized at 100ksps by the LTC1605. The LTC1605 achieves a signal-to-noise and distortion (SINAD) of 87.5dB and very low total harmonic distortion (THD) of −101.7dB.

Figure 350.3 shows an INL error plot for the LTC1605. Guaranteed specifications include ±2.0LSB INL (max) and no missing codes at 16 bits over the industrial temperature range (−40°C to 85°C). The ADC's outstanding accuracy is assured by factory trimming. For the user, this eliminates the software overhead and calibration time delays associated with autocalibrated ADCs.

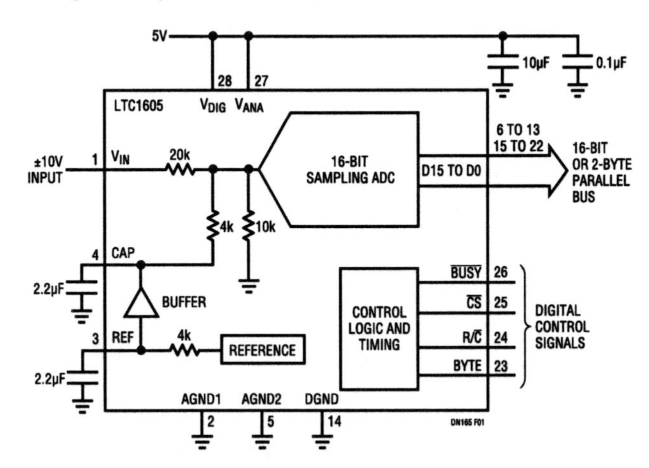

Figure 350.1 • Offering 16-Bit Performance, the LTC1605 Handles ±10V Inputs While Operating on a Single 5V Supply

Analog Circuit Design: Design Note Collection. http://dx.doi.org/10.1016/B978-0-12-800001-4.00350-1

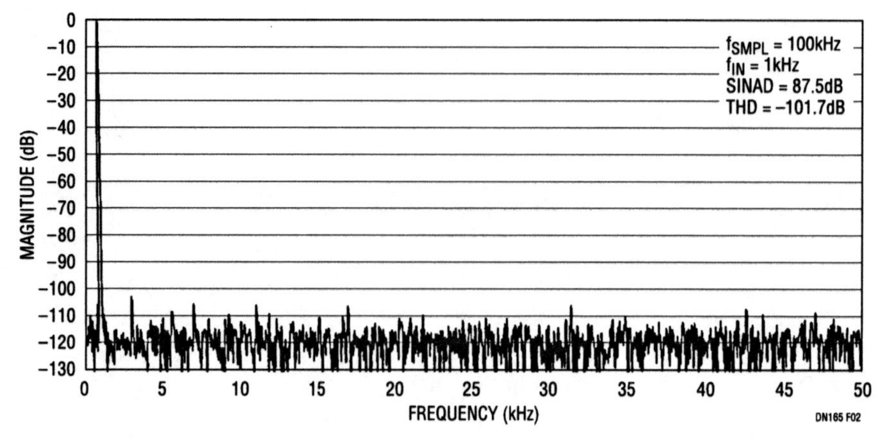

Figure 350.2 • The LTC1605's AC Performance Achieves a 87.5dB SINAD and −101.7dB THD

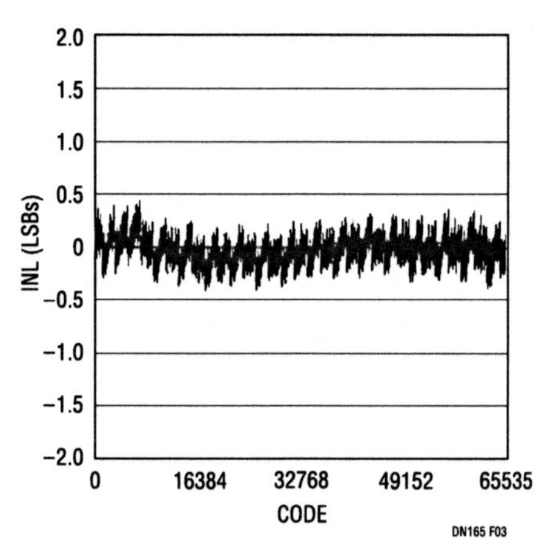

Figure 350.3 • The LTC1605 Achieves Excellent INL Without Cumbersome Autocalibration

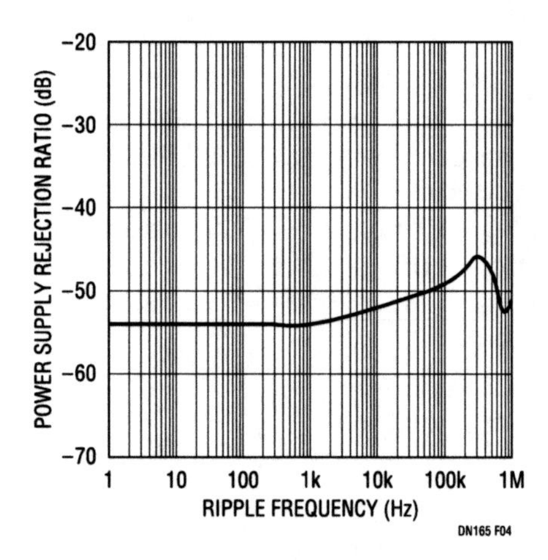

Figure 350.4 • The LTC1605's Differential Architecture Provides Good, Wideband Supply Rejection

One of the benefits of the LTC1605's differential architecture is its high power supply rejection ratio (PSRR). Figure 350.4 shows the power supply rejection ratio of the LTC1605 as a function of frequency.

High PSRR is important to the LTC1605's conversion accuracy and signal-to-noise performance because, at 16-bit quantization, the LSB magnitude is just 305μV (V_{IN}=20V_{P-P}). The LTC1605's high PSRR rejects up to 100mV of power supply noise to below an LSB level.

Applications

With its overvoltage protected ±10V analog input, the LTC1605 fits easily into industrial process control, power management, data acquisition boards and has sufficient speed for multiplexed applications. The LTC1605 is also ideal for wide dynamic range applications that use a PGA and lower resolution ADC. For example, the LTC1605 is a simpler solution and will out perform (in terms of DNL, INL and S/N) a 12-bit ADC converting the output of a PGA with a gain range of 1-to-16.

Conclusion

The LTC1605 is a complete 16-bit ADC with a built-in sample-and-hold and reference. Its wide analog input range, overvoltage protection and DC accuracy make it a good candidate for industrial process control, instrumentation and other high dynamic range applications.

14-bit, 800ksps ADC upgrades 12-bit systems with 81.5dB SINAD, 95dB SFDR

351

Dave Thomas Kevin R. Hoskins

Higher dynamic range ADCs

The new 14-bit, 800ksps LTC1419 enhances new communications, spectral analysis, instrumentation and data acquisition applications by providing an upgrade path to users of 12-bit converters. It provides outstanding 81.5dB SINAD (signal-to-noise and distortion ratio) and 95dB SFDR (spurious-free dynamic range) for frequency-domain applications, and excellent DNL with no missing codes performance for time-domain applications.

LTC1419 features

- Complete 14-Bit, 800ksps ADC
- ±1LSB DNL and ±1.25LSB INL (Max)
- 81.5dB SINAD and 95dB SFDR
- Low Power: 150mW on ±5V Supplies
- Nap/Sleep Power-Down Modes
- Small Footprint: 28-Pin SO or SSOP

The LTC1410's big brother

The new LTC1419 is a 14-bit derivative of the 12-bit LTC1410. It has a similar pinout and function, as shown in the block diagram in Figure 351.1. The wideband differential sample-and-hold (S/H) has a 20MHz bandwidth and can sample either differential or single-ended signals. Unlike some converters, which must be driven differentially to perform well, this ADC operates equally well with single-ended or differential signals.

The switched capacitor SAR architecture yields excellent DC specifications and stability. It is clean, simple to use and delivers an 800ksps conversion rate on just 150mW with ±5V supplies.

The ADC's flexible parallel I/O connects easily to a DSP, microprocessor, ASIC or dedicated logic. Conversions can be started under command of a DSP or microprocessor or from an external sample clock signal. An output disable allows the outputs to be three-stated.

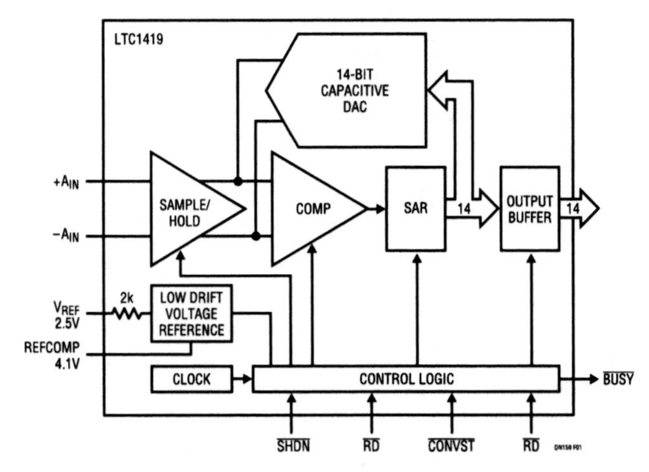

Figure 351.1 • The LTC1419 Complete 800ksps, 14-Bit ADC Has a Fast S/H That Cleanly Samples Wideband Input Signals

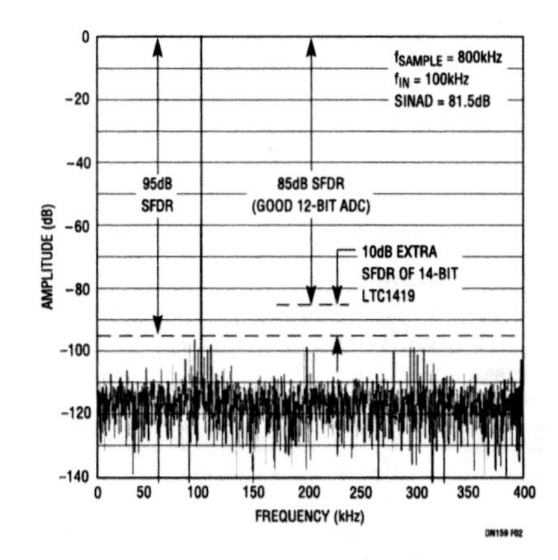

Figure 351.2 • The LTC1419 Gives a 10dB Improvement in Spectral Purity over the Best 12-Bit Devices. This FFT Shows the LTC1419's Outstanding 81.5dB SINAD and 95dB SFDR

Analog Circuit Design: Design Note Collection. http://dx.doi.org/10.1016/B978-0-12-800001-4.00351-3

10dB extra dynamic range for signal applications

Even though the LTC1410 is the cleanest 12-bit ADC on the market (its 72dB SINAD and 85dB SFDR approach the theoretical limit for 12bits) improvements are possible with the LTC1419 (see Figure 351.2). The 14-bit device achieves 81.5dB SINAD and 95dB SFDR. This gives the converter roughly 10dB more resolving power to pick out small signals in communication and spectrum analysis applications. This clean sampling capability is maintained even for wideband inputs. Figure 351.3 shows high effective bits and SINAD for inputs beyond Nyquist.

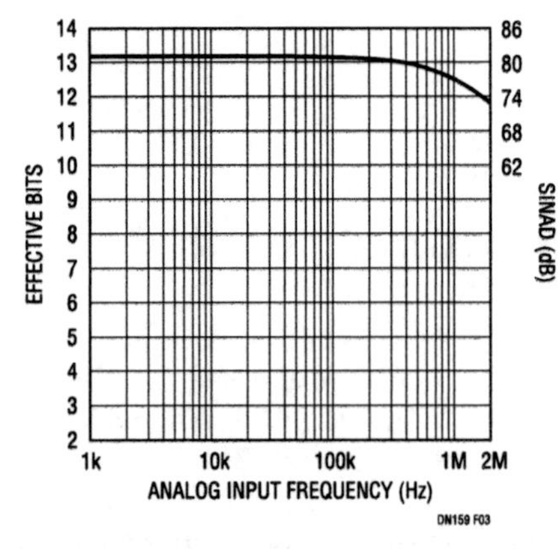

Figure 351.3 • The LTC1419's Essentially Flat SINAD and ENOBs Ensure Spectral Purity at Frequencies to Nyquist and Achieve 12-Bit Performance When Sampling 2MHz Inputs

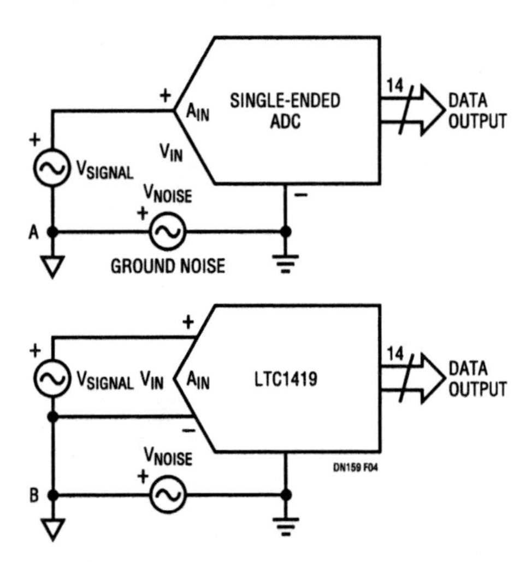

Figure 351.4 • (a) In High Resolution ADC System, Noise Sources Such as Ground Noise and Magnetic Coupling Can Contaminate the ADC's Input Signal. (b) The LTC1419's Differential Inputs Can Be Used to Reject This Noise, Even at High Frequencies

Noise rejecting differential inputs

As the converter's resolution increases and its noise floor drops, other system noises may show up unless they are eliminated. The LTC1419's differential input provides a way to keep noise out. Noise can be introduced in a number of ways including ground bounce, digital noise and magnetic and capacitive coupling (see Figure 351.4a). All of these sources can be reduced dramatically by sampling differentially from the signal source, as in Figure 351.4b. The high CMRR of the differential input (Figure 351.5) allows the LTC1419 to reject resulting common mode noise by over 60dB and maintain a clean signal.

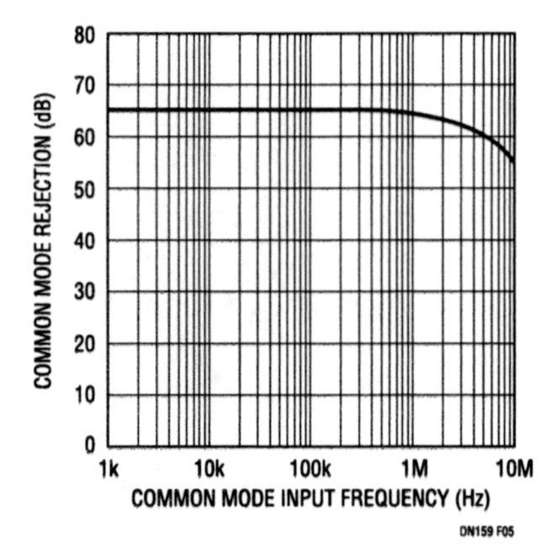

Figure 351.5 • The Common Mode Rejection of the Analog Inputs Rejects Common Mode Input Noise Frequencies to Beyond 10MHz

Other nice features

- Both analog inputs have infinite DC input resistance, which makes them easy to multiplex or AC couple.
- The separate convert-start input pin allows precise control over the sampling instant. The S/H aperture delay is less than 2ns and the aperture jitter is below 5ps RMS.
- Conversion results are available immediately after a conversion and there is no latency in the data (no pipeline delay). This is ideal for both single shot and repetitive measurements.
- The low 150mW power dissipation can be reduced further using the ADC's Nap and Sleep power-down modes. Wake up from Nap mode is instantaneous. Sleep mode wake-up time is several milliseconds.
- The LTC1419 is the industry's smallest high speed 14-bit converter and is packaged in a 28-pin wide SSOP.

Time to upgrade?

The new, low cost LTC1419 is the ideal converter to upgrade 12-bit, high performance designs to 14 bits. Its exceptional dynamic performance gives a 10dB improvement in dynamic range compared to the best 12-bit devices. Low power and flexibility make it useful in a variety of time- and frequency-domain applications. Low cost and ultra-small size make it the ideal candidate for designers who need the next step in ADC performance.

Micropower 4- and 8-channel, 12-bit ADCs save power and space

352

Kevin R. Hoskins

Introduction

Data acquisition applications that require low power dissipation fall into two general areas: products that must use power very efficiently, such as battery-powered portable test equipment, and remotely located data logging equipment. To help meet these requirements, Linear Technology has introduced the LTC1594 and LTC1598.

Micropower ADCs in small packages

The LTC1594/LTC1594L and LTC1598/LTC1598L are micropower 12-bit ADCs that feature a 4- and 8-channel multiplexer, respectively. To cover different system supply voltages, the LTC1594 and LTC1598 operate on 5V and the LTC1594L and LTC1598L operate on 3V. The LTC1594L and the LTC1598L are tested to operate on supplies as low as 2.7V and sample at a maximum of 10.5ksps. The LTC1594 and LTC1598 have a maximum sample rate of 16.8ksps. At full conversion rate, the LTC1594/LTC1598 and LTC1594L/LTC1598L typically draw 320µA and 160µA, respectively. At 1ksps these converters typically draw 20µA. The LTC1594/LTC1594L are available in a 16-pin SO package and the LTC1598/LTC1598L are available in a 24-pin SSOP package.

As shown in Figure 352.1, each converter includes a MUX with separate MUXOUT and ADCIN pins (useful for conditioning an analog input prior to conversion), S/H, 12-bit ADC and a simple, efficient, serial interface that reduces interconnects. Reduced interconnections also reduce board size and allow the use of processors having limited I/O, both of which help reduce system costs.

Conserve power with auto shutdown operation

The LTC1594/LTC1594L and LTC1598/LTC1598L include an auto shutdown feature that reduces power dissipation when the converter is inactive ($\overline{CS} = 1$). Nominal power dissipation while either 5V converter is clocked at 320kHz is typically 1.6mW. The 3V converters dissipate 480µW when clocked at 200kHz. The curve in Figure 352.2 shows the amount of current drawn by this MUXed 12-bit ADC family vs sample rate.

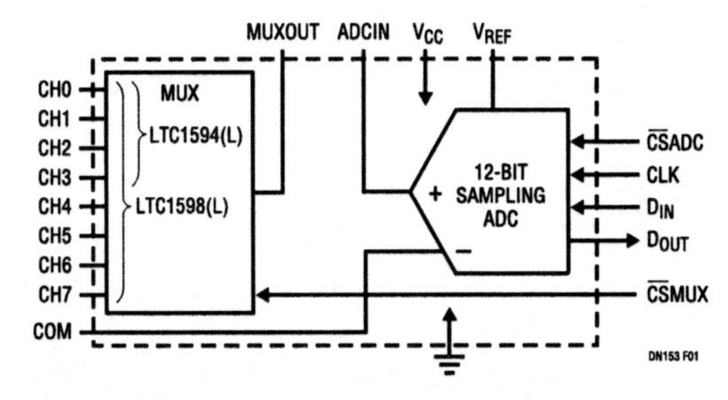

Figure 352.1 • With a 4- or 8-Channel MUX, the LTC1594/LTC1594L and LTC1598/LTC1598L Feature Low Power Dissipation, MUXOUT/ADCIN Connections for External Signal Conditioning and Serial Interface

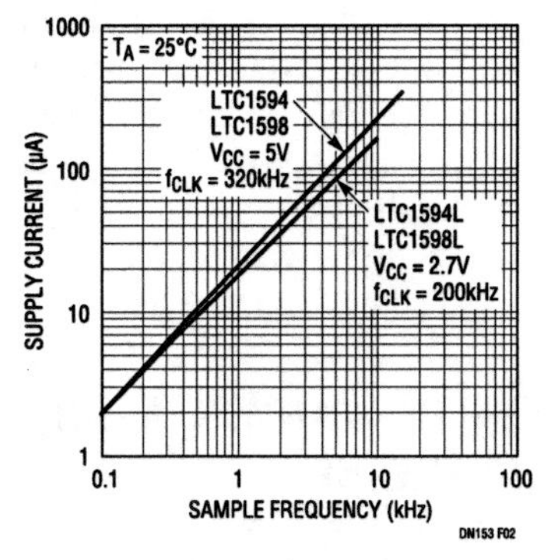

Figure 352.2 • The Auto Shutdown Feature Automatically Reduces Supply Current as Sample Rate Is Reduced. Supply Current Drops to 20µA at 1ksps

Analog Circuit Design: Design Note Collection. http://dx.doi.org/10.1016/B978-0-12-800001-4.00352-5

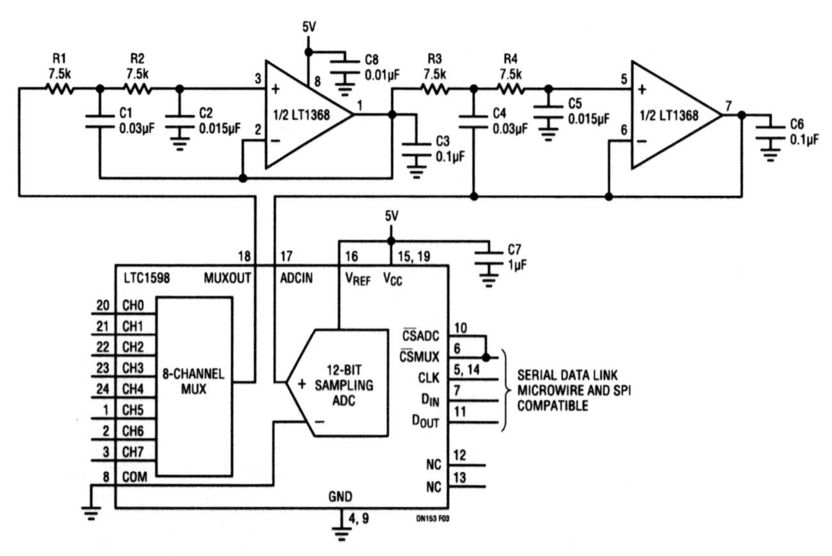

Figure 352.3 • This Data Acquisition System Takes Advantage of the LTC1598's MUXOUT/ADCIN Pins to Filter All Eight Analog Signals with a Single Filter Prior to ADC Conversion

Good DC performance

The DC specs include excellent differential nonlinearity (DNL) of ±3/4LSB, an advantage in pen-screen and other monitoring applications. No missing codes are guaranteed over temperature.

Versatile, flexible serial I/O

The serial interface found on the LTC1594/LTC1594L and LTC1598/LTC1598L is designed for ease of use, flexibility, minimal interconnections and I/O compatibility with QSPI, SPI, MICROWIRE and other serial interfaces. The MUX and the ADC have separate chip select ($\overline{CS}$) and serial clock inputs that can be tied together or used separately, adding versatility. The remaining serial interface signals are data input (D_{IN}) and data output (D_{OUT}). The maximum serial clock frequencies are 320kHz and 200kHz for the 5V and 3V parts, respectively.

Latchup proof MUX inputs

The LTC1594's and LTC1598's input MUXes are designed to handle input voltages that exceed the nominal input range, GND to the supply voltage, without latchup. Although an overdriven unselected channel may affect a selected, correctly driven channel, no latchup occurs and correct conversion results resume when the offending input voltage is removed. The MUX inputs remain latchup proof for input currents up to ±200mA over temperature.

Individual ADC and MUX chip selects enhance flexibility

The LTC1594/LTC1594L and LTC1598/LTC1598L feature separate chip selects for ADC and MUX. This allows the user to select a channel once for multiple conversions. This has the following benefits: first, it eliminates the overhead of sending a D_{IN} word for the same channel each time for each conversion; second, it avoids possible glitches that may occur if a slow-settling anti-aliasing filter is used following the MUX; and third, it sets the gain once for multiple conversions if the MUXOUT/ADCIN pins are used with an op amp to create a programmable gain amplifier (PGA).

MUXOUT/ADCIN economizes signal conditioning

The MUXOUT and ADCIN pins allow the filtering, amplification or conditioning of analog input signals prior to conversion. These input/output connections are also a cost-effective way to perform the conditioning because only one circuit is needed instead of one for each channel. The circuit in Figure 352.3 uses these connections to insert an anti-aliasing filter into the signal path, filtering several analog inputs. The output signal of the selected MUX channel, present on the MUXOUT pin, is applied to R1 of the Sallen-Key filter. The filter band limits the analog signal and its output is applied to ADCIN. When lightly loaded, as in this application, the LT1368 rail-to-rail op amps used in the filter will swing to within 8mV of the positive supply voltage and 6mV of ground. Since only one circuit is used for all channels, each channel sees the same filter characteristics and channel-to-channel matching is ensured.

Conclusion

With their serial interfaces, small packages and auto shutdown, the LTC1594/LTC1594L and LTC1598/LTC1598L achieve very low power consumption while occupying very little circuit board area. Their outstanding DC specifications make them the choice for applications that benefit from low power, battery conserving operation, multichannel inputs and space and component saving signal conditioning.

1.25Msps, 12-bit ADC conserves power and signal integrity on a single 5V supply

353

Kevin R. Hoskins

Introduction

The new LTC1415 expands LTC's family of high speed, low power 12-bit ADCs. Operating on a single 5V supply, it is optimized for applications such as ADSL, HDSL, modems, direct down-conversion, CCD imaging, DSP-based vibration analysis, waveform digitizers and multiplexed systems. The LTC1415's block diagram is shown in Figure 353.1.

Single 5V supply, high speed, lowest power

The new LTC1415 has lower power dissipation and better performance and features than other 12-bit ADCs currently on the market. It samples at 1.25Msps and dissipates just 55mW. NAP and SLEEP modes reduce dissipation even further. Its 28-pin SO and SSOP packages offer the industry's smallest footprint.

A separate output-driver supply pin, OV_{DD}, allows the ADC to interface to either 3V or 5V digital systems.

Tiny package

The LTC1415 is the smallest 1.25Msps 12-bit ADC available. Its 28-pin SSOP occupies just 0.123in^2, 43% of a 28-pin SO. It operates with equally tiny 0.1μF 0805-sized surface mount supply bypass capacitors and a small 10μF 1206-sized ceramic reference bypass capacitor.

Complete ADC with reference and wideband S/H

The LTC1415 includes a 10ppm/°C 2.5V reference that is suitable for use as a system reference. This internal reference sets the LTC1415's full-scale input range from 0V to 4.096V. The circuit in Figure 353.2 trims the full-scale (gain) error without placing additional circuitry in the signal path.

The LTC1415's fully differential S/H has a typical 20MHz input bandwidth and very good CMRR of 60dB or better over a DC to 10MHz bandwidth. AC specifications are measured with single-ended ($-A_{IN}$ grounded) signal drive. Offsetting $-A_{IN}$ above ground eases the drive requirements of rail-to-rail op amps that drive the LTC1415's input.

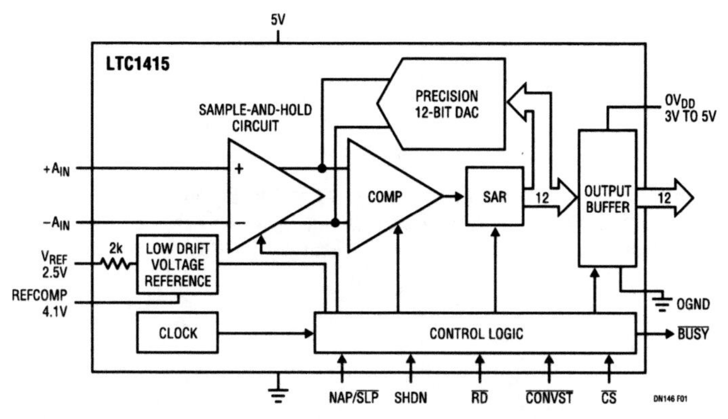

Figure 353.1 • This Complete 5V, 12-Bit, 1.25Msps ADC Comes in a 28-Pin SSOP

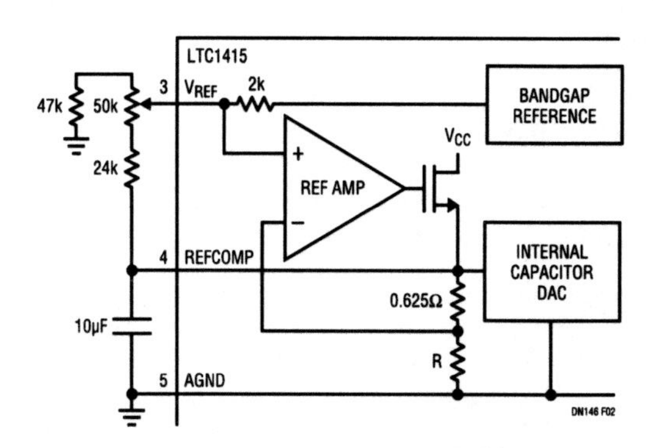

Figure 353.2 • Adding Nothing to the Signal Path, This Simple Circuitry Facilitates Full-Scale Adjustment

Analog Circuit Design: Design Note Collection. http://dx.doi.org/10.1016/B978-0-12-800001-4.00353-7

Benefits

Reduce power with single supply operation and two power saving shutdown modes

The LTC1415 has NAP and SLEEP modes that further reduce the converter's low 55mW power dissipation. NAP mode shuts down all converter circuitry except the 2.5V reference, reducing dissipation by 95%. The converter returns from NAP mode very quickly because the active reference keeps its bypass capacitor charged.

SLEEP mode reduces dissipation to 30μW by shutting down all of the LTC1415's internal circuitry. The converter returns from SLEEP more slowly than from NAP mode because the reference must recharge its bypass capacitor. The converter is ready after a typical settling time of 10ms when using a 10μF bypass capacitor as shown in Figure 353.2.

Wide bandwidth CMRR

The ADC S/H's excellent CMRR eliminates the effects of common mode ground noise. The CMRR is constant over the entire Nyquist bandwidth ($f_S/2$), dropping 3dB at 5MHz. This ability to reject high frequency common mode signals is very helpful in sampling systems where switching transients often cause high frequency noise.

No latency and low bit error rate (BER)

The LTC1415's successive approximation architecture is optimum for high speed data conversion and achieves two important benefits. This first benefit is absence of latency (a time delay of more than one sampling period between sampling an input signal and generating an output code). SAR converters generate data from the current sample before taking the next sample. Thus, unlike high speed converters using pipeline conversion techniques, the LTC1415 is always free of data latency. Absence of latency is advantageous for signal multiplexing, closed-loop control systems, single shot event driven measurements and DSP feedback systems.

The second benefit is reduced BER. The LTC1415 is inherently immune to bit errors or "sparkle codes" (infrequent conversion errors as large as full scale in magnitude), unlike subranging or pipelined converters. The LTC1415's BER is immeasurably low, less than 10^{-15}.

DSP interface

The LTC1415's high speed parallel digital interface can be connected directly to a DSP, or by using the CS and RD pins, data can be retrieved when desired. The edge-triggered CONVST signal precisely controls the sampling instant, essential for processing signals and maintaining signal integrity.

Exemplary AC and DC performance

The LTC1415 handles input signals with great care. Its differential S/H has a wide, 20MHz bandwidth and achieves −75dB THD for a 625kHz, 4.096V_{P-P} input signal. This AC performance is complemented by DC specifications that include ±1LSB maximum differential linearity error (0.25LSB, typ) and integral linearity error (0.2LSB, typ) guaranteed over temperature.

The FFT in Figure 353.3 reveals the LTC1415's 1.25Msps conversion rate performance with a full-scale 100kHz sine wave input signal. The curve shows −88dB THD and 72.8db SINAD (11.8 ENOBs).

Figure 353.4 shows that the 11.8 effective number of bits (ENOBs) remains flat out to an input frequency of 200kHz and maintains 11-bit AC performance at 1MHz.

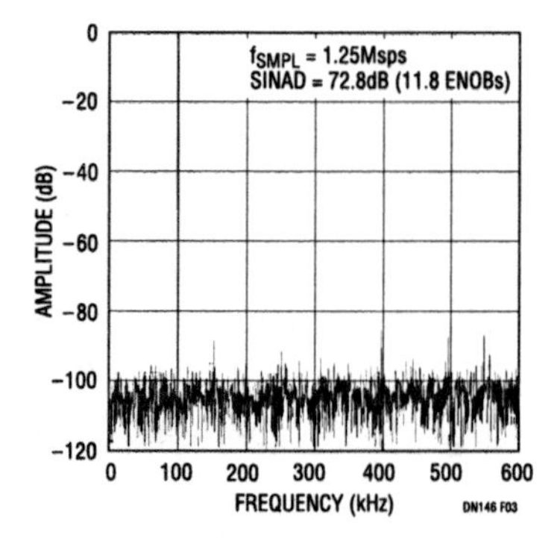

Figure 353.3 • FFT of the LTC1415's Conversion of a Full-Scale 100kHz Sine Wave Shows Excellent Response with a Low Noise Floor While Sampling at 1.25Msps

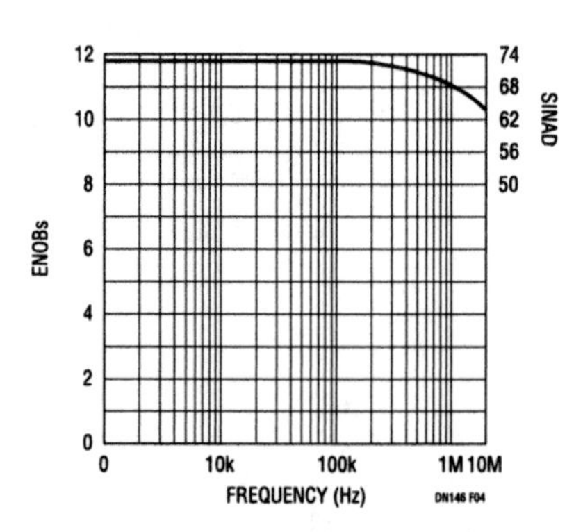

Figure 353.4 • This Curve Shows That the Dynamic Performance of the LTC1415 Remains Robust Out to an Input Frequency of 1MHz, Where It Maintains 11-Bit Performance

Micropower ADC and DAC in SO-8 give PCs a 12-bit analog interface

354

Kevin R. Hoskins

Introduction

Adding two channels of analog input/output to a PC computer is simple, inexpensive, low powered and compact when using the LTC1298 ADC and LTC1446 DAC. The LTC1298 and the LTC1446 are the first SO-8 packaged 2-channel devices of their kind. While the application shown is for PC data acquisition, the two converters provide the smallest, lowest power solutions for many other analog I/O applications.

Small, micropower ADC and DAC

The LTC1298 features a 12-bit ADC, 2-channel multiplexer, and an input S/H. Operating on a 5V supply at maximum conversion rate, the LTC1298 draws just 340μA. A built-in auto shutdown further reduces power dissipation at reduced sampling rates (to 30μA at 1ksps). The serial interface allows remote operation.

The LTC1446 is the first 12-bit dual micropower SO-8 packaged DAC. It features a 1mV/LSB (0V to 4.095V) rail-to-rail output, internal 2.048V reference and a simple, cost-effective, 3-wire serial interface. A D_{OUT} pin allows cascading multiple LTC1446s without increasing the number of serial interconnections. Operating on a supply of 5V, the LTC1446 draws just 1mA (typ).

PC 2-channel analog I/O interface

The circuit shown in Figure 354.1 connects to a PC's serial interface using four interface lines: DTR, RTS, CTS and TX. DTR is used to transmit the serial clock signal, RTS is used to transfer data to the DAC and ADC, CTS is used to receive conversion results from the LTC1298, and the signal on TX selects either the LTC1446 or the LTC1298 to receive input data. The LTC1298's and LTC1446's low power dissipation allow the circuit to be powered from the serial port. The TX and RTS lines charge capacitor C4 through diodes D5 and D6. An LT1021-5 regulates the voltage to 5V. Returning the TX and RTS lines to a logic high after sending data to the DAC or completion of an ADC conversion provides constant power to the LT1021-5.

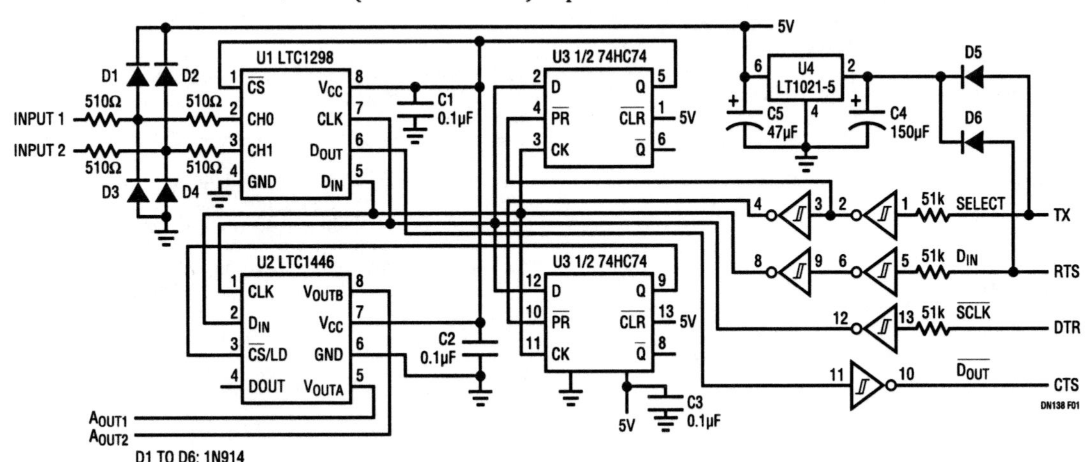

Figure 354.1 • Communicating over the Serial Port, the LTC1298 and LTC1446 in SO-8 Create a Simple, Low Power 2-Channel Analog Interface for PCs

Analog Circuit Design: Design Note Collection. http://dx.doi.org/10.1016/B978-0-12-800001-4.00354-9

Using a 486-33 PC, the throughput was 3.3ksps for the LTC1298 and 2.2ksps for the LTC1446. Your mileage may vary.

Listing 1 is C code that prompts the user to either read a conversion result from the ADC's CH0 or write a data word to both DAC channels.

Conclusion

These SO-8 packaged data converters offer unprecedented space, power and economy to data acquisition system designers. They form a very nice 12-bit analog interface to a wide variety of portable, battery-powered and size constrained products. They are extremely easy to apply and require a minimum of passive support components and interconnections.

Listing 1. Configure Analog Interface with this C Code

```
#define port 0x3FC/* Control register, RS232 */
#define inprt 0x3FE/* Status  reg.  RS232 */
#define LCR  0x3FB/* Line Control Register */
#define high 1
#define low  0
#define Clock 0x01/* pin 4, DTR */
#define Din 0x02/* pin 7, RTS */
#define Dout 0x10/* pin 8, CTS input */
#include<stdio.h>
#include<dos.h>
#include<conio.h>
/* Function module sets bit to high or low */
void set_control(int Port,char bitnum,int flag)
{
        char temp;
        temp = inportb(Port);
        if (flag==high)
    temp |= bitnum; /* set output bit to high */
    else
      temp &= ~bitnum; /* set output bit to low */
       outportb(Port,temp);
}
/* This function brings CS high or low (consult the
schematic) */
void CS_Control(direction)
{
        if (direction)
{
    set_control(port,Clock,low);/* set clock high for Din to
be read */
set_control(port,Din,low); /* set Din low */
set_control(port,Din,low);/* set Din high to make CS goes
high */
    }
    else {
        outportb(port, 0x01);/* set Din & clock low */
        Delay(10);
        outportb(port, 0x03);/* Din goes high to make CS
goes low */
    }
}
/* This function outputs a 24 bit(2x12) digital code to
LTC1446L */
void Din_(long code,int clock)
{
    int x;
    for(x = 0; x<clock; ++x)
    {
        code  <<= 1;/* align the Din bit */
        if (code & 0x1000000)
        {
            set_control(port,Clock,high);\/* set Clock low */
            set_control(port,Din,high);/* set Din bit high */
        }
        else {
         set_control(port,Clock,high); /* set Clock low  */
         set_control(port,Din,low);/* set Din low */
            }
        set_control(port,Clock,low);/* set Clock high for
DAC to latch */
    }
}
/* Read bit from ADC to PC */
Dout_()
{
    int temp, x, volt =0;
    for(x = 0; x<13; ++x)
  {
        set_control(port,Clock,high);
        set_control(port,Clock,low);
        temp = inportb(inprt);/* read status reg. */
        volt <<= 1;/* shift left one bit for serial
transmission */
        if(temp & Dout)
        volt += 1;      /* add 1 if input bit is high */
  }
    return(volt & 0xfff);
}
/* menu for the mode selection */
char menu()
{
    printf("Please select one of the following:\na: ADC\nd:
DAC\nq: quit\n\n");
    return (getchar());
}
void main()
{
    long code;
    char  mode_select;
    int  temp,volt=0;
/* Chip select for DAC & ADC is controlled by RS232 pin
3 TX line.  When LCR's bit 6 is set, the DAC is selected
and the reverse is true for the ADC. */
    outportb(LCR,0x0);/* initialize DAC */
    outportb(LCR,0x64);/* initialize ADC */
    while((mode_select = menu()) != 'q')
  {
   switch(mode_select)
    {
     case 'a':
     {
          outportb(LCR,0x0);/* selecting ADC */
          CS_Control(low);/* enabling the ADC CS */
          Din_(0x680000, 0x5);/* channel selection */
          volt = Dout_();
          outportb(LCR,0x64);/* bring CS high */
          set_control(port,Din,high);/* bring Din signal
high */
          printf("\ncode: %d\n",volt);
          }
        break;
    case 'd':
     {
  printf("Enter DAC input code (0 - 4095):\n");
  scanf("%d", &temp);
  code = temp;
  code += (long)temp << 12;/*  converting 12 bit to 24 bit
word */
  outportb(LCR,0x64);/* selecting DAC */
  CS_Control(low);/* CS enable */
  Din_(code,24);/* loading digital data to DAC */
      outportb(LCR,0x0);/* bring CS high */
      outportb(LCR,0x64);/* disabling ADC */
      set_control(port,Din,high);/* bring Din signal high */
      }
      break;
     }
  }
}
```

Micropower 12-bit ADCs shrink board space

355

Kevin R. Hoskins

Introduction

The LTC1286 and LTC1298 are serial interfaced, micropower 12-bit analog-to-digital converters. In the realm of 12-bit ADCs they bring a new low in power dissipation and the small size of an SO-8 package to low cost, battery-powered electronic products. These micropower devices consume just 250μA (LTC1286) and 340μA (LTC1298) at full conversion speed and feature autoshutdown.

Many portable and battery-powered systems require internal analog-to-digital conversion. Some, such as pen-based computers, have ADCs at their very cores digitizing the pen screen. Other systems use ADCs more peripherally to monitor voltages or other parameters inside the equipment. Regardless of the use, it has been difficult to obtain small ADCs at power levels and prices that are low enough. The LTC1286 and LTC1298 meet these low power dissipation and package size needs.

Table 355.1 LTC Micropower 3V and 5V 12-Bit ADCs

DEVICE	POWER DISSIPATION AT 200ksps	SAMP FREQ	S/(N + D) AT NYQUIST	INPUT RANGE	POWER SUPPLY
LTC1285	12μW 3nW*	7.5ksps	72dB	0V to V_{CC}	2.7V to 6V
LTC1286	25μW 5nW*	12.5ksps	71dB	0V to V_{CC}	4.5V to 9V
LTC1288	12μW 3nW*	6.6ksps	72dB	0V to V_{CC}	2.7V to 6V
LTC1298	12μW 5nW*	11.1ksps	71dB	0V to V_{CC}	4.5V to 5.5V

*5nW and 3nW power dissipation during shutdown.

Micropower and 12-bits in an SO-8 package

The LTC1286 and LTC1298 are the latest members of the growing family of SO-8 packaged parts (Table 355.1). As the first of their kind in SO-8 packages, these are improvements to the 8-bit micropower LTC1096/LTC1098 ADCs. The LTC1286 and LTC1298 use a successive approximation register (SAR) architecture. Both converters contain sample-and-holds and serial data I/O. The LTC1286 has a fully differential analog input and the LTC1298 has a two input multiplexer. While running at a full speed conversion rate of 12.5ksps, the LTC1286 consumes only 250μA from a single 5V supply voltage. The device automatically shuts down to 1nA (typ) when not converting. Figure 355.1 shows how this automatically reduces power at lower sample rates. At a 1ksps conversion rate, the supply current drops to just 20μA (typ). Battery-powered designs will benefit tremendously from this user transparent automatic power dissipation optimization.

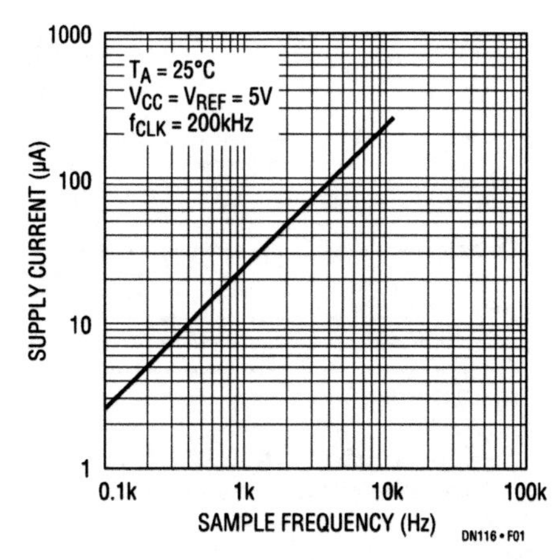

Figure 355.1 • The LTC1286/LTC1298's Autoshutdown Feature Automatically Conserves Power When Operating at Reduced Sample Rates

Analog Circuit Design: Design Note Collection. http://dx.doi.org/10.1016/B978-0-12-800001-4.00355-0

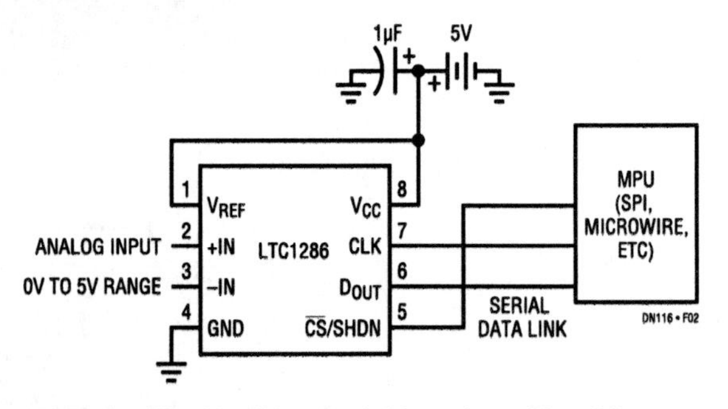

Figure 355.2 • The No-Glue Serial Interface Simplifies Connection to SPI, QSPI or Microwire Compatible Microcontrollers

The DC specifications include an excellent differential linearity error of 0.75LSB (max) and no missing codes. Both are guaranteed over the operating temperature range. Pen screen and other monitoring applications benefit greatly from these tight specifications.

The attractiveness of the LTC1286/LTC1298's small SO-8 design is further enhanced by the use of just one surface mount bypass capacitor (1µF or less). Figure 355. 2 shows a typical connection to a microcontroller's serial port. For ratiometric applications that require no external reference voltage, the LTC1286/LTC1298's reference input is tied to signal source's drive voltage. With their very low supply current requirements, the ADCs can even be powered directly from an external voltage reference. This eliminates the need for a separate voltage regulator.

The LTC1286 and LTC1298 contain everything required except an internal reference (not needed by many applications), keeping systems costs low. The serial interface makes a very space efficient interface and significantly reduces cost of applications requiring isolation. The ADC's high input impedance eliminates the need for buffer amplifiers. All of these features, combined with a very attractive price, make the LTC1286/LTC1298 ideal for new designs.

Resistive touchscreen interface

Figure 355.3 shows the LTC1298 in a 4-wire resistive touchscreen application. Transistor pairs Q1 and Q3, Q2 and Q4 apply 5V and ground to the X axis and Y axis, respectively. The LTC1298 (U1), with its 2-channel multiplexer, digitizes the voltage generated by each axis and transmits the conversion results to the system's processor through a serial interface. RC combinations R1C1, R2C2 and R3C3 form lowpass filters that attenuate noise from possible sources such as the processor clock, switching power supplies and bus signals. Inverter U2A is used to detect screen contact both during a conversion sequence and to trigger its start. Using the single channel LTC1286, 5-wire resistive touchscreens are as easily accommodated.

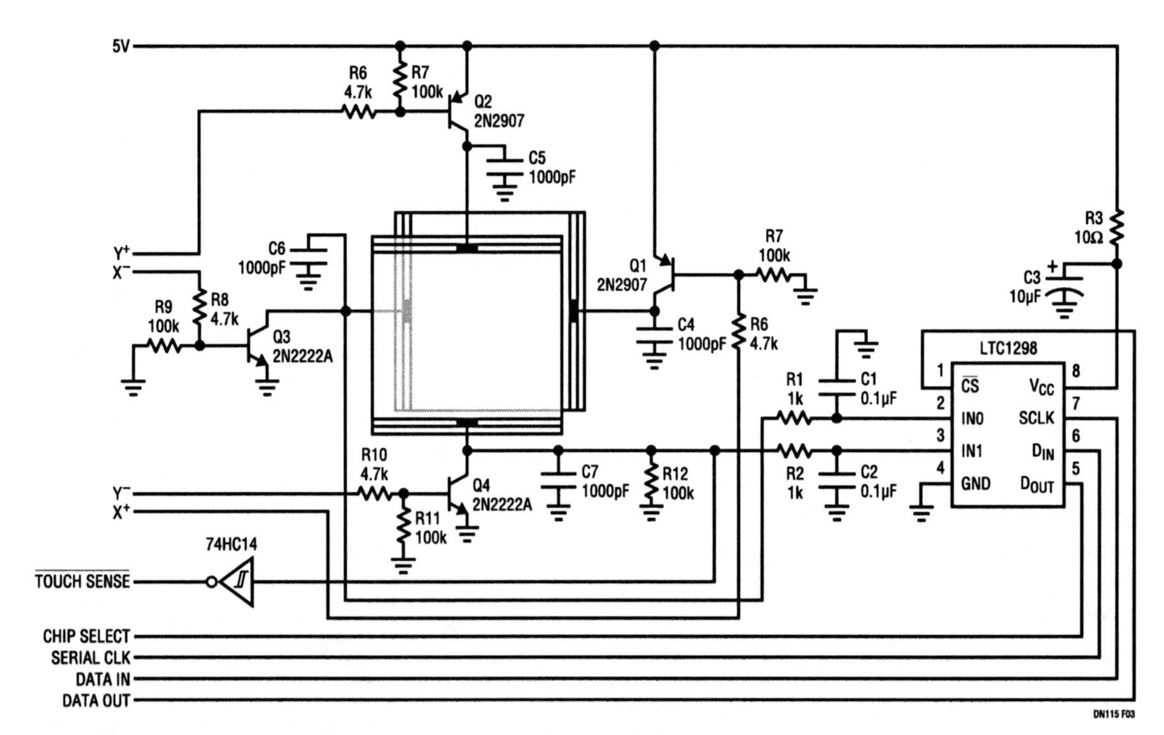

Figure 355.3 • The LTC1298 Digitizes Resistive Touchscreen X and Y Axis Voltages. The ADC's Autoshutdown Feature Helps Maximize Battery Life in Portable Touchscreen Equipment

1.25Msps 12-bit A/D converter cuts power dissipation and size

356

Dave Thomas Kevin R. Hoskins

Introduction

Until now, high speed system designers had to compromise when selecting 1Msps 12-bit A/D converters. While hybrids typically had the best performance, they were big, power hungry (~1W) and costly (>$100). A few manufacturers offered monolithics, but they compromised either AC or DC performance.

That has now changed. The new LTC1410 monolithic 1.25Msps 12-bit ADC performs better than hybrids with the power dissipation, size and cost of monolithics.

Some of the LTC1410's key benefits include:

- 1.25Msps throughput rate
- Fully differential inputs
- 60dB CMRR that remains constant to 1MHz
- Low power: 160mW (typ) from ±5V supplies
- "Instant on" NAP and μpower SLEEP shutdown modes
- Small package: 28-pin SO

The LTC1410's features can increase the performance and decrease the cost of current data acquisition systems and optimize new applications.

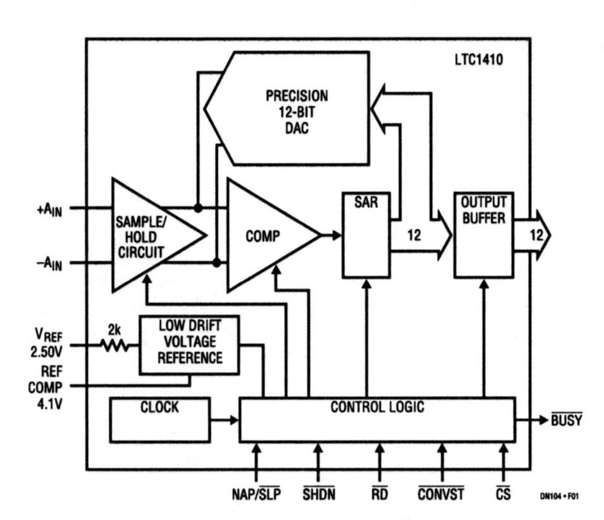

Figure 356.1 • LTC1410 Features a True Differential S/H with Excellent Bandwidth and CMRR

If an application requires an external reference, it can easily overdrive the on-chip reference's 2kΩ output impedance.

High accuracy conversions: AC or DC

In Figure 356.1 the LTC1410 combines a wide bandwidth differential sample-and-hold (S/H) with an extremely fast successive approximation register (SAR) ADC and an on-chip reference. Together they deliver a very high level of both AC and DC performance.

An ADC's S/H determines its overall dynamic performance. The LTC1410's S/H has a very wide bandwidth (20MHz) and generates very low total harmonic distortion (−84db) for the 625kHz Nyquist bandwidth.

Important DC specifications include excellent differential linearity error ≤0.8LSB, linearity error ≤0.5LSB and no missing codes over temperature. The on-chip 10ppm/°C curvature corrected 2.5V bandgap reference assures low drift over temperature.

Important multiplexed applications

The LTC1410's high conversion rate allows very high sample rate multiplexed systems. The S/H's high input impedance eliminates DC errors caused by a MUX's switch resistance. Also, the LTC1410's low input capacitance ensures fast 100ns acquisition times, even with high source impedance.

Ideal for telecommunications

Telecommunications applications such as HDSL, ADSL and modems require high levels of dynamic performance. A key indicator of a sampling ADC's dynamic performance is its signal-to-noise plus distortion ratio (SINAD). The LTC1410's minimum SINAD is 72dB, or 11.67 effective number of bits

Analog Circuit Design: Design Note Collection. http://dx.doi.org/10.1016/B978-0-12-800001-4.00356-2

(ENOB), up to input frequencies of 100kHz. At the Nyquist frequency (625kHz) the SINAD is still a robust 70dB. Figure 356.2 shows that the LTC1410 can under sample signals well beyond the Nyquist rate.

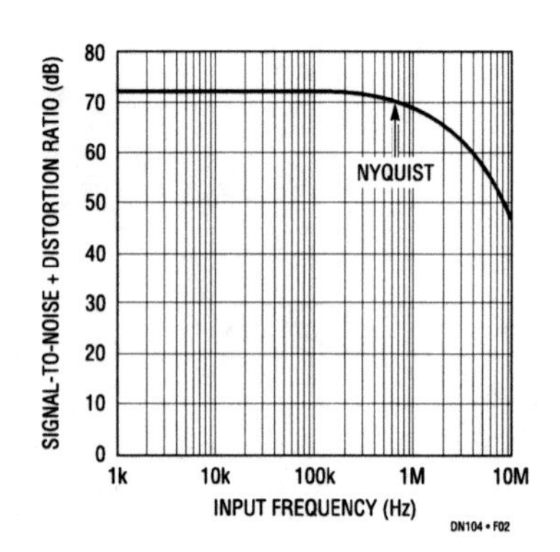

Figure 356.2 • The Wideband S/H Captures Signals Well Beyond Nyquist

Differential inputs reject noise

The LTC1410's differential inputs are ideal for applications whose desired signals must compete with EMI noise. The LTC1410's differential inputs provide a new way to fight noise.

Figure 356.3a shows a single-ended sampling system whose accuracy is limited by ground noise. When a single-ended signal is applied to the ADC's input, the ground noise adds directly to the applied signal. While a filter can reduce this noise, this does not work for in-band noise or common-mode noise at the same frequency as the input signal. However, Figure 356.3b shows how the LTC1410 provides relief. Because of its excellent CMRR, the LTC1410's differential inputs reject the ground noise even if it is at the same frequency as the desired input frequency. Further, the LTC1410's wideband CMRR can eliminate extremely wideband noise, as shown in Figure 356.4.

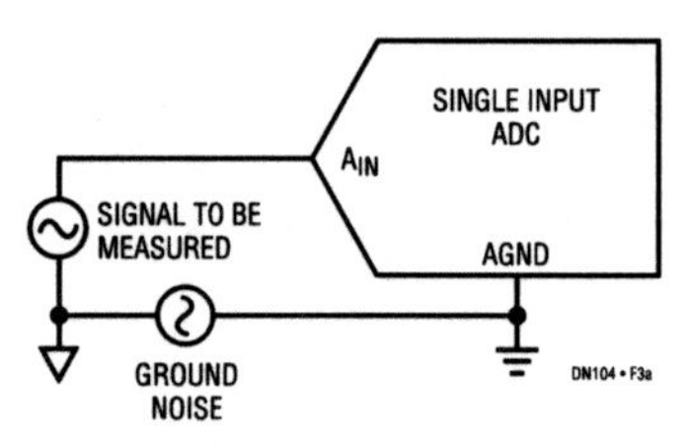

Figure 356.3a • An Input Signal Is Contaminated by Ground Noise with a Single Input ADC

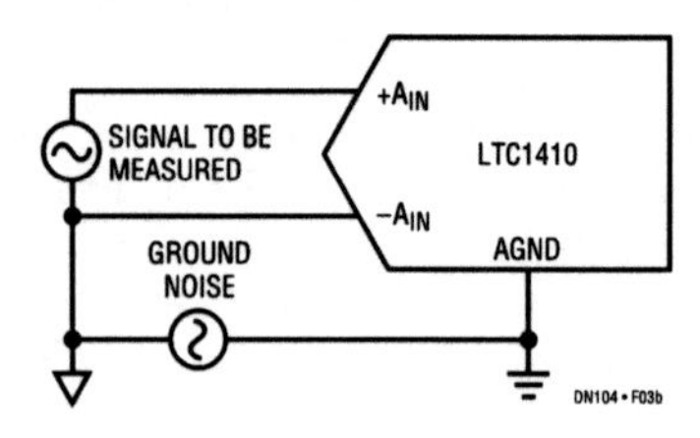

Figure 356.3b • The LTC1410's Differential Inputs Reject Common-Mode Noise and Preserve the Input Signal

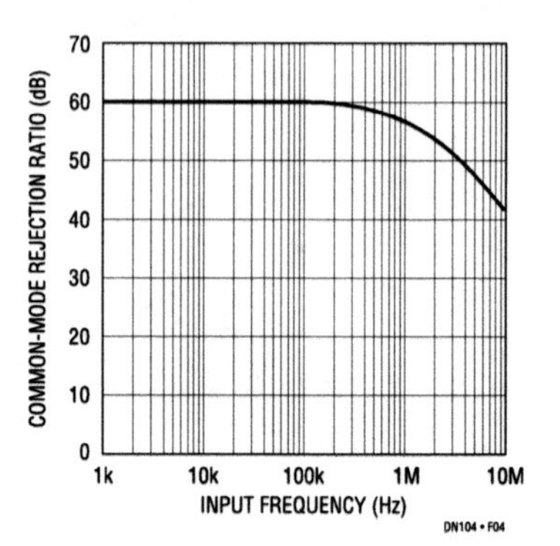

Figure 356.4 • The LTC1410 Rejects Common-Mode Noise Out to 10MHz and Beyond

Low power applications

High speed applications with limited power budgets will greatly benefit from the LTC1410's low 160mW power dissipation. Power can be further reduced by using the power shutdown modes, NAP and SLEEP.

NAP reduces power consumption by 95% (to 7.5mW) leaving only the internal reference powered. The "wake-up" time is a very fast 200ns. The most recent conversion data is still accessible and CS and RD still control the output buffers. NAP is appropriate for those applications that require conversions instantly after periods of inactivity.

SLEEP reduces power consumption to less than 5μW. It is useful for applications that must maximize power savings. SLEEP mode shuts down all bias currents, including the reference. SLEEP mode wake-up time is dependent on the reference compensation capacitor's size. With the recommended 10μF, the wake-up time is 10ms. Typically, NAP mode is used for inactive periods shorter that 10ms and SLEEP is used for longer periods.

Conclusion

Available in 28-pin SO packages, the new LTC1410 is optimized for many high speed dynamic sampling applications including ADSL, compressed video and dynamic data acquisition.

500ksps and 600ksps ADCs match needs of high speed applications

357

Kevin R. Hoskins

Introduction

Combining high speed analog-to-digital conversion with low power dissipation, the 500ksps LTC1278 and 600ksps LTC1279 12-bit ADCs solve major challenges confronting designers of high speed systems: conversion performance, power dissipation, circuit board real estate, complexity, and cost. Applications for the LTC1278/LTC1279 include telecom, communication, PC data acquisition board, and high speed and multiplexed data acquisition systems. In addition to requiring no external references, crystals, or clocks the LTC1278/LTC1279 offer system designers the following significant system-enhancing improvements:

- Power Shutdown: 5mW
- Wakes Up in a Scant 300ns
- Single 5V or ±5V Supply Voltage Operation
- Low Power Dissipation: 150mW (Max), 75mW (Typ)
- Small 24-Pin SO or 24-Pin Narrow DIP Packages

The LTC1278/LTC1279's DC performance includes ±1LSB INL and DNL, no missing codes, and an internal voltage reference with a full-scale drift of only 25ppm/°C.

The AC performance includes 70dB (Min) SINAD, −78dB (Max) THD, and −82dB (Max) spurious-free dynamic range. These specifications were measured at $f_S = 500$ksps (LTC1278) or $f_S = 600$ksps (LTC1279), $f_{IN} = 100$kHz, and are guaranteed over the operating temperature range. The plot of effective number of bits (ENOB) shown in Figure 357.1 clearly indicates that the LTC1278/LTC1279 can accurately sample signals that contain spectral energy beyond the Nyquist frequency.

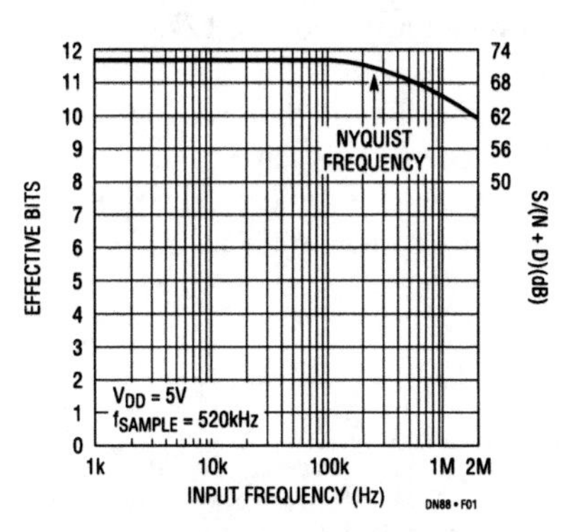

Figure 357.1 • The LTC1278 Can Accurately Digitize Input Signals up to the Nyquist Frequency and Beyond

Table 357.1 LTC's High Speed ADC Family Includes 5V and 3V Devices

DEVICE	SAMPLING FREQUENCY	S/(N + D) @ NYQUIST	INPUT RANGE	POWER SUPPLY	POWER DISSIPATION
LTC1272	250kHz	65dB	0V to 5V	5V	75mW
LTC1273	300kHz	70dB	0V to 5V	5V	75mW
LTC1275	300kHz	70dB	±2.5V	±5V	75mW
LTC1276	300kHz	70dB	±5V	±5V	75mW
LTC1278	**500kHz**	**70dB**	**0V to 5V or ±2.5V**	**5V or ±5V**	**75mW 5mW***
LTC1279	**600kHz**	**70dB**	**0V to 5V or ±2.5**	**5V or ±5V**	**75mW 5mW***
LTC1282	140kHz	68dB	0V to 2.5V or ±1.25V	3V or ±3V	12mW

*5mW power shutdown with instant wake-up.

High speed ADC family members

The LTC1278/LTC1279 are self-contained ADC systems composed of a fast capacitively based charge redistribution successive approximation register (SAR) sampling ADC, internal reference, power shutdown, and internally generated and synchronized conversion clock. The digital interface's

Analog Circuit Design: Design Note Collection. http://dx.doi.org/10.1016/B978-0-12-800001-4.00357-4

flexibility eases connection to external latches, FIFOs, and DSPs. Table 357.1 shows more members of the high speed 12-bit ADC family.

The LTC1278/LTC1279 convert input signals in the 0V to 5V or −2.5V to 2.5V ranges at full speed when operating on a 5V or ±5V supply voltage, respectively. The ±2.5V input range complements the new generation of operational amplifiers that operate on ±5V.

Important applications

The LTC1278's features benefit at least four different application areas: telecom, communication, PC data acquisition boards, and high speed and multiplexed data acquisition.

Telecom digital-data transmission applications such as High-bit-rate Digital Subscriber Line (HDSL) with its high speed T1 data rate benefit from the LTC1278's low power dissipation since these telecom systems usually derive their power from the phone line. While the LTC1278's 500ksps conversion rate easily covers T1 data rates, the LTC1279's 600ksps is ideal for HDSL's faster E1 data rates. Further, these applications use noise and echo cancellation that require excellent dynamic performance from the ADC's sample-and-hold. The LTC1278 satisfies this requirement as indicated by the device's excellent dynamic performance shown in Figure 357.1.

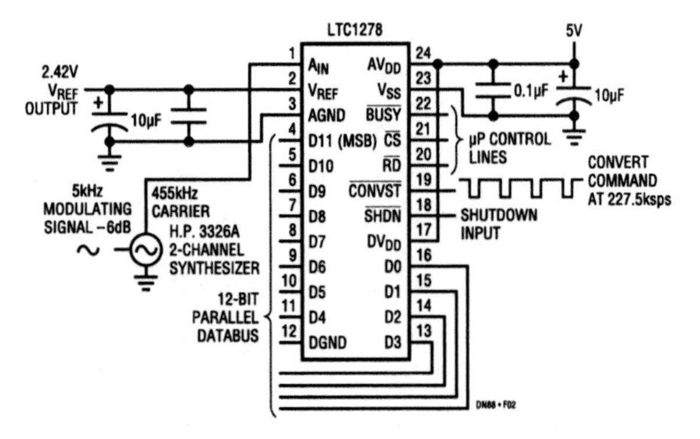

Figure 357.2 • The LTC1278 Undersamples the 455kHz Carrier to Recover the 5kHz Modulating Signal

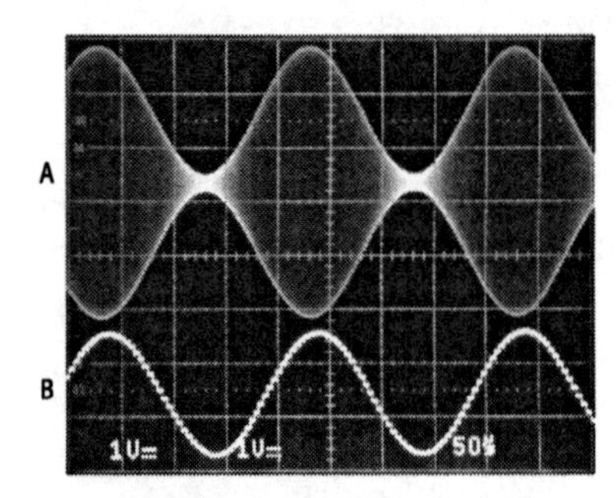

Figure 357.3 • Demodulating an I.F. by Undersampling

Communication applications also benefit from the LTC1278/LTC1279's wide input bandwidth and undersampling capability. The application shown in Figure 357.2 uses the LTC1278 to undersample (at 227.5ksps) a 455kHz I.F. amplitude-modulated by a 5kHz sine wave. Figures 357.3A and 357.3B show, respectively, the 455kHz I.F. carrier and the recovered 5kHz sine wave that results from a 12-bit DAC reconstruction. Figure 357.2 also shows that by taking advantage of surface mount devices, this simple configuration occupies only 0.43in^2 of circuit board real estate.

PC data acquisition cards are another broad application area. The LTC1278's high sampling rate, simple and complete configuration, small outline package, and low cost make this converter ideal for these applications. Additionally, the LTC1278's synchronized internal conversion clock minimizes conversion noise that results when the conversion clock and the sampling command are not synchronized. This internal clock and sampling synchronization overcomes what, in PC environments, can be a cumbersome task.

Both single channel and multiplexed high speed data acquisition systems benefit from the LTC1278/LTC1279's dynamic conversion performance. The 1.6µs and 1.4µs conversion and 200ns and 180ns S/H acquisition times enable the LTC1278/LTC1279 to convert at 500ksps and 600ksps, respectively. Figure 357.4 shows a 500ksps 8-channel data acquisition system. The LTC1278's high input impedance eliminates the need for a buffer amplifier between the multiplexer's output and the ADC's input.

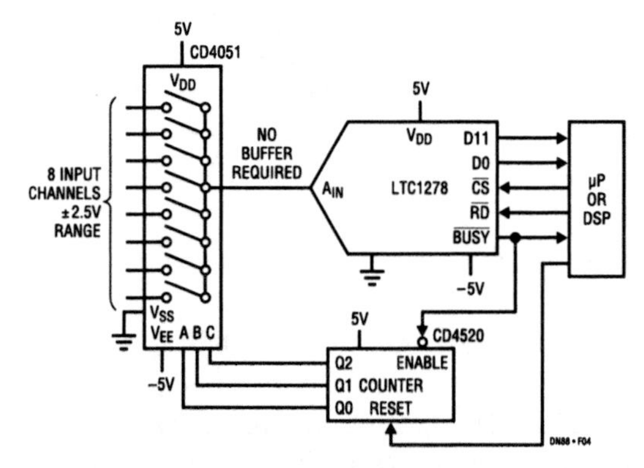

Figure 357.4 • The High Input Impedance of the LTC1278 Allows Multiplexing without a Buffer Amplifier

Conclusion

The LTC1278/LTC1279's new features simplify, improve, and reduce the cost of high speed data acquisition systems. This makes them the converters of choice for telecom, communication, PC data acquisition board, and high speed and multiplexed data acquisition system designers.

5V and 3V, 12-bit ADCs sample at 300kHz on 75mW and 140kHz on 12mW

358

William Rempfer

Four new sampling A/D converters from Linear Technology stand out above the rest of the crowd. These new 5V and 3V 12-bit ADCs offer the best speed/power performance available today (see Figure 358.1). They also provide precision reference, internally trimmed clock, and fast sample-and-hold. With additional features such as single supply operation and high impedance analog inputs, they reduce system complexity and cost. This article will describe the new ADCs, discuss the 5V performance of the LTC1273/5/6 and then the 3V performance of the LTC1282.

Complete ADCs provide lowest power and highest speed on single or dual supplies

The LTC1273/5/6 and LTC1282 provide complete A/D solutions at previously impossible speed/power levels. As shown in Table 358.1, the LTC1273/5/6 all have the same 300kHz maximum sampling rate and 75mW typical power dissipation. The LTC1273 digitizes 0V to 5V inputs from a single 5V rail. The LTC1275 and LTC1276 operate on ±5V rails and digitize ±2.5V and ±5V inputs, respectively.

The LTC1282 samples at 140kHz and typically dissipates only 12mW from either 3V or ±3V supplies. It digitizes 0V to 2.5V inputs from a single 3V supply or ±1.25V inputs from ±3V supplies.

Table 358.1 Four Complete ADCs Offer High Supplies and Low Power on Single or Dual 5V or 3V Supplies

DEVICE	POWER SUPPLIES	INPUT RANGE	SAMPLE RATE	S/(N+D) AT F_{INPUT}	P_{DISS} (TYP)
LTC1273	Single 5V	0V to 5V	300kHz	70dB at 100kHz	75mW
LTC1275/6	±5V	±2.5/±5V	300kHz	70dB at 100kHz	75mW
LTC1282	Single 3V or ±3V	0V to 2.5V or ±1.25	140kHz 140kHz	68dB at 70kHz	12mW

Complete function is provided by the on-chip sample-and-holds, precision references, and internally trimmed clocks. The high impedance analog inputs are easy to drive and can be multiplexed without buffer amplifiers. A single 5V or 3V power supply is all that is needed to digitize unipolar inputs. (Bipolar inputs require ±5V or ±3V supplies but the negative supply draws only microamperes of current). But most significant are the speed/power ratios which are higher than any other ADC in this speed range.

5V ADCs sample at 300kHz on 75mW of power

The LTC1273/5/6 have excellent DC specs, including ±1/2 LSB linearity and 25ppm/°C full-scale drift. In addition, they have excellent dynamic performance. As Figure 358.2 shows, the ADCs provide 72dB of signal-to-noise+distortion (11.7 effective bits) at the maximum sample rate of 300kHz. The S/(N+D) ratio is over 70dB (11.3 effective bits) for input frequencies up to 100kHz.

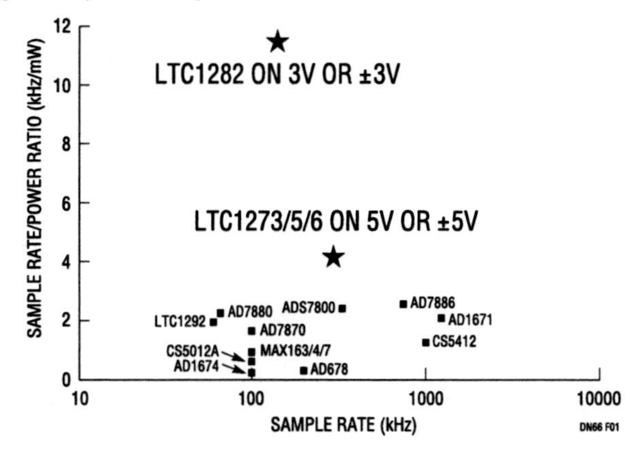

Figure 358.1 • The LTC1273/5/6 and LTC1282 Have Up to 45 Times Higher Speed/Power Ratios Than Competitive ADCs

Analog Circuit Design: Design Note Collection. http://dx.doi.org/10.1016/B978-0-12-800001-4.00358-6

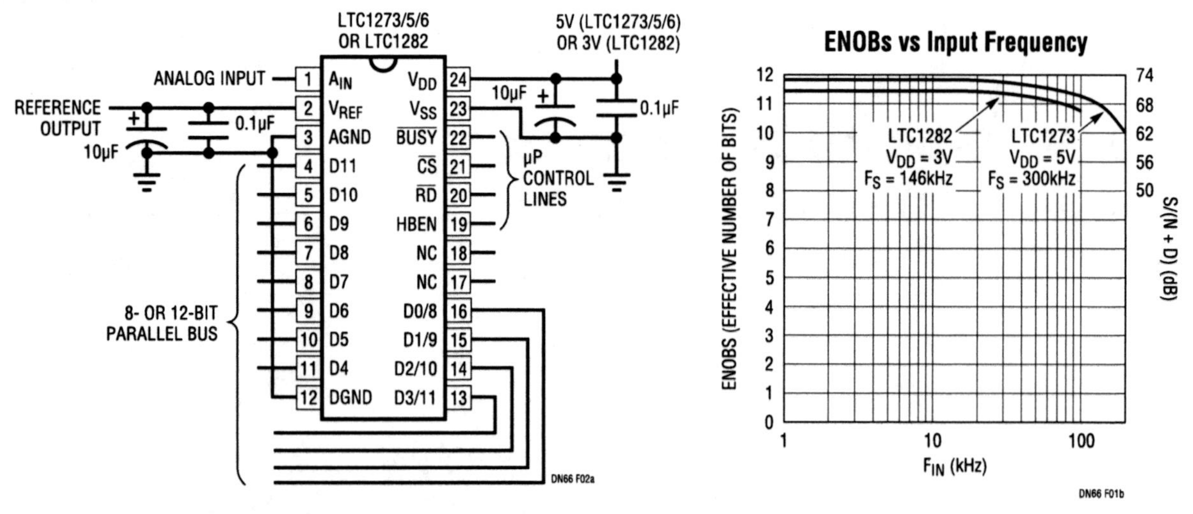

Figure 358.2 • The 300kHz LTC1273 Gives 70dB S/(N+D) with 100kHz Inputs. The 140kHz LTC1282 Gives 68dB S/(N+D) at Nyquist

This 300kHz sample rate and dynamic performance comes at a power level that is more stingy than any other ADC in this speed range. Figure 358.1 shows a graph of speed/power ratio for the competitive ADCs. The speed/power ratio is defined as the maximum sample rate in kHz divided by the typical power dissipation in mW. The 4.0kHz/mW of the LTC1273/5/6 is better than the best competitive ADC.

Even more power savings: 3V ADC samples at 140kHz on 12mW

The low power, 3V LTC1282 provides even more impressive speed/power performance. As fast and dynamically accurate as many power hungry, dual and triple supply ADCs, this complete 3V or ±3V sampling ADC provides extremely good performance on only 12mW of power! DC specs include ±1/2 LSB maximum linearity and the internal reference provides 25ppm maximum full-scale drift. Figure 358.2 shows 11.4 effective bits at 140kHz sample rate with 11 effective bits at the Nyquist frequency of 70kHz. The speed/power ratio, as shown in Figure 358.1, is an outstanding 11.7kHz/mW.

The LTC1282 is ideal for 3V systems but will also find uses in 5V designs where the lowest possible power consumption is required. It interfaces easily to 3V logic but can also talk well to 5V systems. The LTC1282 can receive 5V CMOS levels directly and its 0V to 3V outputs can meet 5V TTL levels and connect directly to 5V systems.

The performance comparison in Table 358.2 shows that using the 3V LTC1282 gives great savings in power with only modest reductions in speed, accuracy and noise. The power dissipation has been reduced 6 times with only a 50% reduction in speed. Linearity and drift do not degrade at all in going to the 3V device. The noise of the LTC1282 is slightly higher, due to the reduced input span and the lower operating current, but the converter still gives more than 70dB S/(N+D).

Table 358.2 5V and 3V Reference Comparison

PARAMETER	LTC1273 ON A 5V SUPPLY	LTC1282 ON A SINGLE 3V OR ±3V SUPPLIES
Power Dissipation (Typ)	75mW	12mW
Sample Rate	300kHz	140kHz
Conversion Time (Max)	2.7μs	6μs
INL (Max)	1/2LSB	±1/2LSB
Typical ENOBs	11.7	11.4
Linear Input Bandwidth (ENOBs>11Bits)	125kHz	70kHz

Conclusion

These 5V and 3V ADCs offer the best speed/power performance available today. They also provide precision reference, internally trimmed clock, and fast sample-and-hold. With additional features such as single supply operation and high impedance analog inputs, they reduce system complexity and cost. For performance, power and cost, these new ADCs must be considered for new designs.

Micropower, SO-8, 8-bit ADCs sample at 1kHz on 3μA of supply current

359

William Rempfer Marco Pan

The LTC1096 and LTC1098 are the lowest power, most compact, sampling analog-to-digital converters in the world. These new 8-bit micropower, sampling ADCs typically draw 100μA of supply current when sampling at 33kHz. Supply current drops linearly as the sample rate is reduced as shown in Figure 359.1. At a 1kHz sample rate, the supply current is only 3μA. The ADCs automatically power down when not performing conversions, drawing only leakage current.

They are packaged in 8-pin SO packages and operate on 3V to 9V supplies or batteries. Both are fabricated on Linear Technology's proprietary LTBiCMOS process.

Two micropower ADCs

The LTC1096 and LTC1098 use a switched-capacitor, successive-approximation (SAR) architecture. Micropower operation is achieved through three design innovations:

1. An architecture which automatically powers up and down as conversions are requested
2. An ultralow power comparator design, and
3. The use of a proprietary BiCMOS process.

Although they share the same basic design, the LTC1096 and LTC1098 differ in some respects. The LTC1096 has a differential input and has an external reference input pin. It can measure signals floating on a DC common-mode voltage and can operate with reduced spans down to 250mV. Reducing the span allows it to achieve 1mV resolution. The LTC1098 has a 2-channel input multiplexer and can convert either channel with respect to ground or the difference between the two.

Longer battery life

Tremendous gains in battery life are possible because of the wide supply voltage range, the low supply current, and the automatic power shut down between conversions. Eliminating the voltage regulator and operating directly off the battery saves the power lost in the regulator. At a sample rate of 1kHz, the 3μA supply current is below the self-discharge rate of many batteries. As an example, the circuit of Figure 359.2, sampling at 1kHz, will run off a Panasonic CR1632 3V lithium coin cell for five years.

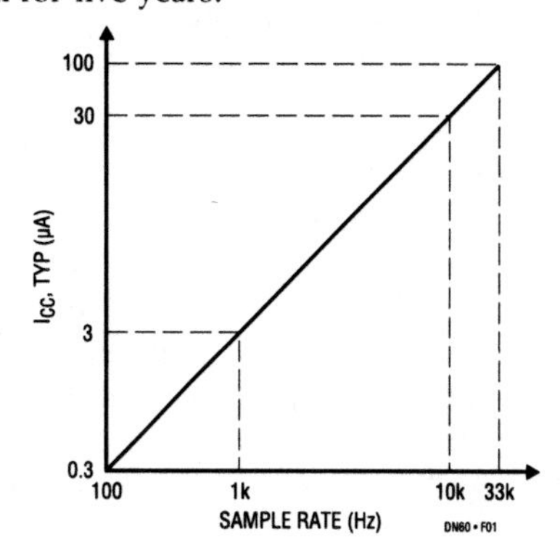

Figure 359.1 • Automatic Power Shutdown between Conversions Allows Power Consumption to Drop with Sample Rate

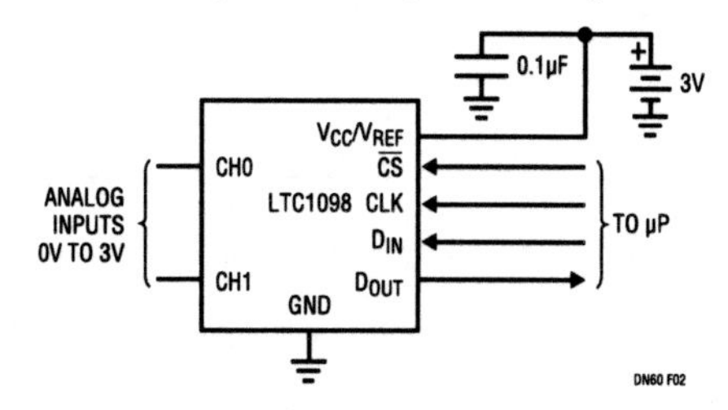

Figure 359.2 • Sampling at 1kHz, This Circuit Draws Only 3μA and Will Run Off a 120mAhr CR1632 3V Lithium Coin Cell for 5 Years

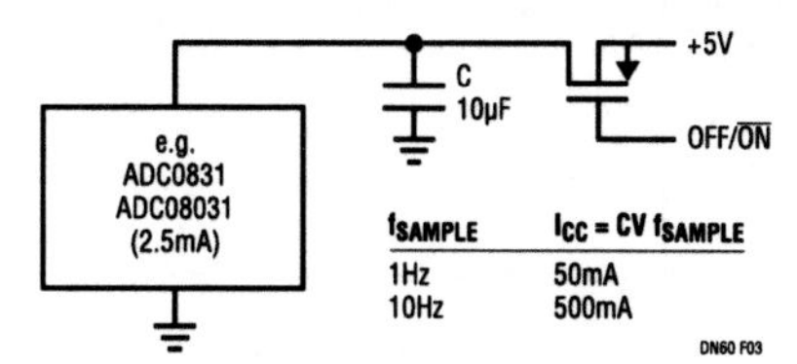

Figure 359.3 • High Side Switching a Power-Hungry ADC Wastes Power. Repeatedly Switching the Required Bypass Capacitor Consumes 500μA Even When Taking Readings at Only 10Hz

Analog Circuit Design: Design Note Collection. http://dx.doi.org/10.1016/B978-0-12-800001-4.00359-8

The automatic shutdown has great advantages over the alternative of high side switching a higher power ADC, shown in Figure 359.3. First, no switching signal or hardware is required. Second, power consumption is orders of magnitude lower with the LTC1096/LTC1098. This is because, when an ADC is high side switched, the current consumed in charging the required bypass capacitor is large, even at very low sample rates. In fact, a 10μF bypass capacitor, high side switched at only 10Hz, will consume 500μA!

A/D conversion for 3V systems

The LTC1096/LTC1098 are ideal for 3V systems. Figure 359.4 shows a 3V to 6V battery current monitor which draws only 70μA from the battery it monitors. The battery current is sensed with the 0.02Ω resistor and amplified by the LT1178. The LTC1096 digitizes the amplifier output and sends it to the microprocessor in serial format. The LT1004 provides the full-scale reference for the ADC. The other half of the LTC1178 is used to provide low battery detection. The circuit's 70μA supply current is dominated by the op amps and the reference. The circuit can be located near the battery and data transmitted serially to the microprocessor.

Smaller instrument size

The LTC1096 and LTC1098 can save board space in compact designs in a number of ways. The SO-8 package saves space. Operating the ADC directly off batteries can eliminate the space taken by a voltage regulator. The LTC1096/LTC1098 can also operate with small, 0.1μF or 0.01μF chip bypass capacitors. The serial I/O requires fewer PC traces and fewer microprocessor pins than a parallel-port ADC. Connecting the ADC directly to sensors can eliminate op amps and gain stages. Finally, the ADCs do not need an external sample-and-hold.

AC and DC performance

The LTC1096/LTC1098 are offered with ±0.5LSB total unadjusted error for applications that require DC accuracy. The ADCs also have a lot to offer in designs that require AC performance.

Figure 359.5 shows remarkable sampling performance for a device that draws only 100μA running at full speed. Dynamic performance of 7.5 effective bits is maintained up to an input frequency of over 40kHz.

In undersampling applications, this 40kHz input bandwidth remains intact as the sample rate (and power consumption) are reduced. A 40kHz waveform can be undersampled at 1kHz with 7.5 bits of accuracy on a supply current of 3μA!

Conclusion

Extremely low power consumption, 3V operation, small size and other benefits will help the LTC1096 and LTC1098 find their way into a variety of micropower, low voltage, battery-powered and compact systems. For more information, refer to the LTC1096/LTC1098 data sheet, Linear Technology Magazine (Volume II Number 1) and application notes.

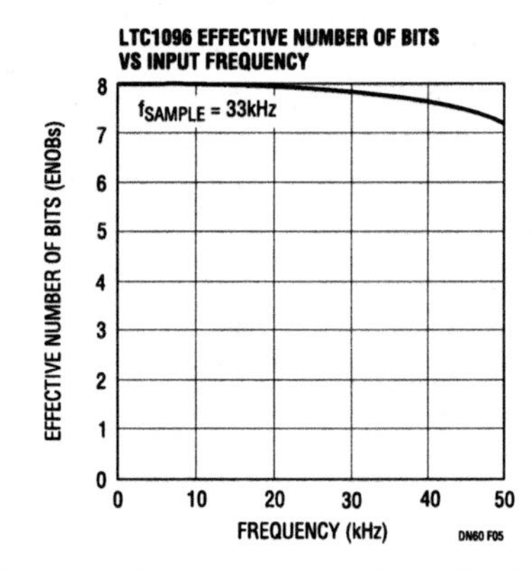

Figure 359.5 • Dynamic Accuracy Is Maintained up to an Input Frequency of over 40kHz

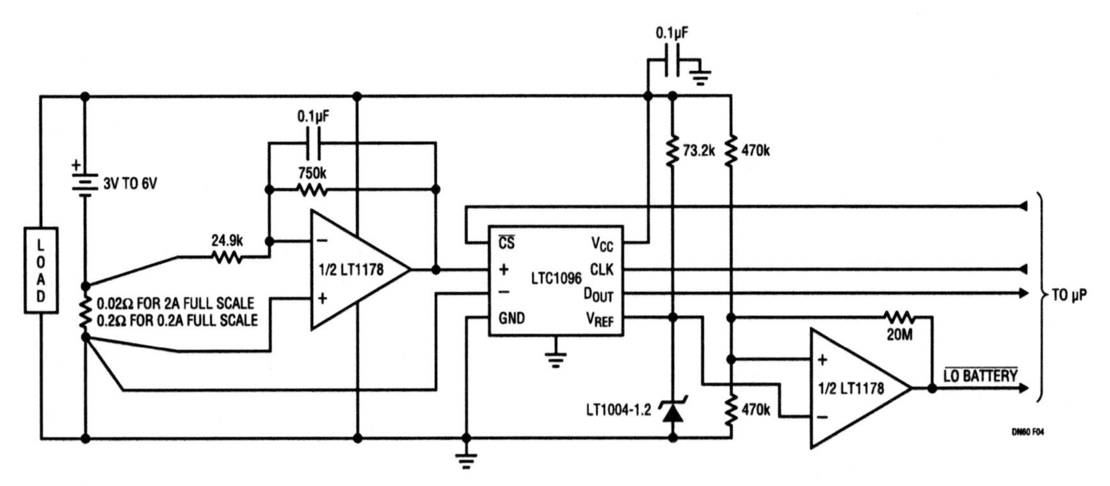

Figure 359.4 • This 0A to 2A Battery Current Monitor Draws Only 70μA from a 3V to 6V Battery

Section 2

Data Conversion: Digital-to-Analog

12-bit DAC in TSOT-23 includes bidirectional REF pin for connection to op amp or external high precision reference

360

Kevin Wrenner   Troy Seman   Mark Thoren

Introduction

The LTC2630's combination of a 12-bit DAC and low-drift integrated reference in a tiny SC-70 package has proven popular for a wide variety of applications. Two new DACs, the LTC2631 and LTC2640, take this winning formula and further expand its reach by adding a bidirectional REF pin and an optional I^2C interface in a tiny TSOT-23.

Like their predecessor, these parts feature 1-bit INL and DNL, offer excellent load regulation driving up to 10mA loads, and can operate rail-to-rail. See Table 360.1 for a list of options.

Applications using REF pin

The bidirectional REF pin can be used as an output, where the accurate 10ppm/°C reference is available to the rest of the application circuit, or it can be used as an input for an external reference.

To configure REF as an output, simply tie the REF_SEL pin high. As an output, the REF pin simplifies pairing the DAC with an op amp. For instance, to achieve an output range centered at 0V, drive the plus input of the op amp, with REF connected to the minus input. Avoid loading the REF pin with DC current; instead, buffer its 500Ω output with an LTC2054 or similar precision op amp.

The LT1991 precision op amp is a superb choice for amplifying or attenuating the DAC output to achieve a desired output range because it requires no precision external resistors. Its integrated, precision resistors are matched to 0.04%, allowing gain to be set by simple pin strapping (see the data sheet for a large variety of gain options). Figure 360.1 shows the configuration for a difference gain of 4, resulting in a ±5V output with 12-bit programmability under I^2C control. Integral nonlinearity, seen in Figure 360.2, is better than 1LSB.

Figure 360.3 shows a negative output system using a similar setup, this time with the LT1991 configured as an inverting amplifier with a gain of −0.25. The 0.1µF capacitor at REF reduces the already low DAC noise by up to 20%.

For applications requiring more accuracy at full scale, the LTC2631 and LTC2640 can be referenced to an external source. Figure 360.4 shows how, using an LT1790 low-dropout reference that's accurate to 0.05%. Tying REF_SEL low configures the REF pin as a reference input. If reset-to-zero is needed, an LTC2640-LZ12 can be substituted. (For that option, pin 8 is rededicated as a $\overline{CLR}$ pin, and, upon powering up, External Reference mode must be selected by software command before the code is changed from zero.)

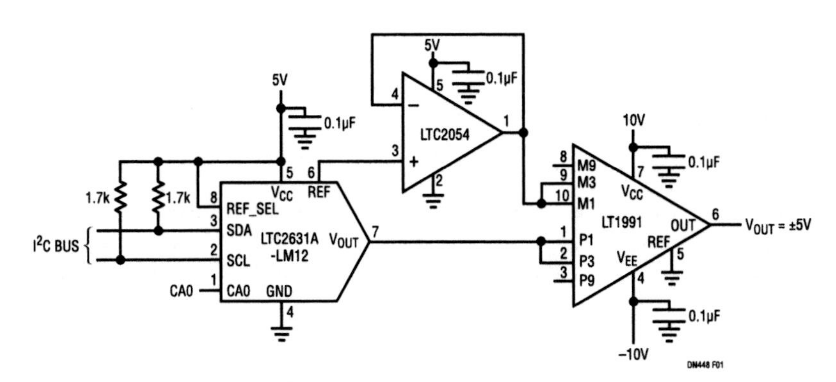

Figure 360.1 • Programmable ±5V Output

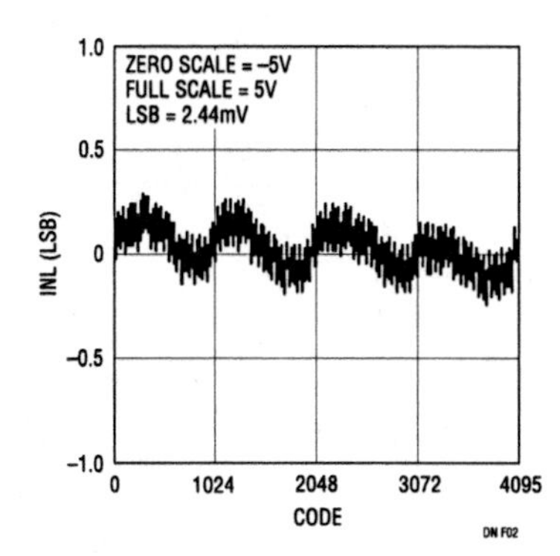

Figure 360.2 • Integral Nonlinearity of Programmable ±5V Output

Analog Circuit Design: Design Note Collection. http://dx.doi.org/10.1016/B978-0-12-800001-4.00360-4

The REF pin enables the LTC2631 and LTC2640 to share their full-scale range with another device, as shown in Figure 360.5. A 16-bit LTC2453 ADC and LTC2631 DAC are referenced to the same 5V full scale. This circuit allows a variety of possible transfer functions to be applied to an input under computer control. It is easy to implement functions such as squaring and square root, or time-dependent functions such as integration or proportional-integral-derivative (PID) control in this manner, resulting in a circuit that is much simpler and more stable than a purely analog circuit.

Conclusion

The LTC2631 and LTC2640 add I^2C capability and a bidirectional REF pin to LTC's family of 12-, 10-, and 8-bit DACs with an integrated reference. For applications requiring a modified output range, the LT1991 op amp with internal precision resistors is an ideal counterpart.

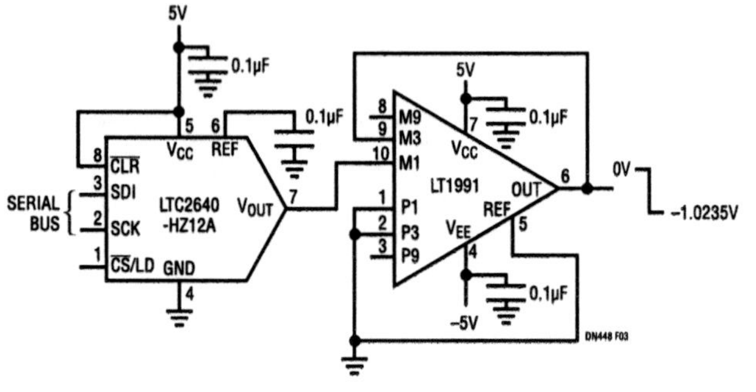

Figure 360.3 • Negative Output, 0V to −1.024V

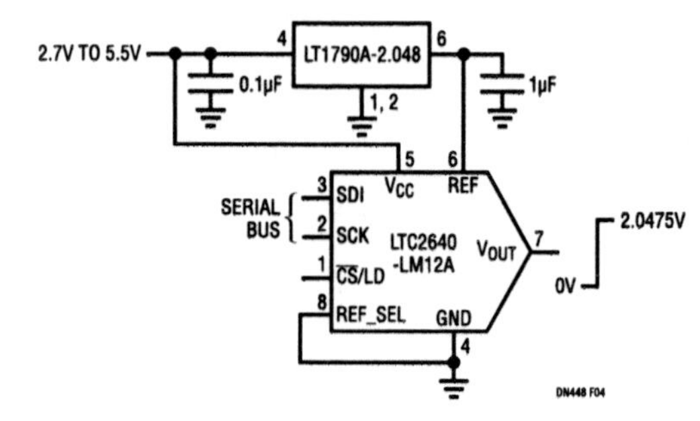

Figure 360.4 • 0V to 2.048V Output Derived from External Reference

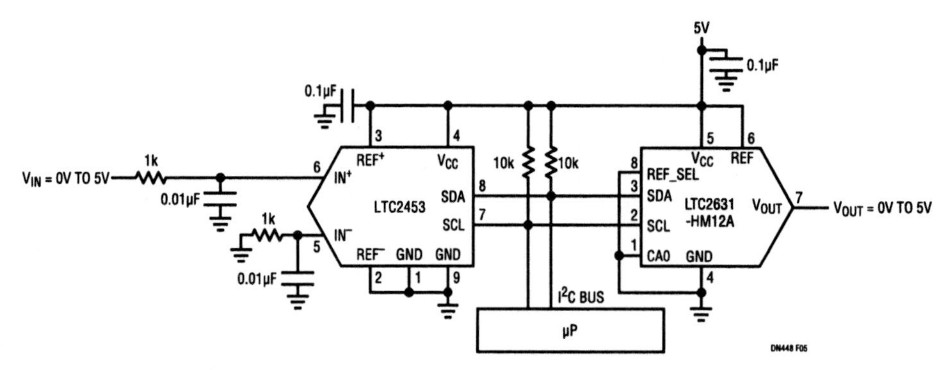

Figure 360.5 • Electronic Transfer Function Generator

Table 360.1 Family Characteristics. Each Part Has a Bidirectional REF Pin and Is Available in 12-, 10-, and 8-Bit Accuracy

PART NUMBER	TYPE	FULL SCALE	POWER-ON RESET CODE	PIN 8 FUNCTION
LTC2631-LM	I^2C	2.5V	Midscale	Select default REF
LTC2631-LZ	I^2C	2.5V	Zero	6 add'l addresses
LTC2631-HM	I^2C	4.096V	Midscale	Select default REF
LTC2631-HZ	I^2C	4.096V	Zero	6 add'l addresses
LTC2640-LM	SPI	2.5V	Midscale	Select default REF
LTC2640-LZ	SPI	2.5V	Zero	DAC Clear
LTC2640-HM	SPI	4.096V	Midscale	Select default REF
LTC2640-HZ	SPI	4.096V	Zero	DAC Clear

Highly integrated quad 16-bit, SoftSpan, voltage output DAC for industrial and control applications

361

Mark Thoren

Introduction

Digital-to-analog converters (DACs) are prevalent in industrial control and automated test applications. General purpose automated test equipment often requires many channels of precisely controlled voltages that span several voltage ranges. The LTC2704 is a highly integrated 16-bit, 4-channel DAC for high-end applications. It has a wide range of features designed to increase performance and simplify design.

Unprecedented integration

The LTC2704 provides true 16-bit performance over six software selectable ranges: 0V to 5V, 0V to 10V, −2.5V to 2.5V, −5V to 5V, −10V to 10V and −2.5V to 7.5V. Four single-range voltage outputs would normally require four current-output DACs, two reference amplifiers and four output amplifiers—seven packages if dual amplifiers are used. Implementing multiple ranges discretely is prohibitive. Design Note 337 explains the difficulty in implementing multiple ranges, including the cost of precision-matched resistors and the performance limitations of analog switches. Control is also complicated, requiring extra digital lines for each DAC and for range control. The LTC2704 integrates all of these functions into a single package with no compromises, and all functions are controlled via an easy-to-use 4-wire SPI bus.

Ease of use

The LTC2704 provides many features to aid system design. The voltage output and feedback are separated, allowing external current booster stages to be added with no loss in accuracy. The C1A, C1B, C1C and C1D pins allow external frequency compensation capacitors to be used, either to allow capacitive loads to be directly driven by the LTC2704's outputs, or to compensate slow booster stages. The V_{OS} pins provide a convenient way to add an offset to the output voltage. The gain from the V_{OS} pin to the output is −0.01, −0.02 or −0.04, depending on the selected range. While this seems like a simple function to perform externally, implementing it inside the LTC2704 eliminates concerns about matching the temperature coefficient of the external offsetting resistor to the internal resistors.

Example circuits

Figure 361.1 shows several ways to use the LTC2704's features. The offset pin of DAC A is driven by an LTC2601 DAC through an LTC1991 amplifier. This provides ±50mV of "system offset" adjustment in the ±2.5V and 0V to 5V ranges, ±100mV of adjustment in the −2.5V to 7.5V, ±5V, and 0V to 10V ranges and ±200mV of adjustment in the ±10V range. The C1 pin is left open for fast settling. An LTC2604 quad DAC can be used to drive all four offset pins, and can share the same SPI bus as the LTC2704.

DAC B drives a 1μF capacitor through a 1Ω resistor, with 2200pF of additional compensation. This is useful for applications where the load has high frequency transients, such as driving the reference pin of an ADC.

DAC C drives an LT3080 low dropout regulator, providing up to 1A of output current. This can be used to power test circuitry directly. Global feedback removes the offset of the regulator, maintaining accuracy at the output.

DAC D is boosted by an LT1970 power op amp, providing 500mA of drive current, either sourcing or sinking. Once again, global feedback preserves DC accuracy.

Conclusion

The LTC2704 provides a highly integrated solution for generating multiple precision voltages. It saves design time, board space and cost compared to implementations using separate DACs and amplifiers.

Analog Circuit Design: Design Note Collection. http://dx.doi.org/10.1016/B978-0-12-800001-4.00361-6

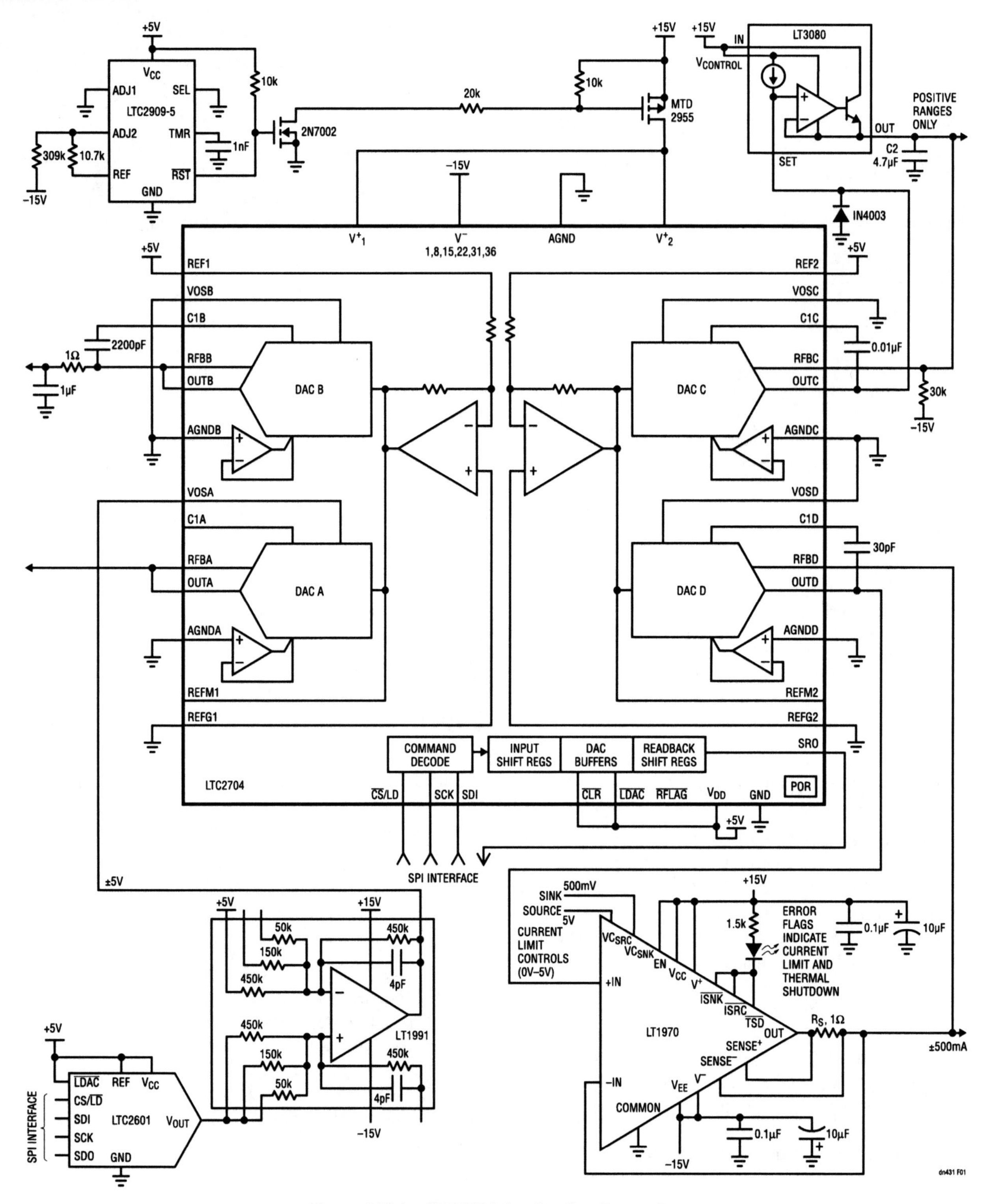

Figure 361.1 • LTC2704 Application Examples

Multiple output range 16-bit DAC design made simple

362

Derek Redmayne

Introduction

Precision 16-bit analog outputs with software-configurable output ranges are often needed in industrial process control equipment, analytical and scientific instruments and automatic test equipment. In the past, designing a universal output module was a daunting task and the cost and PCB real estate associated with this function were problematic, if not prohibitive. Figure 362.1 shows an example of the circuitry formerly required to produce a programmable 16-bit

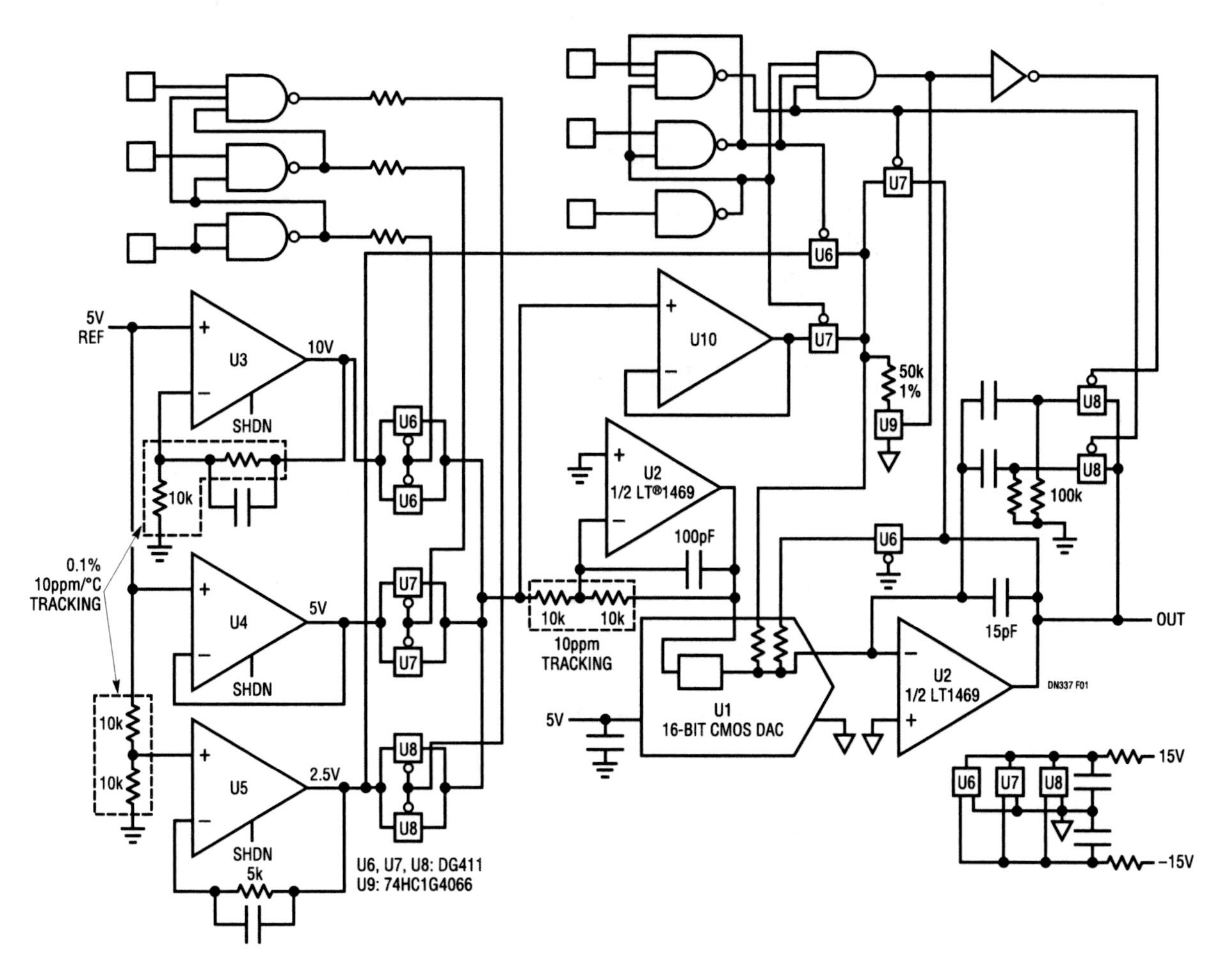

Figure 362.1 • How NOT to Build a Universal 16-Bit Analog Output

Analog Circuit Design: Design Note Collection. http://dx.doi.org/10.1016/B978-0-12-800001-4.00362-8

DAC with a variety of output ranges. However, with the new LTC1592 multiple output range DAC, all of this complexity is unnecessary. Figure 362.2 shows the compact simplicity of an implementation based on the new LTC1592. All the standard industrial ranges (0V to 5V, 0V to 10V, ±5V, ±10V, ±2.5V and −2.5V to 7.5V) are provided, accurately and under software control.

The old way

Figure 362.1 shows a pre-LTC1592 implementation of a multiple output range DAC. The circuit can be made to work, but only with costly components and a lot of PCB real estate. The range switching capability requires the addition of analog switches and precision resistors to the basic DAC. Some of these analog switches are required to compensate for the resistance of switches at other points in the circuit. The circuit as shown, even with its considerable complexity, is a compromise as some of these analog switches are not paired with counterparts in the same package. The analog switches are expensive. They also require PCB real estate, bypassing and decoupling to compensate for poor PSRR to mitigate digital noise. In addition, since they are not switching at virtual ground, the analog switches exhibit on-resistance variation with voltage which will degrade linearity. Leakage can be an issue at high temperature.

Precision matched resistor pairs are shown, as they are available from a number of sources. But unless very expensive devices are used, they will degrade accuracy.

The new, easy way

In contrast is Figure 362.2 where the LTC1592 contains all of the circuitry required to perform these functions—all under processor control. All the ranges are accurate with low drift, fast settling and low glitch operation right out of the box. The LTC1592 incorporates all the switches and precision resistors. A full implementation takes less than 0.5in × 0.5in including the dual operational amplifier, bypass and compensation. This analog output subsystem can be reconfigured in real time and the serial interface makes opto-isolation easy.

Conclusion

Building a precision, multiple output range, software-configurable 16-bit DAC is no longer a complicated, expensive design effort. Now a clean, simple design yields smaller size, lower cost and much better accuracy. The LTC1592 can also be used for embedded or fixed range applications, where its 4-quadrant operation with serial interface make it compelling even if range changing may not be required.

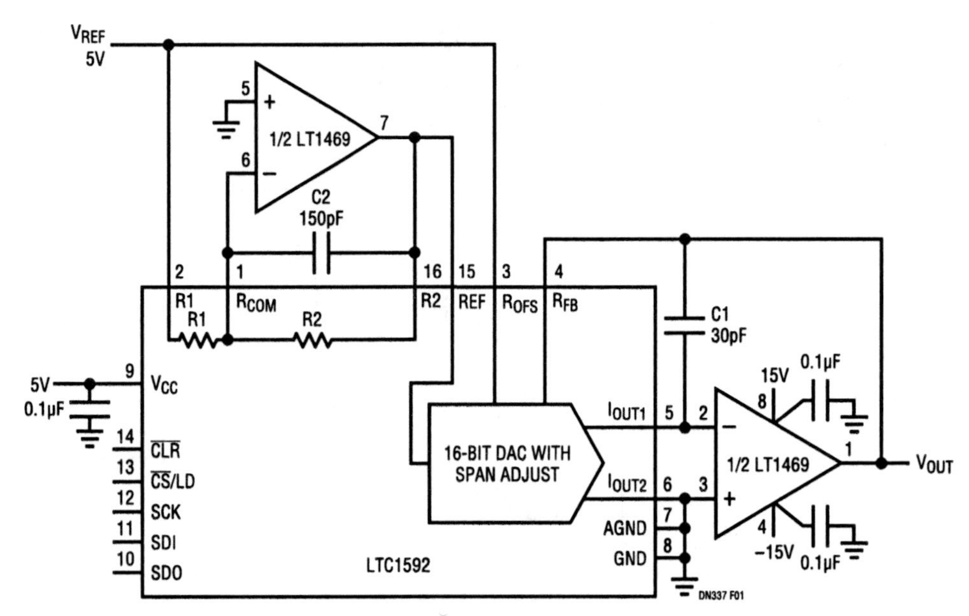

Figure 362.2 • Programmable Output Range 16-Bit SoftSpan DAC

Selecting op amps for precision 16-bit DACs

363

Kevin R. Hoskins Jim Brubaker Patrick Copely

The LTC1597 16-bit current output DAC offers a new level of accuracy and cost efficiency. It has exceptionally high accuracy and low drift: ±1LSB (max) INL and DNL over temperature. It eliminates costly external resistors in bipolar applications because it includes tightly trimmed, on-chip 4-quadrant resistors. To help achieve the best circuit DC performance, this design note offers two sets of easy-to-use design equations for evaluating an op amp's effects on the DAC's accuracy in terms of INL, DNL, offset (unipolar), zero (bipolar), unipolar and bipolar gain errors. With this information, selecting an op amp that gives the accuracy you need in a unipolar or bipolar application is easy.

Figure 363.1 shows a unipolar application that combines the LTC1597 DAC and LT1468 fast settling op amp. The equations for evaluating the effects of the op amp on the DAC's accuracy are shown in Table 363.1. Quick work on Table 363.1's equations with a calculator gives the results shown in Table 363.2. These are the changes the op amp can

Table 363.2 Changes to Figure 363.1's Accuracy in Terms of INL, DNL, Offset and Gain Error When Using the Fast Settling LT1468. $V_{REF} = 10V$

LT1468	TYPICAL	INL (LSB)	DNL (LSB)	OFFSET	GAIN ERROR (LSB)
V_{OS} (mV)	0.03	0.036	0.009	0.2	0.21
I_B (nA)	3	0.0017	0.0005	0.2	0
A_{VOL} (V/V)	9000k	0.0011	0.0003	0	0.015
TOTALS		0.039	0.01	0.4	0.23

Table 363.1 Easy-to-Use Equations Determine Op Amp Effects on DAC Accuracy in Unipolar Applications

OP AMP	INL (LSB)	DNL (LSB)	UNIPOLAR OFFSET (LSB)	UNIPOLAR GAIN ERROR (LSB)
V_{OS} (mV)	$V_{OS} \cdot 1.2 \cdot (10V/V_{REF})$	$V_{OS} \cdot 0.3 \cdot (10V/V_{REF})$	$V_{OS} \cdot 6.6 \cdot (10V/V_{REF})$	$V_{OS} \cdot 6.9 \cdot (10V/V_{REF})$
I_B (nA)	$I_B \cdot 0.00055 \cdot (10V/V_{REF})$	$I_B \cdot 0.00015 \cdot (10V/V_{REF})$	$I_B \cdot 0.065 \cdot (10V/V_{REF})$	0
A_{VOL} (V/V)	$10k/A_{VOL}$	$3k/A_{VOL}$	0	$131k/A_{VOL}$

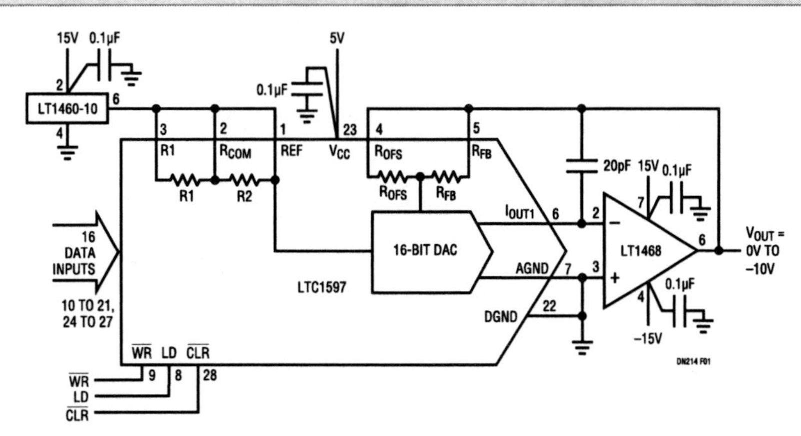

Figure 363.1 • Unipolar V_{OUT} DAC/Op Amp Circuit Swings $10V_{FS}$ (0V to −10V) and Settles to 16 Bits in <2µs. For 0V to 10V Swings, the Reference Can Be Inverted with R1, R2 and Another Op Amp

Analog Circuit Design: Design Note Collection. http://dx.doi.org/10.1016/B978-0-12-800001-4.00363-X

cause to the INL, DNL, unipolar offset and gain error of the DAC. Note, all are substantially less than 1LSB. Thus, the LT1468 is an excellent choice.

The LTC1597's internal design is very insensitive to offset-induced INL and DNL changes. An op amp V_{OS} as large as 0.5mV causes just 0.55LSB INL and 0.15LSB DNL in the output voltage for $10V_{FS}$. So, how does the LT1468 affect, for example, the DAC's DNL? From Table 363.2, the LT1468 adds approximately 0.01LSB of additional DNL to the output. This, along with the LTC1597's typical 0.2LSB DNL, gives a total DAC/op amp DNL of just 0.21LSB (typ). This is well under the −1LSB needed to ensure monotonic operation.

Figure 363.2 shows a bipolar application that combines the LTC1597 and a dual LT1112 op amp. The equations for evaluating the effects of two amplifiers in bipolar operation are just as easy to use and are shown in Table 363.3. A quick application of Table 363.3's equations gives the accuracy changes shown in Table 363.4. Again, each effect is substantially less than 1LSB. This shows that with proper op amp selection, the LTC1597's excellent linearity and precision are not degraded even in bipolar applications.

The totals in Tables 363.2 and 363.4 are algebraic sums of the absolute values, producing a worst-case error.

While not directly addressed by the simple equations in Tables 363.1 and 363.3, temperature effects can be handled just as easily for unipolar and bipolar applications. First, consult an op amp's data sheet to find the worst-case V_{OS} and I_B over temperature. Then, plug these numbers in the V_{OS} and I_B equations from Table 363.1 or Table 363.3 and calculate the temperature induced effects.

Table 363.4 Changes to Figure 363.2's Accuracy in Terms of INL, DNL, Bipolar Zero and Bipolar Gain Error When Using an LT1112, a Precision Dual Op Amp

LT1112	TYPICAL	INL (LSB)	DNL (LSB)	BIPOLAR ZERO ERROR (LSB)	BIPOLAR GAIN ERROR (LSB)
V_{OS1} (mV)	0.02	0.024	0.006	0.198	0.138
I_{B1} (nA)	0.07	0.00004	0.00001	0.0046	0
A_{VOL1}	5000k	0.002	0.0006	0	0.039
V_{OS2} (mV)	0.02	0	0	0.134	0.264
I_{B2} (nA)	0.07	0	0	0.0046	0.009
A_{VOL2}	5000k	0	0	0.013	0.026
TOTALS		0.026	0.007	0.354	0.476

Table 363.3 Easy-to-Use Equations Determine Op Amp Effects on DAC Accuracy in Bipolar Applications

OP AMP	INL (LSB)	DNL (LSB)	BIPOLAR ZERO ERROR (LSB)	BIPOLAR GAIN ERROR (LSB)
V_{OS1} (mV)	$V_{OS1} \cdot 1.2 \cdot (10V/V_{REF})$	$V_{OS1} \cdot 0.3 \cdot (10V/V_{REF})$	$V_{OS1} \cdot 9.9 \cdot (10V/V_{REF})$	$V_{OS1} \cdot 6.9 \cdot (10V/V_{REF})$
I_{B1} (nA)	$I_{B1} \cdot 0.00055 \cdot (10V/V_{REF})$	$I_{B1} \cdot 0.00015 \cdot (10V/V_{REF})$	$I_{B1} \cdot 0.065 \cdot (10V/V_{REF})$	0
A_{VOL1}	$10k/A_{VOL}$	$3k/A_{VOL1}$	0	$196k/A_{VOL1}$
V_{OS2} (mV)	0	0	$V_{OS2} \cdot 6.7 \cdot (10V/V_{REF})$	$V_{OS2} \cdot 13.2 \cdot (10V/V_{REF})$
I_{B2} (nA)	0	0	$I_{B2} \cdot 0.065 \cdot (10V/V_{REF})$	$I_{B2} \cdot 0.13 \cdot (10V/V_{REF})$
A_{VOL2}	0	0	$65k/A_{VOL2}$	$131k/A_{VOL2}$

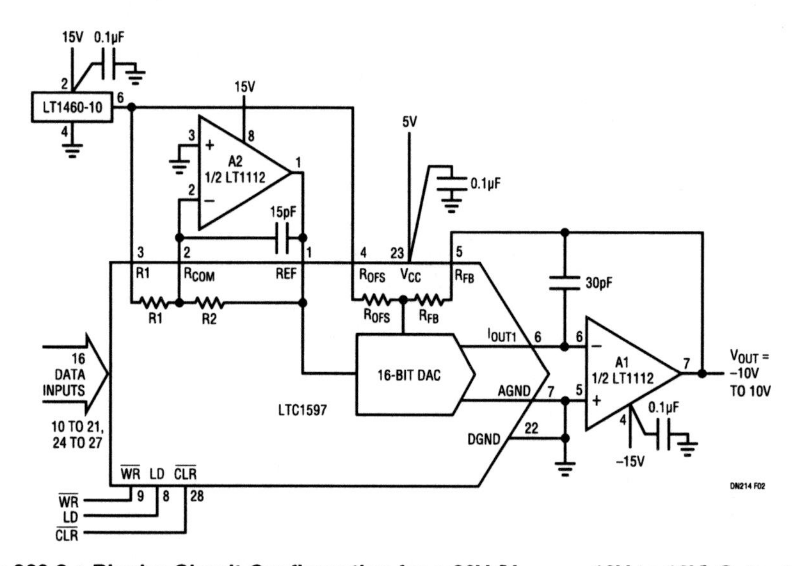

Figure 363.2 • Bipolar Circuit Configuration for a 20V (V_{OUT} = −10V to 10V) Output Swing

Applications versatility of dual 12-bit DAC

364

Kevin R. Hoskins

Introduction

CMOS multiplying digital-to-analog converters (MDACs) make versatile building blocks that go beyond their basic function of converting digital data into analog signals. This design note details some of the other circuits that are possible when using the LTC1590 dual, serial 12-bit current output DAC. This DAC, shown in Figure 364.1 features:

- 2- and 4-Quadrant Multiplying Capability
- Outstanding INL and DNL: 0.1LSB Typ, 0.5LSB Max Over Temperature
- Low Supply Current: 10μA
- 3-Wire, Daisy-Chainable SPI Serial I/O
- Clear Input Resets DACs to Zero Scale
- Small Footprint: Narrow 16-Lead SO Package

The LTC1590 is designed for a wide range of applications, including process control and industrial automation, automatic test equipment, software controlled gain adjustment, digitally controlled filters and power supplies. This DAC is available in 16-lead narrow SO and PDIP packages.

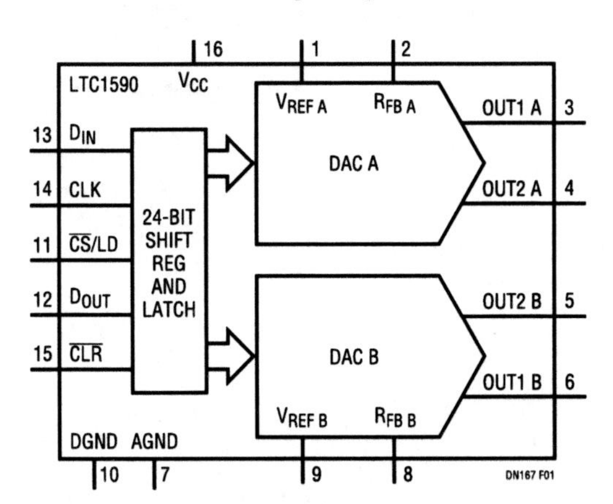

Figure 364.1 • This Single Supply, Dual 12-Bit DAC Features Serial Interface, ±10V Reference Input Range and a CLR Pin That Resets the Output to Zero

Applications

Digitally controlled attenuator and PGA

The circuit shown in Figure 364.2 uses the LTC1590 to create a digitally controlled attenuator using DAC A, and a programmable gain amplifier (PGA) using DAC B. The attenuator's gain is set using the following equation:

$$V_{OUT} = -V_{IN}\frac{D}{2^n} = -V_{IN}\frac{D}{4096}$$

where:

V_{OUT} = output voltage

V_{IN} = input voltage

n = DAC resolution in bits

D = value of code applied to DAC

The attenuator's gain varies from 0 to −4095/4096 in 4096 steps. A code of 0 completely attenuates the input signal.

The PGA's gain is set using the following equation:

$$V_{OUT} = -V_{IN}\frac{2^n}{D} = -V_{IN}\frac{4096}{D}$$

where:

V_{OUT} = output voltage

V_{IN} = input voltage

n = DAC resolution in bits

D = value of code applied to DAC

The gain is adjustable from −4096/4095 to 4096 in 4095 steps. A code of 0 is meaningless, since this results in infinite gain and the amplifier operates open-loop. With either configuration, the attenuator's and PGA's gains are set with 12-bit accuracy.

Amplified attenuator and attenuated PGA

Further modification to the basic attenuator and PGA is shown in Figure 364.3. In this circuit, DAC A's attenuator

Analog Circuit Design: Design Note Collection. http://dx.doi.org/10.1016/B978-0-12-800001-4.00364-1

circuit is modified to give the output amplifier a gain set by the ratio of resistors R3 and R4. The equation for this attenuator with output gain is:

$$V_{OUT} = -V_{IN}\frac{D}{2^n}\frac{R3+R4}{R3} = -V_{IN}\frac{D}{256}$$

With the values shown, the attenuator's gain has a range of 0 to approximately −16. This range is easily modified by changing the ratio of R3 and R4. In the other half of the circuit, an attenuator has been added to the input of DAC B, configured as a PGA. The equation for this PGA with input attenuation is:

$$V_{OUT} = -V_{IN}\frac{2^n}{D}\frac{R1}{R1+R2} = -V_{IN}\frac{256}{D}$$

This sets the gain range from approximately −1/16 to −256. Again, a code of 0 is meaningless. This range can be modified by changing the ratio of R1 and R2.

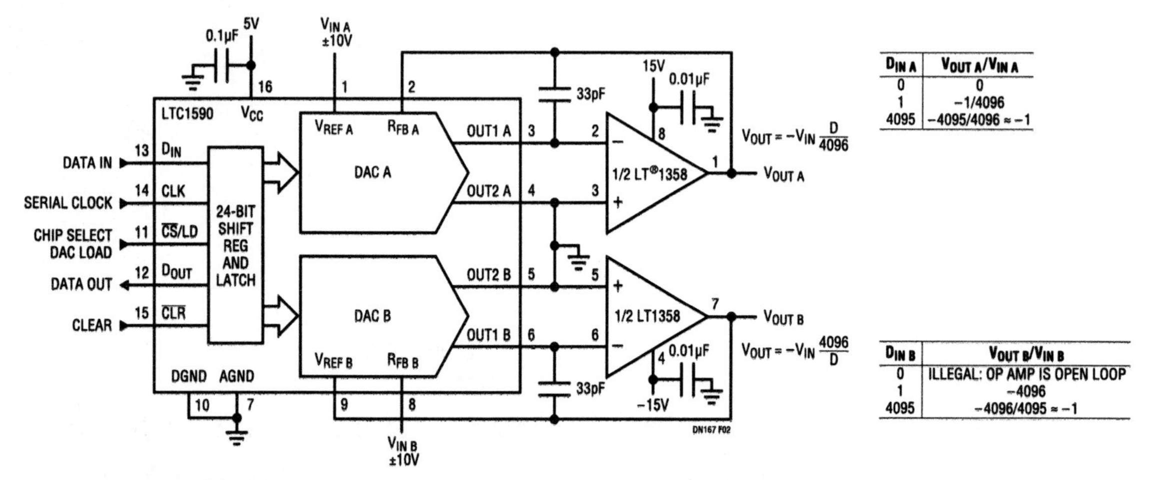

DIN A	VOUT A/VIN A
0	0
1	−1/4096
4095	−4095/4096 ≈ −1

DIN B	VOUT B/VIN B
0	ILLEGAL: OP AMP IS OPEN LOOP
1	−4096
4095	−4096/4095 ≈ −1

Figure 364.2 • The LTC1590's DACs Are Configured as an Attenuator (DAC A) and a Programmable Gain Amplifier (DAC B), Each with a Range of 72dB

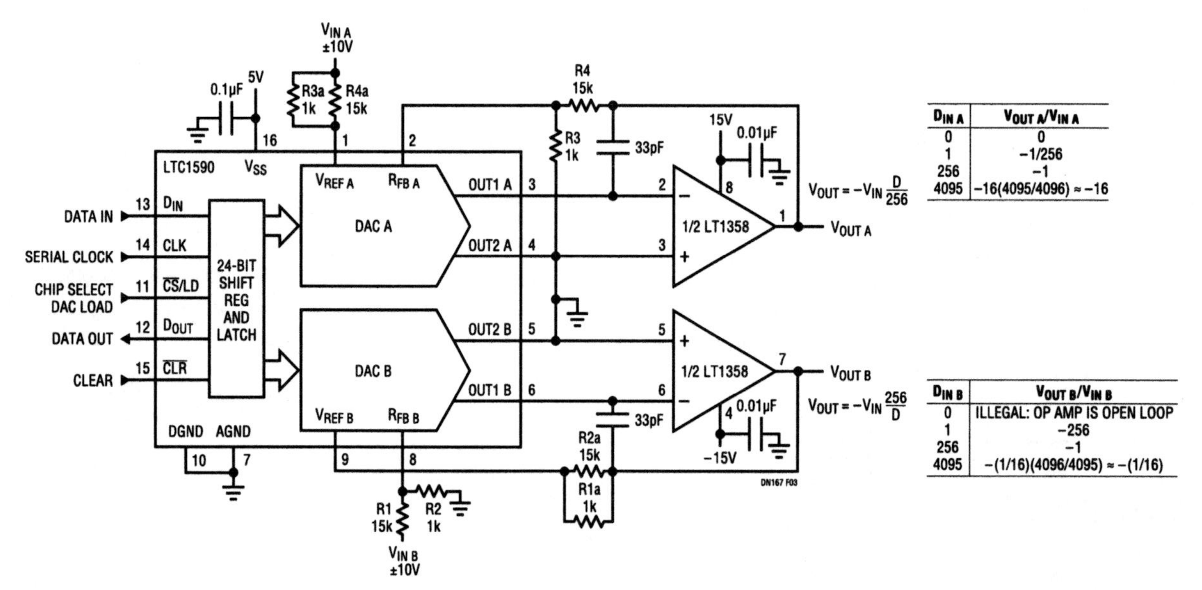

DIN A	VOUT A/VIN A
0	0
1	−1/256
256	−1
4095	−16(4095/4096) ≈ −16

DIN B	VOUT B/VIN B
0	ILLEGAL: OP AMP IS OPEN LOOP
1	−256
256	−1
4095	−(1/16)(4096/4095) ≈ −(1/16)

Figure 364.3 • Minor Additions to Figure 364.1's Circuit Increase Its Flexibility by Adding Gain to the Attenuator (DAC A) and Attenuation to the PGA (DAC B)

365

First dual 12-bit DACs in SO-8

Hassan Malik Kevin R. Hoskins

The LTC1446 and LTC1446L are the first dual, single supply, rail-to-rail voltage output 12-bit DACs. Both parts include an internal reference and two DACs with rail-to-rail output buffer amplifiers, packaged into a space-saving 8-pin SO or PDIP package. The LTC1446's patented architecture is inherently monotonic and has excellent 12-bit DNL, guaranteed to be less than 0.5LSB. These parts have an easy-to-use SPI compatible interface that allows daisy-chaining.

Low power 5V or 3V single supply

The LTC1446 has an output swing of 0V to 4.095V, with each LSB equal to 1mV. It operates from a single 4.5V to 5.5V supply, drawing 1mA. The LTC1446L has an output swing of 0V to 2.5V, operates on a single 2.7V to 5.5V supply and draws 650μA.

Complete standalone performance

Figure 365.1 shows a block diagram of the LTC1446/LTC1446L. The data inputs for both DAC A and DAC B are clocked into one 24-bit shift register. The first 12-bit segment is for DAC A and the second is for DAC B. Each 12-bit segment is loaded MSB first and latched into the shift register on the rising edge of the clock. When all the data has been shifted in, it is loaded into the DAC registers when the signal on the $\overline{CS}$/LD pin changes to a logic high. This updates both 12-bit DACs and internally disables the CLK signal. The D_{OUT} pin allows the user to daisy-chain several DACs together. Power-on reset initializes the outputs to zero scale.

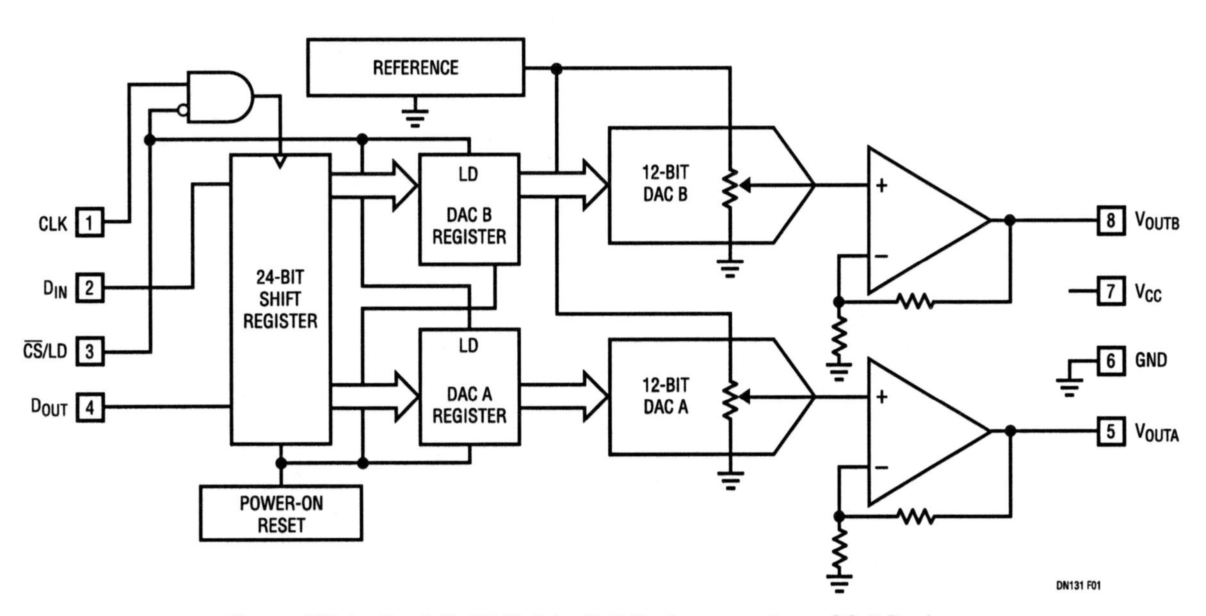

Figure 365.1 • Dual 12-Bit Rail-to-Rail Performance in an SO-8 Package

Analog Circuit Design: Design Note Collection. http://dx.doi.org/10.1016/B978-0-12-800001-4.00365-3

Rail-to-rail outputs

The on-chip output buffer amplifiers can source or sink over 5mA with a 5V supply. More over, they have true rail-to-rail performance. This results in excellent load regulation up to the 4.095V full-scale output with a 4.5V supply. When sinking current with outputs close to zero scale, the effective output impedance is about 50Ω. The midscale glitch on the output is 20nV · s and the digital feedthrough is a negligible 0.15nV · s.

A wide range of applications

Some of the typical applications for these parts include digital calibration, industrial process control, automatic test equipment, cellular telephones and portable battery-powered applications. Figure 365.2 shows how easy these parts are to use.

Figure 365.3 shows how to use one LTC1446 to make an autoranging ADC. The microprocessor adjusts the ADC's reference span and offset by loading the appropriate digital code into the LTC1446. V_{OUTA} controls the common pin for the analog inputs to the LTC1296 and V_{OUTB} controls the reference span by setting the LTC1296's REF^+ pin. The LTC1296 has a Shutdown pin whose output is a logic low in shutdown mode. During shutdown, this logic low turns off the PNP transistor that supplies power to the LTC1446. The resistors and capacitors lowpass filter the LTC1446 outputs, attenuating noise.

Figure 365.4 shows how to use an LTC1446 and an LT1077 to make a wide bipolar output swing 12-bit DAC with a digitally programmable offset. The voltage on DAC A's output (V_{OUTA}) is used as the offset voltage. Figure 365.4 also shows how the circuit's output voltage changes as a function of the input digital code.

Conclusion

The LTC1446 and LTC1446L are the world's only DACs that offer dual 12-bit standalone performance in an 8-pin SO or PDIP package. Along with their amazing density, these DACs do not compromise performance, offering excellent 12-bit DNL, rail-to-rail voltage outputs and very low power dissipation. This allows users to save circuit board space without sacrificing performance.

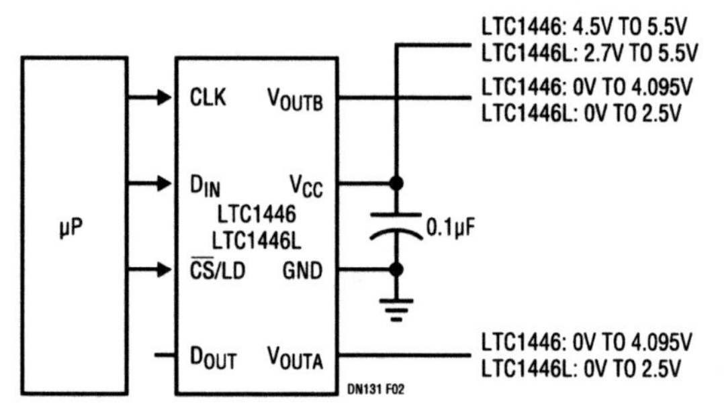

Figure 365.2 • Easy Standalone Application for the LTC1446 or LTC1446L

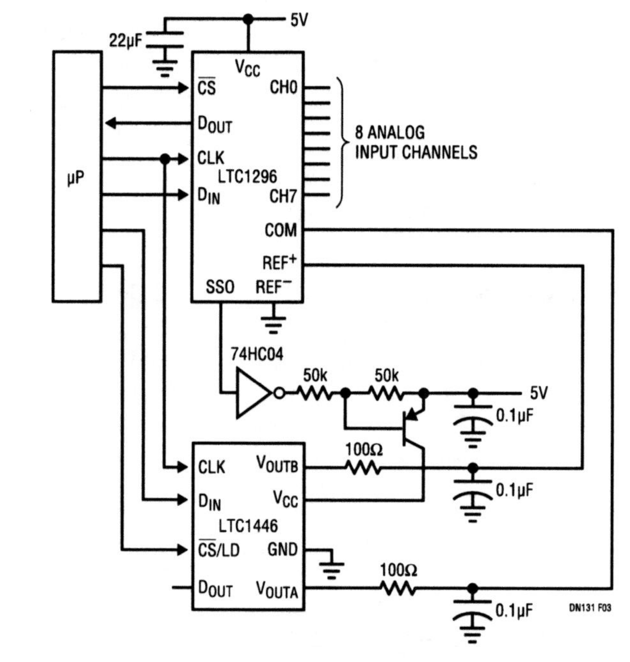

Figure 365.3 • An Autoranging 8-Channel ADC with Shutdown

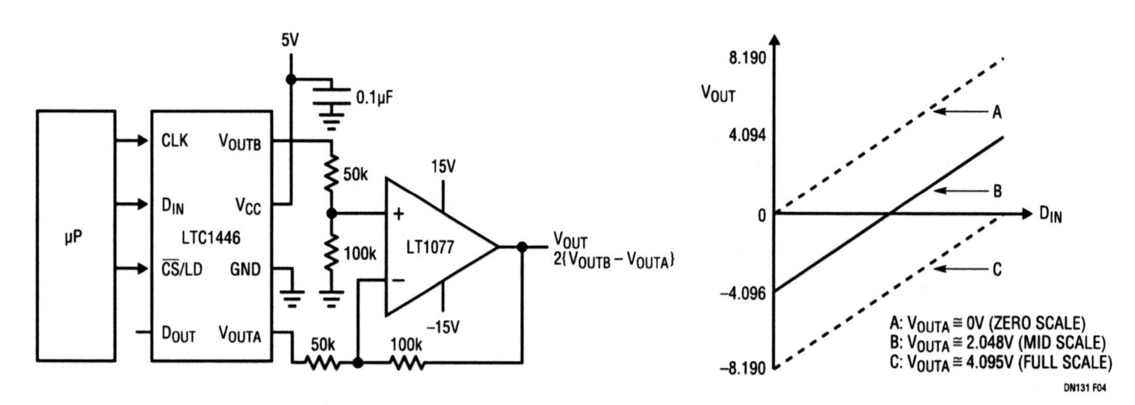

Figure 365.4 • A Wide-Swing, Bipolar Output DAC with Digitally Controlled Offset

3V and 5V 12-bit rail-to-rail micropower DACs combine flexibility and performance

366

Roger Zemke Hassan Malik

The LTC1450 and LTC1450L are complete single supply rail-to-rail voltage output, 12-bit DACs in 24-pin SSOP and PDIP packages. They include an output buffer amplifier, a reference and a double-buffered parallel digital interface. These DACs use a proprietary architecture which guarantees a maximum DNL error of 0.5LSB. A built-in power-on reset ensures that the output of the DAC is initialized to zero scale.

The rail-to-rail buffered output can source or sink 5mA while pulling to within 300mV of the positive supply voltage or ground. The output swings to within a few millivolts of either supply rail when unloaded and has an equivalent output resistance of 40Ω when driving a load to the rails.

Low power, 5V or 3V single supply operation

The LTC1450 draws only 400μA from a 4.5V to 5.5V supply. The DAC can be configured for a 0V to 4.095V or 0V to 2.048V output range. It has a 2.048V internal reference.

The LTC1450L draws 250μA from a 2.7V to 5.5V supply. It can be configured for a 0V to 2.5V or 0V to 1.22V output range. It has a 1.22V internal reference.

Flexibility with standalone performance

The LTC1450 and LTC1450L are complete standalone DACs requiring no external components. Reference output, reference input and gain setting pins provide the user with added flexibility.

Figure 366.1 shows how these DACs may typically be used. REF HI is tied to REF OUT and REF LO and X1/X2 are grounded.

4-quadrant multiplying DAC application

Figure 366.2 shows the LTC1450L configured as a single supply 4-quadrant multiplying DAC. It uses a 5V supply and only one external component, a 5k resistor tied from REF OUT

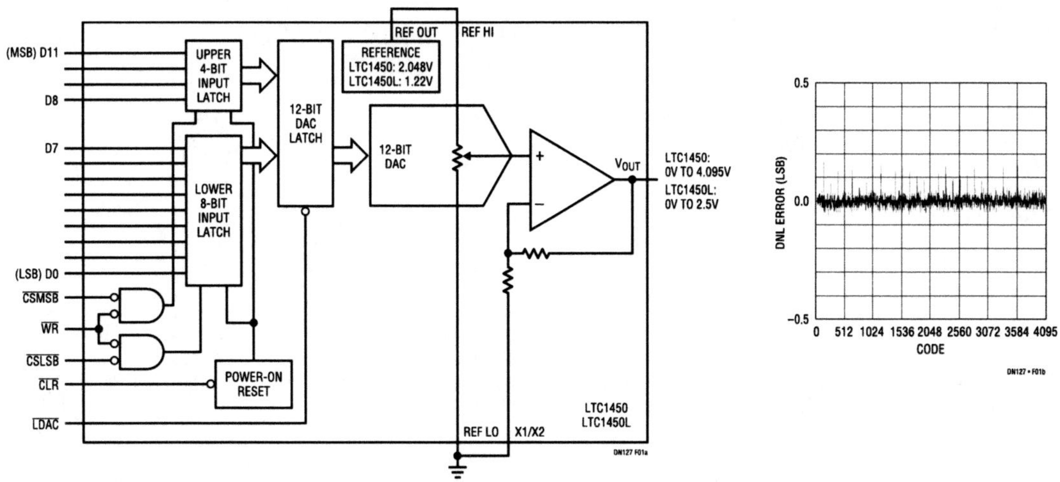

Figure 366.1 • Byte/Parallel Input, Internal/External Reference, Power-On Reset and Gain Select Provide Application Flexibility. The Patented Architecture Guarantees Excellent DNL

Analog Circuit Design: Design Note Collection. http://dx.doi.org/10.1016/B978-0-12-800001-4.00366-5

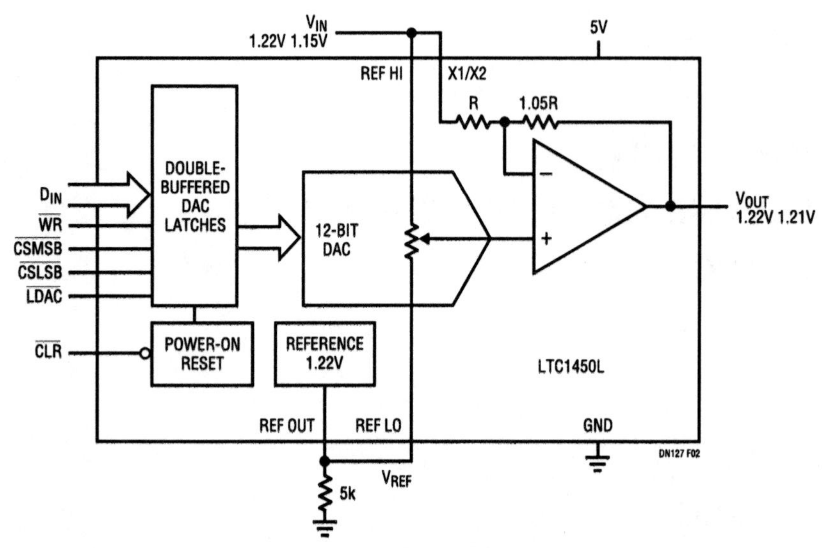

Figure 366.2 • Internal Reference, REF LO/REF HI Pins, Gain Adjust and Wide Supply Voltage Range Allow 4-Quadrant Multiplying on a 5V Single Supply

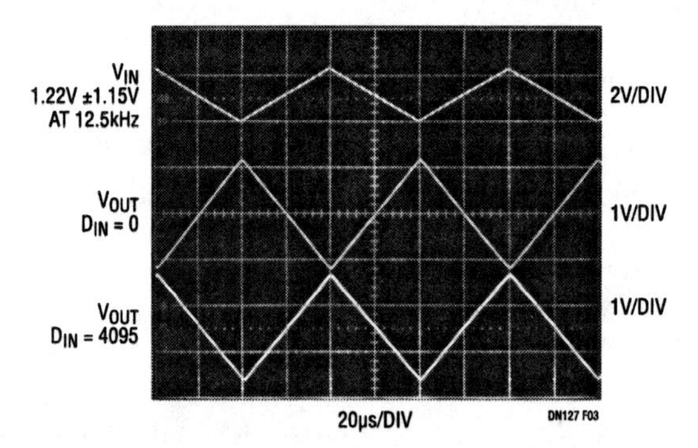

Figure 366.3 • Clean 4-Quadrant Multiplying Is Shown in the Output Waveforms for Zero-Scale and Full-Scale DAC Settings

to ground. (The LTC1450 can be used in a similar fashion.) The multiplying DAC allows the user to digitally change the amplitude and polarity of an AC input signal whose voltage is centered around an offset signal ground provided by the 1.22V reference voltage. The transfer function is shown in the following equations.

$$V_{OUT} = (V_{IN} - V_{REF})\left[\text{Gain}\left(\frac{D_{IN}}{4096} - 1\right) + 1\right] + V_{REF}$$

For the LTC1450L Gain = 2.05 and V_{REF} = 1.22V

$$V_{OUT} = (V_{IN} - 1.22V)\left[2.05\left(\frac{D_{IN}}{4096}\right) - 1.05\right] + 1.22V$$

Table 366.1 shows the expressions for V_{OUT} as a function of V_{IN}, V_{REF} and D_{IN}. Figure 366.3 shows a 12.5kHz, $2.3V_{P-P}$ triangle wave input signal and the corresponding output waveforms for zero-scale and full-scale DAC codes.

Table 366.1 Binary Code Table for 4-Quadrant, Multiplying DAC Application

BINARY DIGITAL INPUT CODE IN DAC REGISTER			ANALOG OUTPUT (V_{OUT})
MSB		LSB	
1111	1111	1111	$(4094/4096)(V_{IN} - V_{REF}) + V_{REF}$
1100	0001	1001	$0.5(V_{IN} - V_{REF}) + V_{REF}$
1000	0011	0010	V_{REF}
0100	0100	1011	$-0.5(V_{IN} - V_{REF}) + V_{REF}$
0000	0110	0100	$-1.0(V_{IN} - V_{REF}) + V_{REF}$
0000	0000	0000	$-1.05(V_{IN} - V_{REF}) + V_{REF}$

12-bit rail-to-rail micropower DACs in an SO-8

367

Hassan Malik Jim Brubaker

The LTC1451, LTC1452 and LTC1453 are complete, single supply, rail-to-rail voltage output 12-bit digital-to-analog (DAC) converters. They include an output buffer amplifier and a space saving SPI compatible three-wire serial interface. There is also a data output pin that allows daisy-chaining multiple DACs. These DACs use a proprietary architecture which guarantees a DNL (differential nonlinearity) error of less than 0.5 LSB. The typical DNL error is about 0.2 LSB as shown in Figure 367.1. There is a built-in power-on reset that resets the output to zero scale. The output amplifier can swing to within 5mV of V_{CC} when unloaded and can source or sink 5mA even at a 4.5V supply. These DACs come in an 8-pin PDIP and SO-8 package.

5V and 3V operation

The LTC1451 has an on-board reference of 2.048V and a nominal output swing of 4.095V. It operates from a single 4.5V to 5.5V supply dissipating 2mW ($I_{CC(TYP)}=400\mu A$).

The LTC1452 is a multiplying DAC with no on-board reference and a full-scale output of twice the reference input. It operates from a single supply that can range from 2.7V to 5.5V. It dissipates 1.125mW ($I_{CC(TYP)}=225\mu A$) at a 5V supply and a mere 0.5mW ($I_{CC(TYP)}=160\mu A$) at a 3V supply.

The LTC1453 has a 1.22V on-board reference and a convenient full scale of 2.5V. It can operate on a single supply with a wide range of 2.7V to 5.5V as shown in Figure 367.2. It dissipates 0.75mW ($I_{CC(TYP)}=220\mu A$) at a 3V supply. The digital inputs can swing above V_{CC} for easy interfacing with 5V logic.

True rail-to-rail output

The output rail-to-rail amplifier can source or sink 5mA over the entire operating temperature range while pulling to within 300mV of the positive supply voltage or ground. The output swings to within a few millivolts of either supply rail when

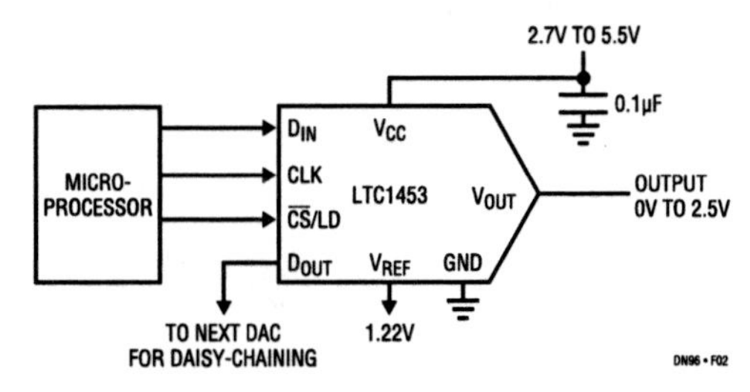

Figure 367.2 • The 3V LTC1453 Is SPI Compatible and Talks to Both 5V and 3V Processors

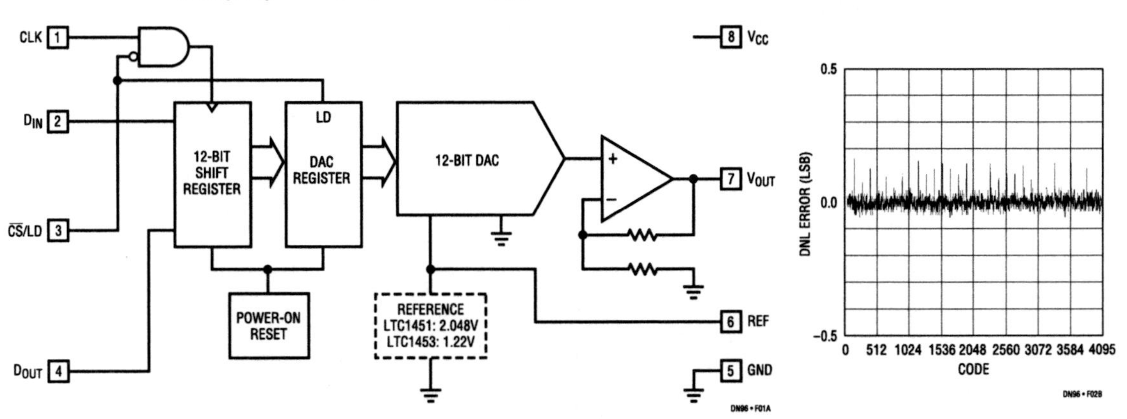

Figure 367.1 • Proprietary Architecture Guarantees Excellent DNL

Analog Circuit Design: Design Note Collection. http://dx.doi.org/10.1016/B978-0-12-800001-4.00367-7

unloaded and has an equivalent output resistance of 50Ω when driving to either rail. The output can drive a capacitive load of up to 1000pF without oscillating.

Wide range of applications

Some of the applications for this family include digital calibration, industrial process control, automatic test equipment, cellular telephones and portable battery-powered applications where low supply current is essential. Figure 367.3 shows how to use an LTC1453 to make an opto-isolated digitally controlled 4mA to 20mA process controller. The controller circuitry, including the opto-isolator, is powered by the loop voltage that can have a wide range of 3.3V to 30V. The 1.22V reference output of the LTC1453 is used for the 4mA offset current and V_{OUT} is used for the digitally controlled 0mA to 16mA current. RS is a sense resistor and the LT1077 op amp modulates the transistor Q1 to provide the 4mA to 20mA current through this resistor. The potentiometers allow for offset and full-scale adjustment. The control circuitry consumes well under the 4mA budget at zero scale.

Flexibility, true rail-to-rail performance and micropower; all in a tiny SO-8

The LTC1451, LTC1452 and LTC1453 are the most flexible micropower, standalone DACs that offer true rail-to-rail performance. This flexibility along with the tiny SO-8 package allows these parts to be used in a wide range of applications where size, power, DNL and single supply operation are important.

Table 367.1 LTC Serial Voltage Output DACs

PART	V_{CC} RANGE	REFERENCE	FULL SCALE	ICC
LTC1451	4.5V to 5.5V	2.048V-Internal	4.095V	400μA at 5V
LTC1452	2.7V to 5.5V	External	2 × REF	225μA at 5V
LTC1453	2.7V to 5.5V	1.22V-Internal	2.5V	250μA at 3V
LTC1257	4.75V to 15.75V	2.048V-Internal (2.5V to 12V-External)	2.048V (2.5V to 12V)	350μA at 5V

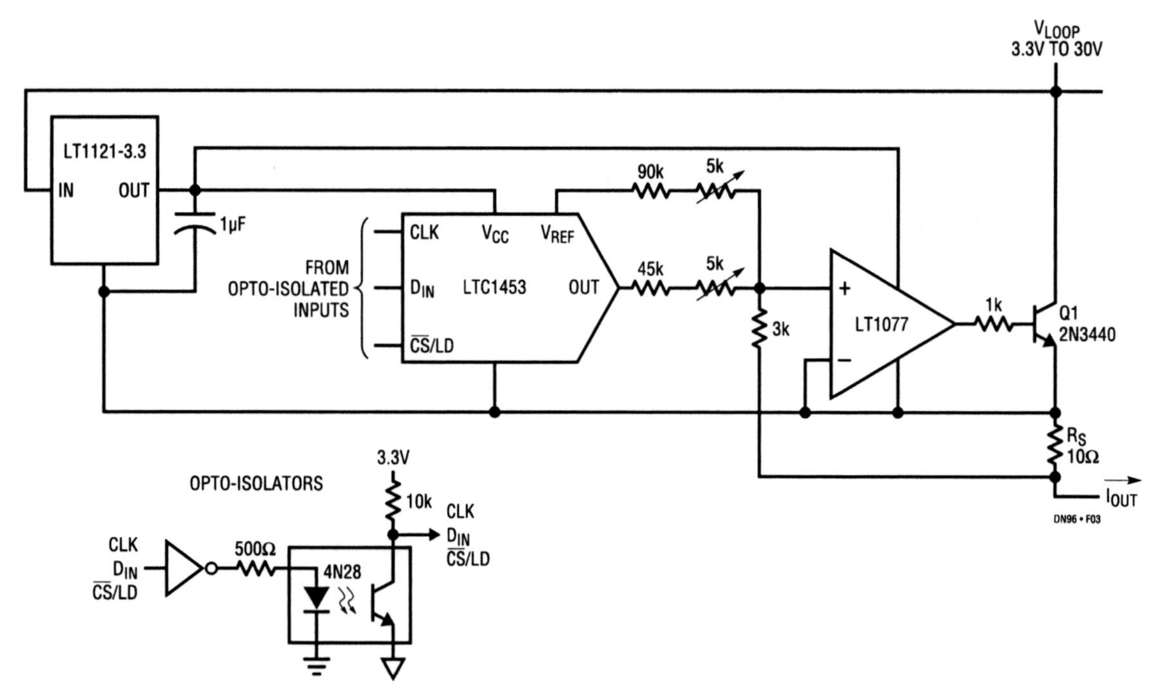

Figure 367.3 • 4mA to 20mA Process Controller Has 3.3V Minimum Loop Voltage

Section 3

Data Acquisition

16-channel, 24-bit ΔΣ ADC provides small, flexible and accurate solutions for data acquisition

368

Sean Wang

Introduction

The LTC2418 is a new 16-channel No Latency ΔΣ ADC that addresses applications where a variety of sensors are monitored. The input can be configured as 16 single-ended, 8 differential, or any combination of differential and single-ended channels to fit the end application. Furthermore, the polarity of a differential channel can be reversed. Each selection maintains the excellent performance of the core 24-bit ADC, has a full-scale differential input range of $-1/2V_{REF}$ to $1/2V_{REF}$ and the input common mode can be anywhere within GND and V_{CC} with DC common mode input rejection better than 140dB.

The converter has an on-chip oscillator that requires no external frequency setting components. Through a single pin, the LTC2418 can provide better than 110dB differential mode rejection at 50Hz or 60Hz ±2%. The LTC2418 communicates through a flexible 4-wire SPI digital interface. No extra register configuration is required except the channel address.

The circuit in Figure 368.1 shows the versatility of the LTC2418. A combination of single-ended, differential, unipolar and bipolar input sources is simultaneously applied to the LTC2418. The low noise (0.2ppm RMS) and high accuracy (total unadjusted error is better than 3ppm) performance enable a wide dynamic input range while near zero drift (0.03ppm/°C full-scale drift, 20nV/°C offset drift) ensures consistent accuracy. Unlike typical delta-sigma converters that require several conversion cycles to settle every time a new channel is switched, the first conversion result is accurate after a new channel is selected (no latency). This feature provides a very simple switching scheme between different channels. The selected channel address and a parity bit are included in the data output to ensure data integrity in noisy isolated environments.

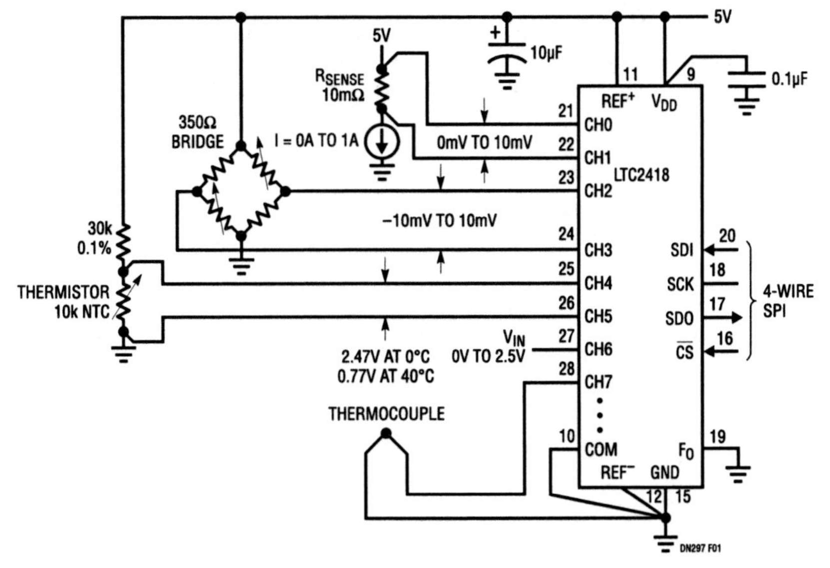

Figure 368.1 • Multiple Measurements with the LTC2418

Analog Circuit Design: Design Note Collection. http://dx.doi.org/10.1016/B978-0-12-800001-4.00368-9

Sensors with common mode output voltages at ground, V_{CC} or anywhere in between may be digitized. Sensors with only a few mV full-scale outputs may be applied directly to the LTC2418. Large sensor offsets and tare voltages are transparently handled by the converter's wide input range, eliminating complex front-end analog processing. Large level input signals ranging from GND to $V_{REF}/2$ (2.5V) may also be applied to the device. The details of each input circuit are described below.

CH0 and CH1 are configured to measure a current sense resistor with a differential input of 0mV to 10mV and common mode near V_{CC}. The current being monitored can have a common mode up to V_{CC} with a wide range of 0A to 1A. The LTC2418 achieves 10,000 counts resolution (0.1mA) without any gain stage or common mode shift.

CH2 and CH3 form a differential input measuring a 350Ω bridge with common mode near $V_{CC}/2$. Typical strain gauge-based bridges deliver 2mV/V of excitation, so the bipolar input is from −10mV to 10mV. The resolution is 1 part in 20,000 without averaging or external gain stages.

CH4 and CH5 measure the thermistor in a half-bridge application. In the example, the output varies from 0.77V to 2.47V over 0°C to 40°C. The LTC2418 can digitize the signal directly due to its excellent common mode rejection and linearity.

CH6 is a large swing single-ended signal that can be from ground up to 2.5V (half of V_{REF}). A reading equivalent to that of a 6-digit DVM is possible due to the accuracy (3ppm total) and low noise ($1\mu V_{RMS}$) of the LTC2418.

CH7 measures a single-ended output of a thermocouple with common mode near ground. Cold junction compensation can be performed to get the accurate absolute temperature using a cold junction sensor similar to the thermistor circuit applied to CH4 and CH5. The temperature measurement may then be used to compensate the temperature effects of the bridge transducers like the one connected to CH2 and CH3.

The above application measures a wide variety of sensors while using just half of the available LTC2418 capacity. The channel address readback and the parity bit included in the output bit stream provide a convenient way to verify which sensor is under measurement and check the digital transmission integrity. Similar circuits can be easily added to the remaining eight channels in order to take more measurements.

Noise reduction

Using its internal oscillator, the LTC2418 can be configured for better than 110dB differential mode rejection at 50Hz or 60Hz ±2% line frequency and harmonics with a single pin set. The unique digital filter design also provides a simultaneous 50Hz and 60Hz rejection if driven by an external clock of 139800Hz. The rejection over 48Hz to 62.4Hz is better than 87dB, as shown in Figure 368.2. In addition to line frequency rejection, the LTC2418 also exhibits excellent noise rejection due to the $Sinc^4$ lowpass filter architecture.

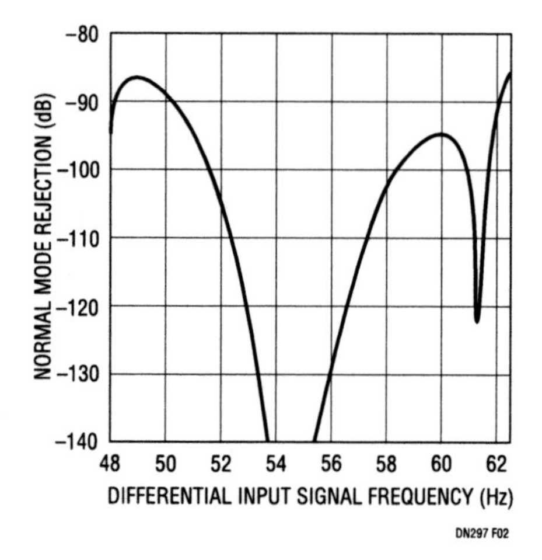

Figure 368.2 • LTC2418 Normal Mode Rejection When Using F_O=139800Hz

Conclusion

The LTC2418 is the multichannel addition to Linear Technology's 24-bit differential delta-sigma family. A pin compatible reduced channel version, the LTC2414, is also available. It can be configured as four differential inputs or eight single-ended inputs. The LTC2414/LTC2418 are available in 28-pin narrow SSOP packages. With high absolute accuracy, ease-of-use and near zero drift, they provide very efficient solutions for multiple channel acquisition applications.

369

A versatile 8-channel multiplexer

Kevin R. Hoskins

Introduction

Available in either 16-pin DIP or narrow body SOIC packages, the CMOS LTC1390 is a high performance 8-to-1 analog multiplexer. It is addressed through a 3-wire digital interface featuring bidirectional serial data.

The LTC1390 features a typical on-resistance of 45Ω, typical switch leakage of 50pA and guaranteed break-before-make switch operation. Charge injection is ±10pC (max). All digital inputs remain logic compatible whether operating on single or dual supplies. The inputs are robust, easily withstanding 100mA fault currents. The LTC1390 operates over a wide power supply voltage range of 3V to ±15V.

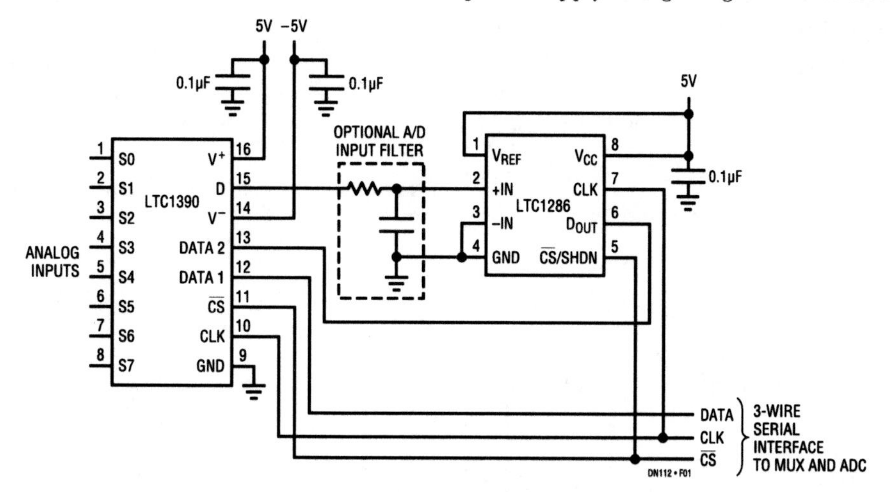

Figure 369.1 • The LTC1390 Expands the 12-Bit Micropower LTC1286 ADC Input Capacity to Eight Channels

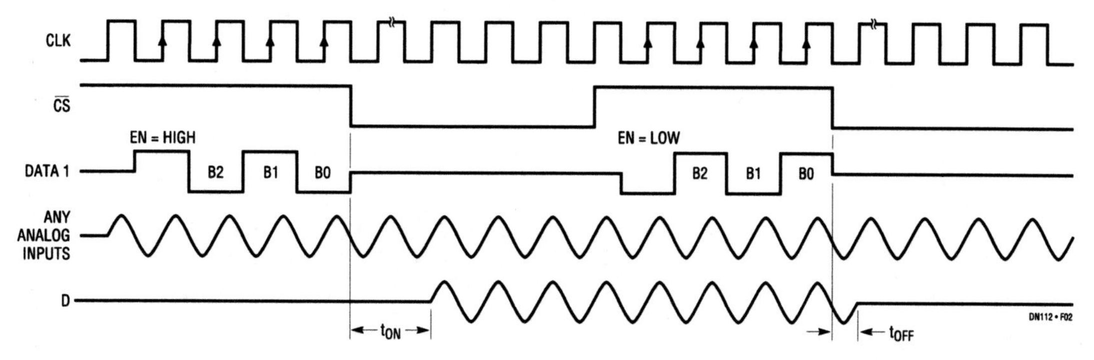

Figure 369.2 • The LTC1390 Clocks in Data While $\overline{CS}$ Is High and Latches the Data, Enabling the Selected Channel, When $\overline{CS}$ Goes Low

Analog Circuit Design: Design Note Collection. http://dx.doi.org/10.1016/B978-0-12-800001-4.00369-0

Low power, daisy-chain serial interface, 8-channel A/D system

Figure 369.1 shows the LTC1390 connected to the LTC1286 12-bit micropower A/D converter. The Clock (CLK) and Chip Select ($\overline{CS}$) signals are connected in parallel to the LTC1390 and LTC1286. Both the LTC1390 and the LTC1286 are designed for serial data transfer. The LTC1390 also includes provisions for daisy-chain operation. This allows full bidirectional communication over a single serial data line. While $\overline{CS}$ is high the LTC1286 is inactive. The serial data used to select a multiplexer channel is applied to the LTC1390 DATA 1 input. Four bits are clocked into the LTC1390. The LTC1286 D_{OUT} is in a high impedance state, ignoring these bits. The LTC1390 latches the four data bits when the $\overline{CS}$ signal goes low, selecting a channel and initiating an A/D conversion from the LTC1286. Subsequent clock cycles shift out LTC1286 conversion data through the LTC1390 (from DATA 1 to DATA 2) to a host microprocessor. Figure 369.2 shows the LTC1390 timing diagram.

It is very easy to expand the width of the multiplexer and take advantage of the LTC1390 daisy-chain capability. Figure 369.3 shows the simple connections needed to add multiplexer channels. The Data 2 on each additional LTC1390 is connected to the Data 1 on the preceding LTC1390. When operating with multiple LTC1390s, the channel selecting data is sent in groups of four bits.

Another useful feature of the LTC1390 is the ability to add analog signal processing between the LTC1390 output and an A/D input. This helps save overall system cost because only one signal processing circuit is needed instead of one circuit per multiplexer input channel.

Figure 369.4 shows an active 2nd order lowpass filter connected between the LTC1390 output and the LTC1286 input. The heart of the lowpass filter is the single supply rail-to-rail precision LT1366 operational amplifier. The filter's cutoff frequency is set to 1kHz.

Conclusion

The LTC1390 provides designers of a serially interfaced data acquisition system with a flexible, low power and cost-effective way to expand the number of A/D channels.

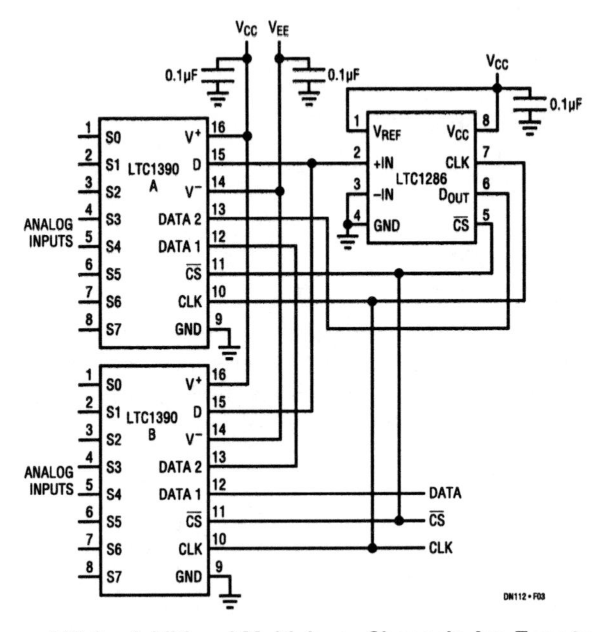

Figure 369.3 • Additional Multiplexer Channels Are Easy to Add without Adding Additional Serial Lines

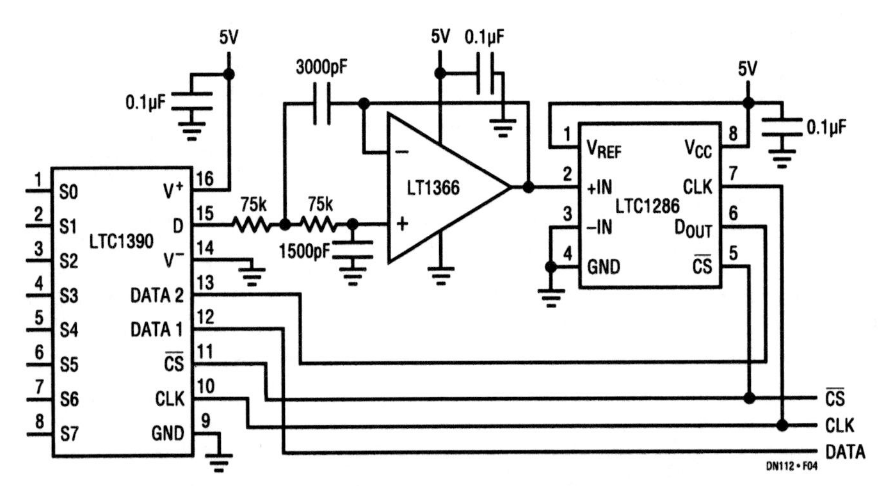

Figure 369.4 • The Connection Between the LTC1390 Output and the LTC1286 Input Is the Perfect Place for Signal Conditioning Circuits Such as the LT1366-Based 2nd Order Active Lowpass Filter

Temperature and voltage measurement in a single chip

370

Ricky Chow Dave Dwelley

Introduction

The LTC1392 is a new micropower, multifunction data acquisition system designed to measure ambient temperature, system power supply voltage and power supply current or differential input voltage. No external components are required for temperature or voltage measurements, and current measurements can be made with a single low value external resistor. An on-board 10-bit A/D converter provides a digital output through a 3- or 4-wire serial interface. Supply current is only 350μA when performing a measurement; this automatically drops to less than 1μA when the chip is not converting. The LTC1392 is designed to be used for PC board temperature and supply voltage/current monitoring or as a remote temperature and voltage sensor for monitoring almost any kind of system. It is available in SO-8 and DIP packages allowing it to fit onto almost any circuit board.

Measurement performance

Wafer level trimming allows the LTC1392 to achieve guaranteed accuracy of ±2°C at room temperature and ±4°C over the entire operating temperature range. The 10-bit A/D converter gives 0.25° resolution over the 0°C to 70°C (LTC1392C) or −40°C to 85°C (LTC1392I) range. Temperature is output as (ADC code/4) −130°C with a theoretical maximum range of −130°C to 125.75°C. Figure 370.1 shows the typical output temperature error of the LTC1392 over temperature.

In supply voltage monitor mode, the A/D converter makes a differential measurement between the 2.42V reference and the actual power supply voltage. Each LSB step is equal to approximately 4.727mV, giving a theoretical measurement range of 2.42V to 7.2V. The LTC1392 has guaranteed accuracy over a voltage range of 4.5V to 6V, with a total absolute error of ±25mV or ±40mV respectively, over the commercial or industrial temperature range. Voltage is output as

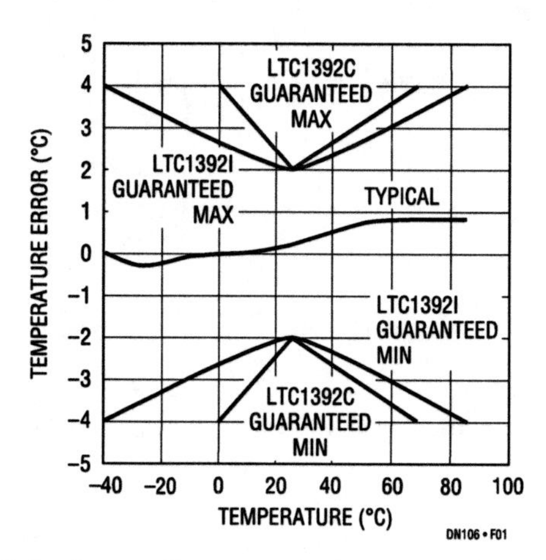

Figure 370.1 • Output Temperature Error

(ADC code × 4.727mV) + 2.42V. Figure 370.2 shows typical integral and differential nonlinearity performance of V_{CC} measurement.

The differential voltage input mode can be configured to operate in either 1V or 0.5V unipolar full-scale mode. Each mode converts the differential voltage between input pins V^+ and V^- directly to bits with the output code equal to ADC code × (full scale/1024). The 1V mode is specified at 8 bits accuracy with the eighth bit accurate to ±0.5LSB or ±2mV, while the 0.5V full-scale mode is specified to 7 bits accuracy ±0.5LSB, giving the same ±2mV accuracy. The differential inputs include a common-mode input range including both power supply rails allowing them to be used to measure the voltage across a sense resistor in either leg of the power supply. They can also be used to make a unipolar differential transducer bridge measurement or to make a single-ended voltage measurement by grounding the V^- pin.

The serial interface in the LTC1392 allows all of its functionality to be implemented in an 8-pin SO or DIP package and makes connection easy to virtually any MPU. Four pins are dedicated to the serial interface: active low chip

Analog Circuit Design: Design Note Collection. http://dx.doi.org/10.1016/B978-0-12-800001-4.00370-7

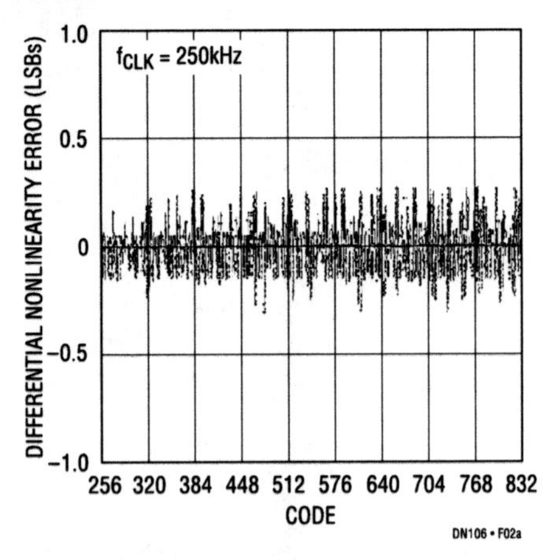

Figure 370.2a • Differential Nonlinearity, Power Supply Voltage Mode

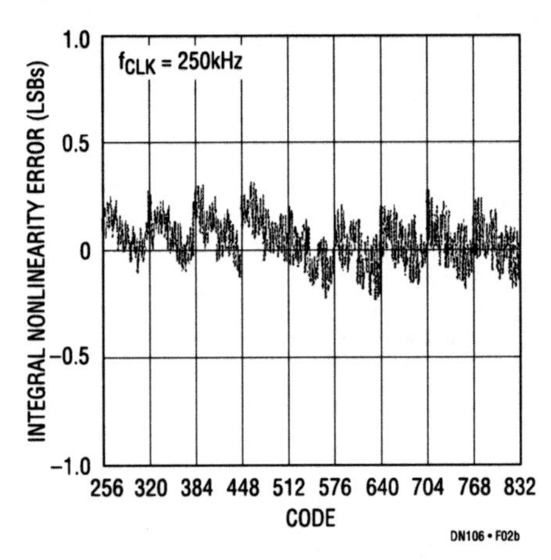

Figure 370.2b • Integral Nonlinearity, Power Supply Voltage Mode

select ($\overline{CS}$), clock (CLK), data input (D_{IN}) and data output (D_{OUT}). The D_{IN} pin is used to configure the LTC1392 for the next measurement and the D_{OUT} pin outputs the A/D conversion data. The D_{IN} pin is disabled after a valid configuration word is received and the D_{OUT} pin is in three-state mode until a valid configuration word is recognized, allowing the two pins to be tied together in a 3-wire system. The serial link allows several devices to be attached to a common serial bus, with separate $\overline{CS}$ lines to select the active chip.

Typical application

A typical LTC1392 application is shown in Figure 370.3. A single point "star" ground is used along with a ground plane to minimize errors in the voltage measurements. The power supply is bypassed directly to the ground plane with a 1μF tantalum capacitor in parallel with a 0.1μF ceramic capacitor.

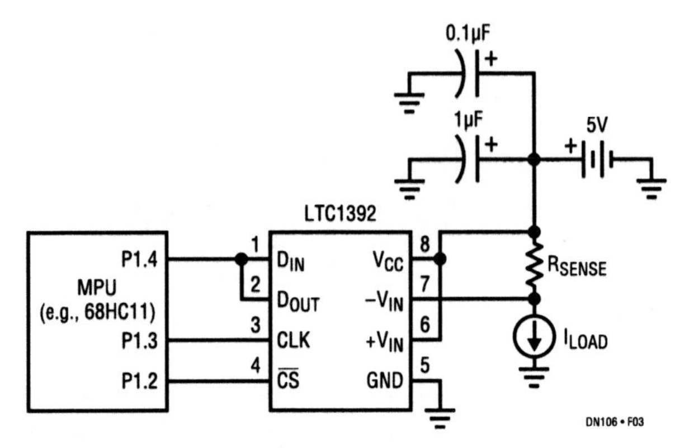

Figure 370.3 • Typical Application

Conclusion

The LTC1392 provides a versatile data acquisition and environmental monitoring system with an easy-to-use interface. Its low supply current, coupled with space-saving SO-8 or DIP packaging make the LTC1392 ideal for systems which require temperature, voltage and current measurement while minimizing space, power consumption and external component count. The combination of temperature and voltage measuring capability on one chip make the LTC1392 unique in the market providing the smallest, lowest power multifunction data acquisition system available.

Applications for a micropower, low charge injection analog switch

371

Guy Hoover William Rempfer Jim Williams

With greater accuracy for both charge and voltage switching, the LTC201A is a superior replacement for the industry standard DG201A. In addition, the micropower LTC201A operates from a single 5V supply, and has lower on-resistance and faster switching speed. These improvements are critical to the operation of the following three circuits.

Micropower V-F converter

Figure 371.1 shows a 100Hz to 1MHz voltage-to-frequency converter. This V-to-F operates from a single supply and draws only 90μA quiescent current, rising to 360μA at 1MHz. Linearity is 0.02% over a 100Hz to 1MHz range.

The circuit consists of an oscillator, a servo amplifier and a charge pump. The oscillator's divided down output is expressed as current (charge per time) by the LTC201A-500pF combination. The input voltage is converted to current by the 220k trimmer pair. The amplifier controls the oscillator frequency to force the net value of the current into A1's summing point to zero.

The 1.5MΩ resistor between V_{IN} and the reference buffer amplifier sums a small input related voltage to the reference,

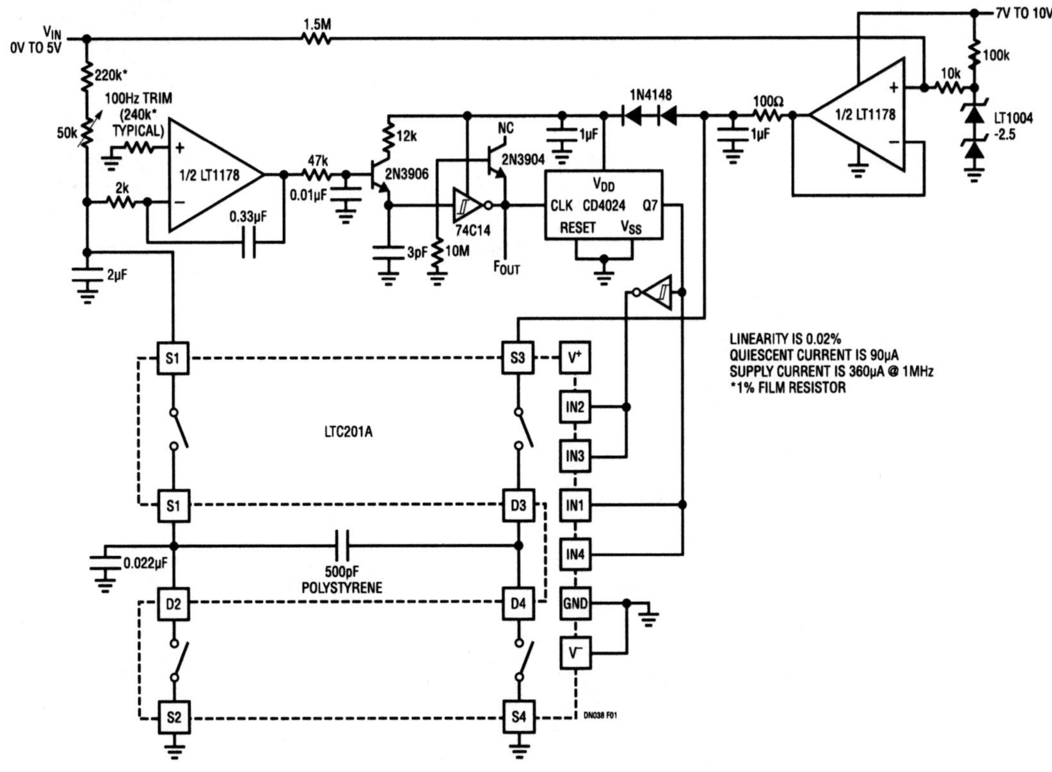

Figure 371.1 • Micropower 100Hz to 1MHz V-to-F Converter

Analog Circuit Design: Design Note Collection. http://dx.doi.org/10.1016/B978-0-12-800001-4.00371-9

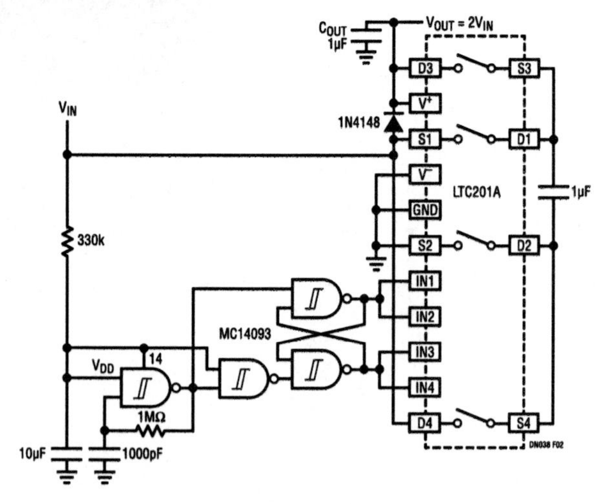

Figure 371.2 • Micropower, 4.5V–15V Input, Voltage Doubler

improving linearity. The 0.022µF capacitor prevents excessive negative transitions at LTC201A D1-D2 pins. The series diodes in the oscillator divider supply the lower supply voltage, decreasing current consumption. The 10MΩ resistor at Q8's collector dominates node leakages ensuring low frequency operation by forcing Q8 to always source current.

Precision voltage doubler

The precision micropower voltage doubler of Figure 371.2 has an input voltage range of 4.5V to 15V. The low supply current of the LTC201A allows it to be powered directly from the input voltage. Total no load supply current of the circuit ranges from 20µA at V_{IN}=4.5V to 130µA at V_{IN}=15V. Output impedance is only 1.2kΩ at V_{IN}=4.5V and reduced to 600Ω at V_{IN}=15V. The accuracy of this circuit is better than 0.2% over the 4.5V to 15V input range.

The MC14093 is used to form an oscillator with complementary non-overlapping outputs. R1 and C1 determine the frequency of oscillation (roughly 1.2kHz at V_{IN}=4.5V). The oscillator outputs drive two sets of switches in the LTC201A and ensure that one pair of switches shuts off before the other set turns on. C_{IN} is alternately charged to V_{IN} and then stacked on top of V_{IN} to charge C_{OUT}. R2 reduces the supply voltage to the MC14093 which keeps current drain low. The diode ensures latch-free power-up for any input rise time condition.

Quad 12-bit sample and hold

Figure 371.3's sample and hold uses the low charge injection of the LTC201A combined with the low offset voltage of the LT1014 to produce a sample to hold offset of only 0.6mV. This makes it accurate enough for 12-bit applications. Acquisition time to 0.6mV is 20µs. Aperture time is 300ns (the off time of the LTC201A). Droop rate is 2mV/ms and is limited by the I_B of the LT1014. The input range is 3.5V to −5V with ±5V supplies.

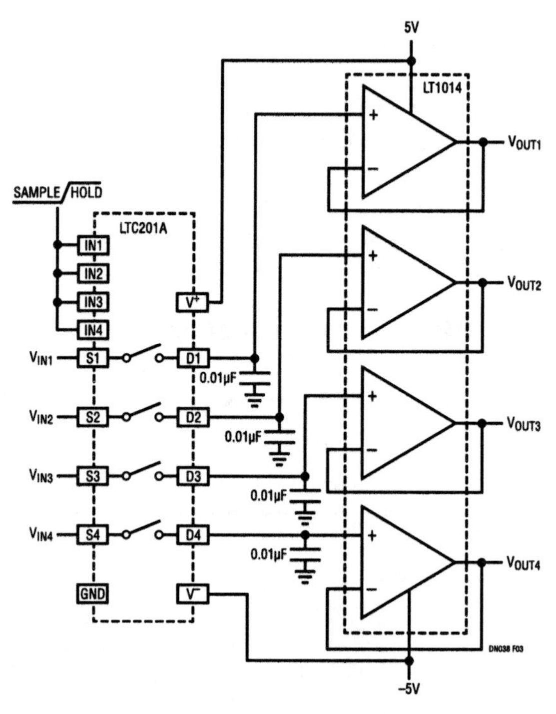

Figure 371.3 • Quad 12-Bit Sample and Hold

12-bit 8-channel data acquisition system interfaces to IBM PC serial port

372

Guy Hoover William Rempfer

IBM PCs collect analog data

IBM PC compatibles can be found just about everywhere. In those instances where a PC is not already in place, battery-operated portables are readily available. This makes the PC a good choice for controlling a data acquisition system. Typically, such data acquisition systems have been expensive. Using dedicated A/D cards or IEEE-488 controllers and instruments, these systems tie up slots in the PC and are not readily transportable from one machine to another. As an alternative, the schematic of Figure 372.1 shows a 12-bit, 8-channel data acquisition system that connects to the serial port of the PC. This system uses an LTC1290, a reference, a handful of other low cost components and requires 12 lines of BASIC to transfer data into the PC. If only ten bits of resolution are required the LTC1290 can be replaced with an LTC1090. Additionally, if the LTC1090 is used, the system can be powered directly from the PC serial port with the option shown.

Two glue chips provide the interface

The control and status lines of the PC serial port are used to send data to and receive data from the LTC1290. Due to incompatible data formats the Rx and Tx lines are not used. The LTC1290 is a 12-bit, 8-channel data acquisition system on a chip. ACLK of the LTC1290 controls the A/D conversion rate while SCLK controls D_{IN} and D_{OUT} data rates. While $\overline{CS}$ is low D_{IN} is clocked into the LTC1290 and D_{OUT} is clocked out in a synchronous full duplex format. While $\overline{CS}$ is high the conversion requested by the last D_{IN} word is performed.

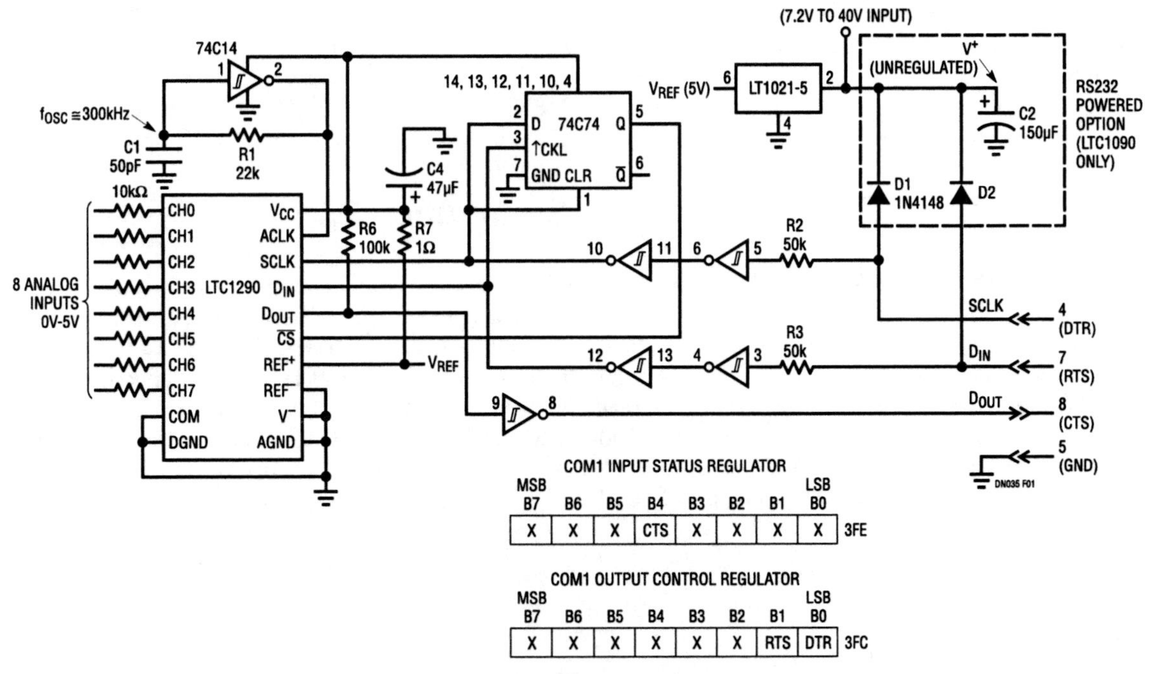

Figure 372.1 • LTC1290 to IBM PC Serial Port Interface

Analog Circuit Design: Design Note Collection. http://dx.doi.org/10.1016/B978-0-12-800001-4.00372-0

```
10   '         LTC1290 TO RS232 IBM PC TRANSFER PROGRAM
20   '                    BY GUY HOOVER
30   '                LINEAR TECHNOLOGY CORP
40   '                       1/4/90
50   '   &H3FC IS THE ADDRESS IN HEX OF THE RS232 OUTPUT CONTROL REGISTER
60   '   &H3FE IS THE ADDRESS IN HEX OF THE RS232 INPUT STATUS REGISTER
66   '    "111101110001"  CH0 \
67   '    "111101110011"  CH1  \
68   '    "111101111001"  CH2   \
69   '    "111101111011"  CH3    \ DIN WORDS FOR CH0-CH7 SINGLE ENDED
70   '    "111101110101"  CH4    / UNIPOLAR, MSB FIRST AND 12 BITS
71   '    "111101110111"  CH5   /
72   '    "111101111101"  CH6  /
73   '    "111101111111"  CH7 /
74  DIN$="111101111111"          'DIN$ IS SENT LSB FIRST.
75                               'THE MSB MUST BE A 1 SO THAT DIN IS NORMALLY HIGH
80                               'THIS  DIN  WORD  CONFIGURES  THE LTC1290 FOR CH7
90                               'WITH RESPECT TO COM, UNIPOLAR, MSB FIRST AND
100                              '12 BITS
110  B=2048                      'B IS SCALE FACTOR FOR DOUT. B=512 FOR LTC1090
120 VOUT=0                       'VOUT IS DECIMAL REPRESENTATION OF LTC1290 DOUT
140 FOR I=1 TO 12                'LOOP TWELVE TIMES
145  OUT &H3FC,(&HFE AND INP(&H3FC))               ' SCLK AND CS GO LOW
150                                                ' DIN IS SHIFTED OUT
160  IF MID$(DIN$,13-I,1) ="0" THEN OUT &H3FC,(&HFD AND INP(&H3FC)) ELSE
     OUT &H3FC,(&H2 OR INP(&H3FC))
180  OUT &H3FC,(&H1 OR INP(&H3FC))                 ' SCLK GOES HIGH
210  IF (INP(&H3FE) AND 16) = 16 THEN D=0 ELSE D=1' READ DOUT
220  VOUT=VOUT+(D*B):B=B/2                         ' SCALE EACH BIT AND SUM BITS
250 NEXT I                                         ' GO THROUGH LOOP AGAIN
260 OUT &H3FC,(&HFD AND INP(&H3FC))                ' DIN GOES LOW
270 OUT &H3FC,(&H2 OR INP(&H3FC))                  ' DIN AND CS GO HIGH
287  'FOR J=1 TO 20: NEXT J                         ' MAKE CS HIGH FOR 52 ACLKS
```

Figure 372.2 • Turbo BASIC Code for LTC1290 to IBM PC Serial Port Interface

A simple RC oscillator is used to generate ACLK. The DTR pin of the PC serial port is used to form SCLK. The DTR signal is also fed into the CLR and D inputs of a 74C74 so that on the first falling SCLK the Q output of the 74C74 drives the $\overline{CS}$ of the LTC1290 low. Between data transfers DTR is held high to charge C2 which provides the unregulated V^+ if the RS232 powered option is used. V^+ is fed into the LT1021 reference which provides a regulated +5V for the LTC1290 and the 74C devices. The RTS pin drives the D_{IN} input of the LTC1290 and the CLK input of the 74C74. During a data transfer, RTS (D_{IN}) changes state only when DTR (SCLK) is low so the 74C74 output ($\overline{CS}$) stays low. After the transfer is completed, RTS is toggled while DTR is high causing the Q output ($\overline{CS}$ of the LTC1290) to go high. D_{OUT} of the LTC1290 goes through an inverter which drives the CTS input of the serial interface. The pull-up resistor on D_{OUT} prevents power consumption in the inverter when D_{OUT} goes into high impedance mode during the conversion.

A few lines of BASIC read the data

The code of Figure 372.2 is written in Turbo BASIC. However, this program will run using GW BASIC at about one-third the transfer rate. The addresses used in this program assume that the interface is connected to COM1 of the PC. The LTC1290 is configured by sending the variable D_{IN}\$ through the RTS line. D_{IN}\$ is a 12-bit string variable which is sent serially LSB first. Bits 11, 10, 9 and 8 are don't cares and bits zero through seven are the actual LTC1290 D_{IN} word as defined in the data sheet. The following loop is executed 12 times. SCLK and $\overline{CS}$ are forced low. D_{IN} is set or reset according to the desired word. SCLK is then set high. D_{OUT} is read one bit at a time and multiplied by a weighting variable B, to produce a variable that ranges from 0 to 4095 (0 to 1023 for the LTC1090). The variable B is initialized to 2048 (512 for the LTC1090) and divided by two after each bit. The last time through the loop SCLK is high and D_{IN} is cycled low then high. This causes $\overline{CS}$ to return high at which time the requested conversion is performed. $\overline{CS}$ must remain high for 52 ACLK cycles, typically 175μs with the RC oscillator shown. This is not a problem except for the fastest of PCs where a simple FOR... NEXT loop as in line 287 can be used to delay execution of the program until the conversion is complete.

Summary

This interface is capable of performing a conversion and shifting the data in 185ms using an XT compatible running at 4.77MHz. Using a 16MHz 386 the same task can be completed in 2.3ms. The code shown is specifically for the IBM PC and compatibles. However, with the proper software the schematic of Figure 372.1 should interface with any RS232 port. For a complete description of the LTC1290 and the LTC1090 please see the desired data sheet.

Auto-zeroing A/D offset voltage

373

Guy Hoover William Rempfer

Introduction

Many A/D converters exhibit low offset errors with large full-scale voltages. However, when the full-scale voltage is decreased the V_{OS}, expressed in LSBs, increases. An A/D converter with 0.5 LSB of offset with a 5V full-scale voltage can have 12.5 LSBs of offset with a 200mV full-scale voltage. With the LTC1090 family of data acquisition systems and a few external components it is now possible to reduce the V_{OS} to only 0.25 LSB even with only a 200mV full-scale voltage. This allows a user to digitize signals from low voltage transducers without the need for a gain stage.

Circuit description

The LTC1090 is a 10-bit data acquisition system with an eight channel multiplexer. The channel to be read is software selectable and all channels can be referred to the COM pin. In the circuit of Figure 373.1, CH0 is used to servo the COM pin giving the use of a seven channel, offset corrected data acquisition system.

Figure 373.2 shows how the processor servos the COM pin to eliminate the A/D offset. CH0 is set to a 0.5 LSB voltage. The COM pin is servoed (by the pulse-width modulated signal on port C2) so that the CH0 reading dithers between 0 and 1 LSB. The 100μF filters the PWM signal at the COM pin. Motorola MC68HC05 code is available from LTC to correct the LTC1090 offset and read the remaining seven channels. This algorithm will work in either unipolar or bipolar mode. (Unipolar is shown. For bipolar the 49.9Ω resistor is changed to 100Ω and the decision block is changed to "CH0 ≤ 0?".)

After initializing the processor, the code sends a D_{IN} word to the LTC1090 requesting CH0 to be read with respect to COM. The next D_{IN} word that is sent will set up the A/D for the desired channel to be read while the CH0 data previously requested is shifted into the processor. If the CH0 D_{OUT} is 0

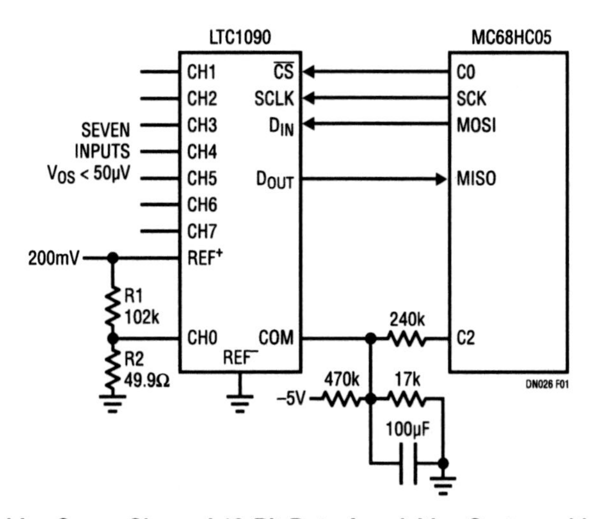

Figure 373.1 • Circuit Provides Seven Channel 10-Bit Data Acquisition System with Less Than 50μV of Offset

Analog Circuit Design: Design Note Collection. http://dx.doi.org/10.1016/B978-0-12-800001-4.00373-2

the C2 is cleared. If the CH0 D_{OUT} is greater than 0 then C2 is set. Another D_{IN} word requesting CH0 data is sent and the D_{OUT} data from the previously requested channel is read into the processor.

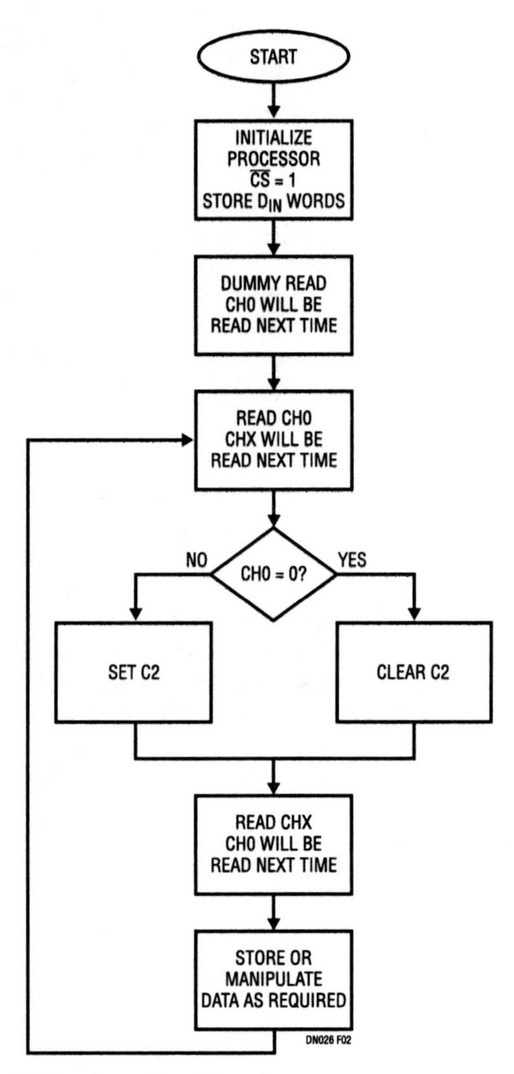

Figure 373.2 • Auto-Zero Flowchart

As can be seen from the LTC1090 data sheet the linearity and full-scale errors with a 200mV full-scale voltage are still within 0.5 LSB. To fully take advantage of the reduced offset of the auto-zero circuit the noise of the LTC1090 must be reduced. This can be done by averaging the data with the processor. Figure 373.3 shows a dynamic cross plot of the output data near half-scale after 64 averages. The top trace is the B9 transition of the LTC1090 while the bottom trace is a binary weight summation of B0 and B1. The horizontal scale is 1 LSB per major division. The averaged noise is much less than 1 LSB.

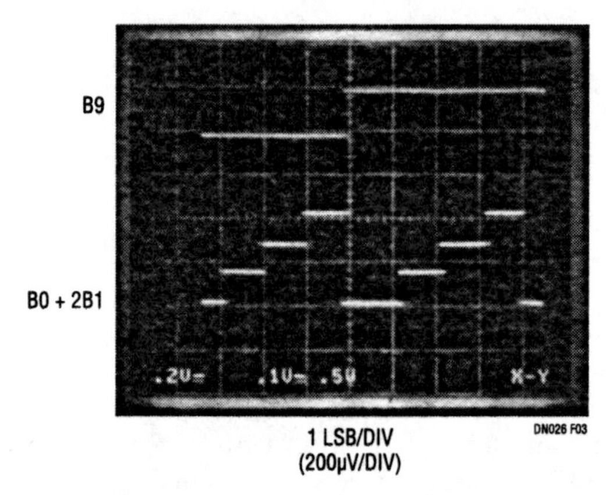

Figure 373.3 • Dynamic Cross Plot Shows Excellent LTC1090 Performance with Only 200mV Full Scale

Complex data acquisition system uses few components

374

Richard Markell

Introduction

Sophisticated filter system designs frequently demand expensive printed circuit boards chock-full of operational amplifiers and precision capacitors. Digital filters require fewer but more expensive devices and a lot of software. However, advances in switched capacitor filters have made the design of elegant filter systems cheaper, easier and much smaller. The system shown in block diagram form in Figure 374.1 is a good example. It is a typical system for filtering transducer signals. Its input is a DC-to-20kHz signal; its output allows signals to be analyzed in three frequency bands.

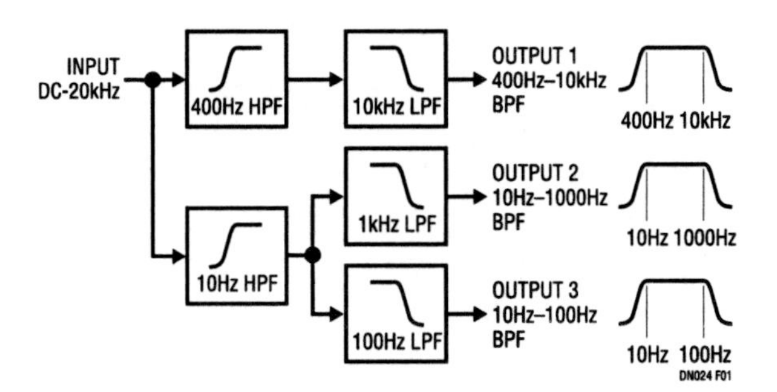

Figure 374.1 • Filter System Block Diagram

Implementation

A system implemented using switched capacitor filters is shown in schematic form in Figure 374.2. This implementation uses two LTC1064 quad switched capacitor building blocks and one LTC1062 5th order Butterworth lowpass filter. The system requires the use of one operational amplifier, an LT1007.

Filter design specifications and test results

Filter 1 – a 400Hz-to-10kHz bandpass filter, with passband ripple of 1dB and passband noise of $200\mu V_{RMS}$, Figure 374.3.

Filter 2 – a 10Hz-to-100Hz bandpass filter, with passband ripple of 1dB and passband noise of $500\mu V_{RMS}$, Figure 374.4.

Filter 3 – a 10Hz-to-1kHz bandpass filter, with passband ripple of 1dB and passband noise of $390\mu V_{RMS}$, Figure 374.5.

These wideband filters are made by cascading 4th order elliptic lowpass and highpass filters. The single exception is the 5th order Butterworth lowpass filter used in the 400Hz-to-10kHz section.

System considerations

The LTC1064 quad switched capacitor filters used are building blocks capable of implementing up to 8th order filters. One LTC1064 implements both a 4-pole elliptic 400Hz highpass filter and a 4-pole elliptic 1kHz lowpass filter. The other LTC1064 implements a 4-pole elliptic 100Hz lowpass filter and a 4-pole elliptic 10Hz highpass filter. The LTC1062 is a 5-pole Butterworth lowpass filter set at 10kHz.

Resistors R_{11A} to R_{H2A} implement the 400Hz elliptic highpass filter in Device A. The 1kHz elliptic lowpass filter in Device A is implemented by R_{13A} to R_{44A}. Resistors R_{11B1} to R_{42B} implement the 10Hz elliptic highpass filter in Device B. The 100Hz elliptic lowpass filter in Device B is implemented by R_{13B} through R_{44B}. The LTC1062 is hardware programmed for 10kHz by R_{50} and C_{50}.

The 8th order LTC1064 devices allow the use of two sections in the 100:1 clock-to-center frequency mode and two

Analog Circuit Design: Design Note Collection. http://dx.doi.org/10.1016/B978-0-12-800001-4.00374-4

sections in the 50:1 mode. (Resistor programming can then be used to further extend the clock-to-center frequency range to 25:1 for two sections and 250:1 for the other two sections.) This allows decade-wide bandpass filters to be built using only one LTC1064 at one clock frequency.

Conclusion

This is only one use of the new switched capacitor building blocks, the LTC1064 family of quad switched capacitor filters. These filters have wide flexibility. For example, the 10Hz-to-100Hz filter could be used at 20Hz-to-200Hz simply by doubling the clock, which sets the filter frequency. Similarly, bands of interest could be inspected by sweeping the clock. The devices work with center frequencies as high as 100kHz in circuits with similar simplicity.

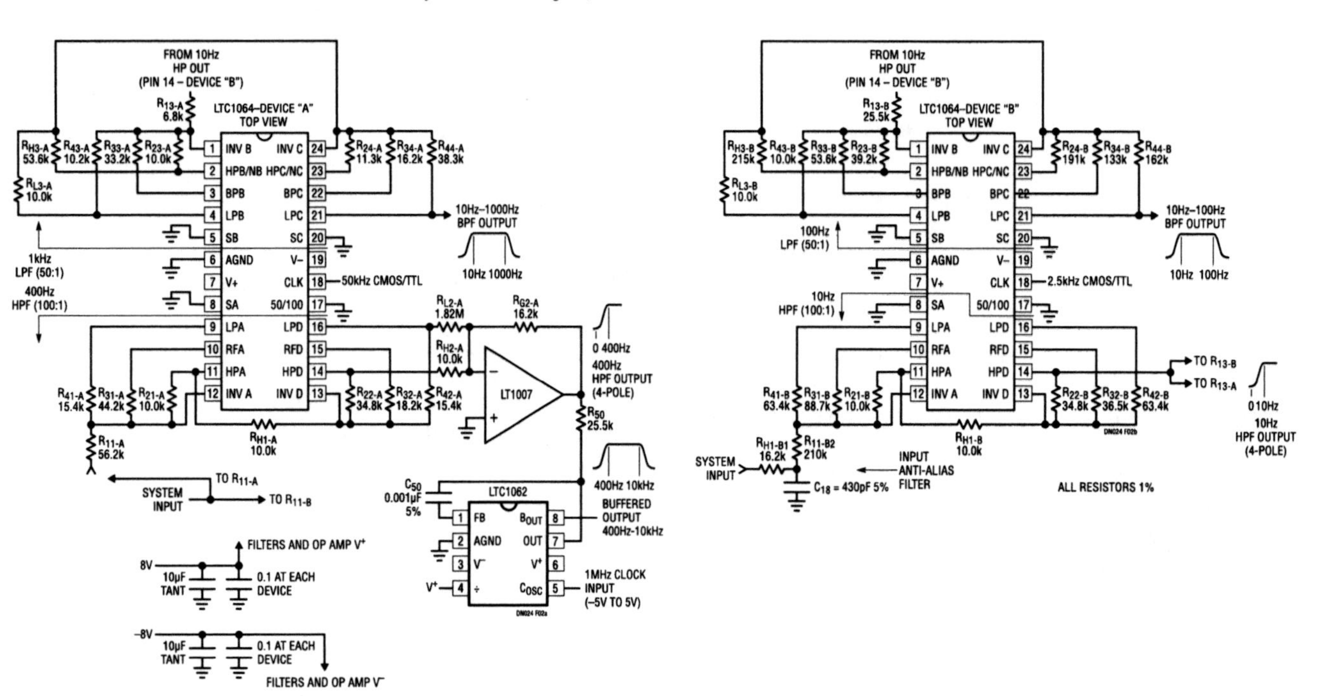

Figure 374.2 • Schematic Diagram

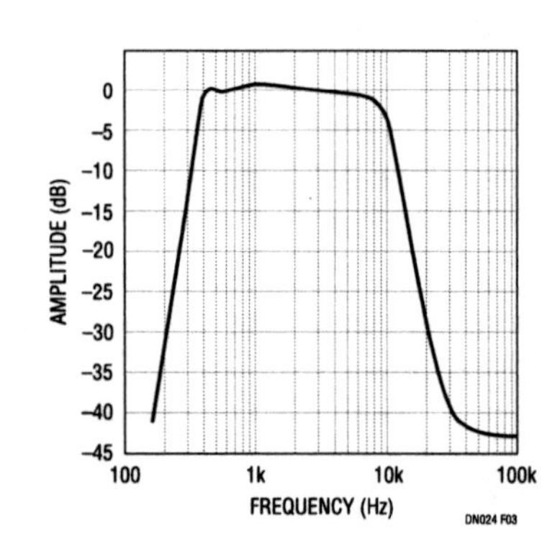

Figure 374.3 • 400Hz–10kHz BP Filter Amplitude Response

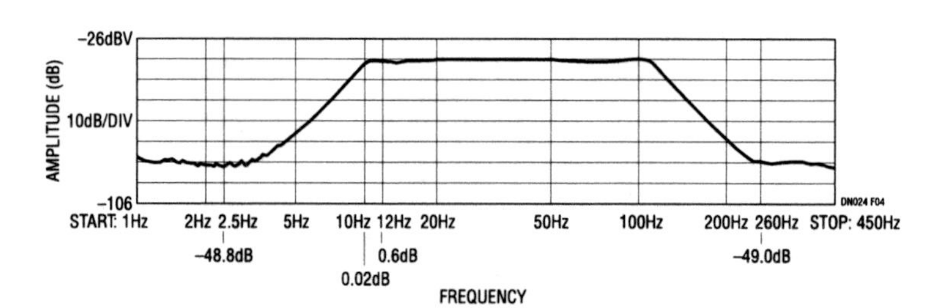

Figure 374.4 • 10Hz–100Hz BPF Amplitude Response

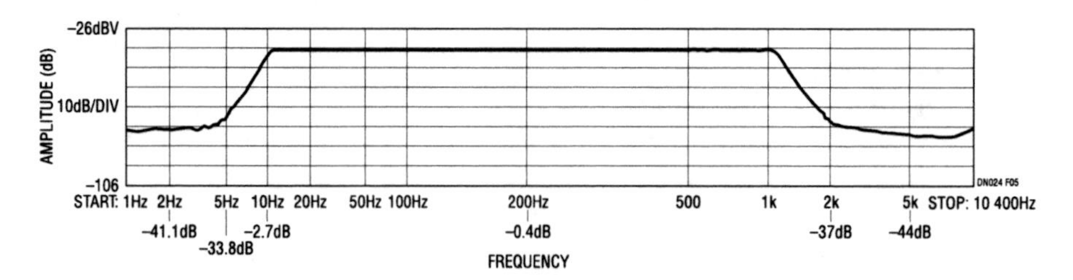

Figure 374.5 • 10Hz–1000Hz BPF Amplitude Response

A two wire isolated and powered 10-bit data acquisition system

375

Guy Hoover William Rempfer

Introduction

For reasons of safety or to eliminate error producing ground loops, it is often necessary to provide electrical isolation between measurement points and the microprocessor. Unfortunately, the isolated side of this measurement system must still be provided with power. One alternative is to power the isolated side of the circuit with batteries. This solution works if power consumption is low, environmental conditions are mild and the batteries are easily accessible. If these conditions are not met a separate isolated supply may be constructed. This can be both difficult and expensive. This design note describes a transformer isolated system in which one small pulse transformer provides both power and a data path.

The circuit of Figure 375.1 is a 10-bit data acquisition system with 700V of isolation. The circuit takes advantage of the serial architecture of the LTC1092 which allows data and power to be transmitted using only one transformer. A 10-bit conversion can be completed and the data transferred to the microprocessor in 100μs. Using standard ribbon cable the isolated side of this circuit has been remotely located as much as 50 feet from the transformer without affecting circuit performance.

Circuit description

In Figure 375.1, a 4μs wide CS pulse clears the 74HC164 shift registers which will hold the D_{OUT} word of the LTC1092. Additionally, the CS signal sends a 15V pulse through the transformer which charges the 1μF capacitor. The CS pulse width must be in the 2–6μs range for the transformer shown, a small Pulse Engineering model. A pulse more than 6μs will saturate the transformer while a pulse width of less than 2μs will not transfer enough energy through the transformer to keep the isolated supply from drooping during the conversion. The CS pulse can be generated with software or hardware. The LT1021-5 produces a regulated 5V at its output once the 1μF capacitor is charged to approximately 7.2V. The LT1021-5 regulated output powers the isolated side of the circuit. Initially several CS pulses may be required to charge the 1μF capacitor to 7.2V. Once charged however, only one CS pulse per cycle is required to keep the isolated supply from drooping as long as the cycle is repeated every 100μs. The 15V CS signal is also attenuated and delayed. This signal is used to reset the clock circuit and begin the conversion of the LTC1092. The delay is required to allow the transformer flyback to die out before transmitting the D_{OUT} word of the LTC192 across it.

The clock circuit is a simple oscillator that is gated by a combination of the CS signal and the 74HC161 counter so that for each CS signal the clock circuit generates 12 pulses and then is gated off. These 12 pulses are used to perform the A/D conversion and shift the D_{OUT} word of the LTC1092 into the 74HC164 shift registers where the data can be acquired by the microprocessor.

The D_{OUT} serial data of the LTC1092 is encoded with the clock, differentiated and sent across the transformer. The encoding circuitry pulse-width modulates the LTC1092 output. The encoding circuitry uses two one shots to combine the data and the clock. For each negative going clock edge a positive pulse is produced at the output of the encoding circuitry. A wide pulse at the output of the encoding circuitry represents a logical 1 and a narrow pulse represents a logical 0 as shown in the timing diagram of Figure 375.2. The Schottky diodes on the output of the 74HC04 capacitor driver are to protect the driver from damage caused by the initial 15V pulse and the resulting flyback.

The differentiated spikes from the transformer are "integrated" by the Schmitt inverters and the 74HC74. Again, Schottky diodes as well as current limiting resistors are used to protect the gates from transformer excursions beyond the supplies at their inputs. The encoded data is decoded by one half of the 74HC221 which reconstructs the clock. The 74HC164s convert the data to parallel format.

The 10kΩ pull-up resistor forces the output of the LTC1092 high when the A/D is in the high impedance state. When the A/D output becomes active a start bit (logic 0) is clocked out.

Analog Circuit Design: Design Note Collection. http://dx.doi.org/10.1016/B978-0-12-800001-4.00375-6

When this initial high to low transition is clocked into bit two of the second 74HC164 it triggers the one shot in the second half of the 74HC221 which provides a load data pulse for the microprocessor.

Summary

The LTC1092 with its simple serial interface is ideally suited for this transformer isolation application. It requires only two isolated lines to transmit 10 bits of data. Additionally, during the time when data is not being transmitted, power for the A/D can be sent through these lines. The circuit provides 700V of isolation and can perform a conversion and shift the data in 100μs. The isolated portion of the circuit can be remotely located up to 50 feet away using ordinary ribbon cable. Possible applications for this circuit are PC-based measurement systems, medical instrumentation, automotive or industrial control loops and other areas where ground loops or large common mode voltages are present.

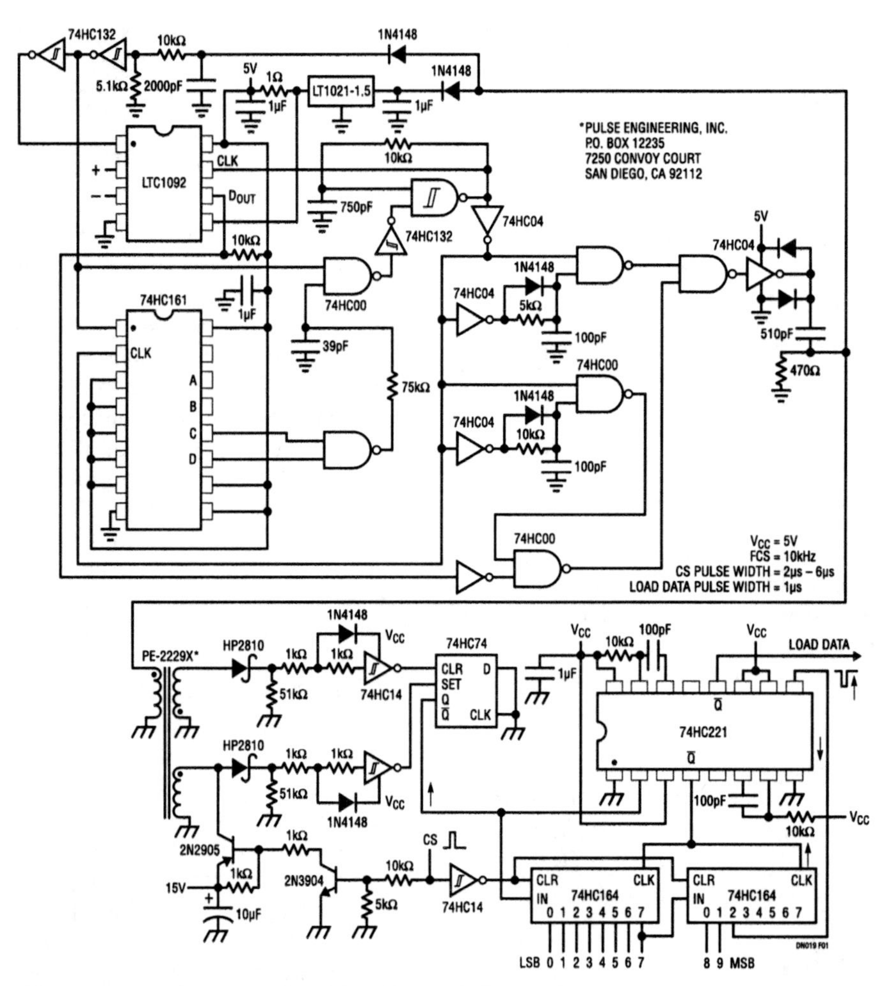

Figure 375.1 • Power and 10-Bit A/D Result Transmitted over Two Isolated Lines

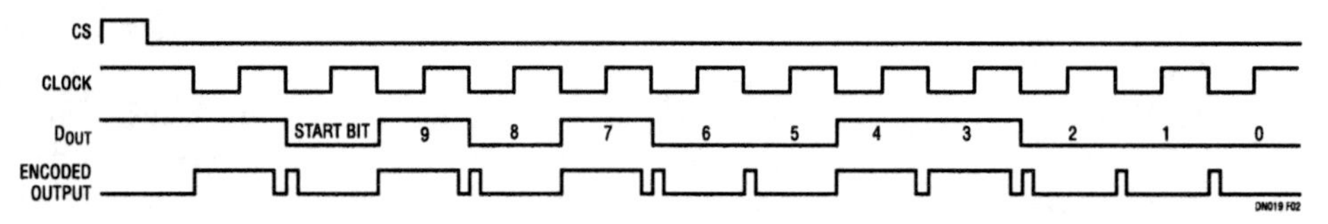

Figure 375.2 • Timing Diagram Shows Pulse Width Coding Technique

Closed loop control with data acquisition systems

376

Guy Hoover William Rempfer

Introduction

The use of microprocessors in process control loops is quite common. A processor-based control loop requires special design considerations as compared to traditional analog loops. Often a single centrally located processor will be used to control several remotely located processes. The outputs of the remote process sensors can be digitized at the sensor location and then be transmitted to the central processor. Unfortunately, transmitting digital signals typically requires one wire for each bit of resolution and requires expensive cabling. Alternatively, the sensor output can be transmitted as an analog signal to the central processor area for digitization. However, transmitting analog signals over distances can introduce errors because of noise and voltage drops in the wires.

The solution to these control loop problems can be found in the LTC1090 series of data acquisition systems. As can be seen in the schematic of Figure 376.2, ten bits of data can be digitized remotely and sent to the processor with only three wires plus ground. The single supply capability and the low DC current drain (1mA typ.) also simplify remote location. The LTC1090 series provides the user with blocks of one, two, six, or eight 10-bit channels which can be chosen according to how many sensors are located in each remote site.

The LTC1090 series is ideally suited for such process control loop applications as position control, temperature control, container filling and tension control.

Circuit description

The circuit of Figure 376.2 is a container filling control loop which has a resolution of .03 pounds with a 30 pound full scale. It was designed to implement an automatic filling station for the model train shown in Figure 376.1. When S1 is closed the MC68HC05 processor reads the LTC1092. If the weight is below the preprogrammed limit in the processor then the motor drive line which controls the pump is turned on. The LTC1092 is continually read by the processor as the truck is filled, until the limit is reached. The motor drive line is then shut off. The limit may be derived in a number of ways. A fixed limit will result in filling to an absolute weight, while relative or tare weight filling can be implemented when the measured empty weight is used in the calculation of the limit. Code for this application is available upon request from Linear Technology Corporation.

The NCI 3220 strain gauge used in this circuit has a linearity specification of .04% which makes it a good match for the .05% linearity of the LTC1092. However, the offset and full scale of the strain gauge are only guaranteed to 10% so trims are required. The circuit is run ratiometrically so an absolute reference is not required. The strain gauge output is amplified by one-half of an LT1013 with the other half being used to buffer the resistor divider that is used for the

Figure 376.1 • A Typical Application. Automatic Filling at a Railroad Siding

Analog Circuit Design: Design Note Collection. http://dx.doi.org/10.1016/B978-0-12-800001-4.00376-8

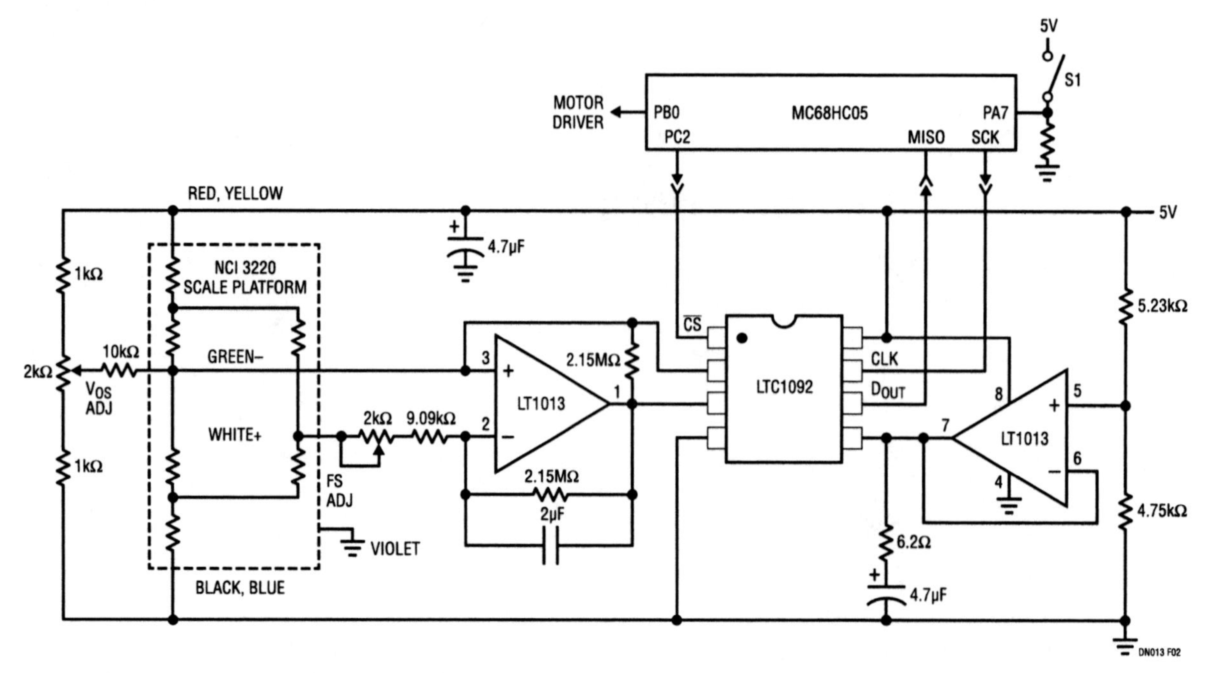

Figure 376.2 • This Circuit Determines Small Weight Changes, Permitting Accurate Filling. Using the Appropriate Transducer, Containers May Range from Perfume Bottles to Railroad Cars

LTC1092's V_{REF} pin. Only one op amp is necessary to amplify the strain gauge output because of the differential inputs of the LTC1092. The 2.15MΩ resistor from pin 1 to 3 of the LT1013 is to balance the load on the strain gauge bridge. With the strain gauge zeroed, both inputs on the LTC1092 are at about 2.5V. As weight is added, the output of the LT1013 into the minus input of the LTC1092 swings toward ground. At the 30 pound full scale the output of the LT1013 is about 100mV above ground which results in a total swing of about 2.4V. The 2µF mylar capacitor filters the LT1013 output eliminating the effects of vibration caused by filling the train car. (As the train car nears the full point, vibration-induced noise can cause the processor to stop the filling too soon.) It is important that the processor monitors the filling process in a timely fashion to prevent overflow. The setup shown relied on a slow fill rate to solve the last problem but with the processor in the loop it is possible to give the fill algorithm some intelligence so that it would run at a high speed to begin with and then run at a slower speed at some preset limit until the final limit is reached.

To calibrate the circuit, offset is first adjusted with no weight on the platform. Next, a known weight near full scale is used to adjust the gain. Once calibrated, variations in the supply voltage within the voltage limits of the LTC1092 should not cause additional errors.

Summary

The LTC1090 series is well suited for use in closed loop control systems. Their low supply current and serial interface make them easy to locate remotely. With a total unadjusted error of .05% over temperature the LTC1090 series is a good match to a wide variety of sensors. The differential inputs of the LTC1090 series can also simplify circuit design while a choice of one, two, six or eight inputs gives the user just the level of complexity that is needed.

Electrically isolating data acquisition systems

377

Guy Hoover William Rempfer

Introduction

In data acquisition systems it is often necessary to electrically isolate the measurement points from the system controller. Reasons for the electrical isolation include the following: to allow floating measurements at high voltages; for safety, to reduce the danger of electrical shock, as might occur in medical applications; and to eliminate ground loops between measurement points and the system controller which can cause errors.

The data transmitted over the isolated lines can be either analog or digital. Analog signals have poor noise immunity and one isolator is required for each signal point. Traditionally, the highly noise immune, digitally encoded signals required many isolated lines for each channel. Now, with the LTC1090 family of serial data acquisition systems, it is possible to transmit eight channels of data with only four isolated lines. Each additional eight channels requires only one additional isolated line.

Both opto-isolators and pulse transformers could be used to isolate the signals. However, since opto-isolators tend to be smaller and less expensive than pulse transformers, they will be the only type of device considered here.

The circuit to be demonstrated is an eight channel data acquisition system with 500V of isolation that uses the LTC1090 and four opto-isolators. With the addition of another opto-isolator, the circuit can be battery operated, drawing only 50μA while taking a reading once every two seconds.

The number of channels can be increased to 16, 24, 32, etc., with one additional opto-isolator used to increase the number of channels in multiples of eight. Up to 24 channels can be powered directly by the LT1021.

Circuit description

The LT1021 powers the analog circuitry and provides an accurate reference. A 1Ω resistor isolates the reference from power supply transients.

The 4N28s in Figure 377.1 are very commonly used opto-isolators. They provide only 500V of isolation, however. If more isolation is desired, up to 2500V of isolation can be obtained by using 4N25s with no other circuit modifications.

PNP transistors were chosen to drive the opto-isolators to optimize signal fall time and clock rate. D_{OUT} of the LTC1090 is transmitted on the falling edge of SCLK. Data is clocked into the processor on the rising edge of SCLK. It is therefore necessary that the falling edge of SCLK have as little delay as possible through the opto-isolator. This insures that D_{OUT} can be output by the LTC1090 in time to be captured by the processor on the rising edge of SCLK. NPNs could be used at slower data rates or if burning more current is not objectionable.

The current limiting resistors in the collectors of the opto drivers are chosen with the current transfer ratio (CTR) of the opto-isolator in mind. The output transistor of the opto-isolator must have enough base current to drive the desired load. The base resistor on the opto output transistor is there to decrease the turn off time of the opto-isolator.

The code written for the Motorola 68HC05 processor is available from Linear Technology. The code powers up the circuit, allows 6.0ms for the 10μF cap to charge to its final value, reads nine channels in 12ms, and then powers down until the next set of readings is required. Nine channel readings are required because, when the LTC1090 is powered up, a dummy read is necessary to initialize the device. The frequency of the ACLK generated by the 74C14 is not too critical. The software must provide a delay of at least two ACLK cycles between the time $\overline{CS}$ goes low and the first SCLK edge is generated. Another delay of 44 ACLKs is required during the time that $\overline{CS}$ is high between data transfers.

Alternatives

The circuit demonstrated in Figure 377.1 is capable of transferring serial data at about a 15kHz rate, which is fast enough for many applications. For higher data transfer rates, pulse

Analog Circuit Design: Design Note Collection. http://dx.doi.org/10.1016/B978-0-12-800001-4.00377-X

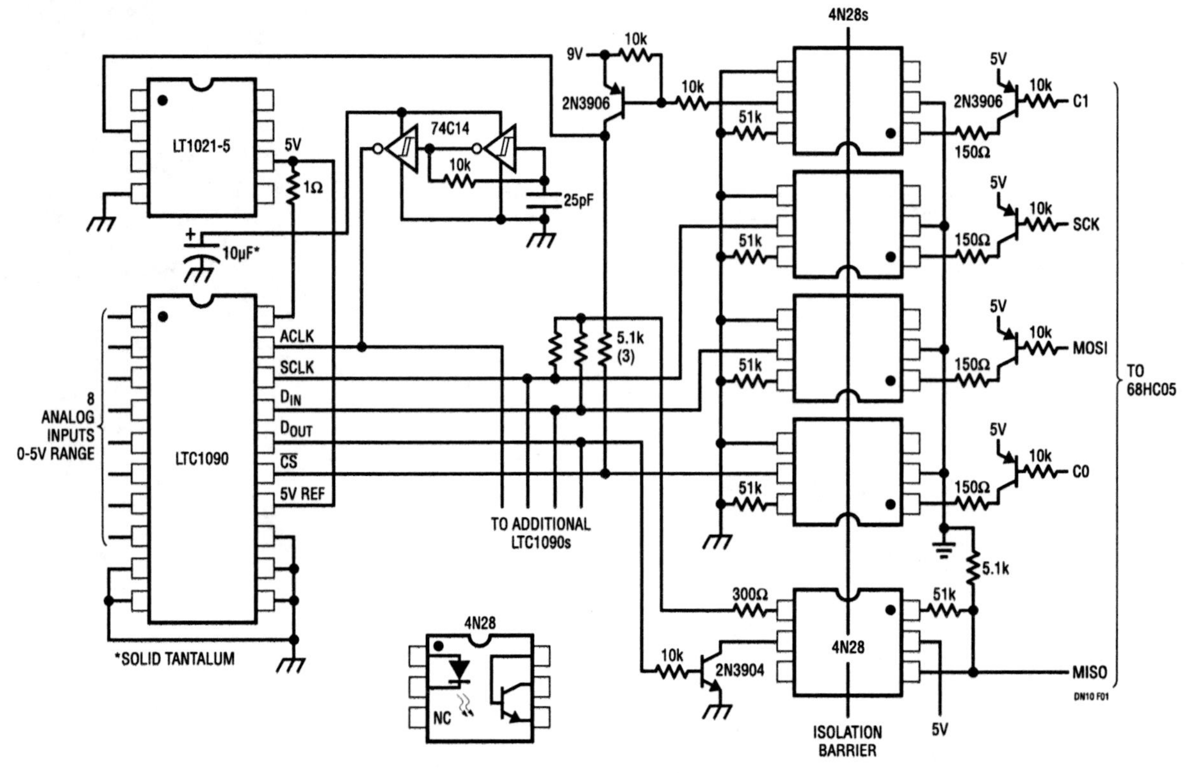

Figure 377.1 • Micropower, 500V Opto-Isolated, Multichannel, 10-Bit Data Acquisition System Is Accessed Once Every Two Seconds

transformers from companies such as Sprague or Pulse Engineering, or high speed opto-isolators such as Hewlett Packard's HCPL2200 are available which can transmit data at the full 1MHz rate that the LTC1090 is capable of.

Summary

The LTC1090 with its serial architecture is ideally suited for isolation applications. It requires only four isolated lines for eight channels of data and can be expanded by adding only one additional isolated line for each additional eight channels of data required. Possible applications for this and similar circuits are PC-based measurement systems, medical instrumentation and other applications where large common mode voltages or ground loops exist.

Temperature measurement using data acquisition systems

378

William Rempfer Guy Hoover

Introduction

Accurate temperature measurement is a difficult and very common problem. Whether recording a temperature, regulating a temperature or modifying a process to accommodate a temperature, the LTC1090 family of data acquisition systems can provide an important link in the chain between the blast furnace temperature and the microcontroller. Features of the LTC1090 family can make temperature measurement easier, cheaper and more accurate.

High DC input resistance and reduced span operation allow direct connection to many standard temperature sensors. Multiplexer options allow one chip to measure up to eight channels of temperature information. Single supply operation, modest power requirements (~5mW) and serial interfaces make remote location possible. Switching power on and off lowers power consumption (560μW) even more for battery applications. Finally, because few sensors have accuracies as good as 0.1%, the 10-bit resolution and 0.05% accuracy of the LTC1090 family are just right for most temperature sensing applications.

Thermocouple systems

The circuit of Figure 378.1 measures exhaust gas temperature in a furnace. The 10-bit LTC1091A gives 0.5°C resolution over a 0°C to 500°C range. The LTC1052 amplifies and filters the thermocouple signal, the LT1025A provides cold junction compensation and the LT1019A provides an accurate reference. The J type thermocouple characteristic is linearized digitally inside the MCU. Linear interpolation between known temperature points spaced 30°C apart introduces less than 0.1°C error. The code for linearizing is available from LTC. The 1024 steps provided by the LTC1091 (24 more than the required 1000) insure 0.5°C resolution even with the thermocouple curvature.

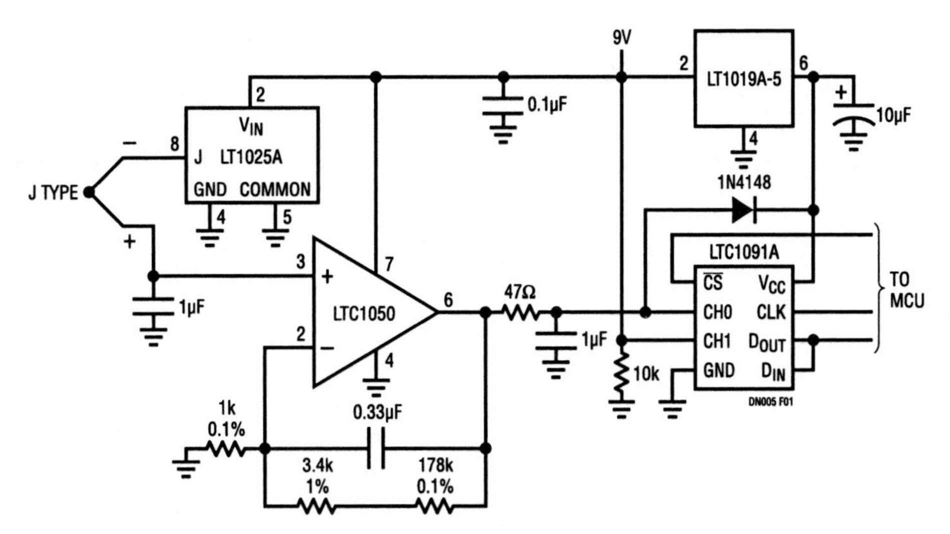

Figure 378.1 • 0°C–500°C Furnace Exhaust Gas Temperature Monitor with Low Supply Detection

Analog Circuit Design: Design Note Collection. http://dx.doi.org/10.1016/B978-0-12-800001-4.00378-1

Offset error is dominated by the LT1025 cold junction compensator which introduces 0.5°C maximum. Gain error is 0.75°C max because of the 0.1% gain resistors and to a lesser extent the output voltage tolerance of the LT1019A and the gain error of the LTC1091A. It may be reduced by trimming the LT1019A or gain resistors. The LTC1091A keeps linearity better than 0.25°C. The LTC1050's 5μV offset contributes negligible error (0.1°C or less). Combined errors are typically 0.5°C or less. These errors do not include the thermocouple itself. In practice, connection and wire errors of 0.5°C to 1°C are not uncommon. With care, these errors can be kept below 0.5°C.

The 20k/10k divider on CH1 of the LTC1091 provides low supply voltage detection (the LT1019A reference requires a minimum supply of 6.5V to maintain accuracy). Remote location is easy, with data transferred from the MCU to the LTC1091 via the 3 wire serial port.

Thermilinear networks

Figure 378.2 shows an eight channel 0°C to 100°C temperature measurement system with 0.1°C resolution. The high DC input resistance and adjustable span of the LTC1090 allow it to measure the outputs of the YSI thermilinear components directly. Accuracy is limited by the sensor repeatability and precision resistors to 0.25°C.

Sensor input voltage (V_{IN}), not critical because of ratiometric operation, is set to around 1.5V to minimize self heating. The zero scale (COM pin) and full scale (REF^+ pin) of the LTC1090 are set by the precision resistor string to directly digitize the roughly 0.2V to 1V sensor output. The LT1006 buffers the 10kΩ reference resistance of the LTC1090. 0°C and 100°C correspond to unipolar output codes of 0 and 1000 (decimal), respectively with an overrange of 102.3°C.

Thermistors

A thermistor is a cheaper alternative to thermilinear components in narrower temperature range applications. In Figure 378.2, CH7 is being used to digitize the output of a 5kΩ thermistor. The resistor shown linearizes the output voltage around the 30°C point. The output remains linear to 0.1°C over a 20°C to 40°C range but gets nonlinear rapidly outside this range. By correcting for the nonlinearity in software this range can be extended to 0°C to 60°C. Beyond that, the repeatability error of the thermistor increases above 0.2°C making correction difficult.

Silicon sensors

Because of its high DC input impedance and reduced span capability, the LTC1090 family can directly measure the output of most industry standard, silicon temperature sensors, both voltage and current mode. Popular sensors of this type include the LM134 and AD590 (current output) and silicon diodes.

Figure 378.3 shows a simple connection between the LTC1092 and industry standard 1μA/K current output sensors. Resolution is 0.25°C and accuracy is limited by the sensor and resistors. Standard 10mV/K voltage output sensors can also be connected directly to the LTC1092 input in a similar manner.

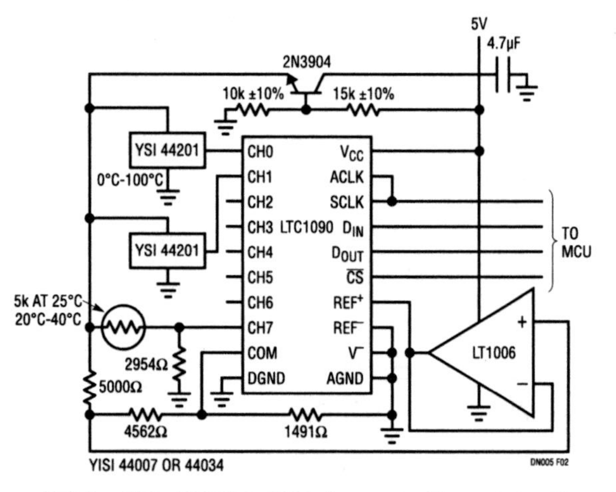

Figure 378.2 • 0°C–100°C 0.25°C Accurate Thermistor-Based Temperature Measurement System

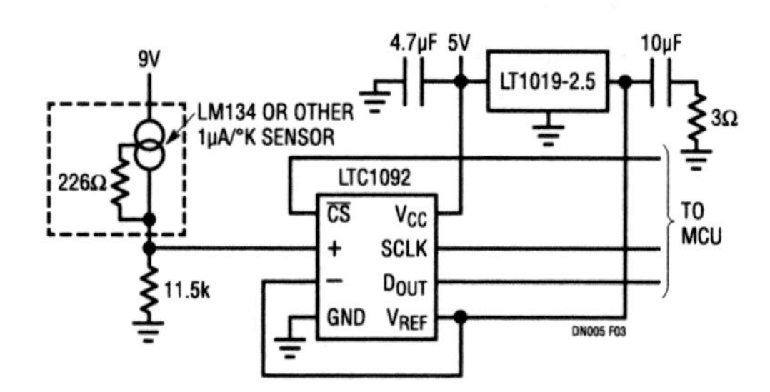

Figure 378.3 • −55°C to +125°C Thermometer Using Current Output Silicon Sensors

379

Sampling of signals for digital filtering and gated measurements

William Rempfer

Introduction

For many signal processing applications a sample-and-hold function is required in a data acquisition system. It is often critical for the processing system to know the exact value of an analog input at an exact time. In DSP applications such as digital filters the usable bandwidth of the system is limited by the Nyquist frequency and the sample-and-hold bandwidth need only be, and is often intentionally limited to, one half the sampling rate. However, another area of application requires infrequently capturing instantaneous values of relatively fast signals, sometimes referred to as gated measurements. In the extreme case of pulse height measurements, only one sample point is required. Here, the sample-and-hold bandwidth should be as high as possible even though the sampling rate is very low.

The LTC1090 excels in both environments. This note shows how the LTC1090 sample and hold can be synchronized to an external event and gives two simple applications: an 8-channel data acquisition system with digital filtering, and the gated measurement of a 1MHz sine wave.

The LTC1090 sample and hold

The LTC1090 provides a sample and hold which is fast, accurate and can be synchronized to an external event. Although the sampling rate is limited by the A/D conversion and data transfer rate to about 30kHz, the signal bandwidth of the sample and hold exceeds 1MHz. The acquisition time is less than 1μs to 0.1% (1LSB). Accuracy is so good, in fact, that it is possible to include all the sample and hold's error contributions (offset, gain, hold step, droop rate, etc.) into the converter specification and still maintain overall system accuracy of ±0.05% (±0.5LSB) over temperature.

Sampling occurs on the failing edge of the last data transfer clock pulse as described in the LTC1090 data sheet. Figure 379.1 shows a typical application which includes circuitry to synchronize sampling to an external sample clock, f_S.

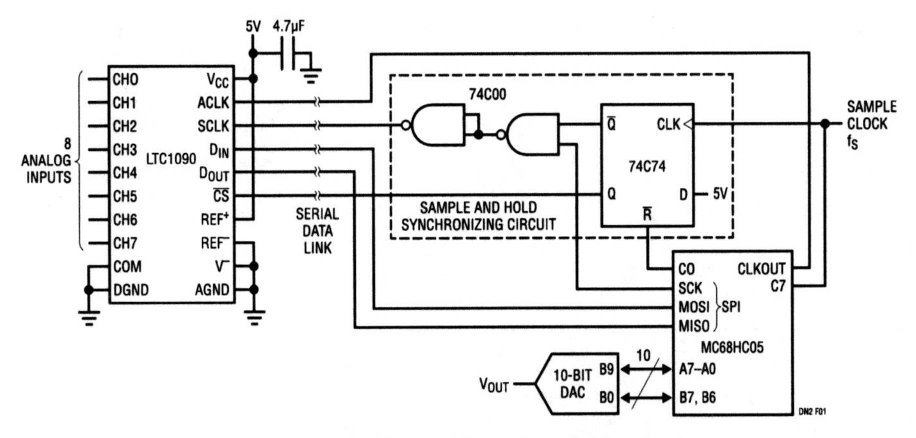

Figure 379.1 • 8-Channel Data Acquisition System Showing Sample-and-Hold Synchronizing Circuitry

Analog Circuit Design: Design Note Collection. http://dx.doi.org/10.1016/B978-0-12-800001-4.00379-3

8-channel data acquisition system with digital filter

The circuit of Figure 379.1 contains an LTC1090 providing multiplexing, sample and hold, A/D conversion and data transfer to the microcontroller (MCU). An MC68HC05C4 is used as the controller (much higher filter performance may be achieved with a dedicated DSP processor). The MCU communicates with the LTC1090 over the serial peripheral interface (SPI), performs the digital filtering algorithm and provides the filtered data on its output port. The DAC provides reconstruction of the filtered waveform for viewing on an oscilloscope or spectrum analyzer. The 74C74 and 74C00 synchronize the sampling of the LTC1090 to the externally applied sample clock, f_S.

In Figure 379.1, the MCU initiates a two byte serial data exchange with the LTC1090. This configures the LTC1090 for the next conversion, simultaneously reads back the previous conversion result and resets the 74C74. The LTC1090 will sample the analog input when the last shift clock (SCLK) pulse fails, so the MCU must end the data transfer by leaving the SCLK in a high state. This inhibits sampling of the selected analog input. When the sample clock, f_S, rises, it clocks the 74C74 which raises the CS and drops the SCLK. This failing SCLK causes the sample to be taken and starts the conversion. After the MCU senses the rising sample clock it waits for the conversion to be completed (44 ACLK cycles) and then initiates another data exchange, preparing the LTC1090 for the next sample. This cycle repeats.

4th order elliptic filter

Using the circuit of Figure 379.1, a 4th order elliptic digital filter was implemented. 10-bit input and output data words and 14-bit coefficients were used with the same coefficients being used for each channel. A direct form II IIR filter was implemented according the following equations:

$$D(n) = [7203 \cdot D(n-1) - 19209 \cdot D(n-2) + 6324 \cdot D(n-3) - 4383 \cdot D(n-4)] \cdot 2^{-14} + X(n)$$

$$Y(n) = [3069 \cdot D(n) + 5505 \cdot D(n-1) + 7824 \cdot D(n-2) + 5504 \cdot D(n-3) + 3066 \cdot D(n-4)] \cdot 2^{-14}$$

where: $X(n)$ = filter input value
$Y(n)$ = filter output value
$D(n)$ = delay node value

The filter frequency response is shown in Figure 379.2. The cutoff frequency is 175Hz, one fourth the sample frequency of 700Hz. The cutoff frequency of the filter can be tuned by varying the frequency of the sample clock.

Because of 68HC05 speed and instruction set limitations, sample rate is limited by the MCU's ability to perform the DSP algorithm. Maximum sample rate was determined to be 700Hz for a single channel filter and 90Hz for eight channels. Using a high performance DSP would allow sample rates approaching the limit of 30kHz for one channel and 3.7kHz for all eight set by the LTC1090. Hopefully, this simple example will encourage the reader to pursue higher order, higher performance applications.

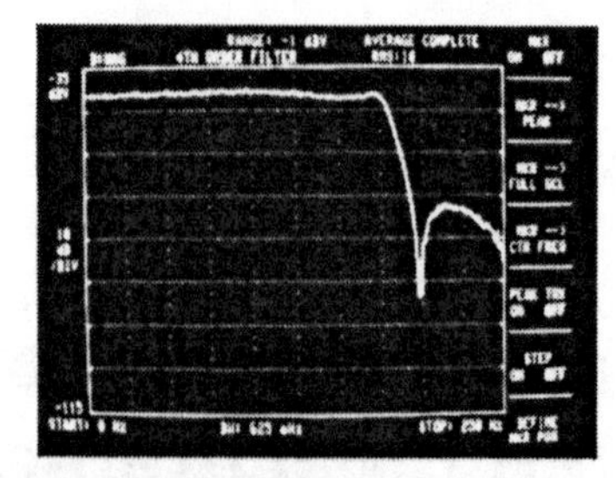

Figure 379.2 • Spectrum of 4th Order Elliptic Digital Filter Used in the Data Acquisition System, f_C = 175Hz

If large amplitude, unwanted AC signals are present on the inputs, a linear filter such as the LTC1062 can be used to remove them and prevent reduction in the dynamic range of the system.

Gated measurements of fast signals

As an example of gated measurements, the circuit of Figure 379.1 was used with no filtering to repetitively sample a $5V_{P-P}$ 1MHz sine wave. The waveform was sampled at 15kHz (approximately one sample every 67 cycles of the 1MHz waveform). A 20ns pulse, triggered off the sample clock, was applied to the z-axis input of a storage scope to illuminate one dot on the CRT per sample. Samples were allowed to accumulate on the storage scope as shown in Figure 379.3. The upper waveform is the sampled input to the LTC1090 and the lower waveform is the sampled output of the DAC. (Remember that the waveforms are not real time: one dot was illuminated only every 67 cycles of the 1MHz sine wave.) With this technique the signal bandwidth of the LTC1090 sample and hold was determined to be 2MHz.

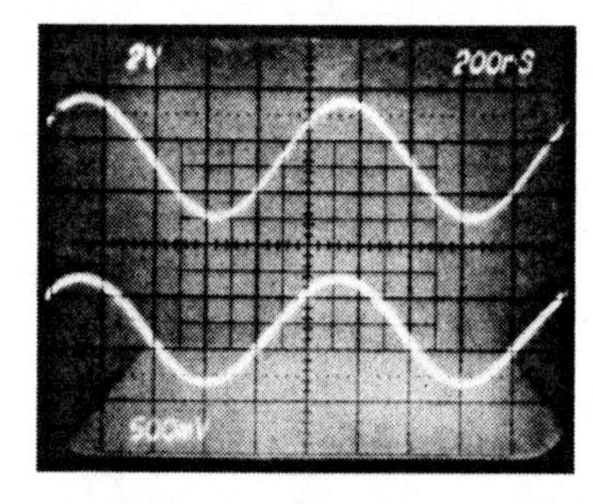

Figure 379.3 • Input and Output Sample Points of a 1MHz Sine Wave Accumulated on a Storage Scope

Using the LTC1090 sample and hold, high speed circuits such as a 1MHz bandwidth AC to DC converter are possible. Because the acquisition time is less than 1µs it is also possible to make a gated measurement of the height of a pulse as narrow as 1µs to 0.1% accuracy.

Data acquisition systems communicate with microprocessors over four wires

380

As board space and semiconductor package pins become more valuable, serial data transfer methods between microprocessors (MPUs) and their peripherals become more and more attractive. Not only does this save lines in the transmission medium, but, because of the savings in package pins, more function can be packed into both the MPU and the peripheral. Users are increasingly able to take advantage of these savings as more MPU manufacturers develop serial ports for their products [1–3]. However, peripherals which are able to communicate with these MPUs must be available in order for users to take full advantage. Also, MPU serial formats are not standardized so not all peripherals can talk to all MPUs.

The LTC1090 family

A new family of 10-bit data acquisition circuits has been developed to communicate over just four wires to the recently developed MPU synchronous serial formats as well as to MPUs which do not have serial ports. These circuits feature software configurable analog circuitry including analog multiplexers, sample and holds, bipolar and unipolar conversion modes. They also have serial ports which can be software configured to communicate with virtually any MPU. Even the lowest grade device features guaranteed ±0.5 LSB linearity over the full operating temperature range. Reduced span operation (down to 200mV), accuracy over a wide temperature range and low power single supply operation make it possible to locate these circuits near remote sensors and transmit digital data back through noisy media to the MPU. Figure 380.1 shows a typical hookup of the LTC1090, the first member of this data acquisition family. For more detail, refer to the 24-page LTC1090 data sheet.

Included are eight analog inputs which can common-mode to both supply rails. Each can be configured for unipolar or bipolar conversions and for single-ended or differential inputs by sending a data input (D_{IN}) word from the MPU to the LTC1090 (Figure 380.1).

Both the power supplies are bypassed to analog ground. The V^- supply allows the device to operate with inputs which swing below ground. In single supply applications it can be tied to ground.

The span of the A/D converter is set by the reference inputs which, in this case, are driven by a 2.5V LT1009 which gives an LSB step size of 2.5mV. However, any reference voltage within the power supply range can be used.

The 4-wire serial interface consists of an active low chip select pin ($\overline{CS}$), a shift clock (SCLK) for synchronizing the data bits, a data input (D_{IN}) and a data output (D_{OUT}). Data is transmitted and received simultaneously (full duplex), minimizing the transfer time required.

The external ACLK input controls the conversion rate and can be tied to SCLK as in Figure 380.1. Alternatively, it can be derived from the MPU system clock (e.g., the 9051 ALE pin) or run asynchronously. When the ACLK pin is driven at 2MHz, the conversion time is 22μs.

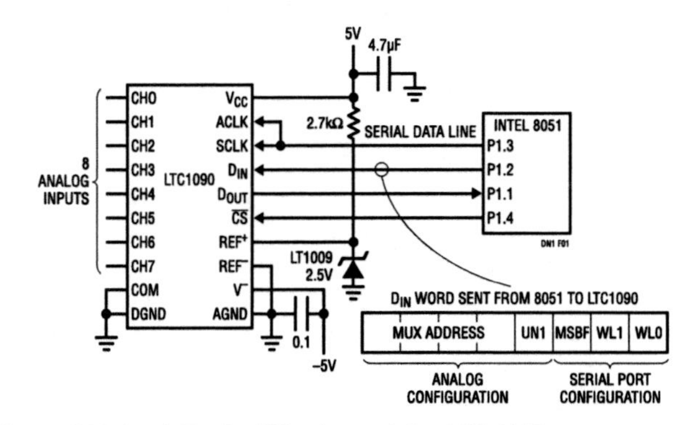

Figure 380.1 • A Typical Hookup of the LTC1090

Advantages of serial communications

The LTC1090 can be located near the sensors and serial data can be transmitted back from remote locations through isolation barriers or through noisy media.

Analog Circuit Design: Design Note Collection. http://dx.doi.org/10.1016/B978-0-12-800001-4.00380-X

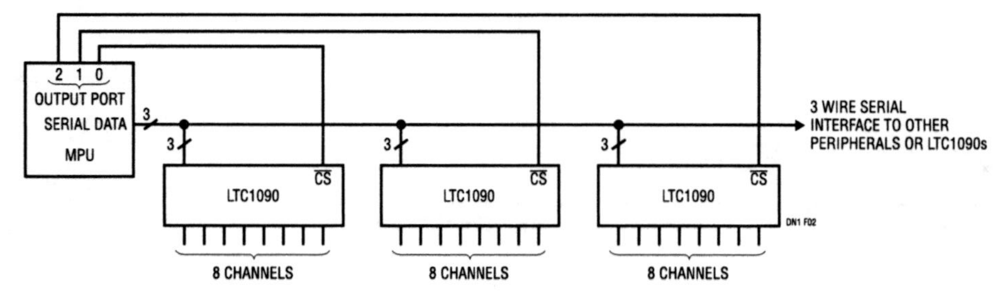

Figure 380.2 • Several LTC1090s Sharing One 3-Wire Serial Interface

Several LTC1090s can share the serial interface and many channels of analog data can be digitized and sent over just a few digital lines (see Figure 380.2). This could, for example, be used to simplify the communications between an instrument and its front panel.

Using fewer pins for communication makes it possible to pack more function into a smaller package. LTC1090 family members are complete systems being offered in packages ranging from 20 pins to 8 pins (e.g., LTC1091).

Speed is usually limited by the MPU

A perceived disadvantage of the serial approach is speed. However, the LTC1090 can transfer a 10-bit A/D result in 10μs when clocked at its maximum rate of 1MHz. With the minimum conversion time of 22μs, throughput rates of 30kHz are possible. In practice, the serial transfer rate is usually limited by the MPU, not the LTC1090. Even so, throughput rates of 20kHz are not uncommon when serial port MPUs are used. For MPUs without serial ports, the transfer time is somewhat longer because the serial signals are generated with software. For example, with the Intel 8051 running at 12MHz, a complete transfer takes 80μs. This makes possible throughput rates of approximately 10kHz.

Talking to serial port MPUs

By accommodating a wide variety of transfer protocols, the LTC1090 is able to talk directly to almost all synchronous serial formats. The last 3 bits of the LTC1090 data input (D_{IN}) word define the serial format. The MSBF bit determines the sequence in which the A/D conversion result is sent to the processor (MSB or LSB first). The two bits WL1 and WL0 define the word length of the LTC1090 data output word. Figure 380.3 shows several popular serial formats and the appropriate D_{IN} word for each. Typically a complete data transfer cycle takes only about 15 lines of processor code.

Talking to MPUs without serial ports

The LTC1090 talks to serial port processors but works equally well with MPUs which do not have serial ports In these cases, $\overline{CS}$, SCLK and D_{IN} are generated with software on three port lines. D_{OUT} is read on a fourth. Figure 380.3 shows the appropriate D_{IN} word for communicating with MPU parallel ports. Figure 380.1 shows a 4-wire interface to the popular Intel 8051. A complete transfer takes only 33 lines of code.

Sharing the serial interface

No matter what processor is used, the serial port can be shared by several LTC1090s or other peripherals (see Figure 380.2). A separate $\overline{CS}$ line for each peripheral determines which is being addressed.

Conclusion

The LTC1090 family provides data acquisition systems which communicate via a simple 4-wire serial interface to virtually any microprocessor. By eliminating the parallel data bus they are able to provide more function in smaller packages, right down to 8-pin DIPs. Because of the serial approach, remote location of the A/D circuitry is possible and digital transmission through noisy media or isolation boundaries is made easier without a great loss in speed.

Hardware and software is available from the factory to interface the LTC1090 to most popular MPUs. The LTC1090 data sheet contains source code for several microprocessors. Further applications assistance is available by calling the factory.

Type of Interface	LTC1090 Data Format	LTC1090 D_{IN} Word: Analog Configuration					MSBF	WL1	WL0
All Parallel Port MPUs	MSB First 10 Bits	X	X	X	X	X	1	0	1
National MICROWIRE	MSB First 12 Bits	X	X	X	X	X	1	1	0
MICROWIRE/PLUS; Morotola SPI	MSB First 16 Bits	X	X	X	X	X	1	1	1
Hitachi Synchronous SCI; TI TMS7000 Serial Port	LSB First 16 Bits	X	X	X	X	X	0	1	1

DN1 F03

Figure 380.3 • The LTC1090 Accommodates Both Parallel and Serial Ports

References

1. Aleaf, Abdul, and Richard, Lazovick, "Microwire/Plus," National Semiconductor, Santa Clara, CA, Wescon '86, Session 21.
2. Derkach, Donald J., "Serial Data Transmission in MCU Systems, " RCA Solid State, Somerville, NJ, Wescon '86, Session 21.
3. Kalinka, Theodore J., "Versatile Serial Peripheral Interface (SPI)," RCA Solid State, Somerville, NJ, Wescon '86, Session 21.

Section 4

Communications Interface Design

Addressable I²C bus buffer provides capacitance buffering, live insertion and nested addressing in 2-wire bus systems

381

John Ziegler

Introduction

In an effort to improve the reliability of large data processing, data storage and communications systems, their subsystems include a growing number of active circuits to monitor parameters such as temperature, fan speed and system voltages. Individual subsystem monitors often communicate with the host system through 2-wire serial busses, such as SMBus or I²C, to a dedicated microcontroller.

As monitoring functions increase in complexity and number, several practical problems arise. First, data bus rise time specifications become difficult to meet. Second, many uninterruptible systems cannot tolerate cycling power whenever a new I/O card is installed. Finally, a device's address is often dictated by the function that it performs. If an existing system already contains a temperature sensor, for example, then inserting a new I/O card with a temperature sensor risks having two devices with the same address. The new LTC4302-1/LTC4302-2, addressable 2-wire bus buffers provide solutions to all of these problems.

Live insertion and removal and capacitance buffering application

The LTC4302 solves bus connect/disconnect problems by creating an active bridge between two physically separate 2-wire busses. The LTC4302's two "input" pins, SDAIN and SCLIN, connect to one 2-wire bus, i.e., a backplane; and its two "output" pins, SDAOUT and SCLOUT, connect to a second 2-wire bus, i.e., an I/O card.

The application shown in Figure 381.1 highlights the live insertion and removal and the capacitance buffering features of the LTC4302-1. Because the LTC4302's SDA and SCL pins default to a high impedance and low capacitance (<10pF) state even when no V_{CC} voltage is applied, an LTC4302 can be inserted onto a live 2-wire bus without corrupting it.

If a staggered connector is available, make CONN the shortest pin and connect a large value resistor from CONN to ground on the I/O card to force the CONN voltage low before it connects to the backplane. This ensures that the

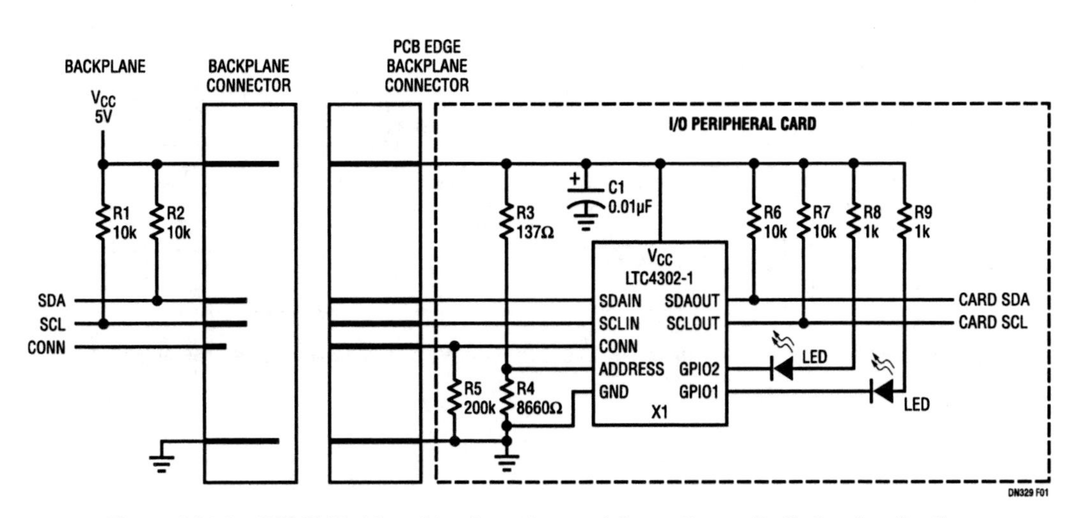

Figure 381.1 • LTC4302-1 in a Live Insertion and Capacitance Buffering Application

Analog Circuit Design: Design Note Collection. http://dx.doi.org/10.1016/B978-0-12-800001-4.00381-1

LTC4302 remains in a high impedance state while SDAIN and SCLIN are making connection during live insertion. With the I/O card firmly connected, drive CONN high to allow masters to communicate with the LTC4302. During live removal, having CONN disconnect first ensures that the LTC4302 enters a high impedance state in a controlled manner before SDAIN and SCLIN disconnect.

Inserting an LTC4302 on the edge of the card isolates the card capacitance from the backplane. As more I/O cards are added to the system, placing an LTC4302 on the edge of each card breaks what would be one large, unmanageable bus into several manageable segments. Moreover, the LTC4302 can be programmed to provide rise time accelerator pull-up currents on all four SDA and SCL pins during rising edges, to further aid in meeting the rise time specification. Figure 381.2 shows the rise time improvement achieved for V_{CC} = 3.3V and bus equivalent capacitances of 50pF and 150pF.

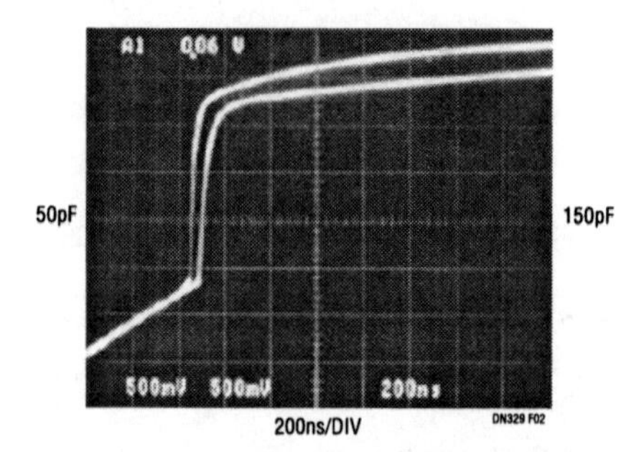

Figure 381.2 • Rise Time Accelerators Reduce Rise Time for C_{BUS} = 50pF, C_{BUS} = 150pF

Nested addressing and 5V to 3.3V level translator application

Figure 381.3 illustrates how the LTC4302 can be used to expand the number of devices in a system by using nested addressing. In this example, each I/O card contains a sensor device having address 1111 111. The LTC4302 isolates the devices on its card from the rest of the system until it is commanded to connect by a master. If masters use the LTC4302s to connect only one I/O card at a time, then each I/O card can have a device with address 1111 111 without any conflicts.

Figure 381.3 also shows the LTC4302-2 providing voltage level translation from the 5V backplane SDA and SCL lines to the 3.3V I/O card lines. The LTC4302-2 functions for voltages ranging from 2.7V to 5.5V on both V_{CC} and V_{CC2}. If either V_{CC} or V_{CC2} supply voltage falls below its UVLO threshold, the LTC4302-2 disconnects the backplane from the card so that the side that is still powered can continue to function.

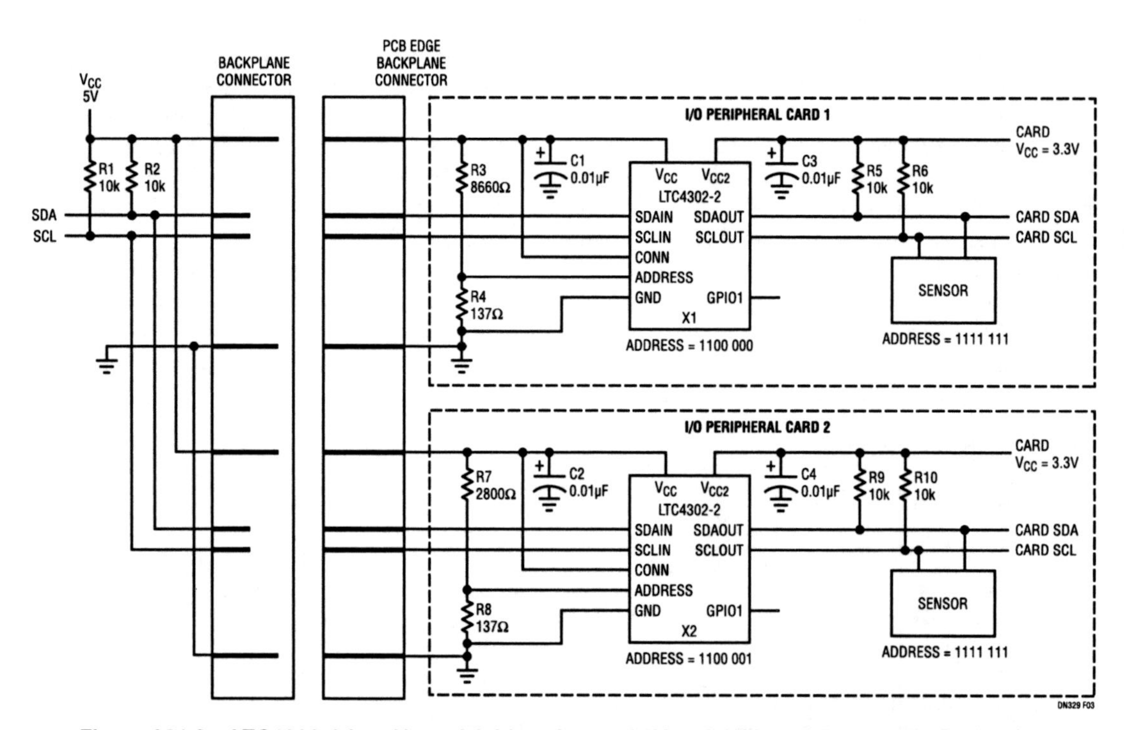

Figure 381.3 • LTC4302-2 in a Nested Addressing and 5V to 3.3V Level Translator Application

Single interface chip controls two smart cards

382

Steven Martin

Introduction

There are considerable challenges to smart card interfacing, including various voltage levels (both input and output) and stringent fault handling requirements. To produce a robust card reading system, designers must comply with extensive and often difficult software as well as hardware standards. Furthermore, there are other complications like in-circuit ESD and pin-to-pin shorts to contend with.

The LTC1955 dual smart card interface provides all of the required power management, control, ESD and fault detection circuitry for two smart cards. Employing a voltage doubling charge pump and two low dropout linear regulators, this device generates two independent levels of either 5V, 3V or 1.8V from a 2.7V to 5.5V input. Both channels have the required pins to support the EMV (Europay, MasterCard, Visa) and the ISO7816 smart card standards. One channel has extra control pins (smart card contact pad locations C4 and C8) to

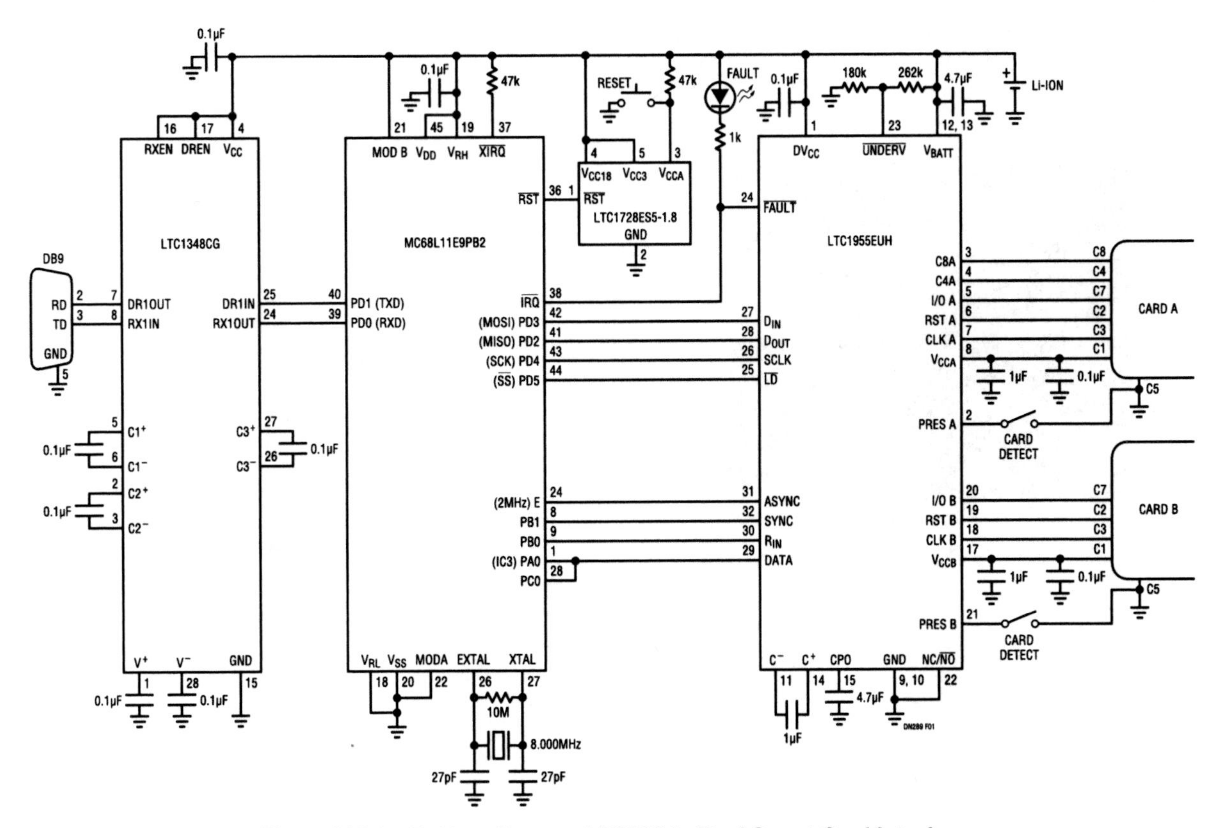

Figure 382.1 • Battery-Powered RS232 to Dual Smart Card Interface

Analog Circuit Design: Design Note Collection. http://dx.doi.org/10.1016/B978-0-12-800001-4.00382-3

support existing memory cards. The entire chip is controlled by a microcontroller-friendly serial interface.

Features

The LTC1955 includes considerable security and functionality, yet remains easy to use. Two independent circuits detect the presence or absence of a smart card. Card insertion is debounced with a 40ms delay to ensure that the contacts are well seated before the card is activated. If a card is removed during a transaction, the LTC1955 automatically deactivates it before its pads leave the connector's contact pins. Figure 382.3 shows the sequencing of the smart card pins during an automatic deactivation.

Providing power to 5V cards from 3V, the charge pump operates in constant frequency mode when heavily loaded and has an autoburst feature for power savings under lightly loaded conditions. The constant frequency operation allows the use of tiny, low profile capacitors. The charge pump is powerful enough to supply both smart cards at rated current requirements.

Internal low dropout linear regulators independently control the voltage of both smart cards. All three smart card classes (1.8V, 3V and 5V) are supported and the smart card signals are shifted to the appropriate level for the cards, independent of the microcontroller supply voltage (which can range from 1.7V to 5.5V).

The data communication pins (I/OX and DATA) are bidirectional and full duplex. This feature allows true acknowledge data to be returned to the microcontroller interface. The bidirectional pins also have special accelerating pull-up sources[1] to ensure fast rise times (see Figure 382.2). These sources are faster than a resistor without dissipating excessive power when the pin is held low. They sense the edge rate on the pin and compare it to a preset limit. If the limit is exceeded, an additional current source is applied to the pin thereby accelerating its rise time. Once the pin reaches its local supply level the acceleration current is disabled. Figure 382.2 shows an example of the data waveforms on the smart card and microcontroller pins.

For the smart card clock pins, special clock divider and synchronization circuitry allows easy interfacing to the microcontroller. Separate clock input pins are available to support either asynchronous smart cards or synchronous memory cards.

Ease of use

Figure 382.1 shows an example of the LTC1955 used in a dual smart card to RS232 application powered by a single Li-Ion battery. A simple 4-wire command and status interface plus a 4-wire smart card communications interface are all that is required. The command/status serial port can be easily daisy-chained and the smart card communications port can be paralleled to expand this application to four or more smart cards while maintaining the same number of wires to the microcontroller.

Conclusion

Requiring a minimum of external components and available in a small 5mm × 5mm × 0.75mm leadless package, the LTC1955 provides a compact, simple and cost-effective solution to the difficult problems facing smart card system designers.

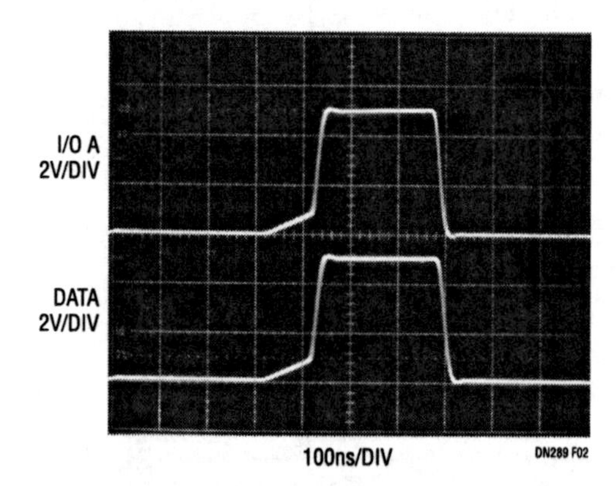

Figure 382.2 • Bidirectional Pin Waveforms with Pull-Up Acceleration

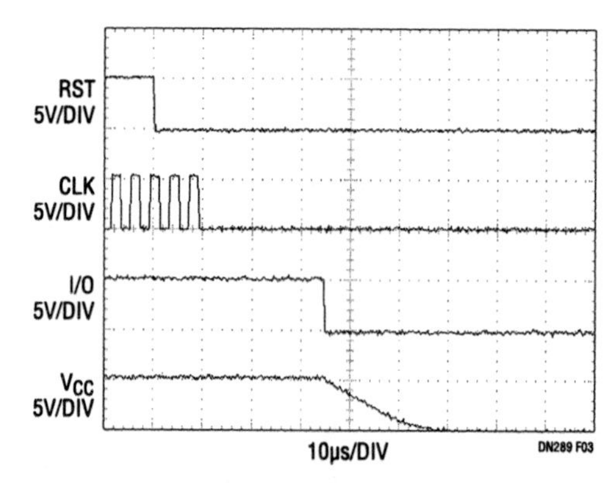

Figure 382.3 • Smart Card Deactivation Sequence

Note 1: Patent pending.

Isolated RS485 transceiver breaks ground loops

383

Mitchell Lee

The RS485 interface standard is designed to handle a −7V to 12V input signal range; however, in practical systems, ground potentials vary widely from node to node, often exceeding the specified range. This can result in an interruption of communications, high current flow through ground loops or worse, destruction of a transceiver. Guarding against large ground-to-ground differentials calls for an isolated interface. A new surface mount device, the LTC1535 isolated RS485 transceiver, provides a one-chip solution for breaking ground loops and achieving a wide input range.

Previously, isolation was achieved by using at least three opto-isolators and a separate isolated power supply. The LTC1535 replaces not only the opto-isolators, but also the power supply, as it includes an on-chip DC/DC converter. Other features include selectable driver slew rate to reduce EMI and susceptibility to reflections, full-duplex pinout and fail-safe detection of open and shorted lines.

The LTC1535 consists of two separate dice assembled on a proprietary, isolated lead frame. The lead frame includes integral coupling capacitors that bridge the isolation barrier and exhibit $2500V_{RMS}$ guaranteed standoff. Data communication takes place via the coupling capacitors, while an on-chip, 400kHz push-pull converter sends power to the isolated side through a small transformer. Total common mode capacitance across the barrier amounts to less than 20pF, with the transformer accounting for about 16pF of the total. Figure 383.1 shows the complete circuit for a fully isolated RS485 port.

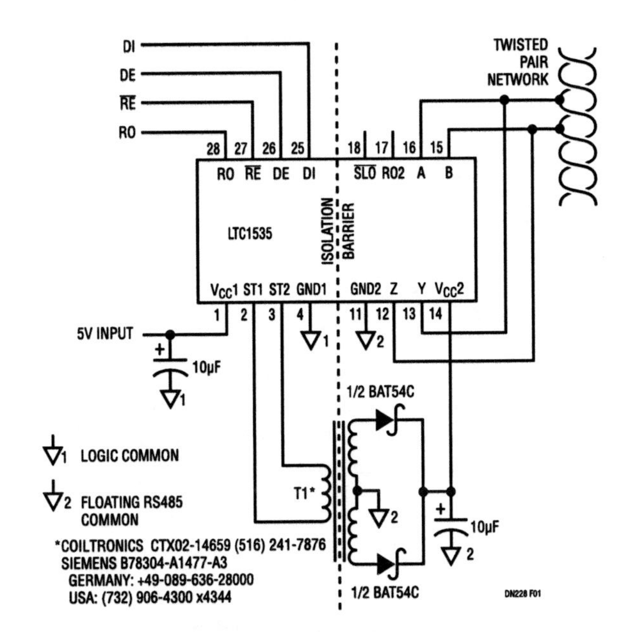

Figure 383.1 • Fully Isolated RS485 Port

Internally, the two halves of the LTC1535 communicate in a ping-pong fashion, first sending transmit data to the isolated side and then sending receive data back to the nonisolated side. The sampling nature of the internal communications link means that some jitter is introduced into the data; this limits the useful baud rate to approximately 500kBd. At 350kBd, the jitter is guaranteed to be less than 10%. Figure 383.2 shows a double pulse propagating through the LTC1535. Waveform (A) is the transmitter data input and waveform (B) is the output of the receiver. The transmitter and receiver are looped back on the isolated side of the chip. The typical jitter is hardly visible. A negative-going double pulse is shown in Figure 383.3. The LTC1535 transceiver is unaffected by the DC average of the data waveform. Total round-trip

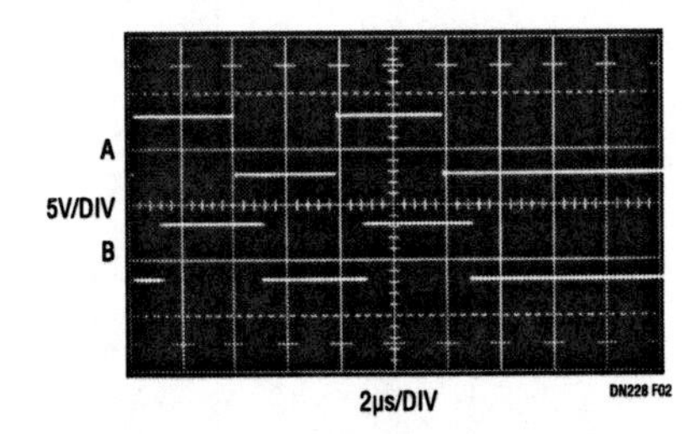

Figure 383.2 • Positive-Going Double-Pulse Behavior: A = Driver Input, B = Receiver Output

Analog Circuit Design: Design Note Collection. http://dx.doi.org/10.1016/B978-0-12-800001-4.00383-5

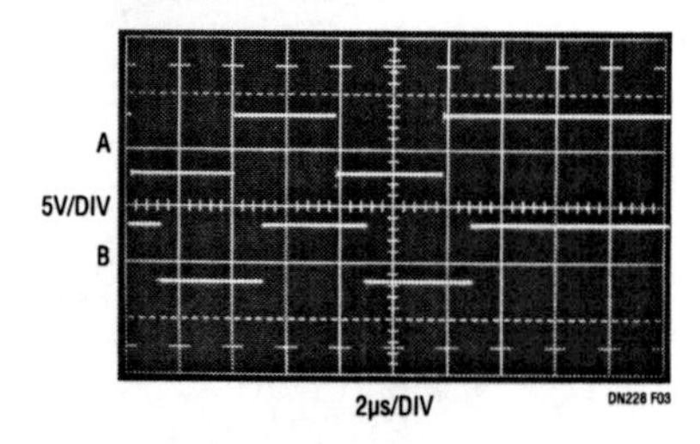

Figure 383.3 • Negative-Going Double-Pulse Behavior: A = Driver Input, B = Receiver Output

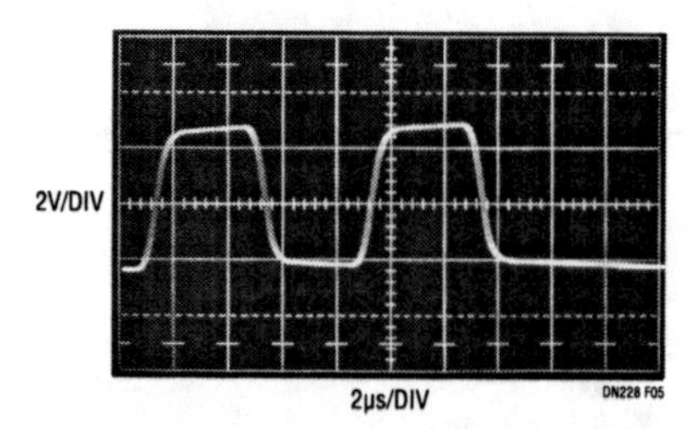

Figure 383.5 • Driver in Slow Slew Mode, Loaded with 5000' of Twice-Terminated Twisted Pair

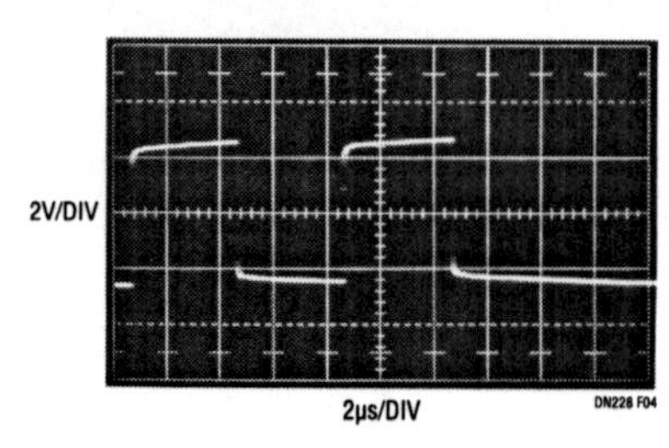

Figure 383.4 • Driver in Fast Slew Mode, Loaded with 5000' of Twice-Terminated Twisted Pair

propagation delay through the LTC1535 is approximately 1μs or roughly equivalent to 328 feet of cable.

Figure 383.4 shows the driver output waveform when loaded by 5000 feet of terminated cable, operating in the fast slew mode (SLO pin pulled high). The effect of the slew pin on the driver output waveform is noticeable in Figure 383.5, where rise and fall times of approximately 1μs result.

The LTC1535 is useful for other types of signal isolation. Figure 383.6 shows a fully isolated, 24-bit differential input A/D converter implemented with the LTC1535 and LTC2402. Power on the isolated side is regulated by an LT1761-5 low noise, low dropout micropower regulator. Its output is suitable for driving bridge circuits and for ratiometric applications.

During power-up, the LTC2402 becomes active at V_{CC} = 2.3V, while the isolated side of the LTC1535 must wait for V_{CC2} to reach its undervoltage lockout threshold of 4.2V. Below 4.2V, the LTC1535's driver outputs Y and Z are in a high impedance state, allowing the 1kΩ pull down to define the logic state at SCK. When the LTC2402 first becomes active it samples SCK; a logic "0" provided by the 1kΩ pull-down invokes the external serial clock mode. In this mode the LTC2402 is controlled by a single clock line from the nonisolated side of the barrier, through the LTC1535's driver output Y. The entire power-up sequence, from the time power is applied to V_{CC1} until the LT1761's output has reached 5V, is approximately 1ms.

Data returns to the nonisolated side through the LTC1535's receiver at RO. An internal divider on receiver input B sets a logic threshold of approximately 3.4V at input A, facilitating communications with the LTC2402's SDO output without the need for any external components. For further details of the LTC2402's logic interface and serial output bit stream, see the LTC2402 data sheet.

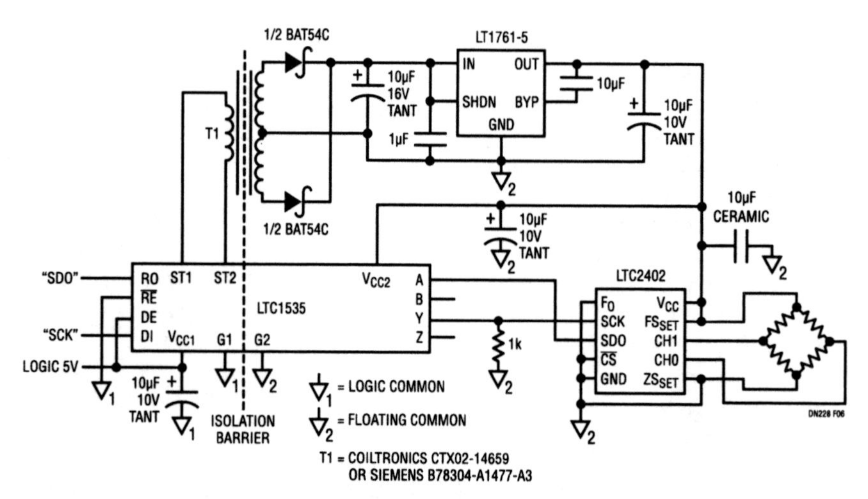

Figure 383.6 • Complete Isolated 24-Bit Data Acquisition System

RS485 transceivers sustain ±60V faults

384

Gary Maulding

Introduction

The LT1785 and LT1791 RS485/RS422 transceivers with ±60V fault tolerance solve a real-world problem of field failures in typical RS485 interface circuits. Modems and other computer peripherals use point-to-point RS422 connections to support higher communication speeds with better noise immunity over greater distances than is possible with RS232 connections. Multipoint RS485 networks are used for LANs and industrial control networks. All of these applications are vulnerable to the unknown, sometimes hostile environment outside of the controlled, shielded environment of a typical electrical equipment chassis. Because the RS485 transceivers are directly in the line of fire, the transceiver chips are often socketed PDIP packages to allow easy field servicing of equipment. Field failures in standard transceiver circuits are caused by data-line voltages exceeding the absolute maximum ratings of the transceiver chips. Installation wiring faults, ground voltage faults and lightning-induced surge voltages are all common causes of overvoltage conditions.

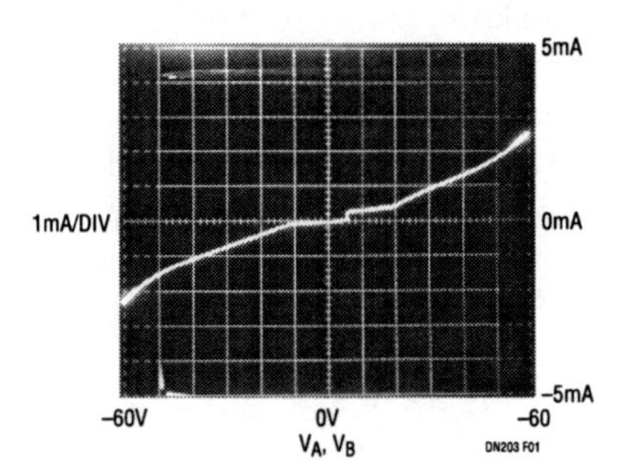

Figure 384.1 • LT1785 Input Current vs V_{IN}

Up to ±60V faults

The electrical standards for RS422 and RS485 signaling reflect the need for tolerance of ground voltage drops in an extended network by requiring receivers to operate with input common mode voltages from −7V to 12V. The RS485 and RS422 transceivers commonly available from various vendors are all vulnerable to damage from fault voltages only slightly outside of the operating envelope. One vendor's RS485 transceivers have absolute maximum voltage ratings of −8V to 12.5V on the data I/O pins. Such narrow margins beyond the required −7V to 12V operating conditions makes such circuits very fragile in a real-world environment. In addition, external protection circuitry is ineffective at protecting these circuits without corrupting normal operating signal levels.

The LT1785 and LT1791, with ±60V absolute maximum ratings on the driver output and receiver input pins are

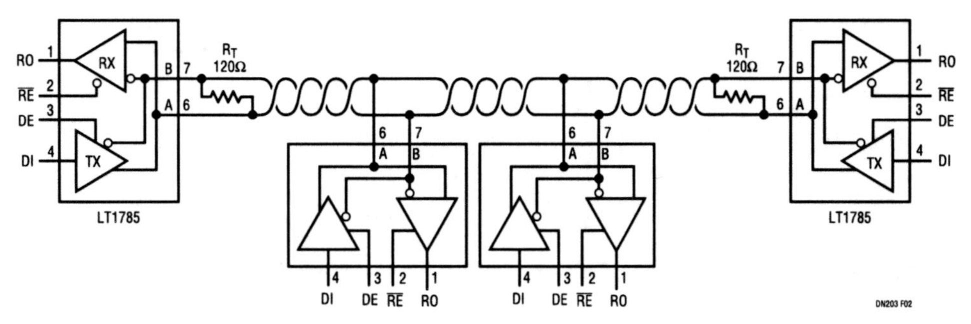

Figure 384.2 • Half-Duplex RS485 Network Operation ±60V DC, ±15kV ESD Protected

Analog Circuit Design: Design Note Collection. http://dx.doi.org/10.1016/B978-0-12-800001-4.00384-7

inherently safe in most environments that will destroy other interface circuits. Standard pinouts in either PDIP or SO packages allow easy upgrades to existing RS422/RS485 networks. Whether the circuit is transmitting, receiving, in standby or powered off, any voltage within ±60V will be tolerated by the chip without damage. Data communication will be interrupted during the fault condition, but the circuit will live to talk another day. Figure 384.1 shows the I-V characteristics at the RS485 input/output pins.

128-node networks at 250kBd

In addition to their unique fault tolerance capabilities, these transceivers feature high input impedance to support extended RS485 networks of up to 128 nodes (Figure 384.2). Controlled slew-rate outputs minimize EMI problems while supporting data rates up to 250kBd (see Figure 384.3). Driver outputs are capable of working with inexpensive telephone cable with characteristic impedance as low as 72Ω with no loss of signal amplitude. "A" grade devices are available that ensure "fail safe" receiver outputs when inputs are open, shorted or no signal is present.

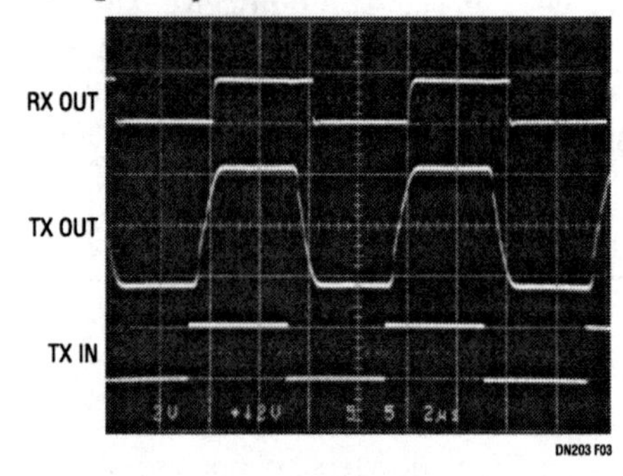

Figure 384.3 • Normal Operation Waveforms at 250kBd

Extending protection beyond ±60V

While ±60V fault tolerance forgives a great number of sins, higher voltage demons may still be lurking. ESD is one such demon, with voltage spikes into the thousands of volts. The LT1785 and LT1791 have on-chip protection to ±15kV air gap ESD transients for other high voltage faults, such as lightning-induced surge voltages or AC line shorts. For such high energy faults, external protection must be used to protect the circuits. Typical protection networks use voltage clamping and current limiting networks. In concept, such networks could be used with normal RS485 circuits to afford extended protection, but in practice, the addition of protection networks would interfere with normal operation of the data network. The voltage clamping Zeners or TransZorbs are not available in tight voltage tolerances, and in addition, their internal impedances cause several volts of additional potential above their nominal breakdown voltage to appear at the protected device's pins. To protect a circuit with a −8V to 12.5V absolute maximum voltage rating would require the use of protection devices with voltage ratings much below the required common mode range of RS485 networks interfering with normal data transmission.

Figure 384.4 gives an example of the use of external clamping and limiting components to extend the LT1785's ±60V tolerance to the peak 120V AC line voltage. 36V TransZorbs are used to clamp the transceiver's line pins below the 60V capability of the transceiver. During a 120V AC line fault, peak surge currents of nearly 3A will flow through the 47Ω limiting resistors and the PolySwitch limiters. The peak current rating and series resistance of the TransZorbs must be considered when selecting the clamp device to ensure that the clamp limiter can withstand the surge and that the peak voltage will remain below the ±60V limitations of the LT1785. At 3A, even high current TransZorbs will exceed their nominal breakdown voltage by several volts, making this protection method ineffective with transceiver circuits with only 1V to 4V margin above their operating ranges. The PolySwitch limiters are thermally activated and increase in resistance by many orders of magnitude in about 10ms. After the PolySwitch transition, fault currents are only a few milliamperes. Carbon composite resistors must be used for limiting the initial surge current before the PolySwitch transition point. Metal film resistors do not have effective surge overload ratings and will fail before the PolySwitch transition drops the currents to sustainable levels.

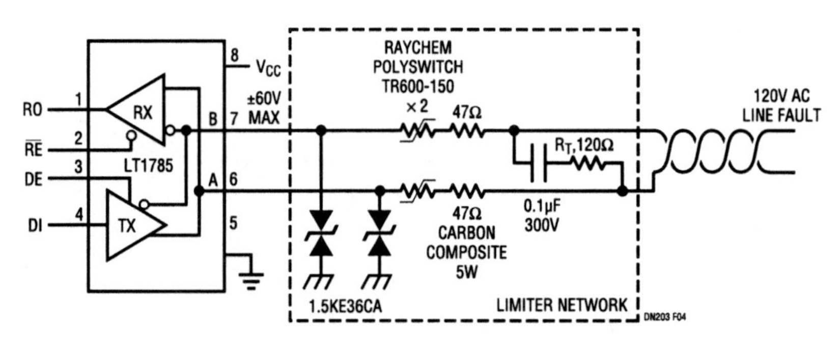

Figure 384.4 • Limiter Network Clamps 120V AC Fault Voltage to Less Than ±60V

SMBus accelerator improves data integrity

385

David Bell   Mark Gurries

Introduction

The System Management Bus (SMBus) is gaining popularity in portable computers as a communication link between smart batteries, the battery charger and the power management microcontroller. The convenience of this simple 2-wire bus is prompting designers to use it for communication with other peripherals such as battery selectors, backlight controllers, temperature monitors, power switches and other devices. Before long, the SMBus will become what its creators envisioned—a general purpose system management bus that connects various low speed peripherals throughout a portable computer (see Figure 385.1).

The SMBus uses the open-drain I^2C protocol for its physical layer, with respecified logic thresholds and pull-up current. Whereas the I^2C bus allows pull-up currents as high as 3mA, the SMBus has been specified with a maximum pull-up current of only 350μA. The maximum 400pF bus capacitance allowed by I^2C is reduced to only *50pF* because of the low pull-up current provided by the SMBus. Although the lower logic thresholds specified for SMBus peripherals mitigate the rise time problem, most SMBus systems include devices with I^2C CMOS logic thresholds that can be as high as $0.8 \cdot V_{CC}$ (microcontrollers are a good example). All it takes is one such peripheral with high logic thresholds, and SMBus rise times can seriously restrict bus capacitance.

The SMBus rise time problem can result in data integrity problems, or in severe cases, cause the bus to stop operating entirely. Because 50pF can easily be exceeded with just a few feet of cable, SMBus systems often fail to operate reliably when simply connected together on the lab bench. The problems of SMBus capacitive loading can become worse when more peripherals are connected by long traces running throughout a portable computer.

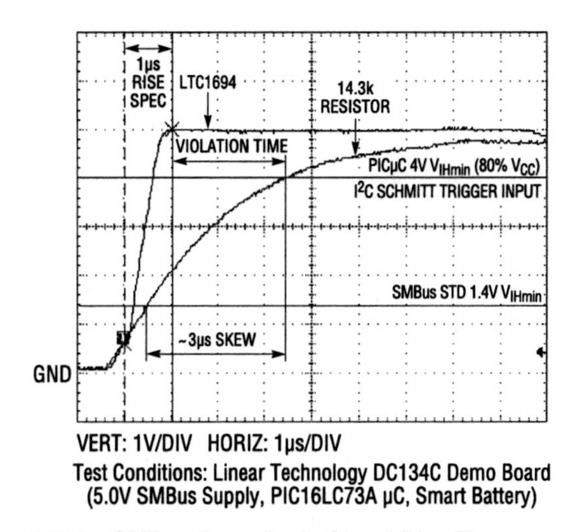

Figure 385.2 • SMBus Open-Drain Signal Rise Times

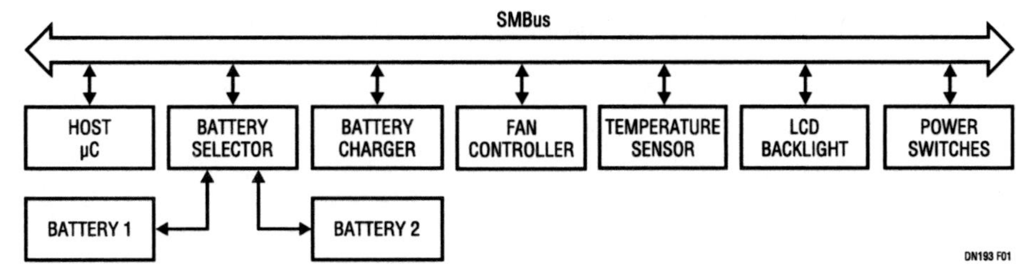

Figure 385.1 • SMBus Applications in a Notebook Computer

Analog Circuit Design: Design Note Collection. http://dx.doi.org/10.1016/B978-0-12-800001-4.00385-9

The solution

Linear Technology developed the LTC1694 SMBus Accelerator[1] active pull-up circuit to alleviate the SMBus rise time problem. This SOT-23 packaged part simply replaces the two external pull-up resistors and reduces the SMBus rise time by a factor of 3× to 4×. Figure 385.2 compares an SMBus signal rise time using a standard resistor pull-up to a signal produced with the SMBus accelerator. SMBus systems that work unreliably (or not at all) with the minimum value pull-up resistor perform as intended with the LTC1694.

Figure 385.3 is a functional block diagram of the LTC1694 SMBus accelerator. Two identical pull-up circuits are contained within the LTC1694: one for the clock line (SCL) and one for the data line (SDA). When both open-drain signals remain high (SMBus idle), the LTC1694 provides a 100μA pull-up current to each line to keep them in the high state. When either signal is pulled low by an SMBus driver, the pull-ups source approximately 275μA. Because this pull-up current is less than the 350μA allowed by the SMBus specification, the logic low (V_{OL}) level on the bus is reduced, resulting in improved low state noise margin.

The biggest improvement occurs when an SMBus driver releases the open-drain signal. If the signal voltage exceeds 0.65V and the positive slew rate exceeds 0.2V/μs, then a 2.2mA pull-up current source is activated in the LTC1694. This 2.2mA current source quickly pulls the signal high until it hits the supply rail. After a short delay the current is reduced to the 100μA level with the signal at a steady state high level. Noise immunity is also built into the slew rate detector to avoid false tripping on narrow noise spikes. In essence, the LTC1694 provides light pull-up when the open-drain signal is static or falling, but accelerates the rising edge once a rising signal is detected.

Making the upgrade

Retrofitting an existing SMBus system is easy—the LTC1694 simply replaces the two pull-up resistors. PCB area is approximately the same, owing to the small 5-pin SOT-23 package. Because SMBus peripherals may operate from either 5V or 3.3V, the LTC1694 is designed to operate equally well from either supply voltage. The SMBus accelerator powers down to only 60μA when both SMBus signals are high, so impact on battery life is also insignificant.

SMBus data integrity is vital, especially since lithium-ion battery charge control may be communicated via the bus. It's not always obvious that "flaky" SMBus operation is the result of excessive rise time, but the test is easy to implement—simply replace your pull-ups with the LTC1694.

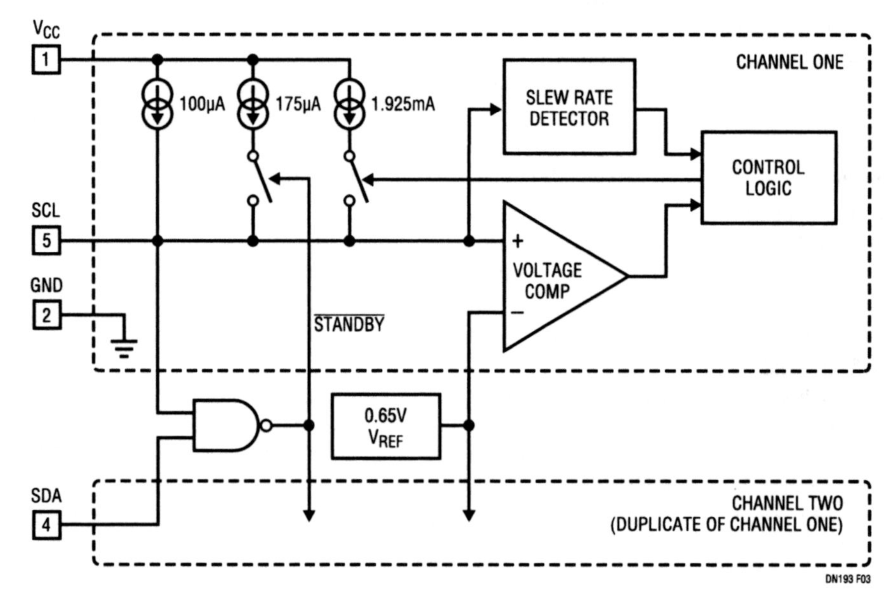

Figure 385.3 • LTC1694 Functional Block Diagram

Note 1: Patent pending.

Providing power for the IEEE1394 "FireWire"

386

Ajmal Godil

Faster microprocessors, more memory and better graphics have fueled the rapid growth of the personal computer industry. However, many of the peripheral connections use older interface technologies that are starting to limit performance and growth for future applications.

The IEEE1394 High Performance Serial Bus ("FireWire") addresses these interface issues by providing a flexible and cost-effective way to share real-time information among data-intensive applications, such as digital camcorders, digital VCRs and digital video disks (DVDs). This serial bus supports data transfer rates of 100Mbps, 200Mbps and 400Mbps. It also provides unregulated 8VDC to 40VDC at up to 1.5A. As many as 16 devices can be connected to the IEEE1394 bus, with cable segments between devices of up to 4.5 meters. The junction or node where a device connects to the bus may be a power source, a power sink or neither. Since there may be more than one power source, all power sources are diode connected. The voltage source with the highest potential is allowed to put power on the bus, while the rest are isolated by reverse-biased diodes (see Figure 386.1).

Figure 386.2 shows an example of a video camera sending digital video data to a monitor and to a computer, which in turn, is connected to both a digital VCR and a printer, via the IEEE1394 bus.

Figure 386.3 is a schematic of a circuit that produces a regulated 5V at 500mA from a FireWire input (8V to 40V).The LT1776 is a high voltage, high efficiency buck converter IC. The IC includes an onboard power switch, oscillator, control

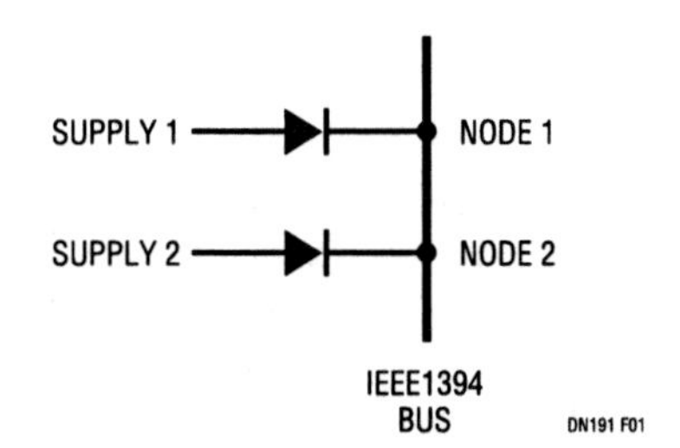

Figure 386.1 • Two Voltage Sources Diode-Connected to the IEEE1394 Bus; the Source with the Highest Potential Provides Power on the Bus

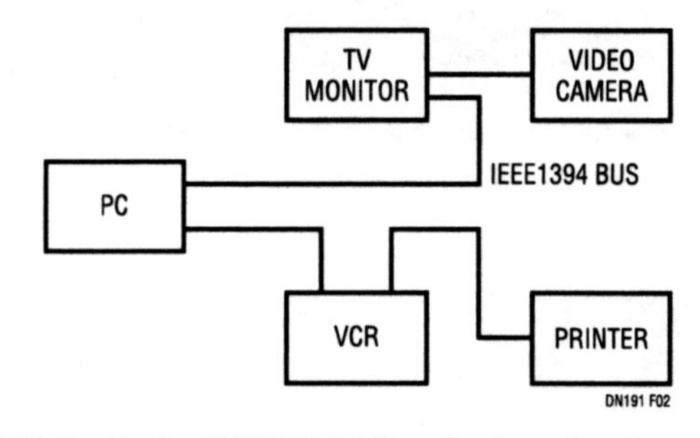

Figure 386.2 • Typical IEEE1394 Bus System Configuration

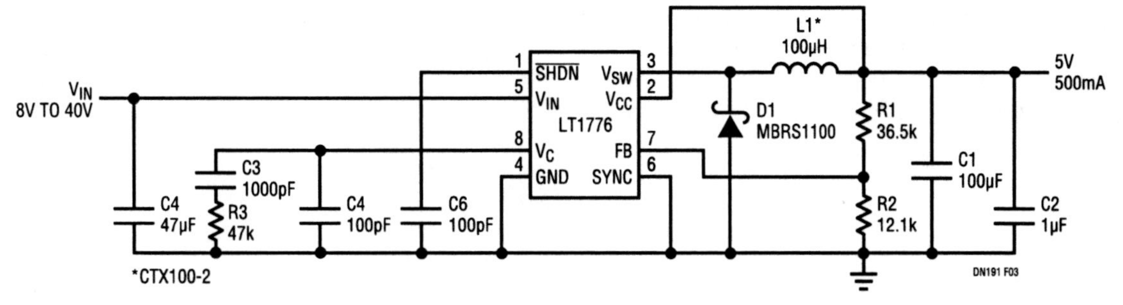

Figure 386.3 • LT1776 Application Circuit for Generating 5V at 500mA

Analog Circuit Design: Design Note Collection. http://dx.doi.org/10.1016/B978-0-12-800001-4.00386-0

and protection circuitry. The part can accept an input voltage as high as 40V and the power switch is rated at 700mA peak current. Current mode control offers excellent dynamic input supply rejection and overcurrent protection. The SO-8 package and 200kHz switching frequency help minimize PC board area requirements. The part can be disabled by connecting the shutdown ($\overline{SHDN}$) pin to ground, thus reducing input current to a few microamperes. In normal operation, decouple the $\overline{SHDN}$ pin with a 100pF capacitor to ground. The part also has a SYNC pin, used to synchronize the internal oscillator to an external clock, which can be anywhere from 250kHz to 400kHz. To use the part's internal 200kHz oscillator, simply connect the SYNC pin to ground. The circuit uses two techniques to maximize efficiency.

The internal control circuitry draws power from the V_{CC} pin and the LT1776 switch circuitry maintains a rapid rise time (see Figure 386.4) at high loads. At light loads, it slows down the rise time (see Figure 386.5) to avoid pulse skipping, thus maintaining constant frequency from heavy to light load. This helps significantly in reducing output ripple voltage and switching noise in the audio frequency spectrum. Figure 386.6 shows typical efficiency curves for various input voltages from 8V to 40V at an output voltage of 5V.

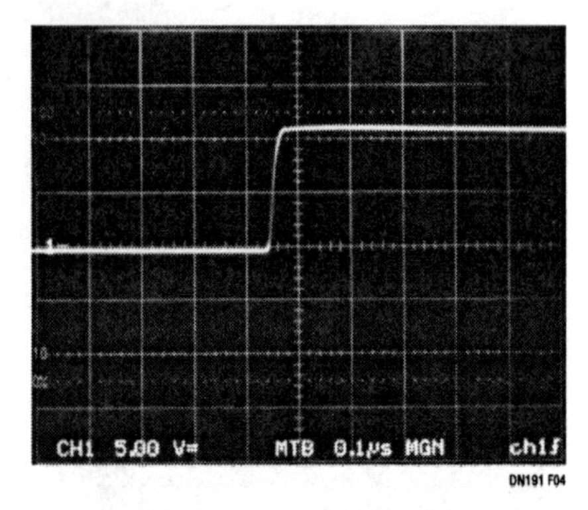

Figure 386.4 • Switch Rise Time at Heavy Loads

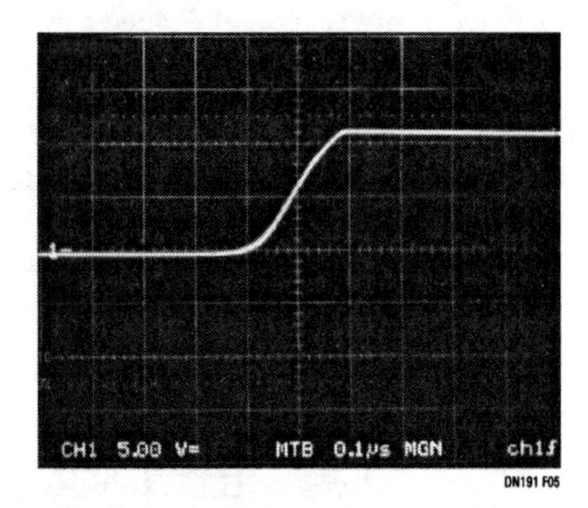

Figure 386.5 • Switch Rise Time at Light Loads

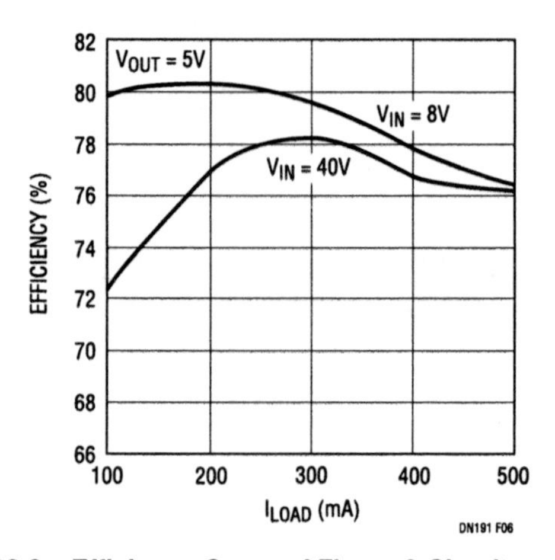

Figure 386.6 • Efficiency Curve of Figure 3 Circuit

Figure 386.7 presents a scheme that takes the unregulated 8V to 40V supply voltage from the IEEE1394 bus and steps it down to the supply voltage for the physical layer (PHY), using an LT1776-based regulator such as the circuit in Figure 386.3. The PHY's input voltage can be 1.25V (min), 3.0V, 3.3V, 5V or any other voltage level, provided that LT1776's input voltage is greater than the desired output voltage.

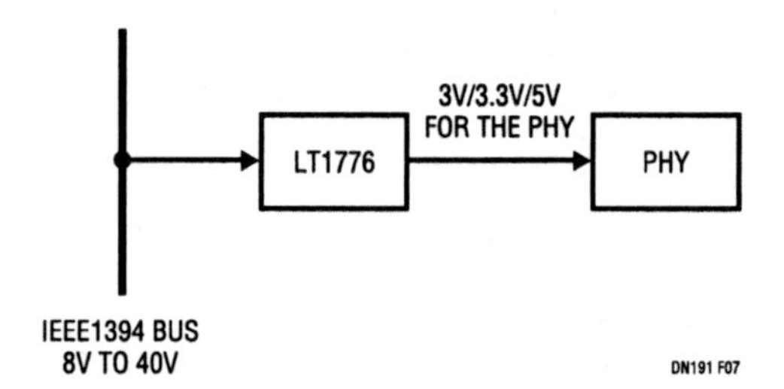

Figure 386.7 • LT1776 Provides Power to Physical Layer Electronics (PHY) from IEEE1394 Bus

Linear Technology also offers the LT1676, which is very similar to the LT1776 except that its maximum input voltage can be as high as 60V. Also, its switching frequency is 100kHz with the option of synchronizing to an external clock in the range of 130kHz to 250kHz. The LT1676's high input voltage range of 7.4V to 60V allows it to be used not only in IEEE1394 "FireWire" applications but also in automotive DC/DC and telecom 48V step-down applications.

5V RS232/RS485 multiprotocol transceiver

387

Y. K. Sim

Introduction

The LTC1387 is a single 5V supply, single-port, logic-configurable RS232 or RS485 transceiver. The LTC1387 features Linear Technology's usual high data rates (120kBd for RS232 and 5MBd for RS485), a loopback mode for self test, a micropower shutdown mode and ±7kV ESD protection at the driver output and receiver inputs. This part is targeted at handheld computers, point-of-sale terminals and applications that require a minimum pin count and software-controlled multiprotocol operation.

RS232 and RS485 interfaces

Most computers communicate with other computers, peripherals or modems through an RS232 interface, a single-ended interconnection standard. The simplest RS232 interface has three wires: a transmit data line, a receive data line and a ground return. EIA-562 is a single-ended electrical standard similar to RS232, with relaxed voltage levels.

RS485, a differential signal interface, is popular because it offers increased noise immunity, longer transmission cable length than RS232 or EIA-562 and allows multiple transceivers on a twisted pair of wires.

Key features

The LTC1387 offers a flexible combination of two RS232 drivers, two RS232 receivers, an RS485 driver, an RS485 receiver and an onboard charge pump to generate boosted voltages for true RS232 voltage levels from a 5V supply. Figures 387.1–387.4 show the LTC1387's versatility in software-controlled RS232/RS485 applications. The RS232 and RS485 transceivers are designed to share the same I/O pins for both single-ended and differential signal communication modes. Both half-duplex and full-duplex communication can be supported. Autodetection of RS232/RS485 interface is feasible via a controller software routine.

The RS232 transceiver supports both RS232 and EIA-562 standards, whereas the RS485 transceiver supports both

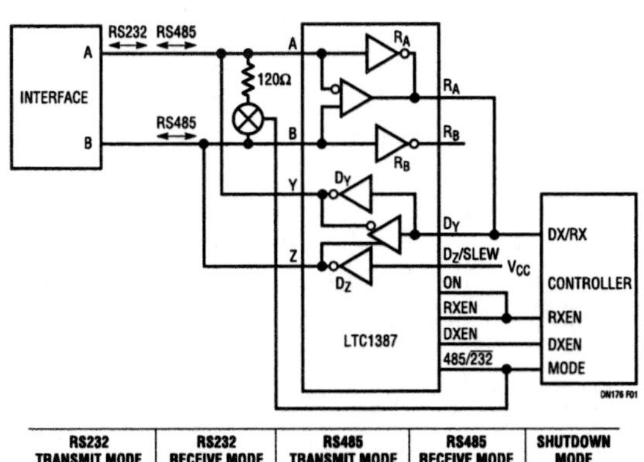

RS232 TRANSMIT MODE	RS232 RECEIVE MODE	RS485 TRANSMIT MODE	RS485 RECEIVE MODE	SHUTDOWN MODE
RXEN = 0	RXEN = 1	RXEN = 0	RXEN = 1	RXEN = 0
DXEN = 1	DXEN = 0	DXEN = 1	DXEN = 0	DXEN = 0
MODE = 0	MODE = 0	MODE = 1	MODE = 1	MODE = X

Figure 387.1 • Half-Duplex RS232 (1-Channel), Half-Duplex RS485

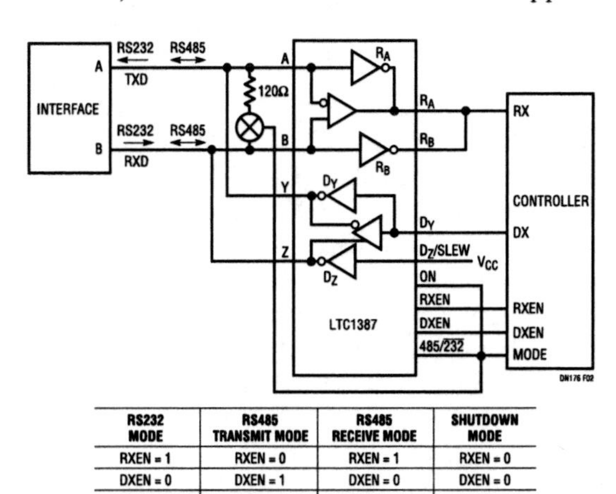

RS232 MODE	RS485 TRANSMIT MODE	RS485 RECEIVE MODE	SHUTDOWN MODE
RXEN = 1	RXEN = 0	RXEN = 1	RXEN = 0
DXEN = 0	DXEN = 1	DXEN = 0	DXEN = 0
MODE = 0	MODE = 1	MODE = 1	MODE = 0

Figure 387.2 • Full-Duplex RS232 (1-Channel), Half-Duplex RS485

Analog Circuit Design: Design Note Collection. http://dx.doi.org/10.1016/B978-0-12-800001-4.00387-2

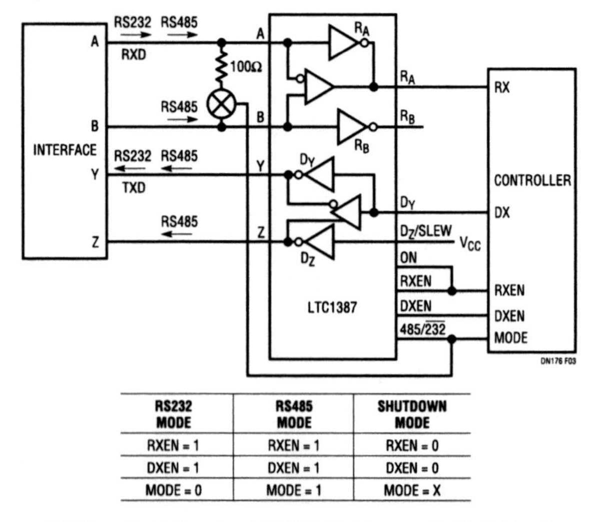

RS232 MODE	RS485 MODE	SHUTDOWN MODE
RXEN = 1	RXEN = 1	RXEN = 0
DXEN = 1	DXEN = 1	DXEN = 0
MODE = 0	MODE = 1	MODE = X

Figure 387.3 • Full-Duplex RS232 (1 Channel), Full-Duplex RS485/RS422

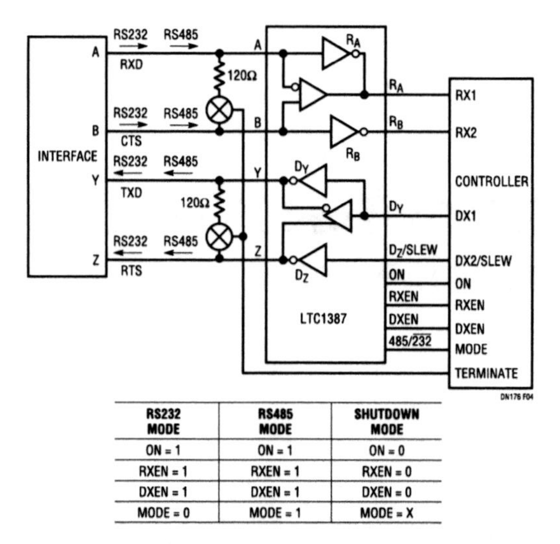

RS232 MODE	RS485 MODE	SHUTDOWN MODE
ON = 1	ON = 1	ON = 0
RXEN = 1	RXEN = 1	RXEN = 0
DXEN = 1	DXEN = 1	DXEN = 0
MODE = 0	MODE = 1	MODE = X

Figure 387.4 • Full-Duplex RS232 (2 Channel), Full-Duplex RS485/RS422 with SLEW and Termination Control

Table 387.1 This Function Table Indicates the Logic Inputs to Configure the LTC1387 for Various RS232/RS485 Modes

SELECT INPUTS				RECEIVER		DRIVER		CHARGE PUMP	LOOPBACK	COMMENTS
ON	RXEN	DXEN	$485/\overline{232}$	RXA	RXB	DXY	DXZ			
1	0	0	0	Hi-Z	Hi-Z	Hi-Z	Hi-Z	ON	OFF	RS232 Mode, DX and RX Off
1	0	1	0	Hi-Z	Hi-Z	ON	ON	ON	OFF	RS232 Mode, DXY and DXZ On, RX Off
1	1	0	0	ON	ON	Hi-Z	Hi-Z	ON	OFF	RS232 Mode, DX Off, RXA and RXB On
1	1	1	0	ON	ON	ON	ON	ON	OFF	RS232 Mode, DXY and DXZ On, RXA and RXB On
0	0	1	0	Hi-Z	Hi-Z	ON	Hi-Z	ON	OFF	RS232 Mode, DXY On, DXZ Off, RX Off
0	1	0	0	Hi-Z	ON	ON	Hi-Z	ON	OFF	RS232 Mode, DXY On, DXZ Off, RXA Off, RXB On
0	1	1	0	ON	ON	ON	ON	ON	ON	RS232 Loopback Mode, DXY and DXZ On, RXA and RXB On
0	0	0	X	Hi-Z	Hi-Z	Hi-Z	Hi-Z	OFF	OFF	Shutdown, RS485 R_{IN}
1	0	0	1	Hi-Z	Hi-Z	Hi-Z	Hi-Z	ON	OFF	RS485 Mode, DX and RX Off
X	0	1	1	Hi-Z	Hi-Z	ON	ON	ON	OFF	RS485 Mode, DX On, RX Off
X	1	0	1	ON	Hi-Z	Hi-Z	Hi-Z	ON	OFF	RS485 Mode, DX Off, RX On
1	1	1	1	ON	Hi-Z	ON	ON	ON	OFF	RS485 Mode, DX On, RX On
0	1	1	1	ON	Hi-Z	ON	ON	ON	ON	RS485 Loopback Mode, DX On, RX On

RS485 and RS422 standards. The logic input (MODE) selects between RS232 and RS485 modes. With three additional control logic inputs (RXEN, DXEN and ON), the LTC1387 adapts easily, as shown in Table 387.1, to various communications needs, including a one-signal-line RS232 I/O mode.

A SLEW input pin available in RS485 mode changes the driver transition between normal and slow-slew-rate modes. In normal slew mode, the twisted-pair cable is terminated at both ends to minimize signal reflection. In slow slew mode, the maximum signal bandwidth is reduced; EMI and signal reflection problems are minimized. Slow slew rate systems can often use incorrectly terminated or unterminated cables with acceptable results. If cable termination is required, external termination resistors can be connected through switches or relays.

The RS485 receiver features an input threshold between 0V and −200mV. The receiver output has a known HIGH output state if both receiver inputs are open, if the cable is shorted or if no driver is active.

Conclusion

The LTC1387 is ideal for point-of-sale terminals, computers, multiplexers, networks or peripherals that must adapt on the fly to various I/O configuration requirements without hardware adjustments.

10Mbps multiple protocol serial chip set: Net1 and Net2 compliance by design

388

David Soo

Introduction

With the increase in multinational computer networks, comes the need for the network equipment to support different serial protocols. When the designer becomes occupied with the details of the interface specification, there is always the possibility that one small detail will be missed. This compliance headache causes designers to seek out a cost-effective, integrated solution.

The LTC1543, LTC1544 and LTC1344A have taken the integrated approach to multiple protocol. By using this chip set, the Net1 and Net2 design work is done (see Figure 388.1). In fact, Detecon Inc. documents compliance in Test Report No. NET2/102201/97. With this chip set, network designers can concentrate on functions that increase the end product value rather than on standards compliance.

Figure 388.1 • The LTC1543, LTC1544 and LTC1344A Multiple Protocol Serial Chip Set

Review of interface standards

The serial interface standards V.28 (RS232), V.35, V.36, RS449, EIA-530, EIA-530A or X.21 specify the electrical characteristics of each signal, the connector type, the transmission rate and the data exchange protocols. In general, the US standards start with RS or EIA and the equivalent European standards start with V or X. The single-ended standard, V.28 (RS232) has a lower data rate than the other differential standards. The current maximum RS232 data rate is 128kbps. As for the V.35, V.36, RS449, EIA-530, EIA-530A and X.21 standards, the maximum data rate is 10Mbps.

Typical application

Like the LTC1343 software-selectable multiprotocol transceiver, the LTC1543 and LTC1544 use the LTC1344A for switching resistive termination. The main difference between these parts is the functional partition: the LTC1343 can be configured as a data/clock chip or a control-signal chip using the CTRL/CLK pin, whereas the LTC1543 is a dedicated data/clock chip and the LTC1544 is a control-signal chip.

Figure 388.2 shows a typical application using the LTC1543, LTC1544 and LTC1344A. By just mapping the chip pins to the connector, the design of the interface port is complete. The chip set supports the V.28 (RS232), V.35, V.36, RS449, EIA-530, EIA-530A or X.21 protocols in either DTE or DCE mode. Shown here is a DCE mode connection to a DB-25 connector.

The mode select pins M0, M1 and M2 are used to select the interface protocol, as summarized in Table 388.1.

Table 388.1 Mode Selection

MODE NAME	M2	M1	M0
Not Used	0	0	0
EIA-530A	0	0	1
EIA-530	0	1	0
X.21	0	1	1
V.35	1	0	0
RS449/V.36	1	0	1
V.28/RS232	1	1	0
No Cable	1	1	1

Analog Circuit Design: Design Note Collection. http://dx.doi.org/10.1016/B978-0-12-800001-4.00388-4

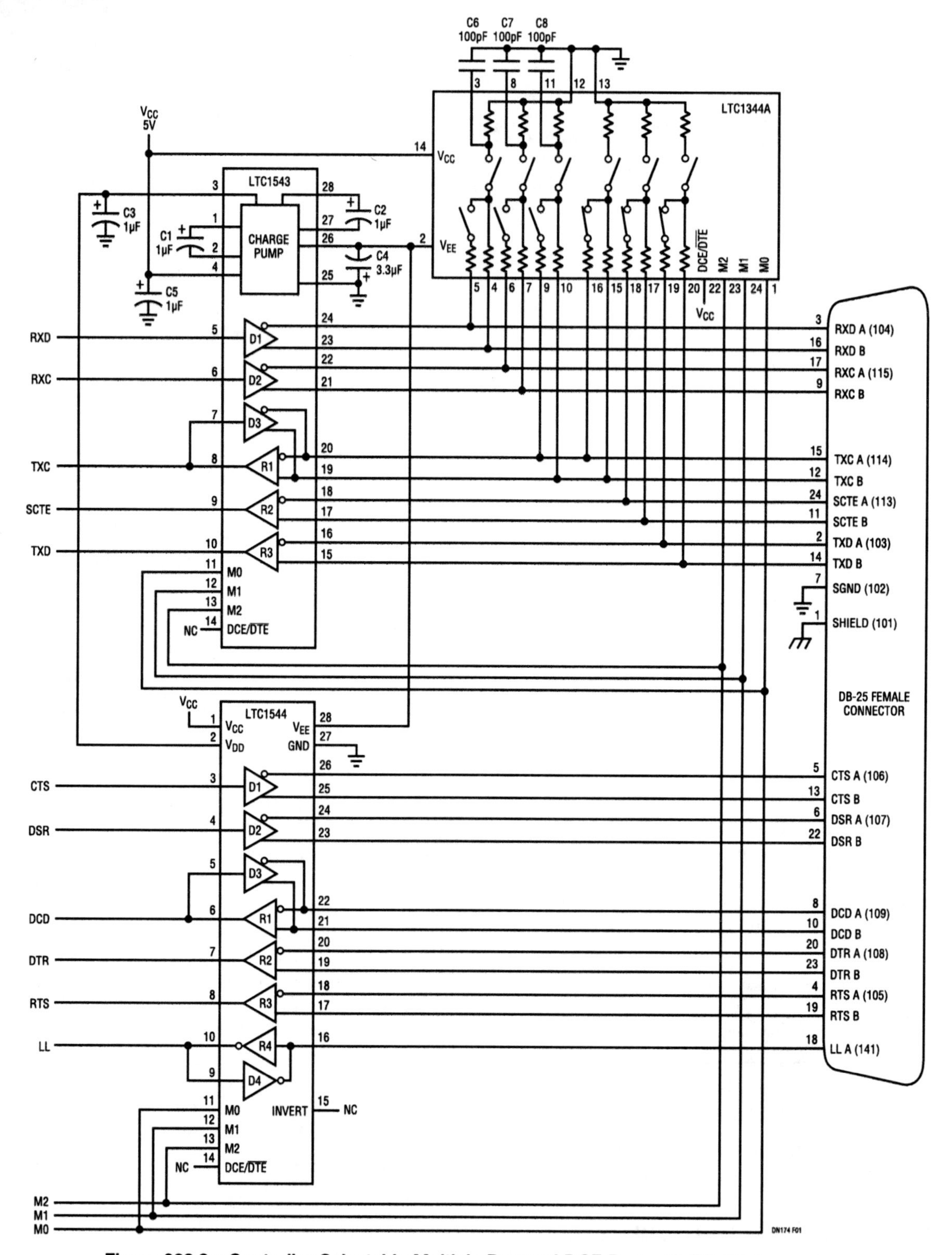

Figure 388.2 • Controller-Selectable Multiple Protocol DCE Port with DB-25 Connector

There are internal 50μA pull-up current sources on the mode select pins, DCE/DTE and the INVERT pins. The protocol may be selected by plugging the appropriate interface cable into the connector. The mode pins can be routed to the connector and are unconnected (logic 1) or wired to ground (logic 0). If all the mode-select pins are not connected (logic 1), the chip set enters the no cable mode in which the chip set lowers the supply current to less than 700μA and three-states the driver and receiver outputs, and the LTC1344A disconnects the termination resistors.

RS485 transceivers operate at 52Mbps over 100 feet of unshielded twisted pair

389

Victor Fleury

The propagation delay of typical RS485 transceivers can vary by as much as 500% over process and temperature. In applications where high speed clock and data waveforms are sent over long distances, propagation delay and skew uncertainties can pose system design constraints and limit the maximum data rate. The LTC1685 high speed RS485 transceiver family addresses this problem by guaranteeing over temperature a precision propagation delay of 18.5ns±3.5ns, a better than ten times improvement over other CMOS transceivers.

The LTC1685 is geared for half-duplex operation, whereas the LTC1686/LTC1687 can operate in full-duplex mode. All include a receiver fail-safe feature, whereby the receiver output remains in a high state over the entire 12V to −7V common mode range when the inputs are left open or shorted together. A novel protection technique permits indefinite short circuiting of the driver and receiver outputs to supply or ground, while limiting the current to 20mA.

High speed differential SCSI (Fast-20/Fast-40 HVD)

The LTC1685's high speed and tight driver/receiver propagation-delay window make the LTC1685 a natural choice as the external transceiver in high speed (40Mbps) differential SCSI applications. Figure 389.1 shows a 100-foot passively terminated category 5 unshielded twisted pair (UTP) connection as used in high speed differential SCSI applications. Figure 389.2 shows a 20ns (50Mbps) pulse propagating down the circuit of Figure 389.1. Note that in order to achieve these speeds at these distances, it is important to use high quality cable, such as category 5 UTP.

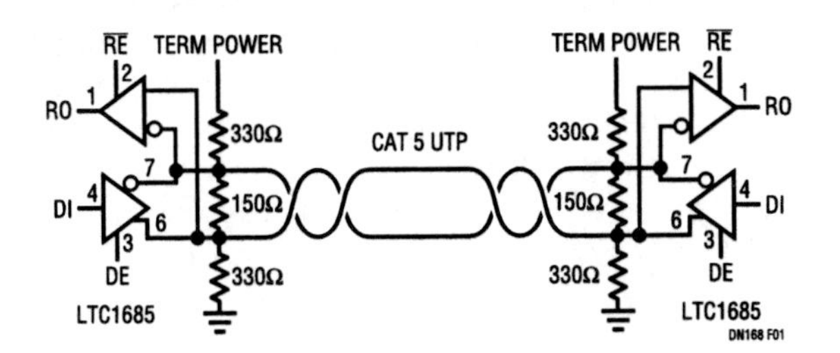

Figure 389.1 • Fast-20/Fast-40 Differential SCSI Application

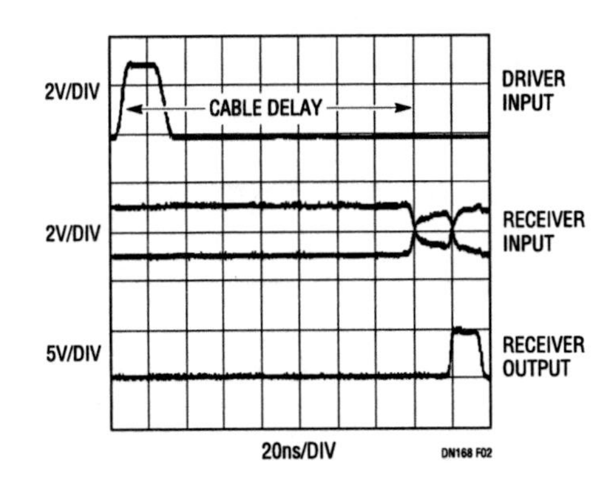

Figure 389.2 • 100 Feet of Category 5 UTP: 20ns Pulse

SCSI applications place strict requirements on the propagation delay variation of its transceivers. For a group of LTC1685 transceivers on the same board, the propagation delay variation will be smaller than the ±3.5ns guaranteed specification. This is because the ±3.5ns propagation delay window covers the entire commercial temperature range, whereas transceivers placed on the same board will have very similar ambient temperatures. Hence, the difference in their propagation delays should be better than the ±3.5ns specification (typically better than ±2ns). This makes the LTC1685 the best choice for high speed SCSI applications.

Transmission over long distances

The LTC1686/LTC1687 can be used as repeaters to extend the effective length of a high speed twisted-pair line. Figure 389.3 shows a 3-repeater configuration using 2000-foot segments of category 5 UTP.

Analog Circuit Design: Design Note Collection. http://dx.doi.org/10.1016/B978-0-12-800001-4.00389-6

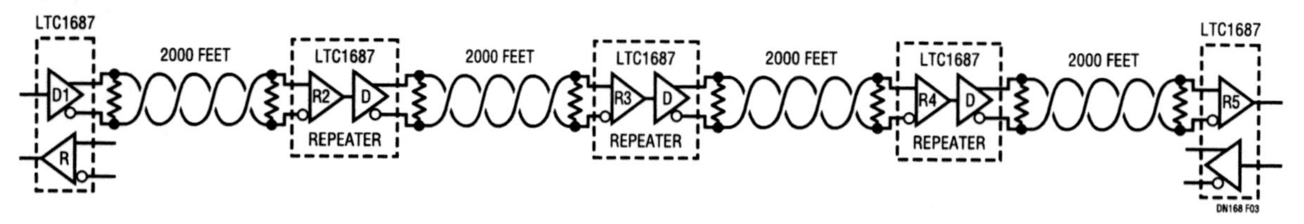

Figure 389.3 • 1.6Mbps, 8000 Feet (1.5 Miles) Using Three Repeaters

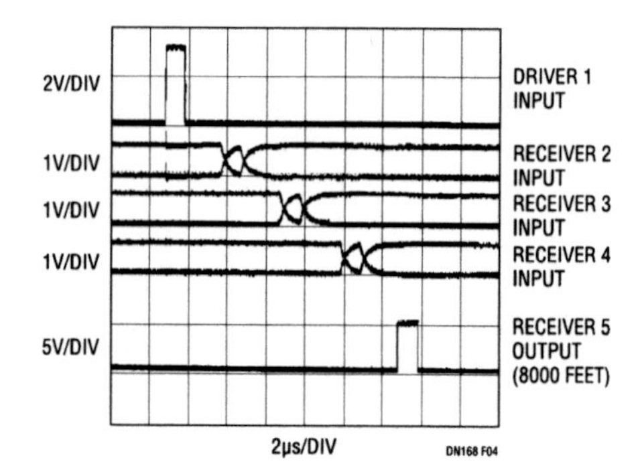

Figure 389.4 • Differential Signals at the Far End of the First Three 2000-Foot Cable Segments

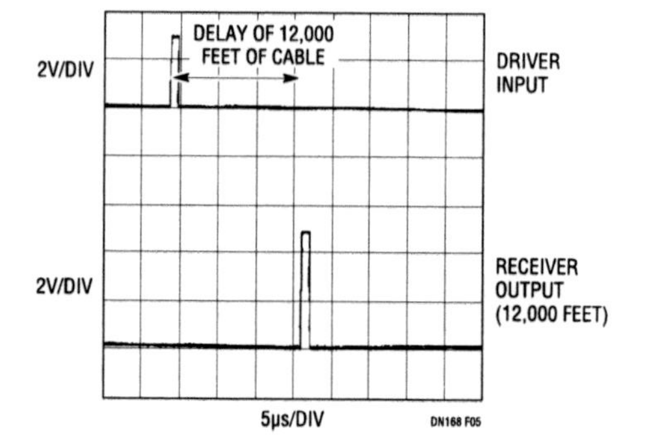

Figure 389.5 • 1µs Pulse over 12,000 Feet Category 5 UTP

1Mbps over 12,000 feet using repeaters

By adding two repeaters to the configuration of Figure 389.3, we are able to propagate a single 1µs pulse (1Mbps) over 12,000 feet of category 5 UTP. At this data rate and cable lengths, one can obtain minimal pulse-width degradation as the signal traverses through the repeater network. Figure 389.4 shows some receiver input and output signals at the far end of the first three 2000-foot cable segments of the network. The DC resistance of 2000 feet of category 5 UTP divides the signal nearly in half. AC losses tend to filter the 1µs pulse. The total attenuation is shown by the middle three traces of Figure 389.4. Note, however, that the output pulse (bottom trace) is nearly the same width as the input pulse (top trace), meaning that the LTC1687 repeaters are able to regenerate the signal with little loss in pulse width. Figure 389.5 shows the waveforms at the near and far end of the entire 12,000-foot network. The imperceptible loss in pulse width implies that we can cascade even more repeater networks and potentially achieve **1Mbps operation at total distances of well over 12,000 feet!**

1.6Mbps over 8000 feet using repeaters

For the same cable distance, high data rates will limit the maximum number of repeaters. Figure 389.6 shows the propagation of a single 600ns pulse through the 3-repeater network of Figure 389.3. The bottom two traces show a 1.6Mbps square wave at the input and output of the network, respectively. Notice that the duty cycle does not noticeably degrade, however, there is a degradation of the pulse width as shown in the second trace. Thus, in order to achieve reliable performance at long distances, a compromise must be reached between cable distance and quality, data rate and number of repeaters.

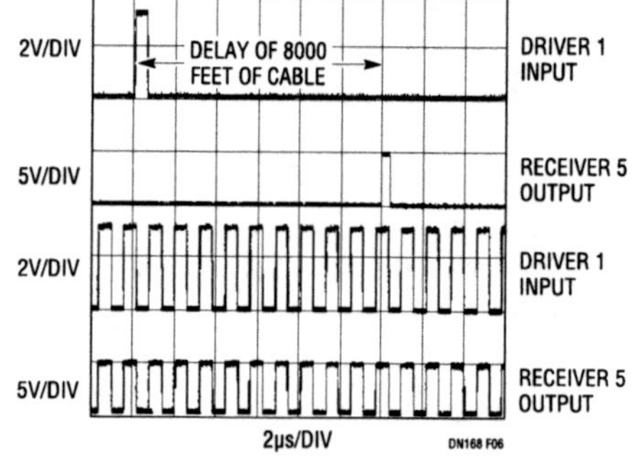

Figure 389.6 • 1.6Mbps Pulse and Square Wave Signals over 8000 Feet Category 5 UTP Using Three Repeaters

Conclusion

The LTC1685 family of high speed RS485 transceivers allows for up to 52Mbps transmission over reasonable distances (100 feet), as well as moderate speed over long distances (1.6Mbps, 2000 feet). Using repeaters can substantially increase the effective length. At 1.6Mbps, three repeaters can carry data a total of 8000 feet and five repeaters can carry 1Mbps data over 12,000 feet of category 5 UTP.

The “smart rock”: a micropower transponder

390

Dale Eager

Introduction

A “smart rock” is a locating device that is buried at a specific site. It is interrogated by a portable source and responds with information about its position, identification number or any data that it has collected since its last interrogation. Ideally, a smart rock, once placed, will wait, listening for its interrogator, for many years, or even for decades. A smart rock buried on a nature trail might send its identification number to a traveler’s handheld transponder, which would decode the identification number and play a message describing the surrounding sights. It could also be used to direct the traveler which way to turn at a trail junction. Smart rocks are sometimes placed along the edges of cliffs so that interrogators built into vehicles, such as bulldozers, will cause them to stop before they get too close to the edge.

The micropower subcircuits

The oscillator

Figure 390.1 shows the LTC1440 implementing a micropower oscillator. This circuit provides the references for both voltage and frequency needed in our rock; it draws only a few microamps of battery current.

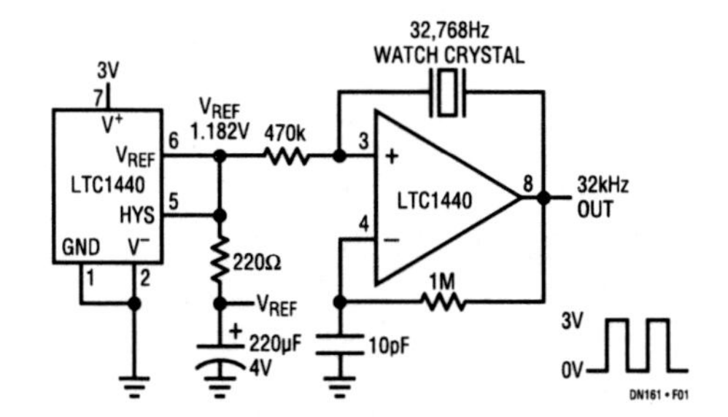

Figure 390.1 • Ultralow Power Crystal Oscillator

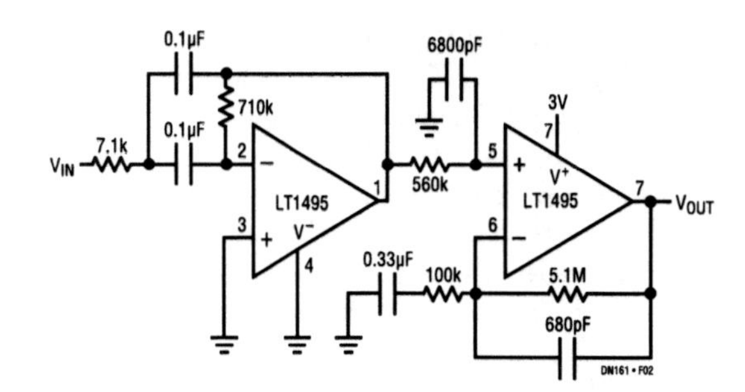

Figure 390.2 • Ultralow Power IF Amplifier (Gain of 2500 at 20Hz)

IF amplifier

Figure 390.2 details the IF amplifier, which has a gain of 2500 at a center frequency of 20Hz. By selecting the LT1495 for our amplifier, we can do this while consuming only 2µA.

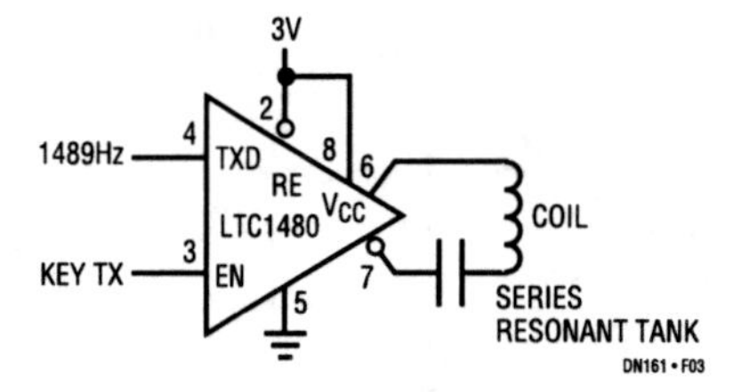

Figure 390.3 • Hefty Driver with Ultralow Sleep Current

Power driver

Figure 390.3 introduces the LTC1480 ultralow standby power RS485 transceiver. In our rock, we only use the LTC1480 in its transmit mode, where it provides currents of about 100mA. The rest of the time the LTC1480 is shut down, drawing a microampere of quiescent current.

Analog Circuit Design: Design Note Collection. http://dx.doi.org/10.1016/B978-0-12-800001-4.00390-2

The smart rock system

Receiver

The 32kHz reference frequency generated by Y1 and U1B (as shown in Figure 390.4) is divided by eleven in U2 and by two in U3A to yield 1489.5Hz, the local oscillator frequency. This LO output is applied to mixer Q3 while Q1 and Q2 are fully enhanced, causing C4 and L1 to act as a parallel resonant antenna. The output of the mixer, Q3, is fed into the IF amplifier created by U5A and U5B, where the signal is multiplied by approximately 2500. When the signal on U5B Pin 7 reaches $1.2V_{P-P}$, Q4 turns on, pulling the START signal line low.

Transmitter

Once the interrogating tone burst is over and the IF amplifier's output has decayed below $1.2V_{P-P}$, Q4 stays off and R11 is allowed to charge C11, raising the voltage on the START node. When the threshold of the clock pin of U3B is crossed, the Q output goes high and the Q signal goes low. D5 quickly discharges C13, pulling the D pin of U3B low and preventing false retriggering of U3B. D3 pulls the START signal low, preventing an early termination of a transmit cycle caused by IF overload. Q1 and Q2 turn off, causing C4 and L1 to form a series resonant circuit connected to the output of U4 (the power driver). At the same time, U4 is enabled and drives the LO frequency into the series resonant tank circuit. This transmitting action continues until R12 discharges C12 to the threshold of the reset pin (Pin 13) of U3B, at which time the flip flop is reset (Q=3V). U4 is disabled and Q1 and Q2 are enhanced.

Blanking

Resetting the flip-flop causes the Q output to toggle to 3V, which causes D3 to go into the blocking state, releasing the START signal from its forced low condition. Because the delay of R13 and C13 is longer than the delay of R11 and C11, U3B is clocked into the off state (Q=3V, the state in which it already exists) as the START signal goes through the clock pin's threshold. Meanwhile, the IF amplifier's output is decaying from the disturbance of transmitting. It decays to well below $1.2V_{P-P}$ before R13 charges C13 to the threshold of the D pin of U3B. This prevents false tripping while waiting to enable the reception of the next interrogation signal.

Conclusion

It is easy to design circuits that, when powered by a single lithium cell, will last for years or even decades while performing real-world significant functions. Linear Technology offers an extensive line of nanopower ICs, including precision operational amplifiers, comparators, voltage references, analog-to-digital converters and line drivers and receivers.

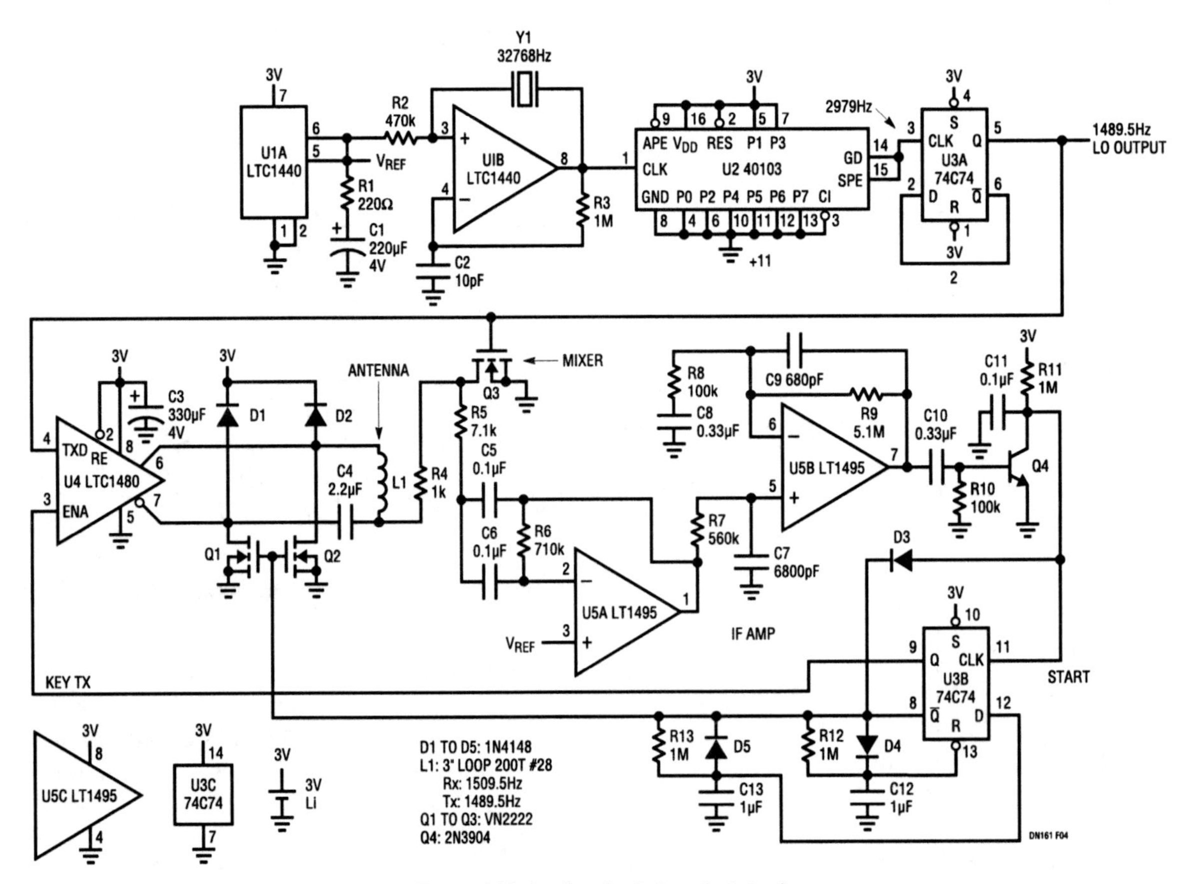

Figure 390.4 • (Lapis Orbus Astutus)

Power supplies for subscriber line interface circuits

391

Eddie Beville

As the demand for worldwide networking grows, so will the need for advanced data transmission products. In particular, ISDN services have become popular because of the recent development of the Internet. ISDN provides higher speed data transmission than standard modems used in PCs. Also, ISDN supports the standard telephone interface (voice and fax), which includes the subscriber line interface circuit (SLIC). A SLIC requires a negative power supply for the interface and the ringer voltages. The power supplies described herein are designed for these applications. Specifically, these designs address the AMD79R79 SLIC device with on-chip ringing.

Circuit descriptions

LT1171 supplies –23.8V at 50mA and –71.5V at 60mA

Figure 391.1 shows a current mode flyback power supply using the LT1171CQ device. This current mode device has a wide input voltage range of 3V to 60V, current limit protection and an on-chip 65V, 0.30Ω bipolar switch. The input voltage range for the circuit is 9V to 18V. This circuit is intended for small wall adapters that power ISDN boxes. The output voltages are −23.8V at 50mA and −71.5V at 60mA.

The circuit shown in Figure 391.1 uses the LT1171 in standard flyback topology. The transformer's turns ratio is 1:1:1:1, where 23.8V appears across each secondary winding and the primary during the switch off time. The remaining secondary windings are stacked in series to develop −47V. The −47V section is then stacked onto the −23.8V section to get −71.5V. This technique provides very good cross-regulation, lowers the voltage rating required on the output capacitors and lowers the RMS currents, allowing the use of cheaper output capacitors. Either the −23.8V output or the −71.5V output can be at full load without effecting the other corresponding output. The circuit's step response is very good; no significant overshoot occurs after either output is shorted and released. Also, the transformer windings are all quadrifilar to lower the leakage inductance and cost.

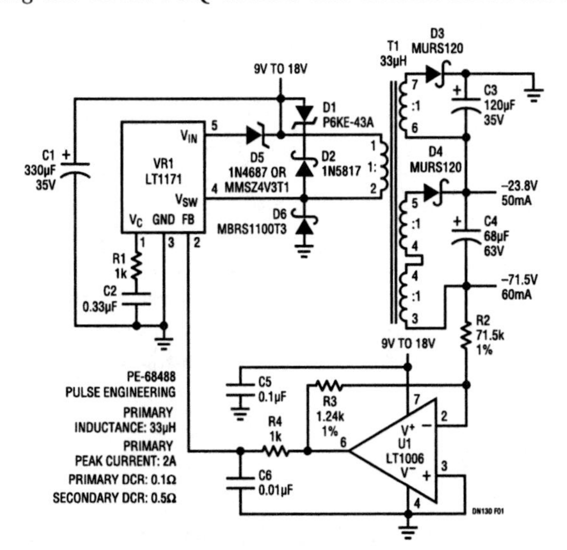

Figure 391.1 • The LT1171 In Standard Flyback Topology

LT1269 supplies –23.5V at 60mA and –71.5V at 120mA from 5V input

Figure 391.2 shows a current mode flyback power supply using the LT1269CQ device. This current mode device has a wide input voltage range, current limit protection and an onboard 60V, 0.20Ω bipolar switch. The input voltage range for the circuit is 5V to 18V. This design provides a wider input voltage range and greater output power than that of Figure 391.1. The output voltages are −23.5V at 60mA and −71.5V at 120mA (8.6W). This circuit is designed to power two SLIC devices. The circuit operation is identical to Figure 391.1, except for a larger switching regulator device (VR1) and a different transformer (T1). These changes allow for 5V operation and higher output power. This circuit is designed for full load on the −71V or −23.5V output. This accommodates the ringing on two SLICs or off hook on two SLICs.

Analog Circuit Design: Design Note Collection. http://dx.doi.org/10.1016/B978-0-12-800001-4.00391-4

Table 1 Bill of Materials

REFERENCE DESIGNATOR	QUANTITY	PART NUMBER	DESCRIPTION	VENDOR	TELEPHONE
C1*	1	ECA-1VFQ331	Capacitor, 330μF, 35V HFQ	Panasonic	
C2, C5	1	0805	Capacitor, 0.33μF Ceramic		
C3	1	UPL1V121MPH	Capacitor, 120μF, 35V Plastic	Nichicon	(708) 843-7500
C4	1	UPL1J680MPH	Capacitor, 68μF, 63V Plastic	Nichicon	
C6	2	0805	Capacitor, 0.01μF		
D1	1	P6KE-43A (MOT), TGL41-43A (GI)	Diode, 0.5W Zener	Motorola or Equiv	
D2	1	1N5817	Diode, 1A Schottky	Motorola or Equiv	
D3, D4	2	MURS120	Diode, Ultrafast	Motorola or Equiv	
D5	1	1N4687, MMSZ4V3T1	Diode, Zener	Motorola or Equiv	
D6	1	MBRS1100T3	Diode	Motorola or Equiv	
R1, R4	2	0805	Resistor, 1k, 5% SMT		
R2	1	0805	Resistor, 71.5k, 1% SMT		
R3	1	0805	Resistor, 1.24k, 1% SMT		
T1*	1	PE-68488	Transformer	Pulse Eng	
U1	1	LT1006S8	IC	LTC	(408) 432-1900
VR1*	1	LT1171CQ	IC	LTC	(408) 432-1900
*Changes and Additions for Figure 391.2's Circuit					
C1	1	205A100M	Capacitor, 100μF, 20V OS-CON	Sanyo	(619) 661-6835
D5, D7	2	1N4001	Diode		
R5, R6	2		Resistor, 50k 0.25W SMT or Through Hole		
T1	1	HM00-96553	Transformer	BI Technology	(714) 447-2656
VR1	1	LT1269CQ	IC	LTC	(408) 432-1900

R5 and R6 are preload resistors for maintaining an accurate −23.5V output at full load with the −71V output at minimum load.

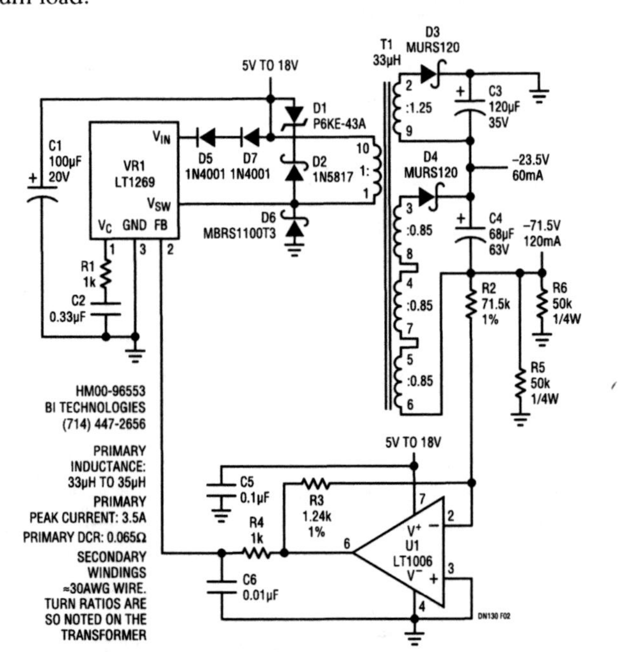

Figure 391.2 • A Current Mode Flyback Power Supply Using the LT1269CQ

Layout and thermal considerations

Printed circuit board layout is an important consideration in the design of switching regulator circuits. A good ground plane is required for all ground connections. The path from the input capacitor to the primary winding of the transformer is a high current path, and requires a short, wide copper trace (0.080″ to 0.1″). The V_{SW} pin connection also needs a short, wide copper trace. R1 and C2 need to be placed close to VR1. The secondary windings can be connected to their associated components with 0.025″ to 0.030″ traces. The feedback circuitry needs to be placed close to the FB pin of VR1. Place C5 close to U1 to decouple the op amp power supply. The LT1171CQ and LT1269CQ are surface mount devices that require about a 1″ copper pad for heat sink mounting. Heat sinking is most critical for the LT1269CQ because of its high output power. Also, via from the copper pad to the internal ground layers are highly recommended.

Bill of materials

A bill of materials has been provided with each schematic (Table 1).

Precision receiver delay improves data transmission

392

Victor Fleury

Moving data from one board to another over a backplane places stringent requirements on channel-to-channel and part-to-part skew and delay. The propagation delay of typical CMOS line receivers can vary as much as 500% over process and temperature. In high speed synchronous systems where clock and data signal timing are critical, transmission rates must often be reduced to minimize the effects of propagation delay and skew uncertainties in the receiver. The LTC1518/LTC1519/LTC1520 family of high speed line receivers solves this problem, reducing propagation delay changes to less than ±17% over production variations and temperature, a better than 10 times improvement over previous solutions.

The LTC1518/LTC1519/LTC1520 family of 50Mbps quad line receivers translate differential input signals into CMOS/TTL output logic levels. The receivers employ a unique architecture that guarantees excellent performance over process and temperature with propagation delay of 18ns ± 3ns. The architecture affords low same channel skew ($|t_{PHL}-t_{PLH}|$ 500ps typ), and low channel-to-channel propagation delay variation (400ps typ). The propagation delay and skew performance are unmatched by any CMOS, TTL or ECL line receiver/comparator.

Circuit description

Figure 392.1 shows a block diagram of the LTC1520 signal path. The input differential pair amplifies the minimum 500mV input signal level. A resistor network expands the input common mode range of the LTC1520 from 0V to 5V, and expands that of the LTC1518/LTC1519 from −7V to 12V while operating on a single 5V supply. A second differential amplifier (g_m) switches a fixed current into its load capacitance. The output of the second stage is a valid logic level that feeds inverters.

To guarantee tight delay and skew performance, delay within each receiver and between channels must be carefully matched. For the LTC1518/LTC1519/LTC1520, the inherent temperature and process tolerance, along with bias and delay trimming, make it possible to guarantee a propagation delay window more than an order of magnitude tighter than that of the typical CMOS line receiver. Since skew is caused by the unequal charging versus discharging of both internal and external capacitances, the first stages are differential to minimize these effects.

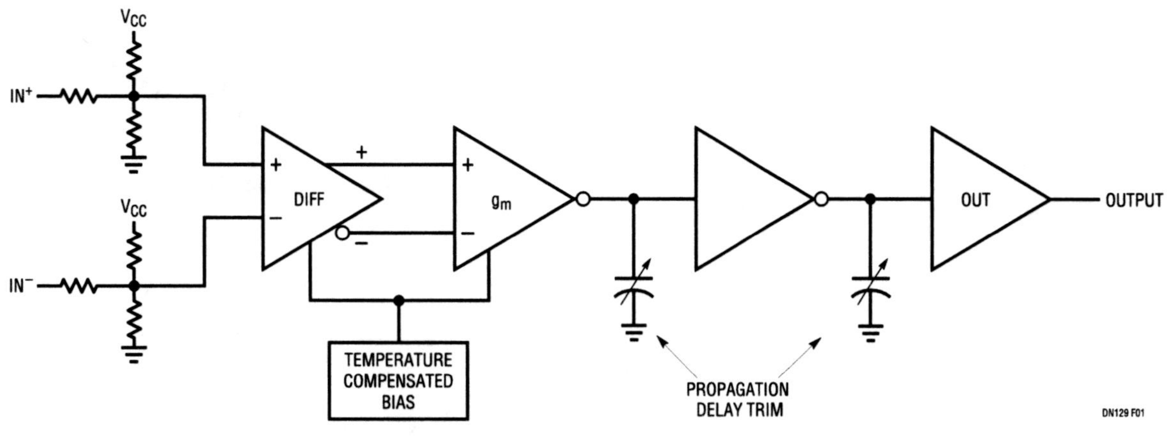

Figure 392.1 • LTC1520 Block Diagram

Analog Circuit Design: Design Note Collection. http://dx.doi.org/10.1016/B978-0-12-800001-4.00392-6

Additional features

Other features include novel short-circuit protection. If the output remains in the wrong state for longer than 60ns, the output current is throttled back to 20mA. When the short is removed and the output returns to the correct state, full output drive is restored. Because of its high input impedance (even when unpowered), the LTC1518/LTC1519/LTC1520 can be "hot swapped" without corrupting backplane signal integrity or causing latchup. The outputs remain high with shorted or floating inputs (LTC1518/LTC1519 only) and can be disabled to a high impedance state. High input resistance (≥18k) also allows multiple parallel receivers.

Applications

The LTC1520 is designed for high speed data/clock transmission over short to medium distances. Its rail-to-rail input common mode range allows it to be driven via long PC board traces, coaxial lines or long (hundreds of feet) twisted pairs. Figure 392.2 shows the LTC1520 in a backplane application. 5V single-ended signals are received with a 2.5V slicing threshold. This configuration can be adapted as a coaxial receiver. In Figure 392.3, the LTC1518 is shown in an RS485-like application, but capable of operating at up to 50Mbps, limited only by cable characteristics.

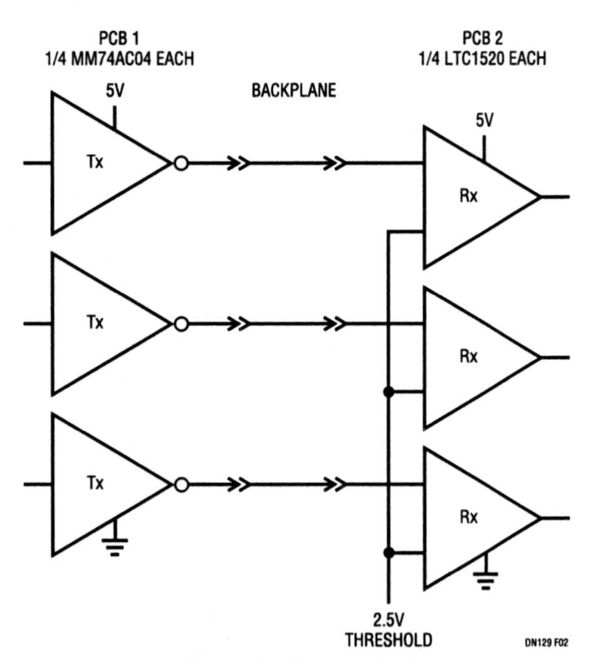

Figure 392.2 • LTC1520 in a Backplane Application

Figure 392.4 shows actual waveforms of the LTC1518 connected in the Figure 392.3 configuration (100 feet of unshielded twisted pair was used). Note that the delay of the twisted pair is almost 200ns versus the receiver's 18ns delay.

The waveforms of Figure 392.5 show 50Mbps operation using the LTC1518 with 100 feet of twisted pair.

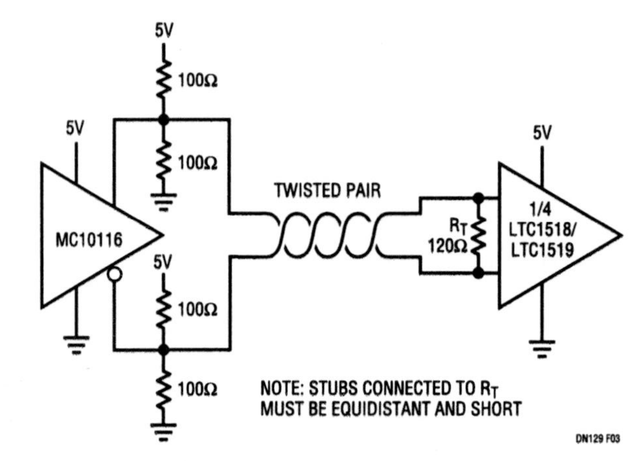

Figure 392.3 • LTC1518 Typical Application

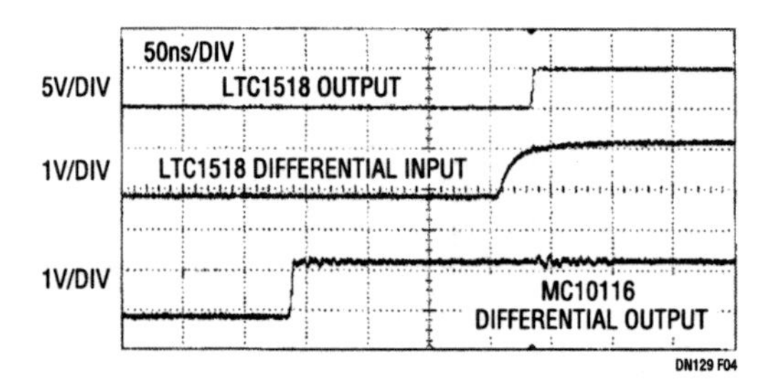

Figure 392.4 • LTC1518 100 Feet Twisted-Pair Connection

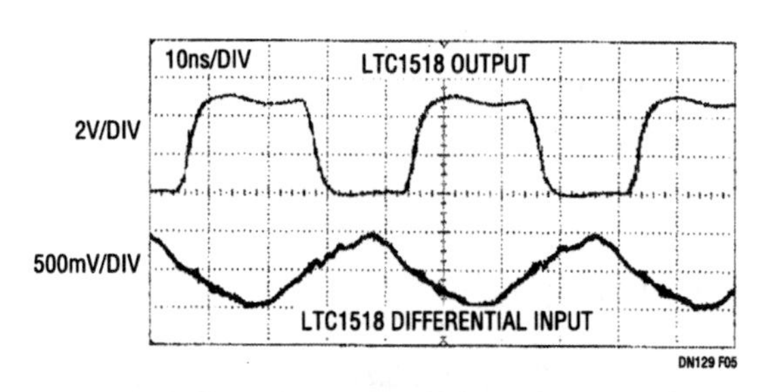

Figure 392.5 • 50Mbps Operation Using 100 Feet of Twisted Pair

RS485 transceivers reduce power and EMI

393

Dave Dwelley Teo Yang Long Yau Khai Cheong Bob Reay

Recent innovations in process and circuit design have enabled the release of three new RS485 transceivers: the LTC1481, LTC1483 and LTC1487. These devices share an improved receiver circuit which features 80μA quiescent current operation (driver disabled) with no loss in AC performance relative to standard RS485 devices, and a new 1μA shutdown mode (Figure 393.1). All three new devices are pin compatible with the industry standard LTC485 pinout, and feature Linear Technology's exclusive ±10kV ESD protection (Human Body Model) at the line I/O pins, eliminating the need for external ESD protection in most cases.

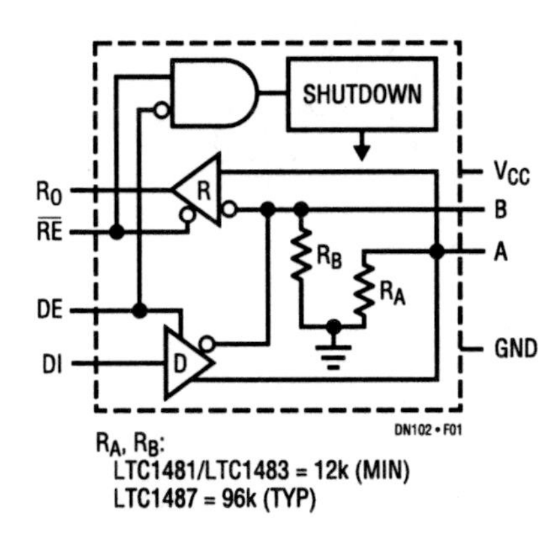

Figure 393.1 • LTC1481/LTC1483/LTC1487 Block Diagram

LTC1481

The LTC1481 provides full 2.5MBd driver and receiver speeds with the low power and improved ruggedness features shared by all three members of the family. Like all Linear Technology RS485 products, it features full RS485 and RS422 compatibility, guaranteed operation over the −7V to 12V common-mode range and a unique driver output circuit that prevents CMOS latch-up and maintains high impedance at the line pins, even when the power is off. An internal driver short-circuit current limit and a thermal overload protection circuit prevent damage under severe fault conditions. The LTC1481 is ideally suited for designs which need to transmit high speed data with minimum power consumption.

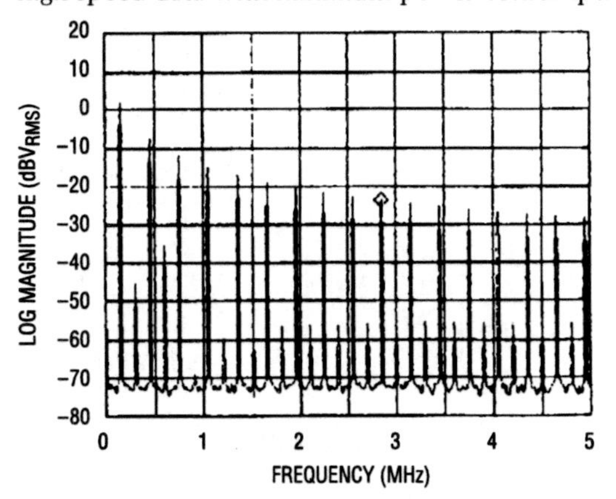

Figure 393.2a • Typical RS485 Driver Output Spectrum Transmitting at 150kHz

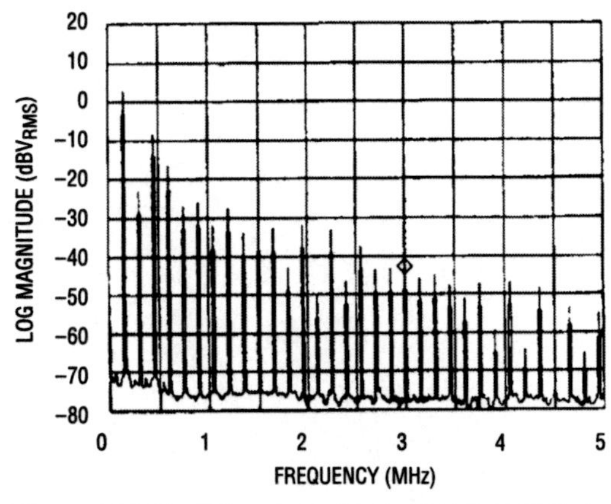

Figure 393.2b • Slew Rate Limited LTC1483 Driver Output Spectrum Transmitting at 150kHz

Analog Circuit Design: Design Note Collection. http://dx.doi.org/10.1016/B978-0-12-800001-4.00393-8

LTC1483

The LTC1483 is a reduced EMI version of the LTC1481 intended for use in systems where electromagnetic interference concerns take precedence over high data rates. The LTC1483 driver slew rate is deliberately limited to reduce the high frequency electromagnetic emissions (Figures 393.2a and 393.2b) while improving signal fidelity by reducing reflections due to misterminated cables. The maximum operating frequency of the LTC1483 driver is limited to 250kBd. All other performance parameters are unchanged from the LTC1481, including the low power receiver operation and the 1μA shutdown mode.

LTC1487

The LTC1487 shares the low power and low slew rate features of the LTC1483. Additionally, the LTC1487 is designed with a high input impedance of 96kΩ (typical) to allow up to 256 transceivers to share a single RS485 differential data line. This exceptionally high input impedance enables additional transceivers to be connected to a single RS485 line, reducing cabling costs and complexity in systems with many nodes.

The RS485 specification requires that a transceiver be able to drive as many as 32 "unit loads." One unit load (UL) is defined as an impedance that draws a maximum of 1mA with up to 12V across it. Most standard RS485 transceivers, including the LTC1481 and LTC1483, have an input resistance of approximately 12k, equivalent to 1UL, which limits a single RS485 bus to 32 nodes. With its high 96kΩ input impedance, the LTC1487 presents only 0.125UL to the line, allowing up to 256 transceivers (32UL/0.125UL=256) to be connected to the data bus line without overloading the driver (Figure 393.3).

Conclusions

The LTC1481, LTC1483 and LTC1487 make up the third generation of the Linear Technology CMOS RS485 transceiver family, all started by the original CMOS RS485 transceiver, the LTC485. These three new devices put exceptional ruggedness features and the lowest power operation available in the industry into three unique niches in the RS485 market: high performance (LTC1481), low EMI (LTC1483) and high input impedance (LTC1487).

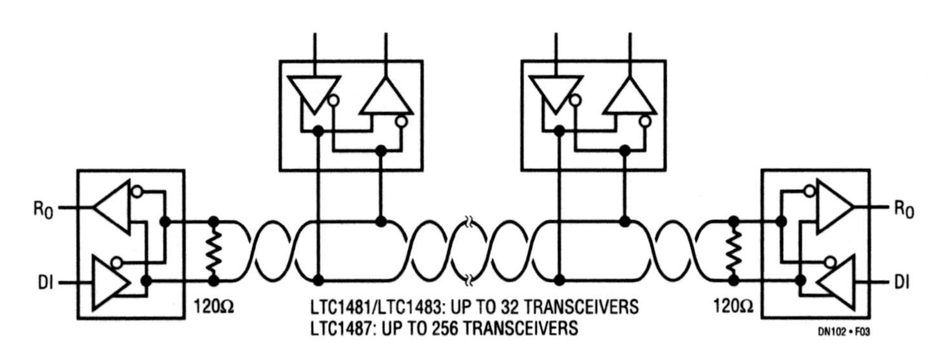

Figure 393.3 • Mulitple Transceivers on One RS485 Bus

Interfacing to V.35 networks

394

Y. K. Sim Robert Reay

What is V.35?

V.35 is a CCITT recommendation for data transmission at 48kbs. The electrical interface between data communication equipment (DCE) and data terminal equipment (DTE) includes a set of balanced differential circuits conforming to Appendix II of Recommendation V.35 and a set of control circuits conforming to Recommendation V.28 (equivalent to RS232). A typical V.35 interface uses five differential signals and five single-ended handshaking signals. The V.35 electrical specifications are summarized in Table 394.1.

Table 394.1 V.35 Electrical Specifications

SPECIFICATION	CONDITION	TRANSMITTER	RECEIVER
Source Impedance	Differential Measurement	50Ω to 150Ω	100Ω ±10Ω
Common-Mode Impedance	Terminals Shorted	150Ω ±15Ω	150Ω ±15Ω
Voltage Swing	100Ω Load	0.55V ±20%	
Common-Mode Swing	100Ω Load	0.6V Max	
Common-Mode Range	Between Dx and Rx Grounds	±4V	±4V

Problems with traditional implementations

The tight tolerance of the transmitter's impedance and voltage swing specifications makes the implementation of the transmitter a little tricky. The traditional approach is to use an RS422 differential driver such as the AM26LS30 (Figure 394.1). Because the chip has a voltage output, the signal must go through a resistive divider to meet the 0.55V swing specification. The problem is that the output voltage is a function of supply voltage, temperature and IC processing, making the 20% tolerance of the swing hard to meet for all conditions. Another problem is meeting the common-mode impedance specification of 150Ω ±15Ω because the output impedance of the driver is not well controlled. For the interface to meet the CCITT specifications under all conditions, another approach is needed.

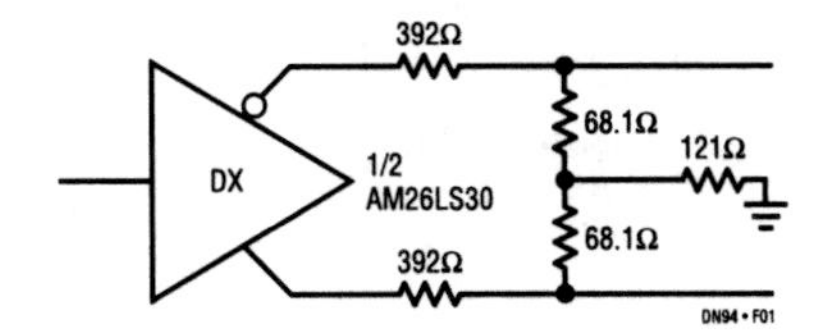

Figure 394.1 • Traditional V.35 Implementation

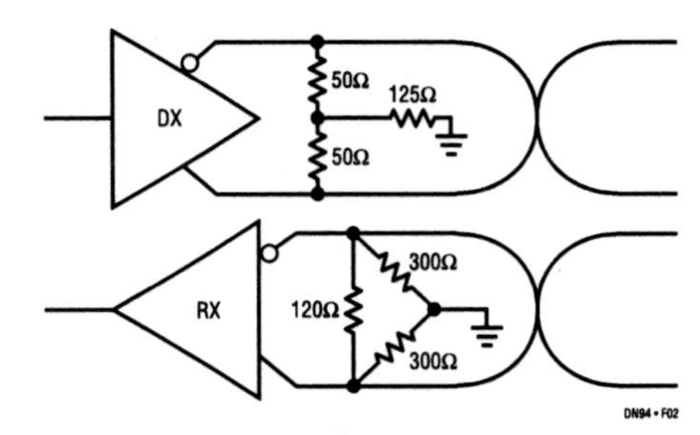

Figure 394.2 • Y and Δ Termination Networks

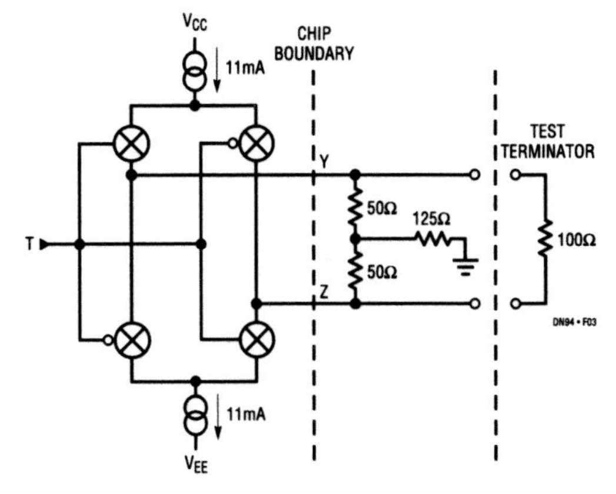

Figure 394.3 • Simplified Transmitter Schematic

Analog Circuit Design: Design Note Collection. http://dx.doi.org/10.1016/B978-0-12-800001-4.00394-X

LTC1345

The LTC1345 is a single 5V V.35 transceiver with three drivers and three receivers that are fully compliant with the V.35 electrical specifications. The chip can be configured for DTE or DCE operation or shutdown using two select pins. In the shutdown mode, supply current is reduced to 1μA.

Each driver or receiver is terminated by an external Y or Δ resistor network (Figure 394.2), guaranteeing compliance with the V.35 impedance specification. The transmitter output consists of complementary switched-current sources of 11mA as shown in Figure 394.3. With a 100Ω test termination resistor, the differential output voltage is set at 0.55V (11mA × 50Ω), and the common-mode voltage is set to 0V, thus meeting the V.35 voltage swing specifications under all conditions. A five Y-termination resistor network in a single narrow body S14 package is available from Beckman Industrial (part number: Beckman 627T500/1250, phone: 714-447-2357).

The differential input receiver has a 40mV input hysteresis to improve noise immunity and a common-mode range from −7V to 12V, allowing the receivers to be used for V.11 (RS422) applications, The output can be forced into high impedance by an output enable (OE) pin or when the receiver is deselected. The negative supply is required to meet the ±4V common-mode operating range requirement. A charge pump generates the negative supply voltage (V_{EE}) with three external 1μF capacitors allowing single 5V operation.

Complete V.35 port

Figure 394.4 shows the schematic of a complete single 5V DTE and DCE V.35 port using three ICs and eight capacitors per port. The LTC1345 with the Beckman 627T500/1250 resistor network is used to transmit and receive the differential clock and data signals. The LT1134 is used to transmit and receive the single-ended control signals.

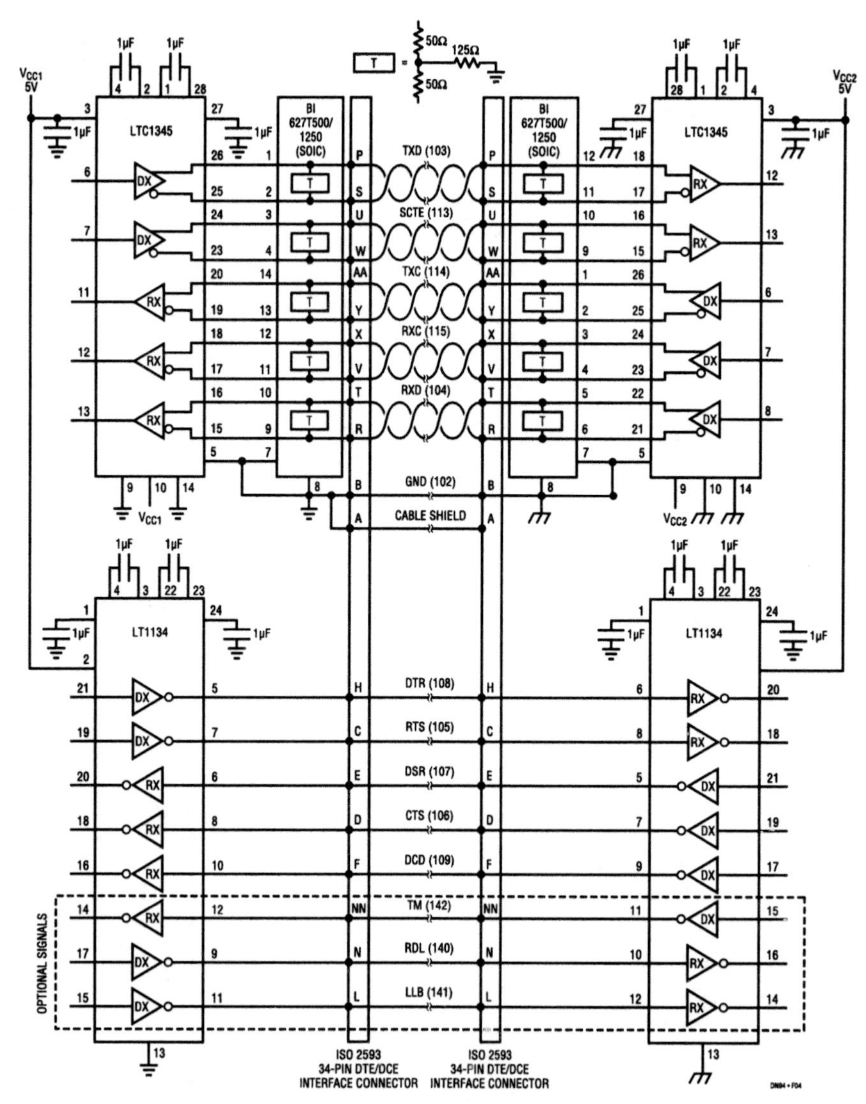

Figure 394.4 • Complete Single 5V V.35 Interface

ESD testing for RS232 interface circuits

395

Gary Maulding

In 1992 Linear Technology introduced the first RS232 interface circuits capable of surviving in excess of ±10kV ESD transients. Since that time, LTC has introduced more than 30 products with this level of protection. The inherent ruggedness of these products eliminates the need to use external protection devices in most applications. Not one unit has been returned from the field to Linear Technology for an ESD related failure analysis since the enhanced ESD protected devices were introduced.

The ±10kV ESD voltage rating is based on the human body ESD model. When evaluated with other standard ESD test methods, the superior ESD ruggedness of LTC's transceivers gives equally impressive results when compared to older conventional designs.

The various ESD test methodologies all share a common configuration as shown in Figure 395.1. A source capacitor is first charged to a high voltage, then the high voltage power supply is disconnected from the capacitor, and the capacitor is connected to the device under test through a limiting resistor. The value of the test capacitor and the limiting resistor differ among the various test standards.

The human body model is the most commonly used ESD test in the United States and is the test method prescribed by Mil-Std-883. This method simulates the ESD discharge waveform seen from human contact to a piece of electronic equipment. The source capacitor is 100pF, limited by 1.5kΩ for the human body model. Linear Technology's RS232 transceivers can withstand in excess of ±10kV when tested with the human body model.

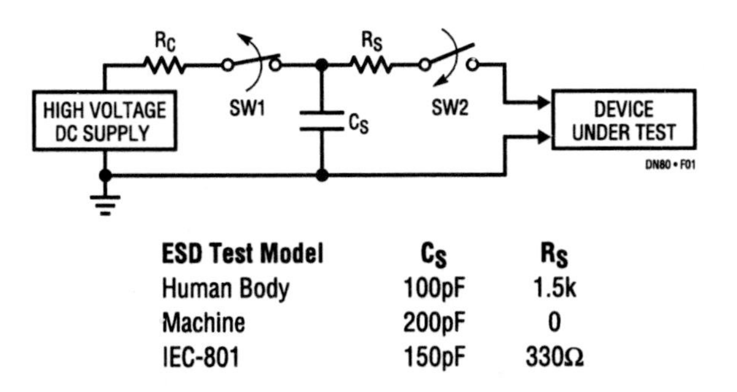

ESD Test Model	C_S	R_S
Human Body	100pF	1.5k
Machine	200pF	0
IEC-801	150pF	330Ω

Figure 395.1 • ESD Test Standards

The machine model, commonly used for ESD testing in Japan, is a more severe ESD test. This model simulates metallic contact between the device under test and a charged body. The source capacitor is 200pF with no limiting resistor. The higher source capacitance and the absence of a limiting resistor causes the device under test to be subjected to more voltage, energy, and current than human body model testing. Therefore failures occur at lower test voltages with machine model than with human body model testing. LTC's RS232 transceivers can withstand ±3.5kV when tested with the machine model.

The IEC-801 test method fits between the human body and machine methods in severity. The source capacitor is 150pF with a 330Ω limiting resistor. LTC's RS232 transceivers pass test voltages of ±7.5kV with the IEC-801 method.

The performance of LTC's 10kV protected RS232 transceivers to each of these test conditions is summarized in Table 395.1. Also included are protection levels achieved to machine model testing by including a simple RC network on the RS232 line pins. The RC network used is a "T" network formed with two 200Ω resistors and a 220pF capacitor to ground. The added resistance and capacitance are small enough to have negligible effect on RS232 signals, but provide a great increase in ESD protection at a lower cost than using TransZorbs with a diode network, which is commonly used for ESD protection. Test voltages higher than those shown in Table 395.1 sometimes cause device damage. The damage seen most commonly is an increase in driver output leakage with functionality failures occurring at even higher voltages.

ESD transients during powered operation

The test methods discussed so far involve testing for permanent damage to the integrated circuit from ESD transients. In today's portable electronics, interconnection of cables to

Analog Circuit Design: Design Note Collection. http://dx.doi.org/10.1016/B978-0-12-800001-4.00395-1

Table 395.1 LTC RS232 Transceiver ESD Test Results

ESD TEST MODEL	DRIVER PIN PROTECTION	RECEIVER PIN PROTECTION
Human Body	±10kV	±10kV
Machine	±3.5kV	±6kV
IEC-801	±7.5kV	±8kV
Machine Model with RC Network on RS232 Pins	±10kV	±10kV

the communications ports may occur while the equipment is operating. This makes it imperative that the circuit can tolerate the ESD transient with minimal disruption of system operation. LTC's RS232 interface circuits can withstand 10kV ESD transients while operating, shut down, or powered down. Disruption of data transfer is unavoidable during the ESD transient event, but data transmission may resume upon the completion of the event.

Figure 395.2 is a scope photograph of the data transmission interruption and recovery seen when a −10kV ESD transient strikes a communications line. The test circuit of Figure 395.3 was used to record this event. The ESD strike is applied to the driver output of an LT1180A and the receiver input of an LT1331. The ESD transient is of too short a duration to be recorded on the photograph, but the effects of the transient can be seen by the corruption of data after the strike. The circuits require about 20μs to recover from the event, after which data transmission continues normally.

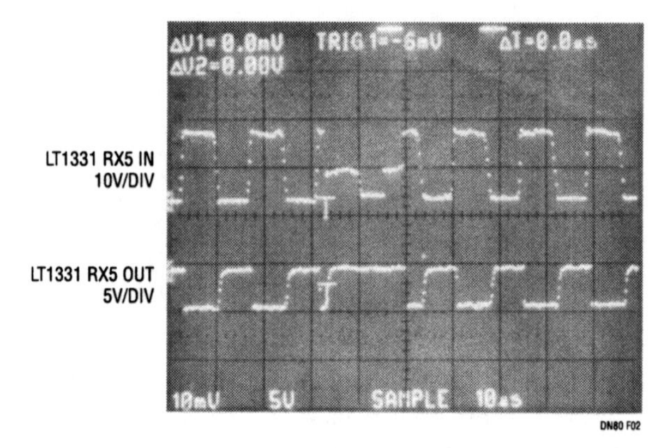

Figure 395.2 • Effects of ESD Transient on Data Transmission Through an LT1331

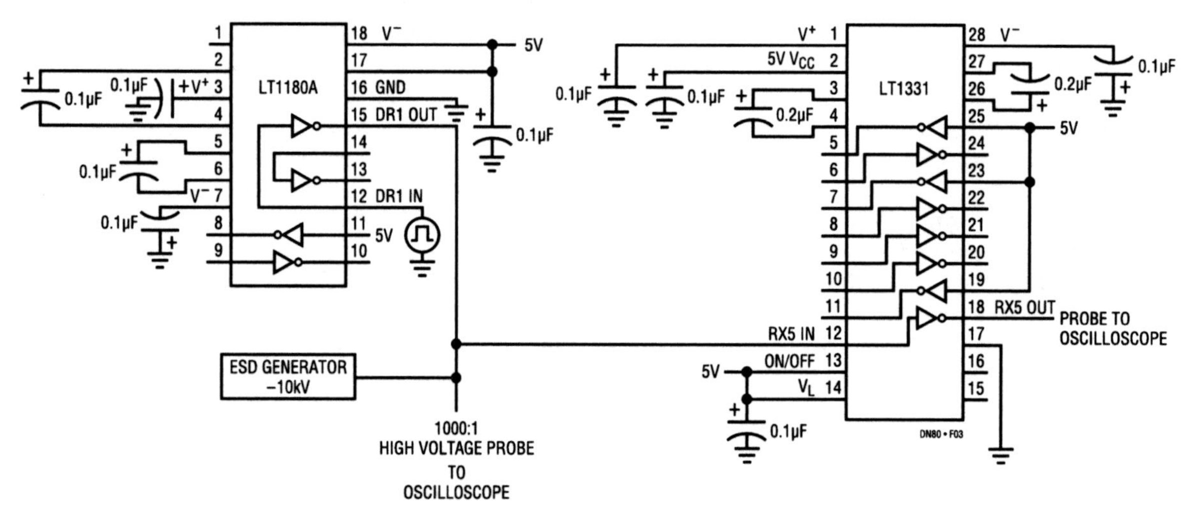

Figure 395.3 • Operating Condition ESD Test Circuit

RS232 interface circuits for 3.3V systems

396

Gary Maulding

The rapid, widespread use of 3.3V logic circuits complicates the selection of RS232 interface circuits. The optimum choice of an interface circuit should be based upon several application dependent factors:

1) Logic circuitry connected to interface chip
2) Power supply voltages available
3) Power consumption constraints
4) Serial interface environment
5) Mouse driving requirements

As Figure 396.1 illustrates, 5V interface circuits cannot be used to directly connect to 3.3V CMOS logic circuits. The receiver output level will forward bias the logic circuit's input protection diode, causing large current flow. In the worst case the CMOS logic circuit may latch up. Resistor voltage dividers or level shift buffers may be used to prevent forward biasing the CMOS input diode, but an RS232 transceiver designed for 3V logic application prevents this problem without extra components or power dissipation.

Many of today's systems have both 5V and 3V power supplies. In these systems, an RS232 interface chip which uses the 5V supply for charge pump and driver operation and the 3V supply for receiver output levels, provides the best performance. The 5V operation of the charge pump and drivers gives full RS232 output levels and sufficient current drive for operating a serial port mouse. The LT1342, LT1330, and LT1331 are all good RS232 transceiver choices for systems with both 5V and 3V power. Typical performance waveforms for the LT1342 operating with $V_{CC}=5V$ and $V_L=3.3V$ are shown in Figure 396.2.

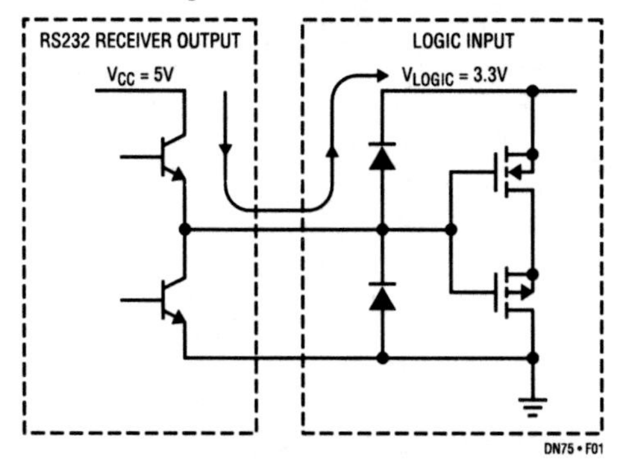

Figure 396.1 • 5V Receiver Forward Biases Logic Input Diode

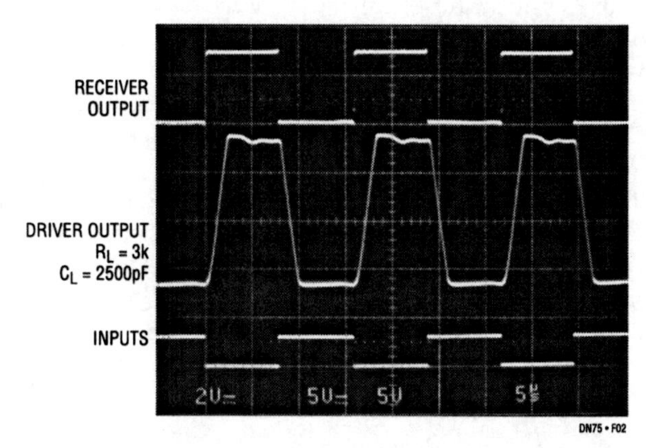

Figure 396.2 • LT1342 Outputs for $V_{CC}=5V$ and $V_L=3.3V$

Systems with only a 3V power supply are unable to use 5V powered RS232 interface circuits. Charge pump triplers (or quadruplers) have losses too great for generating RS232 voltage and current levels from a 3.3V supply. The LT1331 and LTC1327 provide solutions for 3V only systems. The LT1331 circuit is usable in both 5V/3V mixed or 3V only systems. When the charge pump is operated from 3V supplies, it powers the driver circuitry to provide RS562 output levels (see Figure 396.3). RS562 is a newer serial data interface standard than RS232 with lower (±3.7V) driver output levels and extended (64kBd vs 20kBd) data rates. RS562 systems and RS232 systems are universally interoperable.

The LTC1327 also provides RS562 output levels from a 3V supply. This circuit features ultralow 300μA supply current to maximize battery life. An advanced CMOS process makes this low current operation possible without compromising the

Analog Circuit Design: Design Note Collection. http://dx.doi.org/10.1016/B978-0-12-800001-4.00396-3

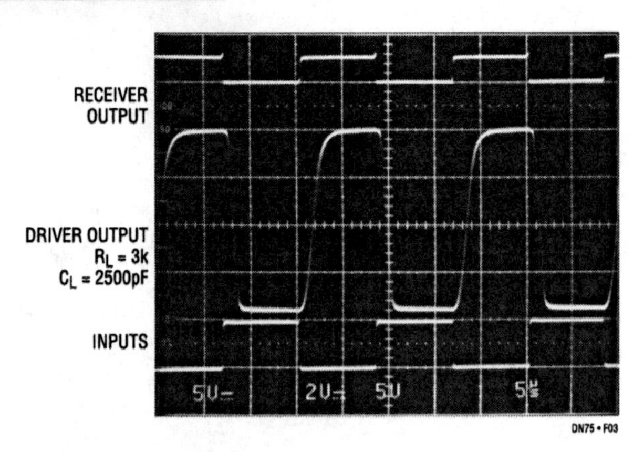

Figure 396.3 • LT1331 Outputs for $V_{CC}=V_L=3.3V$

rugged overvoltage and ESD protection available on Linear Technology's bipolar interface circuits.

VPP switcher drives 3V RS232

When fully RS232 compliant operation or mouse driving is required in a 3V only system, the LT1332 provides the solution. The LT1332 is specifically designed to be used in conjunction with a micropower switching regulator like the LT1109A. The switcher provides 12V needed for flash memory VPP and the RS232 V^+. A capacitor from the switcher's drive pin (V_{SW}) to on-chip diodes in the LT1332 form a charge pump to generate the V^- needed for the RS232 drivers. This two chip solution for VPP generation and RS232 interface is a very economical solution in 3V systems where both these needs coexist. The output driver levels of the LT1332 are fully RS232 compliant and capable of driving serial mice, a capability which cannot be met by other 3V operating circuits.

Battery-powered 3V systems can use the RS232 transceiver's SHUTDOWN and Driver Disable controls to maximize battery charge life. These operating mode controls reduce power consumption when communications needs allow the transceiver to be partially or fully turned off. Keep-alive receivers, available on some transceivers, consume little power (60μA) while monitoring a data line. When data is detected, the system can be fully powered up to accept and process the incoming data.

ESD protection

ESD transient protection of data lines is essential for equipment reliability. Traditional protection measures using TransZorbs and diodes are a large percentage of total interface port component costs. Linear Technology's RS232 and RS562 interface circuits reduce this cost by providing 10kV "Human Body Model" ESD protection on the RS232 data lines without external components. This level of protection is adequate in most applications, but when even higher levels of protection are needed, a simple RC network (see Figure 396.4) may be used. The RC network raises the ESD protection level to 10kV "Machine Model" discharges at a lower cost than TransZorb-based protection networks (see Table 396.1).

Table 396.1 RS232/RS562 Transceivers for 5V/3V and 3V Systems

PART NO.	5V/3V RS232	3V RS562	10kV ESD	COMMENTS
LT1342	✓		✓	LT1137A Pin Compatible
LT1330	✓		✓	Low Power Burst Mode
LT1331	✓	✓	✓	V_{CC} Not Used in SHUTDOWN
LTC1327		✓	✓	300μA Supply Current
LT1332	3V RS232 Used with LT1109A VPP Generator			

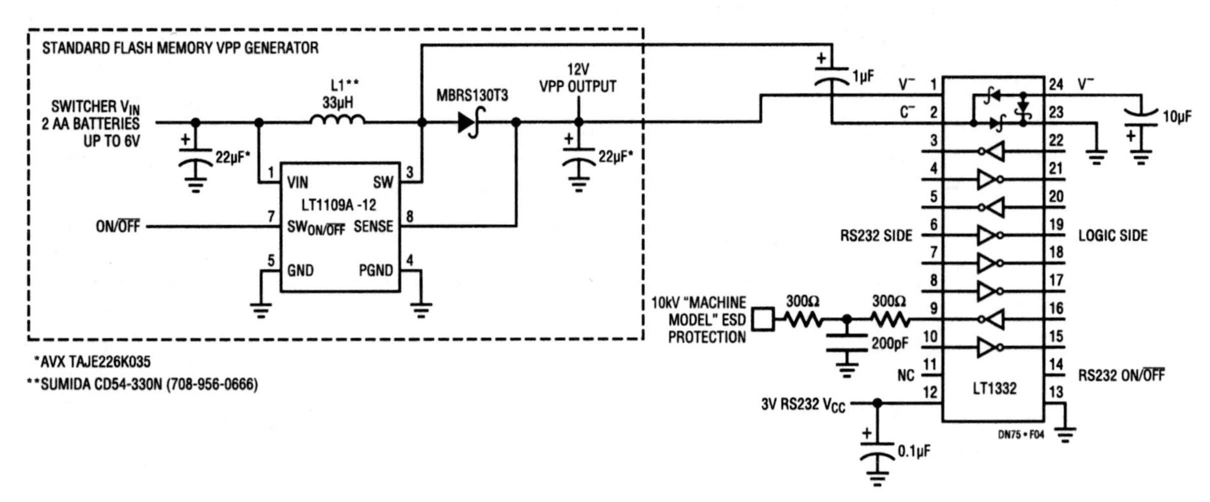

Figure 396.4 • LT1109A-12 and LT1332 Provide VPP Supply and RS232 Interface

RS232 transceivers for handheld computers withstand 10kV ESD

397

Sean Gold

Battery-powered computers and instrumentation are often subjected to severe electrical stress which imposes some stringent demands on serial communication interfaces. As always, operating from a battery mandates minimal power consumption. Transceivers must also tolerate repetitive electrostatic discharge (ESD) pulses because cable connections frequently come in contact with humans and other charged bodies.

Linear Technology's LT1237 addresses the above requirements. The LT1237 is a complete RS232 port, with three drivers, five receivers and a regulated charge pump. Supply current is typically 6mA, but the device can be shut down with two separate logic controls. The driver disable pin shuts off the charge pump and the drivers – leaving all receivers active, $I_{SUPPLY} = 4mA$. The ON/$\overline{OFF}$ pin shuts down all circuitry except for one micropower receiver, $I_{SUPPLY} = 60\mu A$. The active receiver is useful for detecting start-up signals. The LT1237 operates up to 120kBd and is fully compliant with all RS232 specifications. Connections to the RS232 cable are protected with internal ESD structures that can withstand repetitive ±10kV human body model ESD pulses.

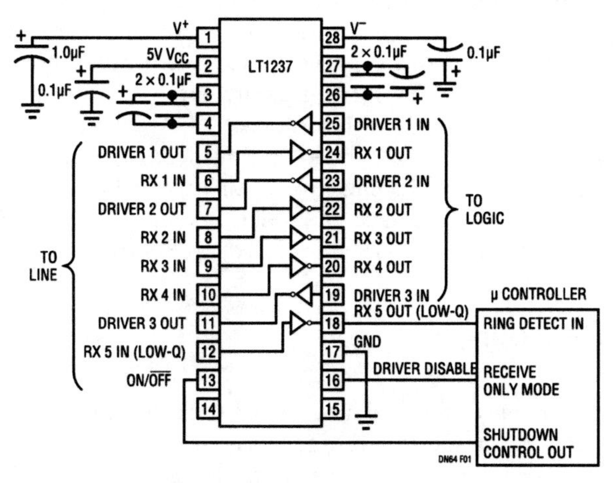

Figure 397.1 • LT1237 Application Circuit

Figure 397.1 shows a typical application circuit. The LT1237's flow through pinout and its ability to use small surface mount capacitors, helps reduce the interface's overall footprint.

Interfacing with 3V logic

Handheld computers are rapidly moving to 3V logic to save power. Yet higher voltage buses are still utilized elsewhere in the system for display driving and other functions. The LT1330 is functionally equivalent to the LT1237 but operates from 5V with a separate logic supply to interface directly with 3V logic. (Figure 397.2)

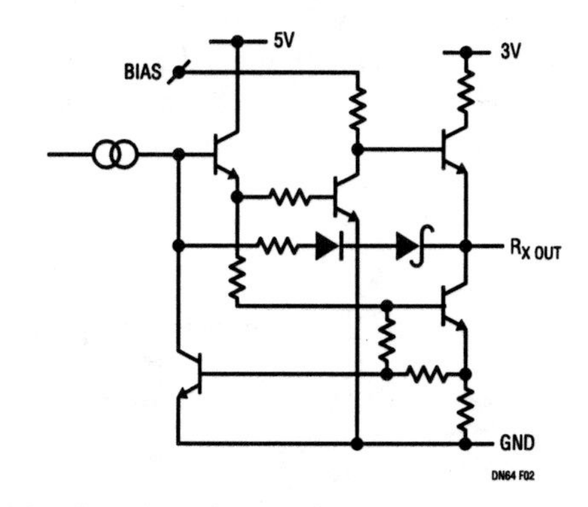

Figure 397.2 • Receiver Output Stages in the LT1330 Are Biased from a Separate Logic Supply to Easily Interface with 3V Systems

ESD protection techniques

Even though the I/O pins on the LT1237 and LT1330 are protected, a basic understanding of electrostatic discharge, its causes and its remedies, is helpful when designing with these circuits.

Analog Circuit Design: Design Note Collection. http://dx.doi.org/10.1016/B978-0-12-800001-4.00397-5

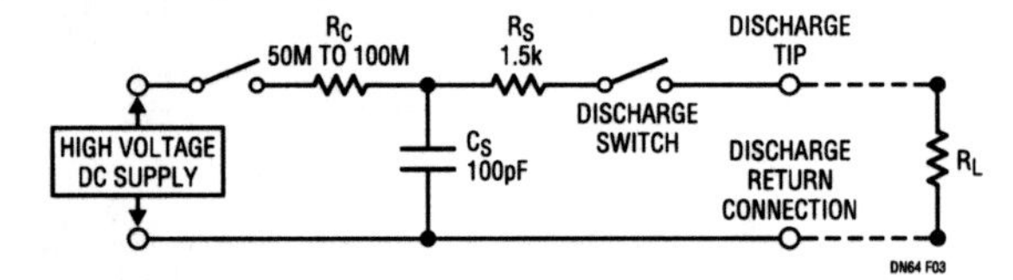

Figure 397.3 • Human Body Circuit Model for ESD Pulses

ESD generated by triboelectric charging of the human body is often the most troublesome problem for portable computers.[1] Energy imparted during a discharge is usually in the form of a rapidly rising high voltage pulse with a slow exponential tail. ESD pulses can be modeled with the switching circuit shown in Figure 397.3. ESD contributes frequency components well into the GHz range. At such frequencies, nearby cables and PC board traces look like receiving antennas for ESD noise.

Circuit damage from ESD can occur as a result of three effects: (1) High current heating, which destroys junctions or metallization. (2) Intense electromagnetic fields, which break down junctions or thin oxides. (3) Radiated noise, which drives the circuit into invalid or locked up states.

Any action which eliminates the charge generator, circumvents charge transfer, or enhances the circuit's ability to absorb energy, will increase a circuit's tolerance of ESD. Eliminating the ubiquitous charge generators and disrupting charge transfer are difficult tasks because they demand strict control of the circuit's operating environment. A more practical approach is to limit ESD entry points by shielding the circuit's enclosure and covering the RS232 port's connector when it is not in use.

Another practical remedy is to increase a transceiver's ability to absorb energy by clamping the RS232 line to ground with fast acting avalanche diodes or dedicated transient suppressors (Figure 397.4). Discrete suppressors are widely available and are extremely effective. Designers are often reluctant to use discrete suppressors because they are expensive. Costing up to $0.40/pin, they can sometimes exceed the cost of the transceiver.

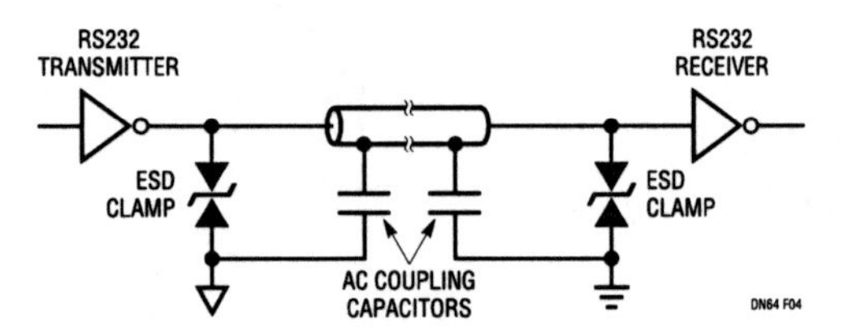

Figure 397.4 • Older Interface Designs Used External ESD Clamps

The LT1237 and LT1330 incorporate the clamps for diverting ESD energy on chip. These active structures quickly respond to positive or negative signals at threshold voltages higher than RS232 signals, yet below destructive levels for the device. The path of high current flow is through large pn junctions which increases the capacity to absorb energy.

When a discharge occurs, the resulting current flow is insignificant when the transceiver is turned off or powered down. When operating, the resulting current may debias internal circuitry and lock up the circuit. Observations have shown these nondestructive errors to be highly dependent upon the logical state of the transceiver. Cycling the power clears the circuit.

When very high levels of ESD protection are required, an external LC filter (Figure 397.5) can be used to drop ESD energy into a range that can be safely dissipated within the transceiver.

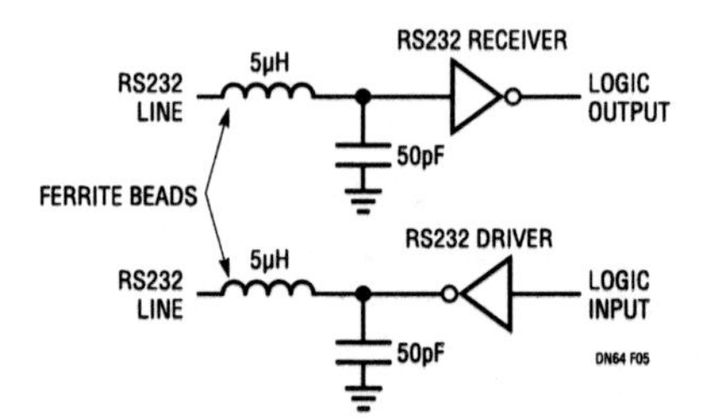

Figure 397.5 • External LC Filters Provide Protection from Very High Levels of ESD Yet Cost Less Than Discrete Suppressors

PC board layout

Energy shunted through an ESD clamp can still cause problems if the impedance of the return path is large enough to create a sizable voltage drop. Such voltage drops may damage unprotected components that share the common return line. Including a low inductance ground plane in the PC board is therefore essential for good ESD protection. For the LT1237 and LT1330, the AC path to ground through V− must also be low impedance. Adding a few hundred picofarads of low ESR capacitance in parallel with the primary storage capacitor provides a good AC ground.

When using discrete transient suppressors or filters, place components as close as possible to the connector with short paths to the return plane. Make the spacing between the circuit board traces as wide as possible. ESD pulses can easily arc from one trace to another when the spacing between traces is narrow. Arcing occurs slowly compared with ESD rise time, so air spark gaps alone will not protect circuitry from ESD. Dedicated spark gaps are effective for limiting ESD energy when used with additional suppression devices.

Do not float the cable shield with respect to local ground. Designers may feel inclined to do this to avoid circulating current due to differences in ground potential. Instead, AC couple the grounds so they are shorted at ESD frequencies.

Conclusion

The techniques described here cannot entirely eliminate ESD problems, but understanding ESD's nature and using careful circuit design, will help protect against its intrusion.

Note 1: Triboelectricity is the charge created as a result of friction between bodies.

Low power CMOS RS485 transceiver

398

Robert Reay

Introduction

The EIA RS485 data transmission standard has become popular because it allows for balanced data transmission in a party line configuration. Users are able to configure inexpensive local area networks and multi-drop communication links using twisted pair wire and the protocol of their choice.

Previous RS485 transceivers have been designed using bipolar technology because the common mode range of the device must extend beyond the supplies and be immune to ESD damage and latchup. Unfortunately, the bipolar devices draw a large amount of supply current and are unacceptable for low power applications. The LTC485 is the first CMOS RS485 transceiver featuring ultralow power consumption ($I_{CC} = 500\mu A$ max.) without sacrificing ESD and latchup immunity.

Proprietary output stage

The LTC485 driver output stage of Figure 398.1 features a common mode range that extends beyond the supplies while virtually eliminating latchup and providing excellent ESD protection. Two Schottky diodes SD3 and SD4 are added to a conventional CMOS inverter output stage. The Schottky diodes are fabricated by a proprietary modification to a standard N-well CMOS process. When the output stage is operating normally, the Schottky diodes are forward biased and have a small voltage drop across them. When the output is in the high impedance state and is driven above V_{CC} or below ground by another driver on the party line, the parasitic diode D1 or D2 will forward bias, but SD3 or SD4 will reverse bias and prevent current from flowing into the N-well or substrate.

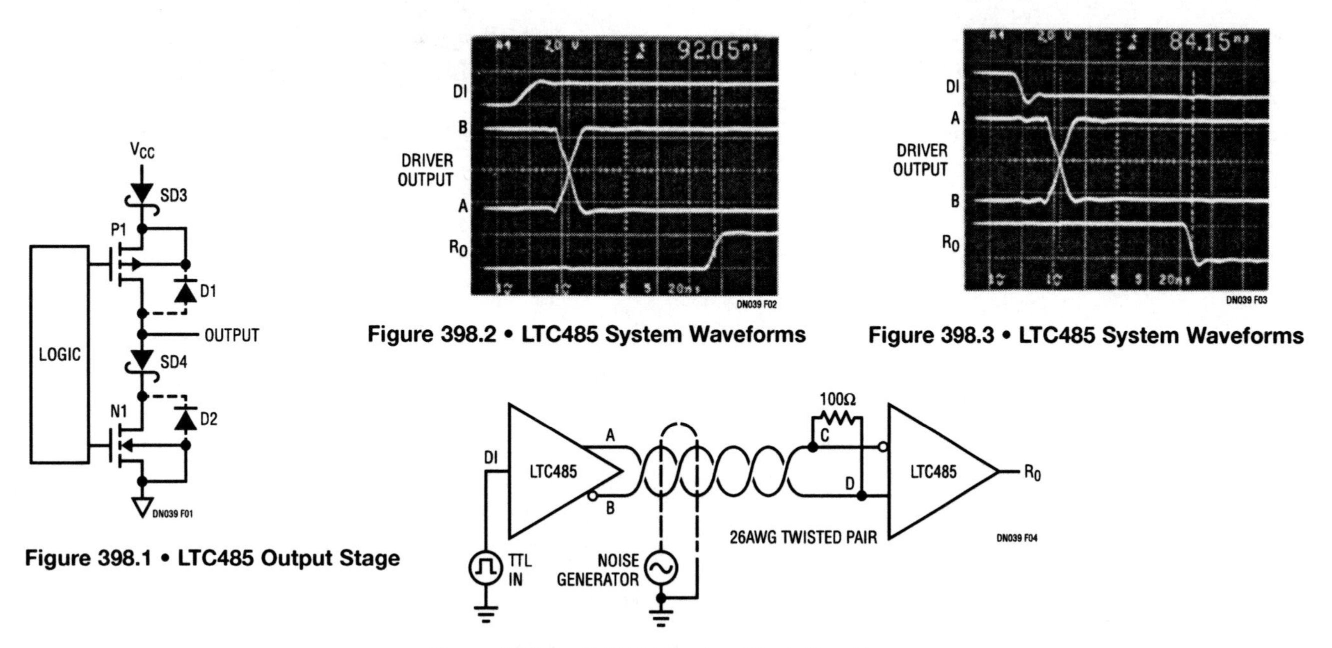

Figure 398.1 • LTC485 Output Stage

Figure 398.2 • LTC485 System Waveforms

Figure 398.3 • LTC485 System Waveforms

Figure 398.4 • LTC485 System Test Circuit

Analog Circuit Design: Design Note Collection. http://dx.doi.org/10.1016/B978-0-12-800001-4.00398-7

Thus, the high impedance state is maintained even with the output voltage beyond the supplies. With no current flow into the N-well or substrate, latchup is virtually eliminated.

Propagation delay

Using the test circuit of Figure 398.4 with only one foot of twisted pair wire, Figures 398.2 and 398.3 show the typical propagation delays.

LTC485 line length vs data rate

The maximum line length allowable for the RS422/RS485 standard is 4000 feet. Using the test circuit of Figure 398.4 with 4000 feet of twisted pair wire, Figure 398.5 and 398.6 show that with ≈20Vp–p common mode noise injected on the line, the LTC485 is able to reconstruct the data stream at the end of the wire.

Figures 398.7 and 398.8 show that the LTC485 is able to comfortably drive 4000 feet of wire at 110kHz.

When specifying line length vs maximum data rate the curve in Figure 398.9 should be used.

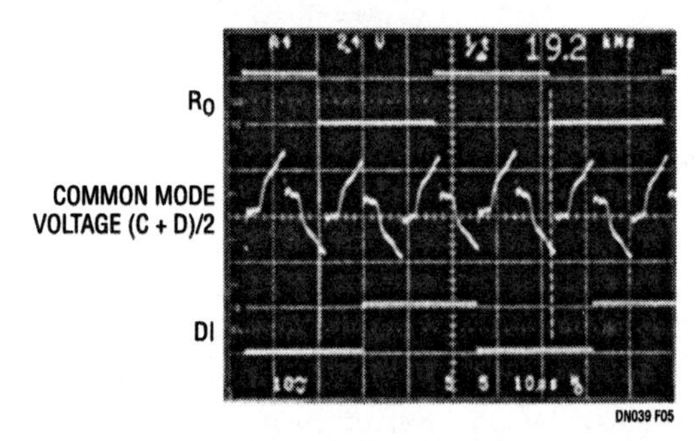

Figure 398.5 • System Common Mode Voltage @ 19.2kHz

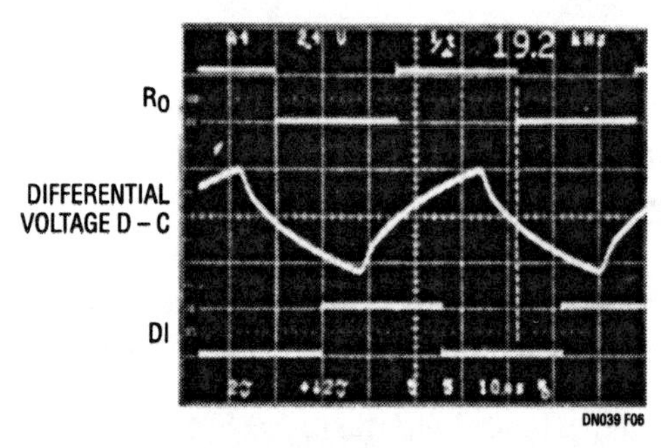

Figure 398.6 • System Differential Voltage @ 19.2kHz

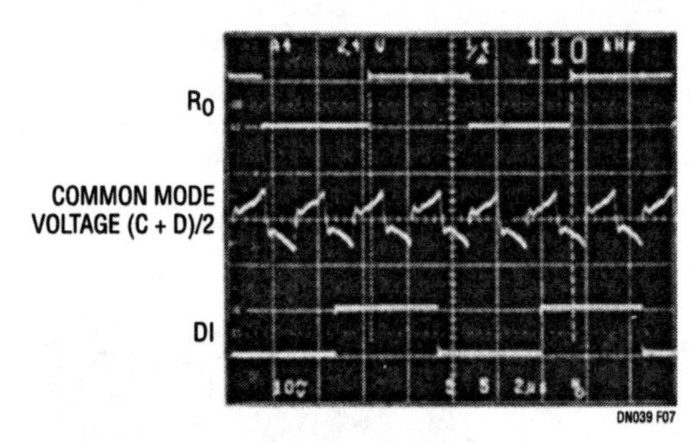

Figure 398.7 • System Common Mode Voltage @ 110kHz

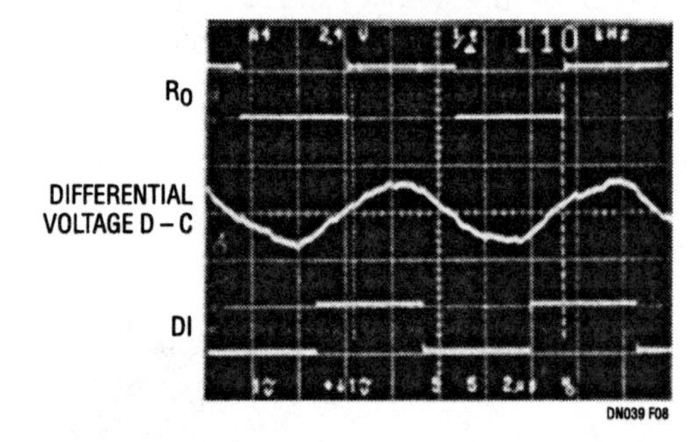

Figure 398.8 • System Differential Voltage @ 100kHz

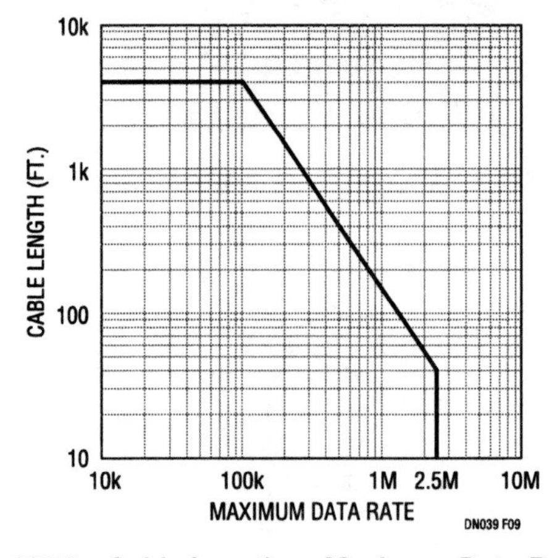

Figure 398.9 • Cable Length vs Maximum Data Rate

399

Active termination for SCSI-2 bus

Sean Gold

Overview of SCSI-2

The SCSI-2 bus[1] is an interface for computers and instrumentation that communicate over small distances—often within the same cabinet. Like GPIB (IEEE 488), SCSI's hardware and software specifications are designed to coordinate independent resources such as disk and tape drives, file servers, printers, and other computers. SCSI-2 is a bidirectional bus, which must be terminated at both ends to 2.85V (Figure 399.1). The terminators are needed because SCSI-2 uses simple open collector output drivers in its transceivers. Terminators link communicating devices to the supplies, and roughly match the transmission line's characteristic impedance. When the load to the bus increases, the role of the termination network becomes more important for maintaining signal integrity at high data rates. An active termination design is now a part of the SCSI-2 standard and is presented here in-depth.

The single-ended SCSI-2 bus is limited to six meters in length, and supports variable speed communication up to 5M transfers/sec. The bus nominally uses 18 data lines which defines the loading requirements for the terminators, because each output driver can sink at most 48mA. Up to eight SCSI devices can access the bus at regular distances along the cable.

Table 399.1 Single-Ended SCSI-2

PARAMETER	VALUE	COMMENTS
Termination Supply	4.25 < TERMPWR < 5.25	0.9A Typical 1.5A Worst Case
Logic Supply	V_{OUT} = 2.85V @0.5A 2.6 < V_{OUT} < 2.9	Per Terminator
Data Rate	5M Transfers/Sec.	Six Meters Max.
Cable Impedance	110Ω 80 < Z_0 < 140	Nominal
Transceivers	TTL Compatible	Negative True Logic 5V = 0, 0V = 1
Signal Levels	0 < V_{OL} < 0.5 2.5 < V_{OH} < 5.25 V_{IL} < 0.8 2.0 < V_{IH} 0.2 < Hysteresis	 –0.4mA < I_{IL} < 0 mA 0.0mA < I_{IH} < 0.1mA
Short Circuit Current	48mA/Transceiver	Based on Old TTL Spec

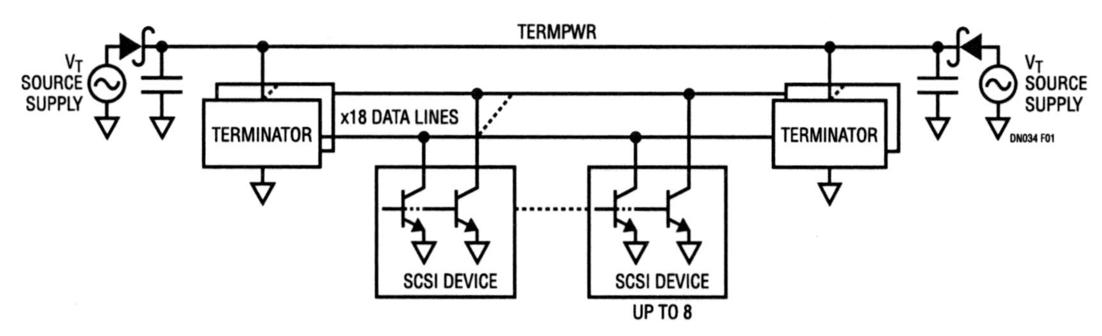

Figure 399.1 • Global View of the SCSI-2 Bus

Note 1: SCSI-2 = Small Computer System Interface Version 2, pronounced "Scuzy-2." The complete specifications standard is available through ANSI #X3T9.2.

Analog Circuit Design: Design Note Collection. http://dx.doi.org/10.1016/B978-0-12-800001-4.00399-9

Any two devices can terminate the cable, but bit error rates are minimized with the terminators attached only at the ends. Local capacitive loading is low under these conditions, making the transmission line more consistent with fewer discontinuities.

SCSI-2's key specifications are repeated from the ANSI standard in Table 399.1.

Shortcomings of passive terminators

The resistive voltage divider shown in Figure 399.2 is commonly used to terminate the SCSI bus. Multiple power sources are allowed to connect to the SCSI cable. Each source is protected with a Schottky diode to prevent damage from reverse currents. The resulting termination power signal, TERMPWR, is not well regulated—subject to variations in source supplies and protection diodes, as well as ohmic losses. Unfortunately, these changes in TERMPWR translate directly to the bus through the resistive divider, which degrades noise margins.

The low values for R1 and R2 reflect a compromise between driver sink current and impedance matching the signal lines. Normally, high resistances would be desirable to minimize driver sink current. Yet, the terminator should match the signal line's 110Ω characteristic impedance, and the bus's quiescent state must be above the TTL logic threshold. It is not possible to meet all of these objectives simultaneously. The SCSI standard suggests R1 = 220Ω and R3 = 330Ω. The resulting bus voltage is 3V with 132Ω impedance, which is mismatched to the nominal 110Ω cable impedance. The Schottky diode aggravates the mismatch because it presents a poor AC ground. In addition to these problems, the small resistors draw 300mA Q-current from TERMPWR, assuming 18 signal lines with the bus inactive.

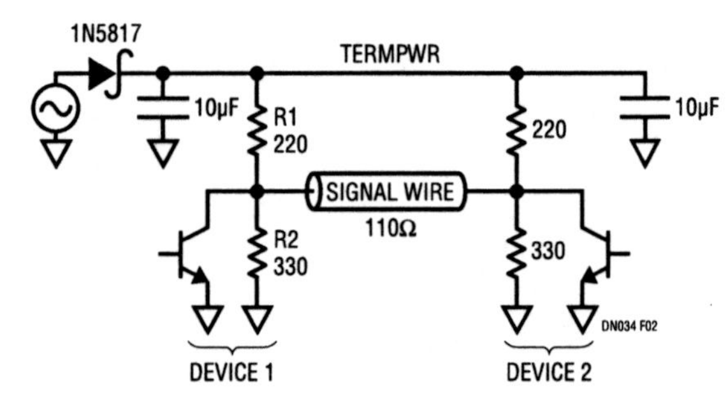

Figure 399.2 • Passive Termination

Active terminators

The active terminator shown in Figure 399.3 uses an LT1117-2.85 low dropout regulator to control the logic supply. The LT1086's line regulation makes the output immune to variations in TERMPWR. After accounting for resistor tolerances and variations in the LT1117's reference, the absolute variation in the 2.85V output is only 4 percent over temperature. When the regulator drops out at TERMPWR – 2.85V = 1.25V, the output linearly tracks the input with a slope of 1V/V. Signal quality is quite good because the 110Ω series resistor closely matches the transmission line's characteristic impedance, and the regulator provides a good AC ground.

In contrast to the passive circuit, two LT1117s require only 20mA quiescent current. For the power levels in this application, the LT1117 does not need a heat sink, and is available in low cost, space saving SOT-223 surface mount packages. Beyond solving basis signal conditioning problems, the LT1117 handles fault conditions with short circuit current limiting, thermal shutdown, and on-ship ESD protection circuitry.

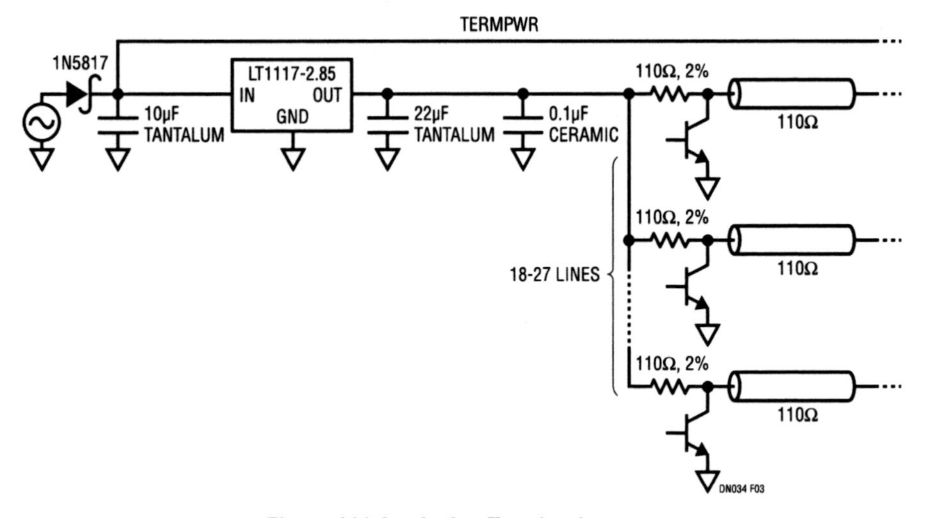

Figure 399.3 • Active Termination

RS232 transceiver with automatic power shutdown control

400

Sean Gold

The LT1180/81 RS232 transceivers with on-chip charge pumps offer some unique features that greatly enhance serial interface performance. Like the LT1080 and LT1130 series transceivers, the LT1180 is fully compliant with all RS232 specifications. The LT1180 is unique since it utilizes a charge pump which oscillates at 150kHz to 200kHz—about twice the frequency of the standard transceivers. In addition to providing excellent current delivery capability, the high speed charge pump can operate with storage capacitors as small as 0.1μF.

Reducing storage capacitor size to 0.1μF shrinks board space, thereby lowering production costs. Small capacitors also shorten the transceiver turn-on time to less than 200μs, which makes the LT1180 ideal for applications which must address the RS232 transceiver quickly. The interface described here takes advantage of fast turn-on to reduce power dissipation.

The circuit shown in Figure 400.1 automatically shuts down when there is no data flow through the interface. A data stream on either the RS232 or logic inputs activates the transceiver. The data must begin with a logic 1 preamble, and the data stream must contain a sufficient number of 1's to keep the transceiver active. The preamble may be as short as 50μs. Alternatively, the input to the Automatic SHUTDOWN circuit could be an RS232 handshake signal, such as Data Set Ready (DSR) or Clear to Send (CTS), which remain high during the data transfer. The LT1180's 200μs turn-on delay does not limit the data rate in the transceiver. Once the

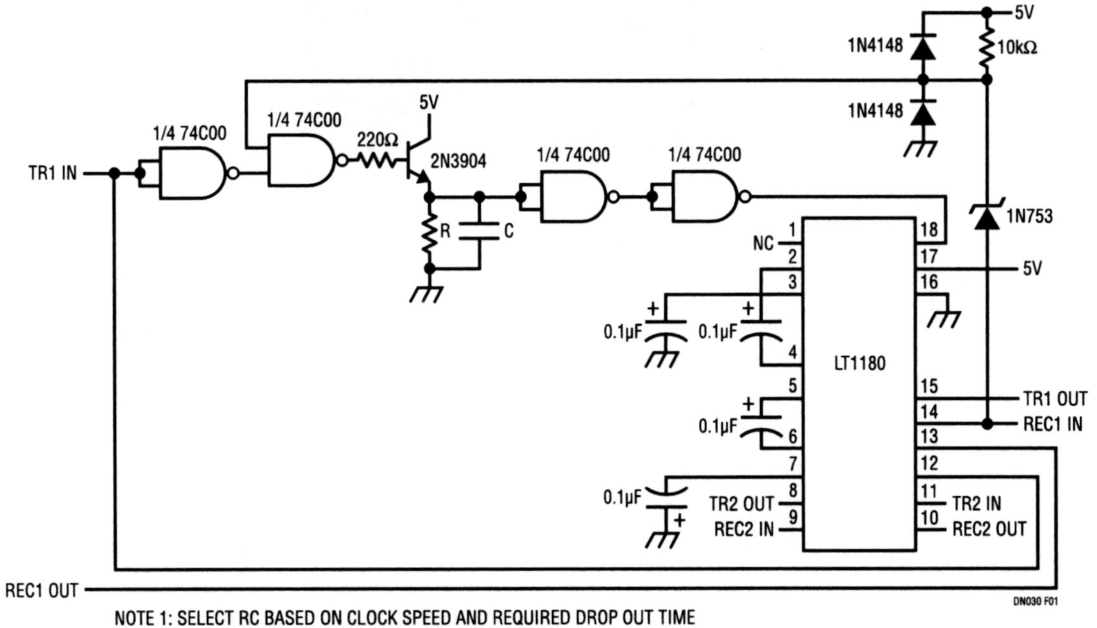

Figure 400.1 • Fast Turn-On Transceiver with Automatic SHUTDOWN Control

Analog Circuit Design: Design Note Collection. http://dx.doi.org/10.1016/B978-0-12-800001-4.00400-2

LT1180 is active, it can process data at the maximum 100kBd data rate.

A peak detector senses data flow. The extra CMOS gates are buffers which ensure the time constant is relatively independent of input signal level. The dropout time, i.e., the duration of inactivity prior to SHUTDOWN, is approximately 0.5RC. More specifically, dropout occurs when the voltage on the peak detector decays from $V_{CC} - 0.7V$ to the logic switch point of $V_{CC}/2$. The RS232 input to the control circuit is clamped to protect the logic inputs. The zener diode, D3, forces the turn-on threshold on the RS232 side to −3.5V, which prevents the transceiver from turning on when the cable is grounded.

Figures 400.2 through 400.4 demonstrate the automatic SHUTDOWN control's response to logic and RS232 signals, as well as zero data flow. The minimum pulse width is 50μs and the dropout time is set to 50ms. The power supply outputs—the lower two traces in Figures 400.2 and 400.3—become active in less than 200μs. When active, the circuit consumes 16mA of quiescent current. In SHUTDOWN state, the Q-current drops to 50μA.

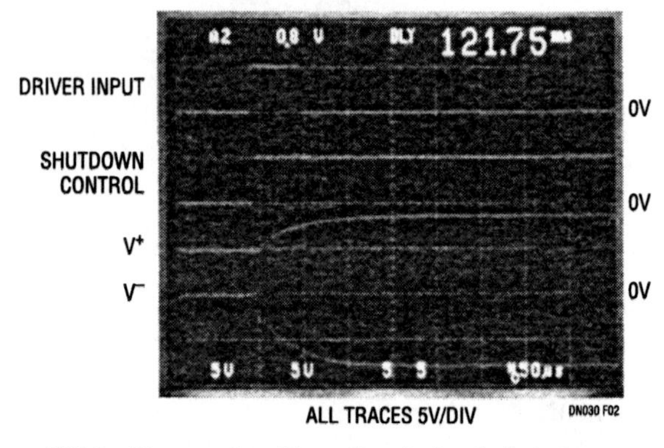

Figure 400.2 • Transceiver Turn-On via Logic Input

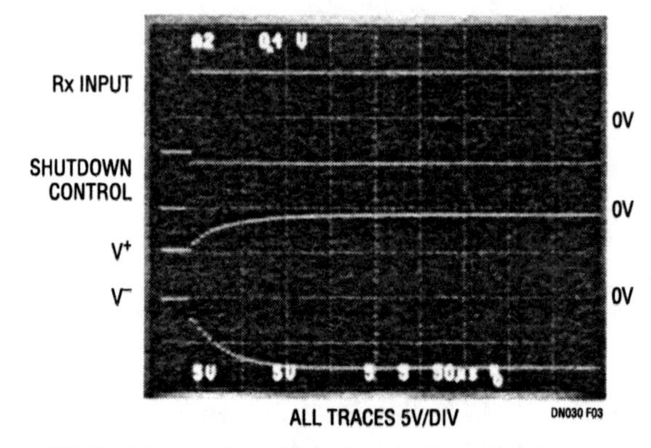

Figure 400.3 • Transceiver Turn-On via Receiver Input

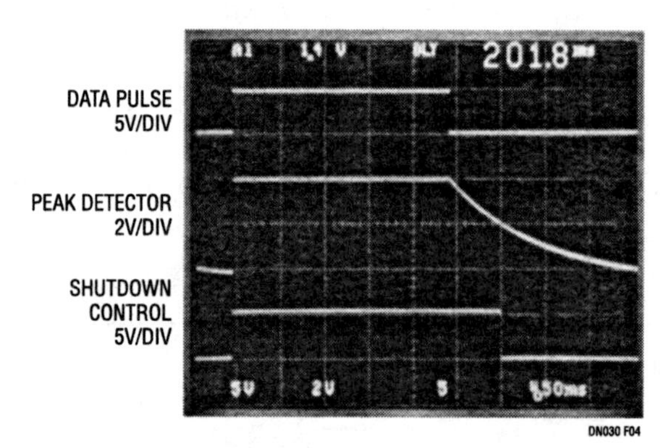

Figure 400.4 • SHUTDOWN After 50ms without Data Transmission

A single supply RS232 interface for bipolar A to D converters

401

Sean Gold

Designing circuitry for single supply operation is often an attractive simplification for reducing production costs. Yet many applications call for just a few additional supplies to solve simple interface problems. The example presented here describes how an advanced RS232 interface can simplify an A to D converter which processes bipolar signals.

The LT1180 RS232 transceiver includes a charge pump which produces low ripple supplies with sufficient surplus current to drive a CMOS A to D converter and precision voltage reference. The circuit in Figure 401.1 operates from a single 5V supply, and draws a total quiescent current of only 37mA. These features make the circuit ideal for applications which must process bipolar signals with minimal support electronics.

The LTC1094 serial A to D converter requires both a low noise supply and reference voltage for accurate operation.[1] These design problems are solved with an LT1021 precision reference, which delivers a stable, low noise, 5V signal from the LT1180's V^+ output. Relatively large storage and filter capacitors must be used with the LT1180 to reduce the noise in the system below 1mV for a 12 bit system. Construction also requires close attention to the layout of the system grounds and other aspects of circuit board design to avoid noise problems.[2]

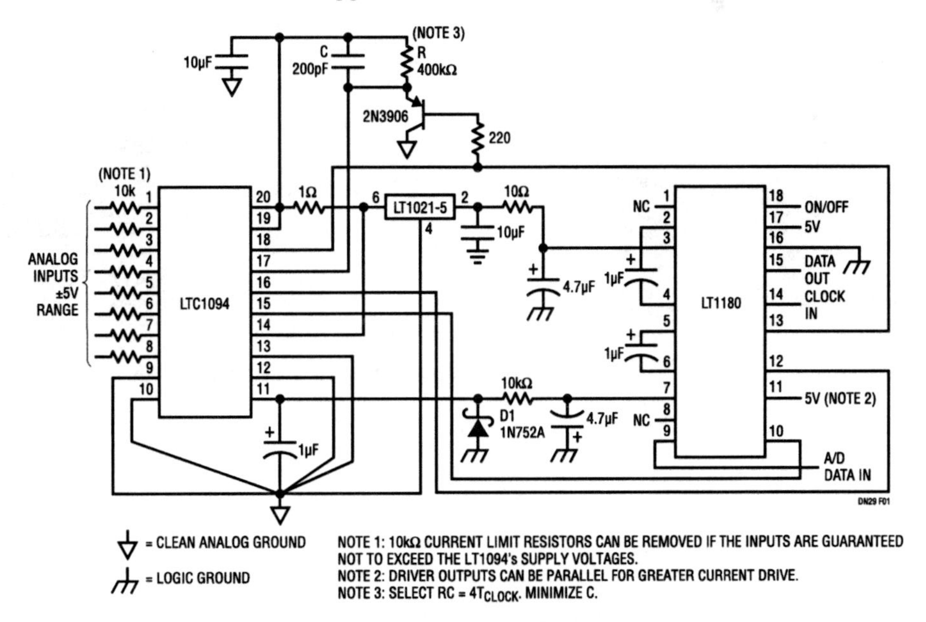

Figure 401.1 • A/D Converter Interface

Note 1: Refer to the data sheets for the LTC1094/LTC1294.

Note 2: An excellent reference on the subject of grounding and low noise circuit design is: "An IC Amplifier User's Guide to Decoupling, Grounding, and Making Things Go Right for a Change," by Paul Brokaw, Analog Devices Application Note AN-202.

Analog Circuit Design: Design Note Collection. http://dx.doi.org/10.1016/B978-0-12-800001-4.00401-4

To accommodate bipolar inputs ($-5 < V_{IN} < 5$), the LTC1094's negative rail must be biased beyond the extreme signal swing, but below absolute maximum ratings for supplies. A 5.6V Zener diode, D1, provides a sufficient bias because the V^- pin draws very little current.

The A to D converter communicates with a remote controller via three wires, which carry the clock, the configuration word, and the output data. The chip select signal, $\overline{CS}$, is generated from the incoming clock with a peak detector, constructed with a single PNP transistor. R and C are designed to hold the $\overline{CS}$ pin low for at least one clock period. Assuming the logic threshold in the LTC1094 is 1.4V, two useful rules of thumb for selecting R and C are: design RC to be at least four times the clock period and select C as small as possible to start the converter quickly. Minor aberrations in the $\overline{CS}$ signal are unimportant because the $\overline{CS}$ pin is level sensitive. The PNP is biased from the clean reference supply so very little noise is coupled into the A to D. Additional buffers are unnecessary because the peak detector drives a CMOS input.

The operating sequence for the LTC1094 is shown in Figure 401.2. The $\overline{CS}$ signal switches to a low state less than 1μs after receiving the system clock, and the configuration word may be transmitted after one clock cycle. After the 18 clock cycles required to complete the conversion, the clock must shut off to allow $\overline{CS}$ to switch to a high state for at least 2μs—the minimum time between conversions. The operating sequence may then be repeated.

A single conversion cycle is shown in Figure 401.3. The LT1180's maximum data rate limits the clock speed to 100kBd. The input voltage is 3.33V which generates a bit pattern of alternating 1's and 0's. Trace B shows the Chip Select signal, and Trace C shows the gating pulse for the system clock. The complete conversion cycle for a 12-bit converter using an LTC1294 is listed in Figure 401.4. For this example, the gating signals are adjusted to allow for the two extra bits of data.[3]

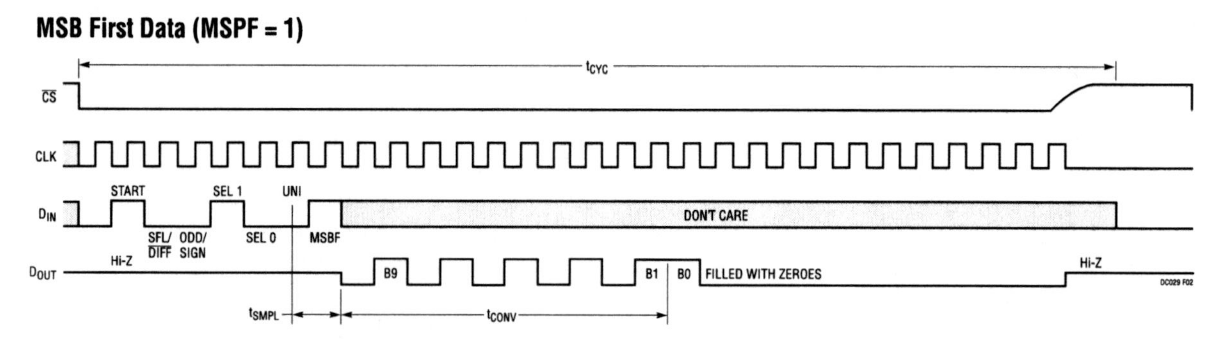

Figure 401.2 • LTC1093/4 Operating Sequence Example: Differential Inputs (CH4+, CH5−), Bipolar Mode

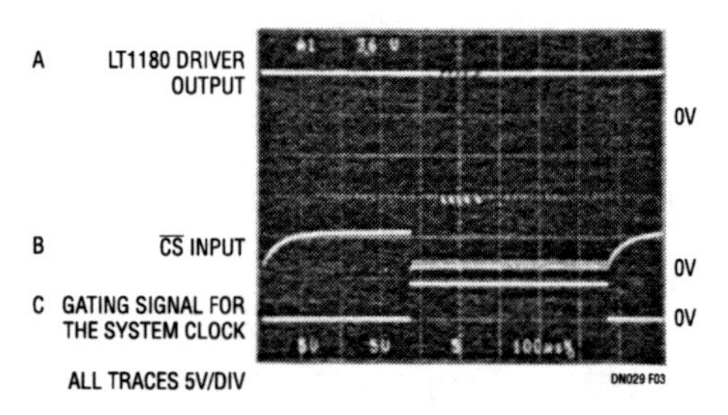

Figure 401.3 • 10-Bit Converter Interface

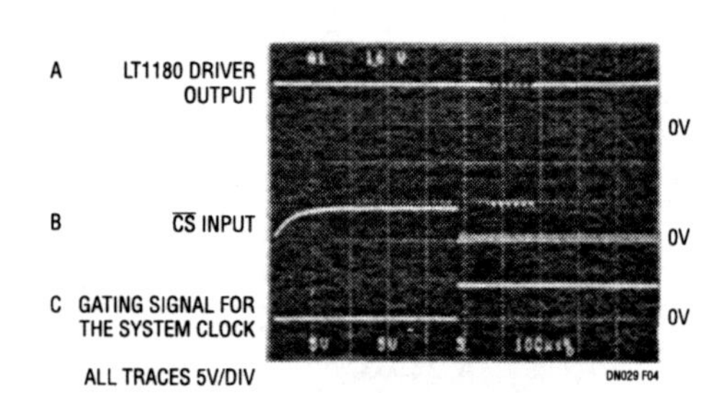

Figure 401.4 • 12-Bit Converter Interface

Note 3: The LTC1094 in Figure 401.1 was directly replaced with an LTC1294, with no changes to the circuit.

Design considerations for RS232 interfaces

402

Sean Gold

Introduction

When designing an RS232 interface, it is necessary to conform to standards published by the Electronics Industry Association, EIA RS232.V28. Some key specifications are summarized in Table 402.1. However, the EIA specifications are often just the beginning of the design. Practical problems such as generating RS232 signal levels, providing sufficient load drive, and ensuring protection against fault conditions must also be considered.

Table 402.1 Key RS232 Transceiver Specifications (EIA RS232C. V28)

SPECIFICATIONS	VALUE	UNITS
Signal Levels	±15 Max; ±5 Min	V
Cable Length	50 Max	Ft
Load Capacitance	2500 Max	pF
Cable Termination	3k<R<7k	Ω
Data Rate	20k Max	Baud
Slew Rate	3<SR<30	V/μs
Fault Conditions	Drivers Must Tolerate:	
	• Conductor to Conductor Shorts	–
	• Line Open Circuit	–
	• ±25V Line Overage	–

Power supply generators

Creating the separate RS232 voltage levels is a common problem in systems which have only a 5V logic supply. Linear Technology has developed a family of transceivers that include an on-chip charge pump to generate the RS232 supplies. These transceivers are available in a wide variety of configurations incorporating up to five drivers and five receivers. Some transceivers have a SHUTDOWN control which turns off the charge pump and places the drivers in a "zero" power–high impedance state.

The charge pump consists of a relaxation oscillator, a capacitive voltage doubler, and a capacitive voltage inverter. The oscillator is designed to operate at a frequency well above the signal frequencies to avoid supply degradation as charge is rapidly removed from the storage capacitors.

The LT1180/LT1181's charge pump oscillator operates at approximately 200kHz, which is two times the frequency of the LT1080 and LT1130 series transceivers. The faster oscillator permits the use of low value capacitors (C > 0.1μF), and shortens the turn-on time from power-off or SHUTDOWN state to less than 200μs. The LT1080 and LT1130 start up in approximately 2ms.

Load driving

It is often desirable to exceed the 20kHz data rate or drive loads greater than 2500pF, e.g., long cables. Slew rate control in the drivers makes this objective possible without

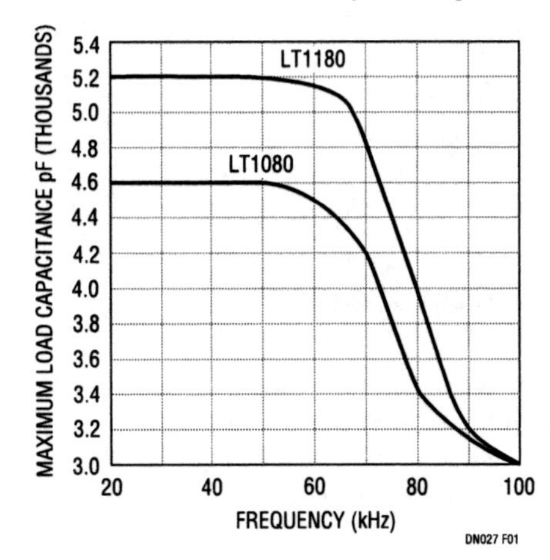

Figure 402.1 • Max Load Capacitance vs Data Rate. Both Transceivers Use 1.0μF Storage Capacitors

Analog Circuit Design: Design Note Collection. http://dx.doi.org/10.1016/B978-0-12-800001-4.00402-6

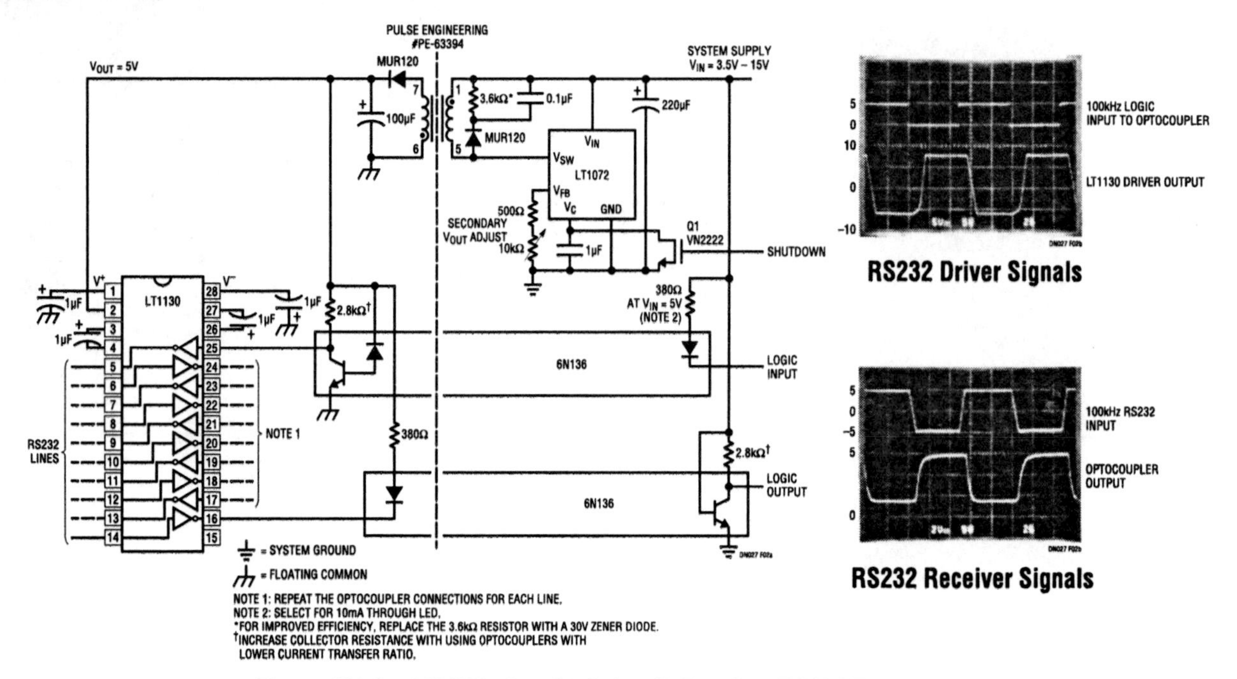

Figure 402.2 • 2500V Isolated 5-Driver/5-Receiver RS232 Transceiver

compromising the remaining specifications. When lightly loaded, the slew rate is set by an internal bias current and compensation capacitor. When heavily loaded, slew rate is limited by the output stage short circuit current and the load capacitance. The plot in Figure 402.1 shows the maximum load capacitance for a given data rate.

Fault conditions

In addition to protecting against all of the fault conditions described in Table 402.1, LTC transceivers are guaranteed for latchup free operation. When the drivers are turned off or SHUTDOWN, the output stage becomes high impedance; even when the output is pulled beyond the supply rails. The small current produced by overvoltage is not directed back into the supplies. High impedance on the driver outputs also eliminates signal feedthrough between the logic inputs and the RS232 lines.

When the device is turned on, overvoltage can, at most, pull the limited short circuit current from the supplies. The receivers are also short circuit current limited to prevent damage to unprotected logic circuitry.

Isolated transceiver

The most frequent cause of failure in interface chips is exposure to extreme fault conditions. Protection against large differences in ground potential, high ground loop currents, or accidental high voltage connections mandates a fully isolated transceiver.

The circuit in Figure 402.2 provides 2500V isolation with optically coupled data lines and an isolated 5V supply. A powered transceiver eliminates the need for three supplies on both sides of the isolation transformer. High speed 6N136 optocouplers permit the LT1130 to operate at its full 100kHz bandwidth. However, slower, less expensive opto-isolators, such as the 4N28, may be used when the data rate is less than 20kBd. The 5V power supply is generated with an isolated LT1072 switching regulator. The LT1072 has no electrical connection to the load; instead, the circuit derives its feedback from the transformers flyback voltage. This technique is often referred to as an isolated flyback regulator.[1] The regulator needs to deliver only modest current levels (200mA max), allowing a physically small isolation transformer. The circuit accepts 3.5V to 15V unregulated inputs which are readily available in most systems. Load regulation is 5% over a 200mA range of output current (50mA–250mA), and efficiency reaches 60% under maximum load conditions. Efficiency may be improved by 10% if the 3.6kΩ snubber resistor is replaced with a 30V Zener diode. Q1 provides shutdown control, which disables the interface to a low power state.

Note 1: Refer to Linear Technology's Application Note 19, pp. 30–34.

New 12-bit data acquisition systems communicate with microprocessors over four wires

403

As board space and semiconductor package pins become more valuable, serial data transfer methods between microprocessors (MPUs) and their peripherals become more and more attractive. Not only does this save lines in the transmission medium, but, because of the savings in the package pins, more function can be packed into both the MPU and the peripheral. Users are increasingly able to take advantage of these savings as more MPU manufacturers develop serial ports for their products [1–3]. However, peripherals which are able to communicate with these MPUs must be available in order for users to take full advantage. Also, MPU serial formats are not standardized so not all peripherals can talk to all MPUs.

The LTC1290 family

A new family of 12-bit data acquisition circuits has been developed to communicate over just four wires to the recently developed MPU synchronous serial formats as well as to MPUs which do not have serial ports. These circuits feature software configurable analog circuitry including analog multiplexers, sample and holds, bipolar and unipolar conversion modes and the ability to shut power completely off. They also have serial ports which can be software configured to communicate with virtually any MPU. Even the lowest grade device features guaranteed ±0.5 LSB linearity over the full operating temperature range. Reduced span operation, accuracy over a wide temperature range and low power single supply operation make it possible to locate these circuits near remote sensors and transmit digital data back through noisy media to the MPU. Figure 403.1 shows a typical hookup of the LTC1290, the first member of this data acquisition family. For more detail, refer to the LTC1290 data sheet.

Included are eight analog inputs which can common-mode to both supply rails. Each can be configured for unipolar or bipolar conversions and for single-ended or differential inputs by sending a data input (D_{IN}) word from the MPU to the LTC1290 (Figure 403.1).

Both the power supplies are bypassed to analog ground. The V^- supply allows the device to operate with inputs which swing below ground. In single supply applications it can be tied to ground.

The span of the A/D converter is set by the reference inputs which, in this case, are driven by a 2.5V LT1009 which gives an LSB step size of 0.61mV. However, any reference voltage within the power supply range can be used.

The 4-wire serial interface consists of an active low chip select pin ($\overline{CS}$), a shift clock (SCLK) for synchronizing the data bits, a data input (D_{IN}) and a data output (D_{OUT}). Data is transmitted and received simultaneously (full-duplex), minimizing the transfer time required.

The external ACLK input controls the conversion rate and can be tied to SCLK as in Figure 403.1. Alternatively, it can be derived from the MPU system clock (e.g., the 8051 ALE pin) or run asynchronously. When the ACLK pin is driven at 4MHz, the conversion time is 13μs.

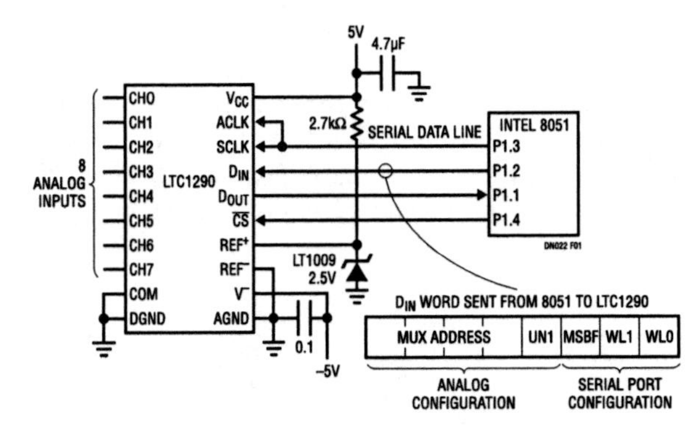

Figure 403.1 • A Typical Hookup of the LTC1290

The LTC1290 can be located near the sensors and serial data can be transmitted back from remote locations through isolation barriers or through noisy media.

Analog Circuit Design: Design Note Collection. http://dx.doi.org/10.1016/B978-0-12-800001-4.00403-8

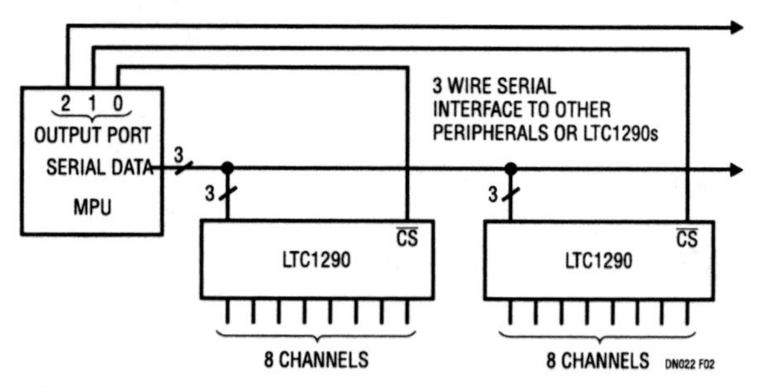

Figure 403.2 • Several LTC1290s Sharing One 3-Wire Serial Interface

Using fewer pins for communication makes it possible to pack more function into a smaller package. LTC1290 family members are complete systems being offered in packages ranging from 20 pins to 8 pins (e.g., LTC1291, 1292, 1293, 1294).

Speed is usually limited by the MPU

A perceived disadvantage of the serial approach is speed. However, the LTC1290 can transfer a 12-bit A/D result in 6μs when clocked at its maximum rate of 2MHz. With the minimum conversion time of 13μs, throughput rates of 50kHz are possible. In practice, the serial transfer rate is usually limited by the MPU, not the LTC1290. Even so, throughput rates of 20kHz are not uncommon when serial port MPUs are used. For MPUs without serial ports, the transfer time is somewhat longer because the serial signals are generated with software. For example, with the Intel 8051 running at 12MHz, a complete transfer takes 96μs. This makes possible throughput rates of approximately 10kHz.

Talking to serial port MPUs

By accommodating a wide variety of transfer protocols, the LTC1290 is able to talk directly to almost all synchronous serial formats. The last 3 bits of the LTC1290 data input (D_{IN}) word define the serial format and power shutdown (see Figure 403.3). The MSBF bit determines the sequence in which the A/D conversion result is sent to the processor (MSB or LSB first). Figure 403.4 shows several popular serial formats and the appropriate D_{IN} word for each. Typically a complete data transfer cycle takes only about 15 lines of processor code.

WL1	WL2	Output Word Length
0	0	8 Bits
0	1	Power ShutDown
1	0	12 Bits
1	1	16 Bits

Figure 403.3 • Word Length and Power Shutdown

Talking to MPUs without serial ports

The LTC1290 talks to serial port processors but works equally well with MPUs which do not have serial ports. In these cases, ($\overline{CS}$), SCLK and D_{IN} are generated with software on three port lines. D_{OUT} is read on a fourth. Figure 403.4 shows the appropriate D_{IN} word for communicating with MPU parallel ports. Figure 403.1 shows a 4-wire interface to the popular Intel 8051. A complete transfer takes only 33 lines of code.

Sharing the serial interface

No matter what processor is used, the serial port can be shared by several LTC1290s or other peripherals (see Figure 403.2). A separate ($\overline{CS}$) line for each peripheral determines which is being addressed.

Conclusions

The LTC1290 family provides data acquisition systems which communicate via simple 4-wire serial interface to virtually any microprocessor. By eliminating the parallel data bus they are able to provide more function in smaller packages, right down to 8-pin DIPs. Because of the serial approach, remote location of the A/D circuitry is possible and digital transmission through noisy media isolation boundaries is made easier without a great loss in speed.

Type of Interface	LTC1290 Data Format	LTC1290 D_{IN} Word: Analog Configuration					MSBF	WL1	WL0
All Parallel Port MPUs	MSB First 12 Bits	X	X	X	X	X	1	1	0
National MICROWIRE	MSB First 12 Bits	X	X	X	X	X	1	1	0
MICROWIRE/PLUS; Morotola SPI	MSB First 16 Bits	X	X	X	X	X	1	1	1
Hitachi Synchronous SCI; TI TMS7000 Serial Port	LSB First 16 Bits	X	X	X	X	X	0	1	1

Figure 403.4 • The LTC1290 Accommodates Both Parallel and Serial Ports

References

[1] Aleaf, Adbul, and Richard Lazovick, "Microwire/Plus," National Semiconductor, Wescon '86, Santa Clara, CA, Session 21(1986).
[2] Derkach, Donald J., "Serial Data Transmission in MCU Systems," RCA Solid State, Wescon '86, Somerville, NJ, Session 21(1986).
[3] Kalinka, Theodore J., "Versatile Serial Peripheral Interface (SOI)," RCA Solid State, Wescon '86, Somerville, NJ, Session 21(1986).

Extending the applications of 5V powered RS232 transceivers

404

High speed operation

Although the EIA RS232 specification is for a relatively slow communications protocol, many applications require RS232 transceivers to operate at higher frequencies. Devices such as the LT1080, LT1081, and the LT1130 series share a common design for the drivers and receivers and are capable of operating over 100kBd.

Although the slew rate is controlled for all of the Linear Technology series of RS232 communications devices, for output levels limited to ±6V the transition time is fast enough to allow high baud rates. With a slew rate of approximately 10V per microsecond, it only takes 1.2 microseconds for a 12V excursion. The two photos (Figures 404.1 and 404.2) show the output waveform and delay associated with a 75kHz square wave input and a 100kHz square wave. Delay times are in the order of 0.5 microseconds and the total slew time is approximately 1.2 microseconds. Output load is 3k. Receivers are much faster and can handle these baud rates with no problem. For higher communication rates, a differential signal is recommended.

Power supply tricks

The power supply generator on 5V powered devices is a charge pump circuit which generates approximately ±9V from a single 5V supply. Parallel operation of the supply charge pumps for 5V powered transceivers is easily achieved to minimize component count. The positive and negative supply have approximately 1μF of holding capacitance for energy storage. If several devices with charge pumps are used in the same system, the output supplies may be paralleled into a single pair of common energy storage capacitors.

Figure 404.3 shows two LT1080s with common power supply capacitors for energy storage. Twice the output current is available for external use. This eliminates two capacitors

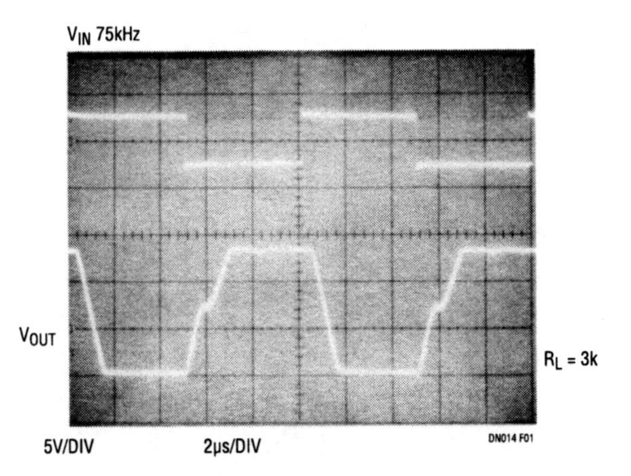

Figure 404.1 • Operation at 75kHz

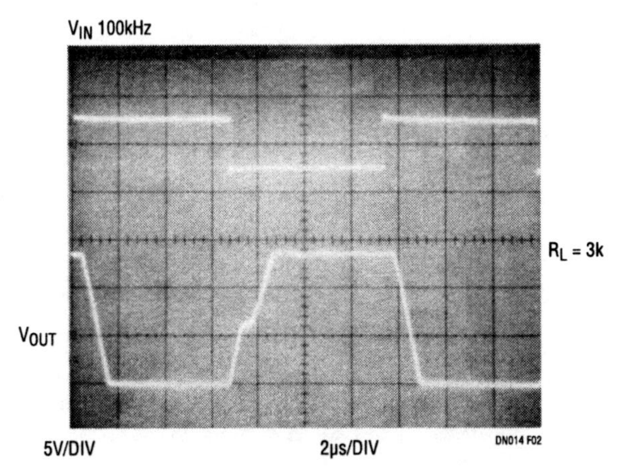

Figure 404.2 • Operation at 100kHz

Analog Circuit Design: Design Note Collection. http://dx.doi.org/10.1016/B978-0-12-800001-4.00404-X

from the system. Individual charge pump capacitors are still needed on each of the devices.

Operation with +5V and +12V supplies

The charge pump circuitry takes the input 5V and doubles it. The doubled voltage is then inverted to obtain a negative output. The only reason for doubling the input is to ensure adequate positive and negative output voltage to meet RS232 specifications. In PC systems, where +12V is available, the internal voltage doubler does not need to be used. The device may be connected directly to a +5V and a +12V supply. The +12V is then inverted to obtain approximately −11V. This eliminates one charge pump capacitor and one holding capacitor for the 12V output. Figure 404.4 shows an LT1080 connected to a 12V and 5V power supply. The +12V is connected into one of the charge pump capacitor pins rather than the 12V output pin. Supply current also decreases to about 9mA.

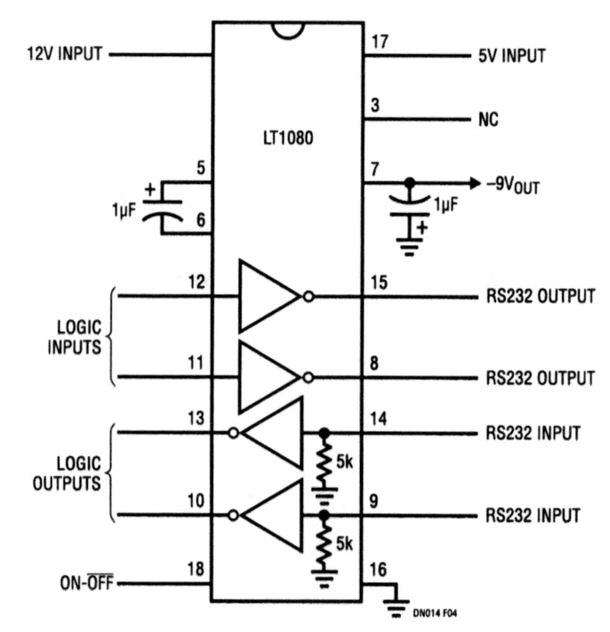

Figure 404.4 • Operation with +12V and +5V Supplies

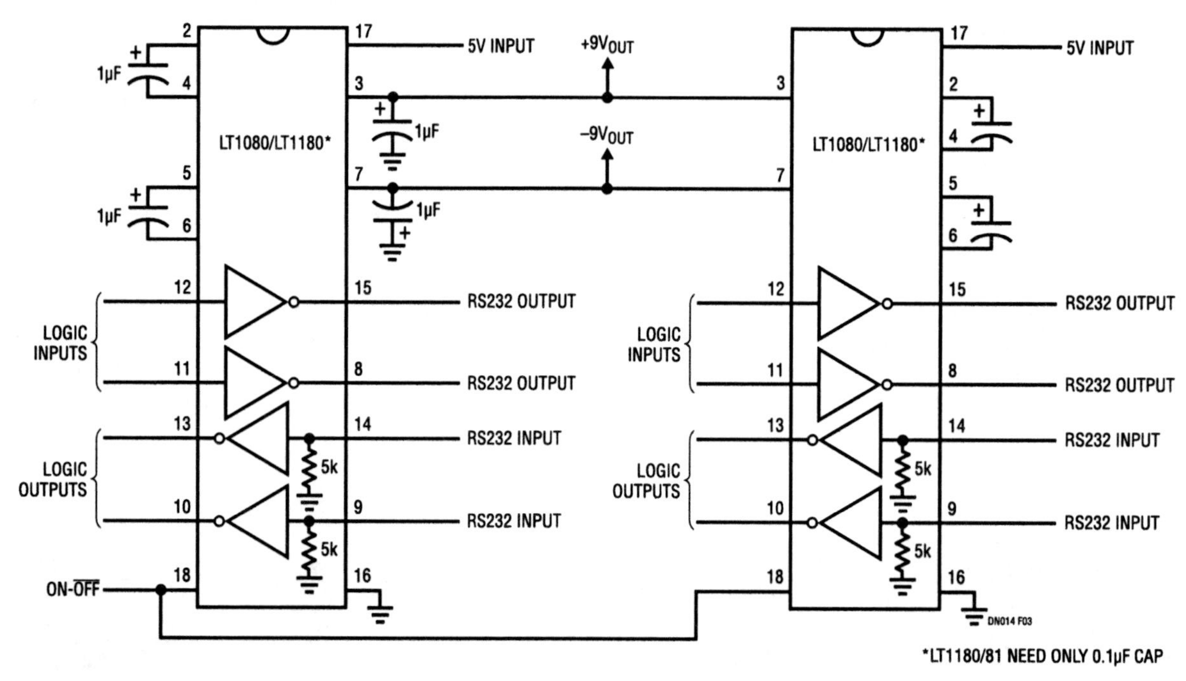

Figure 404.3 • Paralleling Power Outputs

New developments in RS232 interfaces

405

Robert Dobkin

New RS232 interface chips have been developed that offer significant advantages over older devices such as the 1488 and 1489. The new RS232 interface ICs improve speed, power, voltage supply requirements, and protection over older devices. Further, the new chips are easier to use, requiring fewer external components and may be turned off to a "zero" power supply current condition for use in battery-powered systems.

The new RS232 drivers are implemented in a monolithic bipolar technology. A unique output stage was designed that provides large output swings, minimizing power supply voltage requirements, while retaining outstanding overload protection features. The outputs can be driven beyond the power supply voltage without drawing excessive current or forcing current back into the power supplies. Of course, current limiting is included to protect against short-circuit conditions.

Initial consideration of technologies for implementing RS232 interfacing might include CMOS as a possible technology for this type of application. Power supply requirements are low, output voltage swing is high, and higher voltage CMOS technologies are available to allow operation up to ±15V. Consideration of some of the problems associated with CMOS decreases its attractiveness for RS232 drivers.

Inherent in the CMOS structure are diodes between the drain and source of the CMOS devices and the power supplies as is shown in Figure 405.1. A requirement of RS232 interfaces is the ability to withstand voltage applied to the output pins. With a CMOS output stage this is achieved with the inclusion of a 300Ω resistor in series with the output. (The resistor is similar to the resistors included in older drivers.) It protects the interface chip, but still allows damage to other devices powered by the same supply.

A problem occurs when the output of a driver which is powered from the 5V logic supply is connected to an external 12V or 15V source as is allowed by the RS232 specification. External current flows through the 300Ω limiting resistor, through the diodes, which are part of the CMOS structure, and into the power supply. This forces the power supply to 12V and 15V damaging the 5V logic that is connected to the supplies. This problem can even cause latchup if the logic supply is off when external RS232 signals feed voltage into the supply. This problem did not usually exist in the past, because the RS232 interfaces were powered by separate ±12V supplies.

ESD damage is probably the most frequent cause of failure of interface chips. Bipolar devices are relatively rugged but still can be damaged by ESD. System requirements for ESD may be as high as 20kV. No IC can withstand that much voltage without external protection.

A requirement of the RS232 specification is the ability to withstand ±25V input signals. The CMOS LTC1045 which is used as an RS232 receiver has been designed to operate with external resistors in series with the input. These resistors allow very large voltage swings at the input pins and provide ESD protection to the IC. Using on-chip resistors precludes the use of the optimum ESD protection structures, so CMOS devices may be more sensitive to ESD destruction at their inputs.

The output stage of the bipolar drivers is shown in Figure 405.2. Opposed collector NPN and PNP transistors give the widest possible output swings. The PNP transistor

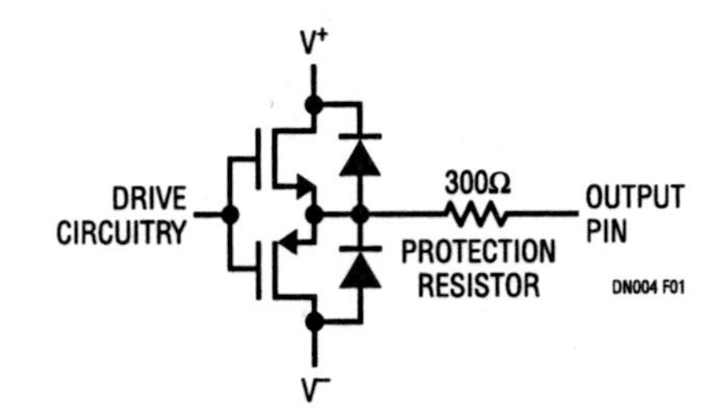

Figure 405.1 • CMOS Line Driver Showing Parasitic Diodes to the Power Supplies

Analog Circuit Design: Design Note Collection. http://dx.doi.org/10.1016/B978-0-12-800001-4.00405-1

will swing to within 200mV of the positive supply while the NPN transistor with its associated Schottky diode will swing within about 900mV of the negative supply. If the output voltage is forced above the positive supply the emitter base junction of the PNP transistor reverse biases, and no current flows into the supply. The device is unaffected by external voltage up to the breakdown voltage of the transistor. If the output is forced below the negative supply, the Schottky diode reverse biases and prevents external current flow into the chip. Capacitor C1 is used to control the output slew rate so that no frequency compensation components are required to meet the RS232 specification 4V/μs to 30V/μs.

Typically the slew rate of these drivers is about 8–10V/μs. This allows them to be used successfully up to about 64kBd. The output slew rate of the bipolar drivers is well controlled by an internal capacitor and relatively independent of load resistance or capacitance. The bipolar receiver is relatively straightforward utilizing a level detector with hysteresis to set the trip point. Nominally the trip point is set at about 1.5V with 200mV of hysteresis. The receivers go into a high output state with an open input. The receivers outputs are both TTL and CMOS compatible.

A recent advance in the drivers and receivers is on-chip power supply generation. Devices like the LT1080 and LT1081 include an oscillator, capacitive voltage doubler, and capacitive inverter to generate ±9V from the 5V power supply. The charge-pump power supply generator requires only four 1μF capacitors to generate RS232 communication levels from a 5V logic supply. Figure 405.3 shows a typical hook-up for the LT1080. The on-chip power supply generators generate excess power over the LT1080 requirements, so another RS232 communication device such as the LT1039 can be powered from the same power supply generator. Table 405.1 gives typical performance of all Linear Technology driver/receiver devices for RS232 communication.

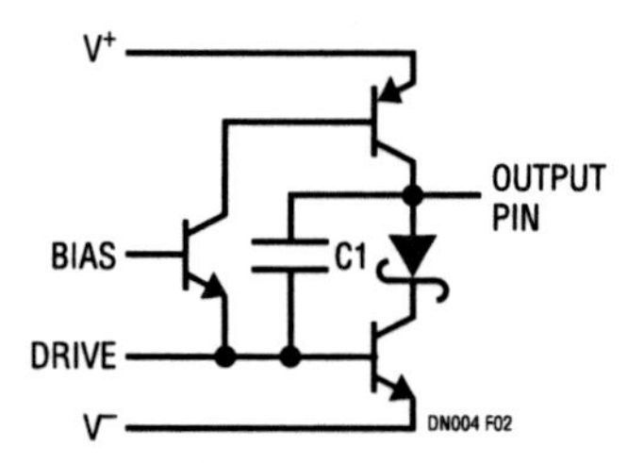

Figure 405.2 • New Bipolar Driver Output Stage

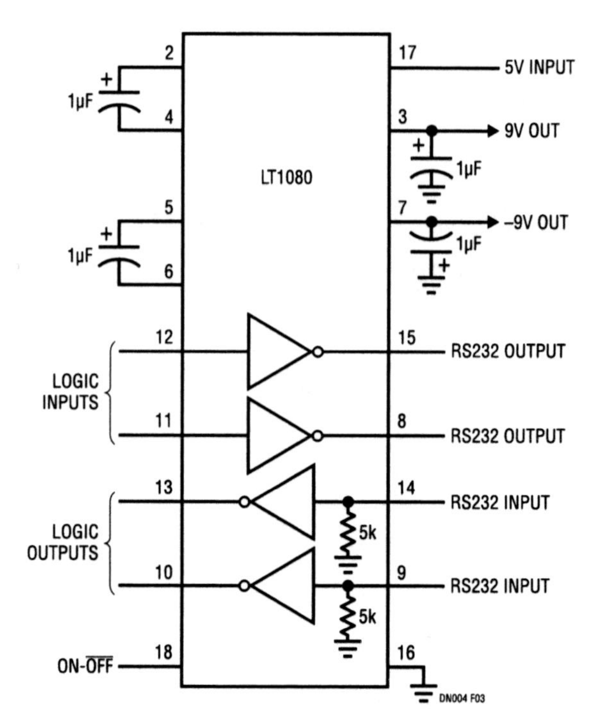

Figure 405.3 • 5V Powered RS232 Driver/Receiver

Table 405.1 New Drivers and Receivers

DEVICE	DRVS	RECS	SHUT-DOWN	SUPPLY GENERATOR	REMARKS
LT1030	4		X		Low Cost
LT1032	4		X		RS423 Compatible
LT1039	3	3	X		
LT1039N16	3	3			MC145406 Compatible
LTC1045		6	X		Micropower
LT1080	2	2	X	X	
LT1081	2	2		X	MAX232 Compatible
LT1130	5	5		X	
LT1131	5	4	X	X	
LT1132	5	3		X	
LT1133	3	5		X	
LT1134	4	4		X	
LT1135	5	3			
LT1136	4	5	X	X	
LT1180	2	2	X		0.1μF Caps
LT1181	2	2			0.1μF Caps

Section 5

Instrumentation Design

System monitor with instrumentation-grade accuracy used to measure relative humidity

406

Leo Chen

Because much can be deduced about a physical system by measuring temperature, it is by far the most electronically measured physical parameter. Selecting a temperature sensor involves balancing accuracy requirements, durability, cost and compatibility with the measured medium. For instance, because of its low cost, a small-signal transistor such as the MMBT3904 is an attractive choice for high volume or disposable sensing applications. Although such sensors are relatively simple, accurate temperature measurement requires sophisticated circuitry to cancel such effects as series resistance.

The LTC2991 system monitor has this sophisticated circuitry built in—it can turn a small-signal transistor into an accurate temperature sensor. It not only measures remote diode temperature to ±1°C accuracy, but it also measures its own supply voltage, single-ended voltages (0 to V_{CC}) and differential voltages (±325mV). While ostensibly designed for system monitor applications, the top shelf performance of the LTC2991 makes it suitable for instrumentation applications as well, such as the accurate psychrometer described here.

A psychrometer: not nearly as ominous as it sounds

A psychrometer is a type of hygrometer, a device that measures relative humidity. A hygrometer uses two thermometers, one dry (dry bulb) and one covered in a fabric saturated with distilled water (wet bulb). Air is passed over both thermometers, either by a fan or by swinging the instrument, as in a sling psychrometer. A psychrometric chart can then be used to calculate humidity by using the dry and wet bulb temperatures. Alternatively, a number of equations exist for this

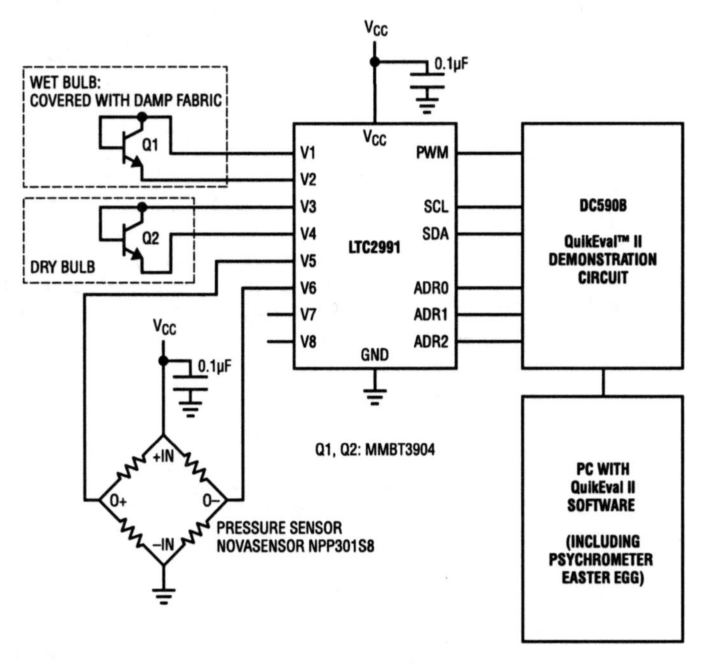

Figure 406.1 • Simple Psychrometer Using the LTC2991

Analog Circuit Design: Design Note Collection. http://dx.doi.org/10.1016/B978-0-12-800001-4.00406-3

purpose. The following equations are used in testing this circuit.

$$A = 6.6 \cdot 10^{-4} \cdot (1 + 1.115 \cdot 10^{-3} \cdot WET)$$

$$ESWB = e^{\left(\frac{16.78 \cdot DRY - 116.9}{WET + 273.3}\right)}$$

where:

$ED = ESWB - A \cdot P \cdot (DRY - WET)$

$HUMIDITY = \frac{ED}{EDSB}$

WET = wet bulb temperature in Celsius

DRY = dry bulb temperature in Celsius

P = pressure in kPa

Figure 406.1 shows an LTC2991-based psychrometer. The two transistors provide the wet bulb and dry bulb temperature readings when connected to the appropriate inputs of the LTC2991.

The equations include atmospheric pressure as a variable, which is determined here via a Novasensor NPP301-100 barometric pressure sensor measured by channels 5 to 6 configured for differential inputs. Full-scale output is 20mV per volt of excitation voltage, at 100kPa barometric pressure (pressure at sea level is approximately 101.325kPa).

The LTC2991 can also measure its own supply voltage, which in our circuit is the same supply rail used to excite the pressure sensor. Thus, it is easy to calculate a ratiometric result from the pressure sensor, removing the error contribution of the excitation voltage.

Error budget

The LTC2991 remote temperature measurements are guaranteed to be accurate to ±1°C. Figure 406.2 shows the error in indicated humidity that results from a 0.7°C error in the worst-case direction, and the error in indicated humidity resulting from a 0.7°C error in the worst-case direction combined with worst-case error from the pressure sensor.

Try it out!

A psychrometer readout is implemented as an Easter egg in the LTC2991 (DC1785A) demonstration software, available as part of the Linear Technology QuikEval software suite (Figure 406.3).

The demo board should be set up as shown in Figure 406.1. To access the readout, simply add a file named tester.txt in the install directory of your DC1785A software. The contents of this file do not matter. On software start-up, the message "Test mode enabled" should be shown in the status bar, and a Humidity option will appear in the Tools menu. Relative humidity readings can then be compared to sensors of similar accuracy grade, such as resistive and capacitive film.

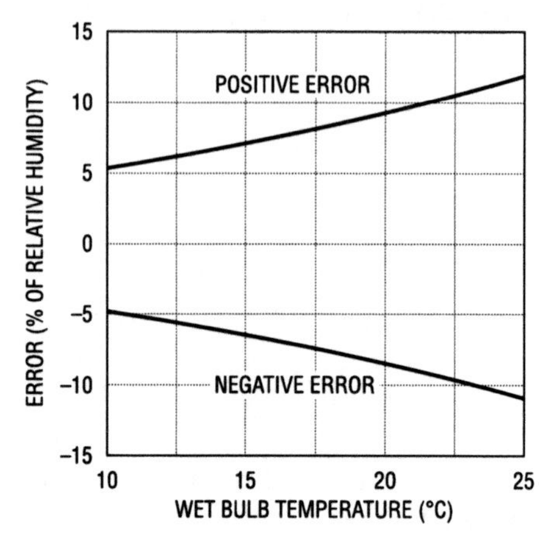

Figure 406.2 • Worst-Case Error

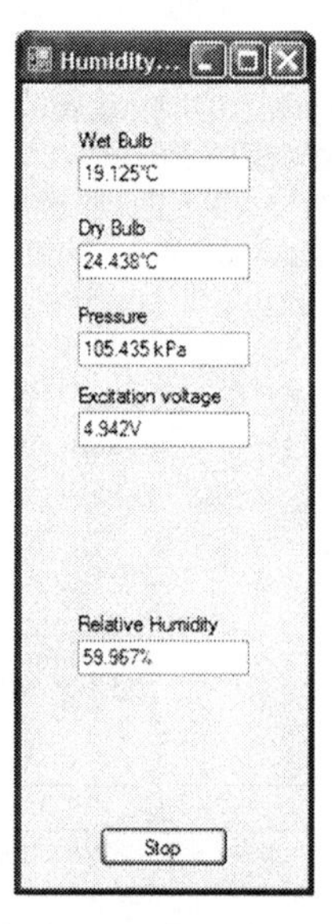

Figure 406.3 • A Psychrometer Readout Is Implemented as an Easter Egg in the LTC2991 (DC1785A) Demonstration Software, Available as Part of Linear's QuikEval Software Suite

6-channel SAR ADCs for industrial monitoring and portable instruments

407

Guy Hoover Steve Logan

The 14-bit LTC2351-14 is a 1.5Msps, low power SAR ADC with six simultaneously sampled differential input channels. It operates from a single 3V supply and features six independent sample-and-hold amplifiers and a single ADC. The single ADC with multiple S/HAs enables excellent range match (1mV) between channels and channel-to-channel skew (200ps).

The versatile LTC2351-14 is ideally suited for industrial monitoring applications such as 3-phase power line monitoring to ensure line voltage compliance, portable power line instrumentation, power factor correction, motor control, and data acquisition. These applications may be battery powered, and it is here that the LTC2351-14's low power and small size are desirable. Power consumption is a mere 16.5mW, which extends battery life. The 3-wire serial interface means fewer pins than parallel output devices, allowing the LTC2351-14 to fit in a 32-pin, 5mm × 5mm QFN package.

Power line monitoring application

Figure 407.1 shows a typical power line monitoring application. Current is sensed by a CR Magnetics CR8348-2500-N current transformer. An LT1790-1.25 biases the output of the transformer to the middle of the LTC2351-14 input range, giving the inputs maximum swing. A 6:1 transformer and 41:1 attenuator scale the line voltage, and the transformer output is similarly biased.

Figure 407.2 shows the AC line voltage in Linear Technology's Mixed Signal lab. The flattened peaks are typical of the

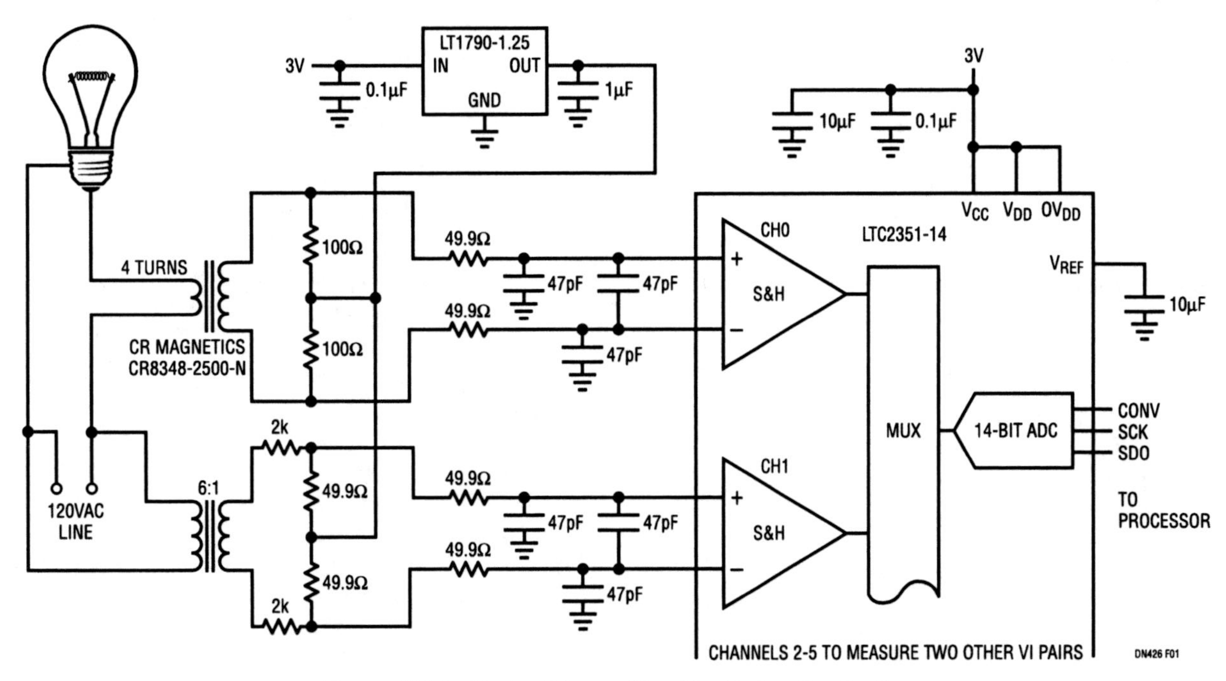

Figure 407.1 • Typical Power Line Monitoring Application

Analog Circuit Design: Design Note Collection. http://dx.doi.org/10.1016/B978-0-12-800001-4.00407-5

voltage in an office building where many of the loads are nonlinear, such as computer power supplies. Figure 407.3 shows the current through a 50W incandescent bulb. Figure 407.4 shows the current through a 15W compact fluorescent bulb, and Figure 407.5 is the current through a 4W LED-based bulb. The 5MHz full linear bandwidth of the LTC2351-14 allows analysis of high frequency components of the line voltage and current, limited in this case by the bandwidth of the sense transformers.

Conclusion

With PCB real estate getting tighter and designers always searching for lower power ICs, fast data acquisition can be a challenge. The LTC2351-14 and other low power SAR converters make it possible to optimize solution size, power and cost.

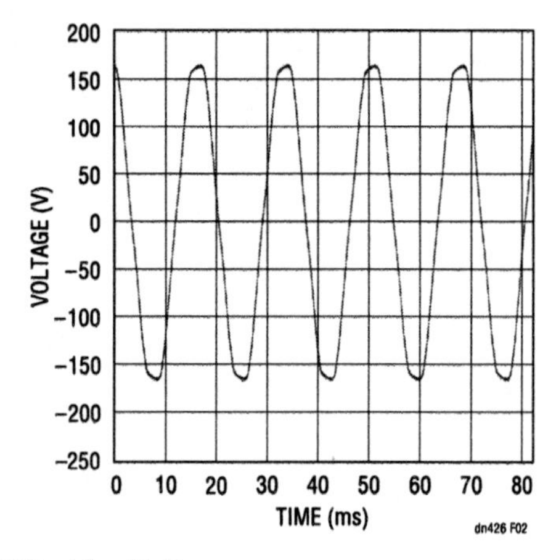

Figure 407.2 • Line Voltage

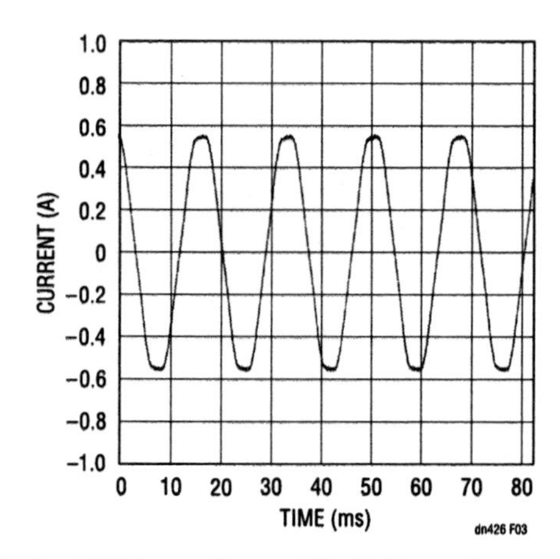

Figure 407.3 • 50W Incandescent Bulb Current

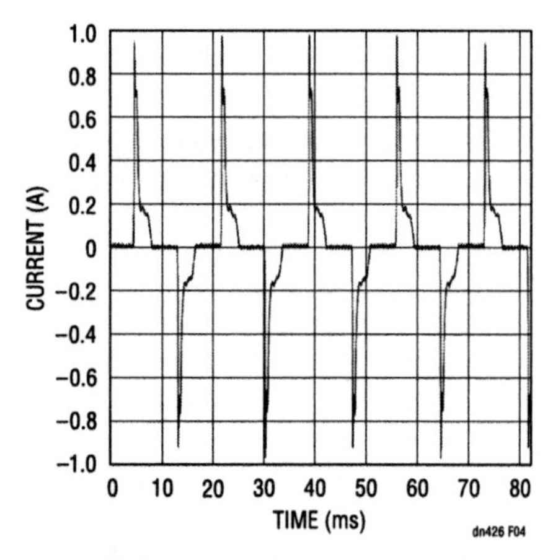

Figure 407.4 • 15W Compact Fluorescent Bulb Current

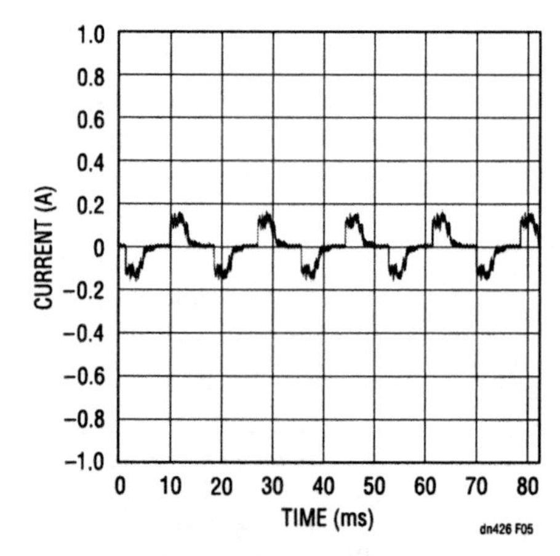

Figure 407.5 • 4W LED Bulb Current

408

Instrumentation amplifiers maximize output swing on low voltage supplies

Glen Brisebois

Introduction

Instrumentation amplifiers suffer from a chronic output swing problem, even when the input common mode range and output voltage swing specifications are not violated. This is because the first stage of an instrumentation amplifier has internal output voltages that can clip at unspecified levels. The clipping itself is invisible to the user, but it affects the output swing adversely, usually causing a gain reduction and thus an invalid result. The new LTC6800 and LT1789-10 both solve this output swing problem, but in two extremely different ways. The LTC6800 incorporates a flying capacitor differential level shifter followed by a rail-to-rail output autozero amplifier. The LT1789-10 is a more classical three op amp instrumentation amplifier with the twist that it takes gain in the final stage.

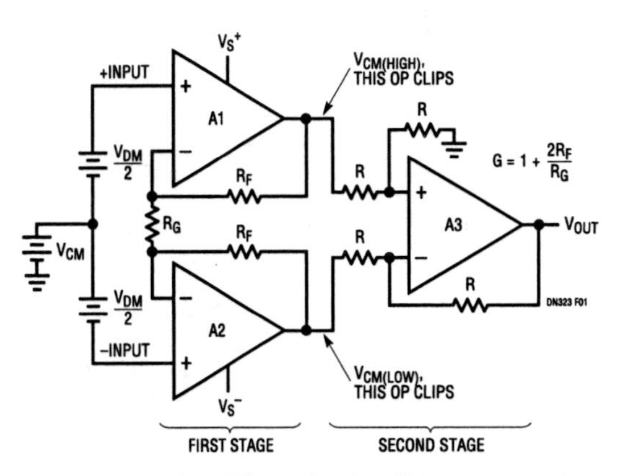

Figure 408.1 • Classical Three Op Amp Instrumentation Amplifier. First Stage Can Have Clipping Problems Depending on V_{CM}. This Reduces the Gain and Gives a False Output Reading

A clearer picture of the problem

Figure 408.1 shows the classical three op amp instrumentation amplifier (IA) topology. Assume that the op amps involved can common mode to V_S^- and have rail-to-rail output stages. This would normally mean that the inputs can be anywhere from V_S^- to about a volt from V_S^+ and that the output can be anywhere within the supply rails. But analysis of the circuit shows that these conditions are not sufficient to ensure a valid output.

For example, assume that the IA is powered on a single 5V supply ($V_S^+ = 5V$, $V_S^- = 0V$), set for a gain of 3 ($R_G = R_F$), and that its inputs are centered at $V_{CM} = 0.5V$. Now, as the differential input voltage is increased around the 0.5V common mode, the output voltages of amplifiers A1 and A2 split apart as well. Note what happens, though, when the differential input voltage (V_{DM}) reaches 1/3V. At that point, the output of A1 goes to 1V and the output of A2 goes to 0V, where it is clipped by the negative supply rail. This happens in spite of the fact that no specified input common mode range or specified output swing has been exceeded.

What can be misleading about this particular error mode is that the gain does not fall to zero, so bench validation tests performed in haste may not catch the problem. The gain is reduced, but there is still a partial signal gain path maintained by A1 and A3 (until A3 clips, of course). Figure 408.2 shows the entire range of valid output swing vs input common mode for an IA similar to that described above powered on a single

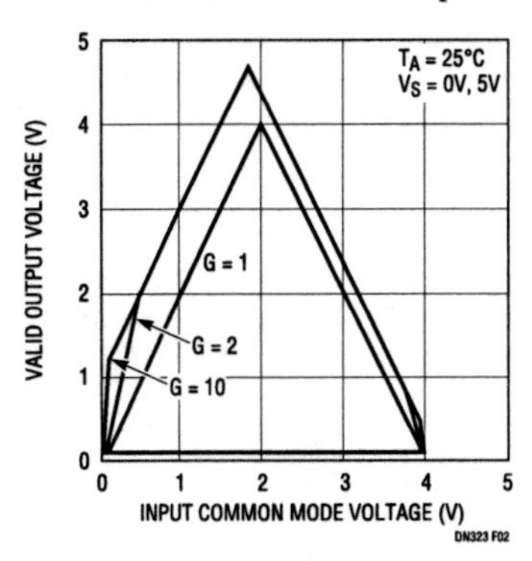

Figure 408.2 • Using Rail-to-Rail Output Op Amps Does Not Guarantee Output Swing Over Input Common Mode

Analog Circuit Design: Design Note Collection. http://dx.doi.org/10.1016/B978-0-12-800001-4.00408-7

5V supply.[1] Note that with the inputs near ground or near 4V, the IA has essentially no valid output swing!

The solutions

Figure 408.3 shows the same plot, this time for an LT1789-10. Note the drastic improvement. A simplified schematic of the LT1789-10 is shown in Figure 408.4. The PNP transistors on the inputs serve to level shift the input voltages up by one V_{BE}, thus ensuring valid small-signal input and output ranges (for A1 and A2) near V_S^-. But the real key to the dramatic improvement in output swing is the gain of 10 in the last stage. By taking gain in the last stage, the outputs of the first stage are not required to swing as much for a given overall gain setting and desired output swing.

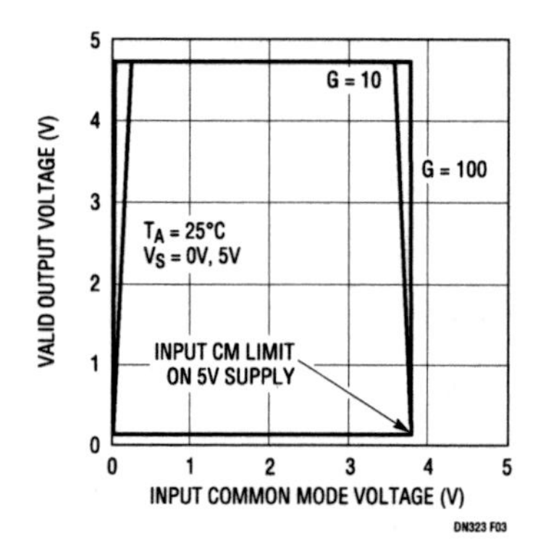

Figure 408.3 • The LT1789-10 Gives Effective Rail-to-Rail Output Validity over Almost the Entire Input Common Mode Range

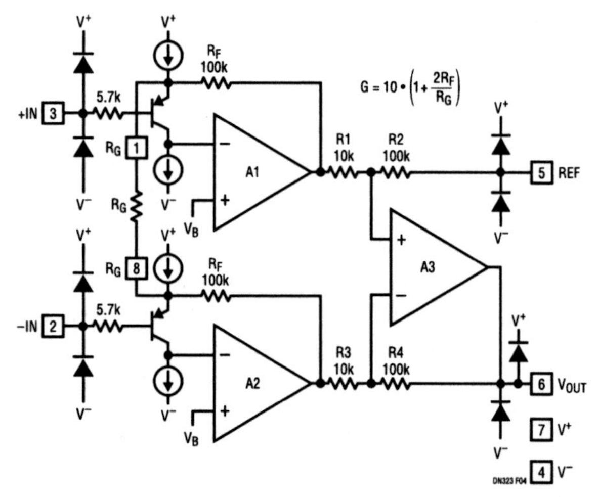

Figure 408.4 • The LT1789-10 Block Diagram. PNP Inputs Level Shift Away from V_S^-. Gain of 10 Around A3 Eases Output Swing Requirements on A1 and A2

[1]This plot is actually taken from an LT1789-1 and incorporates improvements near ground due to a level shifting input PNP stage.

The LTC6800 solution

The LTC6800 achieves similar immunity from the output swing vs input common mode problem, but in a completely different way. The device incorporates a flying capacitor differential level shifter followed by a very precise autozero output op amp as shown in Figure 408.5. The rail-to-rail output op amp is gain configurable in the conventional 2-resistor way, and follows the usual noninverting gain equation $G = 1 + R_F/R_G$.

Figure 408.6 shows the valid output swing vs input common mode for the LTC6800. In a gain of 1, the output validity is clipped to about 3.5V by the input common mode range of the op amp A1. Elsewhere in the plot, the ramp-like limitation characteristics are due to the input referred voltages at the rail-to-rail input switches and capacitors clipping at the supply rails. Like the LT1789-10, the LTC6800 performance represents a dramatic improvement over the classical results of Figure 408.2.

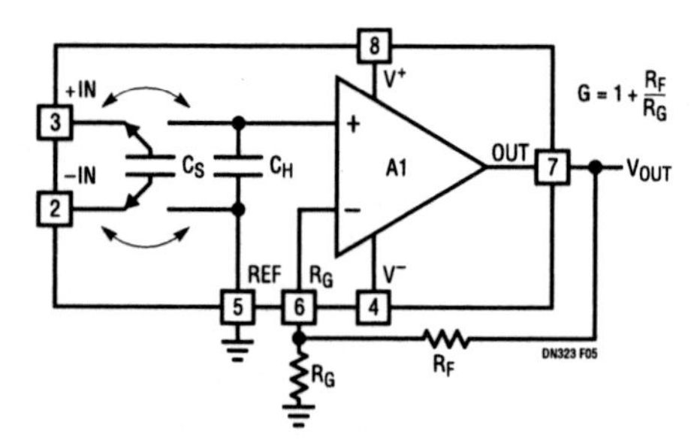

Figure 408.5 • The LTC6800 Block Diagram with External Gain Set Resistors Also Shown

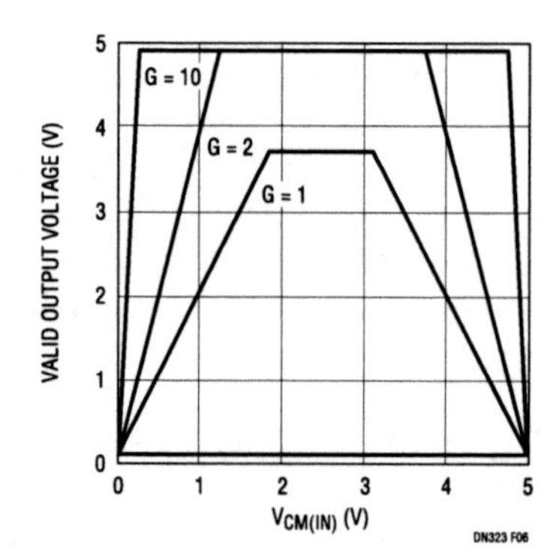

Figure 408.6 • The LTC6800 Output Swing vs Input Common Mode. Drastic Improvement over Classical Architecture

Ultraprecise instrumentation amplifier makes robust thermocouple interface

409

Jon Munson

Introduction

The versatile and precise LTC2053 instrumentation amplifier provides an excellent platform for robust, low power instrumentation products—as exemplified below by the battery-powered thermocouple amplifier circuit. The LTC2053 offers exceptionally low 10μV maximum input offset along with 116dB typical CMRR and PSRR, a result of a combination of switched capacitor and zero-drift op amp technologies. It is optimized for low voltage supplies from 2.7V to 11V single-ended or up to ±5.5V with split supplies. The LTC2053 is ideal for battery-powered instrument applications because of its low 850μA typical current draw. The gain is easily programmed with two resistors, as shown in Figure 409.1, just like a traditional non-inverting op amp. The LTC2053 also features low 1/f noise and rail-to-rail I/O to maximize dynamic range.

The requirements of thermocouple amplification

A robust thermocouple amplification circuit must meet several specific requirements. First, a commonly used type K thermocouple develops 40.6μV/°C, and a standard readout scale is 10mV/°C, so a precision amplifier with a nominal gain of 246 is required. Also, thermocouple leads are generally exposed to the electrical noise of an industrial environment so the fully differential input capability of an instrumentation amplifier helps eliminate errors due to common mode noise pickup. Finally, fault protection against accidental contact of the thermocouple to sources of transients or high voltage is needed but the protection cannot compromise accuracy.

The LTC2053 offers features that help meet all of these requirements. It can withstand a 10mA of fault current in any pin so 10kΩ protection resistors allow ±100V hard faults or Level 4 ESD (8kV contact/15kV air-gap) on the thermocouple junction without damage to the IC. The LTC2053 uses a switched-capacitor input topology, sampling at approximately 2.5kHz. With an internal input sampling capacitance of ~1000pF, the RC transients of the 10kΩ protection resistors settle within the ~180μs sampling window so they do not contribute to offset errors as they might with a typical IA.

A battery-powered thermocouple amplifier

Figure 409.2 shows the LTC2053 used in a battery-powered thermocouple amplifier. The circuit is used as a plug-in adapter for common digital multimeters and is completely portable. This circuit employs the LT1025 thermocouple compensator to improve accuracy over a wide range of ambient conditions and is mounted close to the thermocouple connection points for optimal thermal tracking. It precludes the need to temperature stabilize the thermocouple "cold junctions" and removes the accuracy penalty of a static room temperature correction value.

The output of the LT1025 provides a 10mV/°C correction voltage for the ambient temperature difference from

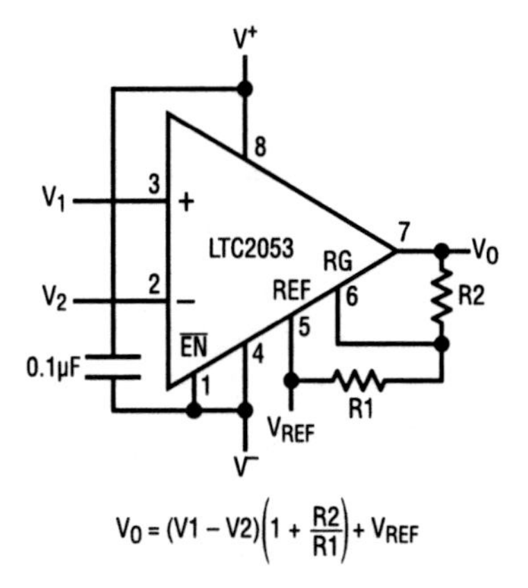

Figure 409.1 • Typical Connection of LTC2053 Instrumentation Amplifier

Analog Circuit Design: Design Note Collection. http://dx.doi.org/10.1016/B978-0-12-800001-4.00409-9

0°C—normally about 250mV at room temperature. The measured probe temperature is the sum of this compensation voltage and the amplified thermocouple voltage. Simple connection of the output of the compensator to the REF input of the LTC2053 is all that is needed to add these two voltages. The only consideration with this configuration is that the correction voltage must be capable of either sourcing or sinking the feedback resistor current that flows. As the LT1025 only sources current, a precision buffer can be used to drive the REF node (e.g., using an LTC2050 zero-drift op amp). The limitation imposed by using a single supply is that both the probe and amplifier unit temperatures must be above 0°C for valid output. If negative temperatures must be accommodated, a simple charge-pump inverter, such as an LTC1046, can be used to develop a minus supply rail. The excellent PSRR of the LTC2053 precludes the need for regulated power supplies, and the additional design and space expense they entail. Four AA alkaline cells supply the ICs in this circuit with 3.5V to 5V, depending on state of charge, yielding a minimum full-scale output of 350°C. The total battery draw is typically only 1.8mA. In a conventional line-powered application, one can use a single LT1025 and buffer amplifier to correct several LTC2053 thermocouple amplifier channels, provided all the thermocouple connections and the LT1025 thermally track.

Filtering and protection

Since the LTC2053 operates by sampling the input ignal, the frequencies of interest are generally below a few hundred hertz so it is useful to roll off the amplifier response by adding 0.1μF in the feedback circuit. The capacitors in the thermocouple input network help absorb RF pickup and suppress sampling artifacts from appearing on the thermocouple leads. The resistors connected to the thermocouple provide a high impedance bias of $V_S/2$ to maximize common mode immunity without inducing voltage drops in the leads. For short thermocouple lead lengths, which minimize common-mode signals, the probe junction may be grounded (note that with split supplies, grounding would be optimal). The 5.1V Zener is used to provide fault-induced supply vervoltage and reverse-battery protection in conjunction with the 560Ω ballast.

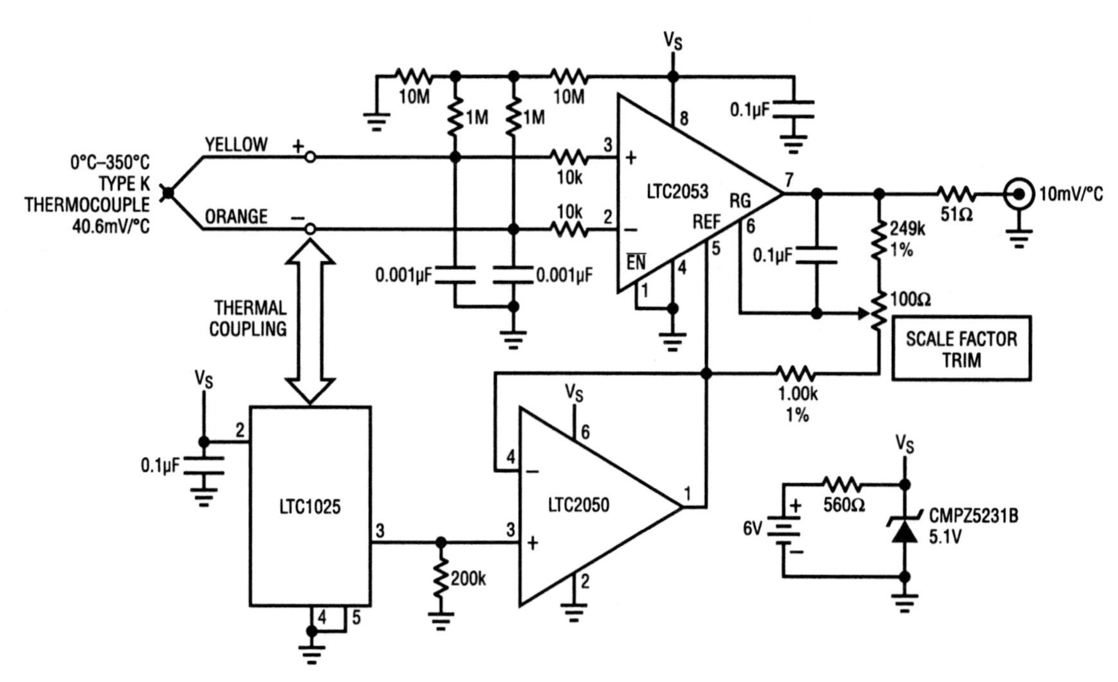

Figure 409.2 • Complete Schematic of the Thermocouple Amplifier

16-bit SO-8 DAC has 1LSB (max) INL and DNL over industrial temperature range

410

Jim Brubaker   William C. Rempfer   Kevin R. Hoskins

The new LTC1595/LTC1596 16-bit DACs from LTC provide the easiest to use, most cost-effective, highest performance solution for upgrading industrial and instrumentation applications from 12 bits to 16 bits. They feature:

- ±1LSB (max) INL and DNL over the industrial temperature range
- Ultralow, 1nV-s glitch impulse
- ±10V output capability
- Small SO-8 package (LTC1595)
- Pin compatible upgrade for industry standard 12-bit DACs (DAC8043/DAC8143 and AD7543)

Nice features of the 16-bit DACs

These new CMOS current output DACs use precision thin-film resistors in a modified R/2R architecture. They have SPI/MICROWIRE compatible serial interfaces and draw only 10μA from a single 5V supply. They generate precision 0V to 10V or ±10V outputs using a single or dual external op amp. The LTC1596 has an asynchronous clear input and both devices have power-on reset.

Because the LTC1595/LTC1596's INL is five times less sensitive to op amp V_{OS} when compared to 12-bit devices, systems using the DAC8043/DAC8143 or AD7543 can be upgraded to true 16-bit resolution and linearity without requiring more precise op amps.

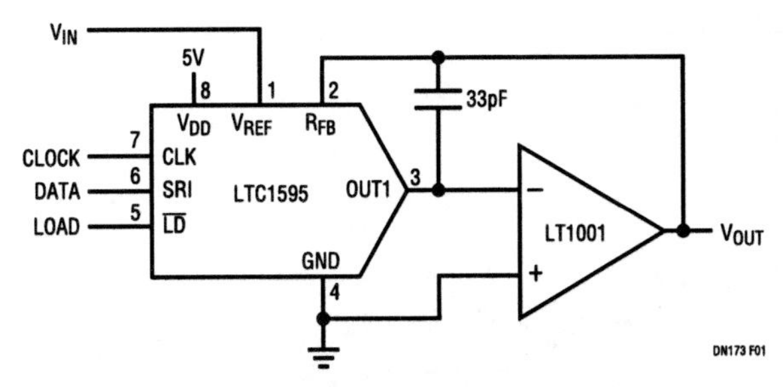

Figure 410.1 • With a Single Op Amp, the 16-Bit DAC Performs 2-Quadrant Multiplication with ±10V Input and Output Ranges. A Fixed −10V Reference Generates a Precision 16-Bit 0V to 10V Unipolar Output

16-bit accuracy over temperature without autocalibration

Autocalibrated DACs achieve their 16-bit accuracy at the cost of additional autocalibration circuitry that increases size and cost, requires cumbersome calibration overhead for the user

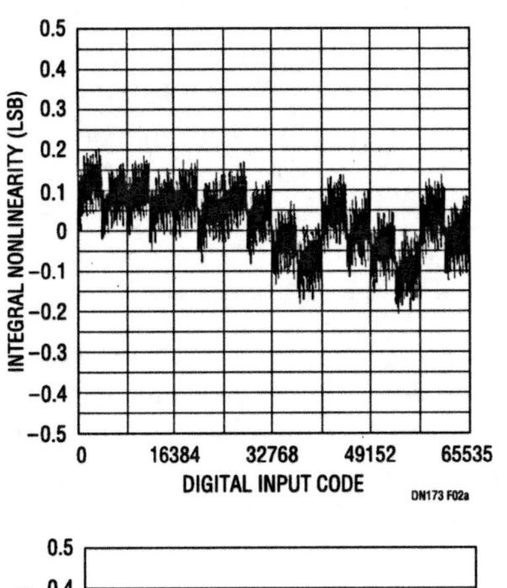

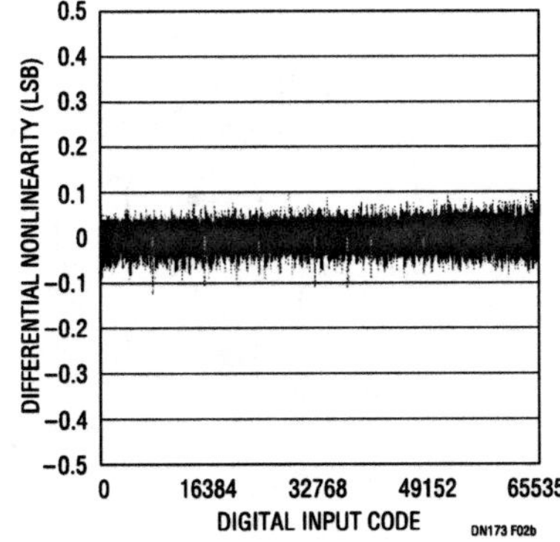

Figure 410.2 • The Outstanding INL and DNL (≤ 0.25LSB Typ) and Very Low Drift Allow a Maximum 1LSB Specification Over Temperature

Analog Circuit Design: Design Note Collection. http://dx.doi.org/10.1016/B978-0-12-800001-4.00410-5

and, because of poor linearity drift, requires DAC recalibration every time the temperature changes.

By eliminating autocalibration, the LTC1595/LTC1596 offer a better choice. Figure 410.2 shows the outstanding 0.25LSB (typ) integral nonlinearity (INL) and differential nonlinearity (DNL). This accuracy and very low drift guarantee a 1LSB (max) INL and DNL specification over the industrial temperature range without autocalibration.

Ultralow 1nV-s glitch

A new proprietary deglitcher brings great benefits to precision applications because it reduces the output glitch impulse to 1nV-s, ten times lower than any other 16-bit industrial DAC, and makes the glitch impulse uniform for any code. Figure 410.3 shows the output glitch for a midscale transition with a 0V to 10V output range.

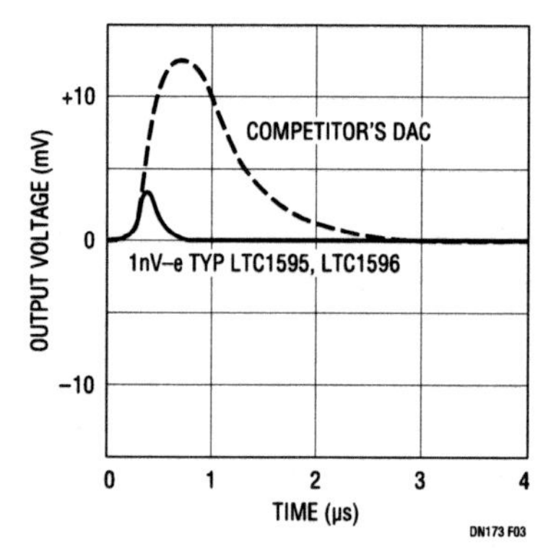

Figure 410.3 • The Output Glitch Is Less Than 1nV-s, Ten Times Less Than the Best 16-Bit Industrial DACs

Precision 0V to10V outputs with one op amp

The LTC1595 can be configured to generate 0V to 10V by applying −10V to the V_{REF} pin as shown in Figure 410.1. This circuit can also perform 2-quadrant multiplication where the reference is driven by a ±10V input signal and V_{OUT} swings from 0V to $-V_{REF}$.

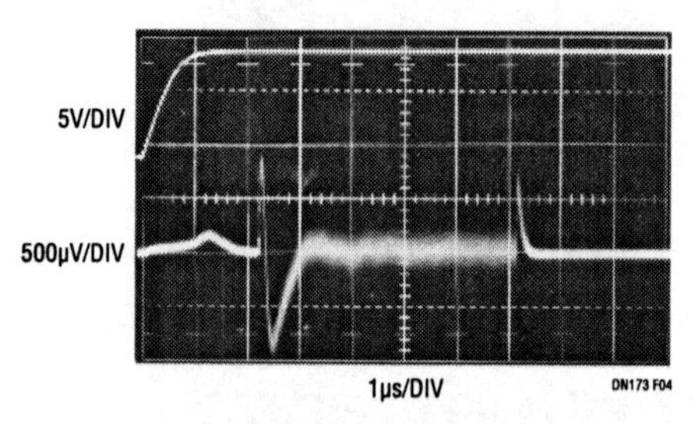

Figure 410.4 • The LTC1595/LTC1596 with an LT1122 Settles in 3μs for a Full-Scale Change. The Top Trace Shows the 0V to 10V DAC Output and the Lower Trace Shows the Gated 3μs Settling Waveform

The LTC1595/LTC1596 allow designers to choose an op amp that optimizes an application's accuracy, speed, power and cost. An LT1001 provides excellent DC precision, low noise and low power dissipation (90mW total for Figure 410.1's circuit). For higher speed, the LT1122 will provide settling to 1LSB in 3μs for a full-scale transition (Figure 410.4).

Precision ±10V outputs with a dual op amp

Figure 410.5 shows a bipolar, 4-quadrant multiplying application. The reference input can vary from −10V to 10V and V_{OUT} swings from $-V_{REF}$ to V_{REF}. Using a fixed 10V reference results in a precision ±10V bipolar output. Use a pack of matched 20k resistors (the 10k resistor is formed using two parallel 20k resistors) for good bipolar gain and offset. Substituting the LT1124 for the LT1112 provides faster settling.

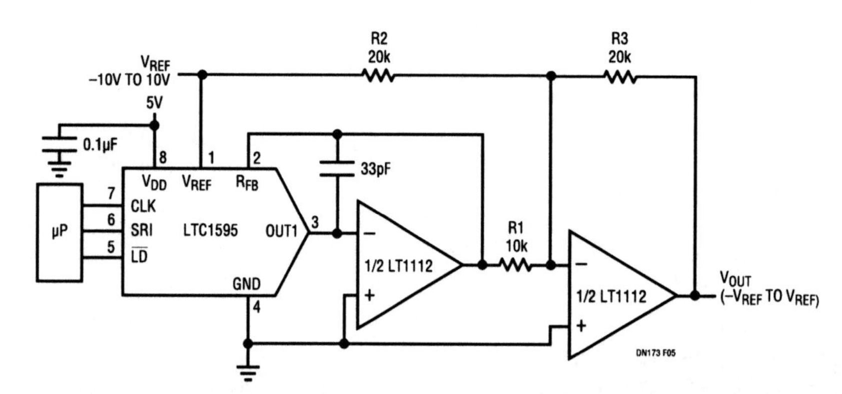

Figure 410.5 • The LTC1595 and the LT1112 Dual Op Amp Achieve 4-Quadrant Multiplication and a 16-Bit ±10V Bipolar Output with 10V Reference

Gain trimming in instrumentation amplifier-based systems

411

Jim Williams

Gain trimming is almost always required in instrumentation amplifier-based systems. Gain uncertainties, most notable in transducers, necessitate such a trim.

Figure 411.1, a conceptual system, shows several points as candidates for the trim. In practice, only one of these must actually be used. The appropriate trim location varies with the individual application.

Figure 411.2 approaches gain trimming by altering transducer excitation. The gain trim adjustment results in changes in the LT1010's output. The LT1027 reference and LT1097 ensure output stability. Transducer output varies with excitation, making this a viable approach. It is important to consider that gain "lost" by reducing transducer drive translates into reduced signal-to-noise ratio. As such, gain reduction by this method is usually limited to small trims, e.g., 5–10%. Similarly, too much gain introduced by this method can cause excessive transducer drive, degrading accuracy. The transducer manufacturer's data sheet should list the maximum permissible drive for rated accuracy.

Figure 411.3 adjusts gain in the instrumentation amplifier stage. The fixed gain LT1101 instrumentation amplifier feeds a second amplifier where the trim occurs. As both cases show,

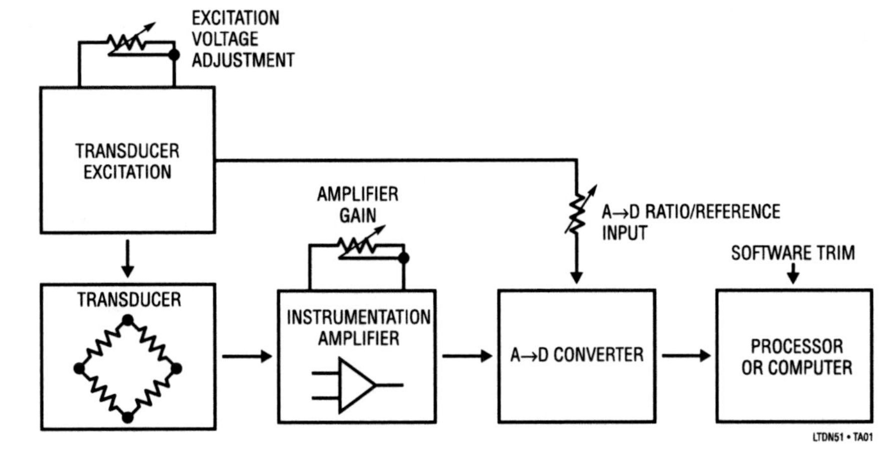

Figure 411.1 • Conceptual Transducer Signal Conditioning Path Showing Gain Trimming Possibilities. In Practice, Only One Adjustment Is Required

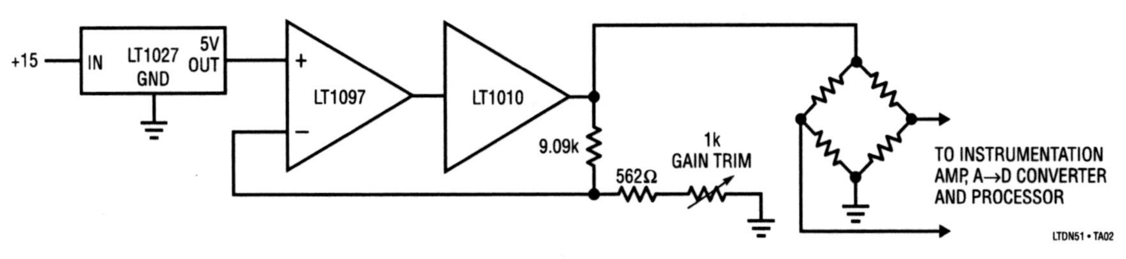

Figure 411.2 • Gain Trimming by Adjustment of Transducer Excitation. This Method Is Useable for Small (5 to 10%) Trims. Large Trims Will Cause Excessive Transducer Power Dissipation or "Starved" Outputs

Analog Circuit Design: Design Note Collection. http://dx.doi.org/10.1016/B978-0-12-800001-4.00411-7

the gain trim may be up or down. A secondary benefit of this trim scheme is that it permits optional offset summing and filtering. Note that either the inverter or follower may be set up for gain addition or reduction. The sole limitation is the signal polarity reversal imposed by the inverter case. This may be corrected by reversing the instrumentation amplifiers' inputs.

A final hardware-based gain trim is shown in Figure 411.4. Here, the A→D reference input is scaled to the appropriate voltage by the op amp and associated components. The op amp input is usually the transducer excitation voltage or, in cases where this is not possible, a reference.

One final way to trim gain is in software. If a processor is involved in the system this is a viable alternative. The software trim does a simple code conversion on the A→D output. When using this approach utilize as much of the analog components' dynamic range as possible to avoid signal-to-noise degradation.

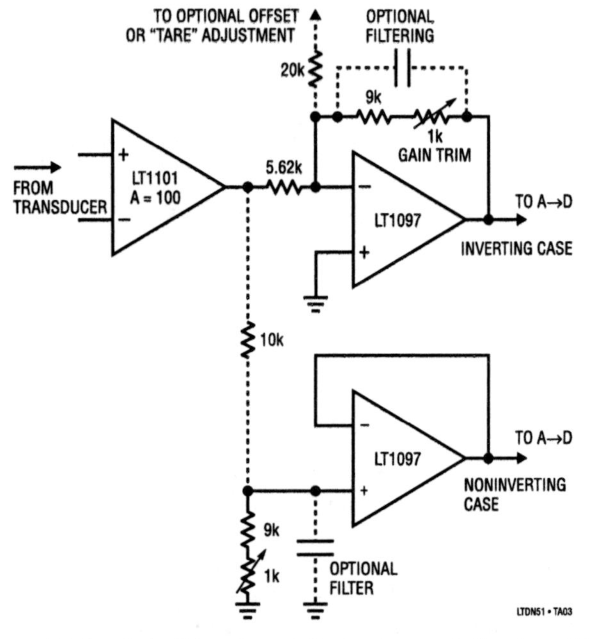

Figure 411.3 • Gain Trimming at the Instrumentation Amplifier. A Second Stage Permits Trimming Gain Up or Down, and Allows Filtering and Offsets to Be Summed In

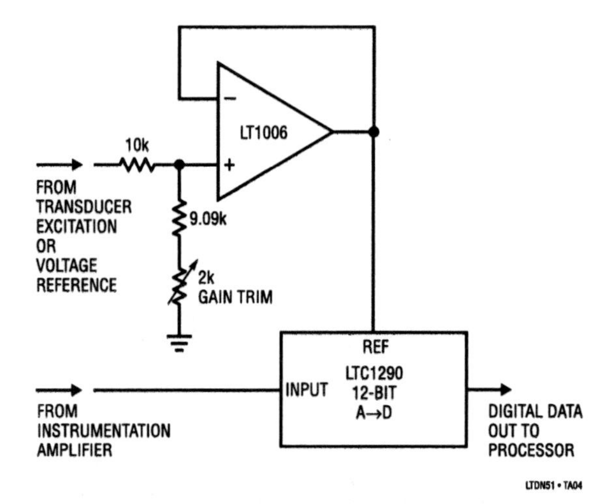

Figure 411.4 • Gain Trimming by Adjustment of the A→D Reference Input Voltage

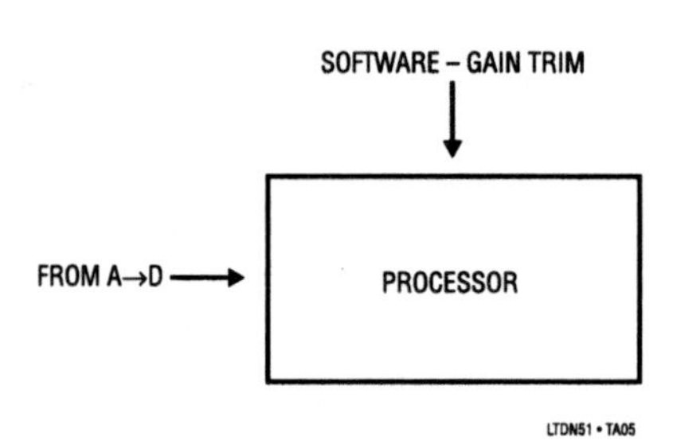

Figure 411.5 • Software Based Trimming

Signal conditioning for platinum temperature transducers

412

Jim Williams

High accuracy, stability, and wide operating range make platinum RTDs (resistance temperature detectors) popular temperature transducers. Signal conditioning these devices requires care to utilize their desirable characteristics. Figure 412.1's bridge-based circuit is highly accurate and features a ground referred RTD. The ground connection is often desirable for noise rejection. The bridge's RTD leg is driven by a current source while the opposing bridge branch is voltage biased. The current drive allows the voltage across the RTD to vary directly with its temperature induced resistance shift. The difference between this potential and that of the opposing bridge leg forms the bridge's output.

A1A and instrumentation amplifier A2 form a voltage controlled current source. A1A, biased by the LT1009 reference, drives current through the 88.7Ω resistor and the RTD. A2, sensing differentially across the 88.7Ω resistor, closes a loop back to A1A. The 2k–0.1μF combination sets amplifier rolloff, and the configuration is stable. Because A1A's loop forces a fixed voltage across the 88.7Ω resistor, the current through Rp is constant. A1's operating point is primarily fixed by the 2.5V LT1009 voltage reference.

The RTD's constant current forces the voltage across it to vary with its resistance, which has a nearly linear positive temperature coefficient. The nonlinearity could cause several degrees of error over the circuit's 0°C to 400°C operating range. The bridge's output is fed to instrumentation amplifier A3, which provides differential gain while simultaneously supplying nonlinearity correction. The correction is implemented

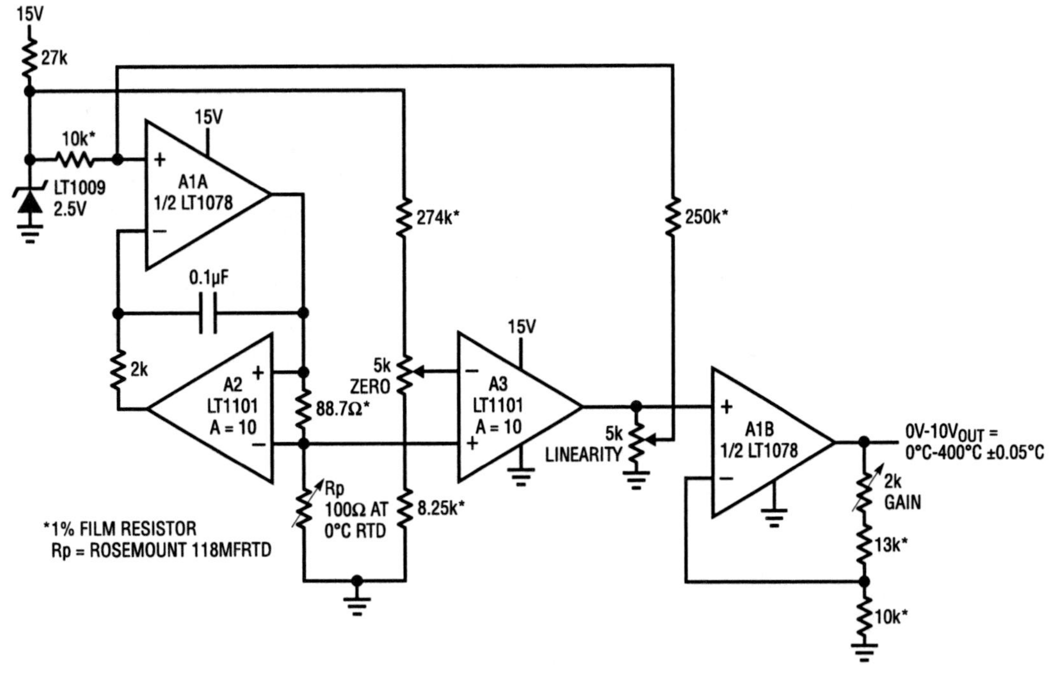

Figure 412.1 • Linearized Platinum RTD Bridge. Feedback to Bridge from A3 Linearizes the Circuit

Analog Circuit Design: Design Note Collection. http://dx.doi.org/10.1016/B978-0-12-800001-4.00412-9

by feeding a portion of A3's output back to A1's input via the 10k to 250k divider. This causes the current supplied to Rp to slightly shift with its operating point, compensating sensor nonlinearity to within ±0.05°C. A1B, providing additional scaled gain, furnishes the circuit output.

To calibrate this circuit, substitute a precision decade box (e.g., General Radio 1432k) for Rp. Set the box to the 0°C value (100.00Ω) and adjust the zero trim for a 0.00V output. Next, set the decade box for a 140°C output (154.26Ω) and adjust the gain trim for a 3.500V output reading. Finally, set the box to 249.0Ω (400.00°C) and trim the linearity adjustment for a 10.000V output. Repeat this sequence until all three points are fixed. Total error over the entire range will be within ±0.05°C. The resistance values given are for a nominal 100.00Ω (0°C) sensor. Sensors deviating from this nominal value can be used by factoring in the deviation from 100.00Ω. This deviation, which is manufacturer specified for each individual sensor, is an offset term due to winding tolerances during fabrication of the RTD. The gain slope of the platinum is primarily fixed by the purity of the material and has a very small error term.

The previous example relies on analog techniques to achieve a precise, linear output from the platinum RTD bridge. Figure 412.2 uses digital corrections to obtain similar results. A processor is used to correct residual RTD nonlinearities. The bridge's inherent nonlinear output is also accommodated by the processor.

The LT1027 drives the bridge with 5V. The bridge differential output is extracted by instrumentation amplifier A1. A1's output, via gain scaling stage A2, is fed to the LTC1290 12-bit A/D. The LTC1290's raw output codes reflect the bridge's nonlinear output versus temperature. The processor corrects the A/D output and presents linearized, calibrated data out. RTD and resistor tolerances mandate zero and full-scale trims, but no linearity correction is necessary. A2's analog output is available for feedback control applications. The complete software code for the 68HC05 processor, developed by Guy M. Hoover, appears in Application Note 43.

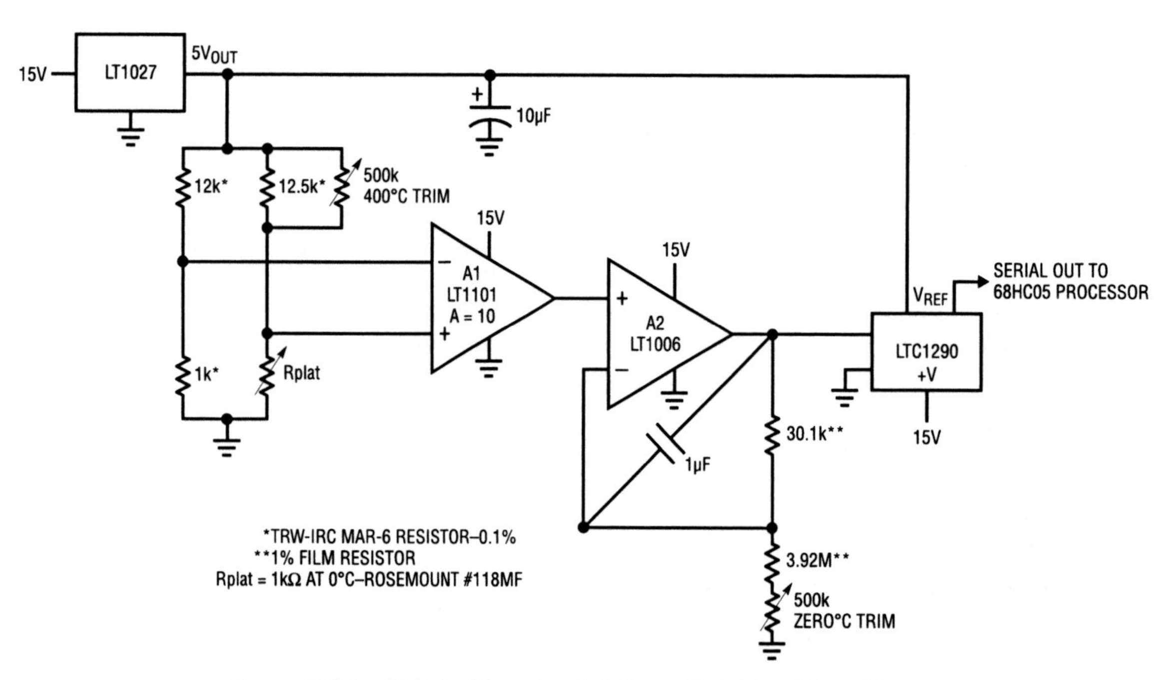

Figure 412.2 • Digitally Linearized Platinum RTD Signal Conditioner

Designing with a new family of instrumental amplifiers

413

Jim Williams

A new family of IC instrumentation amplifiers achieves performance and cost advantages over other alternatives. Conceptually, an instrumentation amplifier is simple. Figure 413.1 shows that the device has passive, fully differential inputs, a single-ended output and internally set gain. Additionally, the output is delivered with respect to the reference pin, which is usually grounded. Maintaining high performance with these features is difficult, accounting for the cost-performance disadvantages previously associated with instrumentation amplifiers.

Figure 413.2 summarizes specifications for the amplifier family. The LTC1100 has the extremely low offset, drift, and bias current associated with chopper stabilization techniques. The LT1101 requires only 105μA of supply current while retaining excellent DC characteristics. The FET input LT1102 features high speed while maintaining precision. Gain error and drift are extremely low for all units, and the single supply capability of the LTC1100 and LT1101 is noteworthy.

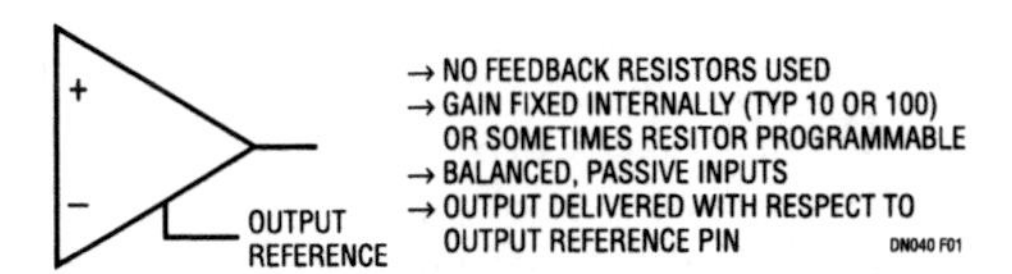

Figure 413.1 • Conceptual Instrumentation Amplifier

PARAMETER	CHOPPER STABILIZED LTC1100	MICROPOWER LT1101	HIGH SPEED LT1102
Offset	10μV	160μV	500μV
Offset Drift	100nV/°C	2μV/°C	2.5μV/°C
Bias Current	50pA	8nA	50pA
Noise (0.1Hz-10Hz)	2μVp-p	0.9μV	2.8μV
Gain	100	10,100	10,100
Gain Error	0.03%	0.03%	0.05%
Gain Drift	4ppm/°C	4ppm/°C	5ppm/°C
Gain Nonlinearity	8ppm	8ppm	10ppm
CMRR	104dB	100dB	100dB
Power Supply	Single or Dual, 16V Max	Single or Dual, 44V Max	Dual, 44V Max
Supply Current	2.2mA	105μA	5mA
Slew Rate	1.5V/μs	0.07V/μs	25V/μs
Bandwidth	8kHz	33kHz	220kHz

Figure 413.2 • Comparison of the New IC Instrumentation Amplifiers

The classic application for these devices is bridge measurement. Accuracy requires low drift, high common mode rejection and gain stability. Figure 413.3 shows a typical arrangement with the table listing performance features for different bridge transducers and amplifiers.

Bridge measurement is not the only use for these devices. They are also useful as general purpose circuit components, in similar fashion to the ubiquitous op amp. Figure 413.4 shows a voltage controlled current source with load and control voltage referred to ground. This simple, powerful circuit produces output current in strict accordance with the sign and magnitude of the control voltage. The circuit's accuracy and stability are almost entirely dependent upon resistor R. A1, biased by V_{IN}, drives current through R (in this case 10Ω)

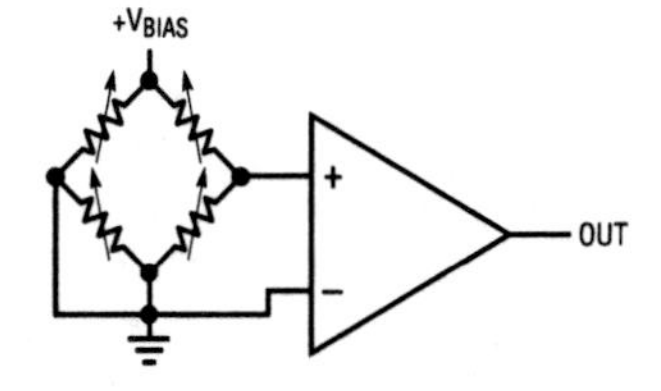

BRIDGE TRANSDUCER	AMPLIFER	V_{BIAS}	COMMENTS
350 Strain Gage (BLH #DHF –350)	LTC1100	10V	Highest Accuracy, 30mA Supply Current
1800Ω Semiconductor (Motorola MPX2200AP)	LT1101	1.2V	Lower Accuracy & Cost. <800μA Supply Current

Figure 413.3 • Characteristics of Some Bridge Transducer-Amplifier Combinations

Analog Circuit Design: Design Note Collection. http://dx.doi.org/10.1016/B978-0-12-800001-4.00413-0

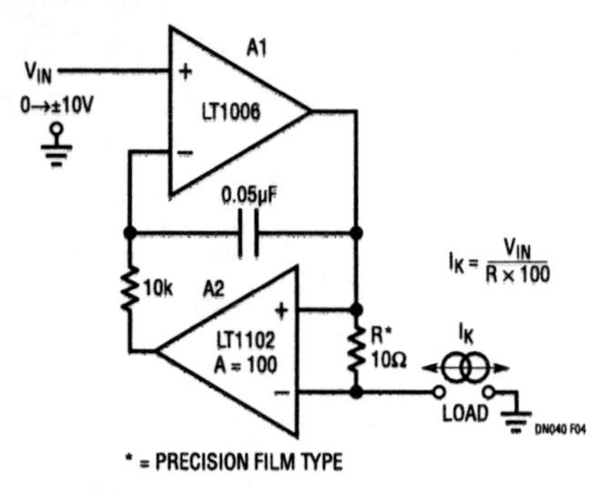

Figure 413.4 • Voltage Programmable Current Source Is Simple and Precise

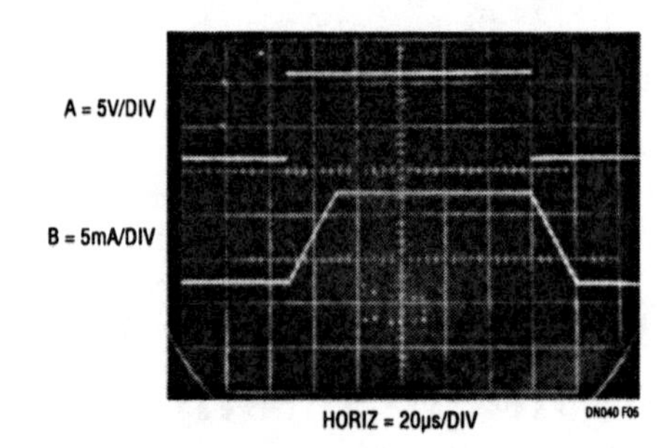

Figure 413.5 • Dynamic Response of the Current Source

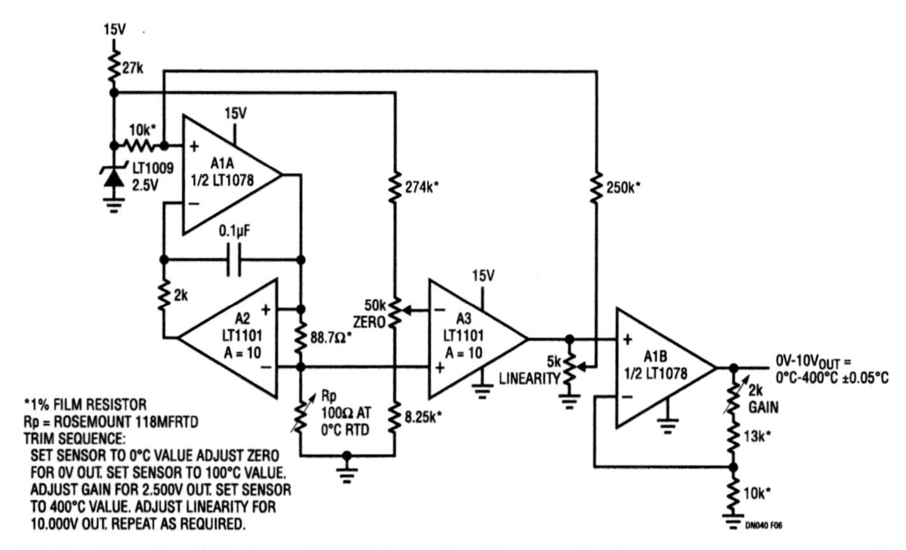

Figure 413.6 • Linearized Platinum RTD Bridge. Feedback to Bridge from A3 Linearizes the Circuit

and the load. A2, sensing differentially across R, closes a loop back to A1. The load current is constant because A1's loop forces a fixed voltage across R. The 10k–0.5μF combination sets rolloff, and the configuration is stable. Figure 413.5 shows dynamic response. Trace A is the voltage control input while trace B is the output current. Response is clean, with no slew residue or aberrations.

A final circuit, Figure 413.6, combines the current source and a platinum RTD bridge to form a complete high accuracy thermometer. A1A and A2 will be recognized as a form of Figure 413.4's current source. The ground referred RTD sits in a bridge composed of the current drive and the LT1009 biased resistor string. The current drive allows the voltage across the RTD to vary directly with its temperature induced resistance shift. The difference between this potential and that of the opposing bridge leg forms the bridge output. The RTD's constant current drive forces the voltage across it to vary with its resistance, which has a nearly linear positive temperature coefficient. The nonlinearity could cause several degrees of error over the circuit's 0°C–400°C operating range. The bridges output is fed to instrumentation amplifier A3, which provides differential gain while simultaneously supplying nonlinearity correction. The correction is implemented by feeding a portion of A3's output back to A1's input via the 10k–250k divider. This causes the current supplied to Rp to slightly shift with its operating point, compensating sensor nonlinearity to within ±0.05°C. A1B, providing additional scaled gain, furnishes the circuit output. To calibrate this circuit, follow the procedure given in Figure 413.6.

Details of these and other instrumentation amplifier circuits may be found in LTC Application Note 43, "Bridge Circuits—Marrying Gain and Balances."

PART 3

Signal Conditioning

Section 1

Operational Amplifier Design Techniques

High voltage CMOS amplifier enables high impedance sensing with a single IC

414

Jon Munson

Introduction

Accurately measuring voltages requires minimizing the impact of the instrument connection to the tested circuit. Typical digital voltmeters (DVMs) use 10M resistor networks to keep loading effects to an inconspicuous level, but even this can introduce significant error, particularly in higher voltage circuits that involve high resistances.

The solution is to use high impedance amplifiers in an electrometer configuration, so only miniscule amplifier input current comes from the test node. To make the input current the lowest possible value, field-effect transistors (FETs) are traditionally employed at the inputs of these circuits. FETs are generally low voltage devices, and introduce voltage offset uncertainty that is difficult to eliminate. Monolithic amplifiers exist that incorporate FET inputs, but these are often very low voltage devices, especially those using typical CMOS fabrication methods, so their utility is limited in high voltage applications. Enter the LTC6090, a CMOS amplifier that can handle over $140V_{P-P}$ signal swings with sub-mV precision, ideally suited to tackle the problem.

The LTC6090 easily solves high voltage sensing problems

The LTC6090 combines a unique set of characteristics in a single device. Its CMOS design characteristics provide the ultimate in high input impedance and "rail-to-rail" output swing, but unlike typical CMOS parts that might be powered by 5V, the LTC6090 can operate with supplies up to ±70V. The device can hold its own in the small-signal regime as well, featuring typical V_{OS} under 500μV and voltage-noise density of $11nV/\sqrt{Hz}$, yielding a spectacular dynamic range. With the high voltage operation comes the possibility of significant power dissipation, so the LTC6090 is available in thermally enhanced SOIC or TSSOP packages. It includes an over temperature output flag and an output disable control that provide flexible protection measures without additional circuitry.

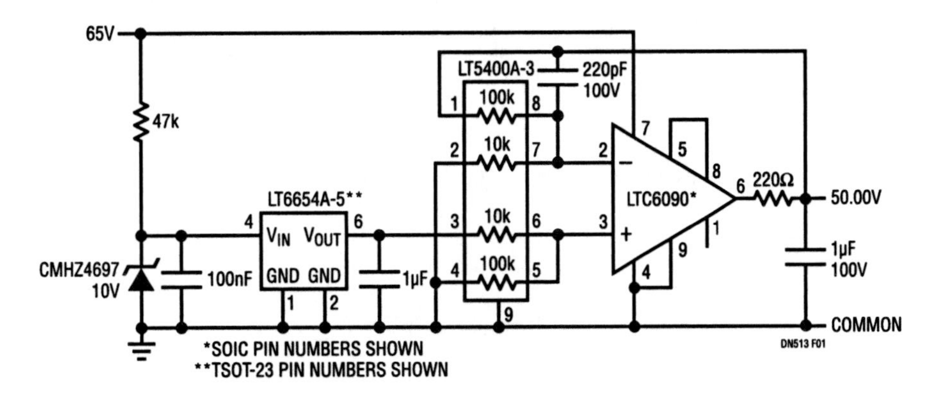

Figure 414.1 • High Voltage Precision Reference

Analog Circuit Design: Design Note Collection. http://dx.doi.org/10.1016/B978-0-12-800001-4.00414-2

Accurate 50.00V reference

The LTC6090 is capable of 140V output levels in single-supply operation, so amplifying a quality 5V reference is a simple matter of using accurate resistor networks to maintain precision. The LT5400 precision resistor array handles voltages up to 80V, so utilizing the 10:1 ratio version for a gain of 10 is an easy way to produce an accurate 50V calibration source without any adjustments required. Figure 414.1 shows a circuit amplifying the LT6654A 5.000V reference to 50.00V with better than 0.1% accuracy. The circuit may be powered from 55V to 140V, with 65V being a useful supply voltage furnished from the optional portable supply shown as part of Figure 414.2.

The LTC6090 is set up with a 1μF output capacitance to provide excellent load-step response. The capacitance is isolated from the op amp with a resistance that forms an effective noise-reduction filter for frequencies above 700Hz. The precision LT5400A-3 resistor network provides 0.01% matched 10k/100k resistances that, along with the absence of loading by the high impedance CMOS op amp inputs, forms a highly accurate amplification factor. Input offset voltage of the LTC6090 contributes <0.03% error, while the LT6654A contributes <0.05%. The entire circuit of Figure 414.1draws about 4mA of quiescent current and can drive 10mA loads.

Simple large-signal buffer

The LTC6090 behaves as an ordinary unity-gain-stable operational amplifier, so constructing an electrometer-grade buffer stage is simply a matter of providing 100% feedback with the classic unity-gain circuit. No discrete FETs or floating biasing supplies are needed.

As shown in Figure 414.2, the LTC6090 can easily be powered with a split supply, such as a small flyback converting battery source. This basic circuit can provide precision measurement of voltages in high impedance circuitry and accurately handle signal swings to within 3V of either supply rail (±62V in this case). With input leakage current typically below 5pA, circuit loading is essentially inconsequential (<V_{OS}) for source impedances approaching a gigaohm. The useful full-swing frequency response is over 20kHz.

Conclusion

The LTC6090 is a unique and versatile high voltage CMOS amplifier that enables simplified high impedance and/or large signal swing, very wide dynamic range amplification solutions.

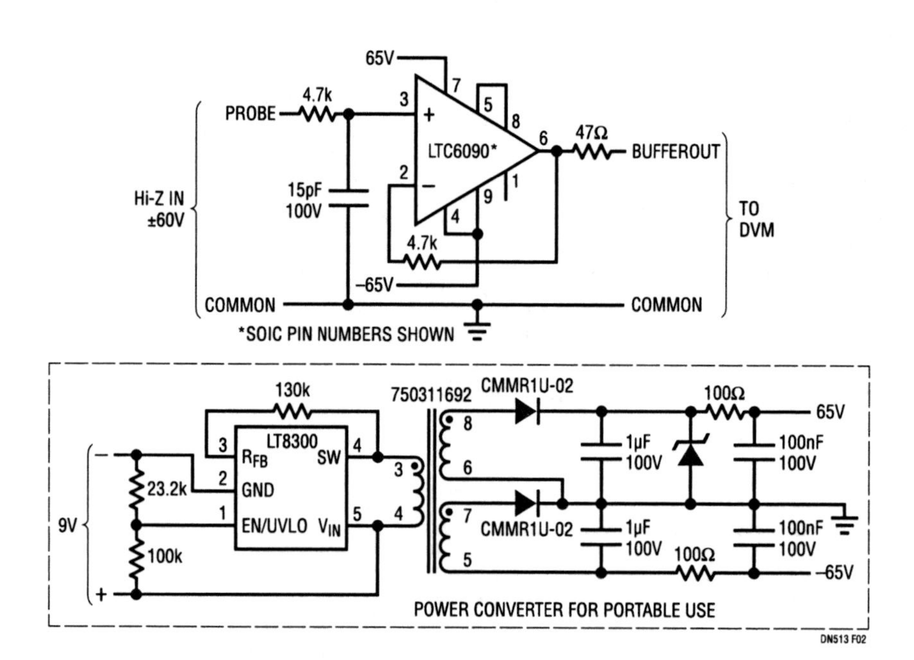

Figure 414.2 • Buffered Probe for Digital Voltmeter

Matched resistor networks for precision amplifier applications

415

Tyler Hutchison

Introduction

Some ideal op amp configurations assume that the feedback resistors exhibit perfect matching. In practice, resistor non-idealities can affect various circuit parameters such as common mode rejection ratio (CMRR), harmonic distortion and stability. For instance, as shown in Figure 415.1, a single-ended amplifier configured to level-shift a ground-referenced signal to a common mode of 2.5V needs a good CMRR. Assuming 34dB CMRR and no input signal, this 2.5V level shifter exhibits an output offset of 50mV, which can even overwhelm the LSB and offset errors of 12-bit ADCs and drivers.

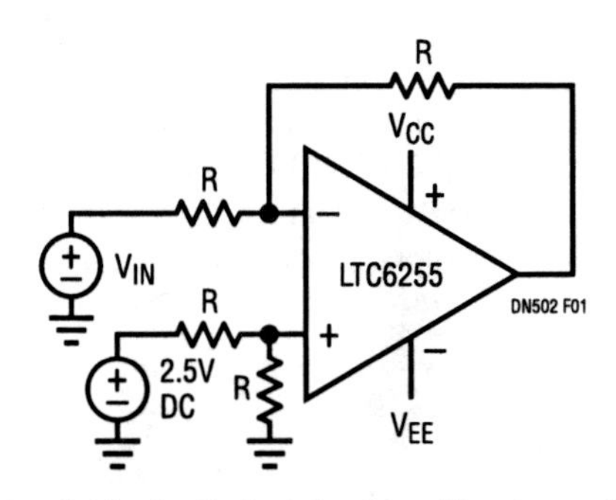

Figure 415.1 • A Single-Ended Op Amp Used as a Level Shifter

For an op amp, 34dB is a less than ideal CMRR. However, a feedback network of 1% tolerance resistors can limit the CMRR to 34dB regardless of the op amp's capabilities. Highly matched resistors, such as those provided by the LT5400, available in 0.01%, 0.025% and 0.05% matching, ensure that the designer can approach or meet amplifier data sheet specifications. This design note compares the LT5400 with thick film, 0402, 1% tolerance surface mount resistors. CMRR, harmonic distortion and stability are considered with these resistors for feedback around an LTC6362 op amp, as shown in Figure 415.2.

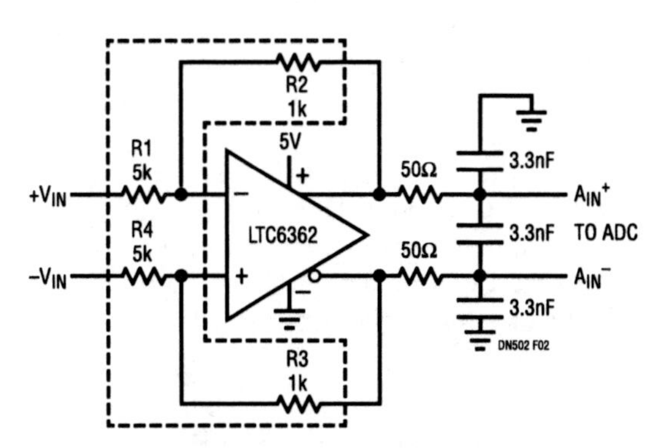

Figure 415.2 • Fully Differential Op Amp Configured for $V_{OUT}/V_{IN} = 0.2$

Common mode rejection ratio

In order to obtain precise measurements in the presence of common mode noise, a high CMRR is important. Input CMRR is defined as the ratio of differential gain ($V_{OUT(DIFF)}/V_{IN(DIFF)}$) to the input common mode to differential conversion ($V_{OUT(DIFF)}/V_{IN(CM)}$).

In ideal single-ended and fully differential amplifiers, only the input differential level affects the output voltage. However, in real circuits, resistor mismatch limits the available CMRR. Consider this circuit configured to attenuate a ±10V signal to a ±2V signal. Using typical surface mount resistors with 2% matching (1% tolerance), the worst-case CMRR contribution from the resistors is 30dB. With 0.01% tolerance, 0.02% matching, the worst-case CMRR contribution from the resistors is 70dB. A limiting factor in the CMRR equation is:

$$\frac{1}{2}\left|\left(\frac{R1}{R2}-\frac{R3}{R4}\right)\frac{R2}{R1}\right|$$

This expression reduces to the resistor matching ratio for typical resistors, but the LT5400 takes an additional step to offer improved CMRR by constraining the matching between resistor pairs R1/R2 and R4/R3. By defining this equation as

Analog Circuit Design: Design Note Collection. http://dx.doi.org/10.1016/B978-0-12-800001-4.00415-4

matching for CMRR, the LT5400 offers accuracy that's better than just the resistor matching ratio. For example, the LT5400A guarantees:

$$\left(\frac{\Delta R}{R}\right)_{CMRR} \leq 0.005\%$$

which improves worst-case CMRR to 82dB.

A bench test of the circuit yielded a CMRR of 50.7dB (highly resistor matching limited) with 1% tolerance resistors, and 86.6dB with the LT5400. In this case, a 2.5V common mode input would result in an offset of 1.5mV with 1% thick-film resistors and an offset of 23μV with the LT5400, making it suitable for 18-bit ADC applications where DC accuracy is critical.

Harmonic distortion

Harmonic distortion is also important when choosing resistors for a precision application. A large-signal voltage across a resistor may significantly change the resistance depending on the size and material. This problem occurs in a number of chip-based resistors, and naturally becomes more severe as the power levels at the resistors increase. Table 415.1 compares the distortion of thick film, through-hole, and the LT5400 resistors based on high power drive and similar power drive. The results suggest that for a given signal, the LT5400 distorts the signal much less than other resistor types.

Stability

Figure 415.3 shows a model of the distributed capacitance between resistors in the LT5400. To achieve high precision matching and tracking in the LT5400, many small SiCr resistors are configured in series and parallel. As a result of the complex interdigitation, the LT5400 resistors can be modeled as a series of infinitesimal resistors with parasitic capacitance between adjacent segments and between individual segments and the exposed pad. In contrast, typical surface mount

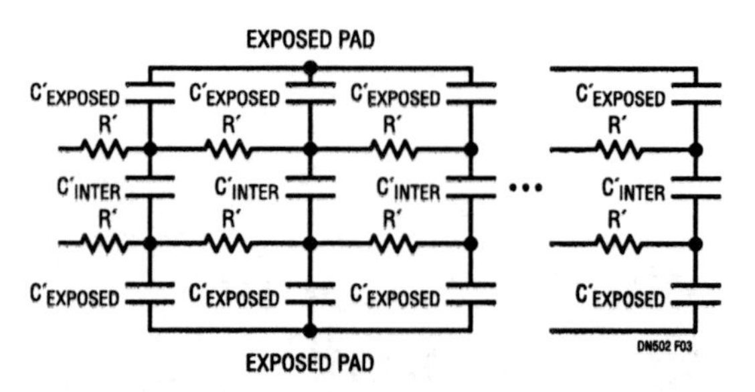

Figure 415.3 • A Simple Model of the Distributed Capacitance in a Matched Resistor IC. The Sum of R′ Components Creates an Equivalent Single Resistor. The Net Effect of C'_{INTER} Is 1.4pF and the Net Effect of $C'_{EXPOSED}$ is 5.5pF

resistors, without the tight layout, typically exhibit significantly less parasitic capacitance.

The effect of inter-resistor capacitance can be mitigated when the exposed pad is grounded. However, even after grounding the exposed pad, this capacitance still affects circuit stability by forming a parasitic pole on the order of the total resistance times the total capacitance.

Since overshoot is inversely proportional to phase margin, minimizing step response overshoot is a good way to ensure circuit stability. The uncompensated LT5400 configuration exhibits 27% compared to 17% overshoot from the 0402 configuration. However, the compensation capacitor necessary to achieve 8% overshoot is approximately the same in both configurations: 18pF with the LT5400; 15pF with 0402 resistors. With nearly identical compensation, the two circuits display similar stability characteristics.

Conclusion

The actual performance of precision amplifiers and ADCs is often difficult to achieve since data sheet specifications assume ideal components. Carefully matched resistor networks, such as those supplied by the LT5400, enable precision matching orders of magnitude better than discrete components, ensuring data sheet specifications are met for precision ICs.

Table 415.1 For a Given Power Level, the LT5400 Behaves More Linearly Than Other Resistor Types

SOURCE HD3	−120.00	AT MAX POWER ($12V_{RMS}$ = 56mW INTO 1kΩ)	
RESISTOR TYPE	**POWER RATING**	**HD3 (56mW POWER)**	**HD3 (1/14TH RATED POWER)**
LT5400	0.8W	−117dBc	−117dBc
5% Through-Hole	0.25W	−100dBc	−114dBc
1% Through-Hole	0.25W	−115dBc	−119dBc
1206 Thick Film	0.25W	−104dBc	−115dBc
0805 Thick Film	0.125W	−93dBc	−117dBc
0603 Thick Film	0.1W	−89dBc	−117dBc
0402 Thick Film	0.068W	−72dBc	−104dBc

Using a differential I/O amplifier in single-ended applications

416

Glen Brisebois

Introduction

Recent advances in low voltage silicon germanium and BiCMOS processes have allowed the design and production of very high speed amplifiers. Because the processes are low voltage, most of the amplifier designs have incorporated differential inputs and outputs to regain and maximize total output signal swing. Since many low-voltage applications are single-ended, the questions arise, "How can I use a differential I/O amplifier in a single-ended application?" and "What are the implications of such use?" This design note addresses some of the practical implications and demonstrates specific single-ended applications using the 3GHz gain-bandwidth LTC6406 differential I/O amplifier.

Background

A conventional op amp has two differential inputs and an output. The gain is nominally infinite, but control is maintained by virtue of feedback from the output to the negative "inverting" input. The output does not go to infinity, but rather the differential input is kept to zero (divided by infinity, as it were). The utility, variety and beauty of conventional op amp applications are well documented, yet still appear inexhaustible. Fully differential op amps have been less well explored.

Figure 416.1 shows a differential op amp with four feedback resistors. In this case the differential gain is still nominally infinite, and the inputs kept together by feedback, but this is not adequate to dictate the output voltages. The reason is that the common mode output voltage can be anywhere and still result in a "zero" differential input voltage because the feedback is symmetric. Therefore, for any fully differential I/O amplifier, there is always another control voltage to dictate the output common mode voltage. This is the purpose of the V_{OCM} pin, and explains why fully differential amplifiers are at least 5-pin devices (not including supply pins) rather than 4-pin devices. The differential gain equation is $V_{OUT(DM)} = V_{IN(DM)} \cdot R2/R1$. The common mode output voltage is forced internally to the voltage applied at V_{OCM}. One final observation is that there is no longer a single inverting input: both inputs are inverting and noninverting depending on which output is considered. For the purposes of circuit analysis, the inputs are labeled with "+" and "−" in the conventional manner and one output receives a dot, denoting it as the inverted output for the "+" input.

Anybody familiar with conventional op amps knows that noninverting applications have inherently high input impedance at the noninverting input, approaching GΩ or even TΩ. But in the case of the fully differential op amp in Figure 416.1, there is feedback to both inputs, so there is no high impedance node. Fortunately this difficulty can be overcome.

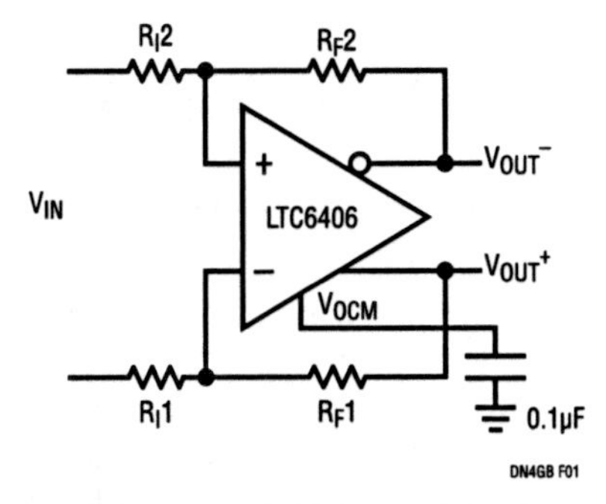

Figure 416.1 • Fully Differential I/O Amplifier Showing Two Outputs and an Additional V_{OCM} Pin

Simple single-ended connection of a fully differential op amp

Figure 416.2 shows the LTC6406 connected as a single-ended op amp. Only one of the outputs has been fed back and only one of the inputs receives feedback. The other input is now

Analog Circuit Design: Design Note Collection. http://dx.doi.org/10.1016/B978-0-12-800001-4.00416-6

high impedance. The LTC6406 works fine in this circuit and still provides a differential output. However, a simple thought experiment reveals one of the downsides of this configuration. Imagine that all of the inputs and outputs are sitting at 1.2V, including V_{OCM}. Now imagine that the V_{OCM} pin is driven an additional 0.1V higher. The only output that can move is V_{OUT}^- because V_{OUT}^+ must remain equal to V_{IN}, so in order to move the common mode output higher by 100mV the amplifier has to move the V_{OUT}^- output a total of 200mV higher. That's a 200mV differential output shift due to a 100mV V_{OCM} shift. This illustrates the fact that single-ended feedback around a fully differential amplifier introduces a noise gain of two from the V_{OCM} pin to the "open" output. In order to avoid this noise, simply do not use that output, resulting in a fully single-ended application. Or, you can take the slight noise penalty and use both outputs.

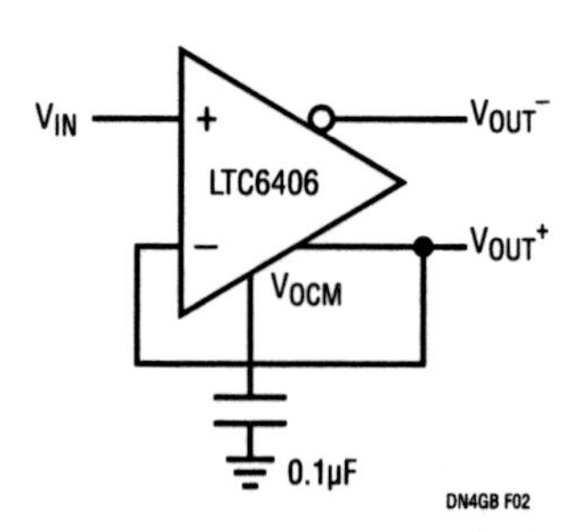

Figure 416.2 • Feedback Is Single-Ended Only. This Circuit Is Stable, with a Hi-Z Input Like the Conventional Op Amp. The Closed Loop Output (V_{OUT}^+ in this Case) Is Low Noise. Output Is Best Taken Single-Ended from the Closed Loop Output, Providing a 3dB Bandwidth of 1.2GHz. The Open Loop Output (V_{OUT}^-) Has a Noise Gain of Two from V_{ocm}, but Is Well Behaved to About 300MHz, Above Which It Has Significant Passband Ripple

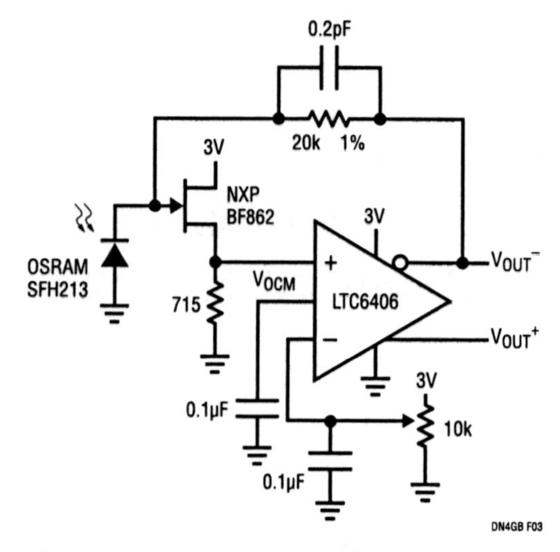

Figure 416.3 • Transimpedance Amplifier. Ultralow Noise JFET Buffers the Current Noise of the Bipolar LTC6406 Input. Trim the Pot for 0V Differential Output under No-Light Conditions

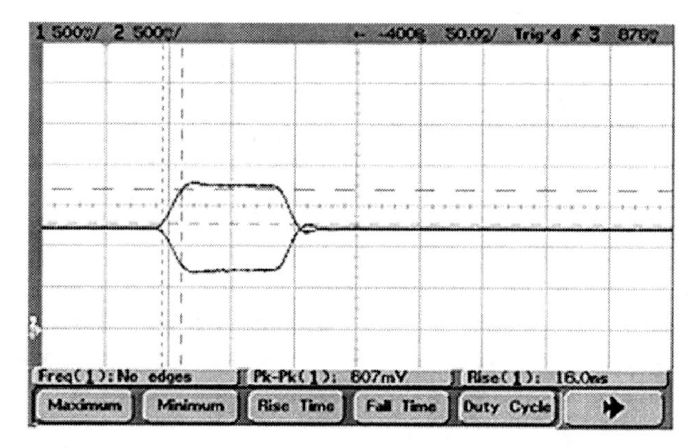

Figure 416.4 • Time Domain Response of Circuit of Figure 416.3, Showing Both Outputs Each with 20kΩ of TIA Gain. Rise Time Is 16ns, Indicating a 20MHz Bandwidth.

A single-ended transimpedance amplifier

Figure 416.3 shows the LTC6406 connected as a single-ended transimpedance amplifier with 20kΩ of transimpedance gain. The BF862 JFET buffers the LTC6406 input, drastically reducing the effects of its bipolar input transistor current noise. The V_{GS} of the JFET is now included as an offset, but this is typically 0.6V so the circuit still functions well on a 3V single supply and the offset can be dialed out with the 10k potentiometer. The time domain response is shown in Figure 416.4. Total output noise on 20MHz bandwidth measurements shows $0.8mV_{RMS}$ on V_{OUT}^+ and $1.1mV_{RMS}$ on V_{OUT}^-. Taken differentially, the transimpedance gain is 40kΩ.

Conclusion

New families of fully differential op amps like the LTC6406 offer unprecedented bandwidths. Fortunately, these op amps can also function well in single-ended and 100% feedback applications.

Single-ended to differential amplifier design tips

417

Philip Karantzalis   Tim Regan

Introduction

A fully differential amplifier is often used to convert a single-ended signal to a differential signal, a design which requires three significant considerations: the impedance of the single-ended source must match the single-ended impedance of the differential amplifier, the amplifier's inputs must remain within the common mode voltage limits and the input signal must be level shifted to a signal that is centered at the desired output common mode voltage.

In all cases, input impedance matching to the source impedance is necessary to prevent high frequency reflections. In designs where the single-ended source is DC coupled to a single supply differential amplifier, then level shifting and the common mode limits are also important considerations. The interaction of these three design parameters is non-trivial—component selection requires spreadsheet analysis using the equations described here.

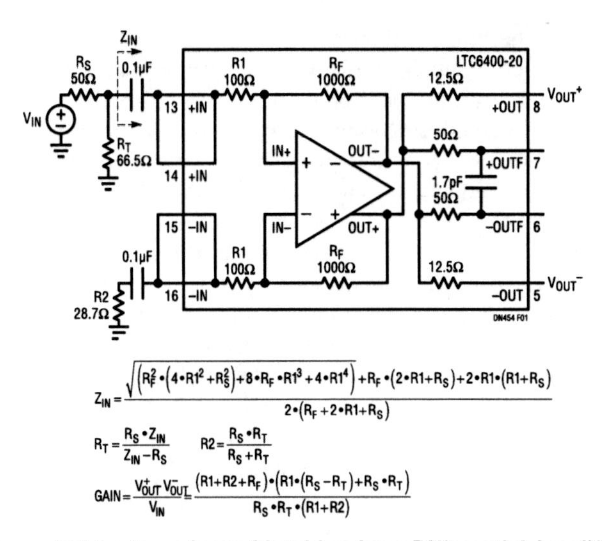

Figure 417.1 • Impedance Matching for a Differential Amplifier with Fixed Gain Integrated Resistors

Input impedance matching

If input AC coupling is used, then impedance matching is the only design issue. Figure 417.1 shows an example of a circuit matching a 50Ω single-ended source to an AC-coupled LTC6400-20 differential amplifier with a gain of 20dB set by internal resistors.

The 66.5Ω resistor, R_T, in parallel with the +IN input impedance, Z_{IN}, matches the circuit input impedance to the 50Ω source. Differential balance is provided with the addition of the 28.7Ω resistor at the −IN input, R2. The balancing resistor assures equivalent feedback factors at the inputs, thus preventing large DC offsets.

To calculate the external resistor values, start by calculating Z_{IN}. Then calculate R_T for impedance matching and the value of the R2 for differential balance. The overall single-ended to differential gain (GAIN) must take into account the input attenuation of the R_S and R_T resistive divider and the effect of adding R2. In this example, the overall gain of the amplifier from signal source to differential output is only 4.44 even though the amplifier has a fixed gain of 10.

By AC coupling at the input, the amplifier's input common mode voltage is equal to its output common mode voltage and the single-ended signal is automatically level shifted to an output differential signal centered on the output common mode voltage.

If the input common mode voltage is not 0V, and the source cannot deliver the DC current into 116.5Ω (50Ω + 66.5Ω), then it is also necessary to AC couple the 66.5Ω resistor.

The DC-coupled differential amplifier

A general purpose, DC coupled, single-ended to differential amplifier circuit with source impedance matching and input level shifting is shown in Figure 417.2. Level shifting is provided by the reference voltage (V_{REF}). If V_{REF} is set to be

Analog Circuit Design: Design Note Collection. http://dx.doi.org/10.1016/B978-0-12-800001-4.00417-8

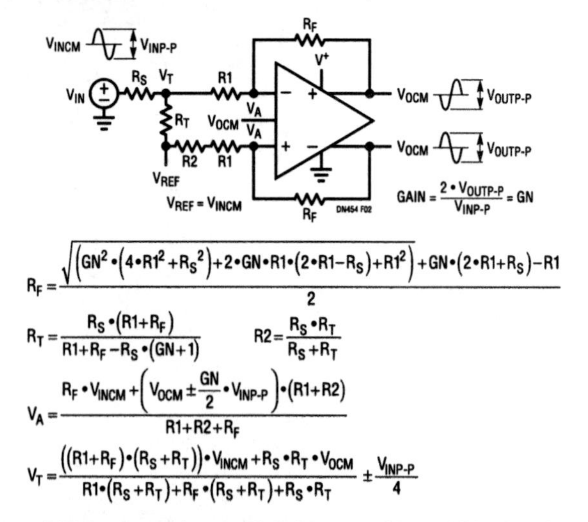

Figure 417.2 • Impedance Matching and Level Shifting for a Differential Amplifier with Gain Set by External Resistors

equal to the input common mode voltage (V_{INCM}) then the single-ended input signal is shifted to a differential signal centered on the output common mode voltage (V_{OCM}).

The design of a single-ended to differential amplifier with external resistors provides an additional design option: specifying the amplifier gain. Figure 417.2 shows the design equations when the R_F and R1 resistors are selectable, not fixed.

The design of this circuit begins with the value of R1. This resistor must be larger than the input source resistance but not so large as to increase circuit noise. Next, calculate the value of the feedback resistor R_F using the desired gain (GN). Then calculate the value of resistors R_T and R2.

Figure 417.3 shows an example of a single-ended to differential amplifier matching a 75Ω source and level shifting from a 2.5V input common mode to a 1.25V output common mode voltage (typical level shifting required from a 5V single-ended circuit to a 3V differential circuit to drive a high speed ADC). The single-ended to differential gain of the Figure 417.3 amplifier is 2 (the $1V_{P-P}$ input signal is amplified into a $2V_{P-P}$ differential output signal, a typical input voltage range of a high speed ADC).

For linear operation, the amplifier's input common mode limits must not be exceeded. Figure 417.2 shows the calculations for the bias voltage (V_T) of the input T-network (R_S, R_T and R1) and the common mode voltage at the differential amplifier's inputs. For example, in Figure 417.3, the 1.99V to 2.44V at the amplifier's inputs (as calculated by the V_A equation) is well within the rail-to-rail input common mode range of the LTC6406 (0V to V^+) (see Table 417.1).

Table 417.1 Sample of LTC High Speed Differential Amplifiers

AMPLIFIER	GBW GHz	SLEW RATE V/µS	VOLTAGE NOISE nV/√Hz	GAIN V/V
LTC6400-26	1.9	6670	1.5	20
LTC6400-20	1.8	4500	2.1	10
LTC6400-14	1.9	4800	2.5	5
LTC6400-8	2.2	3810	3.7	2.5
LTC6401-20	1.3	4500	2.1	10
LTC6401-14	2	3600	2.5	5
LTC6404-1	0.5	450	1.5	R SET
LTC6404-2	0.9	700	1.5	R SET
LTC6405	2.7	690	1.6	R SET
LTC6406	3	630	1.6	R SET

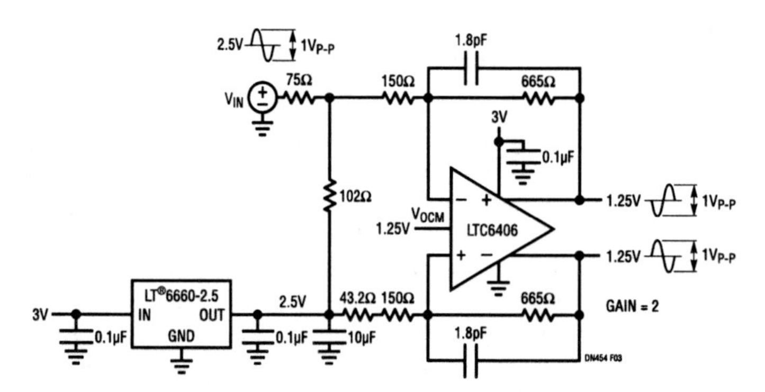

Figure 417.3 • Putting it All Together: A 133MHz Differential Amplifier with External Gain Setting Resistors, Impedance Matching to a 75Ω Source and Shifting from 2.5V to 1.25V

418

Current sense amp inputs work from –0.3V to 44V independent of supply

Glen Brisebois

Introduction

Monitoring current flow in electrical and electromechanical systems is commonly used to provide feedback to improve system operation, accelerate fault detection and diagnosis, and raise efficiency. A current monitoring circuit usually involves placing a sense resistor in series with the monitored conductor and determining the voltage across the sense resistor. To minimize power loss in the sense resistor it is kept as small as possible, resulting in a small differential voltage that must be monitored on top of what may be a fairly large varying common mode voltage. The LT6105 is an ideal current sense amplifier for this application. Just give it any reasonable supply voltage, say 3V, and its inputs can monitor small sense voltages at common modes of –0.3V to 44V and anything in between. The accuracy of the LT6105 over this range is displayed in Figure 418.1.

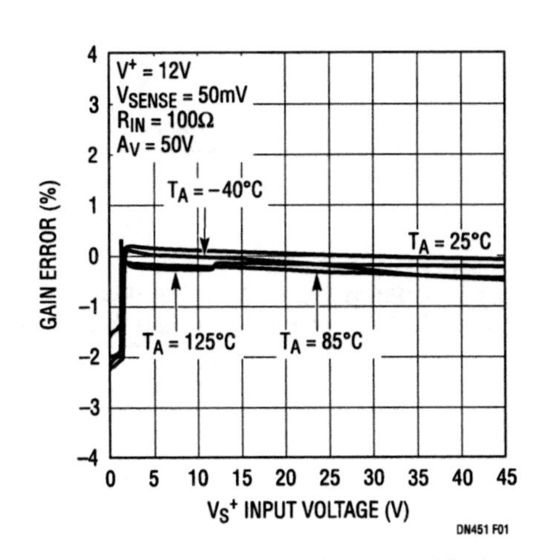

Figure 418.1 • Gain Error vs Input Common Mode

Solenoid monitoring

The large input common mode range of the LT6105 makes it suitable for monitoring currents in quarter, half and full bridge inductive load driving applications. Figure 418.2 shows an example of a quarter bridge. The MOSFET pulls down on the bottom of the solenoid to increase solenoid current. It lets go to decrease current, and the solenoid current freewheels through the Schottky diode. Current measurement waveforms are shown in Figure 418.3. The small glitches occur due to the action of the solenoid plunger, and this provides an opportunity for mechanical system monitoring without an independent sensor or limit switch.

Figure 418.4 shows another solenoid driver circuit, this time with one end of the solenoid grounded and a P-channel MOSFET pulling up on the other end. In this case, the

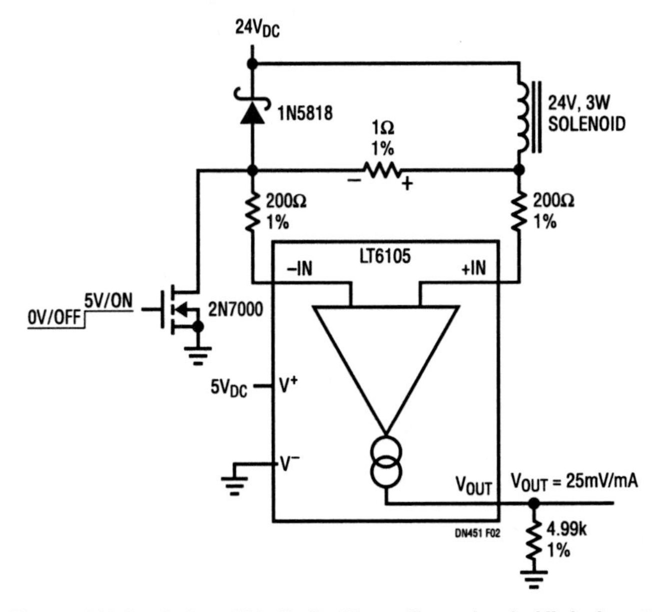

Figure 418.2 • Solenoid Is Pulled Low, Freewheels High. Input Travels from 0V to 24.3V

Analog Circuit Design: Design Note Collection. http://dx.doi.org/10.1016/B978-0-12-800001-4.00418-X

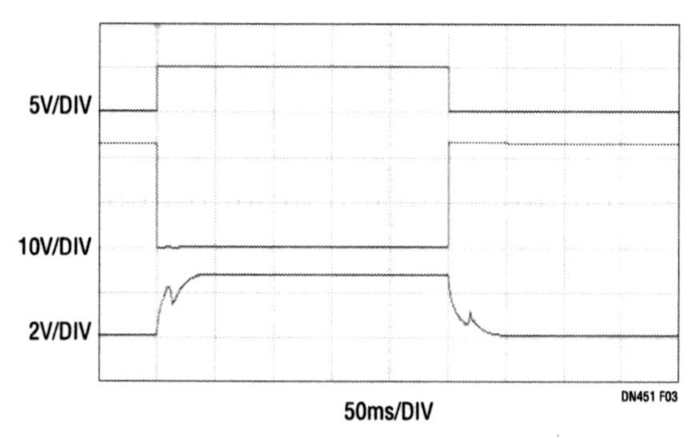

Figure 418.3 • Solenoid Waveforms: MOSFET Gate, Solenoid Bottom and Current Sense Amp Output. Bumps in the Current Result from Plunger Travel

inductor current freewheels around ground, imposing a negative input common mode voltage of one Schottky diode drop. This voltage may exceed the input range of the LT6105. This does not endanger the device, but it severely degrades its accuracy. In order to avoid violating the input range, pull-up resistors can be used as shown in Figure 418.4.

Supply monitoring

The input common mode range of the LT6105 allows it to monitor either positive or negative supplies. Figure 418.5 shows one LT6105 applied as a simple positive supply monitor, and another LT6105 as a simple negative supply monitor. Note that the schematics are practically identical and both have outputs conveniently referred to ground. The only requirement for negative supply monitoring, in addition to the usual constraints of the absolute maximum ratings, is that the negative supply to the LT6105 must be at least as negative as the supply it is monitoring.

Conclusion

Current measurement is popular because it offers improved real-time insight into matters of efficiency, operation and fault diagnosis. The wide input range of the LT6105 and its accuracy over that range make it easy to measure currents in a variety of applications.

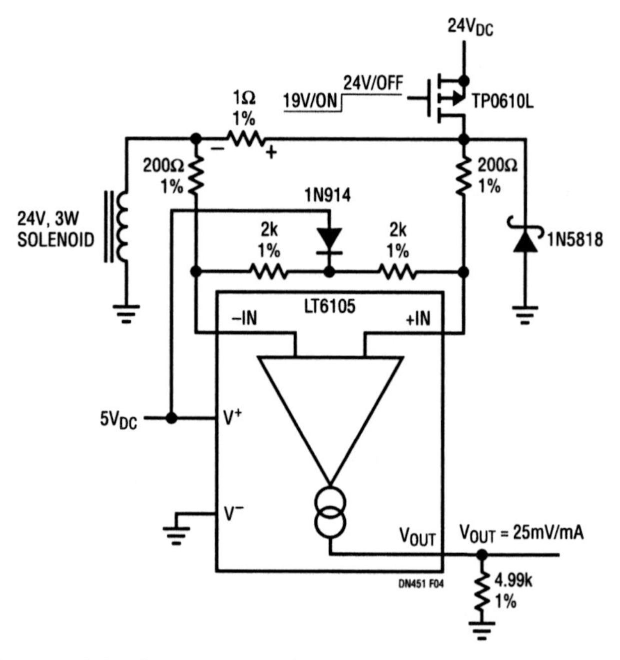

Figure 418.4 • Solenoid Is Pulled High to 24V, Freewheels Low to a Schottky Below Ground. LT6105 Inputs Are Kept within Range by 2k Pull Ups

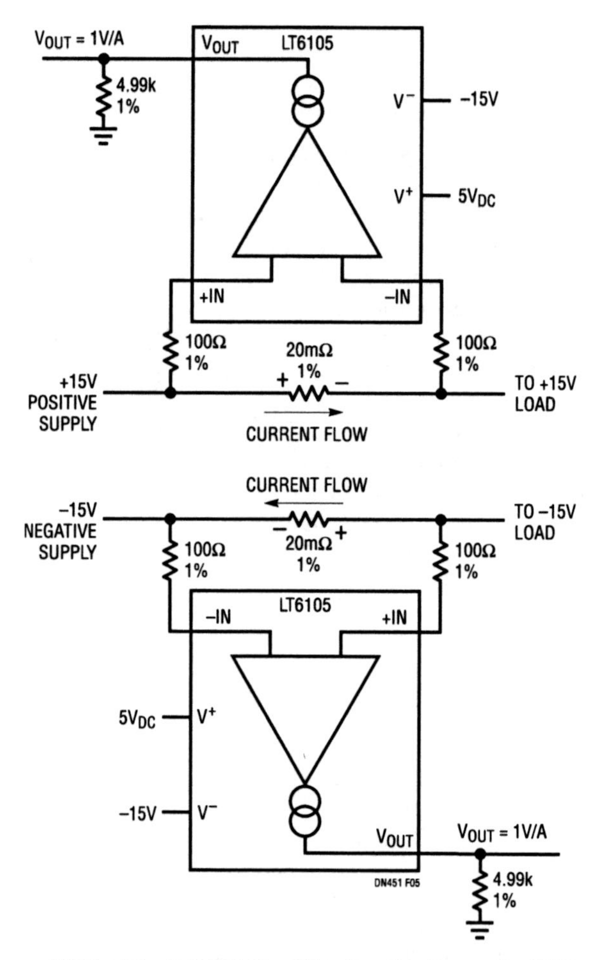

Figure 418.5 • The LT6105 Can Monitor the Current of Either Positive or Negative Supplies, without a Schematic Change. Just Ensure That the Current Flow Is in the Correct Direction

Tiny amplifiers drive heavy capacitive loads at speed

419

Keegan Leary   Brian Hamilton

Introduction

Parasitic capacitance lurks behind every corner of an electronic circuit. FET gates, cabling, ground and power planes all add to the farad bottom line. When the capacitive load gets heavy in high speed circuits, careful op amp selection is paramount for optimizing slew rate, current output capability, power dissipation, and feedback loop stability.

Demanding circuit requirements

For example, consider a 100MHz, $2V_{P-P}$ sine wave signal driving a 350pF capacitive load. The minimum required slew rate without distortion for this scenario is:

$$SR_{MIN} = 2\pi f V_{PK}$$
$$SR_{MIN} = 2\pi(100MHz)(1V)$$
$$\approx 630\frac{V}{\mu s}$$

The slew rate sets the maximum output current—the amplifiers are charging a capacitor, so the maximum output current occurs at maximum slew.

$$I = C\frac{dV}{dt}$$
$$I = (350pF)\left(630\frac{V}{\mu s}\right)$$
$$\approx 220mA$$

Maximum power dissipation is an important consideration. For an op amp operating from ±5V supplies, and assuming the capacitive load starts at 0V and is charged at maximum current, peak power is:

$$P = IV$$
$$P = (220mA)(5V)$$
$$\approx 1.1W$$

With a package that has a thermal resistance of 135°C/W, this much continuous power would result in a 148°C rise in die temperature. If the ambient temperature is 85°C, this brings the die to a package-melting 233°C!

To isolate C_{LOAD} from the amplifier, a design could use a series resistor, R_S. This technique ultimately limits bandwidth when the resistor or capacitive load gets very large. The bandwidth reduction associated with this RC time constant may limit performance. With a current feedback amplifier, increasing the feedback resistor, R_F, is an alternative compensation method to reduce peaking.

Tiny current feedback amplifiers

For the high speed, large capacitive load example above, the 400MHz LT1395/LT1396/LT1397 family of current feedback amplifiers certainly satisfies the slew rate requirement. The LT1395/LT1396/LT1397 can process large signals with speed and 80mA minimum guaranteed output current. However, for the example above, this amplifier family falls short of the 220mA requirement. In this case one may not be enough, but four certainly are. Paralleling these amplifiers satisfies current requirements while maintaining safe power dissipation and stability.

The LT1397 quad was designed to push big loads of current while maintaining good thermal properties. The copper underbelly of the tiny 4mm×3mm DFN package brings the thermal resistance down to 43°C/W, and a die temperature rise above ambient of only 47°C for the given example.

Analog Circuit Design: Design Note Collection. http://dx.doi.org/10.1016/B978-0-12-800001-4.00419-1

Component selection and testing

Without assembling the entire parallel configuration, a single-amplifier test circuit can be constructed to check results into the load capacitance divided by the number of amplifiers to be used, $C_{LOAD}/4$.

The remaining task is to select appropriate values of the feedback resistor (R_F) and series resistor (R_S) to maximize the –3dB bandwidth and sufficiently minimize the amount of peaking in the frequency response. For both R_F and R_S, smaller values result in both additional bandwidth and increased peaking. R_F has a practical lower limit of about 255Ω. As load capacitance increases, R_F and/or R_S values must increase to maintain stability.

Figure 419.2 shows measurement results using the 4-amplifier circuit of Figure 419.1 with various R_F/R_S combinations and 350pF of total load capacitance. Measurements were performed at a gain of 1, so R_G was not used.

The effectiveness of the 4-amplifier circuit topology over a single amplifier can be seen in Figure 419.3. For a more representative effect the load capacitance was tripled to 1000pF. The paralleled 4-amplifier circuit is capable of slewing 4V into 1000pF in under 10ns. This corresponds to a slewing output current of 400mA. The single amplifier current limits at about 140mA, reducing the slew rate into this large capacitive load. The same 4V swing for the single requires 28ns, almost three times longer than the 4-amplifier configuration.

Conclusion

Always consider using all of the amplifiers available in a tiny power-enhanced package to provide the muscle needed to rapidly slew heavy capacitive loads. Also consider current feedback amplifiers such as the LT1397 to make it easy to control a very wide bandwidth circuit.

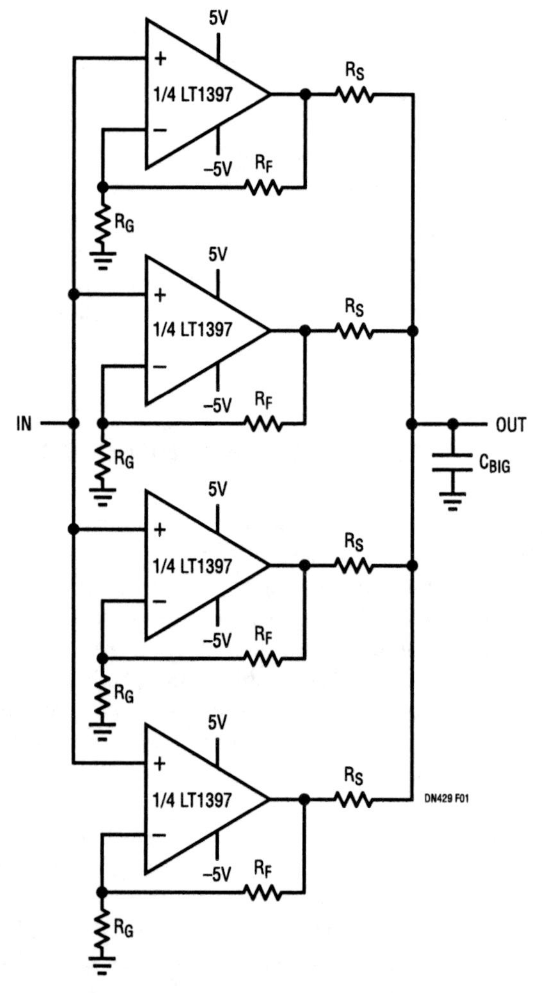

Figure 419.1 • Using All Four Amplifiers of the LT1397 to Drive Large Capacitive Loads

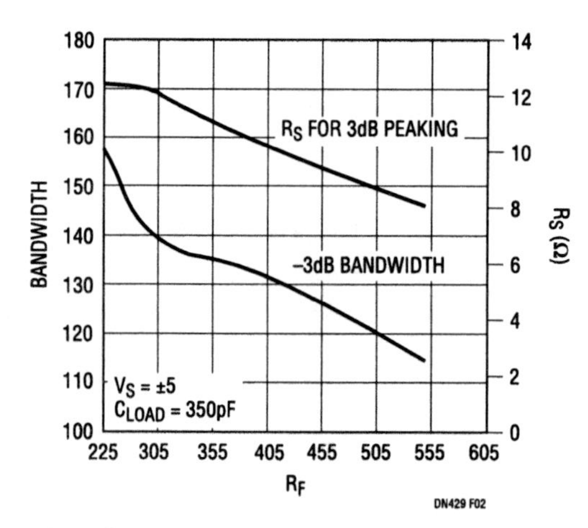

Figure 419.2 • Selecting R_F and R_S to Drive 350pF When Paralleling the Four Amplifiers of the LT1397

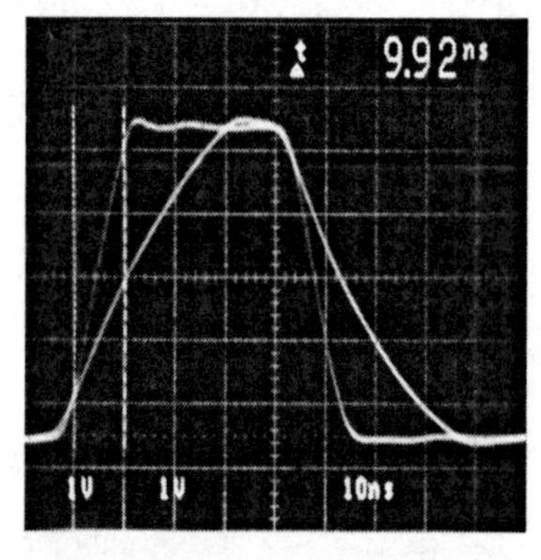

Figure 419.3 • Four Amplifiers Out-Race One Amplifier When Driving a 1000pF Capacitive Load. The Response Time of the Single Amplifier Lags the Quad by a Factor of Three

Micropower op amps work down to 1.8V total supply, guaranteed over temperature

420

Glen Brisebois

Introduction

Micropower op amps extend the run time of battery-powered systems and reduce energy consumption in other energy limited systems. Nevertheless, battery voltages change as they are depleted. To maximize a system's run time, op amps should operate over a wide enough supply range to make use of the complete range of battery voltages, from fully charged to fully depleted. The new LT6000 family of 1μA and 13μA op amps operates on supplies as high as 16V all the way down to 1.8V, guaranteed over temperature.

NiMH and alkaline

A NiMH battery has a nominal cell voltage of 1.2V, but it depletes to 0.9V, below which the voltage rapidly falls off. The LT6000 family of op amps works directly from two series NiMH cells taking full advantage of their entire charge discharge cycle. Likewise, an alkaline battery has a nominal cell voltage of 1.5V, but can deliver energy down to depletion levels of a few hundred millivolts. So, the LT6000 can happily operate from two series alkaline cells, and just as well operate directly from a 9V alkaline battery (6 series cells) from full charge all the way down to very extreme depletions (300mV average cell voltage for 1.8V total). Sure, other low voltage op amps can operate at the depleted end of this battery range, but few of those can also tolerate a 9V supply.

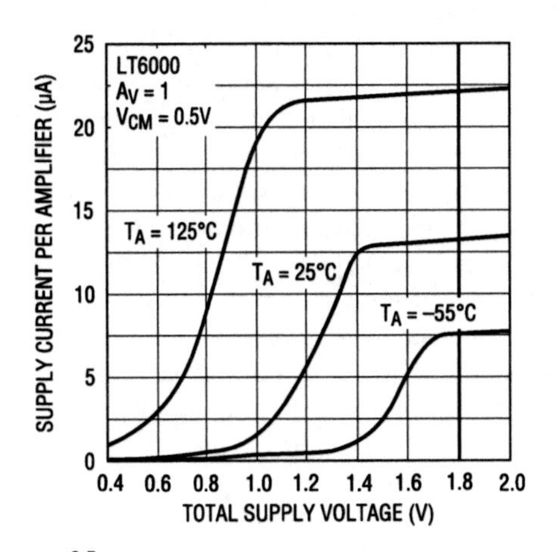

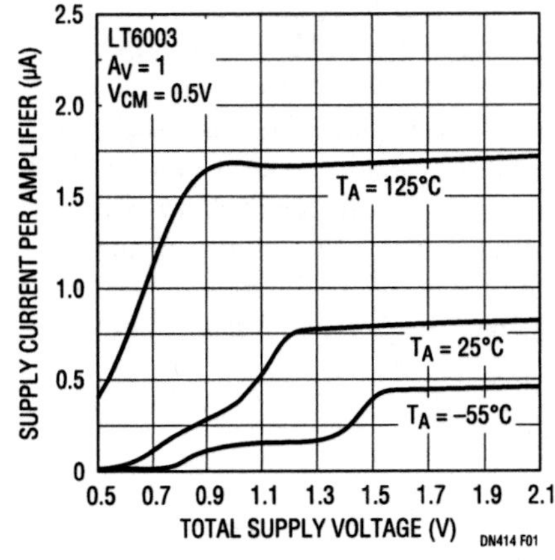

Figure 420.1 • Clean Start-Up Characteristics without Current Spikes

Supply friendliness

Some micropower op amps have annoying properties such as drawing excessive current at start-up (commonly called carrots) or when the output hits a supply rail. These current spikes defeat the purpose of the micropower operation by hastening battery discharge. Worse yet, they may altogether prevent the supply from coming up in the case of a current limited supply, effectively crowbarring the system. Figure 420.1 shows the LT6000 and LT6003 supply current vs applied supply voltage at various temperatures. The LT6000 family eliminates carrots or at least chews them down to stumps.

Analog Circuit Design: Design Note Collection. http://dx.doi.org/10.1016/B978-0-12-800001-4.00420-8

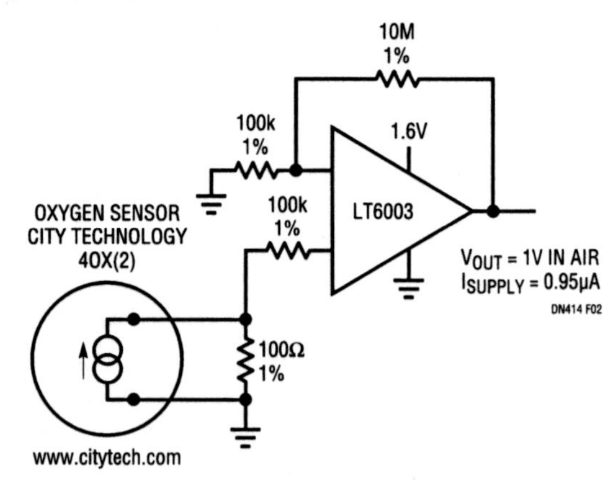

Figure 420.2 • Micropower Oxygen Sensor

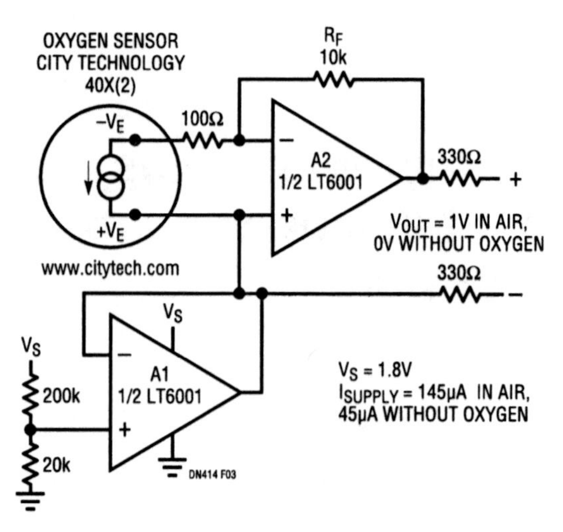

Figure 420.3 • High Accuracy Oxygen Sensor

Portable gas sensor

Figure 420.2 shows the LT6003 applied as an oxygen sensor amplifier. The oxygen sensor acts much like an air-powered battery, and generates 100µA in one atmosphere of fresh air (20.9% oxygen). It is designed to operate into a 100Ω resistor, for a 10mV full-scale reading. The op amp amplifies this voltage with a gain of 100 as shown (101 actually), for a 1V full-scale output. In terms of monitoring environments for adequate human-livable oxygen levels, 18% oxygen content translates to an output voltage of 0.86V. Oxygen contents below this are considered hazardous. Oxygen deprivation in the lungs causes immediate loss of consciousness and bears no resemblance to holding your breath. Total supply current for the circuit is 950nA. The 500µV worst-case input offset voltage at room temperature contributes a 50mV uncertainty in the output reading.

Better low value accuracy can be obtained by implementing a transimpedance approach as shown in Figure 420.3. Op amp A1 provides a buffered reference voltage so the circuit is accurate all the way down to a zero-oxygen environment without clipping at ground. Op amp A2 provides the current-to-voltage function through feedback resistor R_F. The sensor still sees the 100Ω termination, as the manufacturer specifies. The output voltage is still 1V in normal atmosphere, but note that the noise gain is not much higher than unity so the output error due to offset is now 500µV worst case instead of the 50mV of the previous circuit. This considerable improvement in accuracy exacts some price in supply current, because the oxygen sensor current is now provided back through R_F by the op amp output, which necessarily takes it from the supply. The supply current is therefore oxygen-presence dependant. Nevertheless, this solution is still ultralow power when monitoring environments that are oxygen-free by design, such as environments for food storage and those designed to inhibit combustion. It would also be ideal for portable sensors where the detected substance is not oxygen but is rather a hostile substance, which is not normally present and is therefore usually low current.

Conclusion

The LT6000 and LT6003 family of op amps offer 13µA and 1µA micropower operation over a wide supply range from 18V all the way down to 1.8V, guaranteed over temperature. Careful attention was paid during the design phase to minimizing gotchas such as supply current carrots. They are ideal for maximizing battery life in portable applications, operating over a wide range of battery charge levels and environments.

Low noise amplifiers for small and large area photodiodes

421

Glen Brisebois

Introduction

Photodiodes can be broken into two categories: large area photodiodes with their attendant high capacitance (30pF to 3000pF) and smaller area photodiodes with relatively low capacitance (10pF or less). For optimal signal-to-noise performance, a transimpedance amplifier consisting of an inverting op amp and a feedback resistor is most commonly used to convert the photodiode current into voltage. In low noise amplifier design, large area photodiode amplifiers require more attention to reducing op amp input voltage noise, while small area photodiode amplifiers require more attention to reducing op amp input current noise and parasitic capacitances.

Small area photodiode amplifiers

Small area photodiodes have very low capacitance, typically under 10pF and some even below 1pF. Their low capacitance makes them more approximate current sources to higher frequencies than large area photodiodes. One of the challenges of small area photodiode amplifier design is to maintain low input capacitance so that voltage noise does not become an issue and current noise dominates.

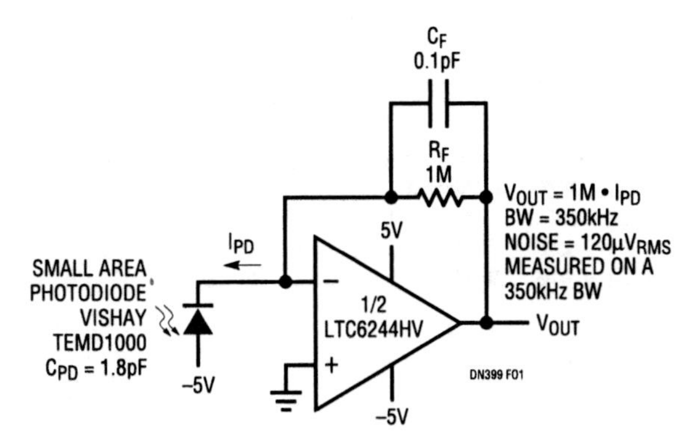

Figure 421.1 • Transimpedance Amplifier for Small Area Photodiode

Figure 421.1 shows a simple small area photodiode amplifier using the LTC6244. The input capacitance of the amplifier consists of C_{DM} (the amplifier's differential mode capacitance) and one C_{CM} (the common mode capacitance at the amplifier's—input only), or about 6pF total. The small photodiode has 1.8pF, so the input capacitance of the amplifier is dominating the capacitance. The small feedback capacitor is an actual component (AVX Accu-F series), but it is also in parallel with the op amp lead, resistor and parasitic capacitances, so the total real feedback capacitance is probably about 0.4pF. This is important because feedback capacitance sets the compensation of the circuit and, with op amp gain bandwidth, the circuit bandwidth. This particular design has a bandwidth of 350kHz, with an output noise of 120μV_{RMS} measured over that bandwidth.

Large area photodiode amplifiers

Figure 421.2a shows a simple large area photodiode amplifier. The capacitance of the photodiode is 3650pF (nominally 3000pF), and this has a significant effect on the noise performance of the circuit. For example, the photodiode capacitance at 10kHz equates to an impedance of 4.36kΩ, so the op amp circuit with 1MΩ feedback has a noise gain of $NG = 1 + 1M/4.36k = 230$ at that frequency. Therefore, the LTC6244 input voltage noise gets to the output as $NG \cdot 7.8nV/\sqrt{Hz} = 1800nV/\sqrt{Hz}$, and this can clearly be seen in the circuit's output noise spectrum in Figure 421.2b. Note that we have not yet accounted for the op amp current noise, or for the $130nV/\sqrt{Hz}$ of the gain resistor, but these are obviously trivial compared to the op amp voltage noise and the noise gain. For reference, the DC output offset of this circuit is about 100μV, bandwidth is 52kHz, and the total noise was measured at 1.7mV_{RMS} on a 100kHz measurement bandwidth.

Analog Circuit Design: Design Note Collection. http://dx.doi.org/10.1016/B978-0-12-800001-4.00421-X

An improvement to this circuit is shown in Figure 421.3a, where the large diode capacitance is bootstrapped by a $1nV/\sqrt{Hz}$ JFET. This depletion JFET has a VGS of about −0.5V, so that RBIAS forces it to operate at just over 1mA of drain current. Connected as shown, the photodiode has a reverse bias of one VGS, so its capacitance will be slightly lower than in the previous case (measured 2640pF), but the most drastic effects are due to the bootstrapping. Figure 421.3b shows the output noise of the new circuit. Noise at 10kHz is now $220nV/\sqrt{Hz}$, and the $130nV/\sqrt{Hz}$ noise thermal noise floor of the 1M feedback resistor is discernible at low frequencies. What has happened is that the $7.8nV/\sqrt{Hz}$ of the op amp has been effectively replaced by the $1nV/\sqrt{Hz}$ of the JFET. This is because the 1M feedback resistor is no longer "looking back" into the large photodiode capacitance. It is instead looking back into a JFET gate capacitance, an op amp input capacitance, and some parasitics, approximately 10pF total. The large photodiode capacitance is across the gate-source voltage of the low noise JFET. Doing a sample calculation at 10kHz as before, the photodiode capacitance looks like 6kΩ, so the $1nV/\sqrt{Hz}$ of the JFET creates a current noise of $1nV/6k = 167fA/\sqrt{Hz}$. This current noise necessarily flows through the 1M feedback resistor, and so appears as $167nV/\sqrt{Hz}$ at the output. Adding the $130nV/\sqrt{Hz}$ of the resistor (RMS wise) gives a total calculated noise density of $210nV/\sqrt{Hz}$, agreeing well with the measured noise of Figure 421.3b. Another drastic improvement is in bandwidth, now over 350kHz, as the bootstrap enabled a reduction of the compensating feedback capacitance. Note that the bootstrap does not affect the DC accuracy of the amplifier, except by adding a few picoamps of gate current.

For more details on photodiode circuits, download the LTC6244 data sheet.

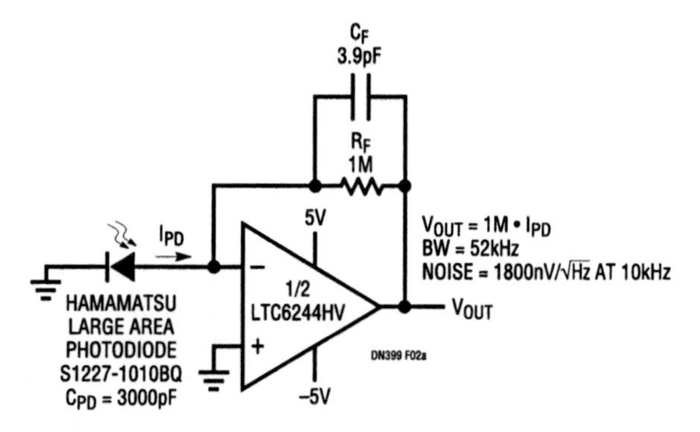

Figure 421.2a • Large Area Photodiode Transimpedance Amp

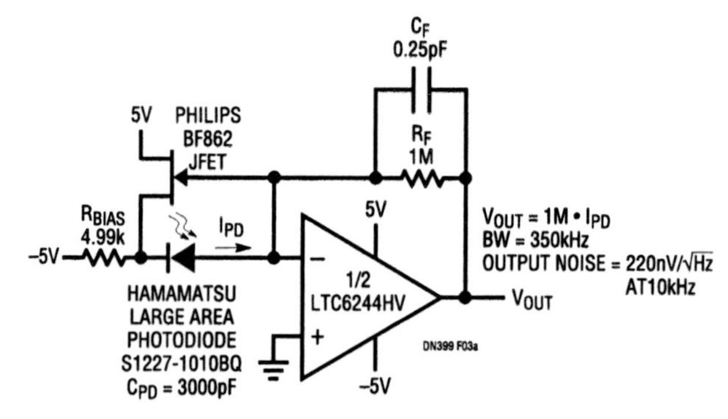

Figure 421.3a • Large Area Diode Bootstrapping

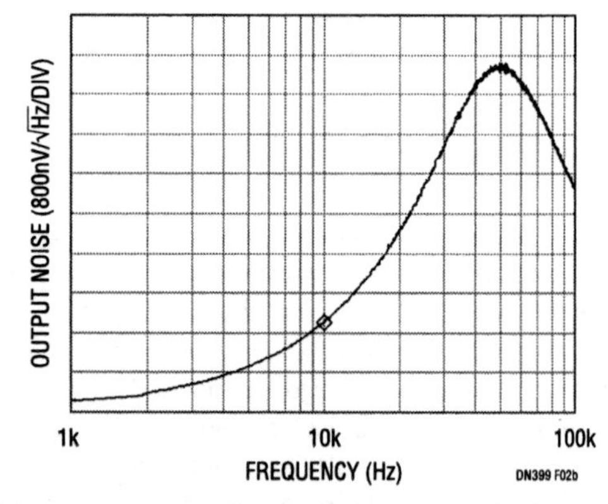

Figure 421.2b • Output Noise Spectral Density of the Circuit of Figure 421.2a. At 10kHz, the $1800nV/\sqrt{Hz}$ Output Noise Is Due Almost Entirely to the 7.8nV Voltage Noise of the LTC6244 and the High Noise Gain of the 1M Feedback Resistor Looking into the High Photodiode Capacitance

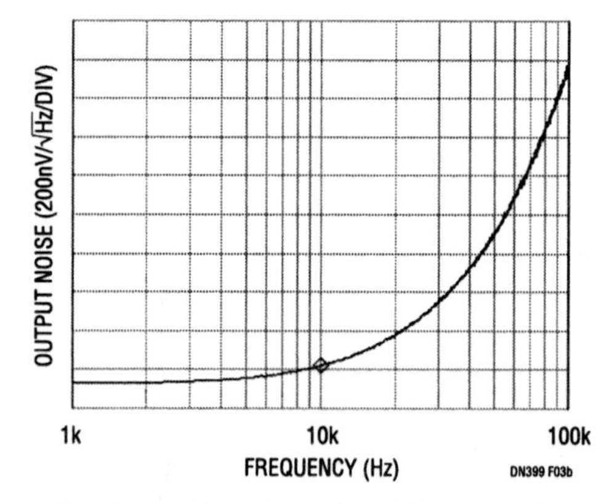

Figure 421.3b • Output Noise Spectral Density of Figure 421.3a. The Simple JFET Bootstrap Improves Noise (and Bandwidth) Drastically. Noise Density at 10kHz Is Now $220nV/\sqrt{Hz}$, About an 8.2x Reduction. This Is Mostly due to the Bootstrap Effect of Swapping the $1nV/\sqrt{Hz}$ of the JFET for the $7.8nV/\sqrt{Hz}$ of the Op Amp

Op amp selection guide for optimum noise performance

422

Glen Brisebois

Introduction

Linear Technology continues to add to its portfolio of low noise op amps. This is not because the physics of noise has changed, but because low noise specifications are being combined with new features such as rail-to-rail operation, shutdown, low voltage and low power operation. Op amp noise is dependent on input stage operating current, device type (bipolar or FET) and input circuitry. This selection guide is intended to help you identify basic noise trade-offs and select the best op amps, new or old, for your application.

Quantifying resistor thermal noise and op amp noise

The key to understanding noise trade-offs is the fact that resistors have noise. At room temperature, a resistor R has an RMS voltage noise density (or "spot noise") of $V_R = 0.13\sqrt{R}$ noise in $nV/\sqrt{Hz}$. So a 10k resistor has $13nV/\sqrt{Hz}$ and a 1M resistor has 130nV/Hz. Rigorously speaking, the noise density is given by the equation $V_R = \sqrt{4\ kTR}$, where k is Boltzman's constant and T is the temperature in kelvin. This dependency on temperature explains why some low noise circuits resort to supercooling the resistors. Note that the same resistor can also be considered to have a noise current of $I_R = \sqrt{4\ kT/R}$ or a noise power density $P_R = 4kT = 16.6 \cdot 10^{-21} W/Hz = 16.6$ zeptowatts/Hz independent of R. Selecting the right amplifier is simply finding which one will add the least amount of noise above the resistor noise.

Don't be alarmed by the strange unit "$/\sqrt{Hz}$." It arises simply because noise power adds with bandwidth (per hertz), so noise voltage adds with the square root of the bandwidth (per root hertz). To make use of the specification, simply multiply it by the square root of the application bandwidth to calculate the resultant RMS noise within that bandwidth.

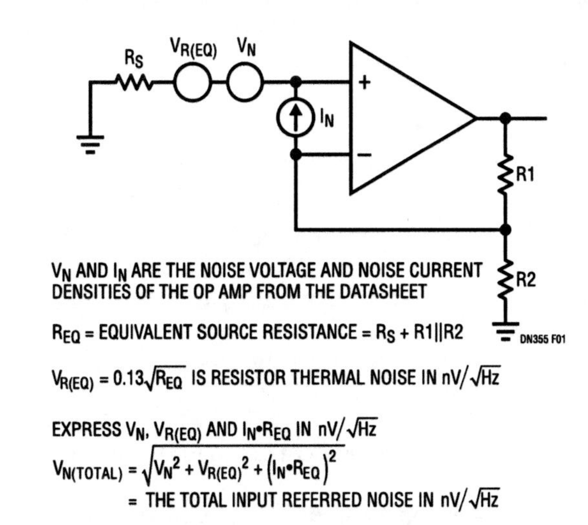

Figure 422.1 • The Op Amp Noise Model. V_N and I_N Are Op Amp Noise Sources (Correlated Current Noise Is Not Shown). $V_{R(EQ)}$ Is the Voltage Noise due to the Resistors

Peak-to-peak noise, as encountered on an oscilloscope for example, will be about 6 times the total RMS noise 99% of the time (assuming Gaussian "bell curve" noise). Do not rely on the op amp to limit the bandwidth. For best noise performance, limit the bandwidth with passive or low noise active filters.

Op amp input noise specifications are usually given in terms of $nV/\sqrt{Hz}$ for noise voltage and $pA/\sqrt{Hz}$ or $fA/\sqrt{Hz}$ for noise current and are therefore directly comparable with resistor thermal noise. Due to the fact that noise density varies at low frequencies, most op amps also specify a typical peak-to-peak noise within a "0.1Hz to 10Hz" or "0.01Hz to 1Hz" bandwidth. For the best ultralow frequency performance, you may want to consider a zero drift amplifier like the LTC2050 or LTC2054.

Analog Circuit Design: Design Note Collection. http://dx.doi.org/10.1016/B978-0-12-800001-4.00422-1

Summing the noise sources

Figure 422.1 shows an idealized op amp and resistors with the noise sources presented externally. The equation for the input referred RMS sum of all the noise sources, $V_{N(TOTAL)}$, is also shown. It is this voltage noise density, multiplied by the noise gain of the circuit ($NG = 1 + R1/R2$) that appears at the output.

From the equation for $V_{N(TOTAL)}$ we can draw several conclusions. For the lowest noise, the values of the resistors should be as small as possible, but since R1 is a load on the op amp output, it must not be too small. In some applications, such as transimpedance amplifiers, R1 is the only resistor in the circuit and is usually large. For low R_{EQ}, the op amp voltage noise dominates (as V_N is the remaining term); for very high R_{EQ}, the op amp current noise dominates (as I_N is the coefficient of the highest order R_{EQ} term). At middle values of R_{EQ}, the resistor noise dominates and the op amp contributes little significant noise. This is the $R_{OPTIMUM}$ of the amplifier and can be found by taking the quotient of the op amp's noise specs: $V_N/I_N = R_{OPT}$.

Selecting the best op amps

Figure 422.2 shows plots of voltage noise density of the source resistance and of various op amps at three different frequencies. Each point labeled by an op amp part number is that part's voltage noise density plotted at its R_{OPT}.

Use the graph with the most applicable frequency of interest. Find your source resistance on the horizontal axis and mark that resistance at the point where it crosses the resistor noise line. This is the "source resistance point." The best noise performance op amps are under that point, the further down the better.

For all candidate op amps, draw a horizontal line from your source resistance point all the way to the right hand side of the plot. Op amps beneath that line will give good noise performance, again the lower the better. Draw another line from the source resistance point down and to the left at one decade per decade. Op amps below that line are also good candidates.

If you still can't find any candidates, then you have a very low source impedance and should use op amps that are closest to the bottom. In such cases, paralleling of low noise op amps is also an option.

Conclusion

Noise analysis can be a daunting task at first and is an unfamiliar territory for many design engineers. The greatest influence on overall noise performance is the source impedance associated with the signal. This selection guide helps the designer, whether novice or veteran, to choose the best op amps for any given source impedance.

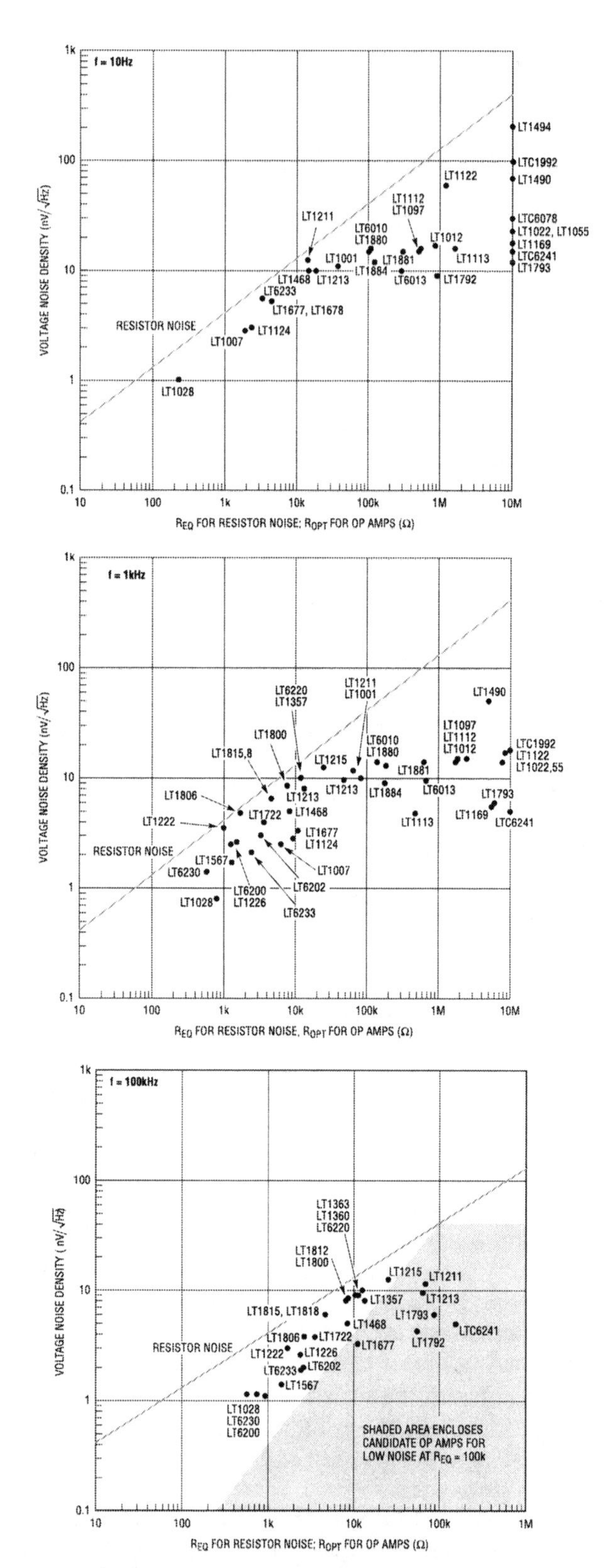

Figure 422.2 • Use These Three Plots to Find the Best Low Noise Op Amps for Your Application

Easy-to-use differential amplifiers simplify balanced signal designs

423

Jon Munson

Introduction

The LTC1992 product family provides simple amplification or level translation solutions for amplifying signals that are intrinsically differential or need to be made differential. In addition to the uncommitted configuration of the base LTC1992, fixed gain versions with space-saving on-chip factory-trimmed resistors are available as the LTC1992-1, LTC1992-2, LTC1992-5 and LTC1992-10, where the nominal gain is indicated by the suffix dash number. Figure 423.1 shows a typical gain-of-ten application where all gain setting components are included in the tiny 8-lead MSOP package.

Easy-to-use circuit topology

The block diagram in Figure 423.2 shows the general configuration of the differential-in/differential-out CMOS amplifier core, along with an output common mode servo. The values of the on-chip gain resistors depend on the dash suffix of the device as indicated. A convenient on-chip voltage-divider resistor network is also provided to support applications where a source of mid-supply potential (V_{MID}) is needed.

The LTC1992 is easy to use. Any signal difference at the inputs (within the input common mode range) are amplified and presented as a voltage difference at the output pins with a gain bandwidth product of about 4MHz. The differential gain, A, is set by resistor values:

$$A = R_F/R_G$$

Any input common mode-induced errors, primarily caused by small mismatches of resistor values, appear at the output as differential error. The common mode (shared offset) of the output pair is $(V_{OUT}^+ + V_{OUT}^-)/2$ and independently governed to track the user-supplied V_{OCM} output common mode control voltage (V_{OCM} may be simply strapped to V_{MID}

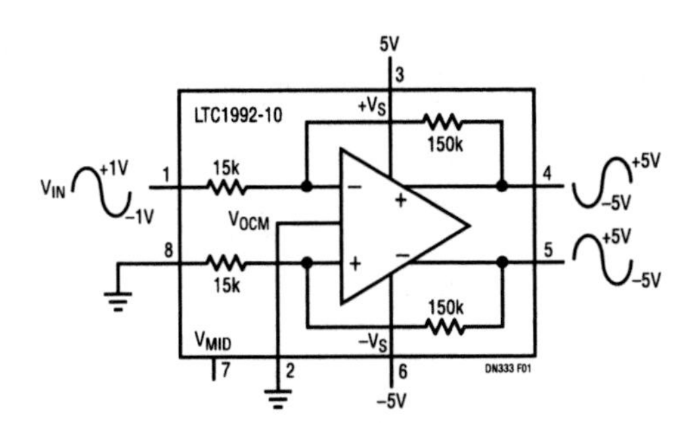

Figure 423.1 • Typical Single-Ended to Differential Conversion

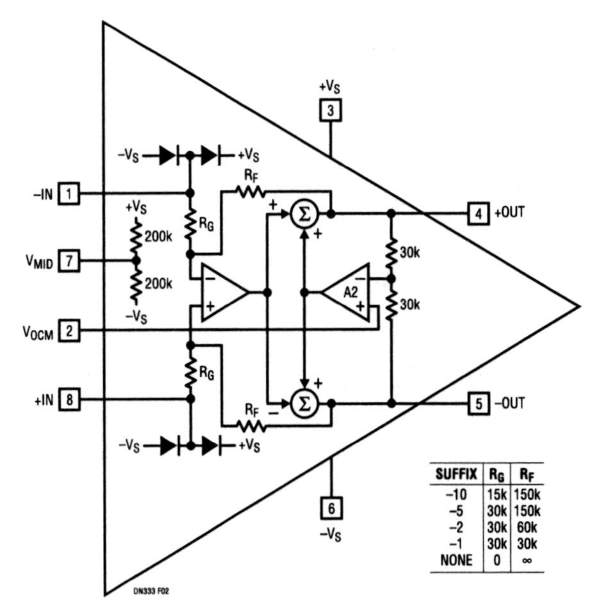

SUFFIX	R_G	R_F
–10	15k	150k
–5	30k	150k
–2	30k	60k
–1	30k	30k
NONE	0	∞

Figure 423.2 • LTC1992 Functional Block Diagram

Analog Circuit Design: Design Note Collection. http://dx.doi.org/10.1016/B978-0-12-800001-4.00423-3

if desired). The uncommitted LTC1992 (no dash suffix) may be user configured for any desired differential gain by selection of external resistors, or configured specifically for other specialized uses.

Common mode range considerations

For a given input common mode voltage (V_{INCM}) and output common mode voltage (V_{OCM}), the designer must verify that voltage appearing at the internal amplifier inputs (V_{ICM}) is within the specified operating range of $-VS\ -0.1V$ to $+V_S\ -1.3V$. With a standard differential amplifier topology having a closed loop gain of A, the following relationship holds:

$$V_{ICM} = (A/(A+1)) \cdot V_{INCM} + (1/(A+1)) \cdot V_{OCM}$$

For example, assume an LTC1992 (no dash) is powered from 5V, configured for a gain of 2.5 with V_{OCM} tied to V_{MID} (i.e. 2.5V), and driven from a source with common mode of 0V. From the relation above, V_{ICM} is $(2.5/3.5)\cdot 0+(1/3.5)\cdot 2.5=0.71V$, which is well within the performance range of the device. In this example, the outputs can swing ±2.5V around the 2.5V V_{OCM} level. Therefore, the differential inputs can swing 1V below ground without clipping effects or the need for a minus rail. The dash suffix versions have an additional input limitation due to the possibility of forward biasing the ESD input protection diodes (shown in Figure 423.2), which limit the maximum allowable signal swings to 0.3V beyond the supply voltages (while the base LTC1992 also includes the ESD diodes, conduction can only occur outside the usable V_{ICM} range).

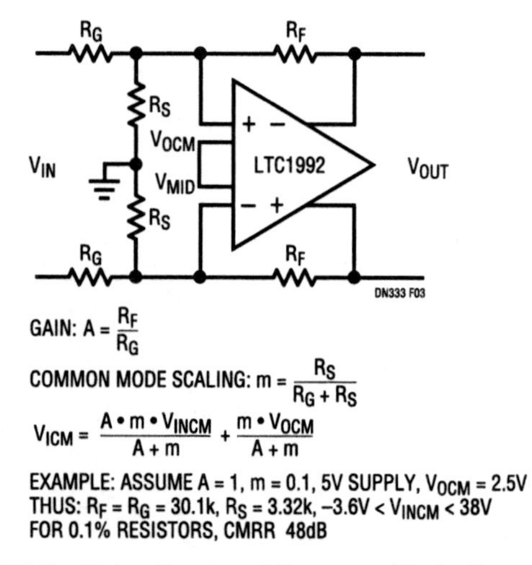

Figure 423.3 • Extending Input Common Mode Range

Common mode input range extension

Use of the non-committed LTC1992 provides the possibility of extending input common mode capability well outside the supply range by operating with a gain below unity and/or introducing common mode shunt resistors (see RS in Figure 423.3). The drawback to the shunt resistor method is that component tolerances of RG and RS become magnified by the common mode improvement of the circuit (approximately), leading to reduced CMRR performance for a given resistor tolerance. For low gain operation, common mode extension of 10× is realizable with the use of high accuracy resistor networks.

Versatile functional block

The LTC1992 family is especially useful for making conversions to or from differential signaling. Analog-to-digital converters (ADCs) are often optimized for differential inputs with a specific common mode input voltage. Use of an LTC1992 amplifier simplifies the ADC interface by using the VOCM control feature to establish the requisite offset. In many cases, the mid-scale potential is provided by the ADC and can be tied directly to the VOCM input. In addition, the source signal input may then be differential or single-ended (by grounding the unused input) or have inverted polarity.

Since it is not necessary to connect to both outputs, one can treat the part as single-ended which provides the useful feature that the V_{OCM} input represents a third algebraic input term (see Figure 423.4). This capability is useful in performing analog addition or simple translation functions.

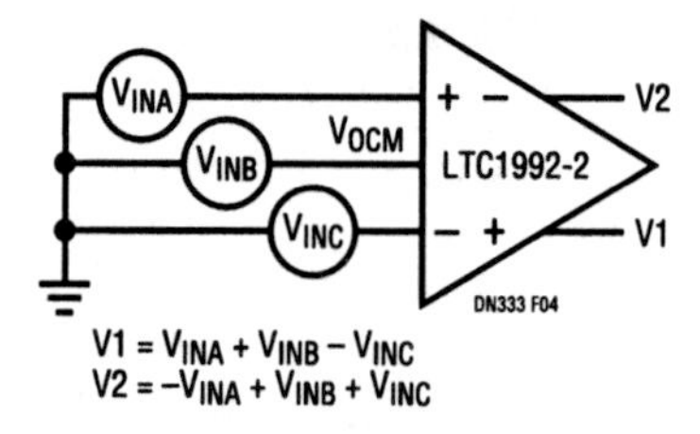

Figure 423.4 • Single-Ended Adder/Subtractor

Conclusion

The LTC1992 family of differential amplifiers offers easy-to-use building blocks that provide simple, minimum component-count solutions to balanced-signal designs. These parts are useful in a wide range of applications, including simple methods of transforming signals to/from differential form to providing component-free gain or DC offset functions.

Dual 25μV micropower op amp fits in 3mm × 3mm package

424

Glen Brisebois

Introduction

Conventional monolithic micropower op amps with a wide supply voltage range require a large die area and, therefore, a large package and footprint. The unconventional LT6011 dual op amp fits 25μV input precision micropower operation and wide 2.7V to 36V supply range in a tiny new package—its 3mm × 3mm DFN package is so small it does not even have leads. The LT6011 also provides rail-to-rail output swing and utilizes superbeta input transistors to achieve picoampere input currents.

Hall sensor amplifier

Figure 424.1 shows the LT6011 applied as a low power Hall sensor amplifier. The magnetic sensitivity of a Hall sensor is proportional to the DC excitation voltage applied across it. With a 1V bias voltage, the sensitivity of this Hall sensor is specified as 4mV/mT of magnetic field. At that level of DC bias, however, the 400Ω bridge consumes 2.5mA. Reducing the excitation voltage would reduce the power consumption, but it would also reduce the sensitivity. This is where the beauty of precision micropower amplification becomes especially apparent.

The LT1790-1.25 micropower reference provides a stable 1.25V reference voltage. The 7.87k:100k resistive ladder attenuates this to about 90mV across the 7.87k and the LT1782 acts as a buffer. When this 90mV is applied as excitation across the Hall bridge, the current is only 230μA. This is less than 1/10 of the original value. (Just imagine if all your batteries could last 10 times longer than they do.) But as mentioned earlier, the sensitivity is now likewise reduced by the same factor, down to 0.4mV/mT.

The way back to high sensitivity is to take gain with a precision micropower amplifier. The LT6011 is therefore configured as an instrumentation amplifier in a gain of 101. Such high gains are permissible and advantageous using an LT6011 because of its exceptional input precision and low drift. The output sensitivity of the circuit is raised to a whopping

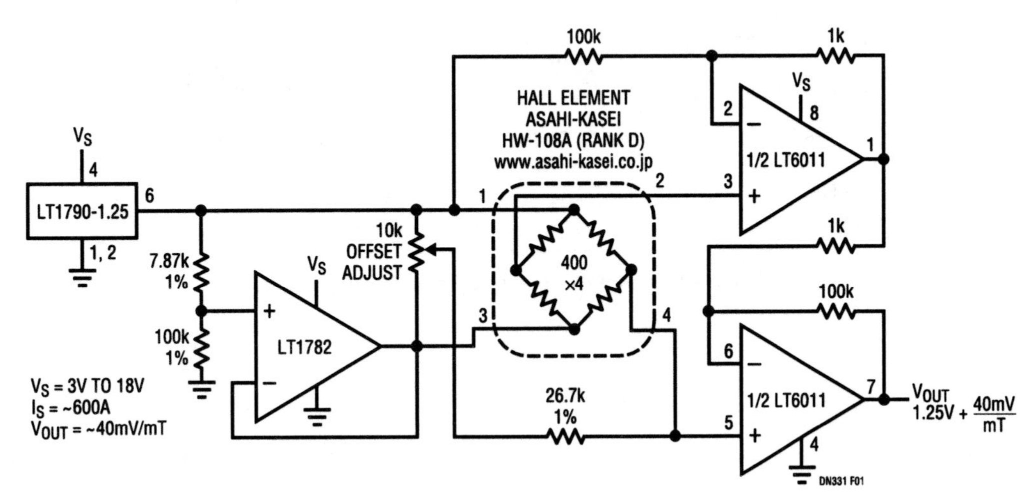

Figure 424.1 • Hall Sensor Amplifier Optimizes Sensitivity vs Supply Current

Analog Circuit Design: Design Note Collection. http://dx.doi.org/10.1016/B978-0-12-800001-4.00424-5

40mV/mT, while consuming a total supply current of only 600μA. To have achieved this sensitivity by increasing the bridge excitation would have required a prohibitive 25mA from the supply! (As an interesting note, here in Milpitas, California, the Earth's 50μT field is about 60° from horizontal and causes a 2mV shift in the circuit's output.)

DAC amplifier

Figure 424.2 shows the LT6011 applied as both a reference amplifier and I-to-V converter with the LTC1592 16-bit DAC. Whereas faster amplifiers such as the LT1881 and LT1469 are also suitable for use with this DAC, the LT6011 is desirable when power consumption is more important than speed. The total supply current of this application varies from 1.6mA to 4mA, depending on code, and is almost entirely dominated by the DAC resistors and the reference.

The DAC itself is powered only from a single 5V supply. Op amp B of the LT6011 inverts the 5V reference using the DAC's internal precision resistors R1 and R2, thus providing the DAC with a negative reference allowing bipolar output polarities. Op amp A provides the I-to-V conversion and buffers the final output voltage. The precision required of the I-to-V converter function is critical because the DAC output resistor network is obviously very code dependent, so the noise gain which the op amp sees is also code dependent. An imprecise op amp in this function would have its input errors amplified somewhat chaotically versus code.

The speed of the circuit is shown in Figure 424.3. Settling is achieved within 250μs. Because the outputs of the LT6011 swing to within 40mV of either supply rail, the supply voltages to the amplifier need to be only barely wider than the desired ±10V output.

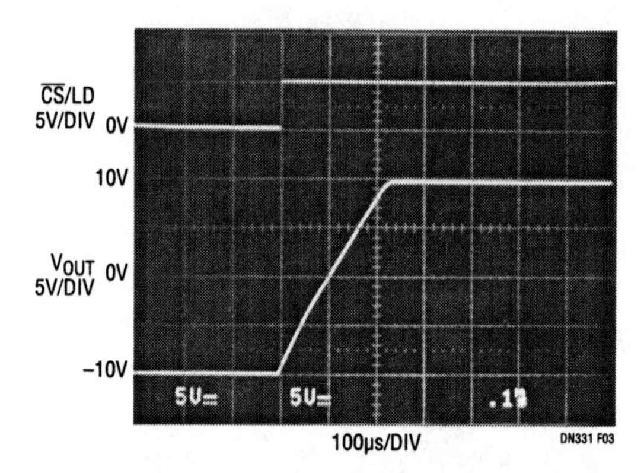

Figure 424.3 • 20V Output Step Time Domain Response

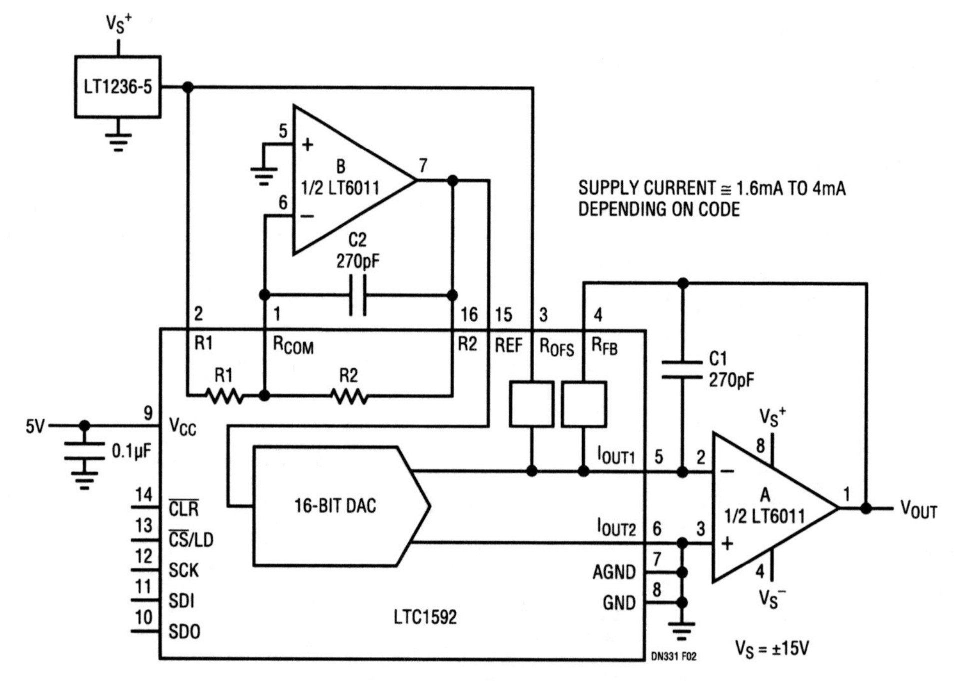

Figure 424.2 • DAC Reference Inverter and I-to-V Converter

100MHz op amp features low noise rail-to-rail performance while consuming only 2.5mA

425

Glen Brisebois

The new LT6202 op amp combines 100MHz gain-bandwidth, $1.9nV/\sqrt{Hz}$ voltage noise and rail-to-rail inputs and outputs, while consuming only 2.5mA. It also features a low $0.75pA/\sqrt{Hz}$ current noise, and it contributes exceptionally low total noise and distortion power in small-signal applications. The device is fully specified on 3V, 5V and ±5V supplies and is available in commercial and industrial temperature grades in both the SOT and SO packages. Dual and quad versions are also available as the LT6203 and LT6204, respectively.

Low power, $2.4nV/\sqrt{Hz}$, photodiode AC transimpedance amplifier outperforms monolithic solutions

You can't optimize for everything. Op amp designs that try to squeeze good JFETs into their high speed monolithic processes inevitably compromise other parameters, usually resulting in high supply currents. Figure 425.1 shows a simple way to get the best of both worlds using the LT6202 and a low noise discrete JFET. The JFET acts as a source follower,

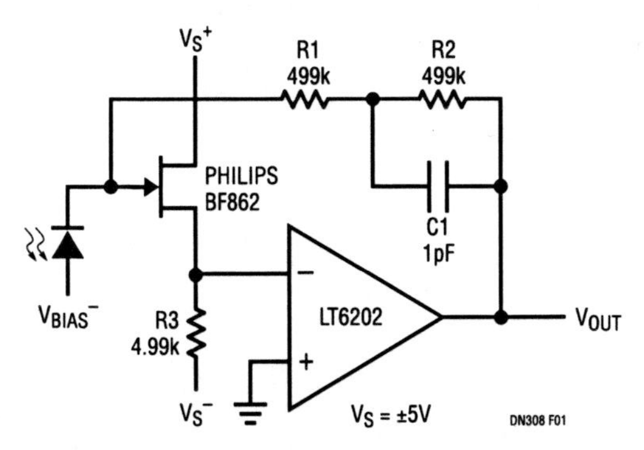

Figure 425.1 • Low Noise, Low Power Photodiode Amplifier Is Better than Monolithics. Low Bias Current, Low Current Noise and Unity Gain Stability Keep the Circuit Operational from 0Ω to Several GΩ

buffering the inverting input of the LT6202 and making it suitable for the high impedance feedback elements R1 and R2. The LT6202 forces the JFET source to 0V, with R3 ensuring that the JFET runs an I_{DRAIN} of 1mA. Because the JFET is run well below its minimum I_{DSS} and has a narrow range of pinchoff voltages, the circuit is guaranteed to self bias just below ground, typically at about −0.5V. Without a photocurrent signal in the photodiode, the LT6202 output sits at the same voltage and tracks it. When the photodiode is illuminated, the current must come from the LT6202 output through R1 and R2 so the output goes up as it would with a normal transimpedance amplifier.

Amplifier input noise density and gain-bandwidth product were measured to be $2.4nV/\sqrt{Hz}$ and 100MHz, respectively, with only 3.8mA supply current. This is unparalleled in the monolithic world where 5 to 6 times the supply current would be expected for similar performance. The 100MHz gain-bandwidth product of the LT6202 is maintained in this circuit because the JFET has a high gm, approximately 1/80Ω, which looks into 4.99kΩ so its loop attenuation is less than 2%. Total circuit input capacitance including board parasitics is measured at 3.5pF. This is less than the specified CGS of the JFET because the JFET source is not grounded but rather looks into R3 and the high impedance op amp input. This fact combined with the low input voltage noise makes the circuit well suited to both large and small photodetectors. Using a small photodiode with 2.5pF junction capacitance and adjusting parasitic feedback capacitance for 4% overshoot in the transient response, closed loop bandwidth is 1.6MHz.

Figure 425.2 shows the LT6202 applied in a manner very similar to that shown in Figure 425.1. In this case however, the JFET is not allowed to dictate the DC bias conditions. Instead of simply grounding the LT6202 noninverting input, an LTC2050 drives it (and therefore the source) exactly to where it needs to be for zero JFET gate voltage. The addition of the LTC2050 increases the total supply current by about 1mA. AC performance is nearly identical to the uncorrected circuit of Figure 425.1, with the additional benefit of

Analog Circuit Design: Design Note Collection. http://dx.doi.org/10.1016/B978-0-12-800001-4.00425-7

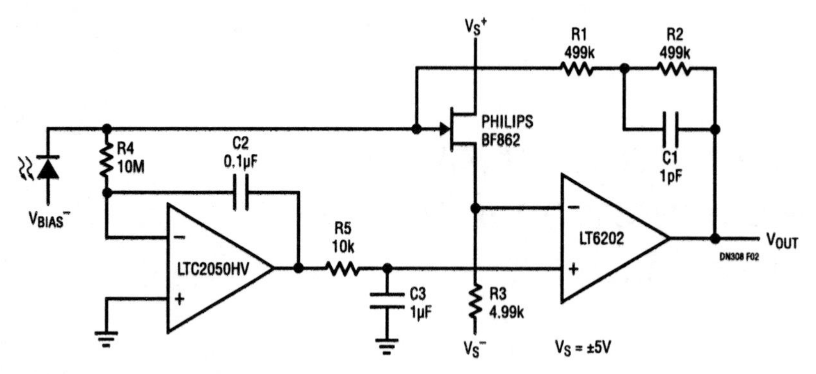

Figure 425.2 • Modified Circuit Provides Similar Noise and AC Performance, with DC Precision Restored by the LTC2050

excellent DC performance. Output offset is 200µV and output noise is 2mV$_{P-P}$ measured in a 20MHz bandwidth.

Single supply 16-bit ADC driver

Figure 425.3 shows the LT6203 driving an LTC1864 unipolar 16-bit, 250ksps A/D converter. The bottom half of the LT6203 is in a gain of 1 and buffers the 0V negative full-scale signal V_{LOW} into the negative input of the LTC1864. The upper half of the LT6203 is in a gain of 10, referenced to the buffered voltage V_{LOW} and drives the positive input of the LTC1864. The input range of the LTC1864 is 0V to 5V so for best results the input range of V_{IN} is from V_{LOW}, about 0.4V to about 0.82V. Figure 425.4 shows an FFT obtained with a 10.1318kHz (coherent) input waveform, with no windowing or averaging. Spurious free dynamic range is seen to be 100dB.

Although the LTC1864 has a sample rate far below the gain bandwidth of the LT6203, using this amplifier is not necessarily a case of overkill. A/D converters have sample apertures that are extremely narrow (infinitesimal as far as mathematicians are concerned) and make demands on upstream circuitry far in excess of what the innocent looking sample rate would imply. In addition, when an A/D converter takes a sample, it applies a small capacitor to its inputs causing a fair amount of glitch energy and expects the voltage on the capacitor to settle to the true value very quickly. Finally, the LTC1864 has a 20MHz analog input bandwidth and can be used in undersampling applications, again requiring a source bandwidth and settling speed higher than the Nyquist criterion would imply.

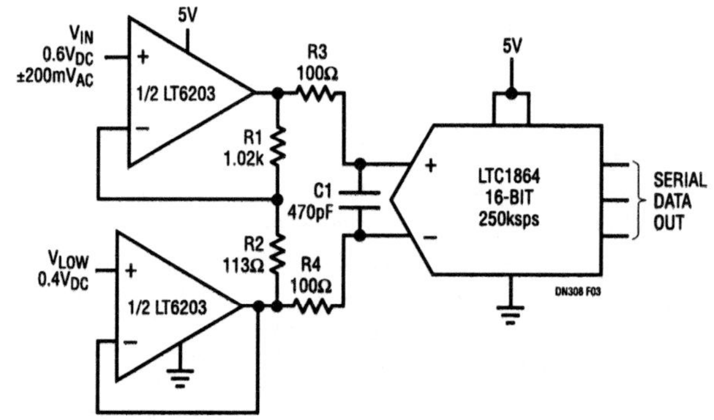

Figure 425.3 • Single Supply 16-Bit ADC Driver

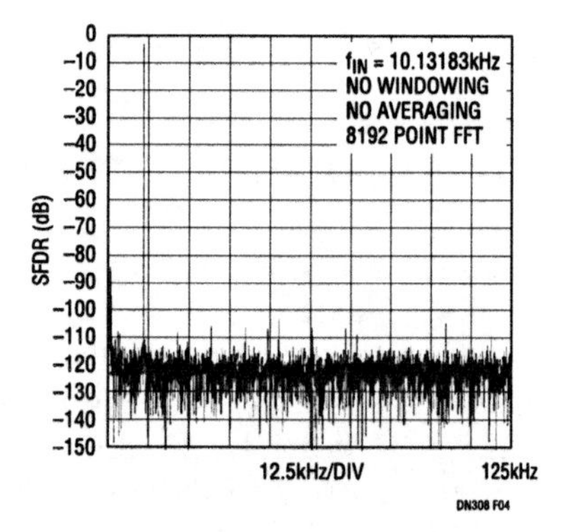

Figure 425.4 • FFT to 10kHz Sine Waves Showing 100dB SFDR

Conclusion

The LT6202, LT6203 and LT6204 are fast, low noise amplifiers which have been optimized for low power consumption. Their rail-to-rail inputs and outputs provide flexibility and ease of use and maximize dynamic range.

High performance op amps deliver precision waveform synthesis

426

Jon Munson

Introduction

With the trend toward ever more precise waveform generation using DSP synthesis and digital-to-analog conversion, such as with the LTC1668 16-bit, 50Msps DAC, increasing demands are being placed on the output amplifier. In some applications, the DAC current-to-voltage function is simply resistive, though this is limited to small-signal situations. The more common solution is to use an amplification or a transimpedance stage to provide larger usable scale factors or level shifting. Figure 426.1 shows one such example, with an LT1722 performing a differential current to single-ended voltage amplification for an LTC1668.

The LT1722, LT1723 and LT1724 low noise amplifiers

The LT1722, LT1723 and LT1724 are single, dual and quadruple operational amplifiers that feature low noise and high speed along with miserly power consumption. The parts are optimized for low voltage operation and draw only 3.7mA (typical) per section from ±5V supplies, yet deliver up to 200MHz GBW and quiet $3.8\text{nV}/\sqrt{\text{Hz}}$, $1.2\text{pA}/\sqrt{\text{Hz}}$ (typical) noise performance. DC characteristics include sub-millivolt input offset precision and output drive greater than 20mA, excellent for cable driving. The LT1722 single is also available in a SOT-23 5-lead package making it easy to fit into PCB layouts.

DAC output amplifier

The circuit in Figure 426.1 provides ±1V at the amplifier output for full-scale DAC currents of 5mA, therefore offering, with the 50Ω series termination shown, a +3dBm sine wave drive into a 50Ω load (~$1V_{P-P}$). In this particular configuration, the LT1722 is operating at a noise-gain of 5, and provides a small-signal bandwidth of about 8MHz (−3dB). The amplifier contribution to output noise is approximately given by $e_n G_n \sqrt{BW}$, or $3.8 \cdot 10^{-9} \cdot 5 \cdot \sqrt{8 \cdot 10^6} = 54\mu V$ for the circuit as shown (resistor noise will increase this to about 75µV). With 16-bit resolution, an LSB increment at the

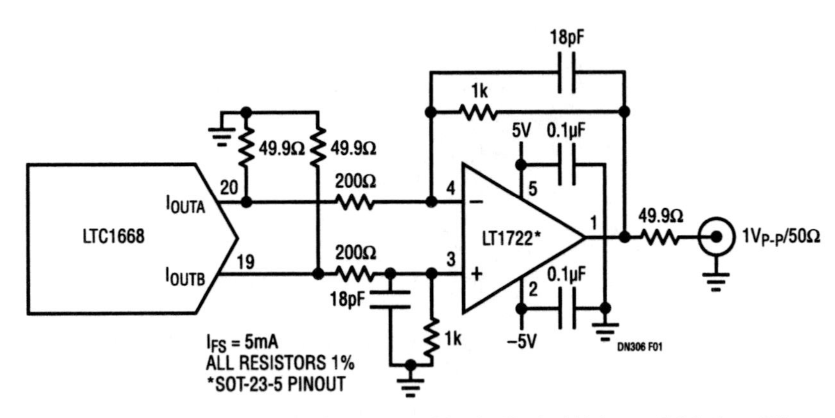

Figure 426.1 • Differential Current to Single-Ended Voltage DAC Amplifier

Analog Circuit Design: Design Note Collection. http://dx.doi.org/10.1016/B978-0-12-800001-4.00426-9

amplifier output is 31μV, therefore the LT1722 amplifier noise will have only minimal impact on the available dynamic range of the converter.

Some applications require amplified differential outputs, such as driving Gilbert-cell mixers (such as the LT5503 I/Q modulator) or RF transformers. For such applications, the LTC1668 differential current outputs can be amplified with twin transimpedance stages as shown in Figure 426.2, which offers the opportunity to reduce the DAC current without loss of signal swing.

The circuit shown has the DAC full-scale currents reduced to 2mA to achieve a substantial power savings over the standard 10mA operation. The scale factor of the transimpedance amplifiers is set to provide $2V_{P-P}$ differentially. Operating at a noise-gain of unity, this circuit provides a small-signal bandwidth of about 12MHz (−3dB). The noise contributed by the LT1723 amplifiers to the differential load is approximately $\sqrt{2} \cdot e_n G_n \cdot \sqrt{BW}$, or $\sqrt{2} \cdot 3.8 \cdot 10^{-9} \cdot 1 \cdot \sqrt{12 \cdot 10^6} = 19\mu V$ for the circuit as shown (the resistors in the circuit will add some additional noise bringing the total to about 24μV). This compares favorably with the nominal 16-bit LSB increment of 31μV, thus barely impacting the converter dynamic range.

The common mode output voltage of the circuit in Figure 426.2 is fixed at 0.5V DC, though some loads may require a different level if DC coupling is to be supported, such as when soft-controlled offset nulling is required. Though not shown here, specific matched currents can easily be introduced to the inverting-input nodes of the two amplifiers to provide common mode output control.

Each of the amplifier circuits presented will deliver +3dBm into 50Ω with harmonic distortion products below −60dBc for a synthesized near-full-scale fundamental of 1MHz as shown in Figures 426.3 and 426.4. The nominal feedback capacitances shown provided ~1% step-response overshoot in the author's prototype configuration, but as with all amplifier circuits, some tailoring may be required to achieve a desired rolloff characteristic in the final printed circuit layout.

Conclusion

When considering candidate devices for DAC post amplification, it is important to consider the noise contribution. The LT1722 family of devices offers the low noise and wide bandwidths demanded by modern 16-bit waveform synthesizers, particularly those used for vector modulation, where high fidelity is paramount.

Additionally, the particular low noise characteristics of the LT1722, LT1723 and LT1724 op amps provide optimal noise performance for external impedances ranging from several hundred ohms to about 12kΩ, making these parts ideal for a variety of precision amplification tasks.

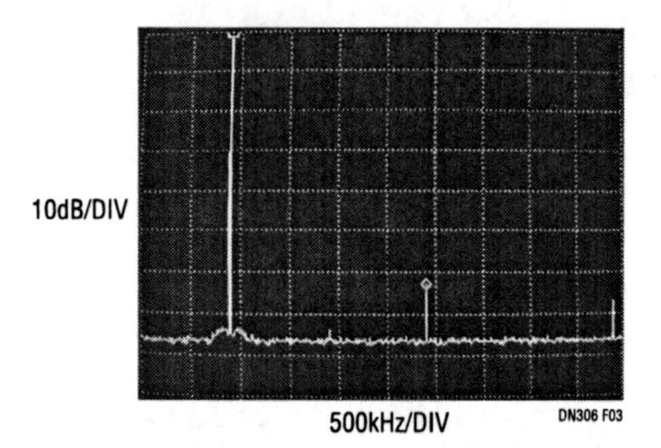

Figure 426.3 • Distortion of LT1722 Circuit (f_O = 1MHz)

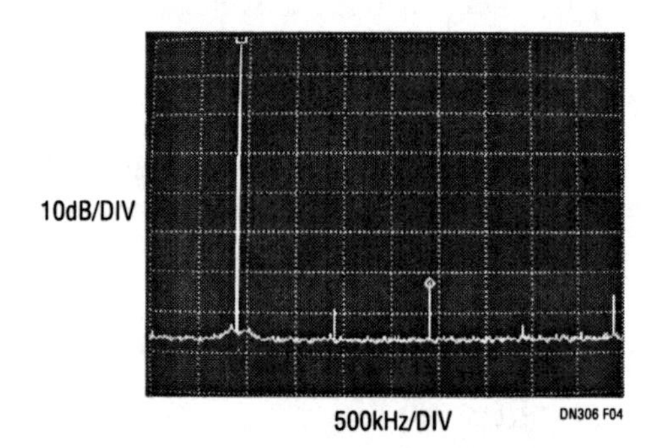

Figure 426.4 • Distortion of LT1723 Circuit (f_O = 1MHz)

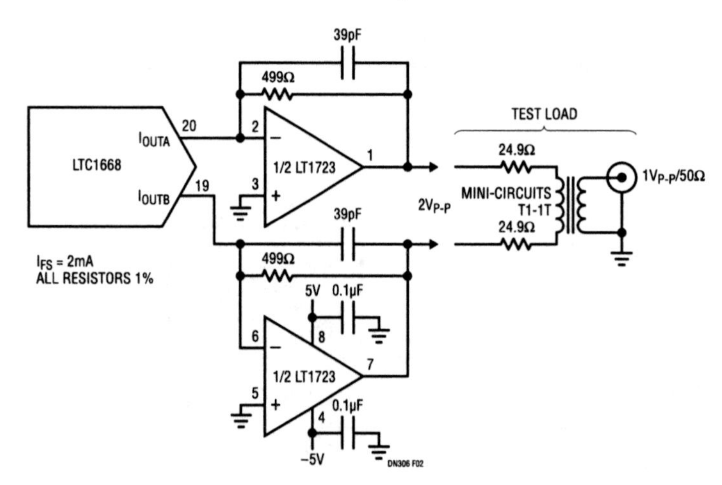

Figure 426.2 • Twin Transimpedance Differential Output DAC Amplifiers

Power op amp provides on-the-fly adjustable current limit for flexibility and load protection in high current applications

427

Tim Regan

Introduction

Many power operational amplifiers offer a built-in current limit where the limit is fixed or programmable through an external resistor. This offers the most basic measure of protection for the load circuitry, and the amplifier itself, under fault conditions. Sometimes, though, there is a need for *on-the-fly* current limiting to satisfy the requirements of different loads. For example, automatic test equipment (ATE) systems use multiple pin drivers to deliver test voltages across a wide range of loads, including faults, to a unit or board to test for continuity or functionality. To protect the load circuitry, the ATE must precisely control the maximum current delivered to each pin. Ideally the maximum current can be controlled on the fly, to accommodate the different loads at each pin.

Introducing the LT1970

The LT1970 op amp can supply ±500mA of output current, with a precise, easy-to-implement current limit. Current limit control is via two simple 0V–5V voltage inputs, V_{CSRC} (for sourcing current) and V_{CSNK} (for sinking current), that set the maximum output current anywhere from 4mA to 500mA. Even at 500mA, the accuracy of the output current limit is guaranteed to a tight 2% (10mA) tolerance. The output current is continuously sensed by a small valued resistor, R_S, connected in series with the load as shown in Figure 427.1. The maximum output current is a function of the control-input voltage and the sense resistor according to the following expression:

$$I_{OUT(LIMIT)} = \frac{V_{LIMIT}}{10 \cdot R_S}$$

By simply changing the voltage between 0V and 5V at the control input, say through a D-to-A converter, the output current limit quickly changes to a new level.

For example, Figure 427.2 shows the output waveforms of the LT1970 driving a 1μF capacitive load in parallel with a 100Ω resistor. The current limit is set to 500mA sourcing (V_{CSRC} = 5V) and 50mA sinking (V_{CSNK} = 500mV). To charge the load capacitance, the amplifier current limits until the output voltage reaches its proper closed loop value. Then, while swinging negative, the sinking current limit prevents the output from going less than −5V.

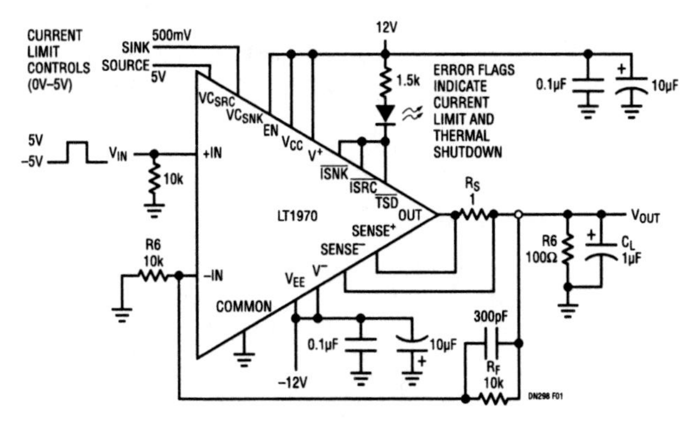

Figure 427.1 • The LT1970: Easy to Use as an Op Amp with an Adjustable Output Current Power Stage

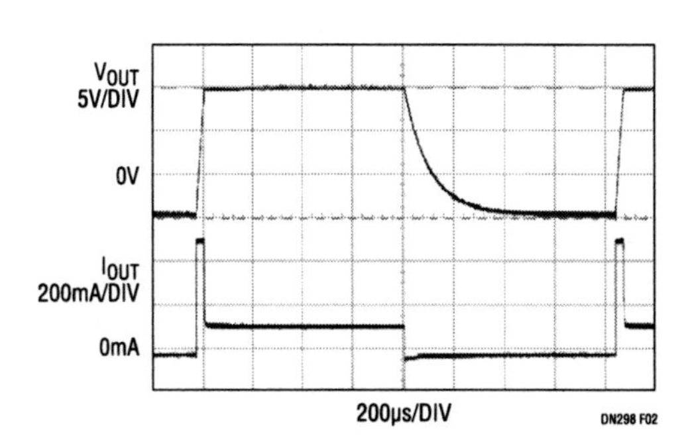

Figure 427.2 • 500mA Source Current Limit and 50mA Sink Current Limit Control Output Response Characteristics

Analog Circuit Design: Design Note Collection. http://dx.doi.org/10.1016/B978-0-12-800001-4.00427-0

The LT1970 also features open-collector error flags. These three outputs indicate that the amplifier is in current limit, either sourcing or sinking, and that the amplifier is in thermal shutdown. Additionally, the Enable input can be used to turn off the amplifier, thus putting the output into a high-impedance, zero output current state. This same input can also be used to simultaneously apply a new set of voltage and current settings to the load. The LT1970 is available in a small 20-pin TSSOP package with exposed underside metal for heat dissipation.

Boosted output current with "snap-back" current limiting

The LT1970 has separate supply pins for the input stage and the power output stage. Only load current flows through the output stage power supplies (V^+ and V^-). These pins can provide gate or base drive to external power transistors to boost the output current capability of the amplifier. A simple power stage, shown in Figure 427.3, increases the output current to ±5A. The same 0V to 5V inputs now set the output current limits a factor of ten higher (to 1A/V) by the use of a smaller current sense resistor, $R_S = 0.1\Omega$.

Externally connected gain setting resistors allow Kelvin sensing at the load. By connecting the feedback resistor right at the load, the voltage placed on the load is exactly what it should be. Any voltage drop across the current sense resistor is inside the feedback loop and thus does not create a voltage error. Figure 427.3 also shows a unique way to use the open-collector error flags to provide extra protection to the load circuitry. When the amplifier enters current limit in either direction, the appropriate error flag goes low. This high impedance to 0V transition can provide a large amount of hysteresis to the current limit control inputs, forcing a drastic reduction in output current. Resistors R1, R2 and R3 set the current limit control feedback at 2V max and 200mV min. Should the load current ever exceed the predetermined maximum limit, the output current snaps back to the min level. The output current remains at this lower level until the signal drops to a point where the load current is less than the minimum set value. When the signal is low enough, the flag output goes open and the current limit reverts to the maximum value. This action simulates an automatically resettable fuse to protect a load. Figure 427.4 shows the action of this feedback with a maximum current limit of 2A snapping back to 200mA when exceeded in either direction.

Conclusion

The LT1970 is a versatile and easy to use power op amp with a built-in precision adjustable current limit, which can protect load circuitry from damage caused by excessive power from the amplifier. This feature is particularly useful in ATE systems where the load is variable (and possibly faulty) at each tested node. Tight control of the output current in these systems is important to prevent damage to the tested unit.

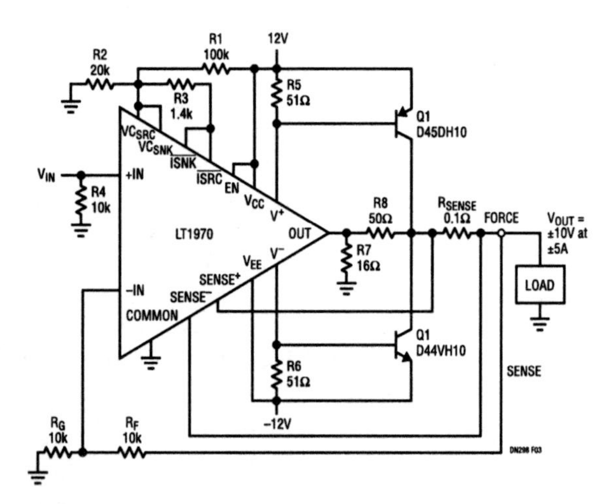

Figure 427.3 • Easily Adjusted Current Limit for a ±5A Boosted Output Current Stage

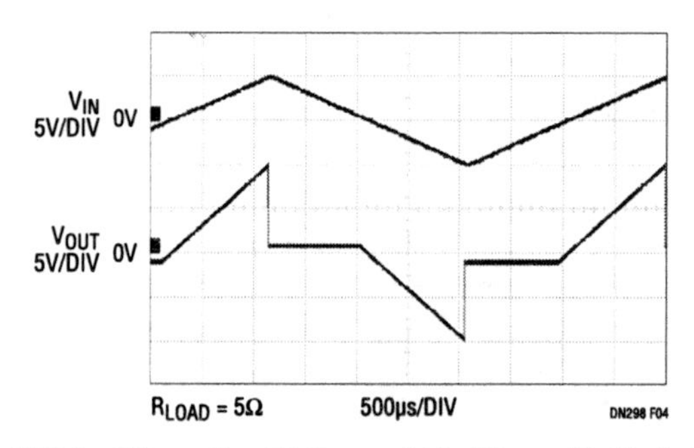

Figure 427.4 • "Snap Back" Current Limiting with Both Source and Sink Current Limit Controlled by a Simple Resistor Network

Fast and accurate 80MHz amplifier draws only 2mA

428

Glen Brisebois William Jett Dahn Tran

Introduction

The 80MHz LT1800 amplifier provides the high speed and DC accuracy required by low voltage signal conditioning and data acquisition systems while consuming a mere 2mA max supply current. The LT1800 operates with supplies from 2.4V to 12V and its rail-to-rail inputs and outputs allow the entire supply range to be used. DC performance is exceptional; the maximum offset voltage is only 350μV and the maximum input bias current is only 250nA. The amplifier is also available in dual and quad versions as the LT1801 and the LT1802, respectively. All are available in commercial and industrial temperature grades. The LT1800 is available in SO and SOT-23 packages, the LT1801 dual in an SO-8 package and the LT1802 quad in an SO-14 package.

Single supply 1A laser driver

Figure 428.1 shows the LT1800 used in a 1A laser driver application. The LT1800 is well suited to this control task because its 2.4V operation ensures that it is awake at power-up and in control before the circuit can cause significant current to flow in the 2.1V threshold laser. Raising the non-inverting input of the LT1800 causes its output to rise, turning on the FMMT619 high current NPN transistor and the SFH495 IR laser.

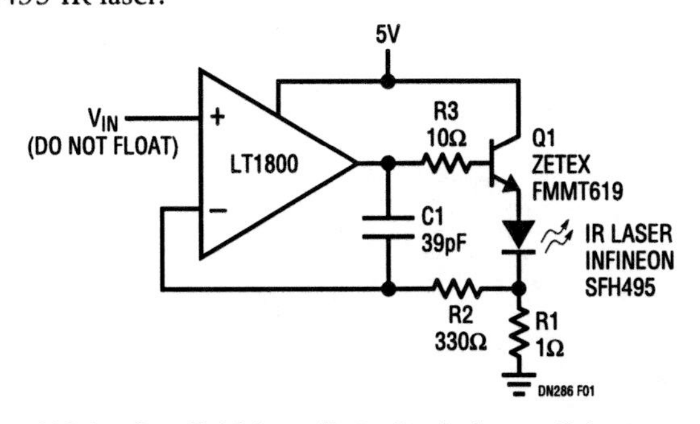

Figure 428.1 • Small 1A Low Duty Cycle Laser Driver

The transistor and laser turn on until the input voltage appears back at the LT1800 inverting input. This voltage therefore also appears across the 1Ω resistor R1. In order for this to occur, a current equal to $V_{IN}/R1$ must exist and the only place it can come from is through the laser. The overall circuit is thus a V-to-I converter with a 1A/V characteristic.

Lower values for R1 may be selected but the designer is reminded to keep series loop traces very short: for example, even 10nH of lead inductance causes a 16MHz pole into 1Ω and a 1.6MHz pole into 0.1Ω! Also when decreasing the value of R1, consider the total of the dynamic impedances of the transistor V_{BE} and the laser. They reduce the feedback voltage to R1 thus increasing circuit noise gain. This has the effect of degrading the DC precision and reducing the achievable bandwidth.

Frequency compensation components R2 and C1 are chosen for fast but zero overshoot time domain response which avoids overcurrent conditions in the laser. Their values may vary from design to design depending upon desired response characteristics, circuit layout, the value of R1 and the actual laser and transistor devices selected. Figure 428.2 shows the time domain response of this circuit, measured at R1 and given a 500mV 230ns input pulse. While the circuit shown is capable of 1A operation, the laser and the transistor are thermally limited and so must be operated at low duty cycles.

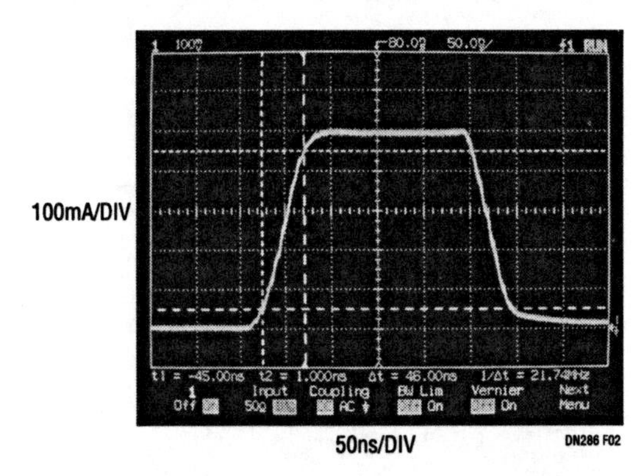

Figure 428.2 • Pulse Response of 1A Laser Driver Circuit Shows Better Than 50ns Rise Time on 500mA Pulse

Analog Circuit Design: Design Note Collection. http://dx.doi.org/10.1016/B978-0-12-800001-4.00428-2

Low power amplifier with 250V output swing

Some recently developed materials have optical characteristics that depend on the presence and strength of a DC electric field. Many applications require a bias voltage applied across such materials, sometimes as high as hundreds of volts, precisely in order to achieve and maintain desired properties in the material. The materials are not conductive and present an almost purely capacitive load.

Figure 428.3 shows the LT1800 used in an amplifier intended for capacitive loads and capable of 250V output swing. When no input signal is present, the op amp output sits at about mid-supply. Transistors Q1 and Q3 create bias voltages for Q2 and Q4 which are forced into a low quiescent current by degeneration resistors R4 and R5. When a transient signal arrives at V_{IN}, the op amp output jumps away from mid-supply and causes current through Q2 or Q4 depending on the signal polarity. The current, limited by the output swing of the LT1800 and the 3kΩ of total emitter degeneration, is level shifted to the high voltage supplies and mirrored into the capacitive load. This causes a voltage slew at V_{OUT} until the feedback loop (through R3) is satisfied.

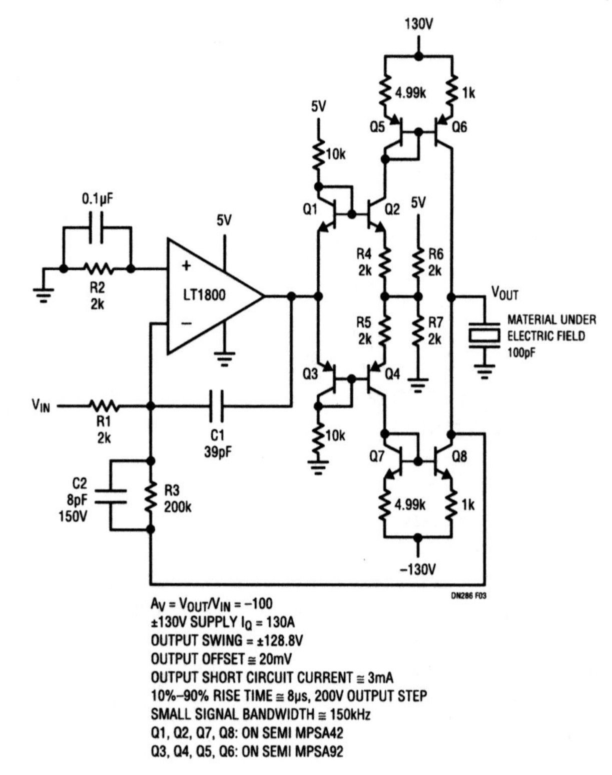

Figure 428.3 • Low Power High Voltage Amplifier

The LT1800 output then returns back to near mid-supply, providing just enough DC output current to maintain the output voltage across R3. The circuit thus alternates between a low current hold state and a higher transient, but limited, current slew state.

Careful attention to current levels minimizes power dissipation allowing for a dense component layout, and also provides inherent output short-circuit protection. To further save power, the LT1800 is operated single supply with its inputs at ground. With the inputs at ground, the LT1800 turns off its internal bias current cancellation and adding R2 externally restores input precision.

Figure 428.4 shows the time domain response of the amplifier providing a ±100V output swing into a 100pF load.

Conclusion

The LT1800 and its LT1801 dual and LT1802 quad derivatives, provide low power solutions to high speed, low voltage signal conditioning. Rail-to-rail input and output maximize dynamic range and can simplify designs by eliminating the negative supply. Circuits that require source impedances of 1k or more, such as filters, benefit from the low input bias currents and low input offset voltage. The combination of speed, DC accuracy and low power makes the LT1800 a top choice for low voltage signal conditioning.

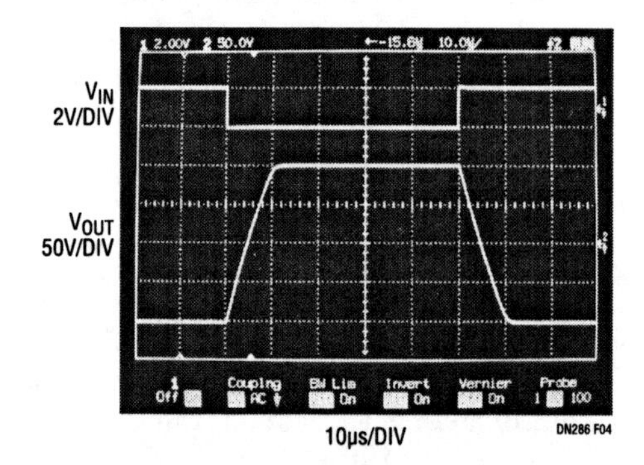

Figure 428.4 • Large-Signal Time Domain Response of the Material Bias Amplifier

SOT-23 superbeta op amp saves board space in precision applications

429

Glen Brisebois

Introduction

The tiny new LT1880 achieves precision unprecedented in a SOT-23 package without resorting to autozeroing techniques. Input offset voltage and drift are typically 40μV and 0.3μV/°C, respectively, with guarantees of 200μV and 1.2μV/°C maximum over temperature. The device operates on total supplies from 2.7V to 40V with rail-to-rail outputs, giving a dynamic range of 120dB. Unlike some competitors' SOT-23 op amps, which claim to maintain good precision, the LT1880 supports its input precision with a high open loop gain of 1.6 million, as well as 135dB CMRR and PSRR. It is available in commercial and industrial temperature grades.

Applications

Getting rail-to-rail operation without rail-to-rail inputs

The LT1880 does not have rail-to-rail inputs, but for most inverting applications and noninverting gain applications, this is largely inconsequential. Figure 429.1 shows the basic op amp configurations, what happens to the op amp inputs, and whether or not the op amp must have rail-to-rail inputs.

The circuit of Figure 429.2 shows an extreme example of the inverting case. The input voltage at the 1M resistor can swing ±13.5V and the LT1880 will output an inverted, divided-by-ten version of the input voltage. The gain accuracy is limited by the resistors to 0.2%. Output referred, this error becomes 2.7mV at 1.35V output. The 40μV input offset voltage contribution, plus the additional error due to input bias current times the ~100k effective source impedance, contribute negligible error.

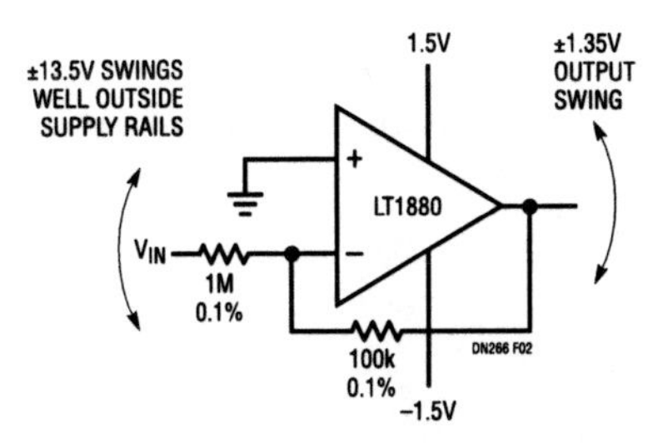

Figure 429.2 • Extreme Inverting Case: Circuit Operates Properly with Input Voltage Swing Well Outside Op Amp Supply Rails

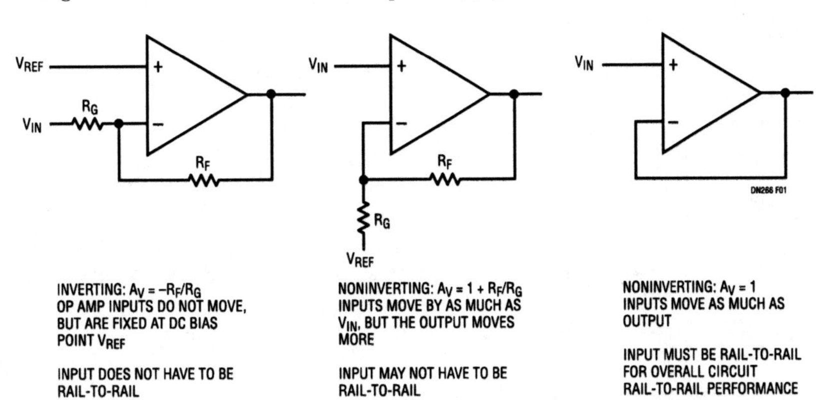

Figure 429.1 • Some Op Amp Configurations Do Not Require Rail-to-Rail Inputs to Achieve Rail-to-Rail Outputs

Analog Circuit Design: Design Note Collection. http://dx.doi.org/10.1016/B978-0-12-800001-4.00429-4

Precision photodiode amplifier

Photodiode amplifiers usually employ JFET op amps because of their low bias current; however, when precision is required, JFET op amps are generally inadequate due to their relatively high input offset voltage and drift. The LT1880 provides a high degree of precision with very low bias current (I_B=150pA typical) and is therefore applicable to this demanding task. Figure 429.3 shows an LT1880 configured as a transimpedance photodiode amplifier. The transimpedance gain is set to 51.1kΩ by R_F. The feedback capacitor, C_F, may be as large as desired where response time is not an issue, or it may be selected for maximally flat response and highest possible bandwidth given a photodiode capacitance C_D. Figure 429.4 shows a chart of C_F and rise time versus C_D for maximally flat response. Total output offset is below 262μV, worst-case, over temperature (0°C to 70°C). With a 5V output swing this implies a minimum 86dB dynamic range, sustained over temperature (0°C to 70°C), and a full-scale photodiode current of 98μA.

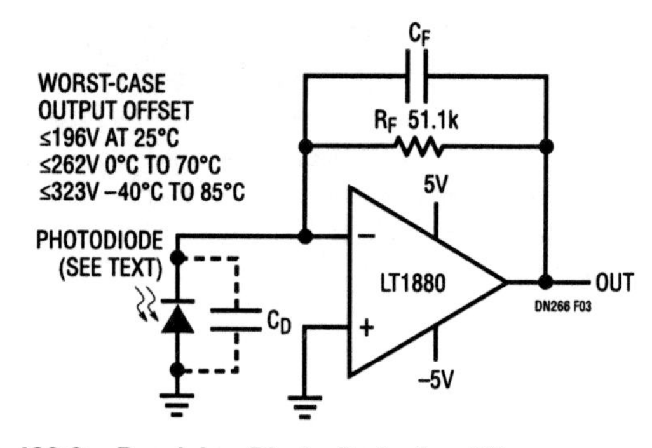

Figure 429.3 • Precision Photodiode Amplifier

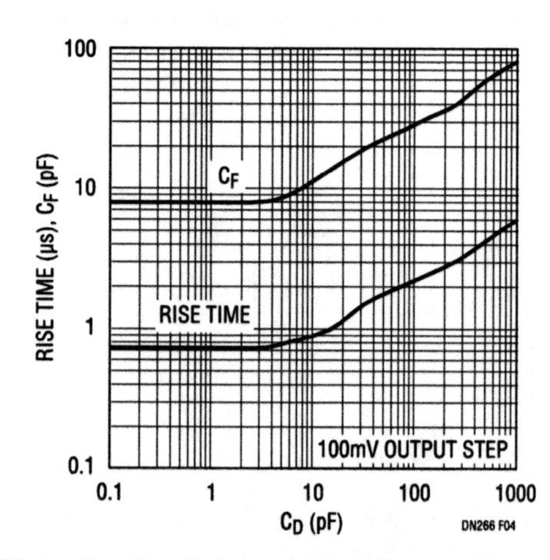

Figure 429.4 • Feedback C_F and Rise Time vs Photodiode C_D

Single supply current source for platinum RTD

The precision, low bias current input stage of the LT1880 makes it ideal for precision integrators and current sources. Figure 429.5 shows the LT1880 providing a simple precision current source for a remote 1kΩ RTD on a 4-wire connection. The LT1634 reference places 1.25V at the noninverting input of the LT1880, which then maintains its inverting input at the same voltage by driving 1mA of current through the RTD and the total 1.25k of resistance set by R1 and R2. Lower precision components R4 and C1 ensure circuit stability, which would otherwise be excessively dependent on the cable characteristics. R5 is also noncritical and is included to improve ESD immunity and decouple any cable capacitance from the LT1880's output. The 4-wire cable allows Kelvin sensing of the RTD voltage while excluding the cable IR drops from the voltage reading. With 1mA excitation, a 1kΩ RTD will have 1V across it at 0°C, and 3.85mV/°C temperature response. This voltage can be easily read in myriad ways, with the best method depending on the temperature region to be emphasized and the particular ADC that will be reading the voltage.

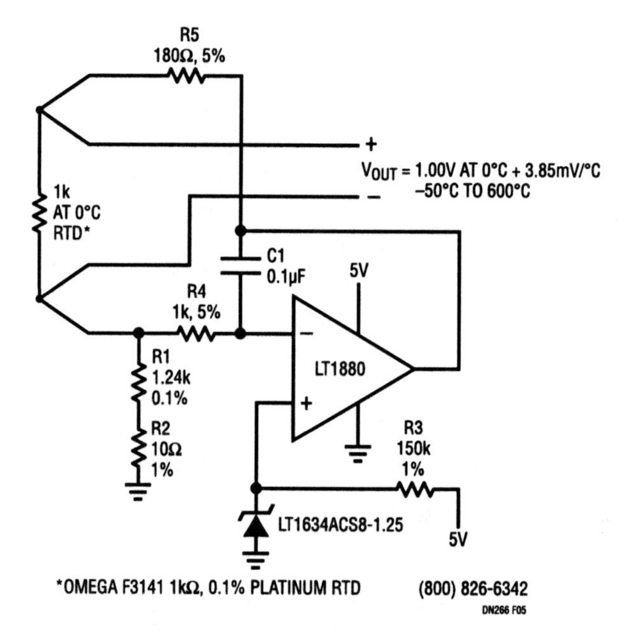

Figure 429.5 • Single Supply Current Source for Platinum RTD

Conclusion

The precision, low bias current input stage of the LT1880 makes it ideal for precision and high impedance circuits. The rail-to-rail output stage renders the op amp capable of driving other devices as simply as possible with extended dynamic range, while the 2.7V to 40V operation means that it will work on almost all supplies. The small SOT-23 package makes it a compelling choice where board space is at a premium or where a composite amplifier is competing against a larger single-chip solution.

430

325MHz low noise rail-to-rail SOT-23 op amp saves board space

Glen Brisebois

The new tiny LT1806 combines 325MHz gain bandwidth, $3.5nV/\sqrt{Hz}$ voltage noise, 100μV input offset voltage and rail-to-rail inputs and outputs in a SOT-23 package. The device is fully specified on 3V, 5V and ±5V supplies with a guaranteed maximum input offset voltage of 850μV at either rail over temperature. It is available in commercial and industrial temperature grades.

1MΩ transimpedance amplifier achieves near theoretical noise performance with large-area photodiodes

The circuit of Figure 430.1 shows the LT1806 in an ultralow noise transimpedance amplifier applied to a large-area high capacitance photodiode. The LT1806 is used for its high gain bandwidth and low noise. The IFN147 ultralow noise JFET[1] operates at its $I_{DSS}(V_{GS}=0V)$ with a typical transconductance of 40mS. With its source grounded, the JFET and its drain resistor, R5,[2] set a voltage gain of about 8. The combination of the ultralow noise JFET gain stage and the LT1806 low noise amplifier achieves the ultralow input noise performance. The circuit's input voltage noise was measured at $0.95nV/\sqrt{Hz}$. Figure 430.2 compares the noise performance of the circuit with that of a competitor's monolithic $6nV/\sqrt{Hz}$ JFET op amp, both with a 680pF capacitive source.

Why is it necessary to have both low voltage noise and low current noise to achieve low total noise in a large-area photodiode transimpedance amplifier? Because the transimpedance circuit's noise gain, which applies to voltage noise but not to current noise or resistor noise, rises drastically with frequency (noise gain $= 1 + Z_F/X_C$). As a sample calculation, a 500pF photodiode has an impedance of 3.2kΩ at 100kHz, giving a 1MΩ transimpedance circuit a noise gain of 314 at

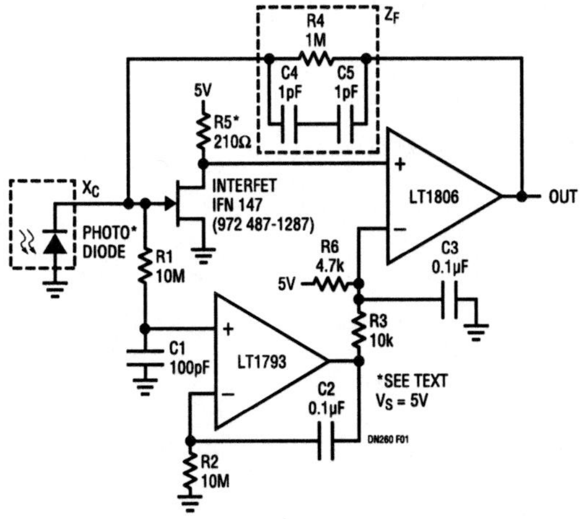

Figure 430.1 • Ultralow Noise, 2.4GHz Gain Bandwidth Large-Area Photodiode Amplifier

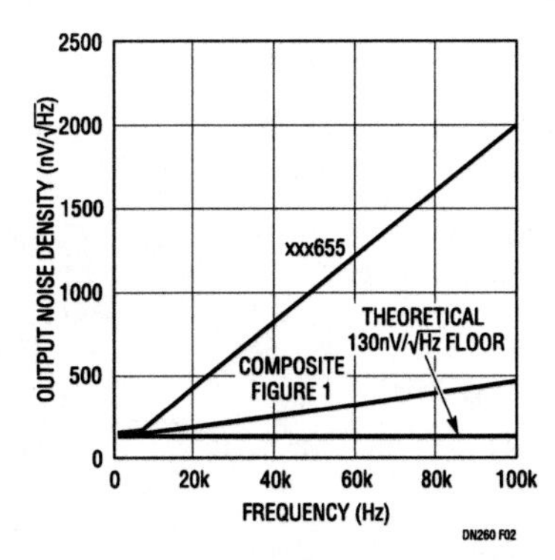

Figure 430.2 • Composite Amplifier vs Competitor Op Amp xxx655

Note 1: Equivalent to Japanese 2SK147.

Note 2: For devices with I_{DSS} higher than about 12mA, R5 should be reduced to 100Ω to avoid saturating the JFET.

Analog Circuit Design: Design Note Collection. http://dx.doi.org/10.1016/B978-0-12-800001-4.00430-0

that frequency. The theoretical noise floor of the 1M resistor is $130nV/\sqrt{Hz}$ (at room temperature), so any input voltage noise above 0.41nV/Hz ($130nV/\sqrt{Hz}/314$) will overtake the resistor noise at 100kHz. Discrete JFETs are available with ultralow voltage noise, but they have high input capacitance (75pF max for the IFN147). Serendipitously, the input capacitance of the JFET is relatively insignificant compared to the 500pF of the example large-area photodiode. Although the capacitance of the JFET does increase the overall noise gain slightly, its much lower input voltage noise is well worth the slight increase in total capacitance and noise gain. Table 430.1 and Figure 430.3 show the bandwidth and noise performance achieved with several large-area photodiodes (and a small-area SFH213 for comparison). Note that large-area detectors also place extra demands on the gain bandwidth of an amplifier. The final case in Table 430.1 shows a 1MΩ transimpedance amplifier with 650kHz bandwidth from a 660pF photodiode. Although this may not seem like much bandwidth, it necessitates a gain bandwidth product of at least 1.8GHz in the amplifier.

The task of the LT1793 is to keep the JFET biased at its I_{DSS} current ($V_{GS}=0V$); it was selected for its low 100pA maximum input offset current over temperature. The LT1793 senses the input voltage at the JFET gate through R1 and nulls this voltage through the LT1806 inverting pin and back around through R4. The time constants formed by R1C1 and R3C3 ensure that the LT1793 noise characteristics do not add to the total noise. C1 shunts the already low LT1793 current noise to ground and R3C3 keeps the LT1793 and resistor thermal noise away from the LT1806 low noise op amp input. Note that with the JFET gate at 0V, there is no reverse bias across the photodiode, eliminating dark current issues.

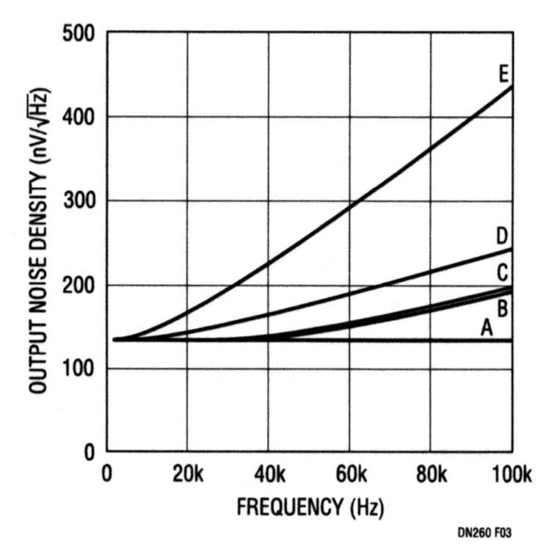

Figure 430.3 • Output Noise Spectra for Various Photodiodes

At first glance, the circuit does not appear stable, since the JFET circuit puts additional gain into the op amp loop and this is usually a recipe for disaster. The reason the circuit is stable (and with quite a bit of margin) is that the gain is greater than unity at frequencies above a few hundred hertz. Because of the relatively high value of the feedback impedances (1MΩ and 0.5pF) and the 75pF minimum input capacitance of the JFET, the gain of the circuit is 150 minimum above 300kHz. The LT1806 is a 325MHz gain bandwidth, unity-gain-stable op amp, so it is quite comfortable maintaining stability above 300kHz in what it sees as about a gain of 19 (150/8). Note that because the JFET circuit has a gain of 8, the gain bandwidth of the composite amplifier is about 2.4GHz. Also of interest are the open-loop gains of 2.4 million ($8 \cdot 300{,}000$) in the fast loop and 350 billion ($3.5\text{ million} \cdot 300{,}000/3$) in the slow loop. These numbers, along with the gain bandwidth and the 1M feedback resistor, determine the impedance that the photodiode sees looking into the amplifier input.

Conclusion

The LT1806 offers exceptional bandwidth and low noise in a SOT-23 package. The rail-to-rail inputs and outputs make the op amp easy to apply and maximize the available dynamic range. The tiny package makes the op amp a compelling choice where PCB real estate is at a premium. The composite photodiode amplifier shown above is just one example where the LT1806 meets a difficult set of requirements. Let's talk about YOUR difficult set of requirements today.

Table 430.1 Performance of the Composite Amplifier with Various Photodiodes

	VENDOR	PART NUMBER	OPTICAL CHARACTER	TYPICAL V = 0 CAPACITANCE	APPROXIMATE BANDWIDTH
A	Siemens/Infineon 408-456-4071	SFH213	Fast IR PIN	11pF	250kHz
B	Siemens/Infineon 408-456-4071	BPW34B	Enhanced Blue PIN	72pF	390kHz
C	Opto-Diode 805-499-0335	ODD45W	Narrow IR GaAlAs	170pF	380kHz
D	Fermionics 805-582-0155	FD1500W	Extended IR InGaAs	300pF	500kHz
E	Siemens/Infineon 408-456-4071	BPW21	Visible Spectrum	660pF	650kHz

Fast op amps operate rail-to-rail on 2.7V

431

Glen Brisebois

The new LT1806 and LT1809 are fast operational amplifiers with rail-to-rail inputs and outputs. The LT1806 is optimized for low noise ($3.5nV/\sqrt{Hz}$) and offset (100μV typical, 550μV max) and has a 325MHz gain bandwidth product. The LT1809 is optimized for low distortion (−90dBc to 5MHz) and has a 180MHz gain bandwidth product. Both amplifiers have 85mA output drive capability and are fully specified over commercial and industrial temperature ranges on 3V, 5V and ±5V supplies.

Parallel composite amplifier achieves low distortion into heavy loads

Achieving low distortion is difficult enough and driving heavy loads only makes the undertaking more difficult. Figure 431.1 shows a parallel composite topology where each amplifier is given a different responsibility. Amplifier U1 is configured in a standard gain of 2 to account for the eventual attenuation caused by the 50Ω output divider. Amplifier U1 alone can drive the load, but its distortion figures would suffer. Therefore, amplifier U2 is added in a slightly higher gain and its output is coupled to the circuit output through the attenuating 10Ω resistor, with the product of the gain and the attenuation set to match U1's gain of 2. In this way, the heavy load current is supplied by U2's output and is not seen by U1. Although the heavy load current on U2 will cause it to generate higher distortion, these distortion products are decoupled by the 10Ω resistor and corrected by U1. Basically, U2 supplies the power and U1 supplies the precision.

While this provides improvement, it is still possible to do better. Normally, lower distortion is achieved by putting the

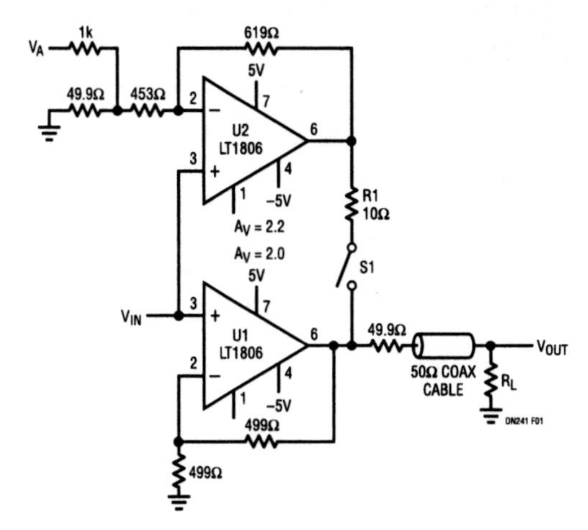

Figure 431.1 • Parallel Composite Amplifier: U2 Provides the Current, Easing the Load on U1. U1 Provides Only a Correction Current

Table 431.1 Distortion of Figure 431.1 at 2.5MHz, V_{SUPPLY} = ±5V

V_{IN}	S1	V_A	R_L	V_{OUT}	HARMONICS		I_{SUPPLY}	NOTE
					2nd	3rd		
$1V_{P-P}$	Open	0V	1M	$2V_{P-P}$	−92dBc	−98dBc	22mA	Low distortion at light load
$1V_{P-P}$	Open	0V	50Ω	$1V_{P-P}$	−80dBc	−84dBc	24mA	Load increases, distortion worse
$1V_{P-P}$	Closed	0V	50Ω	$1V_{P-P}$	−90dBc	−99dBc	25mA	U2 helps, distortion better
$1V_{P-P}$	Closed	1.5V	50Ω	$1V_{P-P}$	−94dBc	−99dBc	28mA	Add offset, distortion better yet
$3V_{P-P}$	Closed	1.5V	50Ω	$3V_{P-P}$	−85dBc	−77dBc	36mA	Results at higher amplitude

Analog Circuit Design: Design Note Collection. http://dx.doi.org/10.1016/B978-0-12-800001-4.00431-2

output into "Class A" operation, in which the bias current is higher than the peak load current. This is often done with a simple load resistor to one of the supply rails. Unfortunately, in cases where load current is already high, additional load current will only make matters worse. However, a kind of "tracking" Class A operation can be achieved by putting a small offset voltage into the power amplifier, U2. This causes a slight DC shift in its output and increases the bias current of U1's output stage. The DC offset is injected at V_A through R7 and can be dynamically adjusted to suit real-time requirements. Table 431.1 shows results achieved with a $1V_{P-P}$ input signal at 2.5MHz with ±5V supplies, with the last entry showing $3V_{P-P}$ throughput. Note that because the circuit shown is noninverting, its input impedance is high and is easy to drive.

Among the benefits of this parallel composite topology over series composite topologies are that it does not require onerous attention to compensation and does not reduce the effective bandwidth of the op amps. The trade-off is that supply current increases, but the designer should remember that a $3V_{P-P}$ signal requires ±30mA of peak current itself, so a 36mA supply current is modest considering the low distortion levels being achieved. This example was shown using two LT1806s for low noise and high bandwidth, but depending on the requirements other amplifiers can be configured in this topology as well.

Rail-to-rail pulse-width modulator using the LT1809

Binary modulation schemes are used in order to improve efficiency and reduce physical circuit size. They do this by reducing the power dissipation in the output driver transistors. In a normal Class A or Class AB amplifier, voltage drop and current flow exist simultaneously in the output transistors and power losses proportional to V·I occur. In a binary modulation scheme, the output transistors, whether bipolar or FET, are switched hard-on and hard-off so that voltage drops do not occur simultaneously with current flow. The circuit of Figure 431.2 shows an example of a binary modulation scheme, in this case pulse-width modulation.

The LT1809 is configured as an integrator in order to generate nice linear rail-to-rail voltage ramps. The polarity of the ramp is determined by the output of comparator A into R4. The heavy hysteresis of R1 around comparator A combined with the feedback of the LT1809 force the devices to perpetually reverse each other, resulting in a 1MHz triangle wave. This constitutes the usual first half of any pulse-width modulator, but the forte of this particular implementation is that it is rail-to-rail allowing a full-scale analog input. Once the triangle wave is achieved, the remainder of the pulse-width modulator is easy, and is constituted by doing a simple comparison using the second half of the LT1714. The triangle wave and the relatively slow moving analog signal (the one to be modulated or to do the modulation, depending on how you look at it) are fed into the inputs of comparator B, whose output is then the PWM representation of the analog input voltage. The higher the analog input voltage, the wider the output pulse. The time averaged output level is thus proportional to the analog input voltage. This binary output can then be fed into power transistors with direct control over motor or speaker winding current, for example, with their inherent lowpass characteristic. Care must be taken to avoid cross conduction in the output power transistors.

The linearity of the pulse-width modulated signal can easily be ascertained by putting a simple 2-pole RC filter at the output (as shown in Figure 431.2). This demodulates the signal which can then be viewed and compared with the original input signal on an oscilloscope. Using a spectrum analyzer and a 1kHz reference signal, this circuit's distortion products were measured as better than −50dBc (0.3%) to about $3.5V_{P-P}$, degrading to −30dBc (3%) as the circuit clips at $5V_{P-P}$ on a single 5V supply.

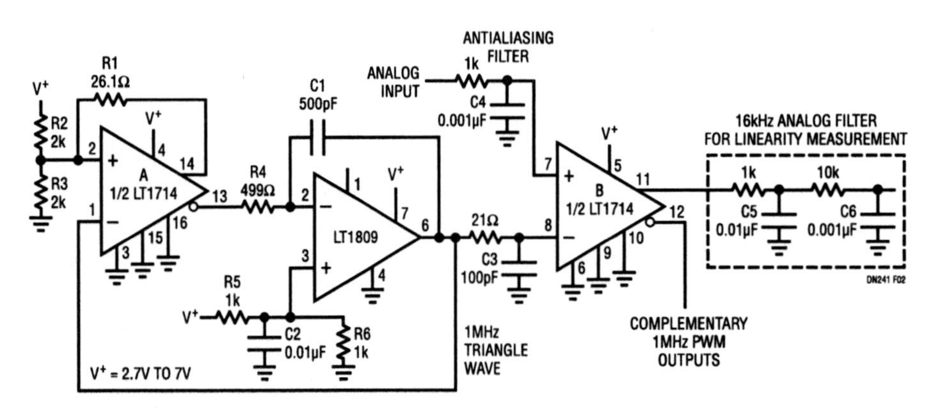

Figure 431.2 • Rail-to-Rail 1MHz Pulse-Width Modulator

Rail-to-rail amplifiers operate on 2.7V with 20μV offset

Glen Brisebois

432

The LT1677 and LT1884 are the latest results of Linear Technology's quest for the "ideal" op amp.[1] Both of them will operate with supplies down to 2.7V, have only 20μV of input offset voltage and have rail-to-rail outputs. The LT1677 features very low noise, $3.2\text{nV}/\sqrt{\text{Hz}}$; the LT1884 features very low 150pA input bias current. Each of the two application circuits shown below take advantage of some particular subset of the features of these amplifiers.

Remote 2-wire Geophone preamp using the low noise LT1677

The LT1677 is optimized for lowest overall noise when looking into transducers of 600Ω to 2700Ω impedance, such as the Geophone shown in Figure 432.1. A low noise amplifier is desired in this application because the seismic signals that must be resolved, whether natural or man made, are extremely small and require high gain. To complicate matters,

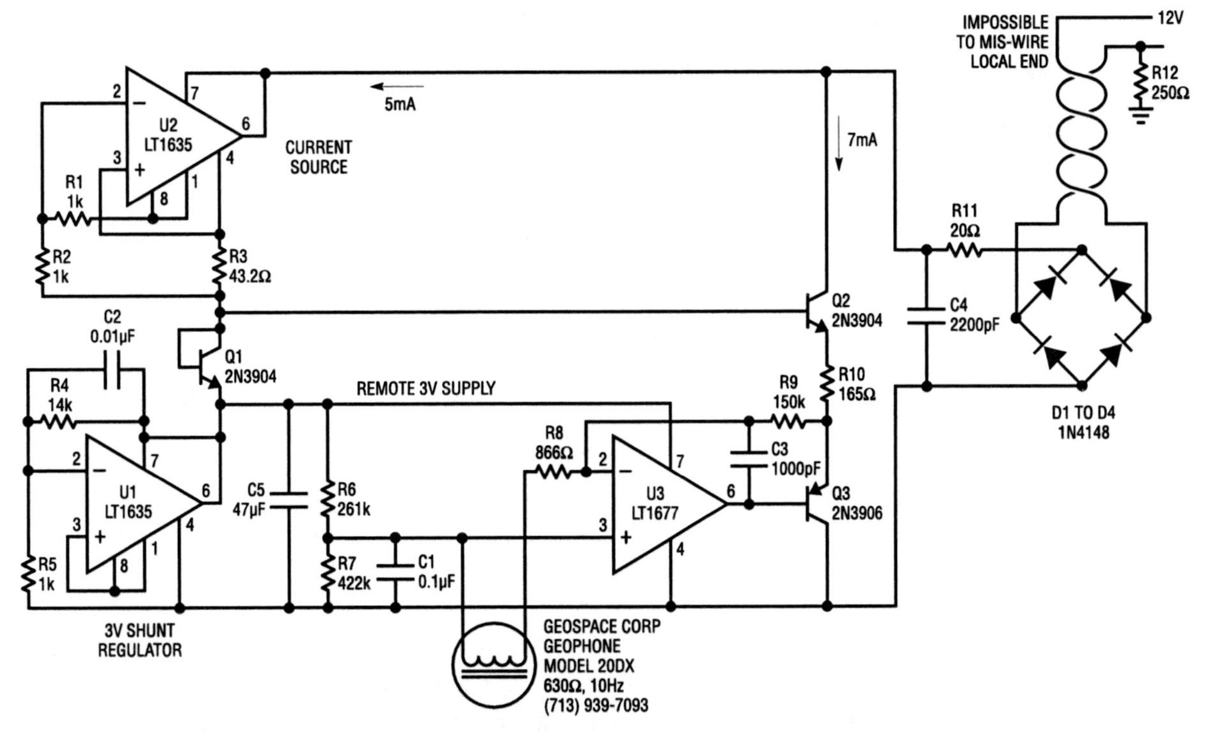

Figure 432.1 • Remote 2-Wire Geophone Preamp

Note 1: Not only would the ideal op amp have zero noise, zero input offset, no parasitic capacitances, infinite gain and bandwidth and supply its own power, but it would do all this for free.

Analog Circuit Design: Design Note Collection. http://dx.doi.org/10.1016/B978-0-12-800001-4.00432-4

Geophones are often buried in order to avoid interference from traffic and other surface effects and so are often necessarily remote.

The circuit in Figure 432.1 applies a gain of ~100 to the Geophone signal and transmits this back to the operator by modulating its own supply current. U2 is an LT1635 configured as a stable current source of 5mA. This then powers the LT1677 as well as another LT1635, this time configured as a 3V shunt regulator. Resistors R6 and R7 set up a DC bias voltage of 1.85V, centering the output swing offset by Q3 and keeping the LT1677 input common mode in its most precise range.[2] This places about 1.15V across R10 thereby pulling an additional 7mA from the main supply through Q2. It is this 7mA that will be modulated by the AC signal. The total current of about 12mA puts 3V across the receiver resistor, with the 7mA allowing a peak signal of ±1.5V about the 3V bias point.

The circuit operates as a current loop and so has good interference immunity, with interference appearing across U2 and Q2 rather than across R12. Q1 temperature compensates Q2. C1 causes a boost in the gain below 10Hz where the Geophone response is falling off. C3 limits the bandwidth to 1kHz. D1 through D4 form a bridge rectifier so that the local wiring is arbitrary. The LT1677 could drive R10 directly, but Q3 is used as an output buffer so that the heavy currents do not eat into the LT1677's high open-loop gain.

Difference amplifier using the LT1884: ±42V CM input range on a single 5V supply without sacrificing differential gain

Measuring small voltages on top of large voltages can be quite difficult. Often, the standard difference amplifier topology is implemented with very high value input resistors

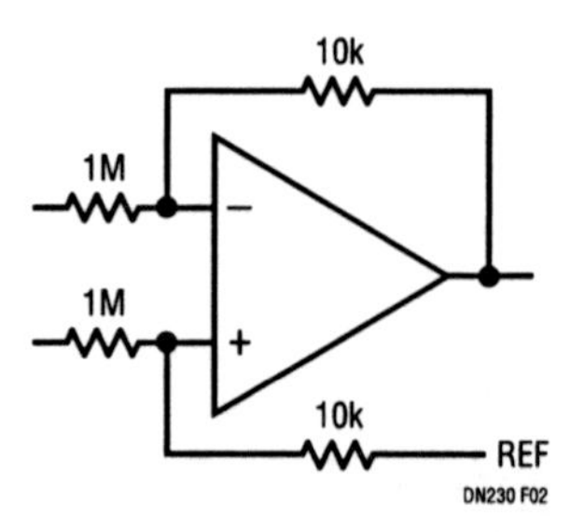

Figure 432.2 • Standard Difference Amplifier Can Handle High Voltage CM Inputs, but at Cost of Differential Gain

and low value divide and feedback resistors, as shown in Figure 432.2. However, this results in significant differential mode attenuation.

The circuit in Figure 432.3 uses an LT1884 to achieve high common mode input range and rejection without sacrificing differential gain. U1B samples the common mode through R5 and R6 and nulls it through R3 and R4. The R3-R1 ratio must be extremely well matched to the R4-R2 ratio to avoid causing a common mode to differential mode translation at this point. Once the common mode is nulled, then the differential mode input voltage is converted to a differential input current and appears unattenuated across R7. The common mode input voltage can theoretically be as high as about 250V (limited by the output of U1B going to ground and the ÷100 ratio maintaining common mode at 2.5V), but is limited in fact by the working voltage of R1 and R2 and by the ratio matching of R1-R3 and R2-R4.

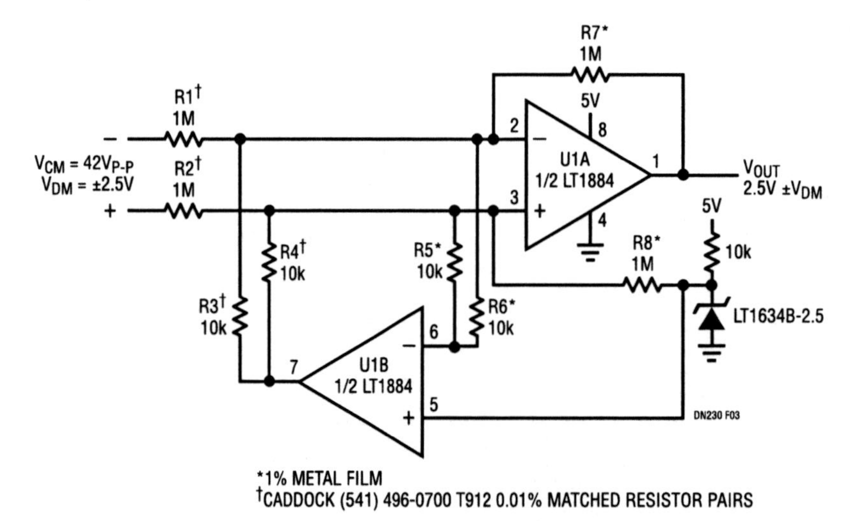

Figure 432.3 • Single Supply Difference Amplifier. U1B Nulls the Common Mode So That U1A Can Concentrate on the Difference Mode

Note 2: Note that the LT1677 has rail-to-rail inputs.

Single resistor sets the gain of the best instrumentation amplifier

433

Alexander Strong Kevin R. Hoskins

Introduction

Linear Technology's next generation LT1167 instrumentation amplifier uses a single resistor to set gains from 1 to 10,000. The single gain-set resistor eliminates expensive resistor arrays and improves V_{OS} and CMRR performance. Careful attention to circuit design and layout, combined with laser trimming, greatly enhances the CMRR, PSRR, gain error and nonlinearity, maximizing application versatility. The CMRR is guaranteed to be greater than 90dB when the LT1167's gain is set at 1. Total input offset voltage (V_{OS}) is less than 60μV at a gain of 10. For gains in the range of 1 to 100, gain error is less than 0.05%, making the gain-set resistor tolerance the dominant source of gain error. The LT1167's gain nonlinearity is unsurpassed when compared to other monolithic solutions. It is specified at less than 40ppm when operating at a gain of 1000 while driving a 2kΩ load. The LT1167 is so robust that it can drive 600Ω loads without a significant linearity penalty. These parametric improvements result in an overall gain error that remains unchanged over the entire input common mode range and is not degraded by supply perturbations or varying load conditions. The LT1167 can operate over a wide ±2.3V to ±18V supply voltage range with only 0.9mA supply current. The LT1167 is offered in 8-pin PDIP and SO packages, saving significant board space compared to multi-op amp designs.

As shown in Figure 433.1, the LT1167's gain is set by the value of one external resistor. A single 0.1% precision resistor sets the gain from 1 to 10, resulting in better than 0.14% accuracy. At very high gains (≥1000), the error is less than 0.2% when using 0.1% precision resistors.

Low input bias current and noise voltage

The LT1167 combines the pA input bias current of FET input amplifiers with the low input noise voltage characteristic of bipolar amplifiers. Using superbeta input transistors, the LT1167's input bias current is only 350pA maximum at room temperature. The LT1167's low input bias current, unlike that of JFET input op amps, does not double for every 10°C. The bias current is guaranteed to be less then 800pA at 85°C. The low noise voltage of 7.5nV√Hz at 1kHz is achieved by idling a large portion of the 0.9mA supply current in the input stage.

Input protection

The inputs of the LT1167 feature low leakage internal protection diodes connected between each input and the supply pins. Their leakage is so low that they do not compromise the low 350pA input bias current. These diodes are rated at 20mA when input voltages exceed the supply rails. Precision and

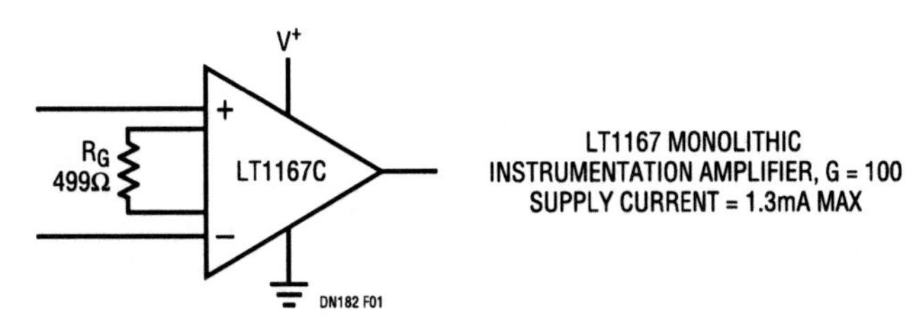

Figure 433.1 • Combining Precision Trimmed Internal Resistors with a Single External Resistor Sets the LT1167 Gain with High Accuracy

Analog Circuit Design: Design Note Collection. http://dx.doi.org/10.1016/B978-0-12-800001-4.00433-6

indestructibility are combined when an external 20k resistor is placed in series with each input. There is little offset voltage penalty because the 320pA offset current from the LT1167 multiplied by the 20k input resistors contributes less than 7μV additional offset. With the 20k resistors, the LT1167 can handle both ±400VDC input faults and ESD spikes over 4kV. This passes the IEC 1000-4-2 level 2 specification.

ADC signal conditioning

In many industrial systems, differential inputs are used to eliminate ground loops and reject noise on long lines. The LT1167 is shown in Figure 433.2 changing a differential signal into a single-ended signal. The single-ended signal is then filtered with a passive 1st order RC lowpass filter and applied to the LTC1400 12-bit analog-to-digital converter (ADC). The LT1167's output stage can easily drive the ADC's small nominal input capacitance, preserving signal integrity. Figure 433.3 shows two FFTs of the amplifier/ADC's output. Figures 433.3a and b show the results of operating the LT1167 at unity gain and a gain of 10, respectively. In both cases, the typical SINAD is 70.6dB.

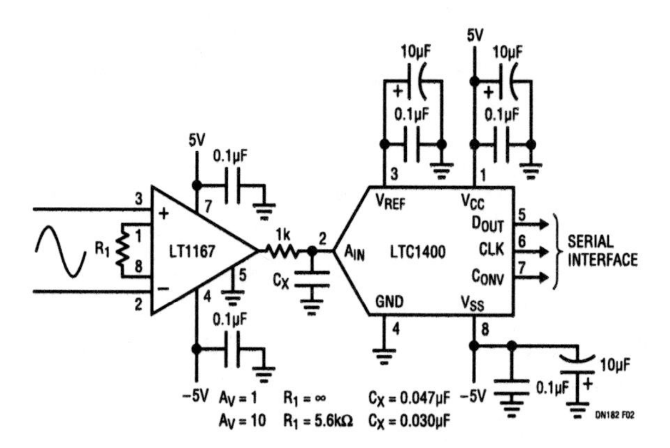

Figure 433.2 • The LT1167's Dynamic Performance Allows It to Convert Differential Signals into Single-Ended Signals for the 12-Bit LTC1400

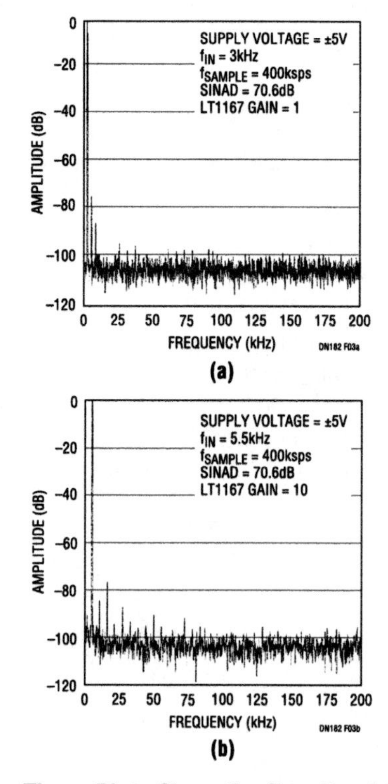

Figure 433.3 • These Plots Show the Results of Operating the LT1167 at Unity Gain (a) and a Gain of 10 (b), Respectively. Each Indicates a Typical SINAD of 70.6dB

Current source

Figure 433.4 shows a simple, accurate, low power programmable current source. The differential voltage across pins 2 and 3 is mirrored across R_G. The voltage across R_G is amplified and applied across R1, defining the output current. For example, opening R_G and setting R1 equal to 1M sets the output current range from 30pA to 10μA for an input voltage of 0V to 10V. The bottom of the range is limited by circuit noise. The circuit can be operated as a current source or sink by applying a positive or negative differential voltage, respectively. The 50μA bias current flowing from the REF pin (Pin 5) is buffered by the LT1464 JFET operational amplifier, increasing the resolution of the current source to 3pA.

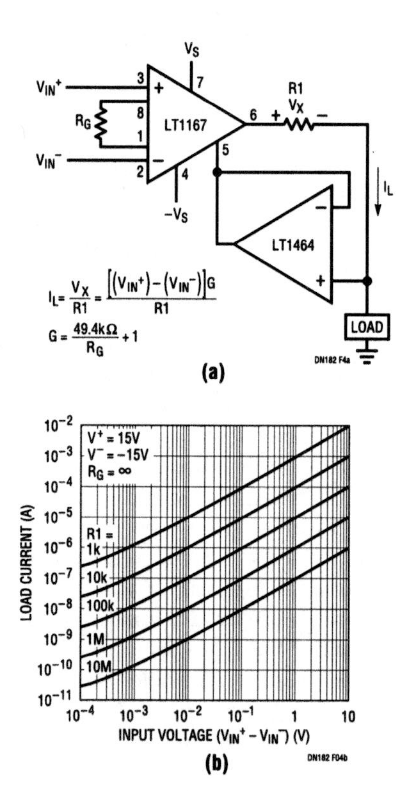

Figure 433.4 • (a) A Simple, Accurate, 3pA Resolution, Low Power Programmable Current Source and (b) Current Source's Output Is Linear Over 8 Decades

Maximize dynamic range with micropower rail-to-rail op amp

434

William Jett

Rail-to-rail amplifiers present an attractive solution for signal conditioning. For battery-powered or other low voltage circuitry, the entire supply voltage can be used by both the input and output signals, maximizing the system's dynamic range. Circuits that require signal sensing near either supply are straightforward to implement using rail-to-rail amplifiers. The LT1466L dual and LT1467L quad combine rail-to-rail input and output operation with precision specifications. Requiring only 75μA of supply current, the LT1466L/LT1467L features a maximum offset of 390μV. Unlike other rail-to-rail amplifiers, the input offset voltage of 390μV maximum is guaranteed across the entire rail-to-rail input range, not just at half supply. The resulting common mode rejection of 83dB minimum is much better than that of other rail-to-rail amplifiers. A minimum open-loop gain of 400V/mV into a 10k load virtually eliminates all gain error.

The following circuits demonstrate the LT1466L's rail-to-rail performance.

Variable current source

The current source shown in Figure 434.1 provides a near 0mA to 50mA output for a 0V to 2.5V control signal. Op amp A1 sets up a variable reference voltage referred to the positive supply; op amp A2 forces the voltage across the sense resistor (R3) to be equal to this voltage. Compliance of the current source is set by the supply voltage; the circuit will operate with voltages from 4V to 16V. The output resistance of the current source is greater than 10V/μA at full scale (50mA). Full-scale accuracy is set primarily by the resistor ratios. The V_{OS} of op amp A2 introduces a maximum error of 80μA or 0.16% of full scale.

High side current sense amplifier

The current sense amplifier shown in Figure 434. 2 amplifies the voltage across a small value sense resistor by the ratio of the current source resistors (R2/R1). The LT1466L controls the low power MOSFET's gate voltage so that the sense voltage appears across current source resistor R1. The resulting current in Q1's drain is converted to a ground-referred voltage at R2. With the values shown, the output can be used to drive an ADC without additional buffering. Conversion gain is 2V/A. Resistor tolerances determine the gain accuracy; the V_{OS} of op amp A1 introduces an error of V_{OS}/R_S (2mA maximum with $R_S=0.2$).

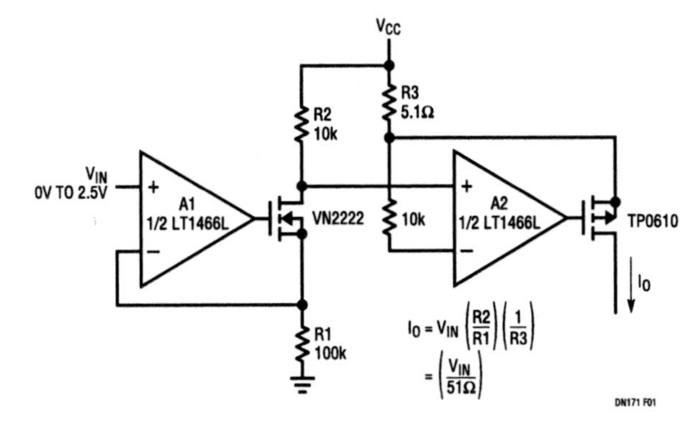

Figure 434.1 • Variable Current Source

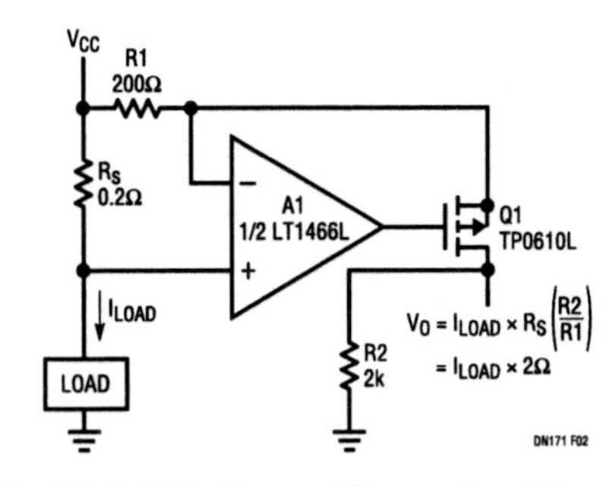

Figure 434.2 • High Side Current Sense Amplifier

Analog Circuit Design: Design Note Collection. http://dx.doi.org/10.1016/B978-0-12-800001-4.00434-8

3.3V, 1kHz, 4th order Butterworth filter

The 4th order Butterworth filter shown in Figure 434.3 operates from supplies as low as 3V and swings rail-to-rail. The circuit has good DC accuracy and low sensitivities for the center frequency and Q. For amplifiers A1 and A3, the common mode voltage is equal to the input voltage, whereas amplifiers A2 and A4 operate in the inverting mode. Component values can be found from the following equations:

$$\omega_0^2 = 1/(R1 \cdot C1 \cdot R2 \cdot C2)$$

where:

$$R1 = 1/(\omega_0 \cdot Q \cdot C1) \text{ and } R2 = Q/(\omega_0 \cdot C2)$$

The DC bias applied to A2 and A4, half supply, is not needed when split supplies are available. The maximum output error is $2 \cdot V_{OS} + I_{OS} \cdot 42k \leq 930\mu V$. Total harmonic distortion (THD) with $V_{IN} = 1V_{RMS}$ and $V_S = 3.3V$ is 0.01% at 100Hz, rising to 0.045% at 1kHz. Figure 434.4 shows the resulting frequency response.

Picoampere input current instrumentation amplifier

The instrumentation amplifier shown in Figure 434.5 features a typical input bias current of 200pA and includes a shield driver. Amplifiers A1A, A1D and A1C form a traditional three op amp configuration, and amplifier A1B both buffers the common mode voltage and cancels the input bias current of A1A and A1D. Input common mode range extends to within 300mV to 400mV of either supply.

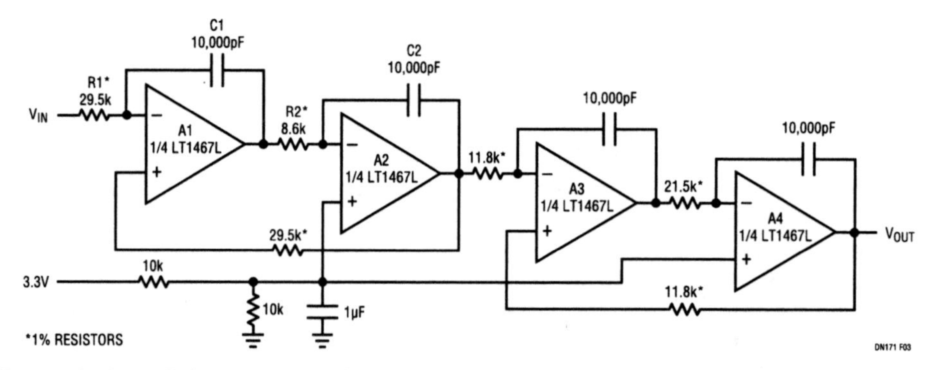

Figure 434.3 • 4-Pole, 1kHz, 3.3V Single-Supply State Variable Filter Using the LT1467L

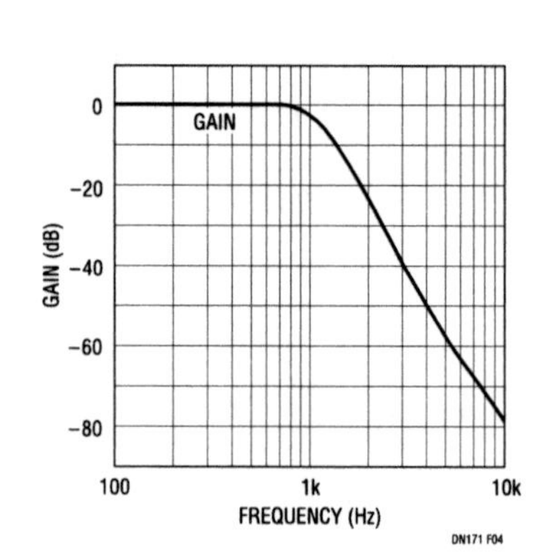

Figure 434.4 • Frequency Response of 4th Order Butterworth Filter

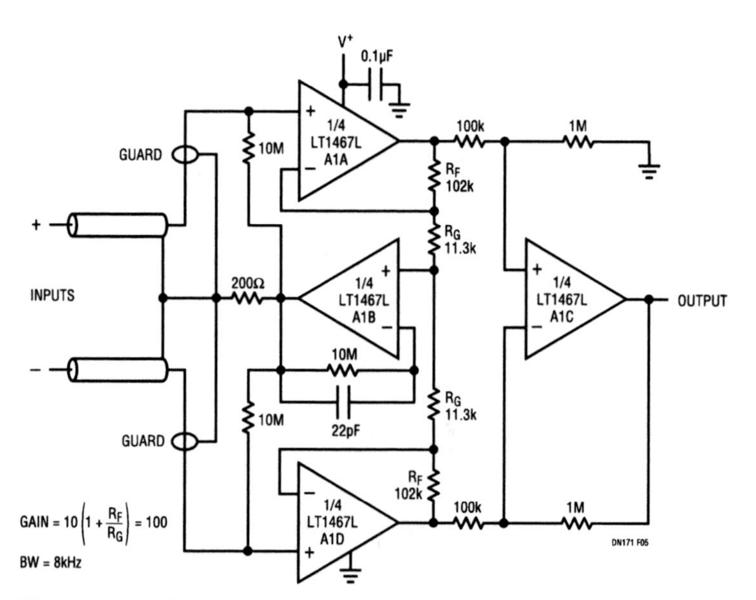

Figure 434.5 • Instrumentation Amplifier

1μA op amp permits precision portable circuitry

Mitchell Lee Jim Williams

A new dual op amp with only 1μA power consumption and precision DC specifications permits high performance portable applications. The LT1495 has 375μV offset, 2μV/°C drift, 1nA bias current and 100dB of open-loop gain. These attributes, combined with careful design, make portable, high performance circuitry possible.

5.5μA, 0.05μV/°C chopped amplifier

Figure 435.1 shows a chopped amplifier requiring only 5.5μA supply current. Offset voltage is 5μV, with 0.05μV/°C drift. Gain exceeding 10^8 affords high accuracy, even at large closed-loop gains.

Micropower comparators C1A and C1B form a biphase 5Hz clock. The clock drives the input-related switches, causing an amplitude modulated version of the DC input to appear at A1A's input. AC-coupled A1A takes a gain of 1000, presenting its output to a switched demodulator similar to the aforementioned modulator.

The demodulator output, a reconstructed, DC amplified version of the circuit's input, feeds DC gain stage A1B. A1B's output is fed back, via gain setting resistors, to the input modulator, closing a feedback loop around the entire amplifier. Amplifier gain is set by the feedback resistor's ratio, in this case 1000.

The circuit's internal AC coupling prevents A1's DC characteristics from influencing overall DC performance, accounting for the extremely low offset errors noted.

The desired micropower operation and A1's bandwidth dictate the 5Hz clock rate. As such, resultant overall bandwidth is low. Full power bandwidth is 0.05Hz with a slew rate of about 1V/s. Clock related noise, about 5μV, can be reduced by increasing C_{COMP}, with commensurate bandwidth reduction.

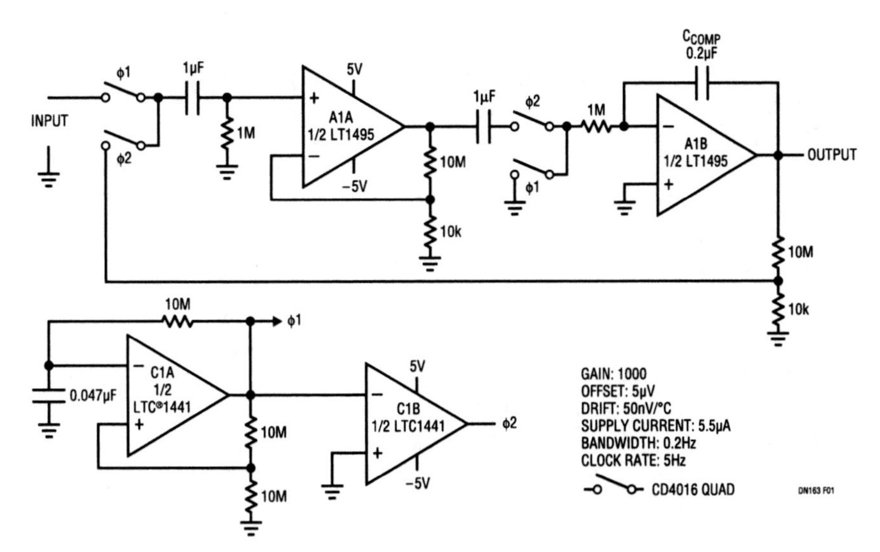

Figure 435.1 • 0.05μV/°C Chopped Amplifier Consumes 5.5μA Supply Current

Analog Circuit Design: Design Note Collection. http://dx.doi.org/10.1016/B978-0-12-800001-4.00435-X

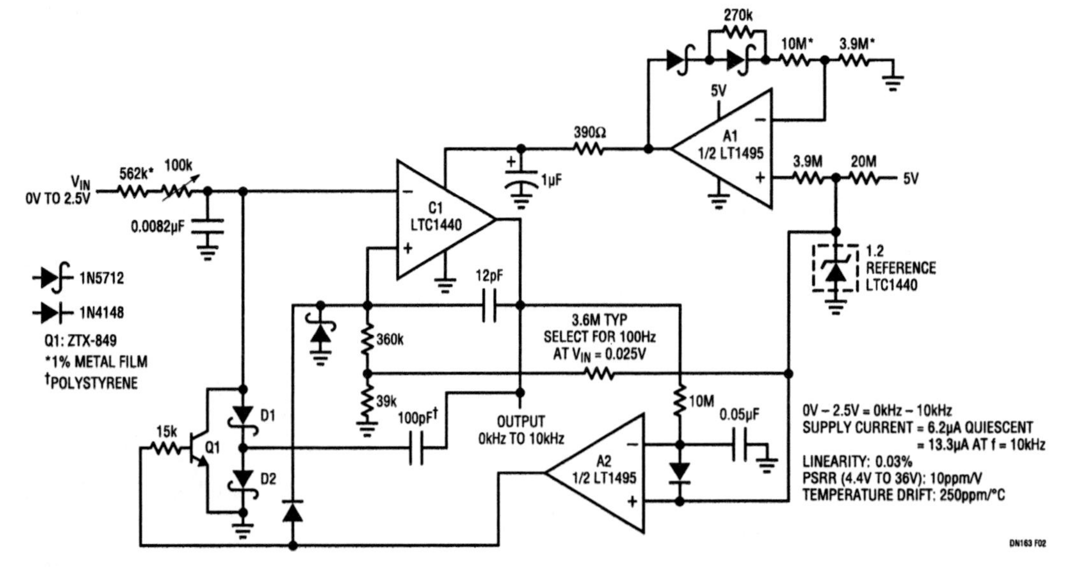

Figure 435.2 • 0kHz to 10kHz Voltage to Frequency Converter Consumes Only 13μA

0.03% linear V/F converter with 13μA power drain

Figure 435.2's voltage-to-frequency converter takes full advantage of the LT1495's low power consumption. A 0V to 2.5V input produces a 0Hz to 10kHz output, with 0.03% linearity, 250ppm/°C drift and 10ppm/V supply rejection. Maximum current consumption is only 13μA, 200 times lower than currently available ICs. Comparator C1 switches a charge pump comprising D1, D2 and the 100pF capacitor to maintain its negative input at 0V. A1 and associated components form a temperature compensating reference for the charge pump. The 100pF capacitor charges to a fixed voltage; hence, the switching repetition rate is the circuit's only degree of freedom to maintain feedback. Comparator C1 pumps uniform packets of charge to its negative input at a repetition rate precisely proportional to the input voltage derived current. This action ensures that circuit output frequency is strictly and solely determined by the input voltage.

Start-up or input overdrive can cause the circuit's AC-coupled feedback to latch. If this occurs, C1's output goes low; A2, detecting this via the 10M/0.05μF lag, goes high. This lifts C1's positive input and grounds the negative input with Q1, initiating normal circuit action.

Portable reference

A final circuit is Figure 435.3's unique portable reference, which draws only 16μA from a pair of AAA alkaline cells. Battery life is five years—equivalent to shelf life.

Two outputs are provided: a buffered, 1.5V voltage output and a regulated 1.5μA current source. The current source compliance ranges from approximately 1V to −43V.

The LT1634A reference is self-biased, completely eliminating line regulation as a concern. Start-up is guaranteed by the LT1495 op amp, whose output initially saturates at 11mV from the negative rail. The 1μA current output is derived from a fraction of the reference voltage impressed across R3.

Note that the portable reference's current output can be pulled well below common, limited only by Q1's 45V breakdown. The 1.5V output can source or sink up to 700μA and is current limited to protect batteries in case of a short circuit.

Once it is powered, there is no reason to turn the circuit off. One AAA alkaline contains 1200mAH capacity, enough to power the circuit through the five year shelf life of the battery.

The voltage output accuracy is about 0.17% and the current output accuracy is about 1.2%. Trim R1 to calibrate voltage (0.1%/kΩ) and R3 to calibrate the output current (0.4%/kΩ).

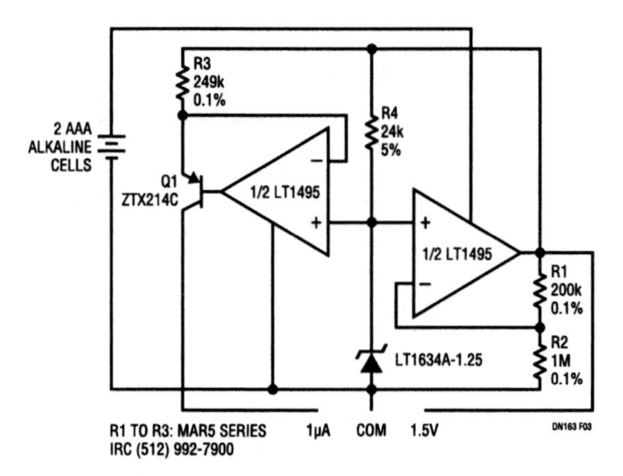

Figure 435.3 • Portable Reference Operates Five Years on One Pair of AAA Cells

Low power, fast op amps have low distortion

436

George Feliz

Introduction

The LT1351/LT1352/LT1353 family of low power operational amplifiers combines a slew rate of 200V/μs with a supply current of 250μA per amplifier. Both input and output stages have been optimized for linearity, achieving outstanding distortion performance with miserly quiescent current. The amplifier is available in single, dual and quad versions, in various packages, including the tiny MSOP package for the LT1351 single amplifier. A summary of key specifications is shown in Table 436.1.

Table 436.1 Key Performance Features of the LT1351/LT1352/LT1353

Feature	Value
Power Supply Range	±2.5V to ±15V
Supply Current (per Amplifer)	250μA
Shutdown Current (LT1351)	10μA
Slew Rate (±15V Supplies) (±5V Supplies)	200V/μs 50V/μs
Gain Bandwidth	3MHz
C-Load Amplifiers Stable	All Capacitive Loads
Maximum Input Offset Voltage	600μV
Maximum Input Bias Current	50nA
Minimum DC Gain, $R_L = 2k$	30V/mV
Input Noise Voltage	14nV/√Hz
Packages: 8-Lead MSOP 8-Lead SO, PDIP 14-Lead SO	LT1351 LT1351,LT1352 LT1353

Buffering data acquisition systems

A low power data acquisition system using the LT1351 as a buffer is shown in Figure 436.1. The LTC1274 is a 12-bit, 2mA, 100ksps, ±2.048V full-scale input ADC. Its input

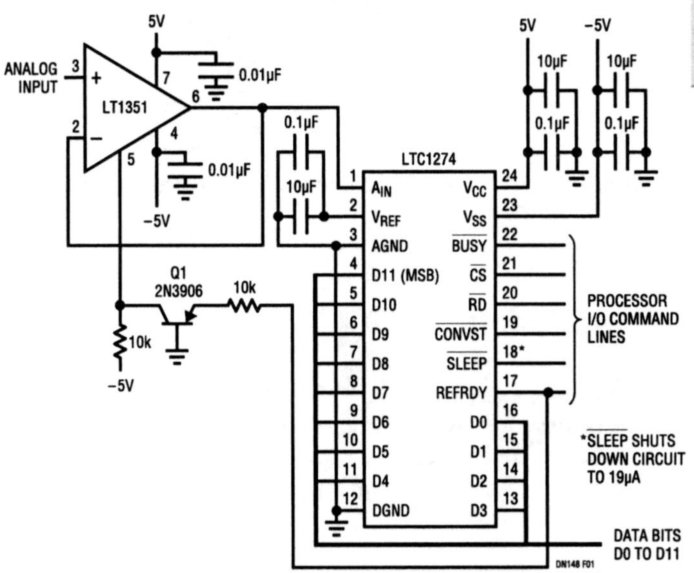

Figure 436.1 • LT1351 as a Buffer for LTC1274 100ksps ADC

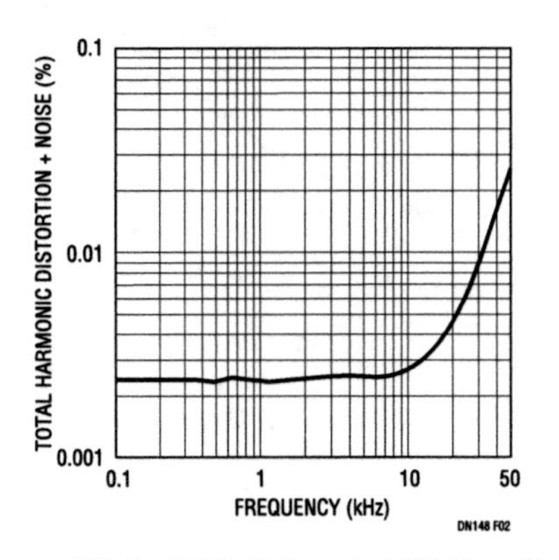

Figure 436.2 • LT1351 $A_V = 1$, ±5V Supplies, $V_{OUT} = 4V_{P-P}$, $R_L = 10k$

Analog Circuit Design: Design Note Collection. http://dx.doi.org/10.1016/B978-0-12-800001-4.00436-1

at Pin 1 can be modeled as a 200Ω switch connected to a 45pF sample cap. This light load presents no problem for the LT1351, which will pass full-scale, 12-bit accurate signals up to 40kHz. Figure 436.2 shows the total harmonic distortion plus noise of the LT1351 configured as a unity-gain buffer driving $4V_{P-P}$ into 10kΩ on ±5V supplies. The buffer settles in under 1.5μs to less than 1mV for a 4V step, thus ensuring acquisition in 2μs. Additionally, the circuit exploits the shutdown feature of the LT1351 to reduce the total supply current to a mere 19μA when Pin 18 is pulled low. Pin 17 signals that the internal ADC voltage reference is valid. When it is high, the amplifier is ready for conversions. Transistor Q1 provides a level shift so that Pin 17 can control the Shutdown pin of the LT1351, which is turned off when its Pin 5 is pulled to V_{EE} (when Pin 17 is low) and is on when Pin 5 is 2V or more above V_{EE} (when Pin 17 is high).

Filters

For large signals, the slew rate of the LT1352 amplifier passes undistorted signals even with a stingy amount of supply current. The 20kHz, 4th order Butterworth filter shown in Figure 436.3 showcases this large-signal performance. The configuration is a standard textbook filter, but the large-signal distortion in Figure 436.4 shows that the $20V_{P-P}$ signals remain below 0.02% THD throughout the passband.

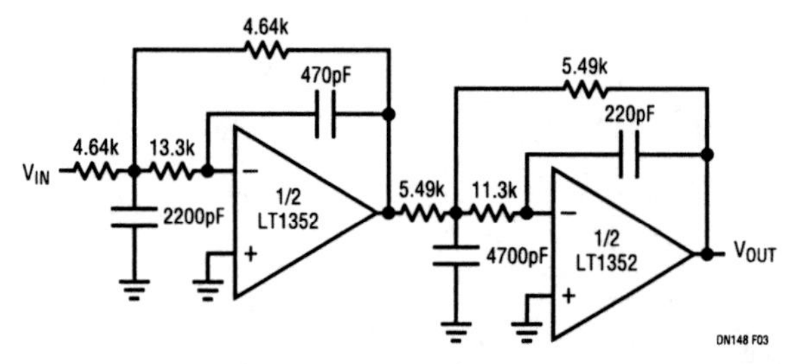

Figure 436.3 • 20kHz, 4th Order Butterworth Filter

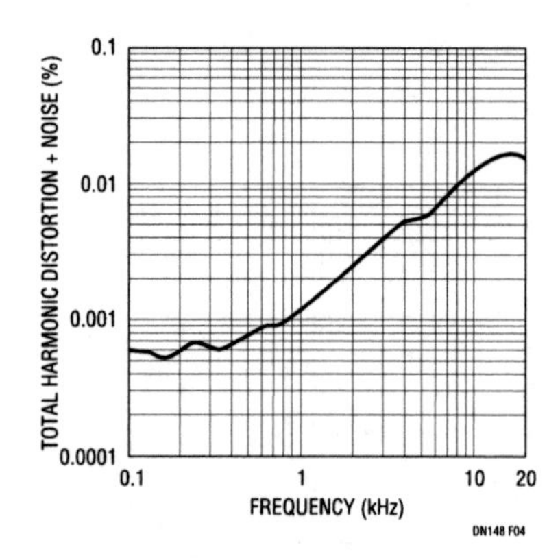

Figure 436.4 • LT1352 20kHz, 4th Order Butterworth Filter ±15V Supplies, $20V_{P-P}$ Signals

This measurement is extraordinary, considering that the circuit draws a mere 500μA of quiescent current.

A two op amp instrumentation amplifier

The two op amp instrumentation amplifier shown in Figure 436.5 has a gain of 102 and a bandwidth of 30kHz. The circuit uses a combination of inversion and summation to cancel the common mode component at the two inputs. Differential gain can be analyzed by calculating the gain from each input with the other input grounded and adding the gains. Figure 436.6 is a plot of total harmonic distortion plus noise for various output levels. The noise of this configuration is a low $370\mu V_{RMS}$ at the output with an 80kHz measurement filter. At output levels below $2.5V_{P-P}$ ($884mV_{RMS}$), noise is the limiting factor in performance. This excellent blend of noise performance and bandwidth is unmatched at this power level.

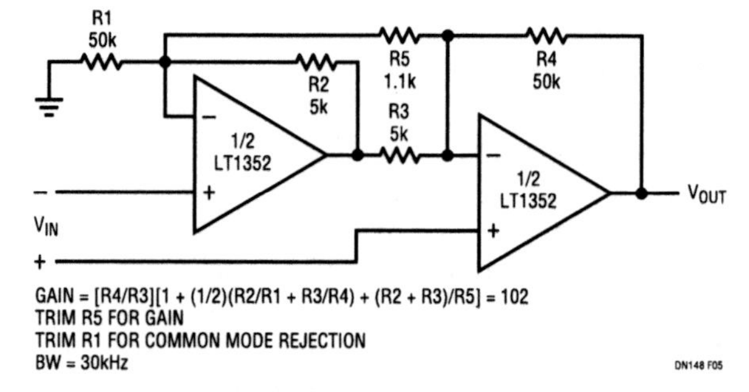

Figure 436.5 • Instrumentation Amplifier

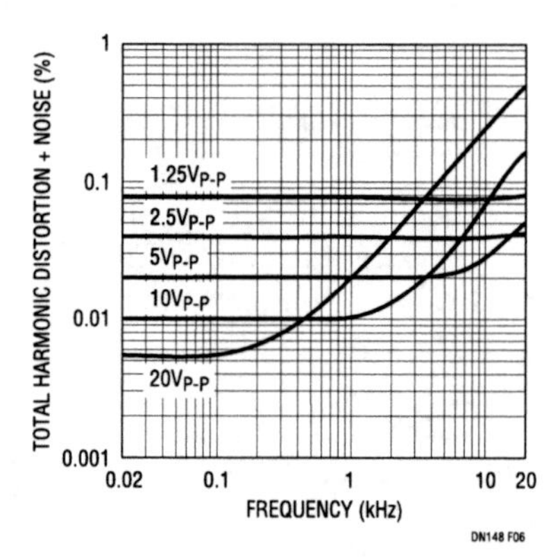

Figure 436.6 • LT1352 2-Amplifier, A_V = 100 Instrumentation Amp, Varying Output Levels

Conclusion

In summary, the LT1351/LT1352/LT1353 family of amplifiers provides low power solutions for low distortion, low noise applications. Even though supply current is only 250μA per amplifier, large-signal performance is outstanding.

Operational amplifier selection guide for optimum noise performance

437

Frank Cox

Eight years ago, George Erdi wrote a very useful Design Note (Chapter 457) that presented information to aid in the selection of op amps for optimum noise performance, in both graphical and tabular form. This chapter is an update of that Design Note. It covers new low noise op amps as well as some high speed op amps. Although a great deal has changed in eight years, especially in electronics, noise is still a critical issue in op amp circuit design and the LT1028 is still the lowest noise op amp for low source impedance applications.

The amount of noise an op amp circuit will produce is determined by the device used, the total resistance in the circuit, the bandwidth of the measurement, the temperature of the circuit and the gain of the circuit. A convenient figure of merit for the noise performance of an op amp is the spectral density or spot noise. This is obtained by normalizing the measurement to a unit of bandwidth. Here the unit is 1Hz and the noise is reported as "$nV/\sqrt{Hz}$." The noise in a particular application bandwidth can be calculated by multiplying the spot noise by the square root of the application bandwidth.

Some other simplifications are made to facilitate comparison. For instance, the noise is referred to the input of the circuit so that the effect of the circuit gain, which will vary with application, does not confuse the issue. Also, the calculations assume a temperature of 27°C or 300K.

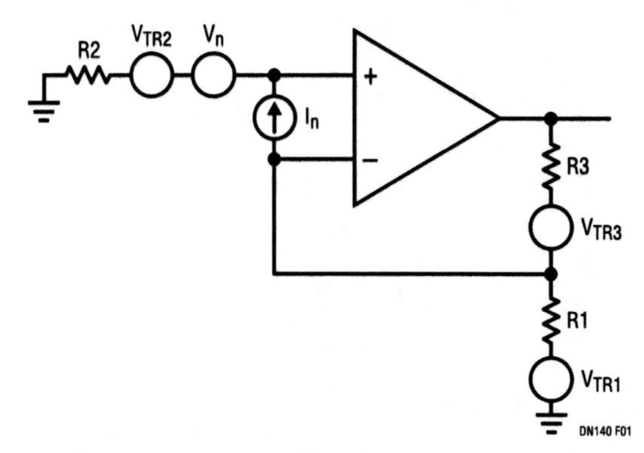

WHERE: V_{TR1}, V_{TR2} and V_{TR3} ARE THERMAL NOISE FROM RESISTORS

$$R_{eq} = R2 + \left(\frac{(R1)(R3)}{R1 + R3}\right)$$

$$4kT = (16.56)(10)^{-21}\ J$$

AND V_n IS THE VOLTAGE SPOT NOISE AND I_n IS THE CURRENT SPOT NOISE OF THE OP AMP AS GIVEN ON THE DATA SHEET.

$$V = \sqrt{(4kT)R_{eq} + V_n^2 + I_n^2(R_{eq}^2)}$$

IS THE INPUT REFERRED SPOT NOISE IN A 1Hz BANDWIDTH.

Figure 437.1 • Circuit Schematic and Formula for Calculating the Spot Noise

The formula used to calculate the spot noise and the schematic of the circuit used are shown in Figure 437.1. Figures 437.2–437.4 plot the spot noise of selected op amps vs the equivalent source resistance. The first two plots show precision op amps intended for low frequency applications, whereas the last plot shows high speed voltage-feedback op amps. There are two plots for the low frequency op amps because at very low frequencies (less than about 200Hz) an additional noise mechanism, which is inversely proportional to frequency, becomes important. This is called 1/f or flicker noise. Figure 437.2 shows slightly higher levels of noise due to this contribution.

Studying the formula and the plots leads to several conclusions. The values of the resistors used should be as small as possible to minimize noise, but since the feedback resistor is a load on the output of the op amp, it must not be too small. For a small equivalent source resistance, the voltage noise dominates. As the resistance increases, the resistor noise becomes most important. When the source resistance is greater than 100k, the current noise dominates because the contribution of the current noise is proportional to Req, whereas the resistor noise is proportional to $\sqrt{Req}$.

For low frequency applications and a source resistance greater than 100k, the LT1169 JFET input op amp is the obvious choice. Not only does the LT1169 have an extremely low current noise of $0.8fA/\sqrt{Hz}$, it also has a very low voltage noise of $6nV/\sqrt{Hz}$. The LT1169 also has excellent DC specifications, with a very low input bias current of 3pA

Analog Circuit Design: Design Note Collection. http://dx.doi.org/10.1016/B978-0-12-800001-4.00437-3

(typical), which is maintained over the input common mode range, and a high gain of 120dB.

High speed op amps, here defined by slew rates greater than 100V/μs, are plotted in Figure 437.4. These op amps come in a wider range of speeds than the precision op amps plotted in Figures 437.2 and 437.3. The faster parts will generally have slightly more spot noise, but because they will most likely be selected on the basis of speed, a selection of parts is plotted. For example, the LT1354–LT1363 (these are single op amps; duals and quads are available) are close in noise performance and consequently cluster close together on the plot, but have a speed range of 12MHz GBW to 70MHz GBW.

The same information is presented in tabular form in Table 437.1.

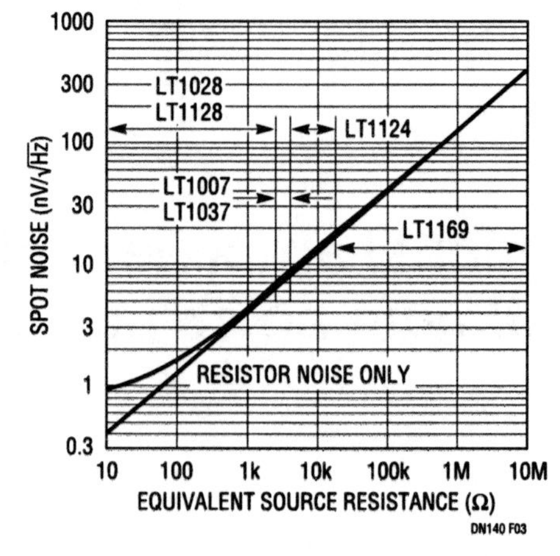

Figure 437.3 • 1kHz Spot Noise vs Equivalent Source Resistance

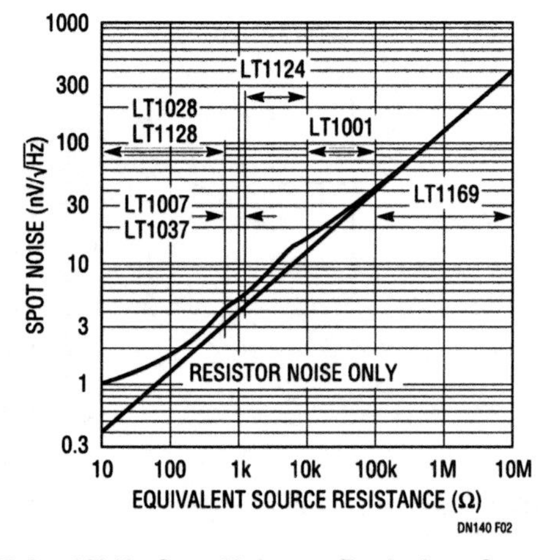

Figure 437.2 • 10kHz Spot Noise vs Equivalent Source Resistance

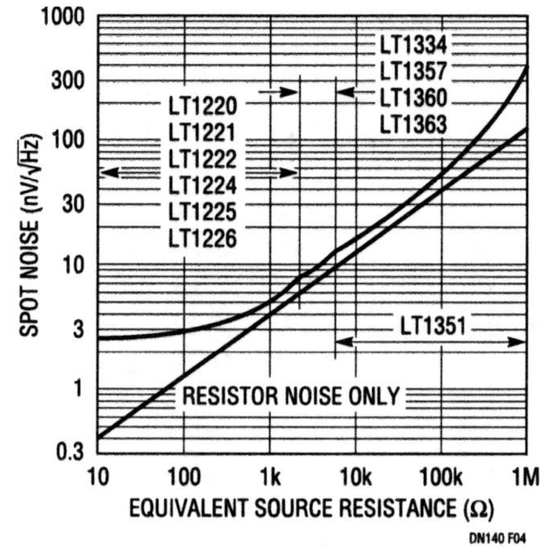

Figure 437.4 • 10kHz Spot Noise vs Equivalent Source Resistance (High Speed Amplifiers)

Table 437.1 Best Op Amp for Lowest Noise vs Source Resistance

SOURCE R (R_{eq})	BEST OP AMP		
	10Hz PRECISION	1000Hz PRECISION	10kHz HIGH SPEED
0Ω to 500Ω	LT1028, LT1115, LT1128	LT1028, LT1115, LT1128	LT1220/21/22/24/25/26
500Ω to 1.5k	LT1007, LT1037	LT1028, LT1115, LT1128	LT1220/21/22/24/25/26
1.5k to 3k	LT1124/25/26/27	LT1028, LT1115, LT1128	LT1220/21/22/24/25/26
3k to 5k	LT1124/25/26/27	LT1007, LT1037	LT1220/21/22/24/25/26
5k to 10k	LT1124/25/26/27	LT1124/25/26/27	LT1354/57/60/63
10k to 20k	LT1001/02	LT1113, LT1124/25/26/27	LT1354/57/60/63
20k to 100k	LT1001/02	LT1055/56/57/58, LT1113, LT1169	LT1351
100k to 1M	LT1022, LT1055/56/57/58, LT1113, LT1122, LT1169	LT1022, LT1055/56/57/58, LT1113, LT1122, LT1169, LT1457	LT1351
1M to 10M	LT1022, LT1055/56/57/58, LT1113, LT1122, LT1169	LT1022, LT1055/56/57/58, LT1113, LT1122, LT1169, LT1457	

Micropower dual and quad JFET op amps feature pA input bias currents and C-Load drive capability

438

Alexander Strong   Kevin R. Hoskins

Introduction

The LT1462/LT1464 duals and the LT1463/LT1465 quads are the first micropower op amps to offer picoampere input bias currents and unity-gain stability when driving capacitive loads. For each amplifier, the LT1462/LT1463 consume only 28μA of supply current; the faster LT1464/LT1465, just 145μA. Low supply current and operation specified at ±5V supplies make these amplifiers appropriate for portable low power applications. The LT1462/3/4/5 family is especially suited for piezo transducer conditioning, strain gauge weight scales, very low droop track-and-holds, wide dynamic range photodiode amplifiers and other applications that benefit from pA input bias current. The LT1462/3/4/5 family also exhibits very low noise current, important to circuits such as low frequency filters. These op amps allow high value resistors to be used with easily obtainable, low value precision capacitors to set a filter's frequency characteristics without compromising noise performance.

Driving large capacitive loads

Though the LT1462/3/4/5 consume very small amounts of supply current, they can easily drive 10nF loads while remaining stable. Instead of increasing their idling current to drive heavy load capacitance, these op amps use a clever compensation technique to lower bandwidth. As load capacitance increases, these op amps automatically reduce their bandwidth by reflecting a portion of the load capacitance back to the gain node, increasing the compensation capacitance. Now instead of a 1MHz op amp trying to drive a large capacitor, a lower bandwidth op amp is able to drive the load capacitance.

Applications

A benefit of the LT1464/LT1465's low power consumption is very low junction leakage current, which in turn, keeps the input bias current below 500fA typically. Track-and-hold

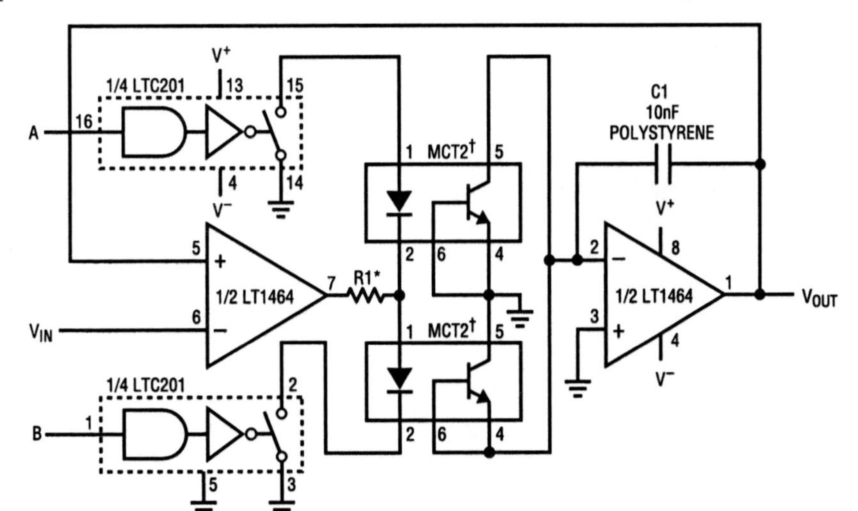

FUNCTION	MODE	IN A	IN B
Track-and-Hold	Track	0	0
	Hold	1	1
Positive Peak Det	Reset	0	0
	Store	0	1
Negative Peak Det	Reset	0	0
	Store	1	0

LTC®201 switch is open for logic "1"

$$\text{TYPICAL DROOP} = \frac{0.5\text{pA}}{10\text{nF}} = 0.05\text{mV/s}$$

TOTAL SUPPLY CURRENT = 460μA MAX

* R1 = 600Ω FOR ±15V SUPPLIES
R1 = 0Ω FOR ±5V SUPPLIES

† MOTOROLA (602) 244-5768

DN136 F01

Figure 438.1 • Low Droop Track-and-Hold/Peak Detector Circuit Takes Advantage of the LT1464's 0.5pA Input Bias Current

Analog Circuit Design: Design Note Collection. http://dx.doi.org/10.1016/B978-0-12-800001-4.00438-5

applications are a natural for this family of op amps. Figure 438.1 is a track-and-hold circuit that uses a low cost optocoupler as a switch. Leakage for these parts is usually in the nA region with 1V to 5V across the output. Since there is less than 0.8mV across the junctions, the leakage is so small that the op amp's I_B dominates. The input signal is buffered by one op amp while the other buffers the stored voltage; this results in a droop of 50μV/s with a 10nF capacitor.

The LT1462/LT1463's low input bias current make it a natural for amplifying low level signals from high impedance transducers. The 1pA input bias current contributes only 0.4 fA/$\sqrt{Hz}$ of current noise or 0.4 nV/$\sqrt{Hz}$ voltage noise with a 1MΩ source impedance. A 1MΩ impedance's thermal noise of 130 nV/$\sqrt{Hz}$ dominates the op amp's noise, showing that even with high source impedances, the LT1462/LT1463 contribute very little to total input-referred noise. Taking advantage of these features, Figure 438.2's photodiode logging amplifier uses two LT1462 duals or an LT1463 quad. Here, the photodiode current is converted to a voltage by the first op amp and D1 and amplified by the first, second and third logarithmic compression amplifiers. A DC feedback path comprising R8, R9, C6 and Q1 is active only for no light conditions. Q1 is off when light is present, isolating the photodiode from C6. When feedback is needed, a small filtered current through R8 prevents the op amp outputs from saturating when no signal is present (see Figure 438.3).

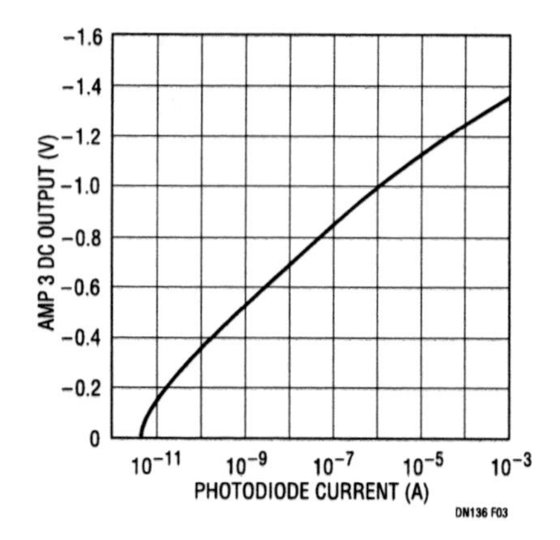

Figure 438.3 • Logging Photodiode Amplifier DC Output

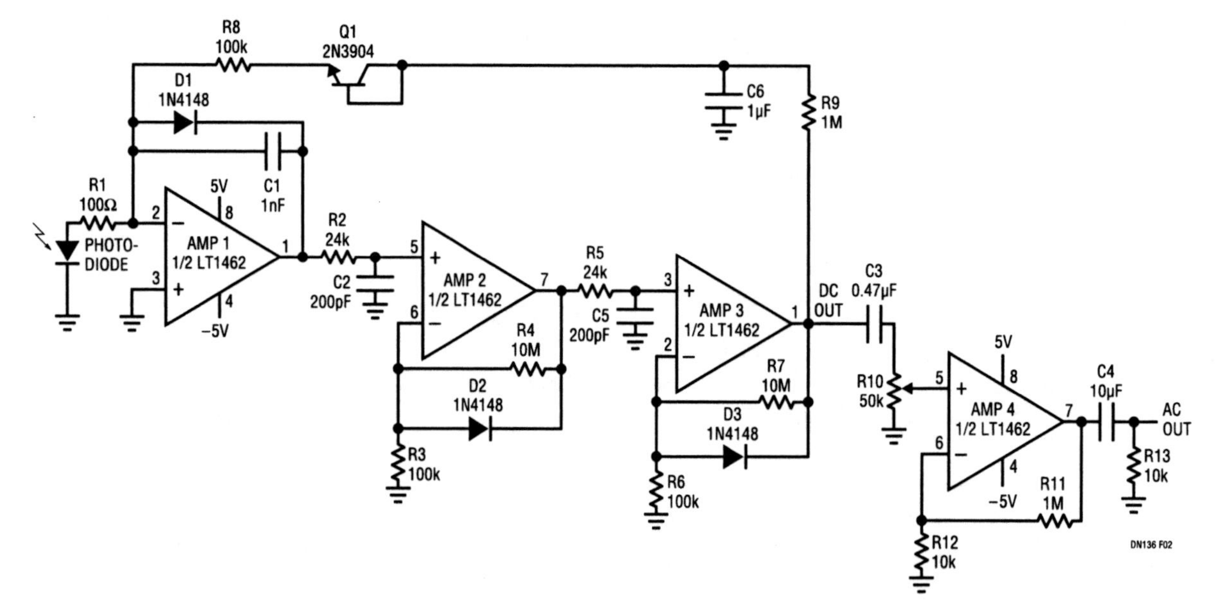

Figure 438.2 • This Logging Photodiode Amplifier Takes Advantage of the LT1462's 1pA Input Bias Current to Amplify the Low Level Signal from the Photodiode's High Source Impedance

Fast current feedback amplifiers tame low impedance loads

439

Sean Gold William Jett

Introduction

Three current feedback amplifiers (CFAs) now available from Linear Technology can considerably ease the task of driving low impedance loads. This design note reviews the capabilities of the LT1206, LT1207 and LT1210 CFAs and addresses some design issues encountered when using them. These CFAs are fast and capable of delivering high levels of current. They can be readily compensated for reactive loads and are fully protected against thermal and short-circuit faults. Table 439.1 summarizes their electrical characteristics.

Driving transformer-coupled loads

Transformer coupling is frequently used to step up transmission line signals. Voltage signals amplified in this way are not constrained by local supply voltages, so the amplifier's rated current rather than its voltage swing usually limits the power delivered to the load. Amplifiers with high output current drive are therefore appropriate for transformer-coupled systems.

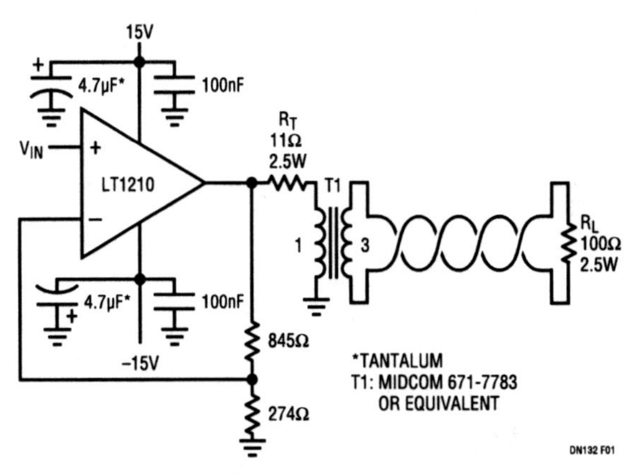

Figure 439.1 • Twisted Pair Driver ADSL. Voltage Gain Is About 6; $5V_{P-P}$ Input Corresponds to Full Output

Figure 439.1 shows a transformer-coupled application for ADSL in which an LT1210 drives a 100Ω twisted pair. The 1:3 transformer turns ratio allows just over 1W to reach the load at full output. Resistor R_T acts as a primary side back-termination and also prevents large DC currents from flowing

Table 439.1 Fast Current Feedback Amplifier Specifications

PART NUMBER	NUMBER OF CFAs	BANDWIDTH (MHz)	RATED OUTPUT CURRENT (A)	SUPPLY RANGE (V)	SLEW RATE (V/µs) (NOTE 1)	THERMAL RESISTANCE θ_{JA} (°C/W) (NOTE 2)	SUPPLY CURRENT	LOW POWER OP/ SHUTDOWN
LT1206	1	60	0.25	±5 to ±15	I_{LIM}/C_{LOAD} to 900	DD = 25, PDIP = 100 SO = 60, TO-220 = 5	20	Yes
LT1207	2	60	0.25	±5 to ±15	I_{LIM}/C_{LOAD} to 900	SO = 40	2 × 20	Yes
LT1210	1	35	1	±5 to ±15	I_{LIM}/C_{LOAD} to 1000	DD = 25, SO = 40 TO-220 = 5	30	Yes

Note 1: Slew rate depends on circuit configuration and capacitive load.
Note 2: θ_{JA} on SO packages measured with part mounted to a 2.5mm thick FR4 2oz copper PC board with 5000mm^2 area.

Analog Circuit Design: Design Note Collection. http://dx.doi.org/10.1016/B978-0-12-800001-4.00439-7

in the coil. The overall frequency response is flat to within 1dB from 500Hz to 2MHz. Distortion products at 1MHz are below −70dBc at a total output power of 0.56W (load plus termination), rising to −56dBc at 2.25W. If R_T is removed, the amplifier will see a load of about 11Ω and the maximum output power will increase to 5W. A DC blocking capacitor should be used in this case.

Bridging can be used to increase the output power transferred to a transformer. Differential operation also promotes the cancellation of even-order distortion. Figure 439.2 shows a differential application using an LT1207 as a bridge driver for HDSL. The dual CFA is configured for a gain of 10, delivering a $10V_{P-P}$ signal to the nominal 35Ω load impedance. The output signal amplitude remains flat over an 8MHz bandwidth.

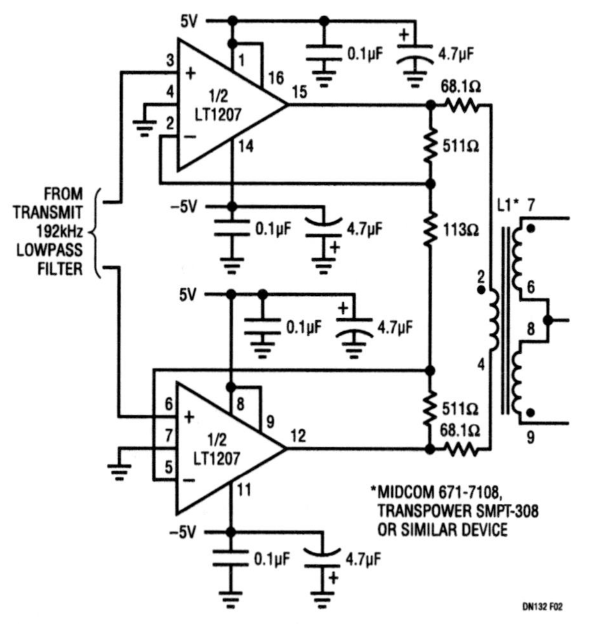

Figure 439.2 • Bridge Driver for HDSL

Driving capacitive loads

The devices in Table 439.1 combine the high output current required to slew large capacitances with appropriate frequency compensation. All of the CFAs described here are C-Load amplifiers and are stable with capacitive loads up to 10,000pF.

A good example of a difficult capacitive load is a clock driver for a charge-coupled device (CCD). These devices require precise multiphase clock signals to initiate the transfer of light-generated pixel charge from one charge reservoir to the next. Noise, ringing or overshoot on the clock signal must be avoided. Two problems complicate clock generation. First, CCDs present an input capacitance (typically 100pF to 3300pF) which is directly proportional to the number of sensing elements (pixels). Second, CCDs often require the clock's amplitude to exceed the logic supply. The amplifying filter in Figure 439.3 addresses these issues. Both CFAs in the LT1207 are configured for a third-order Gaussian lowpass response with 1.6MHz cutoff frequency (one section is shown). This transfer function produces clean clock signals with controlled rise and fall times. Figure 439.4 shows the LT1207's quadrature outputs driving two 3300pF loads that simulate a CCD image sensor. Ringing and overshoot are notably absent from the clock signals, which have rise and fall times of approximately 300ns.

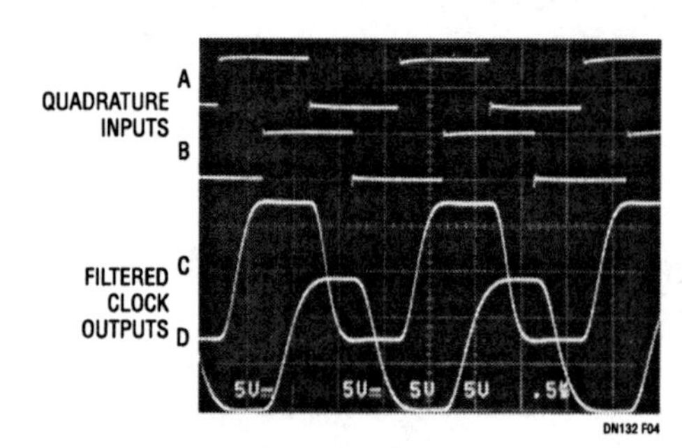

Figure 439.4 • CCD Clock Driver Waveforms

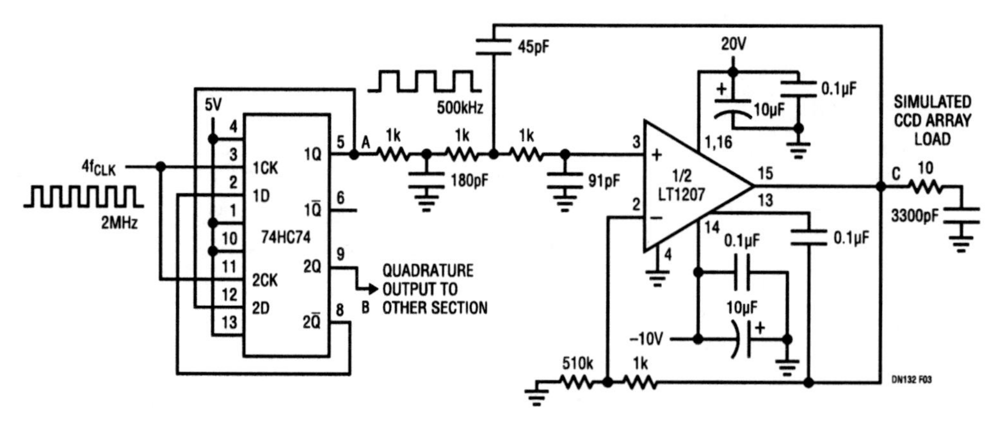

Figure 439.3 • CCD Clock Driver

440

C-Load op amps conquer instabilities

Kevin R. Hoskins

Introduction

Linear Technology Corporation has taken advantage of advances in process technology and circuit innovations to create a series of C-Load operational amplifiers that are tolerant of capacitive loading, including the ultimate, amplifiers that remain stable driving any capacitive load. This series of amplifiers has a bandwidth that ranges from 160kHz to 140MHz. These amplifiers are appropriate for a wide range of applications from coaxial cable drivers to analog-to-digital converter (ADC) input buffer/amplifiers.

Driving ADCs

Most contemporary ADCs incorporate a sample-and-hold (S/H). A typical S/H circuit is shown in Figure 440.1. The hold capacitor's (C1) size varies with the ADC's resolution but is generally in the range of 5pF to 20pF, 10pF to 30pF and 10pF to 50pF for 8-, 10- and 12-bit ADCs, respectively.

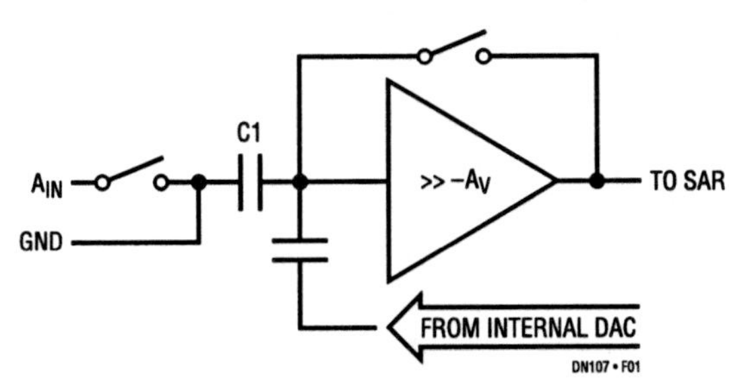

Figure 440.1 • Typical ADC Input Stage Showing Input Capacitors

At the beginning of a conversion cycle, this circuit samples the applied signal's voltage magnitude and stores it on its hold capacitor. Each time the switch opens or closes, the amplifier driving the S/H's input faces a dynamically changing capacitive load. This condition generates current spikes on the input signal. This capacitive load and the spikes produced when they are switched constitutes a very challenging load that can potentially produce instabilities in an amplifier driving the ADC's input. These instabilities make it difficult for an amplifier to quickly settle. If the output of an amplifier has not settled to a value that falls within the error band

Table 440.1 Unity-Gain Stable C-Load Amplifiers Stable with All Capacitive Loads

SINGLES	DUALS	QUADS	GBW (MHz)	I_S/AMP (mA)
–	LT1368	LT1369	0.16	0.375
LT1200	LT1201	LT1202	11	1
LT1220	–	–	45	8
LT1224	LT1208	LT1209	45	7
LT1354	LT1355	LT1356	12	1
LT1357	LT1358	LT1359	25	2
LT1360	LT1361	LT1362	50	4
LT1363	LT1364	LT1365	70	6

Table 440.2 Unity-Gain Stable C-Load Amplifiers Stable with $C_L \leq 10{,}000$pF

SINGLES	DUALS	QUADS	GBW (MHz)	I_S/AMP (mA)
LT1012	–	–	0.6	0.4
–	LT1112	LT1114	0.65	0.32
LT1097	–	–	0.7	0.35
–	LT1457	–	2	1.6

Analog Circuit Design: Design Note Collection. http://dx.doi.org/10.1016/B978-0-12-800001-4.00440-3

of the ADC, conversion errors will result. That is unless the amplifier is designed to gracefully and accurately drive capacitive loads, such as Linear Technology's C-Load line of monolithic amplifiers. Table 440.1 lists Linear Technology's unconditionally stable voltage feedback C-Load amplifiers. Table 440.2 lists other voltage feedback C-Load amplifiers that are stable with loads up to 10,000pF.

Remaining stable in the face of difficult loads

As can be seen in Figure 440.2, an amplifier whose design is not optimized for handling a large capacitive load, has some trouble driving the hold capacitor of the LTC1410's S/H. While the LT1006 has other very desirable characteristics such as very low V_{OS}, very low offset drift, and low power dissipation, it has difficulty accurately responding to dynamically changing capacitive loads and the current glitches and transients they produce (as indicated by the instabilities that appear in the lower trace of Figure 440.2a).

By contrast, Figure 440.2b shows the LT1360 C-Load op amp driving the same LTC1410 input. The photo shows that the LT1360 is an ideal solution for driving the ADC's input capacitor quickly and cleanly with excellent stability. Its wide 50MHz gain-bandwidth and 800V/μs slew rate very adequately complement the LTC1410's 20MHz full power bandwidth. The LT1360 is specified for ±5V operation.

Figure 440.3 shows the circuit used to test the performance of op amps driving the LTC1410's input and measure the input waveforms.

Conclusion

Linear Technology's C-Load amplifiers meet the challenging and difficult capacitive loads of contemporary ADC analog inputs by remaining stable and settling quickly.

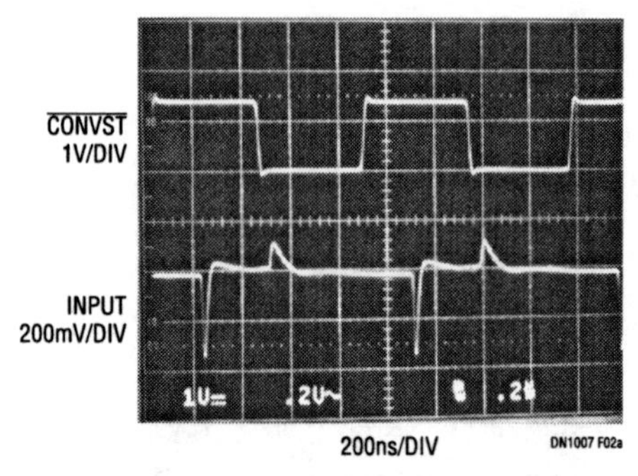

Figure 440.2a • Input Signal Applied to an LTC1410 Driven by an LT1006

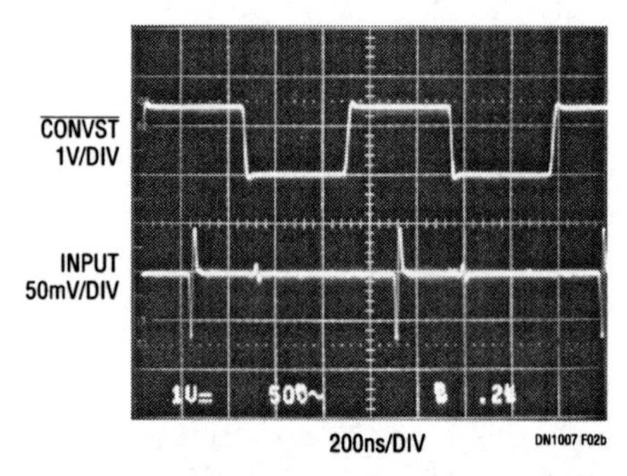

Figure 440.2b • Input Signal Applied to an LTC1410 Driven by an LT1360

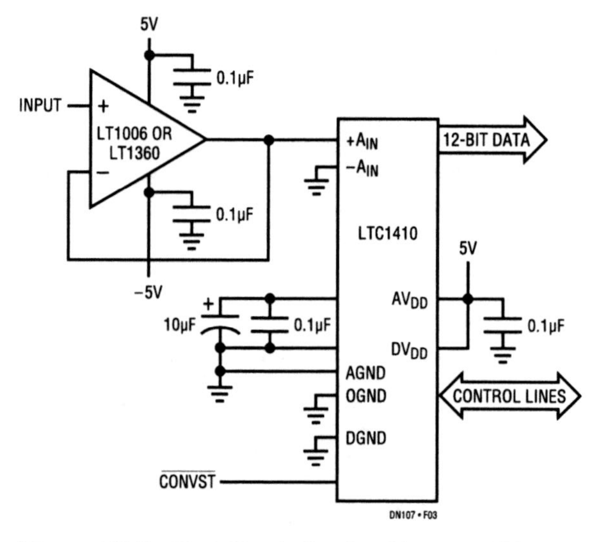

Figure 440.3 • Test Circuit Used to Measure LTC1410 Input Signal Waveform

Applications of a rail-to-rail amplifier

William Jett Sean Gold

441

The LT1366 is Linear Technology's first bipolar dual operational amplifier to combine rail-to-rail input and output operation with precision V_{OS} specifications. The LT1366 maintains precision specifications over a wide range of operating conditions. The device will operate with supply voltages as low as 1.8V and is fully specified for 3V, 5V, and ±15V operation. Offset voltage is typically 200μV when operating from a single 5V supply. Open-loop gain, A_{VOL}, is 2 million driving a 2kΩ load. Supply current is typically 375μA per amplifier. The combination of precision specifications and rail-to-rail operation makes the LT1366 a versatile amplifier, suitable for signal processing tasks that demand the widest possible common-mode range.

Precision low dropout regulator

Microprocessors and complex digital circuits frequently specify tight control of power supply characteristics. The circuit shown in Figure 441.1 provides a precise 3.6V, 1A output from a minimum 3.8V input voltage. The circuit's nominal operating voltage is 4.75V ±5%. The voltage reference and resistor ratios determine output voltage accuracy, while the LT1366's high gain enforces 0.2% line and 0.2% load regulation. Quiescent current is about 1mA and does not change appreciably with supply or load. All components are available in surface mount packages.

The regulator's main loop consists of A1 and a logic level FET, Q1. The output is fed back to the op amp's positive input because of the phase inversion through Q1. The regulator's frequency response is limited by Q1's roll-off and the phase lead introduced by the output capacitor's effective series resistance (ESR). Two pole-zero networks compensate for these effects. The pole formed with R5 and C2 rolls off the gain set with the feedback network, while the pole formed with R7 and C3 rolls off A1's gain directly, which is the dominant influence on settling time. The zeros formed with R6 and C2, and R8 and C3 provide phase boost near the unity-gain crossover, which increases the regulator's phase margin. Although not directly part of the compensation, R9 decouples the op amp's output from the Q1's large gate capacitance.

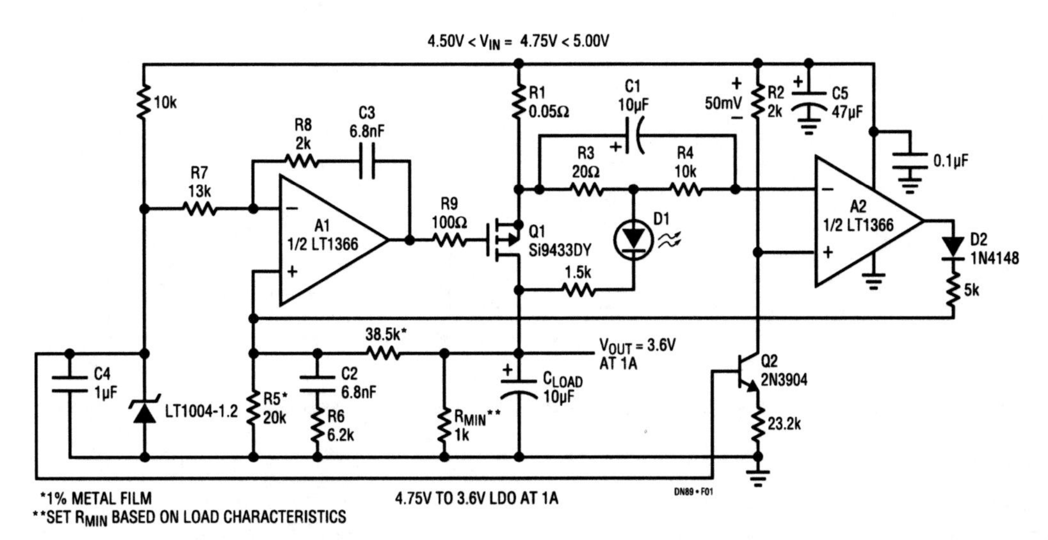

Figure 441.1 • Precision 3.6V, 1A Low Dropout Regulator

Analog Circuit Design: Design Note Collection. http://dx.doi.org/10.1016/B978-0-12-800001-4.00441-5

A second loop provides a foldback current limit. A2 compares the sense voltage across R1 with 50mV referenced to the positive rail. When the sense voltage exceeds the reference, A2's output drives Q2's gate positive via A1. In current limit, the output voltage collapses and the current limit LED (D1) turns on causing about 30mV to drop across R3. A2 regulates Q1's drain current so that the deficit between the 50mV reference and the voltage across R3 is made up across the sense resistor. The reduced sense voltage is 20mV, which sets the current limit to about 400mA. As the supply voltage increases, the voltage across R3 increases, and the current limit folds back to a lower level. The current limit loop deactivates when the load current drops below the regulated output current. When the supply turns on rapidly, C1 bypasses the foldback circuit allowing the regulator to start up into a heavy load.

Q1 does not require a heat sink. When mounted on a type FR4 PC board, Q1 has a thermal resistance of 50°C/W. At 1.4W worst case dissipation, Q1 can operate up to 80°C.

Single supply, 1kHz, 4th order Butterworth filter

The LT1366 is also available as a quad op amp (LT1367), which is used in Figure 441.2 to form a 4th order Butterworth filter. The filter is a simplified state variable architecture consisting of two cascaded 2nd order sections. Each section uses the 360 degree phase shift around the two op amp loop to create a negative summing junction at A1's positive input.[1] The circuit has low sensitivities for center frequency and Q, which are set with the following equations:

$$\omega_0^2 = 1/(R1 \times C1 \times R2 \times C2)$$

where,

$$R1 = 1/(\omega_0 \times Q \times C1) \text{ and } R2 = Q/(\omega_0 \times C2).$$

The DC bias applied to A2 and A4, half supply, is not needed when split supplies are available. The circuit swings rail-to-rail in the passband making it an excellent anti-aliasing filter for A/Ds. The amplitude response is flat to 1kHz then rolls off at 80dB/decade.

Buffering A/D converters

Figure 441.3 shows the LT1366 driving an LTC1288 two-channel micropower A/D. The LTC1288 can accommodate voltage references and input signals equal to the supply rails. The sampling nature of this A/D eliminates the need for an external sample-and-hold, but may call for a drive amplifier because of the A/D's 12µs settling requirement. The LT1366's rail-to-rail operation and low input offset voltage make it well suited for low power, low frequency A/D applications. In addition, the op amp's output settles to 1% in response to a 3mA load step through 100pF in less than 1.5µs.

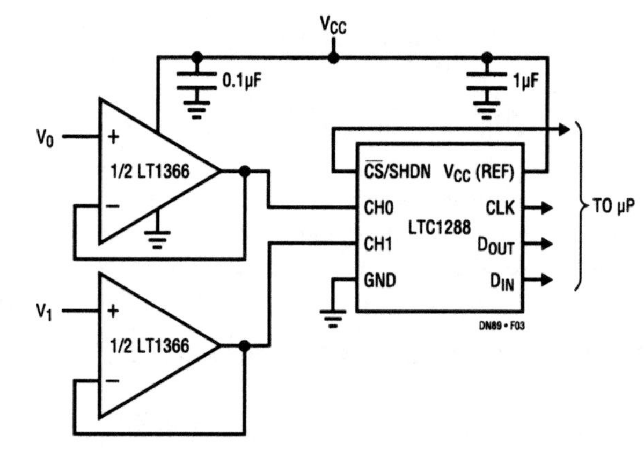

Figure 441.3 • Two-Channel Low Power A/D

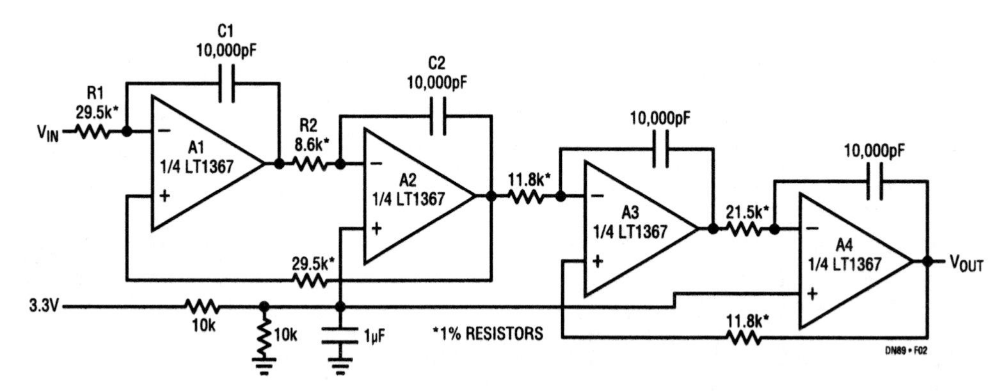

Figure 441.2 • Single Supply Stage Variable Filter Using the LT1367

Note 1: James Hahn, "State Variable Filter Trims Predecessor's Component Count", *Electronics*, April 21, 1982.

Source resistance-induced distortion in op amps

442

William H. Gross

Introduction

Almost all op amp data sheets have typical characteristic curves that show amplifier total harmonic distortion (THD) as a function of frequency. These curves usually show various gains and output levels but almost always the input source resistance is low, typically 50Ω. In some applications, such as active filters, the source impedance will be much larger. If the input impedance of the op amp is nonlinear with voltage, the resulting distortion will be significantly higher than the values indicated in the data sheet.

Test circuit

It is quite easy to evaluate source resistance-induced distortion. Connect the amplifier as a unity-gain buffer operating on ±15V supplies. Feed a low distortion $20V_{P-P}$ signal to the noninverting input through a source resistor and measure the output signal distortion. The setup is shown in Figure 442.1. The readings at 1kHz and 10kHz were recorded for various values of source resistance from 100Ω to 100k. The measured results for several op amps are plotted in Figures 442.2 and 442.3.

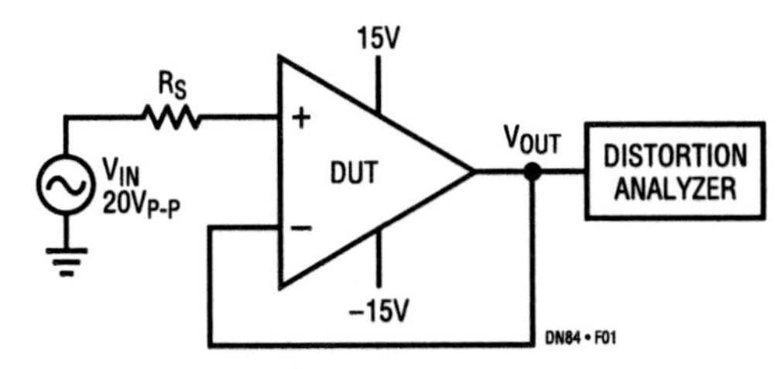

Figure 442.1 • Setup to Evaluate Source Resistance Induced Distortion

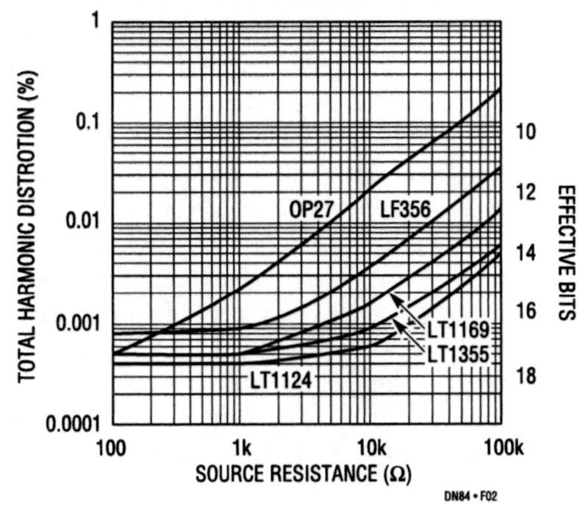

Figure 442.2 • 1kHz Distortion vs Source Resistance

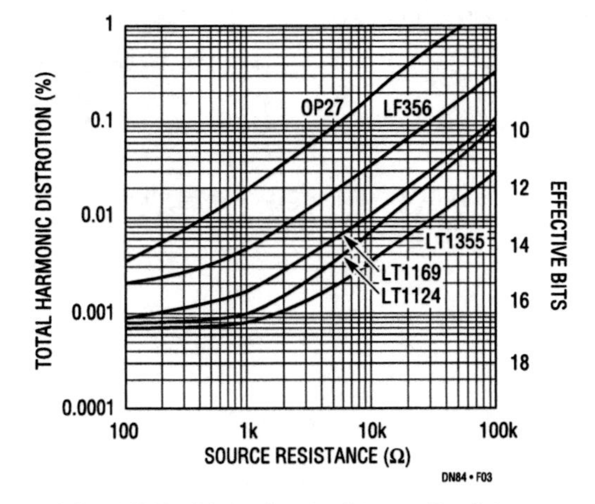

Figure 442.3 • 10kHz Distortion vs Source Resistance

Analog Circuit Design: Design Note Collection. http://dx.doi.org/10.1016/B978-0-12-800001-4.00442-7

Results

Unfortunately there is no easy way to predict which amplifiers will have the lowest source resistance-induced distortion from the data sheets. There are two main causes of the distortion: nonlinear input resistance and nonlinear input capacitance. At first thought, one would not expect the small input capacitance of an op amp to cause distortion at a few kHz. But a 10k source is loaded 0.01% by 1pF at 1.6kHz! Therefore a change in input capacitance of 1pF will cause measurable distortion at 1kHz. For lowest distortion we want an amplifier with low input capacitance as well as very high (and constant) input resistance.

FET input op amps have the highest input resistance but they also have a significant nonlinear input capacitance. The LF356 is a typical FET input op amp; the distortion is 5 to 20 times worse with a 10k source compared with a low source resistance. The LT1169 is a new dual FET input op amp with very low input capacitance (1.5pF) and therefore has about three times lower distortion than the LF356.

The OP27 is a popular high speed precision op amp that has very low distortion when driven from a 50Ω source. Unfortunately the input bias current cancellation circuit works well only at very low frequencies; at 1kHz the input resistance is very nonlinear. The distortion from the OP27 is 50 times worse with a 10k source than with a 100Ω source. The LT1124 is a dual low noise precision op amp that uses a different input bias current cancellation circuit. The LT1124 has the least source resistance induced distortion at 1kH of any of the op amps tested. The LT1355 is a member of a new family of low power, high slew rate op amps that have outstanding high frequency performance. The LT1355 has the least source resistance induced distortion at 10kHz of any op amp tested.

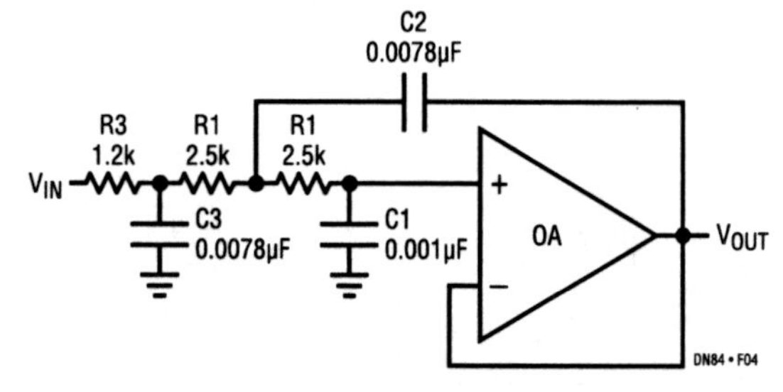

Figure 442.4 • A 20kHz Butterworth Active Filter for Anti-Aliasing or Band Limiting in a Data Acquisition System

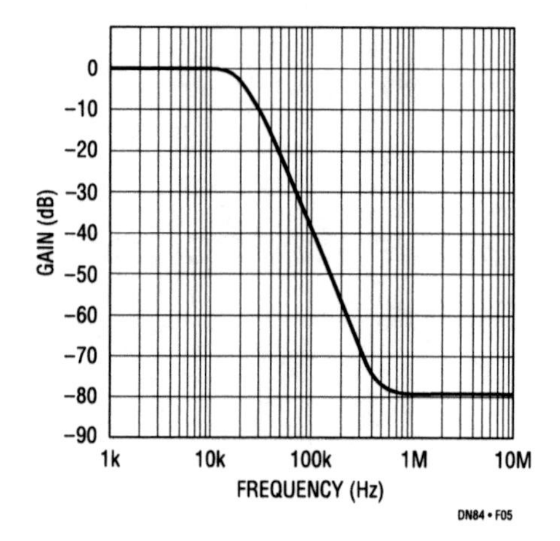

Figure 442.5 • Filter Frequency Response

Figure 442.4 shows a 20kHz Butterworth active filter as might be used for anti-aliasing or band limiting in a data acquisition system. Figure 442.5 shows the frequency response of the circuit. Note that for signals well below the cutoff frequency, the capacitors have no effect and the op amp sees a 6.2k source resistance. Distortion was measured with several op amps in the circuit to confirm the data shown in Figures 442.2 and 442.3. Table 442.1 shows the results of the best op amps.

Table 442.1 Filter Distortion

AMPLIFIER	100Hz	1kHz	2kHz	5kHz	10kHz
LT1124	0.0004%	0.0005%	0.0008%	0.0021%	0.0090%
LT1355	0.0005%	0.0006%	0.0010%	0.0035%	0.0052%
LT1169	0.0005%	0.0012%	0.0024%	0.0080%	0.0100%

Source resistance-induced distortion usually limits the dynamic range of unity-gain RC active filters. An interesting high performance alternative is the LTC1063 and LTC1065. These fifth order, switched-capacitor lowpass filters are not only smaller and easier to use, their distortion is less than 0.01% even with 10k source resistance.

C-Load op amps tame instabilities

443

Richard Markell George Feliz William Jett

Introduction

By taking advantage of advances in process technology and innovative circuit design, Linear Technology Corporation has developed a series of C-Load op amps which are tolerant of capacitive loading, including the ultimate, amplifiers which are stable with any capacitive load. These amplifiers span a range of bandwidths from 1MHz to 140MHz. They are suited for a wide range of applications from coaxial cable drivers to capacitive transducer exciters.

The problem

The cause of the capacitive load stability problem in most amplifiers is the pole formed by the load capacitance and the open-loop output impedance of the amplifier. This output pole increases the phase lag around the loop which reduces the phase margin of the amplifier. If the phase lag is great enough the amplifier will oscillate.

External networks can be used to improve the amplifier's stability with a capacitive load but have serious drawbacks. For instance, most designers are familiar with the use of a series resistor R_S between the load and the amplifier output. The optimum value of R_S depends on the load capacitance, so this approach is not useful for ill-defined loads. Further disadvantages of the external approach include reduced output swing and drive current, and increased component count.

An example

Figure 443.1 shows an example of a competitor's medium speed device which is sensitive to capacitive loading. When 50pF is paralleled with a 5kΩ load, the response exhibits considerable ringing. With a 75pF load the device oscillates. By comparison, the transient responses of the 50MHz LT1360 voltage feedback amplifier (Figure 443.2) shows the improvement in stability achieved in the latest generation of C-Load op amps. In fact the LT1360 maintains a stable transient response for any capacitive load.

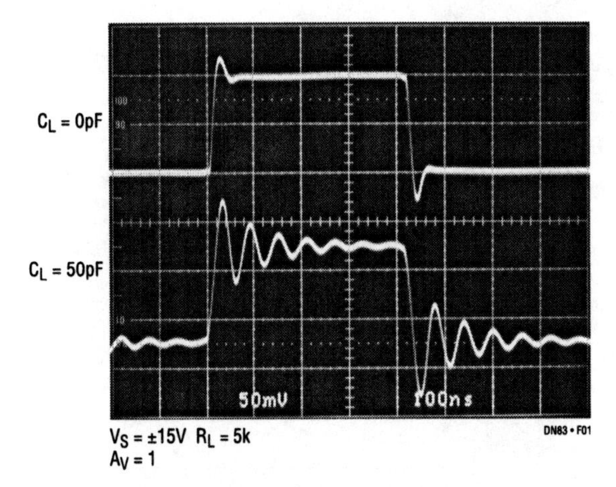

Figure 443.1 • Medium Speed Non-LTC Op Amps

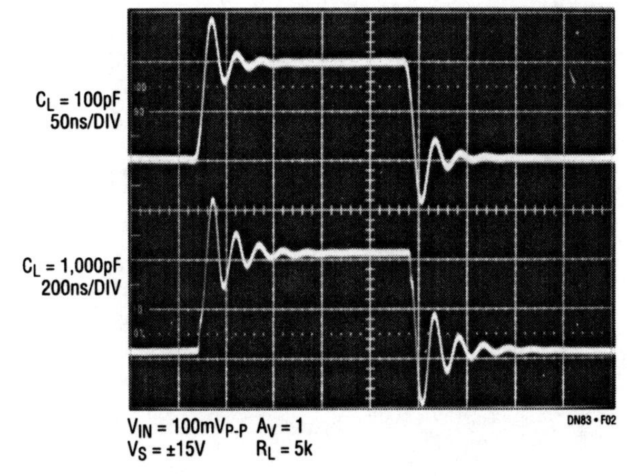

Figure 443.2 • LT1360

Analog Circuit Design: Design Note Collection. http://dx.doi.org/10.1016/B978-0-12-800001-4.00443-9

The solution

LTC's new family of voltage feedback amplifiers adjusts the frequency response of the op amp to maintain adequate phase margin regardless of the capacitive load thus, the amplifiers cannot oscillate. These C-Load amplifiers are great in systems where the load is not fixed or is ill-defined. Examples include driving coaxial cables that may or may not be terminated, driving twisted-pair transmission lines, and buffering the inputs of sampling A/D converters that present time varying impedances.

Table 443.1 Unity-Gain Stable C-Load Amplifiers Stable with All Capacitive Loads

SINGLES	DUALS	QUADS	GBW (MHz)	I_S/AMP (mA)
LT1200	LT1201	LT1202	11	1
LT1220	—	—	45	8
LT1224	LT1208	LT1209	45	7
LT1354	LT1355	LT1356	12	1
LT1357	LT1358	LT1359	25	2
LT1360	LT1361	LT1362	50	4
LT1363	LT1364	LT1365	70	6

Table 443.2 Unity-Gain Stable C-Load Amplifiers Stable with $C_L \leq 10{,}000pF$

SINGLES	DUALS	QUADS	GBW	I_S/AMP
LT1012	—	—	0.6	0.4
—	LT1112	LT1114	0.65	0.32
LT1097	—	—	0.7	0.35
—	LT1457	—	2	1.6

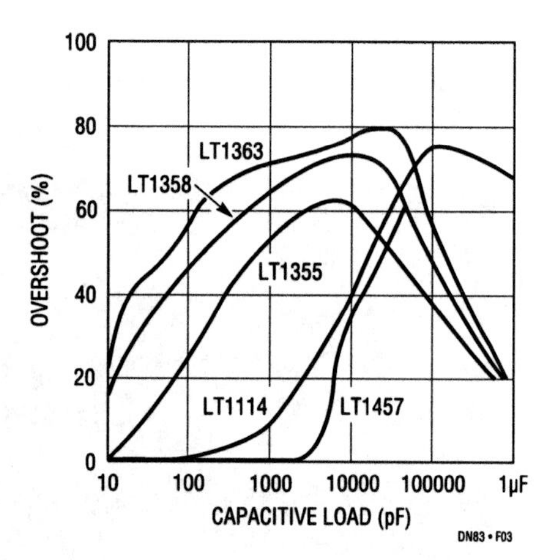

Figure 443.3 • Overshoot vs Capacitive Load

Table 443.1 lists LTC's *unconditionally stable* voltage feedback C-Load amplifiers. Table 443.2 lists other voltage feedback C-Load amplifiers that are stable with loads up to 10,000pF. Figure 443.3 shows overshoot as a function of capacitive load being driven for a wide variety of LTC op amps. Note that the unconditionally stable amplifiers (LT1355, LT1358 and LT1363) have the greatest overshoot for $C_L = 10nF$. Overshoot actually declines as C_L is increased beyond 10nF.

All LTC op amps with adjustable bandwidth can be stabilized for a range of capacitive loads. The bandwidth of current feedback amplifiers is set by the external feedback resistor. Graphs which allow selection of the proper feedback resistor for C_L values to 10,000pF appear in the data sheets of most LTC current feedback amplifiers. As an example, Figure 443.4 shows the LT1206, a 60MHz current feedback amplifier with 250mA output current, driving loads of 1000pF and 10,000pF while remaining stable.

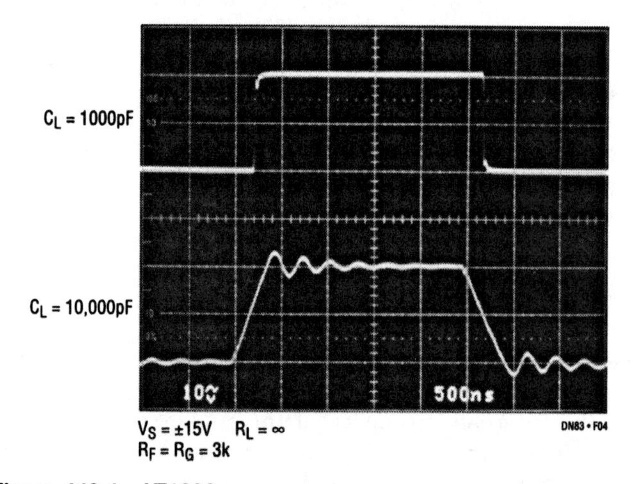

Figure 443.4 • LT1206

Conclusions

Linear Technology has developed families of medium and high speed amplifiers which are much easier to apply than their predecessors. Stable operation with capacitive loads can be achieved without critical external components or loss of output drive. Amplifiers which are stable with any capacitive load are ideal for applications where the load is not well defined. These amplifiers can simplify even low frequency designs by insuring stability under all conditions of loading. For more information on C-Load op amps see the February 1994 issue of Linear Technology Magazine.

A broadband random noise generator

444

Jim Williams

Filter, audio and RF communication testing often requires a random noise source. The circuit in Figure 444.1 provides an RMS amplitude regulated noise source with selectable bandwidth. The RMS output is 300mV with a 1kHz to 5MHz bandwidth selected in decade ranges.

The A1 amplifier, biased from the LT1004 reference, provides optimum drive for D1, the noise source. AC-coupled A2 takes a broadband gain of 100. The A2 output feeds a gain control stage via a simple selectable lowpass filter. The filter's output is applied to LT1228 A3, an operational

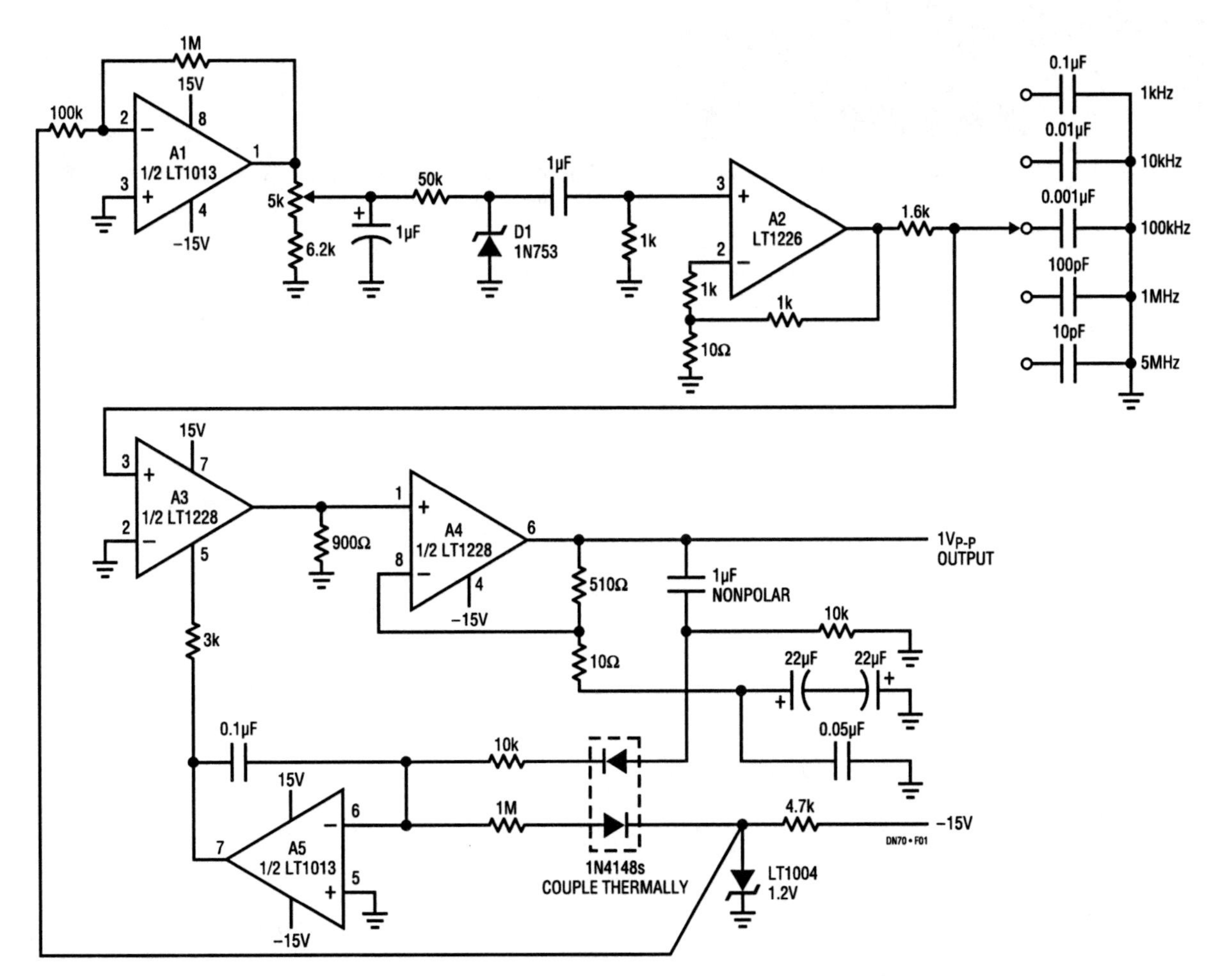

Figure 444.1 • Random Noise Generator with Selectable Bandwidth and RMS Voltage Regulation

Analog Circuit Design: Design Note Collection. http://dx.doi.org/10.1016/B978-0-12-800001-4.00444-0

transconductance amplifier. A3's output feeds LT1228 A4, a current feedback amplifier. A4's output, the circuit's output, is sampled by the A5-based gain control configuration. This closes a gain control loop back at A3. A3's I_{SET} input current controls its gain, allowing overall output level control.

To adjust this circuit, place the filter in the 1kHz position and trim the 5k potentiometer for maximum negative bias at A3, Pin 5.

Figure 444.2 shows noise at a 1MHz bandpass while Figure 444.3 plots amplitude vs RMS noise in the same bandpass. Figure 444.4 plots similar information at full bandwidth. RMS output is essentially flat to 1.5MHz with about ±2dB control to 5MHz before sagging badly.

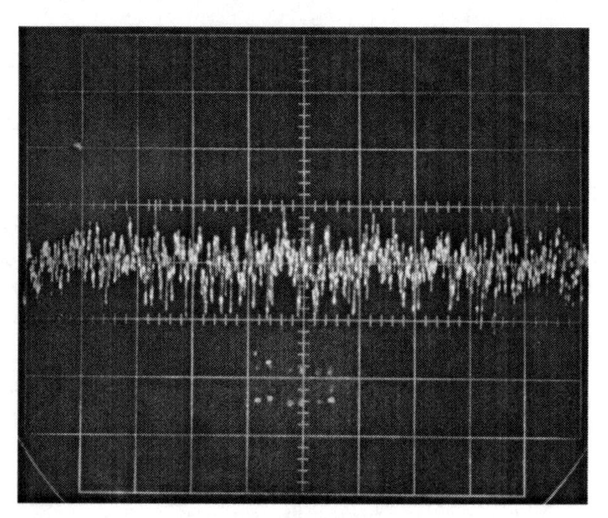
Figure 444.2 • Figure 444.1's Output in the 1MHz Filter Position

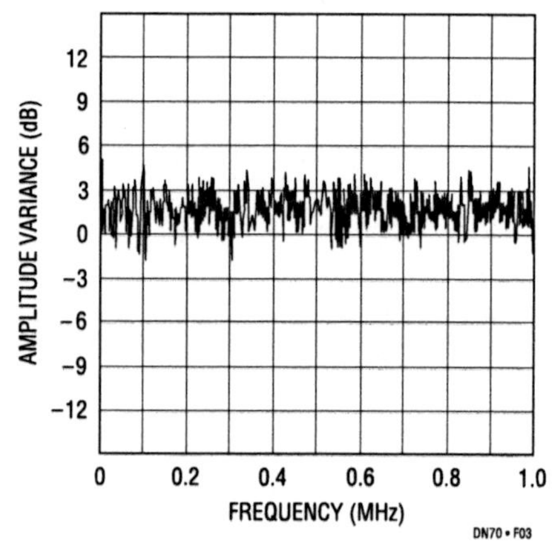

Figure 444.3 • RMS Amplitude vs Frequency for the Random Noise Generator Is Essentially Flat to 1MHz

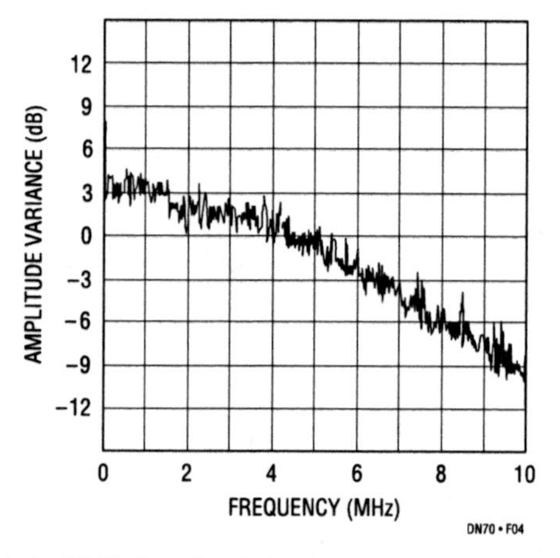

Figure 444.4 • RMS Amplitude Holds within ±2dB Before Sagging Beyond 5MHz

Peak detectors gain in speed and performance

445

John Wright

Introduction

Fast peak detectors place unusual demands on amplifiers. High slew rate is needed to keep the amplifier internal nodes from overracing the output stage. This condition causes either a long overload, or DC accuracy errors. To support the high slew rate at the output, the amplifier must deliver large currents into the capacitive load of the detector. Compounding these problems are issues of amplifier instability with a large capacitive load, as well as the accuracy of the output voltage.

Detecting sine waves

The LT1190 is the ideal candidate for this application, with a high 400V/μs slew rate, large 50mA output current, and a wide 70 degree phase margin. The closed-loop peak detector circuit of Figure 445.1 uses a Schottky diode inside the feedback loop to obtain good accuracy. The 20Ω resistor R_O isolates the 0.01μF load and prevents oscillation. The DC error with a sine wave input is plotted in Figure 445.2 for various input amplitudes. The DC value is read with a DVM. At low frequency, the error is small and dominated by decay of the detector capacitor between cycles. As frequency rises the error increases because capacitor charging time decreases. During this time the overdrive becomes a very small portion of a sine wave cycle. Finally at approximately 4MHz the error rises rapidly due to the slew rate limitation of the op amp. For comparison purposes the error of an LM118 is also plotted for $V_{IN} = 2V_{P-P}$.

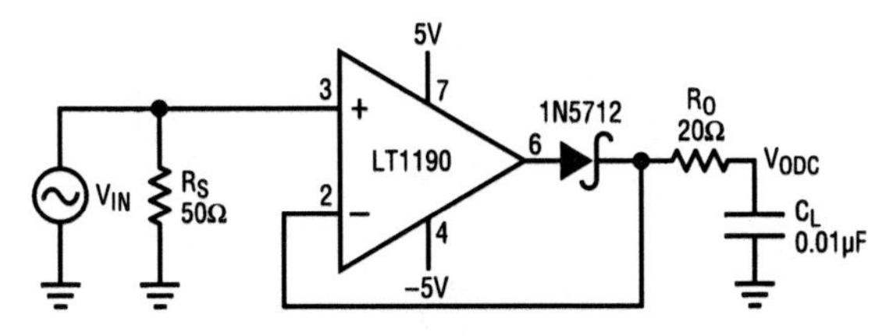

Figure 445.1 • Closed-Loop Peak Detector

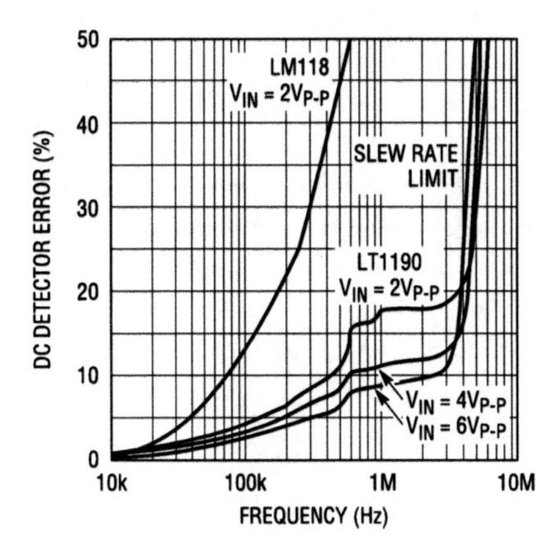

Figure 445.2 • Closed-Loop Peak Detector Error vs Frequency

A fast Schottky diode peak detector can be built with a 1000pF capacitor, and 10k pull down. Although this simple circuit is very fast, it has limited usefulness due to the error of the diode threshold, and its low input impedance. The accuracy of this simple circuit can be improved with the LT1190 circuit of Figure 445.3. In this open-loop design, the detector diode is D1, and a level shifting or compensating diode is D2. A load resistor R_L is connected to −5V, and an identical bias resistor R_B is used to bias the compensating diode. Equal value resistors ensure that the diode drops are equal.

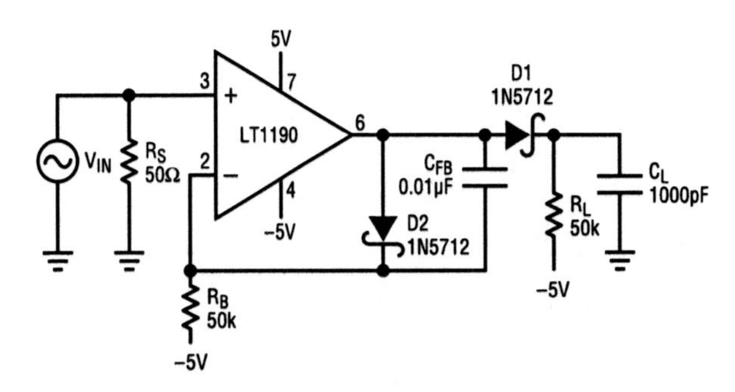

Figure 445.3 • Open-Loop High Speed Peak Detector

Analog Circuit Design: Design Note Collection. http://dx.doi.org/10.1016/B978-0-12-800001-4.00445-2

Low values of R_L and R_B (1k to 10k) provide fast response, but at the expense of poor low frequency accuracy. High values of R_L and R_B provide good low frequency accuracy, but cause the amplifier to slew rate limit, resulting in poor high frequency accuracy. A good compromise can be made by adding a feedback capacitor C_{FB} which enhances the negative slew rate on the (–) input. The DC error with a sine wave input is plotted in Figure 445.4 and is read with a DVM. For comparison purposes the LM118 error is plotted as well as the error of the simple Schottky detector.

Detecting pulses

A fast pulse detector can be made with the circuit of Figure 445.5. A very fast input pulse will exceed the amplifier slew rate and cause a long overload recovery time. Some amount of dV/dt limiting on the input can help this overload condition; however, this will delay the response. Figure 445.6 shows the detector error vs pulse width. Figure 445.7 is the response to a $4V_{P-P}$ input that is 80ns wide. The maximum output slew rate in the photo is 70V/μs. This rate is set by the 70mA current limit driving 1000pF. As a performance benchmark, the LM118 takes 1.2μs to peak detect and settle the same amplitude input. This slower response is due in part to the much lower slew rate and lower phase margin of the LM118.

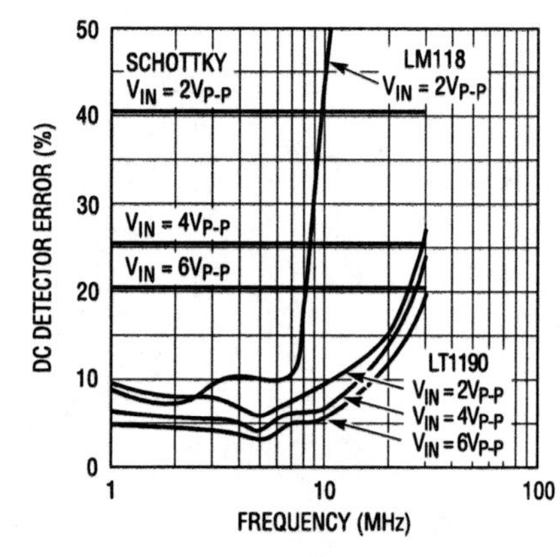

Figure 445.4 • Open-Loop Peak Detector Error vs Frequency

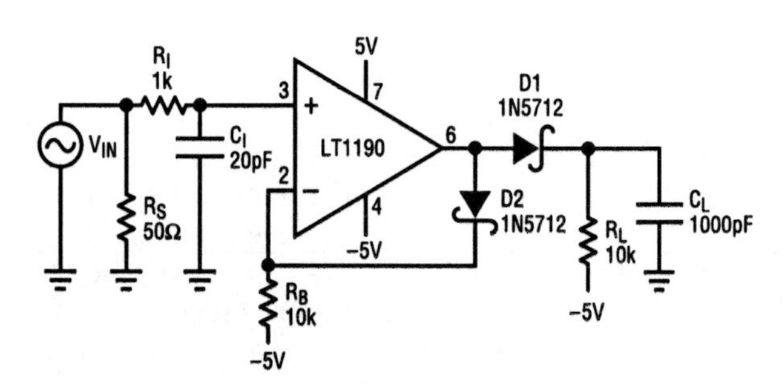

Figure 445.5 • Fast Pulse Detector

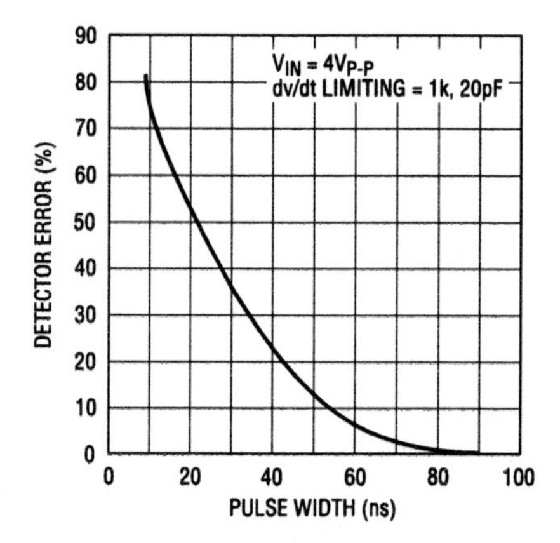

Figure 445.6 • Detector Error vs Pulse Width

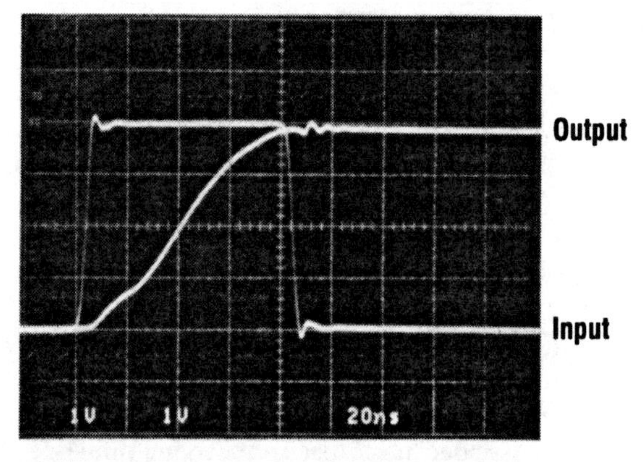

Figure 445.7 • Open-Loop Peak Detector Response

3V operation of Linear Technology op amps

446

George Erdi

The latest trend in digital electronics is the introduction of numerous ICs operating on regulated 3V or 3.3V power supplies. This is a logical development to increase circuit densities and to reduce power dissipation. In addition, many systems are directly powered by two AA cells or 3V lithium batteries. Clearly, analog ICs which work on 3V with good dynamic range to complement these digital circuits are, and will be, in great demand.

Many Linear Technology operational amplifiers work well on a 3V supply. The purpose of this design note is to list these devices and their performance when powered by 3V. The op amps can be divided into two groups: single and dual supply devices. The single supply op amps are optimized for, and fully specified at, a 5V positive supply with the negative supply terminal tied to ground. Input common mode voltage range goes below ground, and the output swings to within a few millivolts of ground while sinking current. Members of the single supply family are the micropower LT1077/LT1078/LT1079 single, dual and quad op amps with 40μA supply current per amplifier, the LT1178/LT1179 dual and quad with

Table 446.1 Single Supply Op Amps: Low Cost Grade Specifications $V_S = 3V, 0V. T_A = 25°C$

PARAMETER		LT1077CN8 LT1078CN8 LT1079CN		LT1178CN8 LT1179CN		LT1006CN8 LT1013CN8 LT1014CN		UNITS
		TYP	MIN/MAX	TYP	MIN/MAX	TYP	MIN/MAX	
Offset Voltage	Single Dual/ Quad	15	80	–	–	35	95	μV
		45	140/170	45	140/170	95	470	μV
Input Voltage Range		−0.3	01.	−0.3	0	−0.3	0	V
		1.8	7	1.9	1.7	1.8	1.7	V
Output Swing	No Load	0.003	0.006	0.006	0.009	0.015	0.025	W
		2.4	2.2	2.4	2.2	2.4	2.2	
	2k to Ground	0.0006	0.0010	0.0002	0.0006	0.007	0.015	V
		2.1	1.9	2.0	1.8	2.3	2.0	V
Voltage Gain	$R_L = 50k$	500	110	180	60	1000	500	V/mV
0.1Hz to 10Hz Noise		0.6	–	1.0	–	0.5	–	μV_{P-P}
Minimum Supply Voltage with 300μV V_{OS} Degradation		–	2.3	–	2.2	–	2.6/25	V
		–	1.8	–	1.7	–	–	V
Gain-Bandwidth Product		160	–	50	–	700	–	kHz

Analog Circuit Design: Design Note Collection. http://dx.doi.org/10.1016/B978-0-12-800001-4.00446-4

13μA per amplifier. The LT1006/LT1013/LT1014 single, dual and quad have faster speed and lower voltage noise, at the expense of 300μA per amplifier.

The performance of these devices at 3V is quite similar to the 5V specs. Clearly, input voltage range and output voltage swing have to be reduced by 2V since the supply is 2V less. Offset voltage change from 5V to 3V is determined by the power supply rejection ratio specs. At 114dB or 2μV/V the degradation in offset voltage is only 4μV (=2V·μV/V). Input bias and offset currents, voltage and current noise, as well as offset voltage drift with temperature, are practically unchanged compared to the 5V specifications.

Table 446.1 summarizes the performance of the low cost grades of these single supply devices at 3V. One note of caution: the minimum operating voltage for the LT1013/LT1014 is 2.95V. All other devices work on lower supplies, ranging from 1.7V to 2.6V.

The LT1101 micropower (=75μA) instrumentation amplifier completes the single supply family. Again, this amp in 8-pin packages is fully specified at 5V. Minimum supply voltage is 1.8V; the performance change in going from 5V to 3V supply is minimal.

The second group of devices are dual supply op amps, i.e., the common mode input voltage and the output swing are limited to a diode voltage (=600mV) above the negative supply terminal for proper operation. In addition, dual supply op amps are traditionally optimized for ±15V operation. Thus, reducing the total supply voltage to 3V represents a significant change. Table 446.2 lists the performance of four op amps: the LT1008 and LT1012 are actually fully tested at reduced supplies. The LT1097 and LT1001 performance is inferred from device evaluation data. Dual versions in 14-pin packages are also available: the LT1002 is a dual LT1001; the LT1024 is a dual version of the LT1012.

In most 3V applications the single supply op amps of Table 446.1 are more flexible and desirable, since no special biasing is needed to shift the input and the output into the operating range. However, the offset voltage drift with temperature performance of the dual supply devices is better. And, most importantly, when picoampere input bias currents are needed, the LT1008/LT1012/LT1097 have no competition. The op amps of Table 446.1 are all at least 6nA. The traditional ways of achieving picoampere bias current are not available either: JFET input or CMOS chopper-stabilized op amps do not function at 3V supply.

Figure 446.1 shows an application using the LT1078 to monitor the condition of the 3V battery. One output warns that the battery voltage is dropping, the other output shuts the system down as the battery voltage falls below the threshold value.

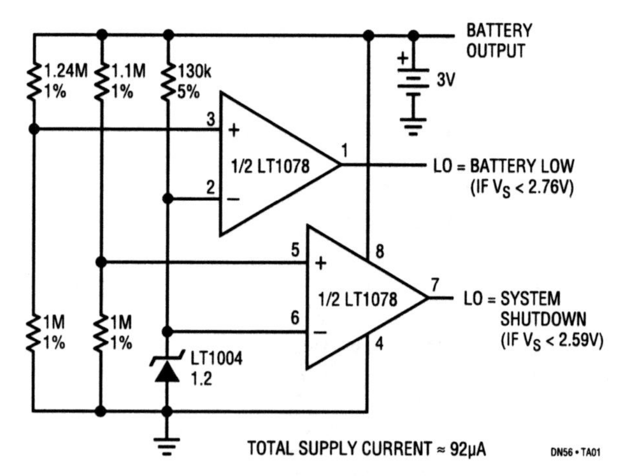

Figure 446.1 • Low Battery Detector with System Shutdown

Table 446.2 Dual Supply Op Amps at V_S=3V, 0V. T_A=25°C. Low Cost Grade Electrical Characteristics

PARAMETER	LT1097CN8		LT1008CN8		LT1012CN8		LT1001CN8		UNITS
	TYP	MIN/MAX	TYP	MIN/MAX	TYP	MIN/MAX	TYP	MIN/MAX	
Offset Voltage	20	100	40	180	25	120	40	150	μV
Drift with Temperature	0.3	1.3	0.3	1.6	0.3	1.3	0.3	1.3	μV/°C
Input Bias Current	40	280	40	150	40	200	600	3500	pA
Input Offset Current	40	260	30	150	30	200	350	3200	pA
Input Voltage Range	0.65	0.80	0.65	0.80	0.65	0.80	0.75	0.90	V
	2.3	2.2	2.3	2.2	2.3	2.2	2.2	2.1	V
Output Swing	0.62	0.8	0.62	0.8	0.62	0.8	0.55	0.7	V
	2.25	2.1	2.25	2.1	2.25	2.1	2.2	2.05	V
Voltage Gain R_L=10k	600	250	500	200	500	200	300	150	V/mV
0.1Hz to 10Hz Noise	0.5	–	0.5	–	0.5	–	0.35	–	$μV_{P-P}$
Minimum Supply Voltage	–	2.4	–	2.4	–	2.4	–	1.9	V
Supply Current	350	560	380	600	380	600	390	550	μA
Gain-Bandwidth Product	500	–	500	–	500	–	600	–	kHz

High frequency amplifier evaluation board

447

Mitchell Lee

Introduction

Demo board DC009 is designed to simplify the evaluation of high speed operational amplifiers. It includes both an inverting and noninverting circuit, and pads are provided to allow the use of board-mounted BNC or SMA connectors. The two circuits are independent, with the exception of shared power supply and ground connections.

High speed layout techniques

Layout is a primary contributor to the performance of any high speed amplifier. Poor layout techniques adversely affect the behavior of a finished circuit. Several important layout techniques, all used in demo board DC009, are described below:

Topside Ground Plane: The primary task of a ground plane is to lower the impedance of ground connections. The inductance between any two points on a uniform sheet of copper is less than the inductance of a narrow, straight trace of copper connecting the same two points. The ground plane approximates the characteristics of a copper sheet and lowers the impedance at key points in the circuit, such as at the grounds of connectors and supply bypass capacitors.

Ground Plane Voids: Certain components and circuit nodes are very sensitive to stray capacitance. Two good examples are the summing node of the op amp and the feedback resistor. Voids are put in the ground plane in these areas to reduce stray ground capacitance.

Input/Output Matching: The width of the input and output traces is adjusted to a stripline impedance of 50Ω. Note that the terminating resistors (R3 and R7) are connected to the end of the input lines — not at the connector. While stripline techniques aren't absolutely necessary for the demo board, they are important on larger layouts where line lengths are longer. The short lines on the demo board can be terminated in 50Ω, 75Ω, or 93Ω without adversely affecting performance.

Separation of Input and Output Grounds: Even though the ground plane exhibits a low impedance, input and output grounds are still separated. For example, the termination resistors (R3 and R7) and the gain-setting resistor (R1) are grounded in the vicinity of the input connector. Supply bypass capacitors (C1, C2, C4, C5, C7, C8, C9, and C10) are returned to ground in the vicinity of the output connectors.

Optional components

The circuit board is designed to accommodate standard 8-pin miniDIP, single operational amplifiers, such as the LT1190 and LT1220 families. Both voltage and current feedback types can be used. Pins 1, 5, and 8 are outfitted with pads for use in adjusting DC offsets, compensation or, in the case of the LT1223 and LT1190/LT1191/LT1192, for shutting down the amplifier.

If a current feedback amplifier such as the LT1223 is being evaluated, omit C3/C6. R4 and R8 are included for impedance matching when driving low impedance lines. If the amplifier is supposed to drive the line directly, or if the load impedance is high, R4 and R8 can be replaced by jumpers. Similarly R10 and R12 can be used to establish a load at the output of the amplifier.

Low profile sockets may be used for the op amps to facilitate changing parts, but performance may be affected above 100MHz.

Supply bypass capacitors

High speed operational amplifiers work best when their supply pins are bypassed with RF-quality capacitors. C1, C5, C8, and C10 should be 10nF disc ceramics with a self-resonant frequency greater than 10MHz. The polarized capacitors (C2, C4, C7, and C9) should be 1μF to 10μF tantalums. Most 10nF ceramics are self-resonant well above 10MHz, and 4.7μF solid tantalums (axial leaded) are self-resonant at 1MHz or below. Lead lengths are critical: the self-resonant frequency of a 4.7μF tantalum drops by a factor of 2 when measured through 2 inch leads. Although a capacitor may become inductive at high frequencies, it is still an effective bypass component above resonance because the impedance is low.

Analog Circuit Design: Design Note Collection. http://dx.doi.org/10.1016/B978-0-12-800001-4.00447-6

Demo DC009 High Frequency Amplifier

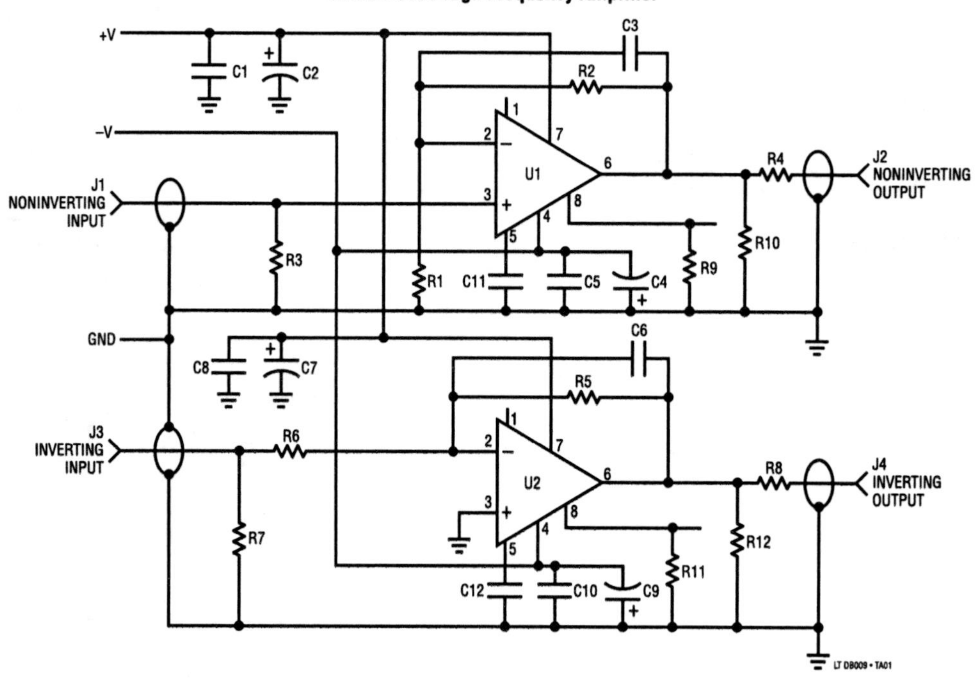

Demo Board DC009 Parts List	
Noninverting Amplifier:	
R1	Gain Setting Resistor
R2	Feedback Resistor
R3	Input Line Termination (51Ω)
R4	Output Line Termination (51Ω)
R9	Shutdown Pin Pull Down
R10	Output Load Resistor
C1	Positive Supply High Frequency Bypass (10nF)
C2	Positive Supply Low Frequency Bypass (4.7μF)
C3	Feedback Capacitor
C4	Negative Supply Low Frequency Bypass (4.7μF)
C5	Negative Supply High Frequency Bypass (10nF)
C11	Compensation Capacitor
J1	Input Connector (AMP 227699-3)
J2	Output Connector (AMP 227699-3)
Inverting Amplifier:	
R5	Feedback Resistor
R6	Gain Setting Resistor
R7	Input Line Termination (51Ω)
R8	Output Line Termination (51Ω)
R11	Shutdown Pin Pull Down
R12	Output Load Resistor
C6	Feedback Capacitor
C7	Positive Supply Low Frequency Bypass (4.7μF)
C8	Positive Supply High Frequency Bypass (10nF)
C9	Negative Supply Low Frequency Bypass (4.7μF)
C10	Negative Supply High Frequency Bypass (10nF)
C12	Compensation Capacitor
J3	Input Connector (AMP 227699-3)
J4	Output Connector (AMP 227699-3)

High Frequency Amplifier, Demo 009A Component Side

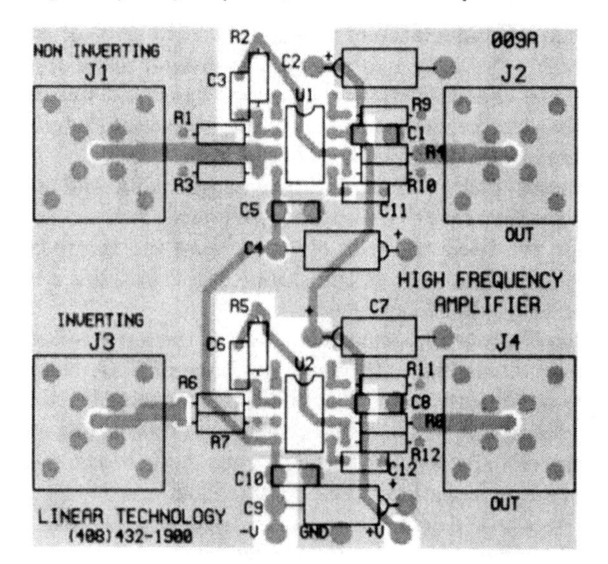

Current feedback amplifier "dos and don'ts"

448

William H. Gross

Introduction

The introduction of current feedback amplifiers, such as the LT1223, has significantly increased the designer's ability to solve difficult high speed amplifier problems. The current feedback architecture has very high slew rate and the small-signal bandwidth is fairly constant for all gains. Current feedback amplifiers are used in broadcast video systems, radar systems, IF and RF stages, RGB distribution systems and many other high speed circuits.

As with any new circuit, there are several new rules that must be kept in mind to prevent problems. Because current feedback amplifiers act very much the same as regular op amps, it is important to note the differences and show how some standard op amp circuits should be implemented.

The most important thing to remember about current feedback amplifiers is that the impedance at the inverting (negative) input sets the bandwidth and therefore the stability of the amplifier. It should be resistive, not capacitive. To slow the amplifier down, increase the resistance driving the inverting input. If the amplifier peaks too much due to capacitive loading or anything else, increase the value of the feedback resistors.

The best way to demonstrate how to use current feedback amplifiers is to show some example circuits. To make it as painless as possible, I will show the traditional op amp implementation next to the current feedback amplifier version.

Op Amp Adjustable Gain Amp

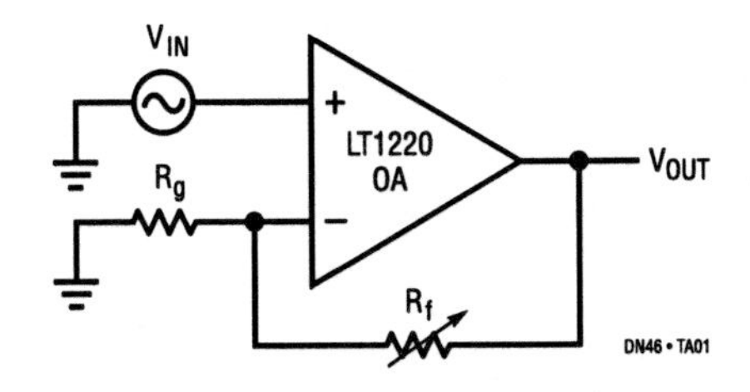

Current Feedback Amp Adjustable Gain Amp

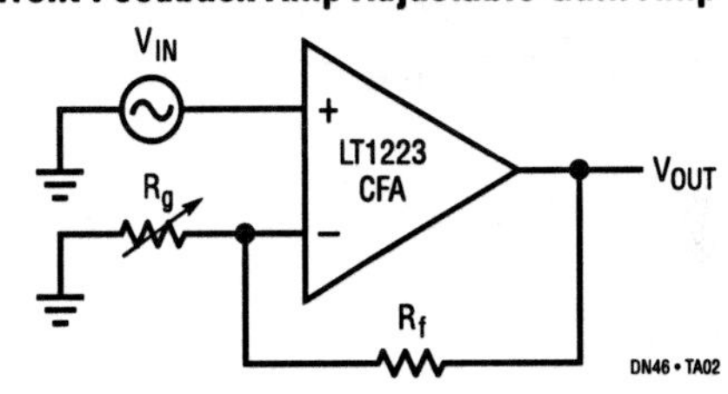

With a standard op amp you can vary the gain of the amplifier with either R_f or R_g. The only real restriction on the values is the loading affect the resistors have on the amplifier output. With a current feedback amplifier the value of R_f should not be varied. Do not make R_f the variable resistor or the bandwidth will be reduced at maximum gain and the circuit will oscillate when R_f is very small.

Op Amp Bandwidth Limiting

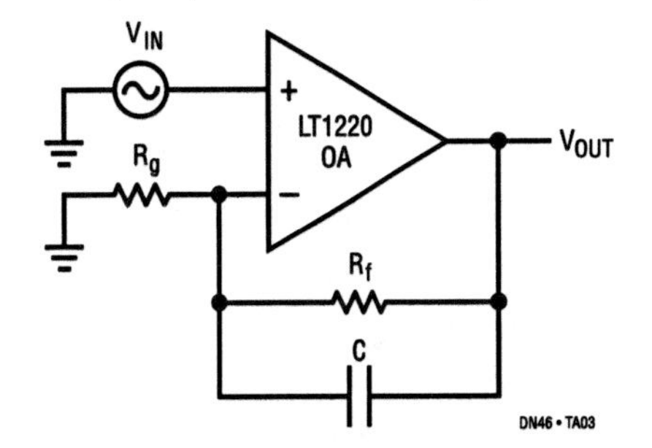

Current Feedback Amp Bandwidth Limiting

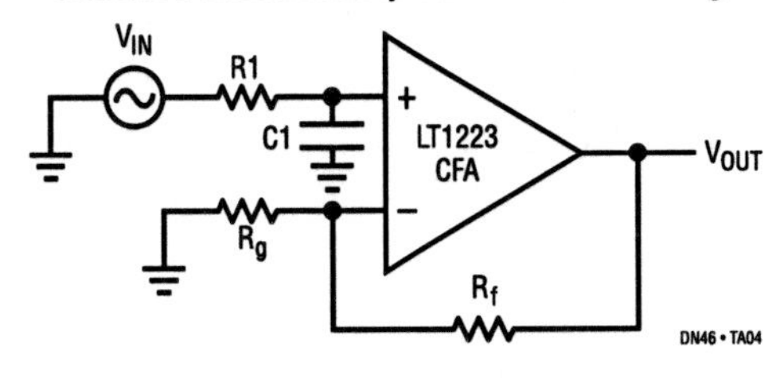

Analog Circuit Design: Design Note Collection. http://dx.doi.org/10.1016/B978-0-12-800001-4.00448-8

It is very common to limit the bandwidth of an op amp by putting a small capacitor in parallel with R_f. This works with all unity-gain-stable op amps; DO NOT PUT A SMALL CAPACITOR FROM THE INVERTING INPUT OF A CURRENT FEEDBACK AMPLIFIER TO ANYWHERE, ESPECIALLY NOT TO THE OUTPUT. The capacitor on the inverting input will cause peaking or oscillations. If you need to limit the bandwidth of a current feedback amplifier, use a resistor and capacitor at the noninverting input (R1 and C1). This technique will also cancel (to a degree) the peaking caused by stray capacitance at the inverting input. Unfortunately, this will not limit the output noise the way it does for the op amp.

Op Amp Integrator

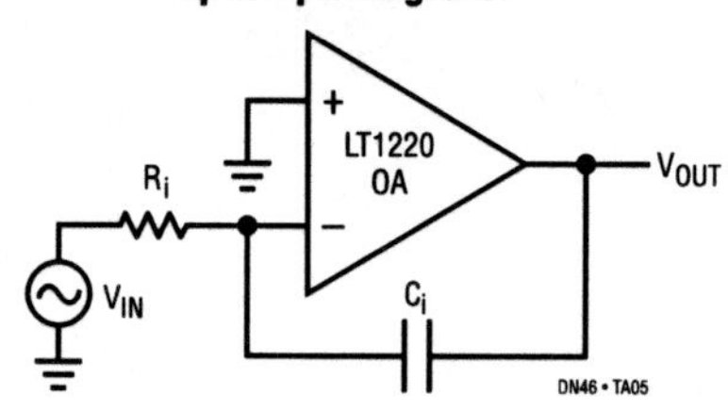

Current Feedback Amplifier Integrator

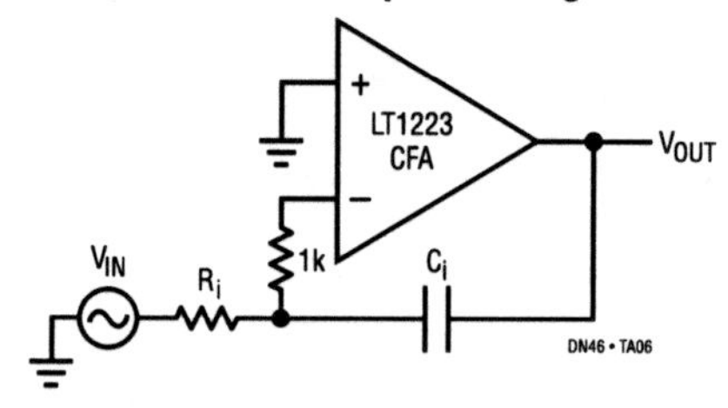

The integrator is one of the easiest circuits to make with an op amp. However, the circuit must be modified before a current feedback amplifier can be used. Since we remember that the inverting input wants to see a resistor, we can add one to the standard circuit. This generates a new summing node where we can apply capacitive feedback. The new current feedback amplifier compatible integrator works just like you would expect; it has excellent large signal capability and accurate phase shift at high frequencies.

Current Feedback Amplifier Summer (DC Accurate)

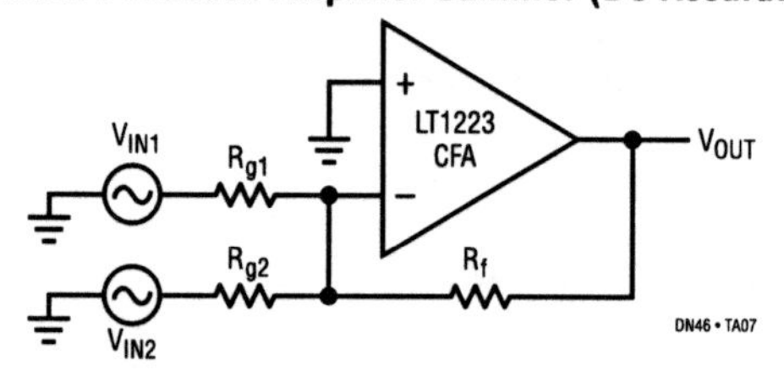

There is no I_{OS} spec on current feedback amplifiers because there is no correlation between the two input bias currents. Therefore we will not improve the DC accuracy of the inverting amplifier by putting an extra resistor in the noninverting input. This is also true of input bias current canceled op amps where the I_{OS} spec is the same as the I_B spec, such as the LT1220.

Two Amplifier Instrumentation Amp

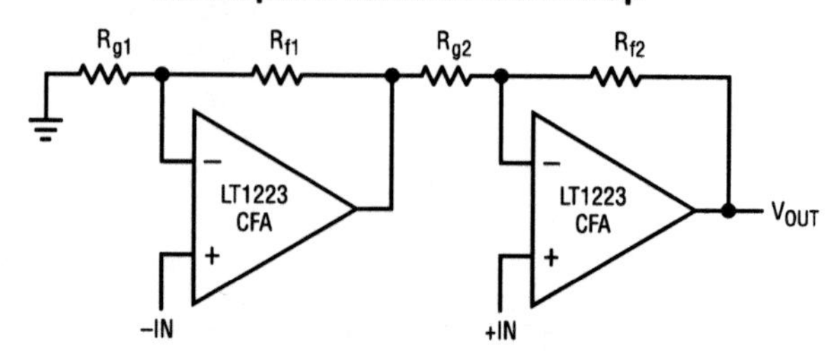

TRIM R_{g2} FOR GAIN, THEN TRIM R_{g1} FOR CMRR. VOLTAGE GAIN, G, IS V_{OUT} DIVIDED BY DIFFERENCE BETWEEN +IN AND –IN.

OP AMP DESIGN EQUATIONS:

$R_{f1} = R_{g2}$; $R_{f2} = (G-1) R_{g2}$; $R_{g1} = R_{f2}$

CURRENT FEEDBACK AMP DESIGN EQUATIONS:

$R_{f1} = R_{f2}$; $R_{g1} = (G-1) R_{f2}$; $R_{g2} = \frac{R_{f2}}{G-1}$

DN46 • TA08

The two amplifier instrumentation amp is easily modified for current feedback amplifiers. The only necessary change is to make the feedback resistor of each amplifier the same and therefore make the gain setting resistors different. This way the bandwidth of both amps is the same and the common mode rejection at high frequencies is better than that of the op amp circuit. In the op amp circuit one amplifier has maximum bandwidth, since it runs at about unity gain, while the other is limited to its gain-bandwidth product divided by the gain.

Cable Driver

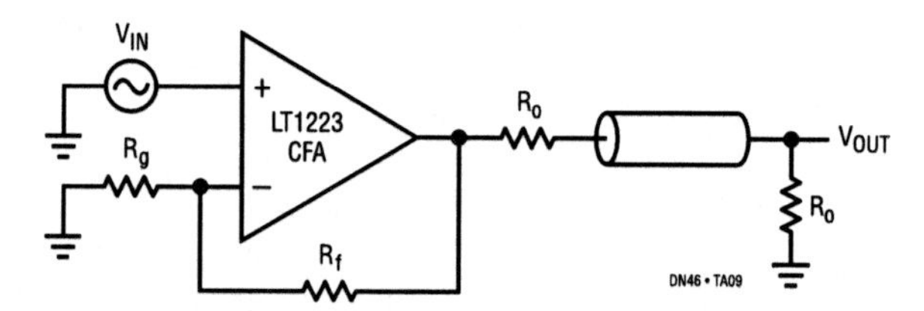

The cable driver circuit is the same for both types of amplifiers. But because most op amps do not have enough output drive current, they are not often used for heavy loads like cables. When driving a cable it is important to properly terminate both ends if even modest high frequency performance is required. The additional advantage of this is that it isolates the capacitive load of the cable from the amplifier so it can operate at maximum bandwidth.

Improved JFET op amp macromodel slews asymmetrically

449

Walt Jung

SPICE macromodels for op amps have been available for some time for both bipolar [1,2] and JFET [3] input stage device types. Interestingly, however, not much attention has been given to the models available to controlled slewing asymmetry. Dependent upon a given amplifier design topology, the large signal characteristics can have various degrees of slew rate (SR) asymmetry. It therefore makes sense to have models which emulate real IC parts in this regard.

A case in point is that of the available P-channel JFET input op amps, many which have a characteristic SR response which is asymmetrical. In fact, popular op amps with topologies like the original 355/356 types are intrinsically faster for negative going output swings than they are for positive. Similar comments apply to such related devices as the OP15, OP16, etc. Since this type of JFET device topology was introduced, the SR specified on the data sheet has typically been the **lower** of two dissimilar rates, i.e., the slower, **positive edge** SR. Thus, given an op amp with a typical SR spec of 14V/μs for positive going edges, the same amp will have a corresponding negative SR of about 28V/μs.

Ironically, this quite common JFET amplifier slewing characteristic has not been well modeled thus far. Most macromodels currently available simply do not address the asymmetric SR issue at all. Others have means of modeling it, but it is seldom found used.

A means of SR control was built into the original Boyle [1] model, and it addresses SR asymmetry for common mode (CM) signals by means of common emitter (source) capacitor, CE (CS, for JFET amps). However, using this capacitor alone for a general SR symmetry control mechanism leaves something to be desired, as the resulting slopes are not consistent. LTC has implemented a new means of modeling SR asymmetry, shown in Figure 449.1.

The circuit as shown here is a simplified Boyle type model with P-channel JFT input devices, J1 and J2. As this type (or similar input structure) of model is typically used, the SR is simply $I_{SS}/C2$, which is symmetrical when CS is zero. When the common source capacitor CS is added, the SR for

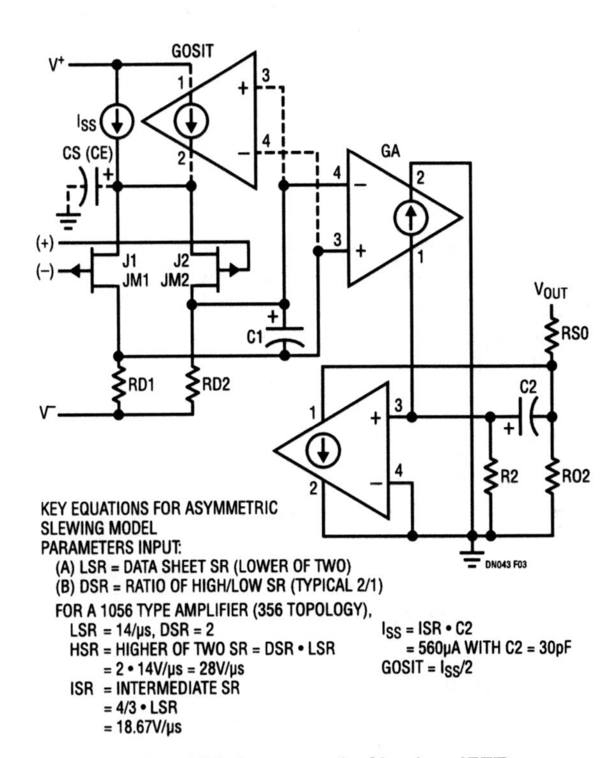

Figure 449.1 • The LTC Asymmetric Slewing JFET Macromodel Has Little Additional Complexity, but Offers Controlled Slewing Response

CM signals can be adapted (corresponds to CE in the Boyle paper). Unfortunately, this strategy works best for CM amplifier inputs, and not as well for inverting inputs.

The LTC method of modeling asymmetrical SR employs an added VCCS (shown dotted), which dynamically modifies the total tail current available to J1/J2.This controlled source, "GOSIT," is driven by the differential output of J1/J2 and produces a current which adds to or subtracts from the fixed current, I_{SS}. The resulting current available to charge/discharge compensation cap C2 is thus higher for one

Analog Circuit Design: Design Note Collection. http://dx.doi.org/10.1016/B978-0-12-800001-4.00449-X

slewing slope than it is for the opposite. This is true regardless of whether the amplifier is operating in an inverting or non-inverting input mode. As an option, CS can still be used for further control of slewing for CM inputs (shown dotted).

In generating a new macromodel with asymmetrical SR, the **lower** of the two slew rates is input from the data sheet. Also input is the **ratio** of the high-to-low SR. Algorithms in the program used by LTC then calculate an appropriate static value for I_{SS} and the gain of VCCS GOSIT, so that the proper slewing characteristic will be produced by the model.

A representative example op amp with these characteristics is the LT1056, a high performance op amp topologically much like the LF156-LF356 and OP-16 types (also produced by LTC, with corresponding macromodels available). Some sample lines of code taken directly from the LT1056 model released in version 2.0 of the LTC library are shown below. These are shown for both the asymmetric form as released, and for an (edited) symmetric case.

Actually, only one SPICE model element is added to produce the asymmetric SR as opposed to symmetric and that is the VCCS GOSIT. The LT1056 example below produces SR of $+14V/\mu s$ and $-28V/\mu s$.

```
**
C1 80 90 1.5000E-11
ISS 7 12 5.6000E-04
GOSIT 7 12 90 80 2.8000E-04
```

When the controlled source GOSIT is omitted, the model reverts to simple symmetric slewing, where the SR will be $\pm(I_{SS})/C2$. This is shown below, with I_{SS} adjusted for a (symmetric) SR of $14V/\mu s$. Those lines of code edited are shown in **bold**.

```
**
C1 80 90 1.5000E-11
* for a (symmetric) SR of 14V/µs
* iss=(1.4e7)*(3e-11)=420µA
ISS 7 12 4/2000E-04
* comment out gosit with first column "*"
* GOSIT 7 12 90 80 2.8000e-04
* intermediate
```

The noninverting mode waveforms of a typical SPICE run using the LT1056 macromodel and parallel lab results with an actual LT1056 device are shown in Figures 449.2A and 449.2B, respectively. As noted, there is quite reasonable correspondence between the two. A complete LT1056 model is contained on the LTC SPICE diskette.

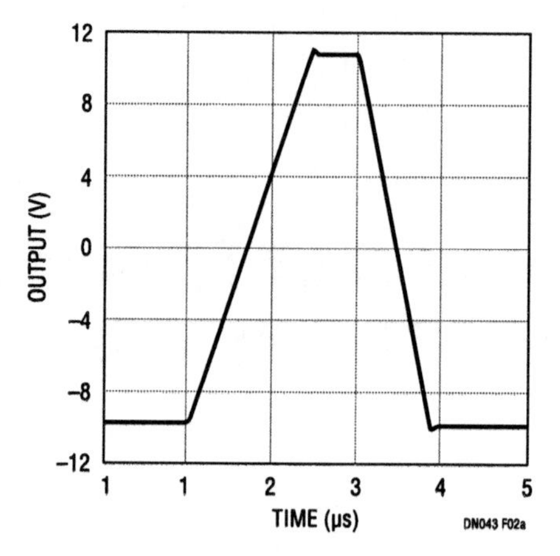

Figure 449.2A • LT1056 SR (+) Mode, Macromodel

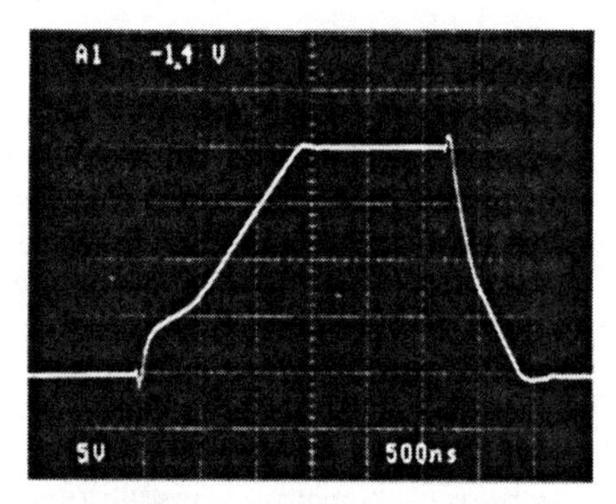

Figure 449.2B • LT1056 SR (+) Mode, Lab Photo

References

Op amp macromodels are now included as part of LTspice, available for download at www.linear.com.

1. Boyle, G.R., Cohn, B.M., Pederson, D.O., and Solomon, J.E., "Macromodeling of Integrated Circuit Operational Amplifiers," IEEE Journal of Solid-State Circuits, Vol. SC-9, #6, December 1974.
2. Solomon, J.E., "The Monolithic Op Amp: A Tutorial Study," IEEE Journal of Solid-State Circuits, Vol. SC-9, #6, December 1974.
3. Krajewska, G. and Holmes, F.E., "Macromodeling of FET/Bipolar Operational Amplifiers," IEEE Journal of Solid-State Circuits, Vol. SC-14, #6, December 1979.

450

Chopper vs bipolar op amps—an unbiased comparison

George Erdi

Over the last few years dozens of new CMOS chopper stabilized and precision bipolar op amps have been introduced. Despite the fact that these two groups compete for the same market, a valid scientific comparison of the merits of choppers and precision bipolar is unavailable. The probable explanation is that most analog IC companies have introduced products in one group or the other but not both. Therefore, articles and news releases have extolled the benefits of one, while knocking the other. Linear Technology is the only company with offerings in both groups with no vested interest in promoting one versus the other. Hence, an attempt will be made for an unbiased comparison.

Table 450.1 Chopper Stabilized vs Precision Bipolar Op Amps

PARAMETER	ADVANTAGE		COMMENTS
	CHOPPER	BIPOLAR	
Offset Voltage	✓		No Contest
Offset Drift	✓		
All Other DC Specs	✓		
Wideband, 20Hz to 1MHz		✓	See Details in Text
Noise		✓	See Details in Text
Output: Light Load	✓		Rail-to-Rail Swing
Heavy Load		✓	2mA Limit on Choppers
Single Supply Application	✓		Inherent to Choppers Needs Special Design Bipolars
±15V Supply Voltage		✓	Except LTC1150
Prejudice/Tradition		✓	Still a Chopper Problem
Cost		✓	Unless DC Performance Needed

Table 450.1 lists the parameters of importance. In all input parameters (except noise) the advantage unquestionably goes to the choppers. 5μV maximum offset voltage, 0.5μV/°C maximum drift are commonly found guaranteed parameters on all Linear Technology choppers. Changes with time and temperature cycling are near zero. These parameters cannot be measured accurately, but can be guaranteed by design; assuming that the auto-zeroing chopper loop, which can be tested independently, is working properly. The best, tightly specified bipolar op amps can only approach this performance, at the cost of great testing and yield expense.

In wideband applications bipolars get the nod. This may seem inconsistent, since typical chopper slew rate is 4V/μs, bandwidth is 2.5MHz—faster than most precision op amps. But choppers have clock frequency spikes, chopping frequency spikes, aliasing errors, millisecond overload recovery, and high wideband noise. All these factors limit the chopper's usefulness as a wideband amplifier.

The noise performance of bipolars is acknowledged to be superior. As shown in Figure 450.1 from 10Hz to 1kHz bipolar noise is nine times better. This comparison is for the industry standard LT1001 and OP-07. Bipolar designs

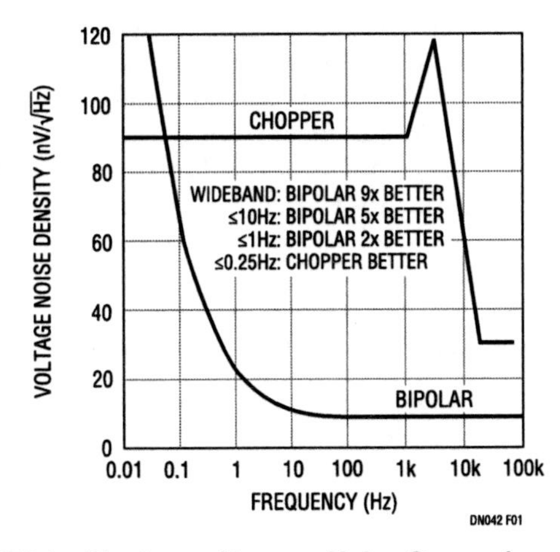

Figure 450.1 • Bipolar vs Chopper Noise Comparison

Analog Circuit Design: Design Note Collection. http://dx.doi.org/10.1016/B978-0-12-800001-4.00450-6

Table 450.2 Chopper Stabilized Op Amps

PART NUMBER	DESCRIPTION	MAX V_{OS} (25°C)	MAX TCV_{OS}	TYPICAL 0.1Hz TO 10Hz NOISE	EXTERNAL CAPS REQUIRED	MAXIMUM SUPPLY VOLTAGE
LTC1049	Single, Micropower	10μV	0.10μV/°C	3.0μVp-p	No	±9V
LTC1050	Single, Low Power	5μV	0.05μV/°C	1.6μVp-p	No	±9V
LTC1051	Dual, Low Power	5μV	0.05μV/°C	1.5μVp-p	No	±9V
LTC1052	Single, 7652 Upgrade	5μV	0.05μV/°C	1.5μVp-p	Yes	±9V
LTC1053	Quad, Low Power	5μV	0.05μV/°C	1.5μVp-p	No	±9V
LTC1150	Single, ±15V Operation	5μV	0.05μV/°C	1.8μVp-p	No	±18V

Table 450.3 Precision Bipolar Op Amps

DESCRIPTION	SINGLE	DUAL	QUAD
Low Cost, Optimum Performance	LT1001	LT1013	LT1014
	LT1012	LT1078	LT1079
	LT1097		
Low Noise, Wideband	LT1007		
	LT1028		
	LT1037		
Low Noise, Audio	LT1115		
Single Supply, Low Power	LT1006	LT1013	LT1014
Single Supply, Micropower	LT1077	LT1078	LT1079
		LT1178	LT1179

optimized for low noise, such as the LT1007, LT1028, LT1037, or LT1115, have 36 to 100 times lower noise than choppers. But choppers do not have 1/f noise, i.e., as frequency decreases bipolar noise increases, while chopper noise stays flat. If the bandwidth is limited chopper noise gets comparatively better. If signal bandwidth is cut off at 0.25Hz—a rather restrictive requirement—chopper noise is actually lower.

Chopper stabilized amplifiers are also limited to ±9V maximum supplies, excluding them from the mainstream ±15V analog applications. The new LTC1150 is the exception. The LTC1150 represents a major breakthrough; it plugs into standard ±15V sockets, yet guarantees the expected 5μV offset and 0.05μV/°C drift.

A non-scientific, yet real, parameter of comparison is prejudice/tradition. Early CMOS circuits have established a reputation of being damaged easily by electrostatic discharge, and latching up under normal operating conditions. Most of the problems were solved years ago, yet the negative impression lingers. Many system designers will not try, and therefore will not use, CMOS choppers.

The cost of precision bipolar op amps is lower than choppers. For example, the 1000 piece price of the LT1097CN8 (50μV max offset voltage, 1μV/°C max drift) is $0.97 versus the LTC1050CN8's $2.10. This, however, is somewhat of an apples to oranges comparison, because the LTC1050CN8's offset and drift performance cannot be obtained at any price on a bipolar op amp.

Table 450.2 summarizes Linear Technology's chopper stabilized op amp offerings. Table 450.3 lists the currently available precision bipolar operational amplifiers.

Ultralow noise op amp combines chopper and bipolar op amps

451

Nello Sevastopoulos

Chopper op amp technology has continuously improved. Contemporary single, dual and quad low noise chopper op amps (LTC1050/51/53), with internal sample-and-hold capacitors, are compatible with industry standard op amp sockets.

Chopper op amps are mainly used to amplify small DC signals and these applications require excellent V_{OS}, V_{OS} drift, low bias current and low noise. The outstanding V_{OS} performance of chopper op amps is well known. However, their low frequency noise compared to precision bipolar op amps is at least an order of magnitude higher. For applications which require both ultralow V_{OS} drift and ultralow noise, neither bipolar nor chopper op amps are optimum. A circuit combining the superior DC performance of the dual chopper LTC1051 with the ultralow noise voltage of the LT1007 precision bipolar op amp appears in Figure 451.1. This composite op amp can, for instance, be used as a strain gauge amplifier. One half of the LTC1051 dual chopper op amp, LTC1051, integrates the small LT1007 input offset voltage and applies a DC correction voltage at its pin 8 via divider (R2, R3). The other half of the LTC1051, buffers the V_{OS} nulling circuitry eliminating loading at input A. Resistors R1, R2, R3 allow the integrator full output swing assuring V_{OS} correction of the

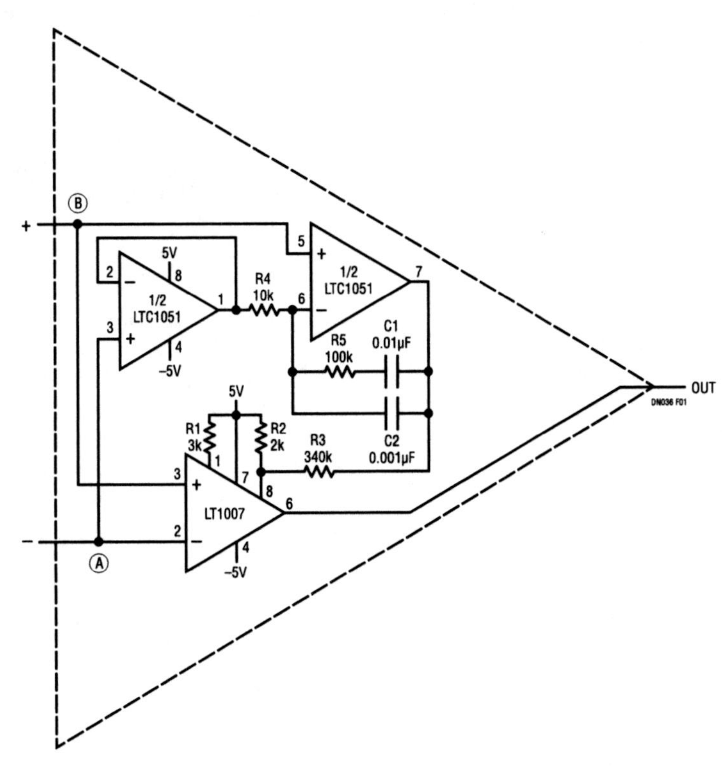

Figure 451.1 • Combining the Low V_{OS} and V_{OS} Drift of the Dual Chopper LTC1051 with the Low Noise of a Precision LT1007 Bipolar Op Amp

Analog Circuit Design: Design Note Collection. http://dx.doi.org/10.1016/B978-0-12-800001-4.00451-8

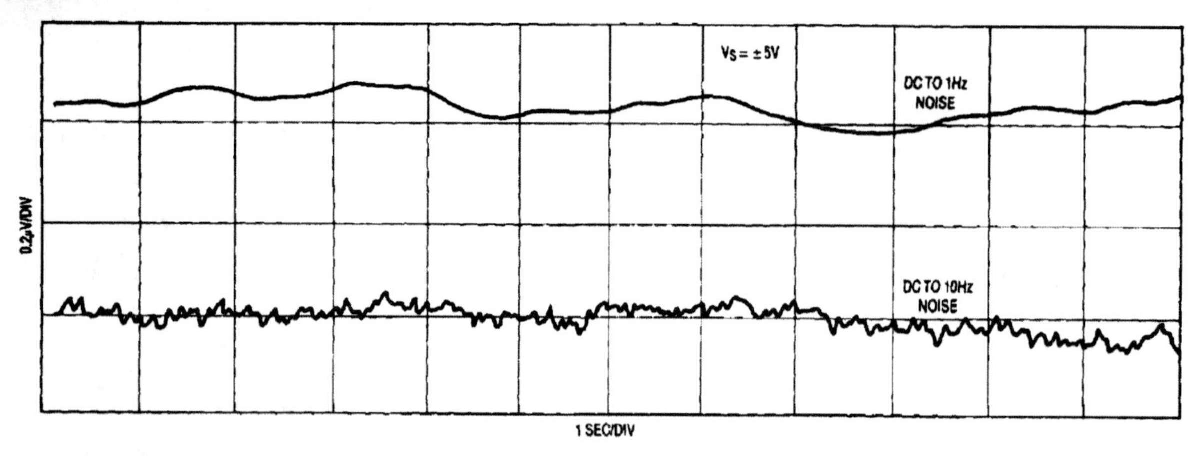

Figure 451.2 • Recorded Peak-to-Peak Noise of the Composite Op Amp, Figure 451.1, during a 10 Sec. Window

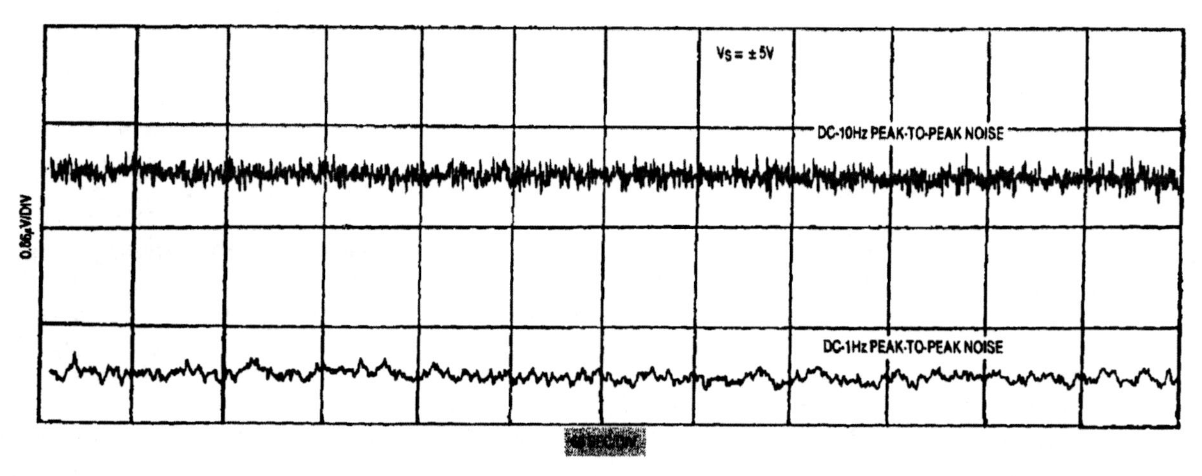

Figure 451.3 • Recorded Peak-to-Peak Noise of the Composite Op Amp, Figure 451.1, during a 10 Minute Window

LT1007. The ratio of R3 to R2 is made as high as possible to limit noise injection of the chopper into pin 8 of the bipolar op amp. The total measured input V_{OS} was 2μV, with a 10nV/°C drift.

Noise measurements

Bipolar op amp data sheets commonly specify 0.1Hz–10Hz peak-to-peak noise measured during a period of 10s. (or "10 sec. window"). The peak-to-peak noise is the difference of the highest positive noise spike to the lowest negative noise spike occurring during the 10 sec. interval. When working with chopper op amps, it is very tempting to apply this 10 sec. window test for noise measured from DC to 10Hz. In fact, the LTC1051 zeros its V_{OS} at a 2.5kHz rate, so it is reasonable to assume its noise spectral density continues to be "flat" below 0.1Hz. Experiments prove this assumption to be quite valid since the 0.1Hz to 10Hz peak-to-peak noise dominates the DC to 0.1Hz noise. For the composite op amp of Figure 451.1, however, the assumption may not be correct. The 0.1Hz to 10Hz noise of the LT1007 is at least an order of magnitude lower than the LTC1051 noise. Assuming that only a minute portion of the LTC1051 noise is injected into the LT1007 offset pin, then most 10 sec. windows should show outstanding noise results. Figure 451.2 shows this is so. Notice that for both DC-1Hz, and DC-10Hz bandwidths the peak-to-peak noise was 100nV!! Any additional noise, therefore, should be contributed by the ultralow frequency "hunting" of the V_{OS} adjusting circuit's integrator loop. Figure 451.3 shows the noise recorded in a 10 minute period. Again, results are quite impressive. The DC to 10Hz peak-to-peak noise is about the same as the DC to 1Hz peak-to-peak noise. The $0.2\mu V_{p-p}$ recorded represents a 7 to 9 times improvement over the LTC1051 DC to 10Hz noise measured under the same conditions, and a 2.5 times improvement of the equivalent DC to 1Hz noise. The turn-on settling time of the circuit is 16 seconds. Once the integrator captures the V_{OS}, the circuit's response time is undistinguishable from a normal amplifier. The circuit of Figure 451.1 should be used with source resistances less than 1kΩ to maintain noise performance.

A SPICE op amp macromodel

452

Walt Jung

Introduction

The Boyle et al. [1] SPICE macromodel for op amps has proven to be quite useful for fast and efficient computer-based IC circuit analysis, used within its limitations. Critics of this type of model point out that it is not optimum for precise transient analysis of amplifiers using complex compensation. On the other hand, the Boyle macromodel may have little match in terms of the computational speed and performance it can achieve, plus how quickly it can be implemented. These virtues are particularly true for lower frequency op amps, or where DC performance parameters are more important.

The Boyle model can be set up to give realistic and quite reasonable working approximations to a variety of IC op amps which use various types of differential transconductance pair front ends. Two fundamental advantages of this model are the relative simplicity and the simulation speed (particularly when a minimum number of junctions are used). Further, the prudent use of the appropriate transistors at the input can simulate real input offset voltage and bias current effects, as well as such IC-unique features as input common-mode clamping [2], making the overall model much more realistic and akin to real-world ICs. While the original Boyle paper used an NPN bipolar input example (the 741), the topology of the macromodel is also readily adaptable to PNP bipolars, as well as JFETs, as design options.

The LT1012

The LT1012 op amp is a popular "universal" high performance internally compensated precision op amp, available in a variety of electrical grades and packages. It uses a rather unique input stage, comprised of a bias current compensated Super-Beta NPN differential pair. This allows the desirably low drift of an NPN pair to be realized, but with typical bias currents of only 30pA, due to the use of both the Super-Beta process and bias current cancellation. Importantly, the low bias currents are *not* achieved at the expense of poor drift and high voltage noise, as the LT1012 (C grade) accomplishes a 1μV/°C (max) drift, and a 14nV/√Hz (typ) voltage noise.

The LT1012 has recently been broadened in terms of performance grades, with the addition of a premium "LT1012A" grade part, featuring 25μV (max) V_{OS}, 0.6μV/°C (max) drift, and 500μA (max) supply current. The added "LT1012D" part has a 140μV (max) V_{OS}, a 1.7μV/°C (max) drift, and an 800μA (max) supply current. All device grades have the unusual combination of performance characteristics which allow use as a low voltage (±1.2V), low supply current micropower op amp, as well as a full ±20V supply range general purpose part. The LT1012 actually exceeds the performance of the industry standard OP-07, doing so at 1/20 the bias/offset currents, and 1/8 of the supply current.

The LT1012 macromodel

While all of the above may be interesting enough to a designer, how the model imitates the real part is more so. The LT1012 macromodel listed in Figure 452.1 has a number of features worthy of mention. Note that it is based on the LT1012C room temperature typical specs, taken from the data sheet. V_{OS}, the input offset voltage of the input pair, is modeled by using two slightly different NPN transistor models, qm1 and qm2. The ratio of their two saturation currents will produce an offset voltage, which is

$$V_{OS} = Kt/q \ \text{In}(Is1/Is2)$$

With the ratio as shown, this produces the typical 10μV offset for the LT1012C.

Bias and offset currents are modeled by using a different Bf for the two input pair halves, as:

$$Bf1 = Ic1/Ib1 \text{ and } Bf2 = Ic2/Ib2$$

Analog Circuit Design: Design Note Collection. http://dx.doi.org/10.1016/B978-0-12-800001-4.00452-X

The Bf values shown for qm1 and qm2 are those which correspond to Ib = 30pA; I_{OS} = 20pA. While the gains listed are enormously high (even for Super-Beta transistors) this is not a problem for SPICE, so bias currents in the range of a typical LT1012C are produced.

Other additions to the generic Boyle macromodel are the optional input diode clamps, ddm1 and ddm2, as in the real part (they can be deleted, if not used). The substitution of a current source, Ip, in the place of the Rp of the original model simulates quiescent power supply current. The LT1012, like many modern day ICs, has a quiescent supply current which is quite constant with supply voltage, thus Ip is more appropriate than a fixed resistor.

```
*
* Linear Technology LT1012 op amp model (with calls for LT1024)
* Written: 08-17-1989 10:16:25 Type: Bipolar npn input, internal comp.
* Typical specs:
* Vos=1.0E-05, Ib=3.0E-11, Ios=2.0E-11, GBP=6.0E+05Hz, Phase mar.= 70 deg,
* SR(+)=2.0E-01V/us, SR(-)=1.9E-01V/us, Av= 126 dB, CMMR= 132 dB,
* Vsat(+)= 1 V, Vsat(-)= 1 V, Isc=+/- 12.5 mA, Iq= 380 uA
* (input differential mode clamp active)
*
* Connections: + - V+V-O
.subckt LT1012 3 2 7 4 6
* input
rc1 7  80 8.842E+03
rc2 7  90 8.842E+03
q1  80 2  10 qm1
q2  90 3  11 qm2
ddm1 2  3  dm2
ddm2 3  2  dm2
c1  80 90 5.460E-12
re1 10 12 2.246E+02
re2 11 12 2.246E+02
iee 12 4  6.000E-06
re  12 0  3.333E+07
ce  12 0  1.579E-12
* intermediate
gcm 0  8  12 0  2.841E-11
ga  8  0  80 90 1.131E-04
r2  8  0  1.000E+05
c2  1  8  3.000E-11
gb  1  0  8  0  1.960E+02
* output
ro1 1  6  1.000E+02
ro2 1  0  9.000E+02
rc  17 0  1.063E-04
gc  0  17 6  0  9.408E+03
d1  1  17 dm1
d2  17 1  dm1
d3  6  13 dm2
d4  14 6  dm2
vc  7  13 1.785E+00
ve  14 4  1.785E+00
ip  7  4  3.740E-04
dsub 4  7  dm2
* models
.model qm1 npn (is=8.000E-16 bf=7.500E+04)
.model qm2 npn (is=8.003E-16 bf=1.500E+05)
.model dm1 d   (is=1.179E-19)
.model dm2 d   (is=8.000E-16)
.ends LT1012
*
.subckt LT1024 3 2 7 4 6
x_LT1024  3 2 7 4 6 LT1012
.ends  LT1024
*
* - - - - - * fini LT1012 family * - - - - - * (oamm vnl 8/89)
```

Figure 452.1 • LT1012 Macromodel

The remaining specifications modeled are shown at the head of the listing, consistent with the LT1012C. The model can also be used for the LT1024 (dual LT1012), if the "x" call is added at the end as shown. A sample small-signal pulse response waveform of the model is shown in Figure 452.2, which can be compared to the similar condition scope photo, from the LT1012 data sheet (p. 7).

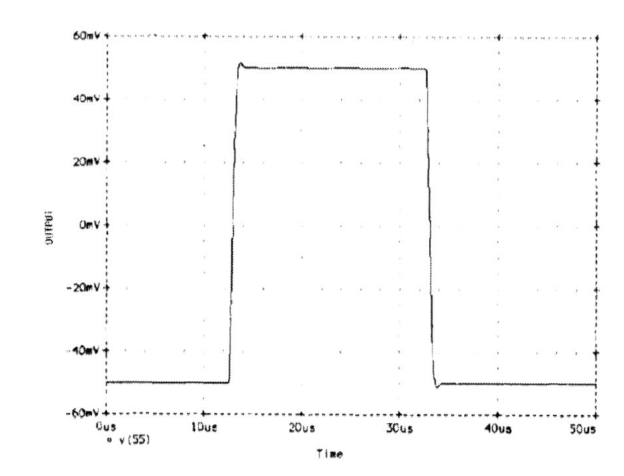

Figure 452.2 • Small-Signal Transient Response

Obtaining this macromodel

Op amp macromodels are now included as part of LTspice, available for download at www.linear.com

References

1. Boyle, G.R., Cohn, B.M., Pederson, D.O., and Solomon, J.E., "Macromodeling of Integrated Circuit Operational Amplifiers," IEEE Journal of Solid-State Circuits, Vol. SC-9, #6, December 1974.
2. Jung, W.G., "An LT1013 Op Amp Macromodel," Linear Technology Design Note Number 13, July 1988 (see Chapter 13).

A single amplifier, precision high voltage instrument amp

453

Walt Jung George Erdi

Instrumentation amplifier (IA) circuits abound in analog systems, in fact virtually any linear applications handbook will show many useful variations on the concept [1]. While this may be somewhat bewildering to a newcomer, all the variations have uses which are differentiated and valuable. A good working knowledge of the alternate forms can be a powerful tool towards designing cost-effective high performance linear circuits.

A case in point is a single amplifier *precision qualified* high voltage IA. This circuit must withstand very high common mode voltages at the input, yet it should still be relatively simple, while at the same time capable of high performance. Whereas dual summing amplifier setups can provide high input-voltage qualifications, a more simple single amp solution is often sought. An IA topology which achieves all the above objectives is shown in Figure 453.1, the "Precision High Voltage IA." The circuit employs the virtues of two key parts in performing its function; the resistor array and the op amp used with it.

Here, the resistor network is a precision high voltage design thin-film system, comprised of R1 through R6. This array, a Vishay type 444, is a thin-film SIP with a 250V/100mW input rating for R1-R3. This high voltage rating allows direct connection to AC or DC line shunts for current monitoring, level shifting from high voltage DC rails, and other such interfacing feats normally uncommon to low voltage IC circuits. The 444 network has a basic common-mode attenuation of 50 times, thus an op amp with an input voltage range of ±10V would allow a theoretical range at input pins 3–7 of ±500V. So, devices with standard ±15V supplies are basically compatible with the network operating parameters. While the network has a CM attenuation of 50 times, the differential signal scaling is nominally unity, with an error of ±0.1%. Functionally then, the differential mode input signal between pins 3–7

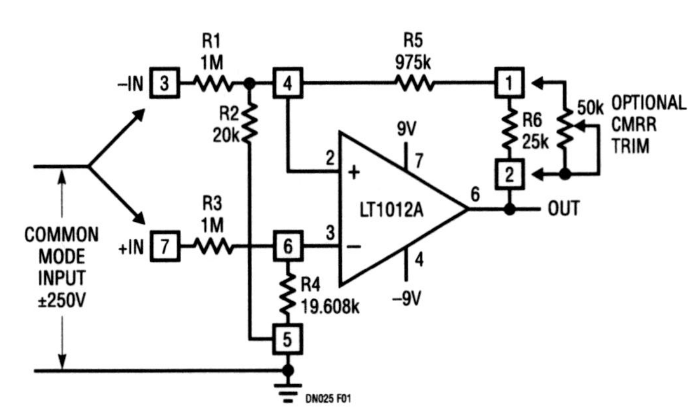

TYPICAL PERFORMANCE:
COMMON MODE REJECTION RATIO = 74dB (RESISTOR LIMITED)
WITH OPTIONAL TRIM = 130dB
OUTPUT OFFSET (TRIMMABLE TO ZERO) = 500µV
OUTPUT OFFSET DRIFT =10µV/°C
INPUT RESISTANCE =1M (CM)
2M (DIFF)
BANDWIDTH = 13kHz
BATTERY CURRENT = 370µA

R1–R6: VISHAY 444 ACCUTRACT THIN-FILM SIP NETWORK
X : VISHAY 444 PIN NUMBERS
VISHAY INTERTECHNOLOGY, INC.
63 LINCOLN HIGHWAY
MANVERN, PA 19355

Figure 453.1 • ±250 Common Mode Range Instrumentation Amplifier ($A_V = 1$)

Analog Circuit Design: Design Note Collection. http://dx.doi.org/10.1016/B978-0-12-800001-4.00453-1

is referred at the output of this circuit to the local ground (pin 5 of the network), with unity gain scaling.

A second keen application point which is a large determining factor towards the overall success of this type of IA is the relative precision of the op amp A1. Indeed, this amplifier is the second "key ingredient" towards high overall performance. Because the circuit basically amplifies the input offset voltage of A1 by the same factor as the CM attenuation, both the initial offset and the drift of A1 can become limitations, as can the CMRR of the device. Here an LT1012A op amp is used, a device with a 25μV(max) input offset voltage; the output offset will then be 1.25mV or less, worst case. The overall CMRR of the circuit has two primary sources for errors, the basic ratio match of the network halves, and to a lesser degree, the CMRR of A1. The LT1012A has a minimum CMRR of 114dB, while the network is factory trimmed to a 0.02% match, corresponding to a 74dB CMRR. For 120dB or more CMRR, a 50k trimmer can be substituted for R6.

While A1 is shown operating from ±9V battery supplies (a feature possible by virtue of the 370μA quiescent current) the LT1012 device family can also be used on standard ±15V supplies, or on lower voltage supplies down to ±1.2V (with reduced CM range, of course). With the 9V supplies shown, input ranges of ±250V or more to the circuit will not tax the network.

For single battery applications (i.e. when pin 4 is grounded), the LT1012A should be replaced by a single supply op amp such as the LT1006 or the LT1077. These devices can handle about −250mV of negative common mode voltage, while manning accuracy. Therefore, the 250V positive common mode range is unchanged, but the negative common mode range is reduced to −12V.

Using an LT1006, bandwidth and battery current are basically unchanged. With the LT1077 micropower op amp, battery current is reduced to 45μA but at the expense of bandwidth (=4.5kHz). Offset voltage and drift specifications are degraded by approximately a factor of two using the LT1006 or the LT1007 compared to the LT1012A.

Reference

1. Jung, W.G. *IC Op Amp Cookbook, 3d Ed.*, Ch 7, "Amplifier Techniques." Indianapolis, IN: Howard W. Sams, 1986.

Micropower, single supply applications: (1) a self-biased, buffered reference (2) megaohm input impedance difference amplifier

454

Walt Jung George Erdi

A self-biased, buffered reference

Voltage reference circuits are common to precision analog designs, in a wide variety of forms. They can be either two or three terminal in basic configuration, and may or may not also provide buffering against line and/or load immunity. Micropower analog circuits are growing in both fashion as well as performance, and micropower voltage references have been available. However, it is not often that a micropower reference combines common features of very low DC errors, and line/load buffering. The circuit of Figure 454.1 is an unusual form of reference circuit, in that it achieves these goals.

The leading virtue of this circuit lies in how it capitalizes on some key operating features for all of the devices used. First, the LT1034, a 1.2V two terminal reference diode allows basic low TC micropower operation, by virtue of its low minimum current requirement of only 20μA. Normally, such a diode would be fed with a simple source resistor to V^+, to maintain the bias current plus the load current. This standard shunt regulator type of use is unbuffered, so for higher load currents, the micropower aspect is lost. It can also be sensitive to line voltage changes.

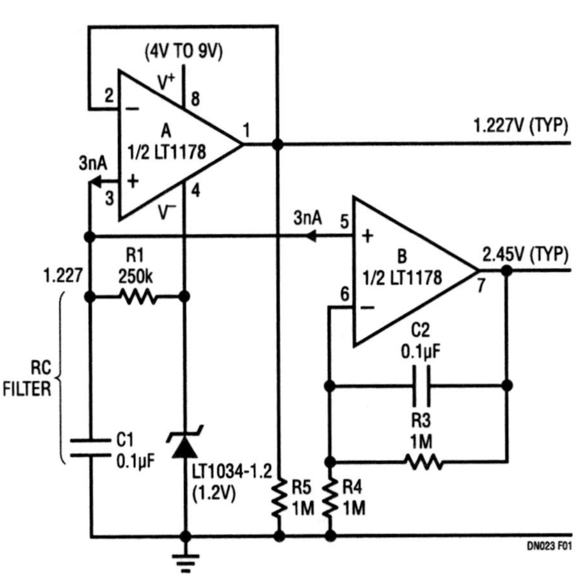

Figure 454.1 • Self-Buffered Micropower Reference

When the LT1178 op amp enters the picture, a "free" and constant bias current source is available – *the 30μA quiescent supply current of the op amp itself!* To allow the op amp to self-bias as well as voltage-buffer the reference diode, the op amp used must have both input and output swings which include the amplifiers V^- pin potential. In the case here, this potential is nominally 1.2V above ground by virtue of the reference diode's terminal voltage. More precisely, this will be 1.225V ±15mV, at the diode cathode. The overall TC of the circuit is essentially that of the LT1034 reference, or 20ppm/°C (maximum for "B" grade).

With an op amp such as the LT1178, whose input and output swing does include the negative rail, a simple follower configuration can be set up to buffer the reference voltage. R1 feeds a filtered version of the reference voltage to the A section op amp's (+) input, which is then replicated with a low source impedance by the DC follower of the A section. The second op amp section is also connected to this node, and is shown here as a precision 2X DC amplifier, providing a buffered +2.45V output. A subtle biasing step is used, where the two amplifier bias currents are combined in R1. This produces a drop of a few mV above the 1.225V, so as to set up the output stage of the A section in a more linear region. The output bleed resistor R2 also helps this biasing, by pulling a constant 0.6μA from the output of this stage.

Overall, the circuit's quiescent current is 30μA, which is essentially the bias current of the dual amplifier, plus the currents in R2 and R3. It can however source several mA of load current, to external loads. For example, the "A" stage output of 1.225V has a typical output impedance of 30μV/mA, for currents of 10mA or less.

Note that current *sinking* types of loads should be used with caution, as the sink current must necessarily flow

Analog Circuit Design: Design Note Collection. http://dx.doi.org/10.1016/B978-0-12-800001-4.00454-3

through the reference diode. While this can be as high as 20mA for the diode itself, the saturation characteristics of the A stage as used here will add some error, proportional to the current. The circuit's greatest application advantage lies with loads which source current, and so allow the true micropower operation. It operates from supplies of 3V greater than the reference voltage, in this case a battery stack of +4V to +9V. Typical line regulation is on the order of 10ppm/V.

More generally, the circuit will also function with the LT1078 op amp, a related micropower dual with a nominal 40μA/channel quiescent current, and input/output ranges similar to the LT1178. It also functions with the LT1004 type 1.2V or 2.5V references, producing proportionally scaled DC outputs, with somewhat greater drift.

If only one of the two reference outputs is needed, the LT1077 single op amp can be substituted for either side A or side B. Supply current is 45μA.

Megaohm input impedance difference amplifier

The usefulness of difference amplifiers is limited by the fact that the input resistance is equal to the source resistance. The picoampere offset current and low current noise of the LT1077 allows the use of 1MΩ source resistors without degradation in performance. In addition, with megaohm resistors micropower operation can be maintained.

Typical performance is:

Bandwidth = 25kHz
Output Offset = 0.7mV
Output Noise = 80μVpp (0.1Hz to 10Hz)
260μV RMS Over Full Bandwidth
Supply Current = 45μA

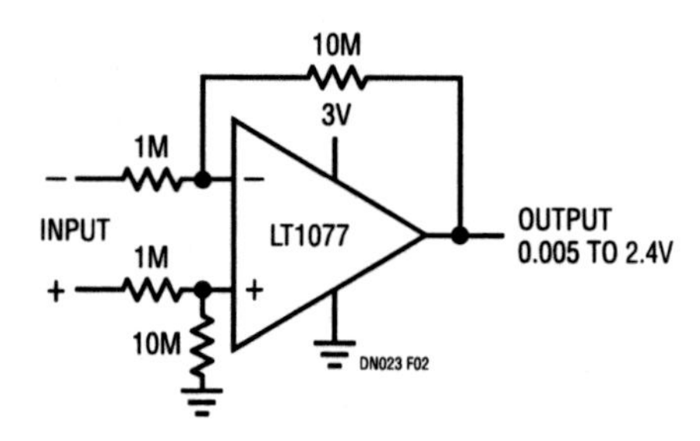

Figure 454.2 • Gain of 10 Difference Amplifier

Although the difference amplifier operates on a single 3V battery, the input common mode range extends to 250mV below ground with proper gain of 10 amplification. As the positive input is pulled further below ground to as low as −1V, the input stage saturates, but the output still stays low because the LT1077 is equipped with a unique phase reversal protection circuit. Using competitive single supply op amps in this application, the output switches high (Figure 454.2).

Another interesting feature of the LT1077 in the differential amplifier configuration is its ability to sink current while swinging to ground. Competitive micropower single supply op amps need a pull down resistor at the output to sink current, the LT1077 does not. When the input common mode voltage is 1.8V, the output has to sink a miniscule 0.16μA. However, competitive devices cannot sink any current, and need a 30k resistor from output to ground to pull the output to 5mV (5mV ≈ 30k × 0.16μA). When the output now swings to 2.4V, 80μA will flow in the pull down resistor, completely dominating the micropower current budget.

Reference

1. Jung, W.G., *IC Op Amp Cookbook, 3rd Ed., Ch 4*, "References." Indianapolis, IN: Howard W. Sams, 1986.

Noise calculations[1] in op amp circuits

455

Alan Rich

Noise calculations in op amp circuits are one of the most confused calculations that an analog engineer must perform.

One cannot just look at noise specifications; the total op amp circuit including resistors and operating frequency range must be included in calculations for circuit noise. A "low" noise amplifier in one circuit will become a "high" noise amplifier in another circuit.

As part of this design note, an IBM-PC[2] or compatible computer program, NOISE, has been written to perform the noise calculations. This program allows the user to calculate circuit noise using LTC op amps, determine the best LTC op amp for a low noise application, display the noise data for LTC op amps, calculate resistor noise, and calculate circuit noise using noise specs for any op amp. At the end of this design note there are detailed operating instructions for the computer program NOISE.

To calculate noise for an op amp circuit, one must consider the op amp voltage and current noise density and 1/f corner frequency, the frequency range of interest, and the resistor noise.

The most comprehensive specification for voltage or current noise is the noise density frequency response curve as shown in Figure 455.1.

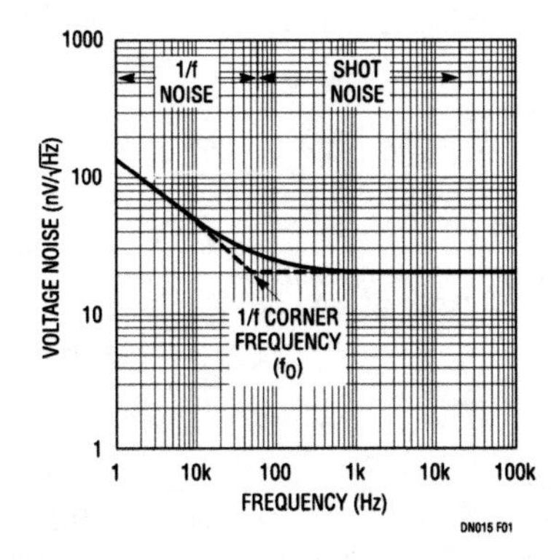

Figure 455.1 • Noise Density Frequency Response Curve

Note 1: Noise calculations are still accurate and appropriate with different op amps. However, the NOISE program has been obsoleted and is no longer supported.
Note 2: IBM made PCs in the 1980s.

There are two distinct regions to consider:

1. The high frequency part of the curve shows the shot noise and is independent of frequency.
2. The low frequency part of the curve is the 1/f noise as shown by a rapidly increasing noise density. In low frequency applications, the 1/f noise limits the minimum level of noise. The point on the curve where the asymptotes of the shot noise and 1/f noise intersect is the 1/f corner frequency.

To calculate the total RMS noise of an op amp over a bandwidth:

$$N = NO \cdot \sqrt{FC \cdot LN(FH/FL) + (FH - FL)} \quad (1)$$

where N is the RMS current or voltage noise measured from a lower frequency FL to an upper frequency FH and NO is the current or voltage shot noise density with a 1/f corner frequency FC.

Consider an audio preamplifier using an LT1037 as a simple inverting circuit (Figure 455.2) and the corresponding noise model (Figure 455.3).

EN is the voltage noise of the op amp, EN1 is the voltage noise developed by the current noise in resistors R1 and R2, EN2 is the voltage noise developed by the current noise in resistor R3, ER1 is the voltage noise of R1 and R2, and E2 is the voltage noise of R3.

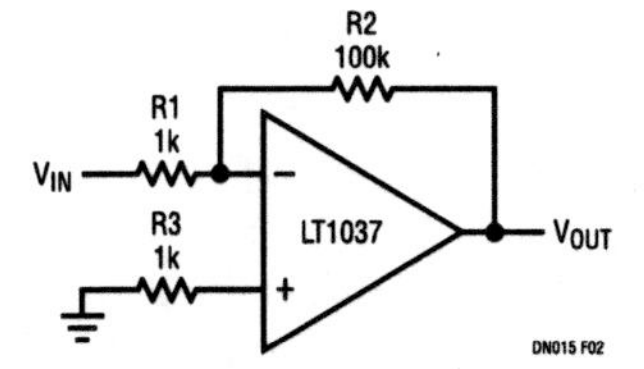

Figure 455.2 • Simple Inverting Circuit

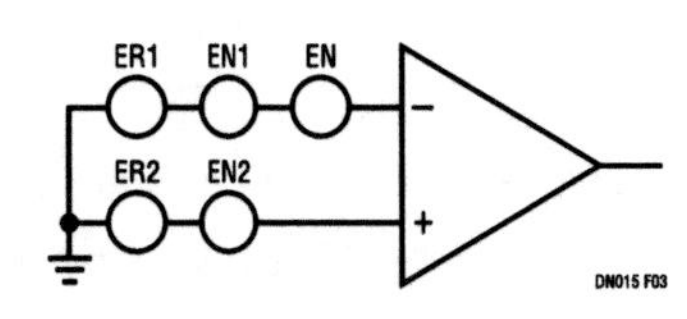

Figure 455.3 • Noise Model

Analog Circuit Design: Design Note Collection. http://dx.doi.org/10.1016/B978-0-12-800001-4.00455-5

Since we are using an LT1037 over the audio frequency range, NO = 2.5nV/$\sqrt{Hz}$, FC = 2.0Hz, FH = 20kHz, FL = 20Hz.

Plugging into Equation (1):

$$EN = 2.5 \cdot \sqrt{2 \cdot LN(20kHz/20Hz) + (20kHz - 20Hz)}$$
$$EN = 354nV, RMS$$

To calculate EN1, first the current noise must be calculated using Equation (1) and a current noise density of 0.57pA/$\sqrt{Hz}$ and 1/f corner frequency of 120Hz.

$$IN = 0.57 \cdot \sqrt{120 \cdot LN(20kHz/20Hz) + (20kHz - 20Hz)}$$
$$IN = 82pA, RMS$$

IN will flow into the parallel combination of R1 and R2.

$$EN1 = 82pA \cdot 1k||100k = 82nV, RMS$$

Similarly, EN2 results from IN flowing in R3.

$$EN2 = 82pA \cdot 1k = 82nV, RMS$$

The voltage noise of the resistors must be calculated next. In general, resistor noise is given by:

$$ER = \sqrt{4 \cdot K \cdot T \cdot R(FH - FL)}$$

where K is Boltzman's Constant, $1.39 \cdot 10^{-23}$, T is temperature (K), R is the resistor value, FH is the upper frequency, and FL is the lower frequency of interest.

At 25°C, this equation reduces to:

$$ER = \sqrt{R \cdot (FH - FL)} \cdot 1.28 \cdot 10^{-10}$$

To calculate ER1 we must consider R1 in parallel with R2,

$$ER1 = \sqrt{(1k||100k) \cdot (20kHz - 20Hz)} \cdot 1.28 \cdot 10^{-10}$$
$$ER1 = 570nV, RMS$$

Similarly, to calculate ER2,

$$ER2 = \sqrt{1k \cdot (20kHz - 20Hz)} \cdot 1.28 \cdot 10^{-10}$$
$$ER2 = 570nV, RMS$$

To calculate the total noise of the audio preamplifier using an LT1037, the RMS sum of the individual terms must be calculated.

$$TOTAL\ NOISE = \sqrt{EN^2 + EN1^2 + EN2^2 + ER1^2 + ER2^2}$$
$$TOTAL\ NOISE = \sqrt{353^2 + 80^2 + 80^2 + 570^2 + 570^2}$$
$$= 880nV, RMS$$

To calculate p–p noise, multiply the RMS noise times 6; the total peak-to-peak noise is 5.3μV for this preamplifier.

It is important to realize this noise is referred to the input of the circuit; to obtain the output noise level, the input noise must be multiplied by the noise gain which can be different from the circuit gain:

$$OUTPUT\ NOISE = TOTAL\ NOISE \cdot NOISE\ GAIN$$
$$OUTPUT\ NOISE = 88nV \cdot 101 = 89\mu V, RMS\ or$$
$$534\mu V,\ peak\text{-}to\text{-}peak$$

It should be noted that design techniques to optimize DC performance will frequently result in higher noise. For example, to minimize DC errors, a balance resistor is often placed in the +Input of an op amp to compensate for an error voltage created by bias current flowing in gain setting resistors connected to the –Input. This resistor will increase the output noise since op amp noise current must flow through the resistor, and thus create a voltage noise generator. For minimum noise levels, the resistor in the +Input should be 0Ω. As a side note, for precision op amps (LT1001, LT1007, OP07) that employ bias current cancellation techniques, this resistor should be 0Ω to minimize DC errors since the bias current equals the offset current.

Instructions for operating NOISE

NOISE is a general purpose computer program to calculate noise in op amp circuits. It will run on any IBM-PC compatible computer with a direct call from DOS.

Noise specifications and data for Linear Technology op amps (LT10XX) are contained in the program's data file. All noise specifications are based on typical specifications at 25°C.

To operate NOISE:

1. Boot the system with DOS and wait for DOS prompt "A>."
2. Insert the NOISE.EXE program disk into the A disk drive.
3. Type "NOISE" and <return>.

Operation in NOISE is menu driven throughout the program with default values on all parameters initially.

Best Op Amp for Lowest Noise vs Source Resistance

SOURCE R (R_{eq})	BEST OP AMP	
	@ LOW FREQ. (10Hz)	@ WIDEBAND (1kHz)
0Ω to 400Ω	LT1028	LT1028
400Ω to 1k	LT1007/37	LT1028
1k to 4k	LT1007/37	LT1028, LT1007/37
4k to 15k	LT1001	LT1007/37
15k to 30k	LT1001	LT1001, LT1007/37
30k to 70k	LT1001, LT1012	LT1001
70k to 150k	LT1012	LT1001, LT1012, LT1055/56/22, LT1057/58
150k to 600k	LT1012, LT1006/13/14	LT1012, LT1006/13/14, LT1055/56/22, LT1057/58
600k to 2M	LT1012, LT1055/56/22, LT1057/58	LT1012, LT1006/13/14, LT1055/56/22, LT1057/58
2M to 10M	LT1055/56/22, LT1057/58	LT1012, LT1055/56/22, LT1057/58
>10M	LT1055/56/22, LT1057/58	LT1055/56/22, LT1057/58

An op amp SPICE macromodel

Walter G. Jung

With the advent of low cost and powerful desktop computers, present day op amp circuit designs can mature more quickly with good simulation tools. One such tool since its inception has been SPICE, the standard analog circuit simulator. However, while PCs and workstations may now be present on more and more desks, a potential bottleneck towards effective simulation has been SPICE models for the more popular parts.

The macromodel approach to simulation of an op amp is viable for many designs, with the great asset of simulation speeds far faster than that of a full device-level circuit. With this design note, Linear Technology Corporation introduces op amp macromodels to its applications library. It is hoped that eventually most op amps in the product line will be developed as macromodels and made available to customers.

The LT1013 and LT1014 devices are popular single supply LTC op amps, and are thus logical candidates for macromodels. While existing macromodels for the generic 358 and 324 types might suffice for some applications, circuit designs which take advantage of the unique precision and functional features of the LT1013 warrant a model which reflects those features. The schematic diagram of the LT1013 and LT1014 macromodel is shown in Figure 456.1, and is applicable to one channel of either device.

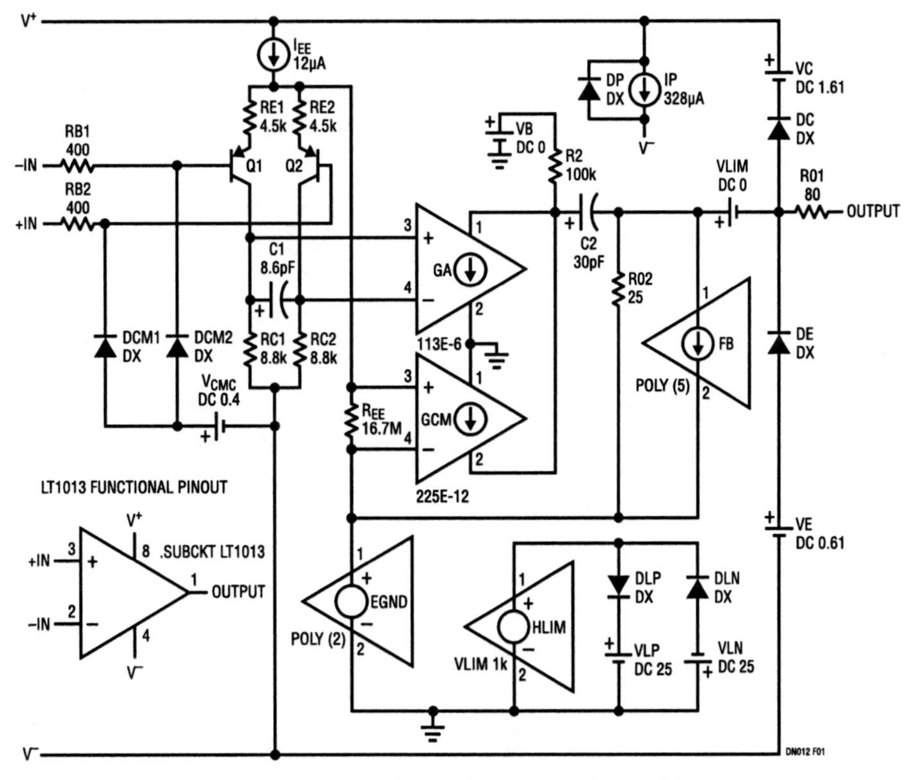

Figure 456.1 • LT1013 Op Amp Macromodel

Analog Circuit Design: Design Note Collection. http://dx.doi.org/10.1016/B978-0-12-800001-4.00456-7

Key op amp specifications for the commercial device are:

Offset voltage = 50μV (offset is not simulated)

Bias current = 8nA

Gain = 1,000,000

Slew rate = 0.4V/μs

Bandwidth = 0.8MHz

Also, the model simulates the input common mode range (which includes ground) and the output characteristics of swinging to ground while sinking current.

This macromodel acts very much like the real LT1013 or LT1014 device which incorporates input common mode clamping, to prevent the sign-reversal errors common to the 358/324 types. For example, comparison responses for the LT1013 and the 358 are shown in Figure 456.2. The common conditions of this test are for an overdriven, +5V single supply follower. In both instances, the input signal is V_{IN}, a −20V to 20V sweep fed through 10k, while the output is V(5).

Note that with the 358, the output reverses sign when the input is overdriven below ground. In contrast, the LT1013 model is well behaved, simply clamping the overdrive at ground level...just like the real LT1013 device does!

The model itself is listed on this page, and can be entered by typing it in (carefully!). Registered users of MicroSim's PSPICE simulator will automatically receive this macromodel as part of the model library update with version 3.07. Interested readers may contact MicroSim at the address or phone number listed at the end of this note for further information.

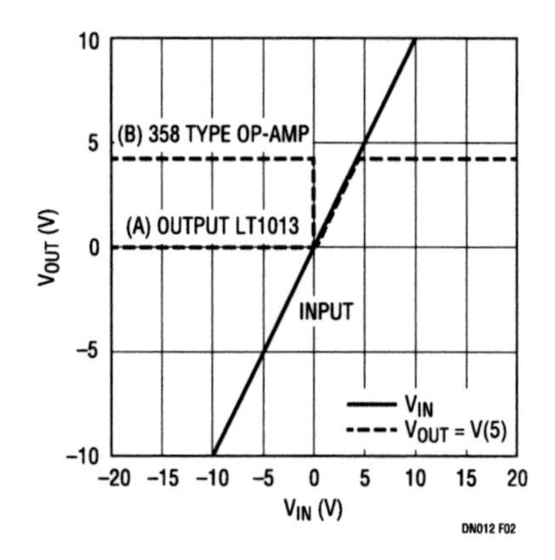

Figure 456.2 • LT1013 Test Circuit: Single Supply (+5V), Overdriven Follower

This LT1013/LT1014 op amp macromodel is being supplied to users as an aid to circuit designs. While it reflects reasonably close similarity to the actual device in terms of performance, it is not suggested as a replacement for breadboarding. Simulation should be used as forerunner or supplement to traditional lab testing.

This more complete macromodel has been adapted from the Parts program generated LT1013/LT1014 model. This version features closer fidelity to the real part, with input common mode clamping and compensated output clamping. It can be used for large signal and/or single supply applications, where the inputs can potentially be overdriven.

SPICE List for LT1013 Macromodel

```
connections:          non-inverting input
*                     |    inverting input
*                     |    |    positive power supply
*                     |    |    |    negative power supply
*                     |    |    |    |    output
*                     |    |    |    |    |
.subckt LT1013        1    2    3    4    5
*
*
c1     11    12                 8.661E-12
c2     6     7                  30.00E-12
dc     8     53    dx
de     54    8     dx
dip    90    91    dx
din    92    90    dx
dp     4     3     dx
egnd   99    0                  poly(2) (3,0) (4,0) 0 .5 .5
fb     7     99                 poly(5) vb vc ve vip vin 0
                                2.475E9 -2E9 2E9 2E9 -2E9
ga     6     0     11    12     113.1E-6
gcm    0     6     10    99     225.7E-12
iee    3     10    dc           12.03E-6
hlim   90    0                  vlim 1k
q1     11    102   13    qx
q2     12    101   14    qx
rb1    2     102   400
rb2    1     101   400
dcm1   105   102   dx
dcm2   105   101   dx
vcmc   105   4     dc           0.4
r2     6     9                  100.0E3
rc1    4     11                 8.841E3
rc2    4     12                 8.841E3
re1    13    10                 4.519E3
re2    14    10                 4.519E3
ree    10    99                 16.63E6
ro1    8     5     80
ro2    7     99    25
ip     3     4                  328E-6; supply current
vb     9     0     dc    0
vc     3     53    dc           1.610
ve     54    4     dc           .61
vlim   7     8     dc    0
vip    91    0     dc    25
vin    0     92    dc    25
.model dx D(Is = 800.0E-18)
.model qx PNP(Is = 800.0E-18 Bf = 400)
.ends
```

PSPICE simulator is available from:
MicroSim
23175 La Cadena Drive
Laguna Hills, CA 92653
(714) 770-3022; (800) 826-8603

Operational amplifier selection guide for optimum noise performance

457

George Erdi

The LT1028 is the lowest noise op amp available today. Its voltage noise is less than that of a 50Ω resistor. In other words, if the LT1028 is operated with source resistors in excess of 50Ω, resistor noise will dominate. If the application requires large source resistors, the LT1028's relatively high current noise will limit performance, and other op amps will provide lower overall noise.

In general, the total noise of any op amp (referred to the input) is given by:

$$\text{total noise} = \sqrt{(\text{voltage noise})^2 + (\text{resistor noise})^2 + \left(\text{current noise} \cdot R_{eq}\right)^2}$$

where,

$$\text{resistor noise} = 0.13\sqrt{R_{eq}} \text{ in nV}\sqrt{\text{Hz}}$$

$$\text{and } R_{eq} = \text{equivalent source resistance} = R2 + R1//R3$$

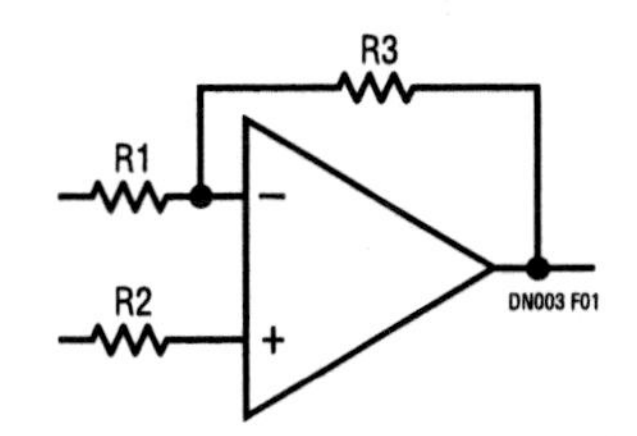

Several conclusions can be reached by inspection of the equation:

(a) To minimize noise, resistor values should be minimized to make the contribution of the second and third terms of the equation negligible. Don't forget, however, that feedback resistor R3 is a load on the output.

(b) Total noise is dominated by:

(i) voltage noise at low R_{eq},

(ii) resistor noise at mid R_{eq},

(iii) current noise at high R_{eq}, because resistor noise is proportional to $\sqrt{R_{eq}}$, while the current noise contribution to total noise is proportional to R_{eq}.

The table below lists which op amp gives minimum total noise for a specified equivalent source resistance. A two step procedure should be followed to optimize noise:

1. Reduce equivalent source resistance to a minimum allowed by the specific application.
2. Enter the table to find the optimum op amp.

Best Op Amp for Lowest Noise vs Source Resistance

SOURCE R (R_{eq})	BEST OP AMP @ LOW FREQ. (10Hz)	BEST OP AMP @ WIDEBAND (1kHz)
0Ω to 400Ω	LT1028	LT1028
400Ω to 1k	LT1007/37	LT1028
1k to 4k	LT1007/37	LT1028, LT1007/37
4k to 15k	LT1001	LT1007/37
15k to 30k	LT1001	LT1001, LT1007/37
30k to 70k	LT1001, LT1012	LT1001
70k to 150k	LT1012	LT1001, LT1012 LT1055/56/22, LT1057/58
150k to 600k	LT1012, LT1006/13/14	LT1012, LT1006/13/14, LT1055/56/22, LT1057/58
600k to 2M	LT1012, LT1055/56/22, LT1057/58	LT1012, LT1006/13/14, LT1055/56/22, LT1057/58
2M to 10M	LT1055/56/22, LT1057/58	LT1012, LT1055/56/22, LT1057/58
>10M	LT1055/56/22, LT1057/58	LT1055/56/22, LT1057/58

Analog Circuit Design: Design Note Collection. http://dx.doi.org/10.1016/B978-0-12-800001-4.00457-9

The table actually has two sets of devices: one for low frequency (instrumentation), one for wideband applications. The slight differences between the two columns occur because voltage and current noise increase at low frequencies (below the so-called 1/f corner) while resistor noise is flat with frequency.

The actual achievable total noise is plotted at 10Hz and 1kHz. The striking feature of these plots is that with the proper selection of op amps total noise is dominated by equivalent source resistor noise over a five decade (100Ω to 10MΩ) range.

100Hz Total Noise vs Equivalent Source Resistance

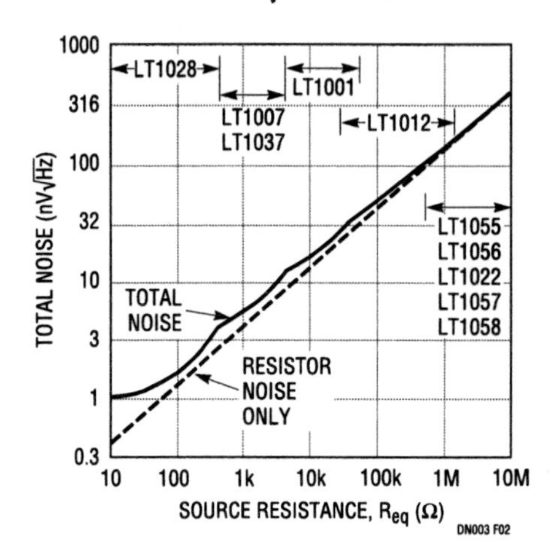

1kHz Total Noise vs Equivalent Source Resistance

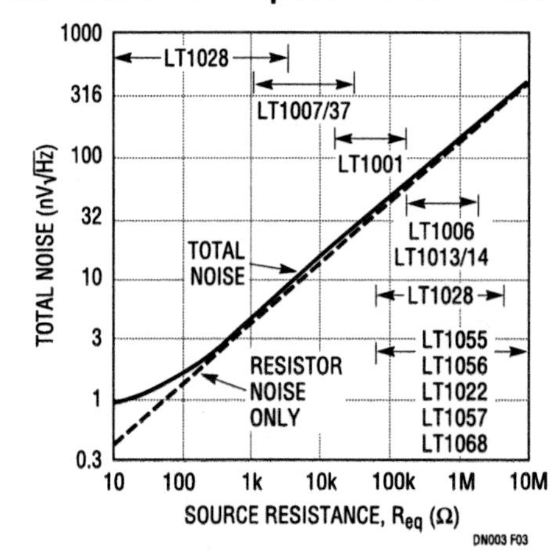

Section 2

Special Function Amplifier Design

Ultraprecise current sense amplifier dramatically enhances efficiency and dynamic range

458

Jon Munson

Introduction

Accurate current measurement is indispensable in many electronic systems. The current is usually measured by amplifying the voltage it generates across a small value resistance. For systems that require a large dynamic measurement range, the sense resistance must be increased or the precision of the amplifier must be improved. Increasing the value of the sense resistor has the detrimental effect of increasing power dissipation. The better option is to improve the precision of the sense amplifier.

Amplifier precision depends a great deal on the input offset voltage of the amplifier. Historically, current sense amplifiers on the market offered input offset voltage performance on the order of hundreds or even thousands of μV. With such parts, achieving a practical dynamic range of 8 to 10 bits can cost more than a watt of power dissipation at full operating current. The LTC6102 ultraprecise current sense amplifier reduces input error to a miniscule 10μV. This dramatic performance enhancement translates directly into a greater measurement dynamic range—16 bits is possible even while *lowering* power dissipation in the sense resistor—thus greatly expanding the gamut of current sensing design options.

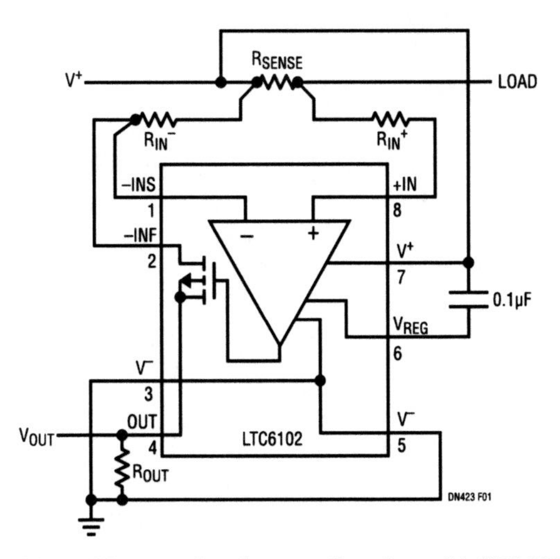

Figure 458.1 • Ultraprecise Current Sensing with LTC6102

Precision buys efficiency

The LTC6102 is easily connected as shown in Figure 458.1.

The input voltage is developed by the sense resistor, and the voltage gain of the amplifier is set by the input and output resistors. The overall scaling is simply:

$$V_{OUT} = I_{LOAD}\left(R_{SENSE} \cdot \frac{R_{OUT}}{R_{IN^-}}\right)$$

The accuracy at small load currents is primarily set by the input offset voltage V_{OS}. The current measurement error I_{OFFSET}, due to the V_{OS}, is given by:

$$I_{OFFSET} = \frac{V_{OS}}{R_{SENSE}}$$

For a given current offset accuracy requirement, it can be seen that with a low V_{OS} that R_{SENSE} may be reduced accordingly, to sub-milliohms in many applications.

In most applications the circuit gain is selected so that V_{OS} translates to about 1LSB (least significant bit) in the analog-to-digital (ADC) acquisition system. Dynamic range is dictated by the maximum signal amplitude that the ADC can handle and how much power the R_{SENSE} resistor is permitted to dissipate.

Consider a comparison between two 8-bit sense amplifier solutions, one using a typical amplifier with $V_{OS} = 500\mu V$ and one using the LTC6102, where $V_{OS} = 10\mu V$. The resolution of each is 20mA. The higher offset part requires a sense resistor of at least 25mΩ, whereas the LTC6102 only requires 500μΩ. At 5A, nearly full-scale current for this example, the R_{SENSE} power loss is 625mW with the higher offset part, but just 13mW with the LTC6102, a 98% reduction in wasted power.

Analog Circuit Design: Design Note Collection. http://dx.doi.org/10.1016/B978-0-12-800001-4.00458-0

Print your own sense resistors

With the ultralow sense resistance capability offered by the LTC6102, the printed circuit foil itself can be used as a practical sensing element. A circuit board using 1oz copper has a nominal sheet resistivity of 500μΩ/square. The value drops proportionally for thicker foils and rises for thinner foils. A trace of width W and length L (in any identical units) has the following resistance:

$$R_{SENSE} \approx 500\mu\Omega \cdot \frac{L}{W}$$

The length of the resistor is simply the spacing between the Kelvin taps along the trace. One ounce copper can generally carry up to about 100mA/mil of trace width (or 4A/mm), which constrains the minimum size of the resistor structure. Another constraint is reproducibility, so the larger, the better. Ultimately the thickness tolerance and tempco of the copper limit the accuracy a printed resistor can have.

Figure 458.2 shows a printed structure for the 5A circuit example discussed previously. In this layout, the L/W factor is set to 1 (for $R_{SENSE} = 500\mu\Omega$) and the size is dictated mainly by the accuracy of printed circuit etching.

Using copper for the sense resistance means that the scaling of the circuit is nearly proportional to absolute temperature, about +0.4%/°C at room temperature. In applications where the current is being monitored for overload protection, the tempco may be convenient, in that a fixed protection threshold will automatically correspond to lower current at higher temperature. For stable measurements, a software calibration and temperature correction approach can be used, or the tempco can be compensated by using a copper-based resistor for R_{IN^-}, such as a small surface mount inductor with known resistance properties (>10Ω readily available, e.g., Vishay IMC series).

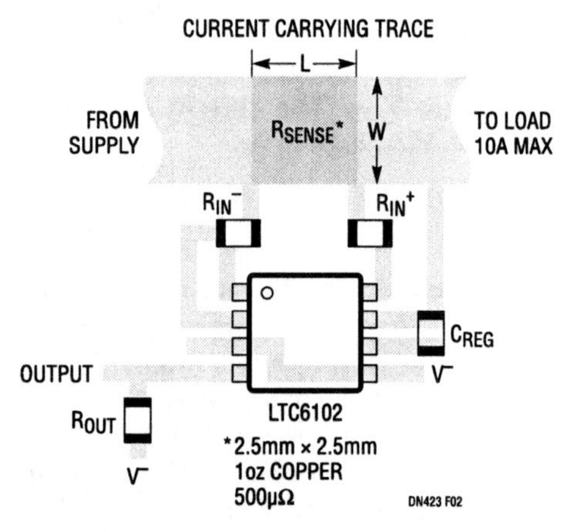

Figure 458.2 • LTC6102 Layout Using Printed Sense Resistance

Design tips and details

If you are not printing your own sense resistors, and need the accuracy of off-the-shelf components, be sure to specify 4-wire (Kelvin) sense resistors for best results. Such resistors are designed so that the resistance is well calibrated between the sensed taps, thus eliminating the error from solder resistance in the load path.

Accurately measuring microvolt level signals raises the real possibility of stray thermocouple effects due to dissimilar metallic interconnections. Figure 458.1 shows the use of an R_{IN^+} that is generally identical to R_{IN^-}. The purpose of this extra resistor is to provide identical metallurgical conditions to both amplifier inputs for minimizing thermocouple effects, as well as to minimize DC bias current imbalance.

The R_{IN^-} value is selected to conduct about 500μA at times of peak measured current I_{PEAK}. The voltage drop on R_{IN^-} is equal to the voltage drop on R_{SENSE}, so:

$$R_{IN^-} \geq \frac{I_{PEAK} \cdot R_{SENSE}}{0.0005}$$

Gain accuracy of the overall circuit is established mainly by the quality of the resistors used. This allows the designer to optimize the cost vs performance trade-off in each specific application.

To minimize copper loss errors in the feedback loop of the LTC6102, the inverting sense input (–INS) and the inverting feedback connection (–INF) have been kept separate so that a Kelvin connection to R_{IN^-} can be made. This connection can also be seen in the suggested layout of Figure 458.2.

Figure 458.2 shows the V^+ connections tied to the load side of R_{SENSE}, whereas Figure 458.1 shows a tie-in to the supply side. The LTC6102 will work in either configuration. The difference is that the Figure 458.2 connection will also include the LTC6102 quiescent supply current (300μA typically) in the measured load current. Supply voltages from 4V to 100V are supported.

Conclusion

The LTC6102 is the industry's highest precision current sense amplifier. The exceptional accuracy allows for dramatic reduction in the R_{SENSE} resistance, thereby improving efficiency, dynamic range and current handling.

Dual current sense amplifiers simplify H-bridge load monitoring

459

Jon Munson

Introduction

The H-bridge power-transistor topology is increasingly popular as a means of driving motors and other loads bidirectionally from a single supply potential. In most cases there is great benefit in monitoring the current delivered to the load and utilizing this information in real time to provide operational feedback to a control system. In most new designs, pulse-width modulation (PWM) techniques are used to provide highly efficient variable power delivery, but this places extremely fast voltage transitions at both terminals of the load and therefore complicates the instrumentation problem. New high side current sense amplifiers from Linear Technology can simplify this problem.

Measuring load current in the H-bridge

The classical approach to load monitoring is to place a small value sense-resistance in series with the load that can develop a measurable voltage drop representing the load current (see Figure 459.1). The difficulty here is that with PWM activity, the common mode voltage at the sense resistor has nasty voltage transitions that can corrupt the sense amplifier operation with high frequency hash. While this hash can be filtered to recover useful low frequency information, the ability to provide fast fault protection is then lost. Additionally, this "flying" sense resistor configuration is unable to monitor switch shoot-through current, leaving many important fault modes undetected or unmanaged (failed switch function, for example).

A far more practical method is to monitor the supply current fed to each half-bridge as shown in Figure 459.2. This scheme provides several benefits that simplify and improve the circuit performance. The main improvement comes from having the sense resistors at a relatively constant common mode voltage (i.e., the power supply voltage) so that fidelity of the PWM current waveform can be preserved. Additionally, by monitoring each half-bridge individually at the supply side, both failed power device operation and load shorts to ground are readily detected and manageable.

By using PWM logic that generates "sign-magnitude" control, one of the half-bridges is in a 100% pull high condition (depending on the direction or polarity of drive). The load current equals the current delivered through the 100% (fully on) switch, unaffected by the duty cycle of the PWM activity

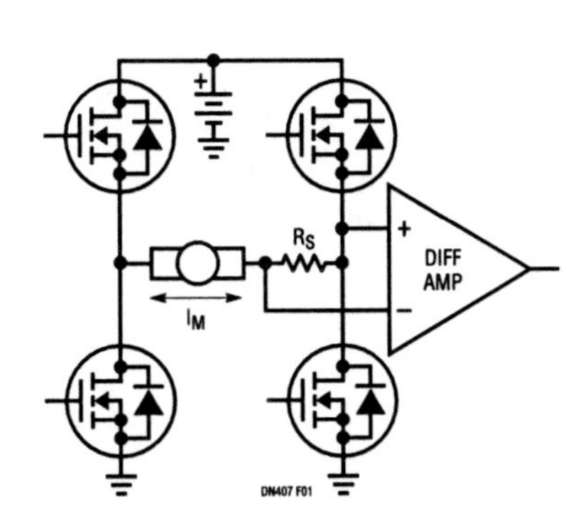

Figure 459.1 • Classical Load Sensing Problematic with PWM

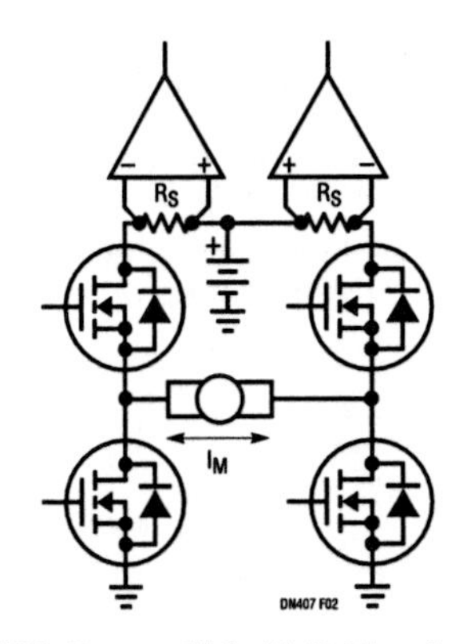

Figure 459.2 • PWM-Compatible H-Bridge Load Sensing

Analog Circuit Design: Design Note Collection. http://dx.doi.org/10.1016/B978-0-12-800001-4.00459-2

on the other half-bridge. This permits simple reconstruction of the load current waveform using suitable high side sense amplification techniques.

The simple solution

The LTC6103 and LTC6104 dual high side sense amplifiers are ideal for performing the H-bridge monitoring function. Both parts include two current sense input channels and furnish either two unidirectional outputs (LTC6103) or a single bidirectional output (LTC6104). Since each current sense channel operates in a unidirectional fashion, only the current from the fully on half-bridge is monitored.

Since the current pulses in the other half-bridge are in the opposite direction, that amplifier channel remains in a cutoff condition and does not impact the reading. This means that the output signals only reflect the fully-on half-bridge current, which is identical to the controlled load current.

With their fast (microsecond level) response times, these parts also offer overload sensing, thereby providing the ability to signal power device protection circuits in the event of fault conditions. Both parts are furnished in tiny MSOP-8 packages for compact layouts and can operate with up to 60V power supply potentials. With their 70V transient capability, the need for additional surge suppression components is eliminated in harsh automotive applications.

The dual outputs of the LTC6103 can be used individually to provide overload detection, and/or may be taken as a differential pair to provide a bidirectional signal to an analog-to-digital converter (ADC) for example. Figure 459.3 shows a typical circuit for a generic H-bridge application. The power devices may be complementary MOSFETs, pure N-MOSFETs, or other switching devices. When the bridge drives the load (a motor assumed in the example shown), one of the LTC6103 outputs rises above ground, while the other remains pulled down to ground, thereby forming an accurate bidirectional differential output with a common mode voltage that never falls below ground. The selection of output resistance (4.99k in the example) can be scaled to satisfy the source-impedance requirement of any ADC.

As an alternative, the output structure of the LTC6104 provides a single bidirectional signal. The output connection can either source or sink current to a load resistance, depending on which input channel is sensing current flow. A negative-going output swing remains linear as long as Pin 4 (V^-) is lower than the lowest expected output level by at least 0.5V. This condition is met if the load resistance is returned to a suitable reference voltage while Pin 4 is grounded (as shown in the Figure 459.4 example). The output resistance could also be returned directly to ground to form a true bipolar output if Pin 4 is tied to a suitable negative supply, such as −3V.

Conclusion

Designing a load current monitor for an H-bridge power driver is not difficult if you have the right amplifier. The LTC6103 and LTC6104 fit the bill. They include dual sense inputs and a choice of two different output configurations—features that reduce complexity and printed circuit area.

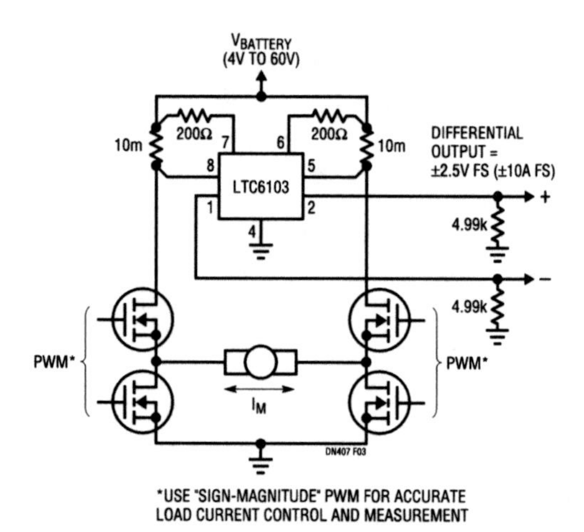

Figure 459.3 • LTC6103 Provides Bidirectional H-Bridge Monitoring with ADC-Friendly Differential Output

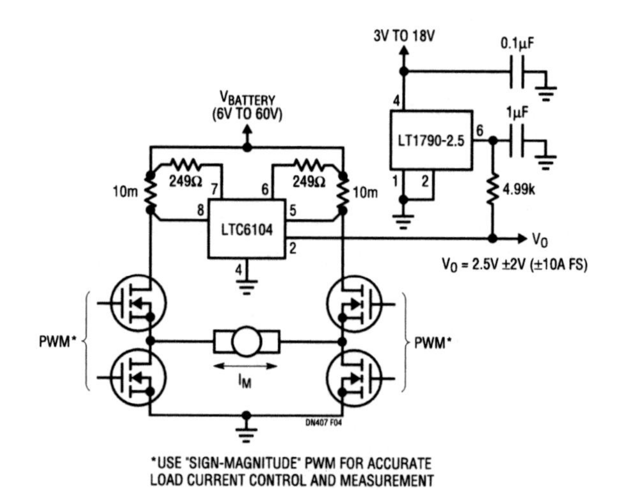

Figure 459.4 • LTC6104 Provides Bidirectional H-Bridge Monitoring with Single-Ended Output

Precise gain without external resistors

460

Glen Brisebois

Introduction

Inventory and manufacturing have associated costs and logistical headaches. "If only we could simplify our stock room and our manufacturing kits." The LT1991 provides so many functions that it may be the last amplifier you ever have to stock. This is *not* a limited-use single-application difference or instrumentation amplifier. This is a flexible part that can be configured into inverting, noninverting, difference amplifiers, and even buffered attenuators, just by strapping its pins. It provides you with internal precisely matched resistors and feedback capacitors, so you can easily configure it into hundreds of different gain circuits without external components.

The LT1991 offers the simplicity of using just one amplifier for unlimited applications—simply hook it up for type and gain and move on. The precisely matched internal resistors, op amp, and feedback capacitor simplify design, reduce external complexity and test time, lower pick and place costs, and make for easy probing. All of this in a small MSOP package.

The resistors: 0.04% worst case

The LT1991 is shown in Figure 460.1. The internal resistors offer 0.04% worst-case matching and 3ppm/°C MAX matching temperature coefficient. They are nominally 50k, 150k and 450k. One end of each resistor is connected to an op amp input, and the other is brought out to a pin. The pins are named "M" or "P" depending on whether its resistor goes to the "minus" or "plus" input, and numbered "M1" M3" or "P9" etc. according to the relative admittance of the resistor. So the "P9" pin has 9 times the admittance (or force) of the "P1" pin. The 450k resistors connected to the M1 and P1 inputs are not diode clamped, and can be taken well outside the supply rails.

The op amp: precision, micropower

The op amp is exceptionally precise, with 15μV typical input offset voltage, 3nA input bias current and 50pA of input offset current. It operates on supplies from 2.7V to 36V with rail-to-rail outputs, and remains stable while driving capacitive loads up to 500pF. Gain bandwidth product is 560kHz, while drawing only 100μA supply current.

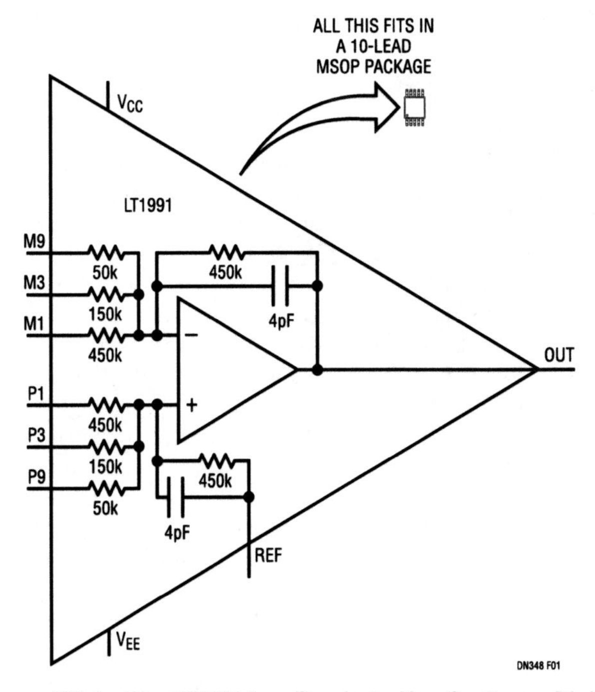

Figure 460.1 • The LT1991 is a Ready-to-Use Op Amp with its Own Resistors and Internal Signal Capacitors, All in a Tiny MSOP Package. Just Wire It Up

Analog Circuit Design: Design Note Collection. http://dx.doi.org/10.1016/B978-0-12-800001-4.00460-9

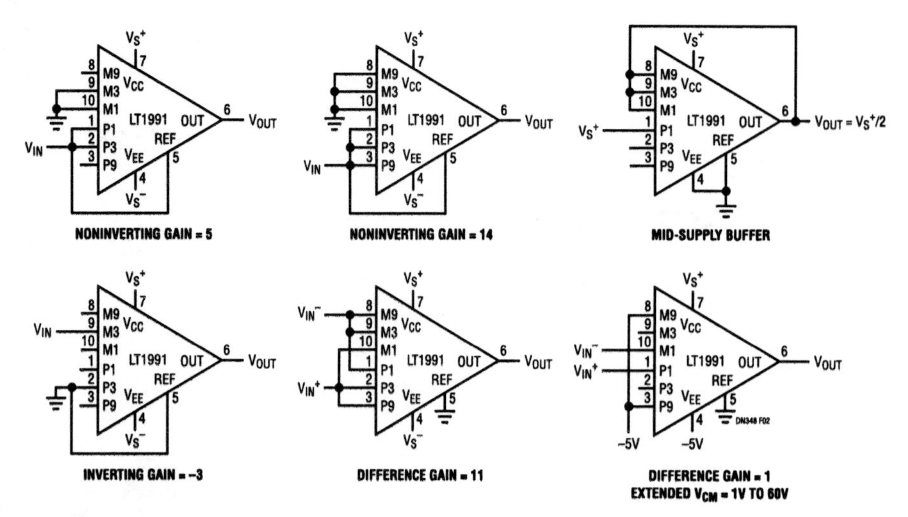

Figure 460.2 • Noninverting, Inverting, Difference Amplifier and Buffered Attenuators: Just a Few Examples Achieved Simply by Connecting Pins on the LT1991. No External Resistors Required

So easy to use

The LT1991 is exceptionally easy to use. Drive or ground or float the P, M, and REF inputs to set the configuration and gain. The M1 and P1 inputs of the LT1991 are not diode clamped, so they can withstand ±60V of common mode voltage. There is a whole series of high input common mode voltage circuits that can be created simply by strapping the pins. Figure 460.2 shows a few examples of different configurations and gains. In fact, there are over 300 unique achievable gains in the noninverting configuration alone. Gains of up to 14 and buffered attenuations down to 0.07 are possible.

Battery monitor circuit

Many batteries are composed of individual cells with working voltages of about 1.2V each, as for example NiMH and NiCd. Higher total voltages are achieved by placing these in series. However, the reliability of the entire battery pack is limited by the weakest cell, so users often like to maintain data on individual cell charge characteristics and histories. Figure 460.3 shows the LT1991 configured as a difference amplifier in a gain of 3, applied across the individual cells of a battery through a dual 4:1 mux. Because of the high valued 150k resistors on its M3 and P3 inputs, the error introduced by the mux impedance is negligible. As the mux is stepped through its addresses, the LT1991 takes each cell voltage, multiplies it by 3 and references it to ground for easiest measurement. Note that worst-case combinations, such as one cell much higher than all the others, can cause the LT1991 output to clip. Connecting the MSB line to the M1 and P1

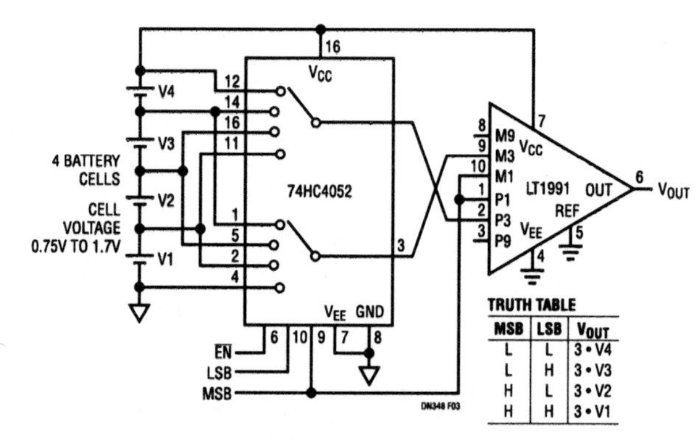

Figure 460.3 • LT1991 Applied as an Individual Battery Cell Monitor for a 4-Cell Battery

inputs helps reduce the effect of the wide input common mode fluctuations from cell to cell. The low supply current of the LT1991 makes it particularly suited to battery-powered applications. With its 110μA maximum supply current specification, it has about the same maximum supply current specification as the CMOS mux!

Conclusion

The precision LT1991 is so simple to use, small, and versatile, it is possible to stock this one amplifier and use it for many varied applications. No external components are needed to achieve hundreds of gains in noninverting, inverting, difference and attenuator configurations. Just strap the pins and go. It's a great way to reduce inventory, ease manufacturing, and simplify a bill of materials.

Sense milliamps to kiloamps and digitize to 12 bits

461

Richard Markell Glen Brisebois

Introduction

The LT1787 high side current sense amplifier provides a precision measurement of current into or out of a power source. The part generates an output voltage that is directly proportional to the current flowing through an external current sense resistor. With a miniscule 40μV (typical) input offset voltage, the LT1787 has a better than 12-bit, zero-cross accurate dynamic range at 250mV full-scale input. A hefty 60V maximum supply voltage specification allows the part to be used not only in low voltage battery applications but also in higher voltage telecom and industrial applications. Independent V_{EE} and V_{BIAS} pins make the application of the LT1787 extremely versatile.

The device is self-powered from the supply that it is monitoring and requires only 60μA of supply current. The power supply rejection ratio of the LT1787 is in excess of 120dB. The part has a fixed voltage gain of 8 from input to output.

Additional LT1787 features include provisions for input noise filtering (both differential and common mode) and the ability to operate over a very wide supply range of 2.5V to 60V. A functional diagram of the part and its theory of operation is detailed in the sidebar. The part is available in both 8-lead SO and MSOP packages.

Operation with an A/D converter

Figure 461.1 shows a detailed schematic of a high resolution (12-bit), bidirectional current-to-bits converter using an LT1787, the LT1783 SOT-23 1.2MHz micropower, rail-to-rail op amp and the LTC1404 SO-8 packaged, 12-bit analog-to-digital converter with shutdown. The circuit, as shown, allows digitization of input current from approximately −3A to approximately 2A with 12-bit resolution.

The LT1787's output voltage is buffered by the LT1783, filtered and applied to the LTC1404's A_{IN} pin. The precision bias voltage applied to the V_{BIAS} pin of the LT1787 is obtained from the LTC1404's reference output and is typically 2.43V. This bias voltage sets the zero-current output of the LT1787 to 2.43V, or approximately mid-range for the A-to-D converter, allowing measurement of current in both the positive and negative directions. Note that the LTC1404's internal reference voltage is 4.096V; thus the output of the converter is 1 count per millivolt of input voltage from the

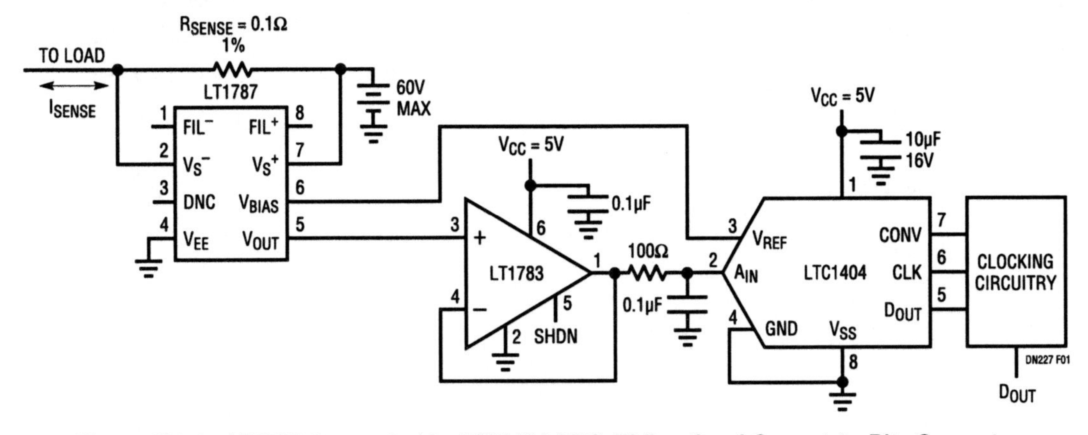

Figure 461.1 • LT1787 Connected to LTC1404 ADC: Bidirectional Current-to-Bits Converter

Analog Circuit Design: Design Note Collection. http://dx.doi.org/10.1016/B978-0-12-800001-4.00461-0

LT1787's output, or 8 counts/mV of input voltage across R_{SENSE}. Zero sense-current translates to 2.43V or 2430 counts from the converter. Full-scale positive current is 4096 counts and full-scale negative current is 0 counts.

Figure 461.2 shows current versus output code for the complete system. The LTC1404 has a digital interface consisting of the CLK and CONV input and the D_{OUT} serial digital output. The signals provide wide flexibility in allowing the part to be interfaced to most microprocessors and DSPs.

The output voltage, V_{OUT}, of the LT1787 is related to the input sense voltage by the following relationships:

$$V_{SENSE} = I_{SENSE} \cdot R_{SENSE}$$
$$V_{OUT} = 8(V_{SENSE}) + V_{BIAS}$$

Although a −3A to 2A range was selected for this illustration, other current ranges can be accommodated by a simple change in value of the sense resistor. The correct R_{SENSE} value is derived so that the product of the maximum sense current and the sense resistor value is equal to the desired maximum sense voltage (250mV for 12-bit resolution). For instance, the value of the sense resistor to sense a maximum current of 10A is 250mV/10A = 0.025Ω. The smallest measurable current is then 10A/4096 counts = 2.44mA/count. If only 10-bit resolution is desired, then the full-scale voltage can be reduced to 60mV and R_{SENSE} reduced to 0.006Ω. Ensure that the power dissipated in the sense resistor, $I^2_{MAX} \cdot R_{SENSE}$, does not exceed the maximum power rating of the resistor.

Conclusion

The LT1787 high side current sense amplifier provides an easy-to-use method of sensing current with 12-bit resolution for a multiplicity of application areas. The part can operate to 60V, making it ideal for higher voltage systems in telecom or industrial applications. Additionally, the part can find application in battery-powered, handheld equipment and computers, where the need for gauging the amount of current consumed and/or the amount of charge remaining in the battery is critical.

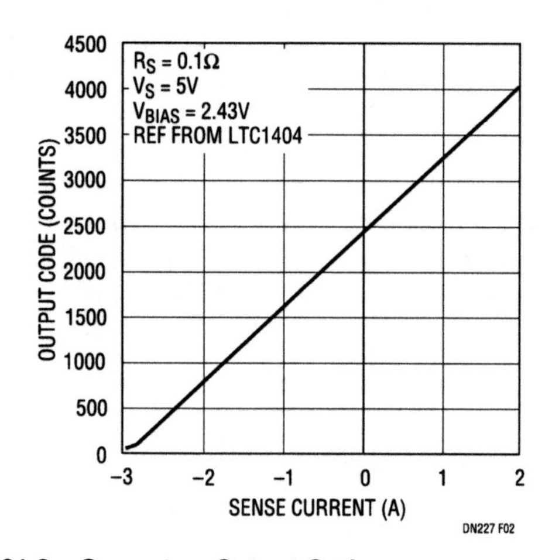

Figure 461.2 • Current vs Output Code

Theory of Operation (See Figure 461.3)

Inputs V_{S+} and V_{S-} apply the sense voltage to matched resistors R_{G1} and R_{G2}. The opposite ends of resistors R_{G1} and R_{G2} are forced to be at equal potentials by the voltage gain of amplifier A1. The currents through R_{G1} and R_{G2} are forced to flow through transistors Q1 and Q2 and are summed at node V_{OUT} by the 1:1 current mirror. The net current from R_{G1} and R_{G2} flowing through resistor R_{OUT} gives a voltage gain of 8. Positive sense voltages result in V_{OUT} being positive with respect to pin V_{BIAS}.

Pins V_{EE}, V_{BIAS} and V_{OUT} may be connected in a variety of ways to interface with subsequent circuitry. Split supply and single supply output configurations are easily supported.

Supply current for amplifier A1 is drawn from the V_{S-} pin. The user may choose to include this current in the monitored current (through R_{SENSE}) by careful choice of connection polarity.

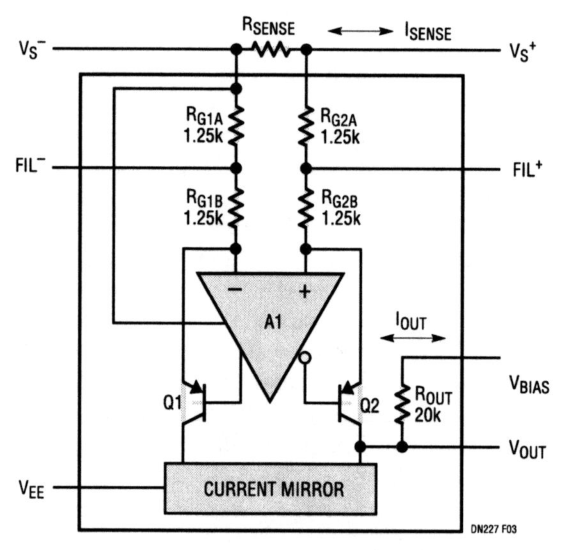

Figure 461.3 • LT1787 Functional Diagram

Op amp, comparator and reference IC provide micropower monitoring capability

462

Jim Williams

Introduction

The LTC1541 combines a micropower amplifier, comparator and 1.2V reference in an 8-pin package. The part operates from a single 2.5V to 12.6V supply with typical supply current of 5μA. Both op amp and comparator feature a common mode input voltage range that extends from the negative supply to within 1.3V of the positive supply. The op amp output stage swings from rail-to-rail. Figure 462.1 lists additional features along with a block diagram of the device. The part's attributes suggest low power monitoring applications and two such circuits are presented here.

Pilot light flame detector with low-battery lockout

Figure 462.2 shows a pilot light flame detector with low-battery lockout. The amplifier ("A"), running open loop, compares a small portion of the reference with the thermocouple-generated voltage. When the thermocouple is hot, the amplifier's output swings high, biasing Q1 on. Hysteresis, provided by the 10M resistor, ensures clean transitions, while the diodes clamp static generated voltages to the rails. The 100k–2.2μF RC filters the signal to the amplifier.

The comparator ("C") monitors the battery voltage via the 2M–1M divider and compares it to the 1.2V reference. A battery voltage above 3.6V holds C's output high, biasing Q2 on and maintaining the small potential at A's negative input. When the battery voltage drops too low, C goes low, signaling a low-battery condition. Simultaneously, Q2 goes off, causing A's negative input to move to 1.2V. This biases A low, shutting off Q1. The low outputs alert downstream circuitry to shut down gas flow.

SUPPLY RANGE: 2.5V TO 12.6V
$I_{QUIESCENT}$: 5μA
OP AMP V_{OS}: 700μV
COMPARATOR V_{OS}: 1mV
COMPARATOR HYSTERESIS: ±3mV
COMMON MODE RANGE: 0V TO (V_{SUPPLY} – 1.3V)
INPUT BIAS CURRENT: 1nA MAX, 10pA (25°C) TYP
REFERENCE: 1.2V ±0.4%

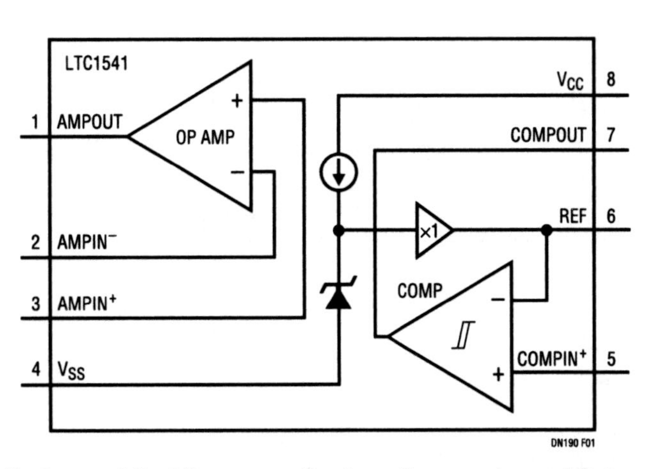

Figure 462.1 • LTC1541 Block Diagram and Features of the Micropower Op Amp, Comparator and Reference

Analog Circuit Design: Design Note Collection. http://dx.doi.org/10.1016/B978-0-12-800001-4.00462-2

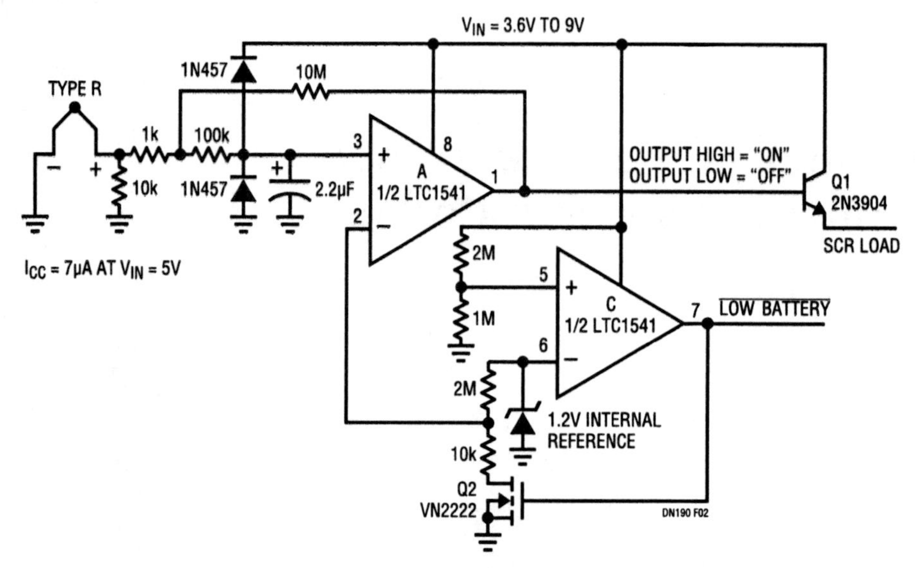

Figure 462.2 • Pilot Light Flame Detector with Low-Battery Lockout

Tip-acceleration detector for shipping containers

Figure 462.3's circuit is a tip-acceleration detector for shipping containers. It detects if a shipping container has been subjected to excessive tipping or acceleration and retains the detected output. The sensitivity and frequency response are adjustable. A potentiometer with a small pendulous mass biases the amplifier ("A"), operating at a gain of 12. Normally, A's output is below C's trip point and circuit output is low. Any tip-acceleration event that causes A's output to swing beyond 1.2V will trip C high. Positive feedback around C will latch it in this high state, alerting the receiving party that the shipped goods have been mishandled. Sensitivity is variable with potentiometer mechanical or electrical biasing or A's gain. Bandwidth is settable by selection of the capacitor at A's input. The circuit is prepared for use by applying power and pushing the button in C's output.

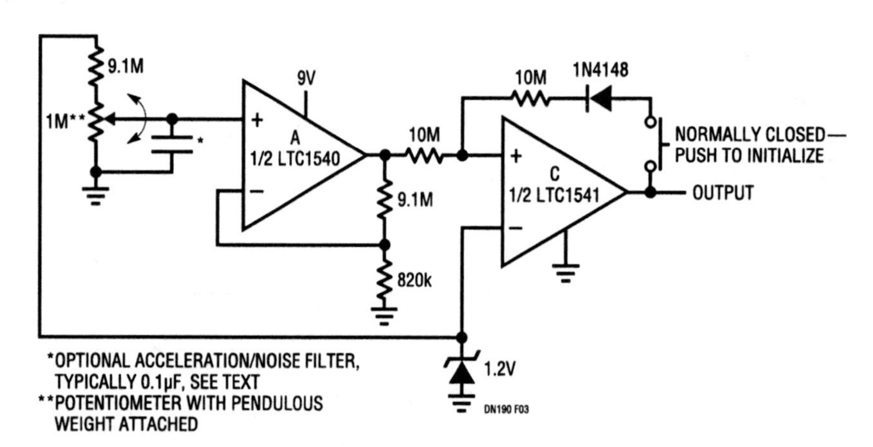

Figure 462.3 • Tip-Acceleration Detector for Shipping Containers Retains Output if Triggered. Sensitivity is Adjustable via Amplifier Feedback Values. Capacitor Sets Acceleration Response Bandwidth

Section 3

Voltage Reference Design

Versatile micropower voltage reference provides resistor programmable output from 0.4V to 18V

463

Jon Munson

Introduction

Voltage reference integrated circuits are widely used to establish accurate and stable voltages in analog circuits. While calibration-grade references are based on buried Zener diode technology (or even more exotic methods), the ubiquitous "band-gap" technique is the workhorse of the general purpose reference offerings. Band-gap references have historically offered fixed 1.2V to 10V outputs, along with a few adjustable models. The highly miniaturized LT6650 extends the scope of band-gap technology to offer the guaranteed ability to operate on a single supply down to 1.4V in ThinSOT packaging and with an output voltage as low as 0.4V. The LT6650 may also be powered by or produce reference voltages up to 18V and operate in either shunt mode or in a low dropout (LDO) series mode. The LT6650 is easy to use, sporting micropower dissipation (about 6μA of quiescent current) and simple 2-resistor voltage programming.

Easy output voltage programming

Figure 463.1 shows the basic connection for developing a fixed 400mV ±1% reference voltage from any supply voltage in the range of 1.4V to 18V. The internal noninverting op amp input port is always driven by a 400mV band-gap derived signal and the inverting op amp port is pinned out as a user connection. In this circuit, the op amp is simply provided with 100% negative feedback, thereby forming a unity-gain buffer for the reference source.

In applications where a reference potential greater than 0.4V is required, the simple addition of a feedback voltage divider programs the buffer op amp to provide gain. Figure 463.2 shows the typical connections for developing a reference voltage above 0.4V with the added feedback components. This configuration provides programmable reference voltages anywhere up to 0.35V below the supply potential used, the dropout voltage. Resistor R_G is chosen in the range from 10k to 100k to set the quiescent loading of the reference, then resistor R_F is simply selected for the required gain. While this illustration indicates fixed component values, the introduction of a variable element can provide a means of dynamically varying the reference output if desired. Figure 463.2 also shows additional input RC filtering which improves rejection of supply noise and a feedback capacitor that serves to both reduce noise gain and improve damping of the load response. The low operating current of the LT6650 and the input series resistor do not impair the low dropout performance significantly.

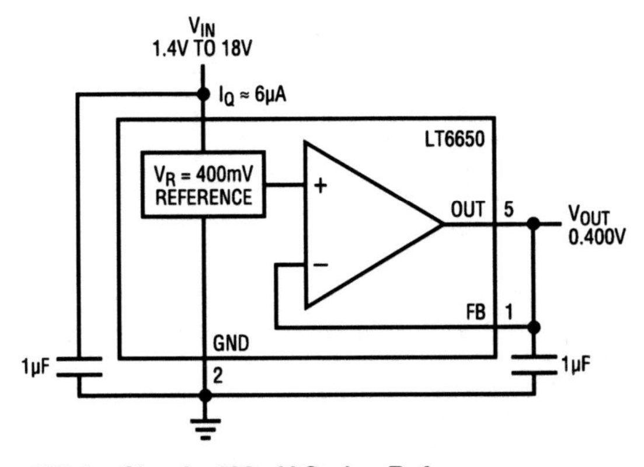

Figure 463.1 • Simple 400mV Series Reference

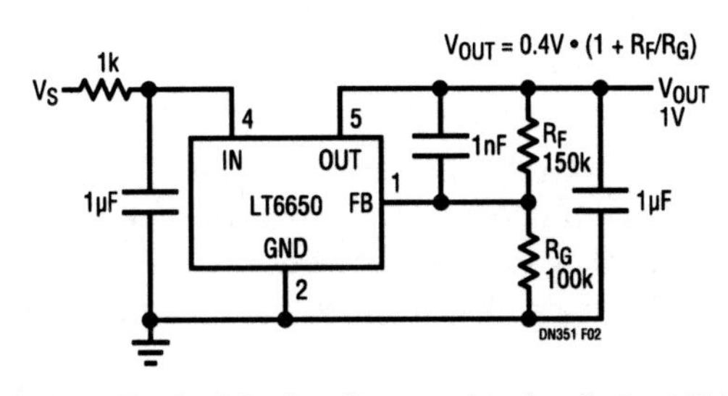

Figure 463.2 • Typical Series Connection for Output Voltages Greater than 0.4V

Analog Circuit Design: Design Note Collection. http://dx.doi.org/10.1016/B978-0-12-800001-4.00463-4

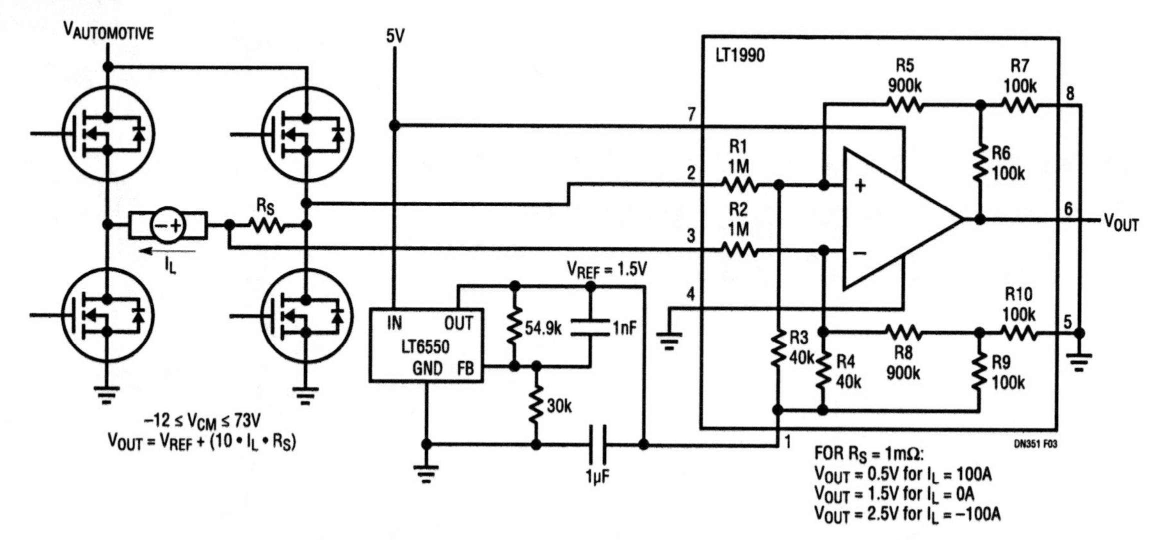

Figure 463.3 • Offsetting a Bidirectional Signal for Unipolar Processing

Create a virtual ground for unipolar processing of bidirectional signals

The LT6650 often finds use in single supply data acquisition circuits where a low voltage offset is needed to provide a shifted "virtual ground." Most ADC inputs can digitize right down to 0V input, but a single-supply input amplifier will not retain its accuracy at that low level, since the output is "saturating" (even with rail-to-rail types). A design solution is to have a voltage reference circuit drive the REF port of input instrumentation amplifiers (IA), thereby introducing a controlled offset mapping that allows the ADC to accurately capture the "zero input" signal level, or even provide a controlled negative signal conversion range within a positive-only input window. Figure 463.3 shows a single supply powered LT1990 difference amp sensing a bidirectional motor current. The LT6650 reference is configured to provide an optimal REF input level for the circuit (1.5V in this example) which both establishes the working common mode input range and introduces an output offset that maps the desired bidirectional signal span into a single supply ADC conversion range. In high accuracy applications, the offset voltage itself may also be digitized so that software algorithms can accurately "auto-zero" the measurements. In multichannel data acquisition systems, a single LT6650 can generally provide offset signaling to an entire IA array.

Shunt mode operation works like precision Zener diode

The LT6650 can easily be configured to behave much like a traditional Zener reference diode, but with far better regulation characteristics and the flexibility to be set to any voltage between 1.4V and 18V. This mode of operation allows the LT6650 to form simple negative references or other precision biasing functions. Figure 463.4 shows a simple negative reference circuit configuration. The programming is done just as with series mode operation, only the load capacitor is increased in value to optimize transient response. The Zener "knee" of the shunt configuration is about 10μA (with R_G set to 100k) and accurate regulation to 200μA is provided.

Conclusion

The LT6650 is an extremely flexible voltage control element, able to form accurate positive, negative or even floating reference voltages. With micropower operation over a wide 1.4V to 18V supply range and miniature ThinSOT packaging, the LT6650 offers design solutions for both portable and industrial applications. For single supply data-acquisition circuitry, the low 400mV output capability offers a simple virtual ground offsetting means that does not unduly sacrifice dynamic range. Thanks to simple resistor-based programming, references of various arbitrary voltages may be produced using a single LT6650 bill-of-material item, thus reducing procurement and inventory costs.

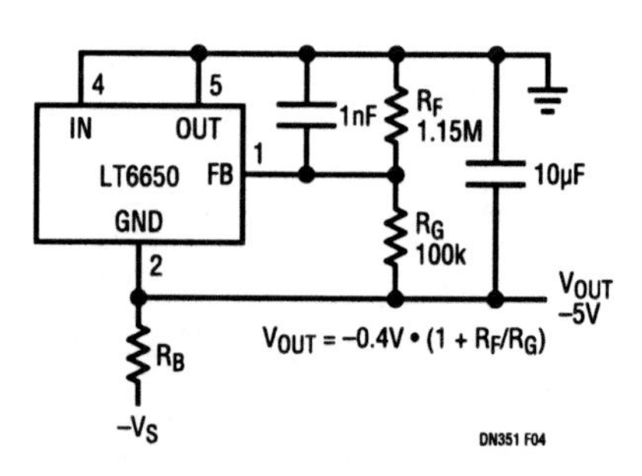

Figure 463.4 • Typical Configuration as −0.4V to −18V Shunt Reference

Don't be fooled by voltage reference long-term drift and hysteresis

464

John Wright

The micropower LT1461 and LT1790 low dropout bandgap voltage references excel not only in temperature coefficient and accuracy, but also in long-term drift and hysteresis (output voltage shift due to temperature cycling). Long-term drift and hysteresis, which are sometimes ignored or wrongly specified by other manufacturers, can be the accuracy limitations of systems. System calibrations can remove TC and initial accuracy errors, but only frequent calibration can remove the long-term drift and hysteresis. Subsurface Zener references, like the LT1236, have the best long-term drift and hysteresis, but they do not offer low output voltage options, low supply current and low operating supplies like these new bandgap references.

Lies about long-term drift

Some manufacturers are now touting phenomenal long-term drift specifications, based on accelerated high temperature testing. **THIS IS A DELIBERATE LIE! Long-term drift cannot be extrapolated from accelerated high temperature testing.** The only way long-term drift can be determined is to measure it over the time interval of interest. The erroneous technique produces numbers that are wildly optimistic and uses the Arrhenius Equation to derive an acceleration factor from elevated temperature readings. The equation is:

$$A_F = e^{\frac{E_A}{K}\left(\frac{1}{T1}-\frac{1}{T2}\right)}$$

where:

E_A = Activation Energy (Assume 0.7)

K = Boltzmann's Constant

T2 = Test Condition in kelvin

T1 = Use Condition Temperature in kelvin

To show how absurd this technique is, compare this calculation to real LT1461 data. Typical 1000 hour long-term drift at 30°C = 60ppm. The typical 1000 hour long-term drift at 130°C is 120ppm. From the Arrhenius Equation the acceleration factor is 767 and the projected "bogus" long-term drift is 0.156ppm/1000hr at 30°C. For a 2.5V reference, this corresponds to a 0.39µV shift after 1000 hours. This is pretty hard to determine (read impossible) if the peak-to-peak output noise is larger than this number. As a practical matter, one of the best laboratory references available has long-term drift of 1.5µV/mo. This performance is only available from the best subsurface Zener references such as the LTZ1000, utilizing specialized heating techniques.

Competitive reference measures 500 times worse than claimed

Long-term drift data was taken with parts that were soldered onto PC boards similar to a "real world" application. These boards were not preconditioned. They were placed into a constant temperature oven with T_A = 30°C and their outputs were scanned regularly and measured with an 8.5 digit

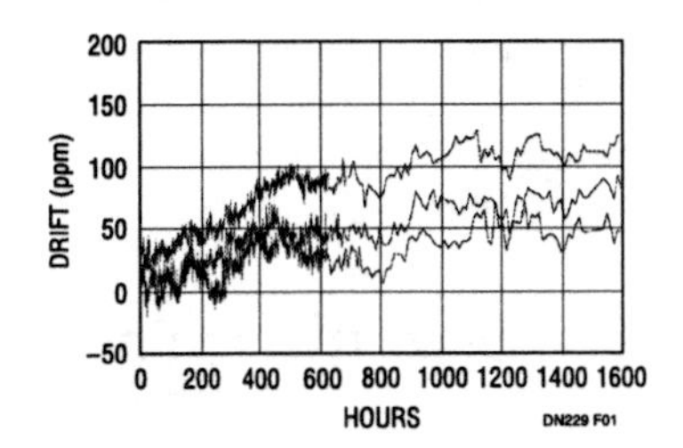

Figure 464.1 • LT1461S8-2.5 Long-Term Drift

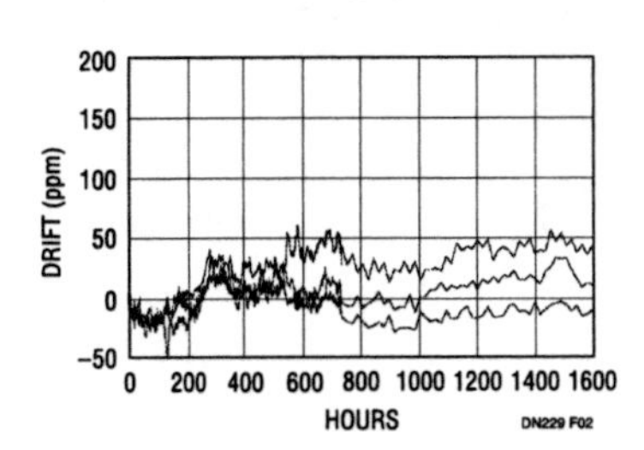

Figure 464.2 • LT1790SOT23-2.5 Long-Term Drift

Analog Circuit Design: Design Note Collection. http://dx.doi.org/10.1016/B978-0-12-800001-4.00464-6

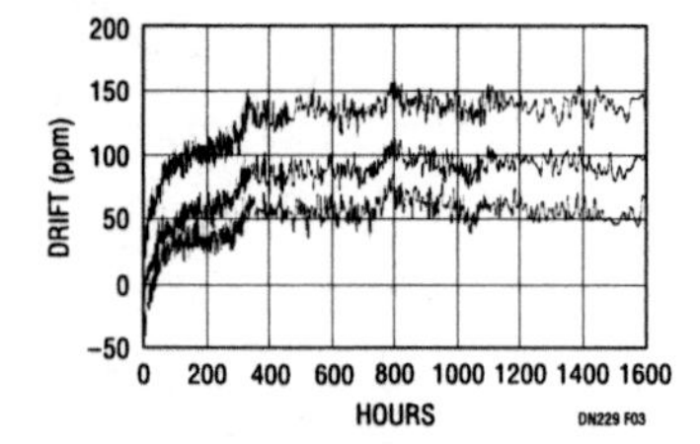

Figure 464.3 • XXX291S8-2.5 Long-Term Drift

DVM. Figures 464.1 and 464.2 show typical long-term drift of the LT1461S8-2.5 and the SOT-23 LT1790S6-2.5. Initially, data was taken every hour where the largest changes occur, but after several hundred hours the frequency was lowered to reduce the large number of data points. Figure 464.3 shows long-term drift of a competitive reference that specifies long-term drift of 0.2ppm/kHr in its data sheet. Measured data shows this reference to have drift between 60ppm/kHr and 150ppm/kHr or 300 to 750 times worse than claimed.

Long-term drift can be reduced by preconditioning the PC board after the reference has been soldered onto the board. Operating the PC board at 25°C or elevated temperature stabilizes initial drifts. This "burn-in" of the PC board eliminates the output shift that occurs in the first several hundred hours of operation. Further changes in output voltage are typically logarithmic and changes after 1000 hours tend to be smaller than before that time. Because of this decreasing characteristic, long-term drift is specified in ppm/$\sqrt{\text{kHr}}$.

Hysteresis limits repeatability

When a reference is soldered onto a PC board, the elevated temperature and subsequent cooling cause stress that influences the output. If the voltage reference is repeatedly temperature cycled, inelastic stress is applied to the chip and the output voltage does not return to the 25°C initial value. The mechanical stress is due to the difference in thermal coefficients of expansion between the silicon chip, plastic package and PC board. This error, known as "thermally induced hysteresis," is expressed in ppm and cannot be trimmed out because it is variable and has memory of previous temperature excursions. Hysteresis is always worse with higher temperature excursions, and differs with die attach and package type.

Hysteresis—often the "missing" spec

Most manufacturers ignore hysteresis specifications, but they can be critical in precision designs. To graphically show hysteresis, many references were IR reflow soldered onto PC boards and the boards underwent a "heat soak" at 85°C (this ensures that they all had the same initializing temperature). The temperature was then cycled multiple times between 85°C, 25°C and −40°C and all 25°C output voltages were recorded. The stabilization time at each temperature was 30 minutes. The worst-case output voltage changes at 25°C are shown in Figures 464.4 and 464.5 for the LT1461S8-2.5 and the SOT-23 LT1790S6-2.5. A competitive reference, which makes no mention of hysteresis on its data sheet, was also measured and is shown in Figure 464.6.

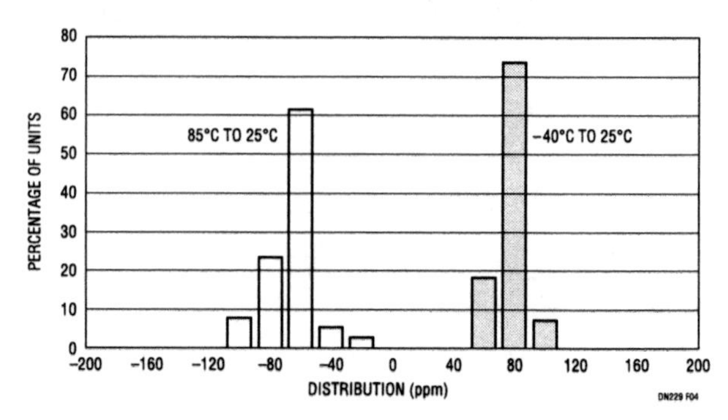

Figure 464.4 • LT1461S8-2.5 Industrial Hysteresis

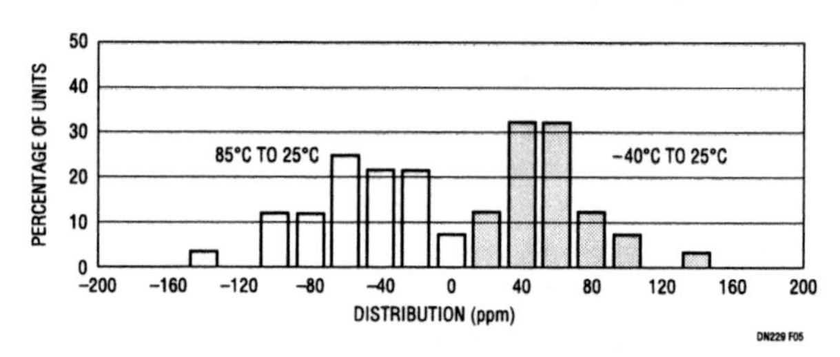

Figure 464.5 • LT1790S6-2.5 Industrial Hysteresis

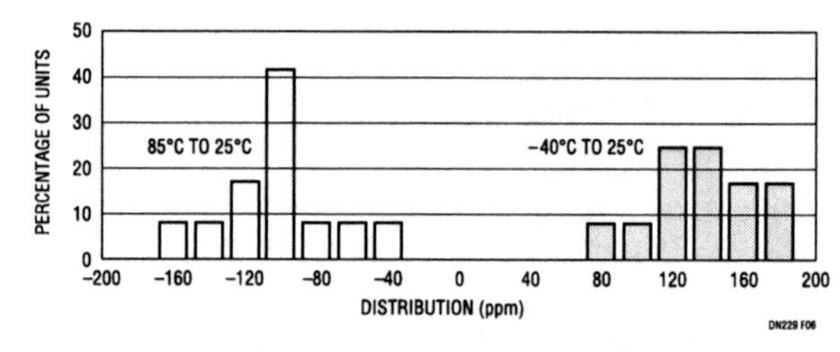

Figure 464.6 • XX780S8-2.5 Industrial Hysteresis

Conclusion

Voltage references from Linear Technology are conservatively and accurately specified, unlike those from other manufacturers that intentionally mislead or eliminate key specifications to cover shortcomings—shortcomings that may cause large errors.

The new LT1461 and LT1790 excel in all specifications that set system precision. There is nothing left out and there is nothing hidden.

Voltage references are smaller and more precise

465

John Wright

Introduction

Two new series voltage references bridge the gap between small package size and high precision. Advances in design, process and packaging have made the introduction of these new voltage references possible. The low power LT1460 is designed for minimum space and is available in all popular voltages, including 2.5V, 5V and 10V. By contrast, the LT1236 is designed for use in 12-bit and higher systems and combines 0.05% accuracy, low noise, low drift and the SO-8 package for difficult industrial temperature range applications.

Longer battery life with precision

The LT1460 low power, series reference has a large advantage over older shunt-style references. Shunt references require a resistor from the power supply to operate. This resistor must be chosen to supply the maximum current that can ever be demanded by the circuit being regulated. When the circuit being controlled is not operating at this maximum current, the shunt reference must always sink this current, resulting in high dissipation and short battery life. Because the LT1460 does not require a current setting resistor, it can operate with any supply voltage from $V_{OUT}+0.9V$ to 20V, while maintaining minimum dissipation and extending battery life. For example, if the 2.5V reference is not delivering load current, it dissipates only 500μW on a 5V supply, yet the same connection can deliver 20mA of load current on demand.

High output accuracy is achieved by the use of trimmed, precision thin-film resistors and curvature compensation is used to reduce drift. The LT1460 family is offered in MSOP, SO-8, PDIP and low cost TO-92 packages.

The small fry

By themselves, surface mount packages are space efficient, but the use of a large output capacitor to stabilize the reference limits the benefits of the small package. The LT1460 is stable with capacitive loads or with no capacitor at all. This can be helpful when fast settling is a goal, because changes in the voltage across load capacitors must recover before the reference's output value is accurate.

The MSOP LT1460 uses so little PC board space that there is a temptation to use the device as a local regulator. This is exactly what the LT1460 was designed to do: deliver substantial load current while maintaining reference-like features. At $I_{OUT}=20mA$, typical load regulation is 70ppm/mA, yet the LT1460 can withstand a short to ground without being destroyed. Additionally, if the power supplies are reversed, the reverse battery protection keeps the reference from conducting current and being damaged Table 465.1 and Figure 465.1.

Table 465.1 Key Specifications of the LT1460-2.5 Voltage Reference (SO-8 Package)

PARAMETER	CONDITIONS	MAX VALUE
Output Voltage Tolerance		
LT1460A		0.075%
LT1460B		0.10%
Temperature Coefficient	$0°C \leq T_A \leq 70°C$	
LT1460A		10ppm/°C
LT1460B		20ppm/°C
Line Regulation	5V to 20V	25ppm/V
Load Regulation Sourcing	$0mA \leq I_{OUT} \leq 20mA$	100ppm/mA
Dropout Voltage	$I_{OUT}=0mA$, $0°C \leq T_A \leq 70°C$	0.9V
Supply Current	$I_{OUT}=0mA$	130μA
Reverse Leakage	$I_{OUT}=0mA$, $V_{IN}=-20V$	10μA

Analog Circuit Design: Design Note Collection. http://dx.doi.org/10.1016/B978-0-12-800001-4.00465-8

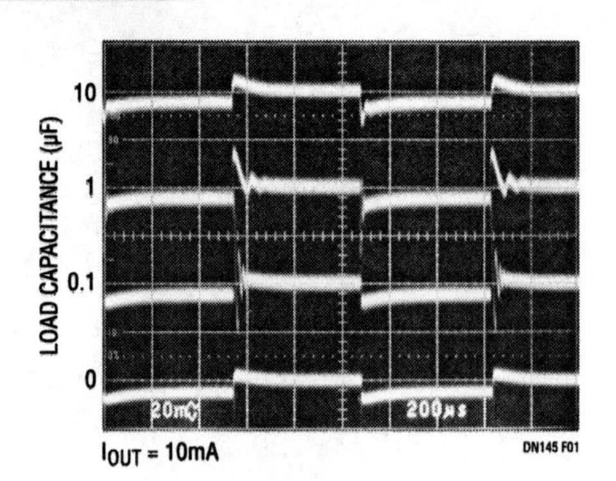

Figure 465.1 • Transient Response

Higher performance, industrial temperature range and surface mount

The LT1236 is a precision reference that combines ultralow drift and noise with excellent long term stability and high output accuracy (Figure 465.2). To address small package requirements, the new reference is available in the SO-8 package and guarantees critical reference parameters from −40°C to 85°C. The LT1236 output will both source and sink 10mA, is almost totally immune to input voltage variations and is stable with any load capacitor. Two output voltages are available: 5V and 10V. The 10V version can be used as a shunt regulator (2-terminal Zener) with the same precision characteristics as the 3-terminal connection. Special care has been taken to minimize thermal regulation effects and temperature-induced hysteresis. The LT1236 is also available in an N8 package.

The LT1236 combines superior accuracy and temperature coefficient specifications without the use of high power, on-chip heaters. The LT1236 references are based on a buried Zener diode structure that eliminates noise and stability problems that plague surface-breakdown devices Table 465.2.

Table 465.2 Key Specifications of the LT1236-10 Voltage Reference (SO-8 Package)

PARAMETER	CONDITIONS	MAX VALUE
Output Voltage Tolerance		
LT1236A		0.05%
LT1236B		0.10%
LT1236C		0.10%
Temperature Coefficient	$-40°C \le T_A \le 85°C$	
LT1236A		5ppm/°C
LT1236B		10ppm/°C
LT1236C		15ppm/°C
Line Regulation	$11.5V \le V_{IN} \le 14.5V$	4ppm/V
Load Regulation Sourcing	$0mA \le I_{OUT} \le 10mA$	25ppm/mA
Output Noise Voltage	$10Hz \le f \le 1kHz$	$6\mu V_{RMS}$
Supply Current	$I_{OUT} = 0mA$	1.7mA

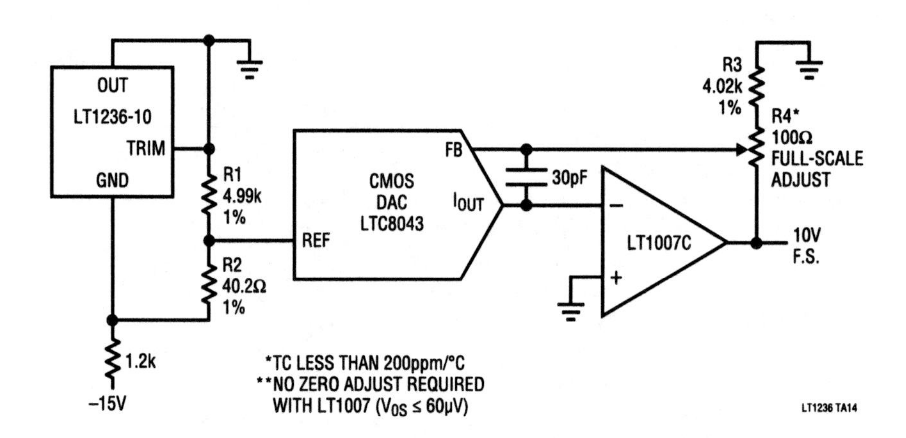

Figure 465.2 • CMOS DAC with Low Drift Full-Scale Trimming**

Section 4

Filter Design

A precision active filter block with repeatable performance to 10MHz

466

Philip Karantzalis

Introduction

The increasing speeds and dynamic ranges of modern communication and control systems make for complex filter requirements that have been, until now, a challenge to satisfy. Signal converters with high sampling rates can alleviate the need for selective filters, but even so, some sort of band limiting is still required. Some wireless systems face yet another problem. Even though the use of oversampling converters has significantly simplified filtering, unexpected signals from neighboring systems can intrude on a converter's allowed bandwidth.

The LT1568 requires only a few external components to satisfy complex filtering requirements, including band rejection. An important feature to note is that the performance of the LT1568 is precisely repeatable from device to device, making it possible to manufacture filters in volume without the need for costly production trimming.

Device description

The LT1568 is a dual 2nd order matched building block which is useful in a variety of ways for different applications. Each individual filter block uses two very low noise ($1.5nV/\sqrt{Hz}$ input voltage noise) operational amplifiers and an inverter op amp output (useful in applications requiring differential outputs or an output with phase reversal).

The LT1568, with a few passive components, can be used to design various filter functions (lowpass, bandpass, highpass, notch or allpass). One of the features of this device is its repeatable AC performance from part to part. This is achieved by trimming all the internal capacitors to a better than 1% tolerance and by trimming the gain bandwidth product (GBW) of all the internal op amps. Thus, any small error caused by the finite speed of the LT1568 active circuitry is highly predictable from part to part.

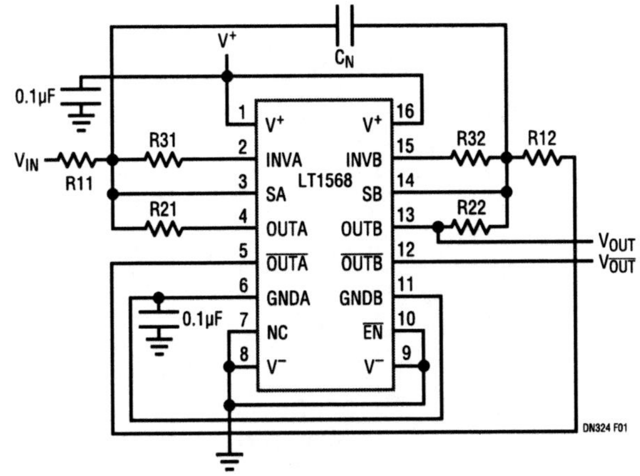

Figure 466.1 • A 4th Order Elliptic Lowpass Filter Circuit (Single Supply Operation)

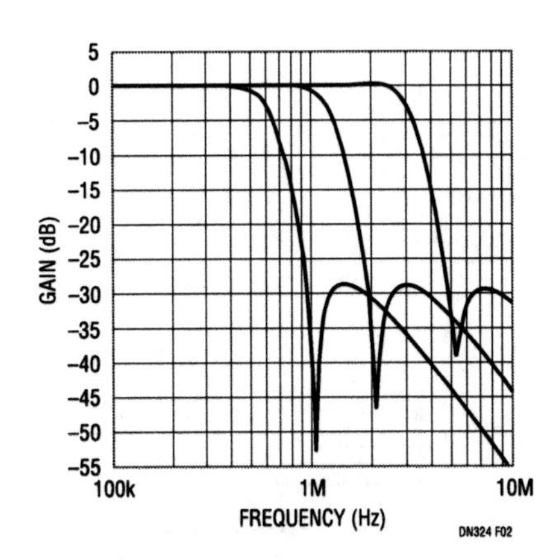

Figure 466.2 • Amplitude Response of Figure 466.1 Circuit for Cutoff Frequencies of 500kHz, 1MHz and 2.5MHz

Analog Circuit Design: Design Note Collection. http://dx.doi.org/10.1016/B978-0-12-800001-4.00466-X

Application examples

The LT1568 is extremely versatile due in part to its generic internal architecture. The values of the external resistors for a few classical filter responses are given in the LT1568 data sheet. The resistor values for a wider selection of filter responses can be determined by using an LT1568 Design Guide found on www.linear.com. The following examples show two LT1568 filter circuits that are implemented using simple design equations.

A 4th order elliptic lowpass filter

The unique architecture of the LT1568 allows the addition of a notch by simply adding an external capacitor between the summing node of the first stage (SA) and the summing node of the second stage (SB), as shown in Figure 466.1. Capacitor C_N and resistor R12 are the only external passive components affecting the accuracy and repeatability of the notch frequency. A 4th order ±0.25dB passband lowpass filter with one stopband notch and equal value resistors is designed as follows:

$C_N = 82pF$;

$R11 = R21 = R31 = R12 = R22 = R32 = 806\Omega$ $(1MHz/f_{CUTOFF})$; f_{CUTOFF} is the ±0.25dB passband

This expression is useful for cutoff frequencies up to 2.5MHz. Cutoff frequencies of 500kHz, 1MHz and 2.5MHz are shown in Figure 466.2.

A 4th order bandpass filter

The LT1568 makes it possible to produce accurate selective bandpass filters in volume which until now has been a difficult task. This is another result of its unique topology and carefully trimmed internal components. The LT1568 also makes the basic design of a bandpass filter extremely easy. Simply replace the input resistor of an LT1568 2nd order lowpass section with a capacitor (C_{IN}) and the lowpass section transforms into a 2nd order bandpass filter. A 4th order bandpass circuit is shown in Figure 466.3. The center frequency of the bandpass function depends on the product R2 · R3 · C1 · C2 (C1 and C2 are the internal LT1568 capacitors) but it also depends on the ratio of $C_{IN}/C2$. If the ratio of $C_{IN}/C2$ is kept small ($C_{IN} \leq C2$), then a five percent variation in the value of C_{IN} keeps the bandpass filter's center frequency variation to around one percent or less.

The component values for a 4th order bandpass filter with unity gain at the center frequency and a −3dB passband equal to $f_{CENTER}/3.5$ are designed as follows:

$C_{IN1} = C_{IN2} = 68pF$;

$$R21 = R31 = R22 = R32 = [(268 \cdot 10^9)/(3.4 \cdot f_{CENTER}^2 + 250 \cdot 10^3 \cdot f_{CENTER})]\Omega$$

where f_{CENTER} is the center frequency in kHz. This expression is useful for cutoff frequencies up to 7MHz.

Figure 466.4 shows the response for a 1MHz bandpass filter.

Conclusion

The LT1568 is the basis for analog filter designs of unprecedented predictability and repeatability. Its unique topology and precision trimmed active and passive components facilitate the implementation of low cost filters in the production of high frequency circuits and systems. The LT1568 is useful as a simple dual matched 2nd or 3rd order lowpass filter in I/Q applications, or with an external five-percent tolerance capacitor, the LT1568 can be used to build selective 4th order elliptic lowpass filters, accurate dual 2nd order or single 4th order bandpass filters.

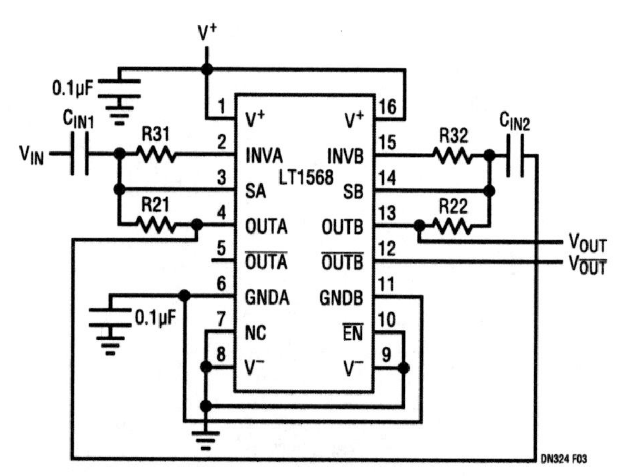

Figure 466.3 • A 4th Order Bandpass Filter Circuit (Single Supply Operation)

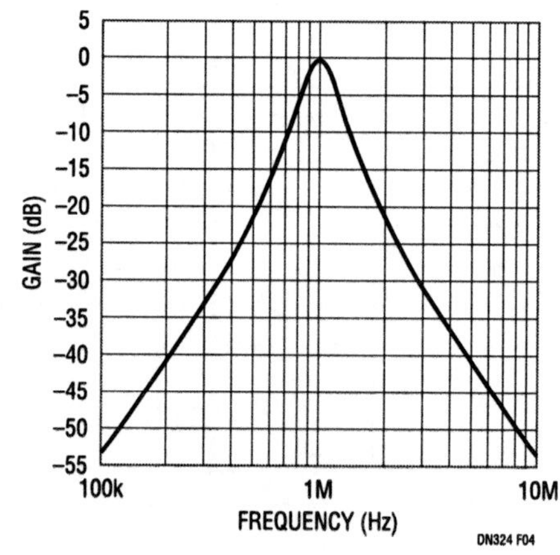

Figure 466.4 • Amplitude Response of Figure 466.3 Circuit for 1MHz Center Frequency

High frequency active anti-aliasing filters

467

Philip Karantzalis

Introduction

High frequency (1MHz or higher) active lowpass filters are now practical alternatives to passive LC filters mainly due to the availability of very high bandwidth (100MHz or higher) integrated amplifiers. Analog signal filtering applications with bandwidths in the megahertz region can be implemented by a discrete active filter circuit using resistors, capacitors and a 400MHz operational amplifier such as the LT1819 or the LT6600-10, a fully integrated lowpass filter. The LT6600-10 has a fixed 10MHz frequency response equivalent to a 4th order flat passband Chebyshev function. An LT1819-based active RC lowpass filter can be designed to have a Chebyshev, Butterworth, Bessel or custom frequency response (up to 20MHz).

The LT6600-10 lowpass filter

The LT6600-10 is a fully integrated, differential, 4th order lowpass filter in a surface mount SO-8 package (Figure 467.1). Two external resistors ranging from 1600Ω to 100Ω set the differential gain in the filter's passband from −12dB to 12dB, respectively. The LT6600-10 passband gain ripple is a maximum of 0.7dB to −0.3dB up to 10MHz and attenuation is typically 28dB at 30MHz and 44dB at 50MHz. The signal to noise ratio (SNR) at the filter's output is 82dB with a $2V_{P-P}$ signal for a passband gain equal to one (a SNR suitable for up to 14 bits of resolution). In addition to lowpass filtering, the LT6600-10 can level shift the input common mode signal. For example, with a single 3V supply, if the input common mode voltage is 0.25V, then the output common mode voltage can be set to 1.5V. The LT6600-10 operates with single 3V or 5V or dual 5V power supplies.

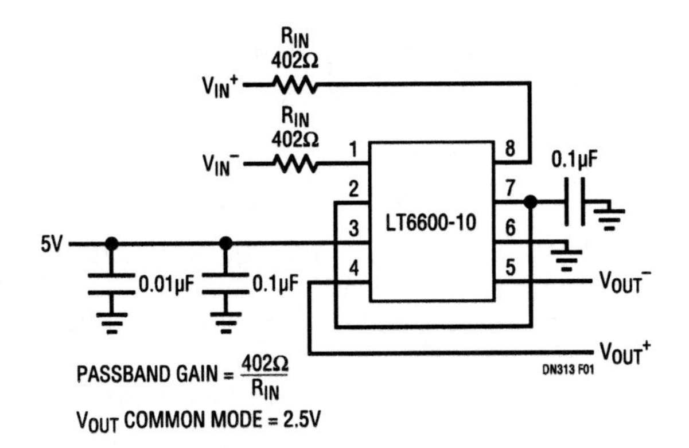

Figure 467.1 • LT6600-10 10MHz, Lowpass Filter Features a Single Supply and Only Two External Resistors

An LT1819-based RC lowpass filter

The LT6600-10 greatly simplifies lowpass filter design because it only requires two external resistors to set the differential gain, but the passband is fixed. For more flexibility, the LT1819 400MHz, high slew rate, low noise and low distortion dual operational amplifier is a good choice. Figure 467.2 shows a differential, 10MHz, 4th order, lowpass filter using two LT1819s. This approach allows for adjustable passband up to 20MHz but at the expense of a large number of passive and active components and high sensitivity to the variation of component values. For example, a component sensitivity analysis of Figure 467.2 shows that in order to maintain a passband ripple similar to the LT6600-10 (±0.5dB up to 10MHz), the component tolerance must not exceed ±0.5% for the resistors and 1% for the capacitors. Also, the LT1819 gain-bandwidth product should not be less than 300MHz. If a Butterworth, Bessel or custom filter response is desired, ±1% resistors and ±5% capacitors are adequate. These filters have lower sensitivity than a "flat" passband Chebyshev filter. The LT1819-based filter operates with single 5V or dual 5V power supplies (for a single 3V power supply filter circuit use an LT1807, a dual 325MHz, rail-to-rail operational amplifier).

Analog Circuit Design: Design Note Collection. http://dx.doi.org/10.1016/B978-0-12-800001-4.00467-1

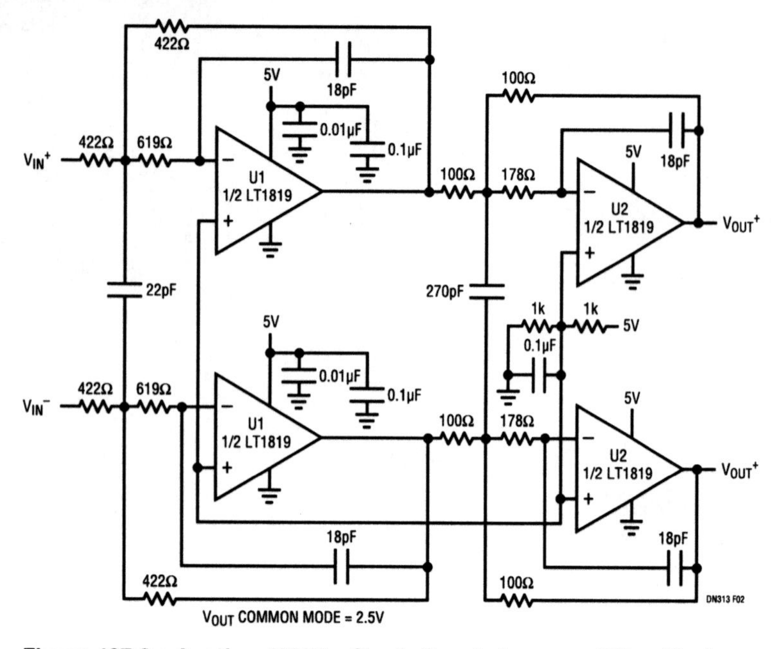

Figure 467.2 • Another 10MHz, Single Supply Lowpass Filter, Similar to Figure 467.1. This Circuit Features the LT1819 Op Amp and Adjustable Bandwidth up to 20MHz

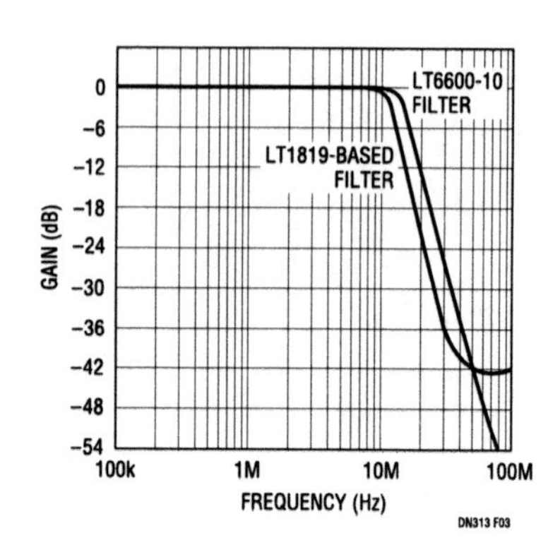

Figure 467.3 • The Frequency Response of the Two 10MHz Anti-aliasing Filters Shown in Figures 467.1 and 467.2

Anti-aliasing 10MHz filters for a differential 50Msps ADC

An LT6600-10 or an LT1819-based 10MHz lowpass filter provides adequate stopband attenuation for reducing aliasing signals at the input of a high speed analog-to-digital converter (ADC) such as the LT1744, a 50Msps, differential input ADC. Figure 467.3 shows the gain response of the LT6600-10 and the LT1819-based 10MHz filters. The LT1819 filter is designed to have higher attenuation at 20MHz than the LT6600-10 filter in order to achieve sufficient stopband attenuation. The stopband attenuation beyond 40MHz of the LT1819 circuit is limited to −42dB by printed circuit stray paths and differential component mismatches that decrease the common mode rejection at very high frequencies. The stopband attenuation of the fully integrated LT6600-10 filter continues increasing beyond 40MHz. Figure 467.4 shows the DC to 10MHz plot of a 1MHz $2V_{P-P}$ differential sine wave processed by an LT6600-10 plus an LT1744 14-bit ADC. The plots are an averaged 4096-point FFT of a 1MHz sine wave digitized at 50 million samples per second. The 1MHz harmonic distortion of Figure 467.4 is virtually the same when an LT1819-based filter is used with an LT1744. In a 10MHz bandwidth, the measured signal-to-noise plus distortion for the LT6600-10 plus LT1744 circuit is 74.5dB and essentially equal to the dynamic range of an LT1744 ADC (a minimum of 75.5dB for a $2V_{P-P}$ signal). The LT1819-based filter is slightly noisier; the measured signal-to-noise ratio plus distortion for the LT1819 plus LT1744 circuit is 71.5dB.

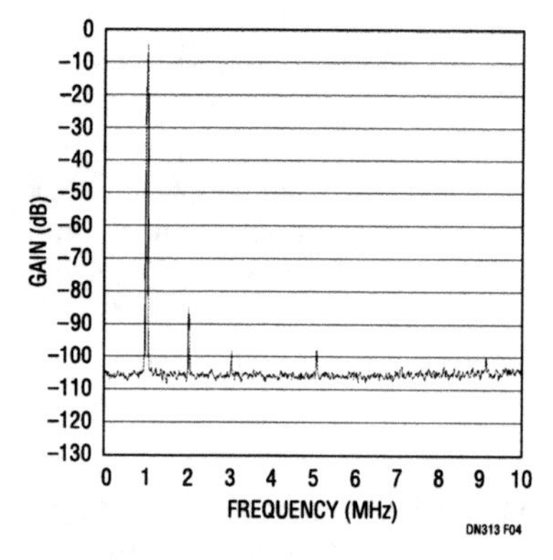

Figure 467.4 • Spectral Plot of a 1MHz, $2V_{P-P}$, Differential Sine Wave Input to an LT6600-10 Filter Plus an LT1744 14-Bit ADC (a DC to 10MHz Plot of a 4096-Point Averaged FFT with a 50MHz Sample Rate)

Conclusion

The LT6600-10 offers a high performance, 10MHz differential filter with gain, in a small package (SO-8) while the LT1819 op amp can be used to create a variety of differential filters up to 20MHz.

Design low noise differential circuits using a dual amplifier building block

468

Philip Karantzalis

Introduction

Many communications systems use differential, low level (400mV to 1V peak-to-peak), analog baseband signals where the baseband circuitry operates from a single low voltage power supply (5V to 3V). Any differential amplifier circuit used for baseband signal conditioning must have very low noise and an output voltage swing that includes most of the power supply range for maximum signal dynamic range. The LT1567, a low noise operational amplifier ($1.4nV/\sqrt{Hz}$ *voltage noise density*) and a unity-gain inverter, is an excellent analog building block (see Figure 468.1) for designing low noise differential circuits. The gain bandwidth of the LT1567 amplifier is 160MHz and its slew rate is sufficient for signal frequencies up to 5MHz. The LT1567 operates from 2.7V to 12V total power supply. The output voltage swing is guaranteed to be 4.4V and 2.6V peak-to-peak, at 1k load with a single 5V and 3V power supply, respectively. The LT1567 is available in an 8-lead MSOP surface mount package.

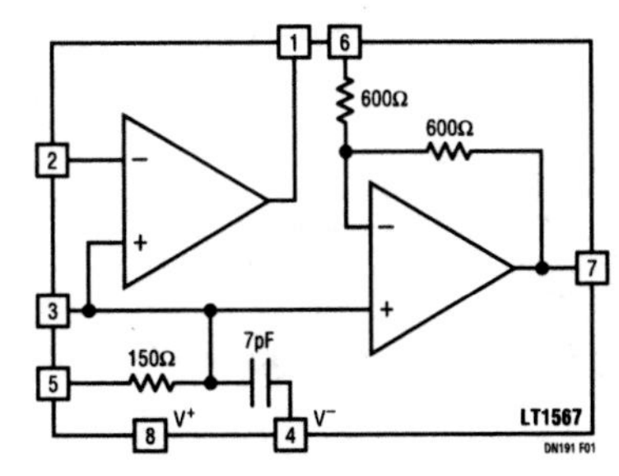

Figure 468.1 • LT1567 Analog Building Block

A single-ended to differential amplifier

Figure 468.2 shows a circuit for generating a differential signal from a single-ended input. The differential output noise is a function of the noise of the amplifiers, the noise of resistors R1 and R2 and the noise bandwidth. For example, if R1 and R2 are each 200Ω, the differential output voltage noise density is $9.5nV/\sqrt{Hz}$ and in a 4MHz noise bandwidth, the total differential output noise is $19\mu V_{RMS}$ (with a low level $0.2V_{RMS}$ differential signal, the signal-to-noise ratio is an excellent 80.4dB). The voltage on Pin 5 (V_{REF}) allows flexible DC bias for the circuit and can be set by a voltage divider or a reference voltage source (with a single 3V power supply, the V_{REF} range is $0.9V \leq V_{REF} \leq 1.9V$). In a single supply circuit, if the input signal is DC coupled, then an input DC voltage (V_{INDC}) is required to bias the circuit within its linear region. If V_{INDC} is within the V_{REF} range, then V_{REF} can be equal to V_{INDC} and the output DC common mode voltage (V_{OUTCM}) at V_{O1} and V_{O2} is equal to V_{REF}. To maximize the unclipped LT1567 output swing however, the DC common mode output voltage must be set at $V^+/2$. The input signal can be AC coupled to the circuit's input resistor R1 and V_{REF} also set to the DC common mode voltage required by any following circuitry (for example the input of an I and Q modulator).

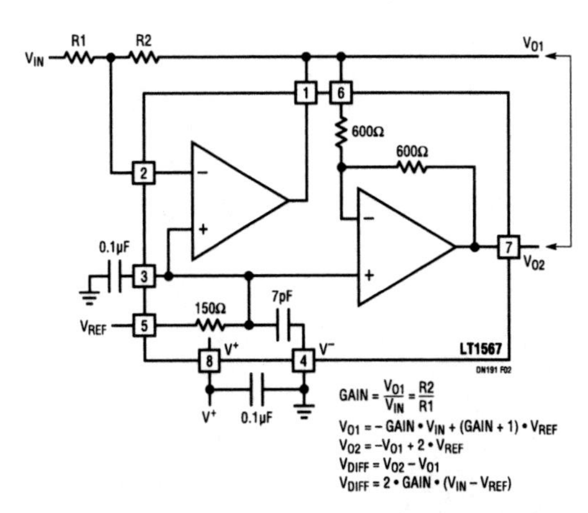

Figure 468.2 • A Single-Ended Input to Differential Output Amplifier

Analog Circuit Design: Design Note Collection. http://dx.doi.org/10.1016/B978-0-12-800001-4.00468-3

A differential buffer/driver

Figure 468.3 shows an LT1567 connected as a differential buffer. The differential output voltage noise density is $7.7\text{nV}/\sqrt{\text{Hz}}$. The differential buffer circuit of Figure 468.3, translates the input common mode DC voltage (V_{INCM}) to an output common mode DC voltage (V_{OUTCM}) set by the V_{REF} voltage ($V_{OUTCM} = 2 \cdot V_{REF} - V_{INCM}$). For example, in a single 5V power supply circuit, if V_{INCM} is 0.5V and V_{REF} is 1.5V then V_{OUTCM} is 2.5V.

A differential to single-ended amplifier

Figure 468.4 shows a circuit for converting a differential input to a single-ended output. For a gain equal to one ($R1 = R2 = 604\Omega$ and $V_{OUT} = V2 - V1$) the input referred differential voltage noise density is $9\text{nV}/\sqrt{\text{Hz}}$ and differential input signal-to-noise ratio is 80.9dB with $0.2V_{RMS}$ input signal in a 4MHz noise bandwidth. The input AC common mode rejection depends on the matching of resistors R1 and R3 and the LT1567 inverter gain tolerance (common mode rejection is at least 40dB up to 1MHz with one percent resistors and two percent inverter gain tolerance). If the differential input is DC coupled, then V_{REF} must be set equal to the input common mode voltage (V_{INCM}). If V_{REF} is greater than V_{INCM} then a peak voltage on Pin 7 may exceed the output voltage swing limit. The DC voltage at the amplifier's output (V_{OUT}, Pin 1) is V_{REF}.

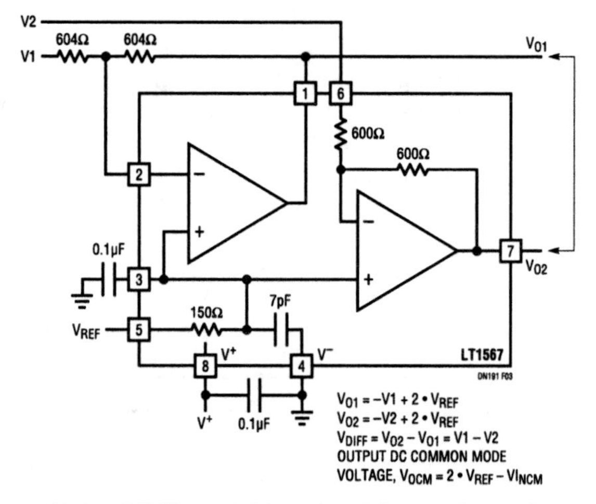

Figure 468.3 • A Differential Input and Output Buffer/Driver

LT1567 free design software

A spreadsheet-based design tool is available at www.linear.com for designing lowpass and bandpass filters using the LT1567.

The simple-to-use spreadsheet requires the user to define the desired corner (or center) frequency, the passband gain and a capacitor value for a choice of second or third order Chebyshev or Butterworth lowpass or second order bandpass filters.

The spreadsheet outputs the required external standard component values and provides a circuit diagram.

Conclusion

With one LT1567 and two or three resistors, it is easy to design low noise, differential circuits for signals up to 5MHz. The LT1567 can also be used to make low noise second and third order lowpass filters and second order bandpass filters with differential outputs.

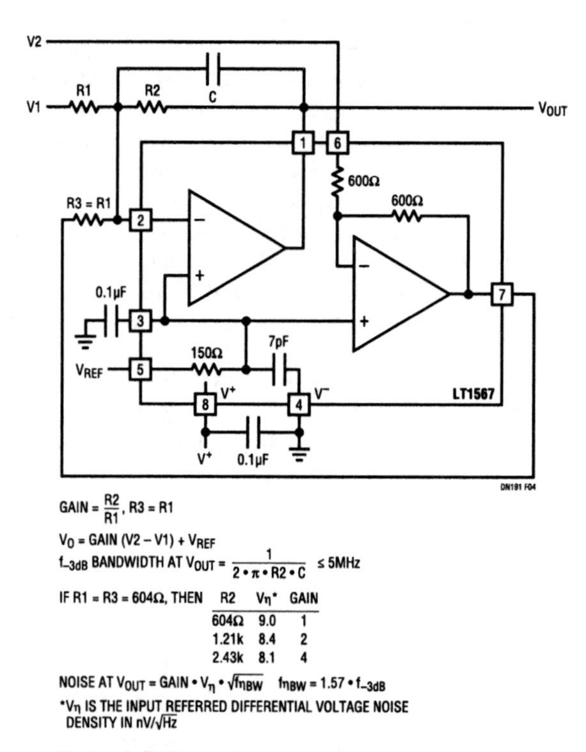

Figure 468.4 • A Differential Input to Single-Ended Output Amplifier

A digitally tuned anti-aliasing/ reconstruction filter simplifies high performance DSP design

469

Max W. Houser   Philip Karantzalis

Introduction

Typically an analog anti-aliasing filter is used to band-limit wideband signals at the input of an analog-to-digital converter. In addition, as the converter's sampling rate changes, an anti-aliasing filter's passband should increase or decrease accordingly. A frequency-tunable analog filter for a high resolution converter requires a large number of expensive precision components. With the LTC1564, designers of data acquisition instruments and digital signal processing (DSP) systems have a low noise, continuous-time, "brick wall" lowpass filter with digital control of the corner frequency f_C, (f_C range 10kHz to 150kHz in 10kHz steps). The LTC1564 also includes a digitally programmable gain amplifier (PGA, 1V/V to 16V/V in 1V/V steps). A simple, on-chip, latching digital interface controls corner frequency and gain settings. The LTC1564 is in a small 16-pin SSOP and operates from a supply voltage of 2.7V to 10.5V total (single or split supplies).

Filtering performance and operation

The LTC1564 is a high resolution filter with a rail-to-rail output. The 8th order lowpass response with two stopband notches gives approximately 100dB attenuation at 2.5 times f_C, making it suitable for high resolution anti-aliasing filtering. Despite the high filter order, the wideband noise is only $33\mu V_{RMS}$ (typical) at a 20kHz corner frequency and unity gain, which is 100dB below the rail-to-rail maximum signal level for ±5V supplies. The output-referred noise rises only slightly at higher gain settings. At the maximum 24dB (16V/V) gain setting, the same 20kHz response just quoted has an output noise level of $40\mu V_{RMS}$ (or an input-referred noise of $2.5\mu V_{RMS}$). Gain control in the LTC1564 is an integral part of the filter, using a proprietary method that deliberately minimizes the total noise. This feature is very difficult to achieve with separate variable gain amplifiers and filter circuits. The LTC1564 satisfies a demand for lowpass filters with roughly "100-100-100" performance: 100dB stopband attenuation, 100dB signal-to-noise ratio (SNR) and 100kHz bandwidth.

You do not have to be a filter expert or analog designer to use the LTC1564. There are only three analog pins: Input, Output and a half-supply reference voltage point, AGND (Figure 469.1). The other pins are digital controls and power supply. The LTC1564 is an instrument in a box with analog input and output jacks and two rotary switches labeled "Frequency" and "Gain." The frequency setting "F" and gain setting "G" are 4-bit codes entered through the F and G digital input pins (Table 469.1). In addition, setting the F code to

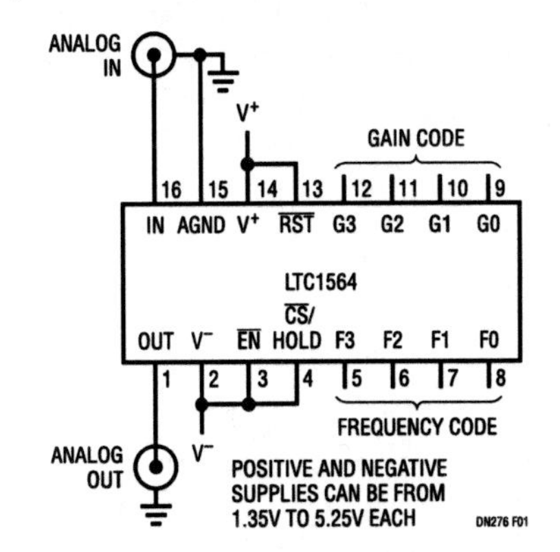

Figure 469.1 • Dual Power Supply Circuit

Table 469.1 Programming the Corner Frequency and Gain of the LTC1564

F3	F2	F1	F0	G3	G2	G1	G0	MODE
0	0	0	1	0	0	0	0	f_C=10kHz, Passband Gain = 1V/V (0dB)
1	1	1	1	0	0	0	0	f_C=150kHz, Passband Gain = 1V/V (0dB)
0	0	0	1	1	1	1	1	f_C=10kHz, Passband Gain = 16V/V (24dB)
0	0	0	0	Don't Care				Mute State, Zero Gain

Analog Circuit Design: Design Note Collection. http://dx.doi.org/10.1016/B978-0-12-800001-4.00469-5

0000 engages a "mute" state where the filter remains fully powered but the gain is a hard zero (typically −100dB). Logic levels for the LTC1564 digital inputs are nominally rail-to-rail CMOS (where a logic 1 is V^+ and a logic 0 is 0V for single 3V or 5V or dual ±5V supply operation).

Application example: 2-chip "universal" DSP front end

In Figure 469.2, an LTC1564 filter drives an LTC1608 16-bit 500ksps analog-to-digital converter (ADC) for a highly flexible, complete 16-bit analog-to-digital signal interface with variable gain, variable sampling rate and variable analog bandwidth up to 150kHz. The LTC1564's frequency-setting "F" code and the rate of sampling controlled by the LTC1608's $\overline{\text{CONVST}}$ input (Pin 31) set signal bandwidth and anti-aliasing filtering.

As an example, with the LTC1564 passband corner (f_C) set to 100kHz, with a sampling rate (f_S) of 500ksps by the LTC1608 ADC, provides 100dB of anti-aliasing protection at the critical analog folding frequency of $f_S/2$, or 250kHz. Another independent option is to sample at a rate (f_S) that is lower than $5 \cdot f_C$. This will move the folding frequency ($f_S/2$) down from $2.5 \cdot f_C$ to somewhere within the analog filter's roll-off band, where the filter's rejection will not be as high as 100dB. This reduces the anti-alias rejection for signals at and above $f_S/2$, but still provides sufficient anti-alias protection in many applications, particularly if, as is often true, the aliasable signals at and above $f_S/2$ have lower levels than the desired signals at and below f_C. The circuit of Figure 469.2 can accommodate either or both of these options by suitable choice of ADC sampling rate and filter F code.

Figure 469.3 shows a measured FFT spectrum of the digital output of Figure 469.2's circuit. A 40kHz, $100mV_{RMS}$ sine wave was preamplified to $4.5V_{P-P}$ by the LTC1564 to nearly span the input range of the LTC1608 ADC. The LTC1564 is set for a cutoff frequency of 50kHz and a gain of 16V/V and the sampling frequency of the ADC is 204.8kHz. Total harmonic distortion (THD) is 86dB down and the dynamic range is 109dB (since the filter noise does not increase with gain, the programmable filter gain extends the dynamic range beyond the unity gain range).

Conclusion

In addition to providing high resolution anti-aliasing, the LTC1564 is useful as a reconstruction filter that eliminates the nonessential high frequency signal spectrum at the output of a digital-to-analog converter (DAC). A simple, compact, economical and high performance digital signal processing and generating system hardware requires only two LTC1564, an ADC and a DAC.

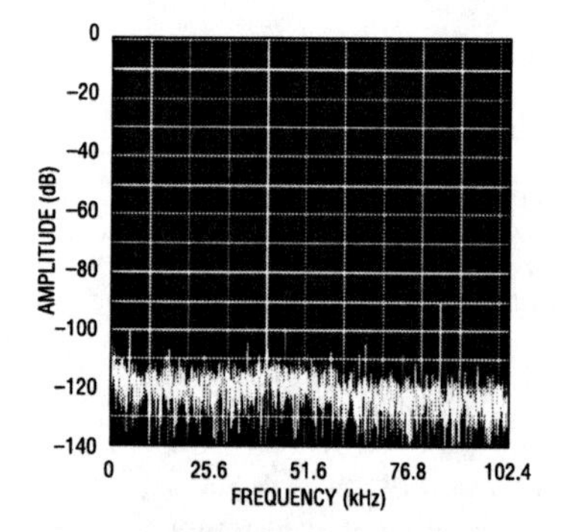

Figure 469.3 • FFT Plot of the Digital Output of Figure 469.2's Circuit

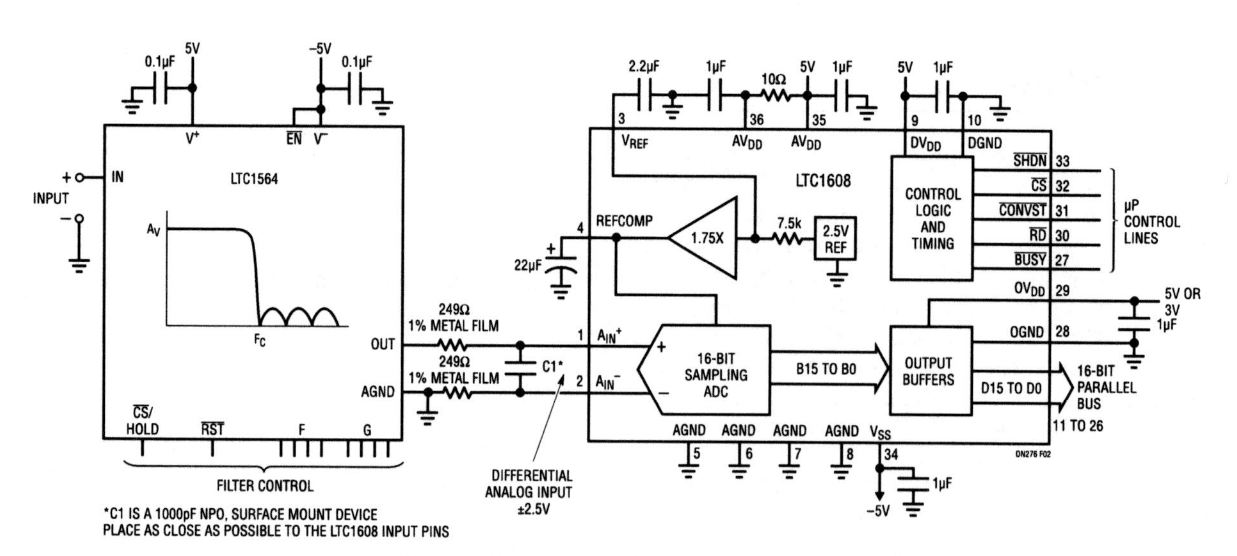

Figure 469.2 • A Universal DSP Front End

Replace discrete lowpass filters with zero design effort, two item BoM and no surprises

470

Doug La Porte

The LTC1563 family of continuous time, monolithic filter products reduces lowpass filter design to a simple resistor value calculation:

$$R = 10k \cdot (256kHz/f_C)$$

where

R = the resistor value in ohms

f_C = the filter's cutoff frequency in hertz ($256Hz \leq f_C \leq 256kHz$)

It's that simple. With the help of any basic calculator (or slide rule), filters are designed in a few seconds.

With the LTC1563, not only is the design process painless, but the end result is an easy-to-build circuit that has been fully specified for its performance as a filter—not just an op amp, a resistor or a capacitor. Additionally, layout sensitivity is minimized and *you only need to specify one passive component—a resistor.*

Lowpass filters—the traditional approach

Filter design is often viewed as a mysterious, intimidating mathematical process. Complex differential equations and elliptic integrals are usually avoided by using filter books with design tables followed by frequency and impedance scaling. Expensive CAD programs can help with much of this effort but these tools are not always available. The final design requires op amps and several resistor and capacitor values.

Once the filter is designed, there are still several component, performance and layout issues to resolve. Which capacitor type is best? How much dynamic range will the circuit have? What are the effects of layout parasitics? All of these questions must be answered to create a finished and robust design.

Lowpass filters—the LTC1563 approach

Designing lowpass filters with the LTC1563 is a snap. There are presently two parts in the LTC1563 family—the LTC1563-2 and the LTC1563-3. The LTC1563-2 is configured to give a unity gain, 4th order Butterworth lowpass filter where the cutoff frequency (f_C) is set by a single resistor value (six resistors are required to form the filter, but they are

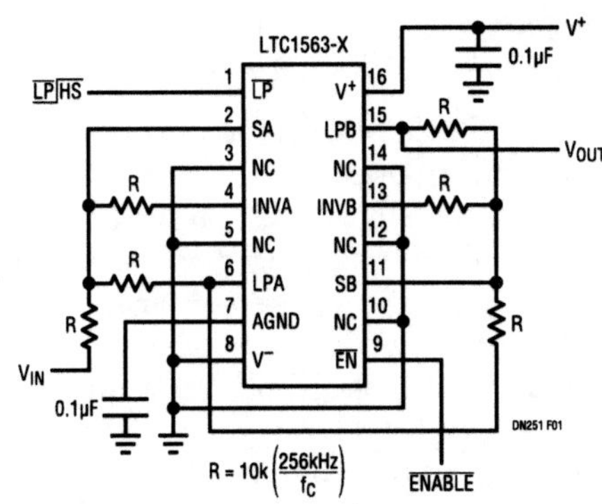

Figure 470.1 • Typical LTC1563-X Single Supply Application

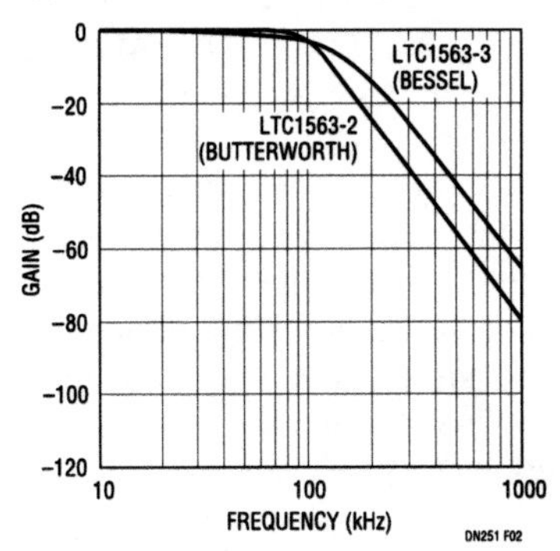

Figure 470.2 • Frequency Response of 100kHz Bessel and Butterworth Filters

Analog Circuit Design: Design Note Collection. http://dx.doi.org/10.1016/B978-0-12-800001-4.00470-1

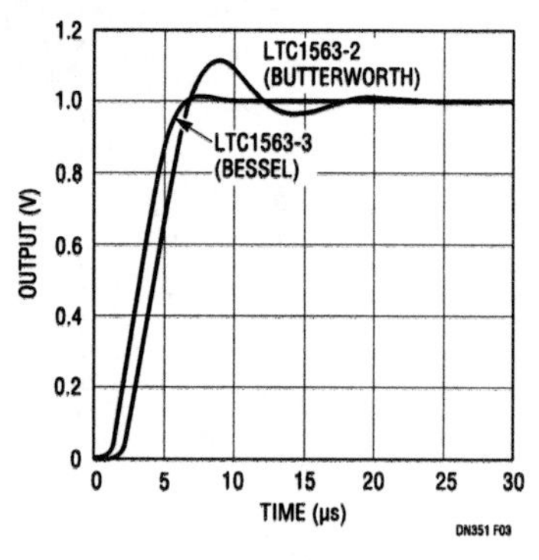

Figure 470.3 • Step Response of 100kHz Bessel and Butterworth Filters

all the same value). Similarly, the LTC1563-3 gives a unity gain, 4th order Bessel lowpass filter where the cutoff frequency is also set by a single resistor value. The resistor value is calculated by the simple formula above.

Figure 470.1 shows the schematic for a typical single-supply application using the LTC1563-X. The frequency responses of the LTC1563-2 (Butterworth) and the LTC1563-3 (Bessel) with a cutoff frequency of 100kHz are shown in Figure 470.2. The step responses are shown in Figure 470.3. The Butterworth response is often chosen for its flat passband gain (from DC to near the cutoff frequency) and reasonably sharp corner. The Bessel trades off the corner sharpness for a perfect step response without overshoot and ringing.

Easy design without sacrificing performance

Just because the LTC1563-X design process is easy does not mean that the filter's performance suffers. These parts are fully characterized filter products, ensuring that the filter's noise, distortion and dynamic range performance are all well known. Depending on the signal level and bandwidth, the LTC1563-X delivers a dynamic range of about 90dB, making these parts suitable for use in 16-bit data acquisition systems. Furthermore, the LTC1563-X supports rail-to-rail input and output operation on supply voltages from 2.7V to ±5V. The LTC1563-X also has a typical output DC offset of 1.5mV with an output DC offset drift of 10µV/°C. Lower frequency applications ($f_C \leq 25.6$kHz) can take advantage of the part's low power mode where supply current is reduced to about 1mA.

The LTC1563-X comes in the narrow 16-lead SSOP package (SO-8 footprint). The part's pinout makes layout easy so that trace parasitics are essentially eliminated. These features, coupled with the LTC1563-X's 2% frequency accuracy, lead to a compact, consistent and robust filter. In the end, you get the filter that you wanted with a known dynamic range and minimal production variation—no surprises.

Also included, Chebyshev filters with gain

Although the configuration of the LTC1563-X leads to the easily designed filters described above, the parts' topology allows for full transfer function freedom by simply using unequal valued resistors.

The LTC1563-X is composed of two cascaded 2nd order sections to form a 4th order filter (the sections are similar but not identical). A 2nd order section is mathematically defined by three parameters: gain, f_O frequency and Q value. Each section of the LTC1563-X requires three resistors to form the 2nd order section. Every gain, f_O and Q triplet corresponds to a unique 3-resistor solution. With this design flexibility, virtually any all-pole filter is obtainable. The LTC1563-X is also cascadable to form 8th order filters. Additionally, with one extra resistor and a capacitor, odd-order filters can be constructed (for example, a 5th order filter with one part).

Figure 470.4 shows the schematic for a 0.1dB ripple Chebyshev filter with a cutoff frequency of 150kHz and a passband gain of 10dB. This is just one example of the wide range of filter designs possible with the LTC1563-X. The design of these other transfer functions (not unity gain, 4th order Butterworth or Bessel) is a bit complex. *The best method of getting a good, solid design is to use Linear Technology's FilterCAD filter design software (version 3.0 or higher).* FilterCAD runs on the Windows operating system and is available free of charge from our web site at www.linear.com.

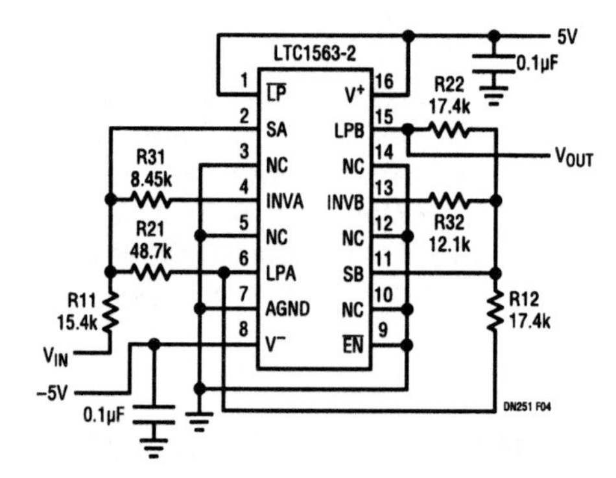

Figure 470.4 • 0.1dB, 150kHz, Chebyshev Lowpass Filter with a DC Gain of 10dB

Conclusion

The LTC1563 family of filter products provides worry-free results with minimal design effort. With its ease of use, consistent results and design flexibility, the LTC1563 may be the only part needed for all of your filter applications.

Free FilterCAD 3.0 software designs filters quickly and easily

471

Philip Karantzalis

Analog filters must satisfy multiple design specifications (i.e., gain, phase and time response) and implementation constraints (i.e., component tolerance, IC package and power supply voltage and current limitations). The design of a complex filter network requires substantial mathematical work and experience in an available technology (i.e., integrated circuits or op amps and discrete components). Linear Technology's FilterCAD 3.0 (FCAD) automates and simplifies the process of analog filter design and implementation. FCAD is a Microsoft Windows-based program. FCAD reduces the lengthy process of selecting and designing an optimal filter to a sequence of mouse clicks. Based on a user's input of filter parameters, FCAD designs a filter response and then implements that response as a practical circuit using a Linear Technology active RC or switched-capacitor filter IC. FCAD can realize lowpass, highpass, bandpass and notch filters with a variety of responses, including Butterworth, Bessel, Chebyshev, elliptic and minimum Q elliptic. In addition, with FCAD, a custom filter can be designed to meet an arbitrary set of specifications. FCAD features two distinct paths, Quick Design/Implement for quick answers to filter questions and Enhanced Design/Implement for comprehensive filter design and implementation choices. FilterCAD can be downloaded from www.linear.com.

Linear phase lowpass filters

In data acquisition and data communication systems, a lowpass filter with a linear phase response in its passband is often required to process a signal with minimum waveform distortion. In classic filter design, a Bessel approximation produces a linear phase lowpass filter. A lowpass Bessel filter has a very low overshoot response to a pulsed input signal. However, a Bessel filter has very low stopband selectivity. Nonstandard linear phase filters can be designed with higher selectivity than a Bessel filter. The following examples will describe how to use FCAD's Quick Design/Implement to obtain standard Bessel and nonstandard linear phase lowpass filter circuits.

Example 1: design a 256kHz linear phase lowpass filter for a single 5V power supply

The first time FCAD is opened, it defaults to the Quick Design window. Clicking on the "Next" button, FCAD advances through a series of windows with simple questions on filter design specifications. In the first window, select Lowpass; in the next, enter 20dB in the Attenuation box, 256kHz in the Passband (f_C) box and twice f_C (512kHz) in the Stopband (f_S) box; in the following two windows, select Linear Phase and 5V Power Supply. One more click on the "Next" button displays a Quick Implement window of Linear Technology filter ICs. The Quick Implement window offers a choice of two linear phase lowpass filters that can provide a 256kHz cutoff while operating with a single 5V power supply: the LTC1569-7 (a self-contained switched capacitor) and the LTC1563-3 (an active RC). Placing the mouse pointer on the device name and clicking causes FCAD to display the filter's gain response. The LTC1563-3 has a standard 4th order Bessel response and the LTC1569-7 has a very selective elliptic response with linearized phase in the passband. Click on the "Next" button to view the device package, then click again to enter a design title; finally, click on the "Done" button to view the filter's schematic, gain and time response to a step input. Selecting the time response window allows for the choice of a filter signal input (step, pulse or sine-burst input). Figure 471.1 shows the filter schematic and 10μs pulse response of the LTC1563-3 and Figure 471.2 shows the same for the LTC1569-7. Choosing the optimum filter IC from the list offered by FCAD depends on the filter application. Specific application requirements can favor one device only or allow a choice of either one. The LTC1569-7 has high selectivity and is tunable (changing an external resistor sets the cutoff frequency), whereas the LTC1563-3 has low pulse overshoot and is not tunable (six equal external resistors must change to set the cutoff frequency). Consult the IC's data

Analog Circuit Design: Design Note Collection. http://dx.doi.org/10.1016/B978-0-12-800001-4.00471-3

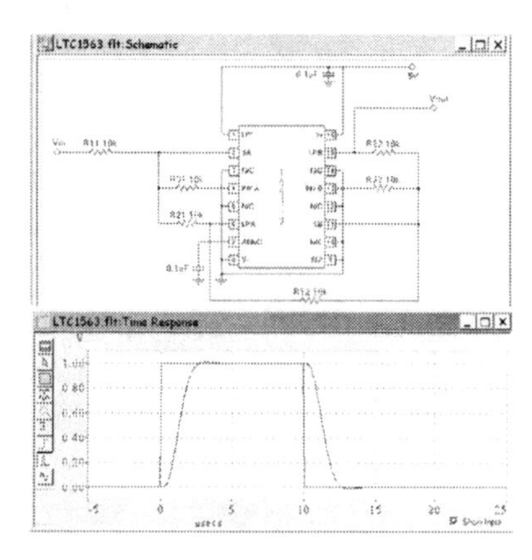

Figure 471.1 • LTC1563-3 Schematic and Pulse Response

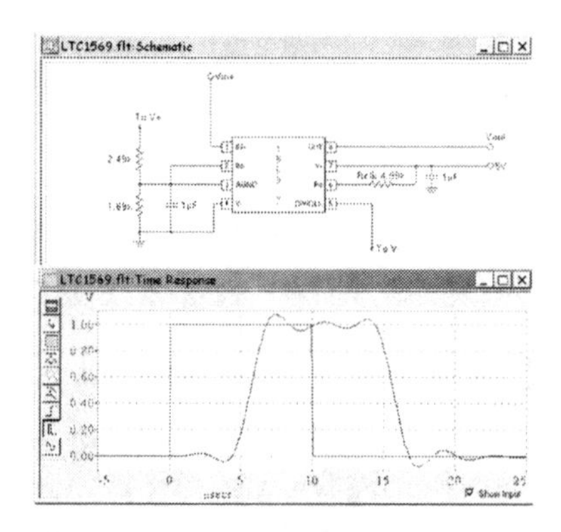

Figure 471.2 • LTC1569-7 Schematic and Pulse Response

sheet for additional information (i.e., the LTC1563-3 has a rail-to-rail input and output voltage range and the LTC1569-7 can be tuned with an external clock to a cutoff frequency as low as 100Hz).

Both the LTC1563-3 and the LTC1569-7 are available in a surface mount package with identical dimensions (the LTC1563-3 in a 16-pin narrow SSOP and the LTC1569-7 in a 8-pin narrow SO package).

Example 2: design a 10kHz low power linear phase lowpass filter for a single 3V power supply

Following the procedure described in Example 1, sequentially select Lowpass, enter 20dB in the Attenuation box, 10kHz in the Passband (f_C) box and twice f_C (20kHz) in the Stopband (f_S) box, then select Linear Phase, 3V Power Supply and Low Power. Click on "Next" to view FCAD's Quick Implement window of three low power linear phase lowpass filters that can operate with a 3V power supply: The LTC1569-6 and LTC1569-7 (switched capacitor) and the LTC1563-3 (active RC). Of the three ICs, the LTC1563-3 has the lowest power supply current (1mA) for cutoff frequencies (f_C) less than 25.6kHz and may be operated to cutoff frequencies as low as 256Hz. The LTC1569-6 has lower power supply current than the LTC1569-7 for cutoff frequencies (f_C) less than 20kHz and it can be clock tuned to a frequency as low as 25Hz.

Example 3: design a 650kHz linear phase lowpass filter for a single 5V power supply

Following the procedure described in Example 1, sequentially select Lowpass, enter 20dB in the Attenuation box, 650kHz in the Passband (f_C) box and twice f_C (1300kHz) in the Stopband (f_S) box, then select Linear Phase and 5V Power Supply. FCAD's Quick Implement window shows the LTC1565-31 as an active RC filter. The LTC1565-31 has differential inputs and outputs. The "dash 31" of LTC1565 denotes a passband frequency equal to 650kHz. For applications requiring specific cutoff frequencies other than that provided by the LTC1565-31, contact LTC Marketing.

FilterCAD 3.0 can provide a variety of filter design solutions. In the event that the program is unable to offer a filter to a set of specifications, Linear Technology's filter applications staff is available to assist in the design of an optimum filter.

The following books are excellent sources in filter design and signal processing:

"Filtering in the Time and Frequency Domains," Herman J. Blinchikoff and Anatol I. Zverev. Robert E. Krieger Publishing Company.

"Filter Design For Signal Processing Using MATLAB and Mathematica." M. D. Lutovac, D. V. Tošic´ and B. L. Evans. Prentice Hall Inc.

SOT-23 micropower, rail-to-rail op amps operate with inputs above the positive supply

472

Raj Ramchandani

Introduction

The only SOT-23 op amps featuring Over-The-Top operation—the ability to operate with either or both inputs above the positive rail—are the 55μA LT1782 and the 300μA LT1783. This feature is important in many current-sensing applications, where the inputs are required to operate at or above the supply. The wide supply voltage range, from 2.7V to 18V, gives the LT1782/LT1783 broad appeal as general purpose amplifiers, while the guaranteed offset voltage of 950μV over temperature is the lowest of any SOT-23 op amp. There is even a shutdown feature for ultralow supply current applications.

Tough general purpose op amps

The LT1782/LT1783 SOT-23 op amps are ideal for general purpose applications that demand excellent performance. These SOT-23 op amps are specified at input common mode voltages as high as 18V, independent of the supply voltage, making them ideal for applications with a wide input range requirement and/or unusual input conditions. In applications that require more bandwidth than the 200kHz LT1782, the LT1783's six-fold increase in supply current gives it six times more bandwidth and slew rate. The parts are available in two pinouts, a 6-lead version with a shutdown feature that reduces supply current to only 5μA, and a standard pinout 5-lead version. Table 472.1 summarizes the performance of these new op amps.

Table 472.1 LT1782/LT1783 SOT-23 Guaranteed Performance, $V_S = 3V/0V$ or $5V/0V$, $T_A = 25°C$

PARAMETER	LT1782	LT1783
Supply Voltage Range	2.7V to 18V	2.7 to 18V
Supply Current (Max)	55μA	300μA
Input Offset Voltage (Max)	800μV	800
Input Bias Current (Max)	15nA	80nA
Input Bias Current, $V^+ = 0V$ (Typ)	0.1nA	0.1nA
Input Offset Current (Max)	2nA	8nA
Open-Loop Gain, $R_L = 10k$ (Min)	200V/mV	200V/mV
PSRR (Min)	90dB	90dB
CMRR (Min)	90dB	90dB
Common Mode Range	0V to 18V	0V to 18V
Output Swing Low ($V_O - V^-$)	8mV	8mV
Output Swing High ($V^+ - V_O$)	90mV	90mV
Slew Rate (Typ)	0.07V/μs	0.42V/μs
Gain Bandwidth Product (Typ)	200kHz	1.25MHz

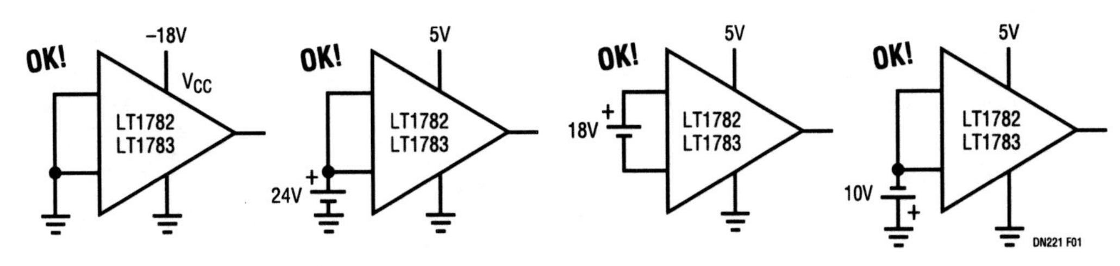

Figure 472.1 • Tough Op Amps

Analog Circuit Design: Design Note Collection. http://dx.doi.org/10.1016/B978-0-12-800001-4.00472-5

Tough op amps

The LT1782/LT1783 are tough op amps that can be exposed to a variety of extreme conditions without being damaged (Figure 472.1). The amplifiers have reverse-battery protection up to 18V. The input pin voltage can extend to 10V below V^- or 24V above V^- without damaging the device. The maximum input differential voltage is 18V, regardless of the supply voltage. All of these features combine to make the LT1782/LT1783 "one tough SOT-23."

Read all of the specs

Common factors that keep most SOT-23 parts from being general purpose amplifiers include low supply voltage range, high input offset voltage, low open-loop voltage gain and poor output stage performance.

The LT1782/LT1783 amplifiers operate on all single and split supplies with a total voltage of 2.7V to 18V. They are stable with capacitive loads up to 500pF under all load conditions. The minimum output current is ±18mA and the unloaded output swing is guaranteed within 8mV of ground and 90mV of the positive rail.

A common problem encountered with other op amps in many applications is that as the output approaches the rail or ground, the gain degrades. The data sheet typically claims the output can swing to within a few millivolts of the rail, but the input overdrive required to achieve this can be quite high. This not the case with the LT1782/LT1783; a few millivolts of input overdrive is enough to swing the outputs to their guaranteed value. Figure 472.2 shows the typical output saturation voltage vs input overdrive.

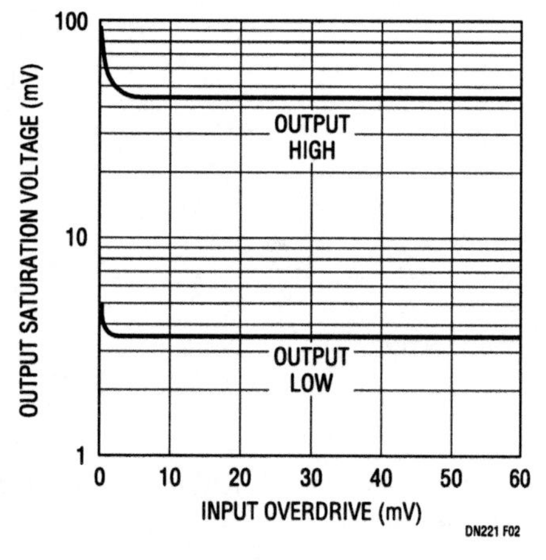

Figure 472.2 • Output Saturation Voltage vs Input Overdrive

Over-The-Top applications

The circuit of Figure 472.3 uses the Over-The-Top capabilities of the LT1783. The 0.2Ω resistor senses the load current while the op amp and NPN transistor form a closed loop, making the collector current of Q1 proportional to the load current. The 2k load resistor converts the current into a voltage. The positive supply rail, V_{BAT}, is not limited to the 5V supply of the op amp and could be as high as 18V. The LT1783 draws no current through the inputs when it is powered down, extending the battery life.

The circuit of Figure 472.4 uses the LT1782 in conjunction with the LT1634 micropower shunt reference. The supply current of the op amp also biases the reference. The drop across resistor R1 is fixed at 1.25V generating an output current equal to 1.25V/R1. Notice that noninverting input is tied to the V_{CC} pin of the op amp.

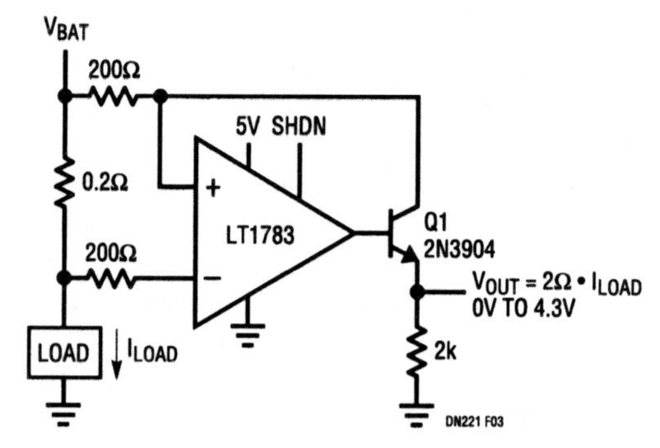

Figure 472.3 • Positive Supply Rail Current Sense

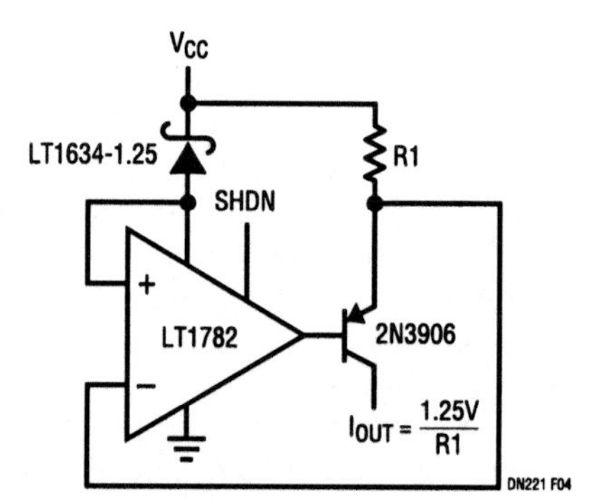

Figure 472.4 • Current Source

Get 100dB stopband attenuation with universal filter family

473

Max W. Hauser

The LTC1562 and LTC1562-2 are compact, high performance, "universal" continuous-time filter products, each containing four 2nd order Operational Filter blocks. These low noise, DC-accurate filters let you tailor their center frequencies (f_0) over a range of roughly 10kHz to 150kHz (LTC1562) or 20kHz to 300kHz (LTC1562-2) and replace several precision capacitors, resistors and op amps. All frequency-setting components are internal and trimmed, except for one resistor per block, which is desensitized (1% error in this external resistor's value contributes only 0.5% error to the programmed f_0). Additional components program each block's Q and gain. A complete application circuit using either the LTC1562 or LTC1562-2 on a surface mount board is about the size of a dime (155mm^2).

Figure 473.1 shows one block, or 2nd order section (each LTC1562 contains four of these), with external resistors to set the standard 2nd order filter parameters f_0, Q and gain. In this example, the section is configured so that the two outputs give lowpass and bandpass responses. Cascades of Figure 473.1 circuits, with appropriate resistor values, can realize any all-pole lowpass or bandpass filter response form such as Chebyshev, Bessel or Butterworth. Adding external capacitors permits highpass forms. These filters can suppress undesired frequencies by 100dB while maintaining low noise and distortion.

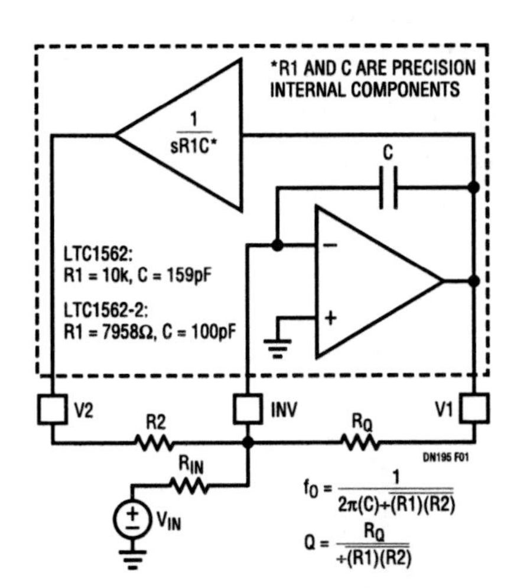

Figure 473.1 • An Operational Filter Block (Inside Dashed Line) Configured with External Resistors for Lowpass (at V2) and Bandpass (at V1) Responses. Each LTC1562 or LTC1562-2 Contains Four Such Blocks

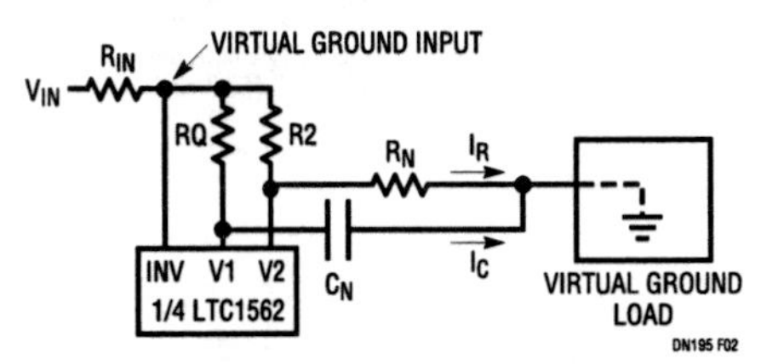

Figure 473.2 • Robust Notch Filtering Using a 1/4 LTC1562 Operational Filter Section. R_N and C_N Control Notch Frequency

Operational Filter blocks, however, have many more creative applications. Each block has a flexible virtual-ground input (INV) and two outputs, V1 and V2. V2 is a time integrated version of V1 and therefore lags V1 by 90° over a very wide range of frequencies. Parallel paths into the virtual-ground input or from the two different outputs permit transfer-function zeroes, of which one of the most useful is the imaginary-axis zero pair, or notch.

Figure 473.2 shows a simple and robust notch-filtering method. A notch filter has zero gain at some frequency f_N. Notch filters are useful not only to remove frequencies *per se* but also to improve selectivity in lowpass or highpass filters by placing notches in the stopband, as illustrated below. (Such responses are broadly called "elliptics" or "Cauers.")

In Figure 473.2, a notch occurs when a 2nd order section drives a virtual-ground input through two paths: one through a capacitor and one through a resistor. The virtual ground can be an op amp input, or as in Figure 473.3, another Operational Filter input. Capacitor C_N adds a further 90° to the 90° phase difference between the V1 and V2 voltages.

Analog Circuit Design: Design Note Collection. http://dx.doi.org/10.1016/B978-0-12-800001-4.00473-7

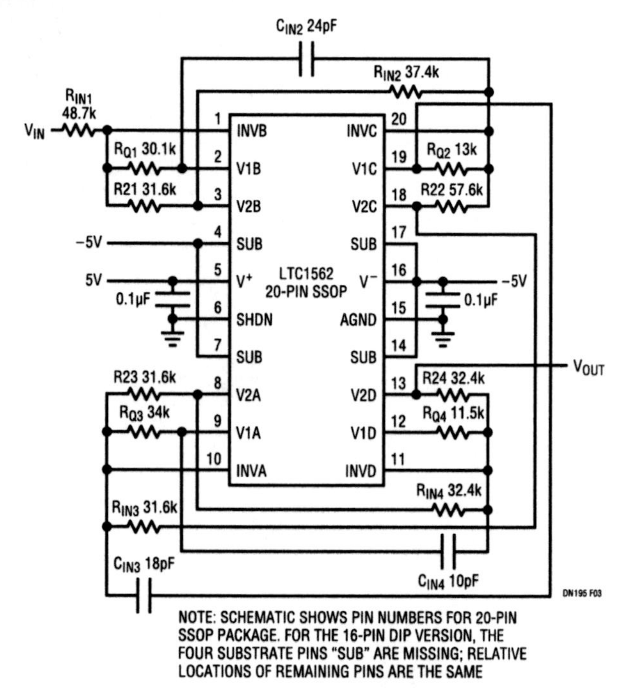

Figure 473.3 • LTC1562 50kHz Elliptic Lowpass Filter with 100dB Stopband Rejection

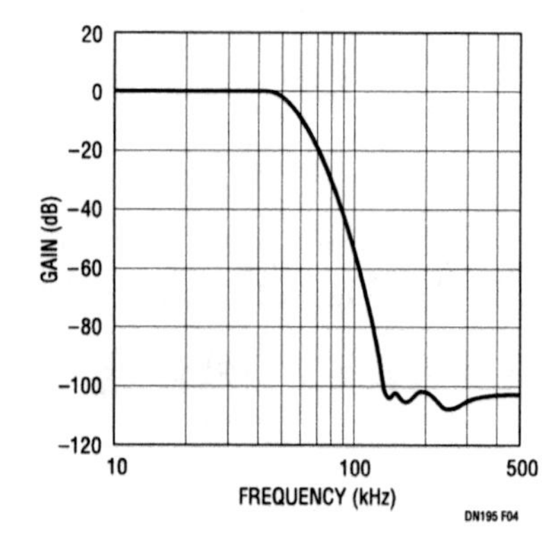

Figure 473.4 • Measured Frequency Response for Figure 473.3

At the frequency where currents I_C and I_R have equal magnitude, the two paths cancel and a 2nd order notch occurs. This frequency is:

$$f_N = \sqrt{\frac{f1}{2\pi(R_N)(C_N)}}$$

Here, f1 is a parameter internally trimmed in each Operational Filter product (100kHz in the LTC1562, 200kHz in the LTC1562-2). The notch frequency, f_N, is independent of the center frequencies, f_0, programmed separately for each 2nd order section, as in Figure 473.1.

A remarkable feature of Figure 473.2 is its inherently deep notch response—the depth does not come from component matching as with other notch-filter circuits. Errors in R_N or C_N values change the notch frequency, f_N, but not the depth of the notch at f_N. Moreover, the square root dependence in the f_N expression desensitizes the notch frequency to errors in the R_N and C_N values.

Figure 473.3 shows an 8-pole modified elliptic response 50kHz lowpass filter using the notch method of Figure 473.2. In this filter, three operational filter blocks ("B," "C" and "A" in the pinout, in sequence) drive RC combinations as in Figure 473.2, giving notches at approximately 133kHz, 167kHz and 222kHz, respectively. A 2nd order lowpass section, per Figure 473.1 with $f_0 = 55.5$kHz, follows (the "D" block in the pinout). Figure 473.4 shows measured frequency response, which falls 100dB in a little more than one octave. The choice of notch frequencies trades off passband flatness against stopband ripple; the user can explore this trade-off via analog filter design software such as FilterCAD for Windows, available free from Linear Technology (1-800-4-LINEAR). The values in Figure 473.3 give stopband attenuations exceeding 100dB above 140kHz. This circuit has output noise (in 500kHz bandwidth) of $60\mu V_{RMS}$ with approximately rail-to-rail input and output swings, or a peak signal-to-noise ratio of 95dB when operating from ±5V supplies. THD is −95dB with $1V_{RMS}$ ($2.8V_{P-P}$) output at 20kHz.

Note that 100dB attenuation at hundreds of kilohertz requires electrically clean, compact construction, with good grounding and supply decoupling, and minimal parasitic capacitances in critical paths (such as the INV inputs). For example, 0.1µF capacitors near the LTC1562 provide adequate decoupling from a clean, low inductance power source. But several inches of wire (i.e., a few microhenrys of inductance) from the power supplies, unless decoupled by substantial (≥10µF) capacitance near the chip, can cause a high-Q LC resonance (at hundreds of kHz) in the LTC1562's supplies or ground reference, impairing SNR and stopband rejection at those frequencies.

Tiny 1MHz lowpass filter uses no inductors

474

Nello Sevastopoulos

The LTC1560-1 is a fully integrated continuous-time filter in an SO-8 package. It provides a 5-pole elliptic response with a pin-selectable cutoff frequency (f_C) of 1MHz or 500kHz. Several features distinguish the LTC1560-1 from other commercially available high frequency, continuous-time monolithic filters:

- 5-pole 0.5MHz/1MHz elliptic in an SO-8 package
- 70dB signal-to-noise ratio (SNR) measured at 0.07% THD
- 75dB signal-to-noise ratio (SNR) measured at 0.5% THD
- 60dB or more stopband attenuation
- No external components required other than supply and ground decoupling capacitors

The LTC1560-1 delivers accurate fixed cutoff frequencies of 500kHz and 1MHz without the need for internal or external clocks. Other cutoff frequencies can be obtained upon demand; please consult LTC marketing. The extremely small size of the part makes it suitable for compact designs that were never before possible using discrete RC active or RLC passive filter designs.

Frequency and time-domain response

Figure 474.1 shows a simple circuit for evaluating the performance of the filter. The LTC1560-1 offers a pin-selectable cutoff frequency of either 500kHz or 1MHz. The filter gain response is shown in Figure 474.2. In the 1MHz mode, the passband gain is flat up to $(0.55)(f_C)$ with a typical ripple of ±0.2dB, increasing to ±0.3dB for input frequencies up to $(0.9)(f_C)$. The stopband attenuation is 63dB starting from $(2.43)(f_C)$ and remains at least 60dB for input frequencies up to 10MHz.

The elliptic transfer function of the LTC1560-1 was chosen as a compromise between selectivity and transient response. Figure 474.3a shows the 2-level eye diagram of the filter. The size of the "eye" opening shows that the filter is suitable for data communications applications. Additional phase equalization can be performed with the help of an external dual op amp and a few passive components. This is shown in Figure 474.4, where a 2nd order allpass equalizer

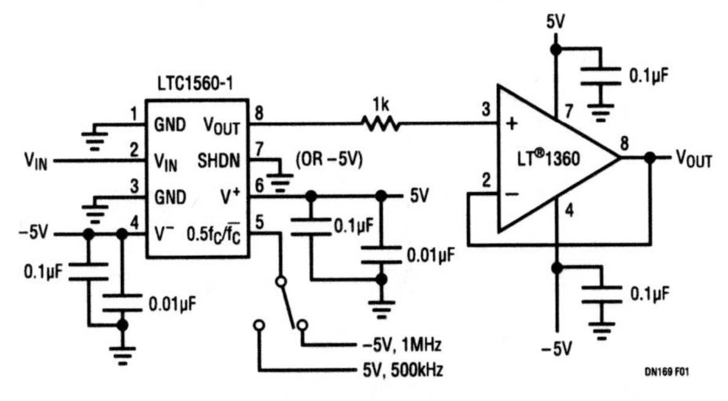

Figure 474.1 • A Typical Circuit for Evaluating the Full Performance of the LT1560-1

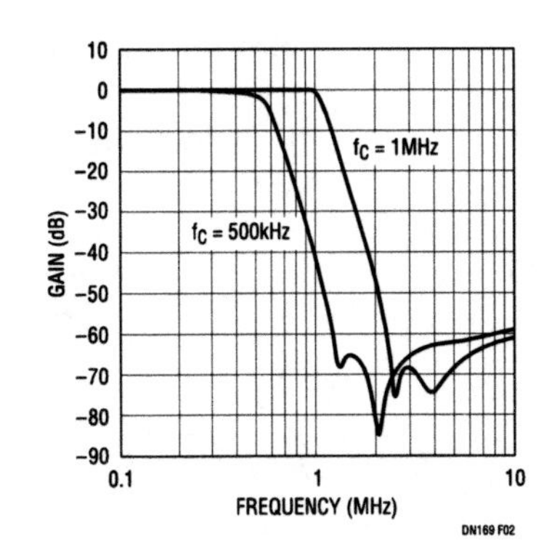

Figure 474.2 • Gain vs Frequency of the 1MHz and 500kHz

Analog Circuit Design: Design Note Collection. http://dx.doi.org/10.1016/B978-0-12-800001-4.00474-9

is cascaded with the IC. The allpass function is achieved through traditional techniques, namely, passing a signal through a low Q inverting bandpass filter and then performing summation with the appropriate gain factors. Figure 474.3b shows the eye diagram of the equalized filter.

DC accuracy

For applications where very low DC offset and DC accuracy are required, the DC offset of the filter can be easily corrected, as shown in Figure 474.5.

The input amplifier stores the DC offset of the IC across its feedback capacitor. The total output DC offset is the input DC offset of the 1/2 LT1364 plus its offset current times the 10k resistor (less than 1.85mV). Upon power-up, the initial settling time of the circuit is dominated by the RC time constant of the DC correcting feedback path; once the DC offset of the LTC1560-1 is stored, the transient behavior of the circuit is dictated by the elliptic filter.

Conclusion

The LTC1560-1 is a 5th order, user friendly, elliptic lowpass filter suitable for any compact design. It is a monolithic replacement for larger, more expensive and less accurate solutions in communications and data acquisition.

Figure 474.3a • 2-Level Eye Diagram of the LTC1560-1 Before Equalization

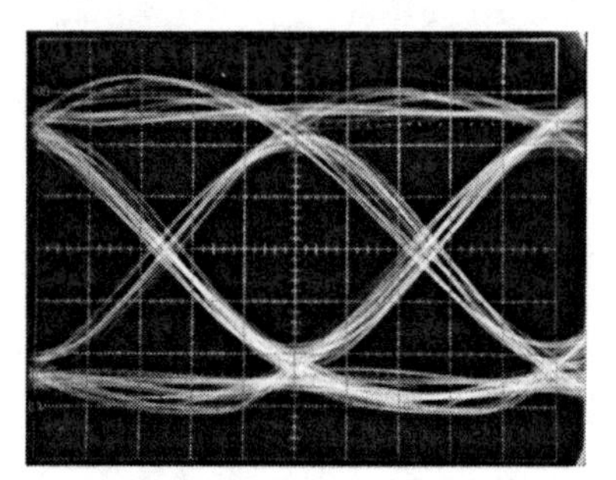

Figure 474.3b • 2-Level Eye Diagram of the Equalized Filter

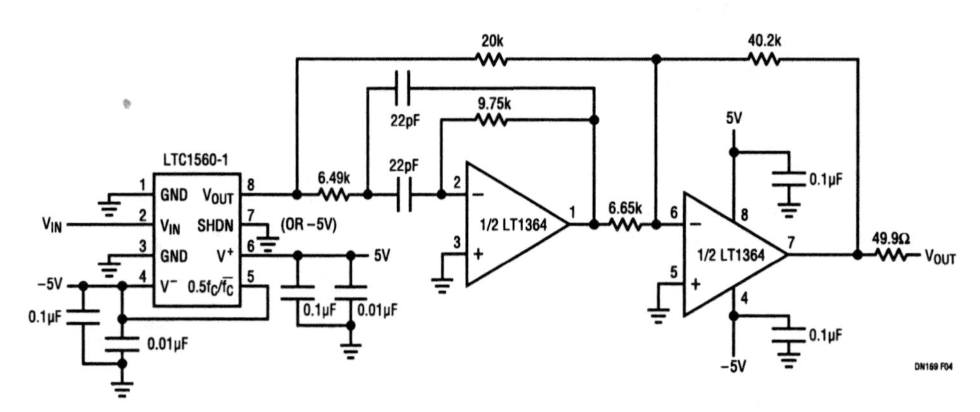

Figure 474.4 • Augmenting the LTC1560-1 for Improved Delay Flatness

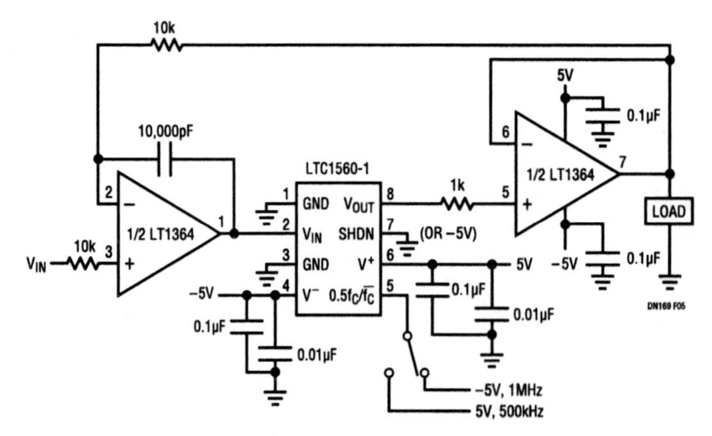

Figure 474.5 • A DC Accurate 500kHz/1MHz Elliptic Filter with an Output Buffer for Driving Cables or Capacitive Loads

A family of 8th order monolithic filters in an SO-8 package

475

Linear's Filter Group

The LTC1069-X family of monolithic filters offers economical solutions for a wide variety of signal processing applications.

Members of the LTC1069-X family are available as standard products or as user-specified, semicustom filters. Both products are fully integrated and are available in small SO-8 packages. They require only power supply decoupling capacitors and a clock source to set the cutoff frequency and form a complete filter.

As semicustom filters, the LTC1069-X can be configured as a single 8th order filter or as two 4th order filters with either identical or independent characteristics. The technology allows the user to specify and optimize the filter response, sampling rate and power consumption for each application. The LTC1069-X can realize lowpass, bandpass, highpass, notch and allpass responses.

The LTC1069-X family also includes standard products, the LTC1069-1, LTC1069-6 and LTC1069-7.

The LTC1069-1 and LTC1069-6 offer either low or very low power, 8th order elliptic lowpass response for dual or single supply applications. The LTC1069-7, although it uses the same technology as the LTC1069-1 and LTC1069-6, has completely different characteristics: it provides lowpass filtering with linear phase response and cutoff frequencies up to 200kHz. All three products have the same pinout.

LTC1069-1: low power elliptic anti-aliasing filter works from single 3.3V to ±5V supplies

The LTC1069-1 amplitude response features a very flat passband up to $0.95f_{CUTOFF}$, and a sharp 20dB attenuation in the vicinity of the cutoff frequency (that is, at $1.2f_{CUTOFF}$). The transition band reaches 52dB attenuation at $1.4f_{CUTOFF}$, and keeps rolling off instead of "bouncing" back as textbook elliptic filters do. Figure 475.1 illustrates this "progressive rolloff."

Figure 475.2 shows typical connections for both single 5V and ±5V supplies. A precision internal resistive divider biases the analog ground pin to one-half the total power supply voltage. For single supply applications, only an external decoupling capacitor is needed to lower the AC impedance of the pin.

The LTC1069-1 cutoff frequency is clock tuned and the clock-to-cutoff frequency ratio and the internal sampling rate

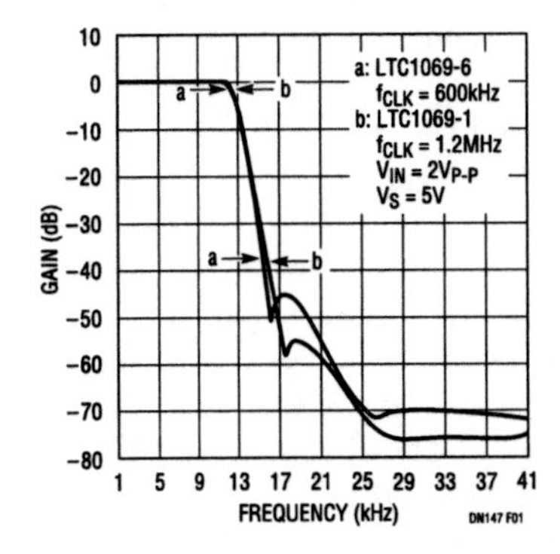

Figure 475.1 • Amplitude Response Comparison: LTC1069-1 vs LTC1069-6

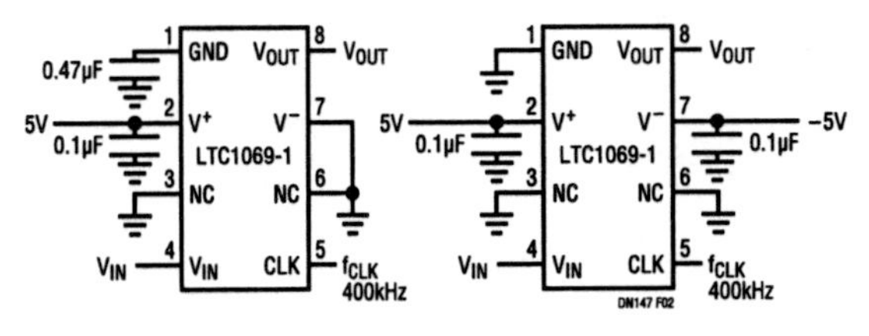

Figure 475.2 • Single 5V or Dual ±5V Supply 4kHz Elliptic Lowpass Filter

Table 475.1 LTC1069-1 Maximum Cutoff Frequency and Resulting Power Consumption

V_S	$I_{SUPPLY(TYP)}$	$F_{CUTOFF(MAX)}$
±5V	3.8mA	12kHz
5V	2.5mA	8kHz
3.3V	1.5mA	4kHz

Analog Circuit Design: Design Note Collection. http://dx.doi.org/10.1016/B978-0-12-800001-4.00475-0

are 100:1. The power consumption and the maximum cutoff frequency are illustrated in Table 475.1.

Despite its low power consumption, the device's dynamic range is not compromised. The widest dynamic range is obtained with ±5V supplies, where a S/(N+THD) ratio of better than 70dB is reached with input voltages between $0.34V_{RMS}$ and $2.5V_{RMS}$.

LTC1069-6: 8th order elliptic lowpass works on single 3V, consumes 1mA

The LTC1069-6 is designed for minimal power consumption and for single 3V and 5V supplies. The low supply current, 1mA for a 3V supply and 1.2mA for a 5V supply, includes the current spent to bias the analog ground pin for optimum dynamic range and for ease of use in single supply applications.

To save power, the clock-to-cutoff frequency range is lowered to 50:1, but the internal sampling rate remains at 100:1. The amplitude response of the filter is also progressive elliptic, as shown in Figure 475.1.

When compared to the LTC1069-1, the LTC1069-6 has a sharper rolloff in the vicinity of its cutoff frequency, the key feature being 42dB attenuation at $1.27f_{CUTOFF}$. Conversely, the LTC1069-1 has better stopband attenuation, a flatter passband and 5dB more dynamic range.

The LTC1069-6 excels in optimizing speed for power consumption. Table 475.2 shows that a cutoff frequency of up to 20kHz can be obtained with a single 5V supply and 1.2mA typical power supply current.

Table 475.2 LTC1069-6 Maximum Cutoff Frequencies and Resulting Power Consumption

V_S	$I_{SUPPLY(TYP)}$	$F_{CUTOFF(MAX)}$	$F_{CLK(MAX)}$
3V	1mA	14kHz	700kHz
5V	1.2mA	20kHz	1MHz

LTC1069-7: linear-phase communication filter delivers up to 200kHz cutoff frequency and symmetrical impulse response

The amplitude response of the IC approximates a "textbook" raised cosine, alpha-of-one filter. The phase of the LTC1069-7 is linearized over the entire passband to provide the fully symmetrical impulse response expected from a raised cosine filter (see Figure 475.3). When compared to a conventional linear phase filter of the same order, such as an LTC1064-3, 8th order Bessel lowpass, the LTC1069-7 has a steeper rolloff in the vicinity of its cutoff frequency, as shown in Figure 475.4. This type of filter is useful in digital communications, where compromises between selectivity and good transient response are always sought.

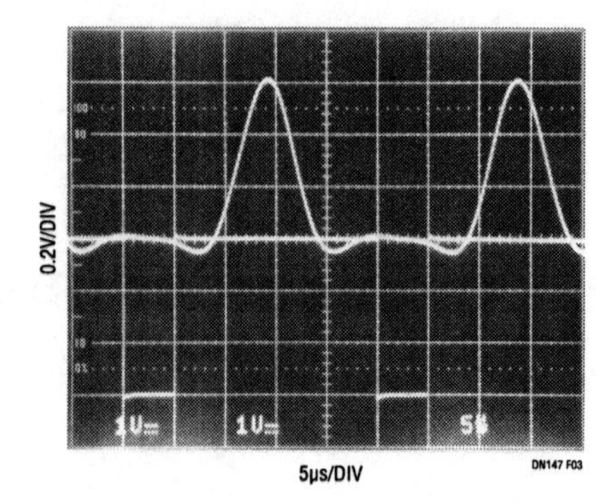

Figure 475.3 • The Pulse Response of the LTC1069-7 is Fully Symmetrical

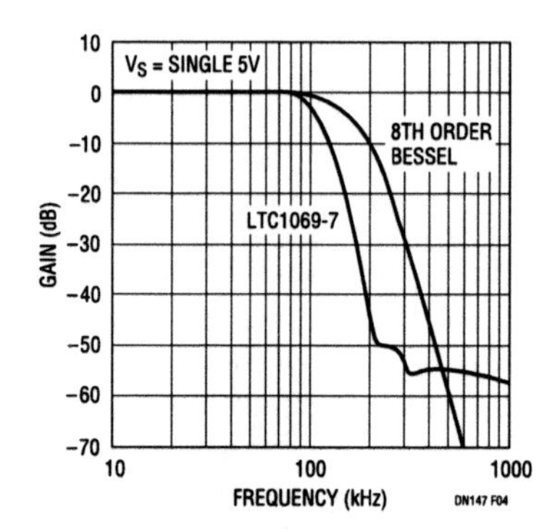

Figure 475.4 • Frequency Response

Note that the raised cosine response is not related to the classical all-pole responses of Butterworth, Chebyshev or Bessel filters. The realization of a raised cosine filter with discrete RC active techniques requires many precision passive and active components.

The LTC1069-7 trades power consumption and sampling rate for speed. The maximum cutoff frequency is 200kHz, the internal sampling rate is 50:1 and the clock-to-cutoff frequency ratio is 25:1. Table 475.3 illustrates the maximum cutoff frequency of the device for single 5V and ±5V supplies.

Table 475.3 LTC1069-7 Maximum Cutoff Frequencies and Resulting Power Consumption

V_S	$I_{SUPPLY(TYP)}$	$F_{CUTOFF(MAX)}$
±5V	18mA	200kHz
5V	12mA	140kHz
3V	8mA	70kHz

Conclusion

A flexible and economic filter technology is available today; it trades power consumption for speed and provides cutoff frequencies ranging from a few hertz up to 200kHz. All of the filters come in SO-8 packages. The filter group of Linear Technology has accumulated years of experience and is always happy to hear from you and talk filters.

A 1mV offset, clock-tunable, monolithic 5-pole lowpass filter

476

Nello Sevastopoulos

The LTC1063 is the first monolithic lowpass filter simultaneously offering outstanding DC and AC performance. It features internal or external clock tunability, cutoff frequencies up to 50kHz, 1mV typical output DC offset, and a dynamic range in excess of 12 bits for over a decade of input voltage.

The LTC1063 approximates a 5-pole Butterworth lowpass filter. The unique internal architecture of the filter allows outstanding amplitude matching from device to device. Typical matching ranges from 0.01dB at 25% of the filter passband to 0.05dB at 50% of the filter passband. This capability is important for multichannel data-acquisition systems where channel-to-channel matching is critical.

Using the filter's internal oscillator

An internal or external clock programs the filter's cutoff frequency. The clock-to-cutoff frequency ratio is 100:1. In the absence of an external clock, the LTC1063's internal precision oscillator can be used. An external resistor and capacitor set the device's internal clock frequency (Figure 476.1). The internal oscillator output is brought out at Pin 4 so that it can be used as a synchronized master clock to drive other LTC1063s. Ten or more filters can be locked together to a single LTC1063 clock output as shown in Figure 476.1.

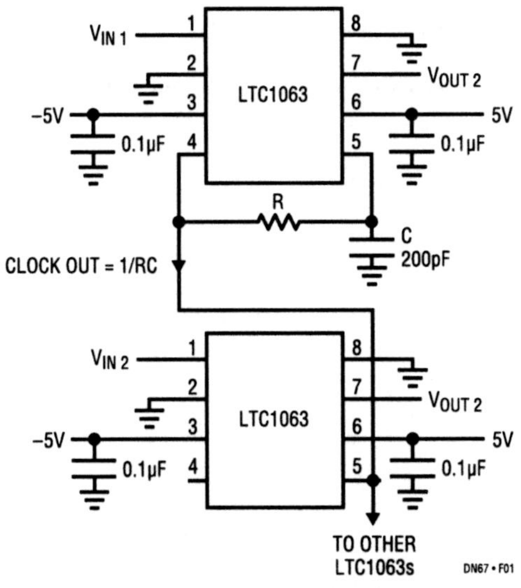

Figure 476.1 • Synchronizing Multiple LTC1063s

DC performance

The LTC1063's output DC offset voltage (typically 1mV or less) is optimized for ±5V supply applications. Output offset is low enough to compete with discrete type RC active filters

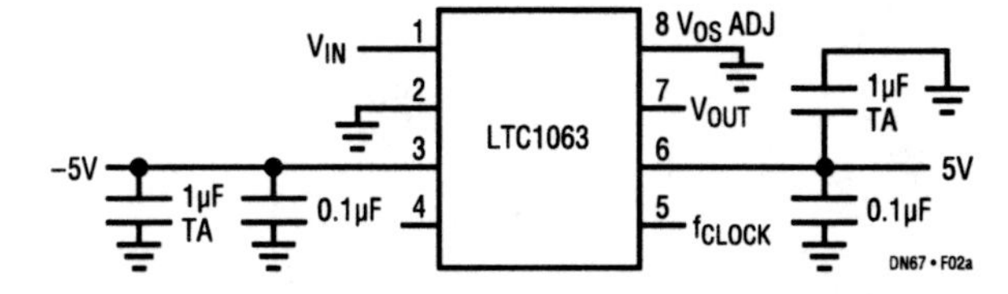

Figure 476.2a • LTC1063 Operating as a Clock-Sweepable Lowpass Filter

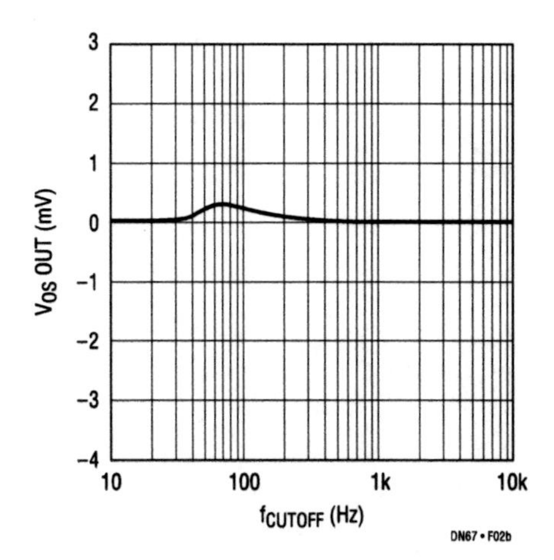

Figure 476.2b • Output DC Offset vs f_{CUTOFF} for Figure 476.2a's Circuit

Analog Circuit Design: Design Note Collection. http://dx.doi.org/10.1016/B978-0-12-800001-4.00476-2

using low offset op amps. Figures 476.2a and 2b show an LTC1063 filter operating as a clock-sweepable lowpass filter exhibiting no more than 200μV of total output offset variation over 3 decades of cutoff frequency.

This measurement was taken with a combination of ceramic and tantalum capacitors and a clean PC board layout. DC offset is also affected by input signal DC voltage. The same circuit, Figure 476.2a, shows a total of 1mV of output offset voltage change with an 8V input change thus, giving a CMRR of 78dB.

Dynamic range

The LTC1063 has both low noise and very low, $50\mu V_{RMS}$, clock feedthrough. The wideband noise is the integral of the noise spectral density; it is usually expressed in μV_{RMS} and is virtually independent of filter cutoff frequency. The LTC1063 has a wideband noise specification of $90\mu V_{RMS}$. This number is clock frequency- and power supply-independent.

The LTC1063's AC design however, is based on optimum dynamic range rather than just wideband noise. Dynamic range measurements take into the account the device's total harmonic distortion. Figure 476.3a shows the typical connection for dynamic range measurement. An inverting buffer is preferred over a unity-gain follower. Large input common-mode signals can severely degrade the distortion performance of noninverting buffers. It is also important to make sure the undistorted op amp swing is equal to or better than that of the filter. Figure 476.3b shows the device's operating distortion plus noise versus input signal amplitude measured with a standard 1kHz pure sine wave input. The THD improves with increased power supply voltage.

Figure 476.3b illustrates how the filter can handle inputs to $4V_{RMS}$ ($11.2V_{P-P}$) with less than 0.02% THD. At this input level, the dynamic range is only limited by distortion and not by wideband noise. The signal-to-noise ratio at $4V_{RMS}$ input is 93dB. Optimum signal-to-noise plus distortion according to Figure 476.3b is 83dB, yet a comfortable 80dB (0.01%) is achieved for input levels between $1V_{RMS}$ and $2.4V_{RMS}$.

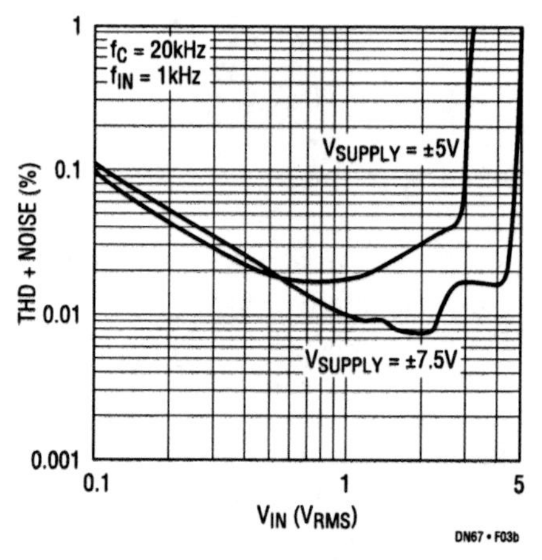

Figure 476.3b • Plot of Distortion + Noise vs V_{IN}

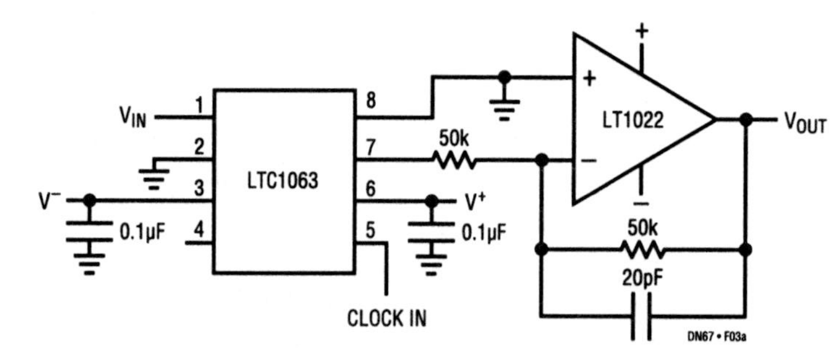

Figure 476.3a • Typical Connection for Measuring Distortion + Noise and Signal to THD + Noise Ratio

High dynamic range bandpass filters for communications

477

Richard Markell

Introduction

Octave or decade wide bandpass filters are non-trivial to design. Wide bandpass filtering requires the designer use a highpass filter at the input in series with a lowpass filter to achieve the desired specifications. This becomes evident if one examines the transfer functions of the state-variable-filter configuration, be it switched capacitor or active RC. Either option limits severely the achievable dynamic range if a wideband bandpass filter is designed using the bandpass output.

Wideband bandpass filters occupy a niche in the communications area of signal processing. The wideband bandpass function is required in receiver IF applications which traditionally use the crystal filter and/or active RCs. Sonar applications also demand steep, wide, low noise bandpass filters to allow analysis of "chunks" of the frequency spectrum, one at a time or in parallel.

Recent improvements in switched capacitor filter technology allow designers the luxury of using switched capacitor filters in these applications. True clock tuning allows variable bandwidth filters to be implemented with only a few parts.

This note details the design of a wideband (10kHz–100kHz) bandpass filter using a single LTC1064 plus an LTC1064-4. The LTC1064 is a quad universal switched capacitor filter while the LTC1064-4 is the same silicon with integrated resistors configured to implement an 8th order elliptic lowpass filter. Both filters have low noise and can operate to frequencies beyond 100kHz. The combination bandpass filter has a set of tough design specifications: total integrated noise in the passband less than 350μV, passband ripple less than 0.4dB or ±0.2dB, and steep rolloffs at the band edges (−70dB at 5kHz, and −70dB at 200kHz).

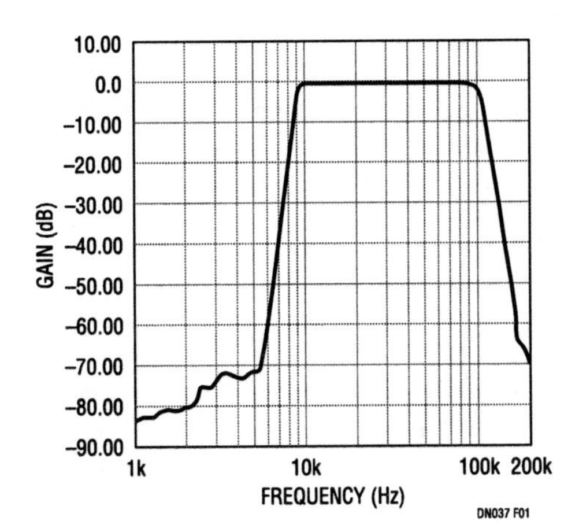

Figure 477.1 • Frequency Response. $V_{IN}=2.2V_{RMS}$, Output Buffer = LT1122

Design

The design ideology combined an LTC1064-4 elliptic lowpass filter with a highpass LTC1064 designed using Linear Technology's FilterCAD filter design software. The LTC1064-4 provides an attenuation of greater than 70dB at 2 times cutoff, while the highpass filter was designed to be almost the mirror image of the lowpass filter. Figure 477.2 shows the schematic of the composite bandpass filter. Note the 74LS90 is used as a divide by 5 whose output clocks the LTC1064 used as the 10kHz highpass filter. In addition to providing both 5MHz and 1MHz clocks, this circuit allows the clocks to be synchronous. This is essential for well behaved operation of the sampled data filters.

Test results

Figure 477.1 shows the overall frequency response of the bandpass filter. Note the steep slopes at the transition regions of the filter. The measured noise in the passband of the filter was $320\mu V_{RMS}$. Passband ripple is shown in Figure 477.3. Figure 477.3 shows excellent ripple specifications of less than ±0.2dB. It should be noted that the values of the resistors

Analog Circuit Design: Design Note Collection. http://dx.doi.org/10.1016/B978-0-12-800001-4.00477-4

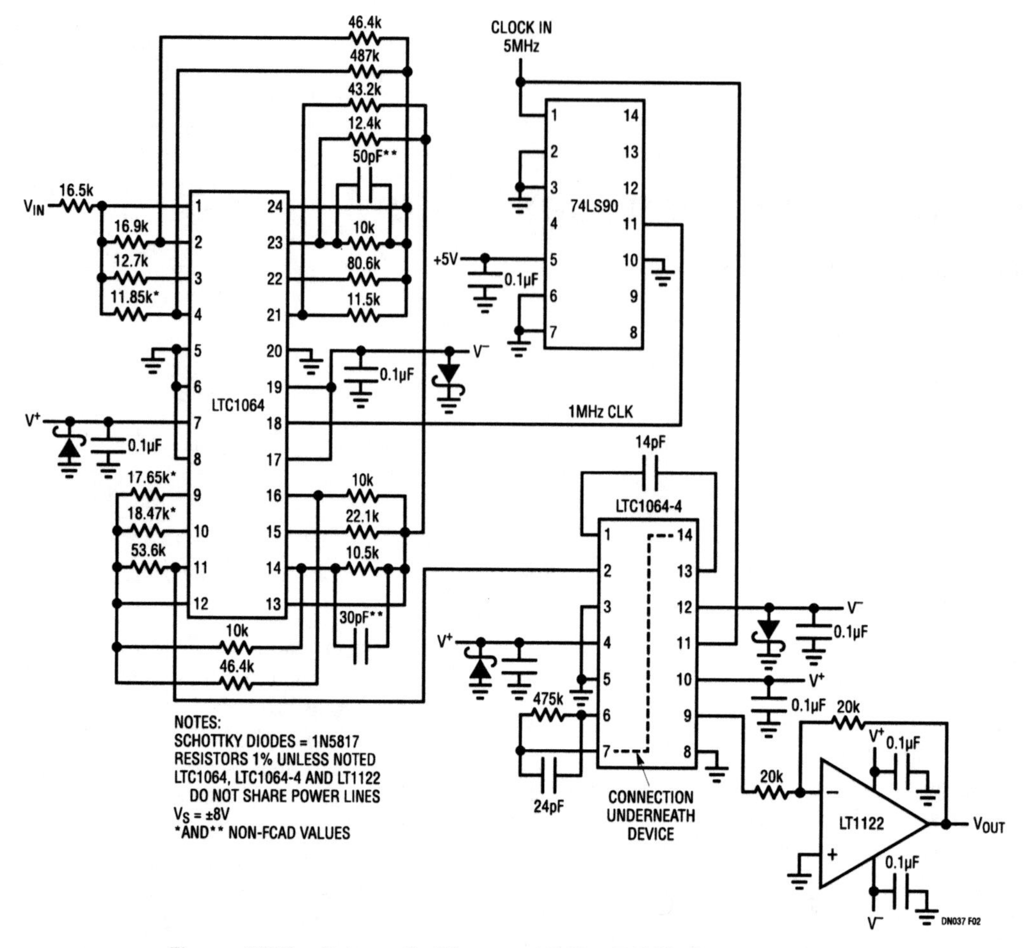

Figure 477.2 • Schematic Diagram 10kHz–100kHz Bandpass Filter

marked with an asterisk were modified slightly from the values given by FilterCAD. Otherwise the passband ripple measured better than ±0.4dB. Additional note should be made of the 50pF and 30pF capacitors shown with double asterisks. These capacitors serve as RC filters on two of the highpass outputs of the LTC1064 to roll off the response of the HPF (well past the passband of the overall filter) to limit noise aliasing back to the filter's passband. Note that these capacitors may limit the overall tunability of the filter or they may need a range changing switch.

Conclusions

A 16th order low noise high quality bandpass filter can be designed with only four active components and a handful of resistors. The filter meets specifications that can only be approached with the most sophisticated hybrid filter solutions. Further, the filter was primarily designed with the FilterCAD program, so that the designer may change parameters with the push of the "Enter" key. The filter is another in the amazing set of achievements available to the system designer by using the new switched capacitor filters from Linear Technology.

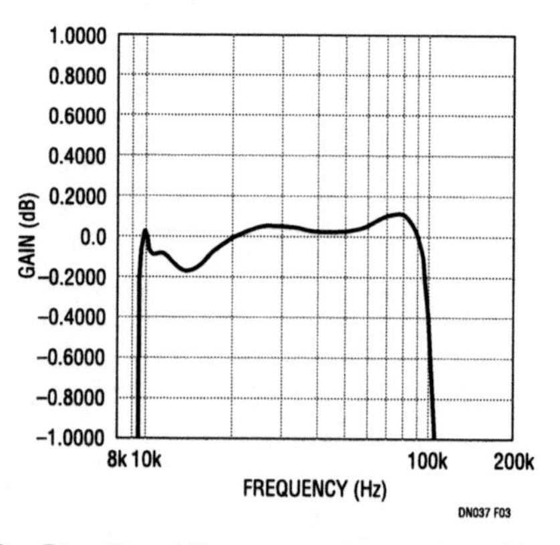

Figure 477.3 • Passband Frequency Response. $V_{IN} = 2.2V_{RMS}$, Output Buffer = LT1122

Switched-capacitor lowpass filters for anti-aliasing applications

478

Richard Markell Nello Sevastopoulos

Introduction

Many signal processing applications require a front end lowpass filter to bandwidth limit the signal of interest. This filter is often crucial to the system designer since it determines the number of bits which the system can resolve by its noise and dynamic range. Until now, the designer rejected the use of switched-capacitor filters as being too noisy, having too much distortion, or because they were not usable at a high enough frequency. The LTC1064-1 8th order Cauer filter can compete directly with the discrete operational amplifier design. Not only that, but the cost and performance advantages are tremendous.

The LTC1064-1 is a complete 8th order, clock-tunable Cauer (also known as elliptic) lowpass switched-capacitor filter *with internal thin film resistors.* The passband ripple is ±0.1dB and the stopband attenuation at 1.5 times the cutoff frequency is 72dB. The device is available in a 14-pin DIP or 16-pin surface mount package.

The LTC1064-1 boasts internal thin film resistors factory adjusted to optimize the Cauer 8th order response. The LTC1064-1 attains wideband noise (2kHz–102kHz) of $150\mu V_{RMS}$ and a total harmonic distortion of 0.03% for $V_{IN} = 3V_{RMS}$. No external components are required for cutoff frequencies up to 20kHz. For cutoff frequencies over 20kHz, two small value capacitors are required to maintain passband flatness.

By way of comparison, older switched-capacitor filters had noise in the millivolts, THD in the percents, and maximum corner frequencies limited to <20kHz.

This note compares the performance of the LTC1064-1 8th order Cauer filter with internal thin film resistors to that of the equivalent filter built with operational amplifiers. The LTC1064-1 quad switched capacitor filter competes favorably with op amp RC designs in most parameters of interest to the designer and wins easily when printed circuit board space is considered. *Since it is tunable*, the LTC1064-1 can replace not just one, but many op amp RC designs, if multi-frequency filtering is required. The specification comparisons become even more favorable to the LTC1064-1 as the frequencies become higher.

Comparing the LTC1064-1 with RC active filters utilizing operational amplifiers

Performance

The Cauer filter has target design specifications as follows: a cutoff frequency of 40kHz, ±0.05dB passband ripple and a −72dB attenuation at 1.5 times the cutoff frequency. This filter is realized with stopband notches and it is considered a quite complex and selective filter realization. Figure 478.1 details the frequency response of this design.

An 8th order active RC was designed using a fully inverting state variable topology. This topology is considered "state-of-the-art" for active filters since all noninverting inputs of the op amps are grounded. The discrete active RC version of the

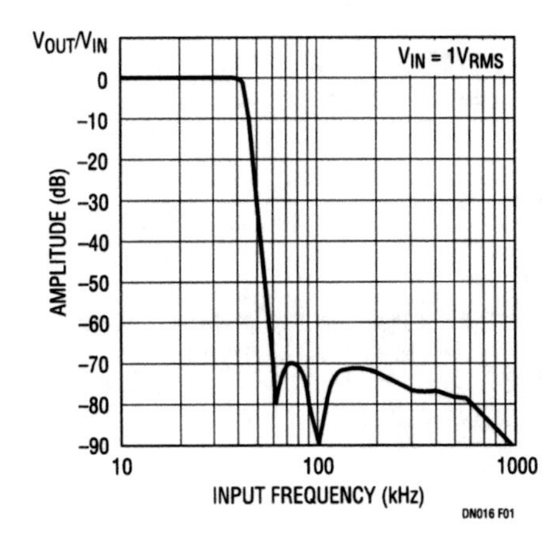

Figure 478.1 • LTC1064-1 Frequency Response

Analog Circuit Design: Design Note Collection. http://dx.doi.org/10.1016/B978-0-12-800001-4.00478-6

Cauer filter is quite complex requiring 16 op amps, 31 resistors and 8 capacitors. The op amps used for this comparison were TL084 quad FET input amplifiers. The circuit topology was optimized to yield the maximum useful input voltage swing.

Test results

Figure 478.1 shows the frequency response of the LTC1064-1 connected as shown in Figure 478.3. The shape of the frequency response of the active RC state variable filter was very similar and its differences cannot be easily shown here. Figure 478.2, curve (a), details the TL084 state variable filter response near the 40kHz cutoff frequency. Laboratory "tweaking" of resistor values could not produce any better response than shown here. This is a passband ripple of approximately ±0.45dB. For comparison, the LTC1064-1 passband ripple is ±0.15dB, as shown in Figure 478.2, curve (b). This is for a clock-to-center frequency ratio of 100:1, or a 4MHz clock. The measured filter amplitude response at 1.5 times the cutoff frequency for the TL084 active RC filter was about −65dB while that of the LTC1064-1 was −68dB. The noise for the TL084 state variable implementation was 111μV$_{RMS}$ while that for the LTC1064-1 was 145μV$_{RMS}$. Second harmonic distortion measurements were also made on both filters and they are included on the summary chart, Table 478.1.

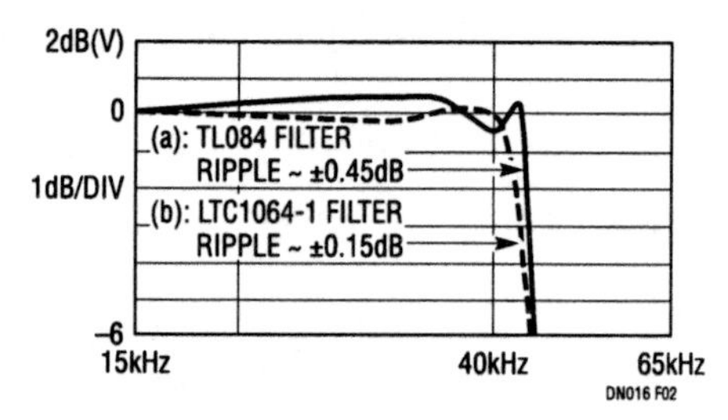

Figure 478.2 • Passband Ripple

Table 478.1 compares the LTC1064-1, the switched-capacitor implementation of the 8th order Cauer lowpass filter, to the active RC. Both circuits operate with dual ±7.5V supplies or a single 15V supply.

System considerations

Not only does the LTC1064-1 compare favorably on individual specifications, but it wins easily when system considerations are evaluated. Suppose four sharp cutoff frequencies are needed. The closest active RC solution is a 7th order single cutoff frequency Cauer filter. Four of these non-tunable devices (each a 2″ × 3″ hybrid) would be required for the four cutoff frequencies. This would be 24 square inches of PC board space. The discrete approach using operational amplifiers requires even more space. Since the LTC1064-1 is tunable, four frequencies can be selected merely by tuning the clock to the LTC1064-1. A complete LTC1064-1 system with tunable clock is estimated to occupy only 4 square inches of board space. This is a whopping savings of 6 times in board area. The LTC1064-1 wins easily in this category.

Summary

In summary it can be seen from Table 478.1 that the LTC1064-1 is the equal of the active RC filter. In the pure specification battle there is no clear winner, but when the amazing difference in hardware complexity, the full clock tunability and the simple method of application of the LTC106-1 device are all considered it is the sure winner.

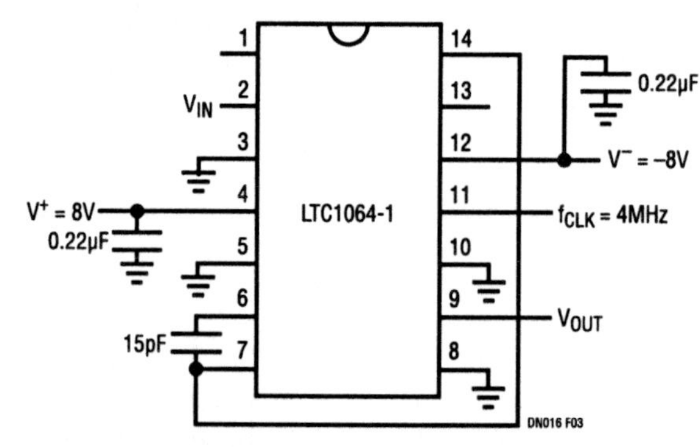

Figure 478.3 • The LTC1064-1, Monolithic 8th Order Cauer Lowpass Filter Operating with a 4MHz Clock and Providing a 40kHz Cutoff Frequency

Table 478.1 8th Order Cauer (Elliptic) LPF with a 40kHz Ripple Bandwidth

	#EXT OP AMPS	#EXT Rs, 1%	#EXT CAPS, 5%	TUNABLE	WIDEBAND NOISE, RMS[4]	DISTORTION V_{IN} = 1V$_{RMS}$, 3V$_{RMS}$ (DB)	V_{OS} OUT (mV)[3]	I_{SUPPLY} (mA)	ATTENUATION AT 60kHz	MEASURED PASSBAND RIPPLE	TRIMMING[2]
RC Active TL084	16	31	8	No	111μV	−87, −87	55	33	65dB	±0.45dB	Yes
LTC1064-1	None[1]	None	1	Yes	145μV	−70, −70	30	18	68dB	±0.15dB	None

[1]An output inverting buffer (LT118) was used for driving cables during measurements.
[2]To obtain the ±0.45dB ripple for the TL084, three resistors were trimmed.
[3]The output offset voltage numbers are as measured by DVM with the input of the filter grounded.
[4]Measurement BW (2kHz–102kHz).

Chopper amplifiers complement a DC accurate lowpass filter

479

Nello Sevastopoulos

Monolithic switched-capacitor lowpass filters, although they offer precise frequency responses, cannot usually be used for DC accurate applications because of their prohibitive DC offsets and poor gain linearity. The LTC1062, however, is quite different from currently available lowpass switched-capacitor filters because it uses an external (R, C) to isolate the IC from the input-signal DC path and to provide anti-aliasing for incoming signals larger than half its clock frequency. The LTC1062 is ideal when used in conjunction with high performance chopper-stabilized op amps.

The LTC1050 is an ultralow offset, low noise chopper with the sampling capacitors internal. It can remove residual clock noise without adding further DC error. Also, the internal capacitor minimizes board area.

Figure 479.1 shows a low cost, 7th order DC accurate, 10Hz lowpass filter where amplitude and phase response closely approximate a Bessel filter. The required clock frequency is 2kHz, thus yielding clock to cutoff frequency ratio of 200:1.

The LTC1050 is configured as a unity gain 2nd order lowpass filter which center frequency is $(1.2\pi R'C') = 1.72 \cdot fCUTOFF = 17.2Hz$ and $Q = 0.5$. Figure 479.2 shows the amplitude response of the filter, and Figure 479.3 shows a well behaved transient response for which Bessel filters are famous. The power supplies used were ±8V to provide a total DC input common-mode range of ±6V. The measured wideband noise was 52μVrms. The clock, and R, C values of Figure 479.1 can be easily modified to provide a 7th order Butterworth 10Hz filter, such as: $f_{CLK} = 1kHz$, $R = 26.7k$, $C = 1\mu F$, $R' = 165k$, $C1 = 0.2\mu F$ and $C2 = 0.047\mu F$. The diode at LTC1062 pin 3 should be used to protect the device from incoming signals above the power supplies.

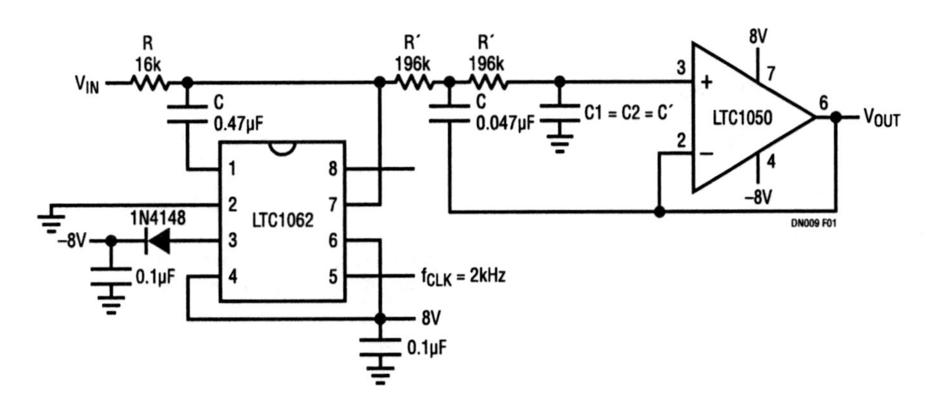

Figure 479.1 • Combining the LTC1050 Chopper Op Amp with the LTC1062 to Provide a 10Hz, DC Accurate Lowpass Bessel Filter

Analog Circuit Design: Design Note Collection. http://dx.doi.org/10.1016/B978-0-12-800001-4.00479-8

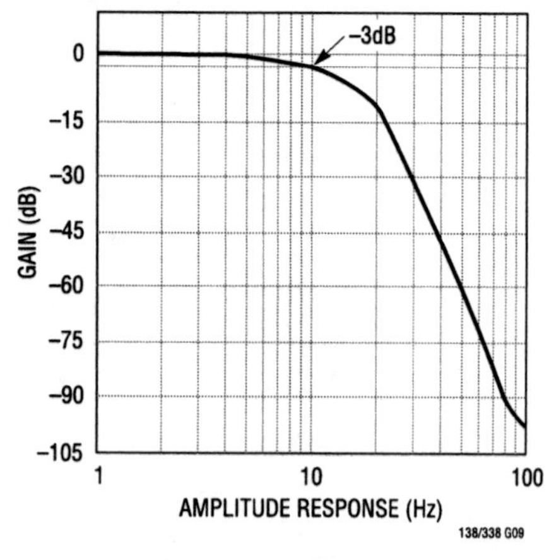

Figure 479.2 • Amplitude Response of Figure 479.1, Providing a Close Approximation to a Bessel Filter

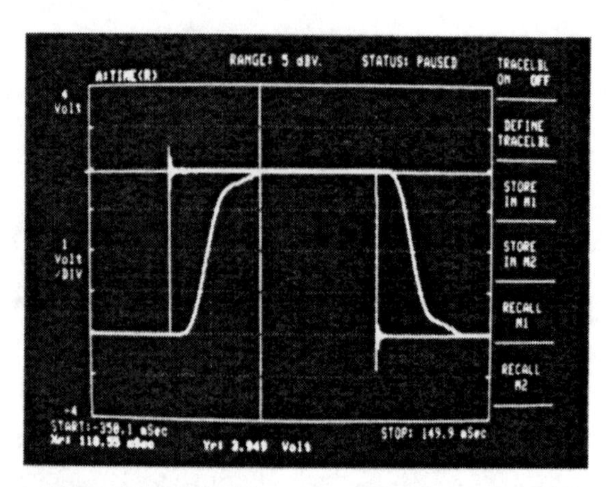

Figure 479.3 • Transient Response of the Bessel Lowpass Filter of Figure 479.1

In Figure 479.4, an external (R1, R2, C1) network is used at the input-output of the internal buffer of the LTC1062 (pins 7 and 8), to provide an additional 2-pole, Q=0.707, highpass filter. The filter output at pin 8 is bandpass, Figure 479.5, whereas the DC accurate Butterworth lowpass filter is still available at the output of the LTC1050. This circuit allows the user to separate the DC and AC components of an incoming signal, V_{IN}. Here, the LTC1050 buffers the lowpass filter section of the overall bandpass filter. For a Q=0.707, the design equation for the highpass sections are straightforward: set R2=2R1; and then the highpass cutoff frequency is, $f_C=(0.707/2\pi R1C1)$. The circuit in Figure 479.4 can be easily operated with a single supply, because resistor R2 and capacitor(s) C1 of the highpass section, also DC bias pin 7 at mid supplies independently of the DC input voltage. If only a single supply is available, simply bias the bottom side of R2 at half supply.

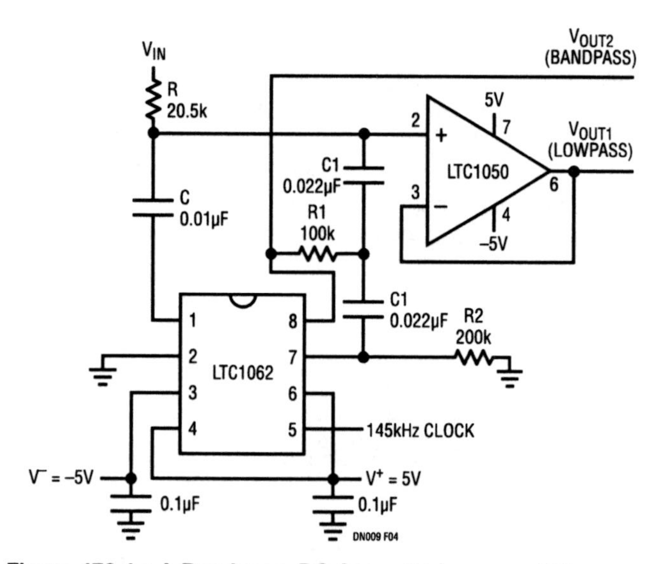

Figure 479.4 • A Bandpass, DC Accurate Lowpass Filter Combination Used to Extract the AC Information from a DC+AC input Signal

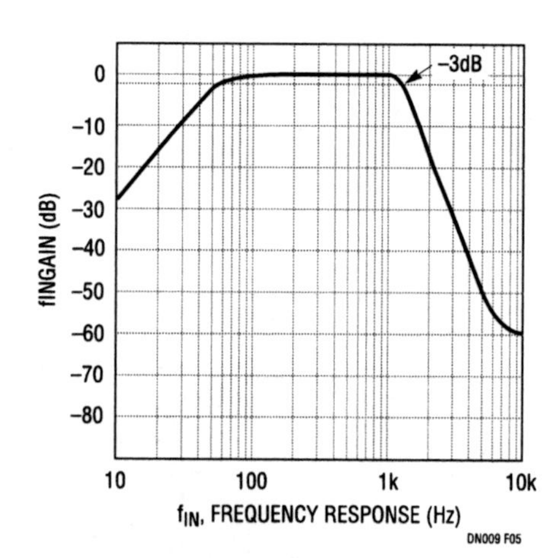

Figure 479.5 • Frequency Response of the Bandpass Filter Output, V_{OUT2}, of Figure 479.4

DC accurate filter eases PLL design

480

Nello Sevastopoulos   Phillip Karantzalis

The LTC1062 is a versatile, DC accurate, instrumentation lowpass filter with gain and phase that closely approximate a 5th order Butterworth filter. The LTC1062 is quite different from presently available lowpass switched capacitor filters because it uses an external (RC) to isolate the IC from the input signal DC path, thus providing DC accuracy. The DC accurate output, pin 7 of Figure 480.1, is buffered by an internal op amp from the switched capacitor network. The output of the switched capacitor network drives the bottom of C1. The input and output appear across an external resistor and, the IC part of the overall filter handles only the AC path of the signal. A buffered output is also provided (Figure 480.1) and its maximum guaranteed offset voltage over temperature is 20mV. Typically the buffered output offset is 0–5mV and drift is 1μV/°C. The use of an input (RC) also provides other advantages, such as lower noise and anti-aliasing.

With commercially available PLLs, the loop filter is designed by the user to optimize the loop performance. For a variety of applications, a 1st or 2nd order lowpass passive or active RC filter will do the job. When minimum output jitter and good transient response are required simultaneously, the design of the loop filter becomes more sophisticated. For instance, a fast transient response implies wide filter bandwidth and a reduced VCO output jitter implies minimum ripple at the VCO input. This is achieved by high outband attenuation of the lowpass filter. The LTC1062 provides the above requirements as well as economy and cutoff frequency programmability to be used advantageously in PLL designs.

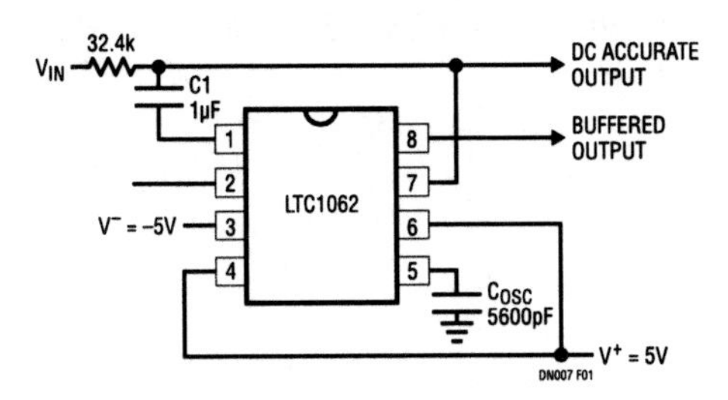

Figure 480.1 • 8Hz 5th Order Butterworth Lowpass Filter

The circuit of Figure 480.2 illustrates the use of the LTC1062 as a loop filter. The power supplies for the circuit are a single 5V for the PLL and ±5V for the LTC1062. The CMOS PLL is a CD4046B. The LYC1062 can also be used with a single 5V with some additional level shifting (see Application Note 20). Phase detector #2 drives a diode-resistor limiter combination to make the voltage at input R of the LTC1062 swing from one diode above ground to one diode below the 5V supply. Additionally, the two 5k resistors establish a maximum AC impedance to keep the LTC1062 in its operating region and to bias the VCO input at its mid point when phase detector #2 switches into a three-state mode.

An empirical design procedure for input frequencies less than 5kHz ($f_{IN} < 5\text{kHz}$, Figure 480.2) is illustrated below:

- Given the minimum input frequency value, the cutoff frequency, f_C of the LTC1062 should be chosen as:

$$1/6\left(f_{IN(MIN)}\right) \leq f_C < 1/4\left(f_{IN(MIN)}\right)$$

  The internal (or external) clock frequency of the LTC1062 should be 150 to 250 times the desired cutoff frequency, f_C.

- The capacitor C_{OSC} setting the LTC1062's internal oscillator should be chosen by:

$$C_{OSC} = \left(\frac{130\text{kHz}}{250 \cdot f_C} - 1\right) \cdot 33\text{pF}$$

  By further decreasing the value of C_{OSC}, the internal clock frequency of the LTC1062 increases and the damping of the loop also increases.

- By letting the value of $C = 0.047\mu\text{F}$, the LTC1062 input resistor R should be:

$$R \simeq \frac{5500\text{k}\Omega}{f_C\ (\text{Hz})}$$

Note: For this application, the loop filter is not required to be maximum flat and, therefore, the (RC) values of the LTC1062 can be within ±5% tolerance.

Analog Circuit Design: Design Note Collection. http://dx.doi.org/10.1016/B978-0-12-800001-4.00480-4

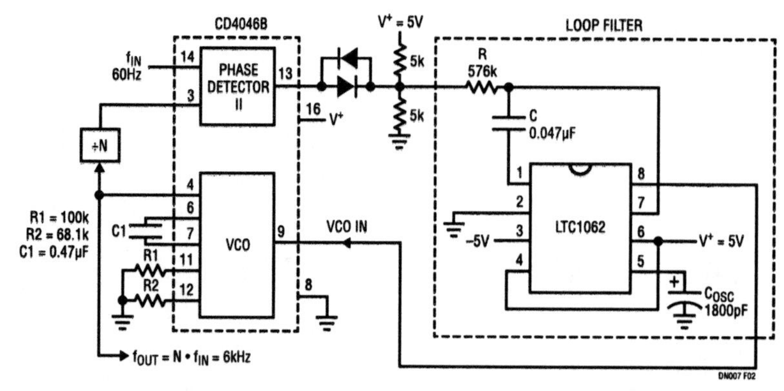

Figure 480.2 • Illustrating Use of the LTC1062 as a Loop Filter

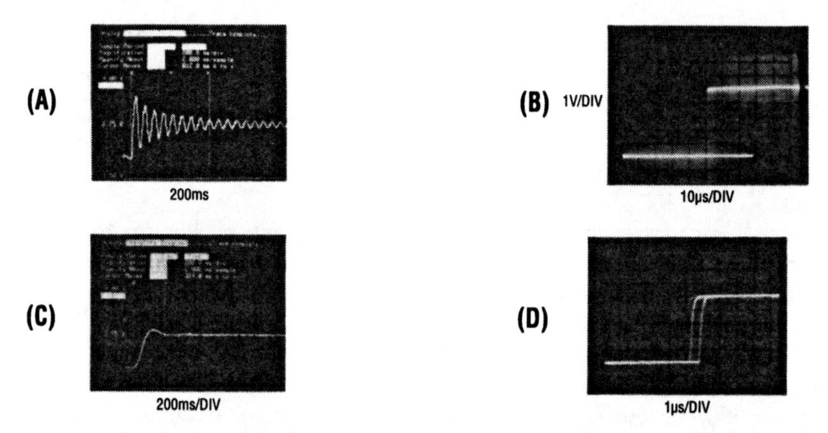

Figure 480.3 • Transient Response (A) and Jitter (B) of the PLL with a Passive RC Loop Filter. The Output Frequency of the VCO Is 6kHz and the ÷N = 100. Transient Response (C) and Jitter (D) of the PLL with the LTC1062 Used as a Loop Filter. The VCO Output Frequency Is 6kHz and the ÷N = 100. The Jitter Is Reduced to the Internal Jitter of the VCO.

To illustrate the performance difference between a lowpass passive RC loop filter and the LTC1062, the circuit of Figure 480.2 was tested for a PLL with a 60Hz ±10% input frequency range with ÷N = 100. Then, the PLL's VCO output could be used to drive the clock input of a precision switched capacitor filter, such as an LTC1060A set up in a 100:1 clock-to-center ratio, and configured as a 60Hz sharp notch or bandpass filter. Figure 480.3A shows the transient response of the loop when a passive RC loop filter, Figure 480.4, is used. The input frequency is shifted from 54Hz to 60Hz and the loop takes 820ms to settle within 5% of its steady stable value. The corner frequency of the RC passive filter is 22Hz. The natural frequency of the loop is approximately 10Hz and the damping factor less than 0.1 Figure 480.3B shows the jitter at the VCO output under the above conditions. A 30µs jitter with f_{OUT} = 6kHz corresponds to 18% instantaneous frequency inaccuracy. This makes the PLL VCO output unusable as a clock generator for a tracking switched capacitor filter. A small improvement in the VCO output jitter could be achieved by further decreasing the filter's cutoff frequency; this however, could further penalize the circuit's setting time.

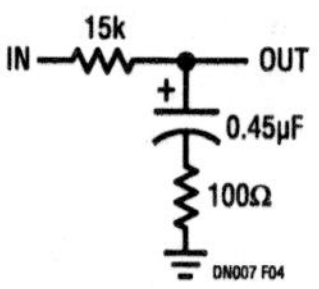

Figures 480.3C and 3D show the PLL performance when an LTC1062 is used as a loop filter. The corner frequency f_C of the LTC1062 was set at 9.5Hz ($\simeq 1/6f_{IN}$) and its internal clock was set for 2.4kHz ($\simeq 252 \cdot f_C$). The settling time of the loop was 320ms and the damping factor was optimally set to 0.7. The 1µs VCO output jitter, f_{OUT} = 6kHz, was measured over five periods and it is attributed to the inherited jitter of the VCO internal circuitry. With the LTC1062 used as a loop filter, the circuit's jitter corresponds to 0.12% equency error. This is quite adequate to drive the clock input of 0.3% accurate switched capacitor filters, such as LTC1059A or LTC1060A.

IN 15k OUT 0.45µF 100Ω DN007 F04

Figure 480.4 • Lowpass RC Filters Used for PLL Example

Section 5

Comparator Design Techniques

Rail-to-rail I/O and 2.4V operation allow UltraFast comparators to be used on low voltage supplies

481

Glen Brisebois

The LT1711 to LT1714 family of UltraFast comparators have full differential rail-to-rail inputs and outputs and operate down to 2.4V, allowing unfettered application on low supply voltages. The LT1711 (single) and LT1712 (dual) are specified at 4.5ns of propagation delay and 100MHz toggle frequency. The lower power LT1713 (single) and LT1714 (dual) are specified at 7ns of propagation delay and 65MHz toggle frequency. All of these comparators are fully equipped to support multiple supply applications and have Latch Enable (LE) pins and complementary outputs like the popular LT1016, LT1116, LT1671 and LT1394. They are available in MSOP and SSOP packages, fully specified over commercial and industrial temperature ranges on 2.7V, 5V and ±5V supplies.

Simultaneous full duplex 75MBd interface with only two wires

The circuit of Figure 481.1 shows a simple, fully bidirectional, differential 2-wire interface that gives good results to 75MBd. using the lower power LT1714. Eye diagrams under conditions of unidirectional and bidirectional communication are shown in Figures 481.2 and 481.3. Although not as pristine as the unidirectional performance of Figure 481.2, the performance under simultaneous bidirectional operation is still excellent. Because the LT1714 input voltage range extends 100mV beyond both supply rails, the circuit works with a full ±3V (one whole V_S up or down) of ground potential difference.

The circuit works well with the resistor values shown, but other sets of values can be used. The starting point is the characteristic impedance, Z_O, of the twisted-pair cable. The input impedance of the resistive network should match the characteristic impedance and is given by:

$$R_{IN} = 2 \cdot R_0 \cdot \frac{R1 \| (R2 + R3)}{R_0 + 2 \cdot [R1 \| (R2 + R3)]}$$

This comes out to 120Ω for the values shown. The Thevenin equivalent source voltage is given by:

$$V_{TH} = V_S \cdot \frac{(R2 + R3 - R1)}{(R2 + R3 + R1)} \cdot \frac{R_0}{R_0 + 2 \cdot [R1 \| (R2 + R3)]}$$

This amounts to an attenuation factor of 0.0978 with the values shown. (The actual voltage on the lines will be cut in half again due to the 120Ω Z_O.) The reason this attenuation factor is important is that it is the key to deciding the ratio between the R2-R3 resistor divider in the receiver path. This divider allows the receiver to reject the large signal of the local transmitter and instead sense the attenuated signal of

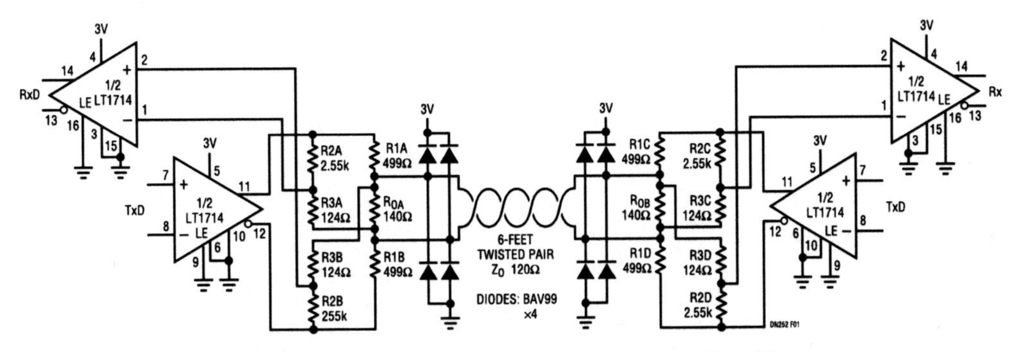

Figure 481.1 • 75MBd Full Duplex Interface on Two Wires

Analog Circuit Design: Design Note Collection. http://dx.doi.org/10.1016/B978-0-12-800001-4.00481-6

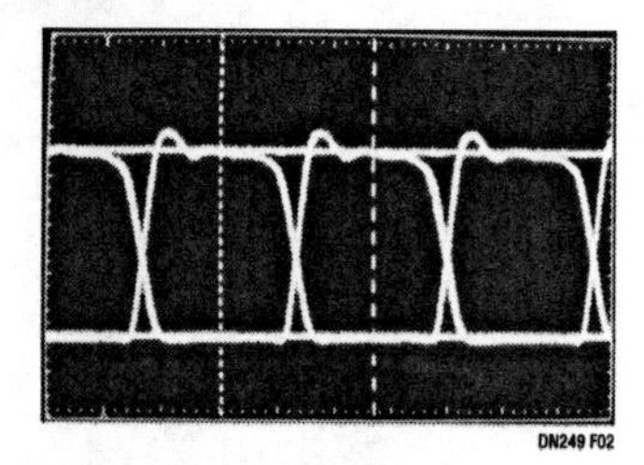

Figure 481.2 • Performance of Figure 481.1's Circuit When Operated Unidirectionally. Eye Is Wide Open

Figure 481.3 • Performance When Operated Simultaneous Bidirectionally (Full Duplex). Crosstalk Appears as Noise. Eye Is Slightly Shut but Performance Is Still Excellent

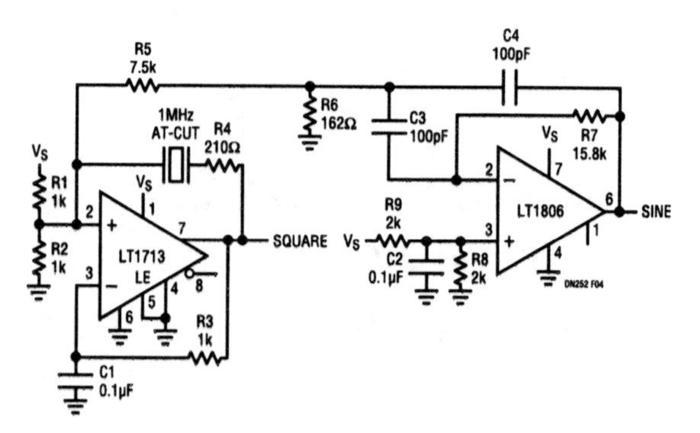

Figure 481.4 • LT1713 Comparator Is Configured as a Series Resonant Xtal Oscillator. LT1806 Op Amp Is Configured in a Q=5 Bandpass with f_C=1MHz

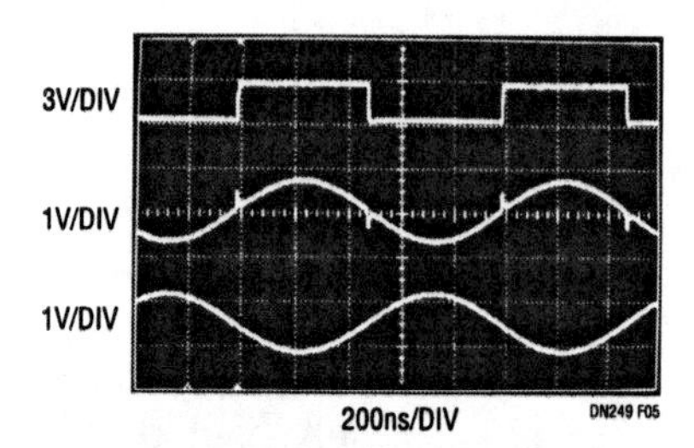

Figure 481.5 • Oscillator Waveforms with V_S=3V. Top Is Comparator Output. Middle Is Xtal Feedback to Pin 2 at LT1713 (Note the Glitches). Bottom Is Buffered, Inverted and Bandpass Filtered with a Q=5 by LT1806

the remote transmitter. Note that in the above equations, R2 and R3 are not yet fully determined because they only appear as a sum. This allows the designer to now place an additional constraint on their values. The R2-R3 divide ratio should be set to equal half the attenuation factor mentioned above[1] or:

$$R3/R2 = 1/2 \cdot 0.0976.$$

Having already designed R2+R3 to be 2.653k (by allocating input impedance across R_O, R1 and R2+R3 to get the requisite 120Ω), R2 and R3 then become 2529Ω and 123.5Ω respectively. The nearest 1% value for R2 is 2.55k and that for R3 is 124Ω.

1MHz series resonant crystal oscillator with square and sinusoid outputs

Figure 481.4 shows a classic 1MHz series resonant crystal oscillator. At series resonance, the crystal is a low impedance and the positive feedback connection is what brings about oscillation at the series resonant frequency. The RC feedback around the other path ensures that the circuit does not find a stable DC operating point and refuse to oscillate.

The comparator output is a 1MHz square wave (top trace of Figure 481.5) with jitter measured at $28ps_{RMS}$ on a 5V supply and $40ps_{RMS}$ on a 3V supply. At Pin 2 of the comparator, on the other side of the crystal, is a clean sine wave except for the presence of the small high frequency glitch (middle trace of Figure 481.5). This glitch is caused by the fast edge of the comparator output feeding back through crystal capacitance. Amplitude stability of the sine wave is maintained by the fact that the sine wave is basically a filtered version of the square wave. Hence, the usual amplitude control loops associated with sinusoidal oscillators are not necessary.[2] The sine wave is filtered and buffered by the fast, low noise LT1806 op amp. To remove the glitch, the LT1806 is configured as a bandpass filter with a Q of 5 and unity-gain center frequency of 1MHz, with its output shown as the bottom trace of Figure 481.5. Distortion was measured at −70dBc and −55dBc on the second and third harmonics, respectively.

Note 1: Using the design value of R2+R3=2.653k rather than the implementation value of 2.55k+124Ω=2.674k.

Note 2: Amplitude will be a linear function of comparator output swing, which is supply dependent and therefore adjustable. The important difference here is that any added amplitude stabilization or control loop will not be faced with the classical task of avoiding regions of nonoscillation versus clipping.

A seven nanosecond comparator for single supply operation

482

Jim Williams

The LT1394—an overview

A new ultra-high speed, single supply comparator, the LT1394, features TTL-compatible complementary outputs and 7ns response time. Other capabilities include a latch pin and good DC input characteristics (see Figure 482.1). The LT1394's outputs directly drive all 5V families, including the higher speed ASTTL, FAST and HC parts. Additionally, TTL outputs make the device easier to use in linear circuit applications where ECL output levels are often inconvenient.

A substantial amount of design effort has made the LT1394 relatively easy to use. It is much less prone to oscillation and other vagaries than some slower comparators, even with slow input signals. In particular, the LT1394 is stable in its linear region. Additionally, output-stage switching does not appreciably change power supply current, further enhancing stability. Finally, current consumption is far lower than that of previous devices. These features make the 200GHz gain bandwidth LT1394 considerably easier to apply than other fast comparators.

This device permits fast circuit functions that are difficult or impractical using other approaches. Two applications are presented here.

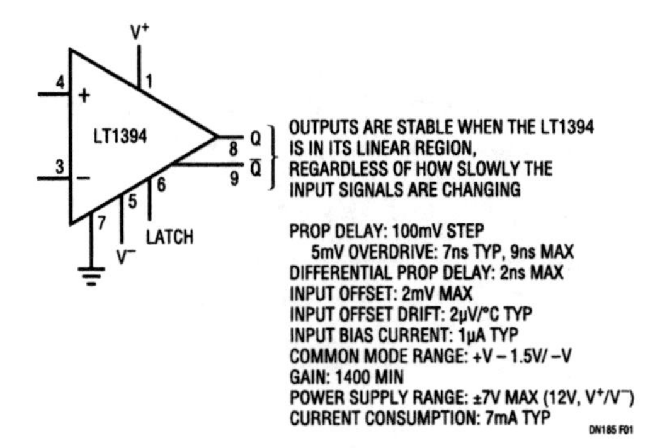

Figure 482.1 • The LT1394 at a Glance

4x NTSC subcarrier tunable crystal oscillator

Figure 482.2, a variant of a basic crystal oscillator, permits voltage tuning the output frequency. Such voltage-controlled crystal oscillators (VCXO) are often employed where slight variation of a stable carrier is required. This example is specifically intended to provide a 4 × NTSC subcarrier tunable oscillator suitable for phase locking.

The LT1394 is set up as a crystal oscillator. The varactor diode is biased from the tuning input. The tuning network is arranged so a 0V to 5V drive provides a reasonably symmetric, broad tuning range around the 14.31818MHz center frequency. The indicated selected capacitor sets tuning bandwidth. It should be picked to complement loop response in phase locking applications. Figure 482.3 is a plot of tuning

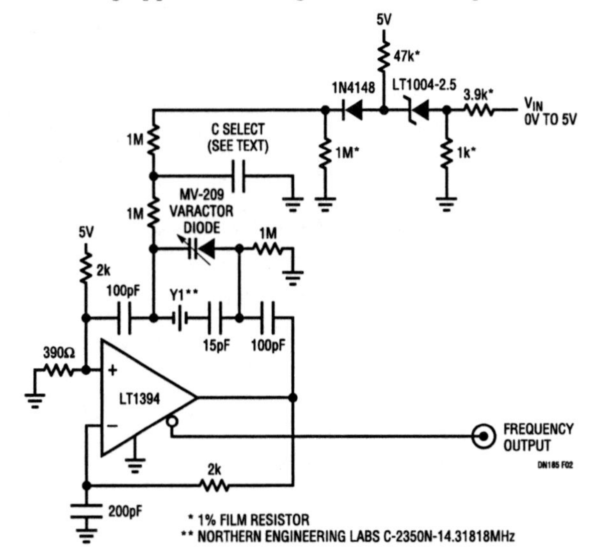

Figure 482.2 • A 4 × NTSC Subcarrier Voltage-Tunable Crystal Oscillator. Tuning Range and Bandwidth Accommodate a Variety of Phase Locked Loops

Analog Circuit Design: Design Note Collection. http://dx.doi.org/10.1016/B978-0-12-800001-4.00482-8

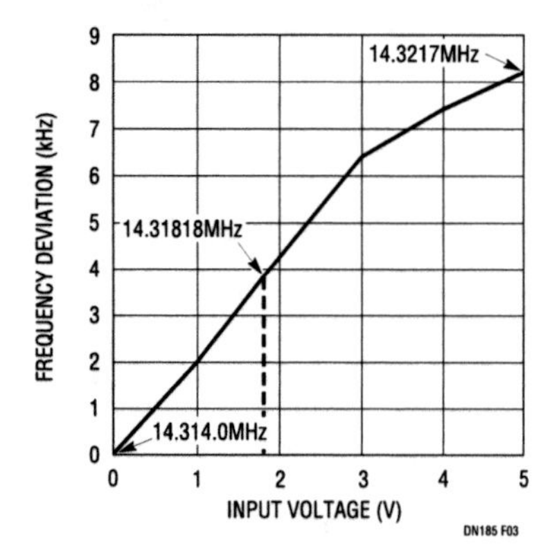

Figure 482.3 • Control Voltage vs Output Frequency for Figure 482.2. Tuning Deviation from Center Frequency Exceeds ±240ppm

input voltage frequency deviation. Tuning deviation from the 4 × NTSC 14.31818MHz center frequency exceeds ±240ppm for a 0V to 5V input.

High speed adaptive trigger circuit

Line and fiber-optic receivers often require an adaptive trigger to compensate for variations in signal amplitude and DC offsets. The circuit in Figure 482.4 triggers on 2mV to 175mV signals from 100Hz to 45MHz while operating from a single 5V rail. A1, operating at a gain of 15, provides wideband AC gain. The output of this stage biases a 2-way peak detector (Q1 through Q4). The maximum peak is stored in Q2's emitter capacitor, while the minimum excursion is retained in Q4's emitter capacitor. The DC value of the midpoint of A1's output signal appears at the junction of the 500pF capacitor and the 3MΩ units. This point always sits midway between the signal's excursions, regardless of absolute amplitude. This signal-adaptive voltage is buffered by A2 to set the trigger voltage at the LT1394's positive input. The LT1394's negative input is biased directly from A1's output. The LT1394's output, the circuit's output, is unaffected by >85:1 signal amplitude variations. Bandwidth limiting in A1 does not affect triggering because the adaptive trigger threshold varies ratiometrically to maintain circuit output.

Figure 482.5 shows operating waveforms at 40MHz. Trace A's input produces Trace B's amplified output at A1. The comparator's output is Trace C.

Additional applications and a tutorial on high speed comparator circuitry can be found in Application Note 72, "A Seven-Nanosecond Comparator for Single Supply Operation."

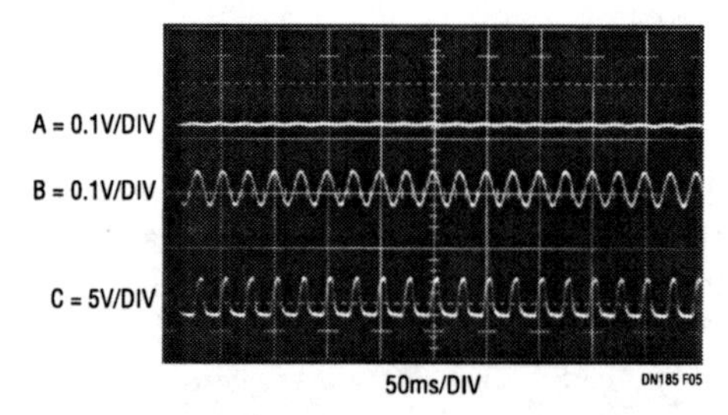

Figure 482.5 • Adaptive Trigger Responding to a 40MHz, 5mV Input. Input Amplitude Variations from 2mV to 175mV Are Accommodated

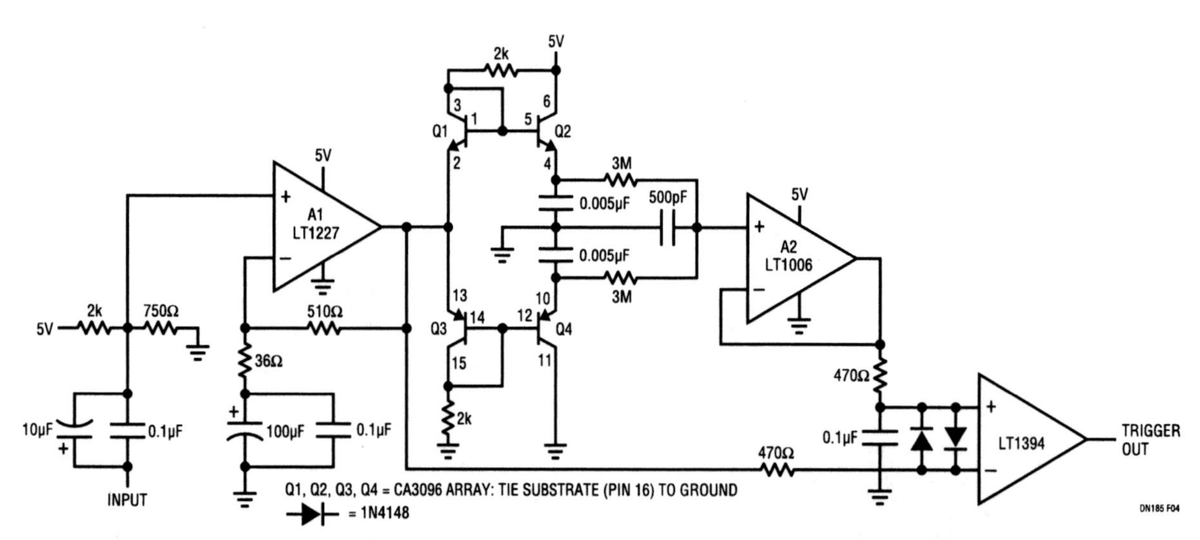

Figure 482.4 • 45MHz Single Supply Adaptive Trigger. Output Comparator's Threshold Varies Ratiometrically with Input Amplitude, Maintaining Data Integrity over >85:1 Input Amplitude Range

Comparators feature micropower operation under all conditions

483

Jim Williams

Some micropower comparators have operating modes that allow excessive current drain. In particular, poorly designed devices can conduct large transient currents during switching. Such behavior causes dramatically increased power drain with rising frequency, or when the inputs are nearly balanced, as in battery monitoring applications.

Figure 483.1 shows a popular micropower comparator's current consumption during switching. Trace A is the input pulse, trace B is the output response and trace C is the supply current. The device, specified for micropower level supply drain, pulls 40mA during switching. This undesirable surprise can upset a design's power budget or interfere with associated circuitry's operation.

The LTC1440 series comparators are true micropower devices. They eliminate current peaking during switching, resulting in greatly reduced power consumption versus frequency, or when the inputs are nearly balanced. Figure 483.2's plot contrasts the LTC1440's power consumption versus

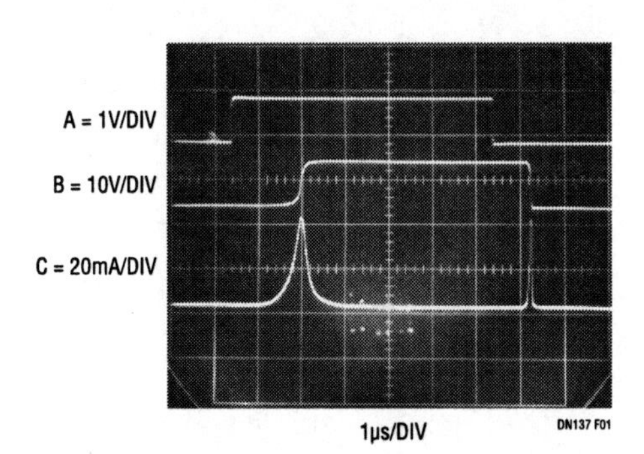

Figure 483.1 • Poorly Designed "Micropower" Comparator Pulls Huge Currents during Transitions. Result Is Excessive Current Consumption with Frequency

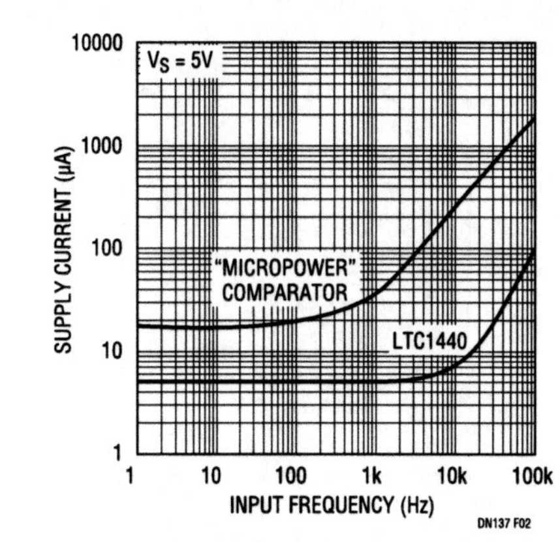

Figure 483.2 • The LTC1440 Family Draws 200 Times Lower Current at Frequency Than Another Comparator

Table 483.1 Some Characteristics of the LTC1440 Family of Micropower Comparators

PART NUMBER	NUMBER OF COMPARATORS	REFERENCE	PROGRAMMABLE HYSTERESIS	PACKAGE	PROP. DELAY (100mV OVERDRIVE)	SUPPLY RANGE	SUPPLY CURRENT
LTC1440	1	1.182V	Yes	8-Lead PDIP, SO	5µs	2V to 11V	4.7µA
LTC1441	2	No	No	8-Lead PDIP, SO	5µs	2V to 11V	5.7µA
LTC1442	2	1.182V	Yes	8-Lead PDIP, SO	5µs	2V to 11V	5.7µA
LTC1443	4	1.182V	No	16-Lead PDIP, SO	5µs	2V to 11V	8.5µA
LTC1444	4	1.221V	Yes	16-Lead PDIP, SO	5µs	2V to 11V	8.5µA
LTC1445	4	1.221V	Yes	16-Lead PDIP, SO	5µs	2V to 11V	8.5µA

Analog Circuit Design: Design Note Collection. http://dx.doi.org/10.1016/B978-0-12-800001-4.00483-X

frequency with that of another comparator specified as a micropower component. The LTC1440 has about 200 times lower current consumption at higher frequencies, while maintaining a significant advantage below 1kHz.

Table 483.1 shows some LTC1440 family characteristics. A voltage reference and programmable hysteresis are included in some versions, with 5µs response time for all devices.

The new devices permit high performance circuitry with low power drain. Figure 483.3's quartz oscillator, using a standard 32.768kHz crystal, starts under all conditions with no spurious modes. Current drain is only 9µA at a 2V supply.

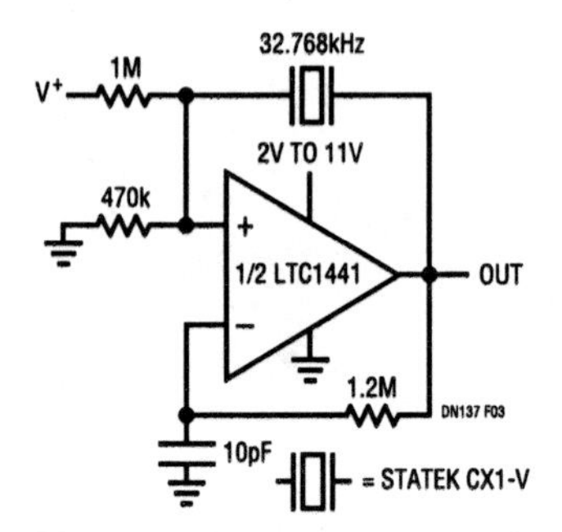

Figure 483.3 • 32.768kHz "Watch Crystal" Oscillator Has No Spurious Modes. Circuit Pulls 9µA at V_S=2V

Figure 483.4's voltage-to-frequency converter takes full advantage of the LTC1441's low power consumption under dynamic conditions. A 0V to 5V input produces a 0Hz to 10kHz output, with 0.02% linearity, 60ppm/°C drift and 40ppm/V supply rejection. Maximum current consumption is only 26µA, 100 times lower than currently available circuits. C1 switches a charge pump, comprising Q5, Q6 and the 100pF capacitor, to maintain its negative input at 0V. The LT1004s and associated components form a temperature-compensated reference for the charge pump. The 100pF capacitor charges to a fixed voltage; hence, the repetition rate is the circuit's only degree of freedom to maintain feedback. Comparator C1 pumps uniform packets of charge to its negative input at a repetition rate precisely proportional to the input voltage derived current. This action ensures that circuit output frequency is strictly and solely determined by the input voltage.

Start-up or input overdrive can cause the circuit's AC-coupled feedback to latch. If this occurs, C1's output goes low; C2, detecting this via the 2.7M/0.1µF lag, goes high. This lifts C1's positive input and grounds the negative input with Q7, initiating normal circuit action.

Figure 483.5 shows the circuit's power consumption versus frequency. Zero frequency current is just 15µA, increasing to only 26µA at 10kHz.

A detailed description of this circuit's operation appears in the August 1996 issue of *Linear Technology* magazine.

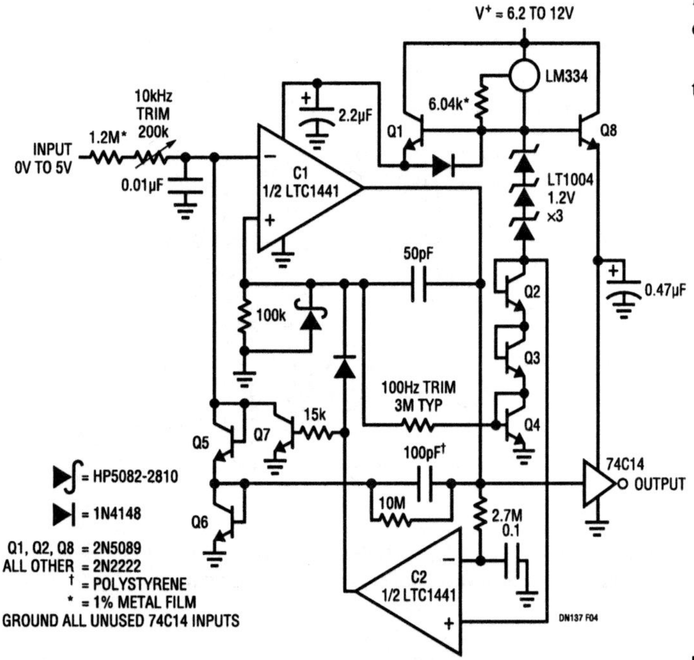

Figure 483.4 • LTC1441-Based 0.02% V/F Converter Requires Only 26µA Supply Current

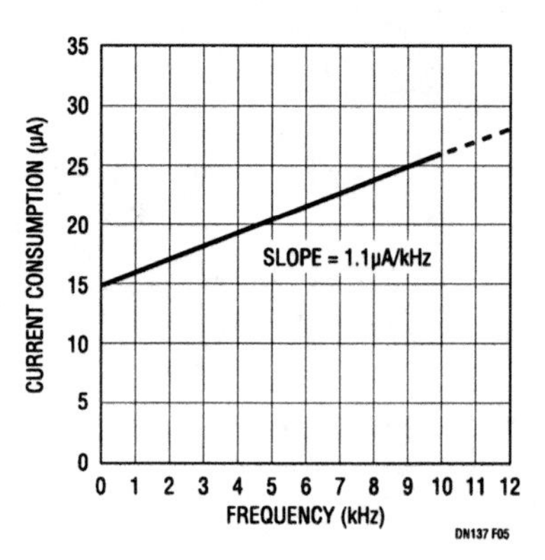

Figure 483.5 • Current Consumption vs Frequency for the V-to-F Converter. Discharge Cycles Dominate 1.1µA/kHz Current Drain Increase

Ultralow power comparators include reference

484

Robert Reay

With the explosion of battery-powered products has come the need for circuits that draw as little supply current as possible in order to extend battery life. Linear Technology's new family of micropower comparators with built-in references is designed to meet that need. Drawing only 1μA of supply current per comparator, the LTC1440–LTC1445 family provides the perfect solution to battery-powered system monitoring problems.

The LTC1440–LTC1445 family features 1μA comparators, adjustable hysteresis, TTL/CMOS outputs that sink and source current and a 1μA reference that can drive a bypass capacitor of up to 0.01μF without oscillation. The parts operate from a 2V to 11V single supply or a ±1V to ±5V dual supply. Each comparator's input voltage range swings from the negative supply rail to within 1.3V of the positive supply. The comparator propagation delay is 12μs with a 10mV overdrive, and the supply current glitches that commonly occur when comparators change logic states have been eliminated. Table 484.1 summarizes the features of each member of the family.

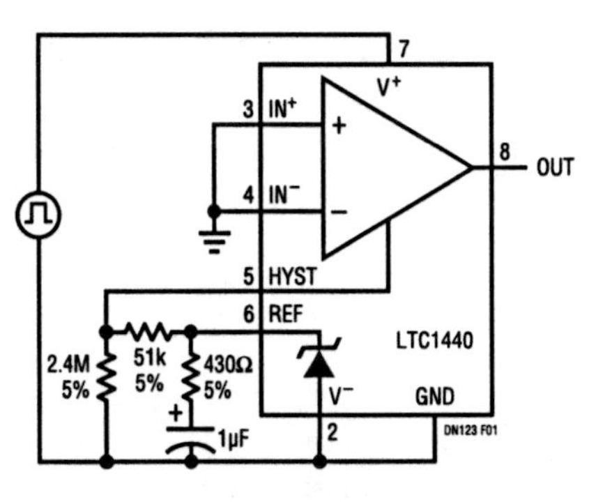

Test Circuit

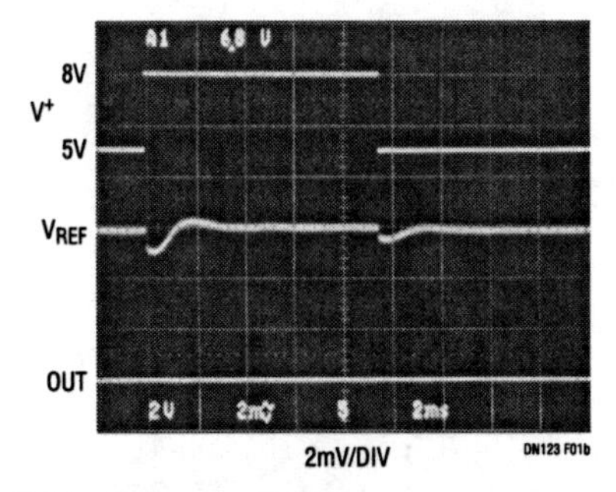

Figure 484.1 • Reference Settling

Voltage reference

The internal bandgap voltage reference has an output voltage of 1.182V above V^- for the LTC1440–LTC1443 and 1.22V ±1% for the LTC1444 and LTC1445. The reference output is

Table 484.1 LTC144x Comparator Family

PART NUMBER	NUMBER OF COMPARATORS	SUPPLY	SUPPLY CURRENT	ADJUSTABLE HYSTERESIS	REFERENCE	COMPARATOR OUTPUT
LTC1440	1	Dual	2.5μA	Yes	1.182V ±1%	CMOS
LTC1441	2	Single	3.5μA	No	1.182V ±1%	CMOS
LTC1442	2	Single	3.5μA	Yes	1.182V ±1%	CMOS
LTC1443	4	Dual	5.0μA	No	1.182V ±1%	CMOS
LTC1444	4	Single	5.0μA	Yes	1.221V ±1%	Open Drain
LTC1445	4	Single	5.0μA	Yes	1.221V ±1%	CMOS

Analog Circuit Design: Design Note Collection. http://dx.doi.org/10.1016/B978-0-12-800001-4.00484-1

capable of sourcing up to 200μA and sinking 15μA. The reference output can directly drive an external bypass capacitor up to 0.01μF without oscillation. By placing a resistor in series with the bypass capacitor, ringing at the reference output can be eliminated and a greater capacitance value can be used. The bypass capacitor prevents reference load transients or power supply glitches from disturbing the reference voltage, which helps eliminate false triggering of the comparators when they are connected to the reference. Figure 484.1 shows the reference voltage settling during a power supply transient.

Undervoltage/overvoltage detector

The LTC1442 can be easily configured as an undervoltage and overvoltage detector as shown in Figure 484.2. R1, R2 and R3 form a resistive divider from V_{CC} so that comparator A goes low when V_{CC} drops below 4.5V, and comparator B goes low when V_{CC} rises above 5.5V. A 10mV hysteresis band is set by R4 and R5 to prevent oscillations near the trip points.

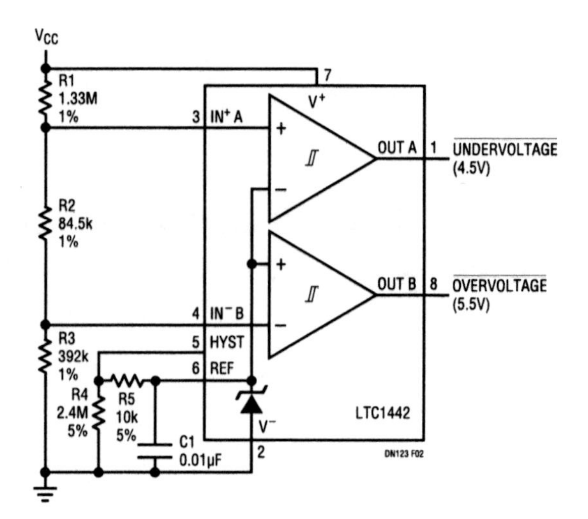

Figure 484.2 • Undervoltage/Overvoltage Detector

Single cell lithium-ion battery supply

Figure 484.3 shows a single cell lithium-ion battery to 5V supply with the low-battery warning, low-battery shutdown and reset functions provided by the LTC1444. The LT1300 micropower step-up DC/DC converter boosts the battery voltage to 5V using L1 and D1. Capacitors C2 and C3 provide input and output filtering.

The voltage monitoring circuitry takes advantage of the LTC1444's open-drain outputs and low supply voltage operation. Comparators A and B, along with R1, R2 and R3, monitor the battery voltage. When the battery voltage drops below 2.6V, comparator A's output pulls low to generate a nonmaskable interrupt to the microprocessor to warn of a low-battery condition. To protect the battery from overdischarge, the output of comparator B is pulled high by R7 when the battery voltage falls below 2.4V. P-channel Q1 and the LT1300 are turned off, dropping the quiescent current to 20μA. Q1 is needed to prevent the load circuitry from discharging the battery through L1 and D1.

Comparators C and D provide the reset input to the microprocessor. As soon as the boost converter output rises above the 4.65V threshold set by R8 and R9, comparator C turns off and R10 starts to charge C4. After 200ms, comparator D turns off and the Reset pin is pulled high by R12.

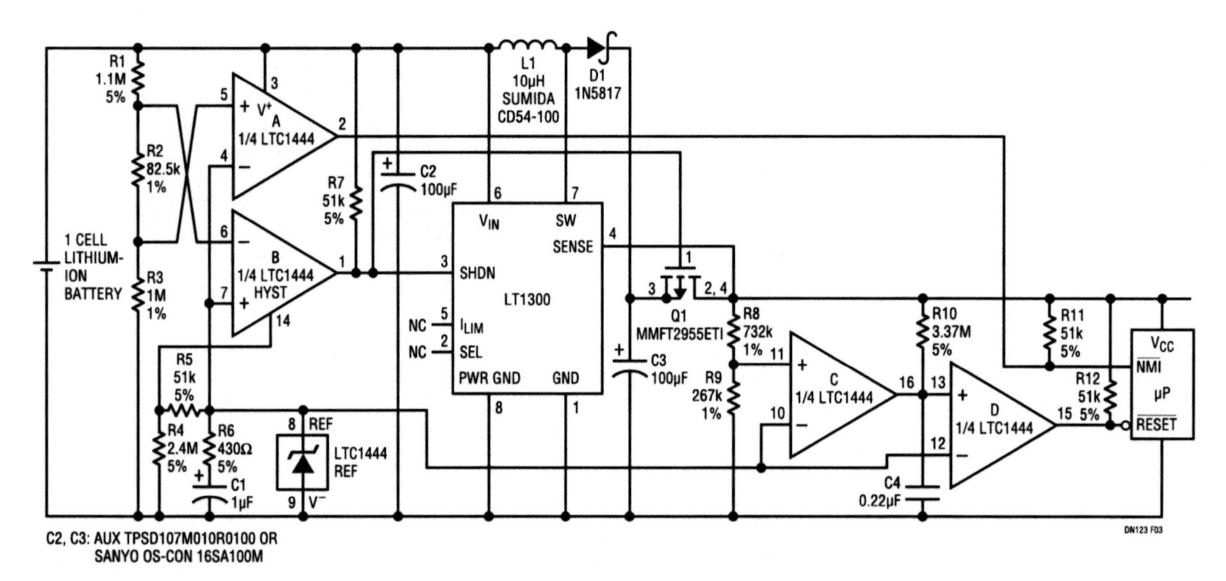

Figure 484.3 • Single Cell to 5V Supply

Section 6

System Timing Design

Using a low power SOT-23 oscillator as a VCO

485

Nello Sevastopoulos

Introduction

The LTC6900 is a precision low power oscillator that is extremely easy to use and occupies very little PC board space. It is a lower power version of the LTC1799.

The output frequency, f_{OSC}, of the LTC6900 can range from 1kHz to 20MHz—programmed via an external resistor, R_{SET}, and a 3-state frequency divider pin, as shown in Figure 485.1.

$$f_{OSC} = \frac{10MHz}{N} \cdot \frac{20k\Omega}{R_{SET}} \quad N = 1, 10, 100 \qquad (1)$$

A proprietary feedback loop linearizes the relationship between R_{SET} and the output frequency so the frequency accuracy is already included in the expression above. Unlike other discrete RC oscillators, the LTC6900 does not need correction tables to adjust the formula for determining the output frequency.

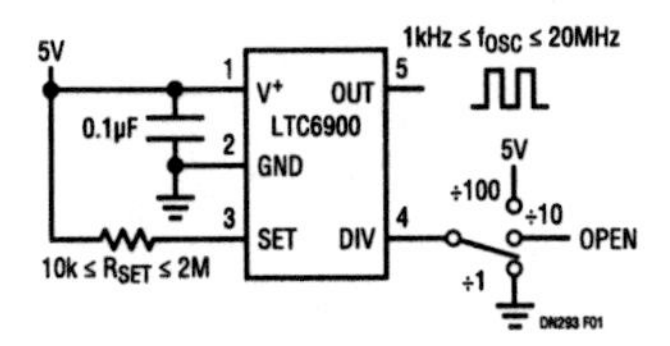

Figure 485.1 • Basic Connection Diagram

Figure 485.2 shows a simplified block diagram of the LTC6900. The LTC6900 master oscillator is controlled by the ratio of the voltage between V^+ and the SET pin and the current, I_{RES}, entering the SET pin. *As long as I_{RES} is precisely the current through resistor R_{SET}*, the ratio of $(V^+ - V_{SET})/I_{RES}$ equals R_{SET} and the frequency of the LTC6900 depends solely on the value of R_{SET}. This technique ensures accuracy, typically ±0.5% at ambient temperature.

As shown in Figure 485.2, the voltage of the SET pin is controlled by an internal bias, and by the gate to source voltage of a PMOS transistor. The voltage of the SET pin (V_{SET}) is typically 1.1V below V^+.

Programming the output frequency

The output frequency of the LTC6900 can be programmed by altering the value of R_{SET} as shown in Figure 485.1 and the accuracy of the oscillator will be as specified. The frequency can also be programmed by steering current in or out of the SET pin, as conceptually shown in Figure 485.3. This

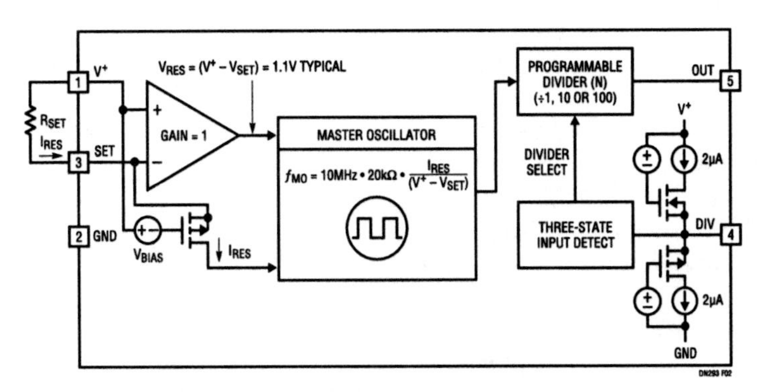

Figure 485.2 • Simplified Block Diagram

Analog Circuit Design: Design Note Collection. http://dx.doi.org/10.1016/B978-0-12-800001-4.00485-3

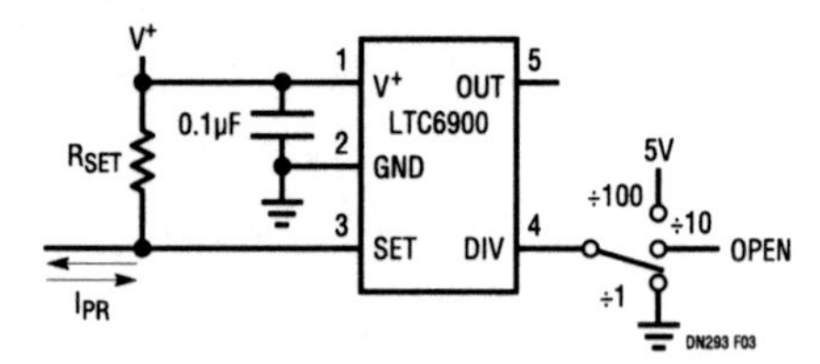

Figure 485.3 • Concept for Programming via Current Steering

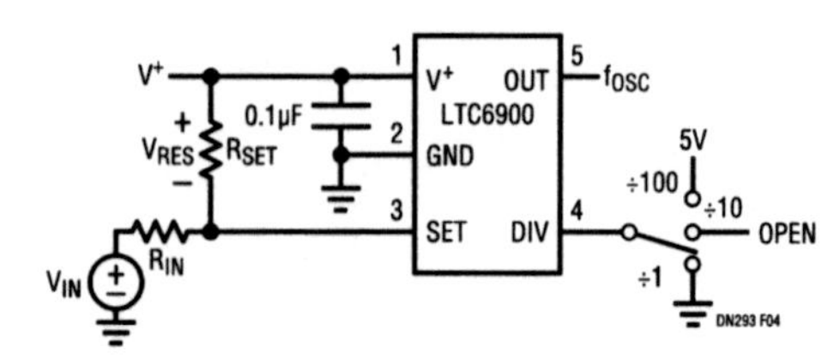

Figure 485.4 • Implementation of the Concept Shown in Figure 485.3

technique can degrade accuracy as the ratio of $(V^+ - V_{SET})/I_{RES}$ is no longer uniquely dependent on the value of R_{SET}, as shown in Figure 485.2. This loss of accuracy will become noticeable when the magnitude of I_{PROG} is comparable to I_{RES}. The frequency variation of the LTC6900 is still monotonic.

Figure 485.4 shows how to implement the concept shown in Figure 485.3 by connecting a second resistor, R_{IN}, between the SET pin and a ground referenced voltage source V_{IN}.

For a given power supply voltage in Figure 485.4, the output frequency of the LTC6900 is a function of V_{IN}, R_{IN}, R_{SET}, and $(V^+ - V_{SET}) = V_{RES}$:

$$f_{OSC} = \frac{10\text{MHz}}{N} \cdot \frac{20\text{k}\Omega}{R_{SET}//R_{IN}} \cdot \left(1 + \frac{V_{IN} - V^+}{V_{RES}} \cdot \frac{1}{1 + R_{IN}/R_{SET}}\right) \tag{2}$$

When $V_{IN} = V^+$ the output frequency of the LTC6900 assumes the highest value and it is set by the parallel combination of R_{IN} and R_{SET}. Also note, the output frequency, f_{OSC}, is independent of the value of $V_{RES} = (V^+ - V_{SET})$ so, the accuracy of f_{OSC} is within the data sheet limits.

When V_{IN} is less than V^+, and especially when V_{IN} approaches the ground potential, the oscillator frequency, f_{OSC}, assumes its lowest value and its accuracy is affected by the change of $V_{RES} = (V^+ - V_{SET})$. At 25°C V_{RES} varies by ±8%, assuming the variation of V^+ is ±5%. The temperature coefficient of V_{RES} is 0.02%/°C. Note that if V_{IN} is the output of a DAC referenced to V^+, the V_{RES} sensitivity to the power supply is eliminated.

By manipulating the algebraic relation for f_{OSC} above, a simple algorithm can be derived to set the values of external resistors R_{SET} and R_{IN}, as shown in Figure 485.4:

1. Choose the desired value of the maximum oscillator frequency, $f_{OSC(MAX)}$, occurring at maximum input voltage $V_{IN(MAX)} \leq V^+$.
2. Set the desired value of the minimum oscillator frequency, $f_{OSC(MIN)}$, occurring at minimum input voltage $V_{IN(MIN)} \geq 0$.
3. Choose $V_{RES} = 1.1\text{V}$ and calculate the ratio of R_{IN}/R_{SET} from the following:

$$\frac{R_{IN}}{R_{SET}} = \frac{\left(V_{IN(MAX)} - V^+\right) - \left(\frac{f_{OSC(MAX)}}{f_{OSC(MIN)}}\right) \cdot \left(V_{IN(MIN)} - V^+\right)}{V_{RES}\left(\frac{f_{OSC(MAX)}}{f_{OSC(MIN)}} - 1\right)} - 1 \tag{3}$$

Once R_{IN}/R_{SET} is known, calculate R_{SET} from:

$$R_{SET} = \frac{10\text{MHz}}{N} \cdot \frac{20\text{k}\Omega}{f_{OSC(MAX)}} \left[\frac{\left(V_{IN(MAX)} - V^+\right) + V_{RES}\left(1 + \frac{R_{IN}}{R_{SET}}\right)}{V_{RES}\left(\frac{R_{IN}}{R_{SET}}\right)}\right] \tag{4}$$

Example 1: In this example, the oscillator output frequency has small excursions. This is useful where the frequency of a system should be tuned around some nominal value.

Let $V^+ = 3\text{V}$, $f_{OSC(MAX)} = 2\text{MHz}$ for $V_{IN(MAX)} = 3\text{V}$ and $f_{OSC(MIN)} = 1.5\text{MHz}$ for $V_{IN} = 0\text{V}$. Solve for R_{IN}/R_{SET} by equation (3), yielding $R_{IN}/R_{SET} = 9.9/1$. $R_{SET} = 110.1\text{k}\Omega$ by equation (4). $R_{IN} = 9.9R_{SET} = 1.089\text{M}\Omega$. For standard resistor values, use $R_{SET} = 110\text{k}\Omega$ (1%) and $R_{IN} = 1.1\text{M}\Omega$ (1%). Figure 485.5 shows the measured f_{OSC} vs V_{IN}. The 1.5MHz to 2MHz frequency excursion is quite limited, so the curve f_{OSC} vs V_{IN} is linear.

Example 2: Vary the oscillator frequency by one octave per volt. Assume $f_{OSC(MIN)} = 1\text{MHz}$ and $f_{OSC(MAX)} = 2\text{MHz}$, when the input voltage varies by 1V. The minimum input voltage is half supply, that is $V_{IN(MIN)} = 1.5\text{V}$, $V_{IN(MAX)} = 2.5\text{V}$ and $V^+ = 3\text{V}$.

Equation (3) yields $R_{IN}/R_{SET} = 1.273$ and equation (4) yields $R_{SET} = 142.8\text{k}\Omega$. $R_{IN} = 1.273R_{SET} = 181.8\text{k}\Omega$. For standard resistor values, use $R_{SET} = 143\text{k}\Omega$ (1%) and $R_{IN} = 182\text{k}\Omega$ (1%).

Figure 485.6 shows the measured f_{OSC} vs V_{IN}. For V_{IN} higher than 1.5V the VCO is quite linear; nonlinearities occur when V_{IN} becomes smaller than 1V, although the VCO remains monotonic.

The VCO modulation bandwidth is 25kHz, that is, the LTC6900 will respond to changes in the frequency programming voltage, V_{IN}, ranging from DC to 25kHz.

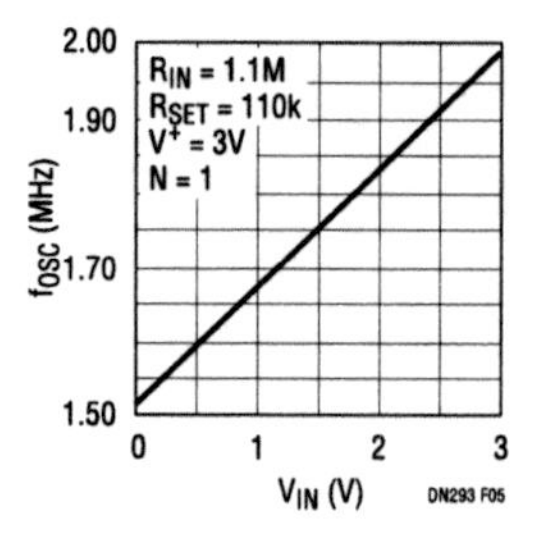

Figure 485.5 • Output Frequency vs Input Voltage

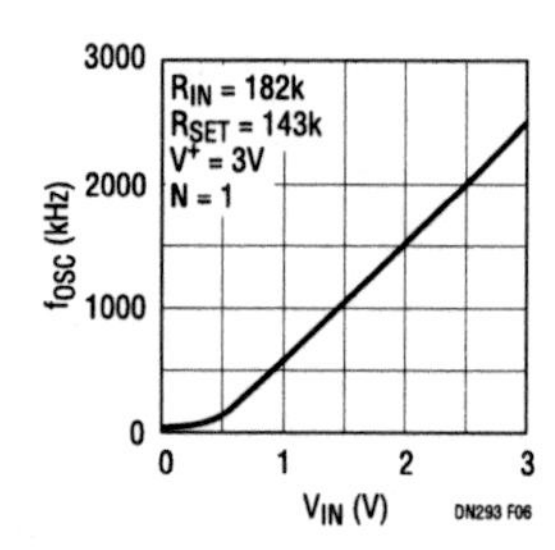

Figure 485.6 • Output Frequency vs Input Voltage

SOT-23 1kHz to 30MHz oscillator with single resistor frequency set

486

Andy Crofts

The LTC1799 resistor-programmed oscillator eliminates the hassle in designing an accurate square-wave frequency reference. A single resistor (R_{SET}) connected between the power supply and an input pin (SET) programs the frequency of a master oscillator to a value between 100kHz and 30MHz. An internal clock divider is programmed using a three-state input pin (DIV) to divide the master oscillator frequency by 1, 10 or 100 before driving the output. This extends the lower limit to 1kHz for a total range of 1kHz to 30MHz. The output frequency is linearly related to R_{SET}, as defined by the frequency-setting equation:

$$f_{OSC} = 10\text{MHz} \cdot \left(\frac{10\text{k}\Omega}{N \cdot R_{SET}}\right), N = \begin{cases} 100, & \text{DIV} = V^+ \\ 10, & \text{DIV} = \text{Open} \\ 1, & \text{DIV} = \text{GND} \end{cases}$$

This simple and accurate relationship is achieved with a proprietary design that linearizes the resistance-to-frequency conversion, eliminating errors such as oscillator propagation delay.

Tiny circuit, big performance

As shown in Figure 486.1, a complete oscillator requires only an LTC1799, a frequency-setting resistor and a bypass capacitor. The circuit's small component count and the LTC1799's diminutive SOT-23 package add up to big savings in PCB space when compared to oscillators built from crystals, ceramic resonators, 555 timers or discrete components.

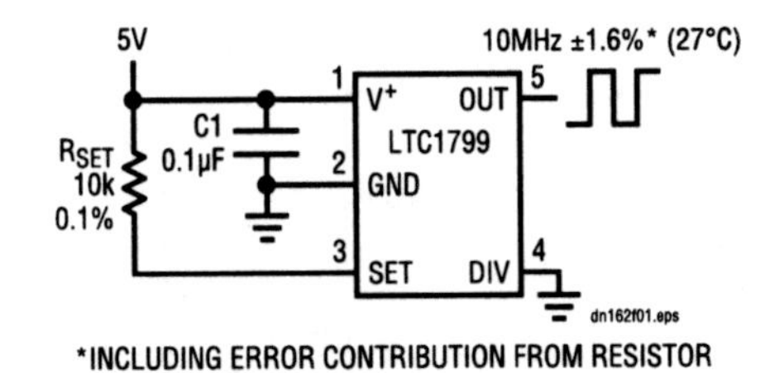

Figure 486.1 • Complete Oscillator Solution

You don't pay a performance penalty for this miniaturization. The LTC1799 has a guaranteed frequency accuracy of ±1.5% (±0.5% typical) at room temperature. This spec applies over the entire 2.7V to 5.5V supply range, made possible by the stingy 0.05%/V typical drift over supply voltage.

The accuracy remains tight over temperature as well. The LTC1799C has a typical temperature drift of ±0.004%/°C, with guaranteed accuracy of ±2% over 0°C to 70°C (LTC1799I guaranteed accuracy is ±2.5% over −40°C to 85°C). Figure 486.2 shows the frequency output of the circuit in Figure 486.1 over the industrial temperature range for three typical parts.

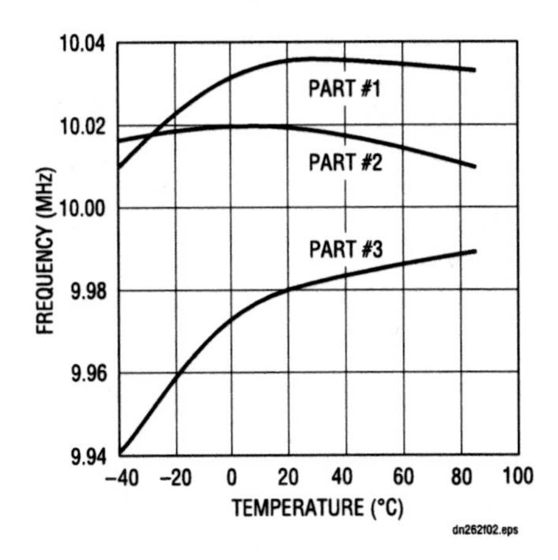

Figure 486.2 • Frequency vs Temperature for Figure 486.1's Circuit

Due to its low sensitivity to supply and temperature variation, the LTC1799 has abilities that no other oscillator can match. Replacing R_{SET} with a potentiometer allows the output frequency to be "tuned" after the circuit is completed.

Analog Circuit Design: Design Note Collection. http://dx.doi.org/10.1016/B978-0-12-800001-4.00486-5

Once set, the LTC1799 will accurately maintain the desired frequency over all operating conditions. Crystals and ceramic resonators cannot be adjusted in this manner; 555 timers and other RC oscillators do not have this level of stability.

Fast start-up time

One common problem designers encounter with crystal oscillators is the long start-up time before the circuit is oscillating at its final frequency. At MHz frequencies, this start-up time is typically 10ms. At frequencies below 100kHz, a crystal oscillator may take up to a second to start up. The LTC1799 takes less than 1ms to settle within 1% of any frequency between 5kHz and 30MHz.

The LTC1799 is immune to vibration and acceleration forces, another problem that plagues crystal oscillators. And its 1mA typical supply current (2.4mA max at 10MHz, 5V supply) is very efficient when compared to the 10mA to 30mA many crystal oscillators consume.

Two-step design process

The LTC1799 combines infinite frequency selectivity (limited only by resistor selection) with incredible ease-of-use. The external resistor R_{SET} determines the frequency of the master oscillator within a 100kHz to 30MHz range. The three-state DIV pin determines whether the master oscillator signal is passed directly to the output, or first divided by 10 or 100. The design process is simple:

1. Use Table 486.1 to determine the proper divider setting.

Table 486.1 Frequency Range vs Divider Setting

DIVIDER SETTING	DIV (Pin 4) CONNECTION	FREQUENCY RANGE
÷1 (N=1)	GND (Pin 2)	>500kHz*
÷10 (N=10)	Floating	50kHz to 1MHz
÷100 (N=100)	VCC (Pin 1)	≤100kHz

*At frequencies above 10MHz (R_{SET} < 10k), the LTC1799 may suffer reduced accuracy on supplies less than 4V.

2. With N known, calculate the best value for R_{SET} using the equation:

$$R_{SET} = 10k \cdot \left(\frac{10MHz}{N \cdot f_{OSC}}\right)$$

It's that simple! Of course, since the LTC1799 converts resistance into frequency, any errors in the value of RSET (due to resistor tolerance or nonideal choice of resistor value) reduce the frequency accuracy. Therefore, 1% or 0.1% resistors are recommended for best performance.

Application: temperature-to-frequency converter

The most straightforward application for the LTC1799 is as a constant-frequency reference. But its resistance-to-frequency conversion architecture allows for a variety of applications. Figure 486.3 shows a temperature-to-frequency converter that is built by simply replacing R_{SET} with a thermistor. The YSI 44011 has a resistance of 100k at 25°C, 333k at 0°C and 16.3k at 70°C, a span that fits nicely into the LTC1799's permitted range for R_{SET}. With its low tempco and high linearity, the LTC1799 adds less than 0.5°C of error over the commercial temperature range. Figure 486.4 plots the output frequency vs temperature.

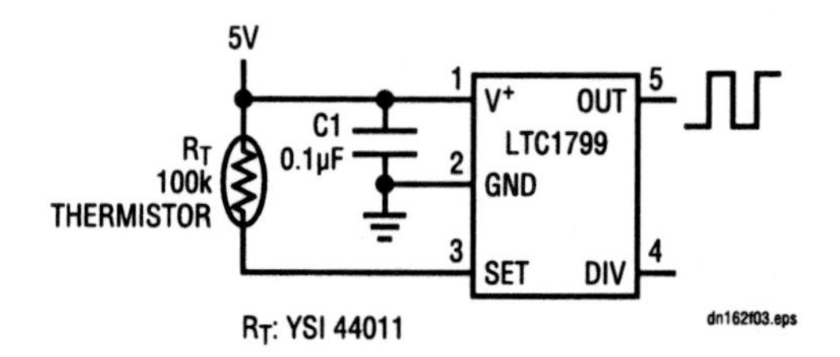

Figure 486.3 • Temperature-to-Frequency Converter

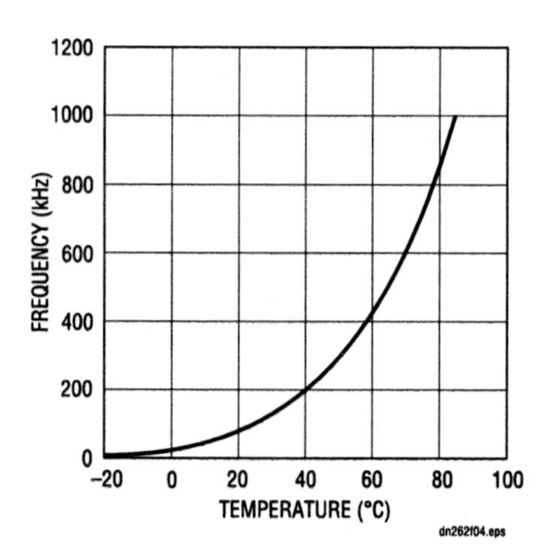

Figure 486.4 • Output Frequency vs Temperature for Figure 486.3's Circuit

Conclusion

The LTC1799 is a tiny, accurate, easy-to-use oscillator that is programmed by a single resistor. With a typical accuracy of better than 0.5% and low temperature and supply sensitivity, the LTC1799's performance approaches that of crystal oscillators and ceramic resonators and yet requires far less PCB space. The resistance-to-frequency conversion architecture allows infinite resolution and design simplicity. The result is a square-wave oscillator with an unprecedented combination of ease-of-use and precision in a tiny SOT-23 package.

Section 7

RMS to DC Conversion

Precision LVDT signal conditioning using direct RMS to DC conversion

487

Cheng-Wei Pei

Introduction

Linear variable differential transformers (LVDTs) are theoretically infinite-resolution displacement measurement devices. An LVDT operates by comparing the magnetic flux coupled into two transformer secondary windings to determine the displacement of a moving transformer core. A low distortion sine wave acts as the input. The amplitude and phase of the output signal across the two secondary windings determine the distance and polarity of the LVDT core with respect to the center.

Proper signal conditioning circuitry allows extremely precise measurements in demanding applications, such as in production manufacturing, fluid level measurements and structural/strain testing. One common application is to use an LVDT at the end of a Bourdon tube to measure minute changes in system or barometric pressure.

The most common method of LVDT signal conditioning is demodulation (i.e., full-wave rectification) and simple lowpass filtering of the rectified sine wave. However, the precision of the demodulation method depends on the accuracy of the phase adjustment. In addition, there are losses associated with demodulation, which usually involve switches and the associated charge injection and timing jitter. It is difficult to achieve 12-bit precision under these circumstances.

A better approach to LVDT signal conditioning is shown in Figure 487.1. The LTC1967 true RMS to DC converter can directly convert an LVDT output sine wave into a precise DC voltage with 0.15% linearity error and 0.3% gain error. The LTC1967's performance is independent of input phase, and

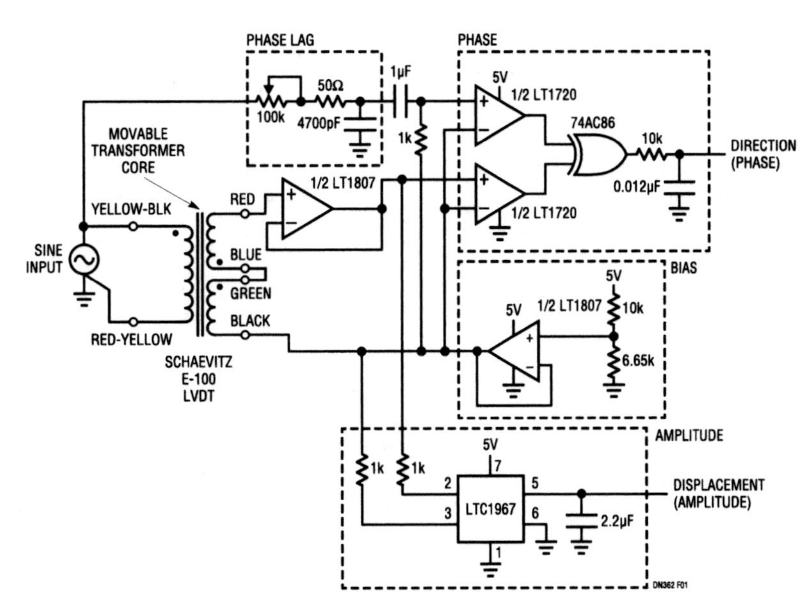

Figure 487.1 • A New Approach to LVDT Signal Conditioning Which Combines an LTC1967 RMS to DC Converter with a Separate Phase Detector Circuit

Analog Circuit Design: Design Note Collection. http://dx.doi.org/10.1016/B978-0-12-800001-4.00487-7

maintains exceptional performance over temperature. A separate circuit determines the phase of the LVDT output, which can be used to determine the polarity of the LVDT core position. True precision performance is achievable with this simple circuit and a minimal amount of calibration.

LVDT operation

LVDTs are driven by a low distortion sine wave in the primary winding of the transformer. In a 12-bit system, the input sine wave needs less than −74dB distortion and better than 0.02% amplitude stability. In the null (center) position, the two secondary windings receive the same amount of magnetic coupling, but the differential voltage across them is not zero due to the flux leakage of the LVDT (see Figure 487.2). When the LVDT core moves in one direction or the other, the differential voltage amplitude increases. The phase of the differential output changes depending on which side of center the LVDT core sits.

Circuit description

The LVDT shown in Figure 487.1, a Schaevitz E-100 with ±2.5mm of linear range, is driven by a low distortion, amplitude-stable $3V_{RMS}$ (the manufacturer's recommended amplitude) sine wave on the primary windings of the transformer. The frequency, 10kHz, is the maximum recommended for the E-100, though LVDTs exist that work well up to hundreds of kilohertz. Use of higher excitation frequencies with the higher frequency LTC1968 RMS to DC converter would result in faster settling times and reduced audio-frequency signals that could cause interference.

To facilitate single supply operation, one-half of an LT1807 amplifier biases the output sine wave DC level to fall within the common mode range of the measurement circuit (approximately 2V on a 5V supply). The other half of the LT1807 buffers the LVDT output for good signal fidelity. The LT1807 was chosen for its high open-loop gain at 10kHz, extremely low distortion and high common mode rejection to maintain the accuracy of the LVDT amplitude signal. The LT1807's buffered output signal is converted by an LTC1967 true RMS to DC converter to a DC signal that is linearly proportional to the displacement of the transformer core.

The phase detector portion of the circuit consists of a phase adjustment network (which provides phase lead or lag, depending on the specific LVDT and excitation frequency), an exclusive-OR (XOR) logic gate and an RC lowpass filter network. The circuit output is high when the LVDT core is on one side of center and low when it moves to the other side. Two LT1720 comparators sense the zero-crossing points of the phase-adjusted input and output sine waves. The XOR gate has a low output when the inputs agree and a high output when the inputs disagree. The RC networks limit the bandwidth of the phase network to 1.3kHz in order to limit the effect of comparator output spikes due to slight phase mismatch. It is recommended that the phase detector band-limiting network be lower in frequency than the LVDT excitation frequency, to minimize the phase output ripple.

Circuit calibration

To calibrate the signal conditioning circuit, first move the LVDT into the null (center) position. The center position is where the amplitude output is at its minimum. Note the output voltage of the phase detector; adjust the phase lead/lag network until the output reaches approximately 2V. Note the amplitude outputs at the extremes which may vary between LVDTs.

Conclusion

Figure 487.2 shows the amplitude and phase outputs of the circuit in Figure 487.1. A novel approach to LVDT signal conditioning yields stable, precise performance and a low IC count. Unlike a synchronous demodulation scheme, the accuracy of the circuit does not hinge on a manual phase adjustment—only on the built-in high precision of the LTC1967 (or LTC1968) true RMS to DC converter. The circuit is robust enough for many industrial and instrumentation applications—it maintains good precision over temperature and it is immune to input signal phase. A separate, easily calibrated phase detector determines which side of center the LVDT core occupies.

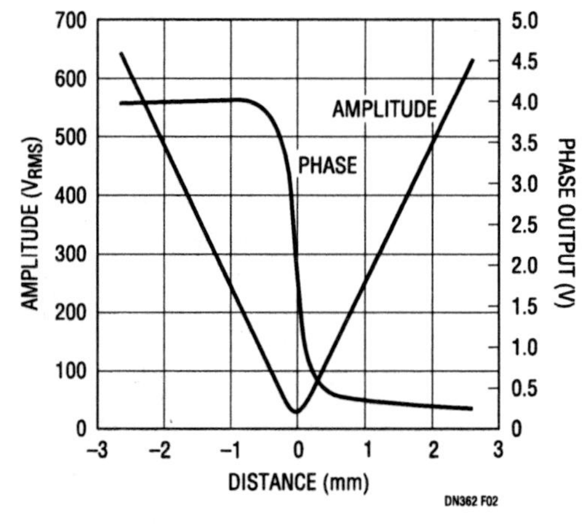

Figure 487.2 • Amplitude and Phase Output versus Position of the LVDT Circuit. The Non-Zero Center Amplitude Is due to Flux Leakage in the LVDT and Is Not Caused by the Measurement Circuit

An autoranging true RMS converter

488

Philip Karantzalis Jim Mahoney

Introduction

The LTC1966 is a true RMS to DC converter that uses a ΔΣ computational technique to make it dramatically simpler to use, significantly more accurate, lower in power consumption and more flexible than conventional log-antilog RMS to DC converters. The LTC1966 RMS to DC converter has an input signal range from $5mV_{RMS}$ to $1.5V_{RMS}$ (a 50dB dynamic range with a single 5V supply rail) and a 3dB bandwidth of 800kHz with signal crest factors up to four.

True RMS voltage detection is most commonly required to measure complex amplitude and time varying signals, such as machine or engine vibration monitoring and complex AC

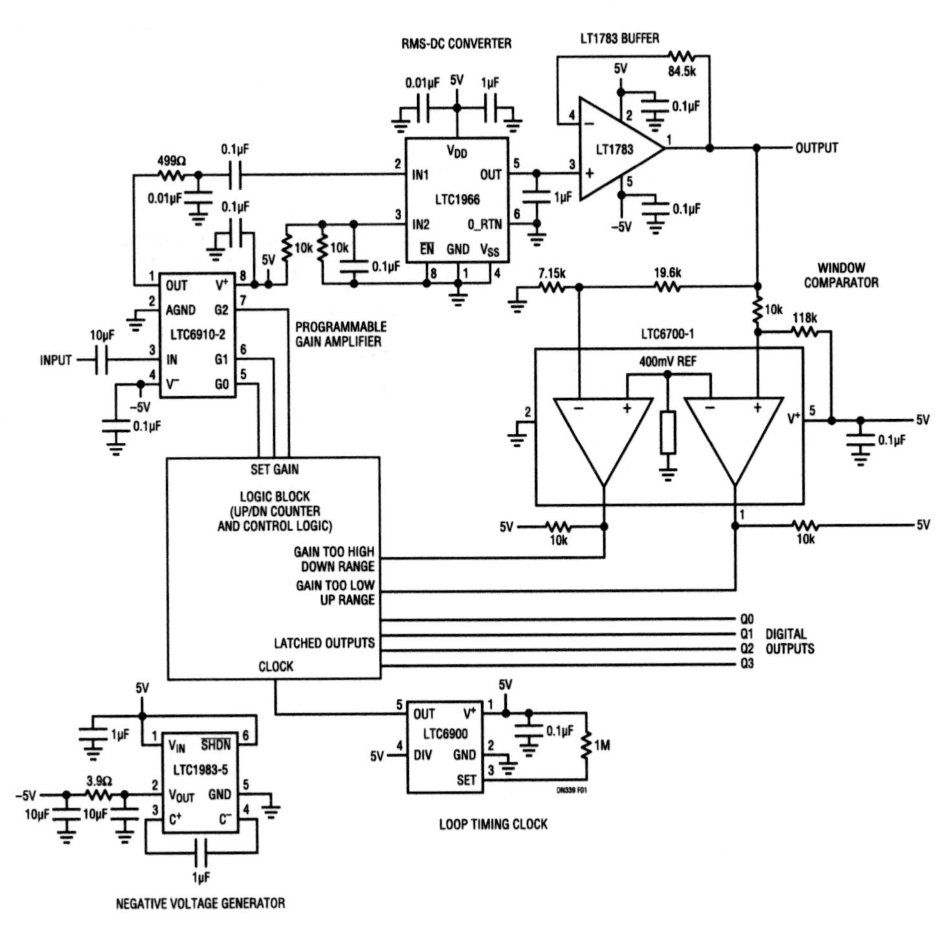

Figure 488.1 • An Autoranging True RMS to DC Converter

Analog Circuit Design: Design Note Collection. http://dx.doi.org/10.1016/B978-0-12-800001-4.00488-9

power line load monitoring. Sometimes these applications require accurate input signal measurement over an extremely wide dynamic range—even more than the 50dB range of the LTC1966. One solution is to add an autoranging function to the LTC1966, thus effectively expanding the dynamic range of the measuring system. Versatility is certainly an advantage of this approach. Figure 488.1 shows a true RMS to DC autoranging converter which has an input signal dynamic range of 80dB, making it suitable to a wide range of applications.

Autoranging expands input dynamic range

The autoranging loop of Figure 488.1 uses an LTC6910-2 programmable gain amplifier (PGA) to provide gain in front of the LTC1966. Under control of a 3-bit input code, the LTC6910-2 provides gain in binary-weighted increments (gain is set to 1, 2, 4, 8, 16, 32 or 64). An LT1783 op amp follower buffers the LTC1966 DC output and drives an LT6700-1-based window comparator (the LT6700-1 combines two micropower, low voltage comparators with a 400mV reference). The window comparator has two logic outputs that go low when the DC output voltage extends beyond or below two preset threshold levels. The comparator outputs enable the clocking of an up/down counter that increases or decreases the front-end gain of the LTC6910-2 as required. An LTC6900 single resistor programmable oscillator controls the response time of the autoranging loop.

Circuit description

The entire circuit is biased from a single 5V supply. The input signal is AC coupled with filtering added in the LTC1966 input. The autoranging true RMS to DC conversion bandwidth is 12Hz to 32kHz. An LTC1983-5 charge pump inverter provides a negative supply for the input PGA and output buffer. This allows their inputs and outputs to operate linearly to zero volts. The thresholds for the window comparator are set to 9.5mV and 1.5V. At power-on it is assumed that there is no input signal present and the PGA gain is set to the maximum value of 64. When an applied signal causes the DC output to exceed the 1.5V down-range threshold, the gain control up/down counter is clocked down by one count. Any gain change is delayed by one second to ensure that the PGA and LTC1966 have plenty of time to settle. The gain continues to clock down until the output signal remains within the window. Conversely when an input signal magnitude is reduced to a level to cause the DC output to fall below the 9.5mV up-range threshold, the gain is clocked up to a higher value. With a maximum front-end gain of 64, signals as low as 150μV_{RMS} are converted. At the minimum gain setting of 1, the input range is 1.5V_{RMS}. For a system to determine the RMS signal level both the DC output and the control code must be read (digital outputs Q3, Q2, Q1 and Q0).

The circuit has three operating conditions, a linear range, an over range and an under range. These three conditions are described as follows:

Linear range: The digital output (Q3, Q2, Q1, and Q0) is in the range 0001 to 0111 and the analog output is within the up-range and down-range voltage range. In the linear range, the input voltage in RMS is equal to the DC output voltage divided by the PGA gain. For example, if the output voltage is 64mV and the digital code is 0111, then the input voltage in RMS is equal to 64mV divided by 64. The circuit's conversion error is less than 1% for an LTC1966 input voltage range of 50mV_{RMS} to 1.5V_{RMS} and increases to 5% for the lowest input of 9.5mV_{RMS}. The 1% error bandwidth is 6kHz.

Over range: The digital output is 0000, the input signal is too high and the auto range circuit cannot provide less gain. The 0001 to 0000 transition indicates an over range signal condition. The PGA gain in this condition is set to 1.

Under range: The digital output is 1000, the input signal is too low and the auto range circuit cannot provide more gain. The transition of the digital output from 0111 to 1000 indicates an under range signal condition. The PGA gain in this condition is set to the maximum of 64.

Conclusion

The autoranging converter shown here expands the dynamic range of the LTC1966 to 80dB, making it extremely versatile. This useful circuit example combines a variety of special function circuits available from Linear Technology. The LTC1966 true RMS to DC converter, the LTC6910-2 programmable gain amplifier, the LT6700-1 window comparator with built-in reference, the LTC6900 resistor programmable oscillator, the LTC1983-5 charge pump voltage inverter and an LT1783 rail-to-rail op amp are all used to handle the analog signal conditioning. The logic block shown on Figure 488.1 can be implemented with discrete logic, a low cost microcontroller or a portion of an FPGA.

489

RMS to DC conversion just got easy

Glen Brisebois Joseph Petrofsky

Introduction

The LTC1966 is a precision, micropower, true RMS to DC converter that utilizes an innovative patented ΔΣ computational technique.[1] The internal delta-sigma circuitry of the LTC1966 makes it simpler to use, more accurate, lower power and dramatically more flexible than conventional log-antilog RMS to DC converters. Unlike previously available RMS to DC converters, the superior linearity of the LTC1966 allows hassle-free system calibration with any input voltage, even DC.

Ease of use

The flexibility of the LTC1966 is illustrated in the typical applications shown in Figures 489.1a–489.1c. The LTC1966 accepts single-ended or differential input signals (for EMI/RFI rejection) and supports crest factors up to 4. Common mode input range is rail-to-rail while the differential input range is $1V_{PEAK}$. The LTC1966 also has a rail-to-rail output with a separate output reference pin providing for flexible level shifting. The LTC1966 operates on a single power supply from 2.7V to 5.5V or dual supplies up to ±5.5V while drawing only 155μA. When the LTC1966 is shut down, supply current is reduced to just 0.1μA.

The trouble with log-antilog

Older RMS to DC converters used log/antilog techniques. The log/antilog function was derived from the logarithmic relationship between the base emitter voltage and collector current of bipolar junction transistors. This method suffers from a variety of problems. BJT transistors match and track well over temperature while operating at the same collector current, for example in op amp differential pair input stages intended to run closed loop. However, their log conformance is NOT very good over wide current variations and they do NOT match and track well when operating at different collector currents in open-loop configurations. This gives rise to the poor linearity and poor temperature rejection characteristic of log-antilog converters and also makes them uncorrectable using simple calibration techniques. In contrast, the LTC1966 gives exceptional accuracy over broad varieties of signal type and temperature, with even better results obtainable via a simple DC calibration. Figure 489.2 compares the

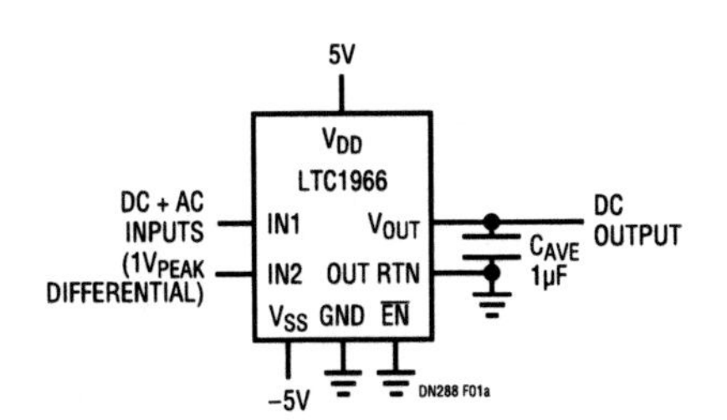

Figure 489.1a • ±5V Supplies, Differential, DC-Coupled RMS to DC Converter

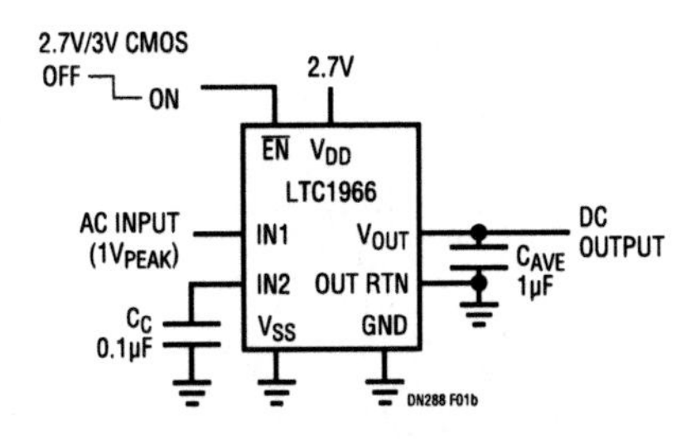

Figure 489.1b • 2.7V Single Supply, Single-Ended, AC-Coupled RMS to DC Converter with Shutdown

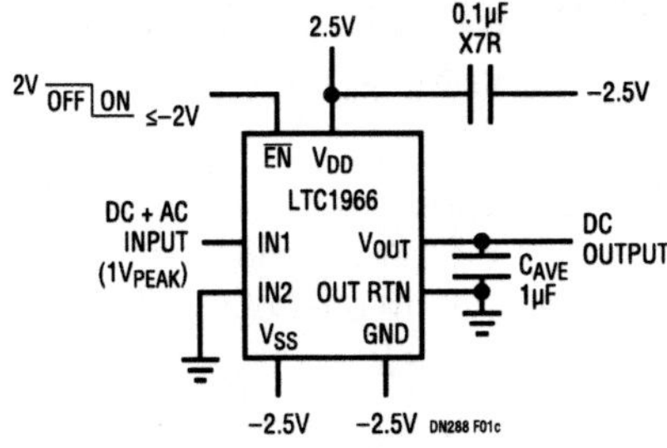

Figure 489.1c • ±2.5V Supplies, Single-Ended, DC-Coupled RMS to DC Converter with Shutdown

Note 1: U.S. Patents numbers 6359576, 6362677, more pending.

Analog Circuit Design: Design Note Collection. http://dx.doi.org/10.1016/B978-0-12-800001-4.00489-0

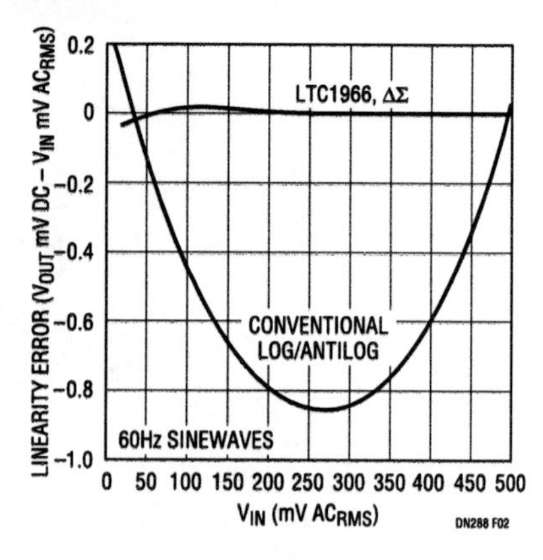

Figure 489.2 • Quantum Leap in Linearity Performance

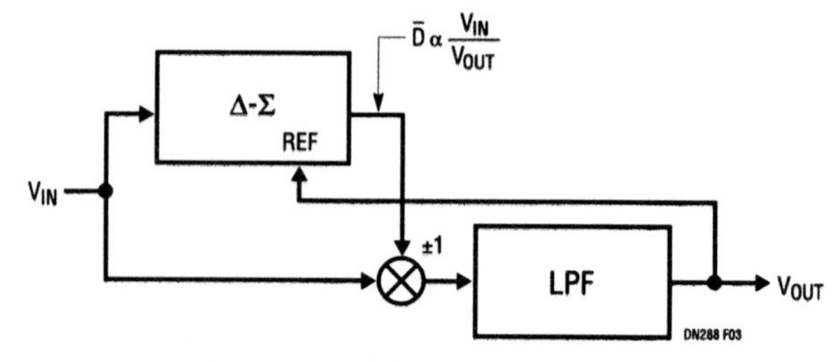

Figure 489.3 • LTC1966 Block Diagram

linearity of the LTC1966 with that of the now inferior log/antilog methods.

Another drawback to log-antilog techniques arises due to the fact that the bandwidth of a BJT depends on how much current flows through it. Thus, log-antilog converters have a bandwidth that varies with signal amplitude. In the extreme, the bandwidth drops to near zero as the signal amplitude drops. To see this effect, take a true RMS meter that employs one of these devices and give it an input signal. Then remove the signal and short the meter inputs. The meter reading will fall fairly quickly at first, but will slow down and keep slowing down and can take as long as a few minutes to get back down to an effective zero. In contrast, the same situation using an LTC1966 gives a true zero reading within seconds.

Still another problem with the log/antilog approach is the need for an absolute value circuit at its front end. Because the input current takes a different path depending on the input polarity, there is a polarity dependant gain error. To see this effect, put an asymmetric signal waveform with 10% to 30% duty cycle into your RMS meter. Now swap the inputs around. You will typically see about a 0.5% difference in the readings. If you don't see that much difference, change the signal amplitude and try again. (Note that this effect will be apparent on DC signals as well, but that most RMS meters are internally AC coupled precluding a DC test.) Because of its symmetric ΔΣ inputs, the LTC1966 does not have an absolute value circuit, and this error is eliminated.

How the LTC1966 RMS to DC converter works

The LTC1966 uses a completely new implementation (Figure 489.3). A ΔΣ modulator acts as the divider and a simple polarity switch is used as the multiplier. Applying V_{OUT} to the ΔΣ reference voltage results in the V_{IN}^2/V_{OUT} function before the lowpass filter and causes the RMS to DC conversion.

The ΔΣ is a 2nd order modulator with excellent linearity. It has a single-bit output whose average duty cycle is proportional to the ratio of the input signal divided by the output. The single-bit output is used to selectively buffer or invert the input signal. Again, this is a circuit with excellent linearity because it operates at only two gains: −1 and +1. The average effective multiplication over time will be on the straight line between these two points.

The lowpass filter performs the averaging of the RMS function and must have a lower corner frequency than the lowest frequency of interest. The LTC1966 needs only one capacitor on the output to implement the lowpass filter. The user selects this capacitor depending on frequency range and settling time requirements, given the 85kΩ output impedance.

This topology is inherently more stable and linear than log-antilog implementations primarily because all of the signal processing occurs in circuits with high gain op amps operating closed loop. Note that the internal scalings are such that the ΔΣ output duty cycle is limited to 0% or 100% only when V_{IN} exceeds $\pm 4 \cdot V_{OUT}$.

Summary

The LTC1966 is a breakthrough in RMS to DC conversion bringing a new level of accuracy to RMS measurements. It is extremely simple to connect and provides excellent accuracy over temperature and time without requiring trims. These features, along with its small size and micropower operation, make the LTC1966 suitable for a wide range of RMS to DC applications, including handheld measurement devices.

PART 4

Wireless, RF & Communications Design

High input IP3 mixer enables robust VHF receivers

490

Andy Mo

Introduction

An increasing number of applications occupy the 30MHz to 300MHz very high frequency (VHF) band. Television and radio broadcasting, navigation controls and amateur radios are a few examples. Modern RF component development is aimed at much higher frequency bands used for voice and data communications systems. Significant advances in circuit techniques and manufacturing processes are required to meet the demanding performance requirements of the next generation of radios. Applying these techniques to lower frequency designs can significantly improve performance.

The LTC5567 is a wideband mixer designed and optimized for performance in the 300MHz to 4GHz frequency band. To create very compact circuit implementations, the LTC5567 contains integrated RF and LO transformers. The Input IP3 linearity performance benchmark is an excellent 30dBm for the LTC5567 in its specified frequency range. Going lower in frequency requires the built-in transformers to maintain this linearity as well as conversion gain.

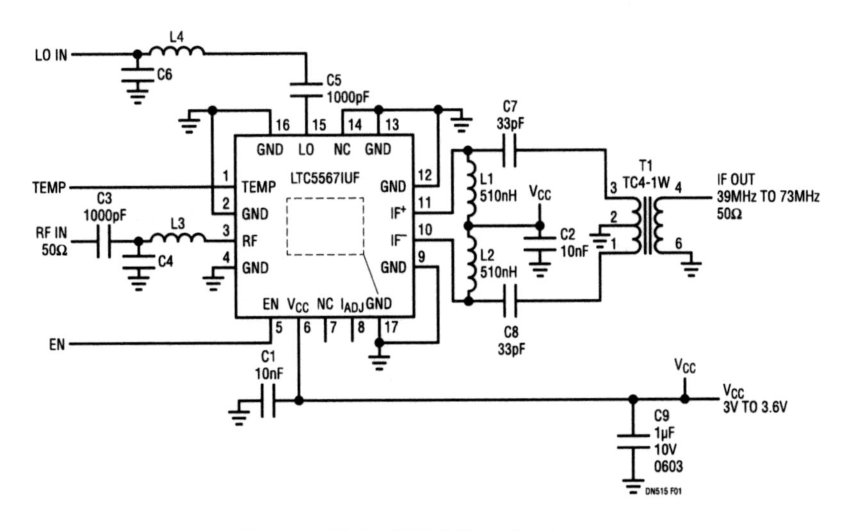

Figure 490.1 • VHF Mixer Design

Table 490.1 VHF Impedance Match Design Values

MATCH	RF INPUT	L3	C4	LO INPUT	L4	C6
A	150MHz	8.2nH	56pF	200MHz	3.9nH	47pF
B	200MHz	6.8nH	39pF	250MHz	2.7nH	33pF
C	250MHz	3.9nH	27pF	300MHz	1.5nH	27pF

Analog Circuit Design: Design Note Collection. http://dx.doi.org/10.1016/B978-0-12-800001-4.00490-7

With such a high level of linearity to start from, it is worthwhile to modify the mixer circuit design and characterize the performance over lower VHF frequencies. The proof of performance is in the testing.

Impedance match design

Figure 490.1 shows an impedance match design with the LTC5567. Table 490.1 shows the design values extending input port match below 300MHz, down to 150MHz, while still achieving outstanding performance. Test results are also provided.

Figure 490.2 shows the LTC5567 mixer gain and input IP3 versus input frequency. The mixer linearity performance improves as input frequency approaches 150MHz. Input, LO and output port return loss measurements are shown in Figures 490.3, 490.4 and 490.5, respectively. The overall performance is maintained in the VHF range compared to higher input frequencies. As a result, the high IP3 and conversion gain yields maximum dynamic range when used in radio designs. Higher dynamic range minimizes adjacent channel interference, improving selectivity. Operating the LTC5567 below 150MHz input is possible with reduced conversion gain, but not recommended, due to the internal transformer becoming lossy.

Conclusion

The LTC5567 offers very high linearity performance at VHF and UHF input frequencies. High IP3 figures and P1dB in (Table 490.2) make it an excellent choice for high performance radio design over a wide range of frequencies.

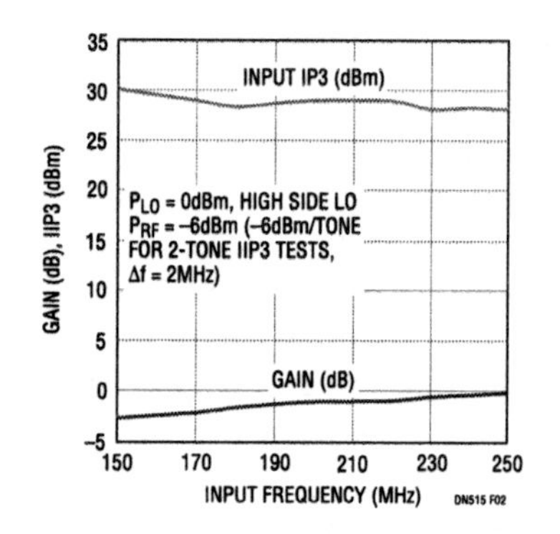

Figure 490.2 • Mixer IIP3 and Gain Performance Results

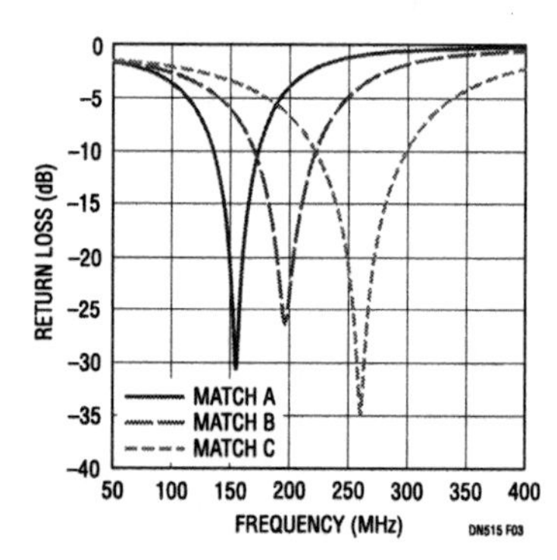

Figure 490.3 • RF Input Return Loss

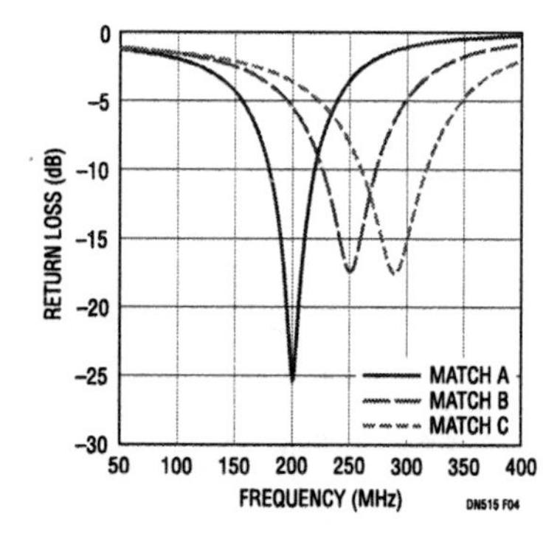

Figure 490.4 • LO Input Return Loss

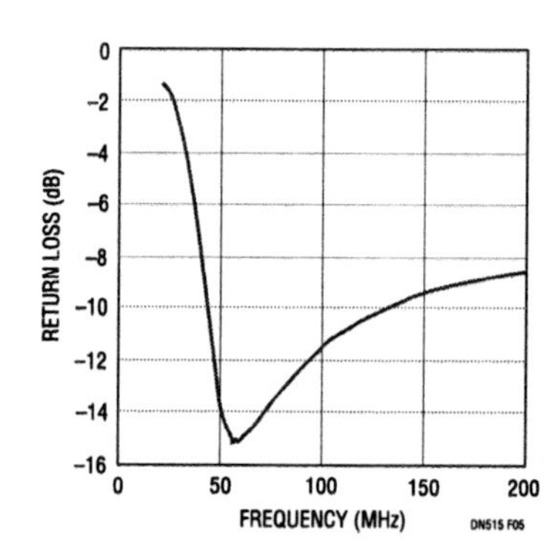

Figure 490.5 • IF Output Return Loss

Table 490.2 P1dB Compression Point and LO Leakage Over Input Frequency. Output Frequency = 50MHz, HSLO

RF INPUT FREQUENCY (MHz)	P1DB (dBm)	LO TO IF LEAKAGE (dBm)
150	12.29	–35
160	12.9	–42
170	12.9	–42
180	12.75	–42
190	12.70	–41.2
200	11.61	–43
210	12.48	–43
220	12.7	–44
230	11.7	–44
240	11.08	–44
250	12.89	–44

A robust 10MHz reference clock input protection circuit and distributor for RF systems

491

Michel Azarian

Introduction

Designing the reference input circuit for an RF system can prove tricky. One challenge is maintaining the phase noise performance of the input clock while meeting the protection, buffering and distribution requirements for the clock. This article shows how to design a 10MHz reference input circuit and optimize its performance.

Design requirements

RF instruments and wireless transceivers often feature an input for an external reference clock, such as the ubiquitous 10MHz reference input port found on RF instruments. Many of these same systems include a provision to distribute the reference clock through the system. Figure 491.1 shows a common scheme, where the reference clock supplies the reference input to two distinct phase-locked loops (PLLs).

A well-designed, robust input would accept both sine and square wave signals over a wide range of amplitudes. It would maintain a constant signal level drive to the destination PLL inputs inside the system, even in the face of varied inputs. The exposed-to-the-world reference input port should have overvoltage/overpower protection. Most importantly, the inevitable degradation in the phase noise performance of the clock signal should be minimized.

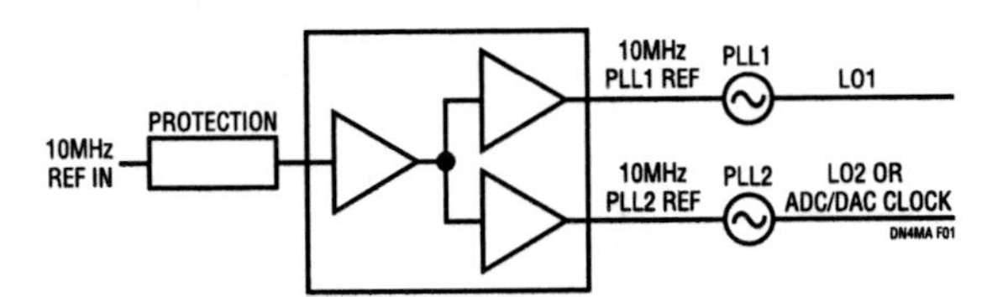

Figure 491.1 • Block Diagram of a Common Reference Input and Reference Distribution for an RF System

Design implementation

The LTC6957 is a very low additive phase noise (or jitter) dual-output clock buffer and logic translator. The input of the LTC6957 accepts a sine or a square wave over a wide range of amplitudes and drives loads at constant amplitude.

The LTC6957 offers various output logic signal options: PECL, LVDS and CMOS (in-phase and complementary), allowing it to drive a wide range of loads. Figure 491.2 shows a 10MHz reference input circuit using the LTC6957-3, which produces two in-phase CMOS outputs.

The transformer shown in Figure 491.2 serves several functions. First, in conjunction with the Schottky diodes following it, it offers input overpower/overvoltage protection. The diodes limit the AC voltage seen by the LTC6957-3. The WBC16-1T can handle up to 0.25W power ($3.5V_{RMS}$ into 50Ω).

The transformer also isolates the connector ground—which is usually tied to the chassis of the RF system—from the internal analog ground of the system.

Furthermore, the transformer applies a voltage gain to the incoming signal, thus steepening the edges seen by the LTC6957-3. This helps reduce AM-to-PM noise conversion, which in turn limits phase noise degradation, especially with small input signals. The WBC16-1T has a voltage gain of four. It is possible to rely on the transformer's voltage gain of four, as opposed to its maximum and ideal power gain of one, because the LTC6957-3 presents a high impedance load to the transformer.

R1 and R2 can be adjusted in combination to match the input port to 50Ω. For small input signals, the diodes are off and the transformer sees a load of 804Ω in Figure 491.2. That load is reflected to the input as approximately 50Ω because of the transformer's primary-to-secondary impedance ratio of 16. For larger input signals, the Schottky diodes turn on, reducing the 604Ω resistance to nearly a short circuit. This degrades the reference input return loss—a problem that can

Analog Circuit Design: Design Note Collection. http://dx.doi.org/10.1016/B978-0-12-800001-4.00491-9

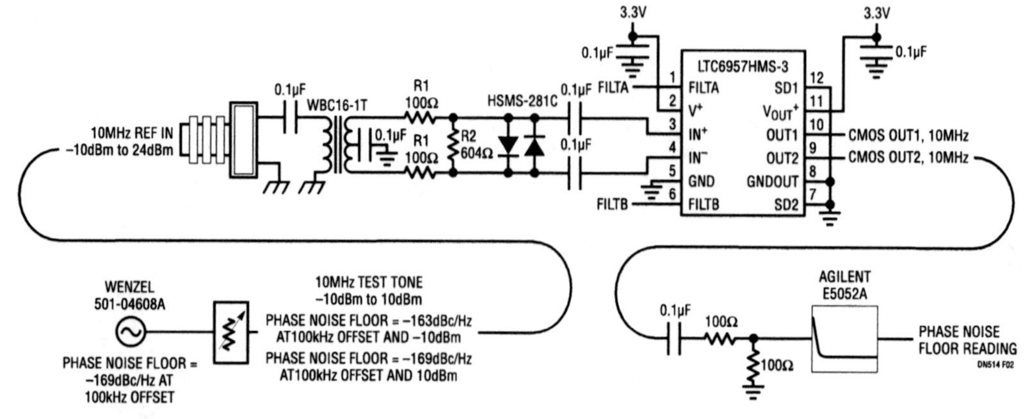

Figure 491.2 • 10MHz Reference Input Circuit Employing the LTC6957-3 with Front-End Protection, Shown with Test Signal and Phase Noise Measurement Set-Up

be avoided by adjusting the values of R1 and R2, but there are trade-offs to doing so.

For large input signals, the input return loss can be improved by increasing R1's value, and reducing R2's value, such that their combined series resistance remains around 800Ω. However, since R1 appears in series with the signal, it adds noise to it. A larger R1 comes in combination with a smaller R2, resulting in a smaller portion of the signal appearing at the LTC6957-3's input, further degrading the phase noise performance. In other words, the designer can trade off phase noise performance for input return loss by playing with the values of R1 and R2. The values shown in Figure 491.2 strike an overall balance of these two performance metrics.

The AC-coupling capacitor separating the connector from the transformer in Figure 491.2 offers input protection from DC sources.

The LTC6957-3 has internal lowpass filters that can be selected via the FILTA and FILTB pins. This option strategically limits the bandwidth of the LTC6957's first amplifier stage, and hence, the additive phase noise of the circuit, especially when the input signal is weak as shown below.

Performance

A 10MHz OCXO is connected to the input of the circuit via a step attenuator as shown in Figure 491.2. The reference input signal is varied between −10dBm and 10dBm while measuring the phase noise floor at the output of the LTC6957-3 with different input filter settings using the Agilent E5052A signal source analyzer. Figure 491.3 shows the phase noise floor of the 10MHz CMOS clock output of the LTC6957-3 measured at a 100kHz offset.

If the amplitude of the externally applied 10MHz reference signal is not known, pulling FILTA low and FILTB high yields good overall phase noise performance as shown in Figure 491.3. Nevertheless, performance can be optimized if the applied signal level at the input is measured and appropriate filter settings are applied.

The R1 and R2 values chosen in Figure 491.2 result in an input return loss of −9dB when the reference input's power is 0dBm into 50Ω. The return loss is better at lower input powers and worse at higher powers.

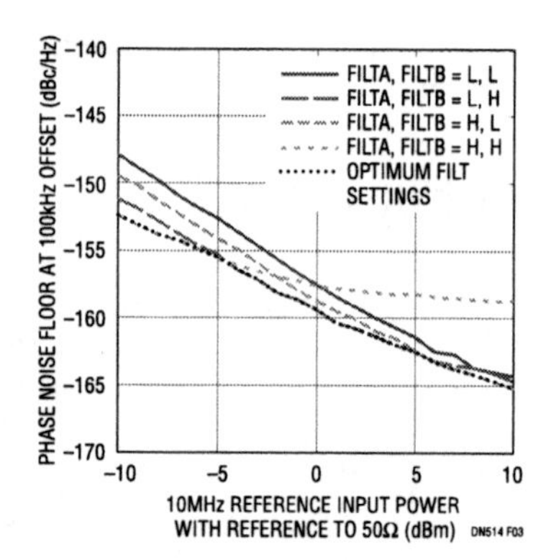

Figure 491.3 • 100kHz Offset Phase Noise Floor at the Output of the LTC6957-3 vs 10MHz Reference Input Power Level for Various LTC6957 Filter Settings

Conclusion

A robust, high performance 10MHz reference input circuit is built around the LTC6957-3. Features include a wide range of input signal type and level compatibility, protection and clock distribution with limited phase noise degradation. The circuit's phase noise and input return loss are evaluated and optimized. The LTC6957-3 simplifies the design process while achieving excellent overall performance.

A low power, direct-to-digital IF receiver with variable gain

492

Walter Strifler

Introduction

Modern communication receivers require an ADC to digitize an incoming analog signal for decoding in a suitable FPGA device. The direct-conversion method of receiver design typically performs a single frequency downconversion and an analog-to-digital conversion (ADC) near baseband. While elegant and simple, this receiver architecture has problems with in-band blockers, out-of-band interferers and LO leakage reflections within the receiver itself.

In the face of these problems, base station receivers often require a robust solution that is achieved using tried-and-true system methods of downconversion to an intermediate frequency (IF) in the range of 70MHz to 240MHz. Demodulating and decoding the IF signal can be performed by various means, but an increasingly popular and cost-effective method is direct-to-digital IF conversion using the recent generation of high speed, low power pipeline data converters available from Linear Technology.

This design note describes a variable gain amplifier plus analog-to-digital converter (VGA + ADC) combination circuit that preserves the IF receiver dynamic range over a 31dB gain adjust range and effectively demodulates and digitizes both the I and Q information in a single step. The combination LTC6412 VGA and LTC2261 14-bit ADC circuit subsamples a 140MHz WCDMA IF channel at 125Msps and provides an equivalent input NF and IP3 that rivals some of the best laboratory spectrum analyzers, while consuming less than 0.5W of power.

IF receiver performance

The performance of a demonstration receiver circuit is shown in Figure 492.1. The inset graph of Figure 492.1 shows the noise-like distribution of the WCDMA signal and is similar to CCDFs of other modern communication signals. At VGA maximum gain, the signal generator power is adjusted to

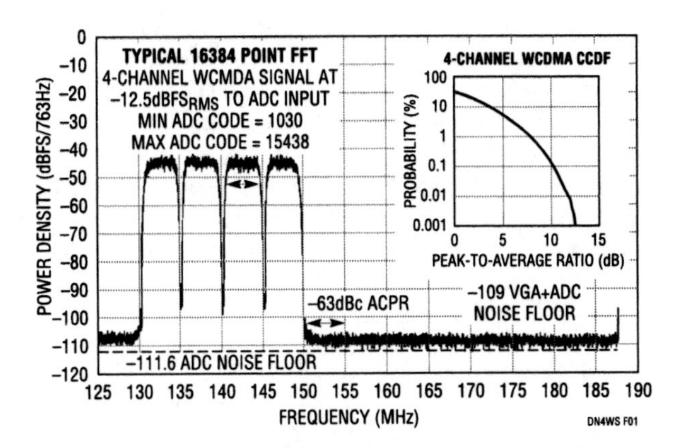

Figure 492.1 • Typical WCDMA Performance at −12.5dBFS$_{RMS}$ over All Gain Adjust Settings. Insert Shows WCDMA CCDF

−12.5dBFS$_{RMS}$ to occupy most of the ADC code range without clipping.

As the input signal power is adjusted higher, the VGA gain is adjusted down to maintain −12.5dBFS$_{RMS}$ and simulate the automatic gain control (AGC) response of a typical receiver. The FFT of the digitized receive signal is plotted over a full Nyquist zone and exhibits a 63dBc ACPR with no measurable spurs and only 2.6dB degradation in the ADC noise floor, *over the full 31dB gain adjust range*. This represents an effective input NF of 13dB and input IP3 of 23dBm for the VGA + ADC pair at 140MHz at maximum gain. The (IP3-NF) delta of 10dBm determines the effective dynamic range of the receive pair and is nearly constant over the entire gain adjust range.

Measurement details and receiver circuit

An Agilent E4436B source generates the multichannel WCDMA test signal with a typical adjacent channel power ratio (ACPR) of 50dBc to 55dBc, perfectly adequate to meet

Analog Circuit Design: Design Note Collection. http://dx.doi.org/10.1016/B978-0-12-800001-4.00492-0

the WCDMA system specifications but insufficient to demonstrate the full quality of this VGA+ADC combination. The test signal is amplified with a high linearity Triquint AH202 and sharply filtered with a SAWTEK 854920 to reduce the test signal's ACPR skirts below 65dBc.

The WCDMA signal is representative of the wideband, noise-like signals found in modern communication systems such as LTE, 802.11g, and WiMAX to name a few. Interestingly, this convergence of statistical signal behavior was predicted over 60 years ago in Claude Shannon's communication theory. He found that the methods to increase spectral efficiency in a modulation format will, by necessity, exercise many degrees of freedom in signal space and approximate the process of additive white Gaussian noise. This was an amazing insight considering the simple AM and FM signals of Shannon's day. This is also a practical insight. One representative noise-like signal can be used to characterize an RF receiver and estimate the performance of other noise-like signals.

Figure 492.2 details a receiver circuit optimized for a 140MHz center frequency and a 20MHz bandwidth typical of a 4-channel WCDMA signal. The filtered test signal feeds to the VGA input balun to perform a single-ended to differential conversion at the input of the LTC6412. The LTC6412 output connects to a simple tank circuit and RC network at the input of the LTC2261. This matching circuit routes bias current to the VGA while performing a low Q impedance transformation to the 100Ω differential load. The matching circuit and RC load also serve to dissipate the differential and common mode charge injections emanating from the sampling switches at the ADC input. This is an important consideration, as these charge impulses need to dampen to better than −85dB during a sampling window (4ns) to preserve the full spur-free dynamic range (SFDR) of the LTC2261. The better damping circuits tend to be small and tight to avoid unnecessary reflection delays and mismatch between the VGA output and ADC input. This particular matching circuit uses 0402 components for most elements and fits inside a board area of 5mm × 10mm.

The balance of connections to the VGA and ADC follow the recommendations of their respective data sheets. The LTC2261 14-bit ADC runs off 1.8V and consumes 127mW at 125Msps. The LTC6412 VGA runs off 3.3V and consumes 360mW for a total power consumption of 490mW.

Conclusion

The LTC6412 VGA drives the LTC2261 14-bit ADC with little compromise in the ADC performance. The VGA buffers the ADC sampling input and provides 31dB of gain adjust to expand the effective dynamic range of the subsampling IF receiver. The LTC2261 is part of a family of 12- and 14-bit low power data converters designed for maximum sampling rates in the range of 80Msps to 125Msps. For complete schematics of this receiver, visit the LTC6412 or LTC2261 product pages at www.linear.com.

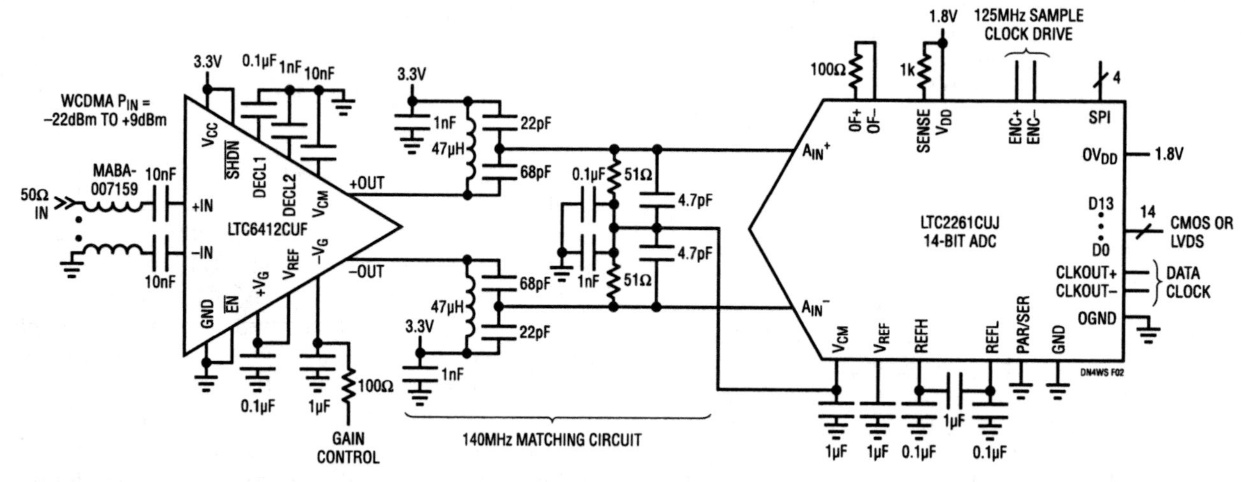

Figure 492.2 • VGA+ADC IF Receiver Circuit. Supply Decoupling Capacitors to the VGA and ADC Omitted for Clarity. For This Measurement, the LVDS Bus Connects to Linear's Data Acquisition Board DC890B for Computer Control and Data Analysis

Fast time division duplex (TDD) transmission using an upconverting mixer with a high side switch

493

Vladimir Dvorkin

Introduction

Many wireless infrastructure time division duplex (TDD) transmit applications require fast on/off switching of the transmitter, typically within one to five microseconds. There are several different ways to implement fast Tx on/off switching, including the use of RF switches in the signal path, or on/off switching of the supply voltage for different stages of the transmitter chain. The advantages of the latter method are low cost, very good performance and power saving during the Tx off-time. In particular, a good place to apply supply switching is at the transmit upconverting mixer because this removes both the transmit signal and all other mixing products from the mixer RF output.

The LT5579 high performance upconverting mixer fits various TDD and Burst Mode transmitter applications with output frequencies up to 3.8GHz. Fast on/off supply voltage (V_{CC}) switching for the LT5579 is as simple as adding an external high side power supply switch (note that this technique is equally effective for the lower frequency upconverting mixer, LT5578).

High side V_{CC} switch for a Burst Mode transmitter using the LT5579 mixer

The high side V_{CC} switch circuit in Figure 493.1 uses a P-channel MOSFET (IRLML6401) with an $R_{DS(ON)}$ of less than 0.1Ω. An N-channel enhancement mode FET (2N7002), connected from the drain of IRLML6401 to ground, further improves fall time. The 2N7002's $R_{DS(ON)}$ is less than 4Ω, which is sufficient for this application.

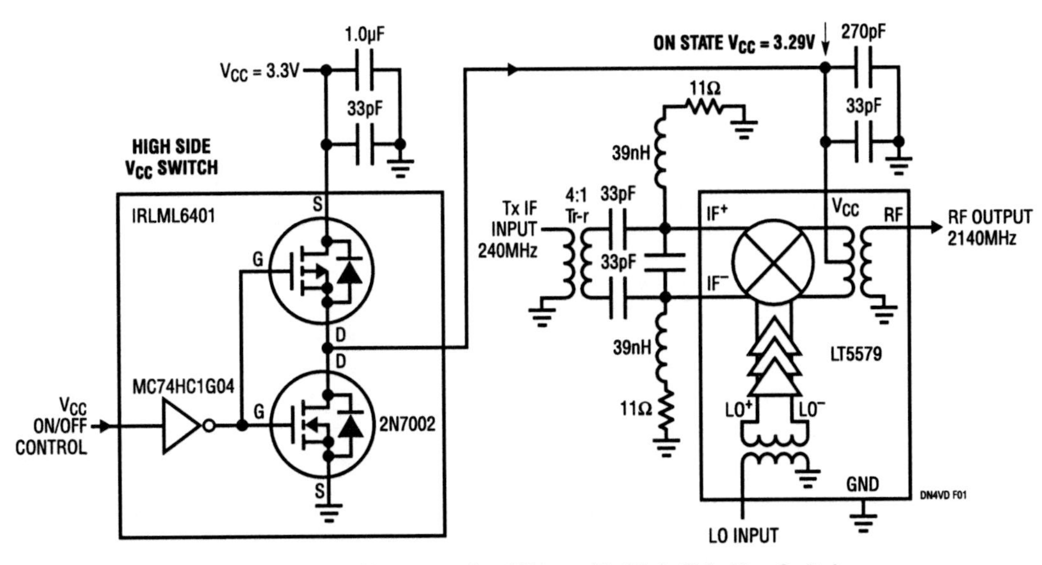

Figure 493.1 • Upconverting Mixer with High Side V_{CC} Switch

Analog Circuit Design: Design Note Collection. http://dx.doi.org/10.1016/B978-0-12-800001-4.00493-2

The input driver for the high side V_{CC} switch is a high speed CMOS inverter (MC74HC1G04) capable of driving capacitive loads. The IRLML6401 input capacitance is typically 830pF and the 2N7002 input capacitance is under 50pF. For faster rise times, two high speed CMOS drivers can be used in parallel. Likewise, for faster fall times, a different N-channel MOSFET with lower on-resistance can be used.

With the LT5579 supply current of 220mA, the power supply voltage drop across the MOSFET is only 11mV. The response time of the high side V_{CC} switch is shown in Figure 493.2. Total turn-on time is only 650ns and total turn-off time is 500ns. These measurements were performed using two RF bypass capacitors at the mixer V_{CC} pin (33pF and 270pF). Higher value RF bypass capacitors can be used, which would result in correspondingly slower rise/fall times.

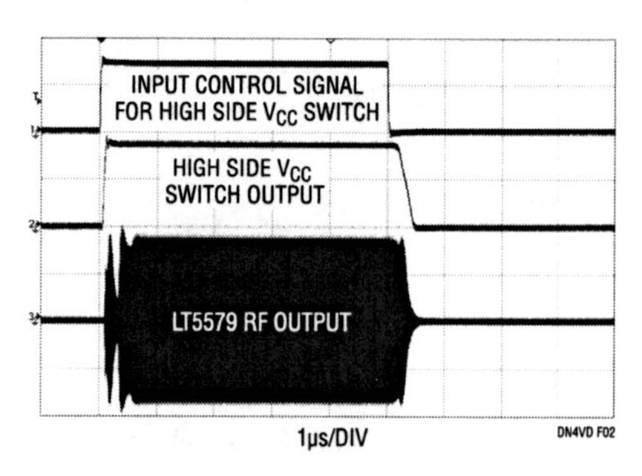

Figure 493.2 • V_{CC} Turn-On and Turn-Off Waveforms

The LT5579 upconverting mixer circuit shown in Figure 493.1 was optimized and tested at an RF output frequency of 2140MHz. The RF output envelope in Figure 493.2 shows a dip about 300ns after the V_{CC} switch turns on, followed by another, smaller dip at about the 500ns point. Both dips represent the mixer's internal feedback circuit reaction to the ramping supply voltage.

LO leakage to the RF output of the LT5579 was measured at −40dBm when V_{CC} is on and −46dBm for V_{CC} off. The LO port of the LT5579 is internally matched and has a return loss of 10dB to 18dB over a frequency range of 1100MHz to 3200MHz.

When the LT5579 mixer is in the off state, the return loss of the LO port is about 3dB to 5dB across the same frequency range of 1100MHz to 3200MHz. It is advisable to use an LO injection VCO with a buffered output for better reverse isolation, and to avoid any VCO pulling while the LO port impedance changes when switching between the on and off states.

Conclusion

LT5579 and LT5578 mixers without an ENABLE pin can be used in TDD applications with external V_{CC} switching. Using only three parts (IRLML6401, 2N7002 and an MC74HC1G04), a high performance high side V_{CC} switch allows turn-on and turn-off in under 1μs.

Precision, matched, baseband filter ICs outperform discrete implementations

494

Philip Karantzalis

Introduction

In digital communication systems, baseband signals must be band-limited in the transmitter or the receiver. Although the bulk of baseband signal shaping and analysis is accomplished using digital signal processing (DSP), analog filtering is used in a number of places along the signal chain. For instance analog filters reduce the imaging of a digital-to-analog converter (DAC), filter out the high frequency noise of an RF demodulator or reduce the aliasing inputs of an analog-to-digital converter (ADC).

Typically, 3G communication systems (CDMA, GSM, UMTS or WiMax) feature a baseband channel bandwidth of 1.25MHz to over 20MHz. In this frequency range, discrete analog filters—those constructed with high speed op amps, resistors and capacitors—are sensitive to PCB layout parasitics, component tolerances and mismatches. The pitfalls of using discrete components can be avoided by using integrated, pin-configurable, precision analog filter ICs, such as the LTC6601-1/-2 and the LTC6605-7/-10/-14. The LTC6601-x is a single 2nd order lowpass filter and the LTC6605-x is a dual, matched filter.

The LTC6601-x lowpass filter

Figure 494.1 shows a block diagram of an LTC6601-x and the 2nd order function it implements. High frequency filters are easily implemented using the LTC6601-x, which integrates a low noise ($1.5nV\sqrt{Hz}$) wideband (600MHz), fully differential amplifier with precision resistors and capacitors. The standard deviation of the on-chip resistors and capacitors is ±0.25% and their matching is ±0.1% (in a differential amplifier, the common mode rejection at high frequencies depends on tight matching of the signal paths). Furthermore, the gain bandwidth (GBW) of the LTC6601-x amplifier is trimmed to ±5%. Note that this level of precision cannot be achieved with discrete analog filters in any practical manufacturing process.

In many LTC6601-x filter implementations, no external components are required. For instance, lowpass filters can be produced by simply hardwiring the input pins for a variety of filter gain and cutoff frequencies from 5MHz to 27MHz.

The product and ratio of the on-chip resistors and capacitors determine the f_0 and Q values of the 2nd order filter function. The f_0 and Q pair sets the filter's cutoff frequency and passband gain peak. The value of the feedback resistor determines the range of the f_0 frequencies. The 400Ω feedback resistors can be shunted by input resistors to increase the f_0 range.

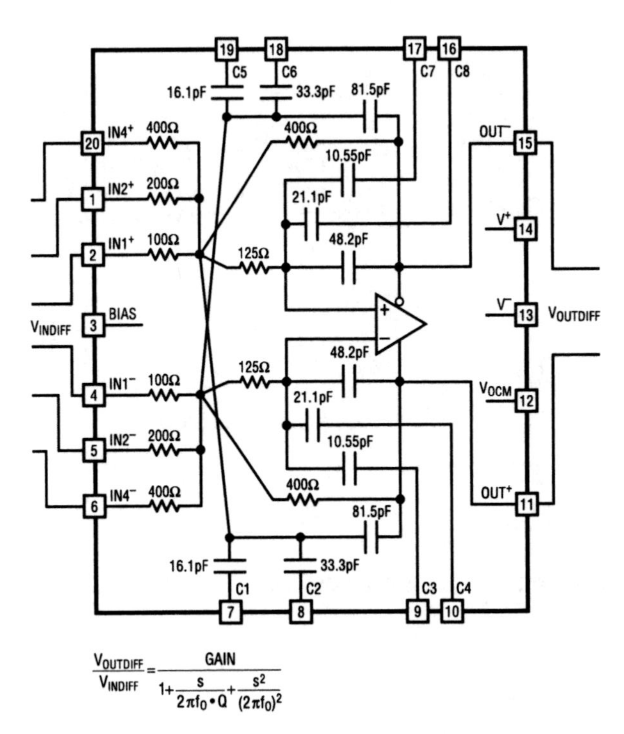

Figure 494.1 • Block Diagram of the LTC6601-x and the 2nd Order Transfer Function It Implements

Analog Circuit Design: Design Note Collection. http://dx.doi.org/10.1016/B978-0-12-800001-4.00494-4

Using an LTC6601-x with input RC or LC filters, 3rd, 4th and 5th order lowpass filters can be implemented (refer to the LTC6601-1 or -2 data sheets for higher order filter options). The LTC6601 family is offered in two options that trade off distortion and noise. Use the LTC6601-1 for low noise and the LTC6601-2 for low distortion at low power.

The LTC6605-x, dual, matched, lowpass filter

Typically, quadrature down conversion is used in a direct conversion or zero-IF receiver. In a quadrature demodulator, the RF signal is split into two paths and mixed with the LO (local oscillator) to produce an I (in phase) and a Q (90° phase or quadrature) signal. In a direct conversion receiver, high image rejection depends on very tight gain and phase matching. High image rejection maximizes the signal-to-noise ratio (SNR) and minimizes the bit error rate.

The LTC6605-x contains two LTC6601 ICs configured and tested as a dual, matched, 2nd order, lowpass filter. There are three LTC6605 versions: the LTC6605-7 with an adjustable $f_{(-3dB)}$ range of 6.5MHz to 10MHz, the LTC6605-10 with an adjustable $f_{(-3dB)}$ range of 9.7MHz to 14MHz and the LTC6605-14 with an adjustable $f_{(-3dB)}$ range of 12.4MHz to 20MHz (the $f_{(-3dB)}$ or $f_{(-1dB)}$ frequency is set by an external resistor).

The maximum gain and phase matching error of an LTC6605-x is ±0.35dB and ±1.2° respectively (equivalent to −32dB of image rejection). This level of gain and phase matching of an LTC6605-x IC is impractical using discrete resistors and capacitors.

The LTC6605-7 or -10 input pins can be hard wired for gains of 1, 4 or 5; the LTC6605-14 for gains of 1, 2 or 3. Either can produce noninteger gains using external input resistors with a slight reduction in gain and phase matching.

Accurate gain and phase matching, when combined with very low noise and distortion, allows for a high dynamic range differential circuit. Figure 494.2 shows an LTC6605-7 driving a dual LTC2205 16-bit ADC and Figure 494.3 shows the FFT plot of the ADC output. The LTC6605-7 is configured for a 12dB gain and a 7MHz, −1dB frequency (suitable application for WiMax).

As with an LTC6601-x, an LTC6605-x can be used with input RC or LC filters, to implement 3rd, 4th and 5th order lowpass filters. The LTC6605-7 is available in a compact 6mm × 3mm, 22-pin leadless DFN package.

The power consumption for an LTC6601-x and for each LTC6605-x 2nd order section, is set by a three-state BIAS pin, allowing a choice between shutdown (I_S=350μA), medium power (I_S=16mA) or full power (I_S=33mA).

Conclusion

The LTC6601-x fully differential, lowpass filter with precision on-chip resistors and capacitors can be configured by hardwiring pins to implement 2nd order filters in a frequency range of 5MHz to 27MHz.

The LTC6601-x is insensitive to PCB parasitics and is a higher performance circuit than a discrete analog filter. The LTC6605-x is a dual, matched, 2nd order lowpass filter for driving dual ADCs in a high performance direct conversion receiver. Using RC or LC filters with an LTC6601-x or LTC6605-x, 3rd, 4th or 5th order lowpass filters can be implemented.

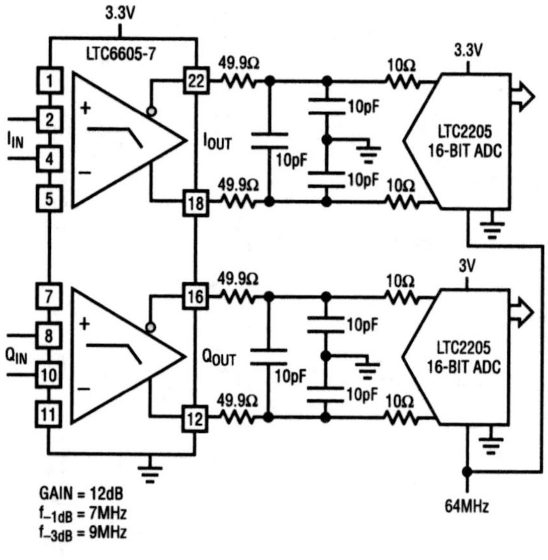

Figure 494.2 • An LTC6605-7 Driving Two LTC2205 ADCs

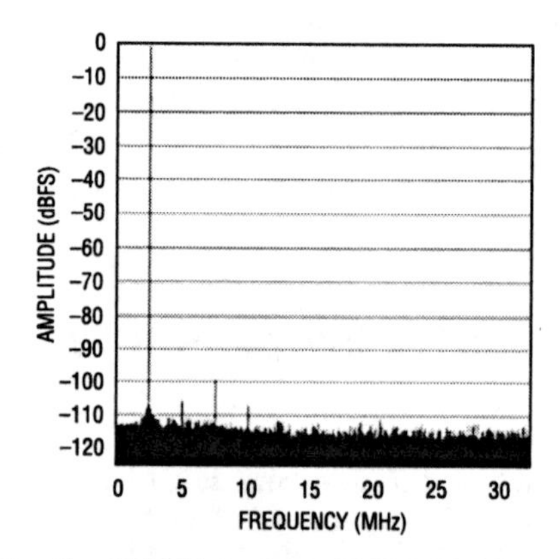

Figure 494.3 • The FFT Plot of the LTC2205's I or Q Output Channels of Figure 494.2 (75dB SNR, −98dB SFDR, 64K-Point FFT, 64Msps, f_{IN}=2.5MHz, −1dBFS)

A complete compact APD bias solution for a 10Gbits/s GPON system

495

Xin Qi

Introduction

Avalanche photo diode (APD) receiver modules are widely used in fiber optic communication systems. An APD module contains the APD and a signal conditioning amplifier, but is not completely self-contained. It still requires significant support circuitry including a high voltage, low noise power supply and a precision current monitor to indicate the signal strength. The challenge is squeezing this support circuitry into applications with limited board space. The LT3482 addresses this challenge by integrating a monolithic DC/DC step-up converter and an accurate current monitor. The LT3482 can support up to a 90V APD bias voltage, and the current monitor provides better than 10% accuracy over four decades of dynamic range (250nA to 2.5mA).

Recent communication design efforts increasingly focus on the 10Gbits/s GPON system, which demands that the transient response of the APD current monitor is less than 100ns for an input current step of two decades of magnitude. A simple compact circuit using the LT3482 is fast enough to meet this challenging requirement.

An APD bias topology with fast current monitor transient response

The circuit in Figure 495.1 shows the LT3482 configured to produce an output voltage ranging from 20V to 45V from a 5V source—capable of delivering up to 2mA of load current. Its operation is straightforward. The LT3482 contains a 48V, 260mA internal switch, which boosts V_{OUT1} to one-half the APD output voltage level. This voltage is doubled through an internal charge pump to generate V_{OUT2}. All boost and charge pump diodes are integrated. V_{OUT2} is regulated by the internal voltage reference and the resistor divider made up of R3 and R4. At this point, V_{OUT2} goes through the integrated high side current monitor (MONIN), which produces a current proportional to the APD current at the MON pin.

The output voltage is available for the APD at the APD pin. The CTRL pin serves to override the internal reference. By tying this pin above 1.25V, the output voltage is regulated with the feedback at 1.25V. By externally setting the CTRL pin to a lower voltage, the feedback and the output voltage follow accordingly.

The $\overline{\text{SHDN}}$ pin not only enables the converter when 1.5V or higher is applied, but also provides a soft-start function to control the slew rate of the switch current, thereby minimizing inrush current. The switching frequency can be set to 650kHz or 1.1MHz by tying the FSET pin to ground or to V_{IN}, respectively. Fixed frequency operation allows for an output ripple that is predictable and easier to filter.

To achieve fast transient response, any time-delay component along the signal path should be minimized. Figure 495.1 shows an APD bias topology with fast current monitor transient response. Unlike the ultralow noise topology with a filter capacitor at the APD pin, the filter capacitor is moved to the MONIN pin of the LT3482. The output sourcing current from the MON pin is directly fed into a transimpedance amplifier.

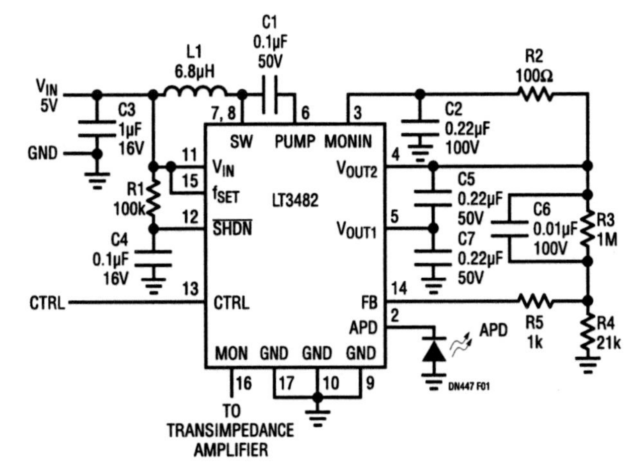

Figure 495.1 • Fast Current Monitor Transient Response ADP Bias Topology

Analog Circuit Design: Design Note Collection. http://dx.doi.org/10.1016/B978-0-12-800001-4.00495-6

A typical measured current monitor transient response consists of the signal generation delay at the APD pin, the built-in current monitor response time and the measurement delay at the MON pin. Thus, every effort should be made to reduce signal generation and the measurement delays.

Figure 495.2 shows the measurement setup. An NPN transistor in common base configuration is used to generate the fast current step representing the APD load. A function generator provides two negative bias voltages at the PWM node that result in two decades current step at the APD pin. At the MON pin, a wideband transimpedance amplifier is implemented using the LT1815. Operating in a shunt configuration, the amplifier buffers the MON output current and dramatically reduces the effective output impedance at the OUT node. Note that there is an inversion and a DC offset present when this measurement technique is used. A regular oscilloscope probe can then be used to capture the fast transient response at the OUT node.

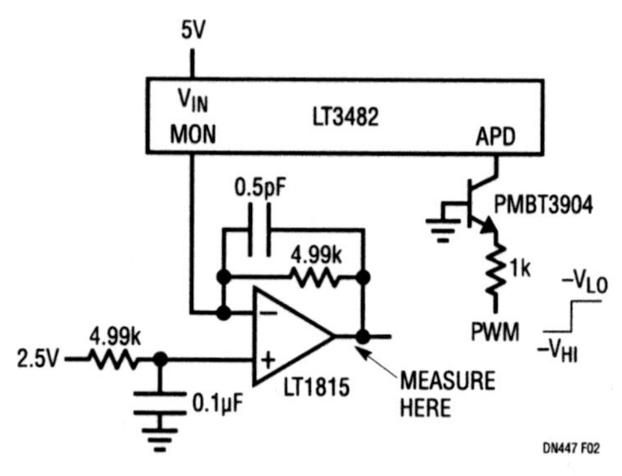

Figure 495.2 • Fast Transient Response Measurement Setup

Figures 495.3 and 495.4 show the measured input signal rising transient response and the measured input signal falling transient response, respectively, where the input current levels are 10µA and 1mA. The PWM input signal levels are selected based on the static measurement results. The APD current is accurately mirrored by the LT3482 with an attenuation of five and sourced from the MON pin. With a 2.5V reference voltage, the OUT node voltage swings between 1.5V (=2.5V – 1mA/5 · 4.99k) and 2.49V(=2.5V – 10µA/5 · 4.99k) responding to the input signal step. The measurements demonstrate less than 50ns transient response time, which exceeds the stringent speed demand of the 10Gbits/s GPON system.

Conclusion

The LT3482 is a complete space-saving solution to APD receiver module support circuitry design. It offers more than just low bias noise and compact solution size; it also features UltraFast current monitor transient speed that addresses the challenges presented in the 10Gbits/s GPON system.

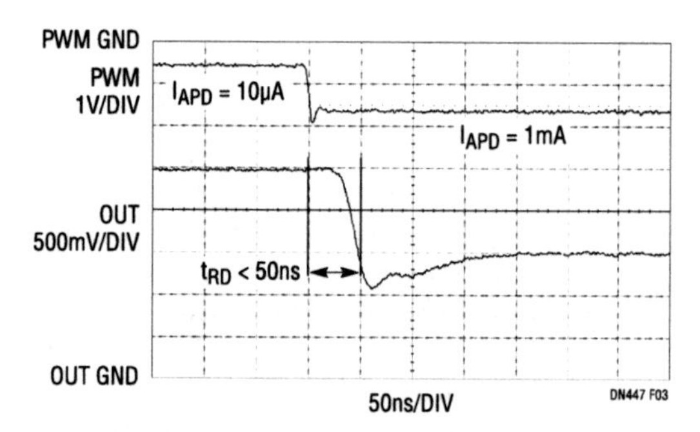

Figure 495.3 • Transient Response on Input Signal Rising Edge (10µA to 1mA)

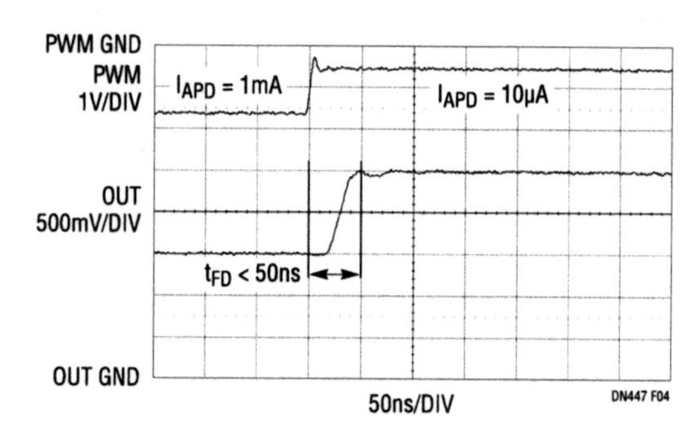

Figure 495.4 • Transient Response on Input Signal Falling Edge (1mA to 10µA)

Signal chain noise analysis for RF-to-digital receivers

496

Cheng-Wei Pei

Introduction

Designers of signal receiver systems often need to perform cascaded chain analysis of system performance from the antenna all the way to the ADC. Noise is a critical parameter in the chain analysis because it limits the overall sensitivity of the receiver. An application's noise requirement has a significant influence on the system topology, since the choice of topology strives to optimize the overall signal-to-noise ratio, dynamic range and several other parameters. One problem in noise calculations is translating between the various units used by the components in the chain: namely the RF, IF/baseband, and digital (ADC) sections of the circuit.

Figure 496.1 shows a simplified system diagram. There is an RF section, an IF/baseband section (represented by an amplifier) and an ADC. The RF section, which includes a mixer or demodulator, is commonly specified using noise figure (NF) in a decibel scale (dB). This may also be specified with a noise power spectral density which is similar to NF in concept (e.g., −160dBm/Hz is equal to an NF of approximately 14dB), so here we use NF. When working in a fixed-impedance (50Ω) environment, using NF simplifies the analysis of an RF signal chain. However, if the assumptions of constant impedance and proper source/load termination are not valid, then NF calculations become less straightforward.

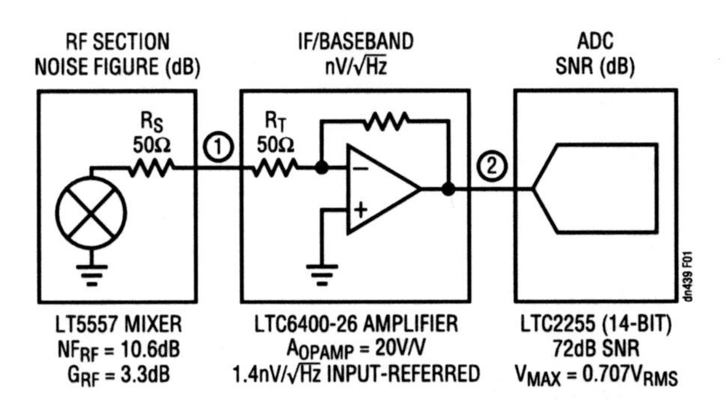

Figure 496.1 • Block Diagram of a Simplified Signal Chain with RF Components (Mixer, LNA, etc.), IF/Baseband Components (Represented by a Simple Amplifier) and an ADC. The Input Resistor of the Amplifier Serves as a Matched Termination for the 50Ω RF Section. A Suggested Product, and Its Specification for Each Section, Are Included

IF/baseband components, such as amplifiers, are typically specified with noise spectral density, which is commonly measured in volts and amps per square-root hertz ($nV/\sqrt{Hz}$ and $pA/\sqrt{Hz}$). The contribution of current noise ($pA/\sqrt{Hz}$) is usually negligible in low impedance environments. ADC noise is primarily specified as a signal-to-noise ratio (SNR) in decibels. SNR is the ratio of the maximum input signal to the total integrated input noise of the ADC. In order to perform a full signal chain analysis, a designer needs to be able to translate between NF, noise density and SNR.

NF to SNR: how much ADC resolution?

The first transition is from the RF section to the IF/baseband section. NF is a convenient unit, but requires constant system impedance. Since noise spectral density is independent of impedance, converting from NF to $nV/\sqrt{Hz}$ makes sense, since in the transition from RF to baseband (node 1 in Figure 496.1), the chain is leaving the fixed 50Ω environment. At node 1, the noise voltage density due to the RF part of the chain can be represented as:

$$e_{N(RF)} = 10^{\left[\frac{(G_{RF}+NF_{RF})}{20}\right]} \cdot e_{N(50)} \cdot 0.5 \left(\frac{nV}{\sqrt{Hz}}\right)$$

where

G_{RF} = cascaded gain of RF component(s) in dB
NF_{RF} = cascaded NF of RF component(s) in dB
$e_{N(50)}$ = noise density of 50Ω ($0.91nV_{RMS}/\sqrt{Hz}$ at 27°C)
0.5 = resistive divider from load termination, equal to 0.5 if R_T and R_S are 50Ω.

Analog Circuit Design: Design Note Collection. http://dx.doi.org/10.1016/B978-0-12-800001-4.00496-8

With the LT5557 shown in Figure 496.1, $e_{N(RF)}$ comes out to $2.25nV/\sqrt{Hz}$. The input-referred voltage noise density of the IF/baseband section (including op amp resistors) can be computed using the op amp data sheet and summed with the contribution from the RF portion (using sum-of-squares addition since the specified values are RMS). Multiplying the result by the amplifier gain (V/V) gives the total noise density at node 2, ignoring the ADC's effective contribution:

$$e_{N2} = A_{OPAMP} \cdot \sqrt{e_{N(OPAMP)}{}^2 + e_{N(RF)}{}^2} \left(\frac{nV}{\sqrt{Hz}}\right)$$

Using the LTC6400-26 amplifier's specifications, e_{N2} comes out to $53nV/\sqrt{Hz}$. The final step is to compute the overall SNR at the ADC. To do so, one must know the total integrated noise at node 2. Assuming the noise spectral density is constant with frequency, one can simply multiply e_{N2} by the square root of the total noise bandwidth. This bandwidth is limited by the amplifier circuit and any ADC anti-alias filtering. Assuming a total bandwidth of 50MHz, the integrated noise in our example is $N2 = 375\mu V_{RMS}$. The total theoretical SNR can be calculated as:

$$SNR_{THEORETICAL} = 20 \cdot \log_{10}\left(\frac{V_{MAX}}{N2}\right) (dB)$$

where

V_{MAX} = maximum sine wave input to the ADC in V_{RMS} ($V_{P-P} \cdot 0.35$)

N2 = total integrated noise at node 2, excluding the ADC, in V_{RMS}

This theoretical SNR, which is 65.5dB in the example, represents the maximum resolution attainable with a perfect ADC. The actual ADC should have an SNR at least 5dB above this number to maintain the performance level down the chain. For example, a practical high performance 14-bit ADC, like Linear Technology's LTC2255 family (or LTC2285 family of dual ADCs), would have an SNR in the 72dB to 74dB range.

SNR to NF

For radio designers, an important consideration in system design is total noise figure, which is affected by all components in the chain. Once the components are selected, one can determine the equivalent input noise figure and the overall sensitivity of the receiver. Assuming that the signal(s) of interest lie within one Nyquist bandwidth of the ADC (a Nyquist bandwidth is $f_{SAMPLE}/2$), the equivalent noise of the ADC is:

$$e_{N(ADC)} = 10^9 \cdot \frac{V_{MAX}}{10^{\left(\frac{SNR_{ADC}}{20}\right)}} \cdot \frac{1}{\sqrt{\frac{f_{SAMPLE}}{2}}} \left(\frac{nV}{\sqrt{Hz}}\right)$$

where

SNR_{ADC} = data sheet SNR at the frequency of interest in dB

f_{SAMPLE} = sample rate of the ADC in hertz

In our example, $e_{N(ADC)}$ comes out to $22.5nV/\sqrt{Hz}$, assuming a 125MHz sample rate. This voltage noise density, $e_{N(ADC)}$, can then be RMS-summed with the amplifier output noise density, e_{N2}, and the result input-referred by dividing by the gain, A_{OPAMP}. To convert back to NF, rearrange the first equation in this article:

$$NF_{TOTAL} = \left\{20\log_{10}\left(\frac{\sqrt{e_{N(ADC)}{}^2 + e_{N2}{}^2}}{A_{OPAMP} \cdot e_{N(50)} \cdot 0.5}\right) - G_{RF}\right\} (dB)$$

This quantity, NF_{TOTAL}, gives the overall input noise figure with the contributions of the RF section, the amplifier and the ADC. In the example, NF_{TOTAL} is 12.7dB for the entire chain of three devices.

Conclusion

When working with an entire system design from RF components to ADC, noise specifications do not always use the same units from component to component. This article addresses the translation between the various nomenclatures. Radio designers can use this information to design their system topology and select components for optimal sensitivity.

497

Programmable baseband filter for software-defined UHF RFID readers

Philip Karantzalis

Introduction

Radio frequency identification (RFID) is an auto-ID technology that identifies any object that contains a coded tag. A UHF RFID system consists of a reader (or interrogator) that transmits information to a tag by modulating an RF signal in the 860MHz–960MHz frequency range. Typically, the tag is passive—it receives all of its operating energy from a reader that transmits a continuous-wave (CW) RF signal. A tag responds by modulating the reflection coefficient of its antenna, thereby backscattering an information signal to the reader.

Tag signal detection requires measuring the time interval between signal transitions (a data "1" symbol has a longer interval than a data "0" symbol). The reader initiates a tag inventory by sending a signal that instructs a tag to set its backscatter data rate and encoding. RFID readers can operate in a noisy RF environment where many readers are in close proximity. The three operating modes, single-interrogator, multiple interrogator and dense-interrogator, define the spectral limits of reader and tag signals. Software programmability of the receiver provides an optimum balance of reliable multitag detection and high data throughput. The programmable reader contains a high linearity direct conversion I and Q demodulator, low noise amplifiers, a dual baseband filter with variable gain and bandwidth and a dual analog-to-digital converter (ADC). The LTC6602 dual, matched, programmable bandpass filter can optimize high performance RFID readers.

The LTC6602 dual bandpass filter

The LTC6602 features two identical filter channels with matched gain control and frequency-controlled lowpass, and highpass networks. The phase shift through each channel is matched to ±1 degree. A clock frequency, either internal or external, positions the pass band of the filter at the required frequency spectrum.

The lowpass and highpass corner frequencies, as well as, the filter bandwidth are set by division ratios of the clock frequency. The lowpass division ratio options are 100, 300 and 600 and the highpass division ratios are, 1000, 2000, 6000. Figure 497.1 shows a typical filter response with a 90MHz internal clock and the division ratios set to 6000 and 600 for the highpass and lowpass, respectively. A sharp 4th order elliptical stopband response helps eliminate out-of-band noise. Controlling the baseband bandwidth permits software definition of the operating mode of the RFID receiver as it adapts to the operating environment.

An adaptable baseband filter for an RFID reader

Figure 497.2 shows a simple LTC6602-based filter circuit that uses SPI serial control to vary the filter's gain and bandwidth to adapt to a complex set of data rates and encoding. (The backscatter link frequency range is 40kHz to 640kHz and the data rate range is 5kbps to 640kbps.)

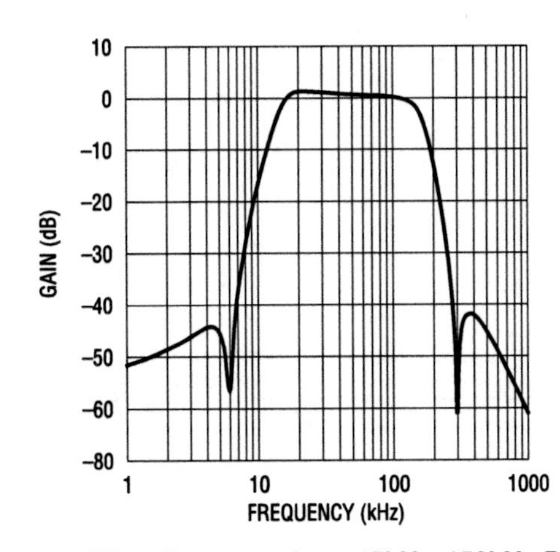

Figure 497.1 • Filter Response for a 15kHz–150kHz Passband

Analog Circuit Design: Design Note Collection. http://dx.doi.org/10.1016/B978-0-12-800001-4.00497-X

For fine resolution positioning of the filter, the internal clock frequency is set by an 8-bit LTC2630 DAC. A 0V to 3V DAC output range positions the clock frequency between 40MHz and 100MHz (234.4kHz per bit). The lowpass and highpass division ratios are set by serial SPI control of the LTC6602. The cutoff range for the highpass filter is 6.7kHz to 100kHz and 66.7kHz to 1MHz for the lowpass filter. The optimum filter bandwidth setting can be adjusted by a software algorithm and is a function of the data clock, data rate and encoding. The filter bandwidth must be sufficiently narrow to maximize the dynamic range of the ADC input and wide enough to preserve signal transitions and pulse widths (the proper filter setting ensures reliable DSP tag signal detection).

Figure 497.3 shows an example of the filter's time domain response to a typical tag symbol sequence (a "short" pulse interval followed by a "long" pulse interval). The lowpass cutoff frequency is set equal to the reciprocal of the shortest interval ($f_{CUTOFF} = 1/10\mu s = 100kHz$). If the lowpass cutoff frequency is lower, the signal transition and time interval will be distorted beyond recognition. The setting of the highpass cutoff frequency is more qualitative than specific. The highpass cutoff frequency must be lower than the reciprocal of the longest interval (for the example shown, highpass $f_{CUTOFF} < 1/20\mu s$) and as high as possible to decrease the receiver's low frequency noise (of the baseband amplifier and the down-converted phase and amplitude noise). The lower half of Figure 497.3 shows the filter's overall response (lowpass plus highpass filter). Comparing the filter outputs with a 10kHz and a 30kHz highpass setting, the signal transitions and time intervals of the 10kHz output are adequate for detecting the symbol sequence (in an RFID environment, noise will be superimposed on the output signal). In general, increasing the lowpass f_{CUTOFF} and/or decreasing the highpass f_{CUTOFF} "enhances" signal transitions and intervals at the expense of increased filter output noise.

Conclusion

The LTC6602 dual bandpass filter is a programmable baseband filter for high performance UHF RFID readers. Using the LTC6602 under software control provides the ability to operate at high data rates with a single interrogator or with optimum tag signal detection in a multiple or dense interrogator physical setting. The LTC6602 is a very compact IC in a 4mm × 4mm QFN package and is programmable with parallel or serial control.

References

1. Dobkin, Daniel M., "The RF in RFID," 9/07, Elsevier Inc.
2. Class-1 Generation-2 UHF RFID Protocol for Communications at 860MHz to 960MHz, Version 1.1.0, www.epcglobalinc.org/standards/specs/.

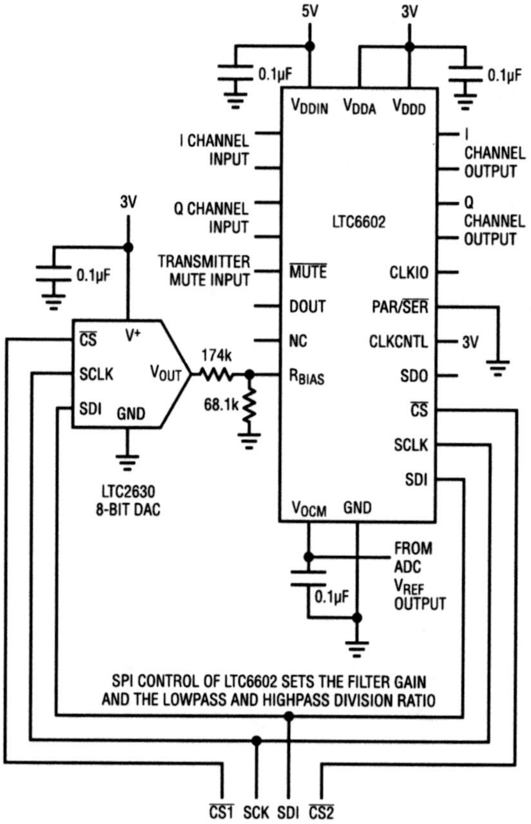

Figure 497.2 • An Adaptable RFID Baseband Filter with SPI Control

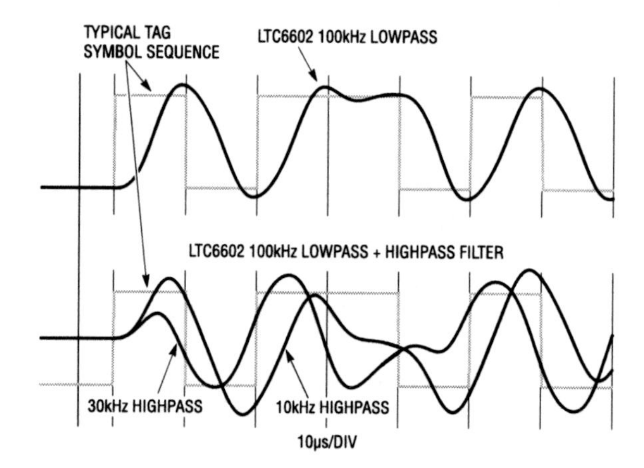

Figure 497.3 • Filter Transient Response to a Tag Symbol Sequence

High linearity components simplify direct conversion receiver designs

498

Cheng-Wei Pei

Introduction

A direct conversion radio receiver takes a high frequency input signal, often in the 800MHz to 3GHz frequency range, and utilizes one mixer/demodulator stage to convert the signal to baseband without going through an intermediate frequency (IF) stage. The resulting low frequency (baseband) signal spectrum has useful information at frequencies from DC to typically a few tens of MHz. Designing these receivers requires the use of very high performance analog ICs. High performance direct conversion radio receiver signal chains for applications such as cellular infrastructure and RFID readers require high linearity, low noise figure (NF), and good matching between the in-phase and quadrature (I and Q) channels.

The right components for the job

Linear Technology's LT5575 direct conversion demodulator has a combination of excellent linearity and noise performance. The most important linearity specification for direct conversion mixers is the 2nd order intercept point (IIP2) due to the 2nd order distortion product falling within the baseband output spectrum, and the LT5575 boasts 54.1dBm at 900MHz (60dBm at 1900MHz). The LT5575 also has high 3rd order linearity and a low noise figure of 12.8dB.

The LTC6406 is a fully differential amplifier with low noise ($1.6nV/\sqrt{Hz}$ at the input) and high linearity (+44dBm OIP3 at 20MHz) in a small 3mm × 3mm QFN package. External resistors set the gain, giving the user maximum

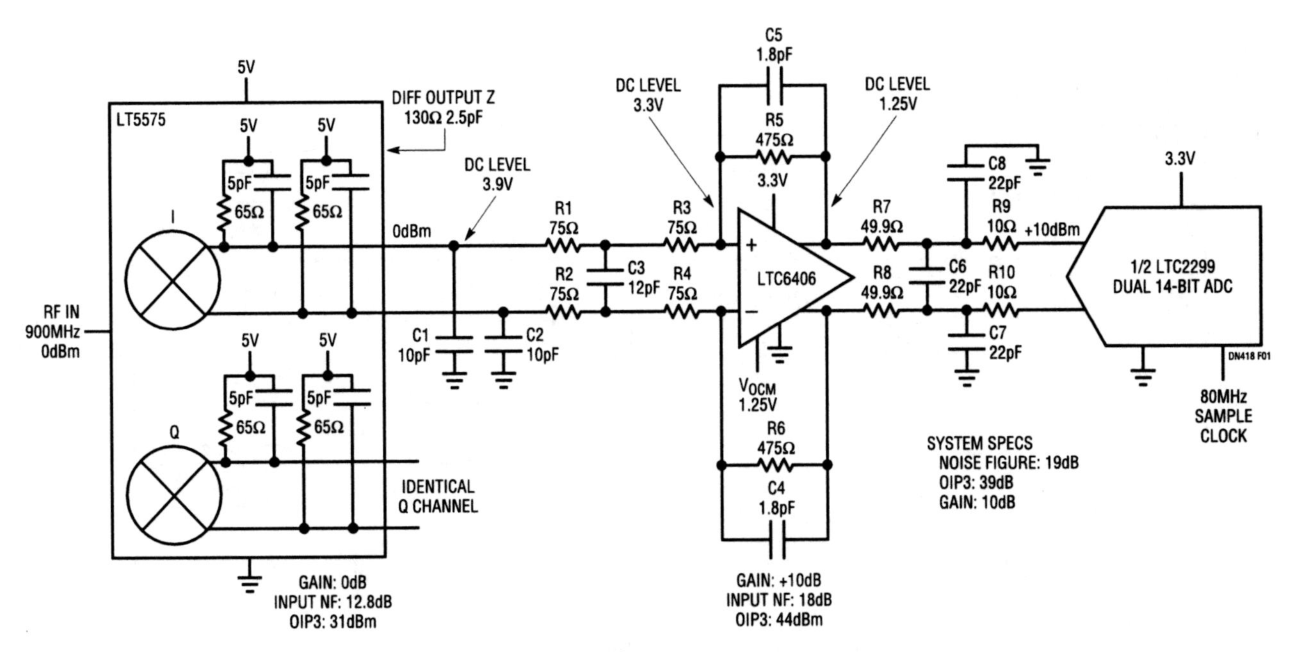

Figure 498.1 • The LT5575 Demodulator and LTC6406 Amplifier Driving an LTC2299 14-Bit ADC. System Bandwidth Is Approximately 40MHz. Overall System OIP3 Was Measured to Be 39dBm

Analog Circuit Design: Design Note Collection. http://dx.doi.org/10.1016/B978-0-12-800001-4.00498-1

design flexibility. The low power consumption (59mW with a 3.3V supply) means that using two amplifiers for I and Q has minimal effect on the system power budget. The LTC6406 maintains high linearity up to 50MHz, which is perfect for WCDMA receivers and other wideband applications.

A basic receiver design

One common design challenge when using active demodulators is level shifting the outputs, which can have a DC level close to V_{CC}, to a usable DC level within the input range of the analog-to-digital converter (ADC). Fortunately, the LTC6406's rail-to-rail inputs make interfacing with the outputs of the LT5575 simple and direct. The LTC6406 also includes an extra feedback loop (controlled by an external V_{OCM} voltage) that independently sets the output common mode DC level, regardless of the input voltage.

Figure 498.1 shows a basic receiver circuit with the LT5575 demodulator and the LTC6406 followed by an LTC2299 14-bit ADC. An RC lowpass filter at the output of the demodulator filters undesired out-of-band signals, and another RC lowpass filter before the ADC anti-aliases and limits noise bandwidth. The DC voltage at the LTC6406 inputs is 3.3V, the same as the supply voltage.

Adding free gain to the system

For signal chains that require more gain, the LTC6401-8 differential amplifier/ADC driver is a good complement to the LT5575 and LTC6406. The LTC6401-8 has higher linearity (50dBm OIP3 at 20MHz) and $2.5nV/\sqrt{Hz}$ of input noise in a 3mm x 3mm QFN package. It contributes gain and linearity without significantly impacting the noise figure. Figure 498.2 adds the LTC6401-8 (also available in 14dB, 20dB and 26dB flavors) to the signal chain to drive the LTC2299. The higher linearity of the LTC6401-8 increases the combined system OIP3 to 45dBm. In addition, 8dB of gain is added with no significant degradation to the noise figure. The 400Ω input impedance of the LTC6401-8 is not a heavy load for the LTC6406, which enables direct coupling of the two amplifiers with minimal signal loss (from series resistors, etc.).

A more selective filter

There are three places where a filter can be implemented in the circuit of Figure 498.2: after the mixer, in between the two amplifier stages, and prior to the ADC. Each has its trade-offs, but the simplest design places the filter after the mixer. This topology attenuates unwanted signals earlier in the signal chain, which preserves the IP3 of the following stages and allows for more gain through the system. An LC filter at the demodulator output minimally affects the distortion and noise figure of the system, whereas LC lowpass filters can present a heavy load impedance to a feedback amplifier output near their resonant frequencies. For reasons outside the scope of this article, it is tricky to design LC networks at the input of a high speed sampling ADC.

A concern when designing the LC network is the need to preserve the I and Q gain/phase matching of the LT5575 (0.04dB/0.4° mismatch), which necessitates using low tolerance LC components (±2% inductors and ±5% capacitors). The frequency response and group delay of the system are almost entirely determined by the LC filter.

Conclusion

Signal chain devices that offer high linearity and excellent noise specifications can greatly simplify the design of high frequency receivers—speeding up entire design cycles.

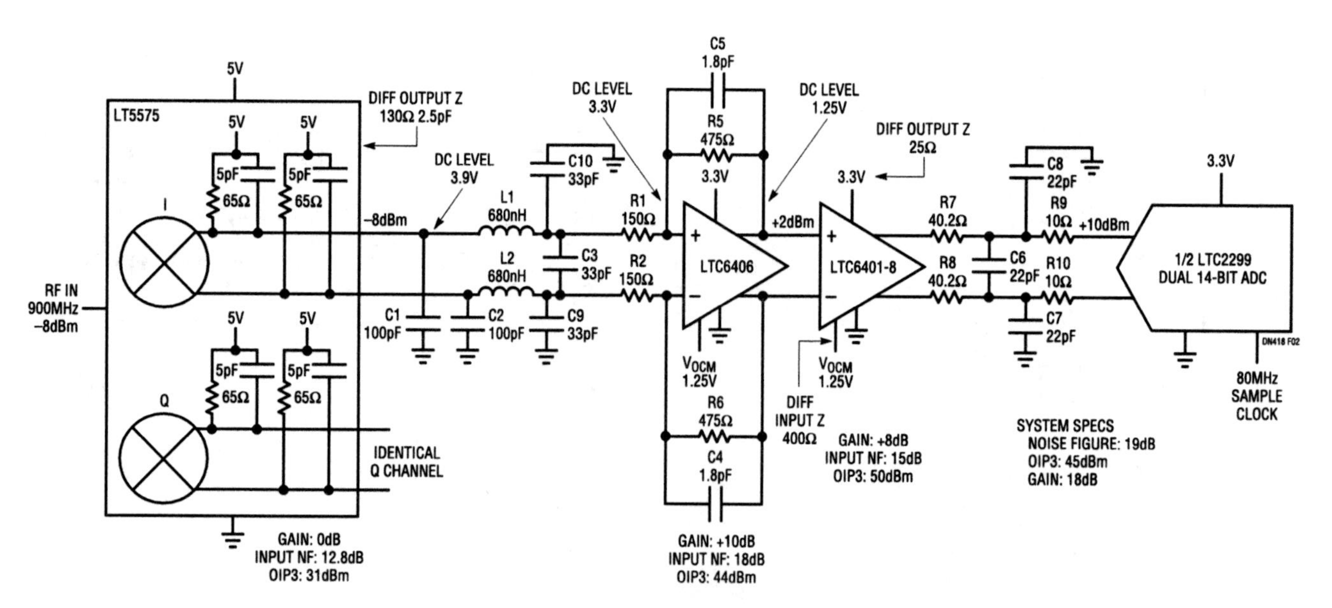

Figure 498.2 • The LT5575 Demodulator with a 20MHz Lowpass Filter Followed by LTC6406 and LTC6401-8. System OIP3 Is Measured to Be 45dBm at 920MHz RF with a 900MHz Local Oscillator. System Noise Figure (NF) Adds Up to Approximately 19dB

Baseband circuits for an RFID receiver

499

Philip Karantzalis

Introduction

Radio frequency identification (RFID) technology uses radiated and reflected RF power to identify and track a variety of objects. A typical RFID system consists of a reader and a transponder (or tag). An RFID reader contains an RF transmitter, one or more antennas and an RF receiver. An RFID tag is simply an uniquely identified IC with an antenna.

Communication between a reader and a tag is via backscatter reflection, similar to a radar system, in the UHF frequencies from 860MHz to 960MHz. This design note describes a high performance RFID receiver.

A direct conversion receiver

Figure 499.1 shows the block diagram of a direct conversion RF receiver—the receiver demodulates an RF carrier directly into a baseband signal without an intermediate frequency down-conversion (a zero IF receiver).

The antenna, shared by both the transmitter and receiver, detects an RF carrier and passes it through a bandpass filter to an LT5516 demodulator's RF input.

The LT5516 direct conversion demodulator frequency range of 800MHz to 1.5GHz includes the UHF range used by RFID readers (860MHz to 960MHz). The excellent linearity of the LT5516 provides for high sensitivity to low level signals, even in the presence of large interfering signals.

The LT6231 low noise dual op amp acts as a differential to single-ended amplifier to drive the single-ended input of the lowpass filter.

Analog baseband filtering is performed by the LT1568, a low noise, precision RC filter building block. The LT1568 filter provides a simple solution for designing lowpass and bandpass filters with cutoff frequencies from 100kHz to 10MHz. These cutoff frequencies are sufficient for the 250kHz to 4MHz signal spectrum typically used in UHF RFID systems.

The differential output of an LT1568 drives the inputs of an LTC2298 ADC. The LTC2298 is a 65Msps, low power (400mW), dual 14-bit analog to digital converter with 74dB

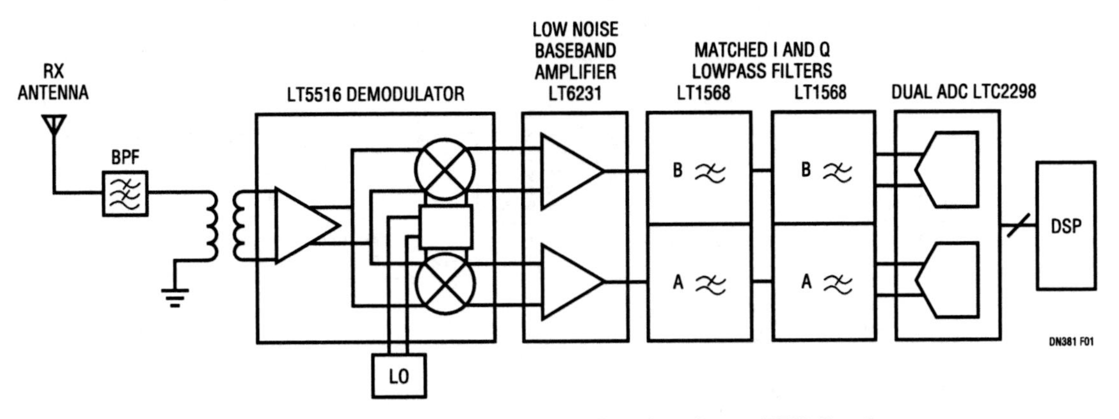

Figure 499.1 • A Direct Conversion Receiver for an RFID Reader

Analog Circuit Design: Design Note Collection. http://dx.doi.org/10.1016/B978-0-12-800001-4.00499-3

signal-to-noise ratio (SNR). The digital signal processor (DSP) that follows the ADC analyzes the received signal from multiple tags and provides additional filtering.

A low noise differential to single-ended amplifier

Figure 499.2 shows an LT6231 difference amplifier used to convert the LT5516 differential I or Q output to a single-ended output. The addition of external 270pF capacitors across the 60Ω resistors limits the demodulator's output to 10MHz to prevent any high frequency interference from reaching the LT6231 amplifier.

AC coupling to the baseband amplifier is used because DC coupling is not necessary for the amplitude shift keying (ASK) RFID signal.

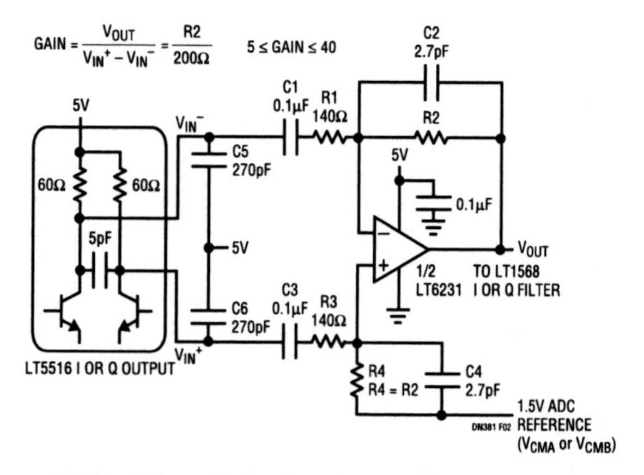

Figure 499.2 • A Low Noise I or Q Baseband Interface

The highpass pole provided by the AC coupling capacitors and the amplifier input resistors is set to 8kHz. The differential amplifier's input resistors are set to 140Ω in order to minimize the input referred noise. The noise floor at the amplifier's output is $4.3\text{nV}/\sqrt{\text{Hz}}$ times the amplifier gain (gain ≥ 5). The 1.5V reference provided by the LTC2298 ADC is used to level shift the amplifier's output to the mid-supply point of the following 3V filter and ADC circuits.

A matched I and Q filter and a dual ADC

Figure 499.3 shows two LT1568 filter building blocks connected as dual, matched, 4th order filters. The LT1568 filter's single-ended input to differential output conversion gain is 6dB. The LT1568 circuit implements an elliptic lowpass filter function with equal resistor values (see Figure 499.3). Stopband attenuation at $2(f_{-3dB})$ is 34dB. I and Q filter matching is assured by the inherent matching of the LT1568s' A and B sides.

The input voltage range of an LTC2298 is adjustable to $2V_{p-p}$ or $1V_{p-p}$.

Conclusion

Using only five ICs (LT5516, LT6231, two LT1568 and an LTC2298), a high performance UHF RFID receiver can be designed with the flexibility to adapt and be optimized to meet the requirements of the present and emerging RFID standards.

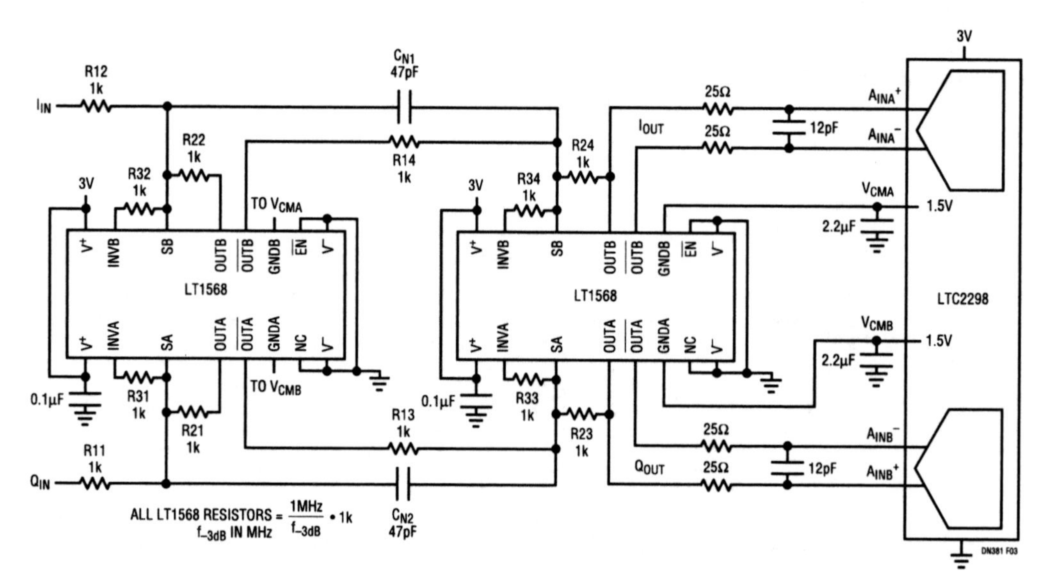

Figure 499.3 • A Matched, 1MHz, 4th Order, I and Q Lowpass Filter and ADC Driver

WCDMA ACPR and AltCPR measurements

500

Doug Stuetzle

Introduction

ACPR (adjacent channel power ratio) and AltCPR (alternate channel power ratio) are important measures of spectral regrowth for digital communication systems that use, for example, WCDMA (wideband code division multiple access) modulation. Both ACPR and AltCPR quantify the ratio of regrowth in a nearby channel to the power in the transmitted channel (Figure 500.1).

To measure ACPR and AltCPR, refer to the test setup shown in Figure 500.2. The DUT (device under test) is the LT5528, which is a high linearity direct I/Q modulator. It accepts complex modulation signals at its baseband inputs and generates a modulated RF signal at the RF output. An accurate measurement of the spectral regrowth of a highly linear device such as the LT5528 is difficult because its dynamic range may rival that of the measurement equipment. Because of this, it is important to account for the noise of the measurement system; i.e., the spectrum analyzer. Refer to Figure 500.3. Some spectrum analyzers offer an ACPR measurement utility. This utility will not, however, give accurate results for highly linear devices, as it does not compensate for the system noise floor.

The spectrum analyzer must have a wide dynamic range. That means a high input 3rd order intercept point, and a low noise floor. The analyzer shown in Figure 500.2 meets both of these requirements.

Note that a free running RF generator provides the LO signal. This type of generator is used because of its superior noise performance. This is critical, as a noisy LO signal may corrupt the ACPR measurement. Its output operating frequency can drift slightly, so manual frequency correction is required.

Also, the baseband source can generate spectral regrowth and noise which may swamp the performance of the DUT. The lowpass filters shown at the baseband generator outputs reduce these impairments to a tolerable level. Filters suggested for this purpose are made by TTE Engineering and

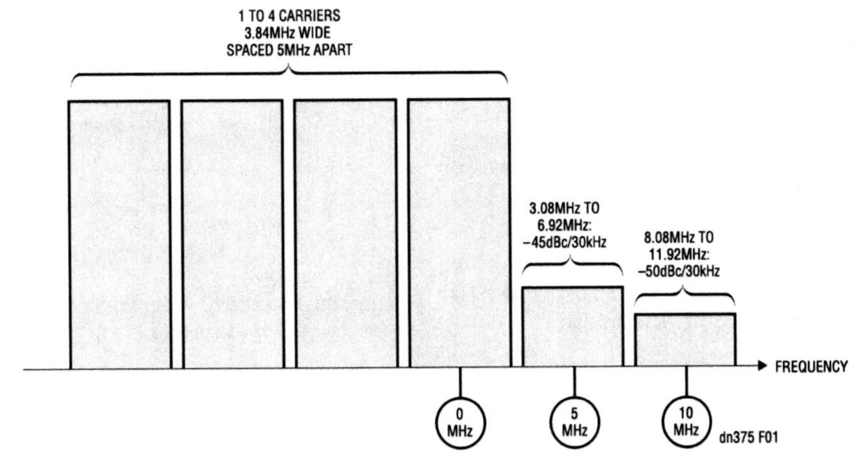

Figure 500.1 • WCDMA ACPR Limits, Per 3GPP TS 25.104, Section 6.6.2.2.1

Analog Circuit Design: Design Note Collection. http://dx.doi.org/10.1016/B978-0-12-800001-4.00500-7

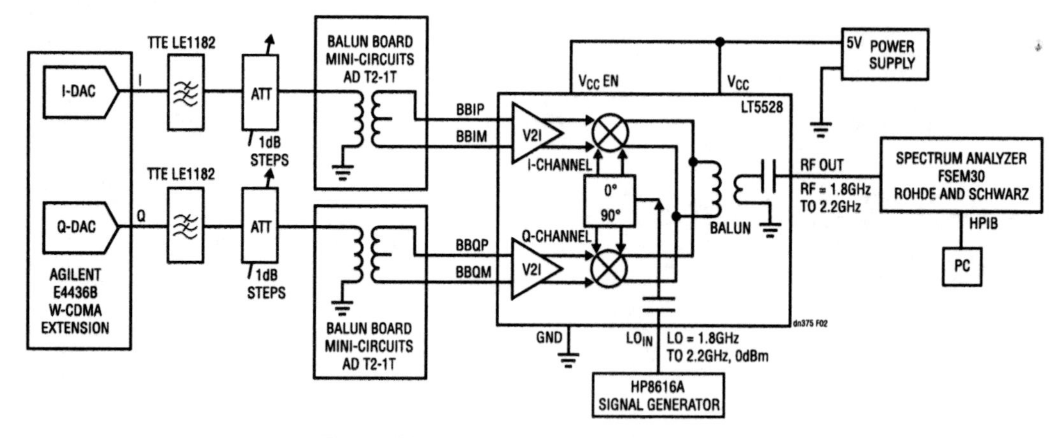

Figure 500.2 • ACPR Measurement Setup

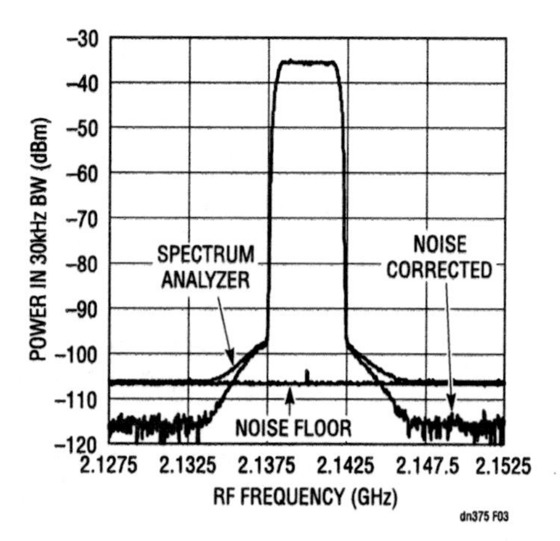

Figure 500.3 • ACPR Spectrum for a Single Carrier WCDMA Signal

offer >20dB rejection at 10.4MHz and >80dB rejection at 13.08MHz.

To start, measure the noise floor of the spectrum analyzer with a 50Ω input termination. The input attenuation of the analyzer is set to minimize the noise figure of the measurement system. A 30kHz resolution bandwidth is used because the spectrum analyzer shown has the lowest noise figure (about 24dB) at that resolution bandwidth. The spectrum analyzer shown includes an RMS display detector mode, which is specifically designed to measure noise-like signals. For spectrum analyzers that do not offer this mode, it is important to set the video bandwidth to at least 3 times the resolution bandwidth; in this case 100kHz. If the ratio of video to resolution bandwidth is too low, the power measurement will be inaccurate. Video averaging helps smooth the result; 100 averages gives good results. Once the settings are correct, use the channel power utility of the analyzer to find the total noise power within a 3.84MHz bandwidth.

Next measure the output spectrum of the DUT using the same settings. For ACPR/AltCPR, center the measurement band 5MHz/10MHz above the center of the highest carrier. To find the true spectral regrowth power, convert the measured spectral power levels to mW and subtract the spectrum analyzer noise floor from the measured DUT power. Reconvert to dBm to get the true spectral regrowth.

The ACPR/AltCPR is equal to the difference in dB between the signal power per carrier and the spectral regrowth.

ACPR and AltCPR vary with output signal level. Figure 500.4 shows the ACPR and AltCPR versus RF output level for a 4-carrier WCDMA signal centered at 2.14GHz. For low RF power levels, these are limited by the output noise floor of the DUT. At high RF output power levels, they are determined by the linearity of the DUT. The maximum ACPR/AltCPR are observed between these extremes, where the spectral regrowth equals the noise floor of the DUT.

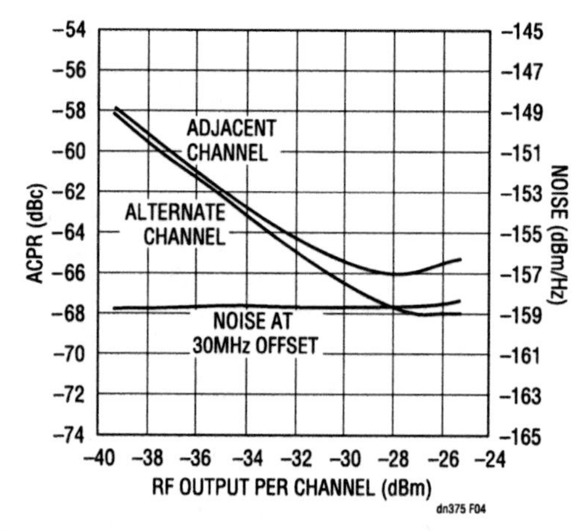

Figure 500.4 • LT5528 4-Channel WCDMA Adjacent and Alternate CPR Measurement vs Channel Power

Low distortion, low noise differential amplifier drives high speed ADCs in demanding communications transceivers

501

Cheng-Wei Pei

Introduction

Today's communications transceivers operate at much higher frequencies and wider bandwidths than those of the past. Combining this with higher resolution requirements, transceiver design can become daunting. For engineers designing these systems, small noise and distortion budgets leave little flexibility when choosing system components.

The LT1993-x is designed to meet the demanding requirements of communications transceiver applications. It can be used as a differential ADC driver or as a general-purpose differential gain block. For single-ended systems, the LT1993-x can replace a transformer in performing single-ended to differential conversion without sacrificing noise or distortion performance.

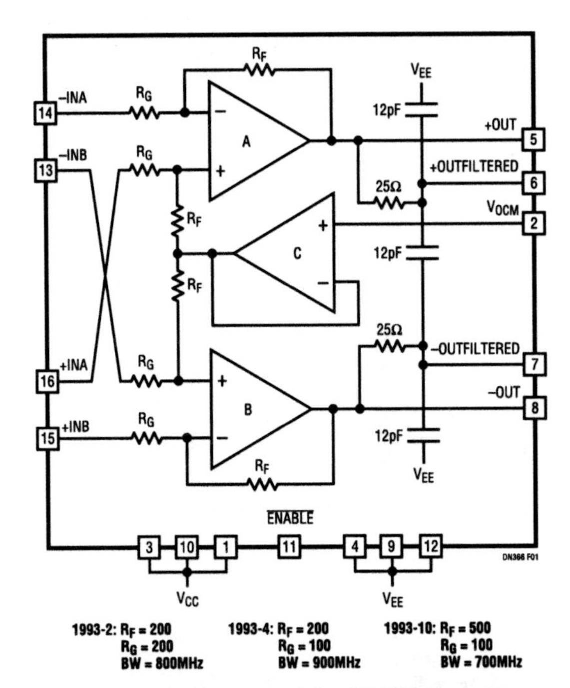

Figure 501.1 • Block Diagram of the LT1993-x and the Differences between the Gain Options. Input Impedance Is 200Ω for the 6dB Version and 100Ω for the Other Two Versions

LT1993-x features

The LT1993-x is a fully differential input and output amplifier with up to 7GHz of gain-bandwidth product and an impressive feature set. There are three fixed-gain options with internal matched resistors: gain of 2 (6dB), gain of 4 (12dB) and gain of 10 (20dB). The LT1993-x is DC-coupled, precluding the need for DC blocking capacitors on the inputs and outputs. The output common mode voltage is independently controlled with an external pin, allowing optimal bias conditions for the ADC inputs. The LT1993-x features two sets of differential outputs: a normal output and a filtered output. The output filter eliminates additional filtering in many applications, but if necessary, additional filtering can be achieved with a few external components. Figure 501.1 shows a block diagram of the LT1993-x.

High speed ADC driving

One of the more challenging tasks in a modern communications transceiver is driving the analog-to-digital converter (ADC). Today's converters sample data at tens to hundreds of megahertz with up to 16 bits of resolution. With each sample cycle, the switching of the internal ADC sample and hold injects charge into the output of the driver which must absorb the charge and settle its output before the next sample is taken. This charge injection is inherent in nearly all high speed, high resolution ADC topologies and must be considered when choosing a suitable driver.

The LT1993-x was designed specifically to drive high speed ADCs to their full potential. With a $3.8\text{nV}/\sqrt{\text{Hz}}$ voltage noise specification and -70dBc of harmonic distortion at 70MHz ($2V_{P-P}$ differential output), the LT1993-x meets and exceeds the requirements for driving high resolution high speed

Analog Circuit Design: Design Note Collection. http://dx.doi.org/10.1016/B978-0-12-800001-4.00501-9

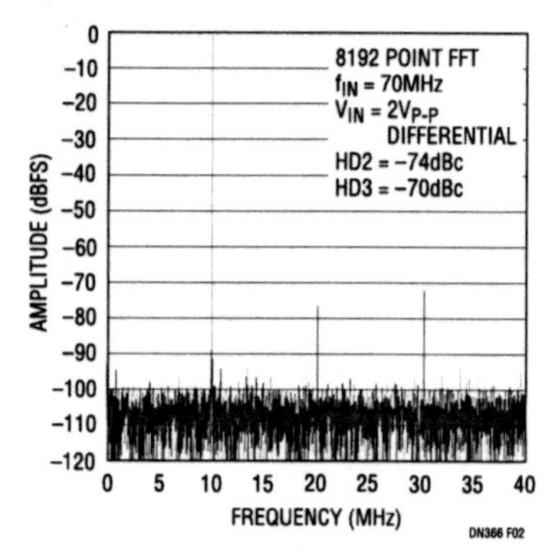

Figure 501.2 • FFT Data Taken Using the LT1993-2 and the LTC2249 ADC Sampling at 80Msps. The Second Harmonic is at –74dBc and the Third Harmonic Is at –70dBc

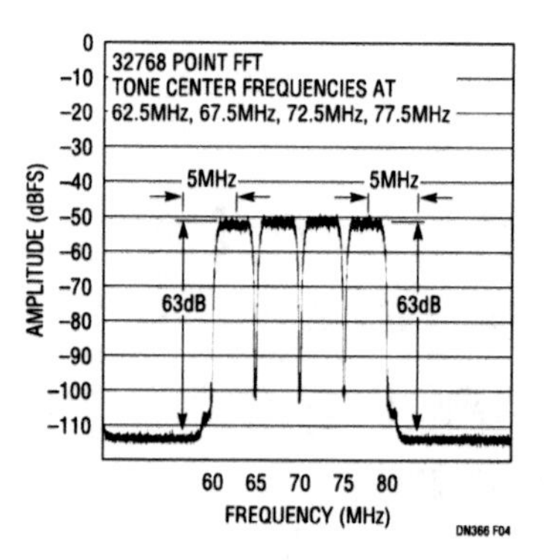

Figure 501.4 • FFT Data Taken from the Output of the LTC2255 ADC. The Low IMD of the LT1993-2 Preserves the Signal-to-Noise Ratio of the WCDMA Channels

ADCs. Figure 501.2 shows an FFT of sampled data taken on a 70MHz input signal with the LT1993-2 driving an LTC2249 sampling at 80Msps.

WCDMA amplifier and ADC driver

Wideband CDMA transceivers often use direct IF sampling, meaning that the ADC samples signals with a 70MHz center frequency and 5MHz of bandwidth per channel. Up to four WCDMA channels are transmitted simultaneously, spaced closely in frequency. This places difficult intermodulation distortion (IMD) and noise requirements on the components in the transceiver, since both raise the noise floor in the closely spaced adjacent channels. The LT1993-2 boasts an exceptional –70dBc IMD and low noise, allowing 63dBc of adjacent channel leakage ratio (ACLR) for WCDMA signals. This figure exceeds most WCDMA manufacturers' ACLR specifications.

Figure 501.3 shows the LT1993-2 driving a LTC2255 14-bit ADC with a 70MHz, 4-channel WCDMA signal. On the output of the LT1993-2 is a simple LC bandpass filter that adds additional out-of-band filtering. Figure 501.4 shows the FFT data from the LTC2255, demonstrating the good ACLR possible with the LT1993-2. The small aberrations on the sides of the WCDMA signals are artifacts of a noisy signal generator, whose output was bandpass filtered prior to reaching the LT1993-2.

Conclusion

The LT1993-x is a flexible, cost saving, and easy-to-use differential amplifier and ADC driver that ensures the best performance in high speed communications transceiver applications. Besides the low noise, low distortion and high speed, the LT1993-x also saves space with its 0.8mm-tall 3mm × 3mm QFN package. Minimal support circuitry is required to operate the LT1993-x under most conditions and output lowpass filtering is included. Three different gain options increase the flexibility of system design and help reduce the gain requirements of noisier system components. The LT1993-x can simplify transceiver designs, reduce component count and reduce product time-to-market.

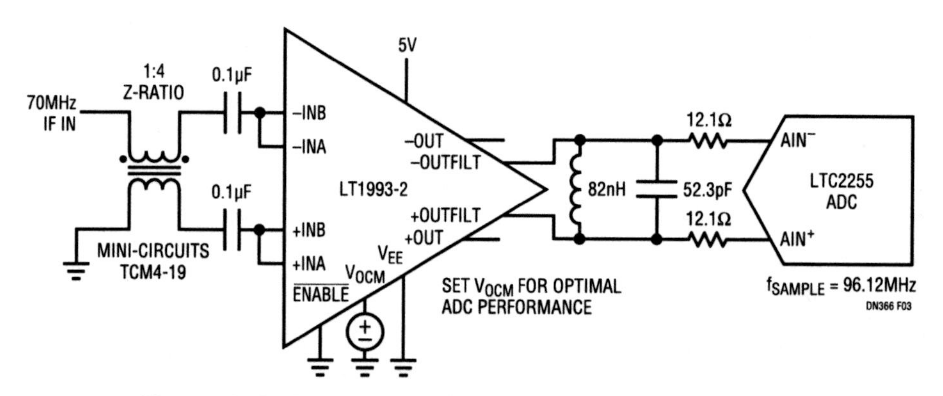

Figure 501.3 • The LT1993-2 Driving an LTC2255 ADC Sampling at 96.12Msps with a 70MHz, 4-Channel WCDMA Signal. The Simple LC Output Network Provides Out-of-Band Filtering

Wideband RF ICs for power detection and control

502

James Wong Vladimir Dvorkin

Introduction

Radio frequency devices are being deployed in ever-increasing numbers, not just in cell phones and cordless telephones. Other applications include 802.11 wireless LAN, RFID (radio frequency identification) tags, inventory monitors, satellite transceivers, fixed wireless access and wireless communications infrastructure. All RF devices must carefully monitor and control their RF power transmission to comply with government regulation and minimize RF interference with other radio devices. For this reason, accurate RF power detection is important in both RF receivers and transmitters.

This article presents some solutions using Linear Technology's versatile family of high frequency Schottky diode detectors. Table 502.1 summarizes the features of this family and lists more applications.

A dual band mobile phone transmitter power control application

Figure 502.1 is a simplified block diagram illustrating transmit power control for a dual band mobile phone (the receiver is not shown here). In this example, a 324Ω, 1% tolerance resistor (R1) followed by a 2.2pF capacitor (C1) form a coupling circuit with 18dB to 20dB coupling factor at 850MHz to 1850MHz, referenced to the LTC5509 RF input pin. C1 is also a DC blocking capacitor. R1 should have a tolerance of 1% while C1 should be 2% to 5%. The coupling circuit (R1 and C1) introduces about 0.15dB to 0.2dB losses into the

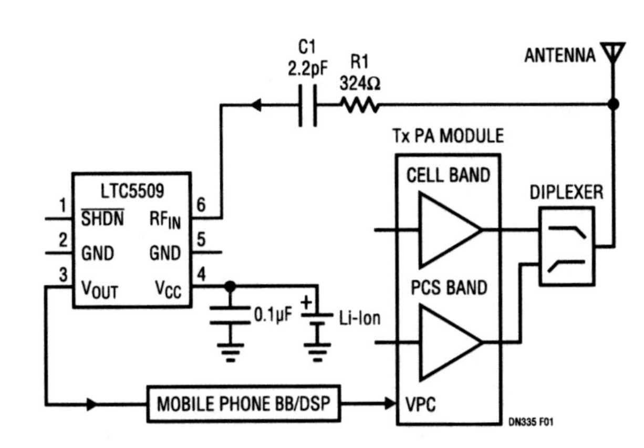

Figure 502.1 • A Dual Band Mobile Phone Transmit Power Control Using a Resistive Tap

Table 502.1 Summary of RF Detector Specifications and Applications

DEVICE	FREQUENCY RANGE	PACKAGE	DYNAMIC RANGE/FEATURES	APPLICATIONS
LTC5505-1	300MHz to 3GHz	ThinSOT	−28dBm to 18dBm★	General Purpose, Phones, ISM
LTC5505-2	300MHz to 3GHz	ThinSOT	−32dBm to 12dBm★	General Purpose, Phones, ISM
LTC5507	100kHz to 1GHz	ThinSOT	−34dBm to 14dBm★	General Purpose LF & Broadband Detection
LTC5508	300MHz to 7GHz	SC–70†	−32dBm to 12dBm★	General Purpose, WLAN, Microwave
LTC5509	300MHz to 3GHz	SC–70†	−30dBm to 6dBm★★	Mobile Phones Tx Power Control
LTC5532	300MHz to 7GHz	ThinSOT	Adjustable Gain & Starting Voltage	Precision RSSI & Envelope Detection

★Gain compression extends the dynamic range with a trade-off of reduced linearity in the transfer characteristic.
★★No gain compression.
†Smallest package.

Analog Circuit Design: Design Note Collection. http://dx.doi.org/10.1016/B978-0-12-800001-4.00502-0

main signal line. R1 should be placed as close as possible to the antenna without forming a "T" connection on the microstrip line and immediately followed by capacitor C1 and the LTC5509. Ideally, C1, R1 and LTC5509 should be placed on the same side of the PCB as the Tx output microstrip line to the antenna. The component values shown here should be used as a reference. In the actual product implementation, component values may differ slightly depending on the output impedance of the Tx PAs, antenna impedance, component placement and PC board parasitics.

An RFID reader application

RFID (radio frequency identification) is a promising technology for many monitoring and tracking applications, including retail store check-out registers, inventory management, vehicle tracking, tire-pressure monitoring and live-stock/agricultural tracking. Common to all of them is the need for well-controlled RF power and a cost-effective means of reliably detecting the received data. Well-regulated RF power allows maximum power transmission to the ID tags while staying within regulatory emission limits. A well-controlled transmitter is possible if an RF detector is used in a closed-loop feedback circuit, similar to the example shown in Figure 502.1. The choice of RF power detector is determined by the RF frequency, as well as by other constraints such as the required dynamic range and sensitivity.

To form a complete RFID reader receiver, an RF Schottky peak detector can also make an excellent low cost data receiver to demodulate ASK or AM modulated signals with data rates up to 3MHz. Because RF detectors such as the LTC5507 can detect RF signals over a wide frequency range, filtering can improve the sensitivity of the receiver. Figure 502.2 shows a data receiver with an input LNA (low noise amplifier) and an input BPF (bandpass filter). The LNA can be a general purpose, low cost gain block that provides fixed gain at the operating frequency of interest. The added gain increases sensitivity and extends the detection range. A lowpass or bandpass filter at the detector output provides additional receiver selectivity, if needed. The RSSI (received signal strength indicator) DC-coupled output provides signal strength information using a lowpass filter (R2 and C5) to filter out the modulation components.

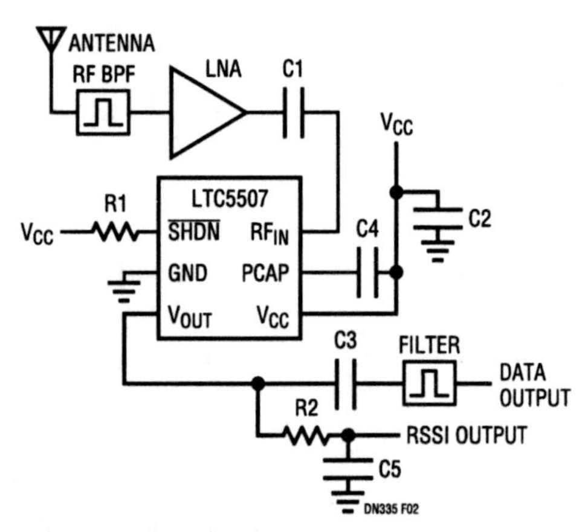

Figure 502.2 • An RFID Reader ASK Receiver with Output Filter

Application of RF power detectors at frequencies above 7GHz

Although the LTC5532 is optimized for an operating frequency range from 300MHz to 7GHz, it can offer useful performance well above this frequency range. The performance at higher frequencies does fall off but gracefully. Figure 502.3 shows a plot of the LTC5532's output voltage versus RF input power characteristics at 12GHz. Figure 502.4 shows the LTC5532's input S11 Smith Chart, extending to 12GHz. Coupling to the LTC5532 at these high frequencies is in principle very similar to lower frequency operation.

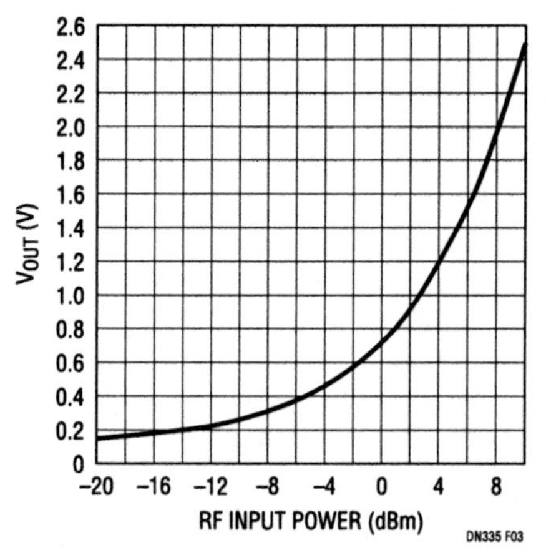

Figure 502.3 • LTC5532 Typical Detector Transfer Characteristics at 12GHz Frequency

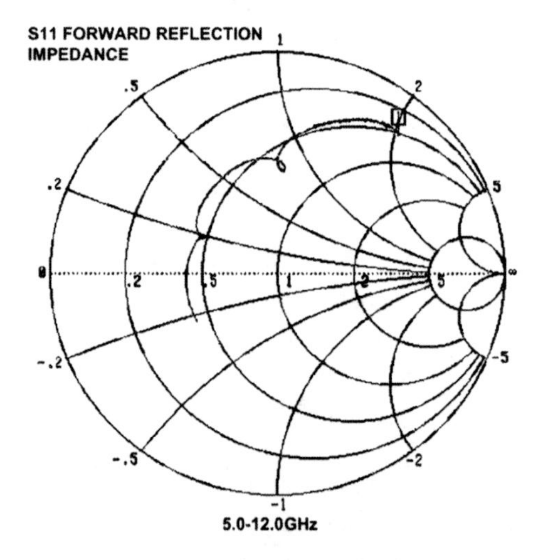

Figure 502.4 • LTC5532 Input S11 Smith Chart

503

Fiber optic communication systems benefit from tiny, low noise avalanche photodiode bias supply

Michael Negrete

Avalanche photodiodes (APDs) are the photo detectors of choice for long-haul fiber optic communication systems because of their high sensitivity and high internal gain. An important characteristic of APDs is that their internal gain is optimal when there is a high voltage reverse bias (30V to 90V) across the APD. Nevertheless, the high gain is all for naught if the sensitivity of the APD is compromised by a noisy bias supply.

Traditionally, such low noise bias supplies required custom circuits that brought with them another problem: large space requirements. Linear Technology's LT1930A 2.2MHz step-up DC/DC converter in a 5-lead SOT-23 package solves these APD bias voltage problems and does so in a compact package suitable for most fiber optic applications.

The LT1930A, a capacitor-diode tripler and an external DAC provide a bias voltage of up to 90V, allowing easy temperature compensation (via the DAC) to optimize internal gain. By running the IC at a switching frequency of 2.2MHz, one can use tiny, low cost capacitors and inductors to keep the circuit footprint under $0.5in^2$. The LT1930A's constant frequency PWM operation keeps output noise low and easy to filter.

Figure 503.1 shows a high voltage, low noise APD bias supply that works from an input range of 2.6V to 6.3V. The DAC, driven from a processor, adjusts the output from 30V to 90V to compensate for temperature dependent APD gain fluctuations. The LT1930A includes a 35V switch making it capable of producing 105V output through a capacitor-diode tripler.

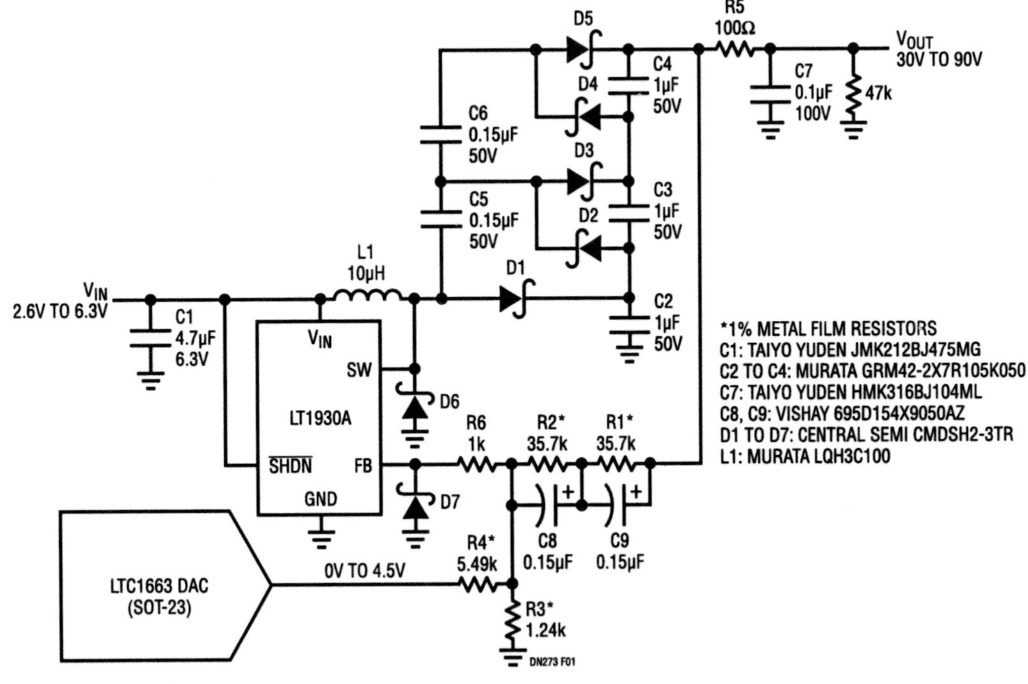

Figure 503.1 • LT1930A-Based Boost Regulator Produces 30V to 90V for Avalanche Photodiode Bias Supplies with Only 200µV_{P-P} Noise

Analog Circuit Design: Design Note Collection. http://dx.doi.org/10.1016/B978-0-12-800001-4.00503-2

To eliminate noise from the internal reference and error amplifier, two 0.15µF tantalum feedback capacitors are used in series. A series connection ensures a sufficient voltage rating of the feedback capacitance. Ceramic feedback capacitors have a piezoelectric response to temperature and low frequency vibrations under 1kHz, which is amplified by the LT1930A internal error amplifier. These should not be used unless noise in that bandwidth is acceptable. To protect the switch pin from negative voltage swings, a clamping diode is tied to ground. An identical diode is placed at the feedback (FB) pin, along with a 1k resistor to protect the part from a sudden short in the load, which would force the feedback capacitor's negative side to the negative value of the output voltage. All other capacitors can be ceramic, which are small and capable of handling the high voltages of the regulator.

Figure 503.2 shows the AC coupled noise of a 50V output with a 5V input. The switching noise is less than 200µV_{P-P}, allowing greater sensitivity and dynamic range than most APD bias solutions. Oscilloscope measurement bandwidth is 100Hz to 10MHz, all probe cables are coaxial and special attention is given to grounding.[1]

Conclusion

The LT1930A exceeds all of the stringent demands of an APD reverse-bias voltage, eliminating the need for custom APD bias supplies. The LT1930A solution not only provides the cleanest output in the industry for APDs, but also achieves this in a fraction of the space required by other solutions.

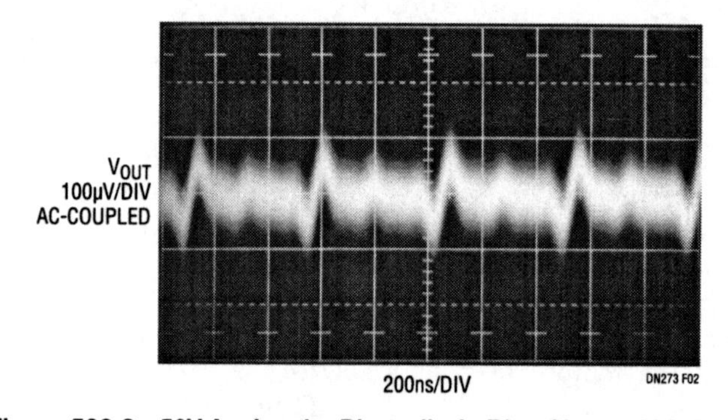

Figure 503.2 • 50V Avalanche Photodiode Bias Shows 200µV_{P-P} Ripple and Noise, Improving Fiber Optic Receiver Sensitivity

Note 1: Discussion of low noise measurement issues is available in "A Monolithic Switching Regulator with 100µV Output Noise," Linear Technology Corporation, Application Note 70 by Jim Williams.

ADSL modems that yield long reach and fast data rates

504

Tim Regan

ADSL technology allows for communication to a remote internet server at data rates nearing 1Mbit/sec, nearly 20 times faster than conventional modems, over normal telephone wiring with a reach of up to 18,000 feet. A key component in making this happen is the power amplifier used to drive the encoded data signal on to the phone line. Ideally the line driver must not introduce any distortion of the transmitted signal. This allows the system to maximize the data rate over the longest wiring reach possible by using all of the available carrier tones for the data transmission. The LT1886 line driver outputs undistorted broadband power into the telephone network.

The LT1886 also maintains very low distortion even when configured for high levels of closed-loop gain. Low supply voltage VLSI digital circuits used to provide the encoded data require significant voltage gain to be taken in the line driver without introducing distortion.

LT1886: low distortion line driver

The LT1886 is a high speed dual power amplifier with excellent high frequency distortion performance. The LT1886 features a 700MHz gain-bandwidth product, 200V/µs slew rate and over 200mA of output current drive in a small SO-8 package. Using only a single 12V supply, the LT1886 is the best choice for the line driver in ADSL CPE modems (customer premises equipment).

Figure 504.1 illustrates a differential line driver with an LT1886 driving a 1:2 turns ratio isolation/coupling transformer. Here it can be seen that both the 2nd and 3rd harmonic components of a 200kHz tone, even with a closed-loop gain of 10, are well below −80dBc with nominal signal power of 13dBm, $4V_{P-P}$ placed on the line. The distortion remains less than −70dBc during large-signal transients, $15V_{P-P}$, which occur often in the complex modulated ADSL transmission signal. The nominal RMS output current from the driver during an ADSL transmission is only 30mA, but the peak driver current capability must be greater than 160mA. To prevent distorting the signal, the line driver cannot current limit.

Traditionally, the power stage is the limiting factor for high frequency performance. The LT1886 is fabricated with an advanced low voltage complimentary bipolar process that combines high frequency amplification with output power handling capability. In addition, to obtain the best distortion performance possible, an amplifier must sustain a very high open-loop gain to as high of a frequency as is practical.

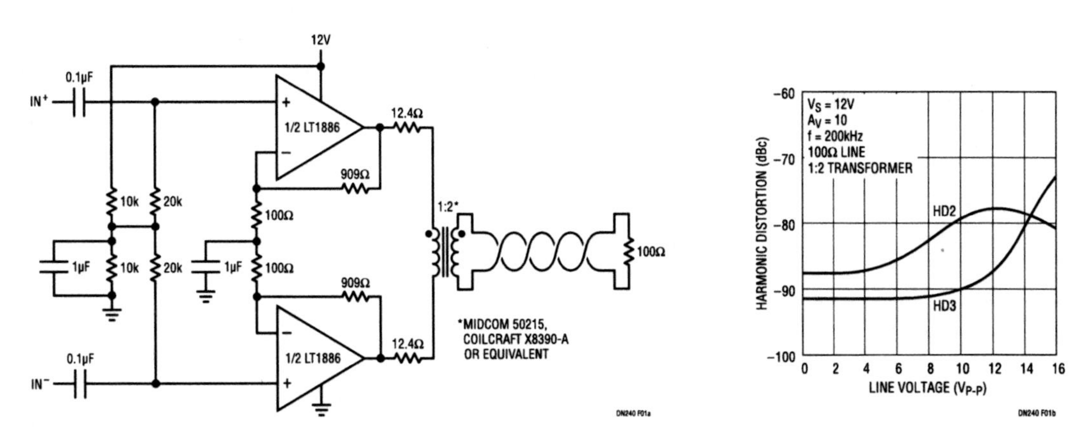

Figure 504.1 • A Single 12V Supply ADSL Modem Line Driver with Low Harmonic Distortion Performance

Analog Circuit Design: Design Note Collection. http://dx.doi.org/10.1016/B978-0-12-800001-4.00504-4

LT1886 frequency response

Figure 504.2 illustrates the gain and phase shift response versus frequency of the LT1886. The amplifier is a voltage feedback design which is compensated for gains greater than 10 and has a conventional single pole open-loop gain roll off characteristic.

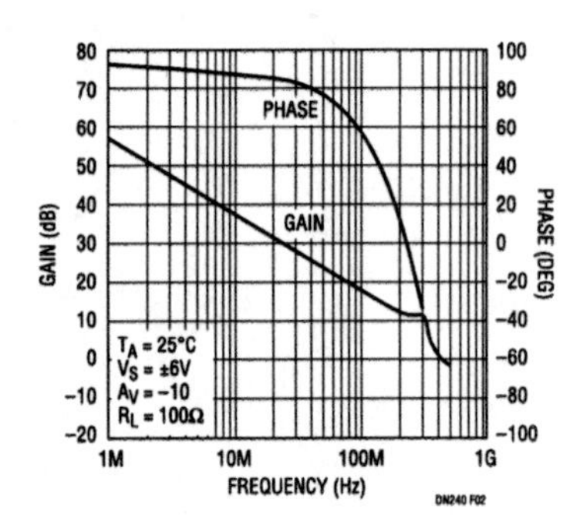

Figure 504.2 • Open-Loop Gain and Phase vs Frequency

The low frequency open-loop gain is over 80dB out to a frequency of 60kHz, a gain-bandwidth product of approximately 700MHz. Notice that with a closed-loop gain of 10 (20dB), the closed-loop bandwidth is 70MHz and the phase margin is nearly 70°, resulting in very stable operation. However, the amplifier is intentionally decompensated to maximize gain bandwidth for distortion performance. At the frequency where the phase margin drops to 0°, approximately 230MHz, the amplifier still has 12dB of gain. With a closed-loop gain this low the amplifier will oscillate.

For stability, it is important for the amplifier to always have a feedback factor of 0.1 or less to maintain a noise gain value of 10 or more. For this reason, in Figure 504.1, the two 100Ω gain setting resistors for the driver are returned to a good AC ground provided by the 1μF bypass capacitor. This ensures that the impedance from the inverting inputs to ground, even beyond 300MHz where the LT1886 still has gain, is controlled and held at a low enough value for stable operation.

A circuit "trick" for a gain of less than 10

In many applications, it may still be desired to operate the driver stage with a gain of less than 10. In this case, a simple circuit "trick" for decompensated op amps can be employed. The addition of a resistor (and an optional capacitor) between the two inputs of the amplifier can provide stable operation at any desired closed-loop gain. Figure 504.3 illustrates these gain compensation components for inverting and noninverting amplifiers.

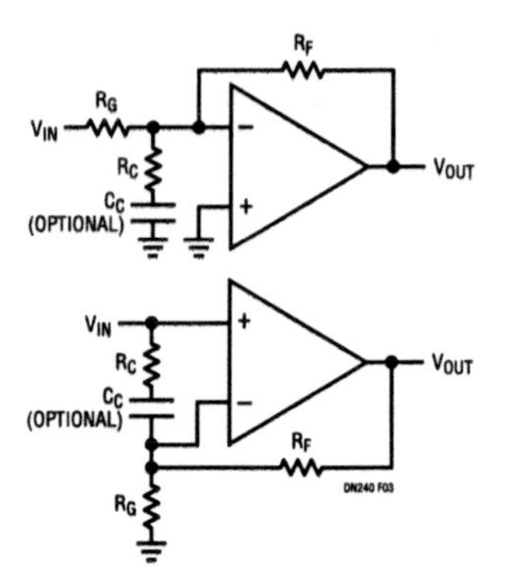

Figure 504.3 • Compensation for Gain < 10

At lower frequencies, where the open-loop gain of the amplifier is very high, there is very little signal between the two inputs. With no signal across the compensating resistor, R_C, it is effectively out of the circuit. The feedback resistor, R_F, and the gain setting resistor, R_G, then determine the closed-loop gain of the amplifier. At higher frequencies, as the open-loop gain reduces, a signal is developed across R_C and it becomes part of the circuit and effectively parallels resistor R_G. This assumes that the impedance to ground at the noninverting input is much lower than the feedback component values. The lower value of R_G reduces the feedback factor and in essence tricks the amplifier into operating at a higher gain. The following equation can be used to determine the value for R_C to ensure stability with lower than required closed-loop gain:

$$R_C = \frac{R_F}{A_{VREQUIRED} - A_{VACTUAL}}$$

Where $A_{VREQUIRED}$ is 10 and $A_{VACTUAL}$ is any gain between 1 and 10. Using this technique the amplifier operates with a stable lower closed-loop gain but has the bandwidth of a gain of 10 amplifier.

The capacitor, C_C, is optional to prevent increased gain to the DC offset of the amplifier. For the LT1886, this capacitor should be sized together with R_C to come into effect at a frequency between 5MHz and 15MHz.

A low power, high output current dual CFA makes xDSL line driving clean and easy

505

Adolfo A. Garcia

Introduction

Driving the balanced lines of high speed modems has become a very popular application for high speed, high output current, dual operational amplifiers. High amplifier speed is required to faithfully process signal passbands of at least 1MHz bandwidth, and high output currents are required to meet peak drive levels into 100Ω or 135Ω termination. Taking into account peak-to-average signal ratios in these applications, Table 505.1 summarizes peak voltage/current drive levels required by current xDSL systems. The peak voltage and current drive values shown are for ADSL systems with 100Ω termination and HDSL/HDSL2 systems with 135Ω termination.

Table 505.1 Peak Drive Levels for Popular xDSL Systems

xDSL	NOMINAL TRANSMIT POWER	PEAK VOLTAGE	PEAK CURRENT
HDSL	13.5dBm	$5.6V_{P-P}$	$42mA_{P-P}$
ADSL DMT Upstream	13dBm	$15V_{P-P}$	$150mA_{P-P}$
ADSL CAP Upstream	13dBm	$8.5V_{P-P}$	$86mA_{P-P}$
ADSL CAP Downstream	20dBm	$18.9V_{P-P}$	$190mA_{P-P}$
HDSL2	16.5dBm	$19.6V_{P-P}$	$146mA_{P-P}$

There are amplifier products available today that can address at least one, but not all, of the xDSL applications listed in Table 505.1. These products are single application specific and suffer from one or more of the following shortcomings: limited supply voltage operation, excessive zero-load power dissipation, large packaging/heat sinking and/or limited dynamic performance (thermal and intermodulation crosstalk).

The LT1497, a 50MHz, 125mA dual CFA from Linear Technology, is the fourth product in our family of high speed, high current dual current feedback amplifiers, capable of addressing any one or all five of the xDSL applications in Table 505.1. It operates from ±2.5V to ±15V supplies, consumes less than 7mA per amplifier and is available in both SO-8 and thermally enhanced S16 packages. With its 125mA output stage and available voltage swing to within 2V of

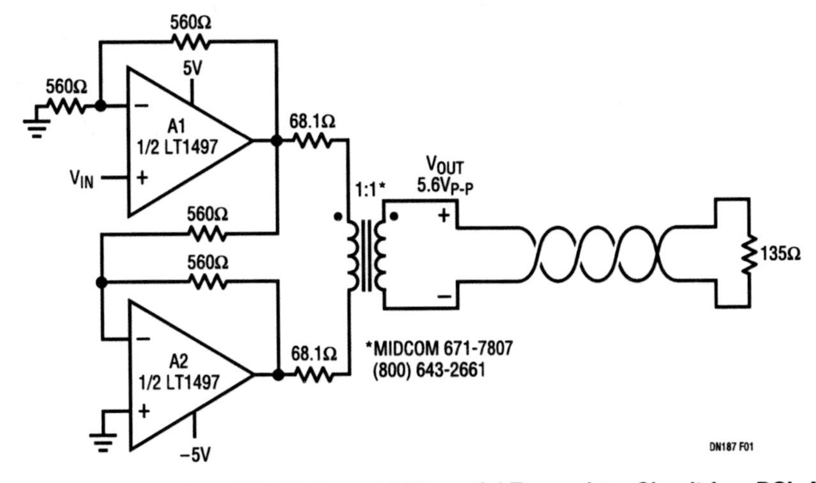

Figure 505.1 • A Low Distortion, LT1497-Based Differential Transmitter Circuit for xDSL Applications

Analog Circuit Design: Design Note Collection. http://dx.doi.org/10.1016/B978-0-12-800001-4.00505-6

either rail, the LT1497 is a good, economical choice because it allows the use of smaller turns-ratio (1:1 and 1:2) transformers to deliver power into the line. These transformers are less expensive than the high turns-ratio transformers required by some amplifiers that operate on a single supply voltage rail. In addition, its low quiescent supply current allows more modem lines per card because less PC board area is required for the device and heat sinking. Its low input offset voltage match (±3.5mV) and low TCV_{OS} (10μV/°C)—not specified for other amplifiers—combine to eliminate output coupling capacitors used to block DC current flow through the transformer's primary windings.

A low distortion HDSL line driver

Figure 505.1 shows an HDSL differential line driver circuit that transmits signals over a 135Ω twisted pair through a 1:1 transformer. The driver amplifiers are configured for gains of two (A1) and minus one (A2) to compensate for the attenuation inherent in the back termination of the line and to provide differential drive to the transformer.

Even though the input circuit configuration is single-ended, the circuit configuration can be very easily manipulated to accommodate differential output analog front ends (AFEs) and/or single supply rail operation for the line driver circuit. For HDSL applications, the LT1497's high output current and voltage swing drive the 135Ω line at the required distortion level of −72dBc. For an HDSL data rate of 1.544Mbps, the fundamental frequency of operation is 392kHz.

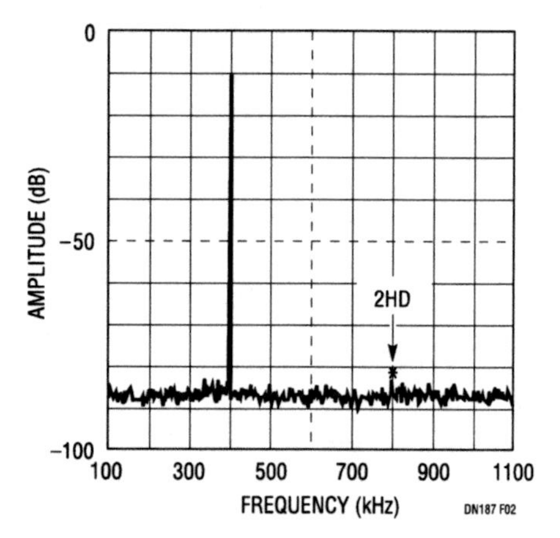

Figure 505.2 • Harmonic Distortion Performance of Figure 505.1's Circuit with a 400kHz Sine Wave at 5.6V_{P-P} into 135Ω

Performance

The circuit of Figure 505.1 was evaluated for harmonic distortion with a 400kHz sine wave and an output level of 5.6V_{P-P} into 135Ω, representing peak drive operation into the HDSL termination. Figure 505.2 shows that the second harmonic is −72.3dB relative to the fundamental for the 135Ω load. Third harmonic distortion is not critical, because received signals are heavily filtered before being digitized by an A/D converter.

With multicarrier applications, such as discrete multitone modulation (DMT), becoming as prevalent as single-carrier applications, another important measure of amplifier dynamic performance is 2-tone intermodulation distortion. This evaluation is a valuable tool for evaluating amplifier linearity when processing more than one tone at a time. For this test, two sine waves at 300kHz and 400kHz were used, with levels set to obtain 5.6V_{P-P} across the 135Ω load. Figure 505.3 shows that the third-order intermodulation products are well below −72dB.

Conclusion

The new LT1497 offers outstanding distortion performance in an SO-8 package with remarkably low power dissipation and very attractive high speed modem system economics. It is ideally suited for single-pair digital subscriber line systems, specifically DMT remote terminal, CAP central office/remote terminal, HDSL or HDSL2 line-driver applications.

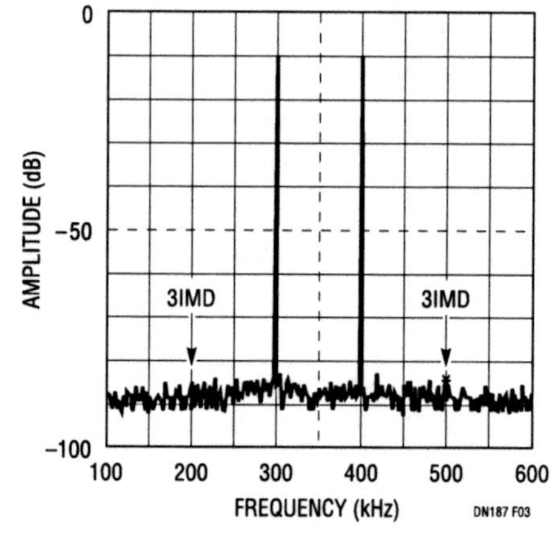

Figure 505.3 • 2-Tone Intermodulation Distortion Performance of Figure 505.1's Circuit. Two Sine Waves at 300kHz and 400kHz Are Used to Obtain 5.6V_{P-P} into 135Ω

A low cost 4Mbps IrDA receiver in MS8 and SO-8 packages

506

Alexander Strong

Introduction

The need for ever increasing data rates required by a vast array of devices, such as notebook computers, printers, mobile phones, pagers and modems, has been satisfied by the technology of infrared data transmission. The Infrared Data Association (IrDA) standard, which covers data rates from 2400bps to 4Mbps, is the overwhelming choice for infrared data transmission. The LT1328 is a photodiode receiver that supports IrDA data rates up to 4Mbps, as well as other modulation methods, such as Sharp ASK and TV remote control.

The LT1328, in the MS8 and SO-8 packages, contains all the necessary circuitry to convert current pulses from an external photodiode to a digital TTL output while rejecting unwanted lower frequency interference. The LT1328 plus five external components is all that is required to make the IrDA-compatible receiver shown in Figure 506.1. An IrDA-compatible transmitter can also be implemented with only six components, as shown in Figure 506.2. Power requirements for the LT1328 are minimal: a single 5V supply and 2mA of quiescent current.

LT1328 functional description

Figure 506.3 is a block diagram of the LT1328. Photodiode current from D1 is transformed into a voltage by feedback resistor R_{FB}. The DC level of the preamp is held at V_{BIAS} by the servo action of the transconductance amplifier's g_m. The servo action only suppresses frequencies below the R_{gm}/C_{FILT} pole. This highpass filtering attenuates interfering signals, such as sunlight or incandescent or fluorescent lamps, and is selectable at Pin 7 for low or high data rates. For high data rates, Pin 7 should be held low. The highpass filter breakpoint is set by the capacitor C4 at $f = 25/(2\pi \cdot R_{gm} \cdot C4)$, where

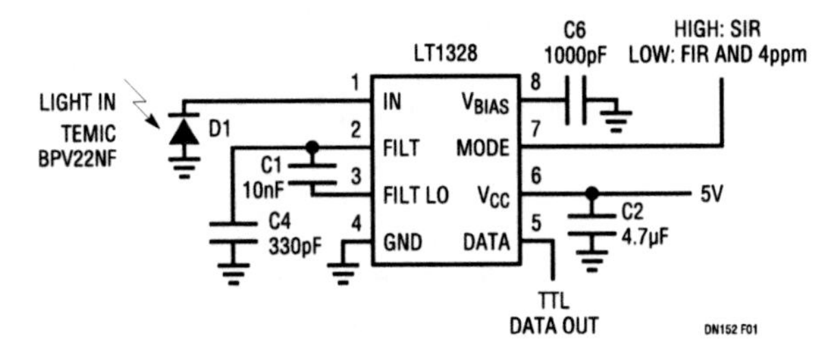

Figure 506.1 • LT1328 IrDA Receiver

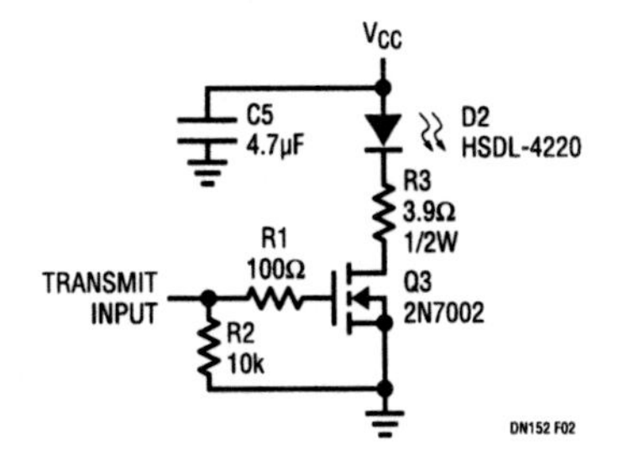

Figure 506.2 • IrDA Transmitter

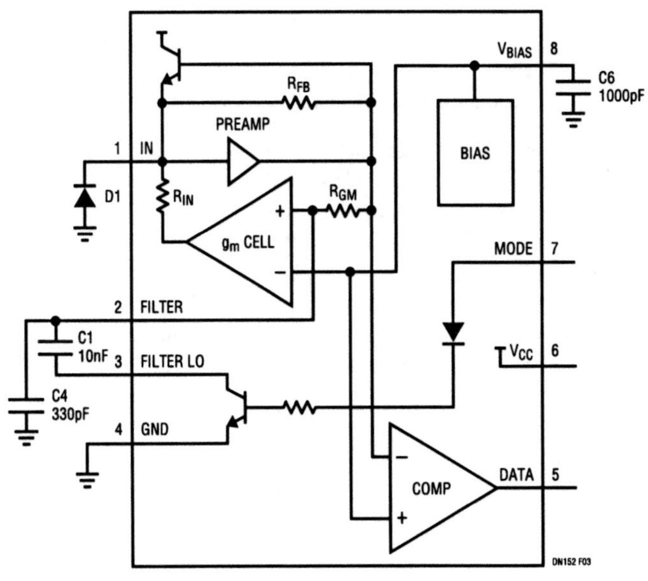

Figure 506.3 • LT1328 Block Diagram

Analog Circuit Design: Design Note Collection. http://dx.doi.org/10.1016/B978-0-12-800001-4.00506-8

R_{gm} = 60k. The 330pF capacitor (C4) sets a 200kHz corner frequency and is used for data rates above 115kbps. For low data rates (115kbps and below), the capacitance at Pin 2 is increased by taking Pin 7 to a TTL high. This switches C1 in parallel with C4, lowering the highpass filter breakpoint. A 10nF capacitor (C1) produces a 6.6kHz corner. Signals processed by the preamp/gm amplifier combination cause the comparator output to swing low.

IrDA SIR

The LT1328 circuit in Figure 506.1 operates over the full 1cm to 1m range of the IrDA standard at the stipulated light levels. For IrDA data rates of 115kbps and below, a 1.6µs pulse width is used for a zero and no pulse for a one. Light levels are 40mW/sr (watts per steradian) to 500mW/sr. Figure 506.4 shows a scope photo for a transmitter input (bottom trace) and the LT1328 output (top trace). Note that the input to the transmitter is inverted; that is, transmitted light produces a high at the input, which results in a zero at the output of the transmitter. The MODE pin (Pin 7) should be high for these data rates.

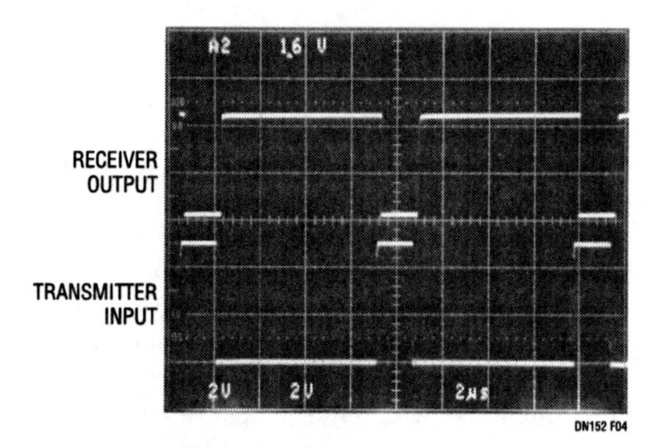

Figure 506.4 • IrDA-SIR Modulation

IrDA FIR

The second fastest tier of the IrDA standard addresses 576kbps and 1.152Mbps data rates, with pulse widths of 1/4 of the bit interval for zero and no pulse for one. The 1.152Mbps rate, for example, uses a pulse width of 217ns; the total bit time is 870ns. Light levels are 100mW/sr to 500mW/sr over the 1cm to 1meter range. A photo of a transmitted input and LT1328 output is shown in Figure 506.5. The LT1328 output pulse width will be less than 800ns wide

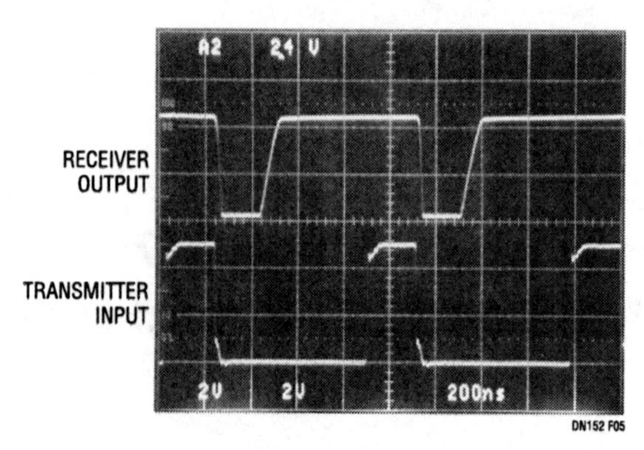

Figure 506.5 • IrDA-FIR Modulation

over all of the above conditions at 1.152Mbps. Pin 7 should be held low for these data rates and above.

4ppm

The last IrDA encoding method is for 4Mbps and uses pulse position modulation, thus its name: 4ppm. Two bits are encoded by the location of a 125ns wide pulse at one of the four positions within a 500ns interval (2 bits · 1/500ns = 4Mbps). Range and input levels are the same as for 1.152Mbps. Figure 506.6 shows the LT1328 reproduction of this modulation.

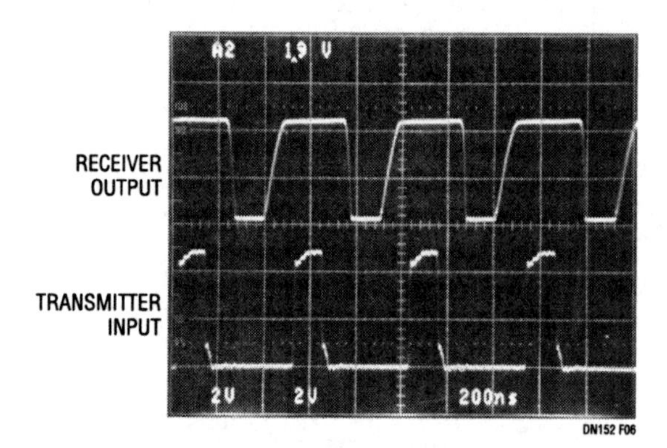

Figure 506.6 • IrDA-4ppm Modulation

Conclusion

In summary, the LT1328 can be used to build a low cost receiver compatible with IrDA standards. Its ease of use and flexibility also allow it to provide solutions to numerous other photodiode receiver applications. The tiny MSOP package saves on PC board area.

507

Telephone ring-tone generation

Dale Eagar

Requirements

When your telephone rings, exactly what is the phone company doing? This question comes up frequently, as it seems everyone is becoming a telephone company. Deregulation opens many new opportunities, but if you want to be the phone company you have to ring bells.

An open-architecture ring-tone generator

Here is a design that you can own, tailor to your specific needs, layout on your circuit board and put on your bill of materials. Finally, you will be in control of the black magic (and high voltages) of ring-tone generation.

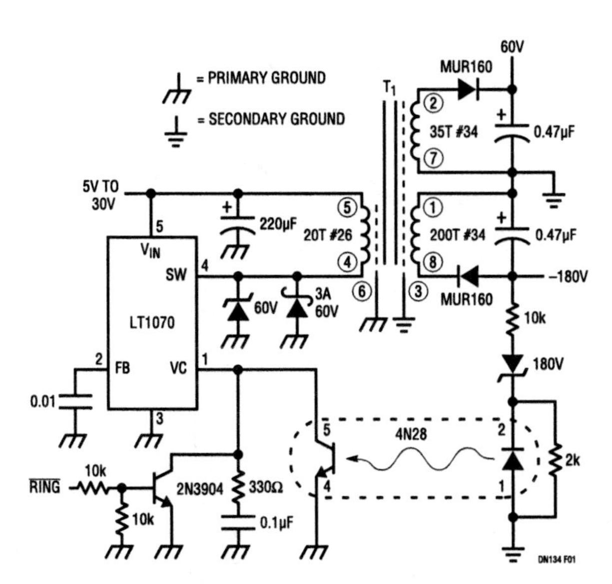

Figure 507.1 • The Switching Power Supply

Not your standard bench supply

Ring-tone generation requires not one but two high voltages, 60VDC and −180VDC (this arises from the need to put $87V_{RMS}$ on −48VDC). Figure 507.1 details the switching power supply that delivers the volts needed to run the ring-tone circuit. This switcher can be powered from any voltage from 5V to 30V and shuts down when not in use. Figure 507.2 is the build diagram of the transformer used in the switching power supply.

Quad op amp rings phones

When a phone rings, it rings with a cadence, a sequence of rings and pauses. The standard cadence is one second ringing followed by two seconds of silence. We use the first 1/4 of

MATERIALS	
2	EFD 20-15-3F3 Cores
1	EFD 20-15-8P Bobbin
2	EFD 20- Clip
2	0.007" Nomex Tape for Gap
Winding 1	Start Pin 1 200T #34
	Term Pin 8
	1 Wrap 0.002" Mylar Tape
Winding 2	Start Pin 2 70T #34
	Term Pin 7
	1 Wrap 0.002" Mylar Tape
Shields	Connect Pin 3 1T Foil Tape Faraday Shield
	1 Wrap 0.002" Mylar Tape
	Connect Pin 6 1T Foil Tape Faraday Shield
	1 Wrap 0.002" Mylar Tape
Winding 3	Start Pin 4 20T #26
	Term Pin 5
	Finish with Mylar Tape

Figure 507.2 • Ring-Tone High Voltage Transformer Build Diagram

Analog Circuit Design: Design Note Collection. http://dx.doi.org/10.1016/B978-0-12-800001-4.00507-X

LT1491 as a cadence oscillator, whose output is at V_{CC} for one second and then at V_{EE} for two seconds. This sequence repeats every three seconds, producing the all-too-familiar pattern. The actual ringing of the bell is done by a 20Hz AC sine wave signal at a signal level of $87V_{RMS}$ superimposed on −48VDC. The 20Hz signal is implemented with the second amplifier in the LT1491 which acts as a gated 20Hz oscillator (see Figure 507.3).

The third amplifier in the LT1491, which is configured as a lowpass filter, converts the square wave output of the oscillator to a sine wave by filtering out unwanted harmonics. Finally, the $87V_{RMS}$ and the −48VDC parts are handled by the fourth amplifier in the LT1491 and its steering of two external 15V regulators.

The rest of what we do, the part that is most difficult to follow, involves the output amplifier. In the output amplifier, the $6V_{P-P}$ signal from the waveform synthesizer is imposed across R12 (see Figure 507.4) into a virtual ground, creating a sine wave signal current. This current is added to the DC current flowing through R15 and the resulting current is imposed across R13. This stage amplifies the sine wave and offsets it to become an $87V_{RMS}$ sine wave imposed on a −48VDC bias. The trick here is that the voltage gain is in the ±15V regulators, not the LT1491 which is merely steering currents.

This complete circuit (Figure 507.4) includes the ring-trip sense circuit to detect when the phone receiver is picked up. This circuit is fully protected for output shorts to any voltage within the power supply window of −180V to 60V.

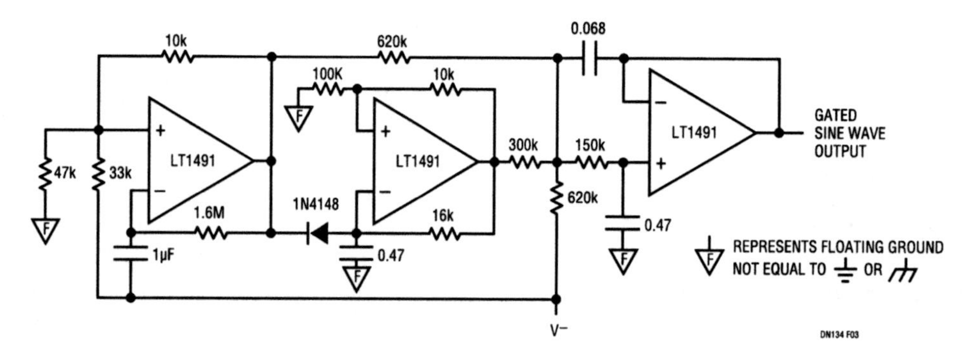

Figure 507.3 • Wave Form Synthesizer

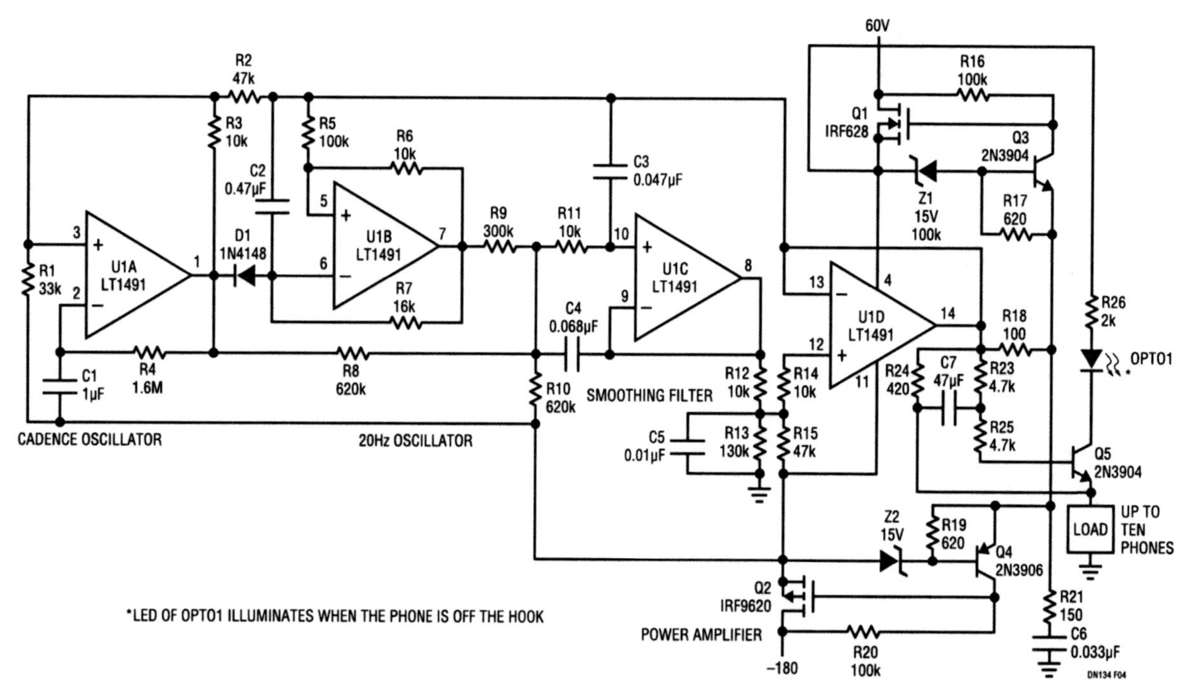

Figure 507.4 • Complete Ring-Tone Generator Circuit

Index

A

B

C

D

E

F

G

H

I

J

K

L

M

N

O

P

Q

R

S

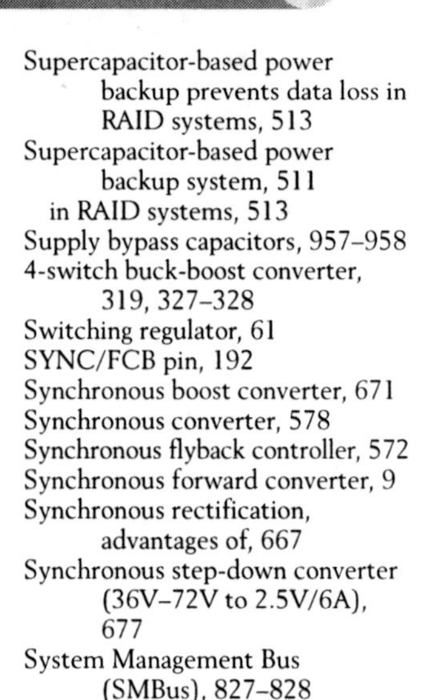

CPSIA information can be obtained at www.ICGtesting.com
Printed in the USA
BVOW06*1141110315

390710BV00004B/2/P

9 780128 000014